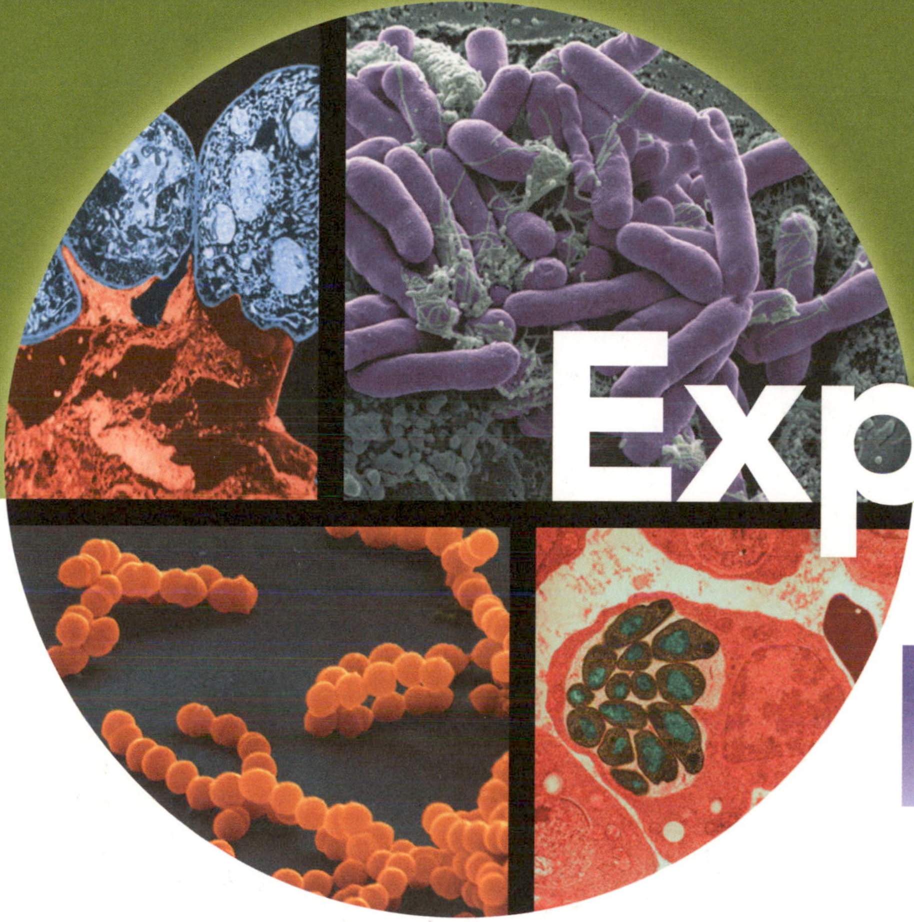

Explore the Invisible

Investigate It

DISEASE IN DEPTH

New **Disease in Depth** spreads visually tell the story of important and representative diseases for each body system, examining the history, present incidents, and potential future developments of specific diseases.

INVESTIGATE IT!

Each **Disease in Depth** feature includes a QR code and Investigate It! question that direct students to a major health website prompting further exploration and critical thinking. New MasteringMicrobiology® assignable Disease in Depth coaching activities encourage students to engage in independent research to apply and test their understanding of key concepts related to the Investigate It! query.

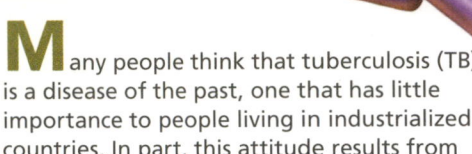

DISEASE IN DEPTH

TUBERCULOSIS

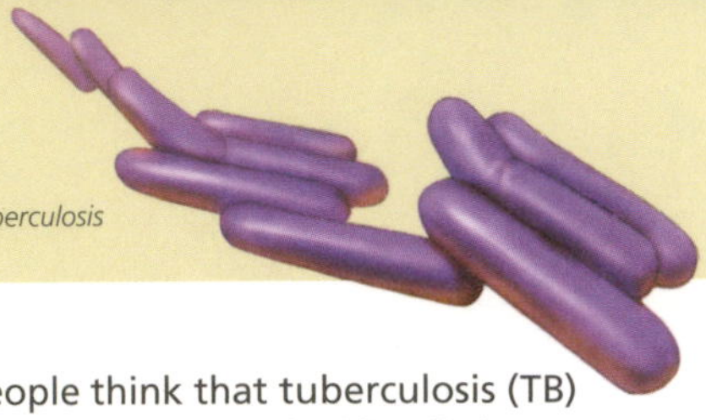

Mycobacterium tuberculosis

SIGNS AND SYMPTOMS

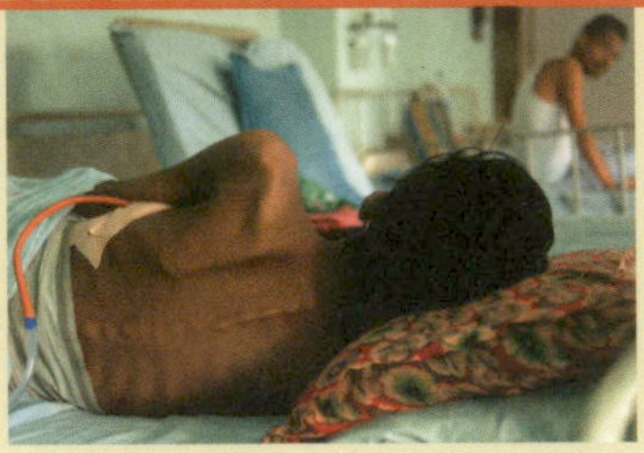

Signs and symptoms of TB are not always apparent, often limited to a minor cough and mild fever. Breathing difficulty, fatigue, malaise, weight loss, chest pain, wheezing, and coughing up blood characterize the disease as it progresses.

Many people think that tuberculosis (TB) is a disease of the past, one that has little importance to people living in industrialized countries. In part, this attitude results from the success health care workers have had in reducing the number of cases. Nevertheless, epidemiologists warn that complacency can allow this terrible killer to reemerge.

PATHOGENESIS

Primary tuberculosis

1 *Mycobacterium* typically infects the respiratory tract via inhalation of respiratory droplets from infected individuals.

2 Macrophages in alveoli phagocytize mycobacteria but are unable to digest them, in part because the bacterium inhibits fusion of lysosomes to endocytic vesicles.

3 Instead, bacteria replicate freely within macrophages, gradually killing the phagocytes. Bacteria released from dead macrophages are phagocytized by other macrophages, beginning the cycle anew.

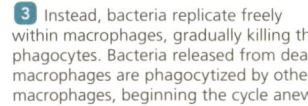

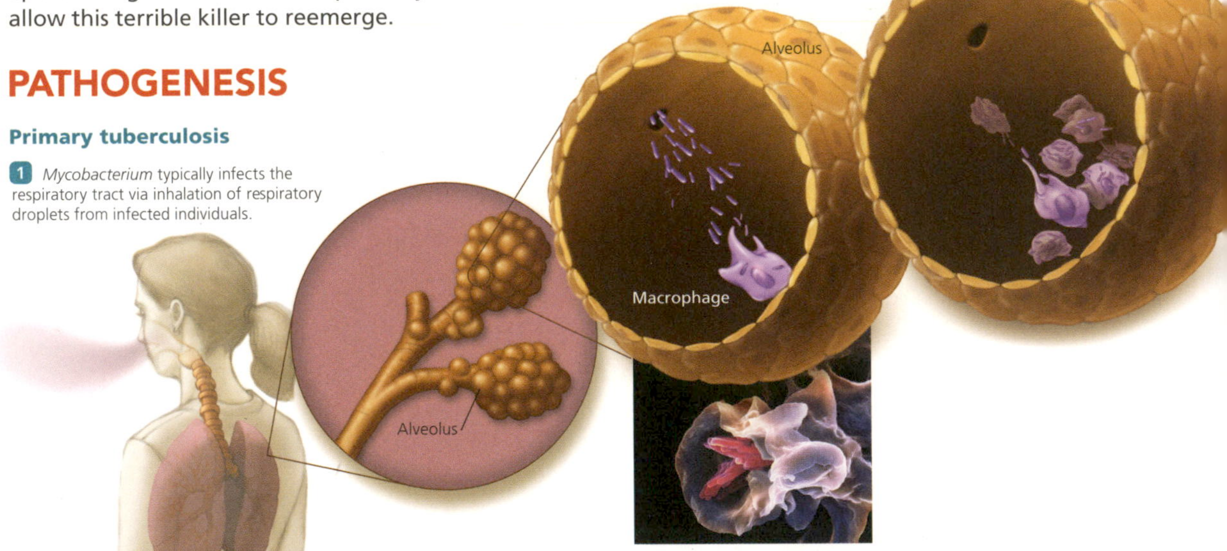

Macrophage engulfing *Mycobacterium.*

INVESTIGATE IT!

What does the development of XDR-TB (extensively drug-resistant strains of Mycobacterium tuberculosis*) portend for the future of the disease?*

Scan this code to visit the Centers for Disease Control and Prevention website to investigate XDR-TB. Then go to MasteringMicrobiology to record your research findings.

EPIDEMIOLOGY

Tuberculosis kills on average four people every minute, mostly in Asia and Africa. TB is on the decline in the U.S., though the CDC estimates that TB may still infect more than 9 million Americans. One third of the world's population is infected, and over 9 million new cases are seen each year.

Left, estimated new TB cases in 2010 per 100,000 (WHO)

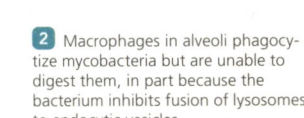

PATHOGEN AND VIRULENCE FACTORS

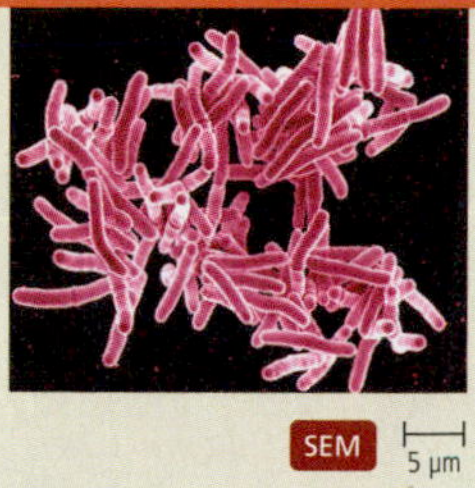

Mycobacterium tuberculosis is a high G + C, aerobic, Gram-positive rod. Virulent strains produce cord factor, a cell wall component that produces strands of daughter cells that remain attached to one another in parallel alignments. Cord factor also inhibits migration of neutrophils and is toxic to mammalian cells. Multi-drug-resistant (MDR-TB) and extensively drug-resistant (XDR-TB) strains of *Mycobacterium* make it more difficult to rid the world of TB.

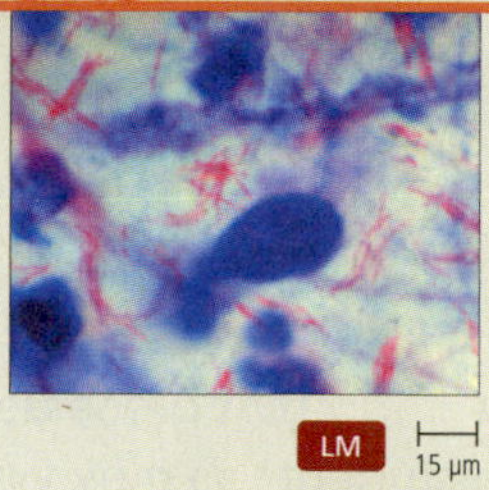

Cell walls contain mycolic acid, a waxy lipid that is responsible for unique characteristics of this pathogen, including slow growth, protection from lysis when cells are phagocytized, intracellular growth, and resistance to Gram staining, detergents, many common antimicrobial drugs, and drying out. (Slow growth is due in part to the time required to synthesize molecules of mycolic acid.)

4 Infected macrophages present antigen to T lymphocytes, which produce lymphokines that attract and activate more macrophages and trigger inflammation. Tightly packed macrophages surround the site of infection, forming a tubercle over a two- to three-month period.

5 Other cells deposit collagen fibers, enclosing infected macrophages and lung cells within the tubercle. Infected cells in the center die, releasing *M. tuberculosis* and producing caseous necrosis—the death of tissue that takes on a cheese-like consistency due to protein and fat released from dying cells. A stalemate between the bacterium and the body's defenses develops.

Secondary/reactivated tuberculosis results when *M. tuberculosis* breaks the stalemate, ruptures the tubercle, and reestablishes an active infection. Reactivation occurs in about 10% of patients; patients whose immune systems are weakened by disease, poor nutrition, drug or alcohol abuse, or by other factors.

Disseminated tuberculosis results when macrophages carry the pathogen via blood and lymph nodes to other sites, including bone marrow, spleen, kidneys, spinal cord, and brain.

Tuberculosis lesions in spleen.

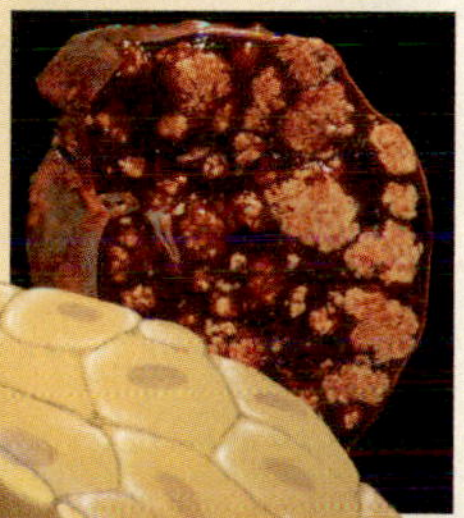

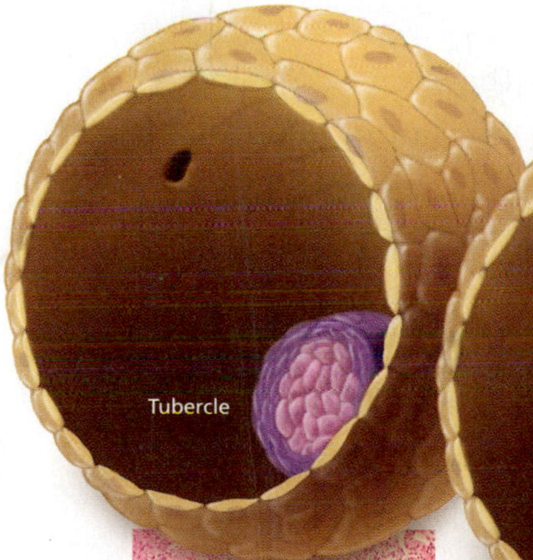

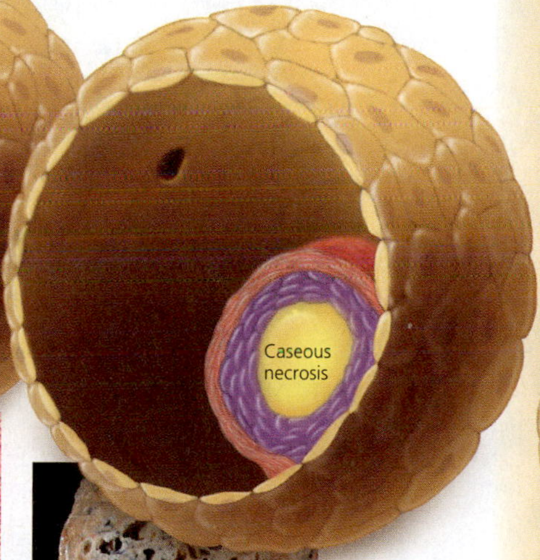

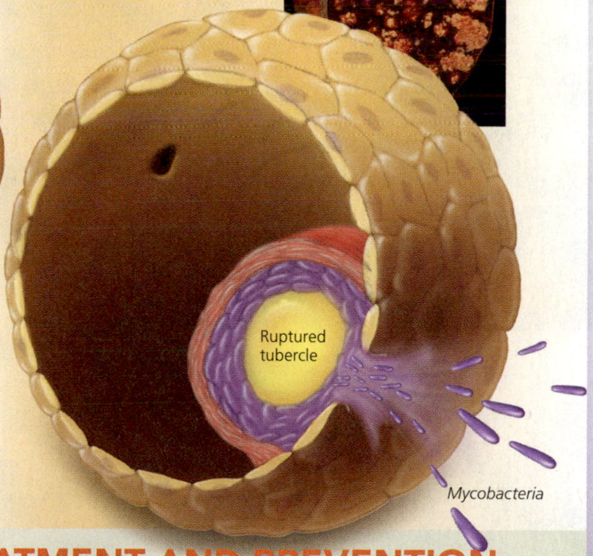

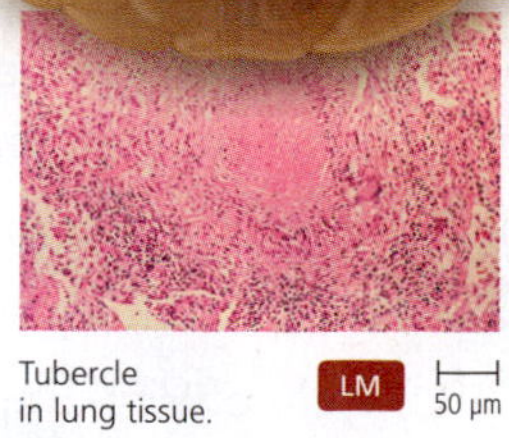

Tubercle in lung tissue.

Lung lesions caused by TB.

DIAGNOSIS

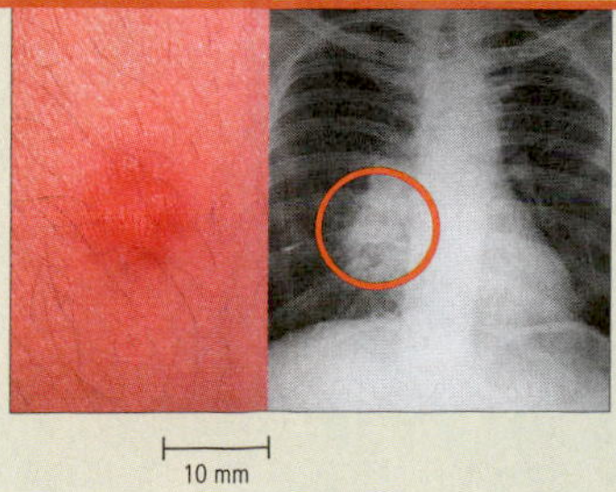

A tuberculin skin test is used to screen patients for TB exposure. A positive reaction is an enlarged, reddened, and raised lesion at the inoculation site. Chest X-ray films can reveal the presence of tubercles in the lungs. Primary TB usually occurs in the lower and central areas of the lung; secondary TB commonly appears higher.

TREATMENT AND PREVENTION

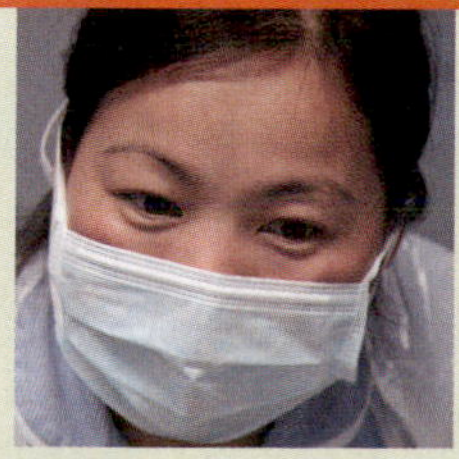

Treatment combines isoniazid, rifampin, and one of several drugs (such as ethambutol, levofloxacin, or streptomycin) for six months. Newly approved bedaquiline is used in combination with other drugs to treat MDR-TB or XDR-TB. In countries where TB is common, health care workers immunize patients with BCG vaccine, which is not recommended for the immunocompromised because it can cause disease. Workers must avoid inhaling respiratory droplets from TB patients.

Make the Invisible Visible

NEW! 18 VIDEO TUTORS

Developed for the Fourth Edition and accessible via QR codes in the text and the student Study Area in MasteringMicrobiology®, new **Video Tutors** by Dr. Robert W. Bauman help students explore important processes and tough topics. These tutorials engage students as they visualize and learn key concepts in microbiology, bringing the textbook art to life. These video tutorials also include assignable multiple-choice questions in MasteringMicrobiology.

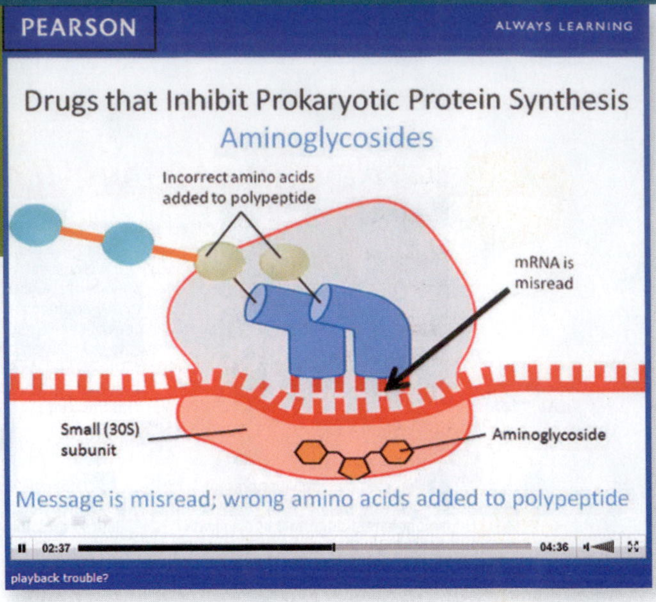

VIDEO TUTOR TOPICS

- The Scientific Method
- The Structure of Nucleotides
- Bacterial Cell Walls
- The Light Microscope
- Electron Transport Chains
- Bacterial Growth Media
- Initiation of Translation
- Action of Restriction Enzymes
- Principles of Autoclaving
- Actions of Some Drugs that Inhibit Prokaryotic Protein Synthesis
- Arrangements of Prokaryotic Cells
- Principles of Sexual Reproduction in Fungi
- The Lytic Cycle of Viral Replication
- Some Virulence Factors
- Inflammation
- Clonal Deletion
- ELISA
- Hemolytic Disease of the Newborn

AMERICAN SOCIETY FOR MICROBIOLOGY

NEW! **Tell Me Why Critical Thinking Questions** end all A-head sections. These questions strengthen the pedagogy and organization of each chapter and consistently provide stop-and-think opportunities for students as they read.

TELL ME WHY

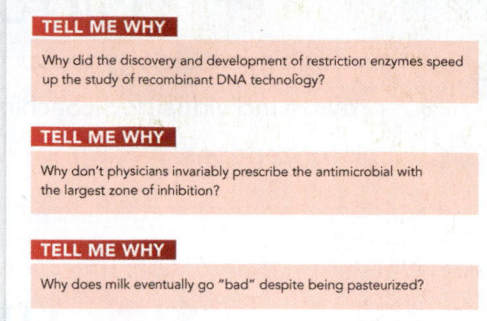

Why did the discovery and development of restriction enzymes speed up the study of recombinant DNA technology?

TELL ME WHY

Why don't physicians invariably prescribe the antimicrobial with the largest zone of inhibition?

TELL ME WHY

Why does milk eventually go "bad" despite being pasteurized?

NEW! **Numbered Learning Outcomes** in the textbook are used to tag Test Bank questions and all Mastering assets. In addition to being tagged to Learning Outcomes, Mastering assessments are tagged to the Global Science Learning Outcomes and Bloom's Taxonomy. The complete Mastering Test Bank is also tagged to ASMCUE recommended outcomes.

NEW! **Expanded Coverage of Helminthes** is provided in new highlight features, and an emphasis on virulence factors is showcased where appropriate in the Fourth Edition's Disease at a Glance and Disease in Depth features.

VISUALIZE IT!

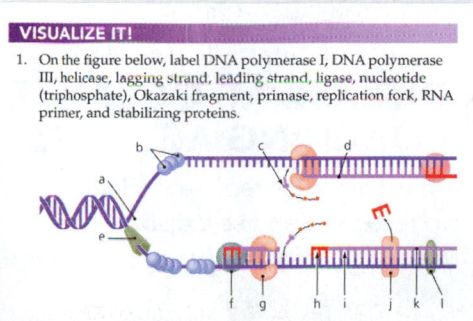

1. On the figure below, label DNA polymerase I, DNA polymerase III, helicase, lagging strand, leading strand, ligase, nucleotide (triphosphate), Okazaki fragment, primase, replication fork, RNA primer, and stabilizing proteins.

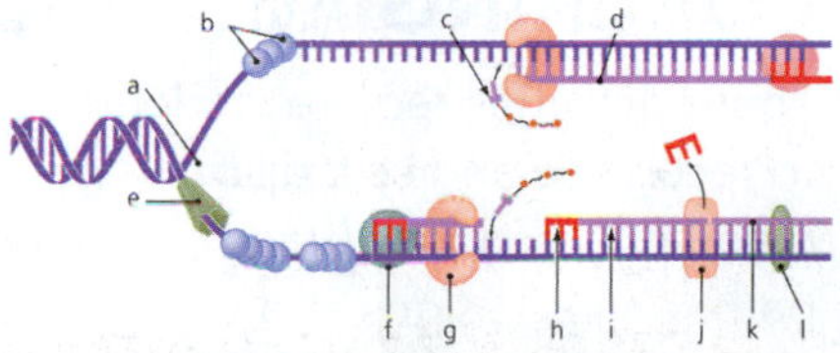

NEW! **VISUALIZE IT!**

Appearing at the end of each chapter, these short-answer or fill-in-the-blank questions are built around illustrations or photos. Visualize It! questions are also assignable as art labeling activities in MasteringMicrobiology.

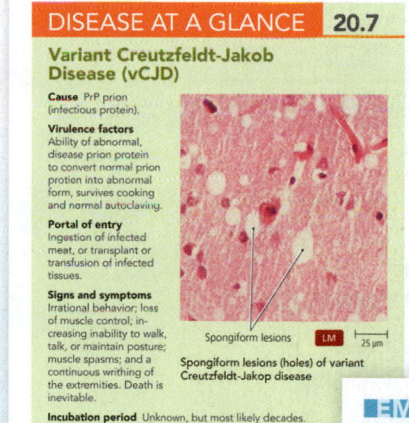

DISEASE AT A GLANCE 20.7

Variant Creutzfeldt-Jakob Disease (vCJD)

Cause PrP prion (infectious protein).

Virulence factors Ability of abnormal, disease prion protein to convert normal prion protien into abnormal form, survives cooking and normal autoclaving.

Portal of entry Ingestion of infected meat, or transplant or transfusion of infected tissues.

Signs and symptoms Irrational behavior; loss of muscle control; increasing inability to walk, talk, or maintain posture; muscle spasms; and a continuous writhing of the extremities. Death is inevitable.

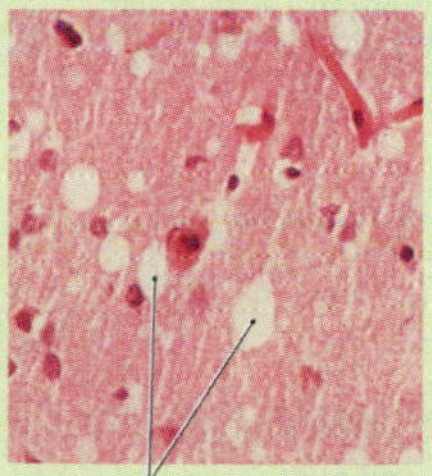

Spongiform lesions (holes) of variant Creutzfeldt-Jakop disease

Incubation period Unknown, but most likely decades.

Susceptibility At greatest risk are individuals who consumed [...] the United Kingdom between 1980 and 1996.

Treatment None.

Prevention Do not eat meat contaminated with infected nerv[e] tissue. Destruction of potentially infected herds is necessary to [...] the spread of the infectious prion.

Additional **Disease at a Glance** features provide more extensive disease coverage.

EMERGING DISEASE CASE STUDY

Microsporidiosis

Darius is sick, which is not surprising for an HIV-infected man. But he is sick in several new ways. Sick of having to stay within 20 feet of a toilet. Sick of the cramping, the gas, the pain, and the nausea. Sick with irregular but persistent, watery diarrhea. He is losing weight because food is passing through him undigested. Most days over the past seven months have been disgusting despite his use of over-the-counter remedies, which provide a few days of intermittent relief. His belief that these normal days signaled the end of the ordeal have kept him from the doctor. But now his eyes have begun to hurt, and his vision is blurry. Whatever it is, it's attacking him at both ends. Time to get stronger drugs from his physician.

Microscopic examination of Darius's stool sample reveals that he is being assaulted by *Encephalitozoon intestinalis*, a member of a group of opportunistic emerging pathogens called microsporidia. The single-celled pathogens are also seen on smears from Darius's nose and eyes. Microsporidia were long thought to be simple single-celled animals, but genetic analysis and comparison with other organisms reveal that they are closer to zygomycete yeasts.

Microsporidia appear to infect humans who engage in unprotected sexual activity, consume contaminated food or drink, or swim in contaminated water. People with active T cells rarely have symptoms; but people with suppressed immunity become easy targets for the fungus.

Microsporidia attack by uncoiling a flexible, hollow filament that stabs into a host cell and serves as a conduit for the microsporidium's cytoplasm to invade. In this way, the pathogens become intracellular parasites. They can destroy the intestinal lining, causing diarrhea, and spread to the eyes, muscles, or lungs.

Fortunately for Darius, an antimicrobial, albendazole, kills the parasite, and the effects of the infection are reversed. Unfortunately for Darius, the loss of helper T cells in AIDS means that another emerging, reemerging, or opportunistic infection is sure to follow.

1. Why are microsporidia considered to be opportunistic pathogens?
2. How could the discovery that microsporidia are fungi rather than animals improve treatment of microsporidiosis?
3. Microsporidia are intracellular pathogens. Which immune cells likely fight off the infection in people with a normal immune system?

NEW! **Critical Thinking Questions** in **Emerging Disease Case Studies** allow students to delve deeper into each case.

Fostering Engagement and Adaptive Learning

Dynamic, Interactive Learning

MasteringMicrobiology® guides students through microbiology topics with assignable, self-paced activities that provide individualized coaching and feedback specific to each student's misconceptions. **www.masteringmicrobiology.com**

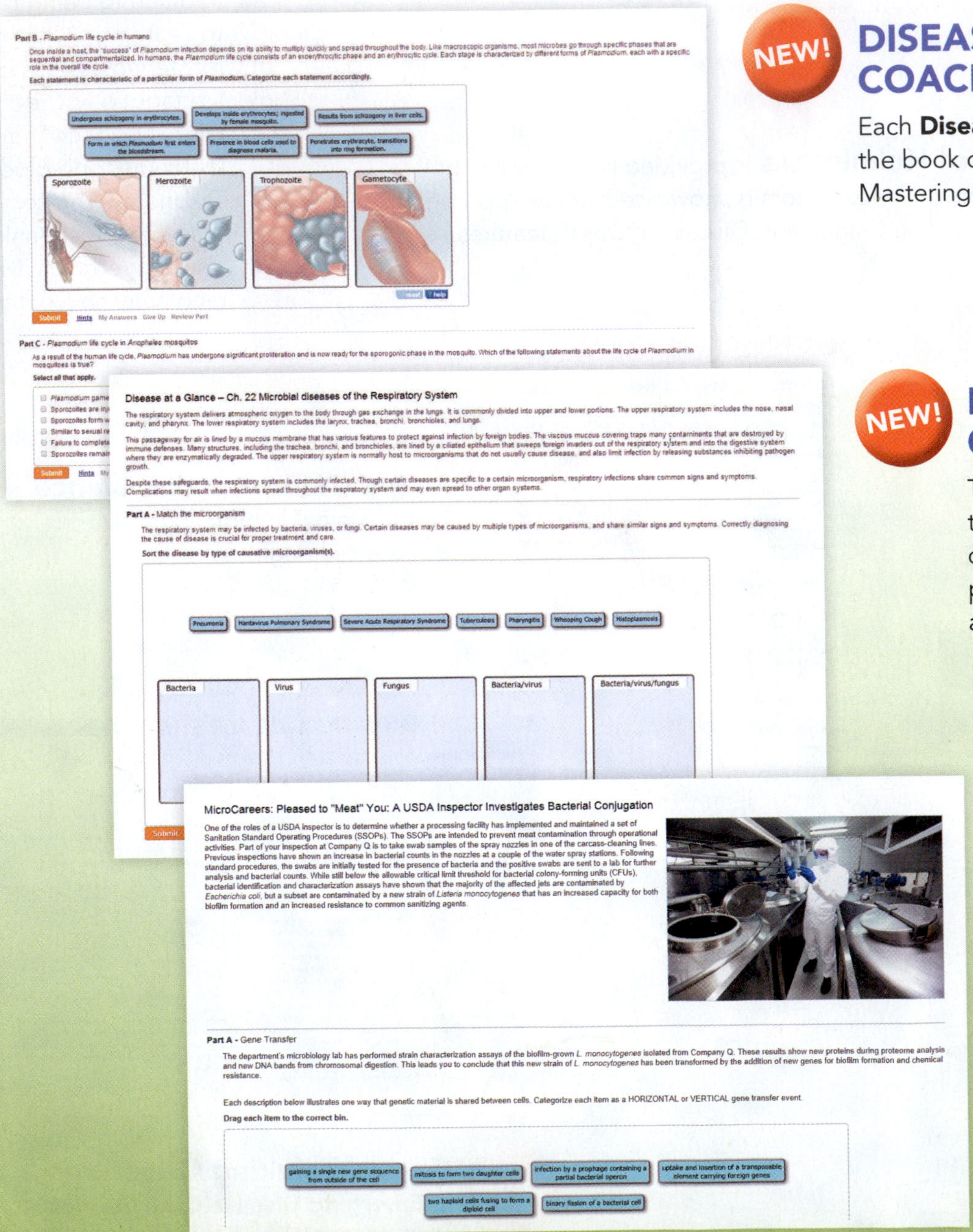

NEW! DISEASE IN DEPTH COACHING ACTIVITIES

Each **Disease in Depth** feature from the book corresponds to an assignable Mastering Coaching activity.

NEW! DISEASE AT A GLANCE COACHING ACTIVITIES

These activities require students to recognize and sort diseases by different categories (transmission type, pathogenesis, signs and symptoms, associated organisms, treatment, etc.).

NEW! MICROCAREERS COACHING ACTIVITIES

Students will learn to think like microbiologists with new MicroCareers coaching activities. These activities offer new opportunities to investigate emerging diseases from different career perspectives and think critically to solve microbiology-related questions.

NEW! CLINICAL CASE STUDY COACHING ACTIVITIES

These activities in MasteringMicrobiology help students connect microbiological theory to real-world disease diagnosis and treatment; they are assignable, and feed directly into the MasteringMicrobiology gradebook.

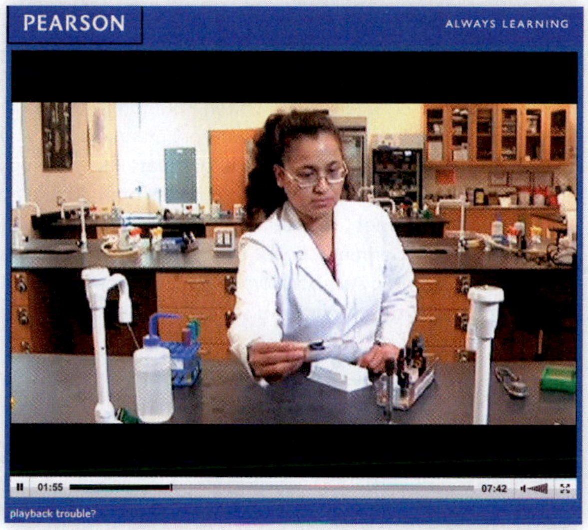

NEW! MICROLAB TUTORS

Helping students get the most out of lab time, each MicroLab Tutor begins with clinical background and a technique video. Select MicroLab Tutors include visually stunning molecular animations, encouraging students to visualize the processes at a molecular level. All 13 Tutors include photomicrographs and video or animation clip hints and feedback designed to assess understanding of lab concepts and techniques outside of formal lecture and lab time.

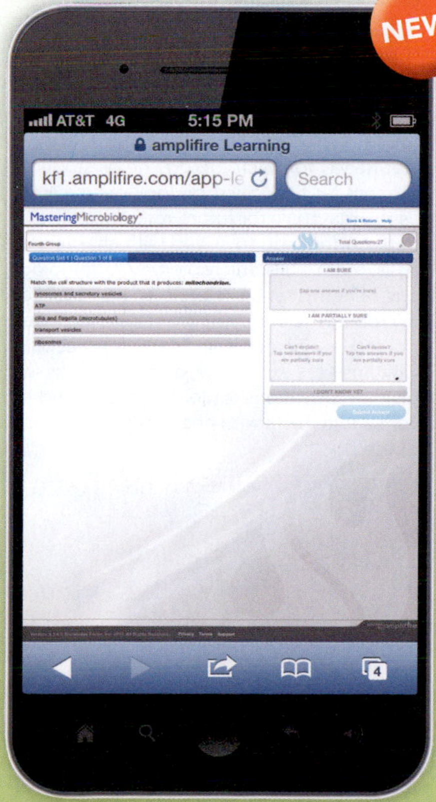

NEW! DYNAMIC STUDY MODULES

MasteringMicrobiology's Dynamic Study Modules, powered by Amplifire, boost knowledge acquisition and retention, fostering more effective study and class time and allowing students to come to class better prepared and ready for higher levels of learning.

NEW! LEARNING CATALYTICS

Now a part of the MasteringMicrobiology suite of powerful resources, this student engagement, assessment, and classroom intelligence system allows students to use their laptops, smartphones, or tablets to respond to questions in class. Learning Catalytics provides meaningful question types and facilitates classroom discussions and activities, supporting active learning in every classroom.

The **Best Support** for **Instructors and Students**

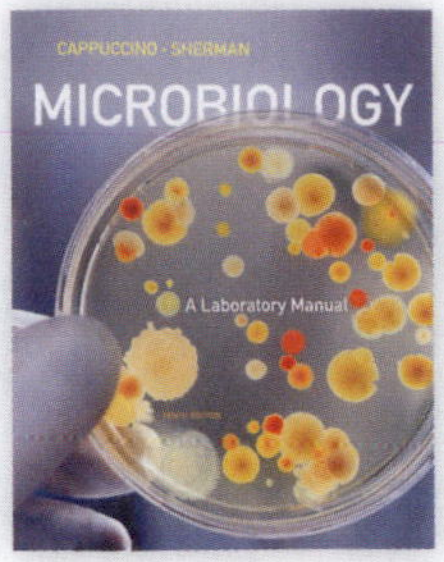

Microbiology: A Laboratory Manual
Tenth Edition
by James Cappuccino and Natalie Sherman
978-0-321-84022-6 ▪ 0-321-84022-4

Versatile, comprehensive, and clearly written, this competitively priced laboratory manual can be used with any undergraduate microbiology text—and now features brief clinical applications for each experiment, MasteringMicrobiology® quizzes that correspond to each experiment, and a new experiment on hand washing. *Microbiology: A Laboratory Manual* is known for its thorough coverage, descriptive and straightforward procedures, and minimal equipment requirements.

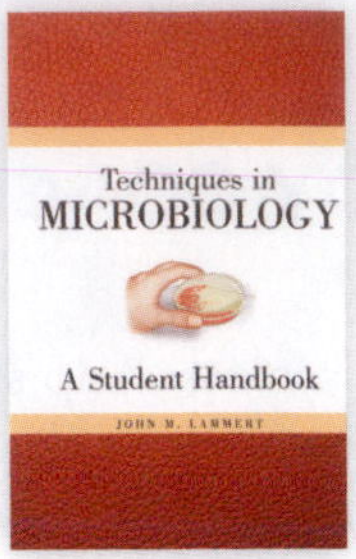

Techniques in Microbiology: A Student Handbook
by John M. Lammert | 978-0-132-24011-6 ▪ 0-132-24011-4

Lammert's approach is visual and incorporates "voice balloons" that keep the student focused on the process described. The techniques are those that will be used frequently for studying microbes in the laboratory, and include those identified by the American Society for Microbiology in its recommendations for the Microbiology Laboratory Core Curriculum.

ALSO AVAILABLE TO HELP YOUR STUDENTS FOR LAB:

Laboratory Experiments in Microbiology Tenth Edition
by Ted R. Johnson and Christine L. Case | 978-0-321-79438-3 ▪ 0-321-79438-9

ADDITIONAL SUPPLEMENTS

FOR INSTRUCTORS

Instructor's Resource DVD
978-0-321-94986-8 ▪ 0-321-94986-2

The Instructor's Resource DVD offers a wealth of instructor media resources, including presentation art, lecture outlines, test items, and answer keys—all in one convenient location. These resources help instructors prepare for class—and create dynamic lectures—in half the time! The IR-DVD includes:

- All figures from the book with and without labels in both JPEG and PowerPoint® formats
- All figures from the book with the Label Edit feature in PowerPoint format
- Select "process" figures from the book with the Step Edit feature in PowerPoint format
- All tables from the book
- Lab Technique Videos, MicroLab Tutors, BioFlix® and MicroFlix™ Animations, Microbiology Animations, and Microbiology Videos
- PowerPoint lecture outlines, including figures and tables from the book and links to the animations and videos
- Clicker Questions
- Quiz Show Questions
- PDF files of Transparency Acetate masters
- The Instructor's Manual as editable Microsoft® Word files
- The Instructor's Manual in PDF format
- The Test Bank as editable Microsoft Word files
- The Test Bank in TestGen® format
- The Instructor's Guide for Cappuccino/Sherman, *Microbiology: A Laboratory Manual*, Tenth Edition in PDF format
- The Preparation Guide for Johnson/Case, *Laboratory Experiments in Microbiology*, Tenth Edition in PDF format

Instructor's Manual / Test Bank
by Nichol Dolby
978-0-321-94984-4 ▪ 0-321-94984-6

This printed guide includes a chapter outline and a detailed chapter summary for each chapter as well as answers to in-text Clinical Case Studies, in-text Critical Thinking questions, and End-of-Chapter Review questions. Each test item in the printed Test Bank has been tagged with its corresponding section title from the textbook as well as book-specific Learning Outcomes and a Bloom's Taxonomy ranking (Knowledge, Comprehension, Application, or Analysis), allowing instructors to test students on a range of learning levels. The Test Bank has been updated with 25% new questions. This supplement is also available in Microsoft Word format on the Instructor's Resource DVD and on the Instructor Resource Center.

COURSE MANAGEMENT OPTIONS

MasteringMicrobiology®—Instant Access
www.masteringmicrobiology.com

Mastering helps instructors maximize class time with easy-to-assign, customizable, and automatically graded assessments that motivate students to learn outside class and arrive prepared for lecture or lab.

Blackboard—Instant Access
www.pearsonhighered.com/elearning

This open-access course management system includes the Pre-Tests, Practice Tests, Microbiology Animations, Microbiology Videos, Microbe Reviews, Flashcards, and the Glossary from the MasteringMicrobiology Study Area (www.masteringmicrobiology.com).

FOR STUDENTS

MasteringMicrobiology® with Pearson eText—Standalone Access Card
978-0-321-95682-8 ▪ 0-321-95682-6

MasteringMicrobiology®—Instant Access
www.masteringmicrobiology.com

See "For Instructors" for full description.

Get Ready for Microbiology Media Update
by Lori K. Garrett and Judy M. Penn
978-0-321-68347-2 ▪ 0-321-68347-1

Get Ready for Microbiology helps students quickly prepare for their microbiology course and provides useful materials for future reference. The workbook gets students up to speed with chapters on study skills, math skills, microbiology terminology, basic chemistry, basic biology, and basic cell microbiology. Each chapter includes a pre-test, guided explanations, interactive practice quizzes with answers explained, quizzes with answers given, motivations for learning, and end-of-chapter cumulative tests with answers given at the back of the book.

FOURTH EDITION

MICROBIOLOGY

WITH DISEASES BY BODY SYSTEM

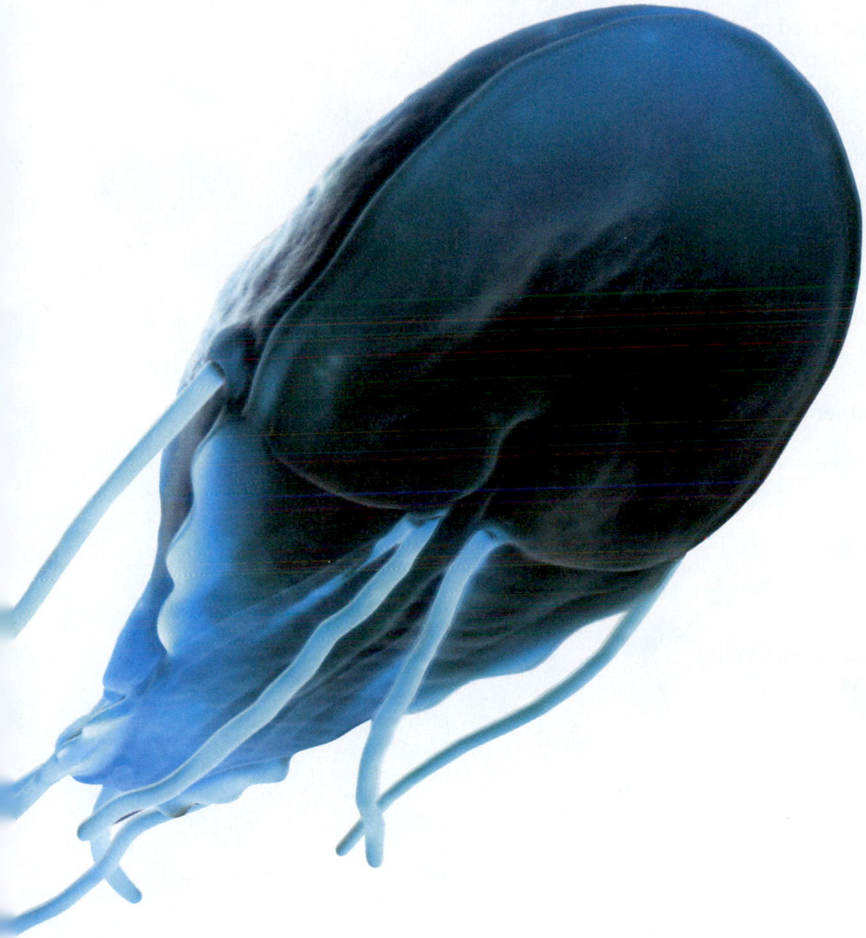

ROBERT W. BAUMAN, Ph.D.
Amarillo College

CLINICAL CONSULTANTS:

Cecily D. Cosby, Ph.D., FNP-C, PA-C
Samuel Merritt College

Janet Fulks, Ed.D.
Bakersfield College

John M. Lammert, Ph.D.
Gustavus Adolphus College

CONTRIBUTIONS BY:

Elizabeth Machunis-Masuoka, Ph.D.
University of Virginia

Jean E. Montgomery, MSN, RN
Austin Community College

PEARSON

Boston Columbus Indianapolis New York
San Francisco Upper Saddle River Amsterdam
Cape Town Dubai London Madrid Milan
Munich Paris Montréal Toronto Delhi
Mexico City São Paulo Sydney Hong Kong
Seoul Singapore Taipei Tokyo

Senior Acquisitions Editor: *Kelsey Churchman*
Associate Editor: *Nicole McFadden*
Director of Development: *Barbara Yien*
Assistant Editor: *Ashley Williams*
Art Development Editor: *Kelly Murphy*
Managing Editor: *Michael Early*
Assistant Managing Editor: *Nancy Tabor*
Project Manager: *Lauren Beebe*
Director, Media Development: *Lauren Fogel*
Assistant Media Producer: *Annie Wang/Natalie Pettry*
Copyeditor: *Sally Peyerfitte*
Design Manager: *Marilyn Perry*
Interior and Cover Designer: *Elise Lansdon*
Illustration: *Precision Graphics*
Associate Director of Image Management: *Travis Amos*
Photo Researcher: *Maureen Spuhler*
Photo Permissions: *PreMedia Global*
Text Permissions Project Manager: *Michael Farmer*
Senior Procurement Specialist: *Stacey Weinberger*
Senior Marketing Manager: *Neena Bali*

Cover Photo Credit: RGB Pictures/Alamy

Credits and acknowledgments for materials borrowed from other sources and reproduced, with permission, in this textbook appear on the appropriate page or on p. CR-1.

Library of Congress Cataloging-in-Publication Data
Bauman, Robert W., author.
Microbiology: with diseases by body system/Robert W. Bauman; clinical consultants, Cecily D. Cosby, Janet Fulks, John M. Lammert ; contributions by Elizabeth Machunis-Masuoka, Jean E. Montgomery. — Fourth edition.
p. ; cm.
ISBN-13: 978-0-321-91855-0
ISBN-10: 0-321-91855-X
I. Title. [DNLM: 1. Microbiological Phenomena. 2. Communicable Diseases—microbiology. 3. Microbiological Techniques—methods. QW 4]
QR41.2
579—dc23

2013044683

ISBN 10: 0-321-91855-X (Student edition)
ISBN 13: 978-0-321-91855-0 (Student edition)
ISBN 10: 0-321-94367-8 (Instructor's Review Copy)
ISBN 13: 978-0-321-94367-5 (Instructor's Review Copy)

PEARSON

www.pearsonhighered.com

1 2 3 4 5 6 7 8 9 10—CKV—16 15 14 13

To Michelle:
My best friend, my closest confidant, my cheerleader, my partner, my love. Thirty-one years! I love you more now than then.

—**Robert**

About the Author

ROBERT W. BAUMAN is a professor of biology and past chairman of the Department of Biological Sciences at Amarillo College in Amarillo, Texas. He teaches microbiology, human anatomy and physiology, and botany. In 2004, the students of Amarillo College selected Dr. Bauman as the recipient of the John F. Mead Faculty Excellence Award. He received an M.A. degree in botany from the University of Texas at Austin and a Ph.D. in biology from Stanford University. His research interests have included the morphology and ecology of freshwater algae, the cell biology of marine algae (particularly the deposition of cell walls and intercellular communication), and environmentally triggered chromogenesis in butterflies. He is a member of the American Society of Microbiology (ASM) where he has held national offices, Texas Community College Teacher's Association (TCCTA), American Association for the Advancement of Science (AAAS), Human Anatomy and Physiology Society (HAPS), and The Lepidopterists' Society. When he is not writing books, he enjoys spending time with his family: gardening, hiking, camping, rock climbing, backpacking, cycling, snowshoeing, skiing, and reading by a crackling fire in the winter and a gently swaying hammock in the summer.

About the Clinical Consultants

CECILY D. COSBY is nationally certified as both a family nurse practitioner and physician assistant. She is a professor of nursing, currently teaching at Samuel Merritt University in Oakland, California, and has been in clinical practice since 1980. She received her Ph.D. and M.S. from the University of California, San Francisco; her BSN from California State University, Long Beach; and her P.A. certificate from the Stanford Primary Care program. She is the Director of Samuel Merritt University's Doctor of Nursing Practice Program.

JANET FULKS is a professor of microbiology at Bakersfield College and a clinical laboratory scientist. She received her M.A. in Biology with an emphasis in microbiology from the University of the Pacific, and her Ed.D. in higher education leadership from Nova Southeastern University. Dr. Fulks and her husband spent six years in Nepal, working with doctors to diagnose diseases and train Nepalese hospital workers. She has also worked at the CDC and at a variety of clinical microbiology labs. Dr. Fulks has taught at Bakersfield College for over 20 years. Her primary research areas are student learning outcomes and assessment, educational data literacy, student success, and educational accountability.

JOHN M. LAMMERT is a professor of biology at Gustavus Adolphus College. He teaches courses in microbiology, immunology, and introductory biology. In 1998, he received the Edgar M. Carlson Award for Distinguished Teaching at Gustavus Adolphus College, and in 2012 he was included in *Princeton Review*'s Best 300 Professors. Dr. Lammert received an M.A. in biology from Valparaiso University and a Ph.D. in immunology from the University of Illinois–Medical Center, Chicago. He is the author of *Techniques in Microbiology: A Student Handbook* and three books on science fair projects (microbes, plants, and the human body).

Preface

The spread of whooping cough, snail fever, spotted fever rickettsiosis, and other emerging diseases; the cases of strep throat, MRSA, and tuberculosis; the progress of cutting-edge research into microbial genetics; the challenge of increasingly drug-resistant pathogens; the continual discovery of microorganisms previously unknown—these are just a few examples of why exploring microbiology has never been more exciting, or more important. Welcome!

I have taught microbiology to undergraduates for over 25 years and witnessed firsthand how students struggle with the same topics and concepts year after year. To address these challenging topics, I have developed and narrated Video Tutors for the first 18 chapters and added full-spread Disease in Depth features to the next six chapters. The Video Tutors and Disease in Depth features walk students through key concepts in microbiology, bringing the art of the textbook to life and important concepts into view. In creating this textbook, my goal was to help students see complex topics of microbiology—especially metabolism, genetics, and immunology—in a way that they can understand, while at the same time presenting a thorough and accurate overview of microbiology. I also wished to highlight the many positive effects of microorganisms on our lives, along with the medically important microorganisms that cause disease.

New to This Edition

In approaching the fourth edition, my goal was to build upon the strengths and success of the previous editions by updating it with the latest scientific and educational research and data available and by incorporating the many terrific suggestions I have received from colleagues and students alike. The feedback from instructors who adopted previous editions has been immensely gratifying and is much appreciated. The Disease at a Glance features have been widely praised by instructors and students, so I, along with art editor Kelly Murphy, developed six new Disease in Depth spreads that use compelling art and photos to provide a detailed overview of a specific disease. Each spread features an Investigate It! question with a QR code directing students to a website, encouraging further, independent research. Another goal for this edition was to provide additional instruction on important concepts and processes. To that end, I developed and narrated the Video Tutors, accessible via QR codes in the textbook and in MasteringMicrobiology®. The result is, once again, a collaborative effort of educators, students, editors, and top scientific illustrators: a textbook that, I hope, continues to improve upon conventional explanations and illustrations in substantive and effective ways.

In this new edition:

- **NEW Disease in Depth** spreads feature important and representative diseases for each body system, extending the visual impact of the art program as well as the highly praised Disease at a Glance features. Each of these six visual spreads contains info-graphics, provides in-depth coverage of the selected disease, and includes a QR code and Investigate It! question that directs students to a major health website, prompting further exploration and critical thinking. New MasteringMicrobiology assignable Disease in Depth coaching activities encourage students to apply and test their understanding of key concepts.

- **NEW Video Tutors** developed and narrated by the author walk students through key concepts in microbiology, bringing the textbook art to life and helping students visualize and understand tough topics and important processes. These 18 video tutorials are accessible via QR codes in the textbook and are accompanied by multiple-choice questions, assignable in MasteringMicrobiology®.
- **NEW Tell Me Why** critical thinking questions end every main section within each chapter. These questions strengthen the pedagogy and organization of each chapter and *consistently* provide stop-and-think opportunities for students as they read.
- **NEW Expanded coverage of helminths** is provided in new highlight features, and an **emphasis on virulence factors** is included in Disease at a Glance and Disease in Depth features.
- **NEW Numbered Learning Outcomes** in the textbook are used to tag Test Bank questions and all Mastering assets. In addition to being tagged to Learning Outcomes, all Mastering assessments are tagged to the Global Science Learning Outcomes and Bloom's Taxonomy. The complete Mastering Test Bank is also tagged to ASMCUE recommended outcomes.
- **NEW Visualize It!** features appear at the end of each chapter. These short-answer or fill-in-the-blank questions are built around illustrations or photos. These are also assignable as art labeling activities in MasteringMicrobiology.
- **The immunology chapters (Chapters 15–18),** which have been and continue to be reviewed in-depth by immunology specialists, reflect the most current understanding of this rapidly evolving field.
- **Over 50 NEW micrographs and photos** enhance student understanding of the text and boxed features.
- **NEW MasteringMicrobiology** includes NEW Disease in Depth and Disease at a Glance coaching activities, NEW Video Tutors with assessments, NEW MicroCareers and Clinical Case Study coaching activities, NEW Visualize It! art labeling activities, and Microbiology Lab Technique videos with assessment and MicroLab Tutor coaching activities. MicroLab Tutors use lab technique videos, 3D molecular animations, and stepped-out tutorials to actively engage students in making the connection between microbiology lecture, lab, and the real world. Disease at a Glance coaching activities ask students to categorize and sort diseases by different concepts, that is, by mode of transmission, signs and symptoms, etc. Additionally, MasteringMicrobiology and the Study Area include NEW MicroLab Practical quizzes, allowing more opportunities to analyze and interpret important lab tests, techniques, and results.

The following section provides a detailed outline of this edition's chapter-by-chapter revisions.

Chapter-by-Chapter Revisions

Every chapter in this edition has been thoroughly revised, and data in the text, tables, and figures have been updated. All Learning Outcomes have been numbered and are tagged to Test Bank questions and Mastering assets. Critical Thinking questions, formerly placed throughout each chapter, are now included in the end-of-chapter content.

The main changes for each chapter are summarized below.

THROUGHOUT THE DISEASE CHAPTERS (19–24)

- Updated disease diagnoses, treatments, and incidence and prevalence data
- Updated immunization recommendations and suggested treatments for all diseases
- Expanded coverage of virulence factors

CHAPTER 1 A BRIEF HISTORY OF MICROBIOLOGY

- Three new Tell Me Why questions
- Four photos replaced for improved pedagogy (Figures 1.5a and b, 1.7b, 1.17)
- One figure revised for improved pedagogy (Figure 1.13)
- Update to CDC-preferred term *healthcare associated infection (HAI)* (formerly *nosocomial infection*)
- New introductory coverage of normal microbiota and agar
- Clarified the use of a control in Pasteur's experiment to disprove spontaneous generation
- Clarified industrial use of microbes in making yogurt and in pest control
- Three new critical thinking questions in the Emerging Disease Case Study: Variant Creutzfeldt-Jakob Disease
- New Clinical Case Study: Can Spicy Food Cause Ulcers?
- New end-of-chapter Short Answer question on healthcare associated (nosocomial) infections
- New Visualize It! question on Pasteur's experiment to disprove spontaneous generation
- New Video Tutor: The Scientific Method

CHAPTER 2 THE CHEMISTRY OF MICROBIOLOGY

- Five new Tell Me Why questions
- Twelve figures revised for improved clarity and pedagogy (Figures 2.2, 2.3, 2.5, 2.7, 2.10–2.12, 2.15, 2.19, 2.20, 2.24, 2.26)
- New figure legend question (Figure 2.3)
- Expanded coverage of term *nucleoside* (nucleoside analogs treat a number of diseases)
- New Visualize It! question on the structure of amino acids
- New Video Tutor: The Structure of Nucleotides

CHAPTER 3 CELL STRUCTURE AND FUNCTION

- Twelve new Tell Me Why questions
- Four new/upgraded photos (Figures 3.7a and b, 3.8, 3.11)
- Five figures revised for improved clarity and pedagogy (Figures 3.9, 3.14, 3.15, 3.20, 3.24)
- Enhanced discussion of bacterial cytoskeletons and of bacterial and archaeal flagella
- Enhanced discussion of the roles of glycocalyces in biofilms
- New Visualize It! question on bacterial flagellar arrangements
- New Video Tutor: Bacterial Cell Walls

CHAPTER 4 MICROSCOPY, STAINING, AND CLASSIFICATION

- Four new Tell Me Why questions
- Four figures revised for improved clarity and pedagogy (Figures 4.2, 4.5, 4.6, 4.17)
- Three new critical thinking questions and one new photo in the Emerging Disease Case Study: Necrotizing Fasciitis
- New Visualize It! question on the light microscope
- New Video Tutor: The Light Microscope

CHAPTER 5 MICROBIAL METABOLISM

- Six new Tell Me Why questions
- Seven figures revised for improved clarity and pedagogy (Figures 5.3, 5.6, 5.10, 5.14, 5.16, 5.17, 5.26)
- Two new figure legend questions (Figures 5.4, 5.12)
- Expanded coverage of vitamins as enzymatic cofactors
- Updated text and figure legends that more clearly explain energy transfer in glycolysis, the Krebs cycle, and electron transport
- Updated text clarifying that glycolysis, the pentose phosphate pathway, and the Krebs cycle supply numerous precursor metabolites for anabolism
- Expanded discussion of bacterial quorum sensing and biofilms
- New end-of-chapter Fill in the Blanks question on anaerobic respiration
- New Visualize It! question on locating glycolysis, the Krebs cycle, and electron transport in eukaryotes
- New Video Tutor: Electron Transport Chains

CHAPTER 6 MICROBIAL NUTRITION AND GROWTH

- Three new Tell Me Why questions
- Two figures revised for improved clarity and pedagogy (Figures 6.1, 6.20)
- Significantly expanded coverage of biofilms and quorum sensing, including a new figure (Figure 6.7)
- Updated Beneficial Microbes: A Nuclear Waste–Eating Microbe?
- New Clinical Case Study about dental caries
- New Clinical Case Study about MRSA infection in a high school
- New Visualize It! question on identifying beta hemolysis
- New Video Tutor: Bacterial Growth Media

CHAPTER 7 MICROBIAL GENETICS

- Four new Tell Me Why questions
- Eleven figures upgraded for greater clarity, accuracy, ease of reading, and better pedagogy (Figures 7.1, 7.4, 7.5, 7.6, 7.9, 7.10, 7.21, 7.24, 7.30, 7.34, 7.37)
- Expanded coverage of the difference between nucleoside and nucleotide (many antimicrobial drugs are analogs of the former, not the latter)
- Clarified section on operons, introduction of the term *polycistronic*, new discussion of quorum-sensing as a trigger for inducible and repressible operons

- Section on regulatory RNA molecules updated for clarity and for inclusion of newly discovered information
- Three new critical thinking questions in Emerging Disease Case Study: *Vibrio vulnificus* Infection
- New Visualize It! question on DNA structure
- New Video Tutor: Initiation of Translation

CHAPTER 8 RECOMBINANT DNA TECHNOLOGY

- Five new Tell Me Why questions
- One new photo (chapter opener)
- Two figures revised for improved pedagogy (Figures 8.2, 8.9)
- New section discussing use of recombinant DNA techniques to address environmental problems, such as the reemergence of dengue fever
- Expanded coverage of the debate concerning genetic modification of agricultural products
- New Highlight: How Do You "Fix" a Mosquito?
- New Highlight: Vaccines on the Menu
- New Visualize It! question on DNA "fingerprinting"
- New Video Tutor: Action of Restriction Enzymes

CHAPTER 9 CONTROLLING MICROBIAL GROWTH IN THE ENVIRONMENT

- Four new Tell Me Why questions
- New photo (Figure 9.9)
- Three figures revised for improved clarity and pedagogy (Figures 9.1, 9.4, 9.13)
- Reorganization of the topics "Methods for Evaluating Disinfectants and Antiseptics" and "Biosafety Levels" for better flow and pedagogy
- New Highlight: Microbes in Sushi?
- Three new critical thinking questions in Emerging Disease Case Study: *Acanthamoeba* Keratitis
- New end-of-chapter critical thinking question on salmonellosis pandemic from smoked salmon
- New Visualize It! question on metal ions as a traditional water disinfectant in India
- New Video Tutor: Principles of Autoclaving

CHAPTER 10 CONTROLLING MICROBIAL GROWTH IN THE BODY: ANTIMICROBIAL DRUGS

- Four new Tell Me Why questions
- One new photo (Figure 10.10)
- Eight figures revised for currency, improved clarity, and pedagogy (Figures 10.2, 10.3, 10.4, 10.6, 10.8, 10.10, 10.15; Emerging Disease Case Study: Community-Associated MRSA map)
- Expanded coverage of the terms *therapeutic index* and *therapeutic window* as applied to antimicrobials
- New coverage on transfer of resistance genes between and among bacteria and on research to discover novel antimicrobials; updated discussion of the efficacy of probiotics
- Updated tables of antimicrobials to include all new antimicrobials mentioned in disease chapters, including antibacterial carbapenems; new antiprotozoan drugs (lumefantrine, nitazoxanide, paromoycin, piperaquine, and tinidazole); the newly approved anti-HIV-1 drug enfuvirtide; the antifungal drug ciclopirox; and antiviral protease inhibitors (boceprevir, darunavir, and telaprevir)
- New end-of-chapter critical thinking question on development of antimicrobial resistance
- Three new critical thinking questions in Emerging Disease Case Study: Community-Associated MRSA
- Nine new Learning Outcomes

- New Visualize It! question on Etest interpretation
- New Video Tutor: Action of Some Drugs that Inhibit Prokaryotic Protein Synthesis

CHAPTER 11 CHARACTERIZING AND CLASSIFYING PROKARYOTES

- Four new Tell Me Why questions
- Fourteen new photos (Figures 11.1, 11.2, 11.7, 11.17, 11.22, 11.23b, 11.24, 11.25b)
- Eight revised figures for improved clarity and pedagogy (Figures 11.1, 11.2, 11.4, 11.5, 11.6, 11.10, 11.21, 11.25)
- Clarified and expanded coverage of "snapping division," which is a distinctive characteristic of corynebacteria, including *C. diphtheriae*
- Updated taxonomy to correspond more completely with current *Bergey's Manual*
- New Beneficial Microbes: Botulism and Botox
- Enhanced discussion of nitrogen fixation, nitrification, and action of *Agrobacterium*
- New Highlight: Your Teeth Might Make You Fat
- Three new critical thinking questions in Emerging Disease Case Study: Pertussis
- Six new Learning Outcomes
- New Visualize It! on endospore identification
- New Video Tutor: Arrangements of Prokaryotic Cells

CHAPTER 12 CHARACTERIZING AND CLASSIFYING EUKARYOTES

- Six new Tell Me Why questions
- Eight new photos (Figures 12.11, 12.13, 12.15a-b, 12.23b, 12.29, 12.30, 12.33e)
- Five revised figures for improved clarity and pedagogy (Figures 12.1, 12.8, 12.11, 12.22, 12.33e)
- Updated algal, fungal, protozoan, water mold, and slime mold taxonomy
- Simplification of the vocabulary in the coverage of the morphology and reproductive strategies of fungi
- New Visualize It! question concerning fungal life cycles
- New Video Tutor: Principles of Sexual Reproduction in Fungi

CHAPTER 13 CHARACTERIZING AND CLASSIFYING VIRUSES, VIROIDS, AND PRIONS

- Four new Tell Me Why questions
- Five new photos (Figures 13.1b, 13.5c, 13.21, 13.23; Beneficial Microbes: Prescription Bacteriophages? photo)
- Four figures revised for improved pedagogy and currency (Figures 13.8, 13.11, 13.13, 13.22)
- Updated viral nomenclature to correspond to changes approved by the International Committee on Taxonomy of Viruses (ICTV)
- New coverage of discovery of *Megavirus*—the largest virus
- Three new critical thinking questions in updated Emerging Disease Case Study: Chikungunya
- New Visualize It! question on recognizing viral shapes in transmission electron micrographs
- New Video Tutor: The Lytic Cycle of Viral Replication

CHAPTER 14 INFECTION, INFECTIOUS DISEASES, AND EPIDEMIOLOGY

- Eight new Tell Me Why questions
- Three new photos (Figures 14.10, 14.6, 14.13)
- Seven figures updated for currency, improved clarity, and pedagogy (Figures 14.8, 14.9, 14.10, 14.14, 14.15, 14.19, 14.20)
- Updated epidemiology charts, tables, and graphs

- Updated list of nationally notifiable infectious diseases
- New discussion of hemolytic uremic syndrome (caused by *E. coli*), provided as an example of an epidemic with reference to an emerging disease (replaces prior discussion of *Hantavirus* pulmonary syndrome)
- New discussion of human West Nile virus infection added to explain the ways epidemiologists report their findings (replaces prior discussion of shigellosis)
- New figure legend questions (Figures 14.15, 14.18)
- Three new critical thinking questions in Emerging Disease Case Study: *Hantavirus* Pulmonary Syndrome
- New Visualize It! question on recognizing viral shapes in transmission electron micrographs
- New Video Tutor: Some Virulence Factors

CHAPTER 15 INNATE IMMUNITY

- Two new Tell Me Why questions
- Six figures revised for improved clarity and pedagogy, including a new rendition to reflect more accurately the sequence of complement cascade and action of complement subunits (Figures 15.6, 15.9, 15.11–14)
- Expanded coverage of the action of antimicrobial peptides (defensins)
- Expanded coverage of NOD receptor proteins and their role in protecting against hepatitis C, AIDS, and mononucleosis
- New Visualize It! question on identification of white blood cells
- New Video Tutor: Inflammation

CHAPTER 16 ADAPTIVE IMMUNITY

- Three new Tell Me Why questions
- Two new photos (Figures 16.1, 16.6)
- Twelve figures revised for improved clarity, pedagogy, and currency (Figures 16.2–16.5, 16.8–16.13, 16.18; Emerging Disease Case Study: Microsporidiosis map)
- Text reorganized to present discussion of T cells, major histocompatibility, antigen processing and presentation, and T cell clonal deletion before the discussion of B cells and B cell clonal deletion
- Three new critical thinking questions in Emerging Disease Case Study: Microsporidiosis
- Revised Learning Outcomes
- New Visualize It! question on major histocompatibility complex proteins
- New Video Tutor: Clonal Deletion

CHAPTER 17 IMMUNIZATION AND IMMUNE TESTING

- Two new Tell Me Why questions
- New photo (Figure 17.10)
- Five figures revised for improved clarity and pedagogy (Figures 17.1–17.3, 17.8, 17.14)
- New CDC 2013 vaccination schedule for children, adolescents, and adults
- Updated table of vaccine-preventable diseases in the United States
- New coverage of quantifying immunoassays—turbidimetry and nephelometry
- New Visualize It! question on interpreting an immunoblot
- New Video Tutor: ELISA

CHAPTER 18 AIDS AND OTHER IMMUNE DISORDERS

- Three new Tell Me Why questions
- New photo (Figure 18.11)
- Two new figures (Figures 18.16, 18.17)
- Three revised figures for improved clarity and pedagogy (Figures 18.8, 18.20, 18.21)
- Updated discussion of AIDS prevalence, transmission, prevention, and treatment
- Updated discussion of HIV attachment, entry, and replication
- New Visualize It! question on recognizing type I, III, and IV hypersensitivities
- New Video Tutor: Hemolytic Disease of the Newborn

CHAPTER 19 MICROBIAL DISEASES OF THE SKIN AND WOUNDS

- Five new Tell Me Why questions
- Ten new photos (Figures 19.7, 19.13, 19.15, 19.17; Disease in Depth and Disease at a Glance figures for *Pseudomonas*, Rocky Mountain spotted fever [RMSF], smallpox, herpes, shingles)
- Three figures revised for improved accuracy, pedagogy, and currency (Figure 19.1; Emerging Disease Case Study: Buruli Ulcer map; Emerging Disease Case Study: Monkeypox map)
- Coverage of spotted fever rickettsioses revised to clarify that Rocky Mountain spotted fever (RMSF) is only one type and to explain that one reason rickettsias are obligate intracellular parasites is their requirement for amino acids and Krebs cycle intermediates
- Updated coverage of chickenpox and shingle vaccine
- Updated treatment regimens for staphylococcal scalded skin syndrome, impetigo, erysipelas, cat scratch disease, cutaneous anthrax, gas gangrene, herpes skin infections, chickenpox, shingles, measles, erythema infectiosum, hand-foot-and-mouth disease, pityriasis versicolor, cutaneous mycoses, chromoblastomycosis, sporotrichosis, and leishmaniasis
- Expanded coverage of methicillin-resistant and vancomycin-resistant *Staphylococcus aureus* (MRSA, VRSA)
- Expanded and updated coverage of action of anthrax toxins
- Three new critical thinking questions in Emerging Disease Case Study: Buruli Ulcer
- Three new critical thinking questions in Emerging Disease Case Study: Monkeypox
- One new end-of-chapter multiple choice question
- Seven new Learning Outcomes
- New Visualize It! question on identification of skin infections
- New Disease at a Glance: *Pseudomonas* Infection
- New Disease in Depth: Necrotizing Fasciitis

CHAPTER 20 MICROBIAL DISEASES OF THE NERVOUS SYSTEM AND EYES

- Six new Tell Me Why questions
- Sixteen new photos (Figures 20.3, 20. 4, 20.14, Highlight: Nipah virus; Clinical Case Studies: Ptosis burnt fingers and N. meningitidis; Disease at a Glance: West Nile Encephalitis; Disease in Depth feature)
- Eight figures revised for currency and improved pedagogy (Figures 20.1, 20.2, 20.10, 20.14, 20.15, 20.16; Emerging Disease Case Study: Melioidosis map, Emerging Disease Case Study: Tick-Borne Encephalitis map)
- Expanded coverage of virulence factors and pathogenesis of diseases, particularly botulism, West Nile virus encephalitis, African sleeping sickness
- Updated treatment regimens for bacterial meningitis, leprosy, foodborne botulism, cryptococcal meningitis, primary amebic meningoencephalopathy, variant Creutzfeldt-Jakob disease, and chlamydial eye infections.
- Three new critical thinking questions in Emerging Disease Case Study: Melioidosis
- Three new critical thinking questions in Emerging Disease Case Study: Tick-Borne Encephalitis
- New Highlight: Nipah Virus: From Pigs to Humans
- New Visualize It! question on lumbar puncture
- New Disease at a Glance: Polio
- New Disease in Depth: Listeriosis

CHAPTER 21 CARDIOVASCULAR AND SYSTEMIC DISEASES

- Four new Tell Me Why questions
- Eighteen new photos (Figures 21.5, 21.13; Beneficial Microbes: Wolbachia; Clinical Case Study: A Tired Freshman, and Man and Cat; Highlight: Malaria; Emerging Disease Case Study: Schistosomiasis; Disease at a Glance: Toxoplasmosis; Disease in Depth feature)
- Thirteen figures revised for currency and improved pedagogy (Figures 21.1, 21.6, 21.9, 21.10, 21.12, 21.16, 21.17, 21.20, 21.21, 21.22; Disease at a Glance: Yellow Fever; Emerging Disease Case Study: Schistosomiasis map; Emerging Disease Case Study: Snail Fever in China map)
- New Clinical Case Study: Nightmare on the Island
- Three new critical thinking questions in Emerging Disease Case Study: Snail Fever in China
- Updated treatment regimens for tularemia, Lyme disease, ehrlichiosis, anaplasmosis, cytomegalovirus disease, malaria, toxoplasmosis, and schistosomiasis
- Two new Learning Outcomes
- New Visualize It! question on Lyme disease
- New Disease at a Glance: Toxoplasmosis
- New Disease in Depth: Malaria

CHAPTER 22 MICROBIAL DISEASES OF THE RESPIRATORY SYSTEM

- Three new Tell Me Why questions
- Twenty-one new photos (chapter opener photo; Figures 22.2, 22.3, 22.4, 22.9, 22.13, 22.17; Disease at a Glance features: Bacterial Pneumonias, Coronavirus Respiratory Syndromes, Respiratory Syncytial Virus Infection, and Histoplasmosis; Clinical Case Study: The Coughing Cousin; Disease in Depth feature)
- Five figures revised for currency and improved pedagogy (Figures 22.1, 22.10, 22.11; Emerging Disease Case Study: Pulmonary Blastomycosis map; Emerging Disease Case Study: H1N1 Influenza map)
- New table comparing and contrasting manifestations of some common respiratory diseases (Table 22.1)
- New discussion of Middle East respiratory syndrome (MERS)
- Expanded discussion of diphtheria, tetanus, pertussis vaccine schedule, and the vaccines' nomenclature
- Introduced new preferred term *rhinosinusitis* to replace *sinusitis*
- Updated treatment regimens for bacterial pneumonia, pneumonic plague, ornithosis, Legionnaires' disease, drug-susceptible tuberculosis (TB), multi-drug-resistant TB (MDR-TB), whooping cough, inhalational anthrax, blastomycosis, and histoplasmosis
- Expanded coverage of multi-drug-resistant tuberculosis (MDR-TB) and extensively drug-resistant TB (XDR-TB)
- Three new critical thinking questions in Emerging Disease Case Study: H1N1 Influenza
- Three new critical thinking questions in Emerging Disease Case Study: Pulmonary Blastomycosis
- New Visualize It! question on bacteria
- New Disease at a Glance: Respiratory Syncytial Virus Infection
- New Disease in Depth: Tuberculosis

CHAPTER 23 MICROBIAL DISEASES OF THE DIGESTIVE SYSTEM

- Four new Tell Me Why questions
- Fifteen new photos (Figures 23.6. 23.11, 23.17b; Disease at a Glance features: Dental Caries, Cholera, and Amebiasis; Disease in Depth feature)
- Five figures revised for currency and improved pedagogy (Figures 23.5, 23.6, 23.14, 23.15, 23.18)
- Updated treatment regimens for peptic ulcers, cholera, shigellosis, traveler's diarrhea, *C. diff* diarrhea/colitis, typhoid fever, oral herpes, hepatitis C, and cryptosporidiosis
- Expanded coverage of Shiga-like toxins, probiotics, oral herpes, hepatitis viruses C and E, the newly approved xTAG Gastrointestinal Pathogen Panel (xTAG GPP) as a way to diagnose causes of gastroenteritis, *Clostridium difficile* diarrhea, and pseudomembranous colitis
- New coverage of the connection between esophageal cancer and the use of antibiotics to treat *Helicobacter* infection
- New coverage of anisakiasis
- New coverage of the reintroduction of the cholera pandemic into North America (Haiti, 2010; Dominican Republic, 2011; Cuba, 2013)
- Three new critical thinking questions in Emerging Disease Case Study: *Norovirus* Gastroenteritis
- One new Learning Outcome
- New Visualize It! question on hepatitis B virus, Dane particles, filamentous particles, and spherical particles
- New Disease at a Glance: Dental Caries
- New Disease in Depth: Giardiasis

CHAPTER 24 MICROBIAL DISEASES OF THE URINARY AND REPRODUCTIVE SYSTEMS

- Seven new Tell Me Why questions
- Twelve new photos (Figures 24.4, 24.12, Beneficial Microbes: Pharmacists of the Future?; Disease at a Glance: Gonorrhea and Genital Warts; Disease in Depth)
- Eight new figures (Figures 23.4, 24.6a, 24.6c, 24.7b, 24.8, 24.13; Disease at a Glance features: Candidiasis, Gonorrhea)
- Five figures revised for currency and improved pedagogy (Figures: 24.3, 24.5, 24.7a, 24.9, 24.11)
- Updated treatment regimens for urinary tract infections, leptospirosis, staphylococcal toxic shock syndrome, lymphogranuloma venereum, gonorrhea, neonatal chlamydial conjunctivitis, and trichomoniasis
- Two new Learning Outcomes
- New Visualize It! question on pathogens of the urinary and reproductive systems
- New Disease at a Glance: Trichomoniasis
- New Disease in Depth: Bacterial Urinary Tract Infections

CHAPTER 25 APPLIED AND ENVIRONMENTAL MICROBIOLOGY

- Four new Tell Me Why questions
- Five new photos (Figures 25.3, 25.6, 25.7, 25.14; Emerging Disease Case Study: Attack in the Lake)
- New figure legend question concerning food sterilization
- Clarification of the terms *unripened* and *ripened* in regard to cheeses and expanded coverage of the processes of cheese-making
- New coverage of biomining—the use of microbes to extract insoluble forms of metals from ore
- New coverage on the presence of significant nitrogen fixation by deep-sea archaea associated in microbial communities with bacteria
- New Emerging Disease Case Study: Attack in the Lake
- New Beneficial Microbes: Oil-Eating Microbes to the Rescue in the Gulf
- New Visualize It! question on nitrogen cycling

Reviewers for the Fourth Edition

I wish to thank the hundreds of instructors and students who participated in reviews, class tests, and focus groups for earlier editions of the textbook. Your comments have informed this book from beginning to end, and I am deeply grateful. For the fourth edition, I extend my deepest appreciation to the following reviewers.

Book Reviewers

Warner B. Bair III
Lone Star College—CyFair

Carrie Burdinski
Delta College

Bradley W. Christian
McLennan Community College

Pamela J. Coker
Pima Community College

Francisco Cruz
Georgia State University

Michael J. Dul
Lakeland Community College and *Fortis College*

Clifton Franklund
Ferris State University

Nicholas Hackett
Moraine Valley Community College

Robert Iwan
Inver Hills Community College

Timothy Johnson
University of Minnesota

James Masuoka
Midwestern State University

Laura Mery
San Antonio College

Jennifer A. Metzler
Ball State University

Ron C. Michaelis
Rutgers University

Karen Persky
College of DuPage

Michael Pressler
Delta College

Nancy Risner
Ivy Tech Community College—Muncie

Ben Rowley
University of Central Arkansas

Debra Scheiwe
Tarrant County College—Northeast

Audra Swarthout
Delta College

Patricia G. Wilber
Central New Mexico Community College

Elizabeth Yelverton
Pensacola State College

Kathy A. Zarilla
Durham Technical Community College

Video Tutor Reviewers

Cheryl Boice
Florida Gateway College

Carroll Weaver Bottoms
Collin College

Teresa G. Fischer
Indian River State College

Leoned Gines
Shoreline Community College

Nicholas Hackett
Moraine Valley Community College

Jennifer Hatchel
College of Coastal Georgia

James B. Herrick
James Madison University

Robert Iwan
Inver Hills Community College

Mary Evelyn B. Kelley
Wayne State University

Denice D. King
Cleveland State Community College

Kevin Mitchell
Northern Essex Community College

Stacy Pfluger
Angelina College

Nancy Risner
Ivy Tech Community College—Muncie

Jennifer Swartz
Pikes Peak Community College

Acknowledgments

As was the case with all the previous editions, this book has truly been a team effort. I am deeply grateful to Kelsey Churchman of Pearson Science and to the team she gathered to produce the fourth edition. Kelsey, dedicated project editor Nicole McFadden, Barbara Yien, project editor of the first two editions, and Robin Pille, project editor of the third edition, helped develop the vision for this fourth edition, coming up with ideas for making it more effective and compelling. As project editor, Nicole also had the unenviable task of coordinating everything and keeping me on track—thank you, Nicole, for being understanding, patient, and lenient with the "dead" in deadline. Thank you, Barbara, for years of support and for introducing me to chocolate truffles. I am excited about your growing adventure! I am grateful to Frank Ruggirello for his unflagging encouragement and support of my work and this book. I am also indebted to Daryl Fox, whose early support for this book never wavered.

Sally Peyrefitte—the eagle-eyed—edited the manuscript thoroughly and meticulously, suggesting important changes for clarity, accuracy, and consistency. Kelly Murphy did an incredibly superb job as art development editor, helping to conceptualize new illustrations and suggesting ways to improve the art overall—thank you, Kelly. My friend Ken Probst is responsible for originally creating this book's amazingly beautiful biological illustrations. My thanks to Precision Graphics for rendering the art in this edition. Nancy Tabor and Lori Bradshaw expertly guided the project through production. Maureen "Mo" Spuhler continued her absolutely incredible job researching photos. I am in your debt, Mo. Rich Robison and Brent Selinger supplied many of the text's wonderful and unique micrographs. Tamara Newman created the beautiful interior design and the stunning cover.

Thanks to Nichol Dolby and Sam Schwarzlose of Amarillo College; Suzanne Long of Monroe Community College; Mindy Miller-Kittrell of University of Tennessee, Knoxville; Jason Andrus of Meredith College; Tiffany Glaven of University of California, Davis; Kathryn Sutton of Clarke College; and Judy Meier Penn of Shoreline Community College for their work on the media and print supplements for this edition. Special thanks are due to Ashley Williams and Denise Wright for managing the supplements, to Shannon Kong in production for her work on the Instructor's Resource DVD, and to Annie Wang for her management of the extraordinary array of media resources for students and instructors, especially MasteringMicrobiology®. Thanks also to Nan Kemp, Corey Webb, Maddie Boston, and Jordan Roeder, RN for their administrative, editorial, and research assistance. Chris Feldman proofread and checked pages—without her help the book would be less useful. I am always grateful to Neena Bali in Marketing and the amazing Pearson sales representatives for continuing to do a terrific job of keeping in touch with the professors and students who provided so many wonderful suggestions for this textbook. You sales representatives inspire and humble me, and your role on the team deserves more praise than I can express here.

I am especially grateful to Phil Mixter of Washington State University, Mary Jane Niles of the University of San Francisco, Bronwen Steele of Estrella Mountain Community College, Jan Miller of American River College, and Jane Reece for their expertise and advice.

I am also indebted to Sam Schwarzlose for his excellent work on the Video Tutor assessments, to Terry Austin for lending his technical expertise to the project, and to all Video Tutor reviewers for their contribution to this great pedagogical tool.

On the home front, "Thank you," Jennie and Nick Knapp, Elizabeth Bauman, Larry Latham, Josh Wood, and Mike Isley. You keep me even-keeled. My wife Michelle deserves more recognition than I can possibly express: "Many have done nobly, but you excel them all. Thank you."

Robert W. Bauman
Amarillo, Texas

Table of Contents

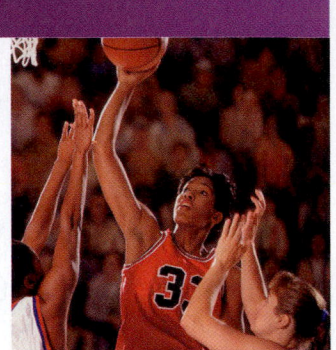

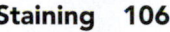

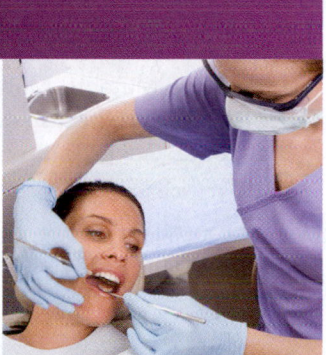

7

Microbial Genetics 196

8

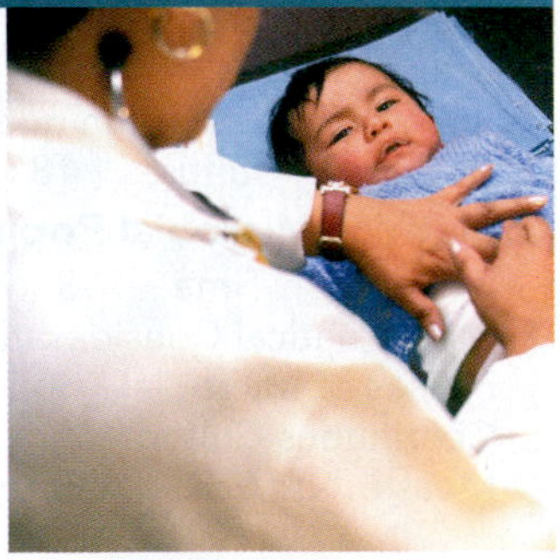

Recombinant DNA Technology 240

9

Controlling Microbial Growth in the Environment 262

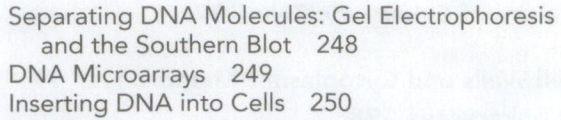

13

Characterizing and Classifying Viruses, Viroids, and Prions 386

14

Infection, Infectious Diseases, and Epidemiology 414

15

Innate Immunity 448

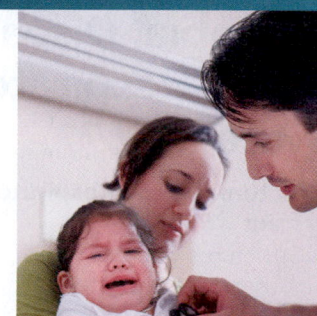

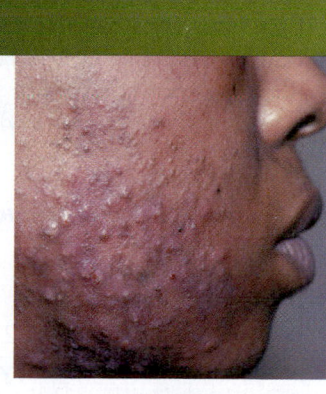

Feature Boxes

BENEFICIAL MICROBES

HIGHLIGHT

EMERGING DISEASE CASE STUDY

CLINICAL CASE STUDY

DISEASE AT A GLANCE

DISEASE IN DEPTHS

1 A Brief History of Microbiology

MICRO IN THE CLINIC

A Simple Case of Traveler's Diarrhea?

Martin is a nurse in Chicago. Every summer, he spends a few weeks in Africa volunteering in a rural village in Zambia. The village has no sanitation system and gets its water from a nearby shallow well. Over time, Martin has gained the villagers' trust and demonstrated handwashing technique, safer food preparation, and other ways to prevent infectious disease. Water purification is especially a challenge: boiling water requires fuel that isn't always available, and chemicals that make water safer to drink are often in short supply.

During the last week of Martin's most recent Africa trip, torrential rains hit the country, causing flash floods and extensive damage to the village. Despite the conditions, Martin manages to return to Chicago on schedule. A day later, he begins experiencing diarrhea. At first, he brushes it off as "traveler's diarrhea," which can be caused by a change in diet and usually goes away quickly. However, over the following days, Martin's symptoms worsen. The diarrhea is much more severe than anything Martin has experienced before; it is milky, with flecks of mucus, and frightening-looking. Martin also develops nausea, vomiting, and muscle cramps. He drinks massive amounts of water and tries over-the-counter diarrhea medicine, but nothing he does relieves the symptoms.

Is Martin suffering from a simple case of "traveler's diarrhea"? Or is something more serious going on? Turn to the end of the chapter (p. 21) to find out.

Explore More: Test your readiness and apply your knowledge with dynamic learning tools at MasteringMicrobiology.

Science is the study of nature that proceeds by posing questions about observations. Why are there seasons? What is the function of the nodules at the base of this plant? Why does this bread taste sour? What does plaque from between teeth look like when magnified? Why are so many crows dying this winter? What causes new diseases?

Many early written records show that people have always asked questions like these. For example, the Greek physician Hippocrates (ca. 460–ca. 377 B.C.) wondered whether there is a link between environment and disease, and the Greek historian Thucydides (ca. 460–ca. 404 B.C.) questioned why he and other survivors of the plague could have intimate contact with victims and not fall ill again. For many centuries, the answers to these and other fundamental questions about the nature of life remained largely unanswered. But about 350 years ago, the invention of the microscope began to provide some clues.

In this chapter we'll see how one man's determination to answer a fundamental question about the nature of life—What does life really look like?—led to the birth of a new science called *microbiology*. We'll then see how the search for answers to other questions, such as those concerning spontaneous generation, the reason fermentation occurs, and the cause of disease, prompted advances in this new science. Finally, we'll look briefly at some of the key questions microbiologists are asking today.

▲ **FIGURE 1.1 Antoni van Leeuwenhoek.** Leeuwenhoek reported the existence of protozoa in 1674 and of bacteria in 1676. *Why did Leeuwenhoek discover protozoa before bacteria?*

Figure 1.1 *Protozoa are generally larger than bacteria.*

The Early Years of Microbiology

The early years of microbiology brought the first observations of microbial life and the initial efforts to organize them into logical classifications.

What Does Life Really Look Like?

LEARNING **OUTCOMES**

1.1 Describe the world-changing scientific contributions of Leeuwenhoek.

1.2 Define microbes in the words of Leeuwenhoek and as we know them today.

A few people have changed the world of science forever. We've all heard of Galileo, Newton, and Einstein, but the list also includes Antoni van Leeuwenhoek (lā´věn-huk; 1632–1723), a Dutch tailor, merchant, and lens grinder, and the man who first discovered the bacterial world **(FIGURE 1.1)**.

Leeuwenhoek was born in Delft, the Netherlands, and lived most of his 90 years in the city of his birth. What set Leeuwenhoek apart from most other men of his generation was an insatiable curiosity coupled with an almost stubborn desire to do everything for himself. His journey to fame began simply enough, when as a cloth merchant he needed to examine the quality of cloth. Rather than merely buying one of the magnifying lenses already available, he learned to make glass lenses of his own **(FIGURE 1.2)**. Soon he began asking, "What does it really look like?" of everything in his world: the stinger of a bee, the brain of a fly, the leg of a louse, a drop of blood, flakes of his own skin. To find answers, he spent hours examining, reexamining, and recording every detail of each object he observed.

Making and looking through his simple microscopes, most really no more than magnifying glasses, became the overwhelming passion of his life. His enthusiasm and dedication are evident from the fact that he sometimes personally extracted the

▲ **FIGURE 1.2 Reproduction of Leeuwenhoek's microscope.** This simple device is little more than a magnifying glass with screws for manipulating the specimen, yet with it, Leeuwenhoek changed the way we see our world. The lens, which is convex on both sides, is about the size of a pinhead. The object to be viewed was mounted either directly on the specimen holder or inside a small glass tube, which was then mounted on the specimen holder.

metal for his microscope from ore. Further, he often made a new microscope for each specimen, which remained mounted so that he could view it again and again. Then one day, he turned a lens onto a drop of water. We don't know what he expected to see, but certainly he saw more than he had anticipated. As he reported to the Royal Society of London[1] in 1674, he was surprised and delighted by

> some green streaks, spirally wound serpent-wise, and orderly arranged. . . . Among these there were, besides, very many little animalcules, some were round, while others a bit bigger consisted of an oval. On these last, I saw two little legs near the head, and two little fins at the hind most end of the body. . . . And the motion of most of these animalcules in the water was so swift, and so various, upwards, downwards, and round about, that 'twas wonderful to see.

Leeuwenhoek had discovered a previously unknown microbial world, which today we know to be populated with tiny animals, fungi, algae, and single-celled protozoa (FIGURE 1.3). In a later report to the Royal Society, he noted that

> the number of these animals in the plaque of a man's teeth, are so many that I believe they exceed the number of men in a kingdom. . . . I found too many living animals therein, that I guess there might have been in a quantity of matter no bigger than the 1/100 part of a [grain of] sand.

From the figure accompanying his report and the precise description of the size of these organisms from between his teeth, we know that Leeuwenhoek was reporting the existence of bacteria. By the end of the 19th century, Leeuwenhoek's "beasties," as he sometimes dubbed them, were called **microorganisms**, and today we also know them as **microbes**. Both terms include all organisms that are too small to be seen without a microscope.

Because of the quality of his microscopes, his profound observational skills, his detailed reports over a 50-year period, and his report of the discovery of many types of microorganisms, Antoni van Leeuwenhoek was elected to the Royal Society in 1680. He and Isaac Newton were the most famous scientists of their time.

How Can Microbes Be Classified?

LEARNING **OUTCOMES**

1.3 List six groups of microorganisms.
1.4 Explain why protozoa, algae, and nonmicrobial parasitic worms are studied in microbiology.
1.5 Differentiate prokaryotic from eukaryotic organisms.

Shortly after Leeuwenhoek made his discoveries, the Swedish botanist Carolus Linnaeus (1707–1778) developed a **taxonomic system**—a system for naming plants and animals and grouping similar organisms together. For instance, Linnaeus and other scientists of the period grouped all organisms into either the animal kingdom or the plant kingdom. Today, biologists still use

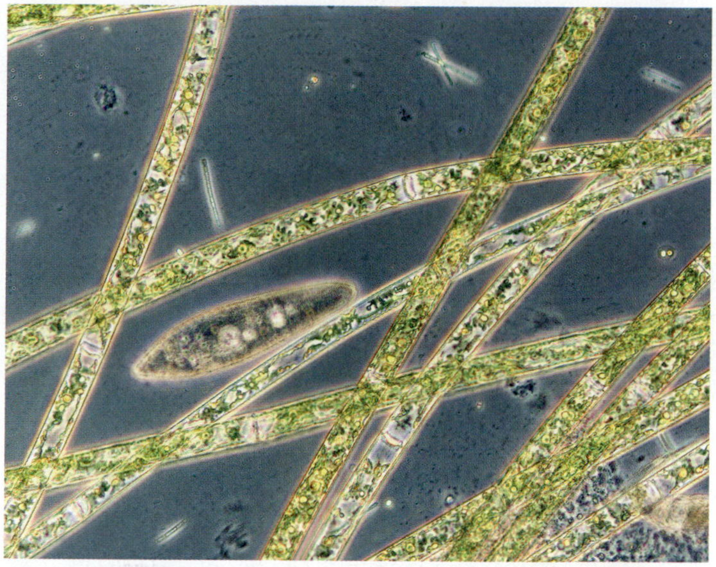
LM 50 μm

▲ **FIGURE 1.3 The microbial world.** Leeuwenhoek reported seeing a scene very much like this, full of numerous fantastic, cavorting creatures.

this basic system, but they have modified Linnaeus's scheme by adding categories that more realistically reflect the relationships among organisms. For example, scientists no longer classify yeasts, molds, and mushrooms as plants but instead as fungi. (We examine taxonomic schemes in more detail in Chapter 4.)

The microorganisms that Leeuwenhoek described can be grouped into six basic categories: bacteria, archaea, fungi, protozoa, algae, and small multicellular animals. The only types of microbes not described by Leeuwenhoek are *viruses*,[2] which are too small to be seen without an electron microscope. We briefly consider organisms in the first five categories in the following sections.

Bacteria and Archaea

because

Bacteria and **archaea** are **prokaryotic**,[3] meaning that they lack nuclei; that is, their genes are not surrounded by a membrane. Bacterial cell walls are composed of a polysaccharide called *peptidoglycan*. (Some bacteria, however, lack cell walls.) The cell walls of archaea lack peptidoglycan and instead are composed of other chemicals. Members of both groups reproduce asexually. (Chapters 3, 4, and 11 examine other differences between bacteria and archaea, and Chapters 19–24 discuss pathogenic [disease-causing] bacteria.)

Most archaea and bacteria are much smaller than eukaryotic cells (FIGURE 1.4). They live singly or in pairs, chains, or clusters in almost every habitat containing sufficient moisture. Archaea are often found in extreme environments, such as the highly saline and arsenic-rich Mono Lake in California, acidic

[1]The Royal Society of London for the Promotion of Natural Knowledge, granted a royal charter in 1662, is one of the older and more prestigious scientific groups in Europe.
[2]Technically, viruses are not "organisms," because they neither replicate themselves nor carry on the chemical reactions of living things.
[3]From Greek *pro*, meaning "before," and *karyon*, meaning "kernel" (which in this case refers to the nucleus of a cell).

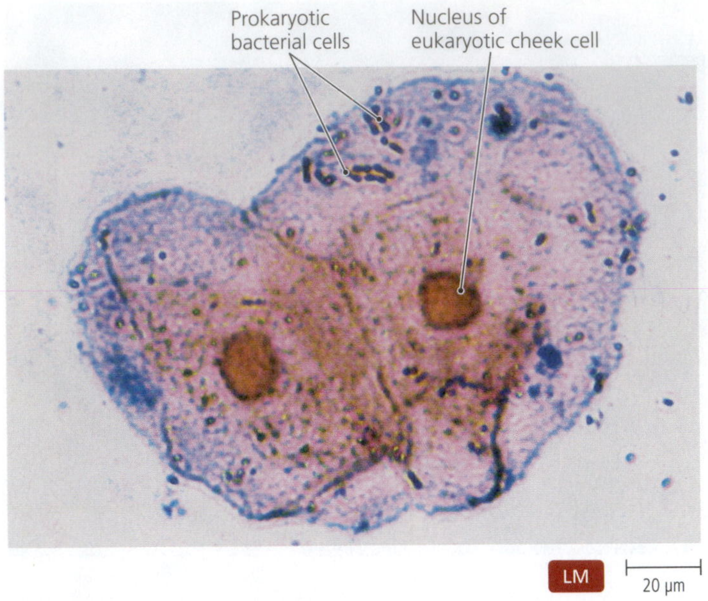

▲ **FIGURE 1.4 Cells of the bacterium *Streptococcus* (dark blue) and two human cheek cells.** Notice the size difference.

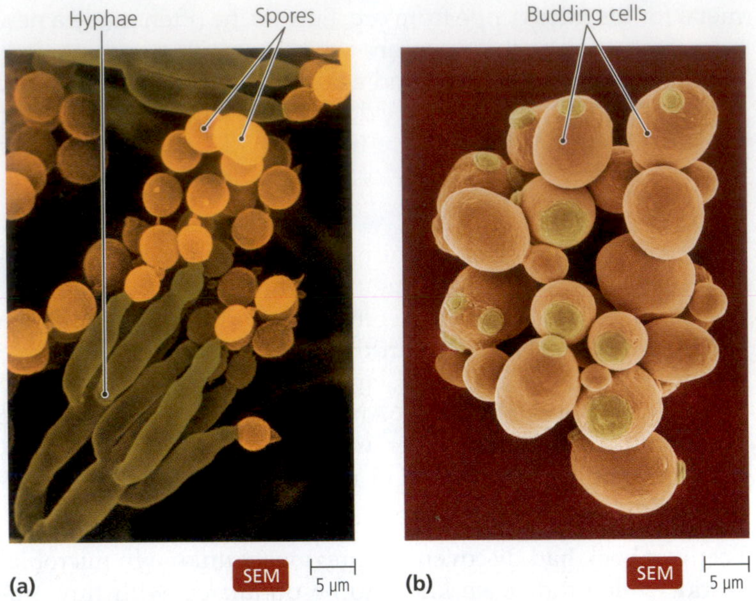

▲ **FIGURE 1.5 Fungi. (a)** The mold *Penicillium chrysogenum*, which produces penicillin, has long filamentous hyphae that intertwine to form its body. It reproduces by spores. **(b)** The yeast *Saccharomyces cerevisiae*. Yeasts are round to oval and typically reproduce by budding.

hot springs in Yellowstone National Park, and oxygen-depleted mud at the bottom of swamps. No archaea are known to cause disease.

Though bacteria may have a poor reputation in our world, the great majority do not cause disease in animals, humans, or crops. Indeed, bacteria are beneficial to us in many ways. For example, bacteria (and fungi) degrade dead plants and animals to release phosphorus, sulfur, nitrogen, and carbon back into the air, soil, and water to be used by new generations of organisms. Without microbial recyclers, the world would be buried under the corpses of uncountable dead organisms. Without beneficial bacteria, our bodies would be much more susceptible to disease.

Fungi

Fungi (fŭn´jī)[4] cells are **eukaryotic**;[5] that is, each of their cells contains a nucleus composed of genetic material surrounded by a distinct membrane. Fungi are different from plants because they obtain their food from other organisms (rather than making it for themselves). They differ from animals by having cell walls.

Microscopic fungi include some molds and yeasts. **Molds** are typically multicellular organisms that grow as long filaments that intertwine to make up the body of the mold. Molds reproduce by sexual and asexual spores, which are cells that produce a new individual without fusing with another cell (**FIGURE 1.5a**). The cottony growths on cheese, bread, and jams are molds. *Penicillium chrysogenum* (pen-i-sil´ē-ŭm krī-so´jĕn-ŭm) is a mold that produces penicillin.

Yeasts are unicellular and typically oval to round. They reproduce asexually by *budding*, a process in which a daughter cell grows off the mother cell. Some yeasts also produce sexual spores. An example of a useful yeast is *Saccharomyces cerevisiae* (sak-ă-rō-mī´sēz se-ri-vis´ē-ī; **FIGURE 1.5b**), which causes bread to rise and produces alcohol from sugar (see **Beneficial Microbes: Bread, Wine, and Beer** on p. 7). *Candida albicans* (kan´did-ă al´bi-kanz) is a yeast that causes most cases of yeast infections in women. (Fungi and their significance in the environment, in food production, and as agents of human disease are discussed in Chapters 12 and 19–24.)

Protozoa

Protozoa are single-celled eukaryotes that are similar to animals in their nutritional needs and cellular structure. In fact, *protozoa* is Greek for "first animals," though scientists today classify them in their own groups rather than as animals. Most protozoa are capable of locomotion, and one way scientists categorize protozoa is according to their locomotive structures: *pseudopods*,[6] *cilia*,[7] or *flagella*.[8] Pseudopods are extensions of a cell that flow in the direction of travel (**FIGURE 1.6a**). Cilia are numerous, short protrusions of a cell that beat rhythmically to propel the protozoan through its environment (**FIGURE 1.6b**). Flagella are also extensions of a cell but are fewer, longer, and more whiplike than cilia (**FIGURE 1.6c**). Some protozoa, such as the malaria-causing *Plasmodium* (plaz-mō´dē-ŭm), are nonmotile in their mature forms.

Protozoa typically live freely in water, but some live inside animal hosts, where they can cause disease. Most protozoa reproduce asexually, though some are sexual as well. (Chapters 12 and 19–24 further examine protozoa and some diseases they cause.)

---

[4]Plural of the Latin *fungus*, meaning "mushroom."
[5]From Greek *eu*, meaning "true," and *karyon*, meaning "kernel."
[6]Plural Greek *pseudes*, meaning "false," and *podos*, meaning "foot."
[7]Plural of the Latin *cilium*, meaning "eyelid."
[8]Plural of the Latin *flagellum*, meaning "whip."

▶ **FIGURE 1.6 Locomotive structures of protozoa.** **(a)** Pseudopods are cellular extensions used for locomotion and feeding, as seen in *Amoeba proteus*. **(b)** Cilia are short, motile, hairlike extrusions, as seen in *Euplotes*. **(c)** Flagella are whiplike extensions that are less numerous and longer than cilia, as seen in *Peranema*. *How do cilia and flagella differ?*

Figure 1.6 *Cilia are short, numerous, and often cover the cell, whereas flagella are long and relatively few in number.*

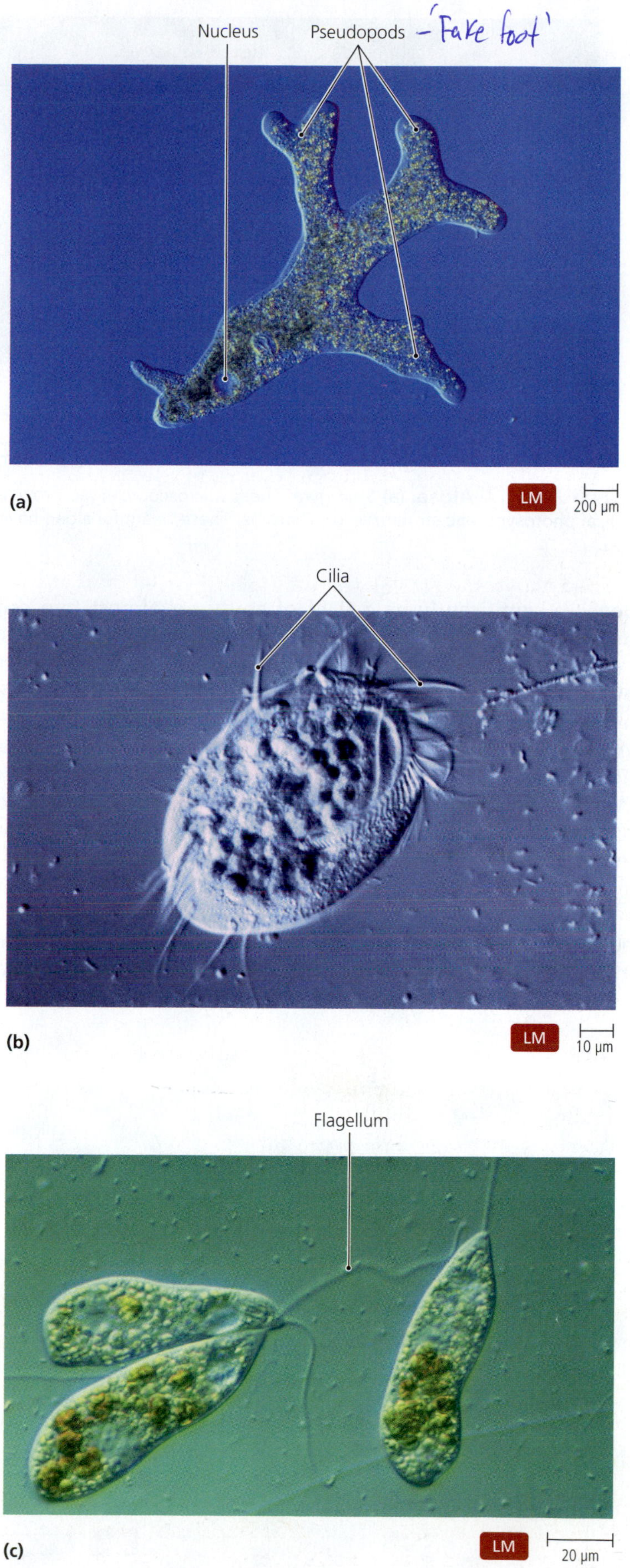

Algae

Algae[9] are unicellular or multicellular *photosynthetic* eukaryotes; that is, like plants, they make their own food from carbon dioxide and water using energy from sunlight. They differ from plants in the relative simplicity of their reproductive structures. Algae are categorized on the basis of their pigmentation and the composition of their cell walls.

Large algae, commonly called seaweeds and kelps, are common in the world's oceans. Chemicals from their gelatinous cell walls are used as thickeners and emulsifiers in many food and cosmetic products as well as in a hardening agent called *agar* in microbiological laboratory media.

Unicellular algae **(FIGURE 1.7)** are common in freshwater ponds, streams, and lakes and in the oceans as well. They are the major food of small aquatic and marine animals and provide most of the world's oxygen as a by-product of photosynthesis. The glasslike cell walls of diatoms provide grit for many polishing compounds. Manufacturers use gelatinous chemicals from the cell walls of some algae as thickeners and emulsifiers in many foods and cosmetics. Scientists use one algae-derived chemical called *agar* to solidify laboratory media. (Chapter 12 discusses other aspects of the biology of algae.)

Other Organisms of Importance to Microbiologists

Microbiologists also study parasitic worms, which range in size from microscopic forms **(FIGURE 1.8)** to adult tapeworms over 7 meters (approximately 23 feet) in length. Even though most of these worms are not microscopic as adults, many of them cause diseases that were studied by early microbiologists. Further, laboratory technicians diagnose infections of parasitic worms by finding microscopic eggs and immature stages in blood, fecal, urine, and lymph specimens. (Chapters 21 and 23 discuss parasitic worms.)

The only type of microbe that remained hidden from Leeuwenhoek and other early microbiologists was the virus, which is much smaller than the smallest prokaryote and is not visible by light microscopy **(FIGURE 1.9)**. Viruses could not be seen until the electron microscope was invented in 1932. All viruses are acellular (not composed of cells) obligatory parasites composed of small amounts of genetic material (either DNA or RNA) surrounded by a protein coat. (Chapter 13 examines the general characteristics of viruses, and Chapters 18–24 discuss specific viral pathogens.)

Leeuwenhoek first reported the existence of most types of microorganisms in the late 1600s, but microbiology did not

[9]Plural of the Latin *alga*, meaning "seaweed."

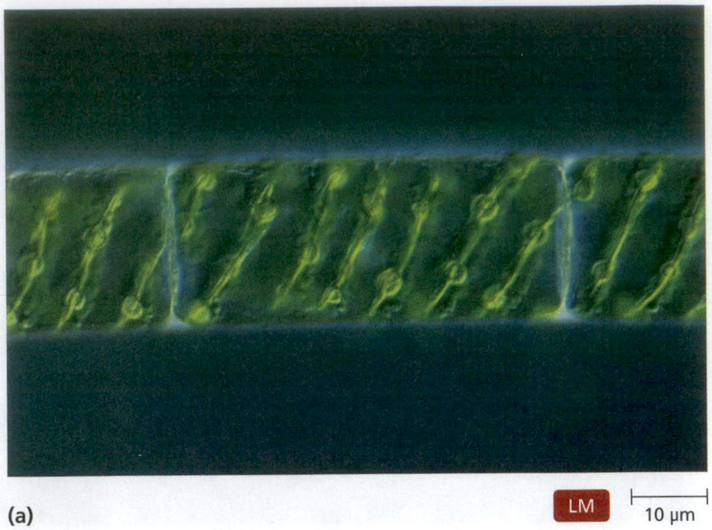

(a)

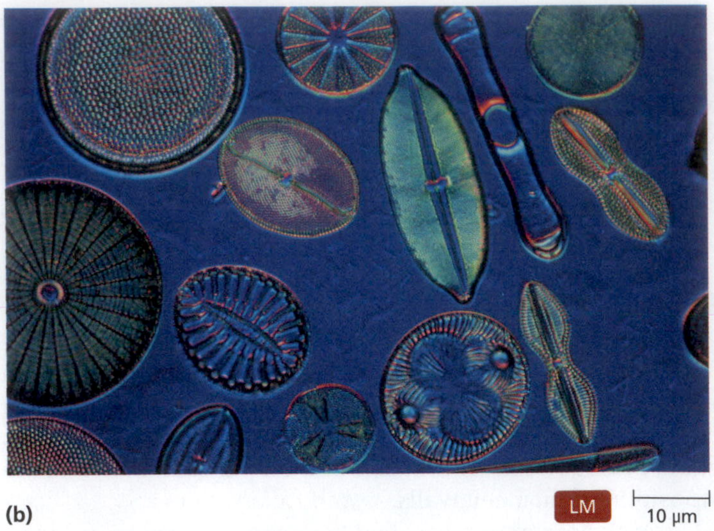

(b)

▲ **FIGURE 1.7 Algae. (a)** *Spirogyra*. These microscopic algae grow as chains of cells containing helical photosynthetic structures. **(b)** Diatoms. These beautiful algae have glasslike cell walls.

develop significantly as a field of study for almost two centuries. There were a number of reasons for this delay. First, Leeuwenhoek was a suspicious and secretive man. Though he built over 400 microscopes, he never trained an apprentice, and he never sold or gave away a microscope. In fact, he never let *anyone*—not his family or such distinguished visitors as the czar of Russia—so much as peek through his very best instruments. When Leeuwenhoek died, the secret of creating superior microscopes was lost. It took almost 100 years for scientists to make microscopes of equivalent quality.

Another reason that microbiology was slow to develop as a science is that scientists in the 1700s considered microbes to be curiosities of nature and insignificant to human affairs. But in the late 1800s, scientists began to adopt a new philosophy, one that demanded experimental evidence rather than mere acceptance of traditional knowledge. This fresh philosophical foundation, accompanied by improved microscopes, new laboratory techniques, and a drive to answer a series of pivotal questions, propelled microbiology to the forefront as a scientific discipline.

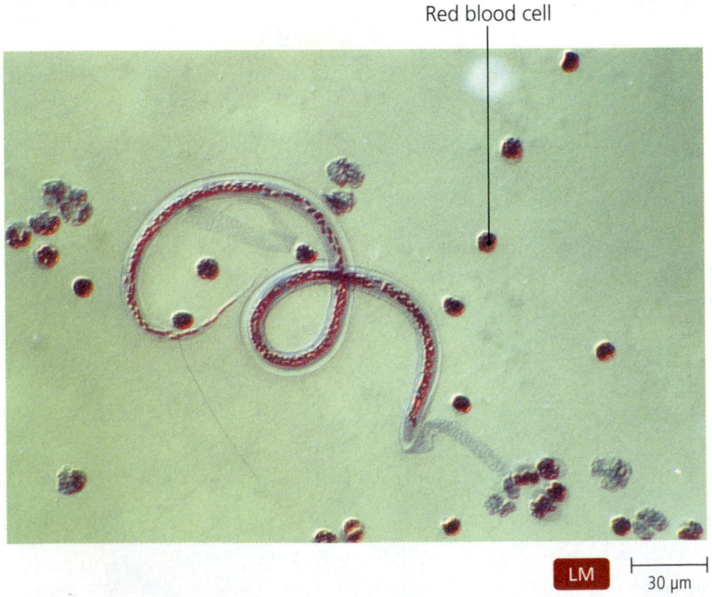

▲ **FIGURE 1.8 An immature stage of a parasitic worm in blood.**

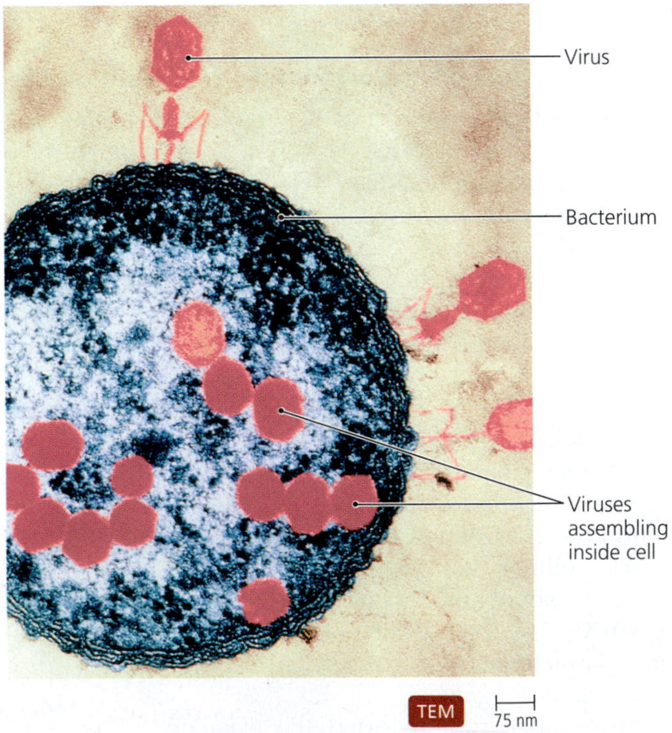

▲ **FIGURE 1.9 A colorized electron microscope image of viruses infecting a bacterium.** Viruses, which are acellular obligatory parasites, are too small to be seen with a light microscope. Notice how small the viruses are compared to the bacterium.

TELL ME WHY

Some people consider Leeuwenhoek the "Father of Microbiology." Explain why this moniker makes sense.

The Golden Age of Microbiology

LEARNING | OUTCOME

1.6 List and answer four questions that propelled research in what is called the "Golden Age of Microbiology."

For about 50 years, during what is sometimes called the "Golden Age of Microbiology," scientists and the blossoming field of microbiology were driven by the search for answers to the following four questions:

- Is spontaneous generation of microbial life possible?
- What causes fermentation?
- What causes disease?
- How can we prevent infection and disease?

Competition among scientists who were striving to be the first to answer these questions drove exploration and discovery in microbiology during the late 1800s and early 1900s. These scientists' discoveries and the fields of study they initiated continue to shape the course of microbiological research today.

In the next sections we consider these questions and how the great scientists accumulated the experimental evidence that answered them.

Does Microbial Life Spontaneously Generate?

LEARNING | OUTCOMES

1.7 Identify the scientists who argued in favor of spontaneous generation.

1.8 Compare and contrast the investigations of Redi, Needham, Spallanzani, and Pasteur concerning spontaneous generation.

1.9 List four steps in the scientific method of investigation.

A dry lake bed has lain under the relentless North African desert sun for eight long months. The cracks in the baked, parched mud are wider than a man's hand. There is no sign of life anywhere in the scorched terrain. With the abruptness characteristic of desert storms, rain falls in a torrent, and a raging flood of roiling water and mud crashes down the dry streambed and fills the lake. Within hours, what had been a lifeless, dry mudflat becomes a pool of water teeming with billions of shrimp; by the next day it is home to hundreds of toads. Where did these animals come from?

Many philosophers and scientists of past ages thought that living things arose via three processes: through asexual

BENEFICIAL MICROBES

Bread, Wine, and Beer

Microorganisms play important roles in people's lives; for example, pathogens have undeniably altered the course of history. However, what may be the most important microbiological event—one that has had a greater impact on culture and society than that of any disease or epidemic—was the domestication of the yeast used by bakers and brewers. Its name, *Saccharomyces cerevisiae*, means "sugar fungus [that makes] beer."

The earliest record of the use of yeast comes from Persia (modern Iran), where archaeologists have found the remains of grapes and wine preservatives in pottery vessels more than 7000 years old. Brewing of beer likely started even earlier, its beginnings undocumented. The earliest examples of leavened bread are from Egypt and show that bread making was routine about 6000 years ago. Before that time, bread was unleavened and flat.

It is likely that making wine and brewing beer occurred earlier than the use of leavened bread because *Saccharomyces* is naturally found on grapes, which can begin to ferment while still on the vine. Historians hypothesize that early bakers may have exposed bread dough to circulating air, hoping that the invisible and inexplicable "fermentation principle" would inoculate the bread. Another hypothesis is that bakers learned to add small amounts of beer or wine to the bread, intentionally inoculating the dough with yeast. Of course, all those years before Leeuwenhoek and Pasteur, no one knew that the fermenting ingredient of wine was a living organism.

Besides its role in baking and in making alcoholic beverages, *S. cerevisiae* is an important tool for the study of cells. Scientists use yeast to delve into the mysteries of cellular function, organization, and genetics, making *Saccharomyces* the most intensely studied eukaryote. In fact, molecular biologists published the complete sequence of the genes of *S. cerevisiae* in 1996—a first for any eukaryotic cell.

Today, scientists are working toward using *S. cerevisiae* in novel ways. For example, some nutritionists and gastroenterologists are examining the use of *Saccharomyces* as a *probiotic*, that is, a microorganism intentionally taken to ward off disease and promote good health. Research suggests that the yeast helps treat diarrhea and colitis and may even help prevent these and other gastrointestinal diseases.

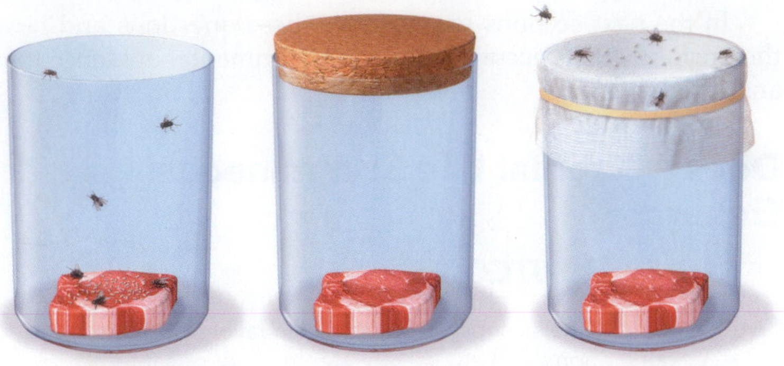

▲ **FIGURE 1.10 Redi's experiments.** When the flask remained unsealed, maggots covered the meat within a few days. When the flask was sealed, flies were kept away, and no maggots appeared on the meat. When the flask opening was covered with gauze, flies were kept away, and no maggots appeared on the meat, although a few maggots appeared on top of the gauze.

reproduction, through sexual reproduction, or from nonliving matter. The appearance of shrimp and toads in the mud of what so recently was a dry lake bed was seen as an example of the third process, which came to be known as *abiogenesis*,[10] or **spontaneous generation**. The theory of spontaneous generation as promulgated by Aristotle (384–322 B.C.) was widely accepted for over 2000 years because it seemed to explain a variety of commonly observed phenomena, such as the appearance of maggots on spoiling meat. However, the validity of the theory came under challenge in the 17th century.

Redi's Experiments

In the late 1600s, the Italian physician Francesco Redi (1626–1697) demonstrated by a series of experiments that when decaying meat was kept isolated from flies, maggots never developed, whereas meat exposed to flies was soon infested **(FIGURE 1.10)**. As a result of experiments such as these, scientists began to doubt Aristotle's theory and adopt the view that animals come only from other animals.

Needham's Experiments

The debate over spontaneous generation was rekindled when Leeuwenhoek discovered microbes and showed that they appeared after a few days in freshly collected rainwater. Though scientists agreed that larger animals could not arise spontaneously, they disagreed about Leeuwenhoek's "wee animalcules"; surely they did not have parents, did they? They must arise spontaneously.

The proponents of spontaneous generation pointed to the careful demonstrations of British investigator John T. Needham (1713–1781). He boiled beef gravy and infusions[11] of plant material in vials, which he then tightly sealed with corks. Some days later, Needham observed that the vials were cloudy, and examination revealed an abundance of "microscopical animals of most dimensions." As he explained it, there must be a "life force" that causes inanimate matter to spontaneously come to

[10]From Greek *a*, meaning "not"; *bios*, meaning "life"; and *genein*, meaning "to produce."
[11]Infusions are broths made by heating water containing plant or animal material.

HIGHLIGHT

"The New Normal": The Challenge of Emerging and Reemerging Diseases

Middle East respiratory syndrome (MERS). Monkeypox. West Nile encephalitis. These and diseases like them are emerging diseases—ones that are diagnosed in a population for the first time. Among them are H1N1 influenza ("swine flu"); Nipah encephalitis, a highly fatal disease carried by pigs; and mosquito-borne chikungunya, which causes severe joint pain and sometimes death. Indeed, unfamiliar diseases have become "the new normal" for health care workers, according to the Centers for Disease Control and Prevention.

Meanwhile, diseases once thought to be near eradication, such as polio, whooping cough, and tuberculosis, have reemerged in troubling outbreaks. Other near-vanquished pathogens such as smallpox or anthrax may become potential weapons in bioterrorist attacks.

How do emerging and reemerging diseases arise? Some are introduced to humans as we move into remote jungles and contact infected animals, some are carried by insects whose range is spreading as climate changes, and some take advantage of the AIDS crisis, infecting immunocompromised patients. In other cases, previously harmless microbes acquire new genes that allow them to be infective and cause disease. Some emerging pathogens spread with the speed of jet planes carrying infected people around the globe, and still others arise when previously treatable microbes develop resistance to our antibiotics.

Workers dumping poultry suspected of harboring avian influenza virus

However they arise, scientists are monitoring emerging and reemerging diseases that may develop into the next generation of high-profile infectious diseases. Throughout this text, you will encounter many boxed discussions of such emerging and reemerging diseases.

life because he had heated the vials sufficiently to kill everything. Needham's experiments so impressed the Royal Society that they elected him a member.

Spallanzani's Experiments

Then, in 1799, the Italian Catholic priest and scientist Lazzaro Spallanzani (1729–1799) reported results that contradicted Needham's findings. Spallanzani boiled infusions for almost an hour and sealed the vials by melting their slender necks closed. His infusions remained clear unless he broke the seal and exposed the infusion to air, after which they became cloudy with microorganisms. He concluded three things:

- Needham either had failed to heat his vials sufficiently to kill all microbes or had not sealed them tightly enough.
- Microorganisms exist in the air and can contaminate experiments.
- Spontaneous generation of microorganisms does not occur; all living things arise from other living things.

Although Spallanzani's experiments would appear to have settled the controversy once and for all, it proved difficult to dethrone a theory that had held sway for 2000 years, especially when so notable a man as Aristotle had propounded it. One of the criticisms of Spallanzani's work was that his sealed vials did not allow enough air for organisms to thrive; another objection was that his prolonged heating destroyed the "life force." The debate continued until the French chemist Louis Pasteur **(FIGURE 1.11)** conducted experiments that finally laid the theory of spontaneous generation to rest.

Pasteur's Experiments

Louis Pasteur (1822–1895) was an indefatigable worker who pushed himself as hard as he pushed others. As he wrote his sisters, "To *will* is a great thing dear sisters, for Action and Work usually follow Will, and almost always Work is accompanied by Success. These three things, Work, Will, Success, fill human existence. Will opens the door to success both brilliant and happy; Work passes these doors, and at the end of the journey Success comes to crown one's efforts." When his wife complained about his long hours in the laboratory, he replied, "I will lead you to fame."

▲ **FIGURE 1.11 Louis Pasteur.** Often called the Father of Microbiology, he disproved spontaneous generation. In this depiction, Pasteur examines some bacterial cultures.

Pasteur's determination and hard work are apparent in his investigations of spontaneous generation. Like Spallanzani, he boiled infusions long enough to kill everything. But instead of sealing the flasks, he bent their necks into an S-shape, which allowed air to enter while preventing the introduction of dust and microbes into the broth **(FIGURE 1.12)**.

Crowded for space and lacking funds, he improvised an incubator in the opening under a staircase. Day after day he

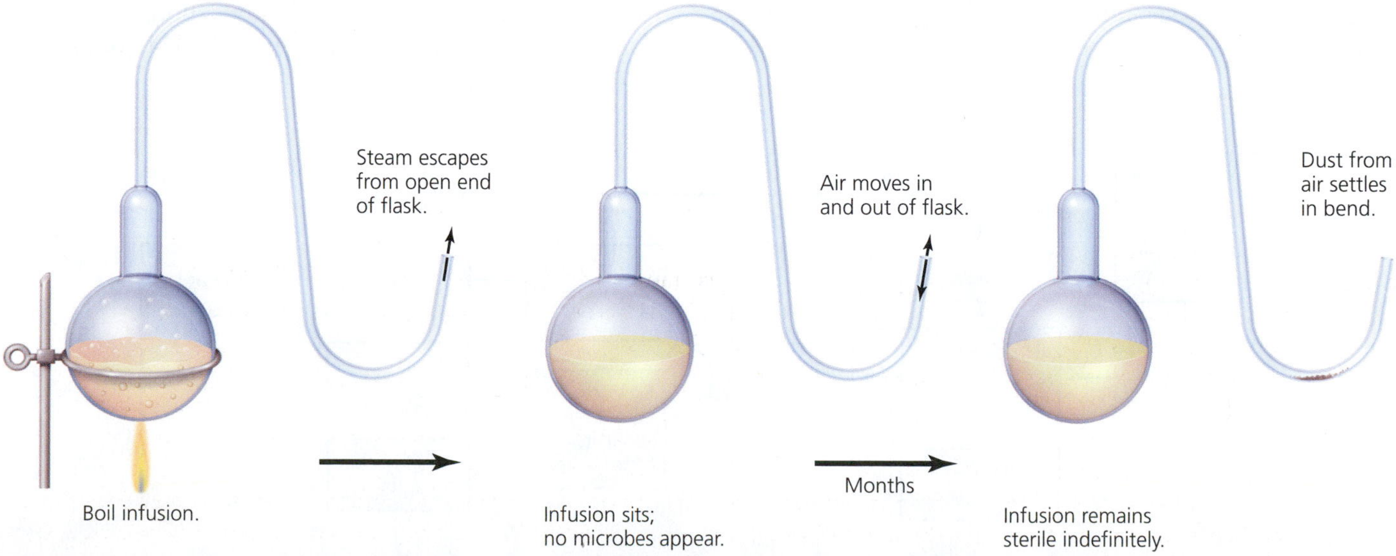

Steam escapes from open end of flask.
Air moves in and out of flask.
Dust from air settles in bend.
Boil infusion.
Months
Infusion sits; no microbes appear.
Infusion remains sterile indefinitely.

▲ **FIGURE 1.12 Pasteur's experiments with "swan-necked flasks."** As long as the flask remained upright, no microbial growth appeared in the infusion.

crawled on hands and knees into this incommodious space and examined his flasks for the cloudiness that would indicate the presence of living organisms. In 1861, he reported that his "swan-necked flasks" remained free of microbes even 18 months later. Because the flasks contained all the nutrients (including air) known to be required by living things, he concluded, "Never will spontaneous generation recover from the mortal blow of this simple experiment."

Pasteur followed this experiment with demonstrations that microbes in the air were the "parents" of Needham's microorganisms. He broke the necks off some flasks, exposing the liquid in them directly to the air, and he carefully tilted others so that the liquid touched the dust that had accumulated in their necks. The next day, all of these flasks were cloudy with microbes. He concluded that the microbes in the liquid were the progeny of microbes that had been on the dust particles in the air.

The Scientific Method

The debate over spontaneous generation led in part to the development of a generalized **scientific method** by which questions are answered through observations of the outcomes of carefully controlled experiments instead of by conjecture or according to the opinions of any authority figure. The scientific method, which provides a framework for conducting an investigation rather than a rigid set of specific "rules," consists of four basic steps **(FIGURE 1.13)**:

1. A group of observations leads a scientist to ask a question about some phenomenon.
2. The scientist generates a hypothesis—that is, a potential answer to the question.
3. The scientist designs and conducts an experiment to test the hypothesis.
4. Based on the observed results of the experiment, the scientist either accepts, rejects, or modifies the hypothesis.

The scientist then returns to earlier steps in the method, either modifying hypotheses and then testing them or repeatedly testing accepted hypotheses until the evidence for a hypothesis is convincing (Figure 1.13). Accepted hypotheses that explain many observations and are repeatedly verified by numerous scientists over many years are called *theories* or *laws*. ▶**VIDEO TUTOR:** ***The Scientific Method***

Note that for the scientific community to accept experiments (and their results) as valid, they must include appropriate *control groups*—groups that are treated exactly the same as the other groups in the experiment except for the one variable that the experiment is designed to test. In Pasteur's experiments on spontaneous generation, for example, his "control flasks" were the sterile infusion composed of all the nutrients living things need as well as air made available through the flasks' "swan necks." His "experimental flasks" for testing his hypothesis were exposed to exactly the same conditions *plus* contact with the dust in the bend in the neck. Because exposure to the dust was the *only* difference between the control and experimental groups, Pasteur was able to conclude that the microbes growing in the infusion arrived on the dust particles.

What Causes Fermentation?

LEARNING | **OUTCOMES**

1.10 Discuss the significance of Pasteur's fermentation experiments to our world today.

1.11 Explain why Pasteur may be considered the Father of Microbiology.

1.12 Identify the scientist whose experiments led to the field of biochemistry and the study of metabolism.

The controversy over spontaneous generation was largely a philosophical exercise among men who conducted research to gain basic scientific knowledge and not to apply the knowledge they gained. However, the second question that moved microbial studies forward in the 1800s had tremendous practical applications.

Our story resumes in 19th-century France, where spoiled, acidic wine was threatening the livelihood of many grape growers. This led to a fundamental question, "What causes the fermentation of grape juice into wine?" This question was so

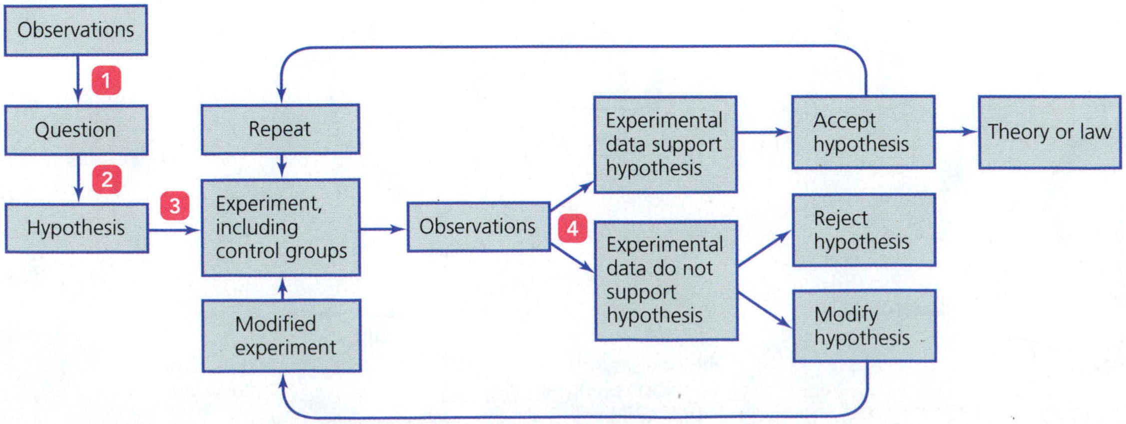

▲ **FIGURE 1.13 The scientific method, which forms a framework for scientific research.**

important to vintners that they funded research concerning fermentation, hoping scientists could develop methods to promote the production of alcohol and prevent spoilage by acid during fermentation.

Pasteur's Experiments

Scientists of the 1800s used the word *fermentation* to mean not only the formation of alcohol from sugar but also other chemical reactions, such as the formation of lactic acid, the putrefaction of meat, and the decomposition of waste. Many scientists asserted that air caused fermentation reactions; others insisted that living organisms were responsible.

The debate over the cause of fermentation reactions was linked to the debate over spontaneous generation. Some scientists proposed that the yeasts observed in fermenting juices were nonliving globules of chemicals and gases. Others thought that yeasts were alive and were spontaneously generated during fermentation. Still others asserted that yeasts not only were living organisms but also caused fermentation.

Pasteur conducted a series of careful observations and experiments that answered the question, "What causes fermentation?" First, he observed yeast cells growing and budding in grape juice and conducted experiments showing that they arise only from other yeast cells. Then, by sealing some sterile flasks containing grape juice and yeast and by leaving others open to the air, he demonstrated that yeast could grow with or without oxygen; that is, he discovered that yeasts are *facultative anaerobes*[12]—organisms that can live with or without oxygen. Finally, by introducing bacteria and yeast cells into different flasks of sterile grape juice, he proved that bacteria ferment grape juice to produce acids and that yeast cells ferment grape juice to produce alcohol (**FIGURE 1.14**).

Pasteur's discovery that *anaerobic* bacteria fermented grape juice into acids suggested a method for preventing the spoilage of wine. His name became a household word when he developed *pasteurization*, a process of heating the grape juice just enough to kill most contaminating bacteria without changing the juice's basic qualities so that it could then be inoculated with yeast to ensure that alcohol fermentation occurred. Pasteur thus began the field of **industrial microbiology** (or **biotechnology**) in which microbes are intentionally used to manufacture products (**TABLE 1.1** on p. 13; see also Chapter 25). Today pasteurization is used routinely on milk to eliminate pathogens that cause such diseases as bovine tuberculosis and brucellosis; it is also used to eliminate pathogens in juices and other beverages.

These are just a few of the many experiments Pasteur conducted with microbes. Although a few of Pasteur's successes can be attributed to the superior microscopes available in the late 1800s, his genius is clearly evident in his carefully designed and straightforward experiments. Because of his many, varied, and significant accomplishments in working with microbes, Pasteur may be considered the Father of Microbiology.

Buchner's Experiments

Studies on fermentation began with the idea that fermentation reactions were strictly chemical and did not involve living organisms. This idea was supplanted by Pasteur's work showing that fermentation proceeded only when living cells were present and that different types of microorganisms growing under varied conditions produced different end products.

In 1897, the German scientist Eduard Buchner (1860–1917) resurrected the chemical explanation by showing that fermentation does not require living cells. Buchner's experiments demonstrated the presence of *enzymes*, which are cell-produced proteins that promote chemical reactions. Buchner's work began the field of **biochemistry** and the study of **metabolism**, a term that refers to the sum of all chemical reactions within an organism.

What Causes Disease?

LEARNING **OUTCOMES**

1.13 List at least seven contributions made by Koch to the field of microbiology.

1.14 List the four steps that must be taken to prove the cause of an infectious disease.

1.15 Describe the contribution of Gram to the field of microbiology.

You are a physician in London, and it is August 1854. It is past midnight, and you have been visiting patients since before dawn. As you enter the room of your next patient, you observe with frustration and despair that this case is like hundreds of others you and your colleagues have attended in the neighborhood over the past month.

A five-year-old boy with a vacant stare lies in bed listlessly. As you watch, he is suddenly gripped by severe abdominal cramps, and his gastrointestinal tract empties in an explosion of watery diarrhea. The voided fluid is clear, colorless, odorless, and streaked with thin flecks of white mucus, reminiscent of water poured off a pot of cooking rice. His anxious mother changes his bedclothes as his father gives him a sip of water, but it is of little use. With a heavy heart you confirm the parents' fear—their child has cholera, and there is nothing you can do. He will likely die before morning. As you despondently turn to go, the question that has haunted you for two months is foremost in your mind: What causes such a disease?

The third question that propelled the advance of microbiology concerned disease, defined generally as any abnormal condition in the body. Prior to the 1800s, disease was attributed to various factors, including evil spirits, astrological signs, imbalances in body fluids, and foul vapors. Although the Italian philosopher Girolamo Fracastoro (1478–1553) conjectured as early as 1546 that "germs[13] of contagion" cause disease, the idea that germs might

[12]From Greek *an*, meaning "not"; *aer*, meaning "air" (i.e., oxygen); and *bios*, meaning "life."
[13]From Latin *germen*, meaning "sprout."

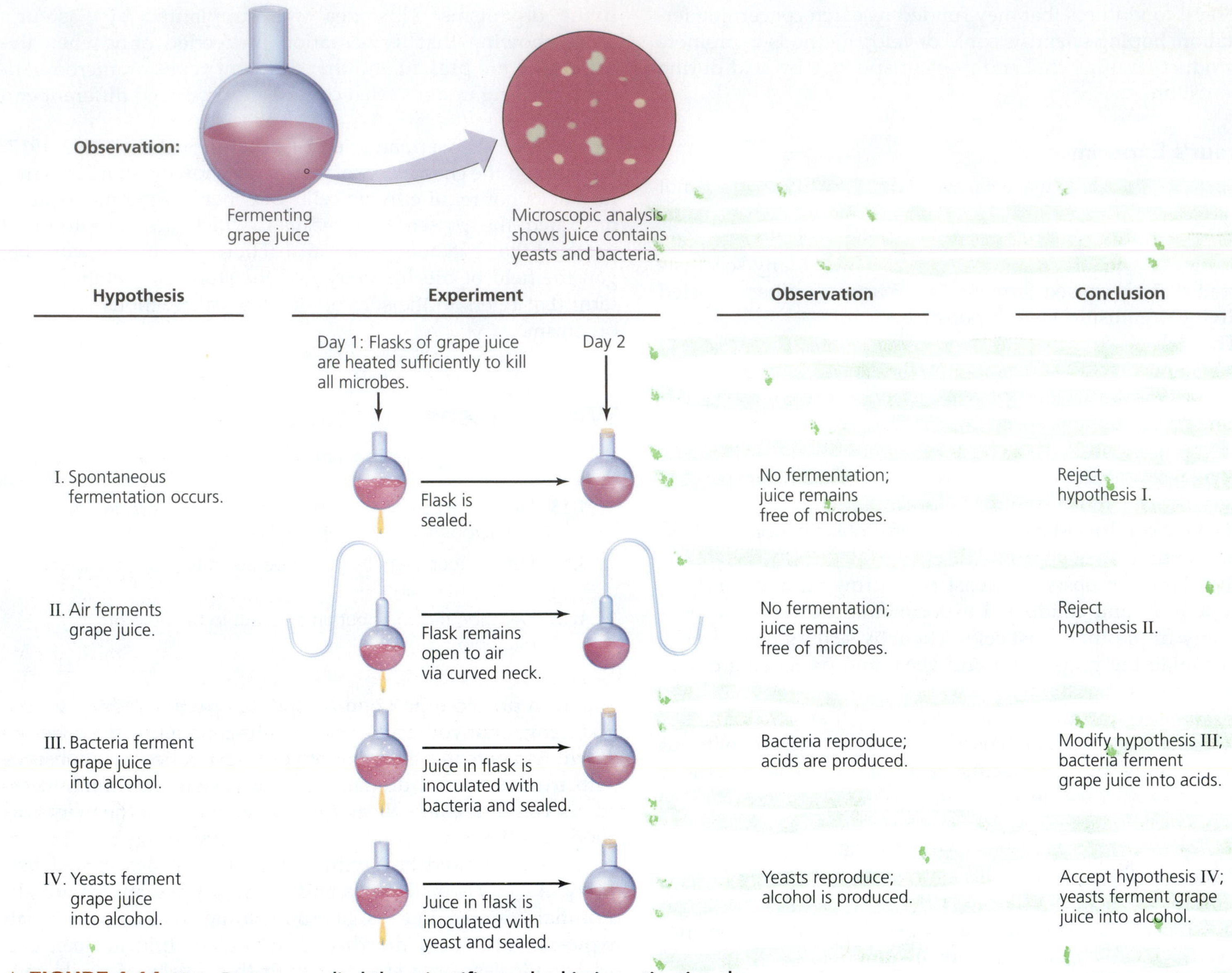
Observation:
Fermenting grape juice
Microscopic analysis shows juice contains yeasts and bacteria.
Hypothesis
Experiment
Observation
Conclusion
Day 1: Flasks of grape juice are heated sufficiently to kill all microbes.
Day 2
I. Spontaneous fermentation occurs.
Flask is sealed.
No fermentation; juice remains free of microbes.
Reject hypothesis I.
II. Air ferments grape juice.
Flask remains open to air via curved neck.
No fermentation; juice remains free of microbes.
Reject hypothesis II.
III. Bacteria ferment grape juice into alcohol.
Juice in flask is inoculated with bacteria and sealed.
Bacteria reproduce; acids are produced.
Modify hypothesis III; bacteria ferment grape juice into acids.
IV. Yeasts ferment grape juice into alcohol.
Juice in flask is inoculated with yeast and sealed.
Yeasts reproduce; alcohol is produced.
Accept hypothesis IV; yeasts ferment grape juice into alcohol.

▲ **FIGURE 1.14 How Pasteur applied the scientific method in investigating the nature of fermentation.** After observing that fermenting grape juice contained both yeasts and bacteria, Pasteur hypothesized that these organisms cause fermentation. On eliminating the possibility that fermentation could occur spontaneously or be caused by air (hypotheses I and II), he concluded that fermentation requires the presence of living cells. The results of additional experiments (those testing hypotheses III and IV) indicated that bacteria ferment grape juice to produce acids and that yeasts ferment grape juice to produce alcohol. *Which of Pasteur's flasks was the control?*

Figure 1.14 *The sealed flask that remained free of microorganisms served as the control.*

be invisible living organisms awaited Leeuwenhoek's investigations 130 years later.

Pasteur's discovery that bacteria are responsible for spoiling wine led naturally to his hypothesis in 1857 that microorganisms are also responsible for diseases. This idea came to be known as the **germ theory of disease**. Because a particular disease is typically accompanied by the same symptoms in all affected individuals, early investigators suspected that diseases such as cholera, tuberculosis, and anthrax are each caused by a specific germ, called a **pathogen**.[14] Today we know that some diseases are genetic and that allergic reactions and environmental toxins cause others, so the germ theory applies only to *infectious*[15] *diseases*.

Just as Pasteur was the chief investigator in disproving spontaneous generation and determining the cause of fermentation, so Robert Koch (1843–1910) dominated **etiology**[16] (disease causation) **(FIGURE 1.15)**.

[14]From Greek *pathos*, meaning "disease," and *genein*, meaning "to produce."
[15]From Latin *inficere*, meaning "to taint" (i.e., with a pathogen).
[16]From Greek *aitia*, meaning "cause," and *logos*, meaning "word" or "study."

TABLE 1.1 Some Industrial Uses of Microbes

Product or Process	Contribution of Microorganism
Foods and Beverages	
Cheese	Flavoring and ripening produced by bacteria and fungi; flavors dependent on the source of milk and the type of microorganism
Alcoholic beverages	Alcohol produced by bacteria or yeast by fermentation of sugars in fruit juice or grain
Soy sauce	Produced by fungal fermentation of soybeans
Vinegar	Produced by bacterial fermentation of sugar
Yogurt	Produced by certain bacteria growing in milk
Sour cream	Produced by bacteria growing in cream
Artificial sweetener	Amino acids synthesized by bacteria from sugar
Bread	Rising of dough produced by action of yeast; sourdough results from bacteria-produced acids
Other Products	
Antibiotics	Produced by bacteria and fungi
Human growth hormone, human insulin	Produced by genetically engineered bacteria
Laundry enzymes	Isolated from bacteria
Vitamins	Isolated from bacteria
Diatomaceous earth (in polishes and buffing compounds)	Composed of cell walls of microscopic algae
Pest control chemicals	Insect pests killed or inhibited by insect-destroying bacteria
Drain opener	Protein-digesting and fat-digesting enzymes produced by bacteria

▲ **FIGURE 1.15 Robert Koch.** Koch was instrumental in modifying the scientific method to prove that a given pathogen caused a specific disease.

Koch's Experiments

Koch was a country doctor in Germany when he began a race with Pasteur to discover the cause of anthrax, which is a potentially fatal disease, primarily of animals, in which toxins produce ulceration of the skin. Anthrax, which can spread to humans, caused untold financial losses to farmers and ranchers in the 1800s.

Koch carefully examined the blood of infected animals, and in every case he identified a rod-shaped bacterium[17] that formed chains. He observed the formation of resting stages (endospores) within the bacterial cells and showed that the endospores always produced anthrax when they were injected into mice. This was the first time that a bacterium was proven to cause a disease. As a result of his successful work on anthrax, Koch moved to Berlin and was given facilities and funding to continue his research.

Heartened by his success, Koch turned his attention to other diseases. He had been fortunate when he chose anthrax for his initial investigations, because anthrax bacteria are quite large and easily identified with the microscopes of that time. However, most bacteria are very small, and different types exhibit few or no visible differences. Koch was puzzled regarding how he was to distinguish among these bacteria.

He solved the problem by taking specimens (e.g., blood, pus, or sputum) from disease victims and then smearing the specimens onto a solid surface such as a slice of potato or a gelatin medium. He then waited for bacteria and fungi present in the specimen to multiply and form distinct colonies (**FIGURE 1.16**). Koch hypothesized that each colony consisted of the progeny of a single cell. He then inoculated samples from each colony into laboratory animals to see which caused disease. Koch's method of isolation is a standard technique in microbiological and medical labs to this day, though *agar* derived from red algae is used instead of gelatin or potato.

Koch and his colleagues are also responsible for many other advances in laboratory microbiology, including the following:

- Simple staining techniques for bacterial cells and flagella
- The first photomicrograph of bacteria
- The first photograph of bacteria in diseased tissue
- Techniques for estimating the number of bacteria in a solution based on the number of colonies that form after inoculation onto a solid surface
- The use of steam to sterilize growth media
- The use of Petri[18] dishes to hold solid growth media
- Laboratory techniques such as transferring bacteria between media using a metal wire that has been heat-sterilized in a flame
- Elucidation of bacteria as distinct species

[17] Now known as *Bacillus anthracis*—Latin for "the rod of anthrax."
[18] Named for Richard Petri, Koch's assistant, who invented them in 1887.

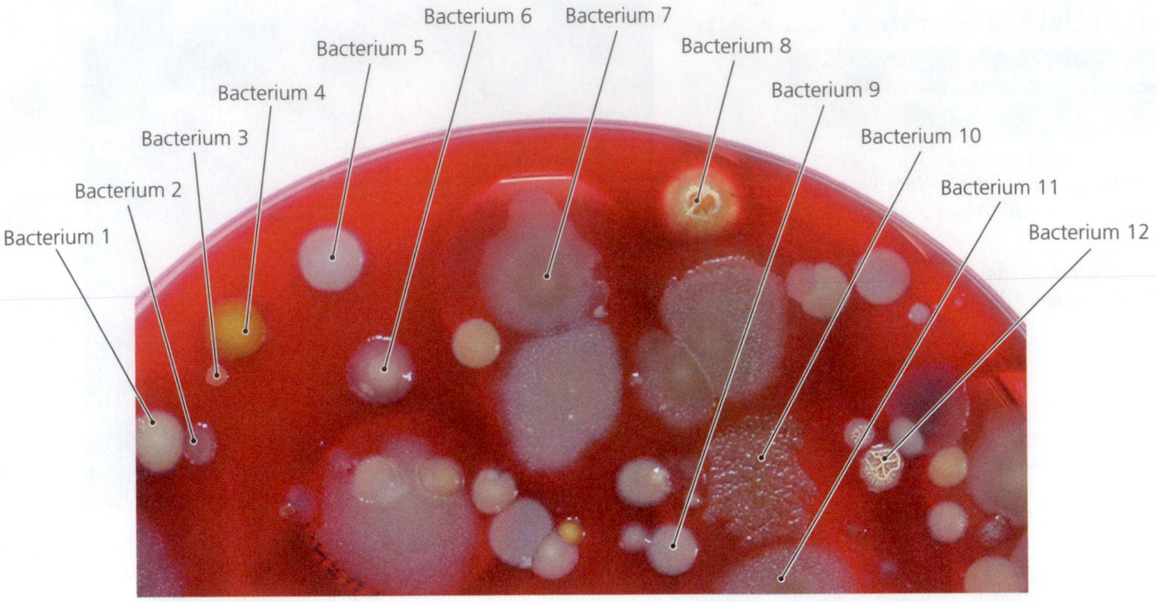

◀ **FIGURE 1.16 Bacterial colonies on a solid surface (agar).** Differences in colony size, shape, and color indicate the presence of different species. Such differences allowed Koch to isolate specific types of microbes that could be tested for their ability to cause disease.

Koch's Postulates

After discovering the anthrax bacterium, Koch continued to search for disease agents. In two pivotal scientific publications in 1882 and 1884, he announced that the cause of tuberculosis was a rod-shaped bacterium, *Mycobacterium tuberculosis* (mī´kō-bak-tēr´ē-ŭm too-ber-kyū-lō´sis). In 1905 he received the Nobel Prize in Physiology or Medicine for this work.

In his publications on tuberculosis, Koch elucidated a series of steps that must be taken to prove the cause of any infectious disease. These steps, now known as **Koch's postulates**, are one of his important contributions to microbiology. His postulates (which we discuss in more detail in Chapter 14) are the following:

1. The suspected causative agent must be found in every case of the disease and be absent from healthy hosts.
2. The agent must be isolated and grown outside the host.
3. When the agent is introduced to a healthy, susceptible host, the host must get the disease.
4. The same agent must be found in the diseased experimental host.

We use the term *suspected causative agent* because it is merely "suspected" until the postulates have been fulfilled, and "agent" can refer to any fungus, protozoan, bacterium, virus, or other pathogen. There are practical and ethical limits in the application of Koch's postulates, but in almost every case they must be satisfied before the cause of an infectious disease is proven.

During microbiology's "Golden Age," other scientists used Koch's postulates as well as laboratory techniques introduced by Koch and Pasteur to discover the causes of most protozoan and bacterial diseases as well as some viral diseases. For example, Charles Laveran (1845–1922) showed that a protozoan is the cause of malaria, and Edwin Klebs (1834–1913) described the bacterium that causes diphtheria. Dmitri Ivanovsky (1864–1920) and Martinus Beijerinck (1851–1931) discovered that a certain disease in tobacco plants is caused by a pathogen that passes through filters with such extremely small pores that bacteria cannot pass through. Beijerinck, recognizing that the pathogen was not bacterial, called it a *filterable virus*. Now such pathogens are simply called *viruses*. As previously noted, viruses could not be seen until electron microscopes were invented in 1932. The American physician Walter Reed (1851–1902) proved in 1900 that viruses can cause such diseases as yellow fever in humans. (Chapter 13 deals with *virology*, and Chapters 18–24 deal with viral diseases.)

A partial list of scientists and the pathogens they discovered is provided in **TABLE 1.2**.

Gram's Stain

The first of Koch's postulates demands that the suspected agent be found in every case of a given disease, which presupposes that minute microbes can be seen and identified. However, because most microbes are colorless and difficult to see, scientists began to use dyes to stain them and make them more visible under the microscope.

Though Koch reported a simple staining technique in 1877, the Danish scientist Hans Christian Gram (1853–1938) developed a more important staining technique in 1884. His procedure, which involves the application of a series of dyes, leaves some microbes purple and others pink. We now label the first group of cells as *Gram positive* and the second as *Gram negative*, and we use the Gram procedure to separate bacteria into these two large groups (**FIGURE 1.17**).

The **Gram stain** is still the most widely used staining technique. It is one of the first steps carried out when bacteria are being identified, and it is one of the procedures you will learn in microbiology lab. (Chapter 4 discusses the full procedure.)

TABLE 1.2 Other Notable Scientists of the "Golden Age of Microbiology" and the Agents of Disease They Discovered

Scientist	Year	Disease	Agent
Albert Neisser	1879	Gonorrhea	*Neisseria gonorrhoeae* (bacterium)
Charles Laveran	1880	Malaria	*Plasmodium* species (protozoa)
Carl Eberth	1880	Typhoid fever	*Salmonella enterica* serotype Typhi (bacterium)
Edwin Klebs	1883	Diphtheria	*Corynebacterium diphtheriae* (bacterium)
Theodor Escherich	1884	Traveler's diarrhea Bladder infection	*Escherichia coli* (bacterium)
Albert Fraenkel	1884	Pneumonia	*Streptococcus pneumoniae* (bacterium)
David Bruce	1887	Undulant fever (brucellosis)	*Brucella melitensis* (bacterium)
Anton Weichselbaum	1887	Meningococcal meningitis	*Neisseria meningitidis* (bacterium)
A. A. Gartner	1888	Salmonellosis (form of food poisoning)	*Salmonella* species (bacterium)
Shibasaburo Kitasato	1889	Tetanus	*Clostridium tetani* (bacterium)
Dmitri Ivanovsky and Martinus Beijerinck	1892 1898	Tobacco mosaic disease	*Tobamovirus tobacco mosaic virus*
William Welch and George Nuttall	1892	Gas gangrene	*Clostridium perfringens* (bacterium)
Alexandre Yersin and Shibasaburo Kitasato	1894	Bubonic plague	*Yersinia pestis* (bacterium)
Kiyoshi Shiga	1898	Shigellosis (a type of severe diarrhea)	*Shigella dysenteriae* (bacterium)
Walter Reed	1900	Yellow fever	*Flavivirus yellow fever virus*
Robert Forde and Joseph Dutton	1902	African sleeping sickness	*Trypanosoma brucei gambiense* (protozoan)

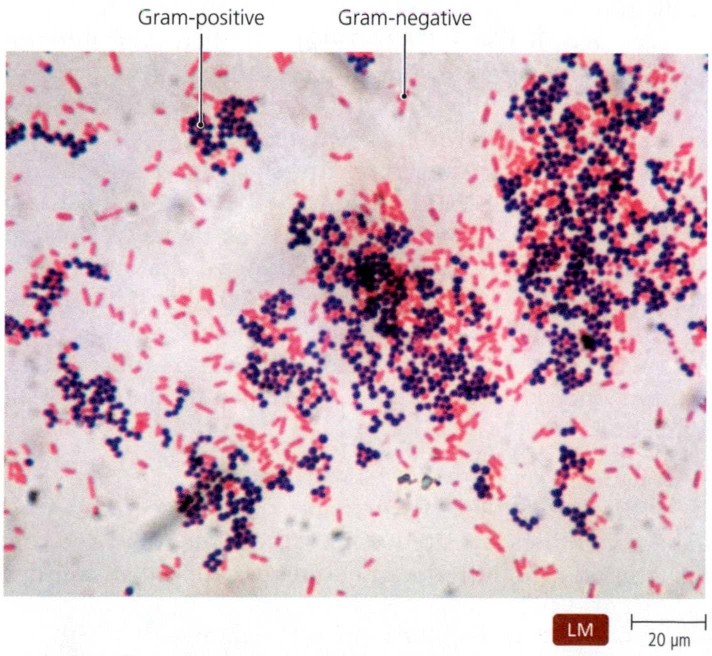

▲ **FIGURE 1.17 Results of Gram staining.** Gram-positive cells (in this case *Staphylococcus aureus*) are purple; Gram-negative cells (in this case *Escherichia coli*) are pink.

How Can We Prevent Infection and Disease?

LEARNING OUTCOMES

1.16 Identify six health care practitioners who did pioneering research in the areas of public health microbiology and epidemiology.

1.17 Name two scientists whose work with vaccines began the field of immunology.

1.18 Describe the quest for a "magic bullet."

The last great question that drove microbiological research during the "Golden Age" was how to prevent infectious diseases. Though some methods of preventing or limiting disease were discovered even before it was understood that microorganisms caused contagious diseases, great advances occurred only after Pasteur and Koch showed that life comes from life and that microorganisms can cause diseases.

In the mid-1800s, modern principles of hygiene, such as those involving sewage and water treatment, personal cleanliness, and pest control, were not widely practiced. Typically, medical personnel and health care facilities lacked adequate

CLINICAL CASE STUDY

Remedy for Fever or Prescription for Death?

In the late 18th century, Philadelphia was one of the larger and wealthier cities in the United States and served as the capital. That changed in 1793. The city had an unusually wet spring, which left behind swamps and stagnant pools that became breeding grounds for mosquitoes. Later, refugees from the slave revolution in Haiti fled to Philadelphia, carrying the yellow fever virus. In late August 1793, a female *Aedes aegypti* mosquito bit an infected refugee and then bit a healthy Philadelphian. This began a yellow fever epidemic that killed 10% of the city's population within three months and forced another 30% to flee for their lives. Victims suffered from high fever, nausea, skin eruptions, black vomit, and jaundice.

The treatment, however, was worse than the disease: physicians administered potions to purge the victims' intestines and drained up to four-fifths of their patients' blood in the mistaken belief the bloodletting would stem fever. These attempted remedies were more harmful than helpful, leaving the patients tired, weak, and unable to fight the infection. Without effective treatments, the epidemic stopped only when the first frost arrived.

1. People who left the city seemed to have milder cases of yellow fever or avoided the infection altogether. Explain why.
2. The story mentions that the coming of the first frost brought an end to the epidemic. Discuss the possible reasons why this would provide at least temporary relief from the epidemic.

cleanliness. **Healthcare associated infections (HAI**, formerly called *nosocomial*[19] *infections*) were rampant. For example, surgical patients frequently succumbed to gangrene acquired while under their doctor's care, and many women who gave birth in hospitals died from puerperal[20] fever. Six health care practitioners who were especially instrumental in changing health care delivery methods were Semmelweis, Lister, Nightingale, Snow, Jenner, and Ehrlich.

Semmelweis and Handwashing

Ignaz Semmelweis (1818–1865) was a physician on the obstetric ward of a teaching hospital in Vienna. In about 1848, he observed that women giving birth in the wing where medical students were trained died from puerperal fever at a rate 20 times higher than the mortality rates of either women attended by midwives in an adjoining wing or women who gave birth at home.

Though Pasteur had not yet elaborated his germ theory of disease, Semmelweis hypothesized that medical students carried "cadaver particles" from their autopsy studies into the delivery rooms and that these "particles" resulted in puerperal fever. Semmelweis gained support for his hypothesis when a doctor who sliced his finger during an autopsy died after showing symptoms similar to those of puerperal fever. Today we know that the primary cause of puerperal fever is a bacterium in the genus *Streptococcus* (strep-tō-kok´ŭs on p. 4; see Figure 1.4), which is usually harmless on the skin or in the mouth but causes severe complications when it enters the blood.

Semmelweis began requiring medical students to wash their hands with chlorinated lime water, a substance long used to eliminate the smell of cadavers. Mortality in the subsequent year dropped from 18.3% to 1.3%. Despite his success, Semmelweis was ridiculed by the director of the hospital and eventually was forced to leave. He returned to his native Hungary, where his insistence on handwashing met with general approval when it continued to produce higher patient survival rates.

Though his impressive record made it easier for later doctors to institute changes, Semmelweis was unsuccessful in gaining support for his method from most European doctors. He became severely depressed and was committed to a mental hospital, where he died from an infection of *Streptococcus*, the very organism he had fought for so long.

Lister's Antiseptic Technique

Shortly after Semmelweis was rejected in Vienna, the English physician Joseph Lister (1827–1912) modified and advanced the idea of *antisepsis*[21] in health care settings. As a surgeon, Lister was aware of the dreadful consequences that resulted from the infection of wounds. Therefore, he began spraying wounds, surgical incisions, and dressings with carbolic acid (phenol), a chemical that had previously proven effective in reducing odor and decay in sewage. Like Semmelweis, he initially met with some resistance, but when he showed that it reduced deaths among his patients by two-thirds, his method was accepted into common practice. In this manner, Lister vindicated Semmelweis, became the founder of antiseptic surgery, and opened new fields of research into antisepsis and disinfection.

Nightingale and Nursing

Florence Nightingale (1820–1910) **(FIGURE 1.18)** was a dedicated English nurse who introduced cleanliness and other antiseptic

[19]From Greek *nosos*, meaning "disease," and *komein*, meaning "to care for" (relating to a hospital).
[20]From Latin *puerperus*, meaning "childbirth."
[21]From Greek *anti*, meaning "against," and *sepein*, meaning "putrefaction."

▲ **FIGURE 1.18 Florence Nightingale.** The founder of modern nursing, she was influential in introducing antiseptic technique into nursing practice.

techniques into nursing practice. She was instrumental in setting standards of hygiene that saved innumerable lives during the Crimean War of 1854–1856. One of her first requisitions in the military hospital was for 200 scrubbing brushes, which she and her assistants used diligently in the squalid wards. She next arranged for each patient's filthy clothes and dressings to be replaced or cleaned at a different location, thus removing many sources of infection. She thoroughly documented statistical comparisons to show that poor food and unsanitary conditions in the hospitals were responsible for the deaths of many soldiers.

After the war, Nightingale returned to England, where she actively exerted political pressure to reform hospitals and implement public health policies. Perhaps her greatest achievements were in nursing education. For example, she founded the Nightingale School for Nurses—the first of its kind in the world.

Snow and Epidemiology

Another English physician, John Snow (1813–1858), also played a key role in setting standards for good public hygiene to prevent the spread of infectious diseases. Snow had been studying the propagation of cholera and suspected that the disease was spread by a contaminating agent in water. In 1854, he mapped the occurrence of cholera cases during an epidemic in London and showed that they centered around a public water supply on Broad Street.

Though Snow did not know the cause of cholera, his careful documentation of the epidemic highlighted the critical need for adequate sewage treatment and a pure water supply. His study was the foundation for two branches of microbiology—**infection control** and **epidemiology**,[22] which is the study of the occurrence, distribution, and spread of disease in humans.

Jenner's Vaccine

In 1796, the English physician Edward Jenner (1749–1823) tested the hypothesis that a mild disease called cowpox provided protection against potentially fatal smallpox. After he intentionally inoculated a boy with pus collected from a milkmaid's cowpox lesion, the boy developed cowpox and survived. When Jenner then infected the boy with smallpox pus, he found that the boy had become immune[23] to smallpox. (Note that experiments that intentionally expose human subjects to deadly pathogens are unethical.) In 1798, Jenner reported similar results from additional experiments, demonstrating the validity of the procedure he named *vaccination* after *Vaccinia virus*,[24] the virus that causes cowpox. Because vaccination stimulates a long-lasting response by the body's protective immune system, the term *immunization* is often used synonymously today. Jenner began the field of **immunology**—the study of the body's specific defenses against pathogens. (Chapters 16–18 discuss immunology.)

Pasteur later capitalized on Jenner's work by producing weakened strains of various pathogens for use in preventing the serious diseases they cause. In honor of Jenner's work with cowpox, Pasteur used the term *vaccine* to refer to all weakened, protective strains of pathogens. He subsequently developed successful vaccines against fowl cholera, anthrax, and rabies.

Ehrlich's "Magic Bullets"

Gram's discovery that stained bacteria could be differentiated into two types suggested to the German microbiologist Paul Ehrlich (1854–1915) that chemicals could be used to kill microorganisms differentially. To investigate this idea, Ehrlich undertook an exhaustive survey of chemicals to find a "magic bullet" that would destroy pathogens while remaining nontoxic to humans. By 1908, he had discovered a chemical active against the causative agent of syphilis, though the arsenic-based drug was toxic to humans. His discoveries began the branch of medical microbiology known as **chemotherapy**.

In summary, the Golden Age of Microbiology was a time when researchers proved that living things come from other living things, that microorganisms can cause fermentation and disease, and that certain procedures and chemicals can limit, prevent, and cure infectious diseases. These discoveries were made by scientists who applied the scientific method to biological investigation, and they led to an explosion of knowledge in a number of scientific disciplines **(FIGURE 1.19)**.

TELL ME WHY

Some people consider Pasteur or Koch to be the Father of Microbiology, rather than Leeuwenhoek. Why might they be correct?

[22]From Greek *epi*, meaning "upon"; *demos*, meaning "people"; and *logos*, meaning "word" or "study."
[23]From Latin *immunis*, meaning "free."
[24]From Latin *vacca*, meaning "cow."

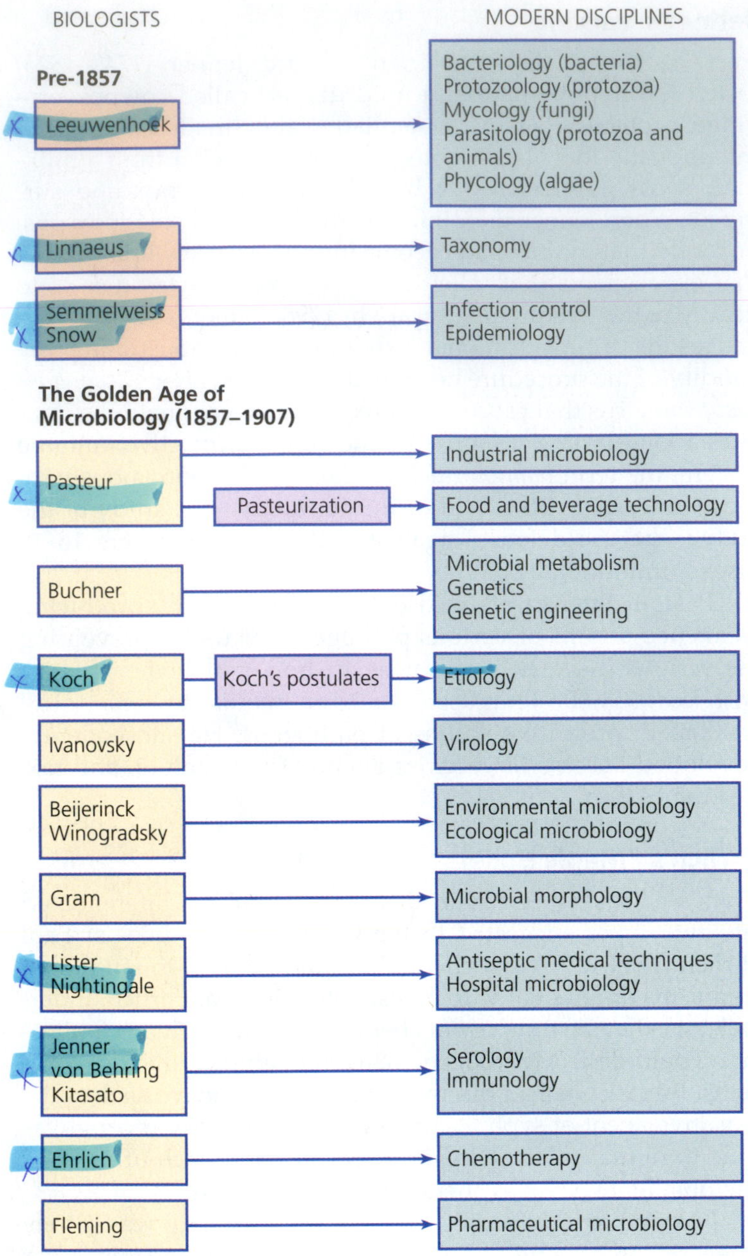

▲ **FIGURE 1.19 Some of the many scientific disciplines and applications that arose from the pioneering work of scientists just before and around the time of the Golden Age of Microbiology.**

TABLE 1.3 Fields of Microbiology

Disciplines	Subject(s) of Study
Basic Research	
Microbe Centered	
Bacteriology	Bacteria and archaea
Phycology	Algae
Mycology	Fungi
Protozoology	Protozoa
Parasitology	Parasitic protozoa and parasitic animals
Virology	Viruses
Process Centered	
Microbial metabolism	Biochemistry: chemical reactions within cells
Microbial genetics	Functions of DNA and RNA
Environmental microbiology	Relationships between microbes and among microbes, other organisms, and their environment
Applied Microbiology	
Medical Microbiology	
Serology	Antibodies in blood serum, particularly as an indicator of infection
Immunology	Body's defenses aginst specific diseases
Epidemiology	Frequency, distribution, and spread of disease
Etiology	Causes of disease
Infection control	Hygiene in health care settings and control of nosocomial infections
Chemotherapy	Development and use of drugs to treat infectious diseases
Applied Environmental Microbiology	
Bioremediation	Use of microbes to remove pollutants
Public health microbiology	Sewage treatment, water purification, and control of insects that spread disease
Agricultural microbiology	Use of microbes to control insect pests
Industrial Microbiology (Biotechnology)	
Food and beverage technology	Reduction or elimination of harmful microbes in food and drink
Pharmaceutical microbiology	Manufacture of vaccines and antibiotics
Recombinant DNA technology	Alteration of microbial genes to synthesize useful products

The Modern Age of Microbiology

LEARNING **OUTCOME**

1.19 List four major questions that drive microbiological investigations today.

The vast increase in the number of microbiological investigations and in scientific knowledge during the 1800s opened new fields of science, including disciplines called environmental science, immunology, epidemiology, chemotherapy, and genetic engineering **(TABLE 1.3)**. Microorganisms played a significant role in the development of these disciplines because microorganisms are relatively easy to grow, take up little space, and are available by the trillions. Much of what has been learned about microbes also applies to other organisms, including humans. In the rest of this text we examine advances made in these branches of microbiology, though it would require thousands of books this size to deal with all that is known.

Once the developing science of microbiology had successfully answered questions about spontaneous generation, fermentation, and disease, additional questions arose in each branch of the new science. Since the early 20th century, microbiologists have worked to answer these new questions. In this section, we briefly consider some of the 20th century's overarching questions in both basic and applied research. The chapter concludes with a look at some of the questions that might propel microbiological research for the next 50 years.

What Are the Basic Chemical Reactions of Life?

Biochemistry is the study of metabolism—that is, the chemical reactions that occur in living organisms. Biochemistry began with Pasteur's work on fermentation by yeast and bacteria and with Buchner's discovery of enzymes in yeast extract, but by the early 1900s, many scientists thought that the metabolic reactions of microbes had little to do with the metabolism of plants and animals.

In contrast, microbiologists Albert Kluyver (1888–1956) and his student C. B. van Niel (1897–1985) proposed that basic biochemical reactions are shared by all living things, that these reactions are relatively few in number, and that their primary feature is the transfer of electrons and hydrogen ions. In adopting this view, scientists could use microbes as model systems to answer questions about metabolism in all organisms. Research during the 20th century validated this approach to understanding basic metabolic processes, but scientists have also documented an amazing metabolic diversity. (Chapter 5 discusses basic metabolic processes, and Chapter 6 considers metabolic diversity.)

Basic biochemical research has many practical applications, including the following:

- The design of herbicides and pesticides that are specific in their action and have no long-term adverse effects on the environment.
- The diagnosis of illnesses and the monitoring of a patient's responses to treatment. For example, physicians routinely monitor liver disease by measuring blood levels of certain enzymes and products of liver metabolism.
- The treatment of metabolic diseases. One example is treating phenylketonuria, a disease resulting from the inability to properly metabolize the amino acid phenylalanine, by eliminating foods containing phenylalanine from the diet.
- The design of drugs to treat leukemia, gout, bacterial infections, malaria, herpes, AIDS, asthma, and heart attacks.

How Do Genes Work?

Genetics, the scientific study of inheritance, started in the mid-1800s as an offshoot of botany, but scientists studying microbes made most of the great advances in this discipline.

Microbial Genetics

While working with the bacterium *Streptococcus pneumoniae* (strep-tō-kok´ŭs nū-mō´nē-ī), Oswald Avery (1877–1955), Colin MacLeod (1909–1972), and Maclyn McCarty (1911–2005) determined that genes are contained in molecules of DNA. In 1958, George Beadle (1903–1989) and Edward Tatum (1909–1975), working with the bread mold *Neurospora crassa* (noo-ros´pōr-ă kras´ă), established that a gene's activity is related to the function of the specific protein coded by that gene. Other researchers, also working with microbes, determined the exact way in which genetic information is translated into a protein, the rates and mechanisms of genetic mutation, and the methods by which cells control genetic expression. (Chapter 7 examines all these aspects of microbial genetics.)

Over the past 40 years, advances in microbial genetics developed into several new disciplines that are among the faster-growing areas of scientific research today, including *molecular biology, recombinant DNA technology,* and *gene therapy.*

Molecular Biology

Molecular biology combines aspects of biochemistry, cell biology, and genetics to explain cell function at the molecular level. Molecular biologists are particularly concerned with *genome*[25] *sequencing.* Using techniques perfected on microorganisms, molecular biologists have sequenced the genomes of many organisms, including humans and many of their pathogens. It is hoped that a fuller understanding of the genomes of organisms will result in practical ways to limit disease, repair genetic defects, and enhance agricultural yield.

The American Nobel Prize winner Linus Pauling (1901–1994) proposed in 1965 that gene sequences could provide a means of understanding evolutionary relationships and processes, establishing taxonomic categories that more closely reflect these relationships, and identifying the existence of microbes that have never been cultured in a laboratory. Two examples illustrate such uses of gene sequencing data:

- In the 1970s, Carl Woese (1928–2012) discovered that significant differences in nucleic acid sequences among organisms clearly reveal that cells belong to one of *three* major groups—bacteria, archaea, or eukaryotes—and not merely two groups (prokaryotes and eukaryotes), as previously thought.
- Scientists showed in 1990 that cat scratch disease is caused by a bacterium that had not been cultured. The bacterium was discovered by recognizing the sequence of a portion of its ribonucleic acid that differs from all other known ribonucleic acid sequences.

Recombinant DNA Technology

Molecular biology is applied in **recombinant DNA technology,**[26] commonly called *genetic engineering,* which was first developed using microbial models. Geneticists manipulate genes in microbes, plants, and animals for practical applications. For instance, once scientists have inserted the gene for human blood-clotting factor into the bacterium *Escherichia coli* (esh-ĕ-rik´ē-ă kō´ lī), the bacterium produces the factor in a pure form.

[25]A genome is the total genetic information of an organism.
[26]Recombinant DNA is DNA composed of genes from more than one organism.

This technology is a benefit to hemophiliacs, who previously depended on a clotting factor isolated from donated blood, which was possibly contaminated by life-threatening viral pathogens.

Gene Therapy

An exciting new area of study is the use of recombinant DNA technology for **gene therapy**, a process that involves inserting a missing gene or repairing a defective one in human cells. In such procedures, researchers insert a desired gene into host cells, where it is incorporated into a chromosome and begins to function normally. (Chapter 8 examines recombinant DNA technology and gene therapy in more detail.)

What Roles Do Microorganisms Play in the Environment?

LEARNING | **OUTCOMES**

1.20 Identify the field of microbiology that studies the role of microorganisms in the environment.

1.21 Name the fastest-growing scientific disciplines in microbiology today.

Ever since Koch and Pasteur, most research in microbiology has focused on pure cultures of individual species; however, microorganisms are not alone in the "real world." Instead, they live in natural microbial communities in the soil, water, the human body, and other habitats, and these communities play critical roles in such processes as the production of vitamins and *bioremediation*—the use of living bacteria, fungi, and algae to detoxify polluted environments.

Microbial communities also play an essential role in the decay of dead organisms and the recycling of chemicals such as carbon, nitrogen, and sulfur. Martinus Beijerinck discovered bacteria capable of converting nitrogen gas (N_2) from the air into nitrate (NO_3), the form of nitrogen used by plants, and the Russian microbiologist Sergei Winogradsky (1856–1953) elucidated the role of microorganisms in the recycling of sulfur. Together these two microbiologists developed laboratory techniques for several important aspects of **environmental microbiology**.

Another role of microbes in the environment is the causation of disease. Although most microorganisms are not pathogenic, in this book (particularly in Chapters 18–25), we focus on pathogenic microbes because of the threat they pose to human health. We examine their characteristics and the diseases they cause as well as the steps we can take to limit their abundance and control their spread in the environment, such as sewage treatment, water purification, disinfection, pasteurization, and sterilization.

How Do We Defend Against Disease?

Why do some people get sick during the flu season while their close friends and family remain well? The germ theory of disease showed not only that microorganisms can cause diseases, but also that the body can defend itself—otherwise, everyone would be sick most of the time.

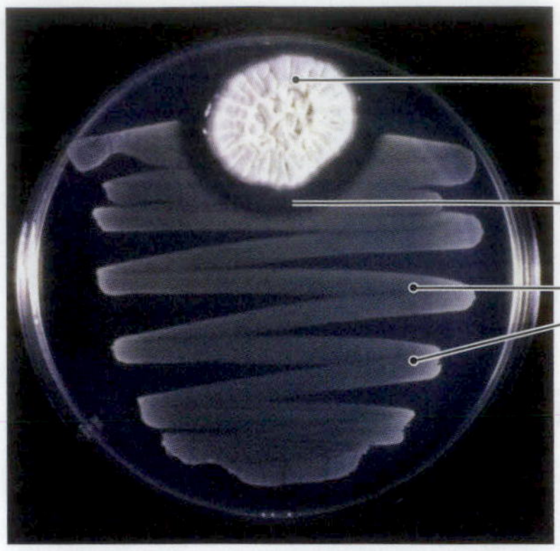

▲ **FIGURE 1.20 The effects of penicillin on a bacterial "lawn" in a Petri dish.** The clear area (zone of inhibition) surrounding the fungus colony, which is producing the antibiotic, is where the penicillin prevented bacterial growth.

The work of Jenner and Pasteur on vaccines showed that the body can protect itself from repeated diseases by the same organism. The German bacteriologist Emil von Behring (1854–1917) and the Japanese microbiologist Shibasaburo Kitasato (1852–1931), working in Koch's laboratory, reported the existence in the blood of chemicals and cells that fight infection. Their studies developed into the fields of *serology,* the study of blood serum[27]—specifically, the chemicals in the liquid portion of blood that fight disease—and *immunology,* the study of the body's defense against specific pathogens. (Chapters 15–18 cover these aspects of microbiology, which are of utmost importance to physicians, nurses, and other health care practitioners.)

Ehrlich introduced the idea of a "magic bullet" that would kill pathogens, but it was not until Alexander Fleming (1881–1955) discovered penicillin **(FIGURE 1.20)** in 1929 and Gerhard Domagk (1895–1964) discovered sulfa drugs in 1935 that medical personnel finally had drugs effective against a wide range of bacteria. (We study chemotherapy and some physical and chemical agents used to control microorganisms in the environment in Chapters 9 and 10.)

What Will the Future Hold?

Science is built on asking and answering questions. What began with the questioning curiosity of a dedicated lens grinder in the Netherlands has come far in the past 350 years, expanding into disciplines as diverse as immunology, recombinant DNA technology, and bioremediation. However, the adage remains true: *The more questions we answer, the more questions we have.*

What will microbiologists discover next? Among the questions for the next 50 years are the following:

- How can we develop successful programs to control or eradicate diseases such as tuberculosis, malaria, and AIDS?
- What is it about the physiology of life forms known only by their nucleic acid sequences that has prevented researchers from growing them in the laboratory?

[27]Latin, meaning "whey." Serum is the liquid that remains after blood coagulates.

- Can bacteria and archaea be used in ultraminiature technologies, such as living computer circuit boards?
- How can an understanding of microbial communities help us understand the positive aspects of microbial action in preventing and curing diseases, recycling nutrients, degrading pollutants, and moderating climate changes?
- How can we reduce the threat from microbes resistant to antimicrobial drugs as well as conquer emerging and reemerging infectious diseases?
- How do bacteria communicate with one another to form *biofilms*—aggregates of cells on a surface—that can have very different properties from the individual cells that compose them?

TELL ME WHY

Why are so many modern questions in microbiology related to genetics?

EMERGING DISEASE CASE STUDY

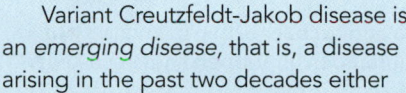

Variant Creutzfeldt-Jakob Disease

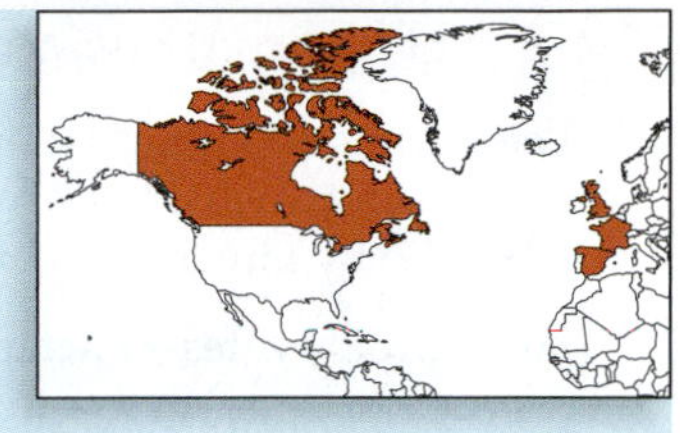

Ellen screamed obscenities as she staggered from the room and collapsed in the hallway, unable to stand and jerking uncontrollably. Her parents were shocked that their kind, considerate, and lovable daughter had changed so drastically during the past year. Sadly, she couldn't even remember her siblings' names.

Ellen had joined the nearly 200 Europeans and one Canadian afflicted with variant Creutzfeldt-Jakob disease (vCJD) (what the media call "mad cow disease" because most humans with the condition acquired the pathogen from eating infected beef). Because vCJD affects the brain by slowly eroding nervous tissue and leaving the brain full of spongelike holes, the signs and symptoms of vCJD are neurological. Ellen's disease started with insomnia, depression, and confusion, but eventually it led to uncontrollable emotional and verbal outbursts, inability to coordinate movements, coma, and death. Typically the disease lasts about a year, and there is no treatment.

Variant Creutzfeldt-Jakob disease is an *emerging disease*, that is, a disease arising in the past two decades either because it is new to a population or because it is newly recognized. Some investigators also include diseases that have been nearly eradicated but are now reemerging. Variant CJD resembles the rare genetic disorder Creutzfeldt-Jakob disease (named for its discoverers), which is caused by a mutation and occurs in the elderly. The difference is that the variant form of CJD results from an acquired infection and often strikes and kills college-aged people, like Ellen in our story. For more about vCJD, see pp. 629–630.

1. The vCJD pathogen is primarily transmitted when a person or animal consumes nervous tissue (brains). How could cattle become infected?
2. Why is vCJD called *variant*?
3. What effect does this pathogen have on cattle?

MICRO IN THE CLINIC FOLLOW-UP

A Simple Case of Traveler's Diarrhea?

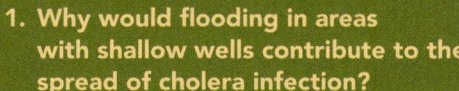

Being a nursing student, Martin suspects he may have contracted *cholera*—a serious illness that, if untreated, can cause severe dehydration and death. He seeks immediate medical care, and a stool culture confirms the diagnosis: infection with *Vibrio cholerae*. Martin is given intravenous fluids and recovers fully.

Outbreaks of cholera occur in areas of South America, Asia, and Africa. If there are people with cholera in an area with poor sanitation, flooding can spread infection, especially in areas with shallow wells. Some people infected with cholera never show any symptoms at all; others experience particularly severe symptoms and die within days or even hours of infection.

1. **Why would flooding in areas with shallow wells contribute to the spread of cholera infection?**
2. **Why do you think cholera is no longer a problem in the United States?**

MM Check your answers to Micro in the Clinic Follow-Up questions in the MasteringMicrobiology Study Area.

Explore the Invisible: The Scientific Method

Practice on-the-go with Dr. Bauman Video Tutors by scanning this QR code with your smart phone. Visit the **MasteringMicrobiology Study Area** to challenge your understanding with practice tests, animation quizzes, and clinical case studies!

MasteringMicrobiology®

CHAPTER SUMMARY

The Early Years of Microbiology (pp. 2–7)

1. Leeuwenhoek's observations of **microbes** introduced most types of **microorganisms** to the world. His discoveries were named and classified by Linnaeus in his **taxonomic system**.
2. Small **prokaryotes**—**bacteria** and **archaea**—live in a variety of communities and in most habitats. Even though some cause disease, most are beneficial.
3. Relatively large microscopic **eukaryotic fungi** include **molds** and **yeasts**.
4. Animal-like **protozoa** are single-celled eukaryotes. Some cause disease.
5. Plant-like eukaryotic **algae** are important providers of oxygen, serve as food for many marine animals, and make chemicals used in microbiological growth media.
6. Parasitic worms, the largest organisms studied by microbiologists, are often visible without a microscope, although their immature stages are microscopic.
7. Viruses, the smallest microbes, are so small they can be seen only by using an electron microscope.

The Golden Age of Microbiology (pp. 7–18)

1. The study of the Golden Age of Microbiology includes a look at the men who proposed or refuted the theory of **spontaneous generation**: Aristotle, Redi, Needham, Spallanzani, and Pasteur (the Father of Microbiology). The **scientific method** that emerged then remains the accepted sequence of study today.

   ▶**VIDEO TUTOR:** *The Scientific Method*
2. The study of fermentation by Pasteur and Buchner led to the fields of **industrial microbiology** (**biotechnology**) and **biochemistry** and to the study of **metabolism**.
3. Koch, Pasteur, and others proved that **pathogens** cause infectious diseases, an idea that is known as the **germ theory of disease. Etiology** is the study of the causation of diseases.
4. Koch initiated careful microbiological laboratory techniques in his search for disease agents. **Koch's postulates**, the logical steps he followed to prove the cause of an infectious disease, remain an important part of microbiology today.
5. The procedure for the **Gram stain** was developed in the 1880s and is still used to differentiate bacteria into two categories: Gram positive and Gram negative.
6. The investigations of Semmelweis, Lister, Nightingale, and Snow are the foundations on which **infection control** and **epidemiology** are built.
7. Jenner's use of a cowpox-based vaccine for preventing smallpox began the field of **immunology**. Pasteur significantly advanced the field.
8. Ehrlich's search for "magic bullets"—chemicals that differentially kill microorganisms—laid the foundations for the field of **chemotherapy**.

The Modern Age of Microbiology (pp. 18–21)

1. Microbiology in the modern age has focused on answering questions regarding **biochemistry**, which is the study of metabolism; microbial genetics, which is the study of inheritance in microorganisms; and **molecular biology**, which involves investigations of cell function at the molecular level.
2. Scientists have applied knowledge from basic research to answer questions in **recombinant DNA technology** and **gene therapy**.
3. The study of microorganisms in their natural environment is **environmental microbiology**.
4. The discovery of chemicals in the blood that are active against specific pathogens advanced immunology and began the field of serology.
5. Advancements in chemotherapy were made in the 1900s with the discovery of numerous substances, such as penicillin and sulfa drugs, that inhibit pathogens.

QUESTIONS FOR REVIEW

Answers to the Questions for Review (except Short Answer questions) begin on p. A-1.

Multiple Choice

1. Which of the following microorganisms are *not* eukaryotic?
   a. bacteria
   b. yeasts
   c. molds
   d. protozoa
2. Which microorganisms are used to make microbiological growth media?
   a. bacteria
   b. fungi
   c. algae
   d. protozoa
3. In which habitat would you most likely find archaea?
   a. acidic hot springs
   b. swamp mud
   c. Great Salt Lake
   d. all of the above
4. Of the following scientists, who first promulgated the theory of abiogenesis?
   a. Aristotle
   b. Pasteur
   c. Needham
   d. Spallanzani
5. Which of the following scientists hypothesized that a bacterial colony arises from a single bacterial cell?
   a. Antoni van Leeuwenhoek
   b. Louis Pasteur
   c. Robert Koch
   d. Richard Petri
6. Which scientist first hypothesized that medical personnel can infect patients with pathogens?
   a. Edward Jenner
   b. Joseph Lister
   c. John Snow
   d. Ignaz Semmelweis
7. Leeuwenhoek described microorganisms as ________________.
   a. animalcules
   b. prokaryotes
   c. eukaryotes
   d. protozoa
8. Which of the following favored the theory of spontaneous generation?
   a. Spallanzani
   b. Needham
   c. Pasteur
   d. Koch
9. A scientist who studies the role of microorganisms in the environment is ________________.
   a. a genetic techologist
   b. an earth microbiologist
   c. an epidemiologist
   d. an environmental microbiologist
10. The laboratory of Robert Koch contributed which of the following to the field of microbiology?
    a. simple staining technique
    b. use of Petri dishes
    c. first photomicrograph of bacteria
    d. all of the above

Fill in the Blanks

Fill in the blanks with the name(s) of the scientist(s) whose investigations led to the following fields of study in microbiology.

1. Environmental microbiology ________________ and ________________
2. Biochemistry ________________ and ________________
3. Chemotherapy ________________
4. Immunology ________________
5. Infection control ________________
6. Etiology ________________
7. Epidemiology ________________
8. Biotechnology ________________
9. Food microbiology ________________

VISUALIZE IT!

1. On the photos below, label *cilium, flagellum, nucleus,* and *pseudopod.*

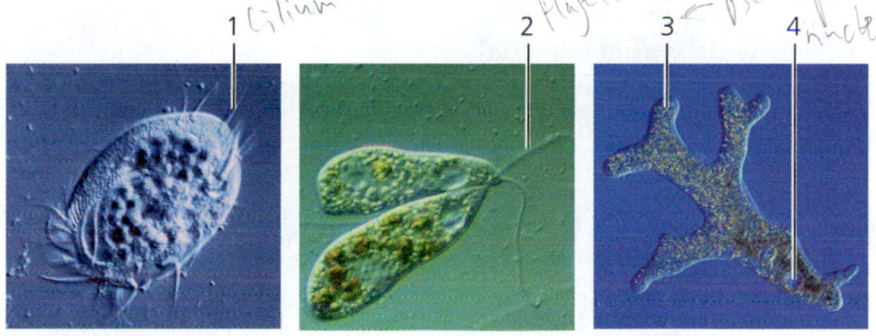

2. Show where microbes ended up in Pasteur's experiment.

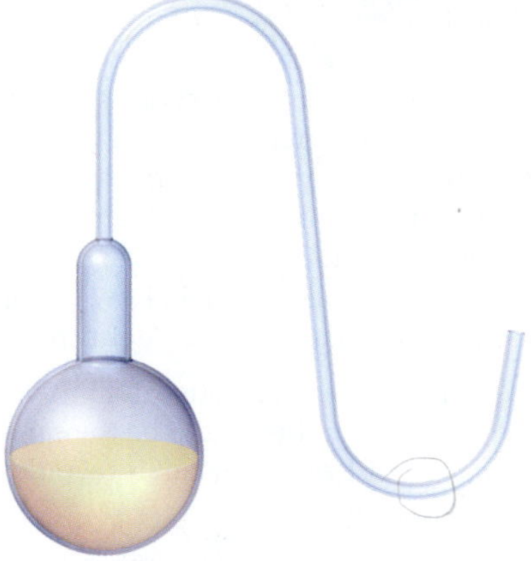

Matching

Match each of the following descriptions with the person it best describes. An answer may be used more than once.

1. J Developed smallpox immunization
2. H First photomicrograph of bacteria
3. C Germ theory of disease
4. CHK Germs cause disease
5. B Sought a "magic bullet" to destroy pathogens
6. A Early epidemiologist
7. C Father of Microbiology
8. E Classification system
9. D Discoverer of bacteria
10. D Discoverer of protozoa
11. I Founder of antiseptic surgery
12. L Developed the most widely used bacterial staining technique

A. John Snow
B. Paul Ehrlich
C. Louis Pasteur
D. Antoni van Leeuwenhoek
E. Carolus Linnaeus
F. John Needham
G. Eduard Buchner
H. Robert Koch
I. Joseph Lister
J. Edward Jenner
K. Girolamo Fracastoro
L. Hans Christian Gram

Short Answer

1. Why was the theory of spontaneous generation a hindrance to the development of the field of microbiology?
2. Discuss the significant difference between the flasks used by Pasteur and Spallanzani. How did Pasteur's investigation settle the dispute about spontaneous generation?
3. List six types of microorganisms.
4. Defend this statement: "The investigations of Antoni van Leeuwenhoek changed the world forever."
5. Why would a *macro*scopic tapeworm be studied in *micro*biology?
6. Describe what has been called the "Golden Age of Microbiology" with reference to four major questions that propelled scientists during that period.
7. List four major questions that drive microbiological investigations today.
8. Refer to the four steps in the scientific method in describing Pasteur's fermentation experiments.
9. List Koch's postulates and explain why they are significant.
10. What does the term *HAI* (nosocomial infection) have to do with patient care?

CRITICAL THINKING

1. If Robert Koch had become interested in a viral disease, such as influenza, instead of anthrax (caused by a bacterium), how might his list of lifetime accomplishments be different? Why?
2. In 1911, the Polish scientist Casimir Funk proposed that a limited diet of polished white rice (rice without the husks) caused beriberi, a disease of the central nervous system. Even though history has proven him correct—beriberi is caused by a thiamine deficiency, which in his day resulted from unsophisticated milling techniques that removed the thiamine-rich husks—Funk was criticized by his contemporaries, who told him to find the microbe that caused beriberi. Explain how the prevailing scientific philosophy of the day shaped Funk's detractors' point of view.
3. *Haemophilus influenzae* does not cause flu, but it received its name because it was once thought to be the cause. Explain how a proper application of Koch's postulates would have prevented this error in nomenclature.
4. Just before winter break in early December, your roommate stocks the refrigerator with a gallon of milk, but both of you leave before opening it. When you return in January, the milk has soured. Your roommate is annoyed because the milk was pasteurized and thus should not have spoiled. Explain why your roommate's position is unreasonable.
5. Design an experiment to prove that microbes do not spontaneously generate in milk.
6. The British General Board of Health concluded in 1855 that the Broad Street cholera epidemic discussed in the chapter (see p. 17) resulted from fermentation of "nocturnal clouds of vapor" from the polluted Thames River. How could an epidemiologist prove or disprove this claim?
7. Compare and contrast the investigations of Redi, Needham, Spallanzani, and Pasteur in relation to the idea of spontaneous generation.
8. If you were a career counselor directing a student in the field of applied microbiology, describe three possible disciplines you could suggest.
9. A few bacteria produce disease because they derive nutrition from human cells and produce toxic wastes. Algae do not typically cause disease. Why not?
10. How might the debate over spontaneous generation have been different if Buchner had conducted his experiments in 1857 instead of 1897?

11. French microbiologists, led by Pasteur, tried to isolate a single bacterium by diluting liquid media until only a single type of bacterium could be microscopically observed in a sample of the diluted medium. What advantages does Koch's method have over the French method?
12. Why aren't Koch's postulates always useful in proving the cause of a given disease? Consider a variety of diseases, such as cholera, pneumonia, Alzheimer's, AIDS, Down syndrome, and lung cancer.
13. Albert Kluyver said, "From elephant to . . . bacterium—it is all the same!" What did he mean?
14. The ability of farmers around the world to produce crops such as corn, wheat, and rice is often limited by the lack of nitrogen-based fertilizer. How might scientists use Beijerinck's discovery to increase world supplies of grain?

CONCEPT MAPPING

Using the following terms, draw a concept map that describes what microbiologists study. You may use some terms more than once. For a sample concept map, see p. 94. Or, complete this and other concept maps online by going to the Mastering-Microbiology Study Area.

Acellular
Algae
Animal-like
Archaea
Bacteria
Eukaryotes
Fungi
Molds
Multicellular
Obligate intracellular parasites
Parasitic worms
Photosynthetic
Prokaryotes
Protozoa
Unicellular (2)
Viruses
Yeasts

2 The Chemistry of Microbiology

MICRO IN THE CLINIC

Can Spicy Food Cause Ulcers?

Ramona is a young mom who takes care of her two children during the day and takes pre-nursing classes at night. Juggling the needs of her family and her studies means a hectic schedule, late nights, very little sleep, and eating on the run. Ramona particularly loves spicy food, and she eats a lot of it. She adds hot sauce to nearly every meal, which tends to be Mexican fast food. She also likes to drink wine with dinner on the weekends and sneaks an occasional cigarette when her children aren't watching.

One night, Ramona notices a burning pain in her upper abdomen. It disappears after a few minutes but then comes back a couple of nights later. Pretty soon she is feeling the pain every night—sometimes accompanied by nausea. She mentions her symptoms to her instructor, Mr. Rowe, who suggests that she might have an ulcer. Mr. Rowe knows about Ramona's love of spicy food and advises her to cut back on the hot sauce to see if that improves her symptoms. Ramona takes his advice, but the pain and nausea continue. Sometimes the symptoms occur during the day, but mostly they flare up during the night.

What do you think? Did eating too much spicy food give Ramona an ulcer? **Turn to the end of the chapter (p. 51) to find out.**

MM **Explore More:** Test your readiness and apply your knowledge with dynamic learning tools at MasteringMicrobiology.

Learning some basic concepts of chemistry will enable you to understand more fully the variety of interactions between microorganisms and their environment—which includes you. If you plan a career in health care, you will find microbial chemistry involved in the diagnosis of disease, the response of the immune system, the growth and identification of pathogenic microorganisms in the laboratory, and the function and selection of antimicrobial drugs. Understanding the fundamentals of chemistry will even help you preserve your own health.

In this chapter we study atoms, which are the basic units of chemistry, and we consider how atoms react with one another to form chemical bonds and molecules. Then we examine the three major categories of chemical reactions. The chapter concludes with a look at the molecules of greatest importance to life: water, acids, bases, lipids, carbohydrates, proteins, nucleic acids, and ATP.

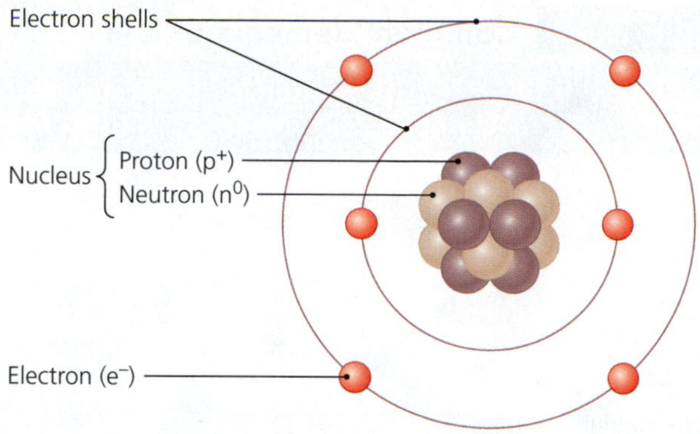

▲ **FIGURE 2.1 An example of a Bohr model of atomic structure.** This drawing is not to scale; for the electrons to be shown in scale with the greatly magnified nucleus, the electrons would have to occupy orbits located many miles from the nucleus. Put another way, the volume of an entire atom is about 100 trillion times the volume of its nucleus.

Atoms

LEARNING **OUTCOME**

2.1 Define *matter*, *atom*, and *element*, and explain how these terms relate to one another.

Matter is defined as anything that takes up space and has mass.[1] The smallest chemical units of matter are **atoms**. Atoms are extremely small, and only the very largest of them can be seen using the most powerful microscopes. Therefore, scientists have developed various models to conceptualize and illustrate the structure of atoms.

Atomic Structure

LEARNING **OUTCOME**

2.2 Draw and label an atom, showing the parts of the nucleus and orbiting electrons.

In 1913, the Danish physicist Niels H. D. Bohr (1885–1962) proposed a simple model in which negatively charged subatomic particles called **electrons** orbit a centrally located nucleus like planets in a miniature solar system (**FIGURE 2.1**). A nucleus is composed of uncharged **neutrons** and positively charged **protons**. (The only exception is the nucleus of a normal hydrogen atom, which is composed of only a single proton and no neutrons.) Protons and neutrons are extremely small. If a meterstick were stretched between the sun and the Earth, a neutron or proton would measure only about the width of a human hair! The number of electrons in an atom typically equals the number of protons, so overall atoms are electrically neutral.

An **element** is matter that is composed of a single type of atom. For example, gold is an element because it consists of only gold atoms. In contrast, the ink in your pen is not an element because it is composed of many different kinds of atoms.

Elements differ from one another in their **atomic number**, which is the number of protons in their nuclei. For example, the atomic numbers of hydrogen, carbon, and oxygen are 1, 6, and 8, respectively, because all hydrogen nuclei contain a single proton, all carbon nuclei have six protons, and all oxygen nuclei have eight protons.

The **atomic mass** of an atom (sometimes called its *atomic weight*) is the sum of the masses of its protons, neutrons, and electrons. Protons and neutrons each have a mass of approximately 1 *atomic mass unit*,[2] which is also called a *dalton*.[3] An electron is much less massive, with a mass of about 0.00054 dalton. Electrons are often ignored in discussions of atomic mass because their contribution to the overall mass is negligible. Therefore, the sum of the number of protons and neutrons approximates the atomic mass of an atom.

There are 93 naturally occurring elements known;[4] however, organisms typically utilize only about 20 elements, each of which has its own symbol that is derived from its English or Latin name (**TABLE 2.1** on p. 28).

Isotopes

LEARNING **OUTCOME**

2.3 List at least four ways that radioactive isotopes are useful.

Every atom of an element has the same number of protons, but atoms of a given element can differ in the number of neutrons in their nuclei. Atoms that differ in this way are called **isotopes**. For example, there are three naturally occurring isotopes of carbon, each having six protons and six electrons (**FIGURE 2.2**). Over 95% of carbon atoms also have six neutrons. Because these atoms have six protons and six neutrons, the atomic mass of this

[1]*Mass* and *weight* are sometimes confused. Mass is the quantity of material in something, whereas weight is the effect of gravity on mass. Even though an astronaut is weightless in space, his mass is the same in space as on Earth.
[2]An atomic mass unit (dalton) is 1/597,728,630,000,000,000,000,000, or 1.673×10^{-24}, grams.
[3]Named for John Dalton, the British chemist who helped develop atomic theory around 1800.
[4]For many years, scientists thought that there were only 92 naturally occurring elements, but natural plutonium was discovered in Africa in 1997.

TABLE 2.1 Common Elements of Life

Element	Symbol	Atomic Number	Atomic Mass[a] (daltons)	Biological Significance
Hydrogen	H	1	1	Component of organic molecules and water; H^+ released by acids
Boron	B	5	11	Essential for plant growth
Carbon	C	6	12	Backbone of organic molecules
Nitrogen	N	7	14	Component of amino acids, proteins, and nucleic acids
Oxygen	O	8	16	Component of many organic molecules and water; OH^- released by bases; necessary for aerobic metabolism
Sodium (Natrium)	Na	11	23	Principal cation outside cells
Magnesium	Mg	12	24	Component of many energy-transferring enzymes
Silicon	Si	14	28	Component of cell wall of diatoms
Phosphorus	P	15	31	Component of nucleic acids and ATP
Sulfur	S	16	32	Component of proteins
Chlorine	Cl	17	35	Principal anion outside cells
Potassium (Kalium)	K	19	39	Principal cation inside cells; essential for nerve impulses
Calcium	Ca	20	40	Utilized in many intercellular signaling processes; essential for muscular contraction
Manganese	Mn	25	54	Component of some enzymes; acts as intracellular antioxidant; used in photosynthesis
Iron (Ferrum)	Fe	26	56	Component of energy-transferring proteins; transports oxygen in the blood of many animals
Cobalt	Co	27	59	Component of vitamin B_{12}
Copper (Cuprum)	Cu	29	64	Component of some enzymes; used in photosynthesis
Zinc	Zn	30	65	Component of some enzymes
Molybdenum	Mo	42	96	Component of some enzymes
Iodine	I	53	127	Component of many brown and red algae

[a]Rounded to nearest whole number.

isotope is about 12 daltons, and it is known as carbon-12, symbolized as ^{12}C. Atoms of carbon-13 (^{13}C) have seven neutrons per nucleus, and ^{14}C atoms each have eight neutrons.

Unlike the first two isotopes, the nucleus of ^{14}C is unstable because of the ratio of its protons and neutrons. Unstable atomic nuclei release energy and subatomic particles such as neutrons, protons, and electrons in a process called *radioactive decay*. Atoms that undergo radioactive decay are *radioactive isotopes*.

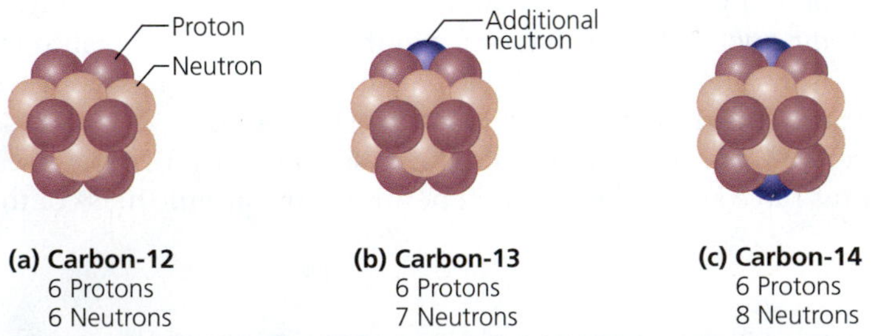

▲ **FIGURE 2.2 Nuclei of the three naturally occurring isotopes of carbon.** Each isotope also has six electrons, which are not shown. *What are the atomic number and atomic mass of each of these isotopes?*

Figure 2.2 The atomic number of all three is 6; their atomic masses are 12, 13, and 14, respectively.

Radioactive decay and radioactive isotopes play important roles in microbiological research, medical diagnosis, the treatment of disease, and the complete destruction of contaminating microbes (sterilization) of medical equipment and chemicals.

Electron Configurations

Although the nuclei of atoms determine their identities, it is electrons that determine an atom's *chemical behavior*. Nuclei of different atoms almost never come close enough together to interact.[5] Typically, only the electrons of atoms interact. Thus, because all of the isotopes of carbon (for example) have the same number of electrons, all these isotopes behave the same way in chemical reactions, even though their nuclei are different.

Scientists know that electrons do not really orbit the nucleus in a two-dimensional circle, as indicated by a Bohr model; instead, they speed around a nucleus 100 quadrillion times per second in three-dimensional *electron shells* or *clouds* that assume unique shapes dependent on the energy of the electrons **(FIGURE 2.3a)**. More accurately put, an electron shell depicts the *probable* locations of electrons at a given time; nevertheless, it is simpler and more convenient to draw electron shells as circles **(FIGURE 2.3b)**.

[5]Except during nuclear reactions, such as occur in nuclear power plants.

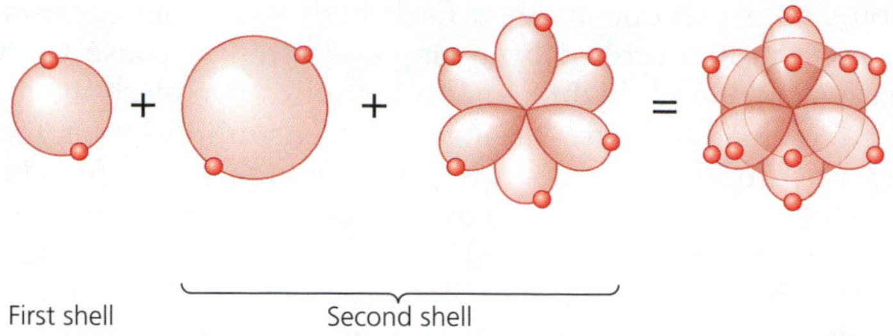

(a) Electron shells of neon: three-dimensional view

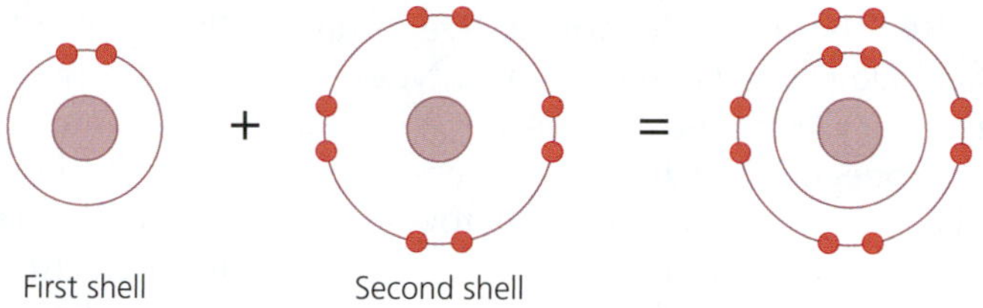

(b) Electron shells of neon: two-dimensional view

▲ **FIGURE 2.3 Electron configurations.** **(a)** Three-dimensional model of the electron shells of neon. In this model, the first shell is a small sphere, whereas the second shell consists of a larger sphere plus three pairs of ellipses that extend from the nucleus at right angles. Larger shells (not shown) are even more complex. **(b)** Two-dimensional model (Bohr diagram) of the electron shells of neon. *How many shells does a sodium atom need to hold its 11 electrons?*

Figure 2.3 *Three shells: two electrons in the first shell, eight in the second, and a single electron in the third.*

Each electron shell can hold only a certain maximum number of electrons. For example, the first shell (the one nearest the nucleus) can accommodate a maximum of two electrons, and the second shell can hold no more than eight electrons. Atoms of hydrogen and helium have one and two electrons, respectively; thus, these two elements have only a single electron shell. A lithium atom, which has three electrons, has two shells.

Atoms with more than 10 electrons require more shells. The third shell holds up to eight electrons when it is the outermost shell, though its capacity increases to 18 when the fourth shell contains two electrons. Heavier atoms have even more shells, but these atoms do not play significant roles in the processes of life.

Electrons in the outermost shell of atoms are called *valence electrons*. **FIGURE 2.4** depicts the electron configurations of atoms of some elements important to microbial life. Notice that except for helium, atoms of all elements in a given column have the same number of valence electrons. Helium is placed in the far right-hand column with the other inert gases because its outer shell is full, though it has two rather than eight valence electrons. Valence electrons are critical for interactions between atoms. Next we consider these interactions, which are called chemical bonds.

TELL ME WHY

Electrons zip around the nucleus at about 5 million miles per hour. Why don't some fly off?

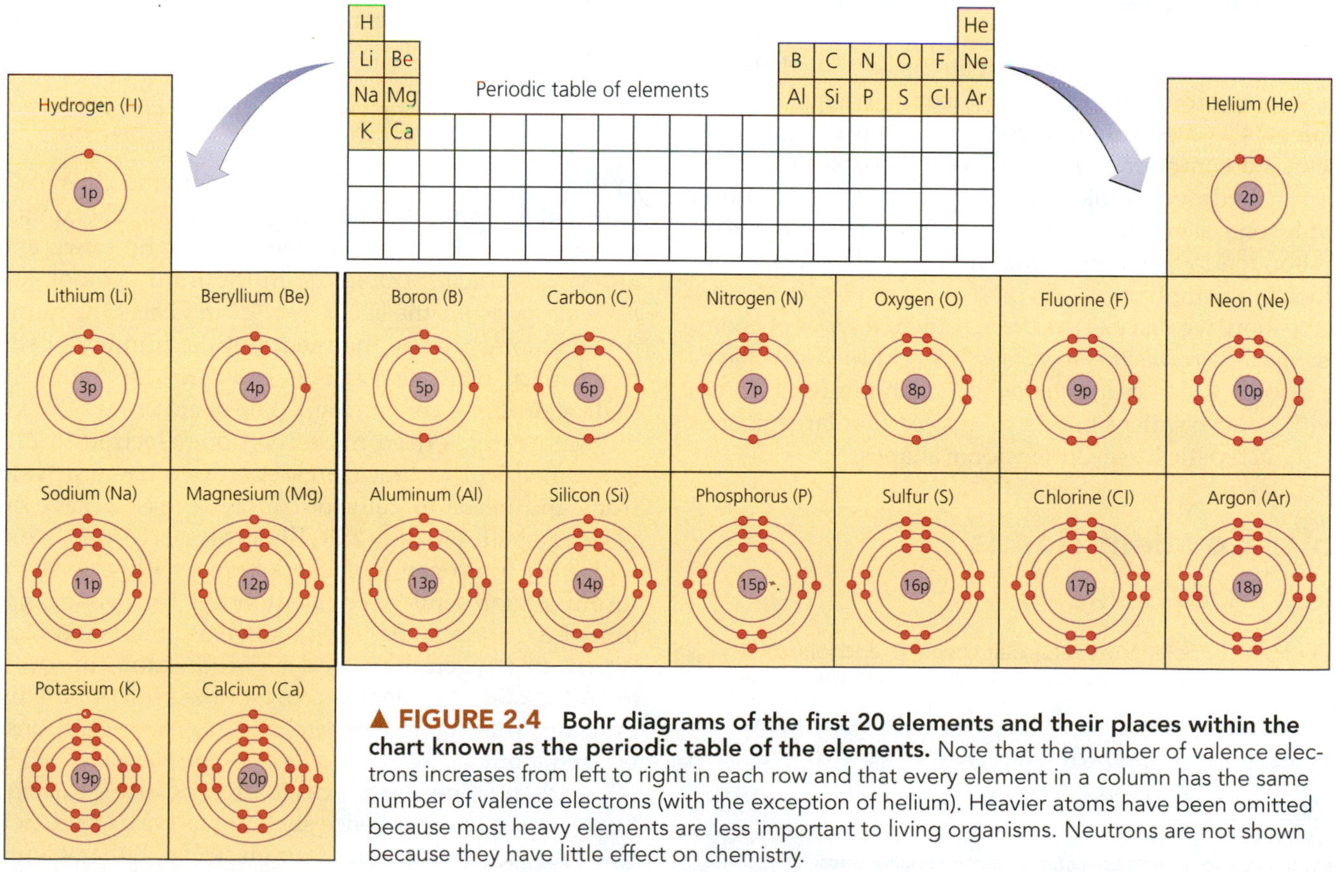

▲ **FIGURE 2.4 Bohr diagrams of the first 20 elements and their places within the chart known as the periodic table of the elements.** Note that the number of valence electrons increases from left to right in each row and that every element in a column has the same number of valence electrons (with the exception of helium). Heavier atoms have been omitted because most heavy elements are less important to living organisms. Neutrons are not shown because they have little effect on chemistry.

Chemical Bonds

LEARNING | OUTCOMES

2.4 Describe the configuration of electrons in a stable atom.
2.5 Contrast molecules and compounds.

Outer electron shells are stable when they contain eight electrons (except for the first electron shell, which is stable with only two electrons, because that is its maximum number). When atoms' outer shells are not filled with eight electrons, they either have room for more electrons or have "extra" electrons, depending on whether it is easier for them to gain electrons or lose electrons. For example, an oxygen atom, with six electrons in its outer shell, has two "unfilled spaces" (see Figure 2.4) because it requires less energy for the oxygen atom to gain two electrons than to lose six electrons. A calcium atom, by contrast, has two "extra" electrons in its outer (fourth) shell because it requires less energy to lose these two electrons than to gain six new ones. When a calcium atom loses two electrons, its third shell, which is then its outer shell, is full and stable with eight electrons.

As previously noted, an atom's outermost electrons are called valence electrons, and thus the outermost shell of an atom is the *valence shell*. An atom's **valence**,[6] defined as its combining capacity, is considered to be positive if its valence shell has extra electrons to give up and to be negative if its valence shell has spaces to fill. Thus, a calcium atom, with two electrons in its valence shell, has a valence of +2, whereas an oxygen atom, with two spaces to fill in its valence shell, has a valence of −2.

Atoms combine with one another by either sharing or transferring valence electrons in such a way as to fill their valence shells. Such interactions between atoms are called **chemical bonds**. Two or more atoms held together by chemical bonds form a **molecule**. A molecule that contains atoms of more than one element is a **compound**. Two hydrogen atoms bonded together form a hydrogen molecule, which is not a compound because only one element is involved. However, two hydrogen atoms bonded to an oxygen atom form a molecule of water (H_2O), which is a compound.

In this section, we discuss the three principal types of chemical bonds: *nonpolar covalent bonds, polar covalent bonds,* and *ionic bonds*. We also consider *hydrogen bonds,* which are weak forces that act with polar covalent bonds to give certain large chemicals their characteristic three-dimensional shapes.

Nonpolar Covalent Bonds

LEARNING | OUTCOME

2.6 Contrast nonpolar covalent, polar covalent, and ionic bonds.

A **covalent**[7] **bond** is the sharing of a pair of electrons by two atoms. Consider, for example, what happens when two hydrogen atoms approach one another. Each hydrogen atom consists of a single proton orbited by a single electron. Because the valence shell of each hydrogen atom requires two electrons to be filled, each atom shares its single electron with the other, forming a hydrogen molecule in which both atoms have full shells **(FIGURE 2.5a)**. Similarly, two oxygen atoms can share electrons, but they must share *two* pairs of electrons for their valence shells to be full **(FIGURE 2.5b)**. Because two pairs of electrons are involved, oxygen atoms form two covalent bonds, or a *double covalent bond,* with one another.

The attraction of an atom for electrons is called its **electronegativity**. The more electronegative an atom, the greater the pull its nucleus exerts on electrons. Note in **FIGURE 2.6**, which displays the electronegativities of atoms of several elements, that electronegativities tend to increase from left to right in the chart. The reason is that elements toward the right of the chart have more protons and thus exert a greater pull on electrons. Electronegativities of elements decrease from top to bottom in the chart because the distance between the nucleus and the valence shell increases as elements get larger.

Atoms with equal or nearly equal electronegativities, such as two hydrogen atoms or a hydrogen and a carbon, share electrons equally or nearly equally. In chemistry and physics, "poles" are opposed forces, such as north and south magnetic poles or positive and negative terminals of a battery. In the case of atoms with similar electronegativities, the shared electrons tend to spend an equal amount of time around each nucleus of the pair, and no poles exist; therefore, the bond between them is a **nonpolar covalent bond**. (All the covalent bonds illustrated in Figure 2.5a–c are nonpolar; formaldehyde is polar.)

A hydrogen molecule can be symbolized a number of ways:

$$\mathrm{H{-}H} \qquad \mathrm{H{:}H} \qquad \mathrm{H_2}$$

In the first symbol, the dash represents the chemical bond between the atoms. In the second symbol, the dots represent the electron pair of the covalent bond. These two symbols are known as *structural formulas*. In the third symbol, known as a *molecular formula,* the subscript "2" indicates the number of hydrogen atoms that are bonded, not the number of shared electrons. Each of these symbols indicates the same thing—two hydrogen atoms are sharing a pair of electrons.

Many atoms need more than one electron to fill their valence shell. For instance, a carbon atom has four valence electrons and needs to gain four more if it is to have eight in its valence shell. **FIGURE 2.5c** illustrates a carbon atom sharing with four hydrogen atoms. As before, a line in the structural formula represents a covalent bond formed from the sharing of two electrons. Two covalent bonds are formed between an oxygen atom and a carbon atom in formaldehyde **(FIGURE 2.5d)**. This fact is represented by a double line, which indicates that the carbon atom shares four electrons with the oxygen atom.

Carbon atoms are critical to life. Because a carbon atom has four electrons in its valence shell, it has equal tendency to either lose four electrons or gain four electrons. Either event produces

[6] From Latin *valentia*, meaning "strength."
[7] From Latin *co*, meaning "with" or "together," and *valentia*, meaning "strength."

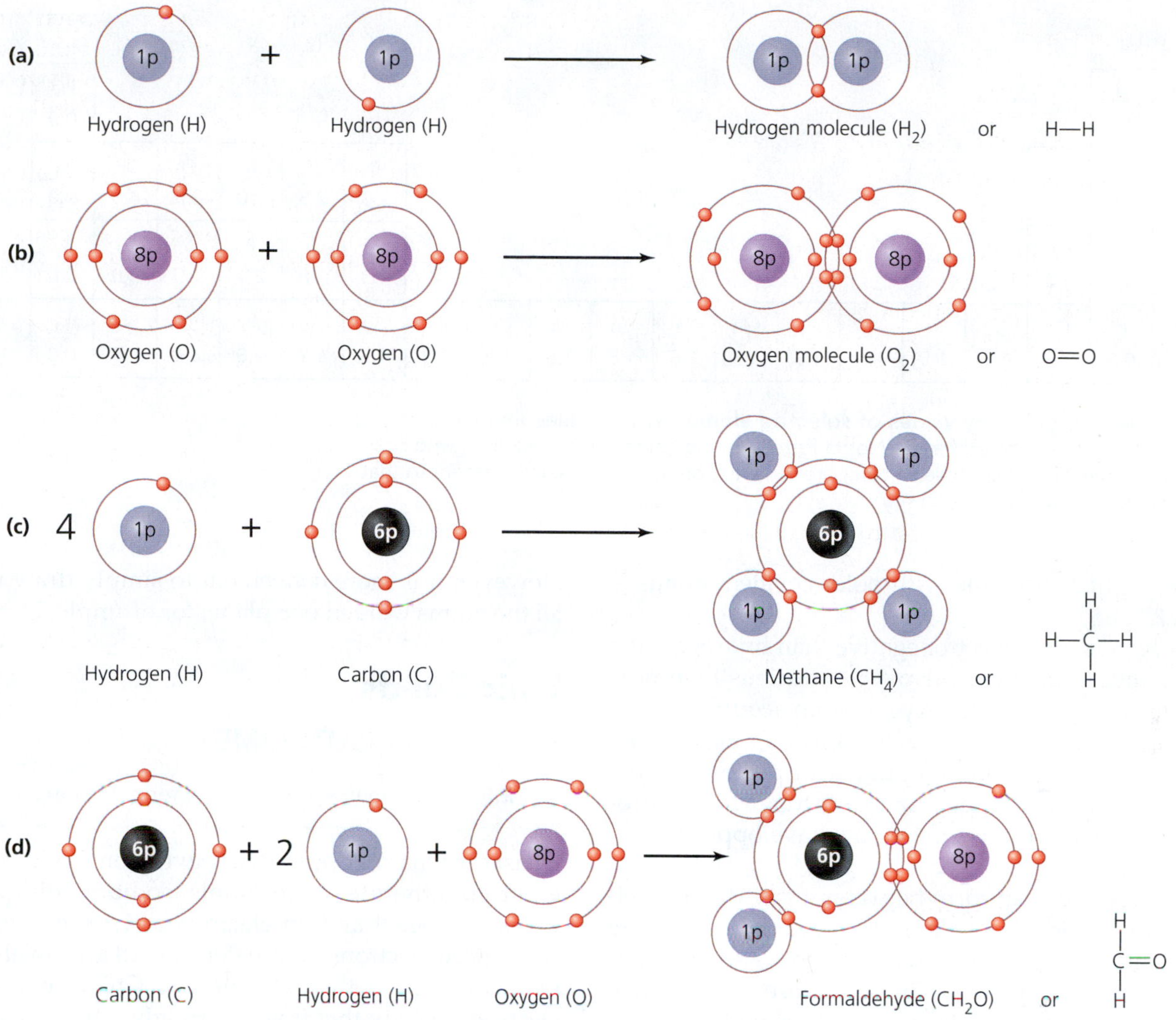

▲ **FIGURE 2.5 Four molecules formed by covalent bonds. (a)** Hydrogen. Each hydrogen atom needs another electron to have a full valence shell. The two atoms share their electrons, forming a covalent bond. **(b)** Oxygen. Oxygen atoms have six electrons in their valence shells; thus, they need two electrons each. When they share with each other, two covalent bonds are formed. Note that the valence electrons of oxygen atoms are in the second shell. **(c)** A methane molecule, which has four single covalent bonds. **(d)** Formaldehyde. The carbon atom forms a double bond with the oxygen atom and single bonds with two hydrogen atoms. *Which of these molecules are also compounds? Why?*

Figure 2.5 *Methane and formaldehyde molecules are also compounds because they are composed of more than one element.*

a full outer shell. The result is that carbon atoms tend to share electrons and form four covalent bonds with one another and with many other types of atoms. Each carbon atom in effect acts as a four-way intersection where different components of a molecule can attach. One result of this feature is that carbon atoms can form very large chains that constitute the "backbone" of many biologically important molecules. Carbon chains can be branched or unbranched, and some even close back on themselves to form rings. Compounds that contain carbon and hydrogen atoms are called **organic compounds**. Among the many biologically important organic compounds are proteins and carbohydrates, which are discussed later in the chapter.

Polar Covalent Bonds

LEARNING **OUTCOME**

2.7 Explain the relationship between electronegativity and the polarity of a covalent bond.

If two covalently bound atoms have significantly different electronegativities, their electrons will not be shared equally. Instead, the electron pair will spend more time orbiting the nucleus of the atom with greater electronegativity. This type of bond, in which there is unequal sharing of electrons, is a **polar covalent**

I	II											III	IV	V	VI	VII	Inert gases
H 2.1																	He 0.0
Li 1.0	Be 1.5											B 2.0	C 2.5	N 3.0	O 3.5	F 4.0	Ne 0.0
Na 0.9	Mg 1.2											Al 1.5	Si 1.8	P 2.1	S 2.5	Cl 3.0	Ar 0.0
K 0.8	Ca 1.0	Sc 1.3	Ti 1.5	V 1.6	Cr 1.6	Mn 1.5	Fe 1.8	Co 1.8	Ni 1.8	Cu 1.9	Zn 1.6	Ga 1.6	Ge 1.8	As 2.0	Se 2.4	Br 2.8	Kr 0.0

▲ **FIGURE 2.6 Electronegativity values of selected elements.** The values are expressed according to the Pauling scale, named for the Nobel Prize–winning chemist Linus Pauling, who based the scale on bond energies. Pauling chose to compare the electronegativity of each element to that of fluorine, to which he assigned a value of 4.0.

bond. An example of a molecule with polar covalent bonds is water **(FIGURE 2.7a)**.

Because oxygen is more electronegative than hydrogen, the electrons spend more time near the oxygen nucleus than near the hydrogen nuclei, and thus the oxygen atom acquires a transient (partial) negative charge (symbolized as δ^-). Each of the hydrogen nuclei has a corresponding transient positive charge (δ^+). The covalent bond between an oxygen atom and a hydrogen atom is called polar because the atoms have opposite partial electrical charges.

Polar covalent bonds can form between many different elements. Generally, molecules with polar covalent bonds are water soluble, and nonpolar molecules are not. The most important polar covalent bonds for life are those that involve hydrogen because they allow hydrogen bonding, which we discuss shortly.

Both nonpolar and polar covalent bonds form angles between atoms such that the distances between electron orbits are maximized. The bond angle for water is shown in **FIGURE 2.7b**.

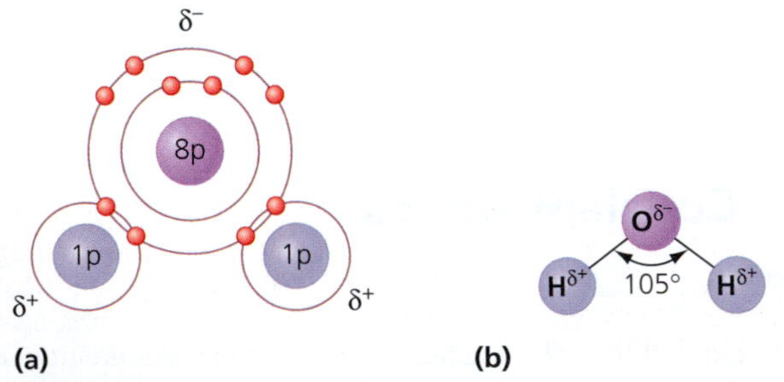

▲ **FIGURE 2.7 Polar covalent bonding in a water molecule.** **(a)** A Bohr model of a water molecule, which has two polar covalent bonds. When the electronegativities of two atoms are significantly different, the shared electrons of covalent bonds spend more time around the more electronegative atom, giving it a transient negative charge (δ^-). Its partner has a transient positive charge (δ^+). **(b)** The bond angle in a water molecule. Atoms maximize the distances between electron orbitals in polar and nonpolar covalent bonds.

However, it is more convenient to simply draw molecules as if all the atoms were in one plane; for example, H—O—H.

Ionic Bonds

LEARNING | **OUTCOME**

2.8 Define *ionization* using the terms *cation* and *anion*.

Consider what happens when two atoms with vastly different electronegativities—for example, sodium, with one electron in its valence shell and an electronegativity of 0.9, and chlorine, with seven electrons in its valence shell and an electronegativity of 3.0—come together **(FIGURE 2.8)**. Chlorine has such a higher electronegativity that it very strongly attracts sodium's valence electron, and the result is that the sodium loses that electron to chlorine 1.

Now that the chlorine atom has one more electron than it has protons, it has a full negative charge, and the sodium atom, which has lost an electron, now has a full positive charge 2. An atom or group of atoms that has either a full negative charge or a full positive charge is called an *ion*. Positively charged ions are called **cations**, whereas negatively charged ions are called **anions**.

Because of their opposite charges, cations and anions attract each other and form what is termed an **ionic bond** 3. They form crystalline compounds composed of metallic and nonmetallic ions known as **salts**, such as sodium chloride (NaCl), also known as table salt, and potassium chloride (KCl, sodium-free table salt). Ionic bonds differ from covalent bonds in that ions do not share electrons. Instead, the bond is formed from the attraction of opposite electrical charges.

The polar bonds of water molecules interfere with the ionic bonds of salts, causing *dissociation* (also called *ionization*) **(FIGURE 2.9)**. This occurs as the partial negative charge on the oxygen atom of water attracts cations, and the partial positive charge on hydrogen atoms attracts anions. The presence of polar bonds interferes with the attraction between the cation and anion.

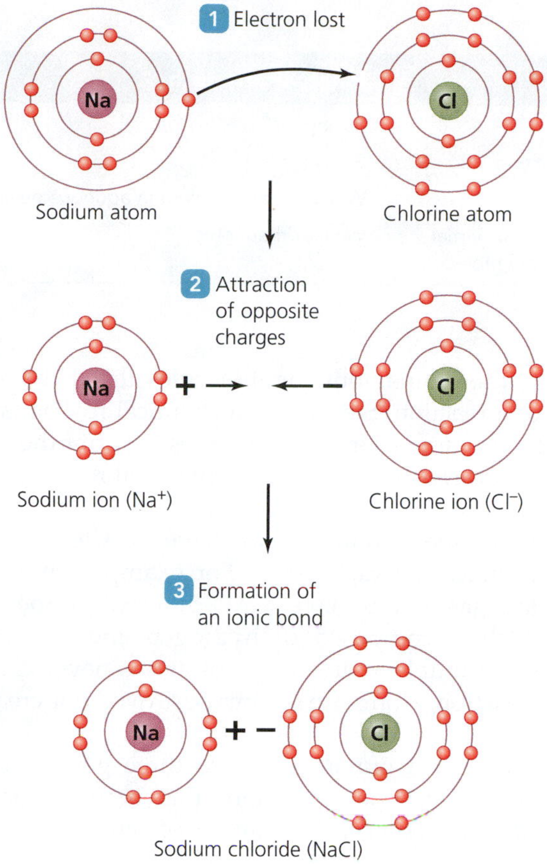

▲ **FIGURE 2.8 The interaction of sodium and chlorine to form an ionic bond.**

When cations and anions dissociate from one another and become surrounded by water molecules (are hydrated), they are called **electrolytes** because they can conduct electricity through the solution. Electrolytes are critical for life because they stabilize a variety of compounds, act as electron carriers, and allow electrical gradients to exist within cells. We examine these functions of electrolytes in later chapters.

In nature, chemical bonds range from nonpolar bonds to polar bonds to ionic bonds. The important thing to remember is that electrons are shared between atoms in covalent bonds and transferred from one atom to another in ionic bonds.

Hydrogen Bonds

LEARNING | **OUTCOME**

2.9 Describe hydrogen bonds and discuss their importance in living organisms.

As we have seen, hydrogen atoms bind to oxygen atoms by means of polar covalent bonds, resulting in transient positive charges on the hydrogen atoms. Hydrogen atoms form polar covalent bonds with atoms of other elements as well.

The electrical attraction between a partially charged hydrogen atom and a full or partial negative charge on either a different region of the same molecule or another molecule is called

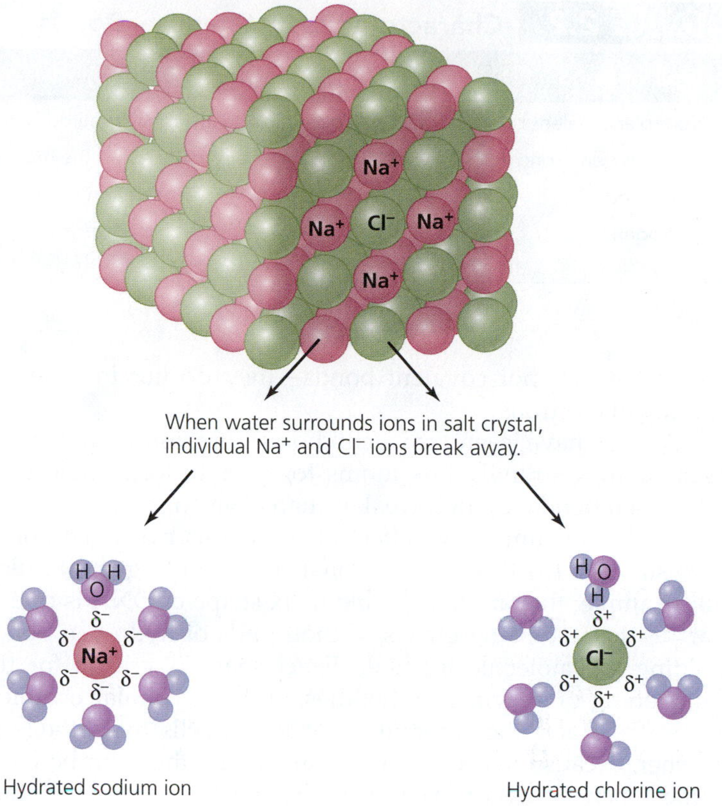

▲ **FIGURE 2.9 Dissociation of NaCl in water.** When water surrounds the ions in a NaCl crystal, the partial charges on water molecules are attracted to charged ions, and the water molecules hydrate the ions by surrounding them. The partial negative charges (δ^-) on oxygen atoms are attracted to cations (in this case, the sodium ions), and the partial positive charges (δ^+) on hydrogen atoms are attracted to anions (the chlorine ions). Because the ions no longer attract one another, the salt crystal dissolves. Hydrated ions are called electrolytes.

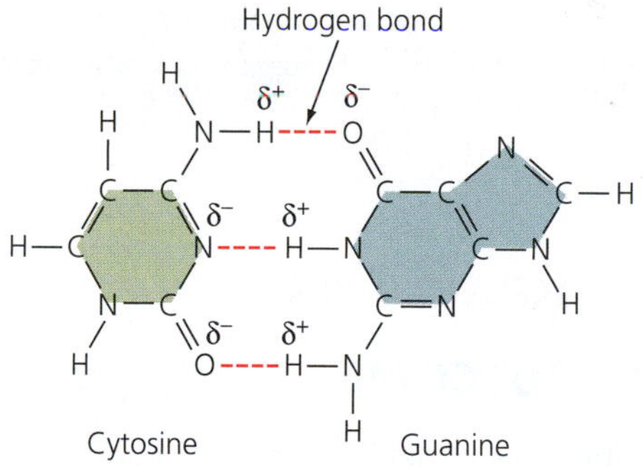

▲ **FIGURE 2.10 Hydrogen bonds.** The transient positive charge (symbolized δ^+) on a hydrogen atom is attracted to a transient negative charge (δ^-) on another atom. Such attraction is a hydrogen bond. Hydrogen bonds can hold together portions of the same molecule or hold two different molecules together. In this case, three hydrogen bonds are holding molecules of cytosine and guanine together.

a **hydrogen bond** (**FIGURE 2.10**). Hydrogen bonds can be likened to weak ionic bonds in that they arise from the attraction of positive and negative charges. Notice also that although they are a consequence of polar covalent bonds between hydrogen atoms and other, more electronegative atoms, hydrogen bonds

TABLE 2.2 Characteristics of Chemical Bonds

Type of Bond	Description	Relative Strength
Nonpolar covalent bond	Pair of electrons is nearly equally shared between two atoms	Strong
Polar covalent bond	Electrons spend more time around the more electronegative of two atoms	Strong
Ionic bond	Electrons are stripped from a cation by an anion	Weaker than covalent in aqueous environments
Hydrogen bond	Partial positive charges on hydrogen atoms are attracted to full and partial negative charges on other molecules or other regions of the same molecule	Weaker than ionic

themselves are not covalent bonds—they do not involve the sharing of electrons.

As we have seen, covalent bonds are essential for life because they strongly link atoms together to form molecules. Hydrogen bonds, though weaker than covalent bonds, are also essential. The cumulative effect of numerous hydrogen bonds is to stabilize the three-dimensional shapes of large molecules. For example, the familiar double-helix shape of DNA is due in part to the stabilizing effects of thousands of hydrogen bonds holding the molecule together. Exact shape is critical for the functioning of enzymes, antibodies, and intercellular chemical messengers and the recognition of target cells by pathogens. Further, because hydrogen bonds are weak, they can be overcome when necessary. For example, the two complementary halves of a DNA molecule are held together primarily by hydrogen bonds, and they can be separated for DNA replication and other processes (see Figure 7.6).

TABLE 2.2 summarizes the characteristics of chemical bonds.

TELL ME WHY

Chlorine and potassium atoms form ionic bonds, carbon atoms form nonpolar covalent bonds with nitrogen atoms, and oxygen forms polar covalent bonds with phosphorus. Explain why these bonds are the types they are.

Chemical Reactions

LEARNING | **OUTCOME**

2.10 Describe three general types of chemical reactions found in living things.

You are already familiar with many consequences of chemical reactions: you add yeast to bread dough, and it rises; enzymes in your laundry detergent remove grass stains; and gasoline burned in your car releases energy to speed you on your way. What exactly is happening in these reactions? What is the precise definition of *chemical reaction*?

We have discussed how bonds are formed via the sharing of electrons or the attraction of positive and negative charges. Scientists define **chemical reactions** as the making or breaking of such chemical bonds. All chemical reactions begin with **reactants**—the atoms, ions, or molecules that exist at the beginning of a reaction. Similarly, all chemical reactions result in **products**—the atoms, ions, or molecules left after the reaction is complete. *Biochemistry* involves the chemical reactions of living things.

Reactants and products may have very different physical and chemical characteristics. For example, hydrogen and oxygen are gases and have very different properties from water, which is composed of hydrogen and oxygen atoms. However, the numbers and types of atoms never change in a chemical reaction; atoms are neither destroyed nor created, only rearranged.

Now let's turn our attention to three general categories of biochemical reactions (reactions that occur in organisms): *synthesis, decomposition,* and *exchange reactions.*

Synthesis Reactions

LEARNING | **OUTCOMES**

2.11 Give an example of a synthesis reaction that involves the formation of a water molecule.

2.12 Contrast endothermic and exothermic chemical reactions.

Synthesis reactions involve the formation of larger, more complex molecules. Synthesis reactions can be expressed symbolically as

$$\text{Reactant} + \text{Reactant} \rightarrow \text{Product(s)}$$

The arrow indicates the direction of the reaction and the formation of new chemical bonds. For example, algae make their own glucose (sugar) using the following reaction:

$$6\,H_2O + 6\,CO_2 \rightarrow C_6H_{12}O_6 + 6\,O_2$$

The reaction is read, "Six molecules of water plus six molecules of carbon dioxide yield one molecule of glucose and six molecules of oxygen." Notice that the total number and kind of atoms are the same on both sides of the reaction.

A common synthesis reaction in biochemistry is a **dehydration synthesis**, in which two smaller molecules are joined together by a covalent bond, and a water molecule is also formed (**FIGURE 2.11a**). The word *dehydration* in the name of this type of reaction refers to the fact that one of the products is a water molecule formed when a hydrogen ion (H^+) from one reactant combines with a hydroxyl ion (OH^-) from another reactant.

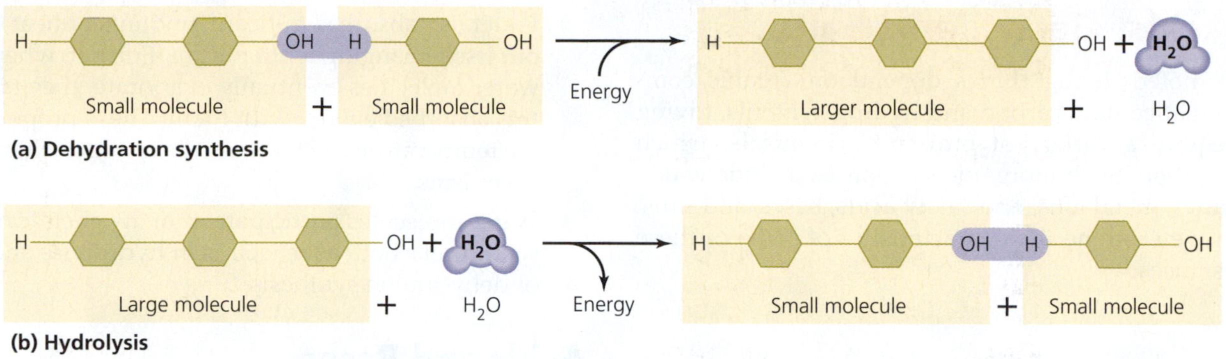

▲ **FIGURE 2.11 Two types of chemical reactions in living things. (a)** Dehydration synthesis. In this energy-requiring reaction, a hydroxyl ion (OH^-) removed from one reactant and a hydrogen ion (H^+) removed from another reactant combine to form hydrogen hydroxide (HOH), which is water. **(b)** Hydrolysis, an energy-yielding reaction that is the reverse of a dehydration synthesis reaction. *What are the scientific words meaning "energy-requiring" and "energy-releasing"?*

Figure 2.11 Endothermic means "energy-requiring," and exothermic means "energy-releasing."

Synthesis reactions require energy to break bonds in the reactants and to form new bonds to make products. Reactions that require energy are said to be **endothermic**[8] **reactions** because they trap energy within new molecular bonds. An energy supply for fueling synthesis reactions is one common requirement of all living things (Chapter 6).

Taken together, all of the synthesis reactions in an organism are called **anabolism**.

Decomposition Reactions

LEARNING | OUTCOME

2.13 Give an example of a decomposition reaction that involves breaking the bonds of a water molecule.

Decomposition reactions are the reverse of synthesis reactions in that they break bonds within larger molecules to form smaller atoms, ions, and molecules. These reactions release energy and are therefore **exothermic reactions**.[9] In general, decomposition reactions can be represented by the following formula:

$$\text{Reactant} \rightarrow \text{Product} + \text{Product}$$

An example of a biologically important decomposition reaction is the aerobic decomposition of glucose to form carbon dioxide and water:

$$C_6H_{12}O_6 + 6\,O_2 \rightarrow 6\,H_2O + 6\,CO_2$$

Note that this reaction is exactly the reverse of the synthesis reaction in algae that we examined previously. Synthesis and decomposition reactions are often reversible in living things.

A common type of decomposition reaction in biochemistry is **hydrolysis**,[10] the reverse of dehydration synthesis **(FIGURE 2.11b)**. In hydrolytic reactions, a covalent bond in a large molecule is broken, and the ionic components of water (H^+ and OH^-) are added to the products.

Collectively, all of the decomposition reactions in an organism are called **catabolism**.

Exchange Reactions

LEARNING | OUTCOME

2.14 Compare exchange reactions to synthesis and decomposition reactions.

Exchange reactions (also called *transfer reactions*) have features similar to both synthesis and decomposition reactions. For instance, they involve breaking and forming covalent bonds, and they involve both endothermic and exothermic steps. As the name suggests, atoms are moved from one molecule to another. In general, these reactions can be represented as either

$$A + BC \rightarrow AB + C$$

or

$$AB + CD \rightarrow AD + BC$$

An important exchange reaction within organisms is the phosphorylation of glucose:

$$\underset{\text{Glucose}}{C_6H_{12}O_6} + \underset{\text{Adenosine triphosphate}}{A{-}Ⓟ{-}Ⓟ{-}Ⓟ} \rightarrow \underset{\text{Glucose phosphate}}{C_6H_{11}O_6{-}Ⓟ} + \underset{\text{Adenosine diphosphate}}{A{-}Ⓟ{-}Ⓟ} + H^+$$

The sum of all of the chemical reactions in an organism, including catabolic, anabolic, and exchange reactions, is called **metabolism**. (We examine metabolism in more detail in Chapter 5.)

TELL ME WHY

Why are decomposition reactions exothermic (release energy)?

[8]From Greek *endon*, meaning "within," and *thermos*, meaning "heat" (energy).
[9]From Greek *exo*, meaning "outside," and *thermos*, meaning "heat" (energy).
[10]From Greek *hydor*, meaning "water," and *lysis*, meaning "loosening."

Water, Acids, Bases, and Salts

As previously noted, living things depend on organic compounds, those that contain carbon and hydrogen atoms. Living things also require a variety of **inorganic chemicals**, which typically lack carbon. Such inorganic substances include water, oxygen molecules, metal ions, and many acids, bases, and salts. In this section we examine the characteristics of some of these inorganic substances.

Water

LEARNING | OUTCOME

2.15 Describe five qualities of water that make it vital to life.

Water is the most abundant substance in organisms, constituting 50% to 99% of their mass. Most of the special characteristics that make water vital result from the fact that a water molecule has two polar covalent bonds, allowing hydrogen bonding between water molecules and their neighbors. Among the special properties of water are the following:

- Water molecules are cohesive; that is, they tend to stick to one another through hydrogen bonding **(FIGURE 2.12)**. This property generates many special characteristics of water, including *surface tension*, which allows water to form a thin layer on the surface of cells. This aqueous layer is necessary for the transport of dissolved materials into and out of a cell.
- Water is an excellent *solvent*; that is, it dissolves salts and other electrically charged molecules because it is attracted to both positive and negative charges (see Figure 2.9).
- Water remains a liquid across a wider range of temperatures than other molecules of its size. This is critical because living things require water in liquid form.
- Water can absorb significant amounts of heat energy without itself changing temperature. Further, when heated water molecules eventually evaporate, they take much of this absorbed energy with them. These properties moderate temperature fluctuations that would otherwise damage organisms.
- Water molecules participate in many chemical reactions within cells both as reactants in hydrolysis and as products of dehydration synthesis.

Acids and Bases

LEARNING | OUTCOME

2.16 Contrast acids, bases, and salts and explain the role of buffers.

As we have seen, the polar bonds of water molecules dissociate salts into their component cations and anions. A similar process occurs with substances known as acids and bases.

An **acid** is a substance that dissociates into one or more hydrogen ions (H^+) and one or more anions **(FIGURE 2.13a)**. Acids can be inorganic molecules, such as hydrochloric acid (HCl) and sulfuric acid (H_2SO_4), or organic molecules, such as amino acids and nucleic acids. Familiar organic acids are found in lemon juice, black coffee, and tea. Of course, the anions of organic acids contain carbon, whereas those of inorganic acids do not.

A **base** is a molecule that binds with H^+ when dissolved in water. Some bases dissociate into cations and *hydroxyl ions* (OH^-) **(FIGURE 2.13b)**, which then combine with hydrogen ions to form water molecules:

$$H^+ + OH^- \rightarrow H_2O$$

Other bases, such as household ammonia (NH_3), directly accept hydrogen ions and become compound ions such as NH_4^+

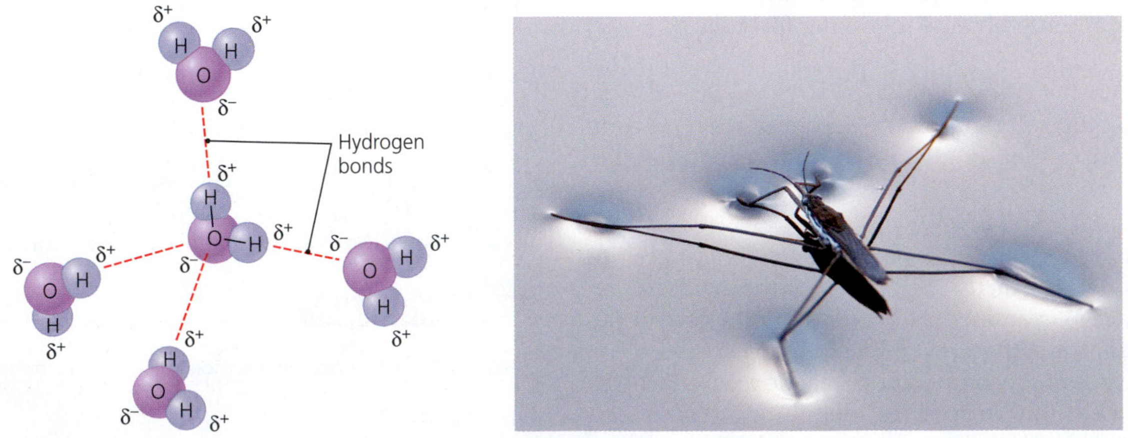

▲ **FIGURE 2.12 The cohesiveness of liquid water. (a)** Water molecules are cohesive because hydrogen bonds cause them to stick to one another. **(b)** One result of cohesiveness in water is surface tension, which can be strong enough to support the weight of insects known as water striders.

BENEFICIAL MICROBES

Architecture-Preserving Bacteria

The Alhambra, a Moorish palace constructed of limestone and marble beginning in the ninth century, was built to last. But not even stone lasts forever. Wind and rain wear away the surface. Acid rain reacts with the calcite crystals in limestone and marble. As years pass, stone slowly crumbles.

Those who would preserve the Alhambra and other historic structures face a dilemma. The microscopic pores that riddle limestone and marble make these materials particularly susceptible to weathering and decay. Sealing the stone's pores can reduce weathering but can also lock in moisture that speeds the stone's decay.

With the help of *Myxococcus xanthus*, a bacterium commonly found in soil, a team of researchers led by mineralogist Carlos Rodríguez-Navarro of the University of Granada may have found a way to protect the stone of structures like the Alhambra.

In many natural environments, bacteria instigate the formation of calcite crystals like the ones in limestone. In tests conducted using samples of the limestone commonly used in historic Spanish buildings, *M. xanthus* formed calcite crystals that lined the stone's pores rather than plugging them. The crystals formed by the bacteria are even more durable than the original stone, offering the potential for long-term protection.

The Alhambra

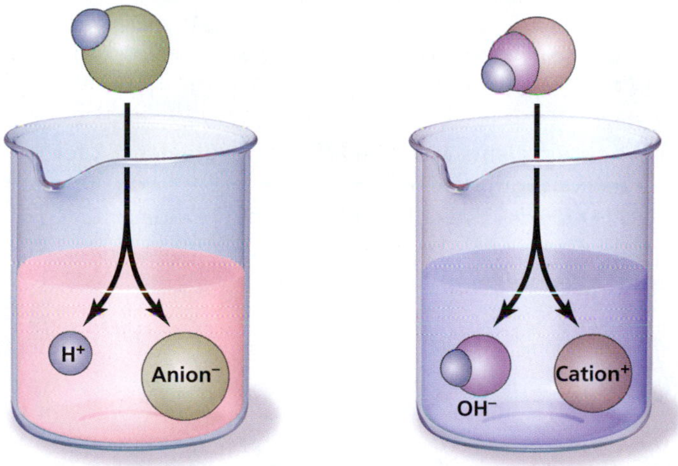

(a) Acid (b) Base

▲ **FIGURE 2.13 Acids and bases. (a)** Acids dissociate in water into hydrogen ions and anions. **(b)** Many bases dissociate into hydroxyl ions and cations.

(ammonium). Another common household base is baking soda (sodium bicarbonate, $NaHCO_3$).

Metabolism requires a relatively constant balance of acids and bases because hydrogen ions and hydroxyl ions are involved in many chemical reactions. Further, many complex molecules such as proteins lose their functional shapes when acidity changes. If the concentration of either hydrogen ions or hydroxyl ions deviates too far from normal, metabolism ceases.

The concentration of hydrogen ions in a solution is expressed using a logarithmic **pH scale** (FIGURE 2.14). The term *pH* comes from *potential hydrogen,* which is the negative of the logarithm of the concentration of hydrogen ions. In this logarithmic scale, it is important to notice that acidity increases as pH values decrease and that each decrease by a whole number in pH indicates a 10-fold increase in acidity (hydrogen ion concentration). For example, a glass of grapefruit juice, which has a pH of 3.0, contains 10 times as many hydrogen ions as the same volume of tomato juice, which has a pH of 4.0. Similarly, tomato juice is 1000 times more acidic than pure water, which has a pH of 7.0 (neutral). Water is neutral because it dissociates into one hydrogen cation and one hydroxyl anion:

$$H_2O \rightarrow H^+ + OH^-$$

Alkaline (basic) substances have pH values greater than 7.0. They reduce the number of free hydrogen ions by combining with them. For bases that produce hydroxyl ions, the concentration of hydroxyl ions is inversely related to the concentration of hydrogen ions.

Organisms can tolerate only a certain, relatively narrow pH range. Fluctuations outside an organism's preferred range inhibit its metabolism and may even be fatal. Most organisms contain natural **buffers**—substances, such as proteins, that prevent drastic changes in internal pH. In a laboratory culture, the metabolic activity of microorganisms can change the pH of microbial growth solutions as nutrients are taken up and wastes are released; therefore, pH buffers are often added to them. One common buffer used in microbiological media is KH_2PO_4 (potassium dihydrogen phosphate), which exists as either a weak acid or a weak base, depending on the pH of its environment. Under acidic conditions, KH_2PO_4 is a base that combines with H^+, neutralizing the acidic environment; in alkaline conditions, however, KH_2PO_4 acts as an acid, releasing hydrogen ions.

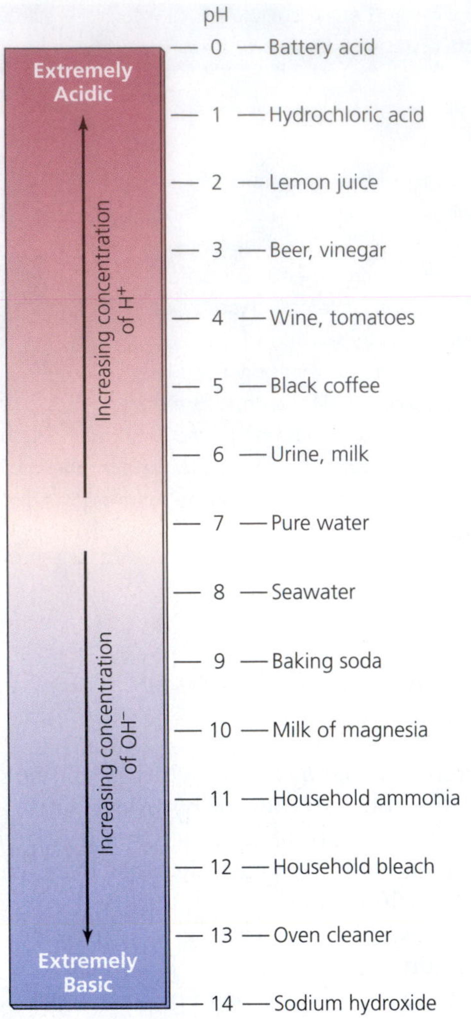

▲ **FIGURE 2.14 The pH scale.** Values below 7 are acidic; values above 7 are basic.

CLINICAL CASE STUDY

Raw Oysters and Antacids: A Deadly Mix?

The highly acidic environment of the stomach kills most bacteria before they cause disease. One bacterium that can slightly tolerate conditions as it passes through the stomach is *Vibrio vulnificus*—a bacterium commonly ingested by eating raw tainted oysters. The bacterium cannot be seen, tasted, or smelled in food or water.

V. vulnificus is an emerging pathogen and a growing cause of food poisoning in the United States: it triggers vomiting, diarrhea, and abdominal pain. The pathogen can also infect the bloodstream, causing life-threatening illness characterized by fever, chills, skin lesions, and deadly loss of blood pressure. About 50% of patients with bloodstream infections die. *V. vulnificus* especially affects the immunocompromised and people with long-term liver disease.

Researchers have discovered that taking antacids may make people more susceptible to becoming ill from *V. vulnificus*. They found that antacids in a simulated gastric environment significantly increased the survival rate of *V. vulnificus*.

1. Why are patients who take antacids at greater risk for infections with *V. vulnificus*?
2. Will antacids raise or lower the pH of the stomach?
3. Other than refraining from antacids, what can people do to reduce their risk of infection?

Reference: Adapted from *MMWR* 45:621–624. 1996.

Microorganisms differ in their ability to tolerate various ranges of pH. Many grow best when the pH is between 6.5 and 8.5. Photosynthetic bacteria known as *cyanobacteria* grow well in more basic solutions. Fungi generally tolerate acidic environments better than most prokaryotes, though acid-loving prokaryotes, called *acidophiles,* require acidic conditions. Some bacteria are tolerant of acid. One such bacterium is *Propionibacterium acnes* (prō-pē-on-i-bak-tēr´ē-ŭm ak´nēz), which can cause acne in the skin and normally has a pH of about 4.0. Another is *Helicobacter pylori* (hel´ĭ-kō-bak´ter pī´lō-rē),[11] a curved bacterium that has been shown to cause ulcers in the stomach, where pH can fall as low as 1.5 when acid is being actively secreted. **Clinical Case Study: Raw Oysters and Antacids: A Deadly Mix?** focuses on how the use of antacids may increase the survival rates of certain disease-causing bacteria in the stomach.

Microorganisms can change the pH of their environment by utilizing acids and bases and by producing acidic or basic wastes. For example, fermentative microorganisms form organic acids from the decomposition of sugar, and the bacterium *Thiobacillus* (thī-ō-bă-sil´ŭs) can reduce the pH of its environment to 0.0. Acid produced by this bacterium in mine water dissolves enough uranium and copper from low-grade ore to make some mines profitable.

Scientists measure pH with a pH meter or with test papers impregnated with chemicals (such as litmus or phenol red) that change color in response to pH. In a microbiological laboratory, changes in color of such pH indicators incorporated into microbial growth media are commonly used to distinguish among bacterial genera.

Salts

As we have seen, a salt is a compound that dissociates in water into cations and anions other than H^+ and OH^-. Acids and hydroxyl-yielding bases neutralize each other during exchange reactions that produce water and salt. For instance, milk of

[11]The name *pylori* refers to the pylorus, a region of the stomach.

magnesia (magnesium hydroxide) is an antacid used to neutralize excess stomach acid. The chemical reaction is

$$\underset{\text{Magnesium hydroxide}}{Mg(OH)_2} + \underset{\text{Hydrochloric acid}}{2\,HCl} \rightarrow \underset{\text{Magnesium chloride (salt)}}{MgCl_2} + \underset{\text{Water}}{2\,H_2O}$$

Cations and anions of salts are electrolytes. A cell uses electrolytes to create electrical differences between its inside and outside, to transfer electrons from one location to another, and as important components of many enzymes. Certain organisms also use salts such as calcium carbonate ($CaCO_3$) to provide structure and support for their cells.

TELL ME WHY

Why does the neutralization of an acid by a base often produce water?

Organic Macromolecules

Inorganic molecules play important roles in an organism's metabolism; however, water excluded, they compose only about 1.5% of its mass. Inorganic molecules are typically too small and too simple to constitute an organism's basic structures or to perform the complicated chemical reactions required of life. These functions are fulfilled by organic molecules, which are generally larger and much more complex.

Functional Groups

LEARNING **OUTCOME**

2.17 Define *functional group* as it relates to organic chemistry.

As we have seen, organic molecules contain carbon and hydrogen atoms, and each carbon atom can form four covalent bonds with other atoms (see Figure 2.5c and d). Carbon atoms that are linked together in branched chains, unbranched chains, and rings provide the basic frameworks of organic molecules.

Atoms of other elements are bound to these carbon frameworks to form an unlimited number of compounds. Besides carbon and hydrogen, the most common elements in organic compounds are oxygen, nitrogen, phosphorus, and sulfur. Other elements, such as iron, copper, molybdenum, manganese, zinc, and iodine, are important in some proteins.

Atoms often appear in certain common arrangements called **functional groups**. For example, —NH_2, the amino functional group, is found in all amino acids, and —OH, the hydroxyl functional group,[12] is common to all alcohols. When a class of organic molecules is discussed, the letter **R** (for *residue*) designates atoms in the compound that vary from one molecule to another. The symbol R—OH, therefore, represents the general formula for an alcohol. **TABLE 2.3** on p. 40 describes some common functional groups of organic molecules.

There is a great variety of organic compounds, but certain basic types are used by all organisms. These molecules—known as *macromolecules* because they are very large—are lipids, carbohydrates, proteins, and nucleic acids.

Lipids

LEARNING **OUTCOMES**

2.18 Describe the structure of a triglyceride molecule, and compare it to that of a phospholipid.

2.19 Distinguish among saturated, unsaturated, and polyunsaturated fatty acids.

Lipids are a diverse group of organic macromolecules not composed of regular subunits. They have one common trait—they are **hydrophobic**;[13] that is, they are insoluble in water. Lipids have little or no affinity for water because they are composed almost entirely of carbon and hydrogen atoms linked by nonpolar covalent bonds. Because these bonds are nonpolar, they have no attraction to the polar bonds of water molecules. To look at it another way, the polar water molecules are attracted to each other and exclude the nonpolar lipid molecules. There are four major groups of lipids in cells: fats, phospholipids, waxes, and steroids.

Fats

Organisms make **fats** via dehydration synthesis reactions that form *esters* between three chainlike fatty acids and an alcohol named glycerol **(FIGURE 2.15a)**. Fats are also called *triglycerides* because they contain three fatty acid molecules linked to a molecule of glycerol.

The three fatty acids in a fat molecule may be identical or different from one another, but each usually has 12 to 20 carbon atoms. An important difference among fatty acids is the presence and location of double bonds between the carbon atoms. When the carbon atoms are linked solely by single bonds, every carbon atom, with the exception of the terminal ones, is covalently linked to two hydrogen atoms. Such a fatty acid is **saturated** with hydrogen and is thus termed a **saturated fatty acid (FIGURE 2.15b)**. In contrast, **unsaturated fatty acids** contain at least one double bond between adjacent carbon atoms and therefore contain at least one carbon atom bound to only a single hydrogen atom. If several double bonds exist in even one fatty acid of a molecule of fat, then it is a **polyunsaturated fat**.

Saturated fats (composed of saturated fatty acids), like those found in animals, are usually solid at room temperature because their fatty acids can be packed closely together. Unsaturated fatty acids, by contrast, are bent at every double bond and so cannot be packed tightly; they remain liquid at room temperature. Most fats in plants are unsaturated or polyunsaturated. **TABLE 2.4** on p. 41 compares the structures and melting points of four common fatty acids.

[12]Note that the hydroxyl functional group is not the same thing as the hydroxyl *ion*, because the former is covalently bonded to a carbon atom.

[13]From Greek *hydor*, meaning "water," and *phobos*, meaning "fear."

TABLE 2.3 Functional Groups of Organic Molecules and Some Classes of Compounds in Which They Are Found

Structure	Name	Class of Compounds
$-OH$	Hydroxyl	Alcohol Monosaccharide Amino acid
$R-CH_2-O-CH_2-R'$	Ether	Disaccharide Polysaccharide
$R-C(=O)-R'$	Internal carbonyl—a carbon atom (in R group) on each side	Ketone Carbohydrate
$R-C(=O)-H$	Terminal carbonyl—a carbon atom (in R group) on only one side	Aldehyde
$R-C(=O)-O-H$	Carboxyl	Amino acid Protein Fatty acid
$R-C(H)-NH_2$	Amino	Amino acid Protein
$R-C(=O)-O-R'$	Ester	Fat Wax
$R-CH_2-SH$	Sulfhydryl	Amino acid Protein
$R-CH_2-O-P(OH)(OH)=O$	Organic phosphate	Phospholipid Nucleotide ATP

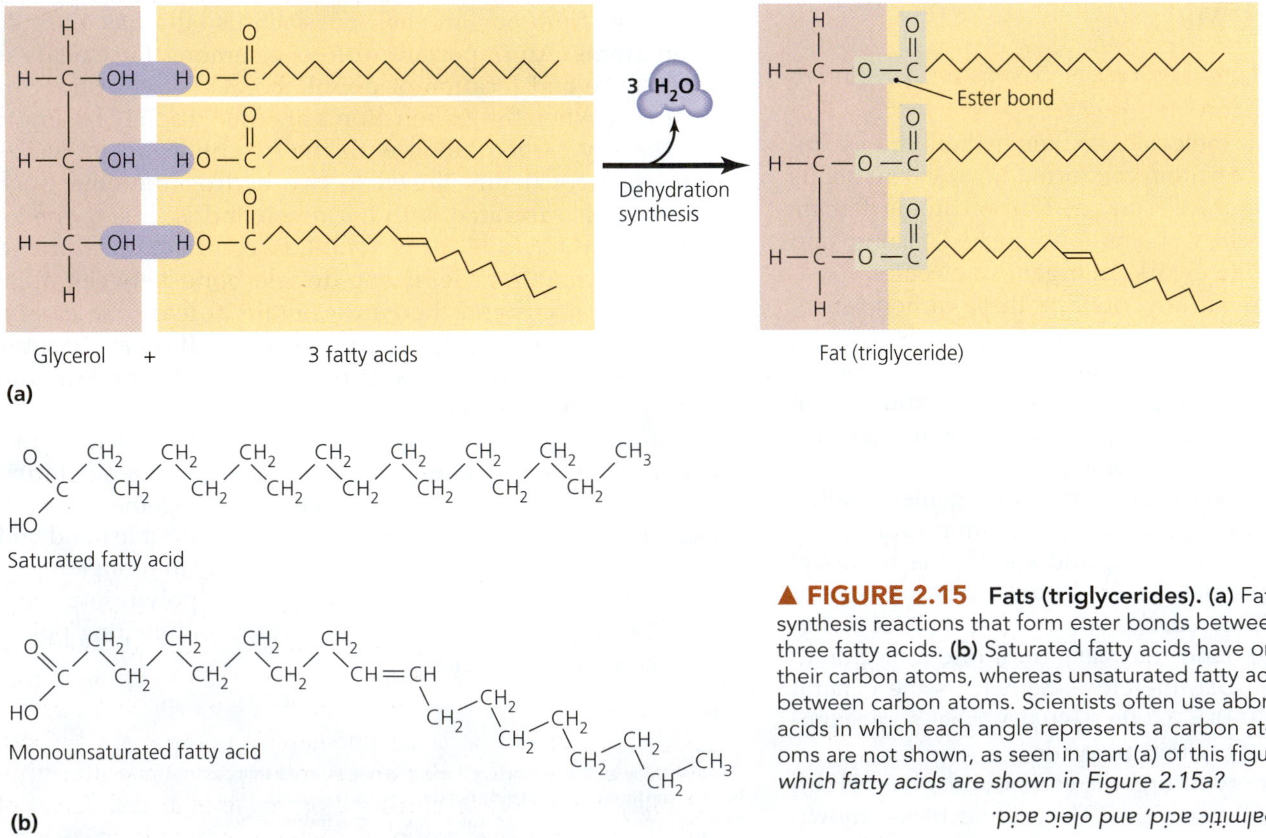

▲ **FIGURE 2.15 Fats (triglycerides). (a)** Fats are made in dehydration synthesis reactions that form ester bonds between a glycerol molecule and three fatty acids. **(b)** Saturated fatty acids have only single bonds between their carbon atoms, whereas unsaturated fatty acids have double bonds between carbon atoms. Scientists often use abbreviated diagrams of fatty acids in which each angle represents a carbon atom and most hydrogen atoms are not shown, as seen in part (a) of this figure. *According to Table 2.4, which fatty acids are shown in Figure 2.15a?*

Figure 2.15 *Stearic acid, palmitic acid, and oleic acid.*

TABLE 2.4 Common Fatty Acids in Fats and Cell Membranes

Numbers of Carbon Atoms: Double Bonds	Type of Fatty Acid	Structure and Formula	Common Name	Melting Point
16:0	Saturated	$CH_3(CH_2)_{14}COOH$ 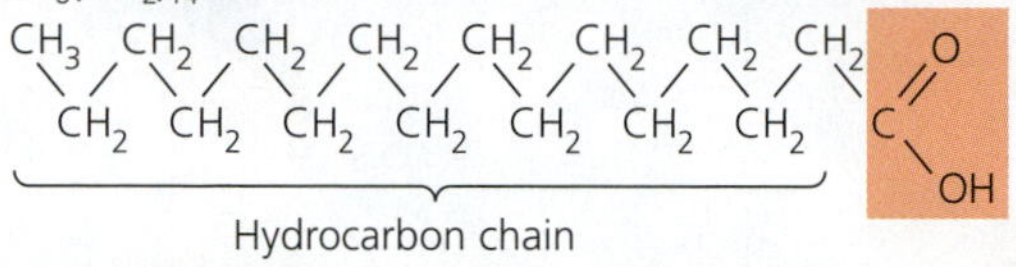	Palmitic acid	63°C
18:0	Saturated	$CH_3(CH_2)_{16}COOH$ 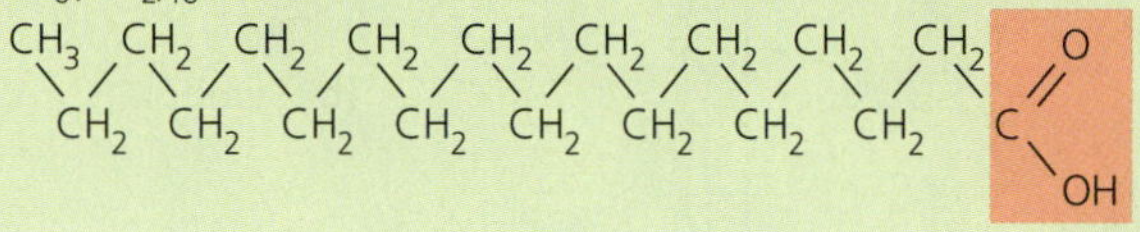	Stearic acid	70°C
18:1	Monounsaturated	$CH_3(CH_2)_7CH{=}CH(CH_2)_7COOH$ 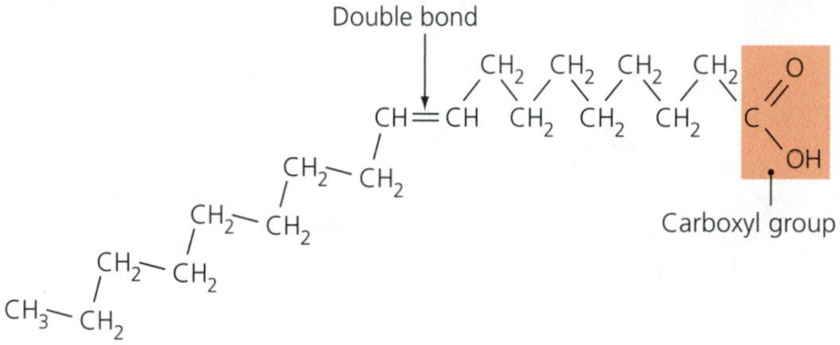	Oleic acid	16°C
18:2	Polyunsaturated	$CH_3(CH_2)_4(CH{=}CHCH_2)_2(CH_2)_6COOH$ 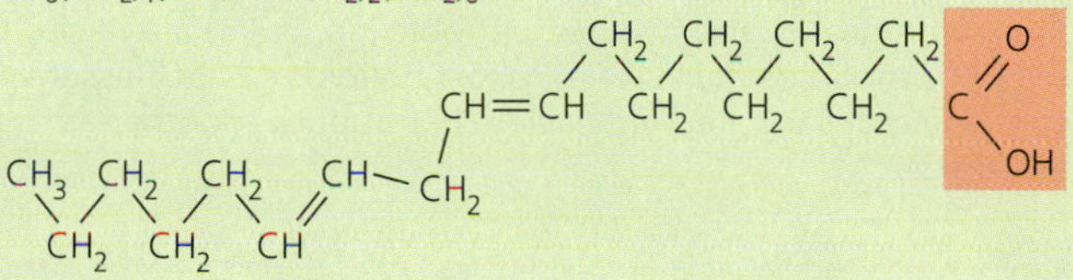	Linoleic acid	−5°C

Fats contain an abundance of energy stored in their carbon-carbon covalent bonds. Indeed, a major role of fats in organisms is to store energy. Fats can be catabolized to provide energy for movement, synthesis, and transport (Chapter 5).

Phospholipids

Phospholipids are similar to fats, but they contain only two fatty acid chains instead of three. In phospholipids, the third carbon atom of glycerol is linked to a phosphate (PO_4) functional group instead of a fatty acid **(FIGURE 2.16a)**. Like fats, different phospholipids contain different fatty acids. Small organic groups linked to the phosphate group provide additional variety among phospholipid molecules.

The fatty acid "tail" portion of a phospholipid molecule is nonpolar and thus hydrophobic, whereas the phospholipid "head" is polar and thus **hydrophilic**.[14] As a result, phospholipids placed in a watery environment will always self-assemble into forms that keep the fatty acid tails away from water. One way they do this is to form a spherical phospholipid bilayer, which resembles a two-ply ball **(FIGURE 2.16b)**.

The fatty acid tails, which are hydrophobic, congregate in the water-free interior of bilayers. The polar phosphate heads orient toward the water because they are hydrophilic. Phospholipid bilayers make up the membranes surrounding cells as well as the internal membranes of plant, fungal, and animal cells.

Waxes

Waxes contain one long-chain fatty acid linked covalently to a long-chain alcohol by an ester bond. Waxes do not have a hydrophilic head; thus, they are completely water insoluble. Certain microorganisms, such as *Mycobacterium tuberculosis* (mī´kō-bak-tēr´ē-ŭm too-ber-kyū-lō´sis), are surrounded by a waxy wall, making them resistant to drying. Some marine microbes use waxes instead of fats as energy storage molecules.

Steroids

A final group of lipids are **steroids**. Steroids consist of four rings (each containing five or six carbon atoms) that are fused to one another and attached to various side chains and functional groups **(FIGURE 2.17a)**. Steroids play many roles in human metabolism. Some act as hormones; another steroid, *cholesterol*, is perhaps familiar to you as an undesirable component of food. However, cholesterol is also an essential part of the phospholipid bilayer membrane surrounding an animal cell. Cells of fungi, plants, and one group of bacteria (mycoplasmas) have

[14]From Greek *philos*, meaning "love."

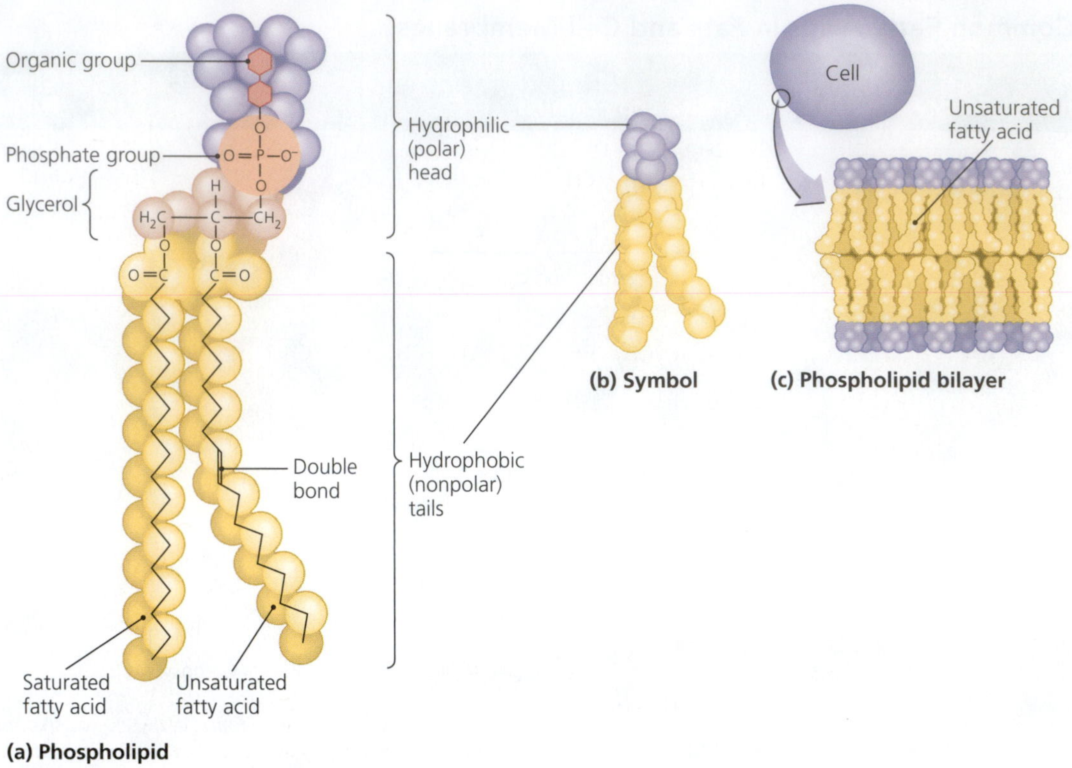

▲ **FIGURE 2.16 Phospholipids.** **(a)** A phospholipid is composed of a hydrophilic (polar) "head," which is composed of glycerol and a phosphate group, and two hydrophobic (nonpolar) fatty acid "tails." **(b)** The symbol used to represent phospholipids. **(c)** In water, phospholipids can self-assemble into spherical bilayers. Phospholipids containing unsaturated fatty acids do not pack together as tightly as those containing saturated fatty acids.

similar sterol molecules in their membranes. Sterols, which are steroids with an —OH functional group, interfere with the tight packing of the fatty acid chains of phospholipids **(FIGURE 2.17b)**. This keeps the membranes fluid and flexible at low temperatures. Without steroids such as cholesterol, the membranes of cells would become stiff and inflexible in the cold.

Carbohydrates, proteins, and nucleic acid macromolecules are composed of simpler subunits known as **monomers**,[15] which are basic building blocks. The monomers of these macromolecules are joined together to form chains of monomers called **polymers**.[16] Some macromolecular polymers are composed of hundreds of thousands of monomers.

Carbohydrates

LEARNING | **OUTCOME**

2.20 Discuss the roles carbohydrates play in living systems.

Carbohydrates are organic molecules composed solely of atoms of carbon, hydrogen, and oxygen. Most carbohydrate compounds contain an equal number of oxygen and carbon atoms and twice as many hydrogen atoms as carbon atoms, so the general formula for a carbohydrate is $(CH_2O)_n$, where n indicates the number of CH_2O units.

Carbohydrates play many important roles in organisms. Large carbohydrates, such as starch and glycogen, are used for the long-term storage of chemical energy, and a smaller carbohydrate molecule—glucose—serves as a ready energy source in most cells. Carbohydrates also form part of the backbones of DNA and RNA, and other carbohydrates are converted routinely into amino acids. Additionally, polymers of carbohydrate form the cell walls of most fungi, plants, algae, and prokaryotes and are involved in intercellular interactions between animal cells. For example, specific carbohydrate-protein combinations found on the surfaces of white blood cells determine which cells interact in immune responses against pathogens.

Monosaccharides

The simplest carbohydrates are **monosaccharides**[17]—simple sugars **(FIGURE 2.18)**. The general names for the classes of monosaccharides are formed from a prefix indicating the number of carbon atoms and from the suffix *-ose*. For example, *pentoses* are sugars with five carbon atoms, and *hexoses* are sugars with six carbon atoms. Pentoses and hexoses are particularly important in cellular metabolism. For example, deoxyribose,

[15]From Greek *mono*, meaning "one," and *meris*, meaning "part."
[16]From Greek *poly*, meaning "many," and *meris*, meaning "part."
[17]From Greek *sakcharon*, meaning "sugar."

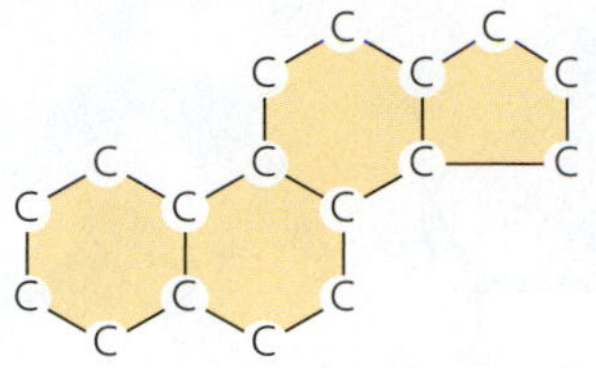

(a)

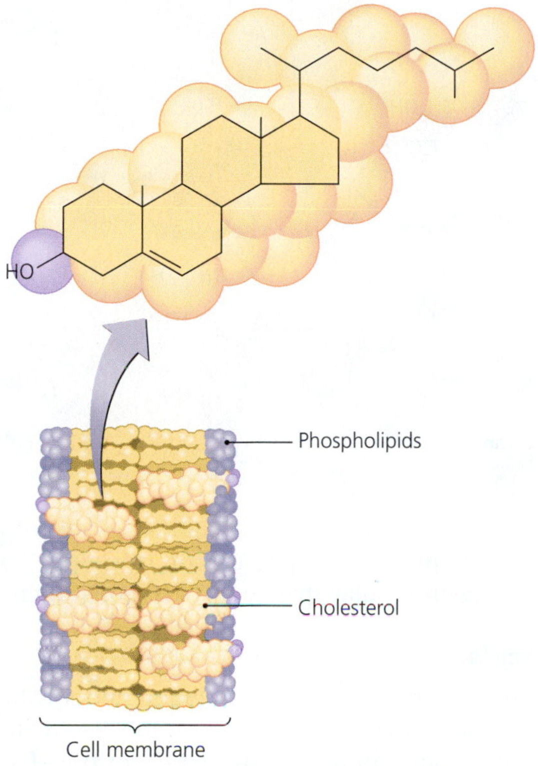

(b)

▲ **FIGURE 2.17 Steroids.** **(a)** Steroids are lipids characterized by four "fused" rings. **(b)** The steroid cholesterol functions in animal and protozoan cell membranes to prevent packing of phospholipids, thereby keeping the membranes fluid at low temperatures.

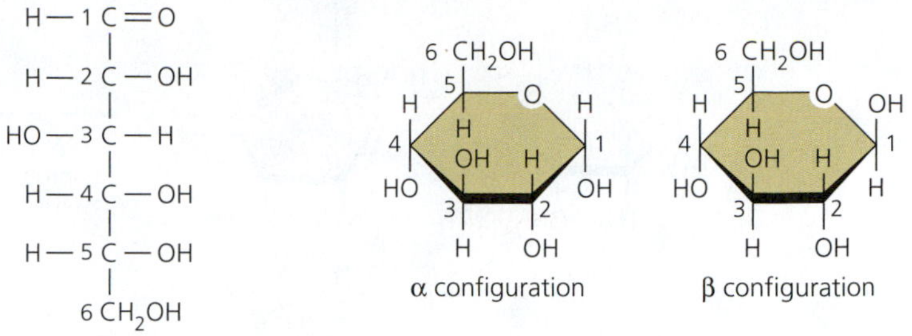

(a) Glucose

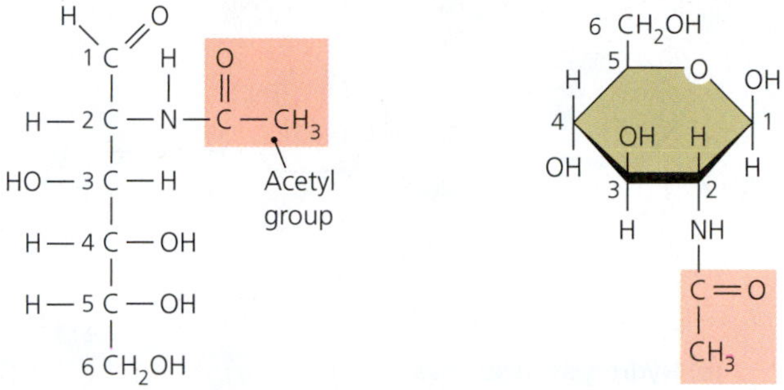

(b) *N*-acetylglucosamine

▲ **FIGURE 2.18 Monosaccharides (simple sugars).** Although simple sugars may exist as either linear molecules (at left) or rings (at right), energy dynamics in the watery cytoplasm of cells generally favor ring forms. **(a)** Glucose, a hexose, is the primary energy source for cellular metabolism and an important monomer in many larger carbohydrates. Chemists number the carbon atoms as shown. Alpha and beta ring configurations differ in the location of oxygen bound to carbon 1. **(b)** *N*-acetylglucosamine (NAG), a monomer in bacterial cell walls.

which is the sugar component of DNA, is a pentose. Glucose is a hexose and the primary energy molecule of cells, and fructose is a hexose found in fruit. Chemists assign numbers to the carbon atoms.

Monosaccharides may exist as linear molecules, but because of energy dynamics, they usually take cyclic (ring) forms. In some cases, more than one cyclic structure may exist. For example, glucose can assume an alpha (α) configuration or a beta (β) configuration (see Figure 2.18a). As we will see, these configurations play important roles in the formation of different polymers.

Disaccharides

When two monosaccharide molecules are linked together via dehydration synthesis, the result is a **disaccharide**. For example, the linkage of two hexoses, glucose and fructose, forms sucrose (table sugar) and a molecule of water **(FIGURE 2.19a)**. Other disaccharides include maltose (malt sugar) and lactose (milk sugar). Disaccharides can be broken down via hydrolysis into their constituent monosaccharides **(FIGURE 2.19b)**.

Polysaccharides

Polysaccharides are polymers composed of tens, hundreds, or thousands of monosaccharides that have been covalently linked in dehydration synthesis reactions. Even polysaccharides that contain only glucose monomers can be quite diverse because they can differ according to their monosaccharide monomer configurations (either alpha or beta) and their shapes (either branched or unbranched). Cellulose, the main constituent of the cell walls of plants and some green algae, is a long unbranched molecule that contains only β-monomers of glucose linked between carbons 1 and 4 of alternating monomers; such bonds are termed β-1,4 bonds **(FIGURE 2.20a)**. Amylose, a starch storage compound in plants, has only α-1,4 bonds and is unbranched **(FIGURE 2.20b)**; glycogen, a storage molecule formed in the liver and muscle cells of animals, is a highly branched molecule with both α-1,4 and α-1,6 bonds **(FIGURE 2.20c)**.

The cell walls of bacteria are composed of *peptidoglycan,* which is made of polysaccharides and amino acids (see Figure 3.15). Polysaccharides may also be linked to lipids to form glycolipids, which can form cell markers such as those involved in the ABO blood typing system in humans.

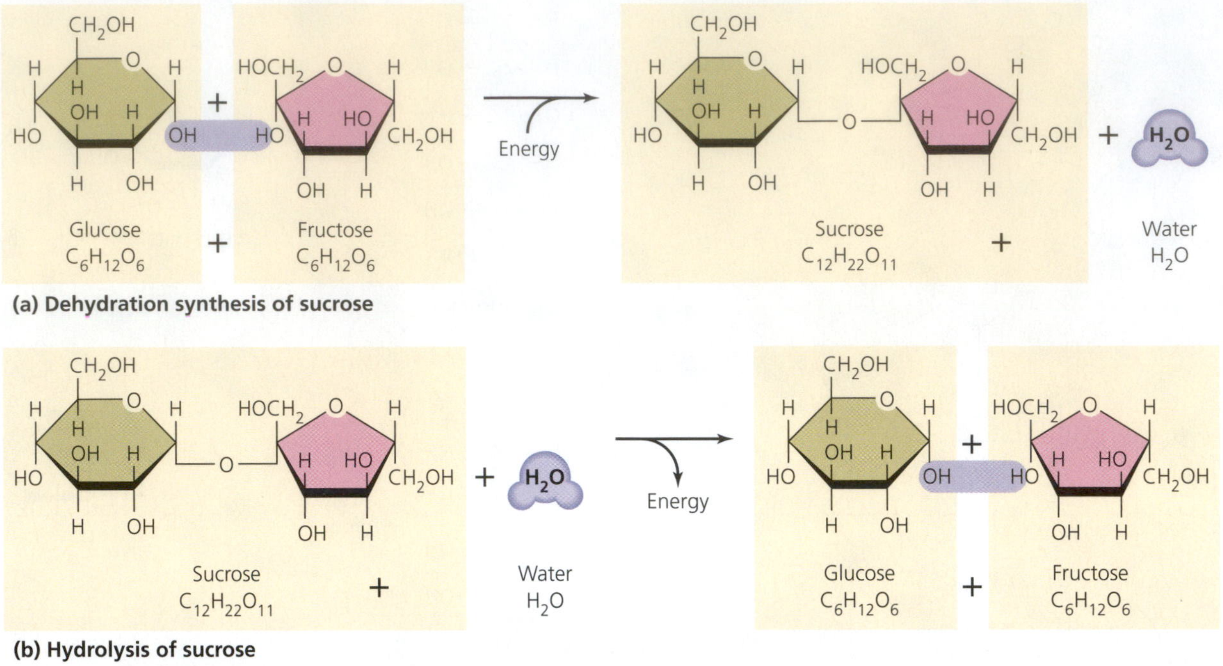

▲ **FIGURE 2.19 Disaccharides. (a)** Formation of the disaccharide sucrose via dehydration synthesis. **(b)** Breakdown of sucrose via hydrolysis.

Proteins

LEARNING **OUTCOMES**

2.21 Describe five general functions of proteins in organisms.
2.22 Sketch and label four levels of protein structure.

The most complex organic compounds are **proteins**, which are composed mostly of carbon, hydrogen, oxygen, nitrogen, and sulfur. Proteins perform many functions in cells, including the following:

- *Structure.* Proteins are structural components found in cell walls, in membranes, and within cells themselves. Proteins are also the primary structural material of hair, nails, the outer cells of skin, muscle, and flagella and cilia (the last two act to move microorganisms through their environment).
- *Enzymatic catalysis.* Catalysts are chemicals that enhance the speed or likelihood of a chemical reaction. Protein catalysts in cells are called *enzymes.*
- *Regulation.* Some proteins regulate cell function by stimulating or hindering either the action of other proteins or the expression of genes. Hormones are examples of regulatory proteins.
- *Transportation.* Certain proteins act as channels and "pumps" that move substances into or out of cells.
- *Defense and offense.* Antibodies and *complement* are examples of proteins that defend your body against pathogens. Some bacteria even produce proteins called *bacteriocins* that kill other bacteria.

A protein's function is dependent on its shape, which is determined by the molecular structures of its constituent parts.

Amino Acids

Proteins are polymers composed of monomers called **amino acids**. Amino acids contain a basic amino group (—NH_2), a hydrogen atom, and an acidic carboxyl group (—COOH). All attach to the same carbon atom, which is known as the α-carbon **(FIGURE 2.21)**. A fourth bond attaches the α-carbon to a side group (—R) that varies among different amino acids. The side group may be a single hydrogen atom, various chains, or various complex ring structures. Hundreds of amino acids are possible, but most organisms use only 21 amino acids in synthesizing proteins.[18] The different side groups affect the way amino acids interact with one another within a given protein as well as how a protein interacts with other molecules. A change in an amino acid's side group may seriously interfere with a protein's normal function.

Because amino acids contain both an acidic carboxyl group and a basic amino group, they have both positive and negative charges and are easily soluble in water. Aqueous solutions of organic molecules such as amino acids and simple sugars bend light rays passing trough the solution. Molecules known as *D forms*[19] bend light rays clockwise; other molecules bend light rays counterclockwise and are known as *L forms.*[20]

Many organic molecules exist as both D and L forms that are *stereoisomers* of one another; that is, they have the same atoms

[18]While 20 amino acids are more common, the genes of most organisms code for a 21st—selenocysteine. The genes of a few prokaryotes code for a 22nd amino acid.
[19]From Latin *dexter*, meaning "on the right."
[20]From Latin *laevus*, meaning "on the left."

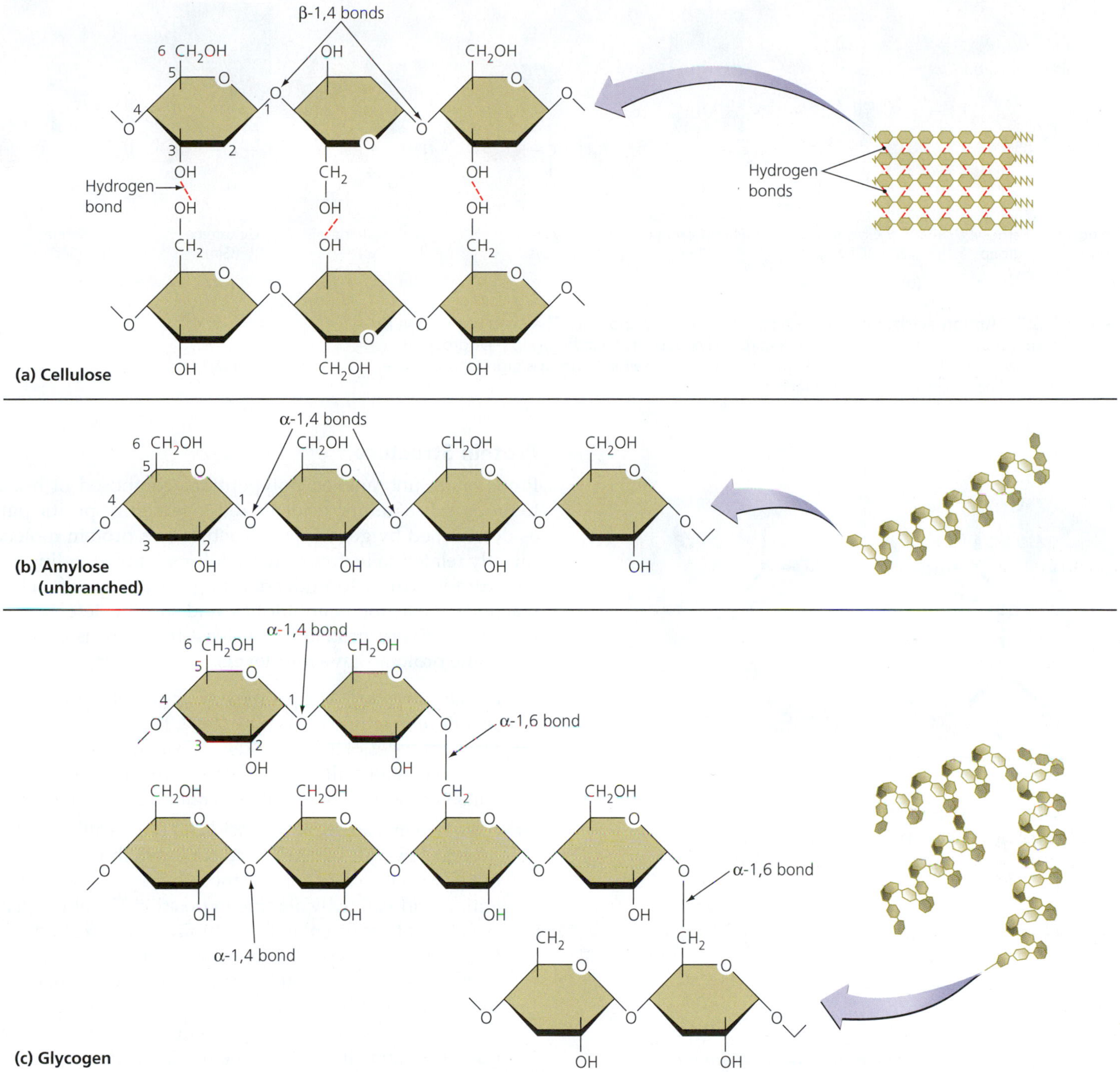

▲ **FIGURE 2.20 Polysaccharides.** All three polysaccharides shown here are composed solely of glucose but differ in the configuration of the glucose monomers and the amount of branching. **(a)** Cellulose, the major structural material in plants, is unbranched and contains only ß-1,4 bonds. **(b)** Amylose is an unbranched plant starch with only α-1,4 bonds. **(c)** Glycogen, a highly branched storage molecule in animals, is composed of glucose monomers linked by α-1,4 or α-1,6 bonds.

and functional groups but are mirror images of each other **(FIGURE 2.22)**. Amino acids in proteins are almost always L forms—except for glycine, which does not have a stereoisomer. Interestingly, organisms almost always use D sugars in metabolism and polysaccharides. Rare stereoisomers—D amino acids and L sugars—do exist in some bacterial cell walls and in some antibiotics.

Peptide Bonds

Cells link amino acids together in chains that somewhat resemble beads on a necklace. By a dehydration synthesis reaction, a covalent bond is formed between the carbon of the carboxyl group of one amino acid and the nitrogen of the amino group of the next amino acid in the chain **(FIGURE 2.23)**. Cells follow the organism's genetic instructions to link amino acids together

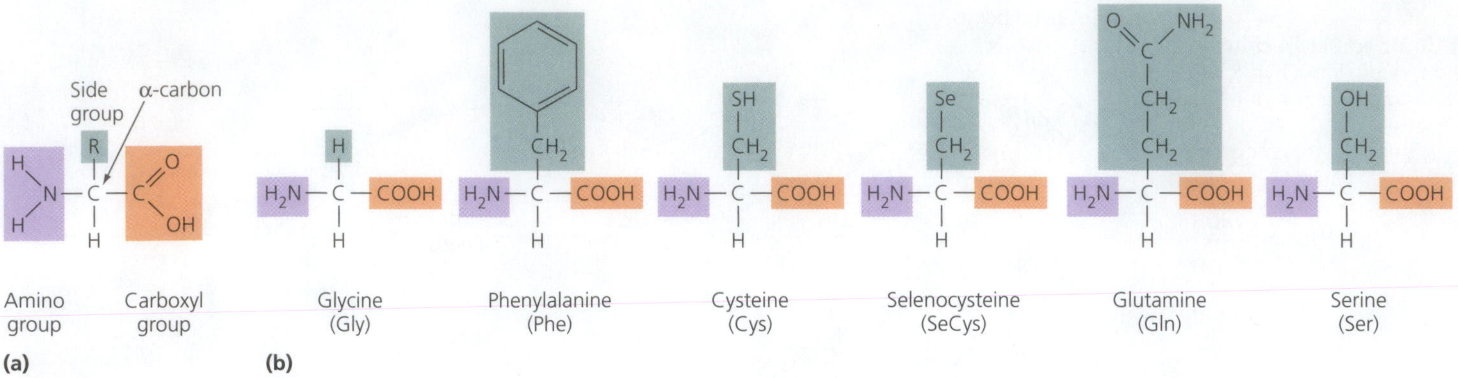

▲ **FIGURE 2.21 Amino acids.** **(a)** The basic structure of an amino acid. The central α-carbon is attached to an amino group, a hydrogen atom, a carboxyl group, and a side group (—R group) that varies among amino acids. **(b)** Some selected amino acids, with their side groups highlighted. Note that each amino acid has a distinctive abbreviation.

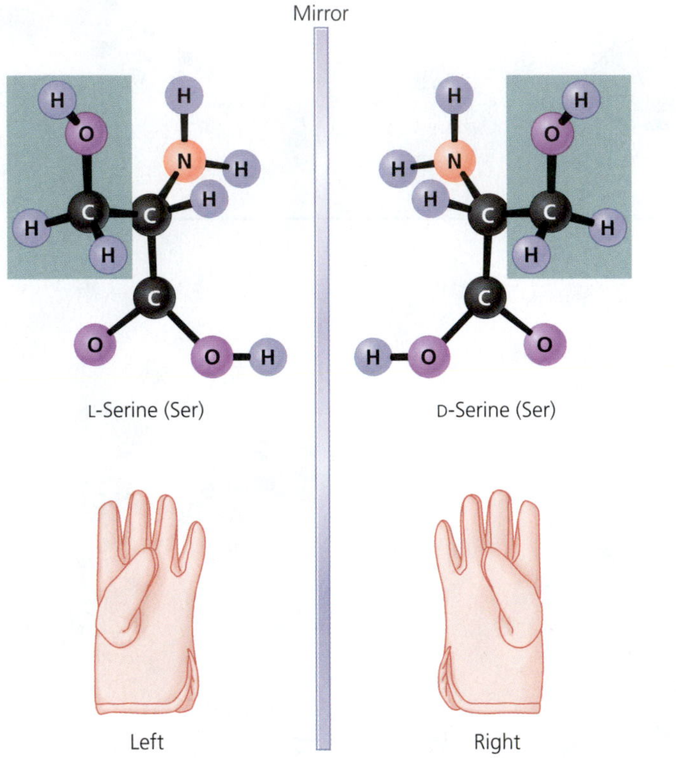

▲ **FIGURE 2.22 Stereoisomers, molecules that are mirror images of one another.** When dissolved in water, D isomers bend light clockwise, and L forms bend light counterclockwise. Just as a right-handed glove does not fit a left hand, so a D stereoisomer cannot be substituted for an L stereoisomer in metabolic reactions.

in precise sequences. (Chapter 7 examines this process in more detail.)

Scientists refer to covalent bonds between amino acids by a special name: **peptide**[21] **bonds**. A molecule composed of two amino acids linked together by a single peptide bond is called a *dipeptide*; longer chains of amino acids are called *polypeptides*.

Protein Structure

Proteins are unbranched polypeptides composed of hundreds to thousands of amino acids linked together in specific patterns as determined by genes. The structure of a protein molecule is directly related to its function; therefore, understanding protein structure is critical to understanding certain specific chemical reactions, the action of antibiotics, and specific defenses against pathogens. Every protein has at least three levels of structure, and some proteins have four levels.

- *Primary structure.* The primary structure of a protein is its sequence of amino acids **(FIGURE 2.24a)**. Cells use many different types of amino acids in proteins, though not every protein contains all types. The primary structures of proteins vary widely in length and amino acid sequence.
- A change in a single amino acid can drastically affect a protein's overall structure and function, though this is not always the case. For instance, the replacement of the amino acid valine by alanine in position 136 of the primary structure of a particular sheep brain protein, called cellular prion (prē´on) protein, may result in a disease called *scrapie.* The altered protein spread into cows, causing *mad cow disease,* and from cows into humans, causing *variant Creutzfeldt-Jakob* (kroytsfelt-yah-kŭp) *disease.*[22] However, numerous substitutions can be made in other, noncritical regions of cellular prion protein, with no ill effects.
- *Secondary structure.* Ionic bonds, hydrogen bonds, and hydrophobic and hydrophilic characteristics cause many polypeptide chains to fold into either coils called α-*helices* or accordion-like structures called β-*pleated sheets* **(FIGURE 2.24b)**. Proteins are typically composed of both α-helices and ß-pleated sheets linked by short sequences of amino acids that do not show such secondary structure. Because of its primary structure, the protein that causes

[21]From *peptone,* the name given to short chains of amino acids resulting from the partial digestion of protein.
[22]Named for the two German neurobiologists who first described the disease.

▶ **FIGURE 2.23 The linkage of amino acids by peptide bonds via a dehydration reaction.** In this reaction, removing a hydroxyl group from amino acid 1 and a hydrogen atom from amino acid 2 produces a dipeptide—which is two amino acids linked by a single peptide bond—and a molecule of water.

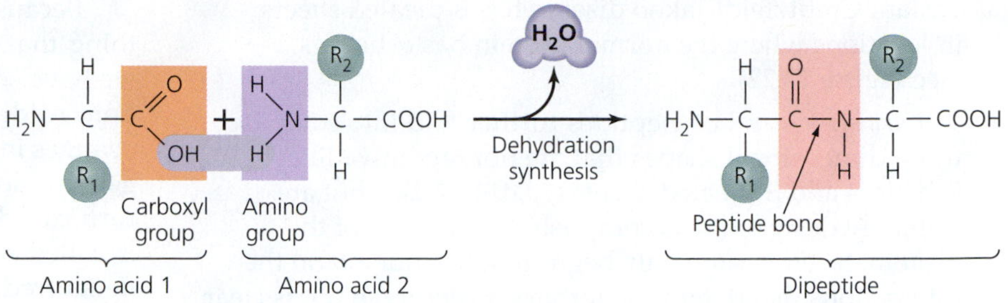

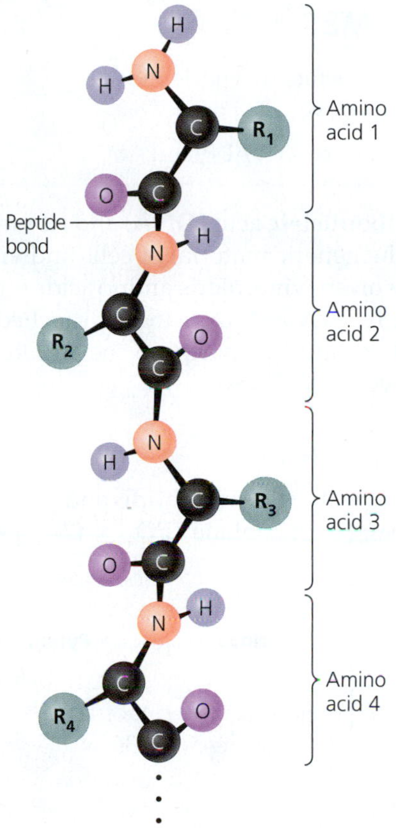

(a) Primary structure

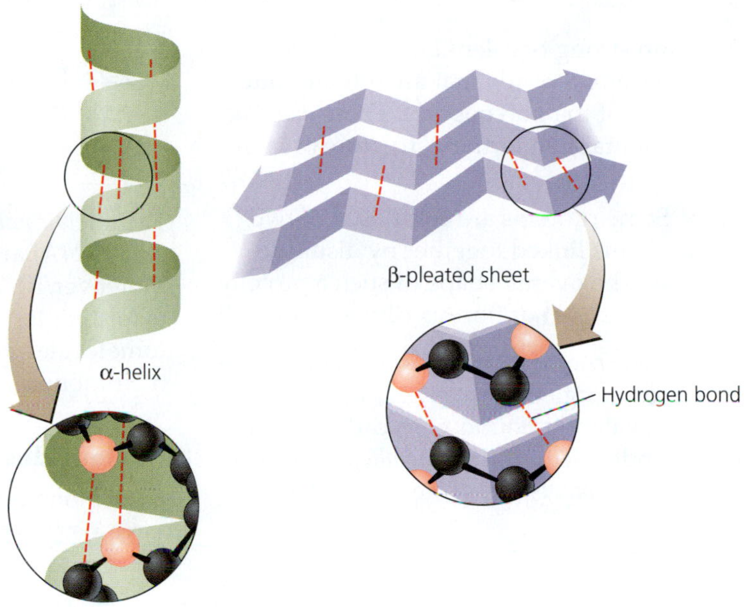

(b) Secondary structure

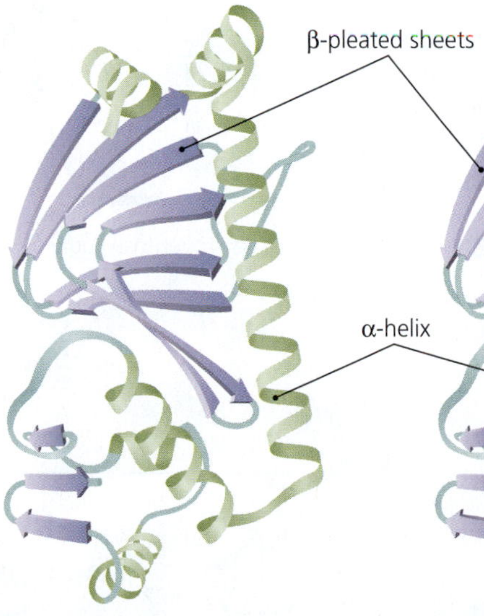

(c) Tertiary structure

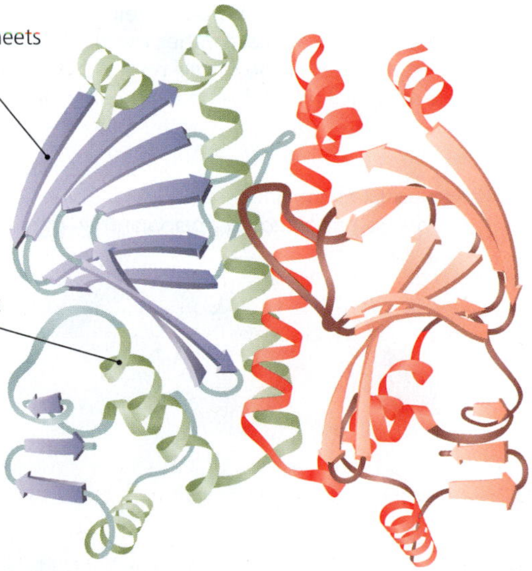

(d) Quaternary structure: two or more polypeptides act together as a single protein

▲ **FIGURE 2.24 Levels of protein structure.** **(a)** A protein's primary structure is the sequence of amino acids in a polypeptide. **(b)** Secondary structure arises as a result of interactions, such as hydrogen bonding, between regions of the polypeptide. Secondary structure takes two basic shapes: α-helices and ß-pleated sheets. **(c)** The more complex tertiary structure is a three-dimensional shape defined by further hydrogen bonding as well as disulfide bridges between neighboring cysteine amino acid molecules. **(d)** Those proteins that are composed of more than one polypeptide chain have a quaternary structure.

variant Creutzfeldt-Jakob disease has ß-pleated sheets in locations where the normal protein has α-helices (see Figure 13.22).

- *Tertiary structure*. Polypeptides further fold into complex three-dimensional shapes that are not repetitive like α-helices and β-pleated sheets **(FIGURE 2.24c)** but are uniquely designed to accomplish the function of the protein. Scientists are only beginning to understand the interactions that determine tertiary structure, but it is clear that covalent bonds between—R groups of amino acids, hydrogen bonds, ionic bonds, and other molecular interactions are important. For instance, nonpolar (hydrophobic) side chains fold into the interior of molecules, away from the presence of water.

  Some proteins form strong covalent bonds between sulfur atoms of cysteine amino acids that are brought into proximity by the folding of the polypeptide. These *disulfide bridges* are critical in maintaining tertiary structure of many proteins.
- *Quaternary structure*. Some proteins are composed of two or more polypeptide chains linked together by disulfide bridges or other bonds. The overall shape of such a protein may be globular **(FIGURE 2.24d)** or fibrous (threadlike).

Organisms may further modify proteins by combining them with other organic or inorganic molecules. For instance, *glycoproteins* are proteins covalently bound with carbohydrates, *lipoproteins* are proteins bonded with lipids, *metalloproteins* contain metallic ions, and *nucleoproteins* are proteins bonded with nucleic acids.

Because protein shape determines protein function, anything that severely interrupts shape also disrupts function. As we have seen, amino acid substitution can alter shape and function. Additionally, physical and chemical factors, such as heat, changes in pH, and salt concentration, can interfere with hydrogen and ionic bonding between parts within a protein. This in turn can disrupt the three-dimensional structure. This process is called **denaturation**. Denaturation can be temporary (if the denatured protein is able to return to its original shape again) or permanent.

Nucleic Acids

LEARNING | **OUTCOMES**

2.23 Describe the basic structure of a nucleotide.
2.24 Compare and contrast DNA and RNA.
2.25 Contrast the structures of ATP, ADP, and AMP.

The nucleic acids **deoxyribonucleic acid (DNA)** and **ribonucleic acid (RNA)** are vital as the genetic material of cells and viruses. Moreover, RNA, acting as an enzyme, binds amino acids together to form polypeptides. Both DNA and RNA are unbranched macromolecular polymers that differ primarily in the structures of their monomers, which we discuss next.

Nucleotides and Nucleosides

Each monomer of nucleic acids is a **nucleotide** and consists of three parts **(FIGURE 2.25a)**: (1) phosphate (PO_4^{3-}); (2) a pentose

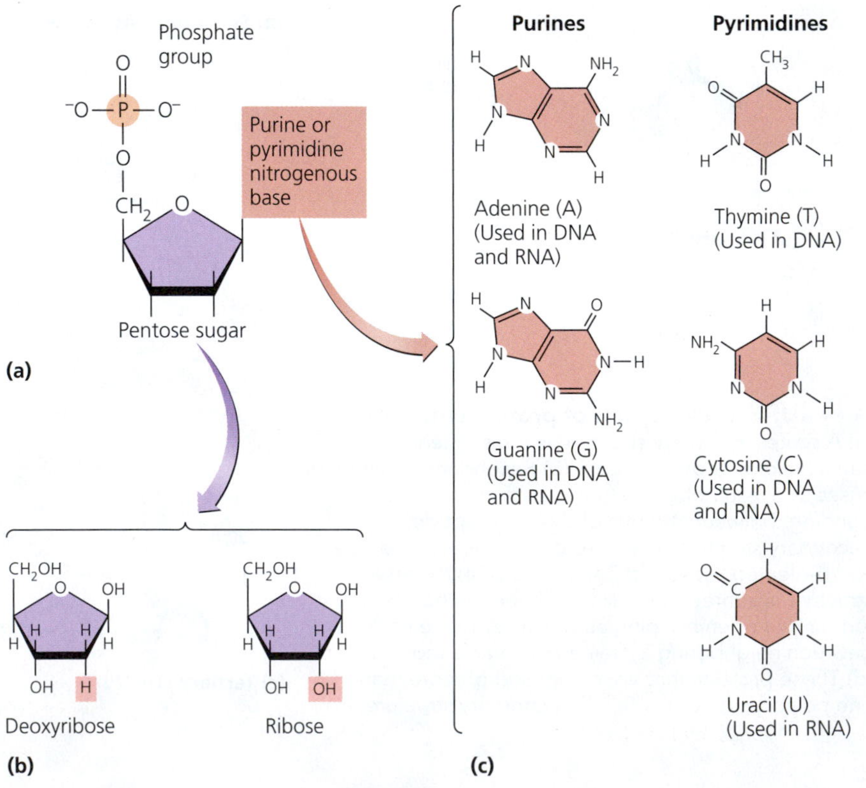

▶ FIGURE 2.25 Nucleotides. (a) The basic structure of nucleotides, each of which is composed of a phosphate, a pentose sugar, and a nitrogenous base. **(b)** The pentose sugars deoxyribose, which is found in deoxyribonucleic acid (DNA), and ribose, which is found in ribonucleic acid (RNA). **(c)** The nitrogenous bases, which are either the double-ringed purines adenine or guanine, or the single-ringed pyrimidines thymine, cytosine, or uracil. *How does a nucleoside differ from a nucleotide?*

Figure 2.25 A nucleoside is composed only of a nitrogenous base and a sugar, whereas a nucleotide has a base, sugar, and phosphate.

sugar, either deoxyribose or ribose **(FIGURE 2.25b)**; and (3) one of five cyclic (ring-shaped) nitrogenous bases: **adenine (A)**, **guanine (G)**, **cytosine (C)**, **thymine (T)**, or **uracil (U)** **(FIGURE 2.25c)**. Adenine and guanine are double-ringed molecules of a class called *purines*, whereas cytosine, thymine, and uracil have single rings and are *pyrimidines*. DNA contains A, G, C, and T bases, whereas RNA contains A, G, C, and U bases. As their names suggest, DNA nucleotides contain deoxyribose, and RNA nucleotides contain ribose. The similarly named **nucleosides** are nucleotides lacking phosphate; that is, a nucleoside is one of the nitrogenous bases attached only to a sugar.

Each nucleotide or nucleoside is also named for the base it contains. Thus, a nucleotide made with ribose, uracil, and phosphate is a uracil RNA nucleotide, which is also called a uracil *ribonucleotide*. Likewise, a nucleoside composed of adenine and deoxyribose is an adenine DNA nucleoside (or adenine *deoxyribonucleoside*). ▶**VIDEO TUTOR:** ***The Structure of Nucleotides***

Nucleic Acid Structure

Nucleic acids, like polysaccharides and proteins, are polymers. They are composed of nucleotides linked by covalent bonds between the phosphate of one nucleotide and the sugar of the next. Polymerization results in a linear spine composed of alternating sugars and phosphates, with bases extending from it rather like the teeth of a comb **(FIGURE 2.26a)**. The two ends of a chain of nucleotides are different. At one end, called the 5′ end[23] (five prime end), carbon 5′ of the sugar is attached to a phosphate group. At the other end (3′ end), carbon 3′ of the sugar is not attached to a phosphate group.

The atoms of the bases in nucleotides are arranged in such a manner that hydrogen bonds readily form between specific bases of two adjacent nucleic acid chains. Three hydrogen bonds form between an adjacent pair composed of cytosine (C) and guanine (G), whereas two hydrogen bonds form between an adjacent pair composed of adenine (A) and thymine (T) in DNA **(FIGURE 2.26b)** or between an adjacent pair composed of adenine (A) and uracil (U) in RNA. Hydrogen bonds do not readily form between other combinations of nucleotide bases; for example, adenine does not readily pair with cytosine, guanine, or another adenine nucleotide.

In cells and most viruses that use DNA as a genome, DNA molecules are double stranded. The two strands of DNA are complementary to one another; that is, the specificity of

[23]Carbon atoms in organic molecules are commonly identified by numbers. In a nucleotide, carbon atoms 1, 2, 3, and so on belong to the base, and carbon atoms 1′, 2′, 3′, and so on belong to the sugar.

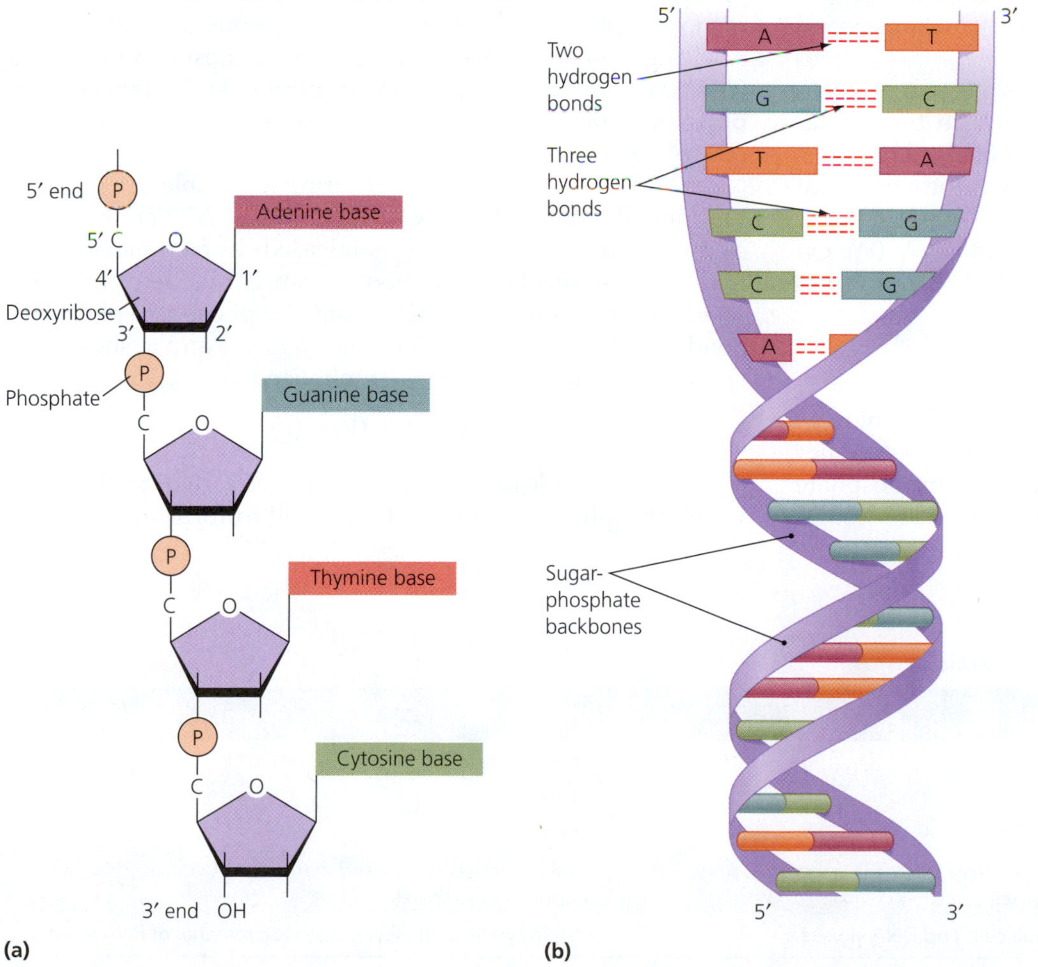

◀ **FIGURE 2.26 General nucleic acid structure.** **(a)** Nucleotides are polymerized to form chains in which the nitrogenous bases extend from a sugar-phosphate backbone like the teeth of a comb. **(b)** Specific pairs of nitrogenous bases form hydrogen bonds between adjacent nucleotide chains to form the familiar DNA double helix. *How can you determine that the molecule in (a) is DNA and not RNA?*

Figure 2.26 *It is DNA because its nucleotides have deoxyribose sugar and because some of them have thymine bases (not uracil, as in RNA).*

nucleotide base pairing ensures that opposite strands are composed of complementary nucleotides. For instance, if one strand has the sequence AATGCT, then its complement has TTACGA.

The two strands are also *antiparallel*; that is, they run in opposite directions. One strand runs from the 3′ end to the 5′ end, whereas its complement runs in the opposite direction, from its 5′ end to its 3′ end. Though hydrogen bonds are relatively weak bonds, thousands of them exist at normal temperatures, forming a stable, double-stranded DNA molecule that looks much like a ladder: the two deoxyribose-phosphate chains are the side rails, and base pairs form the rungs. Hydrogen bonding also twists the phosphate-deoxyribose backbones into a helix. Thus, typical DNA is a double helix. Parvoviruses use single-stranded DNA, which is an exception to this rule.

Nucleic Acid Function

DNA is the genetic material of all organisms and of many viruses; it carries instructions for the synthesis of RNA molecules and proteins. By controlling the synthesis of enzymes and regulatory proteins, DNA controls the synthesis of all other molecules in an organism. Genetic instructions are carried in the sequence of nucleotides that make up the nucleic acid. Even though only four kinds of bases are found in DNA (A, T, G, and C), they can be sequenced in distinctive patterns that create genetic diversity and code for an infinite number of proteins, just as an alphabet of only four letters could spell a very large number of words. Cells replicate their DNA molecules and pass copies to their descendants, ensuring that each has the instructions necessary for life.

Several kinds of ribonucleic acids, such as messenger RNA, transfer RNA, and ribosomal RNA, play roles in the formation of proteins, including catalyzing the synthesis of proteins. RNA molecules also function in place of DNA as the genome of RNA viruses.

TABLE 2.5 compares and contrasts RNA and DNA. (We examine the synthesis and function of DNA and RNA in detail in Chapter 7.)

ATP (Adenosine Triphosphate)

Phosphate in nucleotides and other molecules is a highly reactive functional group and can form covalent bonds with other phosphate groups to make diphosphate and triphosphate molecules. Such molecules made from ribose nucleotides are important in many metabolic reactions. The names of these molecules indicate the nucleotide base and the number of phosphate groups they contain. Thus, cells make adenosine monophosphate (AMP) from the nitrogenous base adenine, ribose sugar, and one phosphate group; adenosine diphosphate (ADP), which has two phosphate groups; and **adenosine triphosphate** (ă-den´ō-sēn trī-fos´fāt) or **ATP**, which has three phosphate groups (**FIGURE 2.27**).

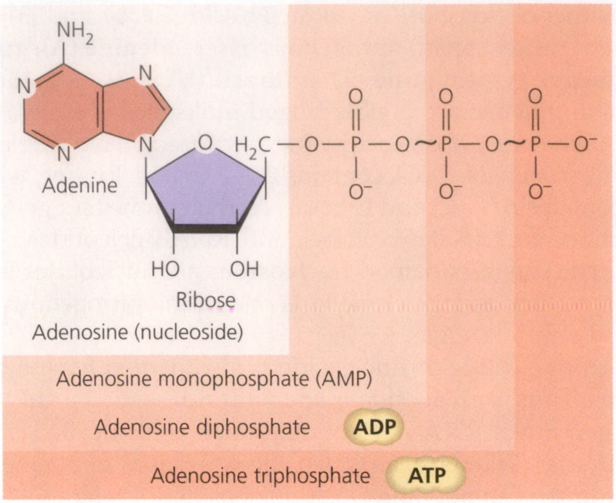

▲ **FIGURE 2.27** **ATP.** Adenosine triphosphate (ATP), the main short-term, recyclable energy supply for cells. Energy is stored in high-energy bonds between the phosphate groups. *What is the relationship between AMP and adenine ribonucleotide?*

Figure 2.27 AMP and adenine ribonucleotide are two names for the same thing.

ATP is the principal, short-term, recyclable energy supply for cells. When the phosphate bonds of ATP are broken, a significant amount of energy is released; in fact, more energy is released from phosphate bonds than is released from most other covalent bonds. For this reason, the phosphate-phosphate bonds of ATP are known as *high-energy bonds,* and to show these specialized bonds, ATP can be symbolized as

$$A—Ⓟ\sim Ⓟ\sim Ⓟ$$

Energy is released when ATP is converted to ADP and when phosphate is removed from ADP to form AMP, though

TABLE 2.5 Comparison of Nucleic Acids

Characteristic	DNA	RNA
Sugar	Deoxyribose	Ribose
Purine nucleotides	A and G	A and G
Pyrimidine nucleotides	T and C	U and C
Number of strands	Double stranded in cells and in most DNA viruses; single stranded in parvoviruses	Single stranded in cells and in most RNA viruses; double stranded in reoviruses
Function	Genetic material of all cells and DNA viruses	Protein synthesis in all cells; genetic material of RNA viruses

the latter reaction is not as common in cells. Energy released from the phosphate bonds of ATP is used for important life-sustaining activities, such as synthesis reactions, locomotion, and transportation of substances into and out of cells.

Cells also use ATP as a structural molecule in the formation of *coenzymes*. Coenzymes such as *flavin adenine dinucleotide, nicotinamide nucleotide,* and *coenzyme A* function in many metabolic reactions (as discussed in Chapter 5).

A cell's supply of ATP is limited; therefore, an important part of cellular metabolism is to replenish ATP stores. (We discuss the important ATP-generating reactions in Chapter 5.)

TELL ME WHY

Why do the cell membranes of microbes living in Arctic water likely contain more unsaturated fatty acids than do membranes of microbes living in hot springs?

MICRO IN THE CLINIC FOLLOW-UP

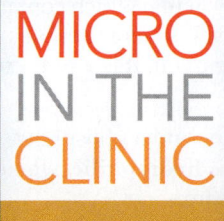

Can Spicy Food Cause Ulcers?

The pain and nausea continue for a month before Ramona finally decides to see a doctor. It turns out that she does, indeed, have an ulcer—but eating spicy food had nothing to do with it. Her doctor explains that most ulcers are caused by bacteria called *Helicobacter pylori*, which thrive in low-pH, highly acidic environments, such as the stomach. *H. pylori* usually lives in the stomach without causing any problems, but occasionally it causes inflammation in the stomach's mucous lining, resulting in an ulcer. The doctor notes that it's a common misconception that spicy food causes ulcers. He adds, however, that alcohol, tobacco, and stress can aggravate ulcers and slow the healing process. He prescribes antibiotics to Ramona to kill the bacteria. Within a short time, Ramona is feeling better—and back to enjoying her spicy food.

1. **Antacids are sometimes taken to treat ulcers. Given what you've learned about the pH environment that *H. pylori* prefers, explain how antacids might help relieve the symptoms of an ulcer.**
2. **What do you think would happen if Ramona consumed a large amount of milk? Would you expect her to feel temporary relief from her ulcer, or would you expect her to feel worse?**

MM Check your answers to Micro in the Clinic Follow-Up questions in the MasteringMicrobiology Study Area.

Explore the Invisible: The Structure of Nucleotides

Practice on-the-go with Dr. Bauman Video Tutors by scanning this QR code with your smart phone. Visit the **MasteringMicrobiology Study Area** to challenge your understanding with practice tests, animation quizzes, and clinical case studies!

MasteringMicrobiology®

CHAPTER SUMMARY

Atoms (pp. 27–29)

1. **Matter** is anything that takes up space and has mass. Its smallest chemical units, **atoms,** contain negatively charged **electrons** orbiting a nucleus composed of uncharged **neutrons** and positively charged **protons.**
2. An **element** is matter composed of a single type of atom.
3. The number of protons in the nucleus of an atom is its **atomic number.** The sum of the masses of its protons, neutrons, and electrons is an atom's **atomic mass,** which is estimated by adding the number of neutrons and protons (because electrons have little mass).
4. **Isotopes** are atoms of an element that differ only in the numbers of neutrons they contain.

Chemical Bonds (pp. 30–34)

1. The region of space occupied by electrons is an electron shell. The number of electrons in the outermost shell, or **valence** shell, of an atom determines the atom's reactivity. Most valence shells hold a maximum of eight electrons. Sharing or transferring valence electrons to fill a valence shell results in **chemical bonds.**
2. A chemical bond results when two atoms share a pair of electrons. The **electronegativities** of each of the atoms, which is the strength of their attraction for electrons, determines whether the bond between them will be a **nonpolar covalent bond** (equal sharing of electrons), a **polar covalent bond** (unequal sharing of electrons), or an **ionic bond** (giving up of electrons from one atom to another).
3. A **molecule** that contains atoms of more than one element is a **compound. Organic compounds** are those that contain carbon and hydrogen atoms.
4. An **anion** is an atom with an extra electron and thus a negative charge. A **cation** has lost an electron and thus has a positive charge. Ionic bonding between the two types of ions makes **salt.** When salts dissolve in water, their ions are called **electrolytes.**
5. **Hydrogen bonds** are relatively weak but important chemical bonds. They hold molecules in specific shapes and confer unique properties to water molecules.

Chemical Reactions (pp. 34–35)

1. **Chemical reactions** result from the making or breaking of chemical bonds in a process in which **reactants** are changed into **products.** Biochemistry involves chemical reactions of life.
2. **Synthesis reactions** form larger, more complex molecules. In **dehydration synthesis,** a molecule of water is removed from the reactants as the larger molecule is formed. **Endothermic reactions** require energy. **Anabolism** is the sum of all synthesis reactions in an organism.
3. **Decomposition reactions** break larger molecules into smaller molecules and are **exothermic** because they release energy. **Hydrolysis** is a decomposition reaction that uses water as one of the reactants. The sum of all decomposition reactions in an organism is called **catabolism.**
4. **Exchange reactions** involve exchanging atoms between reactants.
5. **Metabolism** is the sum of all anabolic, catabolic, and exchange chemical reactions in an organism.

Water, Acids, Bases, and Salts (pp. 36–39)

1. **Inorganic** chemicals typically lack carbon.
2. Water is a vital inorganic compound because of its properties as a solvent, its liquidity, its great capacity to absorb heat, and its participation in chemical reactions.
3. **Acids** release hydrogen ions. **Bases** release hydroxyl anions. The relative strength of each is assessed on a logarithmic **pH scale,** which measures the hydrogen ion concentration in a substance.
4. **Buffers** are substances that prevent drastic changes in pH.

Organic Macromolecules (pp. 39–51)

1. Certain groups of atoms in common arrangements, called **functional groups,** are found in organic macromolecules. **Monomers** are simple subunits that can be covalently linked to form chainlike **polymers.**
2. **Lipids,** which include fats, phospholipids, waxes, and steroids, are **hydrophobic** (insoluble in water) macromolecules.
3. **Fat** molecules are formed from a glycerol and three chainlike fatty acids. **Saturated fatty acids** contain more hydrogen in their structural formulas than **unsaturated fatty acids,** which contain double bonds between some carbon atoms. If several double bonds exist in the fatty acids of a molecule of fat, it is a **polyunsaturated** fat.
4. **Phospholipids** contain two fatty acid chains and a phosphate functional group. The phospholipid head is **hydrophilic,** whereas the fatty acid portion of the molecule is hydrophobic.
5. **Waxes** contain a long-chain fatty acid covalently linked to a long-chain alcohol. Waxes, which are water insoluble, are components of cell walls and are sometimes used as energy storage molecules.
6. **Steroid** lipids such as cholesterol help maintain the structural integrity of membranes as temperature fluctuates.
7. **Carbohydrates** such as **monosaccharides, disaccharides,** and **polysaccharides** serve as energy sources, structural molecules, and recognition sites during intercellular interactions.
8. **Proteins** are structural components of cells, enzymatic catalysts, regulators of various activities, molecules involved in the transportation of substances, and defensive molecules. They are composed of **amino acids** linked by **peptide bonds,** and they possess primary, secondary, tertiary, and (sometimes) quaternary structures that affect their function. **Denaturation** of a protein disrupts its structure and subsequently its function.
9. **Deoxyribonucleic acid (DNA)** and **ribonucleic acid (RNA)** are unbranched macromolecular polymers of **nucleotides,** each composed of either deoxyribose or ribose sugar, ionized phosphate, and a nitrogenous base. Five different bases exist: **adenine, guanine, cytosine, thymine,** and **uracil.** DNA contains A, G, C, and T nucleotides. RNA uses U nucleotides instead of T nucleotides.

   ▶**VIDEO TUTOR:** *The Structure of Nucleotides*
10. The structure of nucleic acids allows for genetic diversity, the correct copying of genes for their passage on to the next generation, and the accurate synthesis of proteins.
11. **Adenosine triphosphate (ATP),** which is related to adenine nucleotide, is the most important short-term energy storage molecule in cells. It is also incorporated into the structure of many coenzymes.

QUESTIONS FOR REVIEW

Answers to the Questions for Review (except Short Answer questions) begin on p. A-1.

Multiple Choice

1. Which of the following structures have *no* electrical charge?
   a. protons
   b. electrons
   c. neutrons
   d. ions
2. The atomic mass of an atom most closely approximates the sum of the masses of all its
   a. protons.
   b. isotopes.
   c. electrons.
   d. protons and neutrons.
3. One isotope of iodine differs from another in
   a. the number of protons.
   b. the number of electrons.
   c. the number of neutrons.
   d. atomic number.
4. Which of the following is *not* an organic compound?
   a. monosaccharide
   b. formaldehyde
   c. water
   d. steroid
5. Which of the following terms most correctly describes the bonds in a molecule of water?
   a. nonpolar covalent bond
   b. polar covalent bond
   c. ionic bond
   d. hydrogen bond
6. In water, cations and anions of salts dissociate from one another and become surrounded by water molecules. In this state, the ions are also called __________.
   a. electrically negative
   b. ionically bonded
   c. electrolytes
   d. hydrogen bonds
7. Which of the following can be most accurately described as a decomposition reaction?
   a. $C_6H_{12}O_6 + 6\ O_2 \rightarrow 6\ H_2O + 6\ CO_2$
   b. glucose + ATP → glucose phosphate + ADP
   c. $6\ H_2O + 6\ CO_2 \rightarrow C_6H_{12}O_6 + 6\ O_2$
   d. A + BC → AB + C
8. Which of the following statements about a carbonated cola beverage with a pH of 2.9 is true?
   a. It has a relatively high concentration of hydrogen ions.
   b. It has a relatively low concentration of hydrogen ions.
   c. It has equal amounts of hydroxyl and hydrogen ions.
   d. Cola is a buffered solution.
9. Proteins are polymers of __________.
   a. amino acids
   b. fatty acids
   c. nucleic acids
   d. monosaccharides
10. Which of the following are hydrophobic organic molecules?
    a. proteins
    b. carbohydrates
    c. lipids
    d. nucleic acids

Fill in the Blanks

1. The outermost electron shell of an atom is known as the ________________ shell.
2. The type of chemical bond between atoms with nearly equal electronegativities is called a(n) ________________ bond.
3. The principal short-term energy storage molecule in cells is ________________.
4. Common long-term storage molecules are ________________, ________________, ________________, and ________________.
5. Groups of atoms such as NH_2 or OH that appear in certain common arrangements are called ________________.
6. The reverse of dehydration synthesis is ________________.
7. Reactions that release energy are called ________________ reactions.
8. All chemical reactions begin with reactants and result in new molecules called ________________.
9. The ________________ scale is a measure of the concentration of hydrogen ions in a solution.
10. A nucleic acid containing the base uracil would also contain ________________ sugar.

VISUALIZE IT!

1. Label a portion of the molecule below where the primary structure is visible; label two types of secondary structure; circle the tertiary structure.

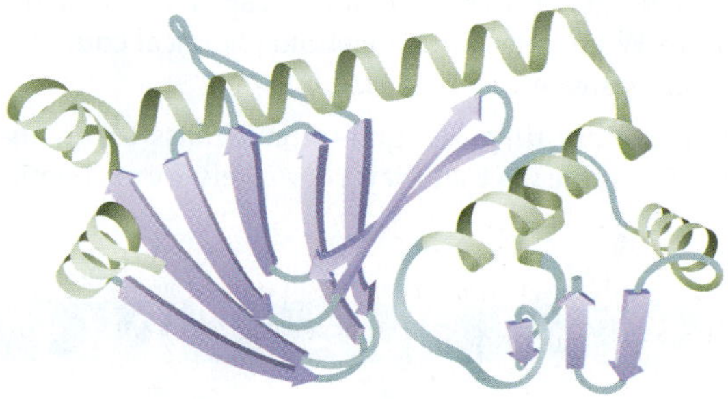

2. Shown is the amino acid tryptophan. Put the letter "C" at the site of every carbon atom. Label the amino group, the carboxyl group, and the side group.

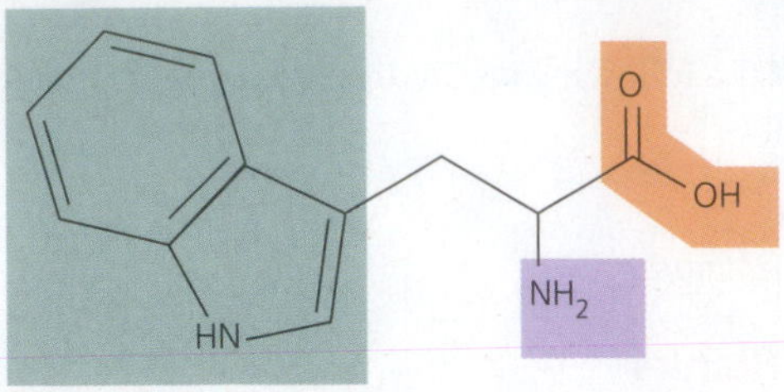

Short Answer

1. List three main types of chemical bonds, and give an example of each.
2. Name five properties of water that are vital to life.
3. Describe the difference(s) among saturated fatty acids, unsaturated fatty acids, and polyunsaturated fatty acids.
4. What is the difference between atomic oxygen and molecular oxygen?
5. Explain how the polarity of water molecules makes water an excellent solvent.

CRITICAL THINKING

1. Anthrax is caused by a bacterium, *Bacillus anthracis*, that avoids the body's defenses against disease by synthesizing an outer glycoprotein covering made from D-glutamic acid. This covering is not digestible by white blood cells that normally engulf bacteria. Why is the covering indigestible?
2. Dehydrogenation is a chemical reaction in which a saturated fat is converted to an unsaturated fat. Explain why the name for this reaction is an appropriate one.
3. Two freshmen disagree about an aspect of chemistry. The nursing major insists that H^+ is the symbol for a hydrogen ion. The physics major insists that H^+ is the symbol for a proton. How can you help them resolve their disagreement?
4. When an egg white is heated, it changes from liquid to solid. When gelatin is cooled, it changes from liquid to solid. Both gelatin and egg white are proteins. From what you have learned about proteins, why can the gelatin be changed back to liquid but the cooked egg cannot?
5. When amino acids are synthesized in a test tube, D and L forms occur in equal amounts. However, cells use only L forms in their proteins. Occasionally, meteorites are found to contain amino acids. Based on these facts, how could NASA scientists determine whether the amino acids recovered from space are evidence of Earth-like extraterrestrial life rather than the result of nonmetabolic processes?
6. The poison glands of many bees and wasps contain acidic compounds. What common household chemical could be used to neutralize this poison?
7. Neon (atomic mass 10) and argon (atomic mass 18) are *inert* elements, which means that they very rarely form chemical bonds. Give the electron configuration of their atoms and explain why these elements are inert.
8. An article in the local newspaper about gangrene states that the tissue-destroying toxin, lecithinase, is an "organic compound." But many people consider "organic" chemicals to mean something is good. Explain the apparent contradiction.
9. The deadly poison hydrogen cyanide has the chemical formula H—C≡N. Describe the bonds between carbon and hydrogen and between carbon and nitrogen in terms of the number of electrons involved.
10. Triple covalent bonds are stronger and more difficult to break than single covalent bonds. Explain why by referring to the stability of a valence shell that contains eight electrons.
11. How can hydrogen bonding between water molecules help explain water's ability to absorb large amounts of energy before evaporating?
12. How can a single molecule of magnesium hydroxide neutralize two molecules of hydrochloric acid?
13. We have seen that it is important that biological membranes remain flexible. Most bacteria lack sterols in their membranes and instead incorporate unsaturated phospholipids in the membranes to resist tight packing and solidification. Examine Table 2.4 on p. 41. Which fatty acid might best protect the membranes of an ice-dwelling bacterium?
14. Why is there no stereoisomer of glycine?
15. A textbook states that only five nucleotide bases are found in cells, but a laboratory worker reports that she has isolated eight different nucleotides. Explain why both are correct.

CONCEPT MAPPING

Using the following terms, draw a concept map that describes nucleic acids. You may use some terms more than once. For a sample concept map, see p. 94. Or, complete this and other concept maps online by going to the MasteringMicrobiology Study Area.

Adenine (2)
Cytosine (2)
Deoxyribonucleotides
Deoxyribose
DNA
Double-stranded
Guanine (2)
Messenger RNA
Nitrogenous bases (2)
Phosphate
Ribonucleotides
Ribose
RNA
Ribosomal RNA
Single-stranded
Thymine
Transfer RNA
Uracil

3 Cell Structure and Function

MICRO IN THE CLINIC

The Big Game

College sophomore Lamara is a star point guard for her school's basketball team. She is excited about the division finals on Friday—she's even heard rumors that a professional scout will be in the stands. On Thursday morning, she wakes up with a sore throat. Her forehead doesn't feel warm, and she is able to eat breakfast without any problems, so she makes it to her Thursday class. However, when Lamara wakes up on Friday morning, her throat is noticeably worse.

Willing herself to be healthy, Lamara tries to attend Friday classes, but by mid-morning she notices a **rash** developing on her face and neck. There are also some tiny bumps on her chest and abdomen. Grumbling, Lamara heads back to the dormitory and checks her temperature—100.8°F. She feels extremely tired, and it is downright painful to swallow. Desperate, she heads to the student health center, where a **nurse** practitioner notices white spots on the back of Lamara's throat and on her tonsils. The divisional basketball game starts in six hours.

Is the same infection causing Lamara's sore throat and the rash? Will she make it to the big game? Turn to the end of the chapter (p. 87) to find out.

Explore More: Test your readiness and apply your knowledge with dynamic learning tools at MasteringMicrobiology.

All living things—including our bodies and the bacterial, protozoan, and fungal pathogens that attack us—are composed of living cells. If we want to understand disease and its treatment, therefore, we must first understand the life of cells. How pathogens attack our cells, how our bodies defend themselves, how current medical treatments assist our bodies in recovering—all of these activities have their basis in the biology of our, and our pathogens', cells.

In this chapter, we will examine cells and the structures within cells. We will discuss similarities and differences among the three major kinds of cells—bacterial, archaeal, and eukaryotic. The differences are particularly important because they allow researchers to develop treatments that inhibit or kill pathogens without adversely affecting a patient's own cells. We will also learn about cellular structures that allow pathogens to evade the body's defenses and cause disease.

Processes of Life

LEARNING | **OUTCOME**

3.1 Describe four major processes of living cells.

Microbiology is the study of particularly small living things. That raises a question: What does living mean; how do we define life? Scientists once thought that living things were composed of special organic chemicals, such as glucose and amino acids, that carried a "life force" found only in living organisms. These organic chemicals were thought to be formed only by living things and to be very different from the inorganic chemicals of nonliving things.

The idea that organic chemicals could come only from living organisms had to be abandoned in 1828, when Friedrich Wöhler (1800–1882) synthesized urea, an organic molecule, using only inorganic reactants in his laboratory. Today we know that all living things contain both organic and inorganic chemicals and that many organic chemicals can be made from inorganic chemicals by laboratory processes. If organic chemicals can be made even in the absence of life, what is the difference between a living thing and a nonliving thing? What is life?

At first, this may seem a simple question. After all, you can usually tell when something is alive. However, defining "life" itself is difficult, so biologists generally avoid setting a definition, preferring instead to describe characteristics common to all living things. Biologists agree that all living things share at least four processes of life: growth, reproduction, responsiveness, and metabolism.

- *Growth*. Living things can grow; that is, they can increase in size.
- *Reproduction*. Organisms normally have the ability to reproduce themselves. Reproduction means that they increase in number, producing more organisms organized like themselves. Reproduction may be accomplished asexually (alone) or sexually with gametes (sex cells). Note that reproduction is an increase in number, whereas growth is an increase in size. Growth and reproduction often occur simultaneously. (We consider several methods of reproduction when we examine microorganisms in detail in Chapters 11–13.)
- *Responsiveness*. All living things respond to their environment. They have the ability to change themselves in reaction to changing conditions around or within them. Many organisms also have the ability to move toward or away from environmental stimuli—a response called *taxis*.
- *Metabolism*. Metabolism can be defined as the ability of organisms to take in nutrients from outside themselves and use the nutrients in a series of controlled chemical reactions to provide the energy and structures needed to grow, reproduce, and be responsive. Metabolism is a unique process of living things; nonliving things cannot metabolize. Cells store metabolic energy in the chemical bonds of *adenosine triphosphate* (ă-den´ō-sēn trī-fos´fāt), or *ATP*. (Major processes of microbial metabolism, including the generation of ATP, are discussed in Chapters 5–7.)

TABLE 3.1 shows how these characteristics, along with cell structure, relate to various kinds of microbes.

Organisms may not exhibit these processes at all times. For instance, in some organisms, reproduction may be postponed or curtailed by age or disease or, in humans at least, by choice. Likewise, the rate of metabolism may be reduced, as occurs in a seed, a hibernating animal, or a bacterial

TABLE 3.1 Characteristics of Life and Their Distribution in Microbes

Characteristic	Bacteria, Archaea, Eukaryotes	Viruses
Growth: increase in size	Occurs in all	Growth does not occur
Reproduction: increase in number	Occurs in all	Host cell replicates the virus
Responsiveness: ability to react to environmental stimuli	Occurs in all	Reaction to host cells seen in some viruses
Metabolism: controlled chemical reactions of organisms	Occurs in all	Viruses use host cell's metabolism
Cellular structure: membrane-bound structure capable of all of the above functions	Present in all	Viruses lack cytoplasmic membrane or cellular structure

endospore,[1] and growth often stops when an animal reaches a certain size. However, microorganisms typically grow, reproduce, respond, and metabolize as long as conditions are suitable. (Chapter 6 discusses the proper conditions for the metabolism and growth of various types of microorganisms.)

TELL ME WHY

The smallest free-living microbe—the bacterium *Mycoplasma*—is nonmotile. Why is it alive, even though it cannot move?

Prokaryotic and Eukaryotic Cells: An Overview

LEARNING **OUTCOME**

3.2 Compare and contrast prokaryotic and eukaryotic cells.

In the 1800s, two German biologists, Theodor Schwann (1810–1882) and Matthias Schleiden (1804–1881), developed the theory that all living things are composed of cells. *Cells* are living entities, surrounded by a membrane, that are capable of growing, reproducing, responding, and metabolizing. The smallest living things are single-celled microorganisms.

There are many different kinds of cells **(FIGURE 3.1)**. Some cells are free-living, independent organisms; others live together in colonies or form the bodies of multicellular organisms. Cells also exist in various sizes, from the smallest bacteria to bird eggs, which are the largest of cells. All cells may be described as either *prokaryotes* (prō-kar´ē-ōts) or *eukaryotes* (yū-kar´ē-ōts).

Scientists categorize organisms based on shared characteristics into groups called *taxa*. "Prokaryotic" is a characteristic of organisms in two taxa—*domain Archaea* and *domain Bacteria*—but "prokaryote" is not itself a taxon. The distinctive feature of **prokaryotes** is that they can make proteins simultaneously to reading their genetic code because a typical prokaryote does not have a membrane surrounding its genetic material (DNA).

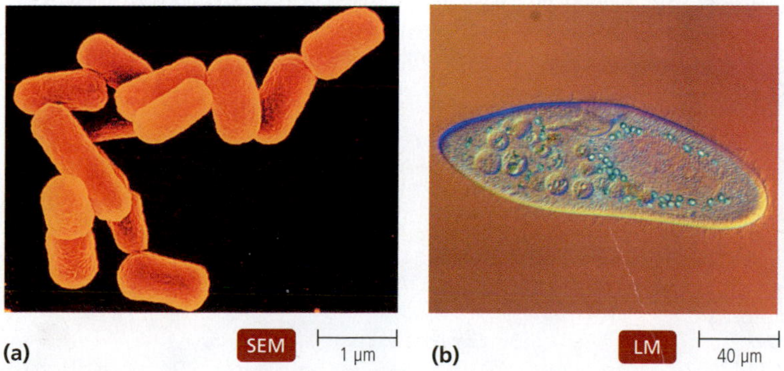

▲ **FIGURE 3.1 Examples of types of cells. (a)** *Escherichia coli* bacterial cells. **(b)** *Paramecium*, a single-celled eukaryote. Note the differences in magnification.

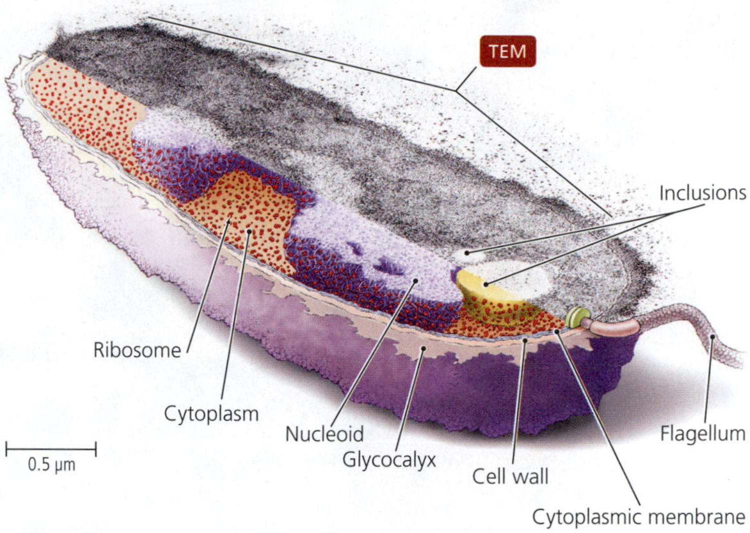

▲ **FIGURE 3.2 Typical prokaryotic cell.** Prokaryotes include archaea and bacteria. The artist has extended an electron micrograph to show three dimensions. Not all prokaryotic cells contain all these features.

In other words, a typical prokaryote does not have a nucleus **(FIGURE 3.2)**. (Researchers have discovered a few prokaryotes with internal membranes that look like nuclei, but further investigation is needed to determine what these structures are.) The word *prokaryote* comes from Greek words meaning "before nucleus." Moreover, electron microscopy has revealed that prokaryotes typically lack various types of internal structures bound with membranes that are present in eukaryotic cells.

Bacteria and archaea differ fundamentally in such ways as the type of lipids in their cytoplasmic membranes and in the chemistry of their cell walls. In many ways, archaea are more like eukaryotes than they are like bacteria. (Chapter 11 discusses archaea and bacteria in more detail.)

Eukaryotes have a membrane called a nuclear envelope surrounding their DNA, forming a nucleus **(FIGURE 3.3)**, which sets eukaryotes in *domain Eukarya*. Indeed, the term *eukaryote* comes from Greek words meaning "true nucleus." Besides the nuclear membrane, eukaryotes have numerous other internal membranes that compartmentalize cellular functions. These compartments are membrane-bound **organelles**—specialized structures that act like tiny organs to carry on the various functions of the cell. Organelles and their functions are discussed later in this chapter. The cells of algae, protozoa, fungi, animals, and plants are eukaryotic. Eukaryotes are usually larger and more complex than prokaryotes, which are typically 1.0 μm in diameter or smaller, as compared to 10–100 μm for eukaryotic cells **(FIGURE 3.4)**.

Although there are many kinds of cells, they all share the characteristic processes of life as previously described, as well as certain physical features. In this chapter, we will distinguish among bacterial, archaeal, and eukaryotic "versions" of physical features common to cells, including (1) external structures,

[1]Endospores are resting stages, produced by some bacteria, that are tolerant of environmental extremes.

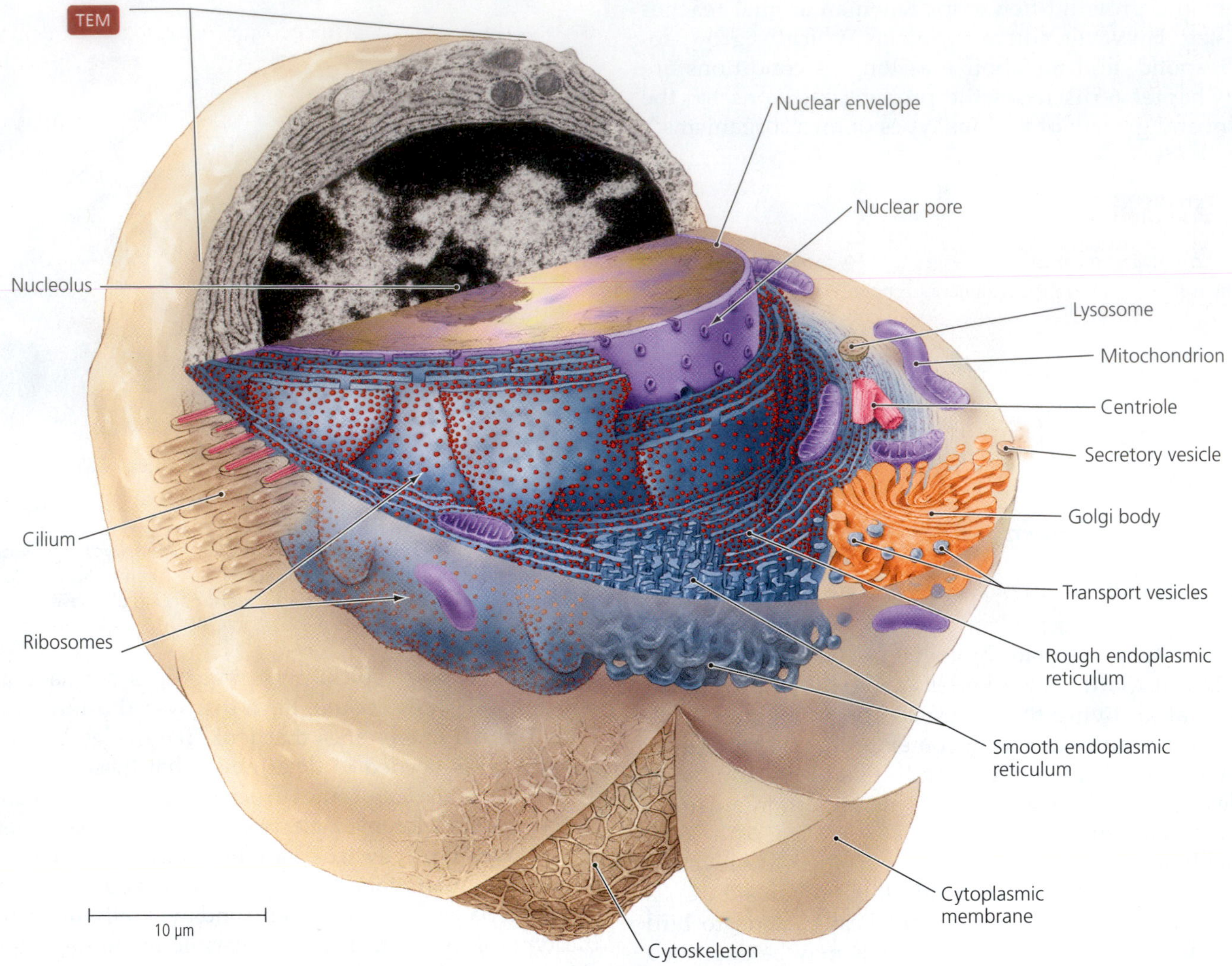

▲ **FIGURE 3.3 Typical eukaryotic cell.** Not all eukaryotic cells have all these features. The artist has extended the electron micrograph to show three dimensions. Note the difference in magnification between this cell and the prokaryotic cell in the previous figure. *Besides size, what major difference between prokaryotes and eukaryotes was visible to early microscopists?*

Figure 3.3 *Eukaryotic cells contain nuclei, which are visible with light microscopes, whereas prokaryotes lack nuclei.*

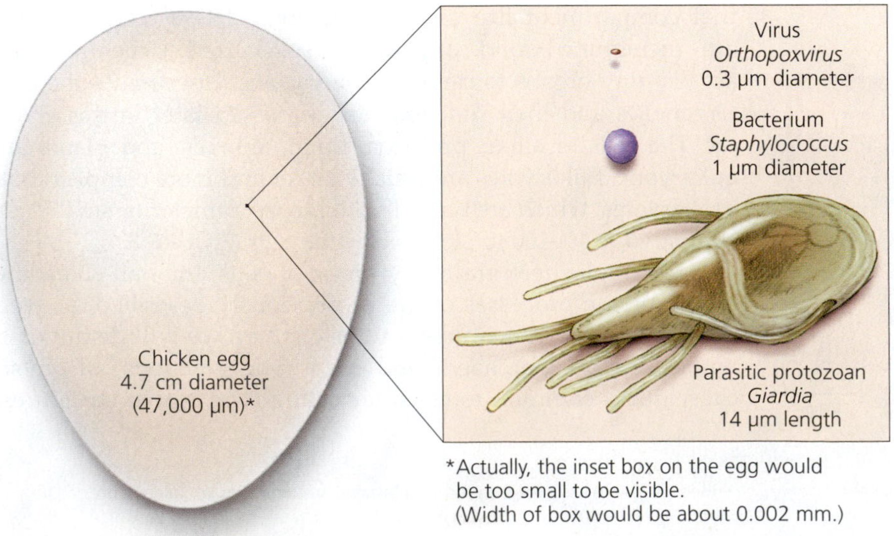

◀ **FIGURE 3.4 Approximate size of various types of cells.** Birds' eggs are the largest cells. Note that *Staphylococcus*, a bacterium, is smaller than *Giardia*, a unicellular eukaryote. A smallpox virus *(Orthopoxvirus)* is shown only for comparison; viruses are not cellular.

(2) the cell wall, (3) the cytoplasmic membrane, and (4) the cytoplasm. We will also discuss features unique to each type. (Chapters 11, 12, and 19–24 examine further details of prokaryotic and eukaryotic organisms, their classification, and their ability to cause disease.)

TELL ME WHY

In 1985, an Israeli scientist discovered a single-celled microbe *Epulopiscium fishelsoni.* This organism is visible with the naked eye. Why did the scientist think *Epulopiscium* is eukaryotic?

What discovery revealed that the microbe is really a giant bacterium?

Next, we explore characteristics of bacterial cells, beginning with external features and working into the cell.

External Structures of Bacterial Cells

Many cells have special external features that enable them to respond to other cells and their environment. In bacteria, these features include glycocalyces, flagella, fimbriae, and pili.

Glycocalyces

LEARNING **OUTCOMES**

3.3 Describe the composition, function, and relevance to human health of glycocalyces.

3.4 Distinguish capsules from slime layers.

Some cells have a gelatinous, sticky substance that surrounds the outside of the cell. This substance is known as a **glycocalyx** (plural: *glycocalyces*), which literally means "sweet cup." The glycocalyx may be composed of polysaccharides, polypeptides, or both. These chemicals are produced inside the cell and are extruded onto the cell's surface.

When the glycocalyx of a bacterium is composed of organized repeating units of organic chemicals firmly attached to the cell's surface, the glycocalyx is called a **capsule** **(FIGURE 3.5a)**. In contrast, a loose, water-soluble glycocalyx is called a **slime layer** **(FIGURE 3.5b)**.

Glycocalyces protect cells from desiccation (drying) and can also play a role in the ability of pathogens to survive and cause disease. For example, slime layers are often sticky and provide one means for bacteria to attach to surfaces as *biofilms*, which are aggregates of many bacteria living together on a surface. Oral bacteria colonize the teeth as a biofilm called dental plaque. The bacteria in the biofilm produce acid and cause *dental caries* (cavities).

The chemicals in many bacterial capsules can be similar to chemicals normally found in the body preventing bacteria from being recognized or devoured by defensive cells of the host. For example, the capsules of *Streptococcus pneumoniae* (strep-tō-kok´ ūs nū-mō´nē-ī) and *Klebsiella pneumoniae* (kleb-sē-el´ă nū-mō´ nē-ī) enable these prokaryotes to avoid destruction by defensive cells in the respiratory tract and to cause pneumonia. Unencapsulated strains of these same bacterial species do not cause disease because the body's defensive cells destroy them.

Flagella

LEARNING **OUTCOMES**

3.5 Discuss the structure and function of bacterial flagella.

3.6 List and describe four bacterial flagellar arrangements.

A cell's motility may enable it to flee from a harmful environment or move toward a favorable environment, such as one where food or light is available. The most notable structures responsible for such bacterial movement are flagella. Bacterial

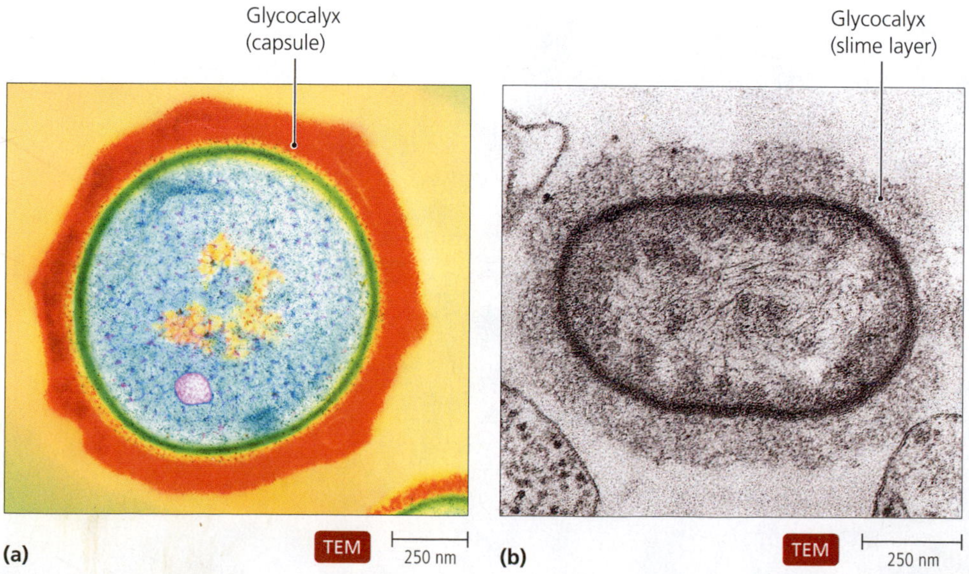

◀ **FIGURE 3.5 Glycocalyces. (a)** Micrograph of a single cell of the bacterium *Staphylococcus*, showing a prominent capsule. **(b)** *Bacteroides*, a common fecal bacterium, has a slime layer surrounding the cell. *What advantage does a glycocalyx provide a cell?*

Figure 3.5 *A glycocalyx provides protection from drying and from being devoured; it may also help attach cells to one another and to surfaces in the environment.*

flagella (singular: *flagellum*) are long structures that extend beyond the surface of a cell and its glycocalyx and propel the cell through its environment. Not all bacteria have flagella, but for those that do, the flagella are very similar in composition, structure, and development. ▶**ANIMATIONS:** ***Motility: Overview***

Structure

Bacterial flagella are composed of three parts: a long, hollow *filament*, a *hook*, and a *basal body* (**FIGURE 3.6**). The filament is a long hollow shaft, about 20 nm in diameter, that extends out into the cell's environment. It is composed of many identical globular molecules of a protein called *flagellin.*

A bacterial flagellum lengthens by growing at its tip. The cell secretes molecules of flagellin through the hollow core of the flagellum, to be deposited in a clockwise helix at the tip. Bacterial flagella react to external wetness, inhibiting their own growth in dry habitats.

No membrane covers the filament of bacterial flagella. At its base, a filament inserts into a curved structure, the hook, which is composed of a different protein. The basal body, which

◀ **FIGURE 3.6 Proximal structure of bacterial flagella.** **(a)** Detail of flagellar structure of a Gram-positive cell. **(b)** Detail of the flagellum of a Gram-negative bacterium. *How do flagella of Gram-positive bacteria differ from those of Gram-negative bacteria?*

Figure 3.6 *Flagella of Gram-positive cells have a single pair of rings in the basal body that function to attach the flagellum to the cytoplasmic membrane. The flagella of Gram-negative cells have two pairs of rings: one pair anchors the flagellum to the cytoplasmic membrane, the other pair to the cell wall.*

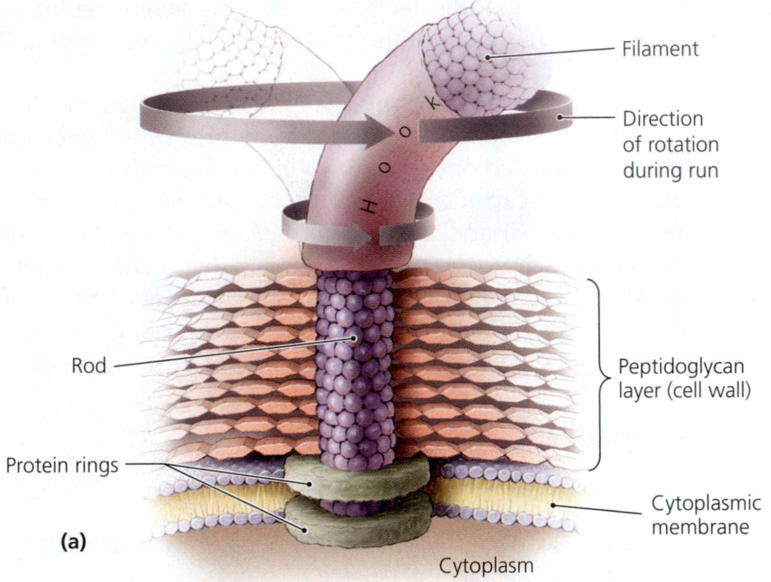

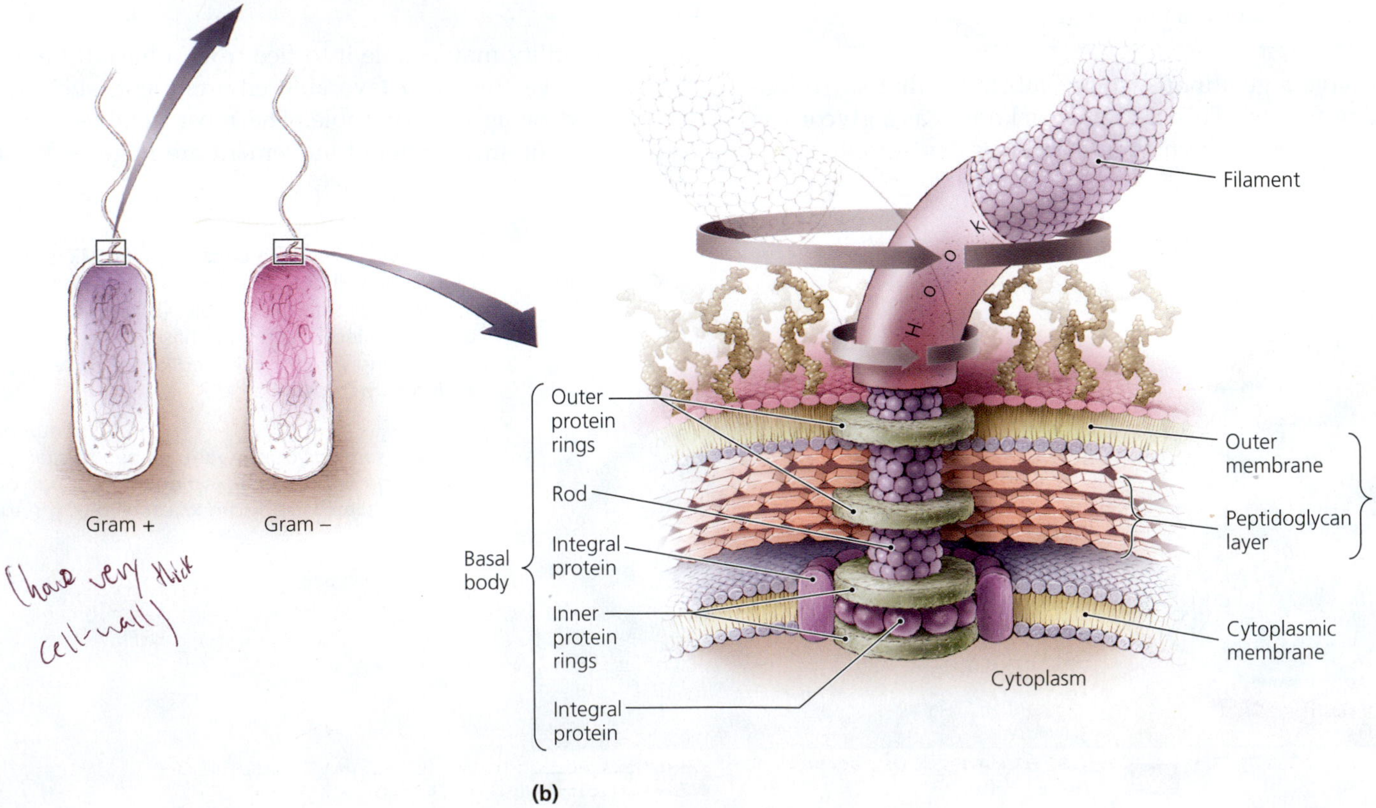

is composed of still different proteins, anchors the filament and hook to the cell wall and cytoplasmic membrane by means of a rod and a series of either two or four rings of integral proteins. Together the hook, rod, and rings allow the filament to rotate 360°. Differences in the proteins associated with bacterial flagella vary enough to allow classification of species into groups (strains) called *serovars*. ▶**ANIMATIONS:** ***Flagella: Structure***

Arrangement

Bacteria may have one of several flagellar arrangements **(FIGURE 3.7)**. Flagella that cover the surface of the cell are termed **peritrichous**;[2] in contrast, **polar** flagella are only at the ends. Some cells have tufts of polar flagella.

Some spiral-shaped bacteria, called *spirochetes* (spī´rō-kēts),[3] have flagella at both ends that spiral tightly around the cell instead of protruding into the surrounding medium. These flagella, called **endoflagella**, form an **axial filament** that wraps around the cell between its cytoplasmic membrane and an outer membrane **(FIGURE 3.8)**. Rotation of endoflagella evidently causes the axial filament to rotate around the cell, causing the spirochete to "corkscrew" through its medium. *Treponema pallidum* (trep-ō-nē´mă pal´li-dŭm), the agent of syphilis, and

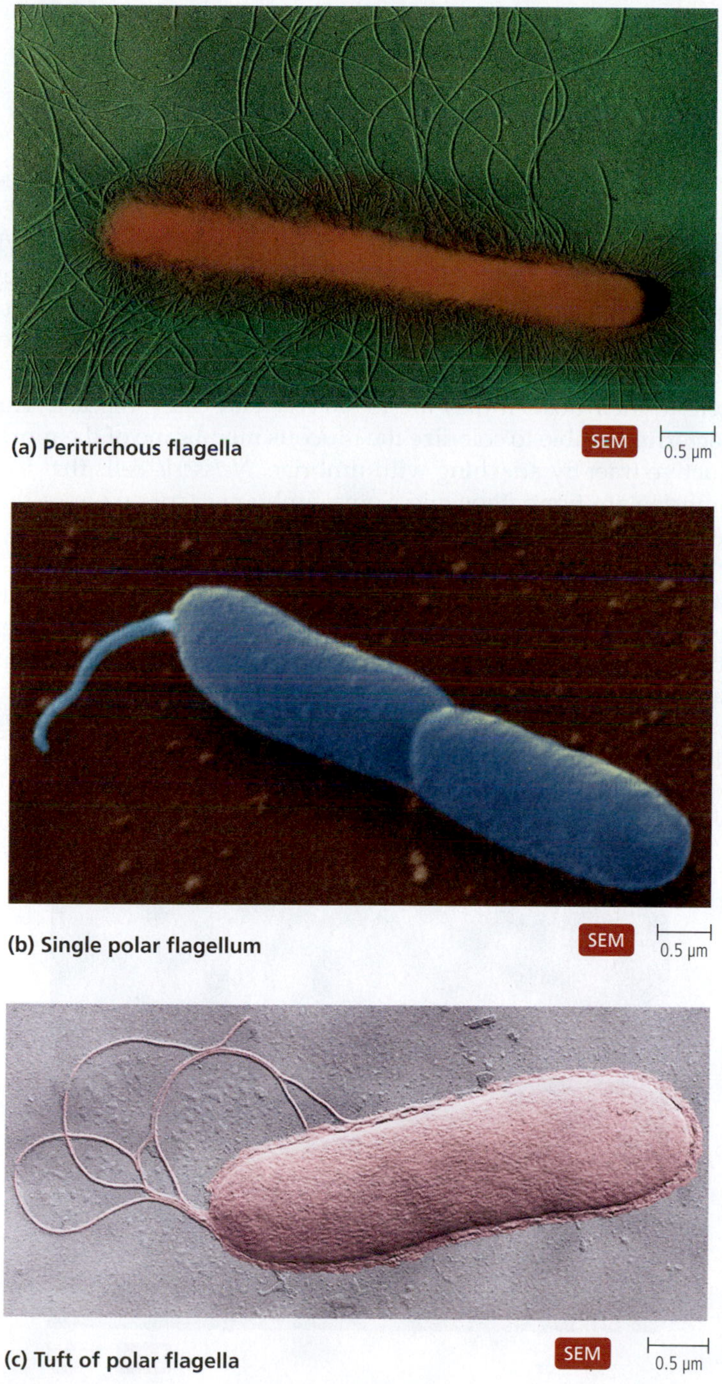

(a) Peritrichous flagella

(b) Single polar flagellum

(c) Tuft of polar flagella

▲ **FIGURE 3.7 Micrographs of basic arrangements of bacterial flagella.**

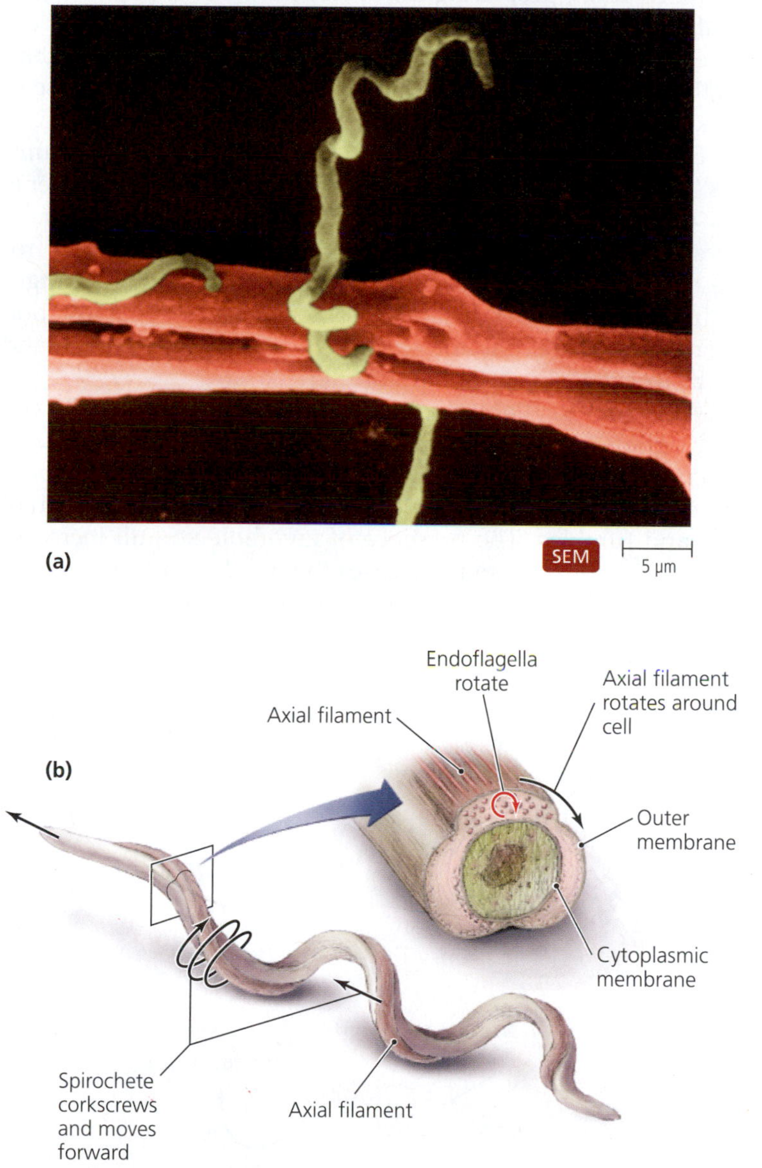

(a)

(b)

▲ **FIGURE 3.8 Axial filament. (a)** Scanning electron micrograph of a spirochete, *Treponema pallidum*. **(b)** Diagram of axial filament wrapped around a spirochete. Cross section reveals that an axial filament is composed of endoflagella.

[2]From Greek *peri*, meaning "around," and *trichos*, meaning a "hair."
[3]From Greek *speira*, meaning "coil," and *chaeta*, meaning "hair."

Borrelia burgdorferi (bō-rē´lē-ă burg-dōr´fer-ē), the cause of Lyme disease, are notable spirochetes. Some scientists think that the corkscrew motility of these pathogens allows them to invade human tissues. ▶**ANIMATIONS:** ***Flagella: Arrangement; Spirochetes***

Function

Although the precise mechanism by which bacterial flagella move is not completely understood, we do know that they rotate 360° like boat propellers rather than whipping from side to side. The flow of hydrogen ions (H^+) or of sodium ions (Na^+) through the cytoplasmic membrane near the basal body powers the rotation, propelling the bacterium through the environment at about 60 cell lengths per second—equivalent to a car traveling at 670 miles per hour! Flagella rotate at more than 100,000 rpm and can change direction from counterclockwise to clockwise.

Bacteria move with a series of "runs" interrupted by "tumbles." Counterclockwise flagellar rotation produces movements of a cell in one direction for some time; this is called a *run*. If more than one flagellum is present, the flagella align and rotate together as a bundle. Tumbles are abrupt, random changes in direction. Tumbles result from clockwise flagellar rotation where each flagellum rotates independently. Both runs and tumbles occur in response to stimuli.

Receptors for light or chemicals on the surface of the cell send signals to the flagella, which then adjust their speed and direction of rotation. A bacterium can position itself in a more favorable environment by varying the number and duration of runs and tumbles. The presence of favorable stimuli increases the duration of runs and decreases the number of tumbles; as a result, the cell tends to move toward an attractant **(FIGURE 3.9)**.

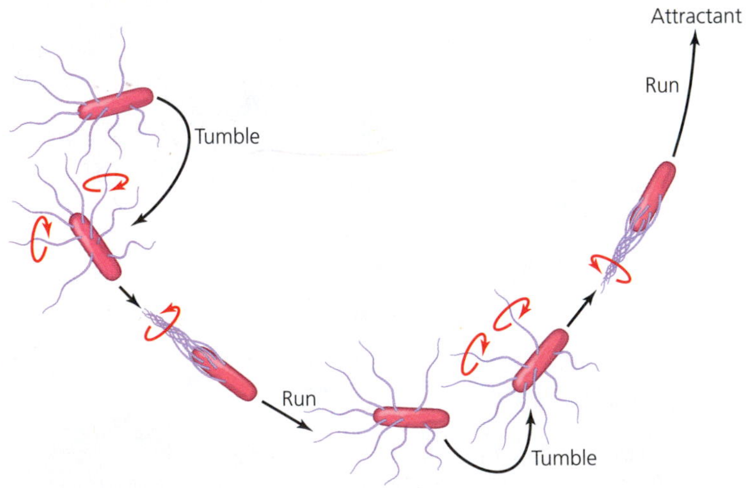

▲ **FIGURE 3.9 Motion of a peritrichous bacterium.** In peritrichous bacteria, runs occur when all of the flagella rotate counterclockwise and become bundled. Tumbles occur when the flagella rotate clockwise, become unbundled, and the cell spins randomly. In positive chemotaxis (shown), runs last longer than tumbles, resulting in motion toward the chemical attractant. *What triggers a bacterial flagellum to rotate counterclockwise, producing a run?*

Figure 3.9 *Favorable environmental conditions induce runs.*

Unfavorable stimuli increase the number of tumbles, which increases the likelihood that it will move randomly in another direction, away from a repellent.

Movement in response to a stimulus is termed **taxis**. The stimulus may be either light **(phototaxis)** or a chemical **(chemotaxis)**. Movement toward a favorable stimulus is *positive taxis*, whereas movement away from an unfavorable stimulus is *negative taxis*. For example, movement toward a nutrient would be *positive chemotaxis*. ▶**ANIMATIONS:** ***Flagella: Movement***

Fimbriae and Pili

LEARNING **OUTCOME**

3.7 Compare and contrast the structures and functions of fimbriae, pili, and flagella.

Many bacteria have rodlike proteinaceous extensions called **fimbriae** (fim´brē-ī; singular: *fimbria*). These sticky, bristlelike projections adhere to one another and to substances in the environment. There may be hundreds of fimbriae per cell, and they are usually shorter than flagella **(FIGURE 3.10)**. An example of a bacterium with fimbriae is *Neisseria gonorrhoeae* (nī-se´rē-ă go-nor-rē´ī), which causes gonorrhea. Pathogens must be able to adhere to their hosts if they are to survive and cause disease. This bacterium is able to colonize the mucous membrane of the reproductive tract by attaching with fimbriae. *Neisseria* cells that lack fimbriae are nonpathogenic. Some fimbriae carry enzymes that render soluble, toxic metal ions into insoluble, nontoxic forms.

Bacteria may use fimbriae to move across a surface via a process similar to pulling an object with a rope. The bacterium extends a fimbria, which attaches at its tip to the surface; then the bacterium retracts the fimbria, pulling itself toward the attachment point.

Fimbriae also serve an important function in **biofilms**, slimy masses of microbes adhering to a substrate by means of

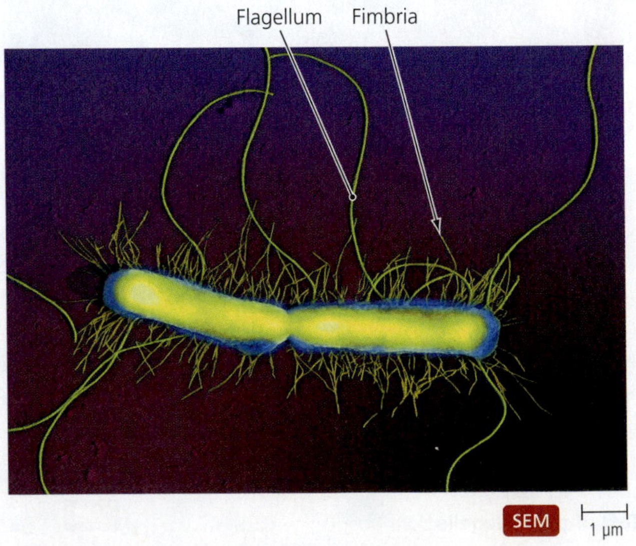

▲ **FIGURE 3.10 Fimbriae.** *Proteus vulgaris* has flagella and fimbriae.

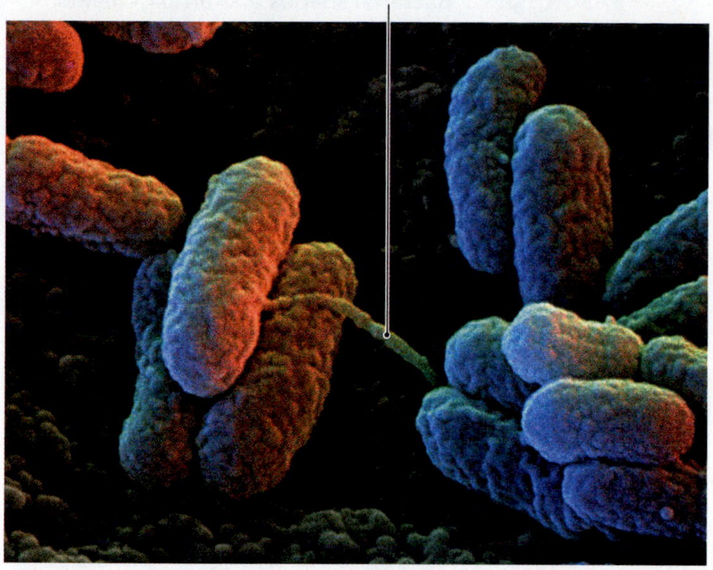

▲ **FIGURE 3.11 Pili.** Two *Salmonella* cells are connected by a pilus. *How are pili different from bacterial flagella?*

Figure 3.11 *Bacterial flagella are flexible structures that rotate to propel the cell; pili are hollow tubes used to transfer DNA from one cell to another.*

fimbriae and glycocalyces. Some fimbriae act as electrical wires, conducting electrical signals among cells in a biofilm. It has been estimated that at least 99% of bacteria in nature exist in biofilms. Researchers are interested in biofilms because of the roles they play in human diseases and in industry. (See **Highlight: Biofilms: Slime Matters.**)

A special type of fimbria is a **pilus** (pī´lus; plural: *pili,* pī´lī), also called *conjugation pilus.* Pili are longer than other fimbriae but usually shorter than flagella. Typically only one to a few pili are present per cell in bacteria that have them. Cells use pili to transfer DNA from one cell to the other via a process termed *conjugation* (**FIGURE 3.11**). (Chapter 7 deals with conjugation in more detail.)

TELL ME WHY

Why is a pilus a type of fimbria, but a flagellum is not?

Bacterial Cell Walls

LEARNING **OUTCOMES**

3.8 Describe common shapes and arrangements of bacterial cells.

3.9 Describe the sugar and peptide portions of peptidoglycan.

3.10 Compare and contrast the cell walls of Gram-positive and Gram-negative bacteria in terms of structure and Gram staining.

HIGHLIGHT

Biofilms: Slime Matters

They form plaque on teeth; they are the slime on rocks in rivers and streams; and they can cause disease, clog drains, or help clean up hazardous waste. They are biofilms—organized, layered systems of bacteria and other microbes attached to a surface. Understanding biofilms holds the key to many important clinical and industrial applications.

Biofilm bacteria communicate via chemical and electrical signals that help them organize and form three-dimensional structures. The architecture of a biofilm provides protection that free-floating bacteria lack. For example, the lower concentrations of oxygen found in the interior of biofilms thwart the effectiveness of some antibiotics.

Furthermore, bacteria in biofilms behave in significantly different ways from individual, free-floating bacteria. For example, as a free-floating cell, the soil bacterium *Pseudomonas putida* propels itself through water with its flagella; however, once it becomes part of a biofilm, it turns off the genes for flagellar proteins and starts synthesizing pili instead. In addition, the genes for antibiotic resistance in *P. putida* are more active within cells in a biofilm than within a free-floating cell.

The more we understand biofilms, the more readily we can reduce their harmful effects or put them to good use. Biofilms account for about two-thirds of bacterial infections in humans, such as the serious lung infections suffered by cystic fibrosis patients. Biofilms are also the culprits in many industrial problems, including corroded pipes and clogged water filters, which cause millions of dollars of damage each year. Fortunately, not all biofilms are detrimental; some show potential as aids in preventing and controlling certain kinds of industrial pollution.

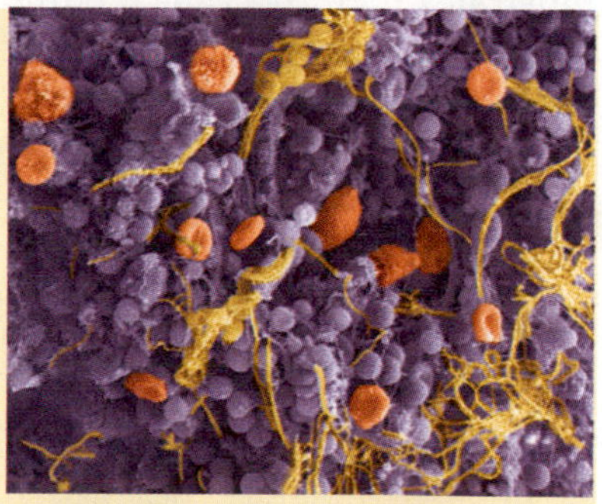

Biofilm on medical tubing

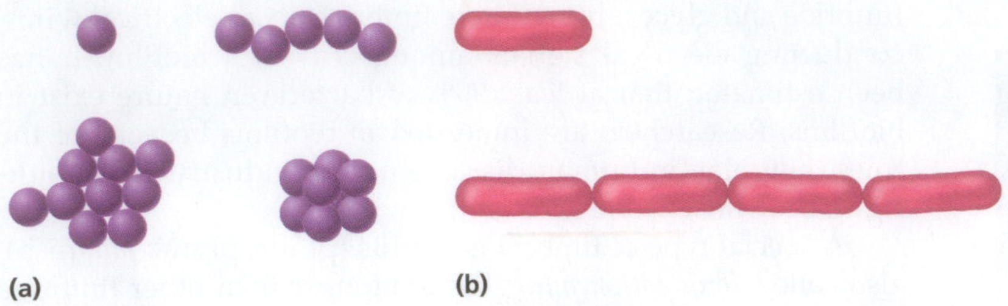

◀ **FIGURE 3.12 Bacterial shapes and arrangements.** **(a)** Spherical cocci may be in arrangements such as single, chains (streptococci), clusters (staphylococci), and cuboidal packets (sarcinae). **(b)** Rod-shaped bacilli may also be single or in arrangements such as chains.

The cells of most prokaryotes are surrounded by a **cell wall** that provides structure and shape to the cell and protects it from osmotic forces. In addition, a cell wall assists some cells in attaching to other cells or in resisting antimicrobial drugs. Note that animal cells do not have walls, a difference that plays a key role in treatment of many bacterial diseases with certain types of antibiotics. For example, penicillin attacks the cell wall of bacteria but is harmless to human cells, which lack walls.

Cell walls give bacterial cells characteristic shapes. Spherical cells, called cocci (kok′sī), may appear in various arrangements, including singly or in chains (streptococci), clusters (staphylococci), or cuboidal packets (sarcinae, sar′si-nī) **(FIGURE 3.12)**, depending on the planes of cell division. Rod-shaped cells, called bacilli (bă-sil′ī), typically appear singly or in chains.

Bacterial cell walls are composed of **peptidoglycan** (pep′ti-dō-glī′kan), a meshlike complex polysaccharide. Peptidoglycan in turn is composed of two types of regularly alternating sugar molecules, called N-*acetylglucosamine (NAG)* and N-*acetylmuramic acid (NAM)*, which are structurally similar to glucose **(FIGURE 3.13)**. Millions of NAG and NAM molecules are covalently linked in chains in which NAG and NAM alternate. These chains are the "glycan" portions of peptidoglycan.

Chains of NAG and NAM are attached to other chains by crossbridges of four amino acids (tetrapeptides). **FIGURE 3.14** illustrates one possible configuration. These peptide cross bridges are the "peptido" portion of peptidoglycan. Depending on the bacterium, tetrapeptide bridges are either bonded to one another or held together by *short connecting chains* of other amino acids, as shown in Figure 3.14. Peptidoglycan covers the entire surface of a cell, which must insert millions of new NAG and NAM subunits if it is to grow and divide.

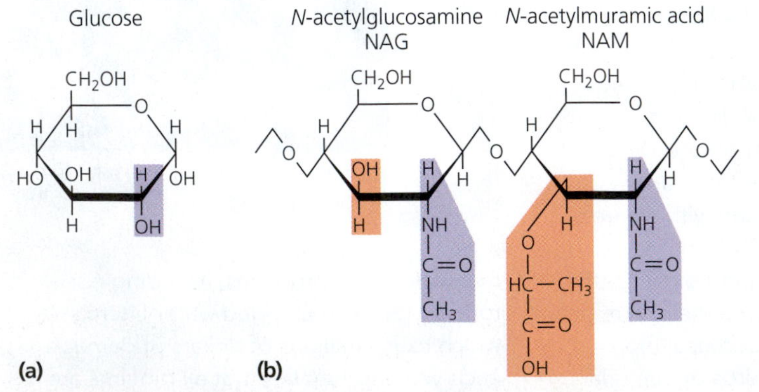

▲ **FIGURE 3.13 Comparison of the structures of glucose, NAG, and NAM.** **(a)** Glucose. **(b)** *N*-acetylglucosamine (NAG) and *N*-acetylmuramic acid (NAM) molecules linked as in peptidoglycan. Blue shading indicates the differences between glucose and the other two sugars. Orange boxes highlight the difference between NAG and NAM.

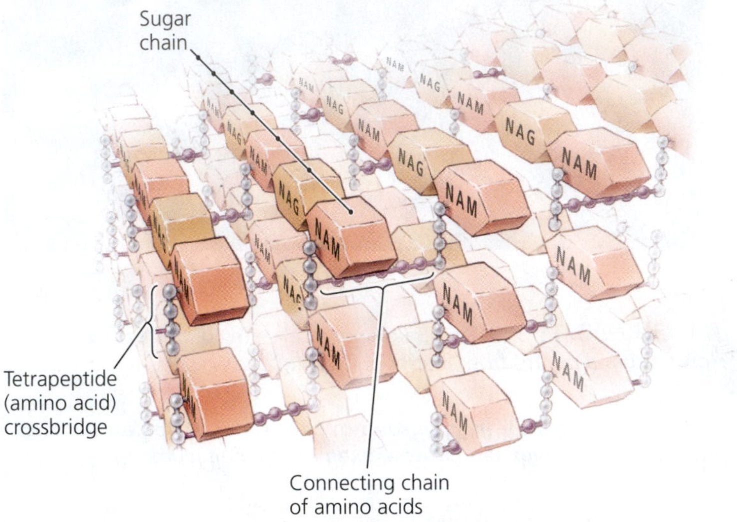

▲ **FIGURE 3.14 Possible structure of peptidoglycan.** Peptidoglycan is composed of chains of NAG and NAM linked by tetrapeptide crossbridges and, in some cases, as shown here, connecting chains of amino acids to form a tough yet flexible structure. The amino acids of the crossbridges differ among bacterial species.

Scientists describe two basic types of bacterial cell walls as *Gram-positive* cell walls or *Gram-negative* cell walls. They distinguish Gram-positive and Gram-negative cells by the use of the Gram staining procedure (described in Chapter 4), which was invented long before the structure and chemical nature of bacterial cell walls were known.

Gram-Positive Bacterial Cell Walls

LEARNING | **OUTCOME**

3.11 Compare and contrast the cell walls of acid-fast bacteria with typical Gram-positive cell walls.

Gram-positive bacterial cell walls have a relatively thick layer of peptidoglycan that also contains unique chemicals called *teichoic* (tī-kō′ik)[4] *acids*. Some teichoic acids are covalently linked to lipids, forming *lipoteichoic acids* that anchor the peptidoglycan

[4]From Greek *teichos*, meaning "wall."

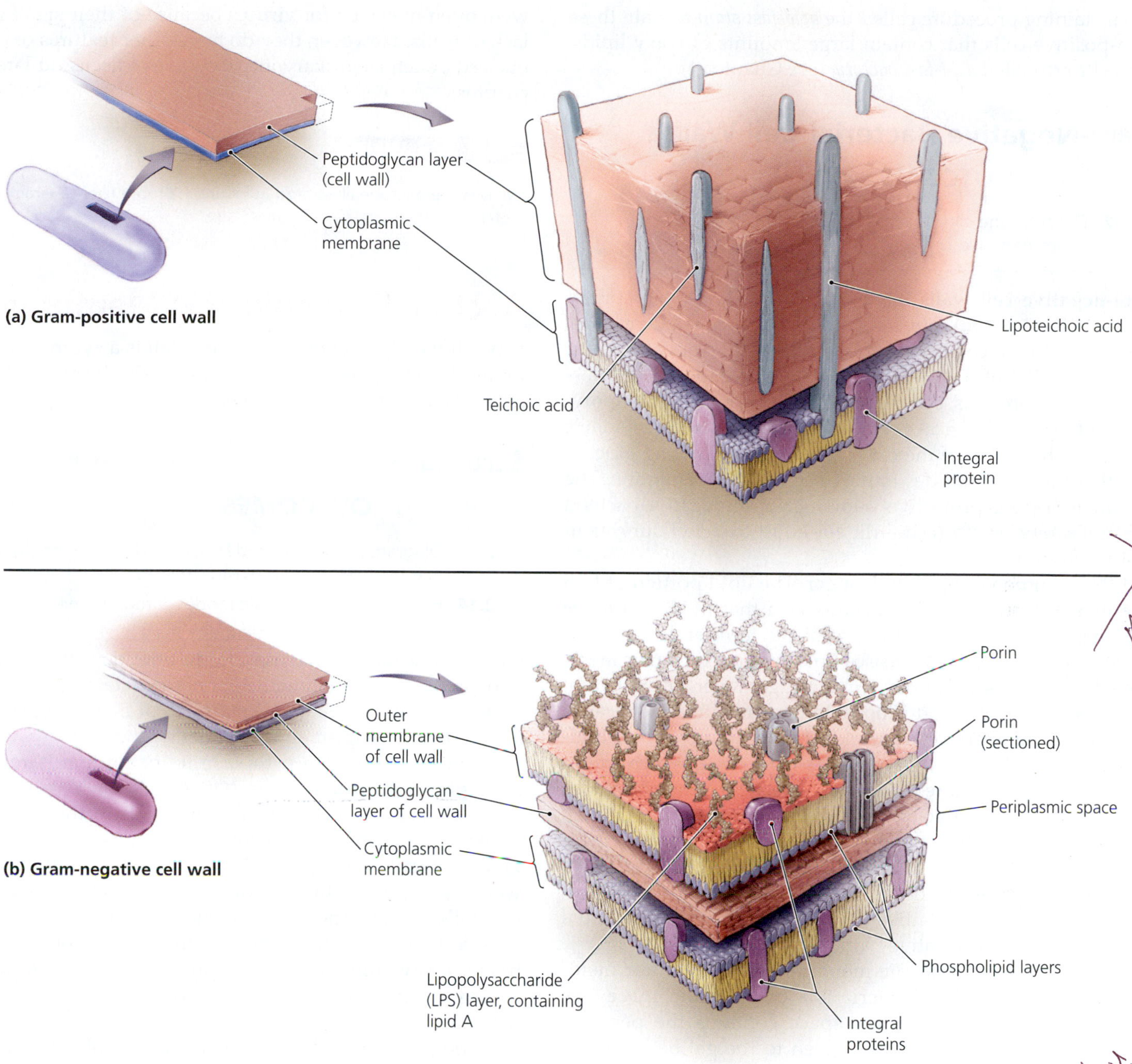

▲ **FIGURE 3.15 Comparison of cell walls of Gram-positive and Gram-negative bacteria.** **(a)** The Gram-positive cell wall has a thick layer of peptidoglycan and lipoteichoic acids that anchor the wall to the cytoplasmic membrane. **(b)** The Gram-negative cell wall has a thin layer of peptidoglycan and an outer membrane composed of lipopolysaccharide (LPS), phospholipids, and proteins. *What effects can lipid A have on human physiology?*

Figure 3.15 Lipid A can cause shock, blood clotting, and fever in humans.

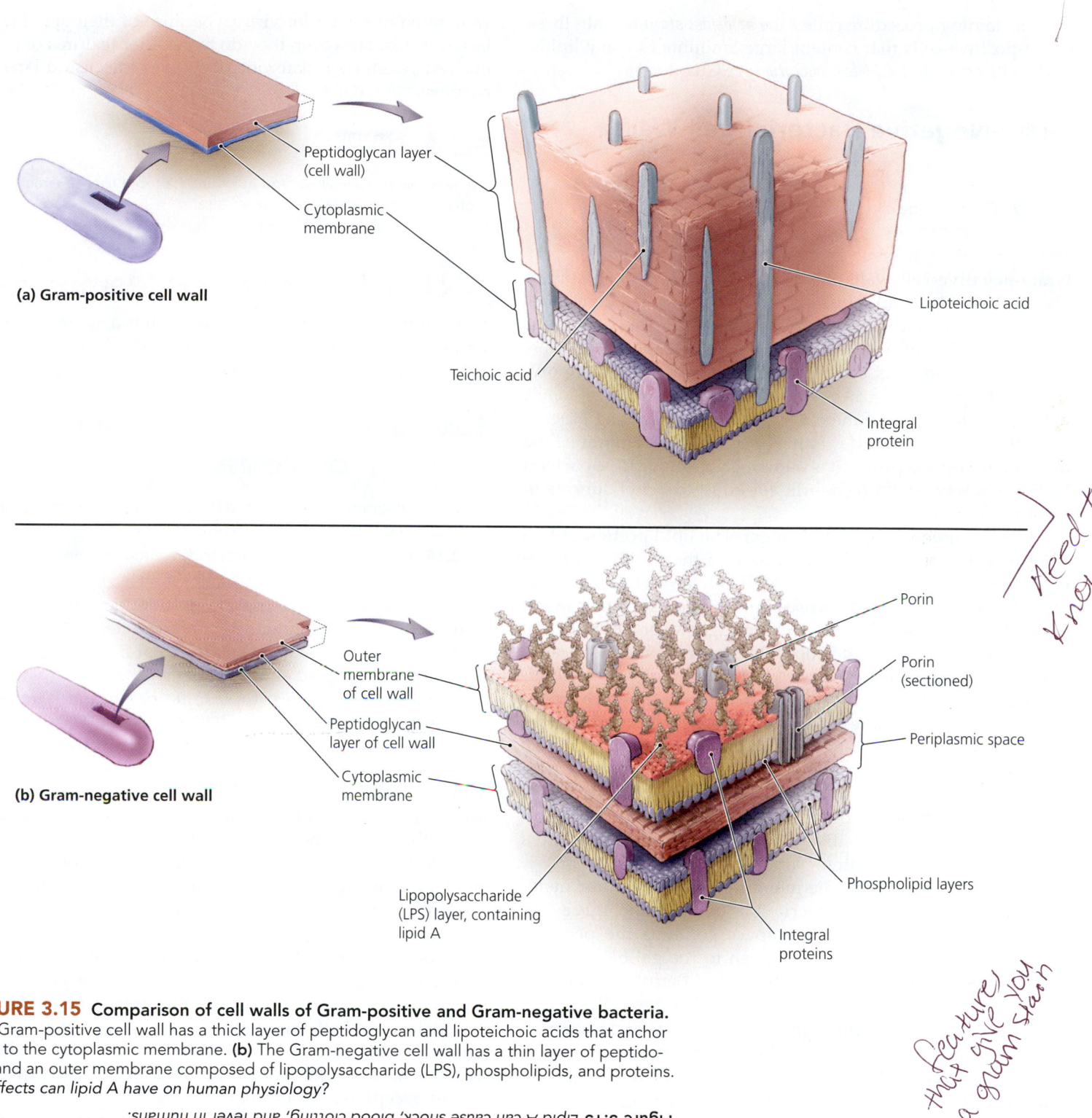

to the cytoplasmic membrane **(FIGURE 3.15a)**. Teichoic acids have negative electrical charges, which help give the surface of a Gram-positive bacterium a negative charge and may play a role in the passage of ions through the wall. The thick cell wall of a Gram-positive bacterium retains the crystal violet dye used in the Gram staining procedure, so the stained cells appear purple under magnification.

Some additional chemicals are associated with the walls of some Gram-positive bacteria. For example, species of *Mycobacterium* (mī′kō-bak-tēr′ē-ŭm), which include the causative agents of tuberculosis and leprosy, have walls with up to 60% mycolic acid, a waxy lipid. Mycolic acid helps these cells survive desiccation (drying out) and makes them difficult to stain with regular water-based dyes. Researchers have developed a special staining procedure called the *acid-fast stain* to stain these Gram-positive cells that contain large amounts of waxy lipids. Such cells are called *acid-fast bacteria* (see Chapter 4).

Gram-Negative Bacterial Cell Walls

LEARNING | OUTCOME

3.12 Describe the clinical implications of the structure of the Gram-negative cell wall.

Gram-negative cell walls have only a thin layer of peptidoglycan **(FIGURE 3.15b)**, but outside this layer is another, outer bilayer membrane composed of two different layers or leaflets. The inner leaflet of the outer membrane is composed of phospholipids and proteins, but the outer leaflet is made of **lipopolysaccharide (LPS)**. Integral proteins called *porins* form channels through both leaflets of the outer membrane, allowing glucose and other monosaccharides to move across the membrane. The outer membrane is protective, allowing Gram-negative bacteria such as *Escherichia coli* (esh-ĕ-rik´ē-ă kō´lī) to better survive in harsh environments.

LPS is a union of lipid with sugar. The lipid portion of LPS is known as **lipid A**. The erroneous idea that lipid A is *inside* Gram-negative cells led to the use of the term **endotoxin**[5] for this chemical. A dead cell releases lipid A when the outer membrane disintegrates, and lipid A may trigger fever, vasodilation, inflammation, shock, and blood clotting in humans. Because killing large numbers of Gram-negative bacteria with antimicrobial drugs releases large amounts of lipid A, which might threaten the patient more than the live bacteria, any internal infection by Gram-negative bacteria is cause for concern.

The Gram-negative outer membrane can also be an impediment to the treatment of disease. For example, the outer membrane may prevent the movement of penicillin to the underlying peptidoglycan, thus rendering the drug ineffectual against many Gram-negative pathogens.

Between the cytoplasmic membrane and the outer membrane of Gram-negative bacteria is a **periplasmic space** (see Figure 3.15b). The periplasmic space contains the peptidoglycan and *periplasm,* the name given to the gel between the membranes of these Gram-negative cells. Periplasm contains water, nutrients, and substances secreted by the cell, such as digestive enzymes and proteins involved in specific transport. The enzymes function to catabolize large nutrient molecules into smaller molecules that can be absorbed or transported into the cell.

Because the cell walls of Gram-positive and Gram-negative bacteria differ, the Gram stain is an important diagnostic tool. After the Gram staining procedure, Gram-negative cells appear pink, and Gram-positive cells appear purple. **▶VIDEO TUTOR:** ***Bacterial Cell Walls***

Bacteria Without Cell Walls

A few bacteria, such as *Mycoplasma pneumoniae* (mī´kō-plaz-mă nū-mō´nē-ī), lack cell walls entirely. In the past, these bacteria were often mistaken for viruses because of their small size and lack of walls. However, they do have other features of prokaryotic cells, such as prokaryotic ribosomes (discussed later in the chapter).

TELL ME WHY

Why is the microbe illustrated in Figure 3.2 more likely a Gram-positive bacterium than a Gram-negative one?

Bacterial Cytoplasmic Membranes

Beneath the glycocalyx and the cell wall is a **cytoplasmic membrane**. The cytoplasmic membrane may also be referred to as the *cell membrane* or a *plasma membrane.*

Structure

LEARNING | OUTCOMES

3.13 Diagram a phospholipid bilayer and explain its significance in reference to a cytoplasmic membrane.

3.14 Explain the fluid mosaic model of membrane structure.

Cytoplasmic membranes are about 8 nm thick and composed of phospholipids (see Figure 2.16) and associated proteins. Some bacterial membranes also contain sterol-like molecules, called *hopanoids,* that help stabilize the membrane.

The structure of a cytoplasmic membrane is referred to as a **phospholipid bilayer (FIGURE 3.16)**. A phospholipid molecule is bipolar; that is, the two ends of the molecule are different. The phosphate-containing heads of each phospholipid molecule are *hydrophilic;*[6] that is, they are attracted to water at the two surfaces of the membrane. The hydrocarbon tails of each phospholipid molecule are *hydrophobic*[7] and huddle together with other tails in the interior of the membrane, away from water. Phospholipids placed in a watery environment naturally form a bilayer because of their bipolar nature.

About half of a bacterial cytoplasmic membrane is composed of *integral proteins* inserted amidst the phospholipids. Some integral proteins penetrate the entire bilayer; others are found in only half the bilayer. In contrast, *peripheral proteins* are loosely attached to the membrane on one side or the other. Proteins of cell membranes may act as recognition proteins, enzymes, receptors, carriers, or channels.

The **fluid mosaic model** describes our current understanding of membrane structure. The term *mosaic* indicates that the membrane proteins are arranged in a way that resembles the tiles in a mosaic, and *fluid* indicates that the proteins and lipids are free to flow laterally within a membrane. **▶ANIMATIONS:** ***Membrane Structure***

[5]From Greek *endo,* meaning "inside," and *toxikon,* meaning "poison."
[6]From Greek *hydro,* meaning "water," and *philos,* meaning "love."
[7]From Greek *hydro,* meaning "water," and *phobos,* meaning "fear."

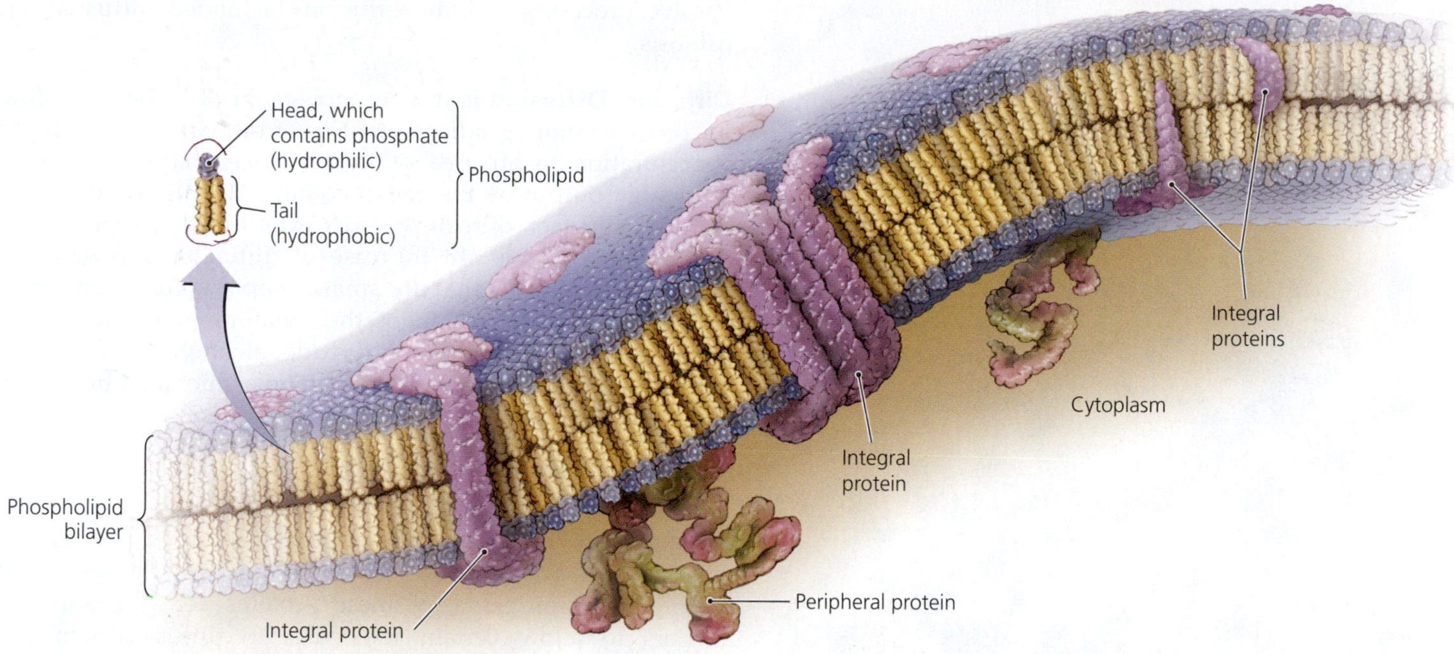

▲ **FIGURE 3.16** **The structure of a prokaryotic cytoplasmic membrane: a phospholipid bilayer.**

Function

LEARNING | **OUTCOMES**

3.15 Describe the functions of a cytoplasmic membrane as they relate to permeability.

3.16 Compare and contrast the passive and active processes by which materials cross a cytoplasmic membrane.

3.17 Define *osmosis*, and distinguish among isotonic, hypertonic, and hypotonic solutions.

A cytoplasmic membrane does more than separate the contents of the cell from the outside environment. The cytoplasmic membrane also controls the passage of substances into and out of the cell. Nutrients are brought into the cell, and wastes are removed. The membrane also functions in producing molecules for energy storage and for harvesting light energy in photosynthetic bacteria. (Energy storage and photosynthesis are discussed in Chapter 5.)

In its function of controlling the contents of the cell, the cytoplasmic membrane is **selectively permeable**, meaning that it allows some substances to cross it while preventing the crossing of others. How does a membrane exert control over the contents of the cell and the substances that move across it? ▶**ANIMATIONS:** ***Membrane Permeability***

A phospholipid bilayer is naturally impermeable to most substances. Large molecules cannot cross through it; ions and molecules with an electrical charge are repelled by it; and hydrophilic substances cannot easily cross its hydrophobic interior. However, cytoplasmic membranes, unlike plain phospholipid bilayers in a scientist's test tube, contain proteins, and these proteins allow substances to cross the membrane by functioning as pores, channels, or carriers.

Movement across the cytoplasmic membrane occurs by either passive or active processes. Passive processes do not require the expenditure of a cell's metabolic energy store, whereas active processes require the expenditure of cellular energy, either directly or indirectly. Active and passive processes will be discussed shortly, but first you must understand another feature of selectively permeable cytoplasmic membranes: their ability to maintain a *concentration gradient*.

Membranes enable a cell to concentrate chemicals on one side of the membrane or the other. The difference in concentration of a chemical on the two sides of a membrane is its **concentration gradient** (also known as a *chemical gradient*).

Because many of the substances that have concentration gradients across cell membranes are electrically charged chemicals, a corresponding **electrical gradient**, or voltage, also exists across the membrane (**FIGURE 3.17**). For example, a greater concentration of negatively charged proteins exists inside the membrane, and positively charged sodium ions are more concentrated outside the membrane. One result of the segregation of electrical charges by a membrane is that the interior of a cell is usually electrically negative compared to the exterior. This tends to repel negatively charged chemicals and attract positively charged substances into cells. ▶**ANIMATIONS:** ***Passive Transport: Principles of Diffusion***

Passive Processes

In passive processes, the electrochemical gradient provides the source of energy; the cell does not expend its energy reserve.

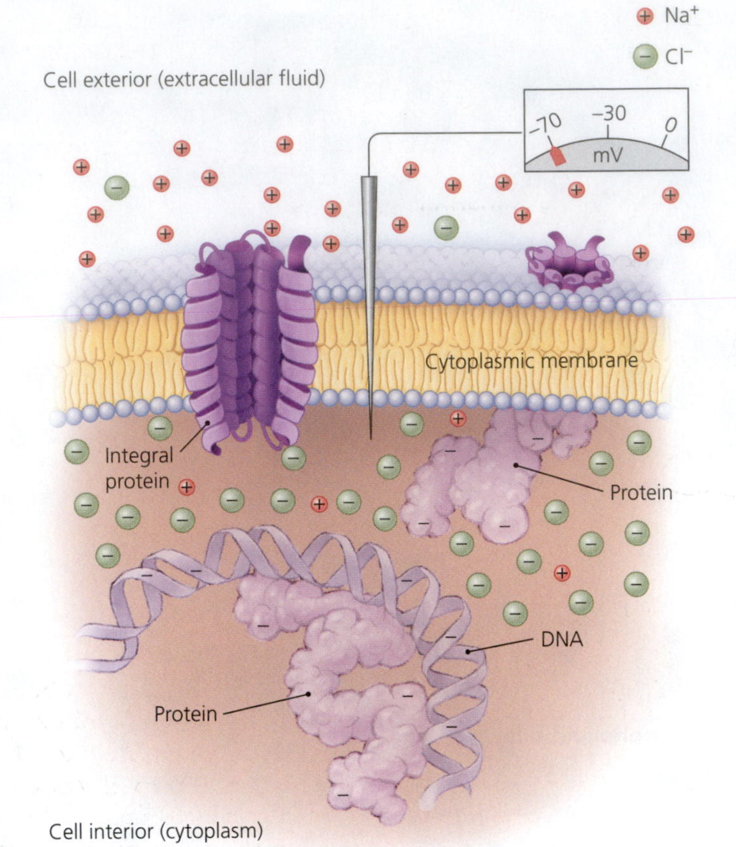

▲ **FIGURE 3.17 Electrical potential of a cytoplasmic membrane.** The electrical potential, in this case −70 mV, exists across a membrane because there are more negative charges inside the cell than outside it.

Passive processes include diffusion, facilitated diffusion, and osmosis.

Diffusion **Diffusion** is the net movement of a chemical down its concentration gradient—that is, from an area of higher concentration to an area of lower concentration. It requires no energy output by the cell, a common feature of all passive processes. In fact, diffusion occurs even in the absence of cells or their membranes. In the case of diffusion into or out of cells, only chemicals that are small or lipid soluble can diffuse through the lipid portion of the membrane (**FIGURE 3.18a**). For example, oxygen, carbon dioxide, alcohol, and fatty acids can freely diffuse through the cytoplasmic membrane, but molecules such as glucose and proteins cannot.

Facilitated Diffusion The phospholipid bilayer blocks the movement of large or electrically charged molecules, so they do not cross the membrane unless there is a pathway for diffusion. As we have seen, cytoplasmic membranes contain integral proteins. Some of these proteins act as channels or carriers to allow certain molecules to diffuse down their concentration gradients into or out of the cell. This process is called **facilitated diffusion** because the proteins facilitate the process by providing a pathway for diffusion. The cell expends no energy in facilitated diffusion; electrochemical gradients provide all of the energy necessary.

Some channel proteins allow the passage of a range of chemicals that have the right size or electrical charge (**FIGURE 3.18b**). Other channel proteins, known as *permeases,* are more specific, carrying only certain substrates (**FIGURE 3.18c**). A permease has a binding site that is selective for one substance.

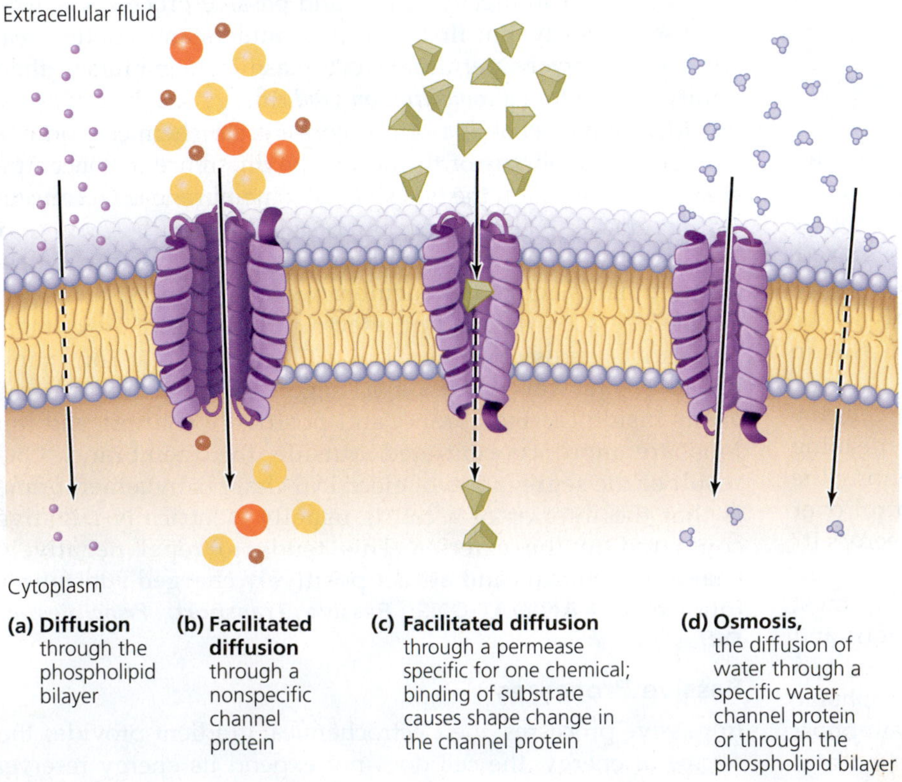

◄ **FIGURE 3.18 Passive processes of movement across a cytoplasmic membrane.** Passive processes always involve movement down an electrochemical gradient.

Osmosis When discussing simple and facilitated diffusion, we considered a solution in terms of the *solutes* (dissolved chemicals) it contains because it is those solutes that move into and out of the cell. In contrast, with osmosis it is useful to consider the concentration of the solvent, which in organisms is always water. **Osmosis** is the special name given to the diffusion of water across a semipermeable membrane—that is, across a membrane that is permeable to water molecules but not to most solutes that are present, such as proteins, amino acids, salts, or glucose **(FIGURE 3.18d)**. Because these solutes cannot freely penetrate the membrane, they cannot diffuse across the membrane, no matter how unequal their concentrations on either side may be. Instead, the water diffuses. Water molecules cross from the side of the membrane that contains a higher concentration of water (lower concentration of solute) to the side that contains a lower concentration of water (higher concentration of solute). In osmosis, water moves across the membrane until equilibrium is reached, or until the pressure of water is equal to the force of osmosis **(FIGURE 3.19)**.

We commonly compare solutions according to their concentrations of solutes. When solutions on either side of a selectively permeable membrane have the same concentration of solutes, the two solutions are said to be **isotonic.**[8] In an isotonic situation, neither side of a selectively permeable membrane will experience a net loss or gain of water **(FIGURE 3.20a)**.

When the concentrations of solutions are unequal, the solution with the higher concentration of solutes is said to be **hypertonic**[9] to the other. The solution with a lower concentration of solutes is **hypotonic**[10] in comparison. Note that the terms *hypertonic* and *hypotonic* refer to the concentration of solute, even though osmosis refers to the movement of the *solvent*, which, in cells, is water. The terms *isotonic, hypertonic,* and *hypotonic* are relative. For example, a glass of tap water is isotonic to another glass of the same water, but it is hypertonic compared to distilled water, and hypotonic when compared to seawater. In biology, the three terms are traditionally used relative to the interior of cells. Most cells are hypertonic to their environments.

Obviously, a hypertonic solution, with its higher concentration of solutes, necessarily means a lower concentration of water; that is, a hypertonic solution has a lower concentration of water than does a hypotonic solution. Like other chemicals, water moves down its concentration gradient from a hypotonic solution into a hypertonic solution. A cell placed in a hypertonic solution will therefore lose water and shrivel **(FIGURE 3.20b)**.

On the other hand, water will diffuse into a cell placed in a hypotonic solution because the cell has a higher solutes-to-water concentration. As water moves into the cell, water pressure against its cytoplasmic membrane increases, and the cell expands **(FIGURE 3.20c)**. One function of a cell wall, such as the peptidoglycan of bacteria, is to resist further osmosis and prevent cells from bursting.

It is useful to compare solutions to the concentration of solutes in a patient's blood cells. Isotonic saline solutions administered to a patient have the same percent dissolved solute (in this case, salt) as do the patient's blood cells. Thus, the patient's intracellular and extracellular environments remain in equilibrium when an isotonic saline solution is administered. However, if the patient is infused with a hypertonic solution, water will move out of the patient's cells, and the cells will shrivel, a condition called *crenation.* Conversely, if a patient is infused with a hypotonic solution, water will move into the patient's cells, which will swell and possibly burst. **▶ANIMATIONS:** ***Passive Transport: Special Types of Diffusion***

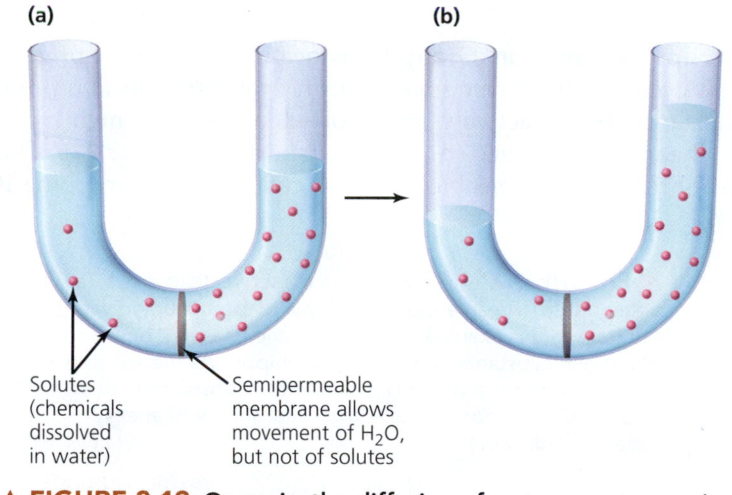

▲ FIGURE 3.19 Osmosis, the diffusion of water across a semipermeable membrane. **(a)** A membrane separates two solutions of different concentrations in a U-shaped tube. The membrane is permeable to water, but not to the solute. **(b)** After time has passed, water has moved down its concentration gradient until water pressure prevented the osmosis of any additional water. *Which side of the tube more closely represents a living cell?*

Figure 3.19 *The right-hand side represents the cell, because cells are typically hypertonic to their environment.*

Active Processes

As stated previously, active processes require the cell to expend energy stored in ATP molecules to move materials across the cytoplasmic membrane against their electrochemical gradient. This is analogous to moving water uphill. As we will see, ATP may be utilized directly during transport or indirectly at some other site and at some other time. Active processes in bacteria include *active transport* by means of carrier proteins and a special process termed *group translocation.* **▶ANIMATIONS:** ***Active Transport: Overview***

Active Transport Like facilitated diffusion, **active transport** utilizes transmembrane permease proteins; however, the functioning of active transport proteins requires the cell to expend ATP to transport molecules across the membrane. Some such proteins are referred to as *gated channels* or *ports* because they are controlled. When the cell is in need of a substance, the protein becomes functional (the gate "opens"). At other times, the gate is "closed."

[8]From Greek *isos,* meaning "equal," and *tonos,* meaning "tone."
[9]From Greek *hyper,* meaning "more" or "over."
[10]From Greek *hypo,* meaning "less" or "under."

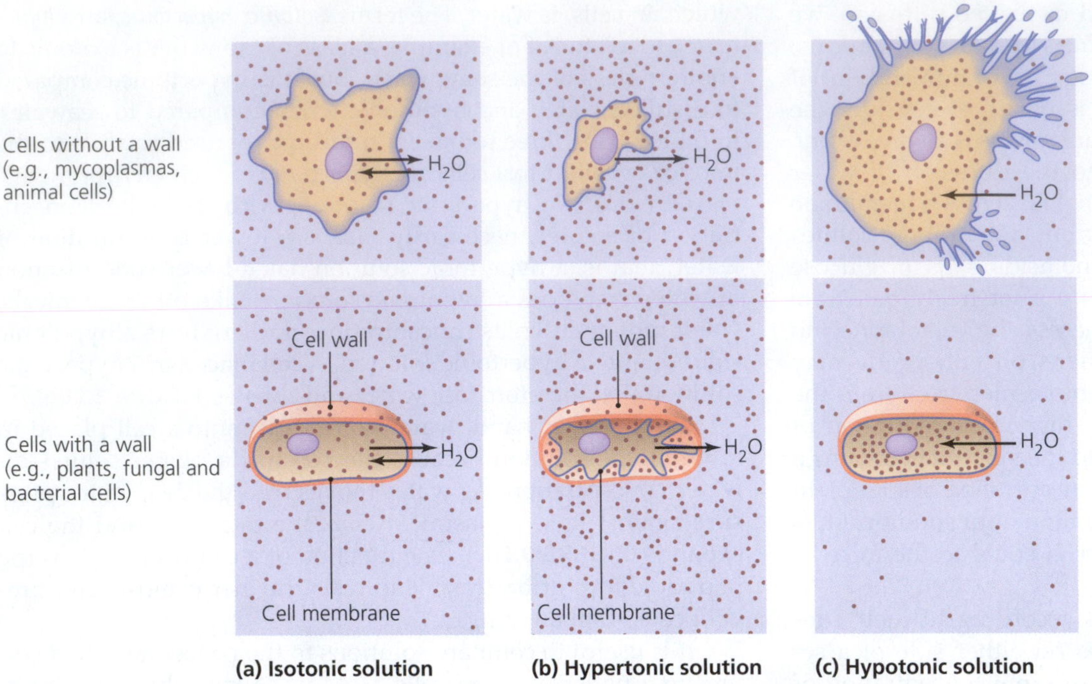

◀ **FIGURE 3.20 Effects of isotonic, hypertonic, and hypotonic solutions on cells.** **(a)** Cells in isotonic solutions experience no net movement of water. **(b)** Cells in hypertonic solutions shrink because of the net movement of water out of the cell. **(c)** Cells in hypotonic solutions undergo a net gain of water. Animal cells burst because they lack a cell wall; in cells with a cell wall, the pressure of water pushing against the interior of the wall eventually stops the movement of water into the cell.

If only one substance is transported at a time, the permease is called a *uniport* **(FIGURE 3.21a)**. In contrast, *antiports* simultaneously transport two chemicals, but in opposite directions; that is, one substance is transported into the cell at the same time that a second substance is transferred out of the cell **(FIGURE 3.21b)**. In other types of active transport, two substances move together in the same direction across the membrane by means of a single carrier protein. Such proteins are known as *symports* **(FIGURE 3.21c)**.

In all cases, active transport moves substances against their electrochemical gradient. Typically, the protein acts as an ATPase—an enzyme that breaks down ATP into ADP and inorganic phosphate during transport, releasing energy that is used to move the chemical against its electrochemical gradient across the membrane.

With symports and antiports, one chemical's electrochemical gradient may provide the energy needed to transport the second chemical, a mechanism called *coupled transport* (Figure 3.21c). For example, H^+ moving into a cell down its electrochemical gradient by facilitated diffusion provides energy to carry glucose into the cell, against the glucose gradient. The two processes are linked by a symport. However, cellular energy may still be utilized for transport because the H^+ gradient can be previously established by the active pumping of H^+ to the outside of the cell by an H^+ uniport. The use of ATP is thus separated in time and space from the active transport of glucose, but ATP was still expended. ▶**ANIMATIONS:** ***Active Transport: Types***

Group Translocation **Group translocation** is an active process that occurs only in some bacteria. In group translocation, the substance being actively transported across the membrane is chemically changed during transport **(FIGURE 3.22)**. The membrane is impermeable to the altered substance, trapping it

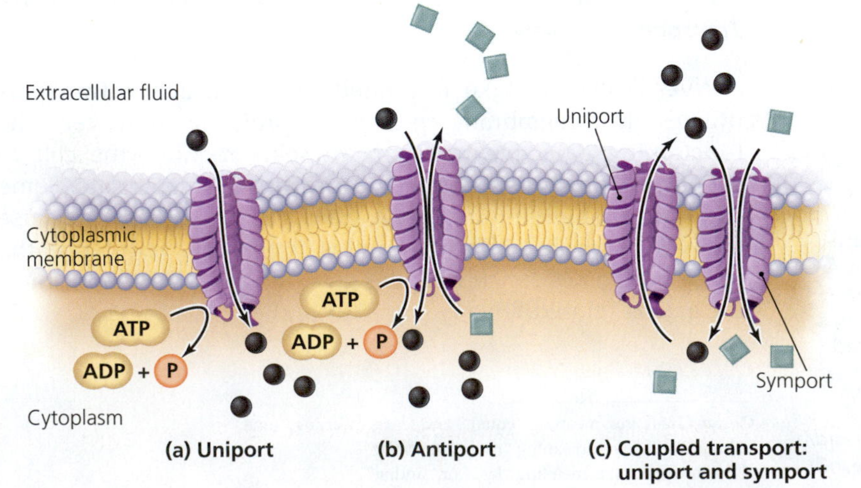

◀ **FIGURE 3.21 Mechanisms of active transport.** **(a)** Via a uniport. **(b)** Via an antiport. **(c)** Via a uniport coupled with a symport. In this example, the membrane uses ATP energy to pump one substance out through a uniport. As this substance flows back into the cell, it brings another substance with it through the symport. *What is the usual source of energy for active transport?*

Figure 3.21 *ATP is the usual source of energy for active transport processes.*

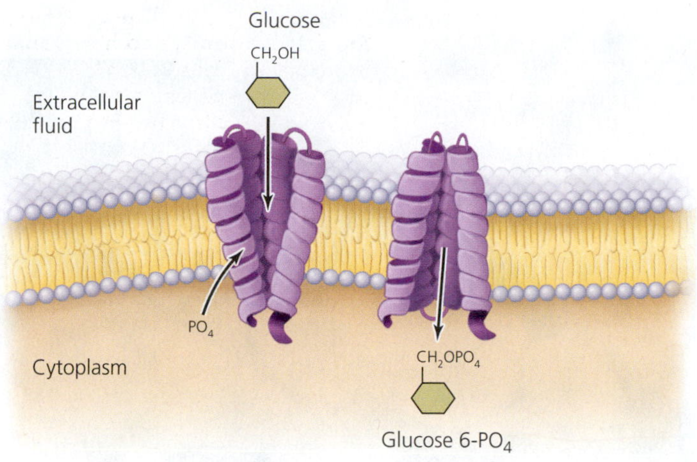

▲ **FIGURE 3.22 Group translocation.** This process involves a chemical change in a substance as it is being transported. This figure depicts glucose being transported into a bacterial cell via group translocation.

inside the cell. Group translocation is very efficient at bringing substances into a cell. It can operate efficiently even if the external concentration of the chemical being transported is as low as 1 part per million (ppm).

One well-studied example of group translocation is the accumulation of glucose inside a bacterial cell. As glucose is transported across the bacterial cell membrane, it is phosphorylated; that is, a phosphate group is added to the glucose. The glucose is changed into glucose 6-phosphate, a sugar that cannot cross back out but can be utilized in the ATP-producing metabolism of the cell. Other carbohydrates, fatty acids, purines, and pyrimidines are also brought into bacterial cells by group translocation. A summary of bacterial transport processes is shown in **TABLE 3.2**.

TELL ME WHY

E. coli grown in a hypertonic solution turns on a gene to synthesize protein that transports potassium into the cell. Why?

Cytoplasm of Bacteria

LEARNING **OUTCOMES**

- **3.18** Describe bacterial cytoplasm and its basic contents.
- **3.19** Define *inclusion*, and give two examples.
- **3.20** Describe the formation and function of endospores.

Cytoplasm is the general term used to describe the gelatinous material inside a cell. Cytoplasm is semitransparent, fluid, elastic, and aqueous. It is composed of cytosol, inclusions, ribosomes, and, in many cells, a cytoskeleton. Some bacterial cells produce internal, resistant, dormant forms called endospores.

Cytosol

The liquid portion of the cytoplasm is called **cytosol**. It is mostly water, but it also contains dissolved and suspended substances, including ions, carbohydrates, proteins (mostly enzymes), lipids, and wastes. The cytosol of prokaryotes also contains the cell's DNA in a region called the **nucleoid**. Recall that a distinctive feature of prokaryotes is lack of a phospholipid membrane surrounding this DNA.

Most bacteria have a single, circular DNA molecule organized as a chromosome. Some bacteria, such as *Vibrio cholerae* (vib′rē-ō kol′er-ī), the bacterium that causes cholera, are unusual in that they have two chromosomes.

TABLE 3.2 Transport Processes Across Bacterial Cytoplasmic Membranes

	Description	Examples of Transported Substances
Passive Transport Processes	Processes require no use of energy by the cell; the electrochemical gradient provides energy.	
Diffusion	Molecules move down their electrochemical gradient through the phospholipid bilayer of the membrane.	Oxygen, carbon dioxide, lipid-soluble chemicals
Facilitated diffusion	Molecules move down their electrochemical gradient through channels or carrier proteins.	Glucose, fructose, urea, some vitamins
Osmosis	Water molecules move down their concentration gradient across a selectively permeable membrane.	Water
Active Transport Processes	Cell expends energy in the form of ATP to move a substance against its electrochemical gradient.	
Active transport	ATP-dependent carrier proteins bring substances into cell.	Na^+, K^+, Ca^{2+}, H^+, Cl^-
Group translocation	The substance is chemically altered during transport; found only in some bacteria.	Glucose, mannose, fructose

The cytosol is the site of some chemical reactions. For example, enzymes within the cytosol function to produce amino acids and degrade sugar.

Inclusions

Deposits, called **inclusions**, are often found within bacterial cytosol. Rarely, a cell surrounds its inclusions with a polypeptide membrane. Inclusions may include reserve deposits of lipids, starch, or compounds containing nitrogen, phosphate, or sulfur. Such chemicals may be taken in and stored in the cytosol when nutrients are in abundance and then utilized when nutrients are scarce. The presence of specific inclusions is diagnostic for several pathogenic bacteria.

Many bacteria store carbon and energy in molecules of glycogen, which is a polymer of glucose molecules, or as a lipid polymer called *polyhydroxybutyrate (PHB)* (**FIGURE 3.23**). Long chains of PHB accumulate as inclusion granules in the cytoplasm. Slight chemical modification of PHB produces a plastic that can be used for packaging and other applications (see **Beneficial Microbes: Plastics Made Perfect?**). PHB plastics are biodegradable, breaking down in a landfill in a few weeks rather than persisting for years as petroleum-based plastics do.

Many aquatic cyanobacteria (blue-green photosynthetic bacteria) contain inclusions called *gas vesicles* that store gases in protein sacs. The gases buoy the cells to the surface and into the light needed for photosynthesis. Other interesting inclusions are small crystals of magnetite stored by *magnetobacteria*. Infoldings of the cytoplasmic membrane surround the magnetite to form membrane-bound sacs.

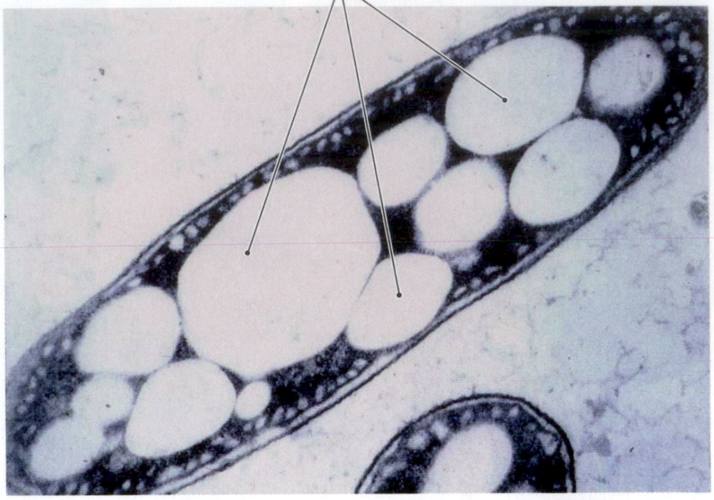

▲ **FIGURE 3.23 Granules of PHB in the bacterium *Azotobacter chroococcum*.**

BENEFICIAL MICROBES

Plastics Made Perfect?

Petroleum-based plastics play a considerable role in modern life, appearing in packaging, bottles, appliances, furniture, automobiles, disposable diapers, and many other synthetic goods. Despite the good that plastic brings to our lives, there are problems with this artificial polymer.

Manufacturers make plastic from oil, exacerbating U.S. dependence on foreign supplies of crude oil. Consumers discard plastic, filling landfills with more than 15 million tons of plastic every year in the United States. Because plastic is artificial, microorganisms do not break it down effectively, and discarded plastic might remain in landfills for decades or centuries. What is needed is a functional "green" plastic—a plastic that is strong and light and that can be shaped and colored as needed, yet is more easily biodegradable.

Enter the bacteria. Many bacterial cells, particularly Gram-negative bacteria, use polyhydroxybutyrate (PHB) as a storage molecule and energy source, much as humans use fat. PHB and similar storage molecules turn out to be rather versatile plastics that are produced when bacteria metabolizing certain types of sugar are simultaneously deprived of an essential element, such as nitrogen, phosphorus, or potassium. The bacteria, faced with such a nutritionally stressed environment, convert the sugar to PHB, which they store as intracellular inclusions. Scientists harvest these biologically created molecules by breaking the cells open and treating the cytoplasm with chemicals to isolate the plastics and remove dangerous endotoxin.

PHB bottle caps. One partially biodegraded in 60 days.

Purified PHB possesses many of the properties of petrochemically derived plastic—its melting point, crystal structure, molecular weight, and strength are very similar. Further, PHB has a singular, overwhelming advantage compared to artificial plastic: PHB is naturally and completely biodegradable—bacteria catabolize PHB into carbon dioxide and water. The positive effect this would have on our overtaxed landfills would be tremendous, and replacing just half of the oil-based plastic used in the United States with PHB could reduce oil imports by more than 250 million barrels per year, improving our foreign trade balance and reducing our dependence on overseas oil suppliers.

Endospores

Some bacteria, notably *Bacillus* (ba-sil´ŭs) and *Clostridium* (klos-trid´ē-ŭm), are characterized by the ability to produce unique structures called **endospores**, which are important for several reasons, including their durability and potential pathogenicity. Though some people refer to endospores simply as "spores," endospores should not be confused with the reproductive spores of actinobacteria, algae, and fungi. A single bacterial cell, called a *vegetative* cell to distinguish it from an endospore, transforms into only one endospore, which then germinates to grow into only one vegetative cell; therefore, endospores are not reproductive structures. Instead, endospores constitute a defensive strategy against hostile or unfavorable conditions.

A vegetative cell normally transforms itself into an endospore only when one or more nutrients (such as carbon or nitrogen) are in limited supply. The process of endospore formation, called *sporulation,* requires 8 to 10 hours and proceeds in eight steps **(FIGURE 3.24)**. During the process, two membranes, a thick layer of peptidoglycan, and a spore coat form around a copy of the cell's DNA and a small portion of cytoplasm. The cell deposits large quantities of dipicolinic acid, calcium, and DNA-binding proteins within the endospore while removing most of the water. Depending on the species, a cell forms an endospore either *centrally, subterminally* (near one end), or *terminally* (at one end). Sometimes an endospore is so large it swells the vegetative cell.

Endospores are extremely resistant to drying, heat, radiation, and lethal chemicals. For example, they remain alive in boiling water for several hours; are unharmed by alcohol, peroxide, bleach, and other toxic chemicals; and can tolerate over 400 rad of radiation, which is more than five times the dose that is lethal to most humans. Endospores are stable resting stages that barely metabolize—they are essentially in a state of suspended animation—and they germinate only when conditions improve. Scientists do not know how endospores are able to resist harsh conditions, but it appears that the double membrane, spore coats, dipicolinic acid, calcium, and DNA-binding proteins serve to stabilize DNA and enzymes, protecting them from adverse conditions.

The ability to survive harsh conditions makes endospores the most resistant and enduring cells. In one case, scientists were able to revive endospores of *Clostridium* that had been sealed in a test tube for 34 years. This record pales, however, beside other researchers' claim to have revived *Bacillus* endospores from inside 250-million-year-old salt crystals retrieved from an underground site near Carlsbad, New Mexico. Some scientists question this claim, suggesting that the bacteria might be recent contaminants that entered through invisible cracks in the salt crystals. In any case, there is little doubt that endospores can remain viable for a minimum of tens, if not thousands, of years.

Endospore formation is a serious concern to food processors, health care professionals, and governments because endospores are resistant to treatments that inhibit other microbes, and because endospore-forming bacteria produce deadly toxins that cause such fatal diseases as anthrax, tetanus, and gangrene. (Chapter 9 considers techniques for controlling endospore formers.)

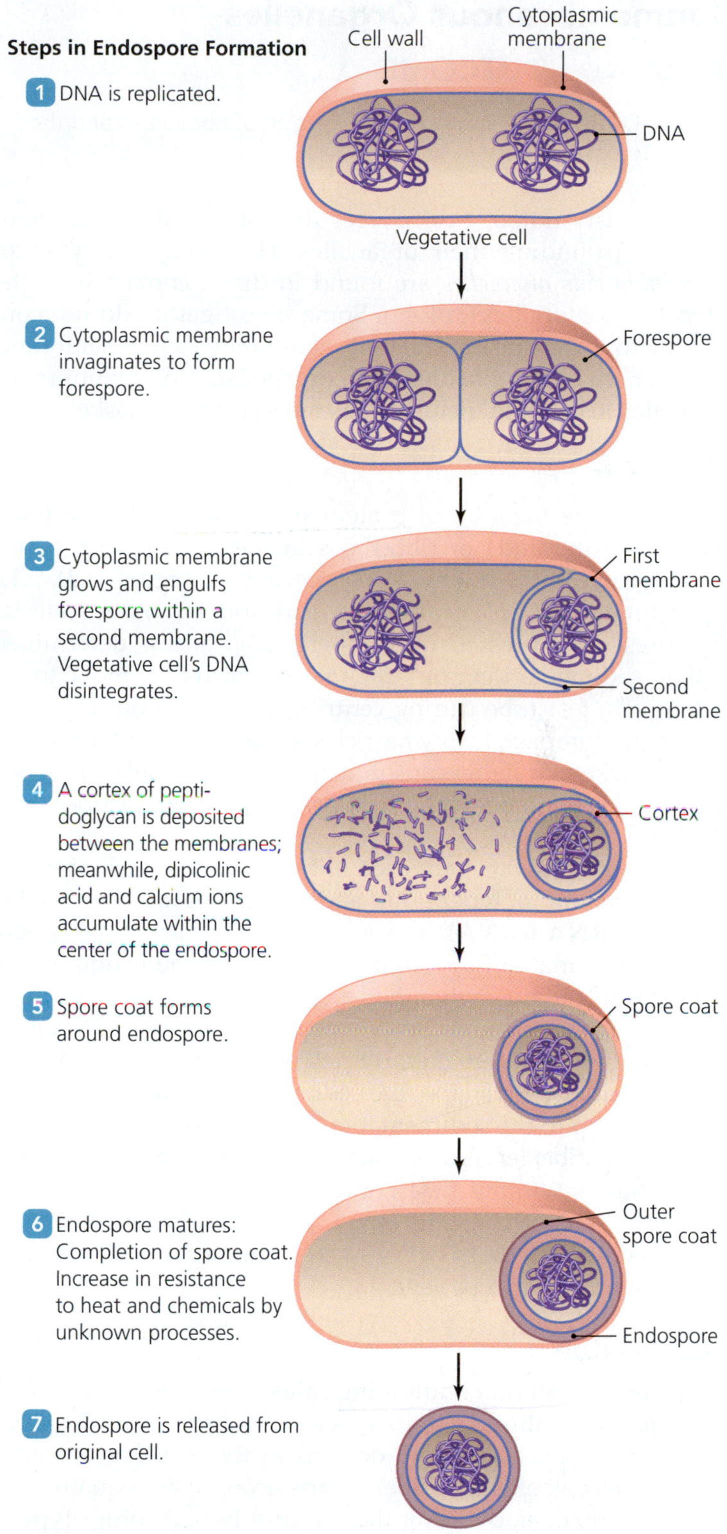
Steps in Endospore Formation
Cell wall
Cytoplasmic membrane
1 DNA is replicated.
DNA
Vegetative cell
2 Cytoplasmic membrane invaginates to form forespore.
Forespore
3 Cytoplasmic membrane grows and engulfs forespore within a second membrane. Vegetative cell's DNA disintegrates.
First membrane
Second membrane
4 A cortex of peptidoglycan is deposited between the membranes; meanwhile, dipicolinic acid and calcium ions accumulate within the center of the endospore.
Cortex
5 Spore coat forms around endospore.
Spore coat
6 Endospore matures: Completion of spore coat. Increase in resistance to heat and chemicals by unknown processes.
Outer spore coat
Endospore
7 Endospore is released from original cell.

▲ **FIGURE 3.24 The formation of an endospore.** The steps depicted occur over a period of 8–10 hours.

Nonmembranous Organelles

LEARNING | OUTCOME

3.21 Describe the structure and function of ribosomes and the cytoskeleton.

As previously noted, prokaryotes do not usually have membranes surrounding their organelles. However, two types of *nonmembranous organelles* are found in direct contact with the cytosol in bacterial cytoplasm. Some investigators do not consider them to be true organelles because they lack a membrane, but other scientists consider them organelles. Nonmembranous organelles in bacteria include ribosomes and the cytoskeleton.

Ribosomes

Ribosomes are the sites of protein synthesis in cells. Bacterial cells have thousands of ribosomes in their cytoplasm, which gives cytoplasm a grainy appearance (see Figure 3.2). The approximate size of ribosomes—and indeed other cellular structures—is expressed in Svedbergs (S)[11] and is determined by their sedimentation rate—the rate at which they move to the bottom of a test tube during centrifugation. As you might expect, large, compact, heavy particles sediment faster than small, loosely packed, or light ones, and are assigned a higher number. Prokaryotic ribosomes are 70S; in contrast, the larger ribosomes of eukaryotes are 80S.

All ribosomes are composed of two subunits, each of which is composed of polypeptides and molecules of RNA called **ribosomal RNA (rRNA)**. The subunits of prokaryotic 70S ribosomes are a smaller 30S subunit and a larger 50S subunit; the 30S subunit contains polypeptides and a single rRNA molecule, whereas the 50S subunit has polypeptides and two rRNA molecules. Because sedimentation rates depend not only on mass and size but also on shape, the sedimentation rates of subunits do not add up to the sedimentation rate of a whole ribosome.

Many antibacterial drugs act on bacterial 70S ribosomes or their subunits without deleterious effects on the larger 80S ribosomes of eukaryotic cells (see Chapter 10). This is why such drugs can stop protein synthesis in bacteria without affecting protein synthesis in a patient.

Cytoskeleton

Cells have an internal scaffolding called a **cytoskeleton**, which is composed of three or four types of protein fibers. Bacterial cytoskeletons play a variety of roles in the cell. For example, one type of cytoskeleton fiber wraps around the equator of a cell and constricts, dividing the cell into two. Another type of fiber forms a helix down the length of some cells **(FIGURE 3.25)**. Such helical fibers appear to play a role in the orientation and deposition of strands of NAG and NAM sugars in the peptidoglycan wall, thereby determining the shape of the cell. Other fibers help keep DNA molecules segregated to certain areas within bacterial cells. An unusual motile bacterium, *Spiroplasma* (spī´ro-plaz-mă), which lacks flagella, uses contractile elements of its cytoskeleton to swim through its environment.

[11]Svedberg units are named for Theodor Svedberg, a Nobel Prize winner and the inventor of the ultracentrifuge.

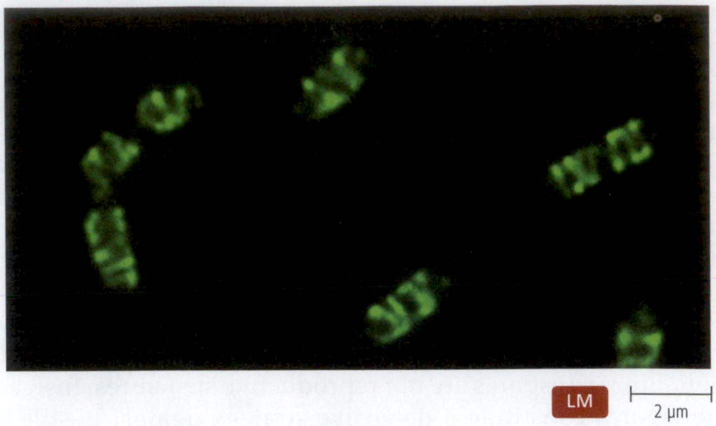

▲ **FIGURE 3.25 A simple helical cytoskeleton.** The rod-shaped bacterium *Bacillus subtilis* has a helical cytoskeleton composed of only a single protein, which has been stained with a fluorescent dye.

TELL ME WHY

The 2001 bioterrorist anthrax attack in the U.S. involved *Bacillus anthracis.* Why is *B. anthracis* able to survive in mail?

We have considered bacterial cells. Next we turn our attention to archaea—the other prokaryotic cells—and compare them to bacterial cells.

External Structures of Archaea

Archaeal cells have external structures similar to those seen in bacteria. These include glycocalyces, flagella, and fimbriae. Some archaea have another kind of proteinaceous appendage called a *hamus.* We consider each of these in order beginning with the outermost structures—glycocalyces.

Glycocalyces

LEARNING | OUTCOME

3.22 Compare the structure and chemistry of archaeal and bacterial glycocalyces.

Like those of bacteria, archaeal glycocalyces are gelatinous, sticky, extracellular structures composed of polysaccharides, polypeptides, or both. Scientists have not studied archaeal glycocalyces as much as those of bacteria, but archaeal glycocalyces function at a minimum in the formation of biofilms—adhering cells to one another, to other types of cells, and to nonliving surfaces in the environment. Organized glycocalyces (capsules) of bacteria and bacterial biofilms are often associated with disease, but researchers have not demonstrated such a link between archaeal capsules or biofilms and disease. Though some research has demonstrated the presence of archaea in some biofilms associated with oral gum disease, no archaeon has been shown conclusively to be pathogenic.

Flagella

LEARNING OUTCOMES

3.23 Describe the structure and formation of archaeal flagella.
3.24 Compare and contrast archaeal flagella with bacterial flagella.

Archaea use flagella to move through their environments, though at a slower speed than bacteria. An archaeal flagellum is superficially similar to a bacterial flagellum: it consists of a basal body, hook, and filament, each composed of protein. The flagellum extends outside the cell and is not covered by a membrane. The basal body anchors the flagellum in the cell wall and cytoplasmic membrane. As with bacterial flagella, archaeal flagella rotate like propellers.

However, scientists have discovered many differences between archaeal and bacterial flagella:

- Archaeal flagella are 10–14 nm in diameter, which is about half the thickness of bacterial flagella.
- Archaeal flagella are not hollow.
- Archaeal flagella lack a central channel; therefore, they grow with the addition of subunits at the base of the filament rather than at the tip.
- The proteins making up archaeal flagella share common amino acid sequences across archaeal species. These are very different from the amino acid sequences common to bacterial flagella.
- Sugar molecules are attached to the filaments of many archaeal flagella, a condition that is rare in bacteria.
- Archaeal flagella are powered with energy stored in molecules of ATP, whereas the flow of hydrogen ions across the membrane powers bacterial flagella.
- Archaeal flagella rotate together as a bundle both when they rotate clockwise and when they rotate counterclockwise. In contrast, bacterial flagella operate independently when rotating clockwise.

These differences indicate that archaeal flagella arose independently of bacterial flagella; they are *analogous* structures—having similar structure without having a common ancestor.

Fimbriae and Hami

LEARNING OUTCOMES

3.25 Compare the structure and function of archaeal and bacterial fimbriae.
3.26 Describe the structure and function of hami.

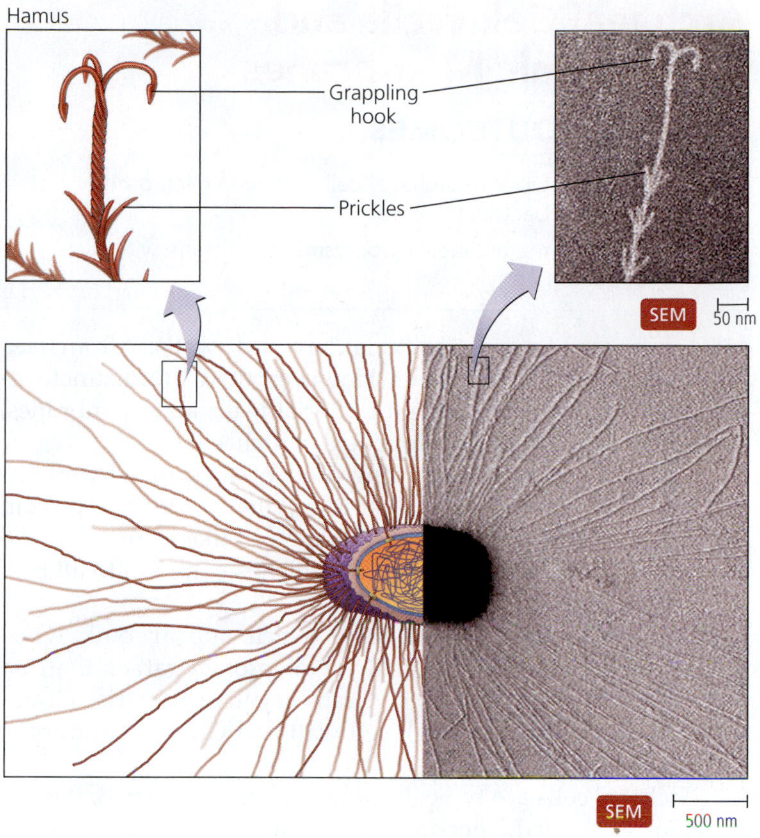

▲ **FIGURE 3.26 Archaeal hami.** Archaea use hami, which are shaped like grappling hooks on barbed wire, to attach themselves to structures in the environment.

Many archaea have fimbriae—nonmotile, rodlike, sticky projections. As with bacteria, archaeal fimbriae are composed of protein and anchor the cells to one another and to environmental surfaces.

Some archaea make unique proteinaceous, fimbriae-like structures called **hami**[12] (singular: *hamus*). More than 100 hami may radiate from the surface of a single archaeon **(FIGURE 3.26)**. Each hamus is a helical filament with tiny prickles sticking out at regular intervals, much like barbed wire. The end of the hamus is frayed into three distinct arms, each of which has a thickened end and bends back toward the cell to make the entire structure look like a grappling hook. Indeed, hami function to securely attach archaea to surfaces.

TELL ME WHY

Why do scientists consider bacterial and archaeal flagella to be analogous rather than evolutionary relations?

[12]From Latin *hamus,* meaning "prickle," "claw," "hook," or "barb."

Archaeal Cell Walls and Cytoplasmic Membranes

LEARNING | OUTCOMES

3.27 Contrast types of archaeal cell walls with each other and with bacterial cell walls.

3.28 Contrast the archaeal cytoplasmic membrane with that of bacteria.

Most archaea, like most bacteria, have cell walls. All archaea have cytoplasmic membranes. However, there are distinct differences between archaeal and bacterial walls and membranes, further emphasizing the uniqueness of archaea.

Archaeal cell walls are composed of specialized proteins or polysaccharides. In some species, the outermost protein molecules form an array that coats the cell like chain mail. All archaeal walls lack peptidoglycan, which is common to all bacterial cell walls.

Gram-negative archaeal cells, which appear pink when Gram-stained, have an outer layer of protein rather than an outer lipid bilayer as seen in Gram-negative bacteria. Gram-positive archaea have a thick cell wall and Gram stain purple, like Gram-positive bacteria.

Archaeal cells are typically spherical or rod shaped, though irregularly shaped, needle-like, rectangular, and flattened square archaea exist (**FIGURE 3.27**).

Archaeal cytoplasmic membranes are composed of lipids that lack phosphate groups and have branched hydrocarbons linked to glycerol by ether linkages rather than the ester linkages seen in bacterial membranes (see Table 2.3 on p. 40). Ether linkages are stronger in many ways than ester linkages, allowing archaea to live in extreme environments such as near-boiling water and hypersaline lakes. Some archaea—particularly those that thrive in very hot water—have a single layer of lipid composed of two glycerol groups covalently linked with branched hydrocarbon chains.

The archaeal cytoplasmic membrane maintains electrical and chemical gradients in the cell. It also functions to control the import and export of substances from the cell using membrane proteins as ports and pumps, just as proteins are used in bacterial cytoplasmic membranes.

TELL ME WHY

Why did scientists in the 19th and early 20th centuries think that archaea were bacteria?

Cytoplasm of Archaea

LEARNING | OUTCOME

3.29 Compare and contrast the cytoplasm of archaea with that of bacteria.

Cytoplasm is the gel-like substance found in all cells, including archaea. Like bacteria, archaeal cells have 70S ribosomes, a fibrous cytoskeleton, and circular DNA suspended in a liquid cytosol. Also like bacteria, they do not have membranous organelles.

However, archaeal cytoplasm differs from that of bacteria in several ways. For example, the ribosomes of archaea have proteins different from those of the ribosomes of bacteria; indeed, archaeal ribosomal proteins are similar to those of eukaryotes. Scientists further distinguish archaea from bacteria in that archaea use different metabolic enzymes to make RNA and use a genetic code more similar to the code used by eukaryotes. (Chapter 7 discusses these genetic differences in more detail.) **TABLE 3.3** contrasts features of archaea and bacteria.

To this point, we have discussed basic features of bacterial and archaeal prokaryotic cells. Next we turn our attention to eukaryotic cells. (Chapter 11 discusses the classification of prokaryotic organisms in more detail.)

TELL ME WHY

Why do some scientists consider archaea, which are prokaryotic, more closely related to eukaryotes than they are to bacteria?

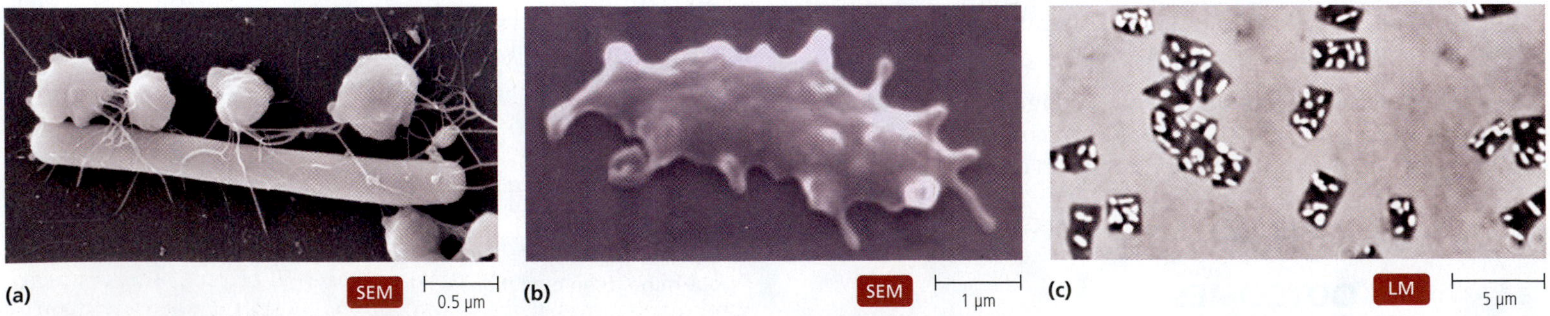

▲ **FIGURE 3.27 Representative shapes of archaea.** **(a)** Cocci, *Pyrococcus furiosus*, attached to rod-shaped *Methanopyrus kandleri*. **(b)** Irregularly shaped archaeon, *Thermoplasma acidophilum*. **(c)** Square archaeon, *Haloquadratum walsbyi*. *What are the stringlike extensions of* Pyrococcus?

Figure 3.27 *The extensions are fimbriae or hami.*

TABLE 3.3 Some Structural Characteristics of Prokaryotes

Feature	Archaea	Bacteria
Glycocalyx	Polypeptide or polysaccharide	Polypeptide or polysaccharide
Flagella	Present in some, 10–14 nm in diameter, grow at base, rotate both counterclockwise and clockwise as bundles	Present in some, about 20 nm in diameter, grow at the tip, rotate counterclockwise in bundles to cause runs, rotate independently clockwise to cause tumbles
Fimbriae	Proteinaceous, used for attachment and in formation of biofilms	Proteinaceous, used for attachment, gliding motility, and in formation of biofilms
Pili	None discovered	Present in some, proteinaceous, used in bacterial exchange of DNA
Hami	Present in some, used for attachment	Absent
Cell walls	Present in most, composed of polysaccharides (not peptidoglycan) or proteins	Present in most, composed of peptidoglycan—a polysaccharide
Cytoplasmic membrane	Present in all, membrane lipids made with ether linkages, some have single lipid layer	Present in all, phospholipids made with ester linkages in bilayer
Cytoplasm	Cytosol contains circular DNA molecule and 70S ribosomes, ribosomal proteins similar to eukaryotic ribosomal proteins	Cytosol contains at least a circular DNA molecule and 70S ribosomes with bacterial proteins

External Structure of Eukaryotic Cells

Some eukaryotic cells have glycocalyces, which are similar to those of prokaryotes.

Glycocalyces

LEARNING **OUTCOME**

3.30 Describe the composition, function, and importance of eukaryotic glycocalyces.

Animal and most protozoan cells lack cell walls, but a cell may have a sticky **glycocalyx**[13] that is anchored to its cytoplasmic membrane via covalent bonds to membrane proteins and lipids. The functions of eukaryotic glycocalyces, which are not as structurally organized as prokaryotic capsules, include helping to anchor animal cells to each other, strengthening the cell surface, providing some protection against dehydration, and functioning in cell-to-cell recognition and communication. Glycocalyces are absent in eukaryotes that have cell walls, such as plants and fungi.

TELL ME WHY

Why are eukaryotic glycocalyces covalently bound to cytoplasmic membranes, and why don't eukaryotes with cell walls have glycocalyces?

[13]From Greek *glykys,* meaning "sweet," and *kalyx,* meaning "cup" or "husk."

Eukaryotic Cell Walls and Cytoplasmic Membranes

LEARNING **OUTCOMES**

3.31 Compare and contrast prokaryotic and eukaryotic cell walls and cytoplasmic membranes.

3.32 Contrast exocytosis and endocytosis.

3.33 Describe the role of pseudopods in eukaryotic cells.

The eukaryotic cells of fungi, algae, and plants have cell walls. Recall that glycocalyces are absent from eukaryotes with cell walls; instead, the cell wall takes on one of the functions of a glycocalyx by providing protection from the environment. The wall also provides shape and support against osmotic pressure. Most eukaryotic cell walls are composed of various polysaccharides but not the peptidoglycan seen in the walls of bacteria.

The walls of plant cells are composed of *cellulose,* a polysaccharide that is familiar to you as paper and dietary fiber. Fungi also have walls of polysaccharides, including cellulose, *chitin,* and/or *glucomannan.* The walls of algae **(FIGURE 3.28)** are composed of a variety of polysaccharides or other chemicals, depending on the type of alga. These chemicals include cellulose, proteins, *agar, carrageenan, silicates, algin, calcium carbonate,* or a combination of these substances. (Chapter 12 discusses fungi and algae in more detail.)

All eukaryotic cells have cytoplasmic membranes **(FIGURE 3.29)**. A eukaryotic cytoplasmic membrane, like those of bacteria, is a fluid mosaic of phospholipids and proteins, which act as recognition molecules, enzymes, receptors, carriers, or channels. Channel proteins for facilitated diffusion are more common in eukaryotes than in prokaryotes. Additionally, within multicellular organisms some membrane proteins serve to anchor cells to each other.

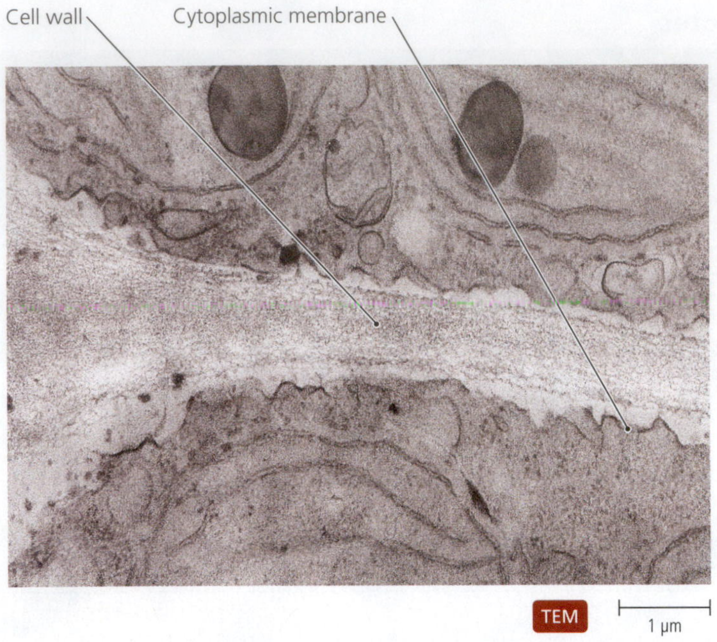

▲ **FIGURE 3.28 A eukaryotic cell wall.** The cell wall of the red alga *Gelidium* is composed of layers of the polysaccharide called agar. *What is the function of a cell wall?*

Figure 3.28 *The cell wall provides support, protection, and resistance to osmotic forces.*

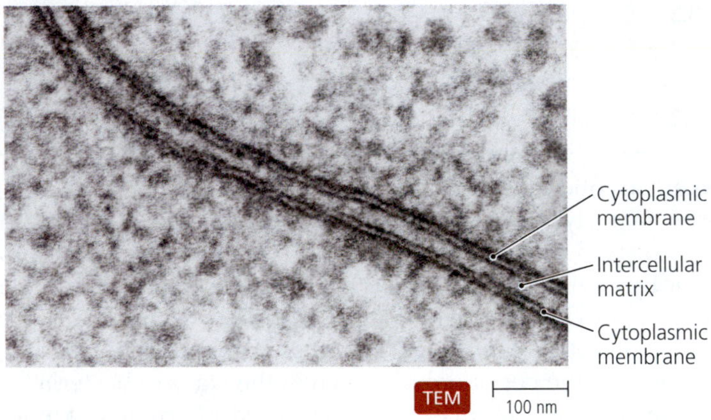

▲ **FIGURE 3.29 Eukaryotic cytoplasmic membrane.** Note that this micrograph depicts the cytoplasmic membranes of two adjoining cells.

Eukaryotic cytoplasmic membranes may differ from prokaryotic membranes in several ways. Eukaryotic membranes contain steroid lipids *(sterols),* such as cholesterol in animal cells, that help maintain membrane fluidity. Paradoxically, at high temperatures sterols stabilize a phospholipid bilayer by making it less fluid, but at low temperatures sterols have the opposite effect—they prevent phospholipid packing, making the membrane more fluid.

Eukaryotic cytoplasmic membranes may contain small, distinctive assemblages of lipids and proteins that remain together in the membrane as a functional group and do not flow independently amidst other membrane components. Such distinct regions are called **membrane rafts**. Eukaryotic cells use membrane rafts to localize cellular processes, including signaling the inside of the cell, protein sorting, and some kinds of cell movement. Some viruses, including those of AIDS, Ebola, measles, and flu, use membrane rafts to enter human cells or during viral replication. Researchers hope that blocking molecules in membrane rafts will provide a way to limit the spread of these viruses.

Eukaryotic cells frequently attach chains of sugar molecules to the outer surfaces of lipids and proteins in their cytoplasmic membranes; prokaryotes rarely do this. Sugar molecules may act in intercellular signaling, cellular attachment, and in other roles.

Like its prokaryotic counterpart, a eukaryotic cytoplasmic membrane controls the movement of materials into and out of a cell. Eukaryotic cytoplasmic membranes use both passive processes (diffusion, facilitated diffusion, and osmosis; see Figure 3.18) and active transport (see Figure 3.20). Eukaryotic membranes do not perform group translocation, which occurs only in some prokaryotes, but many perform another type of active transport—**endocytosis** **(FIGURE 3.30)**, which involves physical manipulation of the cytoplasmic membrane around the cytoskeleton. Endocytosis occurs when the membrane distends to form **pseudopods** (soo´dō-podz; false feet) that surround a substance, bringing it into the cell. Endocytosis is termed **phagocytosis** if a solid is brought into the cell and **pinocytosis** if only liquid is brought into the cell. Nutrients brought into a cell by endocytosis are then enclosed in a *food vesicle.* Vesicles and digestion of the nutrients they contain are discussed in more detail shortly. (The process of phagocytosis is more fully discussed in Chapter 15 as it relates to the defense of the body against disease.)

Some eukaryotes also use pseudopods as a means of locomotion. The cell extends a pseudopod, and then the cytoplasm streams into it, a process called *amoeboid action.*

Exocytosis, another solely eukaryotic process, is the reverse of endocytosis in that it enables substances to be exported from the cell. Not all eukaryotic cells can perform endocytosis or exocytosis.

TABLE 3.4 lists some of the features of endocytosis and exocytosis.

TELL ME WHY

Many antimicrobial drugs target bacterial cell walls. Why aren't there many drugs that act against bacterial cytoplasmic membranes?

Cytoplasm of Eukaryotes

LEARNING | **OUTCOMES**

3.34 Compare and contrast the cytoplasm of prokaryotes and eukaryotes.

3.35 Identify nonmembranous and membranous organelles.

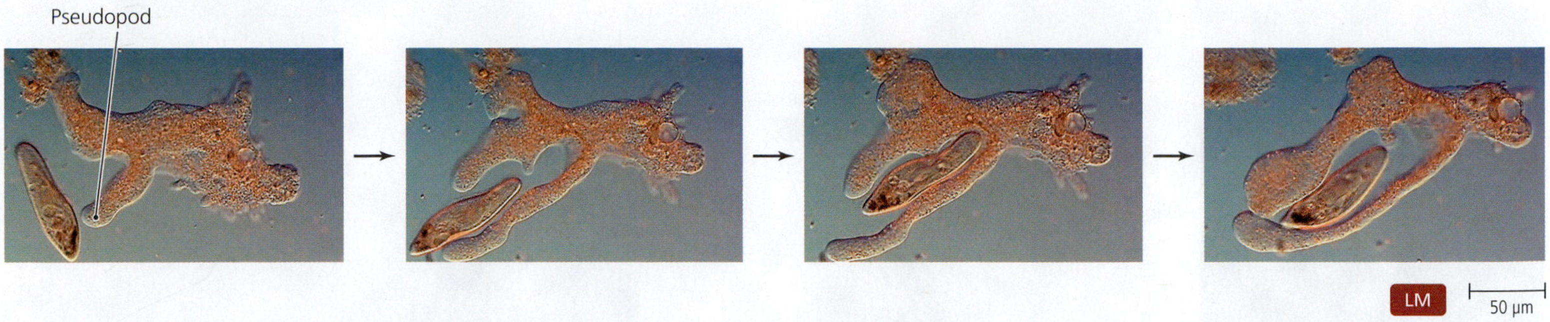

▲ **FIGURE 3.30 Endocytosis.** Pseudopods extend to surround solid and/or liquid nutrients, which become incorporated into a food vesicle inside the cytoplasm. *What is the difference between phagocytosis and pinocytosis?*

Figure 3.30 Phagocytosis is endocytosis of a solid; pinocytosis is endocytosis of a liquid.

TABLE 3.4 Active Transport Processes Found Only in Eukaryotes: Endocytosis and Exocytosis

	Description	Examples of Transported Substances
Endocytosis: phagocytosis and pinocytosis	Substances are surrounded by pseudopods and brought into the cell. Phagocytosis involves solid substances; pinocytosis involves liquids.	Bacteria, viruses, aged and dead cells; liquid nutrients in extracellular solutions
Exocytosis	Vesicles containing substances are fused with cytoplasmic membrane, dumping their contents to the outside.	Wastes, secretions

The cytoplasm of eukaryotic cells is more complex than that of either bacteria or archaea. The most distinctive difference is the presence of numerous membranous organelles in eukaryotes. However, before we discuss these membranous organelles, we will consider organelles of locomotion and other nonmembranous organelles in eukaryotes.

Flagella

LEARNING | OUTCOME

3.36 Compare and contrast the structure and function of prokaryotic and eukaryotic flagella.

Structure and Arrangement

Some eukaryotic cells have whiplike extensions called **flagella**. Flagella of eukaryotes **(FIGURE 3.31a)** differ structurally and functionally from flagella of prokaryotes. First, eukaryotic flagella are within the cytoplasmic membrane; they are internal structures that push the cytoplasmic membrane out around them. Their basal bodies are in the cytoplasm. Second, the shaft of a eukaryotic flagellum is composed of molecules of a globular protein called *tubulin* arranged in chains to form hollow *microtubules*. Nine pairs of microtubules surround two microtubules in the center **(FIGURE 3.31c)**. This "9 + 2" arrangement of microtubules is common to all flagellated eukaryotic cells, whether they are found in protozoa, algae, animals, or plants. The filaments of eukaryotic flagella are anchored in the cytoplasm by a basal body, but no hook connects the two parts, as in prokaryotes. The basal body has *triplets* of microtubules instead of pairs, and there are no microtubules in the center, so scientists say it has a "9 + 0" arrangement of microtubules. Eukaryotic flagella may be single or multiple and are generally found at one pole of the cell.

Function

The flagella of eukaryotes also move differently from those of prokaryotes. Rather than rotating like prokaryotic flagella, those of eukaryotes undulate rhythmically **(FIGURE 3.32a)**. Some eukaryotic flagella push the cell through the medium (as occurs in animal sperm), whereas others pull the cell through the medium (as occurs in many protozoa). Positive and negative phototaxis and chemotaxis are seen in eukaryotic cells, but such cells do not move in runs and tumbles.

Cilia

LEARNING | OUTCOMES

3.37 Describe the structure and function of cilia.
3.38 Compare and contrast eukaryotic cilia and flagella.

Other eukaryotic cells move by means of motile, internal, hairlike structures called **cilia**, which extend the surface of the cell and are shorter and more numerous than flagella **(FIGURE 3.31b)**. No prokaryotic cells have cilia. Like flagella, cilia are composed primarily of tubulin microtubules, which are arranged in a

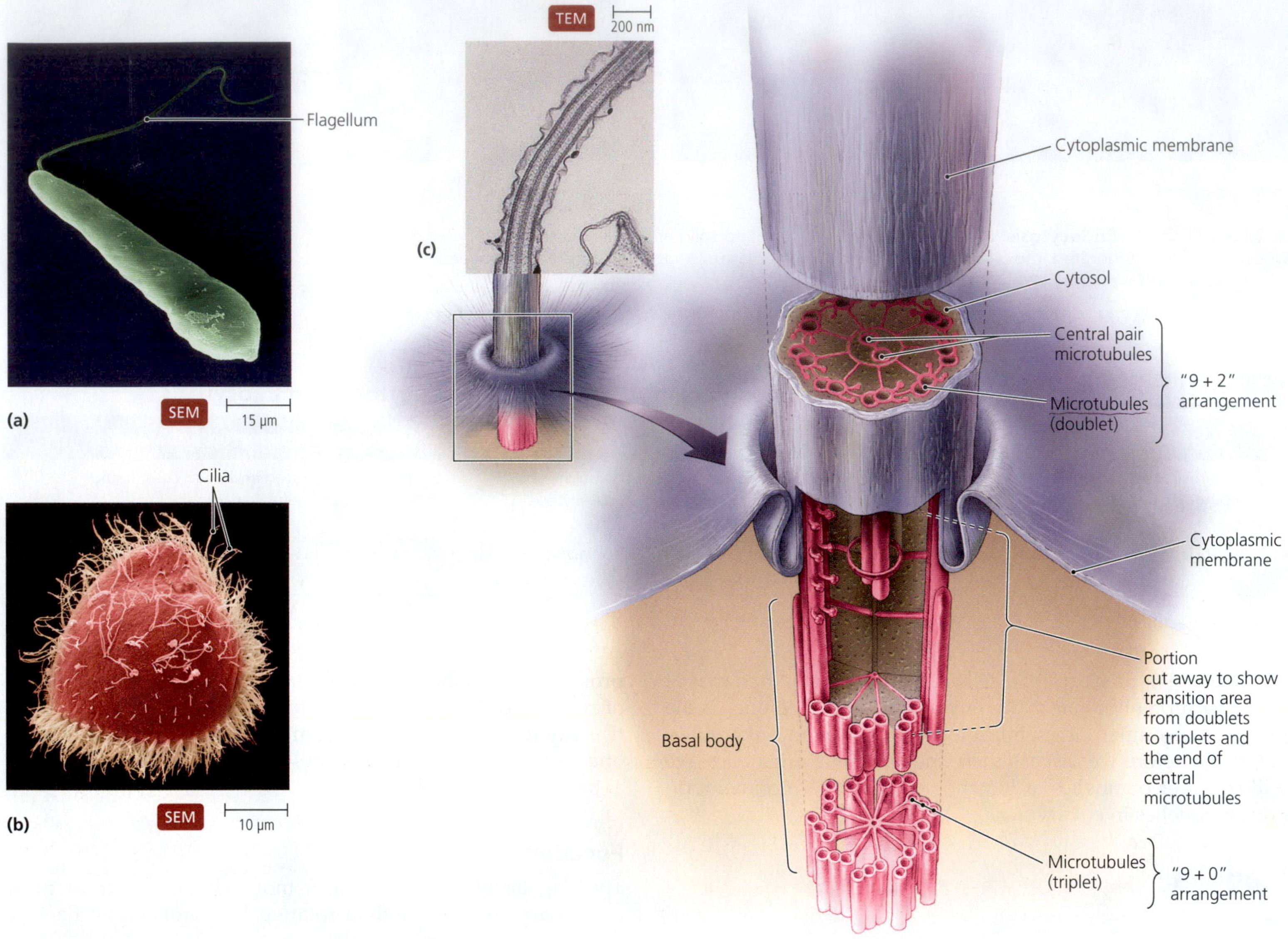

▲ **FIGURE 3.31 Eukaryotic flagella and cilia. (a)** Micrograph of *Euglena*, which possesses a single flagellum. **(b)** Scanning electron micrograph of a protozoan, *Blepharisma*, which has numerous cilia. **(c)** Details of the arrangement of microtubules of eukaryotic flagella and cilia. Both flagella and cilia have the same internal structure. *How do eukaryotic cilia differ from flagella?*

Figure 3.31 *Flagella are longer and less numerous than cilia.*

"9 + 2" arrangement of pairs in their shafts and a "9 + 0" arrangement of triplets in their basal bodies.

A single cell may have hundreds or even thousands of motile cilia. Such cilia beat rhythmically, much like a swimmer doing a butterfly stroke **(FIGURE 3.32b)**. Coordinated beating of cilia propels single-celled eukaryotes through their environment. Cilia are also used within some multicellular eukaryotes to move substances in the local environment past the surface of the cell. For example, such movement of cilia helps cleanse the human respiratory tract of dust and microorganisms.

Other Nonmembranous Organelles

LEARNING | **OUTCOMES**

3.39 Describe the structure and function of ribosomes, cytoskeletons, and centrioles.

3.40 Compare and contrast the ribosomes of prokaryotes and eukaryotes.

3.41 List and describe the three filaments of a eukaryotic cytoskeleton.

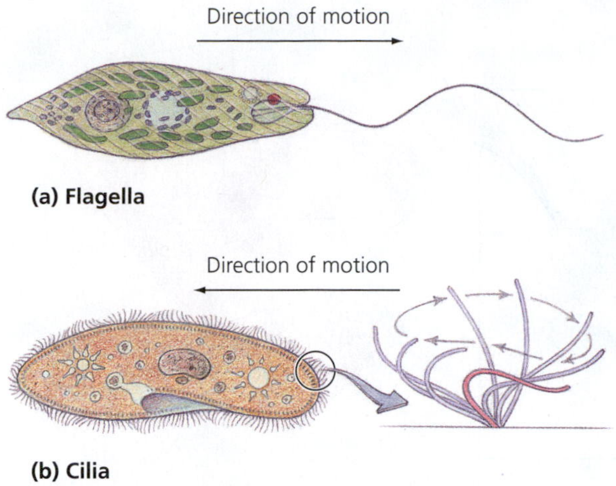

▲ **FIGURE 3.32 Movement of eukaryotic flagella and cilia.** **(a)** Eukaryotic flagella undulate in waves that begin at one end and traverse the length of the flagellum. **(b)** Cilia move with a power stroke followed by a return stroke. In the power stroke a cilium is stiff; it relaxes during the return stroke. *How is the movement of eukaryotic flagella different from that of prokaryotic flagella?*

Figure 3.32 *Eukaryotic flagella undulate in a wave that moves down the flagellum; the flagella of prokaryotes rotate about the basal body.*

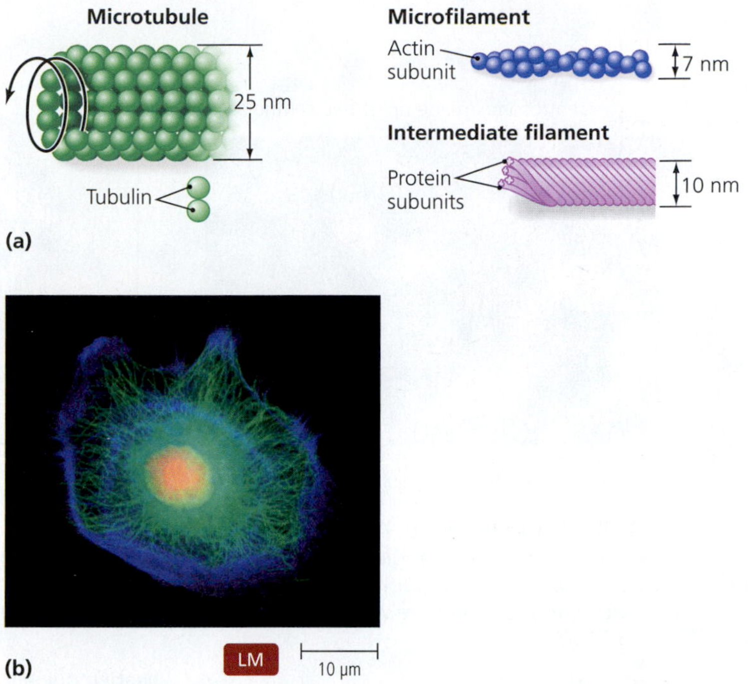

▲ **FIGURE 3.33 Eukaryotic cytoskeleton.** The cytoskeleton of eukaryotic cells serves to anchor organelles, provides a "track" for the movement of organelles throughout the cell, and provides shape to animal cells. Eukaryotic cytoskeletons are composed of microtubules, microfilaments, and intermediate filaments. **(a)** Artist's rendition of cytoskeleton filaments. **(b)** Various elements of the cytoskeleton shown here have been stained with different fluorescent dyes. DNA in the nucleus is stained yellow-orange.

Here we discuss three nonmembranous organelles found in eukaryotes: ribosomes and cytoskeleton (both of which are also present in prokaryotes), and centrioles (which are present only in certain kinds of eukaryotic cells).

Ribosomes

The cytosol of eukaryotes, like that of prokaryotes, is a semitransparent fluid composed primarily of water containing dissolved and suspended proteins, ions, carbohydrates, lipids, and wastes. Within the cytosol of eukaryotic cells are protein-synthesizing **ribosomes** that are larger than prokaryotic ribosomes; instead of 70S ribosomes, eukaryotic ribosomes are 80S and are composed of 60S and 40S subunits. In addition to the 80S ribosomes found within the cytosol, many eukaryotic ribosomes are attached to the membranes of the endoplasmic reticulum (discussed shortly).

Cytoskeleton

Eukaryotic cells contain an extensive **cytoskeleton** composed of an internal scaffolding of fibers and tubules. The eukaryotic cytoskeleton acts to anchor organelles and functions in cytoplasmic streaming and in movement of organelles within the cytosol. Cytoskeletons in some cells enable the cells to contract, move the cytoplasmic membrane during endocytosis and amoeboid action, and produce the basic shapes of the cells.

The eukaryotic cytoskeleton is made up of *tubulin microtubules* (also found in flagella and cilia), thinner *microfilaments* composed of *actin,* and *intermediate filaments* composed of various proteins **(FIGURE 3.33)**.

Centrioles and Centrosome

Animal cells and some fungal cells contain two **centrioles**, which lie at right angles to each other near the nucleus, in a region of the cytoplasm called the **centrosome** **(FIGURE 3.34)**. Plants, algae, and most fungi (and prokaryotes) lack centrioles but usually have a region of cytoplasm corresponding to a centrosome. Centrioles are composed of nine *triplets* of tubulin microtubules arranged in a way that resembles the "9 + 0" arrangement seen at the base of eukaryotic flagella and cilia.

Centrosomes play a role in *mitosis* (nuclear division), *cytokinesis* (cell division), and the formation of flagella and cilia. However, because many eukaryotic cells that lack centrioles, such as brown algal sperm and numerous one-celled algae, are still able to form flagella and undergo mitosis and cytokinesis, the function of centrioles is the subject of ongoing research.

TABLE 3.5 on p. 82 summarizes characteristics of nonmembranous organelles of cells and contrasts them with characteristics of membranous organelles, which we consider next.

Membranous Organelles

LEARNING | **OUTCOMES**

3.42 Discuss the function of each of the following membranous organelles: nucleus, endoplasmic reticulum, Golgi body, lysosome, peroxisome, vesicle, vacuole, mitochondrion, and chloroplast.

3.43 Label the structures associated with each of the membranous organelles.

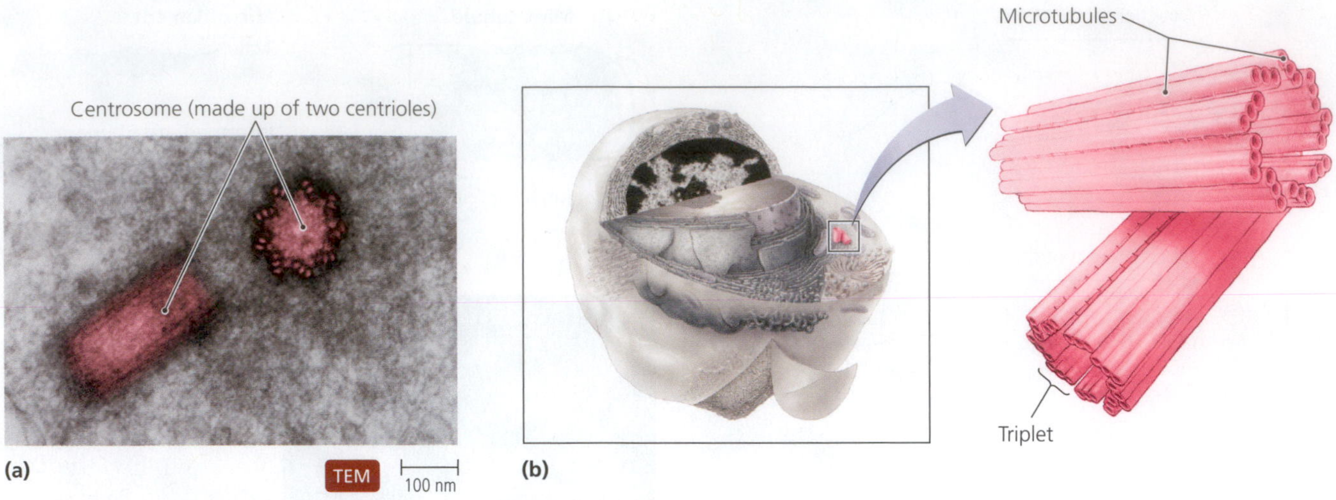

▲ **FIGURE 3.34 Centrosome.** A centrosome is a region of cytoplasm that in animal cells contains two centrioles at right angles to one another; each centriole has nine triplets of microtubules. **(a)** Transmission electron micrograph of centrosome and centrioles. **(b)** Artist's rendition of a centrosome. *How do centrioles compare with the basal body and shafts of eukaryotic flagella and cilia (see Figure 3.31c)?*

Figure 3.34 Centrioles have the same "9 + 0" arrangement of microtubules that is found in the basal bodies of eukaryotic cilia and flagella.

TABLE 3.5 Nonmembranous and Membranous Organelles of Cells

	General Function	Prokaryotes	Eukaryotes
Nonmembranous Organelles			
Ribosomes	Protein synthesis	Present in all	Present in all
Cytoskeleton	Shape in prokaryotes; support, cytoplasmic streaming, and endocytosis in eukaryotes	Present in some	Present in all
Centrosome	Appears to play a role in mitosis, cytokinesis, and flagella and cilia formation in animal cells	Absent in all	Present in animals
Membranous Organelles	Sequester chemical reactions within the cell		
Nucleus	"Control center" of the cell	Absent in all	Present in all
Endoplasmic reticulum	Transport within the cell, lipid synthesis	Absent in all	Present in all
Golgi bodies	Exocytosis, secretion	Absent in all	Present in some
Lysosomes	Breakdown of nutrients, self-destruction of damaged or aged cells	Absent in all	Present in some
Peroxisomes	Neutralization of toxins	Absent in all	Present in some
Vacuoles	Storage	Absent in all	Present in some
Vesicles	Storage, digestion, transport	Absent in all	Present in all
Mitochondria	Aerobic ATP production	Absent in all	Present in most
Chloroplasts	Photosynthesis	Absent in all, though infoldings of cytoplasmic membrane called photosynthetic lamellae have same function in photosynthetic prokaryotes	Present in plants and algae

Eukaryotic cells contain a variety of organelles that are surrounded by phospholipid bilayer membranes similar to the cytoplasmic membrane. These membranous organelles include the nucleus, endoplasmic reticulum, Golgi body, lysosomes, peroxisomes, vacuoles, vesicles, mitochondria, and chloroplasts. Prokaryotic cells lack these structures.

Nucleus

The **nucleus** is usually spherical to ovoid and is often the largest organelle in a cell[14] **(FIGURE 3.35)**. Some cells have a single nucleus; others are multinucleate, while still others lose their nuclei. The nucleus is often referred to as "the control center of the cell" because it contains most of the cell's genetic instructions in the form of DNA. Cells that lose their nuclei, such as mammalian red blood cells, can survive for only a few months.

Just as the semiliquid portion of the cell is called cytoplasm, the semiliquid matrix of the nucleus is called **nucleoplasm**. Within the nucleoplasm may be one or more **nucleoli** (noo-klē´ō-lī; singular: *nucleolus*), which are specialized regions where RNA is synthesized. The nucleoplasm also contains **chromatin**, which is a threadlike mass of DNA associated with special proteins called *histones* that play a role in packaging nuclear DNA. During mitosis (nuclear division), chromatin becomes visible as *chromosomes*. (Chapter 12 discusses mitosis in more detail.)

Surrounding the nucleus is a double membrane called the **nuclear envelope**, which is composed of two phospholipid bilayers, for a total of four phospholipid layers. The nuclear envelope contains **nuclear pores** that function to control the import and export of substances through the envelope.

Endoplasmic Reticulum

Continuous with the outer membrane of the nuclear envelope is a netlike arrangement of flattened hollow tubules called **endoplasmic reticulum (ER) (FIGURE 3.36)**. The ER traverses the cytoplasm of eukaryotic cells. Endoplasmic reticulum functions as a transport system and is found in two forms: **smooth endoplasmic reticulum (SER)** and **rough endoplasmic reticulum (RER)**. SER plays a role in lipid synthesis as well as transport. Rough endoplasmic reticulum is rough because ribosomes adhere to its outer surface. Proteins produced by ribosomes on the RER are inserted into the lumen (central canal) of the RER and transported throughout the cell.

Golgi Body

A **Golgi body**[15] is like the "shipping department" of a cell: it receives, processes, and packages large molecules for export

[14]Historically, the nucleus was not considered an organelle because it is large and not considered part of the cytoplasm.

[15]Camillo Golgi was an Italian histologist who first described the organelle in 1898. This organelle is also known as a *Golgi complex* or *Golgi apparatus* and in plants and algae as a *dictyosome*.

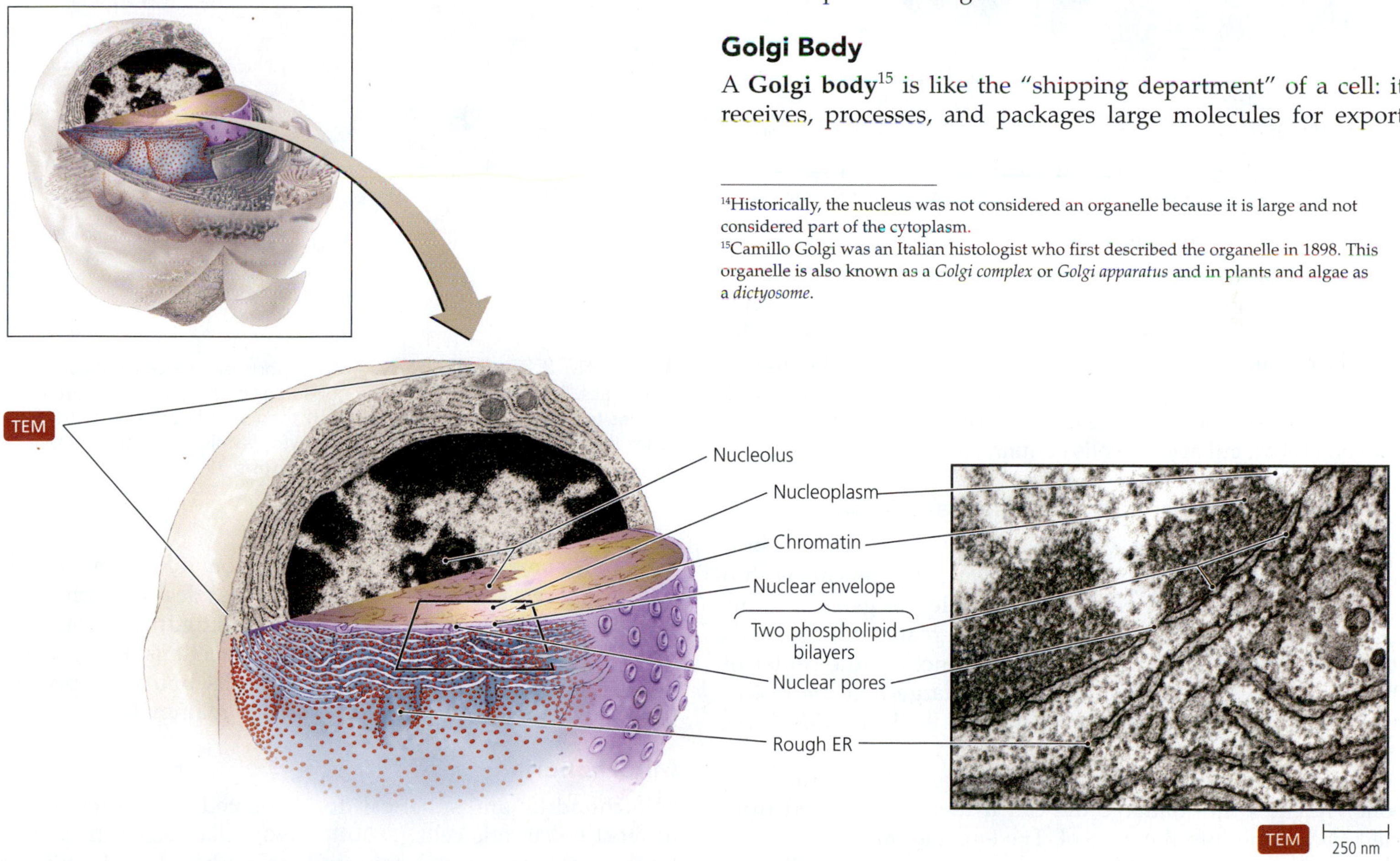

▲ **FIGURE 3.35 Eukaryotic nucleus.** Micrograph and artist's conception of a nucleus showing chromatin, a nucleolus, and the nuclear envelope. Nuclear pores punctuate the two membranes of the nuclear envelope.

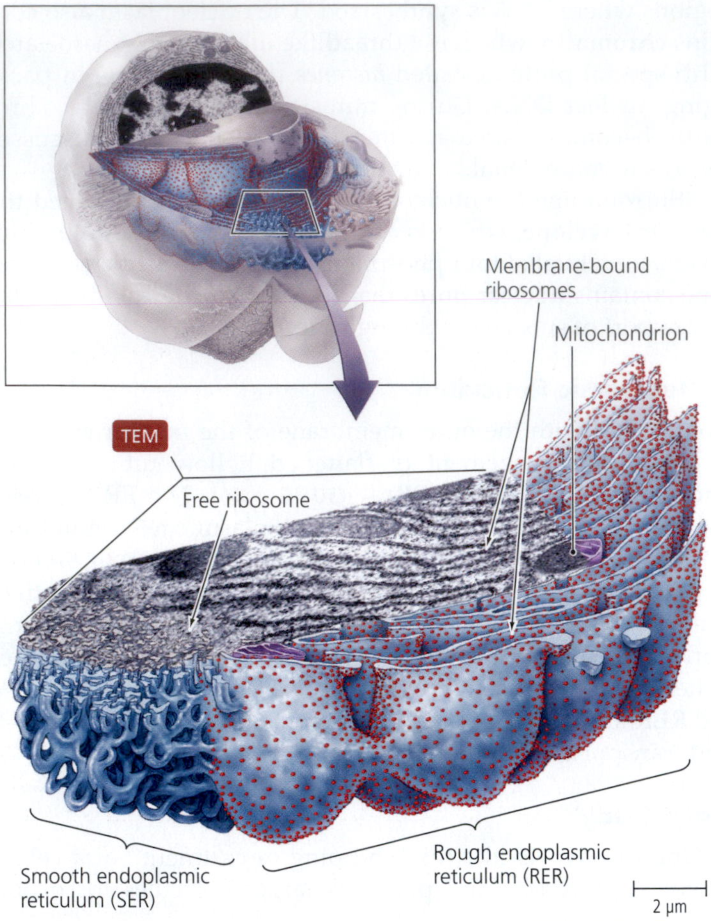

▲ **FIGURE 3.36 Endoplasmic reticulum.** ER functions in transport throughout the cell. Ribosomes are on the surface of rough ER; smooth ER lacks ribosomes.

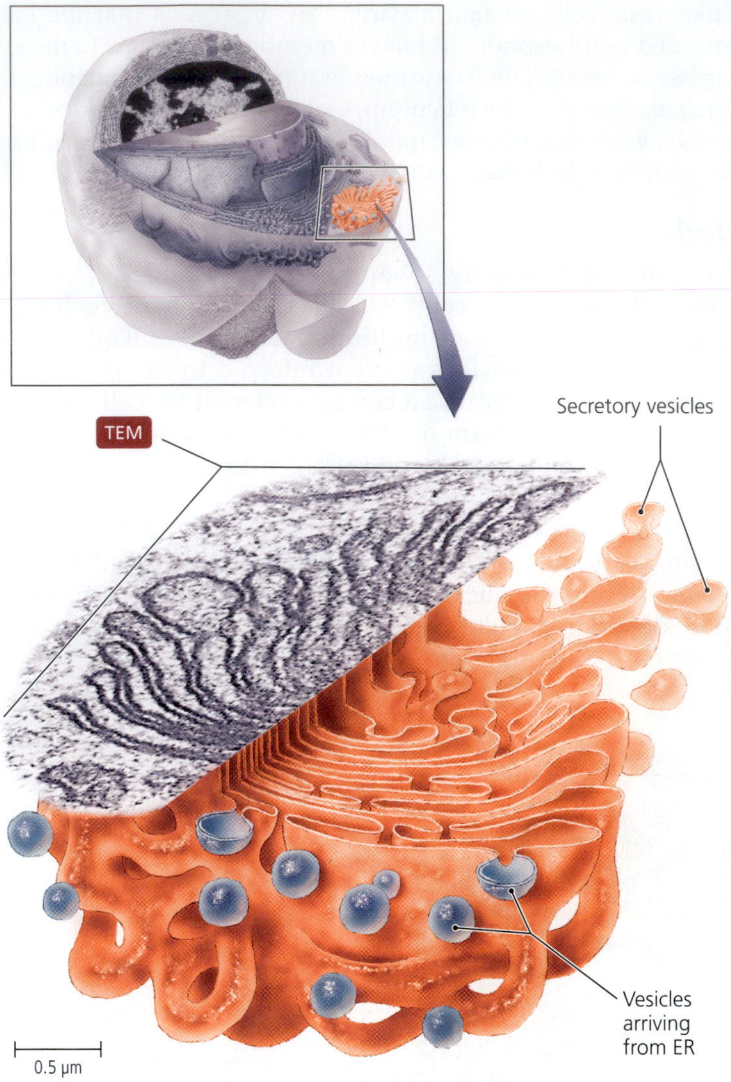

▲ **FIGURE 3.37 Golgi body.** A Golgi body is composed of flattened sacs. Proteins synthesized by ribosomes on RER are transported via vesicles to a Golgi body. The Golgi body then modifies the proteins and sends them via secretory vesicles to the cytoplasmic membrane, where they can be secreted from the cell by exocytosis.

from the cell **(FIGURE 3.37)**. The Golgi body packages secretions in sacs called **secretory vesicles**, which then fuse with the cytoplasmic membrane before dumping their contents outside the cell via exocytosis. Golgi bodies are composed of a series of flattened hollow sacs that are circumscribed by a phospholipid bilayer. Not all eukaryotic cells contain Golgi bodies.

Lysosomes, Peroxisomes, Vacuoles, and Vesicles

Lysosomes, peroxisomes, vacuoles, and vesicles are membranous sacs that function to store and transfer chemicals within eukaryotic cells. Both **vesicle** and **vacuole** are general terms for such sacs. Large vacuoles are found in plant and algal cells that store starch, lipids, and other substances in the center of the cell. Often this central vacuole is so large that the rest of the cytoplasm is pressed against the cell wall in a thin layer **(FIGURE 3.38)**.

Lysosomes, which are found in animal cells, contain catabolic enzymes that damage the cell if they are released from their packaging into the cytosol. The enzymes are used during the self-destruction of old, damaged, and diseased cells and to digest nutrients that have been phagocytized. For example, white blood cells utilize the digestive enzymes in lysosomes to destroy phagocytized pathogens **(FIGURE 3.39)**.

Peroxisomes are vesicles derived from ER. They contain *oxidase* and *catalase*, which are enzymes that degrade poisonous metabolic wastes (such as free radicals and hydrogen peroxide) resulting from some oxygen-dependent reactions. Peroxisomes are found in all types of eukaryotic cells but are especially prominent in the kidney and liver cells of mammals.

Mitochondria

Mitochondria are spherical to elongated structures found in most eukaryotic cells **(FIGURE 3.40)**. Like nuclei, they have two membranes, each composed of a phospholipid bilayer. The inner membrane forms numerous folds called *cristae* that increase the inner membrane's surface area. Mitochondria are often called the "powerhouses of the cell" because their cristae

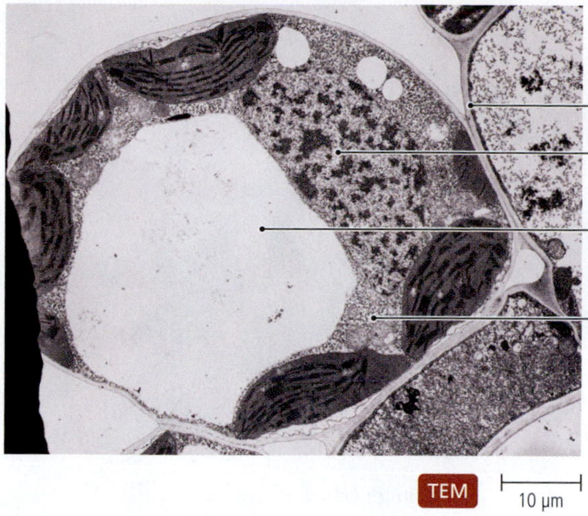

◀ **FIGURE 3.38 Vacuole.** The large central vacuole of a plant cell, which constitutes a storehouse for the cell, presses the cytoplasm against the cell wall.

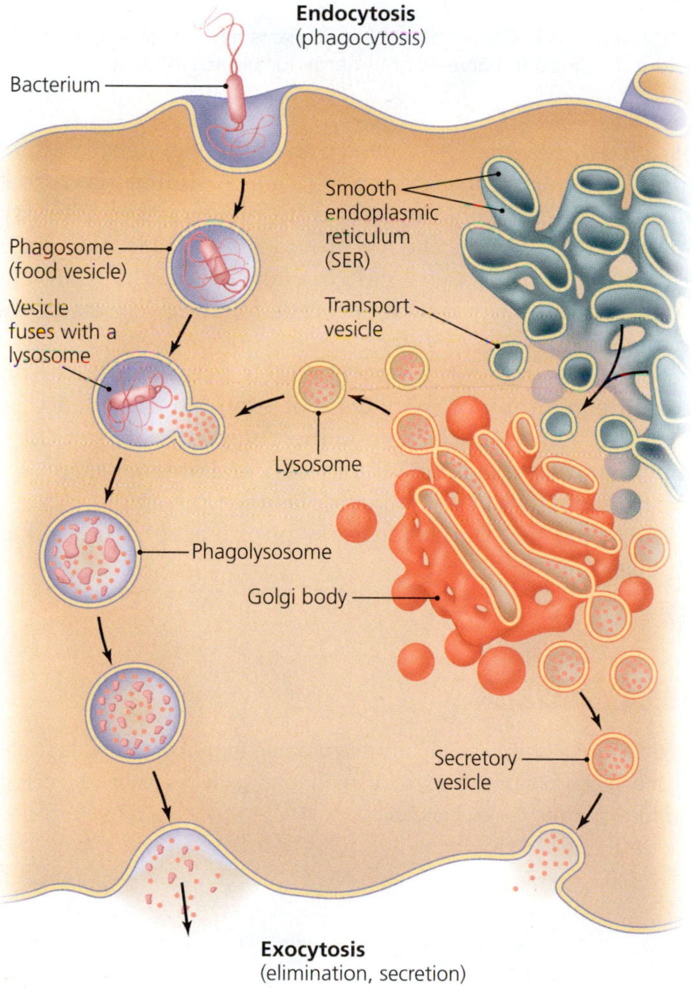

▲ **FIGURE 3.39 The roles of vesicles in endocytosis and exocytosis.** Before endocytosis, vesicles from the endoplasmic reticulum deliver digestive enzymes to a Golgi body, which then packages them into lysosomes. During endocytosis, phagocytized particles (in this case, a bacterium) are enclosed within a vesicle called a phagosome (food vesicle), which then fuses with a lysosome to form a phagolysosome vesicle. Once digestion within the phagolysosome is complete, the resulting wastes can be expelled from the cell via exocytosis. A Golgi body can also form secretory vesicles that deliver secretions outside the cell.

produce most of the ATP in many eukaryotic cells. (The chemical reactions that produce ATP are discussed in Chapter 5.)

The interior matrix of a mitochondrion contains ("prokaryotic") 70S ribosomes and a circular molecule of DNA. This DNA contains genes for some RNA molecules and for a few mitochondrial polypeptides that are manufactured by mitochondrial ribosomes; however, most mitochondrial proteins are coded by nuclear DNA and synthesized by cytoplasmic ribosomes.

Chloroplasts

Chloroplasts are light-harvesting structures found in photosynthetic eukaryotes **(FIGURE 3.41)**. Like mitochondria and the nucleus, chloroplasts have two phospholipid bilayer membranes and DNA. Further, like mitochondria, chloroplasts can synthesize a few polypeptides with their own 70S ribosomes. The pigments of chloroplasts gather light energy to produce ATP and form sugar from carbon dioxide. Numerous membranous sacs called *thylakoids* form an extensive surface area for the biochemical and photochemical reactions of chloroplasts. The fluid between the thylakoids and the inner membrane is called the *stroma.* The space enclosed by the thylakoids is called the *thylakoid space.*

Photosynthetic prokaryotes lack chloroplasts and instead have infoldings of their cytoplasmic membranes called *photosynthetic lamellae.* (Chapter 5 discusses the details of photo synthesis.)

The functions of the nonmembranous and membranous organelles, and their distribution among prokaryotic and eukaryotic cells, are summarized in Table 3.5 on p. 82.

Endosymbiotic Theory

LEARNING | **OUTCOMES**

3.44 Describe the endosymbiotic theory of the origin of mitochondria, chloroplasts, and eukaryotic cells.

3.45 List evidence for the endosymbiotic theory.

Mitochondria and chloroplasts are semiautonomous; that is, they divide independently of the cell but remain dependent on the cell for most of their proteins. As we have seen, both mitochondria and chloroplasts contain a small amount of DNA and 70S ribosomes, and each can produce a few polypeptides with its own ribosomes. The presence of circular DNA, 70S ribosomes, and two bilipid membranes in these semiautonomous organelles led scientists to the **endosymbiotic**[16] **theory** for the formation of eukaryotic cells. This theory suggests that eukaryotes formed from the union of small aerobic[17] prokaryotes with

[16]From Greek *endo,* meaning "inside," and *symbiosis,* meaning "to live with."
[17]*Aerobic* means "requiring oxygen"; *anaerobic* is the opposite.

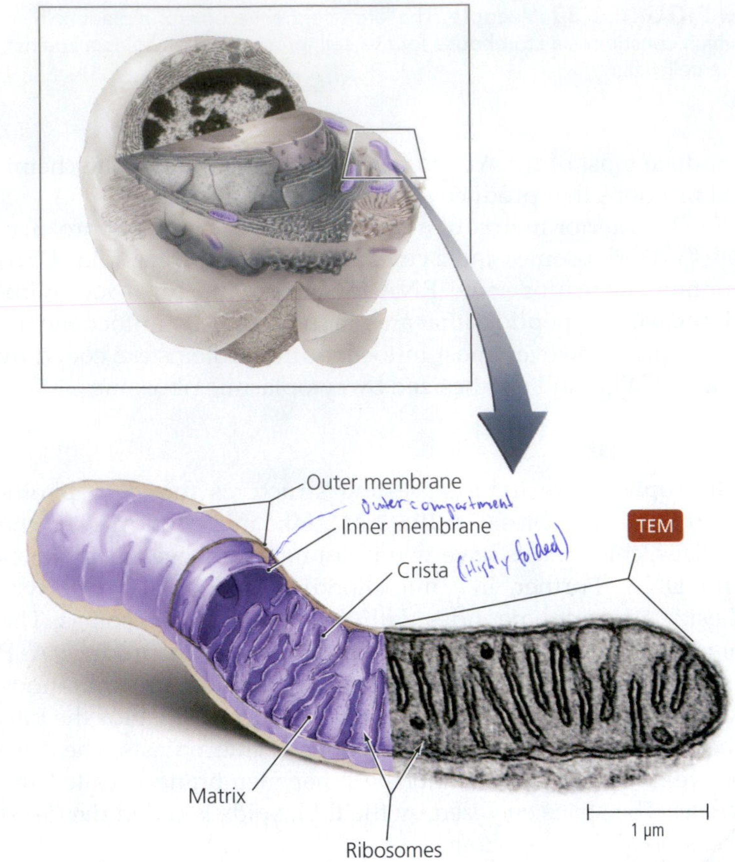

▲ **FIGURE 3.40 Mitochondrion.** Note the double membrane. The inner membrane is folded into cristae that increase its surface area. *What is the importance of the increased surface area of the inner membrane that cristae make possible?*

Figure 3.40 *The chemicals involved in aerobic ATP production are located on the inner membranes of mitochondria. Increased surface area provides more space for more chemicals.*

larger anaerobic prokaryotes. The smaller prokaryotes were not destroyed by the larger cells but instead became internal parasites that remained surrounded by a vesicular membrane of the host.

According to the theory, the parasites eventually lost the ability to exist independently, but they retained a portion of their DNA, some ribosomes, and their cytoplasmic membranes. During the same time, the larger cell became dependent on the parasites for metabolism. According to the theory, the aerobic prokaryotes eventually evolved into mitochondria, and their cytoplasmic membranes became cristae. A similar scenario explains the origin of chloroplasts from phagocytized photosynthetic prokaryotes. The theory provides an explanation for the presence of 70S ribosomes and circular DNA within mitochondria and chloroplasts, and it accounts for the presence of their two membranes.

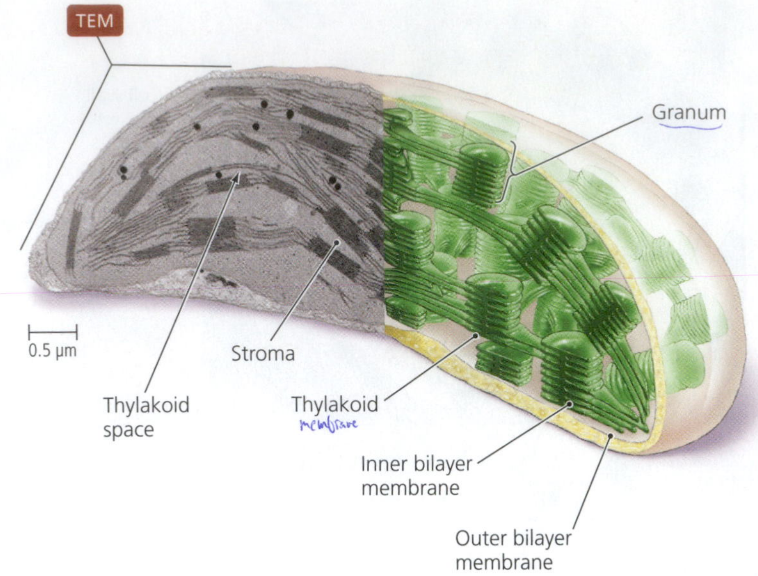

▲ **FIGURE 3.41 Chloroplast.** Chloroplasts have an ornate internal structure designed to harvest light energy for photosynthesis.

The endosymbiotic theory is widely accepted; however, it does not explain all of the facts. For example, the theory provides no explanation for the two membranes of the nuclear envelope, nor does it explain why most of the organelles' proteins come from nuclear DNA and cytoplasmic ribosomes.

TABLE 3.6 summarizes features of prokaryotic and eukaryotic cells.

TELL ME WHY

Colchicine is a drug that inhibits microtubule formation. Why does colchicine inhibit phagocytosis, movement of organelles within the cell, and formation of flagella and cilia?

TABLE 3.6 Comparison of Archaeal, Bacterial, and Eukaryotic Cells

Characteristic	Archaea	Bacteria	Eukaryotes
Nucleus	Absent in all	Absent	Present in all
Free organelles bound with phospholipid membranes	Absent in all	Present in few	Present; include ER, Golgi bodies, lysosomes, mitochondria, and chloroplasts
Glycocalyx	Present	Present as organized capsule or unorganized slime layer	Present, surrounding some animal cells
Motility	Present in some	Present in some	Some have complex undulating flagella and cilia composed of a "9 + 2" arrangement of microtubules; others move with amoeboid action using pseudopods
Flagella	Some have flagella, each composed of basal body, hook, and filament; flagella rotate	Some have flagella, each composed of basal body, hook, and filament; flagella rotate	Present in some
Cilia	Absent in all	Absent in all	Present in some
Fimbriae or pili	Present in some	Present in some	Absent in all
Hami	Present in some	Absent in all	Absent in all
Cell wall	Present in most, lacking peptidoglycan	Present in most; composed of peptidoglycan	Present in plants, algae, and fungi
Cytoplasmic membrane	Present in all	Present in all	Present in all
Cytosol	Present in all	Present in all	Present in all
Inclusions	Present in most	Present in most	Present in some
Endospores	Absent in all	Present in some	Absent in all
Ribosomes	Small (70S)	Small (70S)	Large (80S) in cytosol and on ER, smaller (70S) in mitochondria and chloroplasts
Chromosomes	Commonly single and circular	Commonly single and circular	Linear and more than one chromosome per cell

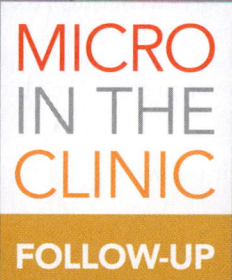

MICRO IN THE CLINIC FOLLOW-UP

The Big Game

The nurse practitioner suspects that Lamara has strep throat. However, because many viral and bacterial throat infections can look the same, testing is needed to determine whether Lamara's illness is strep throat. After the initial exam, the practitioner takes a swab of Lamara's throat for a rapid streptococcal antigen test, which screens for group A, **beta-hemolytic** *Streptococcus pyogenes*. This test responds to the presence of a particular strain of *Streptococcus* and is positive when a patient has strep throat. A throat culture can confirm an infection and help determine what antimicrobial drug is most effective.

The results of the rapid test are ready in few minutes, and Lamara's test comes back positive. Her sore throat and rash are both caused by the streptococcal infection. The nurse practitioner prescribes an antimicrobial and sends Lamara back to her dorm. Lamara will miss the big game, but she will be healthy enough to play again soon.

1. **Strep throat is caused by a Gram-positive bacteria called *Streptococcus pyogenes* that has a capsule. Describe how the capsule may contribute to the pathogens ability to cause disease.**
2. **What are two advantages of rapid identification tools, such as the rapid strep test used in Lamara's diagnosis, over a culture for identifying pathogens and prescribing treatments?**

MM Check your answers to Micro in the Clinic Follow Up questions in the MasteringMicrobiology Study Area.

CHAPTER SUMMARY

Processes of Life (pp. 56–57)

1. All living things have some common features, including **growth**, an increase in size; **reproduction**, an increase in number; **responsiveness**, reactions to environmental stimuli; **metabolism**, controlled chemical reactions in an organism; and cellular structure.
2. Viruses are acellular and do not grow, self-reproduce, or metabolize.

Prokaryotic and Eukaryotic Cells: An Overview (pp. 57–59)

1. All cells can be described as either **prokaryotic** or **eukaryotic**. These descriptive terms help scientists categorize organisms in groups called taxa. Generally, prokaryotic cells make proteins simultaneously to reading the genetic code, and they lack a nucleus and organelles surrounded by phospholipid membranes. Domain Bacteria and domain Archaea are prokaryotic taxa.
2. Eukaryotes (domain Eukarya) have internal, membrane-bound **organelles**, including nuclei. Animals, plants, algae, fungi, and protozoa are eukaryotic.
3. Cells share common structural features. These include external structures, cell walls, cytoplasmic membranes, and cytoplasm.

External Structures of Bacterial Cells (pp. 59–63)

1. The external structures of bacterial cells include glycocalyces, flagella, fimbriae, and pili.
2. **Glycocalyces** are sticky external sheaths of cells. They may be loosely attached **slime layers** or firmly attached **capsules**. Glycocalyces prevent cells from drying out. Capsules protect cells from phagocytosis by other cells, and slime layers enable cells to stick to each other and to surfaces in their environment.
3. A prokaryotic **flagellum** is a long, whiplike protrusion of some cells composed of a basal body, hook, and filament. Flagella allow cells to move toward favorable conditions such as nutrients or light, or move away from unfavorable stimuli such as poisons.
   ▶**ANIMATIONS:** *Motility: Overview; Flagella: Structure*
4. Bacterial flagella may be **polar** (single or tufts) or cover the cell **(peritrichous)**. **Endoflagella**, which are special flagella of a spirochete, form an **axial filament**, located in the periplasmic space.
   ▶**ANIMATIONS:** *Flagella: Arrangement; Spirochetes*
5. **Taxis** is movement that may be either a positive response or a negative response to light **(phototaxis)** or chemicals **(chemotaxis)**.
   ▶**ANIMATIONS:** *Flagella: Movement*
6. **Fimbriae** are extensions of some bacterial cells that function along with glycocalyces to adhere cells to one another and to environmental surfaces. A mass of such bacteria on a surface is termed a **biofilm**. Cells may also use fimbriae to pull themselves across a surface or to conduct signals to neighboring cells.
7. **Pili**, also known as conjugation pili, are hollow, nonmotile tubes of protein that allow bacteria to pull themselves forward and mediate the movement of DNA from one cell to another. Not all bacteria have fimbriae or pili.

Bacterial Cell Walls (pp. 63–66)

1. Most prokaryotic cells have **cell walls** that provide shape and support against osmotic pressure. Cell walls are composed primarily of polysaccharide chains.
2. Cell walls of bacteria are composed of a large, interconnected molecule of **peptidoglycan**. Peptidoglycan is composed of alternating sugar molecules called *N*-acetylglucosamine (NAG) and *N*-acetylmuramic acid (NAM).
3. A **Gram-positive** bacterial cell has a thick layer of peptidoglycan.
4. A **Gram-negative** bacterial cell has a thin layer of peptidoglycan and an external wall membrane with a **periplasmic space** between. This wall membrane contains **lipopolysaccharide (LPS)**, which contains **lipid A**, which is also known as **endotoxin**. During an infection with Gram-negative bacteria, endotoxins can accumulate in the blood, causing shock, fever, and blood clotting.
   ▶**VIDEO TUTOR:** *Bacterial Cell Walls*
5. Acid-fast bacteria have waxy lipids in their cell walls.

Bacterial Cytoplasmic Membranes (pp. 66–71)

1. A **cytoplasmic membrane** is typically composed of phospholipid molecules arranged in a double-layer configuration called a **phospholipid bilayer**. Proteins associated with the membrane vary in location and function and are able to flow laterally within the membrane. The **fluid mosaic model** is descriptive of the current understanding of membrane structure.
   ▶**ANIMATIONS:** *Membrane Structure*
2. The **selectively permeable** cytoplasmic membrane prevents the passage of some substances while allowing other substances to pass through protein pores or channels, sometimes requiring carrier molecules.
   ▶**ANIMATIONS:** *Membrane Permeability*

3. The relative concentrations of a chemical inside and outside the cell create a **concentration gradient**. Differences of electrical charges on the two sides of a membrane create an **electrical gradient** across the membrane. The gradients have a predictable effect on the passage of substances through the membrane.

   ▶**ANIMATIONS:** *Passive Transport: Principles of Diffusion*

4. Passive processes that move chemicals across the cytoplasmic membrane require no energy expenditure by the cell. Molecular size and concentration gradients determine the rate of simple **diffusion. Facilitated diffusion** depends on the electrochemical gradient and carriers within the membrane that allow certain substances to pass through the membrane. **Osmosis** specifically refers to the diffusion of water molecules across a selectively permeable membrane.

5. The concentrations of solutions can be compared. **Hypertonic** solutions have a higher concentration of solutes than **hypotonic** solutions, which have a lower concentration of solutes. Two **isotonic** solutions have the same concentrations of solutes. In biology, comparisons are usually made with the cytoplasm of cells.

   ▶**ANIMATIONS:** *Passive Transport: Special Types of Diffusion*

6. Active transport processes require cell energy from ATP. **Active transport** moves a substance against its electrochemical gradient via carrier proteins. These carriers may move two substances in the same direction at once (symports) or move substances in opposite directions (antiports). **Group translocation** occurs in prokaryotes, during which the substance being transported is chemically altered in transit.

   ▶**ANIMATIONS:** *Active Transport: Overview, Types*

Cytoplasm of Bacteria (pp. 71–74)

1. **Cytoplasm** is composed of the liquid **cytosol** inside a cell plus nonmembranous organelles and inclusions. **Inclusions** in the cytosol are deposits of various substances.
2. Both prokaryotic and eukaryotic cells contain nonmembranous organelles.
3. The **nucleoid** is the nuclear region in prokaryotic cytosol. It has no membrane and usually contains a single circular molecule of DNA.
4. Inclusions include reserve deposits of lipids, starch, or compounds containing nitrogen, phosphate, or sulfur. Inclusions called gas vesicles store gases.
5. Some bacteria produce dormant, resistant **endospores** within vegetative cells.
6. **Ribosomes**, composed of protein and **ribosomal RNA (rRNA)**, are nonmembranous organelles, found in both prokaryotes and eukaryotes, that function to make proteins. The 70S ribosomes of prokaryotes are smaller than the 80S ribosomes of eukaryotes.
7. The **cytoskeleton** is a network of fibers that appears to help maintain the basic shape of prokaryotes.

External Structures of Archaea (pp. 74–75)

1. Archaea form polysaccharide and polypeptide glycocalyces that function in attachment and biofilm formation but are evidently not associated with diseases.
2. Archaeal flagella differ from bacterial flagella. For example, archaeal flagella are thinner than bacterial flagella.
3. Archaeal flagella rotate together as a bundle in both directions and are powered by molecules of ATP.
4. Archaea may have fimbriae and grappling-hook-like **hami** that serve to anchor the cells to environmental surfaces.

Archaeal Cell Walls and Cytoplasmic Membranes (p. 76)

1. Archaeal cell walls are composed of protein or polysaccharides but not peptidoglycan.
2. Phospholipids in archaeal cytoplasmic membranes are built with ether linkages rather than ester linkages, which occur in bacterial membranes.

Cytoplasm of Archaea (p. 76)

1. Gel-like archaeal cytoplasm is similar to the cytoplasm of bacteria, having DNA, ribosomes, and a fibrous cytoskeleton, all suspended in the liquid cytosol.
2. 70S ribosomes of archaea have proteins more similar to those of eukaryotic ribosomes than to bacterial ribosomes.

External Structure of Eukaryotic Cells (p. 77)

1. Eukaryotic animal and some protozoan cells lack cell walls but have **glycocalyces** that prevent desiccation, provide support, and enable cells to stick together.
2. Wall-less eukaryotic cells have glycocalyces, which are not found with eukaryotic cells that have walls.

Eukaryotic Cell Walls and Cytoplasmic Membranes (pp. 77–78)

1. Fungal, plant, algal, and some protozoan cells have cell walls composed of polysaccharides or other chemicals. Cell walls provide support, shape, and protection from osmotic forces.
2. Fungal cell walls are composed of chitin or other polysaccharides. Plant cell walls are composed of cellulose. Algal cell walls contain agar, carrageenan, algin, cellulose, or other chemicals.
3. Eukaryotic cytoplasmic membranes contain sterols such as cholesterol, which act to strengthen and solidify the membranes when temperatures rise and provide fluidity when temperatures fall.
4. **Membrane rafts** are distinct assemblages of certain lipids and proteins that remain together in the cytoplasmic membrane. Some viruses use membrane rafts during their infections of cells.
5. Some eukaryotic cells transport substances into the cytoplasm via **endocytosis**, which is an active process requiring the expenditure of energy from ATP. In endocytosis, **pseudopods**—movable extensions of the cytoplasm and membrane of the cell—surround a substance and move it into the cell. When solids are brought into the cell, endocytosis is called **phagocytosis**; the incorporation of liquids by endocytosis is called **pinocytosis**.
6. **Exocytosis** is the active export of substances out of a cell.

Cytoplasm of Eukaryotes (pp. 78–87)

1. Eukaryotic cytoplasm is characterized by membranous organelles, particularly a nucleus. It also contains nonmembranous organelles and cytosol.

2. Some eukaryotic cells have long, whiplike **flagella** that differ from the flagella of prokaryotes. They have no hook, and the basal bodies and shafts are arrangements of microtubules. Further, eukaryotic flagella are internal to the cytoplasmic membrane.
3. Some eukaryotic cells have **cilia**, which have the same structure as eukaryotic flagella but are much shorter and more numerous. Cilia are internal to the cytoplasmic membrane.
4. The 80S **ribosomes** of eukaryotic cells are composed of 60S and 40S subunits. They are found free in the cytosol and attached to endoplasmic reticulum. The ribosomes within mitochondria and chloroplasts are 70S.
5. The eukaryotic **cytoskeleton** is composed of microtubules, intermediate filaments, and microfilaments. It provides an infrastructure and aids in movement of cytoplasm and organelles.
6. **Centrioles**, which are nonmembranous organelles in animal and some fungal cells only, are found in a region of the cytoplasm called the **centrosome** and are composed of triplets of microtubules in a "9 + 0" arrangement. Centrosomes function in the formation of flagella and cilia and in cell division.
7. The **nucleus**, a membranous structure in eukaryotic cells, contains **nucleoplasm**, in which are found one or more **nucleoli** and **chromatin**. Chromatin consists of the multiple strands of DNA and associated histone proteins that become obvious as chromosomes during mitosis. **Nuclear pores** penetrate the four phospholipid layers of the **nuclear envelope** (membrane).
8. The **endoplasmic reticulum (ER)** functions as a transport system. It can be **rough ER (RER)**, which has ribosomes on its surface, or **smooth ER (SER)**, which lacks ribosomes.
9. A **Golgi body** is a series of flattened hollow sacs surrounded by phospholipid bilayers. It packages large molecules destined for export from the cell in **secretory vesicles**, which release these molecules from the cell via exocytosis.
10. **Vesicles** and **vacuoles** are general terms for membranous sacs that store or carry substances. More specifically, **lysosomes** of animal cells contain digestive enzymes, and **peroxisomes** contain enzymes that neutralize poisonous free radicals and hydrogen peroxide.
11. Four phospholipid layers surround **mitochondria**, site of production of ATP in a eukaryotic cell. The inner bilayer is folded into cristae, which greatly increase the surface area available for chemicals that generate ATP.
12. Photosynthetic eukaryotes possess **chloroplasts**, which are organelles containing membranous thylakoids that provide increased surface area for photosynthetic reactions.
13. The **endosymbiotic theory** explains why mitochondria and chloroplasts have 70S ribosomes, circular DNA, and two membranes. The theory states that the ancestors of these organelles were prokaryotic cells that were internalized by other prokaryotes and then lost the ability to exist outside their host—thus forming early eukaryotes.

QUESTIONS FOR REVIEW

Answers to the Questions for Review (except Short Answer questions) begin on p. A-1.

Multiple Choice

1. A cell may allow a large or charged chemical to move across the cytoplasmic membrane, down the chemical's electrical and chemical gradients, in a process called ________________.
   a. active transport
   b. facilitated diffusion
   c. endocytosis
   d. pinocytosis

   p. 71

2. Which of the following statements concerning growth and reproduction is *false?*
   a. Growth and reproduction may occur simultaneously in living organisms.
   b. A living organism must reproduce to be considered alive.
   c. Living things may stop growing and reproducing yet still be alive.
   d. Normally, living organisms have the ability to grow and reproduce themselves.
3. A "9 + 2" arrangement of microtubules is seen in ________________.
   a. archaeal flagella
   b. bacterial flagella
   c. eukaryotic flagella
   d. all prokaryotic flagella
4. Which of the following is most associated with diffusion?
   a. symports
   b. antiports
   c. carrier proteins
   d. endocytosis
5. Which of the following is *not* associated with prokaryotic organisms?
   a. nucleoid
   b. glycocalyx
   c. cilia
   d. circular DNA
6. Which of the following is true of Svedbergs?
   a. They are not exact but are useful for comparisons.
   b. They are abbreviated "sv."
   c. They are prokaryotic in nature but exhibit some eukaryotic characteristics.
   d. They are an expression of sedimentation rate during high-speed centrifugation.
7. Which of the following statements is true?
   a. The cell walls of bacteria are composed of peptidoglycan.
   b. Peptidoglycan is a fatty acid.
   c. Gram-positive bacterial walls have a relatively thin layer of peptidoglycan anchored to the cytoplasmic membrane by teichoic acids.
   d. Peptidoglycan is found mainly in the cell walls of fungi, algae, and plants.

8. Which of the following is *not* a function of a glycocalyx?
   a. It forms pseudopods for faster mobility of an organism.
   b. It can protect a bacterial cell from drying out.
   c. It hides a bacterial cell from other cells.
   d. It allows a bacterium to stick to a host.

9. Bacterial flagella ________________.
   a. are anchored to the cell by a basal body
   b. are composed of hami
   c. are surrounded by an extension of the cytoplasmic membrane
   d. are composed of tubulin in hollow microtubules in a "9 + 2" arrangement ↓ Eukaryotes

10. Which cellular structure is important in classifying a bacterial species as Gram positive or Gram negative?
    a. flagella
    b. cell wall
    c. cilia
    d. glycocalyx

11. A Gram-negative cell is moving uric acid across the cytoplasmic membrane against its chemical gradient. Which of the following statements is true?
    a. The exterior of the cell is probably electrically negative compared to the interior of the cell.
    b. The acid probably moves by a passive means such as facilitated diffusion.
    c. The acid moves by an active process such as active transport.
    d. The movement of the acid requires phagocytosis.

12. Gram-positive bacteria ________________.
    a. have a thick cell wall, which retains crystal violet dye
    b. contain teichoic acids in their cell walls
    c. appear purple after Gram staining
    d. all of the above

13. Endospores ________________.
    a. are reproductive structures of some bacteria
    b. occur in some archaea
    c. can cause shock, fever, and inflammation
    d. are dormant, resistant cells

14. Inclusions have been found to contain ________________.
    a. DNA
    b. sulfur globules
    c. dipicolinic acid
    d. tubulin

15. Dipicolinic acid is an important component of ________________.
    a. Gram-positive archaeal walls
    b. cytoplasmic membranes in eukaryotes
    c. endospores
    d. Golgi bodies

Matching

1. Match the structures on the left with the descriptions on the right. A letter may be used more than once or not at all, and more than one letter may be correct for each blank.

   D Glycocalyx
   DHI Flagella
   F Axial filaments
   H Cilia
   A Fimbriae
   CG Pili
   E Hami

   A. Bristlelike projections found in quantities of 100 or more
   B. Long whip
   C. Responsible for conjugation
   D. "Sweet cup" composed of polysaccharides and/or polypeptides
   E. Numerous "grappling-hook" projections
   F. Responsible for motility of spirochetes
   G. Extensions not used for cell motility
   H. Made of tubulin in eukaryotes
   I. Made of flagellin in bacteria

2. Match the term on the left with its description on the right. Only one description is intended for each term.

   A Ribosome
   D Cytoskeleton
   F Centriole
   E Nucleus
   I Mitochondrion
   G Chloroplast
   C ER
   H Golgi body
   B Peroxisome

   A. Site of protein synthesis
   B. Contains enzymes to neutralize hydrogen peroxide
   C. Functions as the transport system within a eukaryotic cell
   D. Allows contraction of the cell
   E. Site of most DNA in eukaryotes
   F. Contains microtubules in "9 + 0" arrangement
   G. Light-harvesting organelle
   H. Packages large molecules for export from a cell
   I. Its internal membranes are sites for ATP production

VISUALIZE IT!

1. Label the structures of the following prokaryotic and eukaryotic cells. With a single word or short phrase, explain the function of each structure.

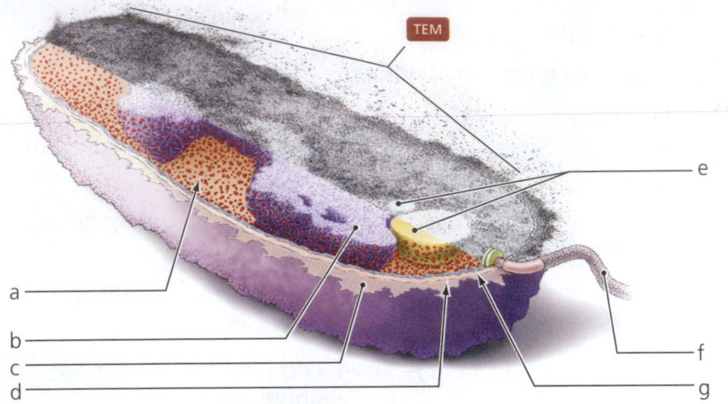

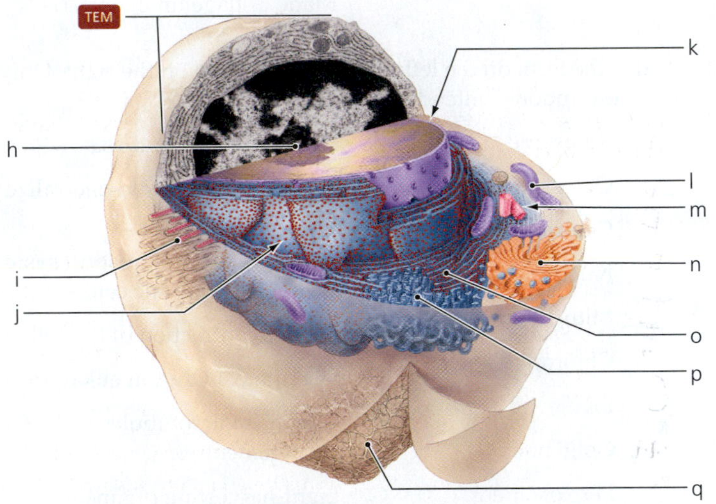

2. Label each type of flagellar arrangement.

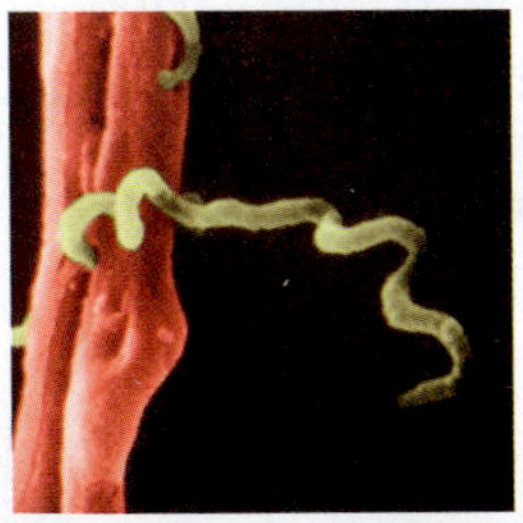
a. ______________

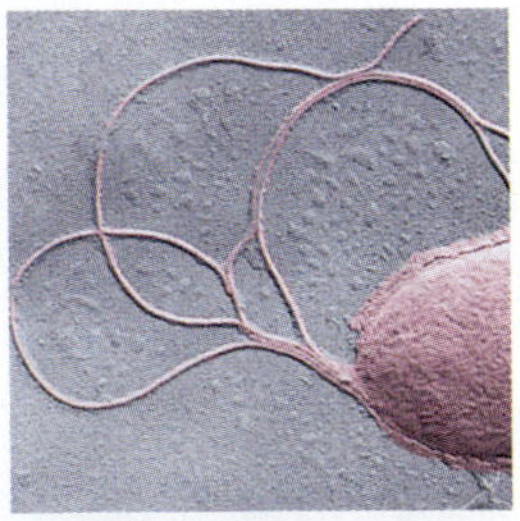
b. ______________

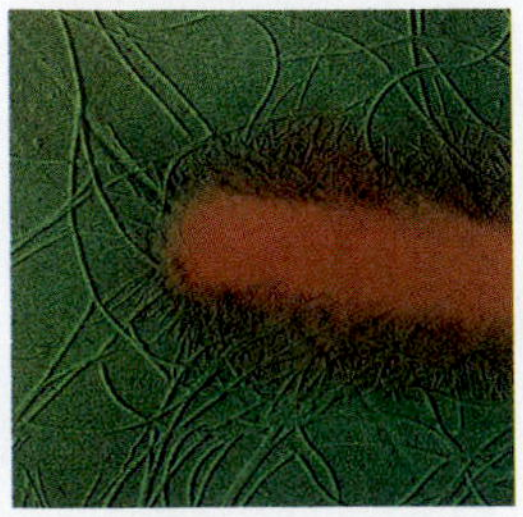
c. ______________

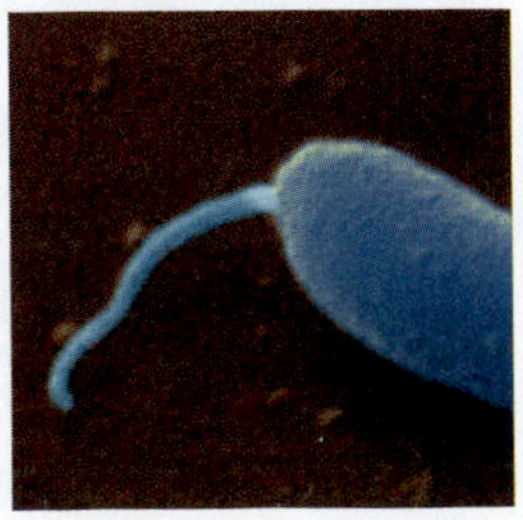
d. ______________

3. A scientist who is studying passive movement of chemicals across the cytoplasmic membrane of *Salmonella enterica* serotype Typhi measures the rate at which two chemicals diffuse into a cell as a function of external concentration. The results are shown in the following figure. Chemical A diffuses into the cell more rapidly than does B at lower external concentrations, but the rate levels off as the external concentration increases. The rate of diffusion of chemical B continues to increase as the external concentration increases.
   a. How can you explain the differences in the diffusion rates of chemicals A and B?
   b. Why does the diffusion rate of chemical A taper off?
   c. How could the cell increase the diffusion rate of chemical A?
   d. How could the cell increase the diffusion rate of chemical B?

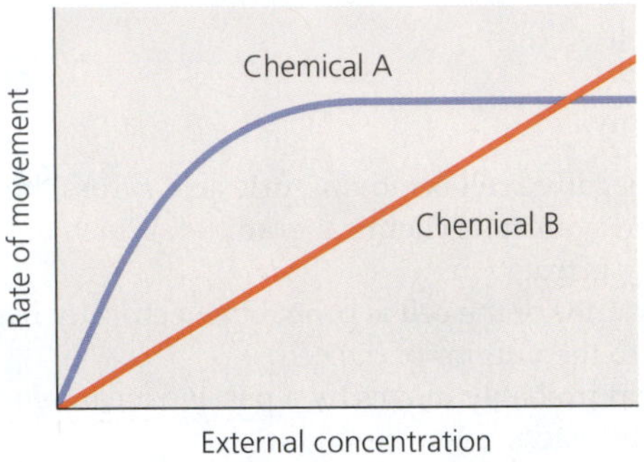

Short Answer

1. Describe (or draw) an example of diffusion down a concentration gradient.
2. Sketch, name, and describe three flagellar arrangements in bacteria.
3. Define *cytosol*.
4. The term *fluid mosaic* has been used in describing the cytoplasmic membrane. How does each word of that phrase accurately describe our current understanding of a cell membrane?
5. A local newspaper writer has contacted you, an educated microbiology student from a respected college. He wants to obtain scientific information for an article he is writing about "life" and poses the following query: "What is the difference between a living thing and a nonliving thing?" Knowing that he will edit your material to fit the article, give an intelligent, scientific response.
6. What is the difference between growth and reproduction?
7. Compare bacterial cells and algal cells, giving at least four similarities and four differences.
8. Contrast a cell of *Streptococcus pyogenes* (a bacterium) with the unicellular protozoan *Entamoeba histolytica*, listing at least eight differences.
9. Differentiate among pili, fimbriae, and cilia, using sketches and descriptive labels.
10. Can nonliving things metabolize? Explain your answer.
11. How do archaeal flagella differ from bacterial flagella and eukaryotic flagella?

12. Contrast bacterial and eukaryotic cells by filling in the following table.

Characteristic	Bacteria	Eukaryotes
Size		
Presence of nucleus		
Presence of membrane-bound organelles		
Structure of flagella		
Chemicals in cell walls		
Type of ribosomes		
Structure of chromosomes		

13. What is the function of glycocalyces and fimbriae in forming a biofilm?
14. What factors may prevent a molecule from moving across a cell membrane?
15. Compare and contrast three types of passive transport across a cell membrane.
16. Contrast the following active processes for transporting materials into or out of a cell: active transport, group translocation, endocytosis, exocytosis.
17. Contrast symports and antiports.
18. Describe the endosymbiotic theory. What evidence supports the theory? Which features of eukaryotic cells are *not* explained by the theory?

CRITICAL THINKING

1. A scientist develops a chemical that prevents Golgi bodies from functioning. Contrast the specific effects the chemical would have on human cells versus bacterial cells.
2. Methylene blue binds to DNA. What structures in a yeast cell would be stained by this dye?
3. A new chemotherapeutic drug kills bacteria but not humans. Discuss the possible ways the drug may act selectively on bacterial cells.
4. Some bacterial toxins cause cells lining the digestive tract to secrete ions, making the contents of the tract hypertonic. What effect does this have on a patient's water balance?
5. A researcher carefully inserts an electrode into an algal cell. He determines that the electrical charge across the cytoplasmic membrane is –70 millivolts. Then he slips the electrode deeper into the cell across another membrane and measures an electrical charge of –90 millivolts compared to the outside. What large organelle is surrounded by the second membrane? Explain your answer.
6. The smallest, single-celled, free-living eukaryote known is a green alga, *Ostreococcus tauris*. What membranous organelles must this photosynthetic cell have?
7. An electron micrograph of a newly discovered cell shows long projections with a basal body in the cell wall. What kind of projections are these? Is the cell prokaryotic or eukaryotic? How will this cell behave in its environment because of these projections?
8. An entry in a recent scientific journal reports positive phototaxis in a newly described species. What condition could you create in the lab to encourage the growth of these organisms?
9. A medical microbiological lab report indicates that a sample contained a biofilm and that one species in the biofilm was identified as *Neisseria gonorrhoeae*. Is this strain of *Neisseria* likely to be pathogenic? Why or why not?
10. A researcher treats a cell to block the function of SER only. Describe the initial effects this would have on the cell.
11. After a man infected with the bacterium *Escherichia coli* was treated with the correct antibiotic for this pathogen, the bacterium was no longer found in the man's blood, but his symptoms of fever and inflammation worsened. What caused the man's response to the treatment? Why was his condition worsened by the treatment?
12. Solutions hypertonic to bacteria and fungi are used for food preservation. For instance, jams and jellies are hypertonic with sugar, and pickles are hypertonic with salt. How do hypertonic solutions kill bacteria and fungi that would otherwise spoil these foods?
13. Following the bioterrorist anthrax attacks in the fall of 2001, a news commentator suggested that people steam their mail for 30 seconds before opening it. Would the technique protect people from anthrax infections? Why or why not?
14. Eukaryotic cells are almost always larger than prokaryotic cells. What structures might allow for their larger size?

CONCEPT MAPPING

Answers to this Concept Map begin on p. A-1.

Using the following terms, fill in the following concept map that describes the bacterial cell wall. Use this map as a guide for chapters in which you are asked to draw your own maps, but know that all maps will be different because there are many ways to draw them. You can also complete this and other concept maps online by going to the MasteringMicrobiology Study Area.

Glycan chains
Gram-negative cell wall
Gram-positive cell wall
Lipid A
Lipopolysaccharide (LPS)
***N*-acetylglucosamine**
Peptidoglycan
Periplasm
Porins
Teichoic acids

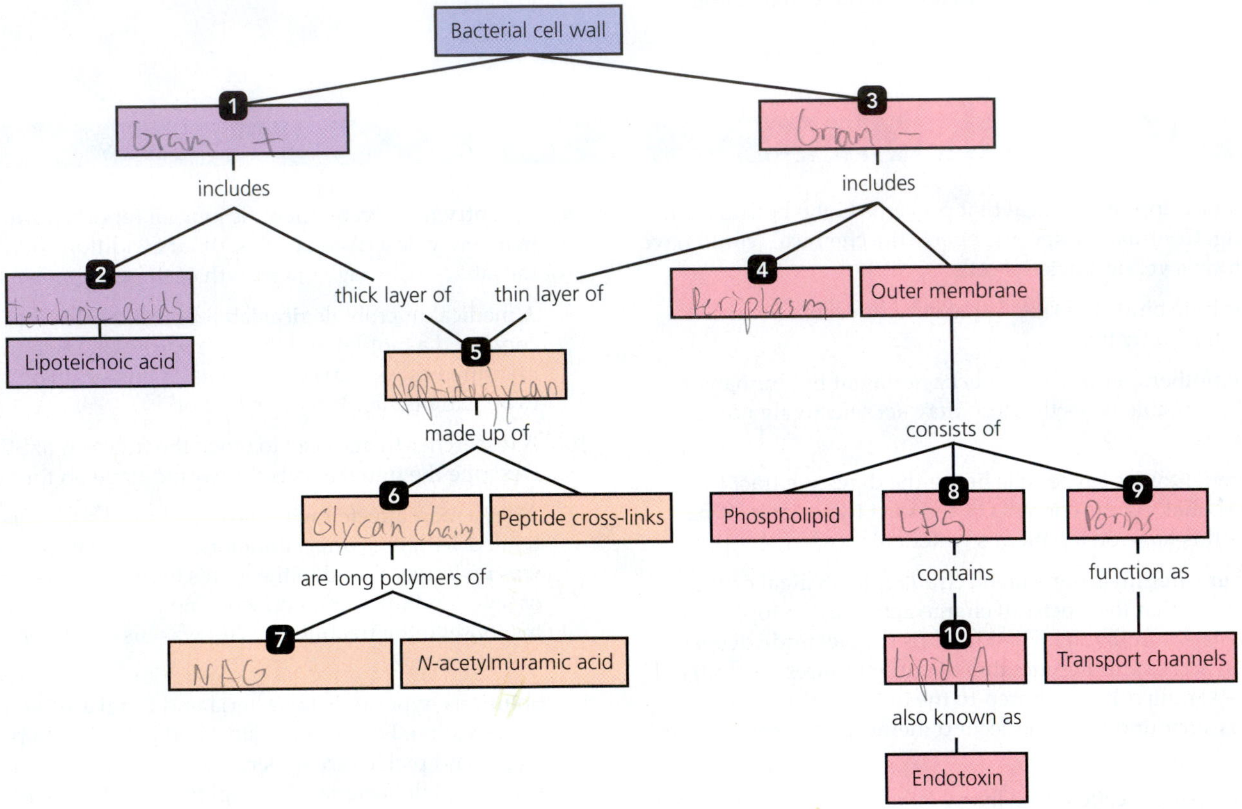

4 Microscopy, Staining, and Classification

MICRO IN THE CLINIC

The Salty Toddler

Lewis is an active three-year-old with a good appetite, though he is significantly smaller than other boys his age. He loves playing with his six-year-old brother, Robert, and is usually able to keep up with him; however, lately Lewis has been coughing a lot and has not been running around as much as usual. He's also had diarrhea, which seems to come and go regardless of his diet. His mother, Dalia, has also noticed a mysterious, distinctively salty taste on Lewis's skin whenever she kisses him on the neck. She wonders whether she should be concerned.

Lewis's symptoms worsen. He starts to wheeze and cough up yellow mucus, sometimes with a little streak of blood in it. He also develops a fever, which prompts Dalia to take him to the hospital. There, she learns that Lewis has an unusual respiratory infection caused by *Pseudomonas*, which does not typically affect healthy children. Why, then, does *Pseudomonas* infect Lewis? Moreover, a lung infection explains only some of Lewis's symptoms. What is causing the persistent diarrhea? Why is Lewis so small for his age?

Is Lewis suffering from more than just an unusual respiratory infection? Could there be a medical explanation for all of his other symptoms? Turn to the end of the chapter (p. 120) to find out.

MM **Explore More:** Test your readiness and apply your knowledge with dynamic learning tools at MasteringMicrobiology.

Either the well was very deep, or she fell very slowly, for she had plenty of time as she went down to look about her, and to wonder what was going to happen next. . . . "Curiouser and curiouser!" cried Alice. . . .

—*Alice's Adventures in Wonderland* by Lewis Carroll

Like Alice falling into Wonderland or traveling through a looking glass, scientists have entered the marvelous microbial world through advances in microscopy. With the invention of new laboratory techniques and the construction of new instruments, biologists are still discovering "curiouser and curiouser" wonders about the microbial world. In this chapter we will discuss some of the techniques microbiologists use to enter that world. We begin with a discussion of metric units as they relate to measuring the size of microbes. We then examine the instruments and staining techniques used in microbiology. Finally, we consider the classification schemes used to categorize the inhabitants of the microbial wonderland. ▶**ANIMATIONS:** ***Microscopy and Staining: Overview***

Units of Measurement

LEARNING | **OUTCOMES**

4.1 Identify the two primary metric units used to measure the diameters of microbes.

4.2 List the metric units of length in order, from meter to nanometer.

Microorganisms are small. This may seem an obvious statement, but it is one that should not be taken for granted. Exactly how small are they? How can we measure the width and length of microbes?

Typically, a unit of measurement is smaller than the object being measured. For example, we measure a person's height in feet or inches, not in miles. Likewise, the diameter of a dime is measured in fractions of an inch, not in feet. So, measuring the size of a microbe requires units that are smaller than even the smallest interval on a ruler marked with English units (typically 1/16 inch). Even smaller units, such as 1/64 inch or 1/128 inch, become quite cumbersome and very difficult to use when we are dealing with microorganisms.

So that they can work with units that are simpler and in standard use the world over, scientists use metric units of measurement. Unlike the English system, the metric system is a decimal system, so each unit is one-tenth the size of the next largest unit. Even extremely small metric units are much easier to use than the fractions involved in the English system.

The unit of length in the metric system is the *meter (m)*, which is slightly longer than a yard. One-tenth of a meter is a *decimeter (dm)*, and one-hundredth of a meter is a *centimeter (cm)*, which is equivalent to about a third of an inch. One-tenth of a centimeter is a *millimeter (mm)*, which is the thickness of a dime. A millimeter is still too large to measure the size of most microorganisms, but in the metric system we continue to divide by multiples of 10 until we have a unit appropriate for use. Thus, one-thousandth of a millimeter is a *micrometer (μm)*, which is small enough to be useful in measuring the size of cells. One-thousandth of a micrometer is a *nanometer (nm)*, a unit used to measure the smallest cellular organelles and viruses. A nanometer is one-billionth of a meter.

TABLE 4.1 presents these metric units and some English equivalents. (Refer to Figure 3.4 for a visual size comparison of a typical eukaryotic cell, prokaryotic cell, and virus particle.)

TELL ME WHY

Why do scientists use metric rather than English units?

TABLE 4.1 Metric Units of Length

Metric Unit (abbreviation)	Meaning of Prefix	Metric Equivalent	U.S. Equivalent	Representative Microbiological Application of the Unit
Meter (m)	—[a]	1 m	39.37 in (about a yard)	Length of pork tapeworm, *Taenia solium* (e.g., 1.8–8.0 m)
Decimeter (dm)	1/10	0.1 m = 10^{-1} m	3.94 in	—[b]
Centimeter (cm)	1/100	0.01 m = 10^{-2} m	0.39 in; 1 in = 2.54 cm	Diameter of a mushroom cap (e.g., 12 cm)
Millimeter (mm)	1/1000	0.001 m = 10^{-3} m	—	Diameter of a bacterial colony (e.g., 2.3 mm); length of a tick (e.g., 5.7 mm)
Micrometer (μm)	1/1,000,000	0.000001 m = 10^{-6} m	—	Diameter of white blood cells (e.g., 5–25 μm)
Nanometer (nm)	1/1,000,000,000	0.000000001 m = 10^{-9} m	—	Diameter of a poliovirus (e.g., 25 nm)

[a]The meter is the standard metric unit of length.
[b]Decimeters are rarely used.

Microscopy

LEARNING | OUTCOME

4.3 Define *microscopy.*

Microscopy[1] refers to the use of light or electrons to magnify objects. The science of microbiology began when Antoni van Leeuwenhoek (lā´ven-hŭk, 1632–1723) used primitive microscopes to observe and report the existence of microorganisms. Since that time, scientists and engineers have developed a variety of light and electron microscopes.

General Principles of Microscopy

LEARNING | OUTCOMES

4.4 Explain the relevance of electromagnetic radiation to microscopy.

4.5 Define *empty magnification.*

4.6 List and explain two factors that determine resolving power.

4.7 Discuss the relationship between contrast and staining in microscopy.

General principles involved in both light and electron microscopy include the wavelength of radiation, the magnification of an image, the resolving power of the instrument, and contrast in the specimen.

Wavelength of Radiation

Visible light is one part of a spectrum of electromagnetic radiation that includes X rays, microwaves, and radio waves (**FIGURE 4.1**). Note that beams of radiation may be referred to as either rays or waves. These various forms of radiation differ in **wavelength**—the distance between two corresponding parts of a wave. The human eye discriminates among different wavelengths of visible light and sends patterns of nerve impulses to the brain, which interprets the impulses as different colors. For example, we see wavelengths of 400 nm as violet and of 650 nm as red. White light, composed of many colors (wavelengths), has an average wavelength of 550 nm.

Electrons are negatively charged particles that orbit the nuclei of atoms. Besides being particulate, moving electrons act as waves, with wavelengths dependent on the voltage of an electron beam. For example, the wavelength of electrons at 10,000 volts (V) is 0.01 nm; that of electrons at 1,000,000 V is 0.001 nm. As we will see, using radiation of smaller wavelengths results in enhanced microscopy.

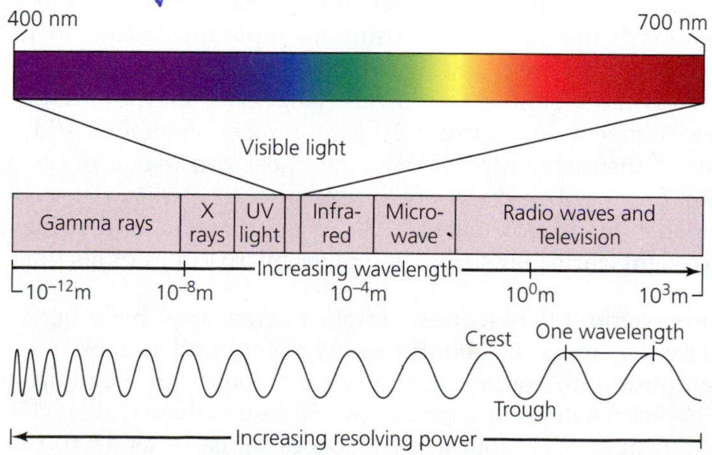

▲ **FIGURE 4.1 The electromagnetic spectrum.** Visible light is made up of a narrow band of wavelengths of radiation. Visible and ultraviolet (UV) light are used in microscopy.

Magnification

Magnification is the apparent increase in size of an object. It is indicated by a number and "×," which is read "times." For example, 16,000× is 16,000 times. Magnification results when a beam of radiation *refracts* (bends) as it passes through a lens. Curved glass lenses refract light, and magnetic fields refract electron beams. Let's consider the magnifying power of a glass lens that is convex on both sides.

A lens refracts light because the lens is *optically dense* compared to the surrounding medium (such as air); that is, light travels more slowly through the lens than through air. Think of a car moving at an angle from a paved road onto a dirt shoulder. As the right front tire leaves the pavement, it has less traction and slows down. Since the other wheels continue at their original speed, the car veers toward the dirt, and the line of travel bends to the right. Likewise, the leading edge of a light beam slows as it enters glass, and the beam bends (**FIGURE 4.2a**). Light also bends as it leaves the glass and reenters the air.

Because of its curvature, a lens refracts light rays that pass through its periphery more than light rays that pass through its center, so that the lens focuses light rays on a *focal point*. Importantly for the purpose of microscopy, light rays spread apart as they travel past the focal point and produce an enlarged, inverted image (**FIGURE 4.2b**). The degree to which the image is enlarged depends on the thickness of the lens, its curvature, and the speed of light through its substance.

Microscopists could combine lenses to obtain an image magnified millions of times, but the image would be faint and blurry. Such magnification is said to be *empty magnification*. The properties that determine the clarity of an image, which in turn determines the useful magnification of a microscope, are *resolution* and *contrast*.

Resolution

Resolution, also called *resolving power*, is the ability to distinguish objects that are close together. An optometrist's eye chart is a test of resolution at a distance of 20 feet (6.1 m). Leeuwenhoek's microscopes had a resolving power of about

[1]From Greek *micro*, meaning "small," and *skopein*, meaning "to view."

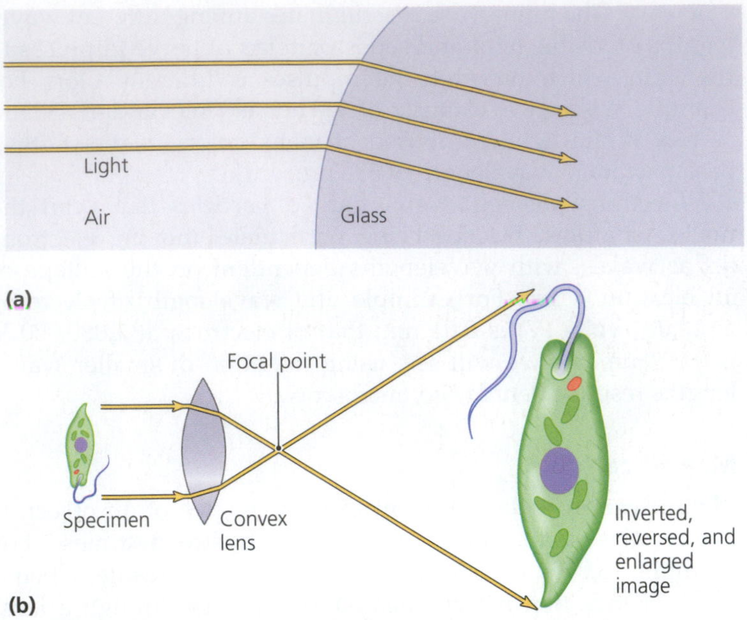

▲ **FIGURE 4.2 Light refraction and image magnification by a convex glass lens. (a)** Light passing through a lens refracts (bends) because light rays slow down as they enter the glass, and light at the leading edge of a beam that strikes the glass at an angle slows first. **(b)** A convex lens focuses light on a focal point. The image is enlarged and inverted as light rays pass the focal point and spread apart.

1 µm; that is, he could distinguish objects if they were more than about 1 µm apart, and objects closer together than 1 µm appeared as a single object. The better the resolution, the better the ability to distinguish two objects that are close to one another. Modern microscopes have fivefold better resolution than Leeuwenhoek's; they can distinguish objects as close together as 0.2 µm. **FIGURE 4.3** illustrates the size of various objects that can be resolved by the unaided human eye and by various types of microscopes.

Why do modern microscopes have better resolution than Leeuwenhoek's microscopes? A principle of microscopy is that resolution distance is dependent on (1) the wavelength of the electromagnetic radiation and (2) the **numerical aperture** of the lens, which refers to the ability of a lens to gather light.

Resolution distance is calculated using the following formula:

$$\text{resolution distance} = \frac{0.61 \times \text{wavelength}}{\text{numerical aperture}}$$

The resolution of today's microscopes is greater than that of Leeuwenhoek's microscopes because modern microscopes use shorter-wavelength radiation, such as blue light or electron beams, and because they have lenses with larger numerical apertures.

Contrast

Contrast refers to differences in intensity between two objects or between an object and its background. Contrast is important in determining resolution. For example, although you can easily distinguish two golf balls lying side by side on a putting green 15 m away, at that distance it is much more difficult to distinguish them if they are lying on a white towel.

Most microorganisms are colorless and have very little contrast whether one uses light or electrons. One way to increase the contrast between microorganisms and their background is to stain them. Stains and staining techniques are covered later in the chapter. As we will see, the use of light that is in *phase*—that is, in which all of the waves' crests and troughs are aligned—can also enhance contrast.

Light Microscopy

LEARNING **OUTCOMES**

4.8 Contrast simple and compound microscopes.
4.9 Compare and contrast bright-field microscopy, dark-field microscopy, and phase microscopy.
4.10 Compare and contrast fluorescence and confocal microscopes.

Several classes of microscopes use various types of light to examine microscopic specimens. The most common microscopes are *bright-field microscopes,* in which the background (or *field*) is illuminated. In *dark-field microscopes,* the specimen is made to appear light against a dark background. *Phase microscopes* use the alignment or misalignment of light waves to achieve the desired contrast between a living specimen and its background. *Fluorescence microscopes* use invisible ultraviolet light to cause specimens to radiate visible light, a phenomenon called *fluorescence.* Microscopes that use lasers to illuminate fluorescent chemicals in a thin plane of a specimen are called *confocal microscopes.* Next we examine each of these kinds of light microscope in turn.

Bright-Field Microscopes

There are two basic types of bright-field microscopes: *simple microscopes* and *compound microscopes.*

Simple Microscopes Leeuwenhoek first reported his observations of microorganisms using a simple microscope in 1674. A **simple microscope,** which contains a single magnifying lens, is more similar to a magnifying glass than to a modern microscope (see Figure 1.2). Though Leeuwenhoek did not invent the microscope, he was the finest lens maker of his day and produced microscopes of exceptional quality. They were capable of approximately 300× magnification and achieved excellent clarity, far surpassing other microscopes of his time.

Compound Microscopes Simple microscopes have been replaced in modern laboratories by compound microscopes. A **compound microscope** uses a series of lenses for magnification **(FIGURE 4.4a)**. Many scientists, including Galileo Galilei (1564–1642), made compound microscopes as early as 1590, but it was not until about 1830 that scientists developed compound microscopes that exceeded the clarity and magnification of Leeuwenhoek's simple microscope.

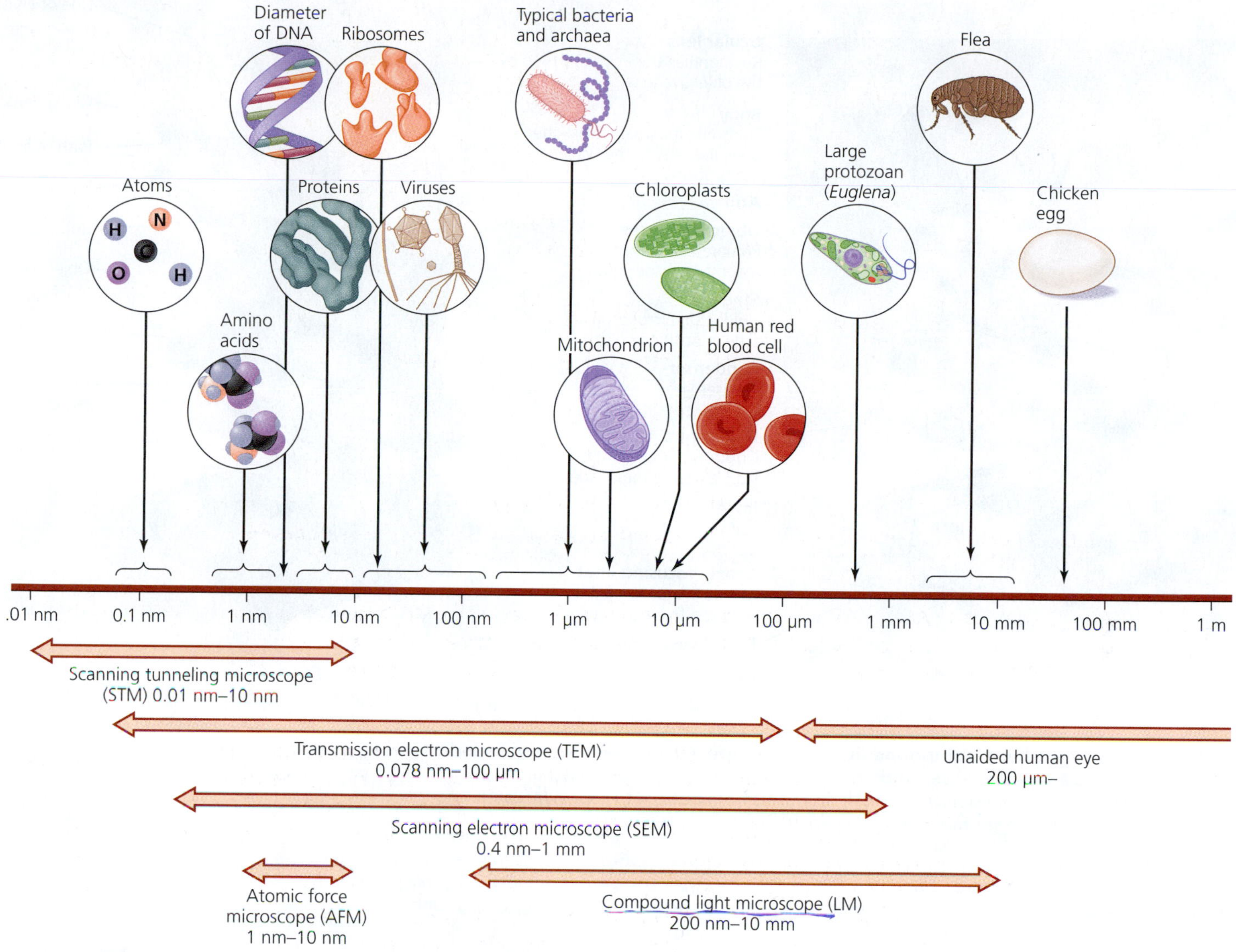

▲ **FIGURE 4.3 The limits of resolution (and some representative objects within those ranges) of the human eye and of various types of microscopes.**

In a basic compound microscope, magnification is achieved as light rays pass through a specimen and into an **objective lens,** which is the lens immediately above the object being magnified **(FIGURE 4.4b)**. An objective lens is really a series of lenses that not only create a magnified image but also are engineered to reduce aberrations in the shape and color of the image. Most light microscopes used in biology have three or four objective lenses mounted on a **revolving nosepiece.** The objective lenses on a typical microscope are *scanning objective lens* (4×), *low-power objective lens* (10×), *high-power lens* or *high dry objective lens* (40×), and *oil immersion objective lens* (100×).

An oil immersion lens increases not only magnification but also resolution. As we have seen, light refracts as it travels from air into glass and also from glass into air; therefore, some of the light passing out of a glass slide is bent so much that it bypasses the lens **(FIGURE 4.5a)**. Placing *immersion oil* between the slide and an *oil immersion objective lens* enables the lens to capture this light because light travels through immersion oil at the same speed as through glass. Because light is traveling at a uniform speed through the slide, the immersion oil, and the glass lens, it does not refract **(FIGURE 4.5b)**. Immersion oil increases the numerical aperture, which increases resolution, because more light rays are gathered into the lens to produce the image. Obviously, the space between the slide and the lens can be filled with oil only if the distance between the lens and the specimen, called the *working distance,* is small.

An objective lens bends the light rays, which then pass up through one or two **ocular lenses,** which are the lenses closest to the eyes. Microscopes with a single ocular lens are *monocular,* and those with two are *binocular.* Ocular lenses magnify the image created by the objective lens, typically another 10×.

The **total magnification** of a compound microscope is determined by multiplying the magnification of the objective lens by the magnification of the ocular lens. Thus, total

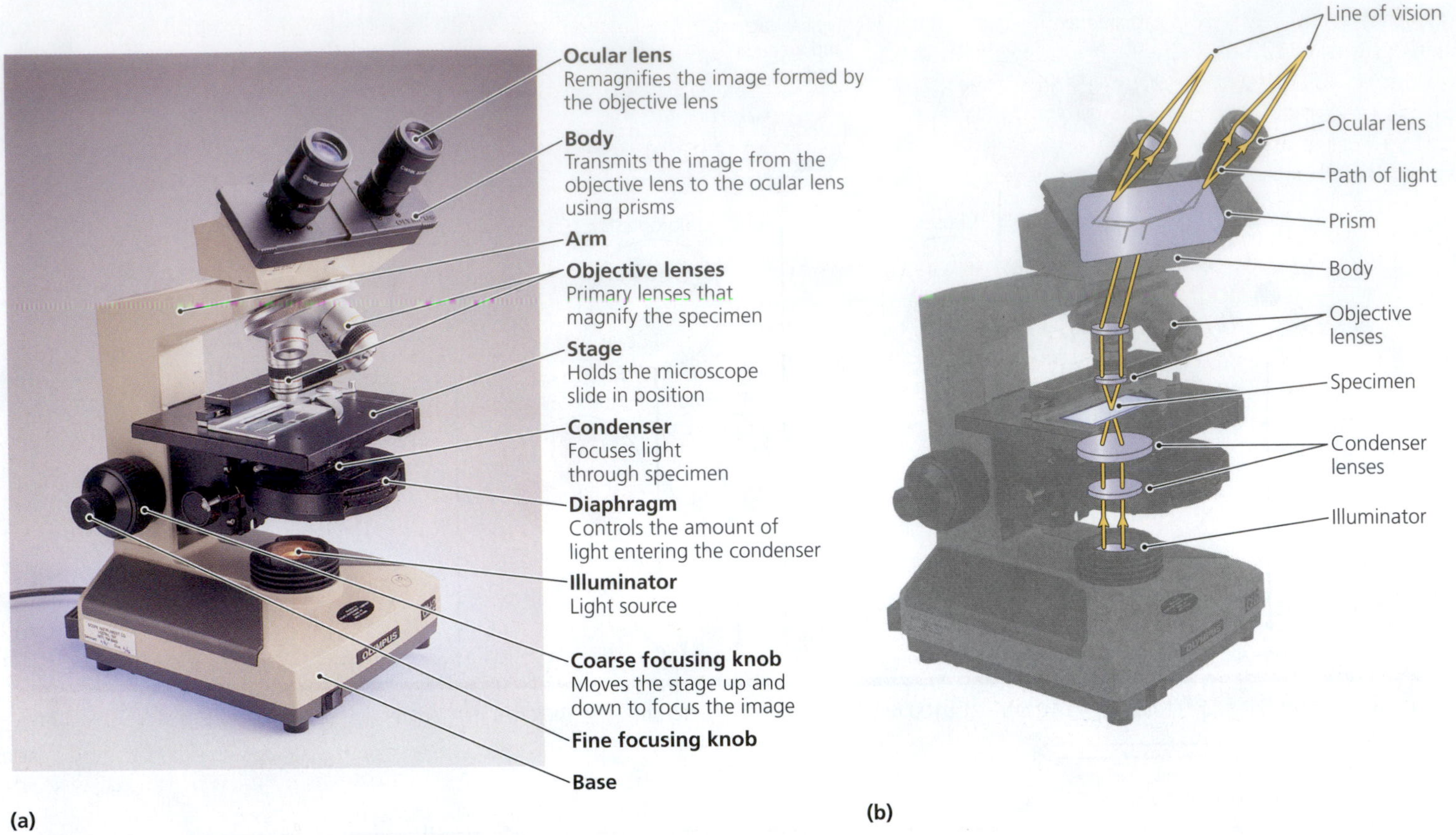

▲ **FIGURE 4.4 A bright-field, compound light microscope. (a)** The parts of a compound microscope, which uses a series of lenses to produce an image at up to 2000× magnification. **(b)** The path of light in a compound microscope; light travels from bottom to top. *Why can't light microscopes produce clear images that are magnified 10,000×?*

Figure 4.4 *Although it is possible for a light microscope to produce an image that is magnified 10,000×, magnification above 2000× is empty magnification because pairs of objects in the specimen are too close together to resolve with even the shortest-wavelength (blue) light.*

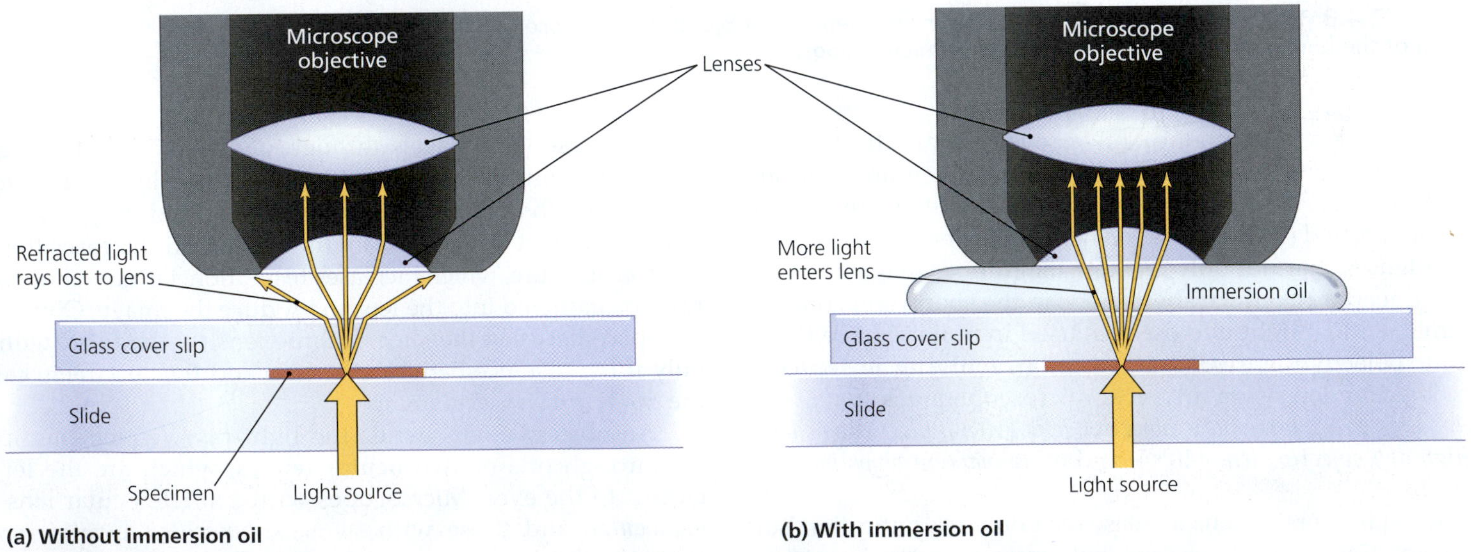

▲ **FIGURE 4.5 The effect of immersion oil on resolution. (a)** Without immersion oil, light is refracted as it moves from the cover glass into the air. Part of the scattered light misses the objective lens. **(b)** With immersion oil. Because light travels through the oil at the same speed as it does through glass, no light is refracted as it leaves the specimen and more light enters the lens, which increases resolution.

magnification using a 10× ocular lens and a 10× low-power objective lens is 100×. Using the same ocular and a 100× oil immersion objective produces 1000× magnification. Some light microscopes, using higher-magnification oil immersion objective lenses and ocular lenses, can achieve 2000× magnification, but this is the limit of useful magnification for light microscopes because their resolution is restricted by the wavelength of visible light.

Modern compound microscopes also have a **condenser lens** (or lenses), which directs light through the specimen, as well as one or more mirrors or prisms that deflect the path of the light rays from an objective lens to the ocular lens (see Figure 4.4b). Some microscopes have mirrors or prisms that direct light to a camera through a special tube. Photographs of such a microscopic image are called light **micrographs.** ▶**VIDEO TUTOR:** *The Light Microscope*

Dark-Field Microscopes

Pale objects are best observed with **dark-field microscopes.** These microscopes utilize a *dark-field stop* in the condenser that prevents light from directly entering the objective lens **(FIGURE 4.6)**. Instead, light rays are reflected inside the condenser, so that they pass into the slide at such an oblique angle that they miss the objective lens. Only those light rays that are scattered by the specimen enter the objective lens and are seen, so the specimen appears light against a dark background. This increases contrast and enables observation of more details than are visible in bright-field microscopy. Dark-field microscopes are especially useful for examining small or colorless cells.

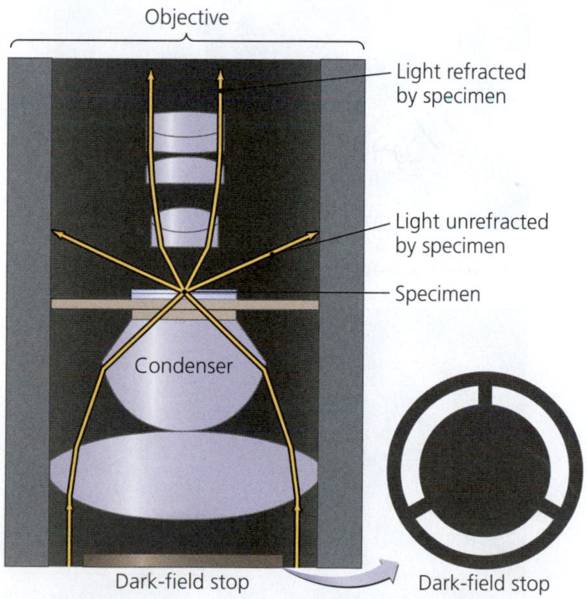

▲ **FIGURE 4.6 The light path in a dark-field microscope.** A dark-field stop prevents light from entering the specimen directly; only light rays that are scattered by the specimen reach the objective lens and can be seen. The resulting high-contrast image—a brightly lit specimen against a dark background—enhances resolution.

Phase Microscopes

Scientists use **phase microscopes** to examine living microorganisms or specimens that would be damaged or altered by attaching them to slides or staining them. Basically, phase microscopes treat one set of light rays differently from another set of light rays.

Light rays are said to be *in phase* when their crests and troughs are aligned and *out of phase* when their crests and troughs are not aligned **(FIGURE 4.7a)**. Light rays that are in phase reinforce one another, producing a brighter image, and light rays that are out of phase interfere with one another, producing a darker image. Light rays passing through a specimen naturally slow down and are shifted about 1/4 wavelength out of phase. A special filter called a *phase plate,* which is mounted in a phase objective lens, retards these rays another 1/4 wavelength, so that they are 1/2 wavelength out of phase with their neighbors. When the phase microscope lens brings the two sets of rays together, troughs of one wave interfere with the crests of the other—because they are out of phase **(FIGURE 4.7b)**—and contrast is created. There are two types of phase microscopes: phase-contrast and differential interference contrast microscopes.

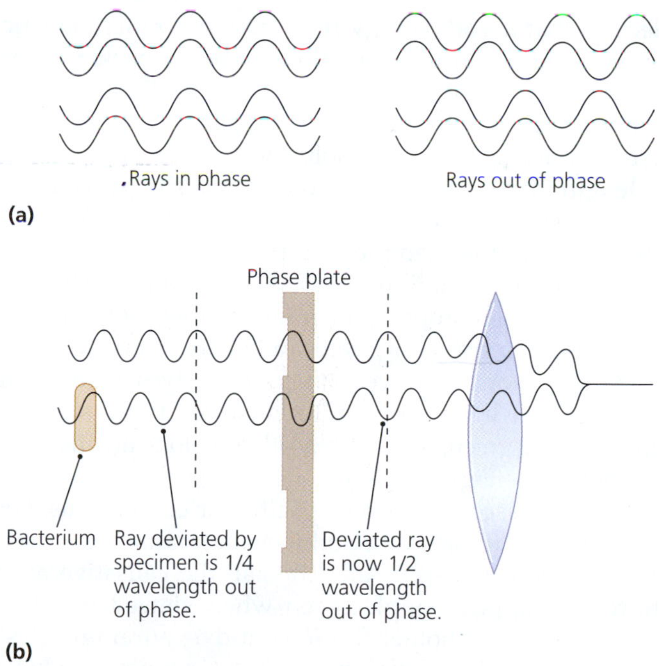

▲ **FIGURE 4.7 Principles of phase microscopy. (a)** Light rays that are in phase are aligned; their crests and troughs reinforce one another to produce a brighter image. Rays that are out of phase interfere with one another and produce a darker image. **(b)** Together, the regions of a specimen and a phase plate built into a phase objective lens slow some of the light rays such that they are 1/2 wavelength out of phase. These rays interfere with light rays that bypass the specimen, which produces contrast between the field and various regions of the specimen (see Figure 4.8c and d).

▶ **FIGURE 4.8 Four kinds of light microscopy.** All four photos show the same human cheek cell and bacteria. **(a)** Bright-field microscopy reveals some internal structures. **(b)** Dark-field microscopy increases contrast between some internal structures and between the edges of the cell and the surrounding medium. **(c)** Phase-contrast microscopy provides greater resolution of internal structures. **(d)** Differential interference contrast (Nomarski) microscopy produces a three-dimensional effect.

Phase-Contrast Microscopes The simplest phase microscopes, **phase-contrast microscopes**, produce sharply defined images in which fine structures can be seen in living cells. These microscopes are particularly useful for observing cilia and flagella.

Differential Interference Contrast Microscopes Differential interference contrast microscopes (also called Nomarski[2] microscopes) create phase interference patterns. They also use prisms that split light beams into their component wavelengths (colors). This significantly increases contrast and gives the image a dramatic three-dimensional or shadowed appearance, almost as though light were striking the specimen from one side. This technique also produces unnatural colors, which enhance contrast.

FIGURE 4.8 illustrates the differences that can be observed in a single specimen when viewed using four different types of light microscopy.

Fluorescence Microscopes

Molecules that absorb energy from invisible radiation (such as ultraviolet light) and then radiate the energy back as a longer, visible wavelength are said to be *fluorescent*. **Fluorescence microscopes** use an ultraviolet (UV) light source to fluoresce objects. UV light increases resolution because it has a shorter wavelength than visible light, and contrast is improved because fluorescing structures are visible against a black background.

Some cells—for example, the pathogen *Pseudomonas aeruginosa* (soo-dō-mō´nas ā-roo-ji-nō´să)—and some cellular molecules (such as chlorophyll in photosynthetic organisms) are naturally fluorescent. Other cells and cellular structures can be stained with fluorescent dyes. When these dyes are bombarded with ultraviolet light, they emit visible light and show up as bright orange, green, yellow, or other colors against a black background (see Figure 3.33b).

Some fluorescent dyes are specific for certain cells. For example, the dye fluorescein isothiocyanate attaches to cells of *Bacillus anthracis* (ba-sil´ŭs an-thrā´sis), the causative agent of anthrax, and appears apple green when viewed in a fluorescence microscope. Another fluorescent dye, auramine O, stains *Mycobacterium tuberculosis* (mī´kō-bak-tēr´ē-ŭm too-ber-kyū-lō´sis; **FIGURE 4.9**).

Fluorescence microscopy is also used in a process called *immunofluorescence*. First, fluorescent dyes are covalently linked to Y-shaped immune system proteins called antibodies

[2]After the French physicist Georges Nomarski, who invented the differential interference contrast microscope.

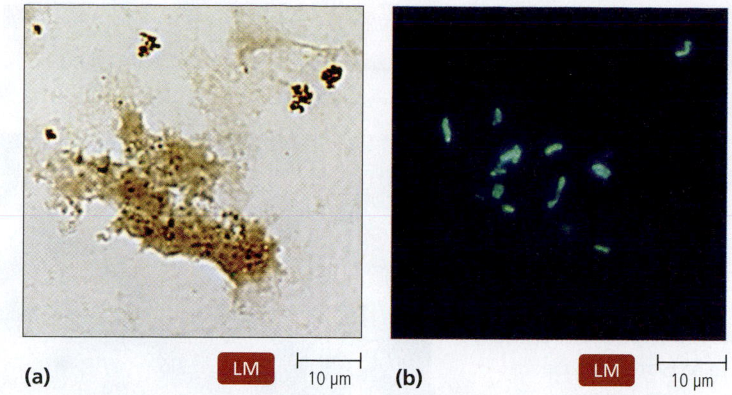

▲ **FIGURE 4.9 Fluorescence microscopy.** Fluorescent chemicals absorb invisible short-wavelength radiation and emit visible (longer-wavelength) radiation. **(a)** When viewed under normal illumination, *Mycobacterium tuberculosis* cells stained with the fluorescent dye auramine O are invisible amid the mucus and debris in a sputum smear. **(b)** When the same smear is viewed under UV light, the bacteria fluoresce and are clearly visible.

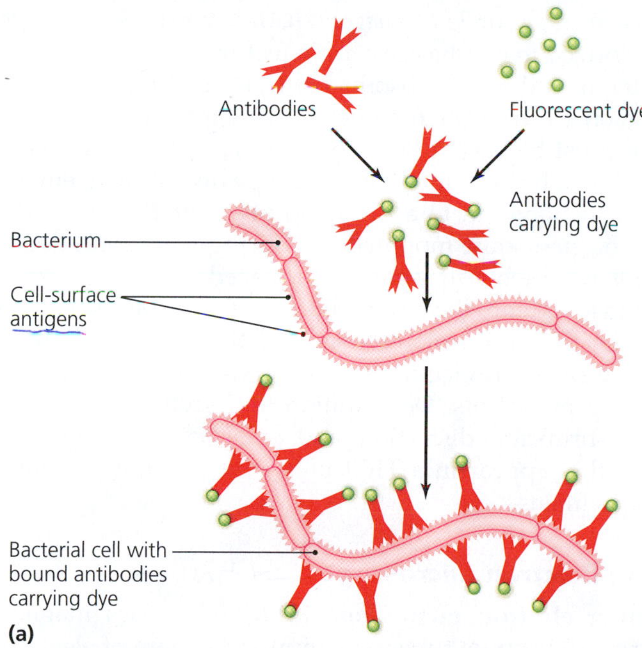

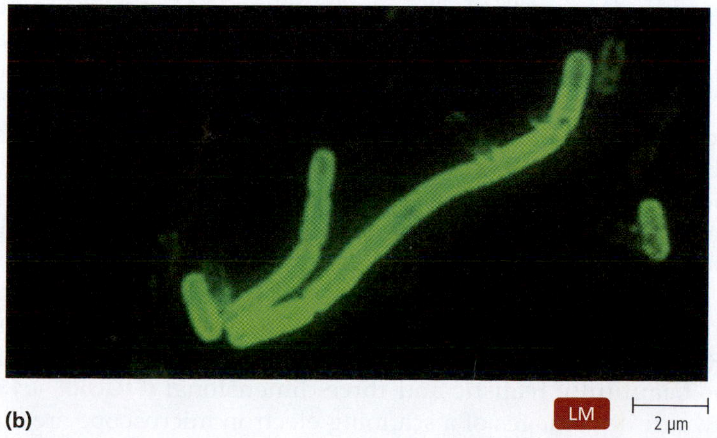

(FIGURE 4.10a). Given the opportunity, these dye-tagged antibodies will bind specifically to complementary-shaped *antigens*, which are portions of molecules that are present, for example, on the surface of microbial cells. When viewed under UV light, a microbial specimen that has bound dye-tagged antibodies becomes visible **(FIGURE 4.10b)**. In addition to identifying pathogens, including those that cause syphilis, rabies, and Lyme disease, scientists can use immunofluorescence to locate and make visible a variety of proteins of interest.

Confocal Microscopes

Confocal[3] **microscopes** also use fluorescent dyes or fluorescent antibodies, but these microscopes use ultraviolet lasers to illuminate the fluorescent chemicals in only a single plane that is no thicker than 1.0 μm; the rest of the specimen remains dark and out of focus. Visible light emitted by the dyes passes through a pinhole aperture that helps eliminate blurring that can occur with other types of microscopes and increases resolution by up to 40%. Each image from a confocal microscope is thus an "optical slice" through the specimen, as if it had been thinly cut. Once individual images are digitized, a computer is used to construct a three-dimensional representation, which can be rotated and viewed from any direction. Confocal microscopes have been particularly useful for examining the relationships among various organisms within complex microbial communities called biofilms (see **Highlight: Studying Biofilms in Plastic "Rocks"** on p. 104). Regular light microscopy cannot produce clear images of structures within a living biofilm, and removing surface layers from a biofilm would change the dynamics of a biofilm community. ▶**ANIMATIONS:** ***Light Microscopy***

Electron Microscopy

LEARNING **OUTCOME**

4.11 Contrast transmission electron microscopes with scanning electron microscopes in terms of how they work, the images they produce, and the advantages of each.

Even with the most expensive phase microscope using the best oil immersion lens with the highest numerical aperture, resolution is still limited by the wavelength of visible light. Because the shortest visible radiation (violet) has a wavelength of about 400 nm, structures closer together than about 200 nm cannot be distinguished using even the best light microscope. By contrast, electrons traveling as waves have wavelengths between 0.01 nm and 0.001 nm, which is one ten-thousandth to one

[3]From *coinciding focal* points of (laser) light.

◀ **FIGURE 4.10 Immunofluorescence. (a)** After a fluorescent dye is covalently linked to an antibody, the dye-antibody combination binds to the antibody's target, making the target visible under fluorescent microscopy. **(b)** Immunofluorescent staining of *Yersinia pestis*, the causative agent of plague. The bacteria are brightly colored against a dark background.

HIGHLIGHT

Studying Biofilms in Plastic "Rocks"

Marine stromatolites are unique rock structures made up of calcium carbonate. The insides of these rock structures are teeming with bacterial communities organized into complex biofilms. These biofilms contain many different microorganisms, including cyanobacteria, aerobic heterotrophic bacteria, and sulfate-reducing bacteria.

To study how the tiny creatures in marine stromatolites interact with each other and their environment, researchers at the University of Southern California pulled samples of the biofilms out of the rock structures and fixed them in nontoxic resin. In some cases, specific sections of the microbial cells were stained using fluorescent probes, and then confocal microscopy was used to study cross sections of the communities. By embedding the bacteria in a resin while maintaining the biofilm structure, the researchers were able to watch the bacterial network in an almost natural state. This technique is versatile and can be used to study other microbial systems to gain further understanding of biofilm mechanics and composition.

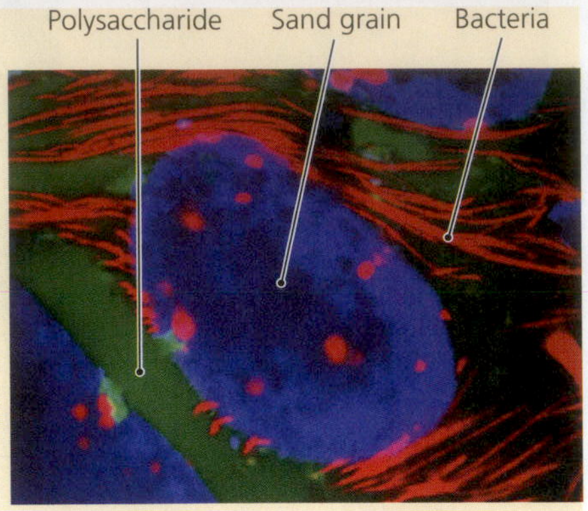

Marine stromatolite CM 25 µm

hundred-thousandth the wavelength of visible light. The resolving power of electron microscopes is therefore much greater than that of light microscopes, and with greater resolving power comes the possibility of greater magnification.

Generally, electron microscopes magnify objects 10,000× to 100,000×, though millions of times magnification with good resolution is possible. Electron microscopes provide detailed views of the smallest bacteria, viruses, internal cellular structures, and even molecules and large atoms. Cellular structures that can be seen only by using electron microscopy are referred to as a cell's *ultrastructure.* Ultrastructural details cannot be made visible by light microscopy because they are too small to be resolved.

There are two general types of electron microscopes: *transmission electron microscopes* and *scanning electron microscopes.*

Transmission Electron Microscopes

A **transmission electron microscope (TEM)** generates a beam of electrons that ultimately produces an image on a fluorescent screen **(FIGURE 4.11a)**. The path of electrons is similar to the path of light in a light microscope. From their source, the electrons pass through the specimen, through magnetic fields (instead of glass lenses) that manipulate and focus the beam, and then onto a fluorescent screen that absorbs electrons, thereby changing some of their energy into visible light **(FIGURE 4.11b)**. Dense areas of the specimen block electrons, resulting in a dark area on the screen. In regions where the specimen is less dense, the screen fluoresces more brightly. As with light microscopy, contrast and resolution can be enhanced through the use of electron-dense stains, which are discussed later. The brightness of each region of the screen corresponds to the number of electrons striking it. Therefore, the image on the screen is composed of light and dark areas, much like a photographic negative.

The screen can be folded out of the way to enable the electrons to strike a photographic film, located in the base of the microscope. Prints made from the film are called *transmission electron micrographs* or *TEM images* **(FIGURE 4.11c)**. Such images can be colorized to emphasize certain features.

Matter, including air, absorbs electrons, so the column of a transmission electron microscope must be a vacuum, and the specimen must be very thin. Before thicker specimens such as whole cells can be examined, they must be dehydrated, embedded in plastic, and cut to a thickness of about 100 nm with a diamond or glass knife mounted in a slicing machine called an *ultramicrotome.* Such thin sections are placed on a small copper grid and inserted into the microscope through an air lock.

Because the vacuum and slicing of the specimens are required, transmission electron microscopes cannot be used to study living organisms. Dehydration and sectioning can also introduce shrinkage, distortion, and other *artifacts,* which are structures that appear in a TEM image but are not present in natural specimens.

Scanning Electron Microscopes

A **scanning electron microscope (SEM)** also uses magnetic fields within a vacuum tube to manipulate a beam of electrons, called primary electrons **(FIGURE 4.12)**. However, rather than passing electrons through a specimen, the SEM rapidly focuses them back and forth across the specimen's surface, which has previously been coated with a metal such as platinum or gold. The primary electrons knock electrons off the surface of the coated specimen, and these scattered secondary electrons pass through a detector and a photomultiplier, producing an amplified signal that is displayed on a monitor. Typically, scanning microscopes are used to magnify up to 10,000× with a resolution of about 20 nm.

One advantage of scanning microscopy over transmission microscopy is that whole specimens can be observed because sectioning is not required. Scanning electron micrographs can be beautifully realistic and three-dimensional **(FIGURE 4.13)**. Two disadvantages of a scanning electron microscope are that

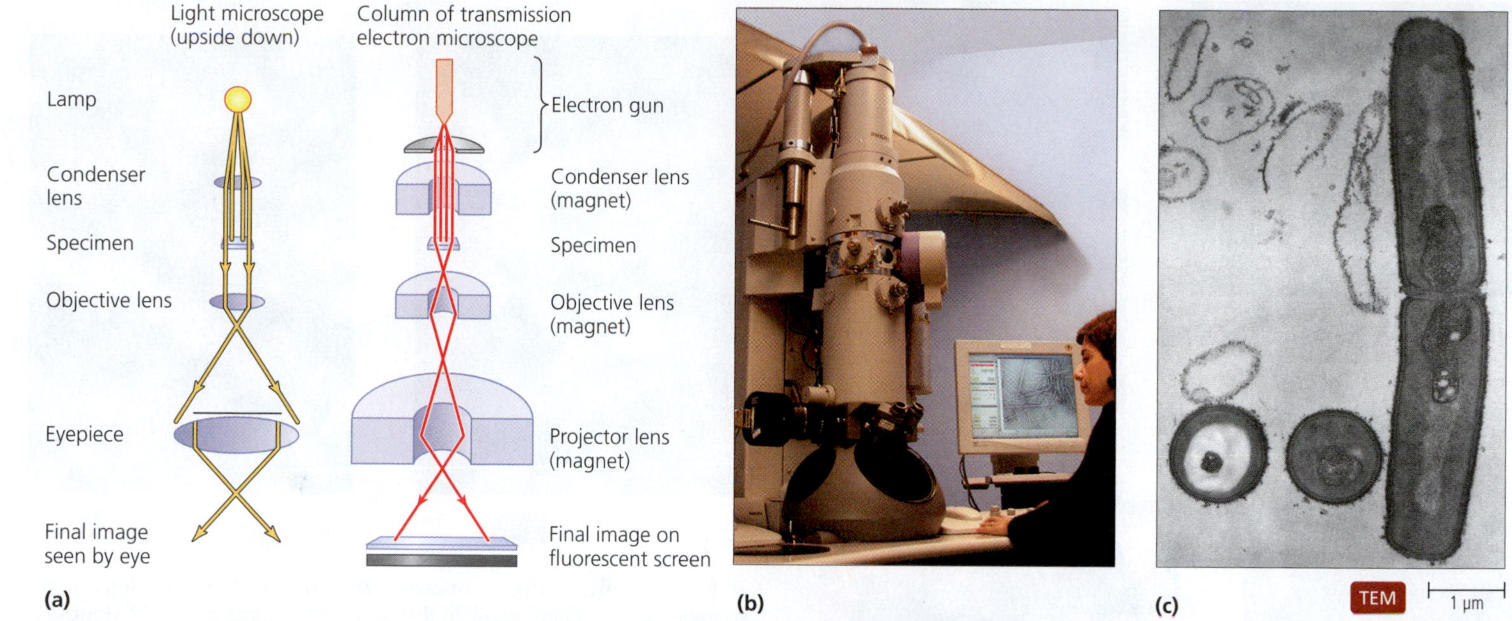

▲ **FIGURE 4.11 A transmission electron microscope (TEM).** **(a)** The path of electrons through a TEM, as compared to the path of light through a light microscope (at left, drawn upside down to facilitate the comparison). **(b)** TEMs are much larger than light microscopes. **(c)** Transmission electron image of a bacterium, *Bacillus subtilis.* A transmission electron micrograph reveals much internal detail not visible by light microscopy. *Why must air be evacuated from the column of an electron microscope?*

Figure 4.11 Air would absorb electrons, so there would be no radiation to produce an image.

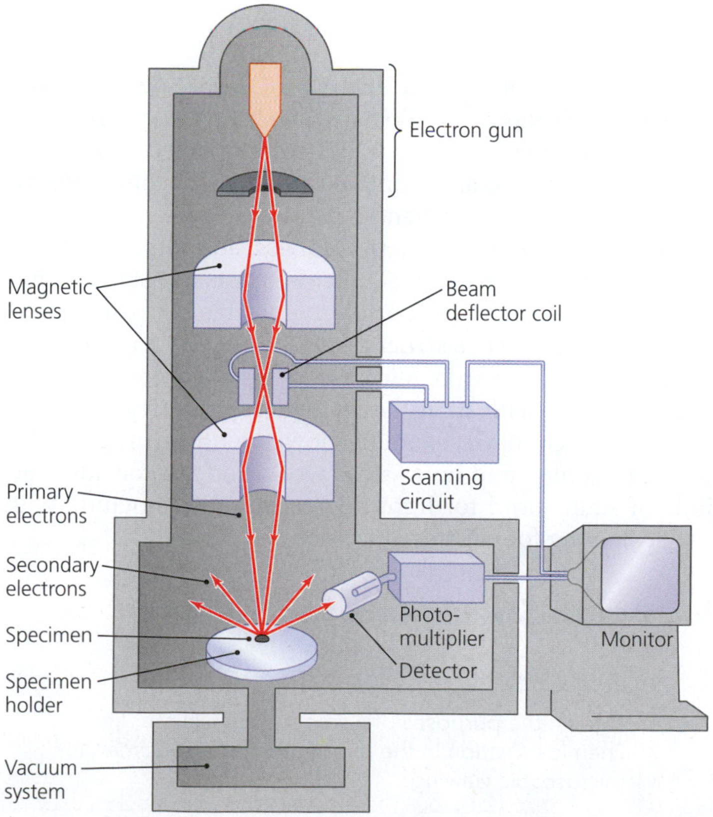

◀ **FIGURE 4.12 Scanning electron microscope (SEM).** The SEM uses magnetic lenses to focus a beam of primary electrons, which are scanned across the metal-coated surface of a specimen. Secondary electrons, knocked off the surface of the specimen by the primary electrons, are collected by a detector, and their signal is amplified and displayed on a monitor.

it magnifies only the external surface of a specimen and that, like TEM, it requires a vacuum and thus can examine only dead organisms. ▶**ANIMATIONS:** ***Electron Microscopy***

Probe Microscopy

LEARNING | **OUTCOME**

4.12 Describe two variations of probe microscopes.

A relatively recent advance in microscopy utilizes minuscule, pointed electronic probes to magnify more than 100,000,000×. There are two variations of probe microscopes: *scanning tunneling microscopes* and *atomic force microscopes.*

Scanning Tunneling Microscopes

A **scanning tunneling microscope (STM)** passes a metallic probe, sharpened to end in a single atom, back and forth across and slightly above the surface of a specimen. Rather than scattering a beam of electrons into a detector, as in scanning electron

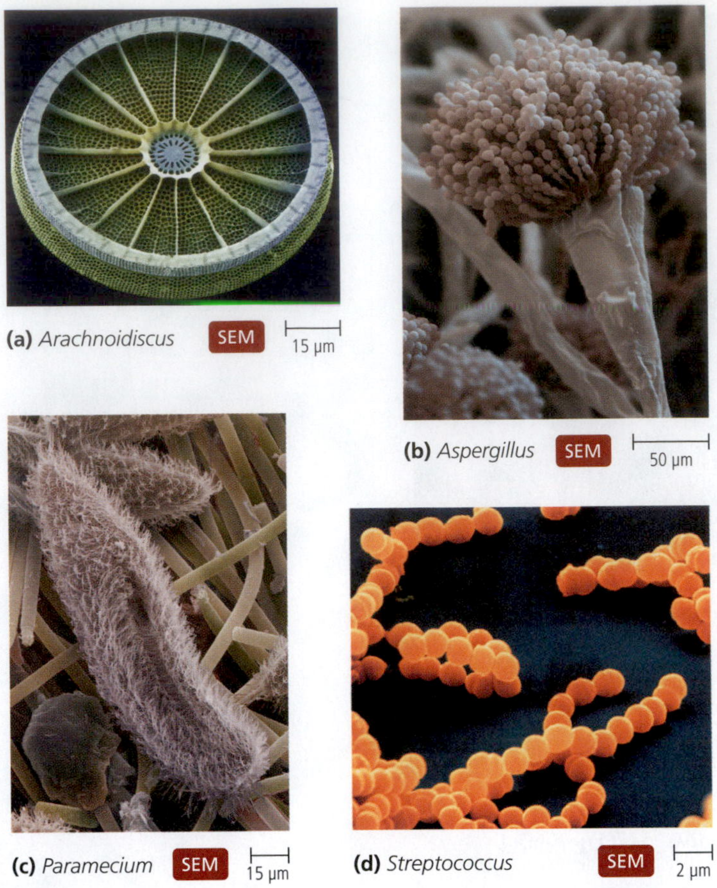

▲ **FIGURE 4.13 SEM images.** **(a)** *Arachnoidiscus*, a marine diatom (alga). **(b)** *Aspergillus*, a fungus. **(c)** *Paramecium*, a unicellular "animal" on top of rod-shaped bacteria. **(d)** *Streptococcus*, a bacterium.

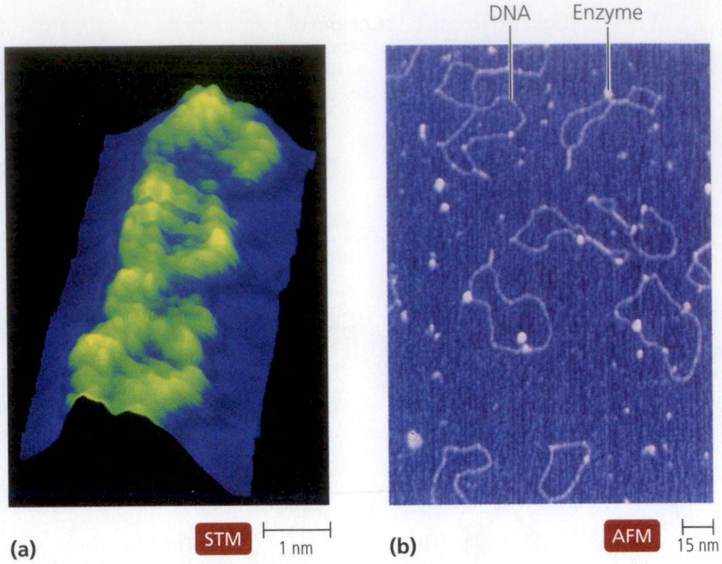

▲ **FIGURE 4.14 Probe microscopy.** **(a)** Scanning tunneling microscopes reveal surface detail; in this case, three turns of a DNA double helix. **(b)** Plasmid DNA being digested by an enzyme, viewed through atomic force microscopy.

microscopy, a scanning tunneling microscope measures the flow of electrons to and from the probe and the specimen's surface. The amount of electron flow, called a *tunneling current*, is directly proportional to the distance from the probe to the specimen's surface. A scanning tunneling microscope can measure distances as small as 0.01 nm and reveal details on the surface of a specimen at the atomic level (**FIGURE 4.14a**). A requirement for scanning tunneling microscopy is that the specimen be electrically conductive.

Atomic Force Microscopes

An **atomic force microscope (AFM)** also uses a pointed probe, but it traverses the tip of the probe lightly on the surface of the specimen rather than at a distance. This might be likened to the way a person reads Braille. Deflection of a laser beam aimed at the probe's tip measures vertical movements, which when translated by a computer reveals the atomic topography.

Unlike tunneling microscopes, atomic force microscopes can magnify specimens that do not conduct electrons. They can also magnify living specimens because neither an electron beam nor a vacuum is required (**FIGURE 4.14b**). Researchers have used these microscopes to magnify the surfaces of bacteria, viruses, proteins, and amino acids. Recent studies using atomic force microscopes have examined single living bacteria in three dimensions while they are dividing.

TABLE 4.2 summarizes the features of the various types of microscopes.

TELL ME WHY

Why is magnification high but color absent in an unretouched electron micrograph?

Staining

Earlier we discussed the difficulty of resolving two distant white golf balls viewed against a white background. If the balls were painted black, they could be distinguished more readily from the background and from one another. This illustrates why staining increases contrast and resolution.

Most microorganisms are colorless and difficult to view with bright-field microscopes. Microscopists use stains to make microorganisms and their parts more visible because stains increase contrast between structures and between a specimen and its background. Electron microscopy also requires that specimens be treated with stains or coatings to enhance contrast.

In this section we examine how scientists prepare specimens for staining and how stains work, and we consider seven kinds of stains used for light microscopy. We conclude with a look at staining for electron microscopy.

Preparing Specimens for Staining

LEARNING | OUTCOME

4.13 Explain the purposes of a smear, heat fixation, and chemical fixation in the preparation of a specimen for microscopic viewing.

TABLE 4.2 Comparison of Types of Microscopes

Type of Microscope	Typical Image	Description of Image	Special Features	Typical Uses
Light Microscopes		Useful magnification 1× to 2000×; resolution to 200 nm	Use visible light; shorter, blue wavelengths provide better resolution	
Bright field	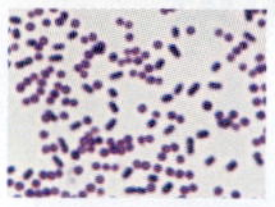	Colored or clear specimen against bright background	Simple to use; relatively inexpensive; stained specimens often required	To observe killed stained specimens and naturally colored live ones; also used to count microorganisms
Dark field		Bright specimen against dark background	Uses a special filter in the condenser that prevents light from directly passing through a specimen; only light scattered by the specimen is visible	To observe living, colorless, unstained organisms
Phase contrast	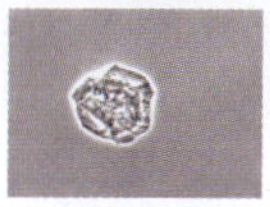	Specimen has light and dark areas	Uses a special condenser that splits a polarized light beam into two beams, one of which passes through the specimen, and one of which bypasses the specimen; the beams are then rejoined before entering the oculars; contrast in the image results from the interactions of the two beams	To observe internal structures of living microbes
Differential interference contrast (Nomarski)	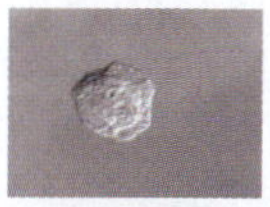	Image appears three-dimensional	Uses two separate beams instead of a split beam; false color and a three-dimensional effect result from interactions of light beams and lenses; no staining required	To observe internal structures of living microbes
Fluorescence	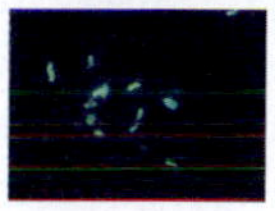	Brightly colored fluorescent structures against dark background	An ultraviolet light source causes fluorescent natural chemicals or dyes to emit visible light	To localize specific chemicals or structures; used as an accurate and quick diagnostic tool for detection of pathogens
Confocal	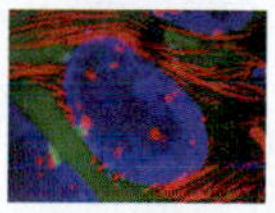	Single plane of structures or cells that have been specifically stained with fluorescent dyes	Uses a laser to fluoresce only one plane of the specimen at a time	Detailed observation of structures of cells within communities
Electron Microscopes		Typical magnification 1000× to 100,000×; resolution to 0.001 nm	Use electrons traveling as waves with short wavelengths; require specimens to be in a vacuum, so cannot be used to examine living microbes	
Transmission	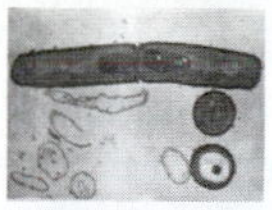	Monotone, two-dimensional, highly magnified images; may be color enhanced	Produces two-dimensional image of ultrastructure of cells	To observe internal ultrastructural detail of cells and observation of viruses and small bacteria
Scanning		Monotone, three-dimensional, surface images; may be color enhanced	Produces three-dimensional view of the surface of microbes and cellular structures	To observe the surface details of structures
Probe Microscopes		Magnification greater than 100,000,000× with resolving power greater than that of electron microscopes	Uses microscopic probes that move over the surface of a specimen	
Scanning tunneling	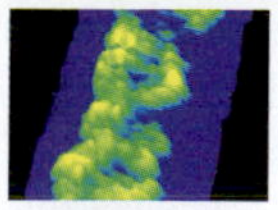	Individual molecules and atoms visible	Measures the flow of electrical current between the tip of a probe and the specimen to produce an image of the surface at atomic level	To observe the surface of objects; provide extremely fine detail, high magnification, and great resolution
Atomic force	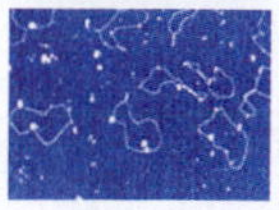	Individual molecules and atoms visible	Measures the deflection of a laser beam aimed at the tip of a probe that travels across the surface of the specimen	To observe living specimens at the molecular and atomic levels

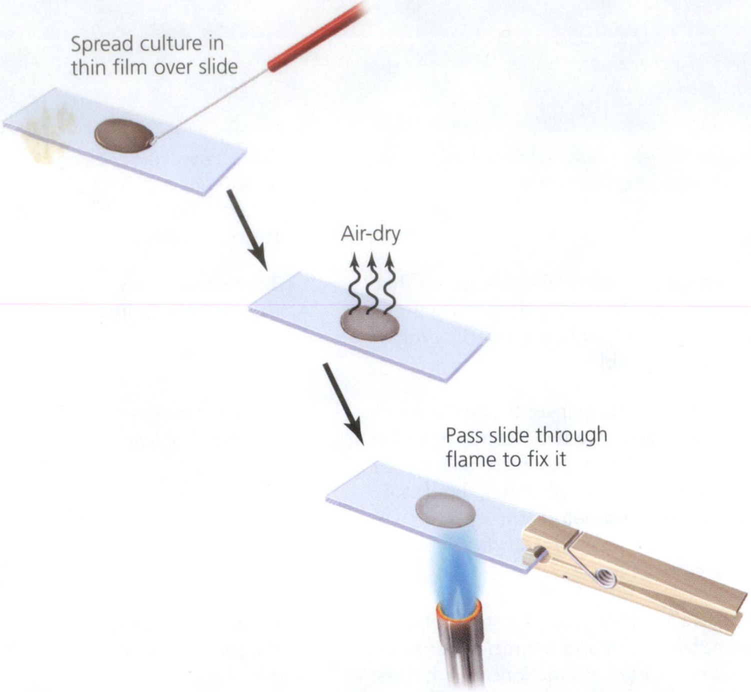
Spread culture in
thin film over slide
Air-dry
Pass slide through
flame to fix it

▲ **FIGURE 4.15 Preparing a specimen for staining.** Microorganisms are spread in liquid across the surface of a slide using a circular motion. After drying in the air, the smear is passed through the flame of a Bunsen burner to fix the cells to the glass. Alternatively, chemical fixation can be used. *Why must a smear be fixed to the slide?*

Figure 4.15 *Fixation causes the specimen to adhere to the glass so that it does not easily wash off during staining.*

Many investigations of microorganisms, especially those seeking to identify pathogens, begin with light microscopic observation of stained specimens. **Staining** simply means coloring specimens with stains, which are also called dyes.

Before microbiologists stain microorganisms, they must place them on and then firmly attach them to a microscope slide. Typically, this involves making a *smear* and *fixing* it to the slide **(FIGURE 4.15)**. If the organisms are growing in a liquid, a small drop is spread across the surface of the slide. If the organisms are growing on a solid surface, such as an agar plate, then they are mixed into a small drop of water on the slide. Either way, the thin film of organisms on the slide is called a **smear.**

The smear is air-dried and then attached or fixed to the surface of the slide. In **heat fixation,** developed more than a hundred years ago by Robert Koch, the slide is gently heated by passing the slide, smear up, through the flame of a Bunsen burner. Alternatively, **chemical fixation** involves applying a chemical such as methyl alcohol to the smear for one minute. Desiccation (drying) and fixation kill the microorganisms, attach them firmly to the slide, and generally preserve their shape and size. It is important to smear and fix specimens properly so that they are not lost during staining.

Specimens prepared for electron microscopy are also dried because water vapor from a wet specimen would stop the electron beam. As we have seen, transmission electron microscopy requires that the desiccated sample also be sliced very thin, generally before staining. Specimens for scanning electron microscopy are coated, not stained.

Principles of Staining

LEARNING | OUTCOME

4.14 Describe the uses of acidic and basic dyes, mentioning ionic bonding and pH.

Dyes used as microbiological stains for light microscopy are usually salts. A salt is composed of a positively charged *cation* and a negatively charged *anion*. At least one of the two ions in the molecular makeup of dyes is colored; this colored portion of a dye is known as the *chromophore*. Chromophores bind to chemicals via covalent, ionic, or hydrogen bonds. For example, methylene blue chloride is composed of a cationic chromophore, methylene blue, and a chloride anion. Because methylene blue is positively charged, it ionically bonds to negatively charged molecules in cells, including DNA and many proteins. In contrast, anionic dyes, for example, eosin, bind to positively charged molecules, such as some amino acids.

Anionic chromophores are also called **acidic dyes** because they stain alkaline structures and work best in acidic (low pH) environments. Positively charged, cationic chromophores are called **basic dyes** because they combine with and stain acidic structures; further, they work best under basic (higher pH) conditions. In microbiology, basic dyes are used more commonly than acidic dyes because most cells are negatively charged. Acidic dyes are used in negative staining, which is discussed shortly.

Some stains do not form bonds with cellular chemicals but rather function because of their solubility characteristics. For example, Sudan black selectively stains membranes because it is lipid soluble and accumulates in phospholipid bilayers.

Simple Stains

LEARNING | OUTCOME

4.15 Describe the simple, Gram, acid-fast, and endospore staining procedures.

Simple stains are composed of a single basic dye, such as crystal violet, safranin, or methylene blue. They are "simple" because they involve no more than soaking the smear in the dye for 30–60 seconds and then rinsing off the slide with water. (A properly fixed specimen will remain attached to the slide despite this treatment.) After carefully blotting the slide dry, the microbiologist observes the smear under the microscope. Simple stains are used to determine size, shape, and arrangement of cells **(FIGURE 4.16)**.

Differential Stains

Most stains used in microbiology are **differential stains,** which use more than one dye so that different cells, chemicals, or

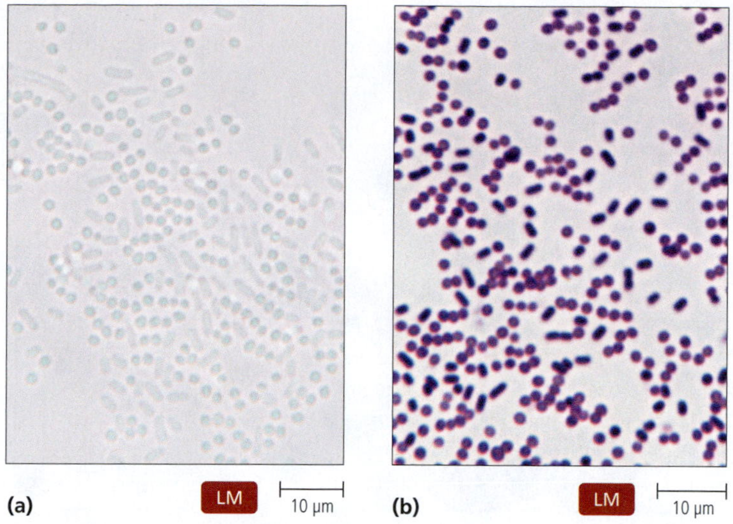

▲ **FIGURE 4.16 Simple stains.** Simply and quickly performed, simple stains increase contrast and allow determination of size, shape, and arrangement of cells. **(a)** Unstained *Escherichia coli* and *Staphylococcus aureus.* **(b)** Same mixture stained with crystal violet. Note that all cells, no matter their type, stain almost the same color with a simple stain because only one dye is used.

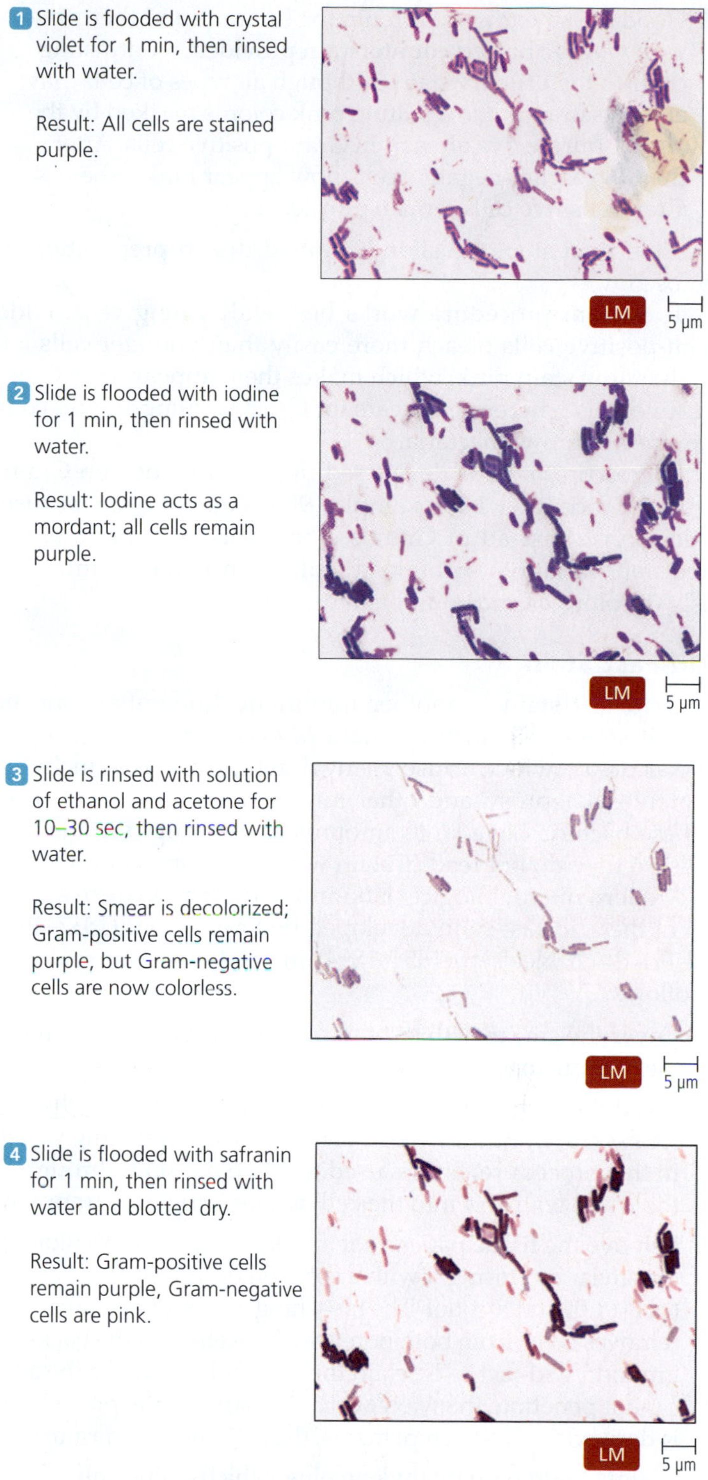

1 Slide is flooded with crystal violet for 1 min, then rinsed with water.
Result: All cells are stained purple.
LM
5 μm
2 Slide is flooded with iodine for 1 min, then rinsed with water.
Result: Iodine acts as a mordant; all cells remain purple.
LM
5 μm
3 Slide is rinsed with solution of ethanol and acetone for 10–30 sec, then rinsed with water.
Result: Smear is decolorized; Gram-positive cells remain purple, but Gram-negative cells are now colorless.
LM
5 μm
4 Slide is flooded with safranin for 1 min, then rinsed with water and blotted dry.
Result: Gram-positive cells remain purple, Gram-negative cells are pink.
LM
5 μm

▲ **FIGURE 4.17 The Gram staining procedure.** A specimen is smeared and fixed to a slide. The classical procedure consists of four steps. Gram-positive cells (in this case, *Bacillus cereus*) remain purple throughout the procedure; Gram-negative cells (here, *Escherichia coli*) end up pink.

structures can be distinguished when microscopically examined. Common differential stains are the *Gram stain*, the *acid-fast stain*, the *endospore stain*, *Gomori methenamine silver stain*, and *hematoxylin and eosin stain*.

Gram Stain

In 1884, the Danish scientist Hans Christian Gram developed the most frequently used differential stain, which now bears his name. The **Gram stain** differentiates between two large groups of microorganisms: purple-staining Gram-positive cells and pink-staining Gram-negative cells. These cells differ significantly in the chemical and physical structures of their cell walls (see Figure 3.15). Typically, a Gram stain is the first step a medical laboratory technologist performs to identify bacterial pathogens.

Let's examine the Gram staining procedure as it was originally developed and as it is typically performed today, over a hundred years later. For the purposes of our discussion, we'll assume that a smear has been made on a slide and heat fixed and that the smear contains both Gram-positive and Gram-negative colorless bacteria. The classical Gram staining procedure has the following four steps **(FIGURE 4.17)**:

1. Flood the smear with the basic dye crystal violet for 1 minute and then rinse with water. Crystal violet, which is called the **primary stain,** colors all cells.
2. Flood the smear with an iodine solution for 1 minute and then rinse with water. Iodine is a **mordant,** a substance that binds to a dye and makes it less soluble. After this step, all cells remain purple.
3. Rinse the smear with a solution of ethanol and acetone for 10–30 seconds and then rinse with water. This solution, which acts as a **decolorizing agent,** breaks down the thin cell wall of Gram-negative cells, allowing the stain and mordant to be washed away; these cells are now colorless. Gram-positive cells, with their thicker cell walls, remain purple.

4 Flood the smear with safranin for 1 minute and then rinse with water. This red **counterstain** provides a contrasting color to the primary stain. Although all types of cells may absorb safranin, the resulting pink color is masked by the darker purple dye already in Gram-positive cells. After this step, Gram-negative cells now appear pink, whereas Gram-positive cells remain purple.

After the final step, the slide is blotted dry in preparation for microscopy.

The Gram procedure works best with young cells. Older Gram-positive cells bleach more easily than younger cells and can therefore stain pink, which makes them appear to be Gram-negative cells. Therefore, smears for Gram staining should come from freshly grown bacteria.

Microscopists have developed minor variations on Gram's original procedure. For example, 95% ethanol may be used to decolorize instead of Gram's ethanol-acetone mixture. In a three-step variation, safranin dissolved in ethanol simultaneously decolorizes and counterstains.

Acid-Fast Stain

The **acid-fast stain** is another important differential stain because it stains cells of the genera *Mycobacterium* and *Nocardia* (nō-kar´dē-ă), which cause many human diseases, including tuberculosis, leprosy, and other lung and skin infections. Cells of these bacteria have large amounts of waxy lipid in their cell walls, so they do not readily stain with the Gram stain.

Modern microbiological laboratories commonly use a variation of the acid-fast stain developed by Franz Ziehl (1857–1926) and Friedrich Neelsen (1854–1894) in 1883. Their procedure is as follows:

1. Cover the smear with a small piece of tissue paper to retain the dye during the procedure.
2. Flood the slide with the red primary stain, carbolfuchsin for several minutes while warming it over steaming water. In this procedure, heat is used to drive the stain through the waxy wall and into the cell, where it remains trapped.
3. Remove the tissue paper, cool the slide, and then decolorize the smear by rinsing it with a solution of hydrochloric acid (pH < 1.0) and alcohol. The bleaching action of acid-alcohol removes color from both non-acid-fast cells and the background. Acid-fast cells retain their red color because the acid cannot penetrate the waxy wall. The name of the procedure is derived from this step; that is, the cells are colorfast in acid.
4. Counterstain with methylene blue, which stains only bleached, non-acid-fast cells.

The Ziehl-Neelsen acid-fast staining procedure results in pink acid-fast cells, which can be differentiated from blue non-acid-fast cells, including human cells and tissue **(FIGURE 4.18)**. The presence of *acid-fast bacilli (AFBs)* in sputum is indicative of mycobacterial infection.

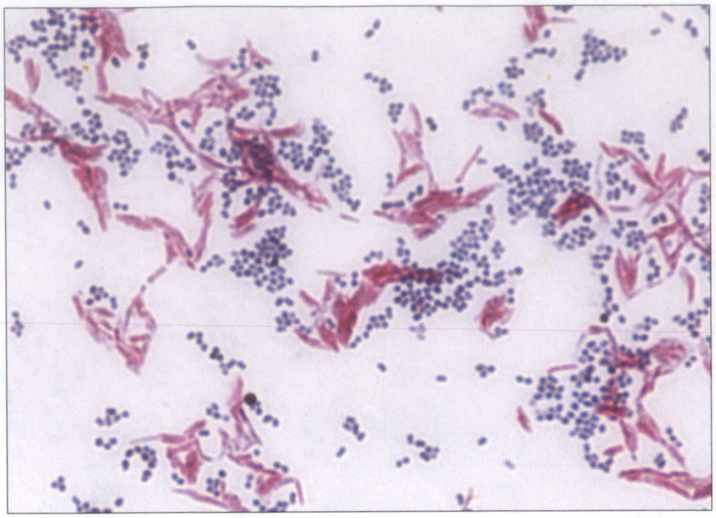

LM 5 µm

▲ **FIGURE 4.18 Ziehl-Neelsen acid-fast stain.** Acid-fast cells such as these rod-shaped *Mycobacterium bovis* cells stain pink or red. Non-acid-fast cells—in this case, *Staphylococcus*—stain blue. *Why isn't the Gram stain utilized to stain* Mycobacterium?

Figure 4.18 *Cell walls of Mycobacterium are composed of waxy materials that repel the water-based dyes of the Gram stain.*

Endospore Stain

Some bacteria—notably those of the genera *Bacillus* and *Clostridium* (klos-trid´ē-ŭm), which contain species that cause such diseases as anthrax, gangrene, and tetanus—produce **endospores.** These dormant, highly resistant cells form inside the cytoplasm of the bacteria and can survive environmental extremes such as desiccation, heat, and harmful chemicals. Endospores cannot be stained by normal staining procedures because their walls are practically impermeable to all chemicals. The **Schaeffer-Fulton endospore stain** uses heat to drive the primary stain, *malachite green,* into the endospore. After cooling, the slide is decolorized with water and counterstained with safranin. This staining procedure results in green-stained endospores and red-colored vegetative cells **(FIGURE 4.19)**.

Histological Stains

Laboratory technicians use two popular stains to stain histological specimens, that is, tissue samples. *Gomori*[4] *methenamine silver (GMS) stain* is commonly used to screen for the presence of fungi and the locations of carbohydrates in tissues. *Hematoxylin and eosin (HE) stain,* which involves applying the basic dye hematoxylin and the acidic dye eosin, is used to delineate many features of histological specimens, such as the presence of cancer cells.

Special Stains

Special stains are simple stains designed to reveal special microbial structures. There are three types of special stains: *negative stains, flagellar stains,* and *fluorescent stains* (which we already discussed in the section on fluorescent microscopy).

[4]Named for George Gomori, a noted Hungarian American histologist.

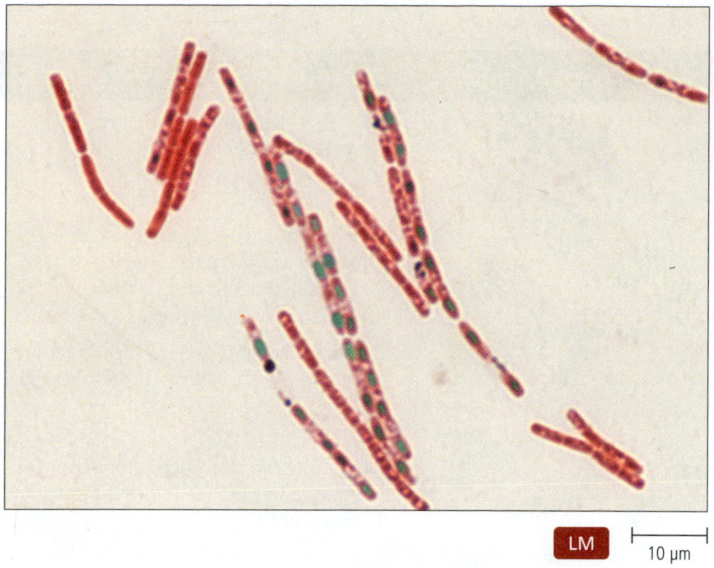

▲ **FIGURE 4.19 Schaeffer-Fulton endospore stain of *Bacillus anthracis*.** The nearly impermeable spore wall retains the green dye during decolorization. Vegetative cells, which lack spores, pick up the counterstain and appear red. *Why don't the spores stain red as well?*

***Figure 4.19** Heat from steam is used to drive the green primary stain into the endospores. Counterstaining is performed at room temperature, and the thick, impermeable walls of the endospores resist the counterstain.*

Negative (Capsule) Stain

Most dyes used to stain bacterial cells, such as crystal violet, methylene blue, malachite green, and safranin, are basic dyes. These dyes stain cells by attaching to negatively charged molecules within them.

Acidic dyes, by contrast, are repulsed by the negative charges on the surface of cells and therefore do not stain them. Such stains are called **negative stains** because they stain the background and leave cells colorless. Eosin and nigrosin are examples of acidic dyes used for negative staining. A counterstain may be added to color the cells.

Negative stains are used primarily to reveal the presence of negatively charged bacterial capsules. Therefore, they are also called **capsule stains.** Encapsulated cells appear to have a halo surrounding them (**FIGURE 4.20**).

Flagellar Stain

Bacterial flagella are extremely thin and thus normally invisible with light microscopy, but their presence, number, and arrangement are important in identifying some species, including some pathogens. Flagellar stains, such as pararosaniline and carbolfuchsin, and mordants, such as tannic acid and potassium alum, are applied in a series of steps. These molecules bind to the flagella, increase their diameter, and change their color, all of which increase contrast and make them visible (**FIGURE 4.21**).

Stains used for light microscopy are summarized in **TABLE 4.3** on p. 112. **Beneficial Microbes: Glowing Viruses** on p. 112 illustrates a unique kind of stain that uses viruses to stain

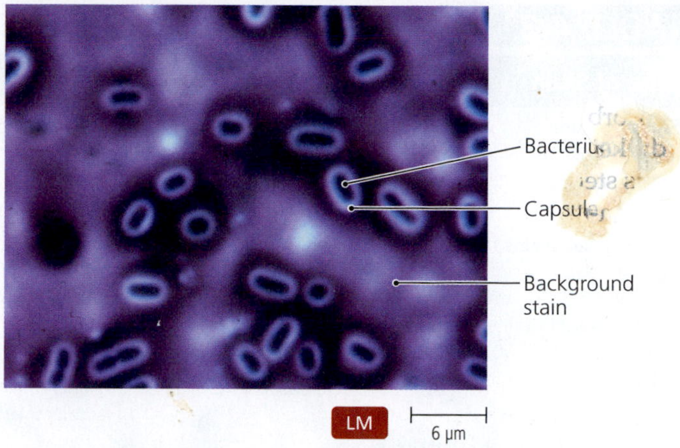

▲ **FIGURE 4.20 Negative (capsule) stain of *Klebsiella pneumoniae*.** Notice that the acidic dye stains the background and does not penetrate the capsule.

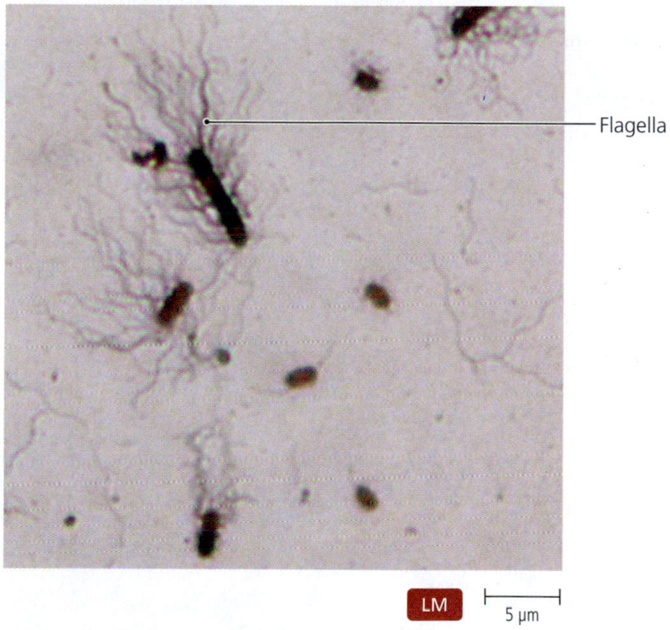

▲ **FIGURE 4.21 Flagellar stain of *Proteus vulgaris*.** Various bacteria have different numbers and arrangements of flagella, features that might be important in identifying some species. *How can the flagellar arrangement shown here be described?*

***Figure 4.21** Peritrichous.*

particular strains of bacteria that are then viewed through a fluorescent microscope.

Staining for Electron Microscopy

LEARNING **OUTCOME**

4.16 Explain how stains used for electron microscopy differ from those used for light microscopy.

Laboratory technicians increase contrast and resolution for transmission electron microscopy by using stains, just as they do for light microscopy. However, stains used for transmission

TABLE 4.3 Some Stains Used for Light Microscopy

Type of Stain	Examples	Results	Typical Images	Representative Uses
Simple stains (use a single dye)	Crystal violet Methylene blue	Uniform purple stain Uniform blue stain	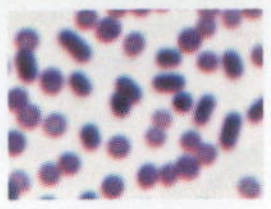	Reveals size, morphology, and arrangement of cells
Differential stains (use two or more dyes to differentiate between cells or structures)	Gram stain	Gram-positive cells are purple; Gram-negative cells are pink	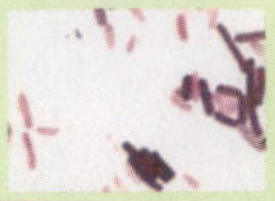	Differentiates Gram-positive and Gram-negative bacteria, which is typically the first step in their identification
	Ziehl-Neelsen acid-fast stain	Pink to red acid-fast cells and blue non-acid-fast cells	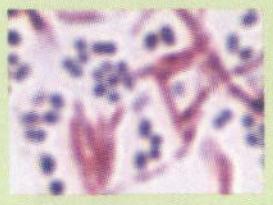	Distinguishes the genera *Mycobacterium* and *Nocardia* from other bacteria
	Schaeffer-Fulton endospore stain	Green endospores and pink to red vegetative cells	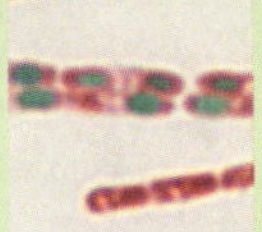	Highlights the presence of endospores produced by species in the genera *Bacillus* and *Clostridium*
Special stains	Negative stain for capsules	Background is dark, cells unstained or stained with simple stain	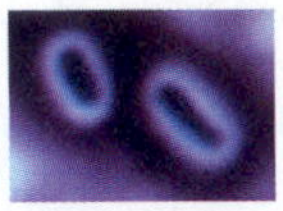	Reveals bacterial capsules
	Flagellar stain	Bacterial flagella become visible	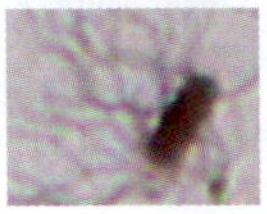	Allows determination of number and location of bacterial flagella

BENEFICIAL MICROBES

Glowing Viruses

A bacteriophage is a virus that inserts its DNA into a bacterium. Commonly called a *phage*, it adheres only to a select bacterial strain for which each phage type has a specific adhesion factor. Many phages are so specialized for their particular bacterial strain that scientists have used phages to identify and classify bacteria. Such identification is called *phage typing*.

Scientists at San Diego State University have taken phage specificity a step further. They successfully linked a fluorescent dye to the DNA of phages of the bacterium *Salmonella* and used the phages to detect and identify *Salmonella* species. Such fluorescent phages rapidly and accurately detect specific strains of *Salmonella* in mixed bacterial cultures.

Fluorescent phages have advantages over fluorescent antibodies: unlike antibodies, phages are not metabolized by bacteria. Phages are also more stable over time and are not as sensitive to vagaries in temperature, pH, and ionic strength. Further, fluorescent phages have a long shelf life; they protect the fluorescent dye inside their phage coat until the dyed DNA is injected.

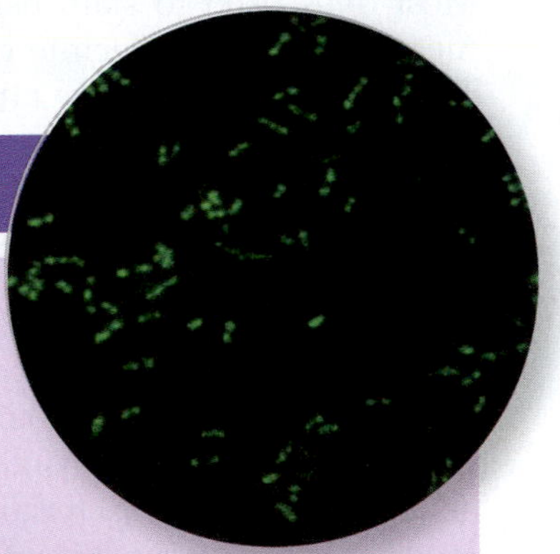

Fluorescent phages light up bacteria. LM 20 µm

There are numerous uses for test kits using fluorescent phages. Environmental scientists could use them to detect bacterial contamination of streams and lakes, food processors could identify potentially fatal *Escherichia coli* strain O157:H7 in meat and vegetables, or homeland security agents could positively establish the presence or absence of bacteria used for biological warfare. Antibody-based kits frequently failed to accurately detect *Bacillus anthracis* used in the 2001 terrorist attacks. Fluorescent phage field kits should be much more robust and precise.

electron microscopy are not colored dyes but instead chemicals containing atoms of heavy metals, such as lead, osmium, tungsten, and uranium, which absorb electrons. Electron-dense stains may bind to molecules within specimens, or they may stain the background. The latter type of negative staining is used to provide contrast for extremely small specimens, such as viruses and molecules.

Stains for electron microscopy can be general in that they stain most objects to some degree, or they may be highly specific. For example, osmium tetroxide (O_sO_4) has an affinity for lipids and is thus used to enhance the contrast of membranes. Electron-dense stains can also be linked to antibodies to provide an even greater degree of staining specificity because antibodies bind only to their specific target molecules. ▶**ANIMATIONS:** ***Staining***

TELL ME WHY

Why is a Gram-negative bacterium colorless but a Gram-positive bacterium purple after it is rinsed with decolorizer?

Classification and Identification of Microorganisms

LEARNING **OUTCOME**

4.17 Discuss the purposes of classification and identification of organisms.

Biologists classify organisms for several reasons: to bring a sense of order and organization to the variety and diversity of living things, to enhance communication, to make predictions about the structure and function of similar organisms, and to uncover and understand potential evolutionary connections. They sort organisms on the basis of mutual similarities into nonoverlapping groups called **taxa**.[5] **Taxonomy**[6] is the science of classifying and naming organisms. Taxonomy consists of *classification,* which is the assigning of organisms to taxa based on similarities; *nomenclature,* which is concerned with the rules of naming organisms; and *identification,* which is the practical science of determining that an isolated individual or population belongs to a particular taxon. In this book we concentrate on classification and identification.

Because all members of any given taxon share certain common features, taxonomy enables scientists both to organize large amounts of information about organisms and to make predictions based on knowledge of similar organisms. For example, if one member of a taxon is important in recycling nitrogen in the environment, it is likely that others in the group will play a similar ecological role. Similarly, a clinician might suggest a treatment against one pathogen based on what has been effective against another pathogen in the same taxon.

[5]From Greek *taxis,* meaning "order."

[6]From *taxis* and Greek *nomos,* meaning "rule."

Identification of organisms is an essential part of taxonomy because it enables scientists to communicate effectively and be confident that they are discussing the same organism. Further, identification is often essential for treating groups of diseases, such as meningitis and pneumonia, which can be caused by pathogens as different as fungi, bacteria, and viruses.

In this section we examine the historical basis of taxonomy, consider modern advances in this field, and briefly consider various taxonomic methods. This chapter also presents a general overview of the taxonomy of prokaryotes (which are considered in greater detail in Chapter 11); of animals, protozoa, fungi, and algae (Chapter 12); and of viruses, viroids, and prions (Chapter 13).

Linnaeus and Taxonomic Categories

LEARNING **OUTCOMES**

4.18 Discuss the difficulties in defining species of microorganisms.

4.19 List the hierarchy of taxa from general to specific.

4.20 Define binomial nomenclature.

4.21 Describe a few modifications of the Linnaean system of taxonomy.

Our current system of taxonomy began in 1753 with the publication of *Species Plantarum* by the Swedish botanist Carolus Linnaeus (1707–1778). Until his time, the names of organisms were often strings of descriptive terms that varied from country to country and from one scientist to another. Linnaeus provided a system that standardized the naming and classification of organisms based on characteristics they have in common. He grouped similar organisms that can successfully interbreed into categories called **species**.

The definition of *species* as "a group of organisms that interbreed to produce viable offspring" works relatively well for more complex, sexually reproducing organisms, but it is not satisfactory for asexual organisms, including most microorganisms. As a result, some scientists define a microbial species as a collection of *strains*—populations of cells that arose from a single cell—that share many stable properties, differ from other strains, and evolve as a group. Alternatively, biologists define a microbial species as cells that share at least 70% common sequences of DNA. Not surprisingly, these definitions sometimes result in disagreements and inconsistencies in the classification of microbial life. Some researchers question whether microbes exist as unique species at all.

In Linnaeus's system, which forms the basis of modern taxonomy, similar species are grouped into **genera**[7] and similar genera into still larger taxonomic categories. That is, genera sharing common features are grouped together to form **families**; similar families are grouped into **orders**; orders are grouped into **classes**; classes into **phyla**;[8] and phyla into **kingdoms** (**FIGURE 4.22**).

[7]Plural of Latin *genus,* meaning "race" or "birth."

[8]Plural of *phylum,* from Greek *phyllon,* meaning "tribe." The term *phylum* is used for animals and bacteria; the corresponding taxon in mycology and botany is called a *division.*

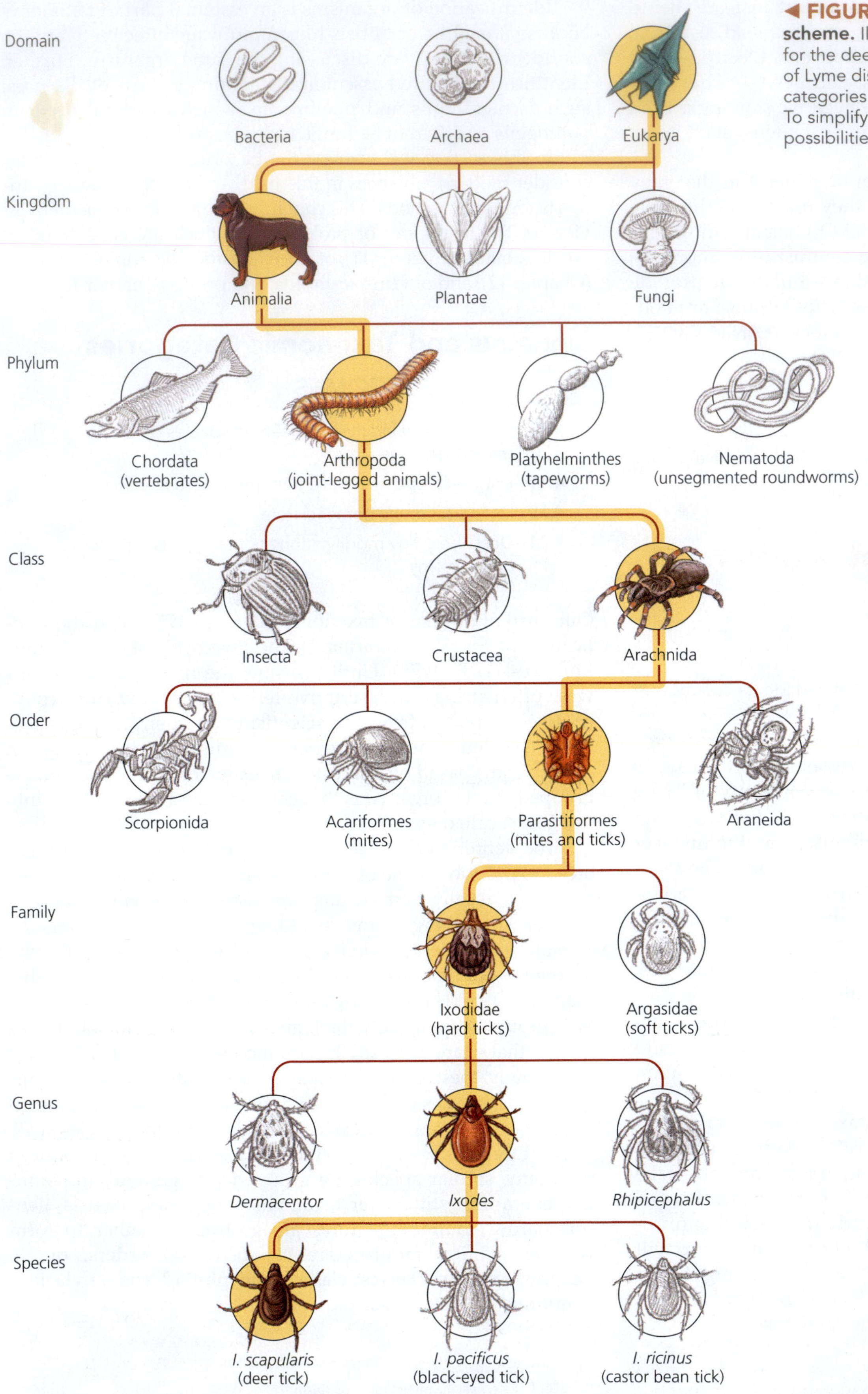

◀ **FIGURE 4.22 Levels in a Linnaean taxonomic scheme.** Illustrated here are the taxonomic categories for the deer tick, *Ixodes scapularis*, the primary vector of Lyme disease. Notice that higher taxonomic categories are more inclusive than lower ones. To simplify the diagram, only selected classification possibilities are shown.

All these categories, including species and genera, are taxa, which are hierarchical; that is, each successive taxon has a broader description than the preceding one, and each taxon includes all the taxa beneath it. The rules of nomenclature require that all taxa have Latin or Latinized names, in part because the language of science during Linnaeus's time was Latin and in part because using Latin ensures that no country or ethnic group has priority in the language of taxonomy. The name *Chondrus crispus* (kon´drŭs krisp´ŭs) describes the exact same algal species all over the world despite the fact that in England its common name is Irish moss, in Ireland it is carragheen, in North America it is curly moss, and it isn't really a moss at all!

When new microscopic, genetic, or biochemical techniques identify new or more detailed characteristics of organisms, a taxon may be split into two or more taxa. Alternatively, several taxa may be lumped together into a single taxon. For example, the genus "*Diplococcus*" has been united (synonymized) with the genus *Streptococcus* (strep-tō-kok´ŭs), and the name *Diplococcus pneumoniae* (nū-mō´nē-ī) has been changed to reflect this synonymy—to *Streptococcus pneumoniae*.

Linnaeus assigned each species a descriptive name consisting of its genus name and a **specific epithet**. The genus name is always a noun, and it is written first and capitalized. The specific epithet always contains only lowercase letters and is usually an adjective. Both names, together called a *binomial*, are either printed in italics or underlined. Because the Linnaean system assigns two names to every organism, it is said to use **binomial**[9] **nomenclature**.

Consider the following examples of binomials. *Enterococcus faecalis* (en´ter-ō-kok´ŭs fē-kă´lis)[10] is a fecal bacterium. Whereas *Enterococcus faecium* (fē-sē´ŭm) is in the same genus, it is a different species because of certain differing characteristics, so it is given a different specific epithet. Humans are classified as *Homo sapiens* (hō´mō sā´pē-enz); the genus name means "man," and the specific epithet means "wise."

Note that even though binomials are often descriptive of an organism, sometimes they can be misleading. For instance, *Haemophilus influenzae* (hē-mof´i-lŭs in-flu-en´zī) does not cause influenza. In some cases, binomials honor people. Examples include *Pasteurella haemolytica* (pas-ter-el´ă hē-mō-lit´i-kă), a bacterium named after the microbiologist Louis Pasteur; *Escherichia coli* (esh-ĕ-rik´ē-ă kō´lī), a bacterium named after the physician Theodor Escherich (1857–1911); and *Izziella abbottiae*, (iz-ē-el´lă ab´ot-tē-ī), a marine alga named after the phycologist and taxonomist Isabella Abbott (1919–2010).

Most scientists still use the Linnaean system today, though significant modifications have been adopted. For example, scientists sometimes use additional categories, such as tribes, sections, subfamilies, and subspecies.[11] Further, Linnaeus divided all organisms into only two kingdoms (Plantae and Animalia), and he did not know of the existence of viruses. As scientists learned more about organisms, they adopted taxonomic schemes to reflect their advances in knowledge. For example, a widely accepted taxonomic approach was based on five kingdoms: Animalia, Plantae, Fungi, Protista, and Prokaryotae. Though widely accepted, this scheme grouped together in the kingdom Protista such obviously disparate organisms as massive brown seaweeds (kelps) and unicellular, animal-like microbes. Further, because it does not address the taxonomy of viruses or all the differences between microorganisms, scientists have proposed other taxonomic schemes that have from 4 to more than 50 kingdoms.

Sometimes students are upset that taxonomists do not agree about all the taxonomic categories or the species they contain, but it must be realized that classification of organisms reflects the state of our current knowledge and theories. Taxonomists change their schemes to accommodate new information, and not every expert agrees with every proposed modification.

Another significant development in taxonomy is a shift in its basic goal—from Linnaeus's goal of classifying and naming organisms as a means of cataloging them to the more modern goal of understanding the relationships among groups of organisms. Linnaeus based his taxonomic scheme primarily on organisms' structural similarities, and whereas such terms as *genus* and *family* may suggest the existence of some common lineage, he and his contemporaries thought of species as divinely created. However, when Charles Darwin (1809–1882) propounded his theory of the evolution of species by natural selection (a century after Linnaeus published his pivotal work on the taxonomy of plants), taxonomists came to consider that common ancestry largely explains the similarities among organisms in the various taxa. Today, most taxonomists agree that a major goal of modern taxonomy is to reflect a *phylogenetic*[12] *hierarchy;* that is, that the ways in which organisms are grouped should reflect their evolution from common ancestors.

Taxonomists' efforts to classify organisms according to their ancestry have resulted in reduced emphasis on comparisons of physical and chemical traits, and in greater emphasis on comparisons of their genetic material. Such work has led to a proposal to add a new, most inclusive taxon: the *domain*.

Domains

LEARNING **OUTCOME**

4.22 List and describe the three domains proposed by Carl Woese.

Carl Woese (1928–2012) labored for years to understand the taxonomic relationships among cells. Morphology (shape) and biochemical tests did not provide enough information to classify organisms fully, so for over a decade Woese painstakingly sequenced the nucleotides of the smaller subunits of ribosomal RNA (rRNA) in an effort to unravel the relationships among these organisms. Because these rRNA molecules are present in all cells and are crucial to protein synthesis, changes in their nucleotide sequences are presumably very rare.

---

[9]From Greek *bi*, meaning "two," and *nomos*, meaning "rule."

[10]From Greek *enteron*, meaning "intestine"; *kokkos*, meaning "berry"; and Latin *faeces*.

[11]Subspecies, which are also called *varieties* or *strains*, differ only slightly from each other. Taxonomists disagree on how much difference between two microbes constitutes a strain versus a species.

[12]From Greek *phyllon*, meaning "tribe," and Latin *genus*, meaning "birth" (i.e., origin of a group).

In 1976 he sequenced rRNA from an odd group of prokaryotes that produce methane gas as a metabolic waste. Woese was surprised when their rRNA did not contain nucleotide sequences characteristic of bacteria. Repeated testing showed that methanogens, as they are called, were not like other prokaryotic or eukaryotic organisms. They were something new to science, a third branch of life, ushering microbiologists into a new and curious wonderland. Woese and his coworkers discovered that there are three basic types of ribosome, leading them to propose a new classification scheme in which a new taxon, called a **domain,** contains the Linnaean taxon of kingdom. The three domains identified by Woese—**Eukarya**, **Bacteria**, and **Archaea**—are based on three basic types of cells as determined by ribosomal nucleotide sequences.

Domain Eukarya includes all eukaryotic cells, all of which contain eukaryotic rRNA sequences. Domains Bacteria and Archaea include all prokaryotic cells. They contain bacterial and archaeal rRNA sequences, respectively, which differ significantly from one another and from those in eukaryotic cells. In addition to differences in rRNA sequences, cells of the three domains differ in many other characteristics, including the lipids in their cell membranes, transfer RNA (tRNA) molecules, and sensitivity to antibiotics. (The taxonomy of organisms within the three domains is discussed in Chapters 11 and 12, which cover prokaryotes and eukaryotes, respectively.)

Ribosomal nucleotide sequences further suggest that there may be at least 50 kingdoms of Bacteria and three kingdoms of Archaea. Further, scientists examining substances such as human saliva, water, soil, and rock regularly discover novel nucleotide sequences. When these sequences are compared to known sequences stored in a computer database, they cannot be associated with any previously identified organism. This suggests that many curious new forms of microbial life have never been grown in a laboratory and still await discovery.

Ribosomal nucleotide sequences have also given microbiologists a new way to define prokaryotic species. Some scientists propose that prokaryotes whose rRNA sequence differs from that of other prokaryotes by more than 3% be classified as a distinct species. Although this definition has the advantage of being precise, not all taxonomists agree with it.

Taxonomic and Identifying Characteristics

LEARNING | **OUTCOME**

4.23 Describe five procedures taxonomists use to identify and classify microorganisms.

Other criteria and laboratory techniques used for classifying and identifying microorganisms are quite numerous and include macroscopic and microscopic examination of physical characteristics, differential staining characteristics, growth characteristics, microorganisms' interactions with antibodies, microorganisms' susceptibilities to viruses, nucleic acid analysis, biochemical tests, and organisms' environmental requirements, including the temperature and pH ranges of their various types of habitats. Clearly, then, microbial taxonomy is too broad a subject to cover in one chapter, and thus the details of the criteria for the classification of major groups are provided in subsequent chapters.

It is important to note that even though scientists may use a given technique to either classify or identify microorganisms, the criteria used to identify a particular organism are not always the same as those that were used to classify it. For example, even though medical laboratory scientists distinguish the genus *Escherichia* from other bacterial genera by its inability to utilize citric acid (citrate) as a sole carbon source, this characteristic is not vital in the classification of *Escherichia.*

Bergey's Manual of Determinative Bacteriology, first published in 1923 and now in its ninth edition (1994), contains information used for the laboratory identification of prokaryotes. *Bergey's Manual of Systematic Bacteriology* (second edition, 2001, 2005, 2009, 2010) is a similar reference work that is used for classification based on ribosomal RNA sequences, which taxonomists use to describe relationships among organisms. Each of these manuals is known as "Bergey's Manual."

Linnaeus did not know of the existence of viruses and thus did not include them in his original taxonomic hierarchy, nor are viruses assigned to any of the five kingdoms or Woese's three domains because viruses are acellular and generally lack rRNA. Virologists do classify viruses into families and genera, but higher taxa are poorly defined for viruses. (Chapter 13 further discusses viral taxonomy.)

With this background, let's turn now to some brief discussions of five types of information that microbiologists commonly use to distinguish among microorganisms: physical characteristics, biochemical tests, serological tests, phage typing, and analysis of nucleic acids.

Physical Characteristics

Many physical characteristics are used to identify microorganisms. Scientists can usually identify protozoa, fungi, algae, and parasitic worms based solely on their *morphology* (shape). Medical laboratory scientists can also use the physical appearance of a bacterial colony[13] to help identify microorganisms. As we have discussed, stains are used to view the size and shape of individual bacterial cells and to show the presence or absence of identifying features such as endospores and flagella.

Linnaeus categorized prokaryotic cells into two genera based on two prevalent shapes. He classified spherical prokaryotes in the genus "*Coccus*,"[14] and he placed rod-shaped cells in the genus *Bacillus*.[15] However, subsequent studies have revealed vast differences among many of the thousands of spherical

[13]A group of bacteria that has arisen from a single cell grown on a solid laboratory medium.

[14]From Greek *kokkos,* meaning "berry." This genus name has been supplanted by many genera, including *Staphylococcus, Micrococcus,* and *Streptococcus.*

[15]From Latin *bacillum,* meaning "small rod."

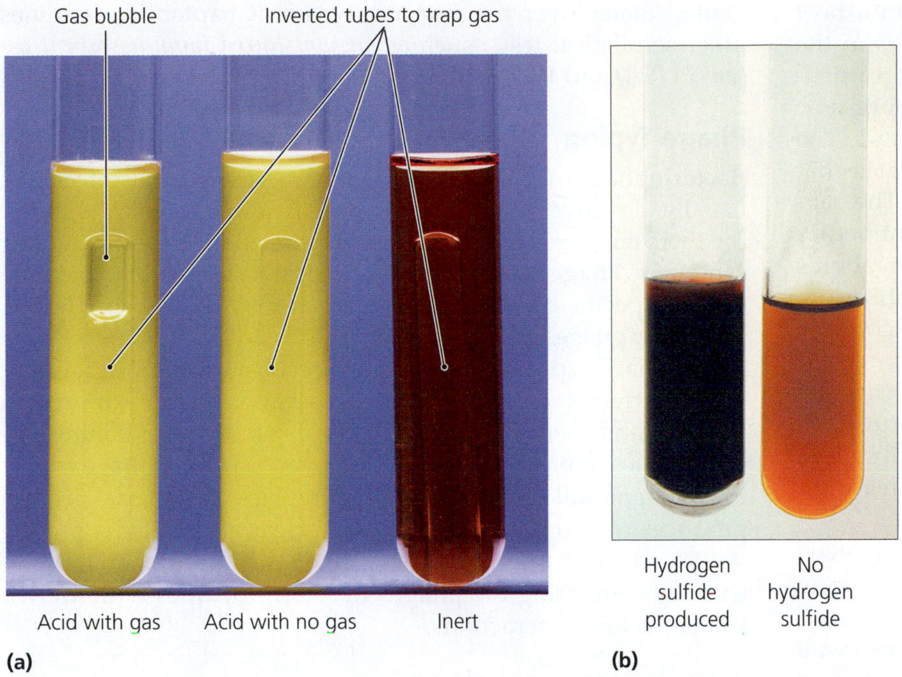

◀ **FIGURE 4.23 Two biochemical tests for identifying bacteria. (a)** A carbohydrate utilization test. At left is a tube in which the bacteria have metabolized a particular carbohydrate to produce acid (which changes the color of a pH indicator, phenol red, to yellow) and gas, as indicated by the bubble. At center is a tube with another bacterium that metabolized the carbohydrate to produce acid but no gas. At right is a tube inoculated with bacteria that are "inert" with respect to this test. **(b)** A hydrogen sulfide (H_2S) test. Bacteria that produce H_2S are identified by the black precipitate formed by the reaction of the H_2S with iron present in the medium.

and rod-shaped prokaryotes, and thus physical characteristics alone are not sufficient to classify prokaryotes. Instead, taxonomists rely primarily on genetic differences as revealed by metabolic dissimilarities and, more and more frequently, on rRNA sequences.

Biochemical Tests

Microbiologists distinguish many prokaryotes that are similar in microscopic appearances and staining characteristics on the basis of differences in their ability to utilize or produce certain chemicals. Biochemical tests include procedures that determine an organism's ability to ferment various carbohydrates; utilize various substrates, such as specific amino acids, starch, citrate, and gelatin; or produce waste products, such as hydrogen sulfide (H_2S) gas **(FIGURE 4.23)**. Differences in fatty acid composition of bacteria are also used to distinguish among bacteria. Obviously, biochemical tests can be used to identify only those microbes that can be grown under laboratory conditions.

Laboratory technicians utilize biochemical tests to identify pathogens, allowing physicians to prescribe appropriate treatments. Many tests require that the microorganisms be *cultured* (grown) for 12 to 24 hours though this time can be greatly reduced by the use of rapid identification tools. Such tools exist for many groups of medically important pathogens, such as Gram-negative bacteria in the family Enterobacteriaceae, Gram-positive bacteria, yeasts, and filamentous fungi. Automated systems for identifying pathogens use the results of a whole battery of biochemical tests performed in a plastic plate containing numerous small wells **(FIGURE 4.24)**. A color change in a well indicates the presence of a particular metabolic reaction, and the machine reads the pattern of colors in the plate to ascertain the identity of the pathogen.

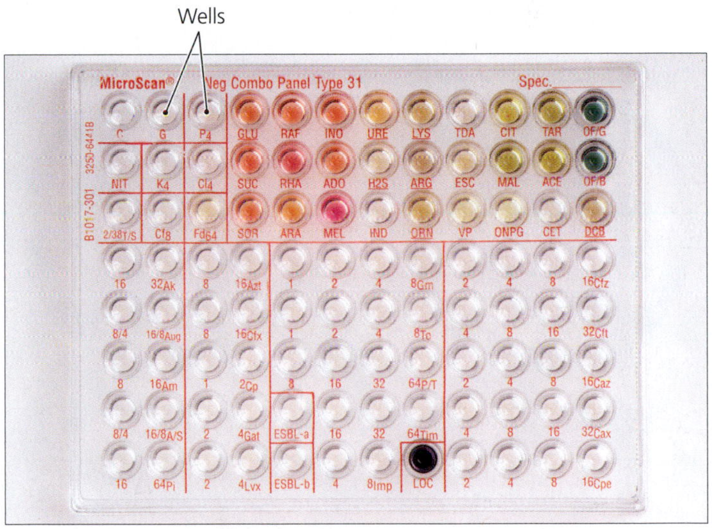

▲ **FIGURE 4.24 One tool for the rapid identification of bacteria, the automated MicroScan system.** A MicroScan panel, a plate containing numerous wells, each the site of a particular biochemical test. The instrument ascertains the identity of the organism by reading the pattern of colors in the wells after the biochemical tests have been performed.

Serological Tests

In the narrowest sense, serology is the study of serum, the liquid portion of blood after the clotting factors have been removed and an important site of antibodies. In its most practical application, serology is the study of antigen-antibody reactions in laboratory settings. Antibodies are immune system proteins that bind very specifically to target antigens (Chapter 16). In this section we briefly consider the use of serological testing to identify microorganisms.

Many microorganisms are *antigenic;* that is, within a host organism they trigger an immune response that results in the production of antibodies. Suppose, for example, that a scientist injects a sample of *Borrelia burgdorferi* (bō-rē´lē-ă burg-dōr´fer-ē), the bacterium that causes Lyme disease, into a rabbit. The bacterium has many surface proteins and carbohydrates that are antigenic because they are foreign to the rabbit. The rabbit responds to these foreign antigens by producing antibodies against them. These antibodies can be isolated from the rabbit's serum and concentrated into a solution known as an **antiserum**. Antisera bind to the antigens that triggered their production.

In a procedure called an **agglutination test,** antiserum is mixed with a sample that potentially contains its target cells. If the antigenic cells are present, antibodies in the antiserum will clump *(agglutinate)* the antigen **(FIGURE 4.25)**. Other antigens, and therefore other organisms, remain unaffected because antibodies are highly specific for their targets.

Antisera can be used to distinguish among species and even among strains of the same species. For example, a particularly pathogenic strain of *Escherichia coli* was classified and is identified by the presence of both antigen number 157 on its cell wall (designated O157 and pronounced "oh one five seven") and antigen number 7 on its flagella (designated H7). This pathogen, known as *E. coli* O157:H7, has caused several deaths in the United States over the past few years. (Chapter 17 examines other serological tests, such as *enzyme-linked immunosorbent assay (ELISA)* and *Western blotting.*)

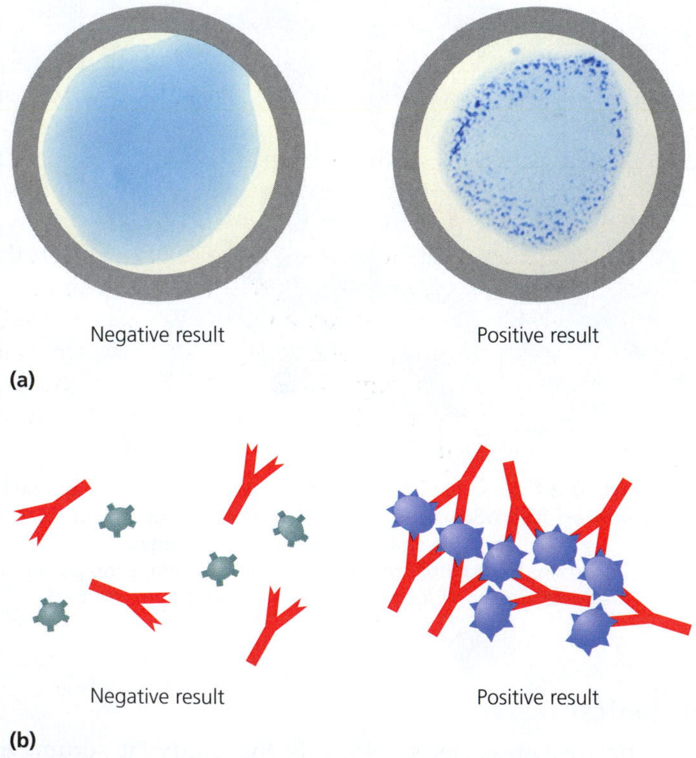

▲ **FIGURE 4.25 An agglutination test, one type of serological test. (a)** In a positive agglutination test, visible clumps are formed by the binding of antibodies to their target antigens present on cells. **(b)** The processes involved in agglutination tests. In a negative result, antibody binding cannot occur because its specific target is not present; in a positive result, specific binding does occur. Note that agglutination occurs because each antibody molecule can bind simultaneously to two antigen molecules.

Phage Typing

Bacteriophages (or simply **phages**) are viruses that infect and usually destroy bacterial cells. Just as antibodies are specific for their target antigens, phages are specific for the hosts they can infect. **Phage typing,** like serological testing, works because of such specificity. One bacterial strain may be susceptible to a particular phage while a related strain is not.

In phage typing, a technician spreads a solution containing the bacterium to be identified across a solid surface of growth medium and then adds small drops of solutions containing different types of bacteriophage. Wherever a specific phage is able to infect and kill bacteria, the resulting lack of bacterial growth produces within the bacterial lawn a clear area called a **plaque** **(FIGURE 4.26)**. A microbiologist can identify an unknown bacterium by comparing the phages that form plaques with known phage-bacteria interactions.

Analysis of Nucleic Acids

As we have discussed, the sequence of nucleotides in nucleic acid molecules provides a powerful tool for classifying and identifying microbes. In many cases, nucleic acid analysis has confirmed classical taxonomic hierarchies. In other cases, as in Woese's discovery of domains, curious new organisms and relationships that were not obvious from classical methodologies have come to light. Techniques of nucleotide sequencing and comparison, such as *polymerase chain reaction (PCR)* (Chapter 8), are best understood after we have discussed microbial genetics (Chapter 7).

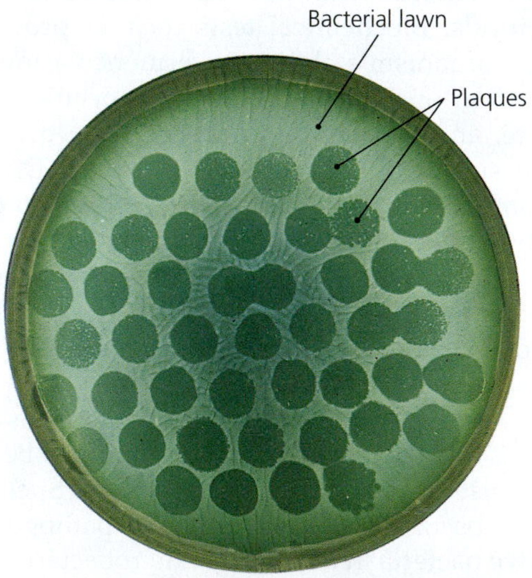

▲ **FIGURE 4.26 Phage typing.** Drops containing type A bacteriophages were added to this plate after its entire surface was inoculated with an unknown strain of *Salmonella*. After 12 hours of bacterial growth, clear zones, called plaques, developed where the phages killed bacteria. Given the great specificity of phage A for infecting and killing its host, the strain of bacterium can be identified as *Salmonella enterica* serotype Typhi.

EMERGING DISEASE CASE STUDY

Necrotizing Fasciitis

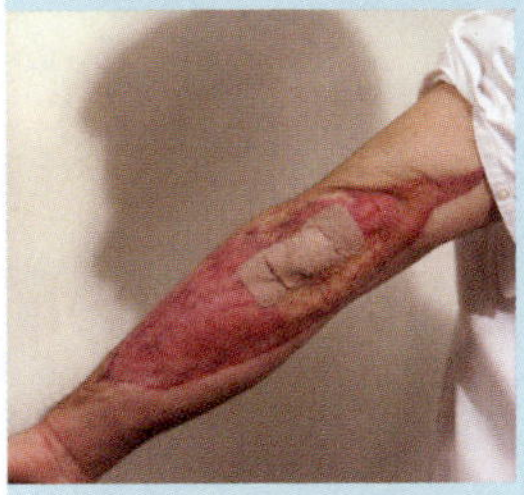

Fever, chills, nausea, weakness, and general yuckiness. Carlos thought he was getting the flu. Further, he had pulled a cactus thorn from his arm the day before, and the tiny wound had swollen to a centimeter in diameter. It was red, extremely hot, and much more painful than such a puncture had a right to be. Everything was against him. He couldn't afford to miss days at work, but he had no choice.

He shivered in bed with fever for the next two days and suffered more pain than he had ever experienced, certainly more than the time he broke his leg. Even more than passing a kidney stone. The red, purple, and black inflammation on his arm had grown to the size of a baseball. It was hard to the touch and excruciatingly painful. He decided it was time to call his brother to take him to the doctor. That decision saved his life.

Carlos's blood pressure dropped severely, and he was unconscious by the time they arrived. The physician immediately admitted Carlos to the hospital, where the medical team raced to treat necrotizing fasciitis, commonly called "flesh-eating" disease. This reemerging disease is caused by group A *Streptococcus*, a serotype of Gram-positive bacteria also known as *S. pyogenes*. Group A strep invades through a break in the skin and travels along the fascia—the protective covering of muscles—producing toxins that destroy human tissues, killing 1500 people each year in the United States.

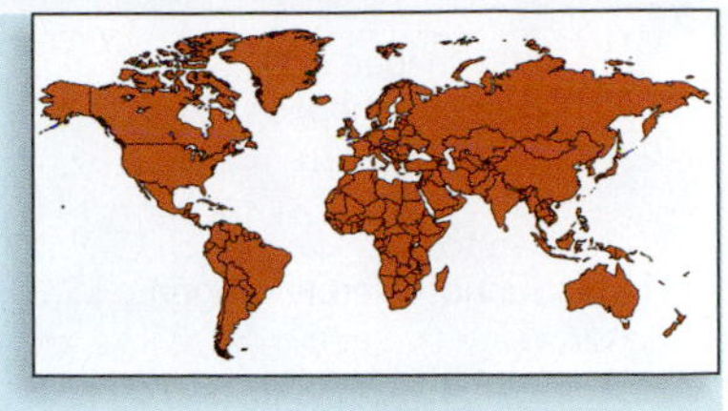

By cutting away all the infected tissue; using high-pressure, pure oxygen to inhibit bacterial growth; and applying antimicrobial drugs to kill the bacterium, the doctors stabilized Carlos. After months of skin grafts and rehabilitation, he returned to work, grateful to be alive. (For more about necrotizing fasciitis, see pp. 568–569.)

1. What color do cells of *S. pyogenes* appear after the Gram staining procedure?

Determining the percentage of a cell's DNA that is guanine and cytosine, a quantity referred to as the cell's *G + C content* (or G + C percentage), has also become a part of prokaryotic taxonomy. Scientists express the content as follows:

$$\frac{G + C}{A + T + G + C} \times 100$$

G + C content varies from 20% to 80% among prokaryotes. Often (but not always) organisms that share characteristics have similar G + C content. Organisms that were once thought to be closely related but have widely different G + C percentages are invariably not as closely related as thought.

Taxonomic Keys

As we have seen, taxonomists, medical clinicians, and researchers can use a wide variety of information, including morphology, chemical characteristics, and results from biochemical, serological, and phage typing tests, in their efforts to identify microorganisms, including pathogens. But how can all these characteristics and results be organized so that they can be used efficiently to identify an unknown organism? All this information is often arranged in **dichotomous keys,** which contain a series of paired statements worded so that only one of two choices applies to any particular organism **(FIGURE 4.27)**. Based on which of the two statements applies, the key either directs the user to another pair of statements or provides the name of the organism in question. Note that more than one key can be created to enable the identification of a given set of organisms, but all such keys involve mutually exclusive, "either/or" choices that send the user along a path that leads to the identity of the unknown organism. **▶ANIMATIONS:** ***Dichotomous Keys: Overview, Sample with Flowchart, Practice***

TELL ME WHY

Why didn't Linnaeus create taxonomic groups for viruses?

1a. Gram-positive cells.............................. Gram-positive bacteria
1b. Gram-negative cells............................. 2

2a. Rod-shaped cells................................. 3
2b. Non-rod-shaped cells........................... Cocci and pleomorphic bacteria

3a. Can tolerate oxygen............................. 4
3b. Cannot tolerate oxygen......................... Obligate anaerobes

4a. Ferments lactose................................. 5
4b. Cannot ferment lactose......................... Non-lactose-fermenters

5a. Can use citric acid as a sole carbon source.................................. 6
5b. Cannot use citric acid alone................... 8

6a. Produces hydrogen sulfide gas............... *Salmonella*
6b. Does not produce hydrogen sulfide gas.... 7

7a. Produces acetoin................................. *Enterobacter*
7b. Does not produce acetoin..................... *Citrobacter*

8a. Produces gas from glucose................... *Escherichia*
8b. Does not produce gas from glucose......... *Shigella*

(a)

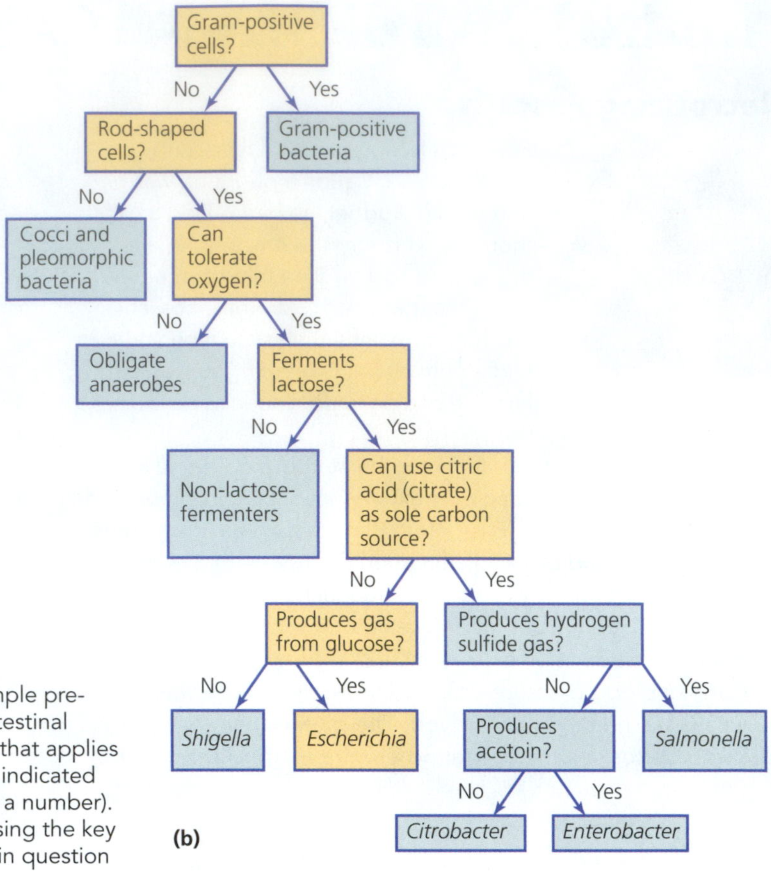

▲ **FIGURE 4.27 Use of a dichotomous taxonomic key.** The example presented here involves identifying the genera of potentially pathogenic intestinal bacteria. **(a)** A sample key. To use it, choose the one statement in a pair that applies to the organism to be identified and then either refer to another key (as indicated by words) or go to the appropriate place within this key (as indicated by a number). **(b)** A flowchart that shows the various paths that might be followed in using the key presented in part (a). Highlighted is the path taken when the bacterium in question is *Escherichia*.

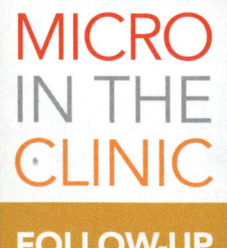

MICRO IN THE CLINIC FOLLOW-UP

The Salty Toddler

Lewis's collection of symptoms—unusual respiratory infection, diarrhea, stunted growth, and salty skin—is a tip-off to the medical staff that Lewis may have cystic fibrosis (CF). CF is a genetic disease that causes the lungs to produce abnormally thick mucus. This mucus makes the lungs vulnerable to secondary infections from bacteria such as *Pseudomonas aeruginosa*. The defective gene that causes CF also compromises the body's ability to move salt into and out of the cells; as a result, excess salt is excreted through sweat glands. Furthermore, children with CF often have good appetites but experience digestive problems and do not gain weight normally.

Lewis's doctors decide to administer a "sweat test": a procedure that measures the amount of salt present in sweat. The test reveals a high level of salt in Lewis's sweat, which confirms the diagnosis of CF. Although there is currently no cure for CF, medical advances have resulted in several treatment options that can extend Lewis's life expectancy and improve his quality of life overall. These include antibiotics, mucus-thinning drugs, procedures to physically loosen and remove mucus from the lungs, and lung transplantation.

1. **What makes biofilms dangerous for patients with cystic fibrosis?**
2. **The Gram stain of Lewis's sputum showed large numbers of the Gram-negative *Pseudomonas* infecting his lungs. Describe the appearance of the *Pseudomonas* at each stage of the Gram staining process.**

MM® Check your answers to Micro in the Clinic Follow-Up questions in the MasteringMicrobiology Study Area.

CHAPTER SUMMARY

▶**ANIMATIONS:** *Microscopy and Staining: Overview*

Units of Measurement (p. 96)

1. The metric system is a decimal system in which each unit is one-tenth the size of the next largest unit.
2. The basic unit of length in the metric system is the meter.

Microscopy (pp. 97–106)

1. **Microscopy** refers to the passage of light or electrons of various **wavelengths** through lenses to **magnify** objects and provide **resolution** and contrast so that those objects can be viewed and studied.
2. Immersion oil is used in light microscopy to fill the space between the specimen and a lens to reduce light refraction and thus increase the **numerical aperture** and resolution.
3. Staining techniques and polarized light may be used to enhance **contrast** between an object and its background.
4. **Simple microscopes** contain a single magnifying lens, whereas **compound microscopes** use a series of lenses for magnification.
5. The lens closest to the object being magnified is the **objective lens,** several of which are mounted on a **revolving nosepiece.** The lenses closest to the eyes are **ocular lenses. Condenser lenses** lie beneath the stage and direct light through the slide.

   ▶**VIDEO TUTOR:** *The Light Microscope*
6. The magnifications of the objective lens and the ocular lens are multiplied together to give **total magnification.**
7. A photograph of a microscopic image is a **micrograph.**
8. **Dark-field microscopes** provide a dark background for small or colorless specimens.
9. **Phase microscopes,** such as **phase-contrast microscopes** and **differential interference contrast** (Nomarski) **microscopes**, cause light rays that pass through a specimen to be out of phase with light rays that pass through the field, producing contrast.
10. **Fluorescence microscopes** use ultraviolet light and fluorescent dyes to fluoresce specimens and enhance contrast.
11. A **confocal microscope** uses fluorescent dyes in conjunction with computers to provide three-dimensional images of a specimen.

    ▶**ANIMATIONS:** *Light Microscopy*
12. A **transmission electron microscope (TEM)** provides an image produced by the transmission of electrons through a thinly sliced, dehydrated specimen.
13. A **scanning electron microscope (SEM)** provides a three-dimensional image by scattering electrons from the metal-coated surface of a specimen.
14. Minuscule electronic probes are used in **scanning tunneling microscopes (STM)** and in **atomic force microscopes (AFM)** to reveal details at the atomic level.

    ▶**ANIMATIONS:** *Electron Microscopy*

Staining (pp. 106–113)

1. Preparing to **stain** organisms with dyes for light microscopy involves making a **smear,** or thin film, of the specimens on a slide and then either passing the slide through a flame **(heat fixation)** or applying a chemical **(chemical fixation)** to attach the specimens to the slide. **Acidic dyes** or **basic dyes** are used to stain different portions of an organism to aid viewing and identification.
2. **Simple stains** involve the simple process of soaking the smear with one dye and then rinsing with water. **Differential stains** such as the **Gram stain, acid-fast stain,** endospore stain, Gomori methenamine silver (GMS) stain, and the hemotoxylin and eosin (HE) stain use more than one dye to differentiate different cells, chemicals, or structures.
3. The Gram stain procedure includes use of a **primary stain,** a **mordant,** a **decolorizing agent,** and a **counterstain** that results in either purple (Gram-positive) or pink (Gram-negative) organisms, depending on the chemical structures of their cell walls.
4. The acid-fast stain is used to differentiate cells with waxy cell walls. **Endospores** are stained by the **Schaeffer-Fulton endospore stain** procedure.
5. Dyes that stain the background and leave the cells colorless are called **negative stains** (or **capsule stains**).

   ▶**ANIMATIONS:** *Staining*

Classification and Identification of Microorganisms (pp. 113–120)

1. **Taxa** are nonoverlapping groups of organisms that are studied and named in **taxonomy.** Carolus Linnaeus invented a system of taxonomy, grouping similar interbreeding organisms into **species,** species into **genera,** genera into **families,** families into **orders,** orders into **classes,** classes into **phyla,** and phyla into **kingdoms.**

2. Linnaeus gave each species a descriptive name consisting of a genus name and **specific epithet.** This practice of naming organisms with two names is called **binomial nomenclature.**

3. Carl Woese proposed the existence of three taxonomic **domains** based on three cell types revealed by rRNA sequencing: **Eukarya, Bacteria,** and **Archaea.**

4. Taxonomists rely primarily on genetic differences revealed by morphological and metabolic dissimilarities to classify organisms. Species or strains within species may be distinguished by using **antisera; agglutination tests;** nucleic acid analysis, particularly G + C content; or **phage typing** with **bacteriophages,** in which unknown bacteria are identified by observing **plaques** (regions of a bacterial lawn where the phage has killed bacterial cells).

5. Microbiologists use **dichotomous keys,** which involve stepwise choices between paired characteristics, to help them identify microbes.

▶**ANIMATIONS:** *Dichotomous Keys: Overview, Sample with Flowchart, Practice*

QUESTIONS FOR REVIEW

Answers to the Questions for Review below (except Short Answer questions) begin on p. A-1.

Multiple Choice

1. Which of the following is smallest?
   a. decimeter
   b. millimeter
   c. nanometer
   d. micrometer

2. A nanometer is ________________ than a micrometer.
   a. 10 times larger
   b. 10 times smaller
   c. 1000 times larger
   d. 1000 times smaller

3. Resolution is best described as ________________.
   a. the ability to view something that is small
   b. the ability to magnify a specimen
   c. the ability to distinguish between two adjacent objects
   d. the difference between two waves of electromagnetic radiation

4. Curved glass lenses ________________ light.
   a. refract
   b. bend
   c. magnify
   d. both a and b

5. Which of the following factors is important in making an image appear larger?
   a. the thickness of the lens
   b. the curvature of the lens
   c. the speed of the light passing through the lens
   d. all of the above

6. Which of the following is different between light microscopy and transmission electron microscopy?
   a. magnification
   b. resolution
   c. wavelengths
   d. all of the above

7. Which of the following types of microscopes produces a three-dimensional image with a shadowed appearance?
   a. simple microscope
   b. differential interference contrast microscope
   c. fluorescent microscope
   d. transmission electron microscope

8. Which of the following microscopes combines the greatest magnification with the best resolution?
   a. confocal microscope
   b. phase-contrast microscope
   c. dark-field microscope
   d. bright-field microscope

9. Negative stains such as eosin are also called ________________.
   a. capsule stains
   b. endospore stains
   c. simple stains
   d. acid-fast stains

10. In the binomial system of nomenclature, which term is always written in lowercase letters?
    a. kingdom
    b. domain
    c. genus
    d. specific epithet

Fill in the Blanks

1. If an objective magnifies 40× and each binocular lens magnifies 15×, the total magnification of the object being viewed is ________________.

2. The type of fixation developed by Koch for bacteria is ________________.

3. Immersion oil ________________ (increases/decreases) the numerical aperture, which ________________ (increases/decreases) resolution because ________________ (more/fewer) light rays are involved.

4. ________________ refers to differences in intensity between two objects.

5. Cationic chromophores such as methylene blue ionically bond to ________________ (positively/negatively) charged chemicals such as DNA and proteins.

VISUALIZE IT!

1. Label each photograph below with the type of microscope used to acquire the image.

a.______________

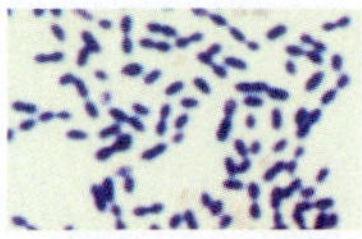

b.______________

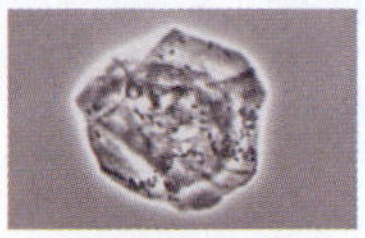

c.______________

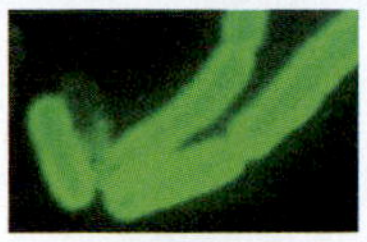

d.______________

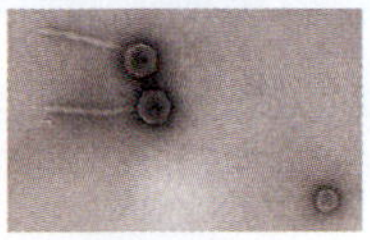

e.______________

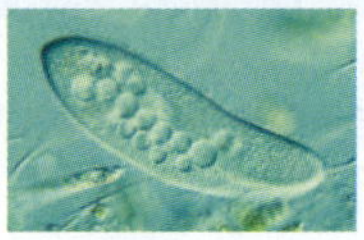

f.______________

2. Label the microscope.

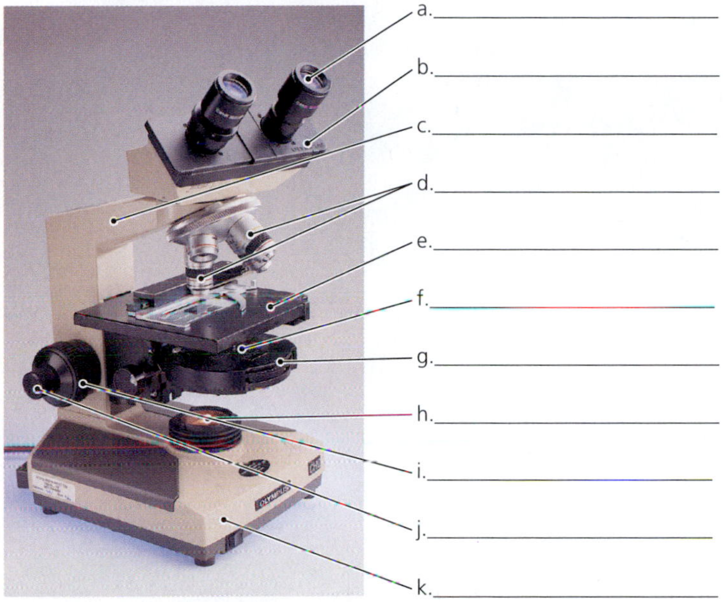

Short Answer

1. Explain how the principle "electrons travel as waves" applies to microscopy.
2. Critique the following definition of *magnification* given by a student on a microbiology test: "Magnification makes things bigger."
3. Why can electron microscopes magnify only dead organisms?
4. Put the following substances in the order they are used in a Gram stain: counterstain, decolorizing agent, mordant, primary stain.
5. Why is Latin used in taxonomic nomenclature?
6. Give three characteristics of a "specific epithet."
7. How does the study of the nucleotide sequences of ribosomal RNA fit into a discussion of taxonomy?
8. An atomic force microscope can magnify a living cell, whereas electron microscopes and scanning tunneling microscopes cannot. What requirement of electron and scanning tunneling microscopes precludes the imaging of living specimens?

CRITICAL THINKING

1. Miki came home from microbiology lab with green fingers and a bad grade. When asked about this, she replied that she was doing a Gram stain but that it never worked the way the book said it should. Christina overheard the conversation and said that she must have used the wrong chemicals. What dye was she probably using, and what structure does that chemical normally stain?
2. Why is the definition of *species* as "successfully interbreeding organisms" not satisfactory for most microorganisms?
3. With the exception of the discovery of new organisms, is it logical to assume that taxonomy as we currently know it will stay the same? Why or why not?
4. A novice microbiology student incorrectly explains that immersion oil increases the magnification of his microscope. What is the function of immersion oil?
5. A light microscope has 10× oculars and 0.3-μm resolution. Using the oil immersion lens (100×), will you be able to resolve two objects 400 nm apart? Will you be able to resolve two objects 40 nm apart?
6. In what ways are the Gram stain and the acid-fast staining procedures similar? In the acid-fast procedure, what takes the place of Gram's iodine mordant?
7. Examine the binomials of the species discussed in this section. What do the genus names and specific epithets indicate about the organisms?
8. Microbiologists have announced the discovery of over 30 new species of bacteria that thrive between the teeth and gums of humans. The bacteria could not be grown in the researchers' laboratories, nor were any of them ever observed via any kind of microscopy.

   If they couldn't culture them or see them, how could the researchers know they had discovered new species? If they couldn't examine the cells for the presence of a nucleus, how did they determine that the organisms were prokaryotes and not eukaryotes?

9. Why is the genus name *"Coccus"* placed within quotation marks but not the genus name *Bacillus*?

10. A clinician obtains a specimen of urine from a patient suspected to have a bladder infection. From the specimen she cultures a Gram-negative, rod-shaped bacterium that ferments lactose in the presence of oxygen, utilizes citrate, and produces acetoin but not hydrogen sulfide. Using the key presented in Figure 4.27, identify the genus of the infective bacterium.

CONCEPT MAPPING

Using the following terms, draw a concept map that describes Gram stain. For a sample concept map, see p. 94. Or, complete this and other concept maps online by going to the MasteringMicrobiology Study Area.

Decolorizer
Ethanol and acetone
Gram-negative bacteria
Gram-positive bacteria
Iodine
Primary stain
Purple
Safranin
Thin peptidoglycan layer

5 Microbial Metabolism

MICRO IN THE CLINIC

What's Lurking in the Fitness Center?

Tom and Kyoko own and operate a spa and fitness center. Tom teaches a variety of yoga classes, while Kyoko runs the organic health food bar. Kyoko serves fruit and vegetable smoothies, fresh salads, and a variety of vegetarian dishes. Her specialty is a traditional Asian noodle dish made with home-prepared fermented tofu (soybean curd). To prepare the dish, Kyoko briefly boils bite-sized chunks of tofu, ferments them at room temperature under plastic wrap for two weeks, and then transfers the tofu chunks into an airtight jar where they are marinated in chili powder, salt, wine, and vegetable oil for three more days. She stores and serves this dish at room temperature, and it is a favorite among their clientele. Between the healthy food and the frequent exercise, Tom and Kyoko are rarely sick.

One day, however, Kyoko wakes up complaining of double vision (diplopia) and dizziness. Tom thinks she may have worked out too hard the day before and just needs to rest. But by the afternoon, Kyoko feels worse. Her arms feel weak, she finds it difficult to swallow, and Tom notices that her speech is slurred and her eyelids are drooping. After two regular spa customers fail to show up for the morning yoga class—and a third customer calls complaining of symptoms similar to Kyoko's—the couple begins to really worry. When Tom also begins to experience the same symptoms, they rush to the emergency room.

Why do you think this health-conscious, physically fit couple is feeling so poorly? Turn to the end of the chapter (p. 159) to find out.

Explore More: Test your readiness and apply your knowledge with dynamic learning tools at MasteringMicrobiology.

How do pathogens acquire energy and nutrients at the expense of a patient's health? How does grape juice turn into wine, and how does yeast cause bread to rise? How do disinfectants, antiseptics, and antimicrobial drugs work? When laboratory personnel perform biochemical tests to identify unknown microorganisms and help diagnose disease, what exactly are they doing?

The answers to all of these questions require an understanding of microbial **metabolism,**[1] the collection of controlled biochemical reactions that takes place within the microbe. Although it is true that metabolism in its entirety is complex, consisting of thousands of chemical reactions and control mechanisms, the reactions are nevertheless elegantly logical and can be understood in a simplified form. In this chapter we will concern ourselves only with central metabolic pathways and energy metabolism.

Your study of metabolism will be manageable if you keep in mind that the ultimate function of an organism's metabolism is to reproduce the organism and that metabolic processes are guided by the following eight elementary statements:

- Every cell acquires *nutrients,* which are the chemicals necessary as building blocks and energy for metabolism.
- Metabolism requires energy from light or from the *catabolism* (kă-tab´ō-lizm; breakdown) of acquired nutrients.
- Energy is often stored in the chemical bonds of *adenosine triphosphate (ATP).*
- Using *enzymes,* cells catabolize nutrient molecules to form elementary building blocks called *precursor metabolites.*
- Using precursor metabolites, other enzymes, and energy from ATP, cells construct larger building blocks in *anabolic* (an-ă-bol´ik; biosynthetic) reactions.
- Cells use enzymes and additional energy from ATP to anabolically link building blocks together to form macromolecules in *polymerization* reactions.
- Cells grow by assembling macromolecules into cellular structures such as ribosomes, membranes, and cell walls.
- Cells typically reproduce once they have doubled in size.

We will discuss each aspect of metabolism in the chapters that most directly apply. For instance, we discussed the first step of metabolism—the active and passive transport of nutrients into cells—in Chapter 3. In this chapter we examine the importance of enzymes in catabolic and anabolic reactions, study the three ways that ATP molecules are synthesized, and show that catabolic and anabolic reactions are linked. We also examine the catabolism of nutrient molecules; the anabolic reactions involved in the synthesis of carbohydrates, lipids, amino acids, and nucleotides; and a few ways that cells control their metabolic activities. Genetic control of metabolism and the polymerization of DNA, RNA, and proteins are discussed in Chapter 7, and the specifics of cell division are covered in Chapters 11 and 12.

[1]From Greek *metabole,* meaning "change."

Basic Chemical Reactions Underlying Metabolism

In the following sections we will examine the basic concepts of catabolism, anabolism, and a special class of reactions called *oxidation-reduction reactions.* The latter involve the transfer of electrons and energy carried by electrons between molecules. Then we will turn our attention briefly to the synthesis of ATP and energy storage before we discuss the organic catalysts called *enzymes,* which make metabolism possible. ▶**ANIMATIONS:** ***Metabolism: Overview***

Catabolism and Anabolism

LEARNING | **OUTCOME**

5.1 Distinguish among metabolism, anabolism, and catabolism.

Metabolism, which is all of the chemical reactions in an organism, can be divided into two major classes of reactions: **catabolism** and **anabolism** (**FIGURE 5.1**). A series of such reactions is called a *pathway.* Cells have *catabolic pathways,* which break

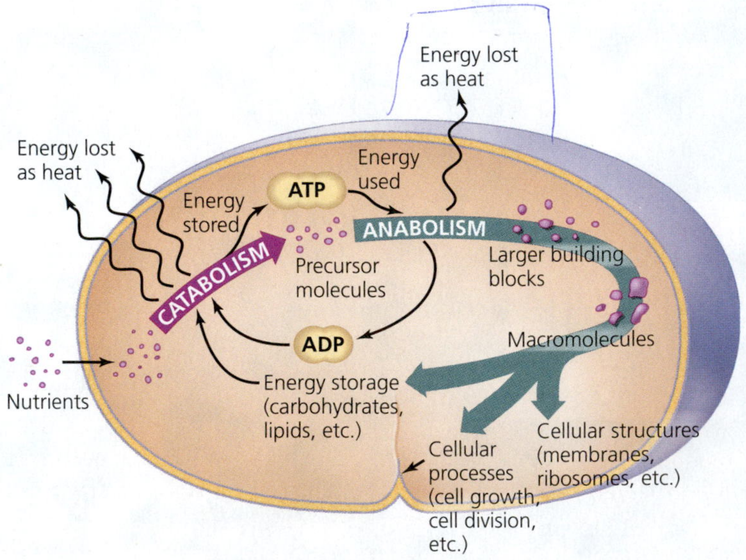

▲ **FIGURE 5.1 Metabolism is composed of catabolic and anabolic reactions.** Catabolic reactions are exergonic—they release energy, some of which is stored in ATP molecules, though most of the energy is lost as heat. Anabolic reactions are endergonic—they require energy, typically provided by ATP. There is some heat loss in anabolism as well. The products of catabolism provide many of the building blocks (precursor metabolites) for anabolic reactions. These reactions produce macromolecules and cellular structures, leading to cell growth and division.

larger molecules into smaller products, and *anabolic pathways*, which synthesize large molecules from the smaller products of catabolism. Even though catabolic and anabolic pathways are intimately linked in cells, it is often useful to study the two types of pathways as though they were separate.

When catabolic pathways break down large molecules, they release energy; that is, catabolic pathways are *exergonic* (ek-ser-gon´ik). Cells store some of this released energy in the bonds of ATP, though much of the energy is lost as heat. Another result of the breakdown of large molecules by catabolic pathways is the production of numerous smaller molecules, some of which are **precursor metabolites** of anabolism. Some organisms, such as *Escherichia coli* (esh-ĕ-rik´ē-ăkō´lī), can synthesize everything in their cells just from precursor metabolites; other organisms must acquire some anabolic building blocks from outside their cells as nutrients. Catabolic *pathways*, but not necessarily *individual* catabolic *reactions*, produce ATP, or metabolites, or both. An example of a catabolic pathway is the breakdown of lipids into glycerol and fatty acids.

Anabolic pathways are functionally the opposite of catabolic pathways in that they synthesize macromolecules and cellular structures. Because building anything requires energy, anabolic pathways are *endergonic* (en-der-gon´ik); that is, they require more energy than they release. The energy required for anabolic pathways usually comes from ATP molecules produced during catabolism. An example of an anabolic pathway is the synthesis of lipids for cell membranes from glycerol and fatty acids.

To summarize, a cell's metabolism involves both catabolic pathways that break down macromolecules to supply molecular building blocks and energy in the form of ATP, and anabolic pathways that use the building blocks and ATP to synthesize macromolecules needed for growth and reproduction.

Oxidation and Reduction Reactions

LEARNING **OUTCOME**

5.2 Contrast reduction and oxidation reactions.

Many metabolic reactions involve the transfer of electrons, which carry energy, from an *electron donor* (a molecule that donates an electron) to an *electron acceptor* (a molecule that accepts an electron). Such electron transfers are called **oxidation-reduction reactions**, or **redox reactions** (**FIGURE 5.2**). An electron acceptor is said to be *reduced*. This may seem backward, but electron acceptors are reduced because their gain in electrons reduces their overall electrical charge (i.e., they are more negatively charged). Molecules that lose electrons are said to be *oxidized* because frequently their electrons are donated to oxygen atoms. An acronym to help you remember these concepts is OIL RIG: oxidation involves loss; reduction involves gain.

Reduction and oxidation reactions always happen simultaneously because every electron donated by one chemical is

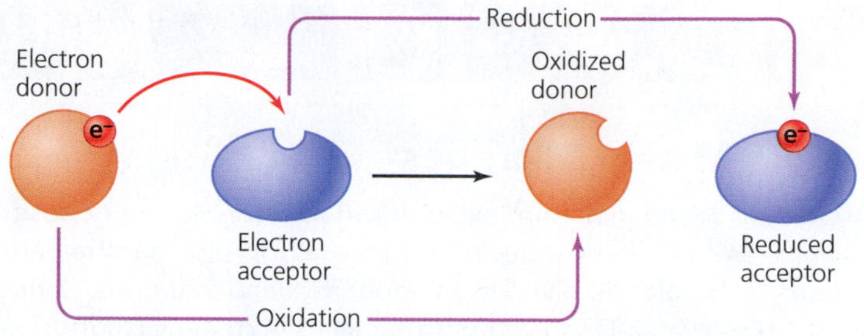

▲ **FIGURE 5.2 Oxidation-reduction, or redox, reactions.** When electrons are transferred from donor molecules to acceptor molecules, donors become oxidized, and acceptors become reduced. *Why are acceptor molecules said to be reduced when they are gaining electrons?*

Figure 5.2 *"Reduction" refers to the overall electrical charge on a molecule. Because electrons have a negative charge, the gain of an electron reduces the molecule's overall charge.*

accepted by another chemical. A chemical may be reduced by gaining either a simple electron or an electron that is part of a hydrogen atom—which is composed of one proton and one electron. **Beneficial Microbes: Gold-Mining Microbes** describes an interesting example of how some prokaryotes are able to reduce gold dissolved in solution.

In contrast, a molecule may be oxidized in one of three ways: by losing a simple electron, by losing a hydrogen atom, or by gaining an oxygen atom. Biological oxidations often involve the loss of hydrogen atoms; such reactions are also called *dehydrogenation* (dē-hī´drō-jen-ā´shŭn) *reactions*.

Electrons rarely exist freely in cytoplasm; instead, they orbit atomic nuclei. Therefore, cells use electron carrier molecules to carry electrons (often in hydrogen atoms) from one location in a cell to another. Three important electron carrier molecules, which are derived from vitamins, are **nicotinamide adenine dinucleotide (NAD$^+$)**, **nicotinamide adenine dinucleotide phosphate (NADP$^+$)**, and **flavin adenine dinucleotide (FAD)**. Cells use each of these molecules in specific metabolic pathways to carry pairs of electrons. One of the electrons carried by either NAD$^+$ or NADP$^+$ is part of a hydrogen atom, forming NADH or NADPH. FAD carries two electrons as hydrogen atoms ($FADH_2$). Many metabolic pathways, including those that synthesize ATP, require such electron carrier molecules. ▶**ANIMATIONS:** ***Oxidation-Reduction Reactions***

ATP Production and Energy Storage

LEARNING **OUTCOME**

5.3 Compare and contrast the three types of ATP phosphorylation.

Nutrients contain energy, but that energy is spread throughout their chemical bonds and generally is not concentrated enough for use in anabolic reactions. During catabolism, organisms

BENEFICIAL MICROBES

Solid gold is gold in its reduced form.

Gold-Mining Microbes

Gold, as found in nature, exists in two forms: gold-ore deposits, which are gold in its reduced form, usually found near the Earth's crust, and gold dissolved in solution, as found in thermal springs and in seawater. Dissolved gold, which is gold in its oxidized forms, is largely useless to humans; it cannot be converted inexpensively into solid gold. Even though gold in either form is toxic when ingested by most living things, scientists have discovered that certain bacteria, such as *Ralstonia metallidurans*, can metabolize oxidized gold. When placed in a solution containing oxidized gold, these microorganisms reduce the gold and encase themselves in solid gold, which is their metabolic waste.

Entrepreneurial minds may wonder whether *Ralstonia* could be potentially profitable. Although it is true that a great deal of dissolved gold is found in thermal springs and oceans, the gold is very dilute—only minute amounts are present in very large volumes of water. Moreover, were someone to perfect a way of using microorganisms to convert dissolved gold to great quantities of solid gold, they would be wise to keep it to themselves: so much solid gold could become available that its market value would plunge dramatically.

release energy from nutrients that can then be concentrated and stored in high-energy phosphate bonds of molecules such as ATP. This happens by a general process called *phosphorylation* (fos´fōr-i-lā´shŭn) in which inorganic phosphate (PO_4^{3-}) is added to a substrate. For example, cells phosphorylate adenosine diphosphate (ADP), which has two phosphate groups, to form adenosine triphosphate (ATP), which has three phosphate groups (see Figure 2.27).

As we will examine in the following sections, cells phosphorylate ADP to form ATP in three specific ways:

- *Substrate-level phosphorylation* (see p. 135), which involves the transfer of phosphate to ADP from another phosphorylated organic compound
- *Oxidative phosphorylation* (see p. 142), in which energy from redox reactions of respiration (described shortly) is used to attach inorganic phosphate to ADP
- *Photophosphorylation* (see pp. 150–152), in which light energy is used to phosphorylate ADP with inorganic phosphate

We will investigate each of these in more detail as we proceed through the chapter.

After ADP is phosphorylated to produce ATP, anabolic pathways use some energy of ATP by breaking a phosphate bond (which re-forms ADP). Thus, the cyclical interconversion of ADP and ATP functions somewhat like rechargeable batteries: ATP molecules store energy from light (in photosynthetic organisms) and from catabolic reactions and then release stored energy to drive cellular processes (including anabolic reactions, active transport, and movement). ADP molecules can be "recharged" to ATP again and again.

The Roles of Enzymes in Metabolism

LEARNING | **OUTCOMES**

5.4 Make a table listing the six basic types of enzymes, their activities, and an example of each.

5.5 Describe the components of a holoenzyme, and contrast protein and RNA enzymes.

5.6 Define *activation energy, enzyme, apoenzyme, cofactor, coenzyme, active site*, and *substrate*, and describe their roles in enzyme activity.

5.7 Describe how temperature, pH, substrate concentration, and competitive and noncompetitive inhibition affect enzyme activity.

Chemical reactions occur when bonds are broken or formed between atoms. In catabolic reactions, a bond must be destabilized before it will break, whereas in anabolic reactions, reactants collide with sufficient energy for bonds to form between them. In anabolism, increasing either the concentrations of reactants or ambient temperatures increases the number of collisions and produces more chemical reactions; however, in living organisms, neither reactant concentration nor temperature is usually high enough to ensure that bonds will form. Therefore, the chemical reactions of life depend on *catalysts*, which are chemicals that increase the likelihood of a reaction but are not permanently changed in the process. Organic catalysts are known as **enzymes**. ▶**ANIMATIONS:** *Enzymes: Overview*

Naming and Classifying Enzymes

The names of enzymes usually end with the suffix "-ase," and the name of each enzyme often incorporates the name of that enzyme's **substrate**, which is the molecule the enzyme acts on. Based on their mode of action, enzymes can be grouped into six basic categories:

- *Hydrolases* catabolize molecules by adding water in a decomposition process known as *hydrolysis*. Hydrolases are used primarily in the depolymerization of macromolecules.
- *Isomerases*[2] rearrange the atoms within a molecule but do not add or remove anything (so they are neither catabolic nor anabolic).
- *Ligases*, or *polymerases*, join two molecules together (and are thus anabolic). They often use energy supplied by ATP.
- *Lyases* split large molecules (and are thus catabolic) without using water in the process.
- *Oxidoreductases* remove electrons from (oxidize) or add electrons to (reduce) various substrates. They are used in both catabolic and anabolic pathways.
- *Transferases* transfer functional groups, such as an amino group (NH_2), a phosphate group, or a two-carbon (acetyl) group, between molecules. Transferases can be anabolic.

TABLE 5.1 summarizes these types of enzymes and gives examples of each.

The Makeup of Enzymes

Many protein enzymes are complete in themselves, but others are composed of both protein and nonprotein portions. The proteins in these combinations are called **apoenzymes** (ap´ō-en-zīms). Apoenzymes are inactive if they are not bound to one or more of the nonprotein substances called **cofactors**. Cofactors are either inorganic ions (such as iron, magnesium, zinc, or copper ions) or certain organic molecules called **coenzymes** (ko-en´zīms). All coenzymes are either vitamins or contain vitamins, which are organic molecules that are required for metabolism but cannot be synthesized by certain organisms (especially mammals). Some apoenzymes bind with inorganic cofactors, some bind with coenzymes, and some bind with both. The binding of an apoenzyme and its cofactor(s) forms an active enzyme, called a **holoenzyme** (hol-ō-en´zīm; **FIGURE 5.3**).

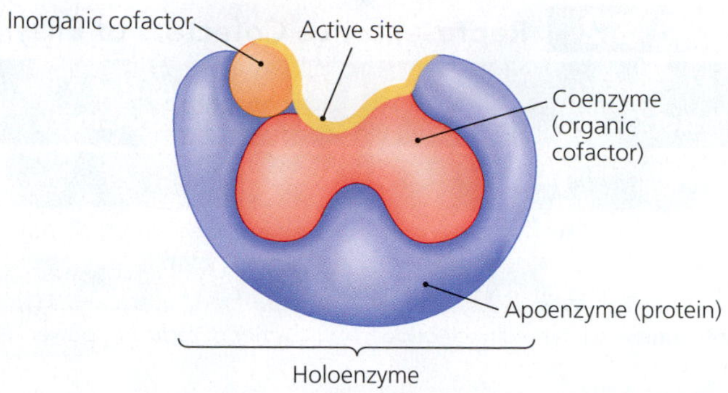

▲ **FIGURE 5.3 Makeup of a holoenzyme.** The combination of a proteinaceous apoenzyme with one or more cofactors forms a holoenzyme, which is the active form. A cofactor is either an inorganic ion or a coenzyme, which is an organic cofactor derived from a vitamin. An apoenzyme is inactive unless it is bound to its cofactors. *Name four metal ions that can act as cofactors.*

Figure 5.3 Iron, magnesium, zinc, and copper ions can act as cofactors.

TABLE 5.2 on p. 130 lists several examples of inorganic cofactors and organic cofactors (coenzymes). Note that three important coenzymes are the electron carriers NAD^+, $NADP^+$, and FAD, which, as we have seen, carry electrons in hydrogen atoms from place to place within cells. We will examine more closely the roles of these coenzymes in the generation of ATP later in the chapter.

Not all enzymes are proteinaceous; some are RNA molecules called **ribozymes**. In eukaryotes, ribozymes process other RNA molecules by removing sections of RNA and splicing the remaining pieces together. Recently, researchers have

[2]An isomer is a compound with the same molecular formula as another molecule, but with a different arrangement of atoms.

TABLE 5.1 Enzyme Classification Based on Reaction Types

Class	Type of Reaction Catalyzed	Example
Hydrolase	Hydrolysis (catabolic)	Lipase—breaks down lipid molecules
Isomerase	Rearrangement of atoms within a molecule (neither catabolic nor anabolic)	Phosphoglucoisomerase—converts glucose 6-phosphate into fructose 6-phosphate during glycolysis
Ligase or polymerase	Joining two or more chemicals together (anabolic)	Acetyl-CoA synthetase—combines acetate and coenzyme A to form acetyl-CoA for the Krebs cycle
Lyase	Splitting a chemical into smaller parts without using water (catabolic)	Fructose-1,6-bisphosphate aldolase—splits fructose 1,6-bisphosphate into G3P and DHAP
Oxidoreductase	Transfer of electrons or hydrogen atoms from one molecule to another	Lactic acid dehydrogenase—oxidizes lactic acid to form pyruvic acid during fermentation
Transferase	Moving a functional group from one molecule to another (may be anabolic)	Hexokinase—transfers phosphate from ATP to glucose in the first step of glycolysis

TABLE 5.2 Representative Cofactors of Enzymes

Cofactors	Examples of Use in Enzymatic Activity	Substance Transferred in Enzymatic Activity	Vitamin Source (of Coenzyme)
Inorganic (Metal Ion)			
Magnesium (Mg^{2+})	Forms bond with ADP during phosphorylation	Phosphate	None
Organic (Coenzymes)			
Nicotinamide adenine dinucleotide (NAD^+)	Carrier of reducing power	Two electrons and a hydrogen ion	Niacin (B_3)
Nicotinamide adenine dinucleotide phosphate ($NADP^+$)	Carrier of reducing power	Two electrons and a hydrogen ion	Niacin (B_3)
Flavin adenine dinucleotide (FAD)	Carrier of reducing power	Two hydrogen atoms	Riboflavin (B_2)
Tetrahydrofolate	Used in synthesis of nucleotides and some amino acids	One-carbon molecule	Folic acid (B_9)
Coenzyme A	Formation of acetyl-CoA in Krebs cycle and beta-oxidation	Two-carbon molecule	Pantothenic acid (B_5)
Pyridoxal phosphate	Transaminations in the synthesis of amino acids	Amine group	Pyridoxine (B_6)
Thiamine pyrophosphate	Decarboxylation of pyruvic acid	Aldehyde group (CHO)	Thiamine (B_1)

discovered that the functional core of a ribosome is a ribozyme; therefore, given that ribosomes make all proteins, ribosomal enzymes make protein enzymes.

Enzyme Activity

Within cells, enzymes catalyze reactions by lowering the **activation energy**, which is the amount of energy needed to trigger a chemical reaction **(FIGURE 5.4)**. Whereas heat can provide energy to trigger reactions, the temperatures needed to reach activation energy for most metabolic reactions are often too high to allow cells to survive, so enzymes are needed if metabolism is to occur. This is true regardless of whether the enzyme is a protein or RNA or whether the chemical reaction is anabolic or catabolic.

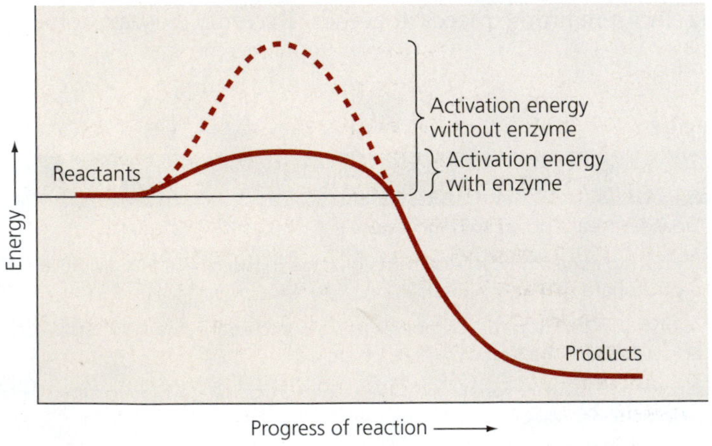

▲ **FIGURE 5.4 The effect of enzymes on chemical reactions.** Enzymes catalyze reactions by lowering the activation energy—that is, the energy needed to trigger the reaction. *Does this graph more closely represent a catabolic or an anabolic reaction?*

Figure 5.4 *Energy of the products is lower than energy of the reactants. Thus, the graph more closely represents catabolism.*

The activity of enzymes depends on the closeness of fit between the functional sites of an enzyme and its substrate. The shape of an enzyme's functional site, called its **active site**, is complementary to the shape of the substrate. Generally, the shapes and locations of only a few amino acids or nucleotides determine the shape of an enzyme's active site. A change in a single component—for instance, through mutation—can render an enzyme less effective or even completely nonfunctional.

Enzyme-substrate specificity, which is critical to enzyme activity, has been likened to the fit between a lock and key. This analogy is not completely apt because enzymes change shape slightly when they bind to their substrate, almost as if a lock could grasp its key once it had been inserted. This latter description of enzyme-substrate specificity is called the **induced-fit model (FIGURE 5.5)**.

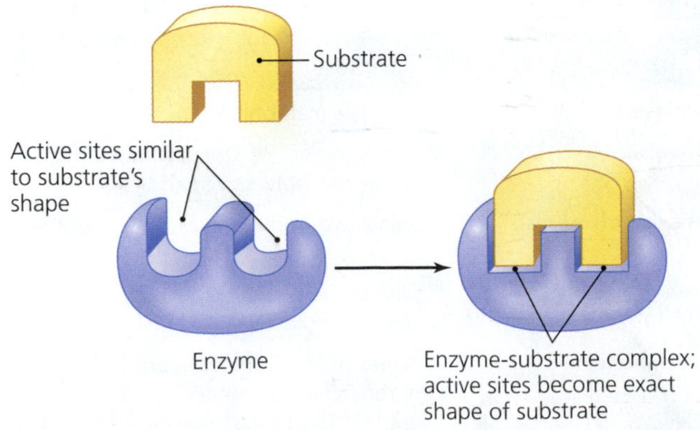

▲ **FIGURE 5.5 Enzymes fitted to substrates.** The induced-fit model of enzyme-substrate interaction. An enzyme's active site is generally complementary to the shape of its substrate, but a perfect fit between them does not occur until the substrate and enzyme bind to form a complex.

In some cases, several different enzymes possess active sites that are complementary to various portions of a single substrate molecule. For example, an important precursor metabolite called phosphoenolpyruvic acid (PEP) is the substrate for at least five enzymes. Depending on the enzyme involved, various products are produced from PEP. In one catabolic pathway, PEP is converted to pyruvic acid, whereas in a particular anabolic pathway, PEP is converted to the amino acid phenylalanine.

Although the exact ways that enzymes lower activation energy are not known, it appears that several mechanisms are involved. Some enzymes appear to bring reactants into sufficiently close proximity to enable a bond to form, whereas other enzymes change the shape of a reactant, inducing a bond to be broken. In any case, enzymes increase the likelihood that bonds will form or break.

The activity of enzymes is believed to follow the process illustrated in **FIGURE 5.6**, which depicts the catabolic lysis of a molecule called fructose 1,6-bisphosphate:

1. An enzyme associates with a specific substrate molecule having a shape that is complementary to that enzyme's active site.
2. The enzyme and its substrate bind to form a temporary intermediate compound called an enzyme-substrate complex. The binding of the substrate induces the enzyme to fit the shape of the substrate even more closely.
3. Bonds within the substrate are broken, forming two (and in some other reactions, more than two) products. (This is a catabolic reaction; in anabolic reactions, reactants are linked together to form products.)
4. The enzyme dissociates from the newly formed molecules, which diffuse away from the site of the reaction, and the enzyme resumes its original configuration and is ready to associate with another substrate molecule.

Many factors influence the rate of enzymatic reactions, including temperature, pH, enzyme and substrate concentrations, and the presence of inhibitors. ▶**ANIMATIONS:** ***Enzymes Steps in a Reaction***

Temperature As mentioned, higher temperatures tend to increase the rate of most chemical reactions because molecules are moving faster and collide more frequently, encouraging bonds to form or break. However, this is not entirely true of enzymatic reactions because the active sites of enzymes change shape as temperature changes. If the temperature rises too high or falls too low, an enzyme is often no longer able to achieve a fit with its substrate.

Each enzyme has an optimal temperature for its activity (**FIGURE 5.7a**). The optimum temperature for the enzymes in the human body is 37°C, which is normal body temperature. Part of the reason certain pathogens can cause disease in humans is that the optimal temperature for the enzymes in those microorganisms is also 37°C. The enzymes of some other microorganisms, however, function best at much higher temperatures; this is the case for *hyperthermophiles*, organisms that grow best at temperatures above 80°C.

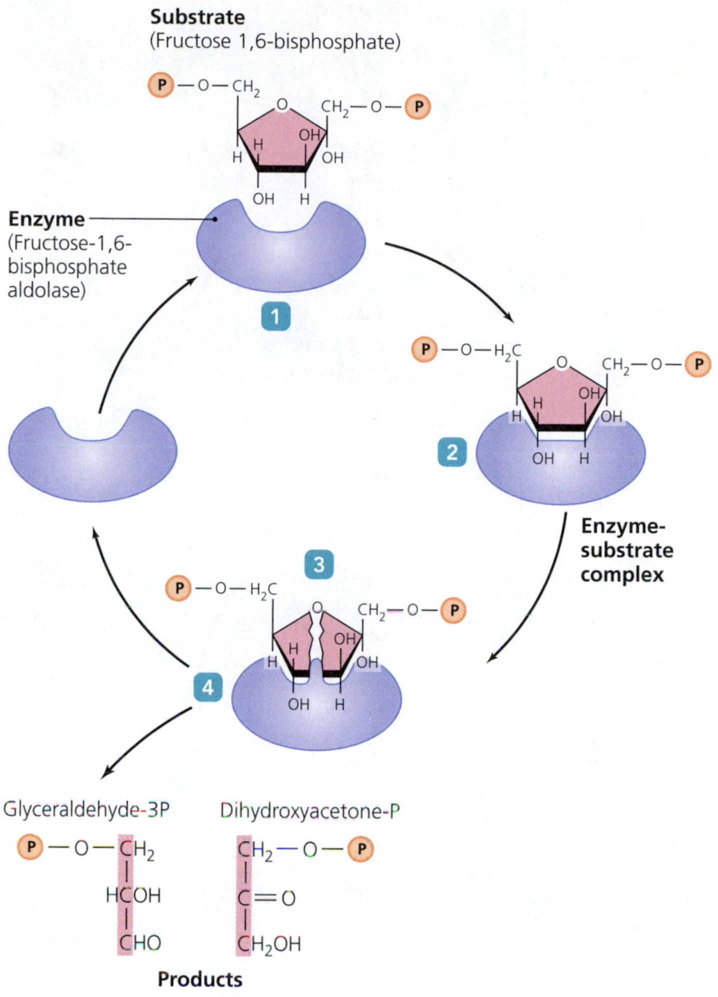

▲ **FIGURE 5.6 The process of enzymatic activity.** Shown here is the lysis of fructose 1,6-bisphosphate by the enzyme fructose-1,6-bisphosphate aldolase (a catabolic reaction). After the enzyme associates with the substrate 1, the two molecules bind to form an enzyme-substrate complex. 2 As a result of binding, the enzyme's active site is induced to fit the substrate even more closely; then, bonds within the substrate are broken 3, after which the enzyme dissociates from the two products 4. The enzyme resumes its initial shape and is then ready to associate with another substrate molecule. This entire process occurs 14 times per second at 37°C (body temperature).

If temperature rises beyond a certain critical point, the noncovalent bonds within an enzyme (such as the hydrogen bonds between amino acids) will break, and the enzyme will **denature** (**FIGURE 5.8**). Denatured enzymes lose their specific three-dimensional structure, so they are no longer functional. Denaturation is said to be *permanent* when an enzyme cannot regain its original three-dimensional structure once conditions return to normal, much like the irreversible solidification of the protein albumin when egg whites are cooked and then cooled. In other cases denaturation is *reversible*—the denatured enzyme's noncovalent bonds re-form on the return of normal conditions.

pH Extremes of pH also denature enzymes when ions released from acids and bases interfere with hydrogen bonding and distort

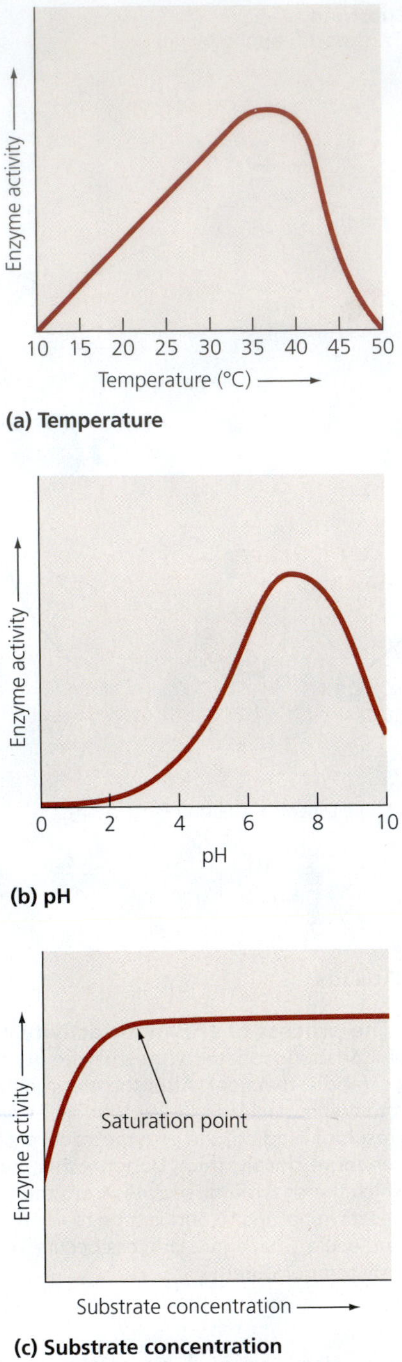

▲ **FIGURE 5.7 Representative effects of temperature, pH, and substrate concentration on enzyme activity.** Effects on each enzyme will vary. **(a)** Rising temperature enhances enzymatic activity to a point, but above some optimal temperature an enzyme denatures and loses function. **(b)** Enzymes typically have some optimal pH, at which point enzymatic activity reaches a maximum. **(c)** At lower substrate concentrations, enzyme activity increases as the substrate concentration increases and as more and more active sites are utilized. At the substrate concentration at which all active sites are utilized, termed the *saturation point*, enzymatic activity reaches a maximum, and any additional increase in substrate concentration has no effect on enzyme activity. *What is the optimal pH of the enzyme shown in part (b)?*

Figure 5.7 *The enzyme's optimal pH is approximately 7.2.*

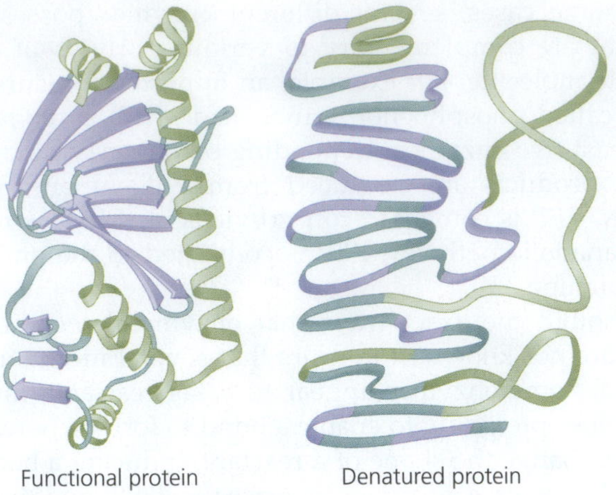

▲ **FIGURE 5.8 Denaturation of protein enzymes.** Breakage of noncovalent bonds (such as hydrogen bonds) causes the protein to lose its secondary and tertiary structure and become denatured; as a result, the enzyme is no longer functional.

and disrupt an enzyme's secondary and tertiary structures. Therefore, each enzyme has an optimal pH **(FIGURE 5.7b)**.

Changing the pH provides a way to control the growth of unwanted microorganisms by denaturing their proteins. For example, vinegar (acetic acid, pH 3.0) acts as a preservative in dill pickles, and ammonia (pH 11.5) can be used as a disinfectant.

Enzyme and Substrate Concentration Another factor that determines the rate of enzymatic activity within cells is the concentration of substrate present **(FIGURE 5.7c)**. As substrate concentration increases, enzymatic activity increases as more and more enzyme active sites bind more and more substrate molecules. Eventually, when all enzyme active sites have bound substrate, the enzymes have reached their saturation point, and the addition of more substrate will not increase the rate of enzymatic activity.

Obviously, the rate of enzymatic activity is also affected by the concentration of enzyme within cells. In fact, one way that organisms regulate their metabolism is by controlling the quantity and timing of enzyme synthesis. In other words, many enzymes are produced in the amounts and at the times they are needed to maintain metabolic activity. (Chapter 7 discusses the role of genetic mechanisms in regulating enzyme synthesis.) Additionally, eukaryotic cells control some enzymatic activities by compartmentalizing enzymes inside membranes so that certain metabolic reactions proceed physically separated from the rest of the cell. For example, white blood cells catabolize phagocytized pathogens using enzymes packaged within lysosomes.

Inhibitors Enzymatic activity can be influenced by a variety of inhibitory substances that block an enzyme's active site. Enzymatic inhibitors, which may be either competitive or noncompetitive, do not denature enzymes.

Competitive inhibitors are shaped such that they fit into an enzyme's active site and thus prevent the normal substrate

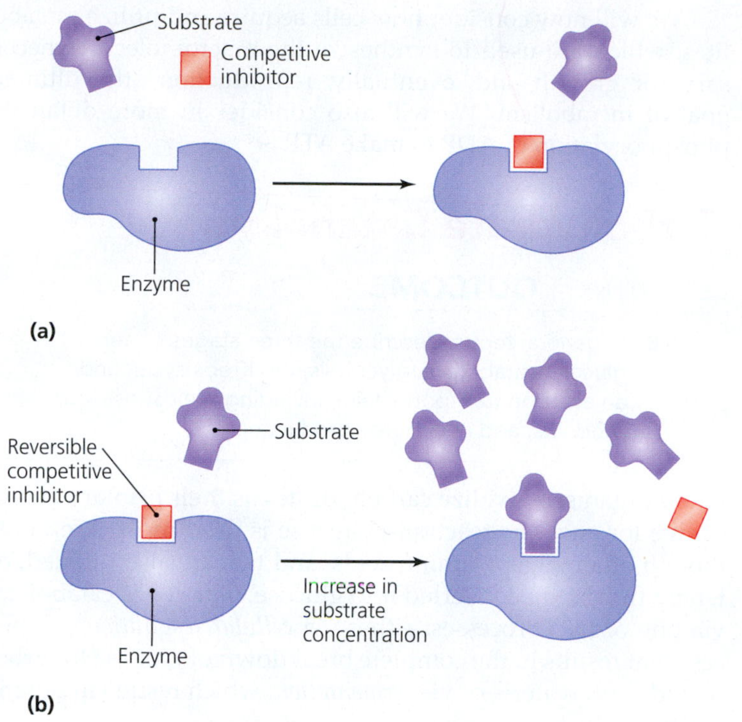

▲ **FIGURE 5.9 Competitive inhibition of enzyme activity.** **(a)** Inhibitory molecules, which are similar in shape to substrate molecules, compete for and block active sites. **(b)** Reversible inhibition can be overcome by an increase in substrate concentration.

from binding **(FIGURE 5.9a)**. However, such inhibitors do not undergo a chemical reaction to form products. Competitive inhibitors can bind permanently or reversibly to an active site. Permanent binding results in permanent loss of enzymatic activity; reversible competition can be overcome by an increase in the concentration of substrate molecules, increasing the likelihood that active sites will be filled with substrate instead of inhibitor **(FIGURE 5.9b)**. ▶**ANIMATIONS:** ***Enzymes: Competitive Inhibition***

An example of competitive inhibition is the action of sulfanilamide, which has a shape similar to that of para-aminobenzoic acid (PABA).

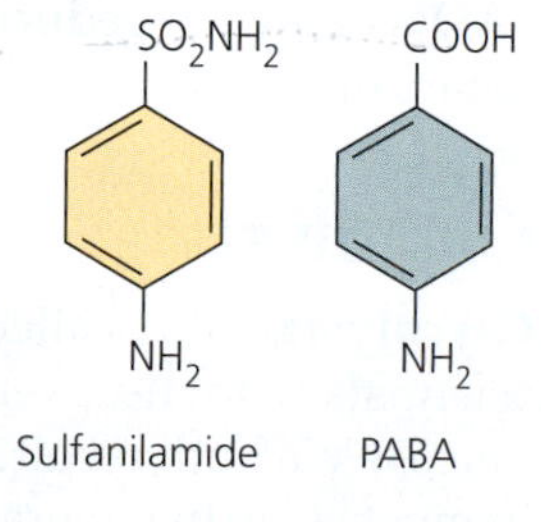

Sulfanilamide has great affinity for the active site of an enzyme required in the conversion of PABA into folic acid, which is essential for DNA synthesis. Once sulfanilamide is bound to the enzyme, it stays bound. As a result, it prevents synthesis of folic acid. Sulfanilamide effectively inhibits bacteria that make folic acid. Humans do not synthesize folic acid—we must acquire it as a vitamin in our diets—so sulfanilamide does not affect us in this way.

Noncompetitive inhibitors do not bind to the active site but instead prevent enzymatic activity by binding to an *allosteric* (al-ō-stār´ik) *site* located elsewhere on the enzyme. Binding at an allosteric site alters the shape of the active site so that substrate cannot be bound. Allosteric control of enzyme activity can take two forms: inhibitory and excitatory. *Allosteric (noncompetitive) inhibition* halts enzymatic activity in the manner just described **(FIGURE 5.10a)**. In *excitatory allosteric activation*, the binding of certain activator molecules (such as a heavy-metal ion cofactor) to an allosteric site causes a change in shape of the active site, which activates an otherwise inactive enzyme **(FIGURE 5.10b)**. Some enzymes have several allosteric sites, both inhibitory and excitatory, allowing their function to be closely regulated. ▶**ANIMATIONS:** ***Enzyme-Substrate Interaction: Noncompetitive Inhibition***

Cells often control the action of enzymes through **feedback inhibition** (also called *negative feedback* or *end-product inhibition*). Allosteric feedback inhibition functions in much the way a thermostat controls a heater. As the room gets warmer, a sensor inside the thermostat changes shape and sends an electrical signal that turns off the heater. Similarly, in metabolic feedback inhibition, the end-product of a series of reactions is an allosteric inhibitor of an enzyme in an earlier part of the pathway **(FIGURE 5.11)**. Because the product of each reaction in the pathway is the substrate for the next reaction, inhibition of the first enzyme in the series inhibits the entire pathway, thereby saving the cell energy. For example, in *Escherichia coli*, the presence of the amino acid isoleucine allosterically inhibits the first enzyme in the anabolic pathway that produces isoleucine. In this manner, the bacterium prevents the synthesis of isoleucine when the amino acid is already available. When isoleucine is depleted, the enzyme is no longer inhibited, and isoleucine production resumes.

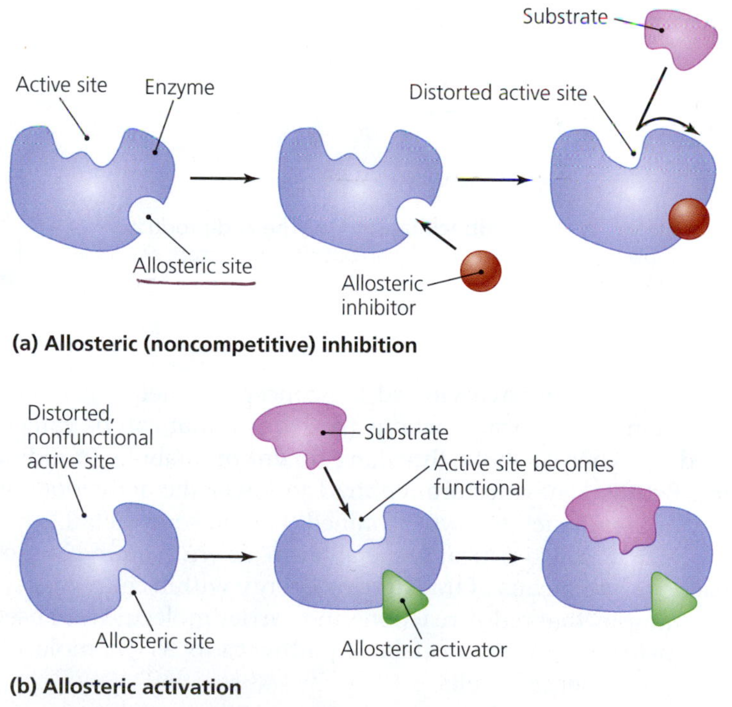

▲ **FIGURE 5.10 Allosteric control of enzyme activity.** **(a)** Allosteric (noncompetitive) inhibition results from a change in the shape of the active site when an inhibitor binds to an allosteric site. **(b)** Allosteric activation results when the binding of an activator molecule to an allosteric site causes a change in the active site that makes it capable of binding substrate.

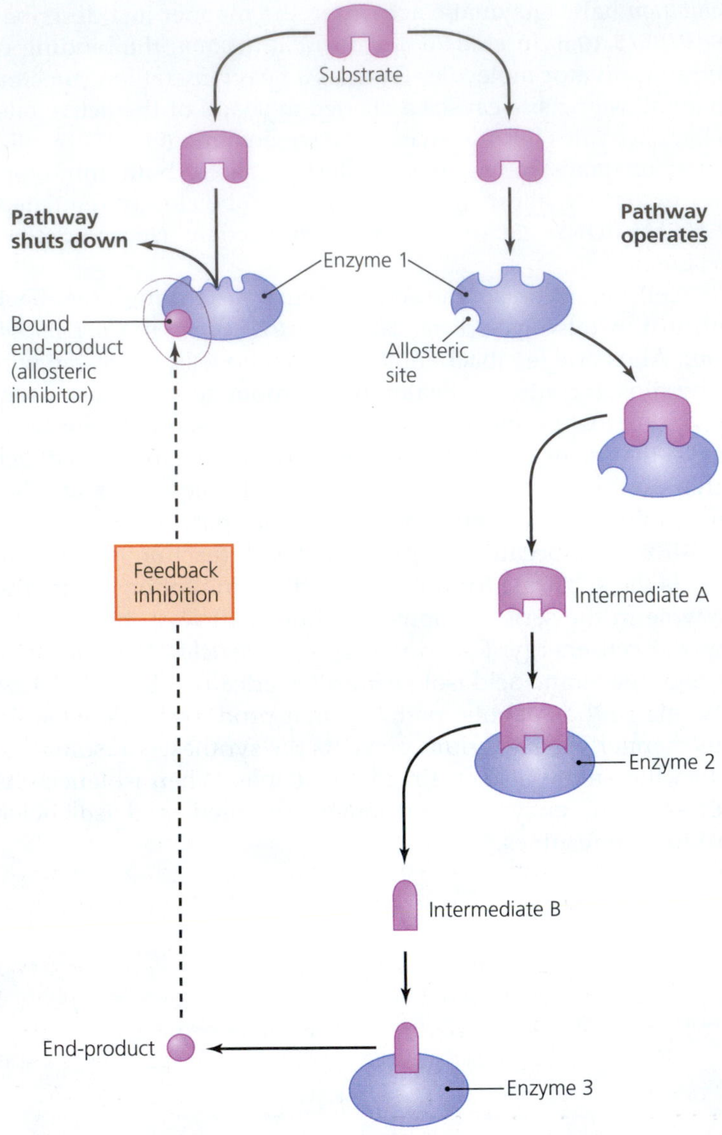

▲ **FIGURE 5.11 Feedback inhibition.** The end-product of a metabolic pathway allosterically inhibits the initial step, shutting down the pathway.

To this point we have viewed the concept of metabolism as a collection of chemical reactions (pathways) that can be categorized as either catabolic (breaking down) or anabolic (building up). Because enzymes are required to lower the activation energy of these reactions, we examined them in some detail.

Energy is also critical to metabolism, so we examined redox reactions as a means of transferring energy within cells. We saw, for example, that redox reactions and carrier molecules are used to transfer energy from catabolic pathways to ATP, a molecule that stores energy in cells.

TELL ME WHY

How can oxidation take place in an anaerobic environment, that is, without oxygen?

We will now consider how cells acquire and utilize metabolites, which are used to synthesize the macromolecules necessary for growth and, eventually, reproduction—the ultimate goal of metabolism. We will also consider in more detail the phosphorylation of ADP to make ATP.

Carbohydrate Catabolism

LEARNING **OUTCOME**

5.8 In general terms, describe the three stages of aerobic glucose catabolism (glycolysis, the Krebs cycle, and an electron transport chain), including their substrates, products, and net energy production.

Many organisms oxidize carbohydrates as their primary energy source for anabolic reactions. Glucose is used most commonly, though other sugars, amino acids, and fats are also utilized, often by first being converted into glucose. Glucose is catabolized via one of two processes: either via *cellular respiration*[3]—a process that results in the complete breakdown of glucose to carbon dioxide and water—or via *fermentation*, which results in organic waste products.

As shown in **FIGURE 5.12**, both cellular respiration and fermentation begin with *glycolysis* (glī-kol′i-sis), a process that catabolizes a single molecule of glucose to two molecules of pyruvic acid (also called pyruvate) and results in a small amount of ATP production. Respiration then continues via the *Krebs cycle* and an *electron transport chain*, which results in a significant amount of ATP production. Fermentation involves the conversion of pyruvic acid into other organic compounds. Because it lacks the Krebs cycle and an electron transport chain, fermentation results in the production of much less ATP than does respiration.

The following is a simplified discussion of glucose catabolism. To help understand the basic reactions in each of the pathways of glucose catabolism, pay special attention to three things: the number of carbon atoms in each of the intermediate products, the relative numbers of ATP molecules produced in each pathway, and the changes in the coenzymes NAD^+ and FAD as they are reduced and then oxidized back to their original forms.

Glycolysis

Glycolysis,[4] also called the *Embden-Meyerhof pathway* after the scientists who discovered it, is the first step in the catabolism of glucose via both respiration and fermentation. Glycolysis occurs in most cells. In general, as its name implies, glycolysis involves the splitting of a six-carbon glucose molecule into two three-carbon sugar molecules. When these three-carbon molecules are oxidized to pyruvic acid, some of the energy released is stored

[3]Cellular respiration is often referred to simply as *respiration*, which should not be confused with breathing, also called respiration.
[4]From Greek *glukus*, meaning "sweet," and *lysis*, meaning to "loosen."

the purpose of Fermentation is to regenerate NADH

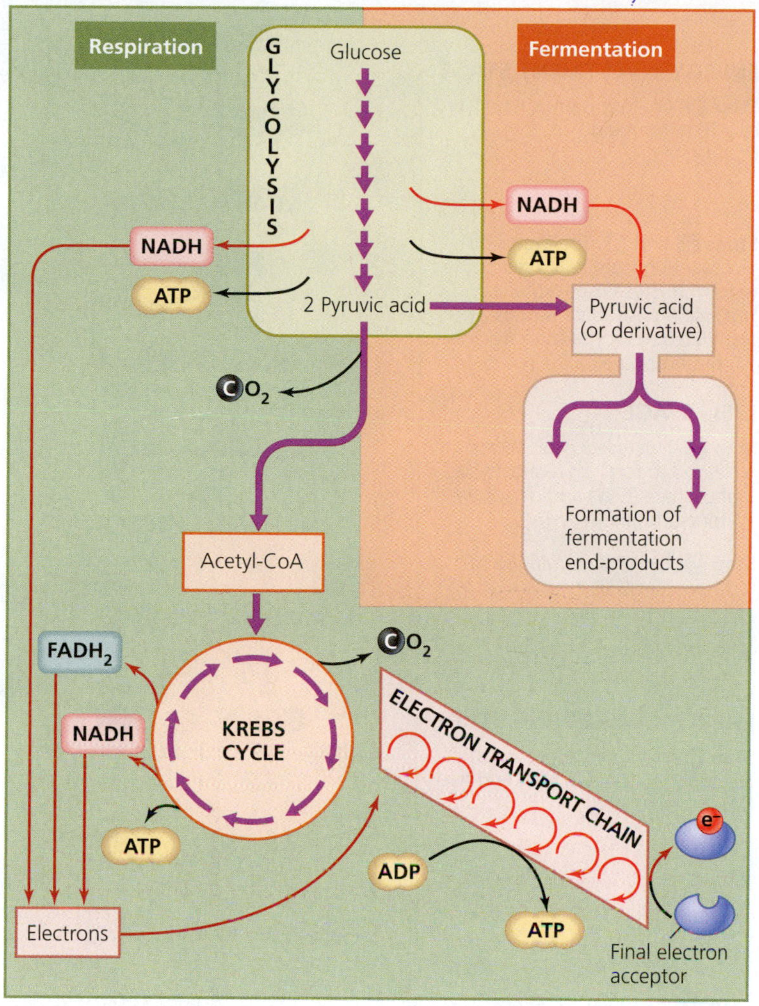

▲ **FIGURE 5.12 Summary of glucose catabolism.** Glucose catabolism begins with glycolysis, which forms pyruvic acid and two molecules of both ATP and NADH. Two pathways branch from pyruvic acid: respiration and fermentation. In aerobic respiration (shown here), the Krebs cycle and the electron transport chain completely oxidize pyruvic acid to CO_2 and H_2O, in the process synthesizing many molecules of ATP. Fermentation results in the incomplete oxidation of pyruvic acid to form organic fermentation products. *Which of the blue ovals represents a reduced chemical?*

Figure 5.12 The upper one, with an additional electron, is reduced.

MM To see a 3-D animation on metabolism, go to the MasteringMicrobiology Study Area and watch the MicroFlix.

in molecules of ATP and NADH. Cells can use many of the intermediate molecules in glycolysis as precursor metabolites. ▶**ANIMATIONS:** ***Glycolysis: Overview***

Glycolysis, which occurs in the cytosol, can be divided into three stages involving a total of 10 steps **(FIGURE 5.13)**, each of which is catalyzed by its own enzyme:

1. *Energy-investment stage* (steps 1–3). As with money, one must invest before a profit can be made. In this case, the energy in two molecules of ATP is invested to phosphorylate a six-carbon glucose molecule and rearrange its atoms to form fructose 1,6-bisphosphate.
2. *Lysis stage* (steps 4 and 5). Fructose 1,6-bisphosphate is cleaved into glyceraldehyde 3-phosphate (G3P)[5] and dihydroxyacetone phosphate (DHAP). Each of these compounds contains three carbon atoms and is freely convertible into the other.
3. *Energy-conserving stage* (steps 6 – 10). G3P is oxidized to pyruvic acid, yielding two ATP molecules. DHAP is converted to G3P and also oxidized to pyruvic acid, yielding another two ATP molecules, for a total of four ATP molecules.

Details of the substrates and enzymes involved are provided in Appendix A on pp. A–6 to A–7. ▶**ANIMATIONS:** ***Glycolysis: Steps***

Our study of glycolysis provides our first opportunity to study *substrate-level phosphorylation* (see steps 1, 3, 7, and 10). Let's examine this important process more closely by considering the 10th and final step of glycolysis.

Each of the two phosphoenolpyruvic acid (PEP) molecules produced in step 9 of glycolysis is a three-carbon compound containing a high-energy phosphate bond. In the presence of a specific holoenzyme (which requires a Mg^{2+} cofactor), the high-energy phosphate in PEP (one substrate) is transferred to an ADP molecule (a second substrate) to form ATP (step 10 and **FIGURE 5.14**); the direct transfer of the phosphate between the two substrates is the reason the process is called **substrate-level phosphorylation**. A variety of substrate-level phosphorylations occur in metabolism. As you might expect, each type has its own enzyme that recognizes both its substrate molecule and ADP.

In glycolysis, two ATP molecules are invested by substrate-level phosphorylation to prime glucose for lysis, and four molecules of ATP are produced by substrate-level phosphorylation. Therefore, a net gain of two ATP molecules occurs for each molecule of glucose that is oxidized to pyruvic acid. Glycolysis also yields two molecules of NADH.

Cellular Respiration

LEARNING **OUTCOMES**

5.9 Discuss the roles of acetyl-CoA, the Krebs cycle, and electron transport in carbohydrate catabolism.

5.10 Contrast electron transport in aerobic and anaerobic respiration.

5.11 Identify four classes of carriers in electron transport chains.

5.12 Describe the role of chemiosmosis in oxidative phosphorylation of ATP.

After glucose has been oxidized via glycolysis or one of the alternate pathways considered shortly, a cell uses the resultant pyruvic acid molecules to complete either cellular respiration or fermentation (which we will discuss in a later section). Our topic here—**cellular respiration**—is a metabolic process that involves

[5]G3P is also known as phosphoglyceraldehyde, or PGAL.

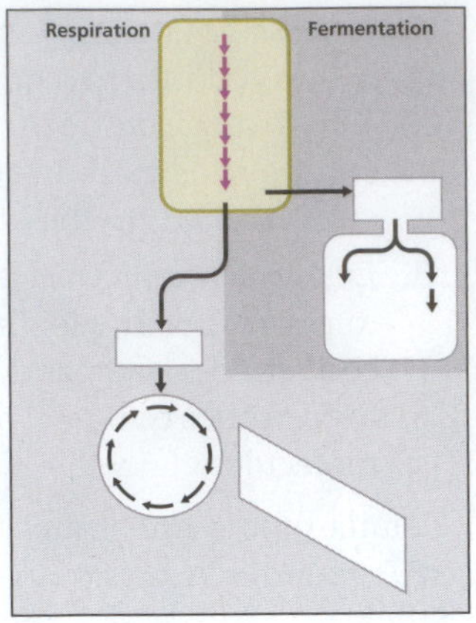

ENERGY-INVESTMENT STAGE

Step 1. Glucose is phosphorylated by ATP to form glucose 6-phosphate.

Steps 2 and 3. The atoms of glucose 6-phosphate are rearranged to form fructose 6-phosphate. Fructose 6-phosphate is phosphorylated by ATP to form fructose 1,6-bisphosphate.

LYSIS STAGE

Step 4. Fructose 1,6-bisphosphate is cleaved to form glyceraldehyde 3-phosphate (G3P) and dihydroxyacetone phosphate (DHAP).

Step 5. DHAP is rearranged to form another G3P.

ENERGY-CONSERVING STAGE

Step 6. Inorganic phosphates are added to the two G3P, and two NAD^+ are reduced.

Step 7. Two ADP are phosphorylated by substrate-level phosphorylation to form two ATP.

Steps 8 and 9. The remaining phosphates are moved to the middle carbons. A water molecule is removed from each substrate.

Step 10. Two ADP are phosphorylated by substrate-level phosphorylation to form two ATP. Two pyruvic acid are formed.

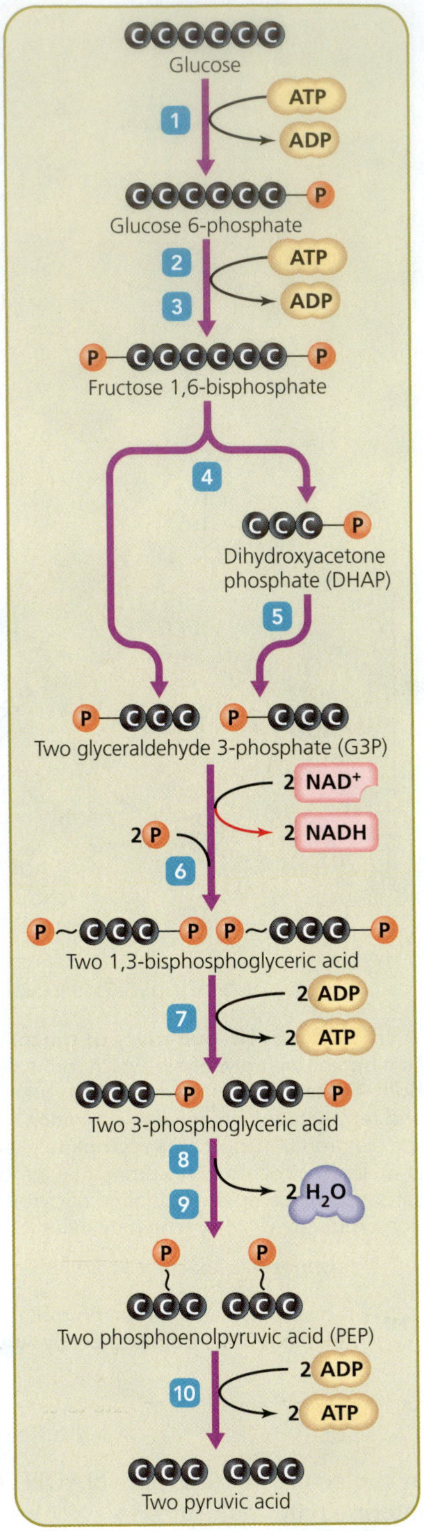

▶ **FIGURE 5.13 Glycolysis.** Glucose is cleaved and ultimately transformed into two molecules of pyruvic acid in this process (also known as the Embden-Meyerhof pathway). Four ATPs are formed and two ATPs are used, so a net gain of two ATPs results. Two molecules of NAD^+ are reduced to NADH. (For simplification, only carbon atoms and phosphate are shown. For details, see Appendix A, pp. A–6 to A–7.)

the complete oxidation of substrate molecules and then production of ATP by a series of redox reactions. The three stages of cellular respiration are (1) synthesis of acetyl-CoA, (2) the Krebs cycle, and (3) a final series of redox reactions, called an electron transport chain, that passes electrons to a chemical not derived from the cell's metabolism.

Synthesis of Acetyl-CoA

Before pyruvic acid (generated by glycolysis or an alternate pathway) can enter the Krebs cycle for respiration, it must first be converted to *acetyl-coenzyme A*, or **acetyl-CoA** (as´e-til kō-ā; see Figure 5.12). Enzymes remove one carbon from pyruvic acid as CO_2 1 and join the remaining two-carbon acetate to *coenzyme A*

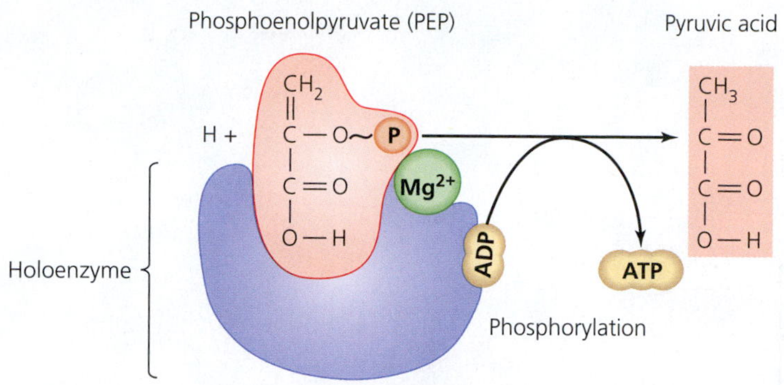

▲ **FIGURE 5.14 Example of substrate-level phosphorylation.** High-energy phosphate bonds are transferred from one substrate to another in this process. *What role does Mg^{2+} play in this reaction?*

Figure 5.14 Mg^{2+} is a cofactor of the enzyme.

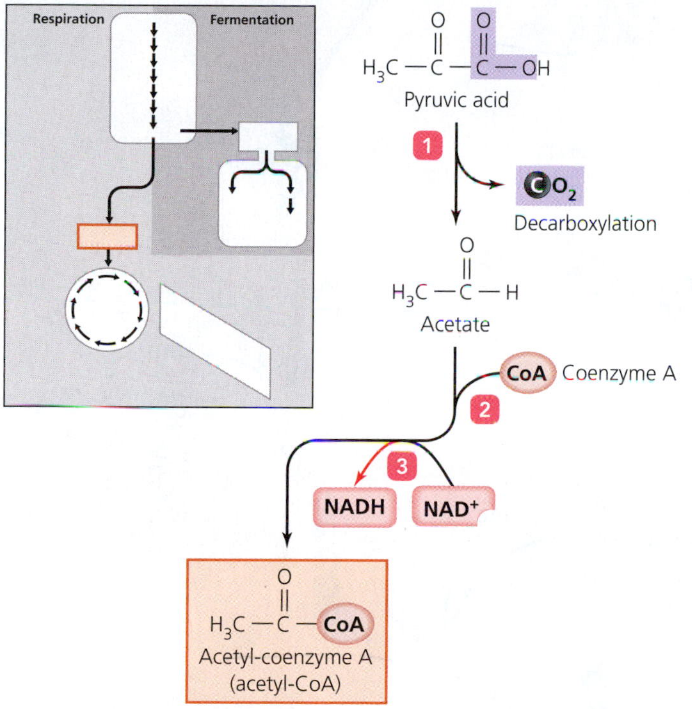

▲ **FIGURE 5.15 Formation of acetyl-CoA.** The responsible enzyme acts in a stepwise manner to 1 remove CO_2 from pyruvic acid, 2 attach the remaining two-carbon acetate to coenzyme A, and 3 reduce a molecule of NAD^+ to NADH. Because glycolysis produces two molecules of pyruvic acid, two molecules of acetyl-CoA are produced.

with a high-energy bond 2 **(FIGURE 5.15)**. The removal of CO_2, called *decarboxylation,* requires a coenzyme derived from the vitamin thiamine. One molecule of NADH is also produced during this reaction 3.

Recall that two molecules of pyruvic acid were derived from each molecule of glucose. Therefore, at this stage, two molecules of acetyl-CoA, two molecules of CO_2, and two molecules of NADH are produced.

The Krebs Cycle

At this point in the catabolism of a molecule of glucose, a great amount of energy remains in the bonds of acetyl-CoA. The **Krebs cycle**[6] is a "circular" series of eight enzymatically catalyzed reactions that transfer much of this stored energy to the coenzymes NAD^+ and FAD (the cycle is diagrammed in **FIGURE 5.16** and presented in more detail on p. A–10). The two carbon atoms in acetate are oxidized, and the coenzymes are reduced. The Krebs cycle occurs in the cytosol of prokaryotes and in the matrix of mitochondria in eukaryotes. It is also known as the *tricarboxylic acid (TCA) cycle,* because many of its compounds have three carboxyl (—COOH) groups, and as the *citric acid cycle,* for the first compound formed in the cycle.

There are six types of reactions in the Krebs cycle:

- Anabolism (of citric acid; step 1)
- Isomerization (step 2)
- Redox reactions (steps 3, 4, 6, and 8)
- Decarboxylations (steps 3 and 4)
- Substrate-level phosphorylation (step 5)
- Hydration (step 7)

In the first step of the Krebs cycle, the splitting of the high-energy bond between acetate and coenzyme A releases enough energy to enable the binding of the freed two-carbon acetate to a four-carbon compound called oxaloacetic acid, forming the six-carbon compound citric acid (step 1).

As you study Figure 5.16, notice that after isomerization (step 2), the decarboxylations of the Krebs cycle release two molecules of CO_2 for each acetyl-CoA that enters (steps 3 and 4). Thus, for every two carbon atoms that enter the cycle, two are lost to the environment. At this juncture in the respiration of a molecule of glucose, all six carbon atoms have been lost to the environment: two as CO_2 molecules produced in decarboxylation of two molecules of pyruvic acid to form two acetyl-CoA molecules, and four in CO_2 molecules produced in decarboxylations in the *two* turns through the Krebs cycle. (One molecule of acetyl-CoA enters the cycle at a time.)

For every two molecules of acetyl-CoA that pass through the Krebs cycle, two molecules of ATP are generated by substrate-level phosphorylation (step 5). A molecule of guanosine triphosphate (GTP), which is similar to ATP, can serve as an intermediary in this process. The two molecules of ATP produced by the Krebs cycle are a small amount compared to the energy available. Most of the energy is still in the electrons carried away by NADH and $FADH_2$.

Redox reactions reduce FAD to $FADH_2$ (step 6) and NAD^+ to NADH (steps 3, 4, and 8) so that for every two molecules of acetyl-CoA that move through the cycle, six molecules of NADH and two of $FADH_2$ are formed. In the Krebs cycle, little energy is captured directly in high-energy phosphate bonds, but much energy is transferred via electrons to NADH and $FADH_2$. These coenzymes are the most important molecules of

[6]Named for biochemist Hans Krebs, who elucidated its reactions in the 1940s.

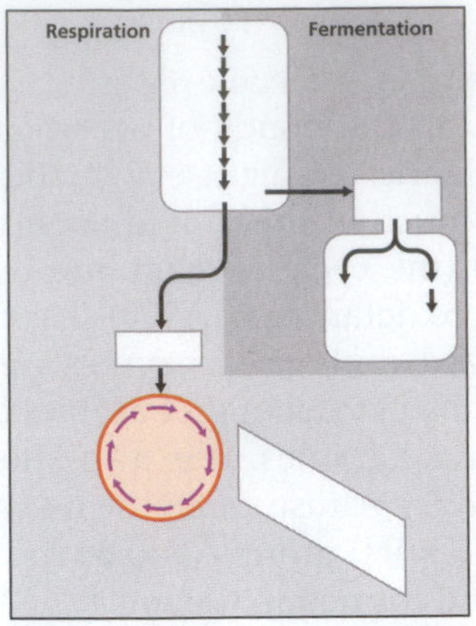

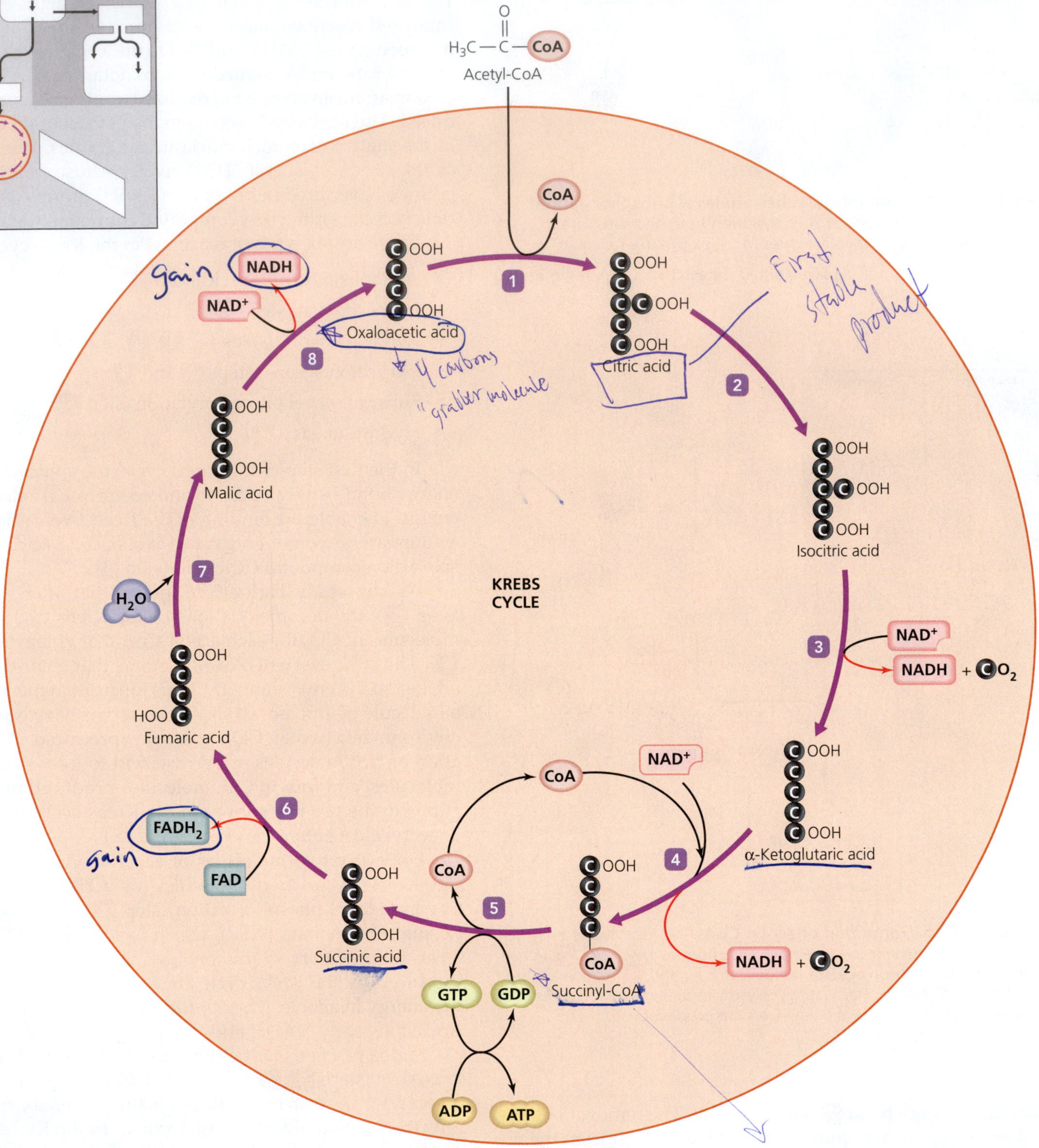

▲ **FIGURE 5.16 The Krebs cycle.** 1 Acetyl-CoA enters the Krebs cycle by joining with oxaloacetic acid to form citric acid, releasing coenzyme A. 2–4 Two oxidations and decarboxylations and the addition of coenzyme A once again yield succinyl-CoA. 5 Substrate-level phosphorylation produces ATP and again releases coenzyme A. 6–8 Further oxidations and rearrangements regenerate oxaloacetic acid, and the cycle can begin anew. Two molecules of acetyl-CoA enter the cycle for each molecule of glucose undergoing glycolysis.

respiration because they carry a large amount of energy that is subsequently used to phosphorylate ADP to ATP. Many of the intermediates of the Krebs cycle are also precursor metabolites; for example, cells can use oxaloacetic acid to make several kinds of amino acids (see Figure 5.31). **▶ANIMATIONS:** ***Krebs Cycle: Overview, Steps***

Electron Transport

Scientists estimate that each day an average human synthesizes his or her own weight in ATP molecules and uses them for metabolism, responsiveness, growth, and cell reproduction. ATP turnover in prokaryotes is relatively as copious. The most significant production of ATP does not occur through glycolysis or the Krebs cycle, but rather through the stepwise release of energy from a series of redox reactions known as an **electron transport chain** **(FIGURE 5.17)**. **▶ANIMATIONS:** ***Electron Transport Chain: Overview***

An electron transport chain consists of a series of membrane-bound carrier molecules that pass electrons from one to another and ultimately to a *final electron acceptor.* Typically, as we have seen, electrons come from the catabolism of an organic molecule such as glucose; however, microorganisms called *lithotrophs* (lith´ō-trōfs) acquire electrons from inorganic sources such as H_2, NO^{2-}, or Fe^{2+}. (Chapter 6 discusses lithotrophs further.) In any case, carrier molecules pass electrons down the chain to the final acceptor like firefighters of old, who passed buckets of water from one to another until the last one threw the water on a fire. In electron transport, NAD^+ is an "empty bucket"; NADH is a "full bucket." As with a bucket brigade, the final step of electron transport is irreversible. Energy from the electrons is used to actively transport (pump) protons (H^+) across the membrane, establishing a *proton gradient* that generates ATP via a process called *chemiosmosis,* which we will discuss shortly.

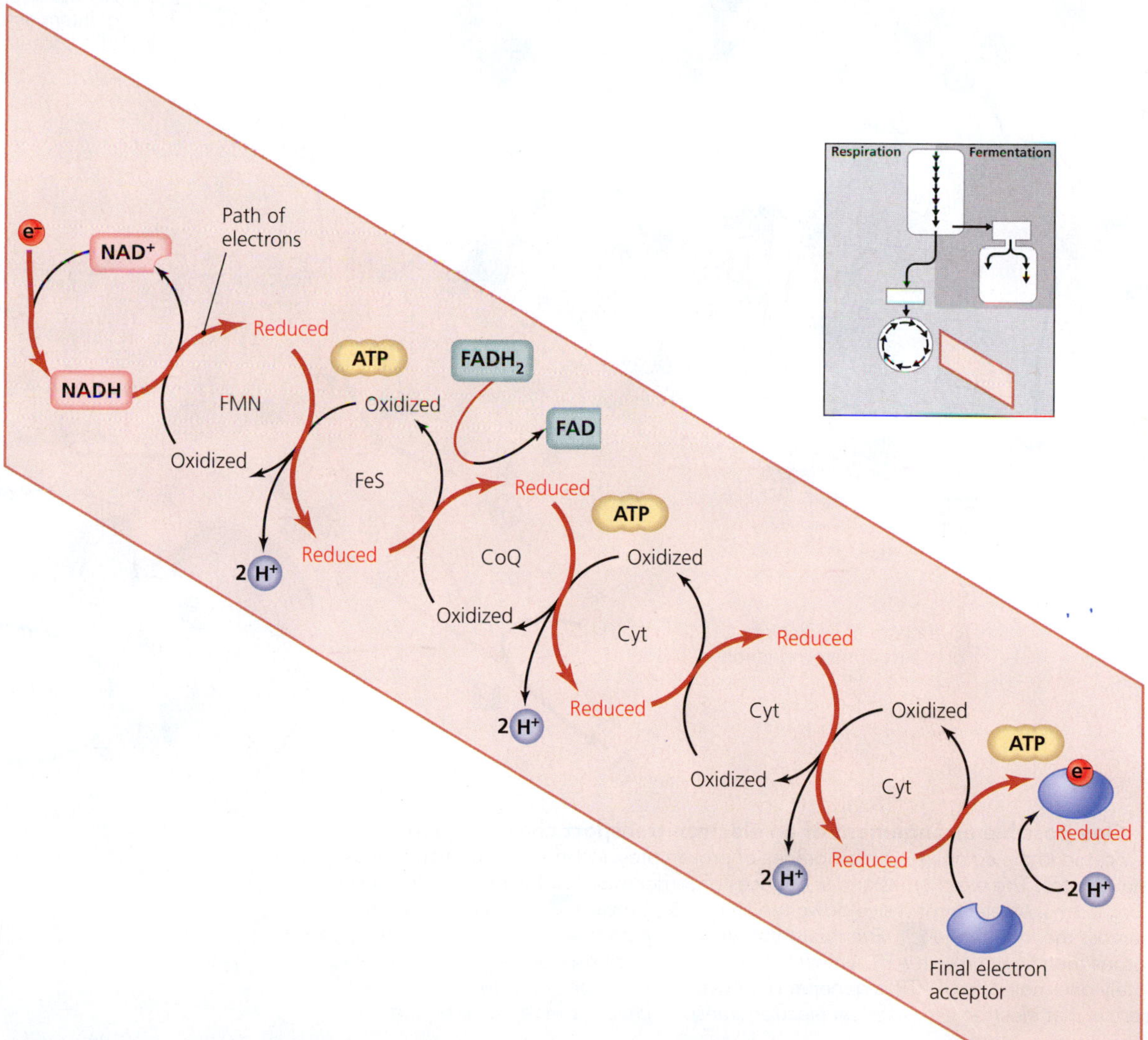

▲ FIGURE 5.17 An electron transport chain. As electrons move down the chain from molecule to molecule (red arrows), energy is released. This energy can be used to synthesize ATP. The ATP icons indicate the approximate point in the chain where enough energy has been released to make an ATP molecule, but the molecules of the chain do not actually synthesize ATP.

To avoid getting lost in the details of electron transport, keep the following critical concepts in mind:

- Electrons pass sequentially from one membrane-bound carrier molecule to another, each time losing some energy. Eventually, they pass to a final acceptor molecule.
- The electrons' energy is used to pump protons across the membrane.

Electron transport chains are located in the inner mitochondrial membranes (cristae) of eukaryotes and in the cytoplasmic membranes of prokaryotes **(FIGURE 5.18)**. Though NADH and

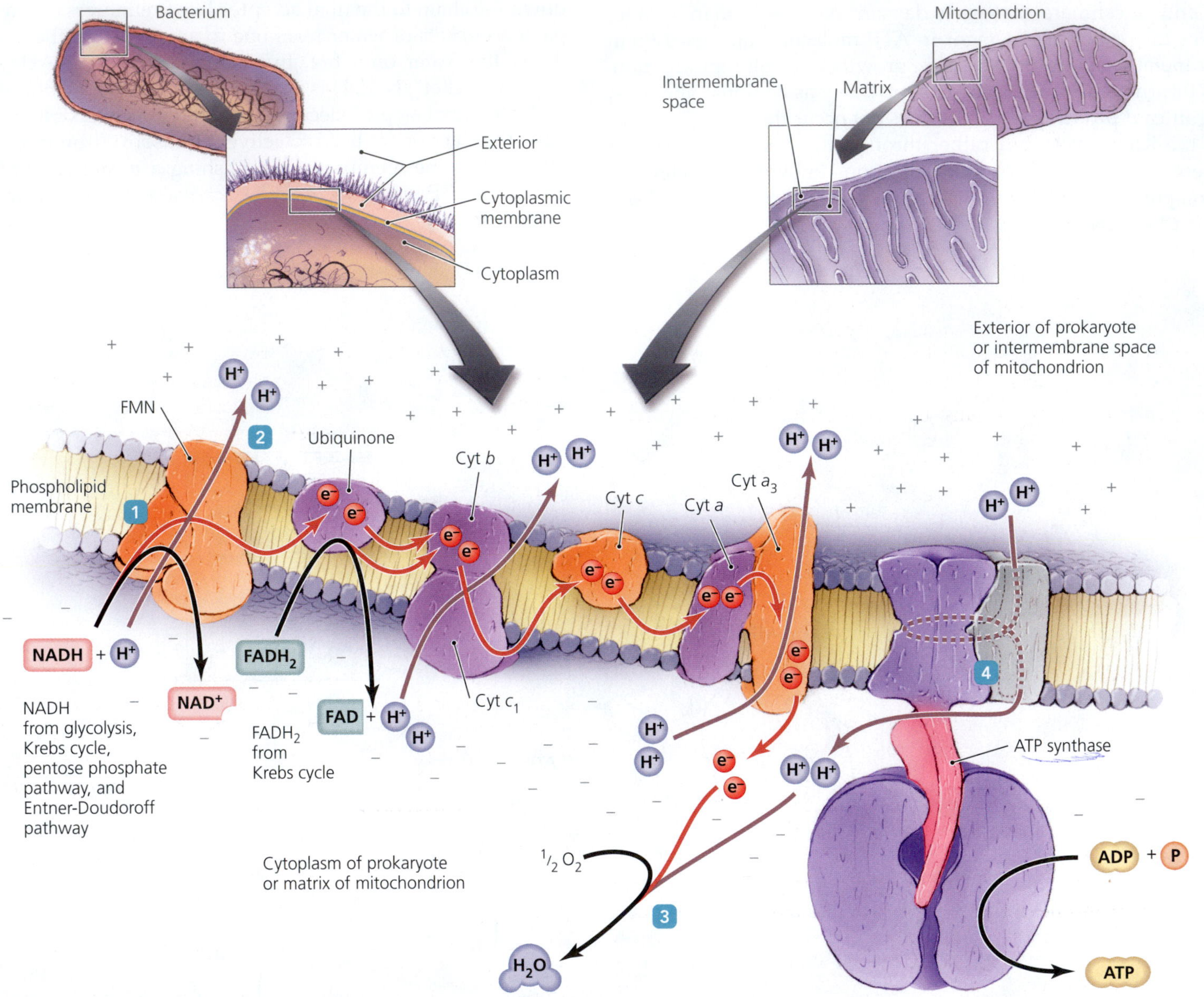

▲ **FIGURE 5.18 One possible arrangement of an electron transport chain.** Electron transport chains are located in the cytoplasmic membranes of prokaryotes and in the inner membranes of mitochondria in eukaryotes. The exact types and sequences of carrier molecules in electron transport chains vary among organisms. As electrons move down the chain (red arrows) 1, their energy is used to pump protons (H^+) across the membrane 2. Eventually the electrons pass to a final acceptor, in this case oxygen 3. The protons then flow through ATP synthase (ATPase), which phosphorylates ADP to make ATP 4. Approximately one molecule of ATP is generated for every two protons that cross the membrane. *Why is it essential that electron carriers of an electron transport chain be membrane bound?*

Figure 5.18 *The carrier molecules of electron transport chains are membrane bound for at least two reasons: (1) Carrier molecules need to be in fairly close proximity so they can pass electrons between each other. (2) The goal of the chain is the pumping of H^+ across a membrane to create a proton gradient; without a membrane, there could be no gradient.*

$FADH_2$ donate electrons as hydrogen atoms (electrons and protons), many carriers pass only the electrons down the chain. There are four categories of carriers in the transport chains:

- *Flavoproteins* are integral membrane proteins, many of which contain flavin, a coenzyme derived from riboflavin (vitamin B_2). One form of flavin is flavin mononucleotide (FMN), which is the initial carrier molecule of electron transport chains of mitochondria. The familiar FAD is a coenzyme for other flavoproteins. Like all carrier molecules in an electron transport chain, flavoproteins alternate between the reduced and oxidized states.
- *Ubiquinones* (yū-bik´wi-nōns) are lipid-soluble, nonprotein carriers that are so named because they are ubiquitous in cells. Ubiquinones are derived from vitamin K. In mitochondria, the ubiquinone is called *coenzyme Q*.
- *Metal-containing proteins* are a mixed group of integral proteins with a wide-ranging number of iron, sulfur, and copper atoms that can alternate between the reduced and oxidized states. Iron-sulfur proteins occur in various places in electron transport chains of various organisms. Copper proteins are found only in electron transport chains involved in photosynthesis (discussed shortly).
- *Cytochromes* (sī´tō-krōms) are integral proteins associated with *heme,* which is the same iron-containing, nonprotein, pigmented molecule found in the hemoglobin of blood. Iron can alternate between a reduced (Fe^{2+}) state and an oxidized (Fe^{3+}) state. Cytochromes are identified by letters and numbers based on the order in which they were identified, so their sequence in electron transport chains does not always seem logical. ▶**ANIMATIONS:** ***Electron Transport Chain: The Process*** ▶**VIDEO TUTOR:** ***Electron Transport Chains***

The carrier molecules in electron transport chains are diverse—bacteria typically have different carrier molecules arranged in different sequences from those of archaea or the mitochondria of eukaryotes. Even among bacteria, the makeup of carrier molecules can vary. For example, the pathogens *Neisseria* (nī-se´rē-ă) and *Pseudomonas* (soo-dō-mō´nas) contain two cytochromes, *a* and a_3—together called *cytochrome oxidase*—that oxidize cytochrome *c*; such bacteria are said to be *oxidase positive.* In contrast, other bacterial pathogens, such as *Escherichia, Salmonella* (sa1´mŏ-nel´ă), and *Proteus* (prō´tē-ŭs) lack cytochrome oxidase and are thus considered to be *oxidase negative.* Some bacteria, including *E. coli,* can even vary their carrier molecules under different environmental conditions.

Electrons carried by NADH enter a transport chain at a flavoprotein, and those carried by $FADH_2$ are introduced via a ubiquinone further down the chain. This explains why more molecules of ATP are generated from NADH than from $FADH_2$. Researchers do not agree on which carrier molecules are the actual proton pumps or on the number of protons that are pumped. Figure 5.18 shows one possibility.

In some organisms, the final electron acceptors are oxygen atoms, which, with the addition of hydrogen ions, generate H_2O; these organisms conduct **aerobic**[7] **respiration** and are called *aerobes.* **Highlight: Glowing Bacteria** on p. 142 describes an unusual aerobic bacterial electron transport system that produces light instead of ATP.

Other organisms, called *anaerobes,*[8] use other inorganic molecules (or rarely an organic molecule) instead of oxygen as the final electron acceptor and perform **anaerobic respiration**. The anaerobic bacterium *Desulfovibrio* (dē´sul-fō-vib´rē-ō), for example, reduces sulfate (SO_4^{2-}) to hydrogen sulfide gas (H_2S), whereas other anaerobes in the genera *Bacillus* (ba-sil´ŭs) and *Pseudomonas* utilize nitrate (NO_3^-) to produce nitrite ions (NO^{2-}), nitrous oxide (N_2O), or nitrogen gas (N_2). Some prokaryotes—particularly archaea called methanogens—reduce carbonate (CO_3^{2-}) to methane gas (CH_4). Medical laboratory scientists test for products of anaerobic respiration, such as nitrite, to help identify some species of bacteria. (Chapter 25). Anaerobic respiration is also critical for the recycling of nitrogen and sulfur in nature. ▶**ANIMATIONS:** ***Electron Transport Chain: Factors Affecting ATP Yield***

In summary, a number of redox reactions in glycolysis or alternate pathways and in the Krebs cycle strip electrons, which carry energy, from glucose molecules and transfer them to molecules of NADH and $FADH_2$. In turn, NADH and $FADH_2$ pass the electrons to an electron transport chain. As the electrons move down the electron transport chain, proton pumps use the electrons' energy to actively transport protons (H^+) across the membrane, creating a proton concentration gradient.

Recall, however, that the significance of electron transport is not merely that it pumps protons across a membrane but also that it ultimately results in the synthesis of ATP. We turn now to the process by which cells synthesize ATP using the proton gradient.

Chemiosmosis

Chemiosmosis is a general term for the use of ion gradients to generate ATP; that is, ATP is synthesized utilizing energy released by the flow of ions down their electrochemical gradient across a membrane. The term should not be confused with osmosis of water. To understand chemiosmosis, we need to review several concepts concerning diffusion and phospholipid membranes (see pp. 67–69 in Chapter 3).

Recall that chemicals diffuse from areas of high concentration to areas of low concentration and toward an electrical charge opposite their own. We call the composite of differences in concentration and charge an *electrochemical gradient.* Chemicals diffuse down their electrochemical gradients. Recall as well that membranes of cells and organelles are impermeable to most chemicals unless a specific protein channel allows their passage across the membrane. A membrane maintains an electrochemical gradient by keeping one or more chemicals in a higher concentration on one side. The blockage of diffusion creates potential energy, like water behind a dam.

Chemiosmosis uses the potential energy of an electrochemical gradient to phosphorylate ADP into ATP. Even though

[7]From Greek *aer,* meaning "air," (i.e., oxygen), and *bios,* meaning "life."
[8]The Greek prefix *an* means "not."

HIGHLIGHT

Glowing Bacteria

Bacteria in the genus *Photobacterium* possess an interesting electron tansport chain that generates light instead of ATP. These organisms can switch the flow of electrons from a standard electron transport chain (composed of cytochromes, proton pumps, and a final electron acceptor) to an alternate chain. Whereas the standard chain establishes a proton gradient that is used to synthesize ATP, the alternate chain uncouples electron transport from ATP synthesis: instead of transferring electrons to proton pumps, the alternate chain shunts the electrons to the coenzyme flavin mononucleotide (FMN). Then, in the presence of an enzyme called luciferase and a long-chain hydrocarbon, the alternate chain emits light as it transfers electrons to O_2 (see the diagram). The exact mechanism of bioluminescence is not known, but both FMN and the hydrocarbon are oxidized as oxygen is reduced.

Interestingly, free-living *Photobacterium* bacteria are not bioluminescent. However, when a multitude of bacteria signal their presence to one another (a situation called *quorum sensing*), primarily when colonizing the tissues of marine animals, the bacteria use their light-generating pathway. The animals gain from the association because the light produced serves as an attractant for mates and a warning against predators.

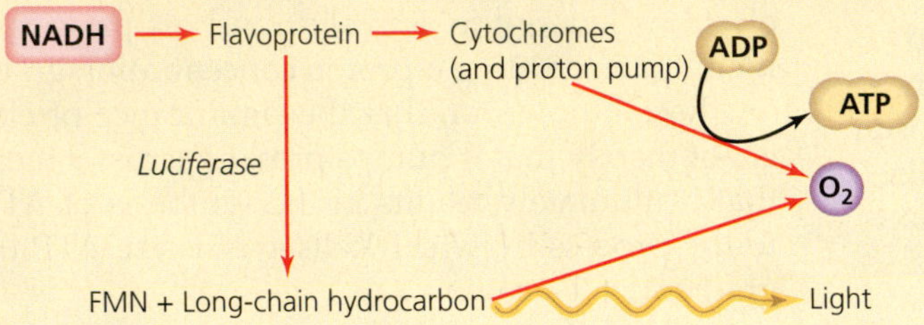

Kryptophanaron alfredi 2 cm

One species, the "flashlight fish" (*Kryptophanaron alfredi*), has a special organ near its mouth that is specially adapted for the growth of luminescent bacteria. Enough light is generated from millions of bacteria that the fish can navigate over coral reefs at night and attract prey to their light. The light organ even has a membrane that descends like an eyelid to control the amount of light emitted.

It is not clear what the bacteria gain from this association. Presumably, the protection and nutrients the bacteria gain from the fish make up for the enormous metabolic cost that the bacteria incur in the form of lost ATP synthesis.

chemiosmosis is a general principle with relevance to both *oxidative phosphorylation* and *photophosphorylation*, here we consider it as it relates to oxidative phosphorylation.

As we have seen, cells use the energy released in the redox reactions of electron transport chains to actively transport protons (H^+) across a membrane. Theoretically, an electron transport chain pumps three pairs of protons for each pair of electrons contributed by NADH, and it pumps two pairs of protons for each electron pair delivered by $FADH_2$. This difference results from the fact that NADH delivers electrons farther up the chain than does $FADH_2$; therefore, energy carried by $FADH_2$ is used to transport one-third fewer protons (see Figure 5.17). Because lipid bilayers are impermeable to protons, the transport of protons to one side of the membrane creates an electrochemical gradient known as a **proton gradient**, which has potential energy known as a *proton motive force*.

Protons, propelled by the proton motive force, flow down their electrochemical gradient through protein channels, called **ATP synthases (ATPases)**, that phosphorylate molecules of ADP to ATP (see Figure 5.18 4). Such phosphorylation is called **oxidative phosphorylation** because the proton gradient is created by the oxidation of components of an electron transport chain.

In the past, scientists attempted to calculate the exact number of ATP molecules synthesized per pair of electrons that travel down an electron transport chain. However, it is now apparent that phosphorylation and oxidation are not directly coupled. In other words, chemiosmosis does not require exact constant relationships among the number of molecules of NADH and $FADH_2$ reduced, the number of electrons that move down an electron transport chain, and the number of molecules of ATP that are synthesized. Additionally, cells use proton gradients for other cellular processes, including active transport and bacterial flagellar motion, so not every transported electron results in ATP production.

Nevertheless, about 34 molecules of ADP per molecule of glucose are oxidatively phosphorylated to ATP via chemiosmosis: three from each of the 10 molecules of NADH generated from glycolysis, the synthesis of acetyl-CoA, and the Krebs cycle and two from each of the two molecules of $FADH_2$ generated in the Krebs cycle. Given that glycolysis produces a net two molecules of ATP by substrate-level phosphorylation and that the Krebs cycle produces two more, the complete aerobic oxidation of one molecule of glucose by a prokaryote can theoretically yield a net total of 38 molecules of ATP (**TABLE 5.3**). The theoretical net maximum for eukaryotic cells is generally given

TABLE 5.3 Summary of Ideal Prokaryotic Aerobic Respiration of One Molecule of Glucose

Pathway	ATP Produced	ATP Used	NADH Produced	$FADH_2$ Produced
Glycolysis	4	2	2	0
Synthesis of acetyl-CoA and Krebs cycle	2	0	8	2
Electron transport chain	34	0	0	0
Total	40	2		
Net total	**38**			

as 36 molecules of ATP because energy from two ATP molecules is required to transport NADH (generated by glycolysis) from the cytoplasm into the mitochondria.

Alternatives to Glycolysis

LEARNING | OUTCOME

5.13 Compare the pentose phosphate pathway and the Entner-Doudoroff pathway with glycolysis in terms of energy production and products.

The initial part of the catabolism of glucose can also proceed via two alternate pathways: the pentose phosphate pathway and the Entner-Doudoroff pathway. Though they yield fewer molecules of ATP than glycolysis, these alternate pathways reduce coenzymes and yield substrate metabolites that are needed in anabolic pathways. Next we briefly examine each of these alternate pathways.

Pentose Phosphate Pathway

The **pentose phosphate pathway** (sometimes called the *phosphogluconate pathway*) is named for the phosphorylated pentose (five-carbon) sugars—ribulose, xylulose, and ribose—that are formed from glucose 6-phosphate by enzymes in the pathway (**FIGURE 5.19**). The pentose phosphate pathway is used primarily for the production of precursor metabolites used in anabolic reactions, including the synthesis of nucleotides for nucleic acids, of certain amino acids, and of glucose by *photosynthesis* (described in a later section). The pathway reduces two molecules of $NADP^+$ to NADPH and nets a single molecule of ATP from each molecule of glucose. NADPH is a necessary coenzyme for anabolic enzymes that synthesize DNA nucleotides, steroids, and fatty acids.

Appendix A, on p. A–8, shows the details of the substrates and enzymes of the pentose phosphate pathway.

Entner-Doudoroff Pathway

Many bacteria use glycolysis and the pentose phosphate pathway, but a few substitute the **Entner-Doudoroff pathway** (**FIGURE 5.20**) for glycolysis. This pathway, named for its discoverers, is a series of reactions that catabolize glucose to pyruvic acid using different enzymes from those used in either glycolysis or the pentose phosphate pathway.

Among organisms, only a very few bacteria use the Entner-Doudoroff pathway. These include the Gram-negative bacterium *Pseudomonas aeruginosa* (ā-roo-ji-nōsă) and the Gram-positive bacterium *Enterococcus faecalis* (en-ter-ō-kok´ŭs fē-kă´lis). Like the pentose phosphate pathway, the Entner-Doudoroff pathway nets only a single molecule of ATP for each molecule of glucose, but it does yield precursor metabolites and NADPH. The latter is unavailable from glycolysis.

Appendix A, on p. A–9 shows the Entner-Doudoroff pathway in more detail.

Fermentation

LEARNING | OUTCOMES

5.14 Describe fermentation and contrast it with respiration.

5.15 List three useful end products of fermentation and explain how fermentation reactions are used to identify bacteria.

5.16 Discuss the use of biochemical tests for metabolic enzymes and products in the identification of bacteria.

Sometimes cells cannot completely oxidize glucose by cellular respiration. For instance, they may lack sufficient final electron acceptors, as in the case, for example, of an aerobic bacterium that lacks oxygen in the anaerobic environment of the colon. Electrons cannot flow down an electron transport chain unless oxidized carrier molecules are available to receive them at the end. Our bucket brigade analogy can help clarify this point.

Suppose the last person in the brigade did not throw the water but instead held onto two full buckets. What would happen? The entire brigade would soon consist of firefighters holding full buckets of water. An analogous situation occurs in an electron transport chain: All the carrier molecules are forced to remain in their reduced states when there is no final electron acceptor. Without the movement of electrons down the chain, protons cannot be transported, the proton motive force is lost, and oxidative phosphorylation of ADP to ATP ceases. Without sufficient ATP, a cell is unable to anabolize, grow, or divide.

ATP could be synthesized in glycolysis and the Krebs cycle by substrate-level phosphorylation. After all, together these pathways produce four molecules of ATP per molecule of glucose. However, careful consideration reveals that glycolysis, formation of acetyl-CoA, and the Krebs cycle require a continual supply of oxidized NAD^+ molecules (see Figures 5.13, 5.15, and 5.16).

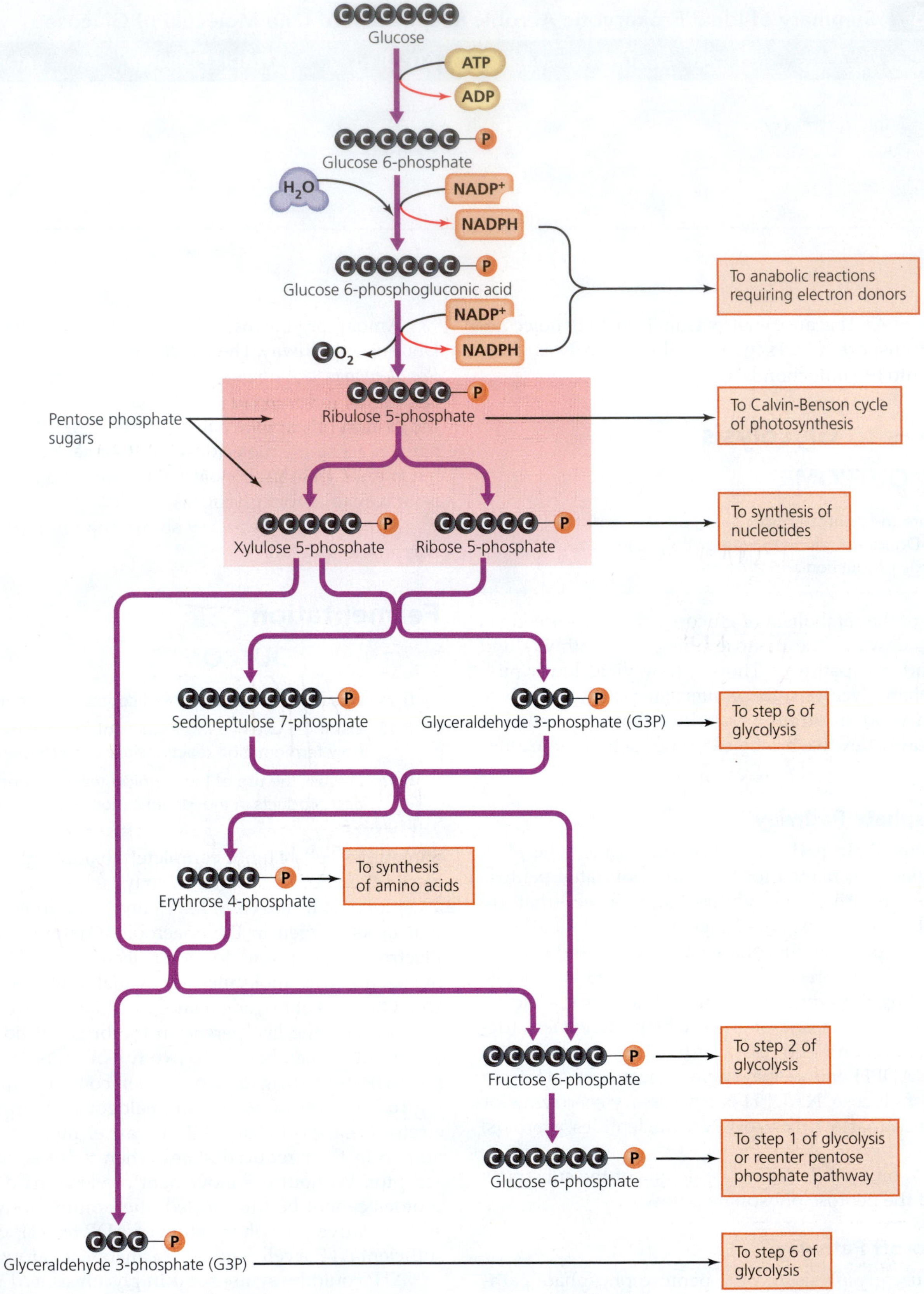

▲ **FIGURE 5.19 The pentose phosphate pathway.** Energy captured during the pentose phosphate pathway is less than that from glycolysis, but the pathway produces ribose 5-phosphate and erythrose 4-phosphate, two metabolites necessary for synthesis of nucleotides and certain amino acids, respectively. Ribulose 5-phosphate, necessary for glucose synthesis in photosynthetic organisms, and NADPH, an electron carrier for anabolic reactions, are also produced.

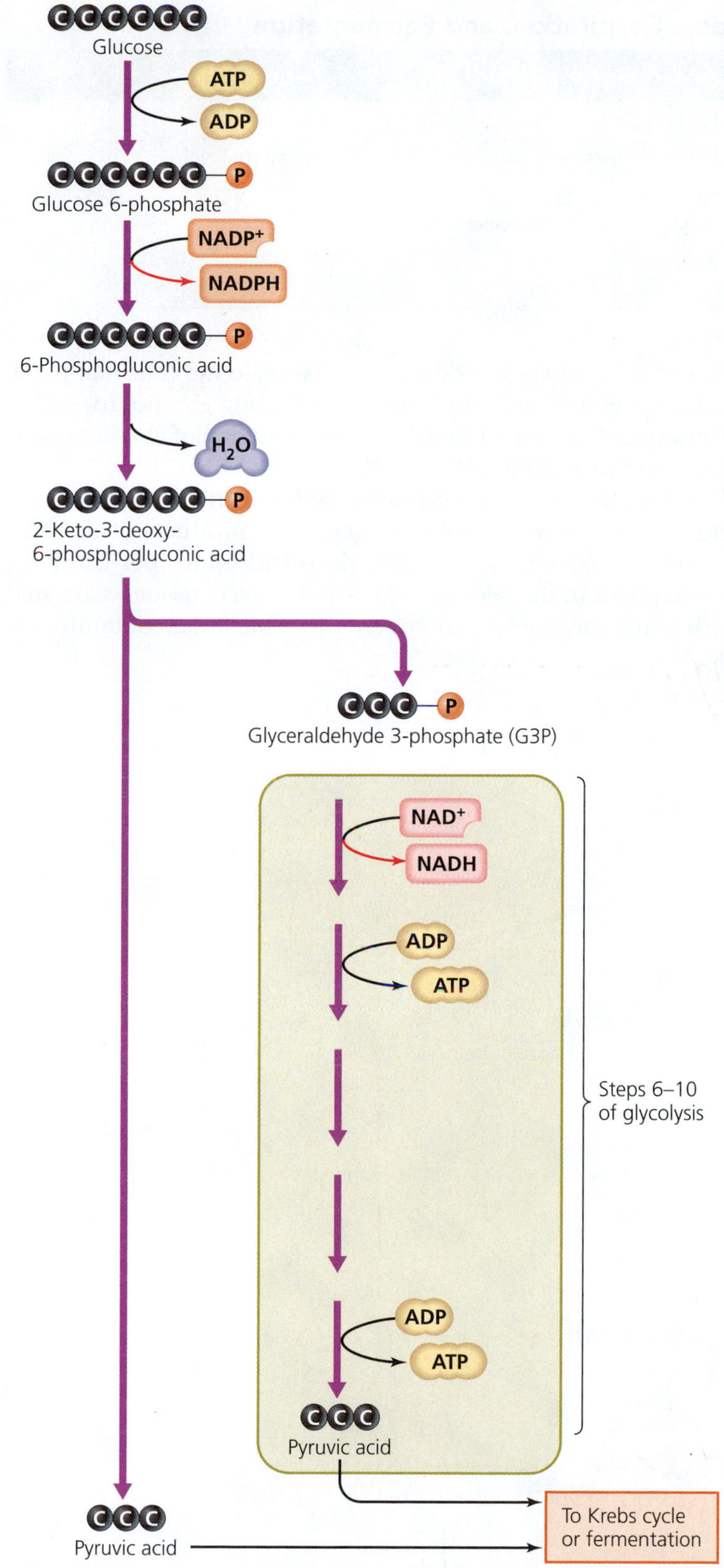

▲ **FIGURE 5.20 Entner-Doudoroff pathway.** A few bacteria use this alternate pathway for the oxidation of glucose to pyruvic acid. For simplicity, only carbon atoms and phosphate are shown. (For details, see Appendix A, p. A–9.) *What potential pathogens use this pathway?*

Figure 5.20 *Pseudomonas aeruginosa* and *Enterococcus faecalis* use the Entner-Doudoroff pathway.

Electron transport produces the required NAD^+ in respiration, but without a final electron acceptor, this source of NAD^+ ceases to be available. A cell in such a predicament must use an alternate source of NAD^+ provided by alternative pathways, called *fermentation pathways*.

In everyday language, fermentation refers to the production of alcohol from sugar, but in microbiology, fermentation has an expanded meaning. **Fermentation** is the partial oxidation of sugar (or other metabolites) to release energy using an organic molecule from within the cell as the final electron acceptor. In other words, fermentation pathways are metabolic reactions that oxidize NADH to NAD^+ while reducing cellular organic molecules. In contrast, respiration reduces externally acquired substances—oxygen in aerobic respiration and, in anaerobic respiration, some other inorganic chemical such as sulfate and nitrate or (rarely) an organic molecule. **FIGURE 5.21** illustrates two common fermentation pathways that reduce pyruvic acid to lactic acid and ethanol, oxidizing NADH in the process.

The essential function of fermentation is the regeneration of NAD^+ for glycolysis so that ADP molecules can be phosphorylated

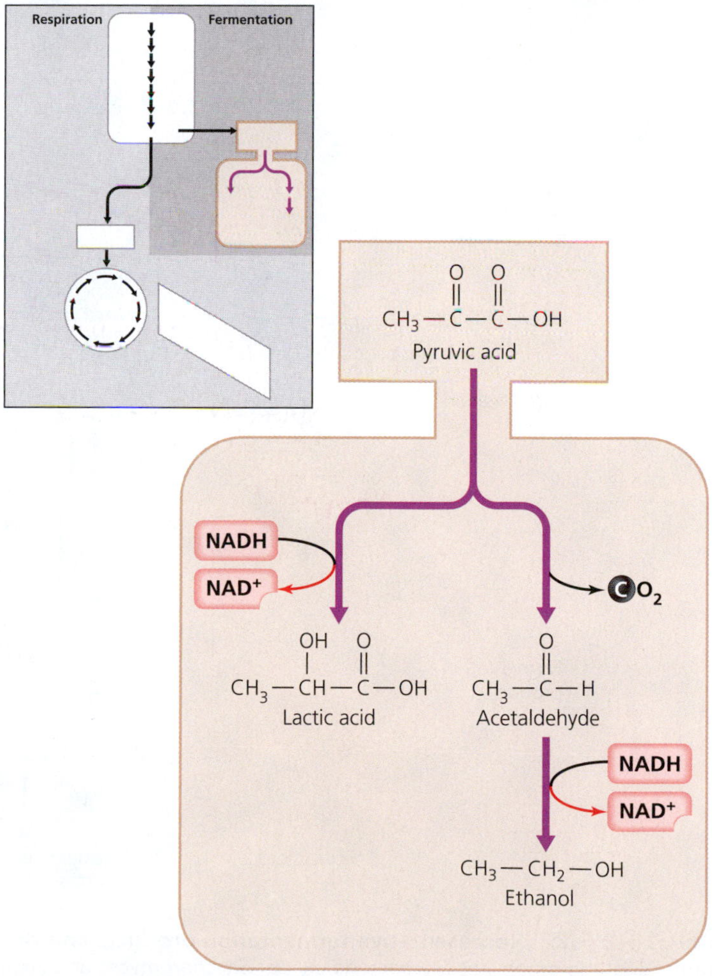

▲ **FIGURE 5.21 Fermentation.** In the simplest fermentation reaction, NADH reduces pyruvic acid to form lactic acid. Another simple fermentation pathway involves a decarboxylation reaction and reduction to form ethanol.

TABLE 5.4 Comparison of Aerobic Respiration, Anaerobic Respiration, and Fermentation

	Aerobic Respiration	Anaerobic Respiration	Fermentation
Oxygen required	Yes	No	No
Type of phosphorylation	Substrate-level and oxidative	Substrate-level and oxidative	Substrate-level
Final electron (hydrogen) acceptor	Oxygen	NO_3^-, SO_4^{2-}, CO_3^{2-}, or externally acquired organic molecules	Cellular organic molecules
Potential molecules of ATP produced per molecule of glucose	38 in prokaryotes, 36 in eukaryotes	2–36	2

to ATP. Even though fermentation pathways are not as energetically efficient as respiration (because much of the potential energy stored in glucose remains in the bonds of fermentation products), the major benefit of fermentation is that it allows ATP production to continue in the absence of cellular respiration. **TABLE 5.4** compares fermentation to aerobic and anaerobic respiration with respect to four crucial aspects of these processes.

Microorganisms produce a variety of fermentation products depending on the enzymes and substrates available to each. Though fermentation products are wastes to the cells that make them, many are useful to humans, including ethanol (drinking alcohol) and lactic acid (used in the production of cheese, sauerkraut, and pickles) **(FIGURE 5.22)**.

Other fermentation products are harmful to human health and industry. For example, fermentation products of the bacterium *Clostridium perfringens* (klos-trid´ē-um per-frin´jens) are involved in the necrosis (death) of muscle tissue associated with gangrene. Pasteur discovered that bacterial contaminants

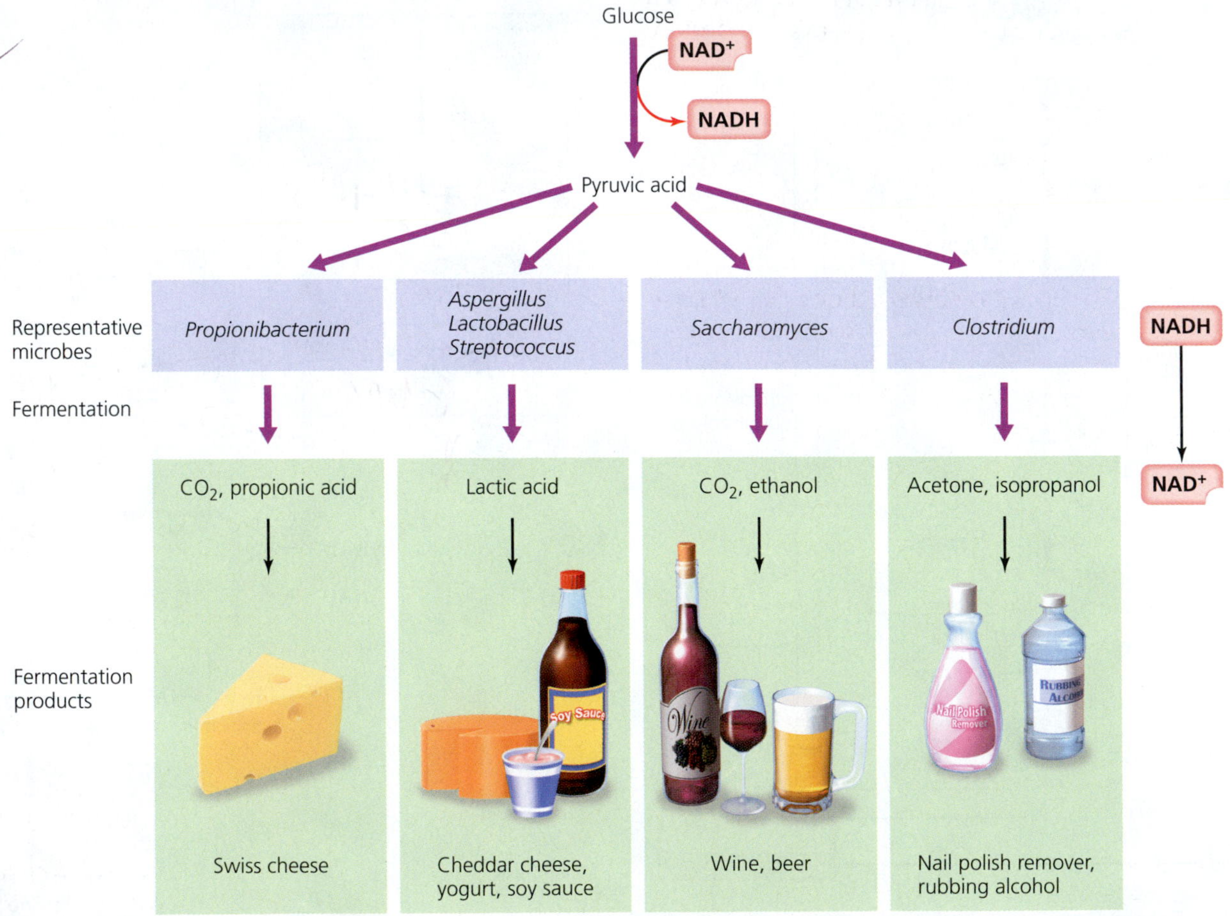

▲ **FIGURE 5.22 Representative fermentation products and the organisms that produce them.** All of the organisms are bacteria except *Saccharomyces* and *Aspergillus*, which are fungi. *Why does Swiss cheese have holes, but cheddar does not?*

Figure 5.22 *Swiss cheese is a product of Propionibacterium, which ferments pyruvic acid to produce CO_2; bubbles of this gas cause the holes in the cheese. Cheddar is a product of Lactobacillus, which does not produce gas.*

in grape juice fermented the sugar into unwanted products such as acetic acid and lactic acid, which spoiled the wine (Chapter 1).

Fermentation products can be used to identify microbes. For example, *Proteus* ferments glucose but not lactose, whereas *Escherichia* and *Enterobacter* ferment both. Further, glucose fermentation by *Escherichia* produces mixed acids (acetic, lactic, succinic, and formic), whereas *Enterobacter* produces 2,3-butanediol. Common fermentation tests contain a carbohydrate and a pH indicator, which is a molecule that changes color as the pH changes. An organism that utilizes the carbohydrate causes a change in pH, causing the pH indicator to change color. ▶**ANIMATIONS:** ***Fermentation***

In this section on carbohydrate catabolism, we have spent some time examining glycolysis, alternatives to glycolysis, the Krebs cycle, and electron transport because these pathways are central to metabolism. They generate all of the precursor metabolites and most of the ATP needed for anabolism. We have seen that some ATP is generated in respiration by substrate-level phosphorylation (in both glycolysis and the Krebs cycle) but that most ATP is generated by oxidative phosphorylation via chemiosmosis utilizing the reducing power of NADH and $FADH_2$. We also saw that some microorganisms use fermentation to provide an alternate source of NAD^+.

TELL ME WHY

Why do electrons carried by NADH allow for production of 50% more ATP molecules than do electrons carried by $FADH_2$?

Thus far we have concentrated on the catabolism of glucose as a representative carbohydrate, but microorganisms can also use other molecules as energy sources. In the next section we will examine catabolic pathways that utilize lipids and proteins.

Other Catabolic Pathways

Lipid and protein molecules contain abundant energy in their chemical bonds and can also be converted into precursor metabolites. These molecules are first catabolized to produce their constituent monomers, which serve as substrates in glycolysis and the Krebs cycle.

Lipid Catabolism

LEARNING **OUTCOME**

5.17 Explain how lipids are catabolized for energy and metabolite production.

The most common lipids involved in ATP and metabolite production are fats, which (as we saw in Chapter 2) consist of glycerol and fatty acids. In the first step of fat catabolism, enzymes called *lipases* hydrolyze the bonds attaching the glycerol to the fatty acid chains (**FIGURE 5.23a**).

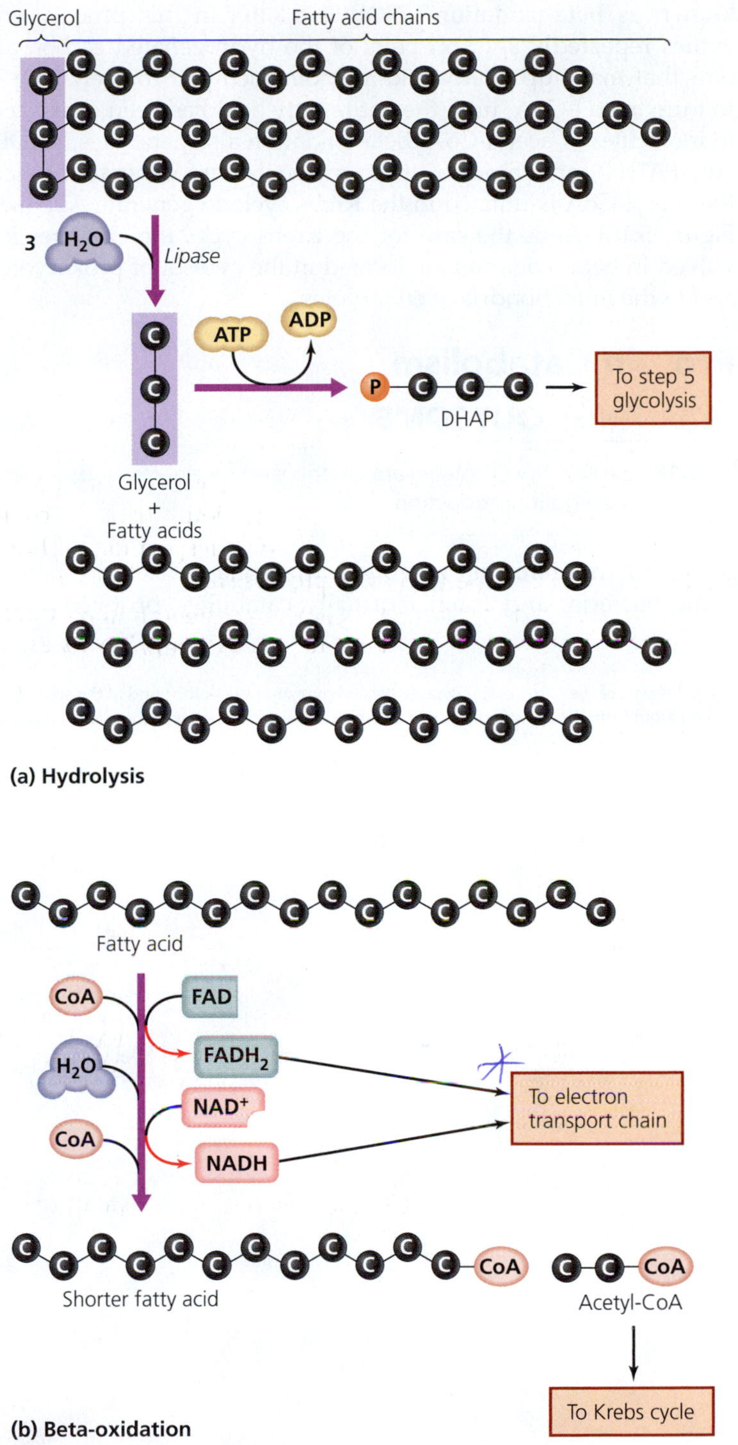

▲ **FIGURE 5.23 Catabolism of a fat molecule. (a)** Lipase breaks fats into glycerol and three fatty acids by hydrolysis. Glycerol is converted to DHAP, which can be catabolized via glycolysis and the Krebs cycle. **(b)** Fatty acids are catabolized via beta-oxidation reactions that produce molecules of acetyl-CoA and reduced coenzymes (NADH and $FADH_2$).

Subsequent reactions further catabolize the glycerol and fatty acid molecules. Glycerol is converted to DHAP, which, as one of the substrates of glycolysis, is oxidized to pyruvic acid (see Figure 5.13 5). The fatty acids are degraded in a catabolic process known as **beta-oxidation**[9] **(FIGURE 5.23b)**. In this process, enzymes repeatedly split off pairs of the hydrogenated carbon atoms that make up a fatty acid and join each pair to coenzyme A to form acetyl-CoA, until the entire fatty acid has been converted to molecules of acetyl-CoA. Beta-oxidation also generates NADH and $FADH_2$, and more of these molecules are generated when the acetyl-CoA is utilized in the Krebs cycle to generate ATP (see Figure 5.16). As is the case for the Krebs cycle, the enzymes involved in beta-oxidation are located in the cytosol of prokaryotes and in the mitochondria of eukaryotes.

Protein Catabolism

LEARNING | OUTCOME

5.18 Explain how proteins are catabolized for energy and metabolite production.

Some microorganisms, notably food-spoilage bacteria, pathogenic bacteria, and fungi, normally catabolize proteins as an important source of energy and metabolites. Most cells catabolize proteins and their constituent amino acids only when carbon sources such as glucose and fat are not available.

Generally, proteins are too large to cross cytoplasmic membranes, so prokaryotes typically conduct the first step in the process of protein catabolism outside the cell by secreting **proteases** (prō´tē-ās-ez)—enzymes that split proteins into their constituent amino acids **(FIGURE 5.24)**. Once released by the action of proteases, amino acids are transported into the cell, where special enzymes split off amino groups in a reaction called **deamination**. The resulting altered molecules enter the Krebs cycle, and the amino groups are either recycled to synthesize other amino acids or excreted as nitrogenous wastes such as ammonia (NH_3), ammonium ion (NH_4^+), or trimethylamine oxide (TMAO). As **Highlight: What's That Fishy Smell?** describes, TMAO plays an interesting role in the production of the odor we describe as "fishy."

TELL ME WHY

Why does catabolism of amino acids for energy result in ammonia and other nitrogenous wastes?

[9]"Beta" is part of the name of this process because enzymes break the bond at the second carbon atom from the end of a fatty acid, and beta is the second letter in the Greek alphabet.

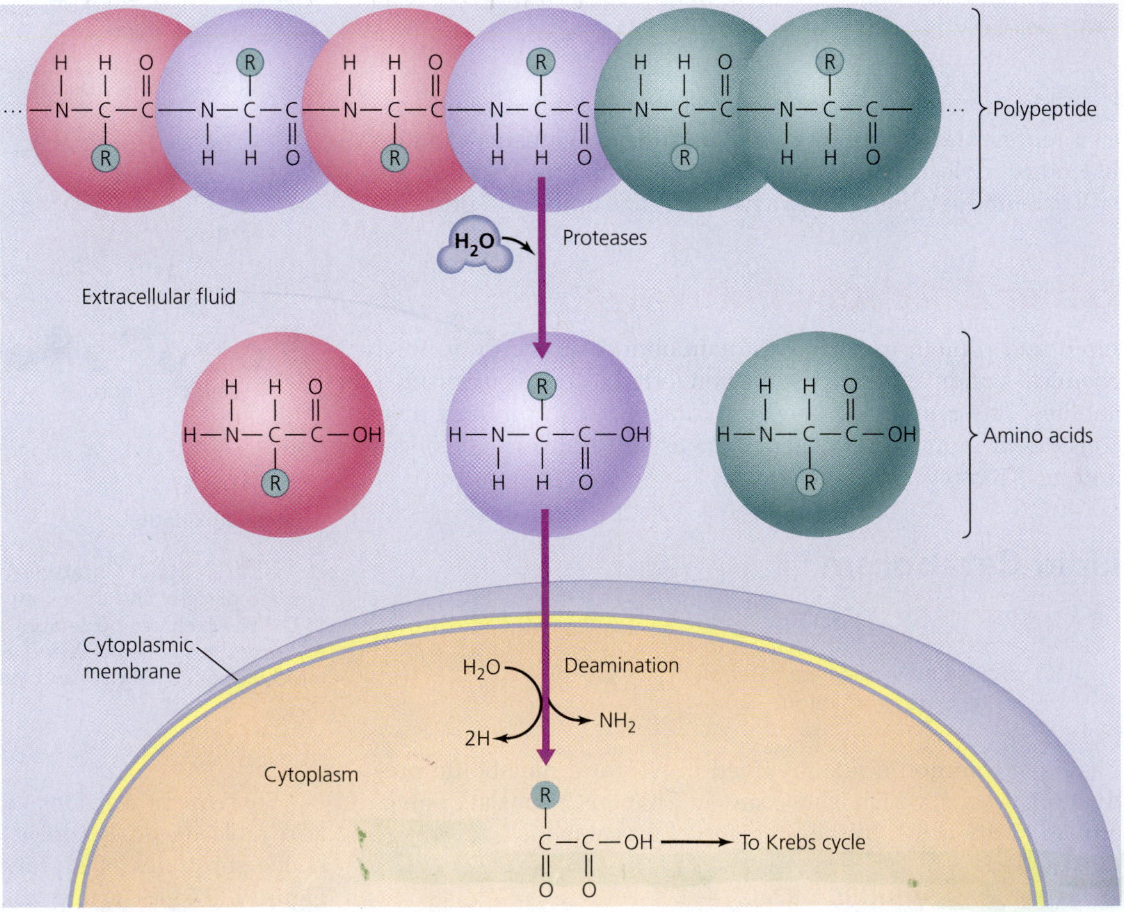

▶ FIGURE 5.24 Protein catabolism. Secreted proteases hydrolyze proteins, releasing amino acids, which are deaminated after uptake to produce molecules used as substrates in the Krebs cycle. R indicates the side group, which varies among amino acids.

HIGHLIGHT

What's That Fishy Smell?

Ever wonder why fish at the supermarket sometimes have that "fishy" smell? The reason relates to bacterial metabolism.

Trimethylamine oxide (TMAO) is an odorless, nitrogenous waste product of fish metabolism used to dispose of excess nitrogen (as amine groups) produced in the catabolism of amino acids. TMAO does not affect the smell, appearance, or taste of fresh fish; however, some bacteria use TMAO as a final electron acceptor in anaerobic respiration, reducing TMAO to trimethyl amine (TMA), a compound with a very definite—and, for most people, unpleasant—"fishy" odor. Thus, the "fishy" odor is really the odor of contaminating bacteria. A human nose is able to detect even a few molecules of TMA. Because bacterial degradation of fish begins as soon as a fish is dead, a fish that smells fresh probably is fresh, or was frozen while still fresh.

Thus far, we have examined the catabolism of carbohydrates, lipids, and proteins. Now we turn our attention to the synthesis of these molecules, beginning with the anabolic reactions of photosynthesis.

Photosynthesis

LEARNING OUTCOME

5.19 Define *photosynthesis.*

Many organisms use only organic molecules as a source of energy and metabolites, but where do they acquire organic molecules? Ultimately every food chain begins with anabolic pathways in organisms that synthesize their own organic molecules from inorganic carbon dioxide. Most of these organisms capture light energy from the sun and use it to drive the synthesis of carbohydrates from CO_2 and H_2O by a process called **photosynthesis.** Cyanobacteria, purple sulfur bacteria, green sulfur bacteria, green nonsulfur bacteria, purple nonsulfur bacteria, algae, green plants, and a few protozoa are photosynthetic. **▶ANIMATIONS:** *Photosynthesis: Overview*

Chemicals and Structures

LEARNING OUTCOME

5.20 Compare and contrast the basic chemicals and structures involved in photosynthesis in prokaryotes and eukaryotes.

Photosynthetic organisms capture light energy with pigment molecules, the most important of which are **chlorophylls.** Chlorophyll molecules are composed of a hydrocarbon tail attached to a light-absorbing *active site* centered around a magnesium ion (Mg^{2+}) **(FIGURE 5.25a)**. The active sites are structurally similar to the cytochrome molecules found in electron transport

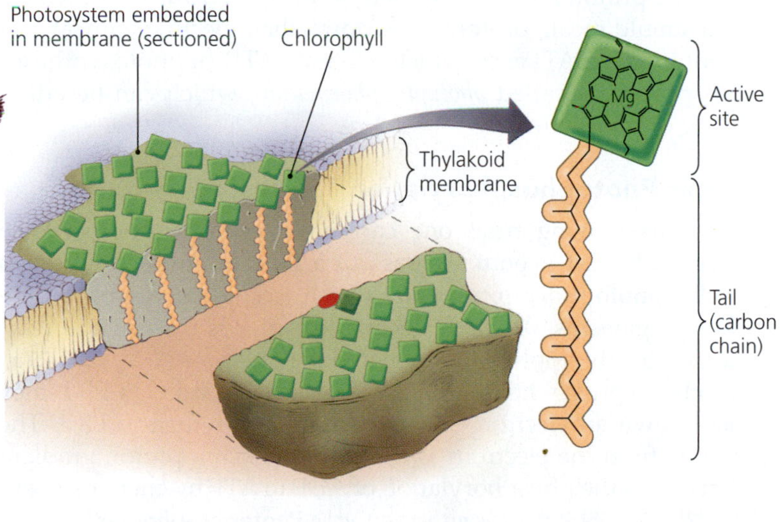

(a)

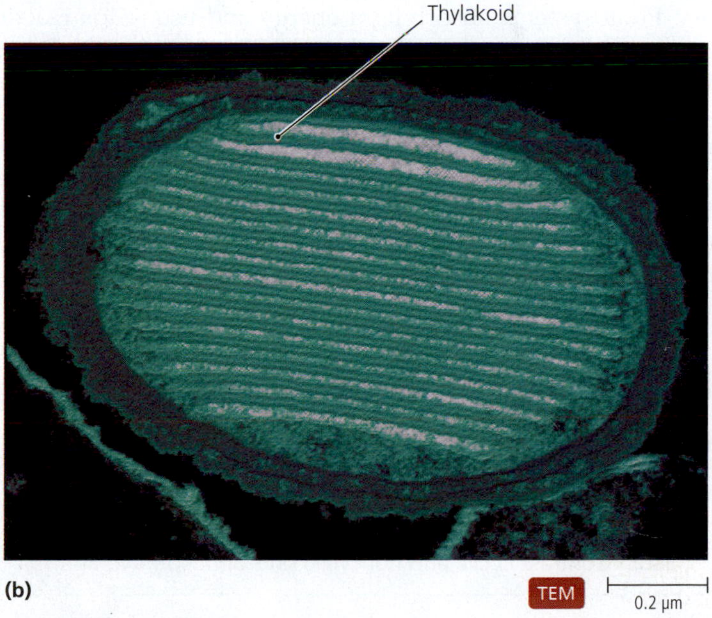

(b)

▲ **FIGURE 5.25 Photosynthetic structures in a prokaryote. (a)** Chlorophyll molecules are grouped together in thylakoid membranes to form photosystems. Chlorophyll contains a light-absorbing active site that is connected to a long hydrocarbon tail. The chlorophyll shown here is bacteriochlorophyll *a*; other chlorophyll types differ in the side chains protruding from the central ring. **(b)** Prokaryotic thylakoids are infoldings of the cytoplasmic membrane, here in a cyanobacterium.

chains, except that chlorophylls use Mg^{2+} rather than Fe^{2+}. Chlorophylls, typically designated with letters—for example, chlorophyll *a*, chlorophyll *b*, and bacteriochlorophyll *a*—vary slightly in the lengths and structures of their hydrocarbon tails and in the atoms that extend from their active sites. Green plants, algae, photosynthetic protozoa, and cyanobacteria principally use chlorophyll *a*, whereas green and purple bacteria use bacteriochlorophylls.

The slight structural differences among chlorophylls cause them to absorb light of different wavelengths. For example, chlorophyll *a* from algae best absorbs light with wavelengths of about 425 nm and 660 nm (violet and red), whereas bacteriochlorophyll *a* from purple bacteria best absorbs light with wavelengths of about 350 nm and 880 nm (ultraviolet and infrared). Because they best use light with differing wavelengths, algae and purple bacteria successfully occupy different ecological niches.

Cells arrange numerous molecules of chlorophyll and other pigments within a protein matrix to form light-harvesting matrices called **photosystems** that are embedded in cellular membranes called **thylakoids**. Thylakoids of photosynthetic prokaryotes are invaginations of their cytoplasmic membranes **(FIGURE 5.25b)**. The thylakoids of eukaryotes appear to be formed from infoldings of the inner membranes of chloroplasts, though thylakoid membranes and the inner membranes are not connected in mature chloroplasts (see Figure 3.40). Thylakoids of chloroplasts are arranged in stacks called *grana*. An outer chloroplast membrane surrounds the grana, and the space between the outer membrane and the thylakoid membrane is known as the *stroma*. The thylakoids enclose a narrow, convoluted cavity called the *thylakoid space*.

There are two types of photosystems, named photosystem I (PS I) and photosystem II (PS II), in the order of their discovery. Photosystems absorb light energy and use redox reactions to store this energy in molecules of ATP and NADPH. Because they depend on light energy, these reactions of photosynthesis are classified as **light-dependent reactions**. Photosynthesis also involves **light-independent reactions** that actually synthesize glucose from carbon dioxide and water. Historically, these reactions were called *light* and *dark reactions*, respectively, but this older terminology implies that light-independent reactions occur only in the dark, which is not the case. We first consider the light-dependent reactions of the two photosystems.

Light-Dependent Reactions

LEARNING **OUTCOMES**

5.21 Describe the components and function of the two photosystems, PS I and PS II.

5.22 Contrast cyclic and noncyclic photophosphorylation.

The pigments of photosystem I absorb light energy and transfer it to a neighboring molecule within the photosystem until the energy eventually arrives at a special chlorophyll molecule called the **reaction center chlorophyll (FIGURE 5.26)**. Light energy from hundreds of such transfers excites electrons in the reaction center chlorophyll, which passes its excited electrons to the reaction center's electron acceptor, which is the initial carrier of an electron transport chain. As electrons move down the chain, their energy is used to pump protons across the membrane, creating a proton motive force. In prokaryotes, protons are pumped out of the cell; in eukaryotes, they are pumped from the stroma into the interior of the thylakoids.

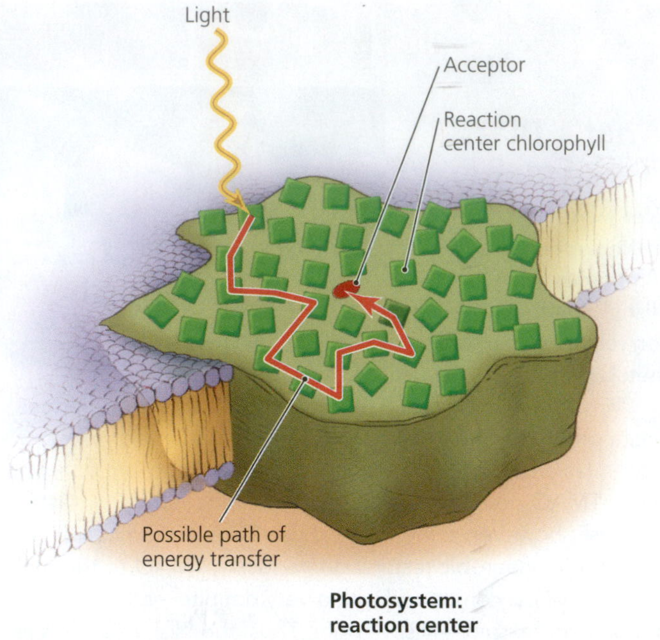

▲ **FIGURE 5.26 Reaction center of a photosystem.** Light energy absorbed by pigments anywhere in the photosystem is transferred to the reaction center chlorophyll, where it is used to excite electrons for delivery to the reaction center's electron acceptor.

The proton motive force is used in chemiosmosis, in which, you should recall, protons flow down their electrochemical gradient through ATPases, which generate ATP. In photosynthesis, this process is called *photophosphorylation*, which can be either cyclic or noncyclic.

Cyclic Photophosphorylation

Electrons moving from one carrier molecule to another in a thylakoid must eventually pass to a final electron acceptor. In **cyclic photophosphorylation**, which occurs in all photosynthetic organisms, the final electron acceptor is the original reaction center chlorophyll that donated the electrons **(FIGURE 5.27a)**. In other words, when light energy excites electrons in PS I, they pass down an electron transport chain and return to PS I. The energy from the electrons is used to establish a proton gradient that drives the phosphorylation of ADP to ATP by chemiosmosis.
▶**ANIMATIONS:** *Photosynthesis: Cyclic Photophosphorylation*

Noncyclic Photophosphorylation

Some photosynthetic bacteria and all plants, algae, and photosynthetic protozoa utilize **noncyclic photophosphorylation** as well. (Exceptions are green and purple sulfur bacteria.) Noncyclic photophosphorylation, which requires both PS I and PS II,

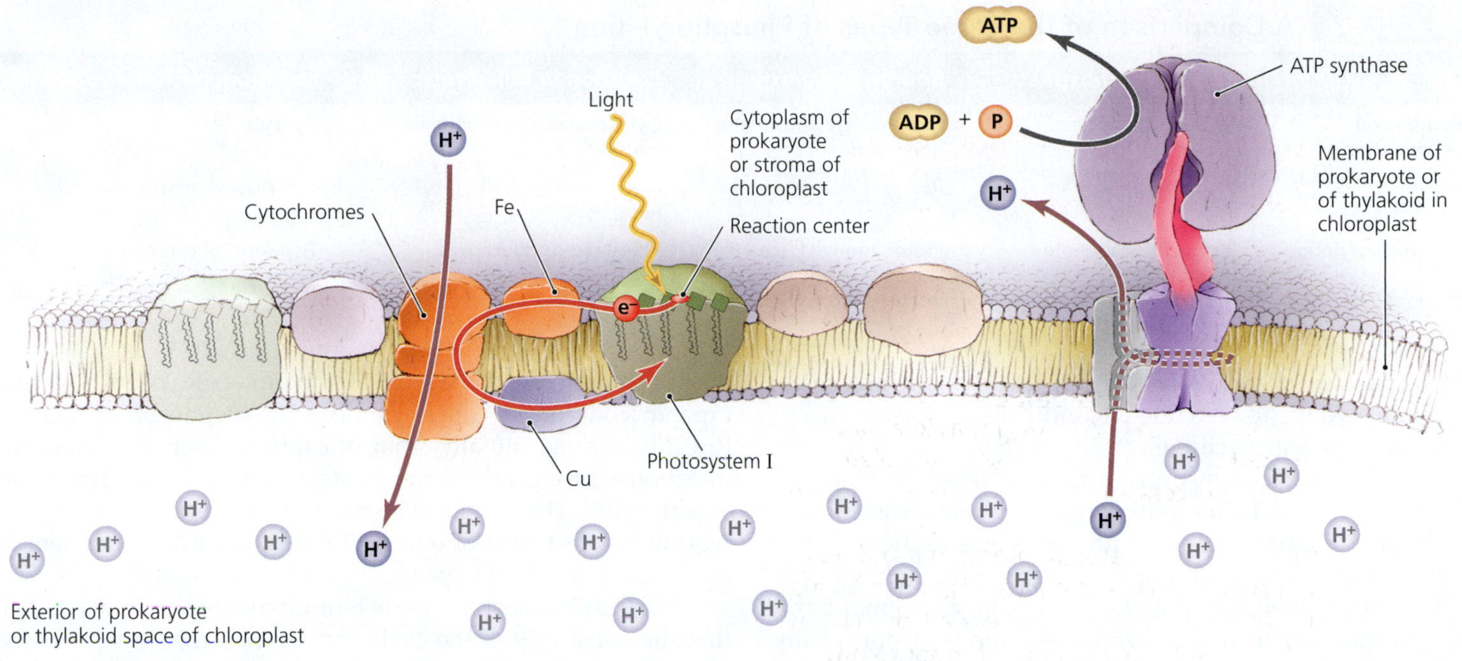

(a) Cyclic photophosphorylation

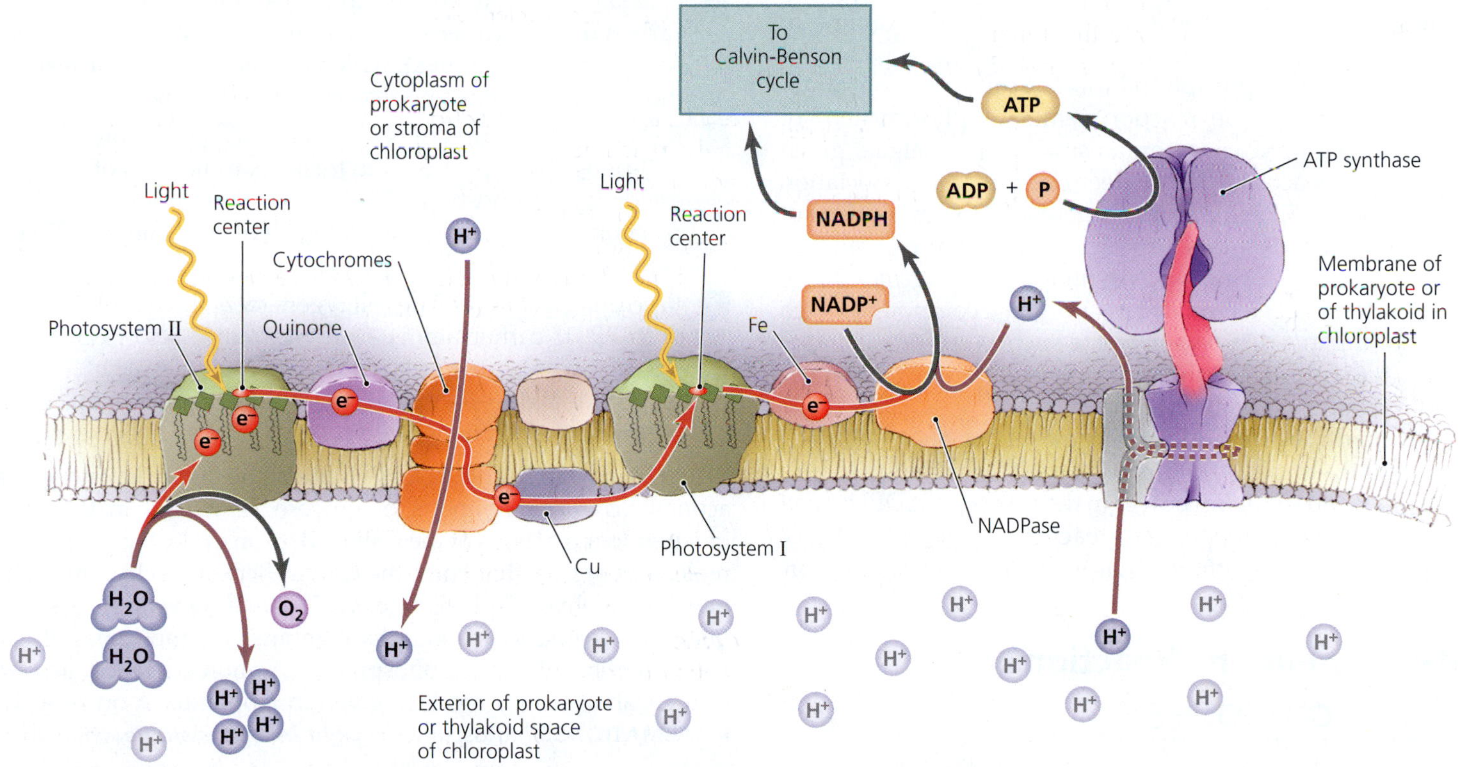

(b) Noncyclic photophosphorylation

▲ **FIGURE 5.27 The light-dependent reactions of photosynthesis: Cyclic and noncyclic photophosphorylation. (a)** In cyclic photophosphorylation, electrons excited by light striking photosystem I travel down an electron transport chain (red arrows) and then return to the photosystem. **(b)** In noncyclic photophosphorylation, light striking photosystem II excites electrons that are passed to photosystem I, simultaneously establishing a proton gradient. Light energy collected by photosystem I further excites the electrons, which are used to reduce $NADP^+$ to NADPH via an electron transport chain. In oxygenic organisms, new electrons are provided by the lysis of H_2O (bottom left). In both types of photophosphorylation, the proton gradient drives the phosphorylation of ADP by ATP synthase (chemiosmosis).

TABLE 5.5 A Comparison of the Three Types of Phosphorylation

	Source of Phosphate	Source of Energy	Location in Eukaryotic Cell	Location in Prokaryotic Cell
Substrate-level phosphorylation	Organic molecule	High-energy phosphate bond of donor	Cytosol and mitochondrial matrix	Cytosol
Oxidative phosphorylation	Inorganic phosphate (PO_4^{3-})	Proton motive force	Inner membrane of mitochondrion	Cytoplasmic membrane
Photophosphorylation	Inorganic phosphate (PO_4^{3-})	Proton motive force	Thylakoid of chloroplast	Thylakoid of cytoplasmic membrane

not only generates molecules of ATP but also reduces molecules of coenzyme $NADP^+$ to NADPH **(FIGURE 5.27b)**.

When light energy excites electrons of PS II, they are passed to PS I through an electron transport chain. (Note that photosystem II occurs first in the pathway and photosystem I second, a result of the naming of the photosystems in the order in which they were discovered, not the order in which they operate.) PS I further energizes the electrons with additional light energy and transfers them through an electron transport chain to $NADP^+$, which is thereby reduced to NADPH. The hydrogen ions added to NADPH come from the stroma or cytosol. NADPH subsequently participates in the synthesis of glucose in the light-independent reactions, which we will examine in the next section.

In noncyclic photophosphorylation, a cell must constantly replenish electrons to the reaction center of photosystem II. *Oxygenic* (oxygen-producing) organisms, such as algae, green plants, and cyanobacteria, derive electrons from the dissociation of H_2O (see Figure 5.27b). In these organisms, two molecules of water give up their electrons, producing molecular oxygen (O_2) as a waste product during photosynthesis. *Anoxygenic* photosynthetic bacteria get electrons from inorganic compounds such as H_2S, resulting in a nonoxygen waste such as sulfur.

TABLE 5.5 compares photophosphorylation to substrate-level and oxidative phosphorylation.

To this point, we have examined the use of photosynthetic pigments and thylakoid structure to harvest light energy to produce both ATP and reducing power in the form of NADPH. Next we examine the light-independent reactions of photosynthesis.

▶**ANIMATIONS:** *Photosynthesis: Noncyclic Photophosphorylation*

Light-Independent Reactions

LEARNING **OUTCOMES**

5.23 Contrast the light-dependent and light-independent reactions of photosynthesis.

5.24 Describe the reactants and products of the Calvin-Benson cycle.

Light-independent reactions of photosynthesis do not require light directly; instead, they use large quantities of ATP and NADPH generated by the light-dependent reactions. The key reaction of the light-independent pathway of photosynthesis is **carbon fixation** by the **Calvin-Benson cycle,**[10] which involves the attachment of molecules of CO_2 to molecules of a five-carbon organic compound called ribulose 1,5-bisphosphate (RuBP). RuBP is derived initially from phosphorylation of a precursor metabolite produced by the pentose phosphate pathway (see Figure 5.19). The Calvin-Benson cycle is very endergonic—it requires a great deal of energy. Life on Earth is dependent on carbon fixation by this cycle.

The Calvin-Benson cycle is reminiscent of the Krebs cycle in that the substrates of the cycle are regenerated. It is helpful to notice the number of carbon atoms during each part of the three steps of the Calvin-Benson cycle **(FIGURE 5.28)**:

1. *Fixation of CO_2.* An enzyme attaches three molecules of carbon dioxide (three carbon atoms) to three molecules of RuBP (15 carbon atoms), which are then split to form six molecules of 3-phosphoglyceric acid (18 carbon atoms).
2. *Reduction.* Molecules of NADPH reduce the six molecules of 3-phosphoglyceric acid to form six molecules of glyceraldehyde 3-phosphate (G3P) (18 carbon atoms). These reactions require six molecules each of ATP and NADPH generated by the light-dependent reactions.
3. *Regeneration of RuBP.* The cell regenerates three molecules of RuBP (15 carbon atoms) from five molecules of G3P (15 carbon atoms). It uses the remaining molecule of glyceraldehyde 3-phosphate to synthesize glucose by reversing the reactions of glycolysis.

In summary, ATP and NADPH from the light-dependent reactions drive the synthesis of glucose from CO_2 in the light-independent reactions of the Calvin-Benson cycle. For every three molecules of CO_2 that enter the Calvin-Benson cycle, a molecule of glyceraldehyde 3-phosphate (G3P) leaves. Glycolysis is subsequently reversed to anabolically combine two molecules of G3P to synthesize glucose 6-phosphate. (A more detailed account of the Calvin-Benson cycle is given in Appendix A on p. A-11.)

▶**ANIMATIONS:** *Photosynthesis: Light-Independent Reaction*

The processes of oxygenic photosynthesis and aerobic respiration complement one another to complete both a carbon cycle and an oxygen cycle. During the synthesis of glucose in oxygenic photosynthesis, water and carbon dioxide are used, and oxygen is released as a waste product; in aerobic respiration, oxygen serves as the final electron acceptor in the oxidation of glucose to carbon dioxide and water.

[10]Named for the men who elucidated its pathways.

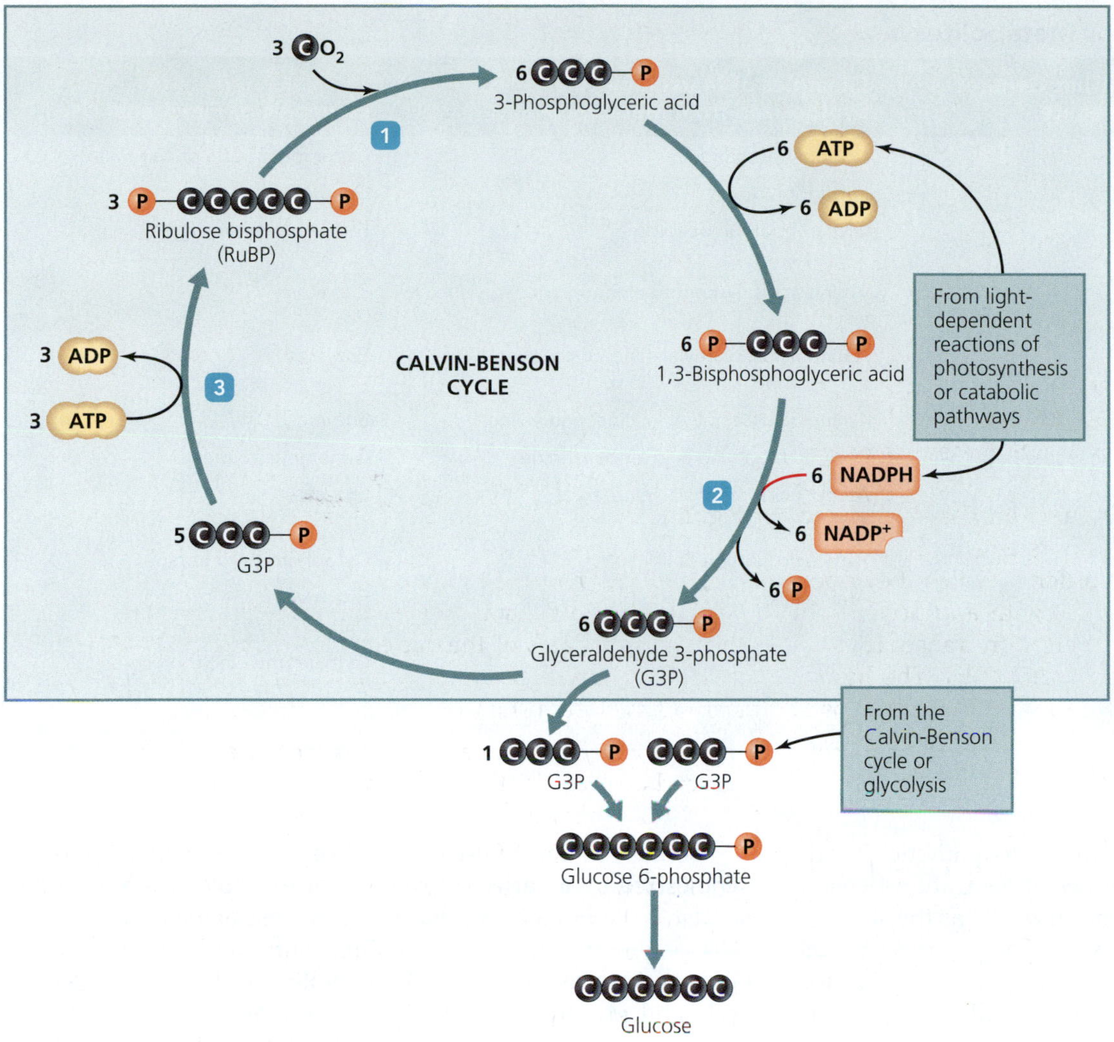

◀ FIGURE 5.28 Simplified diagram of the Calvin-Benson cycle. 1 Fixation of CO_2. Three molecules of RuBP combine with three molecules of CO_2. 2 Reduction. The resulting molecules are reduced to form six molecules of G3P. 3 Generation of RuBP. Five molecules of G3P are converted to three molecules of RuBP, which completes the cycle. Two turns of the cycle also yield two molecules of G3P, which are polymerized to synthesize glucose 6-phosphate.

In this section we examined an essential anabolic pathway for life on Earth—photosynthesis, which produces glucose. Next we will consider anabolic pathways involved in the synthesis of other organic molecules.

TELL ME WHY

An uninformed student describes the Calvin-Benson cycle as "cellular respiration in reverse." Why is the student incorrect?

Other Anabolic Pathways

LEARNING **OUTCOME**

5.25 Define *amphibolic reaction.*

Anabolic reactions are synthesis reactions. As such, they require energy and a source of precursor metabolites. ATP generated in the catabolic reactions of aerobic respiration, anaerobic respiration, and fermentation and in the initial redox reactions of photosynthesis provides energy for anabolism. Glycolysis, the Krebs cycle, and the pentose phosphate pathway provide 12 basic precursor metabolites from which all macromolecules and cellular structures can be made (**TABLE 5.6** on p. 154). Some microorganisms, such as *E. coli*, can synthesize all 12 precursors, whereas other organisms, such as humans, must acquire some precursors in their diets.

Many anabolic pathways are the reversal of the catabolic pathways we have discussed; therefore, much of the material that follows has *in a sense* been discussed in the sections on catabolism. Reactions that can proceed in either direction—toward catabolism or toward anabolism—are said to be **amphibolic reactions**. The following sections discuss the synthesis of carbohydrates, lipids, amino acids, and nucleotides. (Chapter 7 covers anabolic reactions that are closely linked to genetics—the polymerizations of amino acids into proteins and of nucleotides into RNA and DNA.)

Carbohydrate Biosynthesis

LEARNING **OUTCOME**

5.26 Describe the biosynthesis of carbohydrates.

TABLE 5.6 The 12 Precursor Metabolites

	Pathway That Generates the Metabolite	Examples of Macromolecule Synthesized from Metabolite[a]	Examples of Functional Use
Glucose 6-phosphate	Glycolysis	Lipopolysaccharide	Outer membrane of cell wall
Fructose 6-phosphate	Glycolysis	Peptidoglycan	Cell wall
Glyceraldehyde 3-phosphate (G3P)	Glycolysis	Glycerol portion of lipids	Fats—energy storage
Phosphoglyceric acid	Glycolysis	Amino acids: cysteine, selenocysteine, glycine, and serine	Enzymes
Phosphoenolpyruvic acid (PEP)	Glycolysis	Amino acids: phenylalanine, tryptophan, and tyrosine	Enzymes
Pyruvic acid	Glycolysis	Amino acids: alanine, leucine, and valine	Enzymes
Ribose 5-phosphate	Pentose phosphate pathway	DNA, RNA, amino acid, and histidine	Genome, enzymes
Erythrose 4-phosphate	Pentose phosphate pathway	Amino acids: phenylalanine, tryptophan, and tyrosine	Enzymes
Acetyl-CoA	Krebs cycle	Fatty acid portion of lipids	Cytoplasmic membrane
a-Ketoglutaric acid	Krebs cycle	Amino acids: arginine, glutamic acid, glutamine, and proline	Enzymes
Succinyl-CoA	Krebs cycle	Heme	Cytochrome electron carrier
Oxaloacetate	Krebs cycle	Amino acids: aspartic acid, asparagine, isoleucine, lysine, methionine, and threonine	Enzymes

[a]Examples given apply to the bacterium *E. coli*.

As we have seen, anabolism begins in photosynthetic organisms with carbon fixation by the enzymes of the Calvin-Benson cycle to form molecules of G3P. Enzymes use G3P as the starting point for synthesizing sugars, complex polysaccharides such as starch, cellulose for cell walls in algae, and peptidoglycan for cell walls of bacteria. Animals and protozoa synthesize the storage molecule glycogen.

Some cells are able to synthesize sugars from noncarbohydrate precursors such as amino acids, glycerol, and fatty acids by pathways collectively called *gluconeogenesis*[11] (glū´kō-nē-ō-jen´ĕ-sis; **FIGURE 5.29**). Most of the reactions of gluconeogenesis are amphibolic, using enzymes of glycolysis in reverse, but four of the reactions require unique enzymes. Gluconeogenesis is highly endergonic and can proceed only if there is an adequate supply of energy.

Lipid Biosynthesis

LEARNING OUTCOME

5.27 Describe the biosynthesis of lipids.

Lipids are a diverse group of organic molecules that function as energy storage compounds and as components of membranes (as we saw in Chapter 2). *Carotenoids,* which are reddish pigments found in many bacterial and plant photosystems, are also lipids.

Because of their variety, it is not surprising that lipids are synthesized by a variety of routes. For example, fats are synthesized in anabolic reactions that are the reverse of their catabolism—cells polymerize glycerol and three fatty acids **(FIGURE 5.30)**. Glycerol is derived from G3P generated by the Calvin-Benson cycle and glycolysis; the fatty acids are produced by the linkage of two-carbon acetyl-CoA molecules to one another by a sequence of endergonic reactions that effectively reverse the catabolic reactions of beta-oxidation. Other lipids, such as steroids, are synthesized in complex pathways involving polymerizations and isomerizations of sugar and amino acid metabolites.

Mycobacterium tuberculosis (mī´kō-bak-tēr´ē-ŭm too-ber-kyū-lō´sis) the pathogen that causes tuberculosis, makes copious amounts of a waxy lipid called *mycolic acid,* which is incorporated into its cell wall. The arduous, energy-intensive process of making long lipid chains explains why this organism grows slowly and why tuberculosis requires a long course of treatment with antimicrobial drugs.

Amino Acid Biosynthesis

LEARNING OUTCOME

5.28 Describe the biosynthesis of amino acids.

Cells synthesize amino acids from precursor metabolites derived from glycolysis, the Krebs cycle, and the pentose phosphate pathway and from other amino acids. Some organisms (such as *E. coli* and most plants and algae) synthesize all their amino acids from precursor metabolites. Other organisms, including humans, cannot synthesize certain amino acids, called

[11]From Greek *glukus,* meaning "sweet"; *neo,* meaning "new"; and *genesis,* meaning "generate."

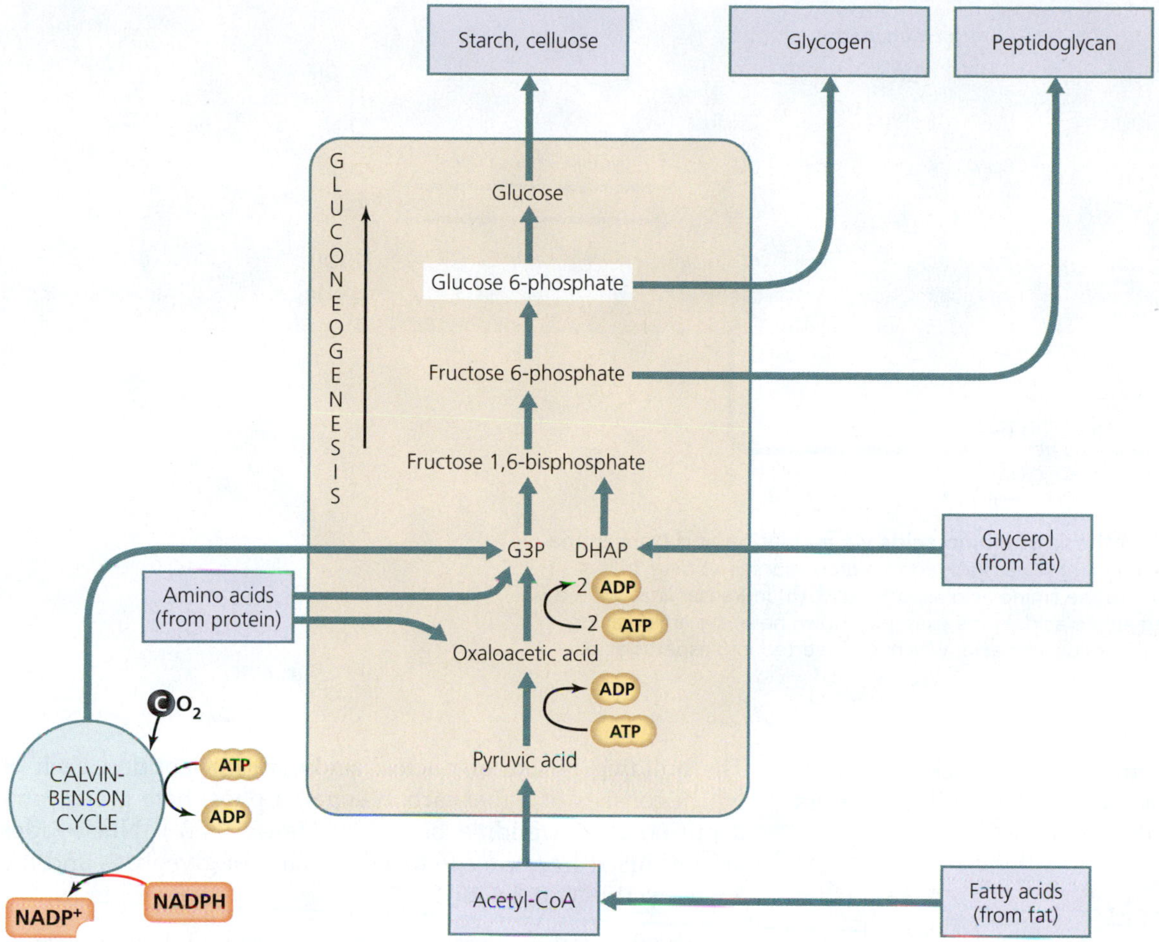

FIGURE 5.29 The role of gluconeogenesis in the biosynthesis of complex carbohydrates. Complex carbohydrates are synthesized from simple sugar molecules such as glucose, glucose 6-phosphate, and fructose 6-phosphate. Starch and cellulose are found in algae; glycogen is found in animals, protozoa, and fungi; and peptidoglycan is found in bacteria.

essential amino acids; that must be acquired in the diet. One extreme example is *Lactobacillus* (lak´tō-bă-sil´ŭs), a bacterium that ferments milk and produces some cheeses. This microorganism cannot synthesize any amino acids; it acquires all of them by catabolizing proteins in its environment.

Precursor metabolites are converted to amino acids by the addition of an amine group. This process is called **amination** when the amine group comes from ammonia (NH_3); an example is the formation of aspartic acid from NH_3 and the Krebs cycle intermediate oxaloacetic acid **(FIGURE 5.31a)**. Amination reactions are the reverse of the catabolic deamination reactions we discussed previously. More commonly, however, a cell moves the amine group from one amino acid and adds it to a metabolite, producing a different amino acid. This process is called **transamination** because the amine group is transferred from one amino acid to another **(FIGURE 5.31b)**. All transamination enzymes use a coenzyme, *pyridoxal phosphate,* which is derived from vitamin B_6.

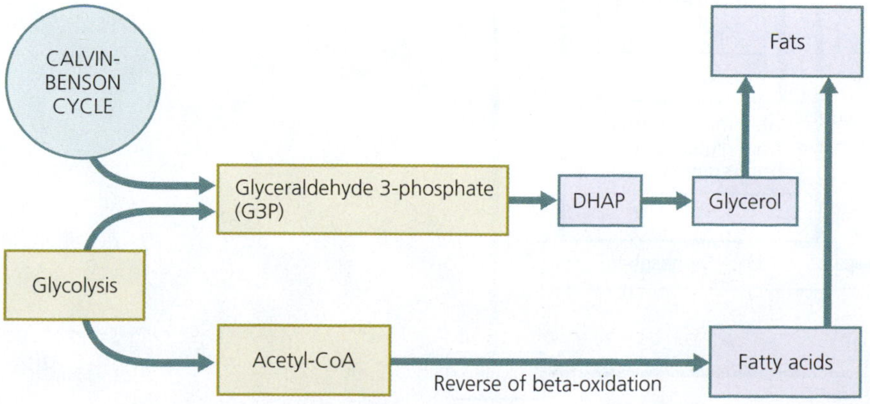

FIGURE 5.30 Biosynthesis of fat, a lipid. A fat molecule is synthesized from glycerol and three molecules of fatty acid, the precursors of which are produced in glycolysis.

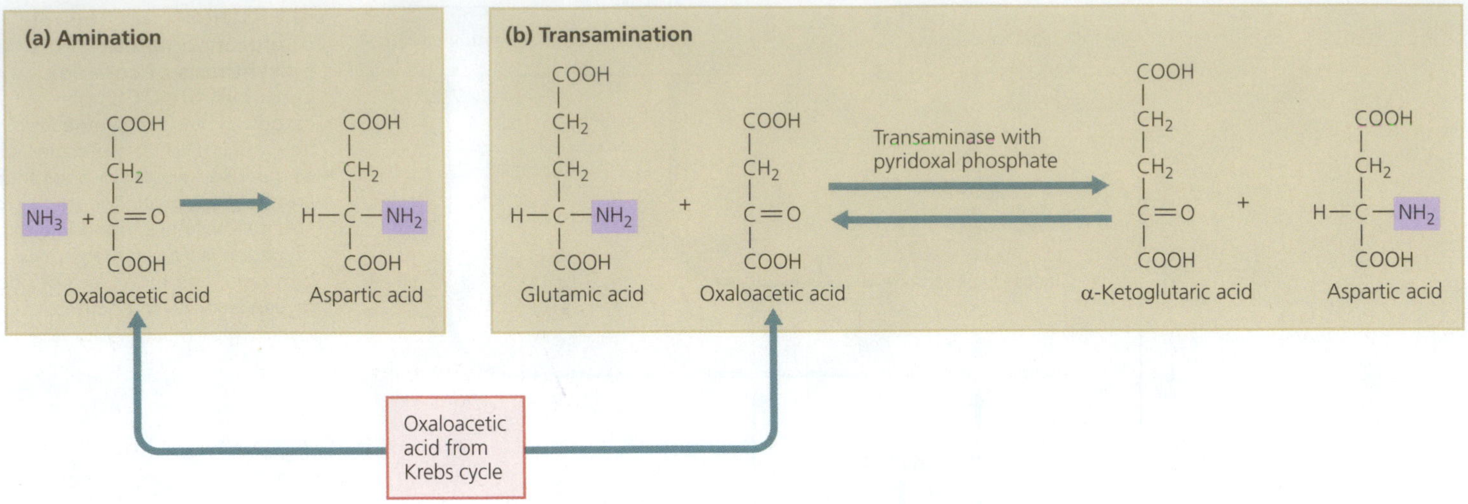

▲ **FIGURE 5.31 Examples of the synthesis of amino acids via amination and transamination. (a)** In amination, an amine group from ammonia is added to a precursor metabolite. In this example, oxaloacetic acid is converted into the amino acid aspartic acid. **(b)** In transamination, the amine group is derived from an existing amino acid. In the example shown here, the amino acid glutamic acid donates its amine group to oxaloacetic acid, which is converted into aspartic acid. Note that transamination is a reversible reaction.

Ribozymes of ribosomes polymerize amino acids into proteins. (Chapter 7 examines this energy-demanding process because it is intimately linked with genetics.)

Nucleotide Biosynthesis

LEARNING **OUTCOME**

5.29 Describe the biosynthesis of nucleotides.

The building blocks of nucleic acids are nucleotides, each of which consists of a five-carbon sugar, a phosphate group, and a purine or pyrimidine base (see Figure 2.25a). Nucleotides are produced from precursor metabolites of glycolysis and the Krebs cycle **(FIGURE 5.32)**:

- The five-carbon sugars—ribose in RNA and deoxyribose in DNA—are derived from ribose 5-phosphate from the pentose phosphate pathway.
- The phosphate group is derived ultimately from ATP.

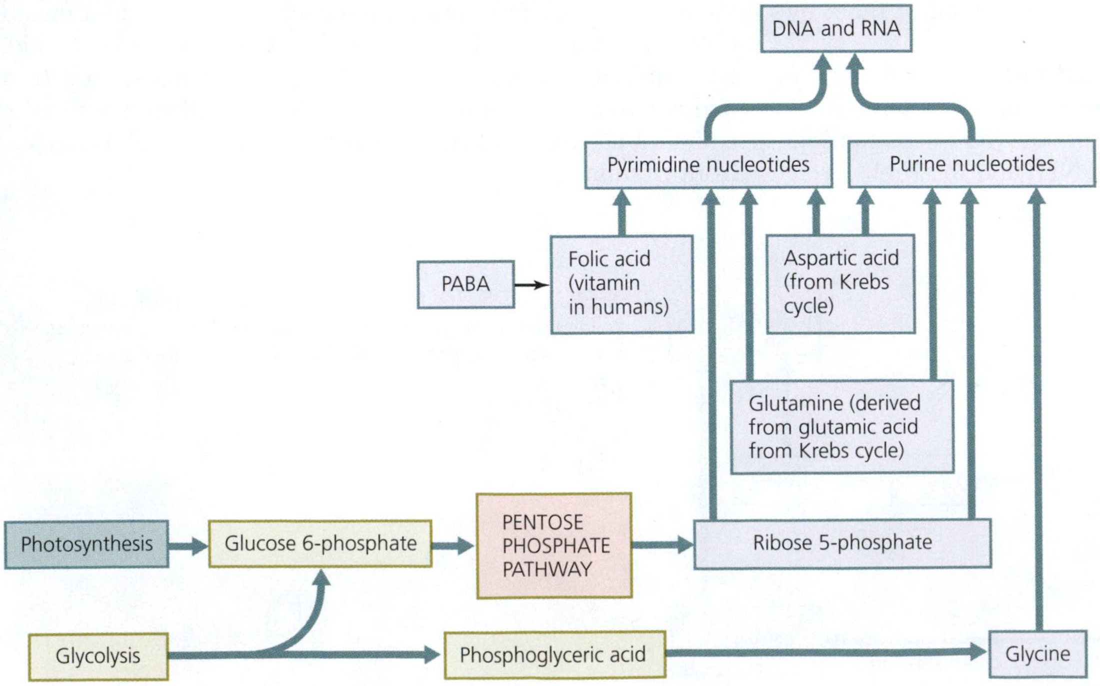

◀ **FIGURE 5.32 The biosynthesis of nucleotides.**

- Purines and pyrimidines are synthesized in a series of ATP-requiring reactions from the amino acids glutamine and aspartic acid derived from Krebs cycle intermediates, ribose 5-phosphate, and *folic acid*. The latter is synthesized by many bacteria and protozoa but is a vitamin for humans.

Chapter 7 examines the anabolic reactions by which polymerases polymerize nucleotides to form DNA and RNA.

TELL ME WHY

Why is nitrogen required for the production fo amino acids by amination?

Integration and Regulation of Metabolic Functions

LEARNING **OUTCOMES**

5.30 Describe interrelationships between catabolism and anabolism in terms of ATP and substrates.

5.31 Discuss regulation of metabolic activity.

As we have seen, catabolic and anabolic reactions interact with one another in several ways. First, ATP molecules produced by catabolism are used to drive anabolic reactions. Second, catabolic pathways produce precursor metabolites to use as substrates for anabolic reactions. Additionally, most metabolic pathways are amphibolic; they function as part of either catabolism or anabolism as needed.

Cells regulate metabolism in a variety of ways to maximize efficiency in growth and reproductive rate. Among the mechanisms involved are the following:

- Cells synthesize or degrade channel and transport proteins to increase or decrease the concentration of chemicals in the cytosol or organelles.
- Cells often synthesize the enzymes needed to catabolize a particular substrate only when that substrate is available. For instance, the enzymes of beta-oxidation are not produced when there are no fatty acids to catabolize.
- If two energy sources are available, cells catabolize the more energy efficient of the two. For example, a bacterium growing in the presence of both glucose and lactose will produce enzymes only for the transport and catabolism of glucose. Once the supply of glucose is depleted, lactose-utilizing proteins are produced.
- Cells synthesize the metabolites they need, but they typically cease synthesis if a metabolite is available as a nutrient. For instance, bacteria grown with an excess of aspartic acid will cease the amination of oxaloacetic acid (see Figure 5.31).
- Eukaryotic cells keep metabolic processes from interfering with each other by isolating particular enzymes within membrane-bound organelles. For example, proteases sequestered within lysosomes digest phagocytized proteins without destroying vital proteins in the cytosol.
- Cells use inhibitory and excitatory allosteric sites on enzymes to control the activity of enzymes (see Figure 5.10).
- Feedback inhibition slows or stops anabolic pathways when the product is in abundance (see Figure 5.11).
- Cells regulate catabolic and anabolic pathways that use the same substrate molecules by requiring different coenzymes for each. For instance, NADH is used almost exclusively with catabolic enzymes, whereas NADPH is typically used for anabolism.

Note that these regulatory mechanisms are generally of two types: *control of gene expression*, in which cells control the amount and timing of protein (enzyme) production, and *control of metabolic expression*, in which cells control the activity of proteins (enzymes) once they have been produced.

FIGURE 5.33 schematically diagrams some of the numerous interrelationships among the metabolic pathways discussed in this chapter. ▶**ANIMATIONS:** ***Metabolism: The Big Picture***

TELL ME WHY

Why is feedback inhibition necessary for controlling anabolic pathways?

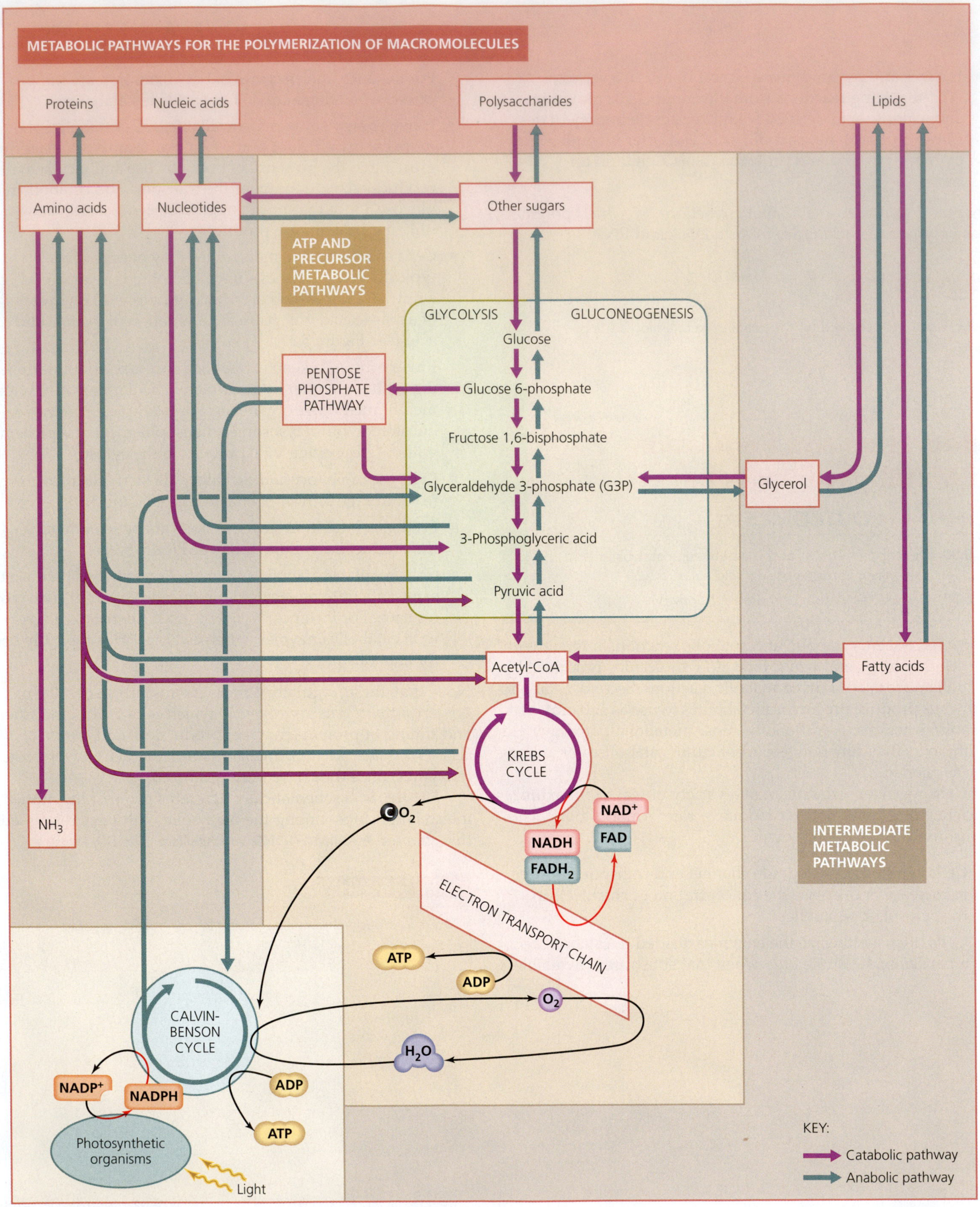

▲ **FIGURE 5.33 Integration of cellular metabolism (shown in an aerobic organism).** Cells possess three major categories of metabolic pathways: pathways for the polymerization of macromolecules (proteins, nucleic acids, polysaccharides, and lipids), intermediate pathways, and ATP and precursor pathways (glycolysis, Krebs cycle, the pentose phosphate pathway, and the Entner-Doudoroff pathway [not shown]). Cells of photosynthetic organisms also have the Calvin-Benson cycle.

MICRO IN THE CLINIC FOLLOW-UP

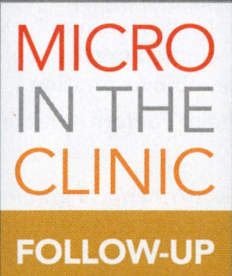

What's Lurking in the Fitness Center?

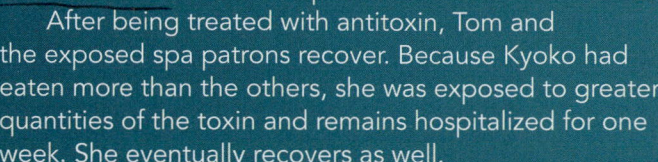

The physician in the emergency room recognizes Tom and Kyoko's symptoms as characteristic of botulinum toxin, a bacterial neurotoxin produced by anaerobic bacteria called *Clostridium botulinum*. The physician immediately notifies the public health department because, although rare, this foodborne intoxication represents a public health emergency. Spa customers are notified, and public health specialists examine the food offerings at the spa. Kyoko, Tom, and the other customers exposed to the toxin are treated with antitoxin and monitored closely.

Meanwhile, the public health department confirms the presence of the botulinum toxin in Kyoko's favorite noodle dish, singling out the home-prepared fermented tofu as the source of the contamination. *C. botulinum* produces highly resistant endospores, which are ubiquitous in nature (refer to Chapter 3, p. 73). Because the tofu was only briefly boiled and then kept for several days in an anaerobic environment, the bacterial endospores were able to germinate, metabolize food sources through anaerobic respiration, and produce the toxin. Although most cases of botulism intoxication in the United States are the result of improperly prepared home-canned foods, botulism in Asia is more commonly the result of homemade fermentation of bean curd products.

After being treated with antitoxin, Tom and the exposed spa patrons recover. Because Kyoko had eaten more than the others, she was exposed to greater quantities of the toxin and remains hospitalized for one week. She eventually recovers as well.

1. **Even though *C. botulinum* endospores are durable and ubiquitous in nature, cases of botulism are not extremely common. Why do you think this is?**
2. **What environment did the plastic wrap and airtight jar create, and how did this contribute to the toxin production?**

MM Check your answers to Micro in the Clinic Follow-Up questions in the MasteringMicrobiology Study Area.

Explore the Invisible: Electron Transport Chains

Practice on-the-go with Dr. Bauman Video Tutors by scanning this QR code with your smart phone. Visit the **MasteringMicrobiology Study Area** to challenge your understanding with practice tests, animation quizzes, and clinical case studies!

MasteringMicrobiology®

CHAPTER SUMMARY

This chapter has MicroFlix. Go to the MasteringMicrobiology Study Area to view movie-quality animations for metabolism.

Basic Chemical Reactions Underlying Metabolism (pp. 126–134)

1. **Metabolism** is the sum of biochemical reactions within the cells of an organism, including **catabolism**, which breaks down molecules and releases energy, and **anabolism**, which synthesizes molecules and uses energy.

   ▶ANIMATIONS: *Metabolism: Overview*

2. **Precursor metabolites**, often produced in catabolic reactions, are used to synthesize all other organic compounds.
3. Reduction reactions are those in which electrons are added. The molecule that donates an electron is oxidized. If the electron is part of a hydrogen atom, an oxidation reaction is also called dehydrogenation. Oxidation and reduction reactions always occur in pairs called **oxidation-reduction (redox) reactions**.

   ▶ANIMATIONS: *Oxidation-Reduction Reactions*

4. Three important electron carrier molecules are **nicotinamide adenine dinucleotide (NAD^+), nicotinamide adenine dinucleotide phosphate ($NADP^+$)**, and **flavin adenine dinucleotide (FAD)**.
5. Phosphorylation is the addition of phosphate to a molecule. Three types of phosphorylation form ATP: **Substrate-level phosphorylation** involves the transfer of phosphate from a phosphorylated organic compound to ADP. In **oxidative phosphorylation**, energy from redox reactions of respiration is used to attach inorganic phosphate (PO_4^{3-}) to ADP. **Photophosphorylation** is the phosphorylation of ADP with inorganic phosphate using energy from light.

6. Catalysts increase the rates of chemical reactions and are not permanently changed in the process. **Enzymes**, which are organic catalysts, are often named for their **substrates**—the molecules on which they act. Enzymes can be classified as hydrolases, isomerases, ligases (polymerases), lyases, oxidoreductases, or transferases, reflecting their mode of action.

   ▶**ANIMATIONS:** *Enzymes: Overview*

7. **Apoenzymes** are the portions of enzymes that may require one or more **cofactors** such as inorganic ions or organic cofactors (also called **coenzymes**). The combination of both apoenzyme and its cofactors is a **holoenzyme**. RNA molecules functioning as enzymes are called **ribozymes**.

8. **Activation energy** is the amount of energy required to initiate a chemical reaction.

9. Substrates fit into the specifically shaped **active sites** of the enzymes that catalyze their reactions.

   ▶**ANIMATIONS:** *Enzymes: Steps in a Reaction*

10. Enzymes may be **denatured** by physical and chemical factors such as heat and pH. Denaturation may be reversible or permanent.

11. Enzyme activity proceeds at a rate proportional to the concentration of substrate molecules until all the active sites are filled.

12. **Competitive inhibitors** block active sites and thereby block enzyme activity. **Noncompetitive inhibitors** attach to an allosteric site on an enzyme, altering the active site so that it is no longer functional.

    ▶**ANIMATIONS:** *Enzymes: Competitive Inhibition; Enzyme-Substrate Interaction: Noncompetitive Inhibition*

13. **Feedback inhibition** (negative feedback) occurs when the final product of a series of reactions is an allosteric inhibitor of some previous step in the series. Thus, accumulation of the end product "feeds back" into the series a signal that stops the process.

Carbohydrate Catabolism (pp. 134–147)

1. **Glycolysis** involves the splitting of a glucose molecule in a three-stage, 10-step process that ultimately results in two molecules of pyruvic acid and a net gain of two ATP and two NADH molecules.

   ▶**ANIMATIONS:** *Glycolysis: Overview, Steps*

2. **Cellular respiration** is a metabolic process that involves the complete oxidation of substrate molecules and the production of ATP following a series of redox reactions.

3. Two carbons from pyruvic acid join coenzyme A to form acetyl-coenzyme A **(acetyl-CoA)**, which then enters the **Krebs cycle**, a series of eight enzymatic steps that transfer electrons from acetyl-CoA to coenzymes NAD^+ and FAD.

   ▶**ANIMATIONS:** *Krebs Cycle: Overview, Steps*

4. An **electron transport chain** is a series of redox reactions that pass electrons from one membrane-bound carrier to another and then to a final electron acceptor. The energy from these electrons is used to pump protons across the membrane.

   ▶**ANIMATIONS:** *Electron Transport Chain: Overview*

5. The four classes of carrier molecules in electron transport systems are flavoproteins, ubiquinones, metal-containing proteins, and cytochromes.

   ▶**ANIMATIONS:** *Electron Transport Chain: The Process*
   ▶**VIDEO TUTOR:** *Electron Transport Chains*

6. Aerobes use oxygen atoms as final electron acceptors in their electron transport chains in a process known as **aerobic respiration**, whereas anaerobes use other inorganic molecules (such as NO_3^-, SO_4^{2-}, and CO_3^{2-}, or rarely an externally acquired organic molecule) as the final electron acceptor in **anaerobic respiration**.

   ▶**ANIMATIONS:** *Electron Transport Chain: Factors Affecting ATP Yield*

7. In **chemiosmosis**, ions flow down their electrochemical gradient across a membrane through **ATP synthase** (**ATPase**) to synthesize ATP.

8. A **proton gradient** is an electrochemical gradient of hydrogen ions across a membrane. It has potential energy known as a proton motive force.

9. Oxidative phosphorylation and photophosphorylation use chemiosmosis.

10. **Fermentation** is the partial oxidation of sugar to release energy using a cellular organic molecule rather than an electron transport chain as the final electron acceptor. End products of fermentation, which are often useful to humans and aid in laboratory identification of microbes, include acids, alcohols, and gases.

    ▶**ANIMATIONS:** *Fermentation*

11. The **pentose phosphate** and **Entner-Doudoroff pathway** are alternative means for the catabolism of glucose that yield fewer ATP molecules than does glycolysis. However, they produce precursor metabolites not produced in glycolysis.

Other Catabolic Pathways (pp. 147–149)

1. Lipids and proteins can be catabolized into smaller molecules, which can be used as substrates for glycolysis and the Krebs cycle.

2. **Beta-oxidation** is a catabolic process in which enzymes split pairs of hydrogenated carbon atoms from a fatty acid and join them to coenzyme A to form acetyl-CoA.

3. **Proteases** secreted by microorganisms digest proteins outside the microbes' cell walls. The resulting amino acids are moved into the cell and used in anabolism, or **deaminated** and catabolized for energy.

Photosynthesis (pp. 149–153)

1. **Photosynthesis** is a process in which light energy is captured by pigment molecules called **chlorophylls** (bacteriochlorophylls in some bacteria) and transferred to ATP and metabolites. **Photosystems** are networks of light-absorbing chlorophyll molecules and other pigments held within a protein matrix on membranes called **thylakoids**.

   ▶**ANIMATIONS:** *Photosynthesis: Overview*

2. The redox reactions of photosynthesis are classified as **light-dependent reactions** and **light-independent reactions**.

3. A **reaction center chlorophyll** is a special chlorophyll molecule in a photosystem in which electrons are excited by light energy and passed to an acceptor molecule of an electron transport chain.

4. In **cyclic photophosphorylation**, the electrons return to the original reaction center after passing down the electron transport chain.

   ▶**ANIMATIONS:** *Photosynthesis: Cyclic Photophosphorylation*

5. In **noncyclic photophosphorylation**, photosystem II works with photosystem I, and the electrons are used to reduce $NADP^+$ to

NADPH. In oxygenic photosynthesis, cyanobacteria, algae, and green plants replenish electrons to the reaction center by dissociation of H_2O molecules, resulting in the release of O_2 molecules. Anoxygenic bacteria derive electrons from inorganic compounds such as H_2S, producing waste such as sulfur.

▶**ANIMATIONS:** *Photosynthesis: Noncyclic Photophosphorylation*

6. In the light-independent pathway of photosynthesis, **carbon fixation** occurs in the **Calvin-Benson cycle**, in which CO_2 is reduced to produce glucose.

▶**ANIMATIONS:** *Photosynthesis: Light-Independent Reaction*

Other Anabolic Pathways (pp. 153–157)

1. **Amphibolic reactions** are metabolic reactions that are reversible—they can operate catabolically or anabolically.
2. Some cells are able to synthesize glucose from amino acids, glycerol, and fatty acids via a process called gluconeogenesis.
3. **Amination** reactions involve adding an amine group from ammonia to a metabolite to make an amino acid. **Transamination** occurs when an amine group is transferred from one amino acid to another.
4. Nucleotides are synthesized from precursor metabolites produced by glycolysis, the Krebs cycle, and the pentose phosphate pathway.

Integration and Regulation of Metabolic Functions (p. 157)

1. Cells regulate metabolism by control of gene expression or metabolic expression. They control the latter in a variety of ways, including synthesizing or degrading channel proteins and enzymes, sequestering reactions in membrane-bound organelles (seen in eukaryotes), and feedback inhibition.

▶**ANIMATIONS:** *Metabolism: The Big Picture*

QUESTIONS FOR REVIEW

Answers to the Questions for Review (except Short Answer questions) begin on p. A-1.

Multiple Choice

For each of the phrases in questions 1–7, indicate the type of metabolism referred to, using the following choices:

a. anabolism only
b. both anabolism and catabolism (amphibolic)
c. catabolism only

1. Breaks a large molecule into smaller ones c
2. Includes dehydration synthesis reactions a
3. Is exergonic c
4. Is endergonic a
5. Involves the production of cell membrane constituents a
6. Includes hydrolytic reactions c
7. Includes metabolism b
8. Redox reactions ________________.
   a. transfer energy
   b. transfer electrons
   c. involve oxidation and reduction
   d. are involved in all of the above
9. A reduced molecule ________________.
   a. has gained electrons
   b. has become more positive in charge
   c. has lost electrons
   d. is an electron donor
10. Activation energy ________________.
    a. is the amount of energy required during an activity such as flagellar motion
    b. requires the addition of nutrients in the presence of water
    c. is lowered by the action of organic catalysts
    d. results from the movement of molecules
11. Coenzymes ________________.
    a. are types of apoenzymes
    b. are proteins
    c. are inorganic cofactors
    d. are organic cofactors
12. Which of the following statements best describes ribozymes?
    a. Ribozymes are proteins that aid in the production of ribosomes.
    b. Ribozymes are nucleic acids that produce ribose sugars.
    c. Ribozymes store enzymes in ribosomes.
    d. Ribozymes process RNA molecules in eukaryotes.
13. Which of the following does *not* affect the function of enzymes?
    a. ubiquinone
    b. substrate concentration
    c. temperature
    d. competitive inhibitors
14. Most oxidation reactions in bacteria involve the ________________.
    a. removal of hydrogen ions and electrons
    b. removal of oxygen
    c. addition of hydrogen ions and electrons
    d. addition of hydrogen ions
15. Under ideal conditions, the fermentation of one glucose molecule by a bacterium allows a net gain of how many ATP molecules?
    a. 2 c. 38
    b. 4 d. 0
16. Under ideal conditions, the complete aerobic oxidation of one molecule of glucose by a bacterium allows a net gain of how many ATP molecules?
    a. 2 c. 38
    b. 4 d. 0

17. Which of the following statements about the Entner-Doudoroff pathway is *false*?
    a. It is a series of reactions that synthesizes glucose.
    b. Its products are sometimes used to determine the presence of *Pseudomonas*.
    c. It is a pathway of chemical reactions that catabolizes glucose.
    d. It is an alternate pathway to glycolysis.

18. Reactions involved in the light-independent reactions of photosynthesis constitute the ________________.
    a. Krebs cycle
    b. Entner-Doudoroff pathway
    c. Calvin-Benson cycle
    d. pentose phosphate pathway

19. The glycolysis pathway is basically ________________.
    a. catabolic
    b. amphibolic
    c. anabolic
    d. cyclical

20. A major difference between anaerobic respiration and anaerobic fermentation is ________________.
    a. in the use of oxygen
    b. that the former requires breathing
    c. that the latter uses organic molecules within the cell as final electron acceptors
    d. that fermentation only produces alcohol

Matching

1. __C__ Occurs when energy from a compound containing phosphate reacts with ADP to form ATP
2. __B__ Involves formation of ATP via reduction of coenzymes in the electron transport chain
3. __E__ Begins with glycolysis
4. __A__ Occurs when all active sites on substrate molecules are filled

A. Saturation
B. Oxidative phosphorylation
C. Substrate-level phosphorylation
D. Photophosphorylation
E. Carbohydrate catabolism

Fill in the Blanks

1. The final electron acceptor in cyclic photophosphorylation is ________________.

2. Two ATP molecules are used to initiate glycolysis. Enzymes generate molecules of ATP for each molecule of glucose that undergoes glycolysis. Thus, a net gain of ________________ molecules of ATP is produced in glycolysis.

3. The initial catabolism of glucose occurs by glycolysis and/or the ________________ and ________________ pathways.

4. ________________ is a cyclic series of eight reactions involved in the catabolism of acetyl-CoA that yields eight molecules of NADH and two molecules of $FADH_2$.

5. The final electron acceptor in aerobic respiration is ________________.

6. Three common inorganic electron acceptors in anaerobic respiration are ________________, ________________, and ________________.

7. Anaerobic respiration typically uses (organic/inorganic) ________________ molecules as final electron acceptors.

8. Complete the following chart:

Category of Enzymes	Description
Hydrolase	Catabolizes substrate by adding water
Isomerase	Rearranges Atoms
Ligase/polymerase	
	Moves functional groups such as an acetyl group
	Adds or removes electrons
Lyase	

9. The use of a proton motive force to generate ATP is __chemiosmosis__.

10. The main coenzymes that carry electrons in catabolic pathways are __NAD+__ and __FAD__.

VISUALIZE IT!

1. Label the diagram below to indicate acetyl-CoA, electron transport chain, $FADH_2$, fermentation, glycolysis, Krebs cycle, NADH, and respiration. Indicate the net number of molecules of ATP that could be synthesized at each stage during bacterial respiration of one molecule of glucose.

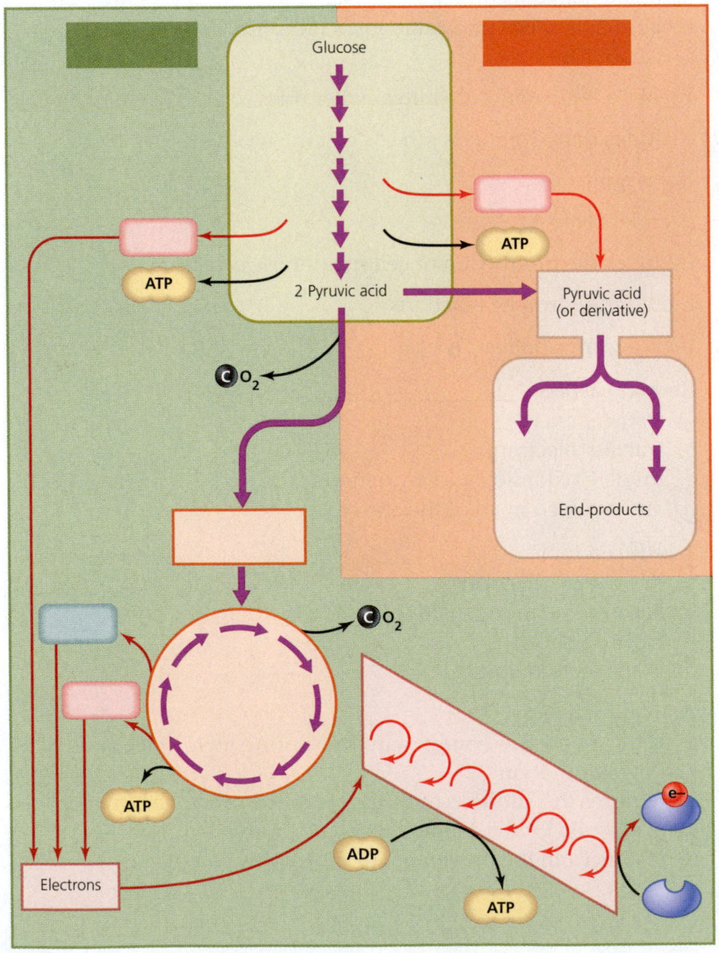

2. Label the electron micrograph to indicate the location of glycolysis, the Krebs cycle, and electron transport chains.

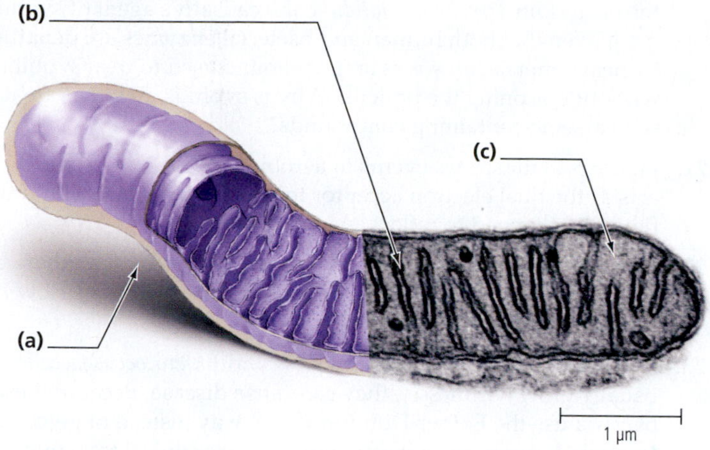

3. Examine the biosynthetic pathway for the production of the amino acids tryptophan, tyrosine, and phenylalanine in the figure. Where do the initial reactants (erythrose 4-phosphate and PEP) originate?

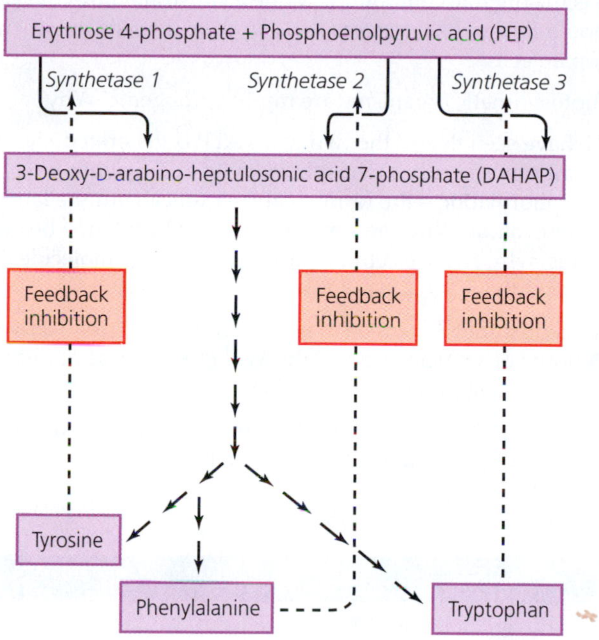

Short Answer

1. How does amination differ from transamination?
2. Why are enzymes necessary for anabolic reactions to occur in living organisms?
3. How do organisms control the rate of metabolic activities in their cells?
4. How does a noncompetitive inhibitor at a single allosteric site affect a whole pathway of enzymatic reactions?
5. Explain the mechanism of negative feedback with respect to enzyme action.
6. Facultative anaerobes can live under either aerobic or anaerobic conditions. What metabolic pathways allow these organisms to continue to harvest energy from sugar molecules in the absence of oxygen?
7. How does oxidation of a molecule occur without oxygen?
8. List at least four groups of microorganisms that are photosynthetic.
9. Why do we breathe oxygen and give off carbon dioxide?
10. Why do cyanobacteria and algae take in carbon dioxide and give off oxygen?
11. What happens to the carbon atoms in sugar catabolized by *Escherichia coli*?
12. How do yeast cells make alcohol and cause bread to rise?
13. Where specifically does the most significant production of ATP occur in prokaryotic and eukaryotic cells?
14. Why are vitamins essential metabolic factors for microbial metabolism?
15. A laboratory scientist notices that a certain bacterium does not utilize lactose when glucose is available in its environment. Describe a cellular regulatory mechanism that would explain this observation.

CRITICAL THINKING

1. Arsenic is a poison that exists in two states in the environment—arsenite ($H_2AsO_3^-$) and arsenate ($H_2AsO_4^-$). Arsenite dissolves in water, making the water dangerous to drink. Arsenate is less soluble and binds to minerals, making this form of arsenic less toxic in the environment. Some strains of the bacterium *Thermus* oxidize arsenic in an aerobic environment but reduce arsenic under anaerobic conditions. In case of arsenic contamination of water, how could scientists use *Thermus* to remediate the problem?
2. Explain why an excess of all three amino acids mentioned in question 1 is required to inhibit production of DAHAP.
3. Why might an organism that uses glycolysis and the Krebs cycle also need the pentose phosphate pathway?
4. Describe how bacterial fermentation causes milk to sour.
5. *Giardia intestinalis* and *Entamoeba histolytica* are protozoa that live in the colons of mammals and can cause life-threatening diarrhea. Interestingly, these microbes lack mitochondria. What kind of pathway must they have for carbohydrate catabolism?
6. Two cultures of a facultative anaerobe are grown in the same type of medium, but one is exposed to air and the other is maintained under anaerobic conditions. Which of the two cultures will contain more cells at the end of a week? Why?
7. What is the maximum number of molecules of ATP that can be generated by a bacterium after the complete aerobic oxidation of a fat molecule containing three 12-carbon chains? (Assume that all the available energy released during catabolism goes to ATP production.)

8. In terms of its effects on metabolism, why is a fever over 40°C often life threatening?
9. Cyanide is a potent poison because it irreversibly blocks cytochrome a_3. What effect would its action have on the rest of the electron transport chain? What would be the redox state (reduced or oxidized) of ubiquinone in the presence of cyanide?
10. How are photophosphorylation and oxidative phosphorylation similar? How are they different?
11. Members of the pathogenic bacterial genus *Haemophilus* require NAD^+ and heme from their environment. For what purpose does *Haemophilus* use these growth factors?
12. Compare and contrast aerobic respiration, anaerobic respiration, and fermentation.
13. Scientists estimate that up to one-third of Earth's biomass is composed of methanogenic prokaryotes in ocean sediments (*Science* 295:2067–2070). Describe the metabolism of these organisms.
14. A young student was troubled by the idea that a bacterium is able to control its diverse and complex metabolic activities even though it lacks a brain. How would you explain its metabolic control?
15. If a bacterium uses beta-oxidation to catabolize a molecule of the fatty acid arachidic acid, which contains 20 carbon atoms, how many acetyl-CoA molecules will be generated?
16. Some desert rodents rarely have water to drink. How do they get enough water for their cells without drinking it?
17. Why do fatty acids typically contain an even number of carbon atoms?
18. We have examined the total ATP, NADH, and $FADH_2$ production in the Krebs cycle for each molecule of glucose coming through Embden-Meyerhof glycolysis. How many of each of these molecules would be produced if the Entner-Doudoroff pathway were used instead of Embden-Meyerhof glycolysis?
19. Explain why hyperthermophiles do not cause disease in humans.
20. In addition to extremes in temperature and pH, other chemical and physical agents denature proteins. These agents include ionizing radiation, alcohol, enzymes, and heavy-metal ions. For example, the first antimicrobial drug, arsphenamine, contained the heavy metal arsenic and was used to inhibit the enzymes of the bacterium *Treponema pallidum,* the causative agent of syphilis.

    Given that both human and bacterial enzymes are denatured by heavy metals, how was arsphenamine used to treat syphilis without poisoning the patient? Why is syphilis no longer treated with arsenic-containing compounds?
21. Figure 5.18 illustrates events in aerobic respiration where oxygen acts as the final electron acceptor to yield water. How would the figure be changed to reflect anaerobic respiration?
22. Suppose you could insert a tiny pH probe into the space between mitochondrial membranes. Would the pH be above or below 7.0? Why?
23. Even though *Pseudomonas aeruginosa* and *Enterococcus faecalis* usually grow harmlessly, they can cause disease. Because these bacteria use the Entner-Doudoroff pathway instead of glycolysis to catabolize glucose, investigators can use clinical tests that provide evidence of the Entner-Doudoroff pathway to identify the presence of these potential pathogens.

    Suppose you were able to identify the presence of any specific organic compound. Name a substrate molecule you would find in *Pseudomonas* and *Enterococcus* cells but not in human cells.
24. Photosynthetic organisms are rarely pathogenic. Why?
25. We have seen that of the two ways ATP is generated via chemiosmosis—photophosphorylation and oxidative phosphorylation—the former can be cyclical, but the latter is never cyclical. Why can't oxidative phosphorylation be cyclical; that is, why aren't electrons passed back to the molecules that donated them?
26. A scientist moves a green plant grown in sunlight to a room with 24 hours of artificial green light. Will this increase or decrease the plant's rate of photosynthesis? Why?
27. What class of enzyme is involved in amination reactions? What class of enzyme catalyzes transaminations?

CONCEPT MAPPING

Using the following terms, draw a concept map that describes aerobic respiration. You may use some terms more than once. For a sample concept map, see p. 94. Or, complete this and other concept maps online by going to the MasteringMicrobiology Study Area.

2 ATP (2)
34 ATP
Chemiosmosis
Electron transport chain
Glycolysis
Krebs cycle
Oxidative phosphorylation
Substrate-level phosphorylation (2)
Synthesis of acetyl-CoA

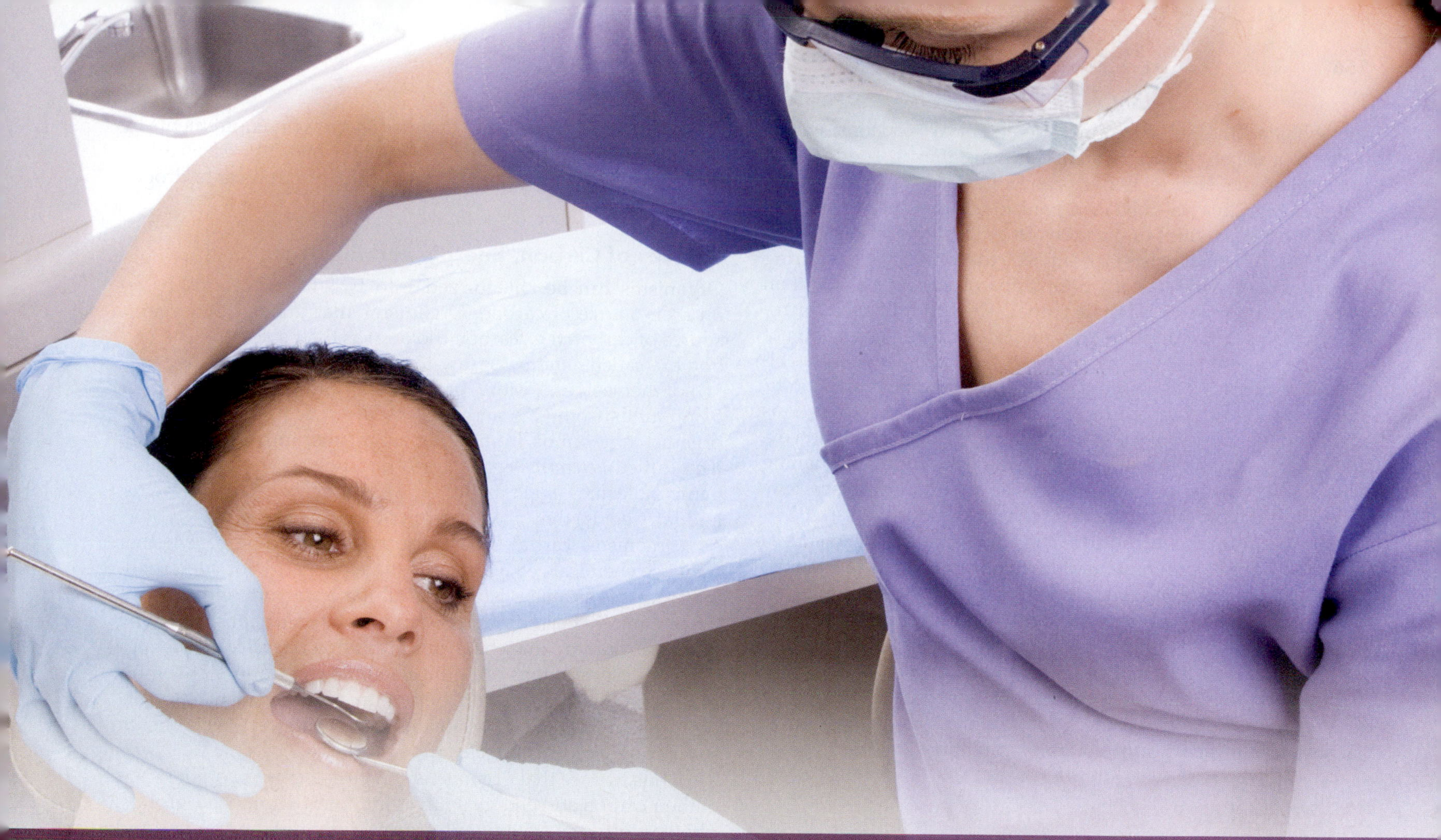

6 Microbial Nutrition and Growth

MICRO IN THE CLINIC

Can a Trip to the Dentist Be Life Threatening?

Betty, a 48-year-old high school teacher, is actually happy about visiting her dentist for her routine teeth cleaning. Since being diagnosed with stage 3 ovarian cancer, life has become surreal, and she actually welcomes the normalcy of a dentist appointment. For months, Betty has been in and out of the hospital, and she is sick of it. First there was the hysterectomy to remove the primary tumor and small secondary tumors that were in her abdominal cavity, and then months of debilitating chemotherapy. She doesn't know yet how effective the chemotherapy has been, but she has been feeling much better.
A trip to the dentist will be easy.

The dentist office is unusually busy, and the hygienist seems rushed, but Betty doesn't mind. She is so thrilled to be doing something ordinary that she hardly notices. With her new toothbrush in hand, Betty leaves the office feeling like she is regaining her normal life back and leaving cancer behind!

Unfortunately, that sense of normalcy is short lived. Two days later, Betty begins to feel sick. She feels weak and achy and has a fever. She immediately goes to the emergency room. The emergency room staff examine her, draw her blood, and call her doctor.

Why does Betty suddenly feel so bad? Are the chemotherapy treatments working? Turn to the end of the chapter (p. 190) to find out.

MM **Explore More:** Test your readiness and apply your knowledge with dynamic learning tools at MasteringMicrobiology.

Metabolism—the set of controlled chemical reactions within cells—is a major characteristic of all living things. The ultimate outcome of metabolic activity is reproduction, an increase in the number of individual cells or organisms. When speaking of the reproductive activities of microbes in general and of bacteria in particular, microbiologists typically use the term *growth,* referring to an increase in the size of a population of microbes rather than to an increase in size of an individual. The result of such microbial growth is either a discrete *colony,* which is an aggregation of cells arising from a single parent cell, or a *biofilm,* which is a collection of microbes living on a surface in a complex community. Put another way, the reproduction of individual microorganisms results in the growth of a colony or biofilm. Further, common expressions such as "The microorganisms *grow* in salt-containing media" are widely understood to mean that the organisms metabolize and reproduce rather than that they increase in size.

In this chapter we consider the characteristics of microbial growth from two different but related perspectives: We examine the requirements of microbes living in their natural settings, including their chemical, physical, and energy requirements, and we explore how microbiologists try to create similar conditions to grow microorganisms in the laboratory so that they can be transported, identified, and studied. We conclude by examining laboratory analysis of bacterial population dynamics and some techniques for measuring bacterial population growth.

Growth Requirements

Organisms use a variety of chemicals—called **nutrients**—to meet their energy needs and to build organic molecules and cellular structures. The most common of these nutrients are compounds containing necessary elements such as carbon, oxygen, nitrogen, and hydrogen. Like all organisms, microbes obtain nutrients from a variety of sources in their environment, and they must bring nutrients into their cells by passive and active transport processes (see Chapter 3). When they acquire their nutrients by living in or on another organism, they may cause disease as they interfere with their hosts' metabolism and nutrition.

Nutrients: Chemical and Energy Requirements

LEARNING | OUTCOMES

6.1 Describe the roles of carbon, hydrogen, oxygen, nitrogen, trace elements, and vitamins in microbial growth and reproduction.

6.2 Compare the four basic categories of organisms based on their carbon and energy sources.

6.3 Distinguish among anaerobes, aerobes, aerotolerant anaerobes, facultative anaerobes, and microaerophiles.

6.4 Explain how oxygen can be fatal to organisms by discussing singlet oxygen, superoxide radical, peroxide anion, and hydroxyl radical and describe how organisms protect themselves from toxic forms of oxygen.

6.5 Define *nitrogen fixation*, and explain its importance.

We begin our examination of microbial growth requirements by considering three things all cells need for metabolism: a carbon source, a source of energy, and a source of electrons or hydrogen atoms.

Sources of Carbon, Energy, and Electrons

Organisms can be categorized into two broad groups based on their source of carbon. Organisms that utilize an inorganic source of carbon (i.e., carbon dioxide) as their sole source of carbon are called *autotrophs*[1] (aw´tō-trōfs), so named because they "feed themselves." More precisely, autotrophs make organic compounds from CO_2 and thus need not acquire carbon from organic compounds from other organisms. In contrast, organisms called *heterotrophs*[2] (het´er-ō-trōfs) catabolize reduced organic molecules (such as proteins, carbohydrates, amino acids, and fatty acids) they acquire from other organisms.

Organisms can also be categorized according to whether they use chemicals or light as a source of energy for such cellular processes as anabolism, intracellular transport, and motility. Organisms that acquire energy from redox reactions involving inorganic and organic chemicals are called *chemotrophs* (kēm´ō-trōfs). These reactions are either aerobic respiration, anaerobic respiration, or fermentation, depending on the final electron acceptor (see Chapter 5). Organisms that use light as their energy source are called *phototrophs*[3] (fō´tō-trōfs).

Thus, we see that organisms can be categorized on the basis of their carbon and energy sources into one of four basic groups: **photoautotrophs, chemoautotrophs, photoheterotrophs,** and **chemoheterotrophs** (**FIGURE 6.1**). Plants, some protozoa, and algae are photoautotrophs, and animals, fungi, and other protozoa are chemoheterotrophs. Bacteria and archaea exhibit greater metabolic diversity than any other group, with members in all four groups.

Additionally, the cells of all organisms require electrons or hydrogen atoms for redox reactions. Hydrogen is the most common chemical element in cells, and it is so common in organic molecules and water that it is never a *limiting nutrient;* that is, metabolism is never interrupted by a lack of hydrogen. Hydrogen is essential for hydrogen bonding and in electron transfer. Heterotrophs acquire electrons (typically as part of hydrogen atoms) from the same organic molecules that provide them carbon and are called **organotrophs** (ōr´-gān-ō-trōfs); alternatively, autotrophic organisms acquire electrons or hydrogen atoms from inorganic molecules (such as H_2, NO^{2-}, H_2S, and Fe^{2+}) and are called **lithotrophs**[4] (lith´ō-trōfs).

Oxygen Requirements

Oxygen is essential for **obligate aerobes** because it serves as the final electron acceptor of electron transport chains, which produce most of the ATP in these organisms. By contrast, oxygen is a deadly poison for **obligate anaerobes.** How can oxygen be essential for one group of organisms and yet be a fatal toxin for others?

[1]From Greek *auto,* meaning "self," and *trophe,* meaning "nutrition."
[2]From Greek *hetero,* meaning "other," and *trophe,* meaning "nutrition."
[3]From Greek *photos,* meaning "light," and *trophe,* meaning "nutrition."
[4]From Greek *lithos,* meaning "rock," and *trophe,* meaning "nutrition."

Carbon source		Energy source: Light (*photo-*)	Energy source: Chemical compounds (*chemo-*)
Carbon source	Carbon dioxide (*auto-*)	***Photoautotrophs*** • Plants, algae, and cyanobacteria use H_2O as an electron source to reduce CO_2, producing O_2 as a by-product • Green sulfur bacteria and purple sulfur bacteria use H_2S as an electron source; they do not produce O_2	***Chemoautotrophs*** • Hydrogen, sulfur, and nitrifying bacteria, some archaea
Carbon source	Organic compounds (*hetero-*)	***Photoheterotrophs*** • Green nonsulfur bacteria and purple nonsulfur bacteria, some archaea	***Chemoheterotrophs*** • Aerobic respiration: most animals, fungi, and protozoa, and many bacteria • Anaerobic respiration: some animals, protozoa, bacteria, and archaea • Fermentation: some bacteria, yeasts, and archaea

◀ **FIGURE 6.1 Four basic groups of organisms based on their carbon and energy sources.** Additionally, organotrophs utilize electrons from organic molecules, and lithotrophs utilize electrons from inorganic molecules.

The key to understanding this apparent incongruity is understanding that neither atmospheric oxygen (O_2) nor covalently bound oxygen in compounds such as carbohydrates and water is poisonous. Rather, the toxic forms of oxygen are those that are highly reactive. They are toxic for the same reason that oxygen is the final electron acceptor for aerobes: They are excellent oxidizing agents, so they steal electrons from other compounds, which in turn steal electrons from still other compounds. The resulting chain of vigorous oxidations causes irreparable damage to cells by oxidizing important compounds, including proteins and lipids.

There are four toxic forms of oxygen:

- **Singlet oxygen ($^1O_2$).** Singlet oxygen is molecular oxygen with electrons that have been boosted to a higher energy state, typically during aerobic metabolism. Singlet oxygen is a very reactive oxidizing agent. Phagocytic cells, such as certain human white blood cells, use it to oxidize pathogens. Because singlet oxygen is also photochemically produced by the reaction of oxygen and light, phototrophic microorganisms often contain pigments called **carotenoids** (ka-rot′e-noyds) that prevent toxicity by removing the excess energy of singlet oxygen.
- **Superoxide radical (O_2^-).** A few superoxide radicals form during the incomplete reduction of O_2 during electron transport in aerobes and during metabolism by anaerobes in the presence of oxygen. Superoxide radicals are so reactive and toxic that aerobic organisms must produce enzymes called *superoxide dismutases* (dis′myu-tās-es) to detoxify them. These enzymes, which have active sites that contain metal ions—Zn^{2+}, Mn^{2+}, Fe^{2+}, Ni^{2+}, or Cu^{2+}, depending on the organism—combine two superoxide radicals and two protons to form hydrogen peroxide (H_2O_2) and molecular oxygen (O_2):

$$2\,O_2^- + 2H^+ \rightarrow H_2O_2 + O_2$$

One reason that anaerobes are susceptible to oxygen is that they lack superoxide dismutase; they die as a result of the oxidizing reactions of superoxide radicals formed in the presence of oxygen.

- **Peroxide anion (O_2^{2-}).** Hydrogen peroxide formed during reactions catalyzed by superoxide dismutase (and during other metabolic reactions) contains peroxide anion, another highly reactive oxidant. It is peroxide anion that makes hydrogen peroxide an antimicrobial agent. Aerobes contain either catalase or peroxidase, enzymes that detoxify peroxide anion.

Catalase converts hydrogen peroxide to water and molecular oxygen:

$$2\,H_2O_2 \rightarrow 2\,H_2O + O_2$$

A simple test for catalase involves adding a sample from a bacterial colony to a drop of hydrogen peroxide. The production of bubbles of oxygen indicates the presence of catalase **(FIGURE 6.2)**.

Peroxidase breaks down hydrogen peroxide without forming oxygen, using a reducing agent such as the coenzyme NADH:

$$H_2O_2 + NADH + H^+ \rightarrow 2\,H_2O + NAD^+$$

Obligate anaerobes either lack both catalase and peroxidase or have only a small amount of them, so they are susceptible to the toxic action of hydrogen peroxide.

- **Hydroxyl radical (OH·).** Hydroxyl radicals result from ionizing radiation and from the incomplete reduction of hydrogen peroxide:

$$H_2O_2 + e^- + H^+ \rightarrow H_2O + OH\cdot$$

▲ **FIGURE 6.2 Catalase test.** The enzyme catalase converts hydrogen peroxide into water and oxygen, the latter of which can be seen as visible bubbles. *Enterococcus faecalis* (above) is catalase negative, whereas *Staphylococcus epidermidis* (below) is catalase positive.

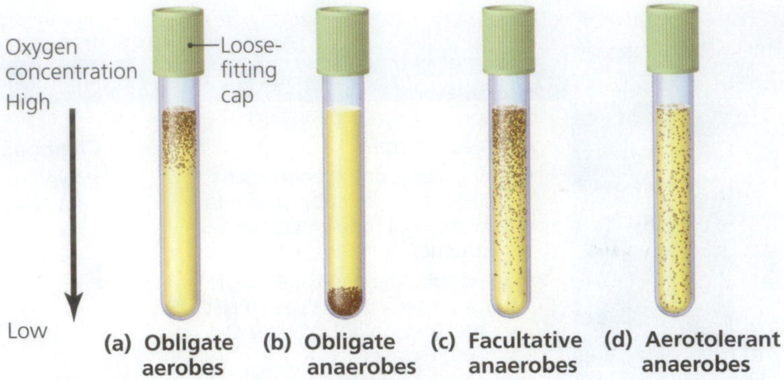

▲ **FIGURE 6.3 Using a liquid thioglycollate growth medium to identify the oxygen requirements of organisms.** The surface is exposed to atmospheric oxygen and is aerobic. Oxygen concentration decreases with depth; the bottom of the tube is anaerobic. **(a)** Obligate aerobes cannot survive below the depth to which oxygen penetrates the medium. **(b)** Obligate anaerobes cannot tolerate any oxygen. **(c)** Facultative anaerobes can grow with or without oxygen, but their ability to use aerobic respiration pathways enhances their growth near the surface. **(d)** Aerotolerant aerobes can grow equally well with or without oxygen; their growth is relatively evenly distributed throughout the medium. *Where in such a test tube would the growth zone be for a microaerophilic aerobe?*

Figure 6.3 *Microaerophiles would be found slightly below the surface but neither directly at the surface nor in the depths of the tube.*

Hydroxyl radicals are the most reactive of the four toxic forms of oxygen, but because hydrogen peroxide does not accumulate in aerobic cells (because of the action of catalase and peroxidase), the threat of hydroxyl radical is virtually eliminated in aerobic cells.

Besides the enzymes superoxide dismutase, catalase, and peroxidase, aerobes use other antioxidants, such as vitamins C and E, to protect themselves against toxic oxygen products. These antioxidants provide electrons that reduce toxic forms of oxygen.

Not all organisms are either strict **aerobes** or **anaerobes;** many organisms can live in various oxygen concentrations between these two extremes. For example, some aerobic organisms can maintain life via fermentation or anaerobic respiration, though their metabolic efficiency is often reduced in the absence of oxygen. Such organisms are called **facultative anaerobes.** *Escherichia coli* (esh-ĕ-rik´ē-ăkō´lī) is an example of a facultatively anaerobic bacterium.

Aerotolerant anaerobes do not use aerobic metabolism, but they tolerate oxygen by having some of the enzymes that detoxify oxygen's poisonous forms. The lactobacilli that transform cucumbers into pickles and milk into cheese are aerotolerant. These organisms can be kept in a laboratory without the special conditions required by obligate anaerobes.

Microaerophiles, such as the ulcer-causing pathogen *Helicobacter pylori*[5] (hel´ĭ-kō-bak´ter pī´lō-rē), require oxygen levels of 2% to 10%. This concentration of oxygen is found in the stomach. Microaerophiles are damaged by the 21% concentration of oxygen in the atmosphere, presumably because they have limited ability to detoxify hydrogen peroxide and superoxide radicals.

Microbial groups contain members with each of the five types of oxygen requirement. Algae, most fungi and protozoa, and many prokaryotes are obligate aerobes. A few yeasts and numerous prokaryotes are facultative anaerobes. Many prokaryotes and a few protozoa are aerotolerant, microaerophilic, or obligate anaerobes. The oxygen requirement of an organism can be identified by growing it in a medium that contains an oxygen gradient from top to bottom (**FIGURE 6.3**).

Nitrogen Requirements

Another essential element is nitrogen, which is contained in many organic compounds, including the amine group of amino acids and as part of nucleotide bases. Nitrogen makes up about 14% of the dry weight of microbial cells.

Nitrogen is often a growth-limiting nutrient for many organisms; that is, their anabolism ceases because they do not have sufficient nitrogen to build proteins and nucleotides. Organisms acquire nitrogen from organic and inorganic nutrients. For example, most photosynthetic organisms can reduce nitrate (NO_3^-) to ammonium (NH_4^+), which can then be used for biosynthesis. In addition, all cells recycle nitrogen from their amino acids and nucleotides.

Though nitrogen constitutes about 79% of the atmosphere, relatively few organisms can utilize nitrogen gas. A few bacteria, notably many cyanobacteria and *Rhizobium* (rī-zō´bē-ŭm), reduce nitrogen gas (N_2) to ammonia (NH_3) via a process called **nitrogen fixation.** Nitrogen fixation is essential for life on Earth because nitrogen fixers provide nitrogen in a usable form to other organisms. (Chapters 11 and 26 discuss nitrogen-fixing prokaryotes as well as nitrifying prokaryotes—those that oxidize nitrogenous compounds to acquire electrons for electron transport.)

Other Chemical Requirements

Together, carbon, hydrogen, oxygen, and nitrogen make up more than 95% of the dry weight of cells; phosphorus, sulfur, calcium,

[5]From the semihelical shape of the cell and from Greek *pyle*, meaning "gate," in reference to *pylorus*, the distal portion of the stomach, which is the gate to the small intestine.

manganese, magnesium, copper, iron, and a few other elements constitute the rest. Phosphorus is a component of phospholipid membranes, DNA, RNA, ATP, and some proteins. Sulfur is a component of sulfur-containing amino acids, which bind to one another via disulfide bonds that are critical to the tertiary structure of proteins, and in vitamins such as thiamine (B_1) and biotin.

Other elements are called **trace elements** because they are required in very small ("trace") amounts. For example, a few atoms of selenium dissolved out of the walls of glass test tubes provide the total requirement for the growth of green algae in a laboratory. Other trace elements are usually found in sufficient quantities dissolved in water. For this reason, tap water can sometimes be used instead of distilled or deionized water to grow microorganisms in the laboratory.

Some microorganisms—for example, algae and photosynthetic bacteria—are lithotrophic photoautotrophs; that is, they can synthesize all of their metabolic and structural needs from inorganic nutrients. They have every enzyme and cofactor they need to produce all their cellular components. Most organisms, however, require small amounts of certain organic chemicals that they cannot synthesize in addition to those that provide carbon and energy. These necessary organic chemicals are called **growth factors** (**TABLE 6.1**). For example, vitamins are growth factors for some microorganisms. Recall that vitamins constitute all or part of many coenzymes. (Note that vitamins are not growth factors for microorganisms that can manufacture them, such as *E. coli*.) Growth factors for various microbes include some amino acids, purines, pyrimidines, cholesterol, NADH, and heme.

TABLE 6.1 Some Growth Factors of Microorganisms and Their Functions

Growth Factor	Function
Amino acids	Components of proteins
Cholesterol	Used by mycoplasmas (bacteria) for cell membranes
Heme	Functional portion of cytochromes in electron transport system
NADH	Electron carrier
Niacin (nicotinic acid, vitamin B_3)	Precursor of NAD^+ and $NADP^+$
Pantothenic acid (vitamin B_5)	Component of coenzyme A
Para-aminobenzoic acid (PABA)	Precursor of folic acid, which is involved in metabolism of one-carbon compounds and nucleic acid synthesis
Purines, pyrimidines	Components of nucleic acids
Pyridoxine (vitamin B_6)	Utilized in transamination syntheses of amino acids
Riboflavin (vitamin B_2)	Precursor of FAD
Thiamine (vitamin B_1)	Utilized in some decarboxylation reactions

In addition to chemical nutrients, organisms have physical requirements for growth, including specific conditions of temperature, pH, osmolarity, and pressure.

Physical Requirements

LEARNING OUTCOME

6.6 Explain how extremes of temperature, pH, and osmotic and hydrostatic pressure limit microbial growth.

Temperature

Temperature plays an important role in microbial life through its effects on the three-dimensional configurations of biological molecules. Recall that to function properly, proteins require a specific three-dimensional shape that is determined in part by temperature-sensitive hydrogen bonds, which are more likely to form at lower temperatures and more likely to break at

HIGHLIGHT

Hydrogen-Loving Microbes in Yellowstone's Hot Springs

If you have ever visited the geothermal springs at Yellowstone National Park, you may recall a "rotten-egg" odor caused by sulfur in the environment. Until recently, it was believed that microorganisms living in these springs used sulfur as their primary source of energy. Researchers at the University of Colorado at Boulder, however, have discovered that most of these microorganisms actually seem to live off hydrogen.

The researchers learned that the gene sequences of bacteria collected from the springs closely matched the gene sequences of other bacteria known to metabolize hydrogen. This was a surprise because many people assumed that the bacteria were sulfur dependent. But the results also made sense given the high-temperature environment of the springs. Sulfur-metabolizing microbes require oxygen, which is poorly soluble at high temperatures. Water temperatures in Yellowstone's springs often surpass 70°C, so it is understandable that microbes living in this environment would rely on hydrogen instead of sulfur.

Thermophilic bacteria, which can be distinguished by their orange-colored carotenoids, surround the Grand Prismatic Spring in Yellowstone National Park.

higher temperatures. When hydrogen bonds break, proteins denature and lose function. Additionally, lipids, such as those that are components of the membranes of cells and organelles, are temperature sensitive. If the temperature is too low, membranes become rigid and fragile; if the temperature is too high, the lipids become too fluid, and the membrane cannot contain the cell or organelle.

Because temperature plays an important role in the three-dimensional structure of many types of biological molecules, different temperatures have different effects on the survival and growth of microbes (**FIGURE 6.4**). The lowest temperature at which an organism is able to conduct metabolism is called the *minimum growth temperature.* Note, however, that many microbes, particularly bacteria, survive (although they do not thrive) at temperatures far below this temperature despite the fact that cell membranes are less fluid and transport processes are too slow to support metabolic activity. The highest temperature at which an organism continues to metabolize is called the *maximum growth temperature;* when the temperature exceeds this value, the organism's proteins are permanently denatured, and it dies. The temperature at which an organism's metabolic activities produce the highest growth rate is the **optimum growth temperature.** Each organism thus survives over a *temperature range* within which its growth and metabolism are supported.

Based on their preferred temperature ranges—the temperatures within which their metabolic activity and growth are best supported—microbes can be categorized into four overlapping groups (**FIGURE 6.5**). **Psychrophiles**[6] (sī´krō-fīls) grow best at temperatures below about 15°C and can even continue to grow at temperatures below 0°C. They die at temperatures much above 20°C. In nature, psychrophilic algae, fungi, archaea, and bacteria live in snowfields, ice, and cold water (**FIGURE 6.6**).

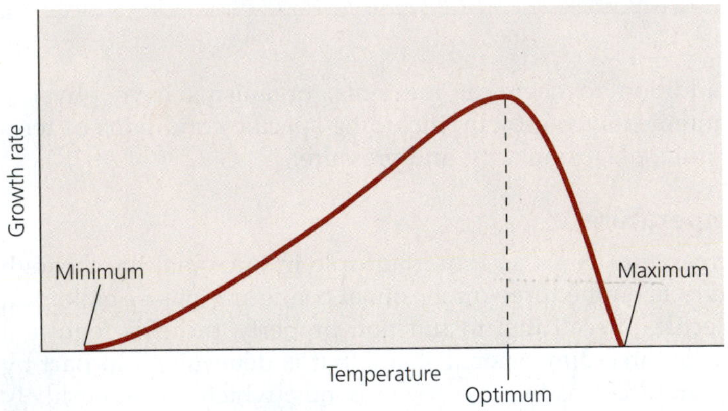

(a)

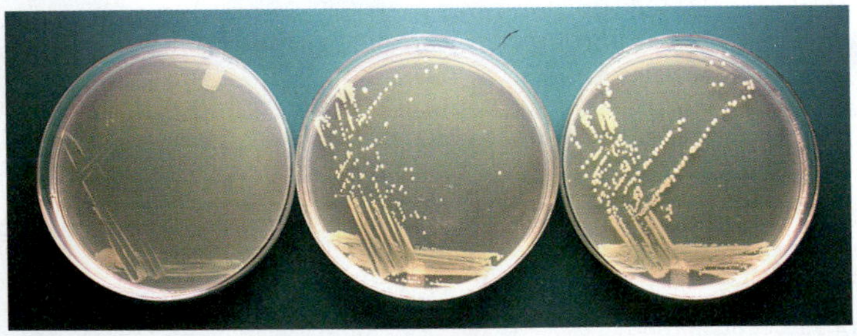

(b)

▲ **FIGURE 6.4 The effects of temperature on microbial growth.** **(a)** Minimum, optimum, and maximum growth temperatures rate plotted against temperature. **(b)** Growth of *Escherichia coli* on nutrient agar after 18 hours of incubation at three different temperatures. *If microorganisms can survive at temperatures lower than their minimum growth temperature, then why is it called "minimum"?*

Figure 6.4 *The minimum growth temperature is defined as the lowest temperature that supports metabolism. Though organisms might survive at lower temperatures, they do not actively metabolize, grow, or reproduce.*

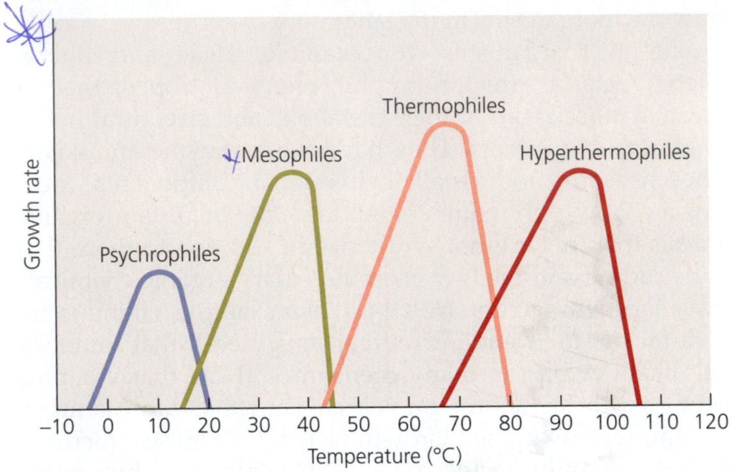

▲ **FIGURE 6.5 Four categories of microbes based on temperature ranges for growth.** *Categorize the bacterium* Vibrio marinus, *which has an optimum growth temperature near 10°C.*

Figure 6.5 Vibrio marinus *is a psychrophile.*

(a)

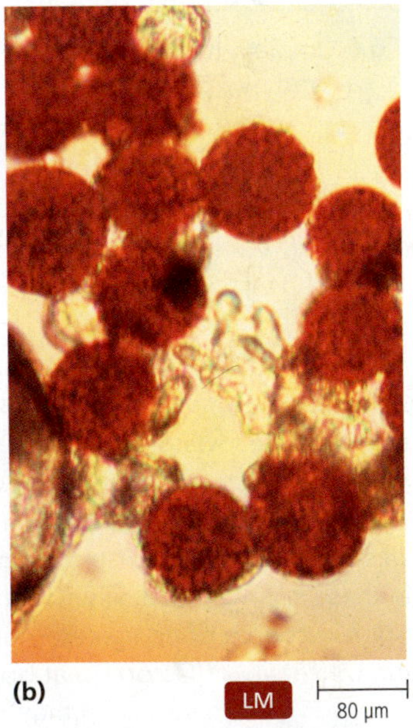

(b)

▲ **FIGURE 6.6 An example of a psychrophile.** **(a)** The alga *Chlamydomonas nivalis* colors this summertime snowbank on Cuverville Island, Antarctica. **(b)** Microscopic view of the red-pigmented spores of *C. nivalis.*

[6]From Greek *psuchros,* meaning "cold," and *philos,* meaning "love."

They do not cause disease in humans because they cannot thrive at body temperature; some do cause food spoilage in refrigerators. Psychrophiles present unique challenges to laboratory investigations because they must be kept cold. For example, microscope stages must be refrigerated, and the air temperatures needed to maintain living psychrophiles are uncomfortably cold for lab personnel.

Mesophiles[7] (mez´ō-fīls) are organisms that grow best in temperatures ranging from 20°C to about 40°C, though they can survive at higher and lower temperatures. Because normal body temperature is approximately 37°C, human pathogens are mesophiles. *Thermoduric*[8] *organisms* are mesophiles that can survive brief periods at higher temperatures. Inadequate heating during pasteurization and canning can result in food spoilage by thermoduric mesophiles.

Thermophiles[9] (ther´mō-fīls) grow at temperatures above 45°C in habitats such as compost piles and hot springs. Some members of the Archaea, called **hyperthermophiles,** grow in water above 80°C; others live at temperatures above 100°C.[10] The current record holder is an archaeon, *Geogemma barossii* (jē´ō-jem-a ba-rōs´ē-ē). *Geogemma* grows and reproduces near submarine hot springs at temperatures between 85°C and 121°C and can survive for at least two hours at 130°C! Thermophiles and hyperthermophiles stabilize their proteins with extra hydrogen and covalent bonds between amino acids. Heat-stable enzymes are useful in industrial, engineering, and research applications. Heat-loving organisms do not cause disease because they "freeze" at body temperature. **Beneficial Microbes: A Nuclear Waste-Eating Microbe?** on p. 172 highlights an unusual thermophile that can also withstand radiation.

pH

Organisms are sensitive to changes in acidity because hydrogen ions and hydroxyl ions interfere with hydrogen bonding within proteins and nucleic acids; as a result, organisms have ranges of acidity that they prefer and can tolerate. pH is a measure of the concentration of hydrogen ions in a solution; that is, it is a measure of the acidity or alkalinity of a substance. A pH below 7.0 is acidic; the lower the pH value, the more acidic a substance is. Alkaline (basic) pH values are higher than 7.0.

Most bacteria and protozoa, including most pathogens, grow best in a narrow range around a neutral pH—that is, between pH 6.5 and pH 7.5, which is also the pH range of most tissues and organs in the human body; such microbes are thus called **neutrophiles** (nū´trō-fīls). By contrast, other bacteria and many fungi are **acidophiles** (ā-sid´ō-phīls), organisms that grow best in acidic habitats. One example of acidophilic microbes are the chemoautotrophic prokaryotes that live in mines and in water that runs through mine tailings (waste rock), habitats that have pHs as low as 0.0. These prokaryotes oxidize sulfur to sulfuric acid, further lowering the pH of their environment. Whereas *obligate acidophiles* require an acidic environment and die if the pH approaches 7.0, *acid-tolerant microbes* merely survive in acid without preferring it.

Many organisms produce acidic waste products that accumulate in their environment until eventually they inhibit further growth. For example, many cheeses are acidic because of lactic acid produced by fermenting bacteria and fungi. The low pH of these cheeses then acts as a preservative by preventing any further microbial growth. Other acidic foods, such as sauerkraut and dill pickles, are also kept from spoiling because most organisms cannot tolerate their low pH.

The normal acidity of certain regions of the body inhibits microbial growth and retards many kinds of infection. At one site, the vaginas of adult women, acidity results from the fermentation of carbohydrates by normal resident bacteria. If the growth of these normal residents is disrupted—for instance, by antibiotic therapy—the resulting higher pH may allow yeasts to grow and lead to a yeast infection. Another site, the stomach, is inhospitable to most microbes because of the normal production of stomach acid. However, the acid-tolerant bacterium *Helicobacter pylori* neutralizes stomach acid by secreting bicarbonate and urease, an enzyme that converts urea to ammonia, which is alkaline. The growth of *Helicobacter* is the cause of most gastric ulcers.

Alkaline conditions also inhibit the growth of most microbes, but **alkalinophiles** live in alkaline soils and water up to pH 11.5. For example, *Vibrio cholerae* (vib´rē-ō kol´er-ī), the causative agent of cholera, grows best outside of the body in water at pH 9.0.

Physical Effects of Water

Microorganisms require water; they must be in a moist environment if they are to be metabolically active. Water is needed to dissolve enzymes and nutrients; also, it is an important reactant in many metabolic reactions. Even though most cells die in the absence of water, some microorganisms—for example, the bacterium *Mycobacterium tuberculosis* (mī´kō-bak-tēr´ē-ŭm too-ber-kyū-lō´sis)—have cell walls that retain water, allowing them to survive for months under dry conditions. Additionally, the spores and cysts of some other single-celled microbes cease most metabolic activity in a dry environment for years; these cells are in essence in a state of suspended animation because they neither grow nor reproduce in their dry condition.

We now consider the physical effects of water on microbes by examining two topics: osmotic pressure and hydrostatic pressure.

Osmotic Pressure *Osmosis* is the diffusion of water across a membrane and is driven by unequal solute concentrations on the two sides of such a membrane. The *osmotic pressure* of a solution is the pressure exerted on a membrane by a solution containing solutes (dissolved material) that cannot freely cross the membrane. Osmotic pressure is related to the concentration of dissolved molecules and ions in a solution. Solutions with greater concentrations of such solutes are *hypertonic* relative to those with a lower solute concentration, which are *hypotonic*.

[7]From Greek *mesos,* meaning "middle," and *philos,* meaning "love."

[8]From Greek *therme,* meaning "hot." The organisms are so named because of their ability to endure or tolerate heat.

[9]From Greek *therme,* meaning "hot," and *philos,* meaning "love."

[10]Water can remain a liquid above 100°C if it has a high salt content or is under pressure, such as occurs in geysers or deep ocean troughs.

BENEFICIAL MICROBES

A Nuclear Waste-Eating Microbe?

Gamma rays emitted by radioactive decay are usually deadly. However, some fungi and bacteria can survive being bombarded with radiation at levels higher than thousands of times what would kill a person. Currently, researchers are studying these organisms in the hope that we can use their biochemistry to develop ways to protect our cells from radiation and use the microbes to clean up nuclear wastes.

Microbes that can survive in extremely hostile environments are called *extremophiles.* Scientists have discovered amazing species of fungi in the genera *Wangiella* and *Cladosporium* inhabiting the walls of the damaged nuclear reactor in Chernobyl, Ukraine, which is considered the site of the worst nuclear accident in history. The fungi absorb radiation with the help of a pigment called *melanin,* the same pigment that colors human skin and hair. Incredibly, the fungi harness radioactive energy so as to grow faster—the first time we have discovered organisms that can use an energy source other than light or chemicals. Since radiation is available in space, perhaps future astronauts will be able to grow fungi as a continual food supply for long space voyages.

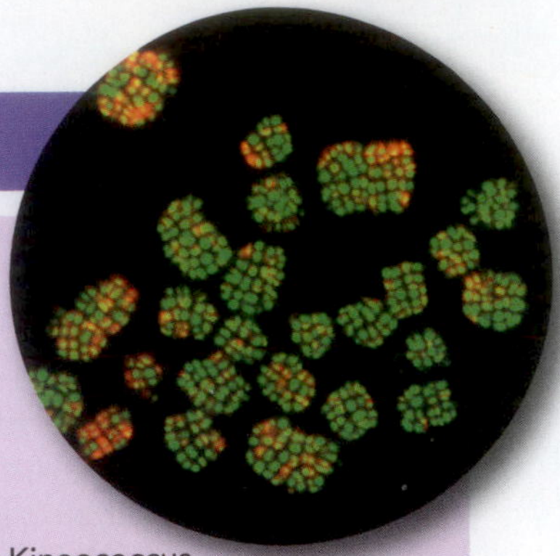

Kineococcus radiotolerans

The bacterium *Kineococcus radiotolerans* is notable among radiation-tolerant extremophiles because it can also break down herbicides, chlorinated compounds, and other toxic substances. Researchers at the U.S. Department of Energy would like to shape *K. radiotolerans* into a biological tool that can clean up environments contaminated with radioactive wastes. Using microbes to break down toxic chemicals in the environment, a process known as bioremediation, is often cheaper, quicker, and more effective than conventional methods. *K. radiotolerans* could potentially slash the cost of nuclear cleanup.

Osmotic pressure can have dire effects on cells. For example, a cell placed in freshwater (a hypotonic solution relative to the cell's cytoplasm) gains water from its environment and swells to the limit of its cell wall. Cells that lack a cell wall—animal cells and some bacterial, fungal, and protozoan cells—will swell until they burst in hypotonic solutions. By contrast, a cell placed in seawater, which is a solution containing about 3.5% solutes and thus hypertonic to most cells, loses water into the surrounding salt water. Such a cell can die from **crenation,** or shriveling of its cytoplasm. Osmotic pressure accounts for the preserving action of salt in jerky and salted fish and of sugar in jellies, preserves, and honey. In those foods, the salt and sugar are solutes that draw water out of any microbial cells that are present, preventing growth and reproduction.

Osmotic pressure restricts organisms to certain environments. Some microbes, called **obligate halophiles,**[11] are adapted to growth under high osmotic pressure such as exists in the Great Salt Lake and smaller salt ponds. They may grow in up to 30% salt and will burst if placed in freshwater. Other microbes are *facultative halophiles;* that is, although they do not require high salt concentrations, they can tolerate them. One potential bacterial pathogen, *Staphylococcus aureus* (staf´i-lō-kok´ŭs o´rē-ŭs), can tolerate up to 20% salt, which allows it to colonize the surface of the skin—an environment that is too salty for most microbes. *S. aureus* causes a number of different skin and mucous membrane diseases ranging from pimples, sties, and boils to life-threatening scalded skin and toxic shock syndromes. (These diseases are covered more fully in Chapters 19 and 24.)

Hydrostatic Pressure Water exerts pressure in proportion to its depth. For every additional 10 m of depth, water pressure increases 1 atmosphere (atm). Therefore, the pressure at 100 m below the surface is 10 atm—10 times greater than at the surface. Obviously, the pressure in deep ocean basins and trenches, which are thousands of meters below the surface, is tremendous. Organisms that live under such extreme pressure are called **barophiles**[12] (bar´ō-fīls). Their membranes and enzymes do not merely tolerate pressure; they also depend on pressure to maintain their three-dimensional, functional shapes. Thus, barophiles brought to the surface quickly die because their proteins denature. Obviously, barophiles cannot cause diseases in humans, plants, or animals that do not live at great depths.

Associations and Biofilms

LEARNING OUTCOME

6.7 Describe how quorum sensing can lead to formation of a biofilm.

Organisms in solitary culture in a laboratory are living very differently from organisms in nature, which live in association with other individuals of their own and different species. The relationships between organisms can be viewed as falling along a continuum stretching from causing harm to providing benefits.

Relationships in which a microbe harms or even kills another organism are considered *antagonistic relationships.* Viruses are especially clear examples of antagonistic microbes; they require cells in which to replicate themselves and almost always kill their cellular hosts.

[11]From Greek *halos,* meaning "salt," and *philos,* meaning "love."
[12]From Greek *baros,* meaning "weight," and *philos,* meaning "love."

Beneficial relationships take at least two forms: synergistic relationships and symbiotic relationships. In *synergistic relationships,* the individual members of an association cooperate such that each receives benefits that exceed those that would result if each lived by itself, even though each member could live separately. In *symbiotic relationships,* organisms live in such close nutritional or physical contact that they become interdependent such that the members rarely (if ever) live outside the relationship. (Symbiotic relationships are discussed in greater detail in Chapter 14, particularly as they relate to the production of disease.)

Biofilms are examples of complex relationships among numerous microorganisms, often different species, attached to surfaces such as teeth (dental plaque), rocks in streams, shower curtains ("soap scum" is really a biofilm), implanted medical devices (e.g., catheters), and mucous membranes of the digestive system. Biofilms are the primary residence of microorganisms in nature. For example, one study showed that more than 10 billion bacteria per square centimeter form the slippery biofilm on rocks in a streambed. The Centers for Disease Control and Prevention estimates that biofilms cause up to 70% of bacterial diseases in industrialized countries, including kidney infections, tooth and gum decay, infections that occur with cystic fibrosis, and health care-associated infections, such as might occur with implantation of medical devices. Cells within biofilms communicate and coordinate with one another, acting similar to a tissue in a multicellular organism.

Biofilm development can involve at least six steps (**FIGURE 6.7**). Free-living cells 1 settle on a surface and attach 2. They develop a gooey, extracellular *matrix,* composed of DNA, proteins, and primarily the tangled fibers of polysaccharides of the cells' glycocalyces 3. This slimy matrix adheres cells to one another, sticks the cells to their substrate, forms microenvironments within the biofilm, sequesters nutrients, forms water channels between groups of cells, and may protect individuals in a biofilm from environmental stresses, including ultraviolet radiation, antimicrobial drugs, and changes in pH, temperature, and humidity.

Biofilms form as a result of a process called **quorum sensing,** in which microorganisms respond to the density of nearby microorganisms. The microbes secrete *quorum-sensing molecules* that act to communicate number and types of cells among members of the biofilm 4. Many cells possess receptors for these signal molecules. When the density of microorganisms increases, the concentration of quorum-sensing molecules also increases such that more and more receptors bind these molecules. Once the binding exceeds a certain threshold amount, the expression of previously suppressed genes is triggered, and the result is that the microorganisms have new characteristics, such as the production of enzymes, changes in cell shape, the formation of mating types, and the ability to form and maintain biofilms. Some researchers estimate quorum sensing may regulate 10% of the genes in a cell.

A matrix not only attaches a biofilm and positions the cells, but also may allow members of the biofilm to concentrate and conserve digestive enzymes, directing them against the underlying structure rather than having them diffuse away in the surrounding medium. New microbes can arrive and the synergistic relationships allowed in a biofilm continue to organize the biofilm community, so that individual members display metabolic and structural traits different from those expressed by the same

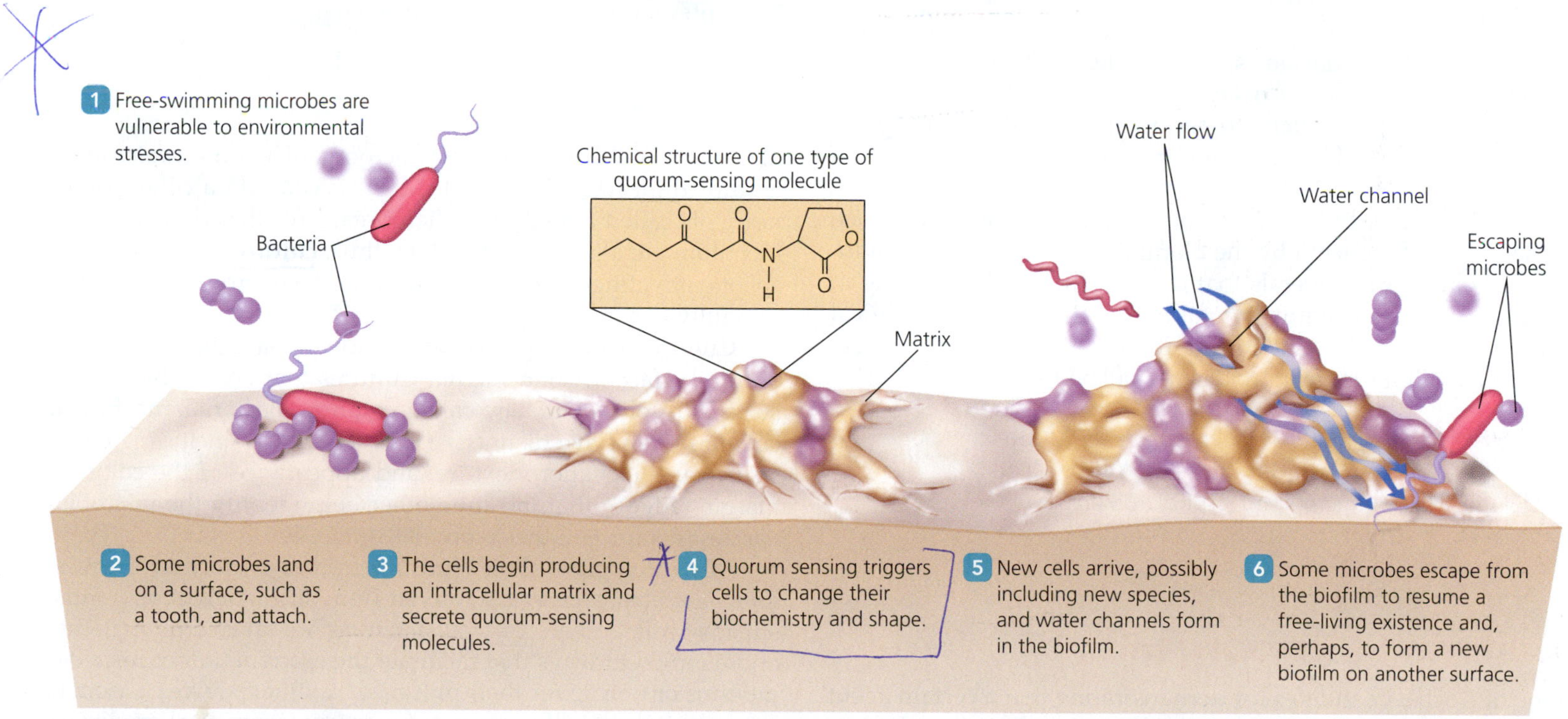
1 Free-swimming microbes are vulnerable to environmental stresses.
Bacteria
Chemical structure of one type of quorum-sensing molecule
Matrix
Water flow
Water channel
Escaping microbes
2 Some microbes land on a surface, such as a tooth, and attach.
3 The cells begin producing an intracellular matrix and secrete quorum-sensing molecules.
4 Quorum sensing triggers cells to change their biochemistry and shape.
5 New cells arrive, possibly including new species, and water channels form in the biofilm.
6 Some microbes escape from the biofilm to resume a free-living existence and, perhaps, to form a new biofilm on another surface.

▲ **FIGURE 6.7 Biofilm development.** Quorum sensing allows microbes to change their behavior in the presence of other microbes to form microbial communities called biofilms.

cells living individually 5. Members assume different roles in different areas of a biofilm, much like cells and tissues of multicellular organisms have different functions in different parts of the body. Some individual cells or groups of cells in *streamers* may leave the biofilm 6.

Given that many microorganisms become more harmful when they are part of a biofilm, scientists are seeking ways to prevent biofilms from forming in the first place. Researchers have learned that drugs that block microbial cell receptors, thereby disrupting communication among cells, can inhibit biofilm formation and thus prevent disease. Such receptor-blocking drugs have successfully blocked biofilm formation and prevented disease in mice and are being considered for use in humans.

Another possible approach to preventing biofilm formation involves artificially amplifying quorum sensing while bacteria are still relatively few in number. Some pathogens hide within capsules and in blood clots, and only after they have multiplied significantly and formed a biofilm do they emerge and become "visible" to the immune system. With amplified quorum sensing, bacteria might produce certain proteins earlier in an infection cycle than normal, thus "revealing" themselves sooner to the immune system, which may then eliminate them before they can cause disease.

Dental plaque is a common biofilm that can lead to dental caries—cavities. Plaque formation usually begins with colonization of the teeth by *Streptococcus mutans* (strep-tō-kok´ŭs mū´ tanz). This bacterium breaks down carbohydrates, particularly the disaccharide sucrose (table sugar), to provide itself with nutrition and a glycocalyx. One of its enzymes catabolizes sucrose into its component monosaccharides—glucose and fructose—which the cell uses as energy sources. A second enzyme releases fructose as an energy source but polymerizes glucose into long, insoluble polysaccharide strands called glucan molecules, which form a sticky glycocalyx matrix around the bacterium. Glucan adheres *S. mutans* to the tooth, provides a home for other species of oral bacteria, and traps food particles. A biofilm has formed.

Bacteria in the biofilm digest nutrients and release acid, which is held against the teeth by the biofilm's matrix. The acid gradually eats away the minerals that compose the tooth, resulting in dental caries and eventually total loss of the teeth. The **Clinical Case Study: Cavities Gone Wild** deals with an especially severe case of a biofilm running amok in a small boy's mouth.

CLINICAL CASE STUDY

Cavities Gone Wild

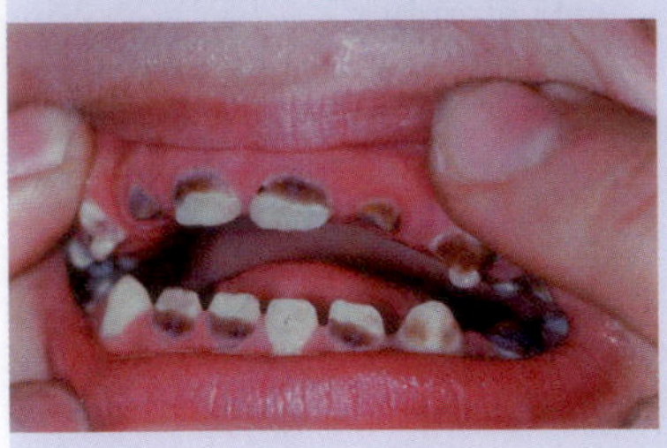

Five-year-old Daniel appears to be shy. He always looks at the floor, has no friends, never plays with the other children, and will rarely speak to adults, and when he does speak, it is difficult to understand his broken enunciation. His skinny frame and the dark circles under his eyes make him appear malnourished. Daniel cries frequently and misses many days of school.

A speech specialist at school finds that only two of Daniel's teeth are healthy; all the others have rotted away to the gum line. The little guy is in constant pain and it hurts to chew. A doctor later determines that bacteria from the cavities in his mouth have entered his bloodstream and infected his heart, causing an irregular heartbeat and poor blood circulation.

1. How does knowledge of biofilms help explain the bulk of Daniel's problems?
2. What can Daniel, his parents, and health care professionals do to cure his diseases?
3. What nutrient should Daniel's parents eliminate from his diet to help prevent a repeat of his condition?
4. A recent study has shown that brushing does more than clean plaque from the teeth; it also disrupts associations between oral bacteria. How does this simple act help prevent the formation of biofilms?

TELL ME WHY

Why should cardiac nurses and respiratory therapists care about biofilms?

Culturing Microorganisms

One of Koch's postulates for demonstrating that a certain agent causes a specific disease requires that microorganisms be isolated and cultivated (Chapter 1). Medical laboratory personnel must also grow pathogens as a step in the diagnosis of many diseases. To cultivate or *culture* microorganisms, a sample called an **inoculum** (plural: *inocula*) is introduced into a collection of nutrients called a **medium.** Microorganisms that grow from an inoculum are also called a *culture;* thus, **culture** can refer to the act of cultivating microorganisms or to the microorganisms that are cultivated.

Cultures can be grown in liquid media called **broths** or on the surface of solid media. Cultures that are visible on the surface of solid media are called **colonies.** Bacterial and fungal colonies often have distinctive characteristics—including color, size, shape, elevation, texture, and appearance of the colony's margin (edge)—that taken together help identify the microbial species that formed the colony **(FIGURE 6.8)**.

Microbiologists obtain inocula from a variety of sources. *Environmental specimens* are taken from such sources as ponds, streams, soil, and air. *Clinical specimens* are taken from patients and handled in ways that facilitate the examination of microorganisms or testing for their presence. Another source of inocula is a culture originally grown from an environmental or clinical specimen and maintained in storage in a laboratory. Next we briefly examine clinical sampling.

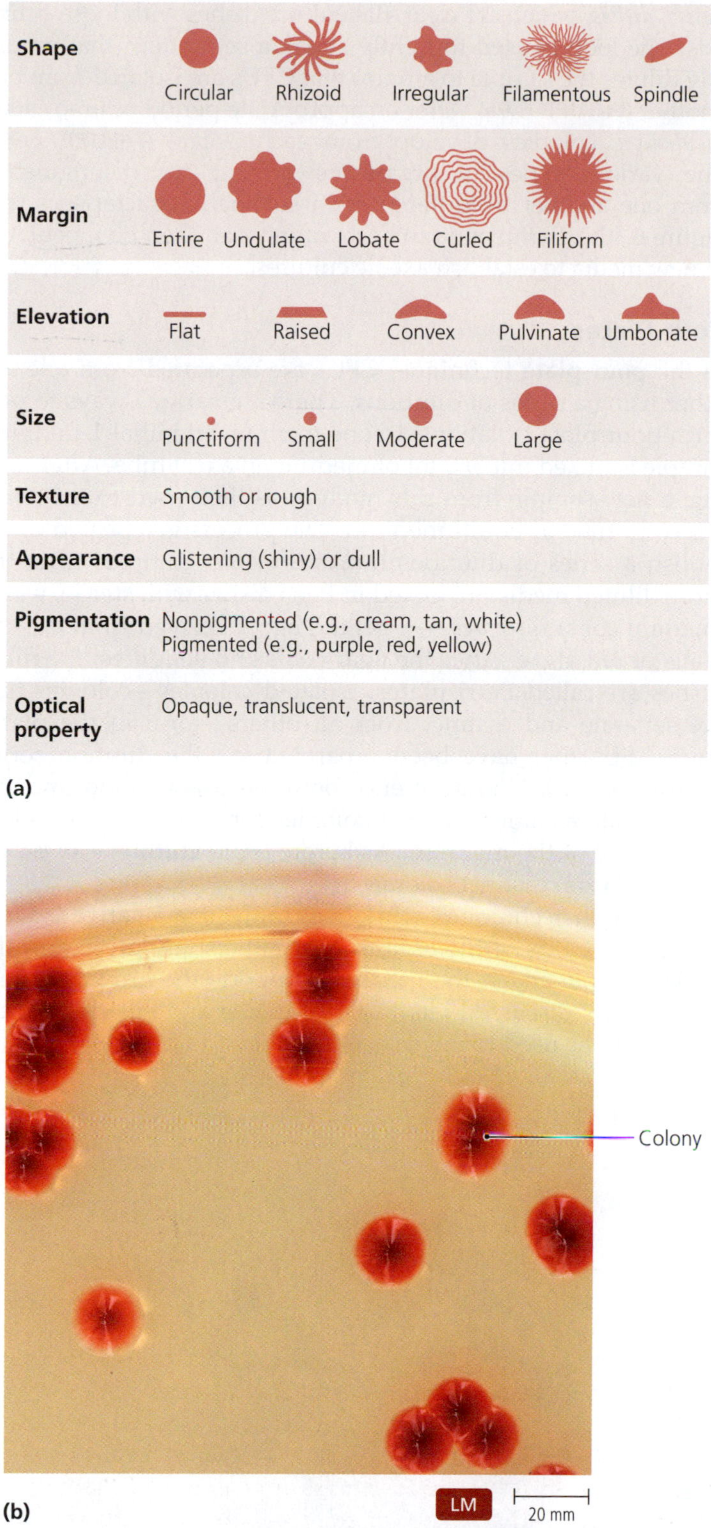

▲ **FIGURE 6.8 Characteristics of bacterial colonies.** **(a)** Shape, margin, elevation (side view), size, texture, appearance, pigmentation (color), and optical properties are described by a variety of terms. **(b)** *Serratia marcescens* growing on an agar surface. These colonies are circular, entire, convex, large, smooth, shiny, red, and opaque.

Clinical Sampling

LEARNING **OUTCOME**

6.8 Describe methods for collecting clinical specimens from the skin and from the respiratory, reproductive, and urinary tracts.

Diagnosing and treating disease often depend on isolating and correctly identifying pathogens. Health care professionals must properly obtain samples from their patients and must then transport them quickly and correctly to a microbiology laboratory for culture and identification. They must take care to prevent contaminating samples with microorganisms from the environment or other regions of the patient's body, and they must prevent infecting themselves with pathogens while sampling. In this regard, the Centers for Disease Control and Prevention has established a set of guidelines, called *standard precautions,* to protect health care professionals from contamination by pathogens.

In clinical microbiology, a **clinical specimen** is a sample of human material, such as feces, saliva, cerebrospinal fluid, or blood, that is examined or tested for the presence of microorganisms. As summarized in **TABLE 6.2**, health care professionals collect clinical specimens using a variety of techniques and equipment. Specimens must be properly labeled and promptly transported to a microbiological laboratory to avoid death of the pathogens and to minimize the growth of microbes normally found at the collection site. Clinical specimens are often transported in special *transport media* that are chemically formulated to maintain the relative abundance of different microbial species or to maintain an anaerobic environment.

TABLE 6.2 Clinical Specimens and the Methods Used to Collect Them

Type or Location of Specimen	Collection Method
Skin, accessible membrane (including eye, outer ear, nose, throat, vagina, cervix, urethra) or open wounds	Sterile swab brushed across the surface; care should be taken not to contact neighboring tissues
Blood	Needle aspiration from vein; anticoagulants are included in the specimen transfer tube
Cerebrospinal fluid	Needle aspiration from subarachnoid space of spinal column
Stomach	Intubation, which involves inserting a tube into the stomach, often via a nostril
Urine	In aseptic collection, a catheter is inserted into the bladder through the urethra; in the "clean catch" method, initial urination washes the urethra, and the specimen is midstream urine
Lungs	Collection of sputum either dislodged by coughing or acquired via a catheter
Diseased tissue	Surgical removal (biopsy)

Obtaining Pure Cultures

LEARNING | OUTCOME

6.9 Describe the two most common methods by which microorganisms can be isolated for culture.

Clinical specimens are collected in order to identify a suspected pathogen, but they also contain *normal microbiota,* which are microorganisms associated with a certain area of the body without causing diseases (see Table 14.2 on p. 417). As a result, the suspected pathogen in a specimen must be isolated from the normal microbiota in culture. Scientists use several techniques to isolate organisms in **pure cultures,** that is, cultures composed of cells arising from a single progenitor. The word *axenic*[13] (ā-zen´ik) is also used to refer to a pure culture. The progenitor from which a particular pure culture is derived may be either a single cell or a group of related cells; therefore, the progenitor is termed a **colony-forming unit (CFU).**

In all microbiological procedures, care must be taken to reduce the chance of contamination, which occurs, for example, when instruments or air currents carry foreign microbes into culture vessels. All media, vessels, and instruments must be **sterile**—that is, free of any microbial contaminants. Sterilization and *aseptic techniques* are designed to limit contamination (discussed in Chapter 9). Now we examine two common isolation techniques: streak plates and pour plates. We consider another method, *serial dilution,* in a later section.

Streak Plates

The most commonly used isolation technique in microbiological laboratories is the **streak-plate** method. In this technique, a sterile inoculating loop (or sometimes a needle) is used to spread an inoculum across the surface of a solid medium in *Petri*[14] *dishes,* which are clear, flat culture dishes with loose-fitting lids. The loop is used to lightly streak a set pattern that gradually dilutes the sample to a point that CFUs are isolated from one another **(FIGURE 6.9a)**. After an appropriate period of time called *incubation,* colonies develop from each isolate **(FIGURE 6.9b)**. The various types of organisms present are distinguished from one another by differences in colonial characteristics (see Figure 6.10b). Samples from each variety can then be inoculated in new media to establish axenic cultures.

Pour Plates

In the **pour-plate** technique, CFUs are separated from one another using a series of dilutions. There are various ways to perform pour-plate isolations. In one method, an initial 1-milliliter sample is mixed into 9.0 ml of medium in a test tube. After mixing, a new sample from this medium is then used to inoculate a second tube of liquid medium. The process is repeated to establish a series of dilutions **(FIGURE 6.10a)**. Samples from the more diluted media are mixed in Petri dishes with sterile, warm medium containing *agar*—a gelling agent derived from the cell walls of red algae. After the agar cools and solidifies, the filled dishes are called **Petri plates.** Isolated colonies—colonies that are separate and distinct from all others—form in the plates from CFUs that have been separated via the dilution series **(FIGURE 6.10b)**. One difference between this method and the streak-plate technique is that colonies form both at and below the surface of the medium. As before, pure cultures can be established from distinct colonies.

Isolation techniques work well only if a relatively large number of CFUs of the organism of interest are present in the initial sample and if the medium supports the growth of that microbe. As discussed later, special media and enriching techniques can be used to increase the likelihood of success.

[13]From Greek *a,* meaning "no," and *xenos,* meaning "stranger."

[14]Named for Richard Petri, Robert Koch's assistant, who invented them in 1887.

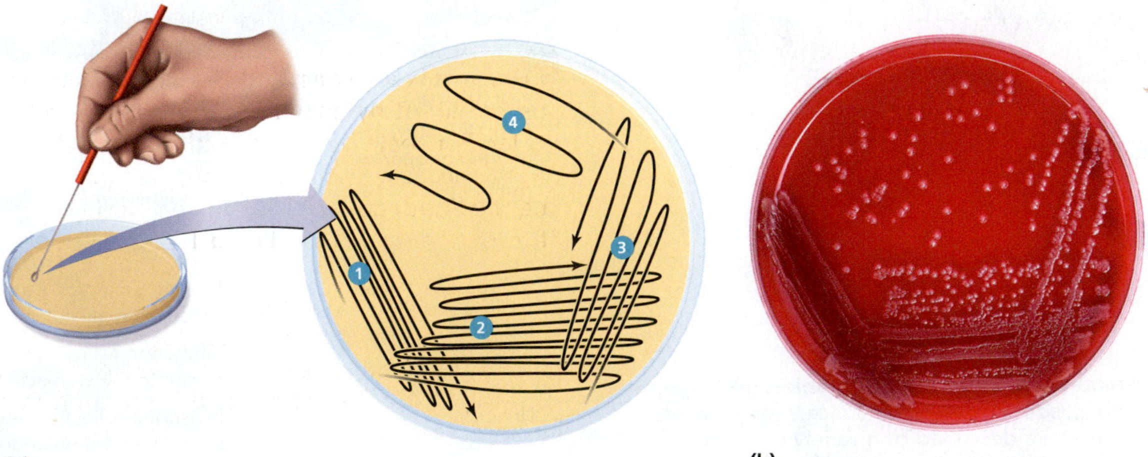

(a) (b)

▲ **FIGURE 6.9 The streak-plate method of isolation. (a)** An inoculum is spread across the surface of an agar plate in a sequential pattern of streaks (as indicated by the numbers and arrows). The loop is sterilized between streaks. In streaks 2, 3, and 4, bacteria are picked up from a previous streak; thus the number of cells is diluted each time. **(b)** A streak plate showing colonies of *Escherichia coli* on blood agar.

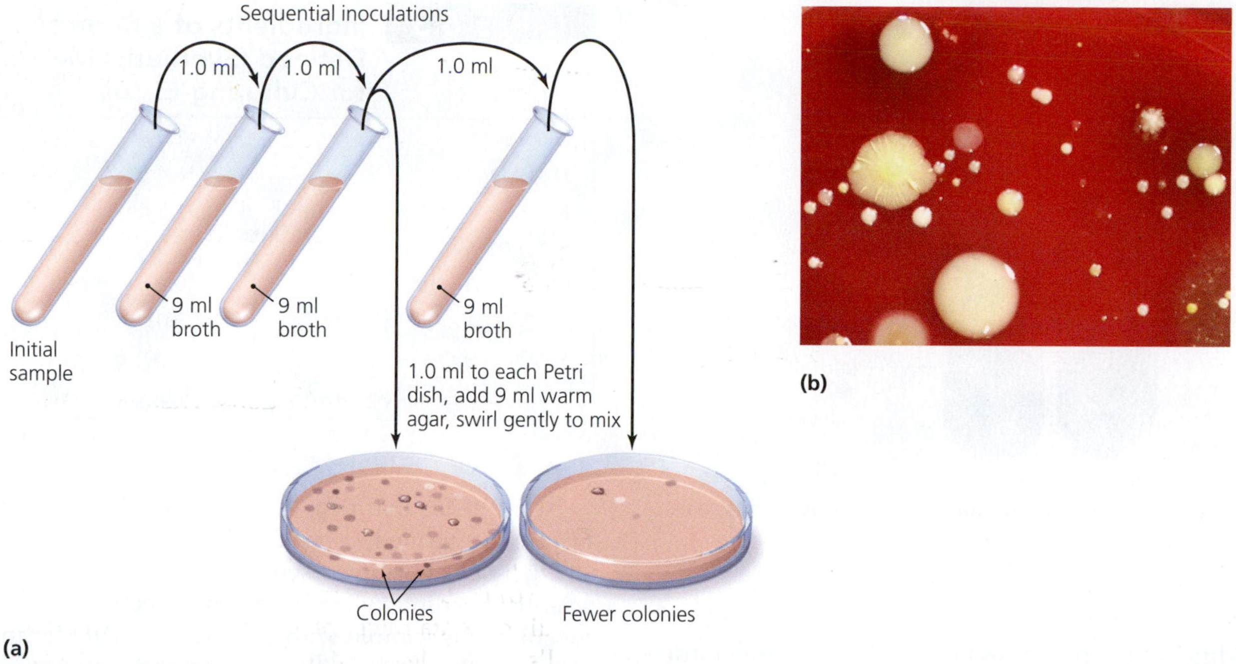

▲ **FIGURE 6.10 The pour-plate method of isolation. (a)** After an initial sample is diluted through a series of transfers, the final dilutions are mixed with warm agar in Petri plates. Individual CFUs form colonies in and on the agar. **(b)** A portion of a plate showing the results of isolation.

Other Isolation Techniques

Streak plates and pour plates are used primarily to establish pure cultures of bacteria, but they can also be used for some fungi, particularly yeasts. Protozoa and motile unicellular algae are not usually cultured on solid media because they do not remain in one location to form colonies. Instead, they are isolated through a series of dilutions but remain in broth culture media. In cases of fairly large microorganisms, such as the protists *Euglena* (yū-glēn´ă) and *Amoeba* (am-ē´bă), hollow tubes with small diameters called *micropipettes* can be used to pick up a single cell that is then used to establish a culture.

Culture Media

LEARNING | OUTCOMES

6.10 Describe six types of general culture media available for bacterial culture.

6.11 Describe enrichment culture as a means of enhancing the growth of less abundant microbes.

Culturing microorganisms can be an exacting science. Although some microbes, such as *E. coli*, are not particular about their nutritional needs and can be grown in a variety of media, bacteria, such as *Neisseria gonorrhoeae* (nī-se´rē-ă go-nor-rē´ī) and *Haemophilus influenzae* (hē-mof´i-lŭs in-flu-en´zī), and many archaea require specific nutrients, including specific growth factors. The majority of most microorganisms have never been successfully grown in any culture medium in part because scientists have concentrated their efforts on culturing commercially important species and pathogens. However, some pathogens, such as the syphilis bacterium *Treponema pallidum* (trep-ō-nē´mă pal´li-dŭm), have never been cultured in any laboratory medium despite over a century of effort.

A variety of media are available for microbiological cultures, and more are developed each year to support the needs of food, water, industrial, and clinical microbiologists. Most media are available from commercial sources and come in powdered forms that require only the addition of water to make broths. A common medium, for example, is *nutrient broth,* which contains powdered beef extract and peptones (short chains of amino acids produced by enzymatic digestion of protein) dissolved in water. For some purposes broths are adequate, but if solid media are needed, dissolving about 1.5% agar into hot broth, pouring the liquid mixture into an appropriate vessel, and allowing it to cool provides a solid surface to support colonial growth. Media made solid by the addition of agar to a broth have the word *agar* in their names; thus, *nutrient agar* is nutrient broth to which 1.5% agar has been added.

Agar, a complex polysaccharide derived from the cell walls of certain red algae, is a useful compound in microbiology for several reasons:

- Most microbes cannot digest agar; therefore, agar media remain solid even when bacteria and fungi are growing on them.
- Powdered agar dissolves in water at 100°C, a temperature at which most nutrients remain undamaged.

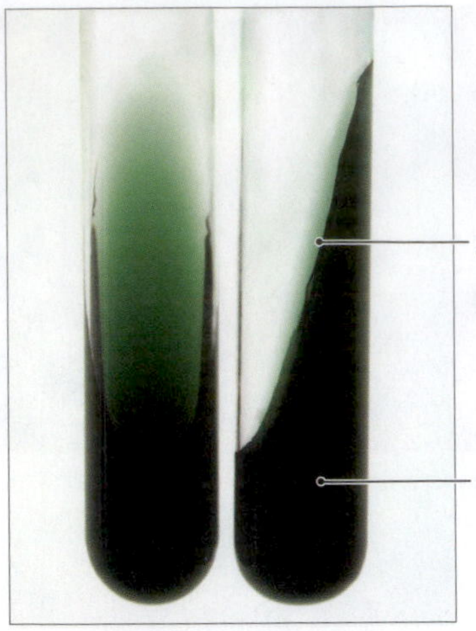

▲ **FIGURE 6.11 Slant tubes containing solid media.** In this case, citrate agar is the medium.

- Agar solidifies at temperatures below 40°C, so temperature-sensitive, sterile nutrients such as vitamins and blood can be added without detriment to cooling agar before it solidifies. Further, cooling liquid agar can be poured over most bacterial cells without harming them. The latter technique plays a role in the pour-plate isolation technique.
- Solid agar does not melt below 100°C; thus, it can even be used to culture some hyperthermophiles.

Still-warm liquid agar media can be poured into Petri dishes to make Petri plates. When warm agar media are poured into test tubes that are then placed at an angle and left to cool until the agar solidifies, the result is **slant tubes,** or **slants** (FIGURE 6.11). The slanted surface provides a larger surface area for aerobic microbial growth while the butt of the tube remains almost anaerobic.

Next we examine six types of general culture media: defined media, complex media, selective media (including enrichment culture), differential media, anaerobic media, and transport media. It is important to note that these types of media are not mutually exclusive categories; that is, in some cases a given medium can belong to more than one category.

Defined Media

If microorganisms are to grow and multiply in culture, the medium must provide essential nutrients (including an appropriate energy source for chemotrophs), water, an appropriate oxygen level, and the required physical conditions (such as the correct pH and suitable osmotic pressure and temperature). A **defined medium** (also called a **synthetic medium**) is one in which the exact chemical composition is known. TABLE 6.3 gives a recipe for one example. Relatively simple defined media containing inorganic salts and a source of CO_2 (such as sodium bicarbonate) are available for autotrophs, particularly cyanobacteria and algae. Chemoheterotrophs require organic molecules, such as glucose, amino acids, and vitamins, which supply carbon and energy or are vital growth factors.

TABLE 6.3 Ingredients of a Representative Defined (Synthetic) Medium for Culturing *E. coli*

Glucose	1.0 g
Na_2HPO_4	16.4 g
KH_2PO_4	1.5 g
$(NH_4)_3PO_4$	2.0 g
$MgSO_4 \cdot 7H_2O$	0.2 g
$CaCl_2$	0.01 g
$FeSO_4 \cdot 7H_2O$	0.005 g
Distilled or deionized water	Enough to bring volume to 1 L

Organisms that require a relatively large number of growth factors are termed *fastidious.* Such organisms may be used as living assays for the presence of growth factors. For example, a scientist needing to know if a sample contains vitamin B_{12} could inoculate the sample with *Euglena granulata* (gran-yū-lă´tă), an organism that requires the vitamin. If the microbe grows, the vitamin is present in the sample. The amount of growth provides an estimate of the amount of vitamin present, scant growth indicating a small amount.

Complex Media

For most clinical cultures, defined media are unnecessarily troublesome to prepare. Most chemoheterotrophs, including pathogens, are routinely grown on **complex media** that contain nutrients released by the partial digestion of yeast, beef, soy, or proteins, such as casein from milk. The exact chemical composition of a complex medium is unknown because partial digestion releases many different chemicals in a variety of concentrations each time it takes place.

Complex media have advantages over defined media. Because a complex medium contains a variety of nutrients, including growth factors, it can support a wider variety of different microorganisms. Complex media are also used to culture organisms whose exact nutritional needs are unknown. Nutrient broth, Trypticase soy agar, and MacConkey agar are some common complex media. Blood is often added to complex media to provide additional growth factors, such as NADH and heme. Such a fortified medium is said to be *enriched* and can support the growth of many fastidious microorganisms.

Selective Media

Selective media typically contain substances that either favor the growth of particular microorganisms or inhibit the growth of unwanted ones. Eosin, methylene blue, and crystal violet dyes as well as bile salts are included in media to inhibit the growth of Gram-positive bacteria without adversely affecting Gram negatives. A high concentration of NaCl (table salt) in a medium selects for halophiles and for salt-tolerant bacteria, such as the pathogen *Staphylococcus aureus*. Sabouraud dextrose agar has a slightly low pH, which by inhibiting the growth of bacteria is selective for fungi (FIGURE 6.12).

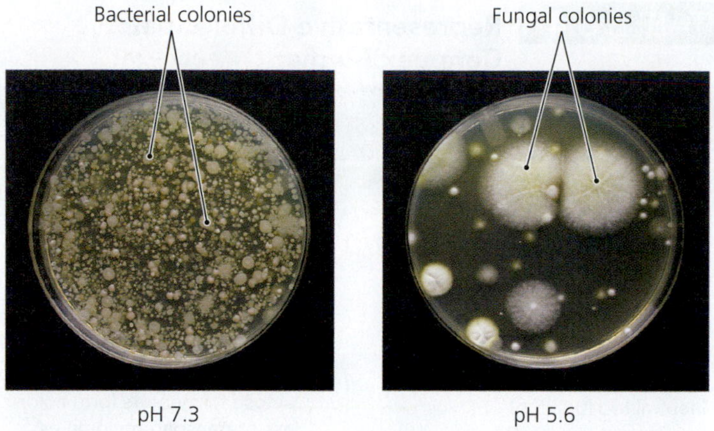

▲ **FIGURE 6.12 An example of the use of a selective medium.** After the medium is inoculated with a diluted soil sample, acidic pH in Sabouraud dextrose agar (right) makes the medium selective for fungi by inhibiting the growth of bacteria. At left for comparison is a nutrient agar plate inoculated with an identical sample.

A medium can also become a selective medium when a single crucial nutrient is left out of it. For example, leaving glucose out of Trypticase soy agar makes the resulting medium selective for organisms that can meet all their carbon requirements by catabolizing amino acids.

Enrichment Culture Bacteria that are present in small numbers may be overlooked on a streak plate or overwhelmed by faster-growing, more abundant strains. This is especially true of organisms in soil and fecal samples that contain a wide variety of microbial species. To isolate potentially important microbes that might otherwise be overlooked, microbiologists enhance the growth of less abundant organisms by a variety of techniques.

In the late 1800s, the Dutch microbiologist Martinus Beijerinck (1851–1931) introduced the most common of these methods, called simply **enrichment culture.** Enrichment cultures use a selective medium and are designed to increase very small numbers of a chosen microbe to observable levels. For example, suppose a microbiologist specializing in environmental cleanup wanted to isolate an organism capable of digesting crude oil to have on hand should it be required to clean an oil-soaked beach. Even though a sample of the beach sand might contain a few such organisms, it would also likely contain many millions of unwanted common bacteria. To isolate oil-utilizing microbes, the scientist would inoculate a sample of the sand into a tube of selective medium containing oil as the sole carbon source and then incubate it. Then a small amount of the culture would be transferred into a new tube of the same medium to be incubated again. After a series of such enrichment transfers, any remaining bacteria will be oil-utilizing organisms. Different species could be isolated by either streak-plate or pour-plate methods.

Cold enrichment is another technique used to enrich a culture with cold-tolerant species, such as *Vibrio cholerae,* the bacterium that causes cholera. Stool specimens or water samples suspected of containing the bacterium are incubated in a refrigerator instead of at 37°C. Cold enrichment works because *Vibrio* cells are much less sensitive to cold than are more common fecal bacteria, such as *E. coli;* therefore, *Vibrio* continues to grow in the cold while the other species are inhibited. The result of cold enrichment is a culture with a greater percentage of *Vibrio* cells than the original sample; the *Vibrio* cells can then be isolated by other methods.

Differential Media

Differential media are formulated such that either the presence of visible changes in the medium or differences in the appearance of colonies help microbiologists differentiate among the kinds of bacteria growing on the medium. Such media take advantage of the fact that different bacteria utilize the ingredients of any given medium in different ways. One example of the use of a differential medium involves the differences in organisms' utilization of red blood cells in blood agar **(FIGURE 6.13)**. *Streptococcus pneumoniae* (nū-mō´nē-ī) partially digests (lyses) red blood cells, producing around its colonies a greenish-brown discoloration denoted *alpha-hemolysis.* By contrast, *Streptococcus pyogenes* (pī-oj´en-ēz) completely digests red blood cells, producing around its colonies clear zones termed *beta-hemolysis. Enterococcus faecalis* (en´ter-ō-kok´ŭs fē-kă´lis) does not digest red blood cells, so the agar appears unchanged, a reaction called *gamma hemolysis* even though no lysis occurs. In some differential media, such as carbohydrate utilization broth tubes, a pH-sensitive dye changes color when bacteria metabolizing sugars produce acid waste products **(FIGURE 6.14)**. Some common differential complex media are described in **TABLE 6.4** on p. 180.

Many media are both selective and differential; that is, they enhance the growth of certain species that can then be

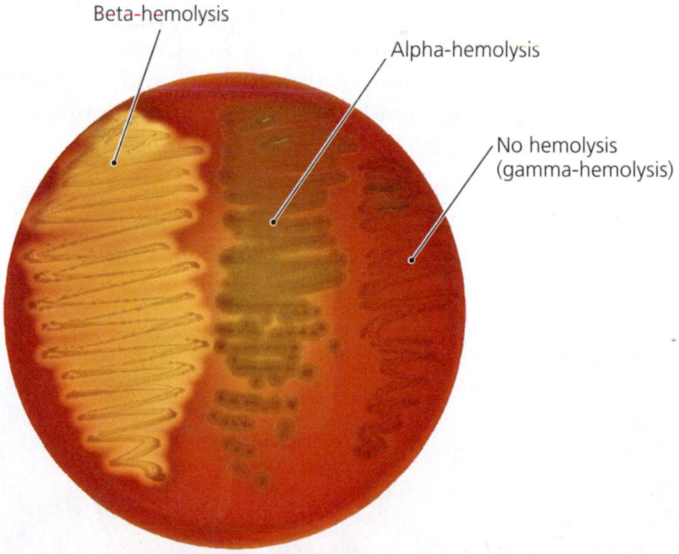

▲ **FIGURE 6.13 The use of blood agar as a differential medium.** *Streptococcus pyogenes* (left) completely uses red blood cells, producing a clear zone termed *beta-hemolysis. Streptococcus pneumoniae* (middle) partially uses red blood cells, producing a discoloration termed *alpha-hemolysis. Enterococcus faecalis* (right) does not use red blood cells; the lack of any change in the medium around colonies is termed *gamma-hemolysis*, even though no red cells are hemolyzed.

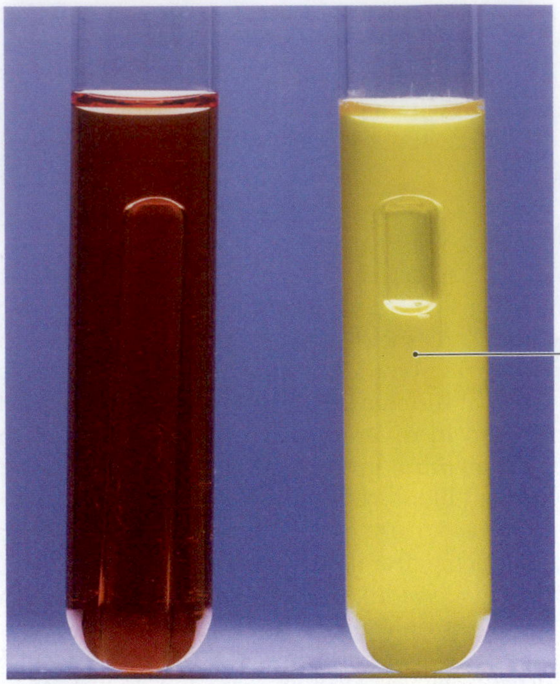

▲ FIGURE 6.14 The use of carbohydrate utilization tubes as differential media. Each tube contains a single kind of simple carbohydrate (a sugar) as a carbon source and the dye phenol red as a pH indicator. *Alcaligenes faecalis* in the tube on the left did not ferment this carbohydrate; because no acid was produced, the medium did not turn yellow. *Escherichia coli* in the tube on the right fermented the sugar, producing acid and lowering the pH enough to cause the phenol red to turn yellow. This bacterium also produced gas, which is visible as a small bubble in the Durham tube.

distinguished from other species by variations in their effect on the medium or by the color of colonies they produce. For example, bile salts and crystal violet in MacConkey agar both inhibit the growth of Gram-positive bacteria and differentiate lactose fermenting from non-lactose-fermenting Gram-negative bacteria **(FIGURE 6.15)**. **▶VIDEO TUTOR:** ***Bacterial Growth Media***

TABLE 6.4 Representative Differential Complex Media

Medium and Ingredients	Use and Interpretation of Results
MacConkey Medium	
Peptone (20.0 g) Agar (12.0 g) Lactose (10.0 g) Bile salts (5.0 g) NaCl (5.0 g) Neutral red (0.075 g) Crystal violet (0.001 g) Water to bring volume to 1 L	For the culture and differentiation of enteric bacteria based on the ability to ferment lactose Lactose fermenters produce red to pink colonies; non-lactose fermenters form colorless or transparent colonies
Blood Agar	
Agar (15.0 g) Pancreatic digest of casein (15.0 g) Papaic digest of soybean meal (5.0 g) NaCl (5.0 g) Water to bring volume to 950.0 ml Sterile blood (50.0 ml, added to medium after autoclaving and cooling)	For culture of fastidious microorganisms and differentiation of hemolytic microorganisms Partial digestion of blood: alpha-hemolysis; complete digestion of blood: beta-hemolysis; no digestion of blood: gamma-hemolysis

Anaerobic Media

Obligate anaerobes require special culture conditions in that their cells must be protected from free oxygen. Anaerobes can be introduced with a straight inoculating wire into the anoxic (oxygen-free) depths of solid media to form a *stab culture,* but special media called **reducing media** provide better anaerobic culturing conditions. These media contain compounds, such as sodium thioglycolate, that chemically combine with free oxygen and remove it from the medium. Heat is used to drive absorbed oxygen from thioglycolate immediately before such a medium is inoculated.

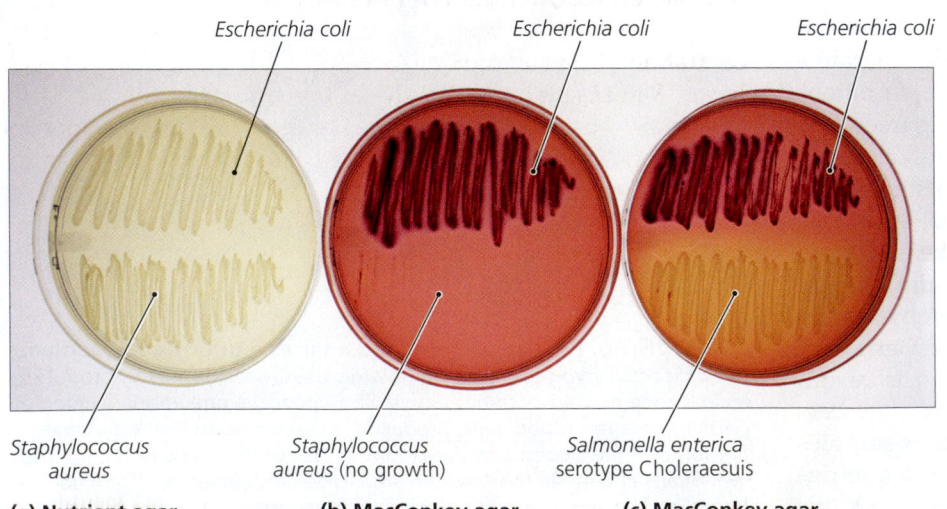

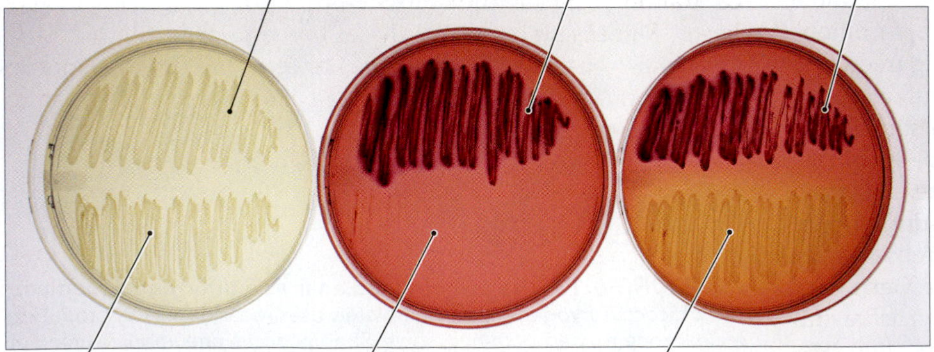

◀ FIGURE 6.15 The use of MacConkey agar as a selective and differential medium. (a) Whereas both the Gram-positive *Staphylococcus aureus* and the Gram-negative *Escherichia coli* grow on nutrient agar, MacConkey agar **(b)** selects for Gram-negative bacteria and inhibits Gram-positive bacteria. **(c)** MacConkey agar also differentiates Gram-negative bacteria based on their ability to ferment lactose. The colonies of the lactose-fermenting *E. coli* are easily distinguished from those of the non-lactose fermenting *Salmonella enterica* serotype Choleraesuis.

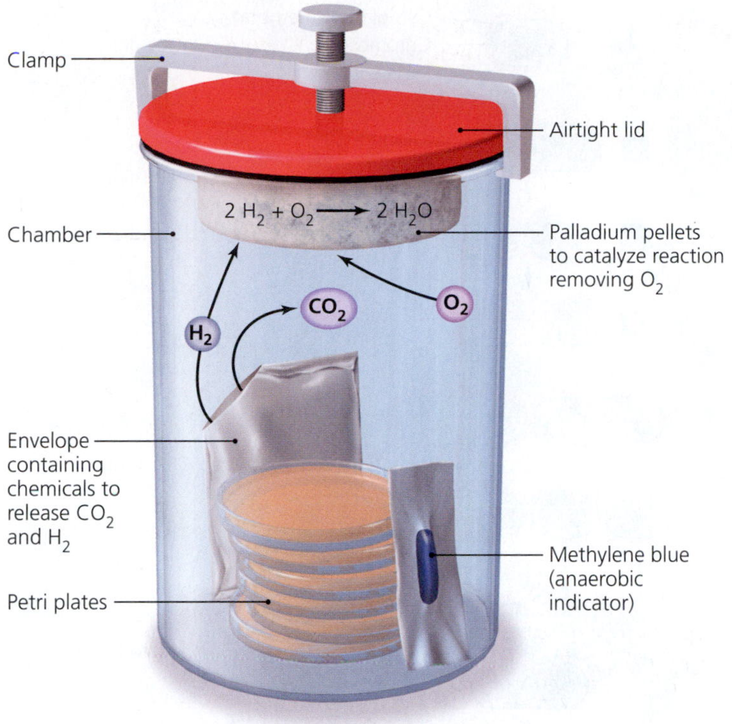

▲ **FIGURE 6.16 An anaerobic culture system.** The system utilizes chemicals to create an anaerobic environment inside a sealable, airtight jar. Methylene blue, which turns colorless in the absence of oxygen, indicates when the environment within the jar is anaerobic.

The use of Petri plates presents special problems for the culture of anaerobes because each dish has a loose-fitting lid that allows the entry of air. For the culture of anaerobes, inoculated Petri plates are placed in sealable containers containing reducing chemicals **(FIGURE 6.16)**. Of course, the airtight lids of anaerobic culture vessels must be sealed so that oxygen cannot enter. Only anaerobes that can tolerate exposure to oxygen can be cultured by this method because inoculation and transfer occur outside the anaerobic environment. Laboratories that routinely study strict anaerobes have large anaerobic glove boxes, which are transparent, airtight chambers with special airtight rubber gloves, chemicals that remove oxygen, and air locks. These chambers allow scientists to manipulate equipment and anaerobic cultures in an oxygen-free environment.

Transport Media

Hospital personnel use special **transport media** to carry clinical specimens of feces, urine, saliva, sputum, blood, and other bodily fluids in such a way as to ensure that people are not infected and that the specimens are not contaminated. Speed in transporting clinical specimens to the laboratory is extremely important because pathogens often do not long survive outside the body. Specimens are transported in buffered media designed to maintain the ratio and life of different microbes. Anaerobic specimens may occasionally be transported for less than an hour inside a syringe, but longer times require the use of anaerobic transport media.

Special Culture Techniques

LEARNING | OUTCOME

6.12 Discuss the use of animal and cell culture and low-oxygen culture.

Not all organisms can be grown under the culture conditions we have discussed. Scientists have developed other techniques to culture many of these organisms.

Animal and Cell Culture

Microbiologists have developed animal and cell culture techniques for growing microbes for which artificial media are inadequate. The causative agents of leprosy and syphilis, for example, must be grown in animals because all attempts to grow them using standard culture techniques have been unsuccessful. *Mycobacterium leprae* (lep´rī) is cultured in armadillos, whose internal conditions (including a relatively low body temperature) provide the conditions this microbe prefers. Rabbits meet the culture needs for *Treponema pallidum*, the bacterium that causes syphilis. Because viruses and small bacteria called rickettsias and chlamydias are obligate intracellular parasites—that is, they grow and reproduce only within living cells—bird eggs and cultures of living cells are used to culture these organisms.

Low-Oxygen Culture

As we have discussed, many types of organisms prefer oxygen conditions that are intermediate between strictly aerobic and anaerobic environments. *Carbon dioxide incubators*, machines that electronically monitor and control CO_2 levels, provide atmospheres that mimic the environments of the intestinal tract, the respiratory tract, and other body tissues and thus are useful for culturing these kinds of organisms. Smaller and much less expensive alternatives to CO_2 incubators are *candle jars*. In these simple but effective devices, culture plates are sealed in a jar along with a lit candle; the flame consumes much of the O_2, replacing it with CO_2. The candle eventually extinguishes itself, creating an environment that is ideal for aerotolerant anaerobes, microaerophiles, and **capnophiles,** which are organisms such as *Neisseria gonorrhoeae* that grow best with a relatively high concentration of carbon dioxide (3–10%) in addition to low oxygen levels. Remaining oxygen in the jar prevents the growth of strict anaerobes. The use of packets of chemicals that remove most of the oxygen from the jar has replaced candles in modern microbiology labs.

Preserving Cultures

LEARNING | OUTCOME

6.13 Contrast refrigeration, deep freezing, and lyophilization as methods for preserving cultures of microbes.

To store living cells, a scientist slows the cells' metabolism to prevent the excessive accumulation of waste products and the exhaustion of all nutrients in a medium. Refrigeration is often the best technique for storing bacterial cultures for short periods of time.

Deep freezing and lyophilization are used for long-term storage of bacterial cultures. **Deep-freezing** involves freezing the cells at temperatures from −50°C to −95°C. Deep-frozen cultures can be restored years later by thawing them and placing a sample in an appropriate medium.

Lyophilization (lī-of′i-li-zā′shŭn; freeze-drying) involves removing water from a frozen culture using an intense vacuum. Under these conditions, ice sublimates (directly becomes a gas) and is removed from cells without permanently damaging cellular structures and chemicals. Lyophilized cultures can last for decades and are revived by adding lyophilized cells to liquid culture media.

TELL ME WHY

Why do clinical laboratory scientists keep many different kinds of culture media on hand?

Growth of Microbial Populations

LEARNING **OUTCOME**

6.14 Describe binary fission as a means of reproduction.

Most unicellular microorganisms reproduce by *binary fission,* a process in which a cell grows to twice its normal size and divides in half to produce two daughter cells of equal size. Binary fission generally involves four steps, as illustrated in **FIGURE 6.17**, for a prokaryotic cell. ▶**ANIMATIONS:** ***Bacterial Growth: Overview***

1. The cell replicates its chromosome (DNA molecule). The duplicated chromosomes are attached to the cytoplasmic membrane. (In eukaryotic cells, chromosomes are attached to microtubules.)
2. The cell elongates and growth between attachment sites pushes the chromosomes apart. (Eukaryotic cells segregate their chromosomes by the process of *mitosis,* described in Chapter 12.)
3. The cell forms a new cytoplasmic membrane and wall (septum) across the midline.
4. When the septum is completed, the daughter cells may remain attached as shown in the figure, or they may separate completely. When the cells remain attached, further binary fission in parallel planes produces a chain. When further divisions are in different planes, the cells become a cluster (as shown in the figure).
5. The process repeats. ▶**ANIMATIONS:** ***Binary Fission***

Other reproductive strategies of prokaryotes and eukaryotes are discussed in Chapters 11 and 12. Here we consider the growth of populations by binary fission, using bacterial cultures as examples. We begin with a brief discussion of the mathematics of population growth.

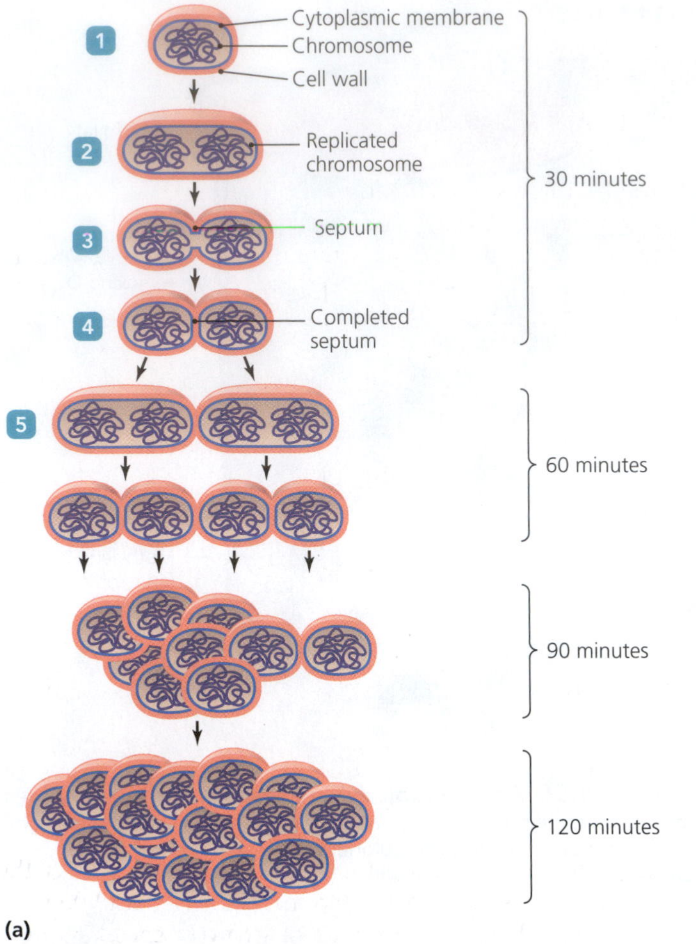

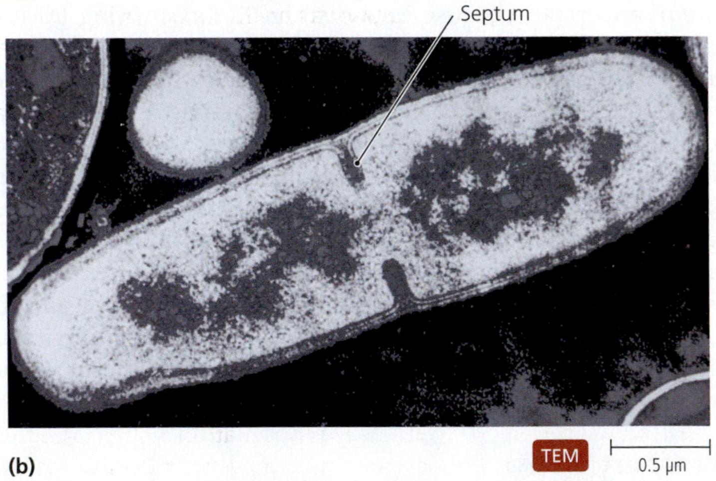

▲ **FIGURE 6.17 Binary fission. (a)** The events in binary fission. All the cells may divide in parallel planes and remain attached to form a chain, or they may divide in different planes to form a cluster (as shown here). **(b)** Transmission electron micrograph of *Bacillus licheniformis* undergoing binary fission.

Generation Time

LEARNING | OUTCOME

6.15 Explain what is meant by the generation time of bacteria.

The time required for a bacterial cell to grow and divide is its **generation time.** Viewed another way, generation time is also the time required for a population of cells to double in number. Generation times vary among populations and are dependent on chemical and physical conditions. Under optimal conditions, some bacteria (such as *E. coli* and *S. aureus*) have a generation time of 20 minutes or less. For this reason, food contaminated with only a few of these organisms can cause food poisoning if not properly refrigerated and cooked. Most bacteria have a generation time of 1 to 3 hours, though some slow-growing species such as *Mycobacterium leprae* require more than 10 days before they double. (Appendix B presents the math required to calculate generation time for a population.)

Mathematical Considerations in Population Growth

LEARNING | OUTCOME

6.16 Describe logarithmic growth.

With binary fission, any given cell divides to form two cells; then each of these new cells divides in two to make four, and then four become eight and so on. This type of growth, called **logarithmic growth** or **exponential growth,** produces very different results from simple addition, known as arithmetic growth. We can compare these two types of growth by considering what would happen over time to two identical hypothetical populations, as shown in **FIGURE 6.18**. In this case we assume that a hypothetical population of species A increases by adding one new cell every 20 minutes, whereas the cells of species B divide by binary fission every 20 minutes. After 20 minutes, each population, which started with a single cell, would have two cells; after 40 minutes, species A would have three cells, whereas species B would have four cells. At this point there is little difference in the growth of the two populations, but after 2 hours, the arithmetically growing species A would have only seven cells, whereas the logarithmically growing species B would have increased to 64 cells. Clearly, logarithmic growth can increase a population's size dramatically—after only 7 hours, species B will have over 2 million cells!

The number of cells arising from a single cell reproducing by binary fission is calculated as 2^n, where n is the number of generations; in other words, multiply 2 times itself n number of times. To calculate the total number of cells in a population, we multiply the original number of cells by 2^n. If, for example, species B had begun with three cells instead of one, then after 2 hours it would have 192 cells ($3 \times 2^6 = 3 \times 64 = 192$).

A visible culture of bacteria may consist of trillions of cells, so microbiologists use scientific notation to deal with the huge numbers involved. One advantage of scientific notation is that large numbers are expressed as powers of 10, making them easier to read and write. For example, consider our culture of species B. After 10 hours (30 generations) it would have 1,073,741,824 (2^{30}) cells. This large number can be rounded off and expressed more succinctly in scientific notation as 1.07×10^9. After 30 more generations, scientific notation would be the only practical way to express the huge number of cells in the culture, which would be 1.15×10^{18} (2^{60}). Such a number, if written out (the digits 115 followed by 16 zeros), would be impractically large. (Appendix B presents scientific notation in more detail.)

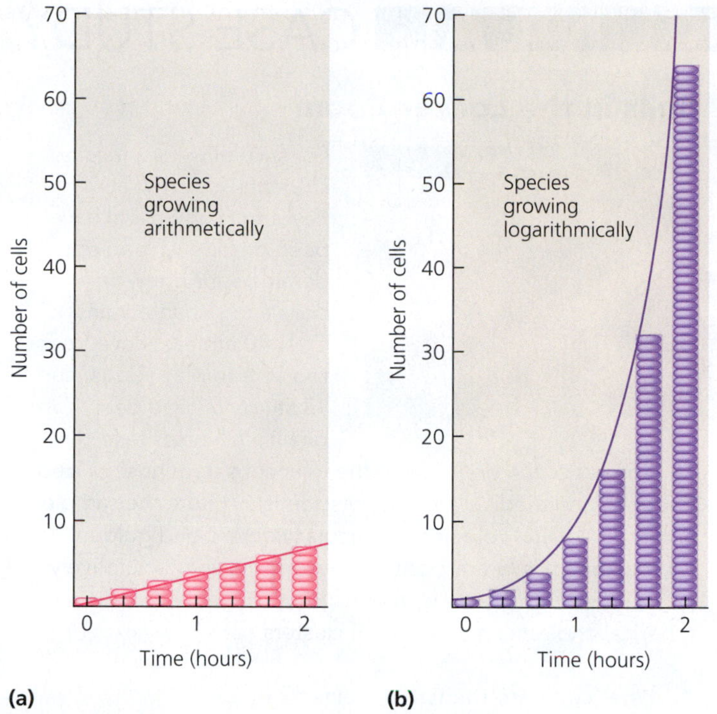

▲ **FIGURE 6.18 A comparison of arithmetic and logarithmic growth.** Given two hypothetical initial populations consisting of a single cell each, after 2 hours an arithmetically growing species will have 7 cells, (a) whereas a logarithmically growing species will have 64 cells (b).

Phases of Microbial Population Growth

LEARNING | OUTCOMES

6.17 Draw and label a bacterial growth curve.

6.18 Describe what occurs at each phase of a population's growth.

A graph that plots the numbers of organisms in a growing population over time is known as a **growth curve.** When drawn using an arithmetic scale on the y-axis, a plot of exponential growth presents two problems (**FIGURE 6.19a**): it is difficult or impossible to distinguish numbers in early generations from the baseline, and as the population grows it becomes impossible to accommodate the graph on a single page.

The solution to these problems is to replace the arithmetic scale on the y-axis with a logarithmic (log) scale (**FIGURE 6.19b**). Such a log scale, in which each division is 10 times larger than the

CLINICAL CASE STUDY

Boils in the Locker Room

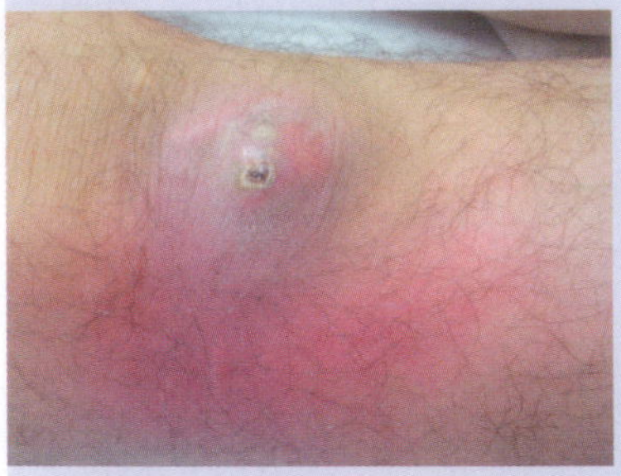

For several weeks, faculty, students, and staff at Rayburn High School have been dealing with an epidemic of unsightly, painful boils and skin infections. Over 40 athletes have been too sore to play sports, and 13 students have been hospitalized. Nurses take clinical samples of the drainage of the infections from hospitalized students, and medical laboratory scientists culture the samples. Isolated bacterial colonies are circular, convex, and golden. The bacterium is Gram-positive, mesophilic, and facultatively halophilic and can grow with or without oxygen. The cells are spherical and remain attached in clusters.

1. What color are the Gram-stained cells?
2. What does the term "facultatively halophilic" mean?
3. What is the scientific description of the bacterium's oxygen requirement?
4. If the bacterium divides every 30 minutes under laboratory conditions, how many cells would there be in a colony after 24 hours?

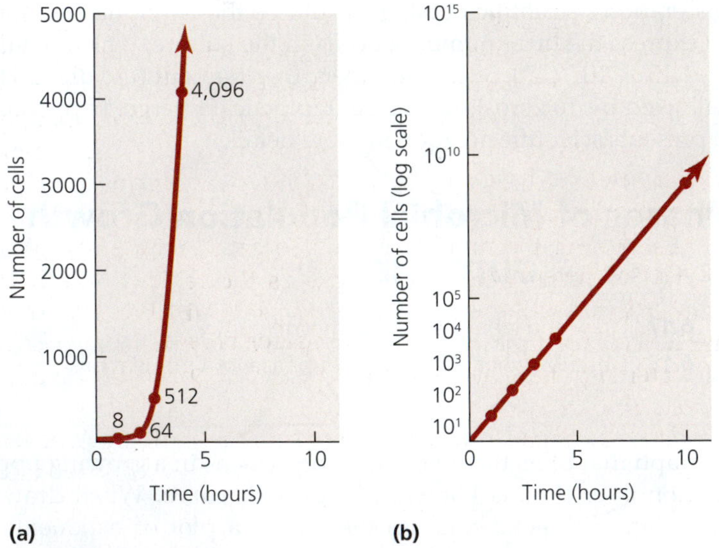

▲ **FIGURE 6.19 Two growth curves of logarithmic growth.** The generation time for this *E. coli* population is 20 minutes. **(a)** An arithmetic graph. Using an arithmetic scale for the *y*-axis makes it difficult to ascertain actual numbers of cells near the beginning and impossible to plot points after only a short time. **(b)** A semilogarithmic graph. Using a logarithmic scale for the *y*-axis solves both of these problems. Note that a plot of logarithmic population growth using a logarithmic scale produces a straight line.

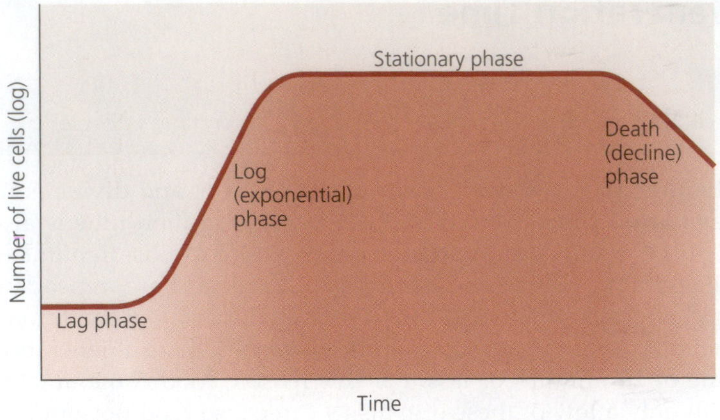

▲ **FIGURE 6.20 A typical microbial growth curve.** The curve shows the four phases of population growth. *Why do cells trail behind their optimum reproductive potential during the lag phase?*

Figure 6.20 *During lag phase, cells are synthesizing the metabolic machinery and chemicals required for optimal reproduction.*

preceding one, can accommodate small numbers at the lower end of the graph and very large numbers at the upper end. This kind of graph is *semilogarithmic* because only one axis uses a log scale.

When bacteria are inoculated into a closed vessel of liquid medium, there are four distinct phases to a population's growth curve: the lag, log, stationary, and death phases **(FIGURE 6.20)**. ▶**ANIMATIONS:** *Bacterial Growth Curve*

Lag Phase

During the **lag phase** the cells are adjusting to their new environment; most cells do not reproduce immediately but instead actively synthesize enzymes to utilize novel nutrients in the medium. For example, bacteria inoculated from a medium containing glucose as a carbon source into a medium containing lactose must synthesize two types of proteins: membrane proteins to transport lactose into the cell and the enzyme lactase to catabolize lactose. The lag phase can last less than an hour or can last for days, depending on the species and the chemical and physical conditions of the medium.

Log Phase

Eventually, the bacteria synthesize the necessary chemicals for conducting metabolism in their new environment, and they then enter a phase of rapid chromosome replication, growth, and reproduction. This is the **log phase,** so named because the population increases logarithmically, and the reproductive rate reaches a constant as DNA and protein syntheses are maximized.

Researchers are interested in the log phase for many reasons. Populations in log phase are more susceptible to antimicrobial drugs that interfere with metabolism, such as erythromycin, and to drugs that interfere with the formation of cell structures, such as the inhibition of cell wall synthesis by penicillin. Populations in log phase are preferred for Gram staining because most cells' walls are intact—an important characteristic for correct staining. Further, because the metabolic rate of individual cells is at a maximum during log phase, this phase may be preferred for industrial and laboratory purposes.

Stationary Phase

If bacterial growth continued at the exponential rate of the log phase, bacteria would soon overwhelm the Earth. This does not occur because as nutrients are depleted and wastes accumulate, the rate of reproduction decreases. Eventually, the number of dying cells equals the number of cells being produced, and the size of the population remains constant—hence the name **stationary phase.** During this phase, the metabolic rate of surviving cells declines.

Death Phase

If nutrients are not added and wastes are not removed, a population reaches a point at which cells die at a faster rate than they are produced. Such a culture has entered the **death phase** (or *decline phase*). Bear in mind that during the death phase, some cells remain alive and continue metabolizing and reproducing, but the number of dying cells exceeds the number of new cells produced so that eventually the population decreases to a fraction of its previous abundance. In some cases all the cells die, whereas in others a few survivors may remain indefinitely. The latter case is especially true for cultures of bacteria that can develop resting structures called *endospores* (see Chapter 3).

Continuous Culture in a Chemostat

LEARNING | OUTCOME

6.19 Explain how a chemostat can maintain a microbial culture in a continuous phase.

Researchers and industrialists can continuously maintain a particular phase of microbial population growth by using a special culture device called a **chemostat (FIGURE 6.21)**. A chemostat is an open system; that is, fresh medium is continuously supplied while an equal amount of old medium (containing microbes) is removed. As nutrients enter the culture vessel of a chemostat, the cells can metabolize, grow, and reproduce but only to the extent that a limiting nutrient (e.g., a particular amino acid) is available. By controlling the amount of limiting nutrient entering the culture vessel, a chemostat maintains a culture in a particular phase, typically log phase. This is impossible in a closed system because as nutrients are depleted and wastes accumulate, the culture enters the stationary phase and then the death phase.

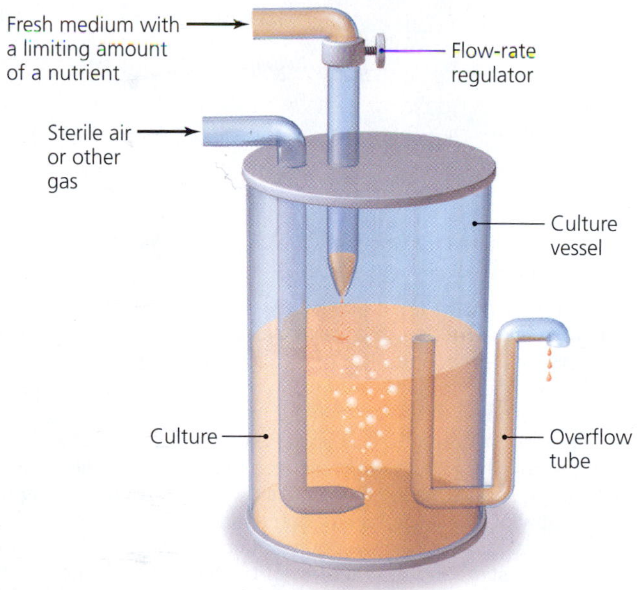

▲ **FIGURE 6.21 Schematic of chemostat.** A microbial culture is continuously maintained in a predetermined phase, typically log phase, by controlling the amount and type of fresh medium added while removing an equal amount of old culture.

Chemostats make possible the study of microbial population growth at steady but low nutrient levels, such as might be found in biofilms. Such studies are essential to understand interactions of microbial species in nature and can reveal aspects of microbial interactions that are not apparent in a closed system. Scientists also use chemostats to maintain log phase population growth for experimental inquiries into aspects of microbial metabolism, such as enzyme activities. Food and industrial microbiologists use chemostats to maintain constant production of useful microbial products.

Measuring Microbial Reproduction

LEARNING | OUTCOME

6.20 Contrast direct and indirect methods of measuring bacterial reproduction.

We have discussed the concepts of population growth and have seen that large numbers result from logarithmic growth, but we have not discussed practical methods of determining the size of a microbial population. Because of each cell's small size and incredible rate of reproduction, it is not possible to actually count every one in a population. For one thing, they grow so rapidly that their number changes during the count. Therefore, laboratory personnel must estimate the number of cells in a population by counting the number in a small, representative sample and then multiplying to estimate the number in the whole specimen. For example, if there are 25 cells in a microliter (μl) sample of urine, then there are approximately 25 million cells in a liter of urine.

Estimating the number of microorganisms in a sample is useful for determining such things as the severity of urinary tract infections, the effectiveness of pasteurization and other methods of food preservation, the degree of fecal contamination of water supplies, and the effectiveness of particular disinfectants and antibiotics.

Microbiologists use either direct or indirect methods to estimate the number of cells. We begin with direct methods of measuring bacterial reproduction.

Direct Methods Not Requiring Incubation

It is possible to directly count cells without having to incubate cultures. Here we consider two such direct methods.

Microscopic Counts Microbiologists can count microorganisms directly through a microscope rather than inoculating them onto the surface of a solid medium. In this method, particularly suitable for stained prokaryotes and relatively large eukaryotes, a sample is placed on a *cell counter* (also called a *Petroff-Hausser counting chamber*), which is a glass slide composed of an etched grid positioned beneath a glass coverslip **(FIGURE 6.22)**. Because the coverslip is 0.02 mm above the grid, the volume of bacterial suspension over a 1 mm^2 portion of the grid is 1 mm × 1 mm × 0.02 mm = 0.02 mm^3. Each 1 mm^2 grid contains 25 large squares, so a microbiologist can count the number of bacteria in several of the large squares and then calculate the mean number of bacteria per square. The number of bacteria per milliliter (cm^3) can be calculated as follows:

$$\text{mean no. of bacteria per square} \times 25 \text{ squares} = \text{no. of bacteria per } 0.02 \text{ mm}^3$$

$$\text{no. of bacteria per } 0.02 \text{ mm}^3 \times 50 = \text{no. of bacteria per mm}^3$$

$$\text{no. of bacteria per mm}^3 \times 1000 = \text{no. of bacteria per cm}^3 \text{ (ml)}$$

This means that one needs only to multiply the mean number of bacteria per square by 1,250,000 (25 × 50 × 1000) to calculate the number of bacteria per milliliter of bacterial suspension.

Direct microscopic counts are advantageous when there are more than 10,000,000 cells per milliliter or when a speedy estimate of population size is required. However, direct counts can be problematic because it is often difficult to differentiate between living and dead cells, and it is difficult to count motile microorganisms.

Electronic Counters A *Coulter*[15] *counter* is a device that directly counts cells as they interrupt an electrical current flowing across a narrow tube held in front of an electronic detector. This device is useful for counting the larger cells of yeast, unicellular algae, and protozoa; it is less useful for bacterial counts because of debris in the media and the presence of filaments and clumps of cells.

Flow cytometry is a variation of counting with a Coulter counter. A cytometer uses a light-sensitive detector to record changes in light transmission through the tube as cells pass. Scientists use this technique to distinguish among cells that have been differentially stained with fluorescent dyes or tagged with fluorescent antibodies. They can count bacteria in a solution and even count host cells that contain fluorescently stained intracellular parasites.

We have considered direct methods not requiring incubation. Now we consider methods with incubation.

Direct Methods Requiring Incubation

Among the many direct techniques are techniques requiring incubation—viable plate counts following dilution, membrane filtration, and the most probable number method. Direct techniques not requiring incubation include microscopic counts and electronic counting. We will consider each in turn.

Serial Dilution and Viable Plate Counts What if the number of cells in even a very small sample is still too great to count? If, for example, a 1-ml sample of milk containing 20,000 bacterial cells per milliliter were plated on a Petri plate, there would be

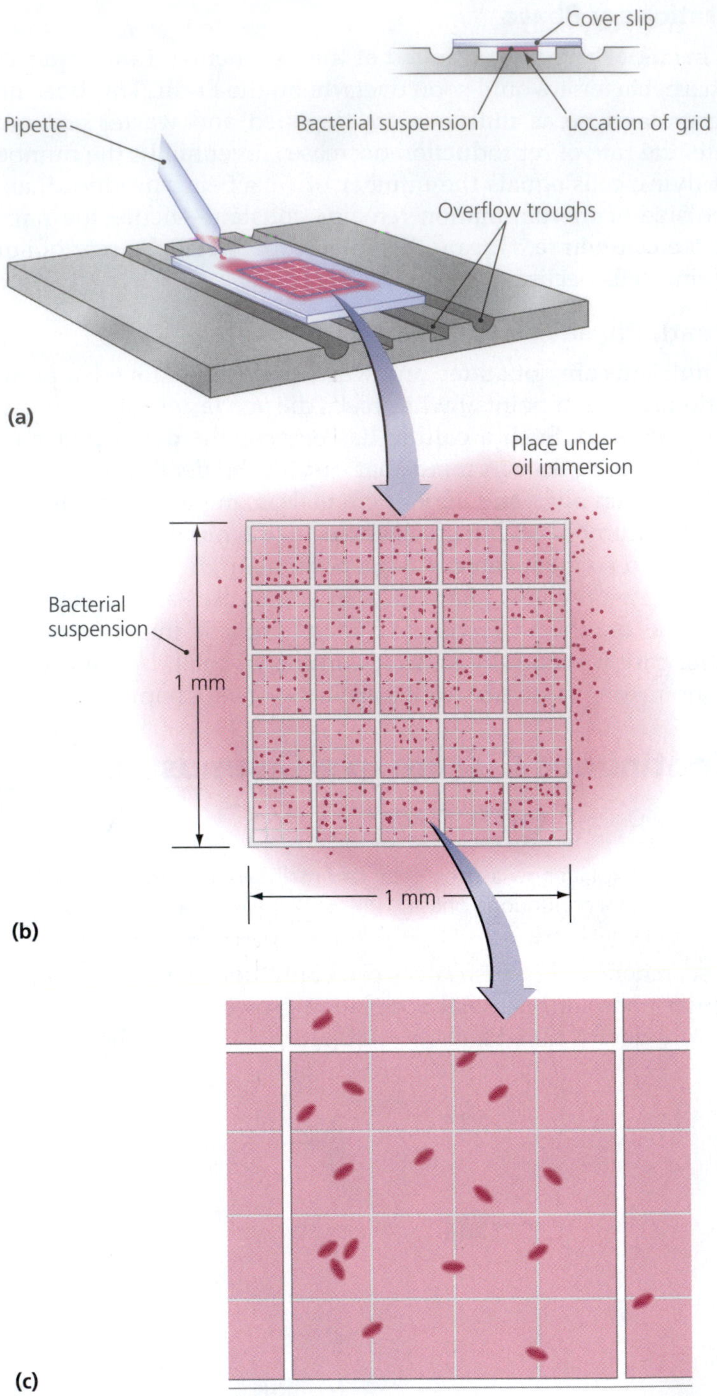

▲ **FIGURE 6.22 The use of a cell counter for estimating microbial numbers. (a)** The counter is a glass slide with an etched grid that is exactly 0.02 mm lower than the bottom of the coverslip. A bacterial suspension placed next to the coverslip through a pipette moves under the coverslip and over the grid by capillary action. **(b)** View of a 1 mm^2 portion of the grid through the microscope. Each square millimeter of the grid has 25 large squares, each of which is divided into 16 small squares. **(c)** Enlarged view of one large square containing 15 cells. The number of bacteria in several large squares is counted and averaged. The calculations involved in estimating the number of bacteria per milliliter (cm^3) of suspension are described in the text.

[15]Named for Wallace Coulter, American inventor.

too many colonies to count. In such cases, microbiologists make a **serial dilution,** which is the stepwise dilution of a liquid culture in which the dilution factor at each step is constant. The scientists plate a set amount of each dilution onto an agar surface and count the number of colonies resulting on a plate from each dilution. They count the colonies on plates with 25 to 250 colonies and multiply the number by the reciprocal of the dilution to estimate the number of bacteria per milliliter of the original culture. This method is called a **viable plate count** (**FIGURE 6.23**).

When a plate has fewer than 25 colonies, it is not used to estimate the number of bacteria in the original sample because the chance of underestimating the population increases when the number of colonies is small. Recall that the number of colonies on a plate indicates the number of CFUs that were inoculated onto the plate. This number differs from the actual number of cells when the CFUs are composed of more than one cell. In such cases, a viable plate count underestimates the number of cells present in the sample.

The accuracy of a viable plate count also depends on the homogeneity of the dilutions, the ability of the bacteria to grow on the medium used, the number of cell deaths, and the growth phase of the sample population. Thoroughly mixing each dilution, inoculating multiple plates per dilution, and using log-phase cultures minimize errors.

Membrane Filtration Viable plate counts allow scientists to estimate the number of microorganisms when the population is very large, but if the population density is very small—as is the case, for example, for fecal bacteria in a stream or lake—microbes are more accurately counted by **membrane filtration** (**FIGURE 6.24**). In this method, a large sample (perhaps as large as several liters) is poured (or drawn under a vacuum) through a membrane filter with pores small enough to trap the cells. The membrane is then transferred onto a solid medium, and the colonies present after incubation are counted. In this case, the number of colonies is equal to the number of CFUs in the original large sample.

Most Probable Number The **most probable number (MPN)** method is a statistical estimation technique based on the fact that the more bacteria are in a sample, the more dilutions are required to reduce their number to zero.

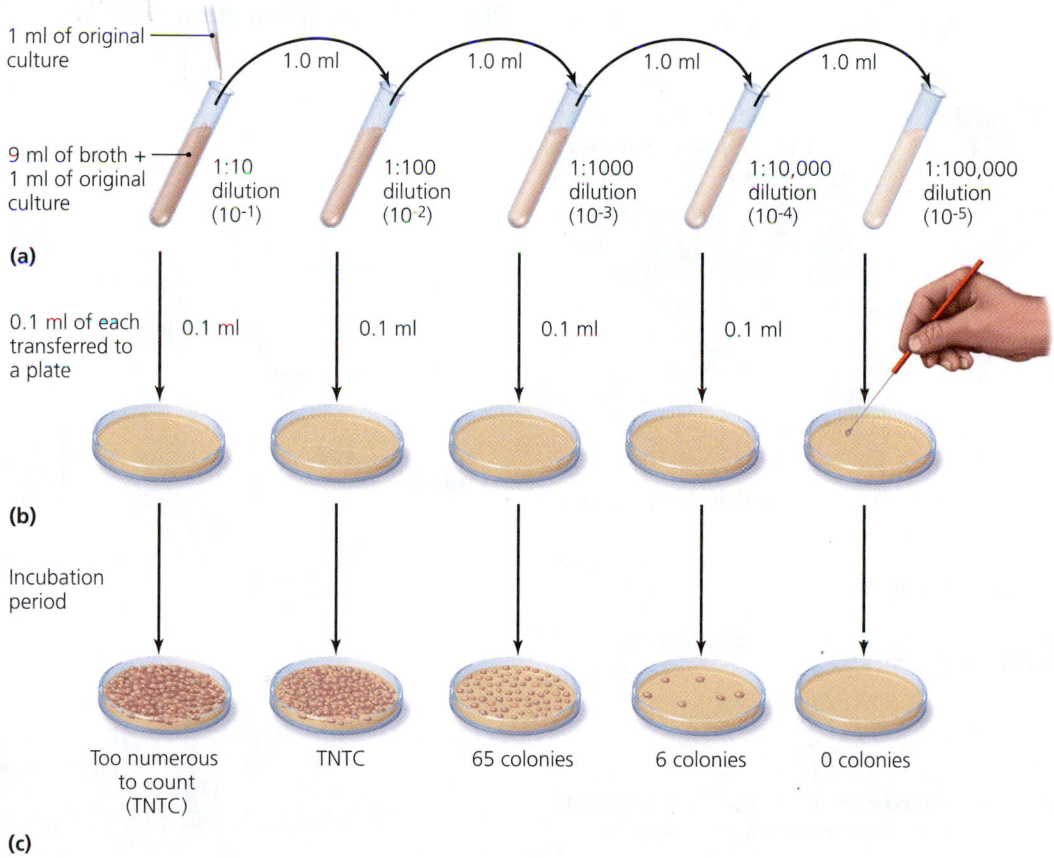

▲ **FIGURE 6.23 A serial dilution and viable plate count for estimating microbial population size. (a)** Serial dilutions. A series of 10-fold dilutions is made. **(b)** Plating. A 0.1-ml sample from each dilution is poured onto a plate and spread with a sterile rod. Alternatively, 0.1 ml of each dilution can be mixed with melted agar medium and poured into plates. **(c)** Counting. Plates are examined after incubation. Some plates may contain so many colonies that they are too numerous to count (TNTC). The number of colonies is multiplied by 10 (because 0.1 ml was plated instead of 1 ml) and then by the reciprocal of the dilution to estimate the concentration of bacteria in original culture—in this case, 65 colonies × 10 × 1000 = 650,000 bacteria/ml.

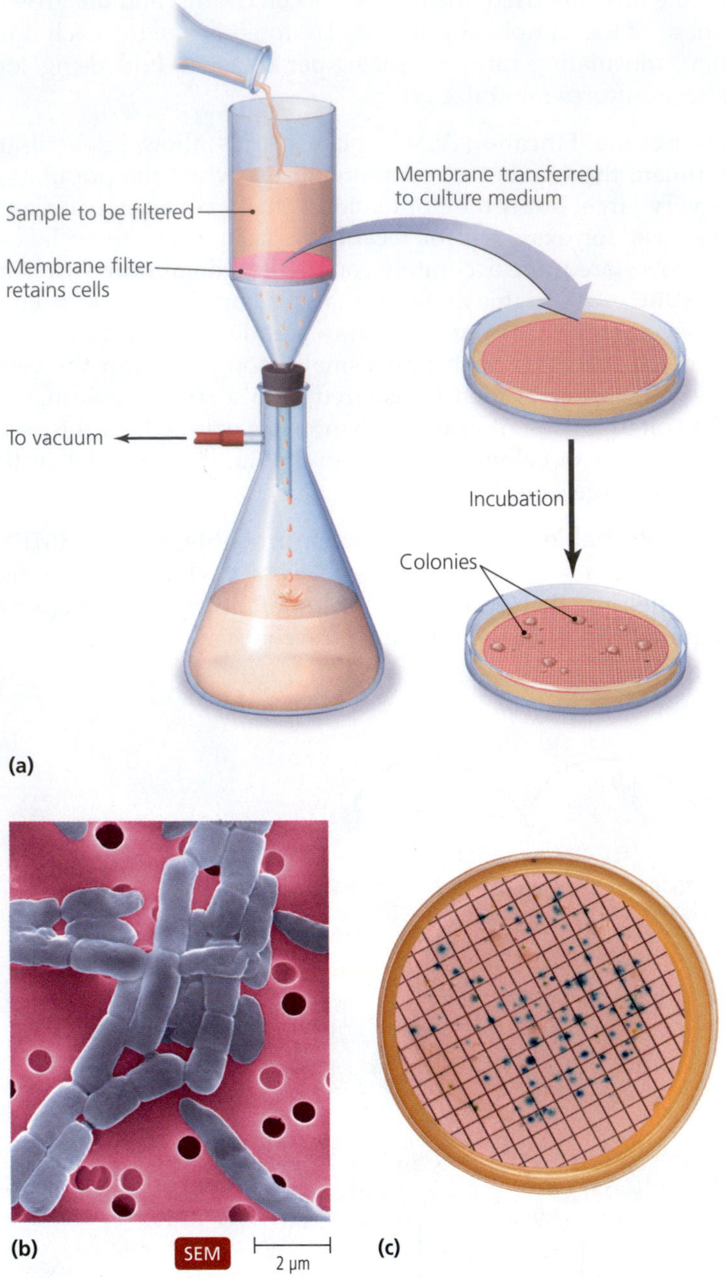

▲ **FIGURE 6.24 The use of membrane filtration to estimate microbial population size. (a)** After all the bacteria in a given volume of sample are trapped on a membrane filter, the filter is transferred onto an appropriate medium and incubated. The microbial population is estimated by multiplying the number of colonies counted by the volume of sample filtered. **(b)** Bacteria trapped on the surface of a membrane filter. **(c)** Colonies growing on a solid medium after being transferred from a membrane filter. Scientists use a superimposed grid to help them count the colonies. *If the colonies in (c) resulted from filtering 2.5 liters of stream water, what is the minimum number of bacteria per liter in the stream?*

Figure 6.24 $\frac{83 \text{ colonies}}{2.5 \text{ L}} = 33.2$ colonies/L

Let's consider an example of the use of the MPN method to estimate the number of fecal bacteria contaminating a stream. A researcher inoculates a set of test tubes of a broth medium with a sample of stream water. The more tubes that are used, the more accurate is the MPN method; even so, accuracy must be balanced against the time and cost involved in inoculating and incubating numerous tubes. Typically, a set of five tubes is inoculated.

The researcher also inoculates a set of five tubes with a 1:10 dilution and another set of five tubes with a 1:100 dilution of stream water. Thus, there are 15 test tubes—the first set of five tubes inoculated with undiluted sample, the second set with a 1:10 dilution, and the third set with a 1:100 dilution **(FIGURE 6.25)**.

After incubation for 48 hours, the researcher counts the number of test tubes in each set that show growth. This generates three numbers in this example—growth occurs in four of the undiluted broth tubes, two of the 1:10 tubes, and in only one of the 1:100 tubes (4, 2, 1). The numbers are compared to the numbers in an MPN table (see **TABLE 6.5**). How statisticians develop MPN tables is beyond the scope of our discussion, but they accurately estimate the number of cells in a solution. In this case the MPN table estimates that there were 26 bacteria per 100 ml of stream water.

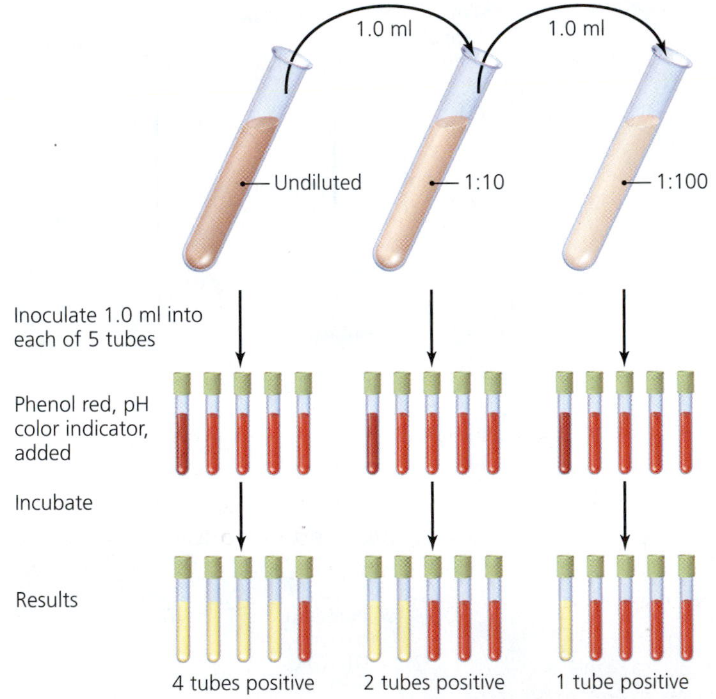

▲ **FIGURE 6.25 The most probable number (MPN) method for estimating microbial numbers.** Typically, sets of five test tubes are used for each of three dilutions. After incubation, the number of tubes showing growth in each set is used to enter an MPN table (see Table 6.5), which provides an estimate of the number of cells per 100 ml of liquid. *If the results were 5, 3, 1, what would be the most probable number of microorganisms in the original broth?*

Figure 6.25 *The MPN is 110/100 ml.*

TABLE 6.5 Most Probable Number Table (partial)

Number Out of Five Tubes Giving Positive Results in Three Dilutions			Most Probable Number of Bacteria per 100 ml
4	0	0	13
4	0	1	17
4	1	0	17
4	1	1	21
4	1	2	26
4	2	0	22
4	2	1	26
4	3	0	27
4	3	1	33
4	4	0	34
5	0	0	23
5	0	1	30
5	0	2	40
5	1	0	30
5	1	1	50
5	1	2	60
5	2	0	50
5	2	1	70
5	2	2	90
5	3	0	80
5	3	1	110
5	3	2	140
5	3	3	170
5	4	0	130
5	4	1	170
5	4	2	220
5	4	3	280
5	4	4	350

The most probable number method is useful for counting microorganisms that do not grow on solid media, when bacterial counts are required routinely, and when samples of wastewater, drinking water, and food samples contain too few organisms to use a viable plate count. The MPN method is also used to count algal cells because algae seldom form distinct colonies on solid media.

Indirect Methods

It is not always necessary to count microorganisms to estimate population size or density. Industrial and research microbiologists use indirect methods that measure such variables as turbidity, metabolic activity, and dry weight instead of counting microorganisms, colonies, or MPN tubes. Scientists can also estimate population size and diversity by analyzing the unique sequences of DNA present in a sample.

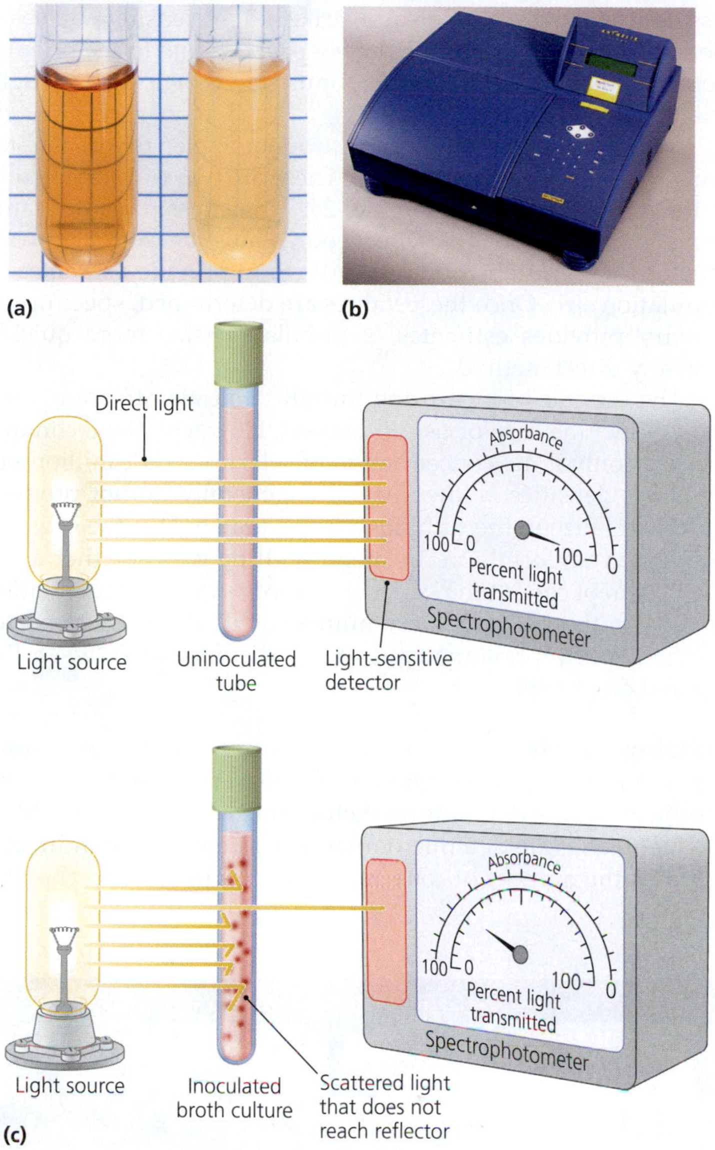

▲ **FIGURE 6.26 Turbidity and the use of spectrophotometry in indirectly measuring population size.** **(a)** Turbidity (right), an increased optical density or cloudiness of a solution. **(b)** A spectrophotometer. **(c)** The principle of spectrophotometry. After a light beam is passed through an uninoculated sample of the culture medium, the scale is set at 100% transmission. In an inoculated sample, the microbial cells absorb and scatter light, reducing the amount reaching the detector. The percentage of light transmitted is inversely proportional to population density.

Turbidity As bacteria reproduce in a broth culture, the broth often becomes *turbid* (cloudy) **(FIGURE 6.26a)**. Generally, the greater the bacterial population, the more turbid a broth will be. An indirect method for estimating the growth of a microbial population involves measuring changes in turbidity using a device called a *spectrophotometer* **(FIGURE 6.26b)**. Researchers most often use this method.

A spectrophotometer measures the amount of light transmitted through a culture under standardized conditions **(FIGURE 6.26c)**. The greater the concentration of bacteria within a broth, the more light will be absorbed and scattered and the less light will pass through and strike a light-sensitive detector. Generally, transmission is inversely proportional to the population size; that is, the larger the population grows, the less light will reach the detector.

Scales on the gauge of a spectrophotometer report *percentage of transmission* and *absorbance*. These are two ways of looking at the same things; for example, 25% transmission is the same thing as 75% absorbance. Direct counts must be calibrated with transmission and absorbance readings to provide estimates of population size. Once these values are determined, spectrophotometry provides estimates of population size more quickly than any direct method.

The benefits of measuring turbidity to estimate population growth include ease of use and speed. However, the technique is useful only if the concentration of cells exceeds 1 million per milliliter; densities below this value generally do not produce turbidity. Further, the technique is accurate only if the cells are suspended uniformly in the medium. If they form either a *pellicle* (a film of cells at the surface) or a *sediment* (an accumulation of cells at the bottom), their number will be underestimated. Further, spectrophotometry does not distinguish between living and dead cells.

Metabolic Activity Under standard temperature conditions, the rate at which a population of cells utilizes nutrients and produces wastes depends on their number. Once they establish the metabolic rate of a microorganism, scientists can indirectly estimate the number of cells in a culture by measuring changes in such things as nutrient utilization, waste production, or pH. Scientists studying environmental water samples often use this method.

Dry Weight The abundance of some microorganisms, particularly filamentous microorganisms, is difficult to measure by direct methods. Instead, these organisms are filtered from their culture medium, dried, and weighed. The *dry weight method* is suitable for broth cultures, but growth cannot be followed over time because the organisms are killed during the process.

Genetic Methods The majority of bacteria and archaea have not been grown in the laboratory, and representatives of most species are too few in number to study by direct observation. How do scientists estimate the number of such uncultured microbes?

Scientists can isolate unique DNA sequences representing uncultured prokaryotic species using genetic techniques such as *polymerase chain reaction (PCR)* and *hybridization* of DNA that codes for ribosomal RNA. For example, one study estimated that more than 100 billion bacteria and archaea, representing more than 10 million different species, are in a single gram of garden soil. (Chapter 8 discusses genetic methods in more detail.)

TELL ME WHY

Some students transfer some "gunk" from a two-week-old bacterial culture into new media. Why shouldn't they be surprised when this "death-phase" sample grows?

MICRO IN THE CLINIC FOLLOW-UP

Can a Trip to the Dentist Be Life Threatening?

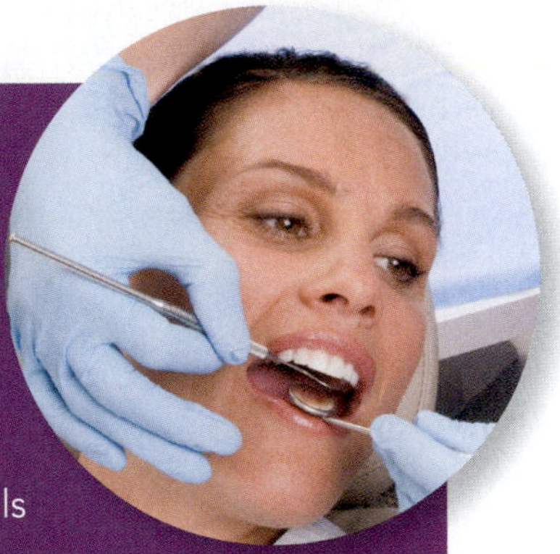

Betty's doctor shows up in the emergency room and asks her a number of questions about what she has been doing since her last treatment. She is surprised but answers his questions, wondering whether her last chemotherapy treatment is related to her current condition. She couldn't be more shocked when her doctor examines the lab results and reveals that Betty's seemingly innocuous dental cleaning is the true culprit. As a result of the procedure, bacteria were actually driven into her blood, causing septicemia, a serious infection of the blood. Because Betty's immune system is compromised (reduced) by the chemotherapy, *Streptococcus mutans*, a harmless bacterium from her teeth, has entered her bloodstream and multiplied.

Because of all the stress she's been under, Betty completely forgot that her doctor had instructed her not to get her teeth cleaned. Even more amazing, the busy hygienist had failed to ask her about chemotherapy. All health care workers should know that the bacterial biofilm typically found on teeth represents a potential threat of infection for immunosuppressed individuals during dental cleaning.

Betty is admitted to the hospital and treated with intravenous antimicrobial drugs. The combination of her quick trip to the hospital, her recovering immune system, and appropriate antimicrobial therapy will contribute to her recovery.

1. **The bacterial biofilm that caused Betty's septicemia is likely to return to the surfaces of the teeth. Why?**
2. **When septicemia is suspected, blood samples are placed into nutritive media. What growth requirements are necessary in the media to ensure that bacteria from the blood specimen will thrive?**

MM Check your answers to Micro in the Clinic Follow-Up questions in the MasteringMicrobiology Study Area.

CHAPTER SUMMARY

Growth Requirements (pp. 166–174)

1. A colony, which is a visible population of microorganisms arising from a single cell or colony-forming unit living in one place, grows in size as the number of cells increases. Most microbes live as biofilms, that is, in association with one another on surfaces.
2. Chemical **nutrients** such as carbon, hydrogen, oxygen, and nitrogen are required for the growth of microbial populations.
3. **Photoautotrophs** use carbon dioxide as a carbon source and light energy to make their own food; **chemoautotrophs** use carbon dioxide as a carbon source but catabolize organic molecules for energy. **Photoheterotrophs** are photosynthetic organisms that acquire energy from light and acquire nutrients via catabolism of organic compounds; **chemoheterotrophs** use organic compounds for both energy and carbon. **Organotrophs** acquire electrons for redox reactions from organic sources, whereas **lithotrophs** acquire electrons from inorganic sources.
4. **Obligate aerobes** require oxygen molecules as the final electron acceptor of their electron transport chains, whereas **obligate anaerobes** cannot tolerate oxygen and must use an electron acceptor other than oxygen.
5. The four toxic forms of oxygen are **singlet oxygen ($^1O_2$),** which is neutralized by pigments called **carotenoids; superoxide radicals (O_2^-),** which are detoxified by superoxide dismutase; **peroxide anion (O_2^{2-}),** which is detoxified by catalase or peroxidase; and **hydroxyl radicals (OH·),** the most reactive of the toxic forms of oxygen.
6. Microbes are described in terms of their oxygen requirements and limitations as strict **aerobes,** which require oxygen; as strict **anaerobes,** which cannot tolerate oxygen; as **facultative anaerobes,** which can live with or without oxygen; as **aerotolerant anaerobes,** which prefer anaerobic conditions but can tolerate exposure to low levels of oxygen; or as **microaerophiles,** which require low levels of oxygen.
7. Nitrogen, acquired from organic or inorganic sources, is an essential element for microorganisms. Some bacteria can reduce nitrogen gas into a more usable form via a process called **nitrogen fixation**.
8. In addition to the main elements found in microbes, very small amounts of **trace elements** are required. Vitamins are among the **growth factors,** which are organic chemicals required in small amounts for metabolism.
9. Though microbes survive within the limits imposed by a minimum growth temperature and a maximum growth temperature, an organism's metabolic activities produce the highest growth rate at the **optimum growth temperature.**
10. Microbes are described in terms of their temperature requirements as (from coldest to warmest) **psychrophiles, mesophiles, thermophiles,** or **hyperthermophiles.**
11. **Neutrophiles** grow best at neutral pH, **acidophiles** grow best in acidic surroundings, and **alkalinophiles** live in alkaline habitats.
12. Osmotic pressure can cause cells to die from either swelling and bursting or from **crenation** (shriveling). The cell walls of some microorganisms protect them from osmotic shock. **Obligate halophiles** require high osmotic pressure, whereas facultative halophiles do not require but can tolerate such conditions.
13. **Barophiles,** organisms that normally live under the extreme hydrostatic pressure at great depth below the surface of a body of water, often cannot live at the pressure found at the surface.
14. **Quorum sensing** is the process by which bacteria respond to changes in microbial density by utilizing signal and receptor molecules. **Biofilms,** which are communities of cells attached to surfaces, use quorum sensing.

Culturing Microorganisms (pp. 174–182)

1. Microbiologists culture microorganisms by transferring an **inoculum** from a clinical or environmental specimen into a **medium** such as **broth** or solid media. The microorganisms grow into a **culture.** On solid surfaces, cultures are seen as **colonies.**
2. A **clinical specimen** is a sample of human material. Standard precautions are the guidelines to protect health care professionals from infection.
3. **Pure cultures** (axenic cultures) contain cells of only one species and are derived from a **colony-forming unit (CFU)** composed of a single cell or group of related cells. To obtain pure cultures, **sterile** equipment and use of aseptic techniques are critical.
4. The **streak-plate** method allows CFUs to be isolated by streaking. The **pour-plate** technique isolates CFUs via a series of dilutions.
5. Petri dishes that are filled with solid media are called **Petri plates. Slant tubes (slants)** are test tubes containing agar media that solidified while the tube was resting at an angle.

6. A **defined medium** (also known as **synthetic medium**) provides exact known amounts of nutrients for the growth of a particular microbe. **Complex media** contain a variety of growth factors. **Selective media** either inhibit the growth of unwanted microorganisms or favor the growth of particular microbes. Microbiologists use **differential media** to distinguish among groups of bacteria. **Reducing media** provide conditions conducive to culturing anaerobes. **Transport media** are designed to move specimens safely from one location to another while maintaining the relative abundance of organisms and preventing contamination of the specimen or environment.

   ▶**VIDEO TUTOR:** *Bacterial Growth Media*

7. Special culture techniques include the use of animal and cell cultures, low-oxygen cultures, **enrichment cultures,** and **cold enrichment** cultures. A **capnophile** grows best with high CO_2 levels in addition to low oxygen levels.
8. Cultures can be preserved in the short term by *refrigeration* and in the long term by **deep-freezing** and **lyophilization.**

Growth of Microbial Populations (pp. 182–190)

1. Bacteria grow by **logarithmic,** or **exponential, growth.**

   ▶**ANIMATIONS:** *Bacterial Growth: Overview, Binary Fission*

2. A population of microorganisms doubles during its **generation time**—the time also required for a single cell to grow and divide.
3. A graph that plots the number of organisms growing in a population over time is called a **growth curve.** When organisms are grown in a broth and the growth curve is plotted on a semilogarithmic scale, the population's growth curve has four phases. In the **lag phase,** the organisms are adjusting to their environment. In the **log phase,** the population is most actively growing. In the **stationary phase,** new organisms are being produced at the same rate at which they are dying. In the **death phase,** the organisms are dying more quickly than they can be replaced by new organisms.

   ▶**ANIMATIONS:** *Bacterial Growth Curve*

4. A **chemostat** is a continuous culture device that maintains a desired phase of microbial population growth by adding limited amounts and kinds of nutrients while removing an equal amount of old medium.
5. Direct methods for estimating population size that do not require incubation are microscopic counts and electronic counters, including flow cytometry. Direct methods requiring incubation—include **serial dilution** with **viable plate counts, membrane filtration,** and the **most probable number (MPN) method**.
6. Indirect methods include measurements of turbidity, metabolic activity, and dry weight and analysis of numbers and kinds of unique genetic sequences.

QUESTIONS FOR REVIEW

Answers to the Questions for Review (except Short Answer questions) begin on p. A-1.

Multiple Choice

1. Which of the following can grow in a Petri plate on a laboratory table?
   a. an anaerobic bacterium
   b. an aerobic bacterium
   c. viruses on an agar surface
   d. all of the above
2. This statement, "In the laboratory, a sterile inoculating loop is moved across the agar surface in a culture dish, thinning a sample and isolating individuals," describes which of the following?
   a. broth culture
   b. pour plate
   c. streak plate
   d. dilution plate
3. Superoxide dismutase ________________.
   a. causes hydrogen peroxide to become toxic
   b. detoxifies superoxide radicals
   c. neutralizes singlet oxygen
   d. is missing in aerobes
4. The most reactive of the four toxic forms of oxygen is ________________.
   a. the hydroxyl radical
   b. the peroxide anion
   c. the superoxide radical
   d. singlet oxygen
5. Microaerophiles that grow best with a high concentration of carbon dioxide in addition to a low level of oxygen are called ________________.
   a. aerotolerant
   b. capnophiles
   c. facultative anaerobes
   d. fastidious
6. Which of the following is *not* a growth factor for various microbes?
   a. cholesterol
   b. water
   c. vitamins
   d. heme
7. Organisms that preferentially may thrive in icy waters are described as ________________.
   a. barophiles
   b. thermophiles
   c. mesophiles
   d. psychrophiles
8. Barophiles ________________.
   a. cannot cause diseases in humans
   b. live at normal barometric pressure
   c. die if put under high pressure
   d. thrive in warm air

9. Which of the following terms best describes an organism that cannot exist in the presence of oxygen?
   a. obligate aerobe
   b. facultative aerobe
   c. obligate anaerobe
   d. facultative anaerobe

10. In a defined medium, ________________.
   a. the exact chemical composition of the medium is known
   b. agar is available for microbial nutrition
   c. blood may be included
   d. organic chemicals are excluded

11. Which of the following is most useful in representing population growth on a graph?
   a. logarithmic reproduction of the growth curve
   b. a semilogarithmic graph using a log scale on the y-axis
   c. an arithmetic graph of the lag phase followed by a logarithmic section for the log, stationary, and death phases
   d. none of the above would best represent a population growth curve

12. Which of the following methods is best for counting fecal bacteria from a stream to determine the safety of the water for drinking?
   a. dry weight
   b. turbidity
   c. viable plate counts
   d. membrane filtration

13. A Coulter counter is ________________.
   a. a statistical estimation using 15 dilution tubes and a table of numbers to estimate the number of bacteria per milliliter
   b. an indirect method of counting microorganisms
   c. a device that directly counts microbes as they pass through a tube in front of an electronic detector
   d. a device that directly counts microbes that are differentially stained with fluorescent dyes

14. Lyophilization can be described as ________________.
   a. freeze drying
   b. deep freezing
   c. refrigeration
   d. pickling

15. Quorum sensing is ________________.
   a. the ability to respond to changes in population density
   b. a characteristic allowing secretion of a matrix
   c. dependent on direct contact among cells
   d. associated with colonies in broth culture

Fill in the Blanks

1. All cells require a source of __electrons__ for redox reactions.
2. A toxic form of oxygen, __singlet__ oxygen, is molecular oxygen with electrons that have been boosted to a higher energy state.
3. All cells recycle the essential element __nitrogen__ from amino acids and nucleotides.
4. __Growth factors__ are small organic molecules that are required in minute amounts for metabolism.
5. The lowest temperature at which a microbe continues to metabolize is called its __minimum growth__. temp.
6. Cells that shrink in hypertonic solutions such as salt water are responding to __osmotic__ pressure.
7. Obligate __halophiles__ exist in salt ponds because of their ability to withstand high osmotic pressure.
8. __Carotenoid__ pigments protect many phototrophic organisms from photochemically produced singlet oxygen.
9. Microbes that reduce N_2 to NH_3 engage in nitrogen __fixation__.
10. A student observes a researcher streaking a plate numerous times, flaming the loop between streaks. The researcher is likely using the __streak plate__ method to isolate microorganisms.
11. Chemolithotrophs *acquire* electrons from (organic/inorganic) __inorganic__ compounds.

VISUALIZE IT!

1. Label each of these thioglycolate tubes to indicate the oxygen requirements of the microbes growing in them.

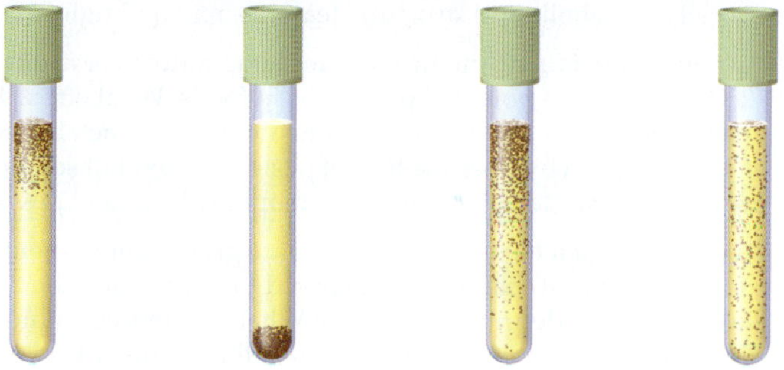

a. ________ b. ________ c. ________ d. ________

2. Describe the type of hemolysis shown by the pathogen *Staphylococcus aureus* pictured here.

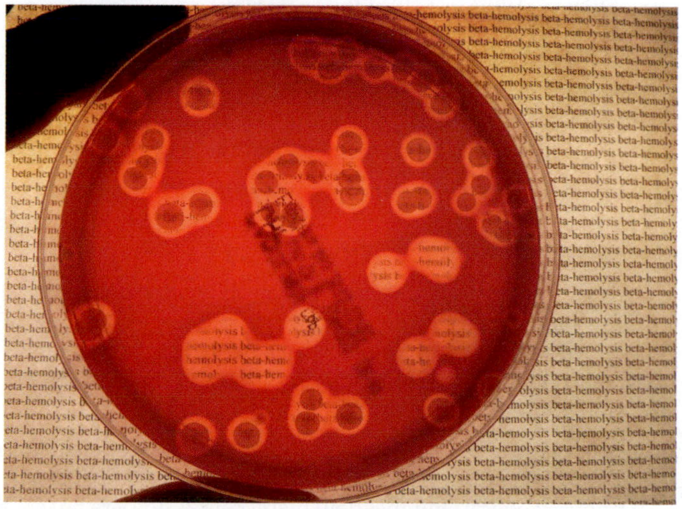

Short Answer

1. High temperature affects the shape of particular molecules. How does this affect the life of a microbe?
2. Support or refute the following statement: Microbes cannot tolerate the low pH of the human stomach.
3. Explain quorum sensing and describe how it is related to biofilm formation.
4. Why must media, vessels, and instruments be sterilized before they are used for microbiological procedures?
5. Why is agar used in microbiology?
6. What is the difference between complex media and defined media?
7. Draw and label the four distinct phases of a bacterial growth curve. Describe what is happening within the culture as it passes through the phases.
8. If there are 47 cells in 1 µl of sewage, how many cells are there in a liter?
9. List three indirect methods of counting microbes.
10. List five direct methods of counting microbes.
11. Explain the differences among photoautotrophs, chemoautotrophs, photoheterotrophs, chemoheterotrophs, organotrophs, and lithotrophs.
12. Contrast the media described in Tables 6.3 and 6.4 on pp. 178, and 180. Why is *E. coli* medium described as defined, whereas MacConkey medium and blood agar are defined as complex?
13. How does a chemostat maintain a constant population size?

CRITICAL THINKING

1. A scientist describes an organism as a chemoheterotrophic, aerotolerant, mesophilic, facultatively halophilic coccus. Describe the cell's metabolic and structural features in plain English.
2. Pasteurization is a technique that uses temperatures of about 72°C to neutralize potential pathogens in foods. What effect does this temperature have on the enzymes and cellular metabolism of pathogens? Why does the heat of pasteurization kill some microorganisms yet fail to affect thermophiles?
3. Two cultures of a facultative anaerobe are grown under identical conditions, except that one was exposed to oxygen and the other was completely deprived of oxygen. What differences would you expect to see between the dry weights of the cultures? Why?
4. Some organisms require riboflavin (vitamin B_2) to make FAD. For what purpose do they use FAD?
5. A scientist inoculates a bacterium into a complex nutrient slant tube. The bacterium forms only a few colonies on the slanted surface but grows prolifically in the depth of the agar. Describe the oxygen requirements of the bacterium.
6. How can regions within biofilms differ in their chemical content?
7. A scientific article describes a bacterium as an obligate microaerophilic chemoorganoheterotroph. Describe the oxygen and nutritional characteristics of the bacterium in everyday language.
8. Microorganisms require phosphorus, sulfur, iron, and magnesium for metabolism. What specifically are these elements used for in microbial metabolism? (Review Chapters 2 and 5.)
9. The bacterium *Desulforudis audaxviator* lives almost 2 miles underground, deriving energy from sulfate, acquiring electrons from hydrogen, and building organic molecules from inorganic carbon found in surrounding rocks. Describe the nutritional classifications of *D. audaxviator*.
10. Starting with 10 bacterial cells per milliliter in a sufficient amount of complete culture medium with a 1-hour lag phase and a 30-minute generation time, how many cells will there be in a liter of medium at the end of 2 hours? At the end of 7 hours?
11. Suppose you perform a serial dilution of 0.1-ml sample from a liter of culture medium as illustrated in Figure 6.23. The 10^{-3} plate gives 440 colonies, and the 10^{-4} plate gives 45 colonies. Calculate the approximate number of bacteria in the original liter.
12. How might the study of biofilms benefit humans?
13. The filamentous bacterium *Beggiatoa* gets its carbon from carbon dioxide and its electrons and energy from hydrogen sulfide. What is its nutritional classification?

    Not all organisms are easy to classify. For instance, the single-celled eukaryote *Euglena granulata* typically uses light energy and gets its carbon from carbon dioxide. However, when it is cultured on a suitable medium in the dark, this microbe utilizes energy and carbon solely from organic compounds. What is the nutritional classification of *Euglena*?
14. Given that *Haemophilus ducreyi* is a chemoheterotrophic pathogen that requires heme as a growth factor, deduce how this bacterium phosphorylates most of its ADP to form ATP. Defend your answer.
15. Examine the graph in Figure 6.4. Note that the growth rate increases slowly until the optimum is reached, and then it declines steeply at higher temperatures. In other words, organisms tolerate a wider range of temperatures below their optimal temperature than they do above the optimum. Explain this observation.
16. Over 100 years ago, doctors infected syphilis victims with malaria parasites to induce a high fever. Surprisingly, such treatment often cured the syphilis infection. Explain how this could occur.
17. Using the terms in Figure 6.8a, describe the shape, margin, pigmentation, and optical properties of two bacterial colonies seen in Figure 6.10b.
18. Why have scientists been unable to axenically culture *Treponema pallidum* in a laboratory medium?
19. Examine the ingredients of MacConkey agar as listed in Table 6.4 on p. 180. Does this medium select for Gram-positive or Gram-negative bacteria? Explain your reasoning.

20. The sole carbon source in citrate medium is citric acid (citrate). Why might a laboratory microbiologist use this medium?
21. Using as many of the following terms as apply—selective, differential, broth, solid, defined, and complex—categorize each of the media listed in Table 6.3 on p. 178 and Table 6.4 on p. 180.
22. Beijerinck used the concept of enrichment culture to isolate aerobic and anaerobic nitrogen-fixing bacteria, sulfate-reducing bacteria, and sulfur-oxidizing bacteria. What kind of selective media could he have used for isolating each of these four types of microbes?
23. Viable plate counts are used to estimate population size when the density of microorganisms is high, whereas membrane filtration is used when the density is low. Why is a viable plate count appropriate when the density is high but not when the density is low?

CONCEPT MAPPING

Using the following terms, draw a concept map that describes culture media. For a sample concept map, see p. 94. Or, complete this and other concept maps online by going to the MasteringMicrobiology Study Area.

Blood agar
Differential
Enriched
Fastidious microorganisms
Fermentation broths (such as phenol red)
General purpose
MacConkey agar
Nutrient broth
Sabouraud agar
Selective
Selective and differential
Trypticase soy agar
Unwanted microorganisms
Visible differences between microorganisms

7 Microbial Genetics

MICRO IN THE CLINIC

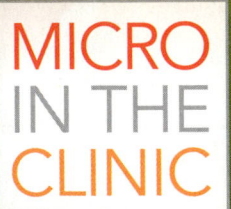

The Case of the Matching Tattoos

Ingrid is an adventurous 30-year-old who loves to travel. Over the years, she's been to France, Italy, Australia, Egypt, Japan, and Thailand. Her motto is "Work hard, play hard," and she is currently saving up for yet another big trip: a backpacking trek through Nepal. Prior to her trip, Ingrid goes to the local blood bank to donate a unit of blood. Two weeks later, Ingrid receives a letter from the blood bank alerting her that she screened positive for hepatitis C. Hepatitis C is typically spread via blood transfusion or injection drug use. Ingrid visits her doctor to confirm the blood test results, and tells him that she's never had a blood transfusion nor tried drugs. He then asks her about the sunburst tattoo on Ingrid's right shoulder.

Ingrid explains that five years ago she went on a trip to Thailand with her two best girlfriends. They had the time of their lives—visiting temples, riding elephants, eating amazing food, and dancing through the night. On the last day of their trip, Ingrid suggested that they all get matching tattoos at a local tattoo parlor, as a symbol of their friendship and as a way to commemorate their travels. Her friends agreed, and all three girls received sunburst tattoos. Ingrid tells her doctor that the tattoo artist assured them the tattoo needles were sterilized.

Could Ingrid's tattoo have anything to do with her abnormal test results? If so, could her girlfriends also be at risk for hepatitis? Turn to the end of the chapter (p. 234) to find out.

MM **Explore More:** Test your readiness and apply your knowledge with dynamic learning tools at MasteringMicrobiology.

Genetics is the study of inheritance and inheritable traits as expressed in an organism's genetic material. Geneticists study many aspects of inheritance, including the physical structure and function of genetic material, mutations, and the transfer of genetic material among organisms. In this chapter, we will examine these topics as they apply to microorganisms, the study of which has formed much of the basis of our understanding of human, animal, and plant genetics.

The Structure and Replication of Genomes

LEARNING OUTCOME

7.1 Compare and contrast the genomes of prokaryotes and eukaryotes.

The **genome** (jē´nōm) of a cell or virus is its entire genetic complement, including both its **genes**—specific sequences of nucleotides that code for RNA or polypeptide molecules—and nucleotide sequences that connect genes to one another. The genomes of cells and DNA viruses are composed solely of molecules of deoxyribonucleic acid (DNA), whereas RNA viruses use ribonucleic acid instead. (We will examine the genomes of viruses in more detail in Chapter 13.) The remainder of this chapter focuses on bacterial genomes—their structure, replication, function, mutation, and repair and how they compare and contrast with eukaryotic genomes and with the genomes of archaea. We begin by examining the structure of nucleic acids.

The Structure of Nucleic Acids

LEARNING OUTCOME

7.2 Describe the structure of DNA, and discuss how it facilitates the ability of DNA to act as genetic material.

Nucleic acids are polymers of basic building blocks called **nucleotides**. Each nucleotide is made up of phosphate attached to a *nucleoside*, which is in turn made up of a pentose sugar (ribose in RNA and deoxyribose in DNA) attached to one of five nitrogenous bases: guanine (G), cytosine (C), thymine (T), adenine (A), or uracil (U) **(FIGURE 7.1)**.

The bases of nucleotides hydrogen-bond to one another in specific ways called complementary **base pairs (bp)**: in DNA, the complementary bases thymine and adenine bond to one another with two hydrogen bonds **(FIGURE 7.1a)**, whereas in RNA, uracil, not thymine, forms two hydrogen bonds with adenine **(FIGURE 7.1b)**. In both DNA and RNA, the complementary bases guanine and cytosine bond to one another with three hydrogen bonds **(FIGURE 7.1c)**.

Deoxyribonucleotides are linked through their sugars and phosphates to form the two backbones of a helical, double-stranded DNA (dsDNA) molecule **(FIGURE 7.1d)**. The carbon atoms of deoxyribose are numbered 1′ (pronounced "one prime") through 5′ ("five prime"). One end of a DNA strand is called the 5′ end because it terminates in a phosphate group attached to a 5′ carbon; the opposite (3′) end terminates with a hydroxyl group bound to a 3′ carbon of deoxyribose. The two strands are constructed similarly but are oriented in opposite directions to each other; one strand runs in a 5′ to 3′ direction, while the other runs 3′ to 5′. Scientists say the two strands are *antiparallel*. The base pairs extend into the middle of the molecule in a way reminiscent of the steps of a spiral staircase.

The lengths of DNA molecules are not usually given in metric units; instead, the length of a DNA molecule is expressed in base pairs. For example, the genome of the bacterium *Carsonella ruddii* (kar-son-el´ă rŭd´ē-ē) is 159,662 bp long, making it the smallest known cellular genome.

The structure of DNA helps explain its ability to act as genetic material. First, the linear sequence of nucleotides carries the instructions for the synthesis of polypeptides and RNA molecules—in much the way a sequence of letters carries information used to form words and sentences. Second, the complementary structure of the two strands allows a cell to make exact copies to pass to its progeny. We will examine the genetic code and DNA replication shortly.

The amount of DNA in a genome can be extraordinary, as some examples will illustrate. The bacterium *Escherichia coli* (esh-ĕ-rik´ē-ă kō´lī) is approximately 2 μm long and 1 μm in diameter, but its genome consists primarily of a 4.6×10^6 bp DNA molecule that is about 1600 μm long—800 times longer than the cell. The human genome has about 6 billion bp in 46 nuclear DNA molecules and numerous copies of a unique mitochondrial DNA molecule, and the entire genome would be about 3 meters (3,000,000 μm) long if all DNA molecules from a single cell were laid end to end. Most of a human cellular genome is packed into a nucleus that is typically only 5 μm in diameter. This is like packing 45 miles of thread into a golf ball and still being able to access any particular section of the thread. To understand how cells package such prodigious amounts of DNA into such small spaces, we must first understand that bacteria, archaea, and eukaryotes package DNA in different ways. We begin by examining the structure of prokaryotic genomes.

The Structure of Prokaryotic Genomes

The DNA of prokaryotic genomes is found in two structures: chromosomes and plasmids.

Prokaryotic Chromosomes

Prokaryotic cells, both bacterial and archaeal, package the main portion of their DNA, along with associated molecules of protein and RNA, as one or two distinct **chromosomes**.[1] Prokaryotic cells have a single copy of each chromosome and are called *haploid* cells.

A typical prokaryotic chromosome **(FIGURE 7.2a)** consists of a circular molecule of DNA localized in a region of the cytoplasm called the **nucleoid**. With few exceptions, no

[1]From Greek *chroma*, meaning "color" (because they typically stain darkly in eukaryotes, where they were first discovered), and *soma*, meaning "body."

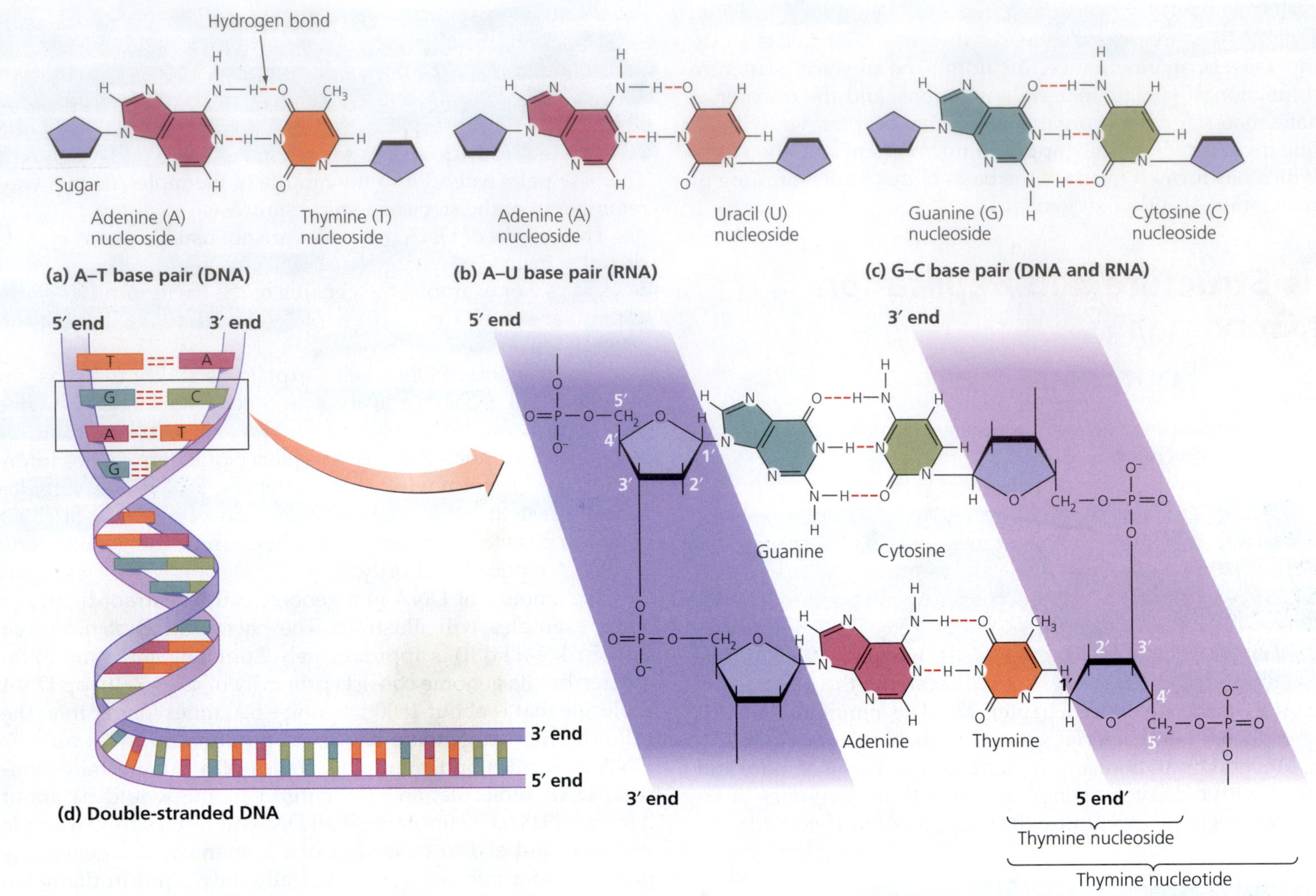

▲ **FIGURE 7.1 The structure of nucleic acids.** Nucleic acids are polymers of nucleotides consisting of nucleosides (a pentose sugar and a nitrogenous base) bound to a phosphate. **(a)** Base pairing between the complementary bases adenine (A) and thymine (T) formed by two hydrogen bonds, found in DNA only. **(b)** Base pairing between adenine and uracil (U), found in RNA only. Notice the structural similarities between thymine and uracil. **(c)** Base pairing between the complementary bases guanine (G) and cytosine (C) formed by three hydrogen bonds, found in both DNA and RNA. **(d)** Double-stranded DNA, which consists of antiparallel strands of nucleotides held to one another by the hydrogen bonding between complementary bases. *What structures do DNA nucleotides and RNA nucleotides have in common?*

Figure 7.1 DNA and RNA nucleotides are each composed of a pentose sugar, a phosphate, and a nitrogenous base.

membrane surrounds a nucleoid, though the chromosome is packed in such a way that a distinct boundary is visible between the nucleoid and the rest of the cytoplasm. Chromosomal DNA is folded into loops that are 50,000 to 100,000 bp long **(FIGURE 7.2b)** held in place by molecules of protein and RNA. Archaeal DNA is wrapped around globular proteins called **histones**. The enzyme *gyrase* further folds and supercoils the entire prokaryotic chromosome like a skein of yarn into a compact mass.

For many years scientists thought that each prokaryote had only a single circular chromosome, but we now know that there are numerous exceptions. For example, *Epulopiscium* (ep´yoo-lō-pis´sē-ŭm), a giant bacterium, has as many as hundreds or thousands of identical chromosomes. Some bacterial species contain two different chromosomes, and at least one member of such a pair may be linear. *Agrobacterium tumefaciens* (ag´rō-bak-tēr´ē-um tū´me-fāsh-enz), a bacterium used to transfer genes into plants, is an example of a prokaryote with two chromosomes, one circular and one linear.

Plasmids

LEARNING | OUTCOME

7.3 Describe the structure and function of plasmids.

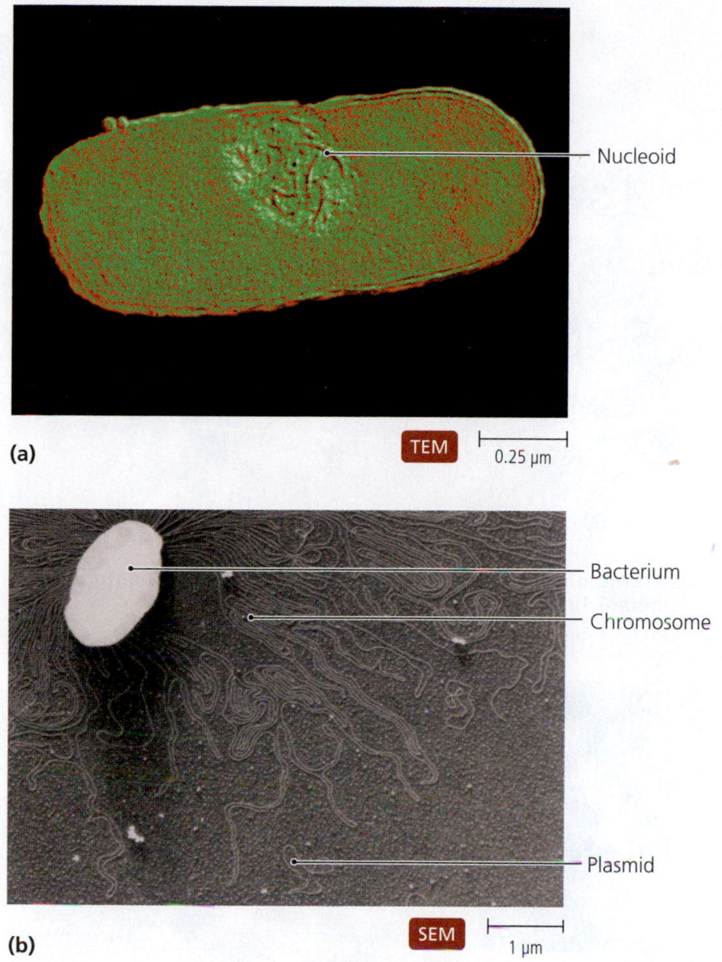

▲ **FIGURE 7.2 Bacterial genome. (a)** Bacterial chromosomes are packaged in a region of the cytosol called the nucleoid, which is not surrounded by a membrane. **(b)** The packing of a circular bacterial chromosome into loops, as seen after the cell was gently broken open to release the chromosome. Extrachromosomal DNA in the form of plasmids is also visible.

In addition to chromosomes, many prokaryotic cells contain one or more **plasmids**, which are small molecules of DNA that replicate independently of the chromosome. Plasmids are usually circular and 1% to 5% of the size of a prokaryotic chromosome (see Figure 7.2b), ranging in size from a few thousand base pairs to a few million base pairs. Each plasmid carries information required for its own replication and often for one or more cellular traits. Typically, genes carried on plasmids are not essential for normal metabolism, for growth, or for cellular reproduction but can confer advantages to the cells that carry them.

Researchers have identified many types of plasmids (sometimes called *factors*), including the following:

- *Fertility (F) plasmids* carry instructions for *conjugation*, a process by which some bacterial cells transfer DNA to other bacterial cells. We will consider conjugation in more detail near the end of this chapter.
- *Resistance (R) plasmids* carry genes for resistance to one or more antimicrobial drugs or heavy metals. By processes we will discuss shortly, certain cells can transfer resistance plasmids to other cells, which then acquire resistance to the same antimicrobial chemicals. One example of the effects of an R plasmid involves strains of *E. coli* that have acquired resistance to the antimicrobials ampicillin, tetracycline, and kanamycin from a strain of bacteria in the genus *Pseudomonas* (soo-dō-mō´nas).
- *Bacteriocin* (bak-tēr´ē-ō-sin) *plasmids* carry genes for proteinaceous toxins called *bacteriocins,* which kill bacterial cells of the same or similar species that lack the plasmid. In this way a bacterium containing this plasmid can kill its competitors.
- *Virulence plasmids* carry instructions for structures, enzymes, or toxins that enable a bacterium to become pathogenic. For example, *E. coli,* a normal resident of the human gastrointestinal tract, causes diarrhea only when it carries plasmids that code for certain toxins.

Now that we have examined the structure of prokaryotic genomes, we turn to the structure of eukaryotic genomes.

The Structure of Eukaryotic Genomes

Eukaryotic genomes consist of both nuclear and extranuclear DNA.

Nuclear Chromosomes

LEARNING **OUTCOME**

7.4 Compare and contrast prokaryotic and eukaryotic chromosomes.

Typically, eukaryotic cells have more than one nuclear chromosome in their genomes, though one species of Australian ant has a single chromosome per nucleus, and some eukaryotic cells, such as mammalian red blood cells, lose their chromosomes as they mature. Eukaryotic cells are often *diploid;* that is, they have two copies of each chromosome.

Eukaryotic chromosomes differ from their typical prokaryotic counterparts in that they are all linear (rather than circular) and are sequestered within a nucleus. A nucleus is an organelle surrounded by two membranes, which together are called the *nuclear envelope.* Given that a typical eukaryotic cell must package substantially more DNA than its prokaryotic counterpart, it is not surprising that nuclear chromosomes are more elaborate than those of prokaryotes.

Most eukaryotic chromosomes are composed of DNA and globular eukaryotic histones,[2] which are similar to archaeal histones. DNA, which has an overall negative electrical charge, wraps around the positively charged histones to

[2]Dinoflagellates, a group of single-celled aquatic microorganisms, are the only eukaryotes without histones.

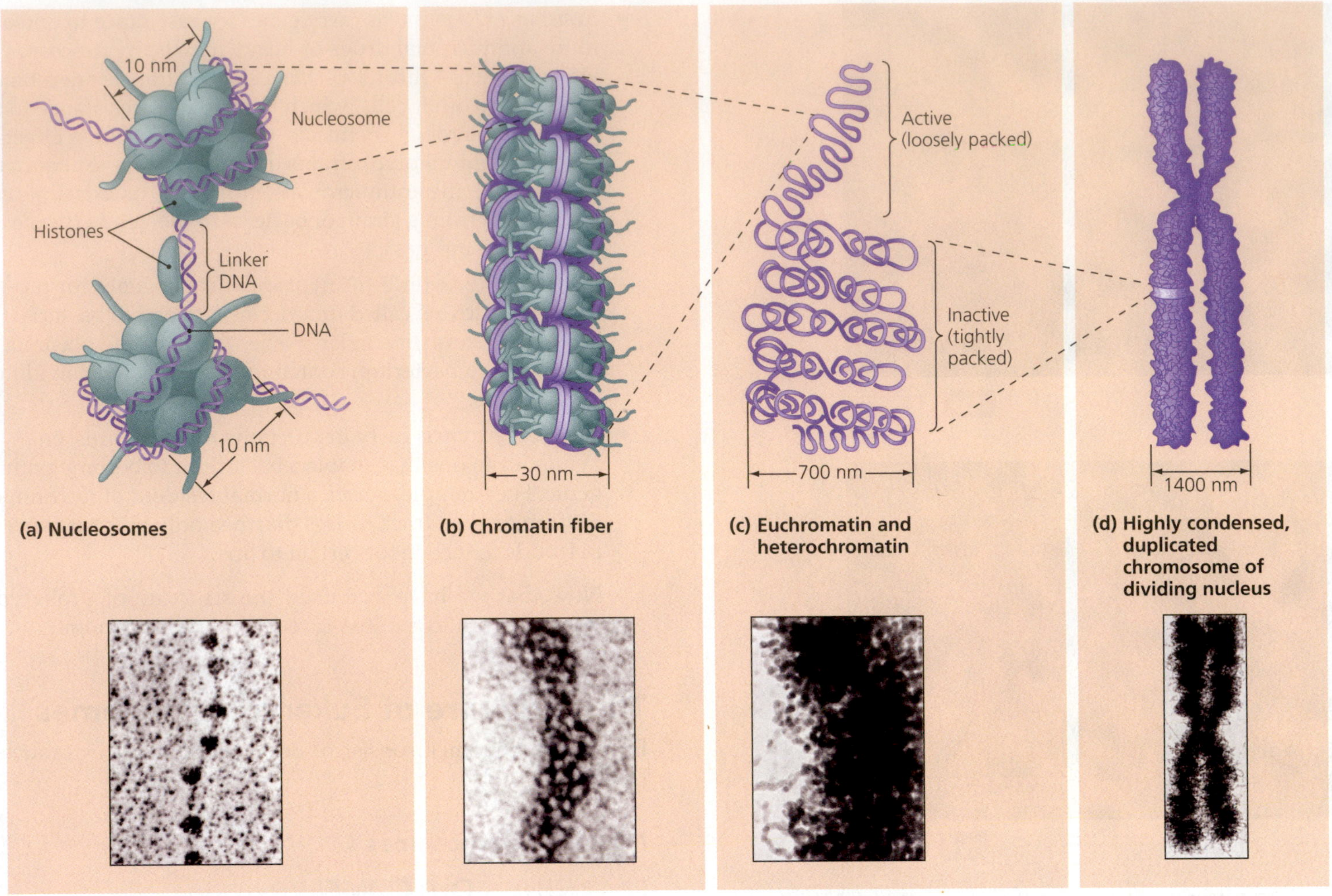

▲ **FIGURE 7.3 Eukaryotic nuclear chromosomal packaging.** **(a)** Histones stabilize and package DNA to form nucleosomes connected by linker DNA. **(b)** Nucleosomes clump to form chromatin fibers. **(c)** Chromatin fibers fold and are organized into active euchromatin and inactive heterochromatin. **(d)** During nuclear division (mitosis), duplicated chromatin fully condenses into a mitotic chromosome that is visible by light microscopy. If the nucleosomes were actually the size shown in the artist's illustration (a), the chromosome in (d) would be 20 m (about 65 feet) long.

form 10-nm-diameter beads called **nucleosomes** **(FIGURE 7.3a)**. Nucleosomes clump with other proteins to form **chromatin fibers** that are about 30 nm in diameter **(FIGURE 7.3b)**. Except during *mitosis* (nuclear division), chromatin fibers are dispersed throughout the nucleus and are too thin to be resolved without the extremely high magnification of electron microscopes. In regions of the chromosome where genes are active, the chromatin fibers are loosely packed to form *euchromatin* (yū-krō´mă-tin); inactive DNA is more tightly packed and is called *heterochromatin* (het´-er-ō-krō´mă-tin) **(FIGURE 7.3c)**.

Prior to mitosis, a cell replicates its chromosomes and then condenses them into pairs of chromosomes visible by light microscopy **(FIGURE 7.3d)**. One molecule of each pair is destined for each daughter nucleus. (Chapter 12 discusses mitosis in more detail.) The net result is that each DNA molecule is packaged as a mitotic chromosome that is 50,000 times shorter than its extended length.

Extranuclear DNA of Eukaryotes

Not all of the DNA of a eukaryotic genome is contained in its nuclear chromosomes; most eukaryotic cells also have mitochondria, and plant, algal, and some protozoan cells have chloroplasts that also contain DNA. DNA molecules of mitochondria and chloroplasts are circular and resemble the circular chromosomes of prokaryotes. Genes located on these "prokaryotic" chromosomes code for about 5% of the RNA and polypeptides required for the organelle's replication and function; nuclear DNA codes for the remaining 95% of the organelle's RNA molecules and polypeptides. Some proteins have a quaternary structure formed from the association of individual polypeptides (see Figure 2.24). Interestingly, polypeptides coded by mitochondrial or chloroplast chromosomes do not alone constitute any functional proteins. Rather, they become functional only when associated with polypeptides coded by nuclear chromosomes.

TABLE 7.1 Characteristics of Microbial Genomes

	Bacteria	Archaea	Eukarya
Number of chromosomes	Single (haploid) copies of one or rarely two	One (haploid)	With one exception, two or more, typically diploid
Plasmids present?	In some cells; frequently more than one per cell	In some cells	In some fungi, algae, and protozoa
Type of nucleic acid	Circular or linear dsDNA	Circular dsDNA	Linear dsDNA in nucleus; circular dsDNA in mitochondria, chloroplasts, and plasmids
Location of DNA	In nucleoid of cytoplasm and in plasmids	In nucleoid of cytoplasm and in plasmids	In nucleus and in mitochondria, chloroplasts, and plasmids in cytosol
Histones present?	No, though chromosome is associated with a small amount of nonhistone protein	Yes	Yes

In addition to the extranuclear DNA in their mitochondria, some fungi, algae, and protozoa carry plasmids. For instance, most strains of the yeast *Saccharomyces cerevisiae* (sak-ă-rō-mi´sēz se-ri-vis´ē-ī) contain about 70 copies of a plasmid known as a *2-μm circle.* Each 2-μm circle is about 6300 bp long and has four protein-encoding genes that are involved solely in replicating the plasmid and confer no other traits to the cell.

In summary, the haploid genome of a prokaryotic cell consists of both chromosomal DNA, which is usually in a single circular chromosome, and all extrachromosomal DNA in the form of plasmids that are present. In contrast, a eukaryotic genome consists of nuclear chromosomal DNA in one or more linear chromosomes, plus all the extranuclear DNA in mitochondria, chloroplasts, and any plasmids that are present. The genomes of prokaryotes and eukaryotes are compared and contrasted in **TABLE 7.1**.

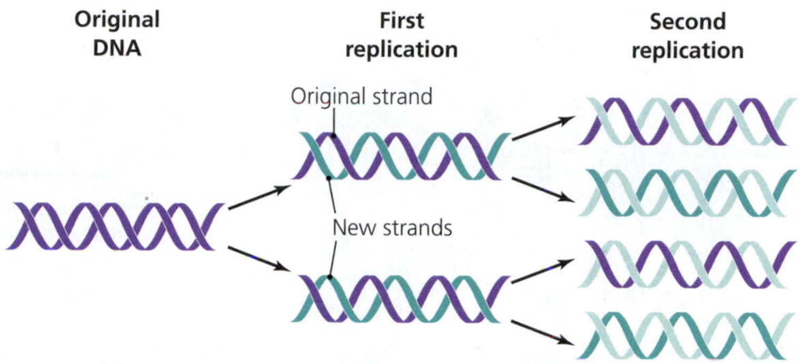

▲ **FIGURE 7.4 Semiconservative model of DNA replication.** Each of the two strands of the original molecule serves as a template for the synthesis of a new, complementary strand.

DNA Replication

LEARNING | **OUTCOMES**

7.5 Describe the replication of DNA as a semiconservative process.

7.6 Compare and contrast the synthesis of leading and lagging strands in DNA replication.

7.7 Contrast bacterial DNA replication with that of eukaryotes.

DNA replication is an anabolic polymerization process that allows a cell to make copies of its genome. Though bacterial, archaeal, and eukaryotic cells package DNA differently, all three types employ similar mechanisms for DNA replication. ▶**ANIMATIONS:** ***DNA Replication: Overview***

The key to DNA replication is the complementary structure of the two strands: Adenine and guanine in one strand bond with thymine and cytosine, respectively, in the other. DNA replication is a simple concept—a cell separates the two original strands and uses each as a template for the synthesis of a new complementary strand. Biologists say that DNA replication is *semiconservative* because each daughter DNA molecule is composed of one original strand and one new strand **(FIGURE 7.4)**.

All polymerization processes require monomers (building blocks) and energy. *Triphosphate deoxyribonucleotides*—DNA nucleotides with three phosphate groups linked together by two high-energy bonds—serve both functions in DNA replication. In other words, the building blocks of DNA carry within themselves the energy required for DNA synthesis. The structure of guanosine triphosphate deoxyribonucleotide (dGTP) **(FIGURE 7.5)** differs from that of cytidine triphosphate (dCTP), thymidine triphosphate (dTTP), and adenosine triphosphate (dATP) only in the kind of base present. dATP has a structure similar to that of the energy storage molecule ATP, except that ATP is a ribonucleotide rather than a deoxyribonucleotide (see Figure 2.27). The following sections focus on bacterial DNA replication and then consider small differences in the process in eukaryotes. Archaeal processes are not as well characterized and are not examined here.

Initial Processes in Bacterial DNA Replication

DNA replication begins at a specific sequence of nucleotides called an *origin.* An enzyme called DNA *helicase* locally "unzips" the DNA molecule by breaking the hydrogen bonds between complementary nucleotide bases, which exposes the bases in a *replication fork* **(FIGURE 7.6a)**. Other protein molecules stabilize

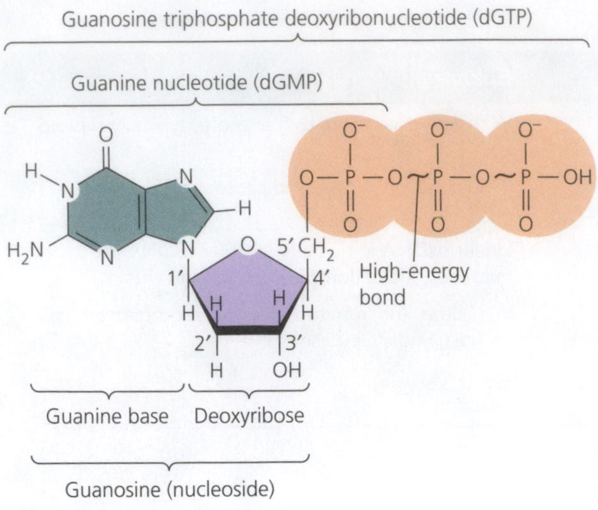

(a)

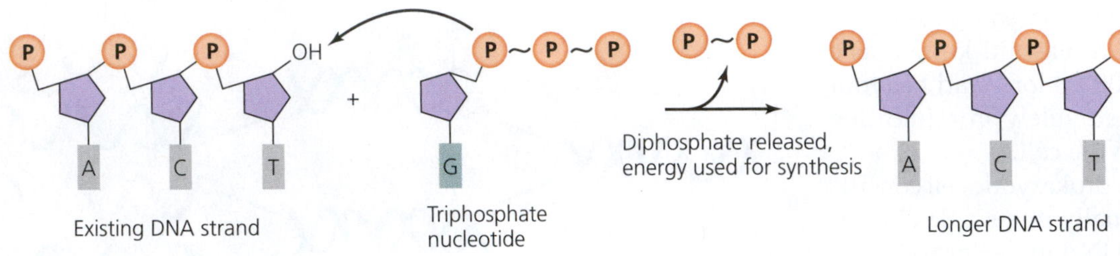

(b)

◀ **FIGURE 7.5 The dual role of triphosphate deoxyribonucleotides as building blocks and energy sources in DNA synthesis. (a)** Guanosine triphosphate deoxyribonucleotide (dGTP), like all the triphosphate monomers of DNA, is a nucleotide to which two additional phosphate groups are attached. **(b)** The energy required for DNA polymerization (the addition of nucleotide building blocks to a DNA strand) is carried by each triphosphate nucleotide in the high-energy bonds between phosphate groups. *What is the difference between dGTP and guanosine triphosphate ribonucleotide (rGTP)?*

Figure 7.5 *This molecule (dGTP) contains deoxyribose; rGTP contains ribose.*

the separated single strands so that they do not rejoin while replication proceeds. ▶**ANIMATIONS:** ***DNA Replication: Forming the Replication Fork***

After helicase untwists and separates the strands, a molecule of an enzyme called *DNA polymerase* (po-lim´er-ās) binds to each strand. Scientists have identified five kinds of bacterial DNA polymerase. These five enzymes vary in their specific functions, but all of them share one important feature—they catalyze synthesis of DNA by the addition of new nucleotides only to a hydroxyl group at the 3′ end of a nucleic acid. **Life in a Hot Tub** on p. 204 describes a type of DNA polymerase that has become a mainstay of genomic investigations. All DNA polymerases replicate DNA by adding nucleotides in only one direction—5′ to 3′—like a jeweler stringing pearls to make a necklace, adding them one at a time, always moving from one end of the string to the other. DNA polymerase III is the usual enzyme of DNA replication in bacteria.

Because the two original (template) strands are antiparallel, cells synthesize new strands in two different ways. One new strand, called the **leading strand**, is synthesized continuously—5′ to 3′—as a single long chain of nucleotides. The other new strand, called the **lagging strand**, is also synthesizd 5′ to 3′ but in short segments that are later joined. We will consider synthesis of the leading strand before examining replication of the lagging strand even though the two processes occur simultaneously. ▶**ANIMATIONS:** ***DNA Replication: Replication Proteins***

Synthesis of the Leading Strand

A cell synthesizes a leading strand toward the replication fork in the following series of five steps, the first three of which are shown in **FIGURE 7.6b**:

1. An enzyme called *primase* synthesizes a short RNA molecule that is complementary to the template DNA strand. This *RNA primer* provides the 3′ hydroxyl group required by DNA polymerase III.
2. Triphosphate deoxyribonucleotides form hydrogen bonds with their complements in the parental strand. Adenine nucleotides bind to thymine nucleotides, and guanine nucleotides bind to cytosine nucleotides.
3. Using the energy in the high-energy bonds of the triphosphate deoxyribonucleotides, DNA polymerase III covalently joins them one at a time to the leading strand. DNA polymerase III can add about 500 to 1000 nucleotides per second to a new strand.
4. DNA polymerase III also performs a proofreading function (not shown). About one out of every 100,000 nucleotides is mismatched with its template; for instance, a guanine might become incorrectly paired with a thymine. DNA polymerase III recognizes most of these errors and removes the incorrect nucleotides before proceeding with synthesis. This role, known as the *proofreading exonuclease function*, acts like the backspace key on a keyboard, removing the

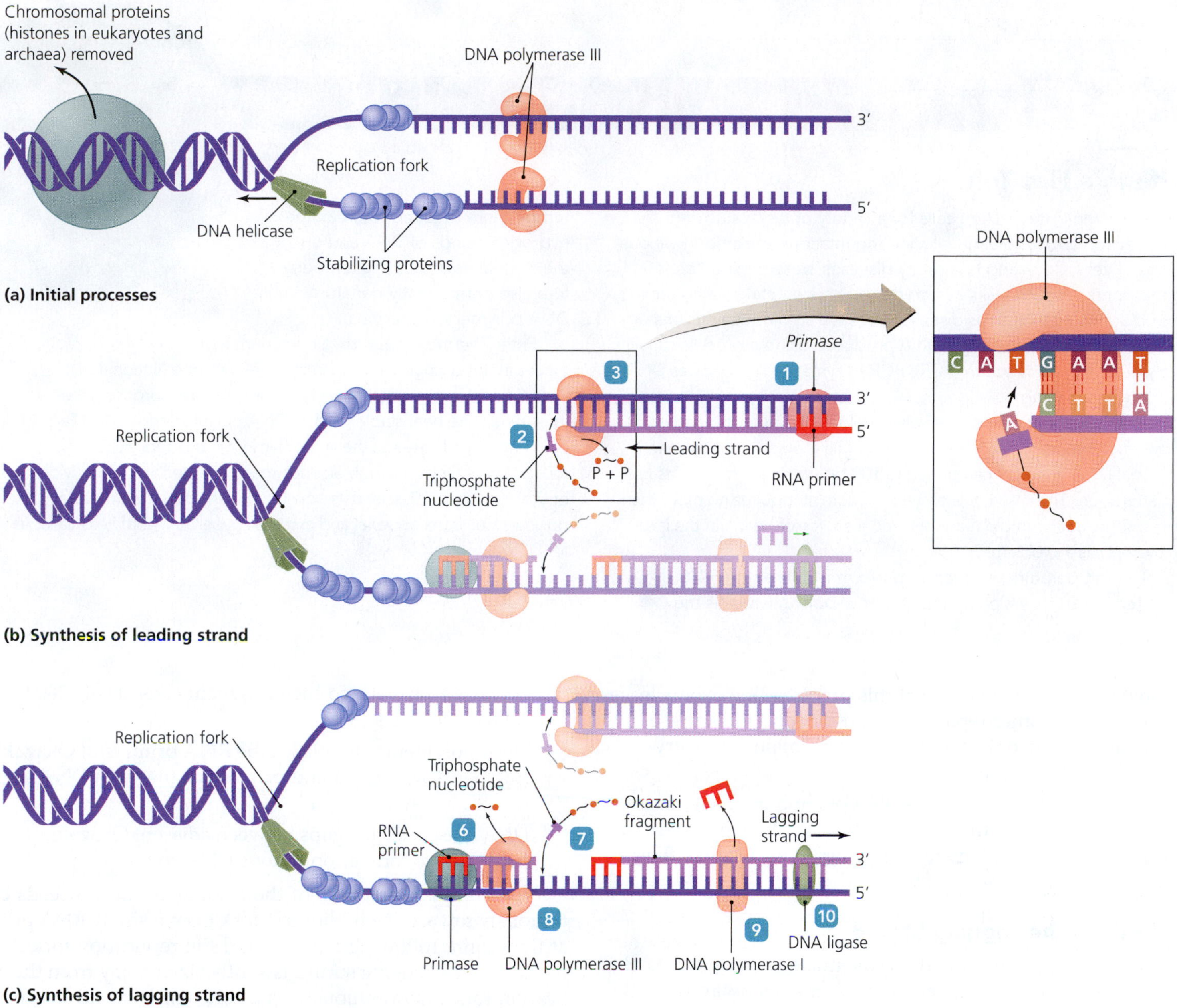

▲ **FIGURE 7.6 DNA replication.** **(a)** Initial processes. The cell removes proteins (called histones in eukaryotes and archaea) from the DNA molecule. Helicase unzips the double helix—breaking hydrogen bonds between complementary base pairs—to form a replication fork. **(b)** Continuous synthesis of the leading strand. DNA synthesis always moves in the 5′ to 3′ direction, so the leading strand is synthesized toward the replication fork. The numbers refer to the steps in the process, which are described in the text. (Steps 4 and 5, the proofreading function of DNA polymerase and the replacement of the RNA primer with DNA, are not shown.) **(c)** Discontinuous synthesis of the lagging strand, which proceeds moving away from the replication fork. Actual Okazaki fragments are about 1000 nucleotides long. *Why is DNA replication termed "semiconservative"?*

Figure 7.6 *"Semiconservative" refers to the fact that each of the daughter molecules retains one parental strand and has one new strand; in other words, each is half new and half old.*

MM® To see a 3-D animation on DNA replication, go to the MasteringMicrobiology Study Area and watch the MicroFlix.

BENEFICIAL MICROBES

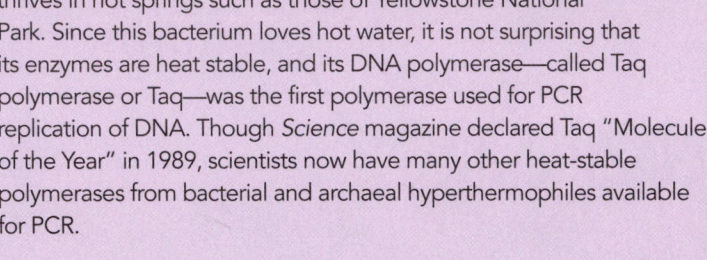

Life in a Hot Tub

Scientists replicate DNA of cells for a variety of tasks: studying gene action and regulation, elucidating relationships among various kinds of cells, detecting hereditary diseases, determining "genetic fingerprints" in such tasks as paternity tests, detecting pathogens, and diagnosing infectious diseases. All such studies use millions or billions of identical copies of DNA produced using a process called polymerase chain reaction (PCR). PCR enzymatically replicates DNA without using living cells.

The concept of PCR is relatively simple. As its inventor wrote in *Scientific American*, "Beginning with a single molecule of the genetic material DNA, the PCR can generate 100 billion similar molecules in an afternoon. The reaction is easy to execute. It requires no more than a test tube, a few simple reagents, and a source of heat." In the latter, however, lies a problem.

The temperature required to perform PCR is about 94°C. This temperature, which is almost that of boiling water, is the temperature required to break the hydrogen bonds of DNA and unzip the double helix, but this temperature also permanently denatures most DNA polymerase enzymes.

Enter *Thermus aquaticus*, a bacterium that thrives in hot springs such as those of Yellowstone National Park. Since this bacterium loves hot water, it is not surprising that its enzymes are heat stable, and its DNA polymerase—called Taq polymerase or Taq—was the first polymerase used for PCR replication of DNA. Though *Science* magazine declared Taq "Molecule of the Year" in 1989, scientists now have many other heat-stable polymerases from bacterial and archaeal hyperthermophiles available for PCR.

most recent error. Because of this proofreading exonuclease function and other repair strategies beyond the scope of this discussion, only about one error remains for every 10 billion (10^{10}) bp replicated.

5. Another DNA polymerase—DNA polymerase I—replaces the RNA primer with DNA (not shown). Note that researchers named DNA polymerase enzymes in the order of their discovery, not the order of their actions.

Synthesis of the Lagging Strand

Because DNA polymerase III adds nucleotides only to the 3′ end of the new strand, the enzyme moves away from the replication fork as it synthesizes a lagging strand. As a result, the lagging strand is synthesized discontinuously and always lags behind the process occurring in the leading strand. The steps in the synthesis of a lagging strand are as follows **(FIGURE 7.6c)**:

6. Primase synthesizes RNA primers, but in contrast to its action on the leading strand, primase synthesizes multiple primers—one every 1000 to 2000 DNA bases of the template strand.
7. Nucleotides pair up with their complements in the template—adenine with thymine and cytosine with guanine.
8. DNA polymerase III joins neighboring nucleotides and proofreads. In contrast to synthesis of the leading strand, however, the lagging strand is synthesized in discontinuous segments called *Okazaki fragments*, named for the Japanese scientist Reiji Okazaki (1930–1975), who first identified them. Each Okazaki fragment uses one of the new RNA primers, so each fragment consists of 1000 to 2000 nucleotides.
9. DNA polymerase I replaces the RNA primers of Okazaki fragments with DNA and proofreads the short DNA segment it has synthesized.
10. *DNA ligase* seals the gaps between adjacent Okazaki fragments to form a continuous DNA strand.

In summary, synthesis of the leading strand proceeds continuously toward the replication fork from a single RNA primer at the origin, following helicase and the replication fork down the DNA. The lagging strand is synthesized away from the replication fork discontinuously as a series of Okazaki fragments, each of which begins with its own RNA primer. All the primers are eventually replaced with DNA nucleotides, and ligase joins the Okazaki fragments.

As noted earlier, DNA replication is semiconservative; each daughter molecule is composed of one parental strand and one daughter strand. The replication process produces double-stranded daughter molecules with a nucleotide sequence identical to that in the original double helix, ensuring that the integrity of an organism's genome is maintained each time it is copied. **▶ANIMATIONS: *DNA Replication: Synthesis***

Other Characteristics of Bacterial DNA Replication

DNA replication is *bidirectional;* that is, DNA synthesis proceeds in both directions from the origin. In bacteria, the process of replication proceeds from a single origin, so it involves two sets of enzymes, two replication forks, two leading strands, and two lagging strands **(FIGURE 7.7)**.

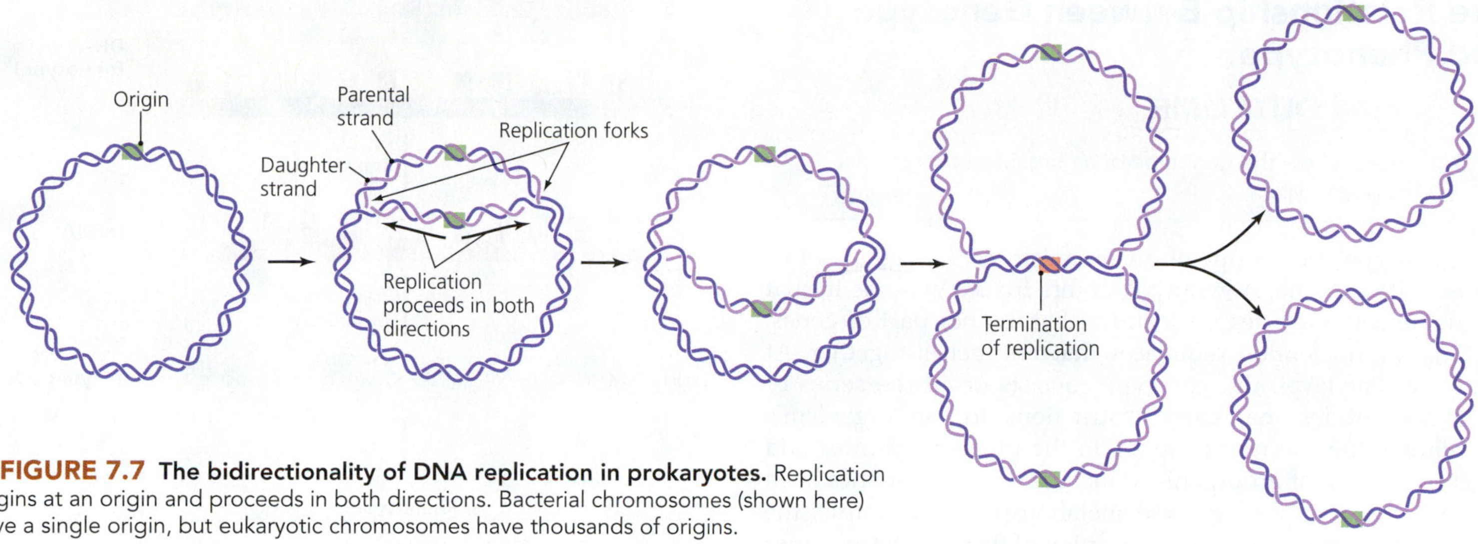

▲ **FIGURE 7.7 The bidirectionality of DNA replication in prokaryotes.** Replication begins at an origin and proceeds in both directions. Bacterial chromosomes (shown here) have a single origin, but eukaryotic chromosomes have thousands of origins.

The unzipping and unwinding action of helicase introduces supercoils into the DNA molecule ahead of the replication forks. Excessive supercoiling creates tension on the DNA molecule—like your grandmother's overwound phone cord—and would stop DNA replication. The enzymes *gyrase* and *topoisomerase* remove such supercoils by cutting the DNA, rotating the cut ends in the direction opposite the supercoiling, and then rejoining the cut ends.

Bacterial DNA replication is further complicated by **methylation** of the daughter strands, in which a cell adds a methyl group ($-CH_3$) to one or two bases that are part of specific nucleotide sequences. Bacteria typically methylate adenine bases and only rarely a cytosine base.

Methylation plays a role in a variety of cellular processes, including the following:

- *Control of genetic expression.* In some cases, genes that are methylated are "turned off" and are not transcribed, whereas in other cases methylated genes are "turned on" and are transcribed.
- *Initiation of DNA replication.* In many bacteria, methylated nucleotide sequences play a role in initiating DNA replication.
- *Protection against viral infection.* Methylation at specific sites in a nucleotide sequence enables cells to distinguish their DNA from viral DNA, which lacks methylation. The cells can then selectively degrade viral DNA.
- *Repair of DNA.* The role of methylation in some DNA repair mechanisms is discussed later in the chapter (pp. 223–224).

Replication of Eukaryotic DNA

Eukaryotes replicate DNA in much the same way as do bacteria; helicases and topoisomerases unwind DNA, protein molecules stabilize single-stranded DNA, and molecules of DNA polymerase synthesize leading and lagging strands simultaneously. However, eukaryotic replication differs from prokaryotic replication in some significant ways:

- Eukaryotic cells use four different DNA polymerases to replicate DNA. DNA polymerase α initiates replication, including synthesis of a primer—the function performed by primase in bacteria. DNA polymerase δ elongates the leading strand, and DNA polymerase ε appears to be responsible for replicating the lagging strand. DNA polymerase γ replicates mitochondrial DNA.[3]
- The large size of eukaryotic chromosomes necessitates thousands of origins per molecule, each generating two replication forks; otherwise, the replication of eukaryotic genomes would take days instead of hours.
- Eukaryotic Okazaki fragments are shorter than those of bacteria—100 to 400 nucleotides long.
- Plant and animal cells methylate cytosine bases exclusively.

TELL ME WHY

DNA replication requires a large amount of energy, yet none of a cell's ATP energy supply is used. Why isn't it?

We have examined the physical structure of cellular genes—the specific sequences of DNA nucleotides—and the way cells replicate their genes. Now we will consider how genes function and how cells control genetic expression.

Gene Function

The first topic we must consider if we are to understand gene function is the relationship between an organism's genotype and its phenotype.

---

[3]Greek letters α, δ, ε, and γ (alpha, delta, epsilon, and gamma) are equivalent to the numbers 1, 4, 5, and 3, corresponding to the order in which the polymerases were elucidated, not the order in which they act.

The Relationship Between Genotype and Phenotype

LEARNING | OUTCOME

7.8 Explain how the genotype of an organism determines its phenotype.

The **genotype**[4] (jēn´ō-tīp) of an organism is the actual set of genes in its genome. A genotype differs from a genome in that a genome also includes nucleotides that are not part of genes, such as the nucleotide sequences that link genes together. At the molecular level, the genotype consists of all the series of DNA nucleotides that carry instructions for an organism's life. **Phenotype**[5] (fē´nō-tīp) refers to the physical features and functional traits of an organism, including characteristics such as structures, morphology, and metabolism. For example, the shape of a cell, the presence and location of flagella, the enzymes and cytochromes of electron transport chains, and membrane receptors that trigger chemotaxis are all phenotypic traits.

Genotype determines phenotype by specifying what kinds of RNA and which structural, enzymatic, and regulatory protein molecules are produced. Though genes do not code *directly* for such molecules as phospholipids or for behaviors such as chemotaxis, ultimately phenotypic traits result from the actions of RNA and protein molecules that are themselves coded by DNA.

Not all genes are active at all times; that is, the information of a genotype is not always expressed as a phenotype. For example, *E. coli* activates genes for lactose catabolism only when it detects lactose in its environment.

The Transfer of Genetic Information

LEARNING | OUTCOME

7.9 State the central dogma of genetics, and explain the roles of DNA and RNA in polypeptide synthesis.

Cells must continually synthesize proteins required for growth, reproduction, metabolism, and regulation. This synthesis requires that they accurately transfer the genetic information contained in DNA nucleotide sequences to the amino acid sequences of polypeptides. However, cells do not transfer the information coded in DNA directly but first make an RNA copy of the gene. In this copying process, called **transcription**,[6] the information in DNA is copied as RNA nucleotide sequences; RNA molecules in ribosomes then synthesize polypeptides in a process called **translation**.[7] These processes make up the **central dogma** of genetics: DNA is transcribed to RNA, which is translated to form polypeptides **(FIGURE 7.8)**.

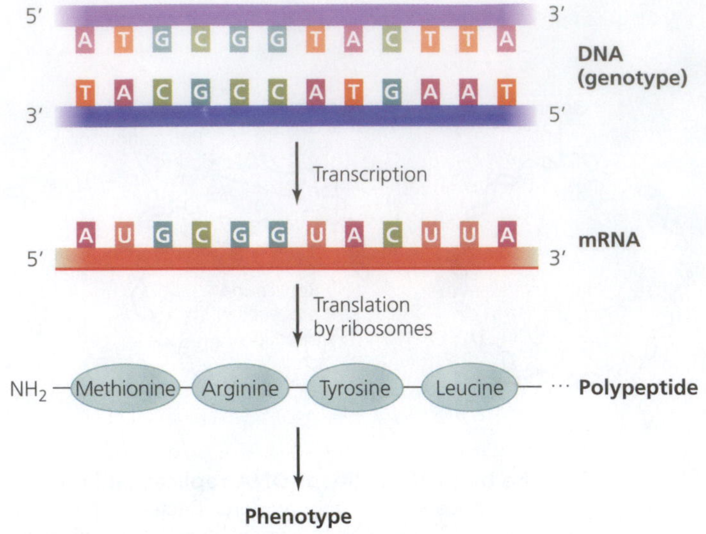

▲ **FIGURE 7.8 The central dogma of genetics.** A cell transcribes RNA from a DNA gene and then translates polypeptides using the code carried by the RNA molecules. Polypeptides determine phenotype by acting as structural, enzymatic, and regulatory proteins.

An analogy serves to illustrate the central dogma. Suppose you were trying to understand the following message (which is a portion of the oath of Hippocrates, written in the Greek alphabet):

> ΔΙΑΙΤΗΜΑΣΙΤΕΧΡΗΣΟΜΑΙΕΠΩΦΕΛΕΙΝ
> ΚΑΜΝΟΝΤΩΝΚΑΤΑΔΥΝΑΜΙΝΚΑΙΚΡΙΣΙΝΕΜΗΝ
> ΕΠΙΔΗΛΗΣΕΙΔΕΚΑΙΑΔΙΚΙΗ

If the Greek alphabet is foreign to you, you might have the Greek characters *transcribed* into the familiar English alphabet as a first step in understanding the message:

> Diaiteimasi te chreisomai ep ophelein
> kamnonton kata dunamin kai krisin emein
> epi deileisei de kai adikiei eirzein

Then you could begin the process of having the Greek words, now expressed in English letters, *translated* into English words:

> I will prescribe treatment to the best of my ability and judgment to help the sick and never for a harmful or illicit purpose

To a ribosome, DNA is like a foreign language written in a foreign alphabet. Thus, a cell must use processes analogous to those just described: It must first *transcribe* the "foreign alphabet" of DNA nucleotides (genes) into the more "familiar alphabet" of RNA nucleotides; then it must *translate* the message formed by these "letters" into the "words" (amino acids) that make up the "message" (a polypeptide). In this way a genotype can be expressed as a phenotype. There are a few exceptions to the central dogma. For example, some RNA viruses transcribe DNA from an RNA template—a process that is the reverse of cellular transcription.

In the following sections, we will examine the processes of transcription and translation. ▶**ANIMATIONS:** ***Transcription Overview; Translation: Overview***

[4] From Greek *genos*, meaning "race," and *typos*, meaning "type."
[5] From Greek *phainein*, meaning "to show."
[6] From Latin *trans*, meaning "across," and *scribere*, meaning "to write"—that is, to transfer in writing.
[7] From Latin *translatus*, meaning "transferred."

The Events in Transcription

LEARNING | OUTCOMES

7.10 Describe three steps in RNA transcription, mentioning the following: DNA, RNA polymerase, promoter, 5′ to 3′ direction, terminator, and Rho.

7.11 Contrast bacterial transcription with that of eukaryotes.

Cells transcribe five main types of RNA from DNA:

- **RNA primer** molecules for DNA polymerase to use during DNA replication
- **messenger RNA (mRNA)** molecules, which carry genetic information from chromosomes to ribosomes
- **ribosomal RNA (rRNA)** molecules, which combine with ribosomal polypeptides to form ribosomes—the organelles that synthesize polypeptides
- **transfer RNA (tRNA)** molecules, which deliver the correct amino acids to ribosomes based on the sequence of nucleotides in mRNA
- **regulatory RNA** molecules, which interact with DNA to control gene expression

We have already considered the role of RNA primer in DNA replication and will more closely examine the functions of the other types of RNA shortly. Next we examine transcription in bacteria and contrast it with eukaryotic transcription; archaeal processes are not as well known.

Transcription occurs in the nucleoid region of the cytoplasm in bacteria. The three steps of RNA transcription are (1) *initiation of transcription*, (2) *elongation of the RNA transcript*, and (3) *termination of transcription*. **FIGURE 7.9** depicts the events in transcription. ▶**ANIMATIONS:** *Transcription: The Process*

Initiation of Transcription

RNA polymerases—the enzymes that synthesize RNA—bind to specific DNA nucleotide sequences called **promoters**, each of which is located near the beginning of a gene and serves to initiate transcription (1a in Figure 7.9a). In bacteria, a polypeptide subunit of RNA polymerase called the *sigma factor* is necessary for recognition of a promoter. Once it adheres to a promoter sequence, an RNA polymerase unzips and unwinds the DNA molecule in the promoter region and then travels along the DNA, unzipping the double helix to form a "bubble" as it moves 1b.

A cell uses different sigma factors and different promoter sequences to provide some control over transcription. RNA polymerases using different sigma factors do not adhere equally strongly to all promoters; there is about a millionfold difference between the strongest attraction and the weakest one. The greater the attraction between a particular sigma factor and a promoter, the more likely that a particular gene will be transcribed; thus, variations in sigma factors and promoters affect the amounts and kinds of polypeptides produced.

Elongation of the RNA Transcript

RNA transcription does not actually begin in the promoter region but, rather, at a spot 10 nucleotides away. There, triphosphate ribonucleotides (rATP, rUTP, rGTP, and rCTP) align opposite their complements in the open DNA "bubble." RNA polymerase links together two adjacent ribonucleotide molecules using energy from the phosphate bonds of the first ribonucleotide (2 in Figure 7.9b). The enzyme then moves down the DNA strand, elongating RNA by repeating the process. Only one of the separated DNA strands is transcribed.

Many molecules of RNA polymerase may concurrently transcribe the same gene **(FIGURE 7.10)**. In this way, a prokaryotic cell simultaneously produces numerous identical copies of RNA from a single gene—much as many identical prints can be made from a single photographic negative.

Like DNA polymerase, RNA polymerase links nucleotides only to the 3′ end of the growing molecule; however, RNA polymerase differs from DNA polymerase in the following ways:

- RNA polymerase unwinds and opens DNA by itself; helicase is not required.
- RNA polymerase does not need a primer.
- RNA polymerase transcribes only one of the DNA strands.
- RNA polymerase is slower than DNA polymerase III, proceeding at a rate of about 50 nucleotides per second.
- RNA polymerase incorporates ribonucleotides instead of deoxyribonucleotides.
- Uracil nucleotides are incorporated instead of thymine nucleotides.
- The proofreading function of RNA polymerase is less efficient, leaving a base-pair error about every 10,000 nucleotides.

Termination of Transcription

Transcription terminates when RNA polymerase and the transcribed RNA are released from DNA (see Figure 7.9c). RNA polymerase is tightly associated with the DNA molecule and cannot be removed easily; therefore, the termination of transcription is complicated. Scientists have elucidated two types of termination processes in bacteria—those that are self-terminating and those that depend on the action of an additional protein called *Rho*. These processes of transcription termination should not be confused with termination of translation examined in a later section.

Self-Termination Self-termination occurs when RNA polymerase transcribes a **terminator** sequence of DNA composed of two symmetrical series: one that is very rich in guanine and cytosine bases, followed by a region rich in adenine bases (see 3a in Figure 7.9c). RNA polymerase slows down during transcription of the GC-rich portion of the terminator because the three hydrogen bonds between each guanine and cytosine base pair make unwinding the DNA helix more difficult. This pause in transcription, which lasts about 60 seconds, provides enough time for the RNA molecule to form hydrogen bonds between its own symmetrical sequences, forming a hairpin loop

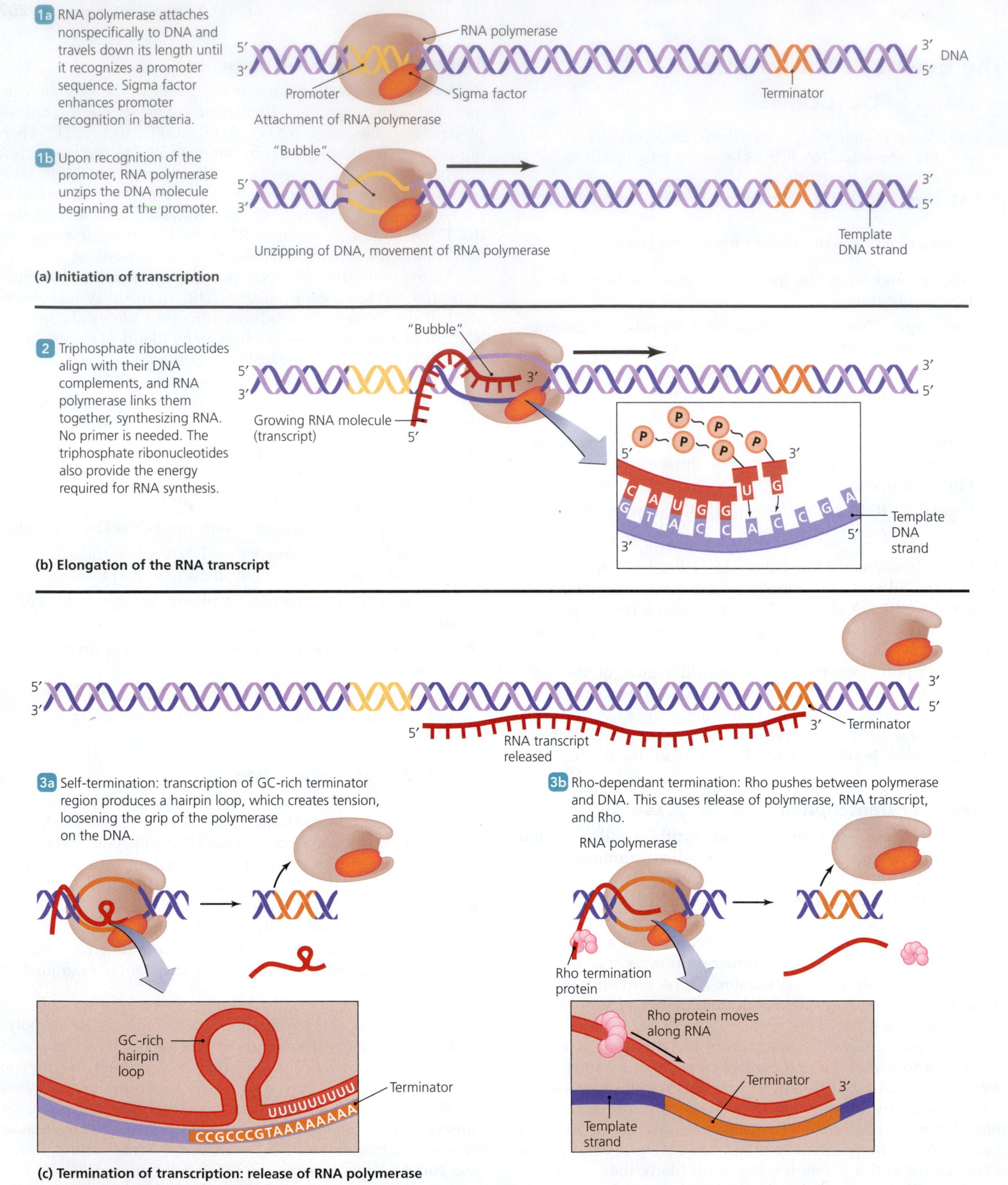
1a RNA polymerase attaches nonspecifically to DNA and travels down its length until it recognizes a promoter sequence. Sigma factor enhances promoter recognition in bacteria.
RNA polymerase
5′
3′
DNA
Promoter
Sigma factor
Terminator
Attachment of RNA polymerase
1b Upon recognition of the promoter, RNA polymerase unzips the DNA molecule beginning at the promoter.
"Bubble"
Template DNA strand
Unzipping of DNA, movement of RNA polymerase
(a) Initiation of transcription
2 Triphosphate ribonucleotides align with their DNA complements, and RNA polymerase links them together, synthesizing RNA. No primer is needed. The triphosphate ribonucleotides also provide the energy required for RNA synthesis.
Growing RNA molecule (transcript)
P
Template DNA strand
(b) Elongation of the RNA transcript
RNA transcript released
Terminator
3a Self-termination: transcription of GC-rich terminator region produces a hairpin loop, which creates tension, loosening the grip of the polymerase on the DNA.
GC-rich hairpin loop
UUUUUUUU
CCGCCCGTAAAAAAAA
Terminator
3b Rho-dependant termination: Rho pushes between polymerase and DNA. This causes release of polymerase, RNA transcript, and Rho.
RNA polymerase
Rho termination protein
Rho protein moves along RNA
Terminator
Template strand
(c) Termination of transcription: release of RNA polymerase

▲ **FIGURE 7.9 The events in the transcription of RNA in prokaryotes.** **(a)** Initiation of transcription. **(b)** Elongation of the RNA transcript. **(c)** Termination of transcription: release of RNA polymerase by one of two methods. *What is the difference between a promoter sequence and an origin?*

Figure 7.9 *A promoter is a DNA sequence that initiates transcription; an origin is a point where DNA replication begins.*

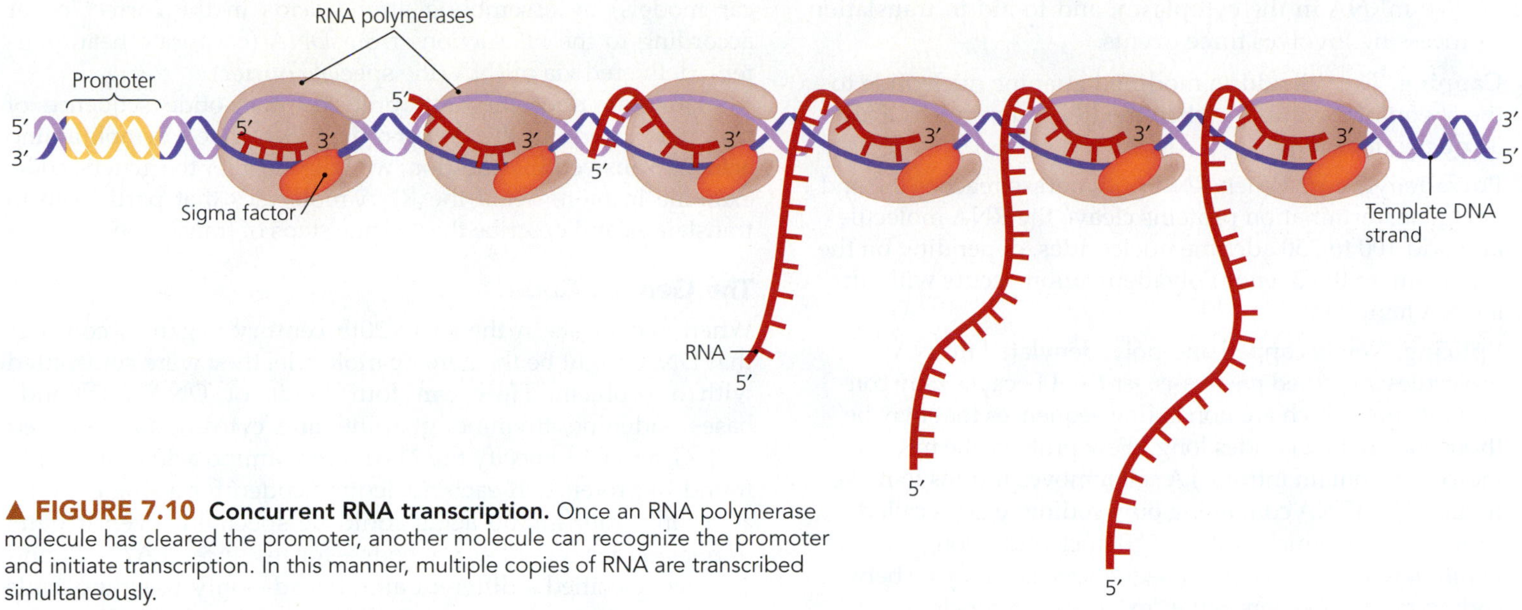

▲ **FIGURE 7.10 Concurrent RNA transcription.** Once an RNA polymerase molecule has cleared the promoter, another molecule can recognize the promoter and initiate transcription. In this manner, multiple copies of RNA are transcribed simultaneously.

structure that puts tension on the union of RNA polymerase and the DNA. When RNA polymerase transcribes the adenine-rich portion of the terminator, the relatively few hydrogen bonds between the adenine bases of DNA and the uracil bases of RNA cannot withstand the tension, and the RNA transcript breaks away from the DNA, releasing RNA polymerase.

Rho-Dependent Termination The second type of termination depends on the termination protein called Rho. Rho binds to a specific RNA sequence near the end of an RNA transcript. Rho moves toward RNA polymerase at the 3′ end of the growing RNA molecule, pushing between RNA polymerase and the DNA strand and forcing them apart; this releases RNA polymerase, the RNA transcript, and Rho (see 3b in Figure 7.9c).

Transcriptional Differences in Eukaryotes

Eukaryotic transcription differs from bacterial transcription in several ways. First, a eukaryotic cell transcribes RNA inside its nucleus, primarily in the region of the nucleus called the nucleolus, as well as inside any mitochondria and chloroplasts that are present. In contrast, transcription in prokaryotes occurs in the cytosol.

Another difference between eukaryotes and bacteria is that eukaryotes have three types of nuclear RNA polymerase—one for transcribing mRNA, one for transcribing the major rRNA gene, and one for transcribing tRNA and smaller rRNA molecules. Mitochondria use a fourth type of RNA polymerase. Further, several separate protein *transcription factors* at a time assist in binding eukaryotic RNA polymerase to promoter sequences, in contrast to a single sigma factor in bacteria. After initiating transcription, eukaryotic RNA polymerases shed most of the transcription factors and recruit another set of polypeptides called *elongation factors*.

Finally, eukaryotic cells must process mRNA before beginning polypeptide translation **(FIGURE 7.11)**. In general, the function of RNA processing is to aid in export from the nucleus,

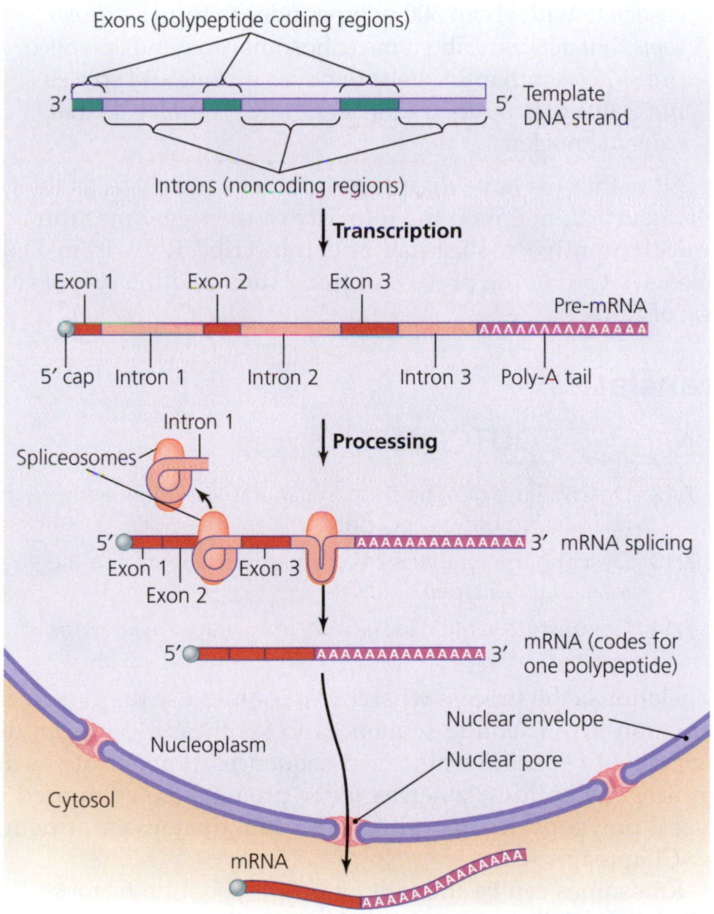

▲ **FIGURE 7.11 Processing eukaryotic mRNA.** Within the nucleus, transcription produces pre-mRNA, which contains coding exons and noncoding introns. Enzymes cap the 5′ end with a modified guanine nucleotide and add hundreds of adenine nucleotides to the 3′ end, a process known as polyadenylation. Ribozymes further process pre-mRNA by removing introns and splicing together exons to form a molecule that codes for a single polypeptide. Eukaryotic mRNA then moves from the nucleus to the cytoplasm.

to stabilize mRNA in the cytoplasm, and to aid in translation. RNA processing involves three events:

1. **Capping.** The cell adds a modified guanine nucleotide to the 5′ end of the mRNA when the RNA molecule is about 30 nucleotides long.
2. **Polyadenylation.** When RNA polymerase reaches the end of a gene, termination proteins cleave the RNA molecule and add 100 to 250 adenine nucleotides, depending on the organism, to the 3′ end. Polyadenylation occurs without a DNA template.
3. **Splicing.** Newly capped and polyadenylated mRNA molecules are called *pre-messenger RNA* because they contain **introns**, which are noncoding sequences that may be thousands of nucleotides long. (Few prokaryotic mRNA molecules contain introns.) A cell removes introns to make functional mRNA containing only coding regions called **exons**, each of which is about 150 nucleotides long. The "in" in *intron* refers to *intervening* sequences (i.e., they lie between coding regions), whereas the "ex" in *exon* refers to the fact that these regions are *expressed*. Five small RNA molecules associate with about 300 polypeptides to form a *spliceosome* that acts as a ribozyme (ribosomal enzyme) to splice pre-mRNA into mRNA—it removes introns and splices the exons to produce a functional mRNA molecule that exits the nucleus.

Now that we have discussed how cells use DNA as the genetic material, maintain the integrity of their genomes through semiconservative replication, and transcribe RNA from DNA genes, we turn to the process of translation and the role of each type of RNA.

Translation

LEARNING | **OUTCOMES**

- **7.12** Describe the genetic code in general, and identify the relationship between codons and amino acids.
- **7.13** Describe the synthesis of polypeptides, identifying the roles of three types of RNA.
- **7.14** Contrast translation in bacteria from that in eukaryotes.

Translation is the process whereby ribosomes use the genetic information of nucleotide sequences to synthesize polypeptides composed of specific amino acid sequences. Some proteins are simple polypeptides, whereas other proteins are composed of several polypeptides bound together in a quaternary structure (see Chapter 2).

Ribosomes can be thought of as "polypeptide factories," so consider the following analogy between translation and a hypothetical automobile factory. Trucks deliver auto parts to the factory at the correct times and in the correct order to manufacture one of a large variety of automobile models, depending on instructions from corporate headquarters delivered by special courier. Similarly, molecules of tRNA (the trucks) deliver preformed amino acids (the parts) to a ribosome (the factory), which can manufacture an infinite variety of polypeptides (the car models) by assembling amino acids in the correct order according to the instructions from DNA (corporate headquarters) delivered via mRNA (the special courier).

How do ribosomes interpret the nucleotide sequence of mRNA to determine the correct order in which to assemble amino acids? To answer this question, we will consider the genetic code, examine in more detail the RNA molecules that participate in translation, and describe the specific steps of translation.

The Genetic Code

When geneticists in the early 20th century began to consider that DNA might be the genetic molecule, they were confronted with a problem: How can four kinds of DNA nucleotide bases—adenine, thymine, guanine, and cytosine (abbreviated A, T, G, and C)—specify the 21 different amino acids commonly found in proteins? If each nucleotide coded for a single amino acid, only four amino acids could be specified. Even if pairs of nucleotides served as the code—for instance, if AA, AT, and TA each specified a different amino acid—only 16 amino acids (i.e., 4^2) could be accommodated. Eventually scientists showed that genes are composed of sequences of three nucleotides that specify amino acids. For example, the DNA nucleotide sequence AAA specifies the amino acid phenylalanine, and GTA codes for histidine. There are 64 possible arrangements of the four nucleotides in triplets (4^3)—more than enough to specify 21 amino acids.

These examples are DNA triplets, but ribosomes do not directly access genetic information on a DNA molecule. Instead, molecules of mRNA carry the code to the ribosomes; therefore, scientists define the genetic code **(FIGURE 7.12)** as triplets of mRNA nucleotides called **codons** (kō′donz), which code for specific amino acids. UUU is a codon for phenylalanine, and CAU is a codon for histidine.

In most cases, 61 codons specify amino acids, and three codons—UAA, UAG, and UGA—instruct ribosomes to stop translating; however, under some conditions, UGA codes for a 21st amino acid, selenocysteine. Codon AUG also has a dual function, acting both as a start signal and as the codon for the amino acid methionine. In bacteria, mitochondria, and chloroplasts, AUG as a start codon codes for *N*-formylmethionine (fMet), a modified amino acid:

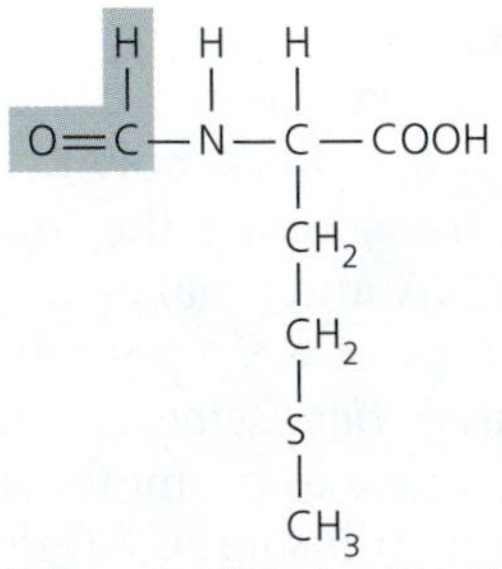

N-formylmethionine (fMet)

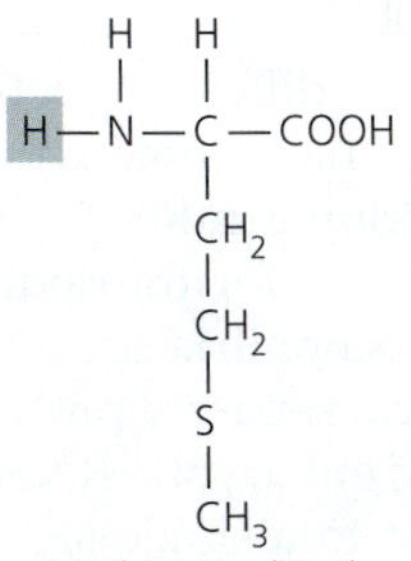

Methionine (Met)

As you examine the genetic code, notice that it is redundant; that is, more than one codon is associated with every amino acid except methionine and tryptophan. With most redundant codons, the first two nucleotides determine the amino acid, and

First nucleotide base (5′ position)	Second nucleotide base: U	C	A	G	Third nucleotide base (3′ position)
U	UUU Phenylalanine (Phe)	UCU Serine (Ser)	UAU Tyrosine (Tyr)	UGU Cysteine (Cys)	U
U	UUC Phenylalanine (Phe)	UCC Serine (Ser)	UAC Tyrosine (Tyr)	UGC Cysteine (Cys)	C
U	UUA Leucine (Leu)	UCA Serine (Ser)	UAA STOP	UGA STOP & Selenocysteine (SeCys)	A
U	UUG Leucine (Leu)	UCG Serine (Ser)	UAG STOP*	UGG Tryptophan (Trp)	G
C	CUU Leucine (Leu)	CCU Proline (Pro)	CAU Histidine (His)	CGU Arginine (Arg)	U
C	CUC Leucine (Leu)	CCC Proline (Pro)	CAC Histidine (His)	CGC Arginine (Arg)	C
C	CUA Leucine (Leu)	CCA Proline (Pro)	CAA Glutamine (Gln)	CGA Arginine (Arg)	A
C	CUG Leucine (Leu)	CCG Proline (Pro)	CAG Glutamine (Gln)	CGG Arginine (Arg)	G
A	AUU Isoleucine (Ile)	ACU Threonine (Thr)	AAU Asparagine (Asn)	AGU Serine (Ser)	U
A	AUC Isoleucine (Ile)	ACC Threonine (Thr)	AAC Asparagine (Asn)	AGC Serine (Ser)	C
A	AUA Isoleucine (Ile)	ACA Threonine (Thr)	AAA Lysine (Lys)	AGA Arginine (Arg)	A
A	AUG START & Methionine (Met)	ACG Threonine (Thr)	AAG Lysine (Lys)	AGG Arginine (Arg)	G
G	GUU Valine (Val)	GCU Alanine (Ala)	GAU Aspartic acid (Asp)	GGU Glycine (Gly)	U
G	GUC Valine (Val)	GCC Alanine (Ala)	GAC Aspartic acid (Asp)	GGC Glycine (Gly)	C
G	GUA Valine (Val)	GCA Alanine (Ala)	GAA Glutamic acid (Glu)	GGA Glycine (Gly)	A
G	GUG Valine (Val)	GCG Alanine (Ala)	GAG Glutamic acid (Glu)	GGG Glycine (Gly)	G

*Also codes for a 22nd amino acid, pyrrolysine, in some prokaryotes.

▲ **FIGURE 7.12 The genetic code.** The table shows the set of mRNA codons and the amino acids for which they code. AUG not only is the start codon but also specifies methionine (Met) in eukaryotes and *N*-formylmethionine (fMet) in prokaryotes, mitochondria, and chloroplasts. Two codons (UAA and UAG) are stop codons that do not typically specify amino acids. UGA functions as a stop codon and also specifies selenocysteine.

the third nucleotide is inconsequential. For example, the codons GUU, GUC, GUA, and GUG all specify the amino acid valine.

Interestingly, the genetic code is nearly universal; that is, with few exceptions, ribosomes in archaeal, bacterial, plant, fungal, protozoan, and animal cells use the same genetic code. Some exceptions are listed in **TABLE 7.2**. ▶**ANIMATIONS:** ***Translation: Genetic Code***

TABLE 7.2 Some Exceptions to the Genetic Code

Codon	Usual Use	Alternative Use
AUA	Codes for isoleucine	Codes for methionine in mitochondria
UAG	STOP	Codes for glutamine in some protozoa and algae and for pyrrolysine, a 22nd amino acid found in some prokaryotes
CGG	Codes for arginine	Codes for tryptophan in plant mitochondria
UGA	STOP, selenocysteine	Codes for tryptophan in mitochondria and mycoplasmas (type of bacteria)

Participants in Translation

As we have discussed, transcription produces messenger RNA, transfer RNA, and ribosomal RNA—each of which is involved in translation. We now discuss each kind of RNA in turn.

Messenger RNA Messenger RNA molecules carry genetic information from chromosomes to ribosomes as triplet sequences of RNA nucleotides (codons) that encode the order of amino acid sequences in a polypeptide. In prokaryotes a basic mRNA molecule contains sequences of nucleotides that are recognized by ribosomes: an AUG start codon, sequential codons for other amino acids in the polypeptide, and at least one of the three stop codons. A single molecule of prokaryotic mRNA often contains start codons and instructions for more than one polypeptide arranged in series **(FIGURE 7.13)**. Because both transcription and the subsequent events of translation occur in the cytosol of prokaryotes, prokaryotic ribosomes can begin translation before transcription is finished.

Eukaryotic mRNA differs from prokaryotic mRNA in several ways:

- As we have seen, eukaryotic cells extensively process pre-mRNA to make mRNA (see Figure 7.11).
- A molecule of eukaryotic mRNA contains instructions for only one polypeptide.

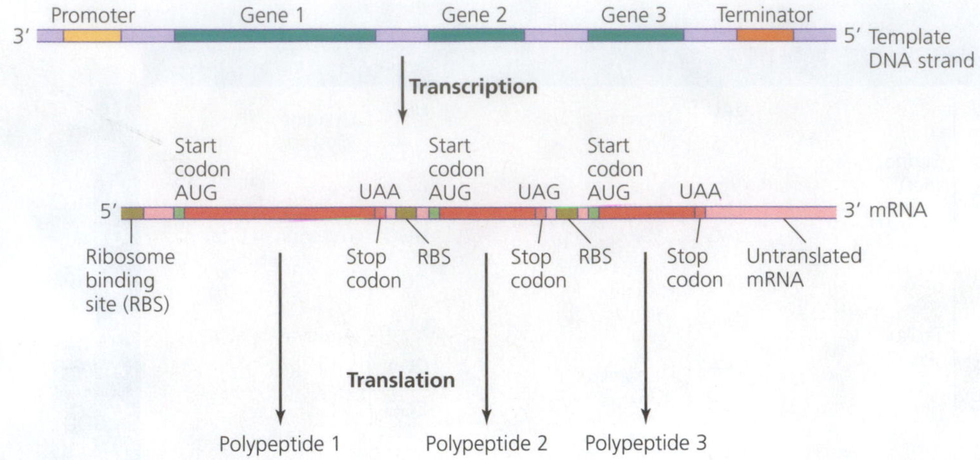

◀ **FIGURE 7.13 A single prokaryotic mRNA can code for several polypeptides.** Prokaryotes typically code for several related polypeptides via a single mRNA molecule. The mRNA molecule shown here has transcripts of three genes encoding three polypeptides. The transcript of each gene begins with a start codon and ends with a stop codon.

- Eukaryotic mRNA is not translated until it is fully transcribed and processed and has left the nucleus. In other words, transcription and translation of a molecule of eukaryotic mRNA do not occur simultaneously because eukaryotic ribosomes are located in the cytoplasm, while transcription occurs in the nucleus.

Transfer RNA A transfer RNA (tRNA) molecule is a sequence of about 75 ribonucleotides that curves back on itself to form three main hairpin loops held in place by hydrogen bonding between complementary nucleotides **(FIGURE 7.14a)**. Although transfer RNA molecules can be modeled simplistically by a cloverleaf structure, their three-dimensional shape is more complex **(FIGURE 7.14b)**. For simplicity, tRNA will be represented in subsequent figures by an icon shaped like the three-dimensional icon in Figure 7.14b.

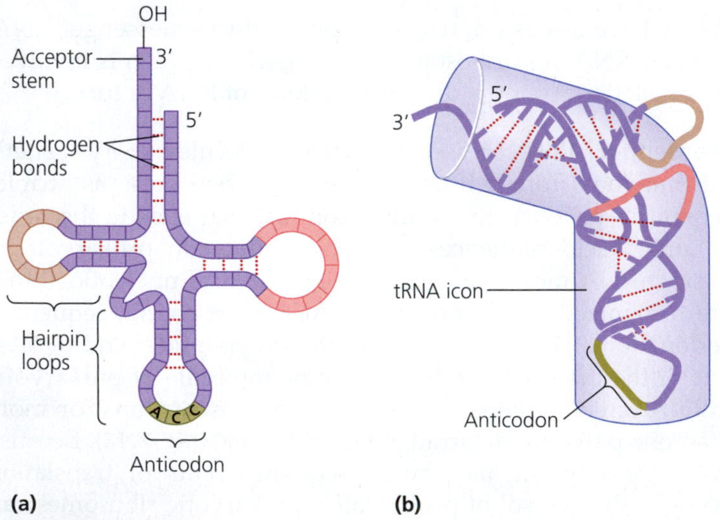

▲ **FIGURE 7.14 Transfer RNA. (a)** A two-dimensional "cloverleaf" representation of tRNA showing three hairpin loops held in place by intramolecular hydrogen bonding. **(b)** A three-dimensional drawing of the same tRNA. A specific amino acid attaches to the 3′ acceptor stem; the anticodon is a nucleotide triplet that is complementary to the mRNA codon for that amino acid. *Which amino acid would be attached to the acceptor stem of this tRNA?*

Figure 7.14 *UGG is the codon for tryptophan.*

A molecule of tRNA transfers the correct amino acid to a ribosome during polypeptide synthesis. To this end, tRNA has an **anticodon** (an-tē-kō´don) triplet in its bottom loop and an *acceptor stem* for a specific amino acid at its 3′ end. Specific enzymes in the cytoplasm *charge* each tRNA molecule; that is, they attach the appropriate amino acid to the acceptor stem.

Anticodons are complementary to mRNA codons, and each acceptor stem is designed to carry one particular amino acid, which varies with the tRNA. In other words, each transfer RNA carries a specific amino acid and recognizes mRNA codons only for that amino acid. A tRNA molecule is designated by a superscript abbreviation of its amino acid. For example, $tRNA^{Phe}$ carries phenylalanine, and $tRNA^{Ser}$ transfers serine.

The fact that 62 codons specify the 21 amino acids used by cells does not mean that there must be 62 anticodons on 62 different types of tRNA because many tRNA molecules recognize more than one codon. *E. coli,* for example, has only about 40 different tRNAs. The variability in codon recognition by tRNA is due to "wobble" of the anticodon's third nucleotide. Wobble, which is a change of angle from the normal axis of the molecule, allows the third nucleotide to hydrogen bond to a nucleotide other than its usual complement. For example, a guanine nucleotide in the third position normally bonds to cytosine, but it can wobble and also pair with uracil; therefore, whether a codon has cytosine or uracil in the third position makes no difference because the same tRNA recognizes either nucleotide in the third position. For example, the codons UUU and UUC specify the amino acid phenylalanine because the anticodon AAG recognizes both of them. Similarly, UCU and UCC code for serine, and UAU and UAC code for tyrosine. This redundancy in the genetic code helps protect cells against the effects of errors in replication and transcription.

Ribosomes and Ribosomal RNA Prokaryotic ribosomes, which are also called 70S ribosomes based on their sedimentation rate in an ultracentrifuge, are extremely complex associations of ribosomal RNAs and polypeptides. Each ribosome is composed of two subunits: 50S and 30S **(FIGURE 7.15a)**. The 50S subunit is in turn composed of two rRNA molecules (23S and 5S) and about 34 different polypeptides, whereas the 30S subunit consists of one molecule of 16S rRNA and

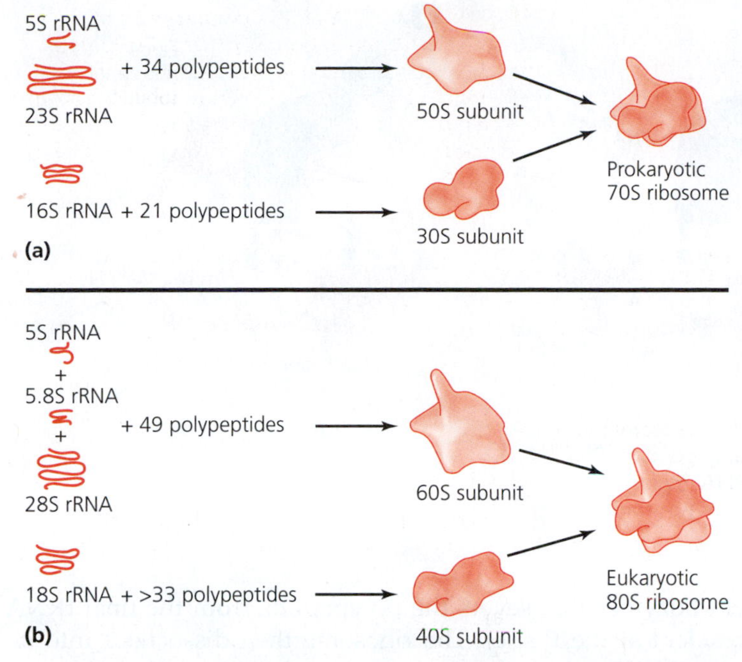

▲ **FIGURE 7.15 Ribosomal structures. (a)** The 70S prokaryotic ribosome, which is composed of polypeptides and three rRNA molecules arranged in 50S and 30S subunits. **(b)** The 80S eukaryotic ribosome, which is composed of molecules of rRNA and polypeptides, arranged in 60S and 40S subunits.

21 ribosomal polypeptides. The ribosomes of mitochondria and chloroplasts are also 70S ribosomes composed of similar subunits and polypeptides.

In contrast, both the cytosol and the rough endoplasmic reticulum (RER) of eukaryotic cells have 80S ribosomes composed of 60S and 40S subunits **(FIGURE 7.15b)**. These subunits contain larger molecules of rRNA and more polypeptides than the corresponding prokaryotic subunits, though researchers do not agree on the exact number of polypeptides. The term *eukaryotic ribosome* is understood to mean only the 80S ribosomes of the cytosol and RER. Because the ribosomes of mitochondria and chloroplasts are 70S, they are "prokaryotic" ribosomes even though they are in eukaryotic cells.

The structural differences between prokaryotic and eukaryotic ribosomes play a crucial role in the efficacy and safety of antimicrobial drugs. Because erythromycin, for example, binds only to the 23S rRNA found in prokaryotic ribosomes, it has no effect on eukaryotic 80S ribosomes and thus little deleterious effect on a patient. (Chapter 10 discusses antimicrobial drugs in more detail.)

The smaller subunit of a ribosome is shaped to accommodate three codons at one time—that is, nine nucleotide bases of a molecule of mRNA. Each ribosome also has three tRNA-binding sites that are named for their function **(FIGURE 7.16)**:

- The **A site** accommodates a tRNA delivering an *amino acid.*
- The **P site** holds a tRNA and the growing *polypeptide.*
- Discharged tRNAs *exit* from the **E site**.

We will now examine translation, the process whereby ribosomes actually synthesize polypeptides using amino acids

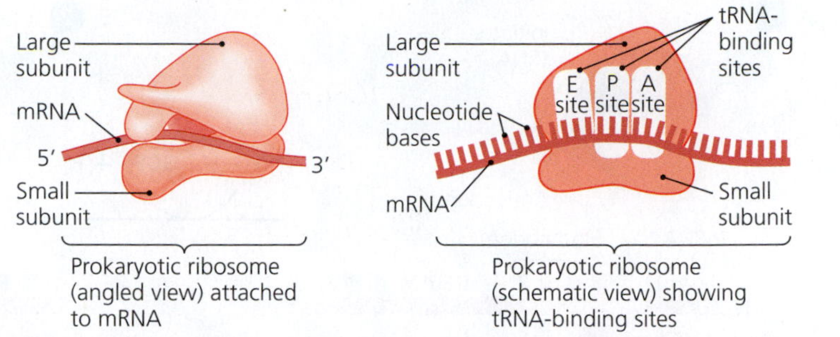

▲ **FIGURE 7.16 Assembled ribosome and its tRNA-binding sites.** The A site accepts the tRNA carrying the next amino acid to be added to the growing polypeptide, whereas the P site holds the tRNA carrying the polypeptide. Empty tRNA molecules exit from the E site.

delivered by tRNAs in the sequence dictated by the order of codons in mRNA.

Events in Translation

Molecular biologists divide translation into three stages: *initiation, elongation,* and *termination.* All three stages require additional protein factors that assist the ribosomes. Initiation and elongation also require energy provided by molecules of the ribonucleotide GTP, which are free in the cytosol (i.e., they are not part of an RNA molecule). Here we consider bacterial translation. **▶ANIMATIONS:** ***Translation: The Process***

Initiation During initiation, the two ribosomal subunits, mRNA, several protein factors, and $tRNA^{fMet}$ form an *initiation complex.* Initiation in prokaryotes may occur while the cell is still transcribing mRNA from DNA. The events of initiation in a bacterium are as follows **(FIGURE 7.17)**:

1. The smaller ribosomal subunit attaches to mRNA at a ribosome-binding site (also known as a Shine-Dalgarno sequence after its discoverers) so as to position a start codon (AUG) at the ribosomal subunit's P site.
2. $tRNA^{fMet}$ (whose anticodon, UAC, is complementary to the start codon, AUG) attaches at the ribosome's P site.
3. The larger ribosomal subunit then attaches to form a complete initiation complex. **▶VIDEO TUTOR:** ***Initiation of Translation***

Elongation Elongation of a polypeptide is a cyclical process that involves the sequential addition of amino acids to a polypeptide chain growing at the P site. **FIGURE 7.18** illustrates several cycles of the process. The steps of each cycle occur as follows:

4. The transfer RNA whose anticodon is complementary to the next codon—in this example, AAA complementary to the codon UUU—delivers its amino acid, in this case, phenylalanine (Phe), to the A site. Proteins called *elongation factors* escort the tRNA along with a molecule of GTP (not shown). Energy from GTP is used to stabilize each tRNA as it binds at the A site.
5. An enzymatic RNA molecule—a ribozyme—in the larger ribosomal subunit forms a peptide bond between the

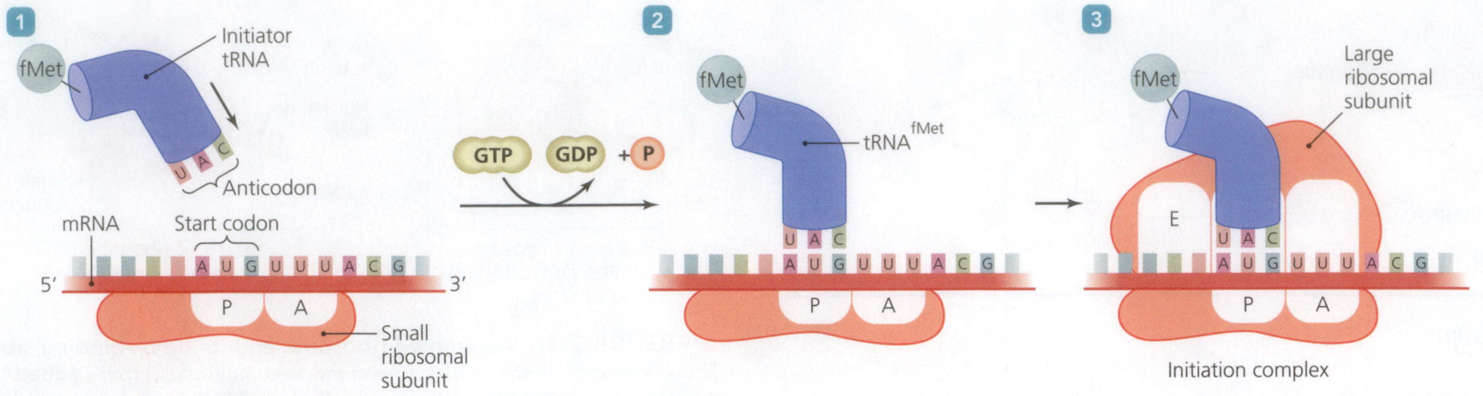

▲ **FIGURE 7.17 The initiation of translation in prokaryotes.** 1 The smaller ribosomal subunit attaches to mRNA at a ribosome-binding site near a start codon (AUG). 2 The anticodon of $tRNA^{fMet}$ aligns with the start codon on the mRNA; energy from GTP is used to bind the tRNA in place. 3 The larger ribosomal subunit attaches to form an initiation complex—a complete ribosome attached to mRNA.

terminal amino acid of the growing polypeptide chain (in this case, *N*-formylmethionine) and the newly introduced amino acid. The polypeptide is now attached to the tRNA occupying the A site.

6 Using energy supplied by more GTP, the ribosome moves one codon down the mRNA. This transfers each tRNA to the adjacent binding site; that is, the first tRNA moves from the P site to the E site, and the second tRNA (with the attached polypeptide) moves to the vacated P site.

7 The ribosome releases the "empty" tRNA from the E site. In the cytosol, the appropriate enzyme recharges the empty tRNA with another molecule of the type of amino acid carried by that tRNA.

8 The cycle repeats, each time adding another amino acid, at a rate of about 15 amino acids per second (in this case, threonine, then alanine, and then glutamine).

As elongation proceeds, ribosomal movement exposes the start codon, allowing another ribosome to attach behind the first one. In this way, one ribosome after another attaches at the start codon and begins to translate identical polypeptide molecules from the same message. Such a group of ribosomes, called a *polyribosome,* resembles beads on a string **(FIGURE 7.19)**.

Termination Termination does not involve tRNA; instead, proteins called *release factors* halt elongation. It appears that release factors somehow recognize stop codons and modify the larger ribosomal subunit in such a way as to activate another of its ribozymes that severs the polypeptide from the final tRNA (resident at the P site). The ribosome then dissociates into its subunits. Termination of translation should not be confused with termination of transcription covered in a previous section. The polypeptides released at termination may function alone as proteins, or they may function with other polypeptides in quaternary protein structures.

Translational Differences in Eukaryotes

Eukaryotic translation is similar to that of bacteria, with some notable differences, including the following:

- Initiation of translation in eukaryotes occurs when the small ribosomal subunit binds to the 5′ guanine cap rather than a specific nucleotide sequence.
- The first amino acid in eukaryotic polypeptides is methionine rather than formylmethionine.
- Ribosomes attached to membranes of endoplasmic reticulum (ER), forming rough ER (RER), can synthesize polypeptides into the cavity of the RER.

Archaeal translation is more similar to that of eukaryotes than to that of bacteria; however, archaea lack ER.

The processes we have examined thus far—how a cell replicates DNA, transcribes RNA, and translates RNA into polypeptides—are summarized in **TABLE 7.3**. Next we examine the way cells control the process of transcription.

TABLE 7.3 Comparison of Genetic Processes

Process	Purpose	Beginning Point	Ending Point (Termination)
Replication	To duplicate the cell's genome	Origin	Origin or the end of a linear DNA molecule
Transcription	To synthesize RNA	Promoter	Terminator
Translation	To synthesize polypeptides	AUG start codon	UAA, UAG, or UGA stop codons

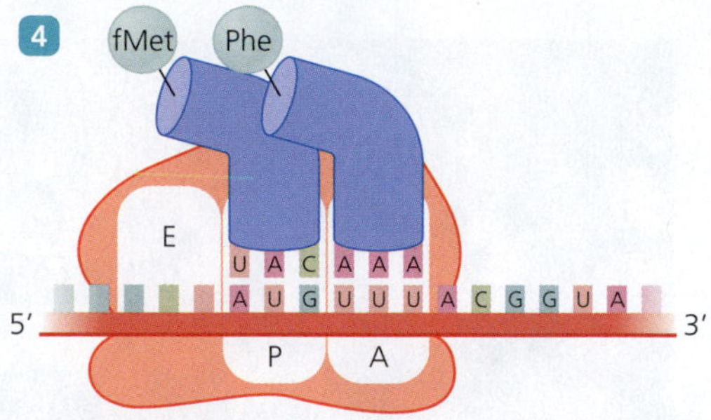

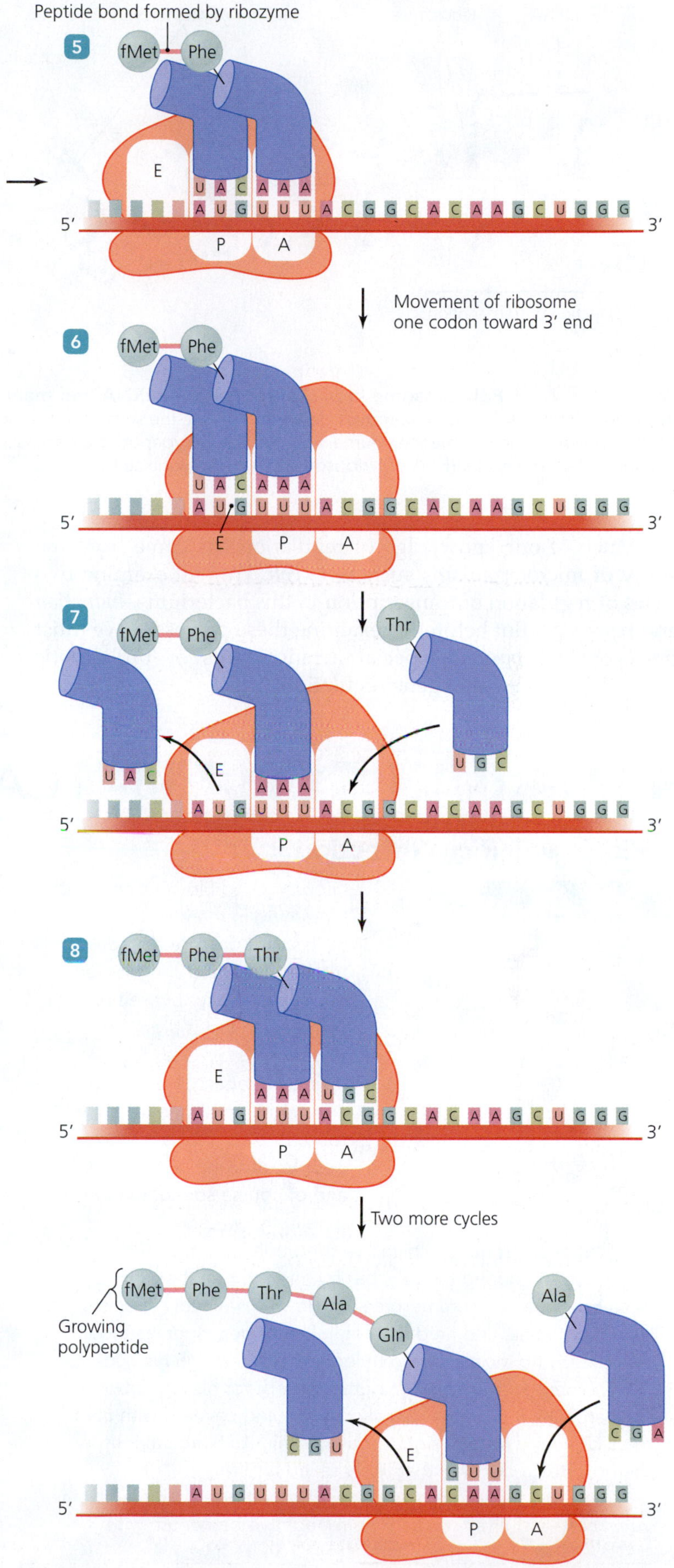

▶ **FIGURE 7.18 The elongation stage of translation.** Transfer RNAs sequentially deliver amino acids as directed by the codons of the mRNA. Ribosomal RNA in the large ribosomal subunit catalyzes a peptide bond between the amino acid at the A site and the growing polypeptide at the P site. The steps in the process are described in the text. *How would eukaryotic translation differ?*

Figure 7.18 *The initial amino acid in eukaryotic polypeptide is methionine.*

Regulation of Genetic Expression

LEARNING **OUTCOMES**

7.15 Explain the operon model of transcriptional control in prokaryotes.

7.16 Contrast the regulation of an inducible operon with that of a repressible operon, and give an example of each.

7.17 Describe the use of microRNA, small interfering RNA, and riboswitches in genetic control.

Most of a bacterium's genes are expressed at all times; that is, they are constantly transcribed and translated and play a persistent role in the phenotype. Such genes code for RNAs and polypeptides that are needed in large amounts by the cell—for example, integral proteins of the cytoplasmic membrane, structural proteins of ribosomes, and enzymes of glycolysis.

Other genes are regulated so that the polypeptides they encode are synthesized only in response to a change in the environment. Protein synthesis requires a large amount of energy, which can be conserved if a cell forgoes production of unneeded polypeptides. For example, *Pseudomonas aeruginosa* (ā-roo-ji-nō´să), a potential pathogen of cystic fibrosis patients, can synthesize harmful proteins. Early in an infection, *Pseudomonas* does not produce the proteins: The body would respond defensively and eliminate the bacterium. Instead, the pathogen uses **quorum sensing**—a process whereby cells secrete quorum-sensing molecules into their environment and other cells detect these signals so as to measure their density. The result is that *Pseudomonas* cells synthesize harmful proteins only after there are numerous bacterial cells, overwhelming the body's defenses. **Emerging Disease Case Study: *Vibrio vulnificus* Infection** on p. 216 describes another case of quorum sensing.

Cells regulate polypeptide synthesis in many ways. They may initiate or stop transcription of mRNA or may stop translation directly.

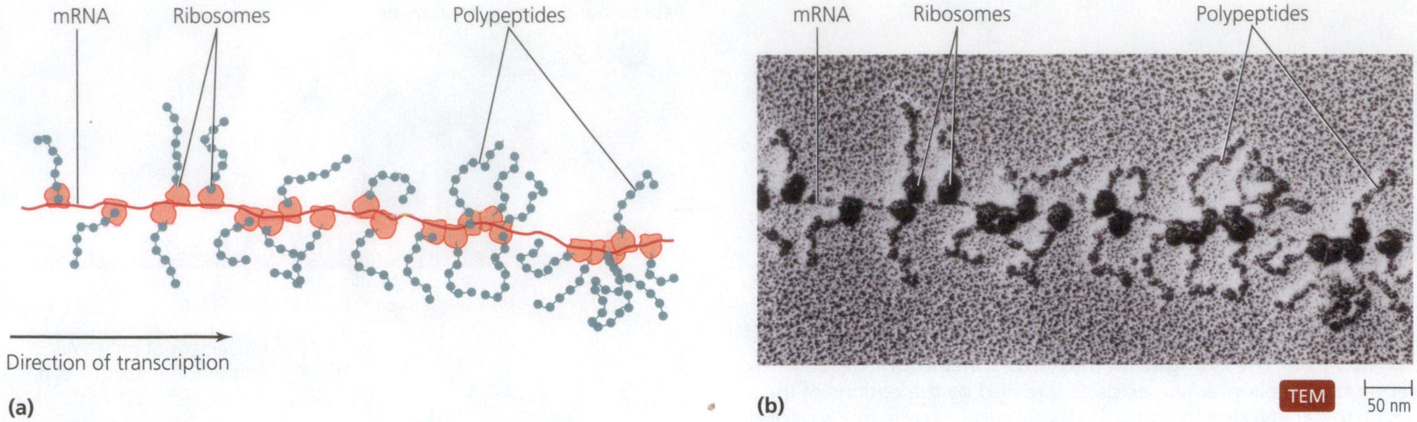

▲ **FIGURE 7.19 Polyribosome in prokaryotes—one mRNA and many ribosomes and polypeptides.** **(a)** As each ribosome moves down the mRNA, the start codon (AUG) becomes available to another ribosome. In this manner, numerous identical polypeptides are translated simultaneously from a single mRNA molecule. **(b)** A polyribosome in a prokaryotic cell.

Much of our knowledge of regulation has come from the study of microorganisms such as *E. coli.* We will examine two types of regulation of transcription in this bacterium—*induction* and *repression.* But before we examine these processes, we must first consider *operons*—special arrangements of prokaryotic genes that play roles in gene regulation.

The Nature of Prokaryotic Operons

As originally described, a prokaryotic **operon** consists of a promoter, a series of genes that code for enzymes and structures (such as channel proteins), and an adjacent regulatory element called an **operator** **(FIGURE 7.20)**, which controls movement of

EMERGING DISEASE CASE STUDY

Vibrio vulnificus Infection

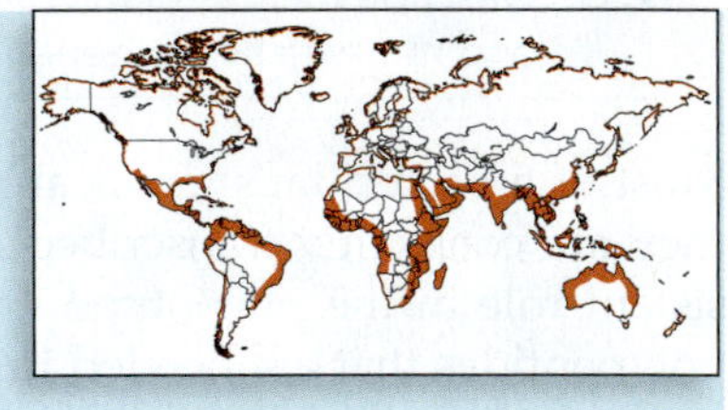

Greg enjoyed Florida's beaches; swimming in the warm water was his favorite pastime. Of course, the salt water did sting his leg where he had cut himself on some coral, but it didn't sting enough to stop Greg from enjoying the beach. He spent the afternoon jogging in the pure sand, throwing a disc, watching people, drinking a few beers, and of course spending more time in the water.

That evening he felt chilled, a condition he associated with too much sun during the day, but by midnight he thought he must have caught a rare summertime flu. He was definitely feverish, extremely weak, and tired. His leg felt strangely tight, as though the underlying muscles were trying to burst through his skin.

The next morning he felt better, except for his leg. It was swollen, dark red, tremendously painful, and covered with fluid-filled blisters. The ugly sight motivated him to head straight for the hospital, a decision that likely saved his life.

Greg was the victim of an emerging pathogen, *Vibrio vulnificus*—a slightly curved, Gram-negative bacterium with DNA similar to that of *V. cholerae* (cholera bacterium). *V. vulnificus* lives in salty, warm water around the globe. Unlike the cholera bacterium, *V. vulnificus* is able to infect a person by penetrating directly into a deep wound, a cut, or even a tiny scratch. In a person, the multiplying bacterium secretes *quorum-sensing molecules*. When the cells sense that there is a certain population size (a quorum), they "turn on" some of their genes, allowing them to thrive in the body and cause disease.

Greg's doctor cut away the dead tissue and prescribed doxycycline and cephalosporin for two weeks. Greg survived and kept his leg. Half of the victims of *V. vulnificus* are not so lucky; they lose a limb or die. Who knew that a beach could be so dangerous? (For more about quorum sensing, see p. 173.)

1. Why is it necessary for *Vibrio vulnificus* to turn on different genes when the microbe invades a human?
2. What does the term "turn on" mean in relation to transcription and translation?
3. Why do you think the related microbe *V. cholerae* is unable to infect through the skin?

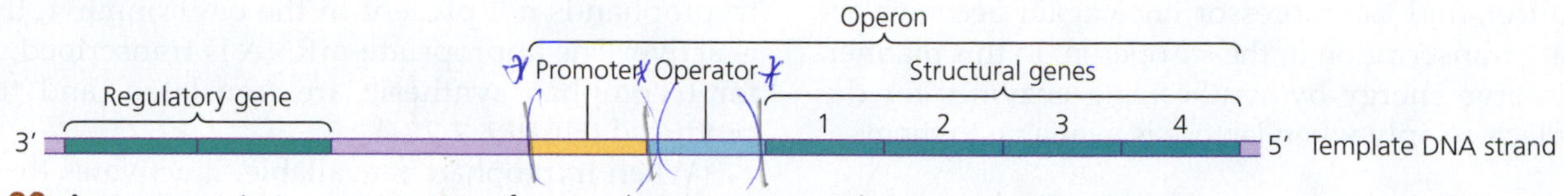

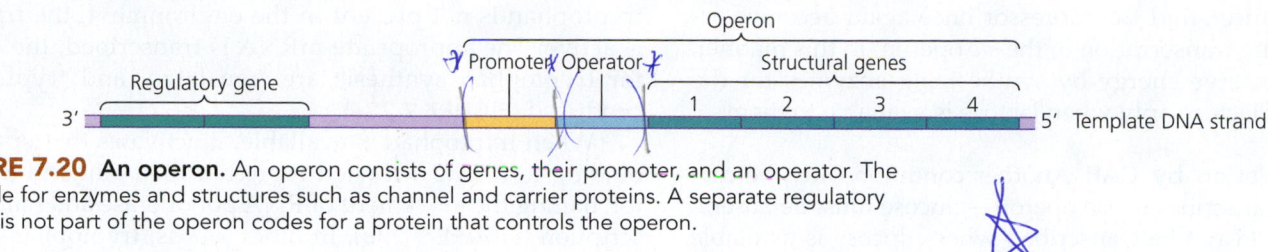

▲ **FIGURE 7.20 An operon.** An operon consists of genes, their promoter, and an operator. The genes code for enzymes and structures such as channel and carrier proteins. A separate regulatory gene that is not part of the operon codes for a protein that controls the operon.

RNA polymerase. Operons typically code for several polypeptides; they are *polygenic*, which is also called *polycistronic*.

Inducible operons are not usually transcribed and must be activated by *inducers*, such as some quorum-sensing polypeptides. **Repressible operons** operate in reverse fashion—they are transcribed continually until deactivated by *repressors*, which bind to the operator and inhibit transcription. To clarify these concepts, let's examine an inducible operon and a repressible operon found in *E. coli*. ▶**ANIMATIONS:** ***Operons: Overview***

The Lactose Operon, an Inducible Operon

The *lactose (lac) operon* of *E. coli* is an inducible operon and the first operon whose structure and action were elucidated. It includes a promoter, an operator, and three genes that encode proteins involved in the transport and catabolism of lactose sugar **(FIGURE 7.21a)**.

Repression and Induction The *lac* operon is controlled by a regulatory gene that is located outside the operon. The regulatory gene is constantly transcribed and translated to produce a repressor protein that attaches to DNA at the *lac* operator 1. This repressor prevents RNA polymerase from binding to the promoter, stopping synthesis of mRNA 2. Thus, the *lac* operon is usually inactive.

Under certain conditions, discussed shortly, *E. coli* takes in lactose whenever it becomes available and converts it to allolactose—an inducer that inactivates the repressor by changing the repressor's quaternary structure so that it can no longer attach to DNA **(FIGURE 7.21b** 3**)**. This allows transcription of the three structural genes to proceed—the operon has been induced and can become active 4. Ribosomes then translate the newly synthesized mRNA to produce enzymes that catabolize lactose. Once the lactose supply has been depleted, there

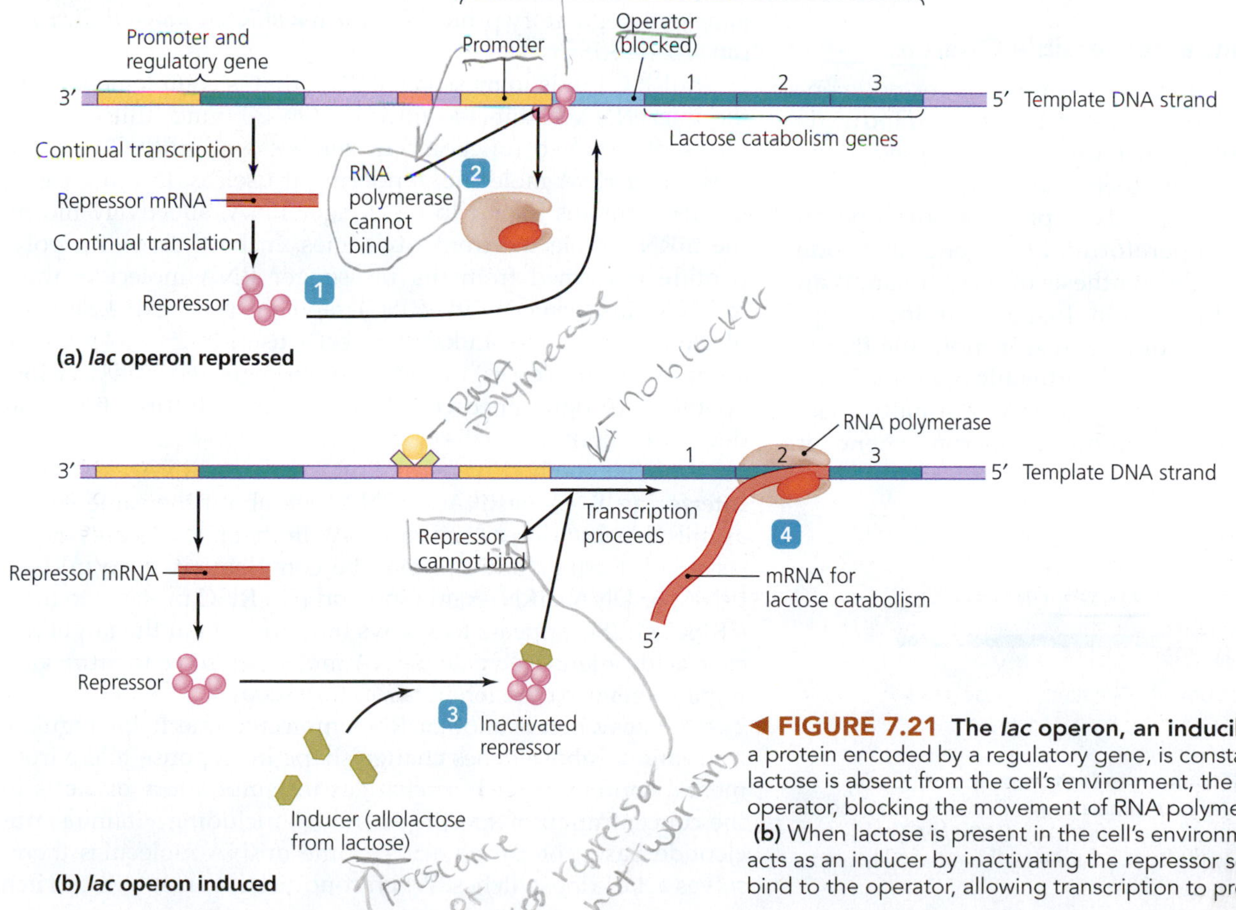

◀ **FIGURE 7.21 The *lac* operon, an inducible operon.** The repressor, a protein encoded by a regulatory gene, is constantly synthesized. **(a)** When lactose is absent from the cell's environment, the repressor binds to the operator, blocking the movement of RNA polymerase and halting transcription. **(b)** When lactose is present in the cell's environment, its derivative, allolactose, acts as an inducer by inactivating the repressor so that the repressor cannot bind to the operator, allowing transcription to proceed.

is no more inducer, and the repressor once again becomes active, suppressing transcription of the *lac* operon. In this manner, *E. coli* cells conserve energy by synthesizing enzymes for the +catabolism of lactose only when lactose is available to them.

Positive Regulation by CAP Another condition must be met before *E. coli* transcribes its *lac* operon—glucose must be absent. *Lac* genes should not be transcribed when glucose is available because glucose is more efficiently catabolized than is lactose. Glucose does not directly inhibit transcription; instead, the small molecule *cyclic adenosine monophosphate (cAMP)* is involved. When glucose is present, the cell does not synthesize much cAMP, so cAMP levels will be low; however, cAMP accumulates in *E. coli* when glucose is absent.

Cyclic AMP binds to an allosteric site of a regulatory protein, *catabolic activator protein (CAP)*, which is then able to bind to a CAP-binding site on DNA near the *lac* operon's promoter. This action is necessary for RNA polymerase to bind effectively. Once bound, RNA polymerase transcribes the *lac* genes **(FIGURE 7.22)**. Thus, CAP-cAMP positively enhances *lac* transcription but only when the operon is also induced by allolactose.

Inducible operons are often involved in controlling catabolic pathways whose polypeptides are not needed unless a particular nutrient is available. Such operons can also be associated with production of harmful proteins (virulence proteins) by pathogens. A different situation occurs with anabolic pathways, such as those that synthesize amino acids. ▶**ANIMATIONS:** ***Operons: Induction***

The Tryptophan Operon, a Repressible Operon

E. coli can synthesize all the amino acids it needs for polypeptide synthesis; however, it can save energy by using amino acids available in its environment. In such cases, *E. coli* represses the genes for a given amino acid's synthetic pathway.

The *tryptophan operon*, which is a polycistronic operon consisting of a promoter, an operator, and five genes that code for the enzymes involved in the synthesis of tryptophan, is an example of such a repressible operon. Just as with the *lac* operon, a regulatory gene codes for a repressor molecule that is constantly synthesized. In contrast to inducible operons, however, the repressor of repressible operons is normally inactive. Thus, in the case of the repressible *trp* operon, whenever

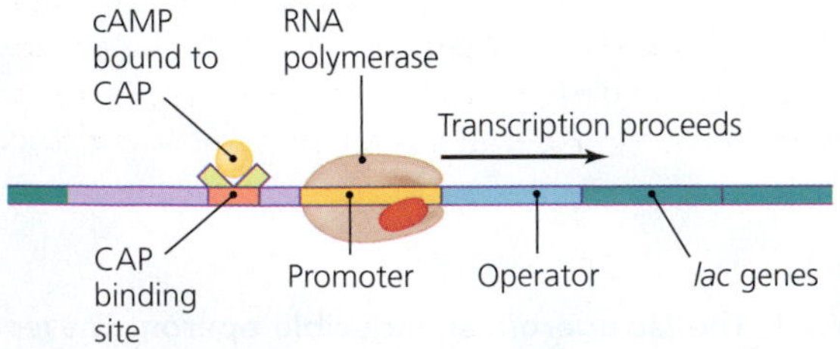

▲ **FIGURE 7.22 CAP-cAMP enhances *lac* transcription.** Cyclic adenosine monophosphate (cAMP), accumulating in *E. coli* when glucose is absent, binds to catabolic activator protein (CAP). CAP-cAMP binds to a CAP-binding site of DNA, allowing RNA polymerase to effectively bind to the *lac* promoter and begin transcription.

tryptophan is not present in the environment, the *trp* operon is active: The appropriate mRNA is transcribed, the enzymes for tryptophan synthesis are translated, and tryptophan is produced **(FIGURE 7.23a)**.

When tryptophan is available, it activates the repressor by binding to it. The activated repressor then binds to the operator, halting the movement of RNA polymerase and halting transcription **(FIGURE 7.23b)**. In other words, tryptophan acts as a *corepressor* of its own synthesis.

Biochemical analyses have revealed numerous variations of repressor-operator regulatory control. For example, scientists have discovered operons that use dual regulatory proteins, multiple operons controlled by a single repressor, and operons with multiple operators. Furthermore, some operons are not merely on-off systems but can be fine-tuned so that transcription rates vary with the concentration of corepressors.

TABLE 7.4 summarizes how the basic characteristics of inducible and repressible operons relate to the regulation of transcription. ▶**ANIMATIONS:** ***Operons: Repression***

RNA Molecules Can Control Translation

Besides inducing and repressing operons, cells can also use molecules of RNA to regulate translation of polypeptides. **Regulatory RNA** molecules include *microRNA, small interfering RNA*, and *riboswitches*.

Eukaryotic cells transcribe single-stranded RNA molecules about 22 nucleotides long called **microRNAs (miRNAs)**. Ribosomes do not translate microRNA molecules; rather, miRNA joins with regulatory proteins to form a *miRNA-induced silencing complex (miRISC)*.

miRISC binds to messenger RNA that is complementary to the microRNA within the miRISC. Once bound, miRISC performs one of two functions: In some cases, miRISC cuts the messenger RNA molecule, rendering it useless. In other cases, miRISC remains bound to messenger RNA, effectively hiding the mRNA molecule from ribosomes. In both cases, no polypeptide is formed from the messenger RNA molecule; thus, miRNAs (associated with RISC) regulate gene expression by blocking translation. Eukaryotic cells use miRISC to regulate a number of processes, including embryogenesis, cell division, apoptosis (programmed cell death), blood cell formation, and development of cancer.

Another method of regulation involving RNA uses **small interfering RNA (siRNA)**. siRNAs are about the same length as miRNAs but differ from miRNAs in that siRNAs are double stranded. Further, siRNAs may be complementary to mRNA, tRNA, or DNA. siRNAs unwind and join RISC proteins to form siRISC. siRISC appears to always bind to and cut the target nucleic acid. Scientists create siRNA molecules so as to artificially regulate gene expression in laboratory studies.

A **riboswitch** is another RNA molecule that helps regulate translation. Riboswitches change shape in response to environmental conditions such as changes in temperature or shifts in the concentration of specific nutrients, including vitamins, nucleotide bases, or amino acids. Some mRNA molecules themselves act as riboswitches. When conditions warrant, riboswitch

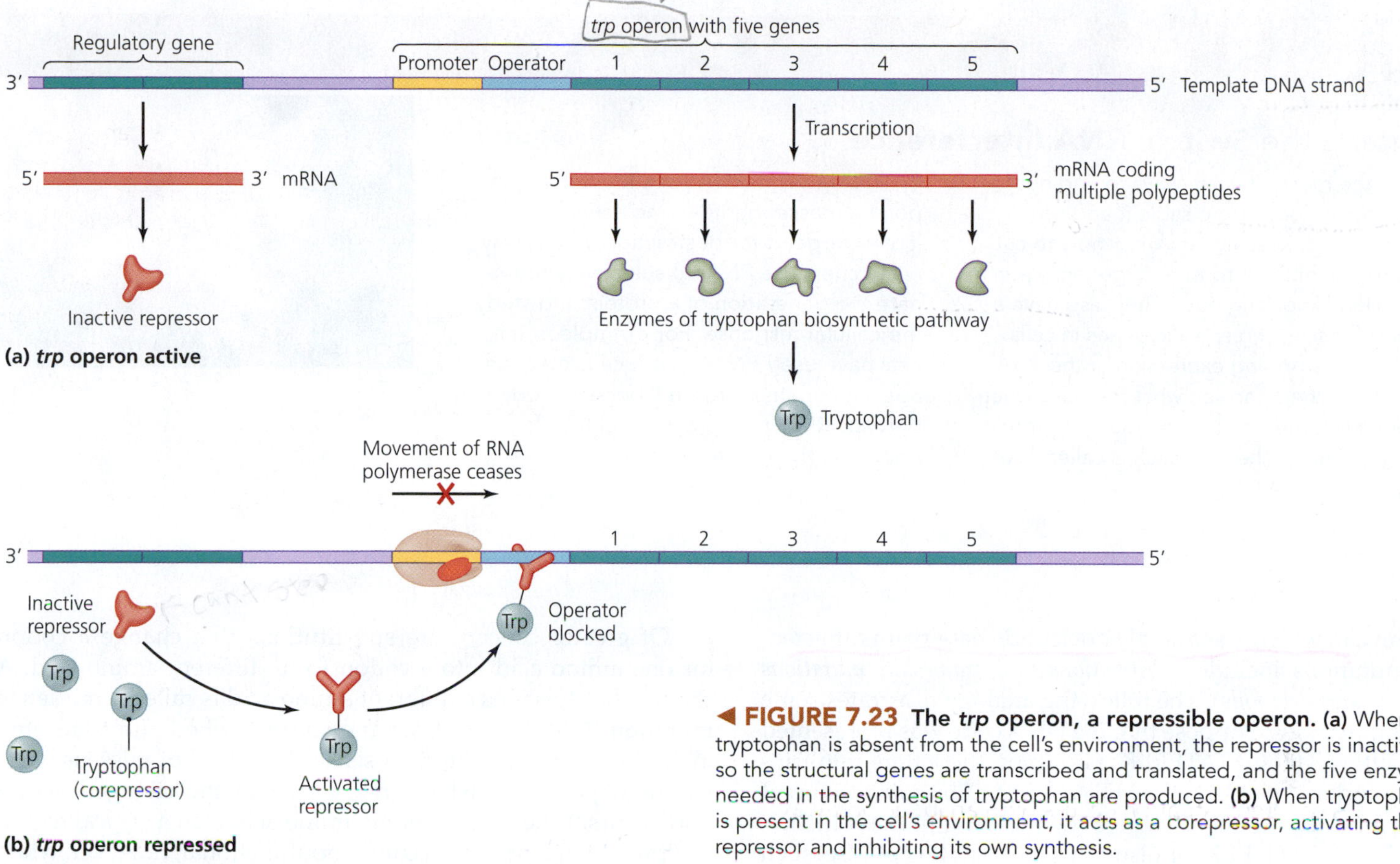

◀ **FIGURE 7.23 The *trp* operon, a repressible operon. (a)** When tryptophan is absent from the cell's environment, the repressor is inactive, so the structural genes are transcribed and translated, and the five enzymes needed in the synthesis of tryptophan are produced. **(b)** When tryptophan is present in the cell's environment, it acts as a corepressor, activating the repressor and inhibiting its own synthesis.

TABLE 7.4 The Roles of Operons in the Regulation of Transcription

Type of Regulation	Type of Metabolic Pathway Regulated	Regulating Condition
Inducible operons	Catabolic pathways; production of virulence proteins	Presence of substrate of pathway, quorum-sensing polypeptides
Repressible operons	Anabolic pathways	Presence of product of pathway, quorum-sensing polypeptides

mRNA folds to either favor or block translation, depending on the need by the cell for the polypeptide it encodes. For example, messenger RNA for the virulence regulator of the plague bacterium *Yersinia pestis* (yer-sin´ē-ă pes´tis) folds in such a way as to prevent translation when the temperature is below human body temperature (37°C). When the bacterium enters a human, the mRNA refolds into a shape that allows translation, and virulence regulator is synthesized; plague ensues.

TELL ME WHY

In bacteria, polypeptide translation can begin even before mRNA transcription is complete. Why can't this happen in eukaryotes?

Mutations of Genes

LEARNING | OUTCOME

7.18 Define *mutation*.

The phenotype of a cell is dependent on both the integrity and accurate control of its genes; however, the nucleotide sequences of genes are not always accurately maintained. A **mutation** is a change in the nucleotide base sequence of a genome, particularly its genes. Mutations of genes are almost always deleterious, though a few make no difference to the organism. Even more rarely a mutation leads to a novel property that improves the ability of an organism and its descendants to survive and reproduce. This is evolution in action. Mutations in unicellular organisms are passed on to the organism's progeny, but mutations in multicellular organisms typically are passed to offspring only if a mutation occurs in gametes (sex cells) or gamete-producing cells.

Types of Mutations

LEARNING | OUTCOME

7.19 Define *point mutation*, and describe three types.

Mutations range from large changes in an organism's genome, such as the loss or gain of an entire chromosome, to **point**

HIGHLIGHT

Flipping the Switch: RNA Interference

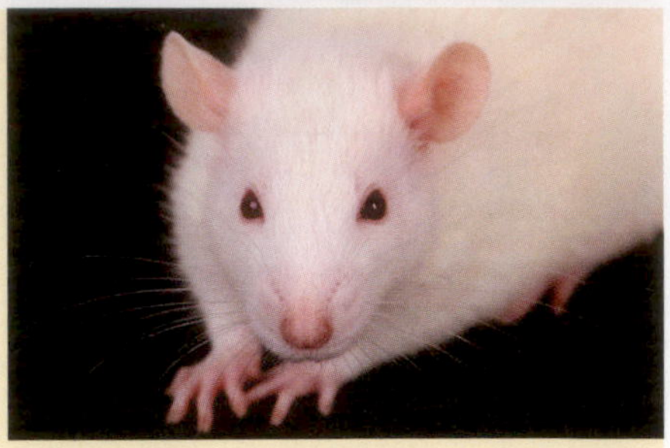

One of the most effective ways to find out what a gene does is to disable it and see what happens. Researchers know how to cut out portions of DNA to turn off genes, a process called "knocking out." They also have a method for silencing specific genes in cells via RNA. By stopping expression at the RNA level, researchers can see what happens when the gene is turned on but its resultant protein is not produced. The approach is called RNA interference (RNAi).

RNAi uses miRNAs or siRNAs that pinpoint a messenger RNA molecule, attach to it, and target it for destruction. In this way, researchers use RNAi to selectively terminate the generation of a protein and study the cellular response. For example, scientists have used RNAi to induce protection against hepatitis virus in laboratory rodents—perhaps RNAi may provide defense for people in the future.

mutations, in which just a single nucleotide base pair is affected. Point mutations include *substitutions* and *frameshift mutations* (*insertions* and *deletions*). The following analogy illustrates some types of mutations. Suppose that the DNA code was represented by the letters THECATATEELK. Grouping the letters into triplets (like codons) yields THE CAT ATE ELK. The substitution of a single letter could either change the meaning of the sentence, as in THE RAT ATE ELK, or result in a meaningless phrase, such as THE CAT RTE ELK. Insertion or deletion of a letter produces more serious changes, such as TRH ECA TAT EEL K (insertion) or TEC ATA TEE LK (deletion). Frameshift mutations can be caused by insertions or deletions because nucleotide triplets following the mutation are displaced, creating new sequences of codons that result in vastly altered polypeptide sequences. Frameshift mutations affect proteins much more seriously than mere substitutions because a frame shift affects all codons subsequent to the mutation.

Mutations can also involve *inversion* (THE ACT ATE KLE), *duplication* (THE CAT CAT ATE ELK ELK), or *transposition* (THE ELK ATE CAT). Such mutations and even larger deletions and insertions are **gross mutations**. ▶**ANIMATIONS:** *Mutations: Types*

Effects of Point Mutations

LEARNING **OUTCOME**

7.20 List three effects of point mutations.

Some base-pair substitutions produce **silent mutations** because redundancy in the genetic code prevents the substitution from altering the amino acid sequence (compare **FIGURE 7.24a** and **b**). For example, when the DNA triplet AAA is changed to AAG, the mRNA codon will be changed from UUU to UUC; however, because both codons specify phenylalanine, there is no change in the phenotype—the mutation is silent because it affects the genotype only.

Of greater concern are substitutions that change a codon for one amino acid into a codon for a different amino acid. A change that specifies a different amino acid is called a **missense mutation** (**FIGURE 7.24c**); what gets transcribed and translated makes sense but not the right sense. The effect of missense mutations depends on where in the protein the changed amino acid occurs. When the different amino acid is in a critical region of a protein, the protein becomes nonfunctional; however, when the different amino acid is in a less important region, the mutation may have no adverse effect.

A third type of mutation occurs when a base-pair substitution changes an amino acid codon into a stop codon. This is called a **nonsense mutation** (**FIGURE 7.24d**). Nearly all nonsense mutations result in nonfunctional proteins.

Frameshift mutations (i.e., insertions or deletions) typically result in drastic missense and nonsense mutations (**FIGURE 7.24e** and **f**) except when the insertion or deletion is very close to the end of a gene.

TABLE 7.5 on p. 222 summarizes the types and effects of point mutations.

Mutagens

LEARNING **OUTCOMES**

7.21 Discuss how different types of radiation cause mutations in a genome.

7.22 Describe three kinds of chemical mutagens and their effects.

Mutations occur naturally during the life of an organism. Such *spontaneous mutations* result from errors in replication and repair as well as from *recombination* in which relatively long stretches of DNA move among chromosomes, plasmids, and viruses, introducing frameshift mutations. For example, we have seen that mismatched base pairing during DNA replication results in one error in every 10 billion (10^{10}) base pairs.

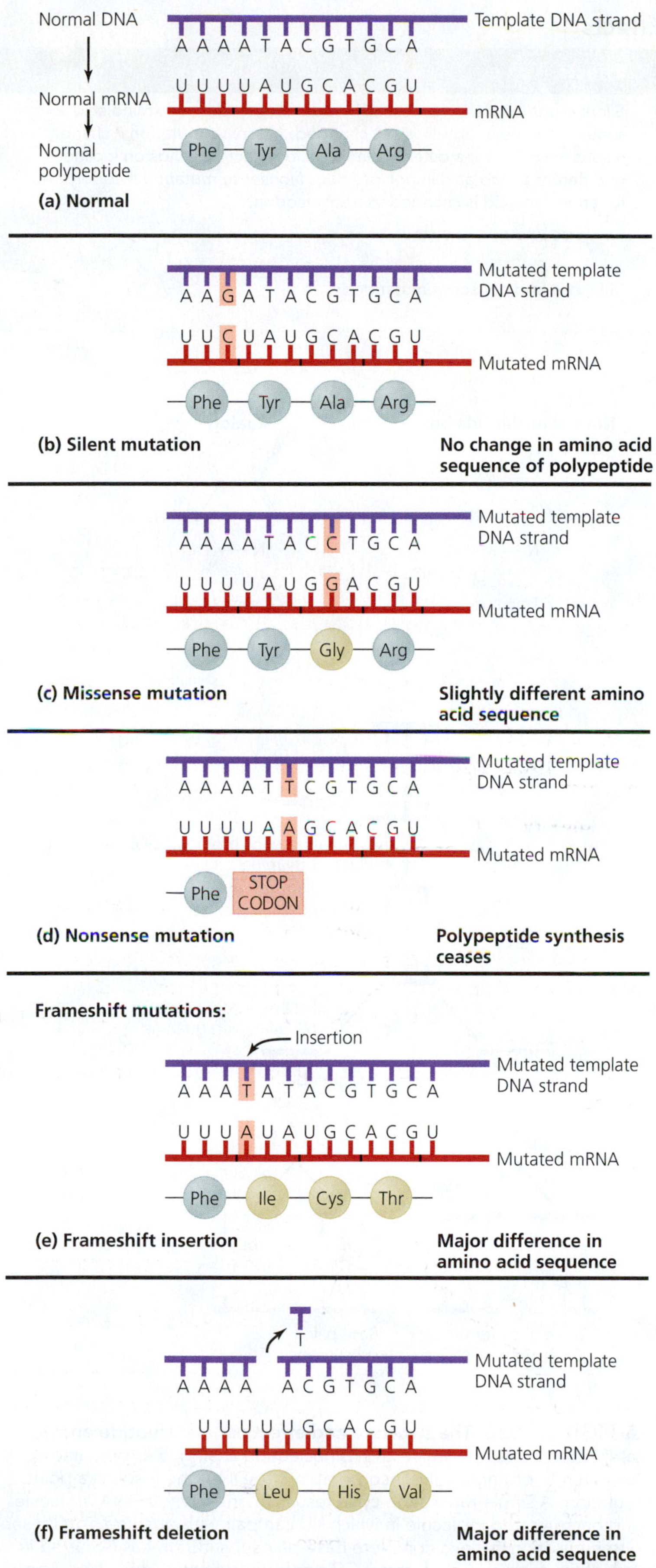

◀ **FIGURE 7.24 The effects of the various types of point mutations.** Normal gene **(a)**. Base-pair substitutions can result in silent mutations **(b)**, missense mutations **(c)**, or nonsense mutations **(d)**. Frameshift insertions **(e)** and frameshift deletions **(f)** usually result in severe missense or nonsense mutations because all codons downstream from the mutation are altered.

Since an average gene has 10^3 base pairs, about one of every 10^7 (10 million) genes contains an error. Further, though cells have repair mechanisms to reduce the effect of mutations, the repair process itself can introduce additional errors. Physical or chemical agents called **mutagens** (myū´tă-jenz), which include radiation and several types of DNA-altering chemicals, induce mutations.
▶**ANIMATIONS:** *Mutagens*

Radiation

In the 1920s, Hermann Muller (1890–1967) discovered that X rays increased phenotypic variability in fruit flies by causing mutations. *Gamma rays* also damage DNA. X rays and gamma rays are *ionizing radiation;* that is, they energize electrons in atoms, causing some of the electrons to escape from their atoms (see Chapter 9). These free electrons strike other atoms, producing ions that can react with the structure of DNA, creating mutations. More seriously, electrons and ions can break the covalent bonds between the sugars and phosphates of a DNA backbone, causing physical breaks in chromosomes and complete loss of cellular control.

Nonionizing radiation in the form of *ultraviolet (UV) light* is also mutagenic because it causes adjacent pyrimidine bases to covalently bond to one another, forming **pyrimidine dimers** (**FIGURE 7.25**). The presence of dimers prevents hydrogen bonding with nucleotides in the complementary strand, distorts the sugar-phosphate backbone, and prevents proper replication and transcription. Cells have several methods of repairing dimers (discussed shortly).

Chemical Mutagens

Here we consider three of the many basic types of mutagenic chemicals.

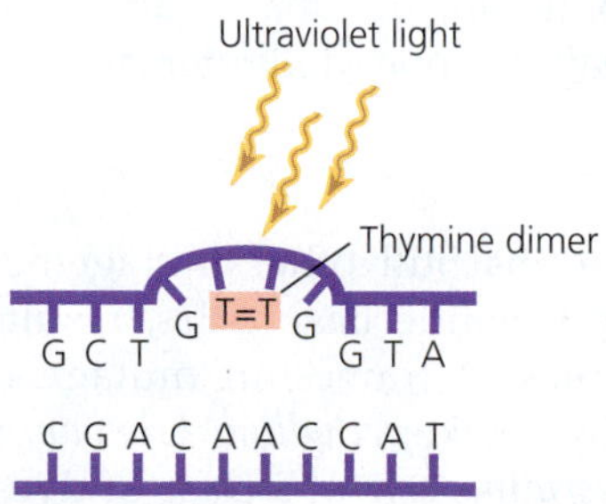

▲ **FIGURE 7.25 A pyrimidine (in this case, thymine) dimer.** Ultraviolet light causes adjacent pyrimidine bases (in this case, thymine bases) to covalently bond to each other, preventing hydrogen bonding with bases in the complementary strand. The resulting distortion of the sugar-phosphate backbone prevents proper replication and transcription.

TABLE 7.5 The Types of Point Mutations and Their Effects

Type of Point Mutation	Description	Effects
Substitution	Mismatching of nucleotides or replacement of one base pair by another	*Silent mutation* if change results in redundant codon, as amino acid sequence in polypeptide is not changed. *Missense mutation* if change results in codon for a different amino acid; effect depends on location of different amino acid in polypeptide. *Nonsense mutation* if codon for an amino acid is changed to a stop codon.
Frameshift (insertion)	Addition of one or a few nucleotide pairs creates new sequence of codons	Missense and nonsense mutations
Frameshift (deletion)	Removal of one or a few nucleotide pairs creates new sequence of codons	Missense and nonsense mutations

Nucleotide Analogs Compounds that are structurally similar to normal nucleotides are called **nucleotide analogs** (FIGURE 7.26a). When nucleotide analogs are available to replicating cells, they may be incorporated into DNA in place of normal nucleotides, where their structural differences either inhibit nucleic acid polymerases or result in mismatched base pairing. For example, when thymine is replaced by 5′-bromouracil, a wrong complement can form—5′-bromouracil can pair with guanine rather than with adenine, resulting in a point mutation (FIGURE 7.26b). (Figure 10.7 illustrates other nucleotide analogs.)

Nucleotide (or nucleoside) analogs make potent antiviral and anticancer drugs because viruses and cancer cells typically replicate faster than normal cells. (Chapter 10 considers the use of analogs as antimicrobial agents.)

Nucleotide-Altering Chemicals Some chemical mutagens alter the structure of nucleotides. For example, a group of nucleotide-altering chemicals, called *aflatoxins,* are produced by *Aspergillus* (as-per-jil´ŭs) molds growing on grains and nuts. Aflatoxins catabolized in the liver can convert guanine nucleotides into thymine nucleotides so that a GC base pair is converted to a TA base pair, resulting in missense mutations and possibly cancer. The Food and Drug Administration prohibits excessive amounts of aflatoxins in human and animal food. Another example is *nitrous acid* (HNO_2), which removes the amine group of adenine, converting adenine into a guanine analog. When a cell replicates DNA containing this analog, an AT base pair is changed to a GC base pair in one daughter molecule—a base-pair-substitution mutation.

Frameshift Mutagens Still other mutagenic chemical agents insert or delete nucleotide base pairs, resulting in frameshift mutations. Examples of frameshift mutagens are *benzopyrene,* which is found in smoke; *ethidium bromide,* which is used to stain DNA; and *acridine,* one of a class of dyes commonly used as mutagens in genetic research. These chemicals are exactly the right size to slip between adjoining nucleotides in DNA, producing a bulge in the molecule (FIGURE 7.27). When DNA polymerase copies the misshapen strands, one or more base pairs may be inserted or deleted in the daughter strand.

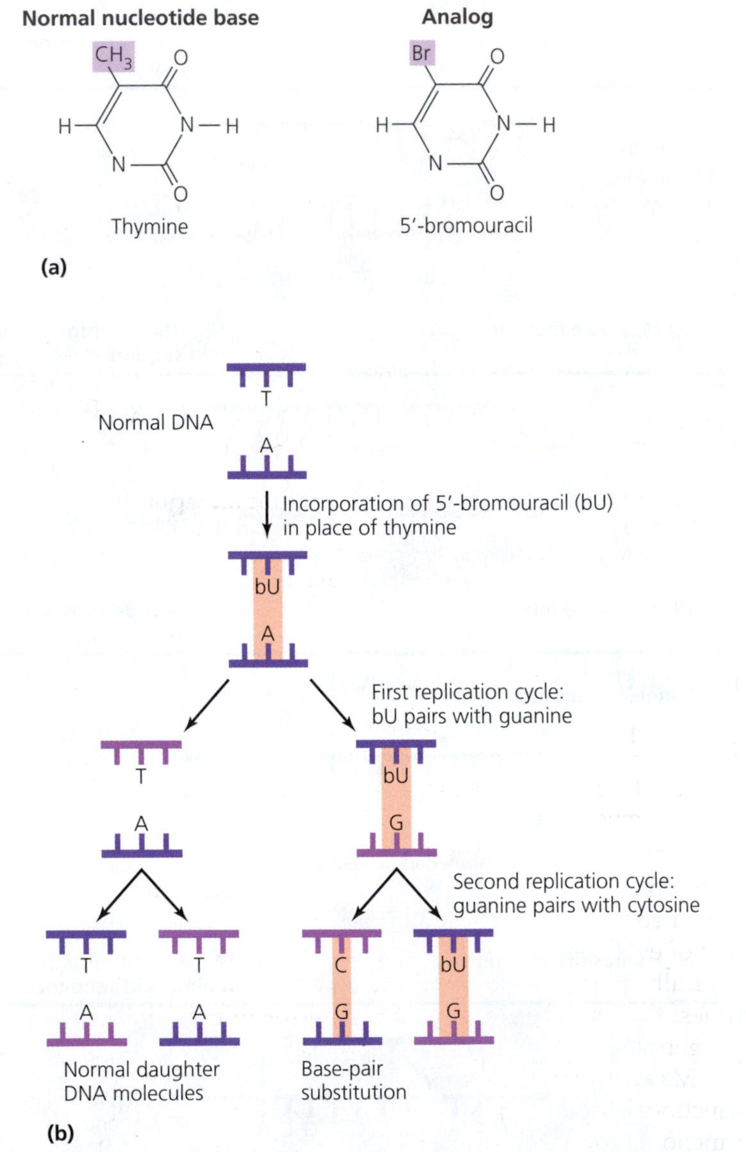

▲ **FIGURE 7.26 The structure and effects of a nucleotide analog.** **(a)** The structures of thymine and its nucleotide analog, 5′-bromouracil (bU). **(b)** When 5′-bromouracil is incorporated into DNA, the result is a point mutation. A single replication cycle results in one normal DNA molecule and one mutated molecule in which bU can pair with guanine. After a second replication cycle, a complete base-pair substitution has occurred in one of the four DNA molecules: CG has been substituted for TA.

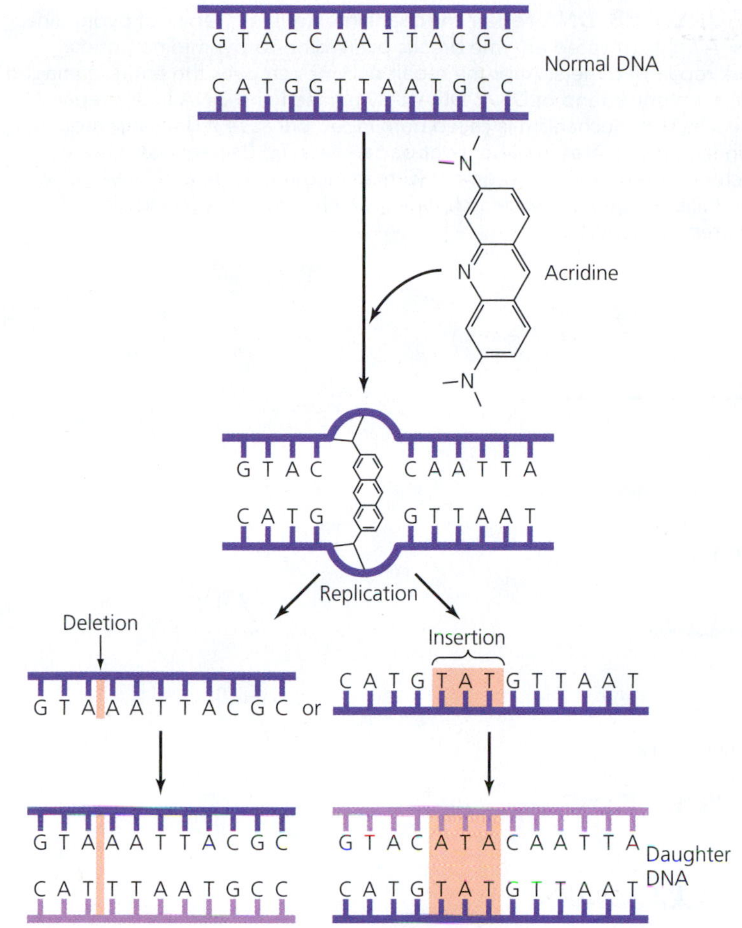

▲ **FIGURE 7.27 The action of a frameshift mutagen.** When DNA polymerase III passes the bulge caused by the insertion of acridine between the two DNA strands, it incorrectly synthesizes a daughter strand that contains either a deletion or an insertion mutation.

Frequency of Mutation

LEARNING | OUTCOME

7.23 Discuss the relative frequency of deleterious and useful mutations.

Mutations are rare events. If they were not, organisms could not live or effectively reproduce themselves. As we have seen, about one of every 10 million (10^7) genes contains an error. Mutagens typically increase the mutation rate by a factor of 10 to 1000 times; that is, mutagens induce an error in one of every 10^6 to 10^4 genes.

Many mutations are deleterious because they code for nonfunctional proteins or stop transcription entirely. Cells without functional proteins cannot metabolize; therefore, deleterious mutations are removed from the population when the cells die. Rarely, however, a cell acquires a beneficial mutation that allows it to survive, reproduce, and pass the mutation to its descendants. This change in gene frequency in a population is evolution. For example, the tuberculosis bacterium has acquired a mutation that confers resistance to the antimicrobial drug rifampin. In patients taking rifampin, bacterial cells without resistance die, but the mutated cell survives and reproduces. As long as rifampin is present, cells with such a mutation have an advantage over cells without the mutation, and the population evolves resistance to rifampin.

DNA Repair

LEARNING | OUTCOME

7.24 Describe light and dark repair of pyrimidine dimers, base-excision repair, mismatch repair, and the SOS response.

We have seen that mutations rarely convey an advantage; most mutations are deleterious. To respond to the dangers mutations pose, cells have numerous methods for repairing damaged DNA, including light and dark repair of pyrimidine dimers, base-excision repair, mismatch repair, and an SOS response. ▶**ANIMATIONS:** ***Mutations: Repair***

Repair of Pyrimidine Dimers

The most common type of mutation is a pyrimidine dimer caused by ultraviolet light. Many cells contain DNA *photolyase*, an enzyme that is activated by visible light to break pyrimidine dimers, reversing the mutation and restoring the original DNA sequence **(FIGURE 7.28a)**. This so-called **light repair** mechanism is advantageous for the cell, but it presents a difficulty to scientists studying UV-induced mutations—they must keep such strains in the dark, or the mutants revert to their normal form.

So-called **dark repair** involves a different repair enzyme—one that doesn't require light. Dark repair enzymes cut the damaged section of DNA from the molecule, creating a gap that is repaired by DNA polymerase I and DNA ligase **(FIGURE 7.28b)**. Though called dark repair, this mechanism operates either in light or in the dark.

Base-Excision Repair

Rarely, DNA polymerase III incorporates an incorrect nucleotide during DNA replication. If the proofreading function of the polymerase does not repair the error, cells may use another enzyme system in a process called **base-excision repair**. This enzyme system excises the erroneous base, and then DNA polymerase I fills in the gap **(FIGURE 7.28c)**.

Mismatch Repair

A similar repair mechanism is called **mismatch repair**. Mismatch repair enzymes scan newly synthesized DNA looking for mismatched bases, which they remove and replace **(FIGURE 7.28d)**. How does the mismatch repair system determine which strand to repair? If it chose randomly, 50% of the time it would choose the wrong strand and introduce mutations. Mismatch repair enzymes, however, do not choose randomly. They distinguish a new DNA strand from an old strand because old strands are methylated. Recognition of an error as far as 1000 bp away from an unmethylated portion

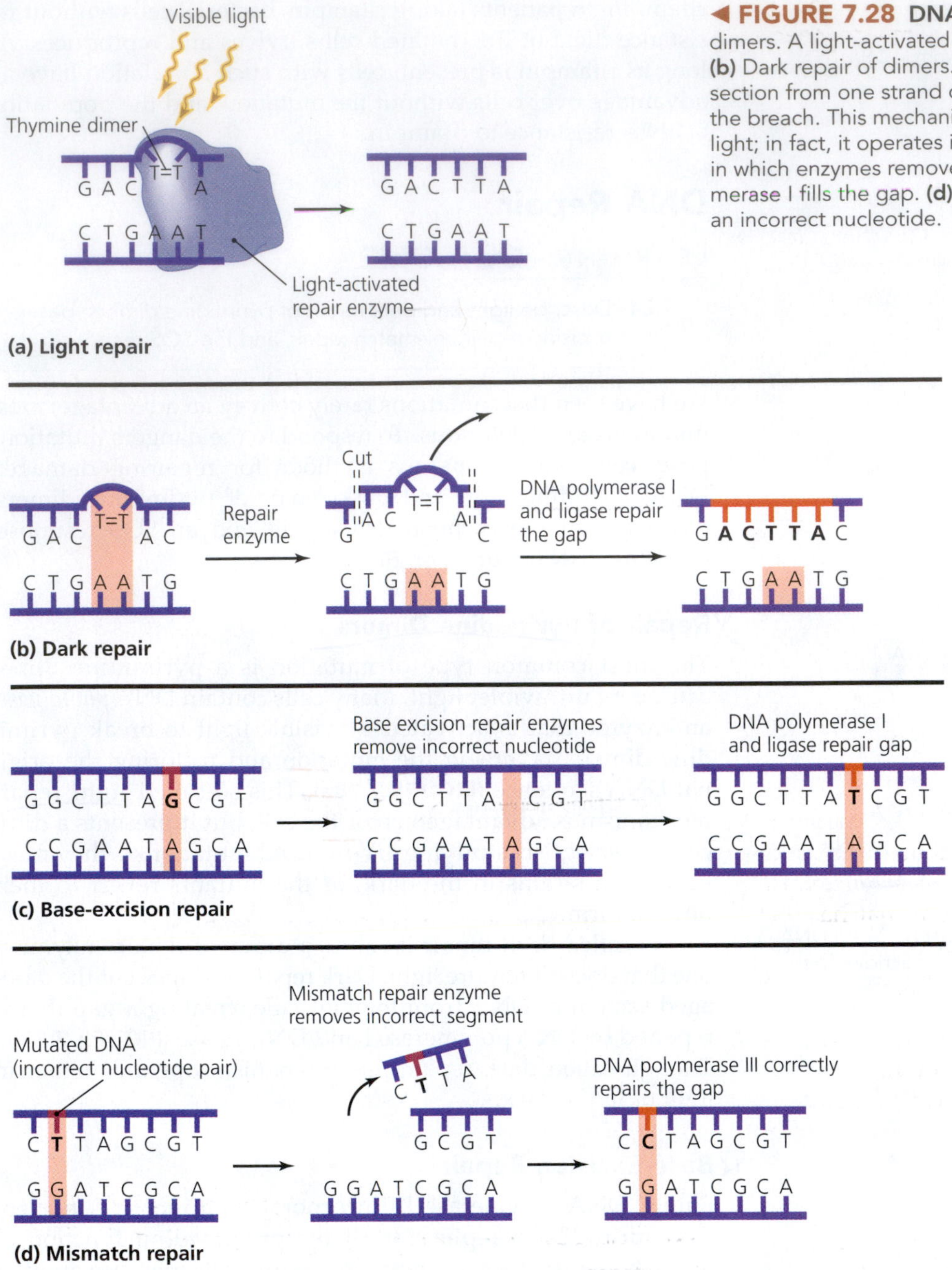

◀ **FIGURE 7.28 DNA repair mechanisms.** **(a)** Light repair of pyrimidine dimers. A light-activated enzyme breaks pyrimidine-to-pyrimidine bonds. **(b)** Dark repair of dimers. After the repair enzyme removes the entire damaged section from one strand of DNA, DNA polymerase I and DNA ligase repair the breach. This mechanism is called dark repair because it does not require light; in fact, it operates in either light or darkness. **(c)** Base-excision repair, in which enzymes remove a segment with an incorrect base and DNA polymerase I fills the gap. **(d)** Mismatch repair, which involves total excision of an incorrect nucleotide.

of DNA triggers the mismatch repair enzymes. Once a new DNA strand is methylated, mismatch repair enzymes cannot correct any errors that remain.

SOS Response

Sometimes damage to DNA is so extreme that regular repair mechanisms cannot cope with the damage. In such cases, bacteria resort to what geneticists call an **SOS response** involving a variety of processes, such as the production of novel DNA polymerases (IV and V) capable of copying less-than-perfect DNA. These polymerases replicate DNA with little regard to the base sequence of the template strand. Of course, this introduces many new and potentially fatal mutations, but presumably SOS repair allows a few offspring of these bacteria to survive.

Identifying Mutants, Mutagens, and Carcinogens

LEARNING | **OUTCOMES**

7.25 Contrast the positive and negative selection techniques for isolating mutants.

7.26 Describe the Ames test, and discuss its use in discovering carcinogens.

If a cell does not successfully repair a mutation, it and its descendants are called **mutants**. In contrast, cells normally found in nature (in the wild) are called **wild-type** cells. Scientists distinguish mutants from wild-type cells by observing or testing for altered phenotypes. Because mutations are rare and nonfatal mutations are even rarer, mutants can easily be "lost in the crowd." Therefore, researchers have developed methods to recognize mutants amidst their wild-type neighbors.

Positive Selection

Positive selection involves selecting a mutant by eliminating wild-type phenotypes. Assume, for example, that researchers want to isolate penicillin-resistant bacterial mutants from a liquid culture. To do so, they spread the liquid medium, which contains mostly penicillin-sensitive cells but also the few penicillin-resistant mutants, onto medium that includes penicillin. Only the penicillin-resistant mutants multiply on this medium and produce visible colonies (**FIGURE 7.29a**).

When a mutagenic agent is added to a liquid culture, it increases the number of mutants (**FIGURE 7.29b**). While the researchers are isolating mutants, they can also determine the rate of mutation by comparing the number of mutant colonies formed after use of the mutagen with the number formed before treatment. The rate of mutation can be calculated as follows.

$$\frac{\text{number of colonies seen with use of mutagen} - \text{number of colonies seen without use of mutagen}}{\text{number of colonies seen without the use of mutagen}} \times 100\%$$

Negative (Indirect) Selection

An organism with nutritional requirements that differ from those of its wild-type phenotype is known as an *auxotroph* (awk´sō-trōf).[8] For example, a mutant bacterium that has lost the ability to synthesize tryptophan is auxotrophic for this amino acid—it must acquire tryptophan from the environment. Obviously, if a researcher attempts to grow tryptophan auxotrophs on media lacking tryptophan, the bacteria will be unable to synthesize all its proteins and will die. Therefore, to isolate such auxotrophs, we must use a technique called **negative (indirect) selection**. The process by which a researcher uses negative selection to culture a tryptophan auxotroph is as follows (**FIGURE 7.30**).

1. The researcher inoculates a sample of a bacterial suspension containing potential mutants onto a plate containing complete media (including tryptophan). The sample is diluted such that the plate receives only about 100 cells.
2. Both auxotrophs and wild-type cells reproduce and form colonies on the plate, but the colonies are indistinguishable.
3. The researcher picks up cells from all the colonies on the plate with a sterile velvet pad by pressing the pad onto the plate.
4. The researcher inoculates two new plates—one containing tryptophan, the other lacking tryptophan—by pressing the pad onto each of them. This technique is called *replica plating*.
5. After the plates have incubated for several hours, the researcher compares the two replica plates. Tryptophan auxotrophs growing on medium containing tryptophan are revealed by the absence of a corresponding colony on the plate lacking tryptophan.
6. The researcher takes cells of the auxotroph colony from the replica plate and inoculates them into a complete medium. The auxotroph is now isolated.

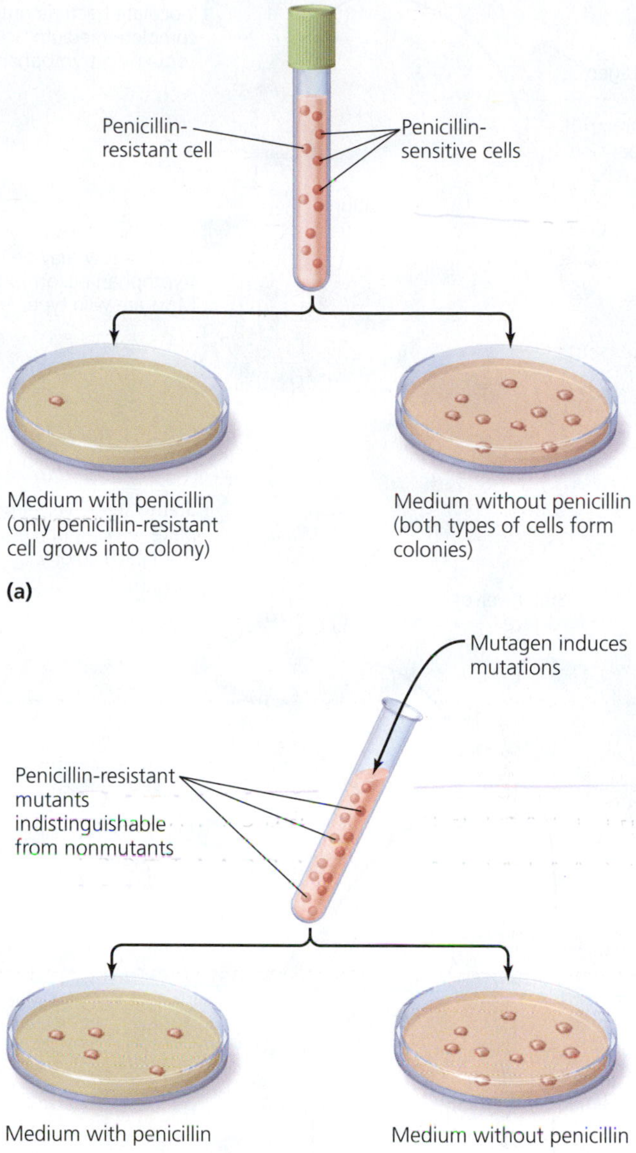

▲ **FIGURE 7.29 Positive selection of mutants.** Only mutants that are resistant to penicillin can survive on the plate containing the antibiotic. **(a)** Normally, a population includes very few mutants. **(b)** Introduction of a mutagen increases the number of mutants and thus the number of colonies that grow in the presence of penicillin. *What is the rate of mutation induced by the mutagen in (b)?*

Figure 7.29 $\frac{5-1}{1} \times 100\% = 400\%$

[8]From Greek *auxein*, meaning "to increase," and *trophe*, meaning "nutrition."

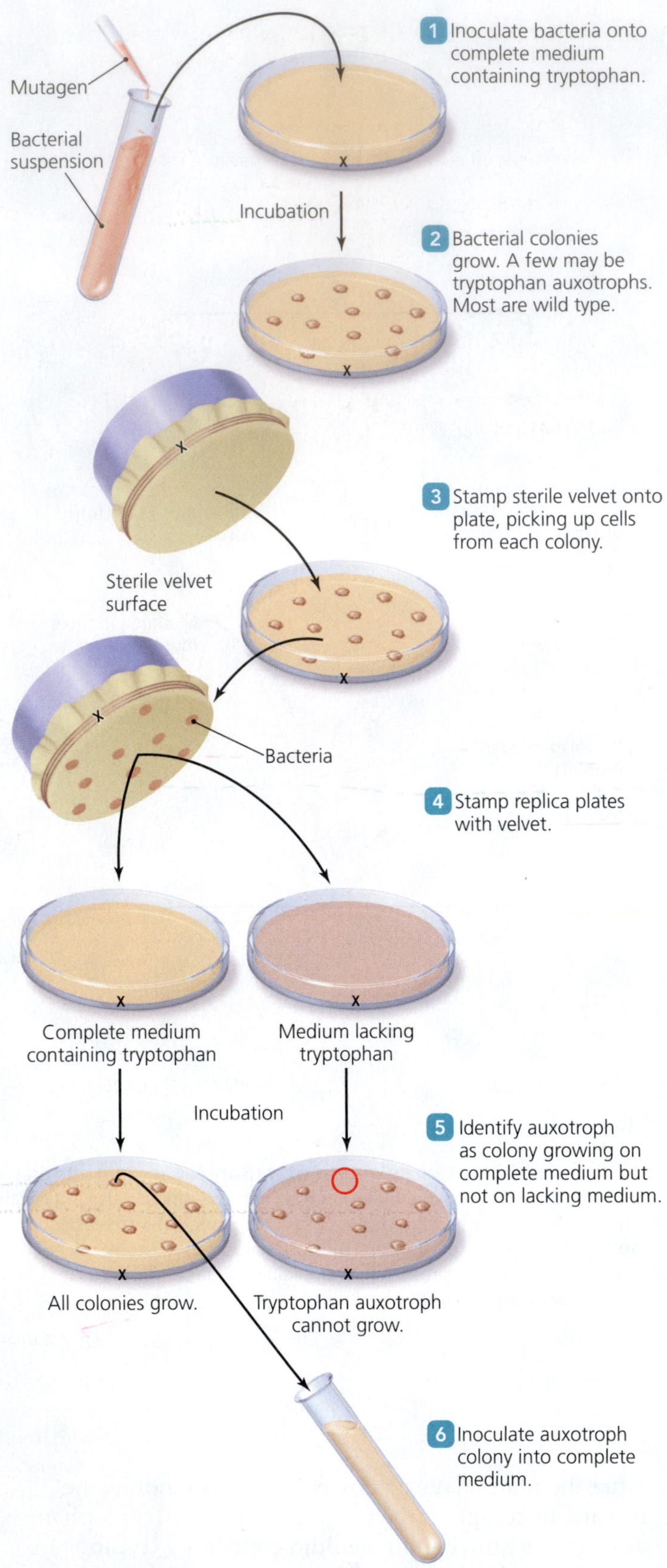
Mutagen
Bacterial suspension
1 Inoculate bacteria onto complete medium containing tryptophan.
Incubation
2 Bacterial colonies grow. A few may be tryptophan auxotrophs. Most are wild type.
3 Stamp sterile velvet onto plate, picking up cells from each colony.
Sterile velvet surface
Bacteria
4 Stamp replica plates with velvet.
Complete medium containing tryptophan
Medium lacking tryptophan
Incubation
5 Identify auxotroph as colony growing on complete medium but not on lacking medium.
All colonies grow.
Tryptophan auxotroph cannot grow.
6 Inoculate auxotroph colony into complete medium.

▲ **FIGURE 7.30 The use of negative (indirect) selection to isolate a tryptophan auxotroph.** The plates are marked (in this case, with an X) so that their orientation can be maintained throughout the procedure. Researchers may have to inoculate hundreds of such plates to identify a single mutant.

The Ames Test for Identifying Mutagens

Numerous chemicals in food, the workplace, and the environment in general have been suspected of being **carcinogenic** (kar´si-nō-jen´ik) mutagens; that is, of causing mutations that result in cancer. Because animal tests to prove that they are indeed carcinogenic are expensive and time consuming, researchers have used a fast and inexpensive method for screening mutagens called an **Ames test**, which is named for its inventor, Bruce Ames (1928–).

An Ames test uses mutant *Salmonella* (sal´mŏ-nel´ă) bacteria possessing a point mutation that prevents the synthesis of the amino acid histidine; in other words, they are histidine auxotrophs, indicated by the abbreviation *his*$^-$. To perform the test, an investigator mixes *his*$^-$ mutants with liver extract and the substance suspected to be a mutagen (**FIGURE 7.31**). The presence of liver extract simulates the conditions in the body under which liver enzymes can turn harmless chemicals into mutagens. The researcher then spreads the treated bacteria on a solid medium lacking histidine. If the suspected substance does in fact cause mutations, some of the mutations will likely reverse the effect of the original mutation, producing revertant

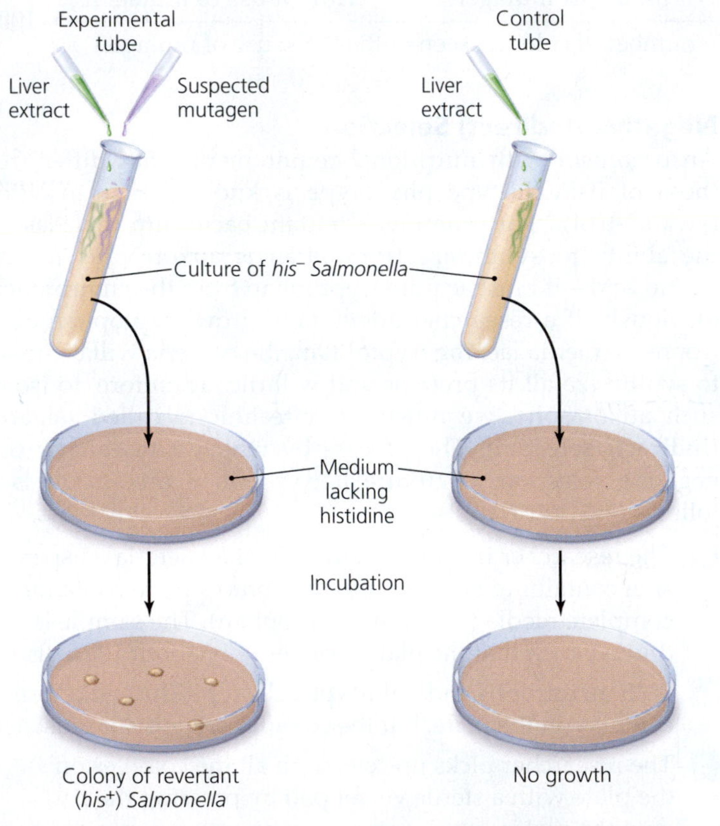
Experimental tube
Liver extract
Suspected mutagen
Control tube
Liver extract
Culture of his⁻ Salmonella
Medium lacking histidine
Incubation
Colony of revertant (his⁺) Salmonella
No growth

▲ **FIGURE 7.31 The Ames test.** A mixture containing *his*$^-$ *Salmonella* mutants, rat liver extract, and the suspected mutagen is inoculated onto a plate lacking histidine. Colonies will form only if a mutagen reverses the *his*$^-$ mutation, producing revertant *his*$^+$ organisms with the ability to synthesize histidine. A control tube that lacks the suspected mutagen demonstrates that reversion did not occur in the absence of the mutagen. *What is the purpose of liver extract in an Ames test?*

Figure 7.31 *Liver extract simulates conditions in the body by providing enzymes that may degrade harmless substances into mutagens.*

cells (designated *his*$^+$) that have regained the ability to synthesize histidine and thus can survive on a medium lacking histidine. Thus, the presence of colonies during an Ames test reveals that the suspected substance is mutagenic in *Salmonella*.

Given that DNA in all cells is very similar, the ability of a chemical to cause mutations in *Salmonella* indicates that it is likely to cause mutations in humans as well. Some mutations cause cancers, so a mutagenic chemical may also be carcinogenic. To prove that the substance can in fact cause cancer, scientists must test it in laboratory animals, a process that usually is warranted only if a chemical is mutagenic in *Salmonella*. Ames testing reduces the cost and time required to assay chemicals for carcinogenicity in animals.

TELL ME WHY

Changes in RNA resulting from poor transcription of RNA to DNA are not as deleterious to an organism as changes to its DNA resulting from mutations. Why is this the case?

Genetic Recombination and Transfer

LEARNING **OUTCOME**

7.27 Define *genetic recombination*.

Genetic recombination refers to the exchange of nucleotide sequences between two DNA molecules and often involves segments that are composed of identical or nearly identical nucleotide sequences called *homologous sequences*. Scientists have discovered a number of molecular mechanisms for genetic recombination, one of which is illustrated simply in **FIGURE 7.32**. In this type of recombination, enzymes nick one strand of DNA at the homologous sequence, and another enzyme inserts the nicked strand into the second DNA molecule. Ligase then reconnects the strands in new combinations, and the molecules resolve themselves into novel molecules. Such DNA molecules that contain new arrangements of nucleotide sequences (and the cells that contain them) are called recombinants. (Chapter 12 discusses crossing over and the formation of gametes in more detail.)

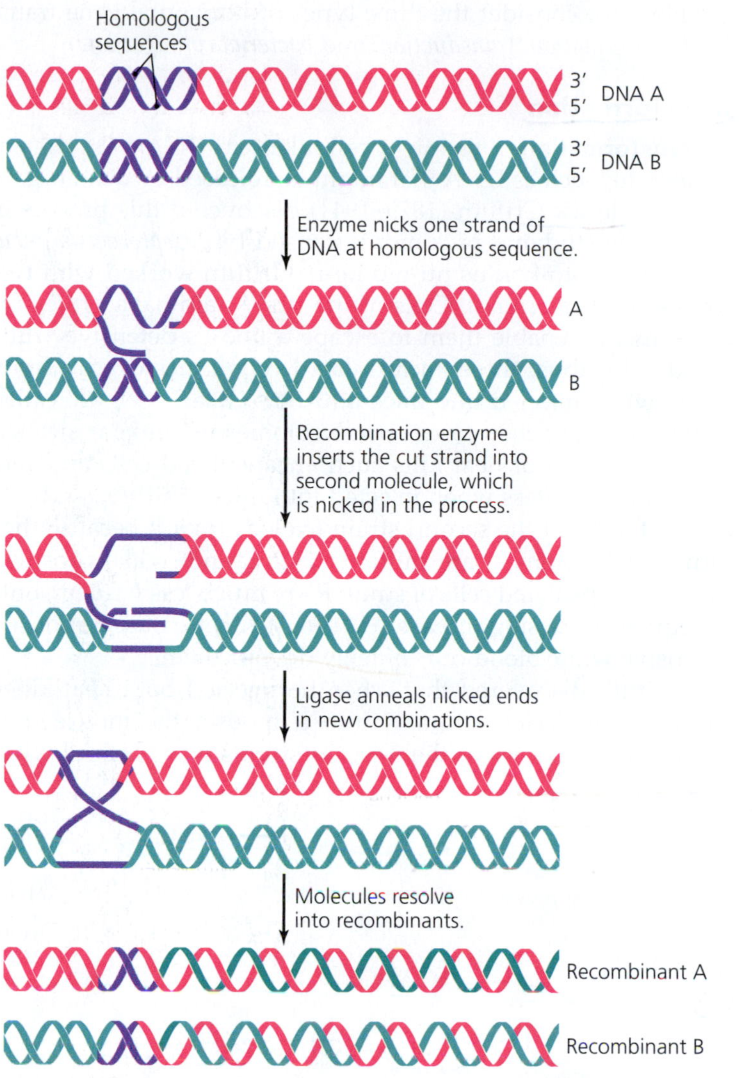

▲ **FIGURE 7.32 Genetic recombination.** Simplified depiction of one type of recombination between two DNA molecules. After an enzyme nicks one strand (here, strand A), a recombination enzyme rearranges the strands, and ligase seals the gaps to form recombinant molecules. *What is the normal function of ligase (during DNA replication)?*

Figure 7.32 *Ligase functions to anneal Okazaki fragments during replication of the lagging strand.*

Horizontal Gene Transfer Among Prokaryotes

LEARNING **OUTCOMES**

7.28 Contrast vertical gene transfer with horizontal gene transfer.

7.29 Explain the roles of an F factor, F$^+$ cells, and Hfr cells in bacterial conjugation.

7.30 Describe the structures and actions of simple and complex transposons.

7.31 Compare and contrast crossing over, transformation, transduction, and conjugation.

As we have discussed, both prokaryotes and eukaryotes replicate their genomes and supply copies to their descendants. This is known as *vertical gene transfer*—the passing of genes to the next generation. In addition, many prokaryotes acquire genes from other microbes of the same generation—a process termed **horizontal (lateral) gene transfer**. In horizontal gene transfer, a **donor cell** contributes part of its genome to a **recipient cell**, which may be of a different species from the donor. Typically, the recipient cell inserts part of the donor's DNA into its own chromosome, becoming a **recombinant** cell. Cellular enzymes then usually degrade remaining unincorporated DNA. Horizontal gene transfer is a rare event, typically occurring in less than 1% of a population of prokaryotes. ▶**ANIMATIONS:** ***Horizontal Gene Transfer: Overview***

Here we consider the three types of horizontal gene transfer: *transformation*, *transduction*, and *bacterial conjugation*.

Transformation

In **transformation**, a recipient cell takes up DNA from the environment, such as DNA that might be released by dead organisms. Frederick Griffith (1879–1941) discovered this process in 1928 while studying pneumonia caused by *Streptococcus pneumoniae* (strep-tō-kok´ŭs nū-mō´nē-ī). Griffith worked with two strains of *Streptococcus*. Cells of the first strain have protective capsules that enable them to escape a body's defensive white blood cells; thus, these encapsulated cells cause deadly septicemia when injected into mice **(FIGURE 7.33a)**. They are called *strain S* because they form *smooth* colonies on an agar surface. The application of heat kills such encapsulated cells and renders them harmless when injected into mice **(FIGURE 7.33b)**. In contrast, cells of the second strain (called *strain R* because they form *rough* colonies) are mutants that cannot make capsules. The unencapsulated cells of strain R are much less virulent, only infrequently causing disease **(FIGURE 7.33c)** because a mouse's defensive white blood cells quickly devour them.

Griffith discovered that when he injected both heat-killed strain S and living strain R into a mouse, the mouse died even though neither of the injected strains was harmful when administered alone **(FIGURE 7.33d)**. Further, and most significantly, Griffith isolated numerous living cells from the dead mouse, cells that had capsules. He realized that harmless, unencapsulated strain R bacteria had been transformed into deadly, encapsulated strain S bacteria. Subsequent investigations showed that transformation, by which cells take up DNA, also occurs *in vitro*[9] **(FIGURE 7.33e)**.

The fact that the living encapsulated cells retrieved at the end of this experiment outnumbered the dead encapsulated cells injected at the beginning indicated that strain R cells were not merely appropriating capsules released from dead strain S cells. Instead, strain R cells had acquired the capability of producing their own capsules by assimilating the capsule-coding genes of strain S cells. In 1944, Oswald Avery (1877–1955), Colin MacLeod (1909–1972), and Maclyn McCarty (1911–2005) extracted various chemicals from S cells and determined that the transforming agent was DNA. This discovery was one of the conclusive pieces of evidence that DNA is the genetic material of cells.

Cells that have the ability to take up DNA from their environment are said to be **competent**. Competence results from

[9]Latin, meaning "within glassware."

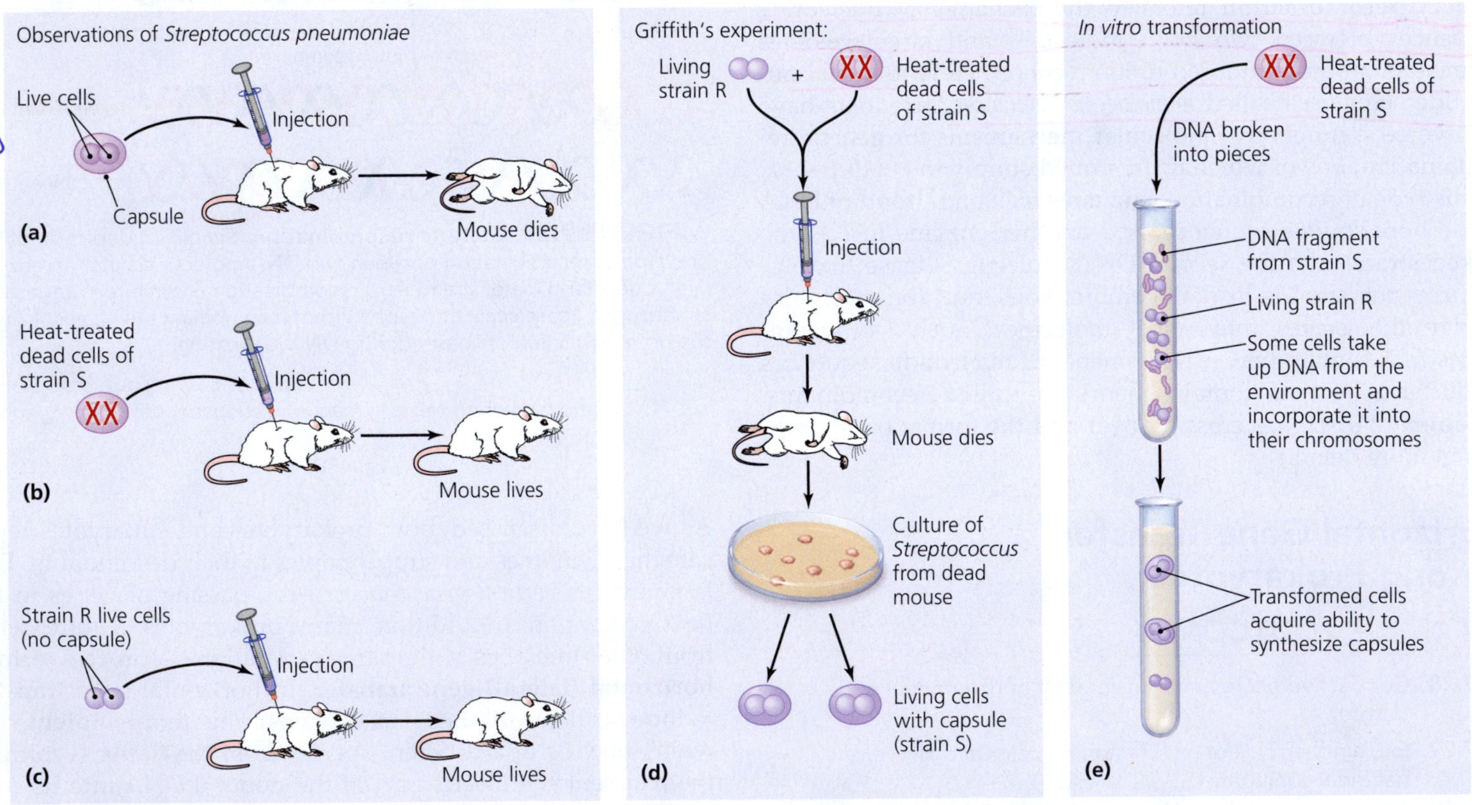

▲ **FIGURE 7.33 Transformation of *Streptococcus pneumoniae*.** Griffith's observations revealed that **(a)** encapsulated strain S killed mice, **(b)** heating renders strain S harmless to mice, and **(c)** unencapsulated strain R did not harm mice. In Griffith's experiment **(d)**, a mouse injected concurrently with killed strain S and live strain R (each harmless) died and was found to contain numerous living, encapsulated bacteria. **(e)** A demonstration that transformation of R cells to S cells also occurs *in vitro*.

alterations in the cell wall and cytoplasmic membrane that allow DNA to enter the cell. Scientists have observed natural competence in a few types of bacteria, including pathogens in *Streptococcus*, *Haemophilus* (hē-mof´i-lŭs), *Neisseria* (nī-se´rē-ă), *Bacillus* (ba-sil´ŭs), *Staphylococcus* (staf´i-lō-kok´ŭs), and *Pseudomonas*. Scientists can also generate competence artificially in *Escherichia* and other bacteria by manipulating the temperature and salt content of the medium. Because competent cells take up DNA from any donor genome, competence and transformation are important tools in recombinant DNA technology, commonly called *genetic engineering*. ▶**ANIMATIONS:** *Transformation*

Transduction

A second method of horizontal gene transfer, called **transduction**, involves the transfer of DNA from one cell to another via a replicating virus. Transduction can occur either between prokaryotic cells or between eukaryotic cells; it is limited only by the availability of a virus capable of infecting both donor and recipient cells. Here we will consider transduction in bacteria.

A virus that infects bacteria is called a **bacteriophage** or simply a **phage** (fāj).[10] The process by which a phage participates in transduction is depicted in **FIGURE 7.34**. To replicate, a bacteriophage attaches to a bacterial host cell and injects its genome into the cell 1. Phage enzymes (enzymes coded by the phage's genome even though translated by bacterial ribosomes) degrade the cell's DNA 2. The phage genome now controls the cell's functions and directs it to synthesize new phage DNA and phage proteins. Normally, phage proteins assemble around phage DNA to form new phage particles, but some phages mistakenly incorporate remaining fragments of bacterial DNA that are about the same length as phage DNA. This forms **transducing phages** 3. Eventually the host cell lyses, releasing daughter and transducing phages. Transduction occurs when a transducing phage injects donor DNA into a new host cell (the recipient) 4. The recipient host cell incorporates the donated DNA into its chromosome by recombination 5.

In *generalized transduction*, the transducing phage carries a random DNA segment from a donor host cell's chromosome or plasmids to a recipient host cell. Generalized transduction is not limited to a particular DNA sequence. In *specialized transduction*, only certain host sequences are transferred (along with phage DNA). In nature, specialized transduction is important in transferring genes encoding for certain bacterial toxins—including those responsible for diphtheria, scarlet fever, and the bloody, life-threatening diarrhea caused by *E. coli* O157:H7—into cells that would otherwise be harmless. (Chapter 8 discusses the use of specialized transduction to intentionally insert genes into cells.) ▶**ANIMATIONS:** *Transduction: Generalized Transduction, Specialized Transduction*

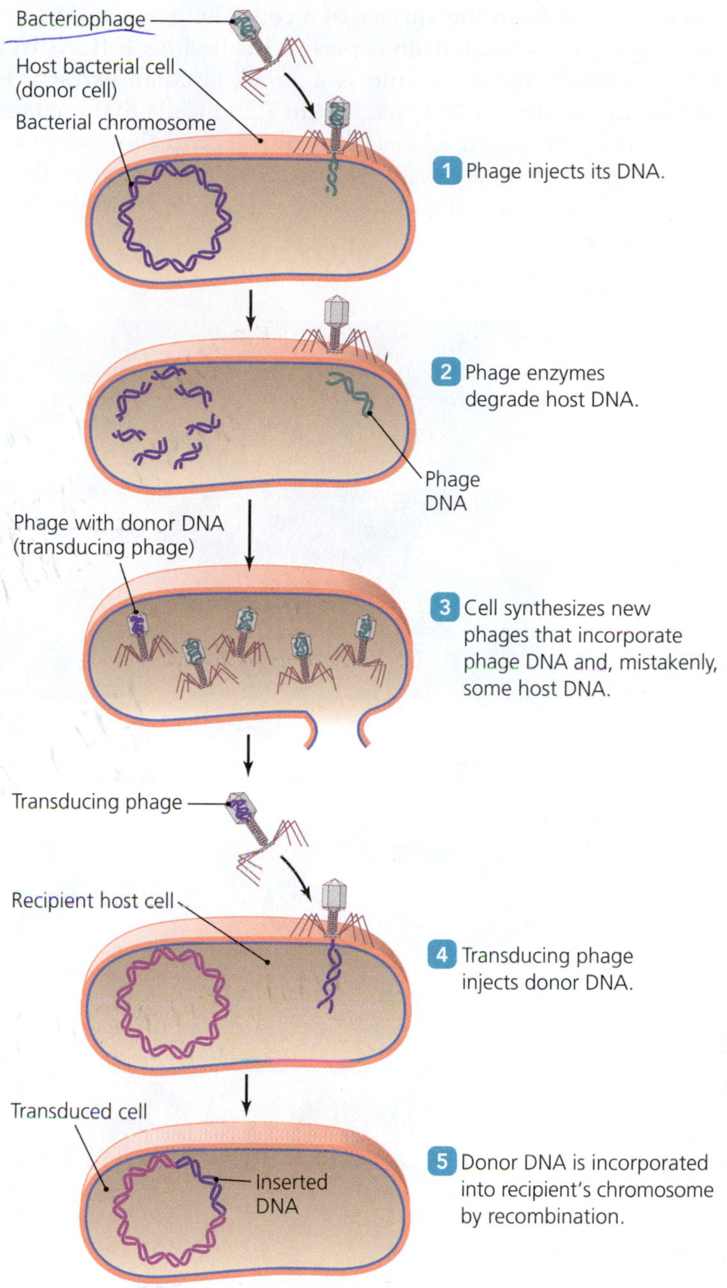

▲ **FIGURE 7.34 Transduction.** After a virus called a bacteriophage (phage) attaches to a host bacterial cell, it injects its genome into the cell and directs the cell to synthesize new phages. During assembly of new phages, some host DNA may be incorporated, forming transducing phages, which subsequently carry donor DNA to a recipient host cell.

Bacterial Conjugation

A third method of genetic transfer in bacteria is **conjugation**.[11] Unlike the typical donor cells in transformation and transduction, a donor cell in conjugation remains alive. Further, conjugation requires physical contact between donor and recipient cells. Scientists discovered conjugation between cells of *E. coli*, and it is best understood in this species. Thus, the remainder of our discussion will focus on conjugation in this bacterium. ▶**ANIMATIONS:** *Conjugation: Overview*

Conjugation is mediated by **pili** (pīlī; singular: *pilus*), also called *conjugation* or *sex pili*,[12] which are thin, proteinaceous

[10]From Greek *phagein*, meaning "to eat."
[11]From Latin *conjugatus*, meaning "yoked together."
[12]Bacterial conjugation may resemble intercourse, but because no gametes are involved, it is not true sex.

tubes extending from the surface of a cell. The gene coding for conjugation pili is located on a plasmid called an **F (fertility) plasmid**. (Recall that a plasmid is a small, circular, extrachromosomal molecule of DNA; see Figure 7.2b.) Cells that contain an F plasmid are called F^+ *cells* ("ef plus"), and they serve as donors during conjugation. Recipient cells are F^-; that is, they lack an F plasmid and therefore have no pili. ▶**ANIMATIONS:** ***Conjugation: F Factor***

The process of bacterial conjugation is illustrated in **FIGURE 7.35**. First, a sex pilus connects a donor cell (F^+) to a recipient cell (F^-) **1**. The pilus may draw the cells together, though DNA transfer may occur when the cells are still more than 10 μm apart **2**. A single strand of the F plasmid DNA transfers to the recipient beginning with a section called the *origin of transfer* **3**. The F^- recipient then synthesizes a complementary strand of F plasmid DNA, becoming an F^+ cell **4**.

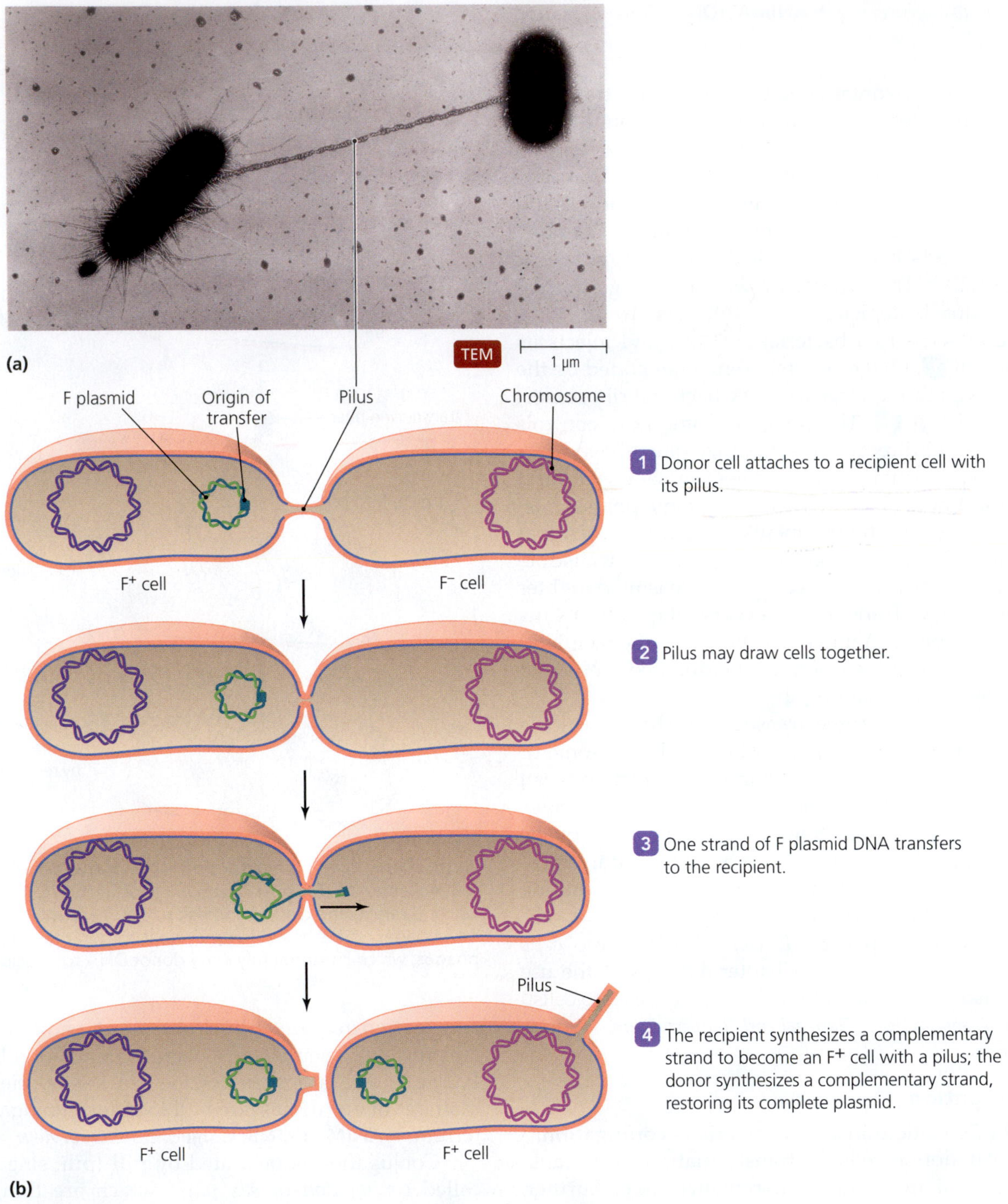
(a)
TEM
1 μm
F plasmid
Origin of transfer
Pilus
Chromosome
F⁺ cell
F⁻ cell
1 Donor cell attaches to a recipient cell with its pilus.
2 Pilus may draw cells together.
3 One strand of F plasmid DNA transfers to the recipient.
Pilus
4 The recipient synthesizes a complementary strand to become an F⁺ cell with a pilus; the donor synthesizes a complementary strand, restoring its complete plasmid.
F⁺ cell
F⁺ cell
(b)

▲ **FIGURE 7.35 Bacterial conjugation.** A conjugation pilus connecting two cells mediates the transfer of DNA between the cells. **(a)** Two *E. coli* cells are connected by a pilus. **(b)** Artist's rendition of the process.

The donor cell also synthesizes a complementary plasmid DNA strand.

In some bacterial cells, an F plasmid does not remain independent in the cytosol but instead integrates at a specific DNA sequence in the cellular chromosome. Such cells, which are called **Hfr (high frequency of recombination) cells**, can conjugate with an F^- cell **(FIGURE 7.36)**. After the F plasmid has integrated **1** and the Hfr and F^- cells join via a sex pilus **2**, DNA transfer begins at the origin of transfer of the F plasmid, carrying with it a copy of the donor's chromosome **3**. In most cases, movement of the cells breaks the intercellular connection before an entire donor chromosome is transferred **4**. Because the recipient receives only a portion of the F plasmid, it remains an F^- cell; however, it also acquires some chromosomal genes from the donor. Recombination can integrate the donor DNA into the recipient's chromosome **5**. The recipient is now a recombinant cell that contains its own genes as well as some donor genes. ▶**ANIMATIONS:** ***Conjugation: Hfr Conjugation***

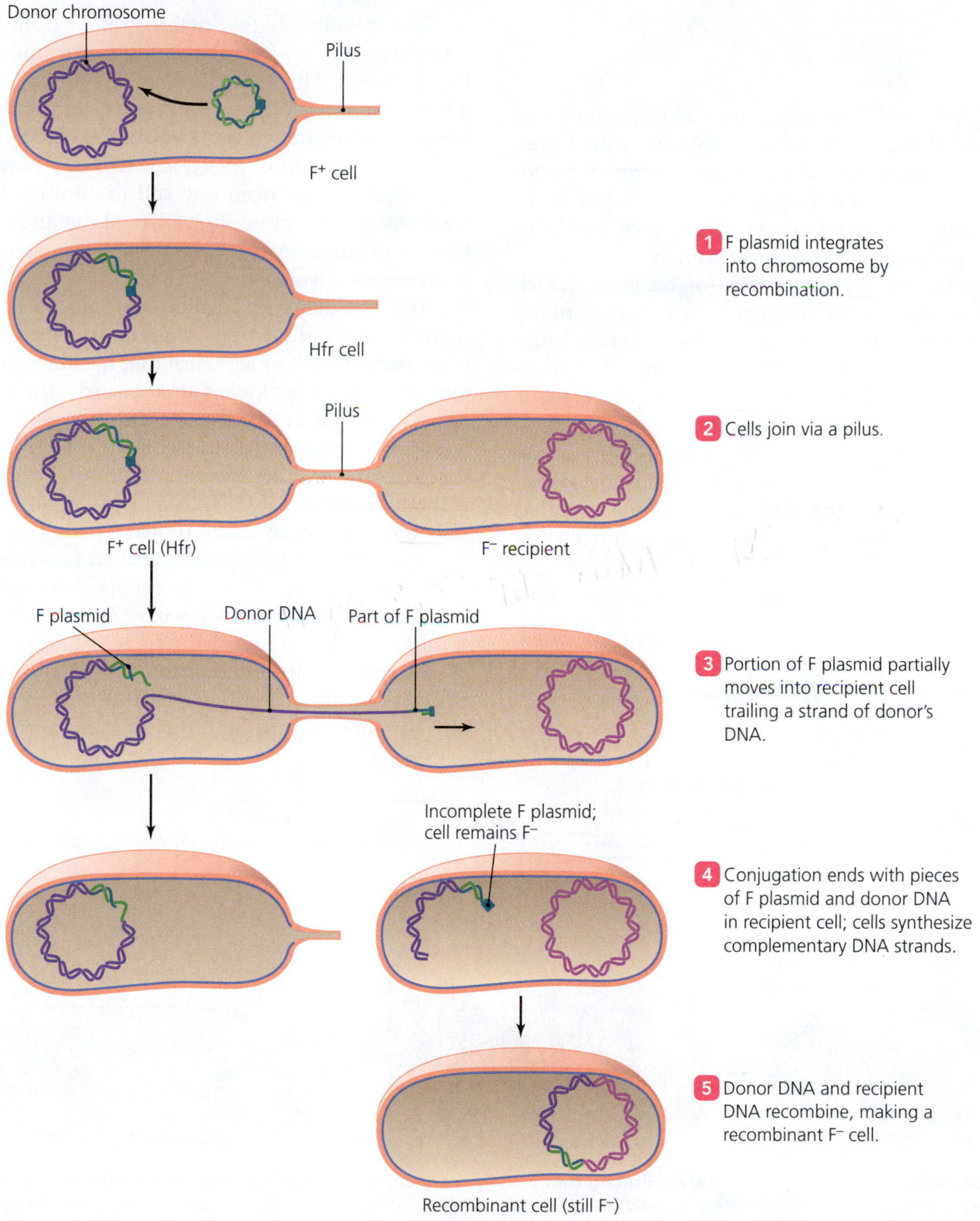
Donor chromosome
Pilus
F⁺ cell
1 F plasmid integrates into chromosome by recombination.
Hfr cell
Pilus
2 Cells join via a pilus.
F⁺ cell (Hfr)
F⁻ recipient
F plasmid
Donor DNA
Part of F plasmid
3 Portion of F plasmid partially moves into recipient cell trailing a strand of donor's DNA.
Incomplete F plasmid; cell remains F⁻
4 Conjugation ends with pieces of F plasmid and donor DNA in recipient cell; cells synthesize complementary DNA strands.
5 Donor DNA and recipient DNA recombine, making a recombinant F⁻ cell.
Recombinant cell (still F⁻)

▲ **FIGURE 7.36 Conjugation involving an Hfr cell.** An Hfr cell is formed when an F^+ cell integrates its F plasmid into its chromosome. Hfr cells donate a partial copy of their DNA and a portion of the F plasmid to a recipient, which is rendered a recombinant cell but remains F^-.

TABLE 7.6 Natural Mechanisms of Horizontal Genetic Transfer in Bacteria

Mechanism	Requirements
Transformation	Free DNA in the environment and a competent recipient
Transduction	Bacteriophage
Conjugation	Cell-to-cell contact and F plasmid, which is either in cytosol or incorporated into chromosome of donor (Hfr) cell

Because an F plasmid integrates into chromosomes at only a few locations, the order in which genes are transferred is consistent. Scientists produced the first gene maps of bacterial chromosomes by noting the time, in minutes, required for a particular gene to transfer from donor cells to recipient cells. ▶**ANIMATIONS:** *Conjugation: Chromosome Mapping*

Conjugation occurs in several species of bacteria, which can be quite promiscuous; that is, conjugation can occur among bacteria of widely varying kinds and even between a bacterium and a yeast cell or between a bacterium and a plant cell. For example, the crown gall bacterium, *Agrobacterium*, transfers some of its genes into the chromosome of its host plant.

The natural transfer of genes by conjugation among diverse organisms heightens some scientists' concerns about the spread of resistance (R) plasmids among pathogens. These scientists note that antibiotic resistance developed by one pathogen can spread to other pathogens.

TABLE 7.6 summarizes the mechanisms of horizontal gene transfer in bacteria.

Transposons and Transposition

LEARNING **OUTCOMES**

7.32 Describe a transposon and a complex transposon.
7.33 Describe transposition and its effects.

Transposons[13] are segments of DNA, 700 to 40,000 bp in length, that transpose (move) themselves from one location in a DNA molecule to another location in the same or a different molecule.

American geneticist Barbara McClintock (1902–1992) discovered these "jumping genes" through a painstaking analysis of the colors of the kernels of corn. She discovered that the genes for kernel color were turned on and off by the insertion of transposons. Subsequent research has shown that transposons are found in many, if not all, prokaryotes and eukaryotes and in many viruses.

The result of the action of a transposon is termed **transposition**; in effect it is a kind of frameshift insertion (see Figure 7.24e). This "illegitimate" recombination does not need a region of homology, unlike other recombination events. Transposition occurs between plasmids and chromosomes and within and among chromosomes **(FIGURE 7.37)**. Further, plasmids can carry transposons from one cell to another. Fortunately, while transposons are common, transposition and the frameshift mutations it causes are relatively rare occurrences. ▶**ANIMATIONS:** *Transposons: Overview*

Transposons vary in their nucleotide sequences, but all of them contain palindromic sequences at each end. A *palindrome*[14] is a word, phrase, or sentence that has the same sequence of letters when read backward or forward—for example, "Madam, I'm Adam." In genetics, a palindrome is a region of DNA in which the sequence of nucleotides is identical to an inverted sequence in the complementary strand. For example, GAATTC is the palindrome of CTTAAG. Such a palindromic sequence is also known as an **inverted repeat (IR)**.

The simplest transposons, called **insertion sequences (IS)**, consist of no more than two inverted repeats and a gene that encodes the enzyme *transposase* **(FIGURE 7.38a)**. Transposase recognizes its own inverted repeat in a target site, cuts the DNA at that site, and inserts the transposon (or a copy of it) into the DNA molecule at that site **(FIGURE 7.38b)**. Such transposition also produces a duplicate copy of the target site.

[13]From *transposable* elements.
[14]From Greek *palin*, meaning "again," and *dramein*, meaning "to run."

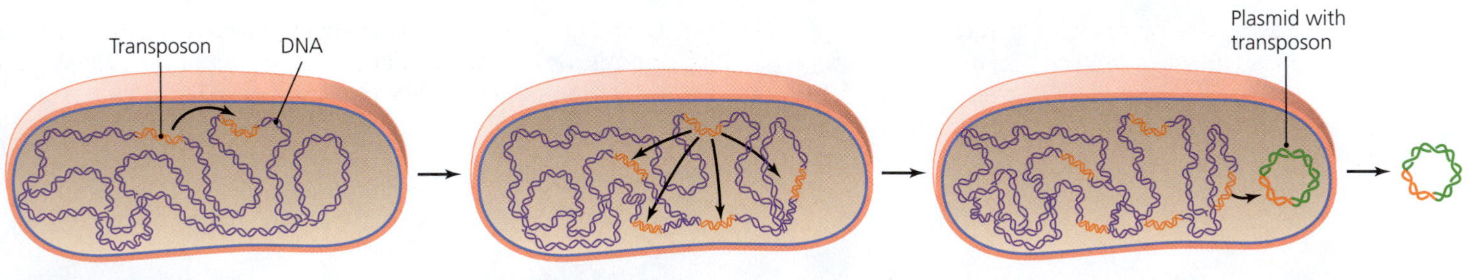

(a) Jumping transposons. Transposons move from one place to another on a DNA molecule.

(b) Replicating transposons. Transposons may replicate while moving, resulting in more transposons in the cell.

(c) Transposons can move onto plasmids. Transposons moving onto plasmids can be transferred to another cell.

▲ **FIGURE 7.37 Transposition.** Transposons move from place to place within, among, and between chromosomes and plasmids. Plasmids can carry transposons to and from cells.

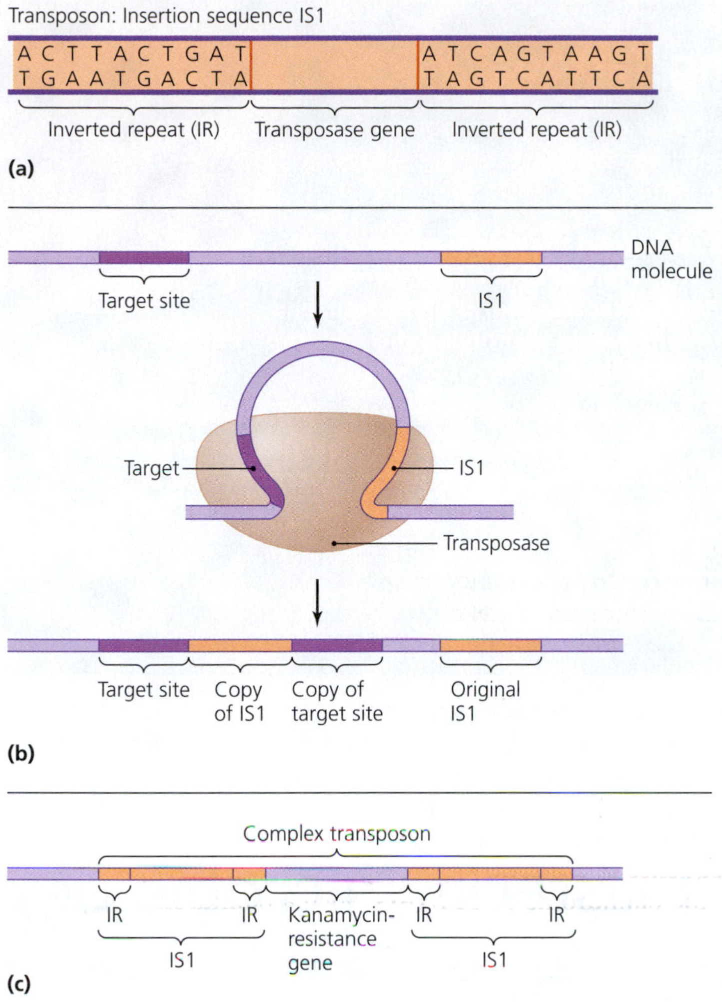

▲ **FIGURE 7.38 Transposons. (a)** A simple transposon, or insertion sequence. Shown here is insertion sequence 1 (IS1), which consists of a gene for the enzyme transposase bounded by identical (though inverted) repeats of nucleotides. **(b)** Transposase recognizes its target site (elsewhere in the same DNA molecule, as shown here, or in a different DNA molecule; then it moves the transposon (or, as in the case of IS1, a copy of the transposon) to its new site. The target site is duplicated in the process. **(c)** A complex transposon, which contains genes not related to transposition. Shown here is Tn5, which consists of a gene for kanamycin resistance between two IS1 transposons.

CLINICAL CASE STUDY

Deadly Horizontal Gene Transfer

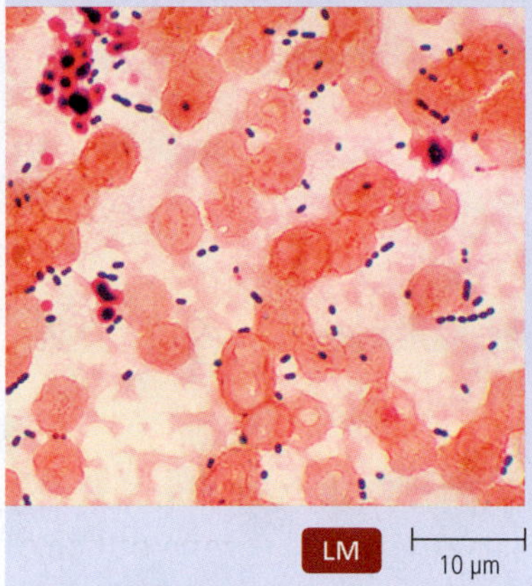

Sarah might have been 85 years old, but her mind was still sharp; it was her body that was failing. After she broke her hip by falling in the bathroom, she had succumbed to a series of complications that ultimately required admittance to a large university-associated hospital in New York City. Treatment necessitated a urinary catheter, a feeding tube, and intensive antibiotic therapy. While hospitalized, she acquired a life-threatening urinary tract infection of multi-drug-resistant (MDR) *Enterococcus faecium.*

Enterococci are normal members of the microbiota of the colon, yet these Gram-positive bacteria have a propensity to acquire genes that convey resistance to antimicrobial drugs. As a result of such horizontal gene transfer, *E. faecium* has become an MDR bacterium—a strain resistant to many different kinds of antimicrobials. Drug-resistant enterococci are a leading cause of healthcare associated infection (nosocomial infection).

Physicians treated Sarah with a series of antimicrobial drugs, including metronidazole, penicillin, erythromycin, ciprofloxacin, and vancomycin, but to no avail. After 90 days of hospitalization, she died.

1. Define *healthcare associated infection.*
2. What is the likely source of infection?
3. List three ways by which *E. faecium* might have acquired genes for drug resistance.
4. How can hospital personnel prevent the spread of resistant *E. faecium* throughout the hospital?

Complex transposons contain one or more genes not connected with transposition, such as genes for antibiotic resistance **(FIGURE 7.38c)**. *R plasmids* often contain transposons. R factors are of great clinical concern because they spread antibiotic resistance among pathogens. ▶**ANIMATIONS:** ***Transposons: Insertion Sequences, Complex Transposons***

TELL ME WHY

Why is the genetic ancestry of microbes much more difficult to ascertain than the ancestry of animals?

Chapter 8 discusses how scientists have adapted the natural processes of transformation, transduction, conjugation, and transposition to manipulate the genes of organisms.

MICRO IN THE CLINIC FOLLOW-UP

The Case of the Matching Tattoos

After undergoing additional tests, Ingrid learns that she does have hepatitis C (HCV). Ingrid's doctor explains that hepatitis C can be contracted from exposure to contaminated needles that are used for tattoos or body piercing. Even if the tattoo artist uses a sterile needle each time, other equipment, such as dye and dye tubs, may be shared with many different customers and transmit infections.

Ingrid also learns that not all hepatitis C viruses are the same. Even though the symptoms for acute viral hepatitis are the same, there are six known genotypes and 50 subtypes of HCV. Determining the specific genotype is important for successful treatment. Her doctor does more testing to determine the genotype and advises Ingrid to encourage her girlfriends who had gotten the matching tattoos to get tested for hepatitis C.

1. **What did the doctor mean when he referred to the HCV genotype causing Ingrid's infection, and why would identifying the genotype be helpful?**
2. **Variations in viral genotypes are the result of mutations, which are permanent changes in the genome. Describe the major types of mutations that occur to produce varying genotypes.**

MM Check your answers to Micro in the Clinic Follow-Up questions in the MasteringMicrobiology Study Area.

Explore the Invisible: Initiation of Translation

Practice on-the-go with Dr. Bauman Video Tutors by scanning this QR code with your smart phone. Visit the MasteringMicrobiology Study Area to challenge your understanding with practice tests, animation quizzes, and clinical case studies!

MasteringMicrobiology®

CHAPTER SUMMARY

MM This chapter has MicroFlix. Go to the MasteringMicrobiology Study Area to view movie-quality animations for DNA replication.

The Structure and Replication of Genomes (pp. 197–205)

1. **Genetics** is the study of inheritance and inheritable traits. **Genes** are composed of specific sequences of nucleotides that code for polypeptides or RNA molecules. A **genome** is the sum of all the genes and linking nucleotide sequences in a cell or virus. Prokaryotic and eukaryotic cells use DNA as their genetic material; some viruses use DNA, and other viruses use RNA.
2. The two strands of DNA are held together by hydrogen bonds between complementary **base pairs (bp)** of **nucleotides**. Adenine bonds with thymine, and guanine bonds with cytosine.
3. Bacterial and archaeal genomes consist of one (or, rarely, two or more) **chromosomes**, which are typically circular molecules of DNA associated with protein and RNA molecules, localized in a region of the cytosol called the **nucleoid**. Archaeal DNA organizes around globular proteins called **histones**. Prokaryotic cells may also contain one or more extrachromosomal DNA molecules called **plasmids**, which contain genes that regulate nonessential life functions, such as bacterial conjugation and resistance to antibiotics.
4. In addition to DNA, eukaryotic chromosomes contain eukaryotic histones and are arranged as **nucleosomes** (beads of DNA) that clump to form **chromatin fibers**. Eukaryotic cells may also contain extranuclear DNA in mitochondria, chloroplasts, and plasmids.
5. DNA replication is semiconservative; that is, each newly synthesized strand of DNA remains associated with one of the parental strands. After helicase unwinds and unzips the original molecule, synthesis of each of the two daughter strands—called the **leading strand** and the **lagging strand**—occurs from 5′ to 3′. Synthesis is mediated by enzymes that prime, join, and proofread the pairing of new nucleotides.

   ▶**ANIMATIONS:** ***DNA Replication: Overview, Forming the Replication Fork, Replication Proteins, Synthesis***
6. After DNA replication, **methylation** occurs. Methylation plays several roles, including the control of gene expression, the initiation of DNA replication, recognition of a cell's own DNA, and repair.

Gene Function (pp. 205–219)

1. The **genotype** of an organism is the actual set of genes in its genome, whereas its **phenotype** refers to the physical and functional traits expressed by those genes.

2. RNA has several forms. These include **RNA primer**; **messenger RNA (mRNA)**, which carries genetic information from DNA to a ribosome; **transfer RNA (tRNA)**, which carries amino acids to the ribosome; **ribosomal RNA (rRNA)**, which, together with polypeptides, makes up the structure of ribosomes; and **regulatory RNA**, which interacts with DNA to control gene expression.
3. The **central dogma** of genetics states that genetic information is transferred from DNA to RNA to polypeptides, which function alone or in conjunction as proteins.
4. The transfer of genetic information begins with **transcription** of the genetic code from DNA to RNA, in which **RNA polymerase** links RNA nucleotides that are complementary to genetic sequences in DNA. Transcription begins at a region of DNA called a **promoter** (recognized by RNA polymerase) and ends with a sequence called a **terminator**. In bacteria, Rho protein may assist in termination, or termination may depend solely on the nucleotide sequence of the transcribed RNA.

   ▶**ANIMATIONS:** *Transcription: Overview, The Process*
5. Eukaryotic mRNA is synthesized as pre-messenger RNA. Before translation can occur, a spliceosome removes noncoding **introns** from pre-mRNA and splices together the **exons**, which are the coding sections.
6. In **translation**, the sequence of genetic information carried by mRNA is used by ribosomes to construct polypeptides with specific amino acid sequences.

   ▶**ANIMATIONS:** *Translation: Overview, Genetic Code, The Process*

   ▶**VIDEO TUTOR:** *Initiation of Translation*
7. The genetic code consists of triplets of mRNA nucleotides, called **codons**. These bind with complementary **anticodons** on transfer RNAs (tRNAs), which are molecules that carry specific amino acids. Ribosomal RNA (rRNA) catalyzes the bonding of one amino acid to another to form a polypeptide. A sequence of nucleotides thus codes for a sequence of amino acids.
8. A ribosome contains three tRNA binding sites: an **A site** (associated with incoming amino acids), a **P site** (associated with elongation of the polypeptide), and an **E site** from which tRNA exits the ribosome.
9. Prokaryotes measure their cellular density by a process called **quorum sensing** in which cells secrete quorum-sensing molecules that neighboring cells detect.
10. An **operon** consists of a series of prokaryotic genes, a promoter, and in some cases an **operator** sequence, all controlled by one regulatory gene. The operon model explains gene regulation in prokaryotes.

    ▶**ANIMATIONS:** *Operons: Overview*
11. **Inducible operons** are normally "turned off" and are activated when the repressor no longer binds to the operator site, whereas **repressible operons** are normally "on" and are deactivated when the repressor binds to the operator site.

    ▶**ANIMATIONS:** *Operons: Induction, Repression*
12. Cells and some viruses use short **microRNAs (miRNAs)** in conjunction with proteins to form RNA silencing complexes (RISC) that attach to mRNA sequences to inhibit polypeptide translation.
13. Genetic expression can also be controlled with **riboswitches** or with **small interfering RNA (siRNA)** associated with protein in RISC.

Mutations of Genes (pp. 219–227)

1. A **mutation** is a change in the nucleotide base sequence of a genome. **Point mutations** involve a change in a single nucleotide base pair and include substitutions and two types of frameshift mutations: insertions and deletions.
2. **Gross mutations** include major changes to the DNA sequence, such as inversions, duplications, transpositions, and large deletions or insertions.
3. Mutations can be categorized by their effects as **silent**, **missense**, or **nonsense mutations**. Most mutations are spontaneous.

   ▶**ANIMATIONS:** *Mutations: Types*
4. Physical or chemical agents called **mutagens** can increase the normal rate of mutation. Physical mutagens include ionizing radiation, such as X rays and gamma rays, and nonionizing ultraviolet light. Ultraviolet light causes adjacent pyrimidine bases to bond to one another to form **pyrimidine dimers**. Mutagenic chemicals include **nucleotide analogs**, chemicals that are structurally similar to nucleotides and can result in mismatched base pairing, and chemicals that insert or delete nucleotide base pairs, producing frameshift mutations.

   ▶**ANIMATIONS:** *Mutagens*
5. Cells repair damaged DNA via **light repair** and **dark repair** of pyrimidine dimers, **base-excision repair**, and **mismatch repair**. When damage is so extensive that these mechanisms are overwhelmed, bacterial cells may resort to an **SOS response**.

   ▶**ANIMATIONS:** *Mutations: Repair*
6. Researchers have developed methods to distinguish **mutants**, which carry mutations, from normal **wild-type** cells. These methods include **positive selection, negative (indirect) selection**, and the **Ames test**, which is used to identify mutagens, which may be potential **carcinogens**.

Genetic Recombination and Transfer (pp. 227–233)

1. Organisms acquire new genes through **genetic recombination**, which is the exchange of segments of DNA. Crossing over occurs during gamete formation, part of sexual reproduction in eukaryotes.
2. Vertical gene transfer is the transmission of genes from parents to offspring. In **horizontal (lateral) gene transfer**, DNA from a **donor cell** is transmitted to a **recipient cell**. A **recombinant** cell results from genetic recombination between donated and recipient DNA. Transformation, transduction, and bacterial conjugation are types of horizontal gene transfer.

   ▶**ANIMATIONS:** *Horizontal Gene Transfer: Overview*
3. In **transformation**, a **competent** recipient prokaryote takes up DNA from its environment. Competence is found naturally or can be created artificially in some cells.

   ▶**ANIMATIONS:** *Transformation*
4. In **transduction**, a virus such as a **bacteriophage**, or **phage**, carries DNA from a donor cell to a recipient cell. Donor DNA is accidentally incorporated in such **transducing phages**.

   ▶**ANIMATIONS:** *Transduction: Generalized Transduction, Specialized Transduction*

5. In **conjugation**, an F^+ bacterium—that is, one containing an **F (fertility) plasmid**—forms a **pilus** (conjugation pilus) that attaches to an F^- recipient bacterium. Plasmid genes are transferred to the recipient, which becomes F^+ as a result.
   ▶**ANIMATIONS:** *Conjugation: Overview, F Factor, Hfr Conjugation, Chromosome Mapping*
6. **Hfr (high frequency of recombination) cells** result when an F plasmid integrates into a prokaryotic chromosome. Hfr cells form conjugation pili and transfer cellular genes more frequently than normal F^+ cells do.
7. **Transposons** are DNA segments that code for the enzyme transposase and have palindromic sequences known as **inverted repeats (IR)** at each end. Transposons move among locations in chromosomes in eukaryotes and prokaryotes—a process called **transposition**. The simplest transposons, known as **insertion sequences (IS)**, consist only of inverted repeats and transposase. **Complex transposons** contain other genes as well.
   ▶**ANIMATIONS:** *Transposons: Overview, Insertion Sequences, Complex Transposons*

QUESTIONS FOR REVIEW

Answers to the Questions for Review (except Short Answer Questions) begin on p. A-1.

Multiple Choice

1. Which of the following is most likely the number of base pairs in a bacterial chromosome?
   a. 4,000,000
   b. 4000
   c. 400
   d. 40
2. Which of the following is a true statement concerning prokaryotic chromosomes?
   a. They typically have two or three origins of replication.
   b. They contain single-stranded DNA.
   c. They are located in the cytosol.
   d. They are associated in linear pairs.
3. A plasmid is ________________.
   a. a molecule of RNA found in bacterial cells
   b. distinguished from a chromosome by being circular
   c. a structure in bacterial cells formed from plasma membrane
   d. extrachromosomal DNA
4. Which of the following forms ionic bonds with eukaryotic DNA and stabilizes it?
   a. chromatin
   b. bacteriocin
   c. histone
   d. nucleoid
5. Nucleotides used in the replication of DNA ________________.
   a. carry energy
   b. are found in four forms, each with a deoxyribose sugar, a phosphate, and a base
   c. are present in cells as triphosphate nucleotides
   d. all of the above
6. Which of the following molecules functions as a "proofreader" for a newly replicated strand of DNA?
   a. DNA polymerase III
   b. primase
   c. helicase
   d. ligase
7. The addition of —CH_3 to a cytosine nucleotide after DNA replication is called ________________.
   a. methylation
   b. restriction
   c. transcription
   d. transversion
8. In translation, the site through which tRNA molecules leave a ribosome is called the ________________.
   a. A site
   b. X site
   c. P site
   d. E site
9. The Ames test ________________.
   a. uses auxotrophs and liver extract to reveal mutagens
   b. is time intensive and costly
   c. involves the isolation of a mutant by eliminating wild-type phenotypes with specific media
   d. proves that suspected chemicals are carcinogenic
10. Which of the following methods of DNA repair involves enzymes that recognize and correct nucleotide errors in unmethylated strands of DNA?
    a. light repair of T dimers
    b. dark repair of P dimers
    c. mismatch repair
    d. SOS response
11. Which of the following is *not* a mechanism of natural genetic transfer and recombination?
    a. transduction
    b. transformation
    c. transcription
    d. conjugation
12. Cells that have the ability to take up DNA from their environment are said to be ________________.
    a. Hfr cells
    b. transposing
    c. genomic
    d. competent
13. Which of the following statements is true?
    a. Conjugation requires a sex pilus extending from the surface of a cell.
    b. Conjugation involves a C factor.
    c. Conjugation is an artificial genetic engineering technique.
    d. Conjugation involves DNA that has been released into the environment.
14. Which of the following are called "jumping genes"?
    a. Hfr cells
    b. transducing phages
    c. palindromic sequences
    d. transposons

15. Although two cells are totally unrelated, one cell receives DNA from the other cell and incorporates this new DNA into its chromosome. This process is ________________.
   a. crossing over of DNA from the two cells
   b. vertical gene transfer
   c. horizontal gene transfer
   d. transposition

16. Transcription produces ________________.
   a. DNA molecules
   b. RNA molecules
   c. polypeptides
   d. palindromes

17. A nucleotide is composed of ________________.
   a. a five-carbon sugar
   b. phosphate
   c. a nitrogenous base
   d. all of the above

18. In DNA, adenine forms ________________ hydrogen bonds with ________________.
   a. three/uracil
   b. two/uracil
   c. two/thymine
   d. three/thymine

19. A sequence of nucleotides formed during replication of the lagging DNA strand is ________________.
   a. a palindrome
   b. an Okazaki fragment
   c. a template strand
   d. an operon

20. Which of the following is *not* part of an operon?
   a. operator
   b. promoter
   c. origin
   d. gene

21. Repressible operons are important in regulating prokaryotic ________________.
   a. DNA replication
   b. RNA transcription
   c. rRNA processing
   d. sugar catabolism

22. Which of the following is part of each molecule of mRNA?
   a. palindrome
   b. codon
   c. anticodon
   d. base pair

23. Ligase plays a major role in ________________.
   a. lagging strand replication
   b. mRNA processing in eukaryotes
   c. polypeptide synthesis by ribosomes
   d. RNA transcription

24. Before mutations can affect a population permanently, they must be ________________.
   a. lasting
   b. inheritable
   c. beneficial
   d. all of the above

25. The *trp* operon is repressible. This means it is usually ________________ and is directly controlled by ________________.
   a. active/an inducer
   b. active/a repressor
   c. inactive/an inducer
   d. inactive/a repressor

Fill in the Blanks

1. The three steps in RNA transcription are ________________, ________________, and ________________.
2. A triplet of mRNA nucleotides that specifies a particular amino acid is called a ________________.
3. Three effects of point mutations are ________________, ________________, and ________________.
4. Insertions and deletions in the genetic code are also called ________________ mutations.
5. An operon consists of ________________, ________________, and ________________ and is associated with a regulatory gene.
6. In general, ________________ operons are inactive until the substrate of their genes' polypeptides is present.
7. A daughter DNA molecule is composed of one original strand and one new strand because DNA replication is ________________.
8. A gene for antibiotic resistance can move horizontally among bacterial cells by ________________, ________________, and ________________.
9. ________________ are nucleotide sequences containing palindromes and genes for proteins that cut DNA strands.
10. ________________ ________________ is a recombination event that occurs during gamete formation in eukaryotes.
11. ________________ RNA carries amino acids.
12. ________________ RNA and ________________ RNA are antisense; that is, they are complementary to another nucleic acid molecule.

Short Answer

1. How does the genotype of a bacterium determine its phenotype? Use the terms *gene, mRNA, ribosome,* and *polypeptide* in your answer.
2. List several ways in which eukaryotic messenger RNA differs from prokaryotic mRNA.
3. Compare and contrast introns and exons.
4. Polypeptide synthesis requires large amounts of energy. How do cells regulate synthesis to conserve energy? Describe one specific example.
5. Describe the operon model of gene regulation.
6. Compare and contrast the structure and components of DNA and RNA in prokaryotes.
7. Besides the fact that it synthesizes RNA, how does RNA polymerase differ in function from DNA polymerase?
8. Describe the formation and function of mRNA, rRNA, and tRNA in prokaryotes and eukaryotes.

9. Describe how DNA is packaged in both prokaryotes and eukaryotes.
10. Explain the central dogma of genetics.
11. Compare and contrast the processes of transformation, transduction, and conjugation.
12. Fill in the following table:

Process	Purpose	Beginning Point	Ending Point
Replication			Origin or end of molecule
Transcription		Promoter	
Translation	Synthesis of polypeptides		

VISUALIZE IT!

1. On the figure below, label DNA polymerase I, DNA polymerase III, helicase, lagging strand, leading strand, ligase, nucleotide (triphosphate), Okazaki fragment, primase, replication fork, RNA primer, and stabilizing proteins.

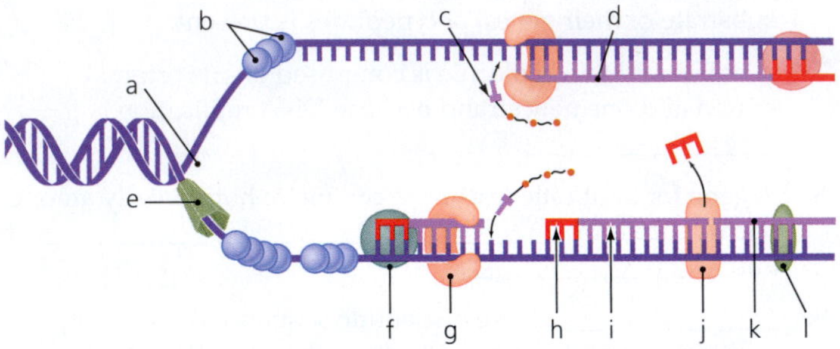

2. This bacteriophage DNA molecule has been warmed. Label the portions that likely have a higher ratio of G-C base pairs and the portions that have a higher ratio of A-T base pairs.

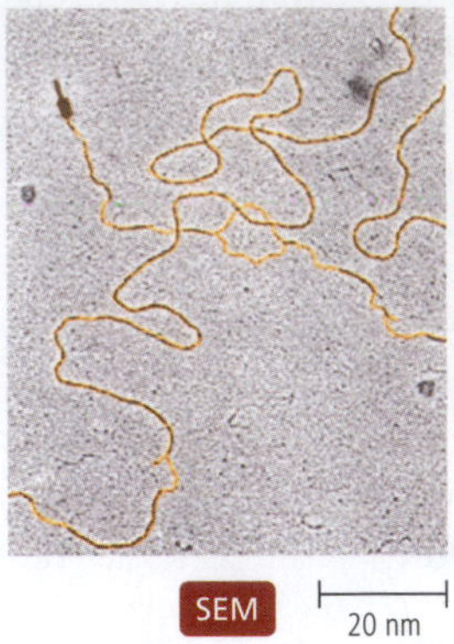

3. The drugs ddC and AZT are used to treat AIDS.

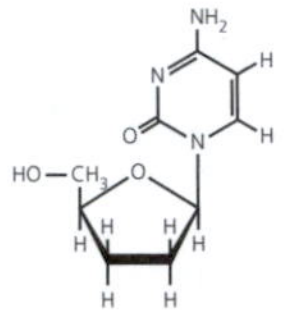

ddC (2′,3′-dideoxycytidine)

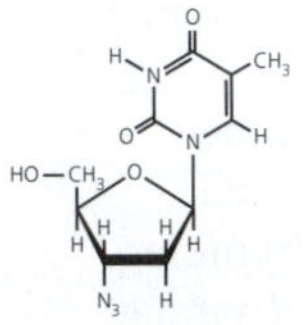

AZT (3′-azido-2′,3′-dideoxythymidine)

Based on their chemical structures, what is their mode of action?

CRITICAL THINKING

1. If molecules of mRNA have the following nucleotide base sequences, what will be the sequence of amino acids in polypeptides synthesized by eukaryotic ribosomes?
   a. AUGGGGAUACGCUACCCC
   b. CCGUACAUGCUAAUCCCU
   c. CCGAUGUAACCUCGAUCC
   d. AUGCGGUCAGCCCCGUGA
2. A scientist uses a molecule of DNA composed of nucleotides containing radioactive deoxyribose as a template for replication and transcription in a nonradioactive environment. What percentage of DNA strands will be radioactive after three DNA replication cycles? What percentage of RNA molecules will be radioactive?
3. Explain why an insertion of three nucleotides is less likely to result in a deleterious effect than an insertion of a single nucleotide.
4. How could scientists use siRNA to turn off a cancer-inducing gene?
5. The chromosome of *Mycobacterium tuberculosis* is 4,411,529 bp long. A scientist who isolates and counts the number of nucleotides in its DNA molecule discovers that there are 2,893,963 molecules of guanine. How many molecules of the other three nucleotides are in the original DNA?
6. Suppose that *E. coli* sustains a mutation in its gene for the *lac* operon repressor such that the repressor is ineffective. What effect would this have on the bacterium's ability to catabolize lactose? Would the mutant strain have an advantage over wild-type cells? Explain your answer.
7. A student claims that nucleotide analogs can be carcinogenic. Another student in the study group insists that nucleotide analogs are used to treat cancer. Explain why both students are correct.
8. Why is DNA polymerase so named?
9. *Corynebacterium diphtheriae,* the causative agent of diphtheria, secretes a toxin that enzymatically inactivates all molecules of elongation factor in a eukaryotic cell. What immediate and long-term effects does this have on cellular metabolism?
10. How can knowledge of nucleotide analogs be useful to a cancer researcher?
11. The endosymbiotic theory states that mitochondria and chloroplasts evolved from prokaryotes living within other prokaryotes (see pp. 85–86). What aspects of the eukaryotic genome support this theory? What aspects do not support the theory?
12. Hydrogen bonds between complementary nucleotides are crucial to the structure of dsDNA because they hold the two strands

together. Why couldn't the two strands be effectively linked by covalent bonds?

13. On average, RNA polymerase makes one error for every 10,000 nucleotides it incorporates in RNA. By contrast, only one base-pair error remains for every 10 billion bp during DNA replication. Explain why the accuracy of RNA transcription is not as critical as the accuracy of DNA replication.
14. A scientist isolates a molecule of mRNA with the following base sequence: CAUGUACGACAUAUGCAUA. What is the sequence of amino acids in the polypeptide synthesized by a bacterial ribosome from this message? What would be different if the message were translated in a mitochondrion instead?
15. We have seen that wobble makes the genetic code redundant in the third position for C and U. After reexamining the genetic code in Figure 7.12, state what other nucleotides in the third position appear to accommodate anticodon wobbling.
16. If a scientist synthesizes a DNA molecule with the nucleotide base sequence TACGGGGGAGGGGGAGGGGGA and then uses it for transcription and translation, what would be the amino acid sequence of the product?
17. What DNA nucleotide triplet codes for codon UGU? Identify a base-pair substitution that would produce a silent mutation at this codon. Identify a base-pair substitution that would result in a missense mutation at this codon. Identify a base-pair substitution that would produce a nonsense mutation at this codon.
18. Suppose you are a scientist who wants to insert into your dog a gene that encodes a protein that protects dogs from heartworms. A dog's cells are not competent, so they cannot take up the gene from the environment; but you have a plasmid, a competent bacterium, and a related (though incompetent) F^+ bacterium that lives as an intracellular parasite in dogs. Describe a possible scenario by which you could use natural processes to genetically alter your dog to be heartworm resistant.

CONCEPT MAPPING

Using the following terms, draw a concept map that describes point mutations. For a sample concept map, see p. 94. Or, complete this and other concept maps online by going to the MasteringMicrobiology Study Area.

Change in DNA sequence
Deleted
Deletion mutation
Effect on amino acid sequence
Frameshift
Incorrect amino acid substituted
Inserted
Insertion mutation
Missense mutation
No change in amino acid sequence
Nonsense mutation
One or few nucleotide base pairs
Premature termination of polypeptide
Silent mutation
Substituted
Substitution mutation

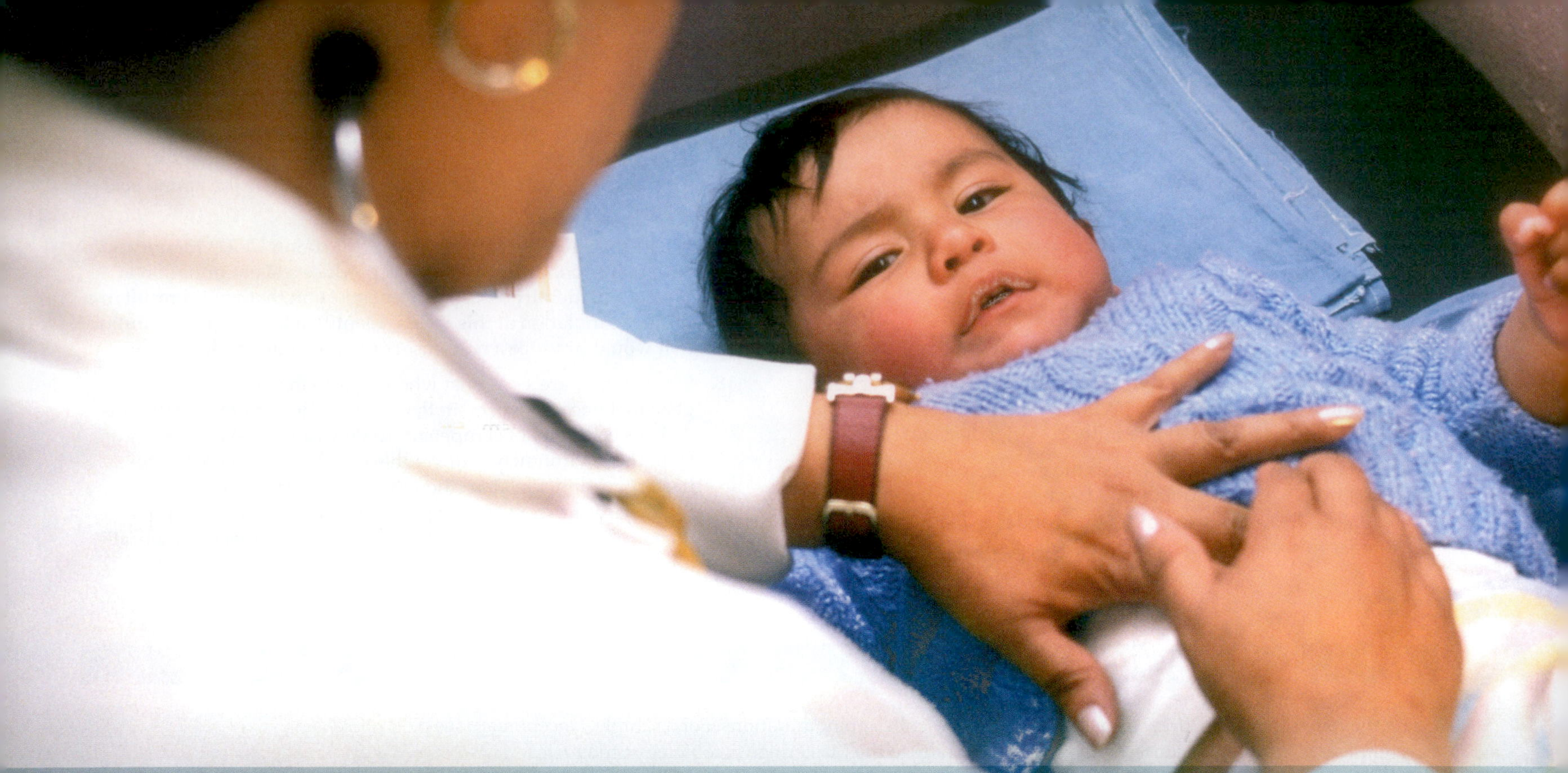

8 Recombinant DNA Technology

MICRO IN THE CLINIC

Can a "Bad" Gene Be Replaced with a "Good" Gene?

Scott and Therese live a quiet life in the suburbs with their children. They have two healthy sons, but their six-month-old daughter, Keesha, has been plagued with health problems since she was born. She suffers from chronic diarrhea, has experienced two bouts of pneumonia, and recently developed mysterious white patches in her mouth. Scott and Therese take Keesha to Children's Hospital, where her pediatrician discovers that she has abnormally low numbers of white blood cells called lymphocytes, and extremely low levels of an enzyme called adenosine deaminase.

Keesha is diagnosed with severe combined immunodeficiency disease (SCID), a genetic disorder that leaves patients with no functional immune system. Lymphocytes are the foot soldiers of the immune system. But Keesha's lymphocytes are dying—because of a defective gene, they are not producing enough adenosine deaminase, which they need to inactivate a toxin that accumulates from the breakdown of DNA. Without treatment, Keesha will remain extremely vulnerable to infections, such as that by *Candida albicans*, which causes the white patches in her mouth. She will likely die before her second birthday.

Keesha's best treatment option approved by the U.S. government is a bone marrow transplant. Unfortunately, tests reveal that none of Keesha's relatives is a compatible donor, and the chances of finding a donor outside the family are slim. After ruling out other options, the family decides to try an experimental gene therapy, in which the "good" gene for adenosine deaminase is isolated using restriction enzymes and polymerase chain reaction. Copies of the good gene are then incorporated into the DNA of Keesha's bone marrow cells.

Will gene therapy cure Keesha? Turn to the end of the chapter (p. 258).

Explore More: Test your readiness and apply your knowledge with dynamic learning tools at MasteringMicrobiology.

This chapter examines how genetic researchers have adopted the natural enzymes and processes of DNA recombination, replication, transcription, transformation, transduction, and conjugation to manipulate genes for industrial, medical, and agricultural purposes. Together these techniques are termed *recombinant DNA technology,* commonly called "genetic engineering." We end the chapter with a discussion of the ethics and safety of these techniques.

The Role of Recombinant DNA Technology in Biotechnology

LEARNING | **OUTCOMES**

8.1 Define *biotechnology* and *recombinant DNA technology.*

8.2 List several examples of useful products made possible by biotechnology.

8.3 Identify the three main goals of recombinant DNA technology.

Biotechnology—the use of microorganisms to make practical products—is not a new field. For thousands of years, humans have used microbes to make products such as bread, cheese, soy sauce, and alcohol. During the 20th century, scientists industrialized the natural metabolic reactions of bacteria to make large quantities of acetone, butanol, and antibiotics. More recently, scientists have adapted microorganisms for use in the manufacture of paper, textiles, and vitamins; to assist in cleaning up industrial wastes, oil spills, and radioactive isotopes; and to aid in mining copper, gold, uranium, and other metals. (Chapter 25 discusses some industrial and environmental applications of biotechnology.)

Until recently, microbiologists were limited to working with naturally occurring organisms and their mutants for achieving such industrial and medical purposes. Since the 1990s, however, scientists have become increasingly adept at intentionally modifying the genomes of organisms, by natural processes, for a variety of practical purposes. This is **recombinant DNA technology** (sometimes called genetic engineering), and it has expanded the possibilities of biotechnology in ways that seemed like science fiction only a few years ago. Today, scientists isolate specific genes from almost any so-called donor organism, such as a human, a plant, or a bacterium, and insert it into the genome of almost any kind of recipient organism.

Scientists who manipulate genomes have three main goals:

- *To eliminate undesirable phenotypic traits in humans, animals, plants, and microbes.* For example, scientists have inserted genes from microbes into plants to make them resistant to pests or freezing, and since 1999 they have cured some children born with a fatal and previously untreatable genetic disorder called severe combined immunodeficiency disease (SCID).
- *To combine beneficial traits of two or more organisms to create valuable new organisms,* such as laboratory animals that mimic human susceptibility to HIV.
- *To create organisms that synthesize products that humans need,* such as paint solvents, vaccines, antibiotics, enzymes, and hormones. For instance, geneticists have successfully inserted the human gene for insulin into bacteria so that the bacteria synthesize human insulin, which is cheaper and safer than insulin derived from animals.

Recombinant DNA technology is not a single procedure or technique but rather a collection of tools and techniques scientists use to manipulate the genomes of organisms. In general, they isolate a gene from a cell, manipulate it *in vitro,*[1] and insert it into another organism. **FIGURE 8.1** illustrates the basic processes involved in recombinant DNA technology.

TELL ME WHY

Why aren't the terms *genetic engineering* and *biotechnology* synonymous?

The Tools of Recombinant DNA Technology

Scientists use a variety of physical agents, naturally occurring enzymes, and synthetic molecules to manipulate genes and genomes. These tools of recombinant DNA technology include *mutagens, reverse transcriptase, synthetic nucleic acids, restriction enzymes,* and *vectors.* Scientists use these molecular tools to create *gene libraries,* which are a time-saving tool for genetic researchers.

Mutagens

LEARNING | **OUTCOME**

8.4 Describe how gene researchers use mutagens.

Mutagens are physical and chemical agents that produce mutations (changes in nucleotide sequence of a genome). Scientists deliberately use mutagens to create changes in microbes' genomes so that the microbes' phenotypes are changed. They then select for and culture cells with characteristics considered beneficial for a given biotechnological application. For example, scientists exposed the fungus *Penicillium* (pen-i-sil´ē-ŭm) to mutagenic agents and then selected strains that produce greater amounts of penicillin. In this manner, they developed a strain of *Penicillium* that secretes over 25 times as much penicillin as did the strain originally isolated by Alexander Fleming (1881–1955). Today, with recombinant DNA techniques (discussed shortly), researchers can isolate mutated genes rather than dealing with entire organisms.

[1]Latin, meaning "within glassware"; that is, in a laboratory.

The Use of Reverse Transcriptase to Synthesize cDNA

LEARNING | OUTCOME

8.5 Explain the function and use of reverse transcriptase in synthesizing cDNA.

Transcription involves the transmission of genetic information from molecules of DNA to molecules of RNA (see Figure 7.9). The discovery of retroviruses, which have genomes consisting of RNA instead of DNA, led to the discovery of an unusual enzyme—**reverse transcriptase.** Reverse transcriptase creates a flow of genetic information in the opposite direction from the flow in conventional transcription: it uses an RNA template to transcribe a molecule of DNA, which is called **complementary DNA (cDNA)** because it is complementary to an RNA template.

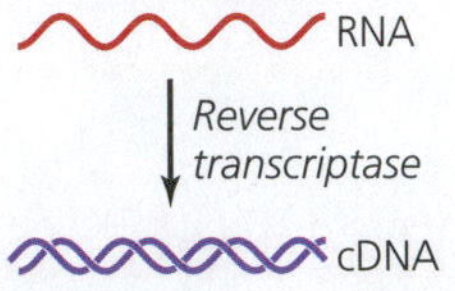
RNA
Reverse transcriptase
cDNA

Because hundreds to millions of copies of mRNA exist for every active gene, it is frequently easier to produce a desired gene by first isolating the mRNA molecules that code for a particular polypeptide and then use reverse transcription to synthesize a cDNA gene from the mRNA template. Further, eukaryotic DNA is not normally expressible by prokaryotic cells, which cannot remove the introns (noncoding sequences) present in eukaryotic pre-mRNA. However, since eukaryotic mRNA has already been processed to remove introns, cDNA produced from it lacks noncoding sequences. Therefore, scientists can successfully insert cDNA into prokaryotic cells, making it possible for the prokaryotes to produce eukaryotic proteins such as human growth factor, insulin, or blood-clotting factors.

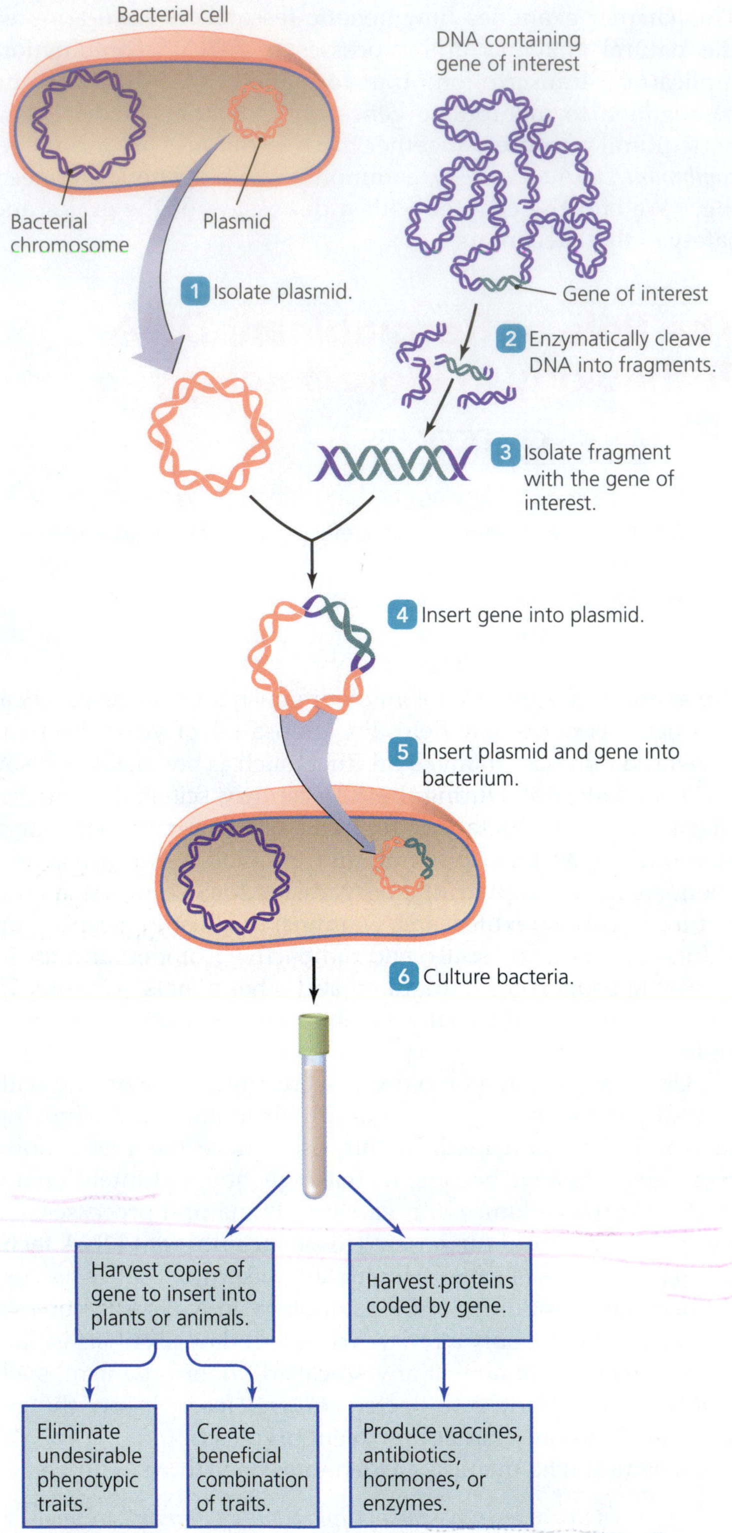
Bacterial cell
DNA containing gene of interest
Bacterial chromosome
Plasmid
1 Isolate plasmid.
Gene of interest
2 Enzymatically cleave DNA into fragments.
3 Isolate fragment with the gene of interest.
4 Insert gene into plasmid.
5 Insert plasmid and gene into bacterium.
6 Culture bacteria.
Harvest copies of gene to insert into plants or animals.
Harvest proteins coded by gene.
Eliminate undesirable phenotypic traits.
Create beneficial combination of traits.
Produce vaccines, antibiotics, hormones, or enzymes.

▲ **FIGURE 8.1 Overview of recombinant DNA technology.**

Synthetic Nucleic Acids

LEARNING | OUTCOMES

8.6 Explain how gene researchers synthesize nucleic acids.
8.7 Describe three uses of synthetic nucleic acids.

The enzymes of DNA replication and RNA transcription function not only *in vivo*[2] but also function *in vitro,* making it possible for scientists to produce molecules of DNA and RNA in cell-free solutions for genetic research. In fact, scientists have so mechanized the processes of nucleic acid replication and transcription that they can produce molecules of DNA and RNA with any nucleotide sequence; all they must do is enter the desired sequence into a synthesis machine's four-letter keyboard. A computer controls the actual synthesis, using a supply of nucleotides and other required reagents. Nucleic acid synthesis machines synthesize molecules over 100 nucleotides long in a few hours, and scientists can join two or more of these molecules end to end with ligase to create even longer synthetic molecules.

Researchers have used synthetic nucleic acids in many ways, including the following:

- **Elucidating the genetic code.** Using synthetic molecules of varying nucleotide sequences and observing the amino

[2]Latin, meaning "in life" (i.e., within a cell).

acids in the resulting polypeptides, scientists elucidated the genetic code. For example, synthetic DNA consisting only of adenine nucleotides yields a polypeptide consisting solely of the amino acid phenylalanine. Therefore, the mRNA codon UUU (transcribed from the DNA triplet AAA) must code for phenylalanine (see Figure 7.12).

- **Creating genes for specific proteins.** Once they know the genetic code and the amino acid sequence of a protein, scientists can create a gene for that protein. In this manner, scientists synthesized a gene for human insulin. Of course, such a synthetic gene likely consists of a nucleotide sequence slightly different from that of its cellular counterpart because of the redundancy in the genetic code (see Figure 7.12).
- **Synthesizing DNA and RNA probes to locate specific sequences of nucleotides. Probes** are nucleic acid molecules with a specific nucleotide sequence that have been labeled with radioactive or fluorescent chemicals so that their locations can be detected. The use of probes to locate specific sequences of nucleotides is based on the fact that any given nucleotide sequence will preferentially bond to its complementary sequence. Thus, a probe constructed with the nucleotide sequence ATGCT will bond to a DNA strand with the sequence TACGA, and the probe's label allows researchers to then detect the complementary site. Probes are essential tools for locating specific nucleic acid sequences such as genes for particular polypeptides.
- **Synthesizing antisense nucleic acid molecules.** Antisense nucleic acid molecules have nucleotide sequences that bind to and interfere with genes and mRNA molecules. Scientists are researching the use of antisense molecules to control genetic diseases.

Restriction Enzymes

LEARNING **OUTCOMES**

8.8 Explain the source and names of restriction enzymes.
8.9 Describe the importance and action of restriction enzymes.

An important development in recombinant DNA technology was the discovery of **restriction enzymes** in bacterial cells. Such enzymes cut DNA molecules and are restricted in their action—they cut DNA only at locations called *restriction sites.* Restriction sites are specific nucleotide sequences, which are usually *palindromes*[3]—they have the same sequence when read forward or backward. In nature, bacterial cells use restriction enzymes to protect themselves from phages by cutting phage DNA into nonfunctional pieces. The bacterial cells protect their own DNA by methylation[4] of some of their nucleotides, hiding the DNA from the restriction enzymes (see Chapter 7).

Researchers name restriction enzymes with three letters (denoting the genus and specific epithet of the source bacterium) and Roman numerals (to indicate the order in which enzymes from the same bacterium were discovered). In some cases, a fourth letter denotes the strain of the bacterium. Thus, *Escherichia coli* (esh-ĕ-rik´ē-ă kō´lī) strain R produces the restriction enzymes *Eco*RI and *Eco*RII. *Hin*dIII is the third restriction enzyme isolated from *Haemophilus influenzae* (hē-mof´i-lŭs in-flū-en´zī) strain Rd.

Scientists have discovered several hundred restriction enzymes and categorize them in two groups on the basis of types of cuts they make. The first type, as exemplified by *Eco*RI, makes staggered cuts of the two strands of DNA, producing fragments that terminate in mortise-like *sticky ends.* Each sticky end is composed of up to four nucleotides that form hydrogen bonds with its complementary sticky end **(FIGURE 8.2a)**. Scientists can use these bits of single-stranded DNA to combine pieces of DNA from different organisms into a single recombinant DNA molecule (the enzyme ligase unites the sugar-phosphate backbones of the pieces) **(FIGURE 8.2b)**. Other restriction enzymes, such as *Hin*dII and *Sma*I (from *Serratia marcescens,* ser-rat´ē-a mar-ses´enz), cut both strands of DNA at the same point, resulting in *blunt ends* **(FIGURE 8.2c)**. It is more difficult to make recombinant DNA from blunt-ended fragments because they are not sticky, but they have a potential advantage—blunt ends are nonspecific. This enables any two blunt-ended fragments, even those produced by different restriction enzymes, to be combined easily **(FIGURE 8.2d)**. In contrast, sticky-ended fragments bind only to complementary, sticky-ended fragments produced by the same restriction enzyme. ▶**VIDEO TUTOR:** ***Action of Restriction Enzymes***

TABLE 8.1 identifies several restriction enzymes and their target DNA sequences. ▶**ANIMATIONS:** ***Recombinant DNA Technology***

TABLE 8.1 Properties of Some Restriction Enzymes

Enzyme	Bacterial Source	Restriction Site[a]
BamHI	*Bacillus amyloliquefaciens* H	G↓GATCC CCTAG↑G
EcoRI	*Escherichia coli* RY13	G↓AATTC CTTAA↑G
EcoRII	*E. coli* R245	CC↓GG GG↑CC
HindII	*Haemophilus influenzae* Rd	GTPy↓PuAC CAPu↑PyTG
HindIII	*H. influenzae* Rd	A↓AGCTT TTCGA↑A
HinfI	*H. influenzae* Rf	G↓ANTC CTNA↑G
HpaI	*H. parainfluenzae*	GTT↓AAC CAA↑TTG
MspI	*Moraxella* sp.	CC↓GG GG↑CC
SmaI	*Serratia marcescens*	CCC↓GGG GGG↑CCC

[a]Arrows indicate sites of cleavage; Py = pyrimidine [either thymine (T) or cytosine (C)], Pu = purine [either adenine (A) or guanine (G)], N = any nucleotide (A, T, G, or C).

[3]From Greek *palin,* meaning "again," and *dramein,* meaning "to run."
[4]Adding a methyl group, —CH_3, to a chemical.

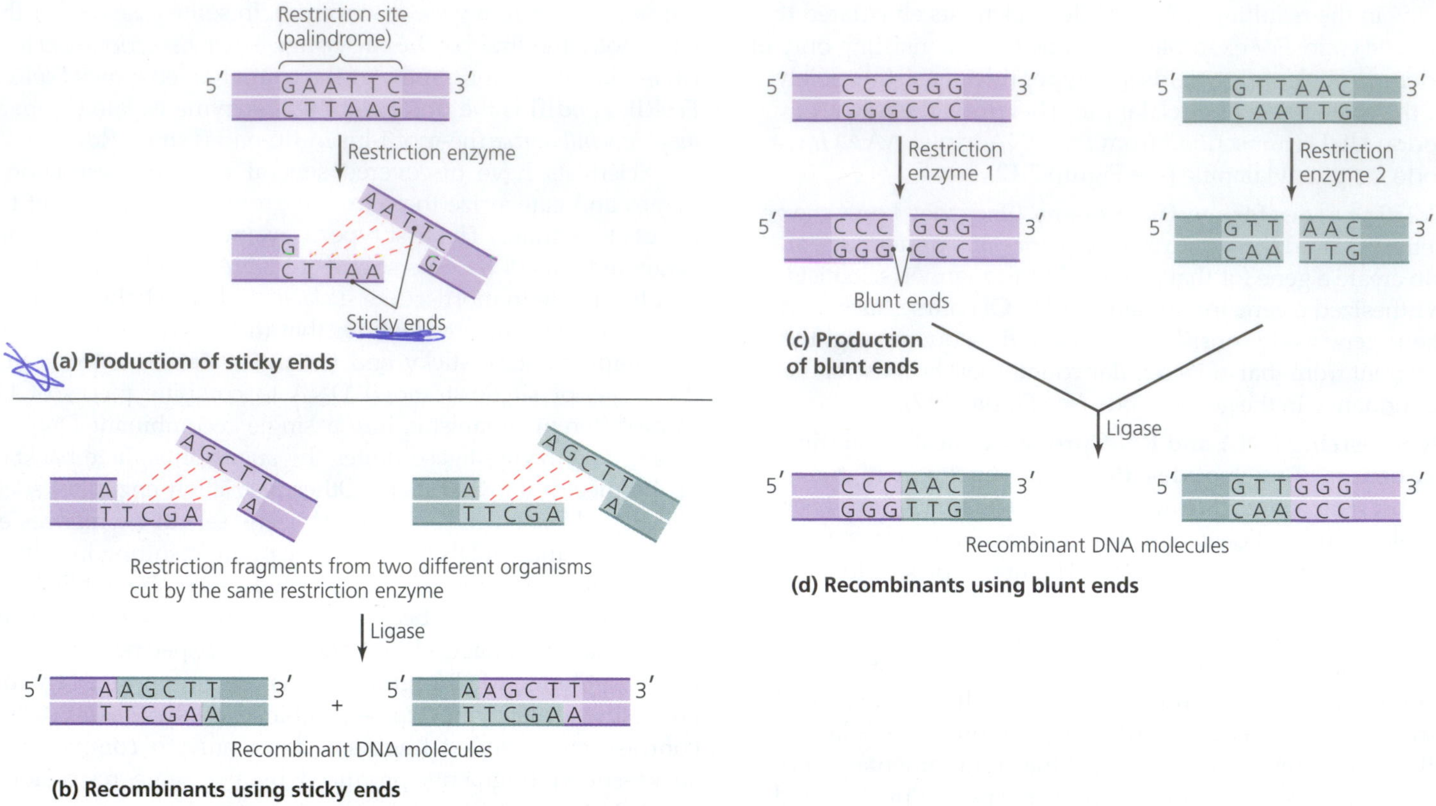

▲ **FIGURE 8.2 Actions of restriction enzymes.** Restriction enzymes recognize and cut both strands of a DNA molecule at a specific (usually palindromic) restriction site. **(a)** Certain restriction enzymes produce staggered cuts with complementary "sticky ends." **(b)** When two complementary sticky-ended fragments come from different organisms, their bonding (catalyzed by ligase) produces recombinant DNA. **(c)** Other restriction enzymes produce blunt-ended fragments. **(d)** A lack of specificity enables blunt-ended fragments produced by different restriction enzymes to be combined easily into recombinant DNA. *Which restriction enzymes act at the restriction sites shown?*

Figure 8.2 *(a)* EcoRI, *(b)* HindIII, and *(c)* SmaI and HpaI.

HIGHLIGHT

How Do You "Fix" a Mosquito?

Dengue is the most common mosquito-borne viral infection of humans in the world; 50 million cases occur every year. Twenty-five thousand people, mostly children, die with intense abdominal pain, severe internal bleeding, and circulatory collapse. Dengue is reemerging in Florida and Hawaii and in France and the Netherlands in Europe. There is no approved vaccine or specific treatment.

Dengue virus spreads through the bite of *Aedes aegypti* mosquitoes that have previously bitten an infected person. Controlling mosquitoes is the way to combat dengue, but this is easier said than done. *Aedes* requires only a thin film of water about the thickness of a dime to breed, and adults often hide inside homes, especially in urban areas.

Recombinant DNA technology may provide a simple and effective way to rid a community of dengue. Scientists have introduced a sterility gene into male mosquitoes, effectively neutering them. When sterile males are released into the wild, and they breed with normal females, their offspring are sterile. This technique reduced by 80% the *Aedes* population in a test area on Grand Cayman, an island in the Caribbean, in 2010. With fewer mosquitoes to carry the dengue virus, the spread of dengue was halted in at least one town on the island.

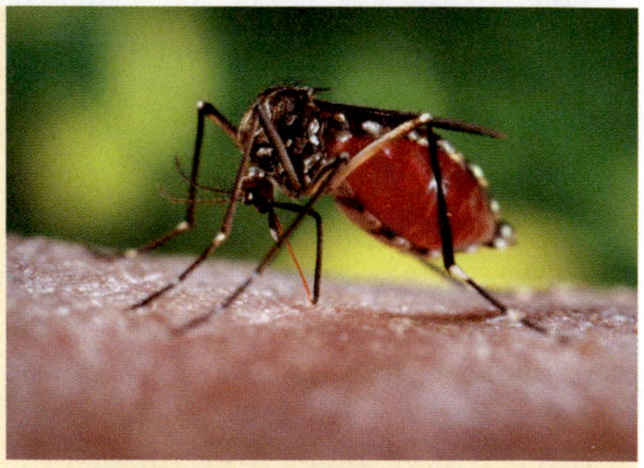

Aedes aegypti **mosquito**

The U.S. Food and Drug Administration (FDA) is considering authorizing the release of gentically modified mosquitoes in Florida.

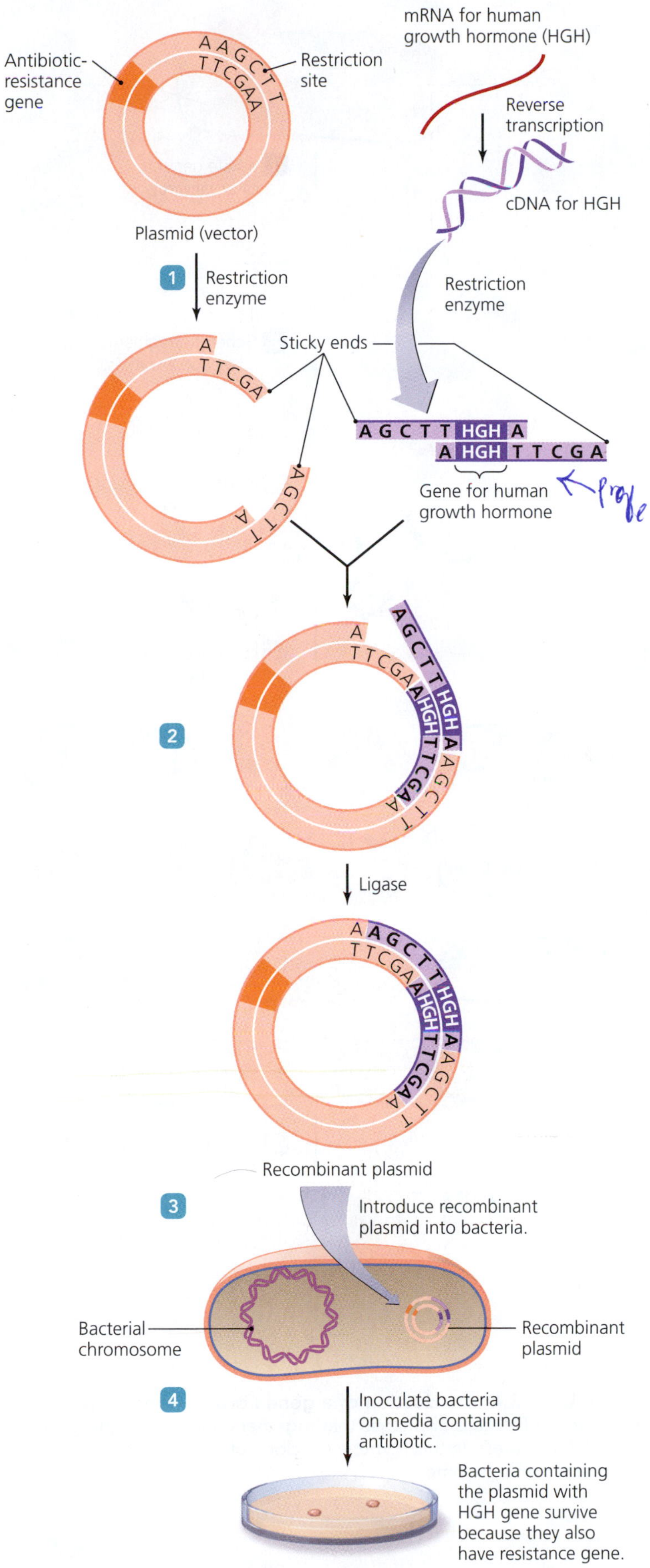

◄ **FIGURE 8.3 An example of the process for producing a recombinant vector.** In this case, the vector is a plasmid. *Which restriction enzyme was used in this example?*

Figure 8.3 *Hin*dIII.

Vectors

LEARNING **OUTCOME**

8.10 Define *vector* as the term applies to genetic manipulation.

One goal of recombinant DNA technology is to insert a useful gene into a cell so that the cell has a new phenotype—for example, the ability to synthesize a novel protein. To deliver a gene into a cell, researchers use **vectors,** which are nucleic acid molecules, such as viral genomes, transposons, and plasmids.

Genetic vectors share several useful properties:

- **Vectors are small enough to manipulate in a laboratory.** Large DNA molecules the size of entire chromosomes are generally too fragile to serve as vectors.
- **Vectors survive inside cells.** Plasmids, which are circular DNA, make good vectors because they are more stable than are linear fragments of DNA, which are typically degraded by cellular enzymes. However, some linear vectors, such as transposons and certain viruses, insert themselves rapidly into a host's chromosome before they can be degraded.
- **Vectors contain a recognizable genetic marker** so that researchers can identify the cells that have received the vector and thereby the specific gene of interest. Genetic markers can either be phenotypic markers, such as those that confer antibiotic resistance or code for enzymes that metabolize a unique nutrient, or radioactive or fluorescent labels.
- **Vectors can ensure genetic expression** by providing required genetic elements such as promoters.

An example of the process used to produce a vector containing a specific gene is depicted in **FIGURE 8.3**. After a given restriction enzyme cuts both the DNA molecule containing the gene of interest (in this example, the human growth hormone gene) and the vector DNA (here a plasmid containing a gene for antibiotic resistance as a marker) into fragments with sticky ends 1, ligase anneals the fragments to produce a recombinant plasmid 2. After the recombinant plasmid has been inserted into a bacterial cell 3, the bacteria are grown on a medium containing the antibiotic 4; only those cells that contain the recombinant plasmid (and thus the human growth hormone gene as well) can grow on the medium.

Generally, viruses and transposons are able to carry larger genes than can plasmids. Researchers are developing vectors from adenoviruses, poxviruses, and a genetically modified form of the human immunodeficiency virus (HIV). HIV in particular might make an excellent vector because HIV inserts itself directly into human chromosomes; however, scientists must ensure that viral vectors do not insert DNA into the middle of a necessary gene, mutating and possibly killing their target cells.

Gene Libraries

LEARNING OUTCOME

8.11 Explain the significance of gene libraries.

Suppose you were a scientist investigating the effects of the genes for 24 different kinds of interleukins (proteins that mediate certain aspects of immunity). Having to isolate the specific genes for each type of interleukin would require much time, labor, and expense. Your task would be made much easier if you could obtain the genes you need from a **gene library,** a collection of bacterial or phage clones—identical descendants—each of which contains a portion of the genetic material of interest. In effect, each clone is like one book in a library in that it contains one fragment (typically a single gene) of an organism's entire genome. Alternatively, a gene library may contain clones with all the genes of a single chromosome or of the full set of cDNA that is complementary to an organism's mRNA.

As depicted in **FIGURE 8.4**, genetic researchers can create each of the clones in a gene library by using restriction enzymes to generate fragments of the DNA of interest and then using ligase to synthesize recombinant vectors. They insert the vectors into bacterial cells, which are then grown on culture media. Once a scientist isolates a recombinant clone and places it in a gene library, the gene that the clone carries becomes available to other investigators, saving them the time and effort required to isolate that gene. Many gene libraries are now commercially available.

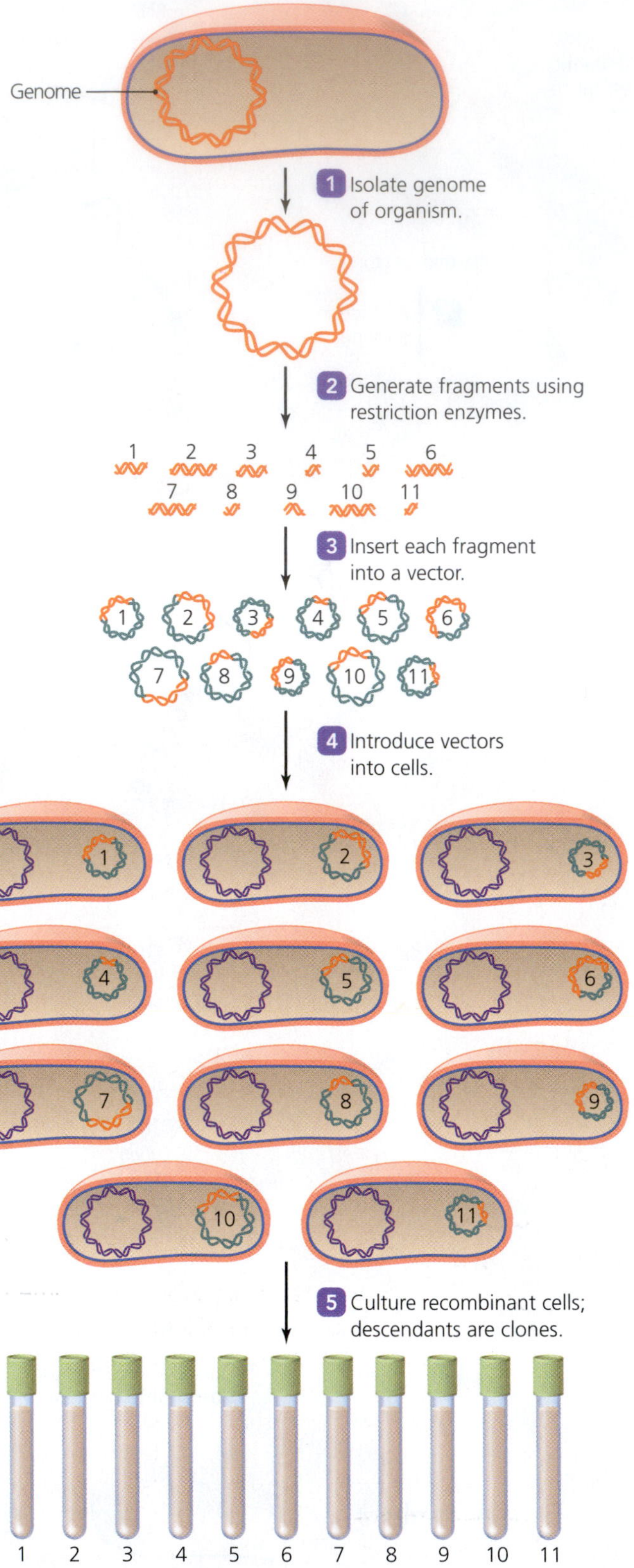

▲ **FIGURE 8.4 Production of a gene library.** A gene library is the population of all cells or phages that together contain all of the genetic material of interest. In this figure, each clone of cells carries a portion of a bacterium's genome.

TELL ME WHY

Why did the discovery and development of restriction enzymes speed up the study of recombinant DNA technology?

Techniques of Recombinant DNA Technology

Scientists use the tools of recombinant DNA technology in a number of basic techniques to multiply, identify, manipulate, isolate, map, and sequence the nucleotides of genes.

Multiplying DNA *In Vitro*: The Polymerase Chain Reaction

LEARNING OUTCOME

8.12 Describe the purpose and application of the polymerase chain reaction.

The **polymerase chain reaction (PCR)** is a technique by which scientists produce a large number of identical molecules of DNA *in vitro.* Using PCR, researchers start with a single molecule of DNA and generate billions of exact replicas within hours. Such rapid amplification of DNA is critical in a variety of situations.

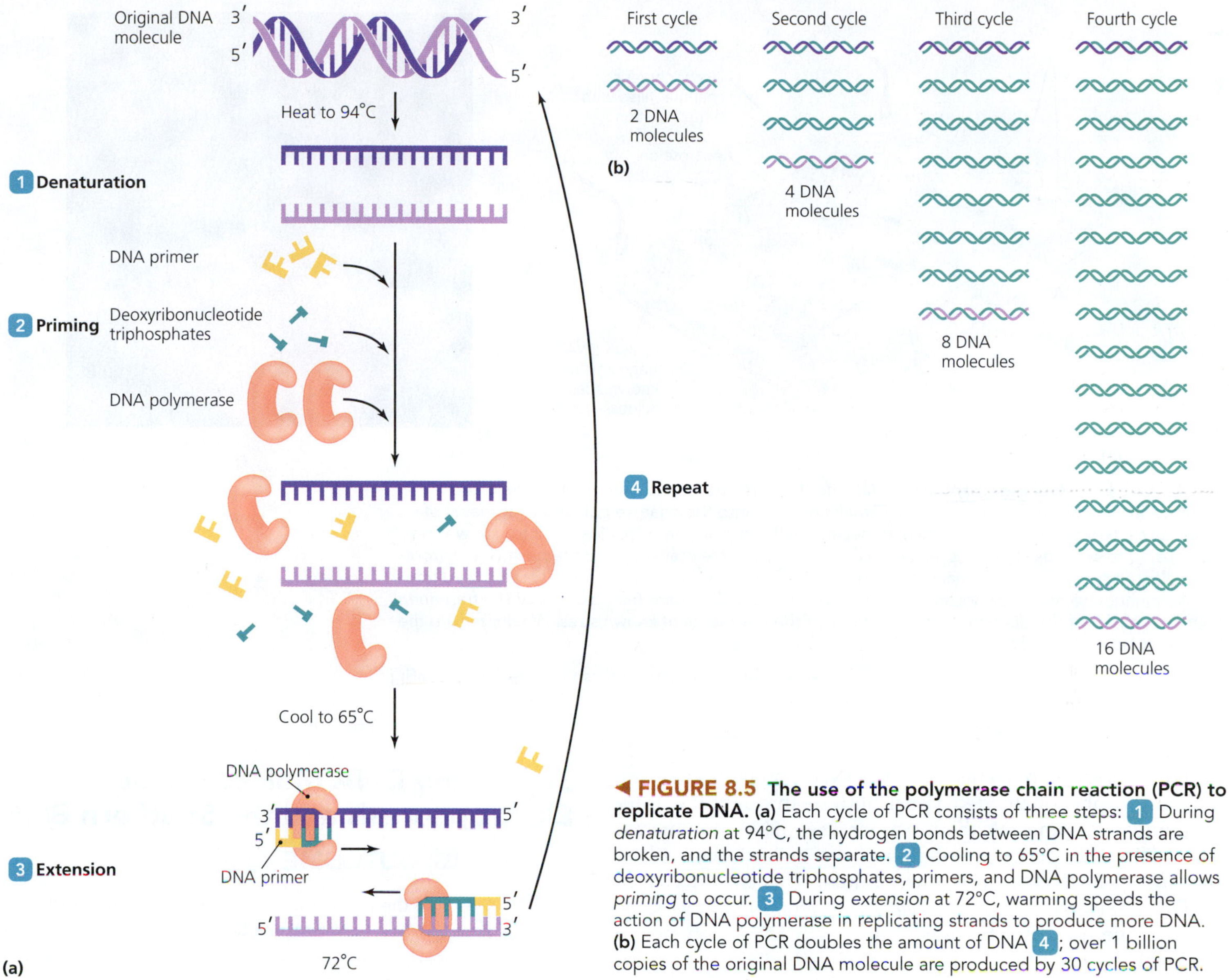

◀ **FIGURE 8.5 The use of the polymerase chain reaction (PCR) to replicate DNA. (a)** Each cycle of PCR consists of three steps: 1 During *denaturation* at 94°C, the hydrogen bonds between DNA strands are broken, and the strands separate. 2 Cooling to 65°C in the presence of deoxyribonucleotide triphosphates, primers, and DNA polymerase allows *priming* to occur. 3 During *extension* at 72°C, warming speeds the action of DNA polymerase in replicating strands to produce more DNA. **(b)** Each cycle of PCR doubles the amount of DNA 4; over 1 billion copies of the original DNA molecule are produced by 30 cycles of PCR.

For example, epidemiologists used PCR to amplify the genome of a previously unknown pathogen that killed people in Hong Kong in 2003 with severe acute respiratory syndrome (SARS). The large number of identical DNA molecules produced by PCR allowed scientists to determine the nucleotide sequence, which was found to be similar to that of coronaviruses, until then thought to cause only mild colds. **▶ANIMATIONS: *Polymerase Chain Reaction (PCR): Overview, Components***

PCR is a repetitive process that alternately separates and replicates the two strands of DNA. Each cycle of PCR consists of the following three steps (**FIGURE 8.5a**):

1 **Denaturation.** Exposure to heat (about 94°C) separates the two strands of the target DNA by breaking the hydrogen bonds between base pairs but otherwise leaves the two strands unaltered.

2 **Priming.** A mixture containing an excess of DNA primers (synthesized such that they are complementary to nucleotide sequences near the ends of the target DNA), DNA polymerase, and an abundance of the four deoxyribonucleotide triphosphates (A, T, G, and C) is added to the target DNA. This mixture is then cooled to about 65°C, enabling double-stranded DNA to re-form. Because there is an excess of primers, single strands are more likely to bind to a primer than to one another. The primers provide DNA polymerase with the 3′ hydroxyl group it requires for DNA synthesis.

3 **Extension.** Raising the temperature to about 72°C increases the rate at which DNA polymerase replicates each strand to produce more DNA. **▶ANIMATIONS: *PCR: The Process***

These steps are repeated over and over 4, so the number of DNA molecules increases exponentially (**FIGURE 8.5b**). After only 30 cycles—which requires only a few hours to complete—PCR produces over 1 billion identical copies of the original DNA molecule.

The process can be automated using a *thermocycler,* a device that automatically performs PCR by continuously cycling

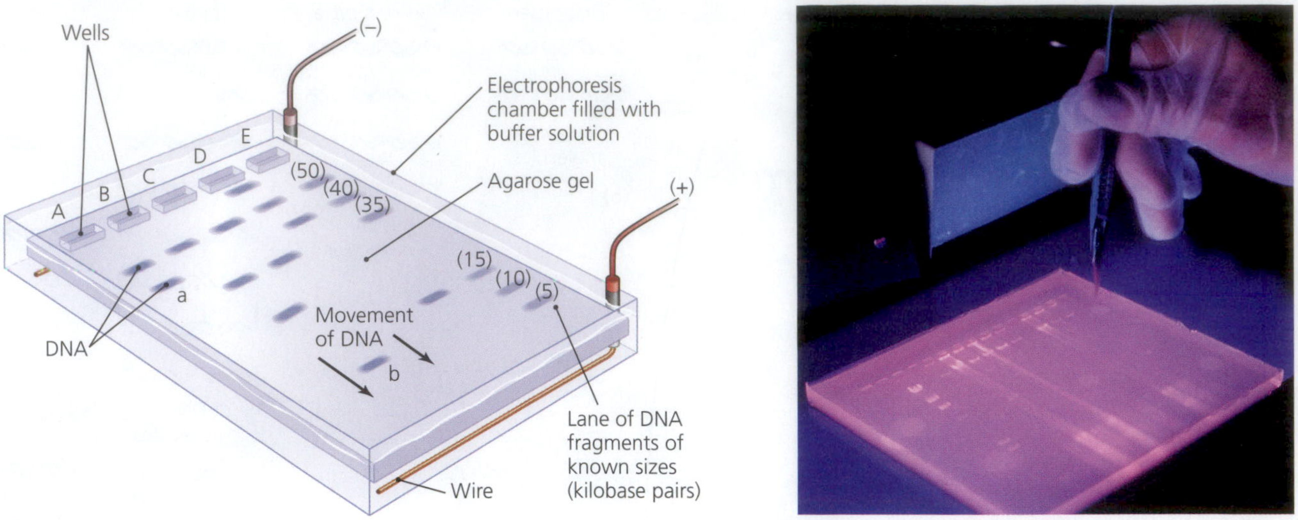

(a) (b)

▲ **FIGURE 8.6 Gel electrophoresis. (a)** After DNA is cleaved into fragments by restriction enzymes, it is loaded into wells, which are small holes cut into the agarose gel. DNA fragments of known sizes, typically in thousands (kilo) of nucleotide base pairs, are often loaded into one well (in this case, E) to serve as standards. After the DNA fragments are drawn toward the positive electrode by an electric current, they are stained with a dye. **(b)** Ethidium bromide dye fluoresces under ultraviolet illumination to reveal the locations of DNA within a gel. *Compare the positions of the fragments in lanes A and B of the diagram to the positions of the fragments of known sizes. What sizes are the fragments labeled a and b?*

Figure 8.6 a, *40 kilobase pairs.* b, *10 kilobase pairs.*

all the necessary reagents—DNA, DNA polymerase, primers, and triphosphate deoxynucleotides—through the three temperature regimes. A thermocycler uses DNA polymerase derived from hyperthermophilic archaea or bacteria, such as *Thermus aquaticus* (ther′mŭs a-kwa′ti-kŭs). This enzyme, called *Taq DNA polymerase* or simply *Taq*, is not denatured at 94°C, so the machine need not be replenished with DNA polymerase after each cycle.

Selecting a Clone of Recombinant Cells

LEARNING | OUTCOME

8.13 Explain how researchers use DNA probes to identify recombinant cells.

Before recombinant DNA technology can have practical application, a scientist must be able to select and isolate recombinant cells that contain particular genes of interest. For example, once researchers have created a gene library, they must find the clone containing the DNA of interest. To do so, scientists use probes—which, as explained earlier in the chapter, bind specifically and exclusively to their complementary nucleotide sequences and have either radioactive or fluorescent markers. Researchers then isolate and culture cells that have the radioactive or fluorescent marker, which also aids in identifying the specific location of the genes of interest, as performed in a technique called *gel electrophoresis*.

Separating DNA Molecules: Gel Electrophoresis and the Southern Blot

LEARNING | OUTCOME

8.14 Describe the process and use of gel electrophoresis, particularly as it is used in a Southern blot.

Electrophoresis (ē-lek-trō-fōr-ē′sis) is a technique that involves separating molecules based on their electrical charge, size, and shape. In recombinant DNA technology, scientists use **gel electrophoresis** to isolate fragments of DNA molecules that can then be inserted into vectors, multiplied by PCR, or preserved in a gene library.

In gel electrophoresis, DNA molecules, which have an overall negative charge, are drawn through a semisolid gel by an electric current toward the positive electrode within an electrophoresis chamber **(FIGURE 8.6)**. The gel is typically composed of a purified sugar component of agar, called *agarose*, which acts as a molecular sieve that retards the movement of DNA fragments down the chamber and separates the fragments by size. Smaller DNA fragments move faster and farther than larger ones. Scientists can determine the size of a fragment by comparing the distance it travels to the distances traveled by standard DNA fragments of known sizes.

As we have seen, DNA probes allow a researcher to find specific DNA sequences such as genes in a cell. Scientists could also use probes to localize specific sequences in electrophoresis gels, but because gels are flimsy and easily broken and deform as they dry, it is difficult to probe gels.

In 1975, Ed Southern (1938–) devised a method, called the **Southern blot,** to transfer DNA from agarose gels to nitrocellulose membranes, which are less delicate. The Southern blot technique begins with the procedures of gel electrophoresis just described (**FIGURE 8.7** 1). The DNA is denatured into single strands with NaOH. Once the DNA fragments have been separated by size, the liquid in the electrophoresis gel is blotted out 2. DNA is transferred and bonded with heat to a nitrocellulose membrane 3. Radioactive probes complementary to DNA sequences of interest are added 4. The probes expose photographic film, revealing the DNA of interest 5. A *northern blot* is a similar technique used to detect specific RNA molecules.

Researchers use Southern blots for a variety of purposes, including genetic "fingerprinting" (discussed shortly) and diagnosing infectious diseases. For example, scientists can detect the presence of genetic sequences unique to hepatitis B virus in a blood sample of an infected patient even before the patient shows symptoms or an immune response.

Scientists also use Southern blotting to demonstrate the incidence and prevalence in an environmental sample of archaea, bacteria, and viruses, particularly those that cannot be cultured.

DNA Microarrays

LEARNING **OUTCOME**

8.15 Describe the manufacture and use of DNA microarrays.

Another tool of biotechnology is a **DNA microarray.** An array consists of molecules of single-stranded DNA, either genetic DNA or cDNA, immobilized on glass slides, silicon chips, or nylon membranes. Robots, similar to those that construct computer chips, deposit PCR-derived copies of hundreds of thousands of different DNA sequences in precise locations on the array (**FIGURE 8.8**). An array may consist of DNA from a single species (e.g., DNA microarrays containing sequences from all the genes of *E. coli* are available commercially), or a DNA array may contain sequences from numerous species. In any case, single strands of fluorescently labeled DNA in a sample washed over an array adhere only to locations on the array where there are complementary DNA sequences.

Scientists use DNA microarrays in a number of ways, including the following:

- **Monitoring gene expression.** One way organisms control metabolism is by controlling RNA transcription. Scientists use DNA microarrays to monitor which genes a cell is transcribing at a particular time by making fluorescently labeled cDNA from mRNA in the cell. These DNA strands bind to complementary DNA sequences on the array, and the location of fluorescence on the array at specific sites reveals which genes the cell was transcribing at the time. Researchers using DNA microarrays can monitor the expression of thousands of genes simultaneously and can compare and contrast genetic expression under different

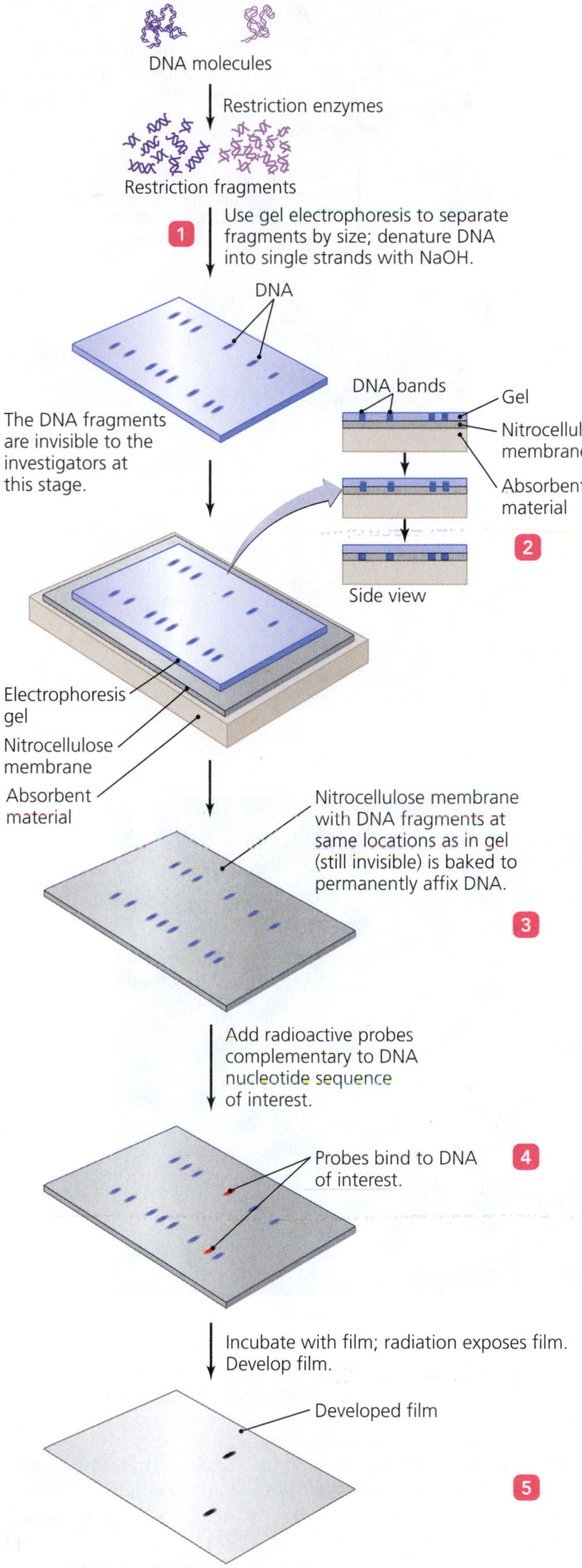

▲ **FIGURE 8.7 The Southern blot technique.** This method enables scientists to locate DNA sequences of interest.

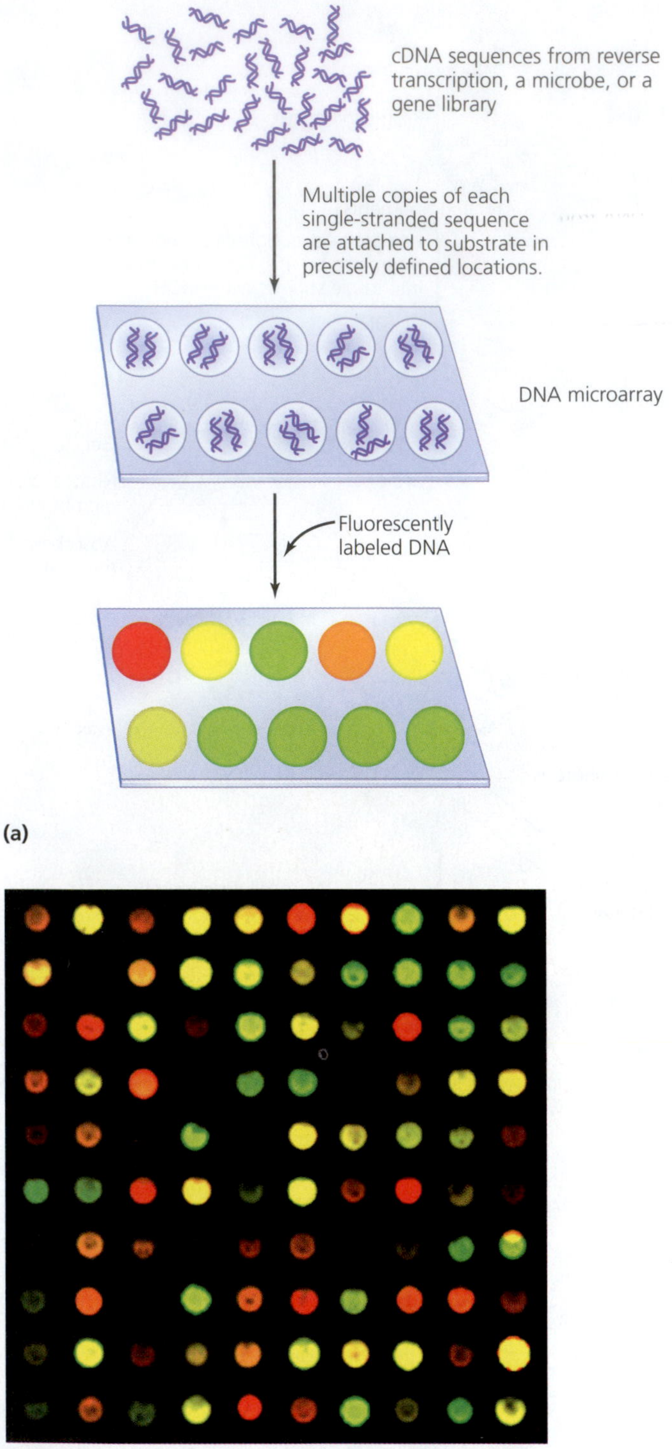
cDNA sequences from reverse transcription, a microbe, or a gene library
Multiple copies of each single-stranded sequence are attached to substrate in precisely defined locations.
DNA microarray
Fluorescently labeled DNA

(a)

(b)

▲ **FIGURE 8.8 DNA microarray. (a)** Construction and use of a microarray. Multiple copies of single-stranded DNA with known sequences are affixed in precise locations on a glass slide, silicon chip, nylon membrane, or other substrate. Fluorescently labeled DNA washed over the microarray binds to complementary strands. **(b)** Photograph of a DNA microarray showing locations of differently labeled cDNA molecules.

Helps with determining genetic disease

conditions. In the latter type of experiment, a different color of fluorescent dye is used to label DNA from microbes grown in each condition.

- **Diagnosing infection.** DNA microarrays made with DNA sequences of numerous pathogens reveal the presence of those pathogens in medical samples.
- **Identifying organisms in an environmental sample.** Microbial ecologists monitor the presence or absence of microbes in an environment by using microarrays of DNA from the organisms.

Inserting DNA into Cells

LEARNING | **OUTCOME**

8.16 List and explain three artificial techniques for introducing DNA into cells.

A goal of recombinant DNA technology is the insertion of a gene into a cell. In addition to using vectors and the natural methods of transformation of competent cells, transduction, and conjugation, scientists have developed several artificial methods to introduce DNA into cells, including the following:

- **Electroporation** **(FIGURE 8.9a)**. Electroporation involves using an electrical current to puncture microscopic holes through a cell's membrane so that DNA can enter the cell from the environment. Electroporation can be used on all types of cells, though the thick-walled cells of fungi and algae must first be converted to *protoplasts*, which are cells whose cell walls have been enzymatically removed. Cells treated by electroporation repair their membranes and cell walls after a time.
- **Protoplast fusion** **(FIGURE 8.9b)**. When protoplasts encounter one another, their cytoplasmic membranes may fuse to form a single cell that contains the genomes of both "parent" cells. Exposure to polyethylene glycol increases the rate of fusion. The DNA from the two fused cells recombines to form a recombinant molecule. Scientists often use protoplast fusion for the genetic modification of plants.
- **Injection.** Two types of injection are used with larger eukaryotic cells. Researchers use a *gene gun* powered by a blank .22-caliber cartridge or compressed gas to fire tiny tungsten or gold beads coated with DNA into a target cell **(FIGURE 8.9c)**. The cell eventually eliminates the inert metal beads. In *microinjection*, a geneticist inserts DNA into a target cell with a glass micropipette having a tip diameter smaller than that of the cell or nucleus **(FIGURE 8.9d)**. Unlike electroporation and protoplast fusion, injection can be used on intact tissues such as in plant seeds.

In every case, foreign DNA that enters a cell remains in a cell's progeny only if the DNA is self-replicating, as in the case of plasmid and viral vectors, or if the DNA integrates into a cellular chromosome by recombination.

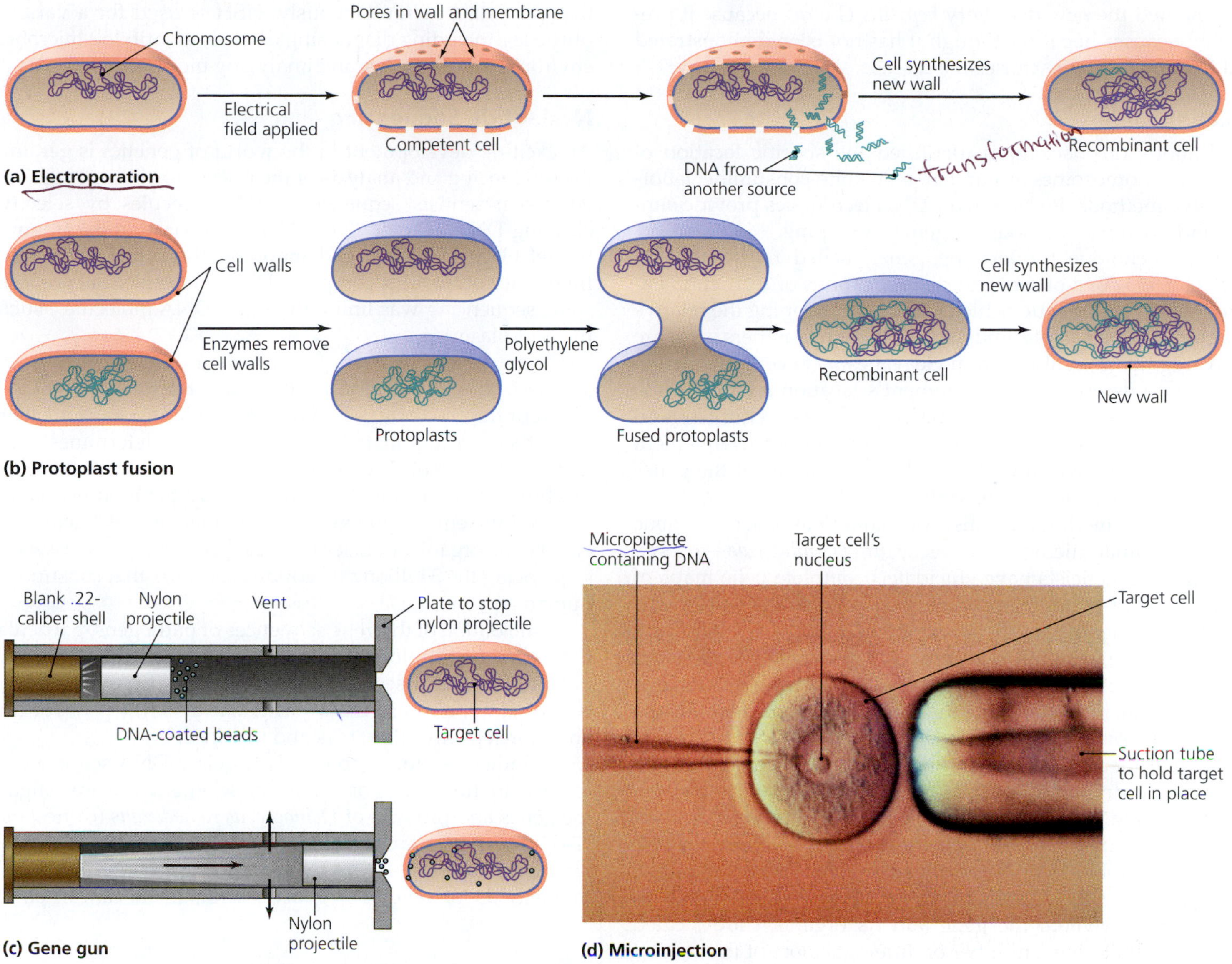

▲ **FIGURE 8.9 Artificial methods of inserting DNA into cells. (a)** Electroporation, in which an electrical current applied to a cell makes it competent to take up DNA. **(b)** Protoplast fusion, in which enzymes digest cell walls to create protoplasts that fuse at a high rate when treated with polyethylene glycol. **(c)** A gene gun, which fires DNA-coated beads into a cell. **(d)** Microinjection, in which a solution of DNA is introduced into a cell through a micropipette.

TELL ME WHY

Why wasn't polymerase chain reaction (PCR) practical before the discovery of hyperthermophilic bacteria?

Applications of Recombinant DNA Technology

The importance of recombinant DNA technology lies not in the novelty, cleverness, or elegance of its procedures but in its wide range of applications. In this section we consider how recombinant DNA technology is used to solve various problems and create research, medical, and agricultural products.

Genetic Mapping

LEARNING **OUTCOME**

8.17 Describe genetic mapping and genomics, and explain their usefulness.

One application of these tools and techniques is **genetic mapping,** which involves locating genes on a nucleic acid molecule. Genetic maps provide scientists with useful facts, including information concerning an organism's metabolism and growth characteristics, as well as its potential relatedness to other microbes. For example, scientists have discovered a virus with a genetic map similar to those of certain hepatitis viruses.

They named the new discovery *hepatitis G virus* because it presumably causes hepatitis, though it has not been demonstrated that the virus actually causes the disease.

Locating Genes

Until about 1970, scientists identified the specific location of genes on chromosomes by cumbersome, time-consuming, labor-intensive methods. Recombinant DNA techniques provide simpler and universal methods for genetic mapping.

One technique for locating genes, called *restriction fragmentation,* was one of the earliest applications of restriction enzymes. In this technique, which is used for mapping the relative locations of genes in plasmids and viruses, researchers compare DNA fragments resulting from cleavages by several restriction enzymes to determine each fragment's location relative to the others. If the researchers know the locations of specific genes on specific fragments, then elucidation of the correct arrangement of the fragments will reveal the relative locations of the genes on the entire DNA molecule.

Using this method, scientists first completed the entire gene map of a cellular microbe—the bacterium *H. influenzae*—in 1995. Since then, geneticists have elucidated complete gene maps of numerous viruses and prokaryotic and eukaryotic organisms.

Often a scientist wants to know where in the environment, clinical sample, or biofilm a particular microbial species is located. When researchers know of a particular gene exclusive to that organism, they can locate the gene and thereby the microbe using *fluorescent* in situ *hybridization (FISH).*

In this method, scientists attach fluorescent chemicals to short, single strands of nucleic acid molecules that are complementary to the gene or its transcribed mRNA. Because complementary strands of nucleic acid best bind one another, these fluorescent probes hybridize with their complementary target. Scientists using fluorescent microscopes to view such probes can determine where the gene and its organism are located **(FIGURE 8.10)**. Using a number of different colors of fluorescent probes, researchers can locate numerous genes and the microbes that carry them simultaneously. FISH is used for a variety of purposes, including diagnosing disease, identifying microbes in environmental samples, and analyzing biofilms.

Nucleotide Sequencing

An exciting development in the world of genetics is **genomics,** the sequencing and analysis of the nucleotide bases of genomes. At first, scientists sequenced DNA molecules by selectively cleaving DNA at A, T, G, or C bases, separating the fragments by gel electrophoresis, and mapping the order in which the fragments occur in a complete DNA molecule. Such cumbersome sequencing was limited to short DNA molecules, such as those of plasmids.

Today, scientists use a faster technique that utilizes cDNA synthesized with nucleotides that have been tagged with four different fluorescent dyes—a different color for each nucleotide base; then an automated DNA sequencer determines the sequence of base colors emitted by the dyes **(FIGURE 8.11)**. Such machines, often running 24 hours a day for months, have sequenced the entire genomes of numerous viruses, bacteria, and eukaryotic organisms. Scientists reached a milestone in 2001 by sequencing the 3 billion nucleotide base pairs that constitute the human genome.

Elucidation of the gene sequences of pathogens, particularly those affecting hundreds of millions of people and those with potential bioterrorist uses, is a current priority of researchers. Scientists hope to use the information to develop novel drugs and more effective therapies and vaccines.

Another use for genomics is to relate DNA sequence data to protein function. For instance, scientists are investigating the genes and proteins of *Deinococcus radiodurans* (dī-nō-kok´ŭs rā-dē-ō-dur´anz), a microorganism that is remarkably resistant to damage of its DNA by radiation. Such studies may lead to methods of reversing genetic damage in cancer patients

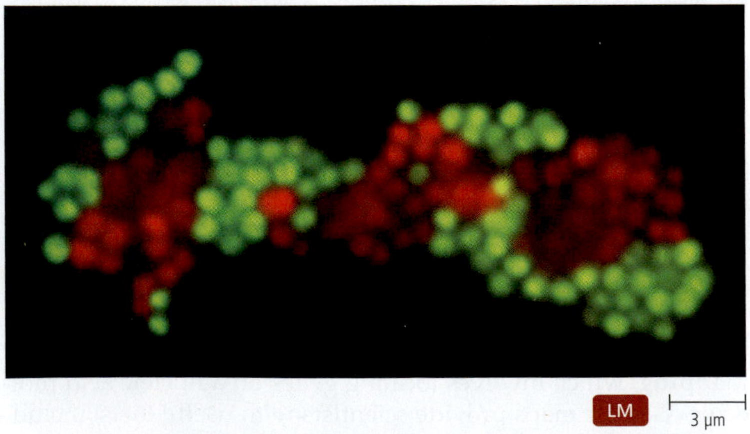

▲ **FIGURE 8.10 Fluorescent *in situ* hybridization (FISH).** Green-fluorescing cells are *Staphylococcus aureus*, whereas red-fluorescing cells are another species of *Staphylococcus*. The absence of fluorescence in some cells indicates that at least one other species is present in this blood sample.

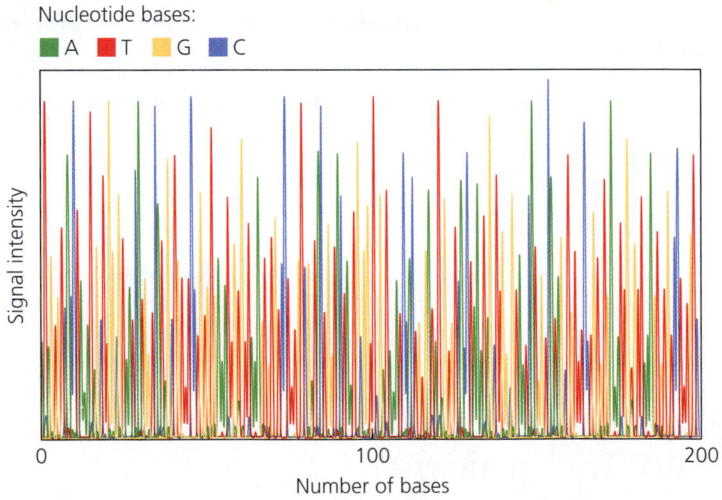

▲ **FIGURE 8.11 Automated DNA sequencing.** Each colored line corresponds to a different nucleotide base; each peak indicates the location of a particular base. *What are the nucleotides in positions 100 and 200?*

Figure 8.11 *Thymine and cytosine.*

TABLE 8.2 Tools and Techniques of Recombinant DNA Technology

Tool or Technique	Description	Potential Application
Mutagen	Chemical or physical agent that creates mutations	Creating novel genotypes and phenotypes
Reverse transcriptase	Enzyme from RNA retrovirus that synthesizes cDNA from an RNA template	Synthesizing a gene using an mRNA template
Synthetic nucleic acid	DNA molecule prepared *in vitro*	Creating DNA probes to localize genes within a genome
Restriction enzyme	Bacterial enzyme that cleaves DNA at specific sites	Creating recombinant DNA by joining fragments
Vector	Transposon, plasmid, or virus that carries DNA into cells	Altering the genome of a cell
Gene library	Collection of cells or viruses, each of which carries a portion of a given organism's genome	Providing a ready source of genetic material
Polymerase chain reaction (PCR)	Produces multiple copies of a DNA molecule	Multiplying DNA for various applications
Gel electrophoresis	Uses electrical charge to separate molecules according to their size	Separating DNA fragments by size
Electroporation	Uses electrical current to make cells competent	Inserting a novel gene into a cell
Protoplast fusion	Fuses two cells to create recombinants	Inserting a novel gene into a cell
Gene gun	Blasts genes into target cells	Inserting a novel gene into a cell
Microinjection	Uses micropipette to inject genes into cells	Inserting a novel gene into a cell
Southern blot	Localizes specific DNA sequences on a stable membrane	Identifying a strain of pathogen
Nucleic acid probes	RNA or DNA molecules labeled with radioactive or fluorescent tags	Localizing specific genes in a Southern blot
Genetic mapping	Uses restriction enzymes to locate relative positions of restriction sites	Locating genes in an organism's genome
DNA sequencing	Determines the sequence of nucleotide bases in DNA	Comparing genomes of organisms
DNA microarray	Reveals presence of specific DNA or RNA molecules in a sample	Diagnosing infection

undergoing radiation therapy. Researchers are also investigating the genetic basis of the enzymes of psychrophiles, which are microorganisms that thrive at temperatures below 20°C. Such enzymes have potential applications in food processing and in the manufacture of drugs.

TABLE 8.2 summarizes the tools and techniques of recombinant DNA technology.

Environmental Studies

It is estimated that more than 99% of microorganisms have never been grown in a laboratory; indeed, scientists know them only by unique DNA patterns in electrophoresis gels and Southern blot membranes. For example, based on such unique DNA sequences, sometimes called signatures or DNA fingerprints, scientists have isolated over 500 species of bacteria from human mouths; however, they have been able to identify only about 150 of these. An understanding of the biology of the other 350 species may lead to a better understanding of tooth and gum decay, diagnosis of disease, and advances in oral health care.

Another application of genetics to environmental studies may have ramifications for global warming. Rice agriculture is possibly the largest human contributor of the so-called greenhouse gas methane to the atmosphere. Rice paddies, concentrated in Asia, contribute 50 million to 100 million metric tons of methane to the environment every year, though neither rice nor humans directly cause this deluge of methane. Scientists, having analyzed the DNA signatures of microbes in the soil, have determined that mud-dwelling, methane-producing archaea feed on carbohydrates released by the rice plants' roots. These archaea have not been isolated or grown in a laboratory. They are known only by their DNA signatures. Thus, the tools and techniques of recombinant DNA technology have revealed the source of a problem. Discovering that these organisms exist is certainly the first step in developing methods to reduce their impact on the environment.

Pharmaceutical and Therapeutic Applications

LEARNING | OUTCOMES

8.18 Describe six potential medical applications of recombinant DNA technology.

8.19 Describe the steps and uses of genetic fingerprinting.

8.20 Define *gene therapy*.

Researchers now supplement traditional biotechnology with recombinant DNA technology to produce a variety of pharmaceutical and therapeutic substances and to perform a host of medically important tasks. Here we explore the use of recombinant DNA technology to synthesize selected proteins, produce vaccines, screen for genetic diseases, match DNA specimens to the organisms from which they came, treat genetic illnesses, and aid in organ transplantation.

Protein Synthesis

Scientists have inserted synthetic genes for insulin, for interferon (a natural antiviral chemical), and for other proteins into bacteria and yeast cells so that the microbes synthesize these proteins in vast quantities. In the past, such proteins were isolated from donated blood or from animals—labor-intensive processes that carry the risk of inducing allergies or of transferring pathogens, such as hepatitis B and HIV. "Genetically engineered" proteins are safer and less expensive than their naturally occurring counterparts.

Vaccines

Vaccines contain *antigens*—foreign substances such as weakened bacteria, viruses, and toxins that stimulate the body's immune system to respond to and subsequently remember these foreign materials. In effect, a vaccine primes the immune system to respond quickly and effectively when confronted with pathogens and their toxins. However, the use of some vaccines entails a risk—they may cause the disease they are designed to prevent.

Scientists now use recombinant DNA technology to produce safer vaccines. Once they have inserted the gene that codes for a pathogen's antigens into a vector, they can inject the recombinant vector or the proteins it produces into a patient. Thus, the patient's immune system is exposed to a subunit of the pathogen—one of the pathogen's antigens—but not to the pathogen itself. Such **subunit vaccines** are especially useful in safely protecting against pathogens that either cannot be cultured or cause incurable fatal diseases. Hepatitis B vaccine is an example of a successful subunit vaccine. Scientists are also pursuing subunit vaccines against HIV.

A promising future approach to vaccination involves introducing genes coding for antigenic proteins of pathogens into common fruits or vegetables, such as bananas or beans. The immune systems of people or animals eating such altered produce would be exposed to the pathogen's antigens and theoretically would develop immunological memory against the pathogen. Such a vaccine would have the advantages of being painless and easy to administer, and vaccination would not require a visit to a health care provider. **Highlight: Vaccines on the Menu** focuses on such vaccines.

Another type of vaccination involves producing a recombinant plasmid carrying a gene from a pathogen and injecting the plasmid into a human whose body then synthesizes polypeptides characteristic of the pathogen. The polypeptides stimulate immunological memory within the human body, readying it to mount a vigorous immune response and prevent infection should it subsequently be exposed to the real pathogen. Clinical trials of such a vaccine against malaria have shown some promise.

Genetic Screening

Genetic mutations cause some diseases, such as inherited forms of breast cancer and Huntington's disease. Laboratory technicians use DNA microarrays to screen patients, prospective parents, and fetuses for such mutant genes. This procedure, called **genetic screening**, can also identify viral DNA sequences in a patient's blood or other tissues. For instance, genetic screening can identify HIV in a patient's cells even before the patient shows any other sign of infection.

DNA Fingerprinting

Medical laboratory technicians and forensic investigators use gel electrophoresis and Southern blotting for so-called **genetic fingerprinting,** or *DNA fingerprinting*—identifying individuals or organisms by their unique DNA sequences.

DNA fingerprinting involves procuring a sample of DNA, making multiple copies of it via PCR, cutting the copies with restriction enzymes, and separating the fragments by gel electrophoresis to produce a unique pattern. The process is analogous to standard fingerprinting in that the pattern resulting from a particular DNA sample is unique, and it must be compared to patterns produced from other DNA molecules **(FIGURE 8.12)**, much like a standard fingerprint must be compared to known fingerprints. For example, the patterns from DNA collected at a crime scene either match or do not match a suspect's or victim's DNA, or the pattern from an environmental sample matches or does not match patterns from known organisms. Genetic fingerprinting is used to determine paternity; to connect blood, semen, or even single skin cells to a particular crime suspect; to identify badly damaged human remains; and to identify pathogens.

Gene Therapy

An exciting use of recombinant DNA technology is **gene therapy,** in which missing or defective genes are replaced with normal

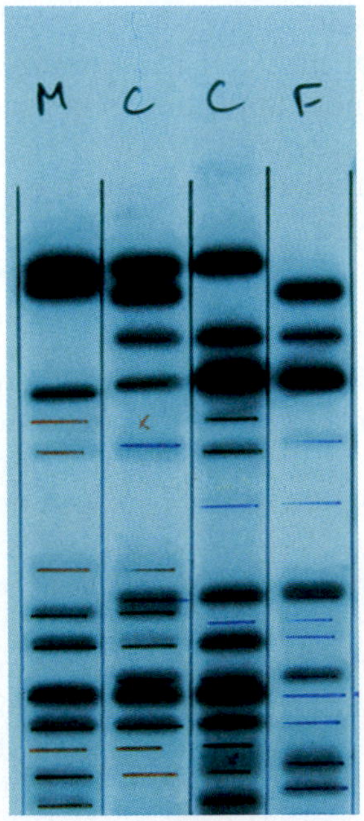

▲ **FIGURE 8.12 DNA fingerprinting.** Shown here is a partial X-ray image of bands of DNA from four family members: a mother (M), father (F), and two children (Cs). Each child shares some bands with each parent, proving they are indeed related. A similar process can be used to compare DNA bands from microbial specimens in order to identify a particular specimen.

HIGHLIGHT

Vaccines on the Menu

Wouldn't it be great if instead of giving your children painful shots, they could be immunized simply by eating a banana that stimulated their immune systems to fight off pathogenic viruses and bacteria? We aren't quite there yet, but scientists are making progress in developing genetically modified banana vaccines that are active against gastrointestinal viruses and bacteria, which together kill about 2 million children each year.

Researchers have isolated genes that code for certain critical proteins of the pathogens and put those genes into the genome of a potato. The edible vaccine protected mice against diarrhea-causing pathogens. Because only a few of the pathogens' genes are expressed, pathogens cannot form, cause disease, or be contagious. The drawback: The mice had to eat the potato raw.

Enter the banana. Bananas are tough and easily grown. The inside of a banana is sweet, sterile, and full of protein, and kids love to eat them raw. Besides bananas, foods under study include tomatoes, rice, wheat, soybean, and corn. Bananas are preferred because they last without refrigeration, their tough skins prevent contamination, and they taste better than raw potatoes or rice.

Potential advantages of edible vaccines are tremendous: Tasty vaccines would reduce children's fears of doctors. People in tropical areas could grow vaccines locally, sustainably, and inexpensively using local farming techniques. Edible vaccines would reduce the need for syringes, which are often in short supply in poorer parts of the world. We would bypass the expenses of transportation, storage, and refrigeration required for many injected vaccines.

Maybe, one day soon, a trip down the produce aisle will provide your family with more than just good nutrition; your immune system will also get a boost.

copies. Scientists remove a few genetically defective cells—for example, cells that produce a defective protein—from a patient, insert normal genes, and replace the cells into the patient, curing the disease. Alternatively, plasmid, bacterial, or viral vectors could deliver genes directly to target cells within a patient.

Unfortunately, gene therapy has proven difficult in practice because of unexpected side effects. Specifically, some patients' immune systems react uncontrollably to the presence of vectors, resulting in the death of these patients. Nevertheless, doctors have successfully treated patients for severe combined immunodeficiency disease (SCID) and a form of blindness. Other diseases that may respond well to gene therapy are cystic fibrosis, sickle-cell anemia, and some types of hemophilia, muscular dystrophy, and diabetes.

Medical Diagnosis

Clinical microbiologists use PCR, fluorescent genetic probes, and DNA microarrays in diagnostic applications. They examine specimens from patients for the presence of gene sequences unique to certain pathogens, such as particular hepatitis viruses, cytomegalovirus, human immunodeficiency virus, or the bacterial pathogens of gonorrhea, tuberculosis, trachoma, and ulcers.

Xenotransplant

Xenotransplants[5] are animal cells, tissues, or organs introduced into the human body. For years physicians have performed xenotransplants, for instance, using valves from pig hearts to repair severely damaged human hearts. However, recombinant DNA technology may expand the possibilities. It is theoretically feasible to insert functional human genes into animals to direct them to produce organs and tissues for transplantation into humans. For example, scientists could induce pigs to produce humanlike cytoplasmic membrane proteins so that entire organs from pigs would not be rejected as foreign tissue by a transplant recipient.

Agricultural Applications

LEARNING | OUTCOME

8.21 Identify six agricultural applications of recombinant DNA technology.

Recombinant DNA technology has been applied to the realm of agriculture to produce **transgenic organisms**—recombinant plants and animals that have been altered for specific purposes by the addition of genes from other organisms. The purposes for which transgenic organisms have been produced are many and varied and include herbicide resistance, tolerance to salty

[5]From Greek *xenos*, meaning "stranger."

soils, resistance to freezing and pests, and improvements in nutritional value and yield. More than 15 million farmers in 29 countries grow transgenic crops on over 2.5 billion acres worldwide—more land than in the entire United States.

Herbicide Tolerance

The biodegradable herbicide *glyphosate* (Roundup) normally kills all plants—weeds and crops alike—by blocking an enzyme that is essential for plants to synthesize several amino acids. After scientists discovered and isolated a gene from *Agrobacterium tumefaciens* (ag´rō-bak-tēr´ē-um tū´me-fāsh-enz) that conveys tolerance to glyphosate, they produced transgenic crop plants containing the gene. As a result of this application of recombinant DNA technology, farmers can now apply glyphosate to a field of transgenic plants to kill weeds without damaging the crop. An added benefit is that farmers do not need to till the soil to suppress weeds during the growing season, reducing soil erosion by 80%. Most of the soybeans, corn, and cotton grown in the United States are genetically modified in this manner to be "Roundup ready." Glyphosate-tolerant rice, wheat, sugar beets, and alfalfa strains are also available, as are soybeans and corn tolerant of other herbicides.

Salt Tolerance

Years of irrigation have resulted in excessive salt buildup in farmland throughout the world, rendering the land useless for farming. Though salt-tolerant plants can grow under these conditions, they are not edible.

Scientists have now successfully removed the gene for salt tolerance from these plants and inserted it into tomato and canola plants to create food crops that can grow in soil so salty that it would poison normal crops. Not only do such transgenic plants survive and produce fruit, but they also remove salt from the soil, restoring the soil and making it suitable to grow unmodified crops as well. Researchers are now attempting to insert the gene for salt tolerance into rice, cotton, wheat, and corn.

Freeze Resistance

Ice crystals form more readily when bacterial proteins (from natural bacteria present in a field) are available as crystallization nuclei. Scientists have modified strains of the bacterium *Pseudomonas* (soo-dō-mō´nas) with a gene for a polypeptide that prevents ice crystals from forming. Crops sprayed with genetically modified bacteria can tolerate mild freezes, so the farmers no longer lose their crops to unseasonable cold snaps.

Pest Resistance

Strains of the bacterium *Bacillus thuringiensis* (ba-sil´ŭs thur-in-jē-en´sis) produce a protein that, when modified by enzymes in the intestinal tracts of insects, becomes **Bt toxin (Bt).** Bt binds to receptors lining the insect's digestive tract and causes the tissue to dissolve. Unlike some insecticides, Bt is naturally occurring, harmful only to insects, and biodegradable. Farmers, particularly organic growers, have used Bt for over 30 years to reduce insect damage to their crops.

Now, genes for Bt toxin have been inserted into a variety of crop plants, including potatoes, cotton, rice, and corn, so that they produce Bt for themselves. Insects feeding on such plants are killed, while humans and other animals that eat them are unharmed. By inserting genes for more than one type of Bt, scientists hope to forestall evolution of resistant insects.

Scientists have also developed crops that are resistant to microbial diseases. Two cases illustrate.

The water mold *Phytophthora infestans* (fī-tof´tho-ră in-fes´tanz) is the most devasting potato pathogen; it caused the great Irish potato famine of the 19th century, resulting in the deaths of at least a million people. The mold still causes many billions of dollars of crop damage each year. Scientists have now cloned genes from potato species resistant to *Phytophthora;* the cloned genes, multiplied by PCR, can be inserted into potato crops to reduce losses to farmers and increase available food for a growing world population.

Papaya ringspot virus causes devastating harm to papayas, mottling and malforming leaves, streaking fruit, and eventually killing the entire plant. Researchers centered at the University of Hawaii saved that state's papaya industry by inserting a gene for ringspot virus capsid into the papaya's genome. When the plant produces the viral protein, it triggers protection by an unknown mechanism against infection by the ringspot virus. Genetically modified plants grow normally **(FIGURE 8.13).**

Improvements in Nutritional Value and Yield

Genetic researchers have increased crop and animal yields in several ways. For example, MacGregor tomatoes remain firm after harvest because the gene for the enzyme that breaks down pectin has been suppressed. This allows farmers to let the tomatoes ripen on the vine before harvesting and increases the tomatoes' shelf life. Scientists suppressed the gene indirectly by inserting a promoter in the noncoding DNA strand that allows transcription of antisense RNA. The antisense RNA then binds to the gene's mRNA, forming double-stranded RNA that makes it impossible to translate the pectin-catabolizing enzyme.

▲ **FIGURE 8.13 Genetically modified papaya plants.** Plants on the left are dying from ringspot virus infection. Those on the right are modified to produce viral protein, which prevents infection by the virus.

Another example of agricultural improvement involves bovine growth hormone (BGH), which when injected into cattle enables them to more rapidly gain weight and produce 10% more milk. Also, use of BGH reduces the fat content in beef. Though BGH can be derived from animal tissue, it is more economical to insert the BGH gene into bacteria so that they produce the hormone, which is then purified and injected into farm and ranch animals.

In yet another application, scientists have improved the nutritional value of rice by adding a gene for beta-carotene, which is a precursor to vitamin A. Vitamin A is required for human embryonic development and for vision in adults, and it is an important antioxidant that plays a role in ameliorating cancer and atherosclerosis (hardening of arteries).

Recombinant DNA technology has progressed to the point that scientists are now considering transplanting genes coding for entire metabolic pathways rather than merely genes encoding single proteins. For instance, researchers are attempting to transfer into corn and rice all the genes that bacteria use to convert atmospheric nitrogen into nitrogenous fertilizer, in effect allowing the recombinant plants to produce their own fertilizer.

Recombinant DNA tools and techniques allow scientists to examine, compare, and manipulate the genomes of microorganisms, plants, animals, and humans for a variety of purposes, including gene mapping and forensic, medical, and agricultural applications. However, as with many scientific advances, concerns arise about the ethics and safety of genetic manipulations. The next section deals with these issues.

TELL ME WHY

Why don't doctors routinely insert genes into their patients to cure the common cold, flu, or tuberculosis?

The Ethics and Safety of Recombinant DNA Technology

LEARNING **OUTCOME**

8.22 Discuss the pros and cons concerning the safety and ethics of recombinant DNA technology.

Recombinant DNA technology provides the opportunity to transfer genes among unrelated organisms, even among organisms in different kingdoms, but how safe and ethical is it? No change in agricultural practices has generated as much controversy. "Frankenfood" and "biological Russian roulette" are some of the terms opponents use to denigrate transgenic agricultural products and gene therapy. Critics of transgenic crops correctly state that the long-term effects of transgenic manipulations are unknown and that unforeseen problems arise from every new technology and procedure. Recombinant DNA technology may burden society with complex and unforeseen regulatory, administrative, financial, legal, social, medical, and environmental problems.

Critics also argue that natural genetic transfer through sexual reproduction and processes such as transformation and transduction could deliver genes from transgenic plants and animals into other organisms. For example, if a herbicide-resistant plant cross-pollinates with a related weed species, we might be cursed with a weed that is more difficult to kill. Opponents further express concern that transgenic organisms could trigger allergies or cause harmless organisms to become pathogenic. Some opponents of recombinant DNA technology desire a ban on all genetically modified products.

The U.S. National Academy of Sciences, the U.S. National Research Council, and 81 research projects conducted between 1985 and 2012 by the European Union have not revealed any risks to human health or the environment from genetically modified agricultural products beyond those found with conventional plant breeding. In fact, the European Union concluded in 2001 that "the use of more precise technology and the greater regulatory scrutiny probably make them [genetically modified foods] even safer than conventional plants and foods." Further, studies have shown that Bt crops spare harmless and beneficial insects that would be killed if pesticides were used. With fewer harmful insects to transmit fungi, Bt corn crops contain less cancer-causing fungal toxin, a potential benefit to human health.

As the debate continues, governments continue to impose standards on laboratories involved in recombinant DNA technology. These are intended to prevent the accidental release of altered organisms or exposure of laboratory workers to potential dangers. Additionally, genetic researchers often design organisms to lack a vital gene so that they should not survive for long outside a laboratory.

Unfortunately, biologists could apply the procedures used to create beneficial crops and animals to create biological weapons that are more infective and more resistant to treatment than their natural counterparts are. Though international treaties prohibit the development of biological weapons, *Bacillus anthracis* (an-thrā´sis) spores were used in bioterrorist attacks in the United States in 2001. Thankfully, the strain utilized was not genetically altered to realize its deadliest potential.

Emergent recombinant DNA technologies raise numerous other ethical issues. Should people be routinely screened for diseases that are untreatable or fatal? Who should pay for these procedures: individuals, employers, prospective employers, insurance companies, health maintenance organizations (HMOs), or government agencies? What rights do individuals have to genetic privacy? If entities other than individuals pay the costs involved in genetic screening, should those entities have access to *all* the genetic information that results? Should businesses be allowed to have patents on and make profits from living organisms they have genetically altered? Should governments be allowed to require genetic screening and then force genetic manipulations on individuals to correct perceived genetic abnormalities that some claim are the bases of criminality, manic depression, risk-taking behavior, and alcoholism? Should HMOs, physicians, or the government demand genetic screening and then refuse to provide services related to the birth or care of supposedly "defective" children?

We as a society will have to confront these and other ethical considerations as the genomic revolution continues to affect people's lives in many unpredictable ways.

TELL ME WHY

Why don't scientists who work with recombinant DNA know all the long-term effects of their work?

MICRO IN THE CLINIC FOLLOW-UP

Can a "Bad" Gene Be Replaced with a "Good" Gene?

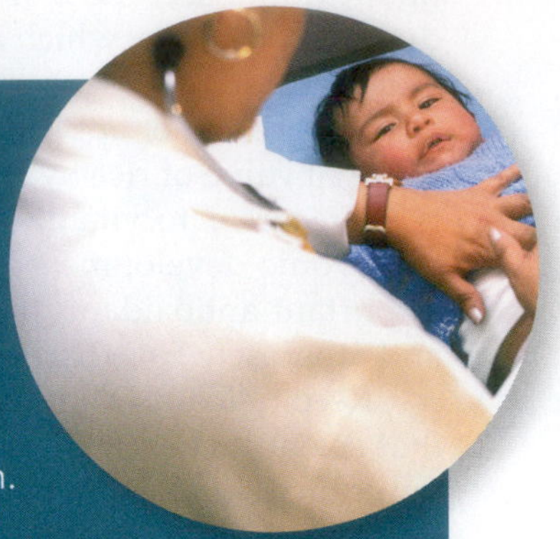

The idea behind Keesha's experimental gene therapy is to deliver a working adenosine deaminase gene to isolated bone marrow cells via a virus vector. In this case, the vector is a crippled, non-disease-causing retrovirus, a virus that integrates its genes into host chromosomes. In the laboratory, the vector is prepared by splicing the working gene into the altered virus. In the laboratory, the vector genome inserts the "good" gene into chromosomes of bone marrow cells, where it then permanently resides. These cells are injected back into Keesha's body, where they repopulate her bone marrow. Every one of their daughter cells has the therapeutic gene in its chromosomes.

One year later, Keesha is feeling much better. She no longer has frequent diarrhea and has not experienced any serious infections. Her lymphocytes can now synthesize adenosine deaminase to detoxify the poisonous by-products of DNA degradation. Keesha can go to the movies, attend school, and play with other children without risk of becoming life-threateningly ill from common microbes.

1. **What is an advantage for choosing a disabled form of retrovirus as a vector for gene therapy?**
2. **What is a risk of using a retrovirus as a vector for gene therapy?**

MM Check your answers to Micro in the Clinic Follow-Up questions in the MasteringMicrobiology Study Area.

Explore the Invisible: Action of Restriction Enzymes

Practice on-the-go with Dr. Bauman Video Tutors by scanning this QR code with your smart phone. Visit the **MasteringMicrobiology Study Area** to challenge your understanding with practice tests, animation quizzes, and clinical case studies!

MasteringMicrobiology®

CHAPTER SUMMARY

The Role of Recombinant DNA Technology in Biotechnology (p. 241)

1. **Biotechnology** is the use of microorganisms to make useful products. Historically these include bread, wine, beer, and cheese.
2. **Recombinant DNA technology** is a new type of biotechnology in which scientists change the genotypes and phenotypes of organisms to benefit humans.

The Tools of Recombinant DNA Technology (pp. 241–246)

1. The tools of recombinant DNA technology include mutagens, reverse transcriptase, synthetic nucleic acids, restriction enzymes, vectors, and gene libraries.
2. **Mutagens** are chemical and physical agents used to create changes in a microbe's genome to effect desired changes in the microbe's phenotype.

3. The enzyme **reverse transcriptase** transcribes DNA from an RNA template; genetic researchers use reverse transcriptase to make **complementary DNA (cDNA)**.
4. Scientists used synthetic nucleic acids to elucidate the genetic code, and they now use them to create genes for specific proteins and to synthesize DNA and RNA **probes** labeled with radioactive or fluorescent markers.
5. **Restriction enzymes** cut DNA at specific (usually palindromic) nucleotide sequences and are used to produce recombinant DNA molecules.

   ▶**VIDEO TUTOR:** *Action of Restriction Enzymes*

   ▶**ANIMATIONS:** *Recombinant DNA Technology*
6. In recombinant DNA technology, a **vector** is a small DNA molecule (such as a viral genome, transposon, or plasmid) that carries a particular gene and a recognizable genetic marker into a cell.
7. A **gene library** is a collection of bacterial or phage clones, each of which carries a fragment (typically a single gene) of an organism's genome.

Techniques of Recombinant DNA Technology

(pp. 246–251)

1. The **polymerase chain reaction (PCR)** allows researchers to replicate molecules of DNA rapidly.

   ▶**ANIMATIONS:** *Polymerase Chain Reaction (PCR): Overview, Components, The Process*
2. **Gel electrophoresis** is a technique for separating molecules (including fragments of nucleic acids) by size, shape, and electrical charge.
3. The **Southern blot** technique allows researchers to stabilize DNA sequences from an electrophoresis gel and then localize them using DNA dyes or probes.
4. **DNA microarrays,** containing nucleotide sequences of thousands of genes, are used to monitor gene activity and the presence of microbes in patients and the environment.
5. Geneticists artificially insert DNA into cells by electroporation, protoplast fusion, or injection.
6. Fluorescent *in situ* hybridization (FISH) uses fluorescent nucleic acid probes to localize specific genetic sequences.

Applications of Recombinant DNA Technology

(pp. 251–257)

1. **Genomics** is the sequencing **(genetic mapping),** analysis, and comparison of genomes. Genetic sequencing has been sped up by an automated machine that distinguishes among fluorescent dyes attached to each type of nucleotide base.
2. Unique DNA sequences reveal the presence of microbes that have never been cultured in a laboratory.
3. Scientists synthesize **subunit vaccines** by introducing genes for a pathogen's polypeptides into cells or viruses. When the cells, the viruses, or the polypeptides they produce are injected into a human, the body's immune system is exposed to and reacts against relatively harmless antigens instead of the potentially harmful pathogen.
4. **Genetic screening** can detect infections and inherited diseases before a patient shows any sign of disease.
5. **Genetic fingerprinting** (DNA fingerprinting), which identifies unique sequences of DNA, is used in paternity investigations, crime scene forensics, diagnostic microbiology, and epidemiology.
6. **Gene therapy** cures various diseases by replacing defective genes with normal genes.
7. In **xenotransplants** involving recombinant DNA technology, human genes would be inserted into animals to produce cells, tissues, or organs for introduction into the human body.
8. **Transgenic organisms** are plants and animals that have been genetically altered by the inclusion of genes from other organisms.
9. Agricultural uses of recombinant DNA technology include advances in herbicide tolerance; salt tolerance; freeze resistance; and pest resistance, such as insertion of the gene for Bt toxin (Bt) into plants; as well as improvements in nutritional value, yield, and shelf life.

The Ethics and Safety of Recombinant DNA Technology (pp. 257–258)

1. Among the ethical and safety issues surrounding recombinant DNA technology are concerns over the accidental release of altered organisms into the environment and the potential for creating genetically modified biological weapons.

QUESTIONS FOR REVIEW

Answers to the Questions for Review (except Short Answer questions) begin on p. A-1.

Multiple Choice

1. Which of the following statements is true concerning recombinant DNA technology?
   a. It will replace biotechnology in the future.
   b. It is a single technique for genetic manipulation.
   c. It is useful in manipulating genotypes but not phenotypes.
   d. It involves modification of an organism's genome.
2. A DNA gene synthesized from an RNA template is ________________.
   a. reverse transcriptase
   b. complementary DNA
   c. recombinant DNA
   d. probe DNA

3. After scientists exposed cultures of *Penicillium* to agents X, Y, and Z, they examined the type and amount of penicillin produced by the altered fungi to find the one that is most effective. Agents X, Y, and Z were probably ______________.
   a. recombinant cells
   b. competent
   c. mutagens
   d. phages

4. Which of the following is *false* concerning vectors in recombinant DNA technology?
   a. Vectors are small enough to manipulate outside a cell.
   b. Vectors contain a recognizable genetic marker.
   c. Vectors survive inside cells.
   d. Vectors must contain genes for self-replication.

5. Which recombinant DNA technique is used to replicate copies of a DNA molecule?
   a. PCR
   b. gel electrophoresis
   c. electroporation
   d. reverse transcription

6. Which of the following would be most useful in following gene expression in a yeast cell?
   a. Southern blot
   b. reverse transcription
   c. DNA microarray
   d. restriction enzymes

7. Which of the following techniques is used regularly in the study of genomics?
   a. Clones are selected using a vector with two genetic markers.
   b. Genes are inserted to produce an antigenic protein from a pathogen.
   c. Fluorescent nucleotide bases are sequenced.
   d. Defective organs are replaced with those made in animal hosts.

8. Restriction enzyme *Hha*I ______________.
   a. recombines DNA
   b. cuts DNA at a specific nucleotide sequence
   c. is likely derived from *Haemophilus influenzae*
   d. all of the above

9. Which application of recombinant DNA technology involves the production of a distinct pattern of DNA fragments on a gel?
   a. genetic fingerprinting
   b. gene therapy
   c. genetic screening
   d. protein synthesis

10. A DNA microarray consists of ______________.
   a. a series of clones containing the entire genome of a microbe
   b. recombinant microbial cells
   c. restriction enzyme fragments of DNA molecules
   d. single-stranded DNA localized on a substrate

Modified True/False

Indicate which of the following are true and which are false. Rewrite any false statements to make them true by changing the underlined words.

1. ____ Restriction enzymes <u>inhibit the movement of</u> DNA.
2. ____ Restriction enzymes act at <u>specific</u> nucleotide sequences within a double-stranded DNA molecule.
3. ____ <u>A thermocycler</u> separates molecules based on their size, shape, and electrical charge.
4. ____ Protoplast fusion is often used in the genetic modification of <u>plants</u>.
5. ____ Gel electrophoresis is used in <u>DNA microarrays</u>.

VISUALIZE IT!

1. Label the reagents and steps of PCR on the figure below. Indicate the temperature of the chemicals at each numbered step.

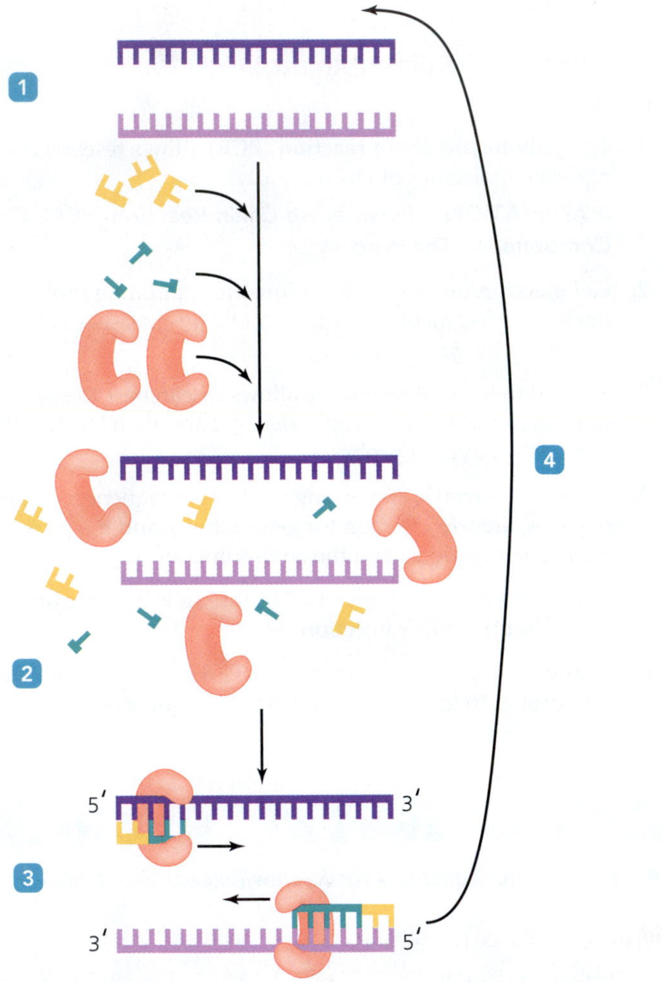

2. Using the "DNA fingerprint" result shown here, which patients can be diagnosed as being infected?

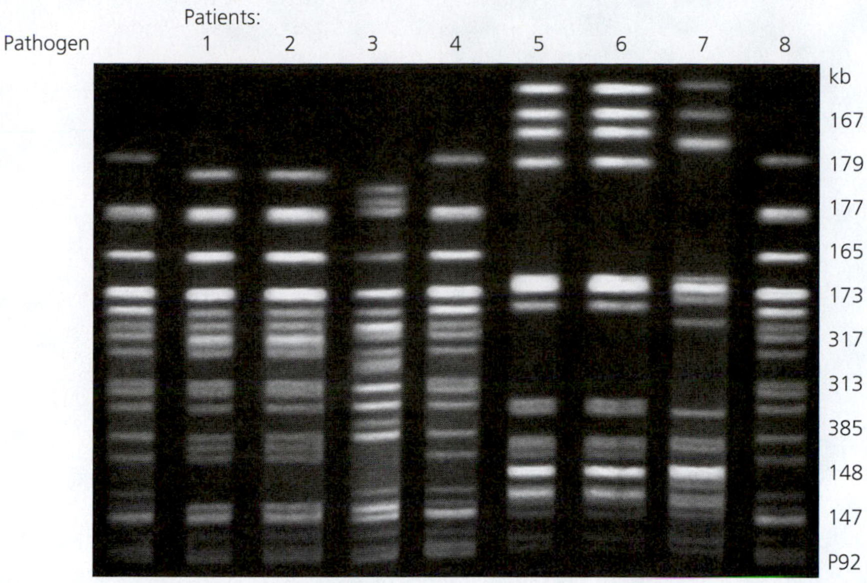

Short Answer

1. Describe three artificial methods of introducing DNA into cells.
2. Why is cloning a practical technique for medical researchers?
3. Describe three ways scientists use synthetic nucleic acids.
4. Describe a gene library and its usefulness.
5. List three potential problems of recombinant DNA technology.

CRITICAL THINKING

1. Examine the restriction sites listed in Table 8.1 on p. 243. Which restriction enzymes produce restriction fragments with sticky ends? Which produce fragments with blunt ends?
2. A cancer-inducing virus, HTLV-1, inserts itself into a human chromosome, where it remains. How can a laboratory technician prove that a patient is infected with HTLV-1 even when there is no sign of cancer?
3. A thermocycler uses DNA polymerase from hyperthermophilic prokaryotes, but it cannot use DNA polymerase derived from *E. coli.* Why not?
4. How is the result of a Southern blot similar to the result of surveillance using a DNA microarray?
5. *Hha*I recognizes and cuts this DNA sequence at the sites indicated:

   ↓
   G-C-G-C
   C-G-C-G
   ↑

   Describe the fragments resulting from the use of this enzyme.
6. PCR replication of DNA is similar to bacterial population growth. If a scientist starts PCR with 15 DNA helices and runs the reaction for 15 cycles, how many DNA molecules will be present at the end? Show your calculations.
7. If a gene contains the sequence TACAATCGCATTGAA, what antisense RNA could be used to stop translation directly?
8. Suppose researchers learn that a particular congenital disease is caused by synthesis of a protein coded by a mutated gene. Describe a way in which recombinant DNA technology might be used to prevent translation of the protein.
9. Even though some students correctly synthesize a fluorescent cDNA probe complementary to mRNA for a particular yeast protein, they find that the probe does not attach to any portion of the yeast's genome. Explain why the students' probe does not work.

CONCEPT MAPPING

Using the following terms, draw a concept map that describes recombinant DNA technology. For a sample concept map, see p. 94. Or, complete this and other concept maps online by going to the MasteringMicrobiology Study Area.

Blunt ends
DNA probes
Electroporation
Fluorescent
Injection
Mutagens
Plasmids
Protoplast fusion
Radioactive
Restriction enzymes
Reverse transcriptase
Southern blot
Transposons

9 Controlling Microbial Growth in the Environment

MICRO IN THE CLINIC

Raw Fish, Raw Deal?

To celebrate her new job as a dental hygienist, Brianne invites her roommate, Sasha, to spend an evening on the town. James, a mutual friend from college, joins the women at a new, low-budget sushi restaurant. They sit at a bar where small plates of colorful sushi rotate on a conveyor belt in front of them. Feeling adventurous, Brianne first tries sashimi—thinly sliced raw fish—which she dips into soy sauce. Then the group selects a variety of sushi rolls, including raw tuna and raw salmon wrapped in rice and seaweed. Sasha, however, sticks to the cooked crab rolls. After a couple of hours of reminiscing about their undergraduate days, the small party breaks up.

At about 4 a.m., Brianne suddenly wakes up with intense stomach pain and nausea. She dashes into the bathroom to vomit. As she washes her face, Brianne looks in the mirror and, to her horror, sees that hives are breaking out all over her body. Her skin starts to itch. Sasha calls a cab and rushes Brianne to a nearby hospital emergency room.

Five days later, James, too, is admitted to the hospital after he doubles over with severe abdominal pain. James yelps when the doctor gently presses on his lower right abdomen. When tests reveal that James's white blood cell count is elevated, the doctor schedules an emergency appendectomy. During the operation, the surgeon examines James's appendix and is surprised to find that it appears normal. However, a short section of James's small intestine is unusually thickened and oozes a bloody fluid. The surgeon decides to consult with a pathologist.

Could bad sushi be responsible for both Brianne's and James's illnesses? Turn to the end of the chapter (p. 283) to find out.

MM **Explore More:** Test your readiness and apply your knowledge with dynamic learning tools at MasteringMicrobiology.

The control of microbes in health care facilities, in laboratories, and at home is a significant and practical aspect of microbiology. In this chapter we study the terminology and principles of microbial control, the factors affecting the efficacy of microbial control, and various chemical and physical means to control microorganisms and viruses. (One important aspect of microbial control—the use of antimicrobial drugs to assist the body's defenses against pathogens—will be considered in Chapter 10.)

Basic Principles of Microbial Control

Scientists, health care professionals, researchers, and government workers should use precise terminology when referring to microbial control in the environment. In the following sections we consider the terminology of microbial control, the concept of microbial death rates, and the action of antimicrobial agents.

Terminology of Microbial Control

LEARNING OUTCOMES

9.1 Contrast sterilization, disinfection, and antisepsis and describe their practical uses.

9.2 Contrast the terms *degerming*, *sanitization*, and *pasteurization*.

9.3 Compare the effects of *-static* versus *-cidal* control agents on microbial growth.

It is important for microbiologists, health care workers, and others to use correct terminology for describing microbial control. Although many of these terms are familiar to the general public, they are often misused.

In its strictest sense, **sterilization** refers to the removal or destruction of *all* microbes, including viruses and bacterial endospores, in or on an object. (The term does not apply to *prions*, which are infectious proteins, because standard sterilizing techniques do not destroy them.)

In practical terms, sterilization indicates only the eradication of harmful microorganisms and viruses; some innocuous microbes may still be present and viable in an environment that is considered sterile. For instance, *commercial sterilization* of canned food does not kill all hyperthermophilic microbes; however, because they do not cause disease and cannot grow and spoil food at ambient temperatures, they are of no practical concern. Likewise, some hyperthermophiles may survive sterilization by laboratory methods (discussed shortly), but they are of no practical concern to technicians because they cannot grow or reproduce under normal laboratory conditions.

The term **aseptic**[1] (ā-sep´tik) describes an environment or procedure that is free of contamination by *pathogens*. For example, vegetables and fruit juices are available in aseptic packaging, and surgeons and laboratory technicians use aseptic techniques to avoid contaminating a surgical field or laboratory equipment.

Disinfection[2] refers to the use of physical or chemical agents known as **disinfectants**, including ultraviolet light, heat, alcohol, and bleach, to inhibit or destroy microorganisms, especially pathogens. Unlike sterilization, disinfection does not guarantee that all pathogens are eliminated; indeed, disinfectants alone cannot inhibit endospores or some viruses. Further, the term *disinfection* is used only in reference to treatment of inanimate objects. When a chemical is used on skin or other tissue, the process is called **antisepsis**[3] (an-tē-sep´sis), and the chemical is called an **antiseptic**. Antiseptics and disinfectants often have the same components, but disinfectants are more concentrated or can be left on a surface for longer periods of time. Of course, some disinfectants, such as steam or concentrated bleach, are not suitable for use as antiseptics.

Degerming is the removal of microbes from a surface by scrubbing, such as when you wash your hands or a nurse prepares an area of skin for an injection. Though chemicals such as soap or alcohol are commonly used during degerming, the action of thoroughly scrubbing the surface may be more important than the chemical in removing microbes.

Sanitization[4] is the process of disinfecting places and utensils used by the public to reduce the number of pathogenic microbes to meet accepted public health standards. For example, steam, high-pressure hot water, and scrubbing are used to sanitize restaurant utensils and dishes, and chemicals are used to sanitize public toilets. Thus, the difference between *disinfecting* dishes at home and *sanitizing* dishes in a restaurant is the arena—private versus public—in which the activity takes place.

Pasteurization[5] is the use of heat to kill pathogens and reduce the number of spoilage microorganisms in food and beverages. Milk, fruit juices, wine, and beer are commonly pasteurized.

So far, we have seen that there are two major types of microbial control—sterilization, which is the elimination of all microbes, and antisepsis or disinfection, which each denote the destruction of vegetative (nonspore) cells and many viruses. Modifications of disinfection include degerming, sanitization, and pasteurization. Some scientists and clinicians apply these terms only to pathogenic microorganisms.

Additionally, scientists and health care professionals use the suffixes *-stasis/-static*[6] to indicate that a chemical or physical agent inhibits microbial metabolism and growth but does not necessarily kill microbes. Thus, refrigeration is bacteriostatic for most bacterial species; it inhibits their growth, but

[1]From Greek *a*, meaning "not," and *sepsis*, meaning "decay."

[2]From Latin *dis*, meaning "reversal," and *inficere*, meaning "to corrupt."

[3]From Greek *anti*, meaning "against," and *sepsis*, meaning "putrefaction."

[4]From Latin *sanitas*, meaning "healthy."

[5]Named for Louis Pasteur, inventor of the process.

[6]Greek, meaning "to stand"—that is, to remain relatively unchanged.

they can resume metabolism when the optimal temperature is restored. By contrast, words ending in *-cide*/*-cidal*[7] refer to agents that destroy or permanently inactivate a particular type of microbe; *virucides* inactivate viruses, *bactericides* kill bacteria, and *fungicides* kill fungal hyphae, spores, and yeasts. *Germicides* are chemical agents that destroy pathogenic microorganisms in general.

TABLE 9.1 summarizes the terminology used to describe the control of microbial growth.

Microbial Death Rates

LEARNING **OUTCOME**

9.4 Define *microbial death rate*, and describe its significance in microbial control.

Scientists define **microbial death** as the permanent loss of reproductive ability under ideal environmental conditions. One technique for evaluating the efficacy of an antimicrobial agent is to calculate the **microbial death rate**, which is usually found to be constant over time for any particular microorganism under a particular set of conditions **(FIGURE 9.1)**. Suppose, for example, that a scientist treats a broth containing 1 billion (10^9) microbes with an agent that kills 90% of them in 1 minute. The most susceptible cells die first, leaving 100 million (10^8) hardier cells after the first minute. After another minute of treatment, another 90% die, leaving 10 million (10^7) cells that have even greater resistance to and require longer exposure to the agent before they die. Notice that in this case, each full minute decreases the number of living cells 10-fold. The broth will be sterile when all the cells are dead. When these results are plotted on a semilogarithmic

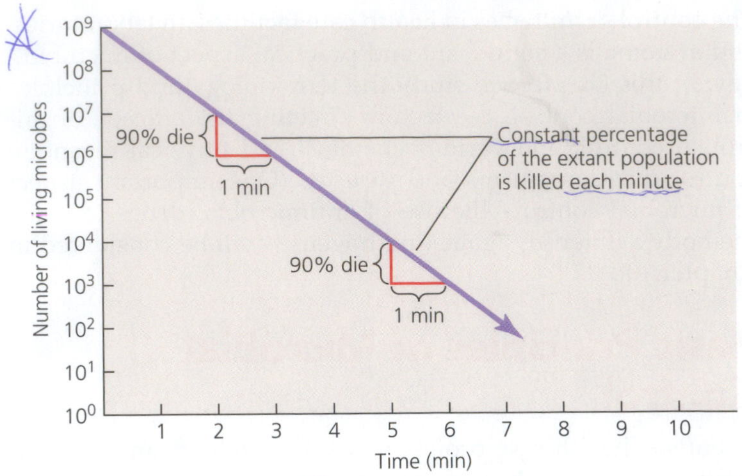

▲ **FIGURE 9.1 A plot of microbial death rate.** Microbicidal agents do not simultaneously kill all cells. Rather, they kill a constant percentage of cells over time—in this case, 90% per minute. On this semilogarithmic graph, a constant death rate is indicated by a straight line. *How many minutes are required for sterilization in this case?*

Figure 9.1 *For this microbe under these conditions, this microbicidal agent requires 9 minutes to achieve sterilization.*

[7]From Latin *cidium*, meaning "a slaying."

TABLE 9.1 Terminology of Microbial Control

Term	Definition	Examples	Comments
Antisepsis	Reduction in the number of microorganisms and viruses, particularly potential pathogens, on living tissue	Iodine; alcohol	Antiseptics are frequently disinfectants whose strength has been reduced to make them safe for living tissues.
Aseptic	Refers to an environment or procedure free of pathogenic contaminants	Preparation of surgical field; hand washing; flame sterilization of laboratory equipment	Scientists, laboratory technicians, and health care workers routinely follow standardized aseptic techniques.
-cide **-cidal**	Suffixes indicating destruction of a type of microbe	Bactericide; fungicide; germicide; virucide	Germicides include ethylene oxide, propylene oxide, and aldehydes.
Degerming	Removal of microbes by mechanical means	Hand washing; alcohol swabbing at site of injection	Chemicals play a secondary role to the mechanical removal of microbes.
Disinfection	Destruction of most microorganisms and viruses on nonliving tissue	Phenolics; alcohols; aldehydes; soaps	The term is used primarily in relation to pathogens.
Pasteurization	Use of heat to destroy pathogens and reduce the number of spoilage microorganisms in foods and beverages	Pasteurized milk and fruit juices	Heat treatment is brief to minimize alteration of taste and nutrients; microbes still remain and eventually cause spoilage.
Sanitization	Removal of pathogens from objects to meet public health standards	Washing tableware in scalding water in restaurants	Standards of sanitization vary among governmental jurisdictions.
-stasis **-static**	Suffixes indicating inhibition but not complete destruction of a type of microbe	Bacteriostatic; fungistatic; virustatic	Germistatic agents include some chemicals, refrigeration, and freezing.
Sterilization	Destruction of all microorganisms and viruses in or on an object	Preparation of microbiological culture media and canned food	Typically achieved by steam under pressure, incineration, or ethylene oxide gas.

graph—in which the *y*-axis is logarithmic and the *x*-axis is arithmetic—the plot of microbial death rate is a straight line; that is, the microbial death rate is constant.

Action of Antimicrobial Agents

LEARNING | **OUTCOME**

9.5 Describe how antimicrobial agents act against cell walls, cytoplasmic membranes, proteins, and nucleic acids.

There are many types of chemical and physical microbial controls, but their modes of action fall into two basic categories: those that disrupt the integrity of cells by adversely altering their cell walls or cytoplasmic membranes and those that interrupt cellular metabolism and reproduction by interfering with the structures of proteins and nucleic acids.

Alteration of Cell Walls and Membranes

A cell wall maintains cellular integrity by counteracting the effects of osmosis when the cell is in a hypotonic solution. If the wall is disrupted by physical or chemical agents, it no longer prevents the cell from bursting as water moves into the cell by osmosis (see Figure 3.20).

Beneath a cell wall, the cytoplasmic membrane essentially acts as a bag that contains the cytoplasm and controls the passage of chemicals into and out of the cell. Extensive damage to a membrane's proteins or phospholipids by any physical or chemical agent allows the cellular contents to leak out—which, if not immediately repaired, causes death.

In enveloped viruses, the envelope is a membrane composed of proteins and phospholipids that is responsible for the attachment of the virus to its target cell. Damage to the envelope by physical or chemical agents fatally interrupts viral replication. The lack of an envelope in nonenveloped viruses accounts for their greater tolerance of harsh environmental conditions, including antimicrobial agents.

Damage to Proteins and Nucleic Acids

Proteins regulate cellular metabolism, function as enzymes in most metabolic reactions, and form structural components in membranes and cytoplasm. As we have seen, a protein's function depends on an exact three-dimensional shape, which is maintained by hydrogen and disulfide bonds between amino acids. When these bonds are broken by extreme heat or certain chemicals, the protein's shape changes (see Figure 5.8). Such *denatured* proteins cease to function, bringing about cellular death.

Chemicals, radiation, and heat can also alter and even destroy nucleic acids. Given that the genes of a cell or virus are composed of nucleic acids, disruption of these molecules can produce fatal mutations. Additionally, that portion of a ribosome that actually catalyzes the synthesis of proteins is a *ribozyme*—that is, an enzymatic RNA molecule. For this reason, physical or chemical agents that interfere with nucleic acids also stop protein synthesis.

Scientists and health care workers have at their disposal many chemical and physical agents to control microbial growth and activity. In the next section we consider the factors and conditions that should be considered in choosing a particular control method as well as some ways to evaluate a method's effectiveness.

TELL ME WHY

Why does milk eventually go "bad" despite being pasteurized?

The Selection of Microbial Control Methods

Ideally, agents used for the control of microbes should be inexpensive, fast acting, and stable during storage. Further, a perfect agent would control the growth and reproduction of every type of microbe while being harmless to humans, animals, and objects. Unfortunately, such ideal products and procedures do not exist—every agent has limitations and disadvantages. In the next section we consider the factors that affect the efficacy of antimicrobial methods.

Factors Affecting the Efficacy of Antimicrobial Methods

LEARNING | **OUTCOMES**

9.6 List factors to consider in selecting a microbial control method.

9.7 Identify the three most resistant groups of microbes and explain why they are resistant to many antimicrobial agents.

9.8 Discuss environmental conditions that can influence the effectiveness of antimicrobial agents.

In each situation, microbiologists, laboratory personnel, and medical staff must consider at least three factors: the nature of the sites to be treated, the degree of susceptibility of the microbes involved, and the environmental conditions that pertain.

Site to Be Treated

In many cases, the choice of an antimicrobial method depends on the nature of the site to be treated. For example, harsh chemicals and extreme heat cannot be used on humans, animals, and fragile objects, such as artificial heart valves and plastic utensils. Moreover, when performing medical procedures, medical personnel must choose a method and level of microbial control based on the site of the procedure because the site greatly affects the potential for subsequent infection. For example, the use of medical instruments that penetrate the outer defenses of the body, such as needles and scalpels, carries a greater potential for infection, so they must be sterilized; however, disinfection may be adequate for items that contact only the surface of a mucous membrane or the skin. In the latter case, sterilization is required only if the patient is immunocompromised.

Relative Susceptibility of Microorganisms

Though microbial death rate is usually constant for a particular agent acting against a single microbe, death rates do vary—sometimes dramatically—among microorganisms and viruses. Microbes fall along a continuum from most susceptible to most resistant to antimicrobial agents. For example, *enveloped* viruses, such as HIV, are more susceptible to antimicrobial agents and heat than are *nonenveloped viruses,* such as poliovirus, because viral envelopes are more easily disrupted than the protein coats of nonenveloped viruses. The relative susceptibility of microbes to antimicrobial agents is illustrated in **FIGURE 9.2**.

Often, scientists and medical personnel select a method to kill the hardiest microorganisms present, assuming that such a treatment will kill more fragile microbes as well. The most resistant microbes include the following:

- *Bacterial endospores.* The endospores of *Bacillus* (ba-sil´ŭs) and *Clostridium* (klos-trid´ē-ŭm) are the most resilent forms of life. They can survive environmental extremes of temperature, acidity, and dryness and can withstand many chemical disinfectants. For example, endospores have survived more than 20 years in 70% alcohol, and scientists have recovered viable endospores that were embalmed with Egyptian mummies thousands of years ago.
- Species of *Mycobacterium.* The cell walls of members of this genus, such as *Mycobacterium tuberculosis* (mī-kō-bak-tēr´ē-ŭm too-ber-kyŭ-lō´sis), contain a large amount of a waxy lipid. The wax allows these bacteria to survive drying and protects them from most water-based chemicals; therefore, medical personnel must use strong disinfectants or heat to treat whatever comes into contact with tuberculosis patients, including utensils, equipment, and patients' rooms.
- *Cysts of protozoa.* A protozoan cyst's wall prevents entry of most disinfectants, protects against drying, and shields against radiation and heat.

Prions, which are infectious proteins that cause degenerative diseases of the brain, are more resistant than any living thing.

The effectiveness of germicides can be classified as high, intermediate, or low, depending on their proficiency in inactivating or destroying microorganisms on medical instruments that cannot be sterilized with heat. *High-level germicides* kill all pathogens, including bacterial endospores. Health care professionals use them to sterilize invasive instruments such as catheters, implants, and parts of heart-lung machines. *Intermediate-level germicides* kill fungal spores, protozoan cysts, viruses, and pathogenic bacteria but not bacterial endospores. They are used to disinfect instruments that come in contact with mucous membranes but are noninvasive, such as respiratory equipment and endoscopes. *Low-level germicides* eliminate vegetative bacteria, fungi, protozoa, and some viruses; they are used to disinfect items that contact only the skin of patients, such as furniture and electrodes.

Environmental Conditions

Temperature and pH affect microbial death rates and the efficacy of antimicrobial methods. Warm disinfectants, for example, generally work better than cool ones because chemicals react faster at higher temperatures **(FIGURE 9.3)**. Acidic conditions enhance the antimicrobial effect of heat. Some chemical disinfectants, such as household chlorine bleach, are more effective at low pH.

Organic materials, such as fat, feces, vomit, blood, and the intercellular matrix of biofilms, interfere with the penetration of heat, chemicals, and some forms of radiation, and in some

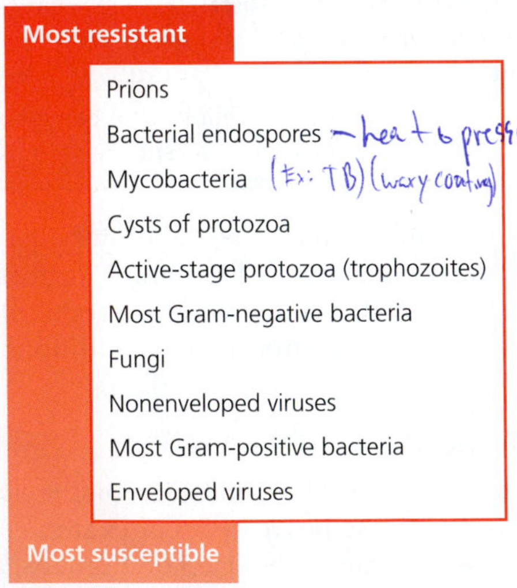

▲ **FIGURE 9.2 Relative susceptibilities of microbes to antimicrobial agents.** *Why are nonenveloped viruses generally more resistant than enveloped viruses?*

Figure 9.2 *A phospholipid envelope is typically more fragile than a protein coat.*

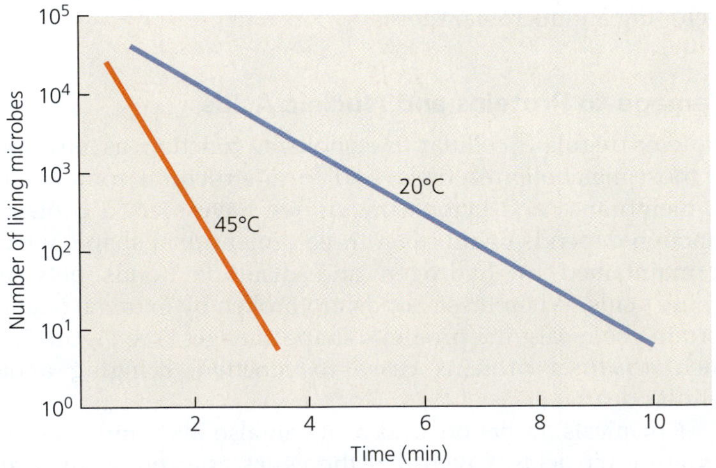

▲ **FIGURE 9.3 Effect of temperature on the efficacy of an antimicrobial chemical.** This semilogarithmic graph shows that the microbial death rate is higher at higher temperatures; to kill the same number of microbes, this disinfectant required only 2 minutes at 45°C but 7 minutes at 20°C.

EMERGING DISEASE CASE STUDY

Acanthamoeba Keratitis

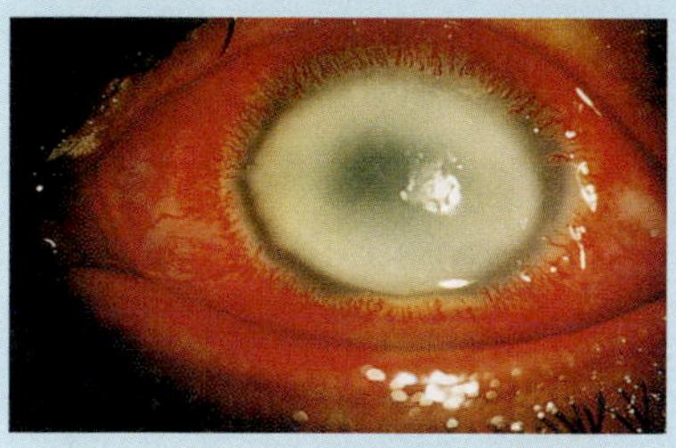

Tim liked the lake; in fact, his girlfriend suggested he was more fish than man. They spent all of their free time swimming, waterskiing, diving, and sunbathing, until Tim met a single-celled amoeba, *Acanthamoeba*.

Tim noticed something was wrong when his right eye began to hurt, turned red, and became so sensitive to light that he could not stand to be outside. Two days later, the pain was excruciating, like nothing Tim had ever experienced. It felt as if someone were pounding pieces of broken glass into the front of his eye and at the same time quickly inserting a thousand tiny needles into the back of the eye. With his eye swollen shut and tears flowing down his face, Tim sought medical aid.

The doctor diagnosed *Acanthamoeba* keratitis, which is inflammation of the covering of the eye (the cornea) caused by the amoeba.

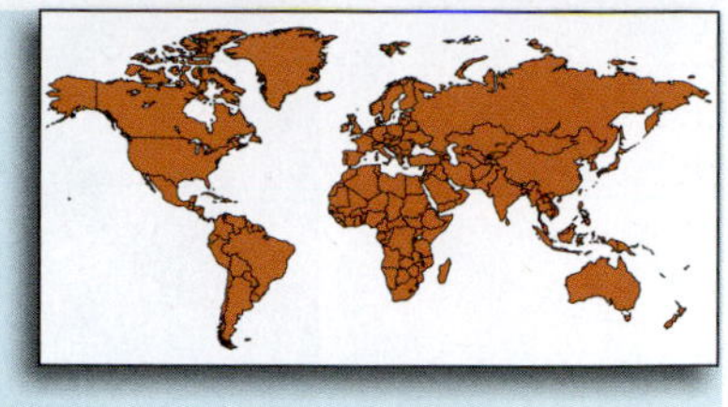

This eukaryotic microbe commonly lives in water, including rivers, hot springs, and lakes. When trapped under a contact lens, the amoeba can penetrate the eye to cause keratitis. Very occasionally, *Acanthamoeba* may also enter the body through the nasal mucous membrane or through a cut in the skin. It has become an emerging menace in our modern society because it can live in hot tubs, pools, shower heads, and sink taps.

The physician prescribed a solution of antiseptic agent that Tim had to drop into his eyes every 30 minutes, day and night, for three weeks. The treatment is painful and time consuming, but at least Tim retained his sight without needing a corneal transplant. He also learned to remove his contact lenses at the lake!

1. **Why are there many more cases of amebic keratitis than there were 100 years ago?**

cases these materials inactivate chemical disinfectants. For this reason, it is important to clean objects before sterilization or disinfection so that antimicrobial agents can thoroughly contact all the object's surfaces.

Biosafety Levels

LEARNING **OUTCOME**

9.9 Describe four levels of biosafety, and give examples of microbes handled at each level.

The Centers for Disease Control and Prevention (CDC) has established guidelines for four levels of safety in microbiological laboratories dealing with pathogens. Each level raises personnel and environmental safety by specifying increasingly strict laboratory techniques, use of safety equipment, and design of facilities.

Biosafety Level 1 (BSL-1) is suitable for handling microbes, such as *Escherichia coli* (esh-ĕ-rik´ē-ă kō´lī), not known to cause disease in healthy humans. Precautions in BSL-1 are minimal and include hand washing with antibacterial soap and washing surfaces with disinfectants.

BSL-2 facilities are similar to those of BSL-1 but are designed for handling moderately hazardous agents, such as hepatitis and influenza viruses and methicillin-resistant *Staphylococcus aureus* (staf´i-lō-kok´ŭs o´rē-ŭs) (MRSA). Access to BSL-2 labs is limited when work is being conducted, extreme precautions are taken with contaminated sharp objects, and procedures that might produce aerosols are conducted within safety cabinets (see Figure 9.11).

BSL-3 is stricter, requiring that all manipulations be done within safety cabinets containing high-efficiency particulate air (HEPA) filters. BSL-3 also specifies special design features for the laboratory. These include entry through double sets of doors and ventilation such that air moves into the room only through an open door. Air leaving the room is HEPA-filtered before being discharged. BSL-3 is designed for experimentation on microbes such as bacteria of tuberculosis and anthrax and viruses of yellow fever and Rocky Mountain spotted fever.

The most secure laboratories are *BSL-4* facilities, designated for working with dangerous or exotic microbes that cause severe or fatal diseases in humans, such as Ebola, smallpox, and Lassa fever viruses. BSL-4 labs are either separate buildings or completely isolated from all other areas of their buildings. Entry and exit are strictly controlled through electronically sealed airlocks with multiple showers, a vacuum room, an ultraviolet light room, and other safety precautions designed to destroy all traces of the biohazard. All air and water entering and leaving the facility are filtered to prevent accidental release. Personnel wear "space suits" supplied with air hoses (**FIGURE 9.4**). Suits and the laboratory itself are pressurized such that microbes are swept away from workers.

TELL ME WHY

Why are BSL-4 suits pressurized? Why not just wear tough regular suits?

Now that we have studied the terminology and general principles of microbial control and biosafety levels, we turn our attention to the actual physical and chemical agents available to scientists, medical personnel, and the general public to control microbial growth.

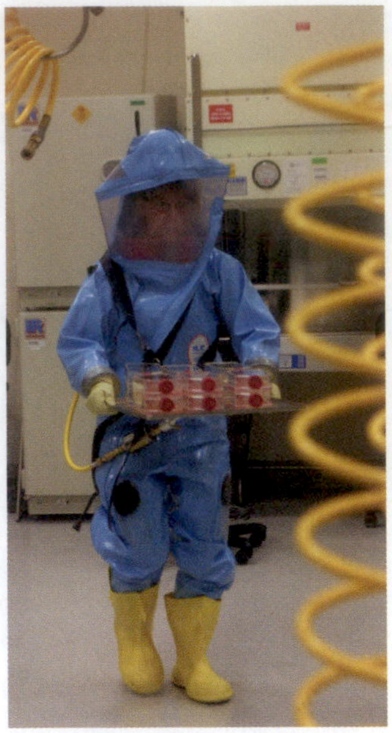

▲ **FIGURE 9.4 A BSL-4 worker carrying Ebola virus cultures.**

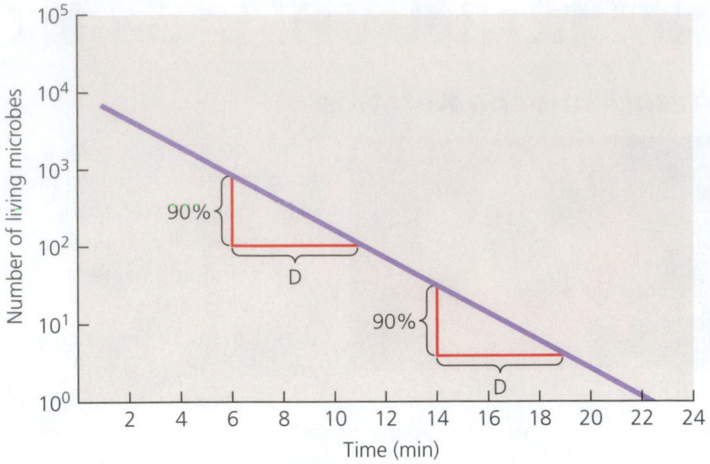

▲ **FIGURE 9.5 Decimal reduction time (D) as a measure of microbial death rate.** D is defined as the time it takes to kill 90% of a microbial population. Note that D is a constant that is independent of the initial density of the population. *What is the decimal reduction time of this heat treatment against this organism? What is the thermal death time?*

Figure 9.5 D = 5 minutes; thermal death time = 22.5 minutes.

Physical Methods of Microbial Control

LEARNING **OUTCOME**

9.10 Describe five types of physical methods of microbial control.

Physical methods of microbial control include exposure of the microbes to extremes of heat and cold, desiccation, filtration, osmotic pressure, and radiation.

Heat-Related Methods

LEARNING **OUTCOMES**

9.11 Discuss the advantages and disadvantages of using moist heat in an autoclave and dry heat in an oven for sterilization.

9.12 Explain the use of *Bacillus stearothermophilus* endospores in sterilization techniques.

9.13 Explain the importance of pasteurization, and describe three different pasteurization methods.

Heat is one of the older and more common means of microbial control. High temperatures denature proteins, interfere with the integrity of cytoplasmic membranes and cell walls, and disrupt the function and structure of nucleic acids. Heat can be used for sterilization, in which case all cells and viruses are deactivated, or for commercial preparation of canned goods. In so-called commercial sterilization, hyperthermophilic prokaryotes remain viable but are harmless because they cannot grow at the normal (room) temperatures in which canned foods are stored.

Though microorganisms vary in their susceptibility to heat, it can be an important agent of microbial control. As a result, scientists have developed concepts and terminology to convey these differences in susceptibility. **Thermal death point** is the lowest temperature that kills all cells in a broth in 10 minutes, and **thermal death time** is the time it takes to completely sterilize a particular volume of liquid at a set temperature.

As we have discussed, cell death occurs logarithmically. When measuring the effectiveness of heat sterilization, researchers calculate the **decimal reduction time (D)**, which is the time required to destroy 90% of the microbes in a sample (**FIGURE 9.5**). This concept is especially useful to food processors because they must heat foods to eliminate the endospores of anaerobic *Clostridium botulinum* (bo-tŭ-lī´num), which could germinate and produce life-threatening botulism toxin inside sealed cans. The standard in food processing is to apply heat such that a population of 10^{12} *C. botulinum* endospores is reduced to 10^0 (i.e., 1) endospore (a 12-fold reduction), which leaves only a very small chance that any particular can of food contains a single endospore. Researchers have calculated that the D value for *C. botulinum* endospores at 121°C is 0.204 minute, so it takes 2.5 minutes (0.204×12) to reduce 10^{12} endospores to 1 endospore.

Moist Heat

Moist heat (which is commonly used to disinfect, sanitize, sterilize, and pasteurize) kills cells by denaturing proteins and destroying cytoplasmic membranes. Moist heat is more effective in microbial control than dry heat because water is a better conductor of heat than air. An example from your kitchen readily demonstrates this: you can safely stick your hand into an oven at 350°F for a few moments, but putting a hand into boiling water at the lower temperature of 212°F would burn you severely.

The first method we consider for controlling microbes using moist heat is boiling.

Boiling Boiling kills the vegetative cells of bacteria and fungi, the trophozoites of protozoa, and most viruses within 10 minutes at sea level. Contrary to popular belief, water at a rapid boil is no hotter than that at a slow boil; boiling water at normal atmospheric pressure cannot exceed boiling temperature (100°C at sea level) because escaping steam carries excess heat away. It is impossible to boil something more quickly simply by applying more heat; the added heat is carried away by the escaping steam. Boiling *time* is the critical factor. Further, it is important to realize that at higher elevations water boils at lower temperatures because atmospheric pressure is lower; thus, a longer boiling time is required in Denver than in Los Angeles to get the same antimicrobial effect.

Bacterial endospores, protozoan cysts, and some viruses (such as hepatitis viruses) can survive boiling at sea level for many minutes or even hours. In fact, because bacterial endospores can withstand boiling for more than 20 hours, boiling is not recommended when true sterilization is required. Boiling is effective for sanitizing restaurant tableware or disinfecting baby bottles.

Autoclaving Practically speaking, true sterilization using heat requires higher temperatures than that of boiling water. To achieve the required temperature, pressure is applied to boiling water to prevent the escape of heat in steam. The reason that applying pressure succeeds in achieving sterilization is that the temperature at which water boils (and steam is formed) increases as pressure increases **(FIGURE 9.6)**. Scientists and medical personnel routinely use a piece of equipment called an *autoclave* to sterilize chemicals and objects that can tolerate moist heat. Alternative techniques (discussed shortly) must be used for items that are damaged by heat or water, such as some plastics and vitamins.

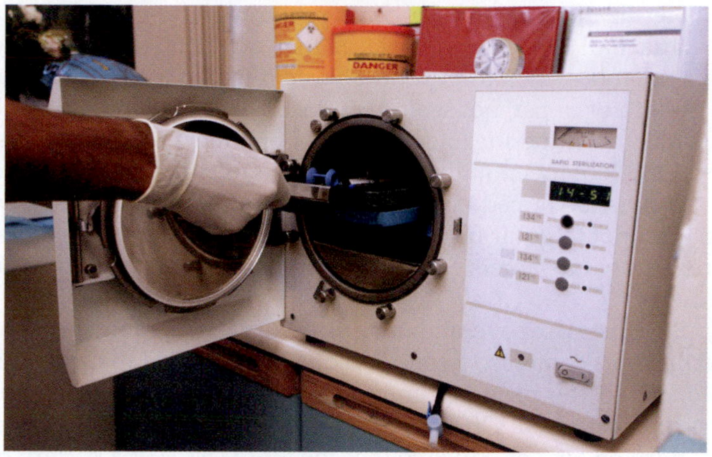

(a)

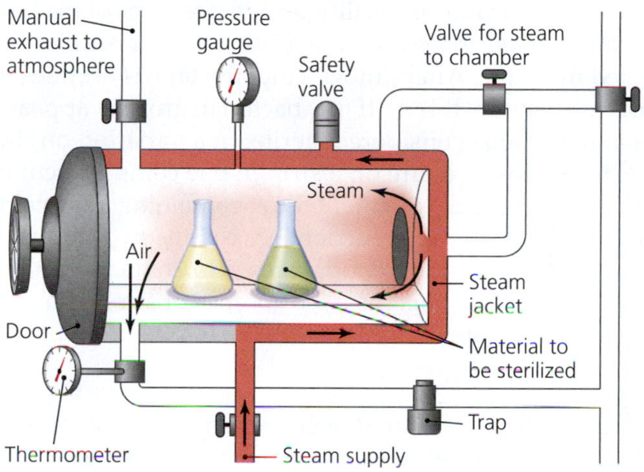

(b)

▲ **FIGURE 9.7 An autoclave. (a)** A photo of a laboratory autoclave. **(b)** A schematic of an autoclave, showing how it functions.

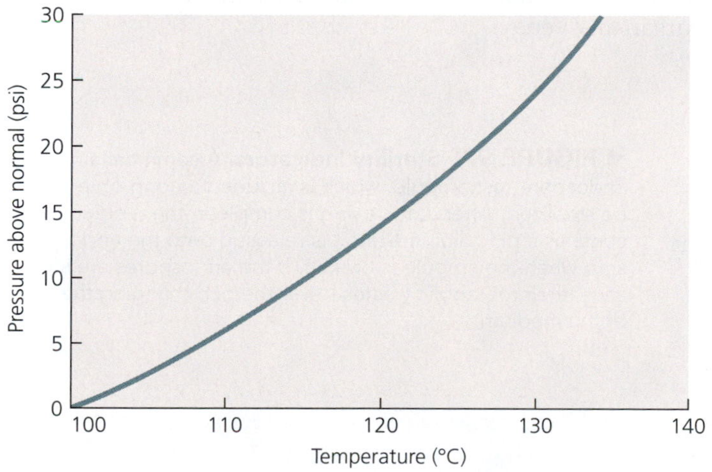

▲ **FIGURE 9.6 The relationship between temperature and pressure.** Note that higher temperatures—and, in consequence, greater antimicrobial action—are associated with higher pressures. *Ultra-high-temperature pasteurization of milk requires a temperature of 134°C; what pressure must be applied to the milk to achieve this temperature?*

Figure 9.6 *29 psi.*

An **autoclave** consists of a pressure chamber, pipes to introduce and evacuate steam, valves to remove air and control pressure, and pressure and temperature gauges to monitor the procedure **(FIGURE 9.7)**. As steam enters an autoclave chamber, it forces air out, raises the temperature of the contents, and increases the pressure until a set temperature and pressure are reached.

Scientists have determined that a temperature of 121°C, which requires the addition of 15 pounds per square inch (psi)[8] of pressure above that of normal air pressure (see Figure 9.6), destroys all microbes in a small volume in about 10 minutes. Typically, an autoclave holds the pressure and temperature for 15 minutes to provide a margin of safety. Sterilizing large volumes of liquids or solids slows the process because they require more time for heat to penetrate. Thus, it requires more time to sterilize 1 liter of fluid in a flask than the same volume of fluid distributed into smaller tubes. Autoclaving solid substances, such as meat, also requires extra time because it takes longer for heat to penetrate to their centers.

[8]The Standard International (SI) equivalent of 1 psi is 6.9×10^3 pascals.

Sterilization in an autoclave requires that steam be able to contact all liquids and surfaces that might be contaminated with microbes; therefore, solid objects must be wrapped in porous cloth or paper, not sealed in plastic or aluminum foil, both of which are impermeable to steam. Containers of liquids must be sealed loosely enough to allow steam to circulate freely, and all air must be forced out by steam. Since steam is lighter than air, it cannot force air from the bottom of an empty vessel; therefore, empty containers must be tipped so that air can flow out of them.

Scientists use several means to ensure that an autoclave has sterilized its contents. A common one is a chemical that changes color when the proper combination of temperature and time have been reached. Often such a color indicator is impressed in a pattern on tape or paper so that the word *sterile* or a pattern or design appears. Another technique uses plastic beads that melt when proper conditions are met.

A biological indicator of sterility uses endospores of the bacterium *Bacillus stearothermophilus* (ba-sil´ŭs ste-rō-ther-ma´fil-ŭs) impregnated into tape. After autoclaving, the tape is aseptically inoculated into sterile broth. If no bacterial growth appears, the original material is considered sterile. In a variation on this technique, the endospores are on a strip in one compartment of a vial that also includes a growth medium containing a pH color indicator. After autoclaving, a barrier between the two compartments is broken, putting the endospores into contact with the medium **(FIGURE 9.8)**. In this case, the absence of a color change after incubation indicates sterility. **▶VIDEO TUTOR:** ***Principles of Autoclaving***

Pasteurization Louis Pasteur developed a method of heating beer and wine just enough to destroy the microorganisms that cause spoilage without ruining the taste. Today, pasteurization is also used to kill pathogens in milk, ice cream, yogurt, and fruit juices. *Brucella melitensis* (broo-sel´lă me-li-ten´sis), *Mycobacterium bovis* (bō´vis), and *Escherichia coli*, the causative agents of undulant fever, bovine tuberculosis, and one kind of diarrhea, respectively, are controlled in this manner.

Pasteurization is not sterilization. *Thermoduric* and *thermophilic*—heat-tolerant and heat-loving—prokaryotes survive pasteurization, but they do not cause spoilage over the relatively short times during which properly refrigerated and pasteurized foods are stored before consumption. In addition, such prokaryotes are generally not pathogenic.

The combination of time and temperature required for effective pasteurization varies with the product. Because milk is the most familiar pasteurized product, we consider the pasteurization of milk in some detail. Historically, milk was pasteurized by the *batch method* for 30 minutes at 63°C, but most milk processors today use a high-temperature, short-time method known as *flash pasteurization*, in which milk flows through heated tubes that raise its temperature to 72°C for only 15 seconds. This treatment effectively destroys all pathogens. *Ultra-high-temperature pasteurization* heats the milk to at least 135°C for only 1 second, but some consumers claim that it adversely affects the taste.

Ultra-High-Temperature Sterilization The dairy industry and other food processors can also use *ultra-high-temperature sterilization*, which involves flash heating milk or other liquids to rid them of all living microbes. The process involves passing the liquid through superheated steam at about 140°C for 1 to 3 seconds and then cooling it rapidly. Treated liquids can be stored indefinitely at room temperature without microbial spoilage, though chemical degradation after months of storage results in flavor changes. Small packages of dairy creamer served in restaurants are often sterilized by the ultra-high-temperature method. **TABLE 9.2** summarizes the dairy industry's use of moist heat for controlling microbes in milk.

Dry Heat

For substances such as powders and oils that cannot be sterilized by boiling or with steam or for materials that can be damaged by repeated exposure to steam (such as some metal objects), sterilization can be achieved by the use of dry heat, as occurs in an oven.

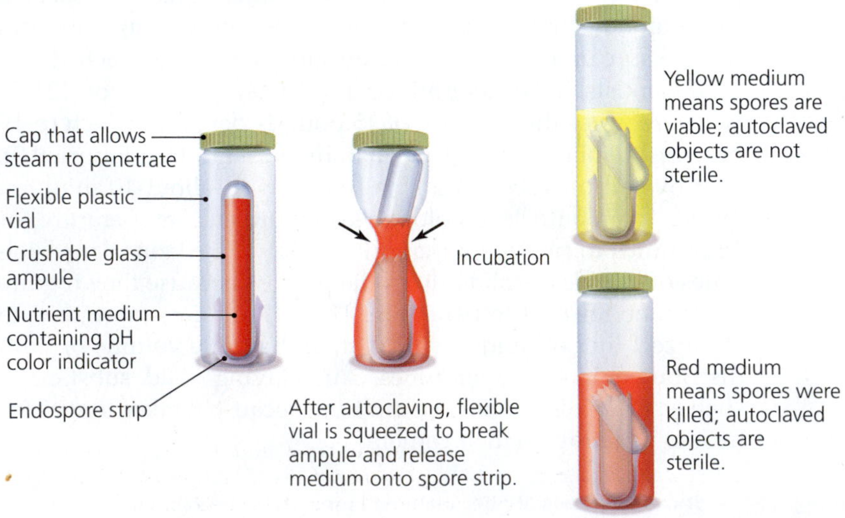
Cap that allows steam to penetrate
Flexible plastic vial
Crushable glass ampule
Nutrient medium containing pH color indicator
Endospore strip
After autoclaving, flexible vial is squeezed to break ampule and release medium onto spore strip.
Incubation
Yellow medium means spores are viable; autoclaved objects are not sterile.
Red medium means spores were killed; autoclaved objects are sterile.

◀ FIGURE 9.8 Sterility indicators. A commercial endospore-test ampule, which is included among objects to be sterilized. After autoclaving is complete, the medium, which contains a pH color indicator, is released onto the endospore strip when the ampule is broken. If the endospores are still alive, their metabolic wastes lower the pH, changing the color of the medium.

TABLE 9.2 Moist Heat Treatments of Milk

Process	Treatment
Historical (batch) pasteurization	63°C for 30 minutes
Flash pasteurization	72°C for 15 seconds
Ultra-high-temperature pasteurization	135°C for 1 second
Ultra-high-temperature sterilization	140°C for 1–3 seconds

Hot air is an effective sterilizing agent because it denatures proteins and fosters the oxidation of metabolic and structural chemicals; however, in order to sterilize, dry heat requires higher temperatures for longer times than moist heat because dry heat penetrates more slowly. For instance, whereas an autoclave needs less than 15 minutes to sterilize an object at 121°C, an oven at the same temperature requires at least 16 hours to achieve sterility. Scientists typically use higher temperatures—171°C for 1 hour or 160°C for 2 hours—to sterilize objects in an oven, but objects made of rubber, paper, and many types of plastic oxidize rapidly (combust) under these conditions.

Complete incineration is the ultimate means of sterilization. As part of standard aseptic technique in microbiological laboratories, inoculating loops are sterilized by heating them in the flame of a Bunsen burner or with an electric heating coil until they glow red (about 1500°C). Health care workers incinerate contaminated dressings, bags, and paper cups, and field epidemiologists incinerate the carcasses of animals that have diseases such as anthrax or bovine spongiform encephalopathy (mad cow disease).

Refrigeration and Freezing

LEARNING OUTCOME

9.14 Describe the use and importance of refrigeration and freezing in limiting microbial growth.

In many situations, particularly in food preparation and storage, the most convenient method of microbial control is either refrigeration (temperatures between 0°C and 7°C) or freezing (temperatures below 0°C). These processes decrease microbial metabolism, growth, and reproduction because chemical reactions occur more slowly at low temperatures and because liquid water is not available at subzero temperatures. Note, however, that psychrophilic (cold-loving) microbes can multiply in refrigerated food and spoil its taste and suitability for consumption.

Refrigeration halts the growth of most pathogens, which are predominantly mesophiles. Notable exceptions are the bacteria *Listeria* (lis-tēr´ē-ă), which can reproduce to dangerous levels in refrigerated food, and *Yersinia* (yer-sin´ē-ă), which can multiply in refrigerated blood products and be passed on to blood recipients. (Chapters 20 and 21 discuss these pathogens in more detail.)

Slow freezing, during which ice crystals have time to form and puncture cell membranes, is more effective than quick freezing in inhibiting microbial metabolism, though microorganisms also vary in their susceptibility to freezing. Whereas the cysts of tapeworms perish after several days in frozen meat, many vegetative bacterial cells, bacterial endospores, and viruses can survive subfreezing temperatures for years. In fact, scientists store many bacteria and viruses in low-temperature freezers at −30°C to −80°C and are able to reconstitute the microbes into viable populations by warming them in media containing proper nutrients. Therefore, we must take care in thawing and cooking frozen food because it can still contain many pathogenic microbes.

Desiccation and Lyophilization

LEARNING OUTCOME

9.15 Compare and contrast desiccation and lyophilization.

Desiccation, or drying, has been used for thousands of years to preserve such foods as fruits, peas, beans, grain, nuts, and yeast (**FIGURE 9.9**). Desiccation inhibits microbial growth because metabolism requires liquid water. Drying inhibits the spread of most pathogens, including the bacteria that cause syphilis, gonorrhea, and the more common forms of bacterial pneumonia and diarrhea. However, most molds can grow on dried raisins and apricots, which have as little as 16% water content.

Scientists use **lyophilization** (lī-of´-i-li-zā´shŭn), a technique combining freezing and drying, to preserve microbes and other cells for many years. In this process, scientists instantly freeze a culture in liquid nitrogen or frozen carbon dioxide (dry ice); then they subject it to a vacuum that removes frozen water through a process called *sublimation*, in which the water is transformed directly from a solid to a gas. Lyophilization prevents the formation of large, damaging ice crystals. Although not all cells survive, enough are viable to enable the culture to be reconstituted many years later.

▲ **FIGURE 9.9 The use of desiccation as a means of preserving apricots in Pakistan.** In this time-honored practice, drying inhibits microbial growth by removing the water that microbes need for metabolism.

Filtration

LEARNING | OUTCOME

9.16 Describe the use of filters for disinfection and sterilization.

Filtration is the passage of a fluid (either a liquid or a gas) through a sieve designed to trap particles—in this case, cells or viruses—and separate them from the fluid. Researchers often use a vacuum to assist the movement of fluid through the filter (**FIGURE 9.10a**). Filtration traps microbes larger than the pore size, allowing smaller microbes to pass through. In the late 1800s, filters were able to trap cells, but their pores were too large to trap the pathogens of such diseases as rabies and measles. These pathogens were thus named *filterable viruses*, which today has been shortened to *viruses*.[9] Now, filters with pores small enough to trap even viruses are available, so filtration can be used to sterilize such heat-sensitive materials as ophthalmic solutions, antibiotics, vaccines, liquid vitamins, enzymes, and culture media.

Over the years, filters have been constructed from porcelain, glass, cotton, asbestos, and diatomaceous earth, a substance composed of the innumerable glasslike cell walls of single-celled algae called diatoms. Scientists today typically use thin (only 0.1 mm thick), circular **membrane filters** manufactured of nitrocellulose or plastic and containing specific pore sizes ranging from 25 μm to less than 0.01 μm in diameter (**FIGURE 9.10b**). The pores of the latter filters are small enough to trap small viruses and even some large protein molecules. Microbiologists also use filtration to estimate the number of microbes in a fluid by counting the number deposited on the filter after passing a given volume through the filter (see Figure 6.24). **TABLE 9.3** lists some pore sizes of membrane filters and the microbes they do not allow through.

Health care and laboratory workers routinely use filtration to prevent airborne contamination by microbes. Medical personnel wear surgical masks to prevent exhaled microbes from contaminating the environment, and cotton plugs are placed in culture vessels to prevent contamination by airborne microbes. Additionally, *high-efficiency particulate air (HEPA) filters* are crucial parts of biological safety cabinets (**FIGURE 9.11**), and HEPA filters are mounted in the air ducts of some operating rooms, rooms occupied by patients with airborne diseases such as tuberculosis, and rooms of immunocompromised patients, such as burn victims and AIDS patients.

[9]Latin, meaning "poisons."

TABLE 9.3 Membrane Filters

Pore Size (μm)	Smallest Microbes That Are Trapped
5	Multicellular algae, animals, and fungi
3	Yeasts and larger unicellular algae
1.2	Protozoa and small unicellular algae
0.45	Largest bacteria
0.22	Largest viruses and most bacteria
0.025	Larger viruses and pliable bacteria (mycoplasmas, rickettsias, chlamydias, and some spirochetes)
0.01	Smallest viruses

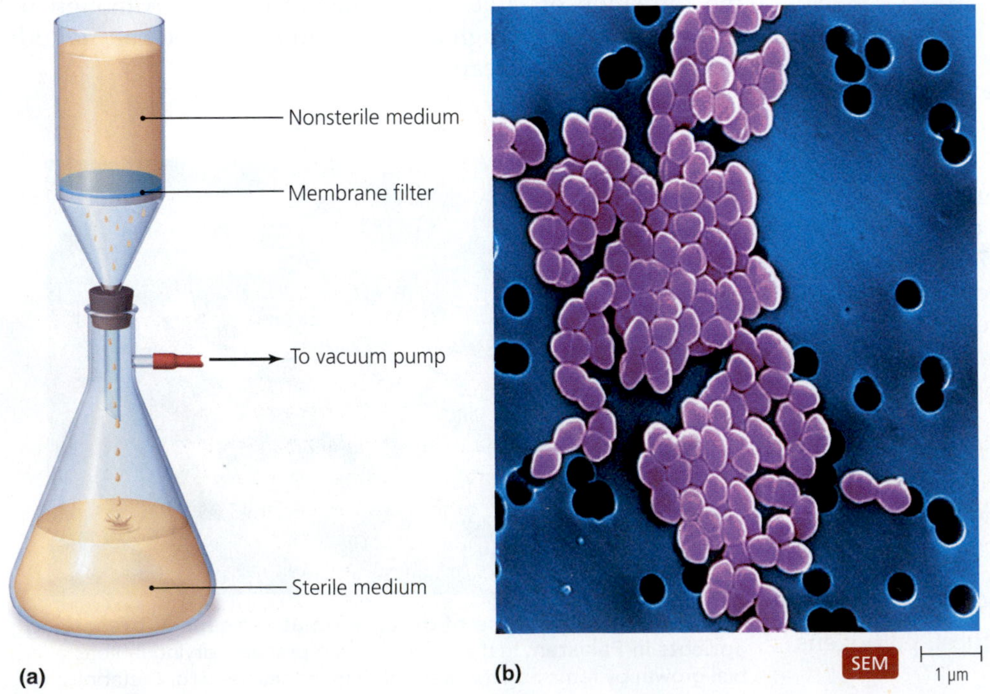

◀ **FIGURE 9.10 Filtration equipment used for microbial control.** **(a)** Assembly for sterilization by vacuum filtration. **(b)** Membrane filters composed of various substances and with pores of various sizes can be used to trap diverse microbes, here a bacterium known as vancomycin-resistant *Enterococcus* (VRE).

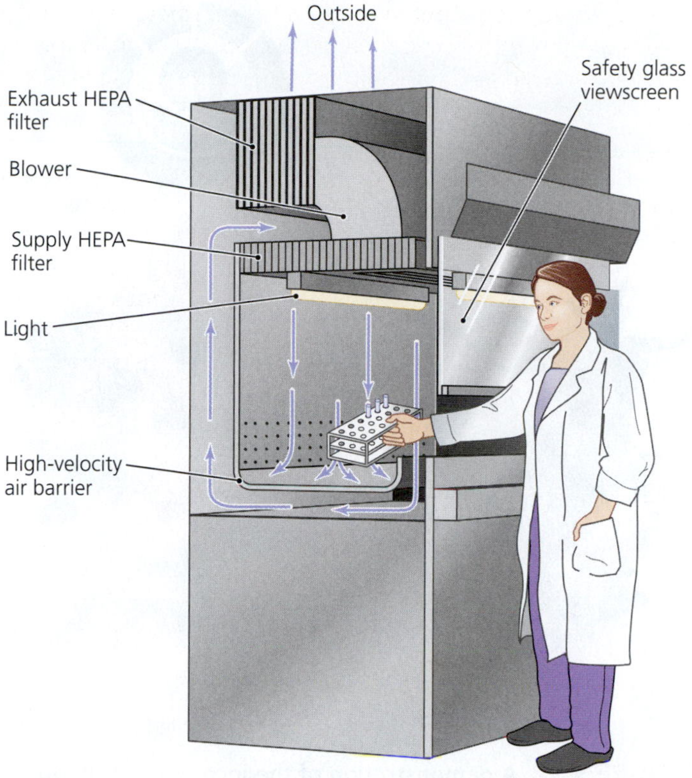

▲ **FIGURE 9.11 The roles of high-efficiency particulate air (HEPA) filters in biological safety cabinets.** HEPA filters protect workers from exposure to microbes (by maintaining a barrier of filtered air across the opening of the cabinet). Hospitals also use HEPA filters in air ducts of operating rooms and of the rooms of highly contagious or immunocompromised patients.

Osmotic Pressure

LEARNING OUTCOME

9.17 Discuss the use of hypertonic solutions in microbial control.

Another ancient method of microbial control is the use of high concentrations of salt or sugar in foods to inhibit microbial growth by **osmotic pressure**. Osmosis is the net movement of water across a semipermeable membrane (such as a cytoplasmic membrane) from an area of higher water concentration to an area of lower water concentration. Cells in a hypertonic solution of salt or sugar lose water, and the cell shrinks (see Figure 3.19b). The removal of water inhibits cellular metabolism because enzymes are fully functional only in aqueous environments. Thus, osmosis preserves honey, jerky, jams, jellies, salted fish, and some types of pickles from most microbial attacks.

Fungi have a greater ability than bacteria to tolerate hypertonic environments with little moisture, which explains why jelly in your refrigerator may grow a colony of *Penicillium* (pen-i-sil´ē-ŭm) mold but is not likely to grow the bacterium *Salmonella* (sal´mŏ-nel´ă).

Radiation

LEARNING OUTCOME

9.18 Differentiate ionizing radiation from nonionizing radiation as they relate to microbial control.

HIGHLIGHT

Microbes in Sushi?

Sushi—it makes you either squirm or salivate! Although technically the term refers to rice, it's generally understood to be bite-sized slices of raw fish served with rice. Long a staple of Japanese cuisine, sushi has become popular throughout the United States. But isn't eating raw fish dangerous? What methods of microbial control are applied in the preparation of sushi?

It's true that raw fish can contain harmful microorganisms. Parasitic roundworms called anisakids are commonly found in sushi and can cause gastrointestinal symptoms in humans. Accordingly, before fish can be served raw, the U.S. Food and Drug Administration (FDA) requires that it be frozen at −20°C for seven days or at −37°C for 15 hours. Unfortunately, freezing does not kill all bacterial or viral pathogens. Consumers should be aware that they always assume some risk of food poisoning caused by such bacteria as *Staphylococcus aureus*, *E. coli*, and species of *Salmonella* and *Vibrio* whenever they eat any kind of raw food, including sushi, raw oysters, ceviche, or carpaccio.

Even though diners tend to fixate on the safety of raw fish, cooked rice left sitting at room temperature is also vulnerable to the growth of pathogens. To counter this potential problem, sushi rice is prepared with vinegar, which acidifies the rice. At pH values below 4.6, rice becomes too acidic to support the growth of most pathogens. Sushi bars can also reduce the risk posed by pathogens by keeping restaurant temperatures cool. Additionally, it is believed that wasabi, the fiery horseradish-like green paste commonly eaten with sushi, contains antimicrobial properties, although its antimicrobial action is not well understood.

With these antimicrobial precautions in place, the vast majority of diners consume sushi safely meal after meal (although pregnant women and people with compromised immune systems should avoid all raw seafood). When properly prepared, sushi is beautiful, low in calories, a source of heart-healthy omega-3 fatty acids—and very delicious.

Another physical method of microbial control is the use of **radiation**. There are two types of radiation: particulate radiation and electromagnetic radiation. Particulate radiation consists of high-speed subatomic particles, such as protons, that have been freed from their atoms. Electromagnetic radiation can be defined as energy without mass traveling in waves at the speed of light (3×10^5 km/sec). Electromagnetic energy is released from atoms that have undergone internal changes. The *wavelength* of electromagnetic radiation, defined as the distance between two crests of a wave, ranges from very short gamma rays; to X rays, ultraviolet light, and visible light; to long infrared rays; and, finally, to very long radio waves (see Figure 4.1). Though they are particles, electrons also have a wave nature, with wavelengths that are even shorter than gamma rays.

The shorter the wavelength of an electromagnetic wave, the more energy it carries; therefore, shorter-wavelength radiation is more suitable for microbial control than longer-wavelength radiation, which carries less energy and is less penetrating. Scientists describe all types of radiation as either *ionizing* or *nonionizing* according to its effects on the chemicals within cells.

Ionizing Radiation

Electron beams, gamma rays, and some X rays, all of which have wavelengths shorter than 1 nm, are **ionizing radiation** because when they strike molecules, they have sufficient energy to eject electrons from atoms, creating ions. Such ions disrupt hydrogen bonding, oxidize double covalent bonds, and create highly reactive hydroxyl radicals (see Chapter 6). These ions in turn denature other molecules, particularly DNA, causing fatal mutations and cell death.

Electron beams are produced by *cathode ray machines.* Electron beams are highly energetic and therefore very effective in killing microbes in just a few seconds, but they cannot sterilize thick objects or objects coated with large amounts of organic matter. They are used to sterilize spices, meats, microbiological plastic ware, and dental and medical supplies, such as gloves, syringes, and suturing material.

Gamma rays, which are emitted by some radioactive elements, such as radioactive cobalt, penetrate much farther than electron beams but require hours to kill microbes. The U.S. Food and Drug Administration (FDA) has approved the use of gamma irradiation for microbial control in meats, spices, and fresh fruits and vegetables **(FIGURE 9.12)**. Irradiation with gamma rays kills not only microbes but also the larvae and eggs of insects; it also kills the cells of fruits and vegetables, preventing both microbial spoilage and overripening.

Consumers have been reluctant to accept irradiated food. A number of reasons have been cited, including fear that radiation makes food radioactive and claims that it changes the taste and nutritive value of foods or produces potentially carcinogenic (cancer-causing) chemicals. Supporters of irradiation reply that gamma radiation passes through food and cannot make it radioactive any more than a dental X ray produces radioactive teeth, and they cite numerous studies that conclude that irradiated foods are tasty, nutritious, and safe.

▲ **FIGURE 9.12 A demonstration of the increased shelf life of food achieved by ionizing radiation.** The circular radura symbol is used in the United States to label irradiated foods.

X rays travel the farthest through matter, but they have less energy than gamma rays and require a prohibitive amount of time to make them practical for microbial control.

Nonionizing Radiation

Electromagnetic radiation with a wavelength greater than 1 nm does not have enough energy to force electrons out of orbit, so it is **nonionizing radiation**. However, such radiation does contain enough energy to excite electrons and cause them to make new covalent bonds, which can affect the three-dimensional structure of proteins and nucleic acids.

Ultraviolet (UV) light, visible light, infrared radiation, and radio waves are nonionizing radiation. Of these, only UV light has sufficient energy to be a practical antimicrobial agent. Visible light and microwaves (radio waves of extremely short wavelength) have little value in microbial control, though microwaves heat food, inhibiting microbial growth and reproduction if the food gets hot enough. The more energetic microwaves produced by some commercial microwave ovens can kill fungal spores, preventing treated food from molding.

UV light with a wavelength of 260 nm is specifically absorbed by adjacent pyrimidine nucleotide bases in DNA, causing them to form covalent bonds with each other rather than forming hydrogen bonds with bases in the complementary DNA strand (see Figure 7.25). Such *pyrimidine* dimers distort the shape of DNA, making it impossible for the cell to accurately transcribe or replicate its genetic material. If dimers remain uncorrected, an affected cell may die.

TABLE 9.4 Physical Methods of Microbial Control

Method	Conditions	Action	Representative Use(s)
Moist heat			
Boiling	10 min at 100°C	Denatures proteins and destroys membranes	Disinfection of baby bottles and sanitization of restaurant cookware and tableware
Autoclaving (pressure cooking)	15 min at 121°C	Denatures proteins and destroys membranes	Autoclave: sterilization of medical and laboratory supplies that can tolerate heat and moisture; pressure cooker: sterilization of canned food
Pasteurization	15 sec at 72°C	Denatures proteins and destroys membranes	Destruction of all pathogens and most spoilage microbes in dairy products, fruit juices, beer, and wine
Ultra-high-temperature sterilization	1–3 sec at 140°C	Denatures proteins and destroys membranes	Sterilization of dairy products
Dry heat			
Hot air	2 h at 160°C or 1 h at 171°C	Denatures proteins, destroys membranes, oxidizes metabolic compounds	Sterilization of water-sensitive materials, such as powders, oils, and metals
Incineration	1 sec at more than 1000°C	Oxidizes everything completely	Sterilization of inoculating loops, flammable contaminated medical waste, and diseased carcasses
Refrigeration	0–7°C	Inhibits metabolism	Preservation of food
Freezing	Below 0°C	Inhibits metabolism	Long-term preservation of foods, drugs, and cultures
Desiccation (drying)	Varies with amount of water to be removed	Inhibits metabolism	Preservation of food
Lyophilization (freeze drying)	−196°C for a few minutes while drying	Inhibits metabolism	Long-term storage of bacterial cultures
Filtration	Filter retains microbes	Physically separates microbes from air and liquids	Sterilization of air and heat-sensitive ophthalmic and enzymatic solutions, vaccines, and antibiotics
Osmotic Pressure	Exposure to hypertonic solutions	Inhibits metabolism	Preservation of food
Ionizing radiation (electron beams, gamma rays, X rays)	Seconds to hours of exposure (depending on wavelength of radiation)	Destroys DNA	Sterilization of medical and laboratory equipment and preservation of food
Nonionizing radiation (ultraviolet light)	Irradiation with 260-nm-wavelength radiation	Formation of thymine dimers inhibits DNA transcription and replication	Disinfection and sterilization of surfaces and of transparent fluids and gases

The effectiveness of UV irradiation is tempered by the fact that UV light does not penetrate well. UV light is therefore suitable primarily for disinfecting air, transparent fluids, and the surfaces of objects, such as barber's shears and operating tables. Some cities use UV irradiation in sewage treatment. By passing wastewater past banks of UV lights, they reduce the number of bacteria without using chlorine, which might damage the environment.

TABLE 9.4 summarizes the physical methods of microbial control discussed in the previous pages.

TELL ME WHY

Why are *Bacillus* endospores used as sterility indicators? (See Figure 9.8.)

Chemical Methods of Microbial Control

LEARNING | **OUTCOME**

9.19 Compare and contrast nine major types of antimicrobial chemicals, and discuss the positive and negative aspects of each.

Although physical agents are sometimes used for disinfection, antisepsis, and preservation, more often chemical agents are used for these purposes. As we have seen, chemical agents act to adversely affect microbes' cell walls, cytoplasmic membranes, proteins, or DNA. As with physical agents, the effect of a chemical agent varies with temperature, length of exposure, and the amount of contaminating organic matter in the environment.

The effect also varies with pH, concentration, and freshness of the chemical. Chemical agents tend to destroy or inhibit the growth of enveloped viruses and the vegetative cells of bacteria, fungi, and protozoa more than fungal spores, protozoan cysts, or bacterial endospores. The latter are particularly resistant to chemical agents, as demonstrated by numerous failed attempts to decontaminate a U.S. Senate office building of anthrax endospores sent there by a bioterrorist in 2001.

In the following sections we discuss nine major categories of antimicrobial chemicals used as antiseptics and disinfectants: *phenols, alcohols, halogens, oxidizing agents, surfactants, heavy metals, aldehydes, gaseous agents,* and *enzymes.* Some chemical agents combine one or more of these. Additionally, researchers and food processors sometimes use antimicrobials—substances normally used to treat diseases—as disinfectants.

Phenol and Phenolics

LEARNING | OUTCOME

9.20 Distinguish between phenol and the types of phenolics, and discuss their action as antimicrobial agents.

In 1867, Dr. Joseph Lister (1827–1912) began using phenol **(FIGURE 9.13a)** to reduce infection during surgery. As stated previously, the efficacy of phenol remains one standard to which the actions of other antimicrobial agents can be compared.

Phenolics are compounds derived from phenol molecules that have been chemically modified by the addition of halogens or organic functional groups **(FIGURE 9.13b)**. For instance, chlorinated phenolics contain one or more atoms of chlorine and have enhanced antimicrobial action and a less annoying odor than phenol. Natural oils, such as pine and clove oils, are also phenolics and can be used as antiseptics.

Bisphenolics are composed of two covalently linked phenolics. An example is *triclosan,* which is incorporated into numerous consumer products, including garbage bags, diapers, and cutting boards **(FIGURE 9.13c)**.

Phenol and phenolics denature proteins and disrupt cell membranes in a wide variety of pathogens. They are effective even in the presence of contaminating organic material, such as vomit, pus, saliva, and feces, and they remain active on surfaces for a prolonged time. For these reasons, phenolics are commonly used in health care settings, laboratories, and households.

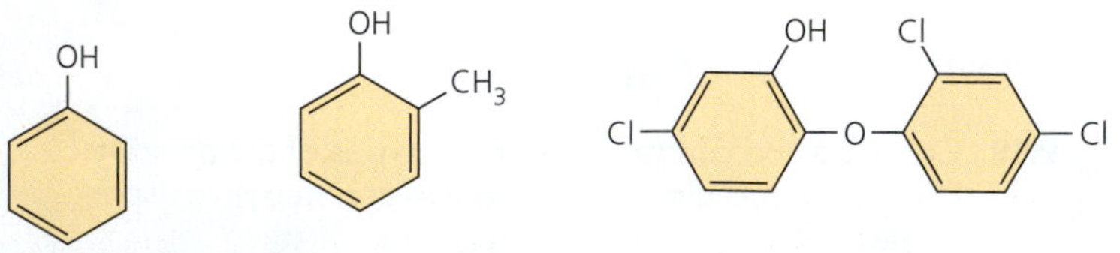

Orthocresol

Triclosan

(a) Phenol **(b) Phenolics** **(c) Bisphenolics**

▲ **FIGURE 9.13 Phenol and phenolics. (a)** Phenol, a naturally occurring molecule that is also called carbolic acid. **(b)** Phenolics, which are compounds synthesized from phenol, have greater antimicrobial efficacy with fewer side effects. **(c)** Bisphenols are paired, covalently linked phenolics.

Negative aspects of phenolics include their disagreeable odor and possible side effects; for example, phenolics irritate the skin of some individuals.

Alcohols

LEARNING | OUTCOME

9.21 Discuss the action of alcohols as antimicrobial agents, and explain why solutions of 70% to 90% alcohol are more effective than pure alcohols.

Alcohols are bactericidal, fungicidal, and virucidal against enveloped viruses; however, they are not effective against fungal spores or bacterial endospores. Alcohols are considered intermediate-level disinfectants. Commonly used alcohols include rubbing alcohol (isopropanol) and drinking alcohol (ethanol):

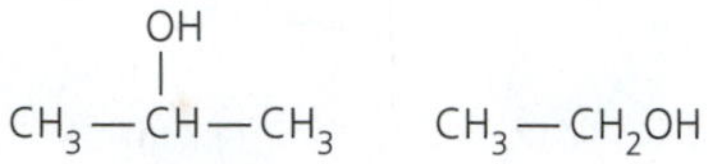

Isopropanol Ethanol

Isopropanol is slightly superior to ethanol as a disinfectant and antiseptic. *Tinctures* (tingk′chūrs), which are solutions of other antimicrobial chemicals in alcohol, are often more effective than the same chemicals dissolved in water.

Alcohols denature proteins and disrupt cytoplasmic membranes. Surprisingly, pure alcohol is not an effective antimicrobial agent because the denaturation of proteins requires water; therefore, solutions of 70% to 90% alcohol are typically used to control microbes. Alcohols evaporate rapidly, which is advantageous in that they leave no residue but disadvantageous in that they may not contact microbes long enough to be effective. Alcohol-based antiseptics are more effective than soap in removing bacteria from hands but not effective against some viruses, such as diarrhea-causing noroviruses. Swabbing the skin with alcohol prior to an injection removes more microbes by physical action (degerming) than by chemical action.

Halogens

LEARNING | OUTCOME

9.22 Discuss the types and uses of halogen-containing antimicrobial agents.

Halogens are the four very reactive, nonmetallic chemical elements: iodine, chlorine, bromine, and fluorine. Halogens are intermediate-level antimicrobial chemicals that are effective against vegetative bacterial and fungal cells, fungal spores, some bacterial endospores and protozoan cysts, and many viruses. Halogens are used both alone and combined with other elements in organic and inorganic compounds. Halogens exert their antimicrobial effect by unfolding and thereby denaturing essential proteins, including enzymes.

BENEFICIAL MICROBES

Hard to Swallow?

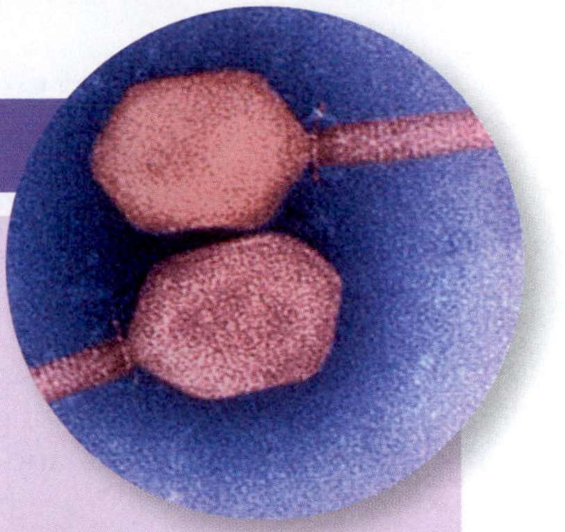

Bacteriophages

Controlling bacteria in the environment is becoming more difficult because they are developing resistance to common disinfectants and antiseptics. So, scientists are turning to natural parasites of bacteria—bacteriophages, also simply called phages—to control bacterial contamination. *Bacteriophages,* which literally means "bacteria eaters," is the term for viruses that specifically attack particular strains of bacteria. Phages are like "smart bombs"; unlike disinfectants and most antimicrobial drugs, phages attack specific bacterial strains and leave neighboring bacteria unharmed. A phage injects its genetic material into a bacterial cell, causing the bacterial cell to produce hundreds of new phages that burst out of the bacterium and kill it. Researchers are developing phage solutions to control bacteria in medical settings, in food, and in patients.

In 2006, the U.S. Food and Drug Administration (FDA) approved the nonmedical use of a phage that specifically kills *Listeria monocytogenes,* which is frequently a bacterial contaminant of cheese and lunch meat. *Listeria* kills about 20% of people who get sick from ingesting it. The approved anti-*Listeria* phage is available in a solution that food processors, delicatessen owners, and consumers can spray on food to reduce the number of *Listeria* cells. Some people find the idea hard to swallow—deliberately contaminating food and equipment with viruses sounds like poor hygiene.

Proponents of using phages point out that an individual consumes millions of phages daily in water and food without ill effect. Indeed, phages in restaurants are more common than mustard and mayonnaise. Medical professionals in the former Soviet Union, particularly the country of Georgia, have used phages to successfully treat disease for over six decades without deleterious side effects.

Iodine is a well-known antiseptic. In the past, backpackers and campers disinfected water with iodine tablets, but experience has shown that protozoan cysts can survive iodine treatment unless the iodine concentration is so great that the water is undrinkable. Knowledgeable campers now filter stream and lake water or carry bottled water.

Medically, iodine is used either as a tincture or as an *iodophor,* which is an iodine-containing organic compound that slowly releases iodine. Iodophors have the advantage of being long lasting and nonirritating to the skin. Betadine is an example of an iodophor used in medical institutions to prepare skin for surgery **(FIGURE 9.14)** and injections and to treat burns.

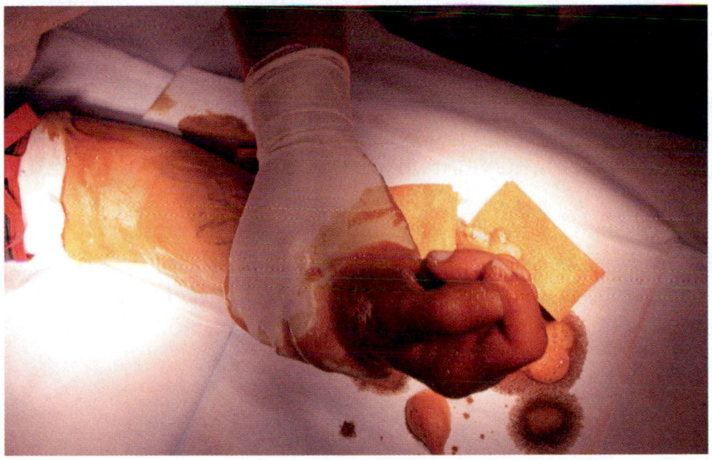

▲ **FIGURE 9.14 Degerming in preparation for surgery on a hand.** Betadine, an iodophor, is the antiseptic used here.

Municipalities commonly use *chlorine* in its elemental form (Cl_2) to treat drinking water, swimming pools, and wastewater from sewage treatment plants. Compounds containing chlorine are also effective disinfectants. Examples include *sodium hypochlorite* ($NaOCl$), which is household chlorine bleach, and *calcium hypochlorite* [$Ca(OCl)_2$]. The dairy industry and restaurants use these compounds to disinfect utensils, and the medical field uses them to disinfect hemodialysis systems. Household bleach diluted by adding two drops to a liter of water can be used in an emergency to make water safer to drink, but it does not kill all protozoan cysts, bacterial endospores, or viruses. *Chlorine dioxide* (ClO_2) is a gas that can be used to disinfect large spaces; for example, it was used in the federal office buildings contaminated with anthrax spores following the 2001 bioterrorism attack. Chloramines—chemical combinations of chlorine and ammonia—are used in wound dressings, as skin antiseptics, and in some municipal water supplies. Chloramines are less effective antimicrobial agents than other forms of chlorine, but they release chlorine slowly and are thus longer lasting.

Bromine is an effective disinfectant in hot tubs because it evaporates more slowly than chlorine at high temperatures. Bromine is also used as an alternative to chlorine in the disinfection of swimming pools, cooling towers, and other water containers.

Fluorine in the form of fluoride is antibacterial in drinking water and toothpastes and can help reduce the incidence of dental caries (cavities). Fluorine works in part by disrupting metabolism in the biofilm of dental plaque.

Oxidizing Agents

LEARNING | OUTCOME

9.23 Describe the use and action of oxidizing agents in microbial control.

Peroxides, ozone, and *peracetic acid* kill microbes by oxidizing their enzymes, thereby preventing metabolism. **Oxidizing agents** are high-level disinfectants and antiseptics that work by releasing oxygen radicals, which are particularly effective against anaerobic microorganisms. Health care workers use oxidizing agents to kill anaerobes in deep puncture wounds.

Hydrogen peroxide (H_2O_2) is a common household chemical that can disinfect and even sterilize the surfaces of inanimate objects such as contact lenses, but it is often mistakenly used to treat open wounds. Hydrogen peroxide does not make a good antiseptic for open wounds because *catalase*—an enzyme released from damaged human cells—quickly neutralizes hydrogen peroxide by breaking it down into water and oxygen gas, which can be seen as escaping bubbles. Though aerobes and facultative anaerobes on inanimate surfaces also contain catalase, the volume of peroxide used as a disinfectant overwhelms the enzyme, making hydrogen peroxide a useful disinfectant. Food processors use hot hydrogen peroxide to sterilize packages such as juice boxes.

Ozone (O_3) is a reactive form of oxygen that is generated when molecular oxygen (O_2) is subjected to electrical discharge. Ozone gives air its "fresh smell" after a thunderstorm. Some Canadian and European municipalities treat their drinking water with ozone rather than chlorine. Ozone is a more effective antimicrobial agent than chlorine, but it is more expensive, and it is difficult to maintain an effective concentration of ozone in water.

Peracetic acid is an extremely effective sporicide that can be used to sterilize surfaces. Food processors and medical personnel use peracetic acid to sterilize equipment because it is not adversely affected by organic contaminants, and it leaves no toxic residue.

Surfactants

LEARNING | OUTCOME

9.24 Define *surfactants,* and describe their antimicrobial action.

Surfactants are "surface active" chemicals. One of the ways surfactants act is to reduce the surface tension of solvents such as water by decreasing the attraction among molecules. One result of this reduction in surface tension is that the solvent becomes more effective at dissolving solute molecules.

Two common surfactants involved in microbial control are soaps and detergents. One end of a soap molecule is hydrophobic because it is composed of fatty acids, and the other end is hydrophilic and negatively charged. When soap is used to wash skin, for instance, the hydrophobic ends of soap molecules are effective at breaking oily deposits into tiny droplets, and the hydrophilic ends attract water molecules; the result is that the tiny droplets of oily material—and any bacteria they harbor—are more easily dissolved in and washed away by water. Thus, soaps by themselves are good degerming agents though poor antimicrobial agents; when household soaps are antiseptic, it is largely because they contain antimicrobial chemicals.

Synthetic **detergents** are positively charged organic surfactants that are more soluble in water than soaps. The most popular detergents for microbial control are **quaternary ammonium compounds**, or **quats**, which are composed of an ammonium cation (NH_4^+) in which the hydrogen atoms are replaced by other functional groups or hydrocarbon chains **(FIGURE 9.15)**. Quats are not only antimicrobial but also colorless, tasteless, and harmless to humans (except at high concentrations), making them ideal for many industrial and medical applications. If your mouthwash foams, it probably contains a quaternary ammonium compound. Examples of quats are benzalkonium chloride (Zephiran) and cetylpyridinium chloride (used in Cepacol mouthwash).

Quats function by disrupting cellular membranes so that affected cells lose essential internal ions, such as potassiumions (K^+). Quats are bactericidal (particularly against Gram-positive

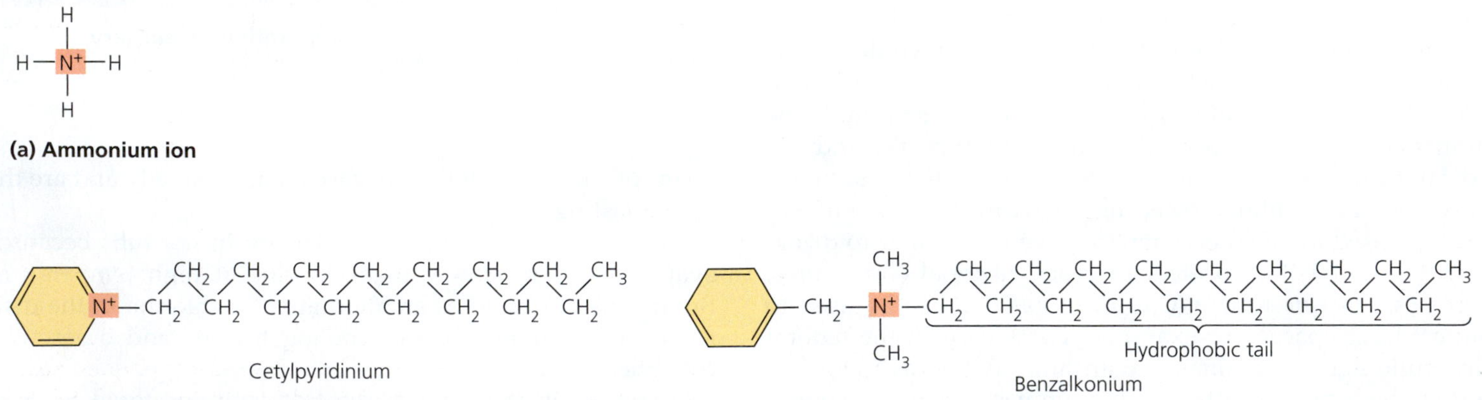

(b) Quaternary ammonium ions (quats)

▲ **FIGURE 9.15 Quaternary ammonium compounds (quats).** Quats are surfactants in which the hydrogen atoms of an ammonium ion **(a)** are replaced by other functional groups **(b)**.

bacteria), fungicidal, and virucidal against enveloped viruses, but they are not effective against nonenveloped viruses, mycobacteria, or endospores. The action of quaternary ammonium compounds is retarded by organic contaminants, and they are deactivated by soaps. Some pathogens, such as *Pseudomonas aeruginosa* (soo-dō-mō´nas ā-roo-ji-nō´să), actually thrive in quats; therefore, quats are classified as low-level disinfectants.

Heavy Metals

LEARNING | **OUTCOME**

9.25 Define *heavy metals*, give several examples, and describe their use in microbial control.

Heavy-metal ions, such as ions of arsenic, zinc, mercury, silver, and copper, are antimicrobial because they combine with sulfur atoms in molecules of cysteine, an amino acid. Such bonding denatures proteins, inhibiting or eliminating their function. Heavy-metal ions are low-level bacteriostatic and fungistatic agents, and with few exceptions their use has been superseded by more effective antimicrobial agents. **FIGURE 9.16** illustrates the effectiveness of heavy metals in inhibiting bacterial reproduction on a Petri plate.

At one time, many states required that the eyes of newborns be treated with a cream containing 1% *silver nitrate* ($AgNO_3$) to prevent blindness caused by *Neisseria gonorrhoeae* (nī-se´rē-ă go-nor-rē´ī), which can enter babies' eyes while they pass through an infected birth canal. Today, silver nitrate has largely been displaced by other antimicrobial ointments that are less irritating and are also effective against other pathogens. Silver still plays an antimicrobial role in some surgical dressings, burn creams, and catheters.

For over 70 years, drug companies used *thimerosal,* a mercury-containing compound, to preserve vaccines. In 1999, the U.S. Public Health Service recommended that alternatives be used because mercury is a metabolic poison, though the very small amount of mercury in vaccines is considered safe. Today only a few adult vaccines contain thimerosal. These include whole-cell pertussis and some vaccines against tetanus, flu, and meningococcal meningitis.

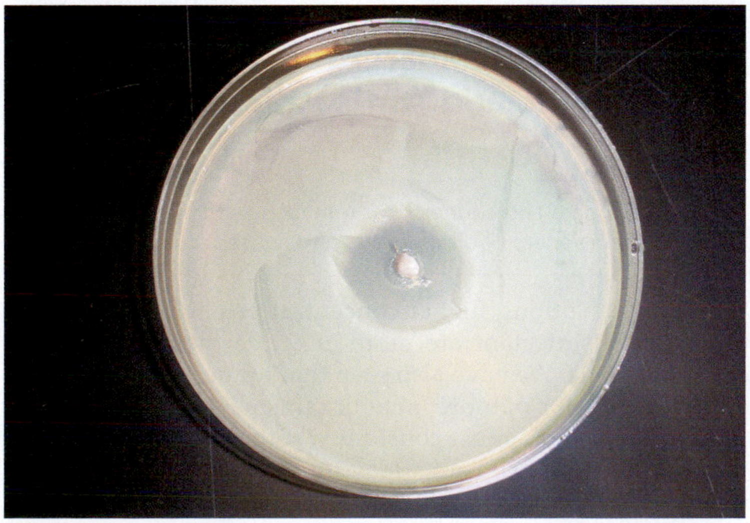

▲ **FIGURE 9.16 The effect of heavy-metal ions on bacterial growth.** Zones of inhibition can form because ions of heavy metals, such as dental amalgam used in fillings and shown in the center of the plate, inhibit bacterial reproduction through their effects on protein function.

Copper, which interferes with chlorophyll, is used to control algal growth in reservoirs, fish tanks, swimming pools, and water storage tanks. In the absence of organic contaminants, copper is an effective algicide in concentrations as low as 1 ppm (part per million). In addition to copper, zinc and mercury are used to control mildew in paint.

Aldehydes

LEARNING | **OUTCOME**

9.26 Compare and contrast formaldehyde and glutaraldehyde as antimicrobial agents.

Aldehydes are compounds containing terminal —CHO groups (see Table 2.3 on p. 40). *Glutaraldehyde,* which is a liquid, and *formaldehyde,* which is a gas, are highly reactive chemicals with the following structural formulas:

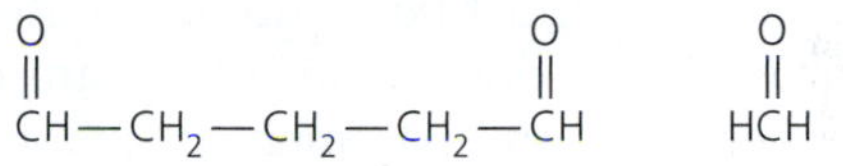

Glutaraldehyde Formaldehyde

Aldehydes function in microbial control by cross-linking amino, hydroxyl, sulfhydryl, and carboxyl organic functional groups, thereby denaturing proteins and inactivating nucleic acids.

Hospital personnel and scientists use 2% solutions of glutaraldehyde to kill bacteria, viruses, and fungi; a 10-minute treatment effectively disinfects most objects, including medical and dental equipment. When the time of exposure is increased to 10 hours, glutaraldehyde sterilizes. Although glutaraldehyde is less irritating and more effective than formaldehyde, it is more expensive as well.

Morticians and health care workers use formaldehyde dissolved in water to make a 37% solution called *formalin.* They use formalin for embalming and to disinfect hospital rooms, instruments, and machines. Formaldehyde must be handled with care because it irritates mucous membranes and is carcinogenic (cancer causing).

Gaseous Agents

LEARNING | **OUTCOME**

9.27 Describe the advantages and disadvantages of gaseous agents of microbial control.

Many items, such as heart-lung machine components, sutures, plastic laboratory ware, mattresses, pillows, artificial heart valves, catheters, electronic equipment, and dried or powdered foods, cannot be sterilized easily with heat or water-soluble chemicals, nor is irradiation always practical for large or bulky items. However, they can be sterilized within a closed chamber containing highly reactive microbicidal and sporicidal gases such as *ethylene oxide, propylene oxide,* and *beta-propiolactone:*

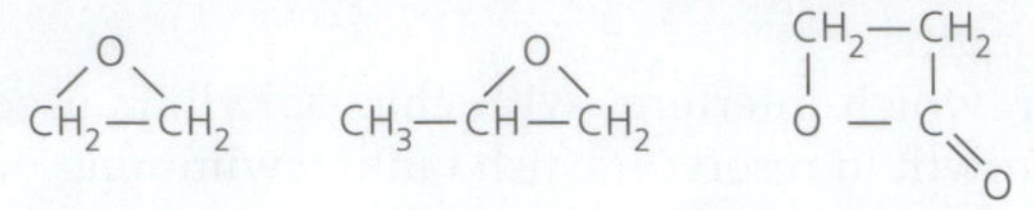

Ethylene oxide Propylene oxide Beta-propiolactone

These gases rapidly penetrate paper and plastic wraps and diffuse into every crack. Over time (usually 4–18 hours), they denature proteins and DNA by cross-linking organic functional groups, thereby killing everything they contact without harming inanimate objects.

Ethylene oxide is frequently used as a gaseous sterilizing agent in hospitals and dental offices, and NASA uses the gas to sterilize spacecraft designed to land on other worlds lest they accidentally export earthly microbes. Large hospitals often use ethylene oxide chambers, which are similar in appearance to autoclaves, to sterilize instruments and equipment sensitive to heat.

Despite their advantages, gaseous agents are far from perfect: They can be extremely hazardous to the people using them. When administered, they must be combined with 10% to 20% nitrogen gas or carbon dioxide because they are often highly explosive. Moreover, they are extremely poisonous, so workers must extensively flush sterilized objects with air to remove every trace of the gas (which adds to the time required to use them). Finally, gaseous agents, especially beta-propiolactone, are potentially carcinogenic.

Enzymes

LEARNING OUTCOME

9.28 Describe the use of an enzyme to remove most bacteria from food and to remove prions from medical instruments.

Many organisms produce chemicals that inhibit or destroy a variety of fungi, bacteria, or viruses. Among these are **antimicrobial enzymes**, which are enzymes that act against microorganisms. For example, human tears contain the enzyme *lysozyme,* which is a protein that digests the peptidoglycan cell walls of bacteria, causing the bacteria to rupture because of osmotic pressure and thus protecting the eye from most bacterial infections.

Scientists, food processors, and medical personnel are researching ways to use natural and chemically modified antimicrobial enzymes to control microbes in the environment, inhibit microbial decay of foods and beverages, and reduce the number and kinds of microbes on medical equipment. For example, food processors use lysozyme to reduce the number of bacteria in cheese, and some vintners use lysozyme instead of poisonous sulfur dioxide (SO_2) to remove bacteria that would spoil wine.

One exciting development is the use of an enzyme to eliminate the prion that causes variant Creutzfeldt-Jakob disease (vCJD), also called mad cow disease. The brain, spinal cord, placenta, eye, liver, kidney, pituitary gland, spleen, lung, and lymph nodes, as well as cerebrospinal fluid, can harbor prions. Medical instruments contaminated by these highly infectious and deadly proteins may remain infectious even after normal autoclaving; boiling; exposure to formaldehyde, glutaraldehyde, or ethylene oxide; or 24 hours of dry heat at 160°C. Until recently, harsh methods, such as autoclaving in sodium hydroxide for 30 minutes or complete incineration, were required to eliminate prions. In 2006, the European Union approved the use of the enzyme *Prionzyme* to safely and completely remove prions on medical instruments. Prionzyme is the first certified, noncaustic chemical to target prions.

Antimicrobials

LEARNING OUTCOME

9.29 Describe the types of antimicrobials and their use in environmental control of microorganisms.

Antimicrobials include antibiotics, semisynthetics, and synthetics. Specifically, *antibiotics* are antimicrobial chemicals produced naturally by microorganisms. When scientists chemically modify an antibiotic, the agent is called a *semisynthetic.* Scientists have also developed wholly *synthetic* antimicrobial drugs. The main difference between these antimicrobials and the chemical agents we have discussed in this chapter is that antimicrobials are typically used for treatment of disease and not for environmental control of microbes. Nevertheless, some antimicrobials are used for control outside the body. For example, the antimicrobials *nisin* and *natamycin* are used to reduce the growth of bacteria and fungi, respectively, in cheese. (Chapter 10 discusses in more detail the nature and use of antimicrobials to treat infectious diseases.)

TABLE 9.5 summarizes the chemical methods of microbial control discussed in this chapter.

Methods for Evaluating Disinfectants and Antiseptics

LEARNING OUTCOME

9.30 Compare and contrast four methods used to measure the effectiveness of disinfectants and antiseptics.

With few exceptions, higher concentrations and fresher solutions of a disinfectant are more effective than more dilute, older solutions. We have also seen that longer exposure times ensure the deaths of more microorganisms. However, anyone using disinfectants must consider whether higher concentrations and longer exposures may damage an object or injure a patient.

Scientists have developed several methods to measure the efficacy of antimicrobial agents. These include the phenol coefficient, the use-dilution test, the Kelsey-Sykes capacity test, and the in-use test.

TABLE 9.5 Chemical Methods of Microbial Control

Method	Action(s)	Level of Activity	Some Uses
Phenol (carbolic acid)	Denatures proteins and disrupts cell membranes	Intermediate to low	Original surgical antiseptic; now replaced by less odorous and injurious phenolics
Phenolics (chemically altered phenol; bisphenols are composed of a pair of linked phenolics)	Denature proteins and disrupt cell membranes	Intermediate to low	Disinfectants and antiseptics
Alcohols	Denature proteins and disrupt cell membranes	Intermediate	Disinfectants, antiseptics, and as a solvent in tinctures
Halogens (iodine, chlorine, bromine, and fluorine)	Presumably denature proteins	Intermediate	Disinfectants, antiseptics, and water purification
Oxidizing agents (peroxides, ozone, and peracetic acid)	Denature proteins by oxidation	High	Disinfectants, antiseptics for deep wounds, water purification, and sterilization of food-processing and medical equipment
Surfactants (soaps and detergents)	Decrease surface tension of water and disrupt cell membranes	Low	Soaps: degerming; detergents: antiseptic
Heavy metals (arsenic, zinc, mercury, silver, copper, etc.)	Denature proteins	Low	Fungistats in paints; silver nitrate cream: surgical dressings, burn creams, and catheters; copper: algicide in water reservoirs, swimming pools, and aquariums
Aldehydes (glutaraldehyde and formaldehyde)	Denature proteins	High	Disinfectant and embalming fluid
Gaseous agents (ethylene oxide, propylene oxide, and beta-propiolactone)	Denature proteins	High	Sterilization of heat- and water-sensitive objects
Enzymes	Denature proteins	High against target substrate	Removal of prions on medical instruments
Antimicrobials (antibiotics, semisynthetics, and synthetics)	Act against cell walls, cell membranes, protein synthesis, and DNA transcription and replication	Intermediate to low	Disinfectants and treatment of infectious diseases

Phenol Coefficient

Joseph Lister introduced the widespread use of phenol (also known as carbolic acid) as an antiseptic during surgery. Since then, researchers have evaluated the efficacy of various disinfectants and antiseptics by calculating a ratio that compares a given agent's ability to control microbes to that of phenol under standardized conditions. This ratio is referred to as the **phenol coefficient**. A phenol coefficient greater than 1.0 indicates that an agent is more effective than phenol, and the larger the ratio, the greater the effectiveness. For example, *chloramine,* a mixture of chlorine and ammonia, has a phenol coefficient of 133.0 when used against the bacterium *Staphylococcus aureus* and a phenol coefficient of 100.0 when used against *Salmonella enterica* (en-ter´i-kă). This indicates that chloramine is at least 133 times more effective than phenol against *Staphylococcus* but only 100 times more effective against *Salmonella.* Measurement of an agent's phenol coefficient has been replaced by newer methods because scientists have developed disinfectants and antiseptics much more effective than phenol.

Use-Dilution Test

Another method for measuring the efficacy of disinfectants and antiseptics against specific microbes is the **use-dilution test**. In this test, a researcher dips several metal cylinders into broth cultures of bacteria and briefly dries them at 37°C. The bacteria used in the standard test are *Pseudomonas aeruginosa*, *Salmonella enterica* serotype Choleraesuis (kol-er-a-su´is), and *S. aureus.* The researcher then immerses each contaminated cylinder into a different dilution of the disinfectants being evaluated. After 10 minutes, each cylinder is removed, rinsed with water to remove excess chemical, and placed into a fresh tube of sterile medium for 48 hours of incubation. The most effective agent is the one that entirely prevents microbial growth at the highest dilution.

The use-dilution test is the current standard test in the United States, though it was developed several decades ago before the appearance of many of today's pathogens, including hepatitis C virus, HIV, and antibiotic-resistant bacteria and protozoa. Moreover, the disinfectants in use at the time were far less powerful than many used today. Some government agencies have expressed concern that the test is neither accurate, reliable, nor relevant; therefore, the Association of Analytical Communities is developing a new standard procedure for use in the United States.

Kelsey-Sykes Capacity Test

The **Kelsey-Sykes capacity test** is the standard alternative assessment approved by the European Union to determine the capacity of a given chemical to inhibit bacterial growth. In this test, researchers add a suspension of a bacterium such as *P. aeruginosa* or *S. aureus* to a suitable concentration of the chemical being tested. Then at predetermined times, they move samples of the mixture into growth medium containing a disinfectant deactivator. After incubation for 48 hours, turbidity in the medium indicates that bacteria survived treatment. Lack of turbidity, indicating lack of bacterial reproduction, reveals the minimum time required for the disinfectant to be effective.

In-Use Test

Though phenol coefficient, use-dilution, and Kelsey-Sykes capacity tests can be beneficial for initial screening of disinfectants, they can also be misleading. These types of evaluation are measures of effectiveness under controlled conditions against one or, at most, a few species of microbes, but disinfectants are generally used in various environments against a diverse population of organisms that are often associated with one another in complex biofilms affording mutual protection.

A more realistic (though more time-consuming) method for determining the efficacy of a chemical is called an **in-use test**. In this procedure, swabs are taken from actual objects, such as operating room equipment, both before and after the application of a disinfectant or an antiseptic. The swabs are then inoculated into appropriate growth media that, after incubation, are examined for microbial growth. The in-use test allows a more accurate determination of the proper strength and application procedure of a given disinfection agent for each specific situation.

Development of Resistant Microbes

Many scientists are concerned that Americans have become overly preoccupied with antisepsis and disinfection, as evidenced by the proliferation of products containing antiseptic and disinfecting chemicals. For example, one can now buy hand soap, shampoo, toothpaste, hand lotion, foot pads for shoes, deodorants, and bath sponges that contain antiseptics, as well as kitty litter, cutting boards, scrubbing pads, garbage bags, children's toys, and laundry detergents that contain disinfectants. There is little evidence that the extensive use of such products adds to human or animal health, but it does promote the development of strains of microbes resistant to antimicrobial chemicals: While susceptible cells die, resistant cells remain to proliferate. Scientists have already isolated strains of pathogenic bacteria, including *M. tuberculosis, P. aeruginosa, E. coli,* and *S. aureus,* that are less susceptible to common disinfectants and antiseptics.

Highlight: Antibacterial Soap: Too Much of a Good Thing? discusses a controversy regarding the use of antibacterial soap.

TELL ME WHY

Many chemical disinfectants and antiseptics act by denaturing proteins. Why does denaturation kill cells?

HIGHLIGHT

Antibacterial Soap: Too Much of a Good Thing?

Although soaps containing antimicrobial drugs are more effective than plain soaps in reducing the presence of microbes, there is concern that overuse of antimicrobials contributes to the evolution of resistant microorganisms: Antibacterial soaps kill off weaker bacteria, leaving stronger, more resistant strains to multiply. The U.S. Centers for Disease Control and Prevention (CDC) has taken a cautious stance, acknowledging the benefits of antimicrobial soaps in certain circumstances while agreeing that we need further research to determine whether such products may actually do more harm than good.

Who knew cleanliness could be so complicated? And in the meantime, what kind of soap should you use? Experts can't seem to agree on a single guideline, but the CDC recommends using mild, regular soap and washing in warm running water for at least 10 to 15 seconds in most cases. Antimicrobial soap should be reserved for limited applications: handling food, caring for newborns, and caring for high-risk patients.

MICRO IN THE CLINIC FOLLOW-UP

Raw Fish, Raw Deal?

The doctors find parasitic roundworms called *anisakids* wriggling on Brianne's stomach lining. They also discover the same parasites in the muscle layer of the surgically removed section of James's small intestine. Occasionally found in raw fish, anisakids are resilient creatures that can survive acids in the stomach as well as digestive enzymes in the small intestine. In Brianne's case, her body began reacting to the presence of the parasites right away. In James's case, the worms did not trigger symptoms until they made their way to the end of his small intestine, where they grew.

Once the worms are removed from Brianne's stomach, she recovers quickly and is discharged with a prescription for an antihistamine to treat her rash, which had been caused by an allergic response to worm proteins. James stays in the hospital for five days to recover from his surgery.

Although raw fish is never 100% safe, sushi restaurants greatly minimize the risk by flash-freezing the fish at −35°C (−31°F) or below for 15 hours, or by keeping it frozen at −20°C (−4°F) for 7 days. Another precautionary measure is to soak raw fish in brine to stop the growth of bacteria that could cause food spoilage or illness. However, neither of these processes can guarantee that all parasites present in fish muscle will be killed.

1. **What is it about the freezing process that can help minimize the risk of microorganisms in sushi?**
2. **How can soaking raw fish in very salty water inhibit the growth of bacteria?**

MM Check your answers to Micro in the Clinic Follow-Up questions in the MasteringMicrobiology Study Area.

Explore the Invisible: Principles of Autoclaving

Practice on-the-go with Dr. Bauman Video Tutors by scanning this QR code with your smart phone. Visit the **MasteringMicrobiology Study Area** to challenge your understanding with practice tests, animation quizzes, and clinical case studies!

MasteringMicrobiology®

CHAPTER SUMMARY

Basic Principles of Microbial Control (pp. 263–265)

1. **Sterilization** is the eradication of microorganisms and viruses; the term is not usually applied to the destruction of prions.
2. An **aseptic** environment or procedure is free of contamination by pathogens.
3. **Antisepsis** is the inhibition/killing of microorganisms (particularly pathogens) on skin or tissue by the use of a chemical **antiseptic**, whereas **disinfection** refers to the use of agents (called **disinfectants**) to inhibit microbes on inanimate objects.
4. **Degerming** refers to the removal of microbes from a surface by scrubbing.
5. **Sanitization** is the reduction of a prescribed number of pathogens from surfaces and utensils in public settings.
6. **Pasteurization** is a process using heat to kill pathogens and control microbes that cause spoilage of food and beverages.
7. The suffixes *-stasis* and *-static* indicate that an antimicrobial agent inhibits microbes, whereas the suffixes *-cide* and *-cidal* indicate that the agent kills or permanently inactivates a particular type of microbe.
8. **Microbial death** is the permanent loss of reproductive capacity. **Microbial death rate** measures the efficacy of an antimicrobial agent.
9. Antimicrobial agents destroy microbes either by altering their cell walls and membranes or by interrupting their metabolism and reproduction via interference with proteins and nucleic acids.

The Selection of Microbial Control Methods (pp. 265–267)

1. Factors affecting the efficacy of antimicrobial methods include the site to be treated, the relative susceptibility of microorganisms, and environmental conditions.

2. The CDC has established four biosafety levels (BSL) for microbiological laboratories. BSL-1 is minimal; BSL-4 requires special suits, rooms, and other precautions.

Physical Methods of Microbial Control (pp. 268–275)

1. **Thermal death point** is the lowest temperature that kills all cells in a broth in 10 minutes, whereas **thermal death time** is the time it takes to completely sterilize a particular volume of liquid at a set temperature. **Decimal reduction time (D)** is the time required to destroy 90% of the microbes in a sample.
2. An **autoclave** uses steam heat under pressure to sterilize chemicals and objects that can tolerate moist heat.

   ▶**VIDEO TUTOR:** *Principles of Autoclaving*
3. Pasteurization, a method of heating foods to kill pathogens and control spoilage organisms without altering the quality of the food, can be achieved by several methods: the historical (batch) method, flash pasteurization, and ultra-high-temperature pasteurization. The methods differ in their combinations of temperature and time of exposure.
4. Under certain circumstances, microbes can be controlled using ultra-high-temperature sterilization, dry-heat sterilization, incineration, refrigeration, or freezing.
5. Antimicrobial methods involving drying are **desiccation**, used to preserve food, and **lyophilization** (freeze drying), used for the long-term preservation of cells or microbes.
6. When used as a microbial control method, **filtration** is the passage of air or a liquid through a material that traps and removes microbes. Some **membrane filters** have pores small enough to trap the smallest viruses. HEPA (high-efficiency particulate air) filters remove microbes and particles from air.
7. The high **osmotic pressure** exerted by hypertonic solutions of salt or sugar can preserve foods such as jerky and jams by removing from microbes the water that they need to carry out their metabolic functions.
8. **Radiation** includes high-speed subatomic particles and even more energetic electromagnetic waves released from atoms. **Ionizing radiation** (wavelengths shorter than 1 nm) produces ions that denature important molecules and kill cells. **Nonionizing radiation** (wavelengths longer than 1 nm) is less effective in microbial control, although UV light causes pyrimidine dimers, which can kill affected cells.

Chemical Methods of Microbial Control (pp. 275–282)

1. **Phenolics**, which are chemically modified phenol molecules, are intermediate- to low-level disinfectants that denature proteins and disrupt cell membranes in a wide variety of pathogens.
2. **Alcohols** are intermediate-level disinfectants that denature proteins and disrupt cell membranes; they are used either as 70% to 90% aqueous solutions or in a tincture, which is a combination of an alcohol and another antimicrobial chemical.
3. **Halogens** (iodine, chlorine, bromine, and fluorine) are used as intermediate-level disinfectants and antiseptics to kill microbes by protein denaturation in water or on medical instruments or skin.
4. **Oxidizing agents** such as hydrogen peroxide, ozone, and peracetic acid are high-level disinfectants and antiseptics that release oxygen radicals, which are toxic to many microbes, especially anaerobes.
5. **Surfactants** include soaps, which act primarily to break up oils during degerming, and **detergents**, such as **quaternary ammonium compounds (quats)**, which are low-level disinfectants.
6. **Heavy-metal ions**, such as arsenic, silver, mercury, copper, and zinc, are low-level disinfectants that denature proteins. For most applications they have been superseded by less toxic alternatives.
7. **Aldehydes** are high-level disinfectants that cross-link organic functional groups in proteins and nucleic acids. A 2% solution of glutaraldehyde or a 37% aqueous solution of formaldehyde (called formalin) is used to disinfect or sterilize medical or dental equipment and in embalming fluid.
8. Gaseous agents of microbial control, which include ethylene oxide, propylene oxide, and beta-propiolactone, are high-level disinfecting agents used to sterilize heat-sensitive equipment and large objects. These gases are explosive and potentially carcinogenic.
9. Many organisms use **antimicrobial enzymes** to combat microbes. Humans use them commercially in food preservation and as a noncaustic, nondestructive way to eliminate prions on medical instruments.
10. **Antimicrobials**, which include antibiotics, semisynthetics, and synthetics, are compounds that are typically used to treat diseases but can also function as intermediate-level disinfectants.
11. Four methods for evaluating the effectiveness of a disinfectant or antiseptic are the **phenol coefficient**, the **use-dilution test**, the **Kelsey-Sykes capacity test**, and the **in-use test**, which provides a more accurate determination of efficacy under real-life conditions.

QUESTIONS FOR REVIEW

Answers to the Questions for Review (except Short Answer questions) begin on p. A-1.

Multiple Choice

1. In practical terms in everyday use, which of the following statements provides the definition of *sterilization*?
   a. Sterilization eliminates organisms and their spores or endospores.
   b. Sterilization eliminates harmful microorganisms and viruses.
   c. Sterilization eliminates prions.
   d. Sterilization eliminates hyperthermophiles.
2. Which of the following substances or processes kills microorganisms on laboratory surfaces?
   a. antiseptics
   b. disinfectants
   c. degermers
   d. pasteurization

3. Which of the following terms best describes the disinfecting of cafeteria plates?
   a. pasteurization
   b. antisepsis
   c. sterilization
   d. sanitization
4. The microbial death rate is used to measure ________________.
   a. the efficiency of a detergent
   b. the efficiency of an antiseptic
   c. the efficiency of sanitization techniques
   d. all of the above
5. Which of the following statements is true concerning the selection of an antimicrobial agent?
   a. An ideal antimicrobial agent is stable during storage.
   b. An ideal antimicrobial agent is fast acting.
   c. Ideal microbial agents do not exist.
   d. all of the above
6. The endospores of which organism are used as a biological indicator of sterilization?
   a. *Bacillus stearothermophilus*
   b. *Salmonella enterica*
   c. *Mycobacterium tuberculosis*
   d. *Staphylococcus aureus*
7. A company that manufactures an antimicrobial cleaner for kitchen counters claims that its product is effective when used in a 50% water solution. By what means might scientists best verify this statement?
   a. disk-diffusion test
   b. phenol coefficient
   c. filter paper test
   d. in-use test
8. Which of the following items functions most like an autoclave?
   a. a boiling pan
   b. an incinerator
   c. a microwave oven
   d. a pressure cooker
9. The preservation of beef jerky from microbial growth relies on which method of microbial control?
   a. filtration
   b. lyophilization
   c. desiccation
   d. radiation
10. Which of the following types of radiation is more widely used as an antimicrobial technique?
    a. electron beams
    b. visible light waves
    c. radio waves
    d. microwaves
11. Which of the following substances would most effectively inhibit anaerobes?
    a. phenol
    b. silver
    c. ethanol
    d. hydrogen peroxide
12. Which of the following adjectives best describes a surgical procedure that is free of microbial contaminants?
    a. disinfected
    b. sanitized
    c. degermed
    d. aseptic
13. Biosafety Level 3 includes ________________.
    a. double sets of entry doors
    b. pressurized suits
    c. showers in entryways
    d. all of the above
14. A sample of *E. coli* has been subjected to heat for a specified time, and 90% of the cells have been destroyed. Which of the following terms best describes this event?
    a. thermal death point
    b. thermal death time
    c. decimal reduction time
    d. none of the above
15. Which of the following substances is least toxic to humans?
    a. carbolic acid
    b. glutaraldehyde
    c. hydrogen peroxide
    d. formalin
16. Which of the following chemicals is active against bacterial endospores?
    a. copper ions
    b. ethylene oxide
    c. ethanol
    d. triclosan
17. Which of the following disinfectants acts against cell membranes?
    a. phenol
    b. peracetic acid
    c. silver nitrate
    d. glutaraldehyde
18. Which of the following disinfectants contains alcohol?
    a. an iodophor
    b. a quat
    c. formalin
    d. a tincture of bromine
19. Which antimicrobial chemical has been used to sterilize spacecraft?
    a. phenol
    b. alcohol
    c. heavy metal
    d. ethylene oxide
20. Which class of surfactant is most soluble in water?
    a. quaternary ammonium compounds
    b. alcohols
    c. soaps
    d. peracetic acids

Short Answer

1. Describe three types of microbes that are extremely resistant to antimicrobial treatment, and explain why they are resistant.
2. Compare and contrast four tests that have been developed to measure the effectiveness of disinfectants.
3. Why is it necessary to use strong disinfectants in areas exposed to tuberculosis patients?
4. Why do warm disinfectant chemicals generally work better than cool ones?
5. Why are Gram-negative bacteria more susceptible to heat than Gram-positive bacteria?
6. Describe five physical methods of microbial control.
7. What is the difference between thermal death point and thermal death time?
8. Defend the following statement: "Pasteurization is not sterilization."
9. Compare and contrast desiccation and lyophilization.
10. Compare and contrast the action of alcohols, halogens, and oxidizing agents in controlling microbial growth.
11. Hyperthermophilic prokaryotes may remain viable in canned goods after commercial sterilization. Why is this situation not dangerous to consumers?
12. Why are alcohols more effective in a 70% solution than in a 100% solution?
13. Contrast the structures and actions of soaps and quats.
14. What are some advantages and disadvantages of using ionizing radiation to sterilize food?
15. How can campers effectively treat stream water to remove pathogenic protozoa, bacteria, and viruses?

VISUALIZE IT!

1. Calculate the decimal reduction time (D) for the two temperatures in the following graph.

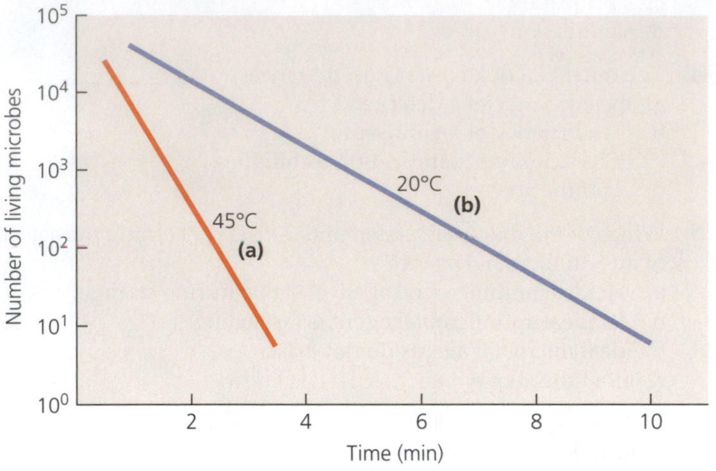

2. Indian tradition holds that storing water in brass pitchers prevents disease. British and Indian scientists have discovered that there is some truth in the tradition. The researchers collected river water samples and found fecal bacterial counts as high as 1 million bacteria per milliliter. However, the scientists could detect no bacteria in the water after it had been stored for two days in traditional brass pitchers. Bacterial levels in plastic or earthenware containers remained high over the same period. How can brass, which is an alloy of copper mixed with zinc, make water safer to drink?

CRITICAL THINKING

1. In 2004, a casino paid $28,000 for a grilled cheese sandwich that was purported to have an image of the Virgin Mary on it. The seller had stored the sandwich in a less-than-airtight box for 10 years without decay or the growth of mold. What antimicrobial chemical and physical agents might account for the longevity of the sandwich?
2. Is desiccation the only antimicrobial effect operating when grapes are dried in the sun to make raisins? Explain.
3. How long would it take to reduce a population of 100 trillion (10^{14}) bacteria to 10 viable cells if the D value of the treatment is 3 minutes?
4. Some potentially pathogenic bacteria and fungi, including strains of *Enterococcus, Staphylococcus, Candida,* and *Aspergillus,* can survive for one to three months on a variety of materials found in hospitals, including scrub suits, lab coats, and plastic aprons and computer keyboards. What can hospital personnel do to reduce the spread of these pathogens?
5. Over 1000 people developed severe diarrhea, and at least four died, from a strain of *Salmonella enterica* in the fall of 2012. Epidemiologists determined that infection resulted from the consumption of contaminated smoked salmon. Based on this chapter, what methods to control microbial growth are available to fishermen, packers, and stores that might have prevented such an outbreak? What other precautions could consumers have taken?
6. An over-the-counter medicated foot powder contains camphor, eucalyptus oil, lemon oil, and zinc oxide. Only one of the ingredients is a proven antimicrobial. Which one? How does it act against fungi?
7. Tsunamis and hurricanes severely contaminate water wells and disrupt water supply lines. What immediate steps should people take to lessen the spread of waterborne illnesses such as cholera?
8. In what ways might it be argued that the widespread commercial use of antiseptics and disinfectants has hurt rather than helped American health?
9. Explain why quaternary ammonium compounds are not very effective against mycobacteria such as *Mycobacterium tuberculosis.*
10. A student inoculates *Escherichia coli* into two test tubes containing the same sterile liquid medium, except that the first tube also contains a drop of a chemical with an antimicrobial effect. After 24 hours of incubation, the first tube remains clear, whereas the second tube has become cloudy with bacteria. Design an experiment to determine whether this amount of the antimicrobial chemical is *bacteriostatic* or *bactericidal* against *E. coli.*
11. Would you expect Gram-negative bacteria or Gram-positive bacteria to be more susceptible to antimicrobial chemicals that act against cell walls? Explain your answer, which you should base solely on the nature of the cells' walls (see Figure 3.15).
12. Where should you place a sterilization indicator within an autoclave? Explain your reasoning.
13. Why is liquid water necessary for microbial metabolism?
14. A virologist needs to remove all bacteria from a solution containing viruses without removing the viruses. What size membrane filter should the scientist use?
15. During the fall 2001 bioterrorist attack in which anthrax endospores were sent through the mail, one news commentator suggested that people should iron all their incoming mail with a regular household iron as a means of destroying endospores. Would you agree that this is a good way to disinfect mail? Explain your answer. Which disinfectant methods would be both more effective and more practical?
16. What common household antiseptic contains a heavy metal as its active ingredient?
17. What is the phenol coefficient of phenol when used against *Staphylococcus?*

CONCEPT MAPPING

Using the following terms, draw a concept map that describes moist heat applications to control microorganisms. You may use some terms more than once. For a sample concept map, see p. 94. Or, complete this and other concept maps online by going to the MasteringMicrobiology Study Area.

100°C, ≥ 10 min.
121°C, 15 psi, ≥ 15 min.
134°C, 1 sec.
140°C, 1–3 sec.
63°C, 30 min.
72°C, 15 sec.
Autoclave
Batch method (classic)
Boiling water
Equivalent treatments
Flash pasteurization
Fungi
Most viruses
Pasteurization
Protozoan trophozoites
Sterilization technique (2)
Thermoduric microorganisms (2)
Ultra-high-temperature pasteurization
Ultra-high-temperature sterilization
Vegetative bacterial cells

10 Controlling Microbial Growth in the Body: Antimicrobial Drugs

MICRO IN THE CLINIC

Battling a Microscopic Enemy

Like his father and grandfather before him, Marine sergeant Ben is proud to serve his country. He is nearing the end of his second tour of duty in Afghanistan and is looking forward to returning home to his wife and son. However, two weeks before he is scheduled to leave Afghanistan, Ben is seriously wounded in a suicide bomber attack and loses his left leg. He initially responds well to treatment, but three days into his recovery, his wound flap becomes infected. His doctors are faced with a difficult situation: They need to act fast, but they don't yet know what bacterium is causing the infection. They decide to administer a course of antimicrobial drugs—trimethoprim-sulfamethoxazole—that is designed to treat multiple species of bacteria commonly involved in wound injuries.

Unfortunately, the antimicrobials fail to improve Ben's condition, and the infection visibly worsens. Ben's doctors decide to try another drug—ampicillin—but after 24 hours, it becomes clear that this has no effect either. On day 6, Ben begins to show signs of *septicemia*—an infection of the blood. The bacteria have now invaded Ben's circulatory system and have access to every part of Ben's body, including his heart, brain, and other vital organs. His doctors must come up with a solution swiftly, or Ben will have survived the suicide bombing attack only to be felled by a tiny bacterial enemy.

How can bacteria survive rigorous antimicrobial therapy? Is there treatment for Ben's life-threatening septicemia? Turn to the end of the chapter (p. 317) to find out.

Explore More: Test your readiness and apply your knowledge with dynamic learning tools at MasteringMicrobiology.

Chemicals that affect physiology in any manner, such as caffeine, alcohol, and tobacco, are called *drugs.* Drugs that act against diseases are called *chemotherapeutic agents.* Examples include insulin, anticancer drugs, and drugs for treating infections—called **antimicrobial agents (antimicrobials),** the subject of this chapter.

In the pages that follow we'll examine the mechanisms by which antimicrobial agents act, the factors that must be considered in the use of antimicrobials, and several issues surrounding resistance to antimicrobial agents among microorganisms. First, however, we begin with a brief history of antimicrobial chemotherapy.

The History of Antimicrobial Agents

LEARNING **OUTCOMES**

10.1 Describe the contributions of Paul Ehrlich, Alexander Fleming, and Gerhard Domagk in the development of antimicrobials.

10.2 Explain how semisynthetic and synthetic antimicrobials differ from antibiotics.

The little girl lay struggling to breathe as her parents stood mutely by, willing the doctor to do something—anything—to relieve the symptoms that had so quickly consumed their four-year-old daughter's vitality. Sadly, there was little the doctor could do. The thick "pseudomembrane" of diphtheria, composed of bacteria, mucus, blood-clotting factors, and white blood cells, adhered tenaciously to her pharynx, tonsils, and vocal cords. He knew that trying to remove it could rip open the underlying mucous membrane, resulting in bleeding, possibly additional infections, and death. In 1902, there was little medical science could offer for the treatment of diphtheria; all physicians could do was wait and hope.

At the beginning of the 20th century, much of medicine involved diagnosing illness, describing its expected course, and telling family members either how long a patient might be sick or when they might expect her to die. Even though physicians and scientists had recently accepted the germ theory of disease and knew the causes of many diseases, very little could be done to inhibit pathogens, including *Corynebacterium diphtheriae* (kō-rī´nē-bak-tēr´ē-ŭm dif-thi´rē-ī), and alter the course of infections. In fact, one-third of children born in the early 1900s died from infectious diseases before the age of five.

It was at this time that Paul Ehrlich (1854–1915), a visionary German scientist, proposed the term *chemotherapy* to describe the use of chemicals that would selectively kill pathogens while having little or no effect on a patient. He wrote of "magic bullets" that would bind to receptors on germs to bring about their death while ignoring host cells, which lacked the receptor molecules.

Ehrlich's search for antimicrobial agents resulted in the discovery of one arsenic compound that killed trypanosome parasites and another that worked against the bacterial agent of syphilis. A few years later, in 1928, the British bacteriologist Alexander Fleming (1881–1955) reported the antibacterial action of penicillin released from *Penicillium* (pen-i-sil´ē-ŭm) mold, which creates a zone where bacteria don't grow (**FIGURE 10.1**).

Though arsenic compounds and penicillin were discovered first, they were not the first antimicrobials in widespread

CLINICAL CASE STUDY

Antibiotic Overkill

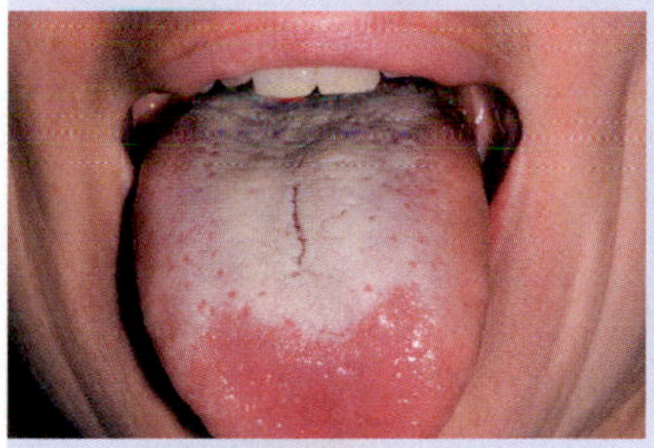

A young woman was taking antibiotic pills for a urinary infection. Several days into her course of medication, she began to experience peculiar symptoms. At first they were hardly noticeable. Very quickly, however, they worsened and became embarrassing and unbearable.

She noticed a white coating on her tongue, bad breath, and an awful taste in her mouth. Despite persistent brushing and mouthwash applications, she was unable to completely remove the film. Furthermore, she had excessive vaginal discharges consisting of a cheeselike white substance. When she began to have vaginal itching, she finally decided it was time to seek help.

Reluctantly she revisited her personal physician and described the symptoms. Her doctor explained the symptoms and provided additional prescriptions to alleviate her distress.

1. What happened to the young woman in this situation?
2. How had her body's defenses been violated?
3. How can she avoid a repeat of this situation?

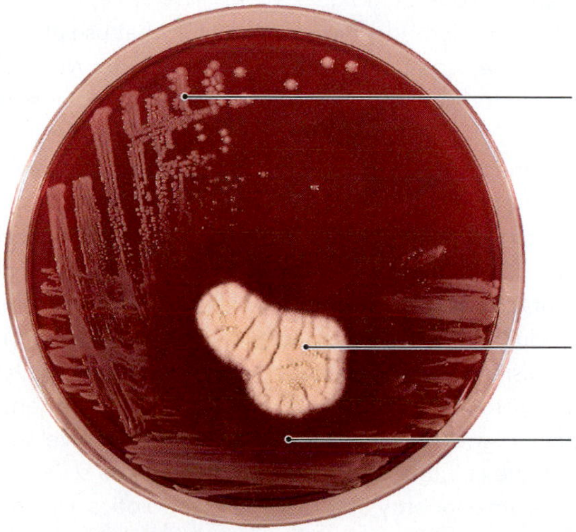

▲ **FIGURE 10.1 Antibiotic effect of the mold *Penicillium chrysogenum*.** Alexander Fleming observed that this mold secretes penicillin, which inhibits the growth of bacteria, as is apparent with *Staphylococcus aureus* growing on this blood agar plate.

use: Ehrlich's arsenic compounds are toxic to humans, and penicillin was not available in large enough quantities to be useful until the late 1940s. Instead, *sulfanilamide*, discovered in 1932 by the German chemist Gerhard Domagk (1895–1964), was the first practical antimicrobial agent efficacious in treating a wide array of bacterial infections.

Selman Waksman (1888–1973) discovered other microorganisms that are sources of useful antimicrobials, most notably species of soil-dwelling bacteria in the genus *Streptomyces* (strep-tō-mī´sēz). Waksman coined the term **antibiotics** to describe antimicrobial agents that are produced naturally by an organism (**Highlight: Microbe Altruism: Why Do They Do It?**). In common usage today, "antibiotic" denotes an antibacterial agent, including synthetic compounds and excluding agents with antiviral and antifungal activity.

Other scientists produced **semisynthetic antimicrobials**—chemically altered antibiotics—that are more effective, longer lasting, or easier to administer than naturally occurring antibiotics. Antimicrobials that are completely synthesized in a laboratory are called **synthetic drugs.** Most antimicrobials are either natural or semisynthetic.

TABLE 10.1 provides a partial list of common antibiotics and semisynthetics and their sources.

TELL ME WHY

Why aren't antibiotics effective against the common cold?

Mechanisms of Antimicrobial Action

LEARNING | OUTCOMES

10.3 Explain the principle of selective toxicity.

10.4 List six mechanisms by which antimicrobial drugs affect pathogens.

As Ehrlich foresaw, the key to successful chemotherapy against microbes is **selective toxicity;** that is, an effective antimicrobial agent must be more toxic to a pathogen than to the pathogen's host. Selective toxicity is possible because of differences in structure or metabolism between the pathogen and its host. Typically, the more differences, the easier it is to discover or create an effective antimicrobial agent.

HIGHLIGHT

Microbe Altruism: Why Do They Do It?

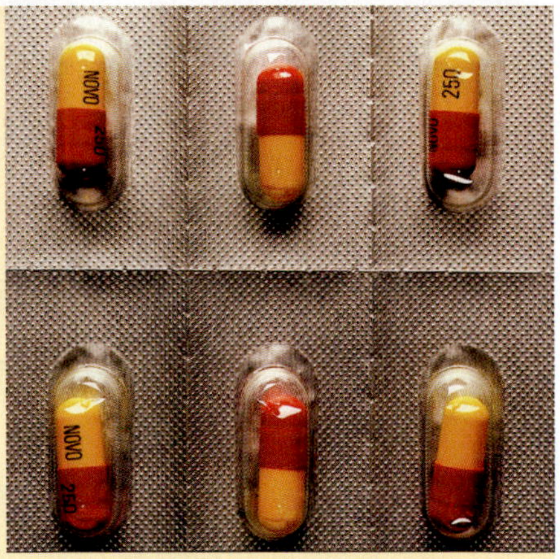

We all know that antibiotics benefit humans, but what good are they to the microorganisms that secrete them? From the viewpoint of evolutionary theory, the answer might seem obvious: Antibiotics are weapons against other microbes; they confer an advantage to the secreting organisms in their struggle for survival. In reality, however, the answer is not so simple.

Antibiotics are members of an extremely diverse group of metabolic products known as *secondary metabolites,* which typically are complex organic molecules that are not essential for normal cell growth and reproduction and are produced only after an organism has already established itself in its environment. The production of secondary metabolites results in a metabolic cost for the cell; that is, producing antibiotics consumes energy and raw materials that the organism could use for growth and reproduction. Tetracycline, for example, is the end result of 72 separate enzymatic steps, and erythromycin requires 28 different chemical reactions—none of which appears to contribute to the normal growth or reproduction of *Streptomyces.* Therefore, the question can be modified: Of what use are metabolically "expensive" antibiotics to organisms that are already secure in their environment?

Adding to the conundrum is the fact that antimicrobials against bacteria have never been discovered in natural soil at high enough concentrations to be inhibitory to neighboring cells. For example, it is almost impossible to detect antibiotics produced by *Streptomyces* except when the bacteria are grown in a laboratory. Minimal and inconsequential quantities of antibiotics hardly give an adaptive edge.

Some scientists have suggested that antibiotics are evolutionary vestiges—leftovers of metabolic pathways that were once useful but no longer have a significant role. However, there should be tremendous selective pressure against the slightest continued manufacture of complex antibiotics if they truly have little purpose for the microorganism. It is more likely that antibiotics are signals used for interbacterial communication within biofilms and that their antimicrobial action is coincidental. Further research is required before we may fully answer the question,"Why do microbes make antibiotics?"

TABLE 10.1 Sources of Some Common Antibiotics and Semisynthetics

Microorganism	Antimicrobial
Fungi	
Penicillium chrysogenum	Penicillin
Penicillium griseofulvum	Griseofulvin
Acremonium[a] *spp.*[b]	Cephalothin
Bacteria	
Amycolatopsis orientalis	Vancomycin
Amycolatopsis rifamycinica	Rifampin
Bacillus licheniformis	Bacitracin
Bacillus polymyxa	Polymyxin
Micromonospora purpurea	Gentamicin
Pseudomonas fluorescens	Mupirocin
Saccharopolyspora erythraea	Erythromycin
Streptomyces griseus	Streptomycin
Streptomyces fradiae	Neomycin
Streptomyces aureofaciens	Tetracycline
Streptomyces venezuelae	Chloramphenicol
Streptomyces nodosus	Amphotericin B
Streptomyces avermitilis	Ivermectin

[a]This genus was formerly called *cephalosporium*.
[b]spp. is the abbreviation for multiple species of a genus.

Because there are many differences between the structure and metabolism of pathogenic bacteria and their eukaryotic hosts, antibacterial drugs constitute the greatest number and diversity of antimicrobial agents. Fewer antifungal, antiprotozoan, and anthelmintic drugs are available because fungi, protozoa, and helminths—like their animal and human hosts—are eukaryotic and thus share many common features. The number of effective antiviral drugs is also limited, despite major differences in structure, because viruses utilize their host cells' enzymes and ribosomes to metabolize and replicate. Therefore, drugs that are effective against viral replication are likely toxic to the host as well.

Although they can have a variety of effects on pathogens, antimicrobial drugs can be categorized into several general groups according to their mechanisms of action (**FIGURE 10.2**):

- Drugs that inhibit cell wall synthesis. These drugs are selectively toxic to certain fungal or bacterial cells, which have cell walls, but not to animals, which lack cell walls.
- Drugs that inhibit protein synthesis (translation) by targeting the differences between prokaryotic and eukaryotic ribosomes.
- Drugs that disrupt unique components of the cytoplasmic membrane.
- Drugs that inhibit general metabolic pathways not used by humans.
- Drugs that inhibit nucleic acid synthesis.
- Drugs that block a pathogen's recognition of or attachment to its host.

In the following sections we examine these mechanisms in turn.
▶**ANIMATIONS:** ***Chemotherapeutic Agents: Modes of Action***

Inhibition of Cell Wall Synthesis

LEARNING **OUTCOME**

10.5 Describe the actions and give examples of drugs that affect the cell walls of bacteria and fungi.

A cell wall protects a cell from the effects of osmotic pressure. Both pathogenic bacteria and fungi have cell walls, which animals and humans lack. First, we examine drugs that act against bacterial cell walls.

Inhibition of Synthesis of Bacterial Walls

The major structural component of a bacterial cell wall is its peptidoglycan layer. Peptidoglycan is a huge macromolecule composed of polysaccharide chains of alternating *N*-acetylglucosamine (NAG) and *N*-acetylmuramic acid (NAM) molecules that are cross-linked by short peptide chains extending between NAM subunits (see Figure 3.14). To enlarge or divide, a cell must synthesize more peptidoglycan by adding new NAG and NAM subunits to existing NAG-NAM chains, and the new NAM subunits must then be bonded to neighboring NAM subunits (**FIGURE 10.3a** and **b**).

Many common antibacterial agents act by preventing the cross-linkage of NAM subunits. Most prominent among these drugs are **beta-lactams** (such as penicillins, cephalosporins, and carbapenems), which are antimicrobials whose functional portions are called *beta-lactam (β-lactam) rings* (**FIGURE 10.3c**). Beta-lactams inhibit peptidoglycan formation by irreversibly binding to the enzymes that cross-link NAM subunits (**FIGURE 10.3d**). In the absence of correctly formed peptidoglycan, growing bacterial cells have weakened cell walls that are less resistant to the effects of osmotic pressure. The underlying cytoplasmic membrane bulges through the weakened portions of cell wall as water moves into the cell, and eventually the cell lyses (**FIGURE 10.3e**).

Chemists have made alterations to natural beta-lactams, such as penicillin G, to create semisynthetic derivatives such as methicillin and imipenem (see Figure 10.3c), which are more stable in the acidic environment of the stomach, more readily absorbed in the intestinal tract, less susceptible to deactivation by bacterial enzymes, or more active against more types of bacteria.

Other antimicrobials such as **vancomycin** (van-kō-mī´sin), which is obtained from *Amycolatopsis orientalis* (am-ē-kō´la-top-sis o-rē-en-tal'is), and **cycloserine,** a semisynthetic, disrupt cell wall formation in a different manner. They directly interfere with particular alanine-alanine bridges that link the NAM subunits in many Gram-positive bacteria. Those bacteria that lack alanine-alanine crossbridges are naturally resistant to these drugs. Still another drug that prevents cell wall formation,

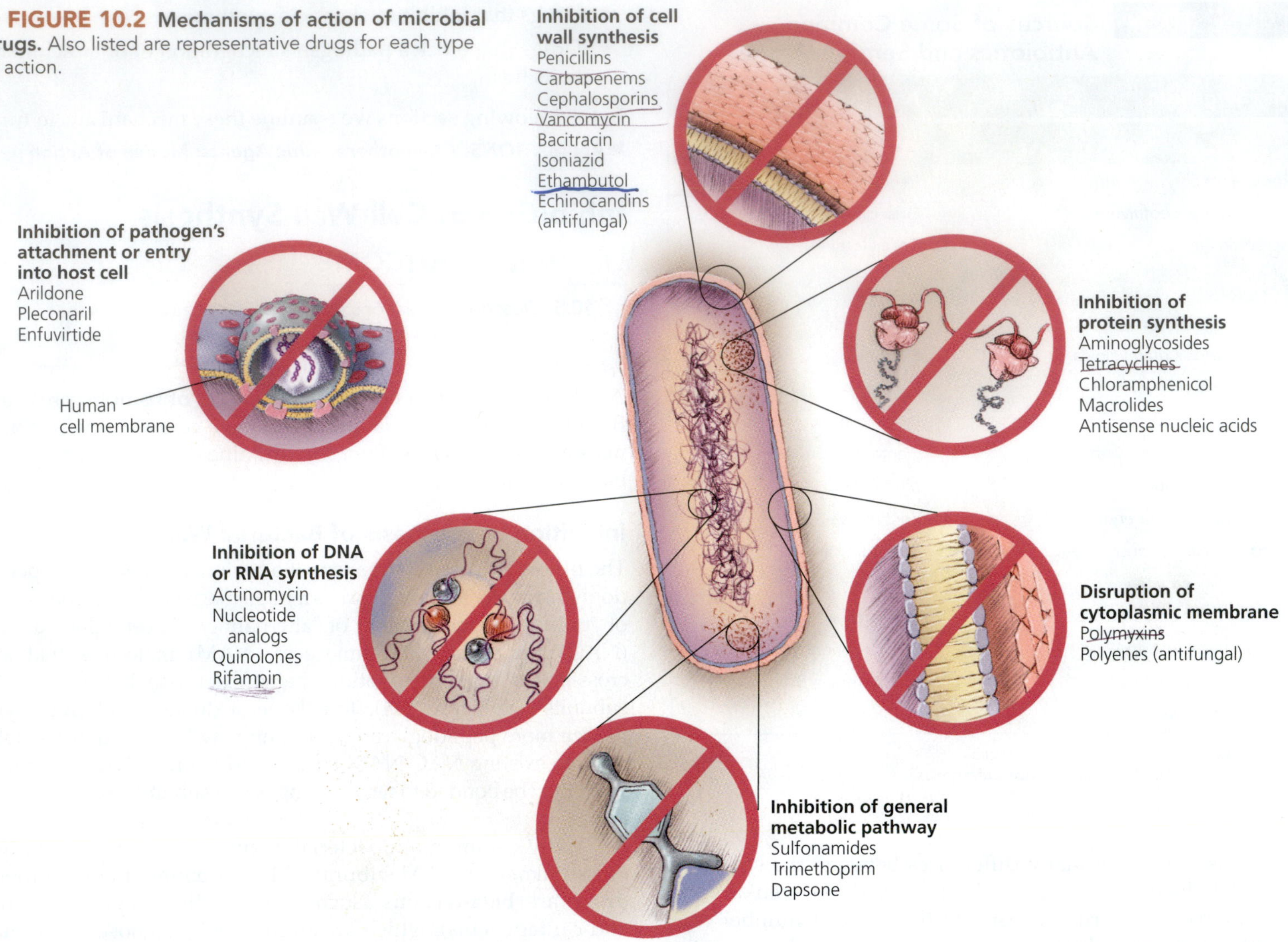

▶ **FIGURE 10.2 Mechanisms of action of microbial drugs.** Also listed are representative drugs for each type of action.

bacitracin (bas-i-trā'sin), blocks the transport of NAG and NAM from the cytoplasm out to the wall. Like beta-lactams, vancomycin, cycloserine, and bacitracin result in cell lysis due to the effects of osmotic pressure.

Since all these drugs prevent bacteria from *increasing* the amount of cell wall material but have no effect on existing peptidoglycan, they are effective only on bacterial cells that are growing or reproducing; dormant cells are unaffected.

Bacteria of the genus *Mycobacterium* (mī'kō-bak-tēr'ē-ŭm), notably the agents of leprosy and tuberculosis, are characterized by unique, complex cell walls that have a layer of arabinogalactan–mycolic acid in addition to the usual peptidoglycan of prokaryotic cells. **Isoniazid** (ī-sō-nī'ă-zid), or **INH**,[1] and **ethambutol** (eth-am'boo-tol) disrupt the formation of this extra layer. Mycobacteria typically reproduce only every 12 to 24 hours, in part because of the complexity of their cell walls, so antimicrobial agents that act against mycobacteria must be administered for months or even years to be effective. It is often difficult to ensure that patients continue such a long regimen of treatment.

[1]From *isonicotinic acid hydrazide*, the correct chemical name for isoniazid.

Inhibition of Synthesis of Fungal Walls

Fungal cell walls are composed of various polysaccharides containing a sugar, 1,3-D-glucan, that is not found in mammalian cells. A new class of antifungal drugs called **echinocandins,** among them *caspofungin,* inhibit the enzyme that synthesizes glucan; without glucan, fungal cells cannot make cell walls, leading to osmotic rupture.

Inhibition of Protein Synthesis

LEARNING | **OUTCOME**

10.6 Describe the actions and give examples of six antimicrobial drugs that interfere with protein synthesis.

Cells use proteins for structure and regulation, as enzymes in metabolism, and as channels and pumps to move materials across cell membranes. Thus, a consistent supply of proteins is vital for the active life of a cell. Given that all cells, including human cells, use ribosomes to translate proteins using information from messenger RNA templates, it is not immediately obvious

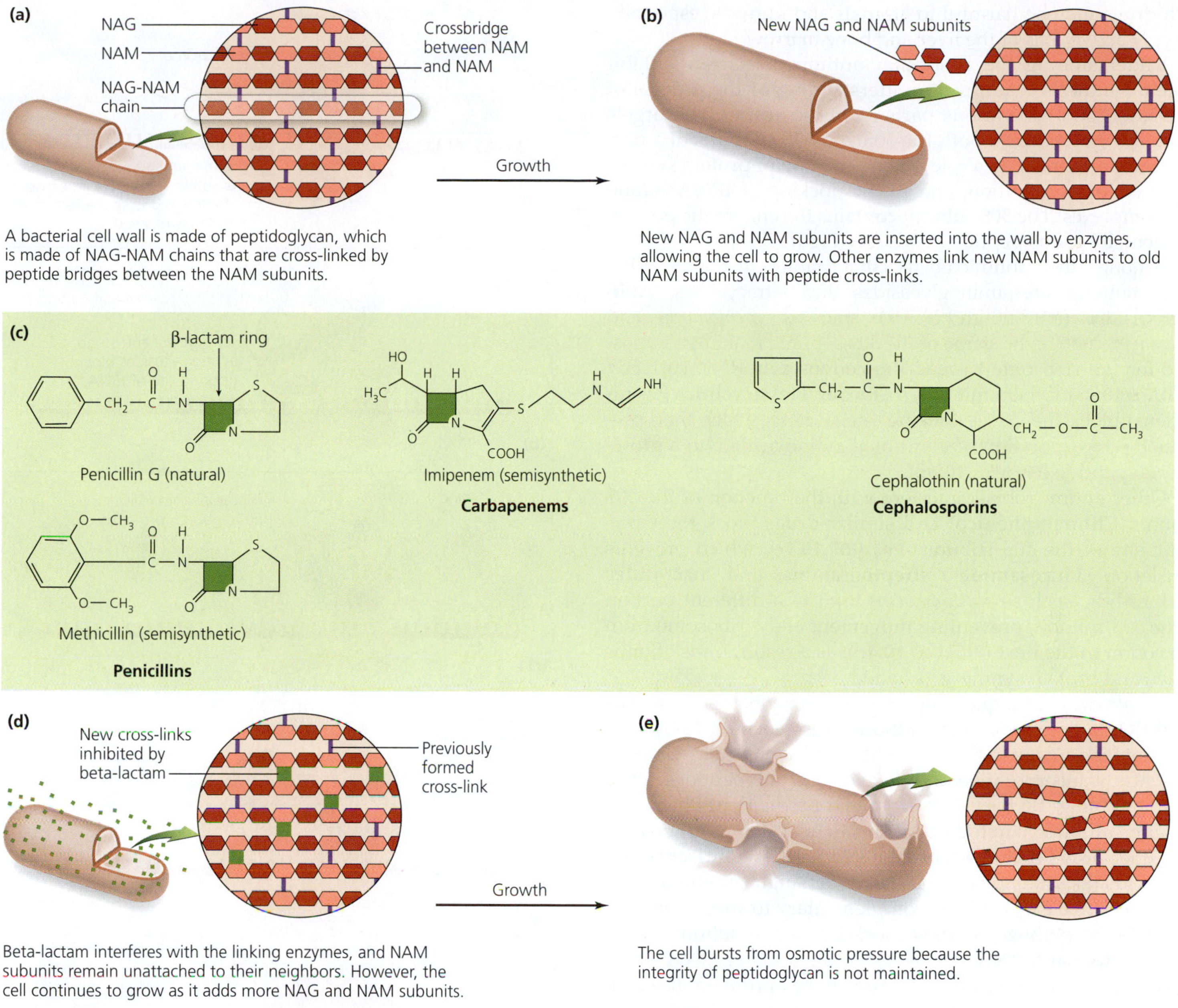
(a)
NAG
NAM
NAG-NAM chain
Crossbridge between NAM and NAM
Growth
A bacterial cell wall is made of peptidoglycan, which is made of NAG-NAM chains that are cross-linked by peptide bridges between the NAM subunits.
(b)
New NAG and NAM subunits
New NAG and NAM subunits are inserted into the wall by enzymes, allowing the cell to grow. Other enzymes link new NAM subunits to old NAM subunits with peptide cross-links.
(c)
β-lactam ring
Penicillin G (natural)
Methicillin (semisynthetic)
Penicillins
Imipenem (semisynthetic)
Carbapenems
Cephalothin (natural)
Cephalosporins
(d)
New cross-links inhibited by beta-lactam
Previously formed cross-link
Growth
Beta-lactam interferes with the linking enzymes, and NAM subunits remain unattached to their neighbors. However, the cell continues to grow as it adds more NAG and NAM subunits.
(e)
The cell bursts from osmotic pressure because the integrity of peptidoglycan is not maintained.

▲ **FIGURE 10.3 Bacterial cell wall synthesis and the inhibitory effects of beta-lactams on it. (a)** A schematic depiction of a normal peptidoglycan cell wall showing NAG-NAM chains and cross-linked NAM subunits. **(b)** A bacterium grows by adding new NAG and NAM subunits and linking new NAM subunits to older ones. **(c)** Structural formulas of some beta-lactam drugs. Their functional portion is the beta-lactam ring. **(d)** A schematic depiction of the effect of penicillin on peptidoglycan in preventing NAM-NAM cross-links. **(e)** Bacterial lysis due to the effects of a beta-lactam drug. *After a beta-lactam weakens a peptidoglycan molecule by preventing NAM-NAM cross-linkages, what force actually kills an affected bacterial cell?*

Figure 10.3 *Osmotic movement of water into the cell causes cells to lyse.*

that drugs could selectively target differences related to protein synthesis. However, prokaryotic ribosomes differ from eukaryotic ribosomes in structure and size: Prokaryotic ribosomes are 70S and composed of 30S and 50S subunits, whereas eukaryotic ribosomes are 80S with 60S and 40S subunits (see Figure 7.15).

Many antimicrobial agents take advantage of the differences between ribosomes to selectively target bacterial protein translation without significantly affecting eukaryotes. Note, however, that because some of these drugs affect eukaryotic mitochondria, which also contain 70S ribosomes like those of prokaryotes, such drugs may be harmful to animals and humans, especially the very active cells of the liver and bone marrow.

Understanding the actions of antimicrobials that inhibit protein synthesis requires an understanding of the process of translation because various parts of ribosomes are the targets of antimicrobial drugs. Both the 30S and 50S subunits of a prokaryotic ribosome play a role in the initiation of protein synthesis, in codon recognition, and in the docking of tRNA–amino acid complexes. The 50S subunit contains the enzymatic portion that actually forms peptide bonds (see Figure 7.18).

Among the antimicrobials that target the 30S ribosomal subunit are aminoglycosides and tetracyclines. **Aminoglycosides** (am´i-nō-glī´kō-sīds), such as *streptomycin* and *gentamicin*, change the shape of the 30S subunit, making it impossible for the ribosome to read the codons of mRNA correctly **(FIGURE 10.4a)**. Other aminoglycosides and **tetracyclines** (tet-ră-sī´klēns) block the tRNA docking site (A site), which then prevents the incorporation of additional amino acids into a growing polypeptide **(FIGURE 10.4b)**.

Other antimicrobials interfere with the function of the 50S subunit. **Chloramphenicol** and similar drugs block the enzymatic site of the 50S subunit **(FIGURE 10.4c)**, which prevents translation. **Lincosamides, streptogramins,** and **macrolides** (mak´rō-līds, such as *erythromycin*) bind to a different portion of the 50S subunit, preventing movement of the ribosome from one codon to the next **(FIGURE 10.4d)**; as a result, translation is frozen and protein synthesis is halted.

Mupirocin is a unique drug that selectively binds to the bacterial tRNA that carries the amino acid isoleucine ($tRNA^{Ile}$). It does not bind to any of the eukaryotic tRNA molecules. Binding prevents the incorporation of isoleucine into polypeptides, effectively crippling the bacterium's protein production. Physicians prescribe mupirocin in topical creams to treat skin infections.

Other drugs that block protein synthesis are **antisense nucleic acids (FIGURE 10.4e)**. These RNA or single-stranded DNA molecules are designed to be complementary to specific mRNA molecules of pathogens. They block ribosomal subunits from attaching to that mRNA with no effect on human mRNA. *Fomivirsen* is the first of this class of drugs to be approved. It inactivates cytomegalovirus and is used to treat eye infections.

Oxazolidinones are antimicrobial drugs that work to stop protein synthesis by blocking initiation of translation **(FIGURE 10.4f)**. Oxazolidinones are used as a last resort in treating infections of Gram-positive bacteria resistant to other antimicrobials,

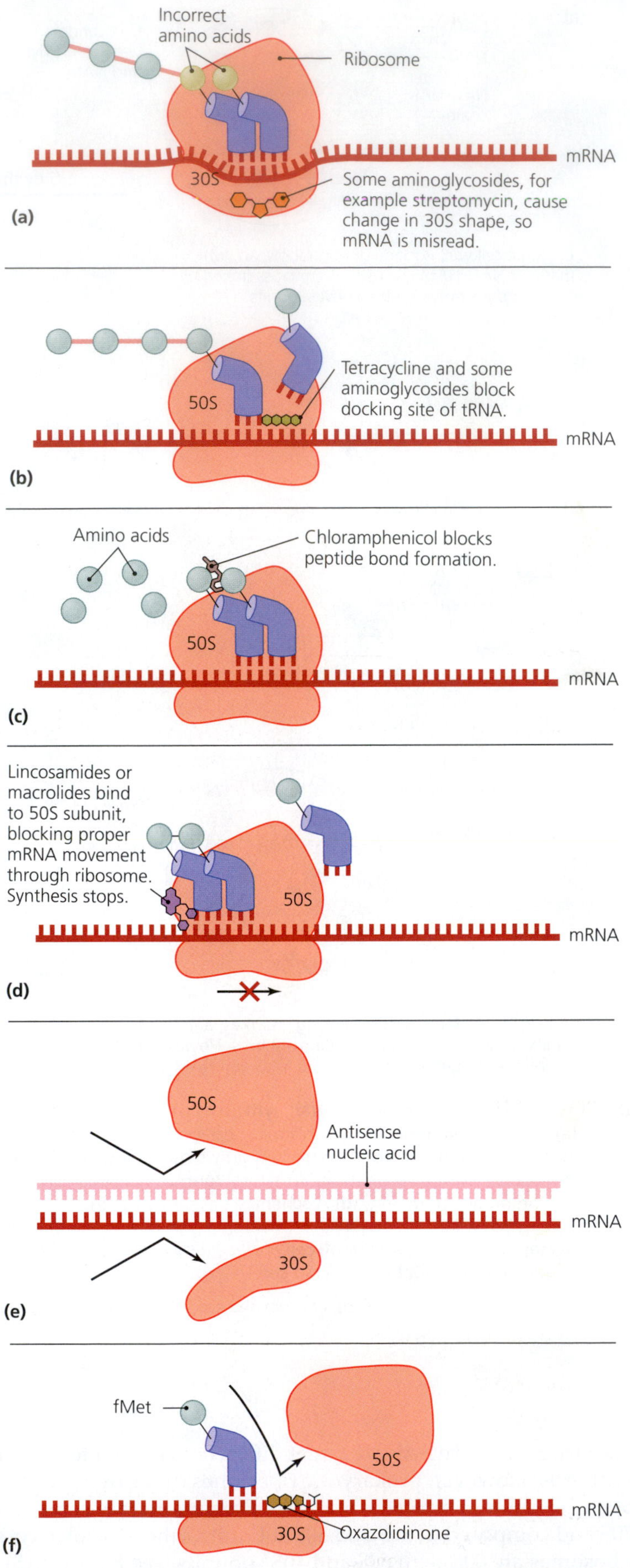

▶ **FIGURE 10.4 The mechanisms by which antimicrobials target prokaryotic ribosomes to inhibit protein synthesis. (a)** Aminoglycosides change the shape of the 30S subunit, causing incorrect pairing of tRNA anticodons with mRNA codons. **(b)** Tetracyclines block the tRNA docking site (A site) on the 30S subunit, preventing protein elongation. **(c)** Chloramphenicol blocks enzymatic activity of the 50S subunit, preventing the formation of peptide bonds between amino acids. **(d)** Lincosamides or macrolides bind to the 50S subunit, preventing movement of the ribosome along the mRNA. **(e)** Antisense nucleic acids bind to mRNA, blocking ribosomal subunits. **(f)** Oxazolidinones inhibit initiation of translation. *Which tRNA anticodon should align with the codon CUG?*

Figure 10.4 *Anticodon GAC is the correct complement for the codon CUG.*

including vancomycin- and methicillin-resistant *Staphylococcus aureus* (staf´i-lō-kok´ŭs o´rē-ŭs). ▶**VIDEO TUTOR:** ***Actions of Some Drugs that Inhibit Prokaryotic Protein Synthesis***

Disruption of Cytoplasmic Membranes

LEARNING | **OUTCOME**

10.7 Describe the action of antimicrobial drugs that interfere with cytoplasmic membranes.

Some antibacterial drugs, such as the short polypeptide *gramicidin,* disrupt the cytoplasmic membrane of a targeted cell, often by forming a channel through the membrane, damaging its integrity. This is also the mechanism of action of a group of antifungal drugs called **polyenes** (pol-ē-ēns´). The polyenes *nystatin* and *amphotericin B* **(FIGURE 10.5a)** are fungicidal because they attach to *ergosterol,* a lipid constituent of fungal membranes **(FIGURE 10.5b)**, in the process disrupting the membrane and causing lysis of the cell. The cytoplasmic membranes of humans are somewhat susceptible to amphotericin B because they contain cholesterol, which is similar to ergosterol, though cholesterol does not bind amphotericin B as well as does ergosterol.

Azoles, such as *fluconazole,* and **allylamines,** such as *terbinafine,* are two other classes of antifungal drugs that disrupt cytoplasmic membranes. They act by inhibiting the synthesis of ergosterol; without ergosterol, the cell's membrane does not remain intact, and the fungal cell dies. Azoles and allylamines are generally harmless to humans because human cells do not manufacture ergosterol.

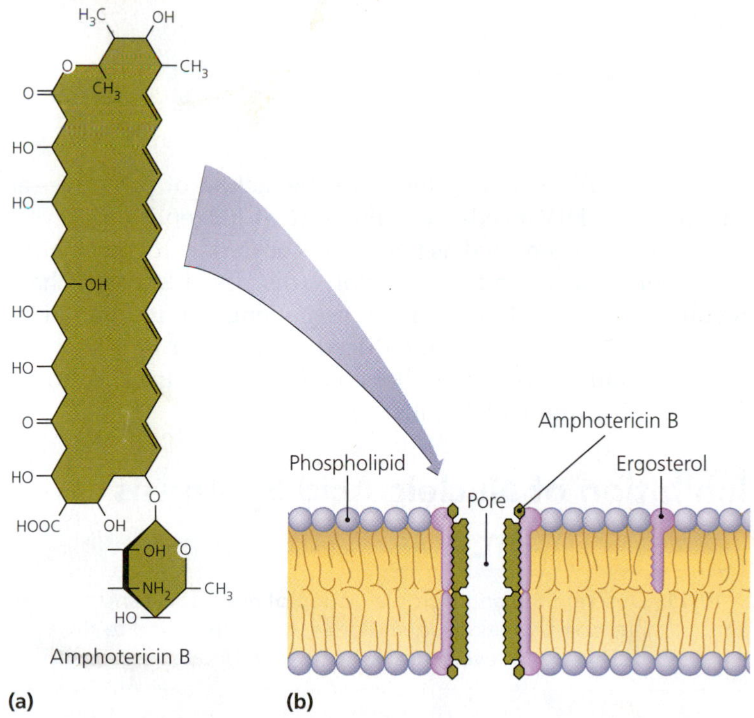

▲ **FIGURE 10.5 Disruption of the cytoplasmic membrane by the antifungal amphotericin B. (a)** The structure of amphotericin B. **(b)** The proposed action of amphotericin B. The drug binds to molecules of ergosterol, which then congregate, forming a pore.

Most bacterial membranes lack sterols, so these bacteria are naturally resistant to polyenes, azoles, and allylamines; however, there are other agents that disrupt bacterial membranes. An example of these antibacterial agents is *polymyxin,* produced by *Bacillus polymyxa* (ba-sil´ŭs po-lē-miks´ă). Polymyxin is effective against Gram-negative bacteria, particularly *Pseudomonas* (soo-dō-mō´nas), but because it is toxic to human kidneys, it is usually reserved for use against external pathogens that are resistant to other antibacterial drugs.

Pyrazinamide disrupts transport across the cytoplasmic membrane of *Mycobacterium tuberculosis* (too-ber-kyū-lō´sis). The pathogen uniquely activates and accumulates the drug. Unlike many other antimicrobials, pyrazinamide is most effective against intracellular, nonreplicating bacterial cells.

Some antiparasitic drugs also act against cytoplasmic membranes. For example, *praziquantel* and *ivermectin* change the permeability of cell membranes of several types of parasitic worms.

Inhibition of Metabolic Pathways

LEARNING | **OUTCOMES**

10.8 Explain the action of antimicrobials that disrupt synthesis of folic acid.

10.9 Define the term *analog* as it relates to antimicrobial drugs.

10.10 Describe the action of antiviral drug that interfere with metabolism.

Metabolism can be defined simply as the sum of all chemical reactions that take place within an organism. Whereas most living things share certain metabolic reactions—for example, glycolysis—other chemical reactions are unique to certain organisms. Whenever differences exist between the metabolic processes of a pathogen and its host, *antimetabolic agents* can be effective.

Various kinds of antimetabolic agents are available, including *atovaquone,* which interferes with electron transport in protozoa and fungi; heavy metals (such as arsenic, mercury, and antimony), which inactivate enzymes; agents that disrupt tubulin polymerization and glucose uptake by many protozoa and parasitic worms; drugs that block the activation of viruses; and metabolic antagonists, such as sulfanilamide, the first commercially available antimicrobial agent.

Sulfanilamide and similar compounds, collectively called **sulfonamides,** act as antimetabolic drugs because they are structural analogs of—that is, are chemically very similar to—*para-aminobenzoic acid* (*PABA;* **FIGURE 10.6a**). PABA is crucial in the synthesis of nucleotides required for DNA and RNA synthesis. Many organisms, including some pathogens, enzymatically convert PABA into dihydrofolic acid and then dihydrofolic acid into tetrahydrofolic acid (THF), a form of folic acid that is used as a coenzyme in the synthesis of purine and pyrimidine nucleotides **(FIGURE 10.6b)**. As analogs of PABA, sulfonamides compete with PABA molecules for the active site of the enzyme involved in the production of dihydrofolic acid **(FIGURE 10.6c)**. This competition leads to a decrease in the production of THF and thus of DNA and RNA. The end result of sulfonamide competition with PABA is the cessation of cell metabolism, which leads to cell death.

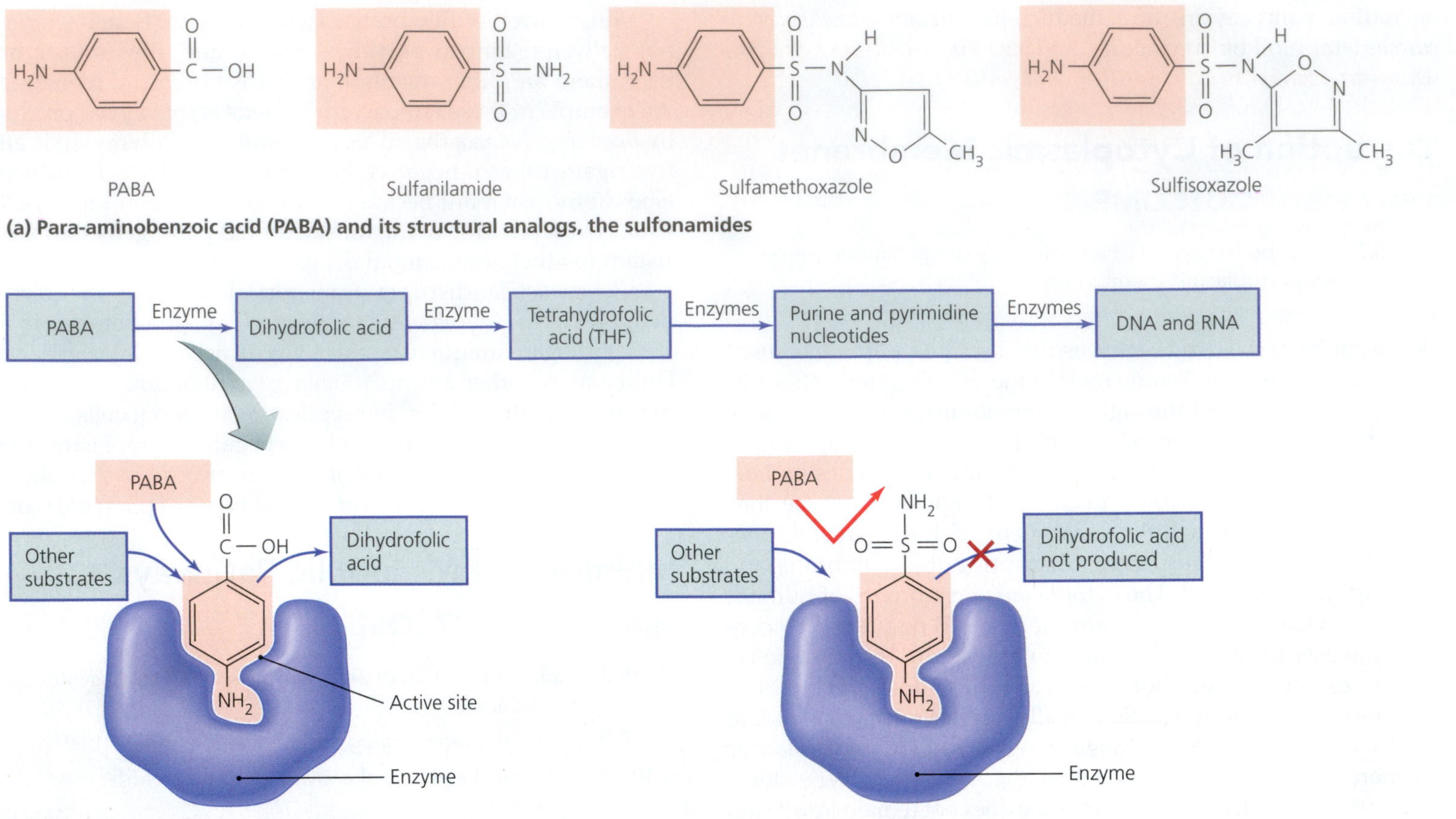

(a) Para-aminobenzoic acid (PABA) and its structural analogs, the sulfonamides

(b) Role of PABA in folic acid synthesis in bacteria and protozoa

(c) Inhibition of folic acid synthesis by sulfonamide

▲ **FIGURE 10.6 The antimetabolic action of sulfonamides in inhibiting nucleic acid synthesis.** **(a)** Para-aminobenzoic acid (PABA) and representative members of its structural analogs, the sulfonamides. The analogous portions of the compounds are shaded. **(b)** The metabolic pathway in bacteria and protozoa by which folic acid is synthesized from PABA. **(c)** The inhibition of folic acid synthesis by the presence of a sulfonamide, which deactivates the enzyme by binding irreversibly to the enzyme's active site.

Note that humans do not synthesize THF from PABA; instead, we take simple folic acids found in our diets and convert them into THF. As a result, human metabolism is unaffected by sulfonamides.

Another antimetabolic agent, *trimethoprim,* also interferes with nucleic acid synthesis. However, instead of binding to the enzyme that converts PABA to dihydrofolic acid, trimethoprim binds to the enzyme involved in the conversion of dihydrofolic acid to THF, the second step in this metabolic pathway.

Some antiviral agents target the unique aspects of the metabolism of viruses. After attachment to a host cell, viruses must penetrate the cell's membrane and be uncoated to release viral genetic instructions and assume control of the cell's metabolic machinery. Some viruses of eukaryotes are uncoated as a result of the acidic environment within phagolysosomes. *Amantadine, rimantadine,* and weak organic bases can neutralize the acid of phagolysosomes and thereby prevent viral uncoating; thus, these are antiviral drugs. Amantadine is used to prevent infections by influenza type A virus.

Protease inhibitors interfere with the action of protease—an enzyme that HIV needs near the end of its replication cycle. These drugs, when used as part of a "cocktail" of drugs including reverse transcriptase inhibitors (discussed shortly), have revolutionized treatment of AIDS patients in industrialized countries. Researchers have reduced the number of pills in the daily cocktail to just a few that contain all the drugs formerly found in 35 or more daily pills.

Inhibition of Nucleic Acid Synthesis

LEARNING **OUTCOME**

10.11 Describe the antimicrobial action of nucleotide and nucleoside analogs, quinolones, drugs that bind to RNA or DNA, and reverse transcriptase inhibitors.

The nucleic acids DNA and RNA are built from purine and pyrimidine nucleotides and are critical to the survival of cells. Several drugs function by blocking either the replication of DNA or its transcription into RNA.

Because only slight differences exist between the DNA of prokaryotes and eukaryotes, drugs that affect DNA replication often act against both types of cells. For example, *actinomycin* binds to DNA and effectively blocks DNA synthesis and RNA transcription not only in bacterial pathogens but in their hosts as well. Generally, drugs of this kind are not used to treat infections, though they are used in research of DNA replication and may be used carefully to slow replication of cancer cells.

One exception is the synthetic drugs called *quinolones*, including *fluoroquinolones*, which are unusual because they are active against prokaryotic DNA specifically. These antibacterial agents inhibit *DNA gyrase*, an enzyme necessary for correct coiling and uncoiling of replicating bacterial DNA; they typically have little effect on eukaryotes or viruses.

Other compounds that can act as antimicrobials by interfering with the function of nucleic acids are **nucleotide[2] analogs** or **nucleoside analogs,** which are molecules with structural similarities to the normal nucleotide building blocks of nucleic acids **(FIGURE 10.7)**. The structures of certain nucleotide or nucleoside analogs, such as the anti-AIDS drug AZT, enable them to be incorporated into the DNA or RNA of pathogens, where they distort the shapes of the nucleic acid molecules and prevent further replication, transcription, or translation. Nucleotide

[2]Nucleotides are composed of a pentose sugar, phosphate, and a nitrogenous base; nucleosides lack the phosphate.

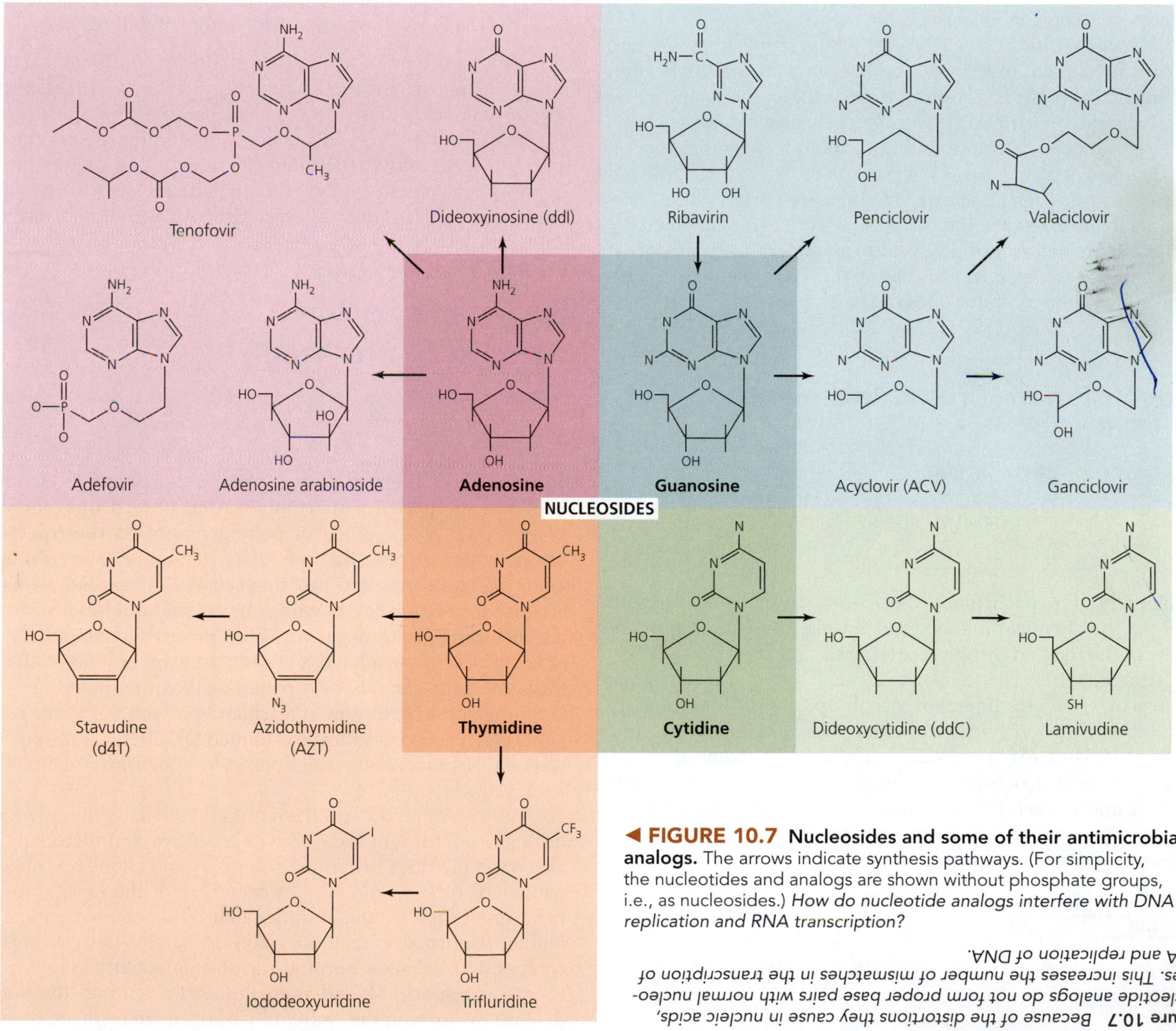

◀ **FIGURE 10.7 Nucleosides and some of their antimicrobial analogs.** The arrows indicate synthesis pathways. (For simplicity, the nucleotides and analogs are shown without phosphate groups, i.e., as nucleosides.) *How do nucleotide analogs interfere with DNA replication and RNA transcription?*

Figure 10.7 Because of the distortions they cause in nucleic acids, nucleotide analogs do not form proper base pairs with normal nucleotides. This increases the number of mismatches in the transcription of RNA and replication of DNA.

and nucleoside analogs are most often used against viruses because viral DNA polymerases are tens to hundreds of times more likely to incorporate these nonfunctional nucleotides into nucleic acids than is human DNA polymerase. Additionally, complete viral nucleic acid synthesis is more rapid than cellular nucleic acid synthesis. These characteristics make viruses more susceptible to nucleotide analogs than their hosts are, though nucleotide analogs can also be effective against rapidly dividing cancer cells.

Other antimicrobial agents function by binding to and inhibiting the action of RNA polymerases during the synthesis of RNA from a DNA template. Several drugs, including *rifampin* (rif'am-pin), bind more readily to prokaryotic RNA polymerase than to eukaryotic RNA polymerase; as a result, rifampin is more toxic to prokaryotes than to eukaryotes. Rifampin is used primarily against *M. tuberculosis* and other pathogens that metabolize slowly and thus are less susceptible to antimicrobials targeting active metabolic processes.

Clofazimine binds to the DNA of *Mycobacterium leprae* (lep´rī), the causative agent of leprosy, and prevents normal replication and transcription. It is also used to treat tuberculosis and other mycobacterial infections. *Pentamidine* and *propamidine isethionate* bind to protozoan DNA, inhibiting the pathogen's reproduction and development.

Reverse transcriptase inhibitors, which are part of AIDS cocktails, act against reverse transcriptase, which is an enzyme HIV uses early in its replication cycle to make DNA copies of its RNA genome (reverse transcription). Since people lack reverse transcriptase, the inhibitor does not harm patients.

Prevention of Virus Attachment and Entry

LEARNING OUTCOME

10.12 Describe the action of antimicrobial attachment antagonists.

Many pathogens, particularly viruses, must attach to their host's cells via the chemical interaction between attachment proteins on the pathogen and complementary receptor proteins on a host cell. Attachment of viruses can be blocked by peptide and sugar analogs of either attachment or receptor proteins. When analogs block these sites, viruses can neither attach to nor enter their host's cells. The use of such substances, called *attachment antagonists,* is an exciting new area of antimicrobial drug development. *Arildone* and *pleconaril* are antagonists of the receptor of polioviruses and some cold viruses. They block attachment of these viruses and deter infections.

TELL ME WHY

Some antimicrobial drugs are harmful to humans. Why can physicians safely use such drugs despite the potential danger?

Clinical Considerations in Prescribing Antimicrobial Drugs

Even though some fungi and bacteria commonly produce antibiotics, most of these chemicals are not effective for treating diseases because they are toxic to humans and animals, are too expensive, are produced in minute quantities, or lack adequate potency. The ideal antimicrobial agent to treat an infection or disease would be one that has all these characteristics:

- Readily available
- Inexpensive
- Chemically stable (so that it can be transported easily and stored for long periods of time)
- Easily administered
- Nontoxic and nonallergenic
- Selectively toxic against a wide range of pathogens

No agent has all of these qualities, so doctors and medical laboratory technicians must evaluate antimicrobials with respect to several characteristics: the types of pathogens against which they are effective; their effectiveness, including the dosages required to be effective; the routes by which they can be administered; their overall safety; and any side effects they produce. We consider each of these characteristics of antimicrobials in the following sections.

Spectrum of Action

LEARNING OUTCOME

10.13 Distinguish narrow-spectrum drugs from broad-spectrum drugs in terms of their targets and side effects.

The number of different kinds of pathogens a drug acts against is known as its **spectrum of action** (**FIGURE 10.8**); drugs that work against only a few kinds of pathogens are **narrow-spectrum drugs,** whereas those that are effective against many different kinds of pathogens are **broad-spectrum drugs.** For instance, because tetracycline acts against many different kinds of bacteria, including Gram negative, Gram positive, chlamydias, and rickettsias, it is considered a broad-spectrum antibiotic. In contrast, penicillin cannot easily penetrate the outer membrane of a Gram-negative bacterium to reach and prevent the formation of peptidoglycan, so its efficacy is limited largely to Gram-positive bacteria. Thus, penicillin has a narrower spectrum of action than tetracycline.

The use of broad-spectrum antimicrobials is not always as desirable as it might seem. Broad-spectrum antimicrobials can also open the door to serious secondary infections by transient pathogens or *superinfections* by members of the normal microbiota unaffected by the antimicrobial. This results because the killing of normal microbiota reduces *microbial antagonism,* the competition between normal microbes and pathogens for nutrients and space. Microbial antagonism reinforces the body's defense by limiting the ability of pathogens to colonize the skin

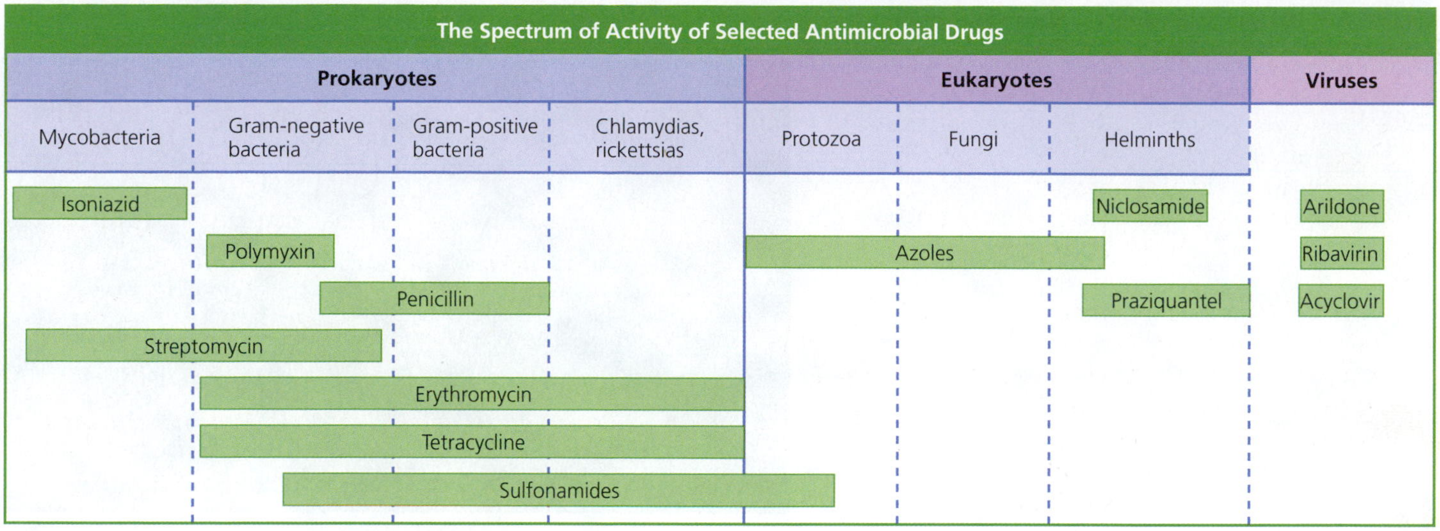

▲ **FIGURE 10.8 Spectrum of action for selected antimicrobial agents.** The more kinds of pathogens a drug affects, the broader its spectrum of action.

and mucous membranes. Thus, a woman using erythromycin to treat strep throat (a bacterial disease) could develop vaginitis resulting from the excessive growth of *Candida albicans* (kan′did-ă al′bi-kanz), a yeast that is unaffected by erythromycin and is freed from microbial antagonism when the antibiotic kills normal bacteria in the vagina.

Effectiveness

LEARNING OUTCOME

10.14 Compare and contrast diffusion susceptibility, Etest, MIC, and MBC tests.

To effectively treat infectious diseases, physicians must know which antimicrobial agent is most effective against a particular pathogen. To ascertain the efficacy of antimicrobials, microbiologists conduct a variety of tests, including diffusion susceptibility tests, the minimum inhibitory concentration test, and the minimum bactericidal concentration test.

Diffusion Susceptibility Test

Diffusion susceptibility tests, also known as *Kirby-Bauer tests,* involve uniformly inoculating a Petri plate with a standardized amount of the pathogen in question. Then small disks of paper containing standard concentrations of the drugs to be tested are firmly arranged on the surface of the plate. The plate is incubated, and the bacteria grow and reproduce to form a "lawn" everywhere except the areas where effective antimicrobial drugs diffuse through the agar. After incubation, the plates are examined for the presence of a **zone of inhibition**—that is, a clear area where bacteria do not grow **(FIGURE 10.9)**. A zone of inhibition is measured as the diameter (to the closest millimeter) of the clear region.

If all antimicrobials diffused at the same rate, then a larger zone of inhibition would indicate a more effective drug.

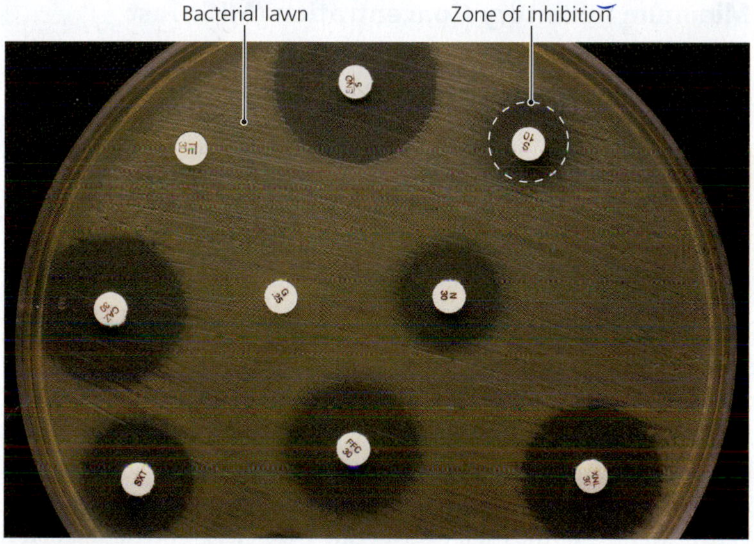

▲ **FIGURE 10.9 Zones of inhibition in a diffusion susceptibility (Kirby-Bauer) test.** In general, the larger the zone of inhibition around disks, which are impregnated with an antimicrobial agent, the more effective that antimicrobial is against the organism growing on the plate. The organism is classified as either susceptible, intermediate, or resistant to the antimicrobials tested based on the sizes of the zones of inhibition. *If all of these antimicrobial agents diffuse at the same rate and are equally safe and easily administered, which one would be the drug of choice for killing this pathogen?*

Figure 10.9 *The drug ENO, a fluoroquinolone found in the uppermost disk, is most effective.*

Instead, drugs with low molecular weights generally diffuse more quickly and so might have a larger zone of inhibition than a more effective but larger molecule. The size of a zone of inhibition must be compared to a standard table for that particular drug before accurate comparisons can be made. Diffusion susceptibility tests enable scientists to classify pathogens as *susceptible, intermediate,* or *resistant* to each drug.

▶ **FIGURE 10.10 Minimum inhibitory concentration (MIC) test in wells.**

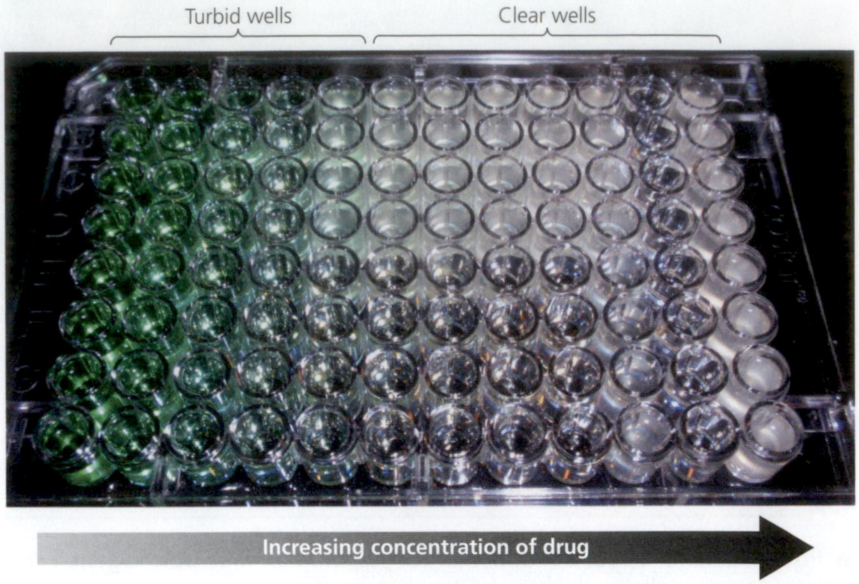

Minimum Inhibitory Concentration (MIC) Test

Once scientists identify an effective antimicrobial agent, they quantitatively express its potency as a **minimum inhibitory concentration (MIC),** often using the unit μg/ml. As the name suggests, the MIC is the smallest amount of the drug that will *inhibit* growth and reproduction of the pathogen. The MIC can be determined via a **broth dilution test,** in which a standardized amount of bacteria is added to serial dilutions of antimicrobial agents in tubes or wells containing broth. After incubation, turbidity (cloudiness) indicates bacterial growth; lack of turbidity indicates that the bacteria were either inhibited or killed by the antimicrobial agent **(FIGURE 10.10)**. Dilution tests can be conducted simultaneously in wells, and the entire process can be automated, with turbidity measured by special scanners connected to computers.

Another test that determines minimum inhibitory concentration combines aspects of an MIC test and a diffusion susceptibility test. This test, called an **Etest,**[3] involves placing a plastic strip containing a gradient of the antimicrobial agent being tested on a plate uniformly inoculated with the organism of interest **(FIGURE 10.11)**. After incubation, an elliptical zone of inhibition indicates antimicrobial activity, and the minimum inhibitory concentration can be noted where the zone of inhibition intersects a scale printed on the strip.

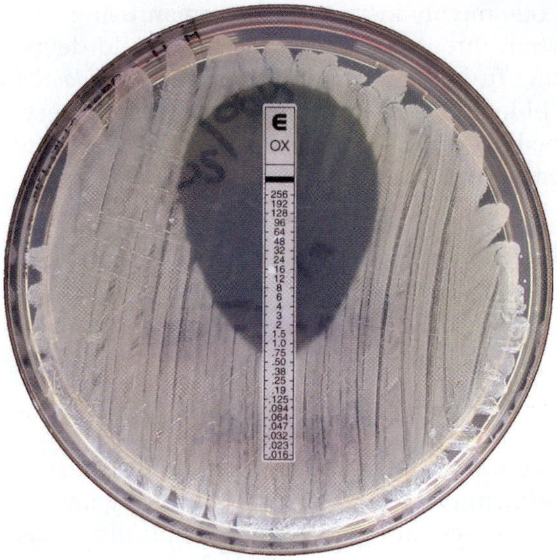

▲ **FIGURE 10.11 An Etest, which combines aspects of Kirby-Bauer and MIC tests.** The plastic strip contains a gradient of the antimicrobial agent of interest. The MIC is estimated to be the concentration printed on the strip where the zone of inhibition intersects the strip. In this example, the MIC is 0.75 μg/ml.

Minimum Bactericidal Concentration (MBC) Test

Similar to the MIC test is a **minimum bactericidal concentration (MBC) test,** though an MBC test determines the amount of drug required to kill the microbe rather than just the amount to inhibit it, as the MIC does. In an MBC test, samples taken from clear MIC tubes (or, alternatively, from zones of inhibition from a series of diffusion susceptibility tests) are transferred to plates containing a drug-free growth medium **(FIGURE 10.12)**. The appearance of bacterial growth in these subcultures after appropriate incubation indicates that at least some bacterial cells survived that concentration of the antimicrobial drug and were able to grow and multiply once placed in a drug-free medium. Any drug concentration at which growth occurs in subculture is *bacteriostatic,* not *bactericidal,* for that bacterium. The lowest concentration of drug for which no growth occurs in the subcultures is the minimum bactericidal concentration (MBC).

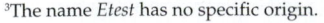

[3]The name *Etest* has no specific origin.

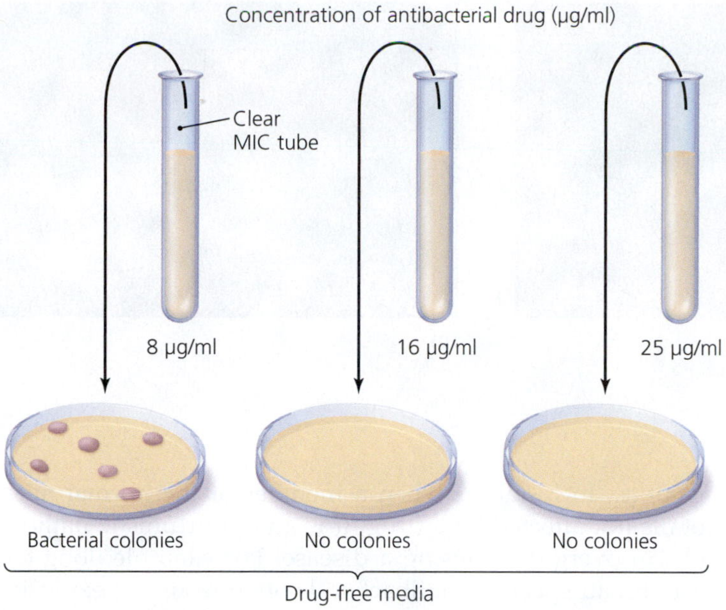

▲ **FIGURE 10.12 A minimum bactericidal concentration (MBC) test.** In this test, plates containing a drug-free growth medium are inoculated with samples taken from zones of inhibition or from clear MIC tubes. After incubation, growth of bacterial colonies on a plate indicates that the concentration of antimicrobial drug (in this case, 8 μg/ml) is bacteriostatic. The lowest concentration for which no bacterial growth occurs on the plate is the minimum bactericidal concentration; in this case, the MBC is 16 μg/ml.

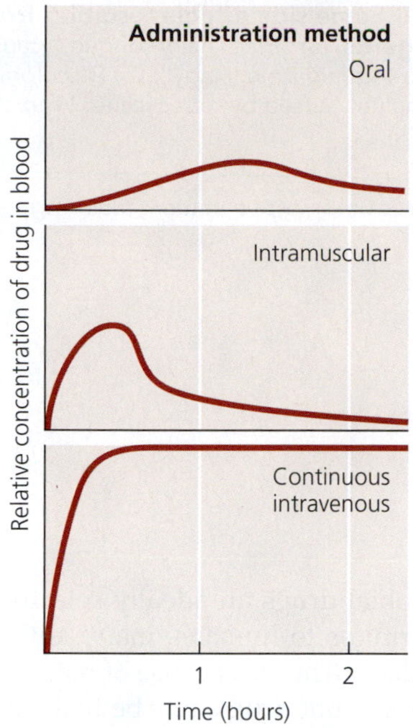

▲ **FIGURE 10.13 The effect of route of administration on blood levels of a chemotherapeutic agent.** Although intravenous (IV) and intramuscular (IM) administration achieve higher drug concentrations in the blood, oral administration has the advantage of simplicity.

Routes of Administration

LEARNING **OUTCOME**

10.15 Discuss the advantages and disadvantages of the different routes of administration of antimicrobial drugs.

An adequate amount of an antimicrobial agent must reach a site of infection if it is to be effective. For external infections such as athlete's foot, drugs can be applied directly. This is known as *topical* or *local* administration. For internal infections, drugs can be administered *orally, intramuscularly (IM),* or *intravenously (IV).* Each route has advantages and disadvantages.

Even though the oral route is simplest (it requires no needles and is self-administered), the drug concentrations achieved in the body are lower than occur via other routes of administration **(FIGURE 10.13)**. Further, because patients administer the drug themselves, they do not always follow prescribed timetables.

IM administration via a hypodermic needle allows a drug to diffuse slowly into the many blood vessels within muscle tissue, but the concentration of the drug in the blood is never as high as that achieved by IV administration, which delivers the drug directly into the bloodstream through either a needle or a catheter (a plastic or rubber tube). The amount of a drug in the blood is initially very high for the IV route, but the concentration can rapidly diminish as the liver and kidneys remove the drug from the circulation, unless the drug is continuously administered. Physicians can administer non-antimicrobial chemicals that prolong an antimicrobial's life span in the body; for example, cilastatin inhibits a kidney enzyme that would destroy imipenem—a type of beta-lactam.

In addition to considering the route of administration, physicians must consider how antimicrobial agents are distributed to infected tissues by the blood. For example, an agent removed rapidly from the blood by the kidneys might be the drug of choice for a bladder infection but would not be chosen to treat an infection of the heart. Finally, given that blood vessels in the brain, spinal cord, and eye are almost impermeable to many antimicrobial agents (because the tight structure of capillary walls in these structures creates the blood-brain barrier), infections there are often difficult to treat.

Safety and Side Effects

LEARNING **OUTCOMES**

10.16 Identify three main categories of side effects of antimicrobial therapy.

10.17 Define *therapeutic index* and *therapeutic range*.

Another aspect of chemotherapy that physicians must consider is the possibility of adverse side effects. These fall into three main categories—toxicity, allergies, and disruption of normal microbiota.

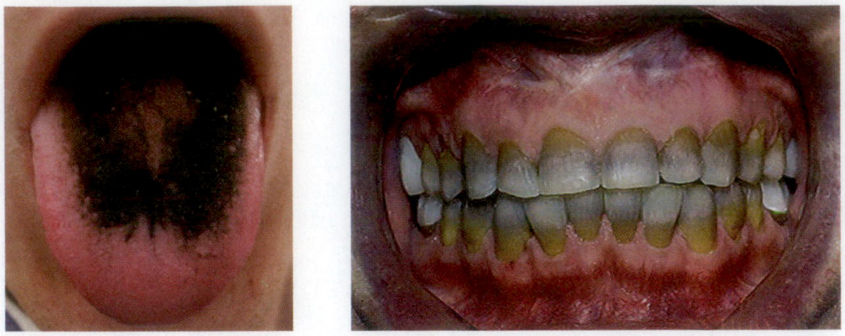

(a) (b)

▶ FIGURE 10.14 Some side effects resulting from toxicity of antimicrobial agents. **(a)** "Black hairy tongue," caused by the antiprotozoan drug metronidazole (Flagyl). **(b)** Discoloration and damage to tooth enamel caused by tetracycline. *Who should avoid taking tetracycline?*

Figure 10.14 Pregnant women and children should not use tetracycline.

Toxicity

Though antimicrobial drugs are ideally selectively toxic against microbes and harmless to humans, many antimicrobials in fact have toxic side effects. The exact cause of many adverse reactions is poorly understood, but drugs may be toxic to the kidneys, the liver, or nerves. For example, polymyxin and aminoglycosides can have fatally toxic effects on kidneys. Not all toxic side effects are so serious. *Metronidazole (Flagyl)*, an antiprotozoan drug, may cause a harmless temporary condition called "black hairy tongue," which results when the breakdown products of hemoglobin accumulate in the papillae of the tongue **(FIGURE 10.14a)**.

Doctors must be especially careful when prescribing drugs for pregnant women, because many drugs that are safe for adults can have adverse affects when absorbed by a fetus. For instance, tetracyclines form complexes with calcium that can become incorporated into bones and developing teeth, causing malformation of the skull and stained, weakened tooth enamel **(FIGURE 10.14b)**.

Researchers are able to estimate the safety of an antimicrobial drug by calculating the drug's **therapeutic index (TI)**, which is essentially a ratio comparing the dose of the drug that a patient can tolerate to the drug's effective dose. The higher the TI, the safer the drug. Clinicians refer to a drug's **therapeutic range (therapeutic window),** which is the range of concentrations of the drug that are effective without being excessively toxic.

Allergies

In addition to toxicity, some drugs trigger allergic immune responses in sensitive patients. Although relatively rare, such reactions may be life threatening, especially in an immediate, violent reaction called *anaphylactic shock*. For example, about 0.1% of Americans have an anaphylactic reaction to penicillin, resulting in approximately 300 deaths per year. However, not every allergy to an antimicrobial agent is so serious. Recent studies indicate that patients with mild allergies to penicillin frequently lose their sensitivity to it over time. Thus, an initial mild reaction to penicillin need not preclude its use in treating future infections. (Chapter 18 discusses allergies in more detail.)

Disruption of Normal Microbiota

Drugs that disrupt normal microbiota and their microbial antagonism of opportunistic pathogens may result in secondary infections. In instances when a member of the normal microbiota is not affected by a drug, it is an opportunistic pathogen and can overgrow, causing a disease. For example, long-term use of broad-spectrum antibacterials often results in explosions in the growth rate of *Candida albicans* in the vagina (vaginitis) or mouth (thrush) and the multiplication of *Clostridium difficile* (klos-trid´ē-ŭm di-fi´sēl) in the colon, causing a potentially fatal condition called *pseudomembranous colitis*. Such opportunistic pathogens are of great concern for hospitalized patients, who are often not only debilitated but also more likely to be exposed to pathogens with resistance to antimicrobial drugs—the topic of the next section.

Beneficial Microbes: Probiotics: The New Sheriff in Town focuses on how normal microbiota may help keep pathogens in check.

TELL ME WHY

Why don't physicians invariably prescribe the antimicrobial with the largest zone of inhibition?

Resistance to Antimicrobial Drugs

Among the major challenges facing microbiologists today are the problems presented by pathogens that are resistant to antimicrobial agents (see **Emerging Disease Case Study: Community-Associated MRSA**). In the sections that follow, we examine the development of resistant populations of pathogens, the mechanisms by which pathogens are resistant to antimicrobials, and some ways that resistance can be retarded.

The Development of Resistance in Populations

LEARNING | **OUTCOMES**

10.18 Describe how populations of resistant microbes can arise.

10.19 Describe the relationship between R plasmids and resistant cells.

BENEFICIAL MICROBES

Probiotics: The New Sheriff in Town

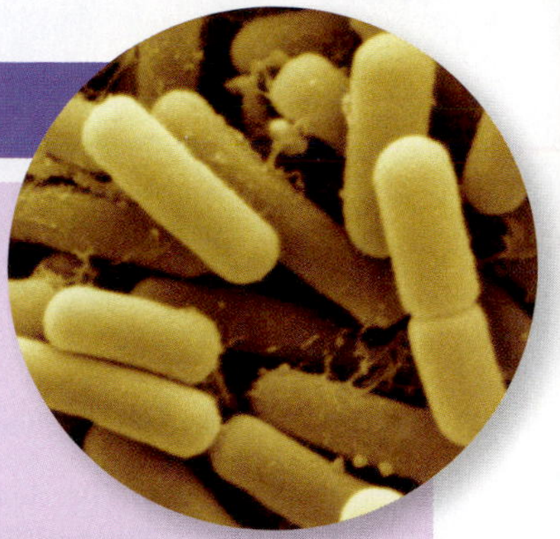

Lactobacillus reuteri

As overuse of antimicrobials has allowed more bacteria that are drug resistant to thrive, scientists are investigating alternative methods of combating microbial infections. One growing field of interest is *probiotics*, the use of microorganisms for health benefits.

Probiotics include bacteria such as *Lactobacillus*, which may help reduce symptoms of diarrhea in children, relieve milk allergies, and alleviate certain respiratory infections. One of the central ideas behind probiotics is that "good" bacteria, such as *Lactobacillus* (which lives naturally and harmlessly in our intestines and other parts of the body), compete with harmful microorganisms for resources, keeping pathogenic microbes in check. Some evidence suggests that probiotics are beneficial, but other studies show no difference in the health of people who use them. Much more research remains to be done.

Not all pathogens are equally sensitive to a given therapeutic agent; a population may contain a few organisms that are naturally either partially or completely resistant. Among bacteria, individual cells can acquire such resistance in two ways: through new mutations of chromosomal genes or by acquiring resistance genes on extrachromosomal pieces of DNA called **R plasmids** (or *R factors*) via the processes of horizontal gene transfer—transformation, transduction, or conjugation. We focus here on resistance in populations of bacteria, but resistance is known to occur among protozoa and viruses.

The process by which a resistant strain of bacteria develops is depicted in **FIGURE 10.15**. In the absence of an antimicrobial drug, resistant cells are usually less efficient than their normal neighbors because they must expend extra energy to maintain

EMERGING DISEASE CASE STUDY

Community-Associated MRSA

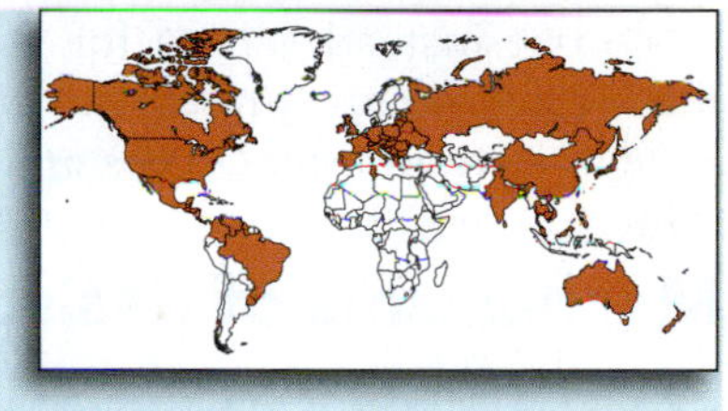

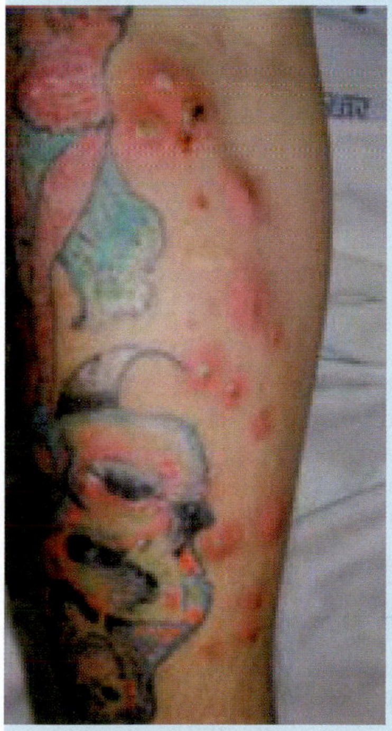

Julie was proud and excited about her new tattoo; it was cool! Jason, her boyfriend, had bought it as a gift for her 18th birthday, and she loved the way it covered her arm. A week later something else covered her arm . . . and his leg.

Julie and Jason had fallen victim to an emerging disease—community-associated, methicillin-resistant *Staphylococcus aureus* (CA-MRSA) infection. Julie's arm was covered with red, swollen, painful, pus-containing lesions, and she had a fever.

Jason thought he had a spider bite on Wednesday, but by Thursday morning a bright red line extended from his ankle to his groin. He could barely walk, and then his condition worsened. By the weekend, he was hospitalized, and his fever hit 107°F during his 10 days there. MRSA had entered his blood (bacteremia) and infected his bones (osteomyelitis). Jason's pain was so severe that he wondered only when he might die to be free of the agony.

For years health care workers have battled healthcare associated MRSA (HA-MRSA), which commonly afflicts patients in hospitals. Now, researchers are concerned that victims who have never been in a hospital are succumbing; MRSA has escaped hospitals and now travels communities worldwide, including among athletes, students in middle schools, and customers of unsafe tattooists. The bacterium can be spread between individuals who share fomites—towels, razors, clothing, or sheets.

Physicians drained Julie's lesions and prescribed oral antimicrobials, including trimethoprim–sulfamethoxazole, doxycycline, and clindamycin. Jason received intravenous vancomycin. Both eventually recovered, but her tattoo is now a reminder of a terrible experience and not a happy birthday. (For more about staphylococcal infections, see Chapter 19.)

1. How were Julie and then Jason likely infected?
2. Why is CA-MRSA considered an "emerging disease," given that MRSA has been around for years?
3. Why doesn't the map show more cases of CA-MRSA in Africa and South America?

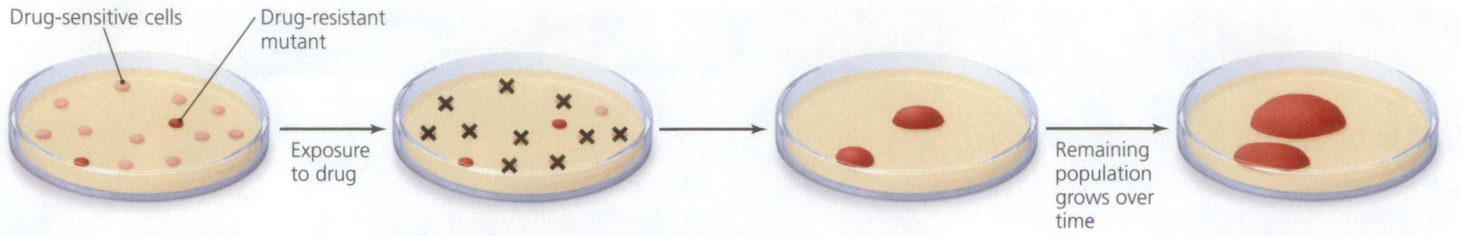

▲ **FIGURE 10.15 The development of a resistant strain of bacteria.** **(a)** A bacterial population contains both drug-sensitive and drug-resistant cells, although sensitive cells constitute the vast majority of the population. **(b)** Exposure to an antimicrobial drug inhibits the sensitive cells; as long as the drug is present, reduced competition from sensitive cells facilitates the multiplication of resistant cells. **(c)** Eventually, resistant cells constitute the majority of the population. *Why do resistant strains of bacteria more often develop in hospitals and nursing homes than in college dormitories?*

Figure 10.15 *Resistant strains are more likely to develop in hospitals and other health care facilities because the extensive use of antimicrobial agents in those places inhibits the growth of sensitive strains and selects for the growth of resistant strains.*

resistance genes and proteins. Under these circumstances, resistant cells remain the minority in a population because they reproduce more slowly. However, when an antimicrobial agent is present, the majority of cells (which are sensitive to the antimicrobial) are inhibited or die while the resistant cells continue to grow and multiply, often more rapidly because they then face less competition. The result is that resistant cells soon replace the sensitive cells as the majority in the population. The bacterium has evolved resistance. It should be noted that the presence of the antimicrobial agent does not *produce* resistance but instead selects for the replication of resistant cells that were already present in the population. ▶**ANIMATIONS:** ***Antibiotic Resistance: Origins of Resistance***

Mechanisms of Resistance

LEARNING | **OUTCOMES**

10.20 List seven ways by which microorganisms can be resistant to antimicrobial drugs.

10.21 List two ways that genes for drug resistance are spread between bacteria.

The problem of resistance to antimicrobial drugs is a major health threat to our world. An "alphabet soup" of resistant pathogens and diseases plague health care professionals: MRSA,[4] VRSA,[5] VISA,[6] VRE,[7] MDR-TB[8], and XDR-TB[9] are just some of these.

How do microbes gain resistance to antimicrobial drugs? Consider the path a typical antimicrobial drug must take to affect a microbe: The drug must cross the cell's wall, then cross the cytoplasmic membrane to enter the cell; there the antimicrobial binds to its target (receptor) molecule. Only then can it inhibit or kill the microbe. Microbes gain resistance by blocking some point in this pathway. There are at least seven mechanisms of resistance:

- Resistant cells may produce an enzyme that destroys or deactivates the drug. This common mode of resistance is exemplified by **beta (β)-lactamases** (penicillinases), which are enzymes that break the beta-lactam rings of penicillin and similar molecules, rendering them inactive **(FIGURE 10.16)**. Many MRSA strains have evolved resistance to penicillin-derived antimicrobials in this manner. Over 200 different lactamases have been identified. Frequently their genes are located on R plasmids.
- Resistant microbes may slow or prevent the entry of the drug into the cell. This mechanism typically involves changes in the structure or electrical charge of the cytoplasmic membrane proteins that constitute channels or pores. Such proteins in the outer membranes of Gram-negative bacteria are called *porins* (see Figure 3.15b). Altered pore proteins result from mutations in chromosomal genes. Resistance against tetracycline and penicillin are known to occur via this mechanism.
- Resistant cells may alter the target of the drug so that the drug either cannot attach to it or binds it less effectively. This form of resistance is often seen against antimetabolites (such as sulfonamides) and against drugs that thwart protein translation (such as erythromycin).
- Resistant cells may alter their metabolic chemistry, or they may abandon the sensitive metabolic step altogether. For example, a cell may become resistant to a drug by producing more enzyme molecules for the affected metabolic pathway, effectively reducing the power of the drug. Alternatively, cells become resistant to sulfonamides by abandoning the synthesis of folic acid, absorbing it from the environment instead.
- Resistant cells may pump the antimicrobial out of the cell before the drug can act. So-called **efflux pumps,** which are

[4]Methicillin-resistant *S. aureus*.
[5]Vancomycin-resistant *S. aureus*.
[6]*S. aureus* with intermediate level of resistance to vancomycin.
[7]Vancomycin-resistant enterococci.
[8]Multi-drug-resistant tuberculosis.
[9]Extensively drug-resistant tuberculosis.

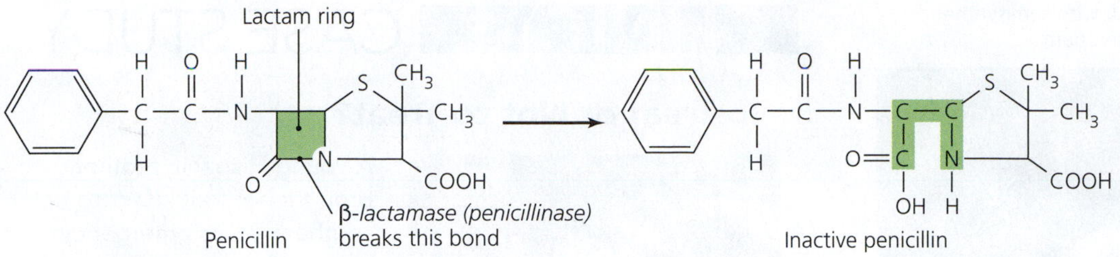

◀ FIGURE 10.16 How β-lactamase (penicillinase) renders penicillin inactive. The enzyme acts by breaking a bond in the lactam ring, the functional portion of the drug.

typically powered by ATP, are often able to pump more than one type of antimicrobial from a cell. Some microbes become multi-drug resistant (perhaps to as many as 10 or more drugs) by utilizing resistance pumps.

- Bacteria within biofilms resist antimicrobials more effectively than free-living cells. Biofilms retard diffusion of the drugs and often slow metabolic rates of species making up the biofilm. Lower metabolic rates reduce the effectiveness of antimetabolic drugs.
- Some resistant strains of the bacterium *Mycobacterium tuberculosis* have a novel method of resistance against fluoroquinolone drugs that bind to DNA gyrase. These strains synthesize an unusual protein that forms a negatively charged, rodlike helix about the width of a DNA molecule. This protein, called *MfpA protein*, binds to DNA gyrase in place of DNA, depriving fluoroquinolone of its target site. This is the first method of antibiotic resistance that involves protecting the target of an antimicrobial drug rather than, say, changing the target or deactivating the drug. MfpA protein probably slows down cellular division of *M. tuberculosis*, but that is better for the bacterium than being killed.
▶ANIMATIONS: ***Antibiotic Resistance: Forms of Resistance***

Despite decades of research, scientists and health care workers still do not fully comprehend what makes certain species, such as *Staphylococcus aureus*, more prone to developing resistance, how and with what frequency resistance to antimicrobials spreads through a population, or whether there are effective ways to limit proliferation of resistance. It has been shown that horizontal gene transfer, most often by *conjugation* and less often by *transformation*, accounts for the spread of antimicrobial resistance among the densely growing cells of biofilms. Often, several resistance genes travel together between cells.

Multiple Resistance and Cross Resistance

LEARNING | OUTCOME

10.22 Define *cross resistance*, and distinguish it from multiple resistance.

A given pathogen can acquire resistance to more than one drug at a time, especially when resistance is conferred by R plasmids, which are exchanged readily among bacterial cells. Such multiresistant strains of bacteria frequently develop in hospitals and nursing homes, where the constant use of many kinds of antimicrobial agents eliminates sensitive cells and encourages the development of resistant strains.

Multiple-drug-resistant pathogens (erroneously called *superbugs* in the popular press) are resistant to three or more types of antimicrobial agents. Multiple-drug-resistant strains of *Staphylococcus, Streptococcus* (strep-tō-kok´ŭs), *Enterococcus* (en´ter-ō-kok´ŭs), *Pseudomonas, Mycobacterium tuberculosis,* and *Plasmodium* (plaz-mō´dē-ŭm) pose unique problems. Caregivers must treat infected patients without effective antimicrobials while taking care to protect themselves and others from infection.

Resistance to one antimicrobial agent may confer resistance to similar drugs, a phenomenon called **cross resistance.** Cross resistance typically occurs when drugs are similar in structure. For example, resistance to one aminoglycoside drug, such as streptomycin, may confer resistance to similar aminoglycoside drugs.

Retarding Resistance

LEARNING | OUTCOME

10.23 Describe four ways to retard development of resistance.

The development of resistant populations of pathogens can be averted in at least four ways. First, sufficiently high concentrations of the drug can be maintained in a patient's body for a long enough time to inhibit the pathogen, allowing the body's defenses to defeat them. Discontinuing a drug too early promotes the development of resistant strains. For this reason, it is important that patients finish their entire antimicrobial prescription and resist the temptation to "save some for another day."

A second way to avert resistance is to use antimicrobial agents in combination so that pathogens resistant to one drug will be killed by other drugs and vice versa. Additionally, one drug sometimes enhances the effect of a second drug in a process called **synergism** (sin'er-jizm) **(FIGURE 10.17)**. For example, the inhibition of cell wall formation by penicillin makes it easier for streptomycin molecules to enter bacteria and interfere with protein synthesis. Synergism can also result from combining an antimicrobial drug and a chemical, as occurs when *clavulanic acid* enhances the effect of penicillin by deactivating β-lactamase. (Not all drugs act synergistically; some combinations of drugs can be *antagonistic*—interfering with each other. For example, drugs that slow bacterial growth are antagonistic to the action of penicillin, which acts only against growing and dividing cells.)

A third way to reduce the development of resistance is to limit the use of antimicrobials to necessary cases. Unfortunately,

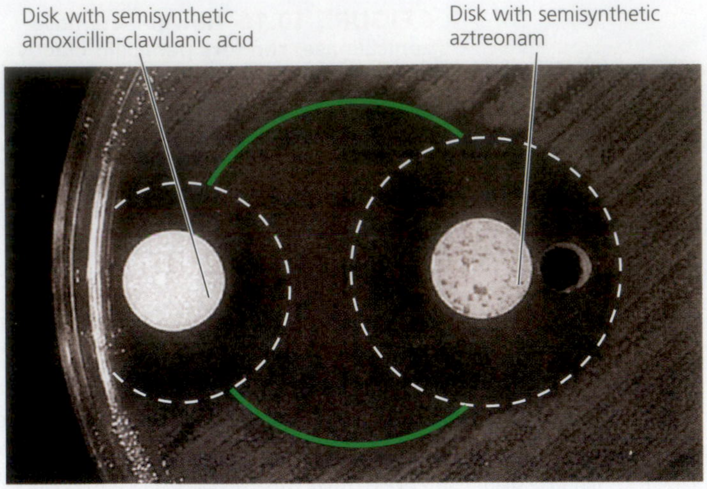

▲ **FIGURE 10.17 An example of synergism between two antimicrobial agents.** The portion of the zone of inhibition outlined in green represents the synergistic enhancement of antimicrobial activity beyond the activities of the individual drugs (outlined in white). *What does clavulanic acid do?*

Figure 10.17 *Clavulanic acid deactivates β-lactamase, allowing penicillins to work.*

many antimicrobial agents are used indiscriminately, both in developed countries and in less developed regions, where many agents are available without a physician's prescription. In the United States, an estimated 50% of prescriptions for antibacterial agents to treat sore throats and 30% of prescriptions for ear infections are inappropriate because the diseases are viral, not bacterial. Likewise, because antibacterial drugs have no effect on cold and flu viruses, 100% of antibacterial prescriptions for treating these diseases are superfluous. The use of antimicrobial agents encourages the reproduction of resistant bacteria by limiting the growth of sensitive cells; therefore, inappropriate use of such drugs increases the likelihood that resistant strains of bacteria will multiply.

Finally, scientists can combat resistant strains by developing new drugs, in some cases by adding different side chains to the original molecule. In this way, scientists develop semisynthetic *second-generation* drugs. If resistance develops to these drugs, *third-generation* drugs may be developed to replace them.

Alternatively, scientists search for new antimicrobials. Researchers explore diverse habitats, such as peat bogs, ocean sediments, garden soil, and people's mouths, for organisms that produce novel antibiotics. Researchers see potential in *bacteriocins*—antibacterial proteins coded by bacterial plasmids. Bacteria use bacteriocins to inhibit other bacterial strains; now we can do the same. Tinkering with these and other antibiotics is yielding promising new semisynthetics.

With the advent of genome sequencing and enhanced understanding of protein folding, some scientists predict that we are moving into a new era of antimicrobial drug discovery. They point out that researchers who know the exact shapes of microbial proteins should be able to design drugs complementary to those shapes—drugs that will inhibit microbial proteins without affecting humans.

CLINICAL CASE STUDY

To Treat or Not to Treat?

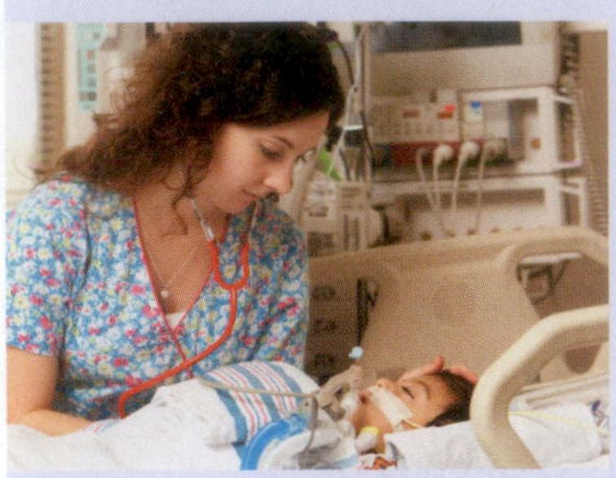

A young Hispanic mother brought her frail infant to a southern Texas emergency room. While she waited, her baby began to have seizures. The staff stabilized the child, while the mother explained that the child had been ill for many days. Feverish at times, the infant had lost weight because she was too short of breath to nurse. Other family members were ill with bad coughs.

The girl was diagnosed with a form of tuberculosis (TB) that affected her brain. She was hospitalized and isolated. Health department officials discovered that two individuals had exposed the baby to TB: an uncle and her father, whom the mother had visited in a Mexican jail. Inmates from the jail carried a strain of TB that was resistant to several standard anti-tuberculosis drugs. It would take many weeks to determine which of the two strains was affecting the baby.

Doctors had to make a tough decision: Should they immediately begin treating the infant for the multi-drug-resistant strain (MDR-TB), which involves using five to seven different drugs with multiple and painful side effects for many months, or treat her for the more normal form of TB using a more typical and less stressful drug regimen for an equal length of time?

1. What issues must be considered to determine the drug therapy for the infant?
2. What treatment would you guess was used for the infant?

Reference: Adapted from *MMWR* 40:373–375, 1991.

Currently scientists are examining ways to inhibit the following: secretion systems, attachment molecules and their receptors, biofilm signaling molecules, and bacterial RNA polymerase.

Despite researchers' best efforts, other scientists wonder if drug developers can stay ahead of the development of resistance by pathogens.

TELL ME WHY

Why is it incorrect to say that an individual bacterium develops resistance in response to an antibiotic?

Selected antimicrobial agents, their modes of action, clinical considerations, and other features are summarized in **TABLES 10.2** to **10.6**. Particular use of antimicrobial drugs against specific pathogens is covered in the relevant chapters of this book.

TABLE 10.2 Antibacterial Drugs

Drug	Description and Mode of Action	Clinical Considerations	Method of Resistance
Antibacterial Drugs That Inhibit Cell Wall Synthesis			
Bacitracin	Isolated from *Bacillus licheniformis* growing on a patient named Tracy; appears to have three modes of action: ■ Interference with the movement of peptidoglycan precursors through the bacterial cell membrane to the cell wall ■ Inhibition of RNA transcription ■ Damage to the bacterial cytoplasmic membrane The latter two modes of action have not been proven definitely	**Spectrum of action:** Gram-positive (G+) bacteria **Route of administration:** Topical **Adverse effects:** Toxic to kidneys	Resistance most often involves changes in bacterial cell membranes that prevent bacitracin from entering the cell
Beta-lactams Representative natural penicillins: Penicillin G Penicillin V Representative semisynthetic penicillin: Amoxicillin Ampicillin Cloxacillin Dicloxacillin Methicillin Nafcillin Oxacillin Representative semisynthetic carbapenem: Imipenem Meropenem Representative natural cephalosporin: Cephalothin Representative semisynthetic cephalosporins: Cefotaxime Ceftriaxone Cefuroxime Cephalexin Representative semisynthetic monobactam: Aztreonam	Large number of natural and semi-synthetic derivatives from the fungi *Penicillium* (penicillins) and *Acremonium* (cephalosporins); bind to and deactivate the enzyme that cross-links the NAM subunits of peptidoglycan Monobactams have only a single ring instead of the two rings seen in other beta-lactams	**Spectrum of action:** Natural drugs have limited action against most Gram-negative (G−) bacteria because they do not readily cross the outer membrane; semisynthetics have broader spectra of action Monobactams have a limited spectrum of action, affecting only aerobic, G− bacteria **Route of administration:** Penicillin V, a few cephalosporins (e.g., cephalexin), and monobactams: oral; penicillin G and many semisynthetics (e.g., methicillin, ampicillin, carbenicillin, cephalothin): IM or IV **Adverse effects:** Allergic reactions against beta-lactams in some adults; monobactams are least allergenic	Develops in three ways in G− bacteria: ■ Change their outer membrane structure to prevent entrance of the drug ■ Modify the enzyme so that the drug no longer binds ■ Synthesize beta-lactamases that cleave the functional lactam ring of the drug; genes for lactamases are often carried on R plasmids
Cycloserine	Analog of alanine that interferes with the formation of alanine-alanine bridges between NAM subunits	**Spectrum of action:** Some G+ bacteria, mycobacteria **Route of administration:** Oral **Adverse effects:** Toxic to nervous system, producing depression, aggression, confusion, and headache	Some G+ bacteria enzymatically deactivate the drug
Ethambutol	Prevents the formation of mycolic acid; used in combination with other antimycobacterial drugs	**Spectrum of action:** Mycobacteria, including *M. tuberculosis* and *M. leprae* **Route of administration:** Oral **Adverse effects:** None	Resistance is due to random mutations of bacterial chromosomes that result in alteration of target site

Continued—

TABLE 10.2 Antibacterial Drugs—*Continued*

Drug	Description and Mode of Action	Clinical Considerations	Method of Resistance
Antibacterial Drugs That Inhibit Cell Wall Synthesis			
Isoniazid (isonicotinic acid hydrazide, INH)	Analog of the vitamins nicotinamide and pyridoxine; blocks the gene for an enzyme that forms mycolic acid	**Spectrum of action:** Mycobacteria, including *M. tuberculosis* and *M. leprae* **Route of administration:** Oral **Adverse effects:** May be toxic to liver	Resistance is due to random mutations of bacterial chromosomes that result in alteration of target site or overproduction of target molecules
Vancomycin	Produced by *Amycolatopsis orientalis;* directly interferes with the formation of alanine-alanine bridges between NAM subunits	**Spectrum of action:** Effective against most G+ bacteria but generally reserved for use against strains resistant to other drugs such as methicillin-resistant *Staphylococcus aureus* (MRSA) **Route of administration:** IV **Adverse effects:** Damage to ears and kidneys, allergic reactions, may cause depression	G− bacteria are naturally resistant because the drug is too large to pass through the outer membrane; some G+ bacteria (e.g., *Lactobacillus*) are naturally resistant because they do not form alanine-alanine bonds between NAM subunits
Antibacterial Drugs That Inhibit Protein Synthesis			
Aminoglycosides Representatives: Amikacin Gentamicin Kanamycin Neomycin Paromomycin Spiramycin Streptomycin Tobramycin	Compounds in which two or more amino sugars are linked with glycosidic bonds; were originally isolated from species of the bacterial genera *Streptomyces* and *Micromonospora.* Inhibit protein synthesis by irreversibly binding to the 30S subunit of prokaryotic ribosomes, which either causes the ribosome to mistranslate mRNA, producing aberrant proteins, or causes premature release of the ribosome from mRNA, stopping synthesis. Also bactericidal by destroying outer membranes of G− bacteria	**Spectrum of action:** Broad: effective against most G− bacteria and some protozoa **Route of administration:** IV; do not traverse blood-brain barrier **Adverse effects:** Toxic to kidneys and to auditory nerves, causing deafness	Uptake of these drugs is energy dependent, so anaerobes (with less ATP available) are less susceptible; aerobic bacteria alter membrane pores to prevent uptake or synthe-size enzymes that alter or degrade the drug once it enters; rarely, bacteria alter the binding site on the ribosome; some bacteria make biofilms when exposed to the drugs
Chloramphenicol	Rarely used drug that binds to the 50S subunits of prokaryotic ribosomes, preventing them from moving along mRNA	**Spectrum of action:** Broad but rarely used except in treatment of typhoid fever **Route of administration:** Oral; traverses blood-brain barrier **Adverse effects:** In 1 of 24,000 patients, causes aplastic anemia, a potentially fatal condition in which blood cells fail to form; can also cause neurological damage	Develops via gene carried on an R plasmid that codes for an enzyme that deactivates drug
Lincosamides Representative: Clindamycin	Binds to 50S ribosomal subunit and stops protein elongation	**Spectrum of action:** Effective against G+ and anaerobic G− bacteria and some protozoa **Route of administration:** Oral or IV; does not traverse blood-brain barrier **Adverse effects:** Gastrointestinal distress, including nausea, diarrhea, vomiting, and pain	Develops via changes in ribosomal structure that prevent drug from binding; resistance genes are same as those of aminoglycosides
Macrolides Representatives: Azithromycin Clarithromycin Erythromycin Natamycin Telithromycin	Group of antimicrobials typified by a macrocyclic lactone ring; the most prescribed is erythromycin, which is produced by *Saccharopolyspora erythraea;* act by binding to the 50S subunit of prokaryotic ribosomes and preventing the elongation of the nascent protein	**Spectrum of action:** Effective against G+ and a few G− bacteria and fungi (natamycin) **Route of administration:** Oral; do not traverse blood-brain barrier **Adverse effects:** Nausea, mild gastrointestinal pain, vomiting; associated with narrowing of the stomach in children under six months old; erythromycin increases risk of cardiac arrest	Develops via changes in ribosomal RNA that prevent drugs from binding, or via R plasmid genes coding for the production of macrolide-digesting enzymes; resistance genes are same as those of lincosamides

TABLE 10.2 —*Continued*

Drug	Description and Mode of Action	Clinical Considerations	Method of Resistance
Antibacterial Drugs That Inhibit Protein Synthesis			
Mupirocin	Produced by *Pseudomonas fluorescens;* binds to bacterial $tRNA^{Ile}$, which prevents delivery of isoleucine to ribosomes, blocking polypeptide synthesis	**Spectrum of action:** Effective primarily against G+ bacteria **Route of administration:** Topical cream **Adverse effects:** None reported	Resistance develops from mutations that change the shape of $tRNA^{Ile}$
Oxazolidinones Representative: Linezolid	Synthetic; inhibits initiation of polypeptide synthesis	**Spectrum of action:** G+ bacteria **Route of administration:** Oral, IV **Adverse effects:** Rash, diarrhea, loss of appetite, constipation, fever	Method of resistance not known
Streptogramin Representatives: Quinupristin Dalfopristin	Binds to 50S ribosomal subunit, stops protein synthesis; the synergistic drugs quinupristin and dalfopristin are taken together	**Spectrum of action:** Broad but reserved for use against multiple-drug-resistant strains **Route of administration:** IV **Adverse effects:** Muscle and joint pain	Not known
Tetracyclines Representatives: Doxycycline Minocycline Tetracycline	Composed of four hexagonal rings with various side groups; prevent tRNA molecules, which carry amino acids, from binding to ribosomes at the 30S subunit's docking site	**Spectrum of action:** Most are broad: effective against many G+ and G− bacteria as well as against bacteria that lack cell walls, such as *Mycoplasma* **Route of administration:** Oral; crosses poorly into brain **Adverse effects:** Nausea, diarrhea, sensitivity to light; forms complexes with calcium, which stains developing teeth and adversely affects the strength and shape of bones	Develops in three ways; bacteria may: ▪ Alter gene for pores in outer membrane such that new pore prevents drug from entering cell ▪ Alter binding site on the ribosome to allow tRNA to bind even in presence of drug ▪ Actively pump drug from cell
Antibacterial Drugs That Alter Cytoplasmic Membranes			
Gramicidin	Short polypeptide that forms pore across cytoplasmic membrane, allowing single-charged cations to cross freely	**Spectrum of action:** G+ bacteria **Route of administration:** Topical **Adverse effects:** Toxic (also forms pores in eukaryotic membranes)	Not known
Nisin	A bacteriocin naturally produced by *Lactococcus lactis*	**Spectrum of action:** G+ and G− bacteria in food **Route of administration:** Used as a preservative in food **Adverse effect:** None reported	Not known
Polymyxin	Produced by *Bacillus polymyxa;* destroys cytoplasmic membranes of susceptible cells	**Spectrum of action:** Effective against G− bacteria, particularly *Pseudomonas,* and some amoebae **Route of administration:** Topical **Adverse effects:** Toxic to kidneys	Results from changes in cell membrane that prohibit entrance of the drug
Pyrazinamide	Disrupts membrane transport and prevents *Mycobacterium* from repairing damaged proteins	**Spectrum of action:** *Mycobacterium tuberculosis* **Route of administration:** Oral **Adverse effects:** Malaise, nausea, diarrhea	Results from point mutations in bacterial gene for enzyme necessary to activate drug

Continued—

TABLE 10.2 Antibacterial Drugs—*Continued*

Drug	Description and Mode of Action	Clinical Considerations	Method of Resistance
Antibacterial Drugs That Are Antimetabolites			
Bedaquiline	Inhibits and ATP synthesis enzyme unique to mycobacteria	**Spectrum of action:** Mycobacteria; reserved for use for multi-drug-resistant tuberculosis **Route of administration:** Oral **Adverse effects:** Disturbance in heart's electrical activity, nausea, headache, joint pain	Not known
Dapsone	Interferes with synthesis of folic acid	**Spectrum of action:** *M. leprae, M. tuberculosis* **Route of administration:** Oral **Adverse effects:** Insomnia, headache, nausea, vomiting, increased heart rate	Not known
Sulfonamides Representatives: Sulfadiazine Sulfadoxine Sulfamethoxazole Sulfanilamide	Synthetic drugs; first produced as a dye; analogs of PABA that bind irreversibly to enzyme that produces dihydrofolic acid; synergistic with trimethoprim	**Spectrum of action:** Broad: effective against G+ and G− bacteria and some protozoa and fungi; however, resistance is widespread **Route of administration:** Oral **Adverse effects:** Rare: allergic reactions, anemia, jaundice, mental retardation of fetus if administered in last trimester of pregnancy	*Pseudomonas* is naturally resistant because of permeability barriers; cells that require folic acid as a vitamin are also naturally resistant; chromosomal mutations result in lowered affinity for the drugs
Trimethoprim	Blocks second metabolic step in the formation of folic acid from PABA; synergistic with sulfonamides, such as sulfamethoxazole	**Spectrum of action:** Broad: effective against G+ and G− bacteria and some protozoa and fungi; however, resistance is widespread **Route of administration:** Oral **Adverse effects:** Allergic reactions or liver damage in some patients	*Pseudomonas* is naturally resistant because of permeability barriers; cells that require folic acid as a vitamin are also naturally resistant; chromosomal mutations result in lowered affinity for the drug
Antibacterial Drugs That Inhibit Nucleic Acid Synthesis			
Clofazimine	Binds to DNA, preventing replication and transcription	**Spectrum of action:** Mycobacteria, especially *M. tuberculosis, M. leprae,* and *M. ulcerans* **Route of administration:** Oral **Adverse effects:** Diarrhea, discoloration of skin and eyes	Not known
Fluoroquinolones Representatives: Ciprofloxacin Levofloxacin Moxifloxacin Ofloxacin	Synthetic agents that inhibit DNA gyrase, which is needed to correctly replicate bacterial DNA; penetrate cytoplasm of cells	**Spectrum of action:** Broad: G+ and G− bacteria are affected **Route of administration:** Oral **Adverse effects:** Tendonitis, tendon rupture	Results from chromosomal mutations that lower affinity for drug, reduce its uptake, or protect gyrase from drug
Nitroimidazoles Representative: Metronidazole	Anaerobic conditions reduce the molecule, which then damages DNA and prevents its correct replication	**Spectrum of action:** Obligate anaerobic bacteria	Not known
Rifamycin Representatives: Rifampin Rifaximin	Natural and semisynthetic derivatives from *Amycolatopsis rifamycinica* that bind to bacterial RNA polymerase, preventing transcription of RNA; used with other antimicrobial bacterial drugs	**Spectrum of action:** Bacteriostatic against aerobic G+ bacteria; bactericidal against mycobacteria **Route of administration:** Oral **Adverse effects:** None of major significance	Results from chromosomal mutation that alters binding site on enzyme; G− bacteria are naturally resistant because of poor uptake

TABLE 10.3 Antiviral Drugs

Drug	Description and Mode of Action	Clinical Considerations	Method of Resistance
Attachment Antagonists			
Arildone Pleconaril	Blocks attachment molecule on host cell or pathogen	**Spectrum of action:** Picornaviruses (e.g., poliovirus, some cold viruses) **Route of administration:** Oral **Adverse effects:** None	Not known
Neuraminidase inhibitors Representatives: Oseltamivir Zanamivir	Prevent influenzaviruses from attaching to or exiting from cells	**Spectrum of action:** Influenzavirus **Route of administration:** Oral (oseltamivir) or aerosol (zanamivir) **Adverse effects:** None	Not known
Antiviral Drugs That Inhibit Viral Entry			
Enfuvirtide	Binds viral protein gp^{41}, preventing fusion of viral envelope with host cell membrane, thereby blocking viral entry	**Spectrum of action:** HIV-1 **Route of administration:** IV **Adverse effects:** Tingling, allergic reaction at injection site	Not known
Antiviral Drugs That Inhibit Viral Uncoating			
Amantadine	Neutralizes acid environment within phagolysosomes that is necessary for viral uncoating	**Spectrum of action:** Influenza A virus **Route of administration:** Oral **Adverse effects:** Toxic to central nervous system; results in nervousness, irritability, insomnia, and blurred vision	Mutation resulting in a single amino acid change in a membrane ion channel leads to viral resistance
Rimantadine	Neutralizes phagolysosomal acid, preventing viral uncoating	**Spectrum of action:** Influenza A virus **Route of administration:** Oral, adults only **Adverse effects:** Toxic to central nervous system; results in nervousness, irritability, insomnia, and blurred vision	Mutation resulting in a single amino acid change in a membrane ion channel leads to viral resistance
Antiviral Drugs That Inhibit Nucleic Acid Synthesis			
Acyclovir (ACV) Representative: Ganciclovir	Phosphorylation by virally coded kinase enzyme activates the drug; inhibits DNA and RNA synthesis	**Spectrum of action:** Viruses that code for kinase enzymes: herpes, Epstein-Barr virus, cytomegalovirus, and varicella viruses **Route of administration:** Oral **Adverse effects:** None	Mutations in genes for kinase enzymes may render them ineffective at drug activation
Adenosine arabinoside	Phosphorylation by cell-coded kinase enzyme activates the drug; inhibits DNA synthesis; viral DNA polymerase more likely to incorporate the drugs than human DNA polymerase	**Spectrum of action:** Herpesvirus **Route of administration:** IV **Adverse effects:** Fatal to host cells that incorporate the drug into cellular DNA; anemia	Results from mutation of viral DNA polymerase

Continued—

TABLE 10.3 Antiviral Drugs—*Continued*

Drug	Description and Mode of Action	Clinical Considerations	Method of Resistance
Antiviral Drugs That Inhibit Nucleic Acid Synthesis			
Nucleotide analogs Representatives (see also Figure 10.7): Adefovir Azidothymidine (AZT) Entecavir Ganciclovir Lamivudine Penciclovir Tenofovir Valaciclovir	Phosphorylation by cell-coded kinase enzyme activates these drugs; inhibits DNA synthesis; viral reverse transcriptase more likely to incorporate these drugs; used in conjunction with protease inhibitor to treat HIV	**Spectrum of action:** HIV, hepatitis B virus **Route of administration:** Oral **Adverse effects:** Nausea, bone marrow toxicity	Results from mutation of viral reverse transcriptase
Ribavirin	Phosphorylation by virally coded kinase enzyme activates the drug; inhibits DNA and RNA synthesis; viral DNA polymerase more likely to incorporate the drugs	**Spectrum of action:** Respiratory syncytial, hepatitis C, influenza A, measles, some hemorrhagic fever viruses **Route of administration:** Oral, aerosol, IV **Adverse effects:** Perhaps harmful to developing fetus	Not known
Antiviral Drugs That Inhibit Protein Synthesis			
Antisense nucleic acids Representative: Fomivirsen	Complementary to mRNA; binding prevents protein synthesis by blocking ribosomes	**Spectrum of action:** Specific to species with complementary mRNA; fomivirsen specific against cytomegalovirus **Route of administration**: Fomivirsen injected weekly into eyes **Adverse effects**: Possible glaucoma	Not known
Antiviral Drugs That Inhibit Viral Proteins			
Protease inhibitors Representatives: Boceprevir Darunavir Fosamprenavir Telaprevir	Computer-assisted modeling of protease enzymes, which are unique to HIV and hepatitis C virus, allowed the creation of drugs that block the enzymes' active sites, often used in conjunction with drugs active against nucleic acid synthesis	**Spectrum of action:** HIV and hepatitis C **Route of administration:** Oral **Adverse effects:** None	Results from mutation in protease gene

TABLE 10.4 Antimicrobials Against Eukaryotes: Antifungal Drugs

Drug	Description and Mode of Action	Clinical Considerations	Method of Resistance
Antifungal Drugs That Inhibit Cell Membranes			
Allylamines Representative: Terbinafine	Antifungal action due to inhibition of ergosterol synthesis	**Spectrum of action:** Fungi **Route of administration:** Oral, IV **Adverse effects:** Headache, nausea, vomiting, diarrhea, liver damage, rash	Not known
Azoles Representatives: Fluconazole Itraconazole Ketoconazole Voriconazole	Antifungal action due to inhibition of synthesis of ergosterol, an essential component of fungal cytoplasmic membranes	**Spectrum of action:** Fungi and protozoa **Route of administration:** Topical, IV **Adverse effects:** Possibly causes cancer in humans	Mutation in gene for target enzyme
Polyenes Representatives: Amphotericin B Nystatin	Associate with molecules of ergosterol, forming a pore through the fungal membrane, which leads to leakage of essential ions from the cell; amphotericin B is produced by *Streptomyces nodosus*	**Spectrum of action:** Fungi, some amoebae **Route of administration:** Amphotericin B: IV; nystatin: topical **Adverse effects:** Chills, vomiting, fever	Rare; decrease in amount or change in chemistry of ergosterol
Other Antifungal Drugs			
Ciclopirox	Synthetic antimetabolite that appears to disrupt DNA repair, inhibit mitosis, and interfere with some aspects of intracellular transport.	**Spectrum of action:** Fungi, especially *Trichophyton* **Route of administration:** Topical **Adverse effects:** Possible burning sensation at application site	Not known
Echinocandins Representative: Caspofungin	Inhibits synthesis of glucan subunit of fungal cell walls	**Specimen of action:** *Candida, Aspergillus* **Route of administration:** IV **Adverse effects:** Rash, facial swelling, respiratory spasms, gastrointestinal distress, toxic to human embryos	Results from mutation in glucan synthase gene
5-Fluorocytosine	Fungi, but not mammals, have an enzyme that converts this drug into 5-fluorouracil, an analog of uracil that inhibits RNA function	**Spectrum of action:** *Candida, Cryptococcus, Aspergillus* **Route of administration:** Oral **Adverse effects:** None	Develops from mutations in the genes for enzymes necessary for utilization of uracil
Griseofulvin	Isolated from *Penicillium griseofulvum;* deactivates tubulin, preventing cytokinesis and segregation of chromosomes during mitosis (see Chapter 12)	**Spectrum of action:** Molds of skin infections **Route of administration:** Topical, oral **Adverse effects:** None	Not known

TABLE 10.5 Antimicrobials Against Eukaryotes: Anthelmintic Drugs

Drug	Description and Mode of Action	Clinical Considerations	Method of Resistance
Anthelmintic Drugs That Are Antimetabolites			
Benzimidazole derivatives Representatives: Albendazole Mebendazole Thiabendazole Triclabendazole	Inhibit microtubule formation and glucose uptake	**Spectrum of action:** Helminths, protozoa **Route of administration:** Oral **Adverse effects:** Possible diarrhea	Not known
Iodoquinol	Halogenated (iodine containing), possibly works by sequestering iron ions required by protozoan	**Spectrum of action:** *Entamoeba* **Route of administration:** Oral **Adverse effects:** Neuropathy and blindness with prolonged use	Not known
Ivermectin **Metrifonate**	Produce flaccid paralysis by blocking neurotransmitters	**Spectrum of action:** Helminths **Route of administration:** Oral **Adverse effects:** Allergic reactions may result from antigens of dead helminths	Not known
Niclosamide	Inhibits oxidative phosphorylation of ATP by mitochondria	**Spectrum of action:** Cestodes **Route of administration:** Oral **Adverse effects:** Abdominal pain, nausea, diarrhea	Not known
Praziquantel	Changes membrane permeability to calcium ions, which are required for muscular contraction; induces complete muscular contraction in helminths	**Spectrum of action:** Cestodes, trematodes **Route of administration:** Oral **Adverse effects:** None	Not known
Pyrantel pamoate **Diethylcarbamazine**	Bind to neurotransmitter receptors, causing complete muscular contraction of helminths	**Spectrum of action:** Nematodes **Route of administration:** Oral **Adverse effects:** None	Not known
Anthelmintic Drugs That Inhibit Nucleic Acid Synthesis			
Niridazole	When partially catabolized by schistosome enzymes, binds to DNA, preventing replication	**Spectrum of action:** *Schistosoma*	Not known
Oltipraz	Possibly acts by reducing the supply of deoxyribonucleotides	**Spectrum of action:** *Schistosoma*	Not known
Oxamniquine	Schistosome enzyme activates drug, which then inhibits DNA synthesis	**Spectrum of action:** *Schistosoma*	Not known

TABLE 10.6 Antimicrobials Against Eukaryotes: Antiprotozoan Drugs

Drug	Description and Mode of Action	Clinical Considerations	Method of Resistance
Antiprotozoan Drugs That Inhibit Protein Synthesis			
Lincosamides Representative: Clindamycin	Binds to 50S ribosomal subunit and stops protein elongation	**Spectrum of action:** Effective against some protozoa and G+ and anaerobic G− bacteria **Route of administration:** Oral or IV; does not traverse blood-brain barrier **Adverse effects:** Gastrointestinal distress, including nausea, diarrhea, vomiting, and pain	Develops via changes in ribosomal structure that prevent drug from binding; resistance genes are same as those of aminoglycosides
Paromomycin	Aminoglycoside that interferes with bacterial 30S ribosomal subunits. Its method of action against protozoa is not known	**Spectrum of action:** Effective against some protozoa, e.g., *Leishmania* and *Entamoeba* and most G− and G+ bacteria **Route of administration:** Oral **Adverse effects:** Allergic reactions in some patients; intestinal blockage	Decreased drug uptake due to altered cytoplasmic membrane
Antiprotozoan Drugs That Are Antimetabolites			
Artemisinins Representative: Artesunate	Derived from Chinese wormwood shrub; interferes with heme detoxification and with Ca^{2+} transport	**Spectrum of action:** *Plasmodium* **Route of administration:** Oral **Adverse effects:** Nausea, vomiting, itching, dizziness	Mutation in Ca^{2+} transporter gene
Atovaquone	Analog of coenzyme Q of several protozoa and of *Pneumocystis;* interrupts electron transport	**Spectrum of action:** Protozoa (especially *Plasmodium*), *Pneumocystis* **Route of administration:** Oral **Adverse effects:** Possible rash, diarrhea, headache	Cells modify the structure of their electron transport chain proteins
Benzimidazole derivatives Representatives: Albendazole Mebendazole	Inhibit microtubule formation and glucose uptake	**Spectrum of action:** Helminths, protozoa **Route of administration:** Oral **Adverse effects:** Possible diarrhea	Not known
Heavy metals (e.g., Hg, As, Cr, Sb) Representatives: Meglumine antimonate Melarsoprol (contains As) Salvarsan (contains As) Sodium stibogluconate (contains Sb)	Deactivate enzymes by breaking hydrogen bonds necessary for effective tertiary structure; drugs containing arsenic were the first recognized selectively toxic chemotherapeutic agents	**Spectrum of action:** Metabolically active cells **Route of administration:** Topical, oral **Adverse effects:** Toxic to active cells, such as those of the brain, kidney, liver, and bone marrow	Not known
Iodoquinol	Mode of action unknown	**Spectrum of action:** Intestinal amoebae **Route of administration:** Oral **Adverse effects:** Fever, chills, rash	Not known
Lumefantrine	Synthetic drug that prevents detoxification of heme released from hemoglobin of damaged red blood cells	**Spectrum of action:** *Plasmodium* **Route of administration:** Oral **Adverse effects:** Headache, dizziness, weakness	Not known
Nifurtimox	Interferes with electron transport	**Spectrum of action:** *Trypanosoma cruzi* **Route of administration:** Oral **Adverse effects:** Abdominal pain, nausea	Not known

Continued—

TABLE 10.6 Antimicrobials Against Eukaryotes: Antiprotozoan Drugs—*Continued*

Drug	Description and Mode of Action	Clinical Considerations	Method of Resistance
Antiprotozoan Drugs That Are Antimetabolites			
Nitazoxanide	Believed to interfere with anaerobic electron transport chain	**Spectrum of action:** Protozoa **Route of administration:** Oral **Adverse effects:** Gastrointestinal distress	Not known
Proguanil **Pyrimethamine**	Block second metabolic step in the formation of folic acid from PABA; synergistic with sulfonamides	**Spectrum of action:** Broad: effective against G+ and G− bacteria and some protozoa and fungi; however, resistance is widespread **Route of administration:** Oral **Adverse effects:** Allergic reactions in some patients	Cells that require folic acid as a vitamin are also naturally resistant; chromosomal mutations result in lowered affinity for the drugs
Sulfonamides Representatives: Sulfadiazine Sulfadoxine Sulfanilamide	Synthetic drugs; first produced as a dye; analogs of PABA that bind irreversibly to enzyme that produces dihydrofolic acid; synergistic with trimethoprim	**Spectrum of action:** Broad: effective against G+ and G− bacteria and some protozoa and fungi; however, resistance is widespread **Route of administration:** Oral **Adverse effects:** Rare: Allergic reactions, anemia, jaundice, mental retardation of fetus if administered in last trimester of pregnancy	Cells that require folic acid as a vitamin are also naturally resistant; chromosomal mutations result in lowered affinity for the drugs
Suramin	Inhibits specific enzymes in some protozoa	**Spectrum of action:** *Trypanosoma brucei* (eastern African variant) **Routine of administration:** Oral **Adverse effects:** None	Not known
Trimethoprim	Block second metabolic step in the formation of folic acid from PABA; synergistic with sulfonamides	**Spectrum of action:** Broad: effective against some protozoa, fungi, and some G+ and G− bacteria; however, resistance is widespread **Route of administration:** Oral **Adverse effects:** Allergic reactions in some patients	Cells that require folic acid as a vitamin are also naturally resistant; chromosomal mutations result in lowered affinity for the drugs
Antiprotozoan Drugs That Inhibit DNA Synthesis			
Eflornithine	Inhibits synthesis of precursors of nucleic acids	**Spectrum of action:** *Trypanosoma brucei gambiense* (western African variant) **Route of administration:** Oral **Adverse effects:** Anemia, inhibition of blood clotting, nausea, vomiting	Not known
Nitroimidazoles Representatives: Benznidazole Metronidazole Tinidazole	Anaerobic conditions reduce the drug, which then appears to damage DNA, preventing correct replication and transcription	**Spectrum of action:** Protozoa **Route of administration:** Oral **Adverse effects:** Metronidazole and tinidazole cause cancer in laboratory rodents	Not known
Pentamidine	Binds to nucleic acids, inhibiting replication, transcription, and translation	**Spectrum of action:** Protozoa, including *Trypanosoma brucei* (western African variant) and *Pneumocystis* (fungus) **Route of administration:** IM, IV **Adverse effects:** Rash, low blood pressure, irregular heartbeat, kidney and liver failure	Not known
Quinolones Representatives: Natural quinine Semisynthetic quinines: Chloroquine Mefloquine Piperaquine Primaquine	Natural and semisynthetic drugs derived from the bark of cinchona tree; inhibit metabolism of malaria parasites by one or more unknown methods	**Spectrum of action:** *Plasmodium* **Route of administration:** Oral **Adverse effects:** Allergic reactions, visual disturbances	Results from the presence of quinoline pumps that remove the drugs from parasite's cells

MICRO IN THE CLINIC FOLLOW-UP

Battling a Microscopic Enemy

Ben is evacuated from the hospital in Afghanistan to Ramstein, a large American military hospital in Europe, where a full analysis can be done. A blood culture reveals that Ben is battling *Acinetobacter baumannii*, a small Gram-negative bacillus that is usually a nonpathogenic, environmental contaminant. Testing identifies the organism as a drug-resistant strain that has been nicknamed "Iraqibacter."

The Ramstein lab conducts antimicrobial sensitivity testing on the bacterium isolated from Ben's blood and finds that it is resistant to most antimicrobials, including fluoroquinolones. In an effort to save Ben's life, the physician prescribes rifampin and imipenem. Imipenem is a newly developed beta-lactam carbapenem-type antimicrobial, administered with cilastatin, which slows down the body's catabolism of imipenem. Imipenem inhibits cell wall formation and is an inhibitor of most Gram-negative bacteria, even those that typically show beta-lactam resistance. Normally, imipenem is quickly excreted in urine, so to prolong its activity in the body, it is combined with cilastatin.

This chemical battle provides just the arsenal for Ben to overcome the infection. After 36 hours of treatment, the signs of septicemia resolve, allowing Ben to reunite with his family and to begin physical therapy to adapt to his limb loss. The specific identification of the bacterium and the determination of antimicrobial sensitivity has saved Ben's life.

1. **What is the mechanism of action for beta-lactams?**
2. **How do bacteria become resistant to beta-lactams?**

MM Check your answers to Micro in the Clinic Follow-Up questions in the MasteringMicrobiology Study Area.

Explore the Invisible: Action of Some Drugs That Inhibit Prokaryotic Protein Synthesis

Practice on-the-go with Dr. Bauman Video Tutors by scanning this QR code with your smart phone. Visit the **MasteringMicrobiology Study Area** to challenge your understanding with practice tests, animation quizzes, and clinical case studies!

MasteringMicrobiology®

CHAPTER SUMMARY

The History of Antimicrobial Agents (pp. 289–290)

1. Chemotherapeutic agents are chemicals used to treat diseases. Among them are **antimicrobial agents (antimicrobials),** which include **antibiotics** (biologically produced agents), **semisynthetic antimicrobials** (chemically modified antibiotics), and **synthetic drugs**.

Mechanisms of Antimicrobial Action (pp. 290–298)

1. Successful chemotherapy against microbes is based on **selective toxicity,** that is, using antimicrobial agents that are more toxic to pathogens than to the patient.
2. Antimicrobial drugs affect pathogens by inhibiting cell wall synthesis, inhibiting the translation of proteins, disrupting cytoplasmic membranes, inhibiting general metabolic pathways, inhibiting nucleic acid synthesis, blocking the attachment of viruses to their hosts, or blocking a pathogen's recognition of its host. ▶**ANIMATIONS:** *Chemotherapeutic Agents: Modes of Action*
3. **Beta-lactams**—penicillins, carbapenems, cephalosporins—have a functional lactam ring. They prevent bacteria from cross-linking NAM subunits of peptidoglycan in the bacterial cell wall during growth. **Vancomycin** and **cycloserine** also disrupt cell wall formation in many Gram-positive bacteria. **Bacitracin** blocks NAG and NAM transport from the cytoplasm. **Isoniazid (INH)** and **ethambutol** block mycolic acid synthesis in the walls of mycobacteria. **Echinocandins** block synthesis of fungal cell walls.

4. Antimicrobial agents that inhibit protein synthesis include **aminoglycosides** and **tetracyclines,** which inhibit functions of the 30S ribosomal subunit, and **chloramphenicol, lincosamides, streptogramins,** and **macrolides,** which inhibit 50S subunits. Mupirocin stops polypeptide synthesis by binding to tRNA molecules that carry isoleucine. **Oxazolidinones** block initiation of translation. **Antisense nucleic acid** molecules also inhibit protein synthesis.

   ▶**VIDEO TUTOR:** *Actions of Some Drugs That Inhibit Prokaryotic Protein Synthesis*

5. **Polyenes, azoles,** and **allylamines** disrupt the cytoplasmic membranes of fungi. Polymyxin acts against the membranes of Gram-negative bacteria.
6. **Sulfonamides** are structural **analogs** of para-aminobenzoic acid (PABA), a chemical needed by some microorganisms but not by humans. The substitution of sulfonamides in the metabolic pathway leading to nucleic acid synthesis kills those organisms. Trimethoprim also blocks this pathway.
7. Drugs that inhibit nucleic acid replication in pathogens include actinomycin, **nucleotide analogs** and **nucleoside analogs,** quinolones, and rifampin.

Clinical Considerations in Prescribing Antimicrobial Drugs (pp. 298–302)

1. Chemotherapeutic agents have a **spectrum of action** and may be classed as either **narrow-spectrum drugs** or **broad-spectrum drugs,** depending on how many kinds of pathogens they affect.
2. **Diffusion susceptibility tests,** such as the Kirby-Bauer test, reveal which drug is most effective against a particular pathogen; in general, the larger the **zone of inhibition** around a drug-soaked disk on a Petri plate, the more effective the drug.
3. The **minimum inhibitory concentration (MIC),** usually determined by either a **broth dilution test** or an **Etest,** is the smallest amount of a drug that will inhibit a pathogen.
4. A **minimum bactericidal concentration (MBC) test** ascertains whether a drug is bacteriostatic and the lowest concentration of a drug that is bactericidal.
5. In choosing antimicrobials, physicians must consider effectiveness, how a drug is best administered—orally, intramuscularly, or intravenously—and possible side effects, including toxicity and allergic responses.
6. A drug's **therapeutic index (TI)** is essentially a ratio of the drug's tolerated dose to its effective dose. The higher the TI, the safer the drug.
7. Clinicians use the term **therapeutic range (therapeutic window)** to indicate the range of concentrations of a drug that are effective without being excessively toxic.

Resistance to Antimicrobial Drugs (pp. 302–316)

1. Some members of a pathogenic population may develop resistance to a drug because of extra DNA pieces called **R plasmids** or the mutation of genes. Microorganisms may resist a drug by producing enzymes such as **β-lactamase** that deactivate the drug, by inducing changes in the cell membrane that prevent entry of the drug, by altering the drug's target to prevent its binding, by altering the cell's metabolic pathways, by removing the drug from the cell with **efflux pumps,** or by protecting the drug's target by binding another molecule to it.

   ▶**ANIMATIONS:** *Antibiotic Resistance: Origins of Resistance, Forms of Resistance*

2. **Cross resistance** occurs when resistance to one chemotherapeutic agent confers resistance to similar drugs. **Multiple-drug-resistant pathogens** are resistant to three or more types of antimicrobial drugs.
3. **Synergism** describes the interplay between drugs that results in efficacy that exceeds the efficacy of either drug alone. Some drug combinations are antagonistic.

QUESTIONS FOR REVIEW

Answers to the Questions for Review (except Short Answer questions) begin on p. A-1.

Multiple Choice

1. Diffusion and dilution tests that expose pathogens to antimicrobials are designed to ________________.
   a. determine the spectrum of action of a drug
   b. determine which drug is most effective against a particular pathogen
   c. determine the amount of a drug to use against a particular pathogen
   d. both b and c
2. In a Kirby-Bauer susceptibility test, the presence of a zone of inhibition around disks containing antimicrobial agents indicates ________________.
   a. that the microbe does not grow in the presence of the agents
   b. that the microbe grows well in the presence of the agents
   c. the smallest amount of the agent that will inhibit the growth of the microbe
   d. the minimum amount of an agent that kills the microbe in question
3. The key to successful chemotherapy is ________________.
   a. selective toxicity
   b. a diffusion test
   c. the minimum inhibitory concentration test
   d. the spectrum of action
4. Which of the following statements is relevant in explaining why sulfonamides are effective?
   a. Sulfonamides attach to sterol lipids in the pathogen, disrupt the membranes, and lyse the cells.
   b. Sulfonamides prevent the incorporation of amino acids into polypeptide chains.
   c. Humans and microbes use PABA differently in their metabolism.
   d. Sulfonamides inhibit DNA replication in both pathogens and human cells.

5. Cross resistance is ________________.
   a. the deactivation of an antimicrobial agent by a bacterial enzyme
   b. alteration of the resistant cells so that an antimicrobial agent cannot attach
   c. the mutation of genes that affect the cytoplasmic membrane channels so that antimicrobial agents cannot cross into the cell's interior
   d. resistance to one antimicrobial agent because of its similarity to another antimicrobial agent

6. Multiple-drug-resistant microbes ________________.
   a. are resistant to all antimicrobial agents
   b. respond to new antimicrobials by developing resistance
   c. frequently develop in hospitals
   d. all of the above

7. Which of the following is most closely associated with a beta-lactam ring?
   a. penicillin
   b. vancomycin
   c. bacitracin
   d. isoniazid

8. Drugs that act against protein synthesis include ________________.
   a. beta-lactams
   b. trimethoprim
   c. polymyxin
   d. aminoglycosides

9. Which of the following statements is *false* concerning antiviral drugs?
   a. Macrolide drugs block attachment sites on the host cell wall and prevent viruses from entering.
   b. Drugs that neutralize the acidity of phagolysosomes prevent viral uncoating.
   c. Nucleotide analogs can be used to stop microbial replication.
   d. Drugs containing protease inhibitors retard viral growth by blocking the production of essential viral proteins.

10. PABA is ________________.
   a. a substrate used in the production of penicillin
   b. a type of β-lactamase
   c. molecularly similar to cephalosporins
   d. used to synthesize folic acid

VISUALIZE IT!

1. Label each figure below to indicate the class of drug that is stopping polypeptide translation.

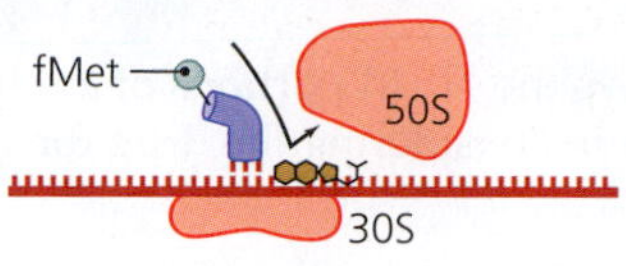

a. ________________
block initiation.

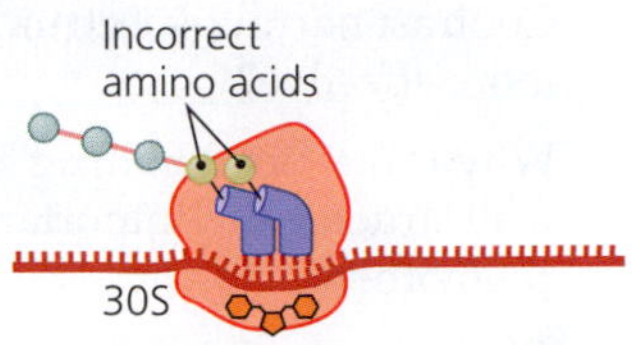

b. ________________
change 30S subunit.

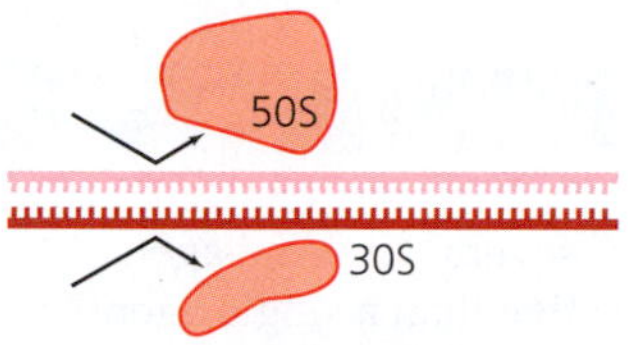

c. ________________
block ribosome attachment.

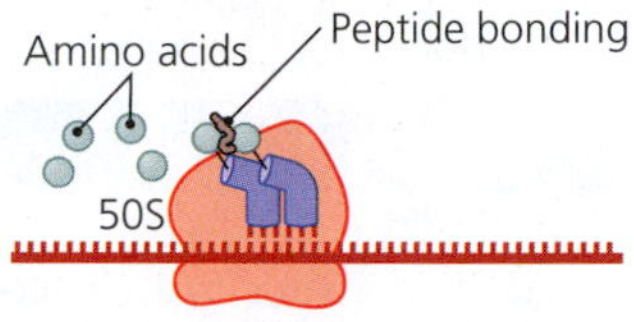

d. ________________
inhibits peptide bonding.

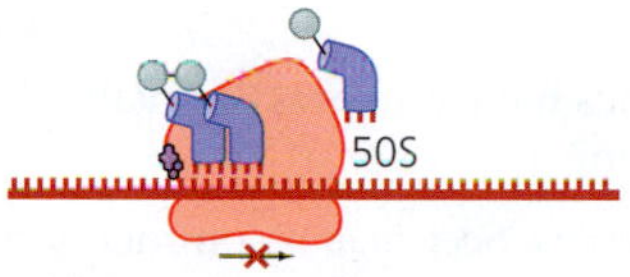

e. ________________
f. ________________
block ribosome movement.

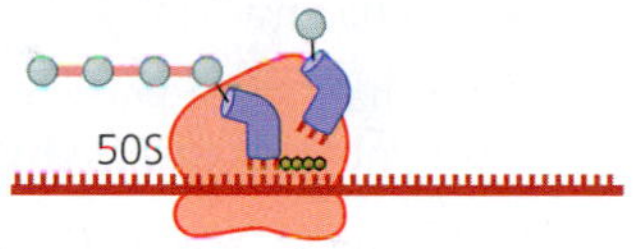

g. ________________
block tRNA docking.

2. What specific test for antimicrobial efficacy is shown? What does this test measure? Draw an oval to predict the size and shape of the zone of inhibition if the drug concentration on the strip were increased twofold.

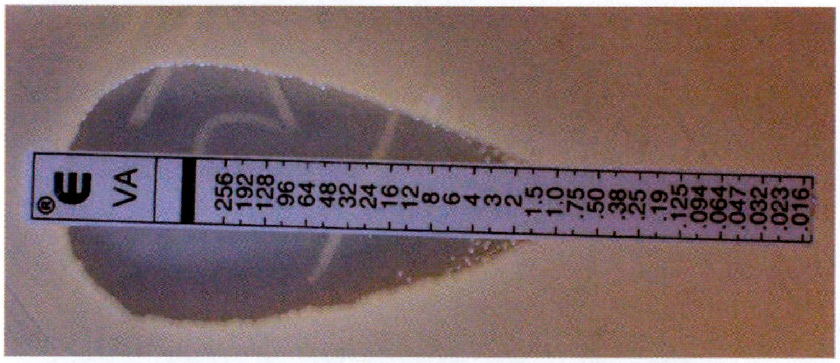

Short Answer

1. What characteristics would an ideal chemotherapeutic agent have? Which drug has these qualities?
2. Contrast narrow-spectrum and broad-spectrum drugs. Which are more effective?
3. Why is the fact that drug Z destroys the NAM portions of a cell's wall structure an important factor in considering the drug for chemotherapy?
4. Given that both human cells and pathogens synthesize proteins at ribosomal sites, how can antimicrobial agents that target this process be safe to use in humans?
5. Support or refute the following statement: antimicrobial agents produce resistant cells.
6. Given that resistant strains of pathogens are a concern to the general health of a population, what can be done to prevent their development?
7. Why are antiviral drugs difficult to develop?
8. A man has been given a broad-spectrum antibiotic for his stomach ulcer. What unintended consequences could arise from this therapy?
9. Compare and contrast the actions of polyenes, azoles, allylamines, and polymyxin.
10. What is the difference in drug action of synergists contrasted with that of antagonists?

CRITICAL THINKING

1. AIDS is treated with a "cocktail" of several antiviral agents at once. Why is the cocktail more effective than a single agent? What is a physician trying to prevent by prescribing several drugs at once?
2. How does *Penicillium* escape the effects of the penicillin it secretes?
3. How might a colony of *Bacillus licheniformis* escape the effects of its own bacitracin?
4. Fewer than 1% of known antibiotics have any practical value in treatment of disease. Why is this so?
5. In an issue of *News of the Lepidopterists' Society*, a recommendation was made to moth and butterfly collectors to use antimicrobials to combat disease in the young of these insects. What are the possible ramifications for human health of such usage of antimicrobials?
6. Even though aminoglycosides such as gentamicin can cause deafness, there are still times when they are the best choice for treating some infections. What laboratory test would a clinical scientist use to show that gentamicin is the best choice to treat a particular *Pseudomonas* infection?
7. Your pregnant neighbor has a sore throat and tells you that she is taking some tetracycline she had left over from a previous infection. Give two reasons why her decision is a poor one.
8. Acyclovir has replaced adenosine arabinoside as treatment for herpes infections. Compare the ways these drugs are activated (see Table 10.3 on pp. 311–312). Why is acyclovir a better choice?
9. Why might amphotericin B affect the kidneys more than other human organs?
10. Antiparasitic drugs in the benzimidazole family inhibit the polymerization of tubulin. What effect might these drugs have on mitosis and flagella?
11. Your cousin reads in a blog that the U.S. Food and Drug Administration (FDA) has approved an antimicrobial called tigecycline. The blog says that the drug is an analog of tetracycline. She asks you what that means and how the drug works. What should you tell your cousin?
12. Scientists have cultured bacteria isolated from within frozen mammoths, which are thousands of years old. Why would it not be surprising if these microbes were to show some resistance to modern antimicrobials that didn't exist when the mammoths died?
13. It would be impractical and expensive for every American to take amantadine during the entire flu season to prevent influenza infections. For what group of people might amantadine prophylaxis be cost effective?
14. Sometimes it is not possible to conduct a susceptibility test because of either a lack of time or an inability to access the bacteria (e.g., from an inner-ear infection). How could a physician select an appropriate therapeutic agent in such cases?
15. *Enterococcus faecium* is frequently resistant to vancomycin. Why might this be of concern in a hospital setting in terms of developing resistant strains of *other* genera of bacteria?

CONCEPT MAPPING

Using the following terms, draw a concept map that describes antimicrobial resistance. For a sample concept map, see p. 94. Or, complete this and other concept maps online by going to the MasteringMicrobiology Study Area.

Altered pore proteins
Altered targets
Antimicrobials
Beta-lactamase
Binding of antimicrobials
Cell division
Conjugation
Efflux pumps
Entry of antimicrobials into cell
Mutation
Pathogen's enzymes
Penicillin
Rapid efflux of antibiotic
Transduction
Transformation

11 Characterizing and Classifying Prokaryotes

MICRO IN THE CLINIC

Can Diabetes Cause a Foot Infection?

Sean is the CEO of a growing bioengineering firm. At 54, he is a workaholic, smokes a pack a day, loves fried foods, and is overweight. A few years ago, Sean was diagnosed with type 2 diabetes, but he has not managed it very well or made any lasting changes to his lifestyle. He gives himself insulin injections to treat the diabetes, but his stubborn refusal to quit burgers and pizza have often sent his blood glucose levels soaring sky high, resulting in several trips to the emergency room. Sean knows he needs to change his ways but can never seem to stick to a healthier lifestyle.

Lately Sean has been noticing an irritating numbness in his feet. At first, he blames it on his new work shoes, but when a discolored, blisterlike lesion appears on the top of his right foot, he decides to visit his doctor. The doctor, alarmed by the sight of the lesion, orders a number of tests. As the nurse debrides the wound (a process of removing damaged and contaminated tissue), a foul smell permeates the doctor's office. When the test results come back, they indicate the presence of white blood cells and endospore-forming, obligately anaerobic Gram-positive bacilli in chains.

Is Sean's foot infection related to his diabetes? What organism is producing the foul-smelling lesion? Turn to the end of the chapter (p. 346) to find out.

MM **Explore More:** Test your readiness and apply your knowledge with dynamic learning tools at MasteringMicrobiology.

Prokaryotes are by far the most numerous and diverse group of cellular microbes. Scientists estimate there are more than 6×10^{31} prokaryotes on Earth. If laid end to end, they would more than encircle the entire Milky Way galaxy. They thrive in various habitats: from Antarctic glaciers to thermal hot springs, from the colons of animals to the cytoplasm of other prokaryotes, from distilled water to supersaturated brine, and from disinfectant solutions to basalt rocks thousands of meters below the Earth's surface. In part because of such great diversity, only a very few prokaryotes have enzymes, toxins, or cellular structures that enable them to colonize humans or cause diseases. In this chapter we will begin by examining general prokaryotic characteristics and conclude with a survey of specific prokaryotic taxa. We will briefly mention human pathogens throughout the chapter. (More detailed discussions of disease-causing microbes appear in Chapters 19–24).

General Characteristics of Prokaryotic Organisms

In this section, we consider prokaryotes' shapes, the ability of some to survive unfavorable conditions by forming resistant endospores, some reproduction strategies, and their spatial arrangement. (Chapters 3 and 5–7 consider other general characteristics of prokaryotic cells, including cellular structure, metabolism, growth, and genetics.)

Morphology of Prokaryotic Cells

LEARNING | OUTCOME

11.1 Identify six basic shapes of prokaryotic cells.

Prokaryotic cells exist in a variety of shapes, or morphologies **(FIGURE 11.1)**. The three basic shapes are **coccus** (kok´ŭs, roughly spherical, plural *cocci*, kok´sī), **bacillus** (ba-sil´ŭs, rod-shaped, plural *bacilli*, bă-sil´ī), and **spiral.** Cocci are not all perfectly spherical; for example, there are pointed, kidney-shaped, and oval cocci. Similarly, bacilli vary in shape; for example, some bacilli are pointed, spindle shaped, or threadlike (filamentous). Spiral-shaped prokaryotes are either **spirilla**, which are stiff, or **spirochetes** (spī´rō-kētz), which are flexible. Curved rods are **vibrios**, and the term **coccobacillus** is used to describe cells that are intermediate in shape between cocci and bacilli, that is, when it is difficult to ascertain if a cell is an elongated coccus or a short bacillus. In addition to these basic shapes, there are star-shaped, triangular, and rectangular prokaryotes as well as prokaryotes that are **pleomorphic**[1] (plē-ō-mōr´fik); that is, they vary in shape and size.

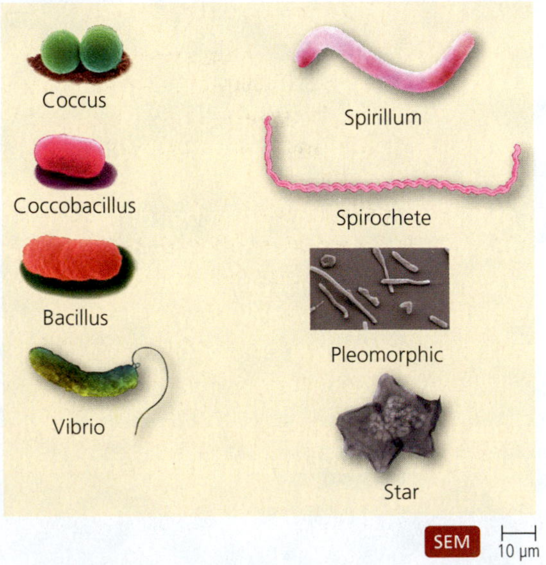

▲ **FIGURE 11.1 Typical prokaryotic morphologies.** *What is one difference between a spirillum and a spirochete?*

Figure 11.1 *Generally, spirilla are stiff, whereas spirochetes are flexible.*

Endospores

LEARNING | OUTCOME

11.2 Describe the formation and function of bacterial endospores.

The Gram-positive bacteria *Bacillus* (ba-sil´ŭs) and *Clostridium* (klos-trid´ē-ŭm) produce **endospores**, which are important for several reasons, including their durability and potential pathogenicity. Endospores constitute a defensive strategy against hostile or unfavorable conditions. They are stable resting stages that barely metabolize and germinate when conditions improve.

Though some people refer to endospores as "spores," endospores should not be confused with reproductive spores, such as those of algae and fungi. A single bacterial cell, called a *vegetative* cell to distinguish it from an endospore, transforms into only one endospore, which then germinates to grow into a single vegetative cell; therefore, endospores are not reproductive structures. No new cells are formed.

The process of endospore formation, called *sporulation*, requires 8 to 10 hours and proceeds in eight steps (see Figure 3.24). Depending on the species, a cell forms endospores either *centrally*, *subterminally* (near one end), or *terminally* (at one end) **(FIGURE 11.2)**.

Food processors, health care professionals, and governments are concerned about endospore formation because endospores can resist our attempts to kill them and because many endospore-forming bacteria produce deadly toxins that cause fatal diseases, such as anthrax, tetanus, and gangrene.

Reproduction of Prokaryotic Cells

LEARNING | OUTCOMES

11.3 List three common types of reproduction in prokaryotes.

11.4 Describe snapping division as a type of binary fission.

All prokaryotes reproduce asexually; none reproduce sexually. The most common method of asexual reproduction is **binary fission**, which proceeds as follows **(FIGURE 11.3)**: **1** The cell replicates its DNA; each DNA molecule is attached to the cytoplasmic membrane. **2** The cell grows, and as the cytoplasmic

[1]From Greek *pleon*, meaning "more," and *morphe*, meaning "form."

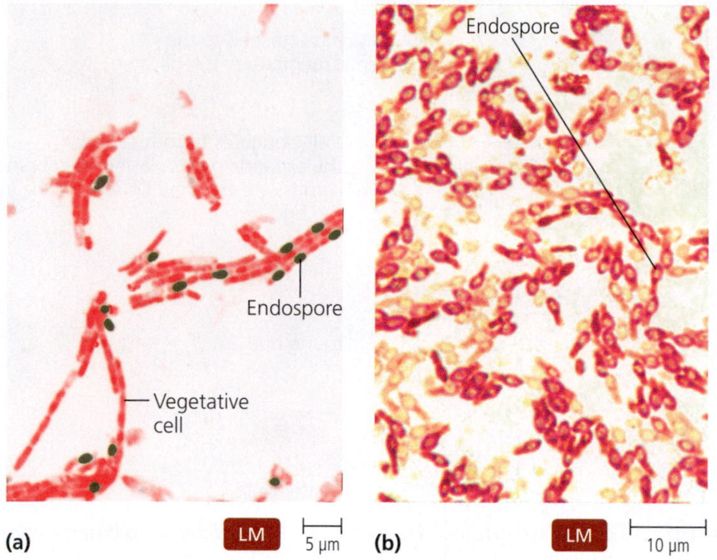

▲ **FIGURE 11.2 Locations of endospores.** **(a)** Central endospores of *Bacillus*. **(b)** Subterminal endospores of *Clostridium botulinum*. The enlarged endospores have swollen the vegetative cells that produced them.

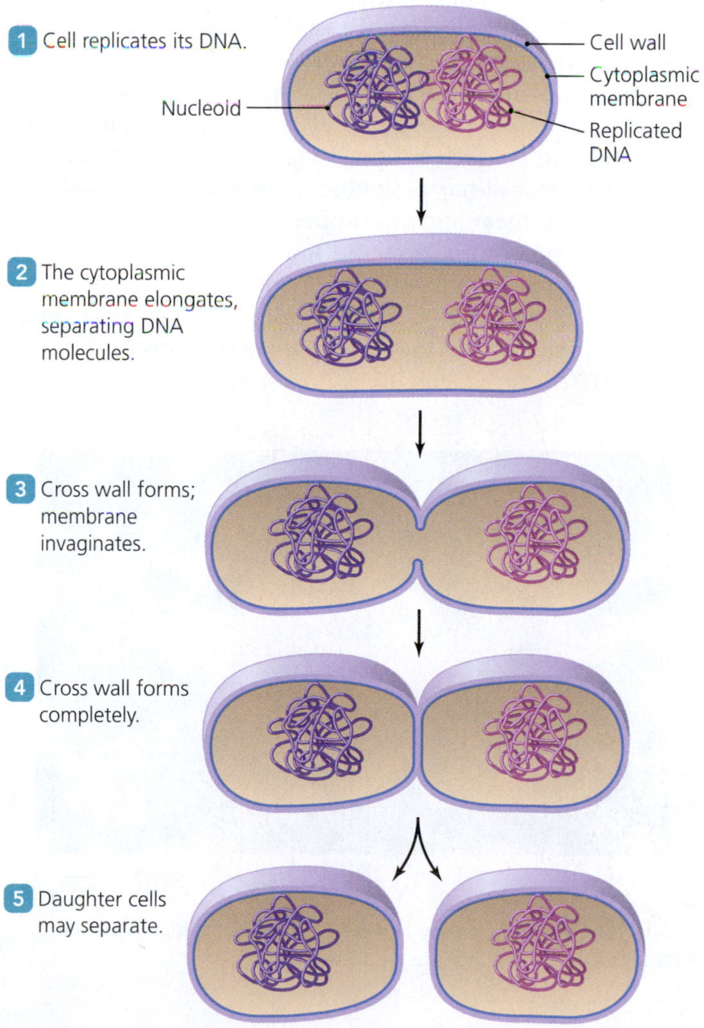

▲ **FIGURE 11.3 Binary fission.**

membrane elongates, it moves the daughter molecules of DNA apart. 3 The cell forms a cross wall, invaginating the cytoplasmic membrane. 4 The cross wall completely divides daughter cells. 5 The daughter cells may or may not separate. The parental cell disappears with the formation of progeny.

▶**ANIMATIONS:** ***Bacterial Growth: Overview***

A variation of binary fission called **snapping division** occurs in some Gram-positive bacilli **(FIGURE 11.4)**. In snapping division, only the inner portion of a cell wall is deposited across the dividing cell. The thickening of this new cross wall puts tension on the outer layer of the old cell wall, which still holds the two cells together. Eventually, as the tension increases, the outer wall breaks at its weakest point with a snapping movement that tears it most of the way around. The daughter cells can then remain hanging together almost side by side being held at an angle by a small remnant of the original outer wall that acts like a hinge.

Some prokaryotes have other methods of reproduction. The parental cell retains its identity during and after these methods. The *actinomycetes* (ak´ti-nō-mī-sētz) produce reproductive cells called **spores** at the ends of their filamentous cells **(FIGURE 11.5)**. These are true spores and should not be confused with endospores. Each spore can develop into a clone of the original organism. Some reproduce by fragmentation into small motile filaments that glide away from the parental strand. Still other prokaryotes, such as the bacterium *Planctomyces* (plank-tō-mī´sēz), reproduce by **budding**, in which an outgrowth of the original cell (a bud) receives a

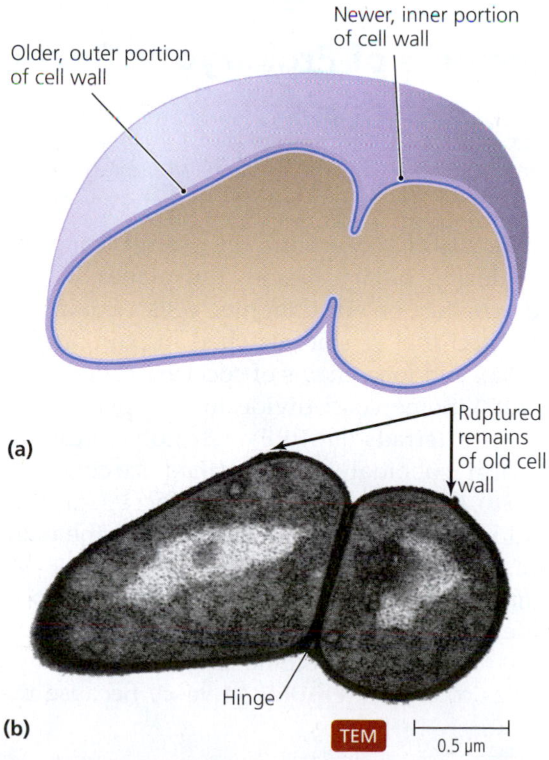

▲ **FIGURE 11.4 Snapping division, a variation of binary fission.** **(a)** Only the inner portion of the cell wall forms a cross wall. **(b)** As the daughter cells grow, tension snaps the outer portion of the cell wall, leaving the daughter cells connected by a hinge of old cell wall material.

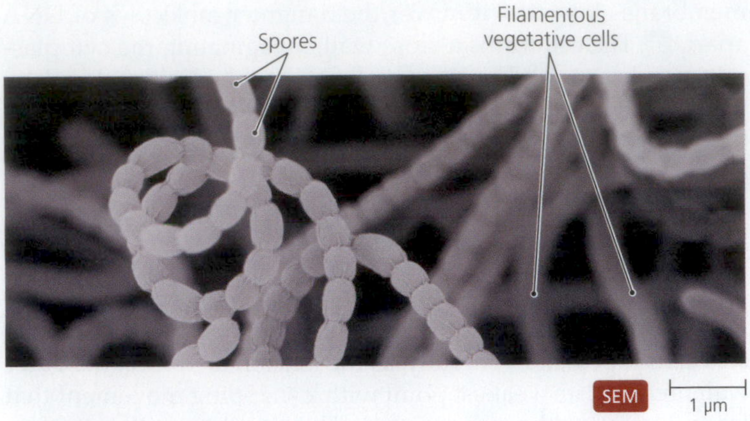
Spores
Filamentous vegetative cells
SEM
1 μm

▲ **FIGURE 11.5 Spores of actinomycetes.** Filamentous vegetative cells produce chains of spores, shown here in *Streptomyces*.

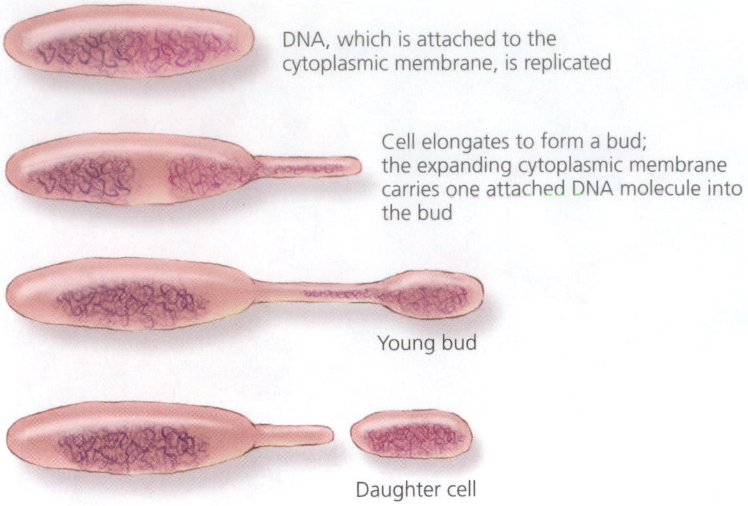
DNA, which is attached to the cytoplasmic membrane, is replicated
Cell elongates to form a bud; the expanding cytoplasmic membrane carries one attached DNA molecule into the bud
Young bud
Daughter cell

▲ **FIGURE 11.6 Budding.** *How does budding differ from binary fission?*

Figure 11.6 *In binary fission, the parent cell disappears with the formation of two equal-sized offspring; in contrast, a bud is often much smaller than its parent, and the parent remains to produce more buds.*

copy of the genetic material and enlarges. Eventually the bud is cut off from the parental cell, typically while it is still quite small **(FIGURE 11.6)**.

Epulopiscium (ep´yoo-lō-pis´sē-ŭm), a giant bacterium that lives inside surgeonfish, and many of its relatives have a truly unique method of reproduction among prokaryotes: They give "birth" to as many as 12 live offspring that emerge from the body of their dead mother cell **(FIGURE 11.7)**. The production of live offspring within a mother is called *viviparity,* and this is the first documented case of viviparous behavior in the prokaryotic world. In these bacteria, formation of internal offspring proceeds in a manner similar to the early stages of endospore formation.

Arrangements of Prokaryotic Cells

LEARNING | **OUTCOME**

11.5 Draw and label five arrangements of prokaryotes.

The arrangements of prokaryotic cells result from two aspects of division during binary fission: the planes in which cells divide and whether or not daughter cells remain attached to each other. Cocci that remain attached in pairs are **diplococci** **(FIGURE 11.8a)**, and long chains of cocci are called **streptococci**[2] **(FIGURE 11.8b)**. Some cocci divide in two planes and remain attached to form **tetrads** **(FIGURE 11.8c)**; others divide in three planes to form cuboidal packets called **sarcinae**[3] (sar´si-nī) **(FIGURE 11.8d)**. Clusters called **staphylococci**[4] (staf´i-lo-kok-sī), which look like bunches of grapes, form when the planes of cell division are random **(FIGURE 11.8e)**.

Bacilli are less varied in their arrangements than cocci because bacilli divide transversely—that is, across their long axis. Daughter bacilli may separate to become single cells or stay attached as either pairs or chains **(FIGURE 11.9a–c)**. Because the cells of *Corynebacterium diphtheriae* (kŏ-rī´nē-bak-tēr´ē-ŭm dif-thi´rē-ī), the causative agent of diphtheria, divide by snapping division, the daughter cells remain attached to form V-shapes and a side-by-side arrangement called a **palisade**[5] **(FIGURE 11.9d)**.

Descriptive words can be used to refer either to a general shape and/or arrangement or to a specific genus. Thus, the characteristic shape of the genus *Bacillus* is a rod-shaped bacterium, and the characteristic arrangement of bacteria in the genus *Sarcina* (sar´si-nă) is cuboidal. In such potentially confusing cases the meaning can be distinguished because genus names are always capitalized and italicized. In other cases, the arrangement uses the plural form, while a genus name is singular; thus, streptococci—spherical cells arranged in a chain—are characteristic of the genus *Streptococcus* (strep-tō-kok´ŭs).
▶**VIDEO TUTOR:** *Arrangements of Prokaryotic Cells*

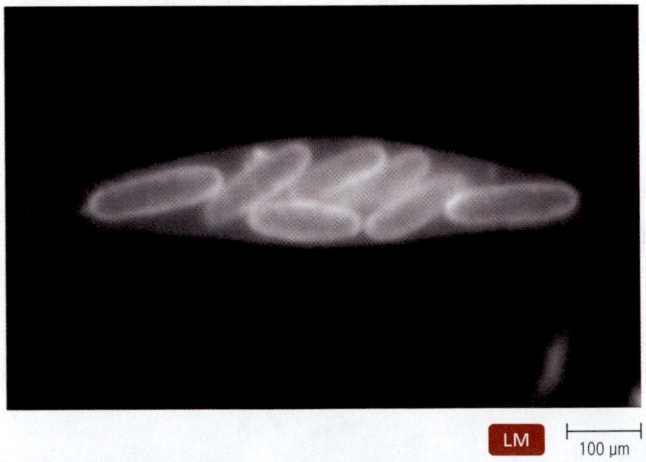
LM
100 μm

▲ **FIGURE 11.7 Viviparity in *Epulopiscium*.** Multiple offspring simultaneously develop inside a mother cell.

[2]From Greek *streptos,* meaning "twisted" because long chains tend to twist.
[3]Latin for "bundles."
[4]From Greek *staphyle,* meaning "bunch of grapes."

[5]From Latin *palus,* meaning "stake," referring to a fence made of adjoining stakes.

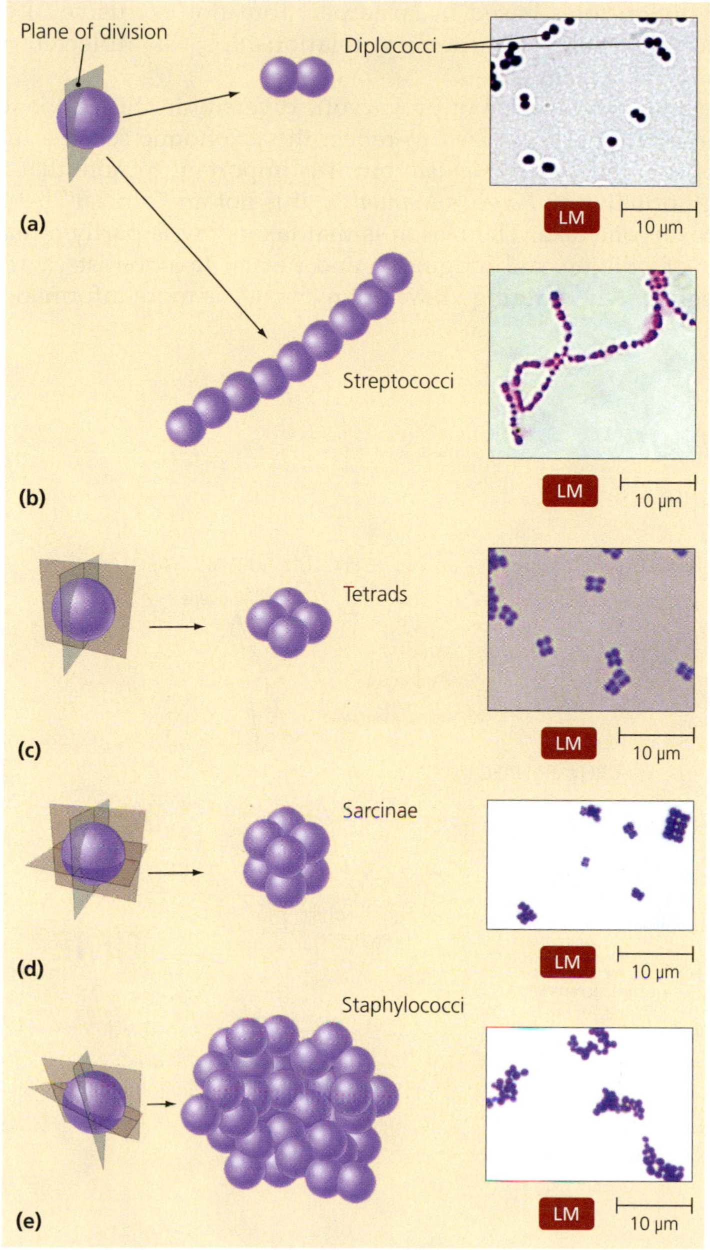

▲ **FIGURE 11.8 Arrangements of cocci. (a)** The diplococci of *Streptococcus pneumoniae.* **(b)** The streptococci of *Streptococcus pyogenes.* **(c)** Tetrads, in this case of *Micrococcus luteus.* **(d)** The genus *Sarcina* is characterized by sarcinae. **(e)** The staphylococci of *Staphylococcus aureus.*

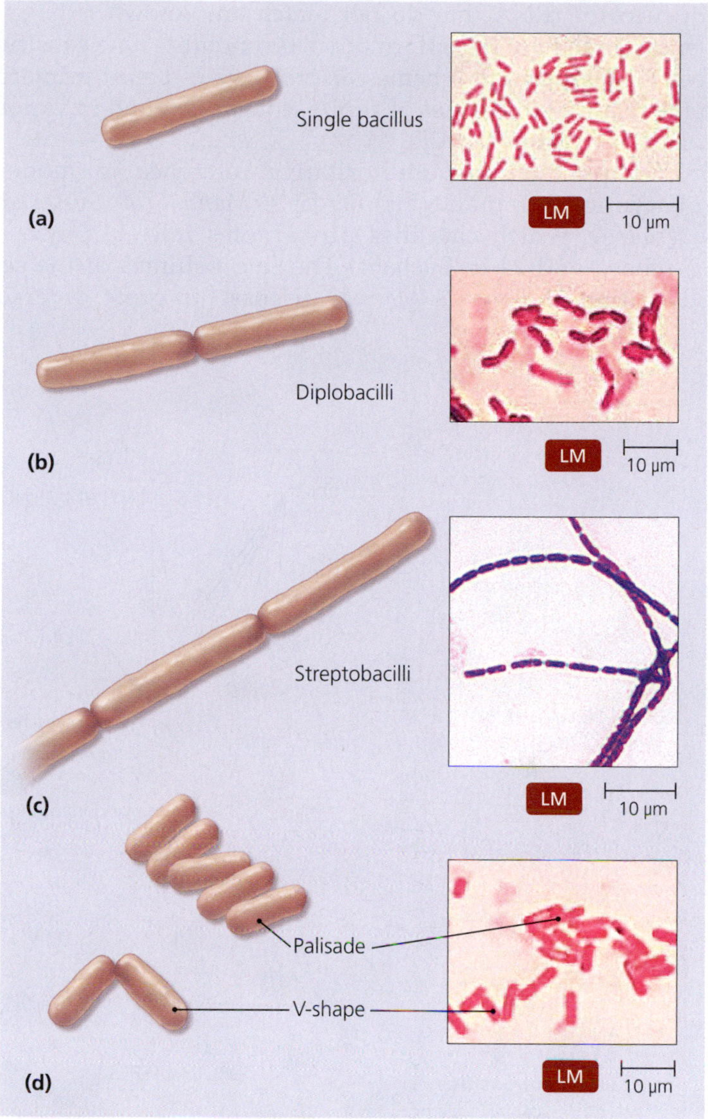

▲ **FIGURE 11.9 Arrangements of bacilli. (a)** A single bacillus of *Escherichia coli.* **(b)** Diplobacilli in a young culture of *Bacillus cereus.* **(c)** Streptobacilli in an older culture of *Bacillus cereus.* **(d)** V-shapes and palisades of *Corynebacterium diphtheriae.*

TELL ME WHY

Why does binary fission produce chains of cocci in some species but clusters of cocci in other species?

Modern Prokaryotic Classification

LEARNING **OUTCOMES**

11.6 Explain the general purpose of *Bergey's Manual of Systematic Bacteriology.*

11.7 Discuss the veracity and limitations of any taxonomic scheme.

Scientists called *taxonomists* group similar organisms into categories called *taxa* (see pp. 113–116). At one time the smallest taxa of prokaryotes (i.e., genera and species) were based solely on growth habits and the characteristics we considered in the previous section, especially morphology and arrangement. More recently, the classification of living things has been based more on genetic relatedness. Accordingly, modern taxonomists place all organisms into three *domains*—Archaea (ar´kē-ă), Bacteria, and Eukarya—which are the largest, most inclusive taxa. Scientists recognize bacterial and archaeal (both prokaryotic) taxa primarily on the basis of similarities in RNA, DNA, and protein sequences.

The vast majority of prokaryotes—perhaps as many as 99.5%, and probably millions of species—have never been isolated or cultured and are known only from their ribosomal (rRNA) "fingerprints"; that is, they are known only from sequences of rRNA that do not match any known rRNA sequences. In light of this information, taxonomists now construct modern classification schemes of prokaryotes based primarily on the relative similarities of rRNA sequences found in various prokaryotic groups **(FIGURE 11.10)**.

Perhaps the most authoritative reference in modern prokaryotic systematics is *Bergey's Manual of Systematic Bacteriology*, which classifies prokaryotes into 26 phyla—2 in Archaea and 24 in Bacteria. The five volumes of the second edition of *Bergey's Manual* discuss the great diversity of prokaryotes based in large part (but not exclusively) on their possible evolutionary relationships, as reflected in their rRNA sequences.

Our examination of prokaryotic diversity in this text is for the most part organized to reflect the taxonomic scheme that appears in *Bergey's Manual*, but it is important to note that as authoritative as *Bergey's Manual* is, it is not an "official" list of prokaryotic taxa. The reason is that taxonomy is partly a matter of opinion and judgment, and not all taxonomists agree. Legitimately differing views often change as more information

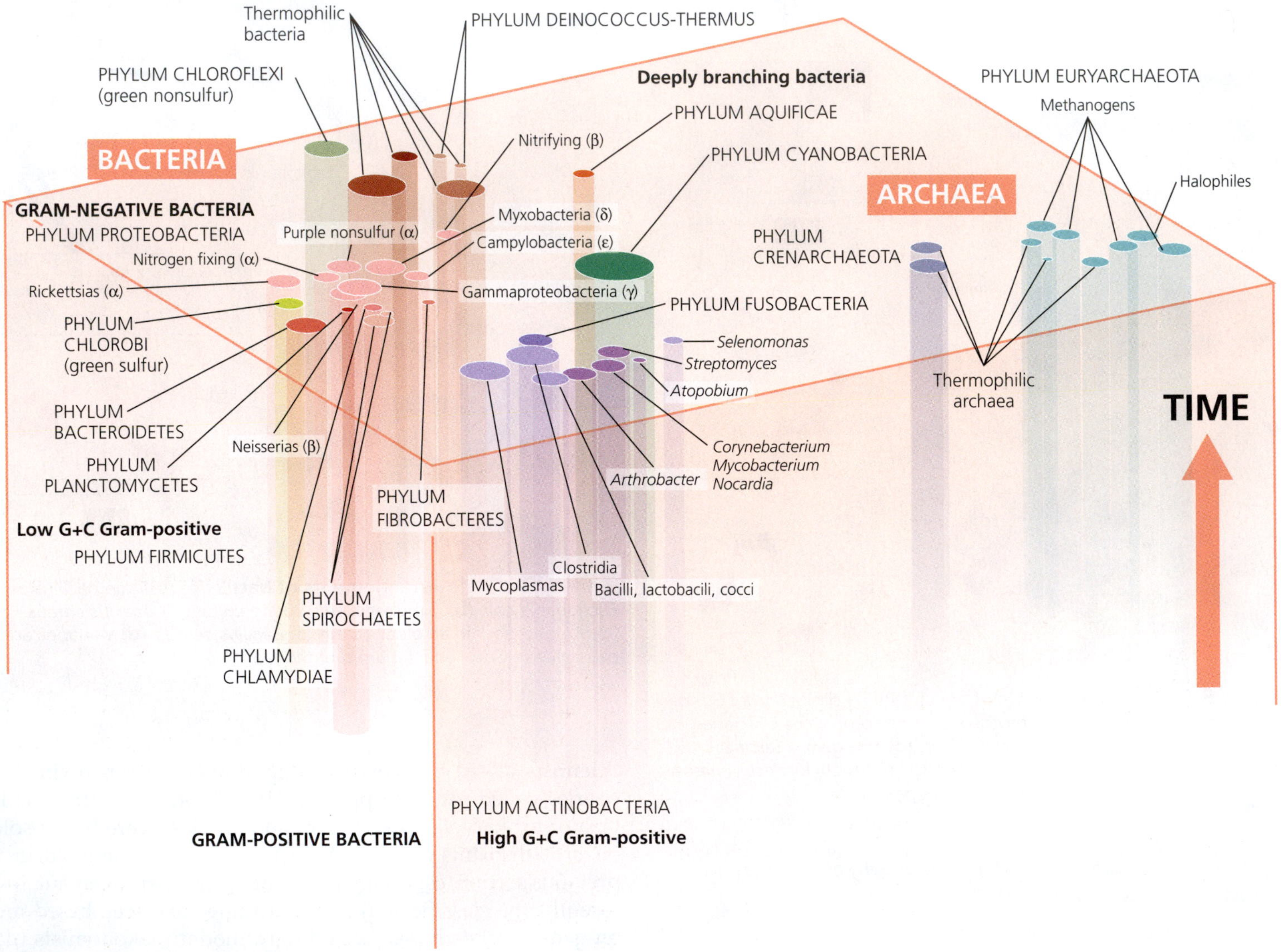

▲ **FIGURE 11.10 Prokaryotic taxonomy.** This scheme is based on relatedness according to rRNA sequences. The closer together the disks, the more similar are the rRNA sequences of the species within the group. The sizes of disks are proportional to the number of species known for that group. The deeper origins and relationships of the groups (evolution) are still uncertain and therefore represented with dashed lines. Note that archaea are distinctly separate from bacteria. The discussion in this chapter is based largely on the scheme depicted in this figure. Not all phyla are shown. (In phylum Proteobacteria, classes are indicated by a Greek letter: α = alpha, β = beta, γ = gamma, δ = delta, and ϵ = epsilon.) (Adapted from *Road Map to Bergey's*. 2002, Bergey's Manual Trust.)

is uncovered and examined, so *Bergey's Manual* is merely a consensus of experts at a given time. More information about *Bergey's Manual* can be found on the textbook website at www.masteringmicrobiology.com.

In the following sections we examine representative prokaryotes if major phyla. We begin our exploration of prokaryotic diversity with a survey of archaea.

TELL ME WHY

Why are taxonomic names and categories in our current taxonomic scheme different from those in 1990?

Survey of Archaea

LEARNING **OUTCOME**

11.8 Identify the common features of microbes in the domain Archaea.

Scientists originally identified archaea as a distinct type of prokaryotes on the basis of unique rRNA sequences. Archaea also share other common features that distinguish them from bacteria:

- Archaea lack true peptidoglycan in their cell walls.
- Their cytoplasmic membrane lipids have branched or ring-form hydrocarbon chains, whereas bacterial membrane lipids have straight chains.
- The initial amino acid in their polypeptide chains, coded by the AUG start codon, is methionine (as in eukaryotes and in contrast to the *N*-formylmethionine used by bacteria).

Archaea are currently classified in two phyla—Crenarchaeota (kren-ar´kē-ō-ta), and Euryarchaeota (ŭ-rē-ar´kē-ō-ta)—based primarily on rRNA sequences. Researchers have discovered RNA from several uncultured archaea that may represent other archaeal phyla, though there is no consensus on these taxa.

Archaea reproduce by binary fission, budding, or fragmentation. Known archaeal cells are cocci, bacilli, spirals, or pleomorphic **(FIGURE 11.11)**. Archaeal cell walls vary among taxa and are composed of a variety of compounds, including proteins, glycoproteins, lipoproteins, and polysaccharides; all lack peptidoglycan. Another interesting feature of archaea is that scientists have not proven that any archaeon causes disease.

Though most archaea live in moderate environmental conditions, some are *extremophiles*, which we discuss next, and *methanogens* (discussed shortly).

Extremophiles

LEARNING **OUTCOME**

11.9 Compare and contrast the two kinds of extremophiles discussed in this section.

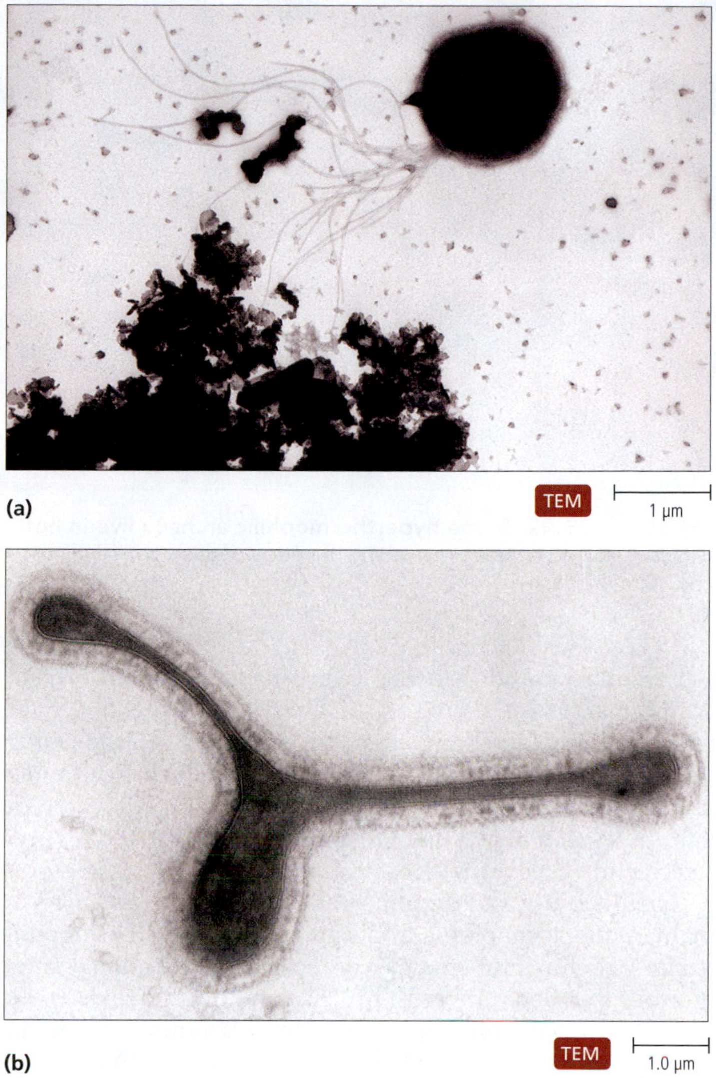

(a) (b)

▲ **FIGURE 11.11 Archaea.** **(a)** *Geogemma*, which has a tuft of flagella. **(b)** *Pyrodictium*, which has disk-shaped cells with filamentous extensions.

Extremophiles are microbes that require what humans consider to be extreme conditions of temperature, pH, and/or salinity to survive. There are extremophilic bacteria as well as archaea. Prominent among the extremophiles are *thermophiles* and *halophiles*.

Thermophiles

Thermophiles[6] are prokaryotes whose DNA, RNA, cytoplasmic membranes, and proteins do not function properly at temperatures lower than about 45°C. Prokaryotes that require temperatures over 80°C are called **hyperthermophiles.** Most thermophilic archaea are in the phylum Crenarchaeota, though some are also found in the phylum Euryarchaeota.

Two representative genera of thermophilic archaea are *Geogemma* (jē´ō-jem-ă) and *Pyrodictium* (pī-rō-dik´tē-um; see Figure 11.10). These microorganisms live in acidic hot springs

[6]From Greek *thermos*, meaning "heat," and *philos*, meaning "love."

▲ **FIGURE 11.12 Some hyperthermophilic archaea live in hot springs.** Orange archaea thrive along the edge of this pool in Yellowstone National Park.

▲ **FIGURE 11.13 The habitat of halophiles: highly saline water.** These are solar evaporation ponds near San Francisco. Halophiles often contain red to orange pigments, possibly to protect them from intense solar energy.

such as those found in deep ocean rifts and similar terrestrial volcanic habitats **(FIGURE 11.12)**. *Geogemma* is the current record holder for surviving high temperatures—it can survive 2 hours at 130°C! The cells of *Pyrodictium,* which live in deep-sea hydrothermal vents, are irregular disks with elongated protein tubules that attach them to grains of sulfur that they use as final electron acceptors in respiration.

Scientists use thermophiles and their enzymes in recombinant DNA technology applications because thermophiles' cellular structure and enzymes are stable and functional at temperatures that denature most proteins and nucleic acids and kill other cells. DNA polymerase from hyperthermophilic archaea makes possible the automated amplification of DNA in thermocyclers (see Chapter 8). Heat-stable enzymes are also ideal for many industrial applications, including their use as additives in laundry detergents.

Halophiles

Halophiles[7] are organisms that inhabit extremely saline habitats, such as the Dead Sea, the Great Salt Lake, and solar evaporation ponds used to concentrate salt for use in seasoning and for the production of fertilizer **(FIGURE 11.13)**. Halophiles can also colonize and spoil such foods as salted fish, sausages, and pork. Halophilic archaea are classified in the phylum Euryarchaeota.

The distinctive characteristic of halophiles is their absolute dependence on a salt concentration greater than 9% to maintain the integrity of their cell walls. Most halophiles grow and reproduce within an optimum range of 17% to 23% salt, and many species can survive in a saturated saline solution (35% sodium chloride). Many halophiles contain red to orange pigments that probably play a role in protecting them from intense sunlight.

The most studied halophile is *Halobacterium salinarium* (hā´lō-bak-tēr´ē-ŭm sal-ē-nar´ē-um), which is an archaeon despite its name. It is a photoheterotroph, using light energy to drive the synthesis of ATP while deriving carbon from organic compounds. *Halobacterium* lacks photosynthetic pigments—chlorophylls and bacteriochlorophylls. Instead, it synthesizes purple proteins, called **bacteriorhodopsins** (bak-tēr´ē-ō-rō-dop´sinz), that absorb light energy to pump protons across the cytoplasmic membrane to establish a proton gradient. Cells use the energy of proton gradients to produce ATP (see Chapter 5). *Halobacterium* also rotates its flagella with energy from the proton gradient so as to position itself at the proper water depth for maximum light absorption.

Methanogens

LEARNING **OUTCOME**

11.10 List at least four significant roles played by methanogens in the environment.

Methanogens are obligate anaerobes that convert CO_2, H_2, and organic acids into methane gas (CH_4). These microbes constitute the largest known group of archaea in the phylum Euryarchaeota. A few thermophilic methanogens are known. For example, *Methanopyrus*[8] (meth´a-nō-pī´rŭs) has an optimum growth temperature of 98°C and grows well in 110°C seawater around submarine hydrothermal vents. Scientists have also discovered halophilic methanogens.

Methanogens play significant roles in the environment by converting organic wastes in pond, lake, and ocean sediments into methane. Other methanogens living in the colons of animals are one of the primary sources of environmental methane. Methanogens dwelling in the intestinal tract of a cow, for example, can produce 400 liters of methane a day. Sometimes the production of methane in swamps and bogs is so great that bubbles rise to the surface as "swamp gas."

[7]From Greek *halos,* meaning "salt."
[8]From Greek *pyrus,* meaning "fire."

Methane is a so-called *greenhouse gas;* that is, methane in the atmosphere traps heat, which adds to global warming. It is about 25 times more potent as a greenhouse gas than carbon dioxide. Methanogens have produced about 10 trillion tons of methane—twice the known amount of oil, natural gas, and coal combined—that lies buried in mud on the ocean floor. If all the methane trapped in ocean sediments were released, it would wreak havoc with the world's climate.

Methanogens also have useful industrial applications. An important step in sewage treatment is the digestion of sludge by methanogens, and some sewage treatment plants burn methane to heat buildings and generate electricity.

Though we have concentrated our discussion on extremophiles and methanogens, most archaea live in more moderate habitats. For example, archaea make up about a third of the prokaryotic biomass in coastal Antarctic water, providing food for marine animals.

TELL ME WHY

Why did scientists formerly think archaea were a type of bacteria?

Survey of Bacteria

As we noted previously, our survey of prokaryotes in this chapter reflects the classification scheme that is featured in the second edition of *Bergey's Manual.* Whereas the classification scheme for bacteria in the first edition of the *Manual* emphasized morphology, Gram reaction, and biochemical characteristics, the second edition bases its classification of bacteria largely on differences in 16S rRNA sequences. We begin our survey of bacteria by considering the deeply branching and phototrophic bacteria.

Deeply Branching and Phototrophic Bacteria

LEARNING **OUTCOMES**

11.11 Provide a rationale for the name "deeply branching bacteria."

11.12 Explain the function of heterocysts in terms of both photosynthesis and nitrogen fixation.

Deeply Branching Bacteria

The **deeply branching bacteria** are so named because their rRNA sequences and growth characteristics lead scientists to conclude that these organisms are similar to the earliest bacteria; that is, they appear to have branched off the "tree of life" at an early stage. For example, the deeply branching bacteria are autotrophic, and early organisms must have been autotrophs because heterotrophs by definition must derive their carbon from autotrophs. Further, many of the deeply branching bacteria live in habitats similar to those some scientists think existed on the early Earth—hot, acidic, anaerobic, and exposed to intense ultraviolet radiation from the sun.

One representative of these microbes—the Gram-negative *Aquifex* (ăk′wē-feks), a bacterium in the phylum Aquificae—is considered to represent the earliest branch of bacteria. It is chemoautotrophic, hyperthermophilic, and microaerophilic, deriving energy and carbon from inorganic sources in very hot habitats containing little oxygen.

Another representative of deeply branching bacteria is *Deinococcus* (dī-nō-kok′ŭs), in phylum Deinococcus-Thermus, which grows as tetrads of Gram-positive cocci. Interestingly, the cell wall of *Deinococcus* has an outer membrane similar to that of Gram-negative bacteria, but the cells stain purple like typical Gram-positive microbes. *Deinococcus* is extremely resistant to radiation because of the way it packages its DNA and the presence of radiation-absorbing pigments, unique lipids within its membranes, and high cytoplasmic levels of manganese that protect its DNA repair proteins from radiation damage. Even when exposed to 5 million rad of radiation, which is enough energy to shatter its chromosome into hundreds of fragments, its enzymes can repair the damage. Not surprisingly, researchers have isolated *Deinococcus* from sites severely contaminated with radioactive wastes.

Phototrophic Bacteria

Phototrophic bacteria acquire the energy needed for anabolism by absorbing light with pigments located in thylakoids called *photosynthetic lamellae.* They lack the membrane-bound thylakoids seen in eukaryotic chloroplasts. Most phototrophic bacteria are also autotrophic—they produce organic compounds from carbon dioxide.

Phototrophs are a diverse group of microbes that are taxonomically confusing. Based on their pigments and their source of electrons for photosynthesis, phototrophic bacteria can be divided into five groups (though classified in four phyla based on 16S rRNA sequencing).

- Blue-green bacteria (phylum Cyanobacteria)
- Green sulfur bacteria (phylum Chlorobi)
- Green nonsulfur bacteria (phylum Chloroflexi)
- Purple sulfur bacteria (phylum Proteobacteria)
- Purple nonsulfur bacteria (phylum Proteobacteria)

We consider them together in the following sections because of their common phototrophic metabolism.

Cyanobacteria **Cyanobacteria** are Gram-negative phototrophs that vary greatly in shape, size, and method of reproduction. They range in size from 1 μm to 10 μm in diameter and are either coccoid or disk shaped. Coccal forms can be single or arranged in pairs, tetrads, chains, or sheets **(FIGURE 11.14a, b)**; disk-shaped forms are often tightly appressed end to end to form filaments that can be either straight, branched, or helical and are frequently contained in a gelatinous glycocalyx called a *sheath* **(FIGURE 11.14c)**. Some filamentous cyanobacteria are motile, moving along surfaces by *gliding.* Cyanobacteria generally reproduce by binary fission, with some species also reproducing by motile fragments or by thick-walled spores called akinetes (ā-kin-ēts′; see Figure 11.14a).

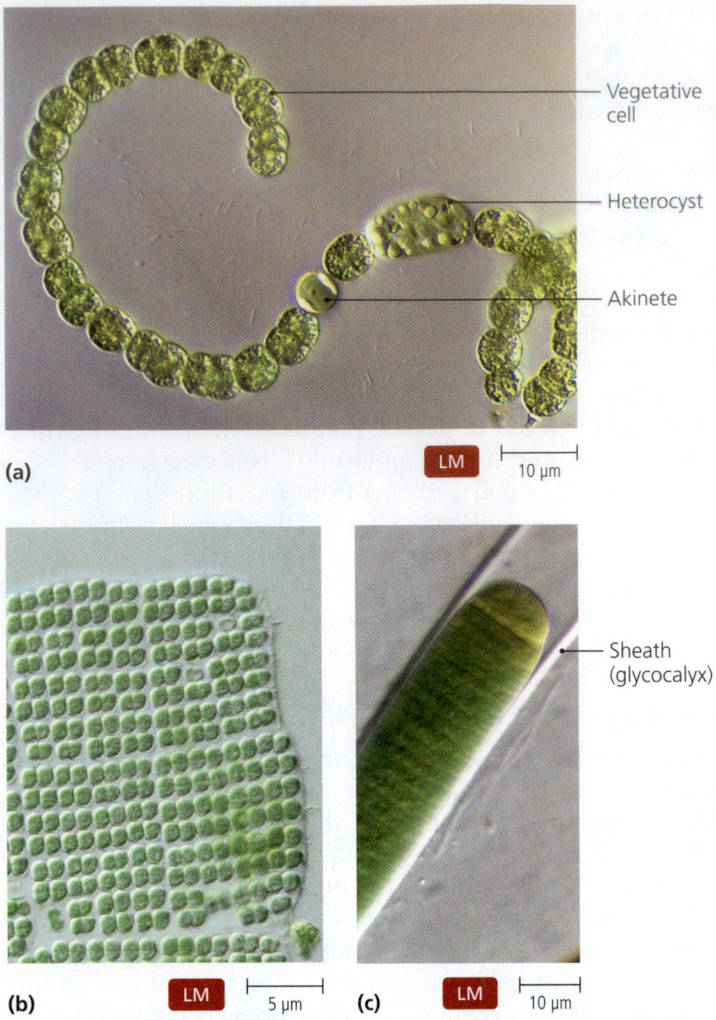

▲ **FIGURE 11.14 Examples of cyanobacteria with different growth habits. (a)** *Anabaena*, which grows as a filament of cocci with differentiated cells. Heterocysts fix nitrogen; akinetes are reproductive cells. **(b)** *Merismopedia*, which grows as a flat sheet of cocci surrounded by a gelatinous glycocalyx. **(c)** *Oscillatoria*, which forms a filament of tightly appressed disk-shaped cells.

Like plants and algae, cyanobacteria utilize chlorophyll *a* and are oxygenic (generate oxygen) during photosynthesis:

$$12\,H_2O + 6\,CO_2 \xrightarrow{\text{light}} C_6H_{12}O_6 + 6\,H_2O + 6\,O_2 \qquad (1)$$

For this reason, cyanobacteria were formerly called *blue-green algae;* however, the name *cyanobacteria* properly emphasizes their true bacterial nature: They are prokaryotic and have peptidoglycan cell walls. Additionally, they lack membranous cellular organelles such as nuclei, mitochondria, and chloroplasts.

Photosynthesis by cyanobacteria is thought to have transformed the anaerobic atmosphere of the early Earth into our oxygen-containing one, and according to the endosymbiotic theory, chloroplasts developed from cyanobacteria. Indeed, chloroplasts and cyanobacteria have similar rRNA and structures, such as 70S ribosomes and photosynthetic membranes.

Nitrogen is an essential element in proteins and nucleic acids. Though nitrogen constitutes about 79% of the atmosphere, relatively few organisms can utilize this gas. A few species of filamentous cyanobacteria as well as some proteobacteria (discussed later in the chapter) reduce nitrogen gas (N_2) to ammonia (NH_3) via a process called **nitrogen fixation**. Nitrogen fixation is essential for life on Earth because nitrogen fixers not only are able to enrich their own growth but also provide nitrogen in a usable form to other organisms.

Because the enzyme responsible for nitrogen fixation is inhibited by oxygen, nitrogen-fixing cyanobacteria are faced with a problem—how to segregate nitrogen fixation, which is inhibited by oxygen, from oxygenic photosynthesis, which produces oxygen. Nitrogen-fixing cyanobacteria solve this problem in one of two ways. Most cyanobacteria isolate the enzymes of nitrogen fixation in specialized, thick-walled, nonphotosynthetic cells called **heterocysts** (see Figure 11.14a). Heterocysts transport reduced nitrogen to neighboring cells in exchange for glucose. A few types of cyanobacteria photosynthesize during daylight hours and fix nitrogen at night, thereby separating nitrogen fixation from photosynthesis in time rather than in space.

Highlight: From Cyanobacteria to Bats to Brain Disease? on p. 332 examines potential negative impacts of cyanobacteria in human health.

Green and Purple Phototrophic Bacteria Green and purple bacteria differ from plants, algae, and cyanobacteria in two ways: They use *bacteriochlorophylls* for photosynthesis instead of chlorophyll *a,* and they are *anoxygenic;* that is, they do not generate oxygen during photosynthesis. Green and purple phototrophic bacteria commonly inhabit anaerobic muds rich in hydrogen sulfide at the bottoms of ponds and lakes. These microbes are not necessarily green and purple in color; rather, the terms refer to pigments in some of the better-known members of the groups.

As previously indicated, the green and purple phototrophic bacteria include both sulfur and nonsulfur forms. Whereas nonsulfur bacteria derive electrons for the reduction of CO_2 from organic compounds such as carbohydrates and organic acids, sulfur bacteria derive electrons from the oxidation of hydrogen sulfide to sulfur, as follows:

$$12\,H_2S + 6\,CO_2 \xrightarrow{\text{light}} C_6H_{12}O_6 + 6\,H_2O + 12\,S \qquad (2)$$

Green sulfur bacteria deposit the resultant sulfur outside their cells, whereas purple sulfur bacteria deposit sulfur within their cells **(FIGURE 11.15)**.

At the beginning of the 20th century, a prominent question in biology concerned the origin of the oxygen released by photosynthetic plants. It was initially thought that oxygen was derived from carbon dioxide, but a comparison of photosynthesis in cyanobacteria (Equation 1) with that in sulfur bacteria (Equation 2) provided evidence that free oxygen is derived from water.

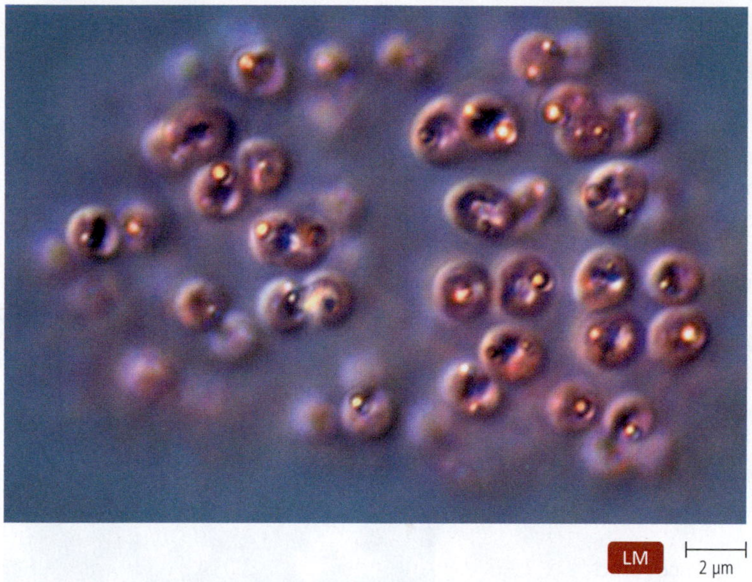

▲ **FIGURE 11.15 Deposits of sulfur within purple sulfur bacteria.** These bacteria oxidize H_2S to produce the granules of elemental sulfur evident in this photomicrograph. *Where do green sulfur bacteria deposit sulfur grains?*

Figure 11.15 Green sulfur bacteria deposit sulfur grains externally.

Whereas green sulfur bacteria are placed in phylum Chlorobi, green nonsulfur bacteria are members of phylum Chloroflexi. The purple bacteria (both sulfur and nonsulfur) are placed in three classes of phylum Proteobacteria, which is composed of Gram-negative bacteria and is discussed shortly. **TABLE 11.1** summarizes the characteristics of phototrophic bacteria. ▶**ANIMATIONS:** ***Photosynthesis: Comparing Prokaryotes and Eukaryotes***

Low G + C Gram-Positive Bacteria

LEARNING | **OUTCOMES**

11.13 Discuss the lack of cell walls in mycoplasmas.

11.14 Identify significant beneficial or detrimental effects of the genera *Clostridium, Bacillus, Listeria, Lactobacillus, Streptococcus,* and *Staphylococcus.*

11.15 Describe the difference between low and high G+C content.

Now we turn our attention to various groups of Gram-positive bacteria, and to a different characteristic of microbes that is used in the classification of Gram-positive bacteria—*G + C content*. This is the percentage of all base pairs in a genome that are guanine-cytosine base pairs—and is a useful criterion in classifying Gram-positive bacteria. Those with G + C content below 50% are considered "low G + C bacteria"; the remainder are considered "high G + C bacteria." Because taxonomists have discovered that Gram-positive bacteria with low G + C content have similar sequences in their 16S rRNA and that those with high G + C content also have rRNA sequences in common, they have assigned low G + C bacteria and high G + C bacteria to different phyla. We discuss the low G + C bacteria first.

The low G + C Gram-positive bacteria are classified within phylum Firmicutes (fer-mik´ū-tēz), which includes three groups: clostridia, mycoplasmas, and other low G + C Gram-positive bacilli and cocci. Next we consider these three groups in turn.

TABLE 11.1 Characteristics of the Major Groups of Phototrophic Bacteria

	Phylum				
	Cyanobacteria	**Chlorobi**	**Chloroflexi**	**Proteobacteria**	**Proteobacteria**
Class	Cyanobacteria	Chlorobia	Chloroflexi	Gammaproteobacteria	Alphaproteobacteria and one genus in betaproteobacteria
Common name(s)	Blue-green bacteria ("blue-green algae")	Green sulfur bacteria	Green nonsulfur bacteria	Purple sulfur bacteria	Purple nonsulfur bacteria
Major photosynthetic pigments	Chlorophyll *a*	Bacteriochlorophyll *a* plus *c*, *d*, or *e*	Bacteriochlorophylls *a* and *c*	Bacteriochlorophyll *a* or *b*	Bacteriochlorophyll *a* or *b*
Types of photosynthesis	Oxygenic	Anoxygenic	Anoxygenic	Anoxygenic	Anoxygenic
Electron donor in photosynthesis	H_2O	H_2, H_2S, or S	Organic compounds	H_2, H_2S, or S	Organic compounds
Sulfur deposition	None	Outside of cell	None	Inside of cell	None
Nitrogen fixation	Some species	None	None	None	None
Motility	Nonmotile or gliding	Nonmotile	Gliding	Motile with polar or peritrichous flagella	Nonmotile or motile with polar flagella

HIGHLIGHT

From Cyanobacteria to Bats to Brain Disease?

The origins of many neurological diseases, such as amyotrophic lateral sclerosis (ALS), Parkinson's disease, and Alzheimer's disease, remain mysterious. These diseases are characterized by paralysis, tremors, and sometimes dementia. In recent years, one of the more intriguing theories concerning their cause indicates that cyanobacteria might be a causative agent because cyanobacteria synthesize an unusual amino acid called β-methylamino-L-alanine (BMAA), which is a known neurotoxin.

Evidence for the hypothesis that BMAA from cyanobacteria is to blame for brain diseases comes from the Chamorro people, who live on the Pacific island of Guam and have had almost a 100-fold higher incidence of ALS than any other group. Chamorros once prized fruit bats as a dietary delicacy, consuming the bats whole—skin, wings, bones, and brains. Fruit bats feed on the seeds and fruits of cycad trees, and BMAA-secreting cyanobacteria inhabit specialized roots of cycads. Researchers hypothesize that the bats concentrate BMAA when they eat cycad seeds and that the Chamorros suffer from the chemical's neurotoxicity. When the Chamorros stopped eating bats (because the bats had become endangered), the number of ALS cases fell to the level seen in other societies.

Researchers have also discovered BMAA in the brains of Alzheimer's and Parkinson's patients. It is presumed that these patients acquire BMAA from drinking water where cyanobacteria live. Although controversial, the idea that cyanobacteria can trigger brain diseases has spawned research around the world.

Clostridia

Clostridia are rod-shaped, obligate anaerobes, many of which form endospores. The group is named for the genus *Clostridium*,[9] which is important both in medicine—in large part because its members produce potent toxins that cause a variety of diseases in humans—and in industry because their endospores enable them to survive harsh conditions, including many types of disinfection and antisepsis. Examples of clostridia include *C. tetani* (te´tan-ē, which causes tetanus), *C. perfringens* (per-frin´jens; gangrene), *C. botulinum* (bo-tū-lī´num; botulism), and *C. difficile* (di-fi´sēl, severe diarrhea). (Chapters 19, 20, and 23 examine these pathogens and the diseases they cause in greater detail.) **Beneficial Microbes: Botulism and Botox** describes how a deadly toxin produced by *C. botulinum* has been put to use for cosmetic purposes.

Microbes related to *Clostridium* include *Epulopiscium*, a giant bacterium that can be seen without a microscope (discussed on p. 324); sulfate-reducing microbes, which produce H_2S from elemental sulfur during anaerobic respiration; and *Selenomonas* (sĕ-lē´nō-mō´nas), a genus of vibrio-shaped bacteria that includes members that live as part of the biofilm (plaque) that forms on the teeth of warm-blooded animals. *Selenomonas* is unusual because even though it has a typical Gram-positive RNA sequence, it has a negative Gram reaction—it stains pink. Researchers have linked one species of *Selenomonas* to obesity (**Highlight: Your Teeth Might Make You Fat** on p. 334).

Mycoplasmas

A second group of low G + C bacteria are the **mycoplasmas**[10] (mī´kō-plaz´mas). These facultative or obligate anaerobes lack cell walls, meaning that they stain pink when Gram stained. Indeed, until their nucleic acid sequences proved their similarity to Gram-positive organisms, mycoplasmas were classified as Gram-negative microbes instead of in phylum Firmicutes with other low G + C Gram-positive bacteria.

Mycoplasmas are able to survive without cell walls in part because they colonize osmotically protected habitats, such as animal and human bodies, and because they have tough cytoplasmic membranes, many of which contain lipids called *sterols* that give the membranes strength and rigidity. Because they lack cell walls, they are pleomorphic. They were named "mycoplasmas" because their filamentous forms resemble the filaments of fungi. Mycoplasmas have diameters ranging from 0.2 μm to 0.8 μm, making them the smallest free-living cells, and many mycoplasmas have a terminal structure that is used for attachment to eukaryotic cells and that gives the

[9]From Greek *kloster*, meaning "spindle."

[10]From Greek *mycos*, meaning "fungus," and *plassein*, meaning "to mold."

BENEFICIAL MICROBES

Botulism and Botox

Clostridium botulinum produces botulinum toxins, some of the deadlier toxins known. When absorbed in the body, botulism toxins interfere with the release of the neurotransmitter that signals muscles to contract. As a result, muscle cells cannot contract, and a progressive paralysis spreads throughout the body. Death occurs if paralysis of respiratory muscles results in respiratory failure. This illness is called botulism (discussed in more detail in Chapter 20).

Purified type A botulinum toxin is marketed as Botox, extremely small doses of which are injected into facial muscles that cause skin wrinkles. The toxin paralyzes the muscles, smoothing the skin. Such treatments may last six months and must be repeated in order to maintain the desired effects.

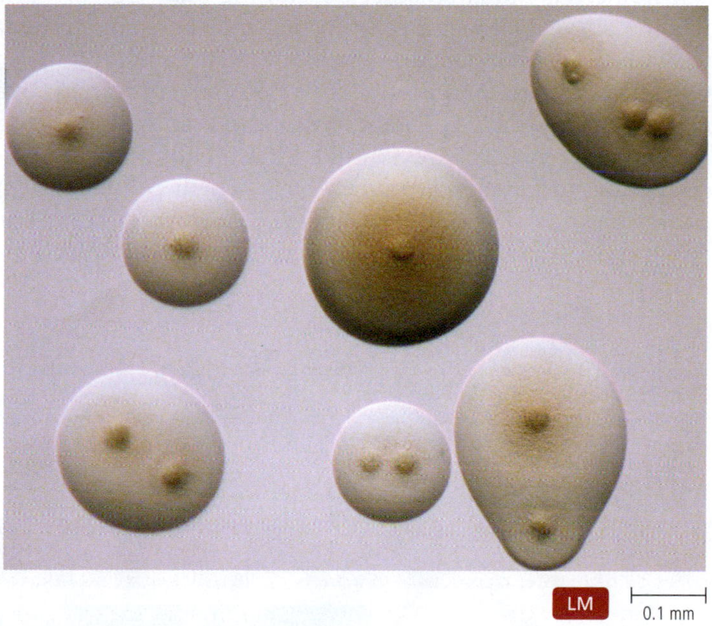

▲ **FIGURE 11.16 The distinctive "fried egg" appearance of *Mycoplasma* colonies.** This visual feature is unique to some species of this genus of bacteria when growing on an agar surface.

bacterium a pearlike shape. They require organic growth factors, such as cholesterol, fatty acids, vitamins, amino acids, and nucleotides, which they acquire from their host or which must be added to laboratory media. When growing on solid media, most species form a distinctive "fried egg" appearance because cells in the center of the colony grow into the agar while those around the perimeter only spread across the surface **(FIGURE 11.16)**.

In animals, mycoplasmas colonize mucous membranes of the respiratory and urinary tracts and are associated with pneumonia and urinary tract infections. (Pathogenic mycoplasmas and the diseases they cause are discussed more fully in Chapters 22 and 24.)

Other Low G + C Bacilli and Cocci

A third group of low G + C Gram-positive organisms is composed of bacilli and cocci that are significant in environmental, industrial, and health care settings. Among the genera in this group are *Bacillus, Listeria* (lis-tēr´ē-ă), *Lactobacillus* (lak´tō-bă-sil´ŭs), *Streptococcus, Enterococcus* (en´ter-ō-kok´ŭs), and *Staphylococcus* (staf´i-lō-kok´ŭs). (Chapters 19–24 discuss the pathogens in these genera in greater detail.)

Bacillus The genus *Bacillus* includes endospore-forming aerobes and facultative anaerobes that typically move by means of peritrichous flagella. The genus name *Bacillus* should not be confused with the general term bacillus. The latter refers to any rod-shaped cell of any genus. Numerous species of *Bacillus* are common in soil.

Bacillus thuringiensis (thur-in-jē-en´sis) is beneficial to farmers and gardeners. During sporulation, this bacterium produces a crystalline protein that is toxic to caterpillars that ingest it **(FIGURE 11.17)**. Gardeners spray *Bt toxin*, as preparations of the bacterium and toxin are known, on plants to protect them from caterpillars. Scientists have achieved the same effect, without the need of spraying, by introducing the gene for Bt toxin into plants' chromosomes. Other beneficial species of *Bacillus* include *B. polymyxa* (po-lē-miks´ă) and *B. licheniformis* (lī-ken-i-for´mis), which synthesize the antibiotics polymyxin and bacitracin, respectively. (Chapter 10 discusses the production and effects of antibiotics in more detail.)

Bacillus anthracis (an-thrā´sis), which causes anthrax, gained notoriety in 2001 as an agent of bioterrorism. Its endospores either are inhaled or enter the body through breaks in the skin. When they germinate, the vegetative cells produce toxins that kill surrounding tissues (see Disease at a Glance on p. 573). Untreated *cutaneous anthrax* is fatal in 20% of patients; untreated inhalational *anthrax* is generally 100% fatal without prompt aggressive treatment. Refrigeration prevents the excessive growth of this and other contaminants (see Chapter 9).

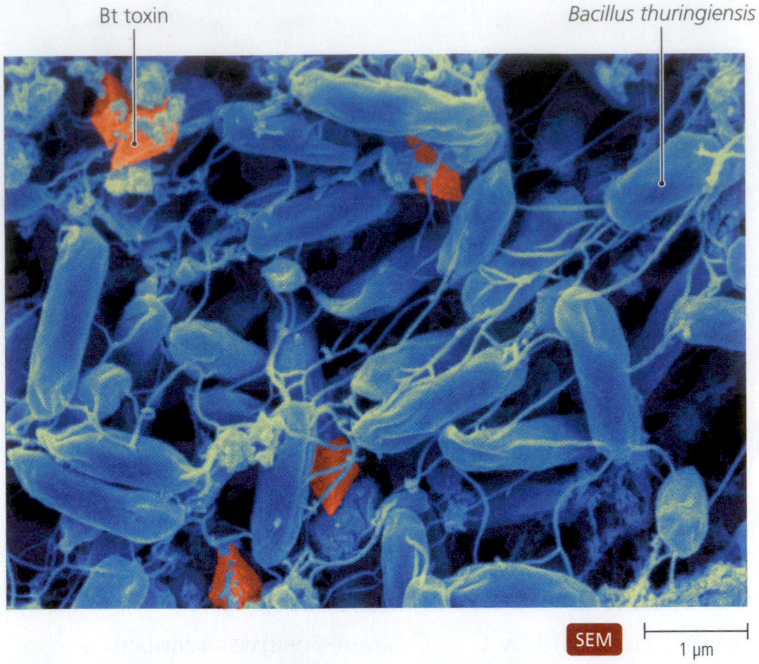

▲ **FIGURE 11.17 Crystals of Bt toxin, produced by the endospore-forming *Bacillus thuringiensis.*** The crystalline protein kills caterpillars that ingest it.

Listeria Another pathogenic low G + C Gram-positive rod is *Listeria monocytogenes* (mo-nō-sī-tah´je-nēz), which can contaminate milk and meat products. This microbe, which does not produce endospores, is notable because it continues to reproduce under refrigeration, and it can survive inside phagocytic white blood cells. *Listeria* rarely causes disease in adults, but it can kill a fetus in an infected woman when it crosses the placental barrier. It also causes meningitis[11] and bacteremia[12] when it infects immunocompromised patients, such as the aged and patients with AIDS, cancer, or diabetes.

Lactobacillus Organisms in the genus *Lactobacillus* are non-spore-forming rods normally found growing in the human mouth, stomach, intestinal tract, and vagina. These organisms rarely cause disease; instead, they protect the body by inhibiting the growth of pathogens—a situation called *microbial antagonism.* Lactobacilli are used in industry in the production of yogurt, buttermilk, pickles, and sauerkraut. (The **Beneficial Microbes: Probiotics** feature on p. 303 discusses the potential role of *Lactobacillus* in promoting human health.)

Streptococcus* and *Enterococcus The genera *Streptococcus* and *Enterococcus* are diverse groups of Gram-positive cocci associated in pairs and chains (see Figure 11.8a and b). They cause numerous human diseases, including pharyngitis, scarlet fever, impetigo, fetal meningitis, wound infections, pneumonia, and diseases of the inner ear, skin, blood, and kidneys. In recent years, health care providers have become concerned over strains of multi-drug-resistant enterococci and streptococci. Of particular concern are so-called flesh-eating streptococci (see Disease in Depth on pp. 568–569), which produce toxins that destroy muscle and fat tissue; and enterococci, which are leading causes of healthcare associated (nosocomial) infections.

[11]Inflammation of the membranes covering the brain and spinal cord; from Greek *meninx*, meaning "membrane."

[12]The presence of bacteria, particularly those that produce disease symptoms, in the blood.

HIGHLIGHT

Your Teeth Might Make You Fat

The Forsyth Institute is a nonprofit organization associated with Harvard Medical School in Boston. Scientists at the institute were instrumental in showing that dental cavities in children are caused by bacteria. This revolutionized oral health care. Now, their researchers may play a role in combating obesity.

No one questions that what you put between your teeth may cause obesity, but scientists at the institute have discovered an intriguing fact: What's on your teeth may be to blame as well. People with an oral bacterium related to the low G + C, Gram-positive clostridia are much more likely to be obese.

In fact, the presence of the bacterium *Selenomonas noxia* is an indicator of obesity 98.4% of the time. Lean people may have the bacterium as a rare member of their oral microbiota, but they have many fewer cells of the species than the obese have. The researchers do not know if the bacterium is a trigger of some pathology that leads to obesity or if the bacterium grows on the teeth as a result of obesity or of a diet leading to obesity. Nevertheless, the presence of *S. noxia* in a child's mouth may serve as a biological warning sign that a child is at risk of developing an overweight condition, allowing parents and health providers to intervene earlier.

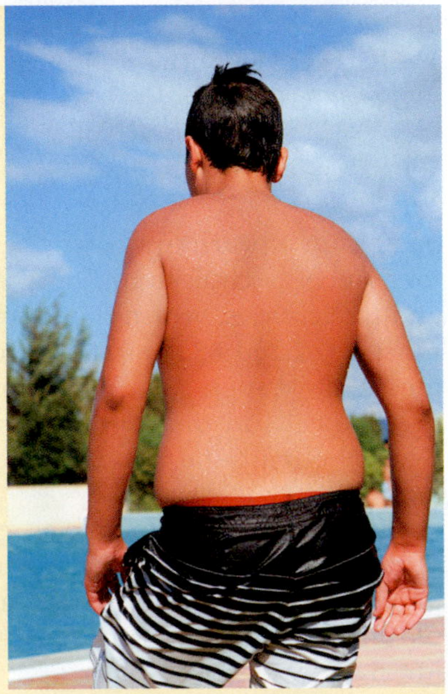

TABLE 11.2 Characteristics of Selected Gram-Positive Bacteria

Phylum/Class	G + C Percentage	Representative Genera	Special Characteristics	Diseases
Firmicutes				
Clostridia	Low (less than 50%)	*Clostridium*	Obligate anaerobic rods; endospore formers	Tetanus Botulism Gangrene
		Epulopiscium	Giant rods	Severe diarrhea
		Selenomonas	Part of oral biofilm on human teeth; stain like Gram-negative bacteria (pink)	Dental caries
Mollicutes	Low (less than 50%)	*Mycoplasma*	Lack cell walls; pleomorphic; smallest free-living cells; stain like Gram-negative bacteria (pink)	Pneumonia Urinary tract infections
Bacilli	Low (less than 50%)	*Bacillus*	Facultative anaerobic rods; endospore formers	Anthrax
		Listeria	Contaminates dairy products	Listeriosis
		Lactobacillus	Produce yogurt, buttermilk, pickles, sauerkraut	Rare blood infections
		Streptococcus	Cocci in chains	Strep throat, scarlet fever, and others
		Staphylococcus	Cocci in clusters	Bacteremia, food poisoning, and others
Actinobacteria				
Actinobacteria	High (greater than 50%)	*Corynebacterium*	Snapping division; metachromatic granules in cytoplasm	Diphtheria
		Mycobacterium	Waxy cell walls (mycolic acid)	Tuberculosis and leprosy
		Actinomyces	Filaments	Actinomycosis
		Nocardia	Filaments; degrade pollutants	Lesions
		Streptomyces	Produce antibiotics	Rare sinus infections

Staphylococcus Among the common inhabitants of humans is *Staphylococcus aureus*[13] (o´rē-ŭs), which is typically found growing harmlessly in clusters in the nasal passages. A variety of toxins and enzymes allow some strains of *S. aureus* to invade the body and cause such diseases as bacteremia, pneumonia, wound infections, food poisoning, toxic shock syndrome, and diseases of the joints, bones, heart, and blood.

The characteristics of the low G + C Gram-positive bacteria are summarized in the first part of **TABLE 11.2**. These bacteria, which are classified into three classes within phylum Firmicutes, include the anaerobic endospore-forming rod *Clostridium,* the pleomorphic *Mycoplasma,* the aerobic and facultative aerobic endospore-forming rod *Bacillus,* the non-endospore-forming rods *Listeria* and *Lactobacillus,* and the cocci *Streptococcus, Enterococcus,* and *Staphylococcus.*

Next we consider Gram-positive bacteria that have high G + C ratios.

High G + C Gram-Positive Bacteria

LEARNING **OUTCOMES**

11.16 Explain the slow growth of *Mycobacterium.*

11.17 Identify significant beneficial or detrimental properties of the genera *Corynebacterium, Mycobacterium, Actinomyces, Nocardia,* and *Streptomyces.*

Taxonomists classify Gram-positive bacteria with a G + C percentage greater than 50% in the phylum Actinobacteria, which includes species with rod-shaped cells (many of which are significant human pathogens) and filamentous bacteria, which resemble fungi in their growth habit and in the production of reproductive spores. Here we examine briefly some prominent high G + C Gram-positive bacteria. (The numerous pathogens in this group are discussed more fully in Chapters 19–24.)

Corynebacterium

Members of the genus *Corynebacterium* are pleomorphic—though generally rod shaped—aerobes and facultative anaerobes. They reproduce by snapping division, which often causes the cells to form V-shapes and palisades (see Figures 11.4 and 11.9d). Corynebacteria are also characterized by their stores of phosphate within inclusions called **metachromatic granules**, which stain differently from the rest of the cytoplasm when the cells are stained with methylene blue or toluidine blue. The best-known species is *C. diphtheriae,* which causes diphtheria.

[13]From Latin *aurum,* meaning "gold" because it produces yellow pigments.

Mycobacterium

The genus *Mycobacterium* (mī´kō-bak-tēr´ē-um) is composed of aerobic species that are slightly curved to straight rods that sometimes form filaments. Mycobacteria grow very slowly, often requiring a month or more to form a visible colony on an agar surface. Their slow growth is partly due to the time and energy required to enrich their cell walls with high concentrations of long carbon-chain waxes called **mycolic acids**, which make the cells resistant to desiccation and to staining with water-based dyes. Microbiologists developed the *acid-fast stain* for mycobacteria because they are difficult to stain by standard staining techniques such as the Gram stain (see Chapter 4). Though some mycobacteria are free living, the most prominent species are pathogens of animals and humans, including *Mycobacterium tuberculosis* (too-ber-kyū-lō´sis) and *Mycobacterium leprae* (lep´rī), which cause tuberculosis and leprosy, respectively. Mycobacteria should not be confused with the low G + C mycoplasmas discussed earlier.

Actinomycetes

Actinomycetes (ak´ti-nō-mī-sētz) are high G + C Gram-positive bacteria that form branching filaments resembling fungi **(FIGURE 11.18)**. Of course, in contrast to fungi, the filaments of actinomycetes are composed of prokaryotic cells. As we have seen, some actinomycetes also resemble fungi in the production of chains of reproductive spores at the ends of their filaments. These spores should not be confused with endospores, which are resting stages and not reproductive cells. Actinomycetes may cause disease, particularly in immunocompromised patients. Among the important actinomycete genera are *Actinomyces* (which gives this group its name), *Nocardia*, and *Streptomyces*.

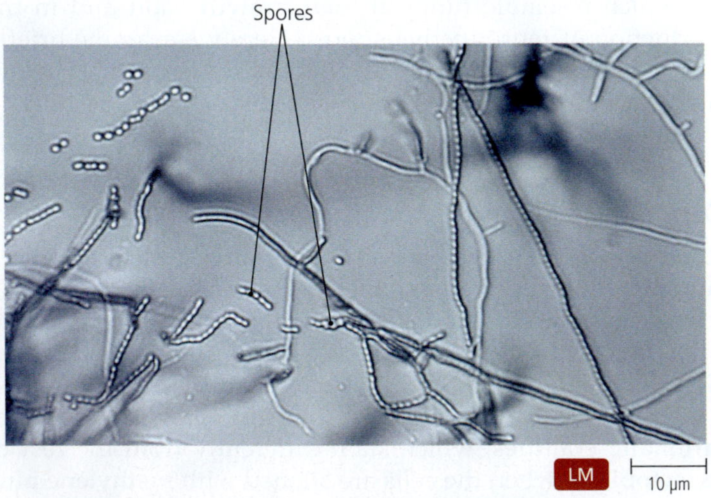

▲ **FIGURE 11.18 The branching filaments of actinomycetes.** This photograph shows filaments of a colony of *Streptomyces* sp. growing on agar. *How do the filaments of actinomycetes compare to the filaments of fungi?*

Figure 11.18 *Filaments of actinomycetes are thinner than those of fungi, and they are composed of prokaryotic cells.*

Actinomyces Species of *Actinomyces* (ak´ti-nō-mī´sēz) are facultative capneic[14] filaments that are normal inhabitants of the mucous membranes lining the oral cavity and throats of humans. *Actinomyces israelii* (is-rā´el-ē-ē) growing as an opportunistic pathogen in humans destroys tissue to form abscesses and can spread throughout the abdomen, consuming every vital organ.

Nocardia Species of *Nocardia* (nō-kar´dē-ă) are soil- and water-dwelling aerobes that typically form aerial and subterranean filaments that make them resemble fungi. *Nocardia* is notable because it can degrade many pollutants of landfills, lakes, and streams, including waxes, petroleum hydrocarbons, detergents, benzene, polychlorinated biphenyls (PCBs), pesticides, and rubber. Some species of *Nocardia* cause lesions in humans.

Streptomyces Bacteria in the genus *Streptomyces* (strep-tō-mī´-sēz) are important in several realms. Ecologically, they recycle nutrients in the soil by degrading a number of carbohydrates, including cellulose, lignin (the woody part of plants), chitin (outer skeletal material of insects and crustaceans), latex, aromatic chemicals (organic compounds containing a benzene ring), and keratin (the protein that forms hair, nails, and horns). The metabolic by-products of *Streptomyces* give soil its musty smell. Medically, *Streptomyces* species produce most of the important antibiotics, including chloramphenicol, erythromycin, and tetracycline (discussed in Chapter 10).

Table 11.2 on p. 335 includes a summary of the characteristics of the genera of high G + C Gram-positive bacteria (phylum Actinobacteria) discussed in this chapter.

To this point we have discussed archaea and deeply branching, phototrophic, and Gram-positive bacteria. Now we turn our attention to the Gram-negative bacteria that are grouped together within the phylum Proteobacteria.

Gram-Negative Proteobacteria

Phylum **Proteobacteria**[15] constitutes the largest and most diverse group of bacteria. Though they have a variety of shapes, reproductive strategies, and nutritional types, they are all Gram negative and share common 16S rRNA nucleotide sequences. The G + C percentage of Gram-negative species is not critical in delineating taxa of most Gram-negative organisms, so we will not consider this characteristic in our discussion of these bacteria.

There are five distinct classes of proteobacteria, designated by the first five letters of the Greek alphabet—alpha, beta, gamma, delta, and epsilon. These classes are distinguished by minor differences in their rRNA sequences. Here we will focus our attention on species with novel characteristics as well as species with practical importance.

[14]Meaning that they grow best with a relatively high concentration of carbon dioxide.
[15]Named for the Greek god *Proteus*, who could assume many shapes.

Class Alphaproteobacteria

LEARNING | OUTCOMES

11.18 Describe the appearance and function of prosthecae in alphaproteobacteria.

11.19 Describe three genera of nitrogen fixing alphaproteobacteria and their association with crops or biofuels.

11.20 Describe nitrification and name a nitrifying bacterium.

11.21 Name two diseases caused by alphaproteobacteria.

Alphaproteobacteria are typically aerobes capable of growing at very low nutrient levels. Many have unusual methods of metabolism, as we will see shortly. They may be rods, curved rods, spirals, coccobacilli, or pleomorphic. Many species have unusual extensions called *prosthecae* (pros-thē´kē), which are composed of cytoplasm surrounded by the cytoplasmic membrane and cell wall **(FIGURE 11.19)**. They use prosthecae for attachment and to increase surface area for nutrient absorption. Some prosthecate species produce buds at the ends of the extensions.

Nitrogen Fixers Two genera of nitrogen fixers in the class Alphaproteobacteria—*Azospirillum* (ā-zō-spī´ril-ŭm) and *Rhizobium* (rī-zō´bē-ŭm)—are important in agriculture. They grow in association with the roots of plants, where they make atmospheric nitrogen available to the plants as ammonia, which is one kind of fixed nitrogen.

$$N_2 + 8H^+ + 8e^- \rightarrow 2NH_3 + H_2 \quad (3)$$

Azospirillum associates with the outer surfaces of roots of tropical grasses, such as sugarcane. In addition to supplying nitrogen to the grass, this bacterium also releases chemicals that stimulate the plant to produce numerous root hairs, increasing a root's surface area and thus its uptake of nutrients.

Rhizobium grows within the roots of leguminous plants, such as peas, beans, and clover, stimulating the formation of nodules on their roots **(FIGURE 11.20)**. *Rhizobium* cells within the nodules make ammonia available to the plant, encouraging growth. Scientists are actively seeking ways to successfully insert the genes of nitrogen fixation into plants such as corn, which requires large amounts of nitrogen.

Rhodopseudomonas palustris (rō-dō´soo-dō-mō´nas pal-us´tris) is another nitrogen-fixing alphaproteobacterium. Scientists are actively studying this bacterium because it can also reduce hydrogen, forming hydrogen gas (H_2), which can be used as a clean-burning fuel. Biofuel producers such as *R. palustris* may help relieve the world's dependence on oil and gas.

Nitrifying Bacteria Organisms need a source of electrons for redox reactions of metabolism (see Chapter 6). Bacteria that derive electrons from the oxidation of nitrogenous compounds are called **nitrifying bacteria**. These microbes are important in the environment and in agriculture, because they convert reduced nitrogenous compounds, such as ammonia (NH_3), into nitrate (NO_3)—a two-step process called **nitrification**. Nitrate moves more easily through soil and is therefore more available to plants.

The first step in nitrification is the oxidation of ammonia into nitrite (NO_2); the second step is further oxidation of nitrite into nitrate.

$$2NH_3 + 3O_2 \rightarrow 2NO_2^- + 2H_2O + 2H^+ \quad (4a)$$
$$2NO_2^- + O_2 \rightarrow 2NO_3^- \quad (4b)$$

No microbe can perform both reactions. Nitrifying alphaproteobacteria include species of *Nitrobacter* (nī-trō-bak´ter), which perform the second step of nitrification. The first step of nitrification is performed either by archaea or by betaproteobacteria.

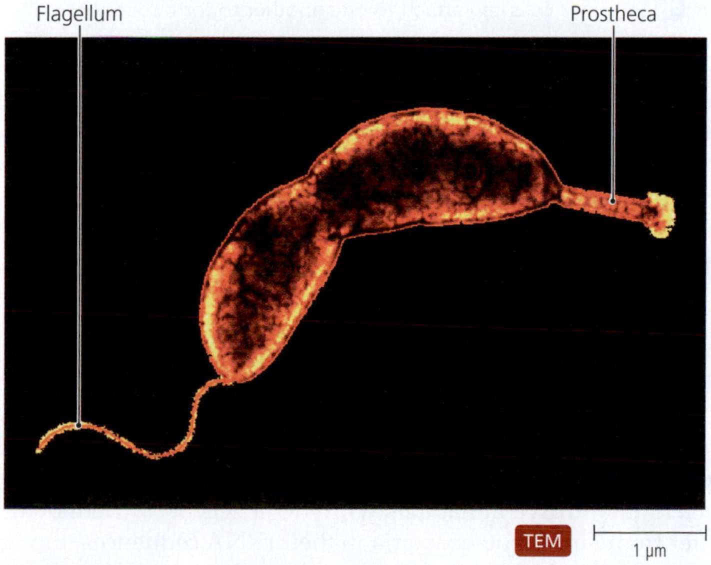

▲ **FIGURE 11.19 A prostheca.** This extension of an alphaproteobacterial cell increases surface area for absorbing nutrients and serves as an organ of attachment. This prosthecate bacterium, *Hyphomicrobium facilis*, also produces buds from its prostheca and has a flagellum.

▲ **FIGURE 11.20 Nodules on pea plant roots.** *Rhizobium*, a nitrogen-fixing alphaproteobacterium, stimulates the growth of such nodules.

Purple Nonsulfur Phototrophs With one exception (the betaproteobacterium *Rhodocyclus*), purple nonsulfur phototrophs are classified as alphaproteobacteria. Purple nonsulfur bacteria grow in the upper layer of mud at the bottoms of lakes and ponds. As we discussed earlier, they harvest light as an energy source by using bacteriochlorophylls, and they do not generate oxygen during photosynthesis. Morphologically, they may be rods, curved rods, or spirals; some species are prosthecate. Refer to Table 11.1 on p. 331 to review the characteristics of all phototrophic bacteria.

Pathogenic Alphaproteobacteria Notable pathogens among the alphaproteobacteria include *Rickettsia* (ri-ket´sē-ă) and *Brucella* (broo-sel´lă).

Rickettsia is a genus of small, Gram-negative, aerobic rods that live and reproduce inside mammalian cells. Outside of a host cell, rickettsias are unstable and quickly die; therefore, they require a vector for transmission from host to host. They are transmitted through the bites of arthropods (fleas, lice, ticks, and mites). Rickettsias cause a number of human diseases, including typhus and Rocky Mountain spotted fever.

Brucella is a coccobacillus that causes *brucellosis*, a disease of mammals characterized by spontaneous abortions and sterility in animals. In contrast, infected humans suffer chills, sweating, fatigue, and fever. *Brucella* is notable because it survives phagocytosis by white blood cells—normally an important step in the body's defense against disease.

Other Alphaproteobacteria Other alphaproteobacteria are important in industry and the environment. For example, *Acetobacter* (a-sē´tō-bak-ter) and *Gluconobacter* (gloo-kon´ō-bak-ter) are used to synthesize acetic acid in the production of vinegar. *Caulobacter* (kaw´lō-bak-ter) is a common prosthecate rod-shaped microbe that inhabits nutrient-poor seawater and freshwater; it can also be found in laboratory water baths.

Caulobacter has a unique reproductive strategy (**FIGURE 11.21**). A cell that has attached to a substrate with its prostheca 1 grows until it has doubled in size, at which time it produces a flagellum at its apex 2; it then divides by asymmetric binary fission 3. The flagellated daughter cell, which is called a swarmer cell, then swims away. The swarmer cell can either attach to a substrate with a new prostheca that replaces the flagellum 4a or attach to other swarmer cells to form a rosette of cells 4b. The process of reproduction repeats about every two hours. **Beneficial Microbes: A Microtube of Superglue** examines an amazing property of *Caulobacter* prosthecae.

Scientists are very interested in the usefulness of another alphaproteobacterium, *Agrobacterium* (ag´rō-bak-tēr´ē-ŭm), which infects plants to form tumors called *galls* (**FIGURE 11.22**). The bacterium inserts a plasmid (an extra chromosomal DNA molecule) that carries a gene for a plant growth hormone through a pilus into a plant cell, which then makes extra growth hormone. Growth hormone causes the cells of the plant to proliferate into a gall and to produce nutrients for the bacterium. Scientists have discovered that they can insert almost any DNA sequence into the plasmid, making it an ideal vector for genetic manipulation of plants.

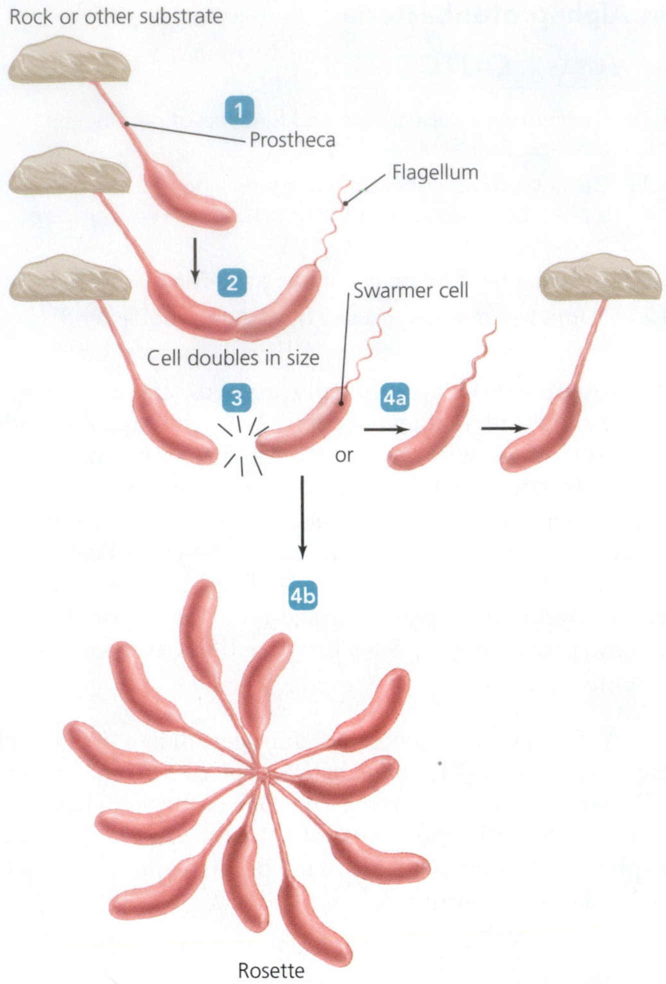

▲ **FIGURE 11.21 Growth and reproduction of *Caulobacter*.** 1 A cell attached to a substrate by its prostheca. 2 Growth and production of an apical flagellum. 3 Division by asymmetrical binary fission to produce a flagellated daughter cell called a swarmer cell. 4a Attachment of a swarmer cell to a substrate by a prostheca that replaces the flagellum. 4b Swarmer cells can attach to one another to form rosettes.

Characteristics of selected members of the alphaproteobacteria are listed in Table 11.4 on p. 345.

Class Betaproteobacteria

LEARNING OUTCOME

11.22 Name three pathogenic, three useful, and one problematic betaproteobacteria.

Betaproteobacteria are another diverse group of Gram-negative bacteria that thrive in habitats with low levels of nutrients. They differ from alphaproteobacteria in their rRNA sequences, though metabolically the two groups overlap. One example of such metabolic overlap is seen with *Nitrosomonas* (nī-trō-sō-mō´nas), an important nitrifying soil bacterium that performs the first

▲ **FIGURE 11.22 A plant gall.** Infecting cells of *Agrobacterium* have inserted into a plant chromosome a plasmid carrying a plant growth hormone gene. The hormone causes the proliferation of undifferentiated plant cells. These gall cells synthesize nutrients for the bacteria.

reaction of nitrification—the conversion of ammonia into nitrite (Equation 4a on p. 337). In this section we will discuss a few other interesting betaproteobacteria.

Pathogenic Betaproteobacteria Species of *Neisseria* (nī-se´rē-ă) are Gram-negative diplococci that inhabit the mucous membranes of mammals and cause such diseases as gonorrhea, meningitis, pelvic inflammatory disease, and inflammation of the cervix, pharynx, and external lining of the eye.

Other pathogenic betaproteobacteria include *Bordetella* (bōr-dĕ-tel´ă), which is the cause of pertussis (whooping cough) (see **Emerging Disease Case Study: Pertussis** on p. 341), and *Burkholderia* (burk-hol-der´ē-ă), which recycles numerous organic compounds in nature. *Burkholderia* commonly colonizes moist environmental surfaces (including laboratory and medical equipment) and the respiratory passages of patients with cystic fibrosis. (Chapters 22 and 24 discuss these pathogens in more detail.)

Other Betaproteobacteria Members of the genus *Thiobacillus* (thī-ō-bă-sil´ŭs) are colorless sulfur bacteria that are important in recycling sulfur in the environment by oxidizing hydrogen sulfide (H_2S) or elemental sulfur (S^0) to sulfate (SO_4^{2-}). Miners use *Thiobacillus* to leach metals from low-grade ore, though the bacterium does cause extensive pollution when it releases metals and acid from mine wastes. (Chapter 25 discusses the sulfur cycle.)

Sewage treatment supervisors are interested in *Zoogloea* (zō´ō-glē-ă) and *Sphaerotilus* (sfēr-ō´til-us), two genera that form *flocs*—slimy, tangled masses of bacteria and organic matter in sewage. *Zoogloea* forms compact flocs that settle to the bottom of treatment tanks and assist in the purification process. *Sphaerotilus*, in contrast, forms loose flocs that do not settle and thus impede the proper flow of waste through a treatment plant.

The characteristics of the genera of betaproteobacteria discussed in this section are summarized in Table 11.4 on p. 345.

BENEFICIAL MICROBES

A Microtube of Superglue

A swarmer cell of the Gram-negative alphaproteobacterium *Caulobacter crescentus* attaches itself to an environmental substrate by secreting an organic adhesive from its prostheca as if it were a tube of glue. This polysaccharide-based bonding agent is the strongest known glue of biological origin, beating out such contenders as barnacle glue, mussel glue, and the adhesion of gecko lizard bristles.

One way scientists gauge adhesive strength is to measure the force required to break apart two glued objects. Commercial superglues typically lose their grip when confronted with a shear force of 18 to 28 newtons (N) per square millimeter.[a] Dental cements bond with strengths up to 30 N/mm^2, but *Caulobacter* glue is more than twice as adhesive. It maintains its grip up to 68 N/mm^2! That is equivalent to being able to hang an adult female elephant on a wall with a spot of glue the size of an American quarter. And remember, this bacterium lives in water, so its glue works even when submerged.

Scientists are researching the biophysical and chemical mechanisms that give this biological glue such incredible gripping power. One critical component of the glue is *N*-acetylglucosamine, one of the sugar subunits of peptidoglycan found in bacterial cell walls. Scientists are struggling to characterize the other molecules that make up the glue. The problem? They cannot pry the glue free to analyze it.

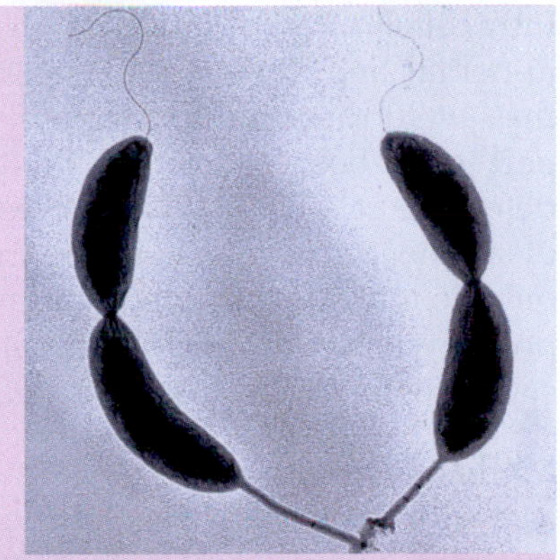

Caulobacter crescentus SEM 1 μm

Some potential applications of such a bacterial superglue include use as a biodegradable suture in surgery, as a more durable dental adhesive, or to stick anti-biofilm disinfectants onto surfaces such as medical devices and ships' hulls.

[a]1 newton is the amount of force needed to accelerate a 1-kg mass 1 meter per second per second.

Class Gammaproteobacteria

LEARNING | OUTCOMES

11.23 Describe the gammaproteobacteria.
11.24 Describe the metabolism of the largest group of gammaproteobacteria.
11.25 Contrast gammaproteobacterial nitrogen fixers with alphaproteobacterial nitrogen fixers.

The **gammaproteobacteria** make up the largest and most diverse class of proteobacteria; almost every shape, arrangement of cells, metabolic type, and reproductive strategy is represented in this group. Ribosomal studies indicate that gammaproteobacteria can be divided into several subgroups:

- Purple sulfur bacteria
- Intracellular pathogens
- Methane oxidizers
- Facultative anaerobes that utilize Embden-Meyerhof glycolysis and the pentose phosphate pathway
- Pseudomonads, which are aerobes that catabolize carbohydrates by the Entner-Doudoroff and pentose phosphate pathways

Here we will examine some representatives of these groups.

Purple Sulfur Bacteria Whereas the purple *non*sulfur bacteria are distributed among the alpha- and betaproteobacteria, **purple sulfur bacteria** are all gammaproteobacteria **(FIGURE 11.23a)**. Purple sulfur bacteria are obligate anaerobes that oxidize hydrogen sulfide to sulfur, which they deposit as internal granules. They are found in sulfur-rich zones in lakes, bogs, and oceans. Some species form intimate relationships with marine worms, covering the body of a worm like strands of a rope **(FIGURE 11.23b)**.

Intracellular Pathogens Organisms in the genera *Legionella* (lē-jŭ-nel′lă) and *Coxiella* (kok-sē-el′ă) are pathogens of humans that avoid digestion by white blood cells, which are normally part of a body's defense; in fact, they thrive inside these defensive cells. *Legionella* derives energy from the metabolism of amino acids, which are more prevalent inside cells than outside. *Coxiella* grows best at low pH, such as is found in the phagolysomes of white blood cells. The bacteria in these genera cause Legionnaires' disease and Q fever, respectively.

Methane Oxidizers Gram-negative bacteria that utilize methane as a carbon source and as an energy source are called **methane oxidizers**. Like archaeal methanogens, bacterial methane oxidizers inhabit anaerobic environments worldwide, growing just above the anaerobic layers that contain methanogens, which generate methane. Methane is one of the so-called greenhouse gases that retains heat in the atmosphere, but methane oxidizers digest most of the methane in their local environment before it can adversely affect the world's climate.

Glycolytic Facultative Anaerobes The largest group of gammaproteobacteria is composed of Gram-negative, facultatively anaerobic rods that catabolize carbohydrates by glycolysis and the pentose phosphate pathway. This group, which is divided into three families (**TABLE 11.3** on p. 342), contains numerous human pathogens. Members of the family Enterobacteriaceae, including *Escherichia coli* (esh-ĕ-rik′ē-ă ko′lī), are frequently used for laboratory studies of metabolism, genetics, and recombinant DNA technology. (Chapter 23 examines pathogenic gammaproteobacteria in detail.)

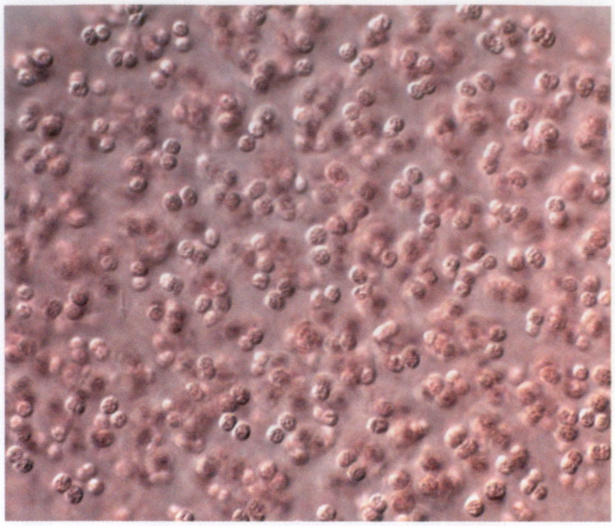

(a)

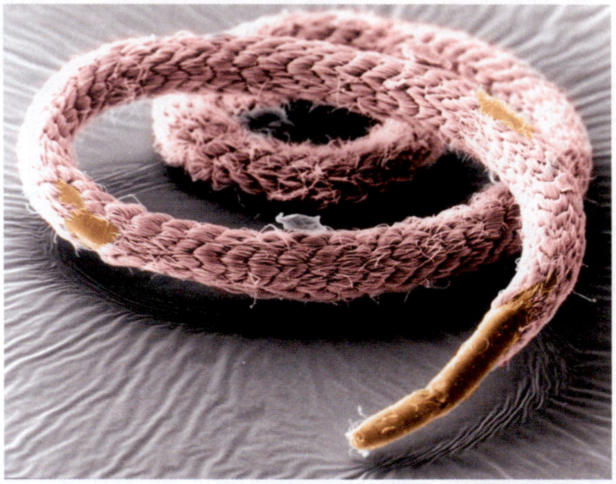

(b)

▲ **FIGURE 11.23 Purple sulfur bacteria. (a)** These bacteria deposit sulfur granules internally. **(b)** Numerous bacteria on the surface of a nematode (roundworm) give the appearance of a rope.

Pseudomonads Bacteria called **pseudomonads** are Gram-negative, aerobic, flagellated, straight to slightly curved rods that catabolize carbohydrates by the Entner-Doudoroff and pentose phosphate pathways. These organisms are noted for their ability to break down numerous organic compounds. Many of them are important pathogens of humans and animals and are involved in the spoilage of refrigerated milk, eggs,

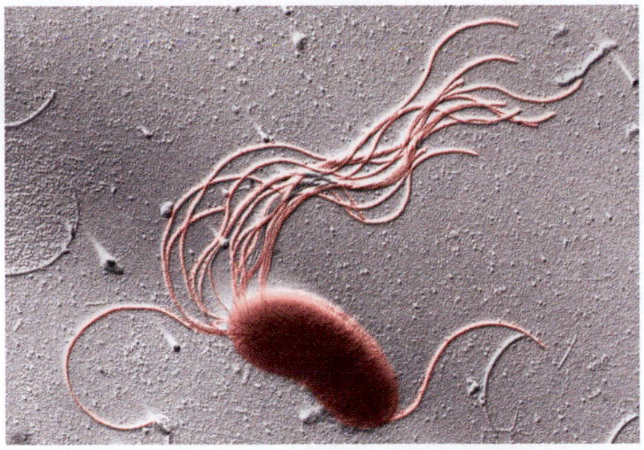

▲ **FIGURE 11.24** ***Pseudomonas* is distinguished by its polar flagella.**

and meat because they can grow and catabolize proteins and lipids at 4°C. The pseudomonad group is named for its most important genus, *Pseudomonas* (soo-dō-mō´nas; **FIGURE 11.24**), which causes diseases such as urinary tract infections, external otitis (swimmer's ear), and lung infections in cystic fibrosis patients. Other pseudomonads, such as *Azotobacter* (ā-zō-tō-bak´ter) and *Azomonas* (ā-zō-mō´nas), are soil-dwelling, nonpathogenic nitrogen fixers; however, in contrast to nitrogen-fixing alphaproteobacteria, these gammaproteobacteria do not associate with the roots of plants.

The characteristics of the members of the gammaproteobacteria discussed in this section are summarized in Table 11.4 on p. 345.

Class Deltaproteobacteria

LEARNING OUTCOME

11.26 List several members of the deltaproteobacteria.

The **deltaproteobacteria** are not a large assemblage, but, like other proteobacteria, they include a wide variety of metabolic types. *Desulfovibrio* (dē´sul-fō-vib´rē-ō) is a sulfate-reducing microbe that is important in recycling sulfur in the environment. It is also an important member of bacterial communities living in the sediments of polluted streams and sewage treatment lagoons, where its presence is often apparent by the odor of hydrogen sulfide that it releases during anaerobic respiration. Hydrogen sulfide reacts with iron to form iron sulfide, so sulfate-reducing bacteria, such as *Desulfovibrio*, play a primary role in the corrosion of iron pipes in heating systems, sewer lines, and other structures.

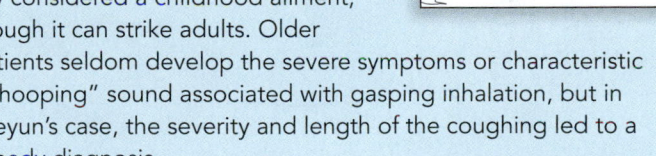

EMERGING DISEASE CASE STUDY

Pertussis

Jeeyun was coughing again. She had been coughing off and on for two weeks. If she had thought of it, she might have noticed that the coughing spells began soon after her flight from New York, where she had visited her grandmother. Within a week of her return to California, she had developed coldlike signs—runny nose, sneezing, and a slight fever—but she didn't remember these things. What she did know was that the coughing was worse; her chest hurt from the constant hacking. Then the coughing broke two ribs.

The surprising pain caused Jeeyun to involuntarily urinate. When the coughing stopped, she vomited and fainted in a heap on the floor. This was no ordinary cough! Jeeyun's roommate called 911. Later, emergency room staff at the hospital bandaged her chest to stabilize the broken ribs and diagnosed the cough as pertussis.

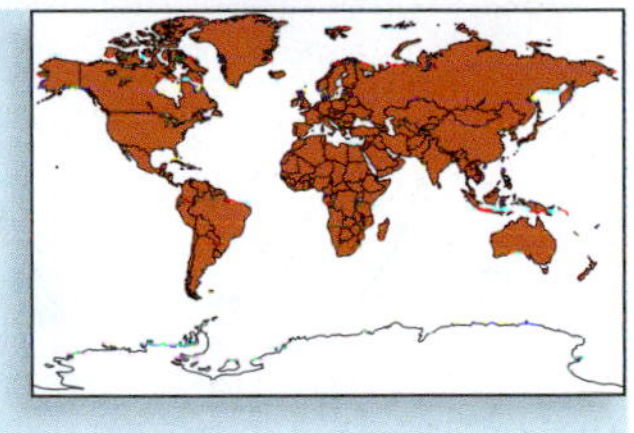

A bacterium, *Bordetella pertussis*, causes pertussis, commonly known as whooping cough. The disease is usually considered a childhood ailment, though it can strike adults. Older patients seldom develop the severe symptoms or characteristic "whooping" sound associated with gasping inhalation, but in Jeeyun's case, the severity and length of the coughing led to a speedy diagnosis.

Immunization is the only way to control pertussis, but adults have been lax in vaccinating children and receiving boosters for themselves. As a result, whooping cough is reemerging as a major problem in the industrialized world. Infected people spread *Bordetella* in respiratory droplets to their neighbors, as happened to Jeeyun, possibly in an airplane's cabin on a long flight. (For more about pertussis, see pp. 691, 694–695.)

1. Why might adults neglect to get a pertussis booster shot for themselves?
2. Why might parents leave their children unimmunized against pertussis?
3. Why is immunization a societal and not merely a personal decision?

TABLE 11.3 Representative Glycolytic Facultative Anaerobes of the Class Gammaproteobacteria

Family	Special Characteristics	Representative Genera	Typical Human Diseases
Enterobacteriaceae	Straight rods; oxidase negative; peritrichous flagella or nonmotile	*Escherichia*	Gastroenteritis
		Enterobacter	(Rarely pathogenic)
		Serratia	(Rarely pathogenic)
		Salmonella	Enteritis
		Proteus	Urinary tract infection
		Shigella	Shigellosis
		Yersinia	Plague
		Klebsiella	Pneumonia
Vibrionaceae	Vibrios; oxidase positive; polar flagella	*Vibrio*	Cholera
Pasteurellaceae	Cocci or straight rods; oxidase positive; nonmotile	*Haemophilus*	Meningitis in children, middle ear infections, pneumonia

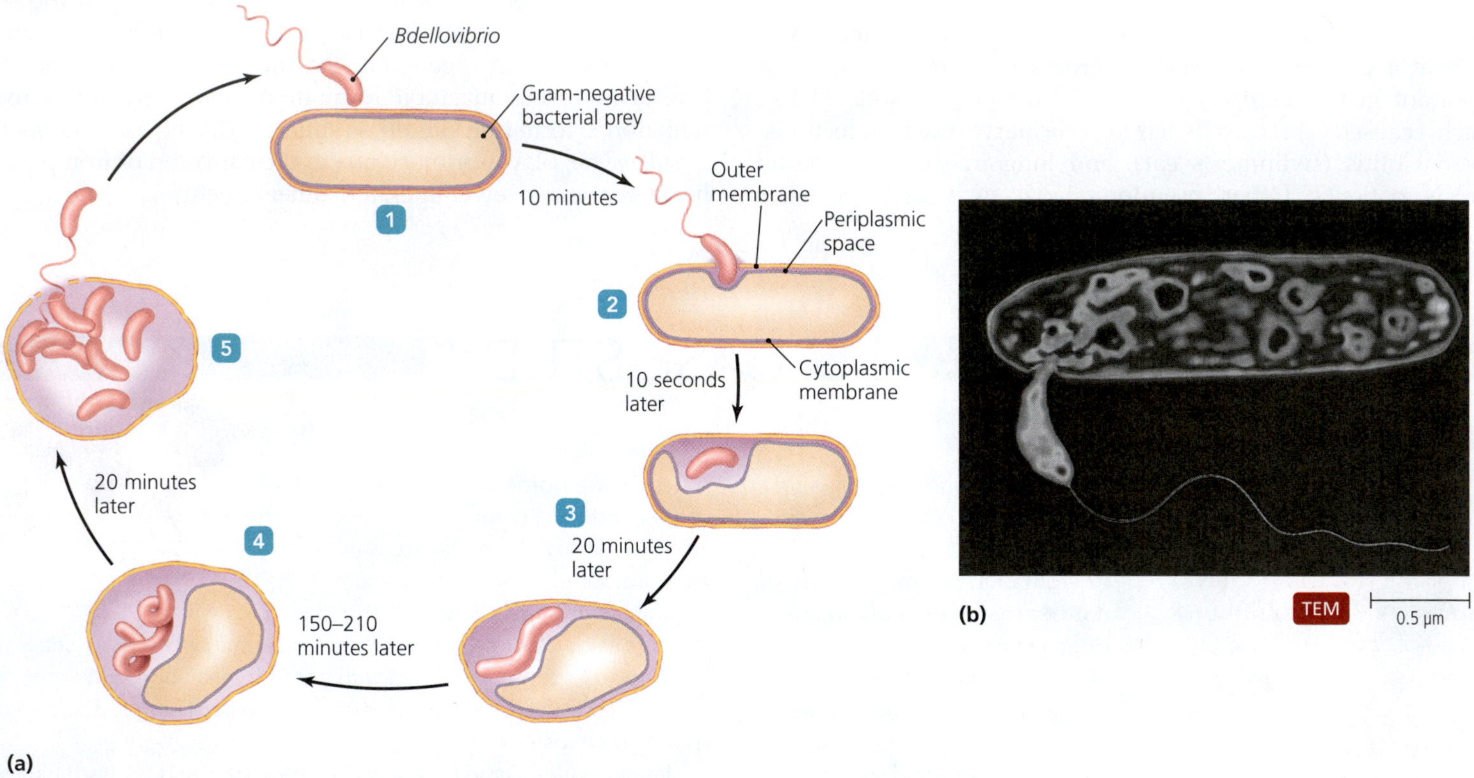

▲ **FIGURE 11.25** ***Bdellovibrio*, a Gram-negative pathogen of other Gram-negative bacteria.**
(a) Life cycle, including the elapsed times between events. **(b)** *Bdellovibrio* invading the periplasmic space of its host.

Bdellovibrio (del-lō-vib´rē-ō) is another deltaproteobacterium. It attacks and destroys other Gram-negative bacteria in a complex and unusual way **(FIGURE 11.25a)**:

1. A free *Bdellovibrio* swims rapidly through the medium until it attaches via fimbriae to a Gram-negative bacterium.
2. It rapidly drills through the cell wall of its prey by secreting hydrolytic enzymes and rotating in excess of 100 revolutions per second **(FIGURE 11.25b)**.
3. Once inside, *Bdellovibrio* lives in the periplasmic space—the space between the cytoplasmic membrane and the outer membrane of the cell wall. It kills its host by disrupting the host's cytoplasmic membrane and inhibiting DNA, RNA, and protein synthesis.
4. The invading bacterium uses the nutrients released from its dying prey and grows into a long filament.
5. Eventually, the filament divides into as many as nine smaller cells at once, each of which, when released from the dead cell, produces a flagellum and swims off to repeat the process. Multiple fissions to produce many offspring is a rare form of reproduction.

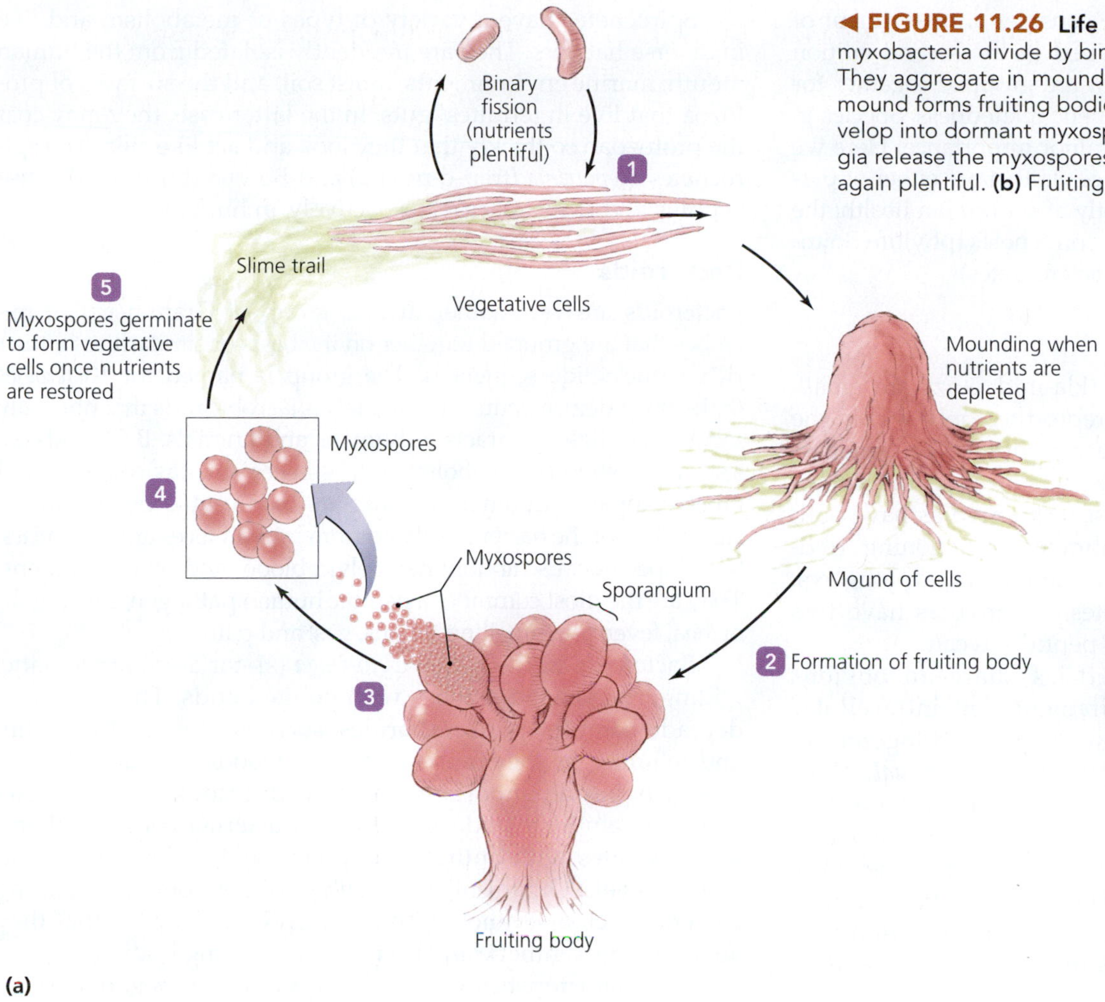

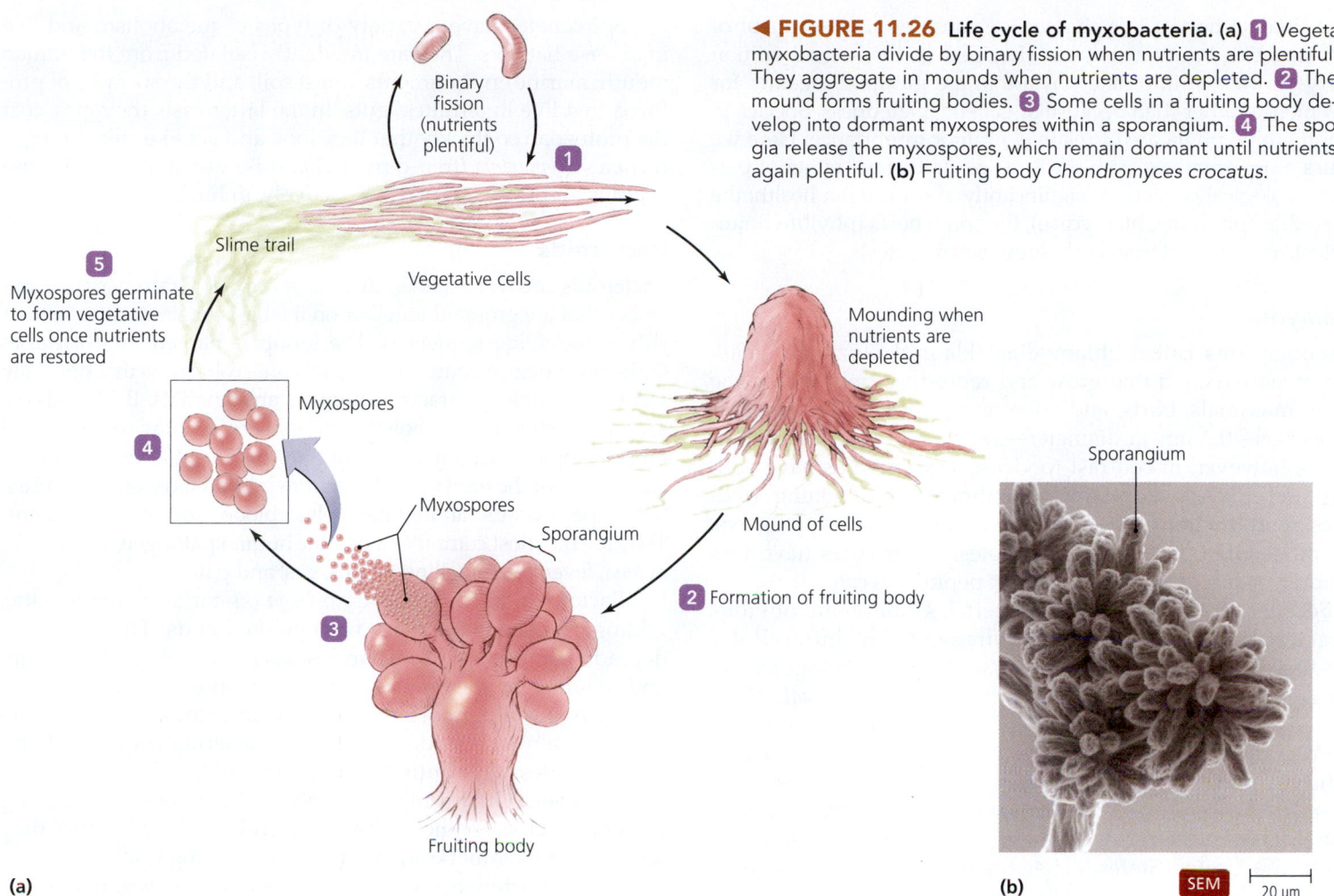

◀ **FIGURE 11.26 Life cycle of myxobacteria. (a)** 1 Vegetative myxobacteria divide by binary fission when nutrients are plentiful. They aggregate in mounds when nutrients are depleted. 2 The mound forms fruiting bodies. 3 Some cells in a fruiting body develop into dormant myxospores within a sporangium. 4 The sporangia release the myxospores, which remain dormant until nutrients are again plentiful. **(b)** Fruiting body *Chondromyces crocatus*.

Myxobacteria are Gram-negative, aerobic, soil-dwelling bacteria with a unique life cycle for prokaryotes in that individuals cooperate to produce differentiated reproductive structures. The life cycle of myxobacteria can be summarized as follows **(FIGURE 11.26a)**:

1. Vegetative myxobacteria glide on slime trails through their environment, digesting yeasts and other bacteria or scavenging nutrients released from dead cells. When nutrients and cells are plentiful, the myxobacteria divide by binary fission; when nutrients are depleted, however, they aggregate by gliding into a mound of cells.
2. Myxobacteria within the mound differentiate to form a macroscopic *fruiting body* ranging in height from 50 μm to 700 μm.
3. Some cells within the fruiting body develop into dormant *myxospores* that are enclosed within walled structures called *sporangia* (singular: *sporangium;* Figure 11.26b).
4. The sporangia release the myxospores, which can resist desiccation and nutrient deprivation for a decade or more.
5. When nutrients are again plentiful, the myxospores germinate and become vegetative cells.

Myxobacteria live worldwide in soils that have decaying plant material or animal dung. Though certain species live in the arctic and others in the tropics, most myxobacteria live in temperate regions.

Class Epsilonproteobacteria

The class **Epsilonproteobacteria** includes Gram-negative rods, vibrios, or spirals. Important genera are *Campylobacter* (kam´pi-lō-bak´ter), which causes blood poisoning and inflammation of the intestinal tract, and *Helicobacter* (hel´ĭ-kō-bak´ter), which causes ulcers. (Chapter 23 examines these pathogens more fully.)

The characteristics of the delta- and epsilonproteobacteria are summarized in Table 11.4 on p. 345.

Other Gram-Negative Bacteria

LEARNING | **OUTCOMES**

11.27 Describe the unique features of chlamydias and spirochetes.

11.28 Describe the ecological importance of bacteroids.

In the final section of this chapter we consider an assortment of Gram-negative bacteria that are classified in the second edition of *Bergey's Manual* into nine phyla that are grouped together for convenience rather than because of genetic relatedness. Species in six of the nine phyla are of relatively minor importance. Here we discuss representatives from the three phyla that are either of particular ecological concern or significantly affect human health: the chlamydias (phylum Chlamydiae), the spirochetes (phylum Spirochaetes), and the bacteroids (phylum Bacteroidetes).

Chlamydias

Microorganisms called **chlamydias** (kla-mid ē-ăz) are small, Gram-negative cocci that grow and reproduce only within the cells of mammals, birds, and a few invertebrates. The smallest chlamydias—0.2 µm in diameter—are smaller than the largest viruses; however, in contrast to viruses, chlamydias have both DNA and RNA, cytoplasmic membranes, functioning ribosomes, reproduction by binary fission, and metabolic pathways. Like other Gram-negative prokaryotes, chlamydias have two membranes, but in contrast they lack peptidoglycan.

Because chlamydias and rickettsias share an obvious characteristic—both have a requirement for intracellular life—these two types of organisms were grouped together in a single taxon in the first edition of *Bergey's Manual*. Now, however, the rickettsias are classified with the alphaproteobacteria, and the chlamydias are in their own phylum. Chlamydias cause neonatal blindness, pneumonia, and a sexually transmitted disease called *lymphogranuloma venereum;* in fact, chlamydias are the most common sexually transmitted bacteria in the United States. (The chlamydial life cycle is illustrated in Figure 24.9).

Spirochetes

Spirochetes are unique helical bacteria that are motile by means of axial filaments composed of flagella that lie within the periplasmic space (see Figure 3.14). When an axial filament rotates, the entire cell corkscrews through the medium.

Spirochetes have a variety of types of metabolism and live in diverse habitats. They are frequently isolated from the human mouth, marine environments, moist soil, and the surfaces of protozoa that live in termites' guts. In the latter case, they may coat the protozoan so thickly that they look and act like cilia. The spirochetes *Treponema* (trep-ō-ne´mă) and Borelia (bō-rē´lē-ā) cause syphilis and Lyme disease, respectively, in humans.

Bacteroids

Bacteroids are yet another diverse group of Gram-negative microbes that are grouped together on the basis of similarities in their rRNA nucleotide sequences. The group is named for *Bacteroides* (bak-ter-oy´dēz), a genus of obligately anaerobic rods that normally inhabit the digestive tracts of humans and animals. Bacteroids assist in digestion by catabolizing substances such as cellulose and other complex carbohydrates that are indigestible by mammals. About 30% of the bacteria isolated from human feces are *Bacteroides.* Some species cause abdominal, pelvic, blood, and other infections. They are the most common anaerobic human pathogen, causing diarrhea, fever, foul-smelling lesions, gas, and pain.

Bacteroids in the genus *Cytophaga* (sī-tof´ă-gă) are aquatic, gliding, rod-shaped aerobes with pointed ends. These bacteria degrade complex polysaccharides, such as agar, pectin, chitin, and cellulose, so they cause damage to wooden boats and piers; they also play an important role in the degradation of raw sewage. Their ability to glide allows these bacteria to position themselves at sites with optimum nutrients, pH, temperature, and oxygen levels. Organisms in *Cytophaga* differ from other gliding bacteria, such as cyanobacteria and myxobacteria, in that they are nonphotosynthetic and do not form fruiting bodies.

The characteristics of these groups of Gram-negative bacteria are listed in **TABLE 11.4**.

TELL ME WHY

Why are bacteria all in the same domain (Bacteria) despite their widely divergent oxygen tolerance, sizes, shapes, and nutritional requirements?

TABLE 11.4 Characteristics of Selected Gram-Negative Bacteria

Phylum/Class	Representative Members	Special Characteristics	Diseases
Proteobacteria			
Alphaproteobacteria	*Azospirillum*	Nitrogen fixer	
	Rhizobium	Nitrogen fixer	
	Nitrobacter	Nitrifying bacterium	
	Purple nonsulfur bacteria	Anoxygenic phototrophs	
	Rickettsia	Intracellular pathogen	Typhus and Rocky Mountain spotted fever
	Brucella	Coccobacillus	Brucellosis
	Acetobacter, Gluconobacter	Synthesize acetic acid	
	Caulobacter	Prosthecate bacterium	
	Agrobacterium	Causes galls in plants; vector for gene transfer in plants	
Betaproteobacteria	*Nitrosomonas*	Nitrifying bacterium	
	Neisseria	Diplococcus	Gonorrhea and meningitis
	Bordetella		Pertussis
	Burkholderia		Lung infection of cystic fibrosis patients
	Thiobacillus	Colorless sulfur bacterium	
	Zoogloea	Used in sewage treatment	
	Sphaerotilus	Blocks sewage treatment pipes	
Gammaproteobacteria	Purple sulfur bacteria		
	Legionella	Intracellular pathogen	Legionnaires' disease
	Coxiella	Intracellular pathogen	Q fever
	Methylococcus	Oxidizes methane	
(Families: Enterobacteriaceae Vibrionaceae Pasteurellaceae)	Glycolytic facultative anaerobes: *Esherichia*, *Enterobacter*, *Serratia*, *Salmonella*, *Proteus*, *Shigella*, *Yersinia*, *Klebsiella*,*Vibrio*, *Haemophilus*	Facultative anaerobes that catabolize carbohydrates via glycolysis and the pentose phosphate pathway	See Table 11.3 on p. 342
	Pseudomonas	Aerobe that catabolizes carbohydrates via Entner-Doudoroff and pentose phosphate pathways	Urinary tract infections, external otitis
	Azotobacter	Nitrogen fixers not associated with plant roots	
	Azomonas		
Deltaproteobacteria	*Desulfovibrio*	Sulfate reducer	
	Bdellovibrio	Pathogen of Gram-negative bacteria	
	Myxobacteria	Reproduces by forming differentiated fruiting bodies	
Epsilonproteobacteria	*Campylobacter*	Curved rod	Gastroenteritis
	Helicobacter	Spiral	Gastric ulcers
Chlamydiae			
Chlamydiae	*Chlamydia*	Intracellular pathogen; lacks peptidoglycan	Neonatal blindness and lymphogranuloma venereum
Spirochaetes			
Spirochaetes	*Treponema*	Motile by axial filaments	Syphilis
	Borrelia	Motile by axial filaments	Lyme disease
Bacteroidetes			
"Bacteroidia"[a]	*Bacteroides*	Anaerobe that lives in animal colons	Abdominal infections
"Sphingobacteria"	*Cytophaga*	Digests complex polysaccharides	

[a]The names of taxa in quotations are not officially recognized.

MICRO IN THE CLINIC FOLLOW-UP

Can Diabetes Cause a Foot Infection?

The doctor diagnoses Sean with diabetic foot syndrome and informs him that his lesion is infected with an obligately anaerobic bacterium called *Clostridium perfringens*. Many patients with this syndrome end up requiring amputation. Luckily, Sean's case is diagnosed early enough to avoid having his foot removed, but it is still difficult to treat. As an insulin-dependent diabetic who has not managed his disease well, Sean has developed nerve damage and reduced blood flow to his foot. Reduced blood flow makes the infection difficult to treat by restricting the ability of antibiotics to reach the infection site, as well as reducing the oxygen content in his tissues, slowing the healing process. The dead and damaged tissue in the lesion further reduces the effectiveness of antimicrobials.

In addition to regular debridement and antibiotics, the doctor recommends hyperbaric treatment. This treatment requires Sean to lie in an enclosed chamber in which oxygen is administered at high pressure to saturate the infected tissues. The treatment is very effective with *C. perfringens* because the anaerobic bacteria cannot tolerate oxygen.

This aggressive treatment will prevent amputation this time. However, diabetic foot problems usually recur. Sean will need to make serious life changes, which include quitting smoking, controlling his blood sugar, and addressing his poor diet and subsequent weight problem. He should also commit to regular foot care and thorough foot examinations at least four times a year. This is a battle Sean must take seriously for the rest of his life, or he will join the large number of diabetes patients who must have limbs amputated because of neuropathy-related infections.

1. **What conditions, caused or complicated by Sean's diabetes, created a unique environment for *C. perfringens* to grow?**
2. **What characteristic of clostridia contributed to the foul smell that permeated the room when the nurse debrided the wound?**

MM Check your answers to Micro in the Clinic Follow-Up questions in the **MasteringMicrobiology** Study Area.

Explore the Invisible: Arrangements of Prokaryotic Cells

Practice on-the-go with Dr. Bauman Video Tutors by scanning this QR code with your smart phone. Visit the **MasteringMicrobiology Study Area** to challenge your understanding with practice tests, animation quizzes, and clinical case studies!

MasteringMicrobiology®

CHAPTER SUMMARY

General Characteristics of Prokaryotic Organisms (pp. 322–325)

1. Three basic shapes of prokaryotic cells are spherical **cocci**, rod-shaped **bacilli**, and **spirals**. Spirals may be stiff (**spirilla**) or flexible (**spirochetes**).
2. Other variations in shapes include **vibrios** (slightly curved rods), **coccobacilli** (intermediate to cocci and bacilli), and **pleomorphic** (variable shape and size).
3. Environmentally resistant **endospores** are produced within vegetative cells of the Gram-positive genera *Bacillus* and *Clostridium*. Depending on the species in which they are formed, the endospores may be terminal, subterminal, or centrally located.
4. Prokaryotes reproduce asexually by **binary fission**, **snapping division** (a type of binary fission), **spore** formation, and **budding**. ▶**ANIMATIONS:** *Bacterial Growth: Overview*

5. Cocci may typically be found in groups, including long chains (**streptococci**), pairs (**diplococci**), foursomes (**tetrads**), cuboidal packets (**sarcinae**), and clusters (**staphylococci**).
6. Bacilli are found singly, in pairs, in chains, or in a **palisade** arrangement.

▶**VIDEO TUTOR:** *Arrangements of Prokaryotic Cells*

Modern Prokaryotic Classification (pp. 325–327)

1. Living things are now classified into three domains—Archaea, Bacteria, and Eukarya—based largely on genetic relatedness.
2. The most authoritative reference in modern prokaryotic systematics is *Bergey's Manual of Systematic Bacteriology,* second edition, which classifies prokaryotes into two phyla of Archaea and 24 phyla of Bacteria. The organization of this text's survey of prokaryotes largely follows Bergey's classification scheme.

Survey of Archaea (pp. 327–329)

1. The domain Archaea includes **extremophiles**, microbes that require extreme conditions of temperature, pH, and/or salinity to survive.
2. **Thermophiles** and **hyperthermophiles** (in the phyla Crenarchaeota and Euryarchaeota) live at temperatures above 45°C and 80°C, respectively, because their DNA, membranes, and proteins do not function properly at lower temperatures.
3. **Halophiles** (phylum Euryarchaeota) depend on high concentrations of salt to keep their cell walls intact. Halophiles such as *Halobacterium salinarium* synthesize purple proteins called **bacteriorhodopsins** that harvest light energy to synthesize ATP.
4. **Methanogens** (phylum Euryarchaeota) are obligate anaerobes that produce methane gas and are useful in sewage treatment.

Survey of Bacteria (pp. 329–345)

1. **Deeply branching bacteria** have rRNA sequences thought to be similar to those of earliest bacteria. They are autotrophic and live in hot, acidic, and anaerobic environments, often with intense exposure to sun.
2. Phototrophic bacteria trap light energy with photosynthetic lamellae. The five groups of phototrophic bacteria are cyanobacteria, green sulfur bacteria, green nonsulfur bacteria, purple sulfur bacteria, and purple nonsulfur bacteria.
3. Many **cyanobacteria** reduce atmospheric N_2 to NH_3 via a process called **nitrogen fixation**. Cyanobacteria must separate (in either time or space) the metabolic pathways of nitrogen fixation from those of oxygenic photosynthesis because nitrogen fixation is inhibited by the oxygen generated during photosynthesis. Many cyanobacteria fix nitrogen in thick-walled cells called **heterocysts**.
4. Green and purple bacteria use bacteriochlorophylls for anoxygenic photosynthesis. Nonsulfur forms derive electrons from organic compounds; sulfur forms derive electrons from H_2S.

   ▶**ANIMATIONS:** *Photosynthesis: Comparing Prokaryotes and Eukaryotes*
5. The phylum Firmicutes contains bacteria with a G + C content (the percentage of all base pairs that are guanine-cytosine base pairs) of less than 50%. Firmicutes includes clostridia, mycoplasmas, and other low G + C cocci and bacilli.
6. Clostridia include the genus *Clostridium* (pathogenic bacteria that cause gangrene, tetanus, botulism, and diarrhea), *Epulopiscium* (which is large enough to be seen without a microscope), and *Selenomonas* (often found in dental plaque).
7. **Mycoplasmas** are Gram-positive, pleomorphic, facultative anaerobes and obligate anaerobes that lack cell walls and therefore stain pink with Gram stain. They are frequently associated with pneumonia and urinary tract infections.
8. Low G + C Gram-positive bacilli and cocci important to human health and industry include *Bacillus* (which contains species that cause anthrax and food poisoning and includes beneficial Bt-toxin bacteria), *Listeria* (which causes bacteremia and meningitis), *Lactobacillus* (used to produce yogurt and pickles), *Streptococcus* (which causes strep throat and other diseases), *Enterococcus* (which can cause endocarditis and other diseases), and *Staphylococcus* (which causes a number of human diseases).
9. High G + C bacteria (*Corynebacterium, Mycobacterium,* and actinomycetes) are classified in phylum Actinobacteria.
10. Bacteria in *Corynebacterium* store phosphates in **metachromatic granules**; *C. diphtheriae* causes diphtheria.
11. Members of the genus *Mycobacterium,* including species that cause tuberculosis and leprosy, grow slowly and have unique, resistant cell walls containing waxy **mycolic acids**.
12. **Actinomycetes** resemble fungi in that they produce spores and form filaments; this group includes *Actinomyces* (normally found in human mouths), *Nocardia* (useful in degradation of pollutants), and *Streptomyces* (produces important antibiotics).
13. Phylum **Proteobacteria** is a very large group of Gram-negative bacteria divided into five classes—the alpha-, beta-, gamma-, delta-, and epsilonproteobacteria.
14. The **alphaproteobacteria** include a variety of aerobes, many of which have unusual cellular extensions called prosthecae. *Azospirillum* and *Rhizobium* are nitrogen fixers that are important in agriculture.
15. Some members of the alphaproteobacteria associate with plant roots and are **nitrifying bacteria**, which oxidize NH_3 to NO_3 via a process called **nitrification**. Nitrifying alphaproteobacteria are in the genus *Nitrobacter.*
16. Most purple nonsulfur phototrophs are alphaproteobacteria.
17. Pathogenic alphaproteobacteria include *Rickettsia* (typhus and Rocky Mountain spotted fever) and *Brucella* (brucellosis).
18. There are many beneficial alphaproteobacteria, including *Acetobacter* and *Gluconobacter,* both of which are used to synthesize acetic acid. *Caulobacter* is of interest in reproductive studies, and *Agrobacterium* is used in genetic recombination in plants.
19. The **betaproteobacteria** include the nitrifying *Nitrosomonas* and pathogenic species, such as *Neisseria* (gonorrhea), *Bordetella* (whooping cough), and *Burkholderia* (which colonizes the lungs of cystic fibrosis patients).
20. Other betaproteobacteria include *Thiobacillus* (ecologically important), *Zoogloea* (useful in sewage treatment), and *Sphaerotilus* (hampers sewage treatment).
21. The **gammaproteobacteria** constitute the largest class of proteobacteria; they include **purple sulfur bacteria**, intracellular

pathogens, facultative anaerobes that utilize glycolysis and the pentose phosphate pathway, and pseudomonads.

22. Both *Legionella* and *Coxiella* are intracellular, pathogenic gammaproteobacteria.
23. **Methane oxidizers** are anaerobic bacteria that use methane for both carbon and energy.
24. Numerous human pathogens are facultatively anaerobic gammaproteobacteria that catabolize carbohydrates by glycolysis.
25. **Pseudomonads**, including pathogenic *Pseudomonas* and nitrogen-fixing *Azotobacter* and *Azomonas,* utilize the Entner-Doudoroff and pentose phosphate pathways for catabolism of glucose.
26. The **deltaproteobacteria** include *Desulfovibrio* (important in the sulfur cycle and in corrosion of pipes), *Bdellovibrio* (pathogenic to bacteria), and **myxobacteria**. The latter form stalked fruiting bodies containing resistant, dormant myxospores.
27. The **epsilonproteobacteria** include some important human pathogens, including *Campylobacter* and *Helicobacter.*
28. **Chlamydias** are Gram-negative cocci typified by the genus *Chlamydia;* they cause neonatal blindness, pneumonia, and a sexually transmitted disease.
29. **Spirochetes** are flexible, helical bacteria that live in diverse environments. *Treponema* (syphilis) and *Borrelia* (Lyme disease) are important spirochetes.
30. **Bacteroids** include *Bacteroides,* an obligate anaerobic rod that inhabits the digestive tract, and *Cytophaga,* an aerobic rod that degrades wood and raw sewage.

QUESTIONS FOR REVIEW

Answers to the Questions for Review (except Short Answer questions) begin on p. A–1.

Modified True/False

For each of the following statements that is true, write "true" in the blank. For each statement that is false, write the word(s) that should be substituted for the underlined word(s) to make the statement correct.

1. ________________ All prokaryotes reproduce <u>sexually</u>.
2. ________________ A <u>bacillus</u> is a bacterium with a slightly curved rod shape.
3. ________________ If you were to view staphylococci, you should expect to see <u>clusters</u> of cells.
4. ________________ Chlamydias <u>have</u> peptidoglycan cell walls.
5. ________________ Archaea are classified into phyla based primarily on <u>tRNA</u> sequences.
6. ________________ <u>Halophiles</u> inhabit extremely saline habitats, such as the Great Salt Lake.
7. ________________ Pigments located in <u>thylakoids</u> in phototrophic bacteria trap light energy for metabolic processes.
8. ________________ Most cyanobacteria form <u>heterocysts</u> in which nitrogen fixation occurs.
9. ________________ A giant bacterium that is large enough to be seen without a microscope is <u>*Selenomonas*</u>.
10. ________________ When environmental nutrients are depleted, <u>myxobacteria</u> aggregate in mounds to form fruiting bodies.

Matching

Match the bacterium on the left with the term with which it is most closely associated

1. _____ *Bacillus anthracis*	A. wood damage
2. _____ *Selenomonas*	B. dental biofilm (plaque)
3. _____ *Clostridium perfringens*	C. gangrene
4. _____ *Clostridium botulinum*	D. botox
5. _____ *Bacillus licheniformis*	E. anthrax
6. _____ *Streptococcus*	F. lymphogranuloma venereum
7. _____ *Streptomyces*	G. leprosy
8. _____ *Corynebacterium*	H. tetracycline
9. _____ *Gluconobacter*	I. vinegar
10. _____ *Bordetella*	J. yogurt
11. _____ *Zoogloea*	K. impetigo
12. _____ *Rhizobium*	L. bacitracin
13. _____ *Desulfovibrio*	M. iron pipe corrosion
14. _____ *Chlamydia*	N. pertussis
15. _____ *Cytophaga*	O. nitrogen fixation
	P. floc formation
	Q. diphtheria

Multiple Choice

1. The type of reproduction in prokaryotes that results in a palisade arrangement of cells is called ________________.
   a. pleomorphic division
   b. endospore formation
   c. snapping division
   d. binary fission
2. The thick-walled reproductive spores produced in the middle of cyanobacterial filaments are called ________________.
   a. akinetes
   b. terminal endospores
   c. metachromatic granules
   d. heterocysts
3. Which of the following terms best describes stiff, spiral-shaped prokaryotic cells?
   a. cocci
   b. bacilli
   c. spirilla
   d. spirochetes
4. Endospores ________________.
   a. can remain alive for decades
   b. can remain alive in boiling water
   c. exist in a state of suspended animation
   d. all of the above

5. How is *Halobacterium salinarium* distinctive?
   a. It is absolutely dependent on high salt concentrations to maintain its cell wall.
   b. It is found in terrestrial volcanic habitats.
   c. It photosynthesizes without chlorophyll.
   d. It can survive 5 million rad of radiation.

6. Photosynthetic bacteria that also fix nitrogen are ________________.
   a. mycoplasmas
   b. spirilla
   c. bacteroids
   d. cyanobacteria

7. Which genus is the most common anaerobic human pathogen?
   a. *Bacteroides*
   b. *Spirochetes*
   c. *Chlamydia*
   d. *Methanopyrus*

8. Flexible spiral-shaped prokaryotes are ________________.
   a. spirilla
   b. spirochetes
   c. vibrios
   d. rickettsias

9. Bacteria that convert nitrogen gas into ammonia are ________________.
   a. nitrifying bacteria
   b. nitrogenous
   c. nitrogen fixers
   d. nitrification bacteria

10. The presence of mycolic acid in the cell wall characterizes ________________.
   a. *Corynebacterium*
   b. *Listeria*
   c. *Nocardia*
   d. *Mycobacterium*

VISUALIZE IT!

1. Label the shapes of these prokaryotic cells.

a. ________________

b. ________________

c. ________________

d. ________________

e. ________________

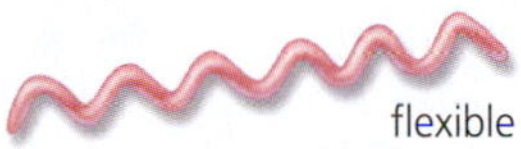

f. ________________

g. ________________

2. Describe the location of these endospores within their cells.

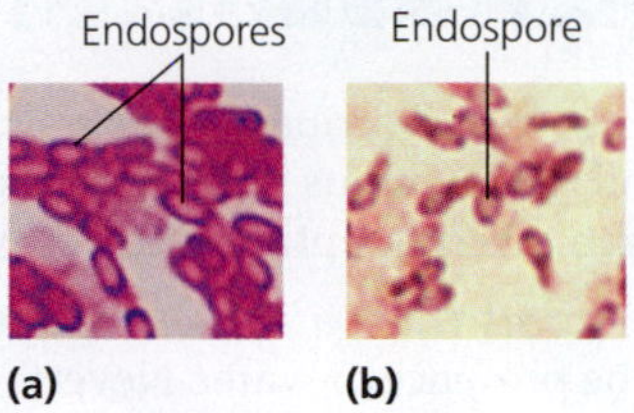

Short Answer

1. Whereas the first edition of *Bergey's Manual* relied on morphological and biochemical characteristics to classify microbes, the new edition focuses on ribosomal RNA sequences. List several other criteria for grouping and classifying bacteria.
2. What are extremophiles? Describe two kinds and give examples.
3. Name and describe three types of bacteria mentioned in this chapter that "glide."
4. Name three groups of low G + C Gram-positive bacteria.
5. Compare and contrast bacterial and archaeal cells.
6. A student was memorizing the arrangements of bacteria and noticed that there are more arrangements for cocci than for bacilli. Why might this be so?
7. How is *Agrobacterium* used in recombinant DNA technology?
8. Name and describe five distinct classes of phylum Proteobacteria.
9. Explain why organisms formerly known as blue-green algae are now called cyanobacteria.
10. Contrast the processes of nitrification and nitrogen fixation.

CRITICAL THINKING

1. A microbiology student described "deeply branching bacteria" as having a branched filamentous growth habit akin to *Streptomyces.* Do you agree with this description? Why or why not?
2. Iron oxide (rust) forms when iron is exposed to oxygen, particularly in the presence of water. Nevertheless, iron pipes typically corrode more quickly when they are buried in moist *anaerobic* soil than when they are buried in soil containing oxygen. Explain why this is the case.
3. Why is it that Gram-positive species don't have axial filaments?
4. Even though *Clostridium* is strictly an anaerobic bacterium, it can be isolated easily from the exposed surface of your skin. Explain how this can be.
5. Louis Pasteur said, "The role of the infinitely small in nature is infinitely large." Explain what he meant by using examples of the roles of microorganisms in health, industry, and the environment.
6. How are bacterial endospores different from the spores of actinomycetes?
7. A scientist who discovers a prokaryote living in a hot spring at 100°C suspects that it belongs to the archaea. Why does she think it might be archaeal? How could she prove that it is not bacterial?
8. Contrast the processes of nitrogen fixation and nitrification.
9. What do the names *Desulfovibrio* and *Bdellovibrio* tell you about the shape of these deltaproteobacteria?

CONCEPT MAPPING

Using the following terms, draw a concept map that describes the domain Archaea. For a sample concept map, see p. 94. Or, complete this and other concept maps online by going to the MasteringMicrobiology Study Area.

>45°C
>80°C
17–25% salt
Acidophiles
Animal colons
Disease
Extremophiles
Great Salt Lake
Halophiles
Hydrothermal vents
Hyperthermophiles
Low pH
Methane
Methanogens
Peptidoglycan
Prokaryotes
Sewage treatment
Thermophiles

12 Characterizing and Classifying Eukaryotes

MICRO IN THE CLINIC

A Souvenir from Paradise

Maria is in the Florida Keys having the vacation of a lifetime. She spends her mornings scuba diving through a wonderland of tropical fish and coral reefs. In the afternoons, she lounges on the beach, listening to music and reading novels. She feels worlds away from her stressful job as an analyst on Wall Street, where she's survived several rounds of layoffs after a tumultuous year. Aside from sunburn and a few insect bites, Maria truly feels like she is relaxing in paradise.

Shortly after returning from vacation, however, Maria comes down with a headache and a high fever. Her joints hurt, a rash appears on her arms, and she feels a stabbing pain behind her eyes. She visits her doctor and says she feels as though she were hit head-on by an 18-wheel truck. After obtaining a complete medical history and hearing a description of Maria's recent travels, the doctor orders a variety of blood tests.

What happened to Maria on her dream vacation? How did she catch this illness? Turn to the end of the chapter (p. 381) to find out.

MM **Explore More:** Test your readiness and apply your knowledge with dynamic learning tools at MasteringMicrobiology.

Eukaryotic microbes include a fascinating and almost bewilderingly diverse assemblage. Eukaryotic microbes include unicellular and multicellular protozoa,[1] fungi,[2] algae,[3] water molds, and slime molds. Additionally, microbiologists study parasitic helminths[4] because they have microscopic stages, and they study arthropod vectors because they are intimately involved in the transmission of microbial pathogens. Eukaryotes include both human pathogens and organisms that are vital for human life. For example, one group of marine algae called diatoms and a set of protozoa called dinoflagellates (dī´nō-flaj´ĕ-lātz) provide the basis for the oceans' food chains and produce most of the world's oxygen. Eukaryotic fungi produce penicillin, and tiny baker's and brewer's yeasts are essential for making bread and alcoholic beverages.

Among the 20 most frequent microbial causes of death worldwide, six are eukaryotic, including the agents of malaria, African sleeping sickness, and amebic dysentery. *Pneumocystis* pneumonia, toxoplasmosis, and cryptosporidiosis—common afflictions of AIDS patients—are all caused by eukaryotic pathogens.

In previous chapters we discussed characteristics of *cells*—their metabolism, growth, and genetics. In this chapter we discuss eukaryotic *organisms* of interest to microbiologists—protozoa (single-celled "animals"), fungi, algae, water molds, and slime molds—and conclude with a brief discussion of the relationship of parasitic helminths and vectors to microbiology.

We begin by discussing general features of eukaryotic reproduction and classification; the following sections survey some representative members of microbiologically important eukaryotic groups, focusing on beneficial, environmentally significant, and unusual species. (Chapters 19–24 discuss fungal and parasitic agents and vectors of human disease in more detail.)

General Characteristics of Eukaryotic Organisms

Our discussion of the general characteristics of eukaryotes begins with a survey of the events in eukaryotic reproduction; then we consider some aspects of the complex matter of classifying the great variety of eukaryotic organisms.

Reproduction of Eukaryotes

LEARNING | **OUTCOMES**

12.1 State four reasons why eukaryotic reproduction is more complex than prokaryotic reproduction.

12.2 Describe the phases of mitosis, mentioning chromosomes, chromatids, centromeres, and spindle.

12.3 Contrast meiosis with mitosis, mentioning homologous chromosomes, tetrads, and crossing over.

12.4 Distinguish among nuclear division, cytokinesis, and schizogony.

A unique characteristic of living things is the ability to reproduce themselves. Prokaryotic reproduction typically involves replication of DNA and binary fission of the cytoplasm to produce two identical offspring. Reproduction of eukaryotes is more complicated and varied than reproduction in prokaryotes for a number of reasons:

- Most of the DNA in eukaryotes is packaged with histone proteins as *chromosomes* in the form of *chromatin* (krō´ma-tin) *fibers* located within nuclei. The remaining DNA in eukaryotic cells is found in mitochondria and chloroplasts, organelles that reproduce by binary fission in a manner similar to prokaryotic reproduction. In this chapter we will discuss only the nuclear portion of eukaryotic genomes.
- Eukaryotes have a variety of methods of asexual reproduction, including binary fission, budding, fragmentation, spore formation, and *schizogony* (ski-zog´ō-nē) (discussed later).
- Many eukaryotes reproduce sexually—that is, via a process that involves the formation of sexual cells called *gametes*, and the subsequent fusion of two gametes to form a cell called a *zygote.*
- Additionally, algae, fungi, and some protozoa reproduce both sexually and asexually. (Animals generally reproduce only one way or the other.)

Eukaryotic reproduction involves two types of division: nuclear division and cytoplasmic division (also called cytokinesis). After we discuss the various aspects of these two types of division, we will consider schizogony.

Nuclear Division

Typically, a eukaryotic nucleus has either one or two complete copies of the chromosomal portion of a cell's genome. A nucleus with a single copy of each chromosome is called a **haploid**,[5] or $1n$, nucleus, and one with two sets of chromosomes is a **diploid**,[6] or $2n$, nucleus. Generally, each organism has a consistent number of chromosomes. For example, each haploid cell of the brewer's yeast *Saccharomyces cerevisiae*[7] (sak-ă-rō-mī´sēz se-ri-vis´ē-ī) has 16 chromosomes.

The cells of most fungi, many algae, and some protozoa are haploid, and the cells of most plants and animals and the remaining fungi, algae, and protozoa are diploid. Typically, gametes are haploid, and a zygote (formed from the union of gametes) is diploid.

A cell divides its nucleus so as to pass a copy of its chromosomal DNA to each of its descendants so that each new generation has the necessary genetic instructions to carry on life. There are two types of nuclear division—*mitosis* (mī-tō´sis) and *meiosis* (mī-ō´sis).

[1]From Greek *protos,* meaning "first," and *zoion,* meaning "animal."
[2]Plural of Latin *fungus,* meaning "mushroom."
[3]Plural of Latin *alga,* meaning "seaweed."
[4]From Greek *helmins,* meaning "worm."
[5]From Greek *haploos,* meaning "single."
[6]From Greek *diploos,* meaning "double."
[7]From Greek *sakcharon,* meaning "sugar," and *mykes,* meaning "fungus," and Latin *cerevisiae,* meaning "beer."

Mitosis Eukaryotic cells have two main stages in their life cycle: a stage called *interphase*,[8] during which the cells grow and eventually replicate their DNA, and a stage during which the cell's nucleus divides. In the type of nuclear division called **mitosis**,[9] which begins after the cell has duplicated its DNA such that there are two exact DNA copies (see Figure 7.6), the cell partitions its replicated DNA equally between two nuclei. Thus mitosis maintains the ploidy of the parent nucleus; that is, a haploid nucleus that undergoes mitosis forms two haploid nuclei, and a diploid nucleus that undergoes mitosis produces two diploid nuclei.

Mitosis has four phases: prophase,[10] metaphase,[11] anaphase,[12] and telophase.[13] The events of mitosis proceed as follows **(FIGURE 12.1a)**:

1. **Prophase.** The cell condenses its DNA molecules into visible threads called *chromatids* (krō´mă-tidz). Two identical chromatids, sister DNA molecules, are joined together in a region called a *centromere* to form one chromosome. Also during prophase, a set of microtubules is constructed in the cytosol to form a *spindle.* In most cells, the nuclear envelope disintegrates during prophase so that mitosis occurs freely in the cytosol; however, many fungi and some unicellular microbes (e.g., diatoms and dinoflagellates) maintain their nuclear envelopes so that mitosis occurs inside their nuclei.
2. **Metaphase.** The chromosomes line up on a plane in the middle of the cell and attach near their centromeres to microtubules of the spindle.
3. **Anaphase.** Sister chromatids separate and crawl along the microtubules toward opposite poles of the spindle. Each chromatid is now called a chromosome.
4. **Telophase.** The cell restores its chromosomes to their less compact, nonmitotic state, and nuclear envelopes form around the daughter nuclei. A cell may divide during telophase, but mitosis is nuclear division, not cell division.

Though certain specific events distinguish each of the four phases of mitosis, the phases are not discrete steps; that is, mitosis is a continuous process, and there are no clear boundaries between succeeding phases—one phase leads seamlessly to the next. For example, late anaphase and early telophase are indistinguishable.

Students sometimes confuse the terms *chromosome* and *chromatid,* in part because early microscopists used the word *chromosome* for two different things. During prophase and metaphase, a chromosome consists of two chromatids (DNA molecules) joined at a centromere. However, during anaphase and telophase, the chromatids separate, and each chromatid is then called a chromosome. In other words, a "chromosome" is a pair of chromatids during the first two phases, whereas "chromatid" and "chromosome" are synonymous terms during the latter two phases of mitosis.

[8]Latin, meaning "between phases."

[9]From Greek *mitos,* meaning "thread," after the threadlike appearance of chromosomes during nuclear division.

[10]From Greek *pro,* meaning "before," and *phasis,* meaning "appearance."

[11]From Greek *meta,* meaning "in the middle."

[12]From Greek *ana,* meaning "back."

[13]From Greek *telos,* meaning "end."

Meiosis In contrast to mitosis, **meiosis**[14] is nuclear division that involves the partitioning of chromatids into four nuclei such that each nucleus receives only half the original amount of DNA. Thus, diploid nuclei use meiosis to produce haploid daughter nuclei. Meiosis is a necessary condition for sexual reproduction (in which nuclei from two different cells fuse to form a single nucleus) because if cells lacked meiosis, each nuclear fusion to form a zygote would cause the number of chromosomes to double, and their number would soon become unmanageable.

Meiosis occurs in two stages known as *meiosis I* and *meiosis II* **(FIGURE 12.1b)**. As in mitosis, each stage has four phases, named prophase, metaphase, anaphase, and telophase. The events in meiosis as they occur in a diploid nucleus proceed as follows:

1. **Early prophase I** (prophase of meiosis I). As with mitosis, DNA replication during interphase has resulted in pairs of identical chromatids, forming chromosomes. But now an additional pairing occurs: *homologous chromosomes*—that is, chromosomes carrying similar or identical genetic sequences—line up side by side. Because these are prophase chromosomes, each of them consists of two identical chromatids; therefore, four DNA molecules are involved in this pairing. An aligned pair of homologous chromosomes is known as a *tetrad.*
2. **Late prophase I.** Once tetrads have formed, the homologous chromosomes exchange sections of DNA in a random fashion via a process called *crossing over.* This results in recombinations of their DNA. It is because of meiotic crossing over that the offspring produced by sexual reproduction have different genetic makeups from their siblings. Prophase I can last for days or longer.
3. **Metaphase I.** Tetrads align on a plane in the center of the cell and attach to spindle microtubules. Metaphase I differs from metaphase of mitosis in that homologous chromosomes remain as tetrads.
4. **Anaphase I.** Chromosomes of the tetrads move apart from one another; however, in contrast to mitotic anaphase, sister chromatids remain attached to one another.
5. **Telophase I.** The first stage of meiosis is completed as the spindle disintegrates. Typically, the cell divides at this phase to form two cells. Nuclear envelopes may form. Each daughter nucleus is haploid, though each haploid chromosome consists of two chromatids.
6. **Prophase II.** Nuclear envelopes disintegrate, and new spindles form.
7. **Metaphase II.** The chromosomes align in the middle of each cell and attach to microtubules of the spindles.
8. **Anaphase II.** Sister chromatids separate as in mitosis.
9. **Telophase II.** Daughter nuclei form. The cells divide, yielding four haploid cells.

[14]From Greek *meioun,* meaning "to make smaller."

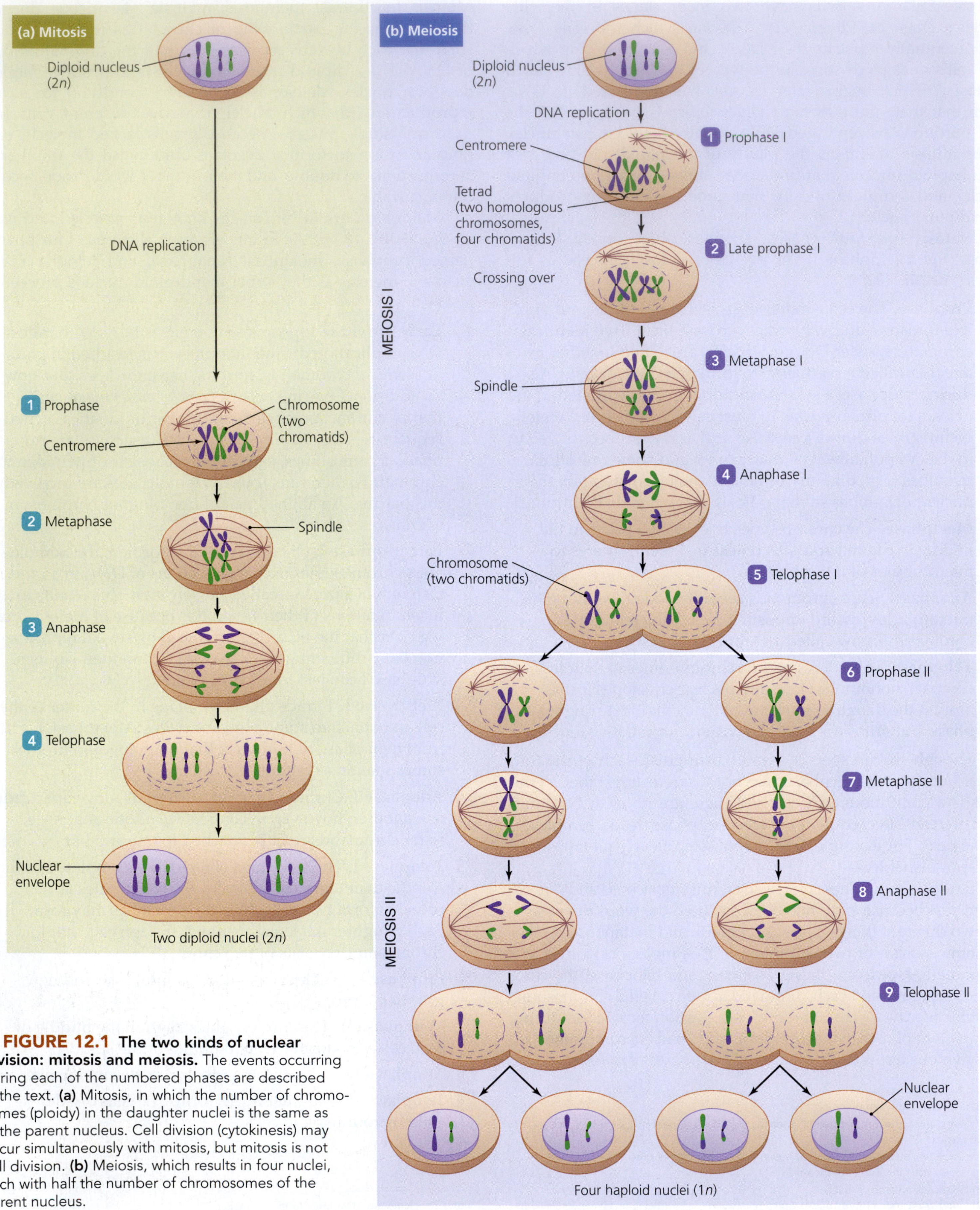

▲ **FIGURE 12.1 The two kinds of nuclear division: mitosis and meiosis.** The events occurring during each of the numbered phases are described in the text. **(a)** Mitosis, in which the number of chromosomes (ploidy) in the daughter nuclei is the same as in the parent nucleus. Cell division (cytokinesis) may occur simultaneously with mitosis, but mitosis is not cell division. **(b)** Meiosis, which results in four nuclei, each with half the number of chromosomes of the parent nucleus.

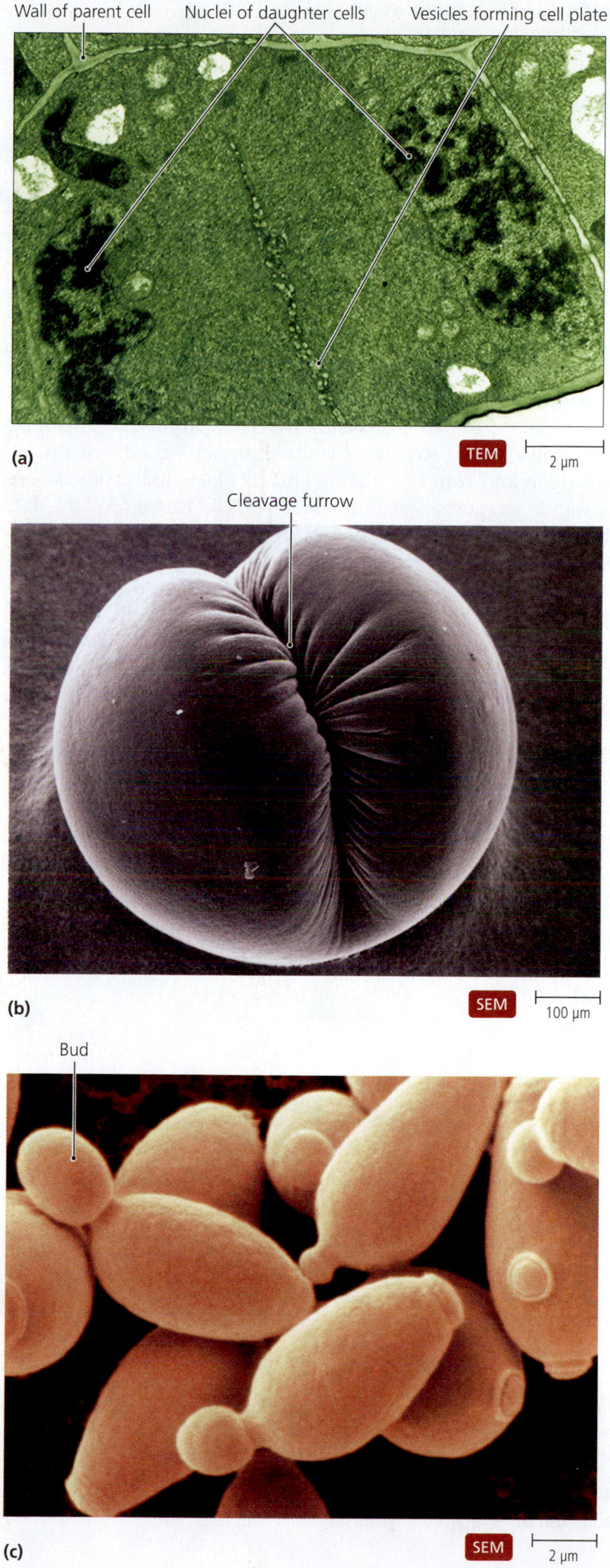

▶ **FIGURE 12.2 Different types of cytoplasmic division. (a)** Cytokinesis in a plant cell, in which vesicles form a cell plate. **(b)** Cytokinesis as it occurs in animals, protozoa, and some fungi. **(c)** Budding in yeast cells.

In summary, meiosis produces four haploid nuclei from a single diploid nucleus. Meiosis can be considered back-to-back mitoses without the DNA replication of interphase between them, though the four phases of meiosis I differ from those of mitosis. The phases of meiosis II are equivalent to those in mitosis. Additionally, crossing over during meiosis I produces genetic recombinations, ensuring that the chromosomes resulting from meiosis are different from the parental chromosomes. This provides genetic variety in the next generation. **TABLE 12.1** on p. 357 compares and contrasts mitotic and meiotic nuclear divisions.

Cytokinesis (Cytoplasmic Division)

Cytoplasmic division—also called **cytokinesis** (sī′tō-ki-nē′sis)—typically occurs simultaneously with telophase of mitosis, though in some algae and fungi it may be postponed or may not occur at all. In these cases, mitosis produces multinucleate cells called **coenocytes** (sē′nō-sītz).

In plant and algal cells, cytokinesis occurs as vesicles deposit wall material at the equatorial plane between nuclei to form a *cell plate*, which eventually becomes a transverse wall between daughter cells **(FIGURE 12.2a)**. Cytokinesis of protozoa and some fungal cells occurs when an equatorial ring of actin microfilaments contracts just below the cytoplasmic membrane, pinching the cell in two **(FIGURE 12.2b)**. Single-celled fungi called *yeasts* form a bud, which receives one of the daughter nuclei and pinches off from the parent cell **(FIGURE 12.2c)**.

Schizogony

Some protozoa, such as *Plasmodium* (plaz-mō′dē-ŭm)—the cause of malaria—reproduce asexually within red blood cells and liver cells via a special type of reproduction called **schizogony** (ski-zog′ō-nē; **FIGURE 12.3**). In schizogony, multiple mitoses form a multinucleate **schizont** (skiz′ont); only then does cytokinesis occur, simultaneously releasing numerous uninucleate daughter cells called *merozoites* (mer-ō-zō′ītz). The body of an infected host responds to the release of huge numbers of merozoites with the cyclic fever and chills characteristic of malaria.

Classification of Eukaryotic Organisms

LEARNING | OUTCOMES

12.5 Briefly describe the major groups of eukaryotes as they were first classified in the late 18th century and as they were classified in the late 20th century.

12.6 List some of the problems involved in the classification of protists in particular.

Historically, the classification of many eukaryotic microbes has been fraught with difficulty and characterized by change. Since

▶ FIGURE 12.3 Schizogony. Sequential mitoses without intervening cytokineses produce a multinucleate schizont, which later undergoes cytokinesis to produce many daughter cells.

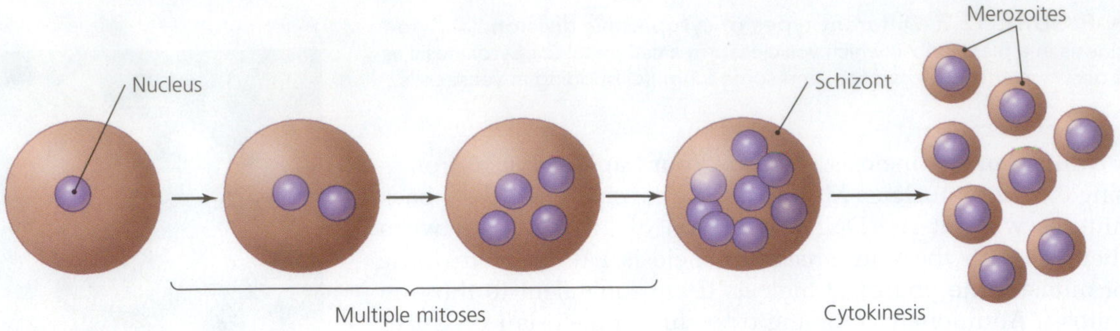

the late 18th century, when Carolus Linnaeus (1707–1778) began modern taxonomy, until near the end of the 20th century, taxonomists grouped organisms together largely according to readily observable structural traits. Linnaeus classified unicellular algae and fungi as plants, and he classified protozoa—as the name suggests—as animals (**FIGURE 12.4a**). By the late 20th century, taxonomists had placed fungi in their own kingdom and grouped protozoa and algae together within the kingdom Protista (**FIGURE 12.4b**), though some taxonomists kept the green algae in the kingdom Plantae.

This scheme was troublesome, in part because the Protista included both large, photosynthetic, multicellular kelps and nonphotosynthetic unicellular protozoa. Adding to the confusion is the fact that taxonomists who classify plants use the term *divisions* to refer to the same taxonomic level that zoologists call *phyla.*

More recently, many taxonomists have abandoned classification schemes that are so strongly grounded in large-scale structural similarities in favor of schemes based on similarities in nucleotide sequences and cellular ultrastructure as revealed by electron microscopy. One of the most evident results of such taxonomic studies is that modern schemes no longer include the taxa "Protozoa" or "Protista"; instead, such eukaryotic microbes belong in several kingdoms.

Though no one classification scheme has garnered universal support and more thorough understanding based on new information will almost certainly dictate changes, many taxonomists favor a scheme similar to the one shown in **FIGURE 12.4c**. In this scheme, on which the discussions of eukaryotic microbes in this chapter are largely based, the organisms we commonly refer to as protozoa are classified in six kingdoms: Parabasala, Diplomonadida, Euglenozoa, Alveolata, Rhizaria, and Amoebozoa; fungi are in the kingdom Fungi; algae are distributed among the kingdoms Stramenopila, Rhodophyta, and Plantae; water molds are in the kingdom Stramenopila; and slime molds are in the kingdom Amoebozoa. As we study the eukaryotic microbes discussed in this chapter, bear in mind that because the relationships

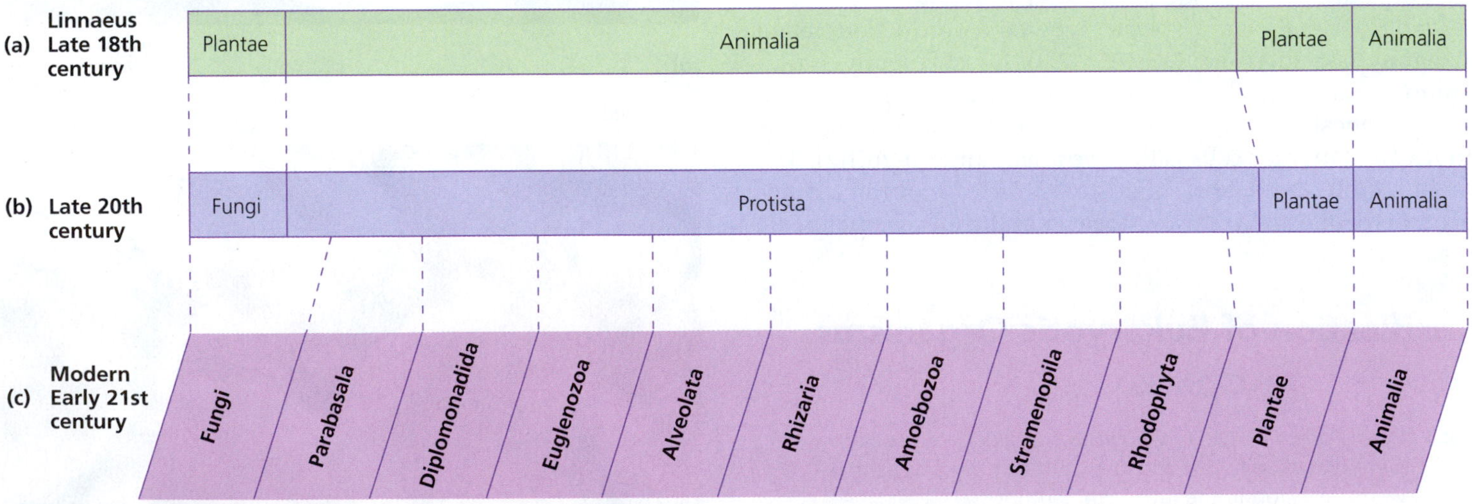

▲ FIGURE 12.4 The changing classification of eukaryotes over the centuries. **(a)** In the late 18th century, Linnaeus classified all organisms as either plants or animals. **(b)** In the late 20th century, taxonomists placed fungi in their own group and recognized a new kingdom, Protista. **(c)** Today, microbial eukaryotes are classified into numerous kingdoms based largely on their genetic relatedness. Though not all taxonomists would agree about every detail of this scheme, it forms the basis for the discussion of eukaryotic organisms in this chapter.

TABLE 12.1 Characteristics of the Two Types of Nuclear Division

	Mitosis	Meiosis
DNA replication	During interphase, before nuclear division	During interphase, before meiosis I begins
Phases	Prophase, metaphase, anaphase, telophase	Meiosis I—prophase I, metaphase I, anaphase I, telophase I Meiosis II—prophase II, metaphase II, anaphase II, telophase II
Formation of tetrads (alignment of homologous chromosomes)	Does not occur	Early in prophase I
Crossing over	Does not occur	Following formation of tetrads during prophase I
Number of accompanying cytoplasmic divisions that may occur	One	Two
Resulting nuclei	Two nuclei with same ploidy as the original	Four nuclei with half the ploidy of the original

among eukaryotic microbes are not fully understood, not all taxonomists agree with this scheme, and new information will shape future alterations in our understanding of the taxonomy of eukaryotic microbes.

We begin our survey of eukaryotic microbes with the group of organisms commonly known as protozoa.

TELL ME WHY

Why is it incorrect to call *mitosis* cell division?

Protozoa

LEARNING **OUTCOME**

12.7 List three characteristics shared by all protozoa.

The microorganisms called **protozoa** (prō-tō-zō´ă) are a diverse group defined by three characteristics: They are eukaryotic, are unicellular, and lack a cell wall. Note that "protozoa" is not a currently accepted taxon. With the exception of one subgroup (called apicomplexans), protozoa are motile by means of cilia, flagella, and/or pseudopodia. By these criteria, protozoa include a diverse assemblage of microbes. The scientific study of protozoa is *protozoology*, and scientists who study these microbes are *protozoologists*.

In the following sections we discuss the distribution, morphology, nutrition, reproduction, and classification of various groups of protozoa.

Distribution of Protozoa

Protozoa require moist environments; most species live worldwide in ponds, streams, lakes, and oceans, where they are critical members of the *plankton*—free-living, drifting organisms that form the basis of aquatic food chains. Other protozoa live in moist soil, beach sand, and decaying organic matter, and a very few are pathogens—that is, disease-causing microbes—of animals and humans.

Morphology of Protozoa

Though protozoa have most of the features of eukaryotic cells (discussed in Chapter 3, pp. 77–87 and illustrated in Figure 3.3), this group of eukaryotic microbes is characterized by great morphological diversity. Indeed, taxonomists once used the variety in locomotory structures as a basis for classification. Locomotory structures no longer figure prominently in the taxonomic classification of protozoa because the presence of a given structure may not indicate evolutionary relatedness.

Some ciliates have two nuclei: a larger *macronucleus*, which contains many copies of the genome (often more than $50n$) and controls metabolism, growth, and sexual reproduction, and a smaller *micronucleus*, which is involved in genetic recombination, sexual reproduction, and regeneration of macronuclei.

Protozoa also show variety in the number and kind of mitochondria they contain. Several groups lack mitochondria, whereas all the others have mitochondria with discoid or tubular cristae rather than the platelike cristae seen in animals, plants, fungi, and many algae. Additionally, some protozoa have *contractile vacuoles* that actively pump water from the cells, protecting them from osmotic lysis **(FIGURE 12.5)**.

All free-living aquatic and pathogenic protozoa exist as a motile feeding stage called a **trophozoite** (trof-ō-zō´ īt), and many have a hardy resting stage called a **cyst**, which is characterized by a thick capsule and a low metabolic rate. Cysts of protozoa are not reproductive structures because one trophozoite forms one cyst, which later becomes one trophozoite. Such cysts allow intestinal protozoa to pass from one host to another and to survive harsh environmental conditions such as desiccation, nutrient deficiency, extremes of pH and temperature, and lack of oxygen.

Nutrition of Protozoa

Most protozoa are chemoheterotrophic; that is, they obtain nutrients by phagocytizing bacteria, decayed organic matter, other protozoa, or the tissues of a host; a few protozoa absorb nutrients from the surrounding water. Because the protozoa called dinoflagellates and euglenids (discussed shortly) are

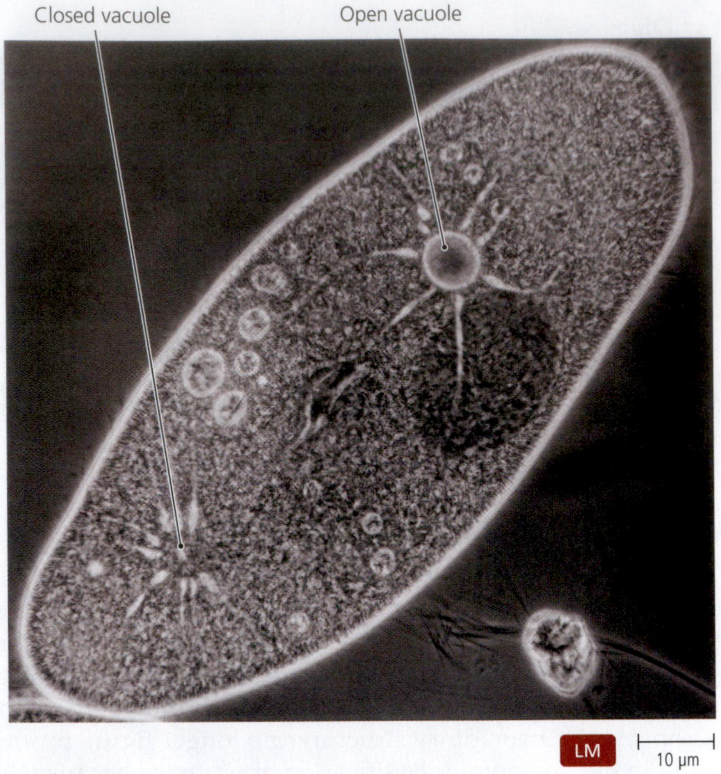

▲ **FIGURE 12.5 Contractile vacuoles.** Many protozoa, such as *Paramecium*, have this prominent feature. Open vacuoles fill with water that entered the cell via osmosis; closed vacuoles are contracted, pumping the water out of the cell. *Is the environment in this case hypertonic, hypotonic, or isotonic to the cell? Explain.*

Figure 12.5 *The environment is hypotonic to the cell; water moves down its concentration gradient—into the cell—by osmosis.*

photoautotrophic, botanists historically classified them as algal plants rather than as protozoa.

Reproduction of Protozoa

Most protozoa reproduce asexually only, by binary fission or schizogony; a few protozoa also have sexual reproduction in which two individuals exchange genetic material. Some sexually reproducing protozoa become **gametocytes** (gametes) that fuse with one another to form a diploid **zygote**. Ciliates, such as *Paramecium* (par-ă-mē´sē-ŭm), reproduce sexually via a complex process called *conjugation* **(FIGURE 12.6)**, which involves the coupling of two compatible mating cells 1, meiosis of diploid micronuclei 2, loss of some haploid micronuclei 3, exchange of micronuclei between the coupled cells 4, uncoupling of the cells 5, fusion of haploid micronuclei to form a diploid micronucleus 6, three mitoses of the micronucleus to form eight micronuclei 7, disintegration of the macronucleus and the subsequent formation of a new macronucleus from four micronuclei 8, and three cytokineses to produce four daughter cells, each with one macronucleus and one micronucleus 9.

Classification of Protozoa

LEARNING | **OUTCOMES**

12.8 Discuss the reasons for the many different taxonomic schemes for protozoa.

12.9 Identify several features of a typical euglenid.

12.10 Compare and contrast three types of alveolate.

12.11 Compare and contrast three types of amoeba.

12.12 Describe the life cycles of plasmodial and cellular slime molds.

12.13 Describe characteristic features of parabasalids, diplomonads, rhizaria, and amoebozoa.

As we have seen, over two centuries ago Linnaeus classified protozoa as animals; later taxonomists grouped protozoa into kingdom Protista. Furthermore, some taxonomists divided the protozoa into four groups based on the organisms' mode of locomotion: Sarcodina (motile by means of pseudopods), Mastigophora (flagella), Ciliophora (cilia), and Sporozoa (nonmotile). Other taxonomists lumped the first two groups together into a single group called Sarcomastigophora. Grouping of the protozoa according to locomotory features is still in common usage for many practical applications.

Taxonomists today recognize that these schemes do not reflect genetic relationships either between protozoa and other organisms or among protozoa. Accordingly, taxonomists continue to revise and refine the classification of protozoa based on 18S rRNA nucleotide sequencing and features made visible by electron microscopy. One such genetic scheme classifies protozoa into the six taxa Parabasala through Amoebozoa shown in Figure 12.4c, which different taxonomists consider kingdoms (as here), subkingdoms, or phyla.

In the following sections we will briefly discuss members of these six taxa of protozoa, formed largely according to similarities in nucleotide sequences and ultrastructure. We begin with parabasalids.

Parabasala

Parabasalids lack mitochondria, but each has a single nucleus and a *parabasal body*, which is a Golgi body–like structure. *Trichonympha* (trik-ō-nimf´ă), a parabasalid with numerous flagella **(FIGURE 12.7)**, inhabits the guts of termites, where it assists in the digestion of wood. Another well-known parabasalid is *Trichomonas* (trik-ō-mō´nas, see p. 776), which lives in the human vagina. When the normally acidic pH of the vagina is raised, *Trichomonas* proliferates and causes severe inflammation that can lead to sterility. It is spread by sexual intercourse and is usually asymptomatic in males.

Diplomonadida

Because members of the group Diplomonadida[15] lack mitochondria, Golgi bodies, and peroxisomes, biologists once thought

[15]From Greek *diploos*, meaning "double," and *monas*, meaning "unit," referring to two nuclei.

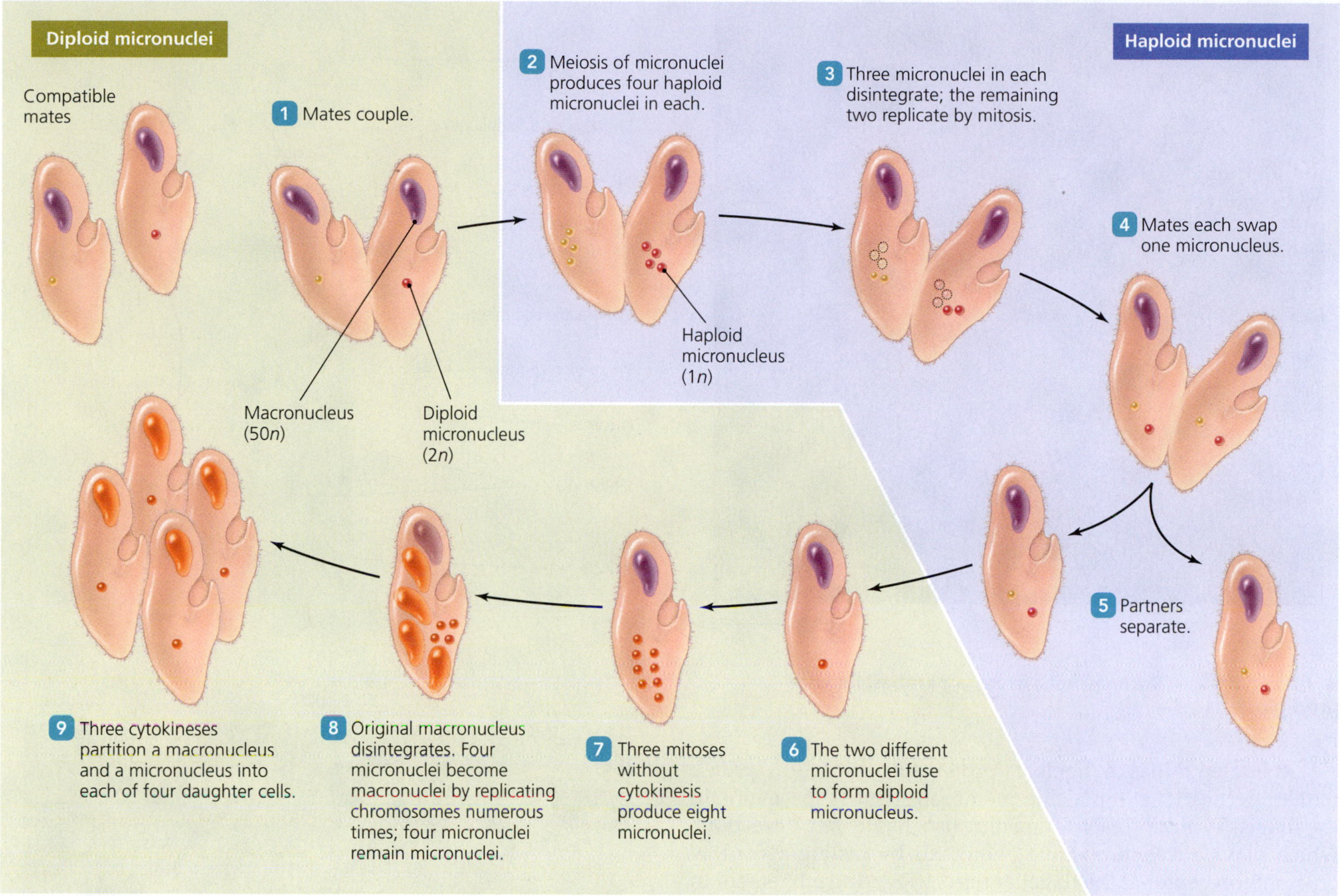

▲ **FIGURE 12.6 Sexual reproduction via conjugation in ciliates.** Shown here for *Paramecium*. Cells with haploid micronuclei are shown against a blue background; those with diploid micronuclei, against a green background.

these organisms were descended from ancient eukaryotes that had not yet phagocytized the prokaryotic ancestors of mitochondria. More recently, however, geneticists have discovered rudimentary *mitosomes* in the cytoplasm and mitochondrial genes in the nuclear chromosomes, a finding that suggests that diplomonads might be descended from typical eukaryotes that somehow lost their organelles.

Diplomonads have two equal-sized nuclei and multiple flagella. A prominent example is *Giardia* (jē-ar´dē-ă), a diarrhea-causing pathogen of animals and humans that is spread to new hosts when they ingest resistant *Giardia* cysts (see Disease in Depth: Giardiasis, pp. 740–741).

Euglenozoa

Part of the reason that taxonomists established the kingdom Protista in the 1960s was to create a "dumping ground" for *euglenids*, eukaryotic microbes that share certain characteristics of both plants and animals. More recently, based on similar 18S rRNA sequences, the presence of a crystalline rod of unknown function in the flagella, and the presence of mitochondria with disk-shaped cristae, some taxonomists have created a new taxon: kingdom Euglenozoa. The euglenozoa include euglenids and some flagellated protozoa called *kinetoplastids*.

Euglenids The group of euglenozoa called **euglenids**, which are named for the genus *Euglena* (yū-glēn´ă; **FIGURE 12.8a**), are photoautotrophic, unicellular microbes with chloroplasts containing light-absorbing pigments—chlorophylls *a* and *b* and carotene. For this reason, botanists historically classified euglenids in the kingdom Plantae. However, one reason for not including euglenids with plants is that euglenids store food as a unique polysaccharide called *paramylon* instead of as starch. Euglenids are similar to animals in that they lack cell walls, have flagella, are chemoheterotrophic phagocytes (in the dark), and move by using their flagella as well as by flowing, contracting, and expanding their cytoplasm. Such a squirming movement, which is similar to amoeboid movement but does not involve pseudopods, is called *euglenoid movement*.

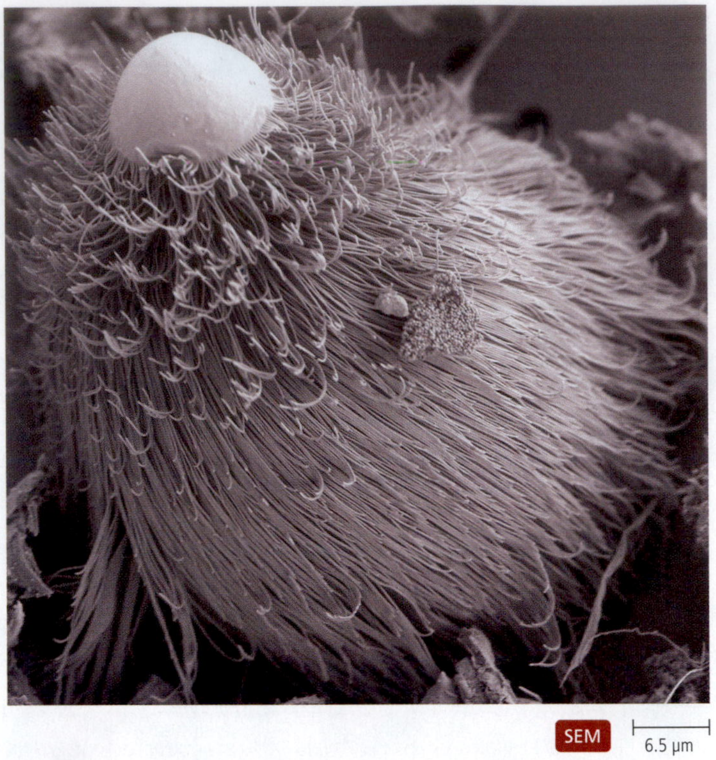

▲ **FIGURE 12.7** ***Trichonympha acuta*, a parabasalid with prodigious flagella.**

A euglenid has a flexible, proteinaceous, helical *pellicle* that underlies its cytoplasmic membrane and helps maintain its shape. Typically each euglenid also has a red "eyespot," which plays a role in positive phototaxis by casting a shadow on a photoreceptor at the flagellar base, triggering movement in that direction. Euglenids reproduce by mitosis followed by longitudinal cytokinesis. They form cysts when exposed to harsh conditions.

Kinetoplastids Euglenozoa called **kinetoplastids (FIGURE 12.8b)** each have a single large mitochondrion that contains a unique region of mitochondrial DNA called a *kinetoplast*. As in all mitochondria, this DNA codes for some mitochondrial polypeptides.

Kinetoplastids live inside animals, and some are pathogenic. Among the latter are the genera *Trypanosoma* (tri-pan´ō-sō-mă) and *Leishmania* (lēsh-man´ē-ă), certain species of which cause potentially fatal diseases of mammals, including humans (see Chapters 19–21).

Alveolates

Alveolates (al-vē´ō-lātz) are protozoa with small membrane-bound cavities called alveoli[16] (al-vē´ō-lī) beneath their cell surfaces **(FIGURE 12.9)**. Scientists do not know the purpose of alveoli. Alveolates share at least one other characteristic—tubular mitochondrial cristae. This group is further divided into three subgroups: *ciliates, apicomplexans,* and *dinoflagellates.*

[16]Latin, meaning "small hollows."

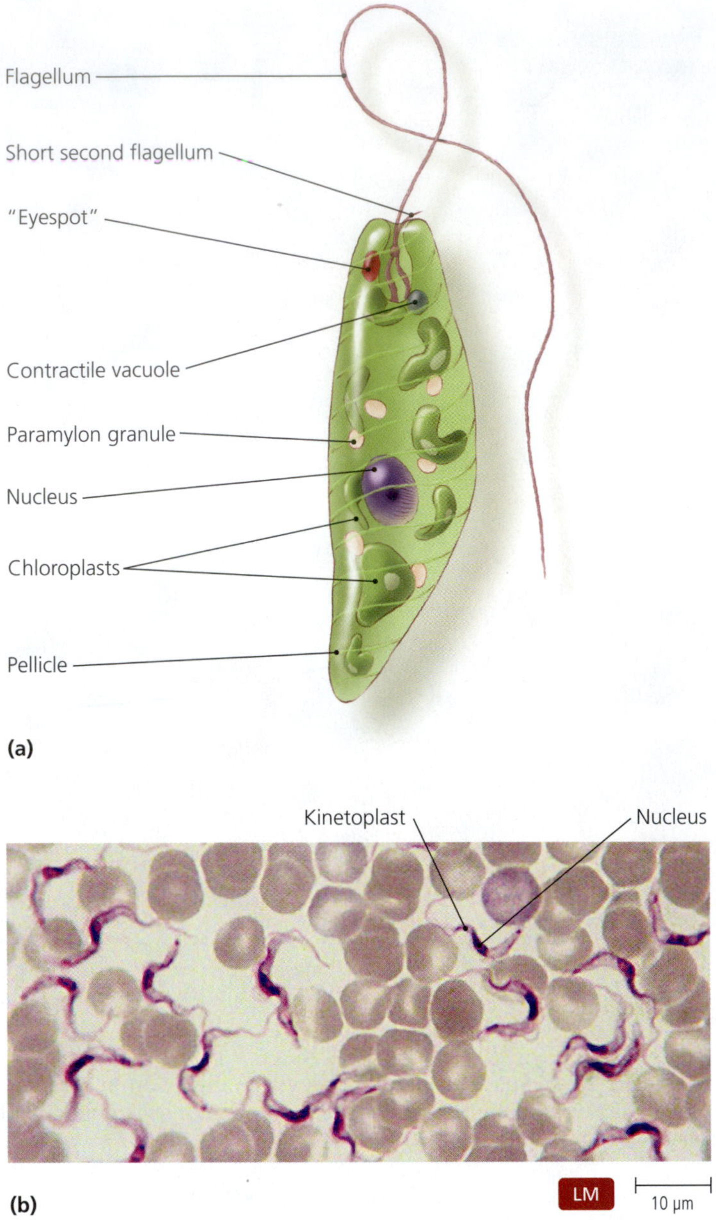

▲ **FIGURE 12.8 Two representatives of the kingdom Euglenozoa. (a)** The euglenid *Euglena*. Euglenids have characteristics that are similar to both plants and animals. **(b)** The kinetoplastid *Trypanosoma*. *What is the function of the kinetoplast in* Trypanosoma?

Figure 12.8 *A kinetoplast is mitochondrial DNA, which codes for some mitochondrial polypeptides.*

Ciliates As their name indicates, **ciliate** (sil´ē-āt) alveolates have cilia by which they either move themselves or move water past their cell surfaces. (The structure and function of cilia are discussed on pp. 79–80). Some ciliates are covered with cilia, whereas others have only a few isolated tufts. All ciliates are chemoheterotrophs and have two nuclei—one macronucleus and one micronucleus. Some taxonomists consider them the sole members of phylum Ciliophora.

Notable ciliates include *Vorticella* (vōr-ti-sel´ă), whose apical cilia create a whirlpool-like current to direct food into its

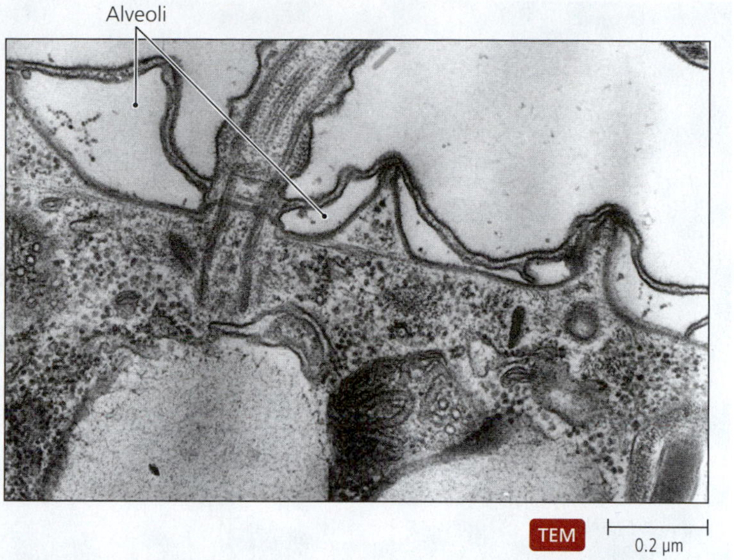

▲ **FIGURE 12.9 Membrane-bound alveoli found in some protozoa.** Even though its function is not yet known, the alveolus is present in eukaryotic microbes with similar 18S rRNA sequences, indicating genetic relatedness and forming the basis of the group of eukaryotes called alveolates.

▲ **FIGURE 12.10 A predatory ciliate, *Didinium* (on left), devouring another ciliate, *Paramecium*.**

"mouth"; *Balantidium* (bal-an-tid´ē-ŭm), which is the only ciliate pathogenic to humans; and the carnivorous *Didinium* (dī-di´nē-ŭm), which phagocytizes other protozoa, such as the well-known pond-water ciliate *Paramecium* **(FIGURE 12.10)**. (Chapter 23 discusses *Balantidium* and its disease in more detail.)

Apicomplexans The alveolates called **apicomplexans** (ap-i-kom-plek´sănz) are all chemoheterotrophic pathogens of animals. The name of this group refers to the *complex* of special intracellular organelles, located at the *apices* of the infective stages of these microbes, that enables them to penetrate host cells. Examples of apicomplexans are *Plasmodium, Cryptosporidium* (krip-tō-spō-rid´-ē-ŭm), and *Toxoplasma* (tok-sō-plaz´mă), which cause malaria, cryptosporidiosis, and toxoplasmosis, respectively. (Chapters 21 and 23 consider representative apicomplexans and the diseases they cause in more detail.)

Dinoflagellates The group of alveolates called **dinoflagellates** are unicellular microbes that have photosynthetic pigments, such as carotene and chlorophylls *a*, c_1, and c_2. Like many plants and algae, their food reserves are starch and oil, and their cells are often strengthened by internal plates of cellulose. Even though botanists have historically classified the dinoflagellates as algae because dinoflagellates are photoautotrophic, taxonomists today note that their 18S rRNA sequences and the presence of alveoli indicate that dinoflagellates are more closely related to ciliates and apicomplexans than they are to either plants or algae. Interestingly, unlike other eukaryotic chromosomes, dinoflagellate chromosomes lack histone proteins.

Dinoflagellates make up a large proportion of freshwater and marine plankton. Motile dinoflagellates have two flagella of unequal length **(FIGURE 12.11)**. The transverse flagellum wraps around the equator of the cell in a groove in the cell wall, and its beat causes the cell to spin; the second flagellum extends posteriorly and propels the cell forward.

Many dinoflagellates are bioluminescent—that is, able to produce light via metabolic reactions. When luminescent dinoflagellates are present in large numbers, the ocean water lights up with every crashing wave, passing ship, or jumping fish.

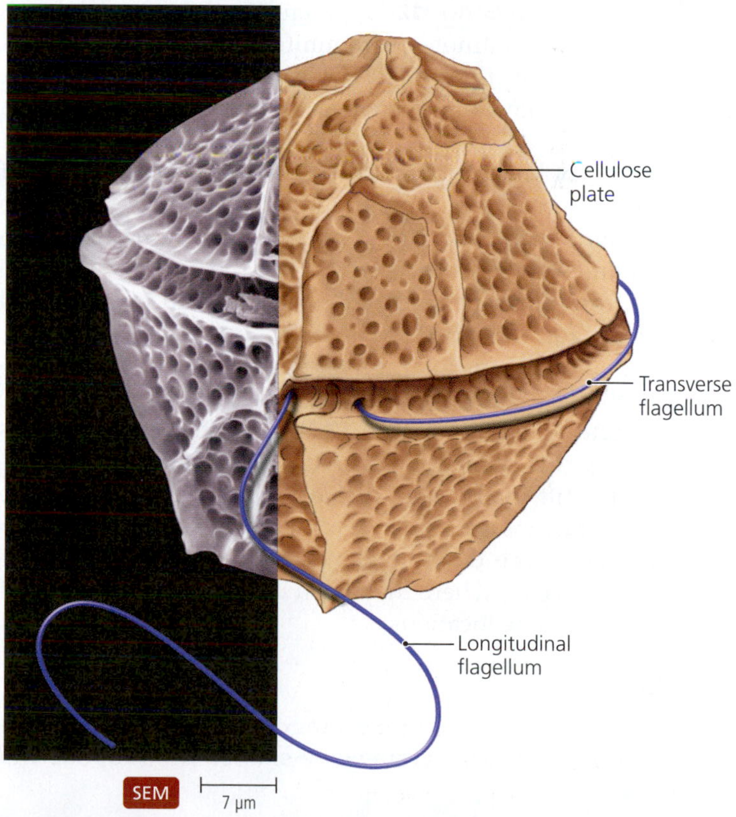

▲ **FIGURE 12.11 *Gonyaulax*, a motile armored dinoflagellate.** Dinoflagellates have two flagella. The transverse flagellum spins the cell; the longitudinal flagellum propels the cell forward.

Other dinoflagellates produce a red pigment, and their abundance in marine water is one cause of a phenomenon called a **red tide**.

Some dinoflagellates, such as *Gymnodinium* (jīm-nō-din′ē-um) and *Gonyaulax* (gon-ē-aw′laks), produce *neurotoxins*—poisons that can act against the human nervous system. Humans can become exposed when they eat shellfish that have ingested planktonic dinoflagellates and concentrated their toxins.

The neurotoxin of another dinoflagellate, *Pfiesteria*[17] (fes-tēr′ē-ă), may be even more potent: It has been claimed that the toxin poisons people who merely handle infected fish or breathe air laden with the microbes, resulting in memory loss, confusion, headache, respiratory difficulties, skin rash, muscle cramps, diarrhea, nausea, and vomiting. The Centers for Disease Control and Prevention (CDC) calls such poisoning *possible estuary-associated syndrome (PEAS).*

SEM 0.25 mm

▲ **FIGURE 12.12 Rhizaria called foraminifera have multichambered, snail-like shells of calcium carbonate.** Pseudopods are not visible.

Rhizaria

Unicellular eukaryotes called **amoebae**[18] are protozoa that move and feed by means of pseudopods (see Figure 3.30). Beyond this common feature and the fact that they all reproduce via binary fission, amoebae exhibit little uniformity. Some taxonomists currently classify amoebae into two kingdoms: Rhizaria and Amoebozoa. We consider them in order.

Rhizaria is a group of amoebae with threadlike pseudopods. A major taxon is composed of armored marine amoebae known as **foraminifera**. A foraminiferan has a porous shell composed of calcium carbonate arranged on an organic matrix in a snail-like manner **(FIGURE 12.12)**. Pseudopods extend through holes in the shell. Commonly, foraminifera live attached to sand grains on the ocean floor. Most foraminifera are microscopic, though scientists have discovered species several centimeters in diameter.

Over 90% of known foraminifera are fossil species, some of which form layers of limestone hundreds of meters thick. The great pyramids of Giza outside Cairo, Egypt, are built of foraminiferan limestone. Geologists correlate the ages of sedimentary rocks from different parts of the world by finding identical foraminiferan fossils embedded in them.

Amoebae called **radiolaria** make up another group of rhizaria, but they have ornate shells composed of silica (SiO_2, the mineral found in opal) **(FIGURE 12.13)** and live unattached as part of the marine plankton. Radiolarians reinforce their pseudopods with stiff internal bundles of microtubules so that the pseudopods radiate from the central body like spokes of a spherical wheel. The dead bodies of radiolarians settle to the bottom of the ocean, where they form ooze that is hundreds of meters thick in some locations.

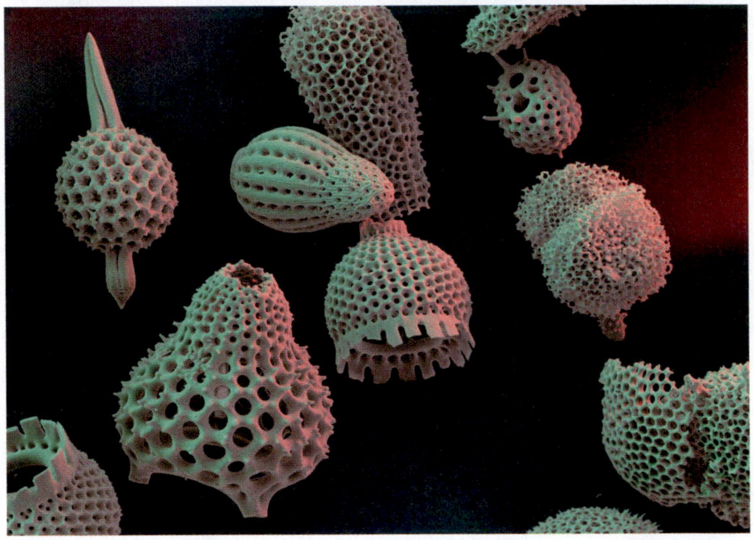

SEM 0.4 mm

▲ **FIGURE 12.13 Rhizaria called radiolarians have ornate shells of silica.** Pseudopods (not present in these dead specimens) extend through the holes.

Amoebozoa

Amoebozoa constitute a second kingdom of amoebae distinguished from rhizaria by having lobe-shaped pseudopods and no shells. Amoebozoa include the normally free-living amoebae *Naegleria* (nā-glē′rē-ă) and *Acanthamoeba* (ă-kan-thă-mē′bă), which can each cause diseases of the eyes or brains of humans and animals that swim in water containing them. Other amoebozoa, such as *Entamoeba* (ent-ă-mē′bă), always live inside animals, where they produce potentially fatal amebic dysentery. (Chapters 20 and 23 examine these pathogenic amoebae in more detail.)

[17]Named for dinoflagellate biologist Lois Pfiester.

[18]From Greek *ameibein*, meaning "to change."

Taxonomists formerly considered another group of amoebozoa—**slime molds**—to be fungi, but the lobe-shaped pseudopods by which they feed and move as well as their nucleotide sequences show that they are amoebozoa. Scientists have identified two types of slime molds: *plasmodial molds* and *cellular slime molds.*

Slime molds differ from true fungi in two main ways:

- They lack cell walls, more closely resembling the amoebae in this regard.
- They are phagocytic rather than absorptive in their nutrition.

Species in the two groups of slime molds differ based on their morphology, reproduction, and 18S rRNA sequences. Slime molds are important to humans primarily as excellent laboratory systems for the study of developmental and molecular biology.

Plasmodial (Acellular) Slime Molds **Plasmodial slime molds**, also known as *acellular slime molds* (e.g., *Physarum,* fi-sar´um), exist as streaming, coenocytic, colorful filaments of cytoplasm that creep as amoebae through forest litter, feeding by phagocytizing organic debris and bacteria. The body, called a *plasmodium,* may contain millions of diploid nuclei and cover many square centimeters (**FIGURE 12.14a**). Nutrients are distributed throughout the plasmodium by cytoplasmic streaming.

When food or water is in short supply, the plasmodium divides into individual masses of cytoplasm, each of which produces a stalked sporangium. Meiosis occurs within the sporangia to generate haploid spores. These spores germinate to produce *myxamoebae,* which look and act like other unicellular amoebae; in the presence of water, however, myxamoebae produce flagella and swim about (not shown). When the water disappears, they become amoeboid again.

Compatible myxamoebae of opposite mating types fuse to form a diploid zygote. The nucleus of the zygote undergoes numerous mitoses—without cytokineses—to form a new coenocytic plasmodium.

Cellular Slime Molds **Cellular slime molds,** such as *Dictyostelium* (dik-tē-ō-stē´lē-um), exist as individual haploid myxamoebae that phagocytize bacteria, yeasts, dung, and decaying vegetation. **FIGURE 12.14b** illustrates their life cycle, all of which is haploid—there is no diploid phase. Myxamoebae reproduce by mitosis and cytokinesis when food is abundant; however, in scarcity, some secrete cyclic adenosine monophosphate (cAMP), which acts as a chemotactic attractant for other myxamoebae. The myxamoebae congregate into a sluglike *pseudoplasmodium,* which can migrate for several days. Unlike the true plasmodium of acellular slime molds, the cells of a pseudoplasmodial slug retain their individuality and can be separated mechanically.

Some cells of a pseudoplasmodium form a stalked sporangium; the remaining cells climb the stalk and become spores. In contrast to the spores of plasmodial slime molds, the spores of cellular slime molds do not result from meiosis and are not enclosed in a common wall.

In summary, protozoa are a heterogeneous collection of single-celled, mostly chemoheterotrophic organisms that lack cell walls. Some taxonomists classify them in Parabasala, Diplomonadida, Euglenozoa, Alveolata, Rhizaria, and Amoebozoa, though their relationships with one another and with other eukaryotic organisms are still unclear. **TABLE 12.2** on p. 365 summarizes the incredible diversity of these microbes.

We next turn our attention to another group of chemoheterotrophs, the fungi, which differ from the protozoa chiefly in that fungi have cell walls.

TELL ME WHY

Why did early taxonomists categorize such obviously different microorganisms as parabasalids, diplomonads, euglenozoa, alveolates, rhizaria, and amoebozoa in a single taxon, Protozoa?

Fungi

LEARNING **OUTCOME**

12.14 Cite at least three characteristics that distinguish fungi from other groups of eukaryotes.

Organisms in the kingdom **Fungi** (fŭn´jī), such as molds, mushrooms, and yeasts, are like most protozoa in that they are chemoheterotrophic; however, unlike protozoa, they have cell walls, which typically are composed of a strong, flexible, nitrogenous polysaccharide called **chitin**. (The chitin in fungi is chemically identical to that in the exoskeletons of insects and other arthropods, such as grasshoppers, lobsters, and crabs.) Fungi differ from plants in that they lack chlorophyll and do not perform photosynthesis; they differ from animals by having cell walls, although genetic sequencing of fungal and animal genomes has shown that fungi and animals are related. The study of fungi is *mycology,*[19] and scientists who study fungi are *mycologists.*

The Significance of Fungi

LEARNING **OUTCOME**

12.15 List five ways in which fungi are beneficial.

Fungi are extremely beneficial microorganisms. In nature, they decompose dead organisms (particularly plants) and recycle their nutrients. Additionally, the roots of about 90% of vascular plants form *mycorrhizae,*[20] which are beneficial associations between roots and fungi that assist the plants to absorb water and dissolved minerals.

Humans use fungi for food (mushrooms and truffles), in religious ceremonies (because of their hallucinogenic properties), and in the manufacture of foods and beverages, including bread, alcoholic beverages, citric acid (the basis of the soft drink

[19]From Greek *mykes,* meaning "mushroom," and *logos,* meaning "discourse."
[20]From Greek *rhiza,* meaning "root."

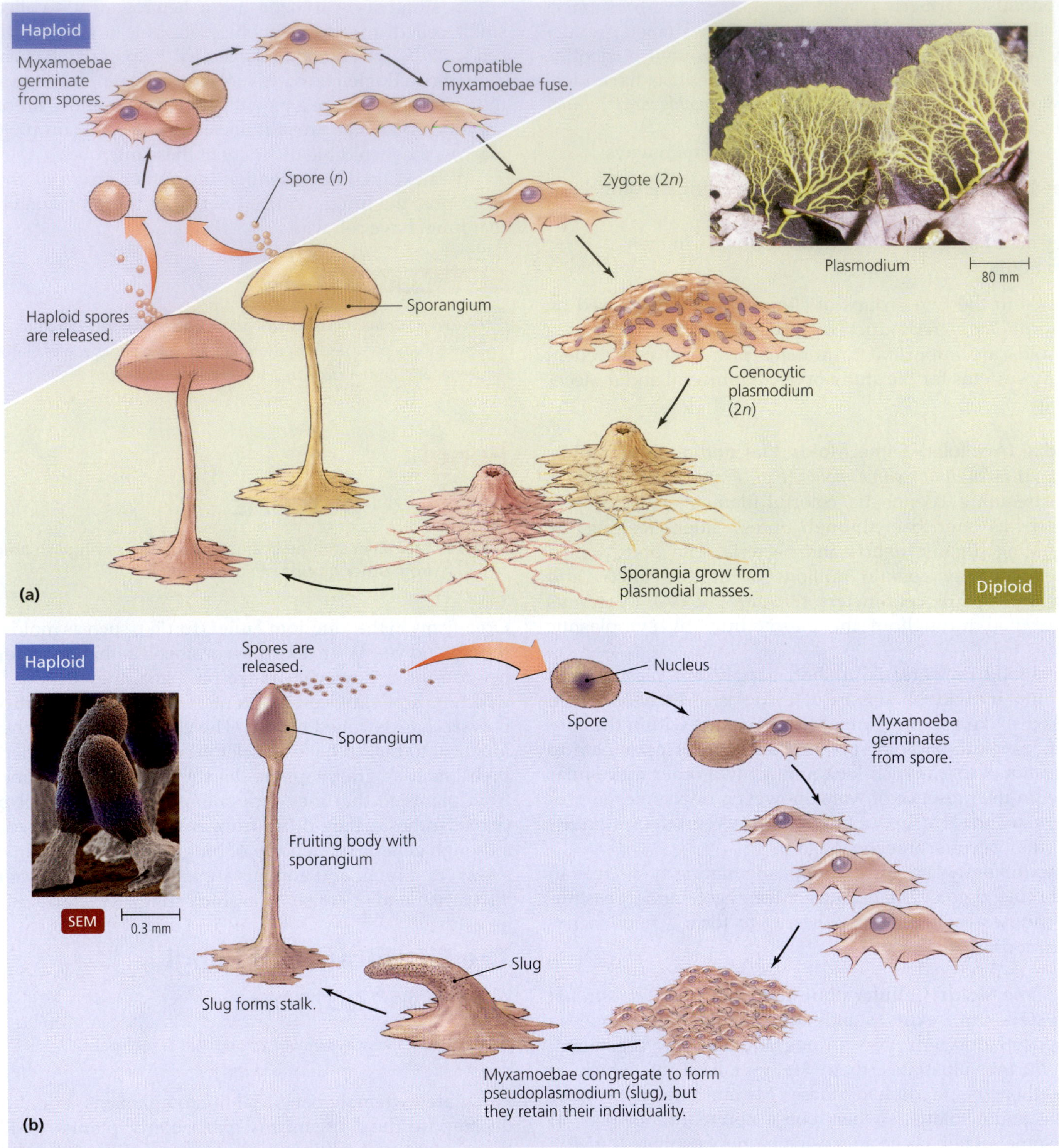

▲ **FIGURE 12.14 Life cycles of slime molds. (a)** A plasmodial (acellular) slime mold. Myxamoebae fuse to form a coenocytic plasmodium that feeds on organic debris and bacteria as the plasmodium creeps through its environment. Under adverse conditions, it produces sporangia that undergo meiosis to produce haploid spores. Compatible spores fuse to form a zygote that undergoes multiple mitoses, but not cytokinesis, to form a new plasmodium. **(b)** A cellular slime mold. Individual myxamoebae congregate during times of starvation to form a multicellular slug, which produces a stalked sporangium. Myxamoebae retain their individuality. The sporangium releases spores that germinate when conditions are more favorable. All phases are haploid.

TABLE 12.2 Characteristics of Protozoa

Category	Distinguishing Features	Representative Genera Mentioned in the Text
Parabasala	Parabasal body; single nucleus; lack mitochondria	*Trichomonas*
Diplomonadida	Two equal-sized nuclei; lack mitochondria, Golgi bodies, and peroxisomes	
Diplomonads	Multiple flagella	*Giardia*
Euglenozoa	Flagella with internal crystalline rod; disk-shaped mitochondrial cristae	
Euglenids	Photosynthesis; pellicle; "eyespot"	*Euglena*
Kinetoplastids	Single mitochondrion with DNA localized in kinetoplast	*Trypanosoma, Leishmania*
Alveolates	Alveoli (membrane-bound cavities underlying the cytoplasmic membrane); tubular cristae in mitochondria	
Ciliates	Cilia	*Balantidium, Paramecium, Didinium*
Apicomplexans	Apical complex of organelles	*Plasmodium, Cryptosporidium, Toxoplasma*
Dinoflagellates	Photosynthesis; two flagella; internal cellulose plates	*Gymnodinium, Gonyaulax, Pfiesteria*
Rhizaria	Threadlike pseudopods	
Foraminifera	Shells of calcium carbonate	
Radiolarians	Threadlike pseudopods, shells of silica	
Amoebozoa	Lobe-shaped pseudopods; no shells	
Free-living and parasitic forms	Do not form aggregates	*Naegleria, Acanthamoeba, Entamoeba*
Plasmodial (acellular) slime molds	Multinucleate body called plasmodium	*Physarum*
Cellular slime molds	Cells aggregate to form pseudoplasmodium but retain individual nature	*Dictyostelium*

industry), soy sauce, and some cheeses. Fungi also produce antibiotics, such as penicillin and cephalosporin; the immunosuppressive drug *cyclosporine,* which makes organ transplants possible; and *mevinic acids,* which are cholesterol-reducing agents.

Fungi are also important research tools in the study of metabolism, growth, and development and in genetics and biotechnology. For instance, based on their work with *Neurospora* (noo-ros´pōr-ă) in the 1950s, George Beadle (1903–1989) and Edward Tatum (1909–1975) developed their Nobel Prize–winning theory that one gene codes for one enzyme. Because of similar research, *Saccharomyces* (brewer's yeast) is the best-understood eukaryote and the first eukaryote to have its entire genome sequenced. (Chapter 25 highlights some uses of fungi in agriculture and industry.)

Not all fungi are beneficial—about 30% of known fungal species produce **mycoses** (mī-kō´sēz), which are fungal diseases of plants, animals, and humans. For example, Dutch elm disease is a mycosis of elm trees, and athlete's foot is a fungal disease of humans. Because fungi tolerate concentrations of salt, acid, and sugar that inhibit bacteria, fungi are responsible for the spoilage of fruit, pickles, jams, and jellies exposed to air.

In the following sections we will consider the basic characteristics of fungal morphology, nutrition, and reproduction before turning to a brief survey of the major groups of fungi.

Morphology of Fungi

LEARNING OUTCOME

12.16 Distinguish among septate hyphae, aseptate hyphae, and mycelia.

Fungi have two basic body shapes. The bodies of *molds* are relatively large and composed of long, branched, tubular filaments called **hyphae**.[21] Hyphae are either **septate** (divided into cells by cross walls called *septa*[22]; **FIGURE 12.15a**) or **aseptate** (not divided by septa; **FIGURE 12.15b**). Aseptate hyphae are *coenocytic* (multinucleate). The bodies of *yeasts* are typically small, globular, and composed of a single cell, which may have buds (**FIGURE 12.15c**).

In response to environmental conditions such as temperature or carbon dioxide concentration, some fungi produce both yeastlike and moldlike shapes (**FIGURE 12.15d**); fungi that produce two types of body shapes are said to be **dimorphic** (which means "two-shaped"). Many medically important fungi are thermally dimorphic; that is, they change growth habits in response to the temperature in their immediate vicinity. Such

[21]From Greek *hyphe,* meaning "weaving" or "web."
[22]Latin, meaning "partitions" or "fences."

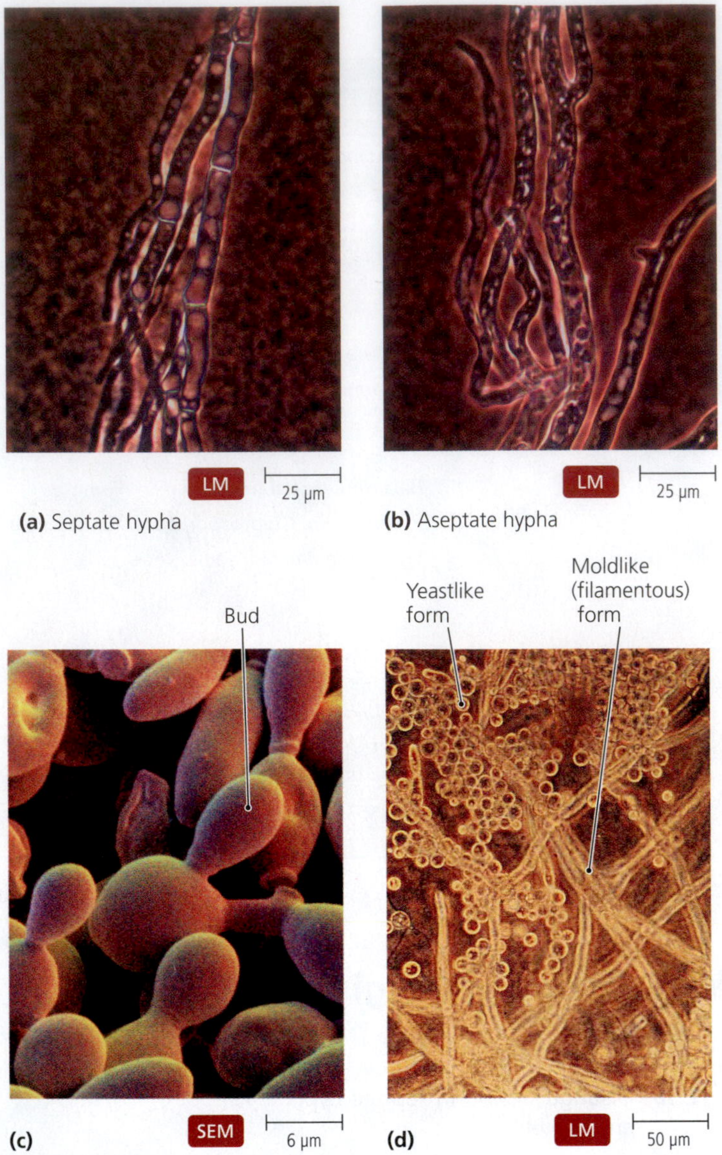

▲ **FIGURE 12.15 Fungal morphology.** **(a)** Septate hyphae have cross walls, **(b)** aseptate hyphae do not. **(c)** Cells of *Saccharomyces* (baker's or brewer's yeast), which are unicellular and spherical to irregularly oval in shape. **(d)** The bodies of a dimorphic fungus, *Mucor rouxii*, showing both yeastlike and moldlike growth in response to environmental conditions.

fungi include *Histoplasma capsulatum* (his-tō-plaz´mă kap-soo-lā´tŭm), which causes a respiratory disease called histoplasmosis, and *Coccidioides immitis* (kok-sid-ē-oy´dēz im´mi-tis), which causes a flulike disease called coccidioidomycosis (kok-sid-ē-oy´dō-mī-kō´sis). Generally, the yeast form of a dimorphic fungal pathogen causes disease, whereas the filamentous form does not.

The body of a mold is composed of hyphae intertwined to form a tangled mass called a **mycelium** (plural: *mycelia;* **FIGURE 12.16**). Mycelia are typically subterranean and thus usually escape our notice, though they can be very large. In fact, the largest known organisms on Earth are fungi in the genus *Armillaria,* the mycelia of which can spread through thousands of acres of forest to a depth of several feet and weigh many hundreds of tons. (In contrast, blue whales, the largest living animals, weigh only about 150 tons.) *Fruiting bodies,* such as puffballs and mushrooms, are the hyphae of the reproductive structures of molds and are only small visible extensions of vast underground mycelia.

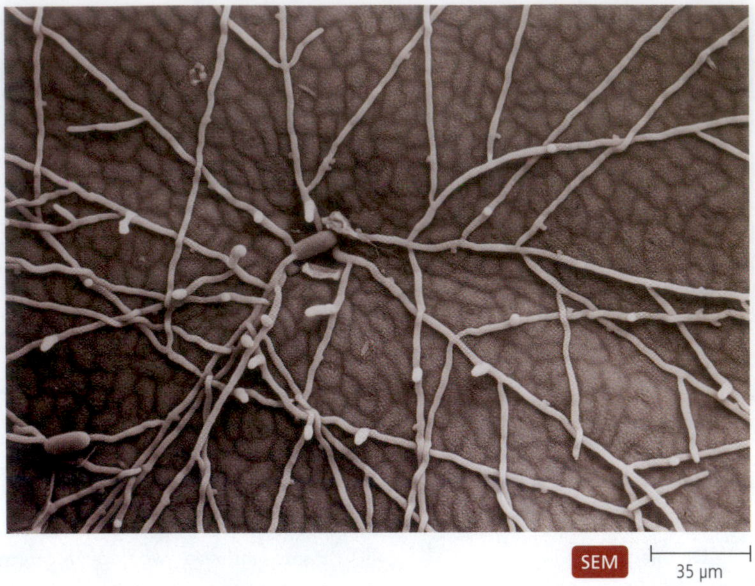

▲ **FIGURE 12.16 A fungal mycelium growing on a leaf.**

Nutrition of Fungi

Fungi acquire nutrients by absorption; that is, they secrete catabolic enzymes outside their bodies to break large organic molecules into smaller molecules, which they then transport throughout their bodies. Most fungi are **saprobes**[23] (sap´rōbz)—they absorb nutrients from the remnants of dead organisms—though some species trap and kill microscopic soil-dwelling nematodes (worms; **FIGURE 12.17**). Fungi that derive their nutrients from living plants and animals usually have modified hyphae called **haustoria**[24] (haw-stō´rē-ă), which penetrate the tissue of the host to withdraw nutrients. Absorptive nutrition is important in the role that fungi play as decomposers and recyclers of organic waste. Cytoplasmic streaming frequently transports nutrients and organelles, including nuclei, throughout a mycelium. Streaming between cells of septate mycelia occurs through pores in the septa.

Recently, scientists have discovered that some fungi use ionizing radiation (radioactivity) as an energy source for metabolism. Many fungi absorb radiation with a black pigment, melanin, but some appear to transform absorbed radiation into chemical energy, which the fungi then use to grow.

Most fungi are aerobic, though many yeasts (e.g., *Saccharomyces*) are facultative anaerobes that obtain energy from fermentation, such as occurs in the reactions that produce alcohol. Anaerobic fungi are found in the digestive systems of many herbivores, such as cattle and deer, where they assist in the catabolism of plant material.

[23]From Greek *sapros,* meaning "rotten," and *bios,* meaning "life."

[24]From Latin *haustor,* meaning "someone who draws water from a well."

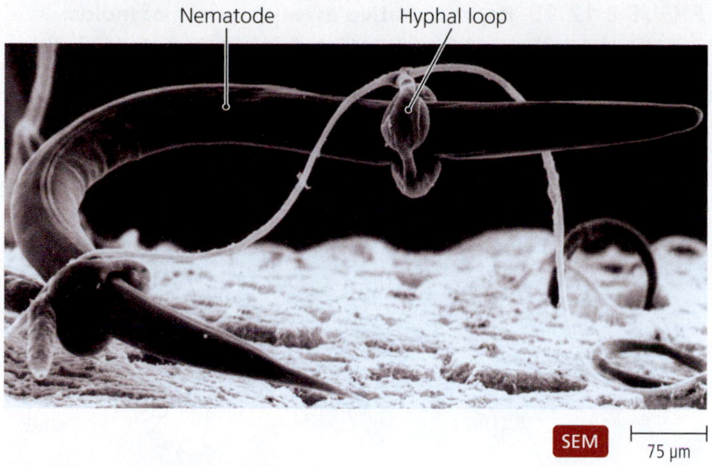

▲ **FIGURE 12.17 Predation of a nematode by the fungus *Drechsierella*.** The fungus produces special looped hyphae that constrict when the worm contacts the inside of the loop. The fungus secretes enzymes that digest the nematode and then absorbs the resulting nutrients. *What is the more typical mode of nutrition found in fungi?*

Figure 12.17 *Most fungi are saprobic.*

Reproduction of Fungi

LEARNING **OUTCOMES**

12.17 Describe asexual and sexual reproduction in fungi.

12.18 List three basic types of asexual spores found in molds.

Whereas all fungi have some means of asexual reproduction involving mitosis followed by cytokinesis, most fungi also reproduce sexually. In the next sections we briefly examine asexual and sexual reproduction in fungi.

Budding and Asexual Spore Formation

Yeasts typically bud in a manner similar to prokaryotic budding. Following mitosis, one daughter nucleus is sequestered in a small bleb (a blisterlike outgrowth) of cytoplasm that is isolated from the parent cell by the formation of a new wall (see Figure 12.15c). In some species, especially *Candida albicans* (kan´did-ă al´bi-kanz), which causes human oral thrush and vaginal yeast infections, a series of buds remain attached to one another and to the parent cell, forming a long filament called a *pseudohypha*. *Candida* invades human tissues by means of such pseudohyphae, which can penetrate intercellular cracks.

Filamentous fungi reproduce asexually by producing lightweight spores, which enable the fungi to disperse vast distances on the wind. Researchers have isolated fungal spores from wind currents many miles above the surface of the Earth. Scientists categorize the asexual spores of molds according to their mode of development. Following are examples of some asexual spores:

- *Sporangiospores* form inside a sac called a *sporangium*,[25] which is often borne on a spore-bearing stalk, called a *sporangiophore*,[26] at either the tips or sides of hyphae **(FIGURE 12.18a)**.
- *Chlamydospores* form with a thickened cell wall inside hyphae **(FIGURE 12.18b)**.
- *Conidiospores* (also called *conidia*) are produced at the tips or sides of hyphae but not within a sac. There are many types of conidia, some of which develop in chains on stalks called *conidiophores* **(FIGURE 12.18c)**.

Medical lab technologists use the presence and type of asexual spores in clinical samples to identify many fungal pathogens.

Sexual Spore Formation

Scientists designate fungal mating types as "+" and "−" rather than as male and female, in part because their bodies are morphologically indistinguishable. The process of sexual reproduction in fungi has four basic steps **(FIGURE 12.19)**:

1. Haploid (n) cells from a + fungus and a − fungus fuse to form a *dikaryon*, a cell containing both + and nuclei. The dikaryotic stage is neither diploid nor haploid but instead is designated ($n + n$).
2. After a period of time that typically ranges from hours to years but can be centuries, a pair of nuclei within a dikaryon fuse to form one diploid ($2n$) nucleus.
3. Meiosis of the diploid nucleus restores the haploid state.
4. The haploid nuclei are partitioned into + and − spores, which reestablish + and − fungi by mitoses and cell divisions.

Fungi differ in the ways they form dikaryons and in the site at which meiosis occurs.

Classification of Fungi

LEARNING **OUTCOMES**

12.19 Compare and contrast the three divisions of fungi with respect to the formation of sexual spores.

12.20 Describe the deuteromycetes, and explain why this group no longer constitutes a formal taxon.

In the following sections we will consider the four major subgroups into which taxonomists traditionally divided the kingdom Fungi. Three of these subgroups, which are taxa called *divisions* that are equivalent to phyla in other kingdoms, are based on the type of sexual spore produced (divisions Zygomycota, Ascomycota, and Basidiomycota); the fourth (the deuteromycetes) was a repository of fungi for which no sexual stage is known. We begin by considering the Zygomycota.

Division Zygomycota

Fungi in the division **Zygomycota** are coenocytic molds called zygomycetes (zī´gō-mī-sēts). Of the approximately 1100 species known, most are saprobes; the rest are obligate parasites of insects and other fungi.

[25]From Greek *spora*, meaning "seed," and *angeion*, meaning "vessel."
[26]From Greek *phoros*, meaning "bearing."

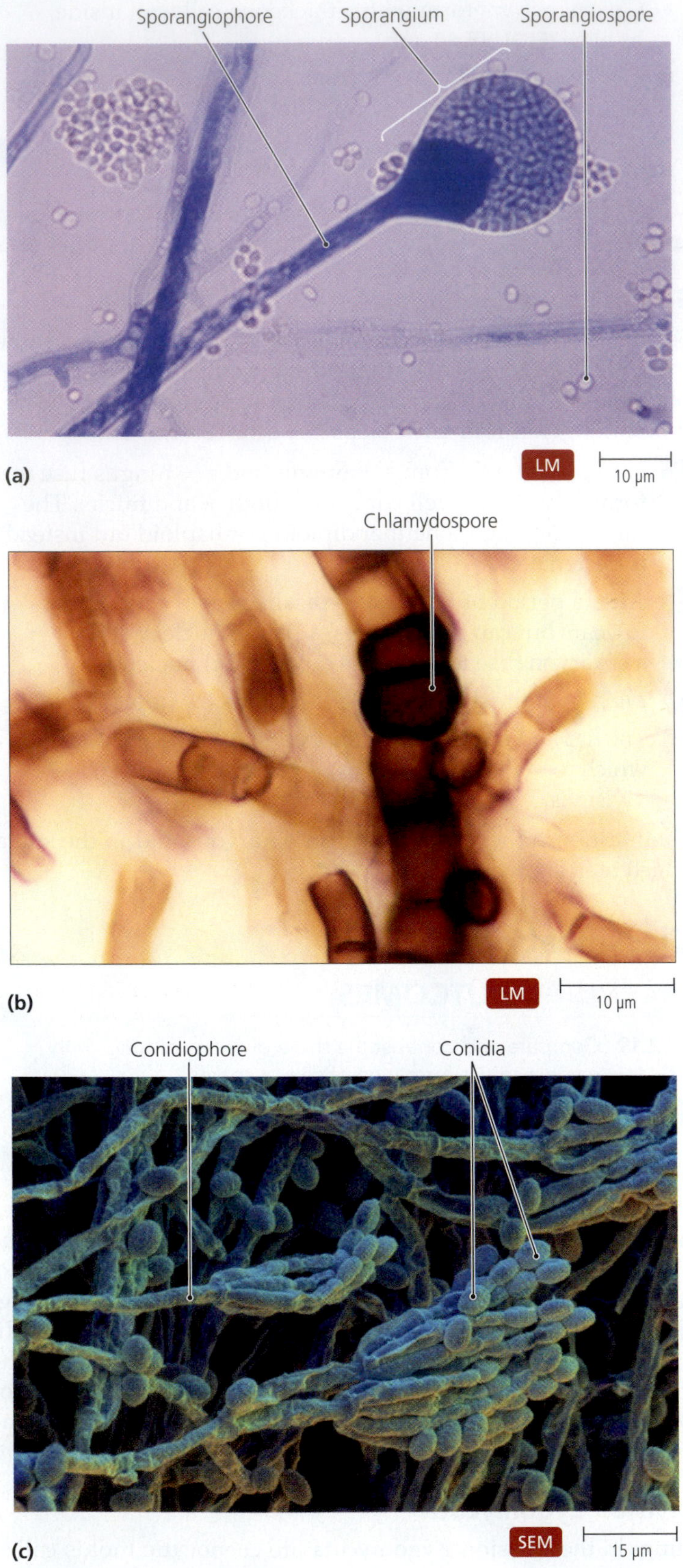

◀ **FIGURE 12.18 Representative asexual spores of molds.** **(a)** Sporangiospores, which develop within a sac called a sporangium that is borne on a sporangiophore, here of *Mucor.* **(b)** Chlamydospores are thick-walled spores that form inside hyphae, here of *Aureobasidium.* **(c)** Conidiospores (conidia) develop at the ends of hyphae, here of *Penicillium. How do conidia differ from sporangiospores?*

Figure 12.18 *Conidia are never enclosed in a sac.*

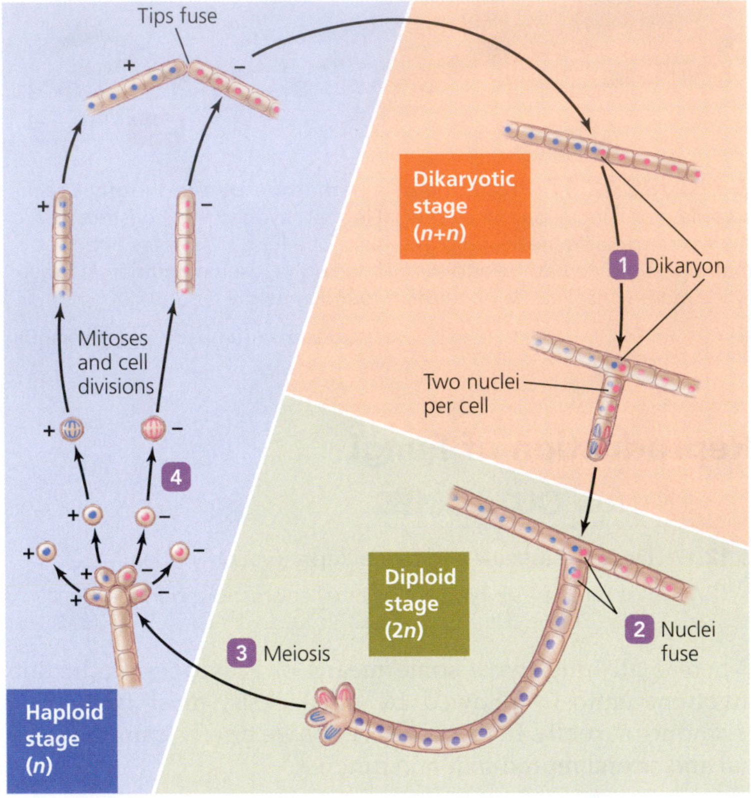

▲ **FIGURE 12.19 The process of sexual reproduction in fungi.** The steps in the process are described in the text. Haploid cells appear against a blue background, diploid cells against a green background, and dikaryotic cells against an orange background.

FIGURE 12.20 illustrates the life cycle of a typical zygomycete: the black bread mold *Rhizopus nigricans* (rī-zō´pŭs ni´gri-kans).[27] Zygomycetes reproduce asexually via sporangiospores 1 to 4, but the distinctive feature of most zygomycetes is the formation of sexual structures called **zygosporangia** (sometimes incorrectly termed zygospores). Zygosporangia of *R. nigricans* are black, rough-walled structures that develop from the fusion of sexually compatible hyphal tips 5 to 8. Like fungal spores, zygosporangia can withstand desiccation and other harsh environmental conditions.

Within a zygosporangium nuclei from one hypha (+) fuse with nuclei from the other hypha (−) to form many diploid nuclei. Each diploid nucleus undergoes meiosis, but only one of the four meiotic daughters of each nucleus survives. The

[27]From Greek *rhiza,* meaning "root," and *pous,* meaning "foot," and Latin *niger,* meaning "black."

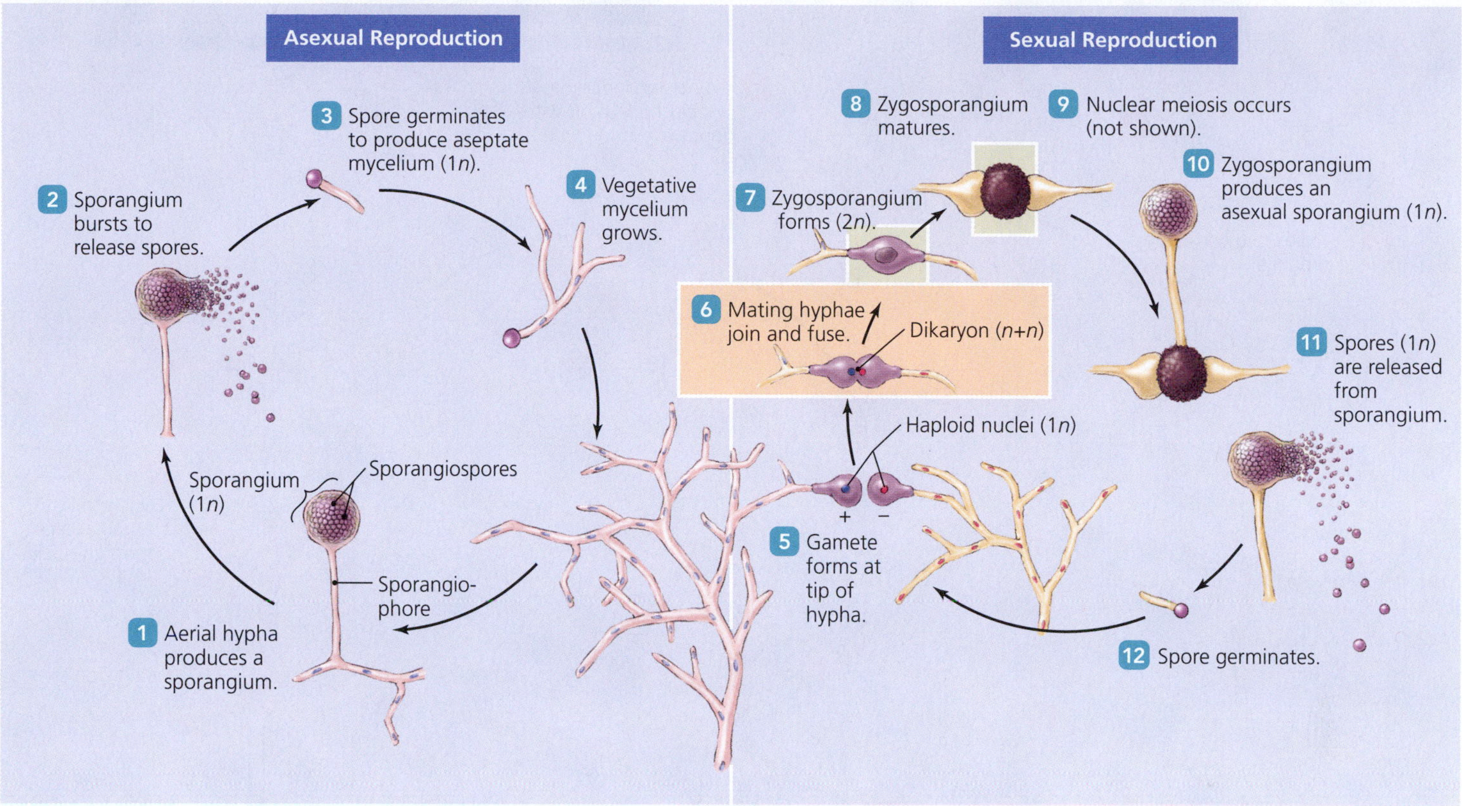

▲ **FIGURE 12.20 Life cycle of the zygomycete *Rhizopus*.** During the asexual cycle, the fungus reproduces via sporangiospores that germinate and produce hyphae. In the sexual cycle, the tips of + and − hyphae fuse and form a diploid zygosporangium that matures, undergoes meiosis, germinates, and produces a sporangium containing haploid sporangiospores, which germinate to form new mycelia.

zygosporangium then produces a haploid sporangium, which is filled with haploid spores (true zygospores). The sporangium releases these spores, each of which germinates to produce either a + or a − mycelium. This completes the life cycle.

Microsporidia **Microsporidia** are small organisms that are difficult to classify. Until 2003, taxonomists thought microsporidia were protozoa, but genetic analysis indicates they are more similar to zygomycetes.

They are obligatory intracellular parasites; that is, organisms that must live within their hosts' cells. Microsporidia spread from host to host as small, resistant spores. An example is *Nosema* (nō-sē´mă), which is parasitic on insects, such as silkworms and honeybees. The Environmental Protection Agency has approved one species of *Nosema* as a biological control agent for grasshoppers. Seven genera of microsporidia, including *Nosema*, *Encephalatizoon* (en-sef-a-lat-e´zō-an), and *Microsporidium* (mī-krō-spor-i´dē-ŭm), are known to cause diseases in immunocompromised patients (see Emerging Disease Case Study: Microsporidiosis in Chapter 16 on p. 497).

Division Ascomycota

The division **Ascomycota** contains about 32,000 known species of molds and yeasts that are characterized by the formation of haploid **ascospores** within sacs called **asci** (singular: **ascus**).[28] Asci occur in fruiting bodies called *ascocarps*, which have various shapes (**FIGURE 12.21**). Ascomycetes (as´kō-mī-sēts), as they are called, also reproduce asexually by conidiospores, as illustrated for a representative ascomycete *Penicillium* in **FIGURE 12.22** 1 to 4.

▲ **FIGURE 12.21 Ascocarps (fruiting bodies) of the common morel, *Morchella esculenta*, a delectable edible ascomycete.** The pits visible in this photograph are lined with asci, sacs that contain numerous ascospores.

[28]From Greek *askus*, meaning "wineskin."

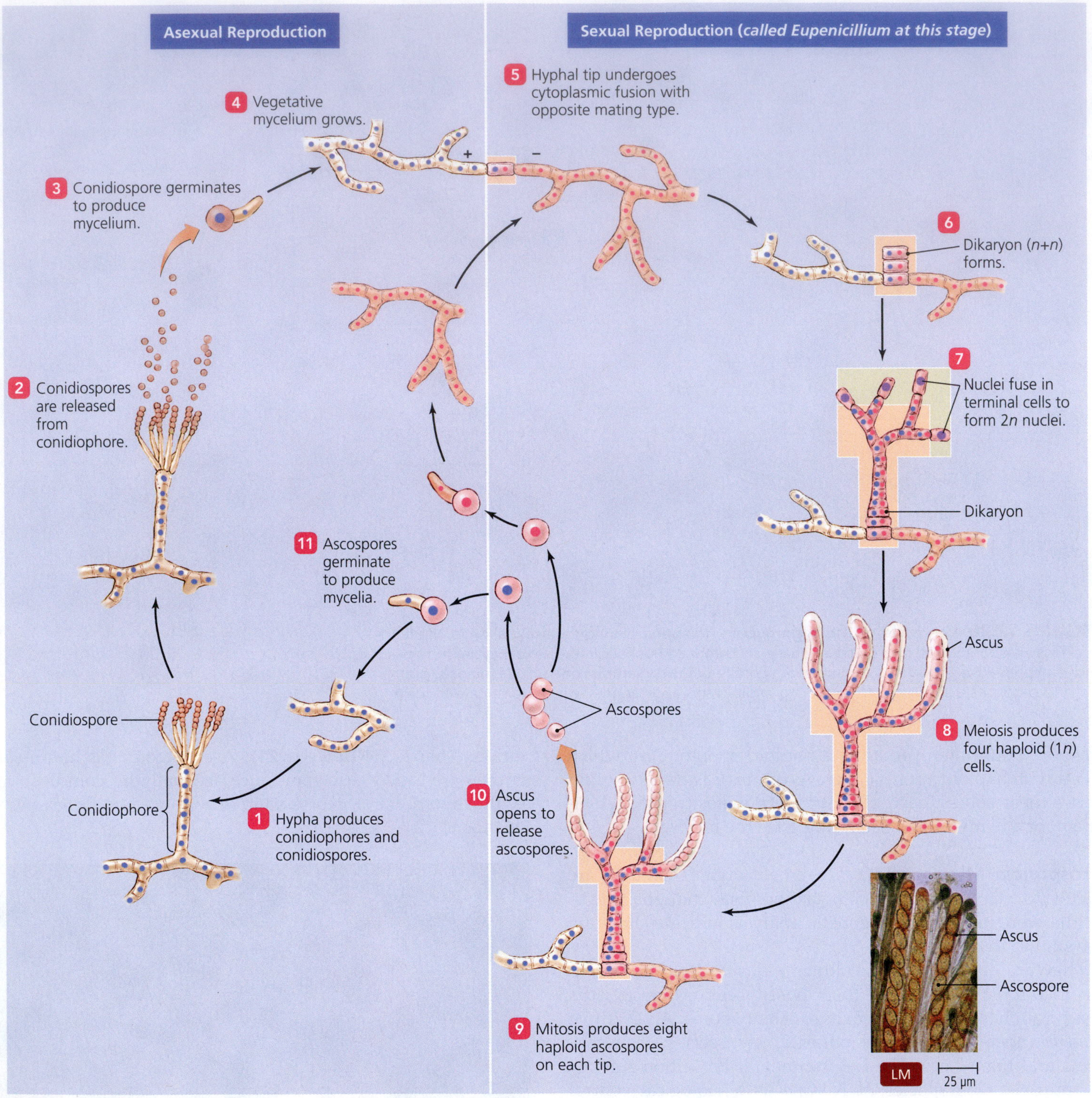

▲ **FIGURE 12.22 Life cycle of an ascomycete.**

Sexual reproduction of an ascomycete proceeds as follows:

5 Multinucleate, hyphal tips of opposite mating types fuse to form a dikaryon.

6 The dikaryon reproduces to form hyphae whose cells are all dikaryotic.

7 In the dikaryon, nuclei of opposite mating types fuse to form a diploid nucleus.

8 Each diploid nucleus undergoes meiosis and cytokinesis to form four haploid cells within an ascus.

9 Each haploid cell may undergo mitosis and cytokinesis to form two haploid ascospores, resulting in eight ascospores, which line up inside the ascus.

10 The asci open to release their ascospores.

11 Each ascospore germinates to produce a + or − hypha.

Ascomycetes are familiar and economically important fungi. For example, most of the fungi that spoil food are ascomycetes. This group also includes plant pathogens, such as the causative agents of Dutch elm disease and chestnut blight, which have almost eliminated their host trees in many parts of the United States. *Claviceps purpurea* (klav´i-seps poor-poo´rē-ă) growing on grain produces *lysergic acid,* which causes abortions in cattle and hallucinations in humans. *Aspergillus* can also infect humans (see **Emerging Disease Case Study: Aspergillosis** on p. 376).

On the other hand, many ascomycetes are beneficial. For example, *Penicillium* (pen-i-sil´ē-ŭm) mold is the source of penicillin; *Saccharomyces,* which ferments sugar to produce alcohol and carbon dioxide gas, is the basis of the baking and brewing industries; and *truffles* (varieties of *Tuber*) grow as mycorrhizae in association with oak and beech trees to form culinary delights (see **Beneficial Microbes: Fungi for $3600 a Pound** on p. 375). As previously noted, another ascomycete, the pink bread mold *Neurospora,* has been an important tool in genetics and biochemistry. Many ascomycetes partner with green algae or cyanobacteria to form *lichens,* which are discussed in more detail shortly.

Division Basidiomycota

A walk through fields and woods in most parts of the world may reveal mushrooms, puffballs, stinkhorns, jelly fungi, bird's nest fungi, or bracket fungi, all of which are the visible fruiting bodies of the almost 22,000 known species of fungi in the division **Basidiomycota**. Poisonous mushrooms are sometimes called *toadstools,* but there is no sure way to always distinguish an edible "mushroom" from a poisonous "toadstool" except by eating them—a truly risky practice!

Mushrooms and other fruiting bodies of *basidiomycetes* (ba-sid´ē-ō-mi-sēts) are called **basidiocarps** (**FIGURE 12.23**). The entire structure of a basidiocarp consists of tightly woven hyphae that extend into multiple, often club-shaped projections called *basidia,* the ends of which produce sexual *basidiospores* (typically four on each basidium). **FIGURE 12.24** illustrates the life cycle of a poisonous mushroom (toadstool), *Amanita muscaria* (am-ă-nī´tă mus-ka´rē-ă).

Besides the edible mushrooms—most notably, cultivated *Agaricus* (a-gār´i-kus)—basidiomycetes affect humans in several ways. Most basidiomycetes are important decomposers that digest chemicals such as cellulose and lignin in dead plants and return nutrients to the soil. Many mushrooms produce toxins or hallucinatory chemicals. An example of the latter is *Psilocybe cubensis* (sil-ō-sī´bē kū-ben´sis), which produces *psilocybin,* a hallucinogen. The basidiomycete yeast *Cryptococcus neoformans* (krip-tō-kok´ŭs nē-ō-for´manz) is the leading cause of fungal meningitis. Other basidiomycetes are *rusts* and *smuts,* which cause millions of dollars in crop loss each year.

Deuteromycetes

As noted previously, the divisions Zygomycota, Ascomycota, and Basidiomycota are based on type of sexual spore produced. Because scientists have not observed sexual reproduction in all fungi, taxonomists in the middle of the 20th century created the division *Deuteromycota* (also called *imperfect fungi*) to contain the fungi whose sexual stages are unknown—either because they do not produce sexual spores or because their sexual spores have not been observed. More recently, however, the analysis of rRNA sequences has revealed that most deuteromycetes in fact belong in the division Ascomycota, and thus modern taxonomists have abandoned Deuteromycota as a formal taxon. Nevertheless, many medical laboratory technologists, health care practitioners, and scientists continue to refer to "deuteromycetes" because it is a traditional name. **▶VIDEO TUTOR:** ***Principles of Sexual Reproduction in Fungi***

Lichens

LEARNING **OUTCOMES**

12.21 Describe a lichen's members.

12.22 List several beneficial roles or functions of lichens.

▶ FIGURE 12.23 Basidiocarps (fruiting bodies). (a) Basidiospores, looking like flattened eggs in birds' nests, develop inside basidiocarps of the bird's nest fungus, *Crucibulum.* **(b)** The familiar shapes of bracket fungi and mushrooms are also basidiocarps of extensive mycelia. Shown here is *Laetiporus sulphureus.*

▶ **FIGURE 12.24 Sexual life cycle of *Amanita muscaria*, a poisonous basidiomycete.** *How can a novice distinguish edible from poisonous mushrooms?*

Figure 12.24 *The only sure way to distinguish edible from poisonous mushrooms is to assume the risk of eating them.*

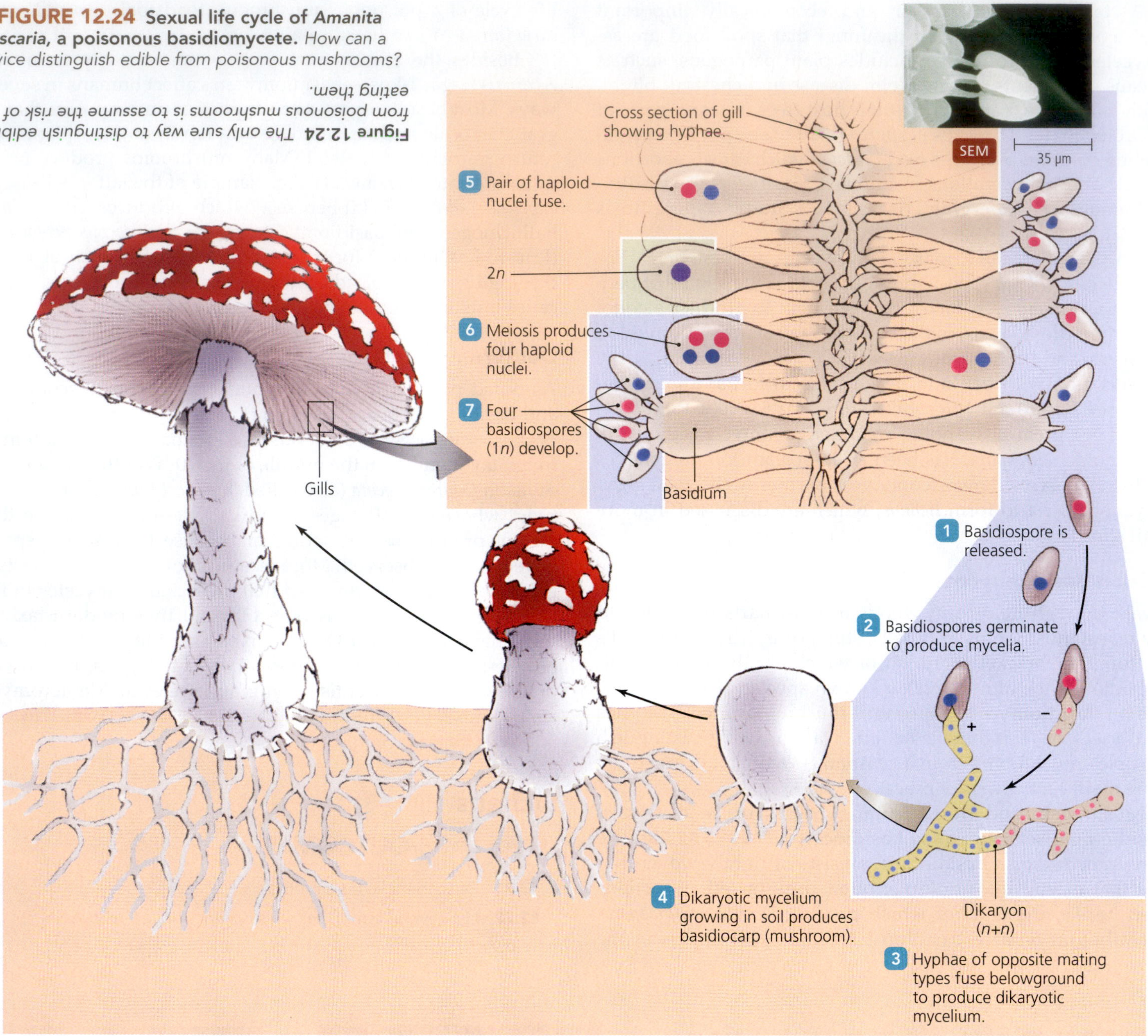

Cross section of gill showing hyphae
SEM
35 μm
5 Pair of haploid nuclei fuse.
2n
6 Meiosis produces four haploid nuclei.
7 Four basidiospores (1n) develop.
Gills
Basidium
1 Basidiospore is released.
2 Basidiospores germinate to produce mycelia.
+
–
4 Dikaryotic mycelium growing in soil produces basidiocarp (mushroom).
Dikaryon (n+n)
3 Hyphae of opposite mating types fuse belowground to produce dikaryotic mycelium.

A discussion of fungi is incomplete without considering **lichens**, which are partnerships between fungi and photosynthetic microbes—commonly cyanobacteria or, less frequently, green algae. In a lichen, the hyphae of the fungus, which is usually an ascomycete, surround the photosynthetic cells **(FIGURE 12.25)** and provide them nutrients, water, and protection from desiccation and harsh light. In return, each alga or cyanobacterium provides the fungus with products of photosynthesis—carbohydrates and oxygen. In some lichens, the phototroph releases 60% of its carbohydrates to the fungus.

The partnership in a lichen is not always mutually beneficial; in some lichens, the fungus produces haustoria that penetrate and kill the photosynthetic member. Such lichens are maintained only because the phototroph's cells reproduce faster than the fungus can devour them.

The fungus of a lichen reproduces by spores, which must germinate and develop into hyphae that capture an appropriate cyanobacterium or alga. Alternatively, wind, rain, and small animals disperse bits of lichen called *soredia*, which contain both phototrophs and fungal hyphae, to new locations where they can establish a new lichen if there is suitable substrate.

Scientists have identified over 14,000 species of lichens, which are abundant throughout the world, particularly in pristine unpolluted habitats, growing on soil, rocks, leaves, tree bark, other lichens, and even the backs of tortoises. Indeed, lichens grow in almost every habitat—from high-elevation alpine tundra to submerged rocks on the oceans' shores, from frozen Antarctic soil to

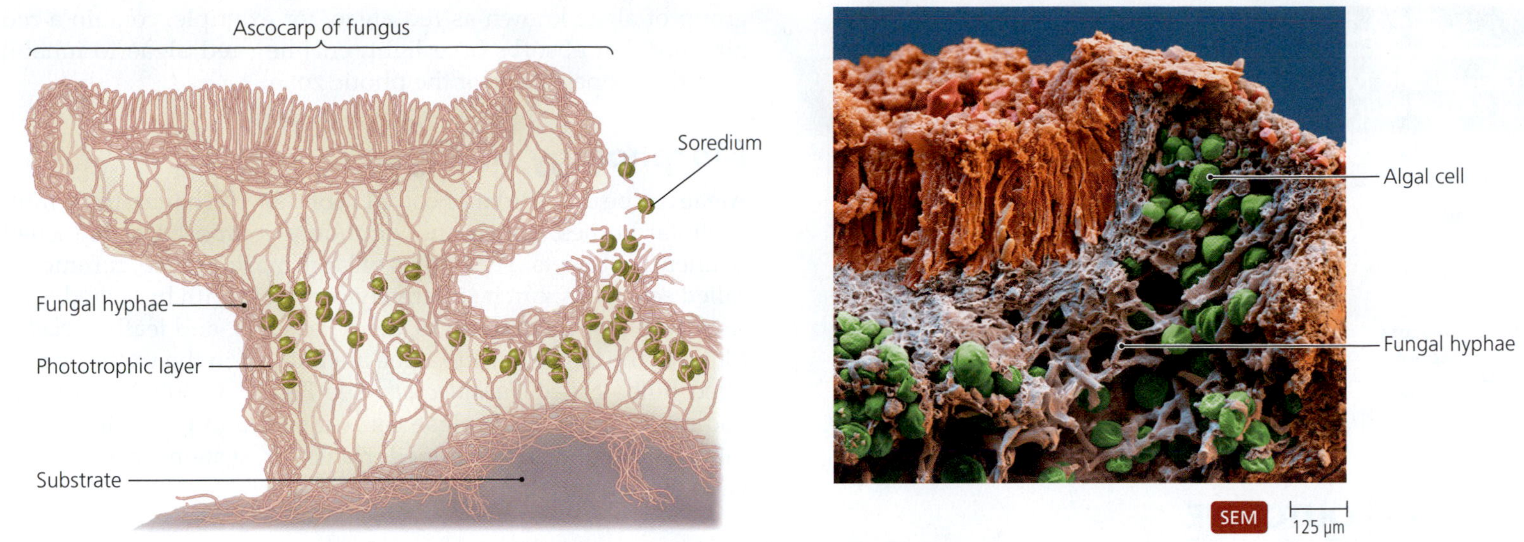

▲ **FIGURE 12.25 Makeup of a lichen.** The hyphae of a fungus (most commonly an ascomycete) constitute the major portion of the body of the lichen. Cells of the photosynthetic member of the lichen are concentrated near the lichen's surface. Soredia, which are bits of fungus surrounding phototrophic cells, are the means by which some lichens propagate themselves. *Members of what two groups of photosynthetic microbes can be part of a lichen?*

Figure 12.25 *Cyanobacteria or green algae can join ascomycetes to form lichens.*

hot desert climes. The only unpolluted places where lichens do not consistently grow are in the dark depths of the oceans and the black world of caves—after all, lichens require light.

Lichens grow slowly, but they can live for hundreds and possibly thousands of years. They occur in three basic shapes (**FIGURE 12.26**):

- *Foliose* lichens are leaflike, with margins that grow free from the substrate.
- *Crustose* lichens grow appressed to their substrates and may extend into the substrate for several millimeters.
- *Fruticose* lichens are either erect or are hanging cylinders.

Lichens create soil from weathered rocks, and lichens containing nitrogen-fixing cyanobacteria provide significant amounts of usable nitrogen to nutrient-poor environments. Many animals eat lichens; for example, reindeer and caribou subsist primarily on lichens throughout the winter. Birds use lichens for nesting materials, and some insects camouflage themselves with bits of living lichen. Humans also use lichens in the production of foods, dyes, clothing, perfumes, medicines, and the litmus of indicator paper. Because lichens will not grow well in polluted environments, ecologists use them as sensitive living assays for monitoring air pollution.

In the preceding sections, we have seen that fungi are chemoheterotrophic yeasts and molds that function primarily to decompose and degrade dead organisms. Some fungi are pathogenic, and others associate with cyanobacteria or algae to form lichens. All fungi reproduce asexually either by budding or via asexual spores, and most fungi also produce sexual spores, by which taxonomists classify them. **TABLE 12.3** on p. 374 summarizes the characteristics of fungi.

▲ **FIGURE 12.26 Gross morphology of lichens.** Crustose forms are flat and tightly joined to the substrate, here a branch; foliose forms are leaflike with free margins; and fruticose forms are either erect (as shown) or pendulant.

TELL ME WHY

Why isn't a fungal dikaryon—with its two haploid (n) nuclei—considered diploid?

TABLE 12.3 Characteristics of Fungi

Division and Type of Sexual Spore	Comments	Representative Genera
Zygomycota **Zygospores**	Coenocytic (aseptate)	*Rhizopus*
Ascomycota **Ascospores**	Septate; some associated with cyanobacteria or green algae to form lichens	*Claviceps, Neurospora, Penicillium, Saccharomyces, Tuber*
Basidiomycota **Basidiospores**	Septate	*Agaricus, Amanita, Cryptococcus*

Algae

LEARNING OUTCOME

12.23 Describe the distinguishing characteristic of algae.

The Romans used the word *alga* (al´ga) to refer to any simple aquatic plant, particularly one found in marine habitats. Their usage thus included the organisms we recognize as algae (al´jē), cyanobacteria, sea grasses, and other aquatic plants. Today, the word *algae* properly refers to simple, eukaryotic, phototrophic organisms that, like plants, carry out oxygenic photosynthesis using chlorophyll *a*. Algae differ from plants such as sea grass in having sexual reproductive structures in which every cell becomes a gamete. In plants, by contrast, a portion of the reproductive structure always remains vegetative.

Algae are not a unified group; rather, they differ widely in distribution, morphology, reproduction, and biochemical traits. Moreover, the word *algae* is not synonymous with any taxon; in the taxonomic scheme shown in Figure 12.4c, algae can be found in kingdoms Alveolata, Euglenozoa, Stramenopila, Rhodophyta, and Plantae. The study of algae is *phycology*,[29] and the scientists who study them are called *phycologists*.

Distribution of Algae

Even though some algae grow in such diverse habitats as in soil and ice, in intimate association with fungi as lichens, and on plants, most algae are aquatic, living in the *photic zone* (penetrated by sunlight) of fresh, brackish, and salt bodies of water. This watery environment provides some benefits and also presents some difficulties for photosynthetic organisms. Whereas most bodies of water contain sufficient dissolved chemicals to provide nutrients for algae, water also differentially absorbs longer wavelengths of light (including red light), so only shorter (blue) wavelengths penetrate more than a meter below the surface. This is problematic for algae because their primary photosynthetic pigment—chlorophyll *a*—captures red light. Thus, to grow in deeper waters, algae must have *accessory photosynthetic pigments* that trap the energy of penetrating, short-wavelength light and pass that energy to chlorophyll *a*. Members of the group of algae known as red algae, for example, contain a red pigment that absorbs blue light, enabling red algae to inhabit even the deepest parts of the photic zone.

[29]From Greek *phykos*, meaning "seaweed," and *logos*, meaning "discourse."

Morphology of Algae

Algae can be unicellular or colonial, or they can have simple multicellular bodies, which are commonly composed of branched filaments or sheets. The bodies of large marine algae, commonly called seaweeds, can be relatively complex, with branched *holdfasts* to anchor them to rocks, stemlike *stipes*, and leaflike *blades*. Many of the larger marine algae are buoyed in the water by gas-filled bulbs called *pneumocysts* (see Figure 12.30). Though the bodies of some marine algae can surpass land plants in length, they lack the well-developed transport systems common to vascular plants.

Reproduction of Algae

LEARNING OUTCOME

12.24 Describe the alternation of generations in algae.

In unicellular algae, asexual reproduction involves mitosis followed by cytokinesis. In unicellular algae that reproduce sexually, each algal cell acts as a gamete and fuses with another such gamete to form a zygote, which then undergoes meiosis to return to the haploid state.

Multicellular algae may reproduce asexually by fragmentation, in which each piece of a parent alga develops into a new individual, or by motile or nonmotile asexual spores. As noted previously, in multicellular algae that reproduce sexually, every cell in the reproductive structures of the alga becomes a gamete—a feature that distinguishes algae from all other photosynthetic eukaryotes.

Many multicellular algae reproduce sexually with an **alternation of generations** of haploid and diploid individuals **(FIGURE 12.27)**. In such life cycles, diploid individuals undergo meiosis to produce male and female haploid spores that develop into haploid male and female bodies, which may look identical to the diploid body. Some haploid algae produce gametes that fuse to form a zygote, which grows into a new diploid alga. Both haploid and diploid algae may reproduce asexually as well.

Classification of Algae

LEARNING OUTCOMES

12.25 List four groups of algae, and describe the distinguishing characteristics of each.

12.26 List several economic benefits derived from algae.

The classification of algae is not yet settled. Historically, taxonomists have used differences in photosynthetic pigments, storage products, and cell wall composition to classify algae into several groups that are named for the colors of their photosynthetic

BENEFICIAL MICROBES

Fungi for $3600 a Pound

Truffles—rare, intensely flavored ascomycetes that grow underground—are one of the most luxurious and expensive foods on Earth, selling on average for more than $800 per pound. There are many different varieties of truffles. The most coveted include *tuber melanosporum*, a black truffle that is also known as the "black diamond," and *Tuber magnatum*, a white truffle that can sell for $3600 per pound!

Because truffles are very difficult to find, truffle hunters often use pigs and dogs trained to sniff them out. (Dogs are preferred because pigs are more likely to eat the truffles.) Dogs may be trained at the University of Truffle Hunting Dogs founded in 1880.

The underground habitat of truffles requires them to form symbiotic relationships with trees for nutrients and with animals, such as squirrels and chipmunks, for spore dispersal—the animals dig up the truffles and thus help spread the spores to other locations.

Incidentally, chocolate truffles derive only their name from their prized fungal counterparts.

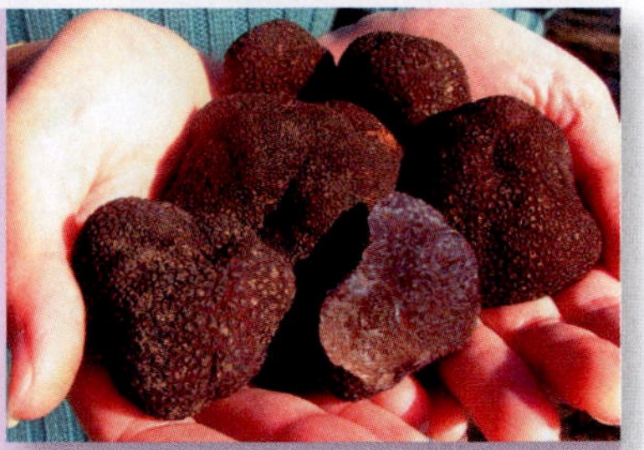

Tuber melanosporum

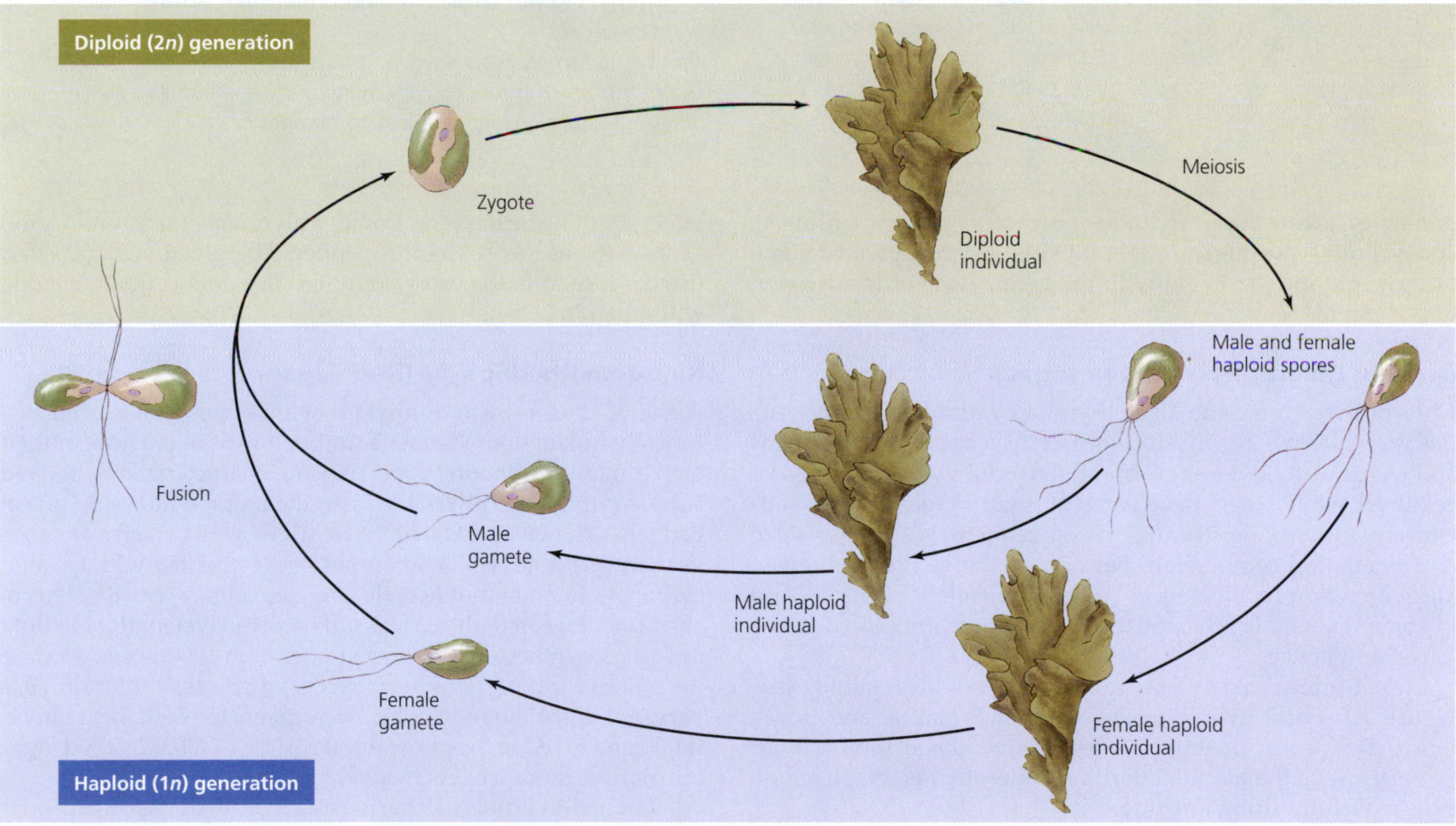

▲ **FIGURE 12.27 Alternation of generations in algae, as occurs in the green alga *Ulva*.** A diploid individual meiotically produces haploid spores that germinate and grow into male and female bodies. These haploid algae produce gametes that fuse to form a diploid zygote, which grows into a new diploid individual. Each generation can also reproduce asexually by fragmentation and by spores.

EMERGING DISEASE CASE STUDY

Aspergillosis

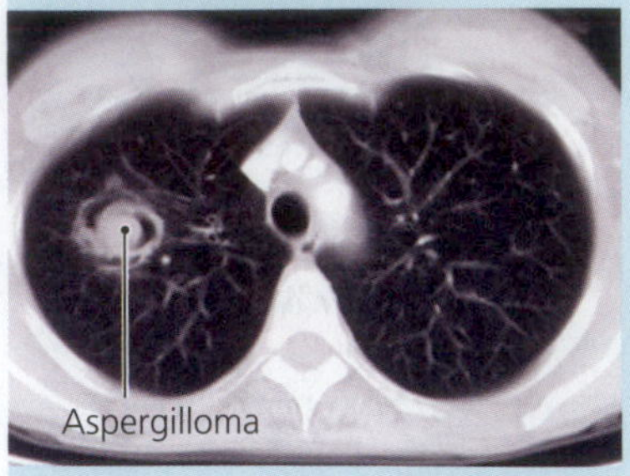

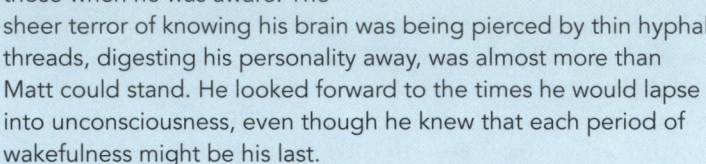

Matt was not in good shape, and it had nothing to do with his physique or time spent at the gym; it had to do with the ball of fungus in his right lung. That fungal sphere, an aspergilloma, had started as a single spore of an ascomycete fungus called *Aspergillus*. The mold had formed the mass of fungal hyphae that was invading the airways of his lung and slowly killing him. Such bronchopulmonary aspergillosis is a rare but increasingly frequent pathogen of the immunocompromised. Besides experiencing difficulty in breathing, fever, and chest pain, Matt most hated coughing up wads of bloody mucus—as if he were expelling the very fabric of his life. In fact, he literally was coughing up pieces of his lung.

Unfortunately for Matt, the disease had not yet peaked. *Aspergillus* had invaded his blood and was even now progressing toward his brain. Soon the signs and symptoms of invasive aspergillosis would be his—extreme tiredness, excessive weakness, severe headaches, and delirium. All would be his daily companions. He might also be paralyzed on one side of his body.

Matt spent four weeks in the hospital. The worst days were those when he was aware. The sheer terror of knowing his brain was being pierced by thin hyphal threads, digesting his personality away, was almost more than Matt could stand. He looked forward to the times he would lapse into unconsciousness, even though he knew that each period of wakefulness might be his last.

A medical miracle in the form of a new antifungal drug (voriconazole) brought Matt back to life. The invasive mold was defeated, and Matt returned home grateful, aware that life is precious and hopeful that no more spores floated his way.

1. How could a microbiologist determine that the fungus in Matt's lung was an ascomycete rather than a basidiomycete?
2. Why is aspergillosis emerging as a "new" disease?
3. Why is a powerful drug like voriconazole effective against fungi but relatively harmless to humans?

pigments: green algae, red algae, brown algae, golden algae, and yellow-green algae. The following sections present some of these groups. We begin with the green algae of the division Chlorophyta.

Division Chlorophyta (Green Algae)

Chlorophyta[30] are green algae that share numerous characteristics with plants—they have chlorophylls *a* and *b* and use sugar and starch as food reserves. Many have cell walls composed of cellulose, while others have walls of glycoprotein or lack walls entirely. In addition, the 18S rRNA sequences of green algae and plants are comparable. Because of the similarities, green algae are often considered to be the progenitors of plants, and in some taxonomic schemes the Chlorophyta are placed in the kingdom Plantae.

Most green algae are unicellular or filamentous (see Figure 1.7a) and live in freshwater ponds, lakes, and pools, where they form green to yellow scum. Some multicellular forms grow in the marine intertidal zone—that is, in the region exposed to air during low tide.

Prototheca (prō-tō-thē´kă) is an unusual green alga in that it lacks pigments, making it colorless. This chemoheterotrophic alga causes a skin rash in sensitive individuals. *Codium* (kō´dē-ŭm) is a member of a group of marine green algae that do not form cross walls after mitosis; thus, its entire body is a single, large, multinucleate cell. Some Polynesians dry and grind *Codium* for use as seasoning pepper. The green alga *Trebouxia* (tre-book´sē-a) is the most common alga found in association with fungi in lichens.

Kingdom Rhodophyta (Red Algae)

Algae of division **Rhodophyta**,[31] which had been placed historically in kingdom Plantae and then Protista, are now in their own kingdom—Rhodophyta. They are characterized by the red accessory pigment **phycoerythrin**; the storage molecule glycogen (also known as *floridean*[32] *starch*); cell walls of **agar** or **carrageenan** (kar-ă-gē´nan), sometimes supplemented with calcium carbonate; and nonmotile male gametes called *spermatia*. Phycoerythrin allows red algae to absorb short-wavelength blue light and photosynthesize at depths greater than 100 meters. Because the relative proportions of phycoerythrin and chlorophyll *a* vary, red algae range in color from green to black in the intertidal zone to red in deeper water **(FIGURE 12.28)**. Most red algae are marine, though a few freshwater genera are known.

The gel-like polysaccharides agar and carrageenan, once they have been isolated from red algae such as *Gelidium* (jel-li´dē-ŭm) and *Chondrus* (kon´drŭs), are used as thickening agents for the production of solid microbiological media and numerous consumer products, including ice cream,

[30]From Greek *chloros*, meaning "green," and *phyton*, meaning "plant."

[31]From Greek *rhodon*, meaning "rose."

[32]Named for a taxon of red algae, Florideophycidae.

▲ **FIGURE 12.28** ***Pterothamnion plumula*, a red alga.**

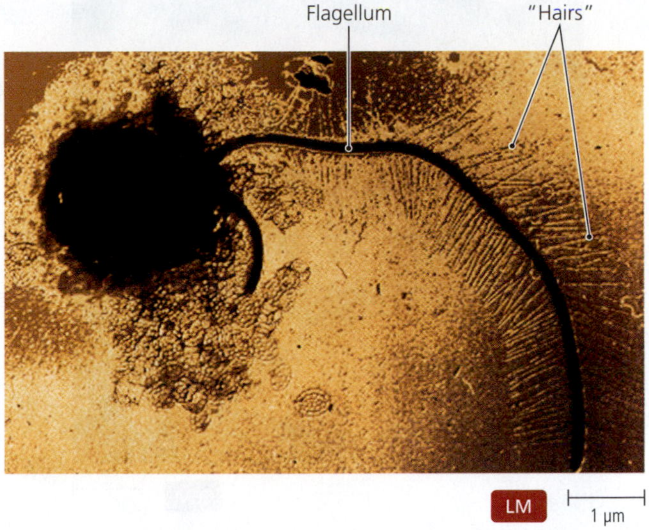

▲ **FIGURE 12.29 Hairy flagellum.** The "hairy" flagellum of brown algae and other stramenopiles is distinctive.

toothpaste, syrup, salad dressings, and snack foods. Some studies suggest that ingested carrageenan can induce inflammation of the colon.

Phaeophyta (Brown Algae)

The **Phaeophyta**[33] are in kingdom Stramenopila[34] based in large part on their gametes being motile by means of two flagella—one whiplike and one with hollow projections giving it a "hairy" appearance **(FIGURE 12.29)**. They have chlorophylls *a* and *c*, carotene, and brown pigments called *xanthophylls* (zan′thō-fils). Depending on the relative amounts of these pigments, brown algae may appear dark brown, tan, yellow-brown, greenish brown, or green. Most brown algae are marine organisms, and some of the giant kelps, such as *Macrocystis* **(FIGURE 12.30)**, surpass the tallest trees in length though not in girth.

Brown algae use the polysaccharide *laminarin* and oils as food reserves and have cell walls composed of cellulose and **alginic acid** (alginate). Alginic acid is used in numerous foods as a thickening agent and in medicine in the preparation of dental impressions.

▲ **FIGURE 12.30 Portion of the giant kelp *Macrocystis*, a brown alga.** A kelp's blades are kept afloat by pneumocysts.

Chrysophyta (Golden Algae, Yellow-Green Algae, and Diatoms)

Chrysophyta[35] is a group of algae that are diverse with respect to cell wall composition and pigments. They are unified in using the polysaccharide *chrysolaminarin* as a storage product. Some additionally store oils. Modern taxonomists group these algae with brown algae and water molds (discussed shortly) in the kingdom Stramenopila based on similarities in nucleotide sequences and flagellar structure. Whereas some chrysophytes lack cell walls, others have ornate external coverings, such as scales or plates. One taxon of chrysophytes, Bacillariophyceae—the **diatoms** (dī′ă-tomz)—are unique in having silica cell walls composed of two halves called *frustules* that fit together like a Petri dish **(FIGURE 12.31)**.

Most chrysophytes are unicellular or colonial. All chrysophytes contain more orange-colored *carotene* pigment than they do chlorophylls *a* and *c*, accounting for the common names of two major classes of chrysophytes—*golden algae* and *yellow-green algae*.

Diatoms are a major component of marine *phytoplankton:* free-floating photosynthetic microorganisms that form the basis of food chains in the oceans. Further, because of their enormous number, diatoms are the major source of the world's oxygen. The silica frustules of diatoms contain minute holes for the exchange of gases, nutrients, and wastes with the environment.

[33]From Greek *phaeo*, meaning "brown."
[34]From Latin *stramen*, meaning "straw," and *pilos*, meaning "hair."
[35]From Greek *chrysos*, meaning "gold."

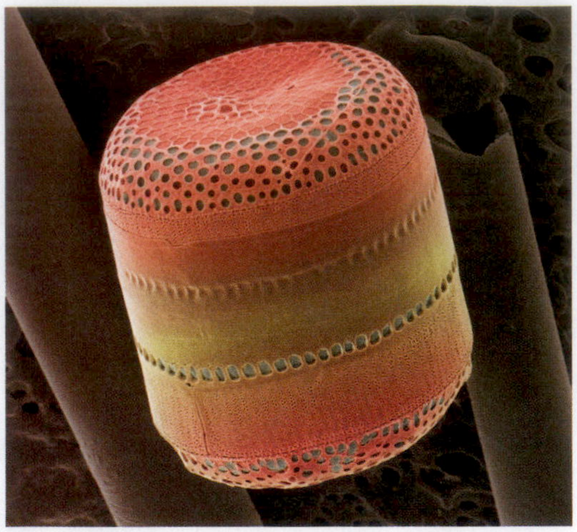

▲ **FIGURE 12.31** ***Coscinodiscus*, a diatom.** Diatoms have frustules, composed of silica and cellulose, that fit together like a Petri dish.

Organic gardeners use *diatomaceous earth,* composed of innumerable frustules of dead diatoms, as a pesticide against harmful insects and worms. Diatomaceous earth is also used in polishing compounds, detergents, and paint removers and as a component of firebrick, soundproofing products, swimming pool filters, and reflective paints.

In summary, algae are unicellular or multicellular photoautotrophs characterized by sexual reproductive structures in which every cell becomes a gamete. The colors produced by the combination of their primary and accessory photosynthetic pigments give them their common names and provide the basis of at least one classification scheme. **TABLE 12.4** summarizes the characteristics of the major groups of algae; the dinoflagellates and euglenids are included in the table because botanists historically classified these phototrophic protozoa as algae.

TELL ME WHY

Why aren't there large numbers of pathogenic algae?

Water Molds

LEARNING **OUTCOME**

12.27 List four ways in which water molds differ from true fungi.

Scientists once classified the microbes commonly known as **water molds** as fungi because they resemble filamentous fungi in having finely branched filaments. However, water molds are not true molds—they are not fungi. Water molds differ from fungi in the following ways:

- They have tubular cristae in their mitochondria.
- They have cell walls of cellulose instead of chitin.
- Their spores have two flagella—one whiplike and one "hairy."
- They have true diploid bodies rather than haploid bodies.

TABLE 12.4 Characteristics of Various Algae

Group (Common Name)	Kingdom	Pigments	Storage Product(s)	Cell Wall Component(s)	Habitat	Representative Genera
Chlorophyta (green algae)	Plantae	Chlorophylls *a* and *b*, carotene, xanthophylls	Sugar, starch	Cellulose or glycoprotein; absent in some	Fresh, brackish, and salt water; terrestrial	*Spirogyra* *Prototheca* *Codium* *Trebouxia*
Rhodophyta (red algae)	Rhodophyta	Chlorophyll *a*, phycoerythrin, phycocyanin, xanthophylls	Glycogen (floridean starch)	Agar or carrageenan, some with calcium carbonate	Mostly salt water	*Chondrus* *Gelidium* *Antithamnion*
Chrysophyta (golden algae, yellow-green algae, diatoms)	Stramenopila	Chlorophylls *a*, c_1, and c_2; carotene; xanthophylls	Chrysolaminarin, oils	Cellulose, silica, calcium carbonate	Fresh, brackish, and salt water; terrestrial; ice	*Stephanodiscus*
Phaeophyta (brown algae)	Stramenopila	Chlorophylls *a* and *c*, carotene, xanthophylls	Laminarin, oils	Cellulose and alginic acid	Brackish and salt water	*Macrocystis*
Pyrrhophyta (dinoflagellates)	Alveolata	Chlorophylls *a*, c_1, c_2; carotene	Starch, oils	Cellulose	Fresh, brackish, and salt water	*Gymnodinium* *Gonyaulax* *Pfiesteria*
Euglenophyta (euglenids)	Euglenozoa	Chlorophylls *a* and *b*, carotene	Paramylon, oils, sugar	Absent	Fresh, brackish, and salt water; terrestrial	*Euglena*

▲ **FIGURE 12.32 An example of the important role of water molds in recycling organic nutrients in aquatic habitats.**

Because water molds have "hairy" flagella and certain similarities in rRNA sequence to sequences of diatoms, other chrysophytes, and brown algae, taxonomists classify all these organisms in kingdom Stramenopila (see Figure 12.4c).

Water molds decompose dead animals and return nutrients to the environment **(FIGURE 12.32)**. Some species are detrimental pathogens of crops, such as grapes, tobacco, and soybeans. In 1845, the water mold *Phytophthora infestans* (fī-tof´tho-ră in-fes´tanz) was accidentally introduced into Ireland and devastated the potato crop, causing the great famine that killed over 1 million people and forced a greater number to emigrate the United States and Canada.

TELL ME WHY

Why are water molds more closely related to brown algae than to true molds?

Other Eukaryotes of Microbiological Interest: Parasitic Helminths and Vectors

LEARNING | **OUTCOMES**

12.28 Explain why microbiologists study large organisms, such as parasitic worms.

12.29 Discuss the inclusion of vectors in a study of microbiology.

Microbiologists are interested also in two other groups of eukaryotes, although they are not microorganisms. The first group are the parasitic *helminths,* commonly called parasitic worms. Microbiologists became interested in parasitic helminths because they observed the microscopic infective and diagnostic stages of the helminths—usually eggs or larvae (immature forms)—in samples of blood, feces, and urine. Thus, microbiologists study parasitic helminths in part because they must distinguish the parasites' microscopic forms from other microbes. (Chapters 21 and 23 discuss helminth parasites of the blood and of the digestive system.)

Microbiologists are also interested in **arthropod vectors**[36]—animals that carry pathogens and have segmented bodies, hard external skeletons, and jointed legs. Some arthropods are *mechanical vectors,* meaning they merely carry pathogens; others are *biological vectors,* meaning they also serve as hosts for microbial pathogens. Given that arthropods are small organisms (so small that we generally don't notice them until they bite us) and given that they produce numerous offspring, controlling arthropod vectors to eliminate their role in the transmission of important human diseases is an almost insurmountable task.

Disease vectors belong to two classes of arthropods: *Arachnida* and *Insecta.* Ticks and mites are arachnoid vectors. (Though spiders are also arachnids, they do not transmit diseases.) Insects account for the greatest number of vectors, and within this group are fleas, lice, flies (such as tsetse flies and mosquitoes), and true bugs (such as kissing bugs). **FIGURE 12.33** shows representatives of major types of arthropod vectors. Most vectors are found on a host only when they are actively feeding. Lice are the only arthropods that may spend their entire lives in association with a single individual host.

Arachnids

LEARNING | **OUTCOMES**

12.30 Describe the distinctive features of arachnids.

12.31 List five diseases vectored by ticks and two diseases vectored by mites.

All adult **arachnids** (ă-rak´nidz) have four pairs of legs. Both **ticks** and **mites** (commonly known as chiggers) go through a six-leg stage when they are juveniles, but they display the characteristic eight legs as adults. Ticks and mites resemble each other morphologically, having disk-shaped bodies. Ticks are roughly the size of a small rice grain, whereas mites are usually the size of sand grains.

Ticks are the most important arachnid vectors. They are distributed worldwide and serve as vectors for bacterial, viral, and protozoan diseases. Ticks are second only to mosquitoes in the number of diseases that they transmit. Hard ticks—those with a hard plate on their dorsal surfaces—are the most prominent tick vectors. They wait on stalks of grasses and brush for their hosts to come by. When a human, for example, walks past them, the ticks leap onto the person and begin searching for exposed skin. They use their mouthparts to cut holes in the skin and attach themselves with a gluelike compound to prevent being dislodged. As they feed on blood, their bodies swell to several times normal size. Some tick-borne diseases are Lyme disease, Rocky Mountain spotted fever, tularemia, relapsing fever, and tick-borne encephalitis.

Parasitic mites of humans live around the world, wherever humans and animals coexist. A few mite species transmit rickettsial diseases (rickettsial pox and scrub typhus) among animals and humans.

[36]From Latin *vectus,* meaning "carried."

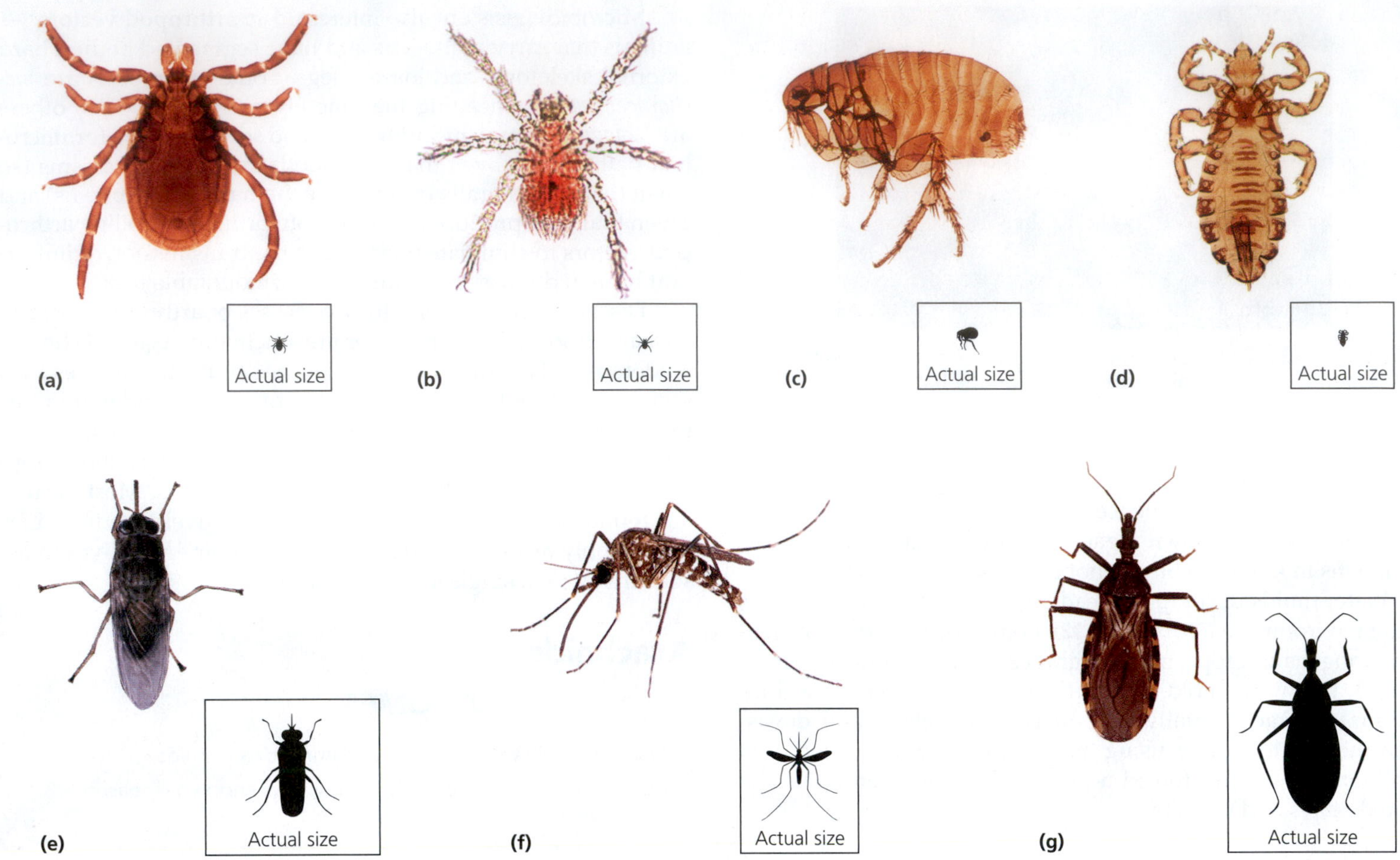

▲ **FIGURE 12.33 Representative arthropod vectors.** Arachnid vectors include **(a)** ticks and **(b)** mites; insect vectors include **(c)** fleas, **(d)** lice, **(e)** true flies, such as this tsetse fly, **(f)** mosquitoes (a type of fly), and **(g)** true bugs.

Insects

LEARNING **OUTCOMES**

12.32 Describe the general physical features of insects and the features specific to the various groups of insect vectors.

12.33 List diseases transmitted by fleas, lice, true flies, mosquitoes, and kissing bugs.

As adults, all **insects** have three pairs of legs and three body regions—head, thorax (chest), and abdomen. Adult insects are, however, far from uniform in appearance. Some have two wings, others have four wings, and others are wingless; some have long legs, some short; and some have biting mouthparts, whereas other have sucking mouthparts. Many insects have larval stages that look very different from the adults, complicating their identification.

The fact that many insects can fly has epidemiological implications. Flying insects have broader home ranges than nonflying insects, and some migrate, making control difficult.

TELL ME WHY

Why are large eukaryotes such as mosquitoes and ticks considered in a microbiology class?

Fleas are small, vertically flattened, wingless arthropods that are found worldwide, though some species have geographically limited ranges. Most are found in association with wild rodents, bats, and birds and are not encountered by humans. A few species, however, feed on humans. Cat and dog fleas are usually just pests (to the animal and its owner), but they can also serve as the intermediate host for a dog tapeworm, *Dipylidium* (dip-ĭ-lid´ē-ŭm). The most significant microbial disease transmitted by fleas is plague, carried by rat fleas.

Lice (singular: *louse*) are parasites that can also transmit disease. They are horizontally flat, soft-bodied, wingless insects found worldwide among humans and their habitations. They live in clothing and bedding and move onto humans to feed. Lice are most common among the poor and those living in severely overcrowded communities. Lice are the vectors involved in epidemic outbreaks of typhus in developing countries.

Flies are among the more common insects, and many different species are found around the world. Flies differ greatly in size, but all have at least two wings and fairly well developed body segments. Not all flies transmit disease, but those that do are usually bloodsuckers. Female sand flies (*Phlebotomus;* fle-bot´ō-mŭs) transmit leishmaniasis in North Africa, the Middle East, Europe, and parts of Asia. Tsetse flies (*Glossina;* glo-sī´nă) are limited geographically to tropical Africa, where they are found in brushy areas and transmit African sleeping sickness.

Mosquitoes are a type of fly, though they are morphologically distinct from other fly species. Female mosquitoes are thin and have wings, elongated bodies, long antennae, long legs, and a long proboscis for feeding on blood. Male mosquitoes do not feed on blood. Mosquitoes are found throughout the world, but particular species are geographically limited. Mosquitoes are the most important arthropod vectors of diseases, and they carry the pathogens that cause malaria, yellow fever, dengue fever, filariasis, viral encephalitis, and Rift Valley fever.

Kissing bugs are relatively large, winged, true bugs with cone-shaped heads and wide abdomens. They are called kissing bugs because of their tendency to take blood meals near the mouths of their hosts. Both sexes feed nocturnally while their victims sleep. Kissing bugs transmit disease in Central and South America. The most important disease they transmit is Chagas' disease.

MICRO IN THE CLINIC FOLLOW-UP

A Souvenir from Paradise

The tests reveal that Maria has *dengue fever*, a blood infection caused by one of four closely related dengue viruses transmitted by the *Aedes* mosquito—making dengue a vector-borne disease. In the past, dengue was primarily a concern for people traveling to areas such as Latin America, Puerto Rico, and Southeast Asia, but cases in the United States have surfaced in *recent* years. The control of vector-borne diseases generally focuses on controlling the vector involved—in this case, mosquitoes. Strategies include eliminating standing water reserves for *mosquito* larvae, wearing long sleeves and long pants to prevent exposure to mosquitoes, and eliminating mosquito access to the house by securely screening doors and windows.

There is no specific *treatment* for dengue fever, and most people fully recover on their own. Maria is fortunate she did not contract dengue hemorrhagic fever, a more serious form of infection that is caused by the same virus and that can be life threatening. Maria's doctor advises her to use acetaminophen or other pain relievers to deal with the ache. After two weeks of a frightening experience, Maria is back to work, dreaming about her next scuba adventure.

1. **What does the term *vector* mean in relation to infectious diseases?**
2. **How are flying arthropod vectors epidemiologically significant?**

MM® Check your answers to Micro in the Clinic Follow-Up questions in the MasteringMicrobiology Study Area.

Explore the Invisible: Principles of Sexual Reproduction in Fungi

Practice on-the-go with Dr. Bauman Video Tutors by scanning this QR code with your smart phone. Visit the **MasteringMicrobiology Study Area** to challenge your understanding with practice tests, animation quizzes, and clinical case studies!

MasteringMicrobiology®

CHAPTER SUMMARY

General Characteristics of Eukaryotic Organisms (pp. 352–357)

1. A typical eukaryotic nucleus may be **haploid** (having a single copy of each chromosome) or **diploid** (having two copies). It divides by **mitosis** in four phases—**prophase**, **metaphase**, **anaphase**, and **telophase**—resulting in two nuclei with the same ploidy as the original.
2. **Meiosis** is nuclear division that results in four nuclei, each with half the ploidy of the original.
3. A cell's cytoplasm divides by **cytokinesis** either during or after nuclear division.
4. **Coenocytes** are multinucleate cells resulting from repeated mitoses but postponed or no cytokinesis.
5. Some microbes undergo multiple mitoses by **schizogony** to form a multinucleate **schizont**, which then undergoes cytokinesis.
6. The classification of eukaryotic microbes is problematic and has changed frequently. Historical schemes based on similarities in morphology and chemistry have been replaced with schemes based on nucleotide sequences and ultrastructural features.

Protozoa (pp. 357–363)

1. **Protozoa** (studied by protozoologists) are eukaryotic, unicellular organisms that lack cell walls. Most of them are chemoheterotrophs.
2. A motile **trophozoite** is the feeding stage of a typical protozoan. A **cyst**, a resting stage that is resilient to environmental changes, is formed by some protozoa.
3. A few protozoa undergo sexual reproduction by forming **gametocytes** that fuse to form a **zygote**.
4. Protozoa may be classified into six groups: parabasalids, diplomonads, euglenozoa, alveolates, rhizaria, and amoebozoa.
5. Parabasalids (e.g., *Trichomonas*) are characterized by a Golgi body–like structure called a parabasal body.
6. Members of the Diplomonadida lack mitochondria, Golgi bodies, and peroxisomes.
7. Unicellular flagellated **euglenids** are euglenozoa that store food as paramylon, lack cell walls, and have eyespots used in positive phototaxis. Because they exhibit characteristics of both animals and plants, they are a taxonomic problem.
8. A **kinetoplastid** is a euglenozoan with a single, large, apical mitochondrion that contains a kinetoplast, which is a region of DNA.
9. Alveolates, with cavities called alveoli beneath their cell surfaces, include **ciliate** alveolates (characterized by cilia), **apicomplexans** (all are pathogenic), and **dinoflagellates** (responsible for **red tides**).
10. Protozoa that move and feed with pseudopods are **amoebae**, which are classified into two kingdoms: Rhizaria and Amoebozoa. Rhizaria include **foraminifera**, which have threadlike pseudopods and calcium carbonate shells, and **radiolaria**, which have threadlike pseudopods and silica shells.
11. Amoebozoa have lobe-shaped pseudopodia. The latter include free-living amoebae, parasitic amoebae, and slime molds. **Slime molds** lack cell walls and are phagocytic in their nutrition.
12. **Plasmodial slime molds** (acellular slime molds) are composed of multinucleate cytoplasm. **Cellular slime molds** are composed of myxamoebae that phagocytize bacteria and yeasts.

Fungi (pp. 363–374)

1. **Fungi** (studied by mycologists) are chemoheterotrophic eukaryotes with cell walls usually composed of **chitin**.
2. Most fungi are beneficial, but some produce **mycoses** (fungal diseases).
3. Molds bodies are composed of tubular filaments called **hyphae**. Yeasts are globular, single cells.
4. Hyphae are described as either **septate** or **aseptate**, depending on the presence of cross walls. A **mycelium** is a tangled mass of hyphae.
5. A **dimorphic** fungus is either moldlike (with hyphae) or yeastlike, depending on environmental conditions.
6. Most fungi are **saprobes**—they acquire nutrients by absorption from dead organisms; others get nutrients from living organisms using **haustoria** that penetrate host tissues.
7. Fungi reproduce asexually either by budding or via asexual spores, which are categorized according to their mode of development. Most fungi also reproduce sexually via sexual spores.

   ▶**VIDEO TUTOR:** *Principles of Sexual Reproduction in Fungi*
8. Most fungi in the division **Zygomycota** produce rough-walled **zygosporangia**. **Microsporidia** are intracellular parasites formerly classified as protozoa but now classed with zygomycetes based on genetic analysis.
9. Fungi in the division **Ascomycota**, a group of economically important fungi, produce **ascospores** within sacs called **asci**.
10. Fungi of the division **Basidiomycota** have fruiting bodies called **basidiocarps** that include mushrooms, puffballs, and bracket fungi. Basidiocarps produce basidiospores at the ends of basidia.
11. Deuteromycetes (imperfect fungi) are an informal grouping of fungi having no known sexual stage.
12. **Lichens** are economically and environmentally important organisms composed of fungi living in partnership with photosynthetic microbes, either green algae or cyanobacteria.

Algae (pp. 374–378)

1. Algae (studied by phycologists) typically reproduce by an **alternation of generations** in which a haploid body alternates with a diploid body.
2. Large algae are multicellular with stemlike stipes, leaflike blades, and holdfasts that attach them to substrates.
3. Division **Chlorophyta** contains green algae, which are metabolically similar to land plants.

4. **Rhodophyta**, red algae, contain the pigment **phycoerythrin**, the storage molecule floridean starch, and cell walls of **agar** or **carrageenan**, substances used as thickening agents.
5. **Phaeophyta**, brown algae, contain xanthophylls, laminarin, and oils. They have cell walls composed of cellulose and **alginic acid**, which is another thickening agent. A brown algal spore is motile by means of one "hairy" flagellum and one whiplike flagellum.
6. **Chrysophyta**—the golden algae, yellow-green algae, and **diatoms**—contain chrysolaminarin as a storage product. The silica cell walls of diatoms are arranged in nesting halves called frustules.

Water Molds (pp. 378–379)

1. **Water molds** have tubular cristae in their mitochondria, cell walls of cellulose, spores having two different flagella, and diploid bodies. They are placed in the kingdom Stramenopila along with chrysophytes and brown algae.

Other Eukaryotes of Microbiological Interest: Parasitic Helminths and Vectors (pp. 379–381)

1. Parasitic helminths are significant to microbiologists in part because their infective stages are usually microscopic.
2. Also important to microbiologists are **arthropod vectors**, animals that carry and transmit pathogens. Mechanical vectors merely carry microbes; biological vectors also serve as microbial hosts.
3. **Ticks** and **mites** (chiggers) are **arachnids**—eight-legged arthropods.
4. **Insect** vectors include **fleas**; **lice**; bloodsucking **flies**, including **mosquitoes**; and **kissing bugs**.

QUESTIONS FOR REVIEW

Answers to the Questions for Review (except Short Answer questions) begin on p. A-1.

Multiple Choice

1. Haploid nuclei ________________.
   a. contain one set of chromosomes
   b. contain two sets of chromosomes
   c. contain half a set of chromosomes
   d. are found in the cytosol of eukaryotic organisms
2. Which of the following sequences reflects the correct order of events in mitosis?
   a. telophase, anaphase, metaphase, prophase
   b. prophase, anaphase, metaphase, telophase
   c. telophase, prophase, metaphase, anaphase
   d. prophase, metaphase, anaphase, telophase
3. Which of the following statements accurately describes prophase?
   a. The cell appears to have a line of chromosomes across the midregion.
   b. The nuclear envelope becomes visible.
   c. The cell constructs microtubules to form a spindle.
   d. Chromatids separate and become known as chromosomes.
4. Multiple nuclear divisions without cytoplasmic divisions result in cells called ________________.
   a. mycoses
   b. coenocytes
   c. haustoria
   d. a pseudohypha
5. Tubular filaments with cross walls found in large fungi are ________________.
   a. septate hyphae
   b. aseptate hyphae
   c. aseptate haustoria
   d. dimorphic mycelia
6. The type of asexual fungal spore that forms within hyphae is called a ________________.
   a. sporangiospore
   b. conidiospore
   c. blastospore
   d. chlamydospore
7. A phycologist studies which of the following?
   a. classification of eukaryotes
   b. alternation of generations in algae
   c. rusts, smuts, and yeasts
   d. parasitic worms
8. The stemlike portion of a seaweed is called its ________________.
   a. thallus
   b. holdfast
   c. stipe
   d. blade
9. Carrageenan is found in the cell walls of which group of algae?
   a. red algae
   b. green algae
   c. dinoflagellates
   d. yellow-green algae
10. Chrysolaminarin is a storage product found in which group of microbes?
    a. dinoflagellates
    b. euglenids
    c. golden algae
    d. brown algae
11. Which of the following features characterizes diatoms?
    a. laminarin and oils as food reserves
    b. protective plates of cellulose in their cells
    c. chlorophylls *a* and *c* and carotene
    d. paramylon as a food storage molecule

12. Amoebae include microbes with ______________.
    a. threadlike pseudopods
    b. eyespots
    c. parabasal bodies
    d. alveoli

13. The motile feeding stage of a protozoan is called ______________.
    a. an apicomplexan
    b. a gametocyte
    c. a cyst
    d. a trophozoite

14. Which of the following is common to mitosis and meiosis?
    a. spindle
    b. crossing over
    c. tetrad of chromatids
    d. cytokinesis

15. Which taxon is characterized by "hairy" flagella?
    a. Apicomplexa
    b. Euglenozoa
    c. Alveolata
    d. Stramenopila

Matching

1.	_____ Mitosis	A.	Cytoplasmic division
2.	_____ Meiosis	B.	Diploid nuclei producing haploid nuclei
3.	_____ Homologous chromosomes	C.	Results in genetic variation
4.	_____ Crossing over	D.	Carry similar genes
5.	_____ Cytokinesis	E.	Diploid nuclei producing diploid nuclei

1.	_____ Chitin	A.	Fungal cell wall component
2.	_____ Basidiospore	B.	Fungus + alga or bacterium
3.	_____ Zygosporangium	C.	Fungal filament
4.	_____ Hypha	D.	Fungal spore formed in a sac
5.	_____ Ascospore	E.	Diploid fungal zygote with a thick wall
6.	_____ Lichen	F.	Fungal spore formed on club-shaped hypha

1.	_____ Chlorophyta	A.	Foraminifera
2.	_____ Rhodophyta	B.	Yellow-green algae
3.	_____ Chrysophyta	C.	Green algae
4.	_____ Phaeophyta	D.	Brown algae
5.	_____ Rhizaria	E.	Red algae

VISUALIZE IT!

1. Label the photos below with the type of fungal spore and indicate whether the spore is asexual or sexual.

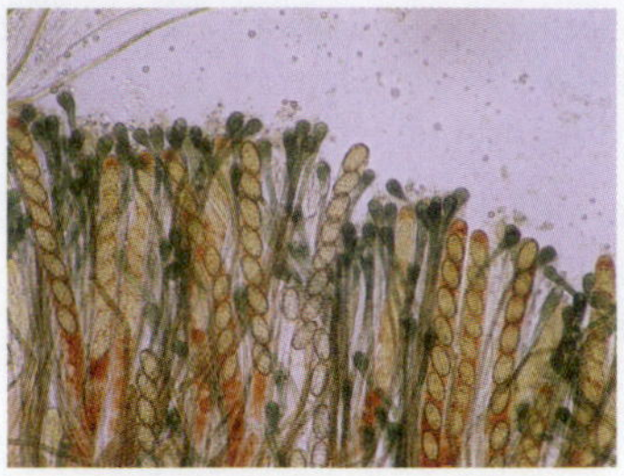

a. ____________, ____________

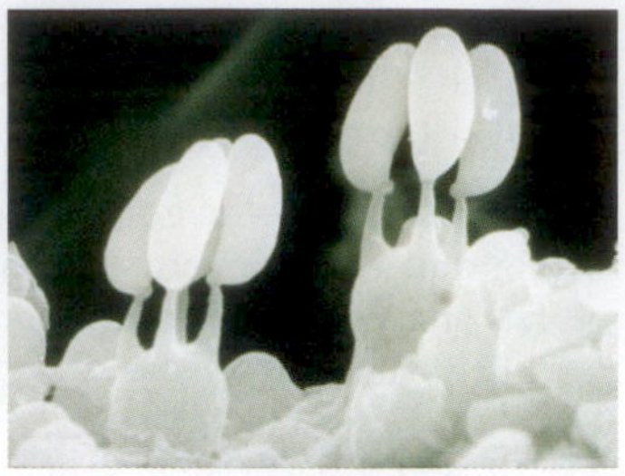

b. ____________, ____________

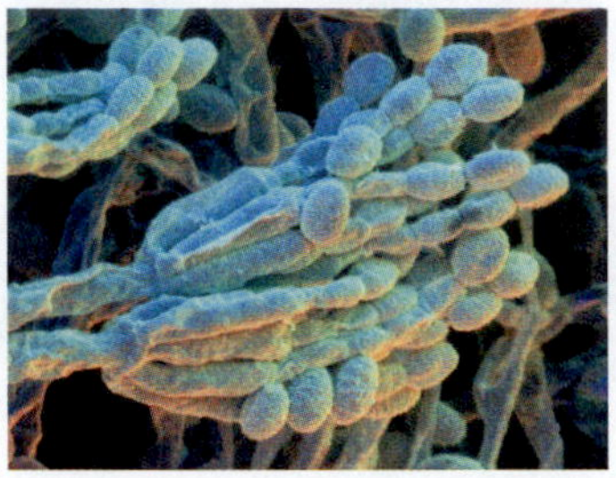

c. ____________, ____________

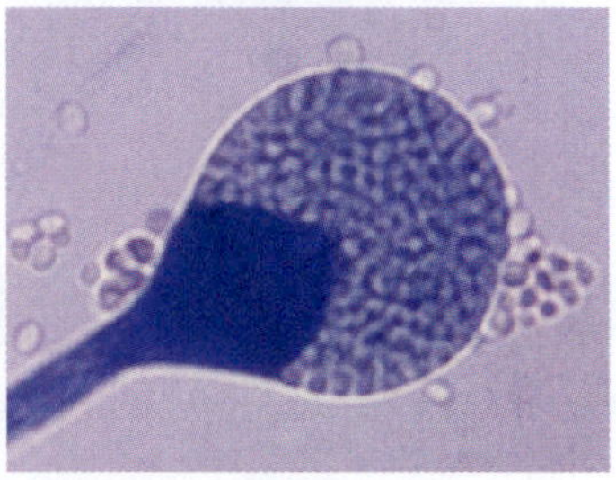

d. ____________, ____________

2. Describe the features of a general fungal life cycle.

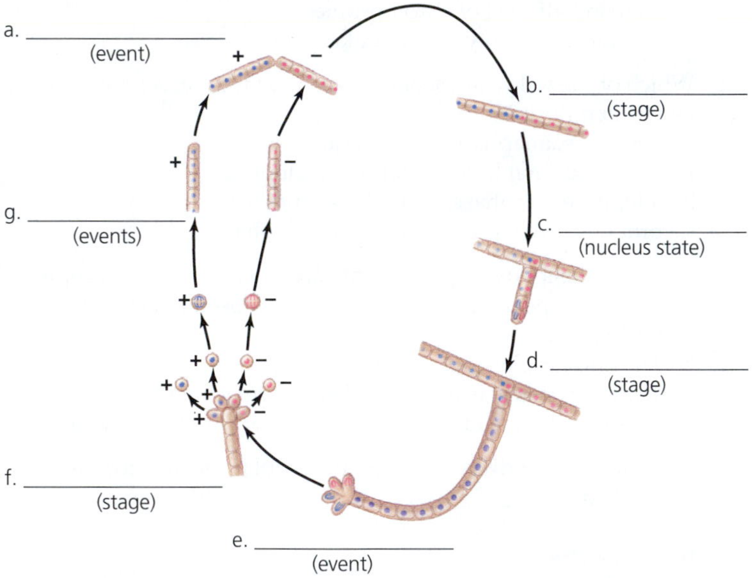

Short Answer

1. Compare and contrast the following closely related terms:
   Chromatid and chromosome
   Mitosis and meiosis II
   Hypha and mycelium
   Algal body and fungal body
   Water mold and slime mold
2. How do fungi transport nutrients?
3. How are lichens useful in environmental protection studies?

4. What are the taxonomic challenges in classifying euglenids?
5. List several economic benefits of algae.
6. Why are relatively large animals, such as parasitic worms, studied in microbiology?
7. Name two ways that slime molds differ from true fungi.
8. What is the role of rRNA sequencing in the classification of eukaryotic microbes?
9. Describe the nuclear divisions that produce eight ascospores in an ascus.

Fill in the Blanks

1. The study of protozoa is called ________________.
2. The study of fungi is called ________________.
3. The study of algae is called ________________.
4. Fungal diseases are called ________________.
5. Amoebae with stiff pseudopods and silica shells are ________________.

CRITICAL THINKING

1. How are cysts of protozoa similar to bacterial endospores? How are they different?
2. The host of a home improvement show suggests periodically emptying a package of yeast into a drain leading to a septic tank. Explain why this would be beneficial.
3. Why doesn't penicillin act against any of the pathogens discussed in this chapter?
4. How can one distinguish a filamentous fungus from a colorless alga?
5. Why do scientists as a group spend more time and money studying protozoa than they do studying algae?
6. Why are there more antibacterial drugs than antifungal drugs?
7. Which type of metabolic pathways are present in protozoa that lack mitochondria (amoebae, diplomonads, and parabasalids)? Which metabolic pathways are absent?
8. Without reference to genetic sequences, mycologists are certain that none of the septate, filamentous deuteromycetes will be shown to make zygospores. How can they be certain when the sexual stages of deuteromycetes are still unknown?
9. Twenty years ago, *Pneumocystis jirovecii*, a pathogen of immunocompromised patients that causes pneumonia, was classified as a protozoan because it is a chemoheterotroph that lacks a cell wall. However, taxonomists today classify *Pneumocystis* as a fungus. Why do you think this pathogen has been reclassified as a fungus?
10. Explain why both dinoflagellates and euglenids were originally classified by zoologists as protozoa and by botanists as algae.
11. Why are cellular slime molds called *cellular*?
12. Fungi tend to reproduce sexually when nutrients are limited or other conditions are unfavorable, but they reproduce asexually when conditions are more ideal. Why is this a successful strategy?
13. Since *Prototheca* is colorless, how do scientists know that it is really a green alga?

CONCEPT MAPPING

Using the following terms, draw a concept map that describes eukaryotic microorganisms. You may use some terms more than once. For a sample concept map, see page 94. Or, complete this and other concept maps online by going to the MasteringMicrobiology Study Area.

Algae
Cell walls (3)
Cellulose
Chitin
Colonial
Cryptosporidium
Fungi
Giardia
Mold
Multicellular (2)
Photosynthetic
Plasmodium
Protozoa
Unicellular (3)
Yeasts

13 Characterizing and Classifying Viruses, Viroids, and Prions

MICRO IN THE CLINIC

Outbreak in the Jungle

Every year fruit bats migrate through the rainforests of the Democratic Republic of Congo—good news for rural residents, who cook and eat the animals as a source of protein. After the hunters in her village bring home a large bounty, 20-year-old Nicia helps prepare the bloody carcasses for a festive meal.

A week later, Nicia suddenly develops a sore throat, high fever, and severe headache. Her symptoms worsen rapidly: she vomits blood and has bloody diarrhea. Then she begins bleeding from her eyes, ears, nose, mouth, and rectum. Soon seizures rack her small body, and her liver and kidneys stop functioning. Five days after her first symptoms appeared, Nicia dies. Nicia's grieving mother, aunt, and 15-year-old sister Odette carefully bathe her blood-covered body and prepare it for burial. The whole village turns out for her funeral.

A few days later, Odette complains of a bad headache and sore throat. Her body temperature soars, and her muscles ache. Soon she, her mother, and her aunt are vomiting blood and experiencing bloody diarrhea even as other villagers start showing the same symptoms. The mysterious killer appears to be spreading through the community.

Terrified, the villagers contact the provincial government in the city of Kananga and plead for help.

What do you think is causing this mysterious illness? How is it spreading among the victims? Turn to the end of the chapter (p. 409) to find out.

Explore More: Test your readiness and apply your knowledge with dynamic learning tools at MasteringMicrobiology.

Not all pathogens are cellular. Many infections of humans, animals, and plants (and even of bacteria) are caused by **acellular** (noncellular) agents, including viruses and other pathogenic particles called viroids and prions. Although these agents are like some eukaryotic and prokaryotic microbes in that they cause disease when they invade susceptible cells, they are simple compared to a cell—lacking cell membranes and composed of only a few organic molecules. In addition to lacking a cellular structure, they lack most of the characteristics of life: They cannot carry out any metabolic pathway; they can neither grow nor respond to the environment; and they cannot reproduce independently but instead must utilize the chemical and structural components of the cells they infect. They must recruit the cell's metabolic pathways in order to increase their numbers.

In this chapter we first examine a range of topics concerning viruses: what their characteristics are, how they are classified, how they replicate, what role they play in some kinds of cancers, how they are maintained in the laboratory, and whether viruses are alive. Then we consider the nature of viroids and prions.

Characteristics of Viruses

Viruses cause most of the diseases that still plague the industrialized world: the common cold, influenza, herpes, and AIDS, to name a few. Although we have immunizations against many viral diseases and are adept at treating the symptoms of others, the characteristics of viruses and the means by which they attack their hosts make cures for viral diseases elusive. Throughout this section, we consider the clinical implications of viral characteristics.

We begin by looking at the characteristics viruses have in common. A **virus** is a minuscule, acellular, infectious agent having one or several pieces of nucleic acid—either DNA or RNA. The nucleic acid is the genetic material (genome) of the virus. Being acellular, viruses have no cytoplasmic membrane (though, as we will see, some viruses possess a membrane-like *envelope*). Viruses also lack cytosol and functional organelles. They are not capable of metabolic activity on their own; instead, once viruses have invaded a cell, they take control of the cell's metabolic machinery to produce more molecules of viral nucleic acid and viral proteins, which then assemble into new viruses via a process we will examine shortly.

Viruses have an extracellular and an intracellular state. Outside a cell, in the extracellular state, a virus is called a **virion** (vir´ē-on). Basically, a virion consists of a protein coat, called a **capsid**, surrounding a nucleic acid core (**FIGURE 13.1a**). Together the nucleic acid and its capsid are also called a *nucleocapsid*, which in many cases can crystallize like crystalline chemicals (**FIGURE 13.1b**). Some virions have a phospholipid membrane called an **envelope** surrounding the nucleocapsid. The outermost layer of a virion (capsid or envelope) provides the virus both protection and recognition sites that bind to complementary chemicals on the surfaces of their specific host cells. Once a virus is inside, the intracellular state is initiated, and the capsid is removed. A virus without its capsid exists solely as nucleic acid but is still referred to as a virus.

Now that we have examined ways in which viruses are alike, we will consider the characteristics that are used to distinguish different viral groups. Viruses differ in the type of genetic material they contain, the kinds of cells they attack, their size, the nature of their capsid coat, their shapes, and the presence or absence of an envelope.

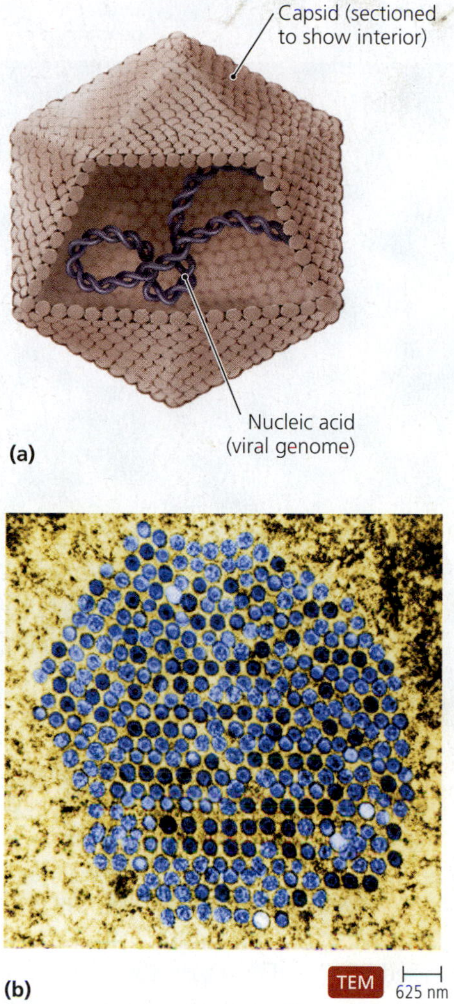

▲ **FIGURE 13.1 Virions, complete virus particles, include a nucleic acid, a capsid, and in some cases an envelope. (a)** A drawing of a nonenveloped polyhedral virus containing DNA. **(b)** A transmission electron microscope image of crystallized tobacco mosaic virus. Like many chemicals and unlike cells, some viruses can form crystals.

Genetic Material of Viruses

LEARNING OUTCOME

13.1 Discuss viral genomes in terms of dsDNA, ssDNA, ssRNA, dsRNA, and number of segments of nucleic acid.

Viruses show more variety in the nature of their genomes than do cells. Whereas the genome of every cell is double-stranded DNA, the genome of a virus may be either DNA or RNA. The primary way in which scientists categorize and classify viruses is based on the type of genetic material that makes up the viral genome.

Some viral genomes, such as those of herpesvirus and chicken pox virus, are double-stranded DNA (dsDNA), like the genomes of cells. Other viruses use either single-stranded RNA (ssRNA),

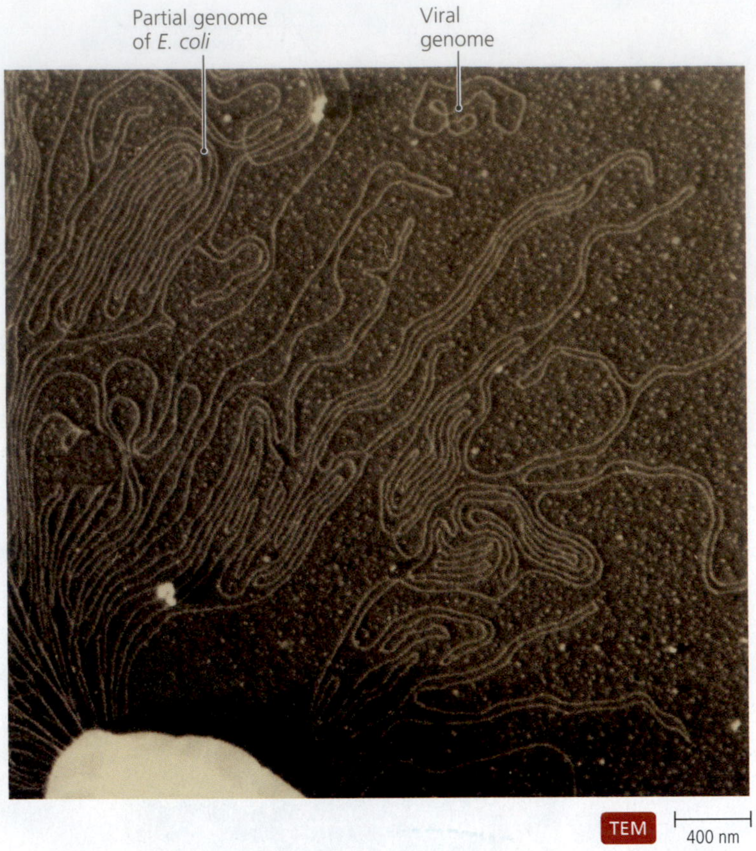

▲ **FIGURE 13.2 The relative sizes of genomes.** A cell of *Escherichia coli*, just visible at lower left, was ruptured to release DNA.

single-stranded DNA (ssDNA), or double-stranded RNA (dsRNA) as their genomes. These molecules never function as a genome in any cell—in fact, ssDNA and dsRNA are almost nonexistent in cells. Further, the genome of any particular virus may be either linear and composed of several molecules of nucleic acid, as in eukaryotic cells, or circular and singular, as in most prokaryotic cells. For example, the genome of an influenzavirus is composed of eight linear segments of single-stranded RNA, whereas the genome of poliovirus is one molecule of single-stranded RNA.

Viral genomes are usually smaller than the genomes of cells. For example, the genome of the smallest bacterium (a species of *Chlamydia*) has almost 1000 genes, whereas the genome of virus MS2 has only three genes. **FIGURE 13.2** compares the genome of a virus with the genome of the bacterium *Escherichia coli* (esh-ĕ-rik´ē-ă kō´lī), which contains over 4000 genes.

Hosts of Viruses

LEARNING | **OUTCOMES**

13.2 Explain the mechanism by which viruses are specific for their host cells.

13.3 Compare and contrast viruses of fungi, plants, animals, and bacteria.

Most viruses infect only a particular host's cells. This specificity is due to the precise affinity of viral surface proteins or glycoproteins for complementary proteins or glycoproteins on the surface of the host cell. Viruses may be so specific that they infect not only a particular host but also a particular kind of cell in that host. For example, HIV (human immunodeficiency virus, the agent that causes AIDS) specifically attacks helper T lymphocytes (a type of white blood cell) in humans and has no effect on, say, human muscle or bone cells. By contrast, some viruses are *generalists;* they infect many kinds of cells in many different hosts. An example of a generalist virus is West Nile virus, which can infect humans, most species of birds, several mammalian species, and some reptiles.

All types of organisms are susceptible to some sort of viral attack. There are viruses that infect archaeal, bacterial, plant, protozoan, fungal, and animal cells **(FIGURE 13.3)**. There is even a tiny virus that attacks a large virus. Most viral research and scientific study has focused on bacterial and animal viruses. A virus that infects bacteria is referred to as a **bacteriophage** (bak-tēr´ē-ō-fāj) or simply a **phage** (fāj). Scientists have determined that bacteriophages outnumber all bacteria, archaea, and eukaryotes put together. We will return our attention to bacteriophages and animal viruses later in this chapter.

Viruses of plants are less well known than bacterial and animal viruses, even though viruses were first identified and isolated from tobacco plants. Plant viruses infect many food crops, including corn, beans, sugarcane, tobacco, and potatoes, resulting in billions of dollars in losses each year. Viruses of plants are introduced into plant cells either through abrasions of the cell wall or by plant parasites, such as nematodes and aphids. After entry, plant viruses follow the replication cycle discussed below for animal viruses.

Fungal viruses have been little studied. We do know that fungal viruses are different from animal and bacterial viruses in that fungal viruses exist only within cells; that is, they seemingly have no extracellular state. Presumably, fungal viruses cannot penetrate a thick fungal cell wall. However, because fusion of cells is typically a part of a fungal life cycle, viral infections can easily be propagated by the fusion of an infected fungal cell with an uninfected one.

Not all viruses are deleterious. The box **Beneficial Microbes: Good Viruses? Who Knew?** on p. 390 illustrates some useful aspects of viruses in the environment.

Sizes of Viruses

In the late 1800s, scientists hypothesized that the cause of many diseases, including polio and smallpox, was an agent smaller than a bacterium. They named these tiny agents "viruses," from the Latin word for "poison." Viruses are so small that only a few can be seen by light microscopy. One hundred million polioviruses could fit side by side on the period at the end of this sentence. Even smaller viruses have a diameter of 24 nm; and the largest virus—*Megavirus*—is about 500 nm in diameter, which is about the diameter of many bacterial cells. **FIGURE 13.4** compares the sizes of selected viruses to *E. coli* and a human red blood cell.

(a)

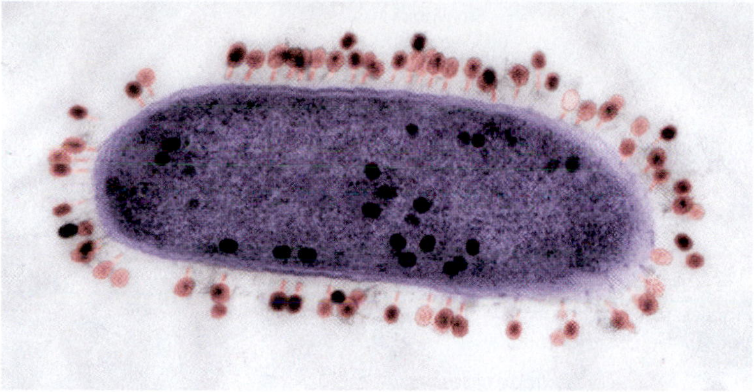

(b) TEM 0.5 μm

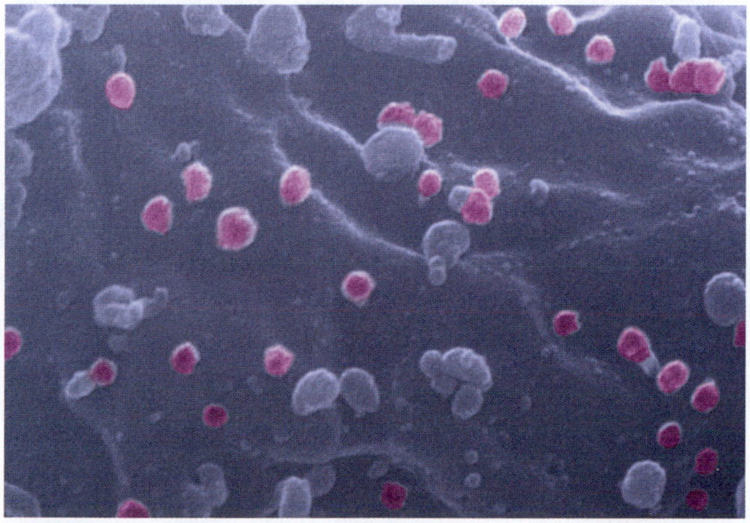

(c) SEM 100 nm

◀ **FIGURE 13.3 Some examples of plant, bacterial, and human hosts of viral infections. (a)** Tobacco mosaic virus—the first virus isolated—causes yellow discolorations of tobacco leaves. **(b)** A bacterial cell (purple) under attack by bacteriophages (pink). **(c)** A human white blood cell's cytoplasmic membrane, to which HIV (pink) is attached.

In 1892, Russian microbiologist Dmitri Ivanowski (1864–1920) first demonstrated that viruses are acellular with an experiment designed to elucidate the cause of tobacco mosaic disease. He filtered the sap of infected tobacco plants through a porcelain filter fine enough to trap even the smallest of bacterial cells. Viruses, however, were not trapped but instead passed through the filter with the liquid, which remained infectious to tobacco plants. This experiment proved the existence of an acellular disease-causing entity smaller than a bacterium. Tobacco mosaic virus (TMV) was isolated and characterized in 1935 by an American chemist, Wendell Stanley (1904–1971). The invention of electron microscopy allowed scientists to finally see TMV and other viruses.

Capsid Morphology

LEARNING OUTCOME

13.4 Discuss the structure and function of the viral capsid.

As we have seen, viruses have capsids—protein coats that provide both protection for viral nucleic acid and a means by which many viruses attach to their hosts' cells. The capsid of a virus is composed of proteinaceous subunits called **capsomeres** (or *capsomers*). Some capsomeres are composed of only a single type of protein, whereas others are composed of several different kinds of proteins. Recall that viral nucleic acid surrounded by its capsid is termed a *nucleocapsid*.

Viral Shapes

The shapes of virions are also used to classify viruses. There are three basic types of viral shapes: helical, polyhedral, and complex **(FIGURE 13.5)**. The capsid of a helical virus is composed of capsomeres that bond together in a spiral fashion to form a tube around the nucleic acid. The capsid of a polyhedral virus is roughly spherical, with a shape similar to a geodesic dome. The most common type of polyhedral capsid is an icosahedron, which has 20 sides.

Complex viruses have capsids of many different shapes that do not readily fit into either of the other two categories. An example of a complex virus is smallpox virus, which has several covering layers (including lipid) and no easily identifiable capsid. The complex shapes of many bacteriophages include icosahedral heads, which contain the genome, attached to helical tails with tail fibers. The complex capsids of such bacteriophages somewhat resemble NASA's lunar lander **(FIGURE 13.6)**.

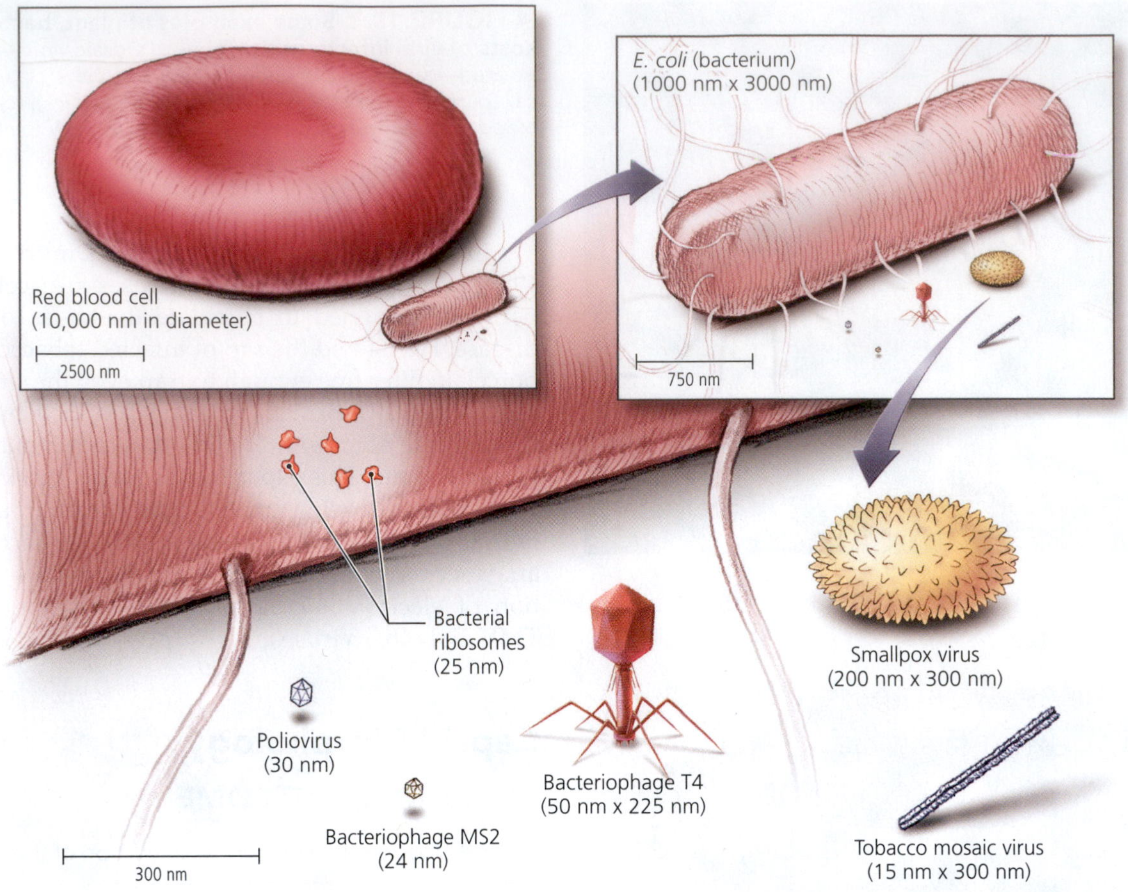

▲ **FIGURE 13.4 Sizes of selected virions.** Selected viruses are compared in size to a bacterium, *Escherichia coli*, and a human red blood cell. *How can viruses be so small and yet still be pathogenic?*

Figure 13.4 *Viruses utilize a host cell's enzymes, organelles, and membranes to complete their replication cycle.*

BENEFICIAL MICROBES

Good Viruses? Who Knew?

An algal bloom off the coast of Seattle, Washington

Viruses, though normally pathogenic to their host cells, do have positive influences, including what appear to be extensive roles in the environment. Recent discoveries by the United Kingdom's Marine and Freshwater Microbial Biodiversity program demonstrate important ways viruses impact our world.

First: Scientists found that a previously unknown virus attacks a tiny marine alga that multiplies to form algal blooms consisting of hundreds of thousands to millions of algal cells per milliliter of water, which are visible from space (see photo). Algal blooms like these can deplete the water of oxygen at night, potentially harming fish and other marine life. The newly discovered virus stops blooms by killing the alga, a result that is good for animal life.

Second: When the algal cells die by this means, they release an airborne sulfate compound that acts to seed clouds. The resulting increased cloudiness noticeably shades the ocean, measurably lowering water temperature. Thus, a marine virus protects wildlife and helps to reduce global warming!

Third: The researchers discovered a bacteriophage of oceanic cyanobacteria that transfers genes for photosynthetic machinery into its hosts' cells so that the cells' photosynthetic rate increases. There are up to 10 million of these viruses in a single milliliter of seawater, so researchers estimate that much of the oxygen we breathe may be attributable to the action of this virus on blue-green bacteria.

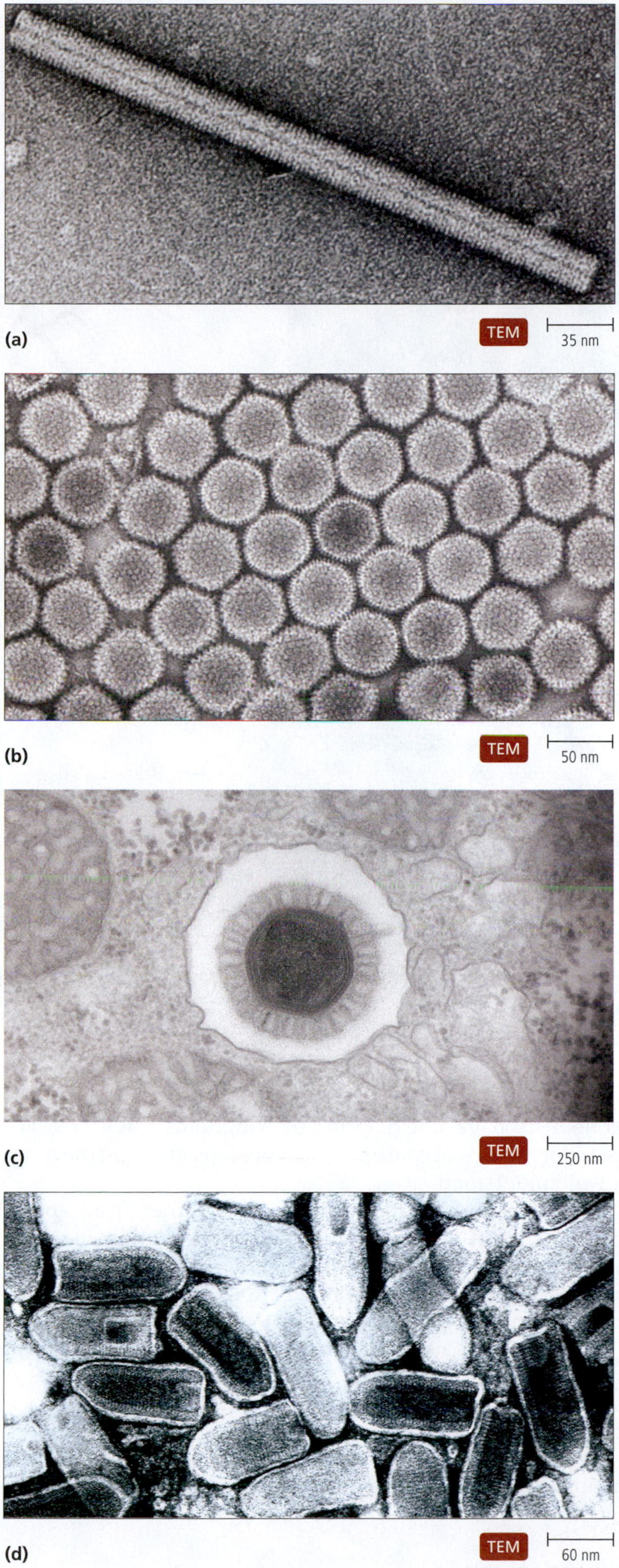

▶ **FIGURE 13.5 The shapes of virions. (a)** A helical virus, tobacco mosaic virus. The tubular shape of the capsid results from the tight arrangement of several rows of helical capsomeres. **(b)** Polyhedral virions of a virus that causes common colds. **(c)** Complex shape of *Megavirus*, which is shown inside a cell's vesicle. **(d)** The complex shape of rabies virus, which results from the shapes of the capsid and bullet-shaped envelope.

The Viral Envelope

LEARNING OUTCOME

13.5 Discuss the origin, structure, and function of the viral envelope.

All viruses lack cell membranes (after all, they are not cells), but some, particularly animal viruses, have an envelope similar in composition to a cell membrane surrounding their capsids. Other viral proteins called *matrix proteins* fill the region between capsid and envelope. A virus with a membrane is an *enveloped virion* **(FIGURE 13.7)**; a virion without an envelope is called a *nonenveloped* or *naked virion*.

An enveloped virus acquires its envelope from its host cell during viral replication or release (discussed shortly). Indeed, the envelope of a virus is a portion of the membrane system of a host cell. Like a cytoplasmic membrane, a viral envelope is composed of a phospholipid bilayer and proteins. Some of the proteins are virally coded glycoproteins, which appear as spikes protruding outward from the envelope's surface (see Figure 13.7). Host DNA carries the genetic code required for the assembly of the phospholipids and some of the proteins in the envelope, while the viral genome specifies the other membrane proteins.

An envelope's proteins and glycoproteins often play a role in the recognition of host cells. A viral envelope does not perform other physiological roles of a cytoplasmic membrane, such as endocytosis or active transport.

TABLE 13.1 on p. 393 summarizes the novel properties of viruses and how those properties differ from the corresponding characteristics of cells. Next we turn our attention to the criteria by which virologists classify viruses.

TELL ME WHY

Why are naked *icosahedral* viruses able to crystallize?

Classification of Viruses

LEARNING OUTCOME

13.6 List the characteristics by which viruses are classified.

The International Committee on Taxonomy of Viruses (ICTV) was established in 1966 to provide a single taxonomic scheme for viral classification and identification. Virologists classify viruses by their type of nucleic acid, presence of an envelope, shape, and size. So far, they have established families for all viral genera, but only three viral orders are described.

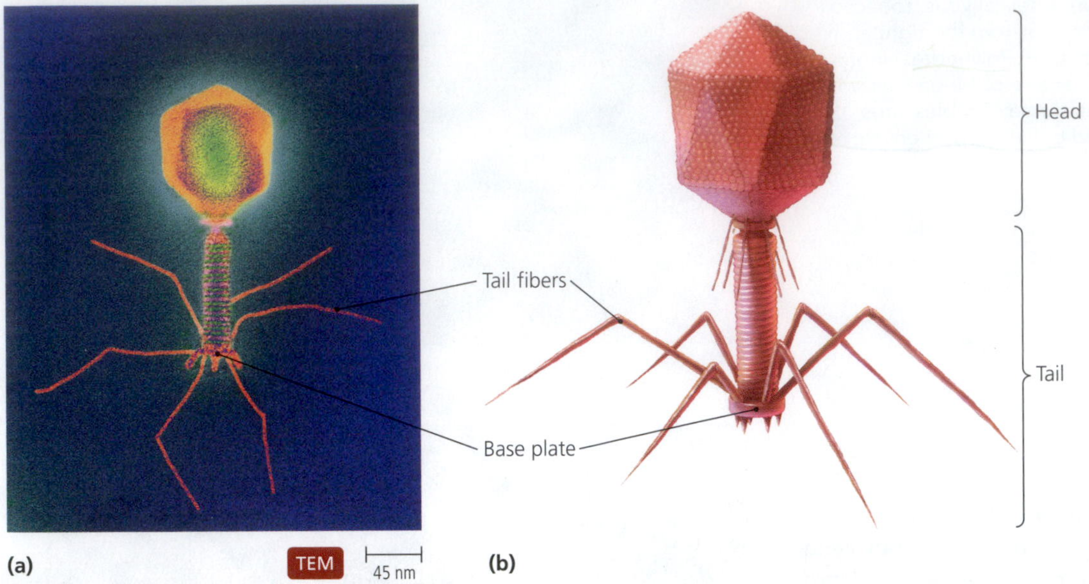

◀ **FIGURE 13.6 The complex shape of bacteriophage T4.** It includes an icosahedral head and an ornate tail that enables viral attachment and penetration.

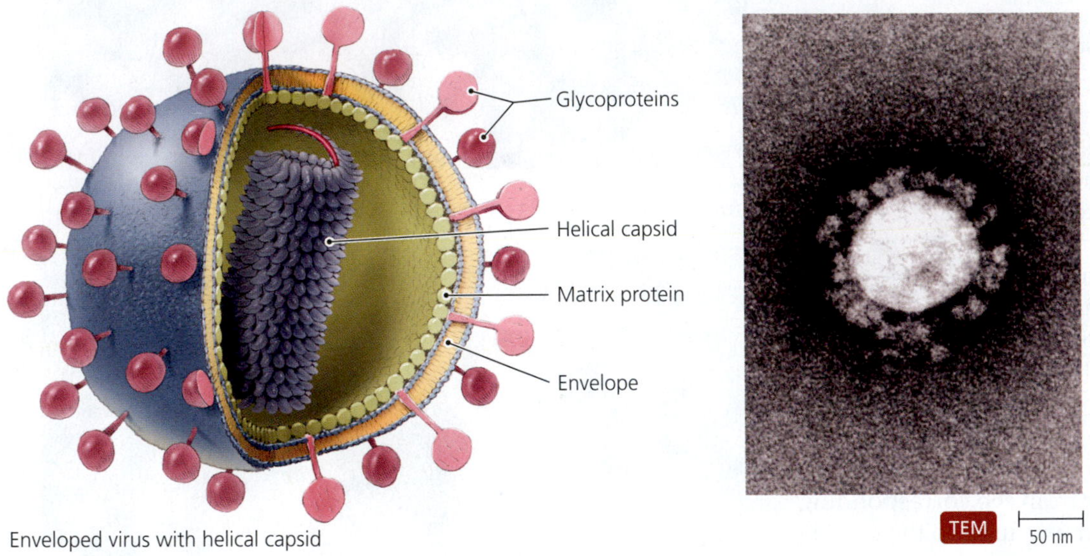

◀ **FIGURE 13.7 Enveloped virion.** Artist's rendition and electron micrograph of severe acute respiratory syndrome (SARS) virus, an enveloped virus with a helical capsid.

Taxonomists have not determined kingdoms, divisions, and classes for viruses because the relationships among viruses are not well understood.

Family names are typically derived either from special characteristics of viruses within the family or from the name of an important member of the family. For example, family *Picornaviridae* contains very small[1] RNA viruses, and *Hepadnaviridae* contains a DNA virus that causes hepatitis B. Family *Herpesviridae* is named for herpes simplex, a virus that can cause genital herpes. **TABLE 13.2** lists the major families of human viruses, grouped according to the type of nucleic acid each contains.

Specific epithets for viruses are their common English designations written in italics. Accordingly, the nomenclature for two important viral pathogens, HIV and rabies virus, is as follows:

	HIV	Rabies Virus
Order	Not yet established	Mononegavirales
Family	*Retroviridae*	*Rhabdoviridae*
Genus	*Lentivirus* (len´ti-ŭvī-rŭs)	*Lyssavirus* (lis´ă-vī-ŭs)
Specific epithet	*human immunodeficiency virus*	*rabies virus*

TELL ME WHY

What characteristics of the genomes of parvoviruses and of reoviruses make them very different from cells?

[1]*Pico* means one-trillionth, 10^{-12}.

TABLE 13.1 The Novel Properties of Viruses

Viruses	Cells
Inert macromolecules outside of a cell but become active inside a cell	Metabolize on their own
Do not divide or grow	Divide and grow
Acellular	Cellular
Obligate intracellular parasites	Most are free living
Contain either DNA or RNA, with few exceptions, such as *Cytomegalovirus* and *Mimivirus*	Contain both DNA and RNA
Genome can be dsDNA, ssDNA, dsRNA, or ssRNA	Genome is dsDNA
Usually ultramicroscopic in size, ranging from 10 nm to 500 nm	200 nm to 12 cm in diameter
Have a proteinaceous capsid around genome; some have an envelope around the capsid	Surrounded by a phospholipid membrane and often a cell wall
Replicate in an assembly-line manner using the enzymes and organelles of a host cell	Self-replicating by asexual and/or sexual means

TABLE 13.2 Families of Human Viruses

Family	Strand Type	Representative Genera (Diseases)
DNA Viruses		
Poxviridae	Double	*Orthopoxvirus* (smallpox)
Herpesviridae	Double	*Simplexvirus* (herpes type 1: fever blisters, respiratory infections; herpes type 2: genital infections); *Varicellovirus* (chicken pox); *Lymphocryptovirus*, Epstein-Barr virus (infectious mononucleosis, Burkitt's lymphoma); *Cytomegalovirus* (birth defects); *Roseolovirus* (roseola)
Papillomaviridae	Double	*Papillomavirus* (benign tumors, warts, cervical and penile cancers)
Polyomaviridae	Double	*Polyomavirus* (progressive multifocal leukoencephalopathy)
Adenoviridae	Double	*Mastadenovirus* (conjunctivitis, respiratory infections)
Hepadnaviridae	Partial single and partial double	*Orthohepadnavirus* (hepatitis B)
Parvoviridae	Single	*Erythrovirus* (erythema infectiosum)
RNA Viruses		
Picornaviridae	Single, +[a]	*Enterovirus* (polio); *Hepatovirus* (hepatitis A); *Rhinovirus* (common cold)
Caliciviridae	Single, +	*Norovirus* (gastroenteritis)
Astroviridae	Single, +	*Astrovirus* (gastroenteritis)
Hepeviridae	Single, +	*Hepevirus* (hepatitis E)
Togaviridae	Single, +	*Alphavirus* (encephalitis); *Rubivirus* (rubella)
Flaviviridae	Single, +	*Flavivirus* (yellow fever, Japanese encephalitis); *Hepacivirus* (hepatitis C)
Coronaviridae	Single, +	*Coronavirus* (common cold, severe acute respiratory syndrome)
Retroviridae	Single, +, segmented	Detaretrovirus (leukemia); *Lentivirus* (AIDS)
Orthomyxoviridae	Single, −[b], segmented	*Influenzavirus* (flu)
Paramyxoviridae	Single, −	*Paramyxovirus* (common cold, respiratory infections); *Pneumovirus* (pneumonia, common cold); *Morbillivirus* (measles); *Rubulavirus* (mumps)
Rhabdoviridae	Single, −	*Lyssavirus* (rabies)
Bunyaviridae	Single, −, segmented	*Bunyavirus* (California encephalitis virus); *Hantavirus* (pneumonia)
Filoviridae	Single, −	*Filovirus* (Ebola hemorrhagic fever); *Marburgvirus* (hemorrhagic fever)
Arenaviridae	Single, −, segmented	*Lassavirus* (hemorrhagic fever)
Reoviridae	Double, segmented	*Orbivirus* (encephalitis); *Rotavirus* (diarrhea); *Coltivirus* (Colorado tick fever)

[a]Positive-sense (+RNA) is equivalent to mRNA; that is, it instructs ribosomes in protein translations.
[b]Negative-sense (−RNA) is complementary to mRNA; it cannot be directly translated.

Viral Replication

As previously noted, viruses cannot reproduce themselves because they lack the genes for all the enzymes necessary for replication; in addition, they do not possess functional ribosomes for protein synthesis. Instead, viruses are dependent on their hosts' enzymes and organelles to produce new virions. Once a host cell falls under control of a viral genome, it is forced to replicate viral genetic material and translate viral proteins, including viral capsomeres and viral enzymes.

The replication cycle of a virus usually results in the death and lysis of the host cell. Because the cell undergoes lysis near the end of the cycle, this type of replication is called a **lytic replication cycle**. In general, a lytic replication cycle consists of the following five stages:

- **Attachment** of the virion to the host cell
- **Entry** of the virion or its genome into the host cell
- **Synthesis** of new nucleic acids and viral proteins by the host cell's enzymes and ribosomes
- **Assembly** of new virions within the host cell
- **Release** of the new virions from the host cell

▶**ANIMATIONS:** *Viral Replication: Overview*

In the following sections we examine the events that occur in the replication of bacteriophages and animal viruses. We begin with lytic replication in bacteriophages, turn to a modification of replication (called lysogenic replication), and then consider the replication of animal viruses.

Lytic Replication of Bacteriophages

LEARNING **OUTCOME**

13.7 Sketch and describe the five stages of the lytic replication cycle as it typically occurs in bacteriophages.

Studies of phages revealed the basics of viral biology. Indeed, bacteriophages make excellent tools for the general study of viruses because they are easier and less expensive to culture than animal or human viruses. **Beneficial Microbes: Prescription Bacteriophages?** on p. 398 is an interesting side note on the potential use of bacteriophages as an alternative to antibiotics.

Here we examine the replication of a much-studied dsDNA phage of *E. coli* called *type 4 (T4)*. T4 virions are complex, having the polyhedral heads and helical tails seen in many bacteriophages. We begin with attachment, the first stage of replication **(FIGURE 13.8)**.

Attachment 1

Because phages, like all virions, are nonmotile, contact with a bacterium occurs by purely random collision, brought about as molecular bombardment and currents move virions through the environment. The structures responsible for the attachment of T4 to its host bacterium are its tail fibers. Attachment is dependent on the chemical attraction and precise fit between attachment proteins on the phage's tail fibers and complementary receptor proteins on the surface of the host's cell wall. The specificity of the attachment proteins for the receptors ensures that the virus will attach only to *E. coli*. Bacteriophages may attach to receptor proteins on bacterial cells' walls, flagella, or pili.

Entry 2

Now that phage T4 has attached to the bacterium's cell wall, it must still overcome the formidable barrier posed by the cell wall and cytoplasmic membrane if it is to enter the cell. T4 overcomes this obstacle in an elegant way. Upon contact with *E. coli*, T4 releases *lysozyme* (lī´sō-zīm), a protein enzyme carried within the capsid that weakens the peptidoglycan of the cell wall. The phage's tail sheath then contracts, forcing an internal hollow tube within the tail through the cell wall and membrane, much as a hypodermic needle penetrates the skin. The phage injects the genome through the tube and into the bacterium. The empty capsid, having performed its task, is left on the outside of the cell looking like an abandoned spacecraft.

After entry, viral enzymes (either carried within the capsid or coded by viral genes and made by the bacterium) degrade the bacterial DNA into its constituent nucleotides.

Synthesis 3

After losing its chromosome, the bacterium stops synthesizing its own molecules and begins synthesizing new viruses under control of the viral genome.

For dsDNA viruses like T4, protein synthesis is straightforward and similar to cellular transcription and translation, except that mRNA is transcribed from viral DNA instead of cellular DNA. Translation by the host cell's ribosomes results in viral proteins, including head capsomeres, components of the tail, viral DNA polymerase (which replicates viral DNA), and lysozyme (which weakens the bacterial cell wall from within, enabling the virions to leave the cell once they have been assembled).

Assembly 4

Scientists do not understand completely how phages are assembled inside a host cell, but it appears that as capsomeres accumulate within the cell, they spontaneously attach to one another to form new capsid heads. Likewise, tails assemble and attach to heads, and tail fibers attach to tails, forming mature virions. Such capsid assembly is a spontaneous process, requiring little or no enzymatic activity. For many years it was assumed that all capsids formed around a genome in just such a spontaneous manner. However, recent research has shown that for some viruses, enzymes pump the genome into the assembled capsid under high pressure—five times that used in a paintball gun. This process resembles stuffing a strand of cooked spaghetti into a matchbox through a single small hole.

Sometimes a capsid assembles around leftover pieces of host DNA instead of viral DNA. A virion formed in this manner is still able to attach to a new host by means of its tail fibers, but instead of inserting phage DNA, it transfers DNA from the first host into a new host. This process is known as *transduction* (described in more detail in Chapter 7).

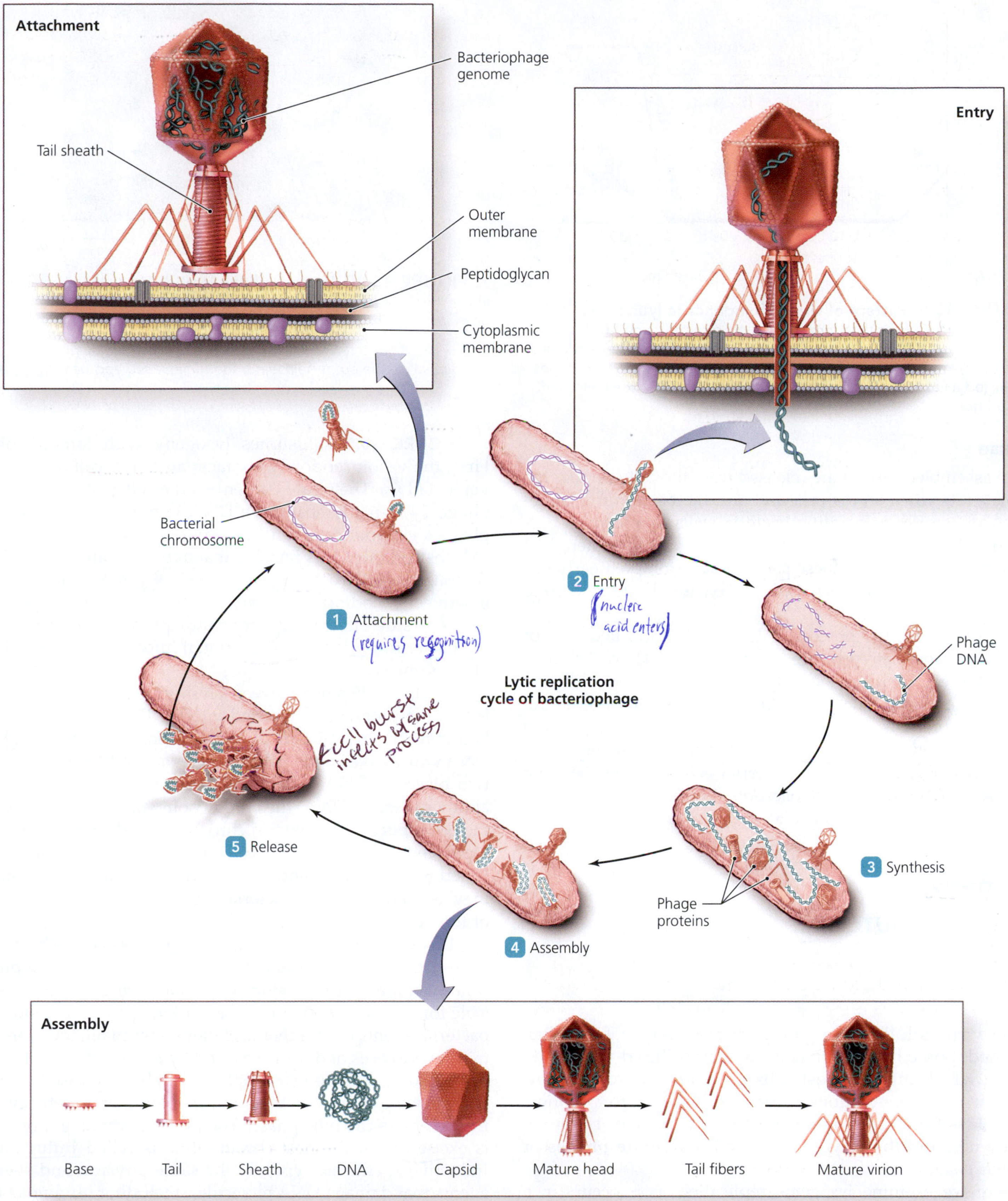

▲ **FIGURE 13.8 The lytic replication cycle in bacteriophages.** The phage shown in this illustration is T4, and the bacterium shown is *E. coli*. The circular bacterial chromosome is represented diagrammatically; in reality, it would be much longer.

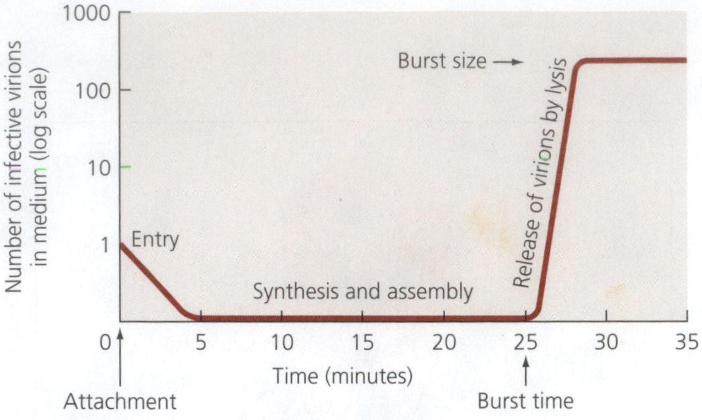

▲ **FIGURE 13.9 Pattern of virion abundance in lytic cycle.** Shown is virion abundance over time for a single lytic replication cycle. New virions are not observed in the culture medium until synthesis, assembly, and release (lysis) are complete, at which time (the burst time) the new virions are released all at once. Burst size is the number of new virions released per lysed host cell.

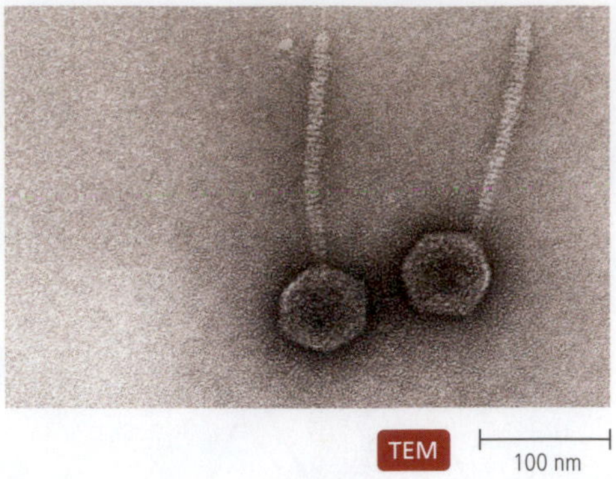

▲ **FIGURE 13.10 Bacteriophage lambda.** Note the absence of tail fibers. *Phage T4 attaches by means of molecules on its tail fibers. How does lambda phage, which lacks fibers, attach?*

Figure 13.10 Lambda has attachment molecules at the end of its tail rather than on its tail fibers.

Release 5

Newly assembled virions are released from the cell as lysozyme completes its work on the cell wall and the bacterium disintegrates. Areas of disintegrating bacterial cells in a lawn of bacteria in a Petri plate look as if the lawn were being eaten, and it was the appearance of these *plaques* that prompted early scientists to give the name *bacteriophage*, "bacterial eater," to these viruses.

For phage T4, the process of lytic replication takes about 25 minutes and can produce as many as 100 to 200 new virions for each bacterial cell lysed (**FIGURE 13.9**). For any phage undergoing lytic replication, the period of time required to complete the entire process, from attachment to release, is called the *burst time*, and the number of new virions released from each lysed bacterial cell is called the *burst size*. ▶**ANIMATIONS:** *Viral Replication: Virulent Bacteriophages* ▶**VIDEO TUTOR:** *The Lytic Cycle of Viral Replication*

Lysogeny

LEARNING **OUTCOME**

13.8 Compare and contrast the lysogenic replication cycle of viruses with the lytic cycle.

Not all viruses follow the lytic pattern of phage T4 we just examined. Some bacteriophages have a modified replication cycle in which infected host cells grow and reproduce normally for many generations before they lyse. Such a replication cycle is called a **lysogenic replication cycle** or **lysogeny** (lī-soj´ĕ-nē), and the phages are called **temperate phages** or *lysogenic phages*.

Here we examine lysogenic replication as it occurs in a much-studied temperate phage, *lambda phage*, which is another parasite of *E. coli*. A lambda phage has a linear molecule of dsDNA in a complex capsid consisting of an icosahedral head attached to a tail that lacks tail fibers (**FIGURE 13.10**).

FIGURE 13.11 illustrates lysogeny with lambda phage. First, the virion randomly contacts an *E. coli* cell and attaches via its tail 1. The viral DNA enters the cell, just as occurs with phage T4, but the host cell's DNA is not destroyed, and the phage's genome does not immediately assume control of the cell. Instead, the virus remains inactive. Such an inactive bacteriophage is called a **prophage** (prō´fāj) 2. A prophage remains inactive by coding for a protein that suppresses prophage genes. A side effect of this repressor protein is that it renders the bacterium resistant to additional infection by other viruses of the same type.

Another difference between a lysogenic cycle and a lytic cycle is that the prophage is inserted into the DNA of the bacterium, becoming a physical part of the bacterial chromosome 3. For DNA viruses like lambda phage, this is a simple process of fusing two pieces of DNA: One piece of DNA, the virus, is fused to another piece of DNA, the chromosome of the cell. Every time the cell replicates its infected chromosome, the prophage is also replicated 4. All daughter cells of a lysogenic cell are thus infected with the quiescent virus. A prophage and its descendants may remain a part of bacterial chromosomes for generations or forever.

Lysogenic phages can change the phenotype of a bacterium, for example from a harmless form into a pathogen—a process called **lysogenic conversion**. Bacteriophage genes are responsible for toxins and other disease-evoking proteins found in the bacterial agents of diphtheria, cholera, rheumatic fever, and certain severe cases of diarrhea caused by *E. coli*.

At some later time a prophage might be excised from the chromosome by recombination or some other genetic event; it then reenters the lytic phase. The process whereby a prophage is excised from the host chromosome is called **induction** 5. Inductive agents are typically the same physical and chemical agents that damage DNA molecules, including ultraviolet light, X rays, and carcinogenic chemicals.

After induction, the lytic steps of synthesis 6, assembly 7, and release 8 resume from the point at which they stopped. The cell becomes filled with virions and breaks open.

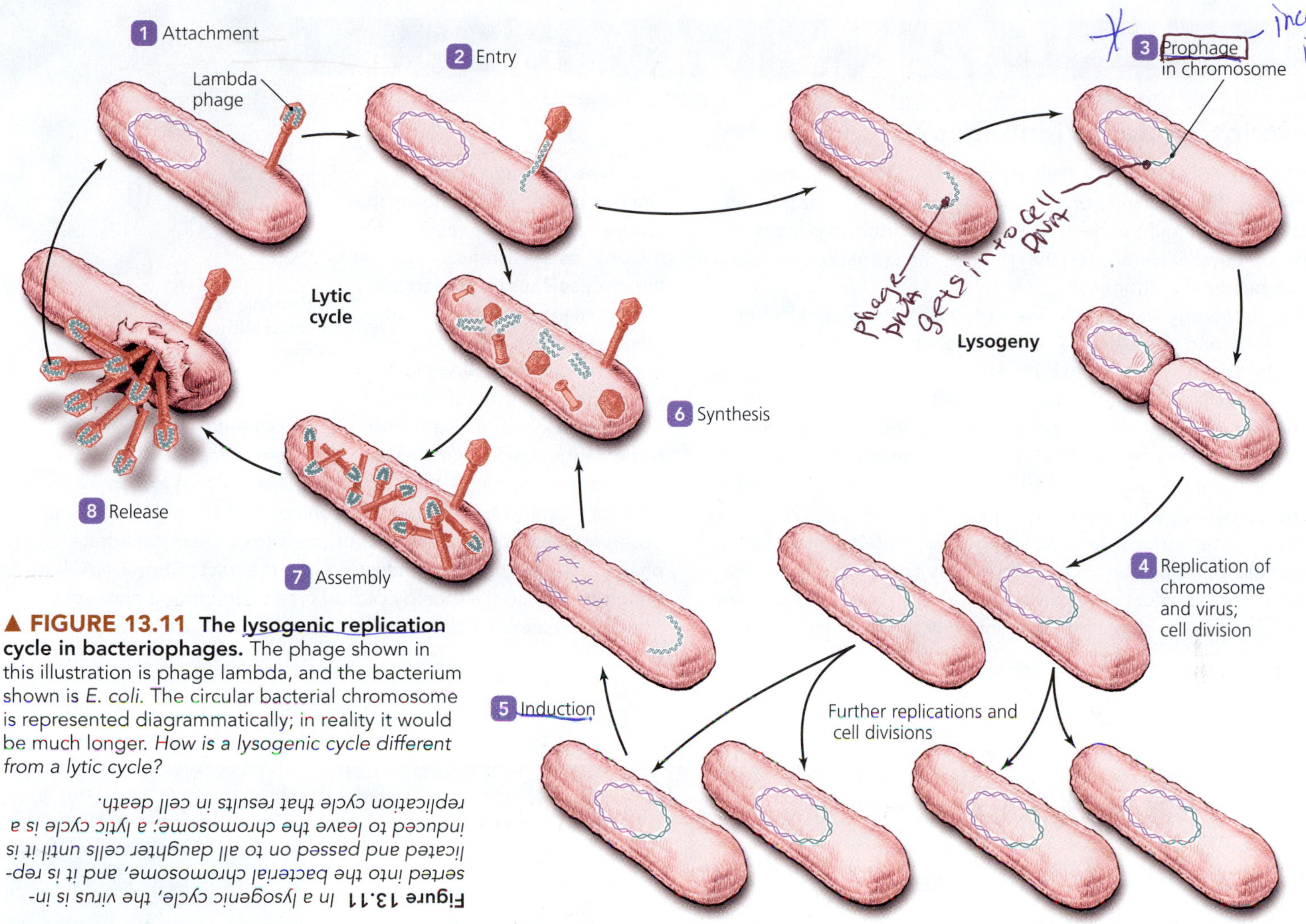

▲ **FIGURE 13.11 The lysogenic replication cycle in bacteriophages.** The phage shown in this illustration is phage lambda, and the bacterium shown is *E. coli*. The circular bacterial chromosome is represented diagrammatically; in reality it would be much longer. *How is a lysogenic cycle different from a lytic cycle?*

Figure 13.11 *In a lysogenic cycle, the virus is inserted into the bacterial chromosome, and it is replicated and passed on to all daughter cells until it is induced to leave the chromosome; a lytic cycle is a replication cycle that results in cell death.*

Bacteriophages T4 and lambda demonstrate two replication strategies that are typical for many DNA viruses. RNA viruses and enveloped viruses present variations on the lytic and lysogenic cycles we have examined. We will next examine some variations seen in replication of animal viruses.
▶**ANIMATIONS:** *Viral Replication: Temperate Bacteriophages*

Replication of Animal Viruses

LEARNING | **OUTCOMES**

13.9 Explain the differences between bacteriophage replication and animal viral replication.

13.10 Compare and contrast the replication and synthesis of DNA, −RNA, and +RNA viruses.

13.11 Compare and contrast the release of viral particles by lysis and budding.

13.12 Compare and contrast latency in animal viruses with phage lysogeny.

Animal viruses have the same five basic steps in their replication pathways as bacteriophages—that is, attachment, entry, synthesis, assembly, and release. However, there are significant differences in the replication of animal viruses that result in part from the presence of envelopes around some of the viruses and in part from the eukaryotic nature of animal cells as well as their lack of a cell wall. **Highlight: The Threat of Avian Influenza** highlights an animal virus that is of great concern to health officials worldwide.

In this section we examine the replication processes that are shared by DNA and RNA animal viruses, compare these processes with those of bacteriophages, and discuss how the synthesis of DNA and RNA viruses differ.

Attachment of Animal Viruses

As with bacteriophages, attachment of an animal virus is dependent on the chemical attraction and exact fit between proteins or glycoproteins on the virion and complementary protein or glycoprotein receptors on the animal cell's cytoplasmic

BENEFICIAL MICROBES

Prescription Bacteriophages?

In 1917, Canadian biologist Felix d'Herelle published a paper announcing the discovery of *bacteriophages,* viruses that prey on bacteria. In fact, half the bacteria on Earth succumb to phages every two days! D'Herelle felt that phages can be natural weapons against bacterial pathogens.

Phage therapy was used in the early 1900s to combat dysentery, typhus, and cholera but was largely abandoned in the 1940s in the United States, eclipsed by the development of antibiotics, such as penicillin. Phage therapy continued in the Soviet Union and Eastern Europe, where research is still centered. Today, motivated by the growing problem of antibiotic-resistant bacteria, scientists in the United States and Western Europe have renewed interest in investigating phage therapy.

A phage reproduces by inserting genetic material into a bacterium, causing the bacterium to build copies of the virus that burst out of the cell to infect other bacteria. A single phage can become 10 trillion phages within 2 hours, killing 99.9% of its host bacteria.

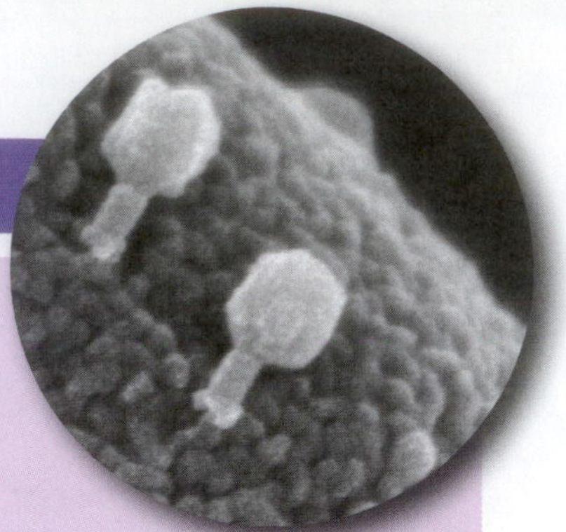

Escherichia coli infected with bacteriophages

Each type of phage attacks a specific strain of bacteria. This means that phage treatment is effective only if the phages are carefully matched to the disease-causing bacterium. It also means that phage treatment, unlike the use of antibiotics, can be effective without killing the body's helpful bacteria.

Introducing an active microbe into a patient does present some dangers, however. Phages can kill bacteria, but some make bacteria more lethal. A strain of *Escherichia coli* that is responsible for a deadly form of food poisoning, for example, has been observed to gain the ability to produce a toxic chemical from a phage genome that integrates itself into the bacterium's DNA. If a phage being used in therapy picked up a toxin-coding gene, the attempted cure could become lethal.

HIGHLIGHT

The Threat of Avian Influenza

In 1997, 18 people in Hong Kong contracted avian influenza, caused by the H5N1 strain of influenzavirus that spreads easily among chickens and other birds. At least half of these people caught the disease directly from birds, something that scientists had previously thought improbable. Human cases have had a death rate of over 60%. When people began dying of the illness, Hong Kong officials slaughtered all 1.5 million chickens within three days. That stopped a potential epidemic in Hong Kong, but it didn't stop the spread of the avian flu.

Avian flu virus very rarely spreads from one person to another. However, the possibility that this may change is of great concern to health officials worldwide. Avian flu viruses mutate quickly and can pick up genes from other flu viruses. If an avian flu virus picks up genes from a human flu virus, it could become a strain that spreads easily from person to person. In a worst-case scenario in this age of jet travel, such a strain of avian/human flu could cause a pandemic killing 2 million to 50 million people worldwide, according to the World Health Organization.

What should be done? In Asia, domesticated fowl have been slaughtered to prevent the virus from spreading. But while governments concentrated on culling domestic poultry, the virus spread to wild birds, such as geese, gray herons, and feral pigeons, and via wild birds it has spread throughout Asia, Europe, and Africa.

Scientists have developed a vaccine that they believe could protect against the H5N1 strain of the virus. Unfortunately, a government could stockpile vaccine for one strain of flu only to have the virus mutate into a new form against which the vaccine is ineffective.

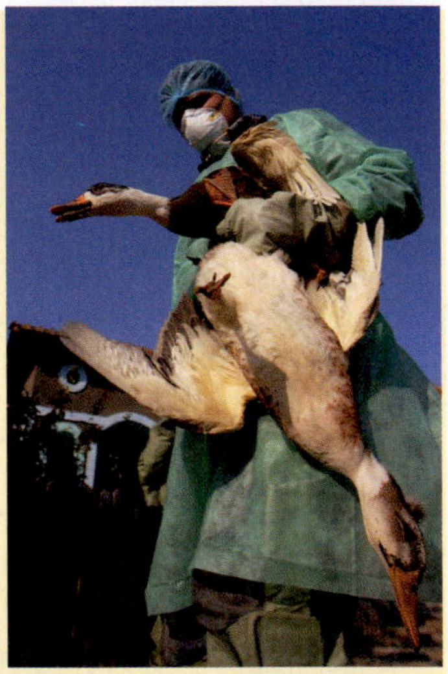

membrane. Unlike the bacteriophages we have examined, animal viruses lack both tails and tail fibers. Instead, animal viruses typically have glycoprotein spikes or other attachment molecules on their capsids or envelopes.

Entry and Uncoating of Animal Viruses

Animal viruses enter a host cell shortly after attachment. Even though entry of animal viruses is not as well understood as entry of bacteriophages, there appear to be at least three different mechanisms: direct penetration, membrane fusion, and endocytosis.

Some naked viruses enter their hosts' cells by *direct penetration*—a process in which the viral capsid attaches and sinks into the cytoplasmic membrane, creating a pore through which the genome alone enters the cell **(FIGURE 13.12a)**. Poliovirus infects host cells via direct penetration.

With other animal viruses, by contrast, the entire capsid and its contents (including the genome) enter the host cell by membrane fusion or endocytosis. With viruses using *membrane fusion*, such as measles virus, the viral envelope and the host cell membrane fuse, releasing the capsid into the cell's cytoplasm and leaving the envelope glycoproteins as part of the cell membrane **(FIGURE 13.12b)**.

Most enveloped viruses and some naked viruses enter host cells by triggering *endocytosis*. Attachment of the virus to receptor molecules on the cell's surface stimulates the cell to endocytize the entire virus **(FIGURE 13.12c)**. Adenoviruses (naked) and herpesviruses (enveloped) enter human host cells via endocytosis.

For those viruses that penetrate a host cell with their capsids intact, the capsids must be removed to release their genomes before the viruses can continue to replicate. The removal of a viral capsid within a host cell is called **uncoating**, a process that remains poorly understood. It apparently occurs via different means in different viruses; some viruses are uncoated within vesicles by cellular enzymes, whereas others are uncoated by enzymes within the cell's cytosol.

Synthesis of DNA Viruses of Animals

Synthesis of animal viruses also differs from synthesis of bacteriophages. Each type of animal virus requires a different strategy for synthesis that depends on the kind of nucleic acid involved—whether it is DNA or RNA and whether it is double stranded or single stranded. DNA viruses typically enter the nucleus, whereas most RNA viruses are replicated in the cytoplasm.

As we discuss the synthesis and assembly of each type of animal virus, consider the following two questions:

- How is mRNA—needed for the translation of viral proteins—synthesized?
- What molecule serves as a template for nucleic acid replication?

dsDNA Viruses Synthesis of new double-stranded DNA (dsDNA) virions is similar to the normal replication of cellular DNA and translation of proteins. The genomes of most dsDNA viruses enter the nucleus of the cell, where cellular enzymes replicate

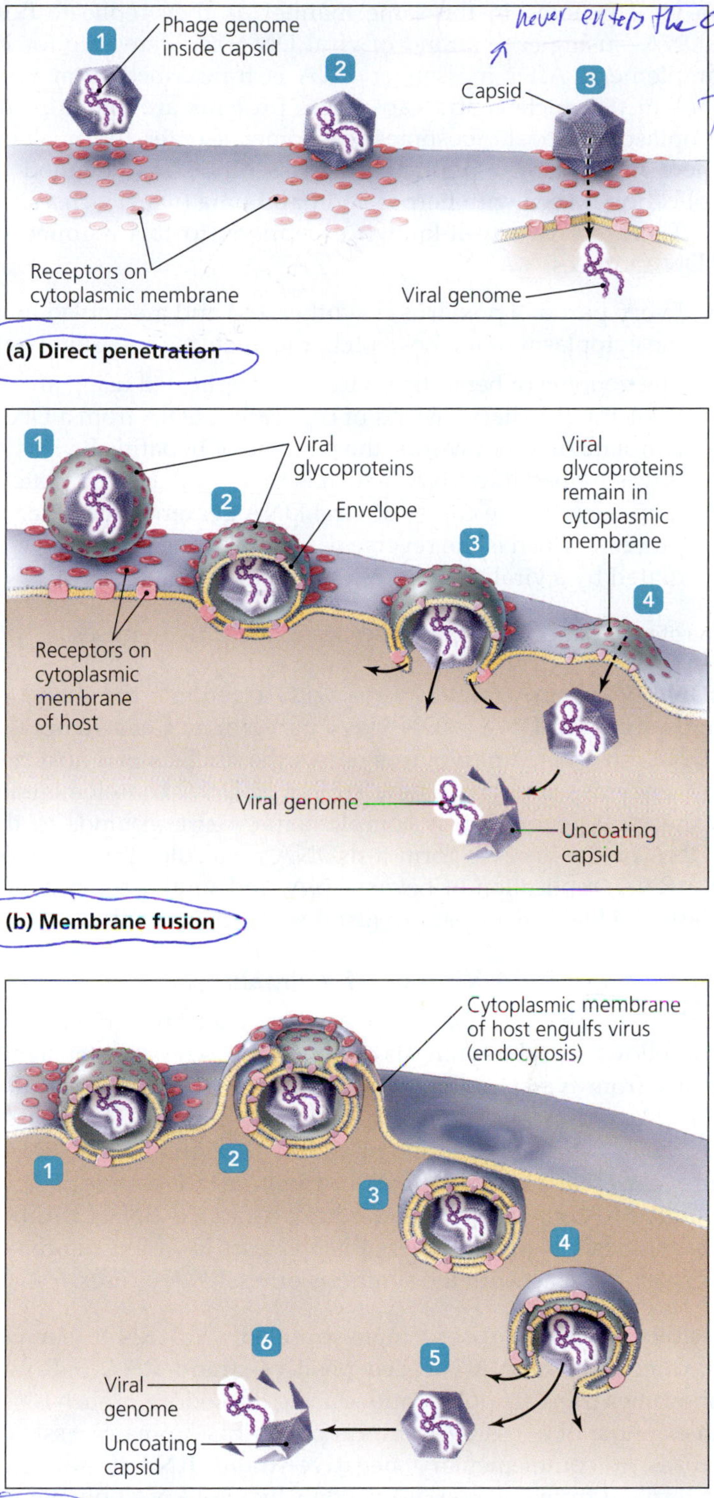

▲ **FIGURE 13.12 Three mechanisms of entry of animal viruses.** **(a)** Direct penetration, a process whereby naked virions inject their genomes into their animal cell hosts. **(b)** Membrane fusion, in which fusion of the viral envelope and cell membrane dumps the capsid into the cell. **(c)** Endocytosis, in which attachment of a naked or an enveloped virus stimulates the host cell to engulf the entire virus. *After penetration, many animal viruses must be uncoated, but bacteriophages need not be. Why is this so?*

Figure 13.12 *Generally, bacteriophages inject their DNA during penetration, so the capsid does not enter the cell.*

the viral genome in the same manner as they replicate host dsDNA—using each strand of viral DNA as a template for its complement. After messenger RNA is transcribed from viral DNA in the nucleus and capsomere proteins are made in the cytoplasm by host ribosomes, capsomeres enter the nucleus, where new virions spontaneously assemble. This method of replication is seen with herpes and papilloma (wart) viruses.

There are two well-known exceptions to this regimen of dsDNA viruses:

- Every part of a poxvirus is synthesized and assembled in the cytoplasm of the host's cell; the nucleus is not involved.
- The genome of hepatitis B viruses is replicated using an RNA intermediary instead of replicating DNA from a DNA template. In other words, the genome of hepatitis B virus is transcribed into RNA, which is then used as a template to make multiple copies of viral DNA genome. The latter process, which is the reverse of normal transcription, is mediated by a viral enzyme, *reverse transcriptase.*

Chapters 19 and 23 discuss diseases of these two viruses.

ssDNA Viruses A human virus with a genome composed of single-stranded DNA (ssDNA) is a parvovirus. Cells do not use ssDNA, so when a parvovirus enters the nucleus of a host cell, host enzymes produce a new strand of DNA complementary to the viral genome. This complementary strand binds to the ssDNA of the virus to form a dsDNA molecule. Transcription of mRNA, replication of new ssDNA, and viral assembly then follow the DNA virus pattern just described.

Synthesis of RNA Viruses of Animals

As previously noted, RNA is not used as genetic material in cells, so it follows that the synthesis of RNA viruses must differ significantly from typical cellular processes and from the replication of DNA viruses as well. There are four types of RNA viruses: positive-sense, single-stranded RNA (designated +ssRNA); retroviruses (a kind of +ssRNA virus); negative-sense, single-stranded RNA (−ssRNA); and double-stranded RNA (dsRNA). The synthesis process for these RNA viruses is varied and rather complex. We start with the synthesis of +ssRNA viruses.

Positive ssRNA Viruses Single-stranded viral RNA that can act directly as mRNA is called **positive-strand RNA (+RNA)**. Ribosomes translate polypeptides using the codons of such RNA. An example of a +ssRNA virus is poliovirus. In many +ssRNA viruses, a complementary **negative-strand RNA (−RNA)** is transcribed from the +ssRNA genome by viral RNA polymerase; −RNA then serves as the template for the transcription of multiple +ssRNA genomes. Such transcription of RNA from RNA is unique to viruses; no cell transcribes RNA from RNA.

Retroviruses Unlike other +ssRNA viruses, the +ssRNA viruses called **retroviruses** do not use their genome as mRNA. Instead, retroviruses use a DNA intermediary that is transcribed from +RNA by reverse transcriptase carried within the capsid. This DNA intermediary then serves as the template for the synthesis of additional +RNA molecules, which act both as mRNA for protein synthesis and as genomes for new virions. Human immunodeficiency virus (HIV) is a prominent retrovirus.

Negative ssRNA Viruses Other single-stranded RNA virions are −ssRNA viruses, which must overcome a unique problem. In order to synthesize a protein, a ribosome can use only mRNA (i.e., +RNA) because −RNA is not recognized by ribosomes. The virus overcomes this problem by carrying within its capsid an enzyme, *RNA-dependent RNA transcriptase,* which is released into the host cell's cytoplasm during uncoating and then transcribes +RNA molecules from the virus's −RNA genome. Translation of proteins can then occur as usual. The newly transcribed +RNA also serves as a template for transcription of additional copies of −RNA. Diseases caused by −ssRNA viruses include rabies and flu.

dsRNA Viruses Viruses that have double-stranded RNA use yet another method of synthesis. The positive strand of the molecule serves as mRNA for the translation of proteins, one of which is an RNA polymerase that transcribes dsRNA. Each strand of RNA acts as a template for transcription of its opposite, which is reminiscent of DNA replication in cells. Double-stranded RNA rotaviruses cause most cases of diarrhea in infants.

FIGURE 13.13 illustrates and **TABLE 13.3** summarizes the various strategies by which animal viruses are synthesized.

Assembly and Release of Animal Viruses

As with bacteriophages, once the components of animal viruses are synthesized, they assemble into virions that are then released from the host cell. Most DNA viruses assemble in and are released from the nucleus into the cytosol, whereas most RNA viruses develop solely in the cytoplasm. The number of viruses produced and released depends on both the type of virus and the size and initial health of the host cell.

Replication of animal viruses takes more time than replication of bacteriophages. Herpesviruses, for example, require almost 24 hours to replicate, as compared to 25 minutes for hundreds of copies of bacteriophage T4.

Enveloped animal viruses are often released via a process called **budding** **(FIGURE 13.14)**. As virions are assembled, they are extruded through one of the cell's membranes—the nuclear, endoplasmic reticulum, or the cytoplasmic membrane. Each virion acquires a portion of membrane, which becomes the viral envelope. During synthesis, some viral glycoproteins are inserted into cellular membranes, and these proteins become the glycoprotein spikes on the surface of the viral envelope.

Because the host cell is not quickly lysed, as occurs in bacteriophage replication, budding allows an infected cell to remain alive for some time. Infections with enveloped viruses in which host cells shed viruses slowly and relatively steadily are called *persistent infections;* a curve showing virus abundance over time during a persistent infection lacks the burst of new virions seen in lytic replication cycles (**FIGURE 13.15**; compare to Figure 13.9).

Naked animal viruses may be released in one of two ways: Either they may be extruded from the cell by exocytosis, in a manner similar to budding but without the acquisition of an

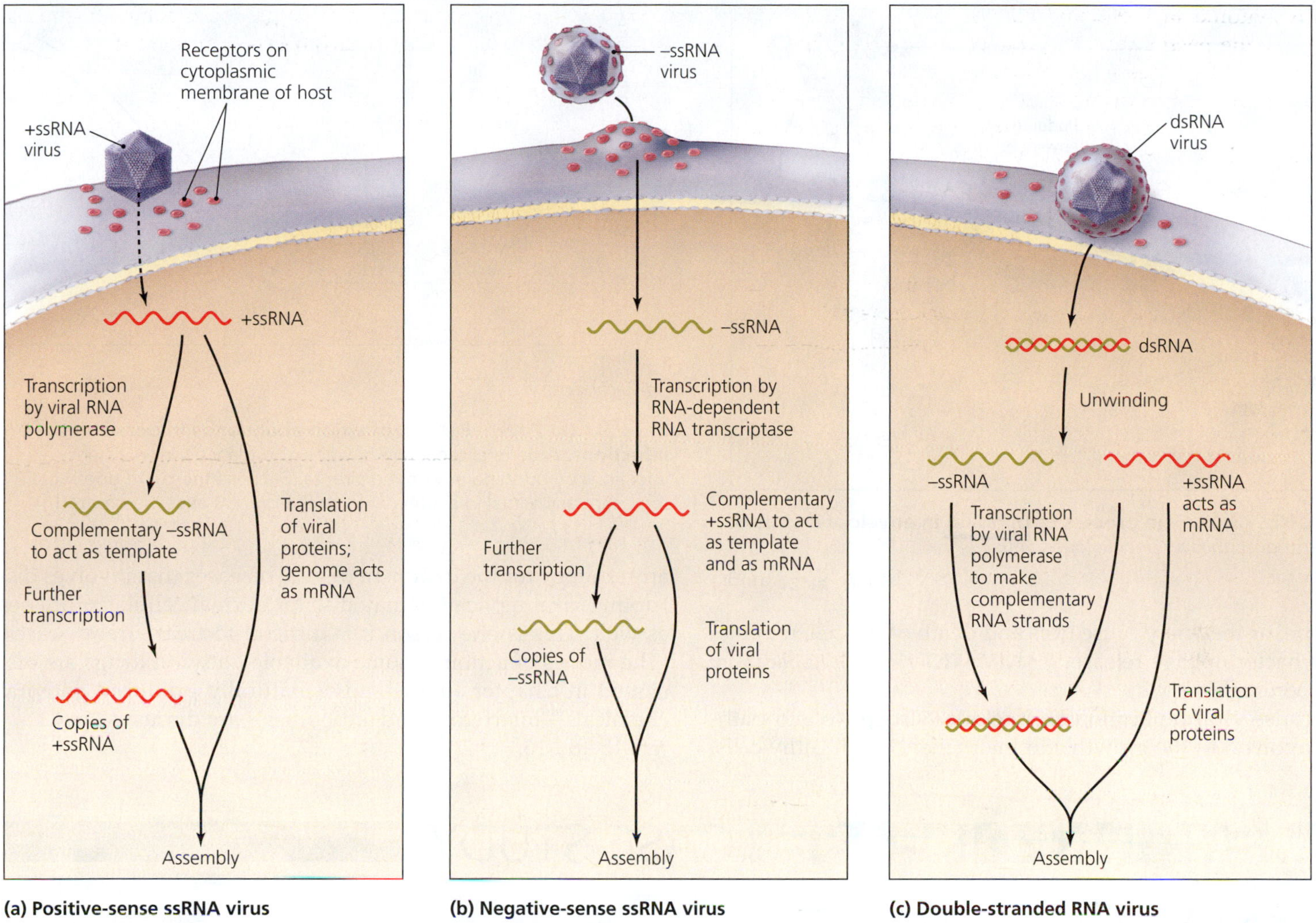

▲ **FIGURE 13.13 Synthesis of proteins and genomes in animal RNA viruses.** **(a)** Positive-sense ssRNA virus, in which +ssRNA acts as mRNA and −ssRNA is the genome template. **(b)** Negative-sense ssRNA virus: transcription forms +ssRNA to serve both as mRNA and as template. **(c)** dsRNA virus genome unwinds so that the positive-sense strand serves as mRNA, and each strand serves as a template for its complement.

TABLE 13.3 Synthesis Strategies of Animal Viruses

Genome	How Is mRNA Synthesized?	What Molecule Is the Template for Genome Replication?
dsDNA	By RNA polymerase (in nucleus or cytoplasm of cell)	Each strand of DNA serves as template for its complement (except for hepatitis B, which synthesizes RNA to act as the template for new DNA)
ssDNA	By RNA polymerase (in nucleus of cell)	Complementary strand of DNA is synthesized to act as template
+ssRNA	Genome acts as mRNA	−RNA complementary to the genome is synthesized to act as template
+ssRNA (*Retroviridae*)	DNA is synthesized from RNA by reverse transcriptase; mRNA is transcribed from DNA by RNA polymerase	DNA
−ssRNA	By RNA-dependent RNA transcriptase	+RNA (mRNA) complementary to the genome
dsRNA	Positive strand of genome acts as mRNA	Each strand of genome acts as template for its complement

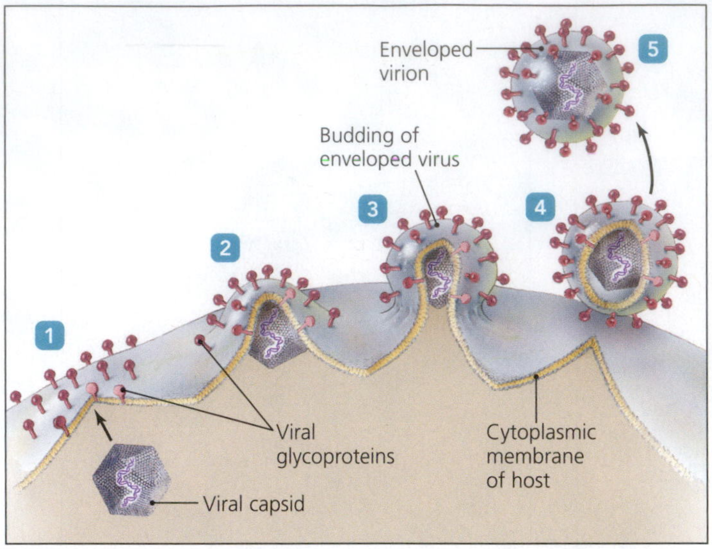

▲ **FIGURE 13.14 The process of budding in enveloped viruses.** *What term describes a nonenveloped virus?*

Figure 13.14 Naked.

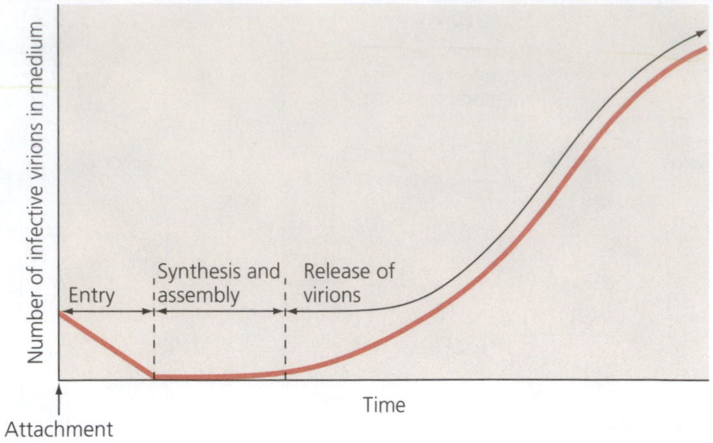

▲ **FIGURE 13.15 Pattern of virion abundance in persistent infections.** A generalized curve of virion abundance for persistent infections by budding enveloped viruses. Because the curve does not represent any actual infection, units for the graph's axes are omitted.

envelope, or they may cause lysis and death of the cell, reminiscent of bacteriophage release. ▶**ANIMATIONS:** ***Viral Replication: Animal Viruses***

Because viral replication uses cellular structures and pathways involved in the growth and maintenance of healthy cells, any strategy for the treatment of viral diseases that involves disrupting viral replication may disrupt normal cellular processes as well. This is one reason it is difficult to treat viral diseases. (The modes of action of some available antiviral drugs are discussed in Chapter 10; the body's naturally produced antiviral chemicals—interferons and antibodies—are discussed in Chapters 15 and 16.)

EMERGING DISEASE CASE STUDY

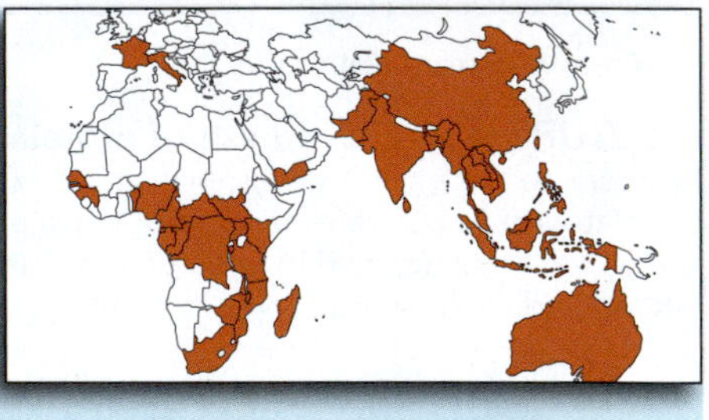

Chikungunya

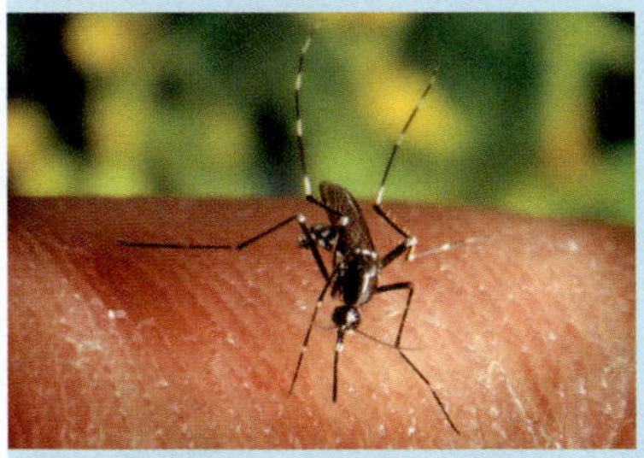

An old man arrived at the doctor's office in Ravenna, Italy, with a combination of signs and symptoms the physician had never heard of: a widespread, severe rash; difficulty in breathing; high fever; nausea; and extreme joint pain. Chikungunya (chik-en-gun´ya) had arrived in Europe.

Though scientists had known of chikungunya virus, which is related to equine encephalitis viruses, for over 50 years, most considered the tropical disease benign—a limited, mild irritation, not a catastrophe. Therefore, few researchers studied chikungunya virus or its disease. Now, they know better.

Over the past decade, chikungunya virus has spread throughout the nations of the Indian Ocean and across Africa. In 2006, officials on the French-owned island of La Réunion in the Indian Ocean reported 47,000 cases of chikungunya in a single week! That same year chikungunya reemerged in India for the first time in four decades with more than 1.5 million reported cases, and in 2010 it emerged in China. Why?

Aedes albopictus (Asian tiger mosquito), which carries the virus, has moved into temperate climates, including Europe and the United States, as the climate has warmed. With the mosquito comes the possibility of viral proliferation—the insects have spread the tropical disease as far north in Europe as France.

And our Italian patient? His crippling pain lasted for months, but he survived. Now that he knows about mosquito-borne chikungunya, he insists that his family and friends use mosquito repellent liberally. Officials in the rest of Europe and in the United States join in his concern: With the coming of *Ae. albopictus*, is incurable chikungunya far behind?

1. **Besides using repellents, how can people protect themselves from mosquito-borne pathogens?**
2. **Why is *Aedes* commonly known as the tiger mosquito?**
3. **Why aren't antibiotics such as penicillin, erythromycin, and ciprofloxacin effective in preventing and treating chikungunya?**

TABLE 13.4 A Comparison of Bacteriophage and Animal Virus Replication

	Bacteriophage	Animal Virus
Attachment	Proteins on tails attach to proteins on cell wall	Spikes, capsids, or envelope proteins attach to proteins or glycoproteins on cell membrane
Penetration	Genome is injected into cell or diffuses into cell	Capsid enters cell by direct penetration, fusion, or endocytosis
Uncoating	None	Removal of capsid by cell enzymes
Site of synthesis	In cytoplasm	RNA viruses in cytoplasm; most DNA viruses in nucleus
Site of assembly	In cytoplasm	RNA viruses in cytoplasm; most DNA viruses in nucleus
Mechanism of release	Lysis	Naked virions: exocytosis or lysis; enveloped virions: budding
Nature of chronic infection	Lysogeny, always incorporated into host chromosome, may leave host chromosome	Latency, with or without incorporation into host DNA; incorporation is permanent

Latency of Animal Viruses

Some animal viruses, including chicken pox and herpes viruses, may remain dormant in cells in a process known as **latency**; the viruses involved in latency are called **latent viruses** or **proviruses**. Latency may be prolonged for years with no viral activity, signs, or symptoms. Though latency is similar to lysogeny as seen with bacteriophages, there are differences. Some latent viruses do not become incorporated into the chromosomes of their host cells, whereas lysogenic phages always do.

On the other hand, some animal viruses (e.g., HIV) are more like lysogenic phages in that they do become integrated into a host chromosome as a provirus. However, when a provirus is incorporated into its host DNA, the condition is permanent; induction does not occur in eukaryotes. Thus, an incorporated provirus becomes a permanent, physical part of the host's chromosome, and all descendants of the infected cell will carry the provirus.

Given that RNA cannot be incorporated directly into a chromosome molecule, how does the ssRNA of HIV become a provirus incorporated into the DNA of its host cell? HIV can become a permanent part of a host's chromosome because it, like all retroviruses, carries reverse transcriptase, which transcribes the genetic information of the +RNA molecule to a DNA molecule—which *can* become incorporated into the host cell's genome.

TABLE 13.4 compares the features of the replication of bacteriophages and animal viruses. Next we turn our attention to the part viruses can play in cancer, beginning with a brief consideration of the terminology needed to understand the basic nature of cancer.

TELL ME WHY

Why are lysogenic and latent viral infections generally longer lasting than lytic infections?

The Role of Viruses in Cancer

LEARNING OUTCOMES

13.13 Define the terms *neoplasia*, *tumor*, *benign*, *malignant*, *cancer*, and *metastasis*.

13.14 Explain in simple terms how a cell may become cancerous, with special reference to the role of viruses.

Under normal conditions, the division of cells in a mature multicellular animal is under strict genetic control; that is, the animal's genes dictate that some types of cells can no longer divide at all and that those that can divide are prevented from unlimited division. In this genetic control, either genes for cell division are "turned off" or genes that inhibit division are "turned on," or some combination of both these genetic events occurs. However, if something upsets the genetic control, cells begin to divide uncontrollably. This phenomenon of uncontrolled cell division in a multicellular animal is called **neoplasia**[2] (nē-ō-plā´zē-ă). Cells undergoing neoplasia are said to be neoplastic, and a mass of neoplastic cells is a **tumor**.

Some tumors are **benign tumors**; that is, they remain in one place and are not generally harmful, although occasionally such noninvasive tumors are painful and rob adjacent normal cells of space and nutrients. Other tumors are **malignant tumors**, invading neighboring tissues and even traveling throughout the body to invade other organs and tissues to produce new tumors—a process called **metastasis** (mĕ-tas´tă-sis). Malignant tumors are also called **cancers**. Cancers rob normal cells of space and nutrients and cause pain; in some kinds of cancer, malignant cells derange the function of the affected tissues, until eventually the body can no longer withstand the loss of normal function and dies.

Several theories have been proposed to explain the role viruses play in the development of cancers. These theories

[2]From Greek *neo*, meaning "new," and *plassein*, meaning "to mold."

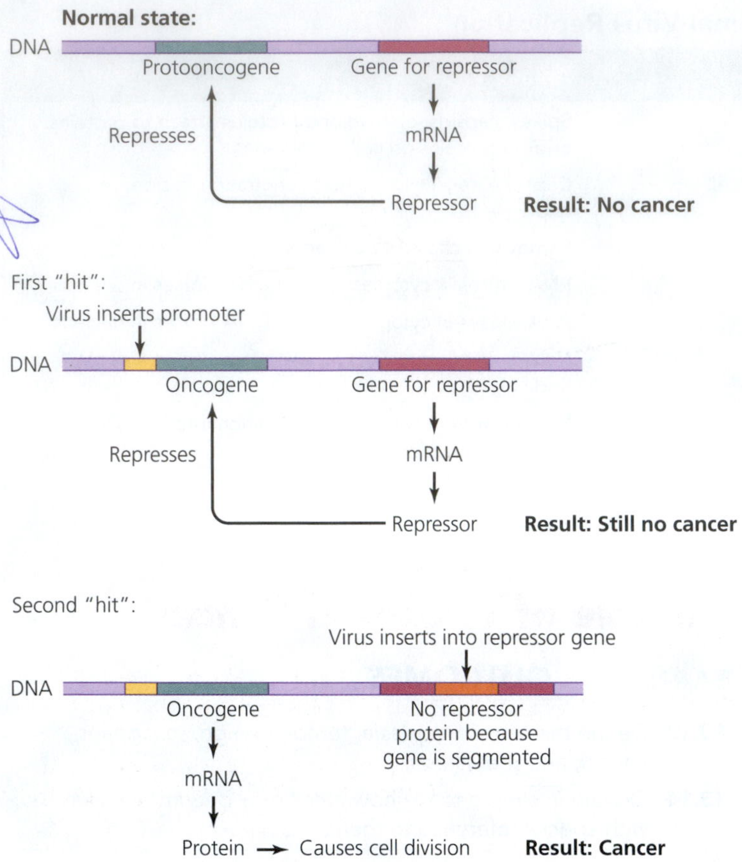

▲ **FIGURE 13.16 The oncogene theory of the induction of cancer in humans.** The theory suggests that more than one "hit" to the DNA (i.e., any change or mutation), whether caused by a virus (as shown here) or various physical or chemical agents, is required to induce cancer.

revolve around the presence of *protooncogenes* (prō-tō-ong´kō-jēnz)—genes that play a role in cell division. As long as protooncogenes are repressed, no cancer results. However, activity of oncogenes (their name when they are active) or inactivation of oncogene repressors can cause cancer to develop. In most cases, several genetic changes must occur before cancer develops. Put another way, "multiple hits" to the genome must occur for cancer to result **(FIGURE 13.16)**.

A variety of environmental factors contribute to the inhibition of oncogene repressors and the activation of oncogenes. Ultraviolet light, radiation, certain chemicals called *carcinogens* (kar-si´nō-jenz), and viruses have all been implicated in the development of cancer.

Viruses cause 20% to 25% of human cancers in several ways. Some viruses carry copies of oncogenes as part of their genomes, other viruses promote oncogenes already present in the host, and still other viruses interfere with normal tumor repression when they insert (as proviruses) into repressor genes.

That viruses cause some animal cancers is well established. In the first decade of the 1900s, virologist F. Peyton Rous (1879–1970) proved that viruses induce cancer in chickens. Though several DNA and RNA viruses are known to cause about 15% of human cancers, the link between viruses and most human cancers has been difficult to document. Among the virally induced cancers in humans are Burkitt's lymphoma, Hodgkin's disease, Kaposi's sarcoma, and cervical cancer. DNA viruses in the families *Adenoviridae, Herpesviridae, Hepadnaviridae, Papillomaviridae,* and *Polyomaviridae* and two RNA viruses in the family *Retroviridae* cause these and other human cancers. (Chapters 18–24 discuss diseases caused by DNA viruses and RNA viruses.)

TELL ME WHY

Why are DNA viruses more likely to cause neoplasias than are RNA viruses?

Culturing Viruses in the Laboratory

LEARNING | OUTCOMES

13.15 Describe some ethical and practical difficulties to overcome in culturing viruses.

13.16 Describe three types of media used for culturing viruses.

Scientists must culture viruses in order to conduct research and develop vaccines and treatments, but because viruses cannot metabolize or replicate by themselves, they cannot be grown in standard microbiological broths or on agar plates. Instead, they must be cultured inside suitable host cells, a requirement that complicates the detection, identification, and characterization of viruses. Virologists have developed three types of media for culturing viruses: media consisting of mature organisms (bacteria, plants, or animals), embryonated (fertilized) eggs, and cell cultures. We begin by considering the culture of viruses in organisms.

Culturing Viruses in Mature Organisms

LEARNING | OUTCOMES

13.17 Explain the use of a plaque assay in culturing viruses in bacteria.

13.18 List three problems with growing viruses in animals.

In the following sections we consider the use of bacterial cells as a virus culture medium before considering the issues involved in growing viruses in living animals.

Culturing Viruses in Bacteria

Most of our knowledge of viral replication has been derived from research on bacteriophages, which are relatively easy to culture because some bacteria are easily grown and maintained. Phages can be grown in bacteria maintained either in liquid cultures or on agar plates. In the latter case, bacteria and phages are mixed with warm (liquid) nutrient agar and poured in a thin layer across the surface of an agar plate. During incubation, bacteria infected by phages lyse and release new

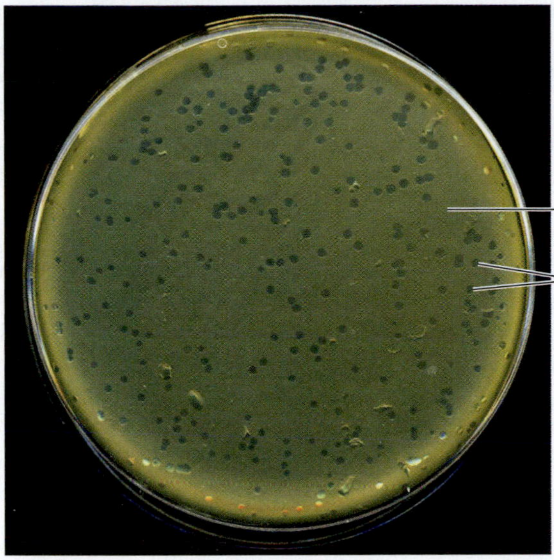

▲ **FIGURE 13.17 Viral plaques in a lawn of bacterial growth on the surface of an agar plate.** *What is the cause of viral plaques?*

Figure 13.17 *Each plaque is an area in a bacterial lawn where bacteria have succumbed to phage infections.*

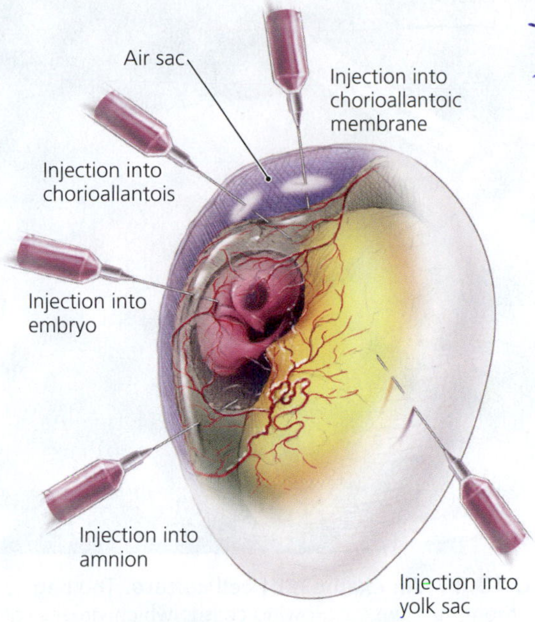

▲ **FIGURE 13.18 Inoculation sites for the culture of viruses in embryonated chicken eggs.** *Why are eggs often used to grow animal viruses?*

Figure 13.18 *Eggs are large, sterile, self-sufficient cells that contain a number of different sites suitable for viral replication.*

phages that infect nearby bacteria, while uninfected bacteria grow and reproduce normally. After incubation, the appearance of the plate includes a uniform bacterial lawn interrupted by clear zones called **plaques**, which are areas where phages have lysed the bacteria **(FIGURE 13.17)**. Such plates enable the estimation of phage numbers via a technique called **plaque assay**, in which virologists assume that each plaque corresponds to a single phage in the original bacterium-virus mixture.

Culturing Viruses in Plants and Animals

Plant and animal viruses can be grown in laboratory plants and animals. Recall that the first discovery and isolation of a virus was the discovery of tobacco mosaic virus in tobacco plants. Rats, mice, guinea pigs, rabbits, pigs, and primates have been used to culture and study animal viruses.

However, maintaining laboratory animals can be difficult and expensive, and this practice raises ethical issues for some. Growing viruses that infect only humans raises additional ethical complications. Therefore, scientists have developed alternative ways of culturing animal and human viruses using fertilized chicken eggs or cell cultures.

Culturing Viruses in Embryonated Chicken Eggs

Chicken eggs are a useful culture medium for viruses because they are inexpensive, are among the largest of cells, are free of contaminating microbes, and contain a nourishing yolk (which makes them self-sufficient). Most suitable for culturing viruses are chicken eggs that have been fertilized and thus contain a developing embryo. Embroyonic tissues (called membranes, which should not be confused with cellular membranes) provide ideal inoculation sites for growing viruses **(FIGURE 13.18)**. Researchers inject samples of virus into embryonated eggs at the sites that are best suited for the particular virus's replication.

Vaccines against some viruses can also be prepared in egg cultures. You may have been asked if you are allergic to eggs before you received such a vaccine because egg protein may remain as a contaminant in the vaccine.

Culturing Viruses in Cell (Tissue) Culture

LEARNING | **OUTCOME**

13.19 Compare and contrast diploid cell culture and continuous cell culture.

Viruses can also be grown in **cell culture**, which consists of cells isolated from an organism and grown on the surface of a medium or in broth **(FIGURE 13.19)**. Such cultures became practical when antibiotics provided a way to limit the growth of contaminating bacteria. Cell culture can be less expensive than maintaining research animals, plants, or eggs, and it avoids some of the moral problems associated with experiments performed on animals and humans. Cell cultures are sometimes called *tissue cultures*, but the term *cell culture* is more accurate because only a single type of cell is used in the culture. (By definition, a tissue is composed of at least two kinds of cells.)

Cell cultures are of two types. The first type, **diploid cell cultures**, are created from embryonic animal, plant, or human cells that have been isolated and provided appropriate

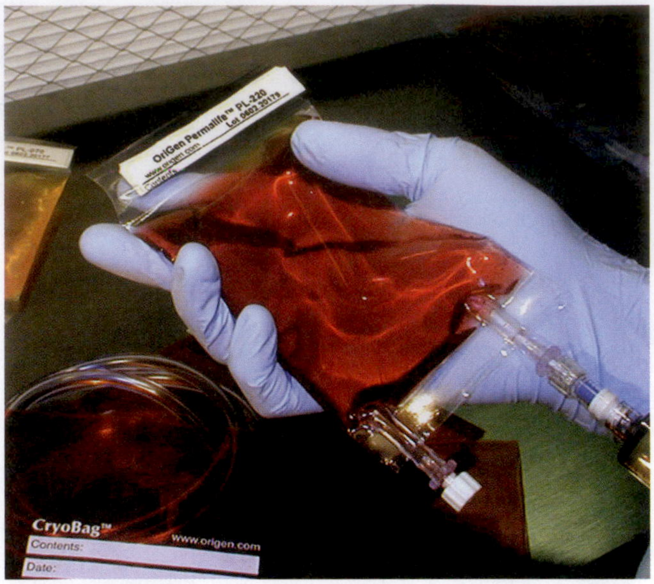

▲ **FIGURE 13.19 An example of cell culture.** The bag contains a colored nutrient medium for growing cells in which viruses can be cultured.

growth conditions. The cells in diploid cell culture generally last no more than about 100 generations (cell divisions) before they die.

The second type of culture, **continuous cell cultures**, are longer lasting because they are derived from tumor cells. Recall that a characteristic of neoplastic cells is that they divide relentlessly, providing a never-ending supply of new cells. One of the more famous continuous cell cultures is of HeLa cells, derived from a woman named *He*nrietta *La*cks, who died of cervical cancer in 1951. Though she is dead, Mrs. Lacks's cells live on in laboratories throughout the world.

It is interesting that HeLa cells have lost some of their original characteristics. For example, they are no longer diploid because they have lost many chromosomes. HeLa cells provide a semistandard[3] human tissue culture medium for studies on cell metabolism, aging, and (of course) viral infection.

TELL ME WHY

HIV replicates only in certain types of human cells, and one early problem in AIDS research was culturing those cells. Why are scientists now able to culture HIV?

Are Viruses Alive?

LEARNING | **OUTCOME**

13.20 Discuss aspects of viral replication that are lifelike and nonlifelike.

Now that we have studied the characteristics and replication processes of viruses, let's ask a question: Are viruses alive?

To be able to wrestle with the answer, we must first recall five characteristics of life: growth, self-reproduction, responsiveness, and the ability to metabolize, all within structures called cells. According to these criteria, viruses lack the qualities of living things, prompting some scientists to consider them nothing more than complex pathogenic chemicals. For other scientists, however, at least three observations—that viruses use sophisticated methods to invade cells, have the means of taking control of their host cells, and possess genomes containing instructions for replicating themselves—indicate that viruses are the ultimate parasites because they use cells to make more viruses. According to this viewpoint, viruses are the least complex living entities.

In any case, viruses are right on the threshold of life—outside cells they do not appear to be alive, but within cells they direct the synthesis and assembly required to make copies of themselves.

TELL ME WHY

Why are viruses seemingly alive and yet not alive?

Other Parasitic Particles: Viroids and Prions

Viruses are not the only submicroscopic entities capable of causing disorders within cells. In this section we will consider the characteristics of two molecular particles that infect cells: viroids and prions.

Characteristics of Viroids

LEARNING | **OUTCOMES**

13.21 Define and describe viroids.
13.22 Compare and contrast viroids and viruses.

Viroids are extremely small, circular pieces of RNA that are infectious and pathogenic in plants **(FIGURE 13.20)**. Viroids are similar to RNA viruses except that they lack capsids. Even though they are circular, viroids may appear linear because of hydrogen bonding within the molecule. Several plant diseases, including some of coconut palm, chrysanthemum, potato, cucumber, and avocado, are caused by viroids, including the stunting shown in **FIGURE 13.21**.

Viroidlike agents—infectious, pathogenic RNA particles that lack capsids but do not infect plants—affect some fungi. (They are not called viroids because they do not infect plants.) No animal diseases are known to be caused by viroidlike molecules, though the possibility exists that infectious RNA may be responsible for some diseases in humans.

[3]HeLa cells are "semistandard" because different strains have lost different chromosomes, and mutations have occurred over the years. Thus, HeLa cells in one laboratory may be slightly different from HeLa cells in another laboratory.

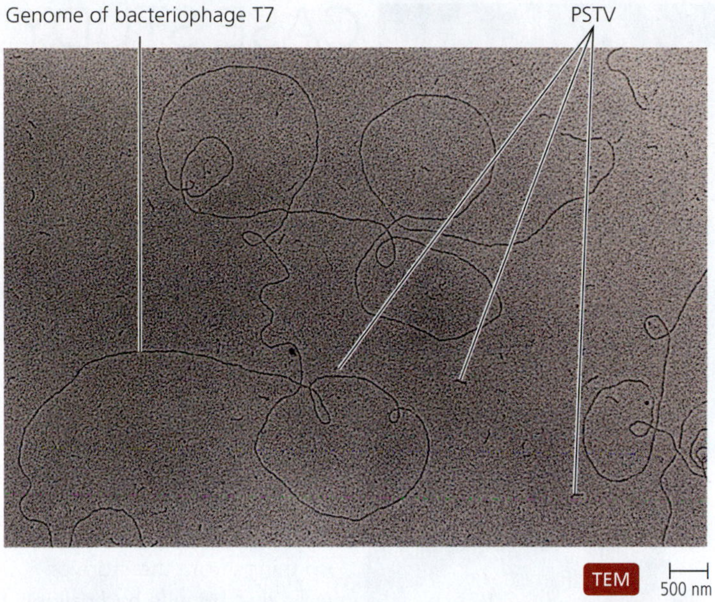

▲ **FIGURE 13.20 The RNA strand of the small potato spindle tuber viroid (PSTV).** Also shown for comparison is the longer DNA genome of bacteriophage T7. *Compare both to the size of a bacterial genome in Figure 13.2. How are viroids similar to and different from viruses?*

Figure 13.20 *Viroids are similar to certain viruses in that they are infectious and contain a single strand of RNA; they are different from viruses in that they lack a proteinaceous capsid.*

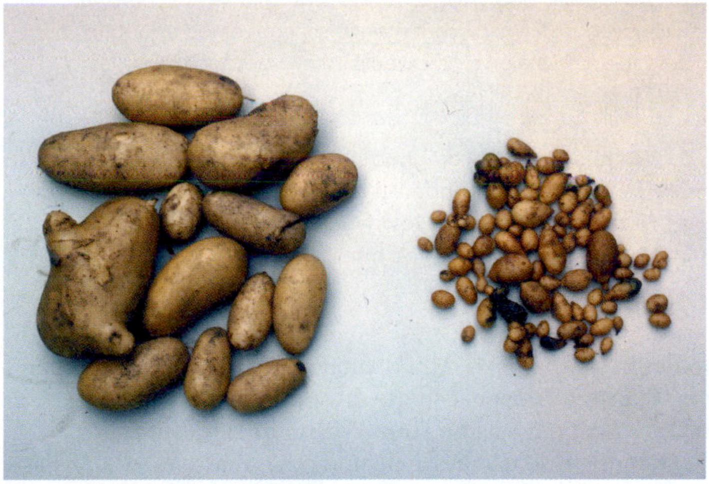

▲ **FIGURE 13.21 One effect of viroids on plants.** The potatoes at right are stunted as the result of infection with PSTV viroids.

Characteristics of Prions

LEARNING | **OUTCOMES**

13.23 Define and describe prions, including their replication process.

13.24 Compare and contrast prions and viruses.

13.25 List four diseases caused by prions.

In 1982, Stanley Prusiner (1942–) described a proteinaceous infectious agent that was different from any other known infectious agent in that it lacked instructional nucleic acid. Prusiner named such agents of disease **prions** (prē´onz), for *pro*teinaceous *in*fective particles. Before his discovery, the diseases now known to be caused by prions were thought to be caused by what were known as "slow viruses," which were so named because 60 years might lapse between infection and the onset of signs and symptoms. Through experiments, Prusiner and his colleagues showed that prions are not viruses because they lack any nucleic acid. **▶ANIMATIONS:** ***Prions: Overview***

Some scientists resist the concept of prions because particles that lack any nucleic acid violate the "universal" rule of protein synthesis—that proteins are translated from a molecule of mRNA. Given that "infectious proteins" lack nucleic acid, how can they carry the information required to replicate themselves?

All mammals make a cytoplasmic membrane protein called *PrP*. PrP is anchored in lipid rafts and plays a role in the normal activity of the brain, though the exact function of PrP is unknown. The amino acid sequence in PrP is such that the protein can fold into two stable tertiary structures: The normal, functional structure of *cellular PrP* has several prominent α-helices, whereas a disease-causing form—*prion PrP*—is characterized by β-pleated sheets (**FIGURE 13.22**).

Scientists have determined that prion PrP acts like a bad influence in a crowd of teenagers, encouraging molecules of normal, cellular PrP to misbehave by refolding into prion PrP molecules, which then clump together. As clumps of prion PrP propagate throughout the brain, neurons stop working properly

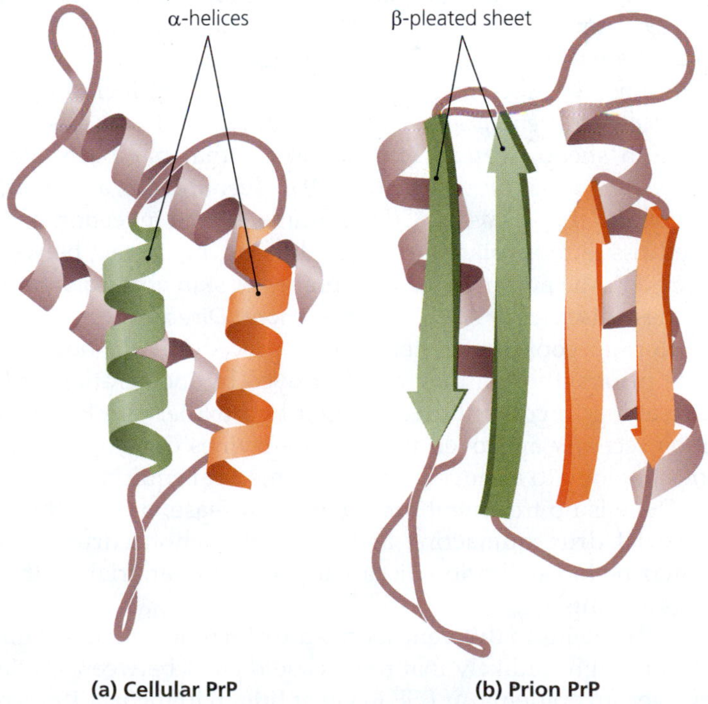

▲ **FIGURE 13.22 The two stable, three-dimensional forms of prion protein (PrP). (a)** Cellular prion protein (normal form) found in functional cells has a preponderance of alpha-helices. **(b)** Prion PrP (abnormal form), which has the same amino acid sequence, is folded to produce a preponderance of beta-pleated sheets.

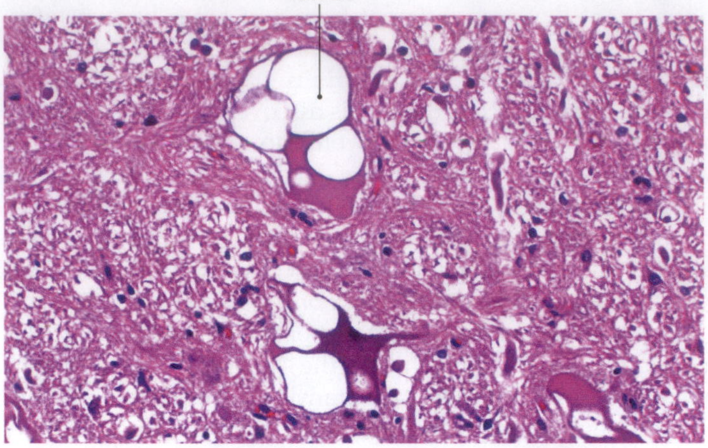

▲ **FIGURE 13.23 A brain showing the large vacuoles and spongy appearance typical in prion-induced diseases.** Shown here is the brain of a sheep with the prion disease called scrapie.

and eventually die, leaving holes and a spongy appearance **(FIGURE 13.23)**. Because of this characteristic, clinicians call prion diseases of the brain *spongiform encephalopathies* (spŭn´ji-fōrm en-sef´ă-lop´ă-thēz). ▶**ANIMATIONS:** ***Prions: Characteristics***

Why don't prions develop in all mammals, given that all mammals have PrP? Under normal circumstances, it appears that other nearby proteins and polysaccharides in lipid rafts force PrP into the correct (cellular) shape. Mutations in the PrP gene can result in the initial formation of prion PrP, but human cellular PrP visually misfolds only if it contains methionine as the 129th amino acid. About 40% of humans have this type of PrP and are thus susceptible to prion disease.

Prions are associated with several diseases, including *bovine spongiform encephalitis* (*BSE,* so-called mad cow disease), *scrapie* in sheep, *kuru* (a human disease that has been eliminated), *chronic wasting disease (CWD)* in deer and elk, and *variant Creutzfeldt-Jakob disease*[4] *(vCJD)* in humans. The ingestion of infected tissue, transplants of infected tissue, or contact between infected tissue and mucous membranes or skin abrasions transmit these diseases. ▶**ANIMATIONS:** ***Prions: Diseases***

Normal cooking or sterilization procedures do not deactivate prions, though they are destroyed by incineration or by autoclaving in concentrated sodium hydroxide. The European Union recently approved the use of enzymes developed using biotechnology to remove prions from medical equipment.

There is no treatment for any prion disease, though the antimalarial drug quinacrine and the antipsychotic drug chlorpromazine forestall prion disease in mice. Human trials of these drugs are ongoing.

PrP proteins of different species are different, and at one time it was thought unlikely that prions could cross between species; however, an epidemic of BSE in Great Britain in the late 1980s resulted in the spread of prions to humans who ate infected beef. To prevent infection, most countries ban the use of animal-derived protein in animal feed. Unfortunately, this step was too late for more than 175 Europeans who developed fatal vCJD. (See **Clinical Case Study: Invasion from Within or Without?**)

Different prions may lie behind other neuronal diseases, such as Alzheimer's disease, Parkinson's disease, Huntington's disease, and amyotrophic lateral sclerosis (ALS). Additionally, some scientists think that prions may cause some cancers and type II diabetes. Research into prions' role in these diseases continues.

[4]"Variant" because it is derived from BSE prions in cattle, as opposed to the regular form of CJD, which is a genetic disease.

CLINICAL CASE STUDY

Invasion from Within or Without?

A 32-year-old father of two small children lived in the midwestern United States. An avid hunter since childhood, the man visited annually with family and friends in Colorado for elk hunting. His job required frequent travel to Europe, where he enjoyed exotic foods.

In 1988, his wife recalls, he began having problems. Frequently he forgot to pick up things from the store or even that his wife had called him. Later that year, he was unable to complete paperwork at his business and had difficulty performing even basic math. In England on business, he had forgotten his home phone number in the United States and couldn't remember how to spell his name for directory assistance.

By September, his wife insisted he seek medical care. All the standard blood tests came back normal. A psychologist diagnosed depression, but a brain scan revealed spongiform changes. He was given six weeks to live because there is no treatment for this disease.

1. What is the likely diagnosis?
2. The man's wife wondered, "Can we catch this disease from my husband?" How would you respond?
3. Where and how was the man probably infected?

TELL ME WHY

Why did scientists initially resist the idea of an infectious protein?

In this chapter we have seen that humans, animals, plants, fungi, bacteria, and archaea are susceptible to infection by acellular pathogens: viruses, viroids, and prions. **TABLE 13.5** summarizes the differences and similarities among these pathogenic agents and bacterial pathogens.

TABLE 13.5 **Comparison of Viruses, Viroids, and Prions to Bacterial Cells**

	Bacteria	Viruses	Viroids	Prions
Width	200–2000 nm	10–400 nm	2 nm	5 nm
Length	200–550,000 nm	20–800 nm	40–130 nm	5 nm
Nucleic acid?	Both DNA and RNA	Either DNA or RNA, never both	RNA only	None
Protein?	Present	Present	Absent	Present (PrP)
Cellular?	Yes	No	No	No
Cytoplasmic membrane?	Present	Absent (though some viruses do have a membranous envelope)	Absent	Absent
Functional ribosomes?	Present	Absent	Absent	Absent
Growth?	Present	Absent	Absent	Absent
Self-replicating?	Yes	No	No	Yes; transform PrP protein already present in cell
Responsiveness?	Present	Some bacteriophages respond to a host cell by injecting their genomes	Absent	Absent
Metabolism?	Present	Absent	Absent	Absent

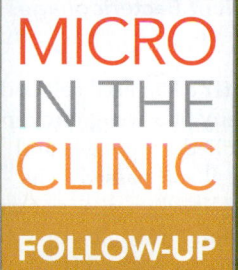

MICRO IN THE CLINIC FOLLOW-UP

Outbreak in the Jungle

When word of the epidemic reaches the government, a medical team quickly mobilizes. With only a foot-path leading to the village, the team carries medical supplies by backpack. When they arrive, half of the villagers are ill and many have died, including Nicia's mother and aunt.

Members of the medical team put on protective gear, take blood samples, and test for pathogens. They also set up a makeshift hospital in the largest building. Because the illness is obviously contagious, everyone showing symptoms is isolated in the "hospital."

Tests identify the killer as Ebola hemorrhagic fever, which is caused by an RNA virus. Humans become infected with Ebola virus through contact with body fluids of infected animals. Apparently, the virus entered Nicia's body through a cut on her finger. Her family became infected when they prepared her body for burial. Friends who took care of the ill also picked up the virus.

By the time the medical team quenches the outbreak, 75% of the villagers have died. Incredibly, Odette is one of the survivors. The medical team teaches the villagers safety guidelines for burying family members and for handling raw animal meat.

1. **Unlike bacteria, which can grow on the dead cells that cover our skin, Ebola virus must enter through breaks in the skin. Why is this?**
2. **Nicia experienced hemorrhaging from traumatic damage to blood vessels. How might replication of Ebola virus contribute to death of the cells that line blood vessels?**

MM Check your answers to Micro in the Clinic Follow-Up questions in the MasteringMicrobiology Study Area.

Explore the Invisible: The Lytic Cycle of Viral Replication

Practice on-the-go with Dr. Bauman Video Tutors by scanning this QR code with your smart phone. Visit the **MasteringMicrobiology Study Area** to challenge your understanding with practice tests, animation quizzes, and clinical case studies!

MasteringMicrobiology®

CHAPTER SUMMARY

1. Viruses, viroids, and prions are **acellular** disease-causing agents that lack cell structure and cannot metabolize, grow, self-reproduce, or respond to their environment.

Characteristics of Viruses (pp. 387–391)

1. A **virus** is a tiny infectious agent with nucleic acid surrounded by proteinaceous **capsomeres** that form a coat called a **capsid**. A virus exists in an extracellular state and an intracellular state. A **virion** is a complete viral particle, including a nucleic acid and a capsid, outside a cell.
2. The genomes of viruses include either DNA or RNA. Viral genomes may be dsDNA, ssDNA, dsRNA, or ssRNA. They may exist as linear or circular and singular or multiple molecules of nucleic acid, depending on the type of virus.
3. A **bacteriophage** (or **phage**) is a virus that infects a bacterial cell.
4. Virions can have a membranous **envelope** or be naked—that is, have no envelope.

Classification of Viruses (pp. 391–393)

1. Viruses are classified based on type of nucleic acid, presence of an envelope, shape, and size.
2. The International Committee on Taxonomy of Viruses (ICTV) has recognized viral family and genus names. With the exception of three orders, higher taxa are not established.

Viral Replication (pp. 394–403)

1. Viruses depend on random contact with a specific host cell type for replication. Typically a virus in a cell proceeds with a **lytic replication cycle** with five stages: **attachment**, **entry**, **synthesis**, **assembly**, and **release**.

   ▶ANIMATIONS: *Viral Replication: Overview*
2. Once attachment has been made between virion and host cell, the nucleic acid enters the cell. With phages, only the nucleic acid enters the host cell. With animal viruses, the entire virion often enters the cell, where the capsid is then removed in a process called **uncoating**.

   ▶ANIMATIONS: *Viral Replication: Animal Viruses*
3. Within the host cell, the viral nucleic acid directs synthesis of more viruses using metabolic enzymes and ribosomes of the host cell.
4. Assembly of synthesized virions occurs in the host cell, typically as capsomeres surround replicated or transcribed nucleic acids to form new virions.
5. Virions are released from the host cell either by lysis of the host cell (seen with phages and animal viruses) or by the extrusion of enveloped virions through the host's cytoplasmic membrane (called **budding**), a process seen only with certain animal viruses. If budding continues over time, the infection is persistent. An envelope is derived from a cell membrane.

   ▶ANIMATIONS: *Viral Replication: Virulent Bacteriophages*

   ▶VIDEO TUTOR: *The Lytic Cycle of Viral Replication*
6. **Temperate phages** (lysogenic phages) enter a bacterial cell and remain inactive in a process called **lysogeny** or a **lysogenic replication cycle**. Such inactive phages are called **prophages** and are inserted into the chromosome of the cell and passed to its daughter cells. **Lysogenic conversion** results when phages carry genes that alter the phenotype of a bacterium. At some point in the generations that follow, a prophage may be excised from the chromosome in a process known as **induction**. At that point the prophage again becomes a lytic virus.

   ▶ANIMATIONS: *Viral Replication: Temperate Bacteriophages*
7. In **latency**, a process similar to lysogeny, an animal virus remains inactive in a cell, possibly for years, as part of a chromosome or in the cytosol. A **latent** virus is also known as a **provirus**. A provirus that has become incorporated into a host's chromosome remains there.
8. With the exception of hepatitis B virus, dsDNA viruses act like cellular DNA in transcription and replication.
9. Some ssRNA viruses have **positive-strand RNA (+RNA)**, which can be directly translated by ribosomes to synthesize protein. From the +RNA, complementary **negative-strand RNA (−RNA)** is transcribed to serve as a template for more +RNA.
10. **Retroviruses**, such as HIV, are +ssRNA viruses that carry reverse transcriptase, which transcribes DNA from RNA. This reverse process (DNA transcribed from RNA) is reflected in the name retrovirus.
11. −ssRNA viruses carry an RNA-dependent RNA transcriptase for transcribing mRNA from the −RNA genome so that protein can then be translated. Transcription of RNA from RNA is not found in cells.

12. In dsRNA viruses, one strand of RNA functions as a genome, and the other strand functions as a template for RNA replication.

The Role of Viruses in Cancer (pp. 403–404)

1. **Neoplasia** is uncontrolled cellular reproduction in a multicellular animal. A mass of neoplastic cells, called a **tumor**, may be relatively harmless **(benign tumor)** or invasive **(malignant tumors)**. Malignant tumors are also called **cancer**. **Metastasis** describes the spreading of malignant tumors. Environmental factors or oncogenic viruses may cause neoplasia.

Culturing Viruses in the Laboratory (pp. 404–406)

1. In the laboratory, viruses must be cultured inside mature organisms, in embryonated chicken eggs, or in cell cultures because viruses cannot metabolize or replicate alone.
2. When a mixture of bacteria and phages is grown on an agar plate, bacteria infected with phages lyse, producing clear areas called **plaques** on the bacterial lawn. A technique called **plaque assay** enables the estimation of phage numbers.
3. Viruses can be grown in two types of **cell cultures**. Whereas **diploid cell cultures** last about 100 generations, **continuous cell cultures**, derived from cancer cells, last longer.

Are Viruses Alive? (p. 406)

1. Outside cells, viruses do not appear to be alive, but within cells, they exhibit lifelike qualities such as the ability to replicate themselves.

Other Parasitic Particles: Viroids and Prions (pp. 406–409)

1. **Viroids** are small circular pieces of RNA with no capsid that infect and cause disease in plants. Similar pathogenic RNA molecules have been found in fungi.
2. **Prions** are infectious protein particles that lack nucleic acids and replicate by converting similar, normal proteins into new prions. Diseases caused by prions are spongiform encephalopathies, which involve fatal neurological degeneration.

▶ANIMATIONS: *Prions: Overview, Characteristics, and Diseases*

QUESTIONS FOR REVIEW

Answers to the Questions for Review (except Short Answer questions) begin on p. A-1.

Multiple Choice

1. Which of the following is *not* an acellular agent?
   a. viroid
   b. virus
   c. rickettsia
   d. prion
2. Which of the following statements is true?
   a. Viruses move toward their host cells.
   b. Viruses are capable of metabolism.
   c. Viruses lack a cytoplasmic membrane.
   d. Viruses grow in response to their environmental conditions.
3. A virus that is specific for a bacterial host is called a ________________.
   a. phage
   b. prion
   c. virion
   d. viroid
4. A naked virus ________________.
   a. has no membranous envelope
   b. has injected its DNA or RNA into a host cell
   c. is devoid of capsomeres
   d. is one that is unattached to a host cell
5. Which of the following statements is *false*?
   a. Viruses may have circular DNA.
   b. dsRNA is found in bacteria more often than in viruses.
   c. Viral DNA may be linear.
   d. Typically, viruses have DNA or RNA but not both.
6. When a eukaryotic cell is infected with an enveloped virus and sheds viruses slowly over time, this infection is ________________.
   a. called a lytic infection
   b. a prophage cycle
   c. called a persistent infection
   d. caused by a quiescent virus
7. Another name for a complete virus is ________________.
   a. virion
   b. viroid
   c. prion
   d. capsid
8. Which of the following viruses can be latent?
   a. HIV
   b. chicken pox virus
   c. herpesviruses
   d. all of the above
9. Which of the following is *not* a criterion for specific family classification of viruses?
   a. the type of nucleic acid present
   b. envelope structure
   c. capsid type
   d. lipid composition
10. A clear zone of phage infection in a bacterial lawn is ________________.
    a. a prophage
    b. a plaque
    c. naked
    d. a zone of inhibition

Matching

Match each numbered term with its description.

1. _____ uncoating
2. _____ prophage
3. _____ retrovirus
4. _____ bacteriophage
5. _____ capsid
6. _____ envelope
7. _____ virion
8. _____ provirus
9. _____ benign tumor
10. _____ cancer

A. dormant virus in a eukaryotic cell
B. a virus that infects a bacterium
C. transcribes DNA from RNA
D. protein coat of virus
E. a membrane on the outside of a virus
F. complete viral particle
G. inactive virus within bacterial cell
H. removal of capsomeres from a virion
I. invasive neoplastic cells
J. harmless neoplastic cells

VISUALIZE IT!

1. Label each step in the bacteriophage replication cycle below.

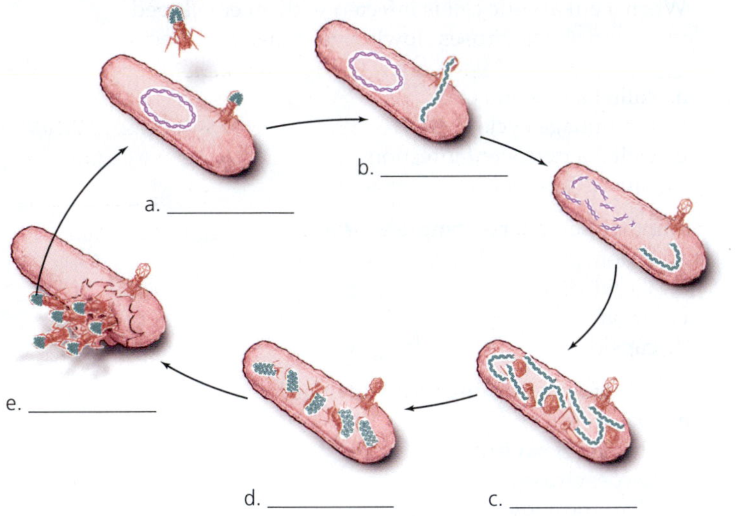

2. Identify the viral capsid shapes.

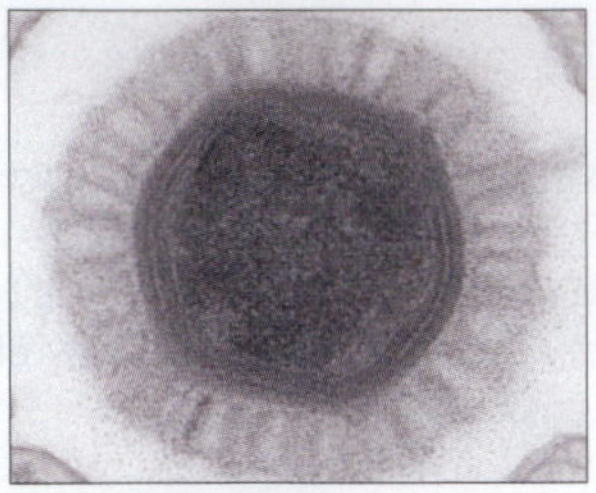

(a)

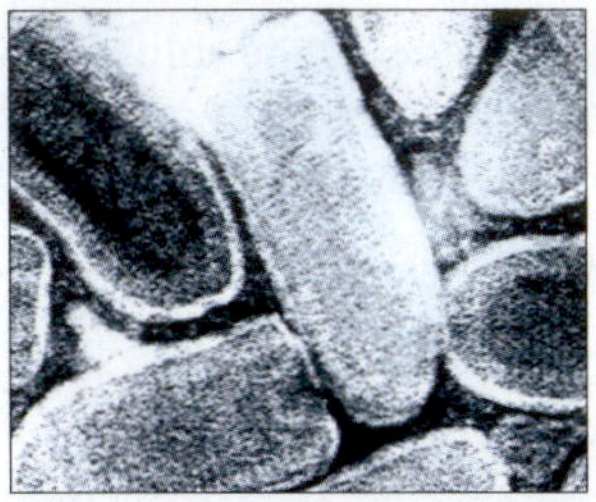

(b)

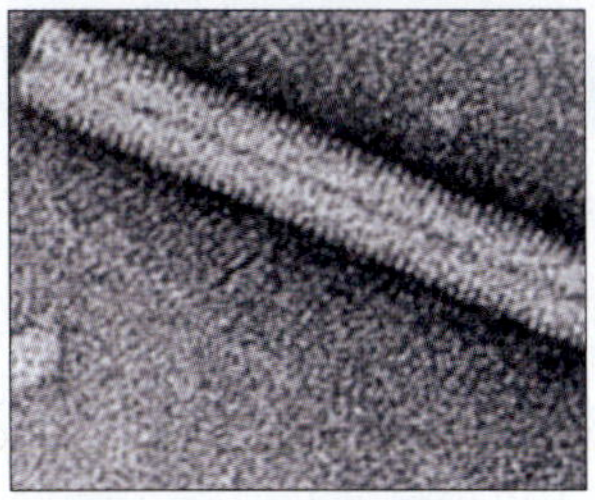

(c)

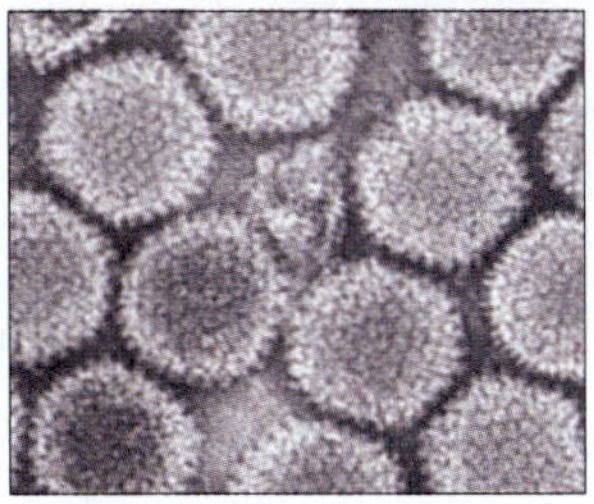

(d)

Short Answer

1. Compare and contrast a bacterium and a virus by writing either "Present" or "Absent" for each of the following structures.

Structure	Bacterium	Virus
Cell membrane		
Functional ribosome		
Cytoplasm		
Nucleic acid		
Nuclear membrane		

2. Describe the five phases of a generalized lytic replication cycle.
3. Why is it difficult to treat viral infections?
4. Describe four different ways that viral nucleic acid can enter a host cell.
5. Contrast lysis and budding as means of release of virions from a host cell.
6. What is the difference between a virion and a virus particle?
7. How is a provirus like a prophage? How is it different?
8. Describe lysogeny.
9. How are viruses specific for their host's cells?
10. Compare and contrast diploid cell culture and continuous cell culture.

CRITICAL THINKING

1. Larger viruses usually have a double-stranded genome, whereas small viruses typically have a single-stranded genome. What reasonable explanation can you offer for this observation?
2. What are the advantages and disadvantages to bacteriophages of the lytic and lysogenic reproductive strategies?
3. How are computer viruses similar to biological viruses? Are computer viruses alive? Why or why not?
4. Compare and contrast lysogeny by a prophage and latency by a provirus.
5. An agricultural microbiologist wants to stop the spread of a viral infection of a crop. Is stopping viral attachment a viable option? Why or why not?
6. Some viral genomes, composed of single-stranded RNA, act as mRNA. What advantage might these viruses have over other kinds of viruses?
7. In some viruses, the capsomeres act enzymatically as well as structurally. What advantage might this provide the virus?
8. Why has it been difficult to develop a complete taxonomy for viruses?
9. If a colony of 1.5 billion *E. coli* cells were infected with a single phage T4 and each lytic replication cycle of the phage produced 200 new phages, how many replication cycles would it take for T4 phages to overwhelm the entire bacterial colony? (Assume for the sake of simplicity that every phage completes its replication cycle in a different cell and that the bacteria themselves do not reproduce.)
10. What differences would you expect in the replication cycles of RNA phages from those of DNA phages? (Hints: Think about the processes of transcription, translation, and replication of nucleic acids. Also, note that RNA is not normally inserted into a DNA molecule.)
11. Although many +ssRNA viruses use their genome directly as messenger RNA, +ssRNA retroviruses do not. Instead, their +RNA is transcribed into DNA by reverse transcriptase. What advantage do retroviruses gain by using reverse transcriptase?
12. If an enveloped virus were somehow released from a cell without budding, it would not have an envelope. What effect would this have on the virulence of the virus? Why?
13. A latent virus that is incorporated into a host cell's chromosome is never induced; that is, it never emerges from the host cell's chromosome to become a free virus. Given that it cannot emerge from the host cell's chromosome, can such a latent virus be considered "safe"? Why or why not?

CONCEPT MAPPING

Using the following terms, draw a concept map that describes the replication of animal viruses. For a sample concept map, see p. 94. Or, complete this and other concept maps online by going to the MasteringMicrobiology Study Area.

+ssRNA
+ssRNA retrovirus
−ssRNA
Assembly
Attachment
Budding
Direct penetration
dsDNA
dsRNA
Endocytosis
Entry
Exocytosis
Fusion
Host cell
Lysis
New virions
Release
ssDNA
Synthesis
Viral genome
Uncoating
Viral nucleic acid
Viral proteins

14 Infection, Infectious Diseases, and Epidemiology

MICRO IN THE CLINIC

Antibiotics: Friend or Foe?

Nancy is an energetic 66-year-old widow who walks three miles every day and does water aerobics three days a week. When she develops an annoying sinus infection, she doesn't let it slow her down; she calls her doctor, gets a prescription for antibiotics, and takes them as prescribed.

After a couple of weeks the sinus infection is gone, but Nancy notices that she has to urinate much more frequently than usual. She has a burning sensation while urinating, and her back aches. Recognizing the common symptoms of a urinary tract infection, Nancy tries her favorite home remedy of drinking lots of cranberry juice—after all, she didn't get to be a healthy 66 by running to the doctor with every problem.

However, her symptoms get worse. The pain spreads to her lower back, she develops a fever, and she feels so weak she can barely get out of bed. Then she notices her urine is reddish and contains what looks like pus. Alarmed, Nancy struggles to the phone and calls her favorite niece, Maya, who's in nursing school. Maya rushes over, examines her aunt quickly, and calls 911. Maya rides with Nancy to the emergency room.

What is making Nancy so ill? Why do you think her urine looks reddish? Turn to the end of the chapter (p. 443) to find out.

Explore More: Test your readiness and apply your knowledge with dynamic learning tools at MasteringMicrobiology.

In this chapter, we examine how microbes directly affect our bodies. Bear in mind that most microorganisms are neither harmful nor expressly beneficial to humans—they live their lives, and we live ours. Relatively few microbes either directly benefit us or harm us.

We first examine the general types of relationships that microbes can have with the bodies of their hosts. Then we discuss sources of infectious diseases of humans, how microbes enter and attach to their hosts, the nature of infectious diseases, and how microbes leave hosts to become available to enter new hosts. Next we explore the ways that infectious diseases are spread among hosts. Finally, we consider *epidemiology,* the study of the occurrence and spread of diseases within groups of humans, and the methods by which we can limit the spread of pathogens within society.

▲ **FIGURE 14.1 Mutualism.** Wood-eating termites in the genus *Reticulitermes* cannot digest the cellulose in wood, but the protozoan *Trichonympha,* which lives in their intestines, can digest cellulose with the help of bacteria. The three organisms maintain a mutualistic relationship that is crucial for the life of the termite. *What benefits accrue to each of the symbionts in this mutualistic relationship?*

Figure 14.1 *The termite gets digested cellulose from the protozoa, which get a constant supply of wood pulp to digest, so the bacteria get a food source as well.*

Symbiotic Relationships Between Microbes and Their Hosts

Symbiosis (sim-bī-o´sis) means "to live together." Each of us has symbiotic relationships with countless microorganisms, although often we are completely unaware of them. The 10 trillion or so cells in your body provide homes for more than 100 trillion bacteria and fungi and even more viruses. We begin this section by considering the types of relationships between microbes and their hosts.

Types of Symbiosis

LEARNING **OUTCOMES**

14.1 Distinguish among the types of symbiosis, listing them in order from most beneficial to most harmful for the host.

14.2 Describe the relationships among the terms *parasite, host,* and *pathogen.*

Biologists see symbiotic relationships as a continuum from cases that are beneficial to both members of a pair to situations in which one member lives at a damaging expense to the other. In the following sections, we examine three conditions along the continuum: mutualism, commensalism, and parasitism.

Mutualism

In **mutualism** (mū´tū-ăl-izm), both members benefit from their interaction. For example, bacteria in your colon receive a warm, moist, nutrient-rich environment in which to thrive, while you absorb vitamin precursors and other nutrients released from the bacteria. In this example, the relationship is beneficial to microbe and human alike but is not required by either. Many of the bacteria could live elsewhere, and you could get vitamins from your diet.

Some mutualistic relationships provide such important benefits that one or both of the parties cannot live without the other. For instance, termites, which cannot digest the cellulose in wood by themselves, would die without a mutualistic relationship with colonies of wood-digesting protozoa and bacteria living in their intestines **(FIGURE 14.1)**. The termites provide wood pulp and a home to the protozoa, while the protozoa break down the wood within the termite intestines using enzymes from the bacteria, which live on the protozoa. The microbes share the nutrients released with the termites. **Beneficial Microbes: A Bioterrorist Worm** on p. 418 describes another mutualistic relationship.

Commensalism

In **commensalism** (kŏ-men´săl-izm), a second type of symbiosis, one member of the relationship benefits without significantly affecting the other. For example, *Staphylococcus epidermidis* (staf´i-lō-kok´us ep-i-der-mid´is) growing on the skin typically causes no measurable harm to a person. However, an absolute example of commensalism is difficult to prove because the host may experience unobserved benefits. In this example, *Staphylococcus* may inhibit pathogenic microbes from colonizing the skin.

Parasitism

Of concern to health care professionals is a third type of symbiosis called **parasitism**[1] (par´ă-si-tizm). A **parasite** derives benefit from its **host** while harming it, though some hosts sustain only slight damage. In the most severe cases, a parasite kills its host, in the process destroying its own home, which

[1]From Greek *parasitos,* meaning "one who eats at the table of another."

TABLE 14.1 The Three Types of Symbiotic Relationships

	Organism 1	Organism 2	Example
Mutualism	Benefits	Benefits	Bacteria in human colon
Commensalism	Benefits	Neither benefits nor is harmed	*Staphylococcus epidermidis* on skin
Parasitism	Benefits	Is harmed	Tuberculosis bacteria in human lung

is not beneficial for the parasite. Therefore, parasites that allow their hosts to survive are more likely to spread. Similarly, hosts that tolerate a parasite are more likely to reproduce. The result over time will be *coevolution* toward commensalism or mutualism. Any parasite that causes disease is called a **pathogen** (path´ō-jen).

A variety of protozoa, fungi, and bacteria are microscopic parasites of humans. Larger parasites of humans include parasitic worms and biting arthropods,[2] including mites (chiggers), ticks, mosquitoes, fleas, and bloodsucking flies.

You should realize that microbes that are parasitic may become mutualistic or vice versa; that is, the relationships between and among organisms can change over time. **TABLE 14.1** summarizes three types of symbiosis.

Normal Microbiota in Hosts

LEARNING | OUTCOME

14.3 Describe the normal microbiota, including resident and transient members.

Even though many parts of your body are *axenic*[3] (ā-zen´ik) environments—that is, sites that are free of any microbes—other parts of your body shelter millions of mutualistic and commensal symbionts. Each square centimeter of your skin, for example, contains more than 3 million bacteria, and your large intestine contains 400 to 1000 kinds of microbes that outnumber your own cells many times. The microbes that colonize the surfaces of the body without normally causing disease constitute the body's **normal microbiota** **(FIGURE 14.2)**, also sometimes called the *normal flora*[4] or the *indigenous microbiota*. The normal microbiota are of two main types: resident microbiota and transient microbiota.

[2]From Greek *arthros*, meaning "jointed," and *pod*, meaning "foot."

[3]From Greek *a*, meaning "no," and *xenos*, meaning "foreigner."

[4]This usage of the term *flora*, which literally means "plants," derives from the fact that bacteria and fungi were considered plants in early taxonomic schemes. *Microbiota* is preferable because it properly applies to all microbes (whether eukaryotic or prokaryotic) and to viruses.

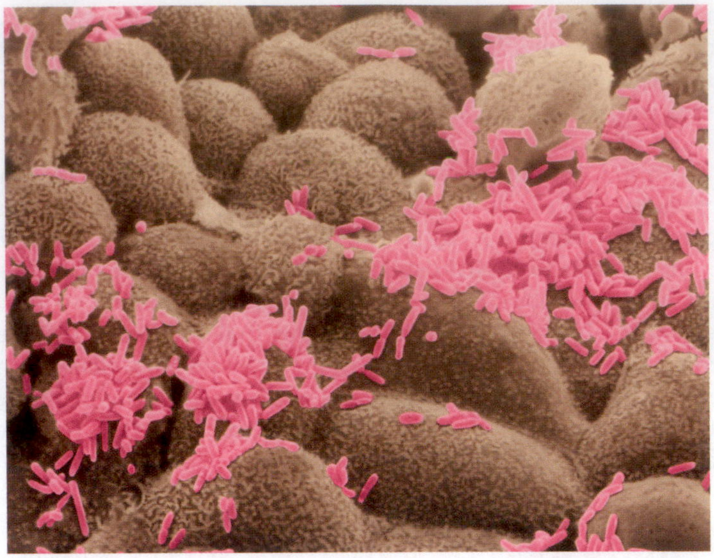

▲ **FIGURE 14.2 An example of normal microbiota.** Bacteria (pink) colonize the human nasal cavity.

Resident Microbiota

Resident microbiota remain a part of the normal microbiota of a person throughout life. These organisms are found on the skin and on the mucous membranes of the digestive tract, upper respiratory tract, distal[5] portion of the urethra,[6] and vagina. **TABLE 14.2** illustrates these environments and lists some of the resident bacteria, fungi, and protozoa that live there. Most of the resident microbiota are commensal; that is, they feed on excreted cellular wastes and dead cells without causing harm.

Transient Microbiota

Transient microbiota remain in the body for only a few hours, days, or months before disappearing. They are found in the same locations as the resident members of the normal microbiota but cannot persist because of competition from other microorganisms, elimination by the body's defense cells, or chemical and physical changes in the body that dislodge them.

Acquisition of Normal Microbiota

You developed in your mother's womb without normal microbiota because you had surrounded yourself with an amniotic membrane and fluid, which generally kept microorganisms at bay, and because your mother's uterus was essentially an axenic environment.

Your normal microbiota began to develop when your surrounding amniotic membrane ruptured and microorganisms came in contact with you during birth. Microbes entered your mouth and nose as you passed through the birth canal, and

[5]*Distal* refers to the end of some structure, as opposed to *proximal*, which refers to the beginning of a structure.

[6]The tube that empties the bladder.

TABLE 14.2 Some Resident Microbiota[a]

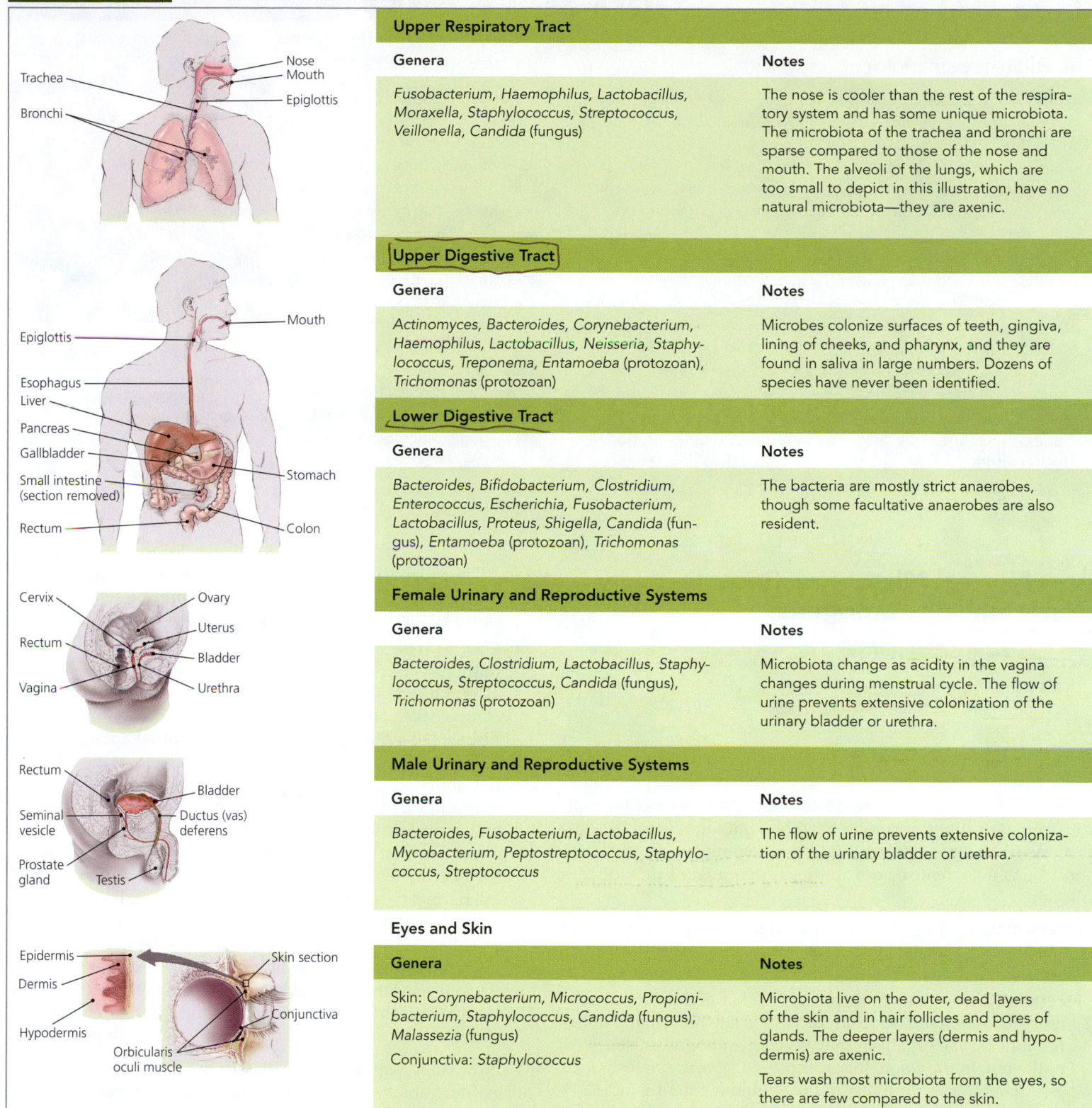

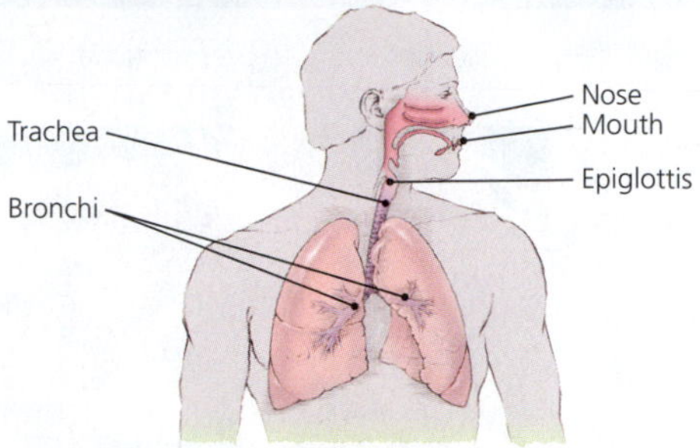

Upper Respiratory Tract

Genera	Notes
Fusobacterium, Haemophilus, Lactobacillus, Moraxella, Staphylococcus, Streptococcus, Veillonella, Candida (fungus)	The nose is cooler than the rest of the respiratory system and has some unique microbiota. The microbiota of the trachea and bronchi are sparse compared to those of the nose and mouth. The alveoli of the lungs, which are too small to depict in this illustration, have no natural microbiota—they are axenic.

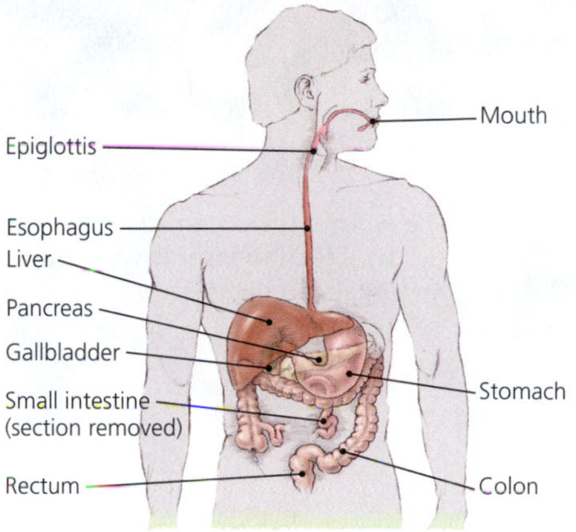

Upper Digestive Tract

Genera	Notes
Actinomyces, Bacteroides, Corynebacterium, Haemophilus, Lactobacillus, Neisseria, Staphylococcus, Treponema, Entamoeba (protozoan), *Trichomonas* (protozoan)	Microbes colonize surfaces of teeth, gingiva, lining of cheeks, and pharynx, and they are found in saliva in large numbers. Dozens of species have never been identified.

Lower Digestive Tract

Genera	Notes
Bacteroides, Bifidobacterium, Clostridium, Enterococcus, Escherichia, Fusobacterium, Lactobacillus, Proteus, Shigella, Candida (fungus), *Entamoeba* (protozoan), *Trichomonas* (protozoan)	The bacteria are mostly strict anaerobes, though some facultative anaerobes are also resident.

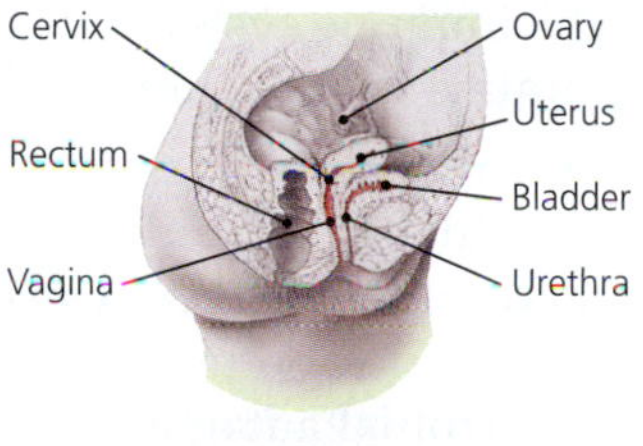

Female Urinary and Reproductive Systems

Genera	Notes
Bacteroides, Clostridium, Lactobacillus, Staphylococcus, Streptococcus, Candida (fungus), *Trichomonas* (protozoan)	Microbiota change as acidity in the vagina changes during menstrual cycle. The flow of urine prevents extensive colonization of the urinary bladder or urethra.

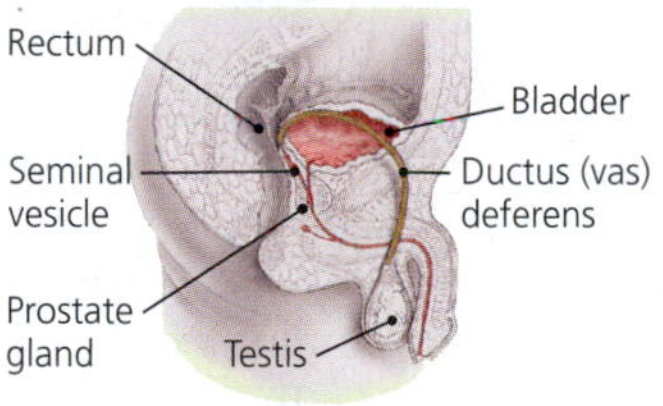

Male Urinary and Reproductive Systems

Genera	Notes
Bacteroides, Fusobacterium, Lactobacillus, Mycobacterium, Peptostreptococcus, Staphylococcus, Streptococcus	The flow of urine prevents extensive colonization of the urinary bladder or urethra.

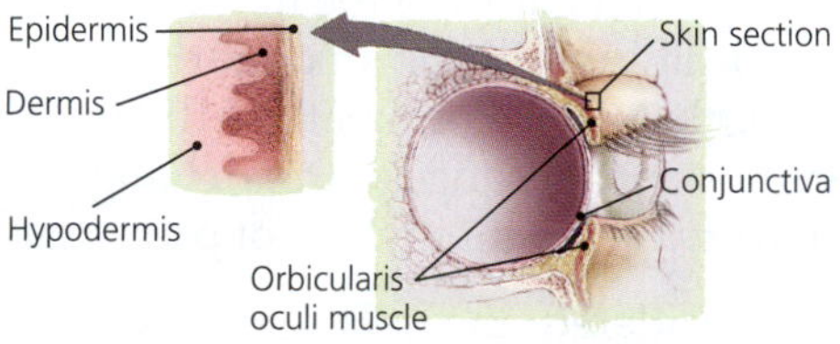

Eyes and Skin

Genera	Notes
Skin: *Corynebacterium, Micrococcus, Propionibacterium, Staphylococcus, Candida* (fungus), *Malassezia* (fungus) Conjunctiva: *Staphylococcus*	Microbiota live on the outer, dead layers of the skin and in hair follicles and pores of glands. The deeper layers (dermis and hypodermis) are axenic. Tears wash most microbiota from the eyes, so there are few compared to the skin.

[a]Genera are bacteria unless noted.

your first breath was loaded with microorganisms that quickly established themselves in your upper respiratory tract. Your first meals provided the progenitors of resident microbiota for your colon, while *Staphylococcus* and other microbes transferred from the skin of both the medical staff and your parents began to colonize your skin. Although you continue to add to your transient microbiota, most of the resident microbiota was initially established during your first months of life.

BENEFICIAL MICROBES

A Bioterrorist Worm

Bioterrorism has been defined as the deliberate release of viruses, bacteria, or other germs to cause illness or death. A nematode worm, *Steinernema*, is by that definition a bioterrorist, releasing a mutualistic bacterial symbiont, *Xenorhabdus*.

Nematodes are microscopic, unsegmented, round worms that live in soil worldwide. *Steinernema* preys on insects, including ants, termites, and the immature stages of various beetles, weevils, worms, fleas, ticks, and gnats. *Steinernema* crawls into an insect's mouth or anus and then crosses the intestinal walls to enter the insect's blood. The worm then releases its symbiotic *Xenorhabdus*. This bacterium inactivates the insect's defensive systems; generates antibacterial compounds, which eliminate other bacteria; produces insecticidal toxins to kill the insect; and secretes digestive enzymes. The bacterial enzymes turn the insect's body into a slimy porridge of nutrients within 48 hours. Meanwhile, *Steinernema* nematodes mature, mate, and reproduce within the insect's liquefying body. The nematodes' offspring feed on the gooey fluid until they are old enough to emerge from the insect's skeleton but not before taking up a supply of bacteria for their own future bioterroist raids on new insect hosts.

This is a true mutualistic symbiosis: The nematode depends on the bacterium to kill and digest the insect host, while the bacterium depends on the nematode to deliver it to new hosts. Both benefit. Farmers, too, can benefit. They can use *Steinernema* and its bacterium to control insect pests without damaging their crops, animals, or people.

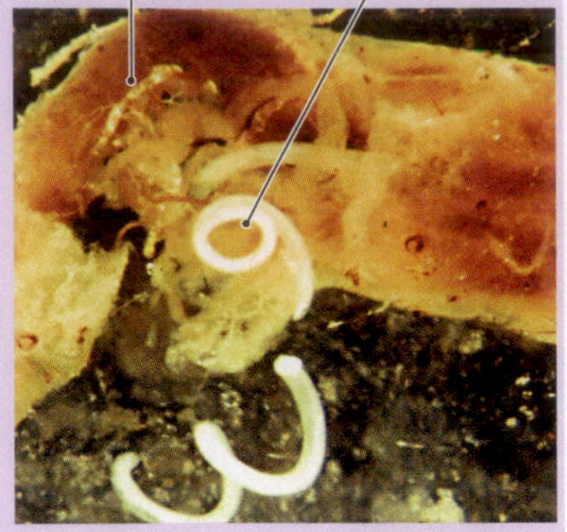

The worm *Steinernema* (white) kills by infecting insects with lethal *Xenorhabdus* bacteria.

How Normal Microbiota Become Opportunistic Pathogens

LEARNING | **OUTCOME**

14.4 Describe three conditions that create opportunities for normal microbiota to cause disease.

Under ordinary circumstances, normal microbiota do not cause disease. However, these same microbes may become harmful if an opportunity to do so arises. In this case, normal microbiota (or other normally harmless microbes from the environment) become **opportunistic pathogens,** or *opportunists.* Conditions that create opportunities for pathogens include the following:

- *Introduction of a member of the normal microbiota into an unusual site in the body.* As we have seen, normal microbiota are present only in certain body sites, and each site has only certain species of microbiota. If a member of the normal microbiota in one site is introduced into a site it normally does not inhabit, the organism may become an opportunistic pathogen. In the colon, for example, *Escherichia coli* (esh-ĕ-rik´ē-ă kō´lī) is mutualistic, but should it enter the urethra, it becomes an opportunist that can produce disease.
- *Immune suppression.* Anything that suppresses the body's immune system—including disease, malnutrition, emotional or physical stress, extremes of age (either very young or very old), the use of radiation or chemotherapy to combat cancer, or the use of immunosuppressive drugs in transplant patients—can enable opportunistic pathogens. AIDS patients often die from opportunistic infections that are typically controlled by a healthy immune system because HIV infection suppresses immune system function.
- *Changes in the normal microbiota.* Normal microbiota use nutrients, take up space, and release toxic waste products, all of which make it less likely that arriving pathogens can compete well enough to become established and produce disease. This situation is known as **microbial antagonism** or **microbial competition.** However, changes in the relative abundance of normal microbiota, for whatever reason, may allow a member of the normal microbiota to become an opportunistic pathogen and thrive. For example, when a woman must undergo long-term antimicrobial treatment for a bacterial blood infection, the antimicrobial may also kill normal bacterial microbiota in the vagina. In the absence of competition from bacteria, *Candida albicans* (kan´did-ă al´bi-kanz), a yeast and also a member of the normal vaginal microbiota, grows prolifically, producing an opportunistic vaginal yeast infection. Other conditions that can disrupt the normal microbiota are hormonal changes, stress, changes in diet, and exposure to overwhelming numbers of pathogens.

Now that we have considered the relationships of the normal microbiota to their hosts, we turn our attention to the movement of microbes into new hosts by considering aspects of human diseases.

TELL ME WHY

Why is an absolute commensalism difficult to prove?

TABLE 14.3 Some Common Zoonoses

Disease	Causative Agent	Animal Reservoir	Mode of Transmission
Helminthic			
Tapeworm infestation	*Dipylidium caninum*	Dogs	Ingestion of larvae transmitted in dog saliva
Fasciola infestation	*Fasciola hepatica*	Sheep, cattle	Ingestion of contaminated vegetation
Protozoan			
Malaria	*Plasmodium* spp.	Monkeys	Bite of *Anopheles* mosquito
Toxoplasmosis	*Toxoplasma gondii*	Cats and other animals	Ingestion of contaminated meat, inhalation of pathogen, direct contact with infected tissues
Fungal			
Ringworm	*Trichophyton* sp. *Microsporum* sp. *Epidermophyton* sp.	Domestic animals	Direct contact
Bacterial			
Anthrax	*Bacillus anthracis*	Domestic livestock	Direct contact with infected animals, inhalation
Bubonic plague	*Yersinia pestis*	Rodents	Flea bites → cause
Lyme disease	*Borrelia burgdorferi*	Deer	Tick bites
Salmonellosis	*Salmonella* spp.	Birds, rodents, reptiles	Ingestion of fecally contaminated water or food
Typhus	*Rickettsia prowazekii*	Rodents	Louse bites
Viral			
Rabies	*Lyssavirus* sp.	Bats, skunks, foxes, dogs	Bite of infected animal
Hantavirus pulmonary syndrome	*Hantavirus* sp.	Deer mice	Inhalation of viruses in dried feces and urine
Yellow fever	*Flavivirus* sp.	Monkeys	Bite of *Aedes* mosquito

Reservoirs of Infectious Diseases of Humans

LEARNING **OUTCOME**

14.5 Describe three types of reservoirs of infection in humans.

Most pathogens of humans cannot survive for long in the relatively harsh conditions they encounter outside their hosts. If these pathogens are to enter new hosts, they must survive in some site from which they can infect new hosts. Sites where pathogens are maintained as a source of infection are called **reservoirs of infection.** In this section we discuss three types of reservoirs: animal reservoirs, human carriers, and nonliving reservoirs.

Animal Reservoirs

Many pathogens that normally infect either domesticated or sylvatic[7] (wild) animals can also affect humans. The more similar an animal's physiology is to human physiology, the more likely its pathogens are to affect human health. Diseases that spread naturally from their usual animal hosts to humans are called **zoonoses** (zō-ō-nō´sēz). Over 150 zoonoses have been identified throughout the world. Well-known examples include yellow fever, anthrax, bubonic plague, and rabies.

Humans may acquire zoonoses from animal reservoirs via a number of routes, including various types of direct contact with animals and their wastes, by eating animals, or via bloodsucking arthropods. Human infections with zoonoses are difficult to eradicate because extensive animal reservoirs are often involved. The larger the animal reservoir (i.e., the greater the number and types of infected animals) and the greater the contact between humans and the animals, the more difficult and costly it is to control the spread of the disease to humans. This is especially true when the animal reservoir consists of both sylvatic and domesticated animals. In the case of rabies, for example, the disease typically spreads from a sylvatic reservoir (often bats, foxes, and skunks) to domestic pets from which humans may be infected. The wild animals constitute a reservoir for the rabies virus, but transmission to humans can be limited by vaccinating domestic pets.

TABLE 14.3 provides a brief view of some common zoonoses. Humans are usually dead-end hosts for zoonotic pathogens—that is, humans do not act as significant reservoirs for the reinfection of animal hosts—largely because the circumstances under which zoonoses are transmitted favor movement from animals to humans but not in the opposite direction. For example, animals do not often eat humans these days, and animals less frequently have contact with human wastes than humans have contact with animal wastes. Zoonotic diseases transmitted via the bites of bloodsucking arthropods are the most likely type to be transmitted back to animal hosts.

[7]From Latin *sylva*, meaning "woodland."

CLINICAL CASE STUDY

A Deadly Carrier

In 1937, a man employed to lay water pipes was found to be the source of a severe epidemic of typhoid fever. The man, an asymptomatic carrier of *Salmonella enterica* serotype Typhi, the bacterium that causes typhoid, habitually urinated at his job site. In the process, he contaminated the town's water supply with bacteria from his bladder. Over 300 cases of typhoid fever developed, and 43 people died before the man was identified as the carrier.

1. How do you think health officials were able to identify the source of this typhoid epidemic?
2. Given that antibiotics were not generally available in 1937, how could health officials end the epidemic short of removing the man from the job site?

Human Carriers

Experience tells you that humans with active diseases are important reservoirs of infection for other humans. What may not be so obvious is that people with no obvious symptoms before or after an obvious disease may also be infective in some cases. Further, some infected people remain both asymptomatic and infective for years. This is true of tuberculosis, syphilis, and AIDS, for example. Whereas some of these **carriers** incubate the pathogen in their body and eventually develop the disease, others remain a continued source of infection without ever becoming sick. Presumably many such healthy carriers have defensive systems that protect them from illness. An example of such a carrier is given in **Clinical Case Study: A Deadly Carrier.**

Nonliving Reservoirs

Soil, water, and food can be **nonliving reservoirs** of infection. Soil, especially if fecally contaminated, can harbor *Clostridium* (klos-trid´ē-ŭm) bacteria, which cause botulism, tetanus, and other diseases. Water can be contaminated with feces and urine containing parasitic worm eggs, pathogenic protozoa, bacteria, and viruses. Meats and vegetables can also harbor pathogens.

TELL ME WHY

Why might animal reservoirs be involved in more diseases than are human reservoirs?

The Invasion and Establishment of Microbes in Hosts: Infection

In this section we examine events that occur when hosts are exposed to microbes from a reservoir, the sites at which microbes can gain entry into hosts, and the ways entering microbes become established in new hosts.

Exposure to Microbes: Contamination and Infection

LEARNING | OUTCOME

14.6 Describe the relationship between contamination and infection.

In the context of the interaction between microbes and their hosts, **contamination** refers to the mere presence of microbes in or on the body. Some microbial contaminants reach the body in food, drink, or the air, whereas others are introduced via wounds, biting arthropods, or sexual intercourse. Several outcomes of contamination by microbes are possible. Some microbial contaminants remain where they first contacted the body (such as the skin or mucous membranes) without causing harm and subsequently become part of the resident microbiota; other microbial contaminants remain on the body for only a short time as part of the transient microbiota. Still others overcome the body's external defenses, multiply, and become established in the body; such a successful invasion of the body by a pathogen is called an **infection.** An infection may or may not result in disease; that is, it may not adversely affect the body.

Next we consider in greater detail the various sites through which pathogens gain entry into the body.

Portals of Entry

LEARNING | OUTCOME

14.7 Identify and describe the portals through which pathogens invade the body.

The sites through which most pathogens enter the body can be likened to the great gates or portals of a castle because those sites constitute the routes by which microbes gain entry. Pathogens thus enter the body at several sites, called **portals of entry** (**FIGURE 14.3**), which are of three major types: the skin, the mucous membranes, and the placenta. A fourth entry point, the so-called *parenteral* (pă-ren´ter-ăl) *route,* is not a portal but a way of circumventing the usual portals. Next we consider each type in turn.

Skin

Because the outer layer of skin is composed of relatively thick layers of tightly packed, dead, dry cells (**FIGURE 14.4**), it forms

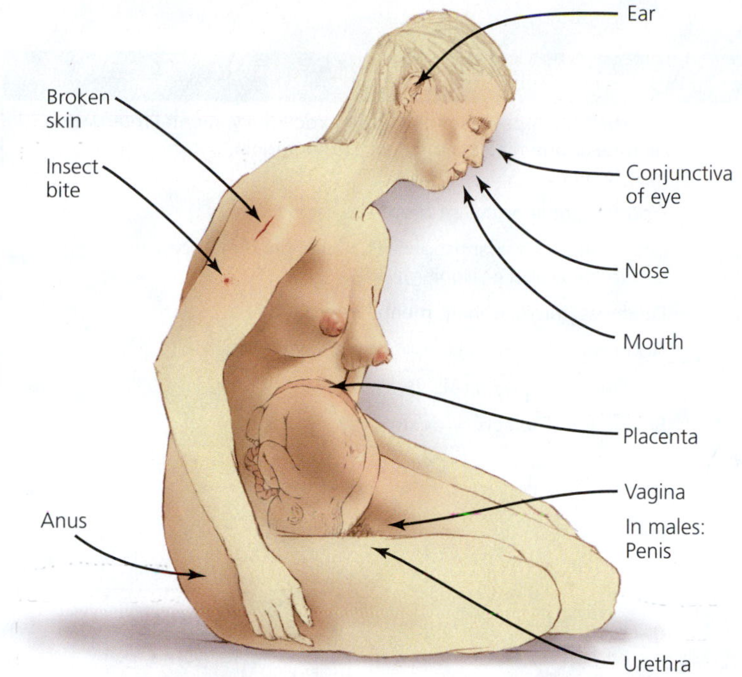

▲ **FIGURE 14.3 Routes of entry for invading pathogens.** The portals of entry include the skin, placenta, conjunctiva, and mucous membranes of the respiratory, gastrointestinal, urinary, and reproductive tracts. The parenteral route involves a puncture through the skin.

a formidable barrier to most pathogens as long as it remains intact. Still, some pathogens can enter the body through natural openings in the skin, such as hair follicles and sweat glands. Abrasions, cuts, bites, scrapes, stab wounds, and surgeries open the skin to infection by contaminants. Additionally, the larvae of some parasitic worms are capable of burrowing through the skin to reach underlying tissues, and some fungi can digest the dead outer layers of skin, thereby gaining access to deeper, moister areas within the body.

Mucous Membranes

The major portals of entry for pathogens are the *mucous membranes,* which line all the body cavities that are open to the outside world. They include the linings of the respiratory, gastrointestinal, urinary, and reproductive tracts as well as the *conjunctiva* (kon-jŭnk-tī′vă), the thin membrane covering the surface of the eyeball and the underside of each eyelid. Like the skin, mucous membranes are composed of tightly packed cells, but unlike the skin, mucous membranes are relatively thin, moist, and warm, and their cells are living. Therefore, pathogens find mucous membranes more hospitable and easier portals of entry.

The respiratory tract is the most frequently used portal of entry. Pathogens enter the mouth and nose in the air, on dust particles, and in droplets of moisture. For example, the bacteria that cause whooping cough, diphtheria, pneumonia, strep throat, and meningitis, as well as some fungi, viruses, and protozoa, enter through the respiratory tract.

Surprisingly, many viruses enter the respiratory tract via the eyes. They are introduced onto the conjunctiva by contaminated

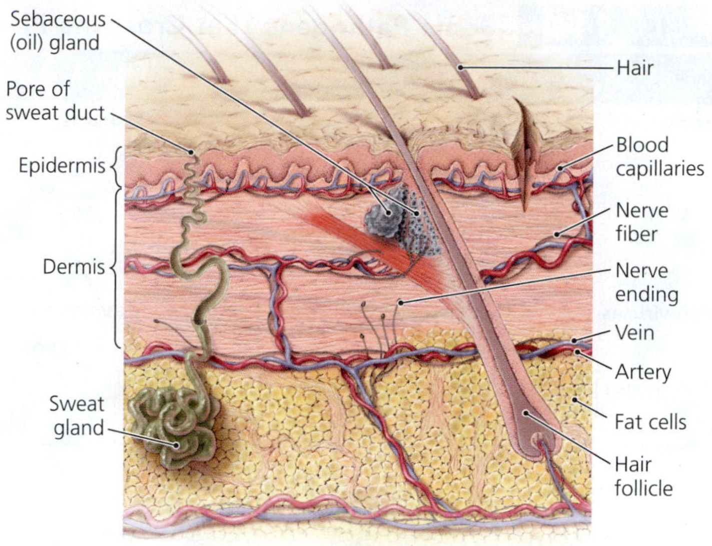

▲ **FIGURE 14.4 A cross section of skin.** The layers of cells constitute a barrier to most microbes as long as the skin remains intact. Some pathogens can enter the body through hair follicles, through the ducts of sweat glands, and parenterally.

fingers and are washed into the nasal cavity with tears. Cold and influenza viruses typically enter the body in this manner.

Some parasitic protozoa, helminths, bacteria, and viruses infect the body through the gastrointestinal mucous membranes. These parasites are able to survive the acidic pH of the stomach and the digestive juices of the intestinal tract. Noncellular pathogens called *prions* enter the body through oral mucous membranes.

Placenta

A developing embryo forms an organ, called the *placenta,* through which it obtains nutrients from the mother. The placenta is in such intimate contact with the wall of the mother's uterus that nutrients and wastes diffuse between the blood vessels of the developing child and of the mother, but because the two blood supplies do not actually contact each other, the placenta typically forms an effective barrier to most pathogens. However, in about 2% of pregnancies, pathogens cross the placenta and infect the embryo or fetus, sometimes causing spontaneous abortion, birth defects, or premature birth. Some pathogens that can cross the placenta are listed in **TABLE 14.4** on p. 422.

The Parenteral Route

The **parenteral route** is not a portal of entry but instead a means by which the portals of entry can be circumvented. To enter the body by the parenteral route, pathogens must be deposited directly into tissues beneath the skin or mucous membranes, such as occurs in punctures by a nail, thorn, or hypodermic needle. Some experts include in the parenteral route breaks in the skin by cuts, bites, stab wounds, deep abrasions, or surgery.

TABLE 14.4 Some Pathogens That Cross the Placenta

	Pathogen	Condition in the Adult	Effect on Embryo or Fetus
Protozoan	*Toxoplasma gondii*	Toxoplasmosis	Abortion, epilepsy, encephalitis, microcephaly, mental retardation, blindness, anemia, jaundice, rash, pneumonia, diarrhea, hypothermia, deafness
Bacteria	*Treponema pallidum*	Syphilis	Abortion, multiorgan birth defects, syphilis
	Listeria monocytogenes	Listeriosis	Granulomatosis infantiseptica (nodular inflammatory lesions and infant blood poisoning), death
DNA viruses	*Cytomegalovirus*	Usually asymptomatic	Deafness, microcephaly, mental retardation
	Erythrovirus	Erythema infectiosum	Abortion
RNA viruses	*Lentivirus* (HIV)	AIDS	Immunosuppression (AIDS)
	Rubivirus	German measles	Severe birth defects or death

The Role of Adhesion in Infection

LEARNING **OUTCOMES**

14.8 List the types of adhesion factors and the roles they play in infection.

14.9 Explain how a biofilm may facilitate contamination and infection.

After entering the body, symbionts must adhere to cells if they are to be successful in establishing colonies. The process by which microorganisms attach themselves to cells is called **adhesion** (ad-hē´zhŭn), or *attachment*. To accomplish adhesion, pathogens use **adhesion factors,** which are either specialized structures or attachment proteins. Examples of such specialized structures are adhesion disks in some protozoa and suckers and hooks in some helminths (see Chapter 23). In contrast, viruses and many bacteria have surface lipoprotein and glycoprotein molecules called *ligands* that enable them to bind to complementary receptors on host cells (**FIGURE 14.5**). Ligands are also called *adhesins* on bacteria and *attachment proteins* on viruses. Adhesins are found on fimbriae, flagella, and glycocalyces of many pathogenic bacteria. Receptor molecules on host cells are typically glycoproteins containing sugar molecules such as mannose and galactose. If ligands or their receptors can be changed or blocked, infection can often be prevented.

The specific interaction of adhesins and receptors with chemicals on host cells often determines the specificity of pathogens for particular hosts. For example, *Neisseria gonorrhoeae* (nī-se´rē-ă go-nor-rē´ī) has adhesins on its fimbriae that adhere to cells lining the urethra and vagina of humans. Thus, this pathogen cannot affect other hosts.

Some bacteria, including *Bordetella* (bōr-dĕ-tel´ă; the cause of whooping cough) have more than one type of adhesin. Other pathogens, such as *Plasmodium* (plaz-mō´dē-ŭm; the cause of malaria), change their adhesins over time, helping the pathogen evade the body's immune system and allowing the pathogen to attack more than one kind of cell. Bacterial cells and viruses that have lost the ability to make

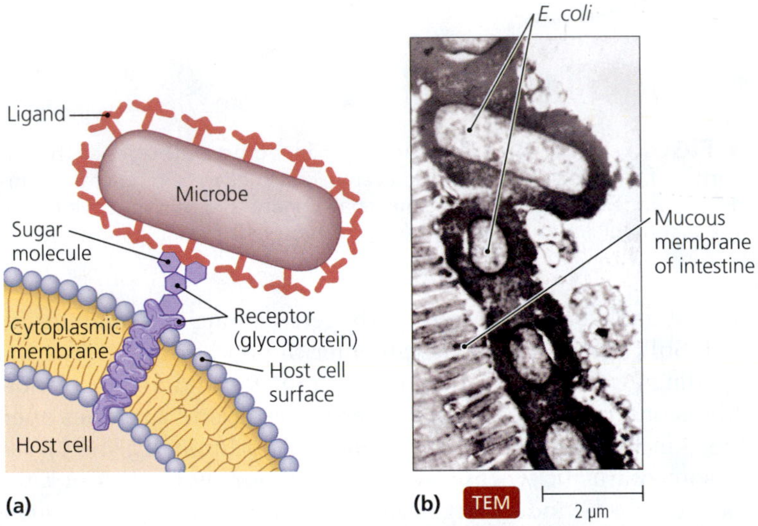

▲ **FIGURE 14.5 The adhesion of pathogens to host cells. (a)** An artist's rendition of the attachment of a microbial ligand to a complementary surface receptor on a host cell. **(b)** A photomicrograph of cells of a pathogenic strain of *E. coli* attached to the mucous membrane of the intestine. Although the thick glycocalyces (black) of the bacteria are visible, the bacteria's ligands are too small to be seen at this magnification.

ligands—whether as the result of some genetic change (mutation) or exposure to certain physical or chemical agents (as occurs in the production of some vaccines)—become harmless, or **avirulent** (ā-vir´ū-lent).

Some bacterial pathogens do not attach to host cells directly but instead interact with each other to form a sticky web of bacteria and polysaccharides called a **biofilm,** which adheres to a surface within a host. A prominent example of a biofilm is dental plaque (**FIGURE 14.6**), which contains the bacteria that cause dental caries (tooth decay).

TELL ME WHY

Why does every infection start with contamination but not every contamination results in an infection?

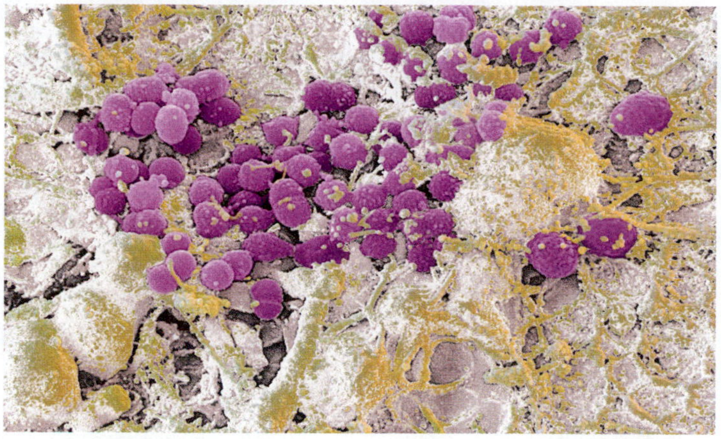

▲ **FIGURE 14.6 Dental plaque.** This biofilm consists of bacteria that adhere to one another and to teeth by means of sticky capsules; extracellular, slimy matrix, and fimbriae.

The Nature of Infectious Disease

LEARNING | OUTCOME

14.10 Compare and contrast the terms *infection, disease, morbidity, pathogenicity,* and *virulence.*

Most infections succumb to the body's defenses (discussed in detail in Chapters 15 and 16), but some infectious agents not only evade the defenses and multiply but also affect body function. By definition, all parasites injure their hosts. When the injury is significant enough to interfere with the normal functioning of the body, the result is termed **disease.** Thus, disease, also known as **morbidity,** is any change from a state of health.

Infection and disease are not the same thing. Infection is the invasion of a pathogen; disease results only if the pathogen multiplies sufficiently to adversely affect the body. Some illustrations will help to clarify this point. While caring for a patient, a nurse may become contaminated with the bacterium *Staphylococcus aureus* (o´rē-ŭs). If the pathogen is able to gain access to his body, perhaps through a break in his skin, he will become infected. Then, if the multiplication of *Staphylococcus* results in the production of a boil, the nurse will have a disease. Another example: A drug addict who shares needles may become contaminated and subsequently infected with HIV without experiencing visible change in body function for years. Such a person is infected but not yet diseased.

Manifestations of Disease: Symptoms, Signs, and Syndromes

LEARNING | OUTCOME

14.11 Contrast symptoms, signs, and syndromes.

Diseases may become manifest in different ways. **Symptoms** are *subjective* characteristics of a disease that can be felt by the patient alone, whereas **signs** are *objective* manifestations of disease that can be observed or measured by others. Symptoms include pain, headache, dizziness, and fatigue; signs include swelling, rash, redness, and fever. Sometimes pairs of symptoms and signs reflect the same underlying cause: Nausea is a symptom and vomiting a sign; chills are a symptom, whereas shivering is a sign. Note, however, that even though signs and symptoms may have the same underlying cause, they need not occur together. Thus, for example, a viral infection of the brain may result in both a headache (a symptom) and the presence of viruses in the cerebrospinal fluid (a sign), but viruses in the cerebrospinal fluid are not invariably accompanied by headache. Symptoms and signs are used in conjunction with laboratory tests to make diagnoses. Some typical symptoms and signs are listed in **TABLE 14.5**.

TABLE 14.5 Typical Manifestations of Disease

Symptoms (Sensed by the Patient)	Signs (Detected or Measured by an Observer)
Pain, nausea, headache, chills, sore throat, fatigue or lethargy (sluggishness, tiredness), malaise (discomfort), itching, abdominal cramps	Swelling, rash or redness, vomiting, diarrhea, fever, pus formation, anemia, leukocytosis/leukopenia (increase/decrease in the number of circulating white blood cells), bubo (swollen lymph node), tachycardia/bradycardia (increase/decrease in heart rate)

A **syndrome** is a group of symptoms and signs that collectively characterizes a particular disease or abnormal condition. For example, acquired immunodeficiency syndrome (AIDS) is characterized by malaise, loss of certain white blood cells, diarrhea, weight loss, pneumonia, toxoplasmosis, and tuberculosis.

Some infections go unnoticed because they have no symptoms. Such cases are **asymptomatic,** or **subclinical,** infections. Note that even though asymptomatic infections by definition lack symptoms, in some cases certain signs may still be detected if the proper tests are performed. For example, leukocytosis (an excess of white blood cells) may be detected in a blood sample from an individual who feels completely healthy.

TABLE 14.6 on p. 424 lists some prefixes and suffixes used to describe various aspects of diseases and syndromes.

Causation of Disease: Etiology

LEARNING | OUTCOMES

14.12 Define *etiology.*

14.13 List Koch's postulates, explain their function, and describe their limitations.

Even though our focus in this chapter is infectious disease, it's obvious that not all diseases result from infections. Some diseases are *hereditary,* which means they are genetically transmitted from parents to offspring. Other diseases, called *congenital diseases,* are diseases that are present at birth, regardless of the cause (whether hereditary, environmental, or infectious). Still

TABLE 14.6 Terminology of Disease

Prefix/Suffix	Meaning	Example
carcino-	Cancer	Carcinogenic: giving rise to cancer
col-, colo-	Colon	Colitis: inflammation of the colon
dermato-	Skin	Dermatitis: inflammation of the skin
-emia	Pertaining to the blood	Viremia: viruses in the blood
endo-	Inside	Endocarditis: inflammation of lining of heart
-gen, gen-	Give rise to	Pathogen: giving rise to disease
hepat-	Liver	Hepatitis: inflammation of the liver
idio-	Unknown	Idiopathic: pertaining to a disease of unknown cause
-itis	Inflammation of a structure	Meningitis: inflammation of the meninges (covering of the brain); endocarditis
-oma	Tumor or swelling	Papilloma: wart
-osis	Condition of	Toxoplasmosis: being infected with *Toxoplasma*
-patho, patho-	Abnormal	Pathology: study of disease
septi-	Literally, *rotting;* refers to presence of pathogens	Septicemia: pathogens in the blood
terato-	Defects	Teratogenic: causing birth defects
tox-	Poison	Toxin: harmful compound

TABLE 14.7 Categories of Diseases[a]

	Description	Examples
Hereditary	Caused by errors in the genetic code received from parents	Sickle-cell anemia, diabetes mellitus, Down syndrome
Congenital	Anatomical and physiological (structural and functional) defects present at birth; caused by drugs (legal and illegal), X-ray exposure, or infections	Fetal alcohol syndrome, deafness from rubella infection
Degenerative	Result from aging	Renal failure, age-related farsightedness
Nutritional	Result from lack of some essential nutrients in diet	Kwashiorkor, rickets
Endocrine (hormonal)	Due to excesses or deficiencies of hormones	Dwarfism
Mental	Emotional or psychosomatic	Skin rash, gastrointestinal distress
Immunological	Hyperactive or hypoactive immunity	Allergies, autoimmune diseases, agammaglobulinemia
Neoplastic (tumor)	Abnormal cell growth	Benign tumors, cancers
Infectious	Caused by an infectious agent	Colds, influenza, herpes infections
Iatrogenic[b]	Caused by medical treatment or procedures; are a subgroup of hospital-acquired diseases	Surgical error, yeast vaginitis resulting from antimicrobial therapy
Idiopathic[c]	Unknown cause	Alzheimer's disease, multiple sclerosis
Nosocomial[d]	Disease acquired in health care setting	*Pseudomonas* infection in burn patient

[a]Some diseases may fall in more than one category.
[b]From Greek *iatros*, meaning "physician."
[c]From Greek *idiotes*, meaning "ignorant person," and *pathos*, meaning "disease."
[d]From Greek *nosokomeion*, meaning "hospital."

other diseases are classified as *degenerative, nutritional, endocrine, mental, immunological,* or *neoplastic.*[8] The various categories of diseases are described in **TABLE 14.7**. Note that some diseases can fall into more than one category. For example, liver cancer, which is usually associated with infection by hepatitis B and D viruses, can be classified as both neoplastic and infectious.

The study of the cause of a disease is called **etiology** (ē-tē-ol´ō-jē). Because our focus here and in subsequent chapters is on infectious diseases, we next examine how microbiologists investigate the causation of infectious diseases, beginning with an examination of the work of Robert Koch.

[8]Neoplasms are tumors, which may either remain in one place (benign tumors) or spread (cancers).

Using Koch's Postulates

In our modern world, we take for granted the idea that specific microbes cause specific diseases, but this has not always been the case. In the past, disease was thought to result from a variety of causes, including bad air, imbalances in body fluids, or astrological forces. In the 19th century, Louis Pasteur, Robert Koch, and other microbiologists proposed the **germ theory of disease,** which states that disease is caused by infections of pathogenic microorganisms (at the time called *germs*).

But which pathogen causes a specific disease? How can we distinguish the pathogen, which causes a disease, from all the other biological agents (fungi, bacteria, protozoa, and viruses) that are part of the normal microbiota and are in effect "innocent bystanders"?

Koch developed a series of essential conditions, or *postulates,* that scientists must demonstrate or satisfy to prove that a particular microbe is pathogenic and causes a particular disease. Using his postulates, Koch proved that *Bacillus anthracis* (ba-sil´ŭs an-thrā´sis) causes anthrax and that *Mycobacterium tuberculosis* (mī-kō-bak-tēr´ē-ŭm too-ber-kyū-lō´sis) causes tuberculosis.

To prove that a given infectious agent causes a given disease, a scientist must satisfy all of **Koch's postulates** (FIGURE 14.7):

1. The suspected agent (bacterium, virus, etc.) must be present in every case of the disease.
2. That agent must be isolated and grown in pure culture.
3. The cultured agent must cause the disease when it is inoculated into a healthy, susceptible experimental host.
4. The same agent must be reisolated from the diseased experimental host.

It is critical that all the postulates be satisfied in order. The mere presence of an agent does not prove that it causes a disease. Although Koch's postulates have been used to prove the cause of many infectious diseases in humans, animals, and plants, inadequate attention to the postulates has resulted in incorrect conclusions concerning some disease causation. For instance, in the early 1900s *Haemophilus influenzae* (hē-mof´i-lŭs in-flū-en´zī) was found in the respiratory systems of flu victims and identified as the causative agent of influenza based on its presence. Later, flu victims who lacked *H. influenzae* in their lungs were discovered. This discovery violated the first postulate, so *H. influenzae* cannot be the cause of flu. Today we know that an RNA virus causes flu and that *H. influenzae* was part of the normal microbiota of those early flu patients. (Later studies, correctly using Koch's postulates, showed that *H. influenzae* can cause an often-fatal meningitis in children.)

Exceptions to Koch's Postulates

Clearly, Koch's postulates are a cornerstone of the etiology of infectious diseases, but using them is not always feasible for the following reasons:

- Some pathogens cannot be cultured in the laboratory. For example, pathogenic strains of *Mycobacterium leprae* (lep´rī), which causes leprosy, have never been grown on laboratory media.
- Some diseases are caused by a combination of pathogens or by a combination of a pathogen and physical, environmental, or genetic cofactors. In such cases, the pathogen alone is avirulent, but when accompanied by another pathogen or the appropriate cofactor, disease results. For example, liver cancer can result when liver cells are infected by both the hepatitis B and hepatitis D viruses but seldom when only one of the viruses infects the cells.
- Ethical considerations prevent applying Koch's postulates to diseases and pathogens that occur in humans only. In such cases the third postulate, which involves inoculation of a healthy susceptible host, cannot be satisfied within ethical boundaries. For this reason, scientists have never attempted to apply Koch's postulates to prove that HIV causes AIDS; however, observations of fetuses that were naturally exposed to HIV by their infected mothers, as well as accidentally infected health care and laboratory workers, have in effect satisfied the third postulate.

Additionally, some circumstances may make satisfying Koch's postulates difficult:

- It is not possible to establish a single cause for such infectious diseases as pneumonia, meningitis, and hepatitis because the names of these diseases refer to conditions that can be caused by more than one pathogen. For these diseases, laboratory technicians must identify the etiologic agent involved in any given case.
- Some pathogens have been ignored. For example, gastric ulcers were long thought to be caused by excessive production of stomach acid in response to stress, but the majority of such ulcers are now known to be caused by a long-overlooked bacterium, *Helicobacter pylori* (hel´ĭ-kō-bak´ter pī´lō-rē).

If Koch's postulates cannot be applied to a disease condition for whatever reason, how can we positively know the causative agent of a disease? *Epidemiological* studies, discussed later in this chapter, can give statistical support to causation theories but not absolute proof. For example, some researchers have proposed that the bacterium *Chlamydophila pneumoniae* (kla-mē-dof´ĭ-lă nū-mō´nē-ī) causes many cases of arteriosclerosis on the basis of the bacterium's presence in patients with the disease. Debate among scientists about such cases fosters a continued drive for knowledge and discovery. For example, we are still searching for the causes of chronic fatigue syndrome, multiple sclerosis, and Alzheimer's disease.

Virulence Factors of Infectious Agents

LEARNING **OUTCOME**

14.14 Explain how microbial extracellular enzymes, toxins, adhesion factors, and antiphagocytic factors affect virulence.

Microbiologists characterize the disease-related capabilities of microbes by using two related terms. The ability of a

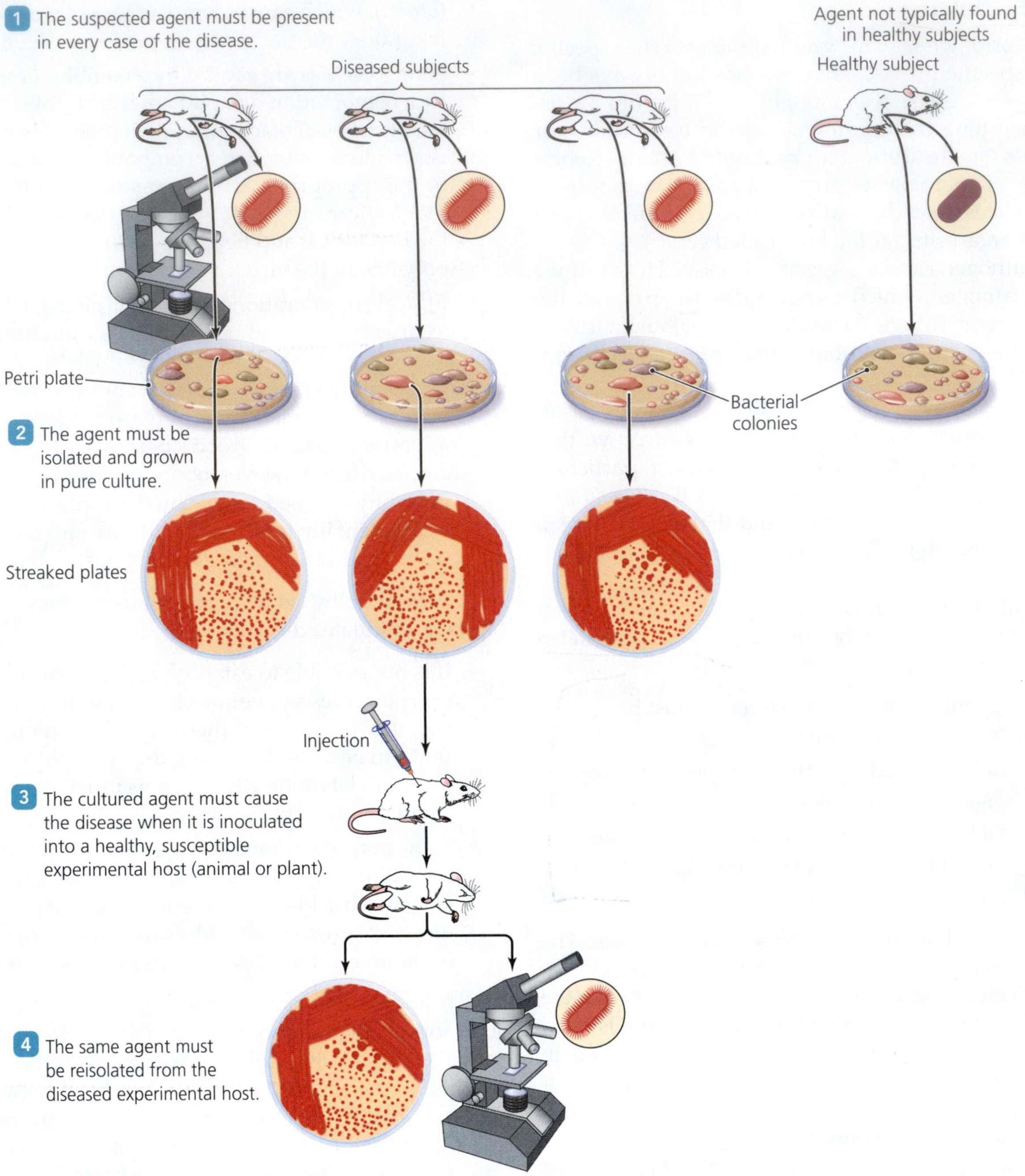

1 The suspected agent must be present in every case of the disease.
Diseased subjects
Agent not typically found in healthy subjects
Healthy subject
Petri plate
Bacterial colonies
2 The agent must be isolated and grown in pure culture.
Streaked plates
Injection
3 The cultured agent must cause the disease when it is inoculated into a healthy, susceptible experimental host (animal or plant).
4 The same agent must be reisolated from the diseased experimental host.

▲ **FIGURE 14.7** **Koch's postulates.**

microorganism to cause disease is termed **pathogenicity** (path´ō-jĕ-nis´i-tē), and the *degree* of pathogenicity is termed **virulence.** In other words, virulence is the relative ability of a pathogen to infect a host and cause disease. Neither term addresses the severity of a disease; the pathogen causing rabbit fever is highly virulent, but the disease is relatively mild. Organisms can be placed along a virulence continuum **(FIGURE 14.8)**; highly virulent organisms almost always cause disease, whereas less virulent organisms (including opportunistic pathogens) cause disease only in weakened hosts or when present in overwhelming numbers.

Pathogens have a variety of traits that interact with a host and enable the pathogen to enter a host, adhere to host cells, gain access to nutrients, and escape detection or removal by the immune system. These traits are collectively called **virulence factors.** Virulent pathogens have one or more virulence factors that nonvirulent microbes lack. We discussed two virulence factors—adhesion factors and biofilm formation—previously; now we examine three other virulence factors: extracellular enzymes, toxins, and antiphagocytic factors. Other virulence factors can be examined online at the MasteringMicrobiology Study Area. ▶**ANIMATIONS:** ***Virulence Factors: Inactivating Host Defenses***

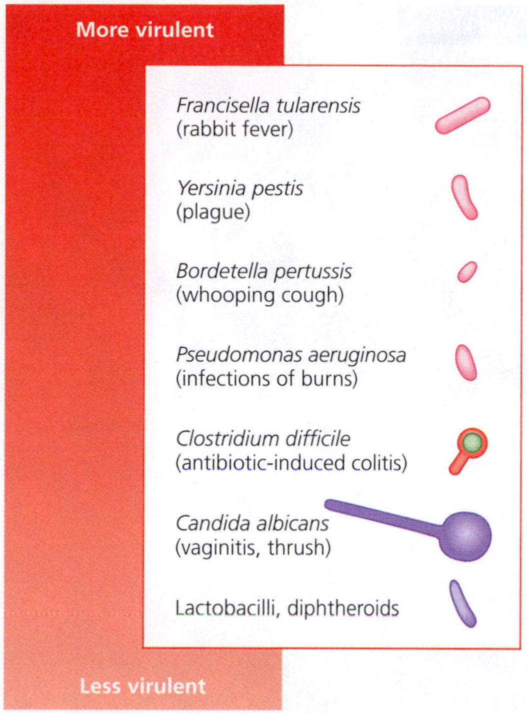

▲ **FIGURE 14.8 Relative virulence of some microbial pathogens.** Virulence involves ease of infection and the ability of a pathogen to cause disease; the term does not indicate the seriousness of a disease.

Extracellular Enzymes

Many pathogens secrete enzymes that enable them to dissolve structural chemicals in the body and thereby maintain an infection, invade further, and avoid body defenses. **FIGURE 14.9a** illustrates the action of some extracellular enzymes of bacteria:

- Hyaluronidase and collagenase degrade specific molecules to enable bacteria to invade deeper tissues. Hyaluronidase digests hyaluronic acid, the "glue" that holds animal cells together, and collagenase breaks down collagen, the body's chief structural protein.
- Coagulase causes blood proteins to clot, providing a "hiding place" for bacteria within a clot.
- Kinases, such as staphylokinase and streptokinase, digest blood clots, allowing subsequent invasion of damaged tissues.

Many bacteria with these enzymes are virulent; mutant strains of the same species that have defective genes for these extracellular enzymes are usually avirulent.

Pathogenic eukaryotes also secrete enzymes that contribute to virulence. For example, fungi that cause "ringworm" produce *keratinase,* which enzymatically digests keratin—the main component of skin, hair, and nails. *Entamoeba histolytica* (ent-ă-mē´bă his-tō-li´ti-kă) secretes *mucinase* to digest the mucus lining the intestinal tract, allowing the amoeba entry to the underlying cells, where it causes amebic dysentery. ▶**ANIMATIONS:** ***Virulence Factors: Penetrating Host Tissues***

Toxins

Toxins are chemicals that either harm tissues or trigger host immune responses that cause damage. The distinction between extracellular enzymes and toxins is not always clear because many enzymes are toxic and many toxins have enzymatic action. In a condition called **toxemia** (tok-sē´mē-ā), toxins enter the bloodstream and are carried to other parts of the body, including sites that may be far removed from the site of infection. There are two types of toxin: exotoxins and endotoxins.

Exotoxins Many microorganisms secrete **exotoxins** that are central to their pathogenicity in that they destroy host cells or interfere with host metabolism. Exotoxins are of three principal types:

- *Cytotoxins,* which kill host cells in general or affect their function **(FIGURE 14.9b)**
- *Neurotoxins,* which specifically interfere with nerve cell function
- *Enterotoxins,* which affect cells lining the gastrointestinal tract

Examples of pathogenic bacteria that secrete exotoxins are the clostridia that cause gangrene, botulism, and tetanus; pathogenic strains of *S. aureus* that cause food poisoning and other ailments; and diarrhea-causing *E. coli, Salmonella enterica* (sal´mŏ-nel´ă en-ter´i-kă), and *Shigella* (shē-gel´lă) species. Some fungi and marine dinoflagellates (protozoa) also secrete exotoxins. Specific exotoxins are discussed in the chapters that examine specific diseases. ▶**ANIMATIONS:** ***Virulence Factors: Exotoxins***

The body protects itself with **antitoxins,** which are protective molecules called *antibodies* that bind to specific toxins and neutralize them. Health care workers stimulate the production of antitoxins by administering immunizations composed of *toxoids,* which are toxins that have been treated with heat, formaldehyde, chlorine, or other chemicals to make them nontoxic but still capable of stimulating the production of antibodies. (Chapters 16 and 17 further discuss antibodies, toxoids, and immunizations.)

Endotoxin Gram-negative bacteria have an outer (wall) membrane composed of lipopolysaccharide, phospholipids, and proteins (see Figure 3.15). **Endotoxin,** also called **lipid A,** is the lipid portion of the membrane's lipopolysaccharide.

Endotoxin can be released when Gram-negative bacteria divide, die naturally, or are digested by phagocytic cells such as macrophages (see Figure 14.9b). Many types of lipid A stimulate the body to release chemicals that cause fever, inflammation, diarrhea, hemorrhaging, shock, and blood coagulation. Although infections by bacteria that produce exotoxins are generally more serious than infections with other bacteria, most Gram-negative pathogens can be potentially life threatening because the release of endotoxin from dead bacteria can produce serious, systemic effects in the host. ▶**ANIMATIONS:** ***Virulence Factors: Endotoxins***

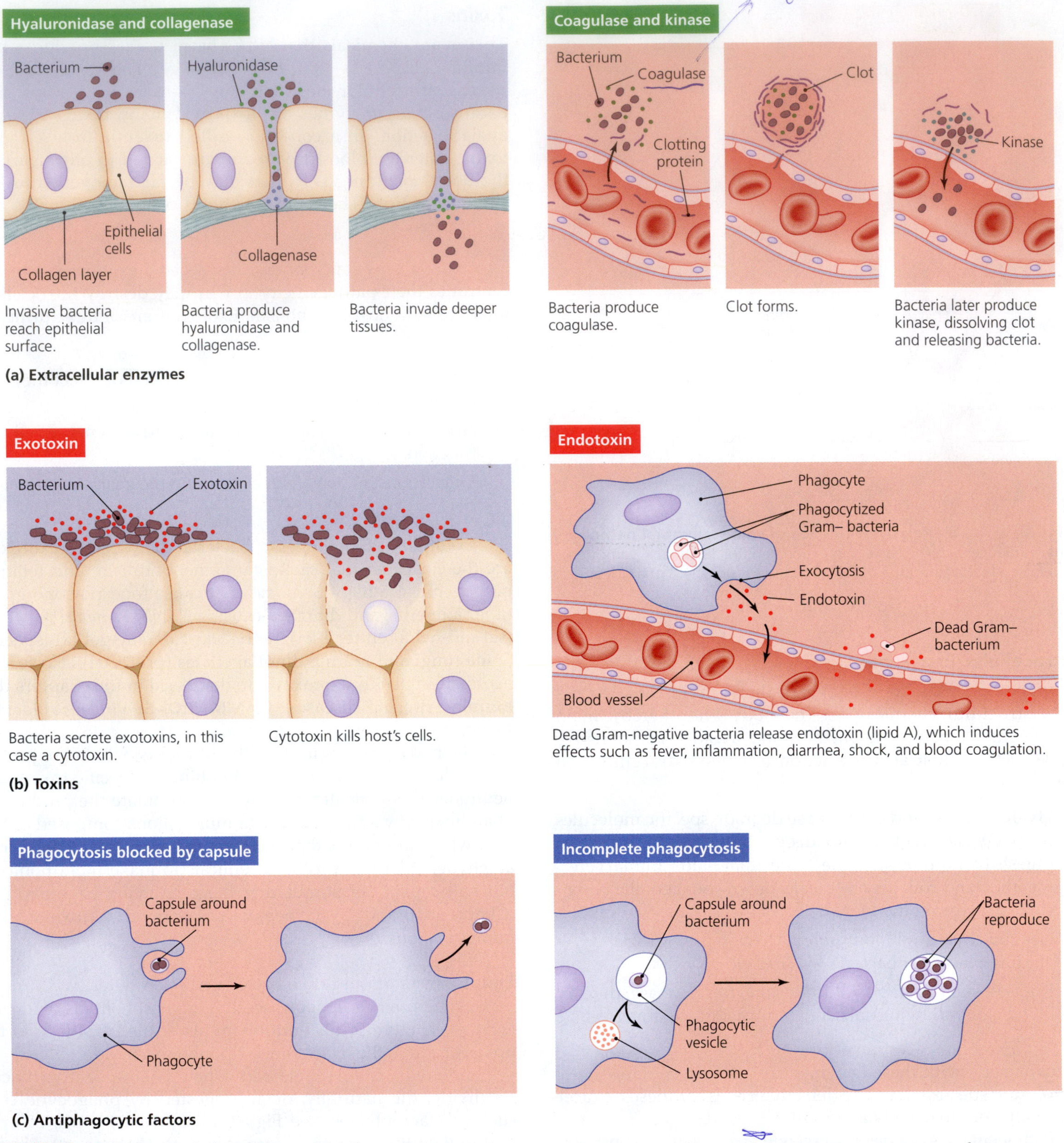

▲ **FIGURE 14.9 Some virulence factors. (a)** Extracellular enzymes. Hyaluronidase and collagenase digest structural materials in the body. Coagulase in effect "camouflages" bacteria inside a blood clot, whereas kinases digest clots to release bacteria. **(b)** Toxins. Exotoxins (including cytotoxin, shown here) are released from living pathogens and harm neighboring cells. Endotoxin is released from many dead Gram-negative bacteria and can trigger widespread disruption of normal body functions. **(c)** Antiphagocytic factors. Capsules, one antiphagocytic factor, can prevent phagocytosis or stop digestion by a phagocyte. *How can bacteria prevent a phagocytic cell from digesting them once they have been engulfed by pseudopods?*

Figure 14.9 *Some bacteria secrete a chemical that prevents the fusion of lysosomes with the phagosome containing the bacteria.*

TABLE 14.8 A Comparison of Bacterial Exotoxins and Endotoxin

	Exotoxins	Endotoxin
Source	Mainly Gram-positive and Gram-negative bacteria	Gram-negative bacteria
Relation to bacteria	Metabolic product secreted from living cell	Portion of outer (cell wall) membrane released upon cell death
Chemical nature	Protein or short peptide	Lipid portion of lipopolysaccharide (lipid A) of outer (cell wall) membrane
Toxicity	High	Low but may be fatal in high doses
Heat stability	Typically unstable at temperatures above 60°C	Stable for up to 1 hour at autoclave temperature (121°C)
Effect on host	Variable, depending on source; may be cytotoxin, neurotoxin, enterotoxin	Fever, lethargy, malaise, shock, blood coagulation
Fever producing?	No	Yes
Antigenicity[a]	Strong: stimulates antitoxin (antibody) production	Weak
Toxoid formation for immunization?	By treatment with heat or fomaldehyde	Not feasible
Representative diseases	Botulism, tetanus, gas gangrene, diphtheria, cholera, plague, staphylococcal food poisoning	Typhoid fever, tularemia, endotoxic shock, urinary tract infections, meningococcal meningitis

[a]Refers to the ability of a chemical to trigger a specific immune response, particularly the formation of antibodies.

TABLE 14.8 summarizes differences between exotoxins and endotoxins.

Antiphagocytic Factors

Typically, the longer a pathogen remains in a host, the greater the damage and the more severe the disease. To limit the extent and duration of infections, the body's phagocytic cells, such as the white blood cells called macrophages, engulf and remove invading pathogens. Here we consider some virulence factors related to the evasion of phagocytosis, beginning with bacterial capsules. ▶**ANIMATIONS:** ***Phagocytosis: Microbes That Evade It***

Capsules The capsules of many pathogenic bacteria (see Figure 3.5a) are effective virulence factors because many capsules are composed of chemicals normally found in the body (including polysaccharides); as a result, they do not stimulate a host's immune response. For example, hyaluronic acid capsules in effect deceive phagocytic cells into treating them and the enclosed bacteria as if they were a normal part of the body. Additionally, capsules are often slippery, making it difficult for phagocytes to surround and phagocytize them—their pseudopods cannot grip the capsule, much as wet hands have difficulty holding a wet bar of soap **(FIGURE 14.9c)**.

Antiphagocytic Chemicals Some bacteria, including the cause of gonorrhea, produce chemicals that prevent the fusion of lysosomes with phagocytic vesicles, allowing the bacteria to survive inside phagocytes (see Figure 14.9c). *Streptococcus pyogenes* (strep-tō-kok´ŭs pī-oj´en-ēz) produces a protein on its cell wall and fimbriae, called *M protein,* that inhibits phagocytosis and thus increases virulence. Other bacteria produce *leukocidins,* which are chemicals capable of destroying phagocytic white blood cells outright. ▶**ANIMATIONS:** ***Virulence Factors: Hiding from Host Defenses*** ▶**VIDEO TUTOR:** ***Some Virulence Factors***

The Stages of Infectious Diseases

LEARNING | **OUTCOME**

14.15 List and describe the five typical stages of infectious diseases.

Following exposure and infection, a sequence of events called the **disease process** can occur. Many infectious diseases have five stages following infection: an incubation period, a prodromal period, illness, decline, and convalescence **(FIGURE 14.10)**.

Incubation Period

The **incubation period** is the time between infection and occurrence of the first symptoms or signs of disease. The length of the incubation period depends on the virulence of the infective agent, the infective dose (initial number of pathogens), the state and health of the patient's immune system, the nature of the pathogen and its reproduction time, and the site of infection. Some diseases have typical incubation periods, whereas for others the incubation period varies considerably. **TABLE 14.9** on p. 430 lists incubation periods for selected diseases.

Prodromal Period

The **prodromal**[9] **period** (prō-drō´măl) is a short time of generalized, mild symptoms (such as malaise and muscle aches) that precedes illness. Not all infectious diseases have a prodromal stage.

Illness

Illness is the most severe stage of an infectious disease. Signs and symptoms are most evident during this time. Typically the patient's immune system has not yet fully responded to the

[9]From Greek *prodromos,* meaning "forerunner."

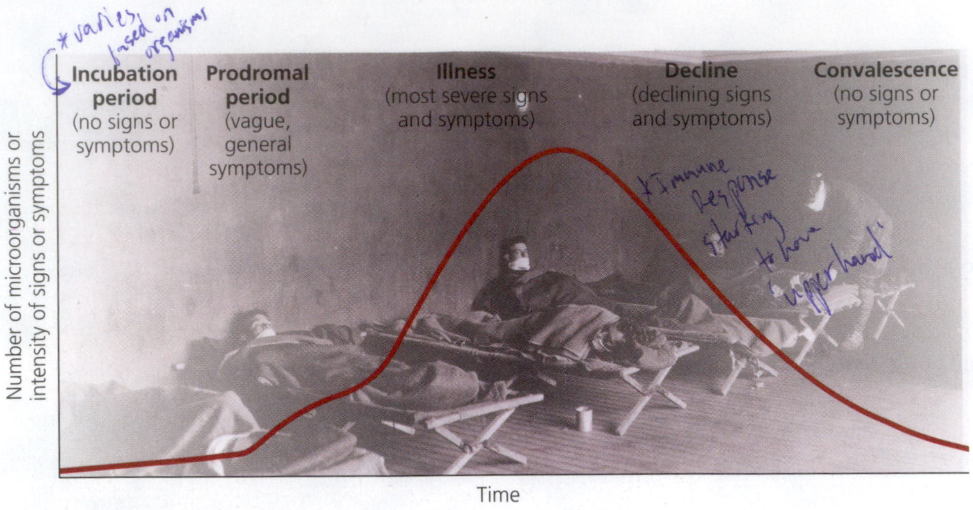

◀ **FIGURE 14.10 The stages of infectious diseases.** Not every stage occurs in every disease.

TABLE 14.9 Incubation Periods of Selected Infectious Diseases

Disease	Incubation Period
Staphylococcus foodborne infection	<1 day
Influenza	About 1 day
Cholera	2 to 3 days
Genital herpes	About 5 days
Tetanus	5 to 15 days
Syphilis	10 to 21 days
Hepatitis B	70 to 100 days
AIDS	1 to >8 years
Leprosy	10 to >30 years

pathogens, and their presence is harming the body. This stage is usually when a physician first sees the patient.

Decline

During the period of **decline,** the body gradually returns to normal as the patient's immune response and/or medical treatment vanquish the pathogens. Fever and other signs and symptoms subside. Normally the immune response and its products (such as antibodies in the blood) peak during this stage. If the patient doesn't recover, then the disease is fatal.

Convalescence

During **convalescence** (kon-vă-les´ens), the patient recovers from the illness; tissues are repaired and returned to normal. The length of a convalescent period depends on the amount of damage, the nature of the pathogen, the site of infection, and the overall health of the patient. Thus, whereas recovery from staphylococcal food poisoning may take less than a day, recovery from Lyme disease may take years.

A patient is likely to be infectious during every stage of disease. Even though most of us realize we are infective during the symptomatic periods, many people are unaware that infections can be spread during incubation and convalescence as well. For example, a patient who no longer has any obvious herpes sores is always capable of transmitting herpesviruses. Good aseptic technique can limit the spread of many pathogens from recovering patients.

TELL ME WHY

Why is mutated *Streptococcus pneumoniae*, which cannot make a capsule, unable to cause pneumonia?

The Movement of Pathogens Out of Hosts: Portals of Exit

Just as infections occur through portals of entry, so pathogens must leave infected patients through **portals of exit** in order to infect others **(FIGURE 14.11)**. Many portals of exit are essentially

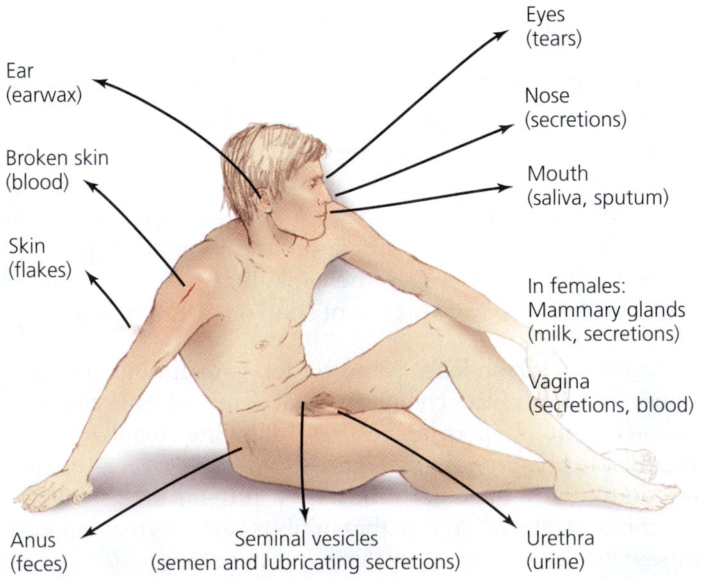

▲ **FIGURE 14.11 Portals of exit.** Many portals of exit are also portals of entry. Pathogens often leave the body via bodily secretions and excretions produced at those sites.

identical to portals of entry. However, pathogens often exit hosts in materials that the body secretes or excretes. Thus, pathogens may leave hosts in secretions (earwax, tears, nasal secretions, saliva, sputum, and respiratory droplets), in blood (via arthropod bites, hypodermic needles, or wounds), in vaginal secretions or semen, in milk produced by the mammary glands, and in excreted bodily wastes (feces and urine). As we will see, health care personnel must consider the portals of entry and exit in their efforts to understand and control the spread of diseases within populations.

TELL ME WHY

Why is the tube emptying the bladder (the urethra) more likely to be a portal of exit than a portal of entry?

Modes of Infectious Disease Transmission

LEARNING | **OUTCOMES**

14.16 Contrast contact, vehicle, and vector transmission of pathogens.
14.17 Contrast *droplet transmission* and *airborne transmission.*
14.18 Contrast *mechanical* and *biological* vectors.

By definition, an infectious disease agent must be transmitted from either a reservoir or a portal of exit to another host's portal of entry. Transmission can occur by numerous modes that are somewhat arbitrarily categorized into three groups: contact transmission, vehicle transmission, and vector transmission. ▶**ANIMATIONS:** *Epidemiology: Transmission of Disease*

Contact Transmission

Contact transmission is the spread of pathogens from one host to another by direct contact, indirect contact, or respiratory droplets.

Direct contact transmission, including *person-to-person spread,* typically involves body contact between hosts. Touching, kissing, and sexual intercourse are involved in the transmission of such diseases as warts, herpes, and gonorrhea. Touching, biting, or scratching can transmit zoonoses such as rabies, ringworm, and tularemia from an animal reservoir to a human. The transfer of pathogens from an infected mother to a developing baby across the placenta is another form of direct contact transmission. Direct transmission within a single individual can also occur if the person transfers pathogens from a portal of exit directly to a portal of entry—as occurs, for example, when people with poor personal hygiene unthinkingly place fingers contaminated with fecal pathogens into their mouths.

Indirect contact transmission occurs when pathogens are spread from one host to another by **fomites** (fōm´i-tēz; singular: *fomes,* fōm´mēz), which are inanimate objects that are inadvertently used to transfer pathogens to new hosts. Fomites include needles, toothbrushes, paper tissues, toys, money, diapers,

▲ **FIGURE 14.12 Droplet transmission.** In this case, droplets are propelled, primarily from the mouth, during a sneeze. By convention, such transmission is considered contact transmission only if droplets transmit pathogens to a new host within 1 meter of their source. *By what portal of entry does airborne transmission most likely occur?*

Figure 14.12 *The most likely portal of entry of airborne pathogens is the respiratory mucous membrane.*

drinking glasses, bedsheets, medical equipment, and other objects that can harbor or transmit pathogens. Contaminated needles are a major source of infection of the hepatitis B and AIDS viruses.

Droplet transmission is a third type of contact transmission. Pathogens can be transmitted within *droplet nuclei* (droplets of mucus) that exit the body during exhaling, coughing, and sneezing **(FIGURE 14.12)**. Pathogens such as cold and flu viruses may be spread in this manner. If pathogens travel more than one meter in respiratory droplets, the mode is considered to be *airborne transmission* (discussed shortly) rather than contact transmission.

Vehicle Transmission

Vehicle transmission is the spread of pathogens via air, drinking water, and food, as well as bodily fluids being handled outside the body.

Airborne transmission involves the spread of pathogens farther than one meter to the respiratory mucous membranes of a new host via an **aerosol** (ār´ō-sol)—a cloud of small droplets and solid particles suspended in the air. Aerosols may contain pathogens either on dust or inside droplets. (Recall that transmission via droplet nuclei that travel less than one meter is considered to be a form of direct contact transmission.) Aerosols can come from sneezing and coughing, or they can be generated by such means as air-conditioning systems, sweeping, mopping, changing clothes or bed linens, or even from flaming inoculating loops in microbiology labs. Dust particles can carry *Staphylococcus, Streptococcus,* and *Hantavirus* (han´tă-vī-rŭs), whereas measles virus and tuberculosis bacilli can be transmitted in dried, airborne droplets. Fungal spores of *Histoplasma* (his-tō-plaz´mă) and *Coccidioides* (kok-sid-ē-oy´dēz) are typically inhaled.

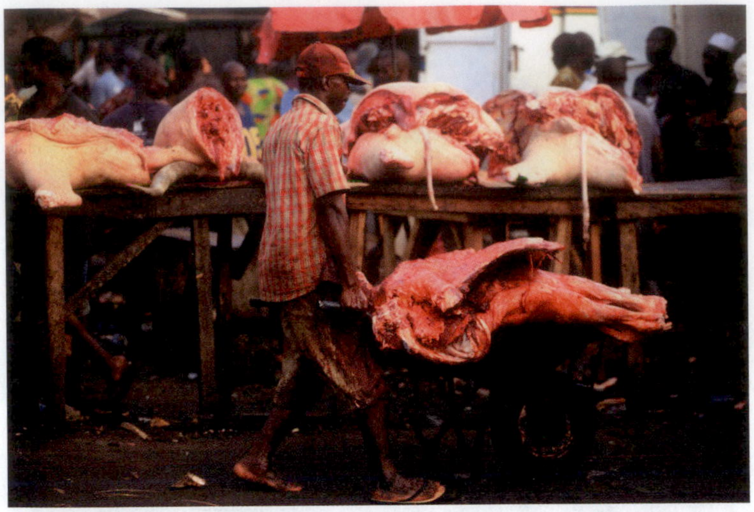

▲ **FIGURE 14.13 Poorly refrigerated foods can harbor pathogens and transmit diseases.**

Waterborne transmission is important in the spread of many gastrointestinal diseases, including giardiasis, amebic dysentery, and cholera. Note that water can act as a reservoir as well as a vehicle of infection. **Fecal-oral infection** is a major source of disease in the world. Some waterborne pathogens, such as *Schistosoma* (skis-tō-sō´mă) worms (see Chapter 21) and enteroviruses (see Chapter 20), are shed in the feces, enter through the gastrointestinal mucous membrane or skin, and subsequently can cause disease elsewhere in the body.

Foodborne transmission involves pathogens in and on foods that are inadequately processed, undercooked, or poorly refrigerated **(FIGURE 14.13)**. Foods may be contaminated with normal microbiota (e.g., *E. coli* and *S. aureus*), with zoonotic pathogens such as *Mycobacterium bovis* (bō´vis) and *Toxoplasma* (tok-sō-plaz´mă), and with parasitic worms that alternate between human and animal hosts. Contamination of food with feces and pathogens such as hepatitis A virus is another kind of fecal-oral transmission. Because milk is particularly rich in nutrients that microorganisms use (protein, lipids, vitamins, and sugars), it would be associated with the transmission of many diseases from infected animals and milk handlers if it were not properly pasteurized.

Because blood, urine, saliva, and other bodily fluids can contain pathogens, everyone—but especially health care workers—must take precautions when handling these fluids in order to prevent **bodily fluid transmission.** Special care must be taken to prevent such fluids—all of which should be considered potentially contaminated with pathogens—from contacting the conjunctiva or any breaks in the skin or mucous membranes. Examples of diseases that can be transmitted via bodily fluids are AIDS, hepatitis, and herpes, which as we have seen can also be transmitted via direct contact.

Vector Transmission

Vectors are animals that transmit diseases from one host to another. Vectors can be either biological or mechanical.

CLINICAL CASE STUDY

TB in the Nursery

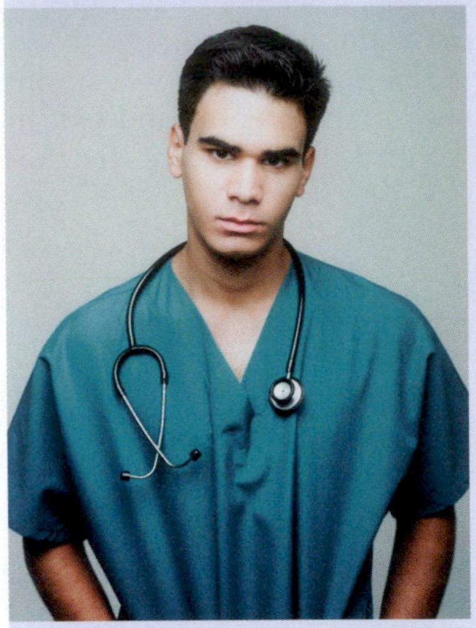

In the early fall, a neonatal nurse in a large metropolitan hospital became ill with a cough and fever. His physician believed he had seasonal allergies and so treated him with cough suppressant, antihistamines, and aerosol steroids. He returned to work in the hospital's nursery.

Three weeks later his condition had worsened; his symptoms were complicated by shortness of breath and bloody sputum. Upon further questioning, his physician noted that he was working in the United States on a work visa and was a native of South Africa. He had a positive skin test for tuberculosis (TB) but had always believed this was his body's natural reaction to the TB vaccine he had received as a child. His chest X-ray films in the past had always been clear of infection.

This time, however, his sputum smear tested positive for acid-fast bacilli. He was diagnosed with active tuberculosis and began a standard drug regimen for TB. He was restricted from work and placed in respiratory isolation for six weeks, but during the three weeks that he had continued to work in the nursery, he exposed over 900 obstetric patients, including 620 newborns, to TB, an airborne infectious disease.

1. How can private physicians quickly assess their clients for the possibility of an infectious disease?
2. What policies should be in place at hospitals to protect patients from exposure to infectious staff members?
3. How could doctors' offices improve public knowledge and protect the public from tuberculosis and other infectious diseases?

Reference: Adapted from *MMWR* 54:1280–1283, 2005.

Biological vectors not only transmit pathogens but also serve as hosts for the multiplication of a pathogen during some stage of the pathogen's life cycle. The biological vectors of diseases affecting humans are typically biting arthropods, including mosquitoes, ticks, lice, fleas, bloodsucking flies, bloodsucking bugs, and mites (see Figure 12.33). After pathogens replicate within a biological vector, often in its gut or salivary gland, the pathogens enter a new host through a bite. The bite site becomes contaminated with the vector's feces, or the vector's bite directly introduces pathogens into the new host.

TABLE 14.10 Selected Arthropod Vectors

	Disease	Causative Agent (bacteria unless otherwise indicated)
Biological Vectors		
Mosquitoes		
Anopheles,	Malaria	*Plasmodium* spp. (protozoan)
Aedes	Yellow fever	*Flavivirus* sp. (virus)
	Elephantiasis	*Wuchereria bancrofti* (helminth)
	Dengue	*Flavivirus* spp. (virus)
	Viral encephalitis	*Alphavirus* spp. (virus)
Ticks		
Ixodes	Lyme disease	*Borrelia burgdorferi*
Dermacentor	Rocky Mountain spotted fever	*Rickettsia rickettsii*
Fleas		
Xenopsylla	Bubonic plague	*Yersinia pestis*
	Endemic typhus	*Rickettsia prowazekii*
Louses		
Pediculus	Epidemic typhus	*Rickettsia typhi*
Bloodsucking flies		
Glossina	African sleeping sickness	*Trypanosoma brucei*
Simulium	River blindness	*Onchocerca volvulus* (helminth)
Bloodsucking bugs		
Triatoma	Chagas' disease	*Trypanosoma cruzi* (protozoan)
Mites (chiggers)		
Leptotrombidium	Scrub typhus	*Orientia tsutsugamushi*
Mechanical Vectors		
Houseflies		
Musca	Foodborne infections	*Shigella* spp., *Salmonella* spp., *Escherichia coli*
Cockroaches		
Blatella, Periplaneta	Foodborne infections	*Shigella* spp., *Salmonella* spp., *Escherichia coli*

Mechanical vectors are not required as hosts by the pathogens they transmit; such vectors only passively carry pathogens to new hosts on their feet or other body parts. Mechanical vectors, such as houseflies and cockroaches, may introduce pathogens such as *Salmonella* and *Shigella* into drinking water and food or onto the skin.

TABLE 14.10 lists some arthropod vectors and the diseases they transmit. **TABLE 14.11** summarizes the modes of disease transmission.

TELL ME WHY

Why can't we correctly say that all arthropod vectors are reservoirs?

TABLE 14.11 Modes of Disease Transmission

Mode of Transmission	Examples of Diseases Spread
Contact Transmission	
Direct contact: e.g., handshaking, kissing, sexual intercourse, bites	Cutaneous anthrax, genital warts, gonorrhea, herpes, rabies, staphylococcal infections, syphilis
Indirect contact: e.g., drinking glasses, toothbrushes, toys, punctures	Common cold, enterovirus infections, influenza, measles, Q fever, pneumonia, tetanus
Droplet transmission: e.g., droplets from sneezing (within one meter)	Whooping cough, streptococcal pharyngitis (strep throat)
Vehicle Transmission	
Airborne: e.g., dust particles or droplets carried more than one meter	Chicken pox, coccidioidomycosis, histoplasmosis, influenza, measles, pulmonary anthrax, tuberculosis
Waterborne: e.g., streams, swimming pools	*Campylobacter* infections, cholera, *Giardia* diarrhea
Foodborne: e.g., poultry, seafood, meat	Food poisoning (botulism, staphylococcal); hepatitis A, listeriosis, tapeworms, toxoplasmosis, typhoid fever
Vector Transmission	
Mechanical: e.g., on bodies of flies, roaches	*E. coli* diarrhea, salmonellosis, trachoma
Biological: e.g., lice, mites, mosquitoes, ticks	Chagas' disease, Lyme disease, malaria, plague, Rocky Mountain spotted fever, typhus fever, yellow fever

Classification of Infectious Diseases

LEARNING | OUTCOMES

14.19 Describe the basis for each of the various classification schemes of infectious diseases.

14.20 Distinguish among acute, subacute, chronic, and latent diseases.

14.21 Distinguish among communicable, contagious, and noncommunicable infectious diseases.

Infectious diseases can be classified in a number of ways. No one way is "*the* correct way"; each has its own advantages. One scheme groups diseases based upon taxonomic groups. A difficulty with this method of classification is that pathogens that share the same taxonomy might still affect different parts of the body. For example, *Staphylococcus* can cause skin diseases, diarrhea, and meningitis.

Another classification system deals with diseases according to the body systems affected. Chapters 18–24 examine diseases based on this approach.

Every disease (not just infectious diseases) can also be classified according to its longevity and severity. If a disease develops rapidly but lasts a relatively short time, it is called an **acute disease.** An example is the common cold. In contrast, **chronic diseases** develop slowly (usually with less severe symptoms) and are continual or recurrent. Infectious mononucleosis, hepatitis C, tuberculosis, and leprosy are chronic diseases. **Subacute diseases** have durations and severities that lie somewhere between acute and chronic. Subacute bacterial endocarditis, a disease of heart valves, is one example. **Latent diseases** are those in which a pathogen remains inactive for a long period of time before becoming active. Herpes is an example of a latent disease.

When an infectious disease comes from another infected host, either directly or indirectly, it is a **communicable disease.** Influenza, herpes, and tuberculosis are examples of communicable diseases. If a communicable disease is easily transmitted between hosts, as is the case for chicken pox or measles, it is also called a **contagious disease**. **Noncommunicable diseases** arise outside hosts or from normal microbiota. In other words, they are not spread from one host to another, and diseased patients are not a source of contamination for others. Tooth decay, acne, and tetanus are examples of noncommunicable diseases. **TABLE 14.12** defines these and other terms used to classify infectious diseases.

Yet another way in which all infectious diseases may be classified is by the effects they have on populations rather than on individuals. Is a certain disease consistently found in a given group of people or geographic area? Under what circumstances is it more prevalent than normal in a given geographic area? How prevalent is "normal"? How is the disease transmitted throughout a population? We next examine issues related to diseases at the population level.

TABLE 14.12 Terms Used to Classify Infectious Diseases

Term	Definition
Acute disease	Disease in which symptoms develop rapidly and that runs its course quickly
Chronic disease	Disease with usually mild symptoms that develop slowly and last a long time
Subacute disease	Disease with time course and symptoms between acute and chronic
Asymptomatic disease	Disease without symptoms
Latent disease	Disease that appears a long time after infection
Communicable disease	Disease transmitted from one host to another
Contagious disease	Communicable disease that is easily spread
Noncommunicable disease	Disease arising from outside hosts or disease from opportunistic pathogen
Local infection	Infection confined to a small region of the body
Systemic infection	Widespread infection in many systems of the body; often travels in the blood or lymph
Focal infection	Infection that serves as a source of pathogens for infections at other sites in the body
Primary infection	Initial infection within a given patient
Secondary infection	Infections that follow a primary infection; often by opportunistic pathogens

TELL ME WHY

Why is an acute disease with a high rate of mortality unlikely to be associated with a pandemic?

Epidemiology of Infectious Diseases

LEARNING **OUTCOME**

14.22 Define *epidemiology.*

Our discussion so far has centered on the negative impact of microorganisms on *individuals.* Now we turn our attention to the effects of pathogens on *populations.* **Epidemiology**[10] (ep-i-dē-mē-ol´ō-jē) is the study of where and when diseases occur and how they are transmitted within populations. During the 20th century, epidemiologists expanded the scope of their work beyond infectious diseases to also consider injuries and deaths related to automobile and fireworks accidents, cigarette smoking, lead poisoning, and other causes; however, we will limit our discussion primarily to the epidemiology of infectious diseases. ▶**ANIMATIONS:** *Epidemiology: Overview*

Frequency of Disease

LEARNING **OUTCOMES**

14.23 Contrast incidence and prevalence.

14.24 Differentiate among the terms *endemic, sporadic, epidemic,* and *pandemic.*

Epidemiologists keep track of the occurrence of diseases by using two measures: incidence and prevalence. **Incidence** is the number of *new* cases of a disease in a given area or population during a given period of time; **prevalence** is the *total number* of cases, both new and already existing, in a given area or population during a given period of time. In other words, prevalence is a cumulative number. Thus, for example, the reported number of new cases—the incidence—of AIDS in adults in the United States in 2009 was 42,959. However, the prevalence of AIDS in 2009 was about 450,000 because more than 400,000 patients who got the disease prior to 2009 still survived with the disease. **FIGURE 14.14** illustrates this relationship between incidence and prevalence.

Epidemiologists report their data in many ways, including maps, graphs, charts, and tables (**FIGURE 14.15**). Why do they report their data in so many different ways? Using a variety of formats enables epidemiologists to observe patterns that may give clues about the causes of or ways to prevent diseases. For example, West Nile virus encephalitis occurs across the United

[10]From Greek *epidemios,* meaning "among the people," and *logos,* meaning "study of."

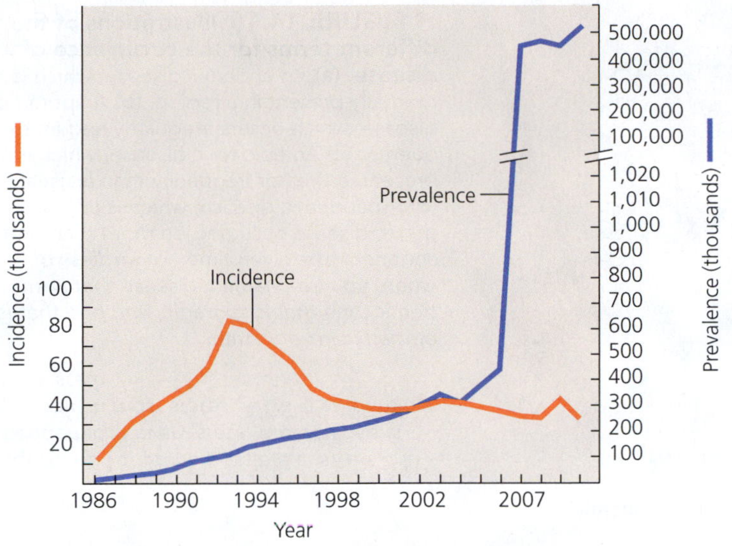

▲ **FIGURE 14.14 Curves representing the incidence and the estimated prevalence of AIDS among U.S. adults.** Note that the scales for the two curves are different. *Why can the incidence of a disease never exceed the prevalence of that disease?*

Figure 14.14 *Because prevalence includes all cases, both old and new, prevalence must always be larger than incidence.*

States, but when data are exported by age group, it becomes obvious that elderly Americans are most at risk.

The occurrence of a disease can also be considered in terms of a combination of frequency and geographic distribution **(FIGURE 14.16)**. A disease that normally occurs continually (at moderately regular intervals) at a relatively stable incidence within a given population or geographical area is said to be **endemic**[11] to that population or region. A disease is considered **sporadic** when only a few scattered cases occur within an area or population. Whenever a disease occurs at a greater frequency than is usual for an area or population, the disease is said to be **epidemic** within that area or population. **▶ANIMATIONS:** ***Epidemiology: Occurrence of Disease***

The commonly held belief that a disease must infect thousands or millions to be considered an epidemic is mistaken. The time period and the number of cases necessary for an outbreak of disease to be classified as an epidemic are not specified; the important fact is that there are more cases than historical statistics indicate are expected. For example, fewer than 70 cases of an emerging disease—hemolytic uremic syndrome caused by a strain of *E. coli*—occurred in Germany in 2011, but because there are typically fewer than five cases annually, the 2011 outbreak was considered an epidemic. The same year, thousands of cases of flu occurred in Germany, but there was no flu epidemic because the number of cases observed did not exceed the number expected. **FIGURE 14.17** illustrates how epidemics are defined according to the number of expected cases and not according to the absolute number of cases.

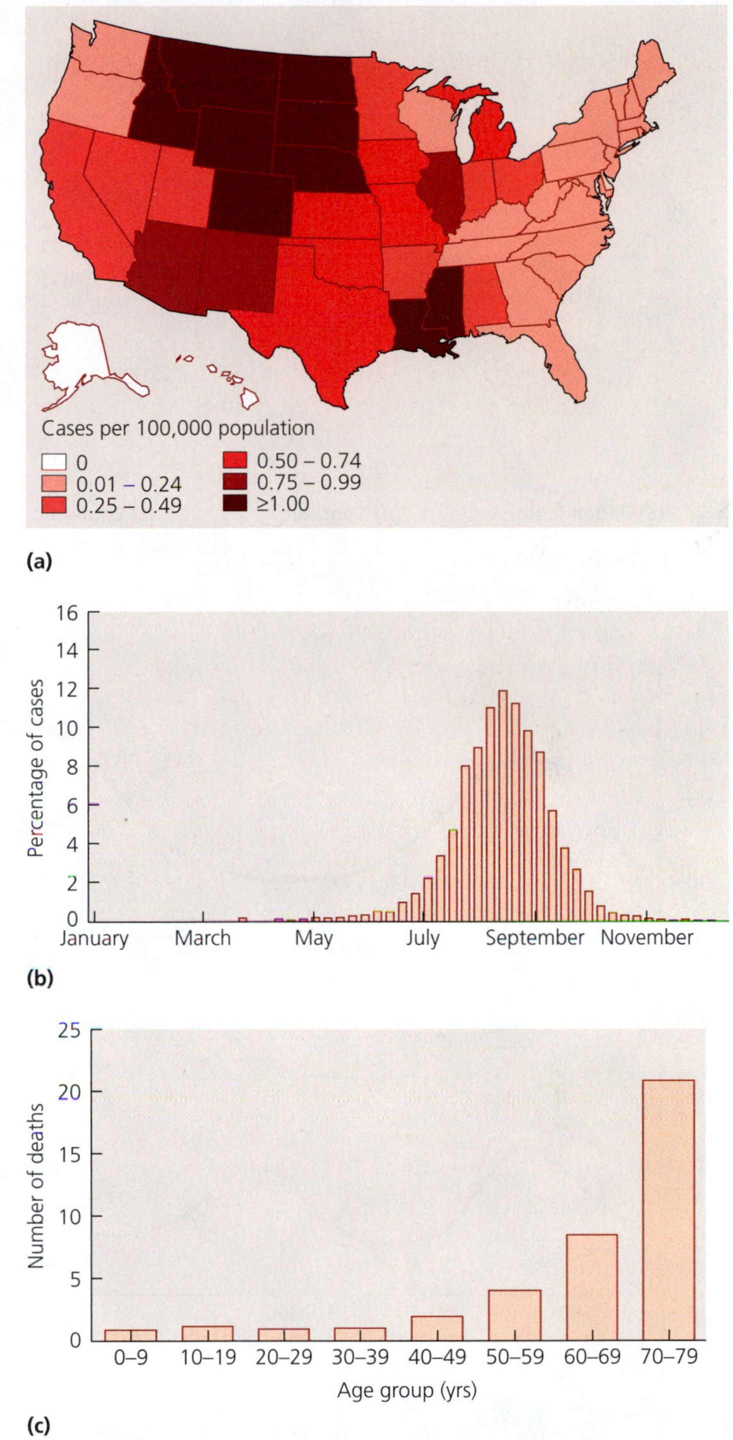

▲ **FIGURE 14.15 Epidemiologists report data in a variety of ways.** Here, incidence of cases of an emerging disease—West Nile virus disease—in which the virus invades the nervous system. Data for the decade 1999–2008 are presented. **(a)** Average annual incidence for the United States is on a map by state. **(b)** Percentage of cases by week of onset. **(c)** Percentage of deaths by age. *How do the data help support the idea that mosquitoes carry West Nile virus?*

Figure 14.15 *Most cases of the disease occur in late summer, when mosquitoes are likely biting and people are more likely to be outdoors.*

[11]From Greek *endemos*, meaning "native."

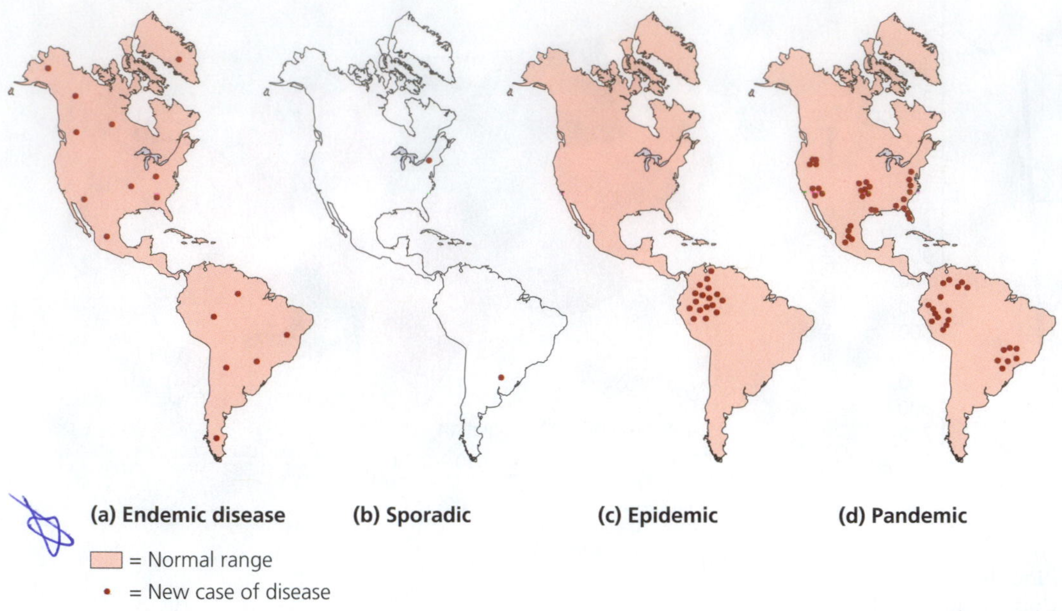

◀ FIGURE 14.16 Illustrations of the different terms for the occurrence of disease. **(a)** An endemic disease, which is normally present in a region. **(b)** A sporadic disease, which occurs irregularly and infrequently. **(c)** An epidemic disease, which is present in greater frequency than is usual. **(d)** A pandemic disease, which is an epidemic disease occurring on more than one continent at a given time. *Regardless of where you live, name a disease that is endemic, one that is sporadic, and one that is epidemic in your state.*

Figure 14.16 *Some possible answers: Flu is endemic in every state; tuberculosis is sporadic in most states; AIDS is epidemic in every state.*

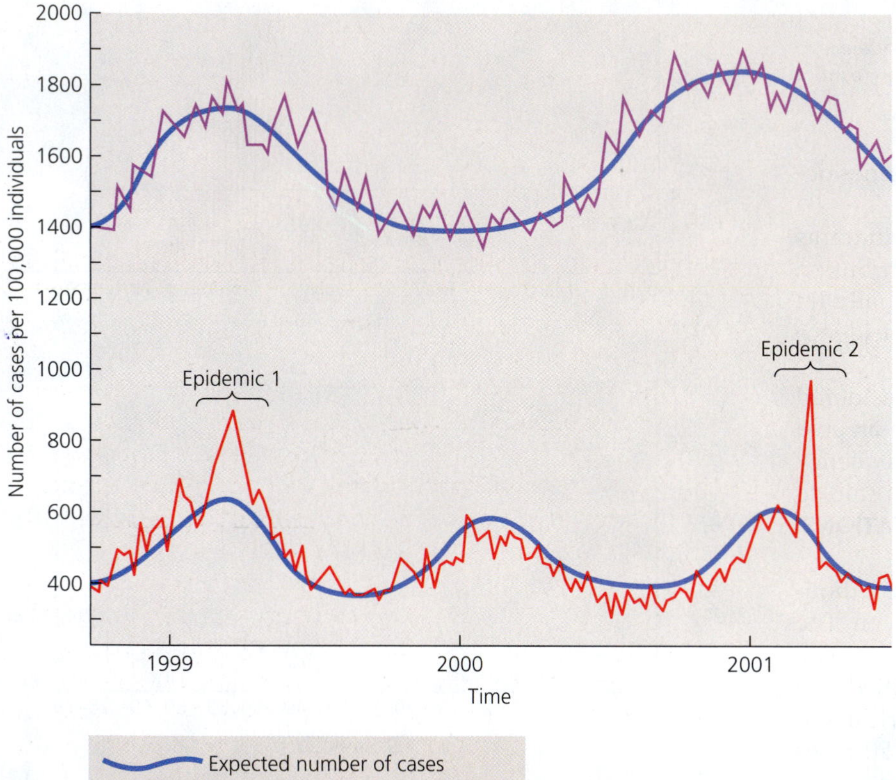

◀ FIGURE 14.17 Epidemics may have fewer cases than nonepidemics. These two graphs demonstrate the independence between the absolute number of cases and a disease's designation as an epidemic. Even though the number of cases of disease A always exceeds the number of cases of disease B, only disease B is considered epidemic—at those times when the number of cases observed exceeds the number of cases expected. *How can scientists know the normal prevalence of a given disease?*

Figure 14.17 *Scientists record every case of the disease so that they will have a "baseline" prevalence, which then becomes the expected prevalence for each disease.*

If an epidemic occurs simultaneously on more than one continent, it is referred to as a **pandemic** (see Figure 14.16d). H1N1 flu, so-called swine flu, became pandemic worldwide in 2009.

Obviously, for disease prevalence to be classified as either endemic, sporadic, epidemic, or pandemic, good records must be kept for each region and population. From such records, incidence and prevalence can be calculated, and then changes in these data can be noted. Health departments at the local and state levels require doctors and hospitals to report certain infectious diseases. Some are also nationally notifiable; that is, their occurrence must be reported to the Centers for Disease Control and Prevention (CDC) in Atlanta, Georgia, which is the headquarters and clearinghouse for national epidemiological research. Nationally notifiable diseases are listed in **TABLE 14.13**. Each week the CDC reports the number of cases of most of the nationally notifiable diseases in the *Morbidity and Mortality Weekly Report (MMWR)* **(FIGURE 14.18)**.

TABLE 14.13 Nationally Notifiable Infectious Diseases[a]

Anthrax	*Haemophilus influenzae*, invasive disease	Mumps	Streptococcal toxic-shock syndrome
Arboviral diseases	Hansen disease (leprosy)	Novel influenza A infections	*Streptococcus pneumoniae*, invasive disease
Babesiosis	*Hantavirus* pulmonary syndrome	Pertussis	Syphilis
Botulism	Hemolytic uremic syndrome, postdiarrheal	Plague	Tetanus
Brucellosis	Hepatitis A	Poliomyelitis	Toxic-shock syndrome, nonstreptococcal
Chancroid	Hepatitis B	Psittacosis	Trichinosis
Chicken pox (varicella)	Hepatitis C	Q fever	Tuberculosis
Chlamydia trachomatis infections	HIV infection	Rabies, animal and human	Tularemia
Cholera	Influenza-related infant deaths	Rubella	Typhoid fever
Coccidioidomycosis	Legionellosis	Rubella, congenital syndrome	Vancomycin-intermediate *Staphylococcus aureus*
Cryptosporidiosis	Listeriosis	Salmonellosis	Vancomycin-resistant *Staphylococcus aureus*
Cyclosporiasis	Lyme disease	Severe acute respiratory syndrome (SARS)	Vibriosis
Dengue virus infections	Malaria	Shiga-toxin-producing *Escherichia coli*	Viral hemorrhagic fever
Diphtheria	Measles	Shigellosis	Yellow fever
Ehrlichiosis/anaplasmosis	Meningococcal disease	Small pox	
Giardiasis		Spotted fever rickettsiosis	
Gonorrhea			

[a]Diseases for which hospitals, physicians, and other health care workers are required to report cases to state health departments and then forward the data to the CDC.

Epidemiological Studies

LEARNING OUTCOME

14.25 Explain three approaches epidemiologists use to study diseases in populations.

Epidemiologists conduct research to study the dynamics of diseases in populations by taking three different approaches, called descriptive, analytical, and experimental epidemiology.

Descriptive Epidemiology

Descriptive epidemiology involves the careful tabulation of data concerning a disease. Relevant information includes the location and time of cases of the disease as well as information about the patients, such as ages, gender, occupations, health histories, and socioeconomic groups. Because the time course and chains of transmission of a disease are an important part of descriptive epidemiology, epidemiologists strive to identify the **index case** (the first case) of the disease in a given area or population. Sometimes it is difficult or impossible to identify an index case because the patient has recovered, moved, or died.

The earliest descriptive epidemiological study was by John Snow (1813–1858), who studied a cholera outbreak in London in 1854. By carefully mapping the locations of the cholera cases in a particular part of the city, Snow found that the cases were clustered around the Broad Street water pump (**FIGURE 14.19**). This distribution of cases, plus the voluminous watery diarrhea of cholera patients, suggested that the disease was spread via contamination of drinking water by sewage.

Analytical Epidemiology

Analytical epidemiology investigates a disease in detail, including analysis of data acquired in descriptive epidemiological studies, to determine the probable cause, mode of transmission, and possible means of prevention of the disease. Analytical epidemiology may be used in situations where it is not ethical to apply Koch's postulates. Thus, even though Koch's third postulate has never been fulfilled in the case of AIDS (because it is unethical to intentionally inoculate a human with HIV), analytical epidemiological studies indicate that HIV causes AIDS and that it is transmitted primarily sexually.

Often analytical studies are *retrospective;* that is, they attempt to identify causation and mode of transmission after an outbreak has occurred. Epidemiologists compare a group of people who had the disease with a group who did not. The groups are carefully matched by factors such as gender, environment, and diet and then compared to determine which pathogens and factors may play a role in morbidity.

Experimental Epidemiology

Experimental epidemiology involves testing a hypothesis concerning the cause of a disease. The application of Koch's postulates is an example of experimental epidemiology. Experimental epidemiology also involves studies to test a hypothesis resulting from an analytical study, such as the efficacy of a preventive measure or certain treatment. For example, analytical epidemiological studies suggested that the bacterium *Chlamydophila pneumoniae* may contribute to arteriosclerosis, resulting in heart attacks. If so, antimicrobial drugs should prevent some heart attacks by killing the bacteria. However, when researchers administered antimicrobial drugs to 7700 patients with a history of

Weekly/Vol.62/No.38 MMWR September 27, 2013

TABLE II. (*Continued*) **Provisional casses of selected notifiable diseases, United States, weeks ending September 21, 2013, and September 22, 2012 (38th week)***

	Giardiasis					Gonorrhea					*Haemophilus influenzae*, invasive† All ages, all serotypes				
	Current	Previous 52 weeks		Cum	Cum	Current	Previous 52 weeks		Cum	Cum	Current	Previous 52 weeks		Cum	Cum
Reporting area	week	Med	Max	2013	2012	week	Med	Max	2013	2012	week	Med	Max	2013	2012
United States	159	256	413	9,343	10,934	3,845	6,170	7,762	225,146	241,673	20	69	190	2,557	2,459
New England	12	23	44	862	1,042	96	123	190	4,644	4,177	—	4	21	154	164
Connecticut	—	4	7	137	173	26	48	102	1,989	1,475	—	1	4	30	43
Maine	1	4	10	135	120	—	5	16	116	321	—	0	2	16	16
Massachusetts	11	12	28	460	514	70	54	100	2,129	1,812	—	2	5	84	81
New Hampshire	—	1	4	46	85	—	2	9	91	106	—	0	3	17	11
Rhode Island	—	0	7	20	40	—	7	25	257	389	—	0	11	—	8
Vermont	—	2	10	64	110	—	1	12	62	74	—	0	1	7	5
Mid. Atlantic	26	57	89	2,047	2,067	607	790	1,084	28,909	32,982	8	12	58	447	464
New Jersey	—	6	21	179	300	49	131	181	4,861	5,582	—	2	10	80	84
New York (Upstate)	—	23	67	875	631	155	129	519	4,552	5,244	1	3	35	114	128
New York City	7	14	27	522	646	156	260	339	9,569	10,891	—	2	6	86	96
Pennsylvania	19	12	25	471	490	247	270	334	9,927	11,265	7	4	13	167	156
E.N. Central	16	36	56	1,268	1,667	267	985	1,439	35,104	42,878	1	10	21	405	422
Illinois	—	5	12	174	256	8	270	428	8,187	13,016	—	3	10	114	114
Indiana	—	3	10	109	159	35	147	180	5,207	5,290	—	2	6	95	81
Michigan	2	9	25	335	422	106	211	378	7,437	9,062	—	2	4	67	61
Ohio	13	10	26	390	427	96	313	391	11,427	12,001	1	3	8	108	115
Wisconsin	1	7	15	260	403	22	84	122	2,846	3,509	—	0	4	21	51
W.N. Central	10	19	62	622	1,244	60	306	415	10,836	12,849	3	5	11	169	174
Iowa	5	4	15	172	191	1	30	51	1,035	1,494	—	0	0	—	—
Kansas	—	2	5	65	103	6	37	66	1,415	1,683	—	1	3	32	21
Minnesota	—	0	26	—	419	—	53	92	1,449	2,067	—	1	3	31	65
Missouri	5	5	13	173	236	53	148	184	5,336	5,863	3	2	7	81	56
Nebraska	—	3	9	105	144	—	25	49	869	1,018	—	0	3	16	20
North Dakota	—	1	4	25	46	—	8	15	273	216	—	0	1	6	12
South Dakota	—	2	9	82	105	—	14	23	459	508	—	0	1	3	—
S. Atlantic	19	43	82	1,679	1,780	1,322	1,379	1,646	51,416	54,294	7	17	37	698	607
Delaware	—	0	2	9	19	16	22	55	934	630	—	0	2	6	6
District of Columbia	—	1	5	39	64	69	42	92	1,643	1,803	—	0	2	10	2
Florida	19	21	56	846	766	270	391	478	14,761	14,308	1	5	13	206	178
Georgia	—	9	36	398	422	243	259	328	9,715	11,502	4	4	12	118	121
Maryland	—	5	9	145	184	99	109	241	3,770	3,858	—	2	4	71	67
North Carolina	N	0	0	N	N	261	266	544	9,779	11,111	2	2	12	109	80
South Carolina	—	2	5	78	93	169	139	207	5,359	5,648	—	2	11	93	54
Virginia	—	4	16	136	193	166	125	214	4,648	4,975	—	1	8	65	71
West Virginia	—	1	5	28	39	29	20	34	807	559	—	0	3	20	28
E.S. Central	—	3	13	113	129	144	501	786	17,584	21,708	—	5	12	193	153
Alabama	—	3	13	113	129	—	157	257	5,533	6,758	—	1	7	60	42
Kentucky	N	0	0	N	N	88	85	143	3,173	3,151	—	0	4	35	29
Mississippi	N	0	0	N	N	—	109	205	3,821	4,980	—	1	2	22	19
Tennessee	N	0	0	N	N	56	154	239	5,057	6,819	—	2	7	76	63
W.S. Central	2	6	15	236	227	650	928	2,555	34,941	34,591	—	3	20	130	136
Arkansas	—	2	8	72	62	79	74	122	2,640	3,273	—	0	3	19	23
Louisiana	2	4	10	164	165	18	116	657	4,365	6,016	—	1	4	30	48
Oklahoma	—	0	0	—	—	—	68	1,325	3,406	2,344	—	2	17	79	64
Texas	N	0	0	N	N	553	654	1,044	24,530	22,958	—	0	1	2	1
Mountain	7	19	36	681	923	208	260	326	9,377	9,742	—	6	14	223	237
Arizona	—	2	5	68	86	81	102	158	3,668	4,208	—	2	7	89	9
Colorado	4	4	18	168	280	52	55	100	1,968	1,999	—	1	4	54	44
Idaho	—	2	9	92	116	—	2	8	58	117	—	0	2	10	13
Montana	—	1	4	48	47	4	3	11	140	70	—	0	1	2	4
Nevada	1	1	6	59	73	52	45	79	1,802	1,629	—	0	1	11	18
New Mexico	—	2	6	61	67	14	32	55	1,201	1,370	—	1	3	28	38
Utah	2	5	8	162	224	5	13	25	525	316	—	0	2	27	28
Wyoming	—	0	6	23	30	—	0	3	15	33	—	0	1	2	3
Pacific	67	51	121	1,835	1,855	491	854	992	32,335	28,352	1	3	9	138	102
Alaska	4	1	7	53	66	6	18	39	714	481	—	0	2	11	10
California	31	32	48	1,183	1,214	381	715	844	27,013	23,989	—	1	5	35	23
Hawaii	—	1	3	30	23	—	14	26	491	574	—	1	2	21	15
Oregon	10	6	17	262	278	42	32	50	1,251	1,069	1	1	5	66	54
Washington	22	10	61	307	274	62	73	154	2,866	2,239	—	0	3	5	—
Territories															
American Samoa	—	0	0	—	—	—	0	0	—	—	—	0	0	—	—
C.N.M.I.	—	—	—	—	—	—	—	—	—	—	—	—	—	—	—
Guam	—	0	1	—	1	—	0	0	—	—	—	0	0	—	—
Puerto Rico	—	0	3	29	2	1	1	6	241	225	—	0	0	—	—
U.S. Virgin Islands	—	0	0	—	—	—	1	8	26	101	—	0	0	—	—

C.N.M.I.: Commonwealth of Northern Mariana Islands.

U: Unavailable. —: No reported cases. N: Not reportable. NN: Not Nationally Notifiable. Cum: Cumulative year-to-date counts. Med: Median. Max: Maximum.

* Case counts for reporting year 2012 are provisional and subject to change. For further information on interpretation of these data, see http://wwwn.cdc.gov/nndss/document/ProvisionalNationaNotifiableDiseasesSurveillanceData20100927.pdf. Data for TB are displayed in Table IV, which appears quarterly.

† Data forH. *influenzae* (age <5 yrs for serotype b, nonserotype b, and unknown serotype) are available in Table I.

◀ **FIGURE 14.18 A page from the *MMWR*.** The CDC's *Morbidity and Mortality Weekly Report (MMWR)* provides epidemiological data state by state for the current week, for the current year to date, and for the previous year to date. The *MMWR* also publishes reports on epidemiological case studies. This page shows the incidence of three diseases in one week in 2013. *Contrast the number of cases of giardiasis in California in 2013 with the numbers in Alaska the same year. Which state has an epidemic of giardiasis?*

Figure 14.18 *An epidemic is defined as more cases of a disease than expected. Since the numbers of cases of giardiasis historically reported in California and Alaska are not given, we cannot classify either as epidemic in 2013.*

heart disease and of infection with *C pneumoniae,* the number of heart attacks was not significantly reduced. Thus, an experimental epidemiological study disproved a hypothesis suggested by an analytical epidemiological analysis.

Clinical Case Study: *Legionella* in the Produce Aisle illustrates the work of epidemiologists.

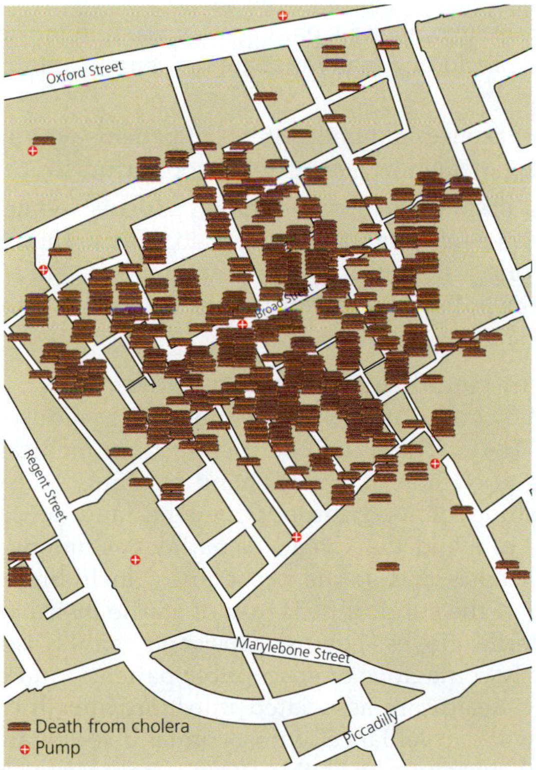

▲ **FIGURE 14.19 A map showing cholera deaths in a section of London, 1854.** From the map he compiled, Dr. John Snow showed that cholera cases centered around the Broad Street pump. Snow's work was a landmark in epidemiological research.

CLINICAL CASE STUDY

Legionella in the Produce Aisle

The Louisiana state health department has received reports of 33 cases of Legionnaires' disease in the town of Bogalusa (population 16,000).

Legionnaires' disease, or legionellosis, is a potentially fatal respiratory disease caused by the growth of a bacterium, *Legionella pneumophila*, in the lungs of patients. The bacterium enters humans via the respiratory portal in aerosols produced by cooling towers, air conditioners, whirlpool baths, showers, humidifiers, and respiratory therapy equipment.

Epidemiologists begin trying to ascertain the source of Bogalusa's outbreak of Legionnaires' disease by interviewing the victims and their relatives to develop complete histories and to identify areas of commonality among the victims that were lacking among nonvictims. Victims include a range of ages, occupations, hobbies, religions, and types and locations of dwellings. Although no significant differences are identified among the lifestyles, ages, or smoking habits of victims and nonvictims, one curious fact is discovered: all the victims did their grocery shopping at the same store. However, healthy individuals also shopped at that store.

The air-conditioning system of the grocery store proves to be free of *Legionella*, but the vegetable misting machine does not. The strain of *Legionella* isolated from the misting machine is identical to the strain recovered from the victims' lungs.

1. Would this outbreak be classified as endemic, epidemic, or pandemic?
2. Is this a descriptive, analytical, or experimental epidemiological study?
3. Knowing the epidemiology and causative agent of Legionnaires' disease, what questions would you ask of the victims or of their surviving relatives?
4. What, as an epidemiologist, would you examine at the store?
5. How did the victims become contaminated? Why didn't everyone who bought vegetables at the store get legionellosis? What could the owners of the store do to limit or prevent future infections?

Reference: Adapted from *MMWR* 39:108–109, 1990.

Hospital Epidemiology: Healthcare Associated (Nosocomial) Infections

LEARNING | OUTCOMES

14.26 Explain how healthcare associated infections differ from other infections.

14.27 Describe the factors that influence the development of healthcare associated infections.

14.28 Describe three types of healthcare associated infections and how they may be prevented.

Of special concern to epidemiologists and health care workers are healthcare associated infections and diseases, which were formerly called nosocomial (nos-ō-kō´mē-ăl) infections and nosocomial diseases. **Healthcare associated infections (HAIs)** are infections acquired by patients or health care workers while they are in health care facilities, including hospitals, dental offices, nursing homes, and doctors' waiting rooms. The CDC estimates that about 10% of American patients acquire an HAI each year. **Healthcare associated diseases** increase the duration and cost of medical care and result in some 100,000 deaths annually in the United States. ▶**ANIMATIONS:** ***Nosocomial Infections: Overview***

Types of Healthcare Associated Infections

When most people think of HAIs, what likely comes to mind are **exogenous** (eks-oj´ĕ-nŭs) **HAIs**, which are caused by pathogens acquired from the health care environment. After all, hospitals are filled with sick people shedding pathogens from every type of portal of exit. However, we have seen that members of the normal microbiota can become opportunistic pathogens as a result of hospitalization or medical treatments such as chemotherapy. Such opportunists cause **endogenous** (en-doj´ĕ-nŭs) **HAIs**; that is, they arise from normal microbiota within the patient that become pathogenic because of factors within the health care setting.

Iatrogenic infections (ī-at-rō-jen´ik; literally meaning "doctor-induced" infections) are a subset of HAIs that ironically are the direct result of modern medical procedures such as the use of catheters, invasive diagnostic procedures, and surgery.

Superinfections may result from the use of antimicrobial drugs that, by inhibiting some resident microbiota, allow others to thrive in the absence of competition. For instance, long-term antimicrobial therapy to inhibit a bacterial infection may allow *Clostridium difficile* (di-fi´sēl), a transient microbe of the colon, to grow excessively and cause a painful condition called pseudomembranous colitis. Such superinfections are not limited to health care settings.

Healthcare associated infections most often occur in the urinary, respiratory, cardiovascular, and integumentary (skin) systems, though surgical wounds can become infected and result in healthcare associated infections in any part of the body.

Factors Influencing Healthcare Associated Infections

HAIs arise from the interaction of several factors in the health care environment, which include the following **(FIGURE 14.20)**:

- Exposure to numerous pathogens present in the health care setting, including many that are resistant to antimicrobial agents
- The weakened immune systems of patients who are ill, making them more susceptible to opportunistic pathogens
- Transmission of pathogens among patients and health care workers—from staff and visitors, to patients, and even from one patient to another via activities of staff members (including invasive procedures and other iatrogenic factors)

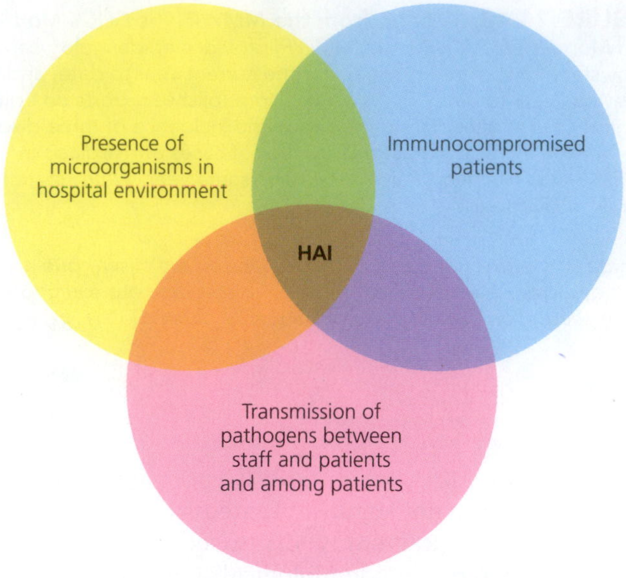

▲ **FIGURE 14.20 The interplay of factors that result in healthcare associated infections (HAIs).** Although HAIs can result from any one of the factors shown, most are the product of the interaction of all three factors.

Control of Healthcare Associated Infections

Aggressive control measures can noticeably reduce the incidence of HAIs. These include disinfection; medical asepsis, including good housekeeping, hand washing, bathing, sanitary handling of food, proper hygiene, and precautionary measures to avoid the spread of pathogens among patients; surgical asepsis and sterile procedures, including thorough cleansing of the surgical field, use of sterile instruments, and use of sterile gloves, gowns, caps, and masks; isolation of particularly contagious or susceptible patients; and establishment of a healthcare associated infection control committee charged with surveillance of nosocomial diseases and review of control measures. ▶**ANIMATIONS:** ***Nosocomial Infections: Prevention***

Numerous studies have shown that the single most effective way to reduce HAIs is effective hand washing by all medical and support staff. In one study, deaths from HAIs were reduced by over 50% when hospital personnel followed strict guidelines about washing their hands frequently.

EMERGING DISEASE CASE STUDY

Hantavirus Pulmonary Syndrome

The deer mouse, *Peromyscus maniculatus*

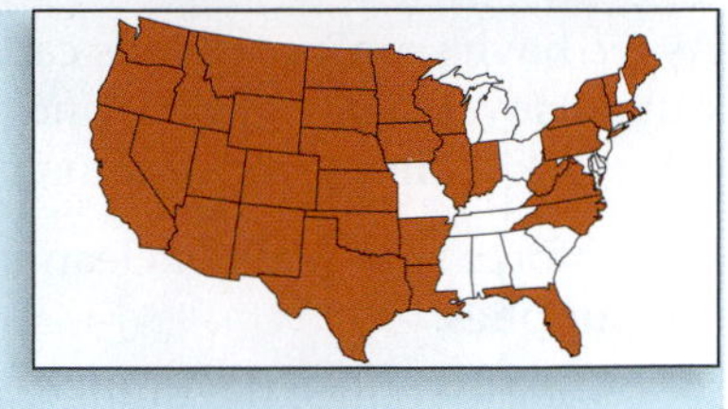

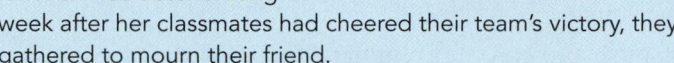

Atsa was excited. Her Navajo basketball team had won the divisional championship for 1993, and she was high point (the high scorer) for the championship game. The thrill of the moment made her forget the argument with her parents earlier in the week about sweeping out the storage shed. They were champions, and life was good.

The next morning, Atsa woke with deep muscle pain that she attributed to the exertions of the game. By noon she had a headache, nausea, and the chill associated with a fever. "The flu," her mother concluded, and sent Atsa to bed with acetaminophen and plenty of fluids. Over the next few days, Atsa's lungs began to congest; she struggled to breathe, and her heart raced. Atsa was drowning in her own bodily fluid.

Desperately worried, Atsa's parents drove her 60 miles to the Navajo medical center, where bewildered doctors provided respiratory and cardiac care while searching for answers. They discovered three patients in the surrounding counties with the same signs and symptoms; all had died. Atsa's prognosis was the same: She was comatose. Her blood platelet level had dropped precipitously. Excessive proteins were in her blood. Her kidneys' function was deteriorating. A week after her classmates had cheered their team's victory, they gathered to mourn their friend.

As heartbreaking as the deaths in this epidemic were, the cases advanced our medical understanding and put to use the power of modern epidemiology and genetic analysis. Within eight days of the initial case, epidemiologists had isolated a suspect virus, sequenced its genes, identified it as a previously unknown species of *Hantavirus*, and shown that the virus was the cause of the condition, which is now known as *Hantavirus* pulmonary syndrome (HPS). Scientists also showed that HPS is acquired when victims inhale the virus in aerosolized deer-mouse urine or feces stirred up by sweeping, a finding that enabled other residents of the region to take preventive measures. (For more about *Hantavirus* pulmonary syndrome, see pp. 702–703.)

1. Why didn't doctors treat Atsa with antibiotics, such as penicillin or tetracycline?
2. Cases of *Hantavirus* pulmonary syndrome are often seen following years with higher-than-average rainfall. Why?
3. What was the likely way that epidemiologists showed that the virus was the cause of the pulmonary condition?

Epidemiology and Public Health

LEARNING OUTCOME

14.29 List three ways public health agencies work to limit the spread of diseases.

As you have likely realized by now, epidemiologists gather information concerning the spread of disease within populations so that they can take steps to reduce the number of cases and improve the health of individuals within a community. In the following sections we will examine how the various public health agencies share epidemiological data, facilitate the interruption of disease transmission, and educate the public about public health issues. Public health agencies also implement immunization programs (immunization is discussed in Chapter 17).

The Sharing of Data Among Public Health Organizations

Numerous agencies at the local, state, national, and global levels work together with the entire spectrum of health care personnel to promote public health. By submitting reports on incidence and prevalence of disease to public officials, physicians can subsequently learn of current disease trends. Additionally, public health agencies often provide physicians with laboratory and diagnostic assistance.

City and county health departments report data on disease incidence to state agencies. Because state laws govern disease reporting, state agencies play a vital role in epidemiological studies. States accumulate data similar to those in the *MMWR* and assist local health departments and medical practitioners with diagnostic testing for diseases such as rabies and Lyme disease.

Data collected by the states are forwarded to the CDC, which is but one branch of the U.S. Public Health Service, the national public health agency. In addition to epidemiological studies, the CDC and other branches of the Public Health Service conduct research in disease etiology and prevention, make recommendations concerning immunization schedules, and work with public health organizations of other countries.

The World Health Organization (WHO) coordinates efforts to improve public health throughout the world, particularly in poorer countries, and the WHO has undertaken ambitious projects to eradicate such diseases as polio, measles, and mumps. Some other current campaigns involve AIDS education, malaria control, and childhood immunization programs in poor countries.

The Role of Public Health Agencies in Interrupting Disease Transmission

As we have seen, pathogens can be transmitted in air, food, and water as well as by vectors and via fomites. Public health agencies work to limit disease transmission by a number of methods:

- Enforce standards of cleanliness in water and food supplies.
- Work to reduce the number of disease vectors and reservoirs.
- Establish and enforce immunization schedules (see Figure 17.3).
- Locate and prophylactically treat individuals exposed to contagious pathogens.
- Establish isolation and quarantine measures to control the spread of pathogens.

A water supply that is **potable** (fit to drink) is vital to good health. Organisms that cause dysentery, cholera, and typhoid fever are just some of the pathogens that can be spread in water contaminated by sewage. Filtration and chlorination processes are used to reduce the number of pathogens in water supplies. Local, state, and national agencies work to ensure that water supplies remain clean and healthful by monitoring both sewage treatment facilities and the water supply.

Food can harbor infective stages of parasitic worms, protozoa, bacteria, and viruses. National and state health officials ensure the safety of the food supply by enforcing standards in the use of canning, pasteurization, irradiation, and chemical preservatives and by insisting that food preparers and handlers wash their hands and use sanitized utensils. The U.S. Department of Agriculture also provides for the inspection of meats for the presence of pathogens, such as *E. coli* and tapeworms.

Milk is an especially rich food, and the same nutrients that nourish us also facilitate the growth of many microorganisms. In the past, contaminated milk has been responsible for epidemics of tuberculosis, brucellosis, typhoid fever, scarlet fever, and diphtheria. Today, public health agencies require the pasteurization of milk, and as a result disease transmission via milk has been practically eliminated in the United States.

Individuals should assume responsibility for their own health by washing their hands before and during food preparation, using disinfectants on kitchen surfaces, and using proper refrigeration and freezing procedures. It is also important to thoroughly cook all meats.

Public health officials also work to control vectors, especially mosquitoes and rodents, by eliminating breeding grounds, such as stagnant pools of water and garbage dumps. Insecticides have been used with some success to control insects.

Public Health Education

Diseases that are transmitted sexually or through the air are particularly difficult for public health officials to control. In these cases, individuals must take responsibility for their own health, and health departments can only educate the public to make healthy choices.

Colds and flu remain the most common diseases in the United States because of the ubiquity of the viral pathogens and their mode of transmission in aerosols and via fomites. Health departments encourage afflicted people to remain at home, use disposable tissues to reduce the spread of viruses, and avoid crowds of coughing, sneezing people. As we have discussed, hand washing is also important in preventing the introduction of cold and flu viruses onto the conjunctiva.

We as a society are faced with several epidemics of sexually transmitted diseases. Syphilis, gonorrhea, genital warts, and sexually transmitted AIDS are completely preventable if the chain of transmission is interrupted by abstinence or mutually faithful monogamy. Their incidence can be reduced but not eliminated by the use of condoms. Based on the premise that "an ounce of prevention is worth a pound of cure," public health agencies expend considerable effort in public campaigns to educate people to make good choices—those that can result in healthier individuals and improved health for the public at large.

TELL ME WHY

Why are all iatrogenic infections healthcare associated, but not all healthcare associated infections are iatrogenic?

MICRO IN THE CLINIC FOLLOW-UP

Antibiotics: Friend or Foe?

The doctor finds *Escherichia coli* in Nancy's urine sample and diagnoses a kidney infection that spread from her urinary tract. *E. coli*, a common cause of *urinary* tract infections, is a normal resident of the intestinal tract but is pathogenic in the urinary tract.

The doctor explains that women are at greater risk for urinary tract infections than men because the female urethra is located closer to the anus. Furthermore, the antibiotics that Nancy took for her sinus infection killed normal microbiota that hold *E. coli* in check, allowing it to multiply in her urinary tract. Nancy is admitted to the hospital for intravenous treatment with another antibiotic.

Unfortunately, after two days in the hospital, Nancy becomes nauseated and develops abdominal tenderness and massive watery diarrhea. A stool test identifies *Clostridium difficile*, a Gram-positive, endospore-forming *bacterium* commonly found in the intestinal tract.

Apparently the antibiotics also upset the normal balance of bacteria in Nancy's intestine. *C. difficile* can overgrow other intestinal microbiota in patients undergoing prolonged antibiotic treatment. *C. difficile* also attacks the lining of the intestine, resulting in excessive diarrhea, which can lead to dehydration and serious complications such as colitis (inflamed colon) or *septicemia* (blood infection). A change of antibiotics eventually controls *C. difficile*, and Nancy recovers thanks to her typically healthy constitution.

1. **Describe the type of symbiotic relationship between *E. coli* and the host's intestinal tract.**
2. **What conditions predisposed Nancy to the infection with *C. difficile*?**

MM Check your answers to Micro in the Clinic Follow-Up questions in the MasteringMicrobiology Study Area.

Explore the Invisible: Some Virulence Factors

Practice on-the-go with Dr. Bauman Video Tutors by scanning this QR code with your smart phone. Visit the **MasteringMicrobiology Study Area** to challenge your understanding with practice tests, animation quizzes, and clinical case studies!

MasteringMicrobiology®

CHAPTER SUMMARY

Symbiotic Relationships Between Microbes and Their Hosts (pp. 415–418)

1. Microbes live with their hosts in **symbiotic** relationships, including **mutualism,** in which both members benefit; **parasitism,** in which a **parasite** benefits while the **host** is harmed; and, more rarely, **commensalism,** in which one member benefits while the other is relatively unaffected. Any parasite that causes disease is called a **pathogen**.
2. Organisms called **normal microbiota** live in and on the body. Some of these microbes are resident, whereas others are transient.
3. **Opportunistic pathogens** cause disease when the immune system is suppressed, when normal **microbial antagonism (competition)** is affected by certain changes in the body, or when a member of the normal microbiota is introduced into an area of the body unusual for that microbe.

Reservoirs of Infectious Diseases of Humans (pp. 419–420)

1. Living and nonliving continuous sources of infectious disease are called **reservoirs of infection.** Animal reservoirs harbor agents of **zoonoses,** which are diseases of animals that may be spread to humans via direct contact with the animal or its waste products or via an arthropod. Humans may be asymptomatic **carriers.**
2. **Nonliving reservoirs** of infection include soil, water, and inanimate objects.

The Invasion and Establishment of Microbes in Hosts: Infection (pp. 420–422)

1. Microbial **contamination** refers to the mere presence of microbes in or on the body or object. Microbial contaminants include harmless resident and transient members of the microbiota as well as pathogens, which after a successful invasion cause an **infection.**
2. **Portals of entry** of pathogens into the body include skin, mucous membranes, and the placenta. These portals may be bypassed via the **parenteral route,** by which microbes are directly deposited into deeper tissues.
3. Pathogens attach to cells—a process called **adhesion**—via a variety of structures or attachment proteins called **adhesion factors.** Some bacteria and viruses lose the ability to make adhesion factors called adhesins and thereby become **avirulent.**
4. Some bacteria interact to produce a sticky web of cells and polysaccharides called a **biofilm** that adheres to a surface.

The Nature of Infectious Disease (pp. 423–430)

1. **Disease,** also known as **morbidity,** is a condition sufficiently adverse to interfere with normal functioning of the body.
2. **Symptoms** are subjectively felt by a patient, whereas an outside observer can observe **signs.** A **syndrome** is a group of symptoms and signs that collectively characterizes a particular abnormal condition.
3. **Asymptomatic,** or **subclinical,** infections may go unnoticed because of the absence of symptoms, even though clinical tests might reveal signs of disease.
4. **Etiology** is the study of the cause of a disease.
5. Nineteenth-century microbiologists proposed the **germ theory of disease,** and Robert Koch developed a series of essential conditions called **Koch's postulates** to prove the cause of infectious diseases. Certain circumstances can make the use of these postulates difficult or even impossible.
6. **Pathogenicity** is a microorganism's ability to cause disease; **virulence** is a measure of pathogenicity. **Virulence factors,** such as adhesion factors, extracellular enzymes, toxins, and antiphagocytic factors, affect the relative ability of a pathogen to infect and cause disease.

   ▶**ANIMATIONS:** *Virulence Factors: Hiding from Host Defenses, Inactivating Host Defenses, Penetrating Host Tissues; Phagocytosis: Microbes That Evade It*

   ▶**VIDEO TUTOR:** *Some Virulence Factors*
7. **Toxemia** is the presence in the blood of poisons called **toxins. Exotoxins** are secreted by pathogens into their environment. **Endotoxin,** also known as **lipid A,** is released from the cell wall of dead and dying Gram-negative bacteria and can have fatal effects.

   ▶**ANIMATIONS:** *Virulence Factors: Exotoxins, Endotoxins*
8. **Antitoxins** are antibodies the host forms against toxins.
9. The **disease process**—the stages of infectious diseases—typically consists of the **incubation period, prodromal period, illness, decline,** and **convalescence.**

The Movement of Pathogens Out of Hosts: Portals of Exit (pp. 430–431)

1. **Portals of exit,** such as the nose, mouth, and urethra, allow pathogens to leave the body and are of interest in studying the spread of disease.

Modes of Infectious Disease Transmission (pp. 431–433)

▶**ANIMATIONS:** *Epidemiology: Transmission of Disease*

1. **Direct contact transmission** of infectious diseases involves person-to-person spread by body contact. Transmission of pathogens via inanimate objects (called **fomites**) is called **indirect contact transmission.**
2. **Droplet transmission** (a third type of contact transmission) occurs when pathogens travel less than one meter in droplets of mucus to a new host as a result of speaking, coughing, or sneezing.
3. **Vehicle transmission** involves **airborne, waterborne,** and **food-borne** transmission. **Aerosols** are clouds of water droplets that, travel more than one meter in airborne transmission. **Fecal-oral infection** can result from drinking sewage-contaminated water or from ingesting fecal contaminants. **Bodily fluid transmission** is the spread of pathogens via blood, urine, saliva, or other fluids.
4. **Vectors** transmit pathogens between hosts. **Biological vectors** are animals, usually biting arthropods, that serve as both host and vector of pathogens. **Mechanical vectors** are not hosts to the pathogens they carry.

Classification of Infectious Diseases (pp. 433–434)

1. There are various ways in which infectious disease may be grouped and studied. When grouped by time course and severity, disease may be described as **acute, subacute, chronic,** or **latent.**
2. When an infectious disease comes either directly or indirectly from another host, it is considered a **communicable disease**. If a communicable disease is easily transmitted from a reservoir or patient, it is called a **contagious disease**. **Noncommunicable diseases** arise either from outside of hosts or from normal microbiota.

Epidemiology of Infectious Diseases (pp. 434–442)

1. **Epidemiology** is the study of where and when diseases occur and of how they are transmitted within populations.

   ▶**ANIMATIONS:** *Epidemiology: Overview*
2. Epidemiologists track the **incidence** (number of new cases) and **prevalence** (total number of cases) of a disease and classify disease outbreaks as **endemic** (usually present), **sporadic** (occasional), **epidemic** (more cases than usual), or **pandemic** (epidemic on more than one continent).

   ▶**ANIMATIONS:** *Epidemiology: Occurrence of Disease*
3. **Descriptive epidemiology** is the careful recording of data concerning a disease; it often includes detection of the **index case**—the first case of the disease in a given area or population. **Analytical epidemiology** seeks to determine the probable cause of a disease. **Experimental epidemiology** involves testing a hypothesis resulting from analytical studies.
4. **Healthcare associated infections (HAIs)** (nosocomial infections) and **healthcare associated diseases** (nosocomial diseases) are acquired by patients or workers in health care facilities. They may be **exogenous** (acquired from the health care environment), **endogenous** (derived from normal microbiota that become opportunistic while in the hospital setting), or **iatrogenic** (induced by treatment or medical procedures).

   ▶**ANIMATIONS:** *Nosocomial Infections: Overview*

5. Health care workers can help protect their patients and themselves from exposure to pathogens by hand washing and other aseptic and disinfecting techniques.

▶**ANIMATIONS:** *Nosocomial Infections: Prevention*

6. Public health organizations, such as the World Health Organization (WHO), use epidemiological data to promulgate rules and standards for clean, **potable** water and safe food, to prevent disease by controlling vectors and animal reservoirs, and to educate people to make healthy choices concerning the prevention of disease.

QUESTIONS FOR REVIEW

Answers to the Questions for Review (except Short Answer questions) begin on p. A-1.

Multiple Choice

1. In which type of symbiosis do both members benefit from their interaction?
   a. mutualism
   b. parasitism
   c. commensalism
   d. pathogenesis

2. An axenic environment is one that ________________.
   a. exists in the human mouth
   b. contains only one species
   c. exists in the human colon
   d. both a and c

3. Which of the following is *false* concerning microbial contaminants?
   a. Contaminants may become opportunistic pathogens.
   b. Most microbial contaminants will eventually cause harm.
   c. Contaminants may be a part of the transient microbiota.
   d. Contaminants may be introduced by a mosquito bite.

4. The most frequent portal of entry for pathogens is ________________.
   a. the respiratory tract
   b. the skin
   c. the conjunctiva
   d. a cut or wound

5. The process by which microorganisms attach themselves to cells is _____.
   a. infection
   b. contamination
   c. disease
   d. adhesion

6. Which of the following is the correct sequence of events in infectious diseases?
   a. incubation, prodromal period, illness, decline, convalescence
   b. incubation, decline, prodromal period, illness, convalescence
   c. prodromal period, incubation, illness, decline, convalescence
   d. convalescence, prodromal period, incubation, illness, decline

7. Which of the following are most likely to cause disease?
   a. opportunistic pathogens in a weakened host
   b. pathogens lacking the enzyme kinase
   c. pathogens lacking the enzyme collagenase
   d. highly virulent organisms

8. The nature of bacterial capsules ________________.
   a. causes widespread blood clotting
   b. allows phagocytes to readily engulf these bacteria
   c. affects the virulence of these bacteria
   d. has no effect on the virulence of bacteria

9. When pathogenic bacterial cells lose the ability to make adhesins, they typically ________________.
   a. become avirulent
   b. produce endotoxin
   c. absorb endotoxin
   d. increase in virulence

10. A disease in which a pathogen remains inactive for a long period of time before becoming active is termed a(n) ________________.
    a. subacute disease
    b. acute disease
    c. chronic disease
    d. latent disease

11. Which of the following statements is the best definition of a pandemic disease?
    a. It normally occurs in a given geographic area.
    b. It is a disease that occurs more frequently than usual for a geographical area or group of people.
    c. It occurs infrequently at no predictable time scattered over a large area or population.
    d. It is an epidemic that occurs on more than one continent at the same time.

12. Which of the following types of epidemiologists is most like a detective?
    a. a descriptive epidemiologist
    b. an analytical epidemiologist
    c. an experimental epidemiologist
    d. a reservoir epidemiologist

13. Consider the following case. An animal was infected with a virus. A mosquito bit the animal, was contaminated with the virus, and proceeded to bite and infect a person. Which was the vector?
    a. animal
    b. virus
    c. mosquito
    d. person

14. A patient contracted athlete's foot after long-term use of a medication. His physician explained that the malady was directly related to the medication. Such infections are termed ________________.
    a. healthcare associated infections
    b. exogenous infections
    c. iatrogenic infections
    d. endogenous infections

15. Which of the following phrases describes a contagious disease?
    a. a disease arising from fomites
    b. a disease that is easily passed from host to host in aerosols
    c. a disease that arises from opportunistic, normal microbiota
    d. both a and b

Fill in the Blanks

1. A microbe that causes disease is called a ________________.
2. Infections that may go unnoticed because of the absence of symptoms are called ________________ infections.
3. The study of the cause of a disease is ________________.
4. The study of where and when diseases occur and how they are transmitted within populations is ________________.
5. Diseases that are naturally spread from their usual animal hosts to humans are called ________________.
6. Nonliving reservoirs of disease, such as a toothbrush, drinking glass, and needle, are called ________________.
7. ________________ infections are those acquired by patients or staff while in health care facilities.
8. The total number of cases of a disease in a given area is its ________________.
9. An animal that carries a pathogen and also serves as host for the pathogen is a ________________ vector.
10. Endotoxin, also known as ________________, is part of the outer (wall) membrane of Gram-negative bacteria.

VISUALIZE IT!

1. Each map below shows the locations (dots) of cases of a disease that normally occurs in the Western Hemisphere. Label each map with a correct epidemiological description of the disease's occurrence.

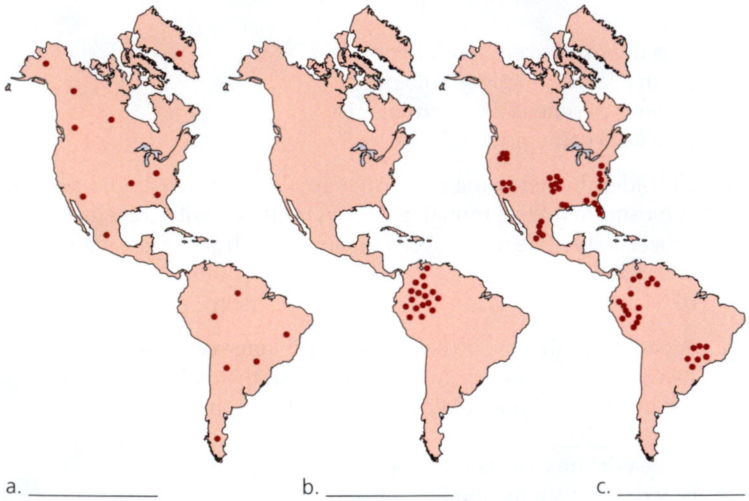

2. Examine the graph of the red epidemic and the blue epidemic. Neither red disease nor blue disease is treatable. A person with either disease is ill for only one day and recovers fully. Both epidemics began at the same time. Which epidemic affected more people during the first three days? What could explain the short time course for the red epidemic? Why was the blue epidemic longer lasting?

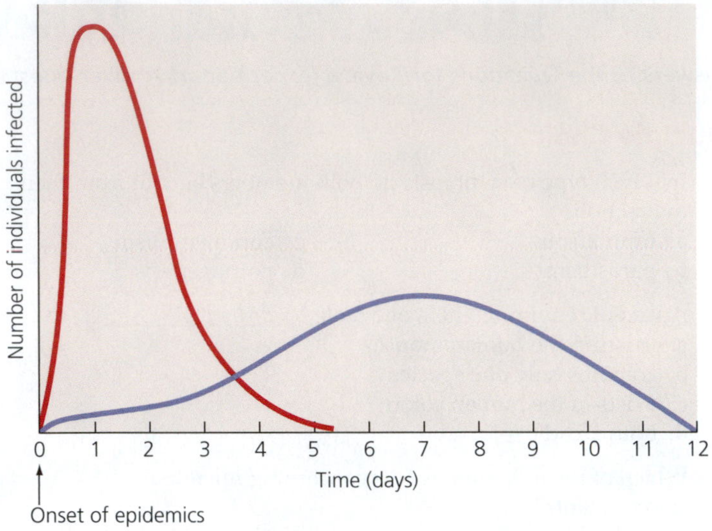

Short Answer

1. List three types of symbiotic relationships, and give an example of each.
2. List three conditions that create opportunities for pathogens to become harmful in a human.
3. List three portals through which pathogens enter the body.
4. List Koch's four postulates, and describe situations in which *not* all may be applicable.
5. List in the correct sequence the five stages of infectious diseases.
6. Describe three modes of disease transmission.
7. Describe the parenteral route of infection.
8. In general, contrast transient microbiota with resident microbiota.
9. Contrast the terms *infection* and *morbidity.*
10. Contrast iatrogenic and healthcare associated diseases.

CRITICAL THINKING

1. Explain why Ellen H., a menopausal woman, may have developed gingivitis from normal microbiota.
2. Will P. died of *E. coli* infection after an intestinal puncture. Explain why this microbe, which normally lives in the colon, could kill this patient.
3. A 27-year-old woman went to her doctor's office with a widespread rash, fever, malaise, and severe muscle pain. The symptoms had begun with a mild headache three days previously. She reported being bitten by a tick one week prior to that.

The doctor correctly diagnosed Rocky Mountain spotted fever (RMSF) and prescribed tetracycline. The rash and other signs and symptoms disappeared in a couple of days, but she continued on her antibiotic therapy for two weeks.

Draw a graph showing the course of disease and the relative numbers of pathogens over time. Label the stages of the disease.

4. Over 30 children younger than three years of age developed gastroenteritis after visiting a local water park. These cases represented 44% of the park visitors in this age group on the day in question. No older individuals were affected. The causative agent was determined to be a member of the bacterial genus *Shigella*. The disease resulted from oral transmission to the children.

   Based only on the information given, can you classify this outbreak as an epidemic? Why or why not? If you were an epidemiologist, how would you go about determining which pools in the water park were contaminated? What factors might account for the fact that no older children or adults developed disease? What steps could the park operators take to reduce the chance of future outbreaks of gastroenteritis?

5. A lichen is an intimate relationship between a fungus and a photosynthetic microbe in which the fungus delivers water and minerals to its partners while the photosynthetic partner delivers sugar to the fungus. What type of symbiosis is involved?

6. Using the data in the **Clinical Case Study: *Legionella* in the Produce Aisle** on p. 439, calculate the incidence of Legionnaires' disease in Bogalusa, Louisiana.

7. Corals are colonial marine animals that feed by filtering small microbes from seawater in tropical oceans worldwide. Biologists have discovered that the cells of most corals are hosts to microscopic algae called zooxanthellae. Design an experiment to ascertain whether corals and zooxanthellae coexist in a mutual, commensal, or parasitic relationship.

8. If a mutation occurred in *Escherichia coli* that deleted the gene for an adhesin, what effect might it have on the ability of *E. coli* to cause urinary tract infections?

CONCEPT MAPPING

Using the following terms, draw a concept map that describes disease transmission. For a sample concept map, see p. 94. Or, complete this and other concept maps online by going to the MasteringMicrobiology Study Area.

Airborne
Arthropods
Biological
Body
Contact transmission
Direct contact
Droplet transmission
Fomites
Foodborne
Indirect contact
Mechanical
Mosquito
Sneezing
Tick
Vector transmission
Vehicle transmission
Waterborne

15 Innate Immunity

MICRO IN THE CLINIC

A Stealth Invader in the Lungs

Tim, a well-known local businessman, is often seen with his ever-present cigar, driving around town, meeting people, and, as he puts it, "investing time with the public." A likeable guy, Tim started smoking in college and has never been able to kick the habit. He jokes about the "smoker's cough" that frequently punctuates his conversations. Recently, Tim has been coughing more than usual—as well as experiencing a scratchy throat, headache, and overall tiredness. Because he doesn't have a fever, he decides not to waste time with a doctor's visit. But the cough starts keeping him up at night, and he increasingly feels as though he can't focus. Frustrated, Tim finally goes to the doctor. Because Tim's cough is dry and he has no fever, the doctor suspects a viral infection. To rule out a bacterial infection, the doctor orders a routine sputum culture (a test of the material coughed up from Tim's lungs). The test doesn't reveal any pathogenic bacteria, although it does note decreased numbers of normal microbiota. Tim goes home, but a week later he's back. Now he's coughing nonstop, and he is short of breath. His chest hurts when he breathes, and he is sweaty and shaking. Tim is exhausted and unable to concentrate or make his "rounds" with the public. At night he lies awake, terrified that he has lung cancer.

On Tim's second visit, his doctor orders a chest X-ray exam and several blood tests.

What is happening to Tim? Will he ever "invest time with the public" again? Turn to the end of the chapter (p. 467) to find out.

MM **Explore More:** Test your readiness and apply your knowledge with dynamic learning tools at MasteringMicrobiology.

A pathogen causes a disease only if it can (1) gain access, either by penetrating the surface of the skin or by entering through some other portal of entry; (2) attach itself to host cells; and (3) evade the body's defense mechanisms long enough to produce harmful changes. In this chapter we will examine the structures, processes, and chemicals that respond in a general way to protect the body from all types of pathogens.

An Overview of the Body's Defenses

LEARNING | **OUTCOMES**

15.1 List and briefly describe the three lines of defense in the human body.

15.2 Explain the phrases *species resistance* and *innate immunity*.

Because the cells and certain basic physiological processes of humans are incompatible with those of most plant and animal pathogens, humans have what is termed **species resistance** to these pathogens. In many cases, the chemical receptors that these pathogens require for attachment to a host cell do not exist in the human body; in other cases, the pH or temperature of the human body is incompatible with the conditions under which these pathogens can survive. Thus, for example, all humans have species resistance both to tobacco mosaic virus and to the virus that causes feline immunodeficiency syndrome in members of the cat family.

Nevertheless, we are confronted every day with pathogens that can cause disease in humans. Bacteria, viruses, fungi, protozoa, and parasitic worms come in contact with your body in the air you breathe, in the water you drink, in the food you eat, and during the contacts you have with other people. Your body must defend itself from these potential pathogens and in some cases from members of your normal microbiota, which may become opportunistic pathogens.

It is convenient to cluster the structures, cells, and chemicals that act against pathogens into three main lines of defense, each of which overlaps and reinforces the other two. The first line of defense is composed chiefly of external physical barriers to pathogens, especially the skin and mucous membranes. The second line of defense is internal and is composed of protective cells, bloodborne chemicals, and processes that inactivate or kill invaders. Together, the first two lines of defense are called **innate immunity** because they are present at birth prior to contact with infectious agents or their products. Innate immunity is rapid and works against a wide variety of pathogens, including parasitic worms, protozoa, fungi, bacteria, and viruses.

By contrast, *lymphocytes*, which are the cells of the third line of defense, respond against unique species and strains of pathogens and alter the body's defenses such that they act more effectively upon subsequent infection with the same specific strain. For this reason, scientists call the third line of defense **adaptive immunity** (see Chapter 16). Now we turn our attention to the two lines of innate immunity. ▶**ANIMATIONS:** *Host Defenses: Overview*

TELL ME WHY

Why aren't the body's skin and mucous membrane barriers significant factors in your resistance to infection by hyperthermophiles?

The Body's First Line of Defense

The body's initial line of defense is made up of structures, chemicals, and processes that work together to prevent pathogens from entering the body in the first place. Here we discuss the main components of the first line of defense: the skin and the mucous membranes of the respiratory, digestive, urinary, and reproductive systems. These structures provide a formidable barrier to the entrance of microorganisms. When these barriers are pierced, broken, or otherwise damaged, they become portals of entry for pathogens. In this section we examine aspects of the first line of defense, including the role of the normal microbiota.

The Role of Skin in Innate Immunity

LEARNING | **OUTCOME**

15.3 Identify the physical and chemical aspects of skin that enable it to prevent the entrance of pathogens.

The skin—the organ of the body with the greatest surface area—is composed of two major layers: an outer **epidermis** and a deeper **dermis**, which contains hair follicles, glands, and nerve endings (see Figure 14.4). Both the physical structure and the chemical components of skin enable it to act as an effective defense.

The epidermis is composed of multiple layers of tightly packed cells. It constitutes a physical barrier to most bacteria, fungi, and viruses. Very few pathogens can penetrate the layers of epidermal cells unless the skin has been burned, broken, or cut.

The deepest cells of the epidermis continually divide, pushing their daughter cells toward the surface. As the daughter cells are pushed toward the surface, they flatten and die and are eventually shed in flakes **(FIGURE 15.1)**. Microorganisms that attach to the skin's surface are sloughed off with the flakes of dead cells. **Beneficial Microbes: What Happens to All That Skin?** describes the fate of lost epidermal cells.

The epidermis also contains phagocytic cells called **dendritic**[1] **cells**. The slender, fingerlike processes of dendritic cells extend among the surrounding cells, forming an almost continuous network to intercept invaders. Dendritic cells both phagocytize pathogens nonspecifically and play a role in adaptive immunity (see Chapter 16).

The combination of the barrier function of the epidermis, its continual replacement, and the presence of phagocytic dendritic cells provides significant nonspecific defense against colonization and infection by pathogens.

The dermis also defends nonspecifically. It contains tough fibers of a protein called collagen. These give the skin strength

[1]From Greek *dendron,* meaning "tree," referring to their branched appearance.

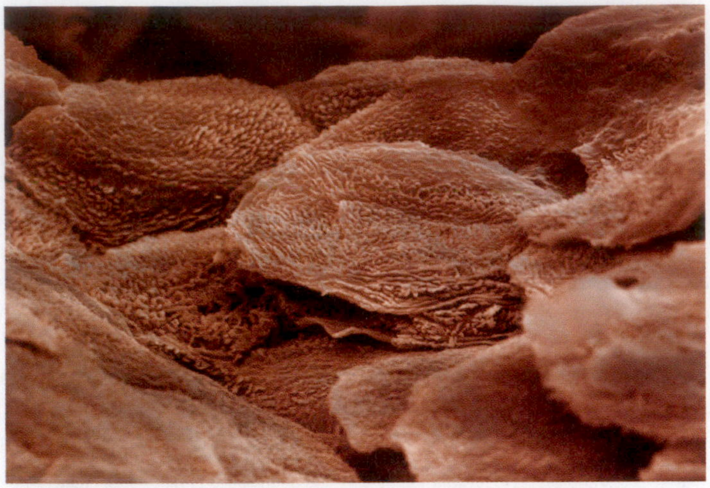

▲ **FIGURE 15.1 A scanning electron micrograph of the surface of human skin.** Epidermal cells are dead and dry and slough off, providing an effective barrier to most microorganisms.

and pliability to prevent jabs and scrapes from penetrating the dermis and introducing microorganisms. Blood vessels in the dermis deliver defensive cells and chemicals, which will be discussed shortly.

In addition to its physical structure, the skin has a number of chemical substances that nonspecifically defend against pathogens. Dermal cells secrete antimicrobial peptides, and sweat glands secrete perspiration, which contains salt, antimicrobial peptides, and lysozyme. Salt draws water osmotically from invading cells, inhibiting their growth and killing them.

Antimicrobial peptides (sometimes called *defensins*) are positively charged chains of 20 to 50 amino acids that act against microorganisms. Sweat glands secrete a class of antimicrobial peptides called *dermcidins.* Dermcidins are broad-spectrum antimicrobials that are active against many Gram-negative and Gram-positive bacteria and fungi. As expected of a peptide active on the surface of the skin, dermcidins are insensitive to low pH and salt. The exact mechanism of dermcidin action is not known.

Lysozyme (lī´sō-zīm) is an enzyme that destroys the cell walls of bacteria by cleaving the bonds between the sugar subunits of the walls. Bacteria without cell walls are more susceptible to osmotic shock and digestion by other enzymes within phagocytes (discussed later in the chapter as part of the second line of defense).

The skin also contains sebaceous (oil) glands, which secrete **sebum** (sē´bŭm), an oily substance that not only helps keep the skin pliable and less sensitive to breaking or tearing but also contains fatty acids that lower the pH of the skin's surface to about pH 5, which is inhibitory to many bacteria.

Although salt, defensins, lysozyme, and acidity make the surface of the skin an inhospitable environment for most microorganisms, some bacteria, such as *Staphylococcus epidermidis* (staf´i-lō-kok´ŭs ep-i-der-mid´is), find the skin a suitable environment for growth and reproduction. Bacteria are particularly abundant in crevices around hairs and in the ducts of glands; usually they are nonpathogenic.

In summary, the skin is a complex barrier that limits access by microbes.

The Role of Mucous Membranes in Innate Immunity

LEARNING | OUTCOMES

15.4 Identify the locations of the body's mucous membranes.

15.5 Explain how mucous membranes protect the body both physically and chemically.

Mucus-secreting (mucous) membranes, a second part of the first line of defense, cover all body cavities that are open to the outside environment. Thus, mucous membranes line the lumens[2] of

[2]A lumen is a cavity or channel within any tubular structure or organ.

BENEFICIAL MICROBES

What Happens to All That Skin?

Your body sheds tens of thousands of skin flakes every time you walk or move, and you shed at only a slightly lower rate when you stand still. That comes to about 10 billion skin cells per day, or 250 grams (about half a pound) of skin every year! What happens to all that skin?

Much of household dust is skin that you and your housemates have shed as you go about your lives. The skin flakes fall to the rug and upholstery, where they become food for microscopic mites that live sedentary and harmless lives waiting patiently for meals to rain down on them from above. They dwell not only in the rug but also in your mattress and pillow and even in the hair follicles of your eyebrows, benefiting you by catching skin cells cascading down your forehead before they can irritate your eyes.

Dust mite

By the way, house dust also contains mite feces and mite skeletons, which can trigger allergies. So after reading this chapter, you just might want to clean your carpet.

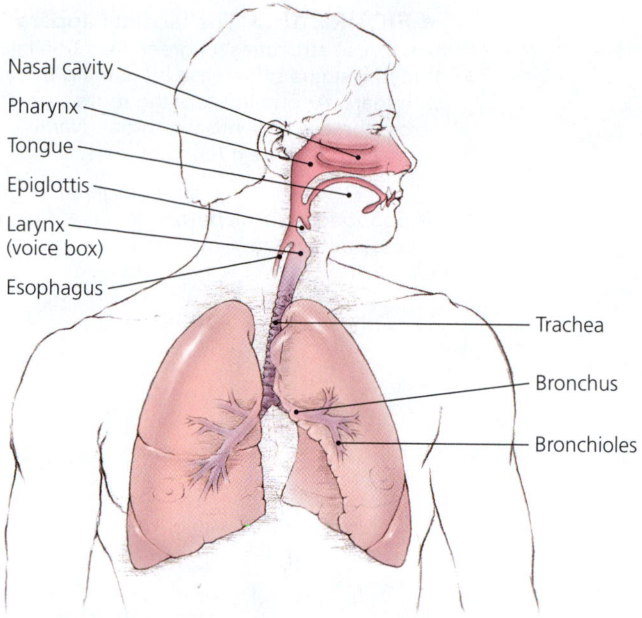

▲ **FIGURE 15.2 The structure of the respiratory system, which is lined with a mucous membrane.** The epithelium of the trachea contains mucus-secreting goblet cells and ciliated cells whose cilia propel the mucus (and the microbes trapped within it) up to the larynx for removal. *What is the function of stem cells within the respiratory epithelium?*

Figure 15.2 *Stem cells in the respiratory epithelium undergo cytokinesis to form both ciliated cells and goblet cells to replace those lost during normal shedding.*

TABLE 15.1 The First Line of Defense: A Comparison of the Skin and Mucous Membranes

	Skin	Mucous Membrane
Number of cell layers	Many	One to a few
Cells tightly packed?	Yes	Yes
Cells dead or alive?	Outer layers: dead; inner layers: alive	Alive
Mucus present?	No	Yes
Relative water content	Dry	Moist
Defensins present?	Yes	With some
Lysozyme present?	Yes	With some
Sebum present?	Yes	No
Cilia present?	No	Trachea, uterine tubes
Constant shedding and replacement of cells?	Yes	Yes

the respiratory, urinary, digestive, and reproductive tracts. Like the skin, mucous membranes act nonspecifically to limit infection both physically and chemically.

Mucous membranes are moist and have two distinct layers: the *epithelium,* in which cells form a covering that is superficial (closest to the surface, in this case the lumen), and a deeper connective tissue layer that provides mechanical and nutritive support for the epithelium. Epithelial cells of mucous membranes are packed closely together, like those of the epidermis, but they form only a thin layer. Indeed, in some mucous membranes, the epithelium is only a single cell thick. Unlike surface epidermal cells, surface cells of mucous membranes are alive and play roles in the diffusion of nutrients and oxygen (in the digestive, respiratory, and female reproductive systems) and in the elimination of wastes (in the urinary, respiratory, and female reproductive systems).

The thin epithelium on the surface of a mucous membrane provides a less efficient barrier to the entrance of pathogens than the multiple layers of dead cells found at the skin's surface. So how are microorganisms kept from invading through these thin mucous membranes? In some cases they are not, which is why some mucous membranes, especially those of the respiratory and reproductive systems, are common portals of entry for pathogens. Nevertheless, the epithelial cells of mucous membranes are tightly packed to prevent the entry of many pathogens, and the cells are continually shed and then replaced by **stem cells**, which are generative cells capable of dividing to form daughter cells of various types. One effect of mucousal shedding is that it carries attached microorganisms away.

Dendritic cells reside below the mucous epithelium to phagocytize invaders. These cells are also able to extend pseudopods between epithelial cells to "sample" the contents of the lumen, helping to prepare adaptive immune responses against particular pathogens that might breach the mucosal barrier (a subject covered more fully in Chapter 16).

In addition, the epithelia of some mucous membranes have still other means of removing pathogens. In the mucous membrane of the trachea, for example, the stem cells produce both *goblet cells,* which secrete an extremely sticky mucus that traps bacteria and other pathogens, and *ciliated columnar cells,* whose cilia propel the mucus and its trapped particles and pathogens up from the lungs **(FIGURE 15.2)**. The effect of the action of the cilia is often likened to that of an escalator. Mucus carried into the throat is coughed up and either swallowed or expelled. Because the poisons and tars in tobacco smoke damage cilia, the lungs of smokers are not properly cleared of mucus, so smokers may develop severe coughs as their respiratory tracts attempt to expel excess mucus from the lungs. Smokers also typically succumb to more respiratory pathogens because they are unable to effectively clear pathogens from their lungs.

In addition to these physical actions, mucous membranes produce chemicals that defend against pathogens. Nasal mucus contains lysozyme, which chemically destroys bacterial cell walls. Mucus also contains antimicrobial peptides (defensins). **TABLE 15.1** compares the physical and chemical actions of the skin and mucous membranes in the body's first line of defense.

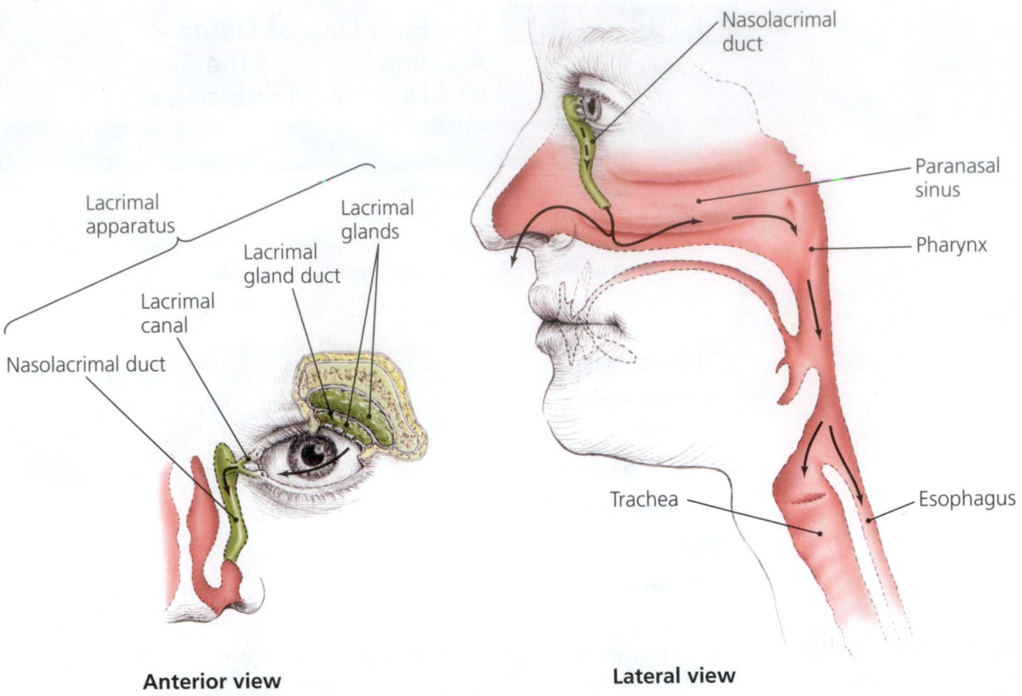

◀ **FIGURE 15.3 The lacrimal apparatus.** These structures (in green) function in the body's first line of defense by bathing the eye with tears. Arrows indicate the route tears take across the eye and into the throat. *Name an antimicrobial protein found in tears.*

Figure 15.3 *Tears contain lysozyme, an antimicrobial protein that acts against the peptidoglycan of bacterial cell walls.*

The Role of the Lacrimal Apparatus in Innate Immunity

LEARNING | **OUTCOME**

15.6 Describe the lacrimal apparatus and the role of tears in combating infection.

The lacrimal apparatus is a group of structures that produce and drain away tears (**FIGURE 15.3**). Lacrimal glands, located above and to the sides of the eyes, secrete tears into lacrimal gland ducts and onto the surface of the eyes. The tears either evaporate or drain into small lacrimal canals, which carry them into nasolacrimal ducts that empty into the nose. There, the tears join the nasal mucus and flow into the pharynx, where they are swallowed. The blinking action of eyelids spreads the tears and washes the surfaces of the eyes. Normally, evaporation and flow into the nose balance the flow of tears onto the eye. However, if the eyes are irritated, increased tear production floods the eyes, carrying the irritant away. In addition to their washing action, tears contain lysozyme, which destroys bacteria.

The Role of Normal Microbiota in Innate Immunity

LEARNING | **OUTCOME**

15.7 Define *normal microbiota*, and explain how they help provide protection against disease.

The skin and mucous membranes of the body are normally home to a variety of protozoa, fungi, bacteria, and viruses. These **normal microbiota** play a role in protecting the body by competing with potential pathogens in a variety of ways, a situation called **microbial antagonism**.

A variety of activities of the normal microbiota make it less likely that a pathogen can compete with them and produce disease. Microbiota consume nutrients, making them unavailable to pathogens. Additionally, normal microbiota can change the pH, creating an environment that is favorable for themselves but unfavorable to other microorganisms.

Further, the presence of microbiota stimulates the body's second line of defense (discussed shortly). Researchers have observed that animals raised in an *axenic*[3] (ā-zēn´ik) environment—that is, one free of all other organisms or viruses—are slower to defend themselves when exposed to a pathogen. Recent studies have shown that members of the normal microbiota in the intestines boost the body's production of antimicrobial substances.

Finally, the resident microbiota of the intestines improve overall health by providing several vitamins, including biotin and pantothenic acid (vitamin B_5), which are important in glucose metabolism; folic acid, which is essential for the production of the purine and pyrimidine bases of nucleic acids; and the precursor of vitamin K, which has an important role in blood clotting.

Other First-Line Defenses

LEARNING | **OUTCOME**

15.8 Describe antimicrobial peptides as part of the body's defenses.

Besides the physical barrier of the skin and mucous membranes, there are other hindrances to microbial invasion. Among these are additional antimicrobial peptides and other processes and chemicals.

[3]From Greek *a*, meaning "no," and *xenos*, meaning "foreigner."

TABLE 15.2 Secretions and Activities That Contribute to the First Line of Defense

Secretion/Activity	Function
Digestive System	
Saliva	Washes microbes from teeth, gums, tongue, and palate; contains lysozyme, an antibacterial enzyme
Stomach acid	Digests and/or inhibits microorganisms
Gastroferritin	Sequesters iron being absorbed, making it unavailable for microbial use
Bile	Inhibitory to most microorganisms
Intestinal secretions	Digests and/or inhibits microorganisms
Peristalsis	Moves gastrointestinal (GI) contents through GI tract, constantly eliminating potential pathogens
Defecation	Eliminates microorganisms
Vomiting	Eliminates microorganisms
Urinary System	
Urine	Contains lysozyme; urine's acidity inhibits microorganisms; may wash microbes from ureters and urethra during urination
Reproductive System	
Vaginal secretions	Acidity inhibits microorganisms; contains iron-binding proteins that sequester iron, making it unavailable for microbial use
Menstrual flow	Cleanses uterus and vagina
Prostate secretion	Contains iron-binding proteins that sequester iron, making it unavailable for microbial use
Cardiovascular System	
Blood flow	Removes microorganisms from wounds
Coagulation	Prevents entrance of many pathogens
Transferrin	Binds iron for transport, making it unavailable for microbial use

Antimicrobial Peptides

As we saw in our examination of skin and mucous membranes, antimicrobial peptides (sometimes called defensins) act against microorganisms. Scientists have discovered hundreds of these antimicrobial peptides in organisms as diverse as silkworms, frogs, and humans. Besides being secreted onto the surface of the skin, antimicrobial peptides are found in mucous membranes and in neutrophils. These peptides act against a variety of potential pathogens, being triggered by sugar and protein molecules on the external surfaces of microbes. Some antimicrobial peptides act only against Gram-positive bacteria or Gram-negative bacteria, others act against both, and still others act against protozoa, enveloped viruses, or fungi.

Researchers have elucidated several ways in which antimicrobial peptides work. Some punch holes in the cytoplasmic membranes of the pathogens, and others interrupt internal signaling or enzymatic action. Some antimicrobial peptides are chemotactic factors that recruit leukocytes to the site, while others assemble to form fibers and microscopic nets that ensnare invading bacteria.

Other Processes and Chemicals

Many other body organs contribute to the first line of defense by secreting chemicals with antimicrobial properties that are secondary to their prime function. For example, stomach acid is present primarily to aid digestion of proteins, but it also prevents the growth of many potential pathogens. Likewise, saliva contains lysozyme as well as a digestive enzyme; further, saliva physically washes microbes from the teeth. The contributions of these and other processes and chemicals to the first line of defense are listed in **TABLE 15.2**.

TELL ME WHY

Some strains of *Staphylococcus aureus* produce exfoliative toxin, a chemical that causes portions of the entire outer layer of the skin to be sloughed off in a disease called scalded skin syndrome. Given that cells of the outer layer are going to fall off anyway, why is this disease dangerous?

The Body's Second Line of Defense

LEARNING **OUTCOME**

15.9 Compare and contrast the body's first and second lines of defense against disease.

When pathogens succeed in penetrating the skin or mucous membranes, the body's second line of innate defense comes into play. Like the first line of defense, the second line operates against a wide variety of pathogens, from parasitic worms to viruses. But unlike the first line of defense, the second line includes no barriers; instead, it is composed of cells (especially phagocytes), antimicrobial chemicals (peptides, complement, interferons), and processes (inflammation, fever). Some cells and chemicals from the first line of defense play additional roles in the second line of defense. We will consider each component of the second line of defense in some detail shortly, but because many of them either are contained in or originate in the blood, we first consider the components of blood.

Defense Components of Blood

LEARNING **OUTCOMES**

15.10 Discuss the components of blood and their functions in the body's defense.

15.11 Explain how macrophages are named.

Blood is a complex liquid tissue composed of cells and portions of cells within a fluid called *plasma*. We begin our discussion of the defense functions of blood by briefly considering plasma.

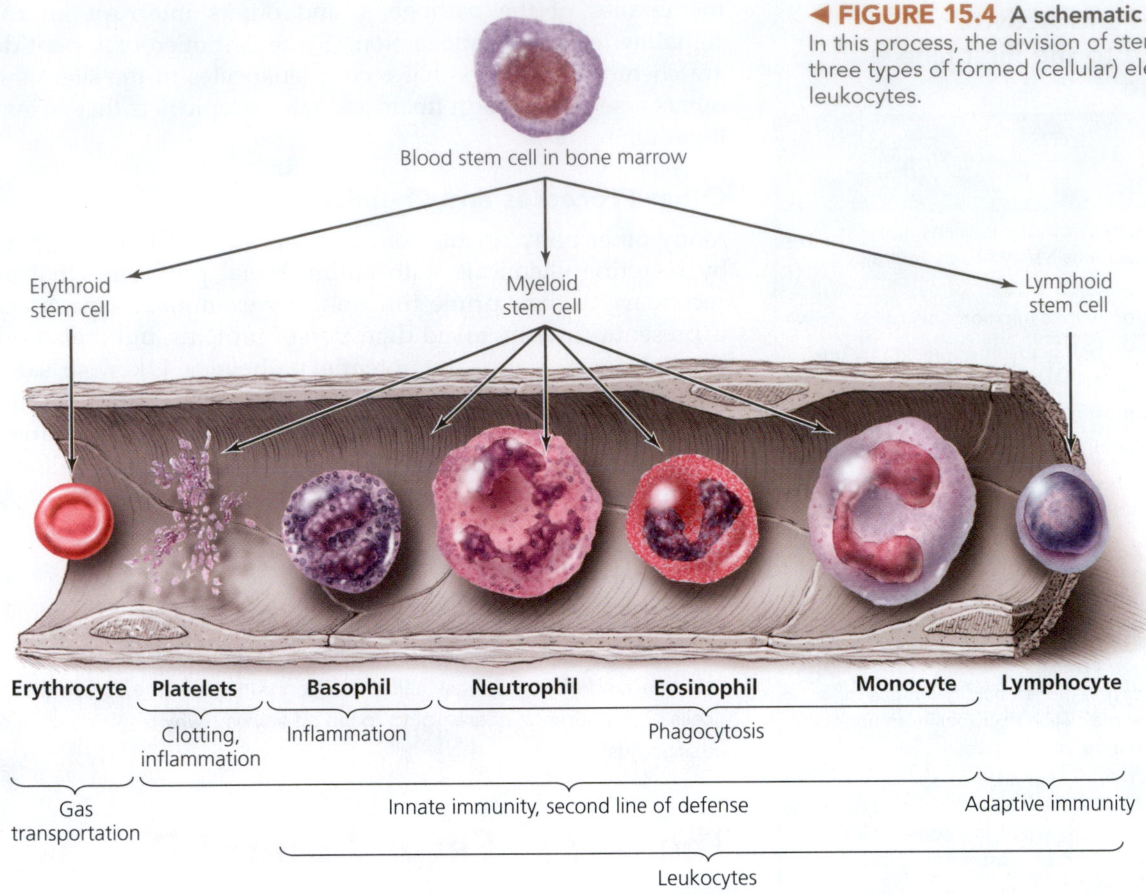

◀ FIGURE 15.4 A schematic representation of hematopoiesis. In this process, the division of stem cells in the bone marrow produces three types of formed (cellular) elements: erythrocytes, platelets, and leukocytes.

Plasma

Plasma is mostly water containing electrolytes (ions), dissolved gases, nutrients, and—most relevant to the body's defenses—a variety of proteins. Some plasma proteins are involved in inflammation (discussed later) and in blood clotting, a defense mechanism that reduces both blood loss and the risk of infection. When clotting factors have been removed from the plasma, as when blood clots, the remaining liquid is called *serum.*

Humans require iron for metabolism: it is a component of cytochromes of electron transport chains, functions as an enzyme cofactor, and is an essential part of hemoglobin—the oxygen-carrying protein of erythrocytes. Because iron is relatively insoluble, in humans it is transported in plasma to cells by a transport protein called *transferrin.* When transferrin-iron complexes reach cells with receptors for transferrin, the binding of the protein to the receptor stimulates the cell to take up the iron via endocytosis. Excess iron is stored in the liver bound to another protein called *ferritin.* Though the main function of iron-binding proteins is transporting and storing iron, they play a secondary, defensive role—sequestering iron so that it is unavailable to microorganisms.

Some bacteria, such as *Staphylococcus aureus* (o´rē-ŭs), respond to a shortage of iron by secreting their own iron-binding proteins called *siderophores.* Because siderophores have a greater affinity for iron than does transferrin, bacteria that produce siderophores can in effect steal iron from the body. In response, the body produces *lactoferrin,* which retakes the iron from the bacteria by its even greater affinity. Thus, the body and the pathogens engage in a kind of chemical "tug-of-war" for the possession of iron.

Some pathogens bypass this contest altogether. For example, *S. aureus* and related pathogens can secrete the protein *hemolysin,* which punches holes in the cytoplasmic membranes of red blood cells, releasing hemoglobin. Other bacterial proteins then bind hemoglobin to the bacterial membrane and strip it of its iron. *Neisseria meningitidis* (nī-se´rē-ă me-nin-ji´ti-dis), a pathogen that causes often fatal meningitis, produces receptors for transferrin and plucks iron from the bloodstream as it flows by.

Another group of plasma proteins, called *complement proteins,* is an important part of the second line of defense and is discussed shortly. Still other plasma proteins, called *antibodies* or *immunoglobulins,* are a part of adaptive immunity, the body's third line of defense.

Defensive Blood Cells: Leukocytes

Cells and cell fragments suspended in the plasma are called **formed elements**. In a process called *hematopoiesis,*[4] blood stem cells located principally in the bone marrow within the hollow cavities of the large bones produce three types of formed elements: **erythrocytes**[5] (ĕ-rith´rō-sītz), **platelets**[6] (plāt´letz), and **leukocytes**[7] (loo´kō-sīts) **(FIGURE 15.4)**. Erythrocytes, the most

[4]From Greek *haima,* meaning "blood," and *poiein,* meaning "to make."
[5]From Greek *erythro,* meaning "red," and *cytos,* meaning "cell."
[6]French for "small plates." Platelets are also called thrombocytes, from Greek *thrombos,* meaning "lump," and *cytos,* meaning "cell," though they are technically not cells but instead pieces of cells.
[7]From Greek *leuko,* meaning "white," and *cytos,* meaning "cell."

numerous of the formed elements, carry oxygen and carbon dioxide in the blood. Platelets, which are pieces of large cells called *megakaryocytes* that have split into small portions of cytoplasm surrounded by cytoplasmic membranes, are involved in blood clotting. Leukocytes, the formed elements that are directly involved in defending the body against invaders, are commonly called white blood cells because they form a whitish layer when the components of blood are separated within a test tube.

Based on their appearance in stained blood smears when viewed under the microscope, leukocytes are divided into two groups: **granulocytes** (gran´ū-lō-sītz) and *agranulocytes* (ā-gran´ū-lō-sītz) **(FIGURE 15.5)**.

Granulocytes have large granules in their cytoplasm that stain different colors depending on the type of granulocyte and the dyes used: **basophils** (bā´sō-fils) stain blue with the basic dye methylene blue, **eosinophils** (ē-ō-sin´ō-fils) stain red to orange with the acidic dye eosin, and **neutrophils** (noo´trō-fils), also known as *polymorphonuclear leukocytes (PMNs)*, stain lilac with a mixture of acidic and basic dyes. Both neutrophils and eosinophils phagocytize pathogens, and both can exit the blood to attack invading microbes in the tissues by squeezing between the cells lining capillaries (the smallest blood vessels). This process is called **diapedesis**[8] (dī´ă-pĕ-dē´sis). As we will see later in the chapter, eosinophils are also involved in defending the body against parasitic worms and are present in large number during many allergic reactions, though their exact function in allergies is disputed. Basophils can also leave the blood, though they are not phagocytic; instead, they release inflammatory chemicals, an aspect of the second line of defense that will be discussed shortly.

The cytoplasm of agranulocytes appears uniform when viewed via light microscopy, though granules do become visible with an electron microscope. Agranulocytes are of two types: **lymphocytes** (lim´fō-sītz), which are the smallest leukocytes and have nuclei that nearly fill the cells, and **monocytes** (mon´ō-sītz), which are large agranulocytes with slightly lobed nuclei. Although most lymphocytes are involved in adaptive immunity, *natural killer (NK) lymphocytes* function in innate defense and thus are discussed later in this chapter. Monocytes leave the blood and mature into **macrophages** (mak´rō-fāj-ĕz), which are phagocytic cells of the second line of defense. Their initial function is to devour foreign objects, including bacteria, fungi, spores, and dust as well as dead body cells.

Macrophages are named for their location in the body. *Wandering macrophages* leave the blood via diapedesis and perform their scavenger function while traveling throughout the body, including extracellular spaces. Other macrophages are fixed and do not wander. These include *alveolar* (al-vē´ō-lăr) *macrophages*[9] of the lungs and *microglia* (mī-krog´lē-ă) of the central nervous system. Fixed macrophages generally phagocytize within specific organs, such as the heart chambers, blood vessels, and lymphatic vessels. (The lymphatic system is discussed in Chapter 16.)

[8]From Greek *dia*, meaning "through," and *pedan*, meaning "to leap."

[9]Alveoli are small pockets at the end of respiratory passages where oxygen and carbon dioxide exchange occurs between the lungs and the blood.

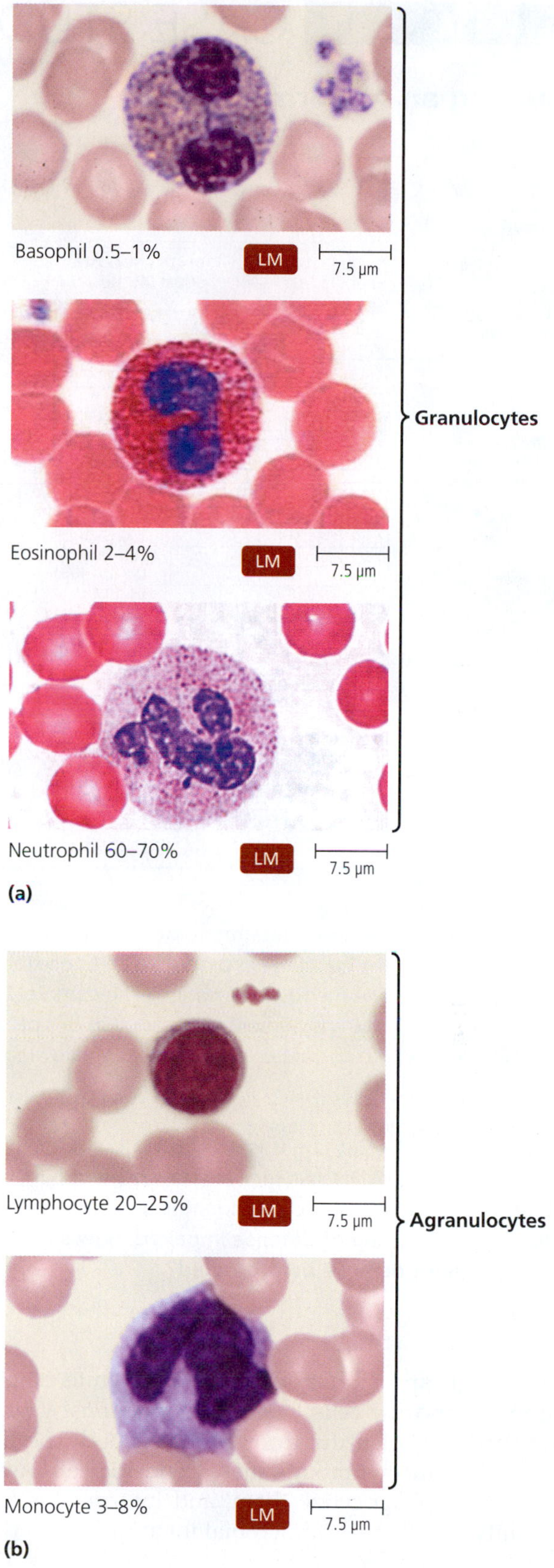

▲ **FIGURE 15.5 Leukocytes as seen in stained blood smears.** **(a)** Granulocytes: basophil, eosinophil, and neutrophil. **(b)** Agranulocytes: lymphocyte and monocyte. The numbers are the normal percentages of each cell type among all leukocytes.

CLINICAL CASE STUDY

Evaluating an Abnormal CBC

CBC Profile

Name: Brown, Roger — Age/Sex: 61/M — Attend Dr: Kevin, Larry
Acct#: 04797747 — Status: ADM IN
Reg: 11/27/13

SPEC #: 0303:AS:H00102T — COLL: 12 / 03 / 13-0620 — STATUS: COMP
RECD: 12 / 03 / 13-0647 — SUBM DR: Kevin, Larry
ENTERED: 12 / 03 / 13-0002 — OTHER DR: NONE, PER PT
ORDERED: CBC W/ MAN DIFF — REQ #: 01797367

Test	Result	Normal range
CBC		
WBC (white blood cells)	0.8	4.8–10.8 K/mm3
RBC (red blood cells)	3.09	4.20–5.40 M/mm3
HGB (hemoglobin)	9.6	12.0–16.0 g/dL
HCT (hematocrit)	28.2	37.0–47.0 %
MCV	91.3	81.0–99.0 fL
MCH	31.1	27.0–31.0 pg
MCHC	34.1	32.0–36.0 g/dl
RDW	17.1	11.5–14.5 %
PLT (platelets)	21	150–450 K/mm3
MPV	8.7	7.4–10.4 fL
DIFF		
CELLS COUNTED	100	#CELLS
SEGS	39	
BAND	4	
LYMPH (lymphocytes)	41	
MONO (monocytes)	15	
EOS (eosinophils)	1	
NEUT# (# neutrophils)	0.3	1.9–8.0 K/mm3
LYMPH#	0.3	0.9–5.2 K/mm3
MONO#	0.1	0.1–1.2 K/mm3
EOS#	0.0	0–0.8 K/mm3
PLATELET EST	DECREASED	

Roger Brown, an African American cancer patient, received a chemotherapeutic agent as a treatment for his disease. The drug used to destroy the cancer also produced an undesirable condition known as bone marrow depression. The complete blood count (CBC) profile shown here indicates that this patient is in trouble. Review the lab values and answer the following questions.

1. Note that the platelet count is very low. How does this affect the patient? Discuss measures to protect him.
2. Note that the white blood cell count is abnormally low. With the second line of defense impaired, how should the first line of defense be protected?

A special group of phagocytes are not white blood cells. These are the dendritic cells, mentioned previously, which are multibranched cells plentiful throughout the body, particularly in the skin and mucous membranes. Dendritic cells await microbial invaders, phagocytize them, and inform cells of adaptive immunity that there is a microbial invasion.

Lab Analysis of Leukocytes Analysis of blood for diagnostic purposes, including white blood cell counts, is one task of medical lab technologists. The proportions of leukocytes, as determined in a **differential white blood cell count**, can serve as a sign of disease. For example, an increase in the percentage of eosinophils can indicate allergies or infection with parasitic worms; bacterial diseases typically result in an increase in the number of leukocytes and an increase in the percentage of neutrophils, whereas viral infections are associated with an increase in the relative number of lymphocytes. The ranges for the normal values for each kind of white blood cell, expressed as a percentage of the total leukocyte population, are shown in Figure 15.5.

Now that we have some background concerning the defensive properties of plasma components and leukocytes, we turn our attention to the details of the body's second line of defense: phagocytosis, nonphagocytic killing by leukocytes, nonspecific chemical defenses, inflammation, and fever.

Phagocytosis

LEARNING **OUTCOME**

15.12 Name and describe the six stages of phagocytosis.

Phagocytosis, which means "eating by a cell," is a way that some microbes obtain nutrients (see Chapter 3), but **phagocytes** (fag´ō-sītz)—phagocytic defense cells of the body—use phagocytosis to rid the body of pathogens that have evaded the body's first line of defense. **▶ANIMATIONS:** ***Phagocytosis: Overview***

Phagocytosis is a complex process that is still not completely understood. For the purposes of our discussion, we will divide the continuous process of phagocytosis into six steps: chemotaxis, adherence, ingestion, maturation, killing, and elimination **(FIGURE 15.6)**.

Chemotaxis

Chemotaxis is movement of a cell either toward a chemical stimulus (positive chemotaxis) or away from a chemical stimulus (negative chemotaxis). In the case of phagocytes, positive chemotaxis involves the use of *pseudopods* (soo´-dō-podz) to crawl toward microorganisms at the site of an infection 1. Chemicals that attract phagocytic leukocytes include microbial components and secretions, components of damaged tissues and white blood cells, and **chemotactic factors** (kem-ō-tak´tik). Chemotactic factors include defensins, peptides derived from complement (discussed later in this chapter), and chemicals called **chemokines** (kē´mō-kīnz), which are released by leukocytes already at a site of infection.

Adherence

After arriving at the site of an infection, phagocytes attach to microorganisms through the binding of complementary chemicals, such as glycoproteins, found on the membranes of cells 2. This process is called **adherence**.

Some bacteria have virulence factors, such as slippery capsules that hinder adherence of phagocytes. Such bacteria are more readily phagocytized if they are pushed up against a surface, such as connective tissue, the wall of a blood vessel, or a blood clot.

All pathogens are more readily phagocytized if they are first covered with antimicrobial proteins, such as complement

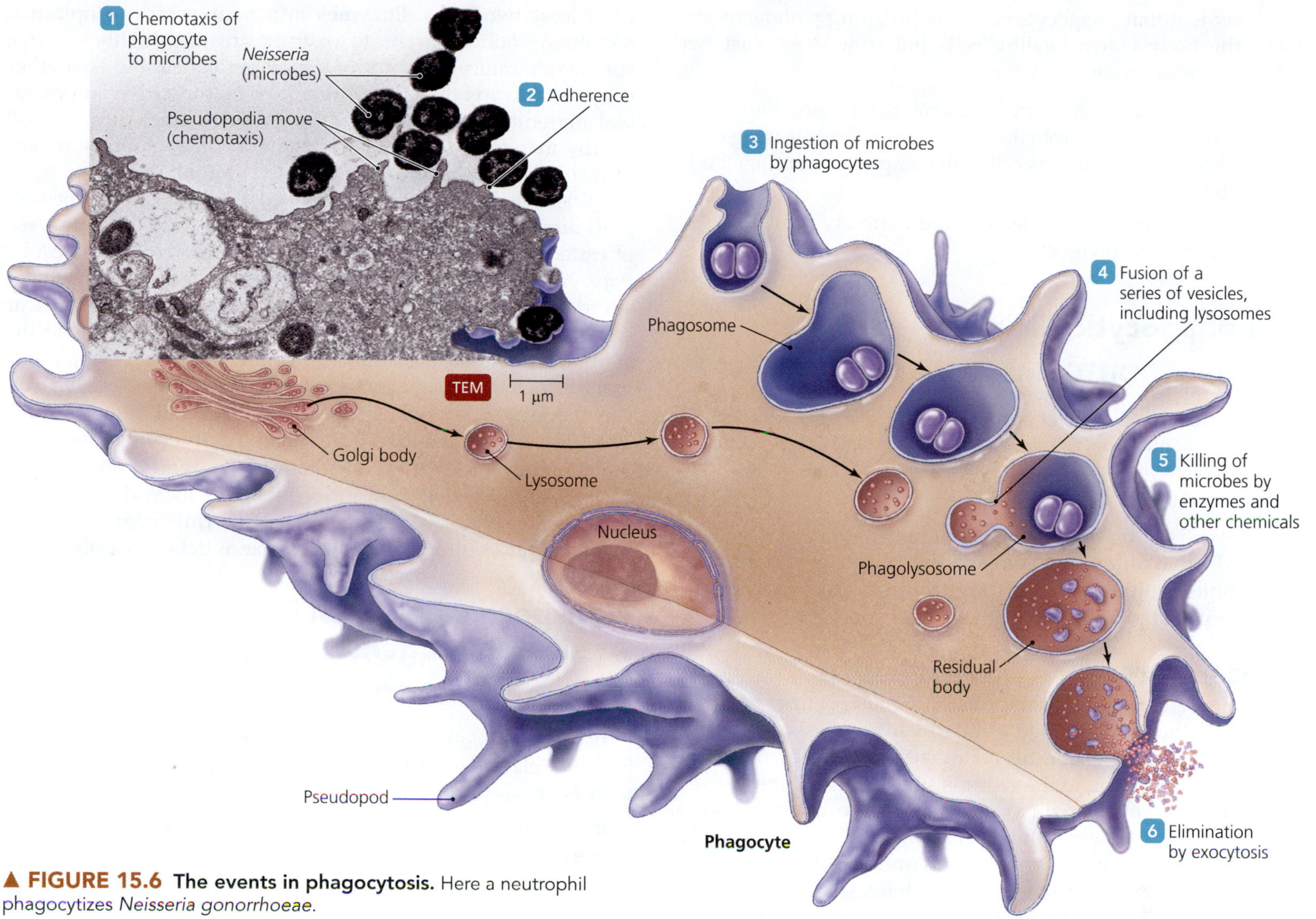

▲ **FIGURE 15.6 The events in phagocytosis.** Here a neutrophil phagocytizes *Neisseria gonorrhoeae*.

proteins (discussed later) or the specific antimicrobial proteins called *antibodies* (discussed in Chapter 16). This coating process is called **opsonization**[10] (op′sŭ-nī-zā′shun), and the proteins are called **opsonins**. Generally, opsonins increase the number and kinds of binding sites on a microbe's surface.

Ingestion

After phagocytes adhere to pathogens, they extend *pseudopods* to surround the microbe 3. The encompassed microbe is internalized as the pseudopods fuse to form a food vesicle called a **phagosome**.

Phagosome Maturation and Microbial Killing

A series of membranous organelles within the phagocyte fuse with newly formed phagosomes to form digestive vesicles. One organelle, the lysosome, adds digestive chemicals to the maturing phagosome, which is now called a **phagolysosome** (fag-ŏ-lī′sō-sōm) 4. Phagolysosomes contain antimicrobial substances, such as highly reactive, toxic forms of oxygen, in an environment with a pH of about 5.5 due to the active pumping of H^+ from the cytosol. These factors, along with 30 or so different enzymes, such as lipases, proteases, nucleases, and a variety of others, destroy the engulfed microbes 5.

Most pathogens are dead within 30 minutes, though some bacteria contain virulence factors (such as waxy cell walls) that resist a lysosome's action. In the end, a phagolysosome is known as a *residual body*.

Elimination

Digestion is not always complete, and phagocytes eliminate remnants of microorganisms via *exocytosis,* a process that is essentially the reverse of ingestion 6. Some microbial components are specially processed and remain attached to the cytoplasmic membrane of some phagocytes, particularly dendritic cells, a phenomenon that plays a role in the adaptive immune response (discussed in Chapter 16). ▶**ANIMATIONS:** ***Phagocytosis: Mechanism***

[10]From Greek *opsonein,* meaning "to supply food," and *izein,* meaning "to cause"; thus, loosely, "to prepare for dinner."

How is it that phagocytes destroy invading pathogens and leave the body's own healthy cells unharmed? At least two mechanisms are responsible for this:

- Some phagocytes have cytoplasmic membrane receptors for various microbial surface components lacking on the body's cells, such as cell wall components or flagellar proteins.
- Opsonins such as complement and antibody provide a signal to the phagocyte.

Nonphagocytic Killing

LEARNING | OUTCOME

15.13 Describe the role of eosinophils, NK cells, and neutrophils in nonphagocytic killing of microorganisms and parasitic helminths.

Phagocytosis involves killing a pathogen once it has been ingested—that is, once it is inside the phagocyte. In contrast, eosinophils, natural killer cells, and neutrophils can accomplish killing without phagocytosis.

Killing by Eosinophils

As discussed earlier, eosinophils can phagocytize; however, this is not their usual mode of attack. Instead, eosinophils secrete antimicrobial chemicals. They attack parasitic helminths (worms) by attaching to the worm's surface, where they secrete extracellular protein toxins onto the surface of the parasite. These weaken the helminth and may even kill it. **Eosinophilia** (ē-ō-sin´ō-fil-e-ă), an abnormally high number of eosinophils in the blood, is often indicative of helminth infestation or allergies.

Besides their attacks against parasitic helminths, eosinophils have recently been discovered to use a never-before-seen tactic against bacteria: Lipopolysaccharide from Gram-negative bacterial cell walls triggers eosinophils to rapidly eject mitochondrial DNA, which combines with previously extruded eosinophil proteins to form a physical barrier. This extracellular structure binds to and then kills the bacteria. This is the first evidence that DNA can have antimicrobial activity, and scientists are investigating exactly how mitochondrial DNA acts as an antibacterial agent.

Killing by Natural Killer Lymphocytes

Natural killer lymphocytes (or **NK cells**) are another type of defensive leukocyte of innate immunity that works by secreting toxins onto the surfaces of virally infected cells and neoplasms (tumors). NK cells identify and spare normal body cells because the latter express membrane proteins similar to those on the NK cells. (The ability to distinguish one's own healthy cells from diseased cells and pathogens is discussed more fully in Chapter 16.)

Killing by Neutrophils

Neutrophils do not always devour pathogens; they can destroy nearby microbial cells without phagocytosis. They can do this in at least two ways. Enzymes in a neutrophil's cytoplasmic membrane add electrons to oxygen, creating highly reactive superoxide radical O_2^- and hydrogen peroxide (H_2O_2). Another enzyme converts these into hypochlorite, the active antimicrobial ingredient in household bleach. These chemicals can kill nearby invaders. Yet another enzyme in the membrane makes nitric oxide, which is a powerful inducer of inflammation.

Scientists have recently discovered another way that neutrophils disable microorganisms in their vicinity. They generate webs of extracellular fibers nicknamed *NETs,* for *neutrophil extracellular traps.* Neutrophils synthesize NETs via a unique form of cellular suicide involving the disintegration of their nuclei. As the nuclear envelope breaks down, DNA and histones are released into the cytosol, and the mixing of nuclear components with cytoplasmic granule membranes and proteins forms NET fibers. Reactive oxygen species—superoxide and peroxide—then kill the neutrophil. The NETs are released from the dying cell as its cytoplasmic membrane ruptures. NETs trap both Gram-positive and Gram-negative bacteria, immobilizing them and sequestering them along with antimicrobial peptides, which kill the bacteria. Thus, even in their dying moments, neutrophils fulfill their role as defensive cells.

Nonspecific Chemical Defenses Against Pathogens

LEARNING | OUTCOMES

15.14 Define *Toll-like receptors,* and describe their action in relation to pathogen-associated molecular patterns.
15.15 Describe the location and functions of NOD proteins.
15.16 Explain the roles of interferons in innate immunity.
15.17 Describe the complement system, including its three activation pathways.

Chemical defenses augment phagocytosis in the second line of defense. The chemicals assist phagocytic cells either by enhancing other features of innate immunity or by directly attacking pathogens. Defensive chemicals include lysozyme and defensins (examined previously) as well as Toll-like receptors, NOD proteins, interferons, and complement.

Toll-Like Receptors (TLRs)

Toll-like receptors (TLRs)[11] are integral proteins of the cytoplasmic membranes of phagocytic cells. TLRs act as an early warning system, triggering your body's responses to a number of molecules that are shared by various bacterial or viral pathogens and are absent in humans. These microbial molecules include peptidoglycan, lipopolysaccharide, flagellin, unmethylated pairs of cytosine and guanine nucleotides from bacteria and viruses, double-stranded RNA, and single-stranded viral RNA. Such microbial components are collectively referred to as **pathogen-associated molecular patterns (PAMPs)**.

[11] *Toll* is a German word meaning "fantastic," originally referring to a gene of fruit flies, mutations of which cause the flies to look bizarre. Toll-like proteins are similar to fruit fly Toll proteins in their amino acid sequence though not in their function.

TABLE 15.3 Toll-Like Receptors and Their Natural Microbial Binding Partners

TLR	PAMP (Microbial Molecule)
In Cytoplasmic Membrane	
TLR1	Bacterial lipopeptides and certain proteins in multicellular parasites
TLR2	Bacterial lipopeptides, lipoteichoic acid (found in Gram-positive cell wall), and cell wall of yeast
TLR4	Lipid A (found in outer membrane of Gram-negative bacteria)
TLR5	Flagellin (bacterial flagella)
TLR6	Bacterial lipopeptides, lipoteichoic acid (found in Gram-positive cell wall), and cell wall of yeast
In Phagosome Membrane	
TLR3	Double-stranded RNA (found only in viruses)
TLR7	Single-stranded viral RNA
TLR8	Single-stranded viral RNA
TLR9	Unmethylated cytosine-guanine pairs of viral and bacterial DNA
Unknown Location	
TLR10	Unknown

Ten TLRs are known for humans. TLRs 1, 2, 4, 5, and 6 are found spanning cytoplasmic membranes, while TLRs 3, 7, 8, and 9 span phagosome membranes. Some TLRs act alone; others act in pairs to recognize a particular PAMP. For example, TLR3 binds to double-stranded RNA from viruses such as West Nile virus, and TLR2 and TLR6 in conjunction bind to lipoteichoic acid—a component of Gram-positive cell walls. **TABLE 15.3** summarizes the PAMPs and the membrane locations of the 10 known TLRs of humans.

Binding of a PAMP to a Toll-like receptor initiates a number of defensive responses, including apoptosis (cell suicide) of an infected cell, secretion of inflammatory mediators or interferons (both discussed shortly), or production of chemical stimulants of adaptive immune responses (discussed in Chapter 16). If TLRs fail, much of immune response collapses, leaving the body open to attack by myriad pathogens.

Scientists are actively seeking ways to stimulate TLRs so as to enhance the body's immune response to pathogens and immunizations. In contrast, methods to inhibit TLRs may provide us with ways to counter inflammatory disorders and some hyperimmune responses.

NOD Proteins

NOD[12] proteins are another set of receptors for microbial molecules, such as PAMPs, but NOD proteins are located inside a cell rather than as part of a cell's cytoplasmic membrane. Scientists have studied NOD proteins that bind to components of Gram-negative bacteria's cell walls and RNA of viruses, such as those that cause AIDS, hepatitis C, and mononucleosis. NOD proteins trigger inflammation, apoptosis, and other innate immune responses against bacterial pathogens, though researchers are still elucidating their exact method of action. Mutations in NOD genes are associated with several inflammatory bowel diseases, including Crohn's disease.

[12]Nucleotide-oligomerization domains, referring to their ability to bind a region of a finite number (oligomer) of DNA nucleotides.

Interferons

So far in this chapter we have focused primarily on how the body defends itself against bacteria and eukaryotes. Now we consider how chemicals in the second line of defense act against viral pathogens.

Viruses use a host's metabolic machinery to produce new viruses. For this reason, it is often difficult to interfere with virus replication without also producing deleterious effects on the host. **Interferons** (in-ter-fēr´onz) are protein molecules released by host cells to nonspecifically inhibit the spread of viral infections. Their lack of specificity means that interferons produced against one viral invader protect somewhat against infection by other types of viruses as well. However, interferons also cause malaise, muscle aches, chills, headache, and fever, which are typically associated with viral infections.

Different cell types produce one of two basic types of interferon when stimulated by viral nucleic acid binding to certain Toll-like receptors (TLR3, TLR7, or TLR8). Interferons within any given type share certain physical and chemical features, though they are specific to the species that produces them. In general, type I interferons—also known as alpha and beta interferons—are present early in viral infections, whereas type II (gamma) interferon appears somewhat later in the course of infection. Because their actions are identical, we examine alpha and beta interferons together before discussing gamma interferon.

Type I (Alpha and Beta) Interferons Within hours after infection, virally infected monocytes, macrophages, and some lymphocytes secrete small amounts of **alpha interferon (IFN-α)**; similarly, fibroblasts, which are undifferentiated cells in such connective tissues as cartilage, tendon, and bone, secrete small amounts of **beta interferon (IFN-β)** when infected by viruses. The structures of alpha and beta interferons are similar, and their actions are identical.

Interferons do not protect the cells that secrete them—these cells are already infected with viruses. Instead, interferons activate natural killer lymphocytes and trigger protective steps in neighboring uninfected cells. Alpha and beta interferons bind to interferon receptors on the cytoplasmic membranes of neighboring cells. Such binding triggers the production of **antiviral proteins (AVPs)**, which remain inactive within these cells until AVPs bind to viral nucleic acids, particularly double-stranded RNA, a molecule that is common among viruses but generally absent in eukaryotic cells **(FIGURE 15.7)**.

At least two types of antiviral proteins are produced: *oligoadenylate synthetase*, the action of which results in the destruction

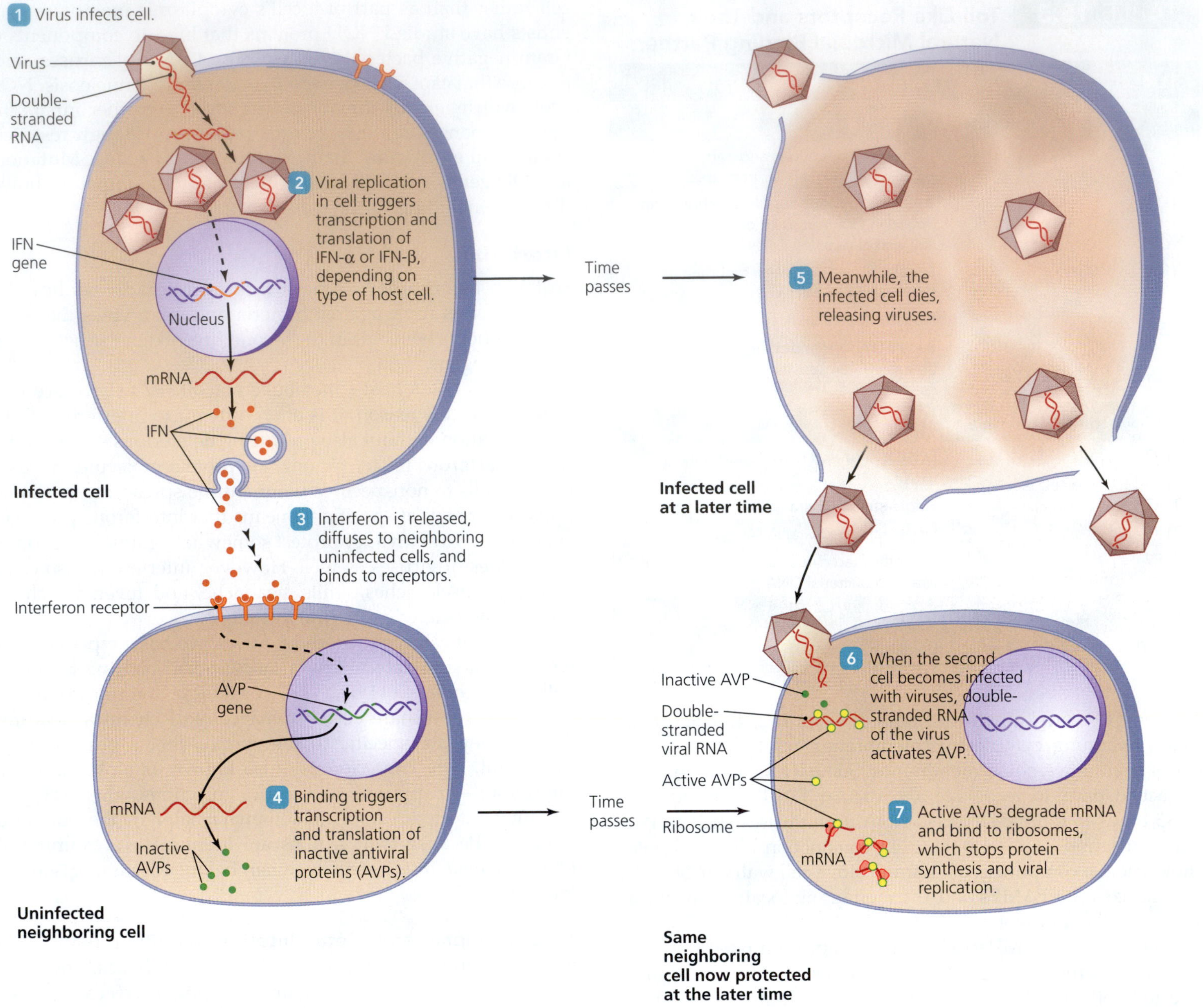
1 Virus infects cell.
Virus
Double-stranded RNA
2 Viral replication in cell triggers transcription and translation of IFN-α or IFN-β, depending on type of host cell.
IFN gene
Nucleus
mRNA
IFN
Infected cell
3 Interferon is released, diffuses to neighboring uninfected cells, and binds to receptors.
Interferon receptor
AVP gene
mRNA
4 Binding triggers transcription and translation of inactive antiviral proteins (AVPs).
Inactive AVPs
Uninfected neighboring cell
Time passes
5 Meanwhile, the infected cell dies, releasing viruses.
Infected cell at a later time
Inactive AVP
Double-stranded viral RNA
Active AVPs
Ribosome
mRNA
6 When the second cell becomes infected with viruses, double-stranded RNA of the virus activates AVP.
7 Active AVPs degrade mRNA and bind to ribosomes, which stops protein synthesis and viral replication.
Same neighboring cell now protected at the later time

▲ **FIGURE 15.7 The actions of alpha and beta interferons.**

of mRNA, and *protein kinase,* which inhibits protein synthesis by ribosomes. Between them, these AVP enzymes essentially destroy the protein production system of the cell, preventing viruses from being replicated. Of course, cellular metabolism is also affected negatively. The antiviral state lasts three to four days, which may be long enough for a cell to rid itself of viruses but still a short enough period for the cell to survive without protein production.

Type II (Gamma) Interferon Gamma interferon (IFN-γ) is produced by activated T lymphocytes and by natural killer (NK) lymphocytes. Because T lymphocytes are usually activated as part of an adaptive immune response (see Chapter 16) days after an infection has occurred, gamma interferon appears later than either alpha or beta interferon. Its action in stimulating the activity of macrophages gives IFN-γ its other name: *macrophage activation factor.* Gamma interferon plays a small role in protecting the body against viral infections; mostly, IFN-γ regulates the immune system, as in its activation of phagocytic activity.

TABLE 15.4 summarizes various properties of interferons in humans.

As scientists learned more about the effects of interferons, many thought these proteins might be a universal weapon against viral infections, but viruses can interfere with the effects of interferon. Many variations of interferons in all three classes have been produced in laboratories using recombinant DNA technology in the hopes that antiviral therapy can be improved.

TABLE 15.4 The Characteristics of Human Interferons

	Type I		Type II
Property	**Alpha Interferon (IFN-α)**	**Beta Interferon (IFN-β)**	**Gamma Interferon (IFN-γ)**
Principal source	Epithelium, leukocytes	Fibroblasts	Activated T lymphocytes and NK lymphocytes
Inducing agent	Viruses	Viruses	Adaptive immune responses
Action	Stimulates production of antiviral proteins	Stimulates production of antiviral proteins	Stimulates phagocytic activity of macrophages and neutrophils
Other names	Leukocyte-IFN	Fibroblast-IFN	Immune-IFN, macrophage activation factor

Complement

The **complement system**—or **complement** for short—is a set of serum proteins designated numerically according to the order of their discovery. These proteins initially act as opsonins and chemotactic factors and indirectly trigger inflammation and fever. The end result of full complement activation is lysis of foreign cells. ▶**ANIMATIONS:** ***Complement: Overview***

Complement is activated in three ways:

- In the *classical pathway,* antibodies activate complement.
- In the *alternative pathway,* pathogens or pathogenic products (such as bacterial endotoxins and glycoproteins) activate complement.
- In the *lectin pathway,* microbial polysaccharides bind to activating molecules.

As **FIGURE 15.8** shows, the three pathways merge. Complement proteins react with one another in an amplifying sequence of chemical reactions in which the product of each reaction becomes an enzyme that catalyzes the next reaction many times over. Such reactions are called *cascades* because they progress in a way that can be likened to a rock avalanche in which one rock dislodges several other rocks, each of which dislodges many others until a whole cascade of rocks is tumbling down the mountain. The products of each step in the complement cascade initiate other reactions, often with wide-ranging effects in the body.

The Classical Pathway Complement got its name from events in the originally discovered "classical" pathway. In this pathway the various proteins act to "complement," or act in conjunction with, the action of antibodies, which we now understand are part of adaptive immunity. As you study the depiction of the classical complement cascade in **FIGURE 15.9**, keep the following concepts in mind:

- Complement enzymes in early events cleave other complement molecules to form *fragments,* which are designated with lowercase letters. For example, inactive complement protein 3 (C3) is cleaved into active fragments C3a and C3b.
- Most fragments have specific and important roles in achieving the functions of the complement system. Some combine to form new enzymes; some act to increase

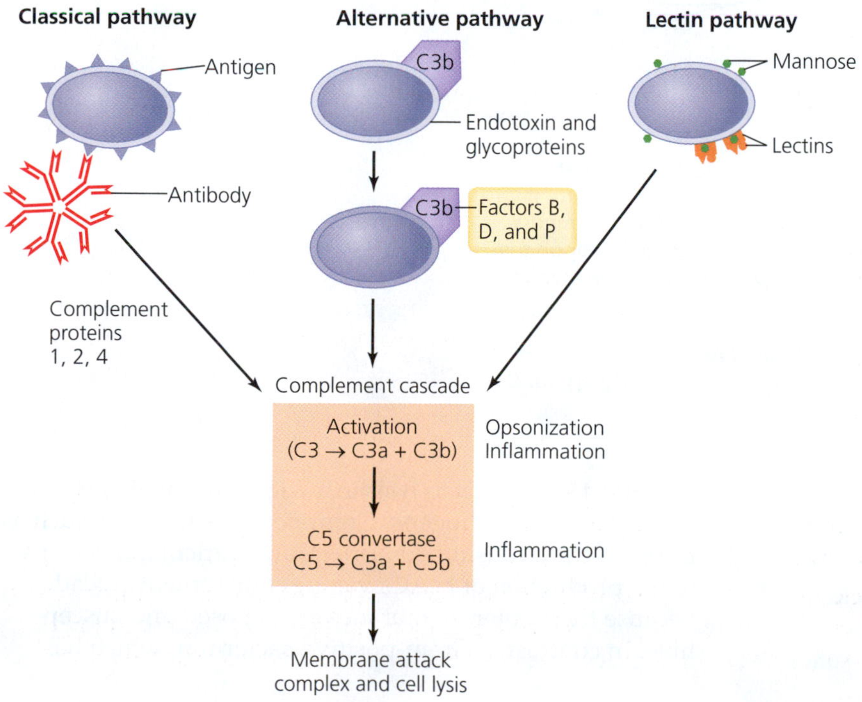

◀ **FIGURE 15.8 Pathways by which complement is activated.** In the classical pathway, the binding of antibodies to antigens activates complement. In the alternative pathway, the binding of factors B, D, and P to endotoxin or glycoproteins in the cell walls of bacteria or fungi activates complement. The lectin pathway activates when lectins bind to microbial carbohydrates. *How did complement get that name?*

Figure 15.8 Complement proteins add to—or complement—the action of antibodies.

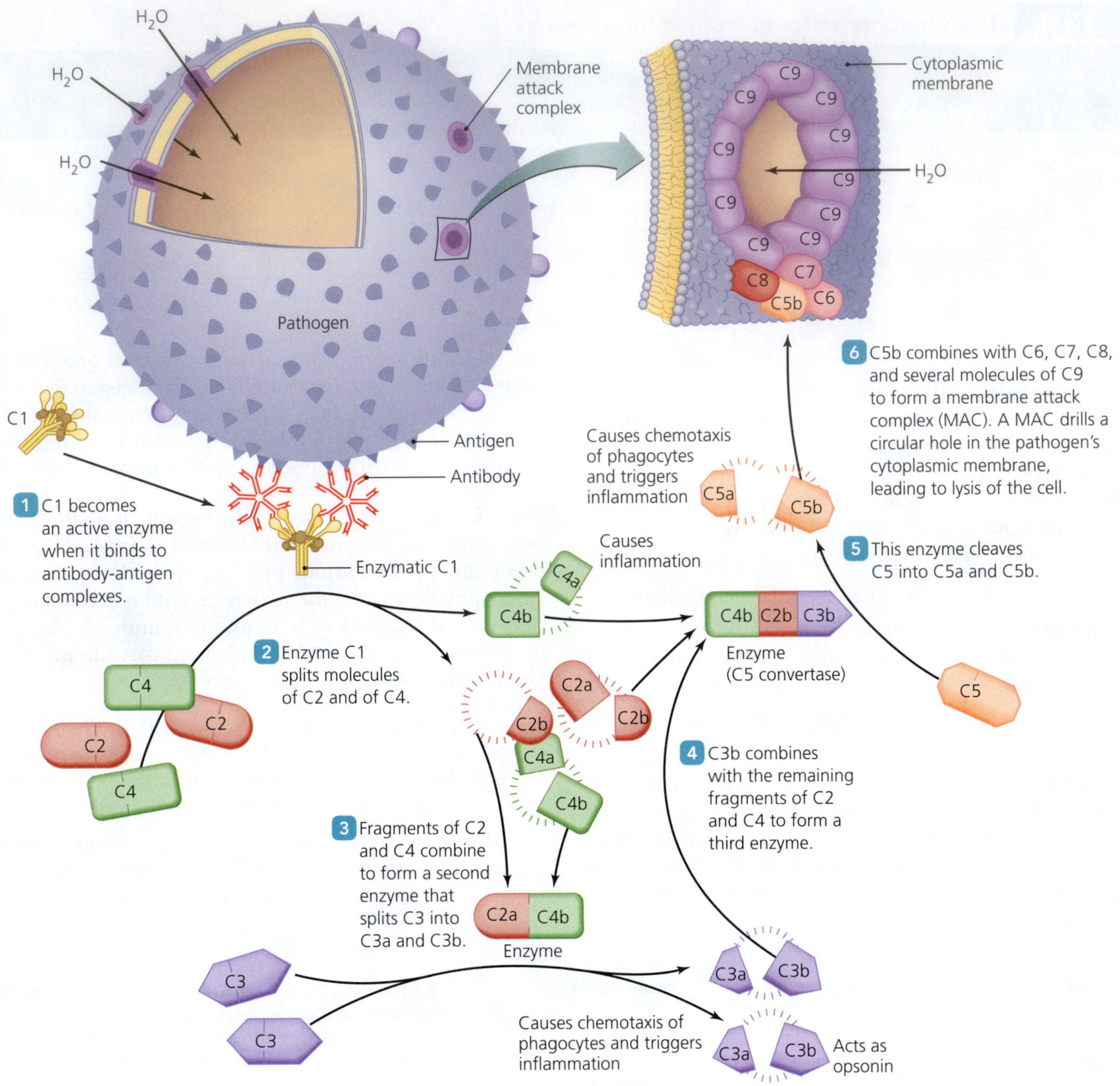

▲ **FIGURE 15.9 The classical pathway and complement cascade.** Steps 2 to 6 are illustrated apart from the pathogen's membrane for clarity; in reality, the enzymes are associated with the site of C1 and antibodies on the membrane. The major functions of complement are opsonization and mediation of chemotaxis and inflammation. A membrane attack complex is a potent antimicrobial weapon that can form against a wide variety of bacterial and eukaryotic pathogens. *What proteins would be involved in activating a complement cascade if this were the alternative pathway?*

Figure 15.9 *Whereas the classical pathway of complement activation involves proteins C1, C2, and C4, the alternative pathway involves factors B, D, and P (properdin).*

vascular permeability, which increases diapedesis; others enhance inflammation; and still others are involved as chemotactic factors, attracting phagocytes, or in opsonization.

- One end product of a full cascade is a **membrane attack complex (MAC)**, which forms a circular hole in a pathogen's membrane. The production of numerous MACs **(FIGURE 15.10)** leads to lysis in a wide variety of bacterial and eukaryotic pathogens. Gram-negative bacteria, such as the bacterium causing gonorrhea, are particularly sensitive to the production of MACs via the complement cascade because their outer membranes are exposed and susceptible. In contrast, a Gram-positive bacterium, which has

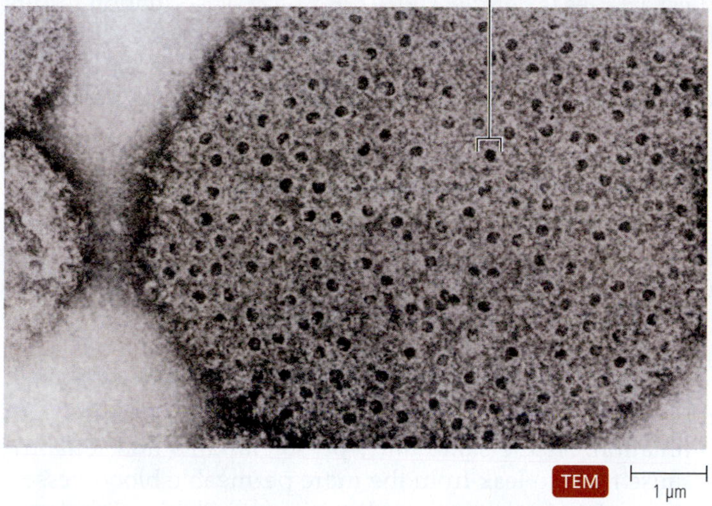

▲ **FIGURE 15.10 Membrane attack complexes.** Transmission electron micrograph of a cell damaged by numerous punctures produced by membrane attack complexes.

a thick layer of peptidoglycan overlying its cytoplasmic membrane, is typically resistant to the MAC-induced lytic properties of complement, though it is susceptible to the other effects of the complement cascade.

In addition to its enzymatic role, fragment C3b acts as an opsonin. Fragment C4b also acts as an opsonin. Fragments C3a and C5a function as chemotactic factors, attracting phagocytes to the site of infection. Along with C4a, they are also inflammatory agents that trigger increased vascular permeability and dilation. The inflammatory roles of these fragments are discussed in more detail shortly.

The Alternative Pathway The alternative pathway was so named because scientists discovered it second. As previously mentioned, antibodies bound to antigens are necessary for the classical activation of complement, whereas activation of the alternative pathway occurs independently of antibodies. The alternative pathway begins with the cleavage of C3 into C3a and C3b. This naturally occurs at a slow rate in the plasma but proceeds no further because C3b is cleaved into smaller fragments almost immediately. However, when C3b binds to microbial surfaces, it stabilizes long enough for a protein called factor B to adhere. Another plasma protein, factor D, then cleaves factor B, creating an enzyme composed of C3b and Bb. This enzyme, which is stabilized by a third protein—factor P (properdin)—cleaves more molecules of C3 into C3a and C3b, continuing the complement cascade and the formation of MACs.

The alternative pathway is useful in the early stages of an infection, before the adaptive immune response has created the antibodies needed to activate the classical pathway.

The Lectin Pathway Researchers have discovered a third pathway for complement activation that acts through the use of *lectins*. Lectins are chemicals that bind to specific sugar subunits of polysaccharide molecules, in this case, to mannose sugar in mannan polysaccharide on the surfaces of fungi, bacteria, or viruses. Mannose is rare in mammals. Lectins bound to mannose act to trigger a complement cascade by cleaving C2 and C4. The cascade then proceeds like the classical pathway (see steps 3 to 6 in Figure 15.9).
▶**ANIMATIONS:** *Complement: Activation, Results*

Inactivation of Complement We have seen that the complement system is nonspecific and that MACs can form on any cell's exposed membrane. How do the body's own cells withstand the action of complement? Membrane-bound proteins on the body's cells bind with and break down activated complement proteins, thereby interrupting the complement cascade before damage can occur.

Inflammation

LEARNING **OUTCOME**

15.18 Discuss the process and benefits of inflammation.

Inflammation is a general, nonspecific response to tissue damage resulting from a variety of causes, including heat, chemicals, ultraviolet light (sunburn), abrasions, cuts, and pathogens. **Acute inflammation** develops quickly, is short lived, is typically beneficial, and results in the elimination or resolution of whatever condition precipitated it. Long-lasting **chronic inflammation** causes damage (even death) to tissues, resulting in disease. Both acute and chronic inflammation exhibit similar signs and symptoms, including redness in light-colored skin (rubor), localized heat (calor), edema (swelling), and pain (dolor).
▶**ANIMATIONS:** *Inflammation: Overview*

It may not be obvious from this list of signs and symptoms that acute inflammation is beneficial; however, acute inflammation is an important part of the second line of defense because it results in (1) dilation and increased permeability of blood vessels, (2) migration of phagocytes, and (3) tissue repair. Although the chemical details of inflammation are beyond the scope of our study, we now consider these three aspects of acute inflammation. ▶**VIDEO TUTOR:** *Inflammation*

Dilation and Increased Permeability of Blood Vessels

Part of the body's initial response to an injury or invasion of pathogens is localized dilation (increase in diameter) of blood vessels in the affected region. The process of blood clotting triggers the conversion of a soluble plasma protein into a nine-amino-acid peptide chain called **bradykinin** (brad-e-kī′nin), which is a potent mediator of inflammation. Patrolling macrophages, using Toll-like receptors and NOD proteins to identify invaders, release other inflammatory chemicals, including **prostaglandins** (pros-tă-glan′dinz) and **leukotrienes** (loo-kō-trī′ēnz). Basophils, platelets, and specialized cells located in connective tissue—called **mast cells**—also release inflammatory mediators, such as **histamine** (his′tă-mēn), when they are exposed to complement fragments C3a or C5a (**FIGURE 15.11**). Recall that these complement peptides were cleaved from larger polypeptides during the complement cascade.

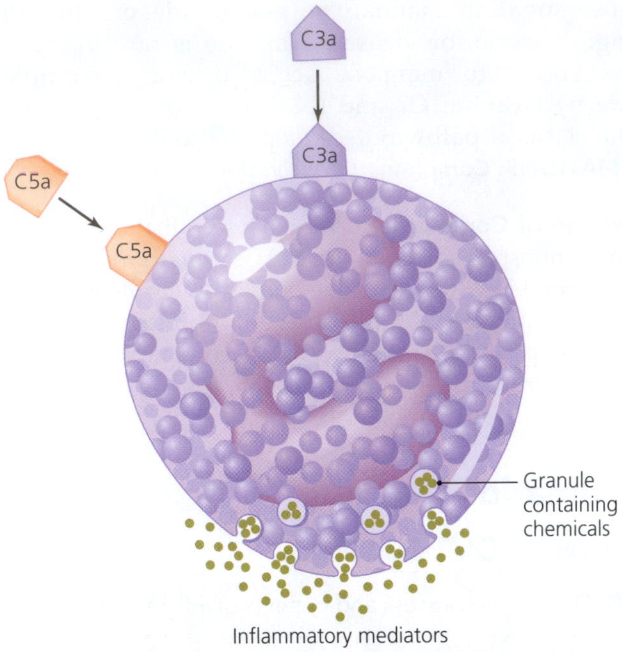

▲ **FIGURE 15.11 The stimulation of inflammation by complement.** Complement fragments C3a and C5a can each bind to platelets, basophils, and mast cells, causing them to release histamine, which in turn stimulates the dilation of arterioles.

Bradykinin and histamine cause vasodilation of the body's smallest arteries (arterioles) (**FIGURE 15.12**). Vasodilation results in delivery of more blood to the site of infection, which in turn delivers more phagocytes, oxygen, and nutrients to the site. Inflammatory mediators cause cells that line blood vessels to make adhesion molecules, which are receptors for leukocytes. Bradykinin, prostaglandins, leukotrienes, and histamine also make small veins more permeable—that is, they cause cells lining the vessels to contract and pull apart, leaving gaps in the walls through which phagocytes can move into the damaged tissue and fight invaders (**FIGURE 15.13**). Increased permeability also allows delivery of more bloodborne antimicrobial chemicals to the site.

Dilation of blood vessels in response to inflammatory mediators results in the redness and localized heat associated with inflammation. At the same time, prostaglandins and leukotrienes cause fluid to leak from the more permeable blood vessels and accumulate in the surrounding tissue, resulting in edema, which is responsible for much of the pain of inflammation as pressure is exerted on nerve endings.

Vasodilation and increased permeability also deliver fibrinogen, the blood's clotting protein. Clots forming at the site of injury or infection wall off the area and help prevent pathogens and their toxins from spreading. One result is the formation of

▶ **FIGURE 15.12 The dilating effect of inflammatory mediators on small blood vessels.** The release of mediators from damaged tissue causes nearby arterioles to dilate. Vasodilation causes capillaries to expand and enables more blood to be delivered to the affected site. The increased blood flow causes the reddening and heat associated with inflammation.

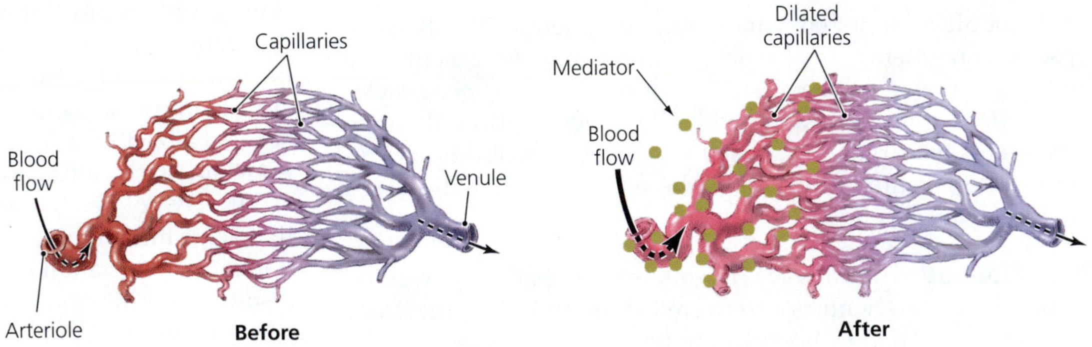

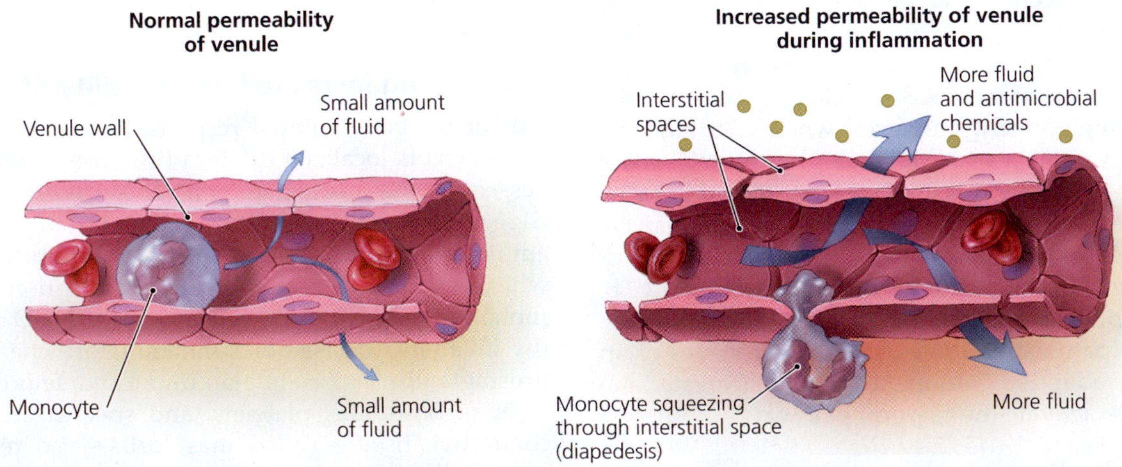

▲ **FIGURE 15.13 Increased vascular permeability during inflammation.** The presence of bradykinin, prostaglandins, leukotrienes, or histamine causes cells lining venules to pull apart, allowing phagocytes to leave the bloodstream and more easily reach a site of infection. The leakage of fluid and cells causes the edema and pain associated with inflammation.

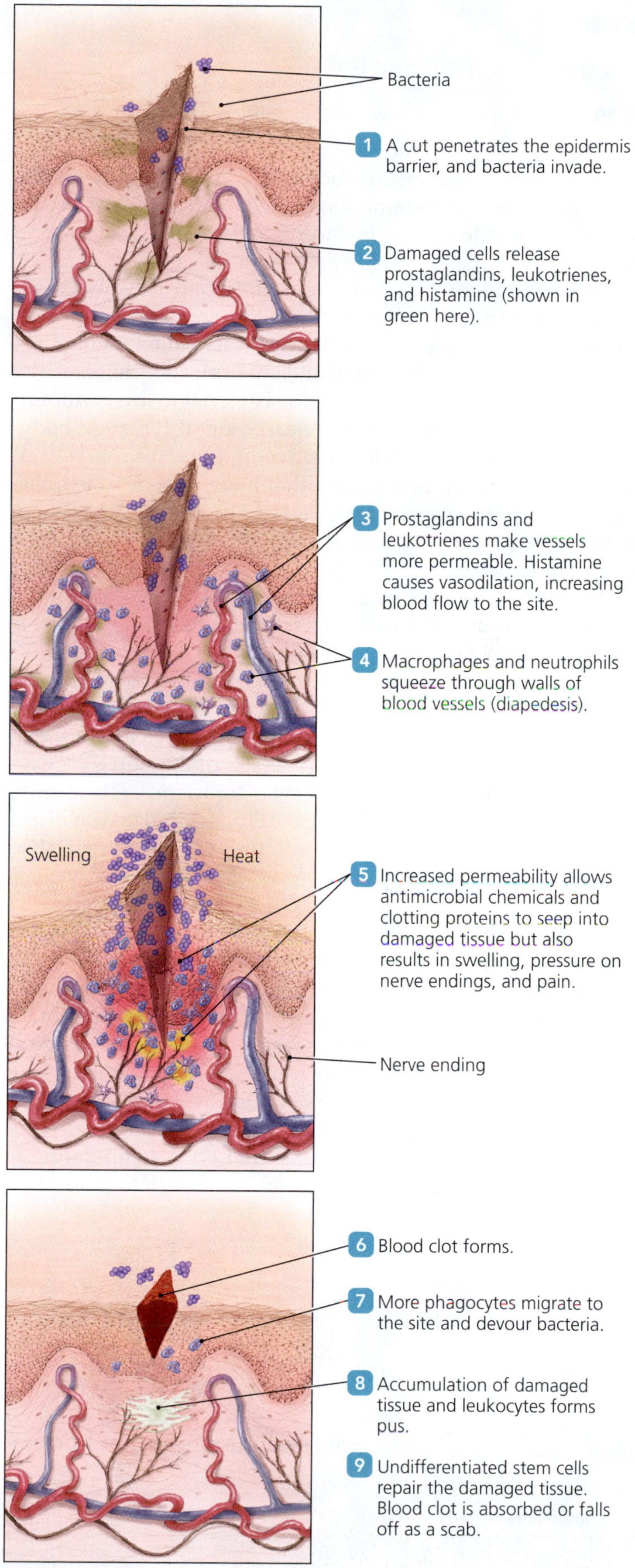
Bacteria
1 A cut penetrates the epidermis barrier, and bacteria invade.
2 Damaged cells release prostaglandins, leukotrienes, and histamine (shown in green here).
3 Prostaglandins and leukotrienes make vessels more permeable. Histamine causes vasodilation, increasing blood flow to the site.
4 Macrophages and neutrophils squeeze through walls of blood vessels (diapedesis).
Swelling
Heat
5 Increased permeability allows antimicrobial chemicals and clotting proteins to seep into damaged tissue but also results in swelling, pressure on nerve endings, and pain.
Nerve ending
6 Blood clot forms.
7 More phagocytes migrate to the site and devour bacteria.
8 Accumulation of damaged tissue and leukocytes forms pus.
9 Undifferentiated stem cells repair the damaged tissue. Blood clot is absorbed or falls off as a scab.

▶ **FIGURE 15.14 An overview of the events in inflammation following a cut and infection.** The process, which is characterized by redness, swelling, heat, and pain, ends with tissue repair. *In general, what types of cells are involved in tissue repair?*

Figure 15.14 *Tissue repair is effected by cells that are capable of cytokinesis and differentiation. If fibroblasts are among them, scar tissue is laid down.*

pus, a fluid containing dead tissue cells, leukocytes, and pathogens in the walled-off area. Pus may push up toward the surface and erupt, or it may remain isolated in the body, where it is slowly absorbed over a period of days. Such an isolated site of infection is called an **abscess**. Pimples, boils, and pustules are examples of abscesses.

The signs and symptoms of inflammation can be treated with antihistamines, which block histamine receptors on blood vessel walls, or with antiprostaglandins. One of the ways aspirin and ibuprofen reduce pain is by acting as antiprostaglandins.

Migration of Phagocytes

Increased blood flow due to vasodilation delivers monocytes and neutrophils to a site of infection. As they arrive, these leukocytes roll along the inside walls of blood vessels until they adhere to the receptors lining the vessels in a process called **margination**. They then squeeze between the cells of the vessel's wall (diapedesis) and enter the site of infection, usually within an hour of tissue damage. The phagocytes then destroy pathogens via phagocytosis.

As mentioned previously, phagocytes are attracted to the site of infection by chemotactic factors, including C3a, C5a, leukotrienes, and microbial components and toxins. The first phagocytes to arrive are often neutrophils, which are then followed by monocytes. Once monocytes leave the blood, they change and become wandering macrophages, which are especially active phagocytic cells that devour pathogens, damaged tissue cells, and dead neutrophils. Wandering macrophages are a major component of pus.

Tissue Repair

The final stage of inflammation is tissue repair, which in part involves the delivery of extra nutrients and oxygen to the site. Areas of the body where cells regularly undergo cytokinesis, such as the skin and mucous membranes, are repaired rapidly. Some other sites are not fully reparable and form scar tissue.

If the damaged tissue contains undifferentiated stem cells, tissues can be fully restored. For example, a minor skin cut is repaired to such an extent it is no longer visible. However, if cells called *fibroblasts* are involved to a significant extent, scar tissue is formed, inhibiting normal function. Some tissues, such as cardiac muscle and parts of the brain, do not replicate, and thus tissue damage cannot be repaired. As a result, these tissues remain damaged following heart attacks and strokes.

FIGURE 15.14 gives an overview of the entire inflammatory process. **TABLE 15.5** on p. 466 summarizes the chemicals involved in inflammation. ▶**ANIMATIONS:** ***Inflammation: Steps***

Fever

LEARNING | OUTCOME

15.19 Explain the benefits of fever in fighting infection.

Fever is a body temperature above 37°C. Fever augments the beneficial effects of inflammation, but like inflammation it also has unpleasant side effects, including malaise, body aches, and tiredness.

The hypothalamus, a portion of the brain just above the brain stem, controls the body's internal (core) temperature. Fever results when the presence of chemicals called **pyrogens**[13] (pī´rō-jenz) trigger the hypothalamic "thermostat" to reset at a higher temperature. Pyrogens include bacterial toxins, cytoplasmic contents of bacteria that are released upon lysis, antibody-antigen complexes formed in adaptive immune responses, and pyrogens released by phagocytes that have phagocytized bacteria. Although the exact mechanism of fever production is not known, the following discussion and **FIGURE 15.15** present one possible explanation.

Chemicals produced by phagocytes 1 cause the hypothalamus to secrete prostaglandin, which resets the hypothalamic thermostat by an unknown mechanism 2. The hypothalamus then communicates the new temperature setting to other parts of the brain, initiating nerve impulses that produce rapid and repetitive muscle contractions (shivering), an increase in metabolic rate, and constriction of blood vessels of the skin 3. These processes combine to raise the body's core temperature until it equals the prescribed temperature setting 4. Because blood vessels in the skin constrict as fever progresses, one effect of inflammation (vasodilation) is undone. The constricted vessels carry less blood to the skin, causing it to appear paler and feel cold to the touch, even though the body's core temperature is higher. This symptom is the *chill* associated with fever.

Fever continues as long as pyrogens are present. As an infection comes under control and fewer active phagocytes are involved, the level of pyrogens decreases, the thermostat is reset to 37°C, and the body begins to cool by perspiring, lowering the metabolic rate, and dilating blood vessels in the skin. These processes, collectively called the *crisis* of a fever, are a sign that the infection has been overcome and that body temperature is returning to normal.

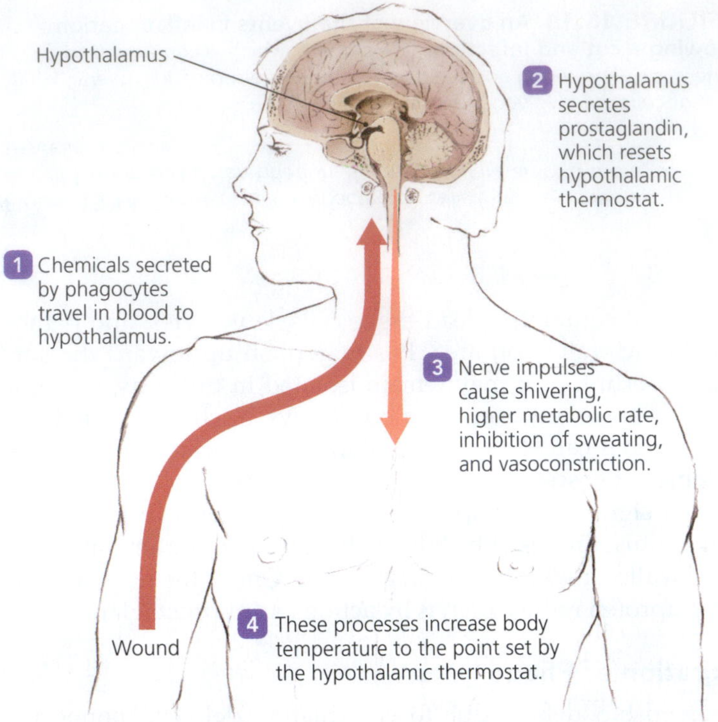

▲ **FIGURE 15.15 One theoretical explanation for the production of fever in response to infection.**

The increased temperature of fever enhances the effects of interferons, inhibits the growth of some microorganisms, and is thought to enhance the performance of phagocytes, the activity of cells of specific immunity, and the process of tissue repair. However, if fever is too high, critical proteins are denatured; additionally, nerve impulses are inhibited, resulting in hallucinations, coma, and even death.

Because of the potential benefits of fever, many doctors recommend that patients refrain from taking fever-reducing drugs unless the fever is prolonged or extremely high. Other physicians believe that the benefits of fever are too slight to justify enduring the adverse symptoms.

TABLE 15.6 summarizes the barriers, cells, chemicals, and processes involved in the body's first two, nonspecific lines of defense.

TELL ME WHY

Why are pathogen-associated molecular patterns (PAMPs) necessary for Toll-like receptors (TLRs) to fully function?

[13]From Greek *pyr*, meaning "fire," and *genein*, meaning "to produce."

TABLE 15.5 Chemical Mediators of Inflammation

Vasodilating chemicals	Histamine, serotonin, bradykinin, prostaglandins
Chemotactic factors	Fibrin, collagen, mast cell chemotactic factors, bacterial peptides
Substances with both vasodilating and chemotactic effects	Complement fragments C5a and C3a, interferons, interleukins, leukotrienes, platelet secretions

TABLE 15.6 A Summary of Some Nonspecific Components of the First and Second Lines of Defense (Innate Immunity)

First Line	Second Line						
Barriers and Associated Chemicals	**Phagocytes**	**Extracellular Killing**	**Complement**	**Interferons**	**Antimicrobial Peptides**	**Inflammation**	**Fever**
Skin and mucous membranes prevent the entrance of pathogens; chemicals (e.g., sweat, acid, lysozyme, mucus) enhance the protection	Macrophages, neutrophils, and eosinophils ingest and destroy pathogens	Eosinophils and NK lymphocytes kill pathogens without phagocytizing them	Components attract phagocytes, stimulate inflammation, and attack a pathogen's cytoplasmic membrane	Increase resistance of cells to viral infection, slow the spread of disease	Interfere with membranes, internal signaling, and metabolism; act against pathogens	Increases blood flow, capillary permeability, and migration of leukocytes into infected area; walls off infected region, increases local temperature	Mobilizes defenses, accelerates repairs, inhibits pathogens

MICRO IN THE CLINIC

FOLLOW-UP

A Stealth Invader in the Lungs

Tim's X-ray film reveals infiltrate in the lungs, an abnormal collection of fluid and cells. A blood test confirms an infection with *Mycoplasma pneumoniae*. This bacterium lacks cell walls and does not Gram stain well. Routine sputum cultures can't detect *M. pneumoniae*. *M. pneumoniae* invades and disrupts the epithelium of the lungs, embedding between the cells and damaging the mucous membranes.

Smokers suffer worse infections by *Mycoplasma* than non-smokers, and they take longer to recover. The irritants in tobacco smoke paralyze the cilia lining the bronchial tubes, allowing mucus and pathogens to accumulate and clog the airways.

Over 2 million cases of pneumonia caused by *M. pneumoniae* occur in the U.S. every year. Among these, about 100,000 patients are hospitalized. Many people with this infection get better on their own, but not Tim—because he is a smoker, his innate immunity is compromised.

Typically, respiratory illnesses are treated with antibiotics that attack the bacterial cell wall, but this won't work with wall-less *Mycoplasma*. Tim's doctor prescribes a macrolide drug called azithromycin and warns him that his recovery will take much longer because of his smoking and the delay in treatment. A check back in five days will determine whether Tim will need a second round of azithromycin.

Tim is relieved, though he has gotten a reality check about the bad state of his lungs. After this experience, he finally decides to quit smoking.

1. **Describe the first line of defense that normally protects the respiratory tract from infection.**
2. **Explain why interferons are not effective in treating *Mycoplasma pneumoniae*.**

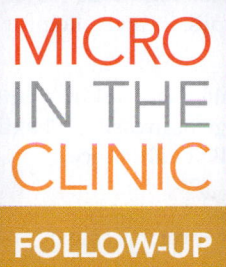

Check your answers to Micro in the Clinic Follow-Up questions in the Study Area at the MasteringMicrobiology Study Area.

CHAPTER SUMMARY

An Overview of the Body's Defenses (p. 449)

1. Humans have **species resistance** to certain pathogens as well as three overlapping lines of defense. The first two lines of defense compose **innate immunity**, which is generally nonspecific and protects the body against a wide variety of potential pathogens. A third line of defense is **adaptive immunity**, which is a specific response to a particular pathogen.

   ▶**ANIMATIONS:** *Host Defenses: Overview*

The Body's First Line of Defense (pp. 449–453)

1. The first line of defense includes the skin, composed of an outer **epidermis** and a deeper **dermis**. **Dendritic cells** of the epidermis devour pathogens. Sweat glands of the skin produce salty sweat containing the enzyme called **lysozyme** and **antimicrobial peptides** (defensins), which are small peptide chains that act against a broad range of pathogens. **Sebum** is an oily substance of the skin that lowers pH, deterring the growth of many pathogens.
2. The mucous membranes, another part of the body's first line of defense, are composed of tightly packed cells that are replaced frequently by **stem cell** division and often coated with sticky mucus secreted by goblet cells.
3. **Microbial antagonism**, the competition between **normal microbiota** and potential pathogens, also contributes to the body's first line of defense.
4. Tears contain antibacterial lysozyme and also flush invaders from the eyes. Saliva similarly protects the teeth. The low pH of the stomach inhibits most microbes that are swallowed.

The Body's Second Line of Defense (pp. 453–467)

1. The second line of defense includes cells (especially **phagocytes**), antimicrobial chemicals (Toll-like receptors, NOD proteins, interferons, complement, lysozyme, and antimicrobial peptides), and processes (phagocytosis, information, and fever).
2. Blood is composed of **formed elements** (cells and parts of cells) within a fluid called **plasma**. Serum is that portion of plasma without clotting factors. The formed elements are **erythrocytes** (red blood cells), **leukocytes** (white blood cells), and **platelets**.
3. Based on their appearance in stained blood smears, leukocytes are grouped as either granulocytes (**basophils**, **eosinophils**, and **neutrophils**) or **agranulocytes** (**lymphocytes**, **monocytes**). When monocytes leave the blood, they become **macrophages**.
4. Basophils function to release histamine during inflammation, whereas eosinophils and neutrophils phagocytize pathogens. They exit capillaries via **diapedesis**.
5. Macrophages, neutrophils, and dendritic cells are phagocytic cells of the second line of defense. Many are named for their location in the body, for example, alveolar macrophages (the lungs) and microglia (the nervous system).
6. A **differential white blood cell count** is a lab technique that indicates the relative numbers of leukocyte types; it can be helpful in diagnosing disease.
7. **Chemotactic factors**, such as chemicals called **chemokines**, attract phagocytic leukocytes to the site of damage or invasion. Phagocytes attach to pathogens via a process called **adherence**.

   ▶**ANIMATIONS:** *Phagocytosis: Overview*
8. **Opsonization**, the coating of pathogens by proteins called **opsonins**, makes those pathogens more vulnerable to phagocytes. A phagocyte's pseudopods then surround the microbe to form a sac called a **phagosome**, which fuses with a lysosome to form a **phagolysosome**, in which the pathogen is killed.

   ▶**ANIMATIONS:** *Phagocytosis: Mechanism*
9. Leukocytes can distinguish the body's normal cells from foreign cells because leukocytes have receptor molecules for foreign cells' components or because the foreign cells are opsonized by complement or antibodies.
10. Eosinophils and **natural killer (NK) lymphocytes** attack nonphagocytically, especially in the case of helminth infections and cancerous cells. **Eosinophilia**—an abnormally high number of eosinophils in the blood—typically indicates such a helminth infection.
11. Microbial molecules called **pathogen-associated molecular patterns (PAMPs)** bind to **Toll-like receptors (TLRs)** on host cells' membranes or to **NOD proteins** inside cells, triggering innate immune responses.
12. **Interferons (IFNs)** are protein molecules that inhibit the spread of viral infections. **Alpha interferons** and **beta interferons**, which are released within hours of infection, trigger **antiviral proteins (AVPs)** to prevent viral reproduction in neighboring cells. **Gamma interferons**, produced days after initial infection, activate macrophages and neutrophils.

13. The **complement system**, or **complement**, is a set of proteins that act as chemotactic attractants, trigger inflammation and fever, and ultimately can effect the destruction of foreign cells via the formation of **membrane attack complexes (MACs)**, which result in multiple, fatal holes in pathogens' membranes. Complement is activated by a classical pathway involving antibodies, by an alternative pathway triggered by bacterial chemicals, or by a lectin pathway triggered by mannose found on microbial surfaces.

   ▶**ANIMATIONS:** *Complement Overview, Activation, Results*

14. **Acute inflammation** develops quickly and damages pathogens, whereas **chronic inflammation** develops slowly and can cause tissue damage that can lead to disease. Signs and symptoms of inflammation include redness, heat, swelling, and pain.

   ▶**ANIMATIONS:** *Inflammation Overview, Steps*

   ▶**VIDEO TUTOR:** *Inflammation*

15. The process of blood clotting triggers formation of **bradykinin**—a potent mediator of inflammation.

16. Macrophages with Toll-like receptors or NOD proteins release **prostaglandins** and **leukotrienes**, which increase permeability of blood vessels. **Mast cells**, basophils, and platelets release **histamine** when exposed to peptides from the complement system. Blood clots may isolate an infected area to form an **abscess**, such as a pimple or boil.

17. When leukocytes rolling along blood vessel walls reach a site of infection, they stick to the wall in a process called **margination** and then undergo diapedesis to arrive at the site of tissue damage. The increased blood flow of inflammation also brings extra nutrients and oxygen to the infection site to aid in repair.

18. **Fever** results when chemicals called **pyrogens**, including substances released by bacteria and phagocytes, affect the hypothalamus in a way that causes it to reset body temperature to a higher level. The exact process of fever and its control are not fully understood.

QUESTIONS FOR REVIEW

Answers to the Questions for Review (except Short Answer questions) begin on p. A-1.

Multiple Choice

1. Phagocytes of the epidermis are called
   a. microglia.
   b. goblet cells.
   c. alveolar macrophages.
   d. dendritic cells.

2. Mucus-secreting membranes are found in
   a. the urinary system.
   b. the digestive cavity.
   c. the respiratory passages.
   d. all of the above

3. The complement system involves
   a. the production of antigens and antibodies.
   b. serum proteins involved in nonspecific defense.
   c. a set of genes that distinguish foreign cells from body cells.
   d. the elimination of undigested remnants of microorganisms.

4. The alternative complement activation pathway involves
   a. factors B, D, and P.
   b. the cleavage of C5 to form C9.
   c. binding to mannose sugar.
   d. recognition of antigens bound to specific antibodies.

5. Complement must be inactivated because if it were *not,*
   a. viruses could continue to multiply inside host cells using the host's own metabolic machinery.
   b. necessary interferons would not be produced.
   c. protein synthesis would be inhibited, thus halting important cell processes.
   d. it could make holes in the body's own cells.

6. The type of interferon present late in an infection is
   a. alpha interferon.
   b. beta interferon.
   c. gamma interferon.
   d. delta interferon.

7. Interferons
   a. do not protect the cell that secretes them.
   b. stimulate the activity of macrophages.
   c. cause muscle aches, chills, and fever.
   d. all of the above

8. Which of the following is *not* targeted by a Toll-like receptor?
   a. lipid A
   b. eukaryotic flagellar protein
   c. single-stranded RNA
   d. lipoteichoic acid

9. Toll-like receptors (TLRs) act to
   a. bind microbial proteins and polysaccharides.
   b. induce phagocytosis.
   c. cause phagocytic chemotaxis.
   d. destroy microbial cells.

10. Which of the following binds iron?
   a. lactoferrin
   b. siderophores
   c. transferrin
   d. all of the above

Modified True/False

Indicate which statements are true. Correct all false statements by changing the underlined words.

1. ________________ The surface cells of the epidermis of the skin are <u>alive</u>.
2. ________________ The surface cells of mucous membranes are <u>alive</u>.
3. ________________ Wandering macrophages experience <u>diapedesis</u>.
4. ________________ <u>Monocytes</u> are immature macrophages.
5. ________________ <u>Lymphocytes</u> are large agranulocytes.
6. ________________ <u>Phagocytes</u> exhibit chemotaxis toward a pathogen.
7. ________________ In phagocytosis, adherence involves the binding between complementary chemicals on a phagocyte and on the membrane of a <u>body cell</u>.
8. ________________ <u>Opsonization</u> occurs when a phagocyte's pseudopods surround a microbe and fuse to form a sac.

9. ________________ Lysosomes fuse with phagosomes to form <u>peroxisomes</u>.

10. ________________ A membrane attack complex drills circular holes in a <u>macrophage</u>.

11. ________________ Rubor, calor, swelling, and dolor are associated with <u>fever</u>.

12. ________________ Acute and chronic inflammation exhibit <u>similar</u> signs and symptoms.

13. ________________ The <u>hypothalamus</u> of the brain controls body temperature.

14. ________________ Defensins are <u>phagocytic parts</u> of the first line of defense.

15. ________________ NETs are webs produced by <u>neutrophils</u> to trap microbes.

Matching

In the blank beside each cell, chemical, or process in the left column, write the letter of the line of defense that first applies. Each letter may be used several times.

1. _____ Inflammation
2. _____ Monocytes
3. _____ Lactoferrin
4. _____ Fever
5. _____ Dendritic cells
6. _____ Alpha interferon
7. _____ Mucous membrane of the digestive tract
8. _____ Neutrophils
9. _____ Epidermis
10. _____ Lysozyme
11. _____ Goblet cells
12. _____ Phagocytes
13. _____ Sebum
14. _____ T lymphocytes
15. _____ Antimicrobial peptides

A. First line of defense
B. Second line of defense
C. Third line of defense

Write the letter of the description that applies to each of the following terms.

1. _____ Goblet cell
2. _____ Lysozyme
3. _____ Stem cell
4. _____ Dendritic cell
5. _____ Cell from sebaceous gland
6. _____ Bone marrow stem cell
7. _____ Eosinophil
8. _____ Alveolar macrophage
9. _____ Microglia
10. _____ Wandering macrophage

A. Leukocyte that primarily attacks parasitic worms
B. Phagocytic cell in lungs
C. Secretes sebum
D. Devours pathogens in epidermis
E. Breaks bonds in bacterial cell wall
F. Phagocytic cell in central nervous system
G. Generative cell with many types of offspring
H. Develops into formed elements of blood
I. Intercellular scavenger
J. Secretes mucus

VISUALIZE IT!

1. Label the steps of phagocytosis.

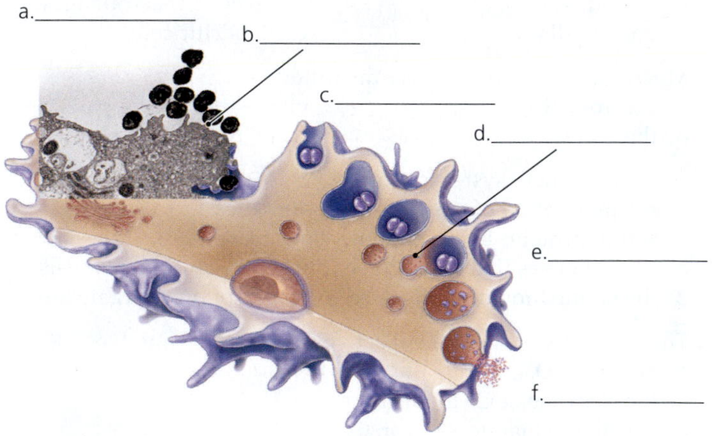

2. Name the cells.

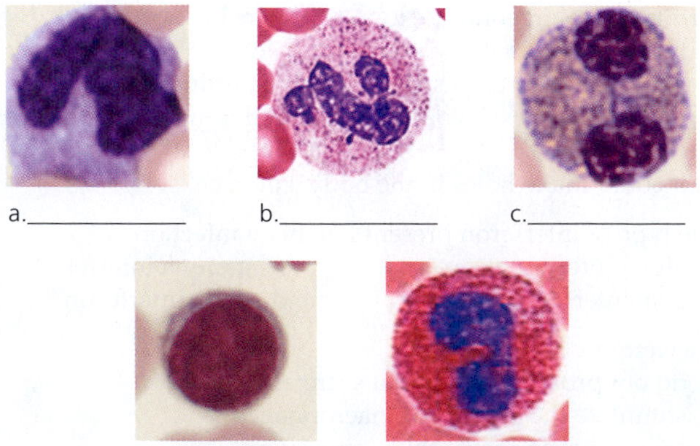

Short Answer

1. In order for a pathogen to cause disease, what three things must happen?
2. How does a phagocyte "know" it is in contact with a pathogen instead of another body cell?
3. Give three characteristics of the epidermis that make it an intolerable environment for most microorganisms.
4. What is the role of Toll-like receptors in innate immune responses?
5. Describe the classical complement cascade pathway from C1 to the MAC.
6. How do NOD proteins differ from Toll-like receptors?

CRITICAL THINKING

1. John received a chemical burn on his arm and was instructed by his physician to take an over-the-counter anti-inflammatory medication for the painful, red, swollen lesions. When Charles suffered pain, redness, and swelling from an infected cut on his foot, he decided to take the same anti-inflammatory drug because his symptoms matched John's symptoms. How is Charles's inflammation like that of John? How is it different? Is it appropriate for Charles to medicate his cut with the same medicine John used?
2. What might happen to someone whose body did not produce C3? C5?
3. Mary, age 65, has had diabetes for 40 years, with resulting damage to the small blood vessels in her feet and toes. Her circulation is impaired. How might this condition affect her vulnerability to infection?
4. A patient's chart shows that eosinophils make up 8% of his white blood cells. What does this lead you to suspect? Would your suspicions change if you learned that the patient had spent the previous three years as an anthropologist living in an African jungle village? What is the normal percentage of eosinophils?
5. There are two kinds of agranulocytes in the blood—monocytes and lymphocytes. Janice noted that monocytes are phagocytic and that lymphocytes are not. She wondered why two agranulocytes would be so different. What facts of hematopoiesis can help her answer her question?
6. A patient has a genetic disorder that prevents him from synthesizing C8 and C9. What effect does this have on his ability to resist bloodborne Gram-negative and Gram-positive bacteria? What would happen if C3 and C5 fragments were also inactivated?
7. Sweat glands in the armpits secrete perspiration with a pH close to neutral (7.0). How does this fact help explain body odor in this area as compared to other parts of the skin?
8. Scientists can raise "germ-free" animals in axenic environments. Would such animals be as healthy as their worldly counterparts?
9. Compare and contrast the protective structures and chemicals of the skin and mucous membranes.
10. Scientists are interested in developing antimicrobial drugs that act like the body's normal antimicrobial peptides. What advantage might such a drug have over antibiotics?
11. A medical laboratory scientist argues that granulocytes are a natural group, whereas agranulocytes are an artificial grouping. Based on Figure 15.4, do you agree or disagree with the lab scientist? What evidence can you cite to justify your conclusion?
12. A patient has a genetic disorder that makes it impossible for her to synthesize complement protein 8 (C8). Is her complement system nonfunctional? What major effects of complement could still be produced?
13. While using a microscope to examine a sample of pus from a pimple, Maria observed a large number of macrophages. Is the pus from an early or a late stage of infection? How do you know?
14. How do drugs such as aspirin and ibuprofen act to reduce fever? Should you take fever-reducing drugs or let a fever run its course?

CONCEPT MAPPING

Using the following terms, draw a concept map that describes phagocytosis. For a sample concept map, see p. 94. Or, complete this and other concept maps online by going to the MasteringMicrobiology Study Area.

Adherence	**Elimination**	**Killing**	**Opsonins**
Chemotactic factors	**Eosinophils**	**Lysosome**	**Phagolysosome**
Chemotaxis	**Exocytosis**	**Macrophages**	**Phagosome**
Dendritic cells	**Ingestion**	**Neutrophils**	

16 Adaptive Immunity

MICRO IN THE CLINIC

A Vaccine Against Cancer?

Kathy, a single mom, has been so busy working and raising her three children that she hasn't had a physical check-up in years. When she finally goes in for a routine gynecological exam, she is stunned when her Pap smear reveals precancerous cells in her cervix. Kathy is also haunted by the memory of her own mother, who died five years ago from breast cancer that was detected too late.

Kathy frequently argues with her daughter, Meghan. Kathy wants Meghan to be vaccinated against the human papillomavirus (HPV), a virus that is strongly linked to cervical cancer. Meghan, however, hates needles. Before starting her freshman year of college, Meghan already received several vaccines protecting against infectious diseases such as diphtheria, tetanus, and meningitis. The last thing she wants is another vaccination and the HPV vaccine involves no fewer than three separate inoculations! She thinks her mother is being overprotective. She is also secretly skeptical of the HPV vaccine. After all, who has ever heard of a vaccine that protects someone from cancer?

Could a vaccine really protect Meghan from cervical cancer? Do Kathy's Pap smear results mean she will definitely develop cancer? Turn to the end of the chapter (p. 498) to find out.

MM **Explore More:** Test your readiness and apply your knowledge with dynamic learning tools at MasteringMicrobiology.

Inborn, or innate, immunity includes two lines of rapid defense against microbial pathogens (see Chapter 15). The first line includes intact skin and mucous membranes, whereas the second line includes phagocytosis, nonspecific chemical defenses (such as complement), inflammation, and fever. Innate defenses do not always offer enough protection in defending the body. Although the mechanisms of innate immunity are readily available and fast acting, they do not adapt to enhance the effectiveness of response to the great variety of pathogens confronting us—a fever is a fever, whether it is triggered by a mild flu virus or the deadly Ebola virus. The body augments the mechanisms of innate immunity with a third line of defense that destroys and targets specific invaders while becoming more effective in the process. This response is called *adaptive immunity.*

Overview of Adaptive Immunity

LEARNING OUTCOMES

16.1 Describe five distinctive attributes of adaptive immunity.

16.2 List the two basic types of white blood cells involved in adaptive immunity.

16.3 List two basic divisions of adaptive immunity and describe their targets.

Adaptive immunity is the body's ability to recognize and then mount a defense against distinct invaders and their products, whether they are protozoa, fungi, bacteria, viruses, or toxins. *Immunologists*—scientists who study the cells and chemicals involved in immunity—are continually refining and revising our knowledge of adaptive immunity. In this chapter, we will examine some of what they have discovered.

Adaptive immunity has five distinctive attributes:

- **Specificity.** Any particular adaptive immune response acts against only one particular molecular shape and not against others. Adaptive immune responses are precisely tailored reactions against specific attackers, whereas innate immunity involves more generalized responses to pathogen-associated molecular patterns (PAMPs)—molecular shapes common to many microbes.
- **Inducibility.** Cells of adaptive immunity are activated in response to the specific pathogens.
- **Clonality.** Once induced, cells of adaptive immunity proliferate to form many generations of nearly identical cells, which are collectively called *clones.*
- **Unresponsiveness to self.** As a rule, adaptive immunity does not act against normal body cells; in other words, adaptive immune responses are self-tolerant. Several mechanisms help ensure that immune responses do not attack the body itself.
- **Memory.** An adaptive immune response has "memory" for specific pathogens; that is, it adapts to respond faster and more effectively in subsequent encounters with a particular type of pathogen or toxin.

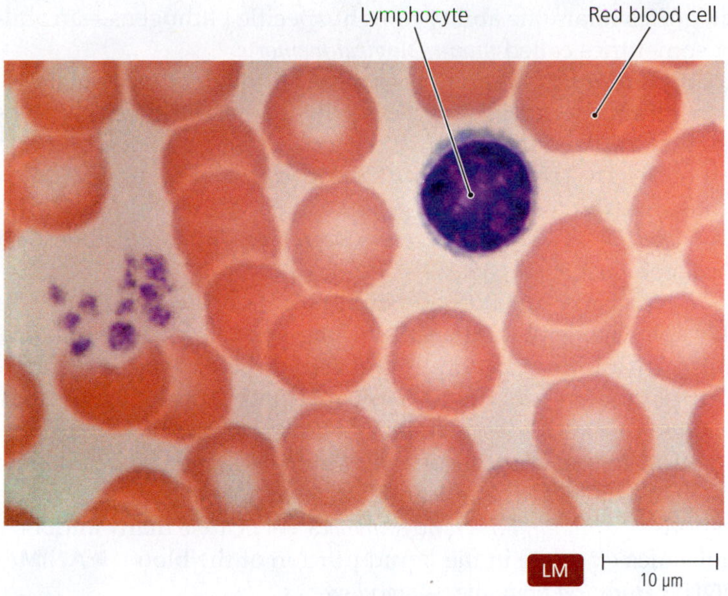

▲ **FIGURE 16.1 Lymphocytes play a central role in adaptive immunity.** A resting lymphocyte is the smallest leukocyte—slightly larger than a red blood cell.

These aspects of adaptive immunity involve the activities of **lymphocytes** (lim´fō-sītz), which are a type of leukocyte (white blood cell) that acts against specific pathogens. Lymphocytes in their resting state are the smallest white blood cells, and each is characterized by a large, round, central nucleus surrounded by a thin rim of cytoplasm **(FIGURE 16.1)**. Initially, lymphocytes of humans form in the *red bone marrow,* located in the ends of long bones in juveniles and in the centers of adult flat bones such as the ribs and hip bones. These sites contain *blood stem cells (hematopoietic stem cells),* which are cells that give rise to all types of blood cells, including lymphocytes (see Figure 15.4).

Although lymphocytes appear identical in the microscope, scientists make distinctions between two main types—*B lymphocytes* and *T lymphocytes*—according to integral surface proteins that are part of each lymphocyte's cytoplasmic membrane. These proteins allow lymphocytes to recognize specific pathogens and toxins by their molecular shapes, and the proteins play roles in intercellular communication among immune cells.

Lymphocytes must undergo a maturation process. **B lymphocytes,** which are also called **B cells,** arise and mature in the red bone marrow of adults. **T lymphocytes,** also known as **T cells,** begin in bone marrow as well but do not mature there. Instead, T cells travel to and mature in the *thymus,* located in the chest near the heart in humans. T lymphocytes are so called because of the role the thymus plays in their maturation.

Adaptive immunity consists of many different immune responses that can be considered under two broad categories: **cell-mediated immune responses** and **antibody immune responses.** Long-lived B and T lymphocytes in each of these categories retain the ability to fight specific pathogens—an ability sometimes called *immunological memory.*

Both types of immune responses stimulate and regulate innate immunity as well as act directly against the specific pathogen that initiated the adaptive response. Cell-mediated immune responses are controlled and carried out by T cells and often act against *intracellular pathogens,* such as viruses replicating inside a cell. ▶**ANIMATIONS:** ***Host Defenses: The Big Picture; Cell-Mediated Immunity: Overview***

B cells carry out antibody immune responses, though T cells play roles in regulating and fulfilling such immune responses. Antibody immune responses are often directed against extracellular pathogens and toxins. An *antibody* is a protective protein secreted by descendants of a B cell that recognize a specific biochemical shape. We will consider antibodies in more detail later. Scientists have also used another term for antibody immune responses—*humoral immune responses*[1]—because many antibody molecules circulate in the liquid portion of the blood. ▶**ANIMATIONS:** ***Humoral Immunity: Overview***

Both cell-mediated and antibody immune responses are powerful defensive reactions that have the potential to severely and fatally attack the body's own cells. Therefore, the body must regulate adaptive immune responses to prevent damage; for example, an immune response requires multiple chemical signals before proceeding, thus reducing the possibility that an immune response will be randomly triggered against uninfected healthy tissue. Autoimmune disorders, hypersensitivities, or immunodeficiency diseases result when regulation is insufficient or overexcited. (Chapter 18 deals with such disorders.)

TELL ME WHY

Why are the activities of B and T cells called *adaptive*?

Elements of Adaptive Immunity

Just as the program at a dramatic presentation might present a synopsis of the performance and introduce the actors and their roles, the following sections present elements of adaptive immunity by describing the "stage" and introducing the "cast of characters" involved in adaptive immunity. We will examine the *lymphatic system*—the organs, tissues, and cells of adaptive immunity; then we consider the molecules, called *antigens,* that trigger adaptive immune responses; next, we take a look at *antibodies;* and finally we examine special *chemical signals* and *mediators* involved in coordinating and controlling a specific immune response.

First, we turn our attention to the "stage"—the lymphatic system, which plays an important role in the production, maturation, and housing of the cells that function in adaptive immunity.

[1]From Latin *humor,* meaning "liquid," referring to bodily fluids, such as blood.

The Tissues and Organs of the Lymphatic System

LEARNING | **OUTCOMES**

16.4 Compare and contrast the flow of lymph with the flow of blood.

16.5 Describe the primary and secondary organs of the lymphatic system.

16.6 Describe the importance to immunity of red bone marrow, the thymus, lymph nodes, spleen, tonsils, and mucosa-associated lymphoid tissue.

The **lymphatic system** is composed of the *lymphatic vessels,* which conduct the flow of a liquid called *lymph,* and lymphatic cells, tissues, and organs, which are directly involved in adaptive immunity **(FIGURE 16.2a)**. Taken together, the components of the lymphatic system constitute a surveillance system that screens the tissues of the body—particularly possible points of entry such as the throat and intestinal tract—for foreign molecules. We begin by examining lymphatic vessels.

The Lymphatic Vessels and the Flow of Lymph

Lymphatic vessels form a one-way system that conducts *lymph* (pronounced *"limf"*) from local tissues and returns it to the circulatory system. Most importantly for immune responses, lymph carries toxins and pathogens to areas where lymphocytes are concentrated.

Lymph is a colorless, watery liquid similar in composition to blood plasma; indeed, lymph arises from fluid that has leaked out of blood vessels into the surrounding intercellular spaces. Lymph is first collected by remarkably permeable *lymphatic capillaries* **(FIGURE 16.2b)**, which are located in most parts of the body (exceptions include the bone marrow, brain, and spinal cord). From the lymphatic capillaries, lymph passes into increasingly larger lymphatic vessels until it finally flows via two large *lymphatic ducts* into blood vessels near the heart. Unlike the cardiovascular system, the lymphatic system has no unique pump and is not circular; that is, lymph flows in one direction. One-way valves ensure that lymph flows only toward the heart as skeletal muscular activity squeezes the lymphatic vessels. Located at various points within the system of lymphatic vessels are about 1000 *lymph nodes,* which house white blood cells including B and T lymphocytes. These lymphocytes recognize and attack foreigners present in the lymph, allowing for immune system surveillance and interactions.

Lymphoid Organs

Once lymphocytes have arisen and matured in the *primary lymphoid organs* of the red bone marrow and thymus, they migrate to *secondary lymphoid organs* and *tissues,* including lymph nodes, spleen, and other less organized accumulations of lymphoid tissue, where they in effect lie in wait for foreign microbes. The hundreds of **lymph nodes** are located throughout the body but concentrated in the cervical (neck), inguinal (groin), axillary (armpit), and abdominal regions. Each lymph node receives

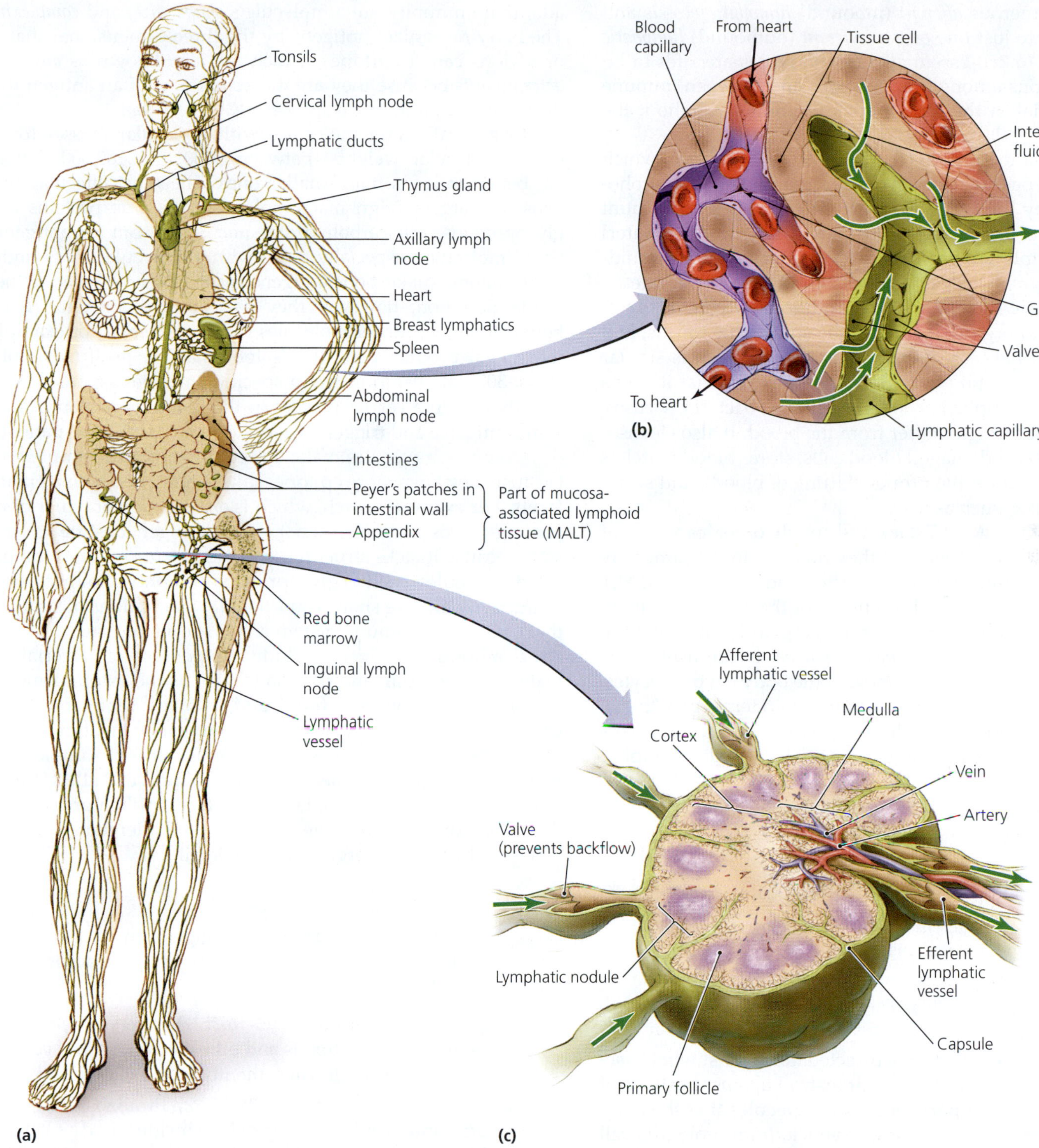

▲ **FIGURE 16.2 The lymphatic system. (a)** The system consists of primary lymphoid organs—bone marrow (not shown) and thymus gland—and secondary lymphoid organs, including lymphatic vessels, lymph nodes, tonsils, and other lymphatic tissue. **(b)** Lymphatic capillaries collect lymph from intercellular spaces. **(c)** Afferent lymphatic vessels carry lymph into lymph nodes; efferent vessels carry it away. The cortex contains primary follicles centers, where B lymphocytes proliferate; the lymphocytes in the medulla encounter foreign molecules. *Why do lymph nodes enlarge during an infection?*

Figure 16.2 *During an infection, lymphocytes multiply profusely in lymph nodes. This proliferation and swelling cause lymph nodes to enlarge.*

lymph from numerous *afferent* (inbound) *lymphatic vessels* and drains lymph into just one or two *efferent* (outbound) *lymphatic vessels* (FIGURE 16.2c). Essentially, lymph nodes are sites to facilitate interactions among immune cells and between immune cells and material in the lymph arriving from throughout the body.

A node has a medullary (central) maze of passages, which filter the lymph passing through and house numerous lymphocytes that survey the lymph for foreign molecules and mount specific immune responses against them. The cortex (outer) portion of a lymph node consists of a tough capsule surrounding *primary follicles*, which is where clones of B cells replicate.

The lymphatic system contains additional secondary lymphoid tissues and organs, including the spleen, the tonsils, and *mucosa-associated lymphoid tissue (MALT)*. The spleen is similar in structure and function to lymph nodes, except that it filters blood instead of lymph. The spleen removes bacteria, viruses, toxins, and other foreign matter from the blood. It also cleanses the blood of old and damaged blood cells, stores blood platelets (which are required for the proper clotting of blood), and stores blood components, such as iron.

The tonsils and MALT lack the tough outer capsules of lymph nodes and the spleen, but they function in the same way by physically trapping foreign particles and microbes. MALT includes the appendix; lymphoid tissue of the respiratory tract, vagina, urinary bladder, and mammary glands; and discrete bits of lymphoid tissue called *Peyer's patches* in the wall of the small intestine. MALT contains most of the body's lymphocytes.

We are considering adaptive immunity in terms of a "play" taking place on the stage of the lymphatic system. Now, let's meet the "villain actors"—the foreign molecules that lymphocytes recognize.

Antigens

LEARNING OUTCOMES

16.7 Identify the characteristics of antigens that stimulate effective immune responses.

16.8 Distinguish among exogenous antigens, endogenous antigens, and autoantigens.

Adaptive immune responses are directed not against whole bacteria, fungi, protozoa, or viruses but instead against portions of cells, viruses, and even parts of single molecules that the body recognizes as foreign and worthy of attack. Immunologists call these biochemical shapes **antigens**[2] (an´ti-jenz). Lymphocytes bind to antigens and can then trigger adaptive immune responses. Antigens from pathogens and toxins are the "villains" of our story.

Properties of Antigens

Not every molecule is an effective antigen. Among the properties that make certain molecules more effective at provoking adaptive immunity are a molecule's *shape, size,* and *complexity.* The body recognizes antigens by the three-dimensional shapes of regions called **epitopes,** which are also known as *antigenic determinants* because they are the actual part of an antigen that determines an immune response (FIGURE 16.3a).

In general, larger molecules with molecular masses (often called molecular weights) between 5000 and 100,000 daltons are better antigens than smaller ones. The most effective antigens are large foreign macromolecules, such as proteins and glycoproteins, but carbohydrates and lipids can be antigenic. Small molecules, especially those with a molecular mass under 5000 daltons, make poor antigens by themselves because they evade detection; however, they can become antigenic when bound to larger, carrier molecules (often proteins). For example, the fungal product penicillin is too small by itself (molecular mass: 302 daltons) to trigger a specific immune response. However, bound to a carrier protein in the blood, penicillin can become antigenic and trigger an allergic response in some patients.

Complex molecules make better antigens than simple ones because they have more epitopes, like a gemstone with its many facets. For example, starch, which is a very large polymer of repeating glucose subunits, is not a good antigen, despite its large size, because it lacks structural complexity. In contrast, complicated molecules, such as glycoproteins and phospholipids, have multiple distinctive shapes and novel combinations of subunits that cells of the immune system recognize as foreign.

Examples of antigens include components of bacterial cell walls, capsules, pili, and flagella as well as the external and internal proteins of viruses, fungi, and protozoa. Many toxins and some nucleic acid molecules are also antigenic. Invading microorganisms are not the only source of antigens; for example, food may contain antigens called *allergens* that provoke allergic reactions, and inhaled dust contains mite feces, pollen grains, dander (flakes of skin), and other antigenic and allergenic particles. (Chapter 18 covers allergies in more detail.)

Types of Antigens

Though immunologists categorize antigens in various ways, one especially important way is to group antigens according to their relationship to the body:

- **Exogenous**[3] **antigens** (FIGURE 16.3b). Exogenous (eks-oj′en-us) antigens include toxins and other secretions and components of microbial cell walls, membranes, flagella, and pili.
- **Endogenous**[4] **antigens** (FIGURE 16.3c). Protozoa, fungi, bacteria, and viruses that reproduce inside a body's cells produce endogenous antigens. The immune system cannot assess the health of the body's cells; it responds to endogenous antigens only if the body's cells incorporate such antigens into their cytoplasmic membranes, leading to their external display.
- **Autoantigens**[5] (FIGURE 16.3d). Antigenic molecules derived from normal cellular processes are autoantigens (or

[2]From *anti*body *gen*erator; antibodies are proteins secreted during an antibody immune response that bind to specific regions of antigens.

[3]From Greek *exo*, meaning "without," and *genein*, meaning "to produce."
[4]From Greek *endon*, meaning "within."
[5]From Greek *autos*, meaning "self."

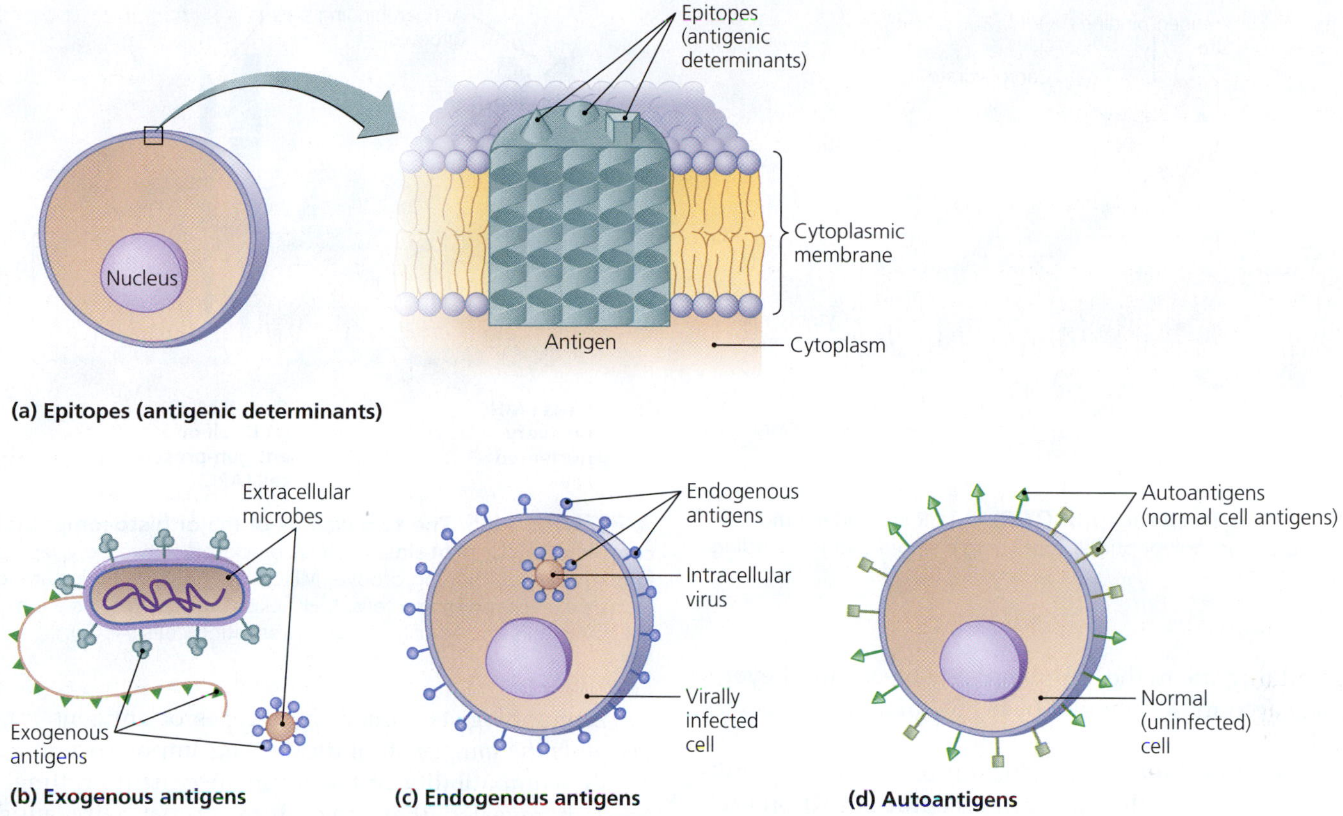

▲ **FIGURE 16.3 Antigens, molecules that provoke a specific immune response. (a)** Epitopes, or antigenic determinants, are three-dimensional regions of antigens whose shapes are recognized by cells of the immune system. **(b–d)** Categories of antigens based on their relationship to the body. Exogenous antigens originate from microbes located outside the body's cells; endogenous antigens are produced by intracellular microbes and are typically incorporated into a host cell's cytoplasmic membrane; autoantigens are components of normal body cells.

self-antigens). As we will discuss more fully in a later section, immune cells that treat autoantigens as if they were foreign are normally eliminated during the development of the immune system. This phenomenon, called *self-tolerance,* prevents the body from mounting an immune response against itself.

So far, we have examined two aspects of immune responses in terms of an analogy to a stage play—the lymphoid organs and tissues provide the stage, and antigens of pathogens are the villains that induce an immune response with their epitopes (antigenic determinants). Next, we examine the activities of lymphocytes in more detail. Recall that T and B lymphocytes act against antigens; they are the "heroes" of the "play." We begin by considering T lymphocytes.

T Lymphocytes (T Cells) and Preparation for an Adaptive Immune Response

LEARNING **OUTCOMES**

16.9 Describe the importance of the thymus to the development of T lymphocytes.

16.10 Describe the basic characteristics common to T lymphocytes.

16.11 Describe the two classes of major histocompatibility complex (MHC) proteins with regard to their location and function.

16.12 Explain the roles of antigen-presenting cells (e.g., dendritic cells and macrophages) and MHC molecules in antigen processing and presentation.

16.13 Contrast endogenous antigen processing with exogenous antigen processing.

16.14 Compare and contrast three types of T cells.

16.15 Describe apoptosis, and explain its role in lymphocyte editing by clonal deletion.

T lymphocytes often act against body cells that harbor intracellular pathogens, such as viruses, and against body cells that produce abnormal cell-surface proteins (such as cancer cells). Because T cells act directly against antigens, T lymphocyte immune activities are sometimes called *cell-mediated immune responses*.

A human adult's red bone marrow produces T lymphocytes (T cells), which are released into the blood. Chemotactic molecules attract T lymphocytes from blood vessels to the thymus, where they adhere and mature under the influence of sticky, adhesive chemicals and molecular signals from the thymus. Following maturation, T cells circulate in the lymph

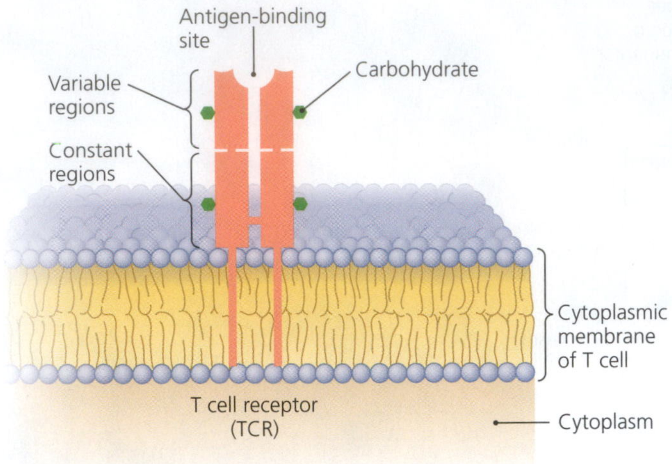

▲ **FIGURE 16.4 A T cell receptor (TCR).** A TCR is a surface molecule composed of two polypeptides containing a single antigen-binding site between them.

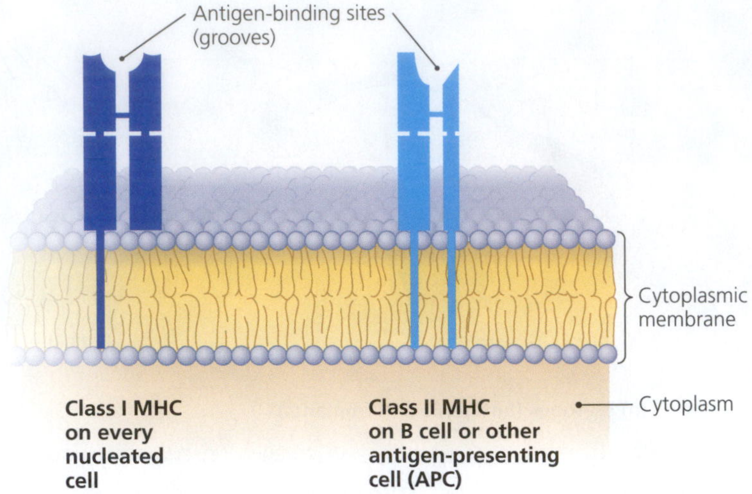

▲ **FIGURE 16.5 The two classes of major histocompatibility complex (MHC) proteins.** Each is composed of two polypeptides that form an antigen-binding groove. MHC class I glycoproteins are found on all cells except red blood cells. MHC class II glycoproteins are expressed only by B cells and special antigen-presenting cells (APCs).

and blood and migrate to the lymph nodes, spleen, and Peyer's patches. They account for about 70% to 85% of all lymphocytes in the blood.

T cell maturation involves production of about half a million copies of a protein called a **T cell receptor (TCR)** on each cell's cytoplasmic membrane. Every T cell randomly chooses and combines segments of DNA from TCR genes to create a new genetic combination specific to that cell, which codes for the cell's unique and specific TCR.

Specificity of the T Cell Receptor (TCR)

The ends of TCRs are composed of variable regions that grant each cell's TCR a specific antigen-binding site (FIGURE 16.4). Because each T cell randomly recombines its TCR genes, every T cell ends up with a differently shaped antigen-binding site, which recognizes and binds to a complementary shape. TCRs do not recognize epitopes directly. Instead, each binds only to an epitope associated with a protein called *MHC protein,* which will be discussed next. In your body, there are at least 10^9 different TCRs—each specific TCR type on a different T cell—enough for every possible epitope–MHC protein shape.

The Roles of the Major Histocompatibility Complex and Antigen-Presenting Cells

When scientists first began grafting tissue from one animal into another to determine a method for treating burn victims, they discovered that if the animals were not closely related, the recipients swiftly rejected the grafts. When they analyzed the reason for such rapid rejection, they found that a graft recipient mounted a very strong immune response against a specific type of antigen found on the cells of unrelated grafts. When the antigen on a graft's cells was sufficiently dissimilar from antigens on the host's cells, as occurred with unrelated animals, the grafts were rejected. This is how scientists came to understand how the body is able to distinguish "self" from "nonself."

Immunologists named these types of antigens *major histocompatibility*[6] *antigens* to indicate their importance in determining the compatibility of tissues in successful grafting. Further research revealed that major histocompatibility antigens are glycoproteins found in the membranes of most cells of vertebrate animals. Major histocompatibility antigens are coded by a cluster of genes called the **major histocompatibility complex (MHC).** In humans, an MHC is located on each copy of chromosome 6.

Because organ grafting is a modern surgical procedure with no counterpart in nature, scientists reasoned that MHC proteins must have some other "real" function. Indeed, immunologists have determined that MHC proteins in cytoplasmic membranes function to hold and position epitopes for presentation to T cells. Each MHC molecule has an *antigen-binding groove* that lies between two polypeptides. Inherited variations in the amino acid sequences of the polypeptides modify the shapes of MHC binding sites and determine which epitopes can be bound and presented. It is important to recall that TCRs recognize only epitopes that are bound to MHC molecules.

MHC proteins are of two classes (FIGURE 16.5). Class I MHC molecules are found on the cytoplasmic membranes of all cells except red blood cells. Special cells called **antigen-presenting cells (APCs)** also have class II MHC proteins. Professional antigen-presenting cells—those that regularly present antigen—are B cells, macrophages, and, most important, **dendritic cells,** which are so named because they have many long, thin cytoplasmic processes called *dendrites*[7] (FIGURE 16.6). Phagocytic dendritic cells are found under the surface of the skin and mucous membranes. Some dendritic cells extend dendrites between skin cells or through a mucous surface to sample antigens, much

[6]From Greek *histos,* meaning "tissue," and Latin *compatibilis,* meaning "agreeable."
[7]From Greek *dendron,* meaning "tree," referring to long cellular extensions that look like branches of a tree.

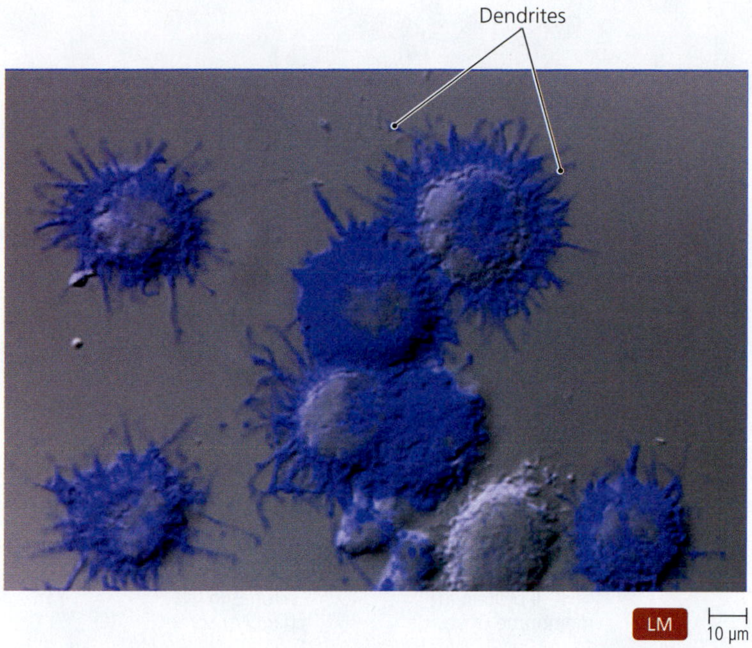
Dendrites
LM
10 μm

▲ **FIGURE 16.6 Dendritic cells.** These important antigen-presenting cells in the body are found in the skin and mucous membranes. They have numerous thin cytoplasmic processes called dendrites.

like a submarine extends a periscope to get a view of surface activity. After acquiring antigens, dendritic cells migrate to lymph nodes to interact with both T and B lymphocytes. Certain other phagocytes, such as *microglia* in the brain and *stellate macrophages* (formerly called *Kupffer cells*) in the liver, may also present antigen under certain conditions. These cells are termed *nonprofessional antigen-presenting cells.*

The cytoplasmic membrane of a professional APC has about 100,000 MHC II molecules, which vary in the epitopes they can bind. Their diversity is dependent on an individual's genotype. If an epitope cannot be bound to an MHC molecule, it typically does not trigger an immune response. Thus, MHC molecules determine which epitopes might trigger immune responses.

Antigen Processing

Before MHC proteins can display epitopes, antigens must be processed. Antigen processing occurs via somewhat different processes according to whether the antigen is endogenous or exogenous. Recall that endogenous antigens come from a cell's cytoplasm or from pathogens living within the cell, whereas exogenous antigens have extracellular sources, such as lymph.
▶ANIMATIONS: ***Antigen Processing and Presentation: Overview***

Processing Endogenous Antigens FIGURE 16.7 illustrates the processing of endogenous antigens. A few molecules of each polypeptide produced within nucleated cells—including some polypeptides produced by intracellular bacteria or polypeptides coded by viruses—are catabolized into smaller pieces containing about 8 to 12 amino acids 1. These pieces, which will include epitopes of the polypeptides, move into the endoplasmic reticulum (ER) and bind onto complementary antigen-binding grooves of MHC class I molecules that were

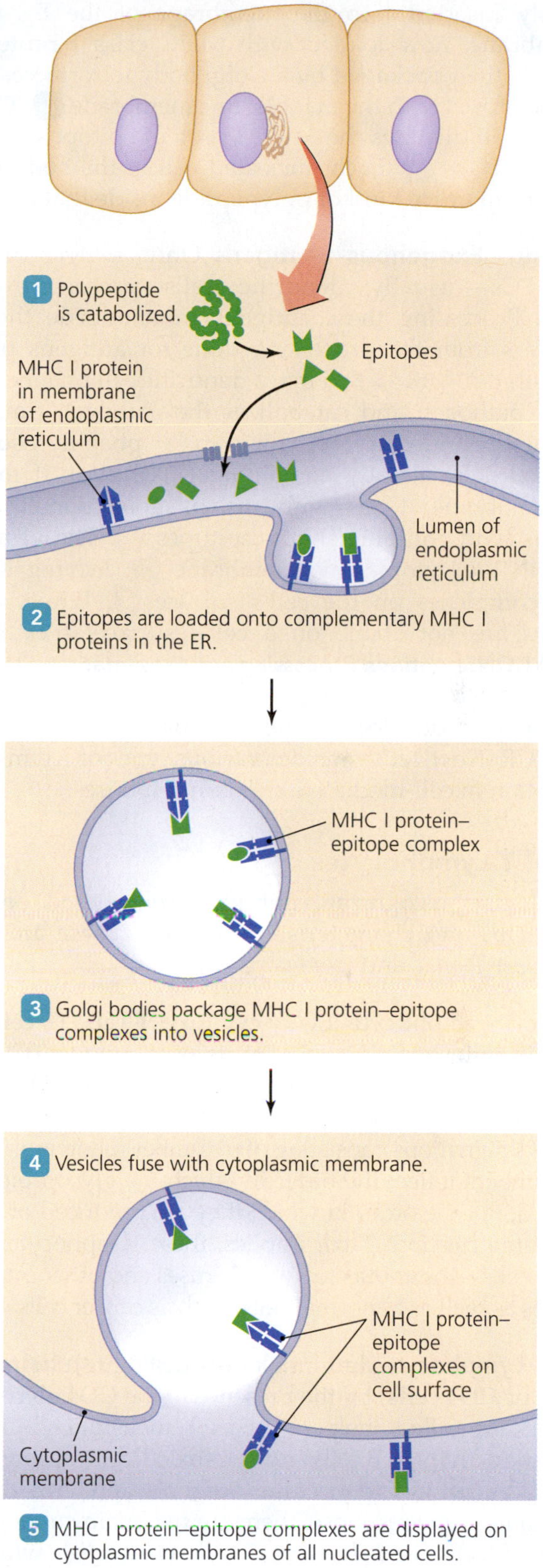
1 Polypeptide is catabolized.
Epitopes
MHC I protein in membrane of endoplasmic reticulum
Lumen of endoplasmic reticulum
2 Epitopes are loaded onto complementary MHC I proteins in the ER.
MHC I protein–epitope complex
3 Golgi bodies package MHC I protein–epitope complexes into vesicles.
4 Vesicles fuse with cytoplasmic membrane.
MHC I protein–epitope complexes on cell surface
Cytoplasmic membrane
5 MHC I protein–epitope complexes are displayed on cytoplasmic membranes of all nucleated cells.

▲ **FIGURE 16.7 The processing of endogenous antigens.** Epitopes from all polypeptides synthesized within a nucleated cell load onto complementary MHC I proteins, which are exported to the cytoplasmic membrane.

previously inserted into the membrane of the ER 2. The ER membrane, now loaded with MHC class I proteins and epitopes, is then packaged by a Golgi body to form vesicles 3. The vesicle fuses with the cytoplasmic membrane 4. The result is that the cell displays the MHC I protein-epitope complex on the cell's surface 5. Each nucleated cell in the body displays epitopes from every kind of polypeptide inside that cell.

Processing Exogenous Antigens Only antigen-presenting cells (APCs)—usually dendritic cells—process exogenous antigens. Processing these antigens from outside the body's cells differs from the processing done for antigens produced within cells (**FIGURE 16.8**). First, a dendritic cell phagocytizes an invading pathogen and catabolizes the pathogen's molecules, producing peptide epitopes within a phagolysosome 1. Another vesicle, already containing MHC class II molecules in its membrane, fuses with the phagolysosome. MHC II molecules bind complementary epitopes 2. The vesicle then fuses with the cytoplasmic membrane 3, leaving MHC II–epitope complexes on the cell's surface 4. Empty MHC II molecules are not stable on a cell's surface; they degrade. ▶**ANIMATIONS:** ***Antigen Processing and Presentation: Steps, MHC***

T lymphocytes develop in the thymus, each synthesizing a unique TCR. Next, let's consider various types of T lymphocytes that function in cell-mediated immune responses.

Types of T Lymphocytes

Immunologists recognize types of T cells based on surface glycoproteins and characteristic functions. These are *cytotoxic T cells, helper T cells,* and *regulatory T cells.*

Cytotoxic T Lymphocyte Every **cytotoxic T cell** (**Tc cell** or **CD8 cell**) is distinguished by copies of its own unique TCR as well as the presence of CD8 cell-surface glycoprotein. CD (for *cluster of differentiation*) glycoproteins are named with internationally accepted designations consisting of a number following the prefix. These numbers reflect the order in which the glycoproteins were discovered, not the order in which they are produced or function. As the name *cytotoxic T cell* implies, these lymphocytes directly kill other cells—those infected with viruses and other intracellular pathogens as well as abnormal cells, such as cancer cells.

Helper T Lymphocyte Immunologists distinguish **helper T cells** (**Th cells or CD4 cells**) by the presence of the CD4 glycoproteins. These cells are called "helpers" because their function is to help regulate the activity of B cells and cytotoxic T cells during immune responses by providing necessary signals and growth factors. During an immune response, there are two main subpopulations of helper T cells: *type 1 helper T cells (Th1 cells),* which assist cytotoxic T cells and stimulate and regulate innate immunity, and *type 2 helper T cells (Th2 cells),* which function in conjunction with B cells. Immunologists distinguish Th1 from Th2 cells on the basis of their secretions and by characteristic cell-surface proteins.

Helper T cells secrete various soluble protein messengers called *cytokines* that regulate the entire immune system, both adaptive and innate portions. We will consider the types and

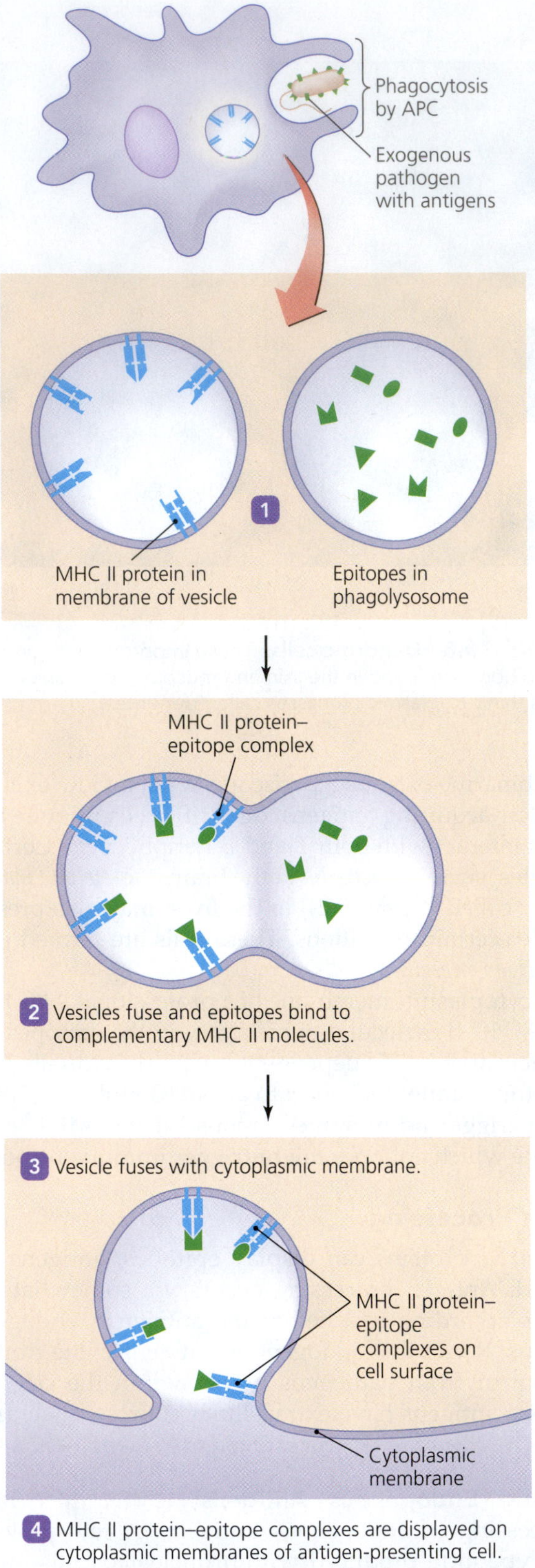

▲ **FIGURE 16.8 The processing of exogenous antigens.** Antigens arising outside the body's cells are phagocytized by an APC and digested to release epitopes, which are loaded into complementary antigen-binding grooves of MHC II molecules. The MHC II–epitopes are then displayed on the outside of the APC's cytoplasmic membrane.

effects of cytokines shortly. **Highlight: The Loss of Helper T Cells in AIDS Patients** describes some of the effects of the destruction of helper T cells in individuals infected with the human immunodeficiency virus (HIV). ▶**ANIMATIONS:** *Cell-Mediated Immunity: Helper T Cells*

Regulatory T Lymphocyte **Regulatory T cells (Tr cells),** previously known as *suppressor T cells,* repress adaptive immune responses and prevent autoimmune diseases. Tr cells express CD4 and CD25 glycoproteins. Scientists have not fully characterized the manner in which Tr cells work, but it is known that they are activated by contact with other immune cells and that they secrete some immunologically active chemicals called *cytokines,* which we examine in a subsequent section.

TABLE 16.1 compares and contrasts the features of various types of T lymphocytes. Now, we turn our attention to the way in which the body eliminates T cells that recognize normal body antigens by means of their TCRs.

Clonal Deletion

Given that T lymphocytes randomly generate the variable region shapes of their receptors, the lymphocytes include numerous cells with receptors complementary to normal body components—the autoantigens mentioned earlier. It is vitally important that adaptive immune responses not be directed against autoantigens; the immune system must be tolerant of "self." When self-tolerance is impaired, the result is an *autoimmune disease* (see Chapter 18).

The body eliminates self-reactive lymphocytes via **clonal deletion,** so named because elimination of a cell deletes its potential offspring (clones). In this process, lymphocytes are exposed to autoantigens, and those lymphocytes that react to autoantigens undergo **apoptosis** (programmed cell suicide) and are thereby deleted from the repertoire of lymphocytes. Apoptosis is the critical feature of clonal deletion and the development of self-tolerance. The result of clonal deletion is that surviving lymphocytes respond only to foreign antigens. In humans, clonal deletion for T lymphocytes occurs in the thymus. ▶**VIDEO TUTOR:** *Clonal Deletion*

Clonal Deletion of T Cells

As we have seen, T cells recognize epitopes when the epitopes are bound to MHC protein. Young T lymphocytes spend about a week in the thymus being exposed to all of the body's natural epitopes through a unique feature of thymus cells. As a group, these cells express all of the body's normal proteins, including proteins that have no function in the thymus. For example, some

HIGHLIGHT

The Loss of Helper T Cells in AIDS Patients

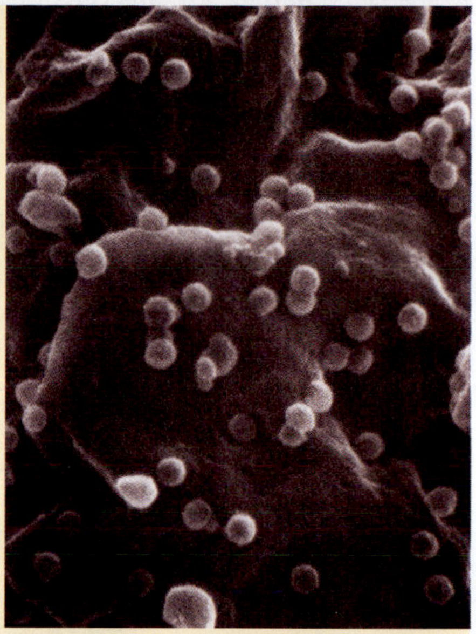

HIV budding from Th cells

Neither B cells nor cytotoxic T cells respond effectively to most antigens without the participation of helper T cells. The reason is that signals are passed more effectively among leukocytes when the CD4 molecules of helper T cells bind to certain leukocytes.

For viruses to enter cells and replicate, they must first attach to a specific protein on their target cells' surface (see Chapter 13). In the case of human immunodeficiency virus (HIV), the causative agent of AIDS, the surface protein to which it attaches is CD4, and its target cells are helper T cells. Once HIV enters a helper T cell, it, like other viruses, takes over the cell's protein-synthesizing machinery and directs the cell to produce more viruses.

Because helper T cells are essential to mounting an effective immune response, the body normally has a surplus of them—usually about three times as many as it needs. As a result, individuals infected with HIV can typically lose up to about two-thirds of their helper T cells before signs of immune deficiency appear. Cell-mediated immunity is usually affected first; even though B cells require the assistance of helper T cells to function optimally, they can still be stimulated directly by large quantities of antigen. Normal human blood typically contains about three CD4 helper T cells for every two CD8 cytotoxic T cells—a CD4-to-CD8 ratio of 3:2. During the course of HIV infection, however, CD4 helper T cells are lost, reducing the CD4-to-CD8 ratio such that individuals with full-blown AIDS may have a ratio lower than 1:7.

TABLE 16.1 Characteristics of T Lymphocytes

Lymphocyte	Site of Maturation	Representative Cell-Surface Glycoproteins	Selected Secretions
Helper T cell type 1 (Th1)	Thymus	CD4, CCR5, and distinctive TCR	Interleukin 2, IFN-γ
Helper T cell type 2 (Th2)	Thymus	CD4, CCR3, CCR4, and distinctive TCR	Interleukin 4
Cytotoxic T cell (Tc)	Thymus	CD8, CD95L, and distinctive TCR	Perforin, granzyme
Regulatory T cell (Tr)	Thymus	CD4, CD25, and distinctive TCR	Cytokines, such as interleukin 10

thymus cells synthesize lysozyme, hemoglobin, and muscle cell proteins, though these proteins are not expressed externally to the cells. Rather, thymus cells process these autoantigens so as to express their epitopes in association with an MHC protein. Since the cells collectively synthesize polypeptides from the body's proteins, together they process and present all the body's autoantigens to young T cells.

Immature T cells undergo one of four fates **(FIGURE 16.9)**:

- Those T cells that do not recognize the body's MHC protein undergo apoptosis, in other words, clonal deletion. Because they do not recognize the body's own MHC protein, they will be of no use identifying foreign epitopes carried by MHC protein. T cells that do recognize the body's MHC protein receive a signal to survive.
- Those that subsequently recognize autoantigen in conjunction with MHC protein mostly die by apoptosis—further clonal deletion.
- A few of these "self-recognizing" T cells remain alive to become regulatory T cells.
- The remaining T cells are those that will recognize the body's own MHC protein in conjunction with foreign epitopes and not with autoantigens. These T lymphocytes become the repertoire of protective T cells, which leave the thymus to circulate in the blood and lymph.

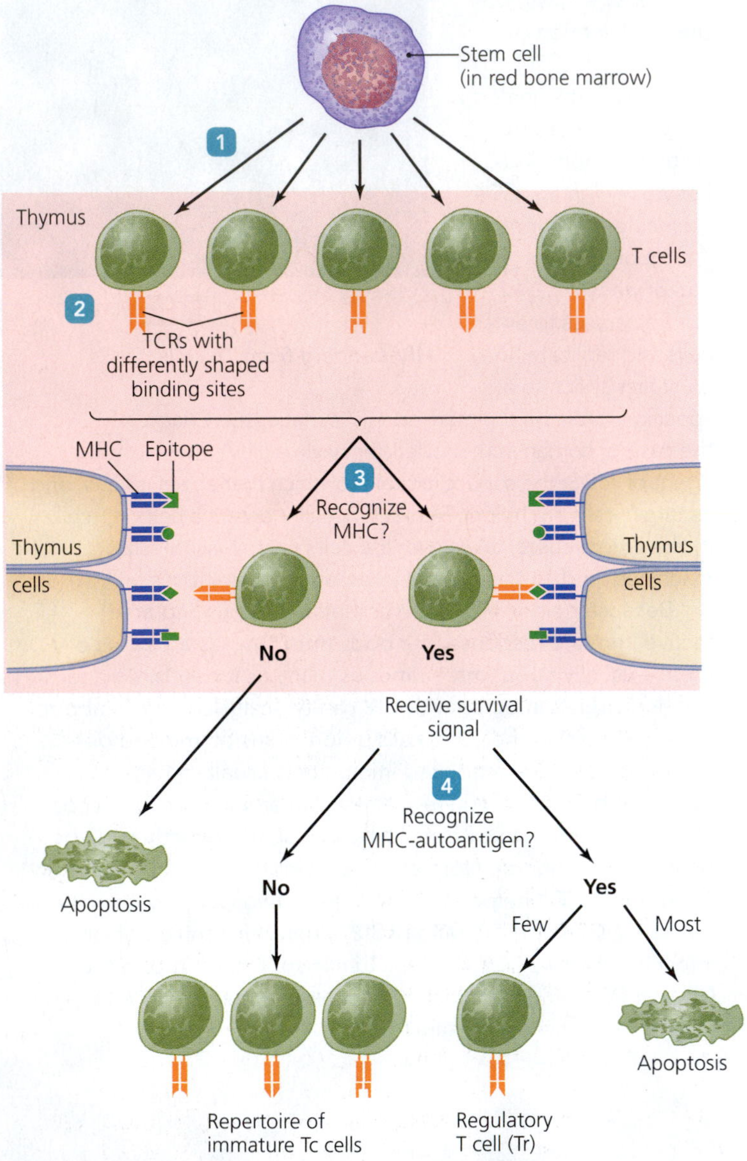

▲ **FIGURE 16.9 Clonal deletion of T cells.** 1 Stem cells in the red bone marrow generate a host of lymphocytes that move to the thymus. 2 In the thymus, each lymphocyte randomly generates a TCR with a particular shape. Note that each cell's TCR binding sites differ from those of other cells. 3 T cells pass through a series of "decision questions" in the thymus: Are their TCRs complementary to the body's MHC protein? If no, they undergo apoptosis—*clonal deletion*. If yes, they survive. 4 Do the surviving cells recognize MHC protein bound to any autoantigen? If no, they survive and become the repertoire of immature T cells. If yes, then most undergo apoptosis (more clonal deletion); a few survive as regulatory T cells (Tr).

Now that we have considered T lymphocytes, major histocompatibility proteins, and antigen-presenting cells, we will consider the other star player in adaptive immune responses—B lymphocytes.

B Lymphocytes (B Cells) and Antibodies

LEARNING **OUTCOMES**

16.16 Describe the characteristic of B lymphocytes that furnishes them specificity.

16.17 Describe the basic structure of an immunoglobulin molecule.

16.18 Contrast the structure, function, and prevalence of the five classes of immunoglobulins.

16.19 Compare and contrast clonal deletion of B cells with clonal deletion of T cells.

B lymphocytes are found in the spleen, in MALT, and in the primary follicles of lymph nodes. A small percentage of B cells circulates in the blood. The major function of B cells is the secretion of soluble antibodies, which we examine in more detail shortly. As we have established, B cells function in antibody immune responses, each of which is against only a particular epitope. As with T cells, such specificity comes from membrane proteins, in this case called *B cell receptors*.

Specificity of the B Cell Receptor (BCR)

The surface of each B lymphocyte is covered with about 500,000 identical copies of a **B cell receptor (BCR).** A BCR is a type of *immunoglobulin* (im´yū-nō-glob´yū-lin; Ig). A simple immunoglobulin contains four polypeptide chains—two identical longer chains called *heavy chains* and two identical shorter *light chains* **(FIGURE 16.10)**. The terms *heavy* and *light* refer to their relative molecular masses. Disulfide bonds, which are covalent bonds between sulfur atoms in two different amino acids, link the light chains to the heavy chains in such a way that a simple immunoglobulin looks like the letter Y. A BCR has two *arms* and a *transmembrane portion*. Each arm's end is a *variable region* because ends of each heavy and each light chain vary in amino acid sequence among B cells. The transmembrane portion

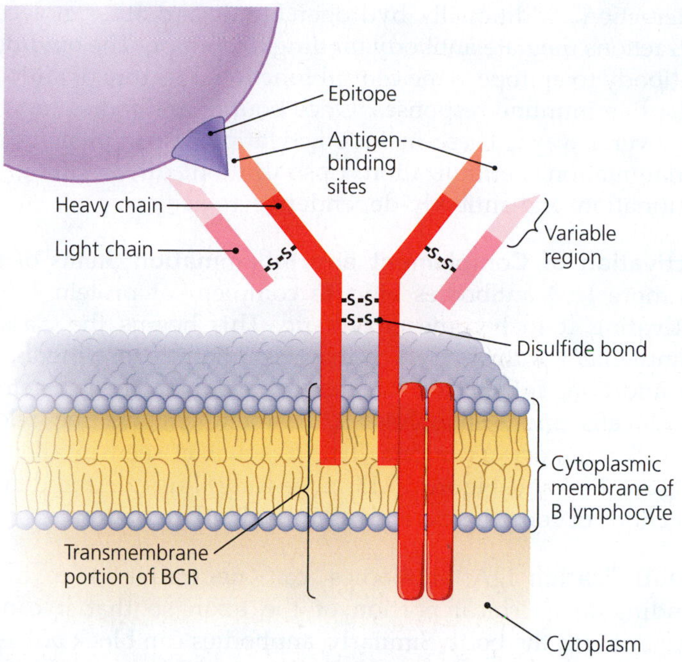

▲ **FIGURE 16.10 B cell receptor (BCR).** The B cell receptor is composed of a symmetrical, epitope-binding, Y-shaped protein in association with two transmembrane polypeptides.

anchors the BCR in the cytoplasmic membrane and consists of the stem of the Y (composed of the tails of the two heavy chains) and two other polypeptides.

Each B cell randomly generates a single BCR gene during its development in the bone marrow. Scientists estimate that there are fewer than 25,000 genes in a human cell, yet there are billions of different BCR proteins in an individual. Obviously, a person cannot have a separate gene for each BCR; instead, a developing B cell randomly recombines segments from three immunoglobulin regions of DNA and combines these segments to develop novel BCR genes. A cell may also randomly change its BCR genes to develop even more diversity. Each newly formed BCR gene codes for a specific and unique BCR. **Highlight: Lymphocyte Receptor Diversity: The Star of the Show** on p. 486 expands on the genetic explanation for the extensive BCR diversity.

All the BCRs of any particular cell are identical because the variable regions of every BCR on a single cell are identical—the two light chain variable regions are identical, and the two heavy chain variable regions are identical. Together the two variable regions form **antigen-binding sites** (see Figure 16.10). Antigen-binding sites are complementary in shape to the three-dimensional shape of an epitope and bind precisely to it. Exact binding between antigen-binding site and epitope accounts for the specificity of an antibody immune response.

Though all of the BCRs on a single B cell are the same, the BCRs of one cell differ from the BCRs of all other B cells, much as each snowflake is distinct from all others. Scientists estimate that each person forms no fewer than 10^9 and likely as many as 10^{13} B lymphocytes—each with its own BCR. Because an antigen (e.g., a bacterial protein) typically has numerous epitopes of various shapes, many different BCRs will recognize any particular antigen's epitopes, but each BCR recognizes only one epitope. BCR genes are randomly generated in sufficient numbers that the entire repertoire of BCRs (each carried by a particular B lymphocyte) is capable of recognizing the entire repertoire of thousands of millions of different epitopes. In other words, at least one BCR is fortuitously complementary to any given specific epitope that the body may or may not encounter.

An analogy will serve to clarify this point. Imagine a locksmith who has a copy of every possible key to fit every possible lock. If a customer arrives at the shop with a lock needing a key, the locksmith can provide it (though it may take a while to find the correct key). Similarly, you have a BCR complementary to every possible epitope in the environment, though you will encounter only some of them. For example, you have lymphocytes with BCRs complementary to epitopes of stingray venom, though it is unlikely you will ever be stabbed by a stingray.

When an antigenic epitope stimulates a specific B cell via the B cell's unique BCR, the B cell responds by undergoing cell division, giving rise to nearly identical offspring that secrete immunoglobulins into the blood or lymph. The immunoglobulins act against the epitope shape that stimulated the B cell. Activated, immunoglobulin-secreting B lymphocytes are called **plasma cells.** They have extensive rough endoplasmic reticulum and many Golgi bodies involved in the synthesis, packaging, and secreting of the immunoglobulins. Next we consider the structure and functions of the secreted immunoglobulins, which are called antibodies.

Specificity and Antibody Structure

Antibodies are free immunoglobulins—not attached to a membrane—and similar to BCRs in shape. Antibodies are secreted and lack most of the transmembrane portions of BCRs **(FIGURE 16.11)**. Thus, a basic antibody molecule is Y-shaped with two identical heavy chains and two identical light chains. The antigen-binding sites of antibodies from a given plasma cell are identical to one another and to the antigen-binding sites of that cell's BCR; thus, antibodies carry the same specificity for an epitope as the BCR of the activated B cell.

Because the arms of an antibody molecule contain antigen-binding sites, they are also known as the *F_{ab} regions (fragment, antigen-binding)*. The angle between the arms and the stem can change because the point at which they join is hinge-like. An antibody stem, which is formed of the lower portions of the two heavy chains, is also called the *F_c region* (because it forms a *fragment* that is *crystallizable*).

There are five basic types of stems (F_c regions), designated by the Greek letters *mu, gamma, alpha, epsilon,* and *delta.* A plasma cell attaches the gene for its heavy chain variable region to one of the five genes to form one of five classes of antibodies known as IgM, IgG, IgA, IgE, and IgD.

Antibody Function

As we have seen, antigen-binding sites of antibodies are complementary to epitopes; in fact, the shapes of the two can match so closely that most water molecules are excluded from the area of contact, producing a strong, noncovalent, hydrophobic

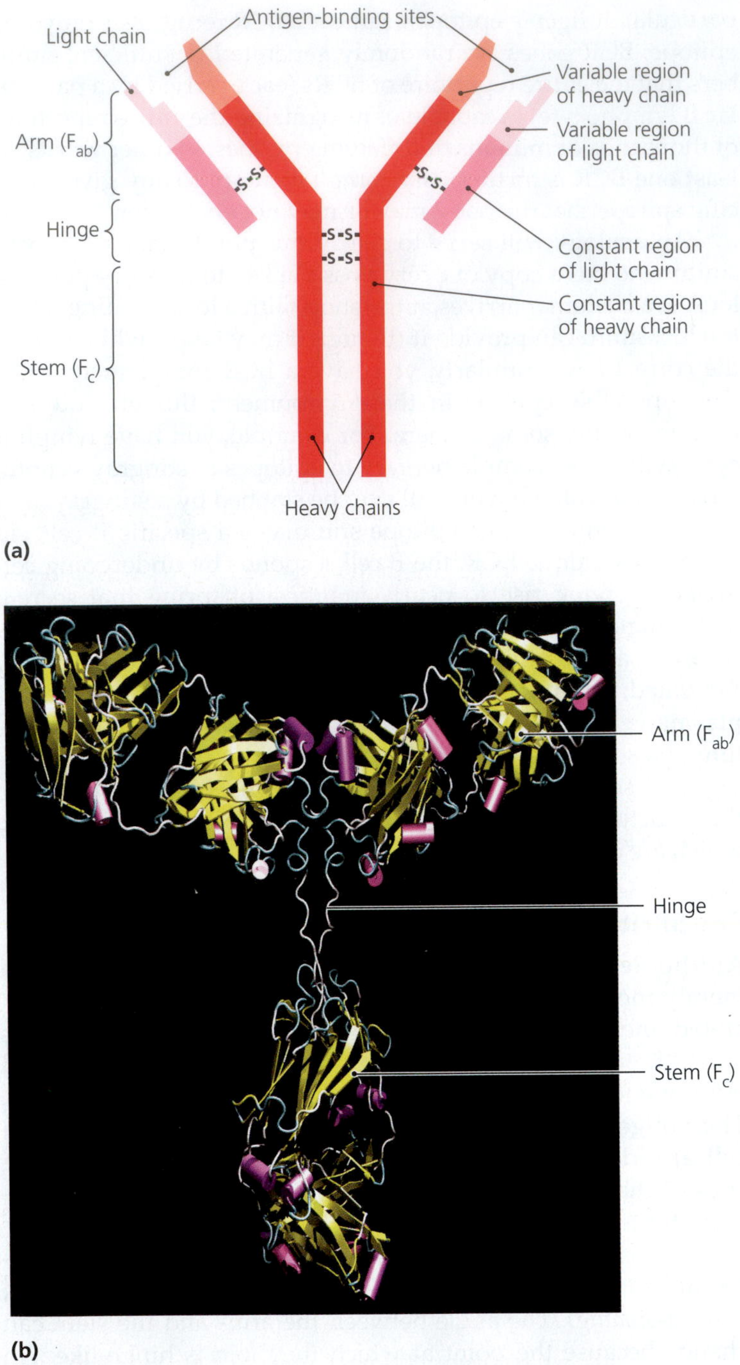

▲ **FIGURE 16.11 Basic antibody structure.** **(a)** Artist's rendition. Each antibody molecule, which is shaped like the letter Y, consists of two identical heavy chains and two identical light chains held together by disulfide bonds. Five different kinds of heavy chains form an antibody's stem (F_c region). The arms (F_{ab} regions) terminate in variable regions to form two antigen-binding sites. The hinge region is flexible, allowing the arms to bend in almost any direction. **(b)** Three-dimensional shape of an antibody based on X-ray crystallography. *Why are the two antigen-binding sites on a typical antibody molecule identical?*

Figure 16.11 The amino acid sequences of the two light chains of an antibody molecule are identical, as are the sequences of the two heavy chains, so the binding sites—each composed of the ends of one light chain and one heavy chain—must be identical.

interaction. Additionally, hydrogen bonds and other molecular attractions mediate antibody binding to epitope. The binding of antibody to epitope is the central functional feature of antibody adaptive immune responses. Once bound, antibodies function in several ways. These include activation of complement and inflammation, neutralization, opsonization, direct killing, agglutination, and antibody-dependent cytoxicity.

Activation of Complement and Inflammation Stems of two or more IgM antibodies bind to complement protein 1 (C1), activating it to become enzymatic. This begins the classical complement pathway, which releases inflammatory mediators. In addition, IgE bound to antigen attaches via its stem to most cells and eosinophils. Attachment triggers the release of inflammatory chemicals. This is what is seen in allergies. (Figure 15.8 and Figure 15.13 more fully illustrate the defense reactions of complement activation and inflammation.)

Neutralization IgA antibodies can **neutralize** a toxin by binding to a critical portion of the toxin so that it can no longer harm the body. Similarly, antibodies can block adhesion molecules on the surface of a bacterium or virus, neutralizing the pathogen's virulence because it cannot adhere to its target cell **(FIGURE 16.12a)**.

Opsonization Antibodies act as **opsonins**[8]—molecules that stimulate phagocytosis. Neutrophils and macrophages have receptors for the stems of IgG molecules; therefore, these leukocytes bind to the stems of antibodies. Once antibodies are so bound, the leukocytes phagocytize them, along with the antigens they carry, at a faster rate compared to antigens lacking bound antibody. Changing the surface of an antigen so as to enhance phagocytosis is called **opsonization** (op´sŏ-nī-zā´shŭn; **FIGURE 16.12b**).

Killing by Oxidation Recently, scientists have shown that some antibodies have catalytic properties that allow them to kill bacteria directly **(FIGURE 16.12c)**. Specifically, antibodies catalyze the production of hydrogen peroxide, ozone, and other potent oxidants that kill bacteria.

Agglutination Because each basic antibody has two antigen-binding sites, each can attach to two epitopes at once. Numerous antibodies can aggregate antigens together—a state called **agglutination** (ă-glū-ti-nā´shŭn; **FIGURE 16.12d**). Agglutination of soluble molecules typically causes them to become insoluble and precipitate. Agglutination may hinder the activity of pathogenic organisms and increases the chance that they will be phagocytized or filtered out of the blood by the spleen. (Chapter 17 examines some uses scientists make of the agglutinating nature of antibodies.)

Antibody-Dependent Cellular Cytoxicity (ADCC) Antibodies often coat a target cell by binding to epitopes all over the target's surface. The antibodies' stems can then bind to receptors on

[8]From Greek *opsonein*, meaning "to supply food," and *izein*, meaning "to cause"; thus, loosely, "to prepare for dinner."

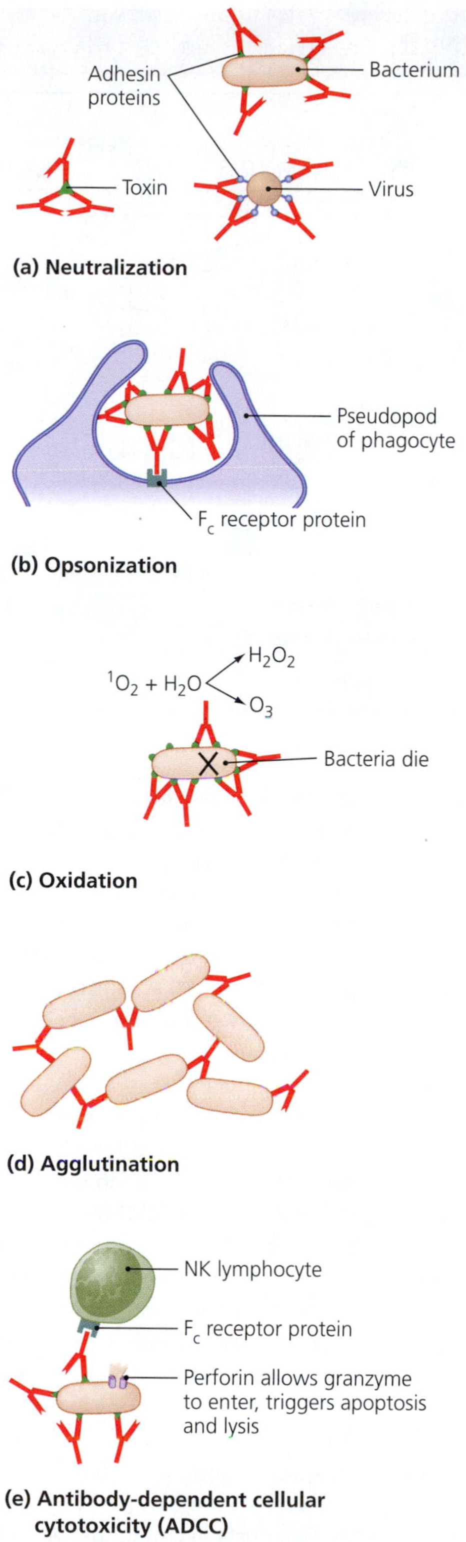

▲ **FIGURE 16.12 Five functions of antibodies.** Drawings are not to scale. **(a)** Neutralization of toxins and microbes. **(b)** Opsonization. **(c)** Oxidation by toxic forms of oxygen such as $^1O_2$ (singlet oxygen), H_2O_2 (hydrogen peroxide), and O_3 (ozone). **(d)** Agglutination. **(e)** Antibody-dependent cellular cytotoxicity. Antibodies have two other functions—activation of complement and participation in inflammation—(illustrated in Chapter 15).

special lymphocytes called *natural killer lymphocytes (NK cells),* which are neither B nor T cells. NK lymphocytes lyse target cells with proteins called **perforin** (per´fōr-in) and **granzyme** (gran´zīm). Perforin molecules form into a tubular structure in the target cell's membrane, forming a channel through which granzyme enters the cell and triggers **apoptosis**[9] (programmed cell suicide; **FIGURE 16.12e**). **Antibody-dependent cellular toxicity (ADCC)** is similar to opsonization in that antibodies cover the target cell; however, with ADCC the target dies by apoptosis, whereas with opsonization the target is phagocytized.
▶**ANIMATIONS:** ***Humoral Immunity: Antibody Function***

Classes of Antibodies

Threats confronting the body can be extremely variable, so it is not surprising that there are several classes of antibody. The class involved in any given antibody immune response depends on the type of invading foreign antigens, the portal of entry involved, and the antibody function required. Here, we consider the structure and functions of the five classes of antibodies.

Every B cell begins by attaching its variable region gene to the gene for the mu stem and thus begins by making class M—**immunoglobulin M (IgM).** Most IgM is secreted during the initial stages of an immune response. A secreted IgM molecule is more than five times larger than the basic Y-shape because secreted IgM is a pentamer, consisting of five basic units linked together in a circular fashion via disulfide bonds and a short polypeptide *joining (J) chain* (see Table 16.2 on p. 488). Each IgM subunit has a conventional immunoglobulin structure, consisting of two light chains and two mu heavy chains. IgM is most efficient at complement activation, which also triggers inflammation, and can be involved in agglutination and neutralization.

In a process called **class switching,** a plasma cell then combines its variable region gene to the gene for a different stem and begins secreting a new class of antibodies. The most common switch is to the gene for heavy chain gamma; that is, the plasma cell switches to synthesizing immunoglobulin G.

Immunoglobulin G (IgG) is the most common and longest-lasting class of antibody in the blood, accounting for about 80% of serum antibodies, possibly because IgG has many functions. Each molecule of IgG has the basic Y-shaped antibody structure.

IgG molecules play a major role in antibody-mediated defense mechanisms, including complement activation, opsonization, neutralization, and antibody-dependent cellular cytotoxicity (ADCC). IgG molecules can leave blood vessels to enter extracellular spaces more easily than can the other immunoglobulins. This is especially important during inflammation because it enables IgG to bind to invading pathogens before they get into the circulatory systems. IgG molecules are also the only antibodies that cross a placenta to protect a developing child.

Immunoglobulin A (IgA), which has alpha heavy chains, is the immunoglobulin most closely associated with various

[9] Greek, meaning "falling off."

HIGHLIGHT

Lymphocyte Receptor Diversity: The Star of the Show

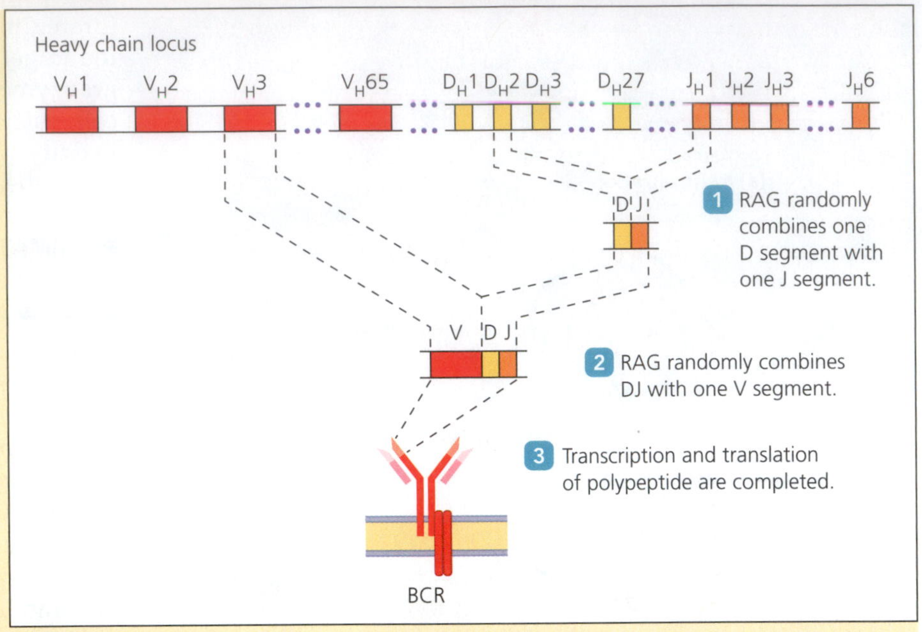

Heavy chain locus
V_H1 V_H2 V_H3 V_H65 D_H1 D_H2 D_H3 D_H27 J_H1 J_H2 J_H3 J_H6
D J
1 RAG randomly combines one D segment with one J segment.
V D J
2 RAG randomly combines DJ with one V segment.
3 Transcription and translation of polypeptide are completed.
BCR

How does your body generate billions of unique lymphocyte receptor proteins (TCRs and BCRs) so as to recognize the thousands of millions of foreign epitopes on pathogens that attack you? There isn't enough DNA in a cell to have individual genes for so many receptors. It would require several thousand times more DNA than you have. The answer to the problem of receptor diversity lies in the ingenious way in which lymphocytes use their relatively small number of diverse receptor genes. Let's consider the diversity generated by B cells.

Genes for BCRs occur in discrete stretches of DNA called *loci* (singular: *locus*). Genes for the constant region are located downstream from loci for the variable regions. We will consider the variable region genes first because it is in the variable region that BCR diversity is greatest.

BCR variable region genes occur in three loci on each of two chromosomes—one locus for the heavy chain variable region and two loci (called *kappa* and *lambda*) for the light chain variable region. Each cell is diploid—having two of each type of chromosome—so there are six loci altogether. However, a developing B cell uses only one chromosome's loci for each of its heavy and light chains. Once a locus on a particular chromosome is used, the other chromosome's corresponding locus is inhibited.

Each locus is divided further into distinct genetic segments coding for portions of its respective chain. The heavy chain variable region locus has three segments called *variable (V_H), diversity (D_H),* and *junction (J_H) segments.* In each of your developing B cells there are 65 variable segment genes, 27 diversity segment genes, and 6 junction segment genes. Each light chain variable region locus has an additional *variable segment (V_L)* and a *junction segment (J_L)*. For the kappa locus, there are 40 variable segment genes and 5 junction genes. The lambda locus is less variable, having only 30 V genes and a single J gene.

A great variety of BCRs form during B cell maturation as the B cell uses an enzyme called RAG—the *recombination activating gene* protein—to randomly combine one of each kind of the various segments to form its BCR gene. An analogy will help your understanding of this concept. Imagine that you have a wardrobe consisting of 65 different pairs of shoes, 27 different shirts, and 6 different pairs of pants that can be combined to create outfits. You can choose any one pair of shoes, any one shirt, and any one pair of pants each day, so with these 98 pieces of clothing, you will potentially have 10,530 different outfits (65 × 27 × 6). Your roommate might have the same numbers of items but in different colors and styles; so between you, you will have twice as many potential outfits—21,060! Similarly, each developing B cell has 10,530 possible combinations of V_H, D_H, and J_H segments on each of its two chromosomes for a total of 21,060 possible heavy chain variable region gene combinations.

The developing B cell also uses RAG to recombine the segments of the light chain variable loci. Because there are 200 (40 × 5) possible kappa genes and 30 (30 × 1) possible lambda genes on each chromosome, there is a total of 460 possible light chain genes on the two chromosomes. Therefore, each B cell can make one of a possible 9,687,600 (21,060 × 460) different BCRs using only 348 genetic segments in the heavy and light chain loci on its two chromosomes.

Still, these nearly 10 million different BCRs do not account for all the variability seen among BCRs. Each B cell creates additional diversity: RAG randomly removes portions of D and J segments before joining the two together. Another enzyme randomly adds nucleotides to each heavy chain VDJ combination, and B cells in the primary follicles of lymph nodes undergo random point mutations in their V regions. The result of RAG recombinations, random deletions, random insertions, and point mutations is tremendous potential variability. Scientists estimate that you may have about 10^{23} B cell receptor possibilities. That's one hundred billion trillion—a number 10 times greater than all the stars in the universe! Your B cells are indeed stars of the show.

body secretions. Some of the body's IgA is a monomer with the basic Y-shape that circulates in the blood, constituting about 12% of total serum antibody. However, plasma cells in the tear ducts, mammary glands, and mucous membranes synthesize **secretory IgA,** which is composed of two IgA molecules linked via a J chain and another short polypeptide (called a *secretory component*). Plasma cells add secretory component during the transport of secretory IgA across mucous membranes. Secretory component protects secretory IgA from digestion by intestinal enzymes.

Secretory IgA agglutinates and neutralizes antigens and is of critical importance in protecting the body from infections arising in the gastrointestinal, respiratory, urinary, and reproductive tracts. IgA provides nursing newborns some protection against foreign antigens because mammary glands secrete IgA into milk. Thus, nursing babies receive antibodies directed against antigens that have infected their mothers and are likely to infect them as well.

Immunoglobulin E (IgE) is a typical Y-shaped immunoglobulin with two epsilon heavy chains. Because it is found in extremely low concentrations in serum (less than 1% of total antibody), it is not critical for most antibody functions. Instead, IgE antibodies act as signal molecules—they attach to receptors on eosinophil cytoplasmic membranes to trigger the release of cell-damaging molecules onto the surface of parasites, particularly parasitic worms. IgE antibodies also trigger mast cells and basophils to release inflammatory chemicals, such as histamine. In developed countries, IgE is more likely associated with allergies than with parasitic worms.

Immunoglobulin D (IgD) is characterized by delta heavy chains. IgD molecules are not secreted but are membrane-bound antigen receptors on B cells that are often seen during the initial phases of an antibody immune response. In this regard, IgD antibodies are like BCRs. Not all mammals have IgD, and animals that lack IgD show no observable ill effects; therefore, scientists do not know the exact function or importance of this class of antibody.

TABLE 16.2 on p. 488 compares the different classes of immunoglobulins and adds some details of antibody structure.

Clonal Deletion of B Cells

Clonal deletion of B cells occurs in the bone marrow in a manner similar to deletion of T cells **(FIGURE 16.13)**, though self-reactive B cells may become inactive or change their BCR rather than undergo apoptosis. In any case, self-reactive B cells are removed from the active B cell repertoire so that the antibody immune response does not act against autoantigens. Tolerant B cells leave the bone marrow and travel to the spleen, where they undergo further maturation before circulating in the blood and lymph.

Surviving B and T lymphocytes move into the blood and lymph, where they form the lymphocyte repertoire that scans for antigens. They communicate among one another and with other body cells via chemical signals called *cytokines.* In terms of our analogy of a stage play, there is dialogue among the cast of characters, but in our story the dialogue consists of chemical signals.

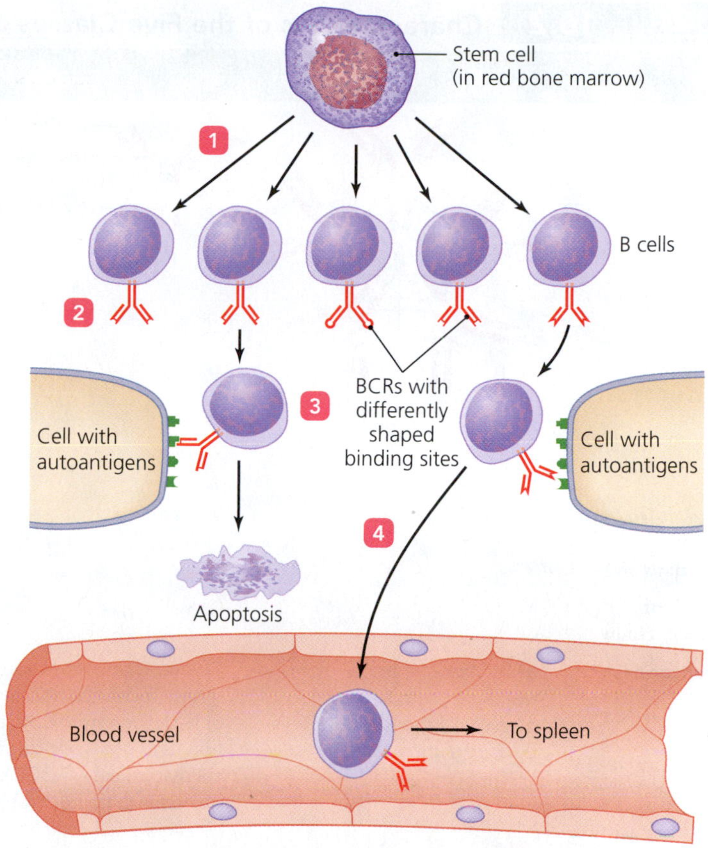

▲ **FIGURE 16.13 Clonal deletion of B cells.** 1 Stem cells in the red bone marrow generate a host of B lymphocytes. 2 Each newly formed B cell randomly generates a BCR with a particular shape. Note that each cell's BCR binding sites differ from those of other cells. 3 Cells whose BCR is complementary to some autoantigen bind with that autoantigen, stimulating the cell to undergo apoptosis. Thus, an entire set of potential daughter B cells (a clone) that are reactive with the body's own cells are eliminated—*clonal deletion.* 4 B cells with a BCR that is not complementary to any autoantigen are released from the bone marrow and into the blood. *Of the B cells shown, which is likely to undergo apoptosis?*

Figure 16.13 *The second and third B lymphocytes from the left will undergo apoptosis; their active sites are complementary to the second and fourth autoantigens from the top, respectively.*

Immune Response Cytokines

LEARNING | **OUTCOME**

16.20 Describe five types of cytokines.

Cytokines (sī´tō-kīnz) are soluble regulatory proteins that act as intercellular messages when released by certain body cells, including those of the kidney, skin, and immune system. Here we are concerned with cytokines that signal among various immune leukocytes. For example, cytotoxic T cells (Tc) do not respond to antigens unless they are first signaled by cytokines.

Immune system cytokines are secreted by various leukocytes and affect diverse cells. Many cytokines are redundant; that is, they have almost identical effects. Such complexity has

TABLE 16.2 Characteristics of the Five Classes of Antibodies

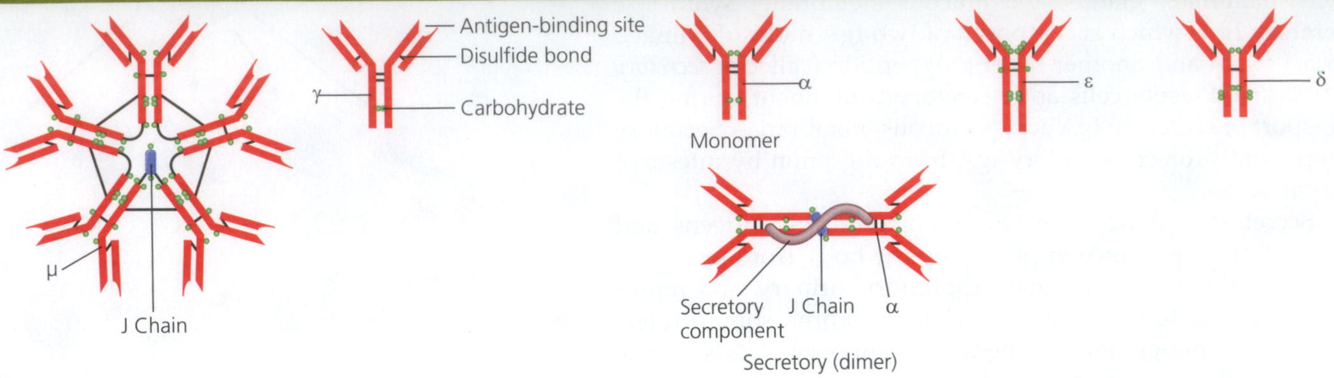

	IgM	IgG	IgA	IgE	IgD
Structure, number of binding sites	Pentamer, 10	Monomer, 2	Monomer, 2 Dimer, 4	Monomer, 2	Monomer, 2
Type of heavy chain	Mu (μ)	Gamma (γ)	Alpha (α)	Epsilon (ε)	Delta (δ)
Functions	Monomer can act as BCR; pentamer acts in complement activation, neutralization, agglutination	Complement activation, neutralization, opsonization, production of hydrogen peroxide, agglutination, and antibody-dependent cellular toxicity (ADCC); crosses placenta to protect fetus	Neutralization and agglutination; dimer is secretory antibody	Triggers release of antiparasitic molecules from eosinophils and of histamines from basophils and mast cells (allergic reactions)	Unknown, but perhaps acts as BCR
Locations	Serum, B cell surface	Serum, mast cell surfaces	Monomer: serum Dimer: mucous membrane secretions (e.g., tears, saliva, mucus); milk	Serum, mast cell surfaces	B cell surface
Approximate half-life (time it takes for concentration to reduce by half) in blood	10 days	20 days	6 days	2 days	3 days
Percentage of serum antibodies	5–10%	80%	10–15%	<1%	<0.05%
Size (mass in kilodaltons)	970	150	Monomer: 160 Dimer: 385	188	184

given rise to the concept of a *cytokine network*—a complex web of signals among all the cell types of the immune system. The nomenclature of cytokines is not based on a systematic relationship among them; instead, scientists named cytokines after their cells of origin, their function, and/or the order in which they were discovered. Cytokines of the immune system include the following substances:

- **Interleukins**[10] (in-ter-lū´kinz; **ILs**). As their name suggests, ILs signal among leukocytes, though cells other than leukocytes may also use interleukins. Immunologists named interleukins sequentially as they were discovered. Currently, scientists have identified about 35 interleukins.
- **Interferons** (in-ter-fēr´onz; **IFNs**). These proteins, which inhibit the spread of viral infections (as discussed in Chapter 15), may also act as cytokines. The most important interferon with such a dual function is gamma interferon (IFN-γ), which is a potent phagocytic activator secreted by type 1 helper T cells.
- **Growth factors.** These proteins stimulate leukocyte stem cells to divide, ensuring that the body is supplied with

[10]From Latin *inter,* meaning "between," and Greek *leukos,* meaning "white."

TABLE 16.3 Selected Immune Response Cytokines

Cytokine	Representative Source	Representative Target	Representative Action
Interleukin 2 (IL-2)	Type 1 helper T (Th1) cell, cytotoxic T (Tc) cell	Tc cell	Cloning of Tc cell
Interleukin 4 (IL-4)	Type 2 helper T (Th2) cell	B cell	B cell differentiates into plasma cell
Interleukin 12 (IL-12)	Dendritic cell	Helper T (Th) cell	Th cell differentiates into Th1 cell
Gamma interferon (IFN-γ)	Th1 cell	Macrophage	Increases phagocytosis
Tumor necrosis factor (TNF)	Macrophages, T cells	Body tissues	Triggers inflammation or apoptosis

sufficient white blood cells of all types. The body can control the progression of an adaptive immune response by limiting the production of growth factors.

- **Tumor necrosis[11] factor (TNF).** Macrophages and T cells secrete TNF to kill tumor cells and to regulate immune responses and inflammation.
- **Chemokines** (kē´mō-kīnz). Chemokines are chemotactic cytokines; that is, they signal leukocytes to move—for example, to rush to a site of inflammation or infection or to move within tissues.

TABLE 16.3 summarizes some properties of selected cytokines.

TELL ME WHY

Why are exogenous epitopes processed in vesicles instead of in endoplasmic reticulum, as endogenous epitopes are?

We have been considering adaptive immunity as a stage play. To this point, we have examined the "stage" (the tissues and organs of the lymphatic system) and the "cast of characters" involved in adaptive immunity. The latter are: antigens with their epitopes (the "villains"); T cells with their TCRs (the "heroes"), and B cells with their BCRs, plasma cells and their antibodies. The dialogue consists of cytokines. Actors who would disrupt the play—those T cells and B cells that recognize autoantigens—have been eliminated by clonal deletion.

We have set the stage for the immune system "play" by examining the preparatory steps of antigen processing and antigen presentation. We have seen that adaptive immunity is specific because of the precise and accurate fit of lymphocyte receptors with their complements; that adaptive immunity involves clones of T and B cells, which are unresponsive to self (because of clonal deletion); and that immune responses are inducible against foreign antigens. Now, we can examine cell-mediated and antibody immune responses in more detail. We begin with cell-mediated immunity.

[11]From Latin *necare*, meaning "to kill."

Cell-Mediated Immune Responses

LEARNING **OUTCOMES**

16.21 Describe a cell-mediated immune response.
16.22 Compare and contrast the two pathways of cytotoxic T cell action.

The body uses cell-mediated immune responses primarily to fight intracellular pathogens and abnormal body cells. Recall that inducibility and specificity are two hallmark characteristics of adaptive immunity. The body induces cell-mediated immune responses only against specific endogenous antigens. Given that many common intracellular invaders are viruses, our examination of cell-mediated immunity will focus on these pathogens. However, cell-mediated immune responses are also mounted against cancer cells, intracellular parasitic protozoa, and intracellular bacteria, such as *Mycobacterium tuberculosis* (mī´kō-bak-tēr´ē-ŭm too-ber-kyū-lō´sis), which causes tuberculosis. **Highlight: Attacking Cancer with Lab-Grown T Cells** on p. 490 describes an experimental use of T cells to treat one form of cancer.

Activation of Cytotoxic T Cell Clones and Their Functions

The body initiates adaptive immune responses not at the site of an infection but rather in lymphoid organs, usually lymph nodes, where antigen-presenting cells interact with lymphocytes. The initial event in cell-mediated immunity is the activation of a specific clone of cytotoxic T cells, as depicted in **FIGURE 16.14**:

1. **Antigen presentation.** A virus-infected dendritic cell (the APC) migrates to a nearby lymph node where it presents virus epitopes in conjunction with the APC's MHC I protein. Because of the vast repertoire of randomly generated T cell receptors (TCRs), at least one cytotoxic T (Tc) cell will have a TCR complementary to the presented MHC I protein–epitope complex. This Tc cell binds to the dendritic cell to form a cell-cell contact site called an *immunological synapse*. CD8 glycoprotein of the Tc cell, which specifically binds to MHC I protein, stabilizes the synapse.
2. **Helper T cell differentiation.** A nearby CD4-bearing helper T (Th) lymphocyte assists by binding to the APC via

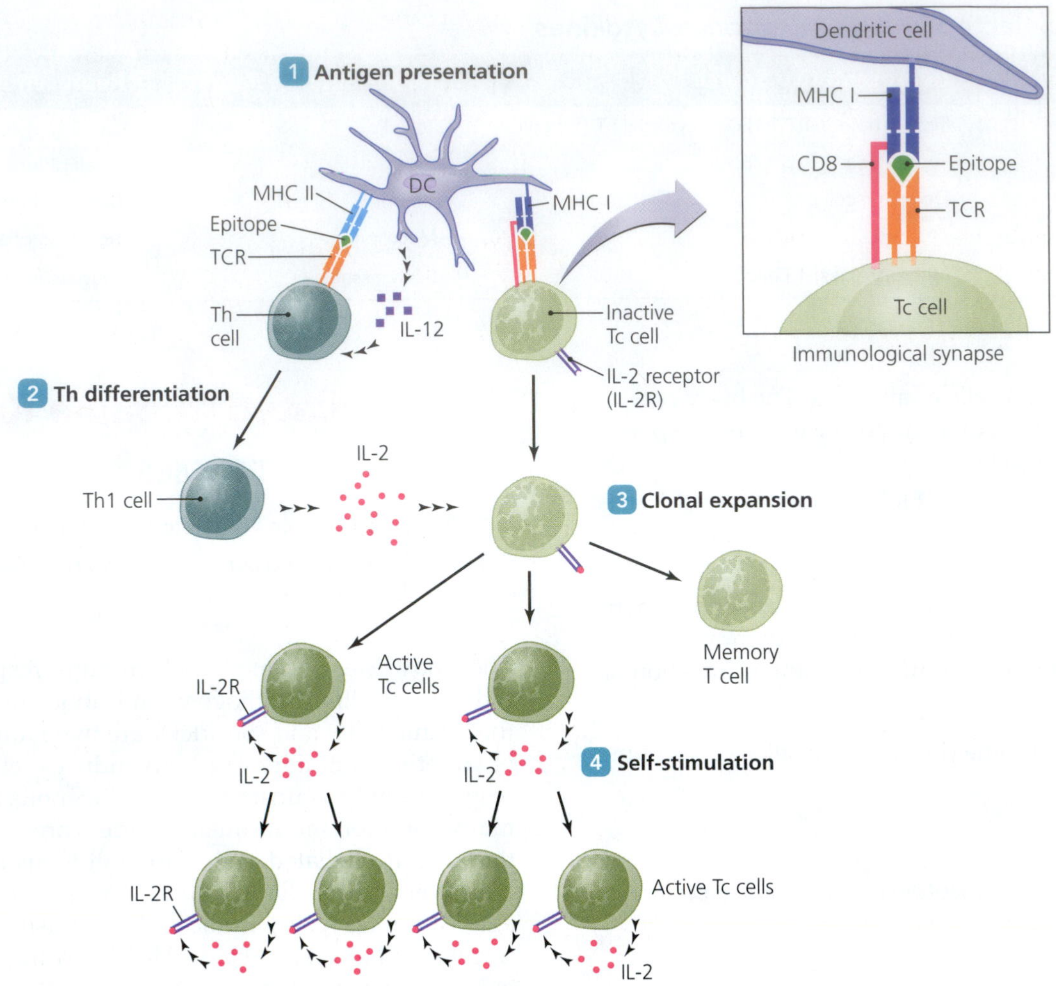

▲ **FIGURE 16.14 Activation of a clone of cytotoxic T (Tc) cells.** 1 Antigen-presenting cells, here a dendritic cell, present epitopes in conjunction with MHC II protein to helper T cells and with MHC I to Tc cells. 2 Infected APCs secrete IL-12, which causes helper T cells to differentiate into Th1 cells. 3 Signaling from the APC and IL-2 from the Th1 cell activates Tc cells that recognize the MHC I protein–epitope complex. IL-2 triggers Tc cells to divide, forming a clone of active Tc cells as well as memory T cells. 4 Active Tc cells secrete IL-2, becoming self-stimulatory.

HIGHLIGHT

Attacking Cancer with Lab-Grown T Cells

The body naturally manufactures T cells that attack cancer cells, but it sometimes does not make enough of them to shrink tumors and effectively halt the tumor's progress. Researchers have taken samples of a patient's own cancer-fighting T cells and cloned them in a laboratory until they numbered in the billions. These billions of identical T cells have then been injected back into the individual that first produced them. Many patients treated this way have become "virtually cancer free," while tumors were substantially reduced in other patients.

Such therapy, called adoptive T cell therapy, remains experimental and is still some time away from becoming a generally accepted cancer treatment. Scientists do not yet understand why the therapy works in some patients and not others. To date, the therapy has been tested primarily against malignant melanoma, but planning to test it against other types of cancer is under way, and some scientists believe that a similar therapy may be effective against viral diseases such as hepatitis.

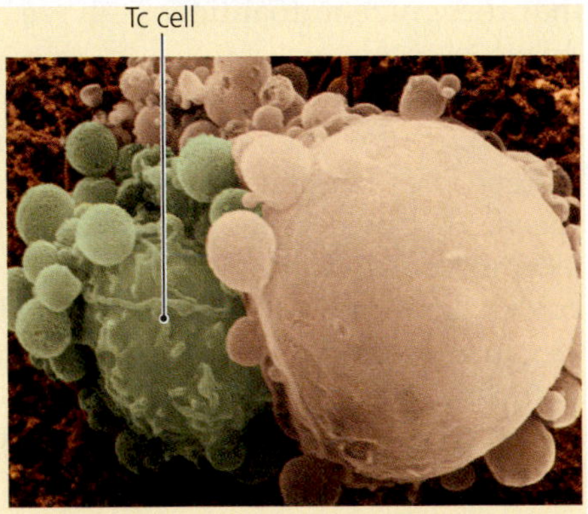

T cell (green) attacking a cancer cell

the TCR of the helper cell (which is complementary to an MHC II protein–epitope complex presented on the APC). This induces the APC to more vigorously signal the Tc cell. Viruses and some intracellular bacteria induce dendritic cells to secrete interleukin 12 (IL-12), which stimulates the helper T cells to become a clone of type 1 helper T (Th1) cells. Th1 cells in turn secrete IL-2. Lacking the assistance of Th cells, an immunological synapse between the APC and the Tc cell fails to progress. This limits improper immune responses.

3 **Clonal expansion.** The dendritic cell imparts a second required signal in the immunological synapse. This signal, in conjunction with any IL-2 from a Th1 cell that may be present, activates the cytotoxic T (Tc) cell to secrete its own IL-2. Interleukin 2 triggers cell division by Tc cells. Activated Tc cells reproduce to form memory T cells (discussed shortly) and more Tc progeny—a process known as **clonal expansion.**

4 **Self-stimulation.** Daughter Tc cells activate and produce both IL-2 receptors (IL-2R) and more IL-2, thereby becoming self-stimulating—they no longer require either an APC or a helper T cell. They leave the lymph node and are now ready to attack virally infected cells.

As previously discussed, when any nucleated cell synthesizes proteins, it displays epitopes from them in the antigen-binding grooves of MHC class I molecules on its cytoplasmic membrane. Thus, when viruses are replicated inside cells, epitopes of viral proteins are displayed on the host cell's surface. An active Tc cell binds to an infected cell via its TCR, which is complementary to the MHC I protein–epitope complex, and via its CD8 glycoprotein, which is complementary to the MHC class I protein of the infected cell **(FIGURE 16.15a)**.

Cytotoxic T cells kill their targets through one of two pathways: the *perforin-granzyme pathway,* which involves the synthesis of special killing proteins, or the *CD95 pathway,* which is mediated through a glycoprotein found on the body's cells.
▶**ANIMATIONS:** *Cell-Mediated Immunity: Cytotoxic T Cells*

The Perforin-Granzyme Cytotoxic Pathway

The cytoplasm of cytotoxic T cells has vesicles containing two key protein cytotoxins—*perforin* and *granzyme,* which are also used by NK cells in conjunction with antibody-dependent cellular cytotoxicity (discussed previously). When a cytotoxic T cell first attaches to its target, vesicles containing the cytotoxins release their contents. Perforin molecules aggregate into a channel through which granzyme enters, activating apoptosis in the target cell **(FIGURE 16.15b)**. Having forced its target to commit suicide, the cytotoxic T cell disengages and moves on to another infected cell.

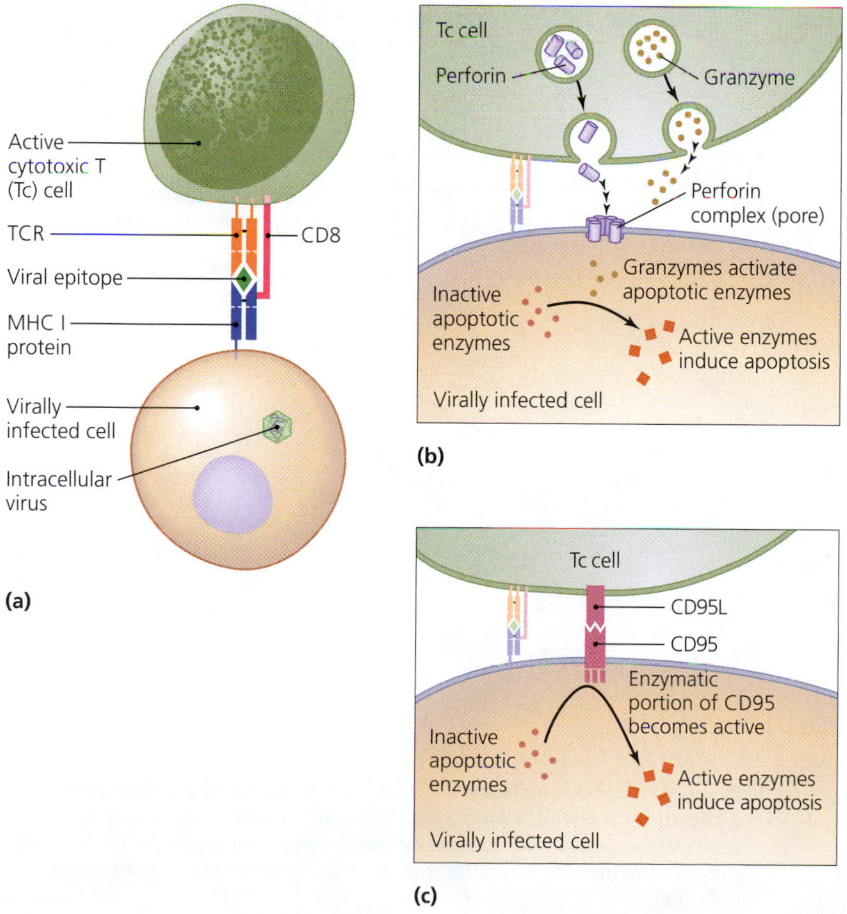

◀ **FIGURE 16.15 A cell-mediated immune response.** **(a)** The binding of a virus-infected cell by an active cytotoxic T (Tc) cell. **(b)** The perforin-granzyme cytotoxic pathway. After perforins and granzymes have been released from Tc cell vesicles, granzymes enter the infected cell through the perforin complex pore and activate the enzymes of apoptosis. **(c)** The CD95 cytotoxic pathway. Binding of CD95L on the Tc cell activates the enzymatic portion of the infected cell's CD95 such that apoptosis is induced.

The CD95 Cytotoxic Pathway

The **CD95 pathway** of cell-mediated cytotoxicity involves an integral glycoprotein called *CD95* that is present in the cytoplasmic membranes of many body cells. Activated Tc cells insert *CD95L*—the receptor for CD95—into their cytoplasmic membranes. When an activated Tc cell comes into contact with its target, its CD95L binds to CD95 on the target, which then activates enzymes that trigger apoptosis, killing the target cells **(FIGURE 16.15c)**.

Memory T Cells

LEARNING | OUTCOME

16.23 Describe the establishment of memory T cells.

Some activated T cells become **memory T cells,** which may persist for months or years in lymphoid tissues. If a memory T cell subsequently contacts an epitope–MHC I protein complex matching its TCR, it responds immediately (without a need for interaction with APCs) and produces cytotoxic T cell clones that recognize the offending epitope. These cells need fewer regulatory signals and become functional immediately. Further, because the number of memory T cells is greater than the number of T cells that recognized the antigen during the initial exposure, a subsequent cell-mediated immune response to a previously encountered antigen is much more effective than a primary response. An enhanced cell-mediated immune response upon subsequent exposure to the same antigen is called a **memory response.**[12]

T Cell Regulation

LEARNING | OUTCOME

16.24 Explain the process and significance of the regulation of cell-mediated immunity.

The body carefully regulates cell-mediated immune responses so that T cells do not respond to autoantigens. As we have seen, T cells require several signals from an antigen-presenting cell to activate. If the T cells do not receive these signals in a specific sequence—like the sequence of numbers in a combination padlock—they will not respond. Thus, when a T cell and an antigen-presenting cell interact in an immunological synapse, the two cell types have a chemical dialogue that stimulates the T cell to fully respond to the antigen. If a T cell does not receive the signals required for its activation, it will "shut down" as a precaution against autoimmune responses.

Regulatory **T** (Tr) cells also moderate cytotoxic T cells by mechanisms that are beyond the scope of our discussion. Suffice it to say that Tr cells provide one more level of control over potentially dangerous cell-mediated immune responses.

[12]Sometimes also called anamnestic responses, from Greek *ana*, meaning "again," and *mimneskein*, meaning "to call to mind."

TELL ME WHY

Why did scientists give the name *perforin* to a molecule secreted by Tc cells?

Antibody Immune Responses

As we have discussed, the body induces antibody immune responses against the antigens of exogenous pathogens and toxins. Recall that inducibility is one of the main characteristics of adaptive immunity: antibody immunity activates only in response to specific pathogens. The following sections examine the activity of B lymphocytes in antibody immune responses.

Inducement of T-Independent Antibody Immunity

LEARNING | OUTCOMES

16.25 Contrast T-dependent and T-independent antigens in terms of size and repetition of subunits.

16.26 Describe the inducement and action of a T-independent antibody immune response.

A few large antigens have many identical, repeating epitopes. These antigens can induce an antibody immune response without the assistance of a helper T cell (Th cell); therefore, these antigens are called *T-independent antigens*, and they trigger response of **T-independent antibody immunity (FIGURE 16.16)**.

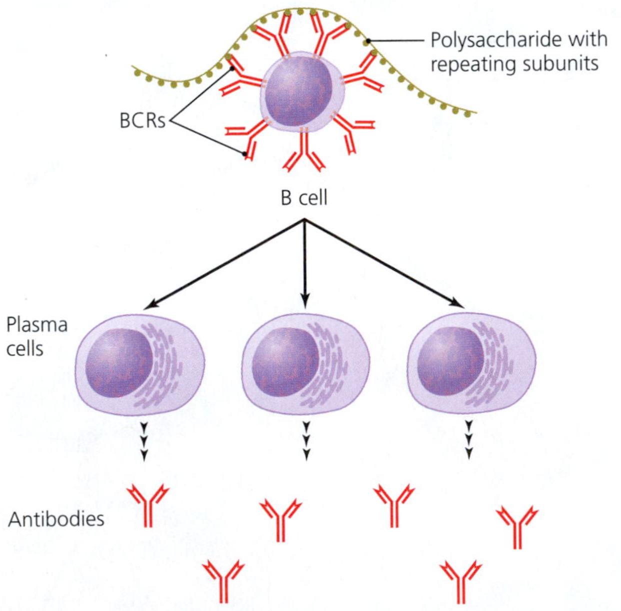

▲ **FIGURE 16.16 The effects of the binding of a T-independent antigen by a B cell.** When a molecule with multiple repeating epitopes (such as the polysaccharide shown here) cross-links the BCRs on a B cell, the cell is activated: it proliferates, and its daughter cells become plasma cells that secrete antibodies.

The repeating subunits of T-independent antigens allow extensive cross-linking between numerous BCRs on a B cell, stimulating the B cell to proliferate. Simultaneous interaction between receptors on the B cell, mediators of innate immunity, and/or bacterial chemicals may facilitate activation of the B cell. Clones of the activated B cell become plasma cells, which have an extensive cytoplasm rich in rough endoplasmic reticulum and Golgi bodies and which secrete antibodies (**FIGURE 16.17**). Though these events occur without the direct involvement of Th cells, cytokines, such as tumor necrosis factor (TNF), are required for some. T-independent antibody immune responses are fast because they do not depend on interactions with Th cells. Their speed is similar to that of innate immunity, but T-independent antibody immunity is relatively weak, disappears quickly, and induces little immunological memory.

T-independent responses are stunted in children, possibly because the repertoire and abundance of B cells is not fully developed in children; therefore, pathogens displaying T-independent antigens can cause childhood diseases, that are rare in adults. An example of such a T-independent antigen and its disease is the capsule of *Haemophilus influenzae* (hē-mof´i-lŭs in-flū-en´zī) type B that causes most cases of meningitis in unvaccinated children. The capsules of other bacteria, lipopolysaccharide of Gram-negative cell walls, bacterial flagella, and the capsids (outer covering) of some viruses constitute other T-independent antigens.

Most antibody immune responses are of the T-dependent type. We consider them next.

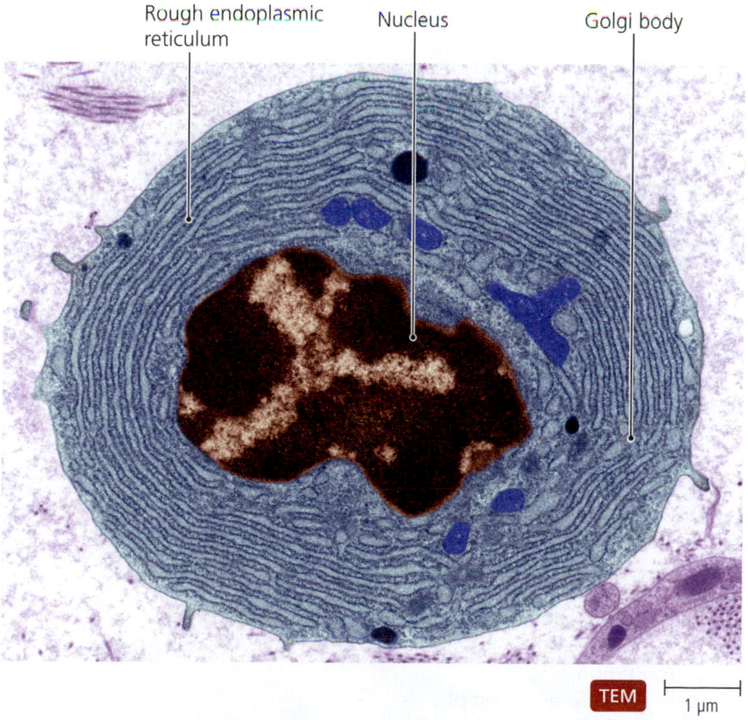

▲ **FIGURE 16.17 A plasma cell.** Plasma cells are almost twice as large as inactive B cells and are filled with rough endoplasmic reticulum and Golgi bodies for the synthesis and secretion of antibodies.

Inducement of T-Dependent Antibody Immunity with Clonal Selection

LEARNING | **OUTCOMES**

16.27 Describe the formation and functions of plasma cells and memory B cells.

16.28 Describe the steps and effect of clonal selection.

T-dependent antigens lack the numerous, repetitive, and identical epitopes and the large size of T-independent antigens, and immunity against them requires the assistance of helper T cells. **T-dependent antibody immunity** begins with the action of an antigen-presenting dendritic cell (APC). After endocytosis and processing of the antigen, the dendritic cell presents epitopes in conjunction with its MHC II proteins. This APC will induce the specific helper T (CD4) lymphocyte with a TCR complementary to the MHC II–epitope complex presented by the APC. The activated Th2 cell must in turn induce the specific B cell that recognizes the same antigen. Lymph nodes facilitate and cytokines mediate interactions among the antigen-presenting cells and lymphocytes, increasing the chance that the appropriate cells find each other.

Thus, a T-dependent antibody immune response involves a series of interactions among antigen-presenting cells, helper T cells, and B cells, all of which are mediated and enhanced by cytokines. Now we will examine each step in more detail (**FIGURE 16.18**):

1. **Antigen presentation for Th activation and cloning.** A dendritic cell, after acquiring antigens in the skin or mucous membrane, moves via the lymph to a local lymph node. The trip takes about a day. As helper T (Th, CD4) cells pass through the lymph node, they survey all the resident APCs for complementary epitopes in conjunction with MHC II proteins. Antigen presentation depends on chance encounters between Th cells and the dendritic cells, but immunologists estimate that every lymphocyte browses the dendritic cells in every lymph node every day; therefore, complementary cells eventually find each other. Once they have established an immunological synapse, CD4 molecules in membrane rafts of the Th cell's cytoplasmic membrane recognize and bind to MHC II, stabilizing the synapse.

   As we saw in cell-mediated immune responses, helper T cells need further stimulation before they activate. The requirement for a second signal helps prevent accidental inducement of an immune response. As before, the APC imparts the second signal by displaying an integral membrane protein in the immunological synapse. This induces the Th cell to proliferate, producing a clone.

2. **Differentiation of helper T cells into Th2 cells.** In antibody immune responses, the cytokine interleukin 4 (IL-4) acts as a signal to the Th cells to become type 2 helper T cells (Th2 cells). Immunologists do not know the source of IL-4, but it may be secreted initially by innate cells, such as mast cells, or secreted later in a response by the Th cells themselves.

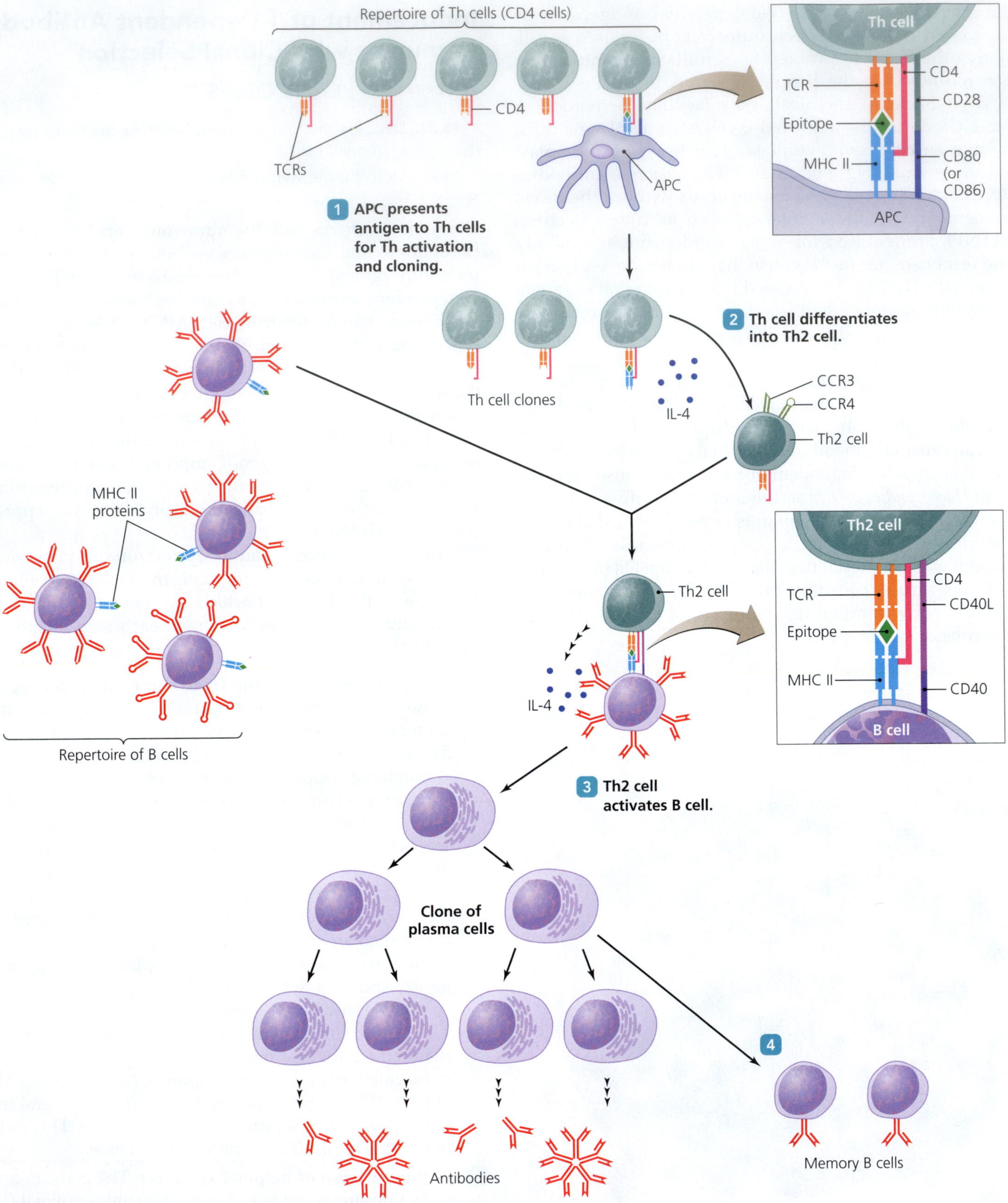

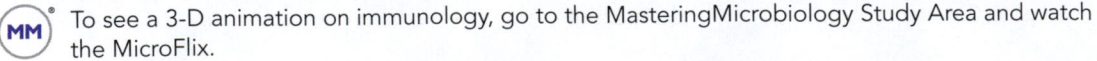

▲ **FIGURE 16.18 A T-dependent antibody immune response.** 1 Antigen presentation, in which an APC, typically a dendritic cell, presents antigen to a complementary Th cell. 2 Differentiation of the Th cell into a Th2 cell. 3 Activation of the B cell in response to secretion of IL-4 by the Th2 cell, which causes the B cell to differentiate into antibody-secreting plasma cells and 4 long-lived memory cells.

MM To see a 3-D animation on immunology, go to the MasteringMicrobiology Study Area and watch the MicroFlix.

3 **Activation of B cell.** The repertoire of B cells and newly formed Th2 cells survey one another. A Th2 cell binds to the B cell with an MHC II protein–epitope complex that is complementary to the TCR of the Th2 cell. CD4 glycoprotein again stabilizes the immunological synapse.

Th2 cells secrete more IL-4, which induces the selected B cell to move to the cortex of the lymph node. A Th2 cell in contact with an MHC II protein–epitope on a B cell is stimulated, expresses new gene products, and inserts a protein called CD40L into its cytoplasmic membrane. CD40L binds to CD40, which is found on B cells. This provides a second signal in the immunological synapse, triggering B cell activation.

4 The activated B cell proliferates rapidly to produce a population of cells (clone) that make up a primary follicle in the lymph node. The clone differentiates into two types of cells—*memory B cells* (discussed shortly) and antibody-secreting *plasma cells.* ▶**ANIMATIONS:** ***Humoral Immunity: Clonal Selection and Expansion***

Most members of a clone become **plasma cells.** The initial plasma cell descendants of any single activated B cell secrete antibodies with binding sites identical to one another and complementary to the specific antigen recognized by their parent cell. However, as the plasma cell clones replicate, the cells slightly modify their antigen-binding-site genes such that they secrete antibodies with slightly different variable regions. Plasma cells that secrete antibodies with a higher affinity for the epitope have a selective survival advantage over plasma cells secreting antibodies with a less good fit; that is, active B cells with BCRs that bind the epitope more closely survive at a higher rate. Thus, as the antibody immune response progresses, there are more and more plasma cells, secreting antibodies whose specificity gets progressively better.

Each plasma cell produces antibody. They begin by secreting IgM and then, through class switching, secrete IgG. Plasma cells are able to secrete their own weight in IgG every day. Some plasma cells later switch a second time and begin secreting IgA or IgE. As discussed previously, antibodies activate complement, trigger inflammation, agglutinate and neutralize antigen, act as opsonins, directly kill pathogens, and induce antibody-dependent cytoxicity.

Individual plasma cells are short lived, at least in part because of their high metabolic rate; they die within a few days of activation, although their antibodies can remain in body fluids for several weeks. Providentially, their descendants persist for years to maintain long-term adaptive responses.

Memory B Cells and the Establishment of Immunological Memory

LEARNING | **OUTCOME**

16.29 Contrast primary and secondary immune responses.

A small percentage of the cells produced during B cell proliferation do not secrete antibodies but survive as **memory B cells**—that is, long-lived cells with BCRs complementary to the specific epitope that triggered their production (Figure 16.18 4). In contrast to plasma cells, memory cells retain their BCRs and persist in lymphoid tissues, surviving for more than 20 years, ready to initiate antibody production if the same epitope is encountered again. Let's examine how memory cells provide the basis for immunization to prevent disease, using tetanus immunization as an example.

Because the body produces an enormous variety of B cells (and therefore BCRs), a few Th cells and B cells bind to and respond to epitopes of *tetanus toxoid* (deactivated tetanus toxin), which is used for tetanus immunization. In a **primary response (FIGURE 16.19a)**, relatively small amounts of antibodies are produced, and it may take days before sufficient antibodies are made to completely eliminate the toxoid from the body. Though some antibody molecules may persist for three

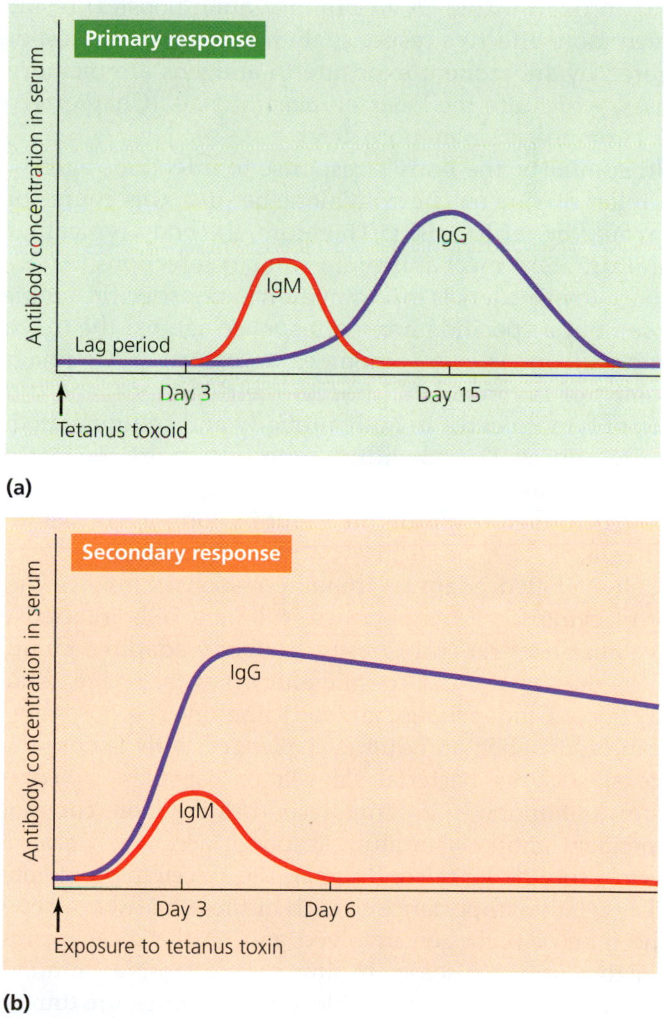

▲ **FIGURE 16.19 The production of primary and secondary antibody immune responses.** This example depicts some events following the administration of a tetanus toxoid in immunization. **(a)** Primary response. After the tetanus toxoid is introduced into the body, the body slowly removes the toxoid while producing memory B cells. **(b)** Secondary response. Upon exposure to active tetanus toxin during the course of an infection, memory B cells immediately differentiate into plasma cells and proliferate, producing a response that is faster and results in greater antibody production than occurs in the primary response.

weeks, a primary immune response basically ends when the plasma cells have lived out their normal life spans. ▶ANIMATIONS: *Humoral Immunity: Primary Immune Response*

Memory B cells, surviving in lymphoid tissue, constitute a reserve of antigen-sensitive cells that become active when there is another exposure to the antigen, in this case, toxin from infecting tetanus bacteria. Exposure may be many years later. Thus, tetanus toxin produced during the course of a bacterial infection will restimulate a population of memory cells, which proliferate and differentiate rapidly into plasma cells. The newly differentiated plasma cells produce large amounts of antibody within a few days (FIGURE 16.19b), and the tetanus toxin is neutralized before it can cause disease. Because many memory cells recognize and respond to the antigen, such a **secondary immune response** is much faster and more effective than the primary response. ▶ANIMATIONS: *Humoral Immunity: Secondary Immune Response*

As you might expect, a third exposure (whether to tetanus toxin or to toxoid in an immunization booster) results in an even more effective response. Enhanced immune responses triggered by subsequent exposure to antigens are memory responses, which are the basis of *immunization.* (Chapter 17 discusses immunization in more detail.)

In summary, the body's response to infectious agents seldom relies on one mechanism alone because this course of action would be far too risky. Therefore, the body typically uses several different mechanisms to combat infections. An initial response to intruders is inflammation (a nonspecific, innate response), but a specific immune response against the invading microorganisms is also sometimes necessary. APCs phagocytize some of the invaders, process their epitopes, and induce clones of lymphocytes in both antibody and cell-mediated immune responses. Key to enduring protection are the facts that adaptive immunity is unresponsive to self and involves immunological memory brought about by long-lived memory B and T cells.

Cell-mediated adaptive immune responses involve the activity of cytotoxic T lymphocytes in killing cells infected with intracellular bacteria and viruses. Antibody adaptive immunity involves the secretion of specific antibodies that have a variety of functions. T-independent antibody immune responses are rare but can occur when an adult is challenged with T-independent antigens, such as bacterial flagella or capsules. In contrast, T-dependent antibody immune responses are more common. A T-dependent antibody immune response occurs when an APC binds to a specific Th cell and signals the Th cell to proliferate.

The relative importance of each of these pathways depends on the type of pathogen involved and on the mechanisms by which they cause disease. In any case, adaptive immune responses are specific, are inducible, involve clones, are unresponsive to self, and give the body long-term memory against their antigenic triggers. ▶ANIMATIONS: *Host Defenses: The Big Picture*

TELL ME WHY

Plasma cells are vital for protection against infection, but memory B cells are not. Why not?

Types of Acquired Immunity

LEARNING | OUTCOME

16.30 Contrast active versus passive acquired immunity and naturally acquired versus artificially acquired immunity.

As we have seen, adaptive immunity is acquired during an individual's life. Immunologists categorize immunity as either naturally or artificially acquired. Naturally acquired immunity occurs when the body mounts an immune response against antigens, such as influenza viruses or food antigens, encountered during the course of daily life. Artificial immunity is the body's response to antigens introduced in vaccines, as occurs with immunization against tetanus and flu. Immunologists further distinguish acquired immune responses as either *active* or *passive;* that is, either the immune system responds actively to antigens via antibody or cell-mediated responses or the body passively receives antibodies from another individual. Next we consider each of four types of acquired immunity.

Naturally Acquired Active Immunity

Naturally acquired active immunity occurs when the body responds to exposure to pathogens and environmental antigens by mounting specific immune responses. The body is naturally and actively engaged in its own protection. As we have seen, once an immune response occurs, immunological memory persists—on subsequent exposure to the same antigen, the immune response will be rapid and powerful and often provides the body complete protection.

Naturally Acquired Passive Immunity

Although newborns possess the cells and tissues needed to mount an immune response, they respond slowly to antigens. If required to protect themselves solely via naturally acquired active immunity, they might die of infectious disease before their immune systems were mature enough to respond adequately. However, they are not on their own; in the womb, IgG molecules cross the placenta from the mother's bloodstream to provide protection, and after birth, children receive secretory IgA in breast milk. Via these two processes, a mother provides her baby with antibodies that protect it during its early months. Because the baby is not actively producing its own antibodies, this type of protection is known as **naturally acquired passive immunity.**

Artificially Acquired Active Immunity

Physicians induce immunity in their patients by introducing antigens in the form of vaccines. The patients' own immune systems then mount active responses against the foreign antigens, just as if the antigens were part of a naturally acquired pathogen. Such **artificially acquired active immunity** is the basis of immunization (see Chapter 17).

EMERGING DISEASE CASE STUDY

Microsporidiosis

Darius is sick, which is not surprising for an HIV-infected man. But he is sick in several new ways. Sick of having to stay within 20 feet of a toilet. Sick of the cramping, the gas, the pain, and the nausea. Sick with irregular but persistent, watery diarrhea. He is losing weight because food is passing through him undigested. Most days over the past seven months have been disgusting despite his use of over-the-counter remedies, which provide a few days of intermittent relief. His belief that these normal days signaled the end of the ordeal have kept him from the doctor. But now his eyes have begun to hurt, and his vision is blurry. Whatever it is, it's attacking him at both ends. Time to get stronger drugs from his physician.

Microscopic examination of Darius's stool sample reveals that he is being assaulted by *Encephalitozoon intestinalis*, a member of a group of opportunistic emerging pathogens called microsporidia. The single-celled pathogens are also seen on smears from Darius's nose and eyes. Microsporidia were long thought to be simple single-celled animals, but genetic analysis and comparison with other organisms reveal that they are closer to zygomycete yeasts.

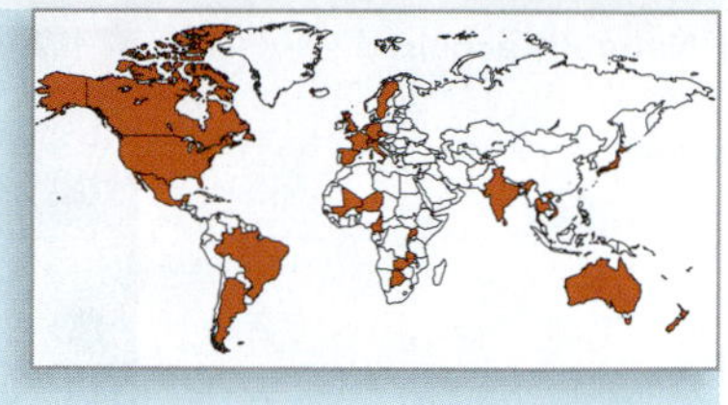

Microsporidia appear to infect humans who engage in unprotected sexual activity, consume contaminated food or drink, or swim in contaminated water. People with active T cells rarely have symptoms; but people with suppressed immunity become easy targets for the fungus.

Microsporidia attack by uncoiling a flexible, hollow filament that stabs into a host cell and serves as a conduit for the microsporidium's cytoplasm to invade. In this way, the pathogens become intracellular parasites. They can destroy the intestinal lining, causing diarrhea, and spread to the eyes, muscles, or lungs.

Fortunately for Darius, an antimicrobial, albendazole, kills the parasite, and the effects of the infection are reversed. Unfortunately for Darius, the loss of helper T cells in AIDS means that another emerging, reemerging, or opportunistic infection is sure to follow.

1. **Why are microsporidia considered to be opportunistic pathogens?**
2. **How could the discovery that microsporidia are fungi rather than animals improve treatment of microsporidiosis?**
3. **Microsporidia are intracellular pathogens. Which immune cells likely fight off the infection in people with a normal immune system?**

Artificially Acquired Passive Immunotherapy

Active immunity usually requires days to weeks to develop fully, and in some cases such a delay can prove detrimental or even fatal. For instance, an active immune response may be too slow to protect against infection with rabies or exposure to rattlesnake venom. Therefore, medical personnel routinely harvest antibodies specific for toxins and pathogens that are so deadly or so fast acting that an individual's active immune response is inadequate. They acquire these antibodies from the blood of immune humans or animals, typically a horse. Physicians then inject such *antisera* or *antitoxins* into infected patients to confer **artificially acquired passive immunotherapy.** (Chapter 17 discusses this type of treatment in greater detail.)

Active immune responses, whether naturally or artificially induced, are advantageous because they result in immunological memory and protection against future infections. However, they are slow acting. Passive processes, in which individuals are provided fully formed antibodies, have the advantage of speed but do not confer immunological memory because B and T lymphocytes are not activated. **TABLE 16.4** on p. 498 summarizes the four types of acquired immunity.

TELL ME WHY

Why is passive immunity effective more quickly than active immunity?

TABLE 16.4 A Comparison of the Types of Acquired Immunity

	Active	Passive
Naturally acquired	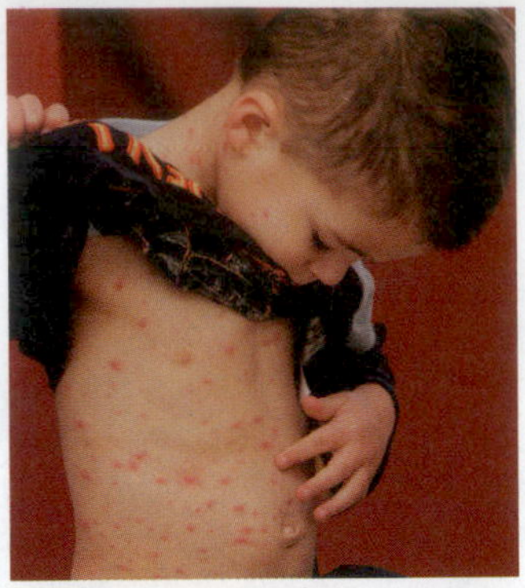 The body responds to antigens that enter naturally, such as during infections.	 Antibodies are transferred from mother to offspring, either across the placenta (IgG) or in breast milk (secretory IgA).
Artificially acquired	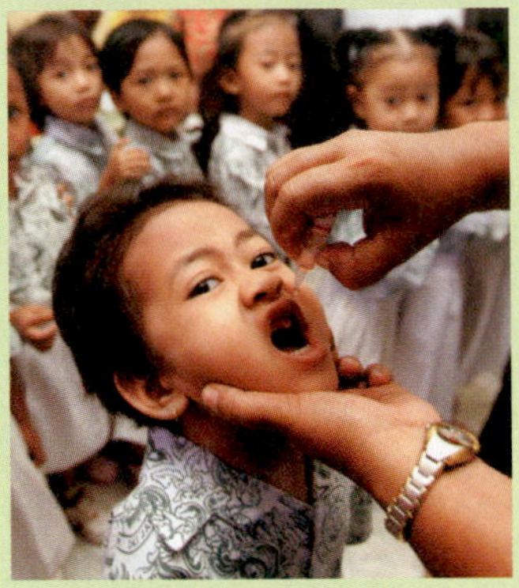 Health care workers introduce antigens in vaccines; the body responds with antibody or cell-mediated immune responses, including the production of memory cells.	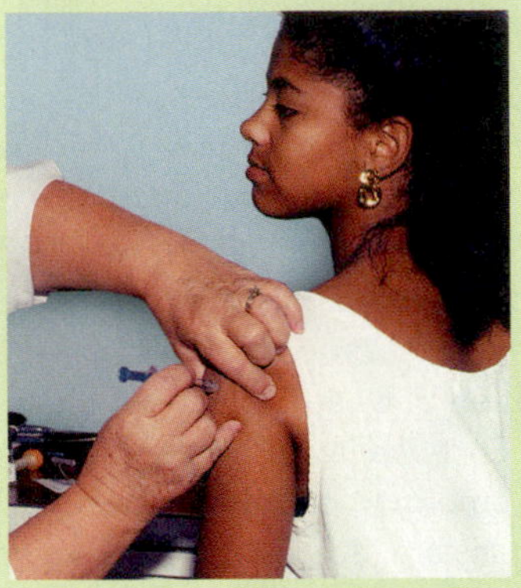 Health care workers give patients antisera or antitoxins, which are preformed antibodies obtained from immune individuals or animals.

MICRO IN THE CLINIC FOLLOW-UP

A Vaccine Against Cancer?

Cervical cancer strikes a half million women every year worldwide, and about half die. Ninety-nine percent of cervical cancer cases start from sexually transmitted human papillomavirus (HPV). HPV causes genital warts, or typically benign tumors called papillomas. Often, a person's immune system prevents HPV from doing harm, and the infections go away without treatment. However, certain HPV strains can trigger the growth of abnormal epithelial cells in the cervix. HPV proteins synthesized within infected cells inhibit cancer suppressor genes, thwarting the immune system that normally detects and destroys cancerous cells as they appear.

Fortunately, Kathy's Pap smear detected abnormal cells before they became cancerous. She is able to schedule a procedure to have the cells removed.

Meanwhile, Meghan does some research and learns that HPV vaccine protects women from the most dangerous strains of HPV. Consisting of only the external viral proteins, the HPV vaccine lacks the internal viral components, so the proteins in the vaccine can't replicate. The external viral proteins stimulate adaptive immunity to make specific antibodies against the HPV virus, preventing sexually transmitted infections and short-circuiting the process that results in cervical cancer. The immunization does not protect against other forms of cancer, but given her mother's recent scare with cervical cancer, Meghan ultimately decides to get the HPV vaccine.

1. **Briefly describe innate protections against viruses in the second line of defense (as discussed in Chapter 15).**
2. **How do antibodies formed against HPV vaccine protect against infection and subsequent cancer?**

MM® Check your answers to Micro in the Clinic Follow-Up questions in the MasteringMicrobiology Study Area.

Explore the Invisible: Clonal Deletion

Practice on-the-go with Dr. Bauman Video Tutors by scanning this QR code with your smart phone. Visit the MasteringMicrobiology Study Area to challenge your understanding with practice tests, animation quizzes, and clinical case studies!

MasteringMicrobiology®

CHAPTER SUMMARY

This chapter has Microflix. Go to the MasteringMicrobiology Study Area for 3-D movie-quality animations on immunology.

Overview of Adaptive Immunity (pp. 473–474)

1. **Adaptive immunity** is the ability of a vertebrate to recognize and defend against distinct species or strains of invaders. Adaptive immunity is characterized by specificity, inducibility, clonality, unresponsiveness to self, and memory.
2. **B lymphocytes (B cells)** attack extracellular pathogens in **antibody immune responses** (also called humoral immune responses), involving soluble proteins called antibodies. **T lymphocytes (T cells)** carry out **cell-mediated immune responses** against intracellular pathogens.

   ▶**ANIMATIONS:** *Host Defenses: The Big Picture; Cell-Mediated Immunity: Overview; Humoral Immunity: Overview*

Elements of Adaptive Immunity (pp. 474–489)

1. The **lymphatic system** is composed of **lymphatic vessels,** which conduct the flow of **lymph,** and lymphoid tissues and organs that are directly involved in specific immunity. The latter include **lymph nodes,** the thymus, the spleen, the tonsils, and mucosa-associated lymphoid tissue (MALT). Lymphocytes originate in the red bone marrow. They mature in the marrow or in the thymus. Mature lymphocytes express characteristic membrane proteins. They migrate to and persist in various lymphoid organs, where they are available to encounter foreign invaders in the blood and lymph.
2. **Antigens** are substances that trigger specific immune responses. Effective antigen molecules are large, usually complex, stable, degradable, and foreign to their host. An **epitope** (or antigenic determinant) is the three-dimensional shape of a region of an antigen that is recognized by the immune system.
3. **Exogenous antigens** are found on microorganisms that multiply outside the cells of the body; **endogenous antigens** are produced by pathogens multiplying inside the body's cells.
4. Ideally, the body does not attack antigens on the surface of its normal cells, called **autoantigens;** this phenomenon is called self-tolerance.
5. T cells have **T cell receptors (TCRs)** for antigens, mature under influence of signals from the thymus, and attack cells that harbor endogenous pathogens during cell-mediated immune responses.
6. Nucleated cells display epitopes of their own proteins and epitopes from intracellular pathogens, such as viruses, on **major histocompatibility complex (MHC)** class I proteins.
7. The initial step in mounting an immune response is that antigens are captured, ingested, and degraded into epitopes by **antigen-presenting cells (APCs),** such as B cells, macrophages, and **dendritic cells.** Epitopes are inserted into major histocompatibility complex (MHC) class II proteins.

   ▶**ANIMATIONS:** *Antigen Processing and Presentation: Overview, Steps, MHC*
8. In cell-mediated immunity, **cytotoxic T cells (Tc cells** or **CD8 cells)** act against infected or abnormal body cells, including virus-infected cells, bacteria-infected cells, some fungus- or protozoan-infected cells, some cancer cells, and foreign cells that enter the body as a result of organ transplantation.
9. Two types of **helper T (Th) cells**—Th1 and Th2—are characterized by **CD4.** They direct cell-mediated and antibody immune responses respectively.

   ▶**ANIMATIONS:** *Cell-Mediated Immunity: Helper T Cells*
10. T cells that do not recognize MHC I protein and most T cells that recognize MHC I protein in conjunction with autoantigens are removed by **apoptosis.** This is **clonal deletion.** A few self-recognizing T cells are retained and become **regulatory T cells (Tr cells).** T cells that recognize MHC I protein but not autoantigens become the repertoire of immature T cells.

    ▶**VIDEO TUTOR:** *Clonal Deletion*
11. B lymphocytes (B cells), which mature in the red bone marrow, make immunoglobulins (Ig) of two types—**B cell receptors (BCRs)** and **antibodies.** Immunoglobulins are complementary to epitopes and consist of two light chains and two heavy chains joined via disulfide bonds to form Y-shaped molecules. BCRs are inserted into the cytoplasmic membranes of B cells via a transmembrane polypeptide, whereas antibodies are secreted.
12. Together the variable regions of a heavy and a light chain form an **antigen-binding site,** and the upper portions of antibody molecules are called F_{ab} regions. Each B cell randomly selects (once in its life) genes for its F_{ab} region; therefore, the F_{ab} regions are called variable regions because they differ from cell to cell. Each basic antibody molecule has two antigen-binding sites and can potentially bind two epitopes.

13. Antibodies function in complement activation, inflammation, **neutralization** (blocking the action of a toxin or attachment of a pathogen), as **opsonins** for **opsonization** (enhanced phagocytosis), direct killing by oxidation, **agglutination,** and **antibody-dependent cellular cytotoxicity (ADCC).**

   ▶**ANIMATIONS:** *Humoral Immunity: Antibody Function*

14. Antibodies are of five basic classes based upon their stems (F_c regions), which differ in their type of heavy chain.
15. **IgM,** a pentamer with 10 antigen-binding sites, is the predominant class of antibody produced first during a primary antibody response. **IgG** is the predominant antibody found in the bloodstream and is largely responsible for defense against invading bacteria. IgG can cross a placenta to protect the fetus. Two molecules of **IgA** are attached via J chains and a polypeptide secretory component to produce **secretory IgA,** which is found in milk, tears, and mucous membrane secretions. **IgE** triggers inflammation and allergic reactions. It also functions during helminth infections. **IgD** is found in cytoplasmic membranes of some animals.
16. Through a process called **class switching,** antibody-producing cells change the class of antibody they secrete, beginning with IgM and then producing IgG and then possibly IgA or IgE.
17. B cells with B cell receptors that respond to autoantigens are selectively killed via apoptosis—further clonal deletion. Only B cells that respond to foreign antigens survive to defend the body.
18. **Cytokines** are soluble regulatory proteins that act as intercellular signals to direct activities in immune responses. Cytokines include **interleukin (ILs), interferons (IFNs), growth factors, tumor necrosis factors (TNFs),** and **chemokines.**

Cell-Mediated Immune Responses (pp. 489–492)

1. Once activated by dendritic cells, cytotoxic T cells (Tc cells) recognize abnormal molecules presented by MHC I protein on the surface of infected, cancerous, or foreign cells. Sometimes cytotoxic T cells require cytokines from Th1 cells.
2. Activated Tc cells reproduce to form memory T cells and more Tc progeny in a process called **clonal expansion.**
3. Cytotoxic T cells destroy their target cells via two pathways: the perforin-granzyme pathway, which kills the affected cells by secreting **perforins** and **granzymes,** or the **CD95 pathway,** in which CD95L binds to CD95 on the target cell, triggering target cell apoptosis. Cytotoxic T cells may also form **memory T cells,** which function in **memory responses.**

   ▶**ANIMATIONS:** *Cell-Mediated Immunity: Cytotoxic T Cells*

Antibody Immune Responses (pp. 492–496)

1. T-independent antigens, such as bacterial capsules, trigger **T-independent antibody immune responses,** which are more common in adults than in children.
2. In **T-dependent antibody immunity,** an APC's MHC II protein–epitope complex activates a helper T cell (Th cell) bearing a complementary TCR. CD4 stabilizes the connection between the cells, which is an example of an immunological synapse. Interleukin 4 (IL-4) then induces the Th cell to become a type 2 helper T cell (Th2).
3. In **clonal selection,** an immunological synapse forms between the Th2 cell and a B cell bearing a complementary MHC II protein–epitope complex. The Th2 cell secretes IL-4, which induces the B cell to divide. Its offspring, collectively called a clone, become **plasma cells** or **memory B cells.**

   ▶**ANIMATIONS:** *Humoral Immunity: Clonal Selection and Expansion*

4. Plasma cells live for only a short time but secrete large amounts of antibodies, beginning with IgM and class switching as they get older. Memory B cells migrate to lymphoid tissues to await a subsequent encounter with the same antigen.
5. The **primary response** to an antigen is slow to develop and of limited effectiveness. When that antigen is encountered a second time, the activation of memory cells ensures that the immune response is rapid and strong. This is a **secondary immune response.** Such enhanced antibody immune responses are memory responses.

   ▶**ANIMATIONS:** *Humoral Immunity: Primary Immune Response, Secondary Immune Response*

Types of Acquired Immunity (pp. 496–498)

1. When the body mounts a specific immune response against an infectious agent, the result is called **naturally acquired active immunity.**
2. The passing of maternal IgG to the fetus and the transmission of secretory IgA in milk to a baby are examples of **naturally acquired passive immunity.**
3. **Artificially acquired active immunity** is achieved by deliberately injecting someone with antigens in vaccines to provoke an active response, as in the process of immunization.
4. **Artificially acquired passive immunotherapy** involves the administration of preformed antibodies in antitoxins or antisera to a patient.

QUESTIONS FOR REVIEW

Answers to the Questions for Review (except Short Answer questions) begin on p. A-1.

Multiple Choice

1. Antibodies function to
   a. directly destroy foreign organ grafts.
   b. mark invading organisms for destruction.
   c. kill intracellular viruses.
   d. directly promote cytokine synthesis.
   e. stimulate T cell growth.

2. MHC class II molecules bind to ________________ and trigger ________________.
   a. endogenous antigens; cytotoxic T cells
   b. exogenous antigens; cytotoxic T cells
   c. antibodies; B cells
   d. endogenous antigens; helper T cells
   e. exogenous antigens; helper T cells

3. Rejection of a foreign skin graft is an example of
   a. destruction of virus-infected cells.
   b. tolerance.
   c. antibody-mediated immunity.
   d. a secondary immune response.
   e. a cell-mediated immune response.
4. An autoantigen is
   a. an antigen from normal microbiota.
   b. a normal body component.
   c. an artificial antigen.
   d. any carbohydrate antigen.
   e. a nucleic acid.
5. Among the key molecules that control cell-mediated cytotoxicity are
   a. perforin.
   b. immunoglobulins.
   c. complement.
   d. cytokines.
   e. interferons.
6. Which of the following lymphocytes predominates in blood?
   a. T cells
   b. B cells
   c. plasma cells
   d. memory cells
   e. All are about equally prevalent.
7. The major class of immunoglobulin found on the surfaces of the walls of the intestines and airways is secretory
   a. IgG.
   b. IgM.
   c. IgA.
   d. IgE.
   e. IgD.
8. Which cells express MHC class I molecules in a patient?
   a. red blood cells
   b. antigen-presenting cells only
   c. neutrophils only
   d. all nucleated cells
   e. dendritic cells only
9. In which of the following sites in the body can B cells be found?
   a. lymph nodes
   b. spleen
   c. red bone marrow
   d. intestinal wall
   e. all of the above
10. Tc cells recognize epitopes only when the latter are held by
   a. MHC proteins.
   b. B cells.
   c. interleukin 2.
   d. granzyme.

Modified True/False

Mark each statement as either true or false. Rewrite false statements to make them true by changing the underlined words.

1. ________________ MHC class II molecules are found on <u>T cells</u>.
2. ________________ <u>Apoptosis</u> is the term used to describe cellular suicide.
3. ________________ Lymphocytes with CD8 glycoprotein are <u>helper</u> T cells.
4. ________________ <u>Cytotoxic T cells</u> secrete immunoglobulin.
5. ________________ Secretion of antibodies by activated B cells is a form of <u>cell-mediated</u> immunity.

Matching

Match each cell in the left column with its associated protein from the right column.

1. ____ Plasma cell	A. MHC II molecule
2. ____ Cytotoxic T cell	B. Interleukin 4
3. ____ Th2 cell	C. Perforin and granzyme
4. ____ Dendritic cell	D. Immunoglobulin

Match each type of immunity in the left column with its associated example from the right column.

1. ____ Artificially acquired passive immunotherapy	A. Production of IgE in response to pollen
2. ____ Naturally acquired active immunity	B. Acquisition of maternal antibodies in breast milk
3. ____ Naturally acquired passive immunity	C. Administration of tetanus toxoid
4. ____ Artificially acquired active immunity	D. Administration of antitoxin

VISUALIZE IT!

1. Label the parts of the immunoglobulin below.

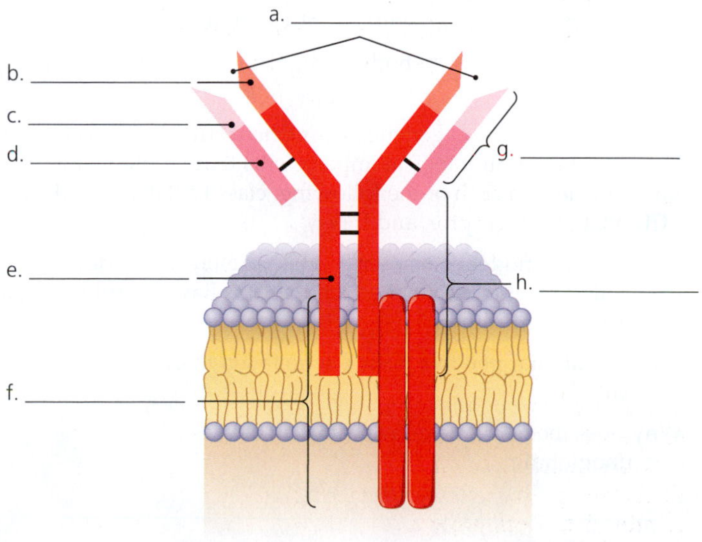

2. This is a transmission electron micrograph of a dendritic cell. Indicate where a scientist could find molecules of MHC I and MHC II. Label a pseudopod and a vesicle.

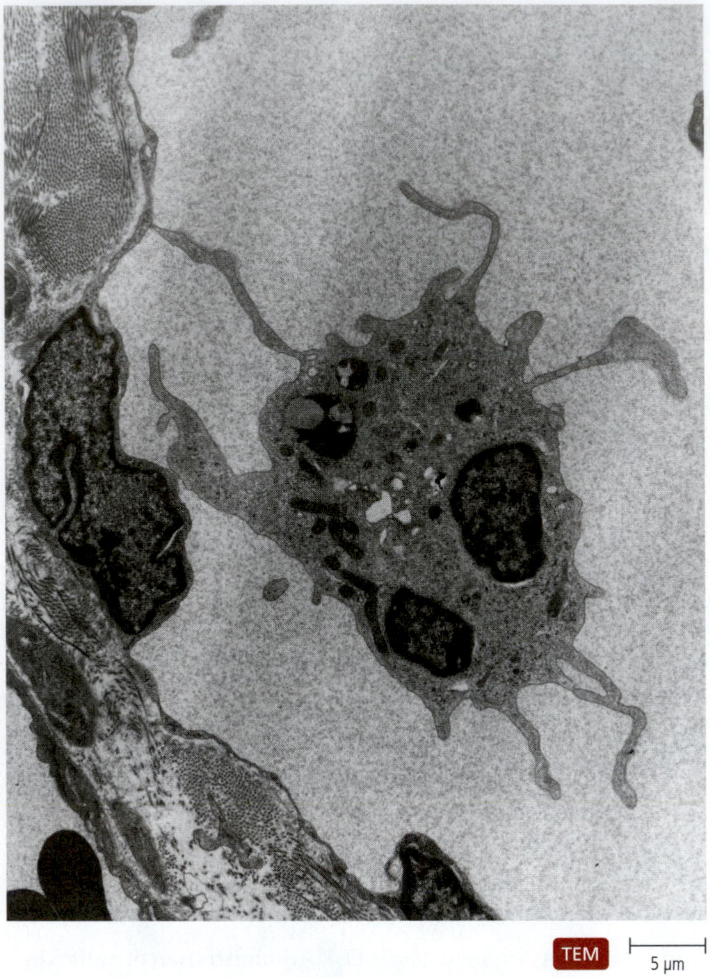

Short Answer

1. When is antigen processing an essential prerequisite for an immune response?
2. Why does the body have both antibody and cell-mediated immune responses?

CRITICAL THINKING

1. Why is it advantageous for the lymphatic system to lack a pump?
2. Contrast innate defenses with adaptive immunity.
3. What is the benefit to the body of requiring the immune system to process antigen?
4. Scientists can develop genetically deficient strains of mice. Describe the immunological impairments that would result in mice deficient in each of the following: class I MHC, class II MHC, TCR, BCR, IL-2 receptor, and IFN-γ.
5. Human immunodeficiency virus (HIV) preferentially destroys CD4 cells. Specifically, what effect does this have on antibody and cell-mediated immunity?
6. What would happen to a person who failed to make MHC molecules?
7. Why does the body make five different classes of immunoglobulins?
8. Some materials, such as metal bone pins and plastic heart valves, can be implanted into the body without fear of rejection by the patient's immune system. Why is this? What are the ideal properties of any material that is to be implanted?
9. What nonmembranous organelle is prevalent in plasma cells? What membranous organelle is prevalent?
10. The cross-sectional area of the afferent lymphatic vessels arriving at a lymph node is greater than the cross-sectional area of the efferent lymphatics exiting the lymph node. The result is that lymph moves slowly through a lymph node. Why is this advantageous?
11. Two students are studying for an exam on the body's defensive systems. One of them insists that complement is part of the nonspecific second line of defense, but the partner insists that complement is part of an antibody immune response in the third line of defense. How would you explain to them that they are both correct?

12. In general, what sorts of pathogens might be able to more successfully attack a patient with an inability to synthesize B lymphocytes?
13. What sorts of pathogens could successfully attack a patient who is unable to produce T lymphocytes?
14. As part of the treatment for some cancers, physicians kill the cancer patients' dividing cells, including the stem cells that produce leukocytes, and then give the patients a bone marrow transplant from a healthy donor. Which cell is the most important cell in such transplanted marrow?

CONCEPT MAPPING

Using the following terms, draw a concept map that describes antibodies. You may use some terms more than once. For a sample concept map, see p. 94. Or, complete this and other concept maps online by going to the MasteringMicrobiology Study Area.

Agglutination
Antigens
Antigen-stimulated B cells
Death by Oxidation
IgA
IgD
IgE
IgG
IgM
Inflammation
Neutralization
Opsonization
Phagocytosis (2)
Plasma cells
Secreted immunoglobulins
Target bacteria
Toxins
Viruses

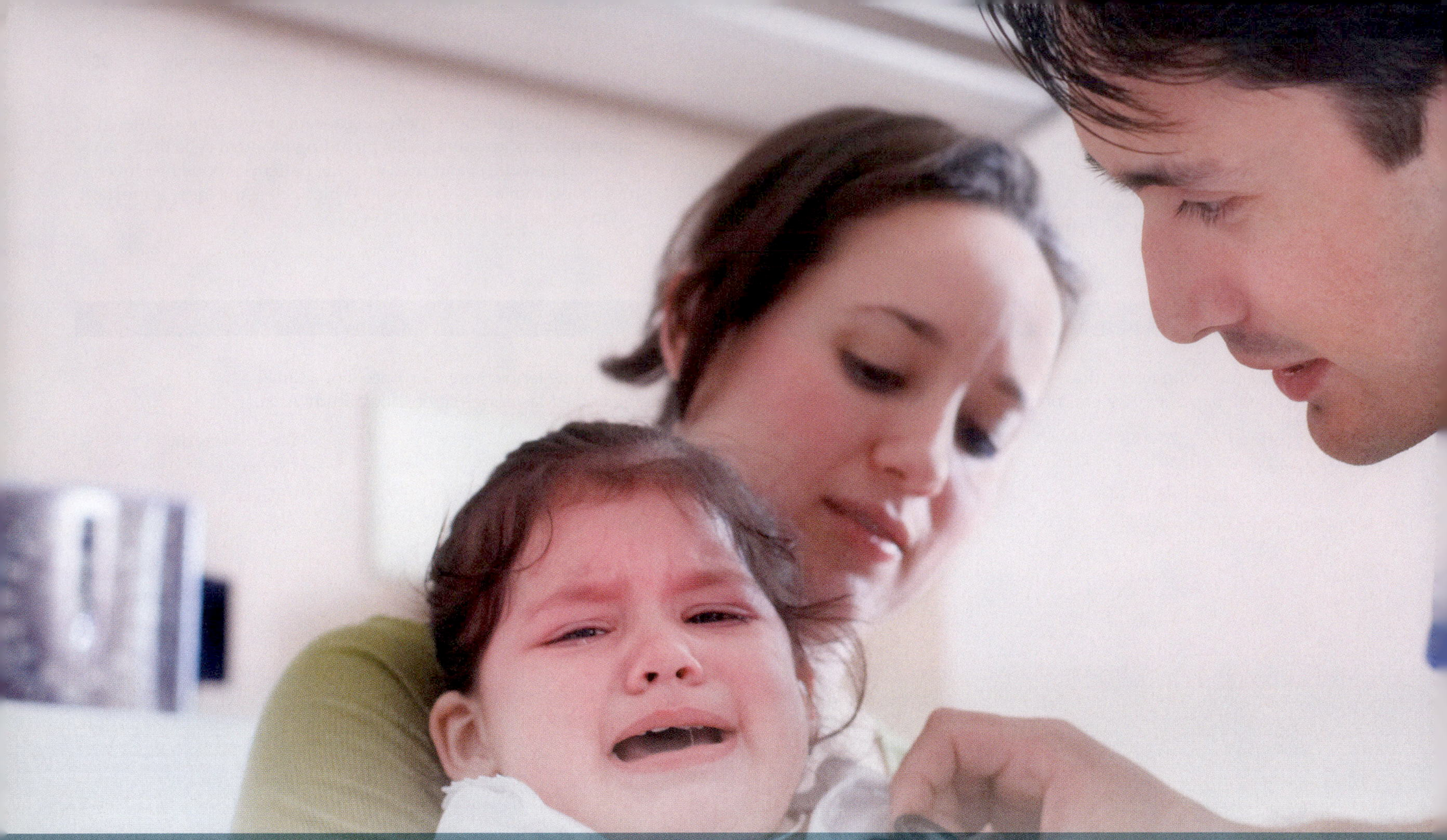

17 Immunization and Immune Testing

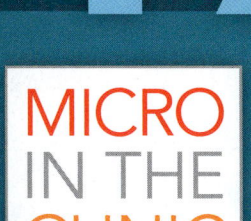

MICRO IN THE CLINIC

Outbreak in the Inner City

Ernesto wants to go to medical school and become a pediatrician. He envisions not only treating children's illnesses but also putting them on a healthy path for life. He has a great GPA and is bolstering his medical school application by volunteering at an inner-city medical clinic.

But this last month at the clinic has been more than Ernesto bargained for. The patient load has increased over 400% because of a mysterious and frightening disease. Hundreds of children are coming in so sick that some do not survive. Most of the casualties are babies less than one year old, and many are Hispanic, which makes Ernesto worry about his own family. Ernesto's language skills, college education, and microbiology classwork make him extremely valuable to the doctors and nurses as a knowledgeable translator. He soon finds himself smack in the middle of the crisis.

One of Ernesto's patients is Maria, a very sick 13-month-old girl. Ernesto's task is to interview the girl's mother and take down Maria's medical history. The interview is constantly interrupted by Maria's forceful coughing, resembling the bark of a sea lion. Each cough is shocking, louder than you would expect from even an adult, followed by a high-pitched gasp for breath. The coughing spasms go on and on, depriving Maria of air until her lips and body turn blue. Eventually, she coughs so hard that she vomits.

Will Maria survive? What is causing this outbreak? Turn to the end of the chapter (p. 521) to find out.

MM **Explore More:** Test your readiness and apply your knowledge with dynamic learning tools at MasteringMicrobiology.

In this chapter we will discuss three applications of immunology: active immunization (vaccination), passive immunotherapy using immunoglobulins (antibodies), and immune testing. Vaccination has proven the most efficient and cost-effective method of controlling infectious diseases. Without the use of effective vaccines, millions more people worldwide would suffer each year from potentially fatal infectious diseases, including measles, mumps, and polio. The administration of immunoglobulins has further reduced morbidity and mortality from certain infectious diseases, such as hepatitis A and yellow fever, in unvaccinated individuals. Medical personnel also make practical use of the immune response as a diagnostic procedure. For example, the detection of antibodies to HIV in a person's blood indicates that the individual has been exposed to that virus and may develop AIDS. The remarkable specificity of antibodies also enables the detection of drugs in urine, recognition of pregnancy at early stages, and the identification or characterization of other biological material. The many tests developed for these purposes are the focus of the discipline of *serology* (sĕ-rol´ō-jē) and are discussed in the second half of this chapter.

Immunization

An individual may be made immune to an infectious disease by two artificial methods: *active immunization,* which involves administering antigens to a patient so that the patient actively mounts an adaptive immune response, and *passive immunotherapy,* in which a patient acquires temporary immunity through the transfer of antibodies formed by other individuals or animals (see Table 16.4 on p. 498).

In the following sections we will review the history of immunization before examining immunization and immunotherapy in more detail.

Brief History of Immunization

LEARNING **OUTCOME**

17.1 Discuss the history of vaccination from the 12th century through the present.

As early as the 12th century, the Chinese noticed that children who recovered from smallpox never contracted the disease a second time. They therefore adopted a policy of deliberately infecting young children with particles of ground smallpox scabs from children who had survived mild cases. By doing so, they succeeded in significantly reducing the population's overall morbidity and mortality from the disease. News of this procedure, called *variolation* (var´ē-ō-lā´shŭn), spread westward through central Asia, and the technique was widely adopted.

Lady Mary Montagu (1689–1762), the wife of the English ambassador to the Ottoman Empire, learned of the procedure, had it performed on her own children, and told others about it upon her return to England in 1721. As a result, variolation came into use in England and in the American colonies. Although effective and usually successful, variolation caused death from smallpox in 1% to 2% of recipients and in people exposed to recipients, so in time the procedure was outlawed.

Thus, when the English physician Edward Jenner demonstrated in 1796 that protection against smallpox could be conferred by inoculation with crusts from a person infected with cowpox—a related but very mild disease—the new technique was adopted. Because cowpox was also called *vaccinia*[1] (vak-sin´ē-ă), Jenner called the new technique **vaccination** (vak´si-nā´shŭn), and the protective inoculum a **vaccine** (vak-sēn´). Today we use the term **immunization** to refer to the administration of any antigenic inoculum, which are all called vaccines. For many years thereafter, vaccination against smallpox was widely practiced, even though no one understood how it worked or whether similar techniques could protect against other diseases.

In 1879, Louis Pasteur conducted experiments on the bacterium *Pasteurella multocida* (pas-ter-el´ă mul-tŏ´si-da) and demonstrated that he could make an effective vaccine against this organism (which causes a disease in birds called fowl cholera). Once the basic principle of vaccine manufacture was understood, vaccines against anthrax and rabies rapidly followed. Once it was discovered that these vaccines provide protection through the actions of antibodies, the technique of transferring protective antibodies to susceptible individuals—that is, *passive immunotherapy* (im´ū-nō-thār´ă-pē)—was developed soon thereafter.

By the late 1900s, immunologists and health care providers had formulated vaccines that significantly reduced the number of cases of many infectious diseases **(FIGURE 17.1)**. We also have successful vaccines against some types of cancer. Health care providers, governments, and international organizations working together have rid the world of naturally occurring smallpox, and we hope for the worldwide eradication of polio, measles, mumps, and rubella. **Highlight: Why Isn't There a Cold Vaccine?** on p. 506 discusses why a vaccine for the common cold does not yet exist.

Even though immunologists have produced vaccines that protect people against many deadly diseases, a variety of political, social, economic, and scientific problems prevent vaccines from reaching all those who need them. In developing nations worldwide, over 3 million children still die each year from vaccine-preventable infectious diseases, primarily because of political obstacles. Additionally, some pathogens, such as the protozoa of malaria and the virus of AIDS, still frustrate attempts to develop effective vaccines against them. Furthermore, the existence of vaccine-associated risks—both medical risks (the low but persistent incidence of vaccine-caused diseases) and financial risks (the high costs of developing and producing vaccines and the risk of lawsuits by vaccine recipients who have adverse reactions)—has in recent years discouraged investment in new vaccines. Thus, although the history of immunization is marked by stunning advancements in public health, the future of immunization poses immense challenges.

Next we take a closer look at active immunization, commonly known as vaccination.

[1]From Latin *vacca,* meaning "cow."

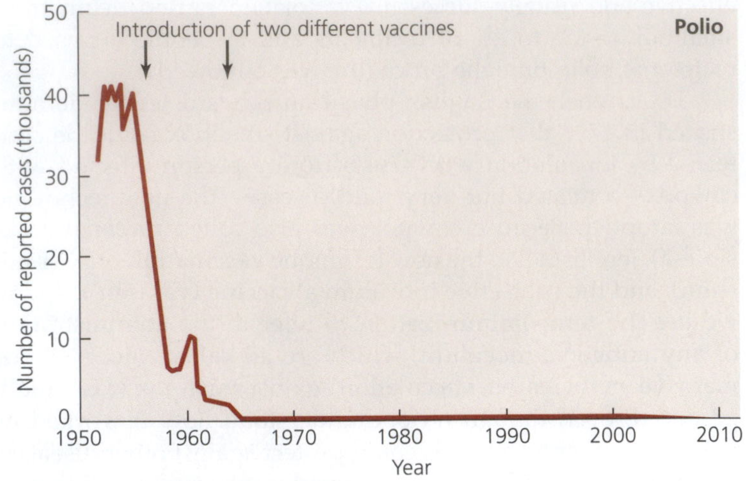

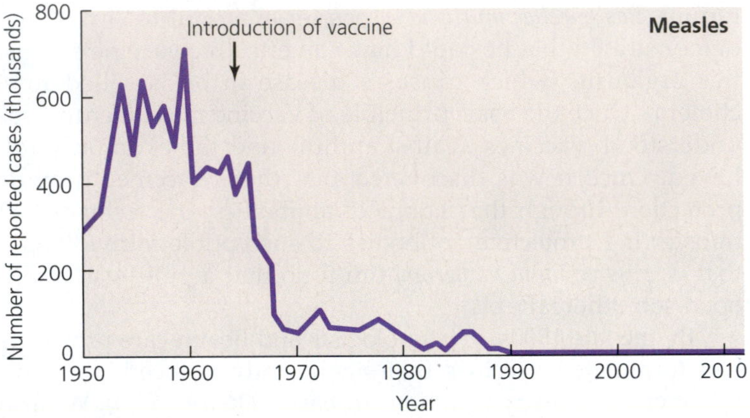

▲ **FIGURE 17.1 The effect of immunization in reducing the prevalence of two infectious diseases in the United States.** Polio is no longer endemic in the United States. Measles is nearly eradicated.

Active Immunization

LEARNING OUTCOMES

17.2 Describe the advantages and disadvantages of five types of vaccines.

17.3 Describe three methods by which recombinant genetic techniques can be used to develop improved vaccines.

17.4 Delineate the risks and benefits of routine vaccination in healthy populations, mentioning contact immunity and herd immunity.

In the following subsections we examine types of vaccines, the roles of technology in producing modern vaccines, and issues concerning vaccine safety.

Vaccine Types

Scientists are constantly striving to develop vaccines of maximal efficacy and safety. In each case, a pathogen is altered or inactivated so that it is less likely to cause illness; however, not all types of vaccines are equally safe or effective. Effectiveness can be checked by measuring the antibody (IgG and IgM) level—called the **titer** (tī´ter)—in the blood. When the titer is low, antibody production can be bolstered by administration of more antigen—a *booster immunization.*

The general types of vaccines, each of which has its own combination of strengths and weaknesses, are attenuated (live) vaccines, killed (or inactivated) vaccines, toxoid vaccines, combination vaccines, and recombinant gene vaccines. Each of these is named for the type of antigen used in the inoculum.

Attenuated (Modified Live) Vaccines Virulent microbes are normally not used in vaccines because they cause disease. Instead, immunologists can reduce virulence so that, although still active, the pathogens no longer cause disease. The process of reducing virulence is called **attenuation** (ă-ten-ū-ā´shŭn).

HIGHLIGHT

Why Isn't There a Cold Vaccine?

We have vaccines for the flu, so why don't we have a vaccine for the common cold? The reason is that whereas strains of only one influenzavirus cause flu, over 200 different adenoviruses, coronaviruses, and rhinoviruses are known to cause the common cold, and each of these viruses has its own distinct antigens and antigenic strains, making it extremely difficult to create a single vaccine to prevent them all. To further complicate matters, viruses can mutate, resulting in changes in their antigens; with over 200 different cold viruses in existence, such mutations create immense logistical challenges in vaccine development. Fortunately, the common cold typically lasts only a few days and is adequately treated with rest and self-care.

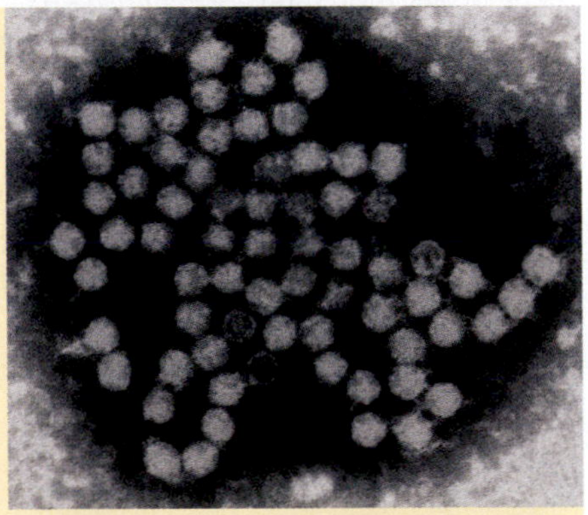

Rhinoviruses TEM 100 nm

A common method for attenuating viruses involves raising them for numerous generations in tissue culture cells until the viruses lose the ability to produce disease. Bacteria may be made avirulent by culturing them under unusual conditions or by using genetic manipulation.

Attenuated vaccines—those containing attenuated microbes—are also called *modified live vaccines.* Because they contain active but avirulent organisms or viruses, these vaccines cause very mild infections but no serious disease under normal conditions. Attenuated viruses in such a vaccine infect host cells and replicate; the infected cells then process endogenous viral antigens. Because modified live vaccines contain active microbes, a large number of antigen molecules are available to stimulate an immune response. Further, vaccinated individuals can infect those around them, providing **contact immunity**—that is, immunity beyond the individual receiving the vaccine.

Although usually very effective, attenuated vaccines can be hazardous because modified microbes may retain enough residual virulence to cause disease in immunosuppressed people. Pregnant women should not receive live vaccines because of the danger that the attenuated pathogen will cross the placenta and harm the fetus. Occasionally, modified viruses actually revert to wild type or mutate to a form that causes persistent infection or disease. For example, in 2000 a polio epidemic in the Dominican Republic and Haiti resulted from the reversion of an attenuated virus in oral polio vaccine to a virulent poliovirus. For this reason, we no longer use oral polio vaccine to immunize children in the United States.

Inactivated (Killed) Vaccines For some diseases, live vaccines have been replaced by **inactivated vaccines**, which are of two types: *whole agent vaccines* are produced with deactivated but whole microbes, whereas *subunit vaccines* are produced with antigenic fragments of microbes. Because neither whole agent nor subunit vaccines can replicate, revert, mutate, or retain residual virulence, they are safer than live vaccines. However, because they cannot replicate, multiple ("booster") doses must be administered to achieve full immunity, and immunized individuals do not stimulate contact immunity. Also, with whole agent vaccines, nonantigenic portions of the microbe occasionally stimulate a painful inflammatory response in some individuals. As a result, whole agent pertussis vaccine is now being replaced with a subunit vaccine called acellular pertussis vaccine.

When microbes are killed for use in vaccines, it is important that their antigens remain as similar to those of living organisms as possible. If chemicals are used for killing, they must not alter the antigens responsible for stimulating protective immunity. A commonly used inactivating agent is *formaldehyde,* which denatures proteins and nucleic acids.

Because the microbes of inactivated vaccines cannot reproduce, they do not present as many antigenic molecules to the body as do live vaccines; therefore, inactivated vaccines are antigenically weak. They are administered in high doses or in multiple doses, or incorporated with materials called **adjuvants**[2] (ad´joo-vănts), substances that increase the effective antigenicity of the vaccine by stimulating immune cell receptors and their actions. Unfortunately, high individual doses and multiple dosing increase the risk of producing allergies, and the use of adjuvants to increase antigenicity may stimulate local inflammation.

Because all types of killed vaccines are recognized by the immune system as exogenous antigens, they stimulate an antibody immune response.

Toxoid Vaccines For some bacterial diseases, notably tetanus and diphtheria, it is more efficient to induce an immune response against toxins than against cellular antigens. **Toxoid vaccines** (tok´soyd) are chemically or thermally modified toxins that are used in vaccines to stimulate active immunity. As with killed vaccines, toxoids stimulate antibody-mediated immunity. Because toxoids have few antigenic determinants, effective immunization requires multiple childhood doses as well as reinoculations every 10 years for life. ▶**ANIMATIONS:** ***Vaccines: Function, Types***

Combination Vaccines The Centers for Disease Control and Prevention (CDC) has approved several **combination vaccines** for routine use. These vaccines combine antigens from several toxoids and inactivated pathogens that are administered simultaneously. Examples include MMR—vaccine against measles, mumps, and rubella—and Pentacel, which is a vaccine against diphtheria, tetanus, pertussis (whooping cough), polio, and diseases of *Haemophilus influenzae* (hē-mof´i-lŭs in-flū-en´zī).

Vaccines Using Recombinant Gene Technology Although live, inactivated, and toxoid vaccines have been highly successful in controlling infectious diseases, researchers are always seeking ways to make vaccines more effective, cheaper, and safer and to make new vaccines against pathogens that have been difficult to protect against. For example, scientists have developed a recombinant DNA vaccine against a fungus, *Blastomyces* (blas-tō-mī´sēz)—the first vaccine against a fungal pathogen. Scientists can also use a variety of genetic recombinant techniques to make improved vaccines. For example, they can selectively delete virulence genes from a pathogen, producing an irreversibly attenuated microbe, one that cannot revert to a virulent pathogen **(FIGURE 17.2a)**.

Scientists also use recombinant techniques to produce large quantities of very pure viral or bacterial antigens for use in vaccines. In this process, scientists isolate the gene that codes for an antigen and insert it into a bacterium, yeast, or other cell, which then expresses and releases the antigen **(FIGURE 17.2b)**. Vaccine manufacturers produce hepatitis B vaccine in this manner using recombinant yeast cells.

Alternatively, a genetically altered microbial cell or virus may express the antigen and act as a live vaccine **(FIGURE 17.2c)**. Experimental recombinant vaccines of this type have used adenoviruses, herpesviruses, poxviruses, and bacteria such as *Salmonella* (sal´mŏ-nel´ă). Vaccinia virus (cowpox virus) is often used because it is easy to administer by dermal scratching or orally and because its large genome makes inserting a new gene into it relatively easy. Another method results in the body's own cells expressing the antigen. The DNA coding for a pathogen's antigen can be inserted into a plasmid vector, which is then injected into the body **(FIGURE 17.2d)**. The body's cells take up

[2]From latin *adjuvo,* meaning "to help."

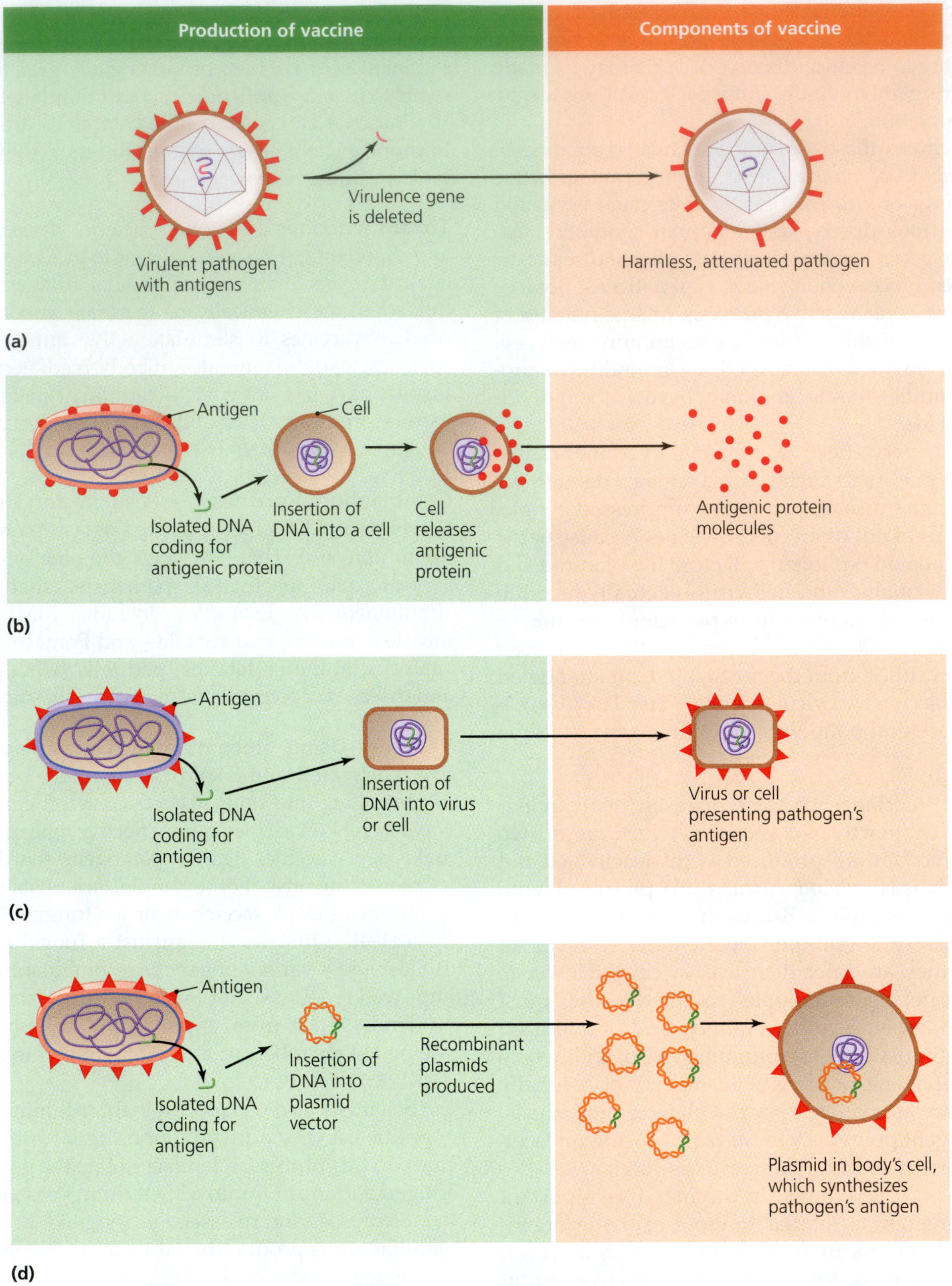

▲ **FIGURE 17.2 Some uses of recombinant DNA technology for making improved vaccines.** **(a)** Deletion of virulence gene(s) to create an attenuated pathogen for use in a vaccine. **(b)** Insertion of a gene that codes for a selected antigenic protein into a cell, which then produces large quantities of the antigen for use in a vaccine. **(c)** Insertion of a gene that codes for a selected antigenic protein into a cell or virus, which displays the antigen. The entire recombinant is used in a vaccine. **(d)** Injection of DNA containing a selected gene (in this case, as part of a plasmid) into an individual. Once some of this DNA is incorporated into the genome of a patient's cells, those cells synthesize and process the antigen, which stimulates an immune response.

the plasmid (with the antigen's DNA) and then transcribe and translate the gene to produce antigen, which triggers an immune response.

Vaccine Manufacture

Manufacturers mass-produce many vaccines by growing microbes in laboratory culture vessels, but because viruses require a host cell to reproduce, they are cultured inside chicken eggs. Availability of sterile eggs is thus critical for manufacturing viral vaccines such as flu vaccines. Because the vaccines are produced in eggs, physicians must withhold such immunizations from patients with egg allergies. Research on gene-based vaccines and development of vaccines in genetically modified plants may result in safer vaccines.

Recommended Immunizations

The CDC and medical associations publish recommended immunization schedules for children, adults, and special populations, such as health care workers and HIV-positive individuals. The recommendations are frequently modified to reflect changes in the relationships between pathogens and the human population. **FIGURE 17.3** highlights the general 2013 immunization

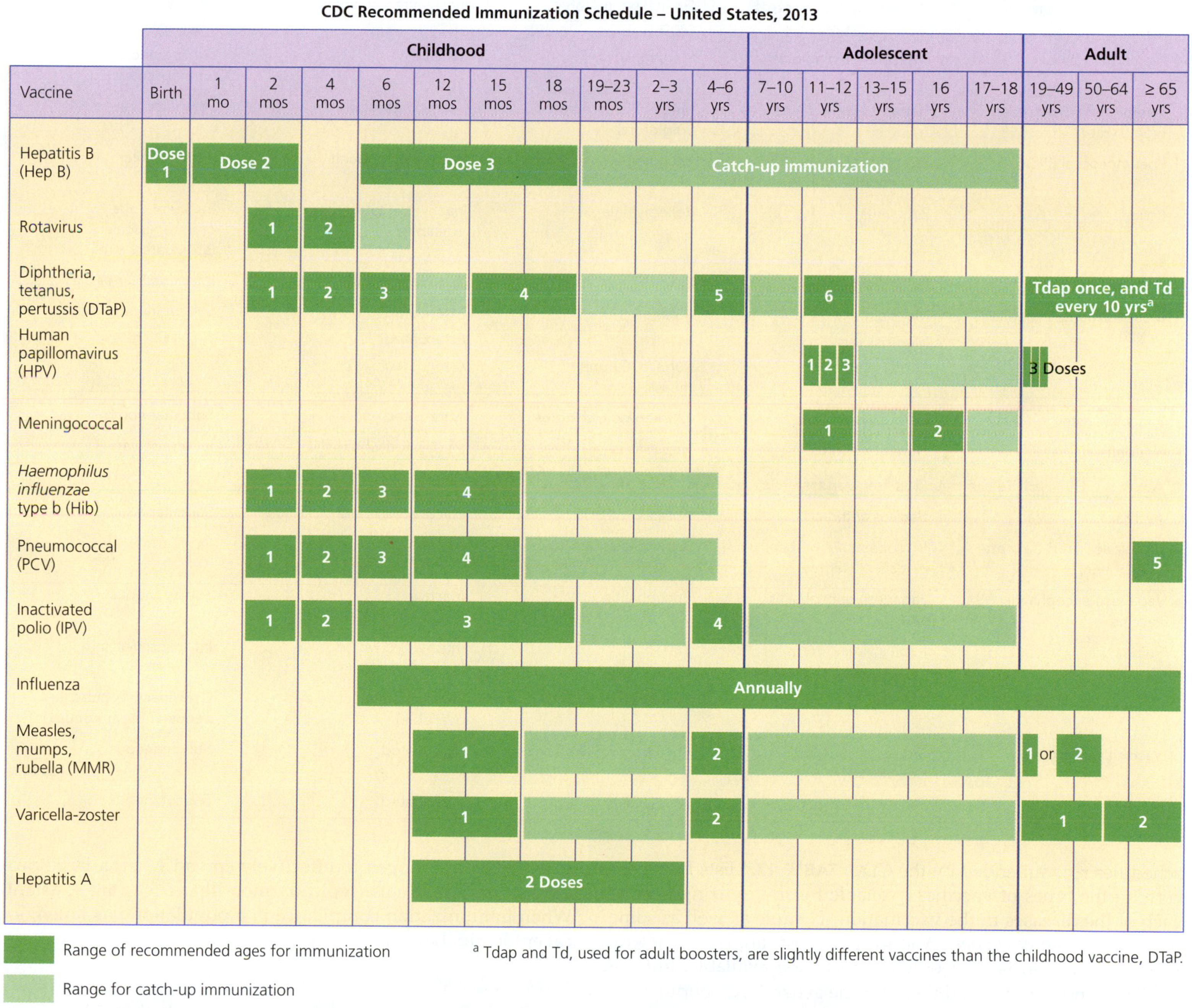

▲ **FIGURE 17.3 The CDC's recommended immunization schedule for the general population.** Adult meningococcal vaccine is recommended for college students who live in dormitories.

TABLE 17.1 Principal Vaccines to Prevent Human Diseases

Vaccine	Disease Agent	Disease	Vaccine Type	Method of Administration
Recommended by CDC				
Hepatitis B	Hepatitis B virus	Hepatitis B	Inactive subunit from recombinant yeast	Intramuscular
Rotavirus	*Rotavirus*	Gastroenteritis	Attenuated, recombinant	Oral
Diphtheria/tetanus/ acellular pertussis (DTaP)	Diphtheria toxin Tetanus toxin *Bordetella pertussis*	Diphtheria Tetanus Whooping cough	Toxoid Toxoid Inactivated subunit (inactivated whole also available)	Intramuscular
Human papillomavirus (HPV)	Human papillomaviruses	Genital warts, cervical cancer	Inactive recombinant	Intramuscular
Meningococcal	*Neisseria meningiditis*	Meningitis	Inactive	Subcutaneous or intramuscular
Haemophilus influenzae type b (Hib)	*Haemophilus influenzae*	Meningitis, pneumonia, epiglottitis	Inactivated subunit	Intramuscular
Pneumococcal (PCV)	*Streptococcus pneumoniae*	Pneumonia	Inactivated subunit	Intramuscular
Polio	Poliovirus	Poliomyelitis	Inactivated (attenuated also available)	Subcutaneous or intramuscular (attenuated: oral)
Influenza	Influenzaviruses	Flu	Inactivated subunit	Intramuscular or oral
Measles/mumps/rubella (MMR)	Measles virus Mumps virus Rubella virus	Measles Mumps Rubella (German measles)	Attenuated Attenuated Attenuated	Subcutaneous
Varicella-zoster	Chicken pox virus	Chicken pox, shingles	Attenuated	Subcutaneous
Hepatitis A	Hepatitis A virus	Hepatitis A	Inactivated whole	Intramuscular
Available but Not Recommended for General Population in the United States				
Anthrax	*Bacillus anthracis*	Anthrax	Inactivated whole	Subcutaneous
BCG (bacillus of Calmette and Guérin)	*Mycobacterium tuberculosis, M. leprae*	Tuberculosis, leprosy	Attenuated	Intradermal
Japanese encephalitis vaccine	Japanese encephalitis virus	Encephalitis	Inactive	Subcutaneous
Rabies	Rabies virus	Rabies	Inactivated whole	Intramuscular or intradermal
Typhoid fever vaccine	*Salmonella enterica*	Typhoid fever	Attenuated (inactive also available)	Oral (inactive: subcutaneous or intramuscular)
Vaccinia (cowpox)	Smallpox virus, monkey pox virus	Smallpox, monkey pox	Attenuated	Subcutaneous
Yellow fever	Yellow fever virus	Yellow fever	Attenuated	Subcutaneous

schedules recommended by the CDC. **TABLE 17.1** lists facts concerning the types of vaccines available for immunizing against each of the diseases in the vaccination schedule as well as some other available vaccines. Vaccines against anthrax, cholera, plague, tuberculosis, and other diseases are available, but the CDC does not recommend them for the general U.S. population.

It is important that patients follow the recommended immunization schedule not only to protect themselves but also to provide society with **herd immunity**. Herd immunity is the protection provided all individuals in a population due to the inability of a pathogen to effectively spread when a large proportion of individuals (typically more than 75%) are resistant. When immunization compliance in a population has fallen, local epidemics have resulted.

Vaccine Safety

Health care providers must carefully weigh the risks associated with vaccines against their benefits. A common vaccine-associated problem is mild toxicity. Some vaccines—especially whole agent vaccines that contain adjuvant—may cause pain at

BENEFICIAL MICROBES

Cowpox: To Vaccinate or Not to Vaccinate?

Vaccinia necrosum

Dr. Edward Jenner developed an early use for a beneficial microbe in medicine. Medical personnel in the United States followed Jenner's example by regularly administering cowpox virus—as the smallpox vaccine—to the general public until 1971, at which time the risk of contracting smallpox was deemed too low to justify required vaccinations. Indeed, in 1980 the World Health Assembly declared smallpox successfully eradicated from the natural world. However, recent concerns about the potential use of smallpox virus as an agent of bioterrorism has sparked debate about whether citizens should once again be vaccinated with cowpox to protect against smallpox.

Although safe and effective for most healthy adults, for others the attenuated cowpox virus can result in serious side effects—even death. Individuals with compromised immune systems (such as AIDS patients or cancer patients undergoing chemotherapy) are considered to be at particularly high risk for developing adverse reactions. Pregnant women, infants, and individuals with a history of the skin condition eczema are also considered poor candidates for the vaccine. Though rare, adverse reactions to the vaccine may also develop in certain otherwise healthy individuals. The more serious side effects include *vaccinia necrosum* (characterized by progressive cell death in the area of vaccination) and encephalitis (inflammation of the brain). Approximately one in every 1 million individuals receiving cowpox virus as a vaccine for the first time develops a fatal reaction to it.

Is the risk of a bioterrorist smallpox attack great enough to warrant the exposure to the known risks of administering cowpox virus 252 vaccine to the general population? If you were a public health official, what would you decide?

the injection site for several hours or days after injection. In rare cases, toxicity may result in general malaise and possibly a fever high enough to induce seizures. Although not usually life threatening, the potential for these symptoms may be sufficient to discourage people from being immunized or having their infants immunized.

A much more severe problem associated with immunization is the risk of *anaphylactic shock*, an allergic reaction that may develop to some component of the vaccine, such as egg proteins, adjuvants, or preservatives. Because people are rarely aware of such allergies ahead of time, recipients should remain for several minutes in the physician's office, where epinephrine is readily available to counter any signs of an allergic reaction.

A third major problem associated with immunization is that of residual virulence, which we previously discussed. Attenuated viruses occasionally cause disease not only in fetuses and immunosuppressed patients but also in healthy children and adults. A good example is the attenuated oral poliovirus vaccine (OPV), which was commonly used in the United States until the late 1990s. Though a very effective vaccine, it causes clinical poliomyelitis in one of every 2 million recipients or their close contacts. Medical personnel in the United States eliminated this problem by switching to inactivated polio vaccine (IPV).

Over the past two decades, lawsuits in the United States and Europe have alleged that certain vaccines against childhood diseases cause or trigger disorders such as autism, diabetes, and asthma. Extensive research has failed to substantiate these allegations. Vaccine manufacturing methods have improved tremendously in recent years, ensuring that modern vaccines are much safer than those in use even a decade ago. The U.S. Food and Drug Administration (FDA) has established a Vaccine Adverse Event Reporting System for monitoring vaccine safety.

The CDC and FDA have determined that the problems associated with immunization are far less serious than the suffering and death that would result if we stopped immunizing people. **Beneficial Microbes: Cowpox: To Vaccinate or Not to Vaccinate?** discusses the issues surrounding the administration of smallpox vaccinations to the general public.

Passive Immunotherapy

LEARNING **OUTCOMES**

17.5 Identify two sources of antibodies for use in passive immunotherapy.

17.6 Compare the relative advantages and disadvantages of active immunization and passive immunotherapy.

Passive immunotherapy (sometimes called *passive immunization*) involves the administration of antibodies to a patient. Physicians use passive immunotherapy when protection against a recent infection or an ongoing disease is needed quickly. Rapid protection is achieved because passive immunotherapy does not require the body to mount a response; instead, preformed antibodies are immediately available to bind to antigen, enabling neutralization and opsonization to proceed without delay. For example, in a case of botulism poisoning (caused by the toxin of *Clostridium botulinum* [klos-trid´ē-ŭm bo-tū-lī´num]), passive immunotherapy with preformed antibodies against the toxin can prevent death.

Antibodies directed against toxins are also called *antitoxins* (an-tē-tok´sinz); *antivenom (antivenin)* used to treat snakebites is an example of an antitoxin. In some cases, infections with certain viruses—hepatitis A and B, measles, rabies, Ebola, chicken

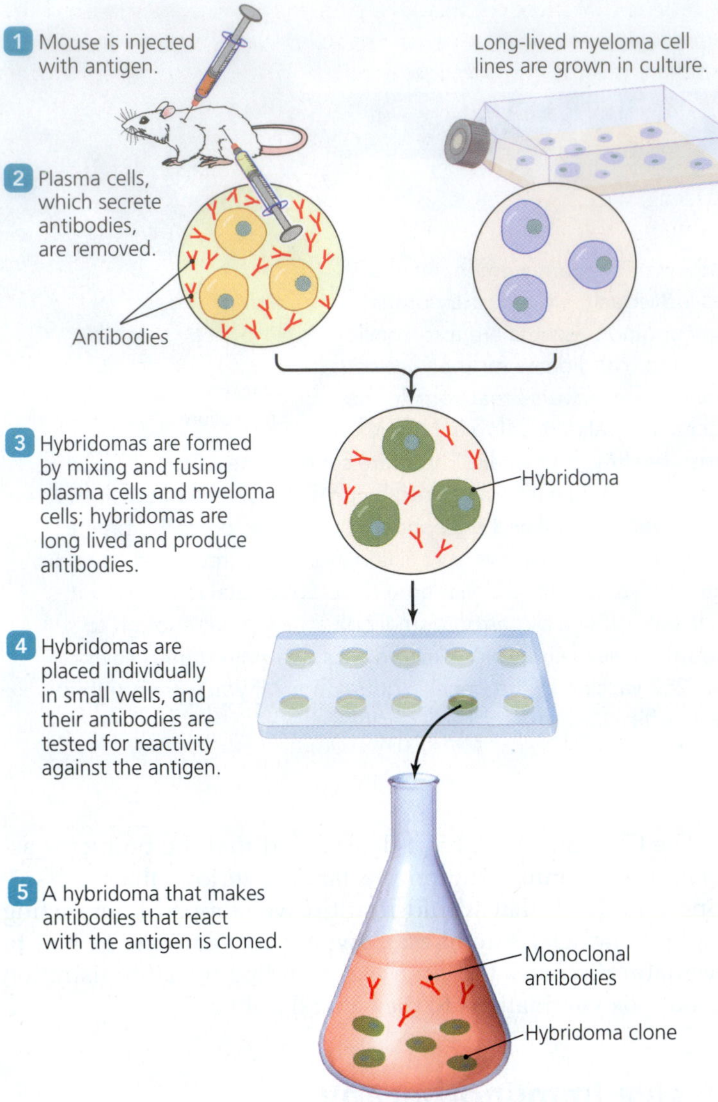

1 Mouse is injected with antigen.
Long-lived myeloma cell lines are grown in culture.
2 Plasma cells, which secrete antibodies, are removed.
Antibodies
3 Hybridomas are formed by mixing and fusing plasma cells and myeloma cells; hybridomas are long lived and produce antibodies.
Hybridoma
4 Hybridomas are placed individually in small wells, and their antibodies are tested for reactivity against the antigen.
5 A hybridoma that makes antibodies that react with the antigen is cloned.
Monoclonal antibodies
Hybridoma clone

▲ **FIGURE 17.4 The production of hybridomas.** After a laboratory animal is injected with the antigen of interest 1, plasma cells are removed from the animal and isolated 2. When these plasma cells are fused with cultured cancer cells called myelomas, hybridomas result 3. Once the hybridomas are cultured individually and the hybridoma that produces antibodies against the antigen of interest is identified 4, it is cloned to produce a large number of hybridomas, all of which secrete identical antibodies called monoclonal antibodies 5.

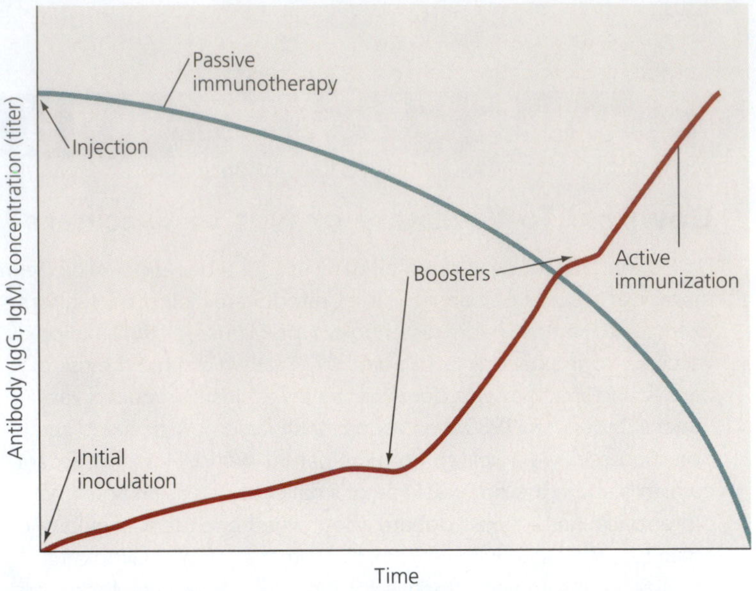

▲ **FIGURE 17.5 The characteristics of immunity produced by active immunization (red) and passive immunotherapy (green).** Passive immunotherapy provides strong and immediate protection, but it disappears relatively quickly. Active immunity takes some time and may require additional booster inoculations to reach protective levels, but it is long lasting and capable of restimulation.

pox, and shingles—are treated with antibodies directed against the causative viruses.

To acquire antibodies for passive immunotherapy, clinicians remove blood cells and clotting factors from the blood of donors; the result is *serum,* which contains a variety of antibodies—particularly gamma globulins. When used for passive immunotherapy, such serum is called **antiserum** (an-tē-sē´rŭm) or sometimes *immune serum.* Antisera are typically collected from human blood plasma donors or from large animals intentionally exposed to a pathogen of interest, because the large blood volume of the animals contains more antibodies than can be obtained from smaller animals. Pooled antisera from a group of human donors can be administered intravenously *(intravenous immunoglobulins, IVIg)* to treat immunodeficiencies and some autoimmune and inflammatory diseases.

Passive immunotherapy has the following limitations:

- Repeated injections of animal-derived antisera can trigger an allergic response called *serum sickness,* in which the recipient mounts an immune response against animal antigens found in the antisera.
- The patient may degrade the antibodies relatively quickly; therefore, protection is not long lasting.
- The body does not produce memory B cells in response to passive immunotherapy, so the patient is not protected against subsequent infections.

Scientists have overcome the limitations of antisera by developing **hybridomas** (hī-brid-ō´măz), which are tumor cells created by fusing antibody-secreting plasma cells with cancerous plasma cells called *myelomas* (mī-ĕ-lō´măz) **(FIGURE 17.4)**. Each hybridoma divides continuously (because of the cancerous plasma cell component) to produce clones of itself, and each clone secretes large amounts of a single antibody molecule. These identical antibodies are called **monoclonal antibodies** (mon-ō-klō´năl) because all of them are secreted by clones originating from a single plasma cell. Once scientists have identified the hybridoma that secretes antibodies complementary to the antigen of interest, they maintain it in tissue culture to produce the antibodies needed for passive immunotherapy. For example, physicians use such a monoclonal antibody to treat newborns infected with respiratory syncytial virus.

Active immunization and passive immunotherapy are used in different circumstances because they provide protection with different characteristics **(FIGURE 17.5)**. As just noted, passive immunotherapy with preformed antibodies is used whenever immediate protection is required. However, because preformed antibodies are removed rapidly from the blood and no memory B cells are produced, protection is temporary, and the recipient becomes susceptible again. Active immunization provides long-term protection that is capable of restimulation. Thus, when initiated before any exposure to *Clostridium tetani* (te´tan-ē) has occurred, active immunization using a tetanus toxoid develops long-lasting protection that is readily available upon exposure to the toxin.

TELL ME WHY

Vaccines have drastically reduced the number of cases of many diseases, such as measles and whooping cough. Why should parents have their children vaccinated given that there are so few cases?

Serological Tests That Use Antigens and Corresponding Antibodies

LEARNING | **OUTCOMES**

17.7 Define *serology*.
17.8 Describe several uses of serological tests.
17.9 In general terms, compare and contrast precipitation, agglutination, neutralization, complement fixation, and labeled antibody testing methods.

The determination of the presence of particular antigens or specific antibodies in blood serum is called **serology** (sě-rol´ō-jē). Scientists have developed a variety of serological tests to identify antigens or antibodies in serum. Serological methods range from simple manual procedures to complex and automated ones.

Serological tests have many uses. Epidemiologists review serological test results to monitor the spread of infection through a population. Physicians order relevant tests, which are conducted by medical laboratory scientists or technologists, to establish diagnoses. For example, when a physician suspects a patient might be infected by both HIV and hepatitis B virus, the doctor orders both an anti-HIV test and a hepatitis B virus surface antigen test. The first determines the presence of antibodies against HIV in the serum—strong evidence that the patient is infected with HIV. The surface antigen test indicates infection with hepatitis B virus.

In the following sections, we examine various types of serological methods: precipitations, turbidimetry, nephelometry, agglutination, neutralization, and labeled antibody tests. Some of these procedures are presented for historical reasons—more accurate and faster modern tests have replaced them. For example, one modern method—*polymerase chain reaction (PCR)*—can amplify copies of genes. Thus, PCR enables testing for the presence of viral genetic material rather than the presence of antibodies against the viruses, allowing infection to be detected before the body produces antibodies.

Precipitation Tests

LEARNING | **OUTCOMES**

17.10 Describe the general principles of precipitation testing.
17.11 Describe the technique of immunodiffusion.

One of the simplest of serological tests relies on the fact that when antigens and antibody are mixed in proper proportions, they form huge, insoluble, lattice-like complexes called precipitates. When, for example, a solution of a soluble antigen, such as that of the fungus *Coccidioides immitis* (kok-sid-ē-oy´dēz im´mi-tis), is mixed with an antiserum containing antibodies against the fungal antigen, the mixture quickly becomes cloudy because of the formation of a precipitate consisting of antigen-antibody complexes, also called **immune complexes**.

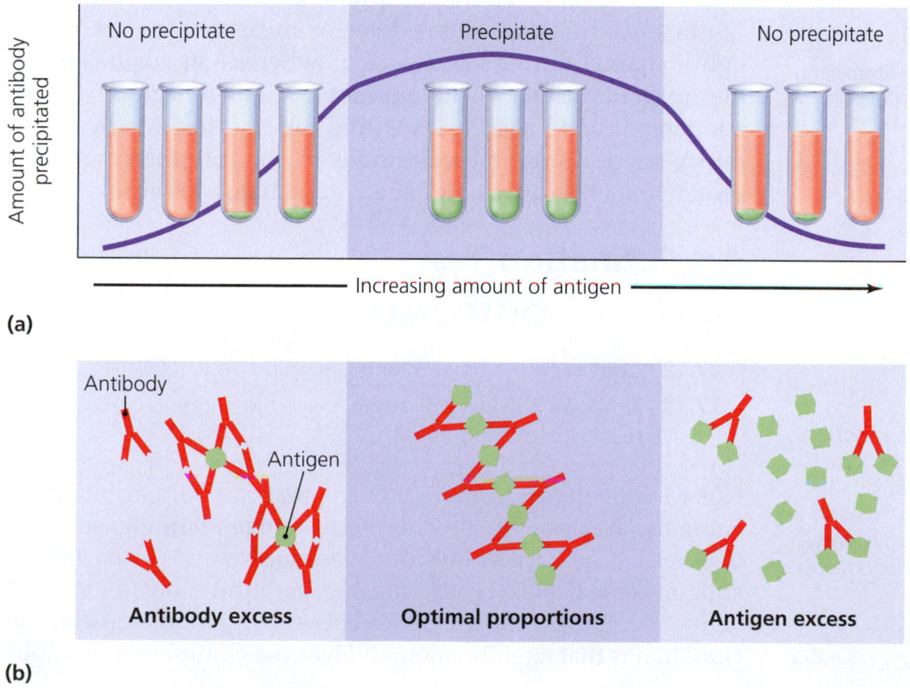

◀ **FIGURE 17.6 Characteristics of precipitation reactions. (a)** When increasing amounts of antigen are placed in tubes containing a constant amount of antibody, precipitate forms only in tubes having moderate amounts of antigen. The overlay graph shows the amount of precipitate formed versus the amount of antigen present. **(b)** In the presence of excess antigen or antibody, immune complexes are small and soluble; only when antigen and antibody are in optimal proportions do large complexes form and precipitate.

When a given amount of antibody is added to each of a series of test tubes containing increasing amounts of antigen, the amount of precipitate increases gradually until it reaches a maximum **(FIGURE 17.6a)**. In test tubes containing still more antigen molecules, the amount of precipitate declines; in fact, in test tubes containing antigen in great excess over antibody, no precipitate at all develops. Thus, a graph of the amount of precipitate versus the amount of antigen has a maximum in the middle values.

The reasons behind this pattern of precipitation reactions are simple. Complex antigens are generally multivalent—each possesses many epitopes—and antibodies have pairs of active sites and therefore can simultaneously cross-link the same epitope on two antigen molecules (see Chapter 16). When there is excess antibody, each antigen molecule is covered with many antibody molecules, preventing extensive cross-linkage and thus precipitation **(FIGURE 17.6b)**. Since there is no precipitation, an observer might conclude that there is no antigen in the solution—a negative test result. This is untrue; antigen is present. Such a *false negative* interpretation is called a *prozone phenomenon*.

When the reactants are in optimal proportions, the ratio of antigen to antibody is such that cross-linking and lattice formation are extensive. As this lattice grows, it precipitates.

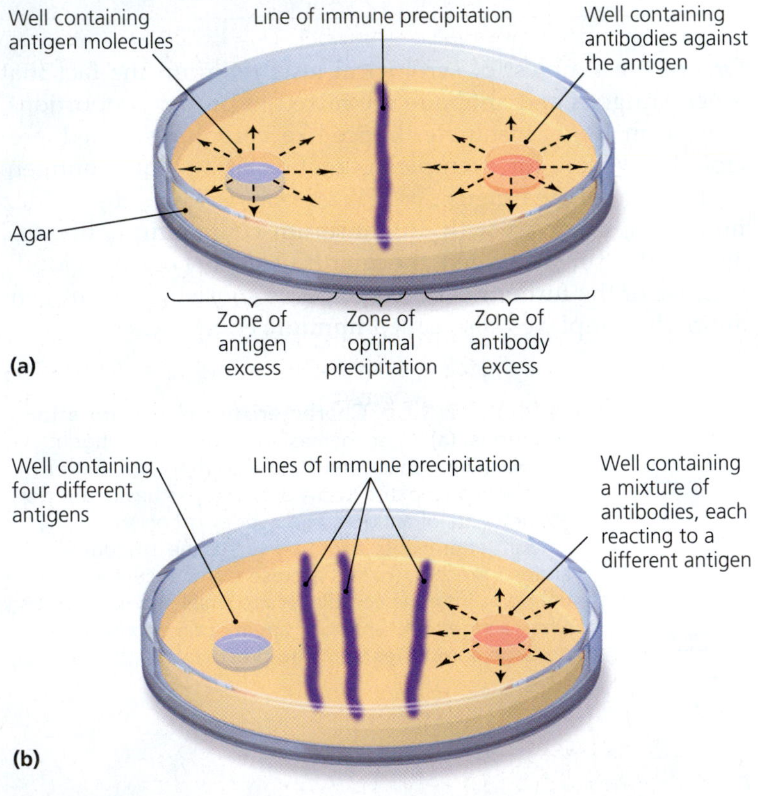

▲ **FIGURE 17.7 Immunodiffusion, a type of precipitation reaction.**
(a) Antigen and antibody placed in wells diffuse out and through the agar; where they meet in optimal proportions, a line of precipitation forms.
(b) When multiple antigens and antibodies are placed in the wells, multiple lines of precipitation mark the sites where different antigen-antibody combinations occurred in optimal proportions. *Why did only three lines of precipitation occur when the antigen well contained four antigens?*

Figure 17.7 *None of the antibodies used in the test was complementary to the fourth antigen.*

In mixtures in which antigen is in excess, little or no precipitation occurs because there are few cross-linkages. Antibody-antigen complexes are small and soluble, so no precipitation occurs.

Because precipitation requires the mixing of antigen and antibody in optimal proportions, it is not possible to perform a precipitation test by combining just any two solutions containing these reagents. To ensure that the optimal concentrations of antibody and antigen come together, scientists historically have used a technique involving movement of the molecules through an agar gel: immunodiffusion.

Immunodiffusion

In the precipitation technique called **immunodiffusion** (im´ū-nō-di-fu´zhŭn), a researcher cuts cylindrical holes called *wells* in an agar plate. One well is filled with a solution of antigen and the other with a solution of antibodies against the antigen. The antigen and antibody molecules diffuse in all directions out of the wells and into the surrounding agar, and where they meet in optimal proportions, a line of precipitation appears **(FIGURE 17.7a)**. If the solutions contain many different antigens and antibodies, each complementary pair of reactants reaches optimal proportions at different positions, and numerous lines of precipitation are produced—one for each interacting antigen-antibody pair **(FIGURE 17.7b)**. Such an immunodiffusion test has been used to indicate exposure to complex mixtures of antigens from fungal pathogens. Only exposed patients have serum antibodies—and show precipitation—against the fungal antigens, so physicians can monitor and treat such patients.

Turbidimetric and Nephelometric Tests

Turbidimitry and *nephelometry* are automated methods that measure the cloudiness of a solution, as occurs when antibodies and antigens are mixed together. As noted previously, when the concentrations of antibodies and antigens are optimal, the initial cloudiness is followed by precipitation.

In turbidimetry, a light detector measures the amount of light passing through a solution, whereas in nephelometry, the machine measures the amount of light reflected from the antigen-antibody complexes within the solution. Medical laboratory scientists use these methods to quantify the amounts of proteins, such as antibodies and complement, in serum.

Agglutination Tests

LEARNING | OUTCOMES

17.12 Contrast agglutination and precipitation tests.
17.13 Describe how agglutination is used in immunological testing, including titration.

Not all antigens are soluble proteins that can be precipitated by antibody. Because of their multiple antigen-binding sites, antibodies can also cross-link particles, such as whole bacteria or antigen-coated latex beads, causing **agglutination** (ă-gloo-ti-nā´shŭn, clumping). The difference between agglutination and precipitation is that agglutination involves the clumping of insoluble

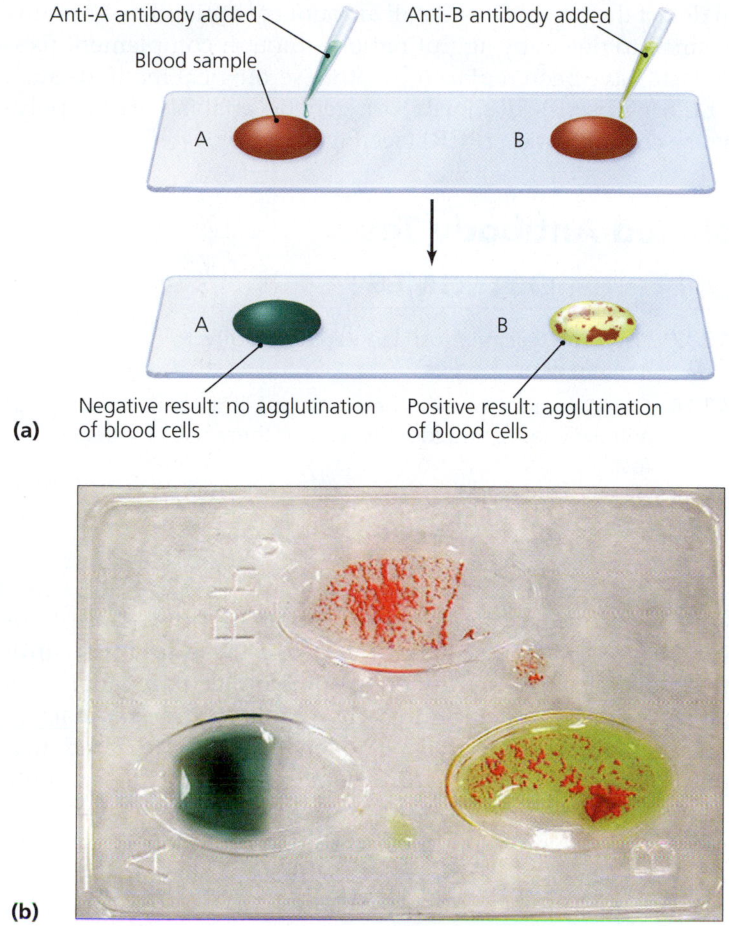

▲ FIGURE 17.8 The use of hemagglutination to determine blood types in humans. **(a)** Antibodies with active sites that bind to either of two surface antigens (antigen A or antigen B) of red blood cells are added to portions of a given blood sample. Where the antibodies react with the surface antigens, the blood cells can be seen to agglutinate, or clump together. **(b)** Photo of actual test, which also includes a test for Rh antigen, to be denoted positive (+) or negative (−). *What is the blood type of the person whose blood was used in this hemagglutination reaction?*

Figure 17.8 *The individual who donated the blood sample has type B+ blood.*

particles, whereas precipitation involves the aggregation of soluble molecules. Agglutination reactions are sometimes easier to see and interpret with the unaided eye.

When the particles agglutinated are red blood cells, the reaction is called *hemagglutination* (hē-mă-gloo´ti-nā´shŭn). One use of hemagglutination is to determine blood type in humans. Blood is considered type A if the red blood cells possess surface antigens called A antigen, type B if they possess B antigens, type AB if they possess both antigens, and type O if they have neither antigen. In a hemagglutination reaction to determine blood type **(FIGURE 17.8)**, two portions of a given blood sample are placed on a slide. Anti-A antibodies are added to one portion and anti-B antibodies to the other; the antibodies agglutinate those blood cells that possess corresponding antigens.

Another use of agglutination is in a type of test that determines the concentration of antibodies in a clinical sample.

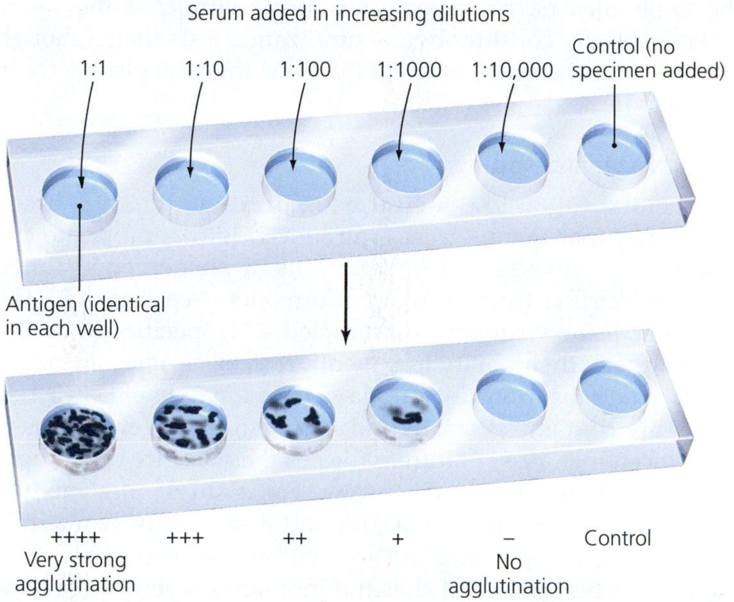

▲ FIGURE 17.9 Titration, the use of agglutination to quantify the amount of antibody in a serum sample. Serial dilutions of serum are added to wells containing a constant amount of antigen. At lower serum dilutions (higher concentrations of antibody), agglutination occurs; at higher serum dilutions, antibody concentration is too low to produce agglutination. The serum's titer is the highest dilution at which agglutination can be detected—in this case, 1:1000.

Although the simple *detection* of antibodies is sufficient for many purposes, it is often more desirable to measure the *amount* of antibodies in serum. By doing so, clinicians can determine whether a patient's antibody levels are rising, as occurs in response to the presence of active infectious disease, or falling, as occurs during the successful conclusion of a fight against an infection. One way of measuring antibody levels in blood sera is by **titration** (tī-trā´shŭn). In titration, the serum being tested undergoes a regular series of dilutions, and each dilution is then tested for agglutinating activity **(FIGURE 17.9)**. Eventually, the antibodies in the serum become so dilute that they can no longer cause agglutination. The highest dilution of serum giving a positive reaction is its **titer**, which is expressed as a ratio reflecting the dilution. Thus, a serum that must be greatly diluted before agglutination ceases (e.g., has been diluted a thousand-fold; that is, has a titer of 1:1000) contains more antibodies than a serum that no longer agglutinates after minimal dilution (has a titer of 1:10).

Neutralization Tests

LEARNING | **OUTCOMES**

17.14 Explain the purpose of neutralization tests.

17.15 Contrast a viral hemagglutination inhibition test with a hemagglutination test.

Neutralization tests work because antibodies can *neutralize* the biological activity of many pathogens and their toxins. For example, combining antibodies against tetanus toxin with a sample of toxin renders the sample harmless to mice because the antibodies have reacted with and neutralized the toxin. Next we briefly consider two neutralization tests that, although not simple to perform, effectively reveal the biological activity of antibodies.

Viral Neutralization

One neutralization test is **viral neutralization**, which is based on the fact that many viruses introduced into appropriate cell cultures will invade and kill the cells, a phenomenon called a *cytopathic effect* (seen in plaque formation; see Figure 13.17). However, if the viruses are first mixed with specific antibodies against them, their ability to kill cultured cells is neutralized. In a viral neutralization test, the lack of cytopathic effects when a mixture containing serum and a known pathogenic virus is introduced into a cell culture indicates the presence of antibodies against that virus in the serum. For example, if a mixture containing an individual's serum and a sample of hantavirus produces no cytopathic effect in a culture of susceptible cells, then it can be concluded that the individual's serum contains antibodies to hantavirus, and these antibodies neutralized the virus. Viral neutralization tests are sufficiently sensitive and specific to ascertain whether an individual has been exposed to a particular virus or viral strain, which may lead a physician to a diagnosis or treatment or to recommendations to prevent future infection or disease.

Viral Hemagglutination Inhibition Test

Because not all viruses are cytopathic—they do not kill their host cell—a neutralization test cannot be used to identify all viruses. However, many viruses (including influenzaviruses) have surface proteins that naturally clump red blood cells. (This natural process, called *viral hemagglutination*, must not be confused with the hemagglutination test we discussed previously—viral hemagglutination is not an antibody-antigen reaction.) Antibodies against influenzavirus inhibit viral hemagglutination; therefore, if serum from an individual stops viral hemagglutination, we know that the individual's serum contains antibodies to that particular strain of influenzavirus. Such **viral hemagglutination inhibition tests** can be used to detect antibodies against influenza, measles, mumps, and other viruses that naturally agglutinate red blood cells.

The Complement Fixation Test

LEARNING **OUTCOME**

17.16 Briefly explain the phenomenon that is the basis for a complement fixation test.

Activation of the classical complement system by antibody leads to the generation of membrane attack complexes (MACs) that disrupt cytoplasmic membranes (see Figure 15.9). This phenomenon is the basis for the **complement fixation test** (kom´plĕ-ment fik-sā´shŭn), which is a complex assay used to detect the presence of specific antibodies in an individual's serum. The test can detect the presence of small amounts of antibody—amounts too small to detect by agglutination—though complement fixation tests have been replaced by other serological methods such as ELISA (discussed shortly) or genetic analysis using polymerase chain reaction (PCR) (see Figure 8.5).

Labeled Antibody Tests

LEARNING **OUTCOMES**

17.17 List three tests that use labeled antibodies to detect either antigen or antibodies.

17.18 Compare and contrast the direct and indirect fluorescent antibody tests, and identify at least three uses for these tests.

17.19 Compare and contrast the methods, purposes, and advantages of ELISA and immunoblotting tests.

A different form of serological testing involves *labeled* (or *tagged*) *antibody tests,* so named because these tests use antibody molecules that are linked to some molecular "label" that enables them to be detected easily. Labels include radioactive chemicals, fluorescent dyes, and enzymes. Automated machines can detect and quantify labels. For example, gamma radiation detectors can count radioactive chemicals, and fluorescence microscopes can measure fluorescent labels. Labeled antibody tests using radioactive or fluorescent labels can be used to detect either antigens or antibodies. In the following sections we will consider fluorescent antibody tests, ELISA, and immunoblotting tests.

Fluorescent Antibody Tests

Fluorescent dyes are used as labels in several serological tests. Some fluorescent dyes can be chemically linked to an antibody without affecting the antibody's ability to bind antigen. When exposed to a specific wavelength of light (as in a fluorescence microscope), the fluorescent dye glows. Fluorescently-labeled antibodies are used in direct and indirect fluorescent antibody tests.

Direct fluorescent antibody tests identify the presence of antigen in a tissue. The test is straightforward: A scientist floods a tissue sample suspected of containing the antigen with labeled antibody, waits a short time to allow the antibody to bind to the antigen, washes the preparation to remove any unbound antibody, and examines it with a fluorescence microscope. If the suspected antigen is present, labeled antibody will adhere to it, and the scientist will see fluorescence. This is not a quantitative test—the amount of fluorescence observed is not directly related to the amount of antigen present.

Scientists use direct fluorescent antibody tests to identify small numbers of bacteria in patient tissues. This technique has been used to detect *Mycobacterium tuberculosis* in sputum and rabies viruses infecting a brain. In one use, medical laboratory scientists employ a direct fluorescent antibody test to detect the presence of fungi in the lungs of a patient, corroborating a diagnosis of fungal pneumonia **(FIGURE 17.10)**.

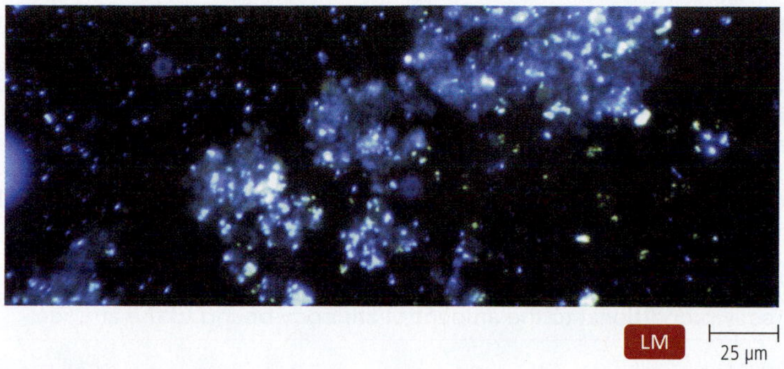

▲ **FIGURE 17.10 A direct fluorescent antibody test.** Fluorescence from labeled antibodies against antigens of the fungal pathogen *Histoplasma capsulatum* in a human lung.

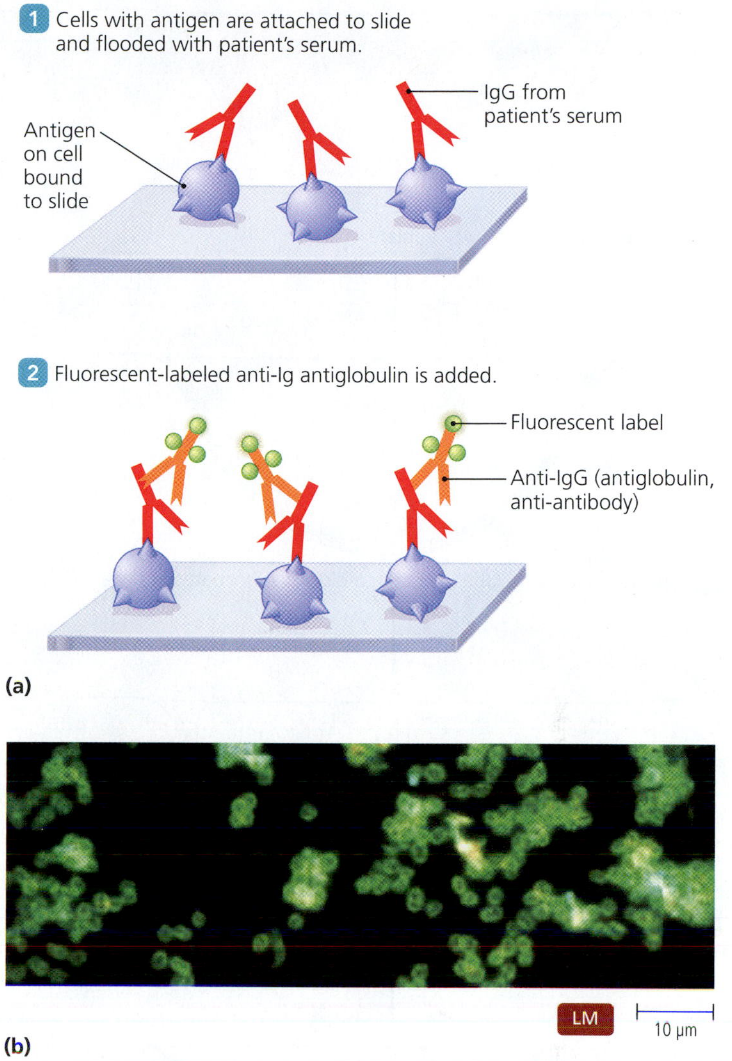

▲ **FIGURE 17.11 The indirect fluorescent antibody test.** This test detects the presence of a specific antibody in a patient's serum. **(a)** The test procedure. 1 Antigen is attached to the slide, which is then flooded with an individual's serum to allow specific antibodies in the serum to bind to the antigen. 2 After anti-antibodies labeled with fluorescent chemical are added and then washed off, the slide is examined with a fluorescence microscope. **(b)** A positive indirect antibody test, in which fluorescence indicates the presence of antibodies in an individual's serum against a particular antigen (here the gonorrhea bacterium).

Indirect fluorescent antibody tests are used to detect the presence of specific antibodies in an individual's serum via a two-step process **(FIGURE 17.11a)**:

1. After an antigen of interest is fixed to a microscope slide, the individual's serum is added for long enough to allow serum antibodies, if present, to bind to the antigen. The serum is then washed off, leaving the antibodies bound to the antigen (but not yet visible).
2. Fluorescently-labeled antibodies against human antibodies (anti-human antibody antibodies) are added to the slide and bind to the antibodies already bound to the antigen. After the slide is washed to remove unbound anti-antibodies, it is examined with a fluorescence microscope.

The presence of fluorescence indicates the presence of the labeled anti-antibodies, which are bound to serum antibodies bound to the fixed antigen; thus, fluoresence indicates that the individual has serum antibodies against the antigen of interest.

Scientists can use indirect fluorescent antibody tests to detect antibodies against many viral, protozoan, or bacterial pathogens, including *Neisseria gonorrhoeae* (nī-se´rē-ă go-nor-rē´ī), the causative agent of gonorrhea **(FIGURE 17.11b)**. The presence of antibodies indicates that the patients have been exposed to the pathogen and may need treatment or counseling on steps to take to lower their risk of future infection.

Scientists routinely identify and separate B and T types of white blood cells, such as lymphocytes, by using specific monoclonal antibodies produced against each cell type. The researchers can attach differently colored fluorescent dyes to the antibodies, allowing them to differentiate types of lymphocytes by the color of the dye attached to each type of antibody. Such identification tests can quantify the numbers and ratios of lymphocyte subsets, information critical in diagnosing and monitoring disease progression and effectiveness of treatment in patients with AIDS, other immunodeficiency diseases, leukemias, and lymphomas.

ELISAs (EIAs)

In another type of labeled antibody test, called an **enzyme-linked immunosorbent assay** (im´ū-nō-sōr´bent as´sā; ELISA), or simply an **enzyme immunoassay (EIA)**, the label is not a dye but instead an enzyme that reacts with its substrate to produce a colored product that indicates a positive test. One form of ELISA is used to detect the presence and quantify the abundance of antibodies in serum—an example of indirect testing—as might be used in diagnosis of Lyme disease, legionellosis (a type of pneumonia), or catscratch disease. An ELISA, which can take place in wells in commercially produced plates, has five basic steps with washes to remove excess chemicals between steps **(FIGURE 17.12)**:

1. Each of the wells in the plate is coated with antigen molecules in solution.
2. Excess antigen molecules are washed off, and another protein (such as gelatin) is added to the well to completely coat any of the surface not coated with antigen.
3. A sample of each of the sera being tested is added to a separate well. Whenever a serum sample contains antibodies

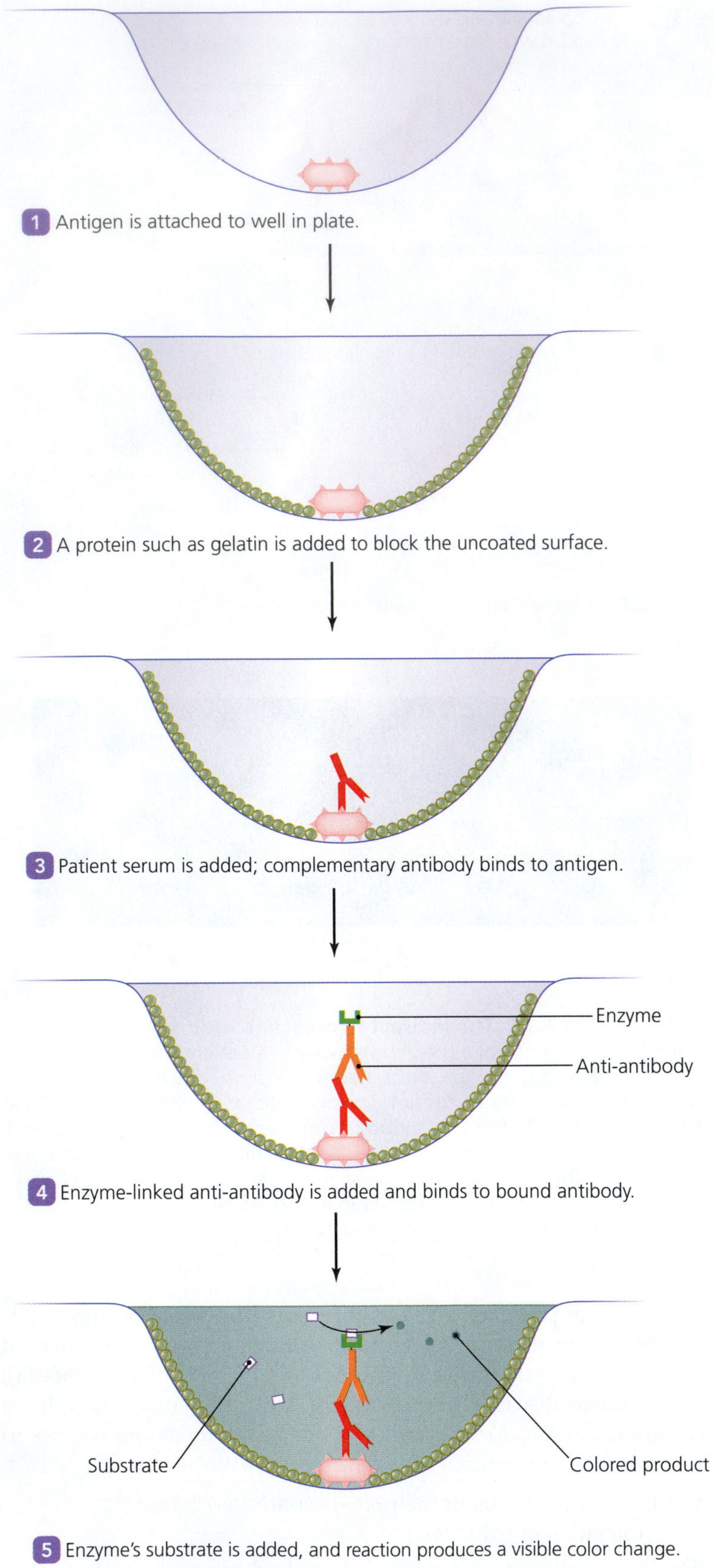

1 Antigen is attached to well in plate.
2 A protein such as gelatin is added to block the uncoated surface.
3 Patient serum is added; complementary antibody binds to antigen.
Enzyme
Anti-antibody
4 Enzyme-linked anti-antibody is added and binds to bound antibody.
Substrate
Colored product
5 Enzyme's substrate is added, and reaction produces a visible color change.

◀ FIGURE 17.12 Enzyme-linked immunosorbent assay (ELISA), also known as enzyme immunoassay (EIA). Shown is one well in a plate. The well is washed between steps. 1 Antigen added to the well attaches to it irreversibly. 2 After excess antigen is removed by washing, gelatin is added to cover any portion of the well not covered by antigen. 3 Test serum is added to the well; any specific antibodies in it bind to the antigen. 4 Enzyme-labeled anti-antibodies are added to the well and bind to any bound antibody. 5 The enzyme's substrate is added to the well, and the enzyme converts the substrate into a colored product; the amount of color, which can be measured via spectrophotometry, is directly proportional to the amount of antibody bound to the antigen.

against the antigen, they bind to the antigen affixed to the plate.

4 Anti-antibodies labeled with an enzyme are added to each well.

5 The enzyme's substrate is added to each well. The enzyme and substrate are chosen because their reaction results in products that cause a visible color change.

A positive reaction in a well, indicated by the development of color, can occur only if the labeled anti-antibody has bound to antibodies attached to the antigen of interest. The intensity of the color, which can be estimated visually or measured accurately using a spectrophotometer, is proportional to the amount of antibody present in the serum.

ELISA has become a test of choice for many diagnostic procedures, such as determination of HIV infection, because of its many advantages:

- Like other labeled antibody tests, ELISA can detect either antibody or antigen.
- ELISAs are sensitive, able to detect very small amounts of antibody (or antigen).
- Unlike some diffusion and fluorescent tests, ELISA can quantify amounts of antigen or antibody. Knowing the amount of antigen or antibody in a patient's serum can provide information concerning the course of an infection or the effectiveness of a treatment.
- ELISAs are easy to perform.
- ELISAs are relatively inexpensive.
- ELISAs can simultaneously test many samples quickly at once.
- ELISAs lend themselves to efficient automation and can be read easily, either by direct observation or by machine.
- Plates coated with antigen and gelatin can be stored for testing whenever they are needed.

A modification of the ELISA technique, called an *antibody sandwich ELISA,* can be used to detect antigen **(FIGURE 17.13)**—an example of direct testing. In testing for the presence of HIV in blood serum, for example, the plates are first coated with antibody against HIV (instead of antigen). Then the sera from individuals being tested for HIV are added to the wells, and any HIV in the sera will bind to the antibody attached to the well. Finally, each well is flooded with enzyme-labeled antibodies specific to the antigen. The name "antibody sandwich ELISA" refers to the fact that the antigen being tested for is

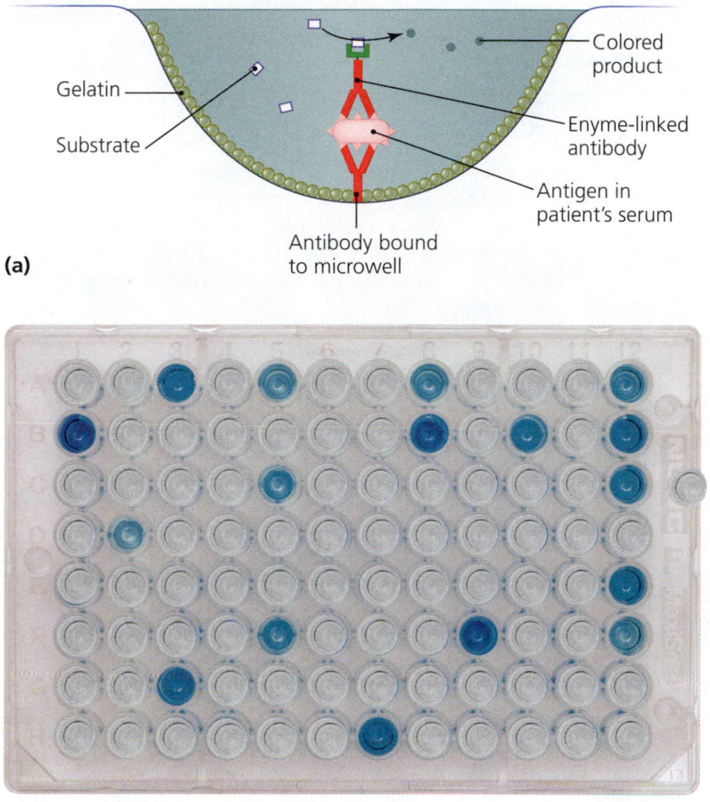

▲ **FIGURE 17.13 An antibody sandwich ELISA.** Because this variation of ELISA is used to test for the presence of antigen, antibody is attached to the well in the initial step. A second antibody sandwiches the antigen. **(a)** Artist's rendition. **(b)** Actual results. *How many wells are positive?*

Figure 17.13 *Seventeen wells are positive.*

"sandwiched" between two antibody molecules. Such tests can also be used to quantify the amount of antigen in a given sample. ▶**VIDEO TUTOR:** ***ELISA***

Immunoblots

An **immunoblot** (also called a *western blot*) is a technique used to detect antibodies against multiple antigens in a complex mixture.

Immunoblotting tests are used to confirm the presence of proteins, including antibodies against pathogens. Physicians use immunoblots to verify the presence of HIV proteins or antibodies against the bacterium of Lymes disease in the blood serum of patients. Immunoblotting involves three steps **(FIGURE 17.14a)**:

1 *Electrophoresis.* Antigens in a solution (in this example, HIV proteins) are placed into wells and separated by gel

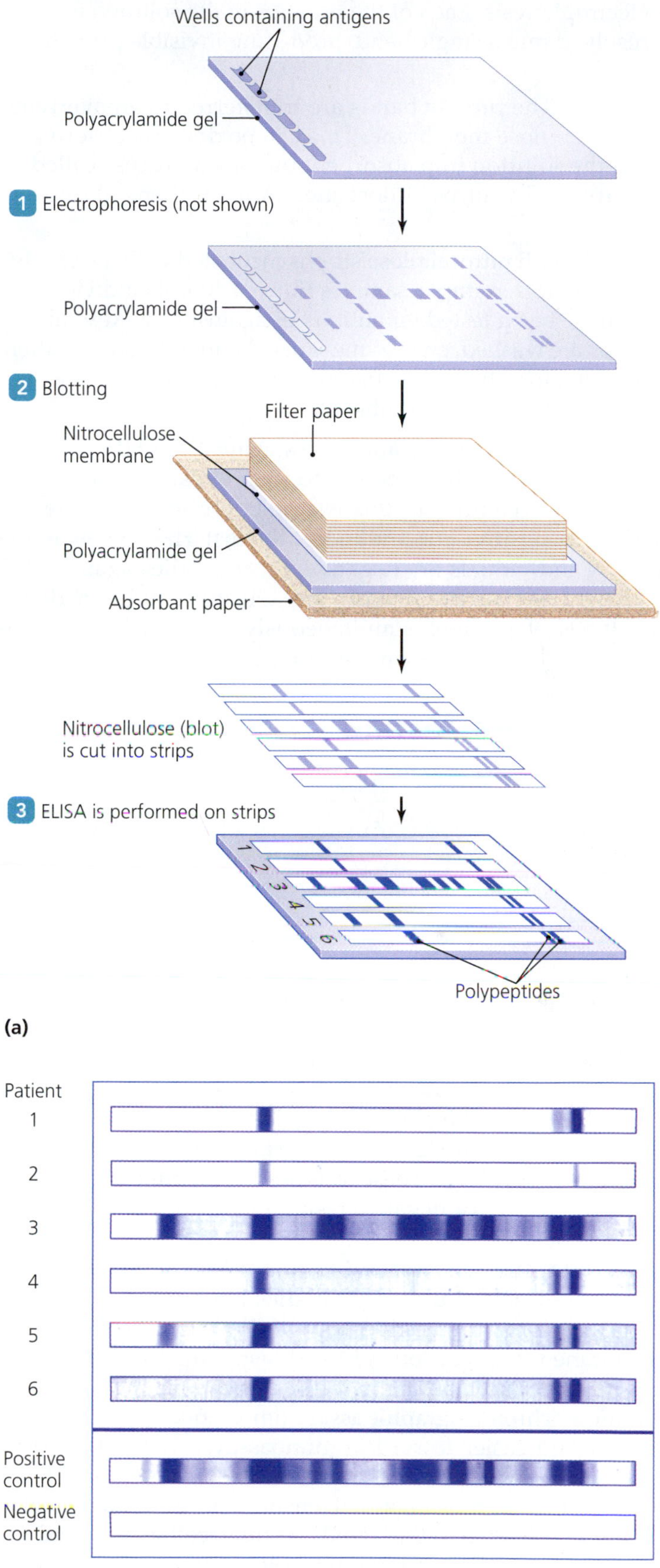

▶ **FIGURE 17.14 Immunoblotting.** This technique demonstrates the presence of antibodies against multiple antigens in a complex mixture. **(a)** Steps in immunoblotting test. 1 Antigens are separated by gel electrophoresis. 2 Separated proteins are transferred to a nitrocellulose membrane. 3 Test solutions, enzyme-labeled anti-antibody, and the enzyme's substrate are added; color changes are detected wherever antibody in the test solutions has bound to proteins. **(b)** Real results of an immunoblot. Patient 3 tested positive.

electrophoresis. Each of the proteins in the solution is resolved into a single band, producing invisible protein bands.

2 *Blotting*. The protein bands are transferred to an overlying nitrocellulose membrane. This can be done by absorbing the solution into absorbent paper—a process called blotting. The nitrocellulose membrane is then cut into strips.

3 *ELISA*. Each nitrocellulose strip is incubated with a test solution—in this example, samples from each of six individuals who are being tested for antibodies against HIV. After the strips are washed, an enzyme-labeled anti-antibody solution is added for a time; then the strips are washed again and exposed to the enzyme's substrate.

Color develops wherever antibodies against the HIV proteins in the test solutions have bound to their substrates, as shown in the positive control. In this example, the individual tested in strip 3 is positive for antibodies against HIV, whereas the other five individuals are negative for antibodies against HIV (**FIGURE 17.14b**). Immunoblots are sensitive and can detect many types of proteins simultaneously. Colored bands common to all patients are normal serum proteins.

Point-of-Care Testing

LEARNING | **OUTCOMES**

17.20 Discuss the benefits of using immunofiltration assays rather than an ELISA.

17.21 Contrast immunofiltration and immunochromatographic assays.

Recent years have seen the development of simple immunoassays that give clinicians useful results within minutes. These assays allow *point-of-care testing*; that is, health care providers do not have to send specimens to a laboratory for testing but can perform the test at the patient's bedside or in a doctor's office. Common point-of-care tests include *immunofiltration* and *immunochromatography assays*. These tests are not quantitative but rapidly give a positive or negative result, making them very useful in arriving at a quick diagnosis.

Immunofiltration assays (im´ū-nō-fil-trā´shŭn) are rapid ELISAs based on the use of antibodies bound to a membrane filter rather than to plates. Because of the large surface area of a membrane filter, reactions proceed faster and assay times are significantly reduced as compared to a traditional ELISA.

Immunochromatographic assays (im´ū-nō-krō´mat-ō-graf´ik) are faster and easier to read immunoassays. In these systems, an antigen solution (such as diluted blood or sputum) flowing through a porous material encounters antibody labeled with either pink colloidal[3] gold or blue colloidal selenium. Where antigen and antibody bind, colored immune complexes form in the fluid, which then flows through a region where

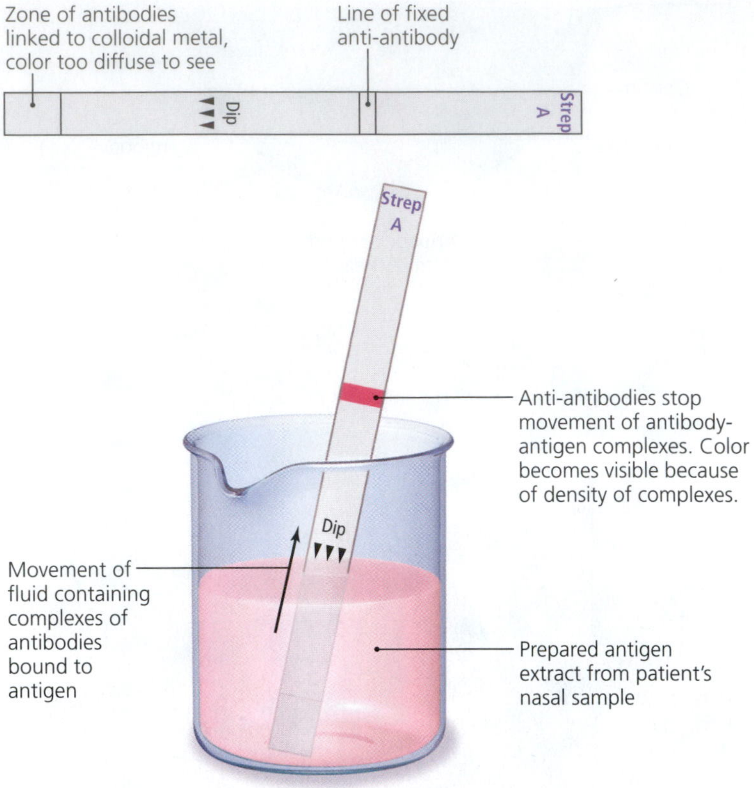

▲ **FIGURE 17.15 Immunochromatographic dipstick.** The dipstick is impregnated with colloidal metal particles linked to movable antibodies against particular antigens at one end and anti-antibodies fixed in a line closer to the other end of the membrane.

the complexes encounter antibody against them, resulting in a clearly visible pink or blue line, depending on the label used. These assays are used for pregnancy testing, which tests for *human chorionic gonadotropin*—a hormone produced only by an embryo or fetus—and for rapid identification of infectious agents such as HIV, *Escherichia coli* (esh-ĕ-rik´ē-ă kō´lī) O157:H7, group A *Streptococcus*, respiratory syncytial virus (RSV), and influenzaviruses. In one adaptation, the antibodies are coated on membrane strips, which serve as dipsticks. At one end, anti-antibodies are fixed in a line so that they cannot move in the membrane. The lower portion of the membrane is coated with antibodies against the antigen in question. These antibodies are linked to a color indicator in the form of a colloidal metal and are free to move in the membrane by capillary action.

FIGURE 17.15 illustrates the procedure used to test for the presence of group A *Streptococcus* in the nasal secretion of a patient. A laboratory scientist prepares a nasal swab from the patient so as to release *Streptococcus* antigens if they are present. She then dips the membrane into the solution. The membrane's antibodies bind to streptococcal antigens, forming complexes. The complexes move up the membrane by capillary action until they reach the line of anti-antibodies, where they bind and must stop because the anti-antibodies are chemically bound to the strip. Previously the complexes were invisible because they were dilute; now they are concentrated at the

[3]*Colloidal* refers to small particles suspended in a liquid or gas.

TABLE 17.2 Antibody-Antigen Immunological Tests and Some of Their Uses

Test	Use
Immunodiffusion (precipitation)	Diagnosis of syphilis, pneumococcal pneumonia
Agglutination	Blood typing; pregnancy testing; diagnosis of salmonellosis, brucellosis, gonorrhea, rickettsial infection, mycoplasma infection, yeast infection, typhoid fever, meningitis caused by *Haemophilus*
Viral neutralization	Diagnosis of infections by specific strains of viruses
Viral hemagglutination inhibition	Diagnosis of viral infections including influenza, measles, mumps, rubella, mononucleosis
Complement fixation	In the past, diagnosis of measles, influenza A, syphilis, rubella, rickettsial infections, scarlet fever, rheumatic fever, infections of respiratory syncytial virus and *Coxiella*
Direct fluorescent antibody	Diagnosis of rabies, infections of group A *Streptococcus*, identification of lymphocyte subsets
Indirect fluorescent antibody	Diagnosis of syphilis, mononucleosis
ELISA	Pregnancy testing; presence of drugs in urine; diagnosis of hepatitis A, hepatitis B, rubella; initial diagnosis of HIV infection
Immunoblot (western blot)	Confirmation of infection with HIV; diagnosis of Lyme disease

line of anti-antibodies and become visible, indicating that this patient has group A *Streptococcus* in his nose. Knowing that the infection is bacterial and not viral, the physician can prescribe antibacterial drugs. The procedure from antigen preparation to diagnosis takes less than 10 minutes.

TABLE 17.2 lists some antibody-antigen immune tests that can be used to diagnose selected bacterial and viral diseases.

TELL ME WHY

A diagnostician used an ELISA to show that a newborn had antibodies against HIV in her blood. However, six months later the same test was negative. How can this be?

MICRO IN THE CLINIC FOLLOW-UP

Outbreak in the Inner City

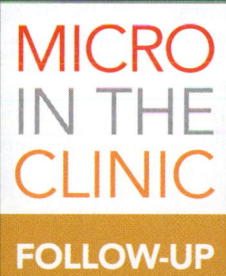

The doctor suspects that Maria has pertussis, a disease caused by *Bordetella pertussis,* which is a Gram-negative coccobacillus. Pertussis is also known as "whooping cough" because of the characteristic gasp for breath after coughing. A throat culture of Maria's sputum confirms the diagnosis. Pertussis is highly contagious. Every cough expels hundreds of thousands of bacteria, which can infect others. Maria and her mother live in an area with a large Hispanic population, and this community was hit especially hard by pertussis.

Until around 1950, *Bordetella* was a leading cause of childhood disease and mortality. Today, immunization with the DTaP (diphtheria, tetanus, acellular pertussis) vaccine can prevent pertussis, but sadly, many parents don't immunize their children. Even many of the adults who were immunized as children don't get the recommended Tdap booster. Rates of pertussis are rising. In inner cities, where immunization compliance is especially low, it's become an epidemic because when immunizations are not regularly maintained, the population loses *herd immunity.*

To control the outbreak, the clinic launches a public education and immunization program offering free DTaP vaccines to children and Tdap boosters to adults. Meanwhile, public health officials are doing molecular typing to rapidly identify bacterial strains and routine antibiotic resistance tests to avoid the growth of erythromycin-resistant strains of *Bordetella.*

As for Maria—the clinic doctor persuades the baby's parents to take her to the hospital immediately, where she is admitted and given intravenous fluids and antibiotics. Thanks to this quick action, Maria survives.

1. **The DTaP vaccine used to prevent pertussis is a combination vaccine protecting against multiple diseases simultaneously. Describe the different types of antigens used in the DTaP vaccine.**
2. **Describe herd immunity, and explain how it is related to the epidemic described in the Clinical Case.**

MM Check your answers to Micro in the Clinic Follow-Up questions in the MasteringMicrobiology Study Area.

Explore the Invisible: ELISA

Practice on-the-go with Dr. Bauman Video Tutors by scanning this QR code with your smart phone. Visit the **MasteringMicrobiology Study Area** to challenge your understanding with practice tests, animation quizzes, and clinical case studies!

MasteringMicrobiology®

CHAPTER SUMMARY

Immunization (pp. 505–513)

1. The first **vaccine** was developed by Edward Jenner against smallpox. He called the technique **vaccination**. **Immunization** is a more general term referring to the use of vaccines against rabies, anthrax, measles, mumps, rubella, polio, and other diseases.
2. Individuals can be protected against many infections by either active immunization or passive immunotherapy.

   ▶**ANIMATIONS:** *Vaccines: Function*
3. Active immunization involves giving antigen in the form of either **attenuated vaccines**, **inactivated vaccines** (killed vaccines), **toxoid vaccines**, or recombinant gene vaccines. Antibody **titer** refers to the amount of antibody produced.

   ▶**ANIMATIONS:** *Vaccines: Types*
4. Pathogens in attenuated vaccines are weakened so that they no longer cause disease, though they are still alive or active and can provide **contact immunity** in unimmunized individuals who associate with immunized people.
5. Inactivated vaccines **are either** whole agent or subunit vaccines and often contain **adjuvants**, which are chemicals added to increase their ability to stimulate active immunity.
6. Toxoid vaccines use modified toxins to stimulate antibody-mediated immunity.
7. A **combination vaccine** is composed of antigens from several pathogens so they can be administered to a patient at once.
8. Having a large proportion of immunized individuals (>75%) in a population interrupts disease transmission, providing protection to unimmunized individuals. Such protection is called **herd immunity**.
9. **Passive immunotherapy** (a type of passive immunization) involves administration of an **antiserum** containing preformed antibodies. Serum sickness results when the patient makes antibodies against the antiserum.
10. The fusion of myelomas (cancerous plasma cells) with plasma cells results in **hybridomas**, the source of **monoclonal antibodies**, which can be used in passive immunization.

Serological Tests That Use Antigens and Corresponding Antibodies (pp. 513–521)

1. **Serology** is the study and use of immunological assays on blood serum to diagnose disease or identify antibodies or antigens. Scientists use antibodies to find an antigen in a specimen and use antigen to find antibodies.
2. The simplest of the serological tests is a precipitation test, in which antigen and antibody meet in optimal proportions to form **immune complexes**, which are often insoluble. Often this test is performed in clear gels, where it is called **immunodiffusion**.
3. Automated light detectors can measure the cloudiness of a solution—an indication of the quantity of protein in the solution. Turbidimetry measures the passage of light through the solution, while nephelometry measures the amount of light reflected by protein in the solution.
4. **Agglutination** tests involve the clumping of antigenic particles by antibodies. The amount of these antibodies, called the **titer**, is measured by diluting the serum in a process called **titration**.
5. Antibodies to viruses or toxins can be measured using a **neutralization test**, such as a **viral neutralization** test. Infection by viruses that naturally agglutinate red blood cells can be demonstrated using a **viral hemagglutination inhibition test**.
6. The **complement fixation test** is a complex assay used to determine the presence of specific antibodies.
7. Fluorescently labeled antibodies—those chemically linked to a fluorescent dye—can be used in a variety of **direct fluorescent antibody** and **indirect fluorescent antibody** tests. The presence of labeled antibodies is visible through a fluorescence microscope.
8. **Enzyme-linked immunosorbent assays (ELISAs)**, or **enzyme immunoassays (EIAs)**, are a family of simple tests that can be readily automated and read by machine. These tests are among the more common serological tests used. A variation of the ELISA is an **immunoblot** (western blot), which is used to detect antibodies against multiple antigens in a mixture.

   ▶**VIDEO TUTOR:** ***ELISA***
9. **Immunofiltration assays** and **immunochromatographic assays** are modifications of ELISA tests that can give much more rapid diagnostic results.

QUESTIONS FOR REVIEW

Answers to the Questions for Review (except Short Answer questions) begin on p. A-1.

Multiple Choice

1. To obtain immediate immunity against tetanus, a patient should receive
   a. an attenuated vaccine of *Clostridium tetani.*
   b. a modified live vaccine of *C. tetani.*
   c. tetanus toxoid.
   d. immunoglobulin against tetanus toxin (antitoxin).
   e. a subunit vaccine against *C. tetani.*
2. Which of the following vaccine types is commonly given with an adjuvant?
   a. an attenuated vaccine
   b. a modified live vaccine
   c. a chemically killed vaccine
   d. an immunoglobulin
   e. an agglutinating antigen
3. Which of the following viruses was widely used in living vaccines?
   a. coronavirus
   b. poliovirus
   c. influenzavirus
   d. retrovirus
   e. myxovirus
4. When antigen and antibodies combine, maximal precipitation occurs when
   a. antigen is in excess.
   b. antibody is in excess.
   c. antigen and antibody are at equivalent concentrations.
   d. antigen is added to the antibody.
   e. antibody is added to the antigen.
5. An anti-antibody is used when
   a. an antigen is not precipitating.
   b. an antibody is not agglutinating.
   c. an antibody does not activate complement.
   d. an antigen is insoluble.
   e. the antigen is an antibody.
6. The many different proteins in serum can be analyzed by
   a. an anti-antibody test.
   b. a complement fixation test.
   c. a precipitation test.
   d. an agglutination test.
   e. an immunodiffusion test.
7. A direct fluorescent antibody test requires which of the following?
   a. heat-inactivated serum
   b. fluorescent serum
   c. immune complexes
   d. heated plasma
   e. antibodies against the antigen
8. An ELISA uses which of the following reagents?
   a. an enzyme-labeled anti-antibody
   b. a radioactive anti-antibody
   c. a source of complement
   d. an enzyme-labeled antigen
   e. an enzyme-labeled antibody
9. A direct fluorescent antibody test can be used to detect the presence of
   a. hemagglutination.
   b. specific antigens.
   c. antibodies.
   d. complement.
   e. precipitated antigen-antibody complexes.
10. Which of the following is a good test to detect rabies virus in the brain of a dog?
    a. agglutination
    b. hemagglutination inhibition
    c. virus neutralization
    d. precipitation
    e. direct fluorescent antibody
11. Attenuation is
    a. the process of reducing virulence.
    b. a necessary step in vaccine manufacture.
    c. a form of variolation.
    d. similar to an adjuvant.
12. An antiserum is
    a. an anti-antibody.
    b. an inactivated vaccine.
    c. formed of monoclonal antibodies.
    d. the liquid portion of blood used for immunization.
13. Monoclonal antibodies
    a. are produced by hybridomas.
    b. are secreted by clone cells.
    c. can be used for passive immunization.
    d. all of the above.
14. The study of antibody-antigen interaction in the blood is
    a. attenuation.
    b. agglutination.
    c. precipitation.
    d. serology.
15. Anti–human antibody antibodies are
    a. found in immunocompromised individuals.
    b. used in direct fluorescent antibody tests.
    c. formed by animals reacting to human immunoglobulins.
    d. an alternative method in ELISA.

True/False

1. ________________ Passive immunotherapy provides more prolonged immunity than active immunization.
2. ________________ It is standard to attenuate killed virus vaccines.
3. ________________ One single serological test is inadequate for an accurate diagnosis of HIV infection.
4. ________________ ELISA is very easily automated.
5. ________________ ELISA has basically replaced immunoblotting.

Matching

Match the characteristic in the first column with the therapy it most closely describes in the second column. Some choices may be used more than once.

1. ____ Induces rapid onset of immunity
2. ____ Induces mainly an antibody response
3. ____ Induces good cell-mediated immunity
4. ____ Increases antigenicity
5. ____ Uses antigen fragments
6. ____ Uses attenuated microbes

A. Attenuated viral vaccine
B. Adjuvant
C. Subunit vaccine
D. Immunoglobulin
E. Residual virulence

VISUALIZE IT!

1. Identify the chemicals represented by this artist's conception of an antibody sandwich ELISA.

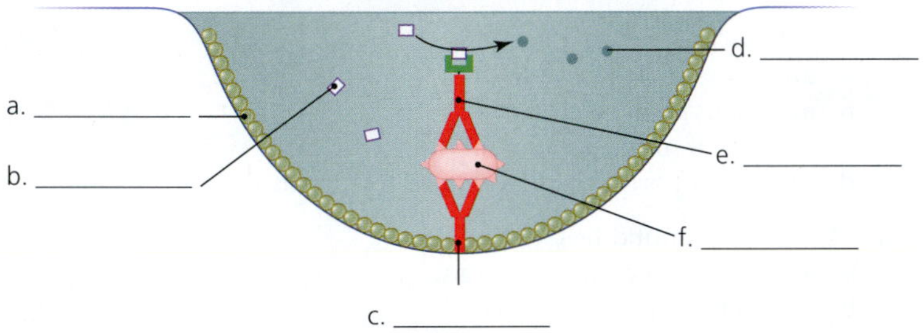

2. The two columns on the left show negative and positive immunoblot results for a particular pathogen. The numbered columns are blots of samples from eleven patients. Which patients are most likely uninfected?

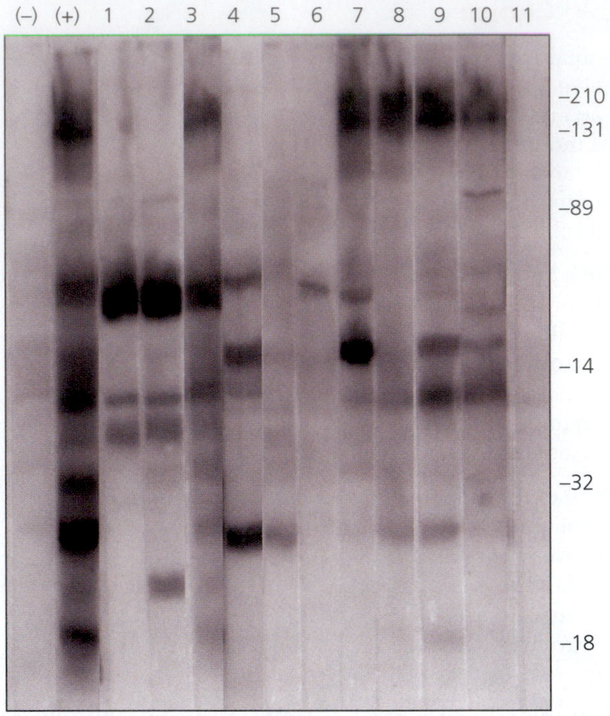

Short Answer

1. Compare and contrast the Chinese practice of variolation with Jenner's vaccination procedure.
2. What are the advantages and disadvantages of attenuated vaccines?
3. Compare the advantages and disadvantages of passive immunotherapy and active immunization.
4. How does precipitation differ from agglutination?
5. Explain how a pregnancy test works at the molecular level.
6. Compare and contrast herd immunity and contact immunity.
7. How does nephelometry differ from turbidimetry?

CRITICAL THINKING

1. Is it ethical to approve the use of a vaccine that causes significant illness in 1% of patients if it protects immunized survivors against a serious disease?
2. Which is worse: to use a diagnostic test for HIV that may falsely indicate that a patient is not infected (false negative) or to use one that sometimes falsely indicates that a patient is infected (false positive)? Defend your choice.
3. Discuss the importance of costs and technical skill in selecting a practical serological test. Under what circumstances does automation become important?
4. What bodily fluids, in addition to blood serum, might be usable for immune testing?
5. Why might a serological test give a false-positive result?
6. Some researchers want to distinguish B cells from T cells in a mixture of lymphocytes. How could they do this without killing the cells?
7. Describe three ways by which genetic recombinant techniques could be used to develop safer, more effective vaccines.
8. How does a toxoid vaccine differ from an attenuated vaccine?
9. Explain why many health organizations promote breast-feeding of newborns. What risks are involved in such nursing?
10. Contrast a hemagglutination test with a viral hemagglutination inhibition test.
11. Sixty years ago, parents would have done almost anything to get a protective vaccine against polio for their children. Now parents fear the vaccine, not the disease. Why?
12. Draw a picture showing, at *both the molecular and the cellular level,* IgM agglutinating red blood cells.

CONCEPT MAPPING

Using the following terms, draw a concept map that describes vaccines. For a sample concept map, see p. 94. Or, complete this and other concept maps online by going to the MasteringMicrobiology Study Area.

Active immunity
Adjuvants
Attenuated vaccines
Hepatitis B vaccine
Immunizations
Inactivated vaccines
Measles vaccine
Microorganisms
Modified toxins
Pertussis
Polio
Subunit vaccines
Tetanus toxoid
Varicella vaccine

18 AIDS and Other Immune Disorders

MICRO IN THE CLINIC

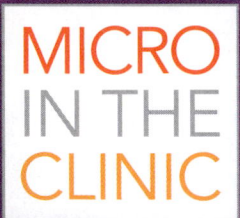

A Simple Case of Bug Bites?

On a sunny, lazy July afternoon at the family lakeside cabin, 10-year-old Ethan is tossing a slobbery tennis ball to Daisy, the family's dog. The dog bounds into the underbrush to retrieve an errant throw. When the game is over, Ethan wraps his arms around Daisy and hugs her.

A week later, Ethan wakes up his mother, Maureen, complaining that he itches. The right side of his face is red and covered with small bumps. Convinced that Ethan is just suffering from bug bites, Maureen gives him an antihistamine to relieve the itching. By morning, though, Ethan is worse—the red angry rash has spread to his neck and arms, his face is swollen, and blisters appear. Maureen applies an over-the-counter anti-itch cream and tells Ethan not to touch the blisters, but he can't help scratching them.

Three days after that, Ethan still feels and looks miserable, and Maureen is really getting worried. Then she notices pus oozing from his blisters. Panicking, she immediately drives Ethan to the doctor's office.

What is wrong with Ethan? Did biting bugs cause Ethan's discomfort? Turn to the end of the chapter (p. 551) to find out.

MM **Explore More:** Test your readiness and apply your knowledge with dynamic learning tools at MasteringMicrobiology.

The immune system is an absolutely essential component of the body's defenses. However, if the immune system functions abnormally—either by overreacting or underreacting—the malfunction may cause significant disease, even death. If the immune system functions excessively, the body develops any of a variety of *immune hypersensitivities,* such as allergies. If the immune system attacks the body's own tissues, *autoimmune diseases* develop. If the immune system fails, *immunodeficiency* (im´ū-nō-dē-fish´en-sē) *disease* result. AIDS is a prime example. In this chapter we discuss each of these three categories of derangements of the immune system.

Hypersensitivities

LEARNING OUTCOME

18.1 Compare and contrast the four types of hypersensitivity.

Hypersensitivity (hī´per-sen-si-tiv´i-tē) may be defined as any immune response against a foreign antigen that is exaggerated beyond the norm. For example, most people can wear wool, smell perfume, or dust furniture without experiencing itching, wheezing, runny nose, or watery eyes. When these symptoms do occur, the person is said to be experiencing a hypersensitivity response. In the following sections we examine each of the four main types of hypersensitivity response, designated as type I through type IV.

Type I (Immediate) Hypersensitivity

LEARNING OUTCOMES

18.2 Describe the two-part mechanism by which type I hypersensitivity occurs.

18.3 Explain the roles of three inflammatory chemicals (mediators) released from mast cell granules.

18.4 Describe three disease conditions resulting from type I hypersensitivity mechanisms.

Type I hypersensitivities are localized or systemic (whole body) reactions that result from the release of inflammatory molecules (such as histamine) in response to an antigen. These reactions are termed *immediate hypersensitivity* because they develop within seconds or minutes following contact with antigens. They are also commonly called **allergies,**[1] and the antigens that stimulate them are called **allergens** (al´er-jenz).

In the next two subsections we will examine the two-part mechanism of a type I hypersensitivity reaction: sensitization upon initial contact with an allergen and the degranulation of sensitized cells.

Sensitization upon Initial Exposure to an Allergen

All of us are exposed to antigens in the environment, against which we typically mount immune responses that result in the production of antibodies of the gamma class (IgG). But when regulatory proteins (cytokines, especially interleukin 4) from type 2 helper T (Th2) cells stimulate B cells in allergic individuals, the B cells become plasma cells that produce class epsilon antibodies (IgE; **FIGURE 18.1a**). Typically, IgE is directed against parasitic worms, which rarely infect most Americans, so IgE is found at low levels in the blood serum. However, in allergic individuals plasma cells produce a significant amount of IgE in response to allergens.

The precise reason that only some people produce high levels of IgE (and thereby suffer from allergies) is a matter of intense research. Many researchers report evidence for the *hygiene hypothesis,* which holds that children exposed to environmental antigens—such as dust mites, molds, parasitic worms, and pet hair—are less likely to develop allergies than children who have been sheltered from common environmental antigens. **Highlight: Can Pets Help Decrease Children's Allergy Risks?** examines the hygiene hypothesis. Conversely, other researchers have discovered evidence that that environmental factors sensitize people, making them hypersensitive and more likely to develop allergies.

[1]From Greek *allos,* meaning "other," and *ergon,* meaning "work."

HIGHLIGHT

Can Pets Help Decrease Children's Allergy Risks?

Several studies suggest that children who grow up in households with two or more cats or dogs may be less likely to develop common allergies than children raised without pets. Researchers suspect that exposure to bacteria harbored by cats and dogs can suppress allergic reactions—not only to pets but also to other common allergens such as dust mites and grass.

In one study, researchers studied 474 children from birth until six to seven years of age, at which point the children were tested for IgE against common allergens. Researchers found that children exposed to two or more dogs or cats during their first year of life were, on average, 66–77% less likely to have antibodies to common allergens than were children exposed to only one or no pets during the same period. Exactly how the pets and/or their bacteria might suppress the allergic response remains unclear; perhaps stimulation of Th1 cells by pet allergens balances or counteracts the production of Th2 cells that would stimulate antibody immune response.

Alternatively, it might be that families with a predisposition to allergies don't keep pets.

In any case, following initial exposure to allergens, the plasma cells of allergic individuals secrete IgE, which binds very strongly with its stem to three types of defense cells—*mast cells, basophils,* and *eosinophils*—sensitizing these cells to respond to future exposures to the allergen.

Degranulation of Sensitized Cells

When the same allergen reenters the body, it binds to the active sites of IgE molecules on the surfaces of sensitized cells (described shortly). This binding triggers a cascade of internal biochemical reactions that causes the sensitized cells to release the inflammatory chemicals from their granules into the surrounding space—an event called *degranulation* (dē-gran-ū-lā´shŭn; **FIGURE 18.1b**). It is these inflammatory mediators that generate the characteristic symptoms of type I hypersensitivity reactions: respiratory distress, rhinitis (inflammation of the nasal mucous membranes commonly called "runny nose"), watery eyes, inflammation, and reddening of the skin.

The Roles of Degranulating Cells in an Allergic Reaction

Mast cells are specialized relatives of white blood cells, deriving from other stem cells in the bone marrow. They are distributed throughout the body in connective tissues other than blood. These large, round cells are most often found in sites close to body surfaces, including the skin and the walls of the intestines and airways. Their characteristic feature is cytoplasm packed with large granules that are loaded with a mixture of potent inflammatory chemicals.

One significant chemical released from mast cell granules is **histamine** (his´tă-mēn), a small molecule related to the amino acid histidine. Histamine stimulates strong contractions in the smooth muscles of the bronchi, gastrointestinal tract, uterus, and bladder, and it also makes small blood vessels dilate (expand) and become leaky. As a result, tissues in which mast cells degranulate become red and swollen. Histamine also stimulates nerve endings, causing itching and pain. Finally, histamine is an effective stimulator of bronchial mucus secretion, tear formation, and salivation.

Other triggers of allergies released by degranulating mast cells include **kinins** (kī´ninz), which are powerful inflammatory chemicals, and **proteases** (prō´tē-ās-ez)—enzymes that destroy nearby cells, activating the complement system, which in turn results in the release of still more inflammatory chemicals. Proteases account for more than half the proteins in a mast cell granule. In addition, the binding of allergens to IgE on

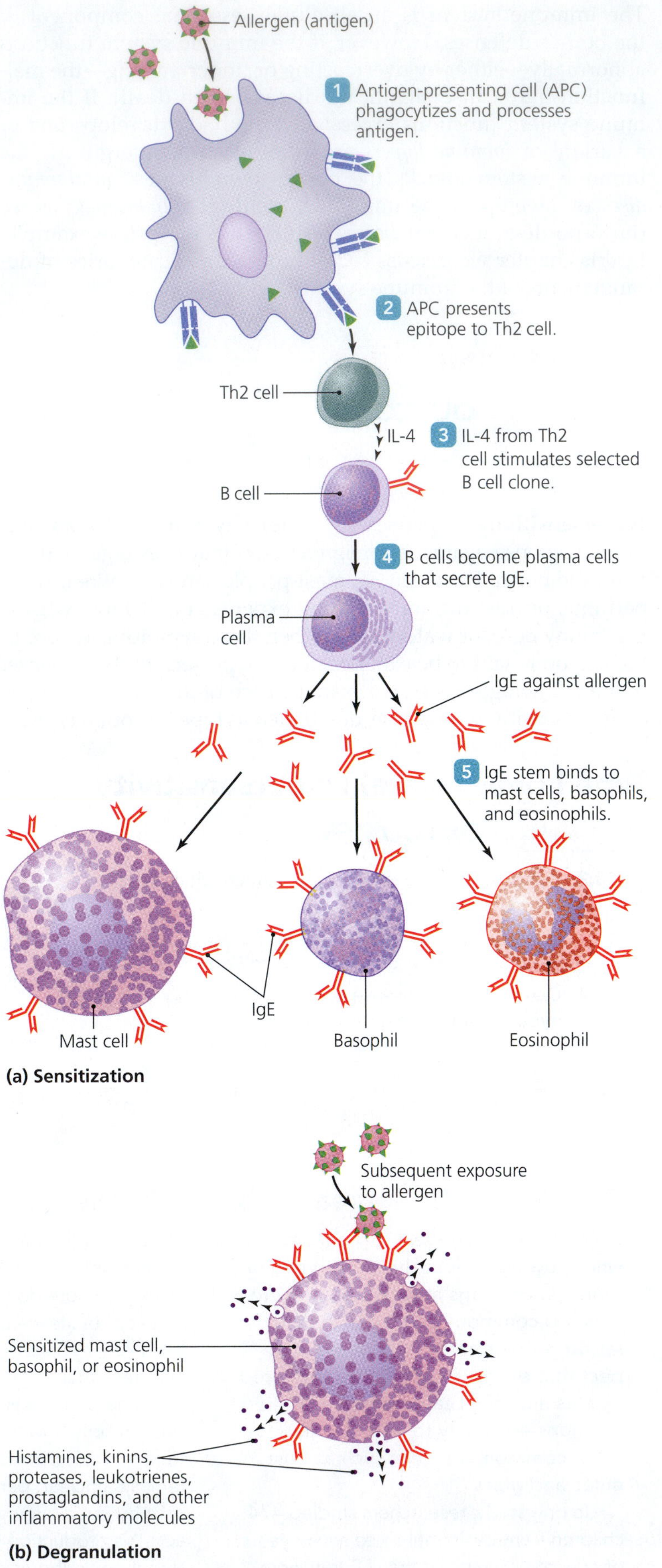

▶ FIGURE 18.1 The mechanisms of a type I hypersensitivity reaction. (a) Sensitization. After normal processing of an antigen (allergen) by an antigen-presenting cell 1, Th2 cells are stimulated to secrete IL-4 2. This cytokine stimulates a B cell 3, which becomes a plasma cell that secretes IgE 4, which then binds to and sensitizes mast cells, basophils, and eosinophils 5. **(b)** Degranulation. When the same allergen is subsequently encountered, its binding to the IgE molecules on the surfaces of sensitized cells triggers rapid degranulation and the release of inflammatory chemicals from the cells.

TABLE 18.1 Inflammatory Molecules Released from Mast Cells

Molecules	Role in Hypersensitivity Reactions
Released During Degranulation	
Histamine	Causes smooth muscle contraction, increased vascular permeability, and irritation
Kinins	Cause smooth muscle contraction, inflammation, and irritation
Proteases	Damage tissues and activate complement
Synthesized in Response to Inflammation	
Leukotrienes	Cause slow, prolonged smooth muscle contraction, inflammation, and increased vascular permeability
Prostaglandins	Some contract smooth muscle; others relax it

mast cells activates other enzymes that trigger the production of **leukotrienes** (loo-kō-trī´ēnz) and **prostaglandins** (pros-tă-glan´dinz), lipid molecules that are powerful inflammatory agents. **TABLE 18.1** summarizes the inflammatory molecules released from mast cells.

Basophils, the least numerous type of leukocyte in blood, contain cytoplasmic granules that stain intensely with basic dyes (see Figure 15.5a) and are filled with inflammatory chemicals similar to those found in mast cells. Sensitized basophils bind IgE and degranulate in the same way as mast cells when they encounter allergens.

The blood and tissues of allergic individuals also accumulate many **eosinophils,** leukocytes that contain numerous cytoplasmic granules that stain intensely with a red dye called eosin and that function primarily to destroy parasitic worms. The process during type I hypersensitivity reactions that results in the accumulation of eosinophils in the blood—a condition termed *eosinophilia*—begins with mast cell degranulation, which releases peptides that stimulate the release of eosinophils from the bone marrow. Once in the bloodstream, eosinophils are attracted to the site of mast cell degranulation, where they themselves degranulate. Eosinophil granules contain unique inflammatory mediators and produce large amounts of leukotrienes, which increase movement from blood vessels and stimulate smooth muscle contraction, thereby contributing greatly to the severity of a hypersensitivity response.

Clinical Signs of Localized Allergic Reactions

Type I hypersensitivity reactions are usually mild and localized. The site of the reaction depends on the portal of entry of the antigens. For example, inhaled allergens may provoke a response in the upper respiratory tract commonly known as **hay fever**—a local allergic reaction marked by a runny nose, sneezing, itchy throat and eyes, and excessive tear production. Fungal (mold) spores; pollens from grasses, flowering plants, and some trees; and feces and dead bodies of house dust mites are among the more common allergens **(FIGURE 18.2)**.

If inhaled allergen particles are sufficiently small, they may reach the lungs. A type I hypersensitivity in the lungs can cause

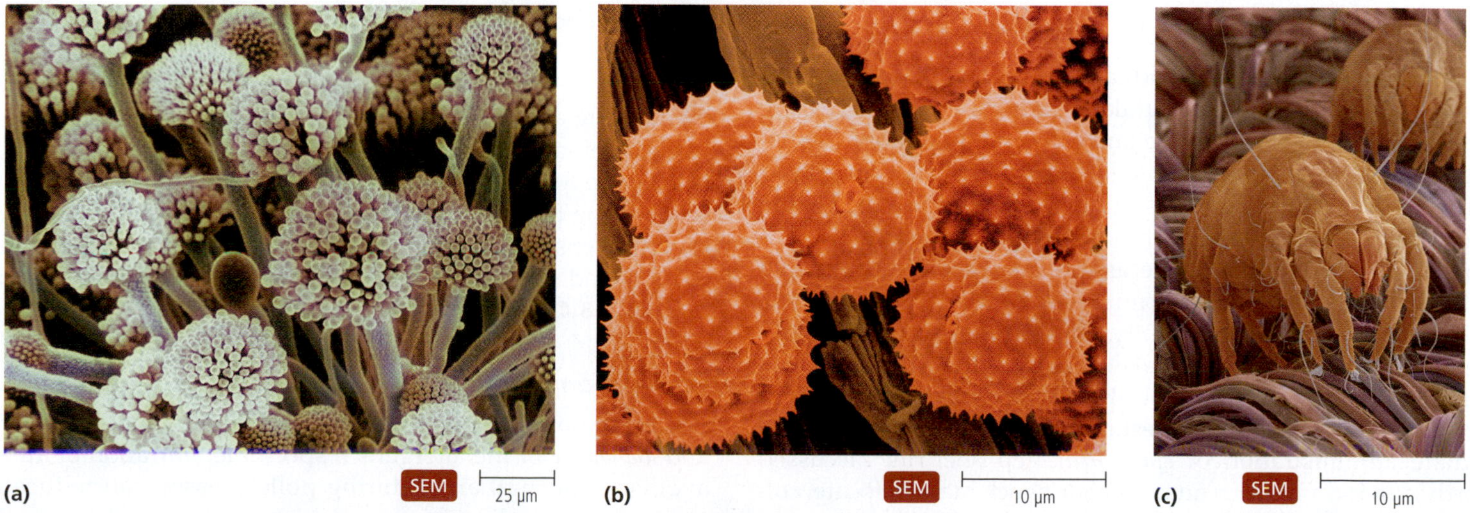

▲ **FIGURE 18.2 Some common allergens. (a)** Spores of the fungus *Aspergillus* on stalks. **(b)** Pollen of *Ambrosia trifida* (ragweed), the most common cause of hay fever in the United States. **(c)** A house dust mite. Mites' fecal pellets and dead bodies may become airborne and trigger allergic responses.

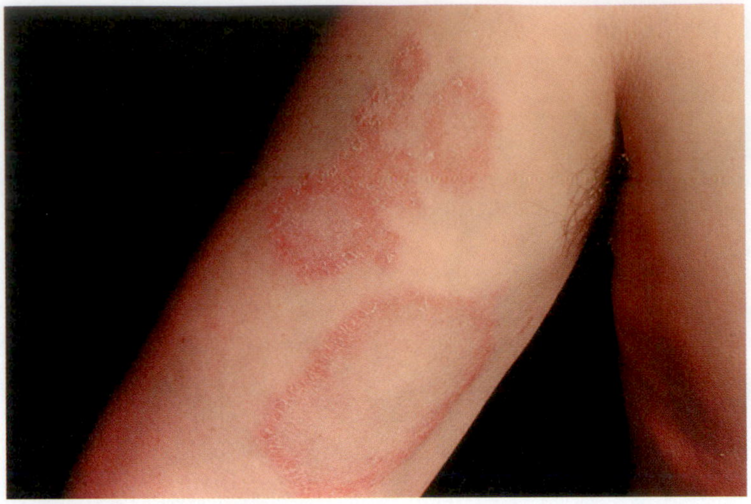

▲ **FIGURE 18.3 Urticaria.** These red, itchy patches on the skin are prompted by the release of histamine in response to an allergen. *What causes fluid to accumulate in the patches seen in urticaria?*

Figure 18.3 *The fluid accumulates because histamine from degranulated mast cells causes blood capillaries to become more permeable.*

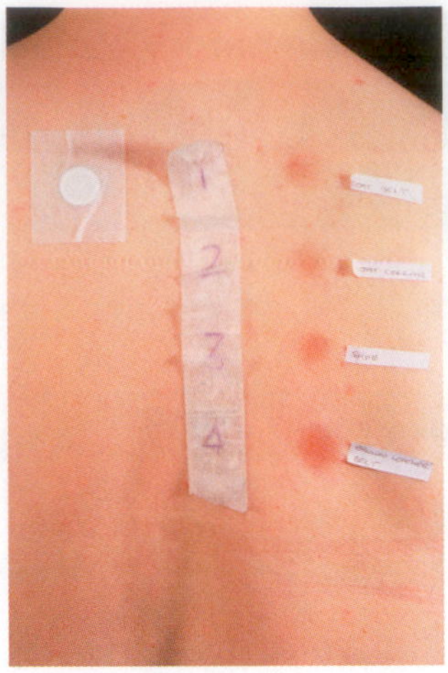

▲ **FIGURE 18.4 Skin tests for diagnosing type I hypersensitivity.** The presence of redness and swelling at an injection site indicates sensitivity to the allergen injected at that site.

an episode of severe difficulty in breathing known as **asthma,** characterized by wheezing; coughing; excessive production of a thick, sticky mucus; and constriction of the smooth muscles of the bronchi. Asthma can be life threatening: without medical intervention, the increased mucus and bronchial constriction can quickly cause suffocation.

Other allergens—including latex, wool, certain metals, and the venom or saliva of wasps, bees, fire ants, deer flies, fleas, and other stinging or biting insects—may cause a localized inflammation of the skin. As a result of the release of histamine and other mediators, plus the ensuing leakage of serum from local blood vessels, the individual suffers raised, red areas called *hives*, or **urticaria**[2] (er´ti-kar´i-ă; **FIGURE 18.3**). These lesions are very itchy because histamine irritates local nerve endings.

Clinical Signs of Systemic Allergic Reactions

Following a sensitized individual's contact with an allergen, many mast cells may degranulate simultaneously, releasing massive amounts of histamine and other inflammatory mediators into the bloodstream. The release of chemicals may exceed the body's ability to adjust, resulting in a condition called **acute anaphylaxis**[3] (an´ă-fī-lak´sis), or **anaphylactic shock.**

The clinical signs of acute anaphylaxis are those of rapid suffocation. Bronchial smooth muscle, which is highly sensitive to histamine, contracts violently. In addition, increased leakage of fluid from blood vessels causes swelling of the larynx and other tissues. The patient also experiences contraction of the smooth muscle of the intestines and bladder. Without immediate administration of *epinephrine* (ep´i-nef´rin; discussed shortly), an individual in anaphylactic shock may suffocate, collapse, and die within minutes.

A common cause of acute anaphylaxis is a bee sting in an allergic and sensitized individual. The first sting produces sensitization and the formation of IgE antibodies, and successive stings—which may occur years later—produce an anaphylactic reaction. Other allergens commonly implicated in acute anaphylaxis are certain foods (notoriously peanuts), vaccines, antibiotics such as penicillin, iodine dyes, local anesthetics, blood products, and certain narcotics such as morphine.

Diagnosis of Type I Hypersensitivity

Clinicians diagnose type I hypersensitivity with a test variously called *ImmunoCAP Specific IgE blood test, CAP RAST,* or *Pharmacia CAP,* in which suspected allergens are mixed with samples of the patient's blood. The specifics of the test are beyond the scope of this chapter, but basically the test detects the amount of IgE directed against each allergen. High levels of a specific IgE indicate a hypersensitivity against that allergen.

Alternatively, physicians diagnose type I hypersensitivity by injecting a very small quantity of a dilute solution of the allergens being tested into the skin. In most cases the individual is screened for more than a dozen potential allergens simultaneously, using the forearms as injection sites. When the individual tested is sensitive to an allergen, local histamine release causes redness and swelling at the injection site within a few minutes **(FIGURE 18.4)**.

Prevention of Type I Hypersensitivity

Prevention of type I hypersensitivity begins with *identification* and *avoidance* of the allergens responsible. Filtration of air and avoidance of rural areas during pollen season can reduce upper respiratory allergies provoked by some pollens. Encasing bed clothes in mite-proof covers, frequent vacuuming, and avoiding home furnishings that trap dust (such as wall-to-wall carpeting and heavy drapes) can reduce the severity of

[2]From Latin *urtica,* meaning "nettle."

[3]From Greek *ana,* meaning "away from," and *phylaxis,* meaning "protection."

HIGHLIGHT

When Kissing Triggers Allergic Reactions

Individuals with particularly severe food allergies need to watch not only what they eat, but also who kisses them. In a study from the University of California at Davis, researchers found that of 316 patients with severe allergies to peanuts, tree nuts, and/or seeds, 20 (about 6%) reported developing allergic reactions after being kissed. In nearly all of the cases, the kisser had eaten nuts to which the allergic individual was sensitive. In some of the cases, the allergic individuals were so sensitive that they developed reactions even though the kissers had brushed their teeth or had consumed the nuts several hours earlier. Food allergists believe that the duration and intensity of a kiss may also be a factor—that is, longer kisses involving exchange of saliva may be more likely to elicit allergic reactions than a quick peck on the cheek. The moral of the story: if you have a severe food allergy, you might want to tell your romantic partner about it!

other household allergies. A dehumidifier can reduce mold in a home. The best way for individuals to avoid allergic reactions to pets is to find out whether they are allergic before they get a pet. If an allergy develops afterward, they may have to find a new home for the animal.

Food allergens can be identified and then avoided by using a medically supervised elimination diet in which foods are removed one at a time from the diet to see when signs of allergy cease. Peanuts and shellfish are among the foods more commonly implicated in anaphylactic shock; allergic individuals must avoid consuming even minute amounts (see **Highlight: When Kissing Triggers Allergic Reactions**). For example, individuals sensitized against shellfish can suffer anaphylactic shock from consuming meat sliced on an unwashed cutting board previously used to prepare shellfish. Some vaccines contain small amounts of egg proteins and should not be given to individuals allergic to eggs.

Health care workers should assess a patient's medical history carefully before administering penicillin, iodine dyes, and other potential allergens. Therapies such as antidotes and respiratory stimulants should be immediately available if a patient's sensitivity status is unknown.

In addition to avoidance of allergens, type I hypersensitivity reactions can be prevented by *immunotherapy* (im´ū-nō-thār´ă-pē), commonly called "allergy shots," which involves the administration of a series of injections of dilute allergen, usually once a week for many months. It is unclear just how allergy shots work, but they may change the helper T cell balance from Th2 cells to Th1 cells, reducing the production of antibodies, or they may stimulate the production of IgG, which binds antigen before the antigen can react with IgE on mast cells, basophils, and eosinophils. Immunotherapy reduces the severity of allergy symptoms by roughly 50% in about two-thirds of patients with upper respiratory allergies; however, the series of injections must be repeated every two to three years. Immunotherapy is not effective in treating asthma.

Treatment of Type I Hypersensitivity

One way to treat type I hypersensitivity is to administer drugs that specifically counteract the inflammatory mediators released by degranulating cells. Thus, **antihistamines** are administered to counteract histamine. However, because histamine is but one of many mediators released in type I hypersensitivity, antihistamines do not completely eliminate all clinical signs. Indeed, because antihistamines are essentially useless in patients with asthma, asthmatics are typically prescribed an inhalant containing *glucocorticoid* (gloo-kō-kōr´ti-koyd) and a *bronchodilator* (brong-kō-dī-lā´ter), which counteract the effects of inflammatory mediators.

The hormone *epinephrine* quickly neutralizes many of the lethal mechanisms of anaphylaxis by relaxing smooth muscle tissue in the lungs, contracting smooth muscle of blood vessels, and reducing vascular permeability. Epinephrine is thus the drug of choice for the emergency treatment of both severe asthma and anaphylactic shock. Patients who suffer from severe type I hypersensitivities may carry a prescription epinephrine kit so that they can inject themselves before their allergic reactions become life threatening.

Type II (Cytotoxic) Hypersensitivity

LEARNING OUTCOMES

18.5 Discuss the mechanisms underlying transfusion reactions.
18.6 Construct a table comparing the key features of the four blood types in the ABO blood group system.
18.7 Describe the mechanisms and treatment of hemolytic disease of the newborn.

The second major form of hypersensitivity results when cells are destroyed by an immune response—typically by the combined activities of complement and antibodies. This *cytotoxic hypersensitivity* is part of many autoimmune diseases, which we will discuss later in this chapter, but the most significant examples of type II hypersensitivity are the destruction of donor red blood cells following an incompatible blood transfusion and the destruction of fetal red blood cells. We begin our discussion by focusing on these two manifestations of type II hypersensitivity.

The ABO System and Transfusion Reactions

Red blood cells have many different glycoprotein and glycolipid molecules on their surface. Some surface molecules of red blood cells, called **blood group antigens,** have various functions, including transportation of glucose and ions across the cytoplasmic membrane.

There are several sets of blood group antigens that vary in complexity. The ABO group system is most famous and consists of just two antigens arbitrarily given the names A antigen and B antigen. Each person's red blood cells have either A antigen, B antigen, both A and B antigens, or neither antigen. Individuals with neither antigen are said to have blood type O.

As you probably know, blood can be transfused from one person to another; however, if blood is transfused to an individual with a different blood type, then the donor's blood group antigens may stimulate the production of antibodies in the recipient. These bind to and eventually destroy the transfused cells. The result can be a potentially life-threatening *transfusion reaction*. Note that it is a blood recipient's own immune system that causes problems; the donated cells merely trigger the response.

Transfusion reactions, the most problematic of which involve the ABO group, develop as follows:

- If the recipient has preexisting antibodies to foreign blood group antigens, then the donated blood cells will be destroyed immediately—either the antibody-bound cells will be phagocytized by macrophages and neutrophils or the antibodies will agglutinate the cells and complement will rupture them, a process called *hemolysis* **(FIGURE 18.5)**. Hemolysis releases hemoglobin into the bloodstream, which may cause severe kidney damage. At the same time, the membranes of the ruptured blood cells trigger blood clotting within blood vessels, blocking them and causing circulatory failure, fever, difficulty in breathing, coughing, nausea, vomiting, and diarrhea. If the patient survives, recovery follows the elimination of all the foreign red blood cells.
- If the recipient has no preexisting antibodies to the foreign blood group antigens, then the transfused cells circulate and function normally, but only for a while—that is, until the recipient's immune system mounts a primary response against the foreign antigens and produces enough antibody to destroy the foreign cells. This happens gradually over a long enough time that the severe symptoms and signs mentioned above do not occur.

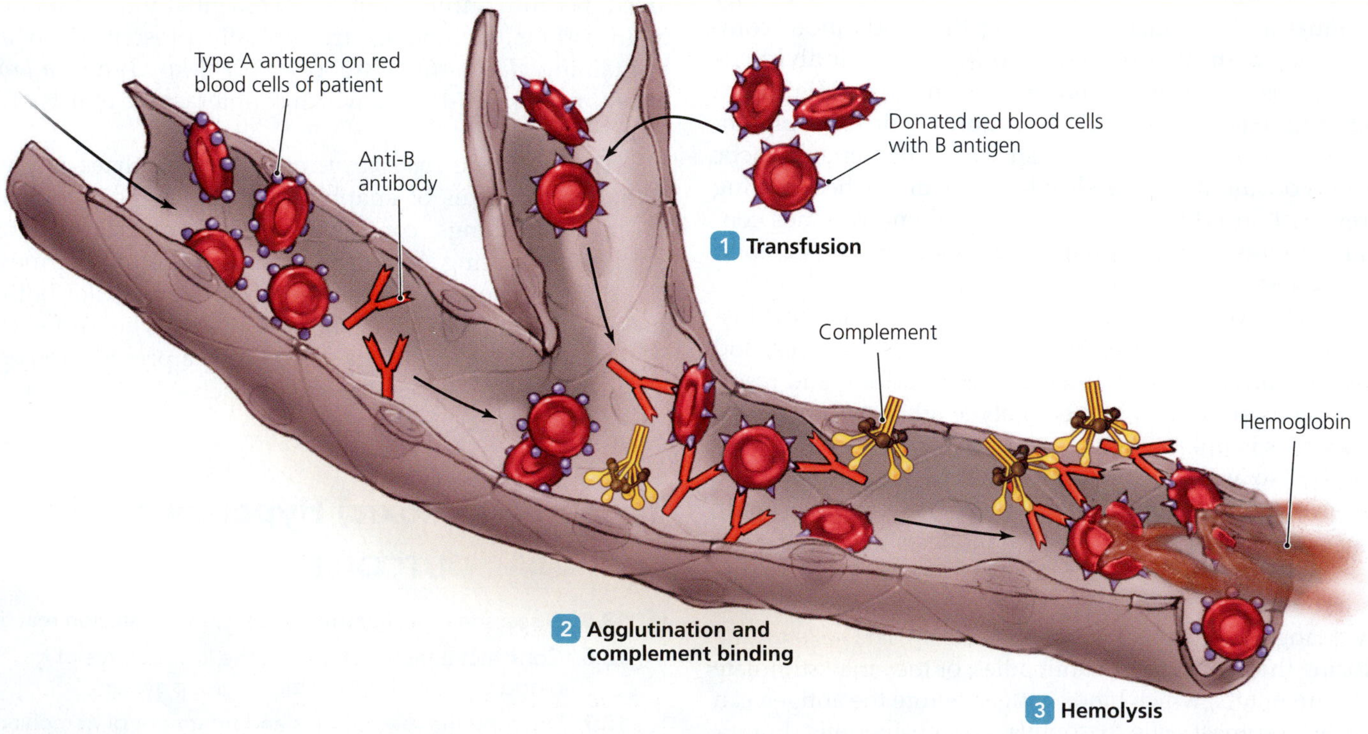

▲ **FIGURE 18.5 Events leading to hemolysis.** One of the negative consequences of a transfusion reaction (here illustrated by a transfusion of type B blood into a patient with type A blood): When antibodies against a foreign ABO antigen combine with the antigen on transfused red blood cells 1, the cells are agglutinated, and the complexes bind complement 2. The resulting hemolysis releases large amounts of hemoglobin into the bloodstream 3 and produces additional negative consequences throughout the body.

TABLE 18.2 ABO Blood Group Characteristics and Donor/Recipient Matches

ABO Blood Group	ABO Antigen(s) Present	Antibodies Present	Can Donate To	Can Receive From
A	A	Anti-B	A or AB	A or O
B	B	Anti-A	B or AB	B or O
AB	A and B	None	AB	A, B, AB, or O (universal recipient)
O	None	Both anti-A and anti-B	A, B, AB, or O (universal donor)	O

To prevent transfusion reactions, laboratory personnel must cross-match ABO blood types between donors and recipients. Before a recipient receives a blood transfusion, the red cells and serum of the donor can be mixed with the serum and red cells of the recipient. If any signs of clumping are seen, then that blood is not used, and an alternative donor is sought. **TABLE 18.2** presents the characteristics of the ABO blood group and the compatible donor/recipient matches that can be made.

In most people, production of antibodies against foreign ABO antigens is stimulated not by exposure to foreign blood cells but instead by exposure to antigenically similar molecules found on a wide range of plants and bacteria. Individuals encounter and make antibodies against these antigens on a daily basis, but only upon receipt of a mismatched blood transfusion do the antigens cause problems.

The Rh System and Hemolytic Disease of the Newborn

Many decades ago, researchers discovered the existence of an antigen common to the red blood cells of humans and rhesus monkeys. They called this antigen, which transports anions and glucose across the cytoplasmic membrane, *rhesus* (rē´sŭs) antigen, or **Rh antigen.** Laboratory analysis of human blood samples eventually showed that the Rh antigen is present on the red blood cells of about 85% of humans; that is, about 85% of the population is *Rh positive* (*Rh+*), and about 15% of the population is *Rh negative* (*Rh–*).

In contrast to the situation with ABO antigens, preexisting antibodies against Rh antigen do not occur. Although a transfusion reaction can occur in an Rh-negative patient who receives more than one transfusion of Rh-positive blood, such a reaction is usually minor because Rh antigen molecules are less abundant than A or B antigens. Instead, the primary problem posed by incompatible Rh antigen is the risk of **hemolytic[4] disease of the newborn** (hē-mō-lit´ik; **FIGURE 18.6**).

This hypersensitivity reaction develops when an Rh– mother is pregnant with an Rh+ baby (who inherited an Rh gene from its father). Normally, the placenta keeps fetal red blood cells separate from the mother's blood, so fetal cells do not enter the mother's bloodstream. However, in 20–50% of pregnancies—-especially during the later weeks of pregnancy, during a clinical or spontaneous abortion, or during childbirth—fetal red blood cells escape into the mother's blood. The Rh– mother's immune system recognizes these Rh+ fetal cells as foreign and initiates an antibody immune response by developing antibodies against the Rh antigen. Initially, only IgM antibodies are produced, and because IgM is a very large molecule that cannot cross the placenta, no problems arise during this first pregnancy.

However, IgG antibodies can cross the placenta, so if during additional pregnancies the fetus is Rh+, anti-Rh IgG molecules produced by the mother cross the placenta and destroy the fetus's Rh+ red blood cells. Such destruction may be limited, or it may be severe enough—especially in a third or subsequent pregnancy—to cause grave problems. One classic feature of hemolytic disease of the newborn is severe *jaundice* from excessive bilirubin (bil-i-roo´bin), which is a yellowish blood pigment released during the degradation of hemoglobin from lysed red blood cells. The liver is normally responsible for removing bilirubin, but because of the immaturity of the fetal liver and overload from the hemolytic process, the bilirubin may instead be deposited in the brain, causing severe neurological damage or death during the last weeks of pregnancy or shortly after the baby's birth.

In the past, when prevention of hemolytic disease of the newborn was impossible, this terrible disease occurred in about 1 of every 300 births. Today, however, physicians can routinely and drastically reduce the number of cases of the disease by administering anti-Rh immunoglobulin (IgG), called *RhoGAM,* to Rh-negative women at 28 weeks into their pregnancy and also within 72 hours following abortion, miscarriage, or childbirth. Any fetal red cells that may have entered the mother's body are destroyed by RhoGAM before the fetal cells can trigger an immune response. As a result, sensitization of the mother does not occur, and future pregnancies are safer. **▶VIDEO TUTOR:** ***Hemolytic Disease of the Newborn***

Drug-Induced Cytotoxic Reactions

Another kind of type II hypersensitivity involves cytotoxic reactions to drugs. Although the molecules of such drugs as quinine, penicillin, or sulfanilamide are too small to trigger an immune response by themselves, they can bind to larger molecules and become antigenic, stimulating production of antibodies.

When such antibodies and then complement bind to drug molecules already bound to blood platelets, the platelets are lysed, producing a disease called **immune thrombocytopenic[5]**

[4]From Greek *haima,* meaning "blood," and *lysis,* meaning "destruction."

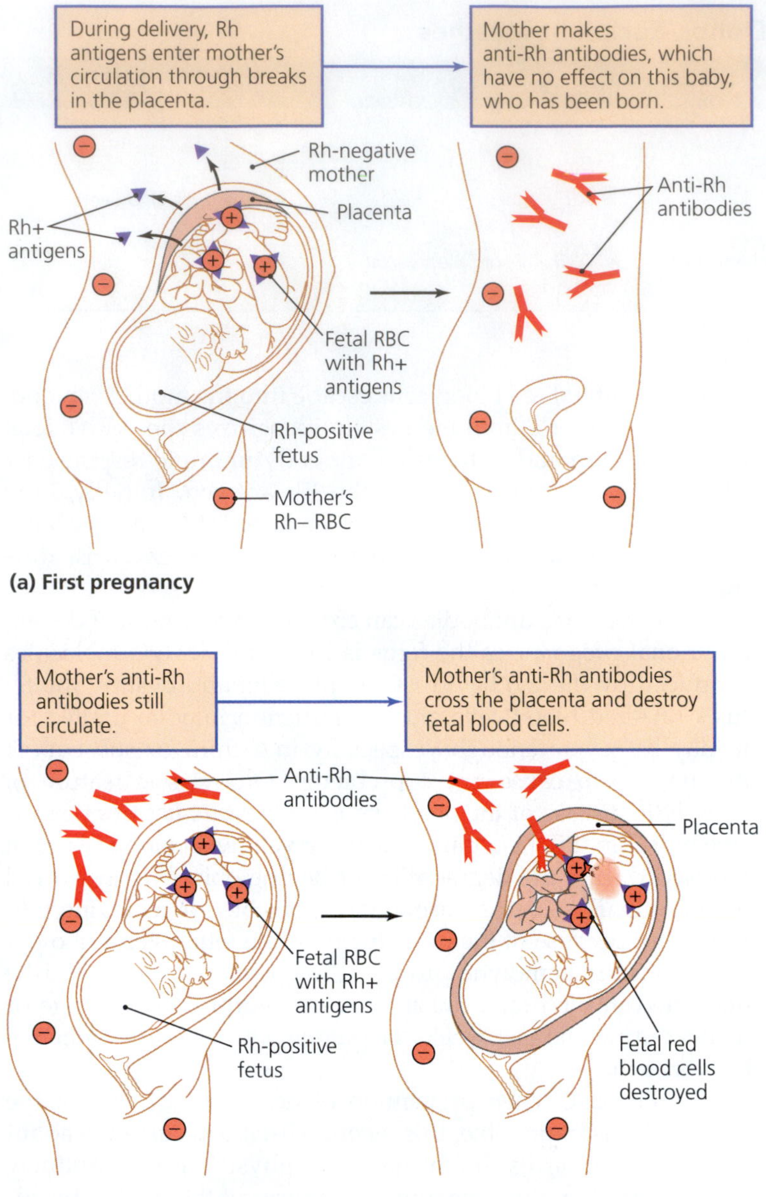
During delivery, Rh antigens enter mother's circulation through breaks in the placenta.
Mother makes anti-Rh antibodies, which have no effect on this baby, who has been born.
Rh-negative mother
Placenta
Rh+ antigens
Fetal RBC with Rh+ antigens
Rh-positive fetus
Mother's Rh– RBC
Anti-Rh antibodies
(a) First pregnancy
Mother's anti-Rh antibodies still circulate.
Mother's anti-Rh antibodies cross the placenta and destroy fetal blood cells.
Anti-Rh antibodies
Placenta
Fetal RBC with Rh+ antigens
Rh-positive fetus
Fetal red blood cells destroyed
(b) Subsequent pregnancy

▲ **FIGURE 18.6 Events in the development of hemolytic disease of the newborn. (a)** During an initial pregnancy, Rh red blood cells from the fetus enter the Rh– mother's circulation, often during childbirth. As a result, the mother produces anti-Rh antibodies. (Antigens and antibodies are not shown to scale.) **(b)** During a subsequent pregnancy with another Rh child, anti-Rh IgG molecules cross the placenta and trigger destruction of the fetus's red blood cells. *How is it possible that the child of an Rh– mother is Rh+?*

Figure 18.6 *The father of the child is Rh positive.*

purpura (throm´bō-sī-tō-pē´nik pŭr´poo-ră; **FIGURE 18.7**). The destruction of the platelets inhibits the ability of the blood to clot correctly, which leads to the production of purple hemorrhages, called *purpura*[6], under the skin. Similar destruction of leukocytes is one form of *agranulocytosis*, and that of red blood cells is called *hemolytic anemia* (ă-nē´mē-ă).

[5]*Thrombocyte* is another name for platelet.

[6]From Latin, meaning "purple."

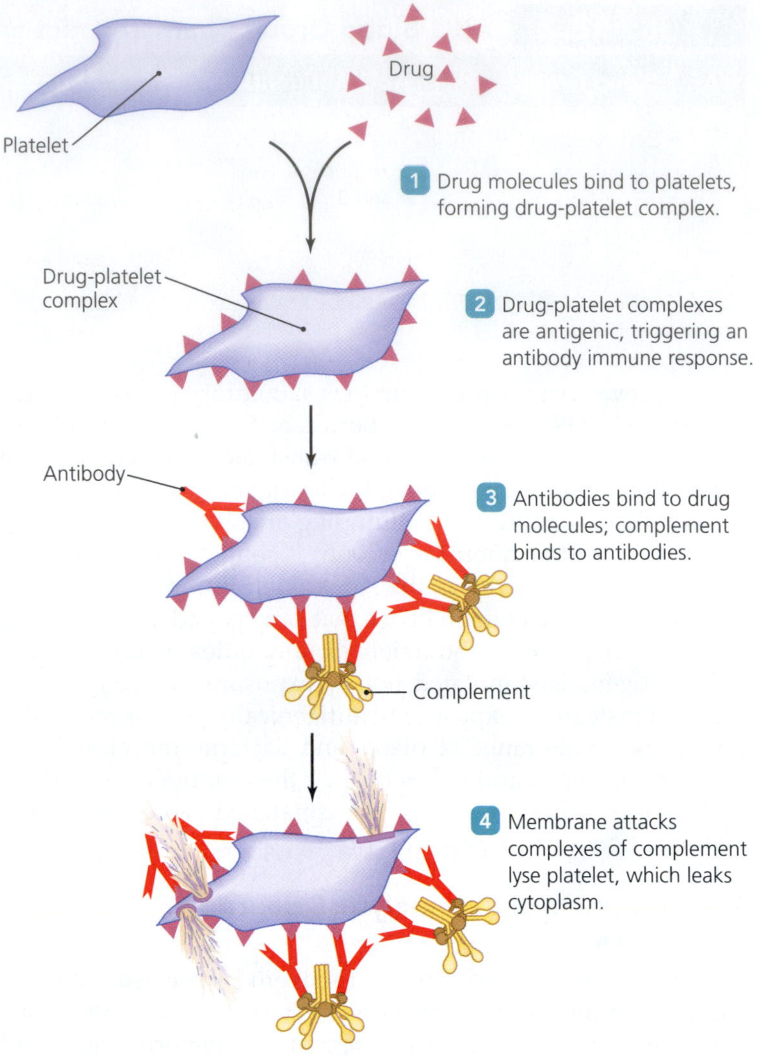
Platelet
Drug
1 Drug molecules bind to platelets, forming drug-platelet complex.
Drug-platelet complex
2 Drug-platelet complexes are antigenic, triggering an antibody immune response.
Antibody
3 Antibodies bind to drug molecules; complement binds to antibodies.
Complement
4 Membrane attacks complexes of complement lyse platelet, which leaks cytoplasm.

▲ **FIGURE 18.7 Events in the development of immune thrombocytopenic purpura.** This disease results from a drug-induced type II (cytotoxic) hypersensitivity reaction. The destruction of platelets through the action of bound antibodies and complement produces the inhibition of blood clotting characteristic of this disease. *What is the similar disease resulting from type II destruction of red blood cells?*

Figure 18.7 *Hemolytic anemia is the type II destruction of red blood cells.*

Type III (Immune Complex–Mediated) Hypersensitivity

LEARNING | **OUTCOMES**

18.8 Outline the basic mechanism of type III hypersensitivity.

18.9 Describe hypersensitivity pneumonitis and glomerulonephritis.

18.10 List the signs and symptoms of rheumatoid arthritis.

18.11 Discuss the cause and signs of systemic lupus erythematosus.

The formation of complexes of antigen bound to antibody, called **immune complexes,** initiates several molecular processes, including complement activation. Normally, immune complexes are removed from the body via phagocytosis. However, in *type III (immune complex–mediated) hypersensitivity* reactions, the immune complexes escape phagocytosis and so circulate in the bloodstream until they become trapped in organs, joints, and tissues (such as the walls of blood vessels). In these sites they trigger mast cells to degranulate, releasing inflammatory chemicals that damage the tissues. **FIGURE 18.8** illustrates the mechanism of type III reactions.

Type III hypersensitivities may be localized or may affect a number of body systems simultaneously. There is no cure for these diseases, though steroids that suppress the immune system may provide some relief. Two localized conditions resulting from immune complex–mediated hypersensitivity are hypersensitivity pneumonitis and glomerulonephritis; two systemic disorders are rheumatoid arthritis and systemic lupus erythematosus.

Hypersensitivity Pneumonitis

Type III hypersensitivities can affect the lungs, causing a form of pneumonia called **hypersensitivity pneumonitis** (noo-mō-nī´tus). Individuals become sensitized when minute mold spores or other antigens are inhaled deep into the lungs, stimulating the production of antibodies. A hypersensitivity reaction occurs when the subsequent inhalation of the same antigen stimulates the formation of immune complexes that then activate complement.

One form of hypersensitivity pneumonitis, called *farmer's lung,* occurs in farmers chronically exposed to spores from moldy hay. Many other syndromes in humans develop via a similar mechanism and are usually named after the source of the inhaled allergen. Thus, *pigeon breeder's lung* arises following exposure to the dust from pigeon feces, *mushroom grower's lung* is a response to the soil or fungal spores encountered in the growing of mushrooms, and *librarian's lung* results from inhaling dust from old books.

Glomerulonephritis

Glomerulonephritis (glō-măr´ū-lō-nef-rī´tis) occurs when immune complexes circulating in the bloodstream are deposited in the walls of *glomeruli,* which are networks of minute blood vessels in the kidneys. Immune complexes damage the glomerular cells, leading to enhanced local production of cytokines that trigger nearby cells to produce more of the proteins that underlie the cells, impeding blood filtration. Sometimes, the immune complexes are deposited in the center of the glomeruli, where they stimulate local cells to divide and compress nearby blood vessels, again interfering with kidney function. The net result is kidney failure; the glomeruli lose their ability to filter wastes from the blood, ultimately resulting in death.

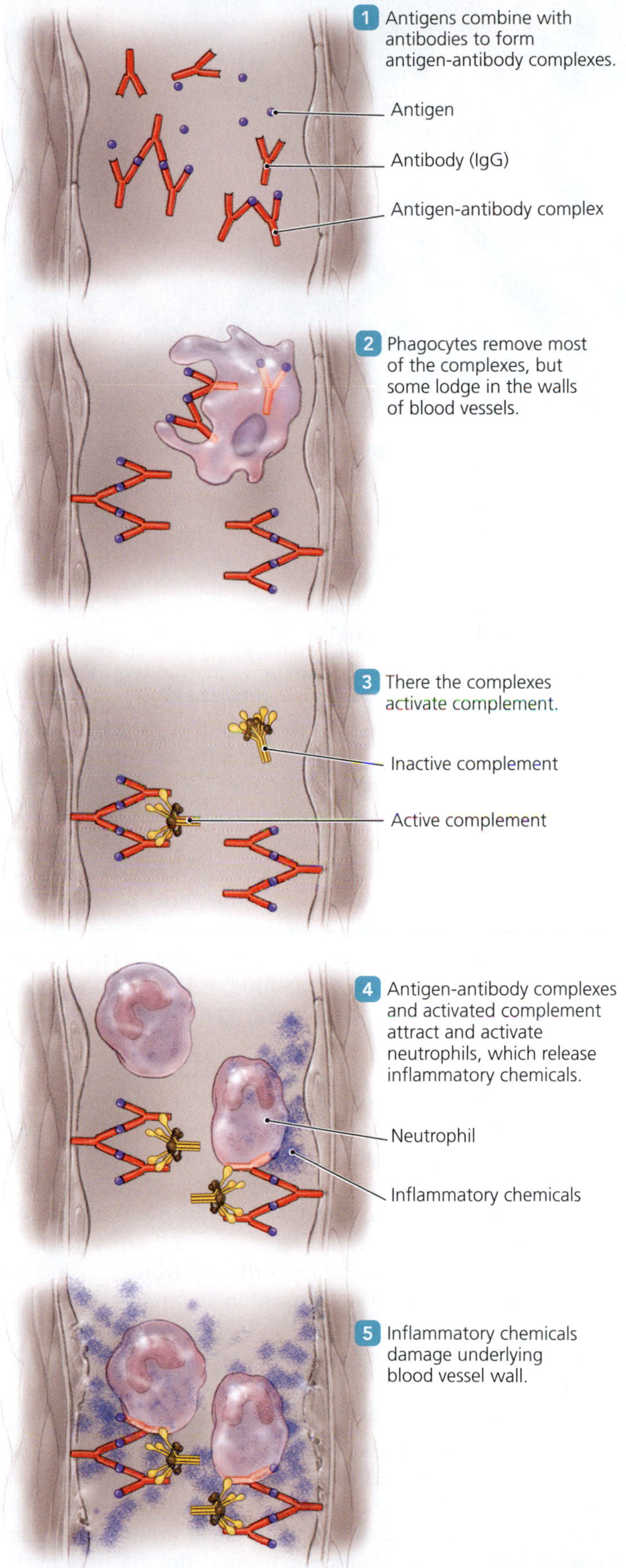
1 Antigens combine with antibodies to form antigen-antibody complexes.
Antigen
Antibody (IgG)
Antigen-antibody complex
2 Phagocytes remove most of the complexes, but some lodge in the walls of blood vessels.
3 There the complexes activate complement.
Inactive complement
Active complement
4 Antigen-antibody complexes and activated complement attract and activate neutrophils, which release inflammatory chemicals.
Neutrophil
Inflammatory chemicals
5 Inflammatory chemicals damage underlying blood vessel wall.

▶ **FIGURE 18.8 The mechanism of type III (immune complex–mediated) hypersensitivity.** After immune complexes form 1, the complexes that are not phagocytized lodge in certain tissues (in this case, the wall of a blood vessel) 2. Trapped complexes then activate complement 3, which attracts and activates neutrophils 4. The release of inflammatory chemicals from the activated neutrophils leads to the destruction of tissue 5.

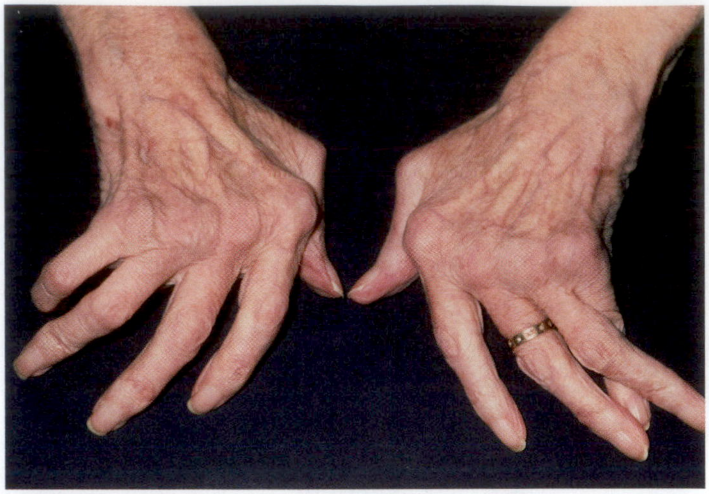

▲ **FIGURE 18.9 The crippling distortion of joints characteristic of rheumatoid arthritis.**

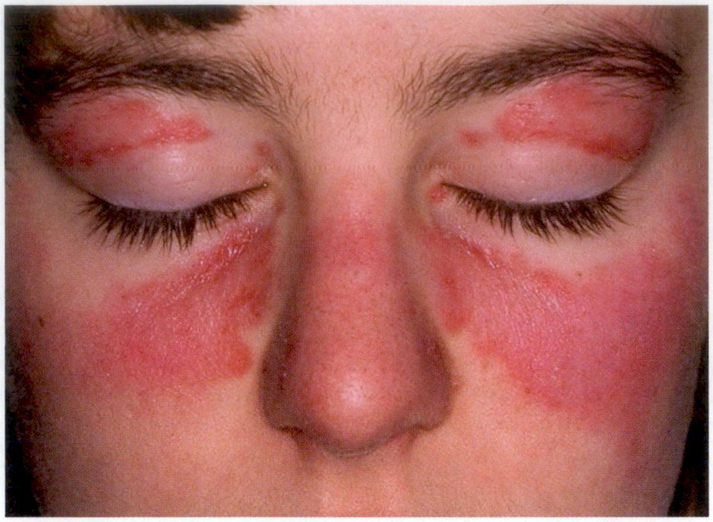

▲ **FIGURE 18.10 The characteristic facial rash of systemic lupus erythematosus.** Its shape corresponds to areas most exposed to sunlight, which worsens the condition.

Rheumatoid Arthritis (RA)

Rheumatoid arthritis (roo´mă-toyd ar-thrī´tis; **RA**) commences when B cells secrete IgM that binds to certain IgG molecules. IgM-IgG complexes are deposited in the joints, where they activate complement and mast cells, which release inflammatory chemicals. The resulting inflammation causes the tissues to swell, thicken, and proliferate into the joint, resulting in severe pain. As the altered tissue extends into the joint, the inflammation further erodes and destroys joint cartilage and the neighboring bony structure until the joints begin to break down and fuse; as a result, affected joints become distorted and lose their range of motion **(FIGURE 18.9)**. The course of rheumatoid arthritis is often intermittent; however, with each recurrence, the lesions and damage get progressively more severe.

The trigger of RA is not well understood. The fact that there are no animal models (because the disease appears to affect humans only) significantly hinders research on its cause. Many cases demonstrate that RA commonly follows an infectious disease in a genetically susceptible individual. Possession of certain immunity (MHC) genes appears to increase susceptibility.

Physicians treat rheumatoid arthritis by administering anti-inflammatory drugs such as ibuprofen to prevent additional joint damage and immunosuppressive drugs to inhibit the antibody immune response.

Systemic Lupus Erythematosus (SLE)

An example of type III hypersensitivity disease that affects multiple organs is **systemic lupus erythematosus** (loo´pŭser-ĭ-thē´mă-tō-sus; **SLE**), often shortened to *lupus*. Patients with this generalized immunological disorder make antibodies against numerous self-antigens found in normal organs and tissues, giving rise to many different pathological lesions and clinical manifestations. One consistent feature of SLE is the development of such self-reactive antibodies, called autoantibodies, against nucleic acids, especially DNA. These autoantibodies combine with free DNA released from dead cells to form immune complexes that are deposited in glomeruli, causing glomerulonephritis and kidney failure. Thus, SLE is an immune complex–mediated hypersensitivity reaction. Immune complexes may also be deposited in joints, where they give rise to arthritis.

The disease's curious name—systemic lupus erythematosus—stems from two features of the disease: *Systemic* simply reflects the fact that it affects different organs throughout the body, *lupus* is Latin for wolf, and *erythematosus* refers to a redness of the skin, so these last two words describe the characteristic red, butterfly-shaped rash that develops on the face of many patients, giving them what is sometimes described as a wolflike appearance **(FIGURE 18.10)**. This rash is caused by deposition of nucleic acid–antibody complexes in the skin and is worse in skin areas exposed to sunlight.

Although autoantibodies to nucleic acids are characteristic of SLE, many other autoantibodies are also produced. Autoantibodies to red blood cells cause hemolytic anemia, autoantibodies to platelets give rise to bleeding disorders, antilymphocyte antibodies alter immune reactivity, and autoantibodies against muscle cells cause muscle inflammation and, in some cases, damage to the heart. Because of its variety of symptoms, SLE can be misdiagnosed.

The trigger of lupus is unknown, though a lupus-like disease can be induced by some drugs. It is likely that SLE has many causes.

Physicians treat lupus with immunosuppressive drugs that reduce autoantibody formation and with glucocorticoids that reduce the inflammation associated with the deposition of immune complexes. Scientists are currently testing a battery of novel drugs for treating lupus.

CLINICAL CASE STUDY

The First Time's Not the Problem

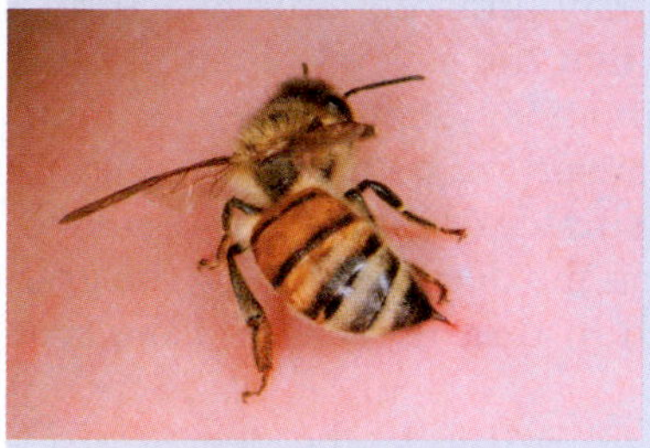

Steven, an eight-year-old boy, is brought to your office Monday morning by his father to have his upper arm checked for a possible infection. Dad is worried because the area of a bee sting on the boy's arm is getting more red, itchy, and tender. The father gave him some children's acetaminophen yesterday, which relieved the discomfort somewhat.

There is no history of medical problems or allergies, and the child takes no regular medication. The child is otherwise feeling well, and his father tells you he is playing and eating normally. There is no previous history of bee stings, and Steven proudly tells you he "hardly even cried" when he got stung.

There is a half-dollar-sized area on his left upper arm that is puffy and red, but there is no streaking or drainage, and the area does not appear to to be infected. Steven's temperature is normal, and his lungs are clear.

1. What type of hypersensitivity reaction is Steven manifesting?
2. What other over-the-counter medication might relieve the itching?
3. What mechanism is causing the signs and symptoms you are seeing?
4. Since the area is not infected, what future health risk for his son should the father be made aware of?
5. What can be done to determine future risk from a bee sting?
6. How would you recognize a severe allergic reaction?
7. What precautions can the family take to protect the boy from future reactions?

Type IV (Delayed or Cell-Mediated) Hypersensitivity

LEARNING | OUTCOMES

18.12 Outline the mechanism of type IV hypersensitivity.
18.13 Describe the significance of the tuberculin test.
18.14 Identify four types of grafts.
18.15 Compare four types of drugs commonly used to prevent graft rejection.

When certain antigens contact the skin of sensitized individuals, they provoke inflammation that begins to develop at the site only after 12–24 hours. Such **delayed hypersensitivity reactions** result not from the action of antibodies but rather from interactions among antigen, antigen-presenting cells, and T cells; thus, a type IV reaction is also called *cell-mediated hypersensitivity*. The delay in this cell-mediated response reflects the time it takes for macrophages and T cells to migrate to and divide at the site of the antigen. We begin our discussion of type IV reactions by considering two common examples: the tuberculin response and allergic contact dermatitis. Then we will consider two type IV hypersensitivity reactions involving the interactions between the body and tissues grafted to (transplanted into) it—graft rejection and graft-versus-host disease—before considering donor-recipient matching and tissue typing.

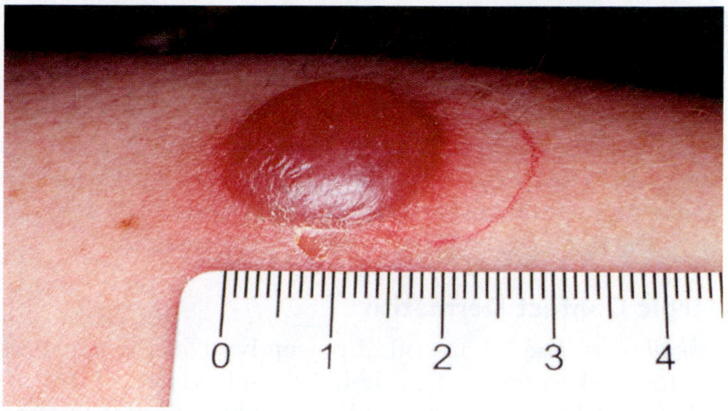

▲ **FIGURE 18.11 A positive tuberculin test, a type IV hypersensitivity response.** The hard, red swelling 10 mm or greater in diameter that is characteristic of the tuberculin response indicates that the individual has been vaccinated against *Mycobacterium tuberculosis* or is now or has been previously infected with the bacterium.

The Tuberculin Response

The **tuberculin response** (too-ber´kyū-lin) is an important example of a delayed hypersensitivity reaction in which the skin of an individual exposed to tuberculosis (TB) or tuberculosis vaccine reacts to a shallow injection of *tuberculin*—a protein solution obtained from *Mycobacterium tuberculosis* (mī´kō-bak-tēr´ē-ŭm too-ber-kyū-lō´sis). Health care providers use the tuberculin test, also called a *Mantoux test* (mahn-too´) after the French physician who perfected it, to diagnose contact with antigens of *M. tuberculosis*.

When tuberculin is injected into the skin of a healthy, never-infected or unvaccinated individual, no response occurs. In contrast, when tuberculin is injected into someone currently or previously infected with *M. tuberculosis* or an individual previously immunized with tuberculosis vaccine, a red, hard swelling (10 mm or greater in diameter) indicating a positive tuberculin test develops at the site (**FIGURE 18.11**). Such inflammation reaches its greatest intensity within 24–72 hours and may persist for several weeks before fading. Microscopic examination of the lesion reveals that it is infiltrated with lymphocytes and macrophages.

A tuberculin response is mediated by memory T cells. When an individual is first infected by or immunized against

M. tuberculosis, the resulting cell-mediated immune response generates memory T cells that persist in the body. When a sensitized individual is later injected with tuberculin, phagocytic cells migrate to the site and attract memory T cells, which secrete a mixture of cytokines that attract still more T cells and macrophages, giving rise to a slowly developing inflammation. The macrophages ingest and destroy the injected tuberculin, allowing the tissues eventually to return to normal.

Allergic Contact Dermatitis

Urushiol (ŭ-rū´shē-ŏl), the oil of poison ivy (*Toxicodendron radicans,* toks´si-kō-den´dron rā´dē-kanz) and related plants, is a small molecule that becomes antigenic when it binds to almost any protein it contacts—including proteins in the skin of anyone who rubs against the plant. The body regards these chemically modified skin proteins as foreign, triggering a cell-mediated immune response and resulting in an intensely irritating skin rash called **allergic contact dermatitis** (der-mă-tī´tis). In severe cases, cytotoxic T lymphocytes (Tc cells) destroy so many skin cells that acellular, fluid-filled blisters develop **(FIGURE 18.12)**.

Other haptens that combine chemically with skin proteins can also induce allergic contact dermatitis. Examples include formaldehyde; some cosmetics, dyes, drugs, and metal ions; and chemicals used in the production of latex for hospital gloves and tubing.

Because T cells mediate allergic contact dermatitis, epinephrine and other drugs used for the treatment of immediate hypersensitivity reactions are ineffective. T cell activities and inflammation can, however, be suppressed by corticosteroid treatment. Good strategies for dealing with exposure to poison ivy include washing the area thoroughly and immediately with a strong soap and washing all exposed clothes as soon as possible.

Graft Rejection

A special case of type IV hypersensitivity is the rejection of **grafts** (or transplants), which are tissues or organs, such as livers, kidneys, or hearts, that have been transplanted, whether between sites within an individual or between a donor and an unrelated recipient. Even though advances in surgical technique enable surgeons to move grafts freely from site to site, grafts perceived as foreign by a recipient may undergo **graft rejection,** a highly destructive phenomenon that can severely limit the success of organ and tissue transplantation. Graft rejection is a normal immune response against foreign major histocompatibility complex (MHC) proteins on the surface of graft cells. The likelihood of graft rejection depends on the degree to which the graft is foreign to the recipient, which in turn is related to the type of graft.

Scientists name graft types according to the degree of relatedness between the donor and the recipient **(FIGURE 18.13)**. A graft is called an **autograft** (aw´tō-graft) when tissues are moved to a different location within the same individual. Autografts do not trigger immune responses because they do not express foreign antigens. Examples of autografts include the grafting of skin from one area of the body to another to cover a burn area or the use of a leg vein to bypass blocked coronary arteries.

Isografts (ī´sō-graftz) are grafts transplanted between two genetically identical individuals, that is, between identical siblings or clones. Because these individuals have identical MHC proteins, the immune system of the recipient cannot differentiate between the grafted cells and its own normal body cells. As a result, isografts are not rejected.

Allografts (al´ō-graftz) are grafts transplanted between genetically distinct members of the same species. Most grafts performed in humans are allografts. Because the MHC proteins of the allograft are different from those of the recipient, allografts typically induce a strong type IV hypersensitivity, resulting in graft rejection. Rejection must be stopped with immunosuppressive drugs if the graft is to survive.

Xenografts (zēn´ō-graftz) are grafts transplanted between individuals of different species. Thus, the transplant of a baboon's

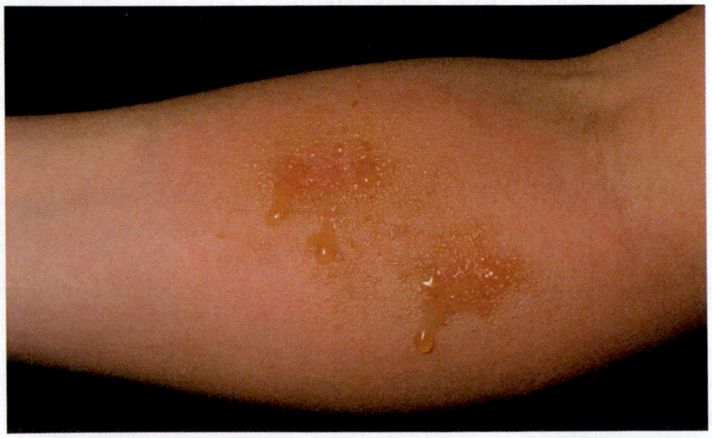

▲ FIGURE 18.12 Allergic contact dermatitis, a type IV hypersensitivity response. The response in this case is to poison ivy. Note the large, acellular, fluid-filled blisters that result from the destruction of skin cells.

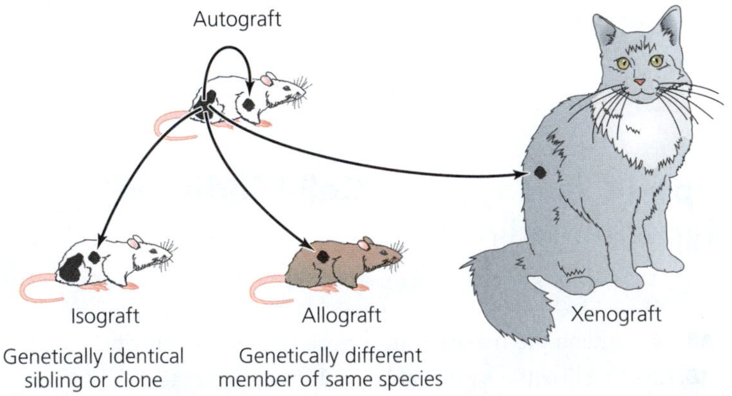

▲ FIGURE 18.13 Types of grafts. The names are based on the degree of relatedness between donor and recipient. Autografts are grafts moved from one location to another within a single individual. In isografts, the donor and recipient are either genetically identical siblings or clones. In allografts, the donor and recipient are genetically distinct individuals of the same species. In xenografts, the donor and recipient are of different species.

TABLE 18.3 The Characteristics of the Four Types of Hypersensitivity Reactions

Descriptive	Name	Cause	Time Course	Characteristic Cells Involved
Type I	Immediate hypersensitivity	Antibody (IgE) on sensitized cells' membranes binds antigen, causing degranulation	Seconds to minutes	Mast cells, basophils, and eosinophils
Type II	Cytotoxic hypersensitivity	Antibodies and complement lyse target cells	Minutes to hours	Red blood cells
Type III	Immune complex–mediated hypersensitivity	Nonphagocytized complexes of antibodies and antigens trigger mast cell degranulation	Several hours	Neutrophils
Type IV	Delayed hypersensitivity	T cells attack the body's cells	Several days	Activated T cells

heart into a human is a xenograft. Because xenografts are usually very different from the tissues of the recipient, both biochemically and immunologically, they usually provoke a rapid, intense rejection that is very difficult to suppress. Therefore, xenografts from mature animals are not commonly used therapeutically.

Graft-Versus-Host Disease

Physicians often use bone marrow allografts as a component of the treatment for leukemias and lymphomas (cancers of leukocytes). In this procedure, physicians use total body irradiation combined with cytotoxic drugs to kill tumor cells in the patient's bone marrow. In the process, the patient's existing leukocytes are also destroyed, which completely eliminates the body's ability to mount any kind of immune response. Physicians then inject the patient with donated bone marrow, which produces a new set of leukocytes. Ideally, within a few months, a fully functioning bone marrow is restored.

Unfortunately, when they are transplanted, donated bone marrow T cells may regard the patient's cells as foreign, mounting an immune response against them and giving rise to a condition called **graft-versus-host disease**. If the donor and recipient mainly differ in MHC class I molecules, the grafted T cells attack all of the recipient's tissues, producing especially destructive lesions in the skin and intestine. If the donor and recipient differ mainly in MHC class II molecules, then the grafted T cells attack the antigen-presenting cells of the host, leading to immunosuppression and leaving the recipient vulnerable to infections. The same immunosuppressive drugs used to prevent graft rejection (discussed shortly) can limit graft-versus-host disease.

Donor-Recipient Matching and Tissue Typing

Although it is usually not difficult to ensure that donor and recipient have identical blood groups, MHC compatibility is much harder to achieve. The reason is the very high degree of MHC variability, which ensures that unrelated individuals differ widely in their MHCs. In general, the more closely donor and recipient are related, the smaller their MHC difference. Given that in most cases an identical sibling is not available, it is usually preferable that grafts be donated by a parent or sibling possessing MHC antigens similar to those of the recipient.

In practice, of course, closely related donors are not always available. And even though a paired organ (such as a kidney) or a portion of an organ (such as liver tissue) may be available from a living donor, unpaired organs (such as the heart) almost always come from an unrelated cadaver. In such cases, attempts are made to match donor and recipient as closely as possible by means of *tissue typing*. Physicians examine the white cells of potential graft recipients to determine what MHC proteins they have. When a donor organ becomes available, it too is typed. Then the individual whose MHC proteins most closely match those of the donor is chosen to receive the graft. Though a perfect match is rarely achieved, the closer the match, the less intense the rejection process and the greater the chance of successful grafting. A match of 50% or less of the MHC proteins is usually acceptable for most organs, but near absolute matches are required for successful bone marrow transplants.

TABLE 18.3 summarizes the major characteristics of the four types of hypersensitivity.

Next we turn our attention to the actions of various types of immunosuppressive drugs used in situations in which the immune system is overactive.

The Actions of Immunosuppressive Drugs

The development of potent immunosuppressive drugs has played a large role in the dramatic success of modern transplantation procedures. These drugs can also be effective in combating certain autoimmune diseases (discussed shortly). In the following discussion we consider four classes of immunosuppressive drugs: glucocorticoids, cytotoxic drugs, cylosporine, and lymphocyte-depleting therapies.

Glucocorticoids, sometimes called *corticosteroids* and commonly referred to as *steroids,* have been used as immunosuppressive agents for many years. Glucocorticoids such as *prednisone* (pred´ni-sōn) and *methylprednisolone* suppress the response of T cells to antigen and inhibit such mechanisms as T cell cytotoxicity and cytokine production. These drugs have a much smaller effect on B cell function.

Cytotoxic drugs inhibit mitosis and cytokinesis (cell division). Given that lymphocyte proliferation is a key feature of specific immunity, blocking cellular reproduction is a powerful although very nonspecific method of immunosuppression. Among the cytotoxic drugs that have been used, the following are noteworthy:

- *Cyclophosphamide* (sī-klō-fos´fă-mīd) cross-links daughter DNA molecules in mitotic cells, preventing their separation and blocking mitosis. It impairs both B cell and T cell responses.

- *Azathioprine* (ā-ză-thī´ō-prēn) is a purine analog; it competes with purines during the synthesis of nucleic acids, thus blocking DNA replication and suppressing both primary and secondary antibody responses.

Three other cytotoxic drugs used for immunosuppression include *mycophenolate mofetil,* which inhibits purine synthesis, and *brequinar sodium* and *leflunomide,* each of which inhibits pyrimidine synthesis and thereby inhibits cellular replication.

Drugs such as **cyclosporine** (sī-klō-spōr´ēn), a polypeptide derived from fungi, prevent production of interleukins and interferons by T cells, thereby blocking Th1 responses. Because cyclosporine acts only on activated T cells and has no effect on resting T cells, it is far less toxic than the nonspecific drugs previously described. When it is given to prevent allograft rejection, only activated T cells attacking the graft are suppressed. Because steroids have a similar effect, the combination of glucocorticoids and cyclosporine is especially potent and can enhance survival of allografts.

Scientists have developed techniques involving relatively specific *lymphocyte-depleting therapies* in an attempt to reduce the many adverse side effects associated with the use of less specific immunosuppressive drugs. One technique involves administering an antiserum called *antilymphocyte globulin,* which is specific for lymphocytes. Another, more specific antilymphocyte technique uses monoclonal antibodies against CD3, which is found only on T cells. An even more specific monoclonal antibody is directed against the interleukin 2 receptor (IL-2R), which is expressed mainly on activated T cells. Such immunosuppressive therapies are effective in reversing graft rejection. Because they target a narrow range of cells, they produce fewer undesirable side effects than less specific drugs.

TABLE 18.4 lists some drugs and their actions for each of the four classes of the immunosuppressive drugs.

TELL ME WHY

During the war in Afghanistan in 2012, an army corporal with type AB blood received a life-saving blood transfusion from his sergeant, who had type O blood. Later the sergeant was involved in a traumatic accident and needed blood desperately. The corporal wanted to help but was told his blood was incompatible. Explain why the corporal could receive blood from but could not give blood to the sergeant.

Autoimmune Diseases

Just as today's military must control its arsenal of sophisticated weapons to avert losses of its own soldiers from "friendly fire," the immune system must be carefully regulated so that it does not damage the body's own tissues. However, an immune system does occasionally produce antibodies and cytotoxic T cells that target normal body cells—a phenomenon called *autoimmunity.* Although such responses are not always damaging, they can give rise to **autoimmune diseases,** some of which are life threatening.

Causes of Autoimmune Diseases

LEARNING OUTCOME

18.16 Briefly discuss eight hypotheses concerning the causes of autoimmunity and autoimmune diseases.

Most autoimmune diseases appear to develop spontaneously and at random. Nevertheless, scientists have noted some common features of autoimmune disease. For example, they occur more often in older individuals, and they are also much more common in women than in men, although the reasons for this gender difference are unclear.

Hypotheses to explain the etiology of autoimmunity abound. They include the following:

- Estrogen may stimulate the destruction of tissues by cytotoxic T cells.
- During pregnancy, some maternal cells may cross the placenta and colonize the fetus. These cells are more likely to survive in a daughter than in a son and might trigger an autoimmune disease later in the daughter's life.
- Conversely, fetal cells may also cross the placenta and trigger autoimmunity in the mother.
- Environmental factors may contribute to the development of autoimmune disorders. Some autoimmune diseases—type 1 diabetes mellitus (dī-ă-bē´tēz mĕ-lī´tēs) and rheumatoid arthritis, for example—develop in a few patients following their recovery from viral infections, though other individuals who develop these diseases have no history of such viral infections.
- Genetic factors may play a role in autoimmune diseases. MHC genes that in some way promote autoimmunity are

TABLE 18.4 The Four Classes of Immunosuppressive Drugs

Class	Examples	Action
Glucocorticoids	Prednisone, methylprednisolone	Anti-inflammatory; kills T cells
Cytotoxic drugs	Cyclophosphamide, azathioprine, mycophenolate mofetil, brequinar sodium, leflunomide	Blocks cell division nonspecifically
Cyclosporine	Cyclosporine	Blocks T cell responses
Lymphocyte-depleting therapies	Antilymphocyte globulin, monoclonal antibodies	Kills T cells nonspecifically, kills activated T cells, inhibits IL-2 reception

found in individuals with autoimmune diseases, whereas MHC genes that somehow protect against autoimmunity may dominate in other individuals. MHC genes may also trigger autoimmune disease by preventing the elimination of some self-reactive T cells in the thymus.

- Some autoimmune diseases develop when T cells encounter self-antigens that are normally "hidden" in sites where T cells rarely go. For example, because sperm develop within the testes during puberty, long after the body has selected its T cell population, men may have T cells that recognize their own sperm as foreign. This is normally of little consequence because sperm are sequestered from the blood. But if the testes are injured, T cells may enter the site of damage and mount an autoimmune response against the sperm, resulting in infertility.
- Infections with a variety of microorganisms may trigger autoimmunity as a result of **molecular mimicry,** which occurs when an infectious agent has an epitope that is very similar or identical to a self-antigen. In responding to the invader, the body produces antibodies that are *autoantibodies* (antibodies against self-antigens), which damage body tissues. For example, children infected with some strains of *Streptococcus pyogenes* (strep-tō-kok´ŭs pī-oj´en-ēz) may produce antibodies to heart muscle and so develop heart disease. Other strains of streptococci trigger the production of antibodies that cross-react with glomerular basement membranes and so cause kidney disease. It is possible that some virally triggered autoimmune diseases may also result from molecular mimicry.
- Other autoimmune responses may result from failure of the normal control mechanisms of the immune system. Thus, even though harmful, self-reactive T lymphocytes are normally destroyed via apoptosis triggered through the cell-surface receptor CD95 (see Figure 16.15), defects in CD95 can cause autoimmunity by permitting abnormal T cells to survive and cause disease.

Examples of Autoimmune Diseases

LEARNING OUTCOME

18.17 Describe a serious autoimmune disease associated with each of the following: blood cells, endocrine glands, nervous tissue, and connective tissue.

Regardless of the specific mechanism that causes an autoimmune disease, immunologists categorize them into two major groups: systemic autoimmune diseases such as lupus (discussed previously) and single-organ autoimmune diseases, which affect a single organ or tissue. Among the many single-organ autoimmune diseases recognized, common examples affect blood cells, endocrine glands, nervous tissue, or connective tissue.

Autoimmunity Affecting Blood Cells

Individuals with **autoimmune hemolytic anemia** produce antibodies against their own red blood cells. These autoantibodies speed up the destruction of the red blood cells, and the patient becomes severely anemic. Different hemolytic anemia patients make antibodies of different classes. Some patients make IgM autoantibodies, which bind to red blood cells and activate the classical complement pathway; the red blood cells are lysed, and degradation products from hemoglobin are released into the bloodstream. Other hemolytic anemia patients make IgG autoantibodies, which serve as opsonins that promote phagocytosis. In this case, red blood cells are removed by macrophages in the liver, spleen, and bone marrow. Even though these latter patients have no hemoglobin in their urine, they are still severely anemic.

The precise causes of all cases of autoimmune hemolytic anemia are unknown, but some cases follow infections with viruses or treatment with certain drugs, both of which alter the surface of red blood cells such that they are recognized as foreign and trigger an immune response.

Autoimmunity Affecting Endocrine Organs

Other common targets of autoimmune attack are the endocrine (hormone-producing) organs. For example, patients can develop autoantibodies or produce T cells against cells of the islets of Langerhans within the pancreas or against cells of the thyroid gland. In most cases, the ensuing autoimmune reaction results in damage to or destruction of the gland and in hormone deficiencies as endocrine cells are killed.

Immunological attack on the islets of Langerhans results in a loss of the ability to produce the hormone *insulin,* which leads to the development of **type 1 diabetes mellitus** (also known as *juvenile-onset diabetes*). As with other autoimmune diseases, the exact trigger of type 1 diabetes is unknown, but many patients endured a severe viral infection some months before the onset of diabetes. Additionally, some patients are known to have a genetic predisposition to developing type 1 diabetes that is associated with the possession of certain class I MHC molecules. Some physicians have been successful in delaying the onset of type 1 diabetes by treating at-risk patients with immunosuppressive drugs before damage to the islets of Langerhans becomes apparent.

An autoimmune response can lead to stimulation rather than to inhibition or destruction of glandular tissue. An example of this is **Graves' disease,** which involves the thyroid gland. This major endocrine gland located in the neck secretes iodine-containing hormones, which help regulate metabolic rate in the body. Like other autoimmune diseases, Graves' disease may be triggered by a viral infection in individuals with certain genetic backgrounds. Affected patients (usually women) make autoantibodies that bind to and stimulate receptors on the cytoplasmic membranes of thyroid cells, which elicits excessive production of thyroid hormone and growth of the thyroid gland. Such patients develop an enlarged thyroid gland—called a goiter—protruding eyes, rapid heartbeat, fatigue, and weight loss despite increased appetite. Physicians treat Graves' disease with antithyroid medicines or radioactive iodine or, in nonresponsive patients, by surgically removing most of the thyroid tissue.

Autoimmunity Affecting Nervous Tissue

Of the group of autoimmune diseases affecting nervous tissue, the most frequent is **multiple sclerosis** (sklĕ-rō'sis; **MS**). The exact cause of MS is unknown, but it appears that a cell-mediated immune response against a bacterium or virus generates cytotoxic T cells that mistakenly attack and destroy the myelin sheaths that normally insulate brain and spinal cord neurons and increase the speed of nerve impulses along the length of the neurons. Consequently, MS patients experience deficits in vision, speech, and neuromuscular function that may be quite mild and intermittent or may ultimately lead to death.

Autoimmunity Affecting Connective Tissue

Rheumatoid arthritis (discussed previously) is another crippling autoimmune disease resulting from a type III hypersensitivity. Autoantibodies are formed against connective tissue in joints.

TELL ME WHY

Why can't scientists use the postulates of Robert Koch to determine the specific cause of Graves' disease?

Immunodeficiency Diseases

LEARNING | **OUTCOME**

18.18 Differentiate primary from acquired immunodeficiencies, and cite one disease caused by each form of immunodeficiency.

You may have noticed that during periods of increased emotional or physical stress you are more likely to have a cold or the flu. You are not imagining this phenomenon; stress has long been known to decrease the efficiency of the immune system. Similarly, chronic defects in the immune system typically first become apparent when affected individuals become sick more often from infections of opportunistic pathogens (or even of organisms not normally considered pathogens). Such opportunistic infections are the hallmarks of *immunodeficiency diseases*, which are conditions resulting from defective immune mechanisms.

Researchers have characterized a large number of immunodeficiency diseases in humans, which are of two general types:

- **Primary immunodeficiency diseases,** which are detectable near birth and develop in infants and young children, result from some genetic or developmental defect.
- **Acquired (secondary) immunodeficiency diseases** develop in later life as a direct consequence of some other recognized cause, such as malnutrition, severe stress, or infectious disease.

Next we consider each general type of immunodeficiency disease in turn.

Primary Immunodeficiency Diseases

Many different inherited defects have been identified in all of the body's lines of defense, affecting first and second lines of defense as well as antibody and cell-mediated immune responses.

One of the more important inherited defects in the second line of defense is **chronic granulomatous disease,** in which children have recurrent infections characterized by the inability of their phagocytes to destroy bacteria. The reason is that these children have an inherited inability to make reactive forms of oxygen, which are necessary to destroy phagocytized bacteria.

Most primary immunodeficiencies are associated with defects in the components of the third line of defense—adaptive immunity. For example, some children fail to develop any lymphoid stem cells whatsoever, and as a result they produce neither B cells nor T cells and cannot mount immune responses. The resulting defects in the immune system cause **severe combined immunodeficiency disease (SCID),** which is discussed in **Highlight: SCID: "Bubble Boy" Disease**.

Other children suffer from T cell deficiencies alone. For example, **DiGeorge syndrome** results from a failure of the thymus to develop. Consequently, there are no T cells. The importance of T cells in protecting against viruses is emphasized by the observation that individuals with DiGeorge syndrome generally die of viral infections while remaining resistant to most bacteria. Physicians treat DiGeorge syndrome with a thymic stem cell transplant.

B cell deficiencies also occur in children. The most severe of the B cell deficiencies, called **Bruton-type agammaglobulinemia**[7] (ā-gam'ă-glob'ū-li-nē'mē-ă), is an inherited disease in which affected babies, usually boys, cannot make immunoglobulins. These children experience recurrent bacterial infections but are usually resistant to viral, fungal, and protozoan infections.

Inherited deficiencies of individual immunoglobulin classes are more common than a deficiency of all classes. Among these, IgA deficiency is the most common. Because affected children cannot produce secretory IgA, they experience recurrent infections in the respiratory and gastrointestinal tracts.

TABLE 18.5 summarizes the major primary immunodeficiency diseases.

Acquired Immunodeficiency Diseases

LEARNING | **OUTCOMES**

18.19 Describe five acquired conditions that suppress immunity.

18.20 Define *AIDS*, and differentiate a disease from a syndrome.

18.21 Describe HIV, including its structure, possible origin, replication cycle, and transmission.

18.22 Describe the relationship of helper the T cell population to the course of AIDS.

18.23 Describe the diagnosis, treatment, and prevention of AIDS, and list four behaviors that increase the risk of infection with HIV.

[7]*Agammaglobulinemia* means "an absence of gamma globulin (IgG)."

HIGHLIGHT

SCID: "Bubble Boy" Disease

As a result of the widely publicized case of David Vetter—a Houston, Texas, resident commonly known as the "bubble boy" because he lived from birth until age 12 in a sterile, plastic-enclosed environment—*severe combined immunodeficiency disease (SCID)* has been known as "bubble boy" disease since the 1970s. Like all SCID patients, David had no immune system—his body produced neither B cells nor T cells—so the slightest infection could be lethal. As a result, contact with him could occur only through a pair of antiseptic rubber gloves built into one of the walls of his plastic enclosure. At age 12, David underwent an experimental bone marrow transplant in the hope that marrow donated by his older sister would enable him to build up an immune system. Unfortunately, the donated cells turned out to be infected with Epstein-Barr virus (a common virus that causes mononucleosis), ultimately killing David.

Today, there is much more hope for patients with SCID. Virus-free bone marrow transplants are now possible. More recently, more than 20 children have undergone gene therapy, in which the gene for an enzyme (adenine deaminase) missing in SCID patients is inserted, via retroviral vectors, into clones of the patient's bone marrow cells. These genetically altered cells are then returned to the child, where they grow and synthesize sufficient enzyme to normalize the immune response. Though successful in several children studied in a French trial, the therapy was halted in 2002 after three children developed leukemia following treatment. Investigations into whether the illness was caused by the gene therapy are under way; however, scientists remain hopeful that gene therapy can achieve a cure for SCID—without side effects—in the near future.

David Vetter

Unlike inherited, primary immunodeficiency diseases, acquired immunodeficiency diseases affect older individuals who had a previously healthy immune system.

Acquired immunodeficiencies result from a number of causes. In all humans, the immune system (but especially T cell production) deteriorates with increasing age; as a result, older individuals normally have less effective immunity, especially cell-mediated immunity, than younger individuals, leading to an increased incidence of both viral diseases and certain types of cancer. Severe stress can also lead to immunodeficiencies by prompting the secretion of increased quantities of corticosteroids, which are toxic to T cells and thus suppress cell-mediated immunity. This is why, for example, cold sores may "break out" on the faces of students during final exams: the causative herpes simplex viruses, which are controlled by a fully functioning cell-mediated immune response, escape immune control in stressed individuals. Malnutrition and certain environmental toxins can also cause acquired immunodeficiency diseases by inhibiting the normal production of B cells and T cells.

Of course, the most significant example of the result of an acquired immunodeficiency is **acquired immunodeficiency syndrome (AIDS)**. From the time of its discovery in 1981 among homosexual males in the United States to its emergence as a worldwide pandemic, no affliction has affected modern life as much.

Signs and Symptoms of AIDS

AIDS is not a single disease but a **syndrome;** that is, a complex of signs, symptoms, and diseases associated with a common cause. Epidemiologists define AIDS as the presence of several

TABLE 18.5 Some Primary Immunodeficiency Diseases

Disease	Defect	Manifestation
Chronic granulomatous disease	Ineffective phagocytes	Uncontrolled infections
Severe combined immunodeficiency disease (SCID)	A lack of T cells and B cells	No resistance to any type of infection, leading to rapid death
Bruton-type agammaglobulinemia	A lack of B cells and thus a lack of immunoglobulins	Death from overwhelming bacterial infections
DiGeorge syndrome	A lack of T cells and thus no cell-mediated immunity	Death from overwhelming viral infections

▶ **FIGURE 18.14 Diseases associated with AIDS. (a)** Disseminated herpes. **(b)** Kaposi's sarcoma, a cancer of blood vessels.

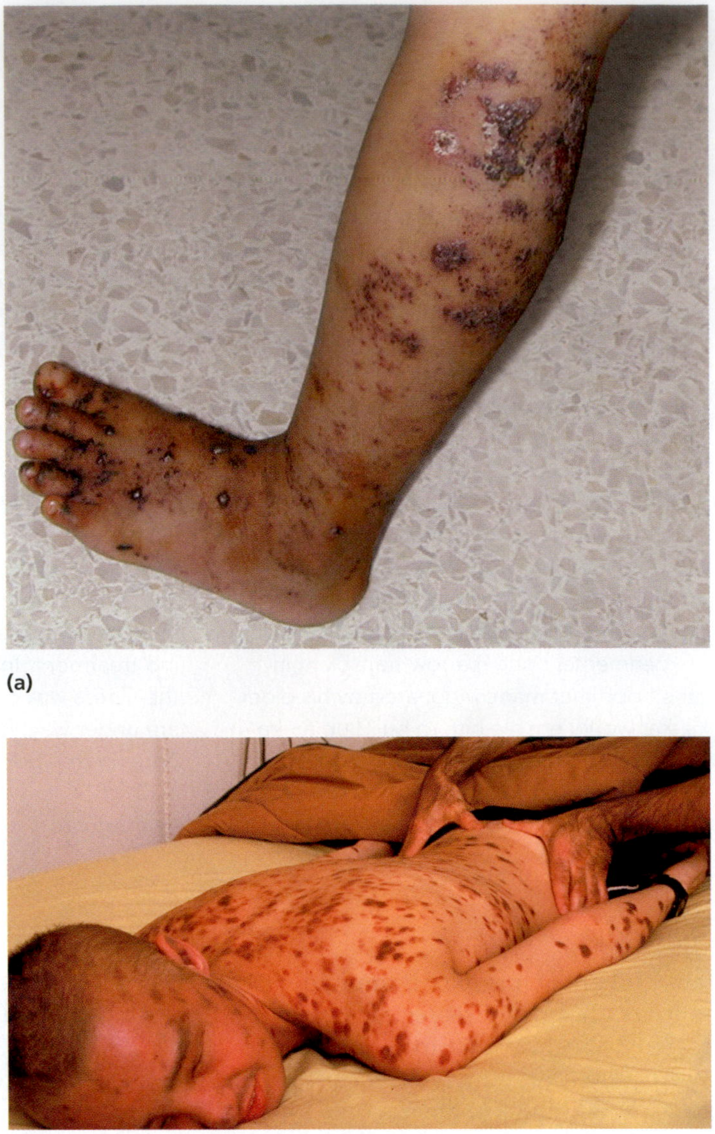

(a)

(b)

opportunistic or rare infections along with infection by human immunodeficiency virus (HIV), or as a severe decrease in the number of lymphocytes called helper T cells (<200 cells/μL blood) and a positive test showing the presence of HIV. Helper T cells are also known as CD4 cells, because the protein CD4 is found on their membranes. The defining diseases and infections of AIDS include diseases of the skin, such as shingles and disseminated (widespread) herpes **(FIGURE 18.14a)**; diseases of the nervous system, including meningitis, toxoplasmosis, and *Cytomegalovirus* (sī-tō-meg´ă-lō-vī´rŭs) disease; diseases of the respiratory system, such as tuberculosis, *Pneumocystis* (nō-mō-sis'tis) pneumonia, histoplasmosis, and coccidioidomycosis (kok-sid-ē-oy´dō-mī-kō-sis); and diseases of the digestive system, including chronic diarrhea, thrush, and oral hairy leukoplakia. A rare cancer of blood vessels called Kaposi's sarcoma is also commonly seen in AIDS patients **(FIGURE 18.14b)**. AIDS often results in dementia during the final stages.

TABLE 18.6 summarizes these and other diseases associated with AIDS; Chapters 19–24 deal with the signs, symptoms, diagnosis, and treatment of specific diseases in more detail.

AIDS Pathogen and Its Virulence Factors

Human immunodeficiency virus (HIV) causes AIDS, though it should be noted that infection with HIV is not AIDS; HIV infection typically leads to AIDS in untreated patients. HIV is an enveloped, positive single-stranded RNA (+ssRNA) virus of the type called *retroviruses,* because it uses **reverse transcriptase** to make a DNA copy of its genome. Luc Montagnier (1932–) and his colleagues at the Pasteur Institute discovered HIV in 1983. Note that the acronym HIV indicates that it is a virus. It is repetitive and incorrect to call this pathogen "HIV virus"; a more appropriate name is "AIDS virus," or

TABLE 18.6 Opportunistic Infections Associated with AIDS

Disease	Causative Agent	Organ Primarily Affected (Chapter Where Covered)
Coccidioidomycosis	*Coccidioides* (fungus)	Lung (22)
Cytomegalovirus disease	*Cytomegalovirus*	Brain (20), liver (23)
Diarrhea (severe and prolonged)	Various bacteria, *Cryptosporidium* (protozoan)	Intestines (23)
Herpes	*Herpesvirus*	Skin (19)
Histoplasmosis	*Histoplasma* (fungus)	Lung (22)
Kaposi's sarcoma	Human herpesvirus 8	Blood vessels (21)
Meningitis	*Cryptococcus* (yeast), *Listeria* (bacterium)	Brain and meninges (20)
Oral hairy leukoplakia	*Lymphocryptovirus* (Epstein-Barr virus)	Tongue (23)
Pneumonia	*Pneumocystis* (fungus)	Lung (22)
Shingles	*Varicellovirus*	Skin (19)
Thrush	*Candida* (yeast)	Mouth and tongue (23), vagina (24)
Toxoplasmosis	*Toxoplasma* (protozoan)	Lungs, liver, heart (21)
Tuberculosis	*Mycobacterium*	Lung (22)

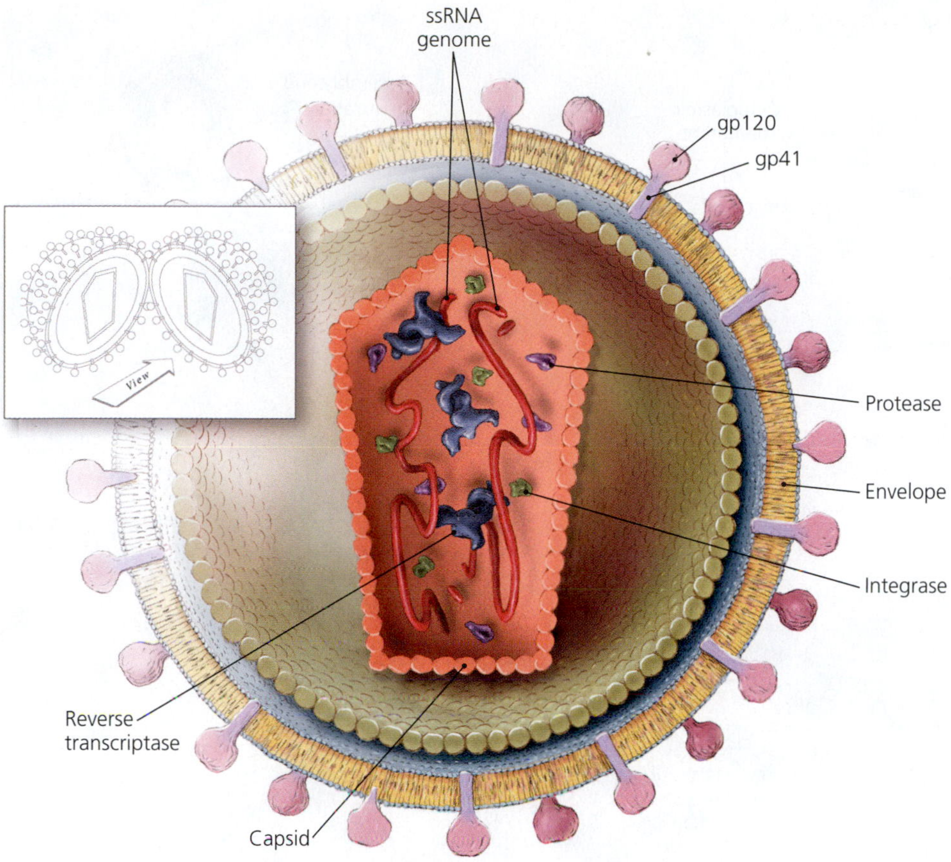

◀ **FIGURE 18.15 Artist's conception of HIV.** The virion is about 90 nm in diameter.

simply "HIV." Taxonomists classify HIV in the genus *Lentivirus,* family *Retroviridae.*

There are two major types of the virus. HIV-1 is more prevalent in the United States and Europe, and HIV-2 is more prevalent in West Africa. HIV-2, which shares about 50% of the nucleic acid sequence of HIV-1, reproduces more slowly than HIV-1. Researchers have studied HIV-1 more, so the following sections focus on this strain.

Structure of HIV HIV is a typical retrovirus in size and shape **(FIGURE 18.15)**. Two antigenic glycoproteins characterize its envelope. The larger glycoprotein, named **gp120**,[8] is the primary attachment molecule of HIV. Its antigenicity changes during the course of prolonged infection, making an effective antibody response against it difficult. The smaller glycoprotein, **gp41**, promotes fusion of the viral envelope to a target cell. The effects of these structural characteristics—antigenic variability and the ability to fuse with host cells—interfere with clearance of HIV from a patient.

Origin of HIV Evidence suggests that HIV arose from mutations of similar viruses—*simian immunodeficiency viruses (SIVs)*—found in African monkeys and chimpanzees; the nucleotide sequence of SIV is similar to that of HIV. Based on mutation rates and the rate of antigenic change in HIV, researchers estimate that HIV emerged in the human population in Africa about 1930. Scientists have identified antibodies against HIV in human blood stored since 1959, though they did not document the first cases of AIDS until 1981.

The relationship among the two types of HIV and strains of SIV is not clear. HIV-1 and HIV-2 may be derived from different strains of SIV, or one type of HIV may be derived from the other. We may never know the exact evolutionary relationships among the immunodeficiency viruses.

Replication of HIV

We can consider replication of HIV as occurring in eight steps **(FIGURE 18.16)**. We first look at the eight-step replication process in general and then examine details of attachment and entry.

1 *Attachment.* HIV primarily attaches to four kinds of cells: helper T cells; cells of the macrophage lineage, including monocytes, macrophages, and microglia (special phagocytic cells of the central nervous system); smooth muscle cells, such as those in arterial walls; and dendritic cells. HIV can also rarely infect nerve cells, liver cells, and some epithelial cells.

Additionally, B lymphocytes adhere to HIV that has been covered with complement proteins. Though HIV does not infect these B cells, the B cells deliver HIV to helper T cells, which then become infected. Similarly, infected macrophages can pass HIV to helper T cells.

[8]gp120 is short for "glycoprotein with a molecular weight of 120,000 daltons."

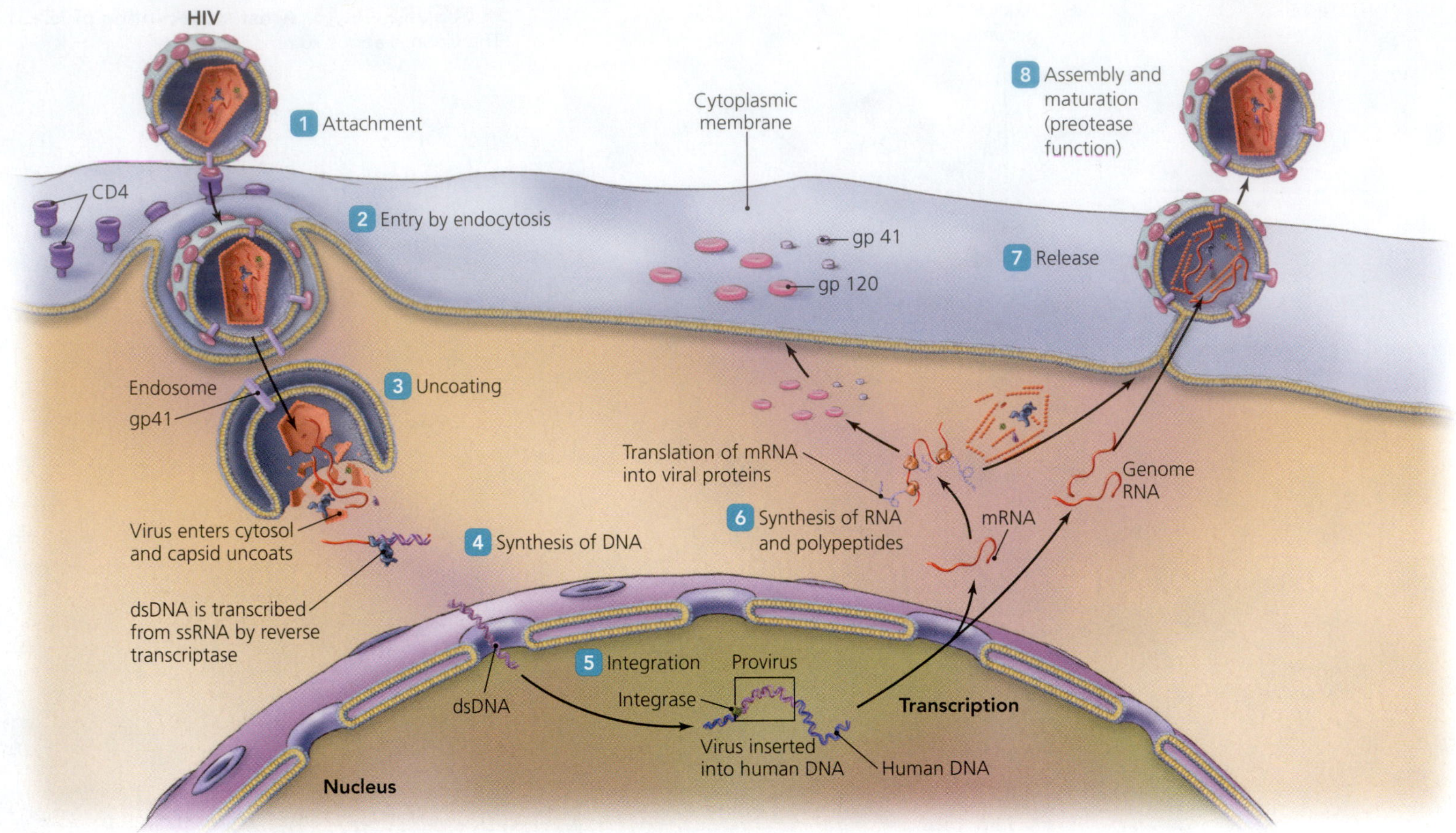

▲ **FIGURE 18.16 The replication cycle of HIV.** The artist's rendition depicts eight steps involved in the replication of the virus within a helper T cell. Photos show attachment and penetration (1–3) and the budding and release of a virion (4–6). *How does the replication of HIV differ from the replication of bacteriophage T4?*

Figure 18.16 Bacteriophage *T4 is a DNA virus; therefore, it does not have reverse transcription. Further, T4 does not integrate into a bacterium's chromosome; it assembles completely before being released from the cell; it has no envelope; and it does not carry enzymes.*

2 ***Entry.*** HIV triggers the cell to endocytose the virus; that is, the cell's cytoplasmic membrane forms a pocket and folds in, surrounding the virus and forming a vesicle with the virus inside.

3 ***Uncoating.*** The viral envelope fuses with the vesicle's membrane, and the intact capsid of HIV enters the cytosol. The virus then uncoats the capsid and releases its two ssRNA molecules from the capsid into the cell's cytoplasm.

4 ***Synthesis of DNA.*** Reverse transcriptase, which has been released from the capsid, synthesizes double-stranded DNA (dsDNA) using viral ssRNA as a template.

5 ***Integration.*** The dsDNA made by reverse transcriptase enters the nucleus and becomes part of a human DNA molecule. There it remains as a part of the cell for life—a condition known as **latency.**

6 ***Synthesis of RNA and polypeptides.*** An infected cell transcribes integrated HIV genes to produce messenger RNA and multiple copies of viral ssRNA that will act as genomes for new viruses. Ribosomes within the infected cell translate mRNA to make viral-encoded polypeptides. These include attachment proteins, integrase, and a large polypeptide composed of inactive reverse transcriptase and capsomeres. The attachment proteins are inserted in the host's cytoplasmic membrane.

7 ***Release.*** Two molecules of genomic RNA, molecules of tRNA, and several viral polypeptides bud from the host's cytoplasmic membrane to form an immature virion.

8 ***Assembly and maturation.*** HIV that buds from a cell is nonvirulent because its capsid is not fully functional, and reverse transcriptase is inactive. **Protease**, a viral enzyme packaged in the virion, cleaves the large polypeptide to release reverse transcriptase and capsomeres. This action of protease, which occurs only after the virus has budded

from the cell, allows final maturation of the viral capsid. HIV is now active.

Now, let us examine HIV replication in more detail.

Details of Attachment, Entry, and Uncoating

HIV uses gp120 and gp41 to attach to and enter target cells **(FIGURE 18.17)**. CD4[9] on a target cell's cytoplasmic membrane is the receptor for gp120; that is, gp120 attaches HIV to CD4 1. The CD4-gp120 complex binds to another membrane receptor, called *fusin* (also known as *CXCR4*), which triggers the cell's membrane to move out and surround the virus—a process called *endocytosis* 2. Endocytosis forms a bubble of membrane with the virus inside. This structure is called an *endosome* 3.

HIV remains intact within an endosome for about 30 minutes. Glycoprotein 41 on the viral envelope evidently then facilitates fusion of the envelope with the endosome membrane. The two fuse 4, and the viral capsid is introduced intact into the cell's cytosol 5. The viral capsid uncoats, releasing viral RNA and proteins 6.

Details of Synthesis and Latency

Reverse transcriptase, which was carried inside the capsid, becomes active in the cytosol. It uses tRNA as a primer to transcribe dsDNA from the ssRNA genome of the virus (see Figure 18.18). Reverse transcriptase is very error prone, making about five errors per genome. This generates multiple antigenic variations of HIV. Billions of variants may develop in a single patient over the course of the syndrome.

HIV is a latent virus. The dsDNA made by reverse transcriptase is known as a *provirus*, and it enters the nucleus. A viral enzyme known as **integrase** inserts the dsDNA provirus into a human chromosome. Once integrated, the provirus permanently remains part of the cellular DNA. It may remain dormant for years or be activated immediately, depending on its location in the human genome and the availability of promoter DNA sequences (see Figure 7.9). Infected macrophages and monocytes are major reservoirs of integrated HIV, and they serve as a means of distribution of the virus throughout the body.

An infected cell replicates integrated DNA every time cellular DNA is replicated. In this way, HIV ends up infecting all progeny of infected cells. An infected cell may also transcribe integrated HIV to make viral messenger RNA molecules as well as entire RNA copies of the whole HIV genome.

Details of Release, Assembly, and Maturation

HIV actively participates in its release from the cell. A viral protein selects a *lipid raft*—a region of regularly packed lipids—in the cytoplasmic membrane as the point of exit. As the virus blebs from the cell, components of the raft become the envelope of the virion. Once outside the cell, capsomeres organize to

[9]Scientists commonly name membrane proteins with letter-number combinations; CD4 stands for the fourth-discovered *cluster of differentiation*.

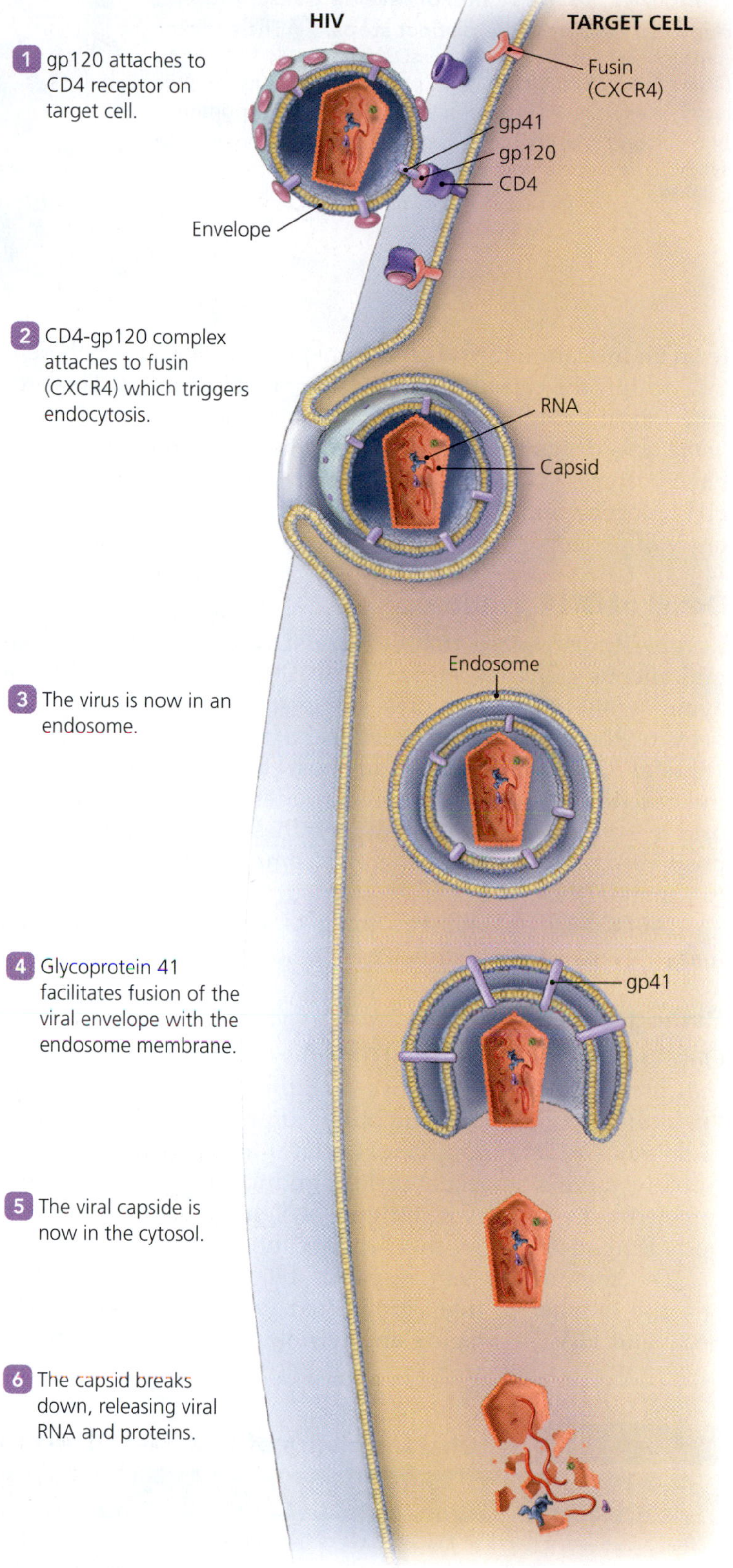
HIV
TARGET CELL
1 gp120 attaches to CD4 receptor on target cell.
Fusin (CXCR4)
gp41
gp120
CD4
Envelope
2 CD4-gp120 complex attaches to fusin (CXCR4) which triggers endocytosis.
RNA
Capsid
Endosome
3 The virus is now in an endosome.
4 Glycoprotein 41 facilitates fusion of the viral envelope with the endosome membrane.
gp41
5 The viral capside is now in the cytosol.
6 The capsid breaks down, releasing viral RNA and proteins.

▲ **FIGURE 18.17 The process by which HIV attaches to and enters a host cell.** 1 Binding of gp120 to a CD4 receptor on the cell membrane. 2 Removal of the CD4-gp120 complex, allowing gp41 to attach to the cytoplasmic membrane. 3 Fusion of the lipid bilayers, introducing the capsid into the cytoplasm.

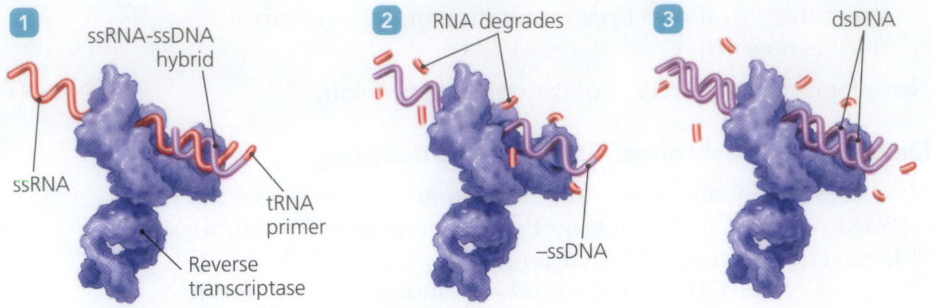

▶ **FIGURE 18.18 Action of reverse transcriptase, depicted here as three distinct steps.** 1 The enzyme transcribes a complementary −ssDNA molecule to form a DNA-RNA hybrid. It uses tRNA brought from a previous host cell as a primer for transcription. 2 The RNA portion of the hybrid is degraded, leaving −ssDNA. 3 The enzyme synthesizes a complementary +ssDNA strand, forming dsDNA.

form an immature capsid, and viral *protease* cleaves a polypeptide within the capsid to release functional proteins. The proteins cause the virus to mature and become infective. *Protease inhibitors*—drugs that interfere with the function of protease—have become a standard therapeutic agent in the treatment of HIV infection and have significantly lengthened the life expectancy of patients.

Detail of DNA Synthesis

Reverse transcriptase transcribes dsDNA in three nearly simultaneous steps, as shown in **FIGURE 18.18**. First, the enzyme synthesizes a negative single-stranded DNA (ssDNA) copy of the viral ssRNA, forming a ssDNA-ssRNA hybrid. Transfer RNA from the previous host cell serves as primer for this reaction. The enzyme then degrades the ssRNA of the hybrid to leave ssDNA and, finally, synthesizes a complementary positive strand of DNA, thereby completing the process of forming dsDNA.

TABLE 18.7 summarizes some of the characteristics that make HIV particularly difficult to combat.

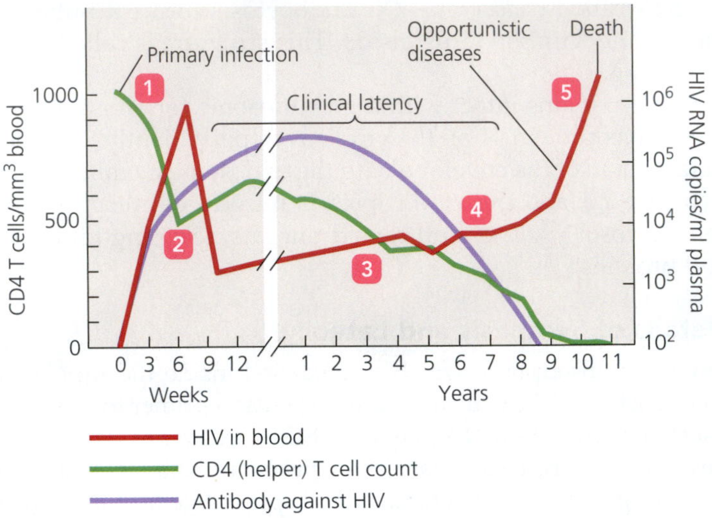

▲ **FIGURE 18.19 The course of AIDS follows the course of helper T cell destruction.** The circled numbers correspond to the steps described in the text.

Pathogenesis of AIDS

Only human cells replicate HIV effectively, and, as its name indicates, the virus destroys a human's immune system. **FIGURE 18.19** illustrates the observation that the destruction of helper T (CD4) cells directly relates to the course of AIDS. Initially, there is a burst of virion production and release from infected cells 1. Fever, fatigue, weight loss, diarrhea, and body aches accompany this primary infection.

The immune system responds by producing antibodies, and the number of free virions (red line) plummets 2. The body and HIV are waging an invisible war with few signs or symptoms. During this period, the body destroys almost a billion virions each day, but the viruses and cytotoxic T cells kill about 100 million CD4 cells. No specific symptoms accompany this stage, and the patient is often unaware of the infection.

Integrated viruses continue to replicate and virions are released into the blood to such an extent that the body cannot make enough helper T cells (green line) 3. Over the course of 5–10 years, the number of helper T cells declines to a level that severely impairs the immune response. The rate of antibody formation (purple line) falls precipitously as helper T cell function is lost.

TABLE 18.7 Characteristics of HIV That Challenge the Immune System

Characteristic	Effect(s)
Retrovirus with a genome that consists of two copies of +ssRNA	Reassortment of viral genes possible; reverse transcription produces much mutation and thus genetic variation; genome integrates into host's chromosome
Targets helper T cells especially, but also macrophages, dendritic cells, and muscle cells, and possibly liver, nerve, and epithelial cells	Permanently infects key cells of host's immune system
Antigenic variability	Numerous antigenic variations due to mutations helps virus evade host's immune response
Induces formation of syncytia	Increases routes of infection; intracellular site helps virus evade immune detection

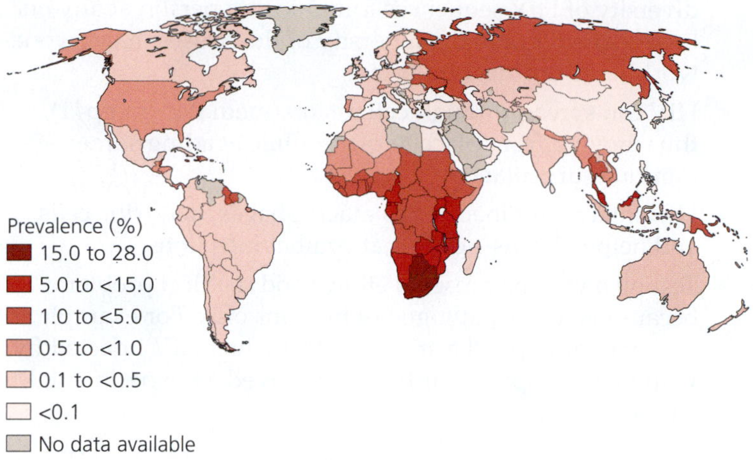

▲ **FIGURE 18.20 The global distribution of HIV/AIDS.** Figures indicate the prevalence of HIV/AIDS among children and adults as of the end of 2010. Note the alarmingly high number of people infected in sub-Saharan Africa.

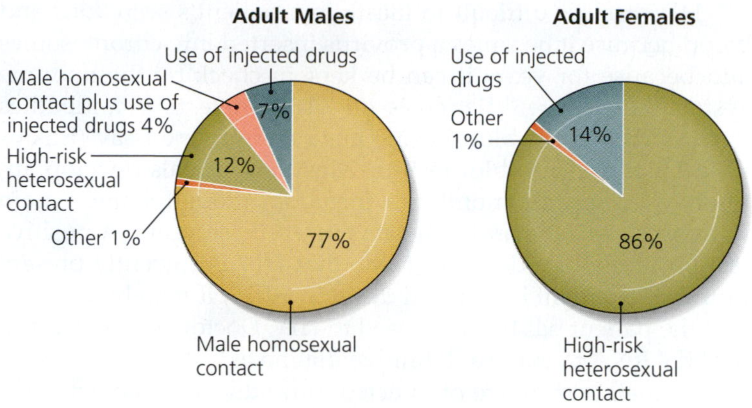

▲ **FIGURE 18.21 Modes of HIV transmission in people over 12 years of age in the United States during 2010.** Percentages represent the estimated proportions of HIV infections that result from each type of transmission. (Estimates are rounded to the nearest 1% and thus do not add up to 100%.)

HIV production climbs 4, and the patient dies 5. Many of the diseases associated with the loss of immune function in AIDS (see Table 18.6 on p. 544) are nonlethal infections in other patients, but AIDS patients cannot effectively resist them. Diseases such as Kaposi's sarcoma, disseminated herpes, toxoplasmosis, and *Pneumocystis* pneumonia occur rarely except in AIDS patients.

Epidemiology of AIDS

Epidemiologists first identified AIDS in young male homosexuals in the United States, but now AIDS is worldwide. The World Health Organization (WHO) estimated as of December 2011 that there were about 34 million HIV-infected people worldwide and that approximately 7000 new infections occur each day, including over 150 new cases in the U.S. daily. About a third of those infected have already developed AIDS; most of the rest will eventually develop the syndrome if they are untreated. The spread of AIDS in sub-Saharan Africa is particularly horrific **(FIGURE 18.20)**. The good news is that new infections, especially in children, are down about 22%.

All body secretions of AIDS patients contain HIV; however, viruses typically exist in sufficient concentration to cause infection only in blood, semen, vaginal secretions, and breast milk. Virions may be free or inside infected leukocytes. Infected blood contains 1000 to 100,000 virions per milliliter, while semen has 10 to 50 virions per milliliter. Other secretions have lower concentrations and are less infective than blood or semen.

Infected fluid must be injected into the body or encounter a tear or lesion in the skin or mucous membranes. Because HIV is only about 90 nm in diameter, a break in the protective membranes of the body allowing entry of HIV may be too small to bleed or to see or feel. Sufficient numbers of virions must be transmitted to target cells to establish an infection, though scientists do not know the exact number of virions required. The number probably varies with the strain of HIV and the overall health of the patient's immune system.

HIV is transmitted primarily via sexual contact (including vaginal, anal, and oral sex, either homosexual or heterosexual) and intravenous drug abuse **(FIGURE 18.21)**. Blood transfusions, organ transplants, tattooing, and accidental medical needlesticks also transmit HIV, though rarely. For example, the risk of infection from a needlestick injury involving an HIV-infected patient is less than 1%. Mothers also transmit HIV to their babies across the placenta and in breast milk; HIV infects approximately one-third of babies born to HIV-positive women.

Behaviors that increase the risk of infection include the following:

- Anal intercourse, especially receptive anal intercourse
- Sexual promiscuity; that is, sex with more than one partner
- Intravenous drug use
- Sexual intercourse with anyone in the previous three categories

The mode of infection is unknown for a small number of AIDS patients, who are often unable or unwilling to answer questions about their sexual and drug abuse history. The CDC has documented a few cases of casual spread of HIV, including infections from sharing razors and toothbrushes and from mouth-to-mouth kissing. In all cases of casual spread, researchers suspect that small amounts of the donor's blood may have entered abrasions in the recipient's mouth.

Diagnosis, Treatment, and Prevention

Physicians diagnose AIDS based on unexplained weight loss, fatigue, fever, and fewer than 200 CD4 lymphocytes per microliter (μl)[10] of blood combined with other signs and symptoms, which vary according to the diseases involved, and the demonstration of antibodies against HIV. Recall that by definition AIDS is the presence of one or more rare diseases and anti-HIV antibodies.

[10] The normal value for CD4 cells is 500–700 cells/μl.

HIV itself is difficult to locate in a patient's secretions and blood because it becomes a provirus inserted into chromosomes and because for years it can be kept in check by the immune system. Therefore, diagnosis involves detecting antibodies against HIV in the blood using enzyme immunoassay (EIA or ELISA), or Western blot testing. Most individuals develop antibodies within six months of infection, though some remain without detectable antibodies for up to three years. A positive test for antibodies does not mean that HIV is currently present or that the patient has or will develop AIDS; it merely indicates that the patient has been exposed to HIV. Doctors use a PCR test for HIV RNA to make a definitive diagnosis.

A small percentage of infected individuals, called *long-term nonprogressors,* do not develop AIDS even years or decades after infection. It appears that either these individuals are infected with defective virions, they have a mutated fusin receptor that does not bind effectively to HIV, or they have unusually well-developed immune systems.

Discovering ever more effective treatments for AIDS is an area of intense research and development. Currently physicians prescribe **antiretroviral therapy (ART),** previously known as *highly active antiretroviral therapy (HAART).* ART is a "cocktail" of three to four antiviral drugs to reduce viral replication, including nucleotide analogs, integrase inhibitors, protease inhibitors (for example, darunavir), attachment inhibitors (such as enfuvirtide), and reverse transcriptase inhibitors. ART is expensive and generally must be taken on a strict schedule. Studies indicate that the therapy stops the replication of HIV because strains of the virus are unlikely to develop resistance to all of the drugs simultaneously. As long as treatment continues, a patient can live a relatively normal life; however, treatment is not a cure, because the infection remains. Some scientists estimate that it would take 60 years on ART for all HIV-infected cells to die. About half the infected patients worldwide have access to ART. In addition to ART, physicians manage and treat individual diseases associated with AIDS on a case-by-case basis. Doctors can use a PCR test for HIV RNA to measure the effectiveness of ART.

Researchers are working to develop new methods that will prevent attachment of HIV to target cells, block the entry of HIV into cells, and halt HIV synthesis and release.

Progress in developing a vaccine against HIV has been disappointing. Among the problems that must be overcome in developing an effective vaccine are the following:

- A vaccine must generate both secretory antibody (IgA) to prevent sexual transmission and infection, and cytotoxic T lymphocytes (to eliminate infected cells).
- Induction of synthesis of gamma class antibodies (IgG), a necessary part of a vaccine, can actually be detrimental to a patient. Because IgG-viral complexes bind to B cells and also remain infective inside phagocytic cells, a vaccine must stimulate cellular immunity more than antibody immunity so as to kill infected cells.
- HIV has a high replication rate and is highly mutable, such that every nucleotide base among the viruses mutates to every other possible base every day. Scientists estimate the diversity of HIV sequences found in one person at any one time is greater than the diversity of flu viruses in everyone worldwide during a full year.
- HIV can spread through cell fusion (mediated by gp41), thus moving from place to place while evading some immune surveillance.
- HIV infects and inactivates macrophages, dendritic cells, and helper T cells—cells that combat infections.
- Testing a vaccine presents ethical and medical problems because HIV is a pathogen of humans only. For example, researchers stopped a study in 1994 when HIV infected five volunteers despite their having received an experimental vaccine.

Scientists have developed a vaccine that protects monkeys from SIV disease but have not been able to translate this success into an effective vaccine for humans, though a vaccine trial concluded in 2013 has shown some efficacy. In the meantime, individuals can slow the AIDS epidemic with personal decisions:

- Abstinence and mutually faithful monogamy between uninfected individuals are the only truly safe sexual behaviors. Studies have shown that whereas condoms reduce a heterosexual individual's risk of acquiring HIV by about 69%, the benefit of condom usage to a population can be undone by an overall increase in sexual activity that may result from a false sense of security that condoms provide safe sex.
- Use of new, clean needles and syringes for all injections, as well as caution in dealing with sharp, potentially contaminated objects, can reduce HIV infection rates. If clean supplies are not available, 10% household bleach deactivates HIV on surfaces uncontaminated by a large amount of organic material (blood, mucus).
- Antiviral drugs given to pregnant women have reduced transfer of HIV across the placenta and in breast milk. Generally, HIV-infected mothers should not breast-feed their infants, though feeding with formula made with contaminated water is often more dangerous, triggering life-threatening diarrhea. WHO recommends formula only when it is "acceptable, feasible, affordable, sustainable, and safe."
- Screening blood, blood products, and organ transplants for HIV and anti-HIV antibodies has virtually eliminated the risk of HIV infection from these sources.
- Proper use of gloves, protective eyewear, and masks can prevent contact with infected blood.
- Researchers have determined that men who are circumcised reduce their risk of infection through sexual activity by at least 60%. Circumcision lowers their partner's risk by 30%.
- For many individuals, pre-exposure prophylaxis (PrEP), which involves taking a daily oral dose of an antiviral drug (tenofovir), prevents sexual acquisition of HIV.

- Women who apply 1% tenofovir gel vaginally before and after sex significantly reduce the chance of infecton from HIV-positive men.

Clinical Case Study: A Case of AIDS examines a clinical case in which a diagnosis of AIDS would be considered.

TELL ME WHY

Why is it unlikely that a medical treatment will be able to cure an HIV-infected person, that is, rid the body's cells of HIV?

CLINICAL CASE STUDY

A Case of AIDS

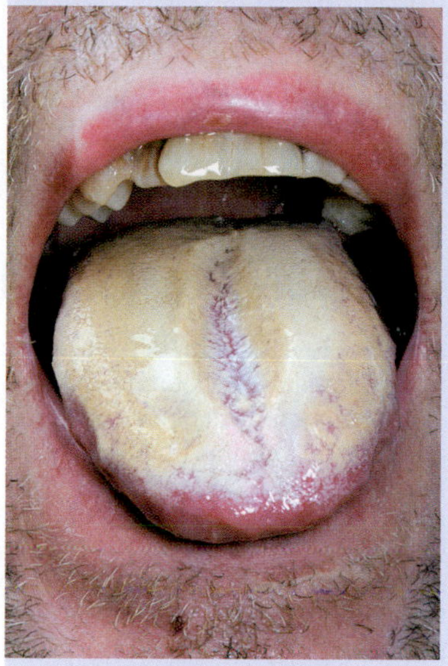

A 25-year-old man is admitted to the hospital with thrush, diarrhea, unexplained weight loss, and difficulty in breathing. Cultures of pulmonary fluid reveal the presence of *Pneumocystis*. The man admits to being a heroin addict and to sharing needles in a "shooting gallery."

1. What laboratory tests could confirm a diagnosis of AIDS?
2. How did the man most likely acquire an HIV infection?
3. What changes to the man's immune system allowed the opportunistic infections of *Pneumocystis* to arise?
4. What precautions should health care providers take in handling blood?

MICRO IN THE CLINIC FOLLOW-UP

A Simple Case of Bug Bites?

As soon as the doctor sees Ethan, she says, "Looks like you had a run-in with poison ivy." She prescribes a glucocorticoid ointment for applying to the skin rash and recommends continuing the antihistamine to control the agonizing itch. The pus that scared Maureen is due not to the poison ivy, but to a secondary bacterial infection caused by Ethan's scratching. Antibiotics clear it up quickly.

What triggers a poison ivy rash? Urushiol, an oily substance in poison ivy sap. When urushiol contacts the skin, it is quickly absorbed and combines with membrane proteins. T lymphocytes respond to urushiol as though pathogenic microbes had invaded. This cell-mediated immune response gives rise to long-lived memory T lymphocytes, on patrol waiting for the next "invasion" by urushiol. Ethan probably became sensitized to urushiol, setting the stage for a delayed-type or cell-mediated hypersensitivity reaction to future encounters. The aggressive attempts of T lymphocytes and macrophages to eliminate urushiol create the inflammation of this allergic contact dermatitis.

Although Ethan did not venture into the poison ivy–rich underbrush that afternoon by the lake, Daisy did. When she fetched the ball, urushiol stuck to her fur and then transferred to Ethan's skin when he hugged her. Daisy didn't get poison ivy because her fur blocked urushiol from reaching her skin. Ethan, as a relatively hairless human, was less fortunate. The doctor explains that it's important to give Daisy a bath after she's been in the woods.

1. **How do memory T lymphocytes sensitized to urushiol cause inflammation?**
2. **Why doesn't epinephrine relieve the inflammation of allergic contact dermatitis?**
3. **Why is the inflammatory response to urushiol delayed for at least 12 to 48 hours after exposure?**

MM Check your answers to Micro in the Clinic Follow-Up questions in the MasteringMicrobiology Study Area.

CHAPTER SUMMARY

1. Immunological responses may give rise to inflammatory reactions called **hypersensitivities.** An immunological attack on normal tissues gives rise to autoimmune disorders. A failure of the immune system to function normally may give rise to immunodeficiency diseases.

Hypersensitivities (pp. 527–540)

1. Type I hypersensitivity gives rise to **allergies.** Antigens that trigger this response are called **allergens.**
2. Allergies result when allergens bind to IgE molecules that are already bound to **mast cells, basophils,** and **eosinophils.** This causes the sensitized cells to degranulate and release **histamine, kinins, proteases, leukotrienes,** and **prostaglandins.**
3. Depending on the amount of these molecules and the site at which they are released, the result can produce various clinical syndromes, including **hay fever, asthma, urticaria** (hives), or various other allergies. When the inflammatory mediators exceed the body's coping mechanisms, **acute anaphylaxis (anaphylactic shock)** may occur. The specific treatment for anaphylaxis is epinephrine.
4. Type I hypersensitivity can be diagnosed by skin testing and can be partially prevented by avoidance of allergens and immunotherapy. Some type I hypersensitivities are treated with **antihistamines.**
5. Type II cytotoxic hypersensitivities, such as incompatible blood transfusions and hemolytic disease of the newborn, result when cells are destroyed by an immune response.
6. Red blood cells have **blood group antigens** on their surface. If incompatible blood is transfused into a recipient, a severe transfusion reaction can result.
7. The most important of the blood group antigens is the ABO group, which is largely responsible for transfusion reactions.
8. Approximately 85% of the human population carries **Rh antigen,** which is also found in rhesus monkeys. If an Rh-negative pregnant woman is carrying an Rh-positive fetus, the fetus may be at risk of **hemolytic disease of the newborn,** in which antibodies made by the mother against the Rh antigen may cross the placenta and destroy the fetus's red blood cells. RhoGAM administered to pregnant Rh-negative women may prevent this disease.
   ▶**VIDEO TUTOR:** *Hemolytic Disease of the Newborn*
9. Drugs bound to blood platelets may subsequently bind antibodies and complement, causing **immune thrombocytopenic purpura,** in which the platelets lyse.
10. In type III hypersensitivity, excessive amounts of **immune complexes** are deposited in tissues, where they cause significant tissue damage. Immune complexes deposited in the lung cause a **hypersensitivity pneumonitis,** of which the most common example is farmer's lung. If large amounts of immune complexes form in the bloodstream, they may be filtered out by the glomeruli of the kidney, causing **glomerulonephritis,** which can result in kidney failure.
11. In **rheumatoid arthritis (RA),** immune complexes result in the growth of inflammatory tissue within joints.
12. **Systemic lupus erythematosus (SLE,** lupus) is a systemic, autoimmune, type III hypersensitivity in which autoantibodies bind to many autoantigens, especially the patient's DNA.
13. Type IV hypersensitivity, also known as **delayed hypersensitivity reaction,** is a T cell–mediated inflammatory reaction that takes 24–72 hours to reach maximal intensity.
14. A good example of a delayed hypersensitivity reaction is the **tuberculin response,** generated when tuberculin, a protein extract of *Mycobacterium tuberculosis,* is injected into the skin of an individual who has been infected with or vaccinated against *M. tuberculosis.*
15. Another example of a type IV hypersensitivity reaction is **allergic contact dermatitis,** which is T cell–mediated damage to chemically modified skin cells. The best-known example is a reaction to poison ivy.
16. An organ or tissue **graft** can be made between different sites within a single individual (an **autograft**), between genetically identical individuals (an **isograft**), between genetically dissimilar individuals (an **allograft**), or between individuals of different species (a **xenograft**). Most surgical organ grafting involves allografts, which, if not treated with immunosuppressive drugs, lead to **graft rejection.**
17. In **graft-versus-host disease,** an organ donor's cells attack the recipient's body.
18. Commonly used immunosuppressive drugs include **glucocorticoids, cytotoxic drugs, cyclosporine,** and lymphocyte-depleting therapies, which involve treatment with antibodies against T cells or their receptors.

Autoimmune Diseases (pp. 540–542)

1. **Autoimmune diseases** may result when an individual begins to make autoantibodies or cytotoxic T cells against normal body components.

2. There are many hypotheses concerning the cause of autoimmune disease. One involves **molecular mimicry,** in which microorganisms with epitopes similar to self-antigens trigger autoimmune tissue damage. Others implicate estrogen, pregnancy, and environmental and genetic factors.
3. One group of autoimmune diseases involve only a single organ or cell type. Examples of such diseases include **autoimmune hemolytic anemia, type 1 diabetes mellitus, Graves' disease, multiple sclerosis (MS),** and rheumatoid arthritis.
4. A second group of autoimmune diseases, such as systemic lupus erythematosus, involves multiple organs or body systems.

Immunodeficiency Diseases (pp. 542–551)

1. Immunodeficiency diseases may be classified as **primary immunodeficiency diseases,** which result from mutations or developmental anomalies and occur in young children, and **acquired (secondary) immunodeficiency diseases,** which result from other known causes such as viral infections.
2. Examples of primary immunodeficiency diseases include **chronic granulomatous disease,** in which a child's neutrophils are incapable of killing ingested bacteria. Inability to produce both T cells and B cells is called **severe combined immunodeficiency disease.** In **DiGeorge syndrome,** the thymus fails to develop. In **Bruton-type agammaglobulinemia,** B cells fail to function.
3. **Acquired immunodeficiency syndrome (AIDS)** is a condition defined by the presence of antibodies against **human immunodeficiency virus (HIV)** in conjunction with certain opportunistic infections or by HIV and a CD4 cound below 200 cells/μL of blood. A **syndrome** is a complex of signs and symptoms with a common cause.
4. Human immunodeficiency viruses (HIV-1 or HIV-2) destroy the immune system. HIV is characterized by glycoproteins such as **gp120** and **gp41,** which enable attachment and have significant antigenic variability. The destruction of the immune system results in vulnerability to any infection.
5. HIV converts its genome into double-stranded DNA, which is integrated permanently into the host DNA—a condition called **latency**.
6. HIV affects helper T cells, macrophages, and dendritic cells. Using gp120, HIV fuses to a cell, uncoats, enters the cell, and transcribes dsDNA to become a provirus, which inserts into a cellular chromosome. After HIV is replicated and released, an internal viral enzyme, **protease,** cleaves a polypeptide, allowing HIV to assemble and become virulent.
7. HIV progressively kills infected cells. The body fights this onslaught but eventually succumbs, allowing opportunistic pathogens to proliferate.
8. Sexual activity and sharing needles are the main ways individuals spread HIV.
9. **Antiretroviral therapy (ART)** is a combination of antiviral drugs for the treatment of AIDS.
10. No effective vaccine against AIDS exists. Prevention of infection primarily depends on faithful mutual monogamy, proper condom usage, use of sterile needles, and screening blood and donated organs for HIV. HIV-positive mothers should not breast-feed their children.

QUESTIONS FOR REVIEW

Answers to the Questions for Review (except Short Answer questions) begin on p. A-1.

Multiple Choice

1. The immunoglobulin class that mediates type I hypersensitivity is
   a. IgA.
   b. IgM.
   c. IgG.
   d. IgD.
   e. IgE.
2. The major inflammatory mediator released by degranulating mast cells in type I hypersensitivity is
   a. immunoglobulin.
   b. complement.
   c. histamine.
   d. interleukin.
   e. prostaglandin.
3. Hemolytic disease of the newborn is caused by antibodies against which major blood group antigen?
   a. MHC protein
   b. MN antigen
   c. ABO antigen
   d. rhesus antigen
   e. type II protein
4. Farmer's lung is a hypersensitivity pneumonitis resulting from
   a. a type I hypersensitivity reaction to grass pollen.
   b. a type II hypersensitivity to red cells in the lung.
   c. a type III hypersensitivity to mold spores.
   d. a type IV hypersensitivity to bacterial antigens.
   e. none of the above.
5. A positive tuberculin skin test indicates that a patient *not* immunized against tuberculosis
   a. is free of tuberculosis.
   b. is shedding *Mycobacterium*.
   c. has been exposed to tuberculosis antigens.
   d. is susceptible to tuberculosis.
   e. is resistant to tuberculosis.
6. Which of the following is an autoimmune disease?
   a. a heart attack
   b. acute anaphylaxis
   c. farmer's lung
   d. graft-versus-host disease
   e. systemic lupus erythematosus

7. When a surgeon conducts a cardiac bypass operation by transplanting a piece of vein from a patient's leg to the same patient's heart, this is
   a. a rejected graft.
   b. an autograft.
   c. an allograft.
   d. a type IV hypersensitivity.
   e. a cardiograft.
8. A deficiency of both B cells and T cells is most likely
   a. a secondary immunodeficiency.
   b. a complex immunodeficiency.
   c. an acquired immunodeficiency.
   d. a primary immunodeficiency.
   e. an induced immunodeficiency.
9. HIV attaches to a cell receptor called
   a. gp41.
   b. an endosome.
   c. CD4.
   d. protease.
   e. gp120.
10. What do medical personnel administer to counteract various type I hypersensitivities?
    a. antihistamine
    b. bronchodilator
    c. corticosteroid
    d. epinephrine
    e. all of the above
11. Which of the following is *not* typically part of an ART cocktail for treating AIDS?
    a. protease inhibitor
    b. corticosteroids
    c. reverse transcriptase inhibitor
    d. nucleotide analog
12. Which of these HIV proteins binds to CD4 on human cells?
    a. integrase
    b. protease
    c. gp41
    d. gp120
13. A provirus is
    a. a virus inserted into its host's chromosome.
    b. any effective virus.
    c. a type of syncytium.
    d. active only when housed in its capsid.
14. Human immunodeficiency viruses most likely arose from
    a. genetically engineered human virus.
    b. reverse transcription.
    c. mutation of protease.
    d. simian immunodeficiency viruses.
15. Protease is an enzyme required by HIV to
    a. integrate the viral genome into the host chromosome.
    b. enter human cells.
    c. replicate the viral genome.
    d. cleave a polypeptide so as to release reverse transcriptase.

Modified True/False

Indicate whether each statement is true or false. If the statement is false, change the underlined word or phrase to make the statement true.

1. _____ <u>Cyclosporine</u> is released by degranulating mast cells.
2. _____ <u>Type III</u> hypersensitivity reactions may lead to the development of glomerulonephritis.
3. _____ ABO blood group antigens are found on <u>nucleated</u> cells.
4. _____ The tuberculin reaction is a <u>type I</u> hypersensitivity.
5. _____ Graft-versus-host disease can follow a bone marrow <u>isograft</u>.

Matching

Match the immune system complication in the first column with the types of hypersensitivities in the second column. Choices may be used more than once or not at all.

1. _____ Acute anaphylaxis	A. Type I hypersensitivity
2. _____ Allergic contact dermatitis	B. Type II hypersensitivity
3. _____ Systemic lupus erythematosus	C. Type III hypersensitivity
4. _____ Allograft rejection	D. Type IV hypersensitivity
5. _____ AIDS	E. Not a hypersensitivity
6. _____ Graft-versus-host disease	
7. _____ Milk allergy	
8. _____ Farmer's lung	
9. _____ Asthma	
10. _____ Hay fever	

Short Answer

1. Why is AIDS more accurately termed a "syndrome" rather than a mere disease?
2. Why is a child born to an Rh+ mother not susceptible to Rh-related hemolytic disease of the newborn?
3. Why is a person who produces a large amount of IgE more likely to experience anaphylactic shock than a person who instead produces a large amount of IgG?
4. Contrast autografts, isografts, allografts, and xenografts.
5. Compare and contrast the functions of four classes of immunosuppressive drugs.

VISUALIZE IT!

1. Label the four types of grafts on the figure below.

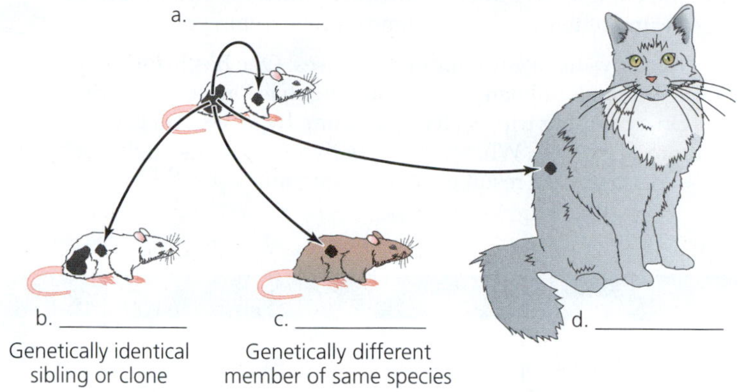

2. Label the replication cycle of HIV.

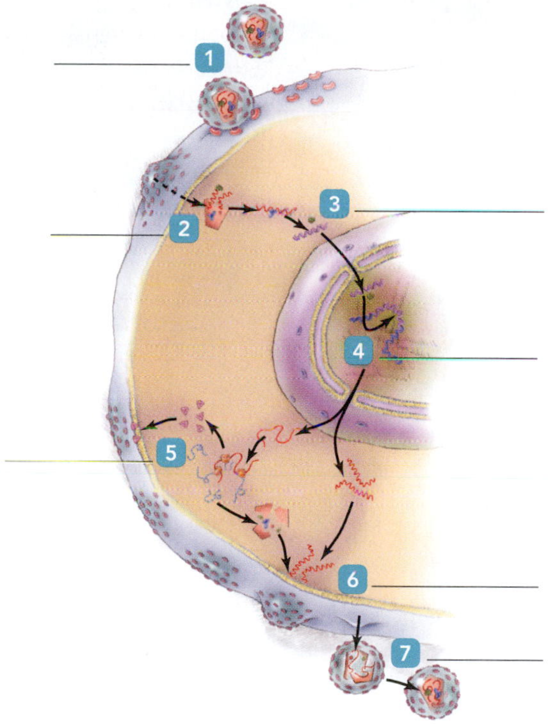

3. Identify the type of hypersensitivity reaction in each photo.

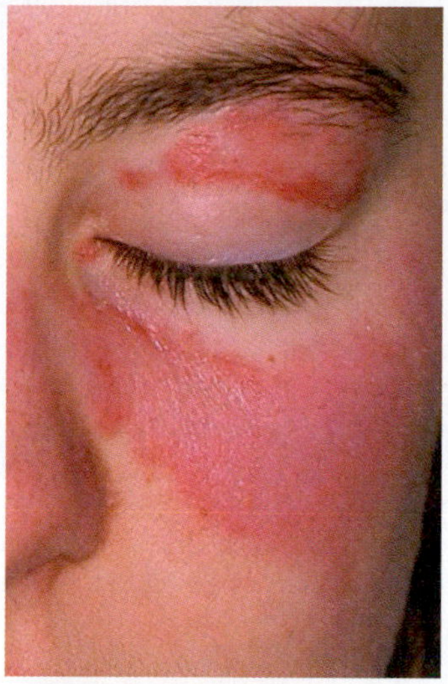

a. ______________________________

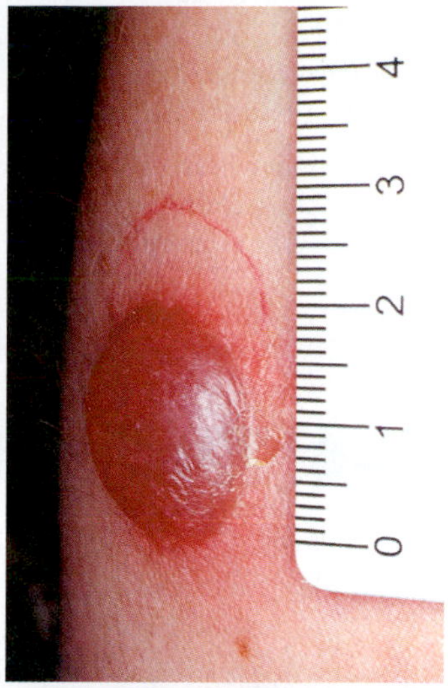

b. ______________________________

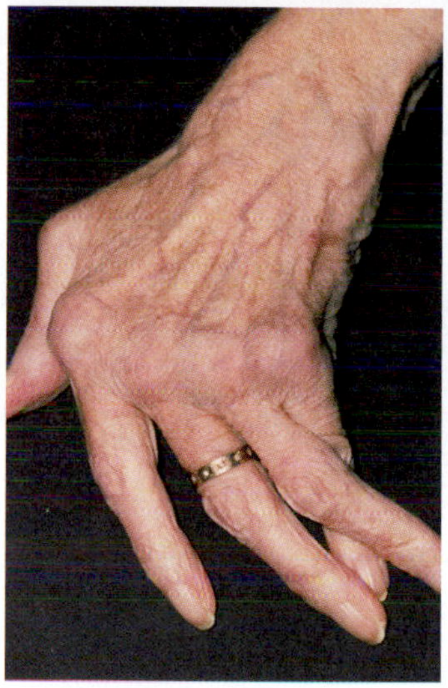

c. ______________________________

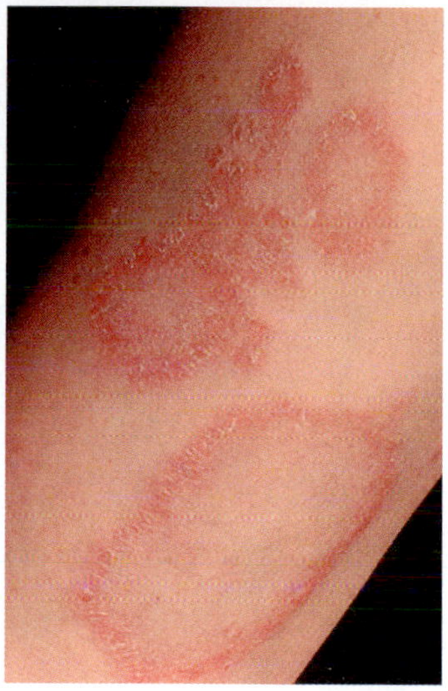

d. ______________________________

CRITICAL THINKING

1. What possible advantages might an individual gain from making IgE?
2. Why can't physicians use skin tests similar to the tuberculin reaction to diagnose other bacterial diseases?
3. In both Graves' disease and juvenile-onset diabetes mellitus, autoantibodies are directed against cytoplasmic membrane receptors. Speculate on the clinical consequences of an autoimmune response to estrogen receptors.
4. In general, people with B cell defects acquire numerous bacterial infections, whereas those with T cell defects get viral diseases. Explain why this is so.
5. What types of illnesses cause death in patients with combined immunodeficiencies or AIDS?
6. Because of the severe shortage of organ donors for transplants, many scientists are examining the possibility of using organs from nonhuman species such as pigs. What special clinical problems might be encountered when these xenografts are used?
7. Why do the blisters of positive tuberculin reactions resemble the blisters of poison ivy?
8. Retroviruses such as HIV use RNA as a primer for DNA synthesis. Why is a primer necessary?

9. Recently scientists synthesized a chemical that inhibits fusin. What effect might this chemical have on infection with HIV?
10. Reverse transcriptase is notoriously sloppy in making DNA copies from RNA templates. How do retroviruses survive when their genomes are subjected to such shoddy replication?
11. A patient arrives at the doctor's office with a rash covering her legs. How could you determine whether the rash is a type I or a type IV hypersensitivity?
12. A 43-year-old woman has been diagnosed with rheumatoid arthritis. Unable to find relief from her symptoms, she seeks treatment from a doctor in another country who injects antibodies and complement into her afflicted joints. Do you expect the treatment to improve her condition? Explain your reasoning.
13. Two boys have autoimmune diseases: One has Bruton-type agammaglobulinemia, and the other has DiGeorge syndrome. On a camping trip, each boy is stung by a bee, and each falls into poison ivy. What hypersensitivity reactions might each boy experience as a result of his camping mishaps?

CONCEPT MAPPING

Using the following terms, draw a concept map that describes immediate hypersensitivity. For a sample concept map, see p. 94.
Or, complete this and other concept maps online by going to the MasteringMicrobiology Study Area.

Allergens
Allergy symptoms
Anaphylactic shock
Anaphylaxis
Antihistamine
Asthma
Basophils
Bee venom
Dust mites
Epinephrine
Hay fever
Histamine
IgE
Increased vascular permeability
Localized
Mast cells
Peanuts
Pollen
Smooth muscle contraction
Systemic
Type I hypersensitivity
Urticaria
Vasodilation

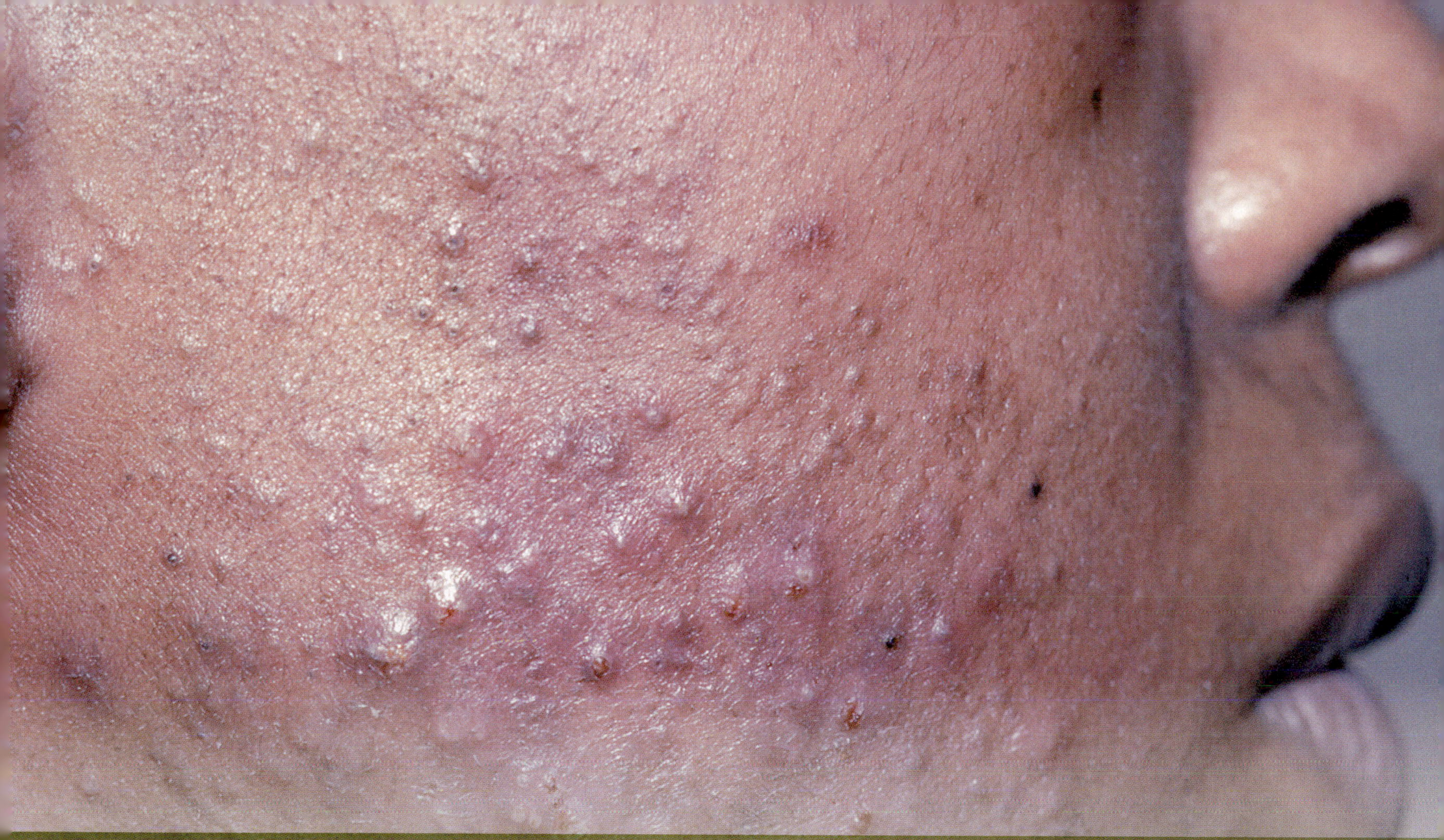

19 Microbial Diseases of the Skin and Wounds

MICRO IN THE CLINIC

A Bad Case of Acne

Thirteen-year-old Rockeem cannot believe the face staring back at him in the mirror. He has pimples everywhere: on his forehead, nose, cheeks, and chin. What's more, the problem isn't limited to his face. His neck and back are inflamed with angry red pustules and cysts as well. At school, the other kids make fun of him, especially during gym class. Junior high school is hard enough—Rockeem is suddenly feeling awkward around girls and drifting apart from some of his best friends from childhood, but the acne problem is the one thing Rockeem worries about the most. Some days, he is so embarrassed by his appearance that he doesn't want to go to school at all.

Rockeem wonders whether his acne is caused by something he is doing. He likes fast food, especially french fries—could grease be causing his pimples? He also likes to play soccer and sweats a lot—could sweat cause acne? When he asks his mom, she tells him not to worry and that he'll grow out of it. As the weeks pass, however, Rockeem grows more and more miserable. Rockeem's mom, worried about her son, schedules an appointment with a dermatologist to see what can be done.

Is there any way to control Rockeem's adolescent acne? Turn to the end of the chapter (p. 594) to find out.

MM **Explore More:** Test your readiness and apply your knowledge with dynamic learning tools at MasteringMicrobiology.

Structure of the Skin

LEARNING | OUTCOME

19.1 Describe the distinctive features of the main two layers of skin and of the hypodermis.

Skin is a marvelous, flexible, tough membrane that is also called the cutaneous membrane. Skin does much more than keep the internal organs together. Your skin prevents excessive water loss, helps regulate body temperature via the production of sweat and the dilation or constriction of blood vessels, assists in the formation of vitamin D, and is involved in sensory phenomena. The skin also has numerous physical and chemical properties that make it a significant barrier against microbial invaders. It limits infection and disease unless it has been burned, broken, cut, or in some other way wounded.

An adult's skin covers about 2 square meters, making it one of the larger organs of the body. Human skin varies in thickness from 0.05 mm on the lips to 4.0 mm thick on the soles of the feet and in calluses, which cover areas of greatest wear and tear.

Skin is composed of two distinct layers—the deeper **dermis**[1] is covered by a superficial (outer) **epidermis**[2] **(FIGURE 19.1)**. The dermis is a tough, leathery structure composed of loosely packed cells, connecting protein fibers, small muscles, sweat glands, sebaceous (oil) glands, blood vessels, nerve endings, and hair follicles, which produce hairs that grow up through the

[1]Greek, meaning "skin."
[2]From Greek *epi*, meaning "on."

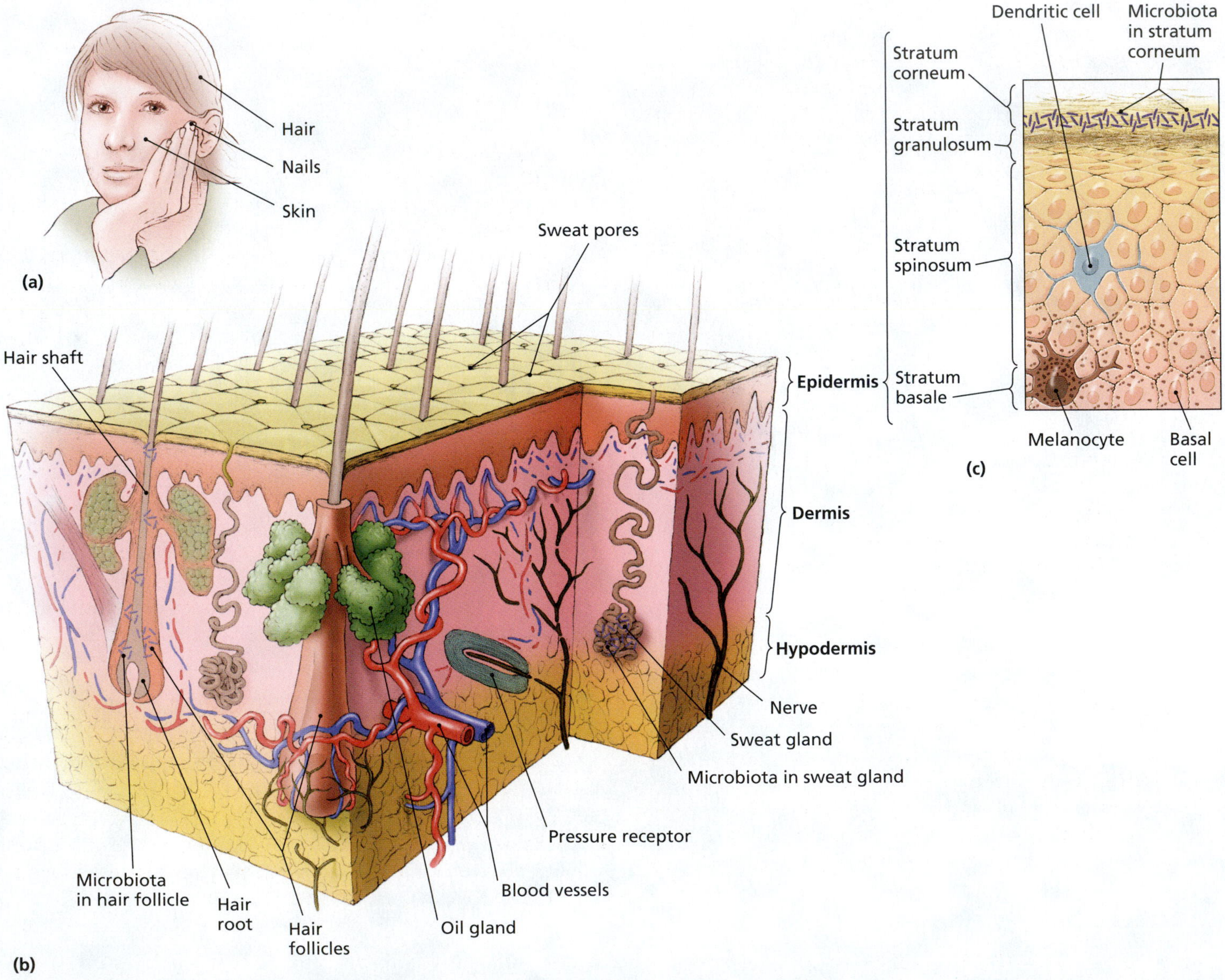

▲ **FIGURE 19.1 The skin. (a)** Hair and nails are accessory structures of the skin. **(b)** Skin is composed of the outer *epidermis* and the deeper *dermis*. In addition, a layer of fat cells called the *hypodermis* lies beneath the dermis. **(c)** Detail of the epidermis.

dermis and epidermis. The dermis provides strength and flexibility and, along with its blood vessels, supports the growth of the overlying epidermis.

The bloodless epidermis consists of four to five layers, each composed of layers of tightly packed cells. *Basal cells* of the epidermis, which adjoin the dermis, divide continuously, pushing their offspring toward the surface. As the daughter cells are pushed away from the dermal blood supply they flatten and die, but not before modifying themselves. They absorb *melanin,* a pigment that is secreted by other cells in the basal layer and that gives skin its color. The more melanin, the darker the skin. Epidermal cells also fill with a waterproofing protein called *keratin.* A hardened form of keratin forms nails and hairs, which are accessory structures to the skin. Between the dying cells, defensive cells called *dendritic cells* phagocytize microbes that penetrate into the deeper layers of the epidermis and deliver microbial antigens to defensive *lymphocytes.*

The surface of the skin is a generally inhospitable environment covered with salt (left behind as sweat evaporates) and *sebum*[3]—an oily lipid secreted by oil (sebaceous) glands in the dermis. Chemicals in sweat and sebum are antimicrobial, preventing the growth of many microorganisms. Thus, the outermost layer of skin consists of flattened, dead, dry, keratinized cells covered with oil and salt—a significant barrier to microbial invasion. Further, the skin sloughs off microbes attached to the outermost skin flakes as epidermal cells are continually pushed up from the basal layer. The body replaces the outer layers of the epidermis about once a month.

The *hypodermis* is a layer of fat cells and fibers lying beneath the dermis. The hypodermis is not technically part of the skin; it is subcutaneous. Stored fat provides cushioning, insulation, and a ready energy source. The fibers in the hypodermis anchor the skin to the underlying tissues.

Wounds are trauma to any tissue of the body. Cuts, abrasions, scrapes, surgery, inoculations, bites, and other penetrating skin wounds, as well as burns, breach the significant mechanical barrier provided by intact epidermis and dermis, allowing microbes to infect the warm, moist, deeper tissues of the body. Dirty wounds provide platforms for the growth of microbes and development of biofilms. Within a wound, microbes can multiply, producing enzymes and toxins that enhance their growth to the detriment of the host.

The body starts to heal a wound by forming a blood clot to temporarily close the breach; then, neighboring connective and epithelial cells multiply and grow into the clot to restore the skin's integrity. In most cases, other body defenses, including phagocytosis, complement, and inflammation, eliminate a wound infection. However, some infections overwhelm the body's defenses, resulting in severe, even fatal, diseases.

TELL ME WHY

Why is the surface of the skin inhospitable for most microbes?

[3]From Latin *sebum,* meaning "grease."

Normal Microbiota of the Skin

LEARNING | **OUTCOMES**

19.2 Define *microbiota,* and describe the major groups found on the skin.

19.3 Describe the beneficial aspects of the microbiota.

Some yeasts and bacteria tolerate (and indeed thrive in) the harsh conditions of the epidermis, in the cavities of the hair follicles, and in the interiors of the sweat ducts. These normally harmless residents of the body make up the **microbiota** (as we saw in Chapter 14). Microbiota compete with potential pathogens for nutrients and space and produce chemicals that interfere with the growth of other microbes, providing further defense against infection. Though vigorous scrubbing may reduce the number of microorganisms on the skin, it can never eliminate them completely—microbes living deep in the hair follicles and sweat ducts soon recolonize the skin's surface. Organisms of the microbiota typically grow in small clusters, particularly in the armpits and between the legs where conditions are more moist. Their waste products produce body odor.

Significant members of the microbiota include small lipophilic[4] yeasts such as *Malassezia* (mal-ă-sē´zē-ă), which digest sebum. Such yeasts are rarely pathogenic, though they can cause disease in immunosuppressed patients. Aerobic, Gram-positive bacteria in the genera *Staphylococcus*[5] (staf´i-lō-kok´ŭs) and *Micrococcus*[6] (mī´krō-kok-ŭs) also grow on the skin of all individuals. These bacteria can tolerate salt concentrations of 5–10%. The most common species is *Staphylococcus epidermidis* (ep-i-der-mid´is).

Diphtheroids are another common type of Gram-positive bacterial microbiota. These pleomorphic bacilli are named for the similarity of their appearance to *Corynebacterium*[7] *diphtheriae* (kŏ-rī´nē-bak-tēr´ē-ŭm dif-thi´rē-ī), which causes diphtheria, though diphtheroids of the microbiota are generally nonpathogenic. A common problematic diphtheroid is *Propionibacterium acnes* (prō-pē-on-i-bak-tēr´ē-ŭm ak´nēz), though it is also protective. This bacterium resides in hair follicles, where it ferments carbohydrates to form propionic acid, which lowers the pH of the skin, acting as a further defense against additional infection.

Despite the inhospitable nature and defensive structures and chemicals of the skin, pathogenic microbes can still produce diseases, particularly if they penetrate the epidermis through wounds, or when the immune system is suppressed. Bacteria, viruses, fungi, protozoa, and arthropods are involved in skin diseases. Diseases of other body systems may also manifest themselves in the skin. We now survey such diseases beginning with diseases caused by bacteria.

[4]From Greek *lipos,* meaning "grease," and *philos,* meaning "love."
[5]From Greek *staphle,* meaning "small bunch of grapes," and *kokkos,* meaning "a berry."
[6]From Greek *micros,* meaning "small."
[7]From Greek *coryne,* meaning "club."

TELL ME WHY

Why is the normal skin microbiota considered beneficial to a person?

Bacterial Diseases of the Skin and Wounds

Bacteria infecting the skin cause diseases that range from mild acne to life-threatening infections. Skin-infecting bacteria include *Staphylococcus, Streptococcus, Propionibacterium, Bartonella, Pseudomonas,* and *Rickettsia.* We begin by examining diseases caused by one of the more common bacterial pathogens, *Staphylococcus.*

Folliculitis

LEARNING **OUTCOMES**

19.4 List and describe four types of folliculitis caused by *Staphylococcus.*

19.5 Discuss the virulence factors of *Staphylococcus* that enable it to be pathogenic, contrasting the virulence of *S. aureus* with that of *S. epidermidis.*

19.6 Describe the diagnosis, treatment, and prevention of folliculitis.

Signs and Symptoms

Folliculitis (fŏ-lik-yū-lī´tis) is an infection of a hair follicle in which the base of the follicle becomes red, swollen, and pus filled. This condition is often called a *pimple.* When it occurs at the base of an eyelid, it is called a *sty.* A **furuncle** (fyu´rŭng-kl), or *boil,* is a large, painful, raised nodular extension of folliculitis resulting from spread of the infection into surrounding tissues. When several furuncles join together—more frequently in areas where the skin is thick, such as at the back of the neck—they form a **carbuncle** (kar´bŭng-kl). In severe cases, the body responds to folliculitis by triggering fever.

Pathogen and Virulence Factors

Staphylococcus—a genus of facultatively anaerobic, Gram-positive bacteria whose spherical cells are typically clustered in grapelike arrangements **(FIGURE 19.2)**—is the most common cause of folliculitis and associated infections of the skin. Staphylococcal cells are salt tolerant: they are capable of growing in media containing up to 10% NaCl, which explains how they tolerate the salty surface of human skin. Additionally, staphylococci are tolerant of drying out, solar radiation, and heat (up to 60°C for 30 min), allowing them to survive on environmental surfaces in addition to skin.

Two species of *Staphylococcus* are typically found on human skin as well as in the upper respiratory, gastrointestinal, urinary, and genital tracts. *Staphylococcus epidermidis,* as its name suggests, is a major member of the microbiota, accounting for up to 90% of bacteria on the skin. The more virulent *Staphylococcus aureus*[8] (o´rē-ŭs), which often grows in nasal passages, produces a variety of disease conditions and symptoms.

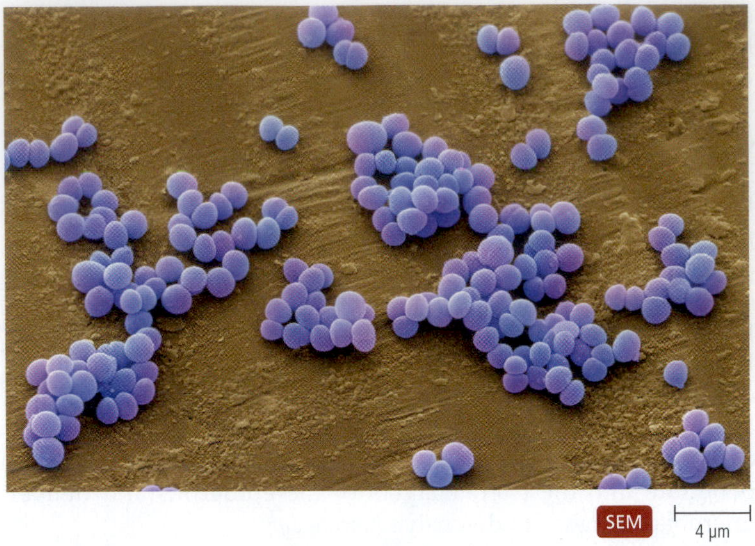

▲ **FIGURE 19.2** ***Staphylococcus.*** *What accounts for the grapelike arrangement of staphylococcal cells?*

Figure 19.2 *The arrangement of staphylococcal cells is due to cell division in random planes.*

Staphylococci have at least three categories of virulence factors that allow them to produce disease: enzymes, structures that enable them to evade phagocytosis, and toxins.

Enzymes Virulent strains of *S. aureus* produce a number of enzymes that contribute to their survival and pathogenicity:

- *Coagulase* clots blood, which may hide the bacterium from phagocytes.
- *Hyaluronidase* breaks down hyaluronic acid, a major component of the matrix between cells, thus enabling the bacterium to spread between cells throughout the body.
- *Staphylokinase* dissolves blood clots, which also allows staphylococci to spread to new locations.
- *Lipases,* which are present in *S. epidermidis* as well, digest lipids, including sebum, providing staphylococci with food on the surface of skin, in hair follicles, and in sebaceous glands.
- *β-Lactamase* plays no role in inhibiting the natural defenses of the body, but it does convey resistance to many beta-lactam antimicrobial drugs such as penicillin and cephalosporin—drugs which β-lactamase inactivates.

Structural Defenses Against Phagocytosis Both *S. aureus* and *S. epidermidis* evade the body's defenses by synthesizing loosely organized polysaccharide slime layers (sometimes called capsules) that inhibit chemotaxis of and phagocytosis by leukocytes. The slime layer also facilitates attachment of staphylococcal biofilms to artificial surfaces such as catheters, shunts, artificial heart valves, and synthetic joints.

[8]Latin, meaning "gold," from the color the colonies sometimes assume when growing on agar.

EMERGING DISEASE CASE STUDY

Buruli Ulcer

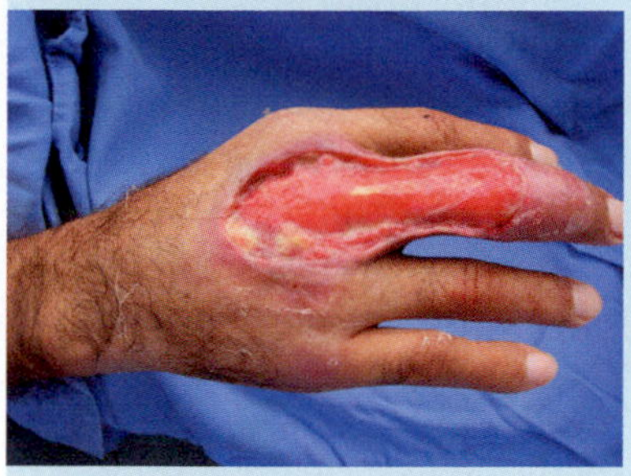

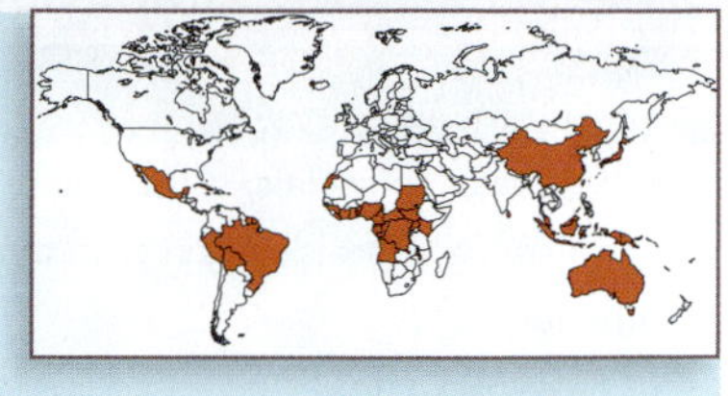

Jacques liked living in the Democratic Republic of the Congo (DRC)—opportunities abounded for exploration, adventure, and wildlife photography, the countryside was beautiful, and he found the people generally friendly. It was on a photographic excursion to the east that Jacques met a not-so-friendly resident of the DRC, an emerging mycobacterial pathogen.

The photographer thought little of the small insect bite he received while documenting wildlife in the swamps along the great Congo River, but he should have been concerned; *Mycobacterium ulcerans* had found a new home in his hand. Jacques would pay a grievous price for his lack of care.

He continued to ignore the infection when it produced a small, painless nodule. He even ignored it when his finger swelled to twice its normal size; it was painless and he could still meet his busy schedule. But the bacterium was producing a potent toxin known as mycolactone that destroys cells below the skin, especially fat and muscle cells. Though his hand continued to swell, making it difficult to work normally, it remained pain free.

After six weeks of this condition, pain began suddenly and excruciatingly. The swollen finger ruptured, and a foul-smelling fluid saturated his camera. It was time to see a doctor.

The physician diagnosed Buruli ulcer, an emerging disease that affects more and more people each year as a result of human encroachment into the swamps where *Mycobacterium ulcerans* lives. After two surgeries to remove dead tissue and bacteria, several skin grafts, and two months of treatment with the antimicrobial drugs rifampicin and streptomycin, Jacques was released with scars that forever remind him of his adventure with *M. ulcerans.*

1. What might be a reason why a Buruli ulcer is initially painless?
2. Why was it necessary to administer antibacterial drugs for two months rather than two weeks?
3. What environmental similarities exist in the endemic countries?

Cells of *S. aureus* are uniformly coated with *protein A,* which binds to the stems of class G antibodies (IgG). Antibodies are opsonins—they enhance phagocytosis—precisely because phagocytic cells bind to antibody stems (recall Chapter 16); therefore protein A, by blocking the stems, effectively inhibits opsonization. Protein A also inhibits the complement cascade, which is triggered by antibody molecules bound to antigen (see Figure 15.9).

Toxins Pathogenic *Staphylococcus aureus* possesses several toxins that contribute to virulence. Cytolytic toxins disrupt the cytoplasmic membranes of a variety of cells. *Leukocidin* kills leukocytes, providing *Staphylococcus* with additional protection against phagocytosis, and *epidermal cell differentiation inhibitor* is a protein that induces large holes in the linings of blood vessels, allowing access for the bacterium to invade body tissues.

Some strains of *S. aureus* also produce *exfoliative toxin* or *toxic shock syndrome toxin*—proteins that cause staphylococcal scalded skin syndrome (discussed shortly) and staphylococcal toxic shock syndrome (see Chapter 24), respectively.

TABLE 19.1 compares and contrasts virulence factors possessed by *S. aureus* and *S. epidermidis.*

TABLE 19.1 Comparison of Virulence Factors of Two Staphylococcal Species

Virulence Factor	*S. aureus*	*S. epidermidis*
Enzymes		
Coagulase	+	–
Staphylokinase	+	–
Lipase	+	+
β-Lactamase	Present in 90% of strains	–
Factors That Inhibit Phagocytosis		
Polysaccharide slime layer	+	+
Protein A on cell surface	+	–
Toxins		
Cytolytic toxins	+	–
Leukocidin	+	–
Epidermal cell differentiation inhibitor	+	–
Exfoliative toxin	Present in some strains	–
Toxic shock syndrome toxin	Present in some strains	–

Pathogenesis

Humans transmit both species of *Staphylococcus* via direct contact between individuals as well as via fomites[9] (fōm´i-tēz),

[9]Latin, meaning "tinder," as in fire-starting material; singular: *fomes* (fō´mēz).

TABLE 19.2 Some Diseases Caused by *Staphylococcus aureus*

Disease	Where Discussed
Skin disease: folliculitis, sty, furuncle, carbuncle	p. 560
Staphylococcal scalded skin syndrome	p. 562
Impetigo	p. 563
Staphylococcal toxic shock syndrome	Chapter 24
Bacteremia	Chapter 21
Endocarditis	Chapter 21
Pneumonia	Chapter 22
Food poisoning	Chapter 23

which are inanimate carriers of infective agents. Contaminated clothing, bedsheets, and medical instruments are examples.

Staphylococcus on the surface of the skin grows into hair follicles and invades sebaceous glands. This triggers fever and inflammation, which are natural responses against infection, and causes the follicle to enlarge and fill with pus composed of leukocytes, dead cells, and bacteria. The infection may spread into the hypodermis to form a furuncle or into neighboring hair follicles to form a carbuncle. *S. aureus* (and infrequently *S. epidermidis*) may also spread into the blood—a condition called *bacteremia*—and be carried to the lining of the heart, lungs, and bones, causing *endocarditis, pneumonia,* and *osteomyelitis,* respectively. **TABLE 19.2** lists some of the many diseases caused by *S. aureus.*

Epidemiology

Staphylococcus epidermidis thrives on almost every square millimeter of human skin. Because it lacks the virulence factors of *S. aureus,* it seldom causes disease, though it can be an opportunistic pathogen in immunocompromised patients or when introduced into the body via intravenous catheters or on prosthetic devices such as artificial heart valves.

Staphylococcus aureus, in contrast, is not a permanent resident but does grow on the skin or mucous membranes of most people at some time in their lives, particularly in the nostrils. About a fifth of the population carries the bacterium with no symptoms for a year or more. The bacterium also intermittently colonizes moist skin folds such as in the armpits and around the groin, often being transferred from the face via the hands.

Diagnosis, Treatment, and Prevention

Diagnosis of staphylococcal folliculitis involves the detection of Gram-positive bacteria in grapelike arrangements isolated from pus. If staphylococci isolated from an infection are able to clot blood, then they are coagulase-positive *S. aureus.* Coagulase-negative staphylococcus is usually *S. epidermidis,* which is a normal part of the microbiota of the skin—its presence in a clinical sample is not normally indicative of staphylococcal disease.

Clinical practice has shown that it is critical to clean and drain abscesses of pus for subsequent topical antibiotic therapy with mupirocin to be effective. Dicloxacillin—a semisynthetic oral form of penicillin not inactivated by β-lactamase—is often the drug of choice for staphylococcal infections.

Unfortunately, drug-resistant *Staphylococcus aureus* has emerged as a major health problem. Strains resistant to many common antimicrobial drugs, including natural penicillin, methicillin, macrolides, aminoglycosides, and cephalosporin, are known. Vancomycin is used to treat such infections. Lately, physicians have become concerned about the increasing prevalence of vancomycin-resistant strains of *S. aureus* (VRSA).

Methicillin-resistant *S. aureus* (MRSA) has become more common in hospitals, so it is imperative that health care workers take precautions against introducing such bacteria into patients. Since *Staphylococcus* is part of the normal microbiota of many people's skin, staphylococcal infections cannot be eliminated completely. Fortunately, because a large inoculum is required to cause disease, proper cleansing of wounds and surgical openings, attention to aseptic use of catheters and indwelling needles, and the appropriate use of antiseptics prevents MRSA infections in most patients.

People with staphylococcal lesions should refrain from working with food, patients with open wounds, immunocompromised patients, and women in labor, and they should not work in nurseries or operating rooms. The most important measures for protecting against healthcare associated infection are proper aseptic techniques, particularly correct hand antisepsis.

Scientists are currently testing a vaccine to protect patients from *S. aureus* infections. Such immunization, if it proves safe and effective, may have a significant effect on nosocomial morbidity and mortality. It would be especially good news for health care providers and their patients who are battling resistant strains of *Staphylococcus.*

Staphylococcal Scalded Skin Syndrome

LEARNING | **OUTCOME**

19.7 Describe staphylococcal scalded skin syndrome, including its diagnosis, treatment, and prevention.

Signs and Symptoms

In **staphylococcal scalded skin syndrome (SSSS)**, cells of the outer epidermis separate from one another and from the underlying tissue **(FIGURE 19.3)**. SSSS involves a reddening and wrinkling of the skin that typically begins near the mouth, spreads over the entire body, and is followed by large blisters that contain clear fluid lacking bacteria or white blood cells. Within two days, the affected outer epidermis peels off in sheets, as though the flesh had been dipped into boiling water.

Pathogen and Virulence Factors

Five percent of strains of *Staphylococcus aureus* secrete one or two distinct **exfoliative toxins** that cause SSSS. Each toxin causes the dissolution of epidermal *desmosomes,* which are

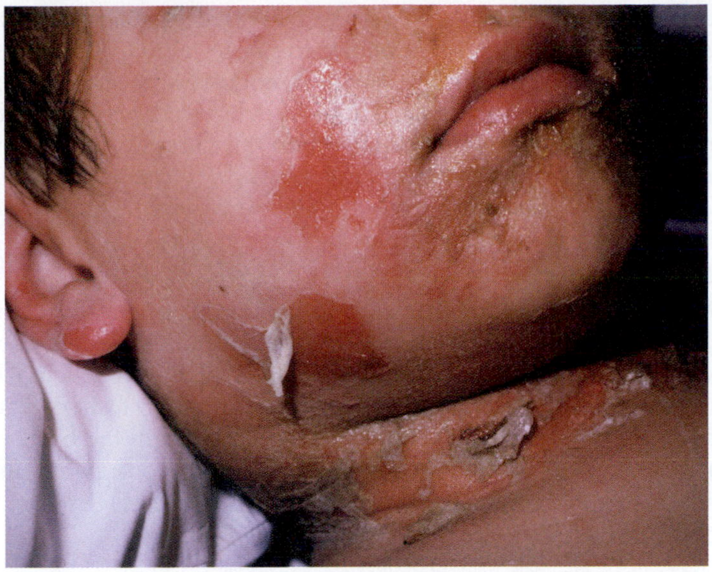

▲ **FIGURE 19.3 Staphylococcal scalded skin syndrome.** Exfoliative toxin, produced by some strains of *Staphylococcus aureus*, causes reddened patches of the epidermis to slough off. *Which patients are most susceptible to exfoliative strains of S. aureus?*

Figure 19.3 Infants, the elderly, and immunosuppressed patients are most susceptible to SSSS.

intercellular bridge proteins that hold together cytoplasmic membranes of adjoining cells. Both toxins affect keratinized cells of the epidermis.

Pathogenesis

The blood carries exfoliative toxins from sites of infection throughout the body. Such circulation of toxins in the blood is called *toxemia.*

The body restores the lost epidermis within 7 to 10 days after protective antibodies circulate in the blood. Though 100% of the skin surface may be afflicted, scarring does not occur because the dermis is not affected by the toxin. Mortality is rare; death, if it occurs, is most often due to secondary infections of skinless areas by yeasts such as *Candida albicans* (kan´did-ă al´bi-kanz) or bacteria such as *Pseudomonas aeruginosa* (soo-dō–mō´nas ā-roo-ji-nō´să).

Epidemiology

SSSS is primarily a disease of infants and children under age five, though it can also affect the elderly and immunosuppressed patients such as those with AIDS; antibodies apparently protect people with well-developed immune systems. Transmission is by person-to-person spread of the bacterium onto skin surfaces, where it penetrates cuts and abrasions.

Diagnosis, Treatment, and Prevention

Diagnosis is made on the basis of the distinctive sloughing of the skin. Fluid in the blisters of SSSS does not contain *S. aureus,* since the disease is mediated by toxins released from a site of infection that may be elsewhere in the body.

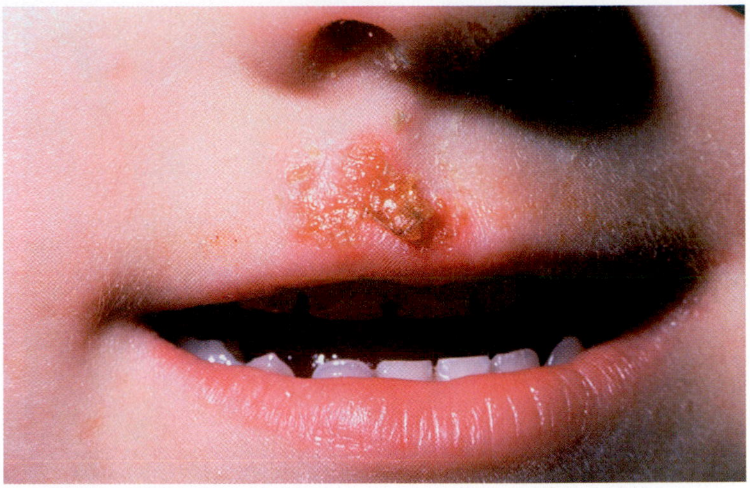

▲ **FIGURE 19.4 Impetigo.** Reddened patches of skin become pus-filled vesicles that eventually form a honey-colored, sticky crust.

Treatment involves intravenous antimicrobial drugs such as cloxacillin. Little can be done to prevent SSSS because *S. aureus* is normally on skin, and susceptible patients lack sufficient immunity.

We have considered disease caused by *Staphylococcus* acting alone. Next we turn our attention to diseases of *Streptococcus* acting alone or in conjunction with *Staphylococcus.*

Impetigo (Pyoderma) and Erysipelas

LEARNING | **OUTCOMES**

19.8 Distinguish the signs, symptoms, and causes of impetigo from those of erysipelas.

19.9 Describe the diagnosis, treatment, and prevention of impetigo and erysipelas.

Signs and Symptoms

Impetigo (im-pe-tī´gō)—also called *pyoderma*[10]—is a contagious disease characterized by small, flattened, red patches that appear primarily on the face and limbs **(FIGURE 19.4)**, particularly of children whose immune systems are incompletely developed. The patches develop into oozing, pus-filled vesicles on a red base. Such vesicles eventually break and form a thick honey-colored, sticky crust that is attached firmly to the skin and may cause intense itching. Numerous vesicles at various stages of development characterize impetigo because bacteria from a vesicle spread to adjacent sites on the skin.

When such an infection of the skin spreads into surrounding lymph nodes and triggers pain and inflammation, the condition is called **erysipelas**[11] (er-i-sip´ĕ-las) **(FIGURE 19.5)**. Erysipelas manifests as reddening of the skin on the face, arms, or legs. The red area has a distinct margin, almost as if it

[10]From Greek *pyon,* meaning "pus," and *derma,* meaning "skin."
[11]From Greek *erythros,* meaning "red," and *pella,* meaning "skin."

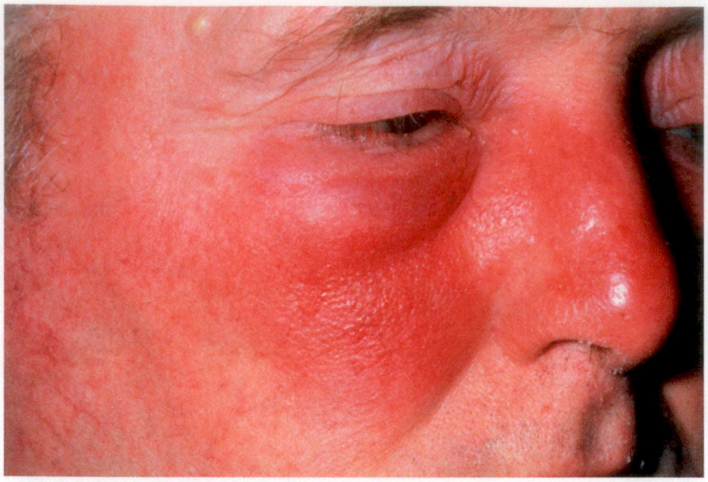

▲ **FIGURE 19.5 Erysipelas.** *What chemicals cause the reddening of the skin in erysipelas?*

Figure 19.5 Proteinaceous pyrogenic toxins released by *Streptococcus pyogenes* cause the red skin seen in erysipelas patients.

were painted onto the skin. Swollen local lymph nodes, pain, fever, chills, and *leukocytosis*—an abnormally large number of leukocytes in the blood—are also seen. Without treatment, erysipelas may be fatal, with mortality ranging from 2% to 17%. The very young, very old, and immunocompromised are more likely to die.

Pathogens and Virulence Factors

S. aureus acting alone causes about 80% of impetigo cases, and about 20% of cases involve *Streptococcus pyogenes* (strep-tō-kok´ŭs pī-oj´en-ēz) alone or in conjunction with *S. aureus. S. pyogenes,* which is synonymously called group A *Streptococcus* after a particular carbohydrate antigen found on the cell's surface, is a Gram-positive coccus whose offspring remain attached in chains following cell division. Besides causing 20% of impetigo cases, *Streptococcus pyogenes* is also the cause of erysipelas.

S. pyogenes has virulence factors that contribute to impetigo:

- *M protein* is a cytoplasmic membrane component that destabilizes complement and interferes with phagocytosis and lysis of the bacterium.
- A hyaluronic acid capsule serves to "camouflage" the bacterium, hiding it from phagocytes since hyaluronic acid is a natural chemical in the body.
- *Pyrogenic*[12] *toxins* (formerly called *erythrogenic*[13] *toxins*) are proteins that stimulate macrophages and helper T lymphocytes to release cytokines that in turn stimulate fever, a widespread rash, and shock.

Pathogenesis

The bacteria of impetigo and erysipelas occasionally colonize the skin and then invade through scratches, abrasions, cold sores, or other wounds where the skin's integrity is compromised. Children are often infected just below the nose, where the skin is abraded by frequent wiping away of mucus. Some strains of *S. pyogenes* may spread from a site of impetigo or erysipelas into the blood (bacteremia) and thence into the kidneys where they cause a disease of the kidneys known as *acute glomerulonephritis.* (Chapter 24 considers this condition in more detail.)

Epidemiology

Both *Staphylococcus* and *Streptococcus* spread among individuals by person-to-person contact or via contaminated fomites such as toys, clothing, bedding, towels, or hairbrushes, particularly in warm, moist, summer conditions. Children two to five years of age are more likely to develop impetigo, though the bacteria can infect older children as well. Epidemics of impetigo in nurseries are of particular concern to hospital workers. Erysipelas occurs most commonly in children and the elderly.

Diagnosis, Treatment, and Prevention

Vesicles of impetigo are diagnostic. Further, the pus found in the blisters is filled with bacteria and white blood cells, which distinguishes impetigo from scalded skin syndrome. The presence of grapelike clusters of Gram-positive cocci indicate staphylococcal impetigo, whereas the presence of chains of Gram-positive cocci characterizes streptococcal impetigo and erysipelas.

Treatment of impetigo involves topical mupirocin and oral clindamycin or amoxicillin. Gentle washing with soap and water twice a day gently removes the crusts and infectious bacteria of impetigo. Physicians treat erysipelas with penicillin.

Good general hygiene and cleanliness help to prevent impetigo and erysipelas. Abrasions and other breaks in the skin of children should be thoroughly cleaned with soap and water, particularly if there has been contact with impetigo patients or contaminated fomites.

Necrotizing Fasciitis

LEARNING | **OUTCOMES**

19.10 Describe the actions of six virulence factors of *Streptococcus pyogenes.*

19.11 Describe the pathogenesis, epidemiology, diagnosis, and treatment of necrotizing fasciitis.

19.12 Explain why preventing necrotizing fasciitis is difficult.

"Bacteria are Eating My Face Off!" screams a tabloid cover; although such a report is sensationalized, the condition is possible. Bacteria can destroy soft body tissues—a condition called "flesh-eating bacteria" by some news media and medically known as *necrotizing fasciitis*[14] (ne´ kro-tī-zing fas-ē-ī´ tis). **Disease in Depth: Necrotizing Fasciitis** on pp. 568–569 visually presents the features of this disease in some detail.

[12]From Greek *pyr,* meaning "fire," and *genein,* meaning "to produce."

[13]From Greek *erythros,* meaning "red," and *genein,* meaning "to produce."

[14]From Greek *nekros,* meaning "corpse"; Latin *fascia* for "band," referring to the bands of tough connective tissue that lie in the skin and surround muscles; and Greek *itis,* meaning "inflammation."

Most patients with necrotizing fasciitis initially report a hot, intensely painful, sunburn-like rash at the site of infection. Subsequently, patients develop fever, tiredness, and muscle aches. As tissue is destroyed, blood pressure drops, and patients become mentally confused and ultimately comatose.

Several bacteria can cause necrotizing fasciitis, but the most common cause is *Streptococcus pyogenes*, also known as group A *Streptococcus*. In addition to the virulence factors listed in association with impetigo, strains of *S. pyogenes* that cause necrotizing fasciitis have factors that allow the bacterium to invade body tissues, resist phagocytosis, and damage cells and tissues. These virulence factors are considered in more detail in the Disease in Depth feature.

The usual route of transmission for group A *Streptococcus* is from person to person through breaks in the skin. Necrotizing fasciitis has developed following surgery, abortion, and seemingly harmless events such as an insect bite or a needlestick to draw blood. Diabetes, cancer, and chickenpox increase the risk for developing necrotizing fasciitis.

Necrotizing fasciitis should be considered a surgical emergency, and physicians must remove dead tissue immediately, repeating the procedure daily until bacterial destruction of tissue is halted. Doctors prescribe intravenous broad-spectrum antimicrobial drugs to curb the severity of disease and inhibit any remaining bacteria. One study has shown that clindamycin and penicillin when used together had a cure rate of 83%, compared to a cure rate of only 41% for penicillin alone.

Acne

LEARNING **OUTCOMES**

19.13 Describe the progression of acne.
19.14 Discuss chemical and physical treatments for acne.

The blackheads and pimples of **acne** are a well-known phenomenon.

Pathogen

The most common causes of acne are propionibacteria, which are small, Gram-positive, rod-shaped diphtheroids commonly found growing on the skin. They are named for the propionic acid that is a by-product of their fermentation of carbohydrates. The species most commonly involved in infections of humans is *Propionibacterium acnes*, which causes acne in 85% of afflicted adolescents and young adults. *Staphylococcus aureus* may also cause acne.

Pathogenesis

FIGURE 19.6 illustrates the development of acne. *Propionibacterium* typically grows on sebum within the sebaceous glands of the skin **1**. Excessive oil production triggered by the hormones of adolescence, particularly testosterone in males, stimulates the growth of the bacteria, which secrete chemicals that attract leukocytes and trigger inflammation. The leukocytes phagocytize the bacteria and release chemicals that stimulate local inflammation. The combination of dead bacteria and dead

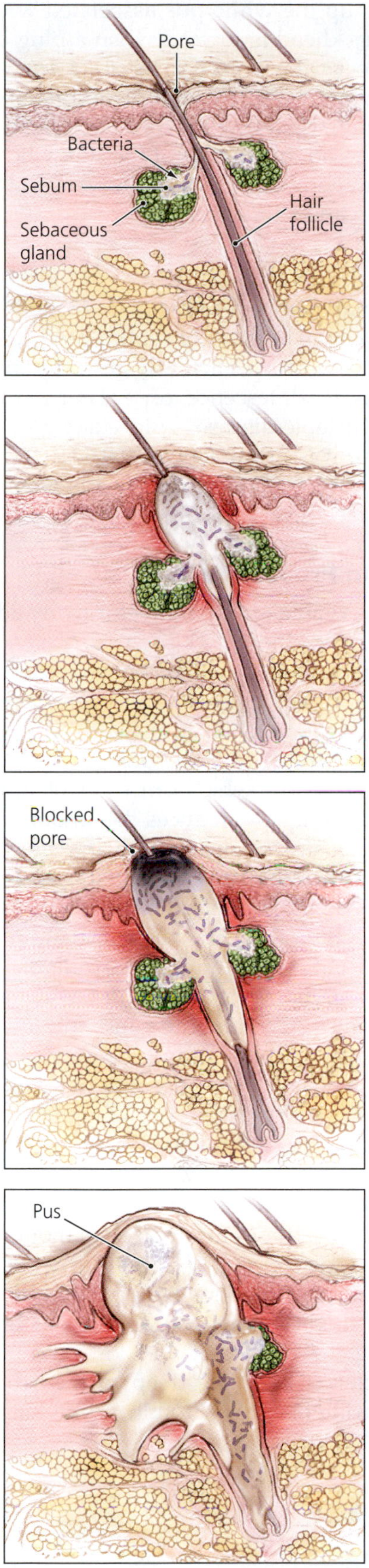

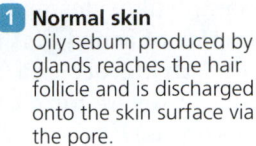

1 Normal skin Oily sebum produced by glands reaches the hair follicle and is discharged onto the skin surface via the pore.

2 Whitehead Inflamed skin swells over the pore when bacteria infect the hair follicle, causing the accumulation of colonizing bacteria and sebum.

3 Blackhead Dead and dying bacteria and sebum form a blockage of the pore.

4 Pustule formation Severe inflammation of the hair follicle causes pustule formation and rupture, producing cystic acne, which is often resolved by scar tissue formation.

▲ **FIGURE 19.6 The development of acne.** *Which bacterial pathogen is a common cause of cystic acne?*

Figure 19.6 *Propionibacterium acnes* is a common cause of acne.

and living leukocytes makes up the white pus associated with the pimples of acne 2. A blackhead is formed when a plug of dead and dying bacteria blocks the pore 3. In *cystic acne,* a particularly severe form of the disease, bacteria form inflamed pustules (cysts) 4 that rupture, triggering the formation of scar tissue. Acne typically develops on those areas of the skin where sebaceous glands are most numerous, including the face, scalp, neck, chest, back, upper arms, and shoulders.

Epidemiology

The bacteria of acne are normal members of the microbiota that grow aggressively due to the large amount of sebum produced as a result of adolescent hormones, particularly testosterone. Though acne typically begins at adolescence, onset later in life is common. Acne is not a reportable disease, so its exact prevalence is not known; however, it is estimated that at least 75% of the world's adolescents develop acne at some time. Severe cystic acne is much rarer.

Diagnosis, Treatment, and Prevention

Diagnosis of acne is easy. It is visible on the skin—for many adolescents, often embarrassingly visible.

In most cases the immune system is able to control *Propionibacterium,* and no treatment is required. For more severe cases, dermatologists may prescribe antimicrobial drugs such as doxycycline, which is highly concentrated in secretions of the skin. They also prescribe benzoyl peroxide, which causes exfoliation of dead skin cells, kills *Propionibacterium,* reduces the amount of lipid on the skin's surface, and has beneficial oxidizing effects. Long-term antimicrobial use can destroy susceptible normal bacterial microbiota, making the individual more vulnerable to opportunistic fungi and to bacteria that are resistant to antimicrobial drugs. Retinoic acid (Accutane), a derivative of vitamin A, inhibits the formation of sebum, reducing the amount available for bacterial metabolism. Because Accutane can cause intestinal bleeding and birth defects, it is prescribed only for severe cases of acne and never for women who are or may become pregnant.

A nonchemical treatment for acne is the use of ultraviolet light (wavelength 315–400 nm). This light, known as UVA, can kill *P. acnes* without causing harm to the body. In one study, two 15-minute exposures a week for a period of four weeks produced a 60% reduction in acne in 80% of patients for up to eight months.

There are many common misconceptions about acne. Scientists have not shown any connection between acne and diet, including chocolate and oily foods—foods do not affect sebum production. Though cleaning the surface of the skin does not remove acne-producing bacteria (they live deep in the sebaceous glands), frequent cleansing may dry the skin, which helps loosen plugs from hair follicles.

Cat Scratch Disease

LEARNING | **OUTCOME**

19.15 Describe the cause, diagnosis, treatment, and prevention of cat scratch disease.

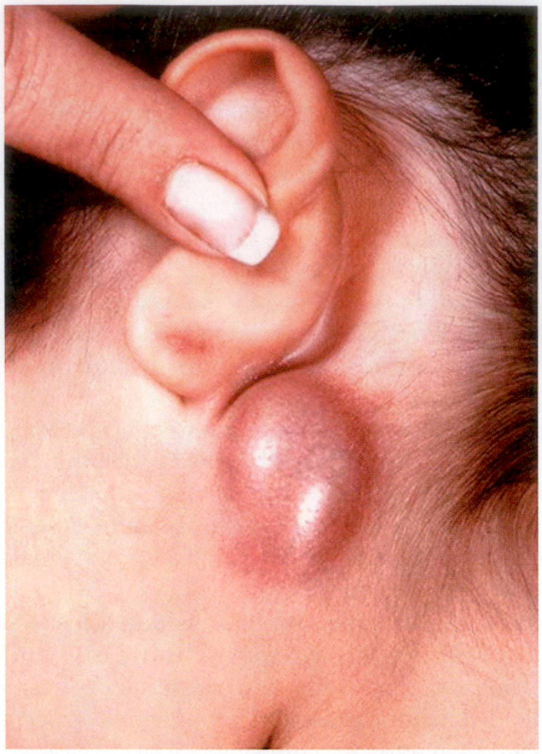

▲ **FIGURE 19.7 Cat scratch disease.** The localized swelling at the site of cat scratches or bites is characteristic.

Signs and Symptoms

Cat scratch disease involves fever for a few days, prolonged malaise, and localized swelling at the site of infection and nearby lymph nodes for several months **(FIGURE 19.7)**.

Pathogen and Virulence Factors

Bartonella henselae (bar-tō-nel´ă hen´sel-ī), a Gram-negative aerobic bacillus, causes cat scratch disease. Its primary virulence factor is *endotoxin (lipid A)* found in the outer membrane of Gram-negative bacteria. Further, *Bartonella* can grow and reproduce inside red blood cells and in cells lining blood vessel walls.

Pathogenesis and Epidemiology

Cat scratches or bites, particularly wounds by kittens, introduce the bacterium into the skin. Blood-sucking arthropods such as fleas may also transmit the bacterium from cats to people. In the skin, the bacterium grows intracellularly. *Bartonella* releases endotoxin when it dies, which can trigger fever, blood clotting, inflammation, and possibly shock (see Figure 21.7).

Though carried by cats, *Bartonella* apparently causes disease only in people; it is not known to cause disease in animals. Cat scratch disease has emerged as a relatively common and occasionally serious infection of children, affecting an estimated 22,000 children annually in the United States.

Diagnosis, Treatment, and Prevention

A positive indirect fluorescent antibody test against *Bartonella* antigens confirms a diagnosis of cat scratch disease in individuals who exhibit the characteristic signs and symptoms following exposure to cats. Physicians prescribe antimicrobials—typically rifampin, ciprofloxacin, or gentamicin—to treat cat scratch disease. Prevention involves avoiding cat-inflicted wounds and adequate cleansing of bites or scratches that do occur.

Beneficial Microbes: New Vessels Made from Scratch? on p. 573 examines a potential positive role for *Bartonella*.

Pseudomonas Infection

LEARNING | **OUTCOMES**

19.16 Describe *Pseudomonas aeruginosa* as an opportunistic pathogen of burn victims.

19.17 Explain the action of nine virulence factors of *Pseudomonas*.

19.18 Explain why diagnosis, treatment, and prevention of *Pseudomonas* infections are problematic.

The skin is a critical barrier to the entrance of many pathogens. This fact is particularly evident in patients whose skin has been burned away by a fire or exposure to steam. In such patients, opportunistic pathogens gain access to the moist, nutrient-rich environment of the fascia and deeper tissues. The most common microorganism seen in burn victims is *Pseudomonas aeruginosa* (soo-dō-mō´nas ā-roo-ji-nō´să). This bacterium may also infect other parts of the body, particularly the lungs. (Cystic fibrosis patients are at particulat risk.)

Signs and Symptoms

When *Pseudomonas aeruginosa* invades the bloodstream, it causes fever, chills, and shock. Massive infections are often readily diagnosed because the bacterium typically produces a blue-green pigment, *pyocyanin,* that colors such infections **(FIGURE 19.8)**.

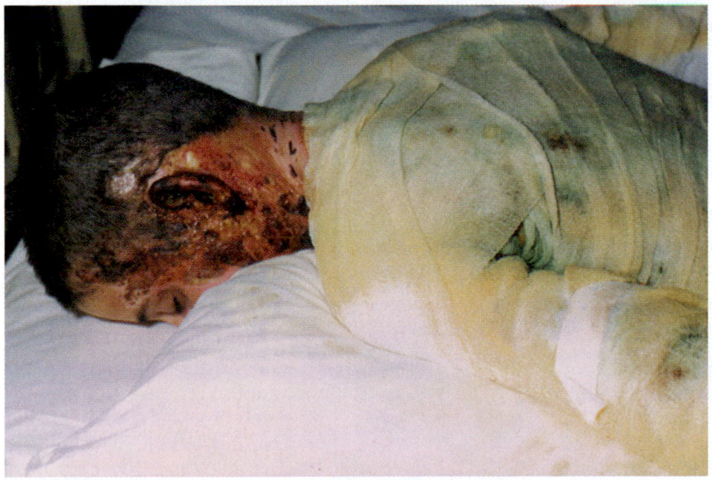

▲ **FIGURE 19.8** ***Pseudomonas aeruginosa* infection.** Bacteria growing under the bandages produce the green color on this burn victim. *What chemical is responsible for the green color of this infection?*

Figure 19.8 *The greenish-blue color of Pseudomonas infections is due to pyocyanin.*

Pathogen and Virulence Factors

Pseudomonas aeruginosa is a Gram-negative, aerobic bacillus that metabolizes a wide range of organic carbon and nitrogen sources. It is almost everywhere in soil, decaying organic matter, and almost every moist environment, including swimming pools, hot tubs, sponges, washcloths, and contact lens solutions. In hospitals, it grows in sinks, moist foods, vases of cut flowers, sponges, toilets, floor mops, dialysis machines, respirators, and humidifiers. Some strains can even grow using small amounts of nutrients left in distilled water.

P. aeruginosa has numerous virulence factors:

- *Fimbriae* and *adhesins* attach to host cells and enable biofilm formation.
- Its *capsule*, composed of a mucoid polysaccharide, plays a role in bacterial attachment and biofilm formation, and it shields the bacterium from phagocytosis.
- *Neuraminidase* modifies host cell receptor molecules to make bacterial attachment to the cells more likely.
- *Elastase* breaks down elastic fiber, degrades complement components, and cleaves immunoglobins A and G (IgA and IgG).
- Endotoxin (lipid A) can trigger fever, blood clotting, inflammation, or possibly shock.
- *Exotoxin A* and *exoenzyme S*, which inhibit eukaryotic protein synthesis, lead to host cell death.
- *Pyocyanin*, the blue-green pigment of *Pseudomonas,* triggers the formation of reactive forms of oxygen (superoxide radical and peroxide anion) that damage host cells.

Pseudomonas aeruginosa is somewhat of a medical puzzle: despite its many virulence factors and ability to live in almost every moist environment using a wide variety of carbon sources, it rarely causes disease. The reason it is only a rare opportunist and not a common and formidable pathogen is that *P. aeruginosa* cannot penetrate the intact structures, cells, and chemical defenses of the skin. This is fortunate, because the ubiquity and inherent resistance of *Pseudomonas* to a wide range of antimicrobial agents would make a more virulent microbe an almost insurmountable challenge to health care professionals.

Pathogenesis

The surface of a burned area provides a warm, moist environment that is quickly colonized by this ubiquitous opportunist. A thick, scablike crust naturally forms over the surface of a severe burn. *Pseudomonas* growing beneath the crust can move into the blood. Once inside the body, it kills cells and destroys tissues, and endotoxin mediates fever, vasodilation, inflammation, shock, and other symptoms. Microbes on the burned area are not readily accessible to medical workers because the crust is generally impermeable to antimicrobial drugs, and blood vessels are absent. Health care workers must remove the crust in a medical procedure termed *debridement* so that antimicrobial drugs can be effective.

DISEASE IN DEPTH

NECROTIZING FASCIITIS

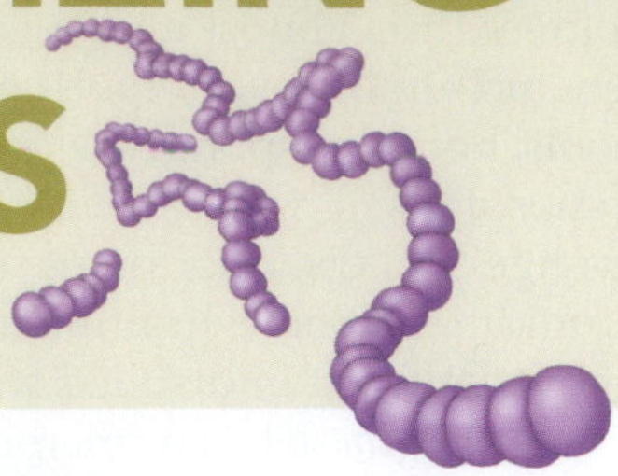

Group A *Streptococcus*

SIGNS AND SYMPTOMS

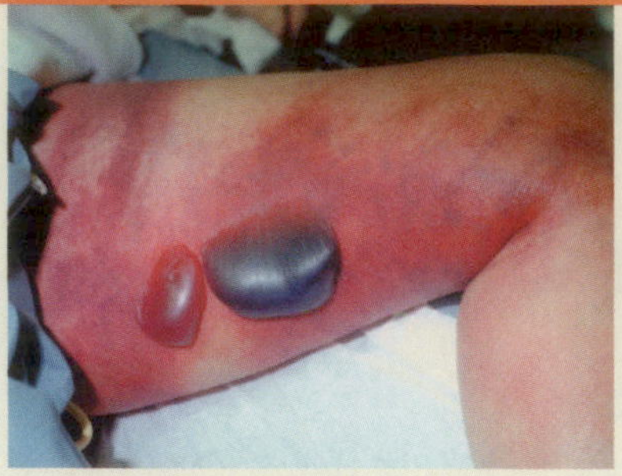

Necrotizing fasciitis is usually characterized by redness, intense pain, and swelling at the site of infection. Initially, the pain does not seem proportionate to the appearance of the infected area. As the bacterium digests muscle fascia—the connective tissue surrounding muscles— and fat tissue, the overlying skin becomes distended and discolored. Patients develop fever, nausea, and malaise, and they may become mentally confused as their blood pressure drops severely.

S*treptococcus pyogenes* can cause necrotizing soft tissue infections, sensationalized in the news as "flesh-eating strep" and medically known as necrotizing fasciitis.

PATHOGENESIS

2 *S. pyogenes* secretes enyzmes that allow the bacterium to invade body tissues.

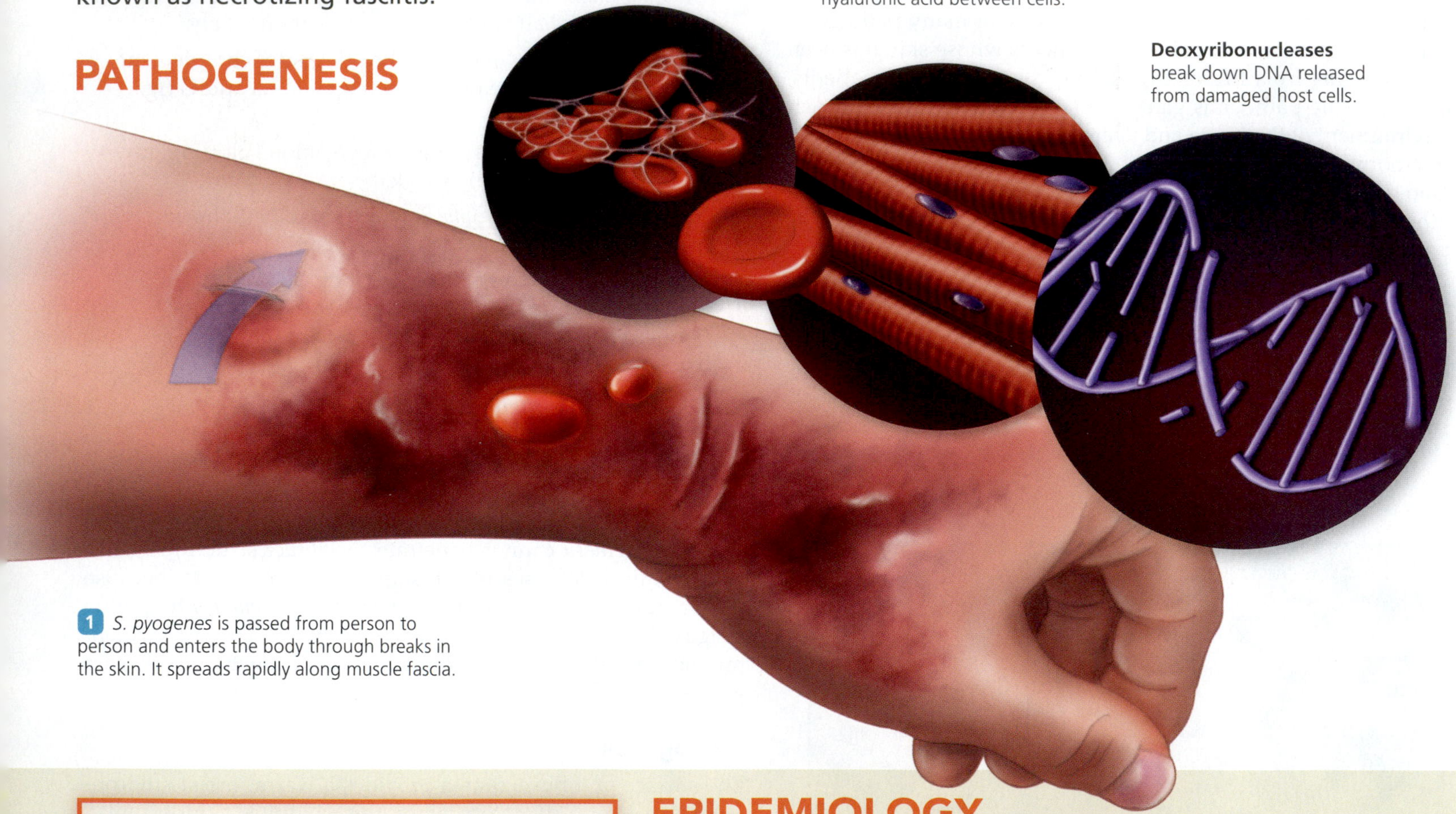

Streptokinases dissolve blood clots.
Hyaluronidase breaks down hyaluronic acid between cells.
Deoxyribonucleases break down DNA released from damaged host cells.

1 *S. pyogenes* is passed from person to person and enters the body through breaks in the skin. It spreads rapidly along muscle fascia.

INVESTIGATE IT!

It can be easy to forget that this disease affects real people.

Scan this code to read the stories of several survivors of necrotizing fasciitis at http://www.nnff.org. What do they have in common? Then go to MasteringMicrobiology to record your research findings.

EPIDEMIOLOGY

Though there has been much recent press coverage, necrotizing fasciitis is not a new disease; it is possible that Hippocrates described the disease, and doctors in China definitively reported cases in 1924. There are about 500,000 cases of necrotizing fasciitis worldwide each year. About 20% of necrotizing fasciitis patients die.

Greek physician Hippocrates (460–370 B.C.) may have described necrotizing fasciitis as early as the 5th century B.C. in *Of the Epidemics.* "…And there were great fallings off (sloughing) of the flesh, sinews, and bones…" (Book II, section III, part 4).

PATHOGEN AND VIRULENCE FACTORS

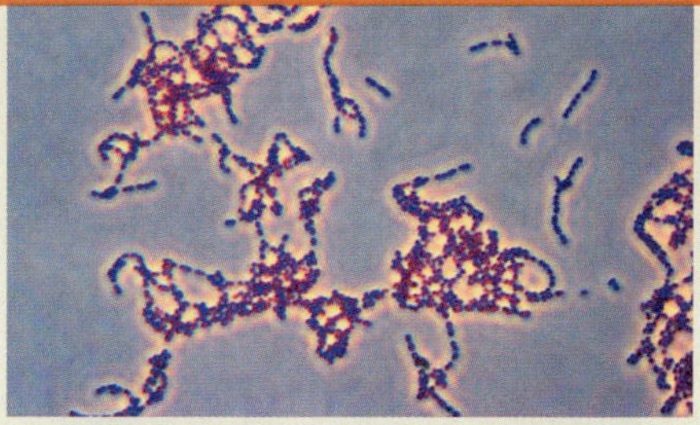

Several bacteria can cause necrotizing fasciitis, including *Staphylococcus aureus*, *Clostridium perfringens*, and *Bacterioides fragilis*, but *Streptococcus pyogenes* (group A *Streptococcus*), shown Gram-stained at left, is the most common cause.

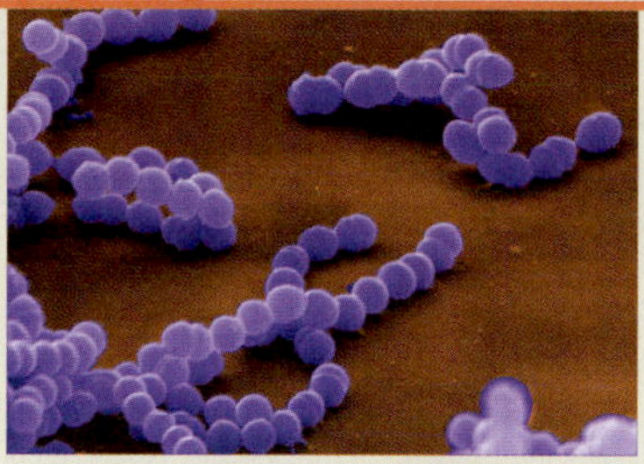

Strains of *S. pyogenes* that cause the condition have enzymes, such as deoxyribonucleases, hyaluronidase, and streptokinases that allow the bacterium to invade body tissues. Streptococcal M protein allows the bacterium to attach to nasopharyngeal cells and to resist phagocytosis. The toxins streptolysin S and exotoxin A damage cells and tissue.

3 Other virulence factors include M protein.

M protein on the surface of streptococcal cells helps the bacterium attach to nose and throat cells, and after entry into the body, M protein allows *S. pyogenes* to survive phagocytosis.

4 *S. pyogenes* also secretes toxins that damage tissue.

Streptolysin S can kill many types of human cells, including neutrophils and erythrocytes.

Exotoxin A triggers an overactive immune response that further damages healthy tissue.

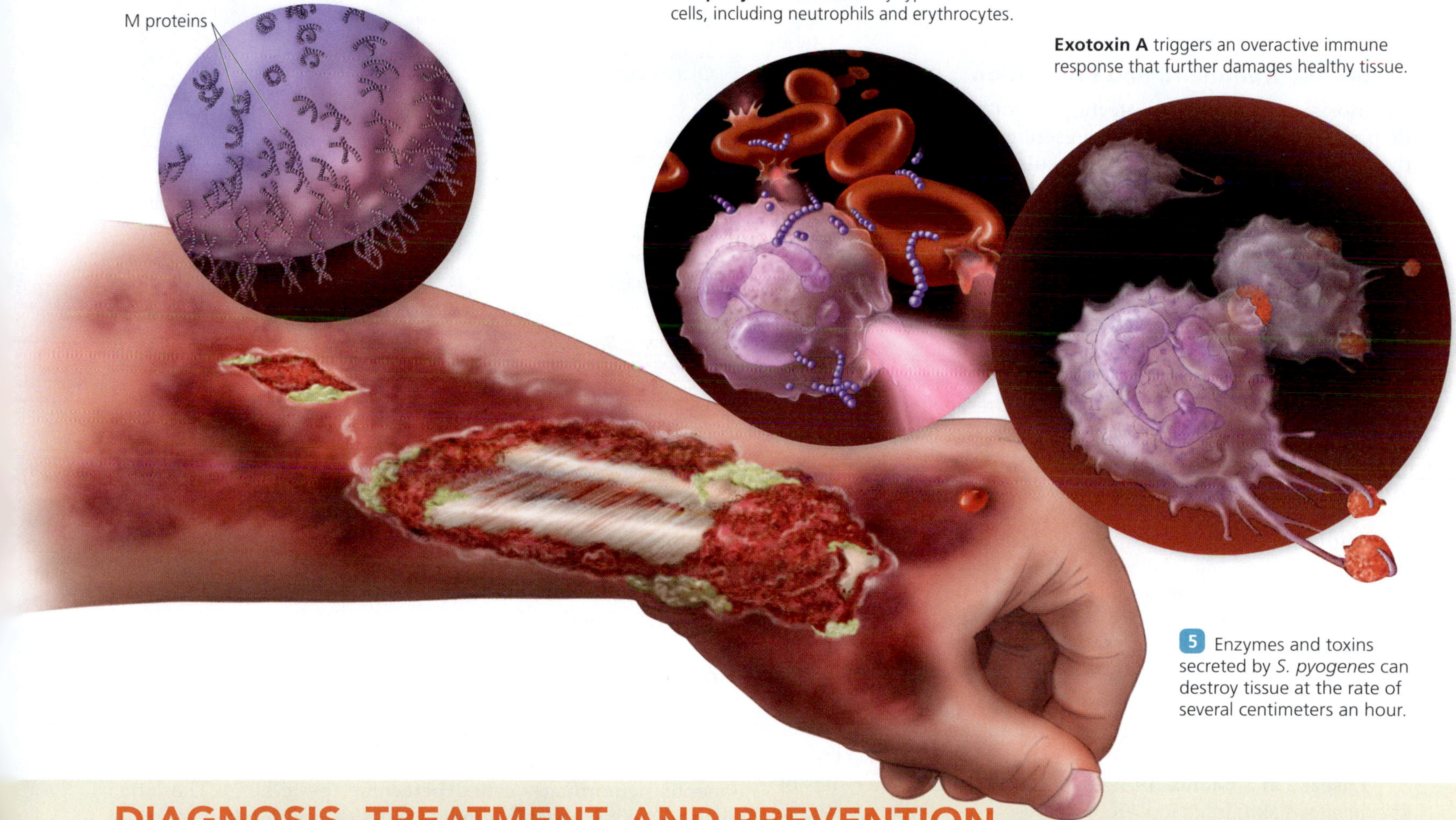

5 Enzymes and toxins secreted by *S. pyogenes* can destroy tissue at the rate of several centimeters an hour.

DIAGNOSIS, TREATMENT, AND PREVENTION

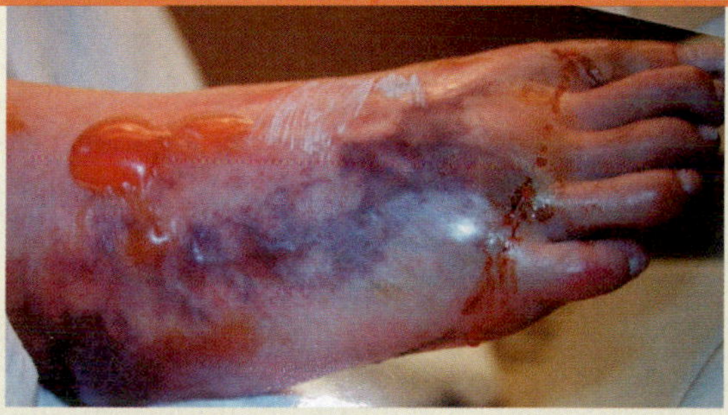

Diagnosis is difficult because early symptoms are general and flulike. Extreme pain that seems out of proportion to an injury should be treated as a possible case of necrotizing fasciitis or infection with *S. pyogenes*. Affected tissue must be removed completely to prevent the spread of bacteria. Intravenous broad-spectrum antimicrobial drugs help curb the severity of the disease. It is difficult to prevent necrotizing fasciitis because *S. pyogenes* is commonly found on humans. Patients with diabetes, cancer, or chickenpox are at a greater risk for necrotizing fasciitis. Any opening in the skin, even a minor cut, is a potential site of invasion by the bacterium; therefore, such openings should be washed thoroughly.

Epidemiology

Although a natural inhabitant of bodies of water and moist soil, *P. aeruginosa* is rarely part of human microbiota. Nevertheless, because of its ubiquity and virulence factors, this opportunistic pathogen is involved in about 10% of healthcare associated infections and colonizes most immunocompromised patients.

Once it breaches the skin or mucous membranes, *P. aeruginosa* can successfully colonize almost any organ and system. Besides skin infections, it can be involved in bacteremia, endocarditis, and urinary, ear, eye, nervous system, gastrointestinal, muscular, and skeletal infections. Infections in cystic fibrosis patients and in burn victims are most common. Almost two-thirds of burn victims develop environmental or nosocomial *Pseudomonas* infections. Swimmers can develop *otitis externa,* also called *swimmer's ear,* which is a painful *Pseudomonas* infection of the external ear canal. The microbe can also infect the hair follicles of people who sit in contaminated hot tubs.

Diagnosis, Treatment, and Prevention

Diagnosis of *Pseudomonas* infection is not always easy because its presence in a culture may represent contamination acquired during collection, transport, or inoculation. Certainly, pyocyanin discoloration of tissues is indicative of massive infection.

Treatment of *P. aeruginosa* can be frustrating. The bacterium is notoriously resistant to a wide range of antibacterial agents, including antimicrobial drugs, soaps, antibacterial dyes, and quaternary ammonium disinfectants. In fact, *Pseudomonas* has been reported to live in solutions of antibacterial drugs and disinfectants.

Resistance is due to the ability of various strains of *Pseudomonas* to metabolize many drugs, to the presence of nonspecific proton/drug antiports that pump many drugs out of the bacterium, and to the ability of its biofilm to resist the penetration of antibacterial chemicals.

Physicians attempt to treat infections with a combination of aminoglycoside and beta-lactam antimicrobials that have first proven efficacious against a particular isolate in a susceptibility test. Polymyxin, a drug that is also toxic to humans, may be used as a last resort.

Pseudomonas aeruginosa is ubiquitous, so it is nearly impossible to prevent exposure; however, the bacterium rarely causes diseases because it does not normally penetrate the skin and mucous membranes, nor does it ultimately evade the body's other defenses. Only in wounded patients does this opportunist thrive.

Disease at a Glance 19.1 summarizes the characteristics of *Pseudomonas* skin infection.

DISEASE AT A GLANCE 19.1

Pseudomonas Infection

Cause *Pseudomonas aeruginosa* (Gram-negative, aerobic, rod-shaped bacterium).

Virulence factors Fimbriae, adhesins, capsule, neuraminidase, elastase, endotoxin (lipid A), exotoxin A, exoenzyme S, pyocyanin.

Portal of entry Contact with wounded skin or deeper tissues following severe burns; inhalation; healthcare associated infections occur from *Pseudomonas* biofilms on endotracheal tubes.

Signs and symptoms Characteristic blue-green color where the bacterium is growing; fever, blood clotting, inflammation, possible shock; with lung infections: breathlessness, coughing, wheezing, rapid breathing, and weight loss.

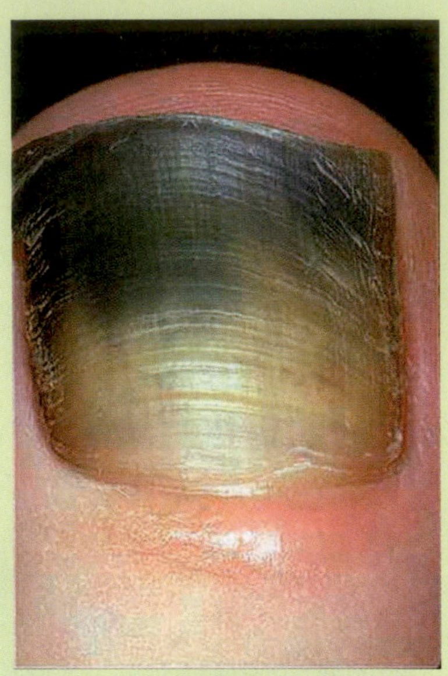

Pseudomonas infection under a nail

Incubation period Eight hours to five days.

Susceptibility Burn victims, cancer patients, immunocompromised patients; cystic fibrosis patients are particularly at risk for lung infection.

Treatment Simultaneous use of both a penicillin and an aminoglycoside.

Prevention The ubiquity of *Pseudomonas* in wet environments makes it nearly impossible to prevent exposure, but the bacterium does not invade intact tissue.

Spotted Fever Rickettsiosis

LEARNING OUTCOMES

19.19 Describe the cause, diagnosis, treatment, and prevention of Rocky Mountain spotted fever.

19.20 Explain why rickettsias cannot live outside another cell.

19.21 Describe the pathogenesis, signs, and symptoms of spotted fever rickettsiosis.

A number of arthropod-borne rickettsias cause rashes in humans, which are called **spotted fever rickettsioses**. The most severe of these is **Rocky Mountain spotted fever (RMSF)**, which we consider here.

Signs and Symptoms

In most cases (90%), RMSF manifests with a non-itchy, spotted rash on the trunk and appendages, including the palms and soles. In about 50% of patients, the rash develops into subcutaneous hemorrhages called **petechiae** (pe-tē´kē-ē). The latter sites are not involved in similar rashes caused by chickenpox or measles viruses. Patients with RMSF also have fever, headache, chills, muscle pain, nausea, and vomiting. In severe cases, the respiratory, central nervous, gastrointestinal, and renal systems fail. Infections of the brain may also occur, producing language disorders, delirium, convulsions, coma, and death.

Pathogen and Virulence Factors

Rickettsia rickettsii (ri-ket´sē-ă ri-ket´sē-ē), the cause of RMSF, is a small (0.3–1 μm), nonmotile, aerobic, Gram-negative, intracellular parasite that possesses a cell wall of peptidoglycan and an outer membrane of lipopolysaccharide. A loosely organized slime layer surrounds the cell.

Rickettsias cannot use glucose as a nutrient; instead, they oxidize amino acids and Krebs cycle intermediates, such as glutamic acid and succinic acid. For this reason, rickettsias are obliged to live inside other cells, where these nutrients are provided.

Rickettsia require a vector for transmission from host to host. For *R. rickettsii*, this vector is a hard tick of the genus *Dermacentor* (der-mă-sen´ter). Once they are in a host, the rickettsia enter the host's cells by stimulating endocytosis. Inside a phagosome, they secrete an enzyme that lyses the phagosome membrane, releasing the bacteria into the host's cytosol before a lysosome fuses with the phagosome. As a result, *Rickettsia* avoids being digested.

The pathogen grows and reproduces slowly, dividing only every 8 to 12 hours. Daughter cells are continually released via exocytosis from long cytoplasmic extensions of host cells.

Pathogenesis

R. rickettsii is typically dormant in the salivary glands of its tick vectors; only when the arachnids feed for several hours is the bacterium infective. The active bacterium is released from a tick's salivary glands into the mammalian host's circulatory system, where it infects a cell lining a small blood vessel. In rare cases, humans become infected following exposure to tick feces or to tissues and fluids from crushed ticks.

R. rickettsii secretes no toxins; instead, disease follows damage to blood vessels, which allows blood to escape, resulting in low blood pressure and insufficient nutrient and oxygen delivery to the body's organs. Almost 5% of patients die, even with treatment. Patients recovering from life-threatening acute Rocky Mountain spotted fever may experience paralysis of the legs, hearing loss, and gangrenous secondary infections with *Clostridium* (klos-trid´ē-ŭm) (as discussed shortly) that necessitate the amputation of fingers, toes, arms, or legs.

Epidemiology

Dermacentor ticks transmit *R. rickettsii* among humans and rodents, which act as reservoirs. Male ticks infect female ticks during mating. Female ticks transmit rickettsia to eggs forming in their ovaries in a process called *transovarian transmission.*

Though the earliest documented cases of RMSF were in the Rocky Mountains, the disease is actually more prevalent in other parts of the U.S. **(FIGURE 19.9)**.

Diagnosis, Treatment, and Prevention

Serological tests such as latex agglutination and fluorescent antibody stains have been used to confirm an initial diagnosis based on a rash on the soles or palms, sudden fever, and headache following exposure to hard ticks. Nucleic acid probes of specimens from rash lesions provide specific and accurate diagnosis, but such tests are expensive and typically are performed only by trained technicians in special laboratories. Early diagnosis is crucial because prompt treatment often makes the difference between recovery and death.

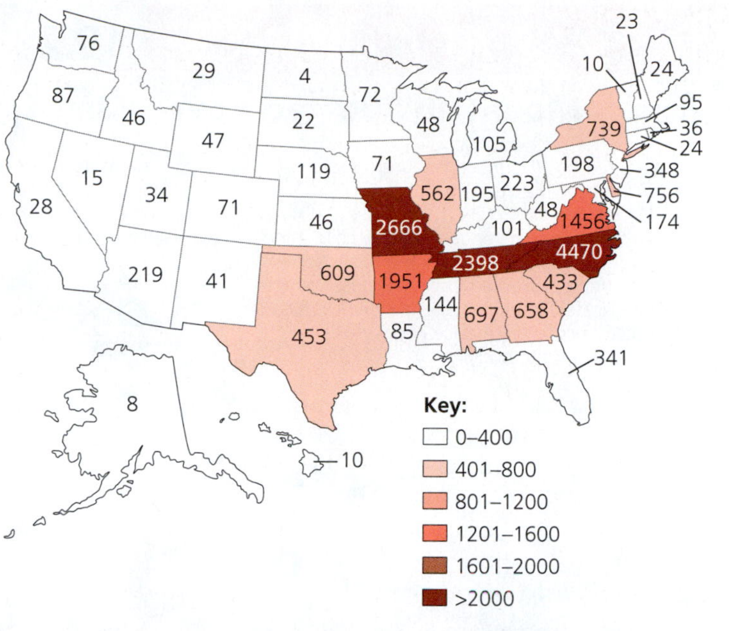

▲ **FIGURE 19.9 Number of cases of Rocky Mountain spotted fever in the United States, 2002–2012.**

Physicians treat Rocky Mountain spotted fever with doxycycline or chloramphenicol. An effective vaccine is not available. Prevention of infection involves wearing tight-fitting clothing, using tick repellants, promptly removing attached ticks, and avoiding tick-infested areas, especially in spring and summer, when ticks are most hungry. It is impossible to eliminate the ticks in the wild, in part because they can survive without feeding for more than four years.

Disease at a Glance 19.2 on p. 572 summarizes the features of Rocky Mountain spotted fever.

Cutaneous Anthrax

LEARNING | **OUTCOME**

19.22 Describe cutaneous anthrax, including its cause and its effect on the skin.

Many bacterial diseases that primarily affect other body systems also affect the skin. For example, a serious systemic disease—Lyme disease (see Chapter 21)—is distinguished initially by a characteristic skin rash. Anthrax is another such disease.

Anthrax has three distinct clinical manifestations—gastrointestinal, inhalation, and cutaneous. Gastrointestinal anthrax is rare in humans. Here we briefly consider cutaneous anthrax. (Chapter 22 discusses the severe, systemic inhalation anthrax in detail.)

Cutaneous anthrax begins when *Bacillus anthracis* (ba-sil´us an-thrā´sis) endospores, often shed by an infected animal, enter a wound in the skin, producing a solid skin nodule. The bacterium's capsule inhibits phagocytosis by white blood cells, and the bacterium synthesizes three toxins that kill cells in the affected area. The nodule spreads to form a painless,

DISEASE AT A GLANCE 19.2

Rocky Mountain Spotted Fever (One Type of Spotted Fever Rickettsiosis)

1. Infected tick *(Dermacentor)* introduces *Rickettsia* while feeding.
2. *Rickettsia* triggers endocytosis by cells lining blood vessels; then it lyses the phagosome's membrane, escaping into the cytosol where it replicates. Infected cells release rickettsia, which once again induce endocytosis throughout the body.
3. Patient develops non-itchy rash on appendages, including palms and soles, that spreads to trunk; other manifestations may include fever, headache, hearing loss, chills, muscle pain, nausea, and vomiting.
4. Damage to cells lining blood vessels allows blood to escape, producing petechiae, low blood pressure, and insufficient delivery of oxygen and nutrients to the body's cells.

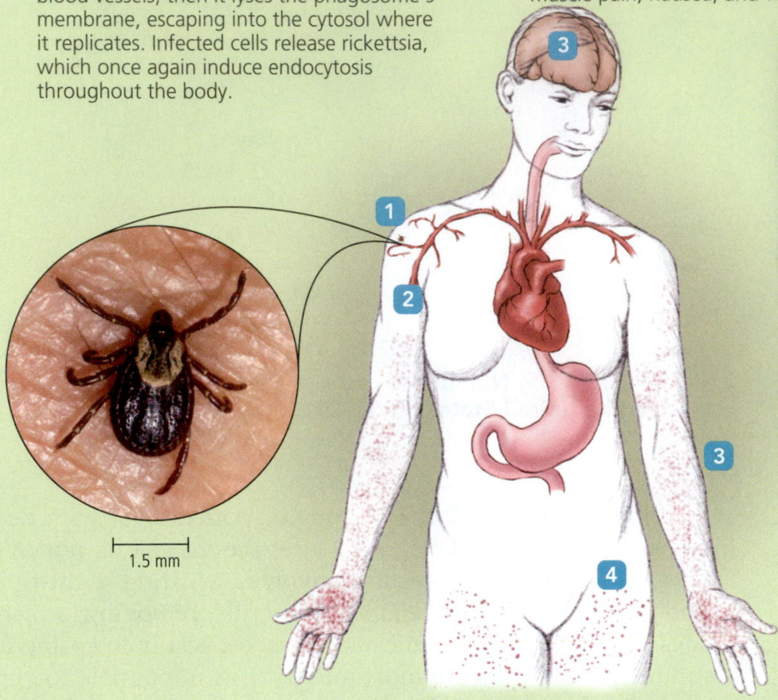

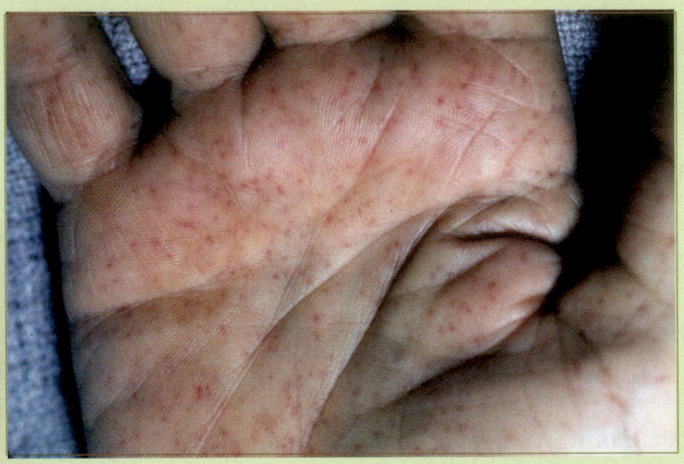

The rash in a case of Rocky Mountain spotted fever

Cause *Rickettsia rickettsii* (aerobic, Gram-negative, obligate intracellular, rod-shaped bacterium).

Virulence factor Escapes phagocytosis by lysing phagosome.

Portal of entry Skin via bite of infected tick.

Signs and symptoms A non-itchy, spotted rash on the trunk and appendages, including the palms and soles; and fever, headache, chills, muscle aches, nausea, vomiting, and petechiae.

Incubation period 5 to 10 days.

Susceptibility People in highly endemic areas are particularly susceptible because of increased exposure.

Treatment Doxycycline is the drug of choice, although chloramphenicol is also effective.

Prevention Wear light-colored, tight-fitting clothes that limit tick exposure. Use tick repellent, promptly remove ticks, examine skin for ticks and bites, and, if possible, avoid tick-infested areas.

swollen, black, crusty ulcer called an **eschar**[15] (es´kar) (**Disease at a Glance 19.3**). Anthrax, which means "charcoal" in Greek, is named for the black color of eschars. Physicians successfully treat cutaneous anthrax with oral antimicrobial drugs such as ciprofloxacin, penicillin, or erythromycin for 60 days, because endospores may take 60 days to germinate. Treated cutaneous anthrax is rarely fatal.

Prevention of naturally occurring anthrax in humans requires control of the disease in animals. Farmers in areas where anthrax is endemic must vaccinate their stock and bury or burn the carcasses of dead animals. Anthrax vaccine has proven effective and safe for humans, but it requires six doses over 18 months plus annual boosters. The Centers for Disease Control and Prevention (CDC) recommends treatment with ciprofloxacin and postinfection immunization, though the vaccine is not licensed for this use.

Gas Gangrene

LEARNING | OUTCOME

19.23 Describe the signs, symptoms, cause, diagnosis, treatment, and prevention of gas gangrene.

When blood supply to a tissue is interrupted, for example by a wound, a condition known as *ischemia* (is-kē´mē-ă)[16] develops.

[15]From Greek *eschara,* meaning a "scab caused by burning."

[16]From Greek *ischo,* meaning "to keep back," and *haima,* meaning "blood."

DISEASE AT A GLANCE 19.3

Cutaneous Anthrax

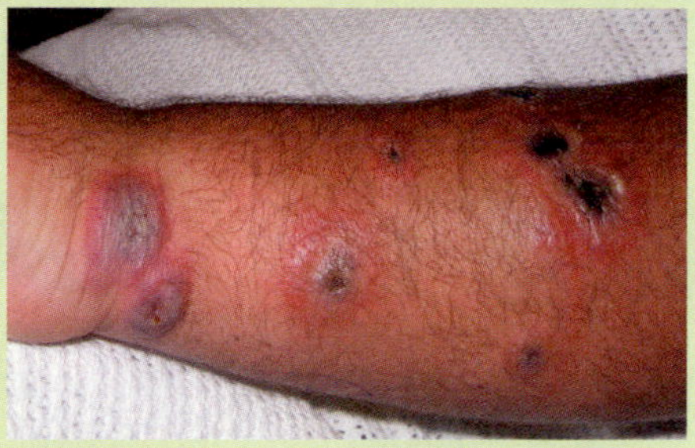

Eschars

Cause *Bacillus anthracis* (facultatively anaerobic, endospore-forming, rod-shaped bacterium).

Virulence factors Endospore, capsule, three anthrax toxims.

Portal of entry Direct contact of endospores with wounded skin, infected animals, or contaminated animal products. (Two other forms of anthrax—gastrointestinal and inhalation—have routes of entry via the gastrointestinal tract and respiratory tract, respectively. More on anthrax is found in Chapter 22.)

Signs and symptoms Localized itching followed by a raised lesion which eventually forms a painless, black eschar (see photo) within 7 to 10 days.

Incubation period Immediate response, within one day.

Susceptibility The most commonly infected individuals are animal handlers.

Treatment Ciprofloxacin is the preferred treatment. There is no treatment for the toxin.

Prevention Vaccinate livestock. The vaccine, considered safe for humans but recommended only for high-risk personnel, requires six doses over an 18-month period and annual boosters.

The tissue becomes anaerobic, and *necrosis* (death) sets in. If the wound is infected with endospores of anaerobic *Clostridium,* then **gas gangrene**[17] (also called simply *gangrene*) develops in the dead tissue.

Signs and Symptoms

Clostridium produces toxins that kill surrounding tissues, providing the bacterium with more nutrients and increasing the size of the anaerobic environment necessary for *Clostridium* to grow. The rapidly growing bacteria then spread into the surrounding tissue, causing the death of muscle and connective tissues.

The immediate result is intense pain at the initial site of infection (often in a foot), followed by gas gangrene, which involves blackening of the infected muscle and skin and the production of bubbles of hydrogen and carbon dioxide gases that may break out with a frothy brownish fluid (see **Disease at a Glance 19.4** on p. 574). The bubbles can be felt under the skin by palpation. Shock, kidney failure, and death can follow, often within a week of infection.

Pathogens and Virulence Factors

Clostridium is an anaerobic, Gram-positive, endospore-forming bacillus that is ubiquitous in soil, water, sewage, and the gastrointestinal tracts of animals and humans. It also grows in the vaginas of 1–9% of healthy women.

Several species of *Clostridium* cause gas gangrene. Pathogenicity is due in great part to the ability of endospores to survive harsh conditions, and to the vegetative cells' secretion of 11 toxins that lyse erythrocytes and leukocytes, increase vascular

[17]From Greek *gangraina,* meaning "an eating sore."

BENEFICIAL MICROBES

New Vessels Made from Scratch?

Bartonella henselae causes cat scratch disease in part by being able to live inside human red blood cells as well as in the cells lining blood vesssels. Scientists at Beth Israel Deaconess Medical Center and the Harvard Medical School in Boston, Massachusetts, discovered recently that the bacterium triggers angiogenesis—the formation of new blood vessels—in infected tissues. Researchers speculate that the pathogen is increasing its food supply and habitat by stimulating the growth of new blood vessels.

How *Bartonella* manages to orchestrate angiogenesis is mysterious. We do know that the bacterium is more efficient in laboratory conditions than is vascular endothelial growth factor—the body's natural angiogenic cytokine. Perhaps it is more efficient in the body as well.

Scientists speculate that a full understanding of *Bartonella*'s method could be harnessed to induce angiogenesis to circumvent blocked arteries in the heart, to prompt tissues to make new blood vessels in damaged limbs, or to speed up wound healing by increasing blood supply to damaged tissues. The investigators continue to probe the genetic basis of the intercellular communication between bacteria and host cells that allow this intriguing phenomenon. Perhaps in the future this pathogenic microbe will benefit patients with blood vessels grown from "scratch."

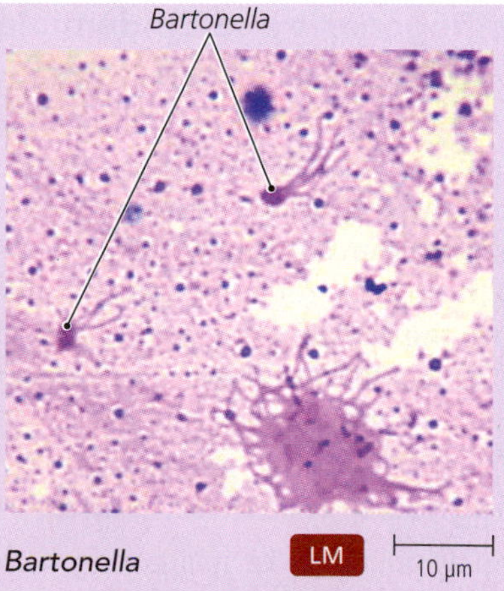

Bartonella

CLINICAL CASE STUDY

A Painful Rash

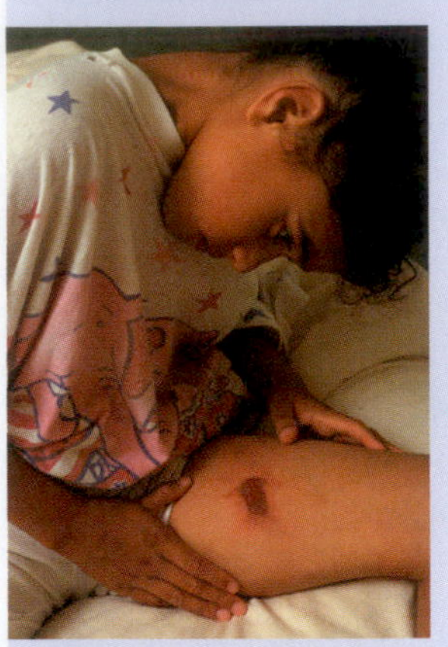

A mother brings her three-year-old daughter to her pediatric office stating that the girl has had fever and chills for three days. The girl also has a large, intensely red patch with a distinct margin on her leg and a nearby swollen lymph node. When the health care provider touches the area it is firm and warm, and the girl screams in pain. Based on these observations, the provider makes a presumptive diagnosis and begins treatment.

1. Is it necessary for the provider to confirm the diagnosis with a lab test? Why or why not?
2. What was the diagnosis? The treatment?
3. How is this different from impetigo?
4. What is the causative agent associated with the girl's condition?
5. How might the girl have contracted the condition?
6. What component(s) of the causative agent stimulated the fever and the lesion?
7. Why is it important for the provider to begin immediate treatment?

DISEASE AT A GLANCE 19.4

Gas Gangrene

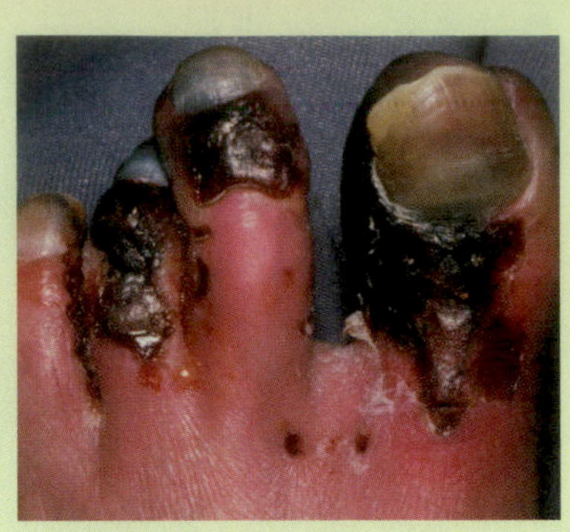

Blackened dead tissue and bubbling of bacterial gas wastes in gas gangrene

Cause *Clostridium* species, particularly *C. perfringens* (anaerobic, Gram-positive, endospore-forming rod).

Virulence factors Endospore, 11 toxins, rapid growth and cell division.

Portal of entry *Clostridium* endospores are introduced into dead tissue via a traumatic event such as surgical incision, puncture, gunshot wound, crushing trauma, abortion, or broken bones punching through the skin.

Signs and symptoms Increased pain and swelling around injury, fever, initially pale skin that turns dusky and then progresses to dark red or purple, foul-smelling drainage from tissues, crepitation (crackly sound due to gas under the skin), and tachycardia. Untreated gas gangrene causes shock, coma, and death.

Incubation period Generally less than 24 hours and no more than three days.

Susceptibility Individuals with deep, lacerating wounds, particularly "dirty" wounds.

Treatment Gas gangrene is a medical emergency, and prompt identification and treatment is crucial. Dead tissue must be surgically removed, and antitoxin, penicillin, and clindamycin administered.

Prevention Proper cleaning of wounds.

permeability, reduce blood pressure, and kill cells, resulting in irreversible damage. *C. perfringens,* the species most frequently isolated from patients, is a large, almost rectangular cell. Although it is nonmotile, its rapid growth and reproduction (cell divisions every 12 minutes) enables it to rapidly colonize wounds.

Pathogenesis and Epidemiology

Clostridium does not affect healthy tissues and is not invasive. A traumatic event, such as a surgical incision, puncture wound, gunshot wound, crushing trauma, abortion, or compound fracture, must introduce endospores into dead tissue. There, in the anaerobic environment of dead tissue, they germinate and grow on nutrients released by the dead cells. Despite therapeutic care, the mortality rate for gas gangrene exceeds 40%. Clostridial toxins kill neighboring cells, and the bacterium spreads into the dead tissue.

Diagnosis, Treatment, and Prevention

The appearance of gas gangrene is usually diagnostic by itself, though the detection of large Gram-positive bacilli is confirmatory. The condition is a medical emergency. A physician must intervene quickly and aggressively to stop the spread of necrosis in gas gangrene by surgically removing dead tissue and administering antitoxin and large doses of intravenous penicillin and clindamycin. Oxygen applied under high pressure may also be effective.

It is difficult to prevent infections of *C. perfringens* because the organism is common in soil. Given that gas gangrene occurs following introduction of endospores into tissues, proper cleaning of wounds can prevent many cases of the disease.

Disease at a Glance 19.4 summarizes the features of gas gangrene.

We have examined some common bacterial pathogens and opportunists that infect or affect the skin. Other bacterial diseases also manifest signs in the skin but have their major effect elsewhere. (Later chapters discuss these diseases. For example, scarlet fever, which manifests with bright red skin, has its most serious effects on other body systems, and is discussed in Chapter 22.) Now we turn our attention to viral pathogens of skin.

TELL ME WHY

Why do most cases of Rocky Mountain spotted fever occur in May, June, and July?

Viral Diseases of the Skin and Wounds

Many viral diseases are systemic in nature and spread by respiratory and oral routes; nevertheless, they manifest signs and symptoms in the skin. These include diseases of poxviruses, herpes infections, warts, chickenpox, shingles, rubella, measles, roseola, erythema infectiosum, and coxsackie viral infections. We begin our survey of viral skin diseases by considering infections of poxviruses.

Diseases of Poxviruses

LEARNING | OUTCOMES

19.24 Describe the progression of lesions in poxvirus infections.
19.25 Discuss the historical importance of poxviruses in immunization.
19.26 Discuss worldwide smallpox eradication.

Human diseases caused by poxviruses include smallpox and three diseases of animals—*orf* (sheep and goat pox), *cowpox*, and *monkeypox*—which rarely affect humans. (Despite its name, chickenpox is not caused by a poxvirus.)

Smallpox and cowpox have played important roles in the history of microbiology, immunization, epidemiology, and medicine. Recall from Chapter 1 that Edward Jenner first demonstrated immunization using the relatively mild cowpox virus to protect against smallpox. He was successful because the antigens of cowpox virus are chemically similar to those of smallpox virus, so that exposure to cowpox results in immunological memory and subsequent resistance to both cowpox and smallpox.

Smallpox was also the first human disease to be eradicated globally in nature. In 1967, the World Health Organization (WHO) began an extensive campaign of identification and isolation of smallpox victims, as well as vaccination of their contacts, with the goal of completely eliminating the disease. The efforts of WHO were successful, and in 1980 natural smallpox was declared eradicated—one of the great accomplishments of 20th-century medicine. Nevertheless, some scientists and government officials are concerned that smallpox could be reintroduced accidentally or through bioterrorism; therefore, knowledge of smallpox is necessary for health care providers.

Signs and Symptoms

All poxviruses produce lesions that progress through a series of stages **(FIGURE 19.10)**. The lesions begin as flat, reddened **macules**[18] (mak´yūlz), which then become raised sores called **papules**[19] (pap´yūlz). When the lesions fill with clear fluid, they are called **vesicles**, which then progress to pus-filled **pustules**[20]

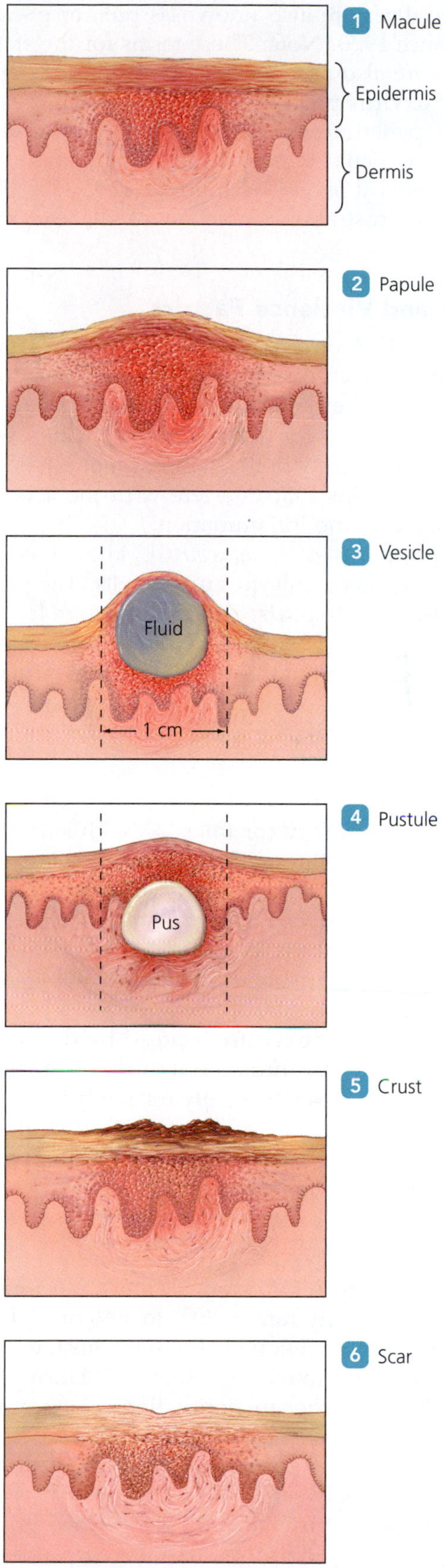

▲ **FIGURE 19.10 The stages of lesions of poxviral skin infections.** Red macules may become hard papules, then fluid-filled vesicles that become pus-filled pustules (pox), which sometimes crust over and leave scars.

[18]From Latin *maculatus*, meaning "spotted."
[19]From Latin *papula*, meaning "pimple."
[20]From Latin *pustule*, meaning "blister."

(pŭs´chūlz), which are also known as **pox**, or pocks (see Disease at a Glance 19.5) (Note: These terms for the stages of poxvirus lesions are also used to describe the lesions of other skin infections.) Poxvirus pustules then dry to form crusts. Because these lesions penetrate the dermis, they may result in characteristic scars, particularly on the face. Additionally, the disease is characterized by fever—as high as 42°C (107°F)—malaise, delirium, and prostration. Infection of the eye may induce blindness.

Pathogens and Virulence Factors

Poxviruses are DNA viruses. They have adhesion molecules allowing them to attach to a host cell's cytoplasmic membrane, where they trigger their own endocytosis. Unlike other DNA viruses, they carry genes for transcription enzymes, which allow them to replicate entirely in the cytoplasm of a cell. The viruses also code for proteins that interfere with the actions of interferon, complement, and inflammation.

Human poxvirus *(Orthopoxvirus)*, known commonly as **variola** virus, attaches only to human cells. The poxviruses of cows *(Vaccinia)*, sheep, goats, and monkeys are less easily contracted by humans because they do not attach as readily to human cells. Transmission of these poxviruses to humans requires prolonged close contact with infected animals, as Edward Jenner observed for milkmaids who contracted cowpox.

Pathogenesis

Close contact is necessary for infection with smallpox viruses. This occurs primarily through inhalation of viruses in droplets or dried crusts. Crusts can remain infective for two years. Smallpox viruses replicate in the respiratory tract and then spread via the blood and lymph throughout the body, causing the characteristic lesions on the skin about 12 days following infection and killing up to 40% of patients.

In contrast, other poxviruses spread by direct contact and do not move extensively throughout the body. Infections of humans with these viruses are usually relatively mild, resulting in pox and scars but little other damage. In the case of cowpox, the advent of milking machines has reduced the number of cases to near zero in the industrialized world.

Epidemiology

Smallpox virus exists in two strains: *variola major* causes severe disease with a mortality rate of 20% to 40% or higher, depending upon the age and general health of the host, whereas *variola minor* has a mortality rate of less than 1%. During the Middle Ages, 80% of the European population contracted smallpox. Later, European colonists unwittingly introduced smallpox into susceptible Native Americans, resulting in the death of as many as 3.5 million people.

Stocks of variola virus are still maintained in laboratories in the United States and in the Russian Federation as research tools for investigations concerning virulence, pathogenicity, protection against bioterrorism, vaccination, and recombinant DNA technology. Some government officials believe that illicit stocks of smallpox virus are being kept in so-called rogue nations such as Iran and North Korea. Now that smallpox vaccination has been generally discontinued, experts are concerned that the world's population is once again susceptible to smallpox epidemics should the virus be accidentally released from storage or used as a biological weapon. For this reason, some scientists advocate the destruction of all smallpox virus stocks; other scientists insist that we must maintain the virus in secure laboratories so that it is available for the development of more effective vaccines and treatments should the virus be disseminated.

An upsurge in the number of monkeypox cases in humans has been seen in the past decade, including an epidemic in the U.S. Midwest in 2003. This upsurge is probably the result of humans moving into monkey habitats in Africa, the importation of pet animals from that continent, and possibly changes in viral antigens. The elimination of smallpox and the cessation of smallpox vaccination may have led to increased human susceptibility to monkeypox, because smallpox vaccination also protects against monkeypox. Some scientists recommend that smallpox vaccination be reinstated in areas where monkeypox is endemic.

Diagnosis, Treatment, and Prevention

Physicians diagnose diseases of poxviruses by their distinctive pox.

Immediate immunization of people exposed to smallpox virus prevents the disease from developing. No treatment exists once smallpox develops.

Smallpox vaccinations ended in 1972 in most locales in the United States, in part because the vaccine caused adverse effects such as inflammation of the heart and, rarely, death. Now, in an effort to ward off potential bioterrorism, military personnel and selected health care providers are immunized. Interestingly, researchers have determined that adverse reactions to the vaccine are currently lower than historical rates, possibly because of better overall health care. They have also determined that previously immunized individuals can be successfully revaccinated with a diluted vaccine containing only 10% of the cowpox viruses in a standard vaccine.

Disease at a Glance 19.5 summarizes the features of smallpox.

Herpes Infections

LEARNING | **OUTCOMES**

19.27 Describe the diseases caused by human herpes viruses 1 and 2.

19.28 List conditions that may reactivate latent herpesviruses.

Signs and Symptoms

The name **herpes**, which is derived from a Greek word meaning "to creep," is descriptive of the slowly spreading skin lesions that are seen with herpesviruses. About a week after infection, the viruses produce painful, itchy skin lesions on the lips (called fever blisters or cold sores) **(FIGURE 19.11)** or genitalia (called

DISEASE AT A GLANCE 19.5

Smallpox

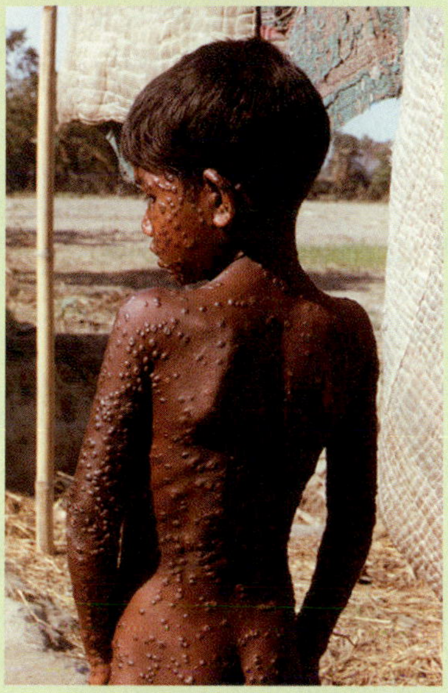

Cause *Orthopoxvirus* (variola virus) (enveloped, double-stranded DNA virus).

Virulence factors Intracellular infection, codes for proteins that inhibit interferon, complement, and inflammation.

Portal of entry Inhalation of viruses in droplets or dried crusts.

Signs and symptoms The prodromal period (two to four days) includes high fever, headaches and body aches, and malaise. A rash begins in the mouth and spreads to the face and downward all over the body. The rash becomes papules, then vesicles, and finally pustules. Eventually the pustules dry to form crusts. Scarring may result.

Incubation period Average 12 to 14 days, but can range from 7 to 17 days. The person is not contagious at this stage.

Susceptibility Currently, there is no naturally occurring smallpox. However, if a bioterrorism event or accidental release were to occur, it is theorized that most people would be susceptible since regular vaccination was discontinued decades ago.

Treatment Immediate vaccination; there is no other approved antiviral treatment for smallpox. Treatment includes supportive therapy.

Prevention Routine vaccination was discontinued in most of the U.S. in 1972. In response to recent world events, the U.S. government ordered new production of the smallpox vaccine and is currently prepared in the event of an outbreak.

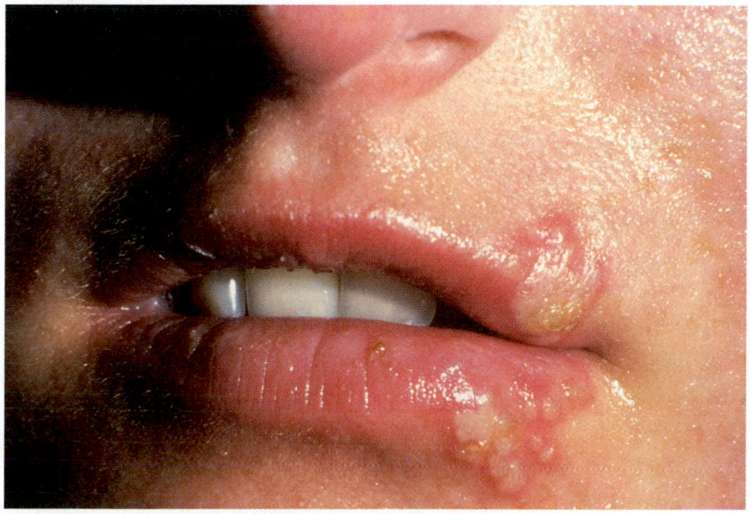

▲ **FIGURE 19.11 Oral herpes lesions.** *What is the most likely pathogen involved in oral herpes?*

Figure 19.11 *Human herpesvirus 1 causes approximately 90% of oral herpes lesions.*

genital herpes). Initial infections may also be accompanied by flulike signs and symptoms, including malaise, fever, and muscle pain. Severe infections in which the lesions extend into the oral cavity, called *herpetic gingivostomatitis*,[21] are most often seen in young patients and in patients with lowered immune function due to disease, chemotherapy, or radiation treatment. An inflamed blister called a *whitlow*[22] (hwit′lō) may occur on a finger if herpesvirus enters a cut or break in the skin of a finger. Athletes may develop herpes lesions in a condition called *herpes gladiatorum* almost anywhere on their skin as a result of contact with herpes lesions during sports.

One of the more distressing aspects of herpes infections is the recurrence of lesions. About two-thirds of patients with herpesvirus will experience recurrences during their lives as a result of the activation of latent viruses. After a primary infection, herpes simplex viruses often enter sensory nerve cells and are carried by cytoplasmic flow to the base of the nerve, where they remain latent in the trigeminal, sacral, or other ganglia,[23] as illustrated in **FIGURE 19.12**. Latent viruses may reactivate later in life as a result of immune suppression caused by stress, fever, trauma, sunlight, menstruation, or disease, and travel down the nerve to produce recurrent lesions as often as every two weeks. Fortunately, recurrent lesions are rarely as severe as those of initial infections because of the existence of immunological memory.

Pathogens and Virulence Factors

Two enveloped viral species of *Simplexvirus* cause herpes: *human herpesvirus 1* (HHV-1) and *human herpesvirus 2* (HHV-2). Historically, HHV-1 has also been known as "above the waist herpes" virus, whereas HHV-2 has been known as "below the waist herpes" virus, though either strain can occur in either location. HHV-2 is also usually the species involved in neonatal herpes. **TABLE 19.3** on p. 578 compares infections caused by human herpesviruses.

Human herpesviruses produce various proteins that act as virulence factors: glycoproteins in the viral envelope mediate viral adhesion and fusion, and other proteins inactivate complement and gamma class antibodies (IgG).

Pathogenesis

A herpesvirus attaches to a host cell through the fusion of its envelope with the cytoplasmic membrane. After the viral genome is replicated and assembled in the cell's nucleus, the virus acquires an envelope from the nuclear membrane and exits the cell via exocytosis or cell lysis.

Active lesions are the usual source of infection, though asymptomatic carriers can shed HHV-2 genitally. After entering the body through cracks or cuts in mucous membranes, viruses reproduce in epithelial cells near the site of infection and produce inflammation and cell death, resulting in painful, localized

[21]From Latin *gingival*, meaning "gum;" Greek *stoma*, meaning "mouth;" and Greek *itis*, meaning "inflammation."

[22]From Scandinavian *whick*, meaning "nail," and *flaw*, meaning "crack."

[23]A ganglion is a collection of enlarged portions of nerve cells that contains the cells' nuclei.

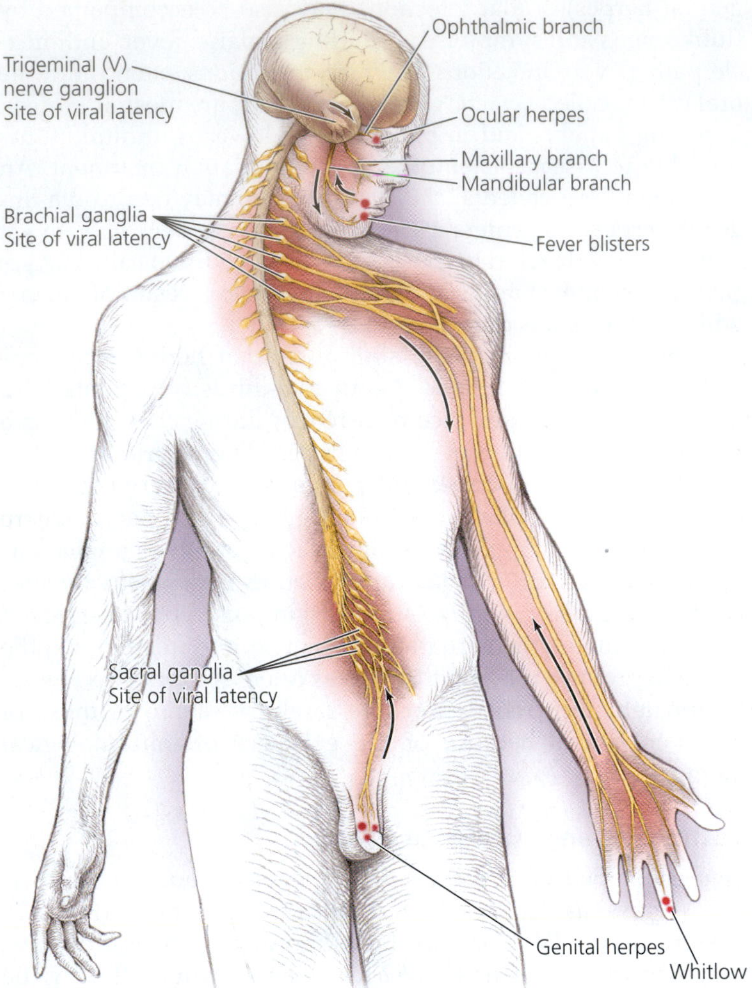

▲ **FIGURE 19.12 Sites of events in herpesvirus infections.** Infections typically occur when viruses invade the mucous membranes of either the lips or the genitalia, or through broken skin of a finger. Viruses may remain latent for years in the trigeminal, brachial, or sacral ganglia before traveling down nerve cells to cause recurrent symptoms in the lips, genitalia, fingers, or eyes (for ocular herpes). *What factors may trigger the reactivation of latent herpesviruses and the recurrence of symptoms?*

Figure 19.12 Stress, trauma, sunlight, menstruation, or diseases such as AIDS may cause recurrent symptoms as a result of immune suppression.

lesions on the skin. By causing infected cells to fuse with uninfected neighboring cells, HHV-1 and HHV-2 can form a structure called a *syncytium,* which is a multinucleate cytoplasmic mass, allowing the viruses to spread and in the process avoid the host's humoral immune system.

Initial infections with herpesviruses are usually age specific. Primary HHV-1 infections typically occur via casual contact during childhood and produce no signs or symptoms; in fact, by age 2, about 80% of children have been asymptomatically infected with HHV-1. Most HHV-2 infections, by contrast, are acquired between the ages of 15 and 29 as a result of sexual activity. (Chapter 24 discusses genital herpes in more detail.)

Epidemiology

Herpesviruses are most often spread between mucous membranes of the mouth and genitals. The viruses may remain inactive inside infected cells, often for years. Inactive viruses may reactivate as a result of aging, chemotherapy, immunosuppression, or physical and emotional stress, causing a recurrence of the manifestations of their diseases.

Whitlow is a hazard for children who suck their thumbs, for patients with genital herpes, and especially for health care workers in the fields of obstetrics, respiratory care, gynecology, and dentistry who come into contact with herpes lesions.

Although herpes infections in adults are painful and unpleasant, they are not life threatening. This is not the case for herpes infections in newborns. A fetus can be infected *in utero* when the virus crosses the placental barrier, but it is more likely that a baby is infected at birth through contact with lesions in the mother's reproductive tract. Also, mothers with oral lesions can infect their babies if they kiss them on the mouth. Neonatal herpes infections can be very severe, with a mortality rate of 30% if the infection is cutaneous or oral, and 80% mortality when the central nervous system is infected.

Diagnosis, Treatment, and Prevention

Physicians diagnose herpes infections by the presence of their characteristic, recurring lesions, especially on the lips and in the genital region. Microscopic examination of infected tissue reveals syncytia. Positive diagnosis is achieved by immunoassay that demonstrates the presence of viral antigen.

TABLE 19.3 Comparative Epidemiology and Pathology of Human Herpesvirus Infections

	HHV-1	HHV-2
Usual diseases	90% of cold sores/fever blisters; whitlow	85% of genital herpes cases
Mode of transmission	Close contact	Sexual intercourse
Site of latency	Trigeminal and brachial ganglia	Sacral ganglia
Locations of lesions	Face, mouth, and rarely trunk	External genitalia, and less commonly thighs, buttocks, and anus
Other complications	15% of genital herpes cases; pharyngitis; gingivostomatitis, ocular/ophthalmic herpes; herpes gladiatorum; 30% of neonatal herpes cases	10% of oral herpes cases; 70% of neonatal herpes cases

DISEASE AT A GLANCE 19.6

Herpes

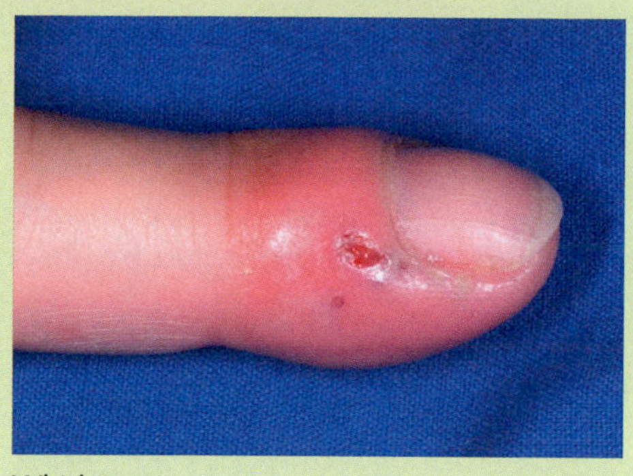
Whitlow

Cause *Human herpesviruses 1* and *2* (enveloped double-stranded DNA viruses).

Virulence factors Glycoproteins that allow viral attachment and entry, intracellular infections, proteins that inactivate complement and IgG.

Portal of entry Usually mucous membranes of the mouth or genitals.

Signs and symptoms A prodromal period of malaise, fever, and/or muscle pain may precede the typical blisters of a herpes infection. The blisters break, leaving a tender ulcer that may take several weeks to heal.

Incubation period 10 to 14 days.

Susceptibility Sexually active people have a greater risk of exposure to the disease, as does a baby during delivery by an infected mother.

Treatment Topical or oral administration of antivirals such as acyclovir, valacyclovir, or famciclovir can shorten outbreaks, but there is currently no cure.

Prevention Sexual abstinence, mutual monogamy, use of condoms, use of latex gloves by health care workers, and cesarean sections in infected mothers.

Herpes infections are among the few viral diseases that can be controlled with chemotherapeutic agents such as acyclovir or its derivatives. Topical applications of the drug limit the duration of the lesions and reduce viral shedding, though the drug does not cure the diseases or free nerve cells of viral infections.

Health care workers can reduce their exposure to infection by wearing latex or nitrile gloves. Since a whitlow is a potential source of infection, symptomatic personnel should not work with patients, especially newborns, until the whitlow has healed.

Disease at a Glance 19.6 summarizes the features of herpes skin infections.

Warts

LEARNING | **OUTCOMES**

19.29 Describe four kinds of warts associated with papillomavirus infections.

19.30 Describe the pathogenesis, treatment, and prevention of warts.

Warts (papillomas) (pap-i-lō´maz) are generally benign growths of the epithelium of the skin or mucous membranes.

Signs and Symptoms

Warts form on many body surfaces, but are most often found on the fingers or toes *(seed warts)*; deep in the soles of the feet *(plantar[24] warts)*; on the trunk, face, elbow, or knees *(flat warts)*; or on the external genitalia *(genital warts)* **(FIGURE 19.13)**. Generally, warts are painless, though they may itch or hurt, particularly plantar warts.

Pathogens and Virulence Factors

Almost 60 different strains of *Papillomavirus* cause warts. Various strains infect either cutaneous or mucosal tissues, causing infected epithelial cells to divide. This produces warts on the skin or mucous membranes, respectively. Some strains of papillomaviruses integrate into host cell chromosomes, potentially triggering the action of oncogenes, which cause cancers.

Pathogenesis

Besides causing infected epithelial cells to divide, papillomaviruses cause the cells to develop distinctive vacuoles surrounding their nuclei and to form dense cytoplasm. The incubation time from infection to the development of a wart is usually three to four months.

Warts are often unsightly, but most are harmless. Nevertheless, papilloma viruses may precipitate various head, neck, anal, vaginal, penile, and oral cancers. (Chapter 24 examines genital warts in more detail.)

Epidemiology

Papillomaviruses infect the skin through cuts or abrasions. They are transmitted via direct contact, including childbirth, and, because they are stable outside the body, via fomites. In a process called autoinoculation, patients can spread wart viruses from one location to another on their own bodies. Viruses that cause genital warts invade the skin and mucous membranes of the penis (particularly when it is uncircumcised), vagina, and anus during sexual intercourse.

Diagnosis, Treatment, and Prevention

Diagnosis of warts is usually a simple matter of observation, though only DNA probes can determine the exact strain of *Papillomavirus* involved. Warts usually regress over time as the cell-mediated immune system recognizes and attacks virally infected cells. Cosmetic concerns and the pain associated with some warts may necessitate removing infected tissue via surgery, freezing, cauterization (burning), laser, or the use of caustic chemicals, as are found in over-the-counter medications. These techniques are not always entirely satisfactory because viruses may remain latent in neighboring tissue and produce new warts at a later time. Laser surgery has the added risk of causing viruses to become airborne—some physicians have developed warts in their noses after inhaling airborne viruses during laser treatment of patients! Some patients have successfully removed

[24]From Latin *planta,* meaning "sole" (of the foot).

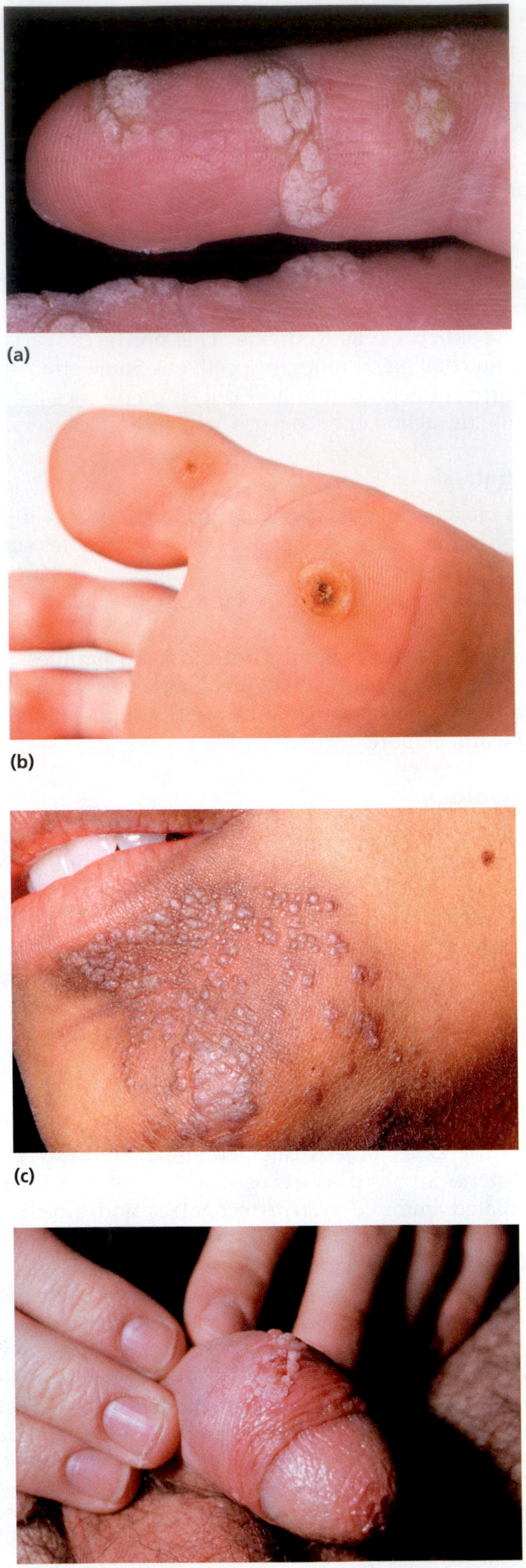
(a) (b) (c) (d)

◀ **FIGURE 19.13 Various kinds of warts—lesions caused by papillomaviruses.** **(a)** Seed wants of the fingers. **(b)** Plantar warts on the sole of a foot. **(c)** Flat warts, which often occur on the trunk, face, elbows, or knees. **(d)** Genital warts can occur on the genitalia in both sexes.

warts by covering them with duct tape. Covered warts usually disappear within two months.

Prevention of most types of warts is difficult. However, genital warts may be prevented by abstinence, mutual monogamy, or vaccine (see Chapter 24).

Chickenpox and Shingles

LEARNING | OUTCOMES

19.31 Compare and contrast the signs, symptoms, and treatment of chickenpox and shingles.

19.32 Explain the relationship between shingles and chickenpox.

Chickenpox is one of numerous childhood diseases that produce skin lesions. The causative virus may become dormant and then years later reactivate to cause shingles.

Signs and Symptoms

Chickenpox, which is medically termed **varicella**, is a highly infectious disease. About two to three weeks after infection with chickenpox virus, the patient develops a slight fever and characteristic skin lesions on the back and trunk that then spread to the face, neck, and limbs. In severe cases, the rash may spread into the mouth, pharynx, and vagina. Chickenpox lesions begin as macules, progress in one to two days to papules, and finally become thin-walled, fluid-filled vesicles on red bases, which have been called "teardrops on rose petals." These vesicles turn cloudy, dry up, and crust over in a few days. Successive crops of lesions appear over a three- to five-day period, so that at any given time all stages can be seen. Usually chickenpox is not life threatening, though it can, rarely, be fatal to newborns.

Chickenpox virus can become latent within sensory nerves and may remain dormant for years. Stress, aging, or immune suppression reactivate the viruses in 15–20% of individuals who have had chickenpox. Reactivated viruses travel down the nerve they inhabit and produce an extremely painful skin rash near the end of the nerve. This rash, known as **shingles** or *herpes zoster*, forms in the band of skin associated with the nerve or in the eye. Shingles lesions are accompanied by a burning sensation, numbness, itchiness, or intense pain.

After the scabs of shingles fall off, most people have no further skin manifestations, though the pain associated with shingles may remain for months or years after the lesions have healed. Some patients report that the pain is so intense they cannot wear clothing over the affected area for months or years.

Pathogen

Both chickenpox and shingles are caused by a type of herpesvirus called **varicella-zoster virus (VZV)** (genus *Varicellovirus*). (Note that chickenpox virus is not a poxvirus but a herpesvirus,

CLINICAL CASE STUDY

A Child with Warts

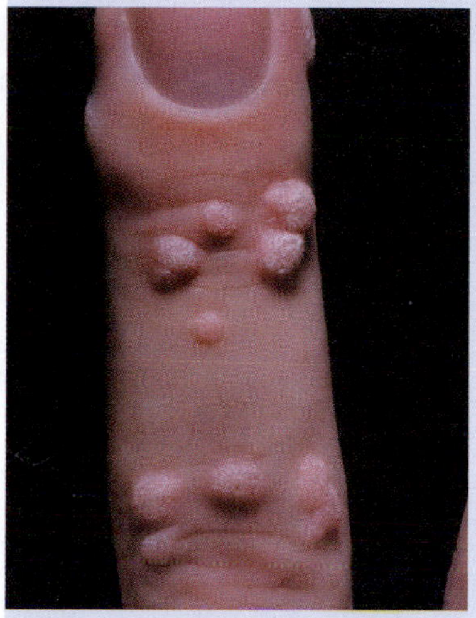

Ten-year-old Rudy has several large warts on the fingers of his hands. They do not hurt, but their unsightly appearance causes him to shy away from people. He is afraid to shake hands, or to play with other children, out of fear that he may transfer the warts to them. Initially, his mother tells him not to worry about them, but Rudy cannot help feeling self-conscious. Furthermore, Rudy fears that the warts may somehow spread on his own body. After consulting with a physician, his mother decides to have the warts surgically removed.

1. Is it possible that Rudy's warts will spread to other parts of his body?
2. Is it possible for someone else to "catch" Rudy's warts by shaking his hand?
3. If Rudy's warts disappear without treatment, are they likely to return to the same sites on his hand?
4. Are the warts on Rudy's finger likely to become cancerous?
5. Following surgical removal of the warts, is there a chance that Rudy will develop warts again?

and its Latin name is spelled differently from its English name.) VZV is named for the medical names of the two diseases.

Pathogenesis

Infection with VZV begins in the mucous membrane of the respiratory tract and then spreads into the liver, spleen, and lymph nodes via the blood and lymph. After two weeks, a second wave of viruses spreads via the blood throughout the body and to the skin. In the dermis, infected cells develop the characteristic rash, which is associated with fever and malaise. Viruses are shed before and during the symptoms through respiratory droplets and the fluid in lesions; dry crusts are not infective. Usually the disease is mild, but it may rarely be fatal, especially if associated with secondary bacterial infections.

The virus becomes latent in nerve ganglia and can reactivate in adults, usually after age 45. Reactivated virus travels down the nerve to manifest the lesions of shingles in the band of skin innervated by the infected nerve.

Epidemiology

Chickenpox is most often seen in children. Before immunization became routine, about 90% of children developed chickenpox acquired through the respiratory route or by the introduction of viruses from vesicles into skin breaks.

Chickenpox in adults is typically more severe than the childhood illness, presumably because much of the tissue damage results from a patient's immune response, and adults have a more developed immune system than children. Adult patients may require hospitalization, since chickenpox infection in adults can lead to fatal pneumonia.

Only 15–20% of adults who had chickenpox as children will develop shingles. Presumably this is because most infected individuals develop complete immunity during their bout with chickenpox. Unlike the multiple recurrences of herpes simplex lesions, shingles usually occurs once; only about 4% of patients ever develop a second case of shingles. The risk of contracting shingles increases with age.

A person who has never had chickenpox can acquire the disease from a shingles patient. This observation proved the two diseases have a common cause. The opposite is not true—a person cannot acquire shingles from a chickenpox victim.

Diagnosis, Treatment, and Prevention

Physicians diagnose chickenpox by the characteristic appearance of the lesions. It is more difficult to distinguish shingles from other types of skin lesions, though the localization within a band of skin on one side of the body is characteristic. Antibody tests are available to verify the diagnosis of both VZV infections.

Uncomplicated chickenpox is typically self-limiting and requires no treatment other than relief of the symptoms with acetaminophen and antihistamines, though the antiviral drug acyclovir may reduce the severity and duration of chickenpox. Aspirin should not be given to children or adolescents with symptoms of chickenpox because of the risk of contracting Reye's syndrome—a condition in which the liver and brain cease to function. Reye's syndrome is associated with aspirin usage during several viral diseases.

Treatment of shingles involves management of the symptoms, bed rest, and oral acyclovir. Loose-fitting clothing and nonadherent dressings may help prevent irritation of the lesions. Acyclovir provides relief from the painful rash for some patients, but it is not a cure.

It is difficult to prevent exposure to VZV because viruses are shed from patients before obvious signs appear. The Centers for Disease Control and Prevention (CDC) recommends immunization against chickenpox with attenuated varicella-zoster virus for all children between 12 and 18 months of age and a second dose before starting school. Since the vaccine provides protection against chickenpox, successfully vaccinated patients will not develop shingles later in life.

The CDC recommends a single dose of a different, more potent, attenuated vaccine for adults aged 60 or older who had chickenpox as children and two doses of this vaccine for everyone older than 19 who has not had chickenpox or successful vaccination in childhood.

Disease at a Glance 19.7 summarizes the features of varicella-zoster virus infections.

DISEASE AT A GLANCE 19.7

Chickenpox and Shingles

1. Inhalation of varicella-zoster virus.
2. Infected cells replicate and release virus into the blood (viremia) and lymph.
3. Viruses infect liver, spleen, and lymph nodes throughout body.
4. Second viremia spreads viruses to skin, causing extensive rash.
5. Viruses shed through respiratory droplets and fluid from skin lesions.
6. Immune system eliminates viruses from blood and most cells.
7. Some viruses become latent (dormant) in nerve ganglia.
8. Upon reactivation, viruses travel inside nerve cells down the nerve to infect skin cells.
9. Rash develops in the band of skin innervated by the infected nerve.

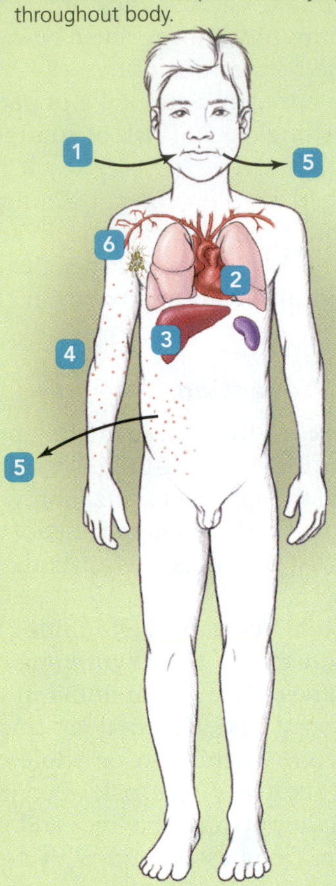

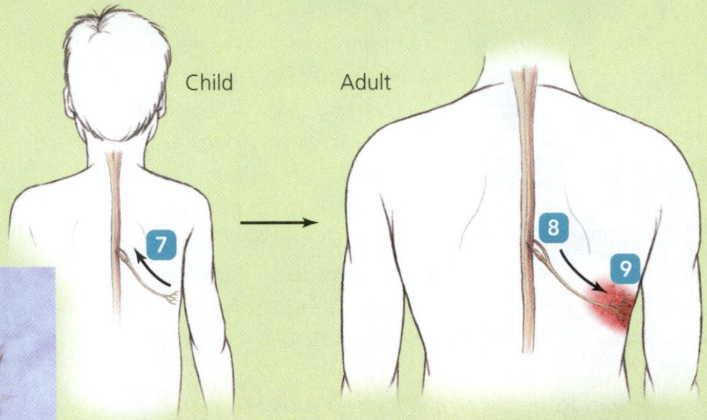

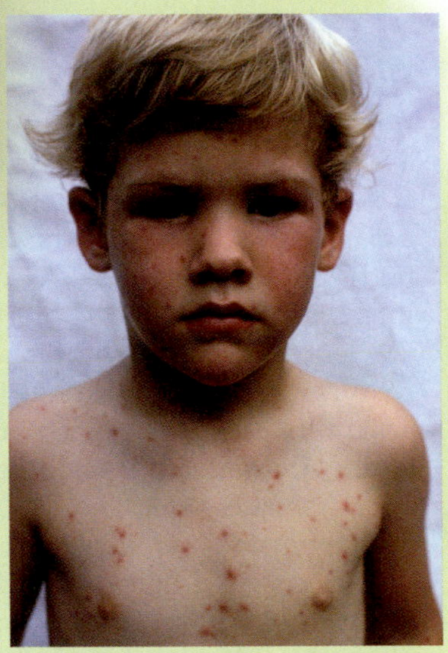

Characteristic chickenpox lesions

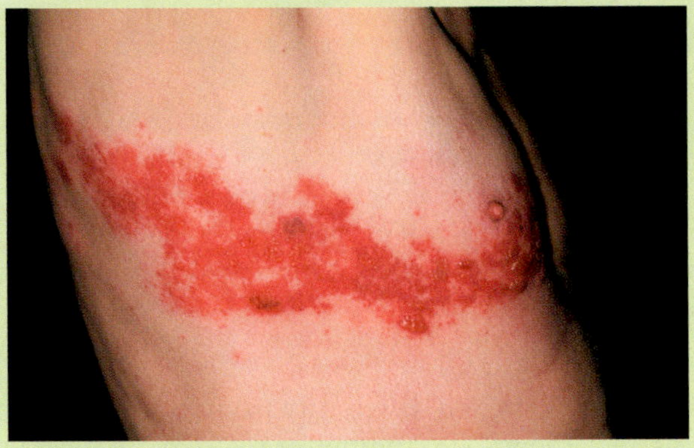

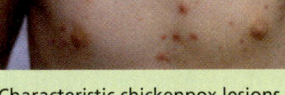

Characteristic shingles rash

Cause *Varicellovirus* (varicella-zoster virus, VZV) (enveloped, double-stranded DNA virus).

Virulence factors Intracellular infection, ability to lie dormant within nerve cells for years.

Portal of entry Respiratory tract.

Signs and symptoms Fever, characteristic rash on back and trunk that spreads to face and extremities. In 15–20% of cases, the virus reactivates years later, causing a painful shingles rash on a band of skin innervated by an infected nerve.

Incubation period Two to three weeks.

Susceptibility Chickenpox is a highly contagious disease to those who are not vaccinated or have not had it before. Persons who have had chickenpox previously are generally immune.

Treatment Treatment is supportive therapy and the antiviral drug acyclovir or its derivatives. Symptoms may be treated with acetaminophen and/or antihistamines. Aspirin should never be given to children with viral infections.

Prevention The CDC recommends an attenuated vaccine for children, a single dose of a more potent attenuated vaccine for adults 19–49 who had chickenpox as children, and two doses of the latter for adults without chickenpox immunity.

Rubella

LEARNING | OUTCOMES

19.33 Describe rubella, including its cause, signs, and symptoms.

19.34 Explain why and how the United States has nearly eradicated rubella.

Rubella[25] (rū-bel′ă), another of the childhood diseases that produces skin lesions, was first distinguished as a separate disease from measles by German physicians. Rubella has been known in the United States as *German measles* or *three-day measles.*

[25]Latin, meaning "little red."

Signs and Symptoms

Rubella is a generally harmless disease in children, causing only slightly swollen lymph nodes and a mild rash of flat, pink to red spots (macules) (see Disease at a Glance 19.8) that lasts about three days. Infections in adults are more severe and may result in arthritis or encephalitis.

Rubella was not seen as a particularly serious disease until 1941, when Dr. Norman Gregg (1892–1966), an Australian eye doctor, recognized that rubella infections of pregnant women resulted in teratogenic[26] birth defects in their babies. These effects include cardiac abnormalities, deafness, blindness, mental retardation, microcephaly,[27] and growth retardation. Death of the fetus is also common. It is now known that the virus can move across the placenta even when the mother is asymptomatic.

Pathogen and Pathogenesis

Rubella virus *(Rubivirus)* is an enveloped, icosahedral, single-stranded RNA virus. Rubella virus infects cells of the upper respiratory tract and spreads from there to lymph nodes, into the blood, and then throughout the body via the blood. Rubella virus does not kill infected cells, but it does cause dividing cells to make errors during DNA replication. The severity of rubella in adults is attributed to cell death resulting from a cell-mediated immune response against virally infected cells.

Epidemiology

Rubella virus spreads through respiratory secretions and infects humans only. Patients shed virions (complete virus particles) in respiratory droplets for approximately two weeks before and two weeks after the rash. The most severe form of rubella is in a fetus under 20 weeks old.

Diagnosis, Treatment, and Prevention

Diagnosis of rubella is usually made by observation, which is confirmed by serological testing for IgM against rubella. No treatment is available, but immunization has proven effective at reducing the incidence of rubella in industrialized countries **(FIGURE 19.14)**. Such immunization is aimed at reducing the number of rubella cases that might serve to introduce rubella to pregnant women. Rubella vaccine (combined with vaccine against measles and mumps as MMR) is made from a live, weakened virus and therefore should never be given to pregnant women or immunocompromised patients.

Disease at a Glance 19.8 summarizes the features of rubella.

DISEASE AT A GLANCE 19.8

Rubella

Cause *Rubivirus* (rubella virus) (enveloped, positive, single-stranded RNA virus).

Virulence factors Attachment proteins, intracellular infection.

Portal of entry Respiratory tract.

Signs and symptoms Mild rash of pink to red macules and swollen lymph nodes in children. Causes severe congenital defects in unborn babies, including deafness, blindness, microcephaly, and mental and growth retardation.

Incubation period 7 to 14 days.

Susceptibility Unvaccinated individuals and unborn babies.

Treatment There is no specific treatment for rubella other than supportive therapy.

Prevention Immunization with the combined MMR (measles, mumps, and rubella) vaccine. Pregnant women should not be given the live, attenuated vaccine.

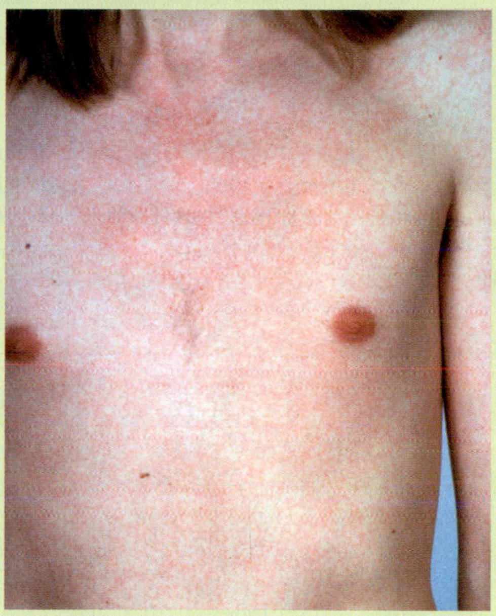

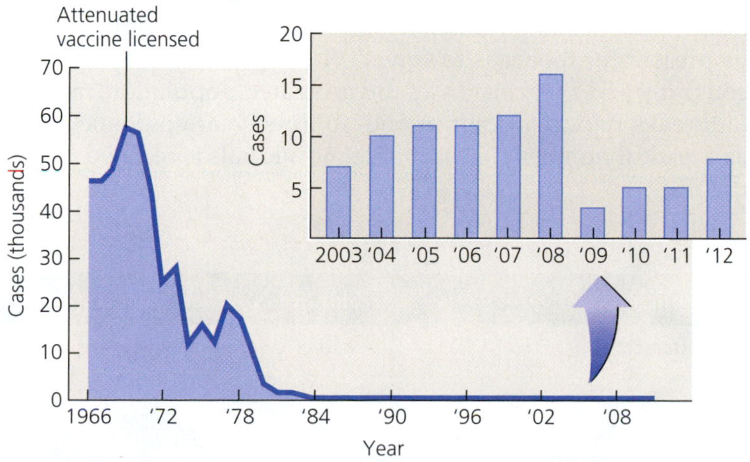

▲ **FIGURE 19.14 The efficacy of immunization against rubella.** The use of a live, attenuated virus vaccine has practically eliminated rubella in the United States.

Measles (Rubeola)

LEARNING | **OUTCOMES**

19.35 Describe the cause, signs and symptoms, and prevention of rubeola.

19.36 Explain how subacute sclerosing panencephalitis (SSPE) develops from defective measles virions.

[26]From Greek *teras,* meaning "monster," and *genein,* meaning "to produce."
[27]From Greek *mikro,* meaning "small," and *kephale,* meaning "to head."

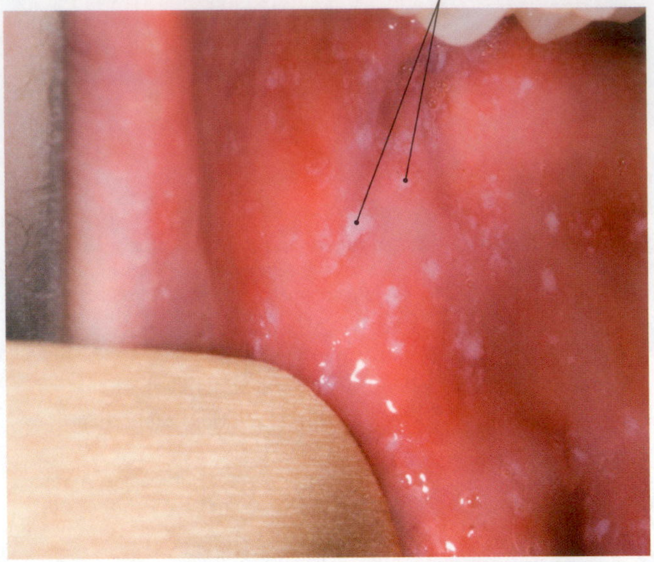

▲ **FIGURE 19.15** Oral lesions of measles (Koplik's spots).

Measles virus is another childhood disease that produces rashes. Measles, also known in the United States as *rubeola*[28] or *red measles,* is one of the more contagious and serious childhood diseases, and it should not be confused with the generally milder childhood rubella (German measles).

Signs and Symptoms

Signs and symptoms of measles include fever, sore throat, headache, dry cough, and conjunctivitis (inflammation of the lining of the eyelid). After two days of illness, lesions called **Koplik's spots** appear on the mucous membrane of the mouth **(FIGURE 19.15)**. These lesions, which have been described as crystals of salt surrounded by a red halo, last one to two days and provide a definitive diagnosis of measles. Red raised (maculopapular) lesions then appear on the head and spread over the body. These later lesions, which initially are similar to those of rubella, are extensive and often fuse to form red patches, which gradually turn brown as the disease progresses.

[28]Rubeola is the name given to German measles (rubella) in some other countries.

Rare complications of measles include pneumonia, encephalitis, and the extremely serious **subacute sclerosing panencephalitis (SSPE)**. SSPE is a slow progressive disease of the central nervous system that involves personality changes, loss of memory, muscle spasms, and blindness. The disease begins one to ten years after the initial infection with measles virus and lasts a few years before resulting in death.

A defective measles virus that cannot make a capsid causes SSPE. The defective virus replicates and moves from brain cell to brain cell via cell fusion, limiting the functioning of infected cells and resulting in the symptoms. SSPE afflicts fewer than 7 measles patients in 1 million and is becoming much rarer as a result of childhood immunizations.

TABLE 19.4 compares and contrasts the more serious measles with the milder rubella (German measles).

Pathogen and Virulence Factors

Measles virus *(Morbillivirus)* is a single-stranded RNA virus with an enveloped helical capsid. Besides adhesion proteins, the envelope has a *fusion protein* that triggers fusion between an infected cell and its neighbors, allowing the virus to pass from cell to cell and avoid antibodies.

Pathogenesis

Measles virus infects cells of the respiratory tract before spreading via lymph and blood throughout the body to infect the conjunctiva, urinary tract, small blood vessels, lymphatics, central nervous system, and skin. Cytotoxic T cells kill infected cells, causing most of the symptoms. Most patients recover within two to three weeks, though 1–5% of infected children die. Secondary infections by bacteria such as *Staphylococcus* or *Streptococcus* can increase the seriousness of measles. Patients with compromised cellular immunity, such as AIDS patients, can have continued measles infection, resulting in death.

Epidemiology

Measles virus is highly contagious. More than 80% of unvaccinated patients exposed to the virus develop symptoms 8–12 days later. Humans are the only host of measles virus, and a numerous, dense population of susceptible individuals must be present for the virus to spread via respiratory droplets from coughing and sneezing. In an unvaccinated population, measles outbreaks typically occur in one- to three-year epidemic cycles as a critical number of susceptible individuals is reached.

TABLE 19.4 A Comparison of Measles and Rubella

Disease	Causative Agent	Primary Patient(s)	Complications	Skin Rash	Koplik's Spots
Rubella (also known as German measles, rubeola, or three-day measles)	*Togaviridae: Rubivirus*	Child, fetus	Birth defects	Mild	Absent
Measles (also known as rubeola or red measles)	*Paramyxoviridae: Morbillivirus*	Child	Pneumonia, encephalitis, subacute sclerosing panencephalitis	Extensive	Present

CLINICAL CASE STUDY

Grandfather's Shingles

The Davises were excited about their newborn twin boys and couldn't wait to take them to see Mr. Davis's father. Grandfather Davis was excited to see his first grandsons as well and thought their visit might help take his mind off the pain of his shingles, which had suddenly appeared only days before.

1. What virus is responsible for Grandfather Davis's shingles?
2. What likely triggered his case of shingles?
3. Are the new parents at risk of catching shingles from Grandfather Davis?
4. Are the newborns at risk of contracting shingles from their grandfather? If Mrs. Davis asked your advice about visiting her father-in-law with the twins, what would you recommend?
5. Could the twins be vaccinated against shingles before the visit?

Every country in the Americas, even the poorest, has halted the spread of measles virus by vaccinating at least 95% of the population. The virus cannot spread, because there are not enough susceptible people. This is not the case in Europe.

Epidemiologists are concerned about the reemergence of measles in Europe—the number of cases has at least quadrupled since 2009. This is directly related to the fact that vaccination rates in Europe have dropped below 90-95% of the population—the rate needed to halt viral transmission. The rate in the United Kingdom dropped to almost 50% following a fraudulent report in 1998 that suggested the vaccine caused autism in children.

Diagnosis, Treatment, and Prevention

The signs of measles, particularly Koplik's spots, are sufficient for diagnosis, but serological testing can confirm the presence of measles antigen in respiratory and blood specimens. Treatment involves supportive therapy and administration of vitamin A, antibodies against measles virus, and the antiviral drug ribavirin.

Scientists introduced an effective, live, attenuated vaccine for measles in 1963. Research has shown that measles elimination requires achieving and sustaining a two-dose immunization

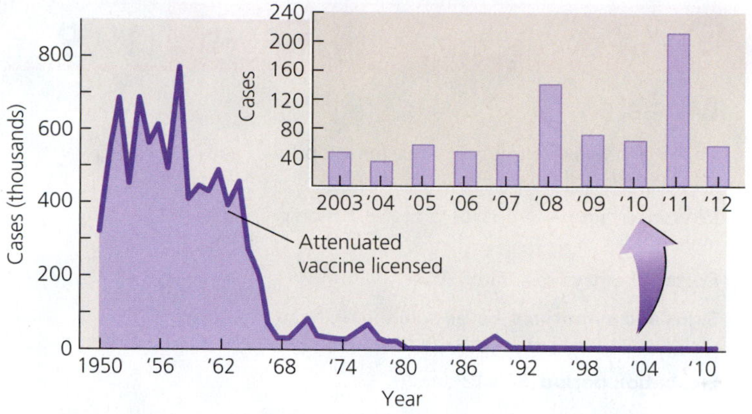

▲ **FIGURE 19.16 Measles cases in the United States since 1950.** The number of cases has declined dramatically since immunization began in 1963. *Why does the measles vaccine pose risks for immunocompromised contacts of vaccinated children?*

Figure 19.16 *Measles vaccine is an attenuated live vaccine; thus, it can replicate and cause disease in immunocompromised patients.*

coverage of at least 95% of a population. Pediatricians administer the two doses of measles vaccine, in combination with vaccines against mumps and rubella (MMR vaccine), to children in the United States at about 12 months and 4 years of age. In this manner, they have nearly eliminated measles as an endemic disease in the United States **(FIGURE 19.16)**, but measles remains a frequent cause of death in other countries.

Disease at a Glance 19.9 summarizes the features of measles.

Other Viral Rashes

LEARNING OUTCOME

19.37 Describe pathogenesis of erythema infectiosum and roseola infection.

Other rare childhood diseases caused by viruses are characterized by rashes. Here we consider three: erythema infectiosum, roseola, and coxsackievirus infection.

Erythema Infectiosum

Erythema infectiosum (er-i-thē´mă in-fek-shē-ō´sŭm) is a respiratory disease caused by a single-stranded DNA (ssDNA) virus called B19 virus in the genus *Erythrovirus*, family *Parvoviridae*. As the name implies, the disease manifests as an infectious reddening of the skin. The disease is also known as *fifth disease* because it was the fifth childhood rash on a list used by pediatricians in the early 1900s. (The others were often listed as scarlet fever, rubella, measles, and a mild scarlet fever–like illness called Duke's disease.)

The rash of erythema infectiosum begins on the cheeks and spreads to the arms, thighs, buttocks, and trunk. The facial rash appears as if the patient has been slapped, a characteristic that child protection workers could misinterpret **(FIGURE 19.17)**. Sunlight aggravates the rash at all stages; earlier lesions fade

DISEASE AT A GLANCE 19.9

Measles

Cause *Morbillivirus* (rubeola virus) (enveloped, negative, single-stranded RNA virus).

Virulence factors Adhesion proteins, fusion proteins, intracellular infection.

Portal of entry Respiratory tract.

Signs and symptoms Fever, sore throat, headache, dry cough, conjunctivitis, Koplik's spots, and maculopapular rash.

Incubation period 8 to 12 days.

Susceptibility Unvaccinated individuals.

Treatment Supportive therapy, ribavirin, vitamin A, and antibodies against the virus.

Prevention Immunization with the combined MMR (measles, mumps, and rubella) vaccine. The vaccine contains live, attenuated measles virus.

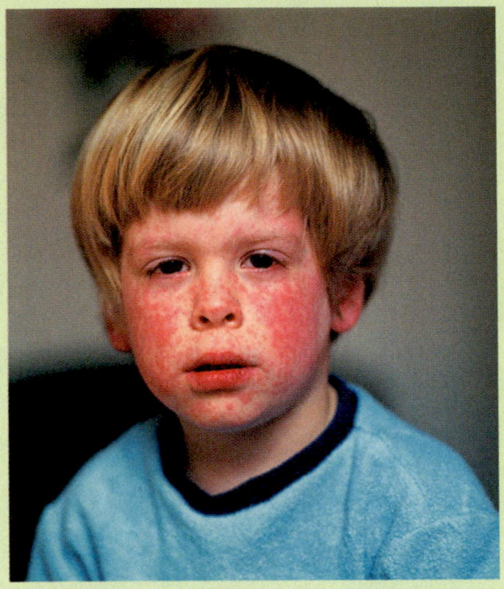

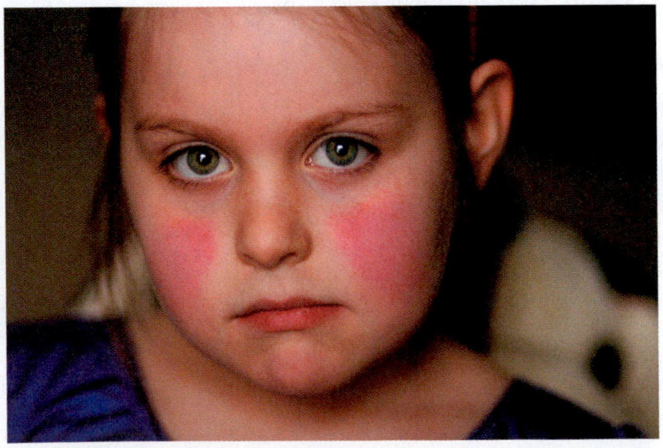

▲ **FIGURE 19.17 A case of erythema infectiosum (fifth disease).** Distinct reddening of facial skin that resembles the result of a slap is characteristic of this parvovirus-caused disease.

as the disease progresses, though they can reappear when the patient is exposed to sunlight. Infected adults who missed the disease in childhood may suffer anemia, joint pain, and rarely miscarriage in addition to the rash. Treatment involves use of nonsteroidal anti-inflammatory drugs (NSAIDs).

Roseola

Roseola (rō-zē-ō´lă) is an endemic illness of children characterized by an abrupt fever, sore throat, enlarged lymph nodes, and a faint pink rash on the face, neck, trunk, and thighs. The name of the disease is derived from its characteristic rose-colored rash. A herpesvirus in the genus *Roseolovirus*, known simply as *human herpesvirus 6* (HHV-6), causes roseola. Some researchers link HHV-6 with the development of multiple sclerosis in adults, and there is some evidence that HHV-6 infection makes individuals more susceptible to HIV infection and AIDS. No pharmaceutical treatment exists.

TELL ME WHY

Why has the number of cases of measles increased dramatically, especially in Europe?

Now that we have examined some bacterial and viral diseases of the skin, we turn our attention to fungal diseases of the skin, hair, and nails.

Mycoses of the Hair, Nails, and Skin

LEARNING | **OUTCOMES**

19.38 Define *mycosis*.

19.39 Describe four general types of mycoses based on location in the body.

Mycoses (fungal diseases) are traditionally classified according to the locations within the body that they affect. Physicians classify mycoses as superficial (on the surface), cutaneous (in the skin), subcutaneous (in the hypodermis and muscles), or systemic (affecting numerous systems). The fungi that cause mycoses are mainly opportunistic.

Superficial and cutaneous mycoses of the hair, nails, and skin are generally not life threatening, though they may cause chronic or recurring diseases. Mycoses affecting the blood and organs of the body may also secondarily affect the skin, but they are considered in later chapters when we discuss the other body systems.

We begin our survey by considering mycoses affecting the hair, nails, and skin.

Superficial Mycoses

LEARNING | **OUTCOME**

19.40 Describe the clinical and diagnostic features of superficial fungi, as exemplified by *Malassezia* infections.

Superficial mycoses, which are the most common fungal infections, are confined to the hair, nails, and outer layers of the skin, all of which are composed of dead cells filled with keratin. Surface mycoses are acquired by direct contact with the hyphae or spores of opportunistic fungi. An example of a superficial mycosis is pityriasis.

Signs and Symptoms

Pityriasis versicolor (pit-i-rī´ă-sis vĕr´si-cŏl-ŏr) is characterized by hypopigmented or hyperpigmented patches of scaly skin resulting from fungal interference with melanin production (**FIGURE 19.18**). This condition, which typically occurs on the trunk, shoulders, and arms, and rarely on the face and neck, is called *versicolor* in reference to the variable pigmentation seen in affected individuals.

Pathogens and Virulence Factors

Malassezia furfur (mal-ă-sē´zē-ă fur´fur), a dimorphic basidiomycete and a normal inhabitant of human skin around the world, causes pityriasis. It feeds on oil produced by the skin, causing infections that tend to be chronic.

Pathogenesis and Epidemiology

Superficial fungi produce *keratinase*, an enzyme that dissolves keratin. These fungi do not penetrate living tissues and do not trigger immune responses, which differentiates them from the cutaneous and subcutaneous pathogens (discussed shortly).

Transmission of the fungi of pityriasis onto the surface of the skin is usually mediated via shared hairbrushes and combs, and several members of a family may be infected at the same time. Adolescents are more likely to be afflicted, possibly due to increased sebum production related to hormonal changes. In severely immunocompromised patients, superficial infections can spread to cover significant areas of skin or invade deeper tissues and become systemic.

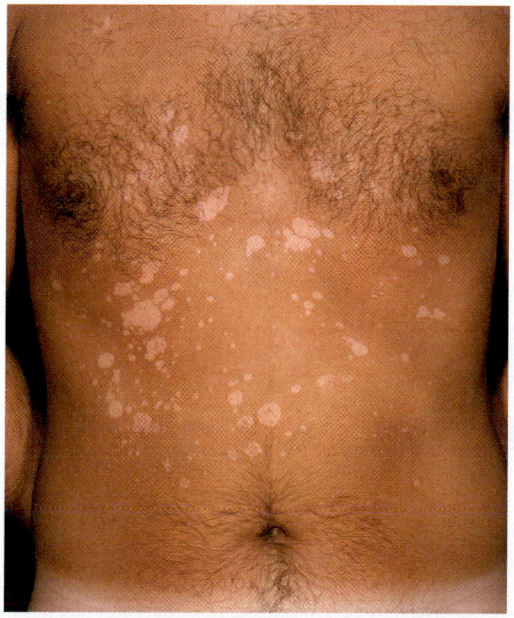

▲ **FIGURE 19.18 Pityriasis versicolor.** The variably pigmented skin patches are caused by *Malassezia furfur.*

Diagnosis, Treatment, and Prevention

The quickest way to diagnose *M. furfur* infections is to examine the patient under ultraviolet illumination with a wavelength of 365 nm; patches of skin infected with *M. furfur* fluoresce pale green. However, fluorescence under ultraviolet is not a definitive diagnosis, because some other fungi also fluoresce. Definitive diagnosis requires microscopic examination. Preparations of skin samples mixed with 10% KOH reveal both masses of budding yeast and short hyphal forms that are diagnostic for *M. furfur.*

Superficial *Malassezia* infections are treated topically with solutions of ciclopirox or antifungal imidazoles, such as ketoconazole shampoos. Alternatively, topical applications of zinc pyrithione, selenium sulfide lotions, or propylene glycol solutions can be used. Oral therapy with ketoconazole may be required to treat extensive infections and infections that do not respond to topical treatments. Relapses are common, and prophylactic topical treatment may be necessary. The skin takes months to regain its normal pigmentation following successful treatment.

Cutaneous Mycoses

LEARNING | **OUTCOMES**

19.41 Describe dermatophytoses in general.

19.42 Explain why cutaneous mycoses are not as common as superficial mycoses.

Some fungi that grow on humans manifest clinically as cutaneous lesions called **dermatophytoses** (der´mă-tō-fī-tō´sēz). They are caused by *dermatophytes,* which are fungi that grow on skin, nails, and hair and invade the body, where they stimulate cell-mediated immune responses that damage deeper tissues. Such triggering of immune responses distinguishes dermatophytoses from superficial fungal infections.

Signs and Symptoms

In the past, dermatophytoses were often called *ringworms* or *tineas* (tin´ē-ăs),[29] because dermatophytes produce circular, scaly patches that made observers think a worm lay just below the surface of the skin (**FIGURE 19.19**). Though these diseases are doubly misnamed—worms are not involved, and the term *phyte* implies plants, not fungi—these common terms remain in use.

Most dermatophytoses are clinically distinctive and so common that they are readily recognized; for example, athlete's foot is a dermatophytosis (**FIGURE 19.20**). Dermatophytoses can have a variety of other clinical manifestations, some of which are summarized in **TABLE 19.5** on p. 588.

[29]Latin, meaning "worms."

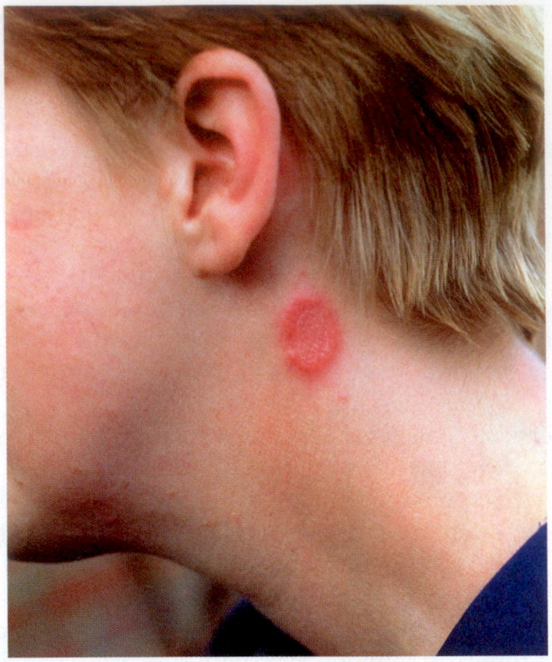

▲ **FIGURE 19.19 Dermatophytosis (ringworm).** This is a cutaneous fungal disease, in this case on the neck. *What three genera cause most dermatophytoses?*

Figure 19.19 Trichophyton, Microsporum, and Epidermophyton are the more common causes of ringworms.

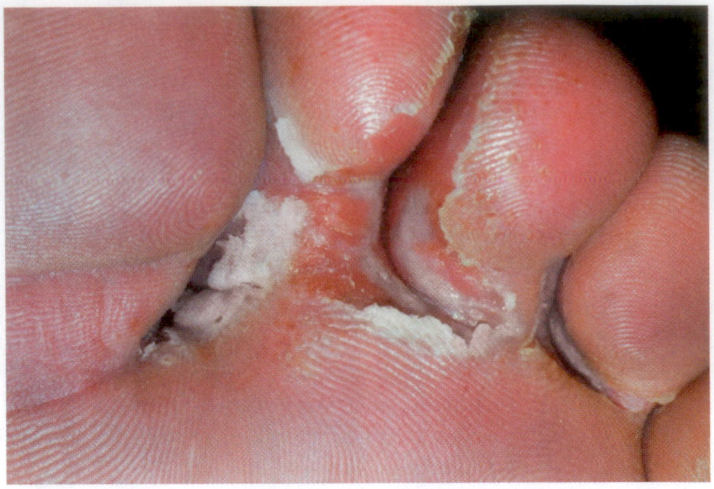

▲ **FIGURE 19.20 Athlete's foot.** Dermatophytosis of skin on the foot.

Pathogens

Three genera of ascomycetes are responsible for most dermatophytoses:

- *Microsporum* (mī-kros´po-rŭm) species
- *Trichophyton* (trik-ō-fī´ton) species
- *Epidermophyton floccosum* (ep´i-der-mof´i-ton flŏk´ō-sŭm)

All three genera cause skin and nail infections, and *Trichophyton* species also infect hair.

Pathogenesis

Dermatophytes use keratin as a nutrient source and thus colonize dead layers of skin, nails, and hair. They also trigger destruction of living cells by the immune system. The fungi involved in cutaneous and subcutaneous mycoses commonly grow on decaying matter in the soil. Infection requires introduction of fungal elements into the living, deeper layers of skin. Thus, exposure is common, but infection is rare.

Most dermatophytoses remain localized to the dermis and hypodermis; infections rarely become systemic.

Epidemiology

Dermatophytes are among the few contagious fungi. Spores and bits of hyphae, along with dead skin cells and hair, are

TABLE 19.5 Common Dermatophytoses

Disease	Agents	Common Signs	Source
Tinea pedis ("athlete's foot")	*Trichophyton rubrum; T. mentagrophytes* var. *interdigitale; Epidermophyton floccosum*	Red, raised lesions on and around the toes and soles of the feet; webbing between the toes is heavily infected	Human reservoirs in toe webbing; carpeting holding infected skin cells
Tinea cruris ("jock itch")	*T. rubrum; T. mentagrophytes* var. *interdigitale; E. floccosum*	Red, raised lesions on and around the groin and buttocks	Usually spreads from the feet
Tinea unguium (onychomycosis)	*T. rubrum; T. mentagrophytes* var. *interdigitale*	Superficial white onychomycosis: patches or pits on the nail surface; Invasive onychomycosis: yellowing and thickening of the distal nail plate, often leading to loss of the nail	Humans
Tinea corporis	*T. rubrum; Microsporum gypseum; M. canis*	Red, raised, ringlike lesions occurring on various skin surfaces (tinea corporis on the trunk, tinea capitis on the scalp, tinea barbae of the beard)	Can spread from other body sites; can be acquired following contact with contaminated soil or animals
Tinea capitis	*M. canis; M. gypseum; T. equinum; T. verrucosum; T. tonsurans; T. violaceum; T. schoenleinii*	Ectothrix invasion: fungus develops arthroconidia on the outside of the hair shafts, destroying the cuticle; Endothrix invasion: fungus develops arthroconidia inside the hair shaft without destruction; Favus: crusts form on the scalp, with associated hair loss	Humans; can be acquired following contact with contaminated soil or animals

constantly shed from infected individuals, making recurrent infections common. Dermatophytes infecting humans are classified according to their natural habitats as follows:

- *Anthropophilic dermatophytes* are associated with humans only and are transmitted either by close human contact or through the sharing of contaminated fomites.
- *Zoophilic dermatophytes* are associated with animals; the fungi are transmitted to humans either by close contact with pets or other animals or through contaminated animal products such as wool.
- *Geophilic dermatophytes* are soil fungi that are transmitted to humans via direct exposure to soil or to dusty animals.

Diagnosis, Treatment, and Prevention

Clinical observation is generally sufficient to diagnose infections with dermatophytes. Potassium hydroxide (KOH) preparations of skin or nail scrapings or hair samples can reveal hyphae and/or conidia (asexual spores), which confirm the diagnosis. When desired, the determination of the specific identity of a dermatophyte requires microscopic examination of a laboratory culture, which may take weeks because these fungi grow slowly in the laboratory.

Limited infections can be treated effectively with topical antifungal agents, but more widespread infections of the scalp or skin, as well as nail infections, must be treated with oral agents. Terbinafine, administered topically for one to four weeks, is effective in most cases. Chronic or stubborn cases are treated with griseofulvin until cured.

CLINICAL CASE STUDY

Is It Athlete's Foot?

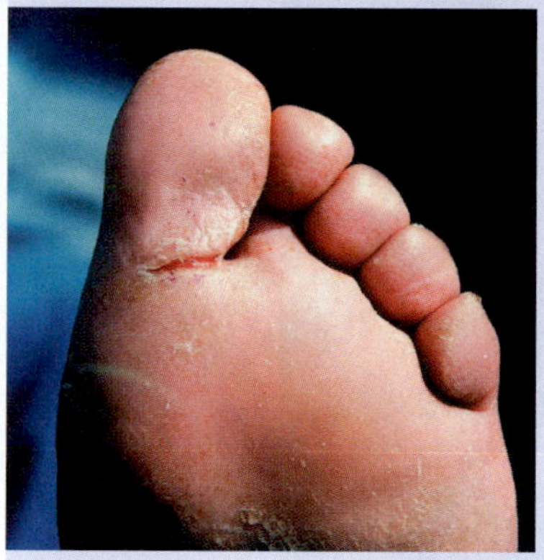

A 30-year-old man comes into a community clinic complaining of persistent redness, itching, and peeling skin on the soles of his feet and between his toes. He states that he works out daily and plays on several sports teams. He concluded he has athlete's foot, but every over-the-counter antifungal agent he has tried has failed to cure his condition. He has finally come to the clinic because his toenails have begun to thicken, yellow, and detach from the nail bed. His condition is pictured.

1. Given the information here, can athlete's foot be diagnosed definitively? If not, what laboratory tests could be performed to positively diagnose the condition and identify the infecting organism?
2. What treatment is likely? Would the recommended treatment be different if two fungal species, and not just one fungal species, were present?

Wound Mycoses

LEARNING **OUTCOMES**

19.43 State the specific difference between chromoblastomycosis and phaeohyphomycosis.

19.44 Define *mycetomas*, and describe their appearance, cause, and treatment.

19.45 Describe how the lesions of lymphocutaneous sporotrichosis correlate with its spread through the lymphatic system.

Some fungi grow on deeper tissues in the body, including the hypodermis and bone, but they do not become systemic; that is, they do not spread throughout the body's systems. Such fungi eventually grow up into the epidermis to produce lesions on the skin's surface. Among these mycoses of wounds are *chromoblastomycosis* and *phaeohyphomycosis,* which are similar mycoses caused by melanin-producing, dark-pigmented fungi; *mycetoma;* and *sporotrichosis.* Here we consider these conditions in order.

Chromoblastomycosis

Four species of ascomycete fungi cause **chromoblastomycosis**[30] (krō´mō-blas´tō-mī-kō´sis): *Fonsecaea pedrosoi* (fon-sē-sē´ă pe-drō´-sō-ē), *F. compacta* (kom-pak´ta), *Phialophora verrucosa* (fī-ă-lof´ō-ră ver-ū-kō´să), and *Cladophialophora carrionii* (klă-dŏf´ē-ă-lŏf-ō-ră kar-rē-ŏn´ē-ē). Initially, chromoblastomycosis uniformly appears as small, itchy but painless, scaly lesions on the skin resulting from fungal growth in subcutaneous tissues near the site of a wound. Over the course of several years the lesions progressively worsen, becoming large, flat to thick, tough, and wartlike. They become tumorlike and extensive if not treated **(FIGURE 19.21)**. Inflammation, development of fibrous tissue, and abscess formation occur in surrounding tissues. The fungus can spread throughout the body.

Microscopic examination of stained skin scrapings reveal the key distinguishing feature of this fungal disease—golden brown bodies that are distinctive and distinguishable from budding yeast. Determining the exact species of fungus relies on laboratory culture, macroscopic examination of colonies on agar, and microscopic examination of spores.

Chromoblastomycosis is difficult to treat, especially in advanced cases, and often the disease cannot be cured. The earlier treatment begins, the more likely it will be successful. For treatment, physicians must remove infected and surrounding tissues—extensive lesions may require amputation—and prescribe a prolonged course of thiabendazole and 5-fluorocytosine.

Despite the worldwide occurrence of the fungi, the overall incidence of chromoblastomycosis is relatively low. People who work daily in the soil with bare feet are at risk due to accidental

[30]From Greek *chroma,* meaning "color;" *blastos,* meaning "germ;" and *mykes,* meaning "fungus."

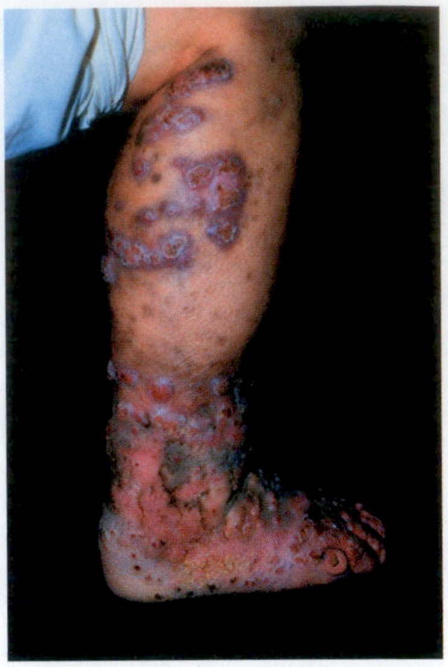

▲ **FIGURE 19.21 A leg with extensive lesions of chromoblastomycosis.** *Fonsecaea pedrosoi* caused this case.

foot wounds. The simple act of wearing shoes greatly reduces the number of infections.

Phaeohyphomycosis

Of the more than 30 genera of fungi that cause **phaeohyphomycosis**[31] (fē´ō-hī´fō-mī-kō´sis), the more common are the ascomycetes *Alternaria* (al-ter-nā´rē-ă), *Exophiala* (ek-sō-fī´ă-lă), *Wangiella* (wang-gē-el´ă), and *Cladophialophora.* Phaeohyphomycosis is acquired when spores, which are prevalent in indoor environments, including hospitals, enter open surgical or traumatic wounds.

Phaeohyphomycoses are more variable in presentation than are chromoblastomycoses. For example, paranasal sinus phaeohyphomycosis involves colonization of the nasal passages and sinuses; it occurs in allergy sufferers and AIDS patients. In cerebral phaeohyphomycosis, the fungus actively invades the brain. Fortunately, this is the rarest form of phaeohyphomycosis and occurs only in the severely immunocompromised. Some cases of phaeohyphomycosis can be treated with itraconazole, but the disease is permanently destructive to tissues.

Microscopic examination of stained skin scrapings, biopsy material, or cerebrospinal fluid (CSF) reveals brown-pigmented hyphae, which are diagnostic. Determining the species of fungus relies on laboratory culture, macroscopic examination of colonies on agar, and microscopic examination of spores.

Mycetomas

Fungal **mycetomas** (mī-sē-tō´maz) are tumorlike infections of the skin, fascia (lining of muscles), and bones of the hands or feet caused by soil fungi of several genera in the division Ascomycota: *Madurella* (mad´ū-rel´ă), *Pseudallescheria* (sood´-al-es-kē´rē-ă), *Exophiala* (ek-sō-fī´ă-lă), and *Acremonium* (ak-rĕ-mō´nē-ŭm). These fungi are distributed worldwide, but infection is most prevalent in countries near the equator.

Prick wounds and scrapes caused by twigs, thorns, or leaves introduce these fungi into humans. As with chromoblastomycosis and phaeohyphomycosis, people who work barefoot in soil are most at risk and wearing protective shoes or clothing can greatly reduce infection.

Infection begins near the site of a wound with the formation of small, hard, subsurface nodules that slowly worsen and spread as time passes. Local swelling occurs, and ulcerated lesions produce pus. Infected areas release an oily fluid containing fungal spores and hyphae. The fungi spread to more tissues, destroying bone and causing permanent deformity **(FIGURE 19.22)**.

A combination of the symptoms and microscopic demonstration of fungi in samples from the infected area is diagnostic for mycetomas. Laboratory culture of specimens produces macroscopic colonies that can be identified down to the level of species.

Treatment involves surgical removal of the mycetoma that may, in severe cases, involve amputation of an entire limb. Surgery is followed by one to three years of antifungal therapy with ketoconazole. Though slowly growing, the fungi invade body tissues; thus, even combinations of surgery and antifungal agents do not always cure the disease.

Sporotrichosis

Sporothrix schenckii (spōr´ō-thriks shen´kē-ē) is a dimorphic ascomycete that causes **sporotrichosis** (spōr´ō-tri-kō´sis), or *rose-gardener's disease,* a subcutaneous infection usually limited to the arms and legs. *S. schenckii* resides in the soil and is most commonly introduced in wounds by thorn pricks or wood splinters. Avid gardeners, farmers, and artisans who work with natural plant materials have the highest incidence of sporotrichosis. Though distributed throughout the tropics and subtropics, most cases occur in Latin America, Mexico, and Africa.

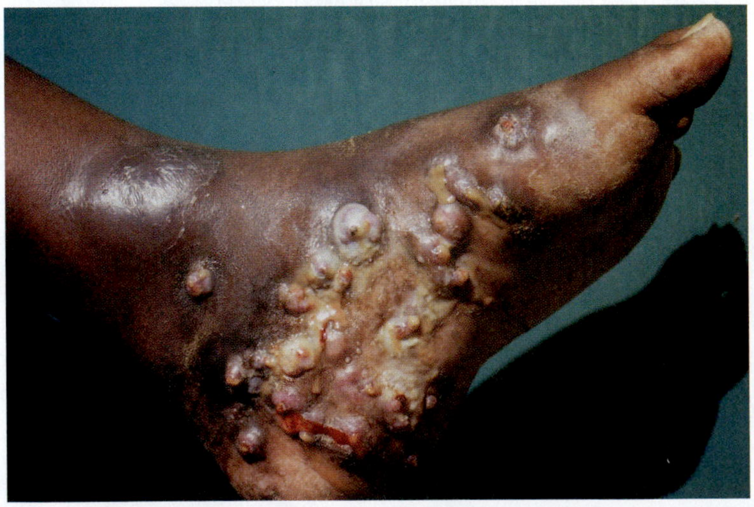

▲ **FIGURE 19.22 A mycetoma of the ankle.** The bone and other tissues were invaded and destroyed by a fungus (in this case, *Madurella mycetomatis).*

[31]From Greek *phaios,* meaning "dusky," *hyphe,* meaning "web," and *mykes,* meaning "fungus."

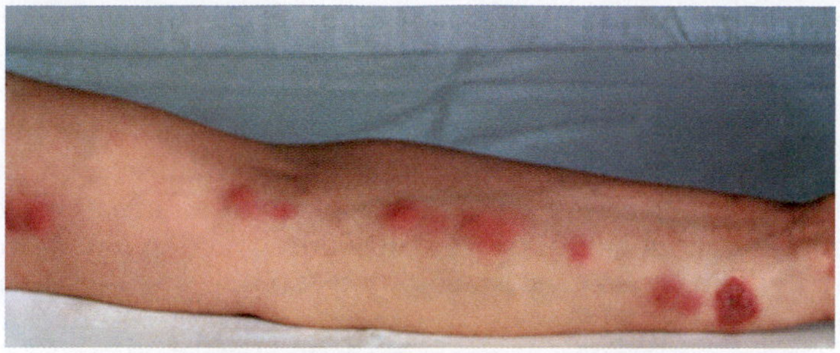

▲ **FIGURE 19.23 Lymphocutaneous sporotrichosis on the arm.** The locations of these secondary, subcutaneous lesions correspond to the course of lymphatic vessels leading from sites of primary, surface lesions. *How is sporotrichosis contracted?*

Figure 19.23 *Sporotrichosis is contracted when this soilborne fungus is inoculated into the skin by thorn pricks and scratches.*

The disease is also common in warm, moist areas of the United States.

Fixed sporotrichosis initially appears as painless, nodular lesions that form around the site of inoculation. With time, these lesions produce a pus-filled discharge, but they remain localized and do not spread. In *lymphocutaneous sporotrichosis,* however, the fungus enters the lymphatic system near the site of a primary lesion, giving rise to secondary lesions on the skin surface along the course of lymphatic vessels **(FIGURE 19.23)**. The fungus remains restricted to subcutaneous tissues and does not enter the blood.

Microscopic observation of pus or biopsy tissue stained with silver can reveal budding yeast forms in severe infections, but often the fungus is present at a low density, making direct examination of clinical samples a difficult method of diagnosis. The patient's history and clinical signs, plus the observation of the dimorphic nature of *S. schenckii* in laboratory culture, are considered diagnostic for sporotrichosis.

Cutaneous lesions can be treated successfully with topical applications of saturated potassium iodide for several months. Itraconazole and amphotericin B are also useful treatments. Prevention requires wearing gloves, long clothing, and shoes to prevent inoculation.

To this point we have examined bacterial, viral, and fungal pathogens that affect the skin. Parasitic protozoa and arthropods can also infest the skin. Next we consider two such parasites.

TELL ME WHY

Onychomycoses (on´i-kō-mī-kō´sēs) are nail infections that can be caused by several species of fungi. Why are these mycoses so difficult to treat? Why is it generally necessary to treat patients with antifungal agents for long periods of time?

Parasitic Infestations of the Skin

Some parasitic protozoa and arthropods infest the skin and cause diseases. Here we examine an example of each, beginning with leishmaniasis.

Leishmaniasis

LEARNING | **OUTCOME**

19.46 Describe leishmaniasis, including its cause, signs and symptoms, pathogenesis, epidemiology, diagnosis, and treatment.

Leishmaniasis (lēsh´mă-nī´ă-sis) is a **zoonosis**—a disease of animals transmitted to humans. This disease of the skin and oral mucous membrane is caused by an intracellular parasitic protozoan.

Signs and Symptoms

The clinical manifestations of leishmaniasis depend on the species of pathogen, the geographical location of hosts and vectors, and the immune response of the infected host. Three clinical forms of leishmaniasis are commonly observed:

- *Cutaneous leishmaniasis* involves large painless skin ulcers that form around bite wounds made by the disease vector. Such lesions often become secondarily infected with bacteria. Scars remain when the lesions heal.
- *Mucocutaneous leishmaniasis* results when skin lesions enlarge to encompass the mucous membranes of the mouth, nose, or soft palate. Damage is severe and permanently disfiguring **(FIGURE 19.24)**. Neither of these forms of leishmaniasis is fatal.
- *Visceral leishmaniasis* (also known as *kala-azar*) is typically fatal in 100% of untreated cases. In this disease, macrophages spread the parasite to the liver, spleen, bone marrow, and lymph nodes. Inflammation, fever, weight loss,

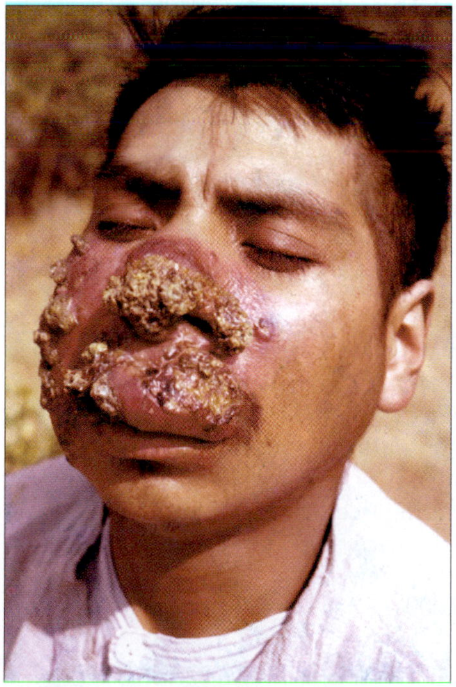

▲ **FIGURE 19.24 Mucocutaneous leishmaniasis.** The large skin lesions are permanently disfiguring.

CLINICAL CASE STUDY

Diagnosis in the Desert

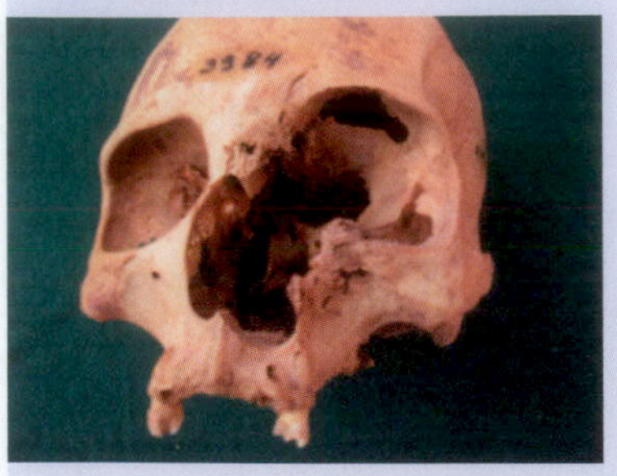

In the 1970s, archeologists unearthed a vast cemetery in the Atacama Desert of Chile. Because the Atacama is the driest desert on Earth, bodies over 800 years old were better preserved than might be expected. Some facial and brain tissues were mummified. The skulls of four women appeared as if the bone had been eaten, though it was obvious to the archeologists that the women had lived with the condition and died from other causes. It wasn't until in 2009 that scientists were able to solve the mystery of these skulls. Genetic analysis of tissue samples revealed that a flagellated protozoan had moved from facial lesions to the women's skulls, slowly eating them away over a 20-year period.

1. What skin pathogen caused the disfigurement?
2. How did the women become infected?
3. If the women were infected today, what treatment would be given?

and anemia increase in severity as the disease progresses. Visceral leishmaniasis is becoming increasingly problematic as an opportunistic infection among AIDS patients.

Pathogen and Virulence Factors

Leishmania (lēsh-man´ē-ă), the cause of leishmaniasis, is a flagellated protozoan commonly hosted by wild and domestic dogs and small rodents. Infected female sand flies of the genera *Phlebotomus* (fle-bot´ō-mŭs) and *Lutzomyia* (lūts-ŏm´yē-a) transmit *Leishmania* from animals to humans through their bites. Of the 30 known species of *Leishmania*, 21 can infect humans.

Pathogenesis and Epidemiology

Upon initial infection, the parasites infect macrophages, which become activated but not sufficiently to kill the intracellular *Leishmania*. Chemicals released by infected macrophages stimulate inflammatory responses that continue to be propagated by the infection of additional macrophages.

Leishmaniasis is endemic in parts of the tropics and subtropics, including Central and South America, central and southern Asia, Africa, Europe, and the Middle East. Of the more than 12 million new cases of leishmaniasis worldwide each year, over 60,000 are fatal.

Diagnosis, Treatment, and Prevention

Microscopic identification of protozoa in samples from cutaneous lesions, the spleen, or bone marrow is diagnostic. Immunoassays using antibodies to detect antigen confirm the diagnosis and identify the strain. Genetic analysis using PCR is needed for definitive speciation.

Most cases of leishmaniasis heal without treatment (though scars may remain) and confer immunity upon the recovered patient. At one time, lesions were purposely encouraged on the buttocks of small children both to induce immunity and to prevent lesions from forming elsewhere and leaving more-visible scars. Treatment, which is required for more serious infections, involves administering amphotericin B or sodium stibogluconate, an antimicrobial drug that contains the heavy metal antimony and which is not licensed for use in the United States.

Prevention is essentially limited to reducing exposure by controlling reservoir host and sand fly populations. For example, rodent nesting sites and burrows can be destroyed around human habitations to reduce contact with potentially infected reservoir populations. In some areas, infected dogs are also destroyed to eliminate this reservoir. Insecticide used around homes reduces the number of sand flies, and personal use of insect repellants, protective clothing, and netting further limits exposure. A vaccine for leishmaniasis does not exist currently.

Scabies

LEARNING OUTCOME

19.47 Describe the cause, signs and symptoms, and treatment of scabies.

A 17th-century Italian physician made one of the earlier connections between microorganisms and human disease by showing that an arachnid causes a common skin disease, **scabies**.

Signs and Symptoms

Scabies is characterized by intense itching, especially at night, and a rash localized to infested areas of the skin. The rash appears as pimple-like irritations or short, slightly raised burrows under the surface of the skin. Lesions are more commonly found in the webbing between fingers, in skin folds of the wrists, elbows, or knees, and around the genitalia, though any area of skin can be affected. Scratching often compounds the symptoms by introducing *Staphylococcus* or *Streptococcus* into the wounds, which triggers bacterial infections, including impetigo.

Pathogen and Virulence Factors

The mite *Sarcoptes scabiei* (sar-kop´tēz skā´bē-ī) causes scabies. This burrowing arachnid has a body about 300 μm long covered with ridges, spines, and hairs **(FIGURE 19.25)**. As it burrows, the mite damages nerve endings (see Figure 19.1a), and its antigens trigger inflammation.

Pathogenesis and Epidemiology

Adult female mites live up to a month in human skin where they burrow to lay their eggs, producing the itching blisters.

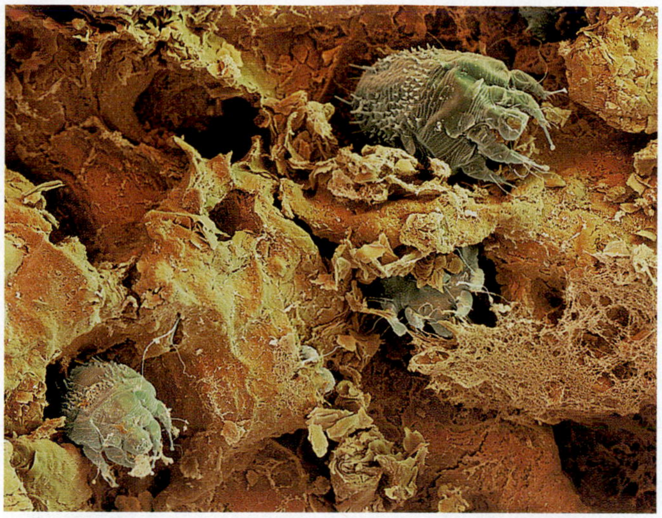

▲ **FIGURE 19.25** **Scabies mites *(Sarcoptes scabiei)* burrowing in human skin.**

Symptoms may take four to six weeks to develop as type IV (delayed) hypersensitivity plays a role in a patient's reaction.

Infested patients spread the mites from one part of their body to another and pass the parasites to others during prolonged bodily contact or by sharing clothing, towels, or bedding, particularly in crowded conditions regardless of personal hygiene. Sexual transmission is common.

Physicians estimate that 300 million individuals develop scabies each year throughout the world; most are children under age 15 and people with weakened immune systems. Epidemics occur among people in crowded conditions such as hospitals, nursing homes, army barracks, day care centers, prisons, and refugee camps.

Diagnosis, Treatment, and Prevention

Dermatologists diagnose scabies by the characteristic burrows of the mites and by finding the microscopic mites, eggs, or fecal matter in skin scrapings. However, since there are typically fewer than 10 mites on the entire body of a patient, clinicians may miss them.

Treatment involves the use of mite-killing lotions applied to the entire body below the neck for eight hours. Since such lotions are absorbed into the blood, frequent use or application on infants or pregnant or lactating women is ill advised. All clothing, bedding, and towels of a patient must be thoroughly washed in hot water and dried in a hot dryer to kill mites and their eggs. Itching may continue for two to three weeks after all mites have been killed.

Immunity to scabies does not develop; the only preventive is good personal hygiene. Patients must avoid contact with infested individuals and their fomites.

EMERGING DISEASE CASE STUDY

Monkeypox

Jacob was excited about the pet he got for his birthday; he was the only boy at school who had a prairie dog! Common dogs were boring; his prairie dog could climb up the drapes, run through plastic tube mazes, and sit on Jacob's shoulder to eat. "Doggy" was the best. In fact, everything would be just about perfect if only Jacob weren't sick in bed.

Two weeks after his birthday Jacob was exhausted and felt terrible. He had fever accompanied by headache and muscle aches, so his mother had kept him home from school. Then the lymph nodes in his neck and armpits had swollen, and he had gotten frightening, large, raised lesions all over his face and hands. He was really sick.

For the next three weeks, the lesions had become crusty and then fallen off, only to be replaced by new lesions. Jacob felt awful and missed playing with Doggy, who had been confined to his cage while Jacob stayed in bed.

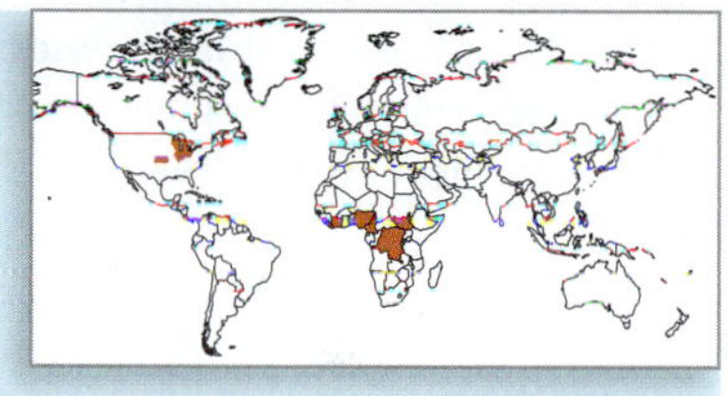

Jacob's doctor had been unable to diagnose the disease, but an infectious disease specialist had finally confirmed monkeypox—a disease named for monkeys who can get the disease. In reality, monkeypox is a disease of rodents, including prairie dogs. Final investigation revealed that the pet wholesaler who had supplied Doggy to Jacob's pet store in Illinois had also supplied African monkeys to stores in Texas. At least one of those monkeys had been infected with monkeypox virus—a virus closely related to smallpox virus. The monkey had infected Doggy, who had then infected Jacob.

Jacob was sick for four weeks, but monkeypox is not as virulent as smallpox, so Jacob recovered with only some scars as a reminder that not all unusual pets are a blessing.

1. Why did Jacob's lymph nodes swell during the course of a skin disease?
2. Why didn't the doctor prescribe penicillin or tetracycline?
3. How can a virus of rodents possibly cause a disease of monkeys and humans?

MICRO IN THE CLINIC FOLLOW-UP

A BAD CASE OF ACNE

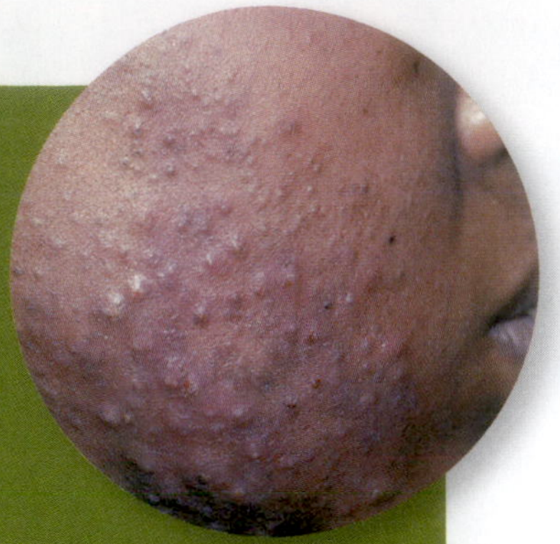

After a full examination of Rockeem's face, neck and back, the dermatologist concludes that Rockeem has typical *acne vulgaris* caused by small Gram-positive bacteria called *Propionibacterium acnes*. She also explains that the acne is not Rockeem's fault—acne is not caused by eating certain foods, how much you sweat, or any other single factor. Over 75% of all adolescents have some form of acne. The doctor, who has great skin now, mentions that she had such a bad case of acne when she was Rockeem's age that it was the major reason she went into dermatology.

The doctor explains that changes in hormones induce excessive oil production during adolescence, which, when combined with dirt and dead skin, creates an anaerobic environment in the skin pores: ideal conditions for acne. *P. acnes* ferments within the pores of the skin, producing propionic acid, inflammation, and pimples. There are several treatments available. In mild cases, washing the affected skin with over-the-counter cleansers containing benzoyl peroxide or salicylic acid can help. For more severe cases, prescription lotions or antibiotics may be necessary. There are also ultraviolet light therapies that target *P. acnes*.

For Rockeem, the dermatologist prescribes an antimicrobial called minocycline. She also instructs Rockeem to use a skin cleanser that contains benzoyl peroxide. Somewhat relieved, Rockeem returns to the high-pressure world of junior high school.

1. **Describe how the skin's first line of defense that usually protects the skin contributes to Rockeem's acne.**
2. **Rockeem was prescribed minocycline. Another acne treatment is retinoic acid, also known as Accutane. What are some potential side effects of Accutane?**

MM Check your answers to Micro in the Clinic Follow-Up questions in the MasteringMicrobiology Study Area.

Explore the Invisible

Visit the **MasteringMicrobiology Study Area** to challenge your understanding with practice tests, animation quizzes, and clinical case studies!

MasteringMicrobiology®

CHAPTER SUMMARY

Structure of the Skin (pp. 558–559)

1. The skin is composed of a **dermis** and an **epidermis**.
2. The leathery dermis, which contains sweat glands, oil glands, and hair follicles, provides strength and resiliency and supports the overlying epidermis. The epidermis, which is continually replaced, is an oily, salty, waterproof barrier against pathogens.
3. A skin wound is trauma to the epidermis or dermis. Pathogens can enter the body through wounds.

Normal Microbiota of the Skin (p. 559–560)

1. A variety of normal **microbiota**, including *Staphylococcus* and **diphtheroids**, live on the skin surface, on hair follicles, and in sweat ducts.

Bacterial Diseases of the Skin and Wounds (pp. 560–574)

1. **Folliculitis** can occur at the base of a hair follicle (called a pimple) or at the base of an eyelid (called a sty). A boil or **furuncle** results when infection spreads to surrounding tissues. When several furuncles join, a **carbuncle** is formed.
2. *Staphylococcus epidermidis* and *S. aureus* are common on the skin. Virulence factors of the latter include enzymes, antiphagocytic capsule, protein A, and toxins.
3. Because resistant strains of *Staphylococcus* have become common, aseptic techniques are especially important in hospitals to protect patients.
4. Some strains of *Staphylococcus aureus* secrete **exfoliative toxins** that dissolve the proteins that hold cell membranes together. The

result is **staphylococcal scalded skin syndrome (SSSS)**, a condition in which the epidermis peels off in sheets.

5. **Impetigo**, also called pyoderma, is a contagious skin disease caused by *Staphylococcus aureus* or *Streptococcus pyogenes* (group A *Streptococcus*). **Erysipelas** results when impetigo infections spread to lymph nodes. Impetigo and erysipelas are more common in children and in the elderly.
6. **Necrotizing fasciitis**, commonly caused by group A *Streptococcus (S. pyogenes)*, is a painful disease characterized by a red painful lesion, flulike symptoms, digestion of connective tissue around muscles, toxemia, and often death.
7. **Acne** pimples are commonly composed of dead *Propionibacterium acnes* or *Staphylococcus aureus* bacteria combined with living leukocytes. Dead bacteria that block pores form blackheads. Cysts can rupture and form scar tissue.
8. *Bartonella henselae* is carried by cats and causes **cat scratch disease** when introduced through a scratch or bite.
9. *Pseudomonas aeruginosa* is a ubiquitous opportunistic pathogen producing pyocyanin, a blue-green pigment. The microbe kills cells, destroys tissues, and triggers shock in immunocompromised patients and burn victims, and it causes otitis externa (swimmer's ear).
10. Ticks carry rickettsias, which cause **spotted fever rickettsioses**, most notably **Rocky Mountain spotted fever (RMSF)** caused by *Rickettsia rickettsii*. Rickettsiosis is characterized by a rash, flulike symptoms, and sometimes subcutaneous hemorrhages called **petechiae**. In severe cases, system failures, encephalitis, and death occur.
11. **Cutaneous anthrax** results when *Bacillus anthracis* infects a cut in the skin, producing a crusty, black ulcer called an **eschar**.
12. Ischemia in a tissue results from interrupted blood supply; without oxygen, necrosis (death) sets in. **Gas gangrene** (gangrene) develops in dead tissue infected with *Clostridium* endospores.

Viral Diseases of the Skin and Wounds (pp. 575–586)

1. Poxvirus causes smallpox, orf, cowpox, and monkeypox. Lesions progress from flat **macules** to raised **papules**, **vesicles**, and pus-filled **pustules** (also called **pox**).
2. Smallpox virus (**variola** virus) exists in two strains: variola major and variola minor. Except for laboratory stocks, worldwide immunization has eradicated smallpox virus.
3. Human herpesviruses 1 and 2 cause oral and genital **herpes**. Herpetic gingivostomatitis is a severe infection in the oral cavity. Sore throats may develop into herpetic pharyngitis. If herpesvirus enters a cut in the finger, a whitlow (blister) may occur. Herpes gladiatorum is the name of lesions acquired by athletes.
4. Inactive herpesviruses in nerve cells may reactivate to cause recurrence of symptoms.
5. **Warts (papillomas)** are growths of epithelium or mucous membranes.
6. **Varicella-zoster virus (VZV)** causes **chickenpox** (also known as **varicella**) in children and adults and a recurrent disease, **shingles** (also known as herpes zoster), in adults.
7. **Rubella** (also called German measles or three-day measles) is caused by *Rubivirus*. The disease is mild in children but more serious in adults and fetuses.
8. **Measles** virus causes rubeola (red measles), a serious, contagious childhood disease characterized by widespread rash and by **Koplik's spots** in the mouth. Rare complications include **subacute sclerosing panencephalitis (SSPE)**, a slow progressive disease of the central nervous system.
9. Viruses also cause **erythema infectiosum** (fifth disease) and **roseola**.

Mycoses of the Hair, Nails, and Skin (pp. 586–591)

1. **Mycoses** are fungal diseases. **Superficial mycoses** are fungal infections that produce the enzyme keratinase, which dissolves keratin of the hair, nails, and skin. **Pityriasis versicolor** results from interference with melanin production in patches of skin.
2. Cutaneous mycoses are **dermatophytoses** (formerly called ringworms or tineas). These infections are caused by fungi that grow on skin, nails, and hair and stimulate immune responses in underlying tissues.
3. Ascomycete fungi that infect wounds cause **chromoblastomycosis**, which progressively manifests as lesions, warts, and tumors.
4. **Phaeohyphomycosis** is typically caused by invasion of ascomycete (fungal) spores in traumatic or surgical wounds.
5. Several genera of soil ascomycetes cause invasive, tumorlike skin infections called **mycetomas**.
6. **Sporotrichosis** is a cutaneous or subcutaneous disease caused by inoculation with *Sporothrix*, often by thorn pricks from roses.

Parasitic Infestations of the Skin (pp. 591–593)

1. *Leishmania* is a parasitic protozoan transmitted to humans by sand flies. It causes **leishmaniasis**, a **zoonosis** (disease of animals transmitted to humans) that can be cutaneous, mucocutaneous, or visceral.
2. The mite *Sarcoptes scabiei* causes **scabies**, a disease of itchy blisters, often on the webbing between the fingers. Scratching can introduce bacteria into the wounds, leading to impetigo.

QUESTIONS FOR REVIEW

Answers to the Questions for Review (except Short Answer questions) begin on page A-1.

Multiple Choice

1. The epidermis
   a. has an intricate network of blood vessels.
   b. consists of layers of loosely packed cells.
   c. contains a waterproofing protein.
   d. has a thin layer of fat cells.

2. The oily lipid secreted by glands in the dermis is called
   a. keratin.
   b. fat.
   c. sebum.
   d. hyaluronic acid.

3. The most severe form of acne is
   a. a blackhead.
   b. cystic acne.
   c. a pimple.
   d. excessive oil production.

4. Which of the following is *not* associated with impetigo?
   a. pus-filled vesicles
   b. erysipelas
   c. *Streptococcus pyogenes*
   d. pyocyanin

5. Which of the following pairs of diseases may be associated with *S. pyogenes*?
   a. impetigo and acne
   b. erysipelas and necrotizing fasciitis
   c. anthrax and blackheads
   d. black piedra and papillomas

6. An eschar is
   a. painless.
   b. black.
   c. a crusty ulcer.
   d. All of the above describe an eschar.

7. The opportunistic pathogen often seen in burn victims is
   a. *Pseudomonas.*
   b. *Streptococcus.*
   c. *Staphylococcus.*
   d. *Dermacentor.*

8. Which of the following ascomycetes causes a cutaneous mycosis?
   a. *Malassezia*
   b. *Sporothrix*
   c. *Phialophora*
   d. *Microsporum*

9. A patient has oral lesions that look like salt crystals surrounded by a red halo. What disease is indicated?
   a. rubella
   b. chickenpox
   c. measles
   d. erysipelas

10. The first disease for which immunization was perfected was
    a. smallpox.
    b. polio.
    c. ringworm.
    d. measles.

11. Which of the following types of warts are found on the soles of the feet?
    a. flat warts
    b. seed warts
    c. genital warts
    d. plantar warts

12. What is a concern for tissue that does *not* receive its needed blood supply?
    a. ischemia
    b. necrosis
    c. gas gangrene
    d. all of the above

13. Which of the following bacteria can have these virulence factors: coagulase, hyaluronidase, lipase, protein A?
    a. *Staphylococcus*
    b. *Streptococcus*
    c. *Bartonella*
    d. *Clostridium*

14. Which of the following is the vector for leishmaniasis?
    a. cat
    b. ringworm
    c. sand fly
    d. tick

15. Subacute sclerosing panencephalitis is caused by
    a. type III hypersensitivity.
    b. seed warts.
    c. measles virus.
    d. a mite.

Matching

Match the disease on the left with the causative pathogen on the right.

1.	_____ Cat scratch disease	A.	*Propionibacterium*
2.	_____ Acne	B.	*Malassezia*
3.	_____ Spotted fever	C.	*Bartonella*
4.	_____ Erysipelas	D.	Dermatophytes
5.	_____ Pityriasis versicolor	E.	Variola major
6.	_____ "Ringworm"	F.	*Rickettsia*
7.	_____ Smallpox	G.	*Simplexvirus*
8.	_____ Herpes	H.	Varicella-zoster virus
9.	_____ Warts	I.	*Rubivirus*
10.	_____ Shingles	J.	*Roseolovirus*
11.	_____ Rubella	K.	*Papillomavirus*
12.	_____ Measles	L.	*Leishmania*
		M.	*Morbillivirus*
		N.	*Streptococcus*

Fill in the Blanks

1. As a group, the normal microbial residents of the body make up the ________________.
2. Staphylococcal scalded skin syndrome is caused by ________________ toxins produced by *Staphylococcus aureus*.
3. Pyoderma is a contagious skin disease also called ________________.
4. "Flesh-eating" disease is formally called ________________.
5. Four types of folliculitis caused by *Staphylococcus* are ________________, ________________, ________________, and ________________.
6. Rocky Mountain spotted fever is transmitted by ticks of the genus ________________.
7. A single-stranded DNA virus causes ________________ (also known as fifth disease) in humans.
8. ________________ causes warts.
9. *Clostridium perfringens* causes most cases of ________________.
10. Most pimples result from infection by ________________, not from eating chocolate, drinking soft drinks, or consuming oily foods.

True/False

1. _____ Vigorous scrubbing of the skin can eliminate microorganisms.
2. _____ Untreated cat scratch disease is fatal.
3. _____ Surface cleansers are highly effective in the treatment of acne.
4. _____ A number of fungi grow on the epidermal layer of skin, but they rarely become systemic infections.
5. _____ Koplik's spots are the lesions of chickenpox.

Short Answer

1. Discuss the pros and cons of using long-term antibiotics to treat acne.
2. Discuss the problems associated with infections in burn victims.
3. Considering that *Pseudomonas aeruginosa* and *Staphylococcus epidermidis* are ubiquitous and each possesses numerous virulence factors, why do they rarely cause disease?
4. What distinguishes the rashes of chickenpox or measles from the rash of Rocky Mountain spotted fever?
5. Maria has patches of black scaly skin on her arm. Her physician suspects that she has *Malassezia furfur.* What is the quickest way to confirm this suspicion?
6. Describe four enzymes produced by *Staphylococcus,* and tell how each contributes to the bacterium's survival and pathogenicity.
7. In 1955, *Staphylococcus* infections were effectively treated with penicillin. Why is this not the drug of choice now?
8. Larry has had flulike symptoms for three days accompanied by extreme shooting pains. What pathogen might be suspected?
9. Why are the terms *dermatophytosis* and *ringworm* inaccurate choices for the names of the superficial fungal infections they are associated with?
10. Mrs. Rathbone called the pediatrician concerning her young daughter Rene, who has had a rosy facial rash and coldlike sniffles for two weeks. What is the most likely cause of Rene's problem?

VISUALIZE IT!

1. Label each lesion seen in poxvirus infections.

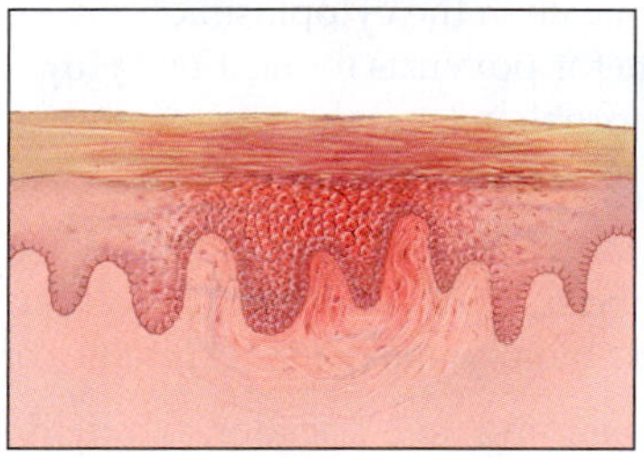

a. ________________

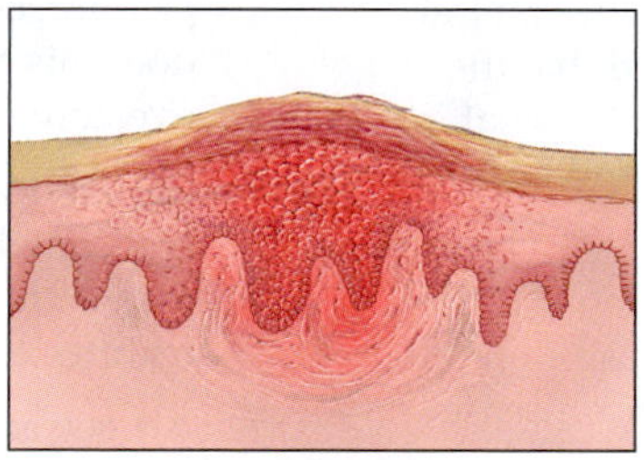

b. ________________

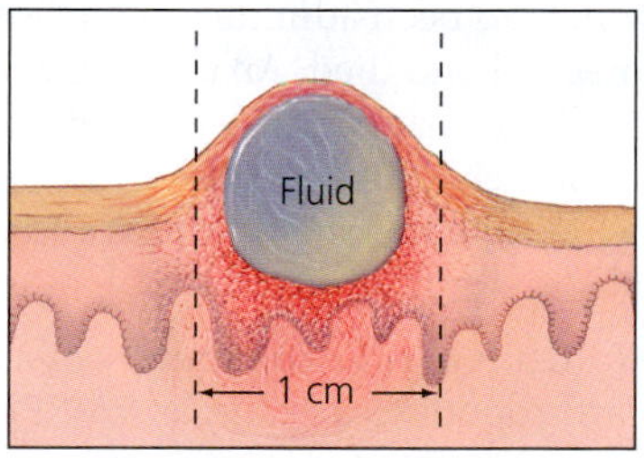

c. ________________

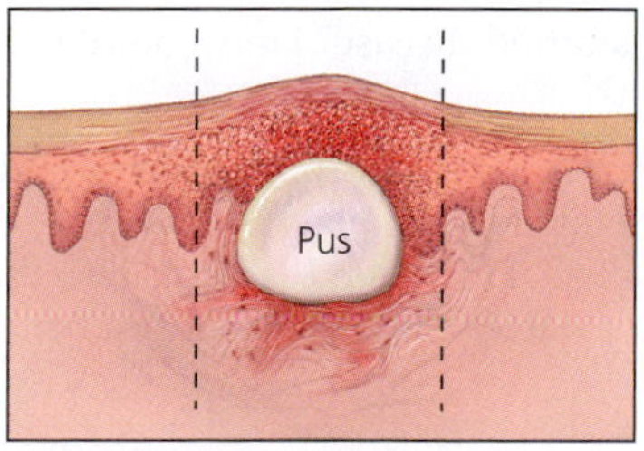

d. ________________

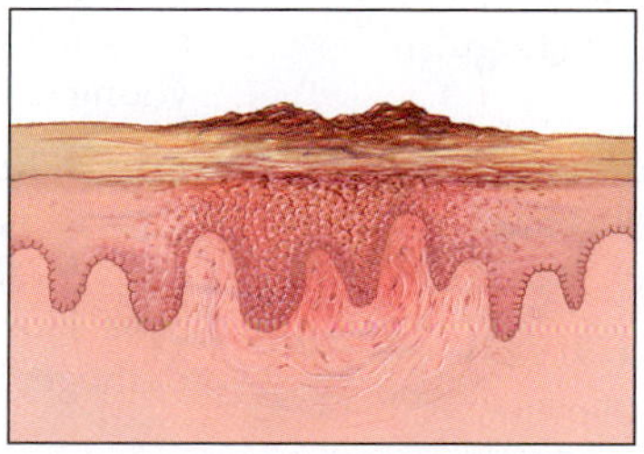

e. ________________

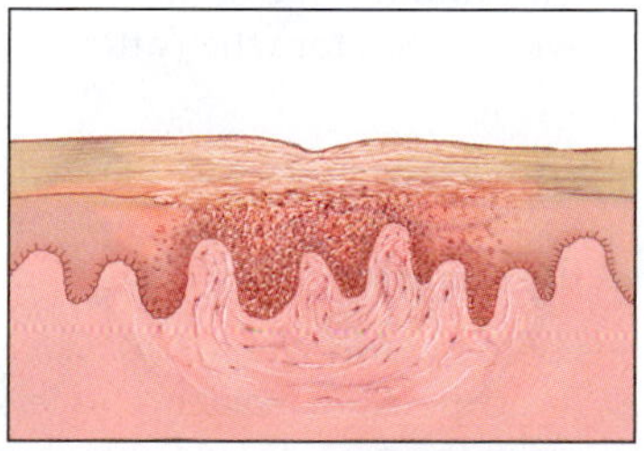

f. ________________

2. Label the lesions of human herpesvirus occurring at each location.

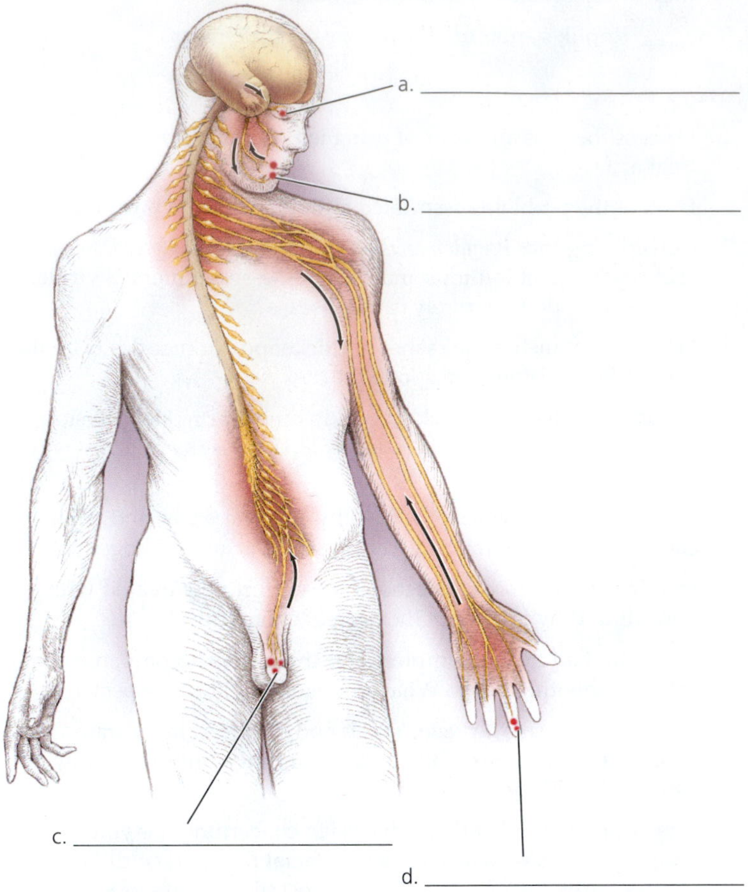

3. Name the skin disease shown in each photo.

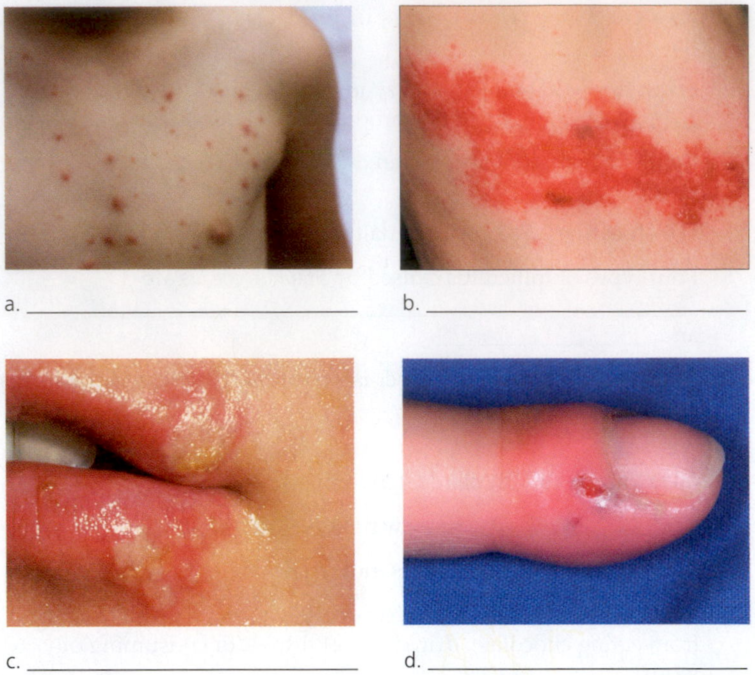

CRITICAL THINKING

1. A few days after the death of a hospitalized child from a methicillin-resistant *Staphylococcus aureus* (MRSA) infection, another child, who had been admitted to the hospital with viral pneumonia, worsened and died. An autopsy revealed that the second child also died from complications of MRSA. By what route was the second child likely infected? What should hospital personnel do to limit the transfer of MRSA and other bacteria among patients?
2. Why is it more difficult to rid a community of a disease transmitted by arthropods (for example, RMSF) than a disease transmitted via contaminated drinking water?
3. Clinically, all fungi that cause subcutaneous mycoses produce lesions on the skin around the site of inoculation. What are some of the things you would look for when attempting to distinguish among them?
4. Given the various predisposing factors that make humans susceptible to opportunistic infections, how can health care providers curtail the rising incidence of such infections?
5. Given the regions of the world where *Leishmania* and HIV are endemic, would you expect the incidence of leishmaniasis to increase or decrease in the next decade? Explain your answer.
6. Most DNA viruses replicate within the nucleus of a host cell, using host enzymes to replicate their DNA. In contrast, poxviruses replicate in the cytoplasm of host cells. What problem does this create for poxvirus replication? How does the virus overcome this problem?
7. Certain features of smallpox viruses allowed them to be eradicated in nature. Which other viruses discussed in this chapter might be suitable candidates for eradication, and what features of their biology make them suitable candidates?
8. A week after spending their vacation rafting down the Colorado River, all five members of the Chen family developed cold sores on their lips. Their doctor told them that the lesions were caused by a herpesvirus. Mr. and Mrs. Chen were stunned: Isn't herpes a sexually transmitted disease? How could it have affected their young children?
9. *Propionibacterium acnes* is a normal member of the skin microbiota that benefits the body by lowering the skin's pH—an antimicrobial effect. However, *P. acnes* is also the leading cause of acne. How can a bacterium be normal and beneficial but also be pathogenic?

10. In 2009, several websites suggested spuriously that measles vaccine causes autism. What effect did these claims have on the number of measles cases in subsequent years?
11. *Sporothrix*, the cause of sporotrichosis (rose-gardener's disease), is dimorphic. What feature characterizes dimorphic fungi?
12. Why do impetigo and erysipelas occur more commonly in children than in young adults?
13. The United States and Russia have repeatedly agreed to destroy their stocks of the smallpox virus, but the deadline for destruction has been postponed numerous times. In the meantime, the entire genome of variola major has been sequenced. What reasons can governments cite for maintaining smallpox viruses? Should all laboratory stores of smallpox viruses be destroyed? Given that the genome of the virus has been sequenced, and that DNA can be reconstructed if the sequence of nucleotides is known, would elimination of all laboratory stocks really be the extinction of the smallpox virus?
14. You learned about the creation and use of dichotomous keys on p. 119. Make a dichotomous key to distinguish among all the fungal mycoses covered in this chapter.

CONCEPT MAPPING

Using the following terms, draw a concept map that describes herpes simplex virus. For a sample concept map, see p. 94. Or, complete this and other concept maps online by going to the MasteringMicrobiology Study Area.

Acyclovir
Cold sores
ds DNA
Envelope
Fever blisters
Finger infection
Genital herpes
Gingivostomatitis
Herpetic whitlow
Human *herpesvirus* 1
Human herpesvirus 2
Latency
Life-threatening
Neonatal herpes
Ocular herpes
Oral herpes
Polyhedral capsid
Primary infection
Recurrences
Simplexvirus

20 Microbial Diseases of the Nervous System and Eyes

MICRO IN THE CLINIC

Under Pressure and Under the Weather

Lin, a freshman in college, is extremely stressed. She is the first person in her family to attend college, which she can afford only because of an academic scholarship. To keep the scholarship, she needs to maintain a B average in all of her classes, including chemistry—a class with a notoriously high failure rate. The final exam is tomorrow, and Lin is not feeling well at all. She is exhausted, feverish, and has a terrible headache. Her mind feels foggy, and her neck has been weirdly stiff all day. She still has hours of studying ahead of her, so she takes an aspirin, hoping to at least cure her headache. A few minutes later, however, Lin runs to the bathroom and vomits. She staggers to bed, lies down, and closes her eyes to rest for a while.

When Lin wakes up, she is disoriented and confused. Why is she in a hospital? The last thing she remembers is lying down in bed in her dorm room. A nurse arrives and explains that Lin's roommate became concerned when Lin couldn't be roused from sleep, so she called 911. The doctors suspect Lin has a serious infectious disease and perform a spinal tap to obtain a sample of Lin's cerebrospinal fluid (CSF). The specimen will be Gram stained, cultured, and examined for blood cells, glucose, and protein. Lin's CSF specimen looks milky-white as the doctors send it to the lab.

What's ailing Lin? What are her doctors testing for? Turn to the end of the chapter (p. 632) to find out.

Explore More: Test your readiness and apply your knowledge with dynamic learning tools at MasteringMicrobiology.

Structure of the Nervous System

LEARNING | OUTCOMES

20.1 Describe the basic divisions, organs, and structures of the nervous system.

20.2 Describe the formation, circulation, absorption, and collection of cerebrospinal fluid.

20.3 Explain how microbes might gain access to the normally axenic central nervous system.

The nervous system is divided into the central nervous system and the peripheral nervous system **(FIGURE 20.1a)**.

Structures of the Central Nervous System

The *central nervous system (CNS)*, which is composed of the brain and spinal cord, functions as the master control center of the body. The brain has several main parts, including the following:

- The *cerebrum*—the largest, upper part of the brain—controls voluntary muscles, perception, and what people commonly call "thinking."
- The lower *cerebellum* controls many involuntary body movements, such as swinging the arms while walking.
- The *brain stem*, which connects the brain to the spinal cord, has a number of functions, including control of breathing, heart rate, and blood pressure.

The spinal cord extends down from the brain stem only as far as the lumbar region (lower back). Below this region, a bundle of nerves called the *cauda equina*[1] extends from the spinal cord. As its name suggests, this bundle resembles the tail of a horse.

Bones of the *cranium*, bones of the *vertebral column*, and three layers of tissue called *meninges*[2] (mĕ-nin´jēz) surround the brain and spinal cord, providing support and protection from external shock. The three meninges vary in their structure and appearance **(FIGURE 20.1b, c)**.

Lying next to the bones is the *dura mater*[3] (dŭ´ră mā´ter), a tough fibrous sheath that provides a strong yet flexible covering for the soft organs of the CNS. It also provides a barrier against the spread of infections from the bones. Deep to the dura mater is the *arachnoid mater*[4] (ă-rak´noyd mā´ter), which contains numerous branching fibers giving the appearance of a spider's web. The cavities between the fibers of the arachnoid mater are collectively called the *subarachnoid space*. The internal layer, which is closely appressed to the spinal cord and brain, is the *pia mater*[5] (pī´ă mā´ter). Blood vessels on top of the pia mater supply the CNS with blood. The walls of these blood vessels are composed of tightly joined cells that form the *blood-brain barrier*, which prevents most microbes and large molecules in the blood from entering the subarachnoid space. Thus, blood infections do not easily spread to the CNS, but unfortunately, neither do many common antimicrobial drugs, such as penicillins, cephalosporins, tetracyclines, and aminoglycosides, making it more difficult to treat infections of the CNS.

Fluid leaks from the blood into the subarachnoid space that lines the brain. This watery fluid—called *cerebrospinal fluid (CSF)*—circulates throughout the subarachnoid space of both brain and spinal cord to bathe both organs. *Arachnoid villi* or *arachnoid granulations*, which are knoblike extensions of the arachnoid mater, extend into a blood-filled cavity at the top of the cranium and return CSF to the blood (see Figure 20.1b).

Cerebrospinal fluid acts as a shock absorber; provides nutrients, electrolytes, and oxygen to the nervous tissues; and removes wastes. In a medical procedure called a **lumbar puncture** *(spinal tap)*, physicians remove a sample of CSF from the region of the subarachnoid space surrounding the cauda equina **(FIGURE 20.1d)**. They insert a needle through the skin between two of the lumbar vertebrae, through the dura mater and subdural space, and through the arachnoid mater to reach the CSF in the subarachnoid space.

Structures of the Peripheral Nervous System

The *peripheral nervous system (PNS)* is composed of nerves that transfer commands from the CNS to muscles and glands throughout the body and provide information to the CNS concerning events in the body. *Cranial nerves* extend from the brain through holes in the cranial bones, and *spinal nerves* extend from the spinal cord through gaps between vertebrae. Branches of nerves often merge together to form a nerve *plexus* (see Figure 20.1a).

Functionally, there are three types of nerves: *Sensory nerves* primarily carry signals toward the CNS. The optic nerves from the eyes are examples. *Motor nerves* carry signals from the CNS to other organs of the body, and *mixed nerves* carry signals both toward and away from the CNS.

Cells of the Nervous System

The entire nervous system is composed of two basic types of cells—*neurons* (nūr´onz) and supportive cells called *neuroglia* (nū-rog´lē-ă). The smaller neuroglia provide a supportive scaffolding, insulation, and nutritive support and phagocytize microbes. The cytoplasmic membrane of a neuron generates an electrical signal called an *action potential* or *nerve impulse*.

The nucleus of a neuron lies in a region of cytoplasm called the *cell body*. Outside the CNS, a collection of many neurons' cell bodies is called a *ganglion*. Two types of fingerlike, cytoplasmic processes extend from a cell body: numerous, perhaps hundreds, of short *dendrites* and a longer single *axon*. Within the cytoplasm of an axon, the cytoskeleton transports substances by a process known as *axonal transport*. Do not confuse dendrites and axons, which are part of individual cells (neurons), with sensory and motor nerves, which are bundles of thousands of cells.

[1]Latin, meaning "horse's tail."
[2]From Greek *meninx*, meaning "membrane."
[3]Latin, *dura*, meaning "hard," and *mater*, meaning "mother."
[4]From Greek *arachne*, meaning "spider," and Latin *mater*, meaning "mother."
[5]Latin, *pia*, meaning "tender," and *mater*, meaning "mother."

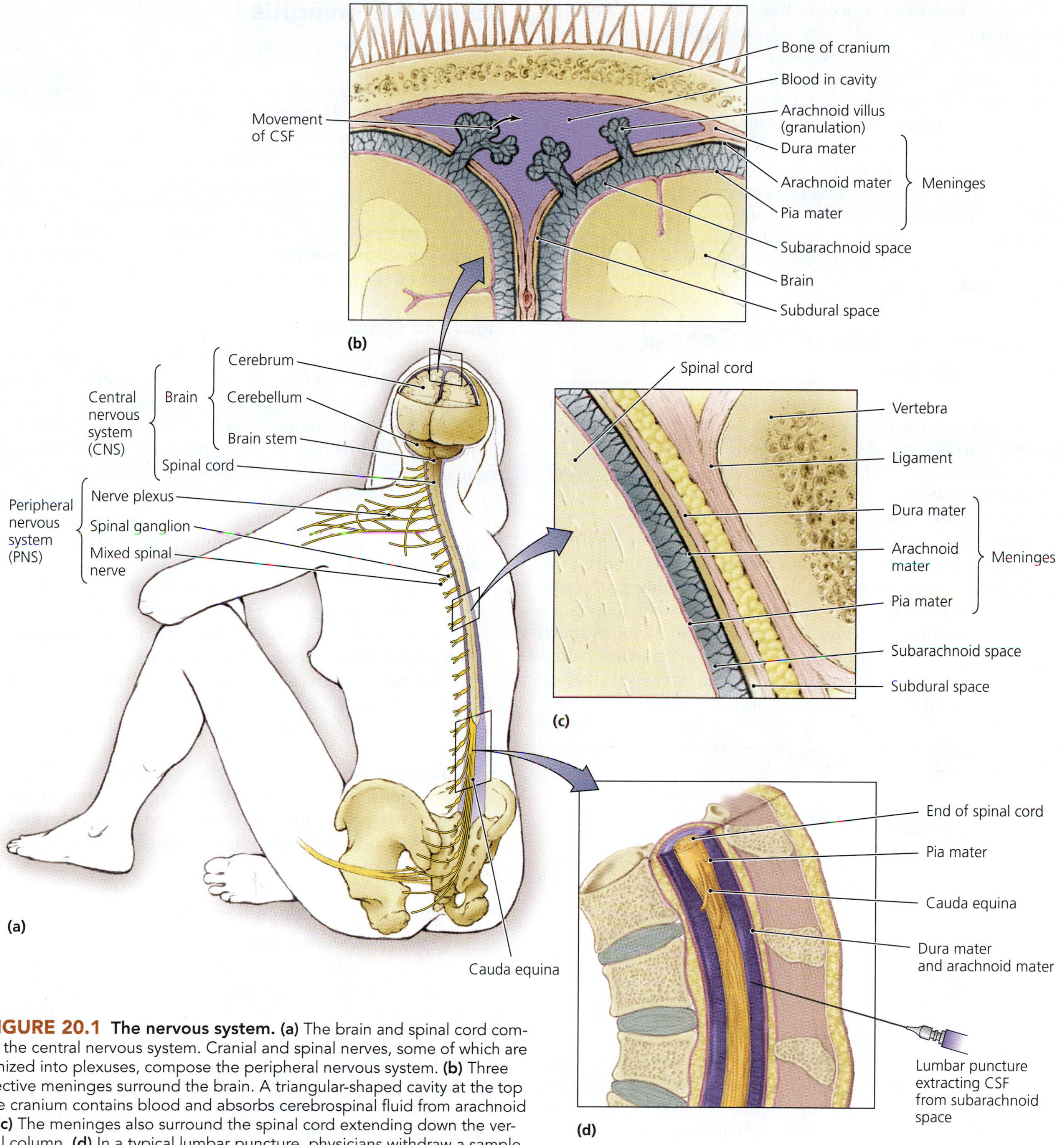

▲ **FIGURE 20.1 The nervous system. (a)** The brain and spinal cord compose the central nervous system. Cranial and spinal nerves, some of which are organized into plexuses, compose the peripheral nervous system. **(b)** Three protective meninges surround the brain. A triangular-shaped cavity at the top of the cranium contains blood and absorbs cerebrospinal fluid from arachnoid villi. **(c)** The meninges also surround the spinal cord extending down the vertebral column. **(d)** In a typical lumbar puncture, physicians withdraw a sample of cerebrospinal fluid from the subarachnoid space in the lumbar region, a region containing the cauda equina instead of the spinal cord. The nerves of the cauda equina are not shown in their entirety. *Why do physicians collect CSF from the lower back and not from higher up the vertebral column?*

Figure 20.1 *The lower vertebral column does not contain the spinal cord; therefore, the physician will not damage the CNS or likely introduce microbes into the CNS. Further, the lumbar vertebrae are large, easy to identify, and have larger spaces between them.*

The terminal ends of axons have thousands of branches that form junctions called *synapses*[6] (sĭ-nap´sēz) with glands, muscles, or other neurons. A synapse mediates transfer of a signal to a neighboring *postsynaptic cell*. In most synapses there is a *synaptic cleft*—an intercellular space about 40 nm wide between the axon terminal and the postsynaptic cell. A synaptic cleft stops the transmission of electrical signals; therefore, the signal between cells is chemical—an axon terminal releases molecules called *neurotransmitters* into the synaptic cleft. A particular neurotransmitter may be stimulatory or inhibitory; that is, it either (1) stimulates a muscle to contract, a gland to secrete, or another neuron to carry a nerve impulse, or (2) inhibits such activities.

Portals of Infection of the Central Nervous System

No openings allow microbial colonization of the central nervous system; therefore, the CNS is an *axenic* (ā-zēn´ik) environment—it has no normal microbiota. Pathogens may access the CNS through breaks in the bones and meninges, through medical procedures such as spinal taps, or by traveling via axonal transport in peripheral neurons to the CNS. Microbes carried in the blood or lymph may penetrate the blood-brain barrier by infecting and killing cells of the meninges, causing **meningitis**[7] (men-in-jī´tis)—inflammation of the meninges. Some pathogens gain access to the CNS when localized inflammation distorts the cells of the blood-brain barrier, changing its permeability; such change is more likely during chronic infection by many pathogens. Circulation of cerebrospinal fluid can carry infective microbes throughout the cranial cavity and spinal column.

We will examine nervous system diseases caused by bacteria, viruses, fungi, protozoa, and prions. We begin by considering bacterial diseases.

TELL ME WHY

Why is it important that the cells forming the blood vessels of the brain and spinal cord be tight against one another, forming a blood-brain barrier?

Bacterial Diseases of the Nervous System

Not only can bacteria infect cells of the nervous system, but toxins released by bacteria growing elsewhere in the body can also affect neurons. In the following sections, we consider disease examples of both types—leprosy, which is a disease of cells found in the PNS, and botulism and tetanus, which involve toxin production. However, the most common bacterial infection of the nervous system is bacterial meningitis, which we consider next.

[6]From Greek *syn*, meaning "together," and *hapto*, meaning "to clasp."

[7]From Greek *meninx*, meaning "membrane," and *itis*, meaning "inflammation."

Bacterial Meningitis

LEARNING | **OUTCOMES**

20.4 Describe the signs, symptoms, diagnosis, treatment, and prevention of bacterial meningitis.

20.5 Compare and contrast the characteristics, including virulence factors, of the five more common causes of bacterial meningitis.

Bacterial meningitis involves inflammatory bacterial infection of the meninges, commonly the pia mater and arachnoid mater and, more rarely, the dura mater.

Signs and Symptoms

Bacterial meningitis is characterized by an increased number of white blood cells in the CSF, sudden high fever, and intense meningeal inflammation. The inflammation accounts for most of the signs and symptoms: Swelling of the meninges retards the normal flow of CSF, putting pressure on the underlying organs. Inflammation of the cranial meninges typically produces severe headache, nausea, vomiting, pain, and in many cases loss of various brain functions, leading to such conditions as drowsiness, confusion, fretfulness, or irritability. Inflammation of the spinal meninges puts pressure on surrounding nerves and muscles, producing stiffness in the neck and affecting sensory input and muscular control. When the brain becomes infected—a condition called *encephalitis*—deafness, blindness, drastic changes in the patient's behavior, coma, or death may result. Signs and symptoms of meningitis may rapidly develop; for example, meningococcal meningitis can kill within six hours of the initial symptoms, allowing little time for treatment.

A lumbar puncture reveals the normally clear CSF to be milky in appearance because of the large number of bacteria and the increased number of white blood cells. **Petechiae** (pe-tē´kē-ē)—small, dark purplish hemorrhages of blood vessels in the skin—are sometimes present.

Pathogens and Virulence Factors

Researchers have shown that more than 50 species of bacteria can cause meningitis. Among these are opportunistic members of the normal microbiota, including species of *Staphylococcus* (staf´i-lō-kok´ŭs) and *Streptococcus* (strep-tō-kok´ŭs) as well as Gram-negative enteric[8] bacteria such as *Escherichia coli* (esh-ĕ-rik´ē-ă ko´lī) and *Klebsiella pneumoniae* (kleb-sē-el´ă nū-mō´nē-ī). However, five other species cause almost 90% of cases of bacterial meningitis. These are *Neisseria meningitidis*, *Streptococcus pneumoniae*, *Haemophilus influenzae*, *Listeria monocytogenes*, and *Streptococcus agalactiae*. All five of these bacteria have virulence factors that allow them to resist phagocytosis and cause disease; the following sections detail these features.

[8]From Greek *entera*, meaning "intestines."

EMERGING DISEASE CASE STUDY

Melioidosis

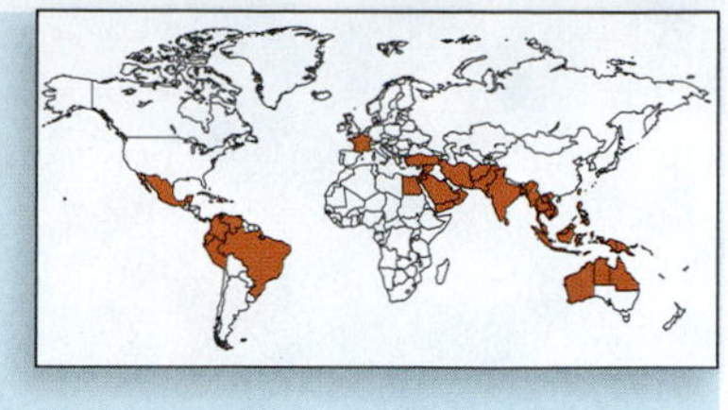

Isabella felt lucky. How many community college students had the opportunity to help a professor do research in northern Australia's Kakadu National Park for a month, get college credit for the experience, and not have to pay for the trip? Though hiking the trails was hot and tiring, she didn't complain. Of course things would have been better if it didn't seem to rain all the time and if the thorns didn't tear at her legs quite so regularly; still, the trip was an adventure.

A week after her return to Houston, Texas, Isabella developed a high fever (39.2°C) and general weakness. All other signs, including the results of blood work, appeared normal. The doctors, suspecting flu, told her to get plenty of rest and sent her home. Two days later Isabella was intermittently drowsy and confused, her breathing was labored, and a small cut on her leg was an inflamed, pus-filled lesion. She was admitted to the hospital and died the next day.

She died of melioidosis, an emerging disease caused by a Gram-negative bacterium, *Burkholderia pseudomallei*, which produces a toxin that inhibits protein synthesis by infected human cells. Melioidosis is endemic to the tropics of Southeast Asia and appears to be spreading into more moderate climes. Isabella had been infected either via inhalation or through the inflamed cut on her leg. Even with treatment, nearly 90% of melioidosis patients die, including Isabella.

1. If a sample of Isabella's CSF were Gram stained, what color would the bacterial cells appear?
2. Why did the doctors suspect flu rather than melioidosis?
3. Why is it unlikely that there will be an epidemic of melioidosis in Sweden or Norway?

Neisseria meningitidis *Neisseria meningitidis* (nī-se´rē-ă me-nin-ji´ti-dis) is one of only two species of Gram-negative cocci that regularly causes disease in humans. Researchers have identified 13 antigenic strains; strains A, B, C, and W135 cause most cases of disease in humans. The cells of all strains of *Neisseria* are nonmotile and are typically arranged as diplococci (pairs) with their common sides flattened in a manner reminiscent of coffee beans **(FIGURE 20.2)**. The bacterium is known as the *meningococcus* and its disease as *meningococcal meningitis*.

Meningococci have fimbriae and polysaccharide capsules, as well as a major cell wall antigen called *lipooligosaccharide* (lĭp´ō-ŏl´ĭ-gō-sak´ă-rīd, LOS), composed of lipid A (endotoxin) and sugar molecules—all of which enable the bacteria to attach to human cells. Cells of *Neisseria* that lack any of these three structural features are avirulent. The polysaccharide capsules also resist lytic enzymes of the body's phagocytes, allowing phagocytized meningococci to survive, reproduce, and be carried throughout the body within neutrophils and macrophages.

Much of the damage caused by *N. meningitidis* results from *blebbing*—a process in which the bacterium sheds extrusions of its outer membrane. The lipid A component of LOS thereby released into the body triggers fever, vasodilation, inflammation, shock, and widespread blood clotting.

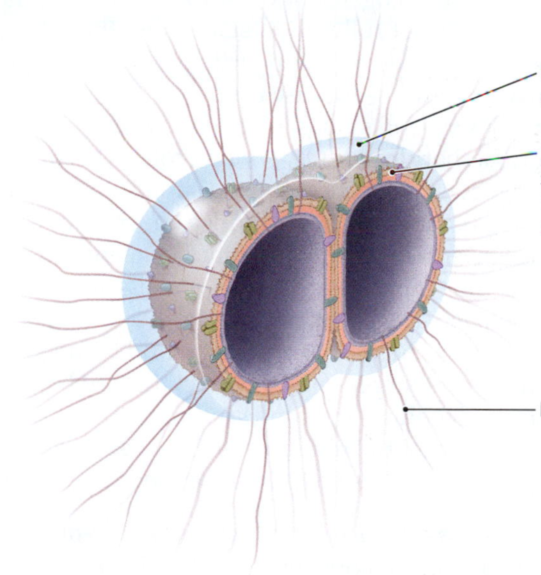

▲ **FIGURE 20.2 Artist's rendition of diplococci of *Neisseria meningitidis*.** *What function of fimbriae contributes to the bacterium's pathogenicity?*

Figure 20.2 *Fimbriae attach the bacterium to target cells.*

Streptococcus pneumoniae Louis Pasteur discovered *Streptococcus pneumoniae* (strep-tō-kok´ŭs nū-mō´nē-ī) in pneumonia patients about 120 years ago. The bacterium is a Gram-positive coccus, which forms short chains or, more commonly, pairs **(FIGURE 20.3)**. In fact, it was once classified in its own genus, "Diplococcus." Ninety-two different strains of *S. pneumoniae*, collectively called pneumococci, are known to infect humans as normal members of the microbiota of the throat that opportunistically grow in the lungs, sinuses, and middle ear

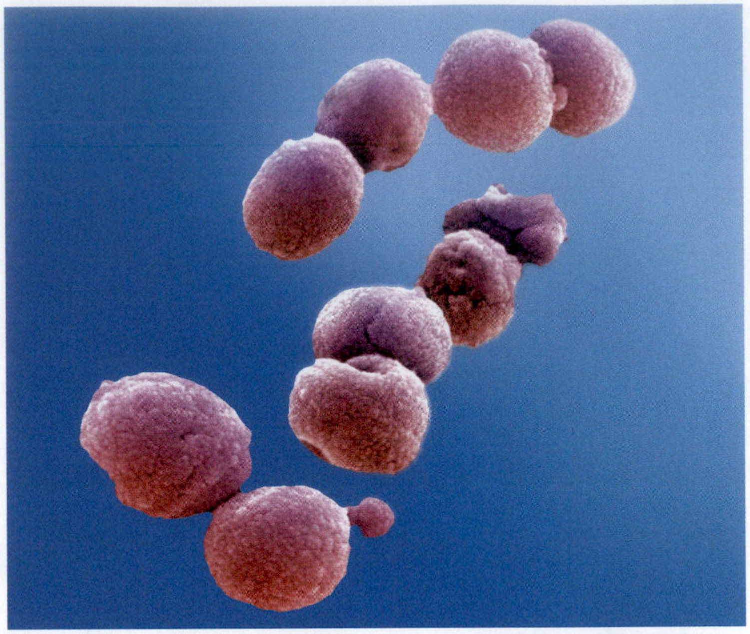

▲ **FIGURE 20.3** **Cells of *Streptococcus pneumoniae* are typically arranged in pairs.**

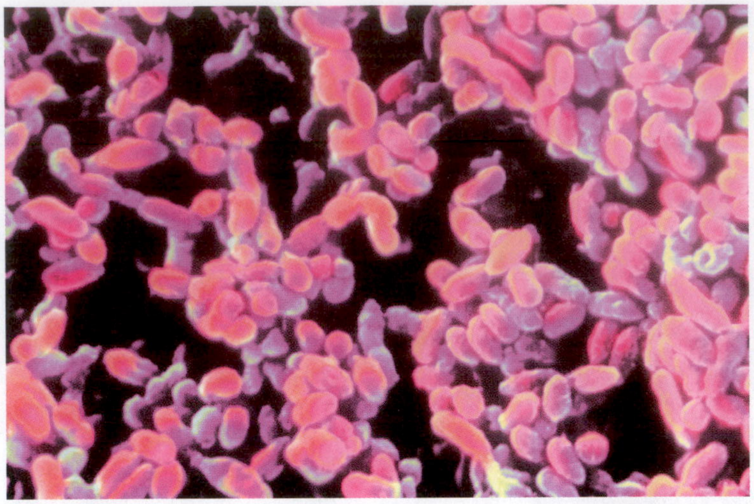

▲ **FIGURE 20.4** **The pleomorphic bacilli of *Haemophilus influenzae* in a Gram-stained smear.**

and from those locations move into the meninges via the blood. *Streptococcus pneumoniae* is a leading cause of meningitis in adults.

Even though microbiologists have studied pneumococci extensively, they still do not fully understand how these bacteria cause disease; nevertheless, certain structural and chemical virulence factors are necessary for disease.

The cells of all virulent strains of *S. pneumoniae* are surrounded by a polysaccharide capsule, which protects them from digestion after phagocytosis. Unencapsulated strains do not cause disease. Pathogenic pneumococci also produce enzymes and toxins that enable the bacteria to counteract immune defenses.

In addition, pathogenic *S. pneumoniae* possess a cell wall chemical called *phosphorylcholine*, which is an adhesin that binds to receptors on cells in the lungs, the meninges, and blood vessel walls. Binding stimulates target cells to endocytize the bacteria. Thus, the phosphorylcholine and polysaccharide capsule together enable pneumococci to "hide" inside body cells. *S. pneumoniae* can then pass across these cells into the blood and brain.

Haemophilus influenzae *Haemophilus*[9] *influenzae* (hē-mof´i-lŭs in-flu-en´zī) is a small pleomorphic bacillus **(FIGURE 20.4)** that requires heme and NAD^+ for growth. As a result, it is an obligate parasite, colonizing mucous membranes of humans and some animals. At one time—as the specific epithet indicates—scientists thought that the bacterium caused the flu pandemics in 1890 and 1918, but *H. influenzae* causes meningitis, not the flu.

Most strains of *H. influenzae* have polysaccharide capsules that resist phagocytosis. Researchers distinguish among six strains of *Haemophilus* by differences in capsular antigens. Before the introduction of an effective vaccine in the 1990s, 95% of *H. influenzae* diseases in the United States were caused by type b.

Listeria monocytogenes *Listeria monocytogenes* (lis-tēr´ē-ă mo-nō-sī-to´je-nēz) is a Gram-positive, non-endospore-forming coccobacillus that enters the body in contaminated food or drink. *Listeria* is rarely pathogenic in healthy adults; infection in pregnant women, fetuses, newborns, the elderly, and immunocompromised patients can result in meningitis.

Disease in Depth on pp. 608–609 examines **listeriosis**, including meningitis, in detail.

Streptococcus agalactiae *Streptococcus agalactiae* (strep-tō-kok´ŭs a-ga-lak´tē-ī), also known as Lancefield[10] group B *Streptococcus* (based on its so-called B antigens), is a normal member of the vaginal microbiota in about a third of all women. The bacterium produces a protective capsule that allows it to evade phagocytosis when it gets into the blood. *S. agalactiae* also causes bacteremia, pneumonia, and meningitis in newborns.

Pathogenesis

Humans inhale *N. meningitidis*, *H. influenzae*, and *S. pneumoniae* in respiratory droplets from infected individuals (who may appear healthy). Babies pick up *S. agalactiae* during passage through an infected birth canal. *Listeria* is transmitted in contaminated food, particularly meat and underpasteurized milk and cheese.

In most cases bacteria spread to the meninges from infections of the lungs, sinuses *(sinusitis)*, or inner ear *(otitis media)* via the blood *(bacteremia)*. Head or neck surgery or trauma may also open a passage into the subarachnoid space of the meninges. The bacteria, somewhat protected by their capsules from phagocytosis, metabolize glucose in the CSF.

[9]From Greek *haima*, meaning "blood," and *philos*, meaning "love."

[10]Named for Rebecca Lancefield (1895–1981), a noted specialist of streptococci.

Epidemiology

Before the 1990's, *H. influenzae* was the leading cause of bacterial meningitis. An effective childhood vaccine has reduced the number of such cases by more than 90%. Today, *S. pneumoniae* (pneumococcus) and *N. meningitidis* (meningococcus) are the more prevalent causes of bacterial meningitis, though *S. agalactiae* is now the leading cause of bacterial meningitis in newborns in the United States, accounting for an estimated 70% of such cases.

S. pneumoniae grows in the mouths and throats of 75% of humans without causing harm; but in some patients, particularly children and the elderly—groups whose immune responses are not fully active—these pneumococci become bloodborne and invade the meninges.

None of the forms of bacterial meningitis are typically spread by casual contact, though *N. meningitidis* can be spread via respiratory droplets to other people who have prolonged contact with a patient. Meningoccocal meningitis is the only type of bacterial meningitis that becomes epidemic, particularly in sub-Saharan Africa, where massive epidemics occur every 5–12 years during the dry season (December to June). For example, in the sub-Saharan nations in 2009, there were almost 90,000 cases.

In the United States, meningococcal meningitis is spread among military personnel in barracks and students in dormitories. In fact, meningococcal disease is 9–23 times more prevalent in students living in dormitories than in the general population. Mortality of meningococcal meningitis approaches 100% in untreated patients, but is about 11% in patients who have been treated appropriately with antimicrobial drugs.

Infant mortality from *S. agalactiae* meningitis has been reduced to about 5% as a result of rapid diagnosis and supportive care, though about 25% of infants surviving this group B streptococcal meningitis have permanent neurological damage, including blindness, deafness, or severe mental retardation.

Listeria infects humans who consume contaminated food. Human-to-human transmission of *Listeria* is limited to the transfer of bacteria from mother to fetus and can result in premature delivery, miscarriage, stillbirth, or meningitis in the newborn.

Diagnosis, Treatment, and Prevention

Symptoms of meningitis should always be considered serious, and a patient should consult a physician immediately. Quick diagnosis is vital. Diagnosis of bacterial meningitis is based on symptoms and culturing bacteria from CSF following a spinal tap. Additionally, serological tests can demonstrate the presence of antibodies against *N. meningitidis*, though strain B of this species is relatively nonimmunogenic and therefore often not revealed by such tests.

Physicians treat bacterial meningitis with any of a number of intravenously administered antimicrobial drugs, including ceftriaxone, cefotaxime, meropenem, vancomycin, or ampicillin. Quick treatment reduces the mortality to below 15% of cases of bacterial meningitis.

Prevention of bacterial meningitis depends on interrupting the transmission of pathogens and their spread in the body. The U.S. Centers for Disease Control and Prevention (CDC) recommends vaccination of children against *Streptococcus pneumoniae, Haemophilus influenzae* type B, and *Neisseria meningitidis* (see Figure 17.3) and administration of penicillin at birth to any child whose mother's vagina is colonized with *Streptococcus agalactiae* (group B *Streptococcus*). Implementation of the latter recommendation in 1996 reduced neonatal meningitis by 70% within five years. Additionally, health care providers prevent the spread of group B streptococcal infection to babies by treating infected pregnant mothers with penicillin or ampicillin, or with vancomycin for patients with penicillin allergy.

The CDC also recommends meningococcal vaccination for all military recruits and college freshmen. Health care providers administer antimicrobials such as sulfonamides, tetracycline, or rifampin to people in prolonged contact with meningococcal patients. Such prophylactic treatment is not recommended following exposure to meningitis caused by other microbes.

Hansen's Disease (Leprosy)

LEARNING **OUTCOMES**

20.6 Describe Hansen's disease, including its diagnosis, treatment, and prevention.

20.7 Distinguish tuberculoid leprosy from lepromatous leprosy.

Hansen's disease is the clinical name for the more dreadful-sounding **leprosy**. It is named after Gerhard Hansen (1841–1912), a Norwegian microbiologist who discovered its bacterial cause in 1873.

Signs and Symptoms

Hansen's disease has two different manifestations depending on the immune response of the patient. Patients with a strong T cell immune response are able to kill cells infected with the bacterium, resulting in a nonprogressive form of the disease, called *tuberculoid leprosy*. Regions of the skin that have lost sensation as a result of nerve damage are characteristic of this form of leprosy.

By contrast, patients with a weak T cell immune response develop *lepromatous leprosy*, in which the bacterium multiplies in skin, mucous membranes, and nerve cells, gradually destroying tissue and leading to the progressive loss of facial features, digits (fingers and toes), and other body structures **(FIGURE 20.5)**. Development of signs and symptoms is very slow; incubation may take years before any are evident. Death from leprosy is rare and usually results from the infection of leprous lesions by other pathogens.

Pathogen and Virulence Factors

Mycobacterium leprae (mī′kō-bak-tēr′ē-ŭm lep′rī) is a high G + C, non-endospore-forming, Gram-positive bacillus with a cell wall containing a large amount of **mycolic** (mī-ko′lik) **acid**—a waxy lipid composed of chains of 60–90 carbon atoms. This unusual cell wall is responsible for several important characteristics of the bacterium:

- Slow growth rate (due to the time required to synthesize numerous molecules of mycolic acid). The generation time varies from hours to several days.
- Protection from lysis once it is phagocytized.

DISEASE IN DEPTH

LISTERIOSIS

Listeria monocytogenes with actin "tails"

SIGNS AND SYMPTOMS

Listeria usually produces no symptoms or only mild flulike symptoms in healthy adults. In contrast, infection in pregnant women, fetuses, newborns, elderly, and immunocompromised patients can be severe, causing bacterial meningitis and possibly death. Human-to-human transmission is limited to the transfer of *Listeria* from pregnant women to fetuses, leading to premature delivery, miscarriage, stillbirth, or meningitis in the newborn.

Researchers have identified more than 50 species of bacteria that can cause bacterial meningitis. One species that is reponsible for many cases is *Listeria monocytogenes. L. monocytogenes* has virulence factors that allow it to resist phagocytosis and cause disease.

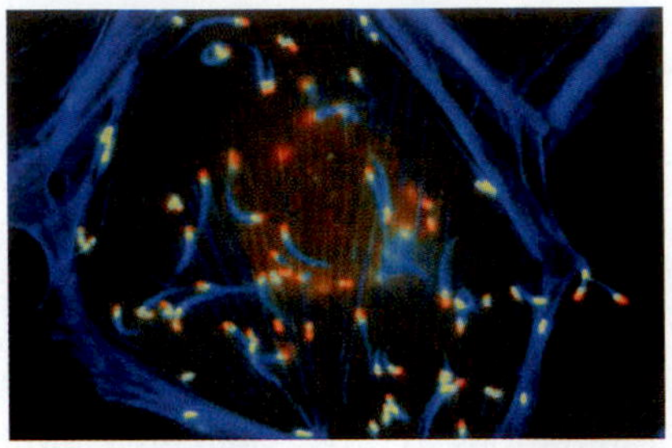

LM
10 μm

Photomicrograph of *Listeria* showing blue-stained actin "tails."

ONE WAY *LISTERIA* AVOIDS HOST'S IMMUNE SYSTEM WHILE INFECTING NEW CELLS

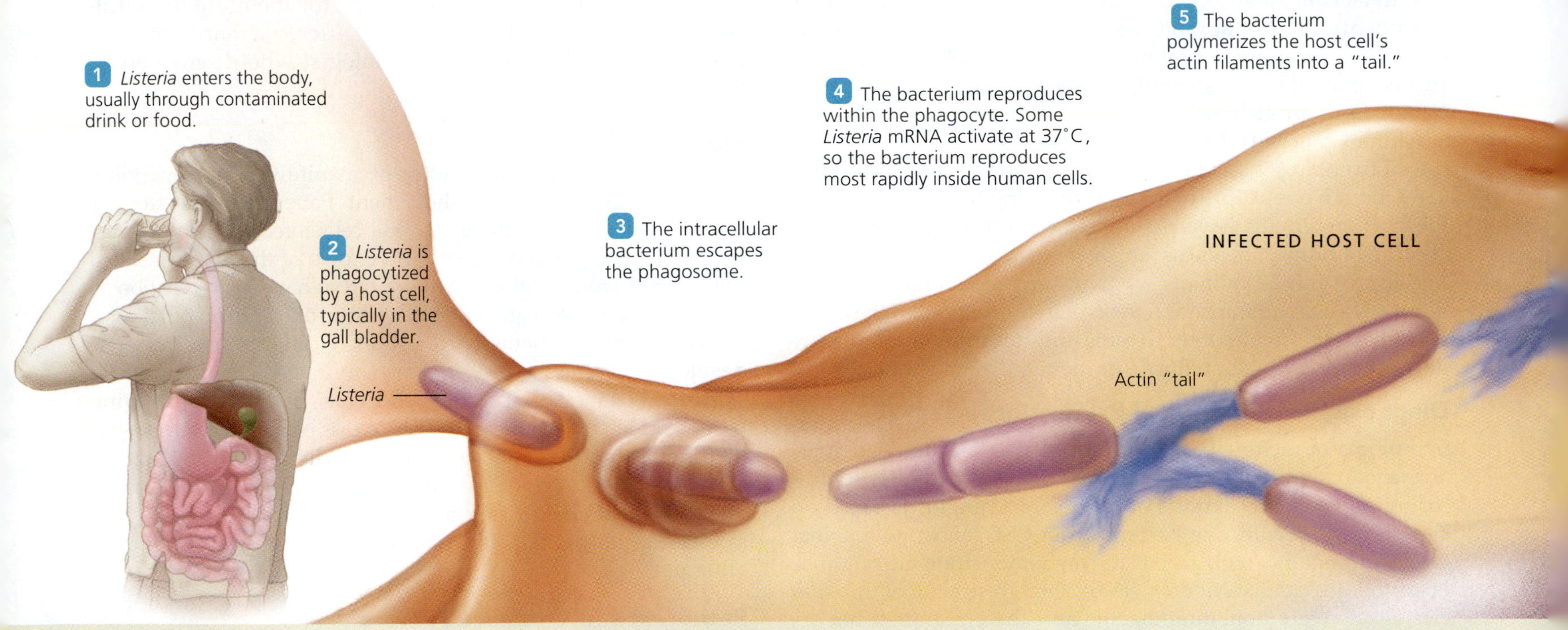

1 *Listeria* enters the body, usually through contaminated drink or food.
2 *Listeria* is phagocytized by a host cell, typically in the gall bladder.
Listeria
3 The intracellular bacterium escapes the phagosome.
4 The bacterium reproduces within the phagocyte. Some *Listeria* mRNA activate at 37°C, so the bacterium reproduces most rapidly inside human cells.
5 The bacterium polymerizes the host cell's actin filaments into a "tail."
INFECTED HOST CELL
Actin "tail"

INVESTIGATE IT!

Consumption of soft cheeses has been associated with listeriosis. How can consumers limit their risk while still continuing to eat soft cheeses?

Scan this code to visit the World Health Organization website to investigate. Then go to MasteringMicrobiology to record your research findings.

EPIDEMIOLOGY

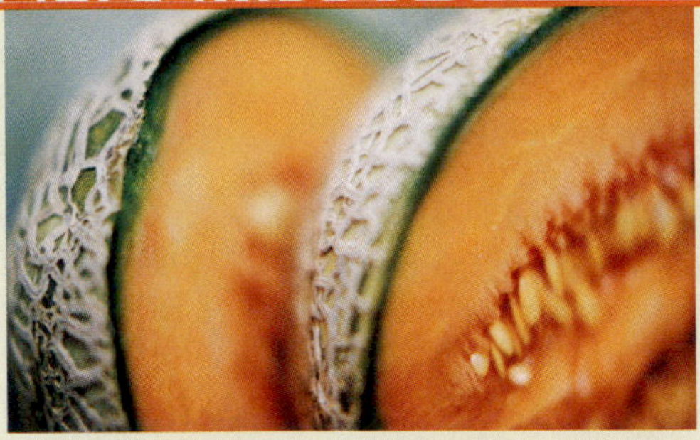

According to the CDC, in 2011, *Listeria*-contaminated cantaloupes "caused the deadliest foodborne disease outbreak in the United States in nearly 90 years," resulting in 30 deaths, including a miscarriage. The death toll could have been higher, but state and local health departments, the FDA, and the CDC worked together to quickly locate the source of the infection (fruit grown on a single farm in Colorado) and to issue a national warning within days of the first reports of the spike in *Listeria* infections.

PATHOGEN AND PATHOGENESIS

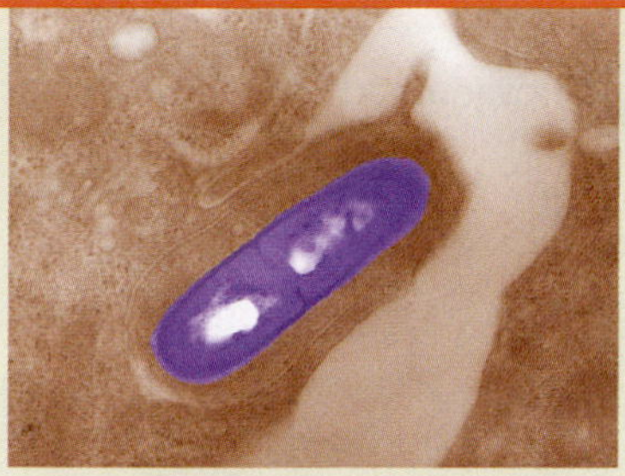

TEM 1 µm

Listeria is a Gram-positive, non-endospore-forming coccobacillus found in soil, water, mammals, birds, fish, and insects. It enters the body through contaminated food and drink, usually deli meats or cheeses.

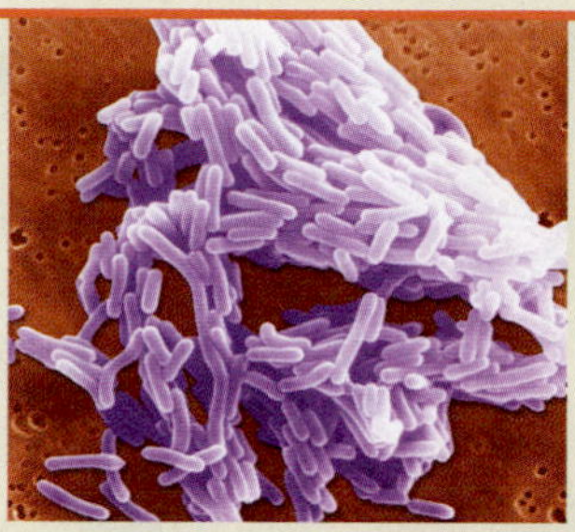

SEM 5 µm

Once ingested, *Listeria* binds to the surface of a macrophage or epithelial cell, triggering its own endocytosis. Inside the cell's phagosome, *Listeria* synthesizes listeriolysin O, an enzyme that breaks open the phagosome before a lysosome can fuse with it; thus, *Listeria* avoids digestion by the cell. *Listeria* then grows and reproduces in the cell's cytosol, sheltered from the B cell immune system.

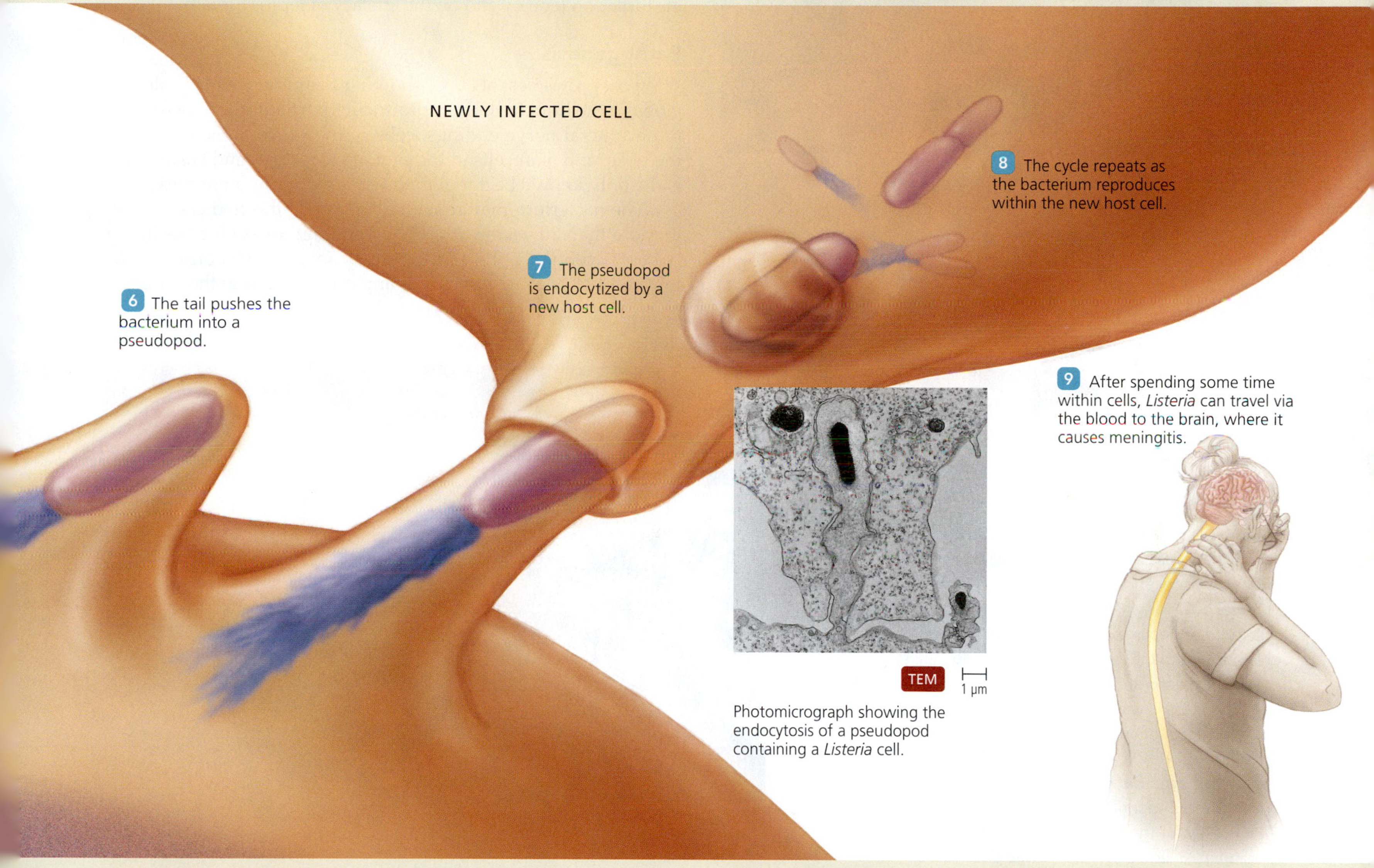
NEWLY INFECTED CELL
8 The cycle repeats as the bacterium reproduces within the new host cell.
7 The pseudopod is endocytized by a new host cell.
6 The tail pushes the bacterium into a pseudopod.
9 After spending some time within cells, Listeria can travel via the blood to the brain, where it causes meningitis.
TEM 1 µm

Photomicrograph showing the endocytosis of a pseudopod containing a *Listeria* cell.

DIAGNOSIS

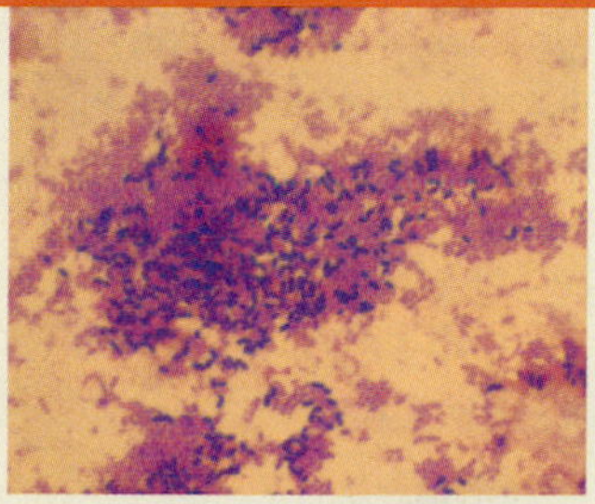

LM 15 µm

Diagnosis can be made by the discovery of *Listeria* in the cerebrospinal fluid of people with symptoms of meningitis. Unfortunately, only a few *Listeria* cells are required to produce disease, so the bacterium is rarely seen in Gram-stained preparations of phagocytes. The bacterium exhibits a characteristic and distinctive end-over-end "tumbling" motility that can be seen at room temperature but not at 37°C.

TREATMENT AND PREVENTION

Most antimicrobial drugs, including penicillin and erythromycin, inhibit *Listeria*, though the bacterium is resistant to tetracycline and trimethoprim. Immunocompromised patients (e.g. AIDS patients) should avoid undercooked meats and vegetables, unpasteurized dairy products, and all soft cheeses.

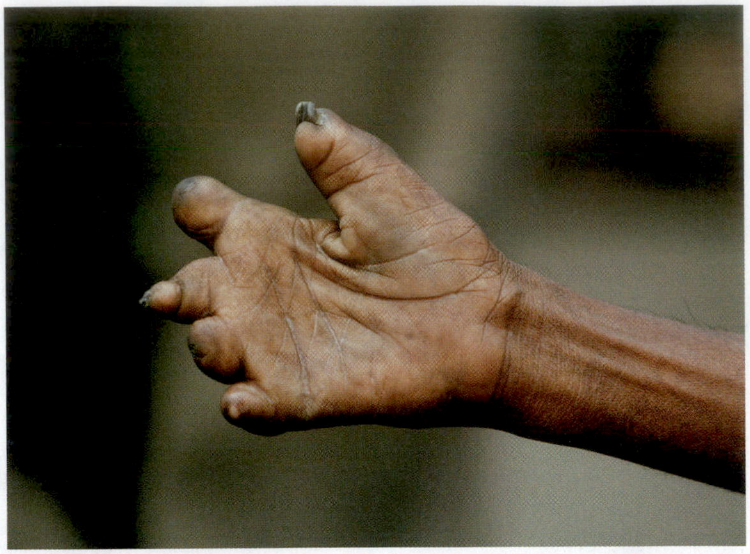

▲ **FIGURE 20.5 Lepromatous leprosy can result in severe deformities.** *What is the medical name for leprosy?*

Figure 20.5 Medically, leprosy is called Hansen's disease.

- Growth within phagocytes.
- Resistance to Gram staining, detergents, many common antimicrobial drugs, and drying. Because mycobacteria stain only Gram stain weakly, the acid-fast staining procedure was developed to differentially stain mycobacteria (see Figure 4.18).

The bacterium has never been grown in cell-free laboratory culture, a fact that has hindered research and diagnostic studies. The nine-banded armadillo, which has a normal body temperature of 30°C, is its only other natural host and has proven valuable in studies on leprosy and on the efficacy of leprosy treatments.

Pathogenesis

M. leprae grows best at 30°C, showing a preference for cooler regions of the human body, reproducing particularly in neuroglia of peripheral nerve endings, cells of the mucous membrane of the nose, and skin cells in the fingers, toes, lips, and earlobes. It is the only known bacterial pathogen of peripheral nerves.

The bacterium may remain alive inside infected cells for 10 to more than 30 years with no obvious sign or symptoms in the patient. Eventually, the body's cellular immune response attacks infected cells, destroying nerves and other tissues in the process.

HIGHLIGHT

Nipah Virus: From Pigs to Humans

Many diseases become known to science either when people encroach on a pathogen's home territory and come into contact with natural hosts or when pathogens are introduced into geographical areas outside their historical range. Diseases resulting under such conditions are called *emerging diseases.* One example of an emerging disease is a new form of encephalitis that has appeared in Southeast Asia, including Malaysia, Singapore, and Bangladesh. The culprit: a never-before-identified RNA virus now known as Nipah virus (genus *Henipavirus*) after the region in Malaysia in which it was discovered.

The victims of Nipah encephalitis experience high fever, severe headache, muscle pain, drowsiness, disorientation, convulsions, and coma. Seventy percent of them die.

Victims contract Nipah virus from infected animals, from other people, from corpses, or from drinking raw date palm sap tainted with bat saliva or feces. In Malaysia, 93% of victims report occupational exposure to pigs.

(a) Collection of date palm sap **(b)** Bat feeding on sap at night

How, then, do pigs become infected with Nipah virus? The prevailing theory involves human encroachment on bat habitats. Certain species of fruit bats are the natural hosts of Nipah virus. Highway workers pushing into these bats' rain forest habitats have disturbed their roosts, driving the bats into proximity with pig farms. The virus then "jumped" species and infected the pigs. Individuals working on pig farms or in slaughterhouses were subsequently infected either via the respiratory route or through breaks in their skin and mucous membranes.

There is no cure or vaccine for Nipah encephalitis.

Epidemiology

Lepromatous leprosy is the more virulent form of the disease, and fortunately it is becoming rare: in the 1990s about 12 million cases were diagnosed annually worldwide, but by 2011 this number had decreased to about 182,000, including 82 cases in the United States. Almost all of the U.S. patients were immigrants. The World Health Organization (WHO) is working with national governments and private organizations to reduce the incidence of leprosy to less than one case per 10,000 people. Only a few endemic countries have yet to achieve this goal.

Human leprosy is transmitted via person-to-person contact, though leprosy is not particularly virulent. Individuals are typically infected only after years of intimate social contact with a victim. Given that the nasal secretions of patients with lepromatous leprosy are loaded with mycobacteria, infection presumably occurs via inhalation of respiratory droplets. Alternatively, infection may occur through breaks in the skin. Industrialized nations no longer quarantine patients with leprosy because the disease is so rarely transmitted and is fully treatable.

Diagnosis, Treatment, and Prevention

Diagnosis of leprosy is based on signs and symptoms of disease—a loss of sensation in skin lesions in the case of tuberculoid leprosy and disfigurement in the case of lepromatous leprosy. Diagnosis is confirmed by the identification of **acid-fast bacilli (AFB)**, also known as *acid-fast rods (AFRs)* in samples from affected sites.

M. leprae quickly develops resistance to single antimicrobial agents, so therapy consists of administering multiple drugs, such as rifampin, clofazimine, and dapsone. Treatment is typically for a minimum of two years and can be lifelong for some patients.

The tuberculosis vaccine (BCG[11]) provides some protection against leprosy, presumably because the bacteria of leprosy and tuberculosis are related and share antigens. Prevention is primarily achieved by limiting exposure to the pathogen and by the prophylactic use of antimicrobial agents when exposure occurs.

Having examined bacterial infections of the meninges and peripheral nerves, we now turn our attention to two neurological diseases caused by bacterial toxins—botulism and tetanus.

Botulism

LEARNING | OUTCOMES

20.8 Describe the signs, symptoms, diagnosis, treatment, and prevention of three kinds of botulism.

20.9 Explain the action of botulism toxin on nerve cells.

Botulism[12] (bot´yū-lizm) is often not an infection, but instead an intoxication (poisoning) caused by a toxin of *Clostridium botulinum* (klos-trid´ē-ūm bo-tū-lī´num) that adversely affects synapses of the peripheral nervous system. Botulism toxin is one of the more powerful natural poisons. Clinicians recognize three manifestations of botulism: foodborne botulism, infant botulism, and wound botulism. Fortunately, all three are rare.

Signs and Symptoms

Patients with **foodborne botulism** become weak and dizzy one to two days following the consumption of botulism toxin in contaminated food, which may not appear or smell spoiled. Blurred vision with fixed dilated pupils, dry mouth, constipation, nausea, vomiting, and abdominal pain are also seen. Progressive paralysis of all voluntary (skeletal) muscles then begins on both sides of the body as peripheral nerves are affected. The patient remains mentally alert throughout the ordeal.

Death, if it occurs, is from paralysis of the diaphragm, which is the major muscle of breathing; the patient cannot inhale. Survivors recover very slowly as their nerve cells grow new endings over the course of months or years, replacing their nonfunctioning tips.

Infant botulism is a disease of infants who are usually under 6 months of age. It differs from foodborne botulism in that the toxin is not ingested. In this disease, the bacterium actually grows in a child's intestinal tract, where it secretes toxin. Infant botulism is characterized by nonspecific symptoms: crying, constipation, and "failure to thrive."

Infants are susceptible to colonization because their GI tracts do not have a sufficient number of benign microbiota to compete with *C. botulinum* for nutrients and space. In adults, *microbial antagonism* by normal intestinal microbiota prevents growth of the bacterium in the intestinal tract.

Wound botulism involves growth of the bacterium in dead tissue following introduction of endospores into wounds. Signs and symptoms are similar to those of the foodborne disease, but the incubation period is longer—four days or more.

Pathogen and Virulence Factors

C. botulinum, the bacterium that produces botulism toxin, is an anaerobic, endospore-forming, Gram-positive bacillus common in soil and water worldwide. Its endospore, which forms at the end of the cell, survives improper canning of non-acidic foods (pH > 4.5) such as meats, eggs, mushrooms, beans, corn, beets, peas, and some cheeses. Endospores germinate to produce vegetative cells that grow and release the debilitating toxin that causes botulism into the jar or can.

Different strains of *C. botulinum* produce one of seven antigenically distinct botulism *neurotoxins*, that is, toxins that affect neurons. Scientists consider botulism toxins the deadliest toxins known—30 grams of pure toxin would be enough to kill every person in the United States. Even a small taste of food contaminated with the powerful toxin, such as licking a spoon, can cause full-blown illness or death.

To learn about a positive use for botulism toxins, see the Beneficial Microbes box on p. 333. Each of the seven toxins is a quaternary protein composed of a single neurologically active polypeptide associated with one or more nontoxic polypeptides that stabilize the toxin and prevent its inactivation by stomach acid.

[11]For bacille Calmette-Guérin, named after the two French developers of this vaccine.

[12]From Latin *botulus*, meaning "sausage."

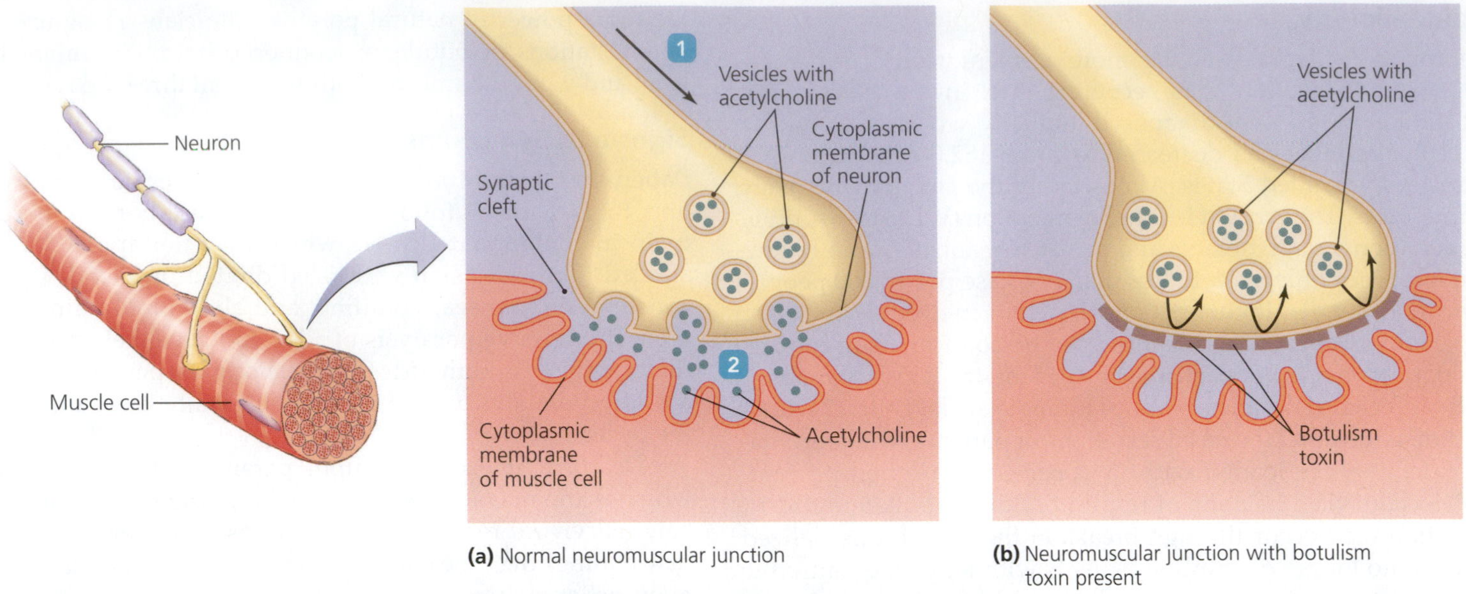

▲ **FIGURE 20.6 How botulism toxin acts at a neuromuscular junction. (a)** Normal function at a neuromuscular junction. A nerve impulse from the central nervous system causes vesicles filled with acetylcholine (ACh) to fuse with the neuron's cytoplasmic membrane 1, releasing ACh into the synaptic cleft 2. The binding of ACh to receptors on the muscle cell's cytoplasmic membrane stimulates a series of events that result in contraction of the muscle cell (not shown). **(b)** Botulism toxin blocks the fusion of ACh vesicles with the neuron's cytoplasmic membrane, thereby preventing secretion of ACh into the cleft; as a result, the muscle cell does not contract.

Pathogenesis

To understand the action of botulism toxins, we must consider the way the nervous system controls muscle contractions. Each of the many ends of a motor neuron forms a synapse with a muscle cell. Such a synapse is called a *neuromuscular junction*. As with most synapses, the two cells do not actually touch; a synaptic cleft remains between them **(FIGURE 20.6a)**. The neuron stores the neurotransmitter *acetylcholine* (as-e-til-kō´lēn; *ACh*) in vesicles near its terminal cytoplasmic membrane. ACh mediates communication between neurons and muscle cells.

When a nerve impulse arrives at the terminus of a motor neuron, ACh vesicles fuse with the neuron's cytoplasmic membrane, releasing ACh into the synaptic cleft. Molecules of ACh then diffuse across the cleft and bind to receptors on the cytoplasmic membrane of the muscle cell. The binding of ACh to the ACh receptor triggers a series of events inside the muscle cell that results in muscle contraction.

Botulism toxins act by binding irreversibly to neuronal cytoplasmic membranes, thereby preventing the fusion of vesicles and the secretion of acetylcholine into the synaptic cleft **(FIGURE 20.6b)**. Thus, botulism neurotoxins prevent muscular contraction, resulting in a flaccid paralysis. Binding of botulism toxin is irreversible; the synapse is forever blocked. Botulism may progress despite aggressive medical care; however, the blocked motor neuron's axon may grow new branches, which form new synapses with muscles.

Epidemiology

About 30 cases each of foodborne and wound botulism occur in the United States each year. Death results in 10% of hospitalized patients.

Infant botulism is now the most common form of botulism in the United States; about 80 cases are reported annually. These cases result from inhalation of endospores in dust or ingestion in food; unpasteurized ("natural") honey is involved in about a third of cases of infant botulism. The mortality rate of hospitalized infants is less than 1%.

Diagnosis, Treatment, and Prevention

The symptoms of botulism are diagnostic; culturing the organism from contaminated food, from feces, or from the patient's wounds confirms the diagnosis. Further, toxin activity may be detected by using a mouse bioassay. In this laboratory procedure, specimens of food, feces, and/or serum are divided into two portions. Botulism antitoxin is mixed with one of the portions, and the portions are then inoculated into two sets of mice. If the mice receiving the antitoxin survive while the other mice die, botulism is confirmed.

Treatment of botulism entails four approaches:

- Vigorous attention to keeping airways open and functional.
- Repeated washing of the intestinal tract to remove *Clostridium*, if the intestinal tract is functioning (as indicated by normal bowel sounds).
- Intravenous administration of botulism immunoglobulin (BIG-IV), which is composed of human antibodies against botulism toxins. BIG-IV neutralizes the neurotoxins in the blood before they can bind to neurons and shortens the length of the disease.
- Administration of antimicrobial drugs to kill the bacterium in wound botulism cases. This approach is not

DISEASE AT A GLANCE 20.1

Infant Botulism

1 Baby inhales or ingests *C. botulinum* endospores, particularly in honey.

2 Endospores germinate in anaerobic environment of intestinal tract. Vegetative cells grow, reproduce, and release botulism toxin.

3 Toxin is absorbed into blood and circulates throughout body.

4 Botulism toxin produces constipation and weak cry.

5 Gradually the toxin paralyzes muscles, including neck muscles and the diaphragm.

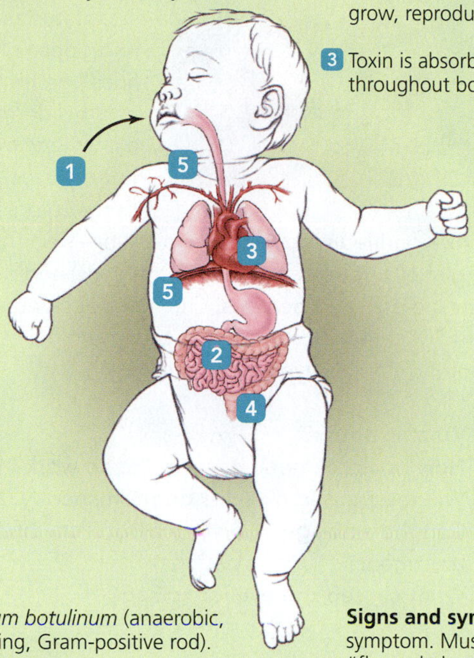

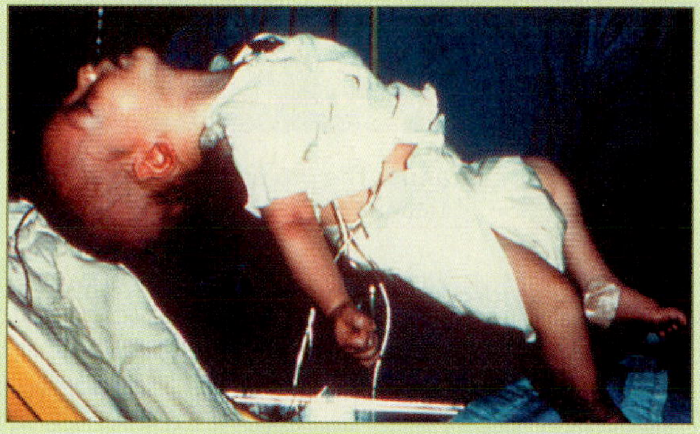

Extreme infantile muscle weakness known as "floppy baby syndrome"

Cause *Clostridium botulinum* (anaerobic, endospore-forming, Gram-positive rod).

Virulence factors Endospores survive improper canning and resist antimicrobial drugs, seven antigenically distinct forms of botulism toxin.

Portal of entry Inhalation of endospores or ingestion of contaminated food.

Signs and symptoms Constipation is the first symptom. Muscle weakness, described as "floppy baby syndrome," causes weak cry, poor feeding (weak suckling), loss of head control, and respiratory distress. Infant may also experience lethargy and descending paralysis.

Incubation period Unknown.

Susceptibility Infants, especially those under six months of age.

Treatment Supportive therapy, including keeping the airway clear and monitoring respiratory functions. Botulism immune globulin (BIG) reduces the length of the disease.

Prevention Do not feed honey to infants less than one year of age.

recommended for treating infant botulism, because the treatment increases toxin release, worsening the disease. Antimicrobials are not effective in treating foodborne botulism, because foodborne botulism results from ingestion of the toxin, not of the bacteria themselves.

Foodborne botulism is prevented by destroying all endospores in contaminated food through proper canning techniques; by preventing endospores from germinating through use of refrigeration or establishment of an acidic environment (pH < 4.5); or by destroying the toxin with heat—food must be heated to at least 80°C for 20 minutes or more. *C. botulinum* produces gas at it grows, so a swollen can of food may indicate contamination. Infant botulism is often associated with the consumption of honey, particularly unpasteurized honey; therefore, pediatricians advise parents not to feed honey to infants until their intestinal microbiota are sufficiently developed to inhibit the germination of *C. botulinum* endospores—usually after one year of age.

Disease at a Glance 20.1 summarizes the features of infant botulism. The accompanying **Clinical Case Study: The Frowning Actor** on p. 614 considers the medical use of botulism toxin.

Tetanus

LEARNING | OUTCOMES

20.10 Describe the action of tetanospasmin.

20.11 Describe the diagnosis, treatment, and prevention of tetanus.

Tetanus[13] is another disease caused by a neurotoxin produced by a species of *Clostridium.*

Signs and Symptoms

Typically the initial and diagnostic sign of tetanus is tightening of the jaw and neck muscles, which is why tetanus is also called *lockjaw*. Other early symptoms include sweating, drooling, grouchiness, and constant back spasms. If the toxin spreads to neurons that control glands and involuntary muscles, then heartbeat irregularities, fluctuations in blood pressure, and

[13]From Greek *tetanus*, meaning "to stretch."

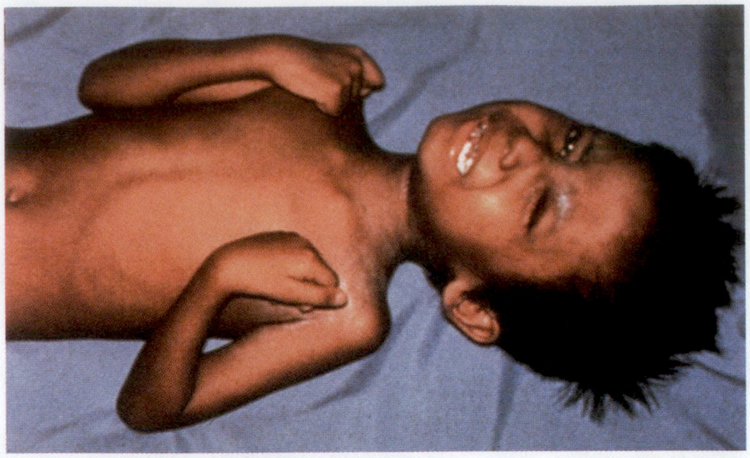

▲ **FIGURE 20.7 A patient with tetanus.** The prolonged muscular contractions are a result of the action of tetanus toxin. *What is the name of tetanus toxin?*

Figure 20.7 *Tetanus toxin is tetanospasmin.*

extensive sweating result. Spasms and contractions may spread to other muscles, becoming so severe that the arms and fists curl tightly, the feet curl down, and the body assumes a stiff backward arch as the heels and back of the head bend toward one another **(FIGURE 20.7)**. Complete, unrelenting contraction of the diaphragm results in a final inhalation—patients die because they cannot exhale.

Pathogen and Virulence Factors

Clostridium tetani (klos-trid´ē-ŭm te´tan-ē) is a small, motile, obligate anaerobe that produces a terminal endospore, giving the cell a distinctive "lollipop" appearance (see Disease at a Glance 20.2 on p. 616). Vegetative cells are extremely sensitive to oxygen and live only in anaerobic environments. In contrast, endospores can survive for decades or longer exposed to air and are nearly universally found in soil, dust, and the intestines of animals and humans.

Cells of *C. tetani* release a potent neurotoxin called **tetanospasmin** (tet´ă-nō-spaz´min) when they die. Tetanospasmin is composed of two polypeptides held together by a disulfide bond. The heavier of the two polypeptides binds to the cytoplasmic membrane of a neuron, triggering the neuron to endocytize the toxin and cleave off the lighter of the two polypeptides. Axonal transport carries the lighter polypeptide to the central nervous system. *C. tetani* remains localized at the site of infection; only the toxin moves to the central nervous system.

Pathogenesis

Contrary to popular belief, deep puncture wounds by rusty nails are not the only (or even primary) source of tetanus. Any break in the skin or mucous membranes can allow endospores of *C. tetani* access to deeper tissues, which lack free oxygen. Wooden splinters, shaving nicks, accidental staple and tack wounds, and drug injections, as well as more serious cuts and punctures of the skin or mucous membranes, have introduced endospores into patients where the endospores germinated, grew, died, and released tetanospasmin.

To understand the action of tetanospasmin, we must further consider events in the central nervous system. As we have seen, nerve impulses traveling down motor neurons cause the motor neurons to secrete acetylcholine onto muscle cells, triggering the muscles to contract. But what tells the motor neuron to generate the nerve impulse?

Two kinds of neurons of the central nervous system act on motor neurons: Stimulatory neurons release a neurotransmitter that excites motor neurons to produce a nerve impulse, resulting in muscle contraction. In contrast, inhibitory neurons release an inhibitory neurotransmitter, hindering motor neurons from producing the nerve impulse, so the muscle remains relaxed. In other words, motor neurons do not directly relax a muscle; rather, inhibitory neurons inhibit the motor neuron, so the muscle is not stimulated to contract **(FIGURE 20.8a)**.

Tetanospasmin blocks the release of inhibitory neurotransmitter. With inhibition blocked, excitation of the motor neurons is unregulated, and the muscle is signaled to contract **(FIGURE 20.8b)**. The result is that muscles contract and do not relax. Contractions can be so severe they break bones.

The incubation period of tetanus ranges from five days to 15 weeks depending on the distance of the infection site in the extremities from the central nervous system.

Epidemiology

The effect of tetanospasmin is not reversible at a synapse, so recovery depends on the growth of new neuronal terminals to

CLINICAL CASE STUDY

The Frowning Actor

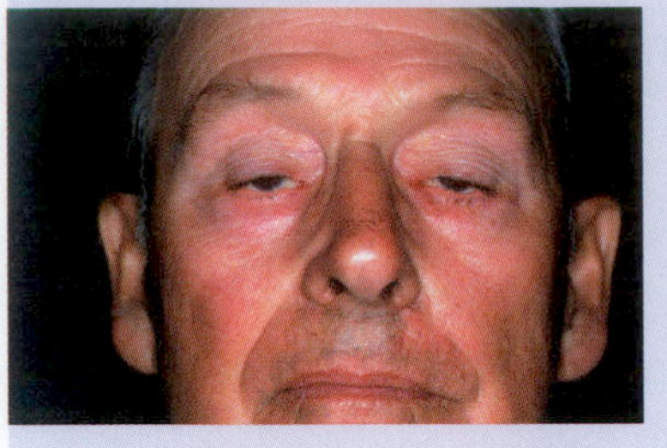

A 55-year-old actor is concerned about maintaining his star status as a leading man and doesn't like the "frown lines" on his forehead. He asks his dermatologist about a cosmetic treatment to relax the facial lines. The dermatologist explains that Botox—purified type A botulism toxin—is approved for use on the glabella (forehead between the eyes) and discusses with the actor the risks and benefits. The actor decides to have a treatment. Two weeks following the treatment, he calls the office reporting that his eyelids are sagging.

1. What is Botox?
2. How does cosmetic botulism toxin work?
3. Will the treated muscles be permanently "frozen"?
4. Can the actor get a systemic botulism infection from the treatment?
5. What caused the eyelids to sag?

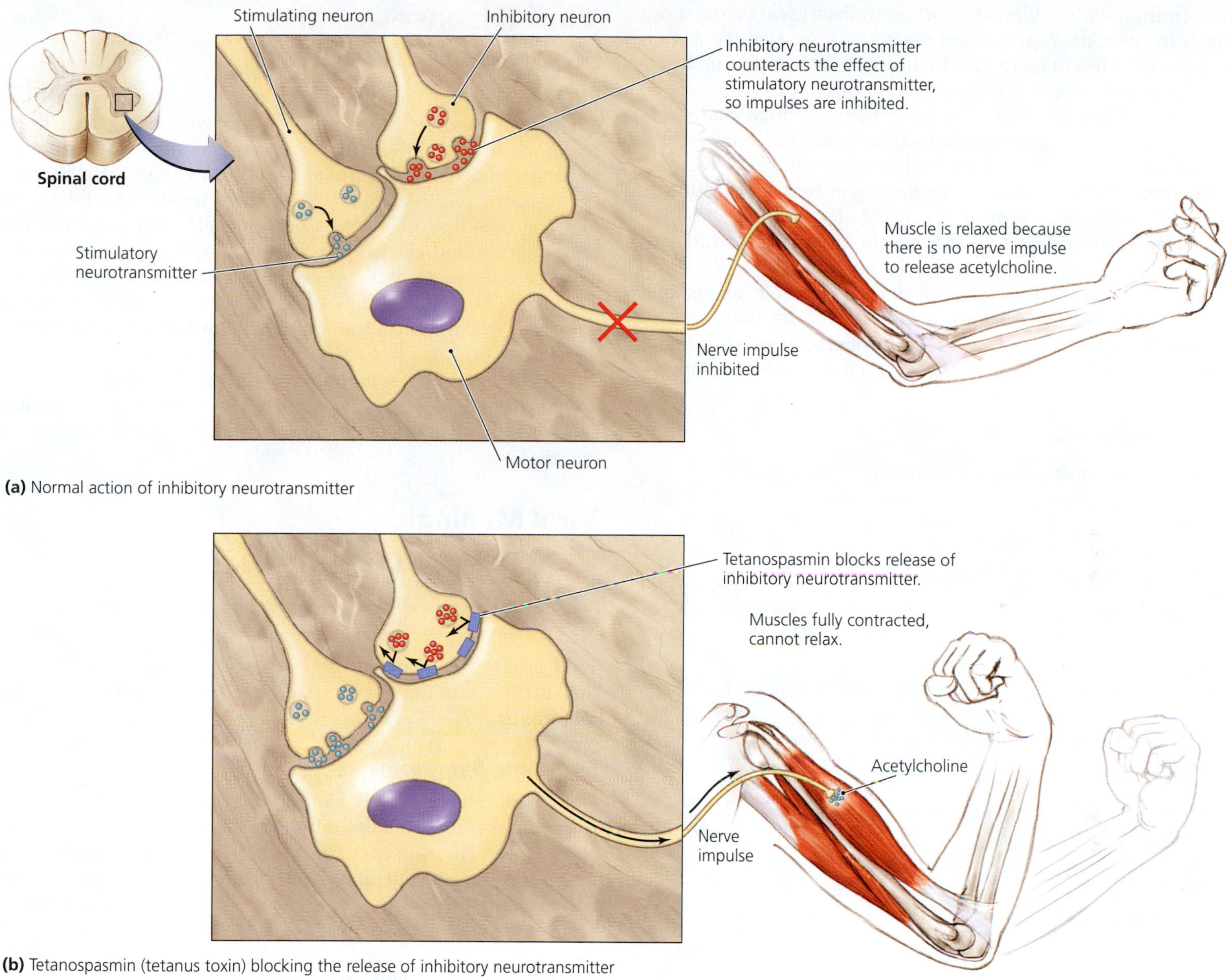
Stimulating neuron
Inhibitory neuron
Inhibitory neurotransmitter counteracts the effect of stimulatory neurotransmitter, so impulses are inhibited.
Spinal cord
Stimulatory neurotransmitter
Muscle is relaxed because there is no nerve impulse to release acetylcholine.
Nerve impulse inhibited
Motor neuron
(a) Normal action of inhibitory neurotransmitter
Tetanospasmin blocks release of inhibitory neurotransmitter.
Muscles fully contracted, cannot relax.
Acetylcholine
Nerve impulse
(b) Tetanospasmin (tetanus toxin) blocking the release of inhibitory neurotransmitter

▲ **FIGURE 20.8 The action of tetanus toxin on a pair of antagonistic muscles.** **(a)** Normally, inhibitory neurotransmitter can block nerve impulses in motor neurons so that muscles can relax. **(b)** Tetanospasmin blocks the inhibitory neurotransmitter. As a result, motor neurons are not inhibited but instead generate nerve impulses that stimulate continuous muscle contraction.

replace those affected. The mortality rate of untreated tetanus is about 50% among all patients, but the mortality of tetanus in newborns, resulting most commonly from infection of the umbilical stump, exceeds 90%.

Incidence of tetanus worldwide has decreased from nearly a million cases annually to fewer than 15,000 cases in 2011, mostly in countries where immunization is unavailable or medical care is inadequate. In the United States, morbidity has declined from 560 reported cases in 1947, the year reporting began, to 36 reported cases in 2012. Widespread immunization has produced a decline in tetanus incidence.

Diagnosis, Treatment, and Prevention

The diagnostic feature of tetanus is the characteristic muscular contraction, which is often noted too late to save the patient. The bacterium itself is rarely isolated from clinical samples because it grows slowly in culture and is extremely sensitive to oxygen.

Treatment involves thorough cleaning of wounds to remove all endospores; immediate passive immunotherapy, in which antibodies directed against tetanospasmin are injected; the administration of antimicrobials such as penicillin; and active immunization to stimulate antibody and memory cell production. Cleansing and antimicrobials eliminate the bacterium, whereas the immunoglobulin binds to and neutralizes toxin before it can attach to neurons. Active immunization stimulates the formation of antibodies to neutralize toxin. Once tetanospasmin binds to a neuron, treatment is limited to supportive care.

The number of cases of tetanus in the United States has steadily declined as a result of effective immunization with *tetanus toxoid*, which is inactivated tetanospasmin. The CDC currently recommends five doses beginning at two months of age, followed by a booster every 10 years for life (see Figure 17.3).

Disease at a Glance 20.2 summarizes the features of tetanus.

We have examined some bacteria and their toxins that cause diseases of the nervous system. Viral infection can also have neurological consequences. In the next section we consider some viruses whose primary effect is on the nervous system.

TELL ME WHY

Why is the incubation period of wound botulism more than twice as long as the incubation period of foodborne botulism?

DISEASE AT A GLANCE 20.2

Tetanus

Cause *Clostridium tetani* (anaerobic, endospore-forming, Gram-positive rod).

Virulence factors Endospores allow bacterium to survive harsh conditions and resist antimicrobial drugs; neurotoxin tetanospasmin blocks action of inhibitory neurons.

Portal of entry Endospores enter through breaks in the skin, such as cuts and punctures.

Signs and symptoms Tightening of the jaw and neck muscles, difficulty swallowing, followed by fever and muscle spasms.

Incubation period Five days to 15 weeks (average of seven days).

Susceptibility Unimmunized individuals with broken skin who contact the microbe.

Treatment Cleansing of wound, administration of human tetanus immunoglobulin (HTIG) and penicillin, and active immunization.

Prevention Tetanus toxoid contained in DTaP, DTP, and Td vaccines.

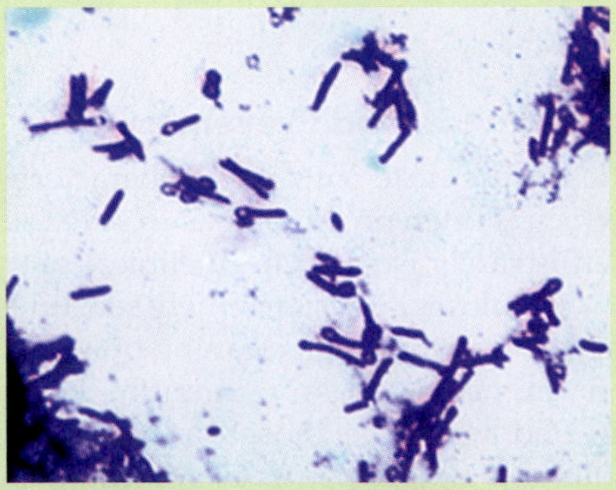

Clostridium tetani cells, some with "lollipop" endospores

Viral Diseases of the Nervous System

Viruses, being smaller than cells, can more readily cross the blood-brain barrier; therefore, it is not surprising that there are more viral infections of the nervous system than bacterial or fungal infections. **Beneficial Microbes: Cocaine No Brainer** suggests one positive aspect of the way viruses enter the brain. Many viruses that attack other body systems can also affect the brain: herpes and chickenpox viruses may remain dormant in nerve cells for years, and measles virus causes *subacute sclerosing panencephalitis*, which is a slow progressive disease of the CNS. Chapter 19 dealt with these pathogens, which primarily affect the skin.

In the following sections we consider viruses that primarily affect the nervous system, causing meningitis, polio, rabies, and encephalitis. We begin our survey by considering viral meningitis.

Viral Meningitis

LEARNING | OUTCOME

20.12 Compare and contrast viral and bacterial meningitis.

Viral meningitis, also known as *aseptic meningitis* to indicate that no bacteria are involved, is the most common form of meningitis.

Signs and Symptoms

Viral meningitis is usually a milder disease than either bacterial or fungal meningitis; although the signs and symptoms—fever, severe headache, stiff neck, drowsiness, confusion, nausea, and vomiting—may be the same, death from viral meningitis is rare. Some meningitis viruses also cause skin rashes, sore throats, and colds.

Pathogens and Virulence Factors

Herpesviruses, mumps virus, and several other viruses may cause viral meningitis, but about 90% of cases result from infections of viruses in the genus *Enterovirus*[14] of the family *Picornaviridae*.[15] As their name indicates, picornaviruses are very small (20–30 nm in diameter), positive, single-stranded RNA (+ssRNA) viruses. They lack envelopes. Enteroviruses are so named because they often spread from person to person via fecal contamination of food, water, or hands. Though ingested enteroviruses attack cells lining the intestinal tract, they do not cause gastrointestinal illnesses; instead, enteroviruses spread via the bloodstream—a condition called **viremia**—to infect other organs, including the meninges.

The common names for the three types of enterovirus that cause most human meningitis are *coxsackie A virus* (named for Coxsackie, New York, where the virus was first isolated),

[14]From Greek *enteron*, meaning "intestine," and Latin *virus*, meaning "poison."
[15]From Italian *pico*, meaning "small," and RNA.

coxsackie B virus, and *echovirus*.[16] (Other viruses, such as West Nile virus, may also affect the meninges, but their primary target is nerve cells. These viruses are considered in later sections.)

Pathogenesis

Enteroviruses primarily attack cells lining the intestinal tract and lungs, in the latter case producing colds. The viruses are cytolytic—they kill their target cells. Damage to cells in the meninges triggers meningitis.

The incubation period of enterovirus infections is between three and seven days, and patients recover completely without treatment after another seven to ten days.

Epidemiology

Viral meningitis appears to be much more common than bacterial or fungal meningitis, but viral meningitis is often mild, and the disease is not reportable, so firm data are not available.

Enteroviruses are contagious, being spread in respiratory droplets and in feces—patients shed viruses in their feces for weeks. For some reason, enteroviruses are more commonly spread in the summer and early fall. The viruses are stable and can survive in chlorinated swimming pools.

Exposure to a patient with meningitis may result in an infection resembling a cold but rarely causes meningitis—fewer than one of every 1000 infected people develop viral meningitis. Patients become contagious when symptoms develop and remain contagious for up to 10 days.

Diagnosis, Treatment, and Prevention

Physicians diagnose viral meningitis based on the characteristic signs and symptoms in the absence of bacteria in CSF obtained with a spinal tap.

No specific treatment exists for viral meningitis; health care providers recommend resting, drinking plenty of fluids, and taking medicine to reduce fever and headache pain.

It is difficult to suppress the spread of enteroviruses because most infected people lack signs or symptoms. Frequent hand antisepsis, avoiding crowded swimming pools, and refraining from bringing contaminated hands near the mouth, nose, or eyes limit the chance of infection.

Poliomyelitis

LEARNING | OUTCOMES

20.13 Describe poliomyelitis with reference to the cause, signs, symptoms, epidemiology, diagnosis, treatment, and prevention.

20.14 Compare and contrast the two polio vaccines.

Older Americans still remember the dreaded **poliomyelitis**, (pō´lē-ō-mī´ĕ-lī´tis) or **polio,** epidemics of the past, when the floors of hospitals were filled with iron lungs **(FIGURE 20.9)** and schoolchildren donated coins to the March of Dimes to develop a polio vaccine. Because polioviruses are stable for prolonged periods in swimming pools and lakes and can be acquired by swallowing contaminated water, parents in the 1930s and 1940s often feared to let their children swim. Those days may soon be over worldwide, as we may soon make polio the second human disease eliminated worldwide **(FIGURE 20.10)**.

Signs and Symptoms

After being ingested and infecting pharyngeal and intestinal cells, poliovirus travels via the lymph and blood to infect cells of the CNS, particularly of the spinal cord. It causes one of the following four conditions:

- *Asymptomatic infections* account for about 90% of all cases.
- *Minor polio* includes nonspecific symptoms such as temporary fever, headache, malaise, and sore throat. Approximately 5% of cases are minor polio.

[16]Derived from *enteric cytopathic human orphan* virus because the virus is acquired intestinally and was not initially associated with disease; it was an "orphan."

BENEFICIAL MICROBES

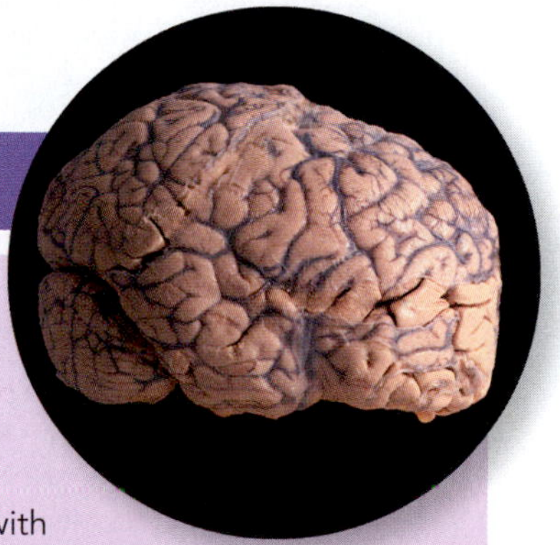

Cocaine No-Brainer

Cocaine addiction is a major health and societal problem in America. The drug allows the accumulation of a neurotransmitter that stimulates an intense sense of joy and feelings of increased confidence. Simply understood, addiction results when people sense a physical or mental need for the effect. Chemical agents have proved disappointing for treating cocaine addiction, but a microbe may change this state of affairs.

Scientists at the Scripps Research Institute in La Jolla, California, have engineered a filamentous bacteriophage that displays cocaine-binding antibodies on its surface. In the central nervous system, the phage absorbs cocaine, preventing the drug from interacting with neurons and mitigating the cocaine's effect. The scientists chose a bacteriophage to deliver the antibodies because bacteriophages can enter directly into the brain through the nose. Perhaps soon addicts can snort a treatment.

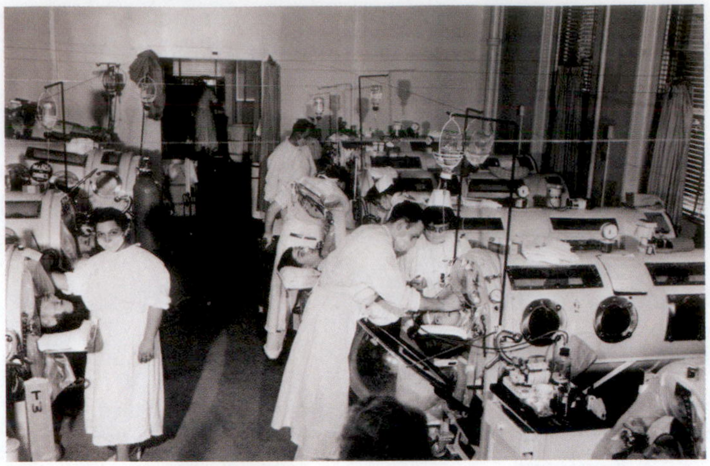

▲ **FIGURE 20.9 A hospital ward filled with "iron lungs."** These mechanical respirators assisted the paralyzed respiratory muscles of polio patients. Such wards were common before 1955.

- *Nonparalytic polio* results from polioviruses invading the meninges and central nervous system, producing muscle spasms and back pain in addition to the general symptoms of minor polio. Nonparalytic polio occurs in about 2% of cases.
- *Paralytic polio* involves viruses invading cells of the spinal cord and the portion of the cerebrum that controls skeletal muscles, producing paralysis by limiting nerve impulse conduction. The degree of paralysis varies with the strain of poliovirus involved, the infective dose, and the health and age of the patient. In a type of paralytic polio called *bulbar poliomyelitis*, the brain stem is infected, resulting in paralysis of respiratory muscles or of muscles in the limbs (see Disease at a Glance 20.3). In the past, iron lungs were used to help victims breathe. In most paralytic cases, complete recovery results after 6–24 months, but in some cases paralysis is lifelong. Paralytic polio occurs in fewer than 2% of infections. Famous victims of paralytic polio include actors Donald Sutherland and Alan Alda.

Postpolio syndrome is a crippling deterioration in the function of polio-affected muscles that occurs in up to 80% of recovered polio patients some 30–40 years after their original bout with poliomyelitis. This condition is not caused by a reemergence of polioviruses, because viruses are not present. Instead, the effects appear to stem from an aging-related aggravation of nerve damage that occurred during the original infection.

Pathogen and Pathogenesis

Poliovirus is another species of *Enterovirus* (family *Picornaviridae*). Poliovirus is relatively stable outside the body and remains infectious in food and water for some time. Scientists distinguish three strains of poliovirus by their antigens. Each of the three can cause all types of polio.

People most often get poliovirus by drinking contaminated water. The viruses are replicated in cells of the throat and small intestine, producing initial symptoms of sore throat and nausea. Viruses then infect lymph nodes and from there enter the blood. In most people, the infection ends there; however, viremia may progress to infection of neurons of the CNS. Paralysis develops following destruction of motor neurons in the upper spinal cord and brain stem; poliovirus does not infect spinal nerves or muscles.

Epidemiology

The near elimination of polio stands as one of the great achievements of 20th-century medicine. Polio currently exists in only a few countries in Africa and Asia (see Figure 20.10).

The last case of naturally occurring poliomyelitis in the Americas occurred in 1979 (though vaccine-induced polio

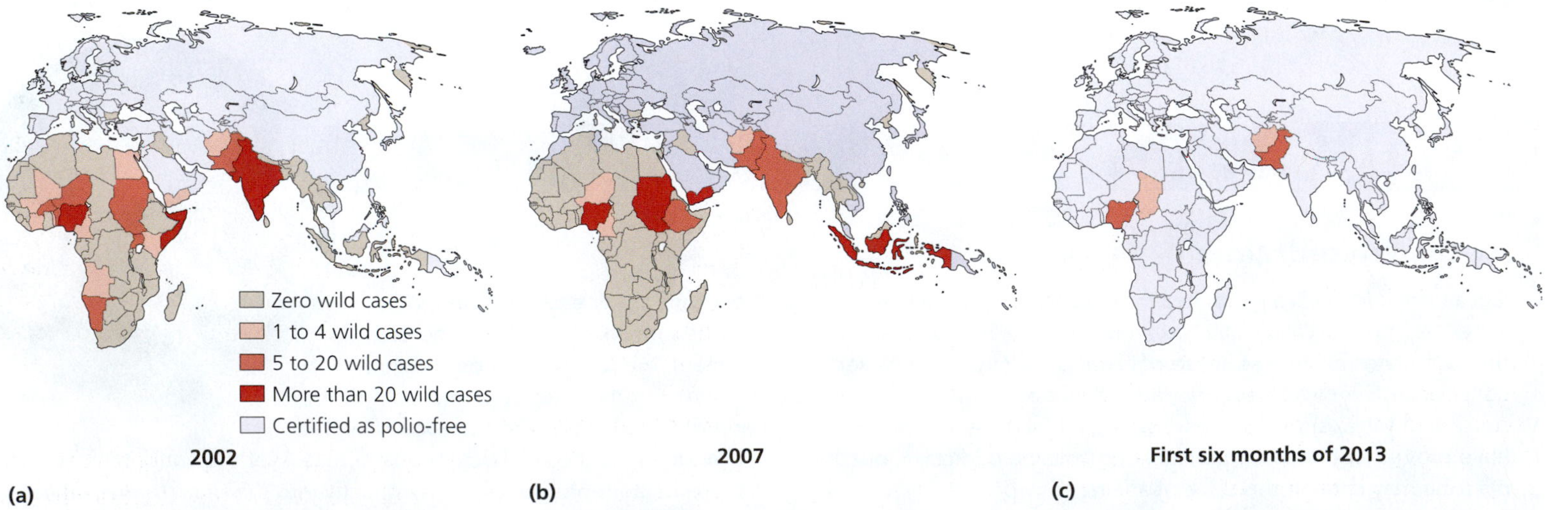

▲ **FIGURE 20.10 Reports of naturally ocurring polio.** **(a)** Cases in 2002. **(b)** Cases in 2007. **(c)** Cases in first six months of 2013. No cases occurred in the Americas. Global eradication of polio is a goal within reach.

TABLE 20.1 Comparison of Polio Vaccines

	Advantages	Disadvantages
Salk vaccine: Inactivated polio vaccine (IPV)	Effective; inexpensive; stable during transport and storage; poses no risk of vaccine-related disease	Requires booster to achieve lifelong immunity; must be injected; requires higher community immunization rate than does OPV
Sabin vaccine: Attenuated, oral polio vaccine (OPV)	Provides lifelong immunity without boosters; triggers secretory antibody response similar to natural infection; easy to administer; results in herd immunity	Less stable than IPV; can mutate to disease-causing form; poses risk of polio developing in immunocompromised contacts of those vaccinated

occurred as recently as 2001). The World Health Organization (WHO) is working with governments and private groups to certify the world polio-free. Then, like smallpox virus, poliovirus would exist only in laboratories. The number of polio cases dropped from 350,000 in 1988 to a low of only 500 in 2001, but the disease has spread in recent years beyond the borders of endemic countries, particularly India and Nigeria (Figure 20.10a). Political and religious tension coupled with poor sanitation and high population density have made it difficult for health care workers to contain the virus. About 1600 cases occurred in 2009.

Diagnosis, Treatment, and Prevention

Diagnosis of polio is based upon demonstration of the virus in throat secretions or feces. There is no specific treatment for polio besides managing the symptoms of infection.

Two effective vaccines make polio eradication possible. Jonas Salk (1914–1995) developed an inactivated polio vaccine (IPV) in 1955. Six years later, it was replaced in the United States by a live, attenuated (weakened), oral polio vaccine (OPV) developed by Albert Sabin (1906–1993). Both vaccines are effective in providing immunity against all three strains of poliovirus. Recently, pediatricians have returned to using IPV for initial immunization of babies because viruses in OPV have occasionally mutated into a virulent form that caused polio. **TABLE 20.1** compares the advantages and disadvantages of the two vaccines.

Disease at a Glance 20.3 summarizes the characteristics of polio.

DISEASE AT A GLANCE 20.3

Polio

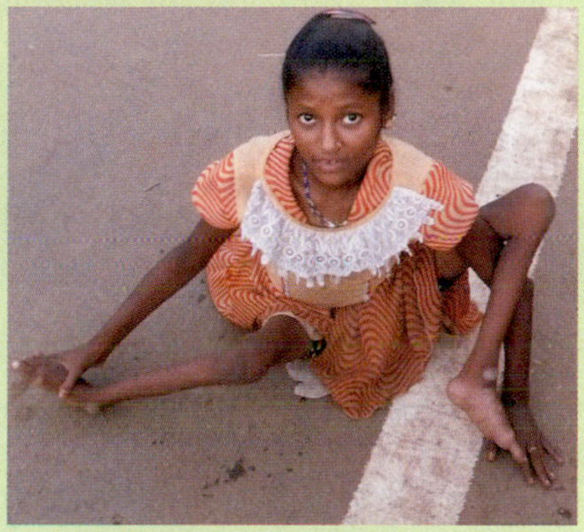

Cause Poliovirus (*Enterovirus*: naked, +ssRNA virus)

Virulence factors Attaches to human intestinal cells; intracellular replication cycle; is relatively stable outside body.

Portal of entry Ingestion in contaminated water.

Signs and symptoms Five possibilities: asymptomatic, minor polio with temporary flulike signs and symptoms; nonparalytic polio with muscle spasms and neck pain; paralytic polio that infects the brain or spinal cord, resulting in paralysis of skeletal muscles; and postpolio syndrome, which is a crippling degeneration that occurs 30–40 years after infection.

Incubation period 6 to 20 days.

Susceptibility Children are most at risk for acute, severe polio.

Treatment Physical therapy for rehabilitation of weakened muscles; noninvasive ventilation to assist the patient's breathing.

Prevention Inactivated (Salk) vaccine or attenuated oral (Sabin) vaccine.

Rabies

LEARNING **OUTCOMES**

20.15 Define *zoonosis*.

20.16 Describe rabies, including its diagnosis, treatment, and prevention.

Although the cry "Mad dog! Mad dog!" may make you apprehensive, in the 19th century—before Pasteur developed a rabies vaccine and treatment—a bite from a rabid dog was a death sentence, and the cry of "Mad dog!" often created panic. Very few people survived **rabies**[17] before Pasteur's important work.

Signs and Symptoms

Initial signs and symptoms of rabies include pain or itching at the site of infection, fever, headache, malaise, and anorexia. Once the virus reaches the central nervous system, neurological manifestations characteristic of rabies develop: hydrophobia[18] (triggered by the pain involved in attempts to swallow water), seizures, disorientation, hallucinations, and paralysis. Death results from respiratory paralysis and other neurological complications.

Pathogen and Virulence Factors

Rabies virus is a negative, single-stranded RNA (-ssRNA) virus in the genus *Lyssavirus*, family *Rhabdoviridae*. Rhabdoviruses

[17]From Latin *rabere*, meaning "rage" or "madness."

[18]From Greek *hydro*, meaning "water," and *phobos*, meaning "fear."

have helical capsids supercoiled into cylinders, which give them a striated appearance, and are surrounded by bullet-shaped envelopes (see Disease at a Glance 20.4). Glycoprotein spikes on the surface of the envelope serve as attachment proteins.

Pathogenesis

Rabies virus attaches to skeletal muscle cells, triggering its own endocytosis. The virus replicates in the cytoplasm of muscle cells. Later, it moves across neuromuscular junctions into neurons and then travels to the central nervous system via axonal transport.

Function of the spinal cord and brain degenerate as a result of infection, though infected cells show little structural damage when examined microscopically. Viruses travel back to the periphery, including the salivary glands, through cranial and spinal nerve cells. Viruses are secreted in the saliva of infected mammals. Transmission of rabies viruses in the saliva of infected animals usually occurs via a bite but can occur through the introduction of viruses into breaks in the skin or mucous membranes or, rarely, through inhalation.

Epidemiology

Rabies is a **zoonosis**; that is, a disease spread from animal reservoirs to humans. Rabies affects mammals, though not all mammals are reservoirs; rodents, for instance, rarely get rabies. The main mammals involved differ from locale to locale and change over time as a result of changes in animal populations and interactions among animals and humans. The primary reservoir of rabies in urban areas is the dog. In the wild, rabies can be found in many animals, including foxes, badgers, raccoons, skunks, cats, bats, and feral dogs **(FIGURE 20.11)**. Bats are the source of most cases of rabies in humans, causing about 75% of cases. Only four cases of rabies in humans occurred in the U.S. in 2009.

Diagnosis, Treatment, and Prevention

The neurological symptoms of rabies are unique and generally sufficient for diagnosis. Tests for antibodies in the blood confirm the diagnosis. Postmortem laboratory tests are often conducted to determine whether a suspected animal in fact carries rabies virus. These tests include antigen detection by immunofluorescence and the identification of aggregates of viruses (called Negri bodies) in the brain **(FIGURE 20.12)**. Unfortunately, by the time symptoms and antibody production occur, it is too late to intervene, and the disease will follow its natural course.

Human rabies vaccine, which is called *human diploid cell vaccine (HDCV)*, is prepared from deactivated rabies viruses cultured in human diploid cells. It is administered intramuscularly on days 0, 3, 7, and 14 after exposure to rabies virus. The vaccine can also be administered before infection to workers who regularly come into contact with animals (veterinarians, zookeepers, and animal control workers) and to people traveling to areas of the world where rabies is prevalent.

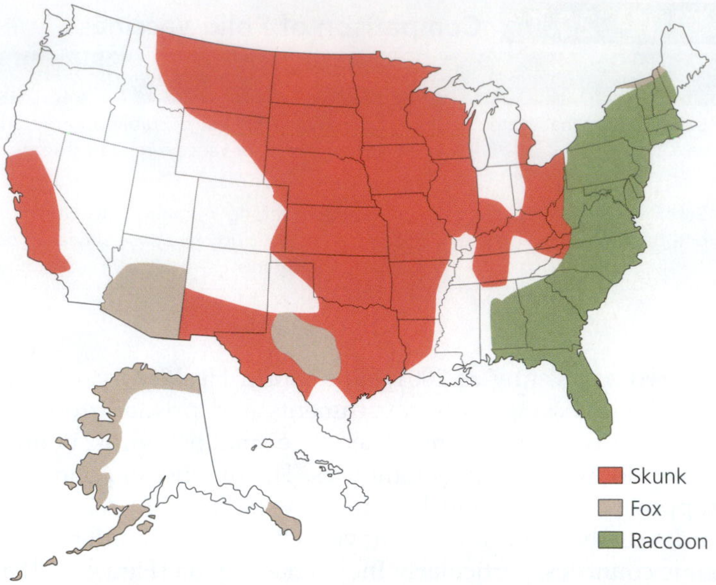

▲ **FIGURE 20.11 Portions of the United States in which skunks, foxes, or raccoons are the predominant wildlife reservoirs for rabies.** Absence of a color does not indicate the absence of rabies in that region, only that no wildlife reservoir is predominant. Bats, for instance, harbor rabies throughout the country but are rarely the predominant reservoir, though they are the major source of human rabies in the United States.

CLINICAL CASE STUDY

A Woman with No Feelings

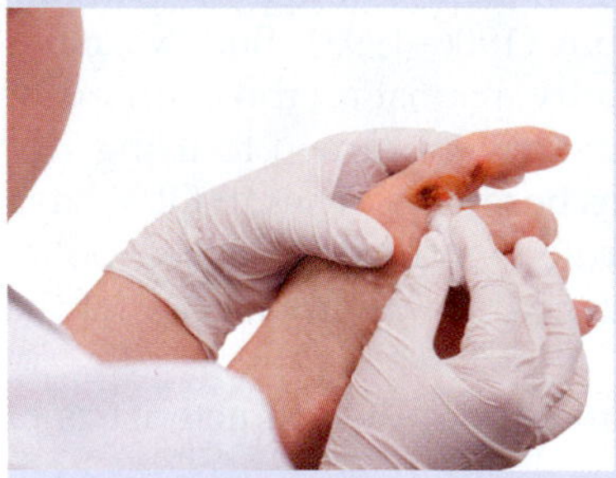

A 41-year-old woman arrives in the hospital emergency department with third-degree burns on the second and third fingers of her left hand. Through a family member who acts as an interpreter, she tells the triage nurse that she had moved to the United States from Brazil barely six months before. This evening she had been sitting in her living room, cigarette in hand, watching the nightly news, when she noticed the odor of something burning. As she rose to investigate, she saw that her fingers were blackened and smoking where the cigarette had burned down. With a jerk she discarded the cigarette and marveled that she felt no pain—it was as if her fingers were not her own.

1. The health care providers suspect the patient has leprosy. How might they verify this?
2. What type of leprosy is this patient more likely to have?
3. What is the appropriate treatment for her condition?
4. Who has the higher risk of contracting the disease: the emergency department personnel or family members with whom she lives?

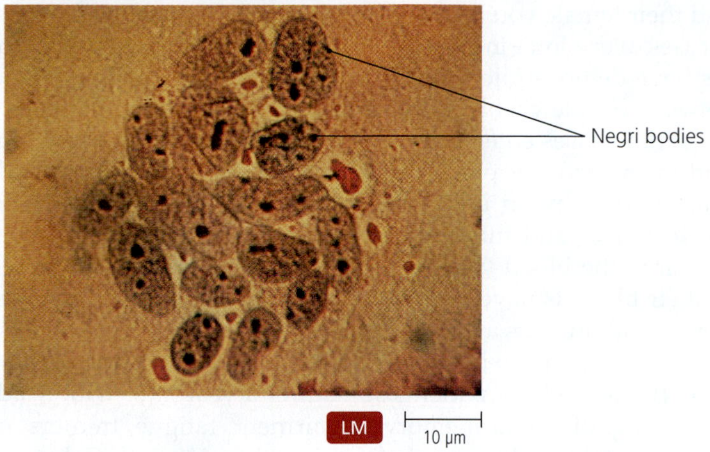

▲ **FIGURE 20.12 Negri bodies, characteristic of rabies infection, in cells of the cerebellum.** *What are Negri bodies?*

Figure 20.12 Negri bodies are stained aggregates of rabies viruses.

Treatment of rabies begins with treatment of the site of infection. The wound should be thoroughly cleansed with water and soap or another substance that deactivates viruses. The World Health Organization recommends anointing the wound with antirabies serum. Initial treatment also involves injection of *human rabies immunoglobulin (HRIG)*. Subsequent treatment involves the four vaccine (HDCV) injections, mentioned previously. Rabies is one of the few infections that can be treated with active immunization because the progress of viral replication and movement to the brain is slow enough to allow effective immunity to develop before disease develops.

The control of rabies involves immunization of domestic dogs and cats and the removal of unwanted strays from urban areas. It is more difficult to eliminate rabies in wild animals because rabies virus can infect so many species, though in the late 1990s an epidemic of rabies was stopped in southern Texas by the successful immunization of the wild coyote population. This was accomplished by lacing meat with an oral vaccine and dropping it from airplanes into the coyotes' range.

Disease at a Glance 20.4 summarizes the features of rabies.

DISEASE AT A GLANCE 20.4

Rabies

Cause Rabies virus (*Lyssavirus*, enveloped, -ssRNA virus).

Virulence factors Glycoprotein spikes on envelope serve as attachement proteins; triggers its own endocytosis by body cells; intracellular replication cycle.

Portal of entry Via the bite or scratch of an infected animal.

Signs and symptoms Pain or itching at site of infection, fever, headache, malaise, anorexia. Neurological manifestations include hydrophobia, seizures, disorientation, hallucinations, paralysis, and death.

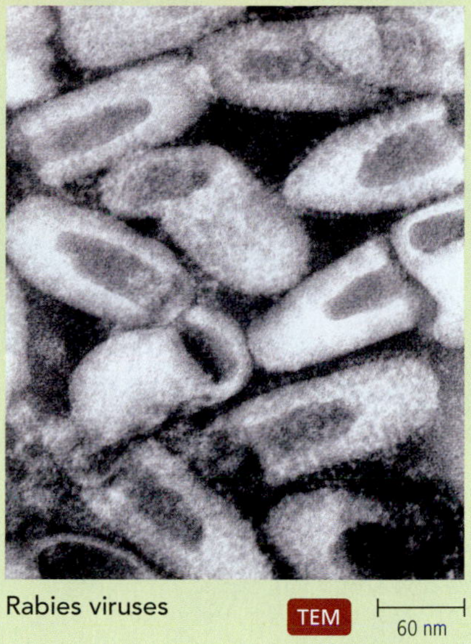

Rabies viruses

Incubation period Variable, generally one to two months, sometimes up to seven years. Dogs have been known to exhibit symptoms within 10 days of infection.

Susceptibility Wild animals are particularly susceptible, but any nonvaccinated mammal may be infected. Animal handlers are at an increased risk of infection.

Treatment Immediately wash the wound with soap and water and begin postexposure prophylaxis (PEP) via administration of one dose of human rabies immune globulin (HRIG) and four doses of rabies vaccine (HDCV) over the course of one month. In cases where PEP was not administered and symptoms develop, treatment is supportive therapy.

Prevention Immunization of dogs and cats. Animal handlers may also be vaccinated.

Arboviral Encephalitis

LEARNING **OUTCOME**

20.17 Compare and contrast Eastern equine, Western equine, Venezuelan equine, West Nile, St. Louis, and California (LaCrosse) encephalitis.

*Ar*thropod-*bo*rne *viruses*—**arboviruses**—are viruses transmitted between hosts by bloodsucking arthropods such as mosquitoes. The term *arbovirus* is a functional name; it is not an official taxonomic term. **Emerging Disease Case Study: Tick-Borne Encephalitis** on p. 624 examines one arboviral disease. Mosquito-borne arboviruses are responsible for various types of **arboviral encephalitis**.[19] These zoonotic diseases typically affect birds, horses, and rodents, and only rarely affect humans.

Signs and Symptoms

Most mosquito-borne arboviruses cause only mild, coldlike symptoms in humans within three to seven days of infection. Occasionally arboviruses in the blood cross the blood-brain barrier to cause arboviral encephalitis, which is characterized by signs and symptoms similar to those of meningitis: high fever, weakness, nausea, vomiting, abrupt headache, and changes in mental state such as confusion, disorientation, and coma. Some patients report body aches and develop a skin rash. Neurological effects may be permanent.

Humans are not the only victims. Arboviruses also attack birds, horses, chimpanzees, cats, dogs, chipmunks, and even alligators. (Scientists are not sure how thick-skinned gators contract the disease.)

[19]From Greek *enkephalos*, meaning "brain," and *itis*, meaning "inflammation."

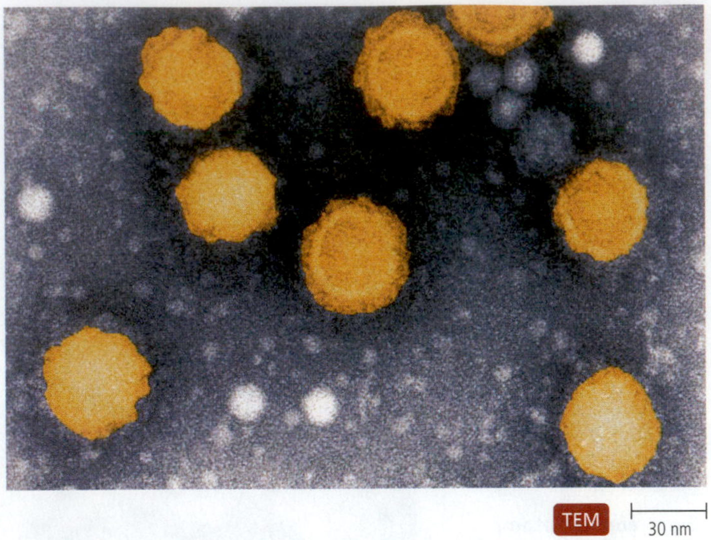

▲ **FIGURE 20.13 Togaviruses.** Each virus has a closely appressed envelope around its capsid.

Of all the arboviruses that affect humans, West Nile virus (WNV) has had the most impact in the United States. Since arriving in 1999, WNV has infected millions of Americans, leading to more than 30,000 cases of severe neurological disease.

Pathogens

Six arboviruses cause most cases of viral encephalitis in Americans. Scientists name these viruses for the geographic regions where the diseases or viruses were first identified. The diseases and their viruses are:

- Eastern equine encephalitis (EEE), Western equine encephalitis (WEE), and Venezuelan equine encephalitis (VEE), all of which were first identified in horses in the eastern U.S., western U.S., and Venezuela, respectively. The viruses causing these conditions are all enveloped, +ssRNA viruses in the family *Togaviridae* **(FIGURE 20.13)**.
- St. Louis encephalitis and West Nile encephalitis, which were first identified in St. Louis, Missouri, and West Nile Province, Uganda (Africa), respectively. Their enveloped, +ssRNA viruses are in the family *Flaviviridae*; they differ from togaviruses in their antigens.
- California (also known as LaCrosse) encephalitis, which is endemic to California but was first described from LaCrosse, Wisconsin. The virus, which is enveloped and has a segmented -ssRNA genome, is in the family *Bunyaviridae*.

Physicians are concerned that new forms of viral encephalitis have appeared in Asia and Australia.

Pathogenesis

As shown in **FIGURE 20.14**, female culex mosquitoes carry the viruses among infected hosts. Mosquitoes remain infected with arboviruses, which they pass to their offspring in eggs; mothers and their female young are a continual source of new infections. Viruses overwinter in eggs or hibernating mosquitoes. Researchers have demonstrated that West Nile virus can be transmitted between people via blood transfusion and transplanted organs.

Arboviruses enter target cells via endocytosis and replicate within them. They produce viremia and can cross the blood-brain barrier by an unknown mechanism to cause encephalitis. In horses and humans, viruses are released from infected cells into the blood (viremia), but the concentration of viruses in their blood is never high enough to infect mosquitoes; thus, horses and humans are "dead-end" hosts for these arboviruses.

Most patients experience severe flulike symptoms but survive, though 50% of patients still suffer a year later with headaches, cognitive and memory impairment, fatigue, tremors, or depression. Mortality ranges from less than 1% with California encephalitis to 35% with EEE, which is the most severe arboviral encephalitis in people.

Epidemiology

The normal host for encephalitis arboviruses is either a bird (viruses of EEE, WEE, St. Louis encephalitis, and West Nile encephalitis) or a rodent (viruses of VEE and California encephalitis). People who engage in work or recreation outdoors in endemic areas are at risk of accidental infection, and those older than 50 are at a higher risk of contracting EEE, St. Louis encephalitis, and West Nile encephalitis. Children are at higher risk for WEE, VEE, and California encephalitis.

As their names suggest, EEE, WEE, and VEE are typically limited to the eastern United States, western United States, and South and Central America, respectively. California encephalitis has been reported in most states west of the Mississippi River; St. Louis and West Nile viral infections occur in the lower continental 48 states.

The incidence of all types of arboviral encephalitis in the United States is seasonal, with peaks that correspond to the times of the year during which adult mosquitoes are active. For example, **FIGURE 20.15** illustrates the seasonal nature of reported cases of encephalitis caused by West Nile virus. Birds spread many types of arboviruses rapidly through an environment. This was dramatically demonstrated by the spread of West Nile virus from New York City in 1999 across the continent in just four years. Species of *Culex* mosquito and 300 species of birds transmit and harbor WNV. House sparrows are an important bird reservoir—they don't die when infected, and their huge population means they outnumber all other bird carriers.

TABLE 20.2 (p. 624) summarizes the distribution, vectors, hosts, and epidemiology of arboviral encephalitis in the United States.

Diagnosis, Treatment, and Prevention

Diagnosis of human arboviral encephalitis is based upon observation of signs and symptoms followed by a positive laboratory test for antibodies against specific arboviruses in the CSF. A commonly used laboratory test measures the concentration of IgM, the antibody class produced early in an infection. The test is positive in most infected people within eight days of onset of symptoms.

◀ FIGURE 20.14 Transmission of six encephalitis arboviruses. Mosquitoes transmit the viruses among birds (EEE, WEE, VEE, St. Louis, and West Nile) or small animals (VEE and California). Horses (EEE, WEE, and VEE) and humans are dead-end hosts. Arboviruses can overwinter in hibernating mosquitoes or their eggs.

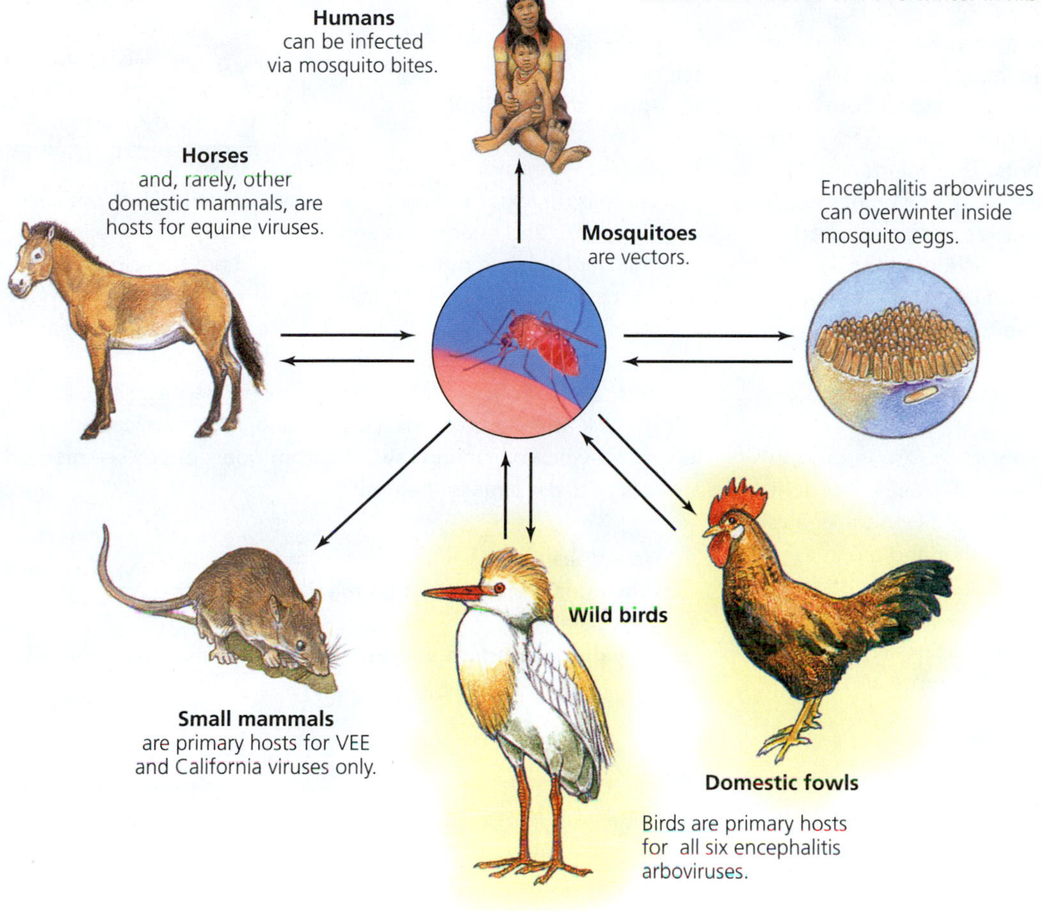

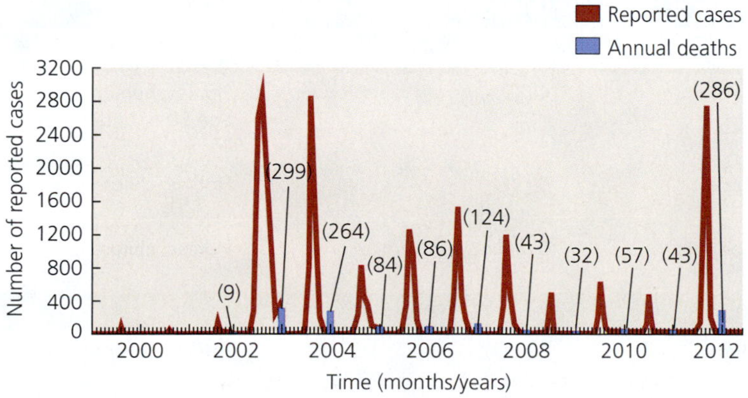

▲ FIGURE 20.15 Human West Nile virus encephalitis in the United States. Note the seasonal nature of the disease. *Why are there more cases of WNV infection in the summer than in the winter?*

Figure 20.15 Mosquitoes transmit West Nile virus among birds and to other animals; in winter it is generally too cold for mosquitoes to be active.

As with viral meningitis, treatment for arboviral encephalitis is supportive: possible hospitalization, administration of intravenous fluids, respiratory support (ventilator), prevention of secondary infections—in other words, good nursing care. Such treatment for a prolonged, serious case of WEE can cost nearly $3 million!

Human disease is prevented by limiting contact with mosquitoes—through the use of netting and insect repellant containing DEET (*N,N*-diethyl-*m*-toluamide)—and by reducing mosquito numbers through the elimination of stagnant water, which is a common breeding site for *Culex*, and the use of insecticides.

Veterinarians can administer effective vaccines against EEE, WEE, VEE, and WNV to horses, but the Food and Drug Administration (FDA) has not approved human vaccines. Scientists have developed an effective human vaccine against West Nile virus; safety trials are ongoing.

Disease at a Glance 20.5 (p. 625) summarizes the features of West Nile encephalitis.

EMERGING DISEASE CASE STUDY

Tick-Borne Encephalitis

Ixodes ricinus tick

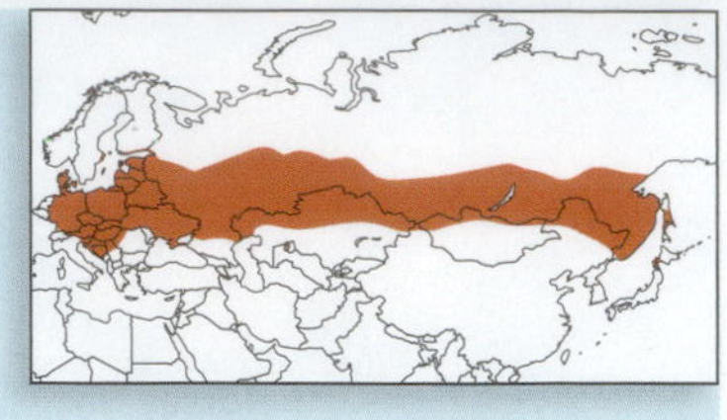

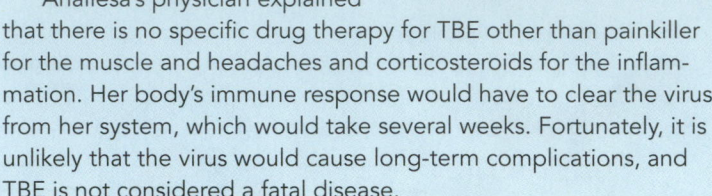

Analiesa's head hurt; rather, her head felt as if it were being pounded from within by gorillas with sledgehammers. This, in addition to last week's fever, nausea, and vomiting, indicated she was more than worn down by her climb two weeks ago to Krimml waterfalls—Austria's highest. There must be something else causing her muscle aches, back pain, and inability to move her shoulders properly.

Indeed, Analiesa is a victim of an emerging disease in Europe and Asia—tick-borne encephalitis (TBE). Several viruses of rodents, transmitted to humans by *Ixodes* ticks, cause TBE. Genetic analysis reveals that the viruses are related to one another and are in the family *Flaviviridae*. Doctors diagnose about 10,000 cases of TBE worldwide each year, but TBE is not a reportable disease, so many more cases likely occur. If global warming allows ticks to survive in higher latitudes and at higher elevations, and as more and more people trek into wilderness areas, scientists expect that tick-borne encephalitis will become more common.

Analiesa's physician explained that there is no specific drug therapy for TBE other than painkiller for the muscle and headaches and corticosteroids for the inflammation. Her body's immune response would have to clear the virus from her system, which would take several weeks. Fortunately, it is unlikely that the virus would cause long-term complications, and TBE is not considered a fatal disease.

And should Analiesa forgo treks in the wilderness? No, but she should take advantage of the TBE vaccine, use chemical tick repellent, and avoid consuming raw milk from goats or cows—infected animals pass the virus in their milk!

1. **How does encephalitis differ from meningitis?**
2. **Why does a brain infection manifest as muscle and joint pain?**
3. **Why don't standard antibiotics provide relief to Analiesa?**

TABLE 20.2 **Characteristics of Arboviral Encephalitis Diseases and Viruses in the United States**

Name of Disease and Virus	Distribution	Vector	Natural Hosts	Number of Human Cases in 2009 in U.S. (Mean Human Mortality)	Special Groups at Risk
Togaviridae					
Eastern equine encephalitis (EEE)	Eastern seaboard, Gulf coast, Great Lakes states	*Aedes* and *Culex* mosquitoes	Birds	4 (35%)	Horses; humans over age 50 or under age 15
Western equine encephalitis (WEE)	States west of Mississippi River	*Culex* and *Culiseta* mosquitoes	Birds	0 (3%)	Horses; children under age 1
Venezuelan equine encephalitis (VEE)	Texas	*Aedes* and *Culex* mosquitoes	Rodents	0 (unknown)	Horses; children
Flaviviridae					
St. Louis encephalitis	Lower 48 states except MA, ME, NH, RI, SC, and VT	*Culex* mosquito	Birds	10 (5%)	Humans over age 50
West Nile encephalitis	Lower 48 states except ME and NH	*Culex* mosquito	Birds	360 (<1%)	Humans over age 50
Bunyaviridae					
California (LaCrosse) encephalitis	Eastern and central states	*Aedes* mosquito	Small mammals	39 (<1%)	Children under age 16

DISEASE AT A GLANCE 20.5

West Nile Encephalitis

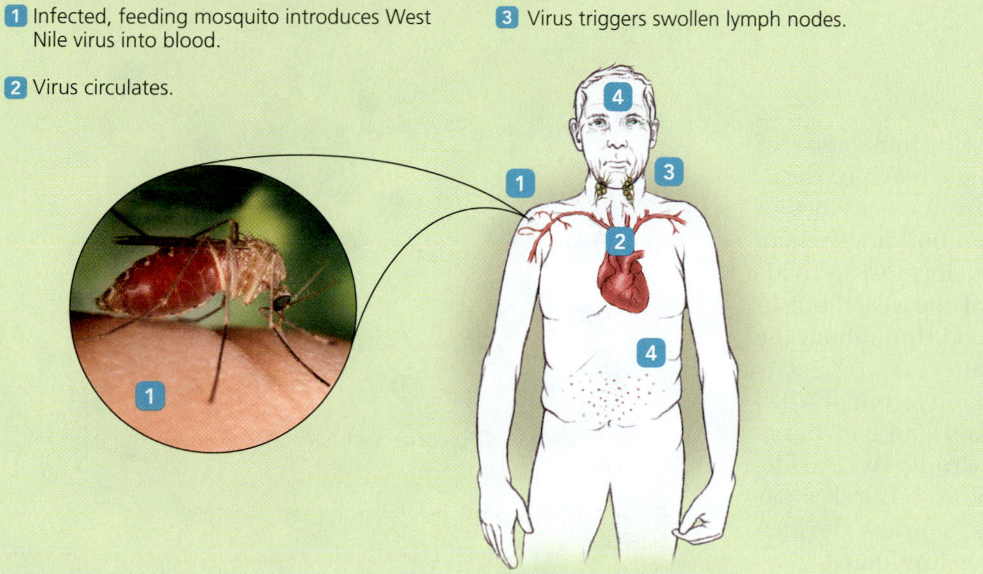
1 Infected, feeding mosquito introduces West Nile virus into blood.
2 Virus circulates.
3 Virus triggers swollen lymph nodes.
4 Body aches, meningitis, and encephalitis are common. Headache, high fever, stiff neck, skin rash, disorientation, coma, and death may occur.
1
2
3
4

West Nile virus encephalitis can be life changing.

Cause West Nile virus (single-stranded RNA virus).

Virulence factors Birds are ubiquitous reservoirs, protein in viral envelope attaches to human cells, virus is intracellular pathogen.

Portal of entry Via the bite of an infected mosquito.

Signs and symptoms A mild case including fever, headache, body aches, and swollen lymph nodes may progress to encephalitis. The manifestations of this more severe form of West Nile virus infection include abrupt headache, high fever, neck stiffness, stupor, disorientation, coma, tremors, convulsions, muscle weakness, paralysis, and possibly a skin rash.

Incubation period Three days to two weeks.

Susceptibility The elderly are more at risk for developing West Nile encephalitis.

Treatment Supportive therapy.

Prevention Limit contact with mosquito populations.

We have examined bacterial and viral diseases of the nervous system. Now we turn our attention to fungal diseases.

TELL ME WHY

The word *enterovirus* literally means "intestine poison," yet enteroviruses do not cause intestinal diseases. Why are these viruses called enteroviruses?

Mycosis of the Nervous System

Fungi rarely infect the CNS, though **mycoses** (fungal diseases) may spread from the lungs via the blood to the CNS, and mushroom poisoning involves toxins that produce neurological dysfunction or hallucinations. (Chapter 22 considers in more detail coccidioidomycosis, histoplasmosis, and blastomycosis, which are diseases of the respiratory system that can have neurological effects.) Here we examine the leading form of fungal meningitis.

Cryptococcal Meningitis

LEARNING | **OUTCOME**

20.18 Describe the features of cryptococcal meningitis.

Cryptococcal meningitis (cryptococcosis) affects people worldwide. Unlike other forms of fungal meningitis, it affects healthy people as well as those who are sick or immunocompromised.

Signs and Symptoms

Cryptococcal meningitis manifests with signs and symptoms common to bacterial meningitis: headache, dizziness, drowsiness, irritability, confusion, nausea, vomiting, and neck stiffness. In late stages of the disease, loss of vision and coma occur. Acute onset of rapidly fatal cryptococcal meningitis occurs in individuals with widespread infection.

Pathogen and Virulence Factors

Cryptococcus neoformans is a basidiomycete yeast; that is, it is a spherical, single-celled fungus that reproduces sexually with basidiospores. It lives in soil, the feces of birds, and the sap of eucalyptus trees.

Scientists recognize two variants of the yeast, both of which are found worldwide. *C. neoformans* var. *gattii* primarily infects immunocompetent individuals, whereas *C. neoformans* var. *neoformans* predominantly infects immunocompromised hosts. Approximately 50% of all cryptococcal infections reported each year are due to the latter variant.

The polysaccharide capsule that surrounds each cryptococcal cell is resistant to phagocytosis by defensive cells of the body. The pathogenesis of *Cryptococcus* is also enhanced by the ability of the yeast to produce melanin, which appears to inhibit phagocytic killing mechanisms. The yeast has a predilection for the central nervous system.

Pathogenesis and Epidemiology

Human infections begin in the lungs following inhalation of spores and/or dried yeast cells made airborne when bird droppings containing the fungus are disturbed by the activities of the birds or humans. People who work around buildings where birds roost are at increased risk of infection. In most individuals, phagocytes in the lungs limit spread of the yeast, but in some patients, the fungus spreads via the blood throughout the body. It infects the meninges and brain tissue.

Cryptococcal infections appear commonly in terminal AIDS patients when little immune function remains and in transplant recipients taking immunosuppressive drugs. Worldwide, the annual incidence in AIDS patients is about 4–12 cases per 100,000 population. For non-AIDS patients, worldwide incidence of *Cryptococcus* infection is 2–9 cases per 1 million.

Diagnosis, Treatment, and Prevention

The preferred method of confirming cryptococcal meningitis is detection of fungal antigen in CSF. In AIDS patients, antigen can be detected in serum as well. Fungal stains revealing the presence of encapsulated yeast in CSF are highly suggestive of cryptococcal meningitis, even if no obvious symptoms are present.

Treatment is with intravenous amphotericin B and oral 5-fluorocytosine administered together for at least four weeks. The two drugs enhance one another, allowing lower doses of amphotericin B, which is toxic to humans, to be used; however, toxicity is not eliminated. Though AIDS patients may appear well following primary treatment, the fungus typically remains and must be actively suppressed by lifelong oral fluconazole treatment.

Because the fungus is a threat to sick individuals, facilities such as hospitals and nursing homes often place devices that deter the roosting of birds near outside air-intake vents in an effort to prevent *Cryptococcus*-contaminated air from entering the building. No vaccine against *Cryptococcus* is available.

Disease at a Glance 20.6 summarizes the features of cryptococcal infection.

Having examined bacterial, viral, and fungal neurological diseases, we turn our attention to representative diseases of the nervous system caused by protozoa.

TELL ME WHY

Why has West Nile virus been able to travel across North America since 1999?

DISEASE AT A GLANCE 20.6

Cryptococcal Meningitis

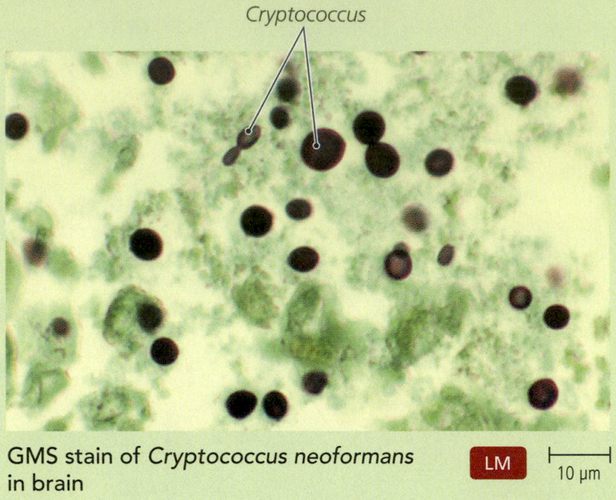

GMS stain of *Cryptococcus neoformans* in brain

Cause *Cryptococcus neoformans* (spherical, single-celled basidiomycete yeast).

Virulence factors Polysaccharide capsule resists phagocytosis; produces melanin, which inhibits intracellular digestion of the yeast.

Portal of entry Inhalation of spores and/or dried yeast cells.

Signs and symptoms Prolonged cough (lasting weeks or months), chest pain, fever, shortness of breath, night sweats, weight loss; proceeds to headache and meningitis.

Incubation period Unknown, but probably two to nine months.

Susceptibility Immunocompromised persons, especially HIV-positive individuals and AIDS patients; people exposed to bird droppings.

Treatment Amphotericin B and 5-fluorocytosine coadministered for 6 to 10 weeks.

Prevention It is difficult to avoid infection by this prevalent fungus; therefore, you should protect your immune system with healthy lifestyle choices.

Protozoan Diseases of the Nervous System

Protozoan infections of the nervous system are relatively rare. Here we examine diseases of two protozoa of the kingdom Euglenozoa: African sleeping sickness, which is caused by a trypanosome, and a form of *meningoencephalitis* (simultaneous meningitis and encephalitis) caused by amoebae.

African Sleeping Sickness

LEARNING OUTCOME

20.19 Describe African sleeping sickness, including its cause, diagnosis, treatment, and prevention.

African sleeping sickness, also known as *African trypanosomiasis*, prevents ranching and extensive human habitation in an area of equatorial Africa larger than that of the lower 48 U.S. states.

Signs and Symptoms

African sleeping sickness progresses through three clinical stages:

1. First, the wound created by the vector's bite becomes a lesion containing necrotic (dead) tissue and rapidly dividing parasites.
2. Next, the presence of parasites in the blood generates fever, swelling of lymph nodes, and headaches.
3. Finally, invasion of the central nervous system results in meningoencephalitis, characterized by headache, abnormal neurological function, and extreme drowsiness. As the name indicates, victims become so tired that they no longer eat, finally succumbing to coma and death. Untreated African trypanosomiasis is invariably fatal.

Cyclical waves of *parasitemia* (parasites in the blood) that occur roughly every 7–10 days characterize the disease. Symptoms develop within months to years after infection, depending on the strain of protozoan involved.

Pathogen and Virulence Factors

Trypanosoma brucei[20] (tri-pan´ō-sō-mă brūs´ē), which is a kinetoplastid[21] protozoan that causes African sleeping sickness, randomly changes its surface glycoprotein antigens when it replicates. The result is that by the time a host's immune system has produced antibodies against a given set of glycoproteins, the parasite has produced a new set, leaving the host's immune system constantly one step behind the parasite—once infected, a patient rarely clears the infection and becomes immune.

Pathogenesis and Epidemiology

Newly hatched bloodsucking tsetse (tset´sē) flies (*Glossina*, glo-sī´nă) spread *T. brucei* among wild and domesicated mammals, and people. **(FIGURE 20.16)** illustrates the life cycle of the parasite. During the course of an infection, some trypanosomes enter the brain, spinal cord, and cerebrospinal fluid; others continue to circulate in the blood, where feeding tsetse flies pick them up.

African sleeping sickness occurs in equatorial and subequatorial savanna and riverine areas of Africa, wherever tsetse flies live.

Diagnosis, Treatment, and Prevention

Microscopic observation of trypanosomes in blood, lymph, spinal fluid, or a tissue biopsy is diagnostic for trypanosomiasis. In mammals, the protozoa are long and thin with a single long flagellum running along the cell and extending past the posterior end (see Figure 12.8b).

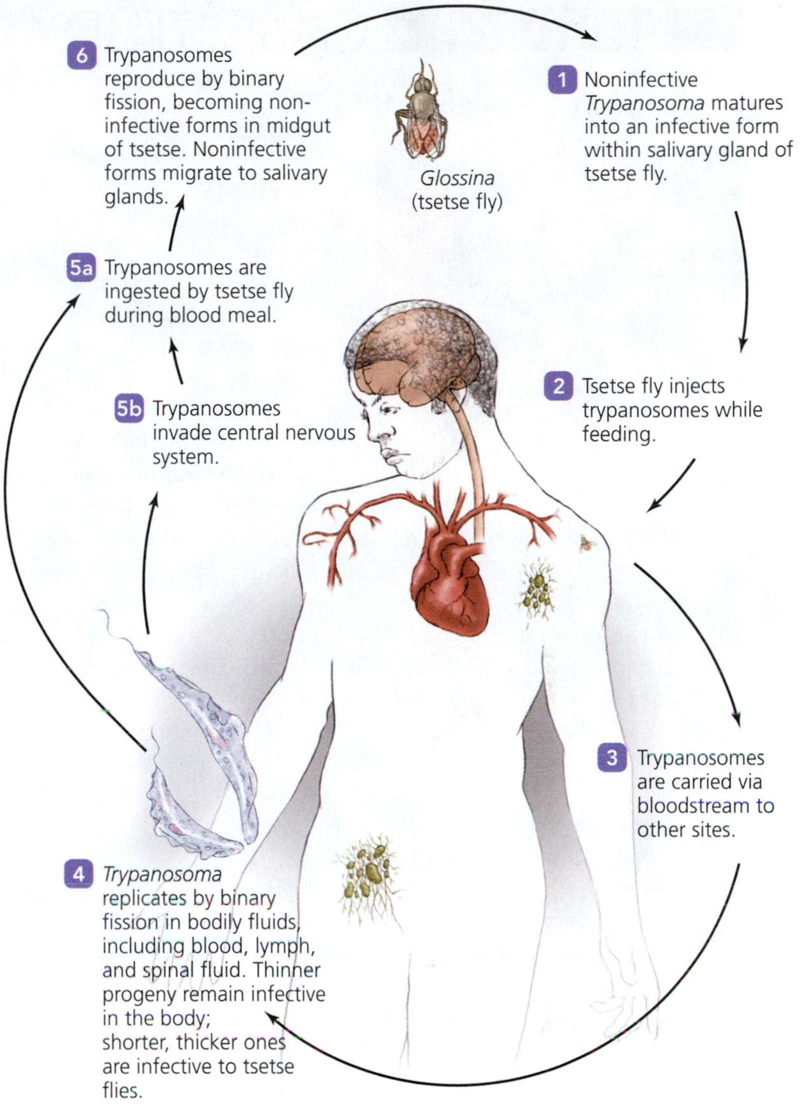
6 Trypanosomes reproduce by binary fission, becoming noninfective forms in midgut of tsetse. Noninfective forms migrate to salivary glands.
Glossina (tsetse fly)
1 Noninfective Trypanosoma matures into an infective form within salivary gland of tsetse fly.
5a Trypanosomes are ingested by tsetse fly during blood meal.
5b Trypanosomes invade central nervous system.
2 Tsetse fly injects trypanosomes while feeding.
3 Trypanosomes are carried via bloodstream to other sites.
4 Trypanosoma replicates by binary fission in bodily fluids, including blood, lymph, and spinal fluid. Thinner progeny remain infective in the body; shorter, thicker ones are infective to tsetse flies.

▲ **FIGURE 20.16 The life cycle of *Trypanosoma brucei*.** This parasite causes African sleeping sickness.

Health care providers treat the early stages of African sleeping sickness with pentamidine and suramin. Melarsoprol is used when the disease progresses to the CNS because this drug crosses the blood-brain barrier. A newer, and expensive, drug called eflornithine is remarkably effective against West African *Trypanosoma*. To be successful, treatment must begin as soon as possible after infection. However, immediate treatment is rare because of poor health care in endemic areas.

Broad application of insecticides has reduced the occurrence of African sleeping sickness in local areas; but large-scale spraying of insecticides is impractical, expensive, and may have disastrous long-term environmental consequences. Personal insecticide use is preferable, as is the use of netting over beds, window screening, and long, loose-fitting clothing. No vaccine currently exists for African sleeping sickness.

[20]From Greek *trypanon*, meaning "auger" (indicating the shape of the protozoan); *soma*, meaning "body"; and British physician Sir David Bruce, who studied the disease.
[21]A kinetoplast is a unique region of mitochondrial DNA found in these microbes.

CLINICAL CASE STUDY

A Threat from the Wild

A man arrived at an emergency room with neurological disorders: uncontrolled facial twitching, anxiety, and feelings of fear. He also had a sore throat, difficulty in swallowing, and complained of itching over his entire body. He remained alive for several days, but became increasingly agitated and refused to drink because of the pain involved in swallowing. He vomited repeatedly, and his temperature rose to 106°F. He died one week after being admitted to the hospital.

An autopsy revealed dark-staining bodies in the cells of his brain and a raised antibody titer in his blood. There were no obvious bites or scratches on his skin, though interviews with friends and family indicated that a bat had landed on the man's face about a month before he was admitted to the hospital.

1. What did the man suffer from?
2. What were the antibodies against?
3. What were the dark-staining bodies in his brain's cells?
4. What preventive measures might health care providers have used to save the man's life?

Reference: Adapted from *MMWR* 52:47–48. 2003.

Primary Amebic Meningoencephalopathy

LEARNING | OUTCOMES

20.20 Describe primary amebic meningoencephalopathy and its pathogens.

20.21 Describe the treatment and prevention of primary amebic meningoencephalopathy.

Now we consider another disease of the CNS caused by protozoa, specifically, two amoebae in the kingdom Euglenozoa.

Signs and Symptoms

Signs and symptoms of **primary amebic[22] meningoencephalopathy** are the same as those for meningitis and encephalitis caused by bacteria, viruses, and fungi—headache, fever, stiff neck, altered mental state, and vomiting. Symptoms progressively worsen over a period of three to seven days until the patient dies.

Pathogens, Pathogenesis, and Epidemiology

Acanthamoeba (ă-kan-thă-mē´bă) and *Naegleria* (nā-glē´rē-ă) each cause this disease. These amoebae are common free-living inhabitants of warm lakes, ponds, puddles, ditches, mud, and moist soil; they are also found in artificial water systems such as swimming pools, air conditioning units, humidifiers, contact lens solution, and dialysis units.

Amoebae enter a host through cuts or scrapes on the skin, through the covering of the eye via abrasions from contact lenses or trauma, or through inhalation of contaminated water while swimming. Replicating amoebae migrate to the brain via cranial nerves.

Primary amebic meningoencephalopathy is rare—fewer than 200 cases were reported from 1962 to 2013 in the United States—but it is nearly always fatal; the CDC reports only three survivors in the United States since records have been kept.

Diagnosis, Treatment, and Prevention

Physicians diagnose *Acanthamoeba* and *Naegleria* infections by detecting amoebae in scrapings from the covering of the eye, cerebrospinal fluid, or biopsy material from the brain. Fluconazole or amphotericin B may have very limited success, but only if administered very early in the course of an infection.

Both amoebae produce resistant cysts that are environmentally hardy; therefore, controlling and preventing infection can be difficult. People should avoid natural waterways in which the organisms are endemic and use only sterile solutions to clean or store contact lenses. Swimming pools should be properly chlorinated and tested periodically to ensure their safety. Air conditioning systems, dialysis units, and humidifiers should be cleaned thoroughly and regularly to prevent amoebae from becoming resident.

TELL ME WHY

Why has the number of cases of primary amebic meningoencephalopathy increased dramatically as societies have become more developed?

Prion Disease

In 1996 a new pathogen leapt to worldwide attention with the revelation that the infectious protein that causes the sheep disease *scrapie* had "jumped species" to cause "mad cow disease" and had then jumped again into humans who ate infected beef. These infectious proteins, called **prions** (prē´onz), may play a role in many neurological diseases such as Alzheimer's and Parkinson's. Here we consider the prion disease of the 1996 epidemic.

[22]Note that in American English, the noun *amoeba* and the adjective *amebic* are spelled dissimilarly.

CLINICAL CASE STUDY

A Protozoan Mystery

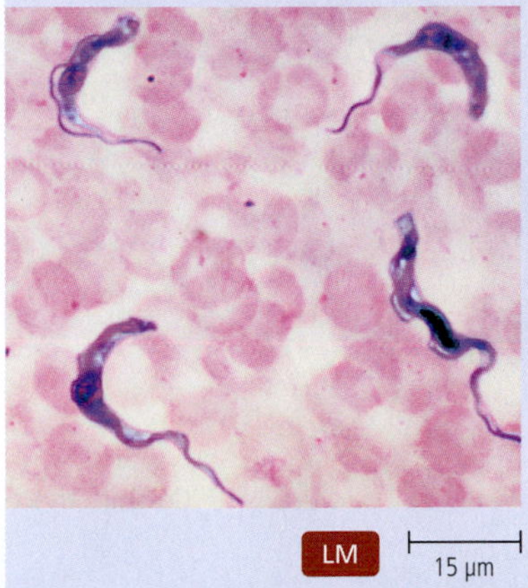

A 20-year-old student was admitted to his college's student health center with fever and headaches shortly after beginning the fall semester. He had spent his summer working with an international aid organization in Nigeria and had returned to the United States only a week earlier. Gross examination revealed numerous insect bites and some swollen lymph nodes. The man had spent most of his summer outdoors in rural areas and had spent some time on African game reserves working with the families of local guides. He could not specifically remember receiving any of the bite wounds on his body, and he did not always use insect repellent in the field. The patient was admitted to the local hospital, where intermittent fever, headache, and swelling continued. Initial blood smears proved negative for malaria.

1. What are some possible protozoan diseases the patient could have contracted in Africa?
2. Can this disease be identified from the symptoms alone?
3. Based on the pictured blood smear, what would you conclude about the cause of the disease?
4. What would the treatment be if the patient had tested positive for malaria?
5. What treatment would you now recommend?
6. What prevention would you have suggested to this individual?
7. Is there a local threat for anyone in America contracting this disease from this student?

Variant Creutzfeldt-Jakob Disease (vCJD)

LEARNING OUTCOME

20.22 Describe the causative agent, signs, symptoms, and prevention of variant Creutzfeldt-Jakob disease.

Scrapie and mad cow disease belong to a class of diseases called *spongiform encephalopathies* because they leave the brains of their victims so full of holes they resemble sponges. German neurologists Hans Creutzfeldt (1885–1964) and Alfons Jakob (1884–1931) described the genetic disease that bears their name as a type of dementia that spontaneously strikes about one person in a million at about age 60. A contagious form of the disease, derived from cattle, is called **variant Creutzfeldt-Jakob** (kroits´felt-ya´kōp) **disease (vCJD)** to distinguish it from the naturally occurring Creutzfeldt-Jakob disease (CJD).

Signs and Symptoms

In vCJD, brain tissue is destroyed and numerous cavities form (see Disease at a Glance 20.7 on p. 630). As the brain deteriorates, victims experience insomnia, weight loss, and memory failure. They also act irrationally, lose control of their muscles, and are unable to speak, walk, or maintain posture. Muscle spasms progressively worsen as the disease progresses. Death usually occurs within 12 months.

Pathogen, Pathogenesis, and Epidemiology

A prion protein causes vCJD. Prions exist in one of two three-dimensional forms (see Figure 13.22): The normal form is anchored in lipid rafts of CNS neurons, membranes and is necessary for normal brain function, though scientists don't understand its exact role. An abnormal prion acts as an enzyme to refold the normal form into a copy of the abnormal form. Newly misfolded prions then proceed to make copies of themselves from surrounding normal prion molecules. In other words, abnormal prions turn normal prion proteins into abnormal prions, which continue the process. The disease progresses unrelentedly, destroying the brain as it does.

Medical procedures such as transplants, blood transfusion from an infected person, use of contaminated surgical instruments, and injection of growth hormones derived from infected pituitary glands can spread the disease. Prions may remain dormant for more than 60 years.

Unlike the genetic form of CJD, variant CJD strikes even young people who consume contaminated nerve tissue, typically in meat products such as sausages. Since its discovery in 1996, vCJD has killed about 200 people, and thousands of others wonder whether they have been infected and will succumb in the future.

Diagnosis, Treatment, and Prevention

The characteristic signs and symptoms of vCJD are diagnostic in the young, but can be confused with other forms of dementia in the elderly. Laboratory tests on samples from the CNS confirm the presence of abnormal prion in the brain.

No treatment is available for vCJD, though interleukins may slow disease progression. The destruction of prions outside the body is problematic—prions survive cooking, freezing, pickling, and even normal autoclaving. Currently, autoclaving in a concentrated solution of sodium hydroxide is recommended for prion destruction, but you would not want to eat a hot dog prepared that way! The European Union has approved an enzyme treatment to remove prions from the environment.

Prevention of vCJD is possible only by remaining free of the prion by not eating contaminated meat, particularly meat cut from bones near the spinal cord and processed meats made from brains or spinal cords of infected animals. As British farmers and ranchers learned, destruction of potentially infected sheep and cattle herds is mandatory to prevent the spread of

DISEASE AT A GLANCE 20.7

Variant Creutzfeldt-Jakob Disease (vCJD)

Cause PrP prion (infectious protein).

Virulence factors Ability of abnormal, disease prion protein to convert normal prion protien into abnormal form, survives cooking and normal autoclaving.

Portal of entry Ingestion of infected meat, or transplant or transfusion of infected tissues.

Signs and symptoms Irrational behavior; loss of muscle control; increasing inability to walk, talk, or maintain posture; muscle spasms; and a continuous writhing of the extremities. Death is inevitable.

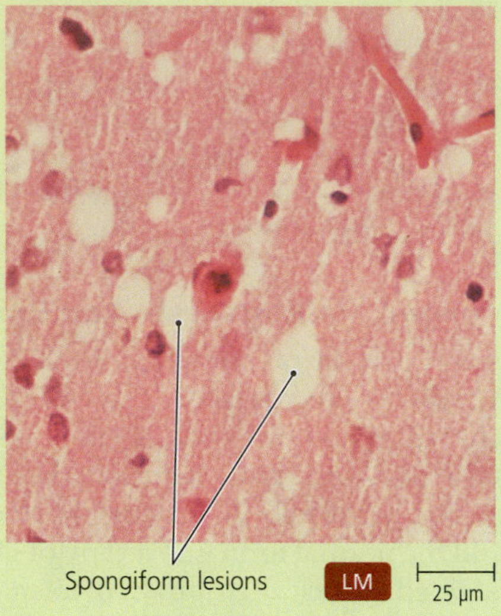

Spongiform lesions (holes) of variant Creutzfeldt-Jakob disease

Incubation period Unknown, but most likely decades.

Susceptibility At greatest risk are individuals who consumed beef in the United Kingdom between 1980 and 1996.

Treatment None.

Prevention Do not eat meat contaminated with infected nervous tissue. Destruction of potentially infected herds is necessary to prevent the spread of the infectious prion.

vCJD. Further, governments have initiated strict laws and inspection procedures to prevent the use of contaminated animal protein in food supplements for herbivores.

Disease at a Glance 20.7 summarizes the features of variant Creutzfeldt-Jakob disease.

TELL ME WHY

Why is infectious CJD called *variant*?

Microbial Diseases of the Eyes

LEARNING | **OUTCOMES**

20.23 Discuss the features of trachoma, including pathogenesis and prevention.

20.24 Define and contrast conjunctivitis and keratitis.

The senses are important parts of the nervous system. Vision is our primary sense, so it is not surprising that nearly half of the function of the cerebrum is dedicated to vision and about 70% of sensory neurons originate in the eyes; in fact, the eyes can be considered extensions of the brain and dura mater.

In the following sections we will briefly consider representative microbial infections of the eyes, but first we consider the structure of an eye.

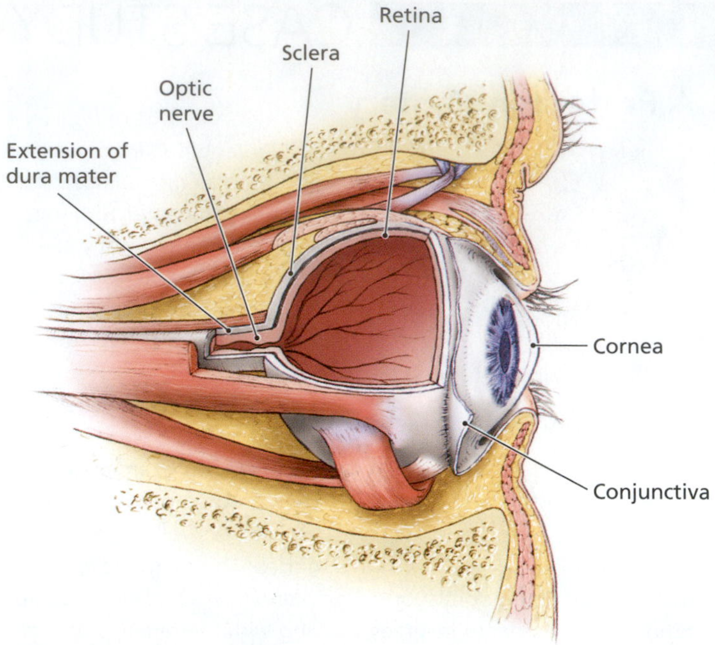

▲ **FIGURE 20.17** **The eye (sagittal section).**

Structure of the Eye

FIGURE 20.17 illustrates the structure of an eye. An eye is a hollow, roughly spherical ball about 2.5 cm in diameter. The outer wall of the eye, called the *fibrous tunic,* is composed of a white *sclera,* which is a direct extension of the dura mater of the brain, and a colorless, clear *cornea,* which covers the front of the eye. The fibrous tunic provides a tough barrier against the penetration of microbes. An extension of the epidermis of the skin, called the *conjunctiva,* lines the backside of the eyelids and all but the center of the cornea.

The interior of the eye is composed of two chambers filled with fluids, called *humors.* The back interior wall of the eye is the *retina,* which contains billions of sensory neurons that respond to light energy and send nerve impulses to the brain down the optic cranial nerve.

In the following sections, we briefly consider several bacterial, fungal, viral, and protozoan diseases of the eye.

Trachoma

Trachoma (trā-kō′mā) is the leading cause of nontraumatic blindness in humans.

Signs and Symptoms

Trachoma is Greek for roughness, which is a good description of this disease in which the conjunctiva and cornea are scarred by bacterial infection, leading to blindness.

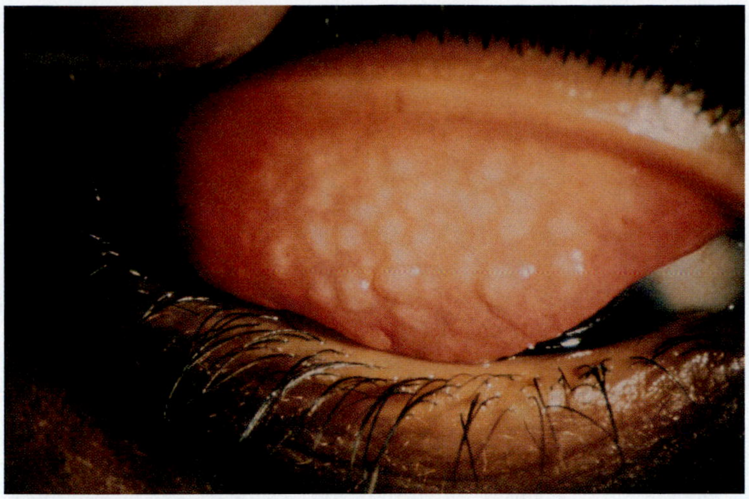

▲ **FIGURE 20.18** An eyelid afflicted with trachoma.

Pathogen, Pathogenesis, and Epidemiology

So-called trachoma strains of the bacterium *Chlamydia trachomatis* (kla-mid´ē-ă tra-kō´ma-tis) causes trachoma. The pathogen multiplies in cells of the conjunctiva and kills them, triggering a large amount of purulent[23] (pus-filled) discharge that scars the conjunctiva. Such scarring in turn causes the patient's eyelids to turn inward, such that the eyelashes scratch, irritate, and scar the cornea.

Corneal scarring triggers an invasion of blood vessels into this normally clear surface of the eye **(FIGURE 20.18)**. A scarred cornea that is filled with blood vessels is no longer transparent, and the eventual result is blindness.

Trachoma is typically a disease of children who have been infected during birth. However, some strains of *C. trachomatis* may produce blindness by a similar process in adults when bacteria from the genitalia are introduced into the eyes via fomites, fingers, or flies.

Diagnosis, Treatment, and Prevention

Diagnosis of chlamydial infection involves demonstration of the bacteria inside cells from the site of infection. Specimens can be obtained from the urethra or vagina by inserting, rotating, and then removing a sterile swab. Stained specimens may reveal bacteria or inclusion bodies within cells, but the most specific method of diagnosis involves amplifying the number of chlamydia by inoculating the specimen into a culture of susceptible cells. Laboratory technicians then demonstrate the presence of *Chlamydia* in the cell culture by means of specific fluorescent antibodies or nucleic acid probes.

Physicians prescribe azithromycin or doxycycline to eliminate genital infections of *C. trachomatis* in adults. Azithromycin is recommended for treatment of pregnant women. Trachoma strains of *C. trachomatis* infecting the eyes of newborns are treated with tetracycline or erythromycin cream for 10–14 days, whereas eye infections in adults are treated with oral tetracycline, doxycycline, or erythromycin for three to six weeks. Surgical correction of eyelid deformities may prevent the scratching, scarring, and blindness that may result from eye infections.

[23]From Latin *pur,* meaning "pus."

Other Microbial Diseases of the Eyes

Bacterial infections of the skin and reproductive tract may also affect the eyes. For example, *Staphylococcus aureus* (staf´i-lō-kok´ŭ s o´rē-ŭs) infections of sebaceous glands in the skin near the eye are called *sties,* and *Neisseria gonorrhoeae* (nī-se´rē-ă go-nor-re´ī), a sexually transmitted bacterium, causes a form of *ophthalmia neonatorum* (of-thal´mē-ă nē´ō-nă-tor´um)—inflammation of the conjunctiva and cornea of a newborn. Babies can be infected during passage through a diseased birth canal.

Conjunctivitis, commonly called *pinkeye,* is inflammation of the conjunctiva, and **keratitis** is inflammation of the cornea. A number of bacteria, fungi, viruses, or protozoa cause the two conditions, which may occur independently or simultaneously. For example, *Haemophilus influenzae* is the most common bacterial cause of conjunctivitis. *Acanthamoeba* can cause both conjunctivitis and keratitis. (See the discussion of *Acanthamoeba keratitis* on p. 267.)

Physicians treat bacterial, fungal, and protozoan conjunctivitis and keratitis with topical antimicrobial drugs. No antiviral drugs are available for these conditions.

TELL ME WHY

Doxycycline—one of the tetracyclines—is the treatment for most adults infected with *Chlamydia trachomatis*; however, it is not recommended for pregnant women or babies. Why not?

CLINICAL CASE STUDY

A Very Sick Sophomore

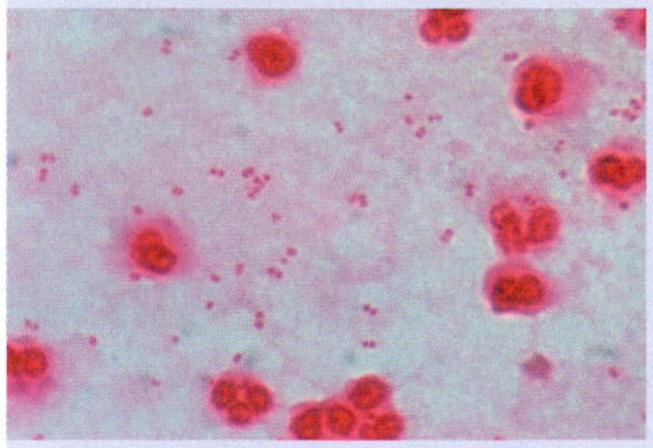

In December, a 19-year-old college student experiences a severe headache, nausea, vomiting, and fever. The student suspects a case of flu and goes to the health center for a diagnosis. By the time he arrives he is feeling worse, with neck stiffness and disorientation. The doctor immediately prepares to perform a spinal tap.

1. What disease does the doctor suspect?
2. What are possible causes of this disease?
3. The spinal tap reveals the presence of Gram-negative cocci in the CSF. What else was present in the CSF of the patient?
4. Given the information in question 3, what is the diagnosis?
5. What might the student's residential assistant in the dorm be concerned about?
6. How might this disease have been prevented?

MICRO IN THE CLINIC FOLLOW-UP

Under Pressure and Under the Weather

The lab report reveals large numbers of white blood cells (neutrophils) in Lin's cerebrospinal fluid (CSF), which explains the specimen's milky appearance. Normal CSF is clear and contains no cells or organisms. Other lab reports reveal that her CSF has abnormally high levels of protein and low levels of glucose. Together, the lab results suggest a diagnosis of meningococcal meningitis—a dangerous and potentially fatal disease.

Several types of bacteria can cause meningitis, but only a few affect healthy young adults like Lin. The Gram stain report describes moderate numbers of Gram-negative diplococci, many of which are intracellular, in the cytoplasm of Lin's neutrophils. The culture takes 24–48 hours but provides the medical team with conclusive evidence that Lin has meningococcal meningitis, caused by *Neisseria meningitidis*. This bacterium can spread between college students living in dormitories, but the risk can be greatly reduced with a preventive vaccine, which Lin did not receive.

Lin begins immediate antibiotic treatment. Because of the rapid diagnosis and treatment, she is able to recover fully, reschedule her chemistry exam, and pass with flying colors.

1. **How did Lin likely contract *Neisseria meningitidis*?**
2. **Review the anatomy of the central nervous system. Why is treating Lin's meningitis a challenge beyond that of normal antibiotic treatment for infections?**

MM Check your answers to Micro in the Clinic Follow-Up questions in the MasteringMicrobiology Study Area.

Explore the Invisible

Visit the **MasteringMicrobiology Study Area** to challenge your understanding with practice tests, animation quizzes, and clinical case studies!

MasteringMicrobiology®

CHAPTER SUMMARY

Structure of the Nervous System (pp. 602–604)

1. The central nervous system (CNS) is composed of the brain and spinal cord surrounded by three layers of meninges: the dura mater (outermost), arachnoid mater, and pia mater.
2. The subarachnoid space is filled with cerebrospinal fluid (CSF), which acts as a shock absorber, provides nutrients and oxygen, and removes wastes.
3. CSF is removed from the lumbar region of the spinal cord in a procedure called a **lumbar puncture**, or spinal tap.
4. Tightly joined cells of blood vessels of the pia mater form the blood-brain barrier, so named because substances in the blood cannot enter the subarachnoid space.
5. The peripheral nervous system (PNS) is composed of cranial nerves and spinal nerves. A plexus is a bundle of branching nerves.
6. Three nerve types are sensory nerves (bundles of nerve cells carrying signals toward the CNS), motor nerves (bundles of nerve cells carrying signals away from the CNS), and mixed nerves.
7. Neuroglia are supportive cells of the nervous system. Neurons are cells which carry nerve impulses. The cell bodies of neurons grouped together form a ganglion.
8. Short cytoplasmic processes that extend from neuron cell bodies and carry impulses toward the cell body are dendrites. Axons are longer processes that carry impulses away from the cell body. Elements of the cytoskeleton carry substances up and down the axon in a process called axonal transport.
9. Synapses are junctions between axons and either glands, muscles, or neurons. In the synaptic cleft (small space between axon and next cell) the axon releases neurotransmitters, which may be stimulatory or inhibitory.
10. **Meningitis** is an infection of the meninges.

Bacterial Diseases of the Nervous System (pp. 604–616)

1. Bacterial meningitis can be caused by many species of bacteria and is characterized by sudden high fever and intense meningeal inflammation, and sometimes **petechiae**, small hemorrhages of blood vessels in the skin.
2. Most cases of bacterial meningitis are caused by *Streptococcus pneumoniae, Neisseria meningitidis, Haemophilus influenzae, Listeria monocytogenes,* and *Streptococcus agalactiae. Listeria* causes **listeriosis**.
3. *Mycobacterium leprae* causes **Hansen's disease (leprosy)**, which manifests as tuberculoid leprosy in patients with a strong immune response or as lepromatous leprosy in those with weak immune response. The bacterial cell wall contains a large amount of **mycolic acid**, which accounts for the slow, resistant growth of the bacteria.
4. Diagnosis of leprosy is based on loss of sensation in skin, disfigurement, and the presence of **acid-fast bacilli (AFB)**.
5. *Clostridium botulinum* produces neurotoxins that cause **botulism** poisoning. It manifests as **foodborne botulism, infant botulism**, and **wound botulism**.
6. Botulism neurotoxins prevent muscular contraction by preventing the secretion of acetylcholine (a necessary chemical) in the neuromuscular junction (the synapse between a motor neuron and a muscle cell).
7. *Clostridium tetani* causes **tetanus** by releasing the neurotoxin **tetanospasmin**, which blocks inhibitory neurotransmitters, resulting in severe muscular contractions.
8. Tetanus toxoid provides effective immunity against tetanus.

Viral Diseases of the Nervous System (pp. 616–625)

1. The most common form of meningitis is viral meningitis. Most cases are caused by viruses of the genus *Enterovirus* of the family *Picornaviridae*, which spread via the blood **(viremia)**.
2. Poliovirus causes **poliomyelitis (polio)**, which is expressed as an asymptomatic infection, minor polio, nonparalytic polio, or paralytic polio.
3. **Postpolio syndrome** is a muscle deterioration that affects polio patients 30–40 years after their original illness.
4. Rabies virus causes **rabies**, a degenerating brain and spinal cord disease that is **zoonotic** (spread by animals to humans).
5. Bloodsucking arthropods transmit **arboviruses**. **Arboviral encephalitis** affects birds, horses, and rodents, and six arboviruses cause encephalitis (rarely) in humans.

Mycosis of the Nervous System (pp. 625–626)

1. Cryptococcal meningitis is a **mycosis** (fungal disease) of the nervous system that can be fatal to AIDS patients and transplant recipients taking immunosuppressive drugs.
2. *Cryptococcus neoformans* is a yeast that normally lives in soil and in bird feces and causes cryptococcal meningitis.

Protozoan Diseases of the Nervous System (pp. 626–628)

1. The protozoan *Trypanosoma brucei* causes **African sleeping sickness** when it enters the body via the bite of an infected tsetse fly.
2. The free-living amoebae *Acanthamoeba* and *Naegleria* cause **primary amebic meningoencephalopathy**, a rare but usually fatal disease.

Prion Disease (pp. 628–630)

1. "Mad cow disease" is caused by a **prion**, an infectious protein. Humans can contract a form of spongiform encephalopathy when they eat meat from infected cattle.
2. **Variant Creutzfeldt-Jakob disease (vCJD)** causes brain deterioration resulting in dementia and death.

Microbial Diseases of the Eyes (pp. 630–631)

1. *Chlamydia trachomatis* infections of the reproductive tract inoculated into a baby's eyes at birth cause **trachoma**, resulting in blindness.
2. **Conjunctivitis** (pinkeye) is an inflammation of the conjunctiva due to infection by any of a variety of microbes.
3. **Keratitis** is the inflammation of the cornea due to microbial infection.

QUESTIONS FOR REVIEW

Answers to the Questions for Review (except Short Answer questions) begin on p. A-1.

Multiple Choice

1. Cerebrospinal fluid is
   a. formed deep within the brain.
   b. found in the subarachnoid space.
   c. withdrawn in a spinal tap.
   d. All of the above are true of the cerebrospinal fluid.
2. The layer of the meninges lying closest to the spinal cord is the
   a. arachnoid mater.
   b. pia mater.
   c. dura mater.
   d. cauda equina.
3. Nerves that primarily carry impulses toward the central nervous system are
   a. sensory nerves.
   b. motor nerves.
   c. mixed nerves.
   d. plexus nerves.
4. A collection of many neuron cell bodies outside the CNS is a
   a. plexus.
   b. dendrite.
   c. ganglion.
   d. axon.

5. Hansen's disease is also known as
   a. meningitis.
   b. leprosy.
   c. rabies.
   d. tetanus.
6. The most common form of meningitis is
   a. bacterial meningitis.
   b. viral meningitis.
   c. cryptococcal meningitis.
   d. septic meningitis.
7. The type of poliomyelitis that accounts for about 90% of the cases is
   a. asymptomatic polio.
   b. minor polio.
   c. nonparalytic polio.
   d. paralytic polio.
8. Which of the following diseases is a zoonosis?
   a. cryptococcal meningitis
   b. rabies
   c. tetanus
   d. tuberculosis
9. Which of the following does *not* play a role in the action of botulism toxin?
   a. acetylcholine in synaptic cleft
   b. inhibition of vesicular fusion
   c. nerve impulse conduction
   d. nerve cell growth
10. Which of the following diseases is *not* a zoonosis?
    a. arboviral encephalitis
    b. rabies
    c. St. Louis encephalitis
    d. viral meningitis
11. Which of the following diseases is caused most commonly by protozoan infections?
    a. variant Creutzfeldt-Jakob disease
    b. trachoma
    c. African sleeping sickness
    d. Hansen's disease
12. Which of the following statements concerning primary amebic meningoencephalitis is true?
    a. Primary amebic meningoencephalopathy is serious but not fatal.
    b. Primary amebic meningoencephalopathy is caused by a prion ingested from infected beef.
    c. A person can become infected by contaminated contact lenses.
    d. The disease is sexually transmitted and can be transmitted to a baby at birth.
13. A bacterium associated with bacteremia, meningitis, and pneumonia in newborns is
    a. *Staphylococcus aureus*.
    b. *Staphylococcus epidermidis*.
    c. *Streptococcus pyogenes*.
    d. *Streptococcus agalactiae*.
14. Which of the following would *not* play a role in preventing infection with amoebae?
    a. sterile drinking water
    b. sterile contact lens solution
    c. avoiding swimming in contaminated lakes
    d. chlorination of swimming pools
15. Human diploid cell vaccine is used to treat what disease?
    a. leprosy
    b. rabies
    c. tetanus
    d. tuberculosis

Fill in the Blanks

1. A ________________ is an intercellular space between an axon and another cell.
2. The cell structure of pneumococci that accounts for virulence is the ________________.
3. A common cause of bacterial meningitis acquired by babies at birth is ________________.
4. Three types of botulism are ________________, ________________, and ________________.
5. The bacterium that causes botulism is ________________.
6. A mycosis of the nervous system that has become more common since the advent of AIDS is ________________.
7. Hydrophobia is characteristic of ________________ infections.
8. The ________________ nervous system is composed of nerves that carry impulses to and from muscles and glands throughout the body.
9. The cytoplasmic processes that carry messages toward the neuron cell body are ________________.
10. ________________ are small, dark purplish hemorrhages of blood vessels in the skin sometimes seen in meningitis patients.
11. The leading cause of nontraumatic blindness in humans is ________________.
12. Pinkeye is an inflammation of the ________________ due to microbial infection.
13. *Staphylococcus aureus* can cause infections of the sebaceous glands near the eye. These infections are called ________________.
14. A sexually transmitted bacterium, ________________, causes ophthalmia neonatorum in newborns who are infected while passing through a diseased birth canal.
15. *Streptococcus agalactiae* is also known as Lancefield ________________ *Streptococcus*.

VISUALIZE IT!

1. Label the neuromuscular junction being acted on by botulism toxin.

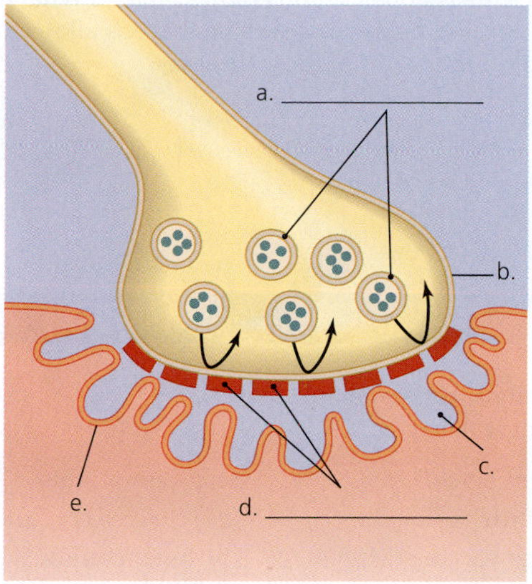

2. Indicate on the drawing where a lumbar puncture would be administered. Label the membranes and spaces the needle would pass through.

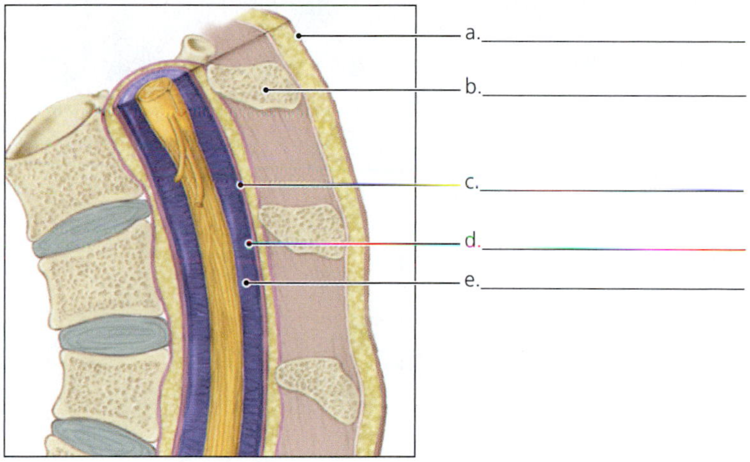

Short Answer

1. Distinguish among the following infections: meningitis, encephalitis, and meningoencephalitis.
2. Explain two ways that pneumococci "hide" inside body cells.
3. A dietician informs Jill that she is at risk for listeriosis. What dietary changes should Jill make?
4. What four effects does mycolic acid in the cell wall of *Mycobacterium leprae* have on its growth rate and resistance to antimicrobial drugs?
5. Describe leprosy with reference to historical control of the disease.
6. Justify this statement: "Botulism is not an infection."
7. Explain why botulism can kill. Include the following terms in your explanation: neurotoxin, neuromuscular junction, acetylcholine, synaptic cleft, flaccid.
8. Explain how tetanus can kill; use the terms *neuromuscular junction, acetylcholine, inhibitory neurotransmitters,* and *tetanospasmin.*
9. List four measures that can be taken to prevent arboviral infections.
10. What feature of the life cycle of *Trypanosoma brucei* makes it difficult to develop a successful vaccine?

CRITICAL THINKING

1. A doctor prescribes a new antimicrobial drug to treat bacterial meningitis in a nursing student. The student asked whether penicillin would kill the pathogen in question. The physician answers, "Perhaps, but. . . ." What is the most probable completion to that sentence?
2. Why would feeding pigeons in the park be more dangerous for a person with AIDS than for a healthy person?
3. Botulism toxin can be used as an antidote for tetanus. Can tetanus toxin be used as an antidote for botulism? Why or why not?
4. A three-year-old boy complains to his day care worker that his head hurts. The worker calls the child's mother, who arrives 30 minutes later to pick up her son. She is concerned that he is now listless and unresponsive, so she drives straight to the hospital emergency room. Though the medical staff immediately treats the boy with penicillin, he dies—only four hours after initially complaining of a headache. What caused the boy's death? Were the day care workers or the hospital staff to blame for his death? What steps should be taken to protect the other children at the day care facility?
5. A 20-year-old woman is brought to a South Carolina hospital's emergency room suffering from seizures, disorientation, hallucinations, and an inordinate fear of water. Her roommates report that she had suffered with fever and headache for several days. What disease does the patient have? Will a vaccine be an effective treatment? What wild animals are the likely sources of infection in South Carolina? What treatment should the patient's friends, classmates, and caregivers receive?
6. How are the actions of botulism and tetanus toxins alike? How are they different?

7. Investigations of tuberculosis have clearly demonstrated that bacteria are transmitted among airline passengers who are breathing recirculated air; however, as of 2012, no cases of in-flight transmission of meningococci had been reported to the Centers for Disease Control and Prevention (CDC). What are some possible explanations for the lack of documented cases of in-flight transmission of *Neisseria meningitidis*?

8. *Haemophilus influenzae* was so named because researchers isolated the organism from flu patients. Specifically, how could a proper application of Koch's postulates have prevented this misnomer?

9. Why does the fact that *M. leprae* grows best at 30°C make it difficult to grow the bacterium in cell culture?

10. Smallpox is the only disease that has been eliminated worldwide, though scientists hope to soon eradicate polio. What features do the smallpox and polio viruses share that has allowed medical science to rid the world of these diseases?

CONCEPT MAPPING

Using the following terms, draw a concept map that describes bacterial meningitis. For a sample concept map, see p. 94. Or, complete this and other concept maps online by going to the MasteringMicrobiology Study Area.

Antimicrobials
Cerebrospinal fluid
Culture
Gram stain
Gram-negative bacilli
Gram-negative diplococci
Gram-positive diplococci
Haemophilus influenzae
Healthy carriers
Hib vaccine
Infants and young children
Lumbar puncture (spinal tap)
Meningococcal vaccine (MCV4)
Neisseria meningitidis
Pneumococcal vaccine (PCV)
Respiratory route
Streptococcus pneumoniae
Third-generation cephalosporins
Vaccines

21 Microbial Cardiovascular and Systemic Diseases

MICRO IN THE CLINIC

Rabbit Fever

Finally, rabbit hunting season in the Arkansas Ozarks has arrived! Fifteen-year-old Jackson stands quietly by a majestic oak tree in the early October morning, barely able to contain his excitement. His patience is rewarded when a cottontail rabbit nervously ventures out from a wild blackberry thicket at the edge of the clearing. Jackson quickly takes aim with his .22 single-shot rifle and shoots the rabbit. As he retrieves the body, the back of his right index finger catches on the prickle of a blackberry bush, drawing a drop of blood.

Back at the family barn, Jackson skins the rabbit. His mother, Mary Ellen, prepares rabbit stew for dinner, and Jackson tans the pelt to preserve it.

Three days later, Jackson wakes up with a high fever, chills, and a headache. His mother applies an over-the-counter antibiotic ointment on the red sore that has developed on his scratched finger. Jackson also takes acetaminophen and feels well enough by the next morning to return to school and football practice. However, two days later, fever, drenching sweats, and aching muscles keep Jackson home again. To make things worse, the lymph nodes under his right arm have become painfully swollen, and the lesion on Jackson's finger has become an open sore. Alarmed that Jackson's symptoms have returned and worsened, Mary Ellen takes her son to their family physician. After Dr. Taylor examines Jackson and learns about his recent rabbit-hunting trip, she doesn't order any laboratory tests. Instead, she prescribes a 10-day series of injections with streptomycin, given twice a day.

Does rabbit hunting have anything to do with Jackson's illness? Why do you think Dr. Taylor prescribed a treatment for Jackson without ordering any tests? Turn to the end of the chapter (p. 670) to find out.

MM **Explore More:** Test your readiness and apply your knowledge with dynamic learning tools at MasteringMicrobiology.

Structures of the Cardiovascular System

LEARNING | OUTCOMES

21.1 Distinguish among the functions of arteries, veins, and capillaries.

21.2 Identify the major arteries and veins.

The heart, blood, and blood vessels compose the *cardiovascular system*, which, along with the lymphatic structures, transports fluids throughout the body. Chapter 16 covers the lymphatic system in detail. In this chapter, we consider the cardiovascular system and some of its more prominent diseases.

The cardiovascular system is a closed system in which the heart pumps blood into *arteries* connected via *capillaries* to *veins*. Veins carry blood back to the heart. The main arteries are the *pulmonary arteries*, which carry blood to the lungs, and the *aorta*, which carries blood to the rest of the body. The main veins are the *pulmonary veins*, which return blood from the lungs to the heart, and the *superior vena cava* and *inferior vena cava*, which carry blood from the upper and lower parts of the body, respectively **(FIGURE 21.1a)**. The heart also has its own coronary arteries and veins.

Blood is a tissue composed of *plasma*—the liquid part of blood containing dissolved nutrients, gases, and proteins—and so-called *formed elements: erythrocytes* (red blood cells), *leukocytes* (white blood cells), and small fragments of cells called *platelets*, which are important in coagulation (blood clotting). *Serum* is the liquid remaining when formed elements and clotting proteins are removed from blood.

Structure of the Heart

LEARNING | OUTCOME

21.3 Describe the structures and functions of the heart and blood vessels.

The heart is an exquisite set of two muscular, parallel pumps that act together to pump blood. Each side of the heart is composed of two chambers: an upper *atrium*[1] and a lower *ventricle*,[2] which are separated by valves[3] **(FIGURE 21.1b)**. The heart is composed of three layers: an outer fibrous *pericardium*,[4] a muscular *myocardium*,[5] and a thin inner *endocardium*.[6] The endocardium lines the atria and ventricles, extends to cover the valves, and is continuous with the linings of the blood vessels.

Movement of Blood and Lymph

LEARNING | OUTCOME

12.4 Trace the flow of fluid through the cardiovascular and lymphatic systems.

Veins called the superior and inferior venae cavae carry blood to the right atrium, which then pumps the blood through the *right atrioventricular (tricuspid) valve* to the right ventricle. The right ventricle pumps blood through the *pulmonary semilunar valve* to the lungs, where oxygen enters the blood and carbon dioxide diffuses out. Oxygenated blood returns via pulmonary veins to the left atrium, which pumps the blood through the *left atrioventricular (mitral) valve* to the left ventricle. The left ventricle, which is the most muscular of the four chambers of the heart, pumps blood through the *aortic semilunar valve* into the aorta, and from there through a series of smaller and smaller branched arteries into capillaries. Heart valves prevent backflow of blood.

Capillaries, which are extremely tiny vessels about 8 μm in diameter, are the part of the vascular system where oxygen and nutrients diffuse into surrounding tissues. Capillaries also leak fluid into spaces between cells, called *interstitial spaces*. Lymphatic vessels pick up this interstitial fluid and return it to the heart as lymph via a one-way system of lymphatic vessels. (Chapter 16 covers lymphatic vessels, the flow of lymph, lymphatic cells, and their roles in immunity.)

Blood is normally germ free (*axenic*[7], ā-zēn´ik); that is, it contains no microbes. However, breaks in the skin or mucous membranes provide a route for microbes to enter the blood. Many of these invaders are benign, but others cause diseases of the cardiovascular and lymphatic systems. The blood and lymph may also spread such microbes throughout the body to cause widespread diseases. The following sections examine such diseases. We begin our survey with cardiovascular and systemic diseases caused by bacteria.

TELL ME WHY

Bacteria infecting the mouth can enter the blood through small cuts resulting from normal brushing and flossing. Why would these bacteria reach the right atrioventricular valve before the other heart valves?

Bacterial Cardiovascular and Systemic Diseases

Bacteria can infect the blood as well as the blood vessels and heart. The following sections examine bacterial effects on the cardiovascular system.

[1]Latin, meaning "halls."
[2]Latin, meaning "little bellies."
[3]From Latin *valva*, meaning "folding door."
[4]From Greek *peri*, meaning "around," and *kardia*, meaning "heart."
[5]From Greek *mys*, meaning "muscle."
[6]From Greek *endon*, meaning "within."

[7]From Greek *a*, meaning "no," and *xenos*, meaning "foreigner."

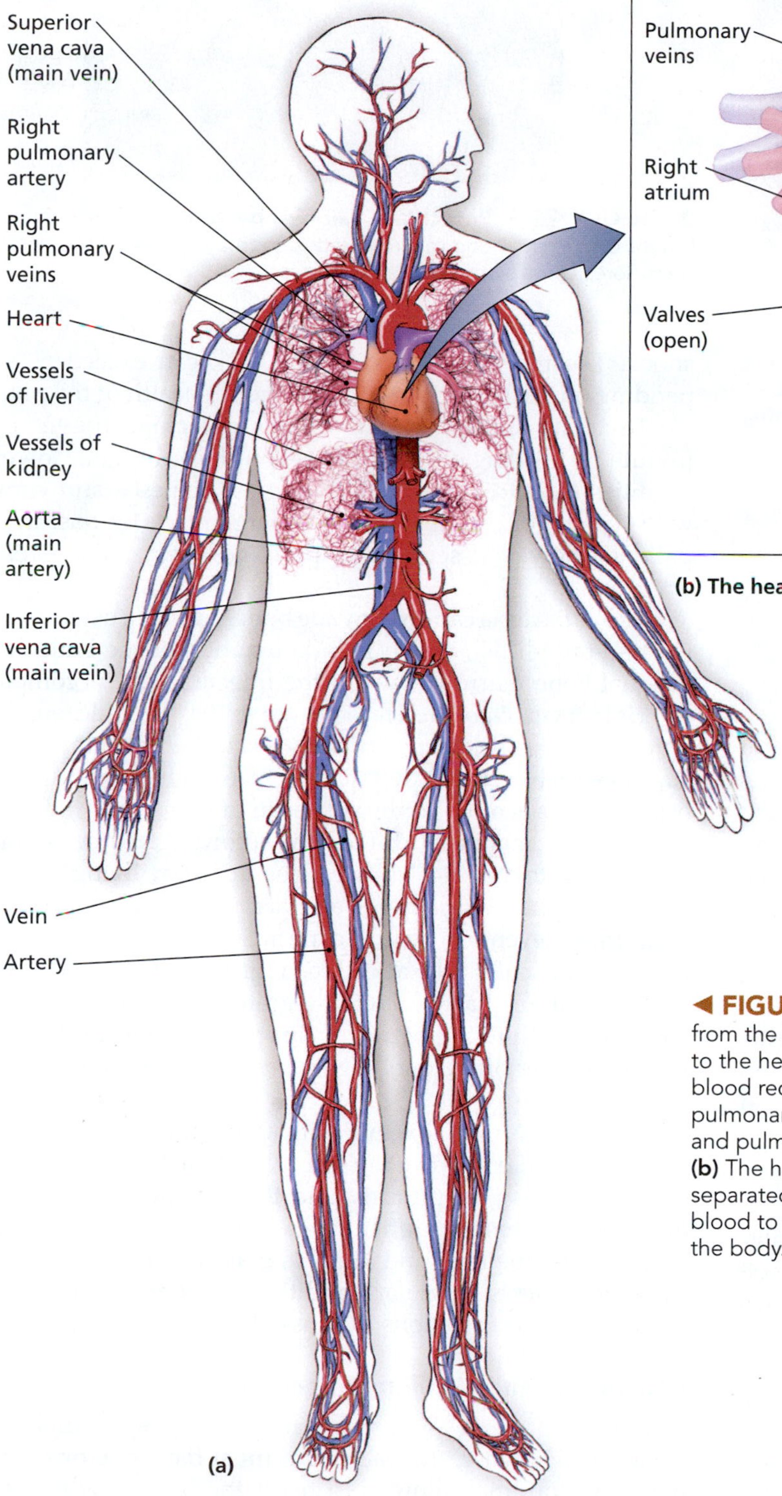

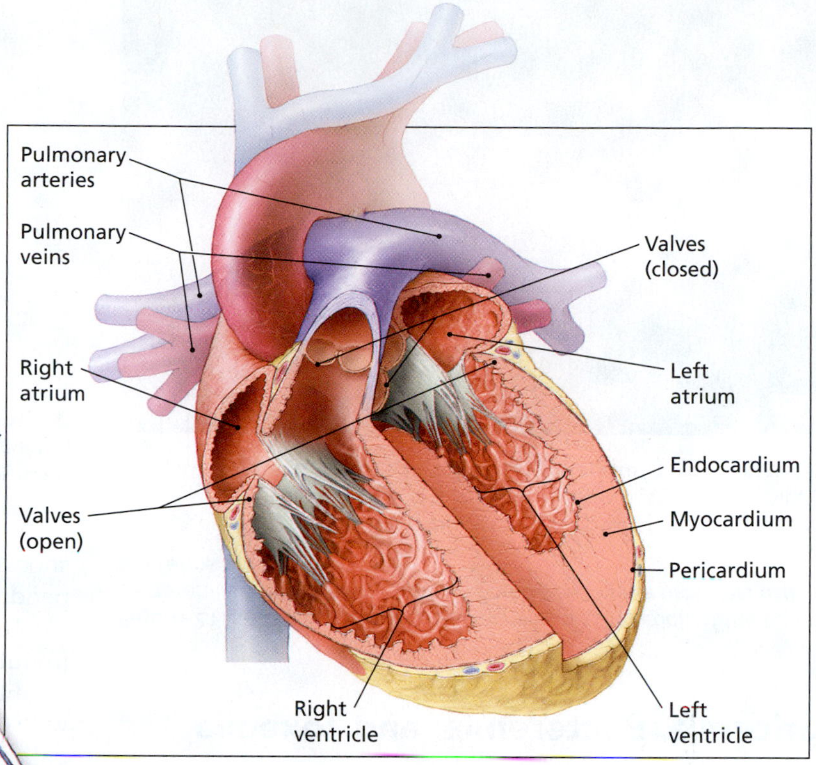

(b) The heart

◀ **FIGURE 21.1 The cardiovascular system. (a)** Arteries carry blood from the heart to the body's organs, tissues, and cells. Veins return blood to the heart. By convention, artists color vessels carrying oxygenated blood red, and vessels carrying unoxygenated blood blue; as a result, pulmonary arteries going from the heart to the lungs are colored blue, and pulmonary veins carrying oxygen from the lungs are colored red. **(b)** The heart is two functional pumps, each composed of two chambers separated by valves. The right side of the heart pumps unoxygenated blood to the lungs; the left side pumps oxygenated blood to the rest of the body.

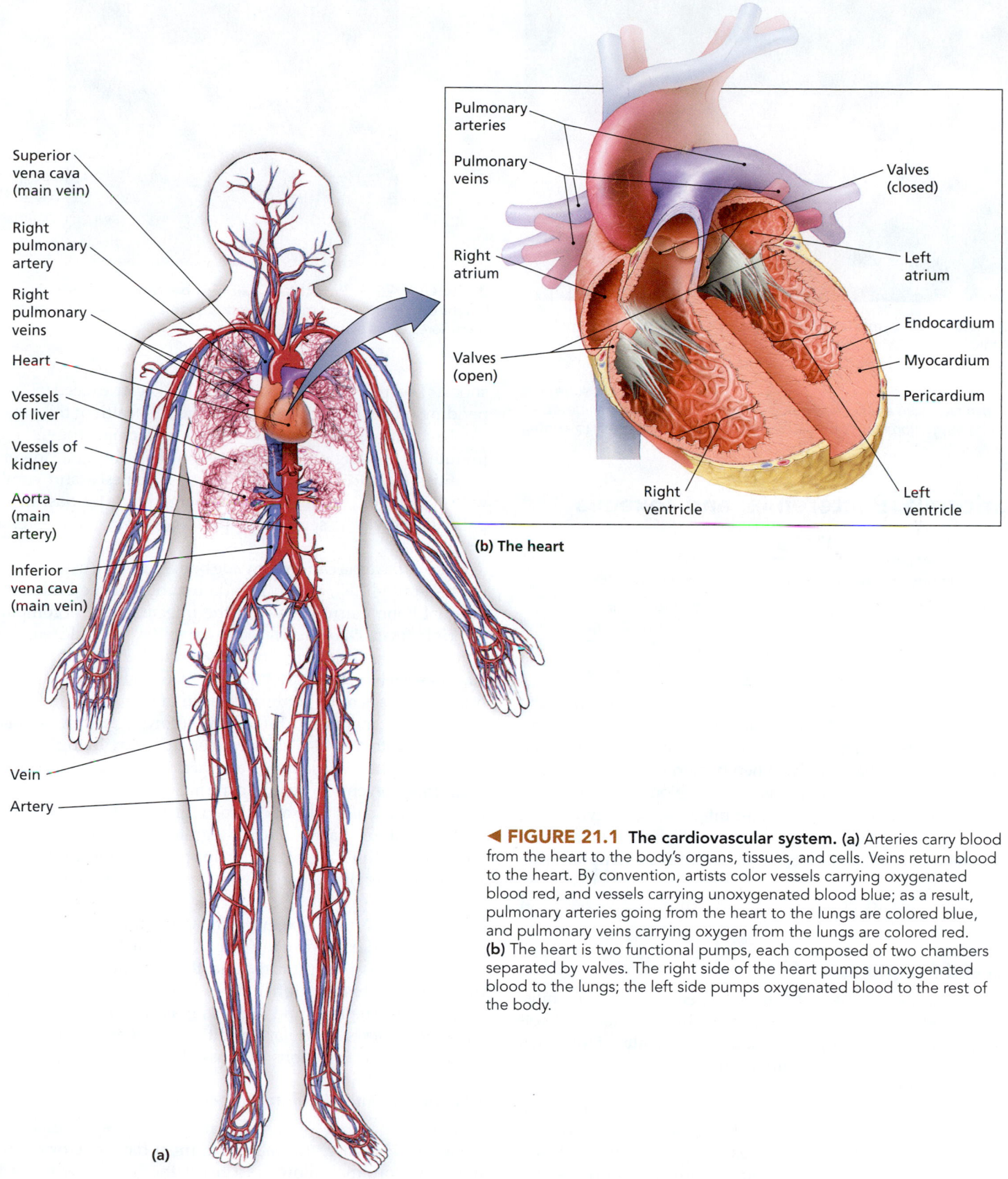

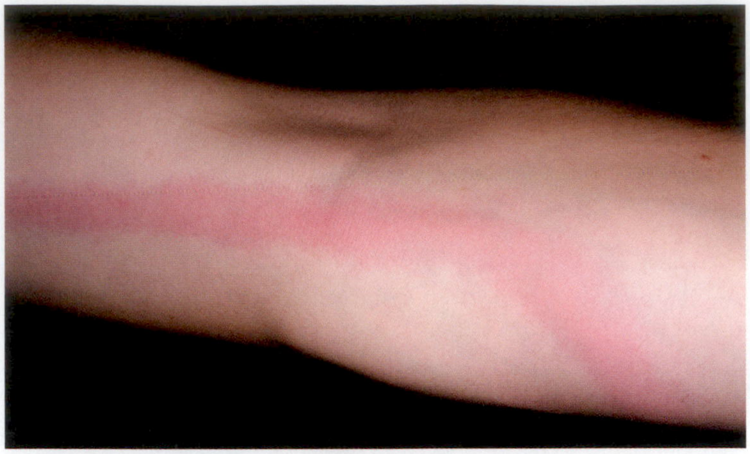

▲ **FIGURE 21.2 Lymphangitis, a sign of septicemia.** As microbes travel through lymphatic vessels from a site of infection, localized inflammation occurs. *How does bacteremia differ from septicemia?*

Figure 21.2 *Septicemia refers to any microbial infection of the blood; bacteremia is bacterial septicemia. Some health care providers use the terms synonymously.*

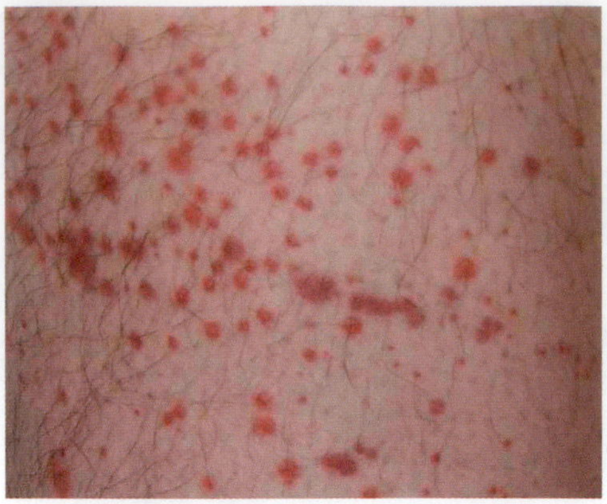

▲ **FIGURE 21.3 Petechiae, a sign of bacteremia.** These skin lesions can be small and relatively diffuse or may coalesce to form large, black sores containing dead cells.

Septicemia, Bacteremia, and Toxemia

LEARNING | **OUTCOMES**

21.5 Distinguish among septicemia, bacteremia, and toxemia.

21.6 Describe the signs, symptoms, causes, diagnosis, treatment, and prevention of septicemia and toxemia.

21.7 Describe the actions of endotoxin (lipid A).

Septicemia[8] refers to the presence of microbial infection of the blood that causes illness. **Bacteremia** refers specifically to bacterial septicemia, though many physicians use the terms *bacteremia* and *septicemia* interchangeably. When bacteria remain fixed at a site of infection but release toxins into the blood, the condition is **toxemia.** Septicemia can lead to an infection of the lymphatic system in which inflamed lymphatic vessels become visible as red streaks under the skin, a condition known as **lymphangitis** (lim-fan-jī´tis) **(FIGURE 21.2)**. Lymphocytes, particularly in lymph nodes, act to limit infection of lymphatic vessels and lymph nodes.

Signs and Symptoms

Septicemia is characterized by fever (over 38°C, 99°F), chills, nausea, vomiting, diarrhea, shortness of breath, malaise (feeling of general discomfort), and changes in mental status such as confusion, anxiety, and an impending feeling of doom. These signs and symptoms can progress rapidly to **septic shock,** a condition of extremely low blood pressure resulting from dilation of blood vessels. Decrease in body temperature, decrease in or absence of urine output, rapid breathing, aberrant blood clotting, increased heart rate, anxiety, and death characterize septic shock. The mortality from septic shock can exceed 50%, depending on the bacterium and the overall health of the patient.

Bacterial septicemia can trigger **petechiae** (pe-tē´kē-ē)—minute hemorrhagic skin lesions—on the trunk and lower extremities **(FIGURE 21.3)**. Petechiae may coalesce and combine with regions of cell death to form large black lesions. In some patients, however, septicemia produces only mild fever and arthritis.

Septic bacteria can also invade bones, causing **osteomyelitis**[9] (os´tē-ō-mī-ĕ-lī´tis), which is inflammation of the bone and its internal bone marrow. Pain in the infected bone accompanied by high fever characterizes osteomyelitis. In children, osteomyelitis typically occurs in areas with well-developed blood supplies such as the growing regions of long bones. In adults, osteomyelitis is more commonly seen in vertebrae.

Toxemia manifests differently according to the toxins involved. **Exotoxins,** which are released from living microbes, include cell-killing *cytotoxins* and *neurotoxins,* such as botulism toxin that prevents muscular contraction and tetanus toxin that prevents muscular relaxation (see Figures 20.6 and 20.8). Dying Gram-negative bacteria disintegrate, releasing **endotoxin,** which is the **lipid A** portion of **lipopolysaccharide (LPS)** from the outer layer of a Gram-negative cell's outer membrane, into the blood.

A severe form of toxemia with septic shock is **streptococcal toxic-shock-like syndrome (TSLS),** in which a patient's blood pressure drops precipitously and the patient may suffer from dizziness, confusion, difficulty in breathing, and a weak and rapid pulse. The liver and kidneys may fail. *Staphylococcal toxic shock syndrome* is a similar condition often associated with reproductive tract infections (discussed in Chapter 24).

Pathogens and Virulence Factors

Numerous bacteria are capable of causing septicemia or toxemia. In every case, the bacterium must have one or more virulence factors that allow it to resist the body's defenses long

[8]From Greek *sepein,* meaning "to become putrid," and *haima,* meaning "blood."

[9]From Greek *osteon,* meaning "bone"; *myelos,* meaning "marrow"; and *itis,* meaning "inflammation."

enough to multiply and upset the body's normal structures and functions, producing signs and symptoms. For example, bacteria that form capsules may resist phagocytosis or intracellular digestion, allowing them to reproduce and metabolize in the blood. Some bacteria have the ability to "steal" iron (which is necessary for bacterial metabolism) by the use of *siderophores* from carrier proteins in the plasma. Other bacteria destroy erythrocytes, releasing iron from hemoglobin into the blood and making it available for bacterial growth. Various Gram-negative bacteria release different kinds of endotoxin molecules. Some endotoxins are relatively harmless, whereas others cause serious signs and symptoms because they stimulate the body to release chemicals that trigger fever, inflammation, diarrhea, hemorrhaging, blood coagulation, or shock.

Frequently, bacteria causing septicemia are *opportunistic*—normal members of the microbiota that become pathogenic. Such organisms can cause healthcare associated diseases. Examples include the Gram-negative aerobes *Pseudomonas aeruginosa* (soo-dō-mō´nas ā-roo-ji-nō´să) and *Neisseria meningitidis* (nī-se´rē-ă me-nin-ji´ti-dis); facultative anaerobes from the intestinal tract such as *Escherichia coli* (esh-ĕ-rik´ē-ă kō´lī) and species of *Salmonella* (sal´mō-nel´ă); and strictly anaerobic species of *Bacteroides* (bac-ter-oy´dēz), which make up a large portion of the normal bacterial microbiota of the colon. Though Gram-negative bacteria are more commonly associated with septicemia, Gram-positive bacteria such as *Staphylococcus aureus* (staf´i-lō-kok´ŭs o´rē-ŭs) and *Streptococcus pneumoniae* (strep-tō-kok´ŭs nū-mō´nē-ī) can also be opportunistic causes of septicemia.

Streptococcus pyogenes (pī-oj´en-ēz) causes streptococcal toxic-shock-like syndrome (TSLS). Streptococcal infections of the skin or wounds rather than of "strep throat" are associated with TSLS. (Streptococcal toxic-shock-like syndrome should not be confused with toxic shock syndrome (TSS), which is associated with tampon use and is caused by *Staphylococcus*; see Chapter 24.)

Pathogenesis and Epidemiology

Septicemia begins with direct inoculation of bacteria into the blood, such as can occur during medical procedures, via nonsterile needle use by drug users, from an infection elsewhere in the body, or through small abrasions in the respiratory or digestive tracts. Septicemia is commonly associated with prolonged venous needle placement, the use of inadequately sterilized kidney dialysis machines, surgical wounds, infected teeth, and urinary tract infections. Healthcare associated infections (HAI), such as biofilms that form on urinary catheters, account for about half of all cases of staphylococcal septicemia.

Healthy people with normal immune systems rarely have septicemia; that is, bacterial blood infections in these patients spontaneously abate. However, people with suppressed immune systems due to alcoholism, other drug abuse, malnutrition, stress, or HIV infection cannot effectively fight bacterial infections; therefore, septicemia is more likely to develop in such patients. Medical procedures that enhance the growth of bacterial strains resistant to antimicrobial drugs or that introduce virulent strains into the blood make septicemia more likely.

Gram-negative bacteria are more likely to cause severe septicemia because lipid A released from the outer membrane of dying Gram-negative bacteria is a potent activator of defensive reactions by the body. Endotoxin activates nearly all nonspecific defensive responses of the body, including complement via the alternative pathway (see Figure 15.8), coagulation (blood clotting), and inflammation. Endotoxin also initiates the release of potent cytokines from macrophages, monocytes, B cells, and other defensive cells when the toxin binds to the cytoplasmic membranes of these cells. Coagulation can be widespread and severe, a condition called **disseminated intravascular coagulation (DIC),** which can be fatal.

Released cytokines, including **tumor necrosis factor (TNF), interleukins (ILs),** and **platelet activating factor (PAF),** normally elicit defensive reactions at localized infections, but they become life threatening when carried to an extreme throughout the body in septicemia. TNF causes tissue damage and is *pyrogenic* (fever producing). IL-1 also triggers fever and causes the bone marrow to release a large number of immature neutrophils, which attach to and damage blood vessel walls. IL-6 and IL-8 damage circulating neutrophils and further injure cells lining blood vessels, allowing plasma to escape the vascular system, which precipitously reduces blood pressure. The drop in blood pressure reduces blood flow and oxygen delivery to vital organs, which can succumb, resulting in death. PAF is another potent trigger for coagulation. **FIGURE 21.4** summarizes the effects of endotoxin on a patient.

The Centers for Disease Control and Prevention (CDC) does not require reporting of cases of septicemia, bacteremia, or toxemia, so the full extent of these conditions in the United States is unknown; however, researchers estimate about 800,000 deaths per year, mostly among the elderly (over 65 years of age), childern under age one, and cancer patients. Morbidity and mortality depend upon the cause; for example, *Haemophilus influenzae* (hē-mof´i-lŭs in-flu-en´zī) in the blood attacks the meninges (covering membranes of the brain and spinal cord), causing meningitis in children, about 4% of whom die; *Streptococcus pneumoniae* is associated with bacteremia that is 0.8% fatal; whereas mortality from either Gram-negative bacterial toxemia or streptococcal toxic-shock-like syndrome can be greater than 50%.

Diagnosis, Treatment, and Prevention

Physicians diagnose septicemia based upon signs and symptoms. Clinicians are able to culture bacteria from the blood of fewer than half of patients with characteristic signs and symptoms of sepsis; the majority of patients therefore have **occult**[10] **septicemia,** a term referring to the fact that the exact bacterial cause is hidden.

Treatment generally involves prompt diagnosis and treatment with appropriate antimicrobial drugs against the specific bacterial cause; elimination of the initial infection (for example, removal of an abscessed tooth); and intravenous fluid replacement to mitigate the fluid loss caused by endotoxin. Prevention

[10]From Latin *occultare,* meaning "to hide."

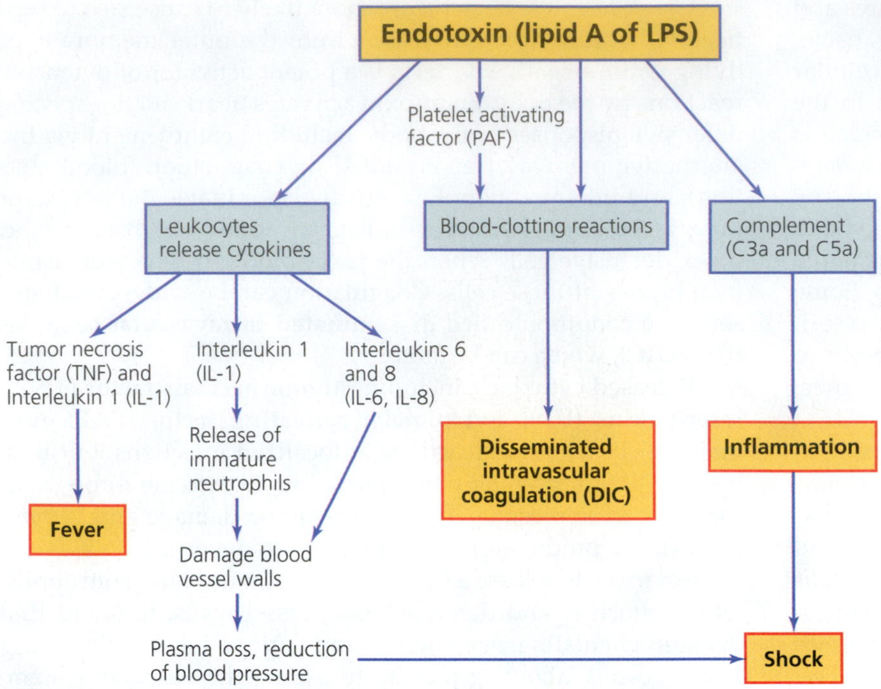

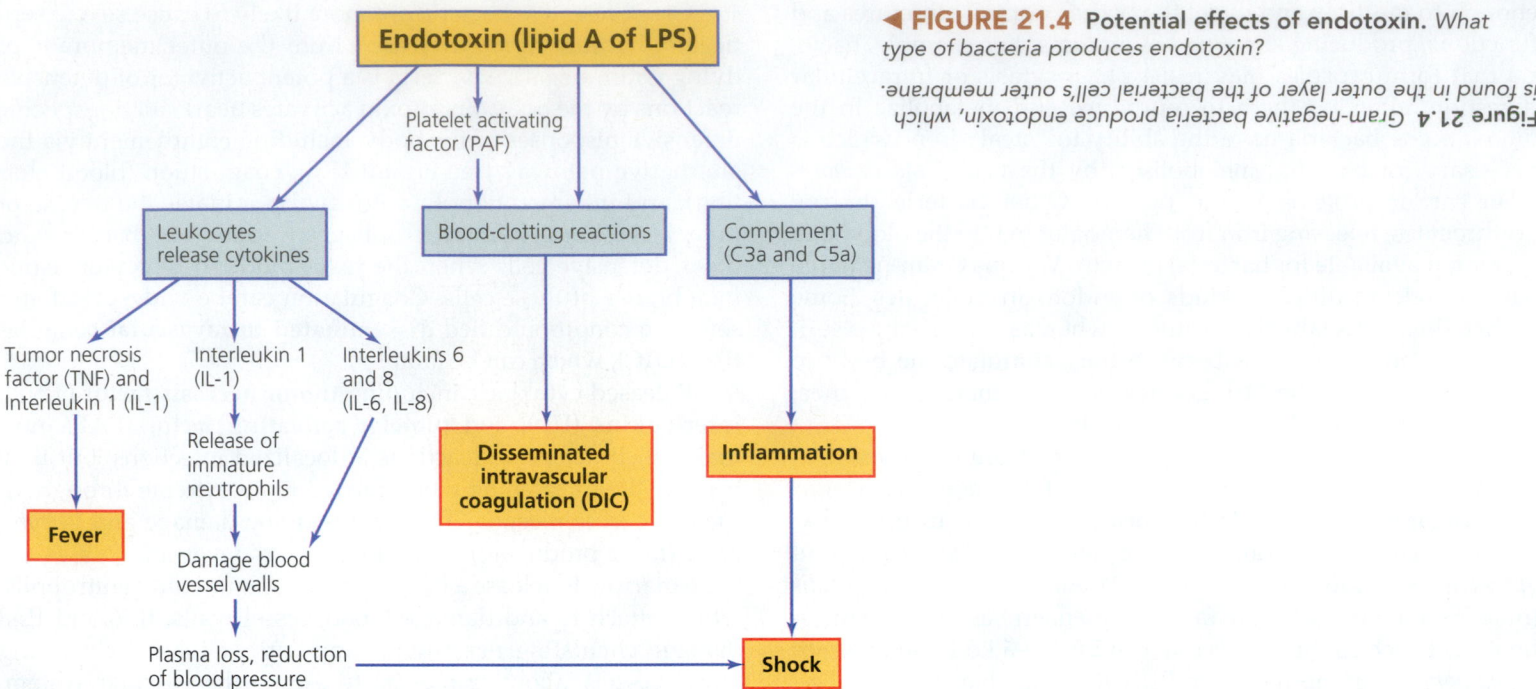

◀ **FIGURE 21.4 Potential effects of endotoxin.** *What type of bacteria produces endotoxin?*

Figure 21.4 Gram-negative bacteria produce endotoxin, which is found in the outer layer of the bacterial cell's outer membrane.

of septicemia depends upon prompt treatment of infections, particularly in patients whose immune system is impaired. If treatment is delayed, bacteria multiply, and the use of antimicrobial drugs can worsen a patient's condition by causing the release of large amounts of endotoxin. Newer treatments utilizing monoclonal antibodies against LPS and TNF show some small success in mitigating the effects of endotoxin.

Endocarditis

LEARNING | OUTCOME

21.8 Describe the signs, symptoms, causes, diagnosis, treatment, and prevention of endocarditis.

Bacteria in the blood can infect the endocardium, which is a thin lining inside the heart chambers, covering the valves, and continuous with the linings of the blood vessels leaving and entering the heart. Bacterial colonization of the endocardium triggers inflammation—**endocarditis**—and the formation of **vegetations,** which are bulky masses of platelets and clotting proteins that surround and bury the bacteria (see Disease at a Glance 21.1 on p. 644), hiding them from defensive cells, antibodies, and antimicrobial drugs.

Signs and Symptoms

Patients with endocarditis typically have fever, extreme fatigue, malaise, and breathing difficulty. A health care provider may detect tachycardia[11] (faster-than-normal heart rate) and murmurs, which are sounds of abnormal blood flow through the heart. Endocarditis may lead to complications such as blood clots, stroke, and the complete destruction of the heart valves, resulting in heart failure. Endocarditis most commonly affects the left atrioventricular (mitral) valve, followed by the aortic semilunar valve. The signs and symptoms of endocarditis may develop slowly over a period of weeks or months—a condition called *subacute endocarditis*—or they may develop quickly *(acute endocarditis)*.

Pathogens

Normal microbiota from elsewhere in the body are the usual causes of bacterial endocarditis; about half the cases are caused by so-called *viridans*[12] *streptococci* (vir´i-danz strep-tō-kok´sī), which are named for the greenish pigment they produce when cultured on blood. Viridans streptococci are not highly invasive but can enter the blood through surgical wounds; small lesions in the lungs during pneumonia; or lacerations of the gums, including undetectable cuts produced by dental procedures, chewing hard candy, or brushing the teeth.

Other opportunistic bacteria are *Staphylococcus epidermidis* (ep-i-der-mid´is) and *S. aureus* from the skin, *Streptococcus pneumoniae* from the pharynx, and *Enterococcus* (en´ter-ō-kok´ŭs) and *Escherichia* from the digestive tract. Dozens of pathogenic and opportunistic bacteria, including *Neisseria, Pseudomonas, Bartonella* (bar-tō-nel´ă), *Mycobacterium* (mī´kō-bak-tēr´ē-ŭm), and *Mycoplasma* (mī´kō-plaz-mă), can also cause endocarditis. "Culture negative" endocarditis is a condition in which the causative agent either has not been cultured or remains unknown.

Pathogenesis and Epidemiology

Most patients with endocarditis have obvious sources of infection such as an infected tooth, skin lesion, or intravascular catheter;

[11]From Greek *tachys,* meaning "quick."

[12]From Latin *viridis,* meaning "green."

BENEFICIAL MICROBES

When a Bacterial Infection Is a Good Thing

Gram-negative bacteria are common opportunistic and nosocomial pathogens of the cardiovascular system, producing bacteremia, toxemia, endocarditis, and other serious conditions. However, Gram-negative bacteria can themselves be the target of bacterial pathogens, specifically cells of *Bdellovibrio* (del-ō-vib′rē-ō) and *Micavibrio* (mī-kă-vib′rē-ō). These Gram-negative predators are voracious devourers of other Gram-negative bacteria; in fact, their Gram-negative cousins are their only diet!

Bdellovibrio latches onto a Gram-negative bacterium such as *Escherichia coli* or the hard-to-treat *Pseudomonas aeruginosa*, enters its prey's periplasm, digests its host, feasts, replicates, and lyses its victim. *Bdellovibrio* daughters quickly attack other cells, reducing the victim's population a hundredfold in short order. *Micavibrio* also attaches to its victim's outer membrane, but remains outside the cell, replicating by binary fission while literally sucking the life (and cytoplasm) from its target. The predators can attack both free-swimming and biofilm-associated Gram-negative bacteria.

Scientists hope to identify, isolate, and utilize the unusual enzymes that allow *Bdellovibrio* and *Micavibrio* to exclusively attach to and kill Gram-negative bacteria. Alternatively, researchers are considering using the bacterial predators as living antimicrobial poultices on skin or wound infections or as living, intravenous, antimicrobial treatments for cardiovascular infections—a patient would be infected to get rid of an infection.

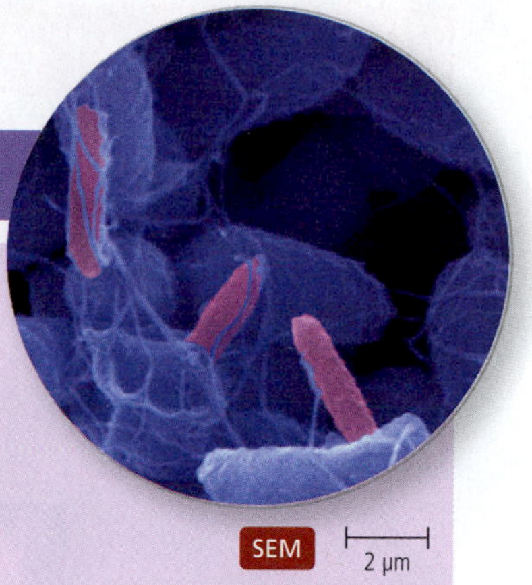

Bdellovibrio (pink) attacks *E. coli* (blue)

however, some patients have no obvious focal site of bacterial infection. Intravenous drug users are at high risk of infection from contaminated syringes. Patients with abnormal hearts, including those with birth defects, scarring from previous bacterial infections, and those with heart valve replacements, are more susceptible to endocarditis because a damaged heart provides sites for bacterial colonization. Vegetations also form roughened spots on the endocardium that provide places for bacteria to lodge and grow, producing additional layers of vegetations.

Fragments of vegetations and blood clots, each called an *embolus*,[13] can break off and travel via the blood to lodge in small blood vessels of the brain, kidneys, lungs, or abdominal organs, interrupting the flow of blood and causing severe damage. A stroke is such an interruption of blood flow through the brain.

Diagnosis, Treatment, and Prevention

Physicians diagnose about 10,000 cases of endocarditis each year in the United States based on symptoms, confirmed by visualization of vegetations via an *echocardiogram*, which is an image of the heart produced by using sound waves, and by culturing the bacterium from the patients' blood. A history of intravenous drug abuse, heart disease, or recent dental work; a new heart murmur; or hemorrhages visible under the nails or on the retina of the eye may indicate that a patient has endocarditis.

Physicians treat endocarditis by administering intravenous antibacterial drugs, which requires hospitalization as sources of infection must be drained of pus and cleansed. Treatment for weeks is common. Physicians choose an antibacterial drug based on susceptibility tests (see Figure 10.9). Early treatment is usually successful; delayed diagnosis and treatment increase the risk of valve damage, heart failure, and the production of emboli. Surgical removal of vegetations and abscesses and repair or replacement of damaged heart valves is requisite.

Patients with known risk factors should ask their physician about preventive use of antimicrobial drugs before dental procedures or surgeries. Good daily dental hygiene is equally important for such patients. Intravenous drug users should seek treatment for their addictions and in the interim always use clean needles.

Disease at a Glance 21.1 summarizes the features of bacterial septicemia and endocarditis.

To this point, we have considered bacteria that infect the blood and the heart. However, blood and lymph can carry pathogens throughout the body to cause **systemic diseases**, that is, diseases that affect the whole body rather than a local area. Next, we examine some bacterial systemic diseases.

Brucellosis

LEARNING OUTCOMES

21.9 Describe the signs, symptoms, and cause of brucellosis.

21.10 Define *zoonosis*, and explain the role of animal husbandry in the spread and prevention of brucellosis.

Brucellosis (brū-sel-ō′sis) is a **zoonosis**, that is, a disease of animals that can transfer to humans under normal conditions. Brucellosis has had a variety of other names, including *Bang's*

[13]From Greek *embolos*, meaning "plug."

DISEASE AT A GLANCE 21.1

Bacteremia/Endocarditis

Causes Numerous bacteria may cause bacteremia or endocarditis. Common opportunistic or nosocomial causes of bacteremia are *Pseudomonas aeruginosa, Neisseria meningitidis, Escherichia coli, Salmonella* species, *Staphylococcus aureus*, and *Streptococcus pneumoniae*. About half the cases of bacterial endocarditis are caused by viridans streptococci. *Bartonella, Mycobacterium*, and *Mycoplasma* species also cause the disease.

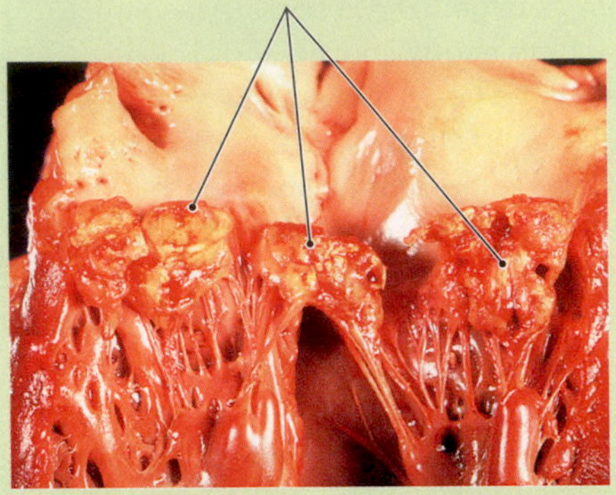

Heart valve

Virulence factors Vary with pathogen but include attachment molecules, capsules, flagella, endotoxin, exotoxins, and antiphagocytic factors.

Portal of entry Inoculation of bacteria into the blood via open wounds.

Signs and symptoms Bacteremia is characterized by fever, chills, nausea, vomiting, diarrhea, shortness of breath, malaise, and confusion or an impending sense of doom. Endocarditis symptoms include fever, extreme fatigue, malaise, difficulty breathing, and tachycardia.

Incubation period Bacteremia symptoms can occur within hours to days following infection. Acute endocarditis can develop over days to a few weeks, whereas subacute bacterial endocarditis develops more slowly, over weeks or months.

Susceptibility People with a suppressed immune system and patients requiring regular or prolonged invasive medical procedures, such as intravenous therapy or kidney dialysis.

Treatment Prompt treatment and identification are important. Administration of appropriate antibiotics, elimination of the initial site of infection, and intravenous fluid replacement are the proper treatment for bacteremia. Treatment for endocarditis is similar: intravenous antibiotic therapy, often for a course of six weeks.

Prevention Patients with known risk factors may be given prophylactic antibiotics before surgeries or dental procedures. Proper sterilization of medical instruments and proper wound care are also standard preventive measures.

disease, after microbiologist Bernhard Bang (1848–1932), who investigated the disease; and *Malta fever, rock fever of Gibraltar*, and *fever of Crete*, all of which were named after localized epidemics in those locales.

Signs and Symptoms

Brucellosis is characterized by a fluctuating fever that spikes at about 40°C (104°F) every afternoon, giving the disease another of its common names—*undulant*[14] *fever*. The patient also has chills, sweating, headache, myalgia[15] (muscle pain), and weight loss.

Pathogen and Virulence Factors

Brucella (broo-sel´lă) is a genus of nonmotile, aerobic, Gram-negative coccobacilli that lack capsules. Although brucellosis in humans has been ascribed to four species of *Brucella*, rRNA analysis has revealed the existence of only a single true species: *Brucella melitensis* (me-li-ten´sis); the other variants are strains of this one species. Nevertheless, many clinicians and veterinarians still use the historical species names—*B. melitensis*, which infects goats and sheep, *B. abortus* (a-bort´us) in cattle, *B. suis* (soo´is) in swine, and *B. canis* (kā´nis) in dogs, foxes, and coyotes.

In animal hosts, the *Brucella* lives as an intracellular parasite in organs such as the uterus and placenta, but these organs are not infected in humans. Infections in animals are either asymptomatic or cause a mild disease, though they sometimes cause sterility or spontaneous abortion.

As a Gram-negative bacterium, *Brucella* has endotoxin that accounts for some of the signs and symptoms of brucellosis. The bacterium also has the ability to grow and multiply inside phagocytic cells, evading antibodies and some antibacterial drugs.

Pathogenesis and Epidemiology

Humans become infected either by consuming unpasteurized contaminated dairy products or through contact with animal blood, urine, or placentas in workplaces such as slaughterhouses, veterinary clinics, and feedlots. The bacterium enters the body through breaks in mucous membranes of the digestive and respiratory tracts. *Brucella* travels inside phagocytic cells to organs of the body, including lymph nodes, spleen, bone marrow, liver, and heart. Arthritis, splenomegaly (enlarged spleen), enlarged testes, endocarditis, meningitis, and encephalitis may result.

Because of immunization of livestock, destruction of infected animals, and quarantine of infected herds, brucellosis is less of a threat in the United States than in years past. In 2012, only 123 cases were reported. About half a million human cases occur worldwide each year.

Diagnosis, Treatment, and Prevention

Daily undulating fever can lead to a diagnosis of brucellosis. Since *Brucella* is difficult to culture in a laboratory, diagnosis is confirmed by serological tests showing a rising level of antibodies against *Brucella*.

Most cases of brucellosis in animals and humans are mild and require no treatment. Physicians treat brucellosis with combinations of antibacterial drugs, including doxycycline and rifampin or streptomycin for several weeks.

An attenuated vaccine for animals exists, but physicians do not administer it to humans because the vaccine sometimes causes brucellosis. With pasteurization of dairy products, immunization of uninfected domesticated animals, and the slaughter of infected ones, the threat of brucellosis for U.S. residents has been reduced. The remaining U.S. cases usually occur in immigrants or people who have traveled outside the country.

[14]From Latin *undula*, meaning "small wave," referring to the wavy appearance of a graph of the patient's temperature over time.

[15]From Greek *mys*, meaning "muscle," and *algos*, meaning "pain."

Tularemia

LEARNING OUTCOME

21.11 Describe the signs, symptoms, cause, diagnosis, treatment, and prevention of tularemia.

Tularemia[16] (tū-lă-rē´mē-ă) is a zoonotic disease.

Signs and Symptoms

Within five days of infection, tularemia patients develop a skin lesion and swollen, tender, pus-filled lymph nodes near the site of infection. This is followed by ascending lymphangitis as bacteria are carried from the site. Patients may also exhibit any or all of the following: fever, chills, headache, malaise, fatigue, shortness of breath, joint stiffness, and myalgia. These general manifestations may last for months or years. Tularemia is fatal in about 5% of untreated patients and 1% of treated patients.

Pathogen and Virulence Factors

Francisella tularensis (fran´si-sel´ă too-lă-ren´sis) is a very small (0.2 μm × 0.2–0.7 μm), nonmotile, strictly aerobic, Gram-negative coccobacillus that is found in temperate regions of the Northern Hemisphere living in water and as an intracellular parasite of animals. The bacterium has an amazingly diverse assortment of hosts. It lives in mammals, birds, fish, and bloodsucking ticks and insects. Indeed, one is hard pressed to find an animal that cannot be its host. The most common reservoirs in the United States are rabbits, muskrats, and ticks, which give the disease two of its common names—*rabbit fever* and *tick fever*.

F. tularensis has a lipid capsule that discourages phagocytosis, but even when phagocytized, the bacterium survives by inhibiting the fusion of lysosomes with phagocytic vesicles. Endotoxin accounts for many of the signs and symptoms of tularemia.

Pathogenesis and Epidemiology

Francisella is incredibly varied in the ways it spreads from its normal animal hosts to humans. *Francisella,* by virtue of its small size, can pass through apparently unbroken skin or mucous membranes. Most often the bite of an infected tick or direct contact with an infected animal introduces the bacterium into the body.

Francisella, an unusually infectious bacterium infection, requires as few as 10 cells when transmitted by a biting arthropod or through skin or mucous membranes. Bloodsucking flies, mosquitoes, and mites also transmit *Francisella,* and humans can be infected by consuming infected meat, drinking contaminated water, or inhaling bacteria in aerosols produced in the laboratory or during animal slaughter. Inhaling only about 50 cells can cause disease, though a person would have to consume 10^8 cells in food or drink to contract the disease via the digestive tract. Fortunately, human-to-human spread does not seem to occur.

Hunters, trappers, taxidermists, and others who come in contact with dead wild animals are at greatest risk for tularemia. Only 152 cases were reported in 2012, but the actual number of infections was probably much higher considering its virulence, prevalence in animals, and multiple modes of transmission. Tularemia frequently remains unsuspected because its signs and symptoms are not notably different from many other bacterial and viral diseases, and because it is difficult to confirm tularemia by using laboratory tests. Though it is typically innocuous, tularemia is fatal to about 5% of untreated patients. The Centers for Disease Control and Prevention removed tularemia from the list of nationally notifiable diseases in 1994, but concern about the possible use of *Francisella* as a bioterrorism weapon led officials to return tularemia to the list in 2000.

[16]Named for Tulare County, California.

CLINICAL CASE STUDY

A Heart-Rending Experience

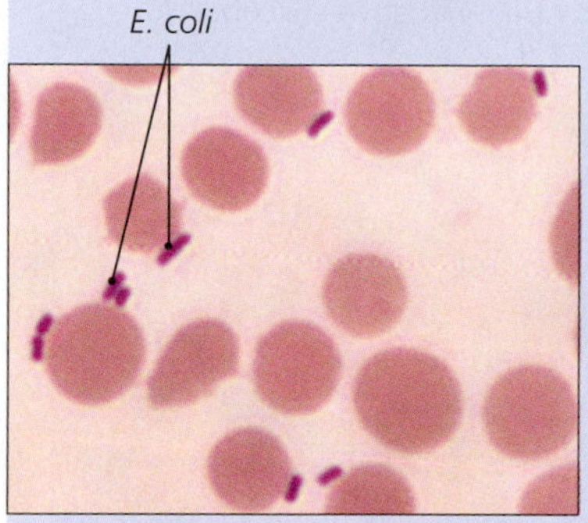

Blood smear LM 7 μm

A construction worker in his mid-30s visited his physician complaining of shortness of breath and a persistent cough for over three weeks. Upon examination, the doctor noted that the client had a history of asthma and intravenous drug use. Believing him to have an upper respiratory disturbance, the physician ordered a steroid to reduce swelling in the airways.

Two weeks later, the construction worker's signs and symptoms were worse. He began to experience periods of dizziness and was extremely fatigued. Worried that something was terribly wrong and getting worse, he sought help at an emergency room. When he arrived, his heart rate was accelerated and he appeared to be dehydrated. The ER staff administered a rapid intravenous infusion of fluids, took blood for analysis, and X-rayed his chest.

Shortly after the infusion, the man's respiratory situation deteriorated to the point that he was placed on a ventilator. The chest X-ray film revealed an abnormally enlarged heart and fluid throughout both lungs. The blood culture contained *Escherichia coli.*

1. How could the client have contracted bacteremia?
2. What had *E. coli* done to the inner structures of his heart?
3. What factors in the medical treatment may have exacerbated the disease? Why?

DISEASE AT A GLANCE 21.2

Tularemia

Cause *Francisella tularensis* (aerobic, Gram-negative coccobacillus).

Virulence factors Intracellular pathogen of many mammals, birds, fish, arachnids, and insects; capsule discourages phagocytosis; inhibits lysosome fusion with phagosome; endotoxin.

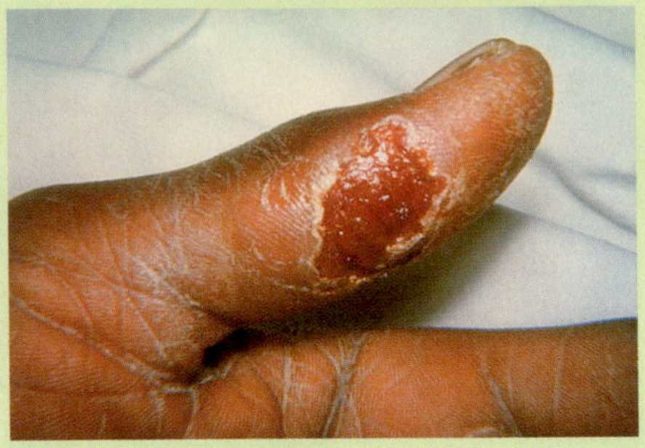

Ulcer of tularemia

Portal of entry Via the bite of an infected tick, consumption of contaminated meat or water, inhalation, or direct contact with an infected animal.

Signs and symptoms Sudden fever, chills, headaches, diarrhea, muscle and joint pain, and progressive weakness and fatigue. Inhalation cases may also exhibit chest pain, bloody sputum, and difficulty breathing. Skin ulcers may develop near the site of an infected tick bite. Tender, swollen lymph nodes appear near the site of infection within five days of infection.

Incubation period Generally three to five days, but can be up to two weeks.

Susceptibility Most risk to animal handlers and rabbit hunters, especially in the endemic areas of the southeastern United States.

Treatment Intramuscularly administered streptomycin and tetracycline.

Prevention Use insect repellent and check for ticks. Wash hands after handling animal carcasses. Cook game thoroughly and ensure water is from a safe source or sterilized.

Diagnosis, Treatment, and Prevention

Diagnosis of tularemia is difficult because the signs and symptoms are general and shared by many diseases. A history of contact with wild animals or of tick bites should prompt suspicion of infection with *F. tularensis.* Serological testing confirms a diagnosis. Culturing is not recommended because the organism is extremely virulent and easily contractible in the laboratory.

F. tularensis produces β-lactamase, so penicillins and cephalosporins are ineffective, but other antimicrobial drugs have been used successfully. Currently, streptomycin or gentamicin are commonly used.

An attenuated vaccine provides some protection for people at risk of exposure. Though the vaccine is not entirely effective, it can lessen the severity of the disease. To prevent infection, people should avoid the major reservoirs of *Francisella* (rabbits, muskrats, and ticks), especially in endemic areas. They can protect themselves from tick bites by wearing long clothing with tight-fitting sleeves, cuffs, and collars, and by using insect repellent containing *DEET* (*N,N*-diethyl-*m*-toluamide). *Francisella* is present in tick saliva and feces, so prompt removal of ticks can reduce the chance of infection. Hikers and hunters should never handle ill-appearing wild animals or their carcasses, and they should wear gloves and masks when field dressing game.

Disease at a Glance 21.2 summarizes the features of tularemia.

Plague

LEARNING **OUTCOMES**

21.12 Describe the cause, diagnosis, treatment, and prevention of plague.

21.13 Compare and contrast the signs and symptoms of bubonic and pneumonic plague.

Plague (plāg) has caused several great pandemics. One major pandemic that lasted from the mid-500s A.D. to the late 700s is estimated to have claimed the lives of more than 40 million people. This devastation was surpassed during a second pandemic in the 14th century that killed about a third of the population of Europe in just five years. A third major pandemic spread from China across Asia, Africa, Europe, and the Americas in the 1860s. The devastation from plague has been so great that, centuries later, the word provokes a sense of dread.

Signs and Symptoms

There are two main clinical manifestations of plague: The sudden appearance of smooth, enlarged, reddened, and painfully inflamed lymph nodes, called **buboes**[17] **(FIGURE 21.5)**, characterizes the first type and gives it its name—**bubonic plague.** Buboes commonly occur in the groin, armpits, or neck near the site of infection. Signs and symptoms of bubonic plague also include fever, chills, malaise, muscular pain, severe headache,

[17]From Greek *boubon,* meaning "groin," referring to a swelling in the groin.

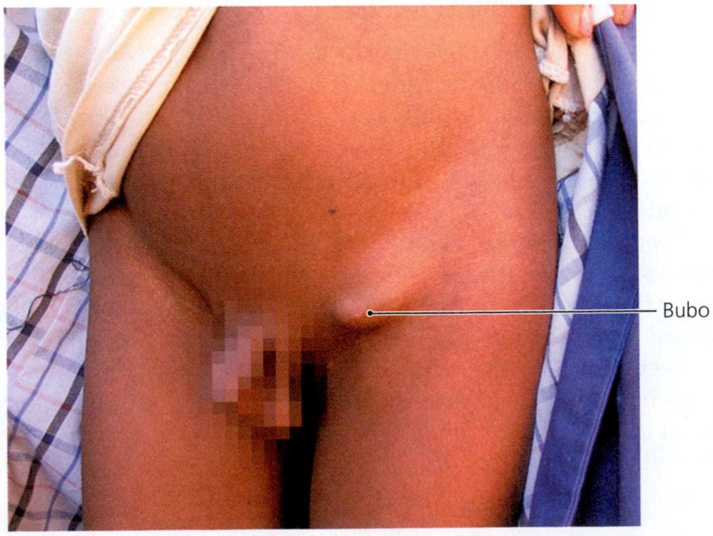

▲ **FIGURE 21.5 Bubo.** Such swollen lymph nodes are a characteristic of bubonic plague.

bacteremia, and septicemia. The latter conditions result in increased heart rate, lowered blood pressure, disseminated intravascular coagulation, subcutaneous hemorrhaging, and death of tissues, which turn black as a result of secondary infection of gangrenous bacteria (discussed in Chapter 19). Because of the extensive darkening of dead skin, plague has been called "Black Death."

The second form of plague—**pneumonic plague**—occurs when plague bacilli in a patient with bubonic plague spread from the bloodstream to the lungs or are inhaled. Pneumonic plague develops very rapidly. Fever, malaise, severe cough, bloody and frothy sputum, and difficulty in breathing can arise within a few hours of infection.

Pathogen and Virulence Factors

Yersinia pestis (yer-sin´ē-ă pes´tis), causative agent, is a Gram-negative, rod-shaped bacterial pathogen of animals. This bacterium carries virulence plasmids that code for *adhesins, type III secretion systems, capsules,* and *antiphagocytic proteins*. Adhesins are molecules that attach pathogens to their target cells, while type III secretion systems inject harmful proteins into the targets. *Yersinia* preferentially injects antiphagocytic proteins into dendritic cells, neutrophils, or macrophages, neutralizing their ability to mount an adaptive immune response or to eliminate the bacterium and thus allowing bacteremia. Endotoxin released from dead *Yersinia* can trigger inflammation, fever, and blood clotting.

Pathogenesis and Epidemiology

Rodents, such as rats, mice, ground squirrels, prairie dogs, and voles, are hosts for the natural endemic cycle of *Y. pestis* **(FIGURE 21.6a)**; they harbor the bacterium but usually do not develop plague. In this cycle, fleas are the vectors for the spread of the bacteria among rodents. As bacteria multiply within a flea, they form a biofilm that blocks the esophagus such that the flea can no longer ingest blood from a host. As the starving flea jumps from host to host vainly seeking to alleviate its hunger with a blood meal, it infects each one with a bite. When other animals—including prairie dogs, rabbits, deer, camels, dogs, and cats—become infected via flea bites, they act as amplifying hosts **(FIGURE 21.6b)**; that is, they support increases in the numbers of bacteria and infected fleas.

Infected fleas that have left their normal animal hosts can spread plague to humans, who can also become infected through direct contact with infected animals or flea feces **(FIGURE 21.6c)**. Buboes develop 2–10 days following a flea bite.

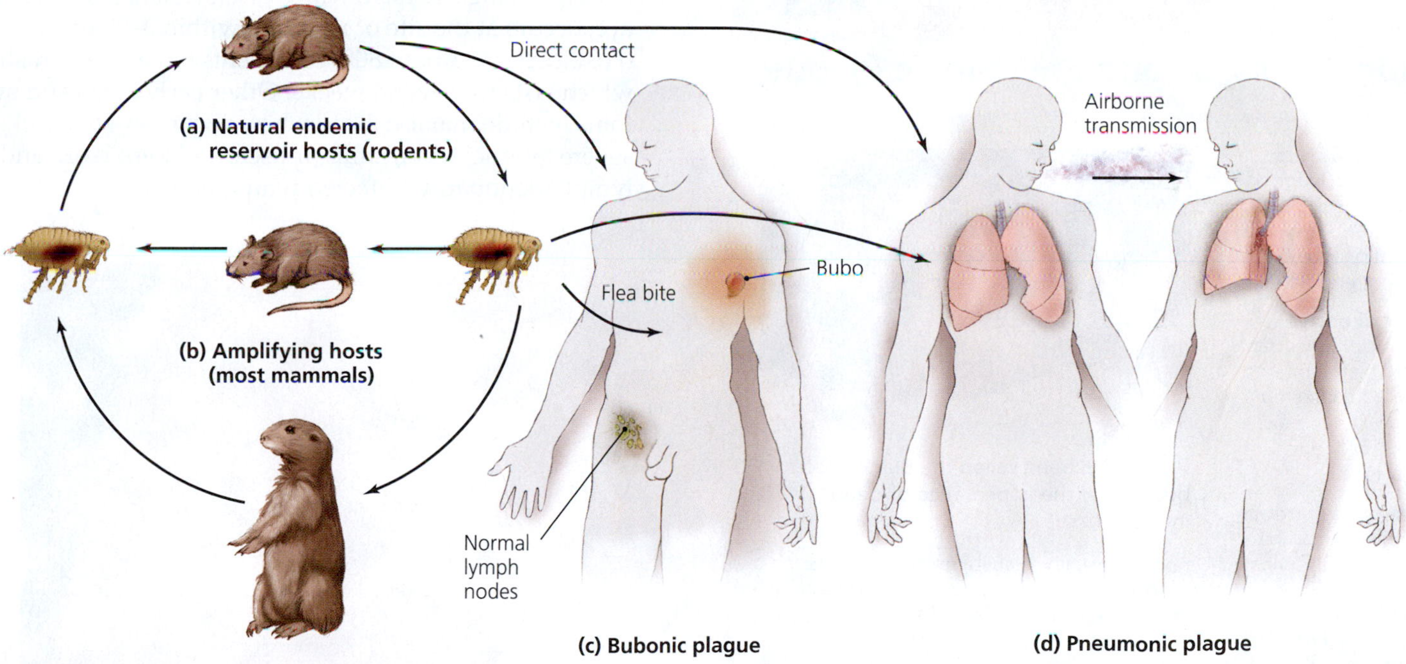

▲ **FIGURE 21.6 The natural history and transmission of *Yersinia pestis*, the bacterium that causes plague. (a)** The natural endemic cycle of *Yersinia* among rodents. **(b)** The cycle involving amplifying mammalian hosts. **(c, d)** Humans develop one of two forms of plague as the result of the bite of infected fleas or direct contact with infected hosts. Bubonic plague involves inflamed lymph nodes, often in neck, armpit, or groin. *Yersinia* can move from the bloodstream into the lungs to cause pneumonic plague, which can be spread between humans via the airborne transmission of the bacteria in aerosols. *Why is plague so much less devastating today than it was in the Middle Ages?*

Figure 21.6 Urban living has reduced contact between people and flea vectors living on wildlife and farm animals; improved hygiene and the use of insecticides have reduced contact with flea vectors living on pets; and antimicrobials are effective against *Y. pestis*.

Untreated bubonic plague is fatal in 50% of cases. Even with treatment, 5% of patients die. In fatal infections, death usually occurs within a week of the onset of symptoms. Bubonic plague is not spread from person to person. Pneumonic plague, by contrast, can spread from person to person through aerosols and sputum **(FIGURE 21.6d)** and if left untreated is fatal in nearly 100% of cases.

Plague is rare in the United States—only three human cases were reported in 2012—but the disease is considered endemic in California, Utah, Arizona, Nevada, and New Mexico. Risk factors include occupational and environmental exposure to rodents and their fleas.

Diagnosis, Treatment, and Prevention

Because plague is deadly, often progresses rapidly, and can spread among people, it is a potential bioterrorism disease. Diagnosis and treatment must be rapid. The characteristic symptoms, especially in patients who have traveled in areas where plague is endemic, are usually sufficient for diagnosis. Although wild animals remain reservoirs, rodent and flea control and better personal hygiene have almost eliminated plague in industrialized countries. Many antibacterial drugs, preferably streptomycin or gentamicin, are effective against *Yersinia*. Researchers are in the final phases of testing a human vaccine for regulatory approval.

Disease at a Glance 21.3 summarizes the features of plague.

DISEASE AT A GLANCE 21.3

Bubonic Plague and Pneumonic Plague

Cause *Yersinia pestis* (Gram-negative coccobacillus).

Virulence factors Plasmids code for adhesins, type III secretion systems, capsule, and antiphagocytic proteins; endotoxin.

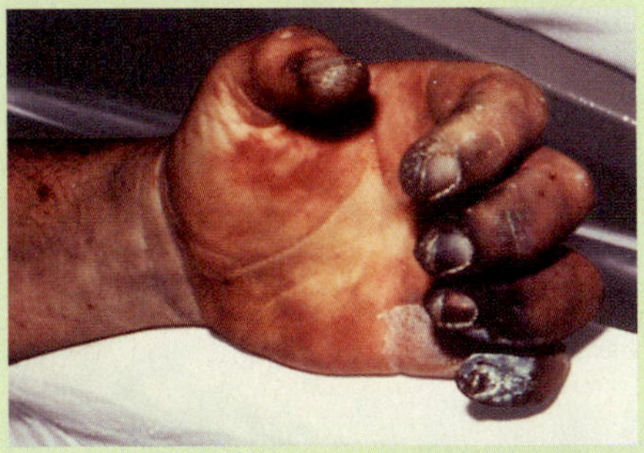

Plague has been called the Black Death because of the appearance of dead infected tissue.

Portal of entry Bubonic plague: transmitted by the bite of a flea from infected rodent. Pneumonic plague; blood to lungs, inhaled.

Signs and symptoms Bubonic plague: buboes (painfully inflamed lymph nodes), fever, chills, malaise, muscle pain, headache, bactermia, septicemia, low blood pressure, disseminated intravascular coagulation, and death of infected tissue, which turns black. Pneumonic plague: fever, malaise, severe cough, bloody sputum, and difficulty in breathing.

Incubation period Bubonic plague: one to seven days. Pneumonic plague: one to four days, though it can be as little as a few hours.

Susceptibility People who enter areas where *Yersinia* is endemic in animal populations, such as prairie dog "towns"; people exposed to others who have pneumonic plague.

Treatment Immediate treatment with antimicrobial drugs such as streptomycin or gentamycin.

Prevention Rodent control; avoidance of endemic areas; possible vaccine.

Lyme Disease

LEARNING | **OUTCOMES**

21.14 Describe the features of the three stages of Lyme disease.

21.15 Discuss the life cycle of the *Ixodes* tick as it relates to Lyme disease.

21.16 Explain how the virulence factors of *Borrelia* allow it to evade the *body's* defenses.

21.17 Elucidate the treatment and prevention of Lyme disease.

In 1975, epidemiologists noted that the incidence of childhood arthritis in Lyme, Connecticut, was over 100 times higher than expected. Upon further investigation, they described a tick-borne zoonosis, **Lyme disease.**

Signs and Symptoms

Lyme disease mimics many other diseases, and its range of signs and symptoms is vast. The disease typically has three phases in untreated patients:

1. An expanding red rash, which often resembles a bull's-eye, occurs at the site of infection within 3–30 days **(FIGURE 21.7)**. About 80% of patients have such a rash, which lasts for several weeks. Other early signs and symptoms include malaise, headaches, dizziness, stiff neck, severe fatigue, fever, chills, muscle and joint pain, and lymphadenopathy (infected lymph nodes).

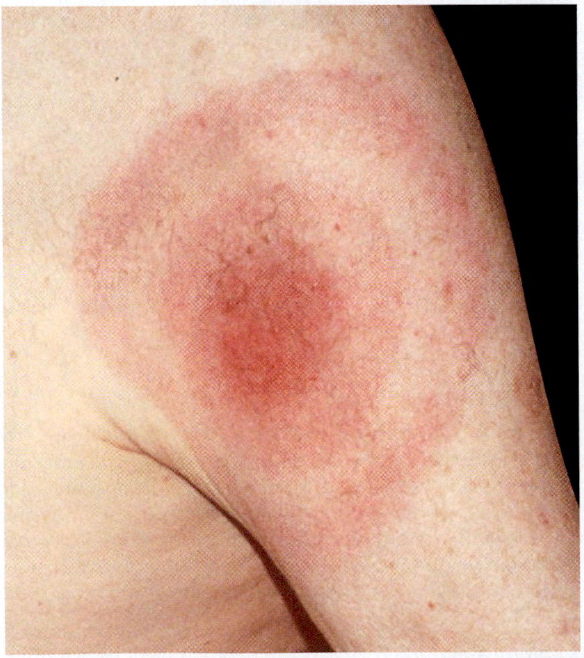

▲ **FIGURE 21.7 Distinctive "bull's-eye" rash of Lyme disease.** Such a rash is often seen in the initial phase of the disease.

CLINICAL CASE STUDY

Nightmare on the Island

Peggy loves her time on The Island each year. Her parents had taken her every year to the resort destination as a girl, and now she was doing the same with her children. Their seaside home near Cape Cod, Massachusetts, was modest in comparison to some of the neighbors', but it had been in Peggy's family for over a hundred years, and she had fond remembrances from every stage of her life. One of Peggy's fondest memories is playing tag with her friends on the lawns of The Island's homes. Now, she smiles as she watches her eight-year-old son, Jacob, help the older son of one of her childhood girlfriends mow the grassy expanse. "Building memories—that's what it's about," she thought. Little does she know that some memories can build nightmares.

Three days later Jacob wakes up complaining of a scratchy throat, headache, and "soreness all over." Peggy is concerned about his dry cough and 103°F temperature. "A summer cold?" She keeps Jacob in bed, which isn't difficult because his breathing becomes more labored and painful. Two days later he begins coughing up blood, and Peggy recognizes that this isn't an ordinary summertime cold.

She rushes Jacob to the local clinic, where the doctor orders immediate intravenous streptomycin and transport to a hospital on the mainland. The physician tells Peggy that Jacob is likely infected with the most virulent bacterium known. He questions her about Jacob's activities on the island: Has the boy touched any animals? Done any outdoor activities? Been bitten by a tick? "No, no, no." Then she recalls that Jacob helped mow the grass earlier in the week.

Within days, Jacob feels better and can answer questions. He tells the doctor that the lawnmower had run over the dried body of a small dead rabbit. The physician suspects the mower had spewed bacteria into the air; Jacob had inhaled a near-fatal dose.

The grassy lawn will no longer recall the fond memories of Peggy's childhood; instead, she will remember men in biohazard suits taking samples, documenting the nightmarish time she almost lost her son.

1. **What bacterium infected Jacob?**
2. **What is a common name of the disease afflicting Jacob?**
3. **What do the laboratory scientists at the hospital determine about the Gram reaction of the bacterium?**
4. **Why didn't the physician use penicillin instead of streptomycin?**

2. Neurological symptoms (for example, meningitis, encephalitis, and peripheral nerve neuropathy[18]) and cardiac dysfunction typify the second phase, which is seen in only 10% of patients.
3. The final phase, seen in almost 80% of patients, is characterized by severe arthritis that can last for years. The pathological conditions of the latter phases of Lyme disease are due in large part to the body's immunological response; rarely is *Borrelia* seen in the involved tissue or isolated in cultures of specimens from sites affected during this last phase. Lyme disease is rarely, if ever, fatal.

Pathogen and Virulence Factors

A large (0.5 μm × 3–20 μm) spirochete—*Borrelia burgdorferi*[19] (bō-rē´lē-ă burg-dōr´fer-ē) **(FIGURE 21.8)**—causes Lyme disease. *B. burgdorferi* has an unusual metabolism in that it does not use iron in its enzymes or its electron transport chains. It uses manganese in place of iron, thereby circumventing one of the body's natural defense mechanisms—the lack of free iron in human tissues and fluids.

The bacterium can change its outer membrane proteins via genetic rearrangement to emerge as antigenically different variants, making it more difficult for immune cells to recognize the pathogen and clear it from the blood. When it dies, *Borrelia* releases an active endotoxin (lipid A) from its outer membrane.

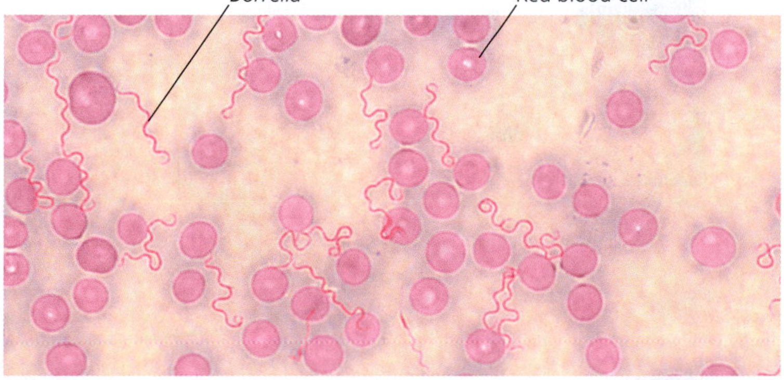

▲ **FIGURE 21.8** ***Borrelia burgdorferi.*** This Gram-negative spirochete causes Lyme disease.

[18]From Greek *neuron*, meaning "nerve," and *pathos*, meaning "suffering."
[19]Named for the French bacteriologist Amedee Borrel and a specialist in tick-borne diseases, Willy Burgdorfer.

Pathogenesis

Hard ticks of the deer tick genus *Ixodes*[20] (ik-sō´dēz) are the vectors of Lyme disease. An understanding of the life cycle of the tick is essential to an understanding of the disease. A deer tick lives for two years, during which it passes through three stages of development: a six-legged larva, an eight-legged nymph, and an eight-legged adult. Once during each stage, it attaches to a vertebrate host for a single blood meal. After each of its three feedings, the tick drops off its host and lives in leaf litter or on brush.

The different stages of *Ixodes* typically feed on different hosts, though all stages of all species of *Ixodes* may feed on humans. Transmission of *Borrelia* from a female tick to her offspring is rare, so larvae that hatch in the spring are usually uninfected 1, becoming infected during their first blood meal **(FIGURE 21.9)** 2. Over the winter, larvae digest their blood meals while *Borrelia* replicates in the ticks' guts 3.

In the spring of their second year, the ticks molt into nymphs 4 and feed a second time, infecting new hosts with *Borrelia* via saliva 5. (Uninfected nymphs can become infected if they feed on an infected host.) Laboratory studies have shown that infected ticks must remain on a host for 36–48 hours in order to transmit enough spirochetes to establish a *Borrelia* infection in that host.

Nymphs drop off their hosts and undergo further development into adults 6. In the fall, adult ticks feed a final time 7, mate, lay eggs 8, and die. Adults infected with *Borrelia* infect their hosts as they feed. Adult ticks are much larger than nymphs are, so humans usually see and remove adults before they can transmit *Borrelia;* thus, nymphs most often infect humans.

Introduced by a tick's bite, Lyme disease spirochetes move from the site of infection through the blood and lymph. They can accumulate in joints and, along with antibodies directed against them, trigger arthritis.

Epidemiology

Even though its discovery was relatively recent, retrospective epidemiological studies have shown that Lyme disease was present in the United States for decades before its discovery. The disease is one of the more reported vector-borne diseases in the United States. Three major events have contributed to an increase in cases of Lyme disease in the United States over the past 30 years **(FIGURE 21.10a)**: the human population has encroached on woodland areas, the deer population has been protected and feeds in suburban yards, and foxes that would

[20]Greek, meaning "sticky."

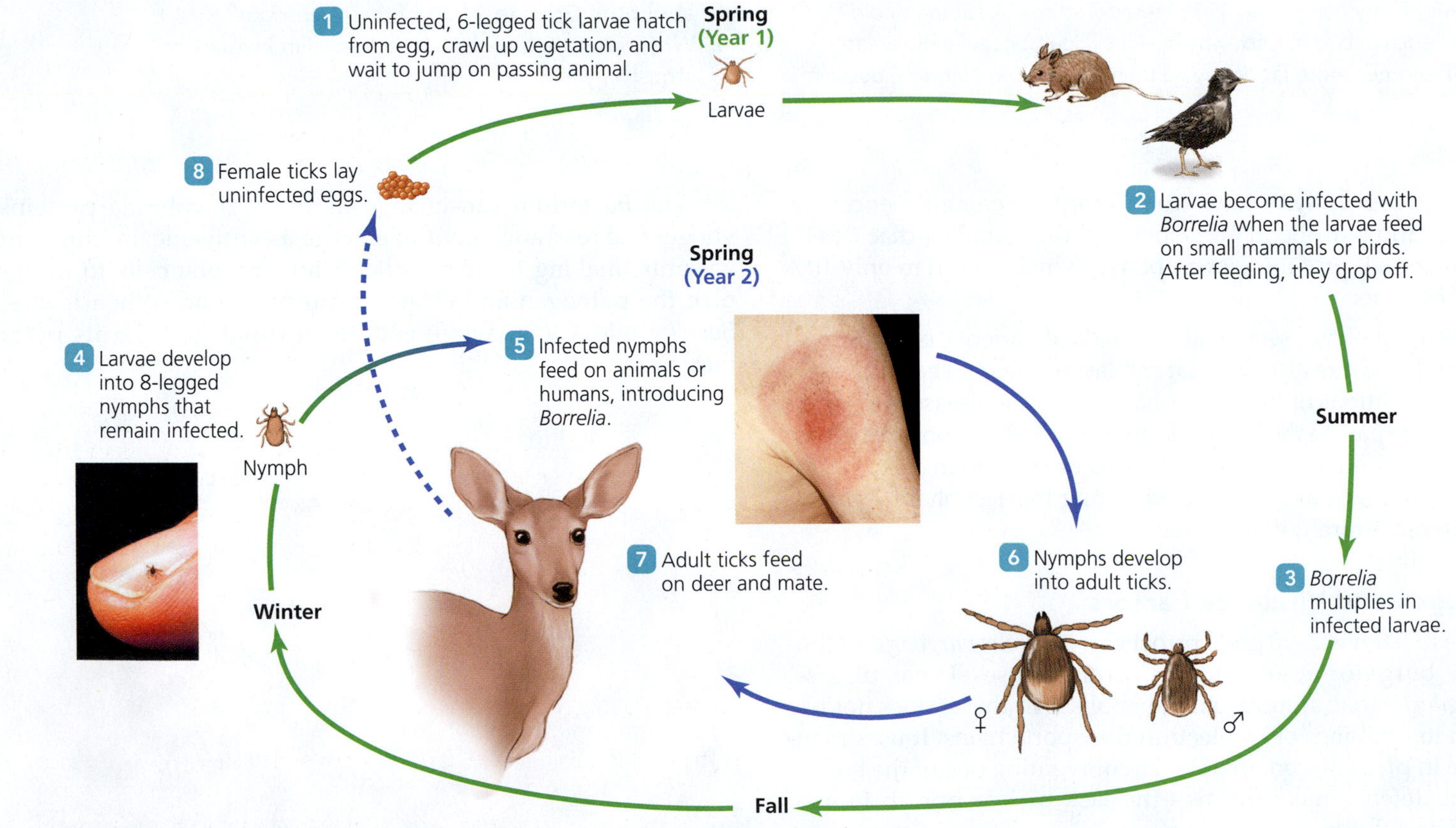
1 Uninfected, 6-legged tick larvae hatch from egg, crawl up vegetation, and wait to jump on passing animal.
Spring (Year 1)
Larvae
2 Larvae become infected with *Borrelia* when the larvae feed on small mammals or birds. After feeding, they drop off.
Summer
3 *Borrelia* multiplies in infected larvae.
Fall
6 Nymphs develop into adult ticks.
♀
♂
7 Adult ticks feed on deer and mate.
Winter
4 Larvae develop into 8-legged nymphs that remain infected.
Nymph
Spring (Year 2)
5 Infected nymphs feed on animals or humans, introducing *Borrelia*.
8 Female ticks lay uninfected eggs.

▲ **FIGURE 21.9 The life cycle of the deer tick *Ixodes* and its role as the vector of Lyme disease.** *Besides the nymph stage, what other stage in the tick's life cycle can infect a human with B. burgdorferi?*

Figure 21.9 Adult ticks infected with B. burgdorferi can also infect humans during a blood meal.

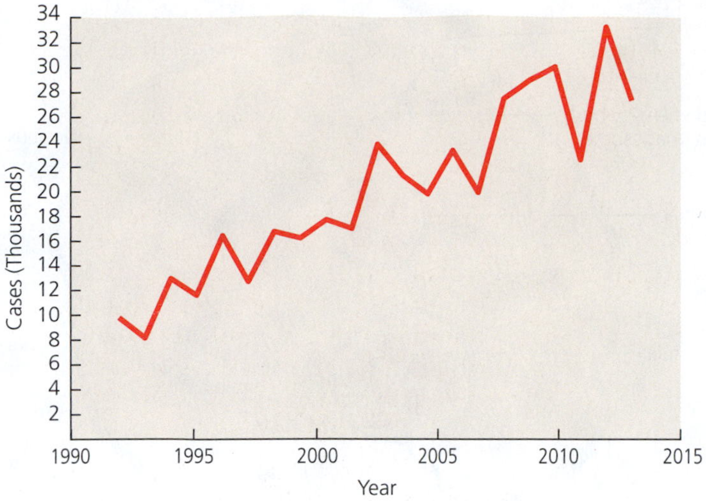

(a)

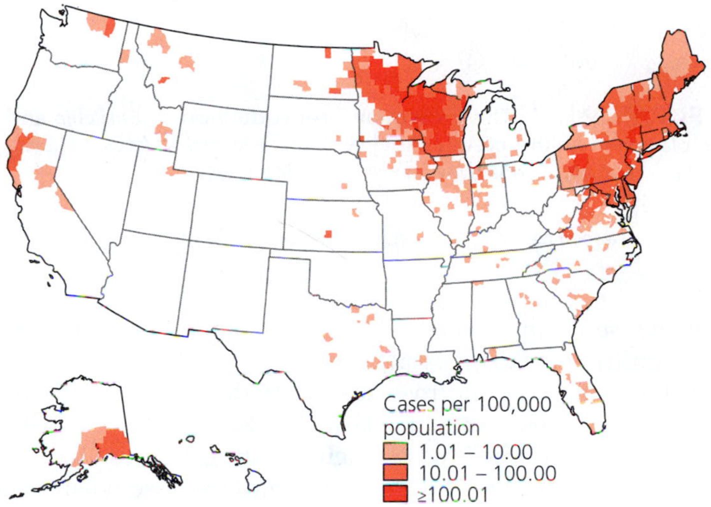

(b)

▲ **FIGURE 21.10 The occurrence of Lyme disease in the United States. (a)** Incidence of cases, 1982–2012. **(b)** The geographic distribution of the disease, 2010.

normally reduce the mouse population have been displaced by coyotes, which rarely feed on mice. Thus, humans have been brought into closer association with ticks infected with *Borrelia* after feeding on mice and deer. Lyme disease occurs in most states, but cases are concentrated in the Northeast and the upper Midwest **(FIGURE 21.10b)**.

Diagnosis, Treatment, and Prevention

A diagnosis of Lyme disease, typically based on observations of its usual signs and symptoms, is rarely confirmed by detecting *Borrelia* itself in blood. Instead, the diagnosis is confirmed with ELISA and western blot tests to demonstrate the presence of antibodies against *Borrelia*.

Treatment with antimicrobial drugs such as doxycycline, cefuroxime, or amoxicillin effectively cures most cases of Lyme disease in the first phase, though prolonged treatment with large doses may be required. Treatment of later phases is more difficult because later symptoms frequently result from immune responses rather than the presence of the spirochetes. Anti-inflammatory drugs act against the arthritic symptoms.

People hiking, picnicking, and working outdoors in areas where Lyme disease is prevalent should take precautions to reduce the chances of infection, particularly during May to July, when nymphs are feeding. People who must be in the woods should wear long-sleeved shirts and long, tight-fitting pants, and should tuck the cuffs of their pants into their socks to deny ticks access to skin. Repellents containing DEET, which is noxious to ticks, should be used. As soon as possible after leaving a tick-infested area, people should thoroughly examine their bodies for ticks or their bites. Attached ticks should be removed using forceps.

Though researchers developed a vaccine against *B. burgdorferi,* it was not widely used because of the ready availability of other preventive measures and because the vaccine produced the symptoms of Lyme disease in some patients. The manufacturer stopped supplying the vaccine in March 2002.

Disease at a Glance 21.4 summarizes the features of Lyme disease.

DISEASE AT A GLANCE 21.4

Lyme Disease

Cause *Borrelia burgdorferi* (Gram-negative spirochete).

Virulence factors Uses magnesium in place of iron; changes outer membrane antigens frequently; endotoxin.

Portal of entry Skin via bite of infected tick.

Signs and symptoms "Bull's-eye" rash, malaise, headache, dizziness, stiff neck, severe fatigue, fever, chills, muscle and joint pain, and lymphadenopathy.

Incubation period 7 to 14 days.

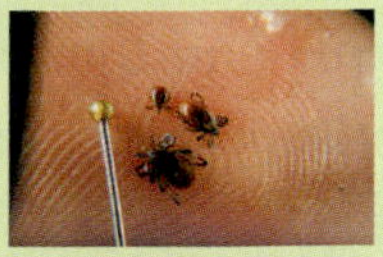

Three stages in the life of the deer tick, *Ixodes*, life size

Susceptibility People in highly endemic areas are particularly susceptible due to increased exposure.

Treatment Doxycycline or amoxicillin antibiotic treatment for three to four weeks if the disease is caught early. Later disease may require intravenous treatment, depending on severity.

Prevention Wear light-colored, tight-fitting clothes outdoors to limit tick exposure, use tick repellent, promptly remove ticks, examine skin for ticks and bites, and avoid tick-infested areas. The Lyme disease vaccine for humans is no longer commercially available.

Ehrlichiosis and Anaplasmosis

LEARNING | **OUTCOMES**

21.18 Describe the manifestations of ehrlichiosis and anaplasmosis.

21.19 Identify the three developmental stages of *Ehrlichia* and *Anaplasma*.

21.20 Discuss difficulties in diagnosing and treating ehrlichiosis and anaplasmosis.

Two emerging, tick-borne human diseases—*human monocytic ehrlichiosis (HME)* and *human granulocytic anaplasmosis (HGA)*—known simply as **ehrlichiosis** and **anaplasmosis,** respectively, were unknown in humans before 1987, though they were known in animals.

Signs and Symptoms

Ehrlichiosis and anaplasmosis manifest similarly, with flulike signs and symptoms of fever, chills, nausea, muscle aches, and headache. *Leukopenia* (a decrease in the number of white blood cells in the blood) and *thrombocytopenia* (a decrease in the number of platelets) also occur. Diarrhea, malaise, or a fine pinhead rash may be present. The symptoms are usually quite general and minimal but may be life threatening depending on the patient's age and general health.

Pathogens and Virulence Factors

Ehrlichia chaffeensis (er-lik´ē-ă chaf-ē-en´sis) causes ehrlichiosis. A related species, *Anaplasma phagocytophilum* (an-ă-plaz´mă fag-ō-sī-tō´fil-ŭm, formerly known as *Ehrlichia equi*), causes anaplasmosis. Both bacteria are rickettsias, which are Gram-negative, highly pleomorphic, obligate intracellular parasites of eukaryotic cells. The bacteria trigger their own endocytosis (internalization by a cell) by white blood cells—monocytes in the case of *Ehrlichia* and neutrophils in the case of *Anaplasma*—and then live inside phagosomes. These bacteria prevent fusion of lysosomes in the phagocyte with their phagosome homes by an unknown mechanism.

Pathogenesis

Hard-shelled ticks, including the Lone star tick (*Ambylomma,* am-bē-lō´ma), the deer tick *(Ixodes),* and the dog tick (*Dermacentor,* der-mă-sen´ter), vector the bacterial agents of ehrlichiosis and anaplasmosis into humans. Once inside blood cells, the bacteria grow and reproduce within phagosomes passing through three developmental stages: an *elementary body,* an *initial body,* and a *morula* **(FIGURE 21.11)**. Release of the bacteria from a morula and into the blood makes them available to feeding ticks, which become infected and pass the bacteria on to a new host.

Epidemiology

Epidemiologists consider ehrlichiosis and anaplasmosis to be emerging diseases because they were unknown before 1987, and the number of reported cases has increased from a few dozen per year in the 1980s to over 1700 per year now. This increase is likely due in part to improved diagnosis and reporting.

Diagnosis, Treatment, and Prevention

Diagnosis of ehrlichiosis and anaplasmosis is difficult because their manifestations may be mild and they resemble other diseases. Physicians consider ehrlichiosis or anaplasmosis in any case of otherwise unexplained acute fever in patients exposed to ticks in endemic areas **(FIGURE 21.12)**. Immunofluorescent antibodies against the bacteria inside white blood cells or polymerase chain reaction (PCR) tests can demonstrate infection, confirming a diagnosis.

Doxycycline is effective against both *Ehrlichia* and *Anaplasma.* Treatment should start immediately, even before diagnosis is confirmed by serological testing, because the complications of infection and mortality rates increase when treatment is delayed.

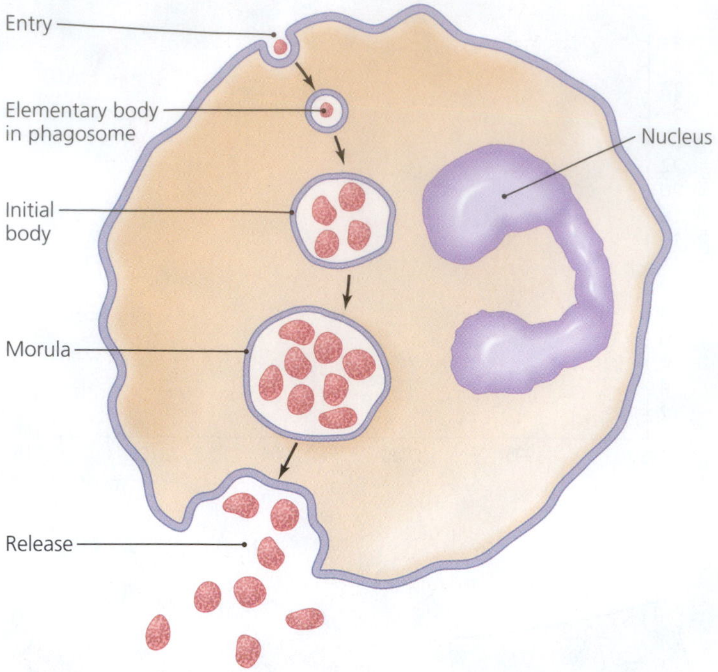

▲ **FIGURE 21.11 The growth and reproduction of *Ehrlichia* and *Anaplasma* in an infected leukocyte.** *If the infected leukocyte shown here is a monocyte, which disease will manifest?*

Figure 21.11 *Ehrlichia chaffeensis* infects monocytes to cause ehrlichiosis.

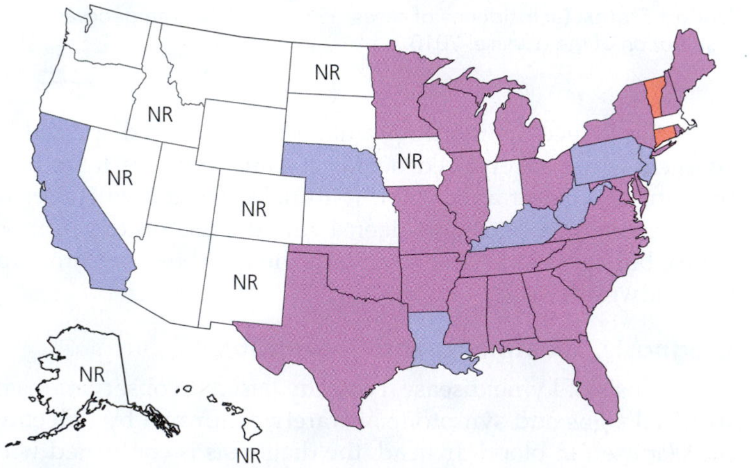

▲ **FIGURE 21.12 The geographical distribution of ehrlichiosis and anaplasmosis in the contiguous United States (2010).** Blue indicates where ehrlichiosis (HME) was reported, orange indicates where anaplasmosis (HGA) was reported, and purple indicates where both diseases were reported.

Prevention involves avoiding tick-infested areas, promptly removing ticks, wearing tight-fitting clothes, and using insect repellents. No vaccines against the two species are available.

We have examined bacterial diseases of the cardiovascular system and bacterial systemic diseases. In the following section, we turn our attention to viral cardiovascular and systemic diseases.

TELL ME WHY

Health departments often recommend that outdoor enthusiasts wear light-colored pants while hiking in areas where *Ixodes* ticks are endemic. Why?

Viral Cardiovascular and Systemic Diseases

Many viral diseases spread through the body via the lymphatic and cardiovascular systems. When viruses infect the blood, the condition is called **viremia** (vī-rē´mē-ā). Here we examine mononucleosis, cytomegalovirus disease, yellow fever, and hemorrhagic fevers.

Infectious Mononucleosis

LEARNING OUTCOMES

21.21 Describe the cause, signs, symptoms, and diagnosis of mononucleosis.

21.22 Explain the role of the immune system in provoking diseases of Epstein-Barr virus.

Infectious mononucleosis, commonly known as *kissing disease* or *mono,* is a condition resulting from interactions of a patient's cellular immune system with infected B lymphocytes.[21]

Signs and Symptoms

Severe sore throat and fever characterize the beginning of infectious mononucleosis. These manifestations are followed by enlarged lymph nodes (especially in the neck), splenomegaly (enlarged spleen), extreme fatigue, nausea, loss of appetite, headache, and a skin rash anywhere on the body. Fatigue and inability to concentrate may last for months.

Pathogen and Virulence Factors

Human herpesvirus 4 (HHV-4), which is also known as Epstein-Barr virus (EBV) after its discoverers, causes infectious mononucleosis. This enveloped, double-stranded DNA virus with an icosahedral capsid is replicated in a host cell's nucleus. It can become latent within cells, resulting in lifelong infection.

[21]Lymphocytes were formerly known as mononuclear leukocytes, from which the name mononucleosis is derived.

CLINICAL CASE STUDY

A Sick Camper

An otherwise healthy 24-year-old woman goes to her doctor complaining of a sudden onset of high fever, chills, uneasiness, and a severe headache. She also shows the doctor a painful swelling she is experiencing in her groin area. The doctor asks her about recent travel. She reports that she returned two days prior from a week-long camping and hiking trip in Texas.

1. How did the woman most likely contract the disease?
2. What are the potential problems associated with diagnosing this disease, and how crucial is prompt diagnosis of this disease?
3. The doctor asks you, as a nursing student rotating through his clinic, your opinion on the disease diagnosis and causative agent. What is your response?
4. How should the patient be treated?
5. Who should be notified once the diagnosis is confirmed? Why?

HHV-4 suppresses **apoptosis** (programmed cell death) of B lymphocytes, causing infected cells to become immortal. Such infected B cells are one source of cancers. Additional factors appear to play a role in the development of such cancers. For example, **Burkitt's lymphoma,** a cancer of the jaw, is almost exclusively limited to young African males exposed previously to malaria parasites.

Although conclusive proof is lacking, the presence of EBV and antibodies against EBV have implicated this herpesvirus as an etiological agent of *chronic fatigue syndrome, B cell lymphomas,* and *oral hairy leukoplakia.* The latter arises in individuals with a T cell deficiency, as occurs in malnourished children, the elderly, AIDS patients, and transplant recipients; it is a precancerous change in the tongue.

Pathogenesis and Epidemiology

Transmission of Epstein-Barr virus usually occurs via saliva, often during the sharing of drinking glasses, while kissing, or from a cough or sneeze. After initially infecting epithelial cells of the throat and salivary glands, EBV enters the blood, where it invades B lymphocytes.

Infectious mononucleosis results from a "civil war" in which cytotoxic T lymphocytes kill infected B lymphocytes. This "war" is responsible for the symptoms and signs of mono.

Ninety-five percent of Americans over the age of 30 have antibodies against EBV, most often acquired during the teen years. The age of the patient at the time of infection is a

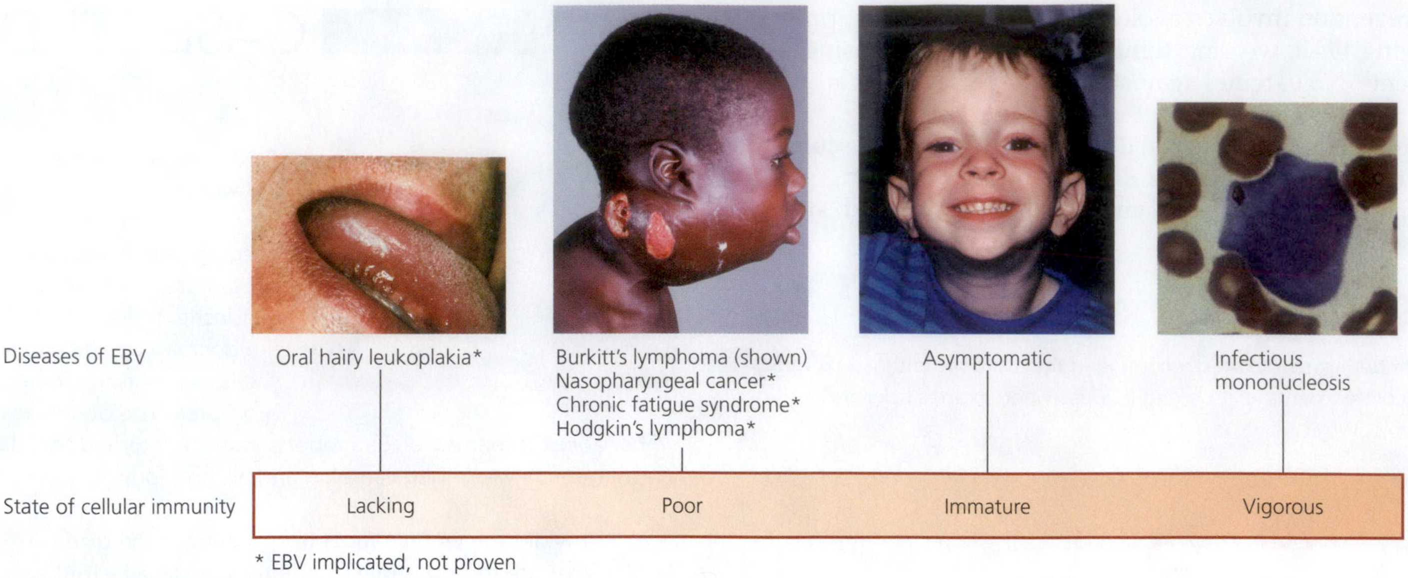

▲ **FIGURE 21.13 Diseases associated with Epstein-Barr virus.** Which disease results from infection with EBV appears to depend on the relative vigor of a host's cellular immune response, which itself can be related to age. Oral hairy leukoplakia, white lesions in the mucous membrane of the tongue, occurs in EBV-infected hosts with severely depressed cellular immunity (for example, AIDS patients). Burkitt's lymphoma occurs primarily in African boys whose immune systems have been suppressed by the malaria parasite. EBV is also implicated in nasopharyngeal cancer and chronic fatigue syndrome, which are associated with impaired immune function. Asymptomatic EBV infections are typical for children exposed at a young age, before the immune response has become vigorous. Infectious mononucleosis, which is characterized by enlarged B lymphocytes with lobed nuclei, results from infection with EBV during adulthood, when the cellular immune response is vigorous.

determining factor in the seriousness of the disease, because the competence of cytotoxic T cells is in part related to age **(FIGURE 21.13)**. Infection during childhood, which is more likely to occur in countries with poor sanitation and inadequate standards of hygiene, is usually asymptomatic because a child's cellular immune system is immature and cannot cause severe tissue damage. Where living standards are higher, childhood infection is less likely, and the postponement of infection until adolescence or later results in a more vigorous cellular immune response that produces the signs and symptoms of mononucleosis in 50% of patients.

Diagnosis, Treatment, and Prevention

Large, lobed B lymphocytes with atypical nuclei and a deficiency of neutrophils (neutropenia) are characteristic of EBV infection. Some diseases associated with EBV, such as hairy leukoplakia and Burkitt's lymphoma, are easily diagnosed by their characteristic signs. Other EBV infections, such as infectious mononucleosis, have symptoms common to many pathogens. Fluorescent antibodies directed against anti-EBV immunoglobulin or ELISA tests provide specific diagnosis.

Care of mono patients involves relief of the symptoms; most patients recover without treatment within two to four weeks, though infection is permanent when EBV becomes latent. Patients should avoid contact sports to reduce the risk of rupturing an enlarged spleen. Burkitt's lymphoma responds well to chemotherapy, and tumors can be removed from affected jaws if surgery is available. There is no effective treatment for other EBV-induced conditions.

Prevention of EBV infection is almost impossible because the viruses are widespread and transmitted readily in saliva. However, only a small proportion of EBV infections result in disease because the immune system is either too immature to cause cellular damage (in children) or because cellular immunity is efficient enough to kill infected B lymphocytes.

Cytomegalovirus Disease

LEARNING | **OUTCOMES**

21.23 Describe the signs, symptoms, and cause of *Cytomegalovirus* disease.

21.24 Discuss treatment and prevention of CMV disease.

Another herpesvirus that affects humans is *Cytomegalovirus* (CMV), so named because cells infected with this virus become enlarged.

Signs and Symptoms

Cytomegalovirus (CMV) disease may result from initial infections or from latent viruses. Most people infected with CMV are asymptomatic, but fetuses, newborns, and immunodeficient patients are susceptible to severe complications of CMV infection. About 10% of congenitally infected newborns develop signs of infection, including enlarged liver and spleen, jaundice, and anemia. CMV may also be *teratogenic*[22] (ter´ă-tō-jen-ik)—that is, cause birth defects—when the virus infects stem cells in an embryo or fetus. In the worst cases, mental retardation, hearing and visual damage, or death may result.

AIDS patients and other immunosuppressed adults, such as transplant recipients, may develop pneumonia, blindness (if the virus targets the retina), or cytomegalovirus mononucleosis, which has signs and symptoms similar to those of infectious mononucleosis caused by Epstein-Barr virus.

Pathogen and Virulence Factors

Cytomegalovirus is an enveloped, double-stranded DNA virus with an icosahedral capsid. Like other herpesviruses, CMV typically infects humans early in their lives, but remains in a latent state until the immune system is compromised. The major exception is prenatal infection across the placenta from mother to child.

Pathogenesis and Epidemiology

Bodily secretions, including saliva, mucus, milk, urine, feces, semen, and cervical secretions, carry CMV. Individual viruses are not highly contagious, so transmission requires intimate contact involving a large exchange of secretion. Transmission usually occurs via sexual intercourse, but can result from *in utero* exposure, vaginal birth, blood transfusions, and organ transplants. CMV infects 7.5% of all neonates, making it the most prevalent viral infection in this age group.

Cytomegalovirus infection is one of the more common infections of humans. Studies have shown that CMV infects about 50% of the adult population of the United States; in some other countries, 100% of the population tests positive for antibodies against CMV. Like other herpesviruses, CMV becomes latent in various cells, and infection by CMV typically lasts for life.

Diagnosis, Treatment, and Prevention

Diagnosis of CMV-induced diseases is dependent on laboratory procedures that reveal the presence of abnormally enlarged cells and inclusions within the nuclei of infected cells **(FIGURE 21.14)**. Viruses and antibodies against them can be detected by ELISA tests and DNA probes. Treatment for fetuses and newborns with complications of CMV infection is difficult because in most cases the damage is done before the infection is discovered.

Treatment of adults is also frustrating. Interferon, anti-CMV gammaglobulin, and nucleotide analogs, such as ganciclovir, slow the release of CMV from adults but do not affect the course of disease. Eye doctors inject *fomivirsen* into the eye to inhibit the replication of CMV in retinal cells. Fomivirsen, which is the first antisense RNA drug, is RNA complementary to CMV mRNA. It binds to CMV mRNA and stops translation of two proteins critical for CMV replication. This stops the replication and spread of CMV but does not cure the disease.

[22]From Greek *teratos*, meaning "monster."

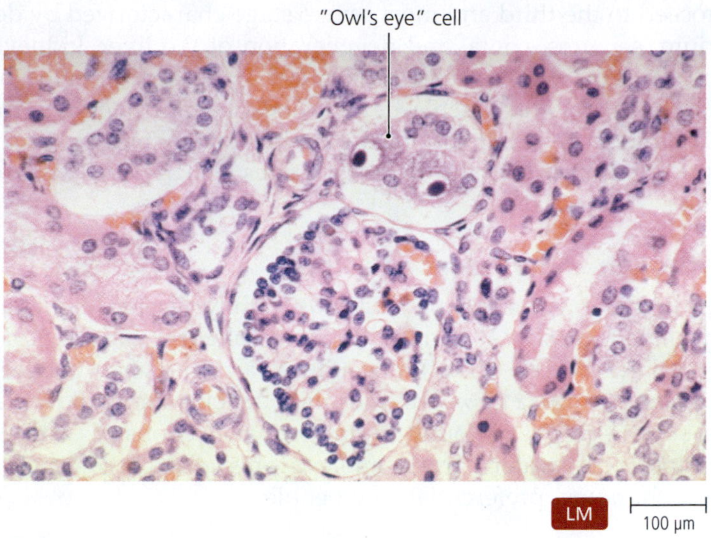

▲ **FIGURE 21.14 An abnormally enlarged "owl's eye" cell indicates *Cytomegalovirus* (CMV) infection.** The cell's enlarged nucleus contains sites of viral assembly called inclusion bodies, which are diagnostic for CMV infection.

Abstinence, mutual monogamy, and use of condoms reduce the chance of infection by CMV. Organs harvested for transplantation can be made safe by treatment with monoclonal antibodies against CMV, which passively reduces the CMV load before the organs are transplanted. There is no vaccine against CMV.

Yellow Fever

LEARNING | **OUTCOME**

21.25 Describe the signs, symptoms, cause, diagnosis, treatment, and prevention of yellow fever.

No disease has influenced the history of the United States as much as **yellow fever.** Slave ships introduced yellow fever and its mosquito vector in the 1600s to the Americas. In 1793, yellow fever killed over 4000 people in Philadelphia—10% of the population of what was then the capital of the United States. Eight years later, an epidemic on the French island colony of Haiti killed 27,000 French troops, discouraging Napoleon and causing him to forfeit goals of a French empire in the West. Instead, he sold the Louisiana Territory to the young United States. More than a century later, yellow fever and malaria further discouraged the French, opening an opportunity for the United States to build the Panama Canal. During the Spanish-American War of 1898, yellow fever killed more American soldiers than did enemy bullets.

Signs and Symptoms

Yellow fever develops in three stages. The early stage involves a slight fever, headache, muscle aches, and vomiting for a few days. This is followed by a period of remission in which the signs and symptoms resolve. Fifteen percent of patients proceed to the third and most severe stage characterized by delirium, seizures, coma, and degeneration of the liver, kidneys, and heart, as well as massive hemorrhaging accompanied by high fever, nausea, nosebleed, and shock. Hemorrhaging in the intestines may result in "black vomit." Liver damage causes jaundice, from which the disease acquires its name and its nickname, "Yellow Jack."

Pathogen and Virulence Factors

Yellow fever virus (genus *Flavivirus*, family *Flaviviridae*) is an enveloped, +ssRNA virus with an icosahedral capsid. *Aedes aegypti*[23] (ā-ē´dēz ē-jip-te´) mosquitoes carry the virus between humans **(FIGURE 21.15)**. This mosquito develops in shallow water such as might be found in domestic collections of water in outdoor pots, water barrels, and old tires. *Aedes* feeds during daylight hours, preferentially on people.

Pathogenesis and Epidemiology

After a bite from an infected *Aedes* mosquito introduces yellow fever virus into the body, the virus travels to the liver where it is replicated rapidly. Signs and symptoms develop three to six days following the mosquito bite. Mortality from severe yellow fever is 20%.

With mosquito control and the development of a vaccine in the 1900s, health care workers eliminated yellow fever from North America, Central America, and much of South America; however, the World Health Organization (WHO) estimates about 200,000 cases still occur annually in South America and Africa combined. Jungle monkeys act as a reservoir for yellow fever virus, making eradication difficult. Laws promulgated in the 1970s to protect the environment from insecticide abuse have allowed *Ae. aegypti* to reestablish itself in the southeastern United States, though yellow fever virus had not been reintroduced into the United States as of 2013.

[23]Greek meaning "unpleasant [thing] from Egypt."

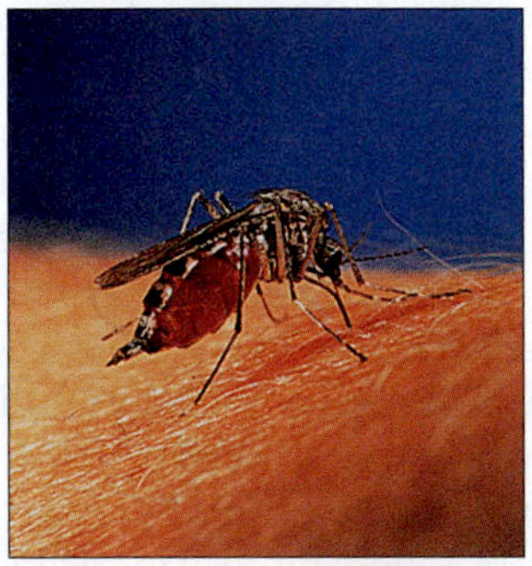

▲ **FIGURE 21.15** ***Aedes aegypti*, the vector of yellow fever and dengue.** *Would you expect more cases of yellow fever and dengue in summer or winter months?*

Figure 21.15 *Mosquitoes, which vector the viruses of yellow fever and dengue, are more active in the summer than winter; therefore, these diseases are more prevalent in summer.*

Diagnosis, Treatment, and Prevention

Though the signs and symptoms of severe yellow fever are obvious, definite diagnosis follows demonstration of viral antigens in the blood via ELISA or viral nucleic acids via PCR. Treatment involves providing supportive care and varies depending upon the systems affected.

People should avoid mosquito bites in geographic areas where yellow fever is endemic. Precautions include use of DEET, wearing protective clothing, and using mosquito netting. The effective, live, attenuated yellow fever vaccine provides protection for 10 years or more. **Disease at a Glance 21.5** on p. 658 summarizes the features of yellow fever.

Dengue Fever and Dengue Hemorrhagic Fever

LEARNING | **OUTCOMES**

21.26 Describe the signs, symptoms, and cause of dengue fever.

21.27 Explain the role of adaptive immunity in causing dengue hemorrhagic fever.

Four other viruses in the genus *Flavivirus* are pathogenic to humans. These antigenically distinct, though related, viruses are known as dengue viruses 1, 2, 3, and 4 and cause diseases that all bear the name **dengue** (den´gā) **fever.**

Signs and Symptoms

Dengue fever usually occurs in two phases separated by 24 hours of remission. First, a patient suffers from fever, weakness, edema (swelling) of the extremities, and severe pain in the head and muscles. The common name for the disease, *breakbone fever,* indicates the severity of this pain. The second phase involves a return of the fever and a bright red rash. Dengue fever is self-limiting and lasts six or seven days. **Dengue hemorrhagic fever (DHF)** is a serious hyperimmune response following reinfection with the dengue virus **(FIGURE 21.16)** in which activated memory T cells release inflammatory lymphokines that trigger rupture of blood vessels, internal bleeding, shock, and possibly death.

Pathogens and Virulence Factors

Dengue viruses are enveloped, +ssRNA viruses with icosahedral capsids about 40–50 nm in diameter. *Aedes* mosquitoes are the vectors for all four strains of dengue viruses. Recall from the previous section that *Aedes* develops in water collected in domestic containers and trash and preferentially feeds on humans.

Pathogenesis and Epidemiology

Usually dengue fever is a mild disease despite the severe pain that can be involved. Lifelong immunity follows recovery; however, since there are four strains of dengue virus, a patient can have dengue fever four times—one case with each serotype. Reinfection with a previously encountered strain does not produce dengue fever, but it can trigger fatal dengue hemorrhagic fever as a result of a hyperimmune response.

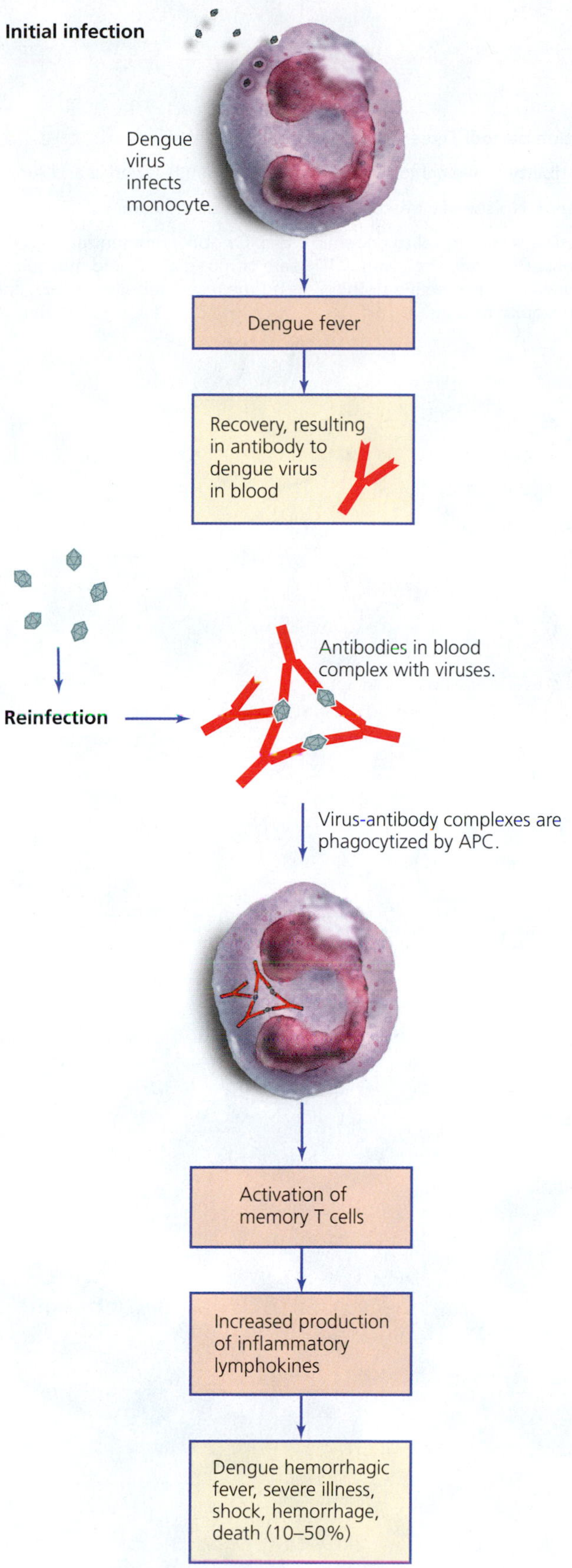
Initial infection
Dengue virus infects monocyte.
Dengue fever
Recovery, resulting in antibody to dengue virus in blood
Reinfection
Antibodies in blood complex with viruses.
Virus-antibody complexes are phagocytized by APC.
Activation of memory T cells
Increased production of inflammatory lymphokines
Dengue hemorrhagic fever, severe illness, shock, hemorrhage, death (10–50%)

FIGURE 21.16 Pathogenesis of dengue hemorrhagic fever (DHF). DHF is a severe hyperimmune response to a second infection with dengue virus. Antibodies from the first infection form complexes with the reintroduced virus. Antigen-presenting cells (APCs) phagocytize the complexes and then activate T memory cells. These release an abundance of lymphokines that induce hemorrhaging and shock.

Health care workers dramatically reduced the number of cases of dengue virus diseases following World War II, but because of lax mosquito control and an increase in world travel, the geographic distribution of dengue diseases is wider than when eradication began in the 1950s. Dengue fever remains an important viral disease of humans. Almost 3 billion people live in areas where *Ae. aegypti* and dengue virus are endemic; DHF is a leading cause of hospitalization and death among children in Southeast Asia. Tens of millions of cases of dengue fever, including cases in northern Mexico, the Florida Keys, and Texas; and hundreds of thousands of cases of dengue hemorrhagic fever occur each year, killing more than 22,000 annually.

Diagnosis, Treatment, and Prevention

Physicians base diagnosis on signs and symptoms in patients who have traveled in endemic areas. Laboratory tests confirm the presence of antigens of dengue virus in a patient's blood.

No specific treatment is available for dengue diseases, but researchers are close to developing an effective safe vaccine. Prevention depends upon control of mosquitoes, a difficult task because the most effective insecticides may harm the environment. **Beneficial Microbes: Eliminating Dengue** on p. 659 describes a pesticide-free alternative.

African Viral Hemorrhagic Fevers

LEARNING **OUTCOME**

21.28 Compare the signs, symptoms, cause, diagnosis, treatment, and prevention of Ebola and Marburg hemorrhagic fevers.

Ebola and **Marburg hemorrhagic fevers,** named for the sites where they were first discovered, are two emerging viral diseases that are of concern to physicians, epidemiologists, and governments.

Signs and Symptoms

Both types of African hemorrhagic fevers begin with fever, fatigue, dizziness, muscle pain, and exhaustion, followed by minor petechiae. This progresses to severe internal hemorrhaging as well as bleeding from other body orifices, including mouth, eyes, and ears. Death results from shock, seizures, or kidney failure.

Pathogens and Virulence Factors

The viruses that cause African hemorrhagic fever are unsegmented −ssRNA viruses. As discussed in Chapter 13, −ssRNA viruses must be converted into mRNA (+ssRNA) before an infected cell can transcribe viral polypeptides. Taxonomists

DISEASE AT A GLANCE 21.5

Yellow Fever

Cause *Flavivirus yellow fever virus* (enveloped, +ssRNA arbovirus with an icosahedral capsid).

Virulence factors Intracellular replication cycle; adhesins.

Portal of entry Injected by the mosquito *Aedes aegypti;* travels in blood to liver.

Signs and symptoms Slight fever, muscle ache, headache, nausea, and vomiting for three to four days; about 20% of patients subsequently develop delirium, seizures, hemorrhaging, nosebleed, shock, and pronounced jaundice and severe fever, which give the disease its name.

Incubation period Three to six days.

Susceptibility Travelers to endemic regions of South America and Africa.

Treatment No specific treatment; supportive nursing care.

Prevention Avoid travel to endemic areas. Or obtain immunization with live attenuated virus, effective for 10 years or more, and avoid mosquito bites by wearing protective clothing, using the insect repellent DEET, and using mosquito netting.

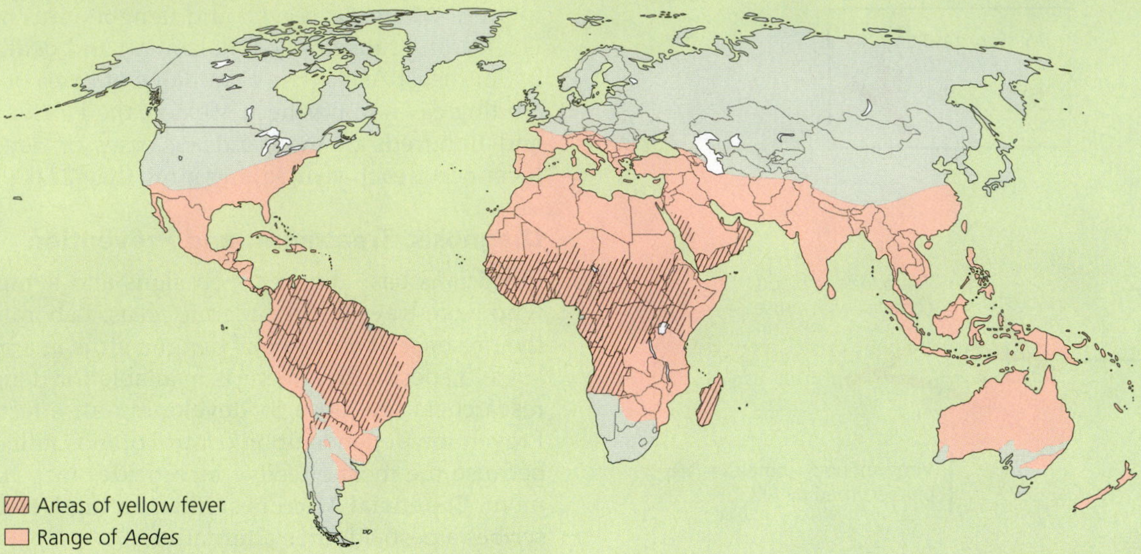

Endemic areas for yellow fever and range of *Aedes*

originally placed them in the family *Rhabdoviridae* but have now assigned them to their own family, *Filoviridae,* primarily based on disease symptoms. These enveloped viruses have long filamentous capsids, which sometimes curve back upon themselves **(FIGURE 21.17)**. The two known genera of filoviruses are *Ebolavirus* and *Marburgvirus.*

Pathogenesis and Epidemiology

African hemorrhagic fevers are acute diseases—there is no carrier state. Signs and symptoms first appear 2–21 days after infection with *Ebolavirus* and within 5–10 days in the case of *Marburgvirus* infections. Hemorrhaging is due to a malfunction of the blood-clotting system, in which infected macrophages trigger localized blood clotting that depletes the serum of clotting proteins, making the body susceptible to massive bleeding.

Viral hemorrhagic fevers occur primarily in Africa, presumably where the natural hosts of the viruses—probably bats—dwell **(FIGURE 21.18)**. Hemorrhagic fever viruses probably initially infect humans via contact with a host, which is thought to be bats, or the host's bodily fluids or wastes. Humans can transmit filoviruses among themselves via contact with bodily

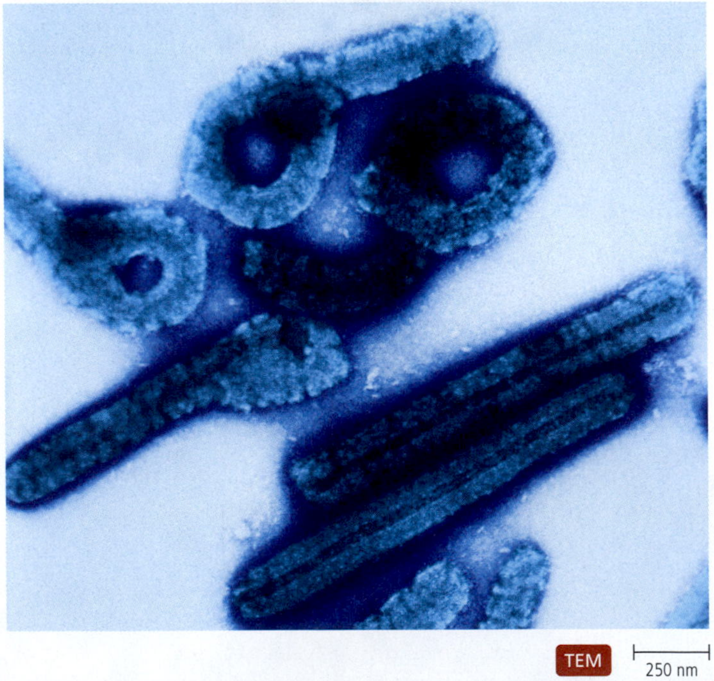

▲ **FIGURE 21.17** Filamentous *Ebolavirus.*

BENEFICIAL MICROBES

Eliminating Dengue

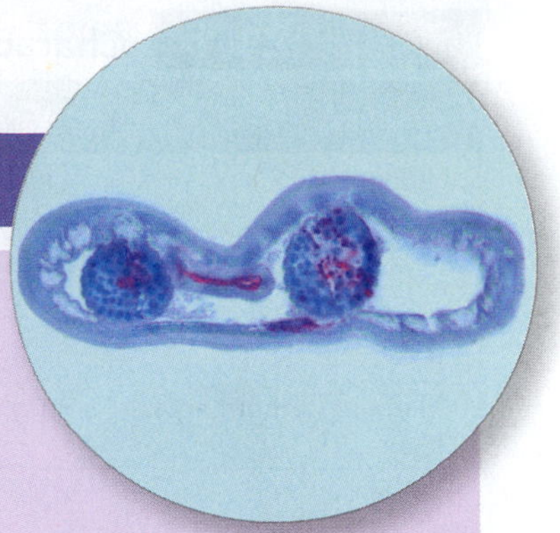

Wolbachia SEM 25 µm

Aedes aegypti has distinctive stripes and a fierce bite. Unlike many mosquitoes, these aggressive bloodsuckers bite during the day. They prefer to live in urban areas, resting in the shade of houses and laying their eggs in modern containers holding a small amount of water, such as tires, cans, and water gutters. Almost 3 billion people share their neighborhoods with *Aedes*. Worst of all, these mosquitoes carry viruses that cause human diseases such as dengue.

There is no vaccine and no treatment for dengue, so prevention involves controlling the mosquitoes, which has had limited success, especially in developing countries. Enter Australian scientist Scott O'Neill and a novel beneficial microbe—a special strain of *Wolbachia pipientis*.

Wolbachia is a Gram-negative, intracellular bacterium that infects about 70% of all insect species, including *Aedes* mosquitoes. Strains of *Wolbachia* often change the biology of their insect hosts, and Dr. O'Neill's *Wolbachia*—strain *w*Mel—is no exception. When male mosquitoes infected with *w*Mel mate with *Wolbachia*-infected females, the females lay eggs normally. All the offspring are infected with *Wolbachia*, which they get from their mother. However, when an infected male mates with an uninfected female, all her eggs are sterile. Thus, *Wolbachia* ensures its own reproductive success and limits mosquito reproduction.

It gets better! It turns out that dengue virus cannot live in infected mosquitoes for some reason. Since dengue virus cannot replicate in infected mosquitoes, the success of *Wolbachia* is the demise of the virus.

O'Neill and his collaborators have released thousands of infected mosquitoes in northern Australia and have successfully altered the mosquito population in the area so that they are unable to transmit dengue. The team plans to release hundreds of thousands of infected mosquitoes into neighborhoods in tropical developing nations with the goal of spreading *Wolbachia pipientis* strain *w*Mel throughout the world. They hope that this will eliminate the scourge of dengue forever.

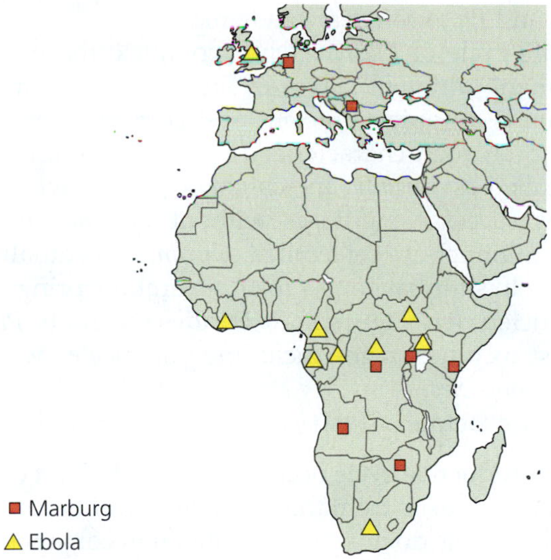

▲ **FIGURE 21.18 Sites in which known locally accquired cases of Marburg and Ebola viruses have occurred.** The Ebola outbreak in Virginia only involved disease in monkeys in a research facility.

fluid, especially blood. Though transmission of filoviruses through the air has been demonstrated in laboratories, such transmission has not been shown in natural settings, hospitals, or mortuaries. *Ebolavirus* is fatal in up to 90% of patients, while *Marburgvirus* is fatal 25% of the time.

Diagnosis, Treatment, and Prevention

Physicians diagnose African hemorrhagic fevers based upon symptoms and demonstration of filoviruses in the blood via ELISA or PCR. Treatment includes supportive care involving fluid and electrolyte replacement. Some physicians treat patients with anticoagulants to stem the initial harmful blood clots. Such treatment has proven efficacious in monkeys, saving the lives of 30% of infected animals. There are no effective antiviral drugs to treat hemorrhagic fevers. Health care workers treat the African hemorrhagic fevers by replacing lost fluids and electrolytes, maintaining blood oxygenation and pressure, replacing lost blood, and treating any secondary infections or complications.

Researchers have developed vaccines that protect monkeys from Ebola hemorrhagic fever and are studying their effectiveness in humans. For now, prevention involves proper procedures in hospitals and morgues to prevent the spread of the viruses.

TABLE 21.1 on p. 660 summarizes features of viral hemorrhagic fevers.

TELL ME WHY

Whereas many doctors are convinced that Epstein-Barr virus causes chronic fatigue syndrome, others deny the association between EBV and the syndrome. Why is the etiology of chronic fatigue syndrome debated even though Epstein-Barr virus is present in patients?

TABLE 21.1 Characteristics of Some Viral Hemorrhagic Fevers

Disease	Viral Genus (Family)	Natural Host(s)	Vector	Geographic Distribution
Yellow fever	*Flavivirus (Flaviviridae)*	Humans, monkeys	*Aedes aegypti* mosquito	Africa, South America
Dengue, dengue hemorrhagic fever	*Flavivirus (Flaviviridae)*	Humans, monkeys	*Aedes aegypti* mosquito	Worldwide, especially tropics
Ebola hemorrhagic fever	*Ebolavirus (Filoviridae)*	Probably bats	None	Central Africa, research facility in the United States
Marburg hemorrhagic fever	*Marburgvirus (Filoviridae)*	Probably bats	None	Central Africa, research facility in Europe

CLINICAL CASE STUDY

A Tired Freshman

An 18-year-old college freshman reports to the campus health clinic. He states that he has felt very fatigued and generally ill for about a week and developed sore throat, fever, and headache the day before. He has been sleeping more than usual and has virtually no appetite. He says his girlfriend, who attends college in a different state and whom he saw last month, has also been feeling very fatigued for the last two weeks or so, but she displays none of the student's other symptoms.

1. What disease should the doctor suspect? What is the infectious agent?
2. How did the student most likely contract the disease?
3. What other symptoms might the student develop?
4. What causes the symptoms of the disease? Why does the student's girlfriend not display the same symptoms?
5. How should the student be treated?

Protozoan and Helminthic Cardiovascular and Systemic Diseases

To this point, we have considered bacterial and viral diseases of the cardiovascular system and several representative systemic diseases. In this section, we examine cardiovascular and systemic diseases caused by protozoa and by a parasitic helminth. We begin with the most prevalent of the protozoan infectious diseases, malaria.

Malaria

LEARNING **OUTCOMES**

21.29 Describe the life cycle of *Plasmodium*, and relate malarial signs and symptoms to stages in the life cycle.

21.30 Describe the diagnosis, treatment, and prevention of malaria.

21.31 Describe *Plasmodium* virulence factors.

21.32 Describe four genetic traits that confer resistance to malaria.

Malaria is a life-threatening disease caused by at least four species of the single-celled parasite *Plasmodium*, which is carried between people by infected *Anopheles* mosquitoes. Over 3 billion people are at risk, because they live in areas where both the mosquito and *Plasmodium* are endemic.

The life cycle of *Plasmodium* is complex; the parasite goes through many different stages in humans and in mosquitoes. *Plasmodium* is an apicomplexan (ap-i-kom-plek´san), which is a type of protozoan characterized by a form of asexual reproduction called *schizogony*. In schizogony, the nucleus of a cell undergoes successive mitoses without cytoplasmic division, forming multinucleate cells called *schizonts*. Eventually, a schizont divides to simultaneously form several offspring, each with a single nucleus (see Figure 12.3). Besides schizonts, *Plasmodium* has at least six other stages, including *gametocytes*, which are infective to *Anopheles*.

Virulence factors of *Plasmodium* include the following:

- The reproductive cycle occurs within red blood cells, which hides the parasite from the immune surveillance since red blood cells do not present antigen in conjunction with major histocompatibility protein.
- A special protein assemblage called the *malaria secretome* injects toxins and enzymes into host cells.
- Adhesins enable infected red blood cells to adhere to certain body tissues such as brain tissue and the linings of blood vessels, which allows the parasite to avoid clearance by the spleen.
- Merozoites form within vesicles of liver cells that are secreted directly into blood vessels, thereby avoiding immune cells of the liver.

- The gametocytes of at least one strain *(P. falciparum)* trigger changes in human body chemistry, presumably breath or body odor, such that *Anopheles* mosquitoes are more attracted; thus *Plasmodium* induces a "bite me" signal in humans carrying the very stage ready to be picked up by a mosquito.

Malaria patients suffer jaundice, severe recurrent fever and chills, headache, vomiting, and diarrhea. Historically, physicians prescribe antimalarial drugs, but strains of *Plasmodium* have developed resistance to many of these drugs.

People living in endemic areas, and their descendants throughout the world, commonly have one or more of the following genetic traits that increase their resistance to malaria:

- *Sickle-cell trait.* Individuals with a sickle-cell gene produce an abnormal type of hemoglobin called hemoglobin S (hemoglobin A is normal). Hemoglobin S causes erythrocytes to become sickle shaped and makes erythrocytes resist *Plasmodium.*
- *Hemoglobin C.* Humans with two genes for hemoglobin C are invulnerable to malaria. The mechanism by which this mutation provides protection is unknown.
- *Genetic deficiency of glucose-6-phosphate dehydrogenase.* In order to synthesize DNA, trophozoites must acquire this enzyme from their host, thus, enzyme deficiency inhibits trophozoite replication.
- *Lack of so-called Duffy antigens on erythrocytes.* Because *P. vivax* requires Duffy antigens to attach to and infect erythrocytes, Duffy-negative individuals are resistant to *P. vivax.*

Antifever medication and blood transfusions may be required as supportive measures. Treatment is usually effective except in severe cases of falciparum malaria.

Draining wetlands and removing standing water reduce mosquito breeding rates but must be balanced against the environmental impact on other wetland-dwelling plants and animals. Personal use of insect repellents containing DEET,[25] use of mosquito netting, and protective clothing reduce mosquito bites.

Researchers are developing and testing several malaria vaccines and alternate antimalarial drugs (**Highlight: In Search of a Malaria Vaccine**). **Disease in Depth: Malaria** (pp. 662–663) examines features of malaria in more detail.

[25] *N,N*-diethyl-*meta*-toluamide.

HIGHLIGHT

In Search of a Malaria Vaccine

A malaria vaccine would protect millions of children from a dread disease.

Researchers are testing malaria vaccines that provide protection by limiting initial infection by sporozoites, vaccines that target merozoites as they are released from liver and blood cells, vaccines against malarial toxins rather than against malaria parasites, and a vaccine against a *Plasmodium* protein that allows the parasite to cross the placenta to invade a developing baby. Exciting developments are also occurring in new ways to deliver malaria vaccines and in the discovery of novel adjuvants—additives to a vaccine that enhance protection. For example, one group of scientists developed a vaccine that is delivered as a nasal spray in conjunction with deactivated cholera toxin as an adjuvant. This vaccine provides 100% protection against malaria infection in mice.

Several groups of researchers are developing vaccines that work inside mosquitoes. Vaccinated mosquitoes cannot spread malaria to humans. Because it is impossible to administer a treatment to every mosquito, scientists vaccinate people, causing them to secrete protective proteins called antibodies into their blood. Mosquitoes then get a stomach full of antibodies with every blood meal. Inside the mosquito, the antibodies attach to chemicals produced by the ookinete stage of the malarial parasite. This prevents an ookinete from becoming an oocyst, breaking the parasite's life cycle permanently.

Although vaccines cannot stop malaria everywhere, they are an important tool in the fight against the dread killer. Using vaccines in conjunction with antimalarial drugs, bed nets to keep mosquitoes at bay, and insecticides to kill mosquitoes, it is possible for the first time in history to realistically contemplate the eradication of malaria from the Earth.

MALARIA

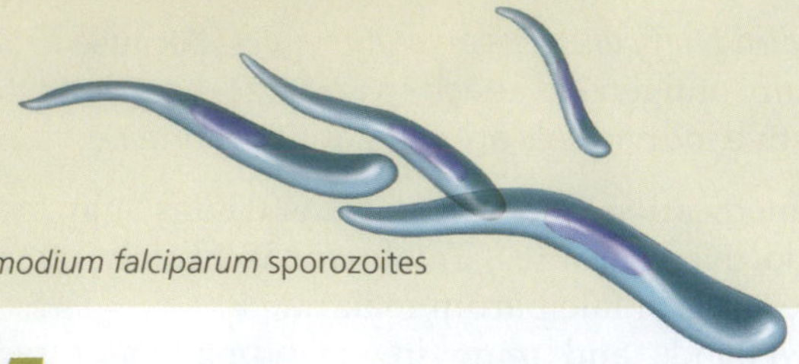

Plasmodium falciparum sporozoites

SIGNS AND SYMPTOMS

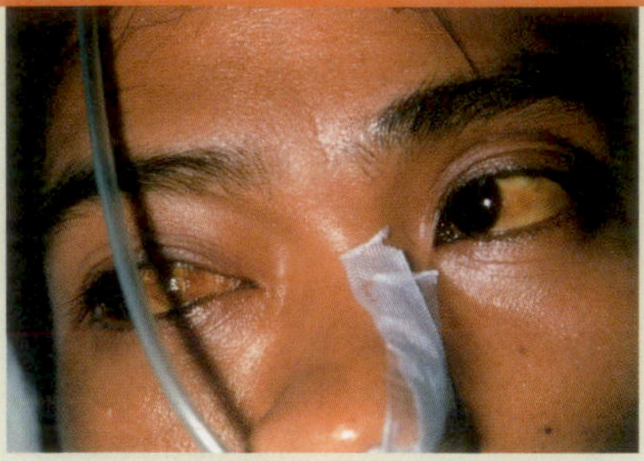

Many of malaria's symptoms are associated with immune response against the parasites, cellular debris, and toxins following the synchronous cycles of the parasite's life within erythrocytes. Two weeks after the cycle begins, enough parasites exist to cause fever, chills, diarrhea, headache, and (occasionally) pulmonary or cardiac dysfunction. Anemia, weakness, and fatigue gradually set in, and patients may become jaundiced, as seen in this malaria patient from Vietnam.

Malaria is the most prevalent and infamous of the protozoan infections. The life cycle of *Plasmodium*, the causative agent of malaria, has three prominent stages: the **exoerythrocytic phase**, the **erythrocytic cycle**, and the **sporogonic phase**.

PLASMODIUM LIFE CYCLE

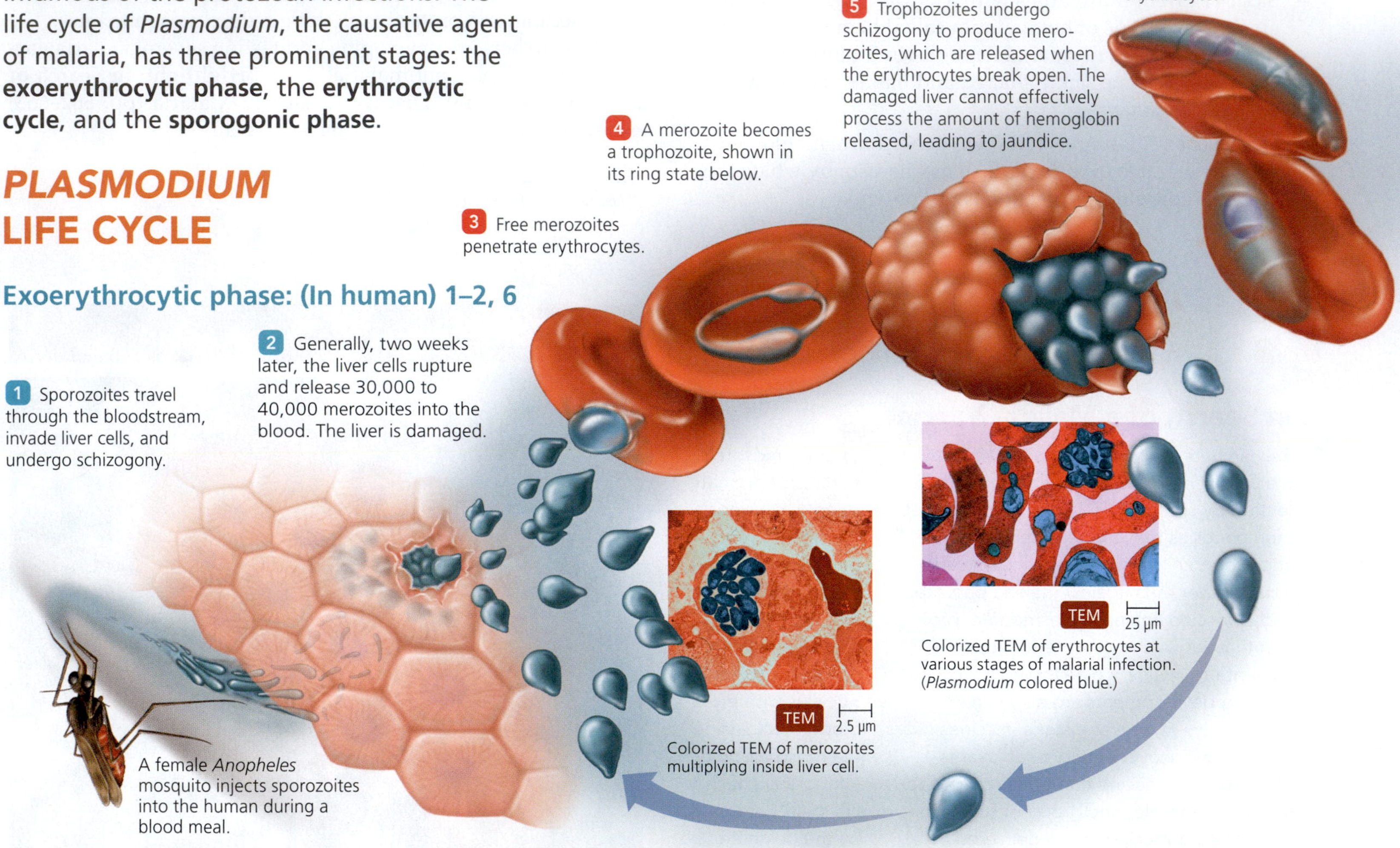
Exoerythrocytic phase: (In human) 1–2, 6
1 Sporozoites travel through the bloodstream, invade liver cells, and undergo schizogony.
2 Generally, two weeks later, the liver cells rupture and release 30,000 to 40,000 merozoites into the blood. The liver is damaged.
Erythrocytic cycle: (In human) 3–5
3 Free merozoites penetrate erythrocytes.
4 A merozoite becomes a trophozoite, shown in its ring state below.
5 Trophozoites undergo schizogony to produce merozoites, which are released when the erythrocytes break open. The damaged liver cannot effectively process the amount of hemoglobin released, leading to jaundice.
6 Some merozoites develop into male and female gametocytes within erythrocytes.
TEM
2.5 μm
TEM
25 μm

A female *Anopheles* mosquito injects sporozoites into the human during a blood meal.

Colorized TEM of merozoites multiplying inside liver cell.

Colorized TEM of erythrocytes at various stages of malarial infection. (*Plasmodium* colored blue.)

INVESTIGATE IT!

Should scientists and government agencies focus more on controlling mosquitoes or developing a malaria vaccine?

Scan this code to visit the World Health Organization website to investigate malaria. Then go to MasteringMicrobiology to record your research findings.

EPIDEMIOLOGY

Malaria is endemic in over 100 countries in the tropics and subtropics, where the parasite's mosquito vector, *Anopheles*, breeds. As many as 300 million people are infected with *Plasmodium*, and an estimated 1.2 million (usually children) die annually. Malaria was prevalent in the U.S. until mosquito eradication programs eliminated the disease decades ago. About 1200 cases seen each year in the U.S. involve immigrants or travelers from endemic areas.

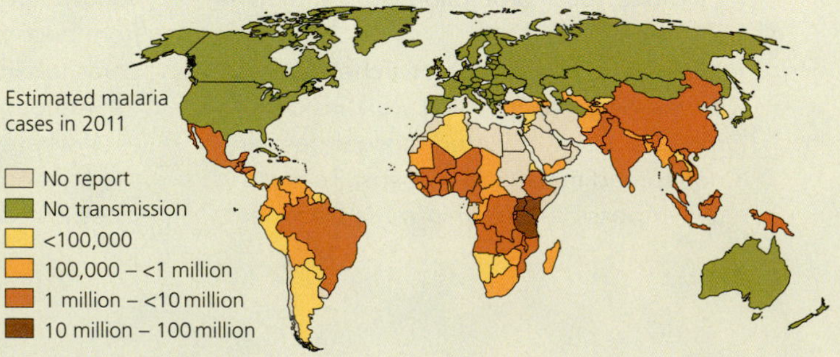
Estimated malaria cases in 2011
No report
No transmission
<100,000
100,000 – <1 million
1 million – <10 million
10 million – 100 million

PATHOGEN AND PATHOGENESIS

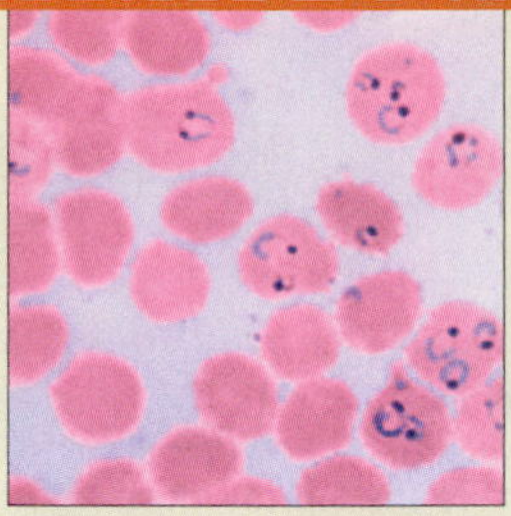

LM 7 µm

At least four species of *Plasmodium* cause malaria in humans: *P. falciparum*, *P. vivax*, *P. ovale*, and *P. malariae* are most significant. Disease severity depends upon the species: *P. ovale* causes mild disease, *P. vivax* results in chronic malaria, and *P. malariae* and *P. falciparum* (left) cause more serious malaria, known as *blackwater fever* because hemoglobin from damaged red blood cells is excreted in the urine. *P. falciparum* malaria can be fatal within 24 hours of the onset of symptoms.

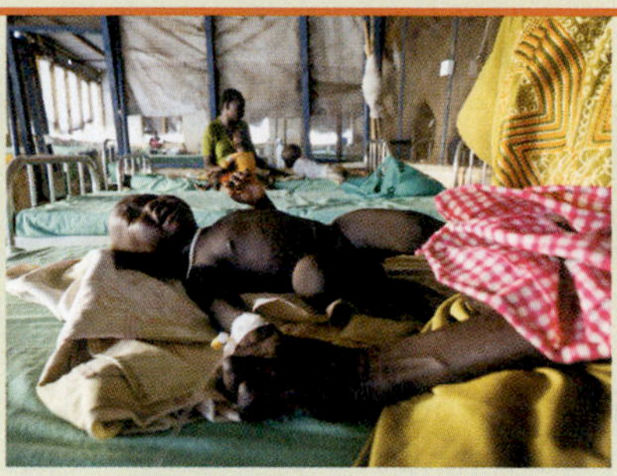

P. falciparum causes fever, erythrocyte lysis, renal failure, and dark urine. Protozoan proteins on the surfaces of erythrocytes cause erythrocytes to adhere to capillary lining, blocking blood flow and leading to small hemorrhages and ultimately tissue death. Children, such as this infant with malaria in South Sudan, are among the most vulnerable.

Sporogonic phase: (In mosquito) 7–11

7 The *Anopheles* mosquito ingests gametocytes during a blood meal.

8 Gametocytes become gametes that fuse to form a zygote.

9 The zygote differentiates into an ookinete and attaches to the gut wall.

10 The ookinete becomes an oocyst in the gut wall. Sporozoites form within the oocyst.

11 Sporozoites exit the oocyst and migrate to the mosquito's salivary glands.

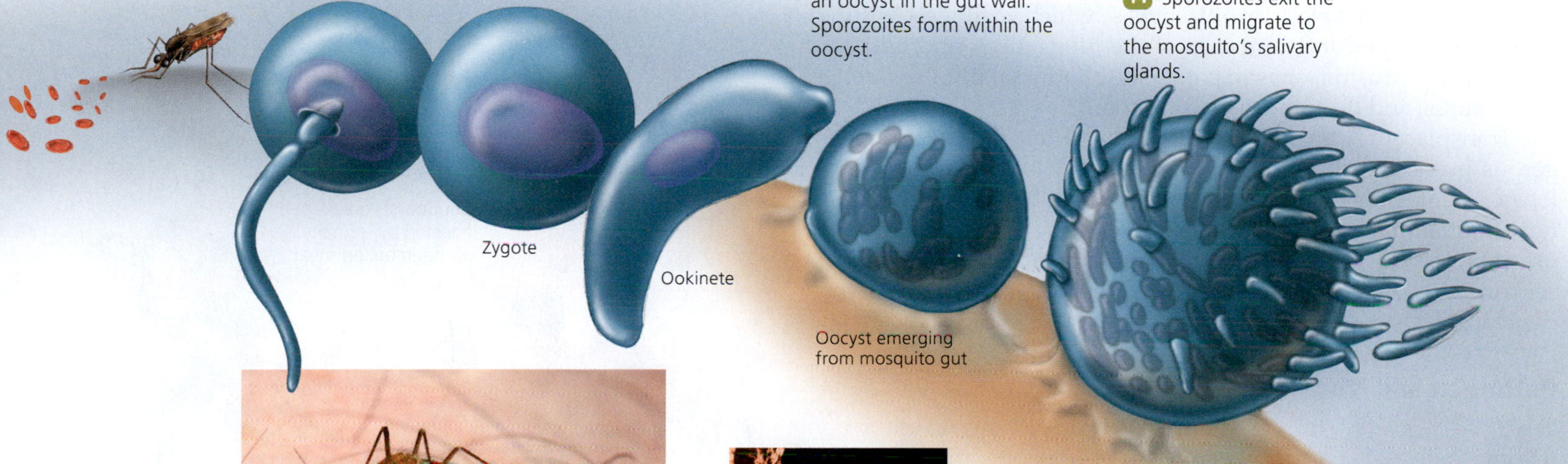

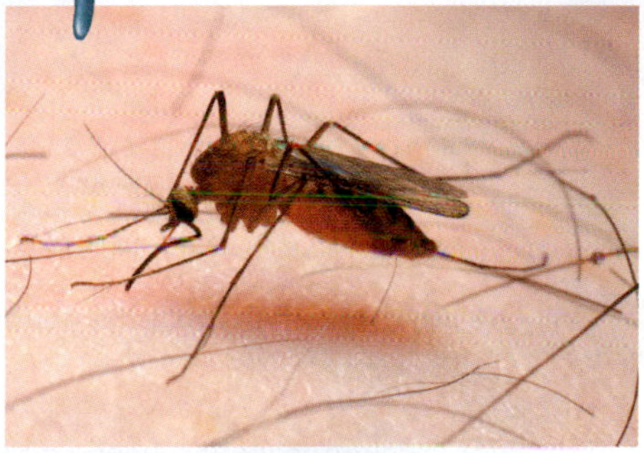

Malarial vector, female *Anopheles* mosquito, during blood meal.

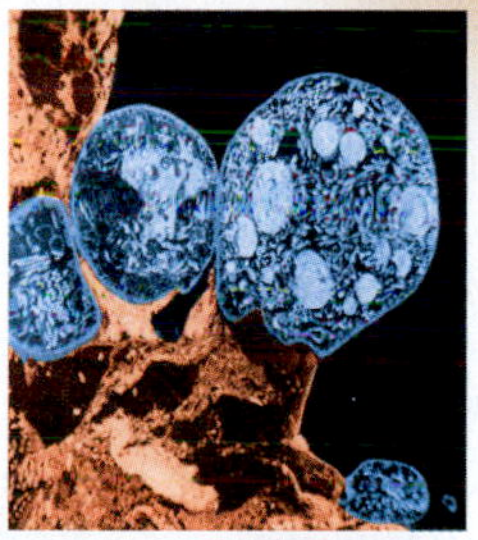

TEM 25 µm

Colorized TEM of oocysts filled with sporozoites.

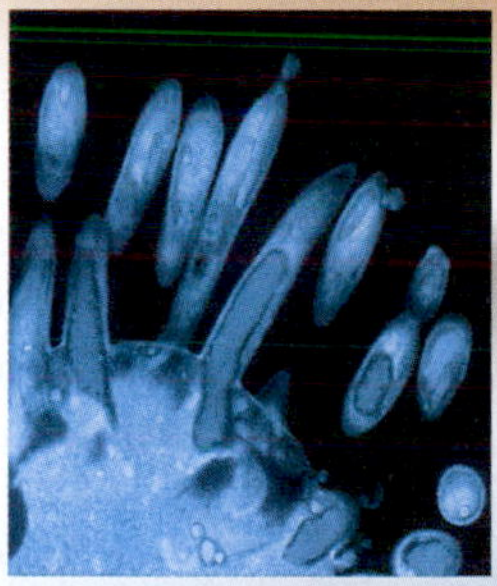

TEM 8 µm

Colorized TEM of sporozoites emerging from oocyst.

12 Sporozoites are injected into a human during a blood meal. The life cycle begins again.

DIAGNOSIS AND TREATMENT

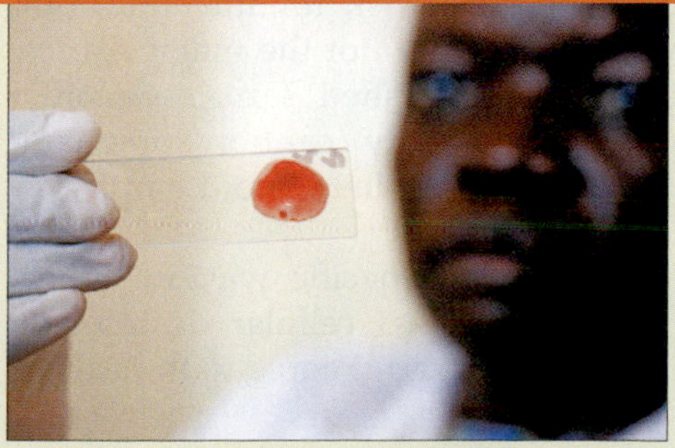

Diagnosis of malaria is commonly based on identification of trophozoites and other stages of *Plasmodium* in blood smears. Standard antimalarial drugs include chloroquine and, in areas where *Plasmodium* is resistant, pyrimethamine with artesunate.

PREVENTION

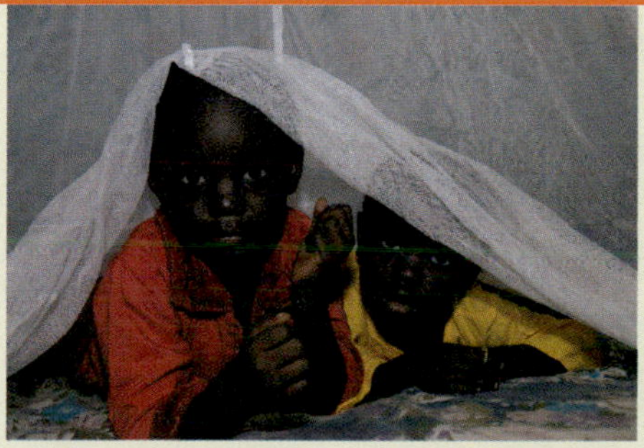

The most cost-effective way to reduce the number of cases is to limit contact with mosquitoes carrying *Plasmodium*. WHO recommends DDT-impregnated mosquito netting. Travelers to endemic countries should take prophylactic antimalarial drugs (such as proguanil), use insect repellent, and wear long-sleeved shirts and long pants.

Toxoplasmosis

LEARNING | OUTCOME

21.33 Describe toxoplasmosis, including its cause, signs, symptoms, pathogenesis, epidemiology, diagnosis, treatment, and prevention.

Toxoplasmosis (tok′sō-plaz-mō′sis), commonly referred to as *toxo,* is a major disease seen in AIDS patients. Unborn children are also at risk.

Signs and Symptoms

Over 80% of patients with toxoplasmosis have no symptoms and no permanent damage, and their infection resolves spontaneously within a few months to a year. Only patients with poor immunity develop toxoplasmosis, which manifests with fever, malaise, and inflammation of the lungs, liver, and heart. Headache, confusion, spastic paralysis, blindness, myocarditis, encephalitis, and death are also common.

Transplacental transfer from mother to fetus is most dangerous in the first trimester of pregnancy; it can result in epilepsy, mental retardation, microcephaly (abnormally small head), inflammation of the retina, blindness, anemia, jaundice, or spontaneous abortion or stillbirth. Ocular infections may remain dormant in children for years, at which time blindness develops.

Pathogen and Virulence Factors

Toxoplasma gondii (tok-sō-plaz′mă gon′dē-ē)—the cause of toxoplasmosis—is an apicomplexan protozoan that lives in the nucleated cells of wild and domestic mammals and birds; cats are its definitive host. In the life cycle of *Toxoplasma* **(FIGURE 21.19)**, male and female *gametes* in the cat's digestive tract fuse to form *zygotes,* which develop into immature oocysts that are shed in the feces **1**. An infected cat shows no signs of infection and is not harmed. Each day the cat can excrete up to 10 million oocysts, which can survive in moist soil for several months. Vegetation growing in soil containing the parasites becomes contaminated with oocysts. As each oocyst matures, it produces internal *sporozoites* **2**. When rodents, livestock, or humans ingest mature oocysts on vegetation **3**, digestion releases the sporozoites, which invade the host's heart, tongue, and diaphragm cells and proliferate asexually to produce *pseudocysts* containing numerous *bradyzoites* **4**. Pseudocysts remain dormant in the tissues of livestock or rodents until humans or cats, respectively, consume these hosts **5**.

Virulence factors include the ability of oocysts and of pseudocysts to parasitize a wide variety of hosts, the ability of *Toxoplasma* to infect many different cells within a host, and the ability of the parasite to survive intracellularly.

Pathogenesis and Epidemiology

Humans typically become infected by ingesting undercooked meat containing the parasite. Over 1 billion people are infected; at greatest risk are butchers, hunters, and anyone who tastes food while preparing it. Ingesting or inhaling contaminated soil can also be a source of infection. The protozoan can also cross a human placenta to infect the fetus. Interestingly, the parasite attacks female fetuses more than males, though scientists do not know why or how *Toxoplasma* selects a fetus by gender. Historically, contact with infected cats and their feces was proposed as a major risk for infection, but recent studies have shown that cats are not the major source of infection for humans because cats shed *Toxoplasma* only briefly. Nevertheless, doctors typically caution pregnant women and immunocompromised people to avoid emptying litter boxes.

Researchers are uncertain of the specific way in which *T. gondii* causes disease. It appears that cellular destruction and the release of cytokines are involved. People with healthy

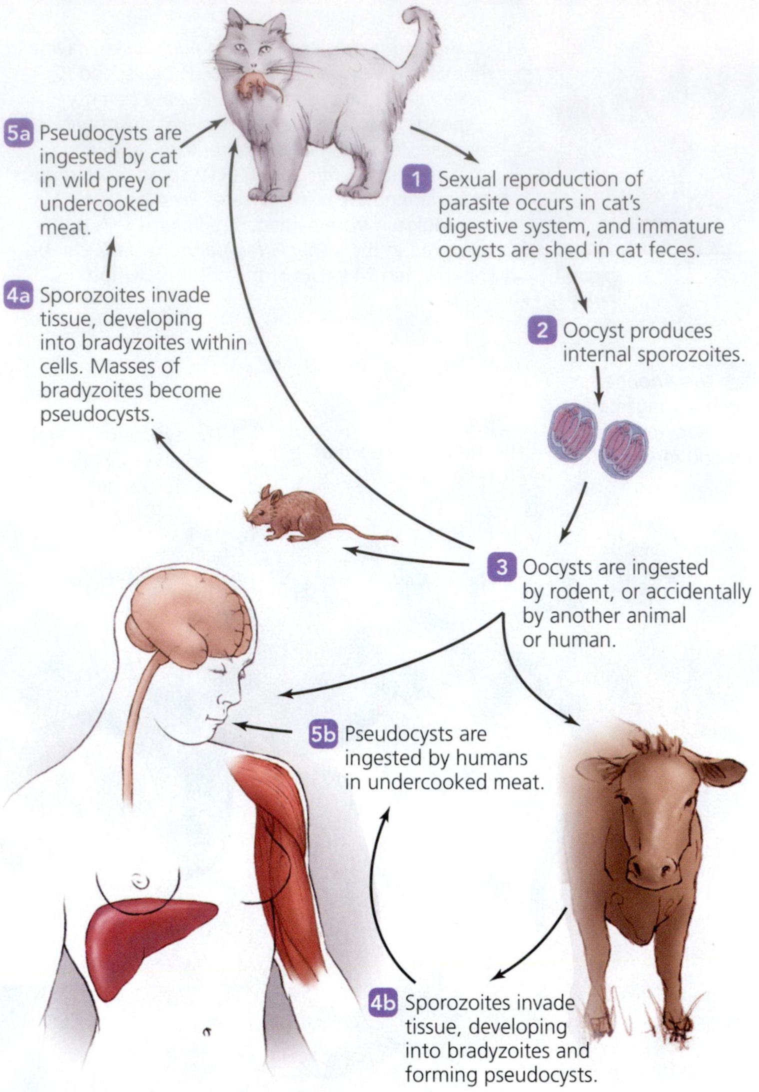

5a Pseudocysts are ingested by cat in wild prey or undercooked meat.
1 Sexual reproduction of parasite occurs in cat's digestive system, and immature oocysts are shed in cat feces.
4a Sporozoites invade tissue, developing into bradyzoites within cells. Masses of bradyzoites become pseudocysts.
2 Oocyst produces internal sporozoites.
3 Oocysts are ingested by rodent, or accidentally by another animal or human.
5b Pseudocysts are ingested by humans in undercooked meat.
4b Sporozoites invade tissue, developing into bradyzoites and forming pseudocysts.

▲ **FIGURE 21.19 The life cycle of *Toxoplasma gondii*.** *Which two groups of humans are at greatest risk from toxoplasmosis?*

Figure 21.19 *AIDS patients and first-trimester fetuses are at greatest risk from toxoplasmosis.*

DISEASE AT A GLANCE 21.6

Toxoplasmosis

Cause *Toxoplasma gondii* (apicomplexans protozoan).

Virulence factors Ability to survive inside many different types of cells in many different hosts.

Portal of entry Ingestion in meat or inhalation of contaminated soil or cat litter.

Signs and symptoms Most patients have no manifestations. In immunocompromised patients: fever; malaise; inflammation of lungs, liver, and heart; headache; confusion; spastic paralysis; blindness; myocarditis; encephalitis. Transplacental transfer to a fetus can cause epilepsy, mental retardation, microcephaly, inflammation of the retina and blindness, anemia, jaundice, or spontaneous abortion or stillbirth.

Incubation period 10 to 23 days.

Susceptibility Immunocompromised and unborn babies.

Treatment Pyrimethamine and sulfonimides.

Prevention Avoid contact with contaminated soil (cat litter); wash all fruits and vegetables; thoroughly cook; freeze, or smoke meat.

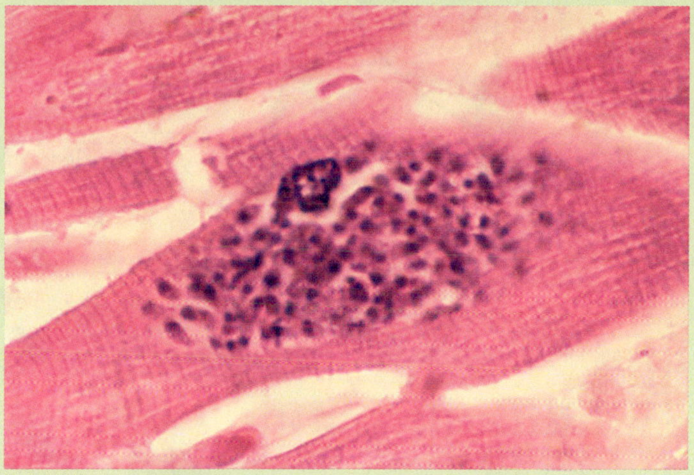

immune systems are able to contain the damage and keep *T. gondii* from spreading beyond the pseudocysts.

Toxoplasma gondii is one of the world's most widely distributed protozoan parasites of humans. The CDC estimates that up to 40% of the U.S. population carry this obligate intracellular parasite. In AIDS patients with a CD4 count under 200, symptoms are thought to result when *Toxoplasma*, resident in the body from a previous infection, reactivates as the immune system fails.

Diagnosis, Treatment, and Prevention

Physicians diagnose *T. gondii* infection by serological detection of organisms in tissue samples, by microscopic identification of parasites in tissue biopsies, or by molecular identification of *T. gondii* genetic material or products in specimens using PCR, Southern blot, or DNA probes. Serology is the most common diagnostic method. Molecular techniques are most useful for congenital toxoplasmosis.

In healthy adults, toxoplasmosis normally resolves without treatment. When treatment is recommended—in AIDS patients, pregnant women, and newborns—physicians prescribe pyrimethamine and sulfonamides, administered together for three to four weeks. Clindamycin can substitute for sulfonamides in patients who are allergic to sulfa. Treatment of pregnant women prevents most transplacental infections but is controversial because the medications are toxic. More aggressive treatment may be needed in AIDS patients, including the addition of steroids to reduce tissue inflammation.

Controlling the incidence of *T. gondii* infection is difficult because so many hosts harbor the apicomplexan. A vaccine for cats is currently under development to reduce the chance of pet-to-owner transmission. The best prevention is to avoid contact with contaminated soil, particularly soil contaminated with cat feces; wash all fruits and vegetables before consumption; and thoroughly cook meat until it reaches an internal temperature of 66°C (150°F); in other words, until a steak is no longer pink in the middle. Smoked meats and meats frozen overnight are considered safe.

Disease at a Glance 21.6 summarizes the characteristics of toxoplasmosis.

Chagas' Disease

LEARNING **OUTCOME**

21.34 Describe the signs, symptoms, causes, diagnosis, treatment, and prevention of Chagas' disease.

Charles Darwin (1809–1882) wrote in *The Voyage of the Beagle*, "At night I experienced an attack (for it deserves no less a name) of . . . the great black bug of the Pampas [Argentina]. It is most disgusting to feel soft wingless insects, about an inch long, crawling over one's body. Before sucking they are quite thin but afterwards they become round and bloated with blood." Following his return to England, Darwin suffered from almost continual ill health and died years later of heart failure. Some researchers hypothesize that Darwin suffered from **Chagas' disease,** named for Brazilian doctor Carlos Chagas (1879–1934). This disease, also called *American trypanosomiasis,* produces manifestations similar to those experienced by Darwin. A parasitic trypanosome, which is transmitted via the bite of the mentioned bug, causes Chagas' disease.

Signs and Symptoms

The initial sign of Chagas' disease is a swelling at the site of infection, followed by general signs and symptoms—fatigue, fever, malaise, and swelling of lymph nodes that drain the site of infection—for four to eight weeks. These initial manifestations occur in only 1% of patients. The disease then enters an intermediate asymptomatic stage for 10–20 years. Subsequently, patients have difficulty in swallowing and develop severe constipation, an irregular heartbeat, and congestive heart failure that is fatal. Not every patient develops chronic manifestations of Chagas' disease.

Pathogen and Virulence Factors

Trypanosoma cruzi (tri-pan´ō-sō-mă kroo´zē) is the flagellated protozoan that causes Chagas' disease. The disease is endemic throughout Central and South America. Opossums and armadillos are the primary reservoirs for *T. cruzi,* but most mammals, including humans, can harbor the organism. Transmission occurs through the bite of insects—true bugs in the genus *Triatoma* (trī-ă-tō´mă). These bloodsucking bugs feed preferentially from blood vessels in the lips, giving the bugs their common name—kissing bugs. *T. cruzi* matures in the hindgut of a kissing bug and enters a mammalian host when the bug's feces are rubbed into the bite wound. *T. cruzi* circulates in the blood and has an intracellular stage, infecting macrophages and heart muscle cells.

T. cruzi evades the immune system in part by living inside host cells. Further, the pathogen has the ability to switch its surface antigens, and it produces a protein that acts to suppress immune cytokines.

FIGURE 21.20 illustrates details of the life cycle of *T. cruzi.* Within the hindgut of a kissing bug, *T. cruzi* becomes infective 1 and is shed in the bug's feces while the bug feeds on a mammalian host 2, usually at night. When the host scratches the itchy wound created by the bug's bite, infective trypanosomes deposited in the feces enter the nearby wounds 3. Alternatively, trypanosomes can enter the body through an open cut, the eyes, or mouth, and infected mothers can pass the parasite to their children in the womb or while breast-feeding. The bloodstream carries the parasites throughout the body, where they penetrate cells, especially macrophages and heart muscle cells, and transform into small, nonflagellated forms 4. Nonflagellated trypanosomes multiply by binary fission 5, and then each develops a flagellum attached to the cell down its length with an undulating membrane 6. These flagellated forms cannot multiply. The host cell bursts, releasing flagellated trypanosomes that either infect other cells 7 or circulate in the bloodstream. Trypanosomes in the blood can be ingested by a kissing bug when it takes a blood meal 8. Within the midgut of the kissing bug, the flagellated cells produce another reproductive form that multiplies by binary fission 9.

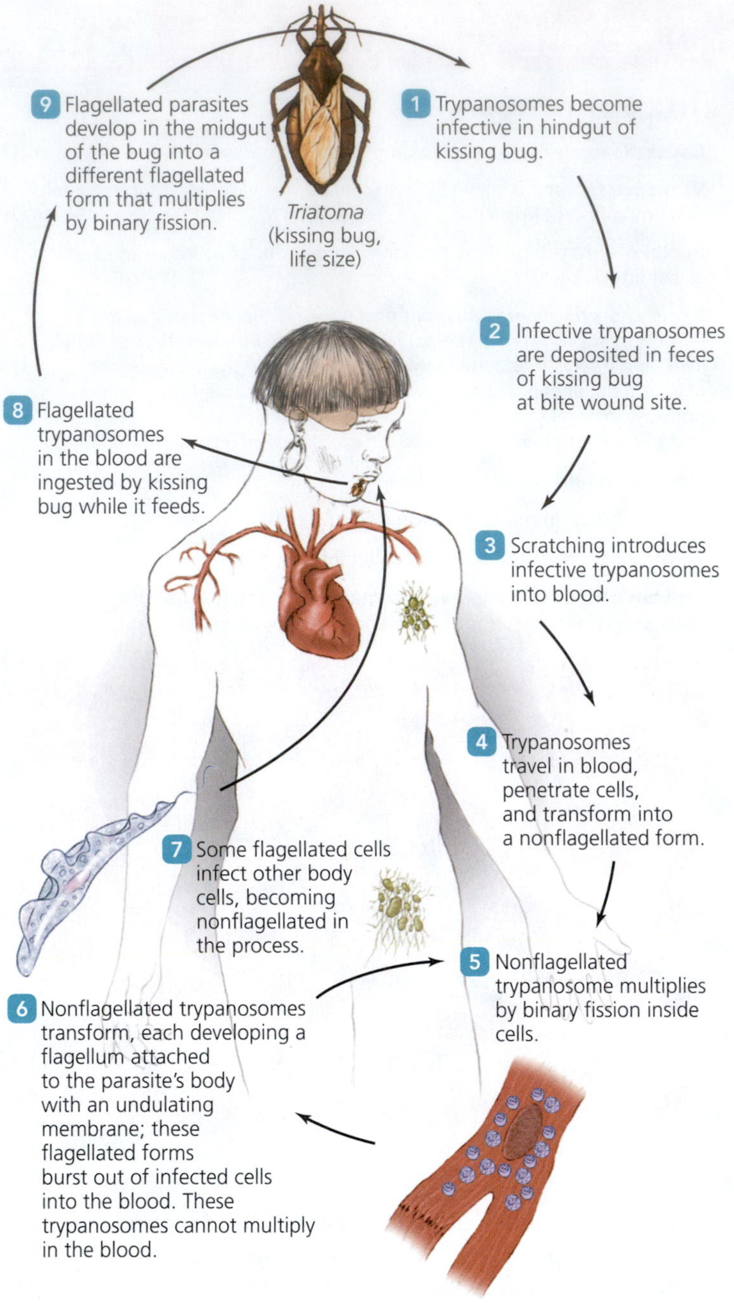

1 Trypanosomes become infective in hindgut of kissing bug.
2 Infective trypanosomes are deposited in feces of kissing bug at bite wound site.
3 Scratching introduces infective trypanosomes into blood.
4 Trypanosomes travel in blood, penetrate cells, and transform into a nonflagellated form.
5 Nonflagellated trypanosome multiplies by binary fission inside cells.
6 Nonflagellated trypanosomes transform, each developing a flagellum attached to the parasite's body with an undulating membrane; these flagellated forms burst out of infected cells into the blood. These trypanosomes cannot multiply in the blood.
7 Some flagellated cells infect other body cells, becoming nonflagellated in the process.
8 Flagellated trypanosomes in the blood are ingested by kissing bug while it feeds.
9 Flagellated parasites develop in the midgut of the bug into a different flagellated form that multiplies by binary fission.
Triatoma (kissing bug, life size)

▲ **FIGURE 21.20 The life cycle of *Trypanosoma cruzi* in a South American.**

Pathogenesis and Epidemiology

Infection follows the bite from infected *Triatoma,* blood transfusion with infected blood, or organ transplantation from an infected donor. Chagas' disease progresses over the course of several months through four stages:

1. The site of the bug's bite swells.
2. The generalized stage is characterized by fever, swollen lymph nodes, myocarditis (inflammation of the heart muscle), splenomegaly, and enlargement of the esophagus and colon.
3. The chronic stage is asymptomatic and can last for years.
4. A final symptomatic stage is characterized primarily by congestive heart failure following the formation of *pseudocysts,* which are clusters of parasites in heart muscle tissue.

An estimated 8–15 million people have Chagas' disease, and about 20,000 die from it each year. Heart failure resulting from infection with *T. cruzi* is one of the leading causes of death in Latin America, especially among children.

Diagnosis, Treatment, and Prevention

Microscopic identification of trypanosomes (see Disease at a Glance 21.7) or their antigens in blood, lymph, spinal fluid, or a tissue biopsy is diagnostic for Chagas' disease. ELISA can demonstrate past infection with *T. cruzi.* Historically, physicians have used a simple and practical diagnostic method called *xenodiagnosis.* In this procedure, the physician allows an uninfected kissing bug to feed on the patient suspected of having a *T. cruzi* infection. Four weeks later, the physician dissects the bug; the presence of parasites within the hindgut of the bug indicates that the patient is infected.

In its earliest stages, Chagas' disease is treated with benznidazole or nifurtimox; however, recall that only 1% of patients develop early manifestations of the disease and thereby know that they have been infected. Most patients do not take antitrypanosome drugs soon enough—late stages of Chagas' disease cannot be treated.

Prevention of Chagas' disease involves replacing thatch and mud building materials, which provide homes for the bugs, with concrete and brick. Alternatively, an insectide-laden paint applied to the walls may provide significant benefit. Tourists staying in well-constructed hotels and resorts are not at risk. The use of insecticides, both personally and in the home, and sleeping under insecticide-impregnated netting can prevent insect feeding. No vaccine exists for Chagas' disease.

Disease at a Glance 21.7 summarizes Chagas' disease.

We have considered protozoan cardiovascular diseases—malaria, toxoplasmosis, and Chagas' disease. Next, we consider a representative blood disease of a parasitic worm, schistosomiasis.

DISEASE AT A GLANCE 21.7

Chagas' Disease

Cause *Trypanosoma cruzi* (flagellated protozoan).

Virulence factors Intracellular life, antigen switching, synthesis of protein that inhibits immunity.

Portal of entry Bite of infected *Triatoma* (kissing) bug.

Signs and symptoms Swelling at site of bite, fever, swollen lymph nodes, myocarditis, splenomegaly, and enlarged esophagus and colon. Following these symptoms a chronic, asymptomatic stage may last for years before congestive heart failure occurs.

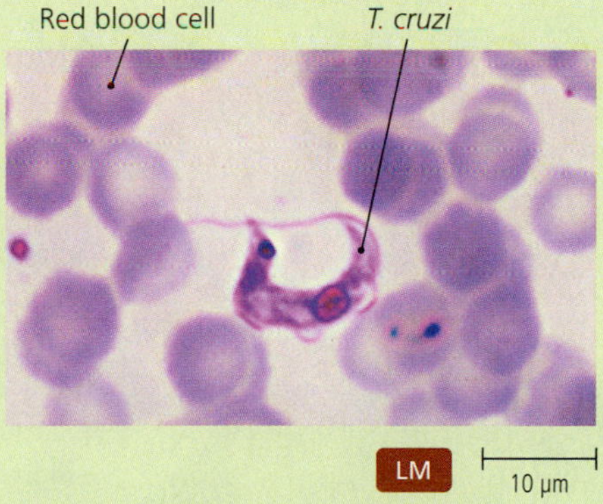

Blood smear containing trypomastigote of *Trypanosoma cruzi*

Incubation period A few days to a few weeks. Congestive heart failure generally occurs 10–20 years following infection.

Susceptibility People visiting or living in endemic areas, especially in poorly constructed dwellings in South and Central America.

Treatment Early stages may be treated with benznidazole or nifurtimox. Late stages of Chagas' disease cannot be treated.

Prevention Avoid sleeping in mud, thatch, or adobe houses in endemic areas, and use insect repellent.

Schistosomiasis

LEARNING **OUTCOME**

21.35 Describe the signs, symptoms, causes, diagnosis, treatment, and prevention of schistosomiasis.

Parasitic blood flukes of the genus *Schistosoma* (skis-tō-sō´mă) are one of the more common parasitic helminths of humans in the world. **Schistosomiasis** (skis´tō-sō-mī´ă-sis) is a potentially fatal disease and one of the major public health problems in the world.

Signs and Symptoms

A transient dermatitis called *swimmer's itch* may occur in the area where infective larvae burrow into the skin. They then migrate throughout the body via the vascular system; such migration produces no signs or symptoms. However, when the worms mature and mate, infections become dangerous as eggs lodge in the liver, lungs, brain, kidneys, and other organs. Trapped eggs die and calcify, leading to renal failure, splenomegaly, increased blood pressure in the pulmonary circulation, or heart failure. Death can result. In some infections, movement of eggs into the bladder and ureters (tubes draining the kidneys into the bladder) causes bladder obstruction and bloody urine, and has been tied to fatal bladder cancer.

Pathogens and Virulence Factors

Three geographically limited species of *Schistosoma* infect humans to cause schistosomiasis:

- *S. mansoni* (man-sō´nē) is common in the Caribbean, Venezuela, Brazil, Arabia, and large areas of Africa.
- *S. haematobium* (hē´mă-tō´bē-ŭm) is found only in Africa and India.
- *S. japonicum* (jă-pon´i-kŭm) occurs in China, Taiwan, the Philippines, and Japan, although infections in Japan are relatively rare.

The freshwater larvae of these worms, called *cercariae*, can burrow through human skin to enter the blood.

Pathogenesis and Epidemiology

Humans are the principal definitive host for most species of *Schistosoma*, though *S. japonicum* infects other mammals. **FIGURE 21.21** illustrates the life cycle of *Schistosoma*: The free-swimming, infective larval stage burrows through human skin 1 that is in contact with contaminated water. Humans become infected while washing clothes or utensils, bathing, swimming, or drinking. Infective larvae enter the vascular system of the liver or bladder (depending on species), where they feed on blood, mature, and mate 2. Females daily lay hundreds of eggs, which have distinctive spines (see Disease at a Glance 21.8),

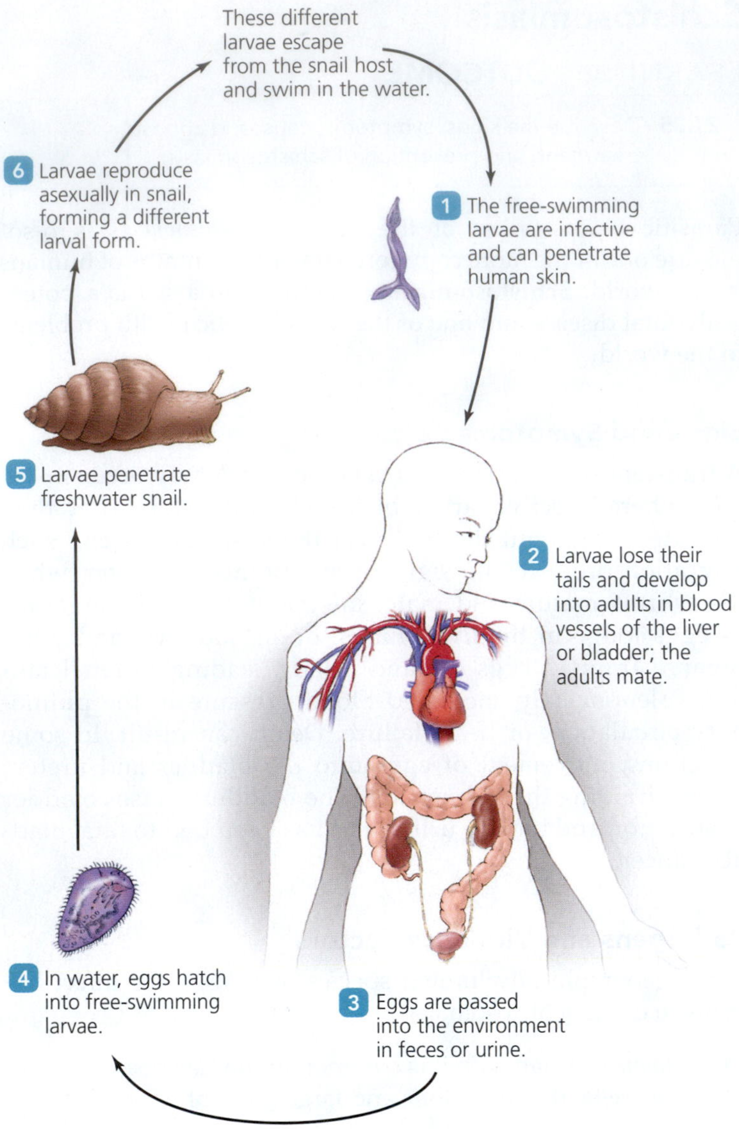
These different larvae escape from the snail host and swim in the water.
6 Larvae reproduce asexually in snail, forming a different larval form.
1 The free-swimming larvae are infective and can penetrate human skin.
5 Larvae penetrate freshwater snail.
2 Larvae lose their tails and develop into adults in blood vessels of the liver or bladder; the adults mate.
4 In water, eggs hatch into free-swimming larvae.
3 Eggs are passed into the environment in feces or urine.

▲ **FIGURE 21.21 The life cycle of *Schistosoma*, a blood fluke.**

in the walls of the blood vessels. The eggs work their way to the lumen of the intestine (*S. mansoni* and *S. japonicum*) or lumens of the urinary bladder and ureters (*S. haematobium*) to be eliminated into the environment 3. In freshwater the eggs hatch, releasing free-swimming larvae 4. Larvae penetrate freshwater snails 5, where they reproduce asexually and develop into another, infective, larval form 6. These larvae escape from the snail to complete the cycle by burrowing into human skin.

The World Health Organization estimates there are about 200 million people infected worldwide with *Schistosoma*, and 600 million at risk. Almost 300 million people die each year in Africa alone. Most cases of schistosomiasis occur in sub-Saharan Africa, but *Schistosoma* is endemic in Asia, South America, and Africa. Schistosomiasis is not found in the United States.

The number of cases of schistosomiasis has increased over the past few decades because of economic stability, improved irrigation systems, and an increase in the number of water reservoirs, which provide more habitats for the schistosomes' intermediate host—snails.

Diagnosis, Treatment, and Prevention

Diagnosis is most effectively made by microscopic identification of spiny eggs in either stool (*S. mansoni* and *S. japonicum*) or urine (*S. haematobium*) samples. The species of blood fluke causing a given infection can be ascertained based on the shape of the egg and the location of its spine. Immunological assays are used to identify antigen when laboratory technicians cannot find eggs in urine or stool samples but infection is suspected.

The drug of choice for treating schistosomiasis is praziquantel. Prevention of infection depends on improved sanitation, particularly sewage treatment, and avoiding contact with contaminated water. A vaccine against *S. haematobium* and another against *S. mansoni* are currently in clinical trials.

Disease at a Glance 21.8 summarizes the features of schistosomiasis.

TELL ME WHY

Why can people who avoid cats get infected by *Toxoplasma* anyway?

CLINICAL CASE STUDY

An Opportunistic Infection

A 43-year-old man, infected with HIV eight years previously, is brought to the hospital because of mental confusion and disorientation, headache, fever, and general discomfort. Further examination also indicates an enlarged liver.

1. What tests might the doctor run to diagnose the disease?
2. What opportunistic infection is likely to be the cause of the man's symptoms?
3. The doctor performs a CD4 count. What is most likely the patient's maximum count?
4. How did the patient most likely become infected?
5. What is the probable outcome for a non-immunocompromised patient infected with the same parasite?

EMERGING DISEASE CASE STUDY

Snail Fever in China

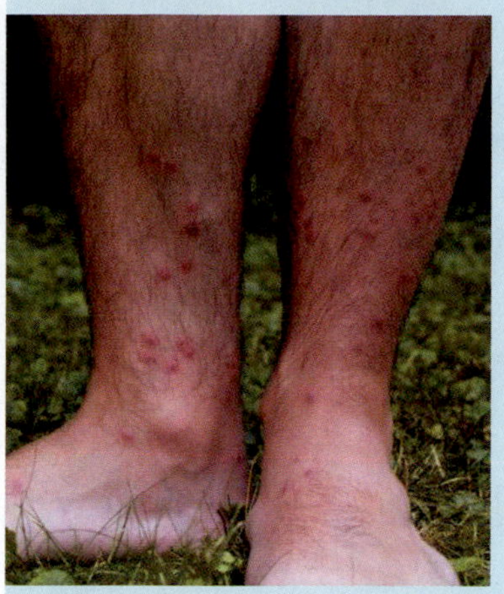

The Year of the Snake was an awful year for Jinhai, a year of too much rain, a year of stress and disturbances in his rural village, and a year that seemed to be filled with nothing but illnesses. His whole family of fishermen and farmers had had fever and chills, coughing, and muscle aches for months. Jinhai and his brothers had had severe rashes that itched mercilessly. His children and nieces and nephews were anemic, tired all the time, and having difficulty learning even the basics of their school lessons. And now, Jinhai had this hugely disfiguring swollen stomach! Yes, the Year of the Snake was a bad one. Jinhai's family was not alone. Schistosomiasis, known in rural China as "snail fever," is reemerging.

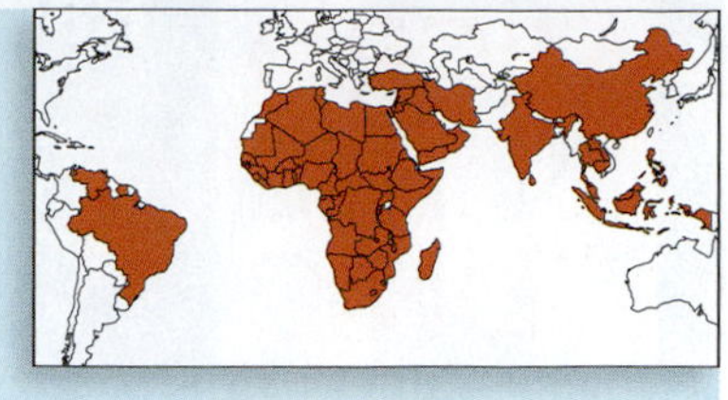

Nearly 900,000 people are infected in China, and an estimated 30 million people are at risk. Why? In the 1950s, the government regularly swept lakes of snails, nearly eradicating schistosomiasis, but with the waning of such control measures today—coupled with frequent floods—snail fever is reemerging in the countryside and even encroaching into cities.

Fortunately for Jinhai and his family and neighbors, the drug praziquantel taken for one to two days can kill the worm in their bodies. Coupled with good sewage treatment and snail suppression, schistosomiasis can be conquered.

1. Why is schistosomiasis called "snail fever"?
2. Why does good sewage treatment help control schistosomiasis?
3. Why is the initial symptom itching skin?

DISEASE AT A GLANCE 21.8

Schistosomiasis

Cause *Schistosoma mansoni* (Caribbean, South America, and Africa), *S. haematobium* (Africa and India), and *S. japonicum* (Asia) (parasitic helminth).

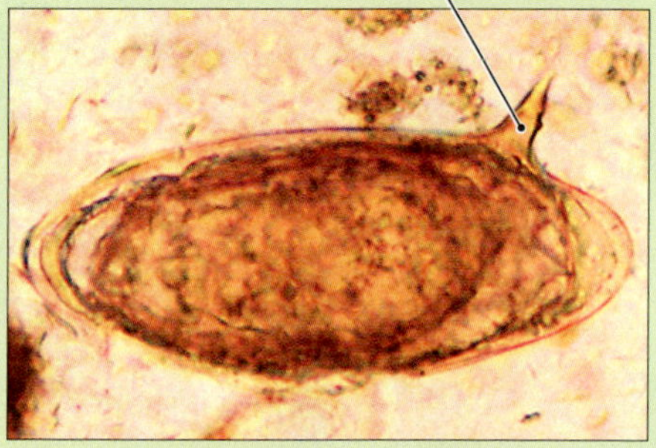

Egg of *Schistosoma mansoni*

Virulence factors Can burrow through human skin.

Portal of entry Larvae burrow through skin.

Signs and symptoms Swimmer's itch may occur at site of infection. There are no signs or symptoms as larvae travel the vascular system, but once the worms mature and mate, fever, chills, cough, and muscle aches occur. As trapped eggs die and calcify, renal failure, splenomegaly, increased blood pressure, bladder obstruction, distended abdomen, and heart failure may occur.

Incubation period A few days for transient dermatitis; one to two months for acute symptoms.

Susceptibility Individuals coming into contact with freshwater (lakes, rivers, or canals) in endemic areas are at risk.

Treatment Praziquantel.

Prevention In endemic areas, avoid swimming and wading in freshwater, boil water for one minute before drinking, and heat bath water to 150°F for five minutes before bathing.

MICRO IN THE CLINIC FOLLOW-UP

Rabbit Fever

Jackson has tularemia, caused by the bacterium *Francisella tularensis*. This microbe is carried by many animals, including those that live in the Ozarks' woods and fields—such as rodents and cottontail rabbits. In fact, tularemia is often referred to as "rabbit fever." Jackson became infected when he handled the freshly killed rabbit. Only a small number of bacterial cells were sufficient to successfully invade Jackson's body through the tiny scratch on his finger. Fortunately, the cooking process killed the bacteria in the rabbit stew, so the rest of the family did not become ill.

Tularemia is very difficult to diagnose and can be confirmed only by serological testing, which can take several weeks—time that a tularemia victim does not have! Dr. Taylor diagnosed Jackson based on his clinical symptoms and his recent contact with a wild rabbit. Her diagnosis proved correct: two days after Jackson begins taking the antibiotics, his fever finally breaks. Eventually, the aches disappear as well. The swollen lymph nodes in his right armpit prevent him from playing football at the homecoming game, but three weeks later the swelling goes away. Tularemia can be fatal, so Jackson is grateful to have survived. Dr. Taylor advises him to always wear gloves and a facial mask when skinning rabbits in the future. She also warns him to beware of ticks while walking in the woods because they, too, are notorious carriers of *F. tularensis*.

1. **How did the rabbit likely become infected with *F. tularensis*?**
2. **Why did Dr. Taylor not ask for a culture of the bacteria growing in the lesion on Jackson's finger?**

MM Check your answers to Micro in the Clinic Follow-Up questions in the MasteringMicrobiology Study Area.

Explore the Invisible

Visit the **MasteringMicrobiology Study Area** to challenge your understanding with practice tests, animation quizzes, and clinical case studies!

MasteringMicrobiology®

CHAPTER SUMMARY

Structures of the Cardiovascular System (p. 638)

1. The main arteries—vessels that carry blood away from the heart—are the pulmonary arteries (to the lungs) and the aorta (to the rest of the body). Arteries connect via capillaries to veins, which carry blood back to the heart. The main veins are the pulmonary veins (from the lungs) and the superior vena cava and inferior vena cava (from the rest of the body except the heart).
2. Blood consists of plasma—the liquid part of blood—and formed elements (blood cells and platelets). When clotting proteins are removed from plasma, serum remains.
3. Each of the paired atria and ventricles of the heart are separated by valves. These prevent backflow of blood into the atria when the ventricles contract.
4. The wall of the heart is composed of an outer, fibrous pericardium, a muscular myocardium, and an inner endocardium.
5. The blood flows through the cardiovascular system in this sequence: venae cavae, right atrium, right atrioventricular (tricuspid) valve, right ventricle, pulmonary semilunar valve, pulmonary arteries, lungs, pulmonary veins, left atrium, left atrioventricular (mitral) valve, left ventricle, aortic semilunar valve, aorta, arteries, capillaries, veins, venae cavae.

Bacterial Cardiovascular and Systemic Diseases (pp. 638–653)

1. **Septicemia** is the presence of pathogens in the blood. When septicemia causes the lymphatic vessels to become inflamed, **lymphangitis** results. **Bacteremia** refers specifically to bacterial septicemia, though physicians often use *bacteremia* and *septicemia* interchangeably.
2. **Toxemia** occurs when bacteria at a fixed site of infection release toxins. Living microbes release **exotoxins,** which can cause host cells to die or fail to function. **Endotoxin** (the **lipid A** portion of **lipopolysaccharide, LPS**) is released from most dying Gram-negative bacteria. Lipid A triggers fever, shock, inflammation, diarrhea, hemorrhaging, and widespread, potentially fatal blood clotting—**disseminated intravascular coagulation (DIC).**
3. Septicemia can progress to a potentially fatal condition called **septic shock.** Bacterial septicemia can trigger **petechiae**—skin lesions—and **osteomyelitis,** which is inflammation of bones. **Streptococcal toxic-shock-like syndrome (TSLS)** is a life-threatening toxemia.
4. Normal members of the microbiota can become pathogenic opportunists particularly in health care situations.
5. Endotoxin can also trigger macrophages to release potent cytokines (including **tumor necrosis factor, interleukins,** and **platelet activating factor**), which are defensive in localized infections but life threatening in septicemia.
6. The exact cause of septicemia in the majority of patients is hidden, a condition named **occult septicemia.**
7. Inflammation of the endocardium is called **endocarditis.** Platelets and clotting proteins surround infecting bacteria and form aggregations called **vegetations,** which resist the body's defenses. Fragments of vegetations and blood clots, each called an embolus, can lodge in small blood vessels, stopping the flow of blood and causing damage such as a stroke.
8. Physicians use sound waves to visualize vegetations in an echocardiogram.
9. Acute endocarditis develops quickly, whereas subacute endocarditis develops slowly.
10. When blood and lymph carry pathogens throughout the body, **systemic diseases** can develop.
11. **Brucellosis** (undulant fever, Bang's disease) is a **zoonosis** caused by the bacterium *Brucella* and characterized by fever that spikes each afternoon.
12. **Tularemia** is a zoonotic disease caused by *Francisella* spp.
13. **Plague** ("Black Death," or **bubonic plague**) is caused by *Yersinia pestis* and is characterized by **buboes** (inflamed lymph nodes), subcutaneous hemorrhaging, and blackened gangrenous tissues. *Yersinia* carries virulence plasmids that code for adhesins and type III secretion systems. The bacterium uses the latter to inject toxins to neutralize macrophages. **Pneumonic plague** occurs when plague bacilli spread from the blood to the lungs and in respiratory aerosols among people.
14. **Lyme disease,** characterized by a "bull's-eye" rash, malaise, neurological symptoms, and arthritis, is caused by *Borrelia burgdorferi,* which is vectored by *Ixodes* ticks.
15. **Ehrlichiosis** and **anaplasmosis** are tick-borne diseases affecting monocytes and neutrophils, respectively, and characterized by fever, chills, headache, and other flulike manifestations. *Ehrlichia chaffeensis* and a related rickettsial species, *Anaplasma phagocytophilum,* cause the diseases.

Viral Cardiovascular and Systemic Diseases (pp. 653–660)

1. **Viremia** is the condition in which viruses infect the blood.
2. **Infectious mononucleosis** (kissing disease) is caused by Epstein-Barr virus (EBV, HHV-4), which suppresses the programmed cell death **(apoptosis)** of infected B lymphocytes. Infected cells produce cancers such as **Burkitt's lymphoma** when certain additional factors are present.
3. EBV is usually transmitted in saliva. Viruses invade B lymphocytes and trigger long-term immune responses, which could possibly be the cause of chronic fatigue syndrome.
4. *Cytomegalovirus* (CMV) disease affects fetuses, newborns, and immunodeficient patients and is transmitted via bodily secretions.
5. **Yellow fever,** caused by the virus *Flavivirus* and carried by *Aedes* mosquitoes, develops in three stages ending in jaundice. The disease is significant in U.S. history.
6. **Dengue fever** (breakbone fever) is characterized by fever, severe pain, weakness, and rash. **Dengue hemorrhagic fever** involves a hyperimmune response to reinfection with dengue virus, resulting in internal bleeding, shock, and possibly death. Dengue viruses are carried by *Aedes* mosquitoes and affect millions of people each year.
7. **Ebola hemorrhagic fever** caused by *Ebolavirus* and **Marburg hemorrhagic fever** caused by *Marburgvirus* are endemic to Africa. The diseases have common features, including progression to severe internal bleeding; bleeding from the mouth, eyes, and ears; and death from shock, seizures, or kidney failure.

Protozoan and Helminthic Cardiovascular and Systemic Diseases (pp. 660–669)

1. **Malaria** is caused by apicomplexans of the genus *Plasmodium,* which infect the liver and erythrocytes, causing recurrent fever and chills, anemia, weakness, fatigue, and jaundice.
2. The life cycle of *Plasmodium* has three stages: The **exoerythrocytic phase** involves *Plasmodium* sporozoites, injected by a female *Anopheles* mosquito, infecting the liver. The **erythrocytic cycle** involves trophozoites cyclically infecting erythrocytes and the development of gametocytes. The **sporogonic phase** involves the development and sexual union of *Plasmodium* in the mosquito.
3. **Toxoplasmosis,** caused by the apicomplexan protozoan *Toxoplasma gondii,* affects AIDS patients and unborn children.
4. **Chagas' disease,** characterized by a swelling at the site of infection, is caused by *Trypanosoma cruzi.* This flagellated protozoan matures in the hindgut of a kissing bug *(Triatoma)* and enters a host when the bug's feces are rubbed into the bite wound. It circulates in the blood and has an intracellular stage, infecting macrophages and heart muscle cells, and can cause fatal heart damage.
5. Chagas' disease can be diagnosed by a xenodiagnosis, a procedure in which a physician allows an uninfected bug to feed on a person suspected of having the disease and then seeks the trypanosome in the bug.
6. Cercariae larvae of the blood fluke *Schistosoma* spp. burrow into the skin, causing swimmer's itch, and migrate through the vascular system. Adults feed on blood, lodge in the organs, and cause **schistosomiasis,** a potentially fatal disease. Snails are the intermediate host.

QUESTIONS FOR REVIEW

Answers to the Questions for Review (except Short Answer questions) begin on p. A-1.

Multiple Choice

1. Blood plasma includes
   a. erythrocytes and leukocytes.
   b. erythrocytes, leukocytes, and platelets.
   c. clotting proteins, erythrocytes, leukocytes, and platelets.
   d. nutrients and gases dissolved in liquid.
2. The aorta is
   a. an artery.
   b. a vein.
   c. a vena cava.
   d. a valve.
3. Another name for the mitral valve is the
   a. aortic semilunar valve.
   b. left atrioventricular valve.
   c. right atrioventricular valve.
   d. aortic valve.
4. *Toxemia* refers to
   a. pathogens in the blood that cause illness.
   b. blood poisoning due to toxins.
   c. relatively harmless bacteria in the blood.
   d. inflammation of the lymphatic vessels due to pathogens in the blood.
5. Which of the following are minute hemorrhagic skin lesions?
   a. petechiae
   b. interleukins
   c. vegetations
   d. emboli
6. Toxin released from cell walls of dying Gram-negative bacteria is
   a. tetanus toxin, for example.
   b. botulism toxin, for example.
   c. called endotoxin.
   d. an opportunistic toxin.
7. Which are more commonly associated with septicemia?
   a. Gram-negative bacteria
   b. Gram-positive bacteria
   c. viruses
   d. protozoa
8. How is septicemia introduced into the body?
   a. food
   b. sexual contact
   c. direct inoculation into the blood
   d. respiratory droplets
9. The type of endocarditis that develops slowly over a period of weeks or months is described as
   a. vegetative endocarditis.
   b. tachycardia endocarditis.
   c. acute endocarditis.
   d. subacute endocarditis.
10. What are vegetations of endocarditis called when released into the blood?
    a. stroke
    b. emboli
    c. ischemia
    d. necrosis
11. A cardiologist examines a patient with history of drug abuse and a recent tooth extraction and notes darkening under the fingernails. What should the diagnosis be?
    a. bacteremia
    b. malaria
    c. black plague
    d. endocarditis
12. Which of the following is important in developing dengue hemorrhagic fever?
    a. dengue virus
    b. immunological memory
    c. previous infection
    d. all of the above
13. Tularemia is also known as
    a. rabbit fever.
    b. mosquito fever.
    c. yellow fever.
    d. camel fever.
14. Which is spread from person to person?
    a. tularemia
    b. pneumonic plague
    c. bubonic plague
    d. Lyme disease

15. Which of the following statements is *false*?
    a. A kissing bug transmits *Trypanosoma cruzi* as it feeds.
    b. *Trypanosoma cruzi* is expelled on the skin of the host in the feces of the kissing bug.
    c. *Trypanosoma cruzi* enters the bloodstream when the host scratches the wound.
    d. *Trypanosoma cruzi* causes Chagas' disease.

16. The presence of lipid A in the outer membranes of Gram-negative bacteria
    a. coagulates blood in the host.
    b. causes these bacteria to be oxidase positive.
    c. triggers the secretion of a protease enzyme to cleave IgA in mucus.
    d. enables enteric bacteria to ferment glucose anaerobically.

17. Following a backpacking trip in Vermont, a hiker experienced flulike symptoms and noticed a round, red rash on her thigh. What is the likely cause of her illness?
    a. *Brucella melitensis*
    b. *Borrelia burgdorferi*
    c. *Francisella tularensis*
    d. *Yersinia pestis*

18. In malaria, which portion of the life cycle occurs in the mosquito?
    a. exoerythrocytic phase
    b. erythrocytic cycle
    c. sporogonic phase
    d. amastigote phase

19. The definitive host for *Toxoplasma gondii* is
    a. humans.
    b. cats.
    c. birds.
    d. mosquitoes.

20. Which of the following is *not* a high risk for endocarditis?
    a. transplant recipient
    b. cat owner
    c. patient with intravenous catheter
    d. IV drug user

21. A horror movie portrays victims of biological warfare with uncontrolled bleeding from the eyes, mouth, nose, ears, and anus. What actual virus causes these symptoms?
    a. *Ebolavirus*
    b. dengue virus
    c. *Cytomegalovirus*
    d. human immunodeficiency virus

22. The intermediate host for the blood fluke *Schistosoma* is the
    a. flea.
    b. mosquito.
    c. snail.
    d. rodent.

23. Streptococcal toxic-shock-like syndrome
    a. results from a hyperimmune response.
    b. is a toxemia.
    c. results from strep throat.
    d. is typically associated with tampon usage.

24. Which of the following is used to treat cases of malaria?
    a. Duffy antigen
    b. mosquito netting
    c. DEET
    d. artemisinin

25. A leading cause of heart failure in South America is
    a. attack by bloodsucking bats
    b. nifurtimox
    c. Chagas' disease
    d. malaria

Matching

Match the disease with the pathogen.

1. _____ Bacterial endocarditis
2. _____ Chagas' disease
3. _____ Bang's disease
4. _____ Tularemia
5. _____ Bubonic plague
6. _____ Lyme disease
7. _____ Infectious mononucleosis
8. _____ Yellow fever
9. _____ Ebola hemorrhagic fever
10. _____ Malaria

A. *Brucella*
B. *Yersinia*
C. Epstein-Barr virus
D. *Trypanosoma*
E. *Flavivirus*
F. *Borrelia*
G. *Francisella*
H. Filovirus
I. *Plasmodium*
J. *Streptococcus*

Match the distinguishing sign or symptom with the disease.

1. _____ "Bull's-eye" rash
2. _____ Bubo
3. _____ "Black vomit"
4. _____ Bleeding from eyes, mouth, nose
5. _____ Swimmer's itch

A. Lyme disease
B. Yellow fever
C. Marburg hemorrhagic fever
D. Black plague
E. Schistosomiasis

Fill in the Blanks

1. Blood leaving the heart on its way to the lungs passes through the ______________ valve.
2. Blood leaving the right atrium on its way to the right ventricle passes through the ______________ valve.
3. The smallest blood vessels where oxygen and nutrients diffuse into surrounding tissues are the ______________.
4. ______________ is the widespread coagulation of blood caused by endotoxin.
5. A patient exhibits fever that spikes daily around 4 p.m. The doctor suspects a systemic disease called ______________.
6. The three parts of the life cycle of *Plasmodium* are the ______________ cycle, the ______________ phase, and the ______________ phase.
7. Toxoplasmosis is caused by the protozoan ______________.
8. *Trypanosoma cruzi* causes ______________ disease.
9. The vector for bubonic plague is the ______________.
10. Small cutaneous hemorrhagic skin lesions are called ______________.

VISUALIZE IT!

1. On the figure of the life cycle of *Plasmodium*, label the three phases (A through C), and the name of the parasite at each stage (a through d).

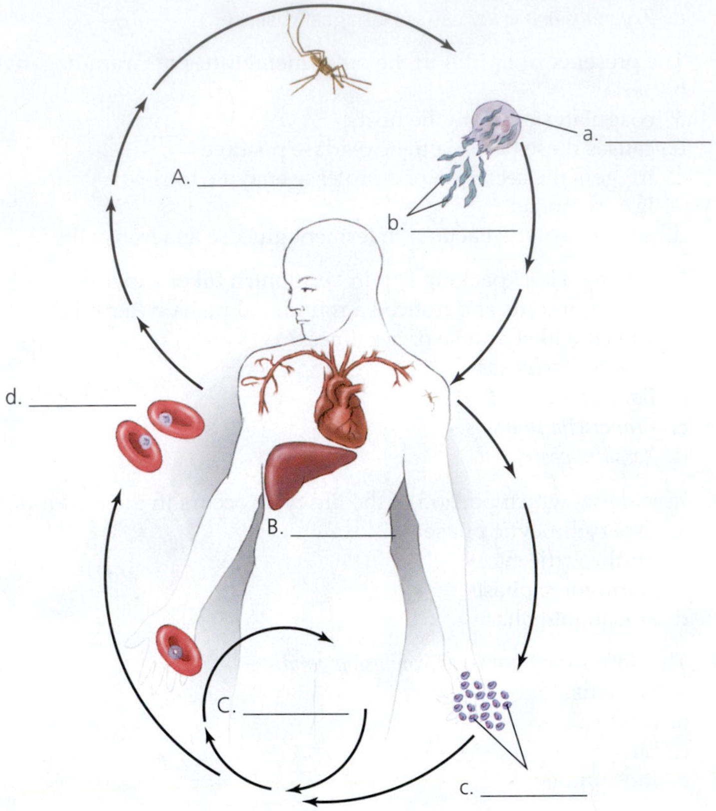
a. __________
A. __________
b. __________
d. __________
B. __________
C. __________
c. __________

2. In the drawing of the life cycle of *Ixodes* and its role in Lyme disease, circle the organisms that might be infected with *Borrelia*.

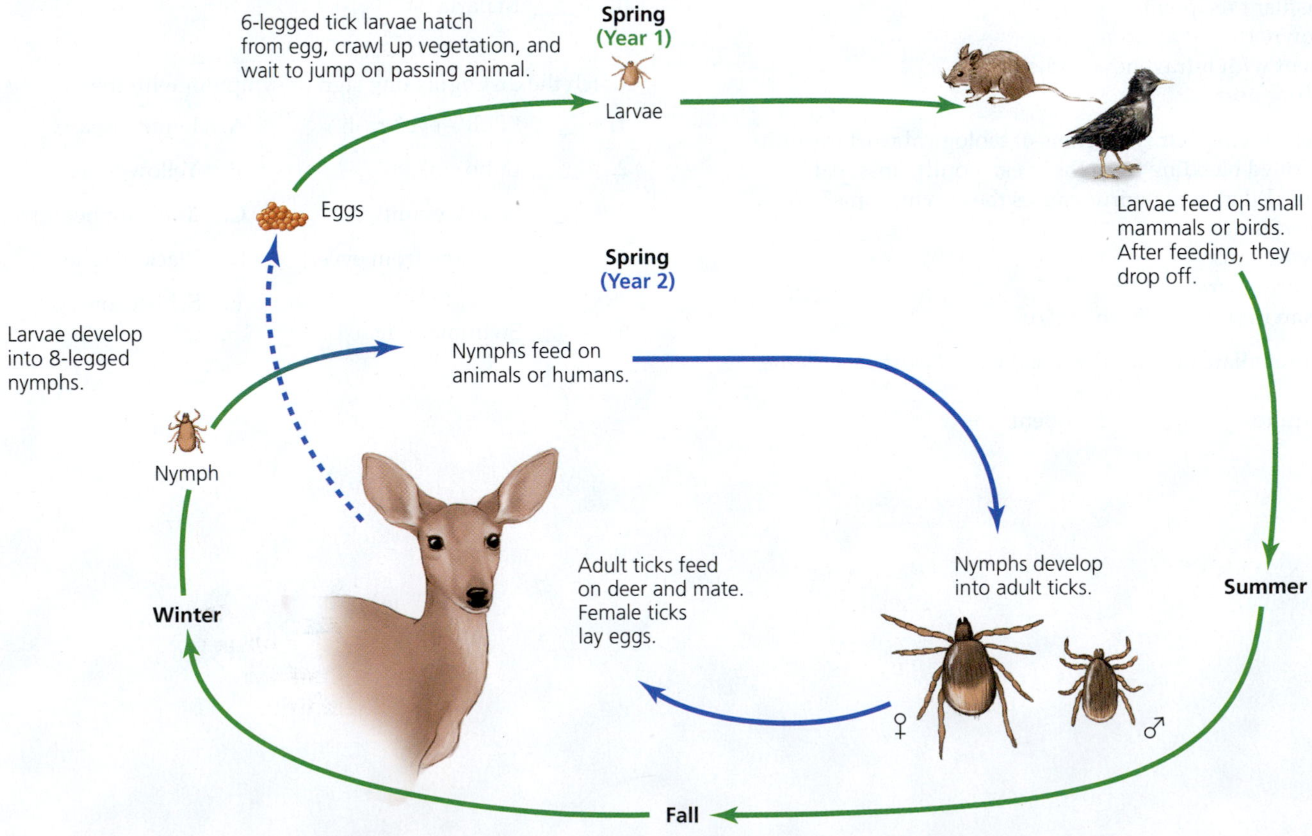
6-legged tick larvae hatch from egg, crawl up vegetation, and wait to jump on passing animal.
Spring (Year 1)
Larvae
Eggs
Larvae feed on small mammals or birds. After feeding, they drop off.
Spring (Year 2)
Larvae develop into 8-legged nymphs.
Nymphs feed on animals or humans.
Nymph
Adult ticks feed on deer and mate. Female ticks lay eggs.
Nymphs develop into adult ticks.
Summer
Winter
♀
♂
Fall

Short Answer

1. List the three layers of the heart in order from the outside.
2. Explain how *Brucella* evades antibodies and some antibacterial drugs.
3. Contrast septicemia, bacteremia, and toxemia.
4. Why is it rare for healthy people with normal immune systems to have bacterial blood infections?
5. Define *occult septicemia.*
6. How do vegetations prolong endocarditis?
7. Explain the situation in which the body's defensive cytokines (specifically tumor necrosis factor, interleukins, and platelet activating factor) can be life threatening.
8. After Tony had been very ill with cardiac problems, his hospital chart listed the cause as "culture negative" endocarditis. What does that mean?
9. Considering that CMV infects 50% of the adult population in the United States and 100% of the population in many other countries, why is there so little *Cytomegalovirus* disease in the world?
10. If 1–3 million people die annually from malaria, how do populations survive in endemic areas?
11. Most of the world's people have been infected with Epstein-Barr virus by age one and show no ill effects, even where medical care is poor. In contrast, individuals in industrialized countries are ordinarily infected after puberty, and these older patients tend to have more severe reactions to infection despite better health and medical care. Explain this paradox.
12. Explain how dead eggs of *Schistosoma* can cause damage in a human.
13. Contrast the severity of malaria caused by each of four species of *Plasmodium.*
14. Why do geneticists say there is a single species of *Brucella,* yet many veterinarians say there are four species?
15. Explain why poor hygiene actually protects against teenage mononucleosis.

CRITICAL THINKING

1. A blood bank refused to accept blood from a potential donor who had just had his teeth cleaned by a dental hygienist. Why did they refuse the blood?
2. An epidemiologist notices a statistical difference in the fatality rates between cases of Gram-positive bacteremia treated with antimicrobial drugs and treated cases of Gram-negative bacteremia—that is, patients with Gram-positive bacteremia are much more likely to respond to treatment and survive. Explain one reason why this might be so.
3. Why is it more difficult to rid a community of a disease transmitted by arthropods (for example, Lyme disease or malaria) than a disease transmitted via contaminated drinking water (for example, cholera)?
4. Compare and contrast the life cycles of *Trypanosoma brucei* (Chapter 20) and *T. cruzi.*
5. Explain how each of the following could lead to the reemergence of malaria in the United States: (a) global warming, (b) increased travel of individuals from endemic regions to the United States, (c) increased immigration of individuals from endemic regions to the United States, and (d) laws protecting wetlands.
6. Discuss why sickle-cell trait is advantageous to people living in malaria endemic areas, but is not advantageous in malaria-free areas.
7. Why are dengue, yellow fever, and malaria reemerging in areas where they have long been eradicated?
8. Why are ehrlichiosis and anaplasmosis considered emerging diseases given that they have been known for decades?
9. Why do physicians substitute trimethoprim and sulfamethoxazole for tetracycline when treating children and pregnant women infected with *Brucella*?
10. Most cases of tularemia in the United States occur in the late spring and summer months; very few cases ever occur in January. Why might this be so?
11. In 1861, yellow fever struck the Mississippi Valley, hitting Memphis, Tennessee, most forcefully. Half the citizens fled the city, and one-quarter of the remaining population died. Historical records indicate that the epidemic ceased only after winter frosts arrived. Why did frost stop the epidemic? Could yellow fever epidemics reappear in the United States?
12. Suppose scientists developed a vaccine for dengue that induced the production of memory T cells. After reviewing the characteristics of infection and reinfection, would you argue for or against the use of such a vaccine? Why?
13. Ebola hemorrhagic fever belongs to a group of diseases called "emerging diseases"—diseases that were previously unidentified or had never been identified in human populations. Emerging diseases are often first seen in developing countries such as the Democratic Republic of the Congo. What factors may explain the emergence of new diseases in these countries?
14. Propose some methods in addition to those in the text that public health officials could take to prevent transmission of *Schistosoma* to humans.

CONCEPT MAPPING

Using the following terms, draw a concept map that describes Lyme disease. For a sample concept map, see p. 94.
Or, complete this and other concept maps online by going to the MasteringMicrobiology Study Area.

Antibodies against 75–80% of patients *Borrelia burgdorferi*
Antimicrobials
Arthritis
Borrelia burgdorferi
Bull's-eye rash
Cardiac symptoms
Deer
Deer ticks
ELISA
Field mice
First phase
Genus: *Ixodes*
Mild flulike symptoms
Neurological symptoms
Reservoirs
Second phase
Spirochete
Symptoms
Third phase
Vectors

22 Microbial Diseases of the Respiratory System

MICRO IN THE CLINIC

This Cough Can Kill

The jumbo jet finally takes off from the airport in Johannesburg, South Africa, and Lance relaxes in his business-class seat. The next stop, 16 hours later, is Atlanta, Georgia. A second-year law student, Lance has just spent six weeks working as a volunteer for a South African law firm that specializes in human rights issues. His work took him to clients living in crowded shacks in some South African townships.

Now, Lance's thoughts turn to the upcoming fall term, when he will practically live in a study carrel, law books piled on the table. The constant, deep coughing of a woman seated beside him interrupts his reverie. He glances at her when he goes to the restroom. She is very thin and huddles against the armrest, looking weak and tired. When the flight attendant offers the woman food, she turns it down. Her loud, incessant coughing continues for the entire flight, and often she rubs her chest as if it hurts.

Four weeks later, Lance opens a letter from the state department of health and learns that he was exposed to a potentially deadly bacterium on that flight from South Africa. The letter advises Lance to have his physician administer a skin test to determine whether he has been infected, but it also tells him to wait for eight more weeks before getting this test.

What bacterium was Lance exposed to on the flight? Why does he need to wait so long before getting tested? **Turn to the end of the chapter (p. 709) to find out.**

MM **Explore More:** Test your readiness and apply your knowledge with dynamic learning tools at MasteringMicrobiology.

Structures of the Respiratory System

The respiratory system serves the vital function of exchanging gases between the atmosphere and the blood. Anatomists commonly divide its structures into two divisions—the upper respiratory system and the lower respiratory system **(FIGURE 22.1a)**. We begin our examination of respiratory anatomy by considering the upper respiratory system and the organs associated with it.

Structures of the Upper Respiratory System, Sinuses, and Ears

LEARNING **OUTCOMES**

22.1 Describe the structures of the upper respiratory system.

22.2 Describe the anatomical relationship between the pharynx and the middle ears and sinuses.

The upper respiratory system collects air; filters dust, pollen, microorganisms, and other contaminants from the air; and delivers it to the lower respiratory organs. The upper respiratory system includes:

- The *nose*, which is the only external part of the respiratory system.
- The *nasal cavity*, which is lined with hairs and a ciliated mucous membrane, receives air from the nose. The hairs filter large dust particles and organisms from the air, while the sticky mucus traps smaller particles and microbes. Ciliary action moves nasal mucus and its contents down, into the throat. *Sinuses*, which are air-filled, hollow regions of bones in the skull, often share fluids—and infecting microorganisms—with the nasal cavity.
- The *pharynx*, which is shared with the digestive system, is lined with a ciliated mucous membrane that propels mucus and contaminants into the digestive system. (Note that *mucus* is a noun, and the spelling *mucous* is an adjective.) A flap extending from the roof of the mouth called the *uvula* partially closes the opening between the nasal cavity and the pharynx during swallowing.

Ducts from the eyes carry contaminants (see Figure 15.3) into the pharynx. *Auditory* (*eustachian*, yū-stā´shŭn) *tubes* from the ears to the pharynx **(FIGURE 22.1b)** allow equalization of air pressure against the inner surfaces of the eardrums. Groups of lymphoid tissue called *tonsils* or *adenoids* are located near the junction of the nasal cavity, pharynx, and auditory tubes. The tonsils contain cells and chemicals to combat microbes in this frequent portal of entry.

The mucus of the upper respiratory system contains antimicrobial chemicals, including defensins, lactoferrin, and lysozyme. Defensins are antimicrobial peptides that act against Gram-positive and Gram-negative bacteria, fungi, and some viruses. Lactoferrin sequesters iron, making this important nutrient unavailable to microbial contaminants. Lysozyme breaks down peptidoglycan in the bacterial cell wall.

Structures of the Lower Respiratory System

The lower respiratory system consists of a series of tubes—the *larynx* (voice box), *trachea* (windpipe), *bronchi*, *bronchioles*, and smaller respiratory tubes—that lead to hundreds of millions of microscopic air sacs called *alveoli* (al-vē´ō-lī) **(FIGURE 22.1c and d)**, hich make up the *lungs*. Protective membranes called *pleurae* (plūr´ē) surround the lungs. The major respiratory muscle is the *diaphragm*, located below the lungs.

Because the structures of the lower respiratory system resemble an upside-down tree with branches that gradually decrease in diameter while increasing in number, anatomists refer to it as the *respiratory tree*. In this analogy, the trachea is the trunk, the bronchi and smaller tubes are the branches, and the alveoli represent leaves. When the diaphragm contracts, the lungs inflate, and air flows from the nose through the pharynx and into the respiratory tree.

The larynx contains the vocal cords, which vibrate as air flows over them. A cartilaginous flap called the *epiglottis* folds over the opening of the larynx during swallowing to prevent food and liquids from entering the lower respiratory organs.

Air flows from the larynx through the trachea, through the bronchi and bronchioles, and into the alveoli of the lungs. In the alveoli, oxygen enters the blood by passing through the thin walls of the alveoli and blood capillaries. Carbon dioxide diffuses from the capillaries into the alveoli to be exhaled. Relaxation of the diaphragm, accompanied by contraction of a different set of small muscles attached to ribs, allows the lungs to deflate, and air flows out.

A ciliated mucous membrane lines the trachea, bronchi, and bronchioles. The cilia beat synchronously about 1000 times per minute to carry mucus and trapped contaminants up to the pharynx. Physiologists refer to this action as a *ciliary escalator*. The mucus and its contents pass into the digestive system, where digestive juices destroy them. Further protection from pathogens is provided by *alveolar macrophages*, which enter the alveoli from blood capillaries and devour microorganisms. Secretory antibodies (IgA), which are present in tears, saliva, and respiratory mucus, also provide protection from many pathogens.

Normal Microbiota of the Respiratory System

LEARNING **OUTCOME**

22.3 Describe the normal microbiota of the upper and lower respiratory tracts.

The lower respiratory system normally lacks microorganisms because the ciliary escalator, secretory antibodies, and phagocytic cells clear the organs of contaminants. In contrast, many types of microorganisms live in the upper respiratory system. The nose, which is cooler than the rest of the respiratory system, supports the growth of bacteria such as species of *Haemophilus* (hē-mof´i-lus) and *Veillonella* (vī-lō-nĕl´ă). Gram-positive *Staphylococcus aureus* (staf´i-lō-kok´ŭs o´rē-ŭs) lives in the nasal

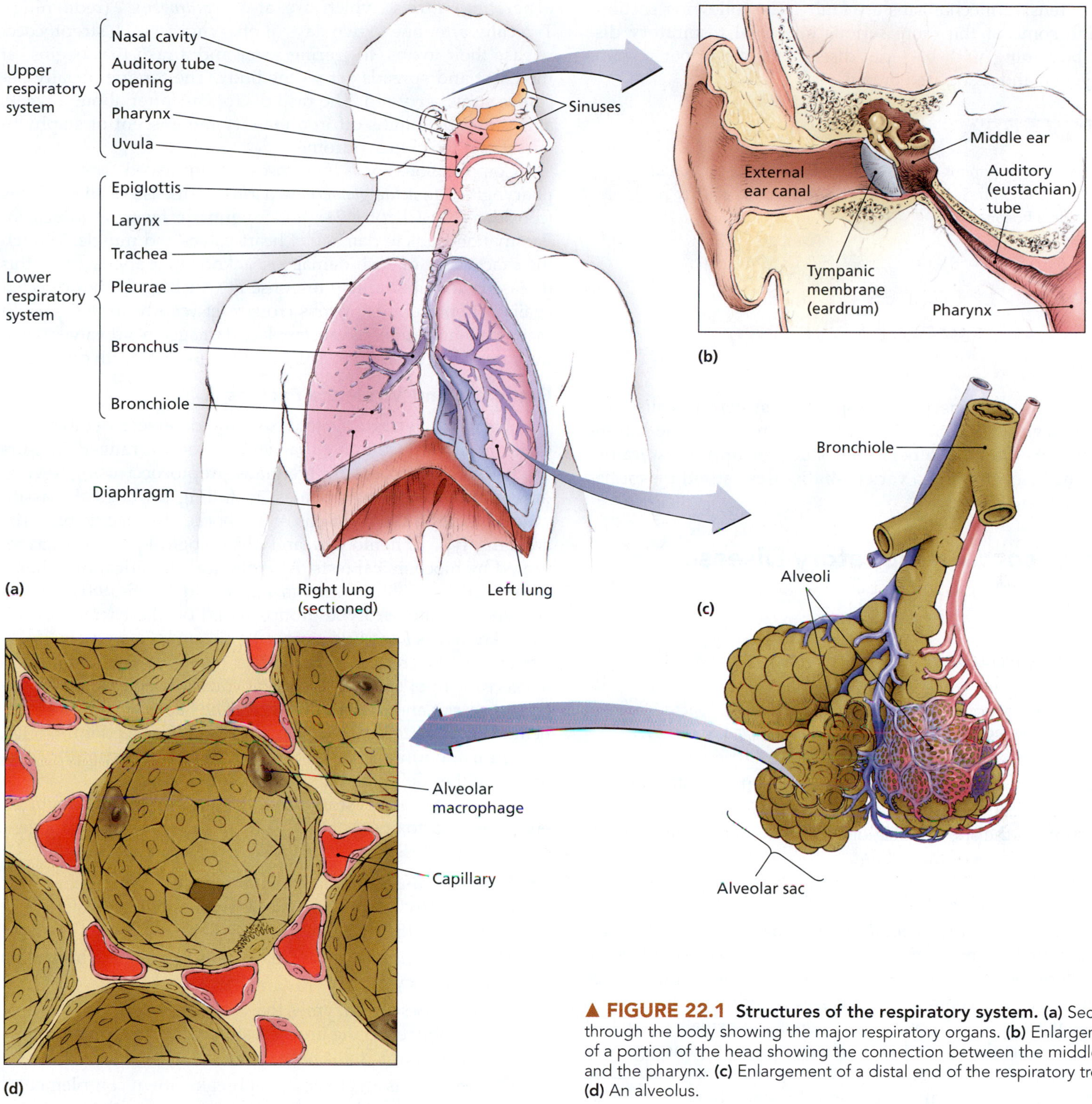

▲ **FIGURE 22.1 Structures of the respiratory system. (a)** Section through the body showing the major respiratory organs. **(b)** Enlargement of a portion of the head showing the connection between the middle ear and the pharynx. **(c)** Enlargement of a distal end of the respiratory tree. **(d)** An alveolus.

cavities of about one-third of healthy Americans without causing disease but can be an opportunistic pathogen. So-called **diphtheroids** (dif´thĕ-royds) also commonly colonize the nose and nasal cavity. These harmless Gram-positive bacteria resemble the bacterium that cause a respiratory disease called diphtheria (discussed shortly).

Gram-negative cocci, diphtheroids, opportunistic *Staphylococcus* spp., and alpha-hemolytic streptococci, including *Streptococcus pneumoniae* (strep-tō-kok´ŭs nū-mō´nē-ī), colonize the upper regions of the pharynx. The latter opportunistic pathogen causes most cases of pneumonia. Organisms in the pharynx can infect the middle ears and sinuses.

The normal microbiota of the upper respiratory system limit infection and disease by removing nutrients and releasing substances that inhibit the growth of pathogens. However, some normal microbiota can cause opportunistic diseases when other defensive mechanisms are faulty. The following sections examine some of the more serious microbial respiratory diseases, beginning with bacterial diseases of the upper respiratory system and associated organs.

TELL ME WHY

Why do patients with methicillin-resistant *Staphylococcus aureus* (MRSA) as part of their normal nasal microbiota pose a risk to other patients in a hospital?

Bacterial Diseases of the Upper Respiratory System, Sinuses, and Ears

Bacteria can infect the upper respiratory system to cause diseases such as sore throat. They can also spread into the sinuses and auditory tubes. We begin our study of upper respiratory infections by examining a variety of diseases caused by species of *Streptococcus*.

Streptococcal Respiratory Diseases

LEARNING **OUTCOMES**

22.4 Describe four respiratory diseases caused by *Streptococcus*.

22.5 Identify the structures, enzymes, and toxins of group A *Streptococcus (S. pyogenes)* that enable this bacterium to survive against the body's defenses and cause disease.

Physicians recognize a variety of diseases of the respiratory system and associated organs caused by species of *Streptococcus*, depending on the site of infection, the strain of bacteria, and the immune responses of the patient.

Signs and Symptoms

"Strep throat," or **streptococcal pharyngitis,** (strep´tō-kok´ăl far-in-jī´tis), is an inflammation of the pharynx caused by streptococci. The back of the pharynx appears red, with swollen lymph nodes and purulent (pus-containing) abscesses covering the tonsils (see Disease at a Glance 22.1 on p. 682). Pain during swallowing, bad breath, fever, malaise,[1] and headache accompany pharyngitis. If bacteria spread into the lower respiratory tract, they may cause inflammation of the larynx or bronchi, conditions known as **laryngitis** and **bronchitis,** respectively. Laryngitis manifests as hoarseness; bronchitis reduces airflow, encourages mucus accumulation in the lungs, and triggers coughing.

The disease **scarlet fever,** also known as *scarlatina*, can accompany pharyngitis caused by a strain of *Streptococcus* carrying a lysogenic bacteriophage that codes for *pyrogenic*[2] (fever generating) *toxins*, which are also *erythrogenic*[3] (reddening). Typically, after one to two days of pharyngitis, such streptococci release their toxins, triggering fever and a rash that begins on the chest and spreads across the body. The tongue usually becomes strawberry red. The rash disappears after about a week, and the skin sloughs off in a manner reminiscent of staphylococcal scalded skin syndrome.

Complications of some cases of untreated streptococcal pharyngitis are *acute glomerulonephritis*, a disease of the kidneys (considered in Chapter 24), and **rheumatic fever,** in which inflammation leads to damage of heart valves and muscle. Though the exact cause of such damage is unknown, it appears that this disease is an autoimmune response in which antibodies directed against streptococcal antigens cross-react with heart antigens. In many patients, surgeons must replace damaged heart valves when the patient reaches middle age. Heart failure and death can occur.

Pathogen and Virulence Factors

The bacterial genus *Streptococcus* is a diverse assemblage of Gram-positive, facultatively anaerobic cocci arranged in pairs or chains. Researchers differentiate streptococci using several different, overlapping schemes, including serological classification based on the reactions of antibodies to specific bacterial antigens, type of hemolysis, and physiological properties as revealed by biochemical tests. A serological classification scheme developed in 1938 by Rebecca Lancefield (1895–1981) divides streptococci into serotype groups based on the bacteria's antigens (known as *Lancefield antigens*).

Lancefield group A *Streptococcus* (synonymously known as *S. pyogenes*, pī-oj´en-ēz) is the major cause of bacterial pharyngitis and scarlet and rheumatic fevers. The bacterium shows beta-hemolysis after 24 hours on blood agar plates (see Figure 6.13). In contrast, harmless streptococci of the upper respiratory system are either nonhemolytic or alpha-hemolytic.

Strains of group A streptococci have a number of structures, enzymes, and toxins that enable them to survive as pathogens in the body. These include the following:

- *M protein* causes inhibition of complement component C3b, thereby interfering with opsonization and lysis (see Figure 15.9).
- The *hyaluronic acid capsule* may "camouflage" the bacterium from phagocytes.
- *Streptokinases* are enzymes that break down blood clots, presumably enabling group A streptococci to spread rapidly through damaged tissues.
- *C5a peptidase* is an enzyme that breaks down complement protein C5a, which is a chemotactic factor. With this enzyme, *S. pyogenes* decreases the movement of leukocytes into the site of infection.
- *Pyrogenic* (also called *erythrogenic*) *toxins* stimulate leukocytes to release cytokines that in turn stimulate fever, rash, and shock.
- *Streptolysins* lyse erythrocytes, leukocytes, and platelets.

[1]French, meaning "discomfort."

[2]From Greek *pyr*, meaning "fire," and *genein*, meaning "to produce."

[3]From Greek *erythros*, meaning "red," and *genein*, meaning "to produce."

One strain of group C *Streptococcus* (also called *S. equisimilis*, ek-wi-si´mi-lis), is also a pathogenic beta-hemolytic bacterium that causes some cases of streptococcal pharyngitis. However, unlike group A strep throat, this kind of group C pharyngitis does not lead to scarlet or rheumatic fevers.

Pathogenesis

Streptococci cause a variety of illnesses depending on the virulence factors present in the various strains. *S. pyogenes* frequently infects the pharynx, but the resulting disease is usually temporary, lasting only until adaptive immune responses against bacterial antigens (particularly M protein and streptolysins) clear the pathogen, usually within a week. Typically, strep throat and streptococcal bronchitis occur only when normal, competing microbiota are missing, when a large inoculum enables the bacterium to gain a rapid foothold before antibodies are formed against it, or when adaptive immunity is impaired. *S. pyogenes* can invade deeper tissues and organs through a break in a mucous membrane to cause *necrotizing fasciitis* (examined in Chapter 19).

Epidemiology

People spread *S. pyogenes* via respiratory droplets. Most cases of streptococcal pharyngitis occur during the winter and spring among elementary and middle school children, probably because of crowded conditions such as those in classrooms and day care centers. One person can spread sufficient bacteria to cause disease by coughing or sneezing within a radius of about 1.5 meters (5 feet) of another person.

Group A *Streptococcus* formerly claimed the lives of millions, but because it is sensitive to antimicrobial drugs, its significance as a deadly pathogen has declined. Nevertheless, the bacterium still sickens thousands of Americans annually.

The incidence of rheumatic fever has also declined significantly, from 7491 cases in 1964 to 112 cases in 1994, the last year rheumatic fever was nationally reportable. Epidemiologists do not understand fully the reason for this decline, but the introduction of antimicrobial drugs, which limit the growth of the bacterium and thereby the severity of streptococcal pharyngitis, likely played a role. It also appears that there has been a decline in the strains of *Streptococcus* that cause rheumatic fever.

Diagnosis, Treatment, and Prevention

Microbiologists estimate that fewer than 50% of patients diagnosed with strep throat actually have it; the rest have viral pharyngitis. Given that the manifestations of bacterial and viral pharyngitis are nearly identical, a sure diagnosis requires serological testing. Correct diagnosis is essential because bacterial pharyngitis is treatable with antibacterial drugs, whereas viral pharyngitis is not.

Because alpha-hemolytic and nonhemolytic streptococci are normally in the pharynx, the presence of streptococci in a respiratory sample is of little diagnostic value. Physicians should use immunological tests to identify the presence of antigens of beta-hemolytic strains of *Streptococcus*.

TABLE 22.1 Manifestations of Some Respiratory Diseases

Ailment	Manifestations
Common cold (viral)	Sneezing, rhinorrhea, congestion, sore throat, headache, malaise, cough
Influenza (viral)	Fever, rhinorrhea, headache, body aches, fatigue, dry cough, pharyngitis, congestion
"Strep" throat (bacterial)	Fever, red and sore throat, swollen lymph nodes in neck
Viral pneumonia	Fever, chills, mucus-producing cough, headache, body aches, fatigue
Bacterial pneumonia	Fever, chills, congestion, cough, chest pain, rapid breathing, and possible nausea and vomiting
Bronchitis (viral or bacterial)	Mucus-producing cough, wheezing
Inhalation anthrax (bacterial)	Fever, malaise, cough, chest discomfort, vomiting
Coronavirus respiratory syndromes (SARS, MERS)	High fever (>38°C), cough, shortness of breath

Oral penicillin is very effective against both *S. pyogenes* and *S. equisimilis*. Physicians prescribe erythromycin or another macrolide to treat penicillin-sensitive patients. Antibodies against M protein provide long-term protection against *S. pyogenes*; however, antibodies directed against the M protein of one strain provide no protection against other strains. For this reason, a person can have strep throat more than once.

Disease at a Glance 22.1 on p. 682 summarizes the manifestations of streptococcal pharyngitis. **TABLE 22.1** compares and contrasts strep throat and bronchitis with other respiratory diseases.

Diphtheria

LEARNING **OUTCOMES**

22.6 Discuss the transmission of *Corynebacterium diphtheriae* and the effect of diphtheria toxin.

22.7 Describe diphtheria.

Physicians have brought the deadly childhood disease **diphtheria** (dif-thēr´ē-ă) under control in industrialized countries using effective immunization. The disease is still a major threat to children living in less-developed regions of the world.

Signs and Symptoms

Diphtheria manifests as sore throat, localized pain, fever, pharyngitis, and the oozing of a fluid in the throat composed of intracellular fluid, blood clotting factors, leukocytes, bacteria, and the remains of dead pharyngeal and laryngeal cells. The fluid thickens into a thick *pseudomembrane* (**FIGURE 22.2**), which gives the disease its name: *diphthera* is a Greek word meaning

DISEASE AT A GLANCE 22.1

Streptococcal Pharyngitis (Strep Throat)

1 *Streptococcus pyogenes* in respiratory droplets from a nearby cough or sneeze enter the body.

2 Pharyngitis results; the back of the pharynx reddens, lymph nodes swell, and tonsils abscess. Fever, malaise, and headache are typical.

3 If bacteria spread to larynx, laryngitis may result.

4 Infection of the bronchi causes bronchitis.

5 Erythrogenic toxins trigger scarlet fever, a rash that spreads from the chest, a strawberry-red tongue, headache, chills, and muscle aches.

6 Rheumatic fever may develop, with pain in heart and joints.

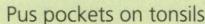

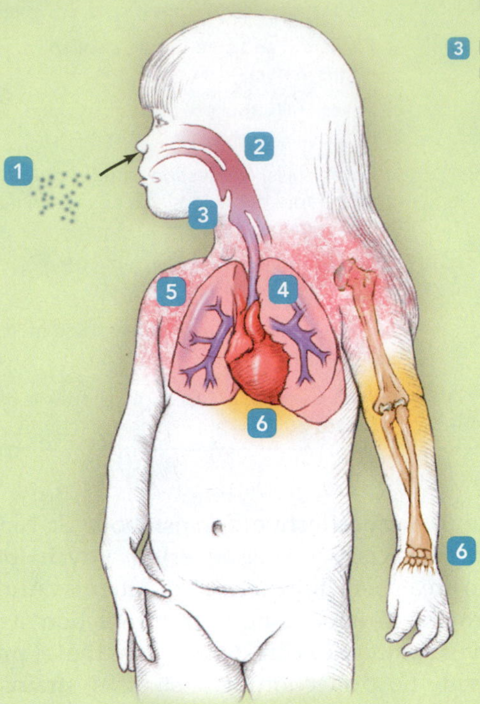

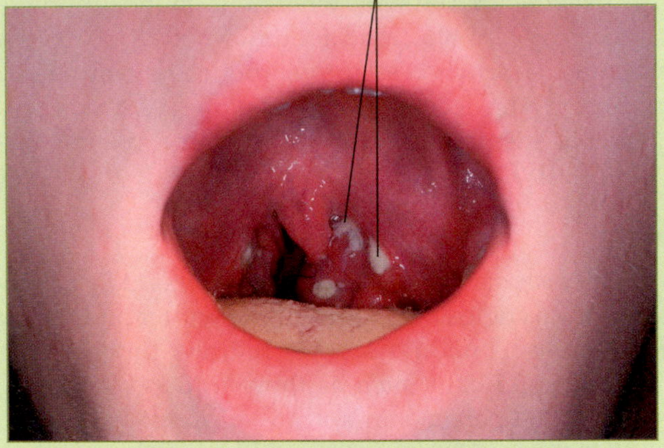

The reddened appearance and pus pockets of a throat with pharyngitis

Cause Group A streptococci (*Streptococcus pyogenes*).

Virulence factors Capsule, M protien, streptokinase, deoxyribonucleases, C5a peptidase, hyaluronidase, pyrogenic toxins, erythrogenic toxin, and streptolysins.

Portal of entry Upper respiratory tract.

Signs and symptoms Sore, red throat; difficulty swallowing; sudden fever; malaise; and loss of appetite. Can develop into scarlet fever, characterized by a "sandpapery" rash that first appears on the neck and chest and spreads all over the body, red "strawberry" tongue, headache, chills, and muscle aches. Rheumatic fever may also develop; symptoms include fever, joint pain (knees, ankles, elbows, and wrists), joint swelling, and possible cardiac problems (chest pain, shortness of breath).

Incubation period Strep throat: 3–5 days; scarlet fever: 1–2 days following strep symptoms; rheumatic fever: up to 20 days.

Susceptibility Children are generally the most susceptible. Rheumatic fever is more common in children ages 6–15.

Treatment A throat culture to diagnose bacterial cause is important. Once streptococcal infection is confirmed, the standard treatment is oral penicillin. Erythromycin may be given to those sensitive to penicillin. Scarlet fever and rheumatic fever are treated similarly.

Prevention Sick individuals are contagious for at least two days following antibiotic therapy and should be kept at home so as not to infect others. Sore throats in children should be taken seriously and treated so that rheumatic fever does not develop.

"leather." A pseudomembrane can adhere so tightly to the tonsils, uvula, roof of the mouth, pharynx, and larynx that it cannot be dislodged without ripping the underlying tissue and causing bleeding. In severe cases, the pseudomembrane can completely block the respiratory passages, resulting in death by suffocation.

Pathogen and Virulence Factors

Corynebacterium diphtheriae (kŏ-rī′nē-bak-tēr′ē-ŭm dif-thi′rē-ī) is a species of pleomorphic, non-endospore-forming, Gram-positive bacteria, which is ubiquitous in animals and humans, colonizing the skin and the respiratory, gastrointestinal, urinary, and genital tracts. The bacterium divides via a type of binary fission called snapping division, in which daughter cells remain attached to form characteristic V-shapes and side-by-side palisade arrangements **(FIGURE 22.3)**. Although other diphtheroids can be pathogenic, the agent of diphtheria is the most common pathogen of this genus.

Virulent *C. diphtheriae* contains a lysogenic phage that codes for *diphtheria toxin.* The phage and its toxin gene are directly responsible for the signs and symptoms of diphtheria. Like many bacterial toxins, diphtheria toxin consists of two polypeptides. One polypeptide binds to human growth factor receptor on many types of human cells, triggering endocytosis of the toxin.

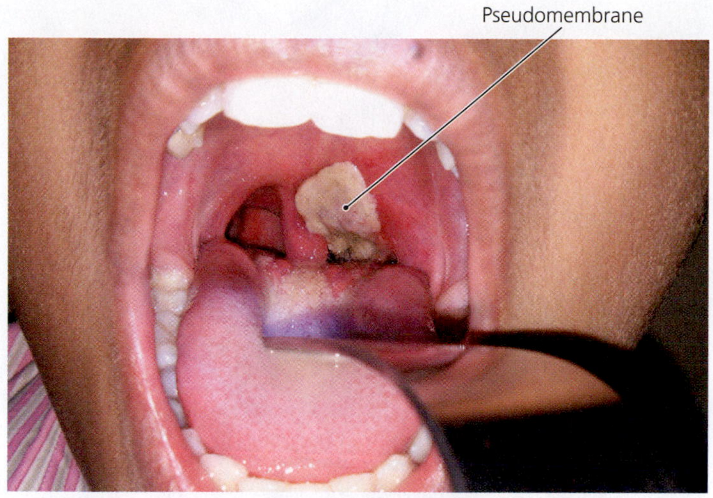

▲ **FIGURE 22.2** **A pseudomembrane, characteristic of diphtheria.**

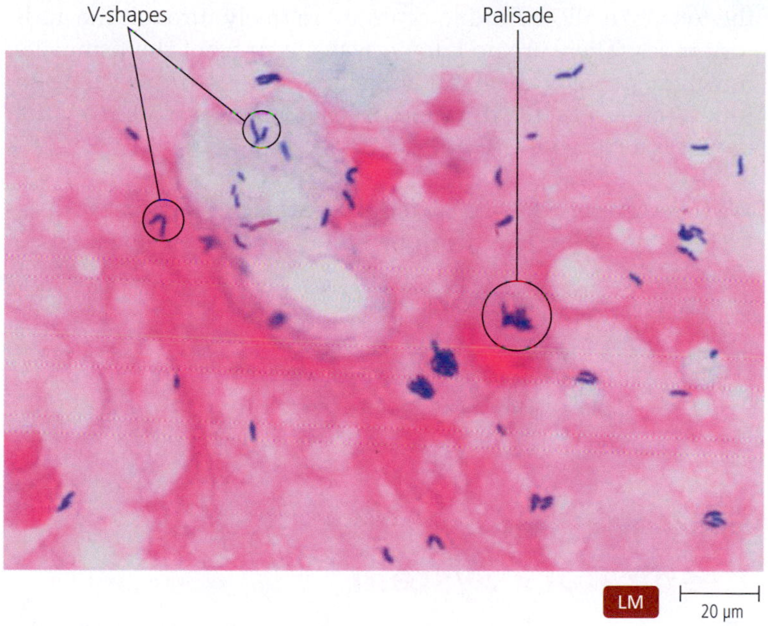

▲ **FIGURE 22.3** **Characteristic arrangements of Gram-stained cells of *Corynebacterium diphtheriae.*** *What process produces these cellular arrangements?*

Figure 22.3 V-shapes and palisade arrangements result from the type of binary fission called snapping division.

Once inside a cell, proteolytic enzymes split the toxin molecule, releasing the second toxin polypeptide into the cytosol. This polypeptide enzymatically destroys a eukaryotic elongation factor—a protein required for translation of polypeptides. Because the action of the toxin is enzymatic, a single molecule of toxin sequentially destroys every molecule of elongation factor in a cell, completely blocking all polypeptide synthesis and resulting in cell death. Diphtheria toxin is thus one of the more potent toxins known.

Pathogenesis and Epidemiology

C. diphtheriae is transmitted from person to person via respiratory droplets or skin contact. Infections with *C. diphtheriae* have different effects depending on a host's immune status and the site of infection. Infections in immune individuals are asymptomatic, whereas infections in immunocompromised individuals result in a mild respiratory disease. Respiratory infections of nonimmune people are most severe, resulting in the sudden and rapid signs and symptoms of diphtheria. Diphtheria is a leading cause of childhood death among the unimmunized.

Before immunization, hundreds of thousands of cases of diphtheria occurred in the United States each year. In contrast, health care workers reported only 20 cases total from 1992 to 2012.

Diagnosis, Treatment, and Prevention

Initial diagnosis of diphtheria is based on the presence of a pseudomembrane. Laboratory examination of the membrane or of tissue collected from the site of infection does not always reveal bacterial cells because the effects are due largely to the action of diphtheria toxin and not the cells directly. Diagnosis is based on observation of the pseudomembrane and an immunodiffusion assay, called an *Elek test*, in which antibodies against the toxin react with the toxin in a sample of fluid from the patient.

The most important aspect of treatment is the administration of antitoxin (immunoglobulins against the toxin) to neutralize diphtheria toxin before it binds to cells; once the toxin binds to a cell, it enters via endocytosis and kills the cell. Penicillin or erythromycin kills *Corynebacterium,* preventing the synthesis of more toxin. In severe cases, a blocked airway must be opened surgically or bypassed with a tracheostomy tube.

Because humans are the only known host for *C. diphtheriae,* the most effective way to prevent diphtheria is immunization. Immunization involves the DTaP[4] (diphtheria, tetanus, acellular pertussis) vaccine. This combines diphtheria and tetanus toxoids (deactivated toxins) with antigens of the pertussis (whooping cough) bacterium. It is administered at 2, 4, 6, and 15–18 months, and 4–6 years of age, followed by booster immunizations without pertussis antigens (Td[5] vaccine) every 10 years, and a single dose of Tdap once in adulthood.

Rhinosinusitis and Otitis Media

LEARNING **OUTCOME**

22.8 Describe the causes, manifestations, diagnosis, treatment, and the possibilities for preventing rhinosinusitis and otitis media.

Bacteria resident in the pharynx can infect the nose and sinuses or middle ears via their connections with the throat, causing **rhinosinusitis** or **otitis media**[6] (earache).

[4]Capital letters indicate full-strength vaccine.
[5]Lowercase letters indicate a reduced amount of vaccine.
[6]From Greek *ous,* meaning "ear," and *itis,* meaning "inflammation."

Signs and Symptoms

In the past, pain and pressure in the region of an inflamed sinus was called simply *sinusitis*, but sinusitis rarely occurs without also involving the nasal mucous membrane; so, the condition is now called *rhinosinusitis*. Malaise typically accompanies the headache and inflamed nasal passages.

Otitis media is a common and painful disease of early childhood that manifests with severe pain in the ears, which may end abruptly when the eardrum ruptures, releasing the pressure. Pressure on the eardrums may interfere with hearing and delay speech development in young children. Rarely, fever and vomiting may be present.

Pathogens and Virulence Factors

A number of bacteria that are normally part of the respiratory microbiota cause otitis media. These include *Streptococcus pneumoniae,* an alpha-hemolytic streptococcus that lacks specific Lancefield antigens (about 35% of cases); *Staphylococcus aureus* (1–2% of cases); *Haemophilus influenzae* (20–30% of cases); and *Moraxella catarrhalis* (10–15% of cases). These bacteria also cause most cases of rhinosinusitis. Additionally, *Streptococcus pyogenes* infections of the pharynx can spread into the sinuses and ears. There is some evidence that damage to the mucous membranes of the upper respiratory system and auditory tubes resulting from viral infections, cigarette smoke, and other irritants allows normal microbiota to become opportunistic pathogens.

Pathogenesis and Epidemiology

Infective agents spread from the pharynx into the sinuses via their connections with the throat. Similarly, the middle ears are infected via the auditory tubes. Inflammation, triggered by an infection, is responsible for the signs and symptoms of these diseases.

Rhinosinusitis is more common in adults than in children, presumably because an adult's sinuses are developed more fully. In contrast, otitis media is more common in children, because a child's auditory tubes are more horizontal and have smaller diameters; thus, they are invaded and blocked more easily. More than 85% of children develop otitis media, accounting for almost half of all visits to pediatricians, but as children's heads grow and they develop specific immunity to the various bacteria, otitis media becomes less common.

Diagnosis, Treatment, and Prevention

In most cases, physicians presume that the signs and symptoms of otitis media indicate bacterial infection, and they treat the condition with antibacterial drugs such as penicillin. Epidemiologists calculate that immunizations against influenzaviruses and *S. pneumoniae* could reduce the number of cases of childhood otitis media by more than 1 million per year.

Physicians may take drastic measures to treat and limit recurrent otitis media. These include lancing the eardrum of the infected ear to relieve pressure, installing plastic tubes through the eardrum to allow drainage of fluid and pus, and removing the tonsils to allow fluid to drain more freely through the auditory tubes. There are no known ways to prevent rhinosinusitis; however, flushing the nasal and sinus cavities with a saline solution can reduce the duration of the symptoms. One irrigation device is a Neti pot **(FIGURE 22.4)**.

▲ **FIGURE 22.4 Neti pot.** Flushing the nasal cavities with saline solution can reduce the duration of rhinosinusitis.

We have examined some bacterial diseases of the upper respiratory system. Now we turn our attention to viral infections of these organs.

TELL ME WHY

Why must diphtheria immunization be boosted every 10 years?

Viral Diseases of the Upper Respiratory System

Viral respiratory diseases, such as the common cold and influenza, are among the more common human diseases. In the following sections, we examine the primary viral disease of the upper respiratory system—the common cold.

Common Cold

LEARNING OUTCOME

22.9 Describe the manifestations and characteristics of common colds.

The common cold is named well—colds are among the most common of human diseases; an adult averages two colds each year. However, there is not a single common cause of the common cold—numerous viruses cause colds.

Signs and Symptoms

Everyone is familiar with the sneezing, rhinorrhea,[7] congestion, sore throat, malaise, and cough of a cold. Fever does not occur during a cold unless there is secondary bacterial infection. Signs and symptoms usually last a week, though sometimes a mild cough persists for several weeks.

Pathogens and Virulence Factors

Over 200 different *serotypes* (strains) of various viruses cause colds. The most common cold viruses—over 115 serotypes—are small viruses with naked polyhedral capsids in the genus *Enterovirus* of the family *Picornaviridae*[8] **(FIGURE 22.5)**. Other cold viruses include numerous serotypes of coronaviruses, over 30 different adenoviruses, several reoviruses, and a few paramyxoviruses.

Enteroviruses infecting the nose are commonly called *rhinoviruses* though this term is no longer an official viral taxon. Rhinoviruses are among the smallest of viruses, about 25 nm in diameter; 500 million rhinoviruses could sit side by side on the head of an ordinary straight pin. Almost all rhinoviruses attach to a human protein named *ICAM-1*, which is found on the cytoplasmic membranes of cells lining the nasal cavities. The complementary binding sites on the viruses lie at the bottom of deep, narrow clefts only 1.3–3 nm wide on the viral capsids. Such deep, narrow sites are protected from human antibodies and antiviral drugs; thus, prevention of common colds still eludes us.

All cold viruses reproduce most effectively at about 33°C, which is the temperature of the nasal cavity. Cold viruses cannot infect the lower respiratory system because of the system's higher temperature. Acid in the stomach and warmth inhibit cold viruses in the gastrointestinal tract.

[7]From Greek *rhis*, meaning "nose," and *rhoia*, meaning "to flow" (i.e., a runny nose).
[8]From Latin *pico*, meaning "small," and RNA virus.

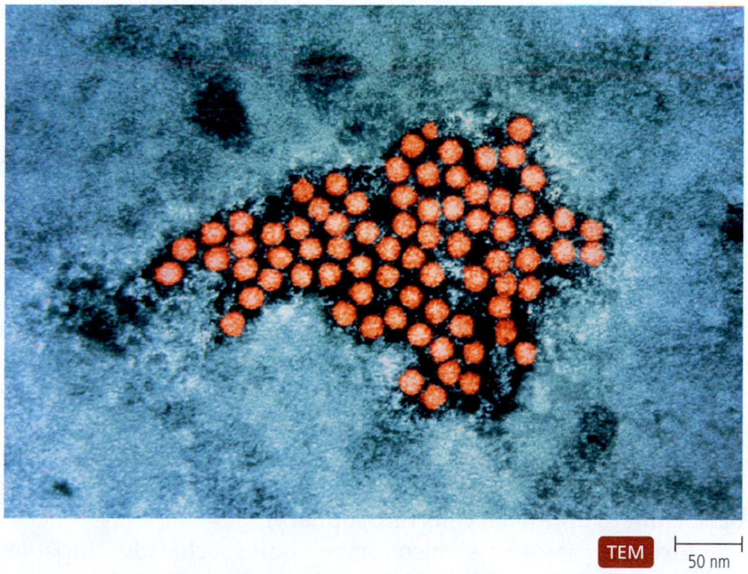

▲ **FIGURE 22.5 Rhinoviruses, the most common cause of colds.** The viruses (stained red) are able to infect microvilli of nasal cells despite a layer of mucus. *What other types of viruses cause common colds?*

Figure 22.5 Besides rhinoviruses (in the family Picornaviridae), adenoviruses, coronaviruses, reoviruses, and paramyxoviruses cause colds.

Pathogenesis

After attaching to cells of the nasal mucous membrane, cold viruses cause the cells to synthesize many more viruses, then kill the cells. The new viruses are released to infect still more cells. When cold symptoms are most severe, over 100,000 virions/ml of nasal mucus may be present. They remain infective for hours outside the body.

Infected cells lose their ciliary action and slough off when they die. These events trigger the release of inflammatory chemicals and stimulate nerve cells, triggering mucus production, sneezing, and localized inflammation of nasal tissue. Inflammation blocks nasal cavities, resulting in congestion.

Epidemiology

Rhinoviruses are extremely infective—a single virus is sufficient to cause a cold in 50% of infected individuals. Symptomatic or not, an infected person can spread viruses in aerosols produced by coughing or sneezing, via *fomites* (fō´mi-tēz, nonliving carriers of pathogens, such as door knobs), or via hand-to-hand contact. Epidemiologists do not agree on the most common method for transmitting cold viruses, but it appears that self-inoculation by touching the mucous membranes of the eyes, where tears can wash the viruses into the nasal cavity, is common. Studies have shown that explosive sneezes rarely transmit colds.

Although people of all ages are susceptible to rhinoviruses, they can acquire some immunity against serotypes that have infected them in the past. For this reason, children typically have six to eight colds per year, younger adults have two to four, and adults over age 60 have one or fewer. An isolated population may acquire a certain amount of *herd immunity* by sharing infections of specific strains of rhinoviruses; however, new serotypes introduced into a population by outsiders or by mutations ensure that no population is free of all colds.

Besides causing colds, adenoviruses can also cause pharyngitis. For some reason, epidemics of respiratory adenoviruses occur on military bases but rarely under the similar conditions found in college dormitories.

Diagnosis, Treatment, and Prevention

The manifestations of a cold are usually diagnostic. Laboratory tests are required only if the actual cause of infection is to be identified.

Although there are many home remedies and over-the-counter medicines for treating colds, none prevents the disease or provides a cure. A prescription drug—pleconaril—taken at the onset of symptoms reduces the seriousness and duration of rhinovirus disease. Adenovirus infection can be treated during the early stages with interferon. Antihistamines, decongestants, pain relievers, rest, and drinking fluids relieve cold symptoms and allow the body to mount an effective immune response.

However, they do not reduce the duration of the disease; a cold still lasts about a week.

Because there are hundreds of differing cold viruses, an effective vaccine against all colds is not practical; such a vaccine would have to protect against all viral serotypes. Nevertheless, a live, attenuated vaccine is available against adenoviruses, though the vaccine is used currently only for military recruits. Because some adenoviruses of animals are known to be oncogenic (cancer causing), there is concern that widespread use of adenovirus vaccine would increase the number of cancer cases. Recently, scientists have discovered an antigen common to rhinoviruses, opening the possibility for developing a vaccine against rhinoviral colds.

Hand antisepsis is probably the most important preventive measure against colds, especially if you have touched the hands of an infected person. Disinfection of fomites is somewhat effective in limiting the spread of cold viruses.

Table 22.1 on p. 681 compares and contrasts common colds with other respiratory diseases.

TELL ME WHY

Why is it inappropriate to treat a cold with penicillin, erythromycin, or ciprofloxacin?

Bacterial Diseases of the Lower Respiratory System

The lower respiratory organs are axenic;[9] that is, they are normally devoid of microorganisms. When bacteria successfully surmount the defenses of the respiratory system or when disease or stress weakens those defenses, life-threatening diseases can result. These include bacterial pneumonias, Legionnaires' disease, pertussis (whooping cough), and tuberculosis.

Bacterial Pneumonias

LEARNING | **OUTCOMES**

- **22.10** Define *pneumonia*, *empyema*, and *pleurisy*.
- **22.11** Describe pneumococcal, primary atypical (mycoplasmal), and *Klebsiella* pneumonias.
- **22.12** List five other bacterial species that can cause pneumonia.
- **22.13** Explain why scientists in different eras have classified mycoplasmas as viruses, Gram-negative bacteria, and Gram-positive bacteria.
- **22.14** Describe the features of mycoplasmal pneumonia.

The term **pneumonia** (nū-mō′nē-ă) describes an inflammation of the lungs in which the alveoli and bronchioles become filled with fluid. In some patients this fluid is pus—a condition known as *empyema*[10] (em-pī-ē′mă). When the pleurae become inflamed, a painful condition called *pleurisy* results. An estimated 2–5 million cases of pneumonia occur annually.

Physicians describe pneumonias according to the affected region of the lungs, the organism causing the disease, or the location of acquisition. For example, *lobar pneumonia* involves entire lobes of the lungs, and *mycoplasmal pneumonia* is pneumonia caused by the bacterium *Mycoplasma* (mī′kō-plaz-mă). *Healthcare associated pneumonia (HAP),* that is, pneumonia acquired in any health care setting, is a common illness among the elderly and immunosuppressed patients. One significant variety of HAP is *ventilator associated pneumonia (VAP)*. A ventilator is a machine that provides oxygen to patients through a tube inserted through the nose, mouth, or a hole in the larynx. Bacterial biofilms can form in and on the tubes, causing VAP.

A number of bacteria, viruses, and fungi cause pneumonia; bacterial pneumonias are the more serious and in adults the more common. In the following sections, we examine some of the more common bacterial pneumonias, beginning with the most prevalent—pneumococcal pneumonia.

Pneumococcal Pneumonia

Pneumonia caused by *Streptococcus*—also called *pneumococcal pneumonia*—is the most common type of bacterial pneumonia, accounting for most cases of *community acquired pneumonia (CAP)*.

Signs and Symptoms Pneumococcal pneumonia is usually lobar, affecting one or more lobes of the lungs. Signs and symptoms include fever, chills, congestion, cough, chest pain, which results in short, rapid breathing, and possibly nausea and vomiting. Blood frequently enters the lungs, causing coughed-up sputum to be rust colored. Neutrophils are present in the patient's sputum smear.

Pathogen and Virulence Factors Louis Pasteur discovered *S. pneumoniae,* and microbiologists have studied the bacterium extensively over the last 125 years, including sequencing the entire genomes of more than 10 strains; nevertheless, scientists still do not fully understand its pathogenicity. The bacterium rarely reaches the lungs because the ciliary escalator sweeps it away.

The bacterium is a Gram-positive coccus that is a normal member of the microbiota of the mouths and pharynges of 75% of humans without causing harm. Commonly known as **pneumococcus**, *S. pneumoniae* forms short chains or, more frequently, pairs. Pathogenic pneumococci secrete an attachment molecule, which is a poorly defined protein that mediates binding of the bacterium to epithelial cells of the pharynx.

Virulent serotypes also have polysaccharide capsules **(FIGURE 22.6)** that protect them from lysis by phagocytes. A capsule is required for virulence, as observed by Griffith during his experiments concerning bacterial transformation

[9] From Greek *a*, meaning "no," and *xenos*, meaning "foreigner."

[10] From Greek *en*, meaning "in," and *pyon*, meaning "pus."

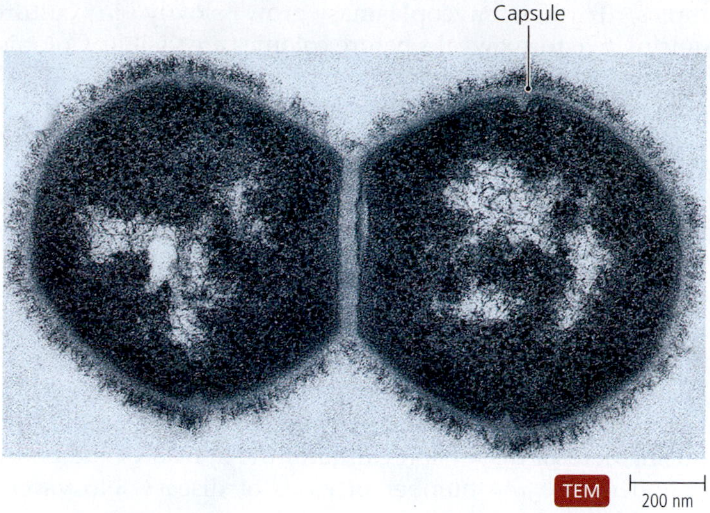

▲ **FIGURE 22.6** ***Streptococcus pneumoniae*, the most common cause of bacterial pneumonia.** The cells are commonly paired and covered with a polysaccharide capsule.

(see Figure 7.33). Unencapsulated variants are avirulent because alveolar macrophages clear them from the lungs. In addition, *S. pneumoniae* inserts into its cell wall a chemical called *phosphorylcholine,* which by binding to receptors on cells in the lungs stimulates endocytosis of the bacterium.

Pneumococci also secrete a cytotoxin called *pneumolysin,* which binds to cholesterol in the cytoplasmic membranes of ciliated epithelial cells, producing transmembrane pores that result in the lysis of the cells. Pneumolysin also suppresses the digestion of phagocytized bacteria by interfering with the action of lysosomes inside phagocytes.

Pathogenesis and Epidemiology Pneumococci are inhaled occasionally from the pharynx into lungs damaged either by a previous viral disease, such as influenza or measles, or by other conditions, such as alcoholism, congestive heart failure, or diabetes mellitus. Phosphorylcholine triggers endocytosis by lung cells, the capsule protects the bacterium, and thereby pneumococci live in and eventually kill lung cells. From their intracellular "hiding place," *S. pneumoniae* can pass into the blood and brain to cause bacteremia and meningitis.

As the bacteria multiply in the alveoli, they damage the lining of the alveoli, allowing erythrocytes, leukocytes, and blood plasma to enter the lung. This fluid fills the alveoli, reducing the lung's ability to transfer oxygen to the blood and causing the pneumonia. Leukocytes attack the bacteria, in the process secreting inflammatory and fever-producing chemicals, which add to the manifestations of the disease.

The body acts to limit migration of bacteria throughout the lungs by binding the microbes with the active sites of secretory IgA. The rest of the antibody molecule then binds to mucus, enabling mucus-enveloped bacteria to be swept from the airways by the action of ciliated epithelium. Pneumococcus counteracts this defense by secreting *secretory IgA protease,* which destroys IgA.

Pneumococcal pneumonia constitutes about 85% of all cases of bacterial pneumonia, including about 30% of healthcare associated pneumonias. It occurs most frequently in fall and winter in children, the elderly, alcohol and drug abusers, diabetics, and AIDS patients—groups whose immune responses are not fully active.

Diagnosis, Treatment, and Prevention Medical laboratory technologists can quickly identify diplococci in Gram stains of sputum smears (see Disease at a Glance 22.2 on p. 690). Because the bacteria are sensitive to most antimicrobial drugs, health care workers must collect samples for smearing before antibacterial therapy has begun. Historically, laboratories confirm the presence of pneumococci with anticapsular antibodies that cause pneumococcal capsules to swell—a so-called *quellung reaction.*

Penicillin has long been the drug of choice against *S. pneumoniae,* though about a third of pneumococcal isolates are now penicillin resistant. Cephalosporin, erythromycin, clindamycin, vancomycin, or fluoroquinolones are effective alternative treatments.

The Centers for Disease Control and Prevention (CDC) recommends a vaccine against pneumococcus administered at 2, 4, 6, and 12–15 months of age and for all adults over age 65.

Primary Atypical (Mycoplasmal) Pneumonia

Mycoplasmal pneumonia is the leading type of pneumonia in children and young adults.

Signs and Symptoms Early symptoms of mycoplasmal pneumonia, including fever, malaise, headache, sore throat, and excessive sweating, are not typical of other types of pneumonia; thus mycoplasmal pneumonia is called **primary atypical pneumonia**. The body responds to infection with a persistent, unproductive cough in an attempt to clear the lungs of the pathogen and accumulated mucus. Primary atypical pneumonia may last for several weeks, but it is usually not severe enough to require hospitalization or to cause death. Because symptoms can be mild, the disease is also sometimes called *walking pneumonia.*

Pathogen and Virulence Factors *Mycoplasma pneumoniae* is a strictly aerobic, encapsulated mycoplasma. Mycoplasmas lack cell walls, allowing them to have a variety of shapes—they are *pleomorphic* (**FIGURE 22.7**). Further, mycoplasmas have lipids in their cytoplasmic membranes called *sterols,* a feature lacking in other prokaryotes.

Mycoplasmas are the smallest free-living microbes; that is, microbes that can grow and reproduce independently of other cells. Their diameters range from 0.1 μm to 0.8 μm. Originally, scientists thought mycoplasmas were viruses because their small size and flexibility enable them to squeeze through the pores of filters that were then commonly used to remove bacteria from solutions; however, mycoplasmas contain both RNA and DNA, and they divide by binary fission—traits that viruses lack.

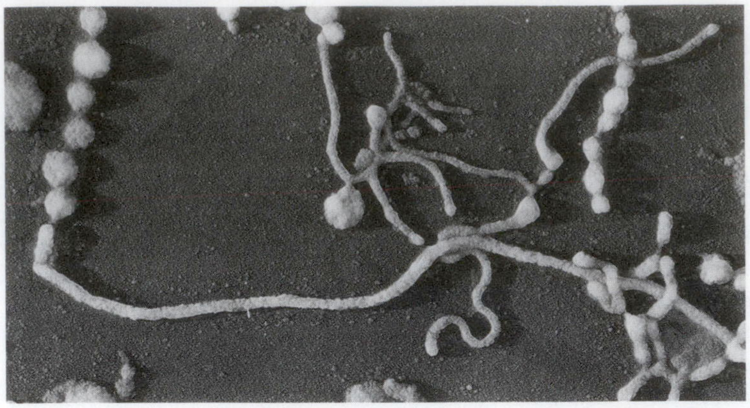

▲ **FIGURE 22.7 Pleomorphic forms of *Mycoplasma*.** *Why do these Gram-positive bacteria appear pink when Gram stained?*

Figure 22.7 Mycoplasmas lack cell walls; therefore, the decolorizing step of the Gram procedure removes crystal violet from the cells.

Before analysis of mycoplasmal rRNA sequences revealed that they are similar to Gram-positive organisms, mycoplasmas were classified as Gram-negative bacteria. Modern taxonomists and the second edition of *Bergey's Manual of Systematic Bacteriology* now categorize mycoplasmas as low G + C, Gram-positive bacteria in the phylum Firmicutes. Despite being Gram-positive, they appear pink when Gram stained, because they lack cell walls.

M. pneumoniae is one of the few mycoplasmas that causes human disease. This is due partly to its production of an adhesive protein that attaches specifically to receptors located at the bases of cilia of epithelial cells lining the respiratory tracts of humans and to its capsule, which provides protection from phagocytosis.

Pathogenesis Attachment of *M. pneumoniae* to the base of cilia causes the cilia to stop beating, and colonization eventually kills the epithelial cells. This interrupts the normal removal of mucus from the respiratory tract by the ciliary escalator, allowing colonization by other bacteria and causing a buildup of mucus that irritates the respiratory tract.

Epidemiology Nasal secretions spread *M. pneumoniae* among people in close contact, such as classmates, family members, and dormitory residents. The disease is uncommon in children under age 5 or in older adults, but it is the most common form of pneumonia seen in high school and college students. However, because primary atypical pneumonia is not a reportable disease and is difficult to diagnose, the actual incidence of infection is unknown.

Primary atypical pneumonia occurs throughout the year. This lack of seasonality is in contrast to pneumococcal pneumonia, which is more commonly seen in the fall and winter.

Diagnosis, Treatment, and Prevention Diagnosis of primary atypical pneumonia is difficult because mycoplasmas are small and difficult to detect in clinical specimens or tissue samples. Further, mycoplasmas grow slowly in culture, requiring two to six weeks before colonies are visible. Colonies of *M. pneumoniae* have a grainy appearance when the bacteria grow on solid surfaces, unlike the "fried-egg" appearance of colonies of other species of *Mycoplasma*. Complement fixation, hemagglutination, and immunofluorescent tests confirm a diagnosis, but such tests are nonspecific and are not by themselves diagnostic.

Physicians treat primary atypical pneumonia with antimicrobials such as erythromycin or doxycycline. Prevention is difficult because patients are often infective for long periods without signs or symptoms, and they remain infective even while undergoing antimicrobial treatment. Nevertheless, frequent handwashing, avoidance of contaminated fomites, and reducing aerosol dispersion can limit the spread of the pathogen and reduce the number of cases of disease. No vaccine against *M. pneumonia* is available.

Klebsiella Pneumonia

Gram-negative bacteria are the leading cause of nosocomial infections, and pneumonias caused by Gram-negative bacteria are among the leading causes of nosocomial deaths. *Klebsiella* pneumonia is one type of Gram-negative bacterial pneumonia.

Signs and Symptoms Besides the common signs and symptoms of bacterial pneumonia—coughing, fever, and chest pain—*Klebsiella* pneumonia often involves destruction of alveoli, resulting in the production of thick, bloody sputum. Further, *Klebsiella* pneumonia patients often have recurrent chills. Mortality rates are higher than with pneumococcal or mycoplasmal pneumonias.

Pathogen and Virulence Factors *Klebsiella pneumoniae* (kleb-sē-el´ă nū-mō´nē-ī) is an opportunistic pathogen that infects the respiratory systems of humans and animals following inhalation. It is a nonmotile, Gram-negative rod that produces a prominent capsule **(FIGURE 22.8)**, giving *Klebsiella* colonies

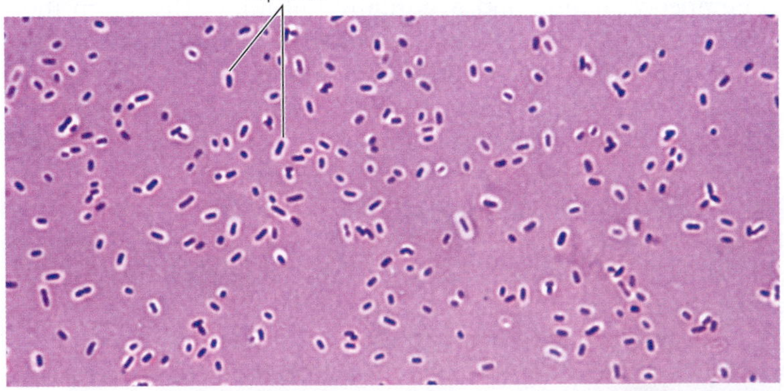

▲ **FIGURE 22.8 The prominent capsule of *Klebsiella pneumoniae*.** *How does the capsule function as a virulence factor?*

Figure 22.8 Capsules inhibit phagocytosis and intracellular digestion by phagocytic cells.

a mucoid appearance and protecting the bacterium from phagocytosis. Besides causing pneumonia, *K. pneumoniae* may also be involved in meningitis, wound infections, and urinary tract infections.

Pathogenesis and Epidemiology *K. pneumoniae* kills alveolar cells and often invades the blood, resulting in bacteremia. When the bacterial cells die, they release endotoxin, which can trigger shock and disseminated intravascular coagulation, leading to death.

Alcoholics and other patients with compromised immunity such as the elderly, people with AIDS, and very young children are at greater risk of pulmonary disease, including *Klebsiella* pneumonia, because of their poor ability to clear aspirated oral secretions from their respiratory tracts.

Diagnosis, Treatment, and Prevention Physicians diagnose *Klebsiella* pneumonia from its signs and symptoms and by culturing *Klebsiella* from sputum specimens. There is no specific treatment for the disease other than supportive care, rest, and fever-reducing drugs. Antimicrobial drugs, such as cephalosporin, imipenem, and quinolones, may be used against *Klebsiella*. Because of tissue damage and the release of endotoxin, damage to the lungs is often permanent and can be fatal despite treatment. No vaccine is available, and prevention involves good aseptic technique by health care workers.

Other Bacterial Pneumonias

Other species of bacteria also cause pneumonia. Normal respiratory microbiota such as the pleomorphic Gram-negative *Haemophilus influenzae* and Gram-positive *Staphylococcus aureus* cause pneumonia with manifestations and treatment similar to those for pneumococcal pneumonia. *S. aureus* produces pus in a chest cavity in about 3% of patients.

Fever, chills, cough, difficulty breathing, and frothy bloody sputum characterize a form of pneumonia called **pneumonic plague,** which is produced by the bubonic plague bacillus, *Yersinia pestis* (yer-sin´ē-ă pes´tis). This bacterium can enter the lungs in respiratory droplets or via the blood (see Figure 21.6d) and can cause pneumonia in just a few hours. Rapid shock and death result if a pneumonic plague patient is not treated promptly, but mortality is reduced to about 5% with treatment with streptomycin or gentamicin. (Chapter 21 discusses plague in more detail.)

Chlamydias, which are Gram-negative, obligate intracellular parasites, can also cause respiratory diseases. Chlamydias form extremely small (0.2–0.4 μm diameter), resistant structures called *elementary* bodies that act as infectious agents. Here we examine pneumonia caused by two chlamydias in the genus *Chlamydophila* (formerly known as *Chlamydia*).

Chlamydophila psittaci[11] (kla-mē-dof´ĭ-lă sit´ă-sē) causes **ornithosis**[12] (ōr-ni-thō´sis), which is a disease of birds that can be transmitted to humans, in whom it typically causes flulike symptoms, though in some patients severe pneumonia occurs. Physicians treat ornithosis with tetracycline, erythromycin, or a fluoroquinolone.

Chlamydophila pneumoniae, spread in respiratory droplets, causes pneumonia as well as bronchitis and rhinosinusitis. Most infections with this bacterium are mild, producing only malaise and a chronic cough, and do not require specific treatment, though erythromycin, fluoroquinolones, and tetracycline can be pescribed. The prevalence of chlamydial pneumonia is unknown because most cases are never diagnosed or reported.

Disease at a Glance 22.2 on p. 690 summarizes the features of eight types of bacterial pneumonia. Table 22.1 on p. 681 compares and contrasts bacterial pneumonia with other respiratory diseases.

Another bacterial disease of the lower respiratory system is also a pneumonia, but physicians and clinicians have given it a specific name—legionellosis.

Legionnaires' Disease

LEARNING | OUTCOME

22.15 Describe the features of Legionnaires' disease.

In 1976, joyous celebration of the 200th anniversary of the Declaration of Independence was curtailed in Philadelphia when over 200 American Legion members attending a convention were stricken with severe pneumonia; 29 died. After extensive epidemiological research, a new disease was identified and dubbed **Legionnaires' disease,** or *legionellosis.* A previously unknown pathogen, *Legionella* (lē-jŭ-nel´lă), causes this disease.

Signs and Symptoms

Legionnaires' disease is characterized by common features of pneumonia—fever, chills, a dry nonproductive cough, and headache; pleurisy—inflammation of the pleurae—may also develop. Complications involving the gastrointestinal tract, central nervous system, liver, and kidneys are common. If Legionnaires' disease is not treated promptly, pulmonary function rapidly decreases, resulting in the death of up to 50% of patients.

Pathogen and Virulence Factors

Cells of *Legionella* are aerobic, slender, pleomorphic, Gram-negative bacteria that are classified in the taxon Gammaproteobacteria. They are extremely fastidious in their nutrient requirements, and laboratory media for culture of *Legionella* must be enhanced with iron salts and the amino acid cysteine. **FIGURE 22.9** shows colonies growing on one special medium—buffered charcoal yeast extract agar. Nineteen species are known to cause disease in humans, but 85% of all infections in humans are caused by *L. pneumophila*[13] (noo-mō´fi-lă).

[11]From Greek *psittakos,* meaning "parrot."
[12]From Greek *ornith,* meaning "bird."
[13]From Greek *pneuma,* meaning "breath," and *philos,* meaning "love."

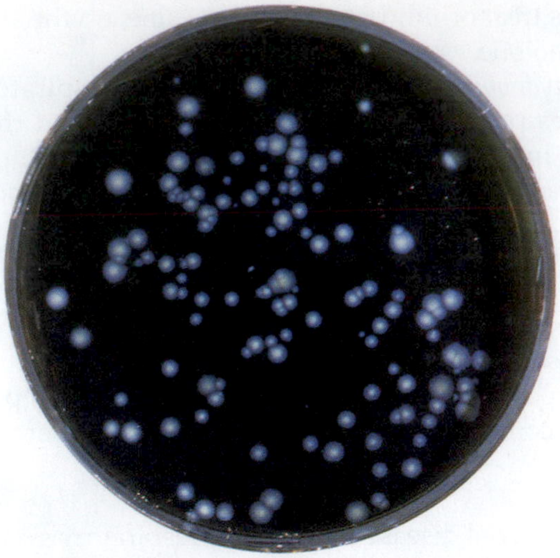

▲ **FIGURE 22.9** ***Legionella pneumophila*** **growing on buffered charcoal yeast extract agar.** This bacterium cannot be grown in culture without such special media.

Pathogenesis

L. pneumophila presented a problem for early investigators: the bacterium is nearly ubiquitous in moist environmental samples, yet it cannot grow on common laboratory media. In the original epidemic, for example, *Legionella* was cultured from condensation in hotel air conditioning ducts, an environment that seems unsuitable for such a fastidious microorganism. Investigations revealed that *Legionella* invades freshwater protozoa, typically amoebae, and reproduces inside phagocytic vesicles; thus, the bacterium survives in the environment as an intracellular parasite. Protozoa release *Legionella*-filled vesicles, and humans acquire the disease by inhaling the vesicles.

Within human cells, *L. pneumophila* lives much as it does in the environment—as an intracellular parasite of macrophages and other cells. The bacterium kills human cells, causing tissue destruction and triggering inflammation of the lungs. For unknown reasons, *Legionella* rarely spreads outside the lungs.

Epidemiology

Legionella tolerates heat and chlorination, so it can live in water pipes and other domestic water sources; in fact, legionellosis was not a common disease until modern devices provided a suitable means of transmitting the bacteria to humans: showers, vaporizers, spa whirlpools, hot tubs, air conditioning systems, and cooling towers produce aerosols containing *Legionella*. The Clinical Case Study on p. 439 describes an epidemic. Epidemiologists have never documented person-to-person spread of legionellosis.

The CDC estimates that about 18,000 people contract Legionnaires' disease each year, but most cases are so mild and isolated that they are never diagnosed or reported. Smokers, the elderly, patients with chronic respiratory diseases, and the immunocompromised are at greatest risk for infection and disease. Most documented cases occur in summer to early fall.

DISEASE AT A GLANCE 22.2

Bacterial Pneumonias

Causes *Streptococcus pneumoniae* (Gram-positive diplococcus also known as pneumococcus), *Mycoplasma pneumoniae* (Gram-positive, but pink staining, pleomorphic bacterium), *Klebsiella pneumoniae* (Gram-negative bacillus), *Haemophilus influenzae* (Gram-negative pleomorphic coccobacillus), *Staphylococcus aureus* (Gram-positive coccus), *Yersinia pestis* (Gram-negative rod), and *Chlamydophila psittaci* and *C. pneumoniae* (obligate intracellular pleomorphic bacteria).

Virulence factors Vary among pathogens but include attachment molecules, capsules, inhibitors of phagocytosis, and lipid A.

Portal of entry Inhalation, also via blood in case of *Yersinia*.

Signs and symptoms Dry, unproductive cough, headache, fever, chills, and chest pain. Bloody mucoid sputum with *Klebsiella;* frothy bloody sputum with *Yersinia*.

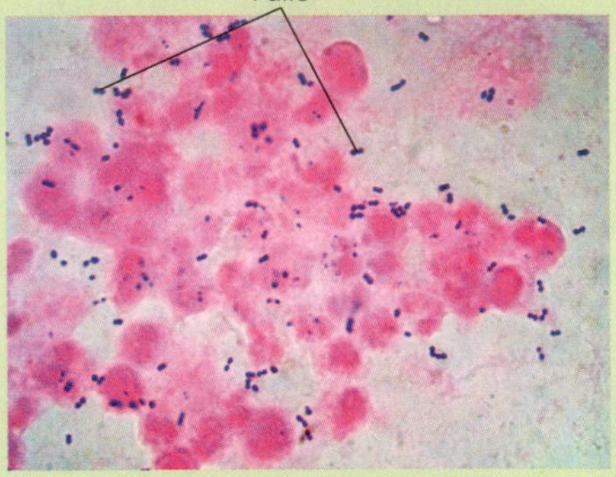

Streptococcus pneumoniae

Incubation period Pneumococcus: 1–3 days; *M. pneumoniae:* 1–4 weeks; *K. pneumoniae:* 1–3 days; *H. influenzae:* 2–4 days; *S. aureus:* variable, usually days; *Y. pestis:* a few hours to 2 days; *Chlamydophila* spp.: 1–4 weeks.

Susceptibility Pneumococcus: immunocompromised individuals; *M. pneumoniae:* high school and college students; *K. pneumoniae:* hospitalized individuals; *H. influenzae:* infants and young children; *S. aureus:* very young, patients with respiratory diseases, such as cystic fibrosis patients; *Y. pestis:* people exposed to bubonic plague; *C. pneumoniae:* most common in school-aged children; *C. psittaci:* individuals in close contact with birds.

Treatment Pneumococcus: penicillin G, vancomycin; *Mycoplasma:* tetracycline, erythromycin; *K. pneumoniae* and *H. influenzae:* cephalosporins; *S. aureus:* vancomycin; *Y. pestis:* tetracycline, streptomycin, chloramphenicol; *Chlamydophila* spp.: tetracycline, erythromycin.

Prevention Use general precautions, such as washing hands frequently. Stop smoking, and, in the case of *C. psittaci*, avoid infected birds. Vaccines against *S. pneumoniae* and *H. influenzae* are available.

Diagnosis, Treatment, and Prevention

Physicians diagnose Legionnaires' disease with fluorescent antibody staining or serological testing that reveals the presence of *Legionella* or antibodies against the bacterium in clinical samples. Quinolones or azithromicin are the antimicrobial drugs of choice for treating Legionnaires' disease.

Eliminating *Legionella* from water supplies is not feasible, because chlorination and heating are only moderately successful against the microorganism. However, the bacterium is not highly virulent, so merely reducing its number is typically a successful control measure.

Tuberculosis

LEARNING | **OUTCOMES**

22.16 Identify two effects of cord factor of *Mycobacterium tuberculosis.*

22.17 Describe the transmission and pathogenesis of *M. tuberculosis* and its effects on the body.

22.18 Discuss the epidemiology, diagnosis, treatment, and prevention of tuberculosis.

22.19 Define and contrast multi-drug-resistant and extensively drug-resistant tuberculosis.

Tuberculosis (TB) is the leading disease killer in the world, though its importance to people in industrialized countries has declined as a result of successful surveillance and the use of effective antimicrobial drugs. Nevertheless, epidemiologists warn us that complacency can allow this terrible killer to resurge, as it did during the late 1900s in the United States after health departments shifted funds from TB eradication programs to other areas.

Tuberculosis is caused by *Mycobacterium tuberculosis* (mī´kō-bak-tēr´ē-ŭm too-ber-kyū-lō´sis), which is a Gram-positive rod with cell walls containing an abundance of **mycolic** (mi-kol-ic) **acid**. This waxy lipid is composed of chains of 60–90 carbon atoms and is directly responsible for many unique characteristics of *M. tuberculosis* and other mycobacteria. Specifically, mycobacteria:

- Grow slowly (in part because of the time required to synthesize numerous molecules of mycolic acid). The generation time varies from hours to several days.
- Are protected from lysis once they are phagocytized.
- Are capable of intracellular growth.
- Are resistant to Gram staining, detergents, many common antimicrobial drugs, and drying out. Because mycobacteria stain only weakly with the Gram procedure (if at all), the acid-fast staining procedure was developed to stain them (see Figure 4.18).

Virulent strains of *M. tuberculosis* produce **cord factor**, a cell wall component that produces strands of daughter cells that remain attached to one another in parallel alignments and inhibits migration of neutrophils and is toxic to mammalian cells. Mutant mycobacteria that are unable to synthesize cord factor do not cause disease.

Disease in Depth: Tuberculosis on pp. 692–693 examines aspects of tuberculosis in more detail.

In most patients, the immune system reaches a stalemate with the bacterium: the immune system is able to prevent further spread of the pathogen and stop the progression of the disease, but it is not able to rid the body of all mycobacteria. *M. tuberculosis* may remain dormant for decades within macrophages and in the centers of tubercles.

Multi-drug-resistant (MDR) strains of *M. tuberculosis*—strains resistant to at least isoniazid (INH) and rifampin—have arisen in several countries. Health care officials are even more concerned about the emergence of **extensively drug-resistant (XDR) tuberculosis,** which is defined as disease caused by *M. tuberculosis* that is resistant *in vitro* to at least INH, rifampin, and three or more other antitubercular drugs. In one South African hospital, 52 of 53 XDR-TB patients died from the disease. Physicians administer levofloxacin combined with three to five other antibacterial drugs to treat MDR-TB. Physicians are heartened that *bedaqiline*—the first new anti-TB drug in over 40 years—was approved by the FDA in 2013. Bedaquiline specifically inhibits ATP synthesis in mycobacteria. Its use is restricted to multi-drug therapy against MDR-TB or XDR-TB.

MDR and XDR strains make it more difficult to rid the world of tuberculosis, so officials encourage strict adherence to antitubercular drug regimens. The World Health Organization (WHO) and the CDC recommend a strategy of drug delivery called *Directly Observed Treatment, Shortcourse (DOTS)*, in which health care workers observe patients to ensure they take their medications on schedule. Effective treatment of cases of MDR and XDR tuberculosis is thereby quite expensive. Researchers are developing a subcutaneous implant that would gradually deliver drugs directly into the body over a period of months so that direct observation is not required.

Renewed efforts to detect and stop infection and the implementation of DOTS have reduced the number of reported U.S. cases of TB. The CDC predicts that the current downward trend in case number indicates that a TB-free country is attainable in the near future.

Pertussis (Whooping Cough)

LEARNING | **OUTCOME**

22.20 Describe the characteristics of pertussis.

What could be more agonizing to a parent than observing one's child being tortured by persistent, powerful coughing spells that leave the child exhausted, blue, and with ruptured blood vessels in the eyes? **Pertussis,** commonly called **whooping cough,** produces these manifestations.

Signs and Symptoms

The beginning signs and symptoms of pertussis resemble those of a common cold—rhinorrhea, slight cough, and mild fever—if they occur at all. After one to two weeks, the signature symptom of severe coughing occurs. Vomiting, diarrhea, and choking can accompany this stage. Oxygen exchange in the lungs may be so severely limited that *cyanosis* develops (the patient turns blue), and death results.

Pathogen and Virulence Factors

Bordetella pertussis (bōr-dē-tel´ă per-tus´is), a small, aerobic, nonmotile, Gram-negative coccobacillus in the class Betaproteobacteria is responsible for pertussis. Various adhesins and toxins

TUBERCULOSIS

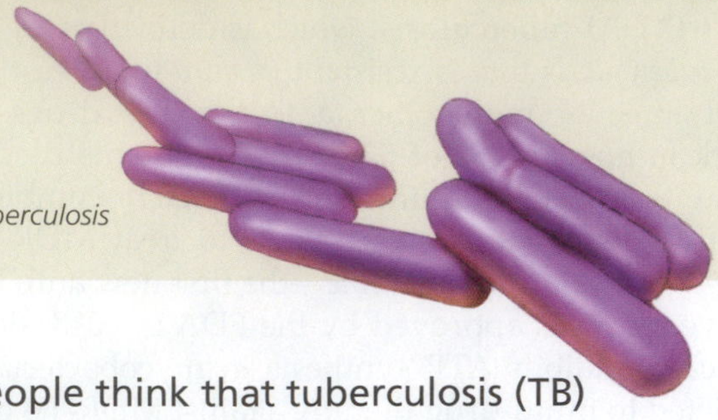

Mycobacterium tuberculosis

SIGNS AND SYMPTOMS

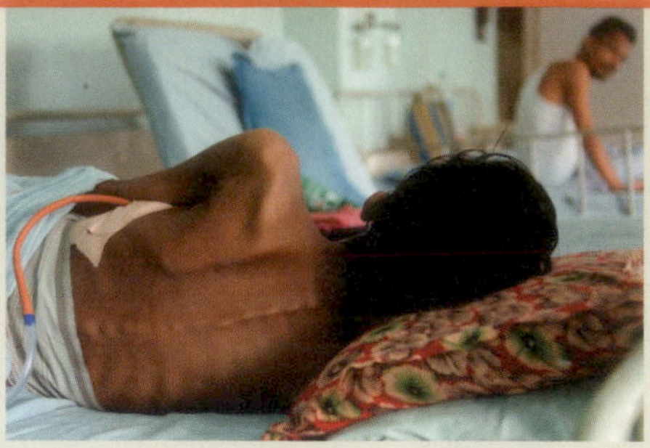

Signs and symptoms of TB are not always apparent, often limited to a minor cough and mild fever. Breathing difficulty, fatigue, malaise, weight loss, chest pain, wheezing, and coughing up blood characterize the disease as it progresses.

Many people think that tuberculosis (TB) is a disease of the past, one that has little importance to people living in industrialized countries. In part, this attitude results from the success health care workers have had in reducing the number of cases. Nevertheless, epidemiologists warn that complacency can allow this terrible killer to reemerge.

PATHOGENESIS

Primary tuberculosis

1 *Mycobacterium* typically infects the respiratory tract via inhalation of respiratory droplets from infected individuals.

2 Macrophages in alveoli phagocytize mycobacteria but are unable to digest them, in part because the bacterium inhibits fusion of lysosomes to endocytic vesicles.

3 Instead, bacteria replicate freely within macrophages, gradually killing the phagocytes. Bacteria released from dead macrophages are phagocytized by other macrophages, beginning the cycle anew.

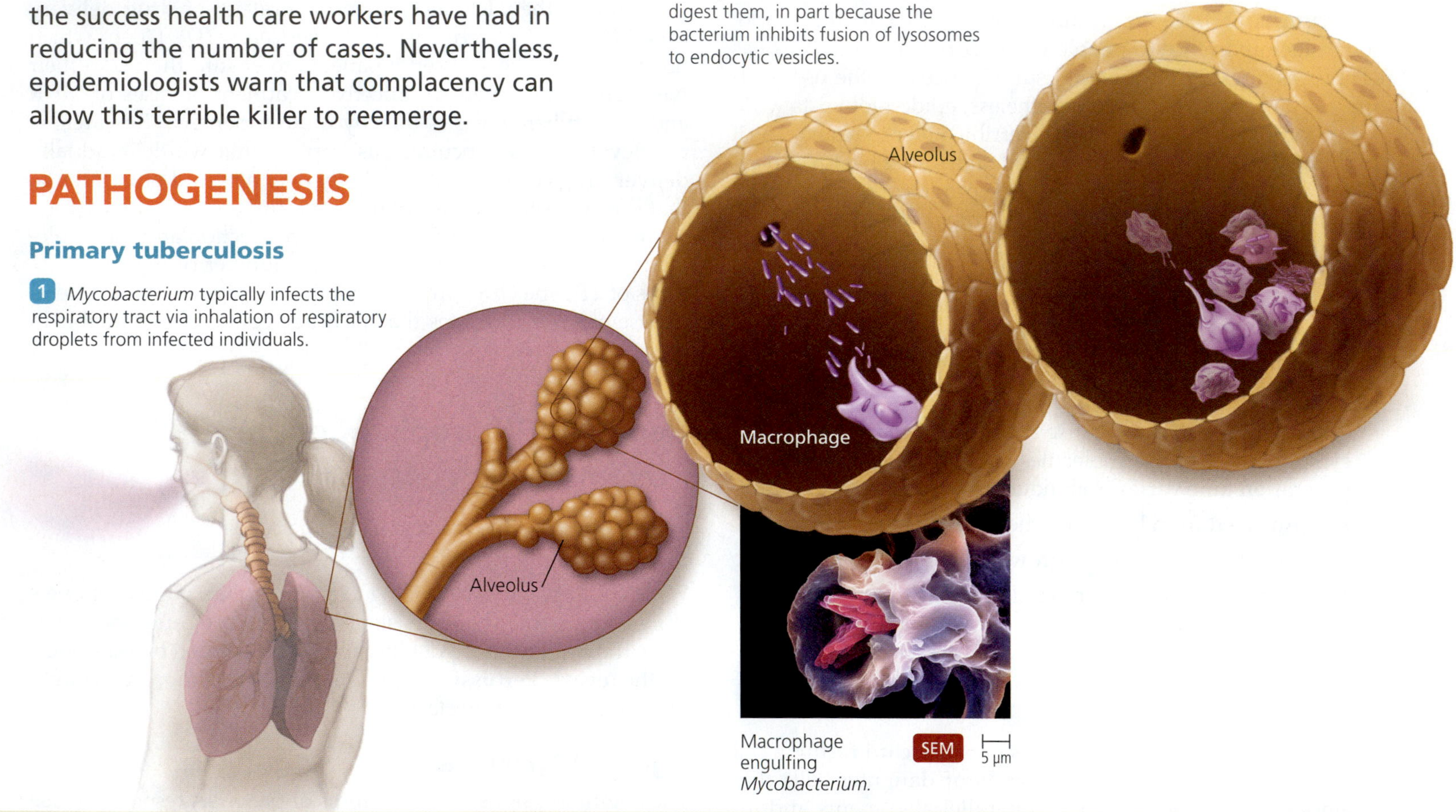

Macrophage engulfing *Mycobacterium.*

INVESTIGATE IT!

What does the development of XDR-TB (extensively drug-resistant strains of Mycobacterium tuberculosis*) portend for the future of the disease?*

Scan this code to visit the Centers for Disease Control and Prevention website to investigate XDR-TB. Then go to MasteringMicrobiology to record your research findings.

EPIDEMIOLOGY

Tuberculosis kills on average four people every minute, mostly in Asia and Africa. TB is on the decline in the U.S., though the CDC estimates that TB may still infect more than 9 million Americans. One third of the world's population is infected, and over 9 million new cases are seen each year.

Left, estimated new TB cases in 2010 per 100,000 (WHO)

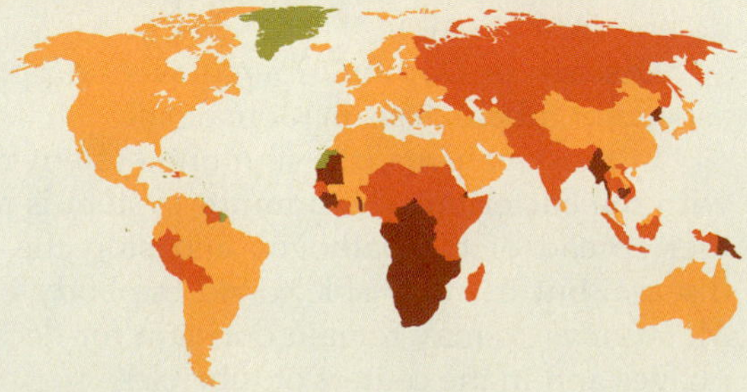

PATHOGEN AND VIRULENCE FACTORS

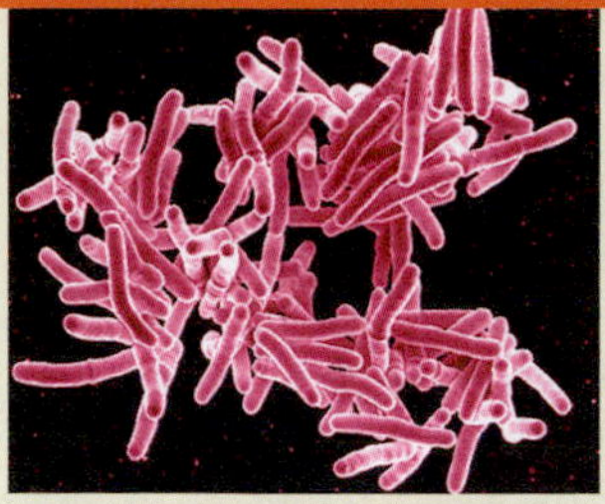

SEM 5 μm

Mycobacterium tuberculosis is a high G + C, aerobic, Gram-positive rod. Virulent strains produce cord factor, a cell wall component that produces strands of daughter cells that remain attached to one another in parallel alignments. Cord factor also inhibits migration of neutrophils and is toxic to mammalian cells. Multi-drug-resistant (MDR-TB) and extensively drug-resistant (XDR-TB) strains of *Mycobacterium* make it more difficult to rid the world of TB.

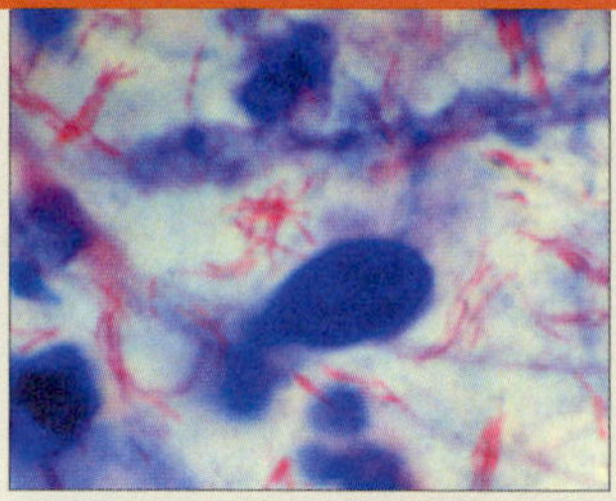

LM 15 μm

Cell walls contain mycolic acid, a waxy lipid that is responsible for unique characteristics of this pathogen, including slow growth, protection from lysis when cells are phagocytized, intracellular growth, and resistance to Gram staining, detergents, many common antimicrobial drugs, and drying out. (Slow growth is due in part to the time required to synthesize molecules of mycolic acid.)

4 Infected macrophages present antigen to T lymphocytes, which produce lymphokines that attract and activate more macrophages and trigger inflammation. Tightly packed macrophages surround the site of infection, forming a tubercle over a two- to three-month period.

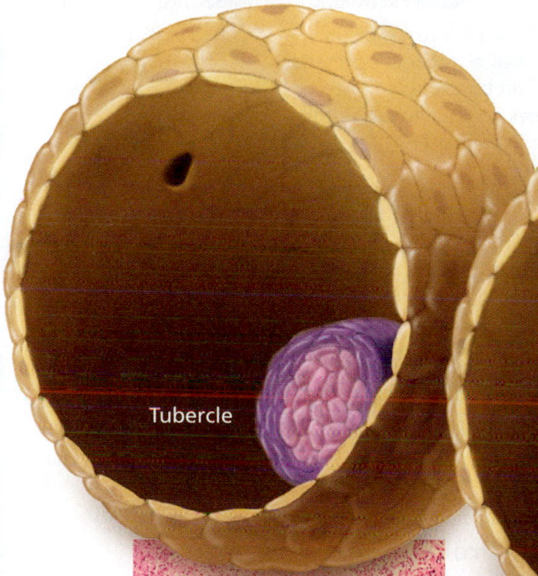

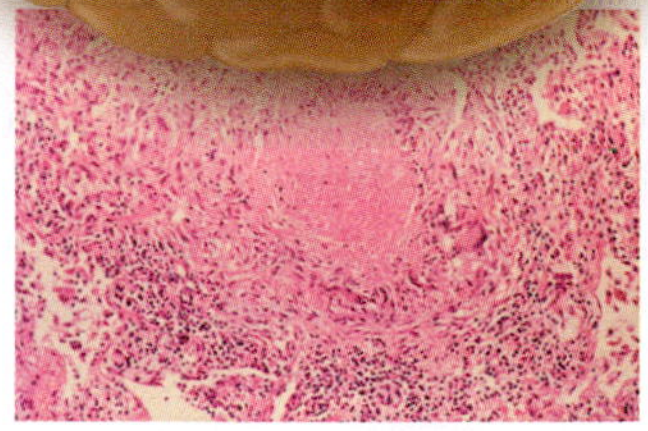

Tubercle in lung tissue. LM 50 μm

5 Other cells deposit collagen fibers, enclosing infected macrophages and lung cells within the tubercle. Infected cells in the center die, releasing *M. tuberculosis* and producing caseous necrosis—the death of tissue that takes on a cheese-like consistency due to protein and fat released from dying cells. A stalemate between the bacterium and the body's defenses develops.

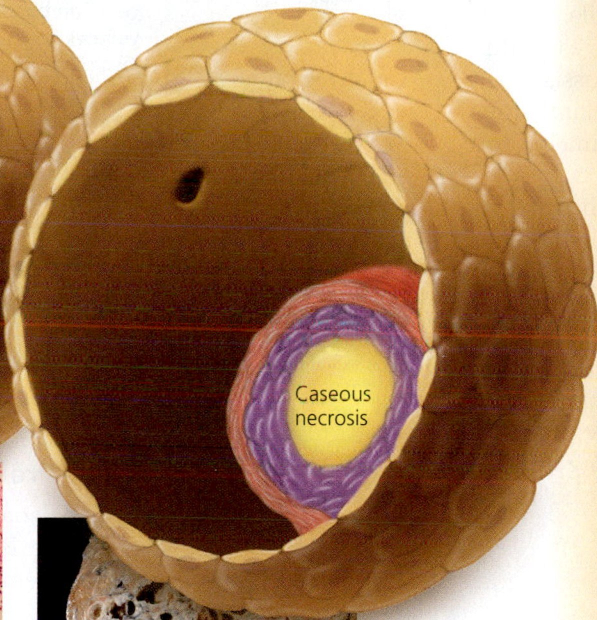

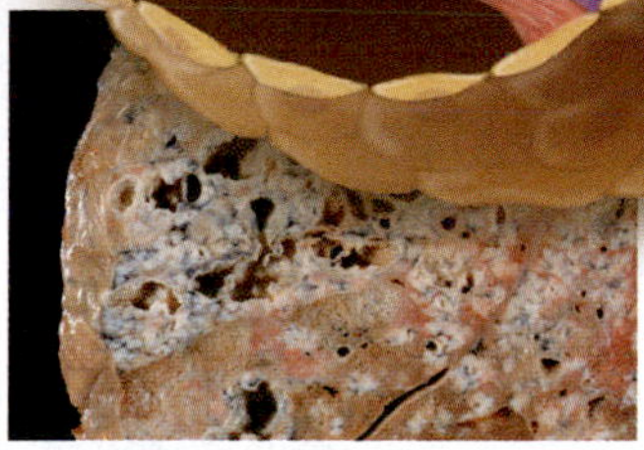

Lung lesions caused by TB.

Secondary/reactivated tuberculosis results when *M. tuberculosis* breaks the stalemate, ruptures the tubercle, and reestablishes an active infection. Reactivation occurs in about 10% of patients; patients whose immune systems are weakened by disease, poor nutrition, drug or alcohol abuse, or by other factors.

Disseminated tuberculosis results when macrophages carry the pathogen via blood and lymph nodes to other sites, including bone marrow, spleen, kidneys, spinal cord, and brain.

Tuberculosis lesions in spleen.

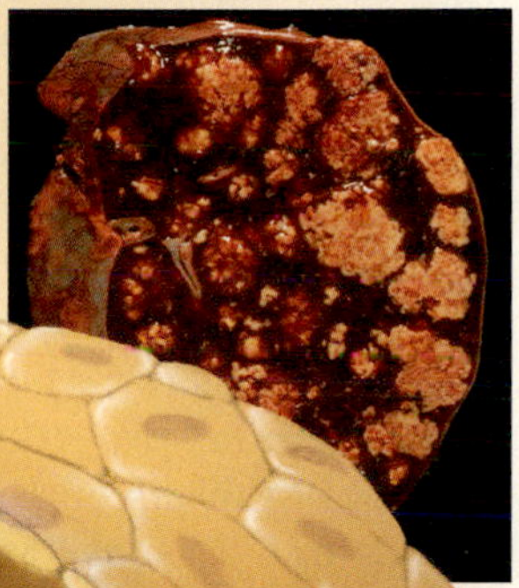

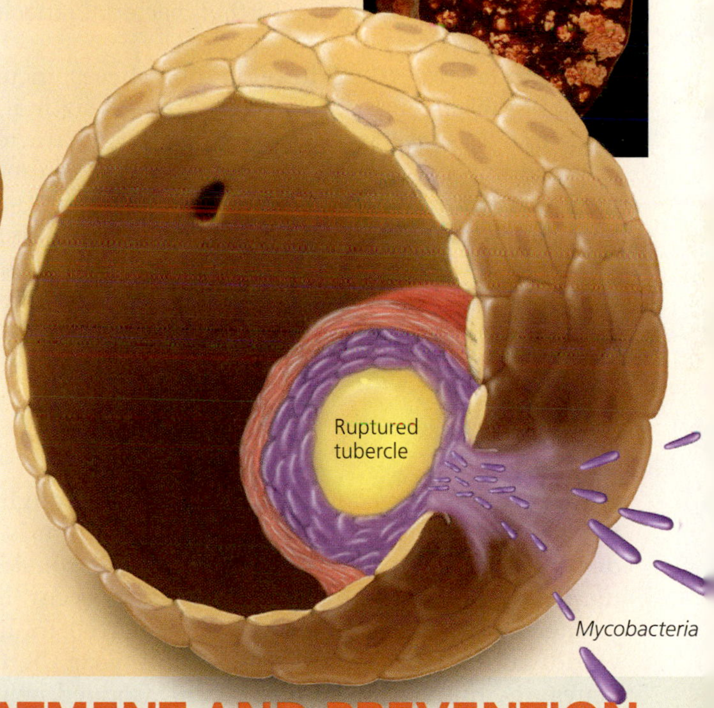

DIAGNOSIS

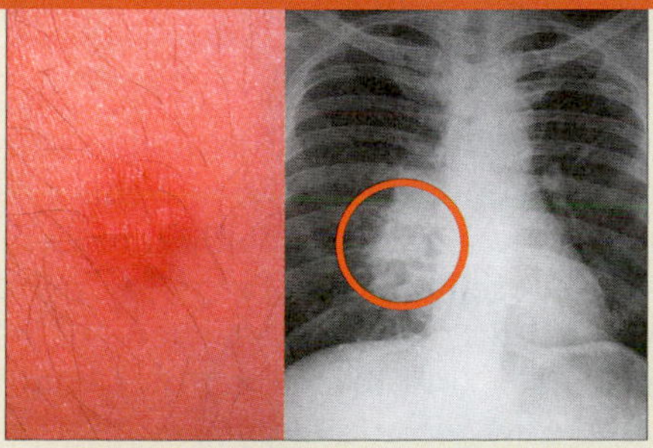

10 mm

A tuberculin skin test is used to screen patients for TB exposure. A positive reaction is an enlarged, reddened, and raised lesion at the inoculation site. Chest X-ray films can reveal the presence of tubercles in the lungs. Primary TB usually occurs in the lower and central areas of the lung; secondary TB commonly appears higher.

TREATMENT AND PREVENTION

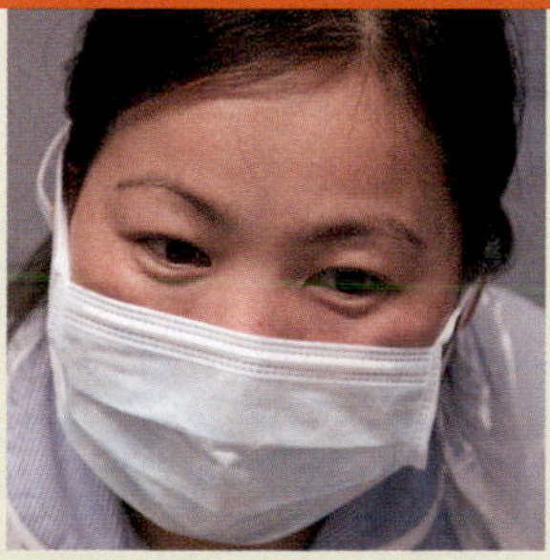

Treatment combines isoniazid, rifampin, and one of several drugs (such as ethambutol, levofloxacin, or streptomycin) for six months. Newly approved bedaquiline is used in combination with other drugs to treat MDR-TB or XDR-TB. In countries where TB is common, health care workers immunize patients with BCG vaccine, which is not recommended for the immunocompromised because it can cause disease. Workers must avoid inhaling respiratory droplets from TB patients.

mediate the disease. Two adhesins are *filamentous hemagglutinin* and *pertussis toxin.* Four toxins are:

- *Pertussis toxin,* a portion of which interferes with ciliated epithelial cells' metabolism, resulting in increased mucus production. (Note that pertussis toxin is both an adhesin and a toxin.)
- *Adenylate cyclase toxin,* which triggers increased mucus production and inhibits leukocyte movement, phagocytosis, and killing.
- *Dermonecrotic toxin,* which causes localized constriction and hemorrhage of blood vessels, resulting in cell death and tissue destruction.
- *Tracheal cytotoxin,* which at low concentrations inhibits the movement of cilia on respiratory cells.

Pathogenesis

Via its adhesins, inhaled *B. pertussis* binds to cilia in the trachea (see Disease at a Glance 22.3), interfering with their action and stopping the ciliary escalator. Filamentous hemagglutinin also binds to the cytoplasmic membranes of neutrophils, initiating endocytosis of the bacterium. *B. pertussis* survives within the phagocytes, evading the immune system. Pertussis toxin causes infected cells to produce more receptors for filamentous hemagglutinin, leading to further bacterial attachment and phagocytosis.

Pertussis progresses through four phases: incubation, catarrhal[14] (kă-tah´răl), paroxysmal[15] (par-ok-siz´măl), and convalescent **(Disease at a Glance 22.3)**. The characteristic sign of whooping cough occurs during the two to four weeks of the paroxysmal phase: to clear mucus from the lungs, the patient coughs two or three times without inhalation, followed by the characteristic "whoop" of inhalation through a congested trachea. Each day a patient may experience 40–50 of these coughing spells, which often end with vomiting and exhaustion. Coughing may be so brutal that blood vessels in the eyes burst.

Epidemiology

Pertussis is highly contagious, spreading through the air in respiratory droplets; fortunately, the bacterium is fragile—it doesn't survive long outside the body. It is considered a pediatric disease, as life-threatening cases occur in children younger than five years old. More than 60 million people worldwide suffer from pertussis each year. In the United States, the number of reported cases dropped below 1250 in 1981, but pertussis has become a *reemerging disease*—there were almost 42,000 reported cases in 2012. All these figures considerably underestimate the actual number of cases, because patients with chronic coughs are not routinely tested for infection with *Bordetella,* and because the disease in older children and adults is typically less severe and can be misdiagnosed as a cold or influenza.

DISEASE AT A GLANCE 22.3

Pertussis (Whooping Cough)

Cause *Bordetella pertussis* (Gram-negative coccobacillus).

Virulence factors
Adhesins that attach to tracheal cells; survives within phagosome; pertussis toxin causes human cells to make more receptors for *Bordetella*.

Portal of entry
Inhalation.

Signs and symptoms
Catarrhal stage begins with signs and symptoms resembling the common cold that last one to two weeks. The following paroxysmal stage lasts two to four weeks and is characterized by persistent, violent coughing spells that consist of three or four coughs without a breath, followed by a high-pitched, wheezing inhalation or "whoop." The convalescent stage lasts several weeks as the cough subsides.

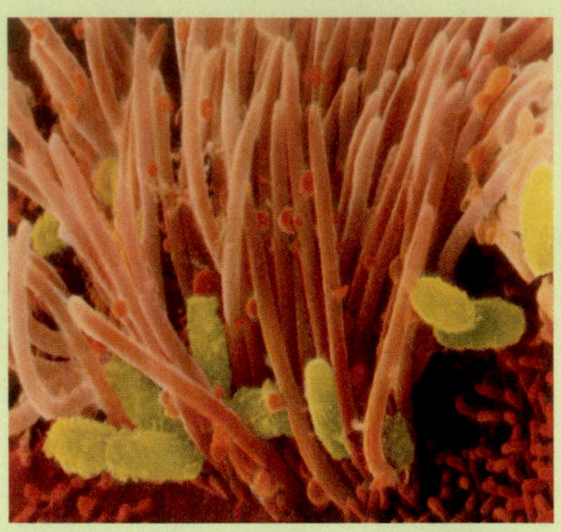

Bordetella pertussis bound to cilia of a tracheal epithelial cell. The bacteria (colored yellow) eventually cause the loss of the cells. SEM 1 µm

Incubation period 6–20 days, with an average of 7–10 days.

Susceptibility Unimmunized children.

Treatment Supportive care; erythromycin has only little effect.

Prevention DTaP vaccine.

Diagnosis, Treatment, and Prevention

The symptoms of pertussis are usually diagnostic. Even though health care workers may isolate *B. pertussis* from respiratory specimens, the bacterium is extremely sensitive to drying out, so specimens must be inoculated at the patient's bedside onto *Bordet-Gengou medium,* which is specially designed to support the growth of this bacterium.

Treatment for pertussis is primarily supportive. By the time the disease is recognized, the immune system has often already "won the battle." Recovery depends on regeneration of the tracheal epithelium, not on the number of bacteria; therefore, antibacterial drugs have little effect on the course of the disease, though they may reduce the patient's infectivity. Regimens include daily erythromycin for two weeks.

Given that *B. pertussis* has no reservoir and that an effective vaccine (the "P" of DTP vaccine) has been available since 1949, whooping cough could be eradicated. Despite this possibility, pertussis remains a problem in the United States. This is due in part to the refusal of parents to immunize their children following publicity concerning adverse reactions to the original attenuated vaccine (DTP), and to the fact that immunity is not lifelong. The CDC now recommends *DTaP vaccine,*

[14]From Greek *katarheo,* meaning "to flow down."
[15]From Greek *paroxysmos,* meaning "to irritate."

which contains an acellular pertussis (aP) component. DTaP has fewer side effects than DTP vaccine but is equally effective in engendering an immune response.

Inhalational Anthrax

LEARNING | OUTCOME

22.21 Describe inhalational anthrax.

The government classifies the bacillus of **anthrax** as one of a handful of potential biological terror agents. In 2001, a terrorist spread the disease via the U.S. postal system by sending letters filled with endospores. Twenty-two people sickened, and five died. There are three forms of anthrax: *cutaneous anthrax*, which is discussed in Chapter 19, manifests on the skin; *gastrointestinal anthrax* is a rare human disease of the digestive system; and **inhalational anthrax**—the most severe form of anthrax in humans—is a respiratory disease, which is examined here.

Signs and Symptoms

The initial symptoms of inhalational anthrax resemble those of a common cold or flu—sore throat, mild fever, myalgia (muscle aches), mild cough, and malaise. After several days, the symptoms progress to include more severe coughing, nausea, vomiting, fainting, confusion, lethargy, shock, and death.

Pathogen and Virulence Factors

Bacillus anthracis, a Gram-positive, endospore-forming, aerobic, rod-shaped bacterium, causes anthrax. Resistant endospores allow the bacillus to survive in the environment indefinitely. In the body, *B. anthracis* forms a protective capsule of glutamic acid, which inhibits phagocytosis by alveolar macrophages. The bacterium also secretes *anthrax toxin,* which kills human cells and triggers edema (swelling due to fluid accumulation).

Pathogenesis and Epidemiology

An infective dose of *B. anthracis* endospores involves inhalation of at least 8000 to 50,000 endospores. In the lungs, the endospores germinate, and vegetative cells secrete anthrax toxin, which impairs respiratory function, initiates toxemia, and often results in death.

Anthrax does not spread from person to person; rather, people must acquire *B. anthracis* from infected animals either by contact or via inhalation of endospores in dust or on animal hides or wool. Most patients with inhalational anthrax die, even with full supportive care and timely use of antimicrobial drugs.

Diagnosis, Treatment, and Prevention

Clinicians can readily identify the large, Gram-positive cells of *Bacillus* in the sputum of patients, but endospores are seen rarely. Unlike harmless environmental species of *Bacillus,* laboratory-cultured *B. anthracis* colonies are sticky and nonhemolytic when grown on blood agar. Serological, DNA, and biochemical testing confirms the presence of *B. anthracis.*

Many antimicrobials, including penicillin, doxycycline, levofloxacin, and ciprofloxacin, are effective against *B. anthracis*; however, damage to the lungs and toxemia can be so severe and rapid that treatment is often ineffectual. During the bioterrorism attack of 2001, physicians learned that early and aggressive treatment of inhalation anthrax with antimicrobial drugs accompanied by persistent drainage of fluid from around the lungs increased the survival rate from less than 1% to almost 50%.

An efficacious vaccine is available to select military personnel, researchers, people who work closely with animals, and health care professionals with anthrax patients.

Table 22.1 on p. 681 compares and contrasts inhalational anthrax with some other respiratory diseases.

TELL ME WHY

Mycoplasma pneumoniae is resistant to penicillin, though *Mycoplasma* does not synthesize an enzyme to break down penicillin. Explain why *Mycoplasma* is resistant to penicillin.

Viral Diseases of the Lower Respiratory System

We have considered important bacterial and viral diseases of the upper respiratory system and bacterial diseases of the lower respiratory organs. Now we turn our attention to viral diseases of the lower respiratory system, beginning with influenza.

Influenza

LEARNING | OUTCOMES

22.22 Describe the general characteristics of influenza.

22.23 Describe the roles of hemagglutinin and neuraminidase in the replication cycle of influenzaviruses and in the origin of new influenzaviruses.

Imagine being the only elementary schoolchild of your sex in a midsized Swedish town because all of your peers died six winters ago; or imagine returning to college after a break, only to learn that half of your fraternity brothers had died during the previous two months. These stories are not fictional; they happened to relatives of this author during the great flu pandemic in the winter of 1918–19 (**FIGURE 22.10**). During that winter, half the world's population was infected with a new, extremely virulent strain of influenzavirus, and approximately 40 million died during that one flu season. In some U.S. cities, 10,000 people died each week for several months. Could it happen again? In this section we will learn about the characteristics of influenzaviruses that enable flu—a common disease, second in prevalence only to common colds—to produce such devastating epidemics, and we will examine some ways to protect ourselves.

Signs and Symptoms

Following infection, influenza has an incubation period of about one day. The signs and symptoms of **influenza**[16] (in-flū-en´ză,

[16]*Influenza,* which is Italian for "influence," derives from the mistaken idea that the alignment of celestial objects caused or influenced the disease.

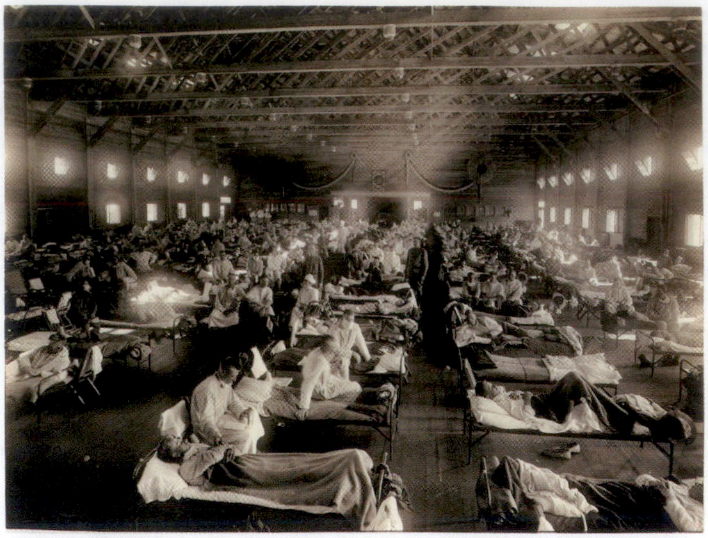

▲ **FIGURE 22.10 A scene from the flu pandemic of 1918–19.** Influenza afflicted so many people in the United States that gymnasiums were used as hospital wards.

or **flu**) usually include sudden fever between 39°C and 41°C (102–106°F), pharyngitis, congestion, dry cough, malaise, headache, and myalgia. Fever distinguishes flu from a common cold. Most people recover in one to two weeks.

Pathogens and Virulence Factors

Two species of viruses, designated types A and B, cause influenza. Each flu virion is segmented, needing eight different −ssRNA molecules, and is surrounded by a pleomorphic envelope studded with prominent glycoprotein spikes composed of either **hemagglutinin**[17] (hē-mă-glū´ti-nin; **HA**) or **neuraminidase** (nūr-ă-min´i-dās; **NA**) **(FIGURE 22.11)**. Both HA and NA play roles in attachment: NA spikes provide the virus access to cell surfaces by hydrolyzing mucus in the lungs, whereas HA spikes actually bind to pulmonary epithelial cells and trigger endocytosis. Because influenzaviruses rarely attack cells outside the lungs, so-called *stomach flu* is probably caused by other viruses or bacteria.

The genomes of flu viruses are extremely variable, especially with respect to the genes that code for HA and NA. Mutations in the genes coding for these glycoprotein spikes are responsible for the production of new strains of influenzavirus, via processes known as antigenic drift and antigenic shift.

Antigenic drift (FIGURE 22.12a) refers to the accumulation of hemagglutinin and neuraminidase gene mutations within a single strain of virus in a given geographic area. Because of relatively minor changes in the virus population's antigens, localized increases in the number of seasonal flu infections occur about every two years. The slow, gradual change in the viral antigens gives the name drift to the process.

[17]The word *hemagglutinin* refers to these spikes' ability to attach to and clump (agglutinate) erythrocytes.

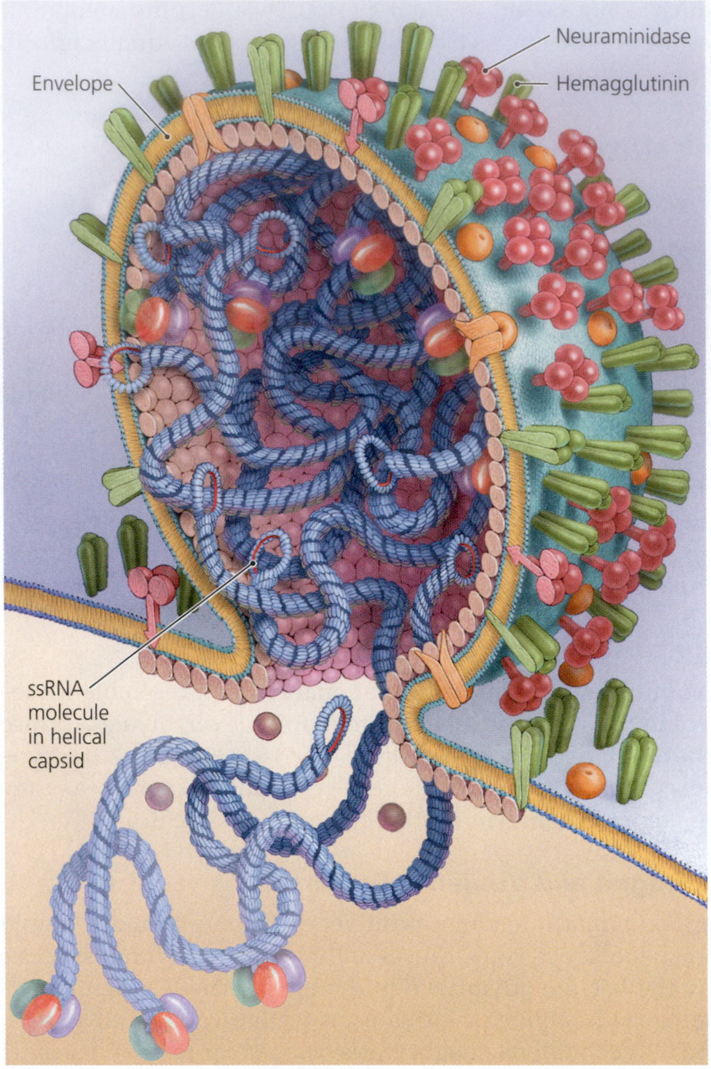

▲ **FIGURE 22.11 Artist's rendition of a cross-sectioned influenzavirus budding from a cell.** The viral genome consisting of eight segments of ssRNA, each of which is enclosed in a helical capsid. *What are the types of glycoprotein spikes, and what is their relationship to viral strain?*

Figure 22.11 *Glycoprotein spikes on influenzavirus are neuraminidase (NA) and hemagglutinin (HA); variations in these glycoproteins determine the strain of the virus.*

In contrast, **antigenic shift (FIGURE 22.12b)** is a major antigenic change that results from the reassortment of genes among different influenza A viruses infecting the same host cell (either human cells or animal cells, including cells of birds and pigs). On average, antigenic shift of influenza A occurs every 10 years. Influenza B does not undergo antigenic shift.

Strains of influenza are named by type (A or B), location and date of original identification, and type of antigens (HA and NA). For example, A/Singapore/1/80 (H1N2) is influenza type A, isolated in Singapore in January 1980, that contains HA and NA antigens of type 1 and type 2, respectively. If the virus is isolated from an animal, the animal name is appended to the location. From these names, common names such as "Hong Kong flu" or "swine flu" arise. The number of possible flu strains is

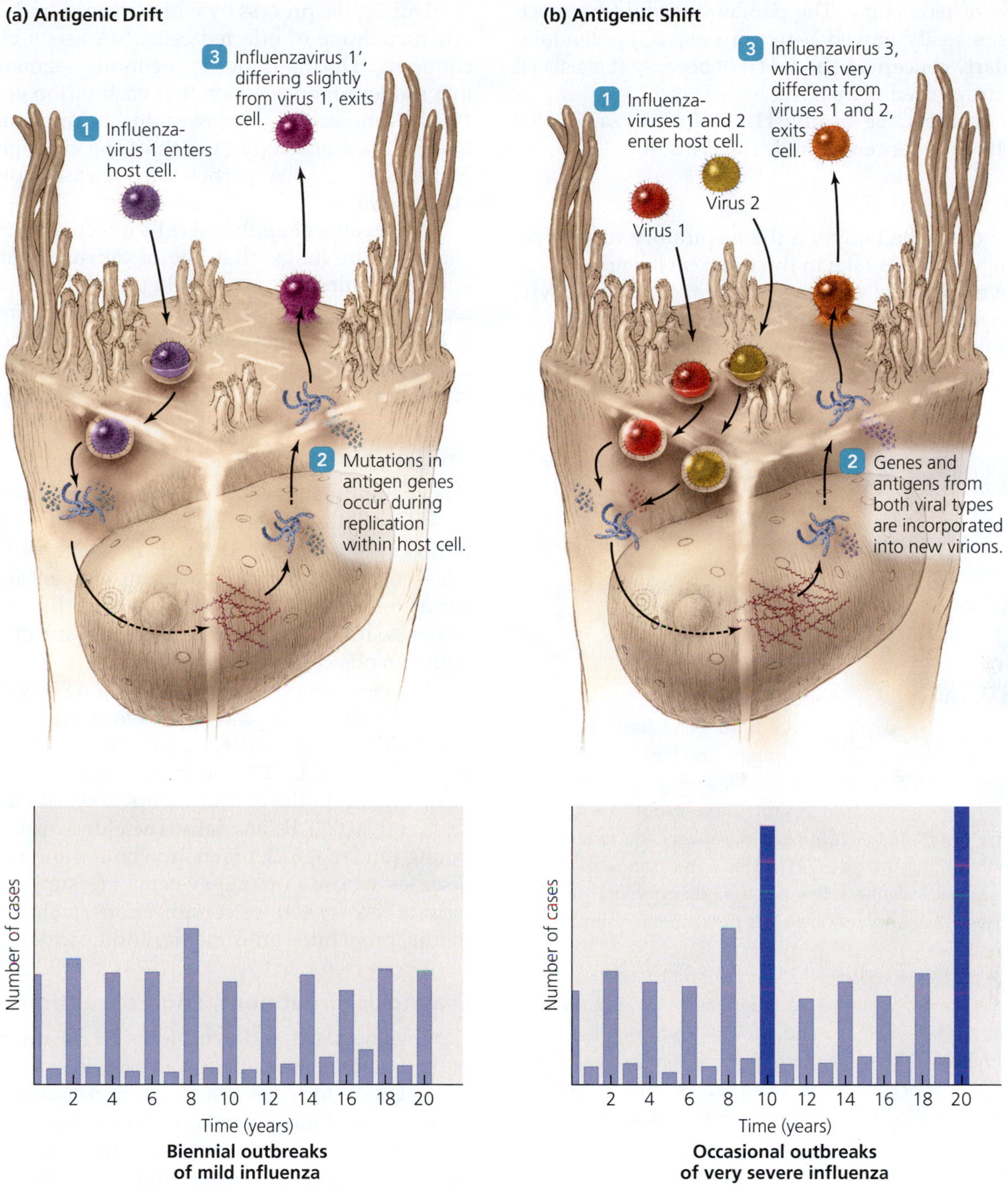
(a) Antigenic Drift
1 Influenzavirus 1 enters host cell.
2 Mutations in antigen genes occur during replication within host cell.
3 Influenzavirus 1′, differing slightly from virus 1, exits cell.
Number of cases
2 4 6 8 10 12 14 16 18 20
Time (years)
Biennial outbreaks of mild influenza
(b) Antigenic Shift
1 Influenzaviruses 1 and 2 enter host cell.
Virus 1
Virus 2
2 Genes and antigens from both viral types are incorporated into new virions.
3 Influenzavirus 3, which is very different from viruses 1 and 2, exits cell.
Number of cases
2 4 6 8 10 12 14 16 18 20
Time (years)
Occasional outbreaks of very severe influenza

▲ **FIGURE 22.12 The development of new strains of flu viruses. (a)** Antigenic drift, which results from variation in the NA and HA spikes of a single strain of influenzavirus, either A or B. **(b)** Antigenic shift, which occurs when RNA molecules from two or more strain A influenzaviruses infecting a single cell are incorporated into a single virion. Because antigenic shift produces significantly greater antigenic variability than occurs in antigenic drift, antigenic shifts can result in major epidemics.

almost infinite. Asia is a major site of antigenic shift, and the source of most pandemic strains, because of the continent's high population densities of humans, ducks, chickens, and pigs—all of which serve as influenzavirus hosts.

Epidemiologists are particularly concerned about type A influenzaviruses of aquatic birds that carry antigens similar to those of major pandemics in 1918–19, 1957–58, and 1968–69. One virus of concern is H5N1, which kills more than 60% of people who contract the virus from infected birds. Fortunately, this virus does not move from person to person; people get it only from birds. Another virus, H1N1, caused a slow, worldwide pandemic beginning in 2009 in Mexico. H1N1 killed an estimated 284,400 that year. Most of its victims were under 65 years of age, which is unusual for a flu virus. In 2013, scientists became concerned about a rapidly expanding pandemic caused by another deadly influenzavirus, H7N9, which affects mainly the elderly, killing over 50% of its victims. The pandemic would be much worse if the virus easily moved between people. Epidemiologists are particularly concerned about H7N9 because it resists all FDA-approved drugs used to treat flu.

Emerging Disease Case Study: H1N1 Influenza follows one patient's 2009 experience with a deadly bird flu.

Pathogenesis

Influenzaviruses enter the body via the respiratory route. Epithelial cells lining the lungs take in the viruses via endocytosis, and the viral envelopes fuse with the membranes of phagocytic vesicles. Flu viruses multiply using positive-sense RNA molecules both for translation of viral proteins and as templates for transcription of −ssRNA genomes.

CLINICAL CASE STUDY

The Coughing Cousin

Twenty-one-year-old Marjorie comes to the family practice office for evaluation of a dry cough that has not gotten better over the last seven days. Before she started coughing, she had a mild fever, runny nose, and sneezing for 10 days, but she was able to carry out her usual routine. She reports "coughing spells" frequently during the day and recalls one episode in which she gagged and vomited. She has a mild coughing episode while in the office, and although it is paroxysmal, there is no "whoop" noted.

Marjorie is a part-time college student attending evening classes and helps her aunt with her younger cousins during the day. Her cousins are five years old and six weeks old and also come to the same office for health care. Both cousins appear well, without any signs or symptoms, and a review of their charts reveals they are current for all scheduled vaccines. Marjorie's last vaccine was at age 16.

1. Why should pertussis be something to consider?
2. What are the three clinical stages of pertussis?
3. If this is pertussis, why doesn't Marjorie have the classic "whoop"?
4. What are the CDC recommendations for adults ages 19–64?
5. Which family members are at the highest risk in this situation?
6. What test can her provider do to confirm the suspicion that Marjorie has pertussis?
7. Is treatment recommended at this time?
8. How is pertussis spread, and how could Marjorie prevent the spread of her infection to others?

During the process by which virions bud from the cytoplasmic membrane of infected cells, NA keeps viral proteins from clumping. During budding, genomic segments are enveloped in a random manner, such that each virion ends up with about 11 RNA molecules. However, to be functional, a virion must have at least one copy of each of the eight genomic segments. Numerous defective particles are released for every functional virion formed.

The deaths of epithelial cells infected with influenzaviruses eliminate the lungs' first line of defense against infection, the epithelial lining. As a result, flu patients are more susceptible to secondary bacterial infection. One common infecting bacterium is *Haemophilus influenzae,* which was so named because it was found in many flu victims. Cytokines released as part of the immune response induce the typical signs and symptoms of flu.

Epidemiology

The changing antigens of flu viruses guarantee that there will be susceptible people, especially children, each season. Infection occurs primarily through inhalation of airborne viruses released by coughing or sneezing, but self-inoculation can occur as people transfer viruses on their fingers to their mouths or noses. Influenzaviruses can remain infective for up to eight hours on objects outside the body.

Infected people are contagious one day before the disease is manifested and for a week or more after signs and symptoms begin. Some individuals are carriers—infected people who are not sick. About 15–20% of the U.S. population gets the flu each year, usually in the winter. About 200,000 victims are hospitalized, and 30,000 die annually. The elderly (over age 65), the very young (under age 2), pregnant women, and people with chronic diseases whose immune systems are suppressed are typically most at risk for serious complications, including bacterial pneumonia, bronchitis, otitis media, and heart failure.

Diagnosis, Treatment, and Prevention

Having the signs and symptoms of flu during a community-wide outbreak is sufficient for an initial diagnosis of influenza. Laboratory tests such as immunofluorescence, ELISA, polymerase chain reaction, and commercially available rapid antigen testing can distinguish strains of flu virus. Early and accurate diagnosis is important because antiviral therapy must begin promptly to be effective; further, early diagnosis can forestall the writing of inappropriate prescriptions for antibacterial drugs.

The CDC recommends the use of either of two drugs to treat influenza: Oseltamivir pills or inhaled zanamivir mist inhibit type A and type B neuraminidase, blocking the release of virions from infected cells. These prescription drugs must be taken during the first 48 hours of infection in order to be effective, because they cannot prevent later manifestations of the disease. As of January 2000, the CDC discouraged using two drugs active against type A influenzaviruses only—amantadine and rimantadine—because the viruses are growing resistant to both. The CDC is evaluating the drugs' effectiveness to see whether this recommendation should remain in effect.

Hundreds of millions of dollars are spent each year in the United States on antihistamines and pain relievers to alleviate the symptoms of the flu. Aspirin and aspirin-like products should not be used to treat the symptoms of flu in children and teenagers because of increased risk of Reye's syndrome, which is a potentially fatal syndrome that sometimes follows viral infections, especially in children who have taken aspirin.

The greatest success in controlling flu epidemics has come from immunization with multivalent vaccines; that is, vaccines that contain several antigens at once. The CDC has personnel in Asia who detect changes in the HA and NA antigens of flu viruses. Antigens of emerging viruses are then used to create a flu vaccine in advance of the next flu season in the United States. In 2013, the U.S. Food and Drug Administration (FDA) approved the first flu vaccine produced from viruses grown in animal cell cultures rather than in eggs.

Flu vaccines are at least 70% effective, but only against the viral antigens they contain. A flu vaccine usually provides protection against the strains included in the vaccine for three years or less. Natural active immunity following infection lasts much longer but is probably not lifelong. In 2003, the U.S. Food and Drug Administration (FDA) approved a flu vaccine administered by means of a nasal spray. This vaccine is intended for nonpregnant patients between 2 and 49 years of age.

Clinical Case Study: Influenza on p. 700 examines some aspects of influenza infection and immunization. Table 22.1 on p. 681 compares and contrasts influenza with some other respiratory diseases.

Coronavirus Respiratory Syndromes

LEARNING | **OUTCOME**

22.24 Describe coronavirus respiratory syndromes.

In the spring of 2003, *severe acute respiratory syndrome (SARS)* literally changed the face of China as people donned masks to prevent the spread of this coronavirus disease. Similarly, another coronavirus disease, *Middle East respiratory syndrome (MERS)*, is changing the face of the Middle East **(FIGURE 22.13)**.

Signs and Symptoms

Coronavirus respiratory syndromes manifest with a high fever (greater than 38.0°C) accompanied by shortness of breath, difficulty in breathing, malaise, and body aches. About 10–20% of patients have diarrhea. After about a week, SARS patients develop a dry cough and pneumonia, and about 10% die. More than 50% of MERS patients die.

EMERGING DISEASE CASE STUDY

H1N1 Influenza

Middle school is supposed to be a time of exploration, learning, and fun, but Maria was too sick to enjoy it right now. Her 104°F fever had lasted two very unpleasant days, but at last it had gone. For another two days she ached all over and was dizzy; her head felt as if it could explode at any moment; and the nausea, frequent vomiting, and extreme tiredness were unrelenting. For over a week her tired red eyes stared listlessly as she struggled to cope with a constantly running nose, severe sore throat, and a dry hacking cough. Would this onslaught never end? A newly emerging strain of influenza type A had a victim in its grasp.

Flu viruses infect birds, pigs, horses, and humans. These viruses normally mutate, producing slightly different strains every year—a process called antigenic drift. About once a decade, however, different strains of influenzaviruses infect a single cell, and within it they exchange major pieces of RNA, producing a new, quite different strain of virus—a process called antigenic shift. The virus attacking Maria was such a newly emerged influenzavirus, a strain that contained RNA from influenzaviruses of humans, birds, and swine in a novel combination. The virus, commonly called swine flu virus, is officially 2009A(H1N1).

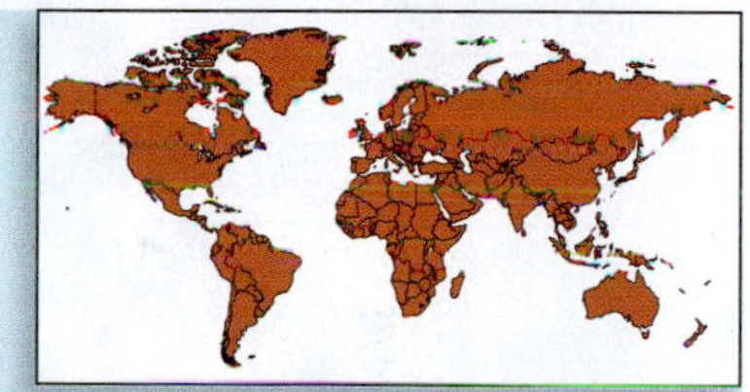

This new strain can have devastating and sometimes fatal effects on hosts because it has antigenic combinations that the adaptive immune system has never seen before, necessitating a prolonged defensive response before the body can conquer the infection. During this period, the well-known signs and symptoms of flu play out. When H1N1 flu cases reached epidemic levels on multiple continents by June 2009, the World Health Organization declared a pandemic—the first flu pandemic since 1968.

Maria was exhausted and weak for another two weeks, but she survived the flu, though many other patients were not so fortunate.

1. Why can people who are immunized still get the flu?
2. What do the letters A, H, and N stand for in the name 2009A(H1N1)?
3. Why are epidemiologists particularly concerned about bird influenzaviruses?

▲ **FIGURE 22.13 The face of coronavirus respiratory syndromes.** A newly identified strain of coronavirus has struck in the Middle East.

CLINICAL CASE STUDY

Influenza

A 26-year-old man reports to his physician in late October, complaining of a sudden onset of fever, a dry cough, headache, and body aches. The man states that he received his flu shot 10 days prior and must have gotten the flu from the immunization. He also states that he had just returned two days before from a weeklong trip to Hong Kong. He mentions that a highlight of his trip was a visit to the farmers' market filled with fresh produce and livestock. A culture confirms the patient is infected with influenza virus.

1. How should the physician counter the patient's assertion that he "got the flu" from the vaccine?
2. Explain the occurrence of this influenza case before the onset of the recognized flu season.
3. Why might the patient have been infected with influenzavirus even after receiving a vaccine?
4. The culture indicates that this is a drastically different flu strain from those seen in recent years. What phenomenon explains this?
5. Will prescription drugs likely be effective in this patient's case? Why or why not?

Pathogen and Virulence Factors

Only four weeks after SARS was recognized as a newly emerging disease, scientists had identified and sequenced the genome of the novel coronavirus—an unprecedented epidemiological and genomic accomplishment.

Coronaviruses are enveloped, +ssRNA viruses with helical capsids whose envelopes form crownlike (*corona*=crown) halos around the capsids, giving the viruses their name (see Disease at a Glance 22.4). Because diseases of coronaviruses are usually mild—coronaviruses are the second most common causes of common colds—the fatalities in the 2003 SARS and 2013 MERS pandemics were particularly alarming.

Pathogenesis and Epidemiology

Coronaviruses enter the body via respiratory droplets and adhere to lung cells. The virus destroys these cells, triggering the respiratory symptoms, and then spreads via the bloodstream to the heart and kidneys.

Epidemiologists accomplished a notable task in tracking the spread of SARS in only six months from Guangdong Province, China, to an apartment building in Hong Kong, and from there to more than two dozen nations in Asia, Europe, and the Americas. WHO reported that 8096 people became sick and 774 died from SARS (9.6% mortality). In the years following that first SARS pandemic, physicians have reported only about a dozen cases of SARS in China, including several cases among workers in SARS laboratories.

Epidemiologists first became aware of MERS in the spring of 2013 in Saudi Arabia. By June, the disease had spread between people across the Middle East and into Africa and Europe. The death rate from MERS is about 55%. Camels appear to be a reservoir for the MERS virus.

Diagnosis, Treatment, and Prevention

Physicians diagnose coronavirus respiratory syndromes based upon signs and symptoms, particularly in patients in endemic areas, and confirm it by isolating the virus or antibodies against the virus in the patient's blood. These techniques are not rapid; therefore, scientists are developing polymerase chain reaction (PCR) tests that will allow rapid diagnosis, which is necessary if future epidemics are to be avoided.

Physicians provide supportive care for patients. As of 2012, no antiviral drug has proven universally effective against these viruses, though antibodies against the virus reduce viral replication. Scientists have developed recombinant DNA vaccines against SARS virus and are testing the vaccines for safety and effectiveness.

Despite the speed with which researchers identified and characterized the SARS virus using modern genetic and immunological techniques, the pandemic was brought under control using centuries-old methods of isolation and quarantine of SARS patients and their contacts. The combination of historical and modern epidemiological practices rendered the SARS pandemic much less severe than many epidemiologists and health officials first feared.

DISEASE AT A GLANCE 22.4

Coronavirus Respiratory Syndromes

Cause SARS- and MERS-associated coronavirus (+ssRNA helical viruses).

Virulence factors Intracellular replication cycle evades immune surveillance.

Portal of entry Respiratory droplets enter through mucous membranes via close person-to-person contact.

Signs and symptoms High fever, headache and body aches, malaise, dry cough developing into pneumonia.

Incubation period Generally 2–7 days, up to 10 days in some cases.

Susceptibility Studies suggest that some people are genetically more susceptible than others are.

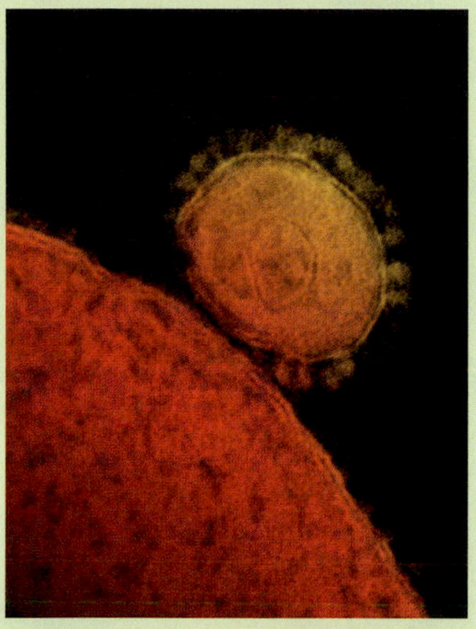

MERS virus, a coronavirus TEM 40 nm

Treatment Treatment is mostly supportive, athough studies show antibodies against SARS may be effective in reducing viral replication.

Prevention Limit travel to endemic areas, use handwashing precautions, quarantine infected persons and their contacts.

Disease at a Glance 22.4 summarizes the characteristics of coronavirus respiratory syndromes. Table 22.1 on p. 681 compares and contrasts manifestations of some common, respiratory diseases.

Respiratory Syncytial Virus (RSV) Infection

LEARNING **OUTCOMES**

22.25 Describe the features of respiratory syncytial virus infection.

22.26 Compare the effects of RSV infection in infants and adults.

Respiratory syncytial virus (RSV) infection is the most common childhood respiratory disease in newborns and young children. Annual community-wide outbreaks lasting four to six months are common during late fall, winter, and early spring.

Signs and Symptoms

About four to six days following infection, RSV triggers fever, rhinorrhea (runny nose), cyanosis (blue cast to skin), coughing, and sometimes wheezing in babies and the immunocompromised. RSV is the leading cause of *bronchiolitis* (inflammation of the bronchioles) and pneumonia among children less than one year of age and the leading respiratory killer of infants worldwide. Infections of RSV in older children and adults with normal immune systems results in coldlike symptoms. Some children develop *tracheobronchitis,* known commonly as **croup** (krūp), which is inflammation of the trachea and bronchi, resulting in breathing difficulty accompanied by a barking cough.

Pathogen

RSV (genus *Pneumovirus,* family *Paramyxoviridae*) is an enveloped, helical, −ssRNA virus. It is relatively unstable outside the body, surviving only about five hours in the environment or two hours on skin or used facial tissues. Soap and water as well as disinfectants deactivate RSV.

Pathogenesis

As its name indicates, the virus causes **syncytia** to form in the lungs. A syncytium is a giant, multinucleated cell formed from the fusion of virally infected cells to neighboring cells **(FIGURE 22.14)**. Plugs of mucus, fibrin, and dead cells in the bronchioles make it difficult to breathe. The action of cytotoxic T cells and other specific immune responses to RSV infection further damages the lungs.

Epidemiology

RSV spreads easily during close contact with infected persons, such as through kissing, touching, and shaking hands, and via contact with recently contaminated fomites. Spread via respiratory droplets is less frequent. Immunocompromised older patients and babies, especially those who are premature, immune

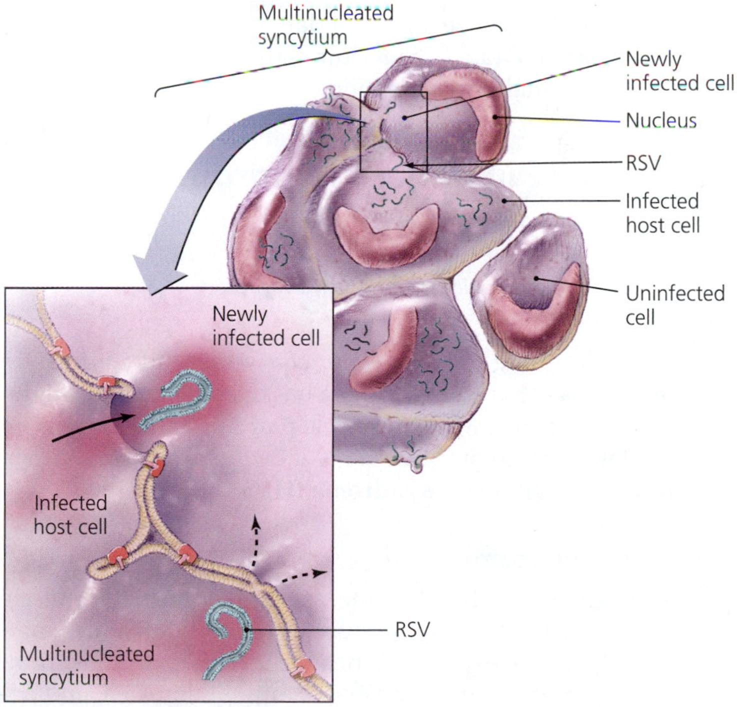

▲ **FIGURE 22.14 A syncytium forms when RSV triggers infected cells to fuse with uninfected cells.** Virions moving through these large multinucleated cells can infect new cells, all the while evading the host's immune system.

impaired, exposed to tobacco smoke, who attend day care, or who have older school-aged siblings, are most at risk. RSV is prevalent in the United States; epidemiological studies reveal that about 98% of children in day care centers are infected by age three. Up to 125,000 of these children require hospitalization each year, and 2000 die. Additionally, RSV kills about 14,000 elderly patients annually.

Diagnosis, Treatment, and Prevention

Prompt diagnosis is essential if infected infants are to get the care they need. The signs of respiratory distress provide some diagnostic clues, but verification of RSV infection is made by immunoassay. Specimens of respiratory fluid may be tested by immunofluorescence, ELISA, or complementary nucleic acid probes.

Older children and most adults require no treatment because their disease is mild. For younger children, supportive treatment includes administration of oxygen, intravenous fluids, drugs to reduce fever, and antibiotics to reduce secondary bacterial infections. Immunoglobulins against RSV, derived from blood donations, have proven effective for treating severe cases. Ribavirin via inhalation in mist form is used to treat extreme cases in premature and immunocompromised infants.

Control of RSV is limited to attempts to delay infection of susceptible infants through proper aseptic techniques, especially by health care workers and day care employees. Handwashing and the use of gowns, goggles, masks, and gloves are important measures to reduce healthcare associated infections. Attempts at developing a vaccine with deactivated RSV have proven difficult because the vaccine enhances the severity of the cellular immune response and lung damage.

Disease at a Glance 22.5 summarizes characteristics of respiratory syncytial virus infection.

DISEASE AT A GLANCE 22.5

Respiratory Syncytial Viral Infection

Cause Respiratory syncytial virus (RSV) (*Pneumovirus*, enveloped, −ssRNA virus).

Virulence factors Attaches to and enters human cells; intracellular replication cycle evades immune system; virus causes infected cells to fuse with their neighbors, so virus spreads without entering blood.

Portal of entry Inhalation.

Signs and symptoms Fever, rhinorrhea, coughing, cyanosis.

Incubation period Four to six days.

Susceptibility Babies and the immunocompromised are most at risk of serious infection.

Treatment Supportive respiratory care; antibodies against RSV and inhaled ribavirin may slow the disease.

Prevention Delay infection of newborns by proper aseptic technique, particularly handwashing.

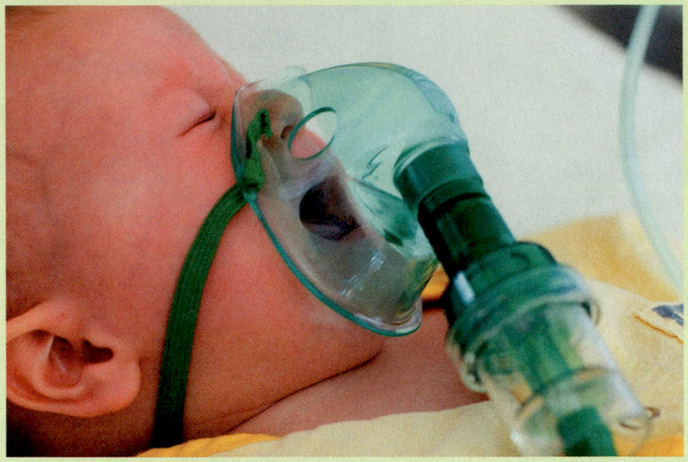

Hantavirus Pulmonary Syndrome (HPS)

LEARNING **OUTCOME**

22.27 Describe *Hantavirus* pulmonary syndrome.

True story: A young Native American scores an impressive 44 points for his tribal team during a basketball game one night. After the game, he is aware of muscle pains but attributes them to his athletic endeavors; five days later he is dead, a victim of ***Hantavirus* pulmonary syndrome (HPS).**

Signs and Symptoms

Early symptoms of HPS include fever, fatigue, and muscle aches, particularly in the large muscles of the thighs, hips, and back. Some patients experience headache, chills, and gastrointestinal symptoms, such as nausea, vomiting, diarrhea, and abdominal pain. An elevated leukocyte count and a low or falling platelet count are indicative of HPS.

Four to ten days after the initial manifestations, the patient begins coughing, goes into shock, and has difficulty breathing as the lungs rapidly fill with fluid. Fifty percent of diagnosed patients die, drowning in their own fluids, despite intensive medical care.

Pathogen

Hantavirus (han´tā-vī-rŭs) is a genus of enveloped, segmented, −ssRNA viruses in the family *Bunyaviridae* **(FIGURE 22.15)** that infect various species of mice, particularly deer mice (see Disease at a Glance 22.6), without causing disease. Two American strains of *Hantavirus* are transmitted via inhalation in dried mouse urine, feces, or saliva to infect the lungs of humans and cause HPS. A viral protein inhibits cellular responses to interferon.

Pathogenesis

Following inhalation, *Hantavirus* enters the blood via an unknown mechanism and travels throughout the body infecting the cells that make up blood capillary walls, particularly in the lungs. The body responds with inflammation, which causes the capillaries to leak fluid into the surrounding tissue. Blood pressure drops precipitously, and about 50% of patients die from pneumonia and shock.

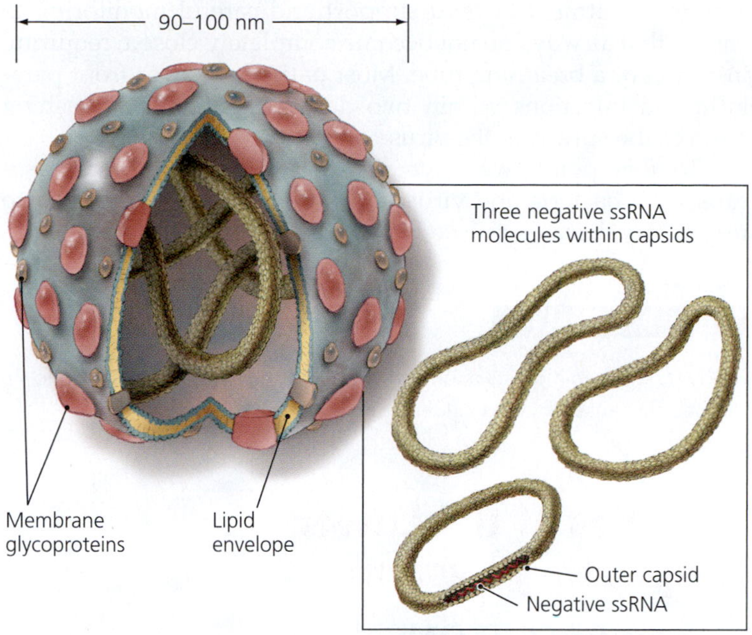

▲ **FIGURE 22.15** ***Hantavirus* is an enveloped, segmented, −ssRNA bunyavirus.** *What does the term* segmented *mean in reference to a viral genome?*

Figure 22.15 *A segmented genome has more than one nucleic acid molecule.*

DISEASE AT A GLANCE 22.6

Hantavirus Pulmonary Syndrome

Cause *Hantavirus* (enveloped, segmented, −ssRNA virus).

Virulence factors Intracellular replication helps avoid immune surveillance, proteins inhibit interferon responses.

Portal of entry Inhalation of infected rodent excrement or saliva.

The deer mouse, *Peromyscus maniculatus*

Signs and symptoms Sudden onset of fever, fatigue, muscle aches, headache, chills, abdominal pain; followed by coughing and severe difficulty in breathing.

Incubation period 14–30 days.

Susceptibility Those exposed to rodents in endemic areas.

Treatment Supportive therapy for symptoms must occur under hospital supervision.

Prevention Avoid rodent droppings, camp on ground sheet, carefully clean out potential rodent nesting sites at home, and keep home clean.

Epidemiology

Epidemiologists have reported several hundred cases in the United States since HPS was first recognized during an epidemic in the Four Corners[18] area of the United States in May 1993. Nevertheless, the syndrome has probably been around as long as mice and mankind have shared domiciles. The number of infections increases when abundant rainfall stimulates plant growth, providing abundant food material for the host mouse species. As the mouse population increases dramatically, humans are more likely to contact mice and their excrement and saliva. Despite a large number of virions in the blood vessels of infected humans, few viruses pass out of the lungs; therefore, person-to-person spread does not occur.

Diagnosis, Treatment, and Prevention

Physicians diagnose *Hantavirus* pulmonary syndrome based on typical manifestations of the disease—low platelet count, sudden onset of fever, and muscle aches in the major muscles of the legs and trunk—in patients who have contacted mice. Diagnosis is confirmed by detection of anti-hantavirus IgM, a rising titer of similar IgG, or demonstration of hantaviral RNA in clinical specimens using polymerase chain reaction.

No pharmacological treatment exists for HPS. Supportive care may include pulmonary intubation, fever-reducing drugs, pain medication, and supplemental oxygen. No vaccine against *Hantavirus* exists. Rodent control is necessary to prevent infection. This includes keeping food in sealed containers, blocking openings into homes (deer mice can squeeze through a hole the size of a man's shirt button), sleeping on a ground cover when camping, and removing mouse feces, urine, and nesting material from the home.

Disease at a Glance 22.6 summarizes the features of *Hantavirus* pulmonary syndrome.

Other Viral Respiratory Diseases

LEARNING **OUTCOME**

22.28 Describe human *Metapneumovirus* and parainfluenzavirus infections.

Other viruses cause common respiratory illnesses in children, elderly adults, and the immunocompromised. Among these are *Cytomegalovirus*, which is examined in more detail in Chapter 21, *Metapneumovirus* (MPV), and *parainfluenzaviruses*.

Metapneumovirus and parainfluenzaviruses are enveloped, unsegmented, −ssRNA viruses in the family *Paramyxoviridae*. Scientists discovered MPV in 2001 using nucleic acid probes which bound to genetic sequences that did not match the sequences of any known organism or virus. Since then, researchers have found that antibodies against the virus form in all children by age five, and they estimate that MPV is second only to rhinoviruses as a common cause of viral respiratory disease. As with most viral diseases, there is no treatment.

Three strains of parainfluenzaviruses cause croup and viral pneumonia, particularly in young children. There is no specific

[18]Four Corners is the geographic area where Arizona, Colorado, Utah, and New Mexico meet.

CLINICAL CASE STUDY

A Blue Baby

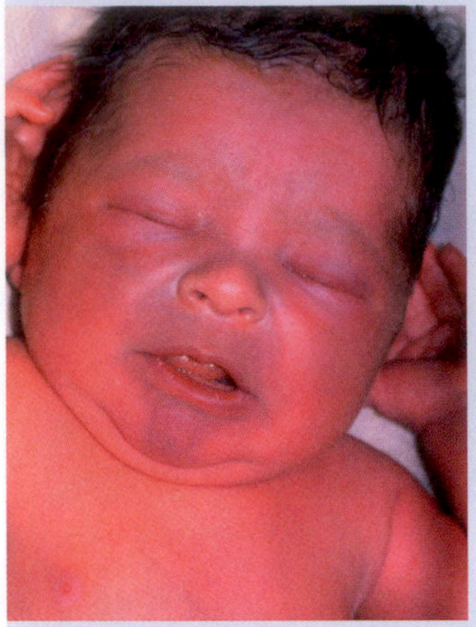

A woman reports to the emergency room in January with her cyanotic (blue-colored) 10-month-old child. The mother reports that the infant has had a runny nose, fever, and slight cough for a day and has had increasing trouble breathing. The child does not have a history of bronchial disease and was not premature. The mother also states that the infant's five-year-old brother is recovering from symptoms that resemble a cold.

1. What is the presumptive diagnosis?
2. How can the doctor confirm the diagnosis?
3. Describe the possible treatment for the child.
4. Were the parents irresponsible for not immunizing their child?
5. Is it likely that the infant caught the disease from his older brother? If so, why did the older child not display signs of respiratory distress?

antiviral treatment beyond support and careful monitoring to ensure that airways do not become completely closed, requiring insertion of a breathing tube. Most patients recover from parainfluenza infections within two days. Frequent handwashing reduces the spread of the virus.

To this point, we have considered respiratory diseases caused by bacteria and viruses. Next, we turn our attention to respiratory diseases caused by fungi.

TELL ME WHY

Why do epidemiologists think that there will be a major flu pandemic in people caused by bird influenzaviruses?

Mycoses of the Lower Respiratory System

LEARNING **OUTCOME**

22.29 Define *systemic mycosis.*

The number of cases of **mycoses** (diseases caused by fungi) has increased over the last two decades, mostly because AIDS patients are susceptible to fungal infections. The following sections examine three *systemic mycoses*—fungal infections that spread throughout the body—seen in North America **(FIGURE 22.16)** and a type of fungal pneumonia common to AIDS patients. We begin with a systemic mycosis caused by *Coccidioides.*

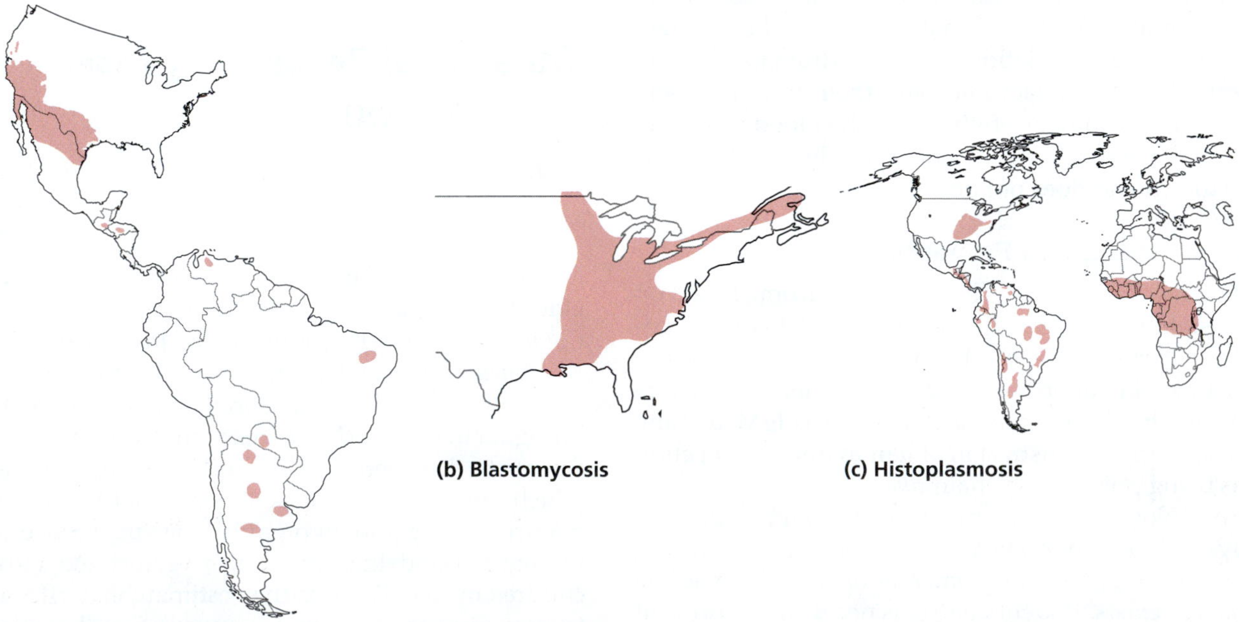

(a) Coccidioidomycosis (b) Blastomycosis (c) Histoplasmosis

▲ **FIGURE 22.16** **The geographic distributions of three systemic fungal diseases endemic to North America.**

Coccidioidomycosis

LEARNING | OUTCOME

22.30 Describe coccidioidomycosis.

Coccidioidomycosis (kok-sid′ē-oy′dō-mī-kō′sis), commonly known as *valley fever,* occurs primarily in the San Joaquin Valley of California.

Signs and Symptoms

The major manifestation of coccidioidomycosis is pulmonary, initially resembling pneumonia or tuberculosis, though about 60% of patients experience only mild, unremarkable respiratory symptoms that typically resolve on their own. Other patients develop more severe infections characterized by fever, cough, chest pain, difficulty breathing, coughing up or spitting blood, headache, night sweats, weight loss, and pneumonia; in some individuals, a diffuse rash may appear on the trunk.

In less than 1% of cases, generally in patients who are severely immunocompromised, the fungus spreads from the lungs to various other sites. Invasion of the central nervous system may result in meningitis, headache, nausea, and emotional disturbance. The fungus can also spread to the bones, joints, and subcutaneous tissues; subcutaneous lesions are inflamed masses of granular material **(FIGURE 22.17)**.

Pathogen and Virulence Factors

Coccidioides immitis (kok-sid-ē-oy′dēz im′mi-tis) is a dimorphic soil fungus in the division Ascomycota. In the warm and dry summer and fall months, particularly in drought cycles, *C. immitis* grows as a mycelium and produces sturdy chains of asexual spores called *arthroconidia.* Mature arthroconidia germinate into mycelia. The fungus assumes a pathogenic yeast form at human body temperature.

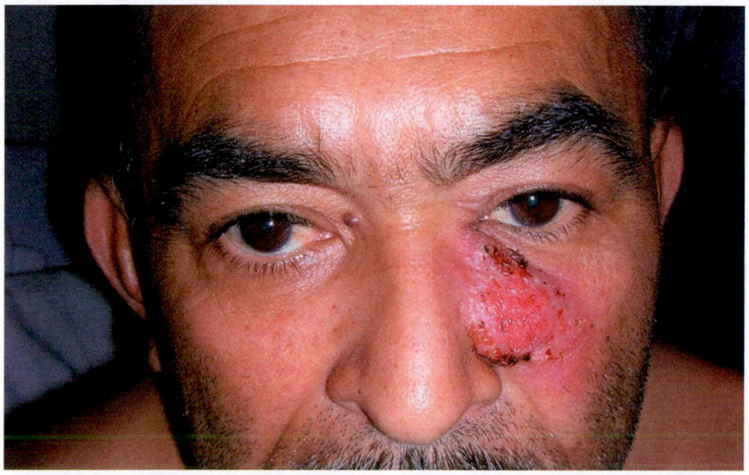

▲ **FIGURE 22.17 Coccidioidomycosis lesions in subcutaneous tissue.** Painless lesions result from the spread of *Coccidioides immitis* from the lungs.

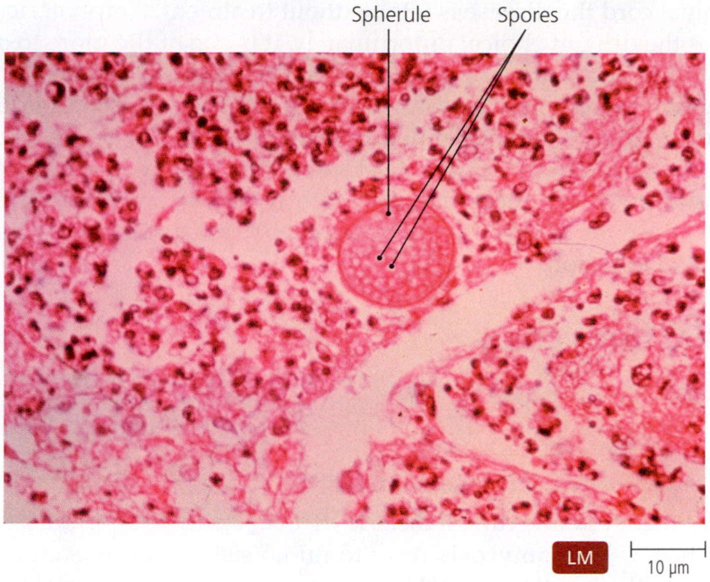

▲ **FIGURE 22.18 Spherules of *Coccidioides immitis.*** Note the numerous spores within a spherule.

Pathogenesis

Coccidioides enters the body through inhalation of arthroconidia from the soil. Disease begins as a generalized pulmonary infection that then spreads to the rest of the body. Arthroconidia germinate in the alveoli into a form called a *spherule* (sfer′ool) **(FIGURE 22.18)**. As each spherule matures, it enlarges and generates a large number of spores via multiple divisions, until it ruptures and releases the spores into the surrounding tissue. Each spore then forms a new spherule to continue the cycle of division and release. This type of spreading growth accounts for the seriousness of coccidioidomycosis.

Epidemiology

Coccidioides is found almost exclusively in the southwestern United States and northern Mexico (see Figure 22.16a). Small endemic areas also exist in semiarid parts of Central and South America. About 3% of people living in endemic areas develop the disease each year.

Any activity that disrupts the soil can disseminate arthroconidia into the air. Local epidemics have occurred among archeologists, model plane enthusiasts practicing their hobby in the desert, and drivers of off-road vehicles. Windstorms and earthquakes can disturb large tracts of contaminated soil, spreading arthroconidia for miles. In 1978, there were many cases of coccidioidomycosis in Sacramento, California, 500 miles north of the endemic area, following a severe dust storm in southern California.

Diagnosis, Treatment, and Prevention

Diagnosis of coccidioidomycosis is based on the identification of spherules in clinical specimens; diagnosis is confirmed by injecting antigen beneath the skin and observing an inflammatory response.

Although infections in otherwise healthy patients generally resolve on their own, when the fungus spreads to the brain and spinal cord the disease is fatal without treatment. Amphotericin B is the drug of choice; unfortunately, it is one of the more toxic antifungal agents to humans. In AIDS patients, maintenance therapy with other antifungal drugs, such as itraconazole or fluconazole, is recommended to prevent relapse or reinfection. The wearing of protective masks in endemic areas can prevent exposure to arthroconidia, although it may be impractical for daily use for all but those whose occupations put them at clear risk of infection.

Blastomycosis

LEARNING | OUTCOME

22.31 Describe blastomycosis.

Another systemic fungal disease that begins as a respiratory infection is **blastomycosis** (blas´tō-mī-kō´sis), which is endemic across the southeastern United States north to Canada (see Figure 22.16b).

Signs and Symptoms

Blastomycosis begins with flulike signs and symptoms. The fungus may then spread to cause cutaneous blastomycosis (in 60–70% of cases), which manifests as generally painless lesions on the face or upper body **(FIGURE 22.19)**. In roughly 30% of cases, purulent (pus-filled) lesions may develop and expand in the bones, prostate, testes, or other organs as the yeast multiplies, resulting in necrosis (death of tissues) and cavity formation.

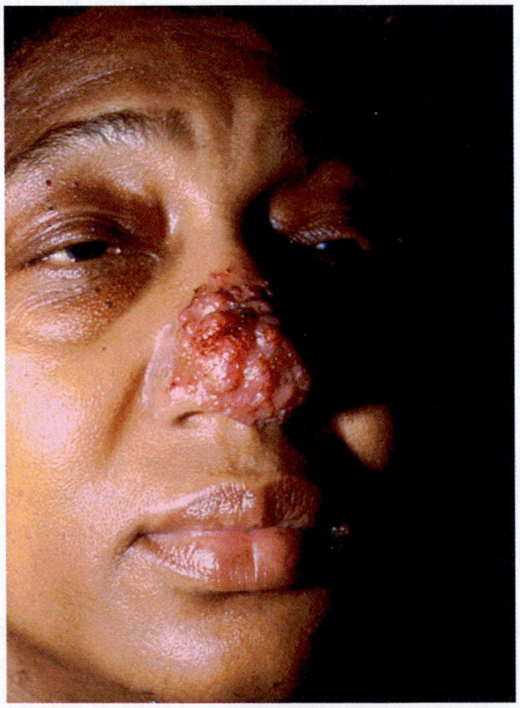

▲ **FIGURE 22.19 Cutaneous blastomycosis in an American woman.** The lesions result from the dissemination of *Blastomyces dermatitidis* from the lungs to the skin.

Pathogen

A dimorphic, pathogenic ascomycete—*Blastomyces dermatitidis* (blas-tō-mī´sēz der-mă-tit´i-dis)—causes blastomycosis. The fungus normally grows in soil rich in organic material such as decaying vegetation and animal wastes, where cool, damp conditions favor growth and sporulation. The fungus assumes a yeast form in the higher temperature of the human body.

Pathogenesis and Epidemiology

Inhalation of dust carrying fungal spores infects the lungs. In the lungs, spores germinate to form yeasts and multiply. Initial pulmonary lesions are asymptomatic in most individuals. In otherwise healthy people, pulmonary blastomycosis and minor skin lesions typically resolve, though the disease may become chronic and fatal. Respiratory failure and death occur at a high frequency among immunocompromised patients.

Epidemiologists have reported blastomycoses in Latin America, Africa, Asia, and Europe. One to two cases occur annually per 100,000 population in endemic areas. As **Emerging Disease Case Study: Pulmonary Blastomycosis** relates, the incidence of human infection is increasing.

Diagnosis, Treatment, and Prevention

Diagnosis relies on identification of *B. dermatitidis* in culture or direct examination of various samples such as sputum, bronchial washings, biopsies, cerebrospinal fluid, or skin scrapings. Observation of dimorphism in laboratory cultures coupled with microscopic examination is diagnostic.

Physicians treat blastomycosis with amphotericin B for 10 weeks, though longer treatment may be necessary. Oral itraconazole may be used as an alternative but must be administered for six months. Relapse is common in AIDS patients, and suppressive maintenance therapy with itraconazole is recommended.

Scientists have developed a live recombinant DNA vaccine against *Blastomyces* that provides protection in mice.

Histoplasmosis

LEARNING | OUTCOME

22.32 Describe histoplasmosis.

Histoplasmosis (his´tō-plaz-mō´sis) is the most common fungal systemic disease affecting humans.

Signs and Symptoms

In almost 95% of individuals, histoplasmosis is asymptomatic, subclinical, and resolves without damage. About 5% of patients develop clinical histoplasmosis, which is characterized by severe coughing with blood-tinged sputum or skin lesions. An AIDS patient with histoplasmosis often rapidly develops an enlarged spleen and liver, which can be fatal. Some patients' bodies mount a type I hypersensitivity reaction against the fungus in the eyes, producing inflammation and redness.

EMERGING DISEASE CASE STUDY

Pulmonary Blastomycosis

Phil was glad he had spent the summer fishing, hiking, and camping in western Ontario. Canada is beautiful, and he had needed time off from his two jobs and full schedule of nursing classes. Now he was back in college, and remembrances of Canada were helping him cope with the stress.

However, Phil didn't feel well. He was probably coming down with the flu—fever and chills, shivering and coughing, muscle aches, tiredness, a general feeling of yuckiness, and no runny nose; yes, it must be the flu. Or was it? He knew from microbiology class that many diseases of bacteria, protozoa, and viruses have flulike symptoms. Phil had forgotten fungi.

Phil used every over-the-counter remedy, to no avail. He tried the health clinic on campus, but the antibiotics he received made things worse, not better. He was losing weight, and pus-filled, raised sores had appeared on his face, neck, and legs. More alarming: his testes were swollen and aching. This was getting serious.

Blastomyces, an emergent, dimorphic fungus, was attacking Phil. Inhaled spores from hyphae growing on wet leaves primarily in Wisconsin and Ontario had germinated in Phil's lungs. The yeast phase was multiplying and spreading throughout his body, producing skin lesions that lasted for months, finally resolving into raised, wartlike scars.

Researchers don't know why blastomycosis is becoming more prevalent. Perhaps it has to do with better diagnosis and reporting; perhaps it has to do with a growing number of AIDS patients who are susceptible to infection; or perhaps it has to do with more people like Phil adventuring into the wilderness.

The good news is that with proper diagnosis, Phil got the treatment he needed—itraconazole for six months. He graduated, and now he is a nurse who knows that flulike symptoms can sometimes indicate serious fungal infection.

1. *Blastomyces* is dimorphic; what does this adjective mean?
2. Why does AIDS increase the incidence of blastomycosis?
3. Given that *Blastomyces* is a mold that grows on dead leaves, how does it cause disease in people?

Pathogen

Histoplasma capsulatum (his-tō-plaz´mă kap-soo-lā´tŭm), the causative agent of histoplasmosis, is a dimorphic ascomycete that is found in moist soils containing high levels of nitrogen such as from the droppings of bats and birds, especially chickens, starlings, and blackbirds. The fungus becomes a pathogenic yeast at human body temperature (37°C). Besides changing morphology, *Histoplasma* at this temperature produces several proteins that prevent full activation of macrophages and abrogate host defenses such as production of free radicals and antimicrobial peptides.

Pathogenesis and Epidemiology

H. capsulatum is an intracellular parasite that, upon inhalation, first attacks alveolar macrophages in the lungs. Infected macrophages disperse the fungus beyond the lungs via the blood and lymph. Cell-mediated immunity eventually develops, clearing the organism from healthy patients.

Histoplasmosis is particularly prevalent in the eastern United States along the Ohio River Valley (see Figure 22.16c), but endemic areas also exist in Africa and Central and South America. People inhale spores that have become airborne when soil containing the fungus is disturbed by wind or by human activities.

Diagnosis, Treatment, and Prevention

Diagnosis of histoplasmosis is based on the identification of budding yeast within macrophages or in stained samples of skin scrapings, sputum, cerebrospinal fluid, or various tissues; the diagnosis is confirmed by the observation of dimorphism in cultures grown from such samples. Cultured *H. capsulatum* produces distinctively spiny spores that are also diagnostic (see Disease at a Glance 22.7 on p. 708). Antibody tests are not useful indicators of *Histoplasma* infection because many people have been exposed without contracting disease. In the endemic regions of the United States, close to 90% of the population have antibodies against *H. capsulatum.*

Infections in immunocompetent individuals typically resolve without treatment. When symptoms do not resolve, physicians can prescribe fluconazole, itraconazole, or amphotericin B. Maintenance therapy for AIDS patients is recommended.

Disease at a Glance 22.7 on p. 708 summarizes the features of histoplasmosis.

Pneumocystis Pneumonia (PCP)

LEARNING OUTCOME

22.33 Describe *Pneumocystis* pneumonia.

Before the AIDS epidemic, ***Pneumocystis*** (nū-mō-sis´tis) **pneumonia (PCP[19])** was observed only in malnourished, premature

[19]Originally stood for the old name of the disease: *Pneumocystis carinii* pneumonia.

DISEASE AT A GLANCE 22.7

Histoplasmosis

1. Spores of *Histoplasma capsulatum* are present in moist, nitrogen-rich soils, especially at sites of bat and bird droppings.
2. Airborne spores are inhaled.
3. Spores attack alveolar macrophages. Symptoms in 95% of infections are mild: coughing, aches, and pains.
4. Infected macrophages carry spores through circulatory and lymphatic systems.
5. Patient may develop chronic pulmonary histoplasmosis with symptoms similar to those of tuberculosis.
6. Patient may develop chronic cutaneous histoplasmosis, characterized by ulcers.
7. Immunocompromised patients may develop systemic histoplasmosis characterized by enlargement of spleen and liver; death may result.
8. Ocular histoplasmosis may develop. Inflamed, red eyes indicate type I hypersensitivity to the fungus.

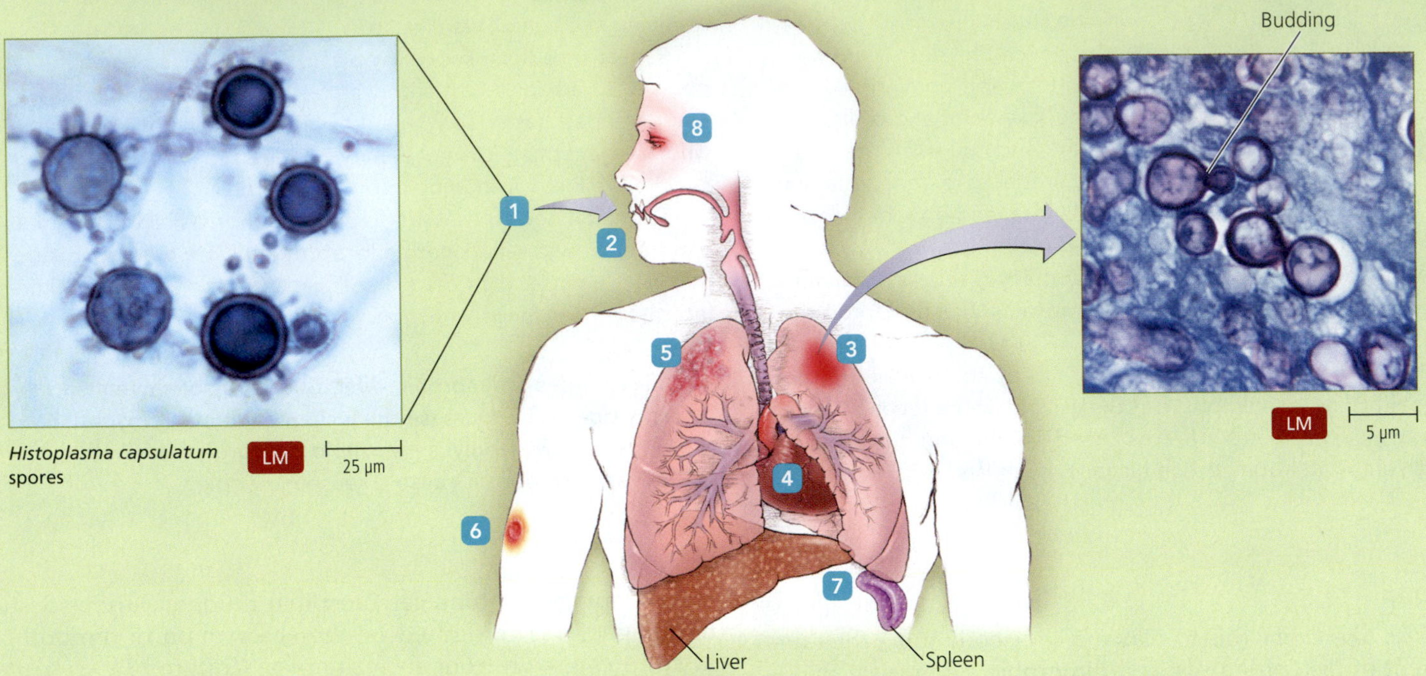

Cause *Histoplasma capsulatum* (dimorphic ascomycete fungus).

Virulence factors Ability to change to infective, yeast form at 37°C; production of proteins that inhibit macrophages and other immune responses.

Portal of entry Inhalation.

Signs and symptoms Dry, unproductive cough, shortness of breath, chest pain, fever, chills, headache, and malaise.

Incubation period Approximately 10 days.

Susceptibility Anyone, but more common in children up to 15 years old and those exposed to the soil in endemic areas.

Treatment Itraconazole, ketoconazole, or amphotericin B.

Prevention Minimize exposure to soil, especially near chicken coops or bat caves, or wear a mask.

infants and weak elderly patients; now, the disease is almost diagnostic for AIDS.

Signs and Symptoms

Signs and symptoms of PCP include increasing difficulty in breathing, mild anemia, hypoxia (low tissue oxygen), and fever. A nonproductive cough occurs in some cases. Rarely, extrapulmonary lesions develop in the lymph nodes, spleen, liver, and bone marrow. If left untreated, PCP involves more and more lung tissue until death occurs.

Pathogen

Pneumocystis jirovecii (nū-mō-sis´tis jē-rō-vĕt´zē-ē), which is a normal member of the respiratory microbiota, is an obligate parasitic ascomycete formerly known as *"P. carinii."* Originally it was considered a protozoan because of its morphology and development, but scientists have reclassified it as a fungus based on rRNA nucleotide sequences and biochemistry.

Pathogenesis and Epidemiology

P. jirovecii cannot survive on its own in the environment; therefore, transmission most likely occurs through inhalation of droplet nuclei containing the fungus. In normal people, infection is asymptomatic, and generally clearance of the fungus from the body is followed by lasting immunity. However, some individuals may remain infected for years; in such carriers, the organism remains in the alveoli and can be passed in sputum and presumably in respiratory droplets. Once the fungus enters the lungs of an immunocompromised patient, it multiplies rapidly, extensively colonizing the lungs.

P. jirovecii is distributed worldwide in humans. Based on serological confirmation of antibodies, 75% of healthy children have been exposed to the fungus by the age of five, but disease results only in immunocompromised patients. PCP is one of the more common diseases seen in AIDS patients.

Diagnosis, Treatment, and Prevention

Diagnosis of PCP relies on clinical and microscopic findings. Chest X rays usually reveal abnormal lung features. Stained specimens of fluid from the lungs or from biopsies can reveal distinctive morphological forms of the fungus **(FIGURE 22.20)**. The use of fluorescent antibody on samples taken from patients is more sensitive and provides a more specific diagnosis.

Although classified as a fungus, *Pneumocystis* does not respond to antifungal drugs; rather, both primary treatment and maintenance therapy are with an oral or intravenous combination of trimethoprim and sulfamethoxazole known as TMP-SMX. It is virtually impossible to prevent infection because *P. jirovecii* is ubiquitous in humans; however, the fungus produces disease only in immunocompromised individuals, so steps to ensure a healthy immune system such as stopping smoking, good nutrition, and prevention of HIV infection can prevent PCP in most people.

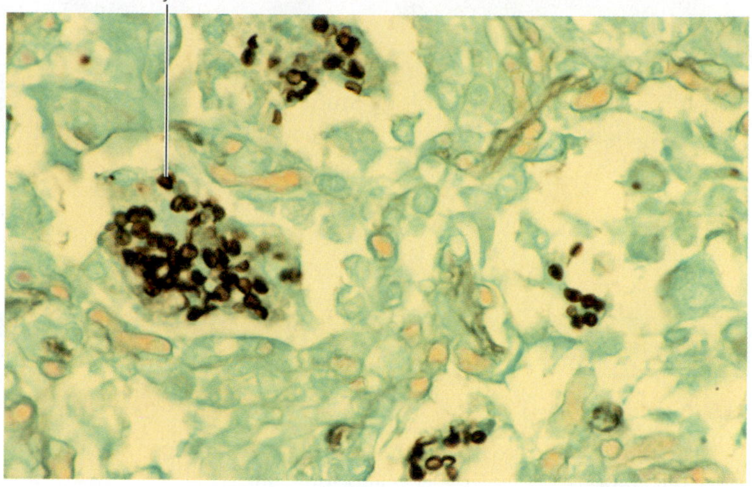

▲ **FIGURE 22.20 Cysts of *Pneumocystis jirovecii* in lung tissue.** Such microscopic findings are diagnostic for PCP.

TELL ME WHY

Outbreaks of blastomycosis have occurred in Latin America even though the organism itself is not normally found there. Why might a few cases of blastomycosis appear outside endemic areas?

MICRO IN THE CLINIC FOLLOW-UP

This Cough Can Kill

Lance learns that the woman seated behind him on the airplane from South Africa was diagnosed with tuberculosis (TB), caused by *Mycobacterium tuberculosis*. She was hospitalized one week after arriving in Atlanta, when tests showed that her sputum contained many acid-fast bacilli. She died five days later from extensive bleeding in her lungs. To complicate the clinical picture, the infecting strain of *M. tuberculosis* is a multi-drug-resistant (MDR) strain that does not respond to common anti-TB medicines.

Why does Lance have to wait so long to be tested? Any TB bacilli he may have inhaled from the respiratory droplets spewed by the woman's coughing will take up to 12 weeks to grow in his lungs. The cell-mediated hypersensitivity response of the tuberculin skin test reflects the body's immune reaction to the bacteria. Although Lance tries to forget about it for the next two months, he can't help worrying a little, and he has a hard time concentrating on school.

Lance's tuberculin test is negative: he does not have MDR *M. tuberculosis*. He's lucky, because a study found that passengers on a flight that lasts more than 8.5 hours and seated within two rows of a person with active tuberculosis are at risk of becoming infected. Had Lance's tuberculin skin test been positive, he would face a 6- to 12-month regimen of daily doses of several anti-TB medicines to reduce his risk of developing active tuberculosis. These medicines can have serious side effects, including liver damage, and could have interfered with his studies.

1. **An elected U.S. government official claims that funding for TB eradication programs should be greatly reduced because antibiotics are available to treat the disease. What evidence would you cite to counter his assertions?**
2. **How did Lance's legal work with people living in the shanties of the South African townships also place him at risk of contracting TB?**

MM Check your answers to Micro in the Clinic Follow-Up questions in the MasteringMicrobiology Study Area.

CHAPTER SUMMARY

Structures of the Respiratory System (pp. 678–680)

1. The upper respiratory system—consisting of the nose, nasal cavity, and pharynx—collects and filters air.
2. Cilia move mucus with trapped microorganisms to the pharynx to be swallowed.
3. Chemicals and cells of the tonsils fight microbial contaminants.
4. The lower respiratory system resembles an upside-down tree, composed of the larynx, trachea, bronchi, bronchioles, and alveoli of the lungs. They carry air through progressively smaller tubes to eventually allow exchange of gases with the blood capillaries of the lungs.
5. Pathogens cause inflammation of the respiratory tubes, resulting in restricted air flow.
6. Microbes stuck in mucus are carried by a ciliary escalator up and out of the bronchioles, bronchi, and trachea.
7. Alveolar macrophages and secretory antibodies (IgA) provide protection from many pathogens.
8. Harmless **diphtheroids** and other microbiota are commonly found in the nose and nasal cavity. Many may become opportunistic pathogens.

Bacterial Diseases of the Upper Respiratory System, Sinuses, and Ears (pp. 680–684)

1. **Streptococcal pharyngitis**—caused by streptococcal inflammation of the pharynx—is commonly known as strep throat.
2. Pharyngitis can progress into **scarlet fever** (scarlatina) when *Streptococcus* releases toxins that trigger fever and a bright red skin rash.
3. Untreated streptococcal pharyngitis can spread to cause complications in the kidneys (acute glomerulonephritis) and **rheumatic fever,** which may result in heart damage.
4. Inflammation of the larynx, **laryngitis,** causes hoarseness, and inflammation of the bronchi—**bronchitis**—results in restricted airflow to the lungs.
5. Group A *Streptococcus (S. pyogenes)* is the major cause of bacterial pharyngitis, scarlet fever, and rheumatic fever. Its survival in the body is aided by M protein, a hyaluronic acid capsule, pyrogenic toxins, streptolysins, and enzymes that allow the pathogen to spread in a variety of ways.
6. *S. pyogenes* from one individual travels via respiratory droplets to other individuals.
7. *Corynebacterium diphtheriae* produces a toxin that causes **diphtheria,** a disease characterized by the formation of a thick pseudomembrane on the surfaces of the upper respiratory tract.
8. **Rhinosinusitis** (inflammation of the nasal passages and sinuses) and **otitis media** (earache) are usually caused by *S. pneumoniae, Haemophilus influenzae,* or *Moraxella catarrhalis,* which spread from the pharynx to the sinuses or ears.

Viral Diseases of the Upper Respiratory System (pp. 684–686)

1. Of the over 200 different serotypes of viruses that cause colds, the most common are in the genus *Rhinovirus* (family *Picornaviridae*). Some coronaviruses, adenoviruses, reoviruses, and paramyxoviruses also cause colds.
2. Cold viruses are sensitive to warm temperatures and low pH, so they do not infect the lower respiratory tract.
3. Rhinoviruses are extremely infective and are transmitted via respiratory droplets, skin contact, and fomites.

Bacterial Diseases of the Lower Respiratory System (pp. 686–695)

1. **Pneumonia** results from pulmonary inflammation and the accumulation of fluid in the alveoli and bronchioles of the lungs. Physicians give different names to pneumonias, which are derived from the affected region of the lungs or from the causative organism.
2. The most common type of pneumonia is pneumococcal pneumonia, which is caused by *Streptococcus pneumoniae* (commonly known as **pneumococcus**).
3. Mycoplasmal pneumonia, caused by *Mycoplasma pneumoniae,* is also called **primary atypical pneumonia** or walking pneumonia.
4. *Klebsiella* pneumonia produces thick, bloody sputum and recurrent chills.
5. **Pneumonic plague** is a form of pneumonia caused by *Yersinia pestis,* the bubonic plague bacterium.
6. *Chlamydophila psittaci* causes **ornithosis** in birds, which is a disease that manifests as severe pneumonia in humans.
7. *Chlamydophila pneumoniae* causes bronchitis, pneumonia, and rhinosinusitis.
8. **Legionnaires' disease,** or legionellosis, is a pneumonia caused by *Legionella,* a genus of bacteria that are transmitted in aerosols.

9. The cell walls of *Mycobacterium tuberculosis* contain **mycolic acid,** which protects the pathogen when it infects the lungs, where it forms nodules called **tubercles** and the disease **tuberculosis (TB)**. **Multi-drug-resistant (MDR)** and **extensively drug-resistant (XDR)** strains of *Mycobacterium* pose a significant challenge to health care workers.
10. **Cord factor** is another cell wall component of *M. tuberculosis* that inhibits migration of neutrophils and kills cells.
11. A skin test determines if a person has been exposed to *M. tuberculosis* antigens. Chest X rays reveal tubercles in the lungs. Directly Observed Treatment, Shortcourse (DOTS) is a strategy to ensure treatment of TB patients.
12. *Bordetella pertussis* causes **pertussis (whooping cough)** by interfering with ciliated epithelial cells of the trachea via adhesins and toxins.
13. Pertussis progresses in four stages: incubation, catarrhal phase, paroxysmal phase, and convalescent phase.
14. *Bacillus anthracis* causes the most lethal form of **anthrax—inhalational anthrax,** which progresses from the manifestations of a common cold to lethargy and shock.

Viral Diseases of the Lower Respiratory System (pp. 695–704)

1. So-called A and B strains of orthomyxoviruses are surrounded by lipid envelopes with glycoprotein spikes composed of **hemagglutinin (HA)** or **neuraminidase (NA)** that play roles in the attachment of the viruses to the cells of the lungs to cause **influenza (flu)**.
2. The accumulation of HA and NA mutations in a geographic area results in an increase in cases of flu and is called **antigenic drift.**
3. **Antigenic shift** is a major antigenic change resulting from the reassortment of genomes from different influenzavirus A strains. Such shifts may result in pandemics.
4. **Coronavirus respiratory syndromes (SARS and MERS)** are emerging diseases that destroy lung cells and spreads via the bloodstream to the heart and kidneys.
5. **Respiratory syncytial virus (RSV) infection** is a common childhood respiratory disease. RSV causes giant, multinucleated cells **(syncytia)** to form in the lungs, leading to bronchiolitis or tracheobronchitis, commonly known as **croup.**
6. ***Hantavirus* pulmonary syndrome (HPS)** is a potentially fatal respiratory disease caused by *Hantavirus,* which is transmitted in dried mouse excreta.
7. Parainfluenzavirus strains 1, 2, and 3 are associated with croup and viral pneumonia.
8. *Metapneumovirus* (MPV) is the second most common cause of respiratory tract disease in children.

Mycoses of the Lower Respiratory System (pp. 704–709)

1. **Mycoses** are diseases caused by fungi; spread throughout the body they are called systemic mycoses.
2. *Coccidioides immitis* causes **coccidioidomycosis,** also known as valley fever. The initial manifestation resembles pneumonia or tuberculosis. In immunocompromised patients the manifestations include meningitis, headache, nausea, and emotional disturbance.
3. *Blastomyces dermatitidis* causes **blastomycosis,** a systemic fungal disease that begins as a flulike infection. The fungus may spread to cause lesions on the upper body and destroy tissues in bone, prostate, testes, and other organs.
4. *Histoplasma capsulatum* causes **histoplasmosis**—the most common fungal systemic disease affecting humans. About 5% of patients develop clinical histoplasmosis.
5. *Pneumocystis jirovecii* causes ***Pneumocystis* pneumonia (PCP)** in malnourished, premature, debilitated elderly, and immunocompromised patients.

QUESTIONS FOR REVIEW

Answers to the Questions for Review (except Short Answer questions) begin on p. A-1.

Multiple Choice

1. The movement of mucus from lungs to pharynx is due to
   a. epiglottal flow.
   b. a ciliary escalator.
   c. sneezing.
   d. pharyngeal reflux.
2. Compared to the upper respiratory system, the lower respiratory system
   a. provides an environment conducive to the growth of microorganisms.
   b. is normally devoid of microorganisms.
   c. provides an ideal environment for diphtheroids.
   d. is several degrees cooler.
3. The major cause of bacterial pharyngitis is
   a. group A *Streptococcus*.
   b. group B *Streptococcus*.
   c. *Mycobacterium*.
   d. *Bordetella*.
4. The glycoprotein spikes on influenzaviruses are composed of
   a. cord factor.
   b. hemagglutinin or neuraminidase.
   c. streptokinase and hyaluronic acid.
   d. M protein.
5. Group A *Streptococcus* is camouflaged from phagocytes by
   a. M protein.
   b. a hyaluronic acid capsule.
   c. pus resulting from the action of streptokinase.
   d. streptolysin.

6. The action of streptolysin results in
   a. breaking the hyaluronic acid capsule around cells.
   b. the inhibition of complement protein and decrease in the number of leukocytes at the site of infection.
   c. breaks in the membranes of erythrocytes, leukocytes, and platelets.
   d. the destruction of streptococcal bacteria.
7. Which pathogenic fungus is found in the droppings of bats, chickens, and blackbirds?
   a. *Histoplasma capsulatum*
   b. *Blastomyces dermatitidis*
   c. *Coccidioides immitis*
   d. *Parainfluenzavirus*
8. Which term associated with tuberculosis refers to protein and fat with a cheese-like consistency in the lungs?
   a. tubercles
   b. tuberculin tests
   c. tuberculous cavities
   d. caseous necrosis
9. The phase of whooping cough in which the characteristic "whoop" is obvious is the
   a. paroxysmal phase.
   b. catarrhal phase.
   c. convalescent phase.
   d. incubation phase.
10. Which of the following is associated with the fusion of neighboring cells?
    a. neuraminidase
    b. respiratory syncytial virus
    c. tuberculin
    d. agglutinin

Fill in the Blanks

1. The medical name for the inflammation of the throat known as strep throat is ________________.
2. A thick, leathery membrane in the throat is a sign of ________________.
3. RSV is characterized by the formation of giant multinucleated cells in the lungs called ________________.
4. A drug commonly used to treat systemic fungal diseases is ________________.
5. A condition in adolescents associated with taking aspirin to treat viral infections is ________________ syndrome.

Modified True/False

Modify each false statement to make it true by changing the underlined word(s).

1. _____ A normal cold produces fever in most cases.
2. _____ The number of fungal infections has increased over the last 20 years because of damp climatic conditions.
3. _____ Antigenic shift accounts for an increase in flu infections at a locality every two years.
4. _____ The formation of a hard, red lesion at the site of a tuberculosis skin test is a conclusive indication of the presence of the tuberculosis bacterium.
5. _____ Death by pneumonia would be similar to drowning.

VISUALIZE IT!

1. Color each map to show the general areas where each disease is endemic.

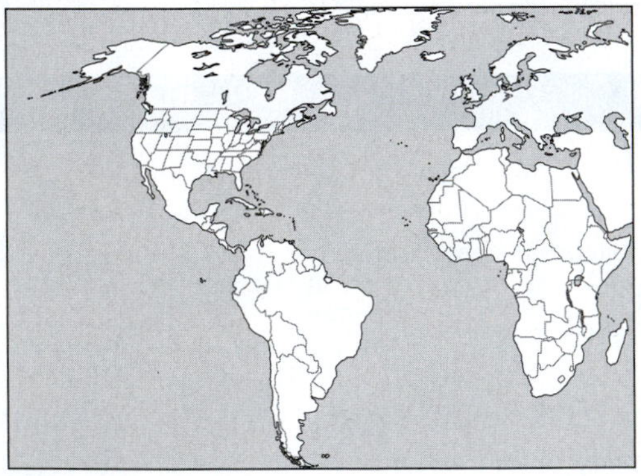

Blastomycosis

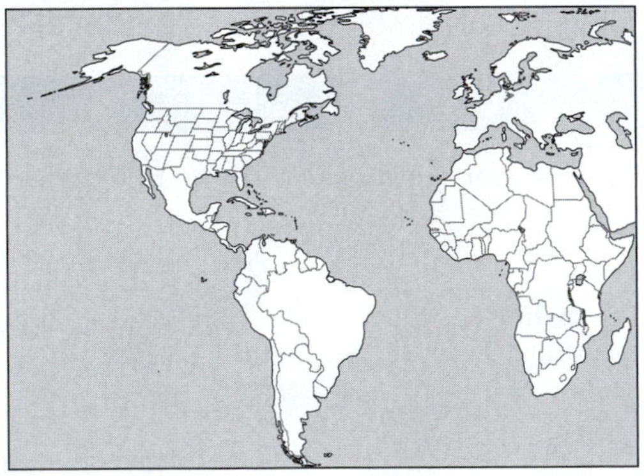

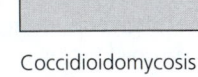

Coccidioidomycosis

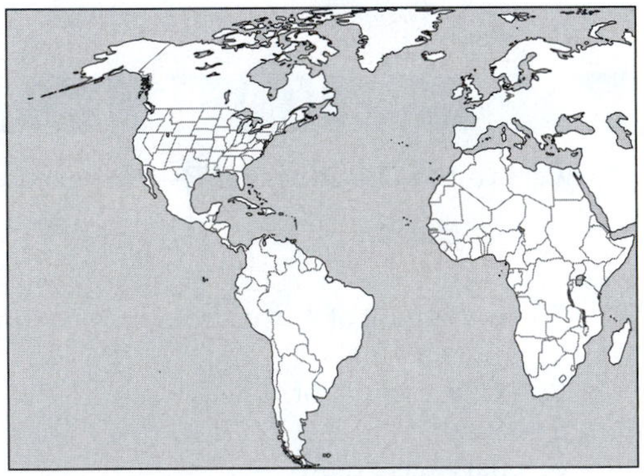

Histoplasmosis

2. Identify the following bacteria discussed in this chapter.

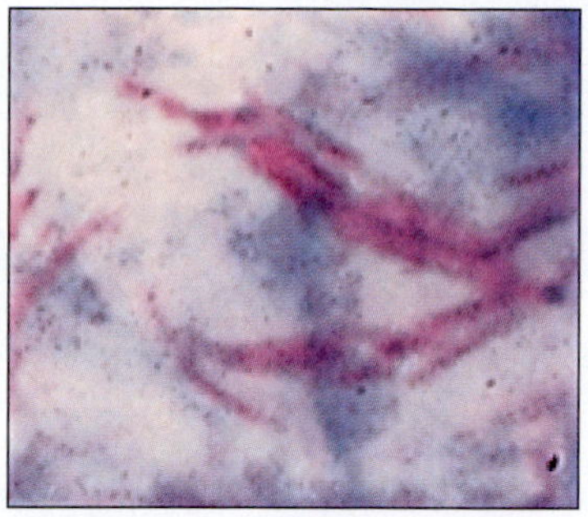
(a)

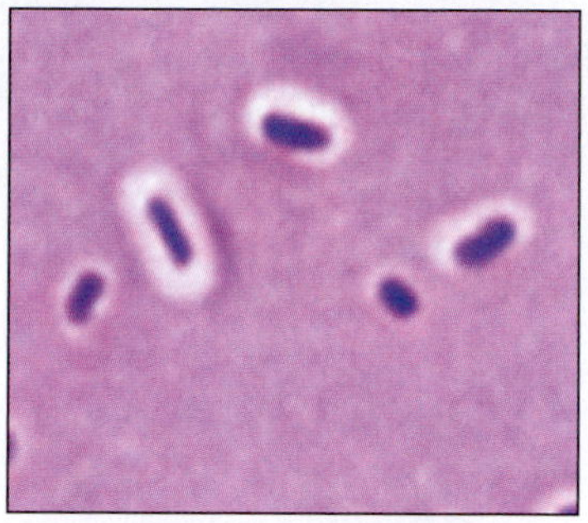
(b)

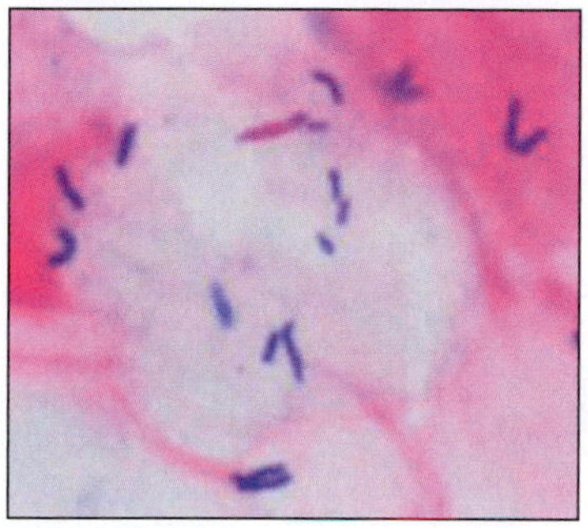
(c)

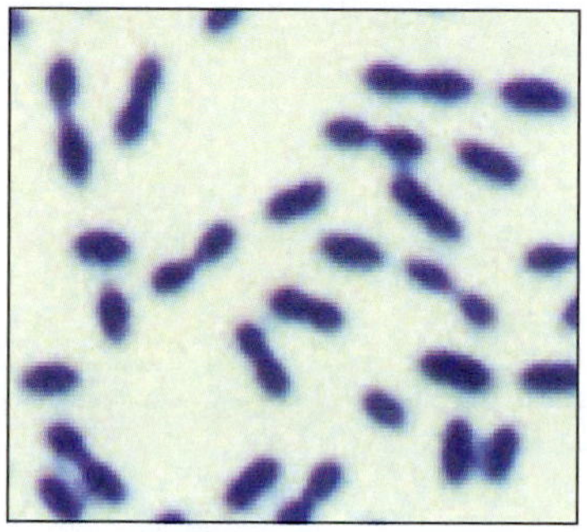
(d)

Short Answer

1. What is the function of adenoids?
2. Explain why the lower respiratory system is called the "respiratory tree."
3. The microorganisms that cause diphtheria and pneumonia are frequently found in the upper pharynx of healthy individuals. Why do relatively few individuals contract these two illnesses?
4. After listening to a lecture on diseases caused by *Streptococcus* spp., a student exclaimed, "Do you mean to say that *Streptococcus* can make a person kill his own heart?" To what was the student referring?
5. Describe the action of diphtheria toxin.
6. Give two reasons why there is no prevention for the common cold.
7. Give two reasons why cold viruses are more prolific in the winter.
8. On what basis is the decision made concerning the antigens selected for the flu vaccine in the U.S. each year?
9. Contrast the way infections of *Chlamydophila psittaci* spread from the way *C. pneumoniae* spreads.
10. What does a positive tuberculin skin test reveal?

CRITICAL THINKING

1. Explain the following identification label from a laboratory vial of influenza virus: B/Kuwait/6/97/(H1N3).
2. An elderly man is admitted to the hospital with severe pneumonia, from which he eventually dies. What bacterial species is the most likely cause of his demise? What antimicrobial drug is effective against this species? How could the man have been protected from infection? Is the hospital staff at significant risk of infection from the man? Which groups of patients would be at risk if the man had visited their rooms before he died?
3. Compare and contrast viral pneumonia with bacterial pneumonia in terms of cause, prevention, and potential seriousness.
4. Laboratory observation of *Pneumocystis jirovecii* led one researcher to conclude that it was a protozoan. Why has it been classified as a fungus?
5. A patient is admitted to a hospital in Ohio with a respiratory fungal infection. He had returned the previous month from a trip to Arizona, where he bought some dusty old blankets and two pots from a roadside vendor. What diseases might he have?
6. Compare and contrast antigenic drift and antigenic shift in influenza A virus.
7. In mid-November, a worried couple brought their 29-day-old newborn to their small-town doctor's office. The weak infant had been coughing severely over the past five days, so much so that she was choking on her formula. During her examination, the infant became blue and breathless. The baby girl tested positive on a DNA amplification test for *Bordetella pertussis.* (Adapted from *MMWR* 54:71–72. 2005.) Why is this disease so dangerous for small infants? Why did this baby become ill when this disease is preventable with vaccination?
8. As you have probably noticed, colds occur more frequently in the fall and winter. One explanation is that more people are crowded together in buildings when school starts and the weather cools. Design an experiment or epidemiological survey to test the hypothesis that crowded conditions explain the prevalence of colds in the fall and winter.
9. Why don't physicians try to prevent the spread of TB by simply administering prophylactic antimicrobial drugs to everyone living in endemic areas?
10. Could a traveler with coccidioidomycosis establish an endemic region of the disease in northern Russia?
11. Statistically, men are more likely than women to contract histoplasmosis. What might explain this fact?

CONCEPT MAPPING

Using the following terms, draw a concept map that describes tuberculosis. For a sample concept map, see p. 94.
Or, complete this and other concept maps online by going to the MasteringMicrobiology Study Area.

Acid-fast stain of sputum
Antibiotic resistance
Blood-tinged sputum
Chest X ray
Cough
Culture on special media
Disseminated tuberculosis
Dormant for life
Dormant infection
Ethambutol
Extensive lung damage
For 6–12 months
Isoniazid (INH)
MDR-TB
Mycobacterium tuberculosis
Other body locations
Primary infection
Rifampin
Strong immune response
Tubercular disease
Tuberculin skin test
XDR-TB

23 Microbial Diseases of the Digestive System

MICRO IN THE CLINIC

Trouble at the Rec Center

Andrea recently completed her nursing degree, got a great job, and moved with her husband and two toddlers to southern California. Their new community is a suburban paradise with excellent schools, shopping, and a recreational center that includes an artificial lake with swimming areas, a wading pool for very young children, and a beach.

In early August, many children at the rec center suddenly begin complaining of diarrhea. The diarrhea comes and goes and, aside from the number of children suddenly experiencing it, does not seem cause for concern. But in late August, Andrea's kids develop frequent, frothy diarrhea that smells like rotten eggs and sometimes contains mucus. Recalling the earlier complaints, Andrea suspects the lake water may be contaminated. She contacts the director of the rec center, who insists that the water is fine and that everyone just has the flu. However, the symptoms continue, and before long Andrea and her husband also come down with the same watery, foul-smelling diarrhea. They also suffer from abdominal cramps, bloating, and nausea.

Andrea informs the rec center director that if he doesn't call the public health department, she will. The director finally agrees, and public health officials come out to take water samples from the lake and wading pool. While waiting to hear the results, Andrea's entire family visits the doctor.

What is afflicting this suburban paradise? Turn to the end of the chapter (p. 747) to find out.

MM **Explore More:** Test your readiness and apply your knowledge with dynamic learning tools at MasteringMicrobiology.

Structures of the Digestive System

LEARNING | OUTCOMES

23.1 Describe the structure and function of the major parts of the gastrointestinal tract.

23.2 List the accessory digestive organs, and describe their functions in digestion.

Anatomists often divide the structures of the digestive system into two groups: those of the *gastrointestinal (GI)* (or *digestive*) *tract*—the tubular path from the mouth to the anus—and those of the *accessory digestive organs,* which either grind food or inject digestive secretions. We begin our survey by examining the GI tract.

The Gastrointestinal Tract

The gastrointestinal tract is a long tube lined with mucous membrane and composed of the mouth, esophagus, stomach, small intestine, large intestine, rectum, and anus **(FIGURE 23.1)**. The GI tract processes food into nutrients, absorbs nutrients and water into the blood, and eliminates waste. A membranous covering called the *peritoneum* (not shown) surrounds and protects most organs of the GI tract. The peritoneal cavity is the space between the organs and the peritoneum.

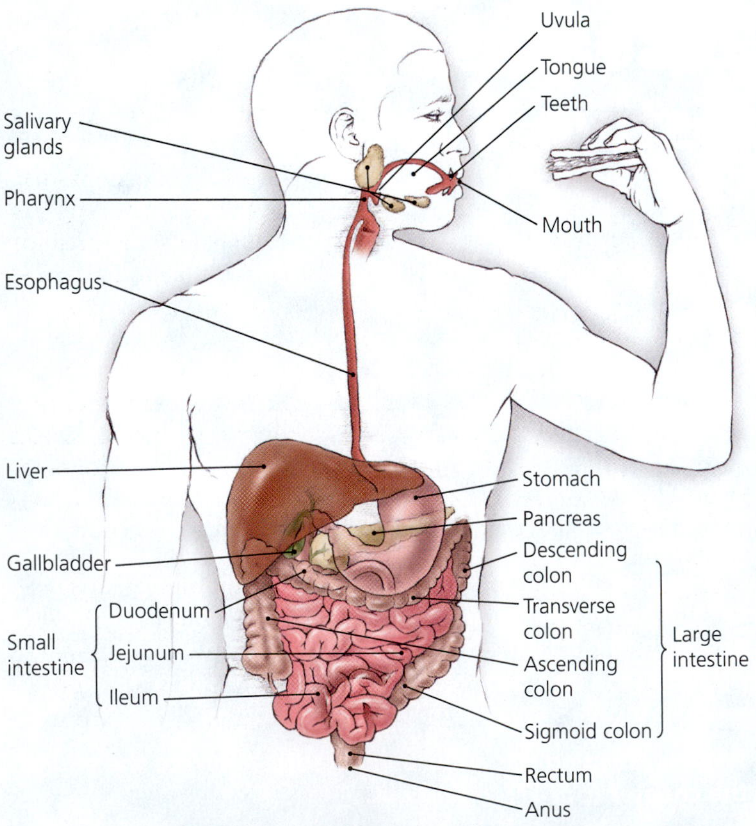

▲ **FIGURE 23.1 Major structures of the digestive system.**

When we eat, we chew and moisten the food in the *mouth* before swallowing. Digestion begins here with salivary enzymes. Muscle contractions, called *peristalsis,* move moistened food down the *esophagus,* which is the muscular tube at the back of the throat, to the *stomach.* The stomach secretes hydrochloric acid and a protein-catabolizing enzyme called *pepsin.* These chemicals further the chemical digestion of food as it is held in the stomach. The stomach gradually moves partially digested food into the small intestine.

The *small intestine* is so named because it is only about 3 cm in diameter, though it is about 6 m long. This portion of the GI tract has three subdivisions—the *duodenum* (doo-od´ĕ-nŭm), the *jejunum* (jĕ-joo´nŭm), and the *ileum* (il´ē-ŭm)—which are responsible for most of the digestion and absorption of nutrients. To this end, the internal surface of the small intestine folds into millions of fingerlike projections called *villi,* each of which is lined with cells having a cytoplasmic membrane convoluted into *microvilli,* giving the small intestine an absorptive surface area estimated to be about 2 million cm^2 (about 2150 ft^2)—the size of an average two-story American house!

Intestinal peristalsis moves remaining undigested and unabsorbed material into the *large intestine* (also known as the *colon*), which is about 7 cm in diameter and 1.5 m long. Anatomists name the regions of the colon for their location or shape—the *ascending colon,* the *transverse colon,* the *descending colon,* and the *sigmoid*[1] *colon.* The colon completes absorption of nutrients and water.

The remaining undigested materials, called *feces,* are mostly fiber. Feces pass into the *rectum,* which stores them until they are eliminated through the *anus,* a process called *defecation.*

The Accessory Digestive Organs

Accessory digestive organs include the tongue, teeth, liver, gallbladder, and pancreas. The *teeth* and *tongue* are important accessory organs of the mouth that masticate (chew) food into small bits, while the *salivary glands* secrete saliva that lubricates the food for swallowing. Saliva also contains *salivary amylase* that begins the digestion of starch.

Teeth have two functions in chewing. The *incisors* and *canines* at the front of the mouth tear food, and *molars* near the back of the mouth grind it. The surface of a tooth is *enamel*—a hard calcium phosphate mineral **(FIGURE 23.2)**. A softer material called *dentin* composes the body of a tooth, which extends as one or more roots into the *gingiva* (jin´ji-vă) (gums) and bone of the jaw. The interior of a tooth contains soft *pulp* with blood vessels and nerves.

The *liver* serves several major functions in the body, including production of *bile* to aid digestion and neutralization or removal of harmful substances from the blood. Bile, a yellow-greenish solution, is concentrated and stored in the *gallbladder.* Bile moves into the duodenum, where it emulsifies fat, that is, helps turn large fat droplets into millions of smaller ones, making them more accessible to digestive enzymes. The liver

[1]Named for the Greek letter sigma, which is S-shaped.

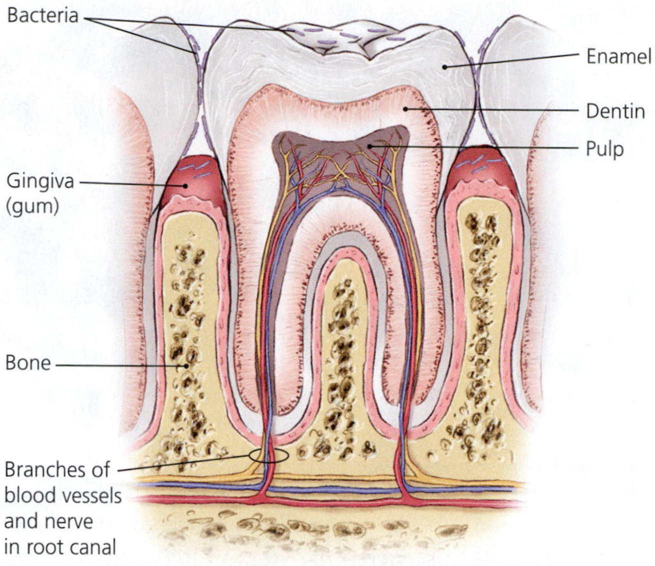

▲ **FIGURE 23.2 Detailed structure of teeth and socket.**

breaks down, excretes, or stores toxins, including *bilirubin*, a breakdown product of hemoglobin from dead erythrocytes. Liver damage leads to toxin buildup in the blood and jaundice from bilirubin buildup.

The *pancreas* produces *pancreatic juice*—digestive enzymes and bicarbonate buffer—which is released into the duodenum of the small intestine. The buffer neutralizes stomach acid as it enters the intestine and the enzymes are responsible for food digestion.

TELL ME WHY

Why is the digestive system an important *portal of entry* for microorganisms?

Normal Microbiota of the Digestive System

LEARNING **OUTCOME**

23.3 Describe the types and locations of normal microbiota in the mouth and intestines.

The esophagus, stomach, and duodenum are almost free of microbes. Peristalsis helps prevent the accumulation of food particles and microorganisms in the esophagus, and stomach acid (about pH 2.0) is antimicrobial. Further, the relatively rapid transport of food through the stomach and duodenum prevents most microbes from colonizing these regions.

However, microorganisms colonize the tongue, teeth, jejunum, ileum, colon, and rectum. The mouth and pharynx provide numerous microscopic pits and crevices as well as food for bacteria, fungi, and a few protozoa that colonize all oral surfaces. Each milliliter of saliva contains millions of bacteria, and scientists have discovered more than 700 species in oral biofilms. Most prevalent of the oral microbes are species of *Streptococcus* (strep-tō-kok´us) known as **viridans**[2] **streptococci**. Viridans streptococci are alpha-hemolytic, are Gram-positive, and lack so-called Lancefield carbohydrates; thus, they do not fall into any of the Lancefield classifications. Each strain of viridans streptococci has an adhesion factor allowing it to attach to a specific chemical on the gingiva, lining of the cheeks, tongue, pharynx, intestine, or teeth. *S. mutans* (mū´tanz) is the species that grows specifically on teeth.

The lower small intestine and colon are home to an estimated 100 trillion (10^{14}) bacteria, and over 10^{11} bacteria are in every gram of feces, accounting for approximately 40% of total fecal mass. Most of the bacteria are species of Gram-negative anaerobes of the genus *Bacteroides* (bak-ter-oy´dēz), followed in predominance by Gram-positive *Lactobacillus* (lak´tō-bă-sil´ŭs) and facultative enterobacteria such as *Escherichia* (esh-ĕ-rik´-ē-ă), *Enterobacter* (en´ter-ō-bak´ter), *Proteus* (prō´tē-ŭs), and *Klebsiella* (kleb-sē-el´ă). Fungi such as the yeast *Candida* (kan´did-ă) and protozoa such as *Entamoeba* (ent-ă-mē´bă) also live in the colon. Normal intestinal microbiota feed on the partially digested and indigestible contents of the colon. The mucous membrane lining the GI tract prevents most microbes from entering the blood, though some microbes can affect the body negatively without invading by changing the chemical nature of ingested chemicals, increasing toxicity, or producing cancer-causing chemicals.

Generally, intestinal microbiota serve to protect the body by outcompeting pathogens—a situation called **microbial antagonism**. The metabolism of intestinal microorganisms also produces vitamins, including vitamin B_{12}, folic acid, biotin, and vitamin K, in addition to a daily 500 ml of *flatus*—intestinal gas composed of nitrogen gas, carbon dioxide, hydrogen gas, and quite odorous dimethyl sulfide and methane.

Oral antimicrobials, taken for any disorder, can inhibit intestinal microbiota, thus undermining their defensive properties. Their loss during long-term antimicrobial therapy can allow colonization by pathogenic microbes. **Beneficial Microbes: Microbes to the Rescue?** on p. 718 examines potential benefits of changing the makeup of the microbiota.

Not all microbes in the GI tract are helpful or harmless members of the protective microbiota; numerous bacteria, fungi, viruses, protozoa, and parasitic worms cause diseases of the digestive system. The following sections examine some of the more common diseases, beginning with diseases caused by bacteria.

TELL ME WHY

Why does use of antibacterial drugs over an extended time increase the likelihood of oral candidiasis (thrush) and so-called *C. diff* diarrhea?

[2]From Latin *viridis*, meaning "green," for the pigment produced when grown on blood media.

BENEFICIAL MICROBES

Microbes to the Rescue?

The digestive tract is home to viruses, bacteria, protozoa, fungi, and parasitic helminths. The normal microbiota helps protect the body by competing with pathogens for nutrients and space. Recent scientific studies indicate that microbes added to people's diets improve and maintain their health. Many researchers, nutritionists, and health care professionals are excited about the possibilities.

Beneficial microbes help ward off bowel problems, such as irritable bowel syndrome, long-term inflammation of the end of the ileum (Crohn's disease); reduce incidence of yeast infection; alleviate symptoms of gastroenteritis; and may shorten the duration of colds by 36 hours. And, the benefits don't stop with the digestive system. One study indicates that *Lactobacillus* growing in the vagina inhibits attachment of the gonorrhea bacterium. Many probiotics, as such microbes are called, are bacteria used to ferment food, particularly species of *Lactobacillus* or the related genus *Bifidobacterium.* Despite favorable anecdotal evidence and some positive laboratory studies for the benefits of probiotics, some scientists remain skeptical that consumers can successfully change the makeup of their intestinal microbiota.

Other research suggests that the makeup of the microbiota mediates whether a person is obese or lean. Obese people have a greater percentage of Gram-positive bacteria in their colons, whereas lean people have a larger proportion of Gram-negative bacteria called bacteroids. Gram-positive bacteria break down indigestible polysaccharides, releasing sugars that can be absorbed and add to weight gain. In lean people with bacteroids, the polysaccharides remain undigested and pass from the body. When scientists changed the gut microbiota of mice, obese mice became thin even though they ate the same amount and kind of food and exercised the same amount.

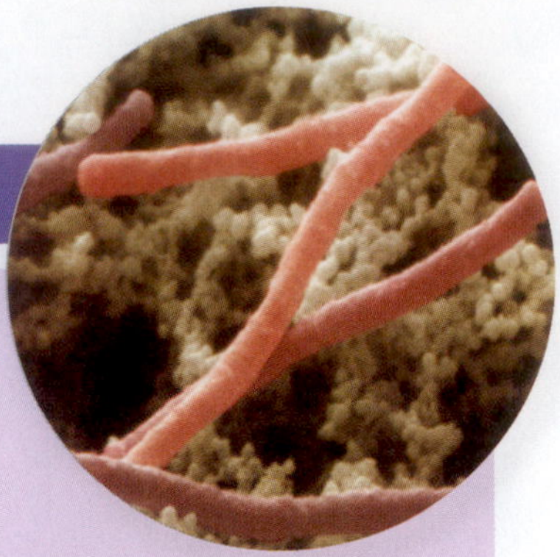

Lactobacillus, a potential probiotic

Could probiotics of the future permanently change the makeup of the microbiota of the digestive tract and alleviate disease and reduce obesity? Researchers continue working to answer these questions.

Bacterial Diseases of the Digestive System

In the United States and other developed nations, bacterial diseases of the digestive system are generally viewed as an annoyance, but in many areas of the world, these diseases are fatal. We begin by examining bacterial diseases of the mouth.

Dental Caries, Gingivitis, and Periodontal Disease

LEARNING **OUTCOMES**

- **23.4** Explain the process of dental caries formation.
- **23.5** Describe the progression to gingivitis and more severe periodontal disease.
- **23.6** Describe the treatment and prevention of cavities and periodontal disease.

Dental **caries**[3] (kār´ēz, *tooth decay* or *cavities*) are second only to common colds in frequency. They occur in people of all age groups, though they usually form during childhood. **Gingivitis**—inflammation of the gums—is a form of **periodontal disease**, which is inflammation and infection of the tissues surrounding and supporting the teeth.

Signs and Symptoms

Caries generally appear as holes or pits in the teeth, particularly in the later stages of disease, and can result in tooth loss. Initially, cavities are painless, but as they continue to develop, toothaches occur, often happening with the consumption of sweet, hot, or cold foods or drinks. Fractured teeth or an inability to bite down on a tooth without pain is also a sign of caries.

General symptoms of periodontal disease include swollen and/or bleeding gums, gums that are tender to the touch and appear shiny, or gums that appear bright red or red-purple in color. In the case of advanced periodontal disease, loose teeth and a foul breath odor occur. An extreme and rare type of periodontal diseases is *acute necrotizing ulcerative gingivitis (ANUG),* which was called *trench mouth* during World War I. Patients with ANUG show the manifestations of other periodontal diseases with the addition of craterlike ulcers between the teeth, extensive gum bleeding, a foul taste in the mouth, and a grayish biofilm that appears on the gums.

Pathogens, Virulence Factors, and Pathogenesis

Dental caries begins when bacteria, particularly *Streptococcus mutans,* produce an insoluble, sticky, polysaccharide slime called *dextran* from sucrose (table sugar). Dextran and adhesion

[3]Latin, meaning "decay."

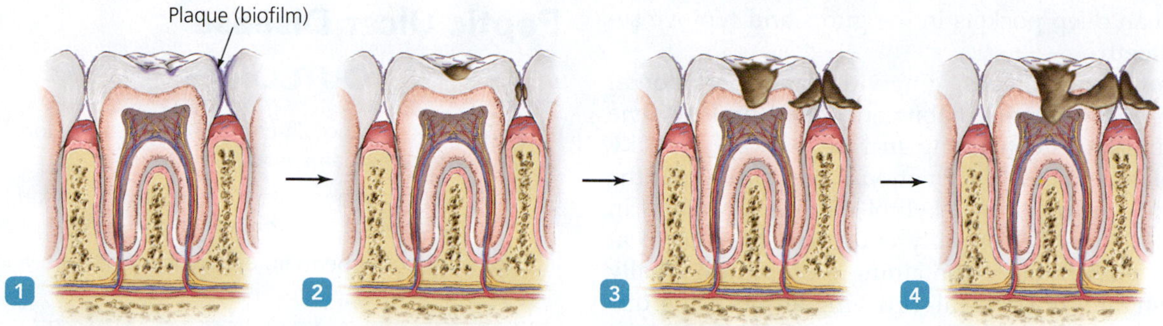

▲ **FIGURE 23.3 The process of tooth decay.** *Streptococcus mutans* produces dextran from sucrose, allowing plaque formation in pits and crevices on enamel 1. *Lactobacillus* and other bacteria ferment sugars to acids, which dissolve enamel 2. Decay continues into the dentin 3 and pulp 4.

factors, such as fimbriae, allow bacteria to form a biofilm known as *dental plaque* on tooth enamel (see Figure 14.6). Dental plaque may be more than 500 bacterial cells thick and contain nearly 10 billion bacteria.

S. mutans and other bacteria in plaque, such as *Lactobacillus,* ferment sugars to acid (about pH 5), which dissolves tooth enamel and allows bacterial invasion of the dentin and pulp inside the tooth **(FIGURE 23.3)**. The bacteria and their secretions can destroy dentin, pulp, and eventually the nerves and blood vessels of a tooth, possibly leading to loss of the tooth.

Hard deposits called *tartar* or *dental calculus* form when calcium salts mineralize plaque. Tartar trapped at the base of the teeth triggers the initial form of periodontal disease—gingivitis. Swelling of the gums, plaque, and tartar can also form oxygen-free pockets that become colonized by anaerobic bacteria, such as *Porphyromonas gingivalis* (pōr-fir-ō-mōn´ăs jin´ji-val-is), compounding the infection and producing a condition called *periodontitis. P. gingivalis* produces five protein-digesting enzymes that break down gingival tissue. Further destruction occurs as bacteria invade the bone, causing *osteomyelitis* (inflammation of bone marrow and bone), and teeth become loose and fall out.

Scientists do not know the exact cause of trench mouth, but they suspect an overabundance of anaerobes and spirochetes, including nonculturable species of *Treponema* (trep-ō-nē´mă).

Epidemiology

About 6% of personal health care expenditures in the United States are for dental services. Of American adults 20–64 years of age, 92% have experienced dental caries in their permanent teeth, with 42% of children aged 2–11 having at least one cavity. Diets high in sucrose increase the risk of tooth decay by increasing bacterial dextran and acid production. Continual snacking stimulates continual microbial activity, decreasing the pH further. Foods high in natural acid content, such as citrus fruit, can contribute to the problem.

Plaque buildup, injury or trauma to the gums, misaligned teeth, rough edges to fillings, ill-fitting dental appliances (such as dentures), and cavities contribute to gingivitis. Additional contributory factors to gingivitis include pregnancy, uncontrolled diabetes, general poor health and diet, poor dental hygiene, use of birth control pills, and lead poisoning. Gingivitis occurs frequently in many people and to varying degrees over the course of one's lifetime, though it usually first appears during puberty and early adulthood. If not treated, it leads to recurrent gingivitis or periodontal disease, which occurs in about 70% of the U.S. population, with up to 20% experiencing bone loss.

Diagnosis, Treatment, and Prevention

Dentists diagnose dental caries by visual inspection or physical examination during routine checkups. Probing with sharp dental instruments and X-ray exams can reveal soft spots before observable pits form. Dentists diagnose gingivitis by observing excessive plaque and tartar near the gum line, swollen red to purple gums that are sensitive to touch, receding gums with enlarged pockets around the bases of the teeth, or loose teeth. Dental X-ray exams reveal loss of bone structure that occurs in advanced periodontal disease.

Dentists can treat cavities that are diagnosed early with minimal expense and pain. They fill small cavities by removing softened tissue and filling the resulting hole with silver alloy, gold, porcelain, or resin. When cavities have destroyed a significant portion of a tooth, dentists remove soft material and overlay the tooth with a *crown* (cap) of gold, silver alloy, porcelain, or porcelain overlaid on metal. Decay resulting in death of the nerves feeding the tooth necessitates a *root canal,* in which a dentist removes the decayed material, pulp, nerves, and blood vessels from the core of the tooth and replaces them with a sealant.

Dental hygienists and dentists initially treat gingivitis and other periodontal diseases with *scaling*—the physical removal of plaque and tartar from the teeth in order to reduce inflammation. Scaling may result in bleeding, discomfort, or pain, which can be treated with over-the-counter pain relievers. They also repair dental irritants, such as loose-fitting dentures and misaligned teeth, so they no longer rub against the gums. Scaling must be followed by strict and effective daily oral hygiene along with regular professional cleanings, or the problems will return. Dentists may prescribe antibacterial mouth rinses. In the case of advanced periodontal disease, surgery may be required to expose and clean deep pockets in the gums and remove severely damaged teeth.

Prevention of dental caries, gingivitis, and periodontal disease relies on healthy eating habits and good oral hygiene practices. Avoiding foods containing sucrose, as well as sticky foods, reduces plaque and acid formation. Consumption of the naturally occurring sugar alcohol xylitol reduces *S. mutans* in the mouth. People should floss daily and brush their teeth at least twice a day with fluoride-containing toothpastes; ideally they should brush after every meal. Brushing and flossing disrupt the biofilm (plaque) on teeth, between teeth, and from the gums. Rinsing the mouth with water after a meal or snack can reduce plaque buildup when brushing is not possible. Teeth should be cleaned regularly in a dental office as part of routine care. Additionally, dentists can seal teeth with a plastic resin to deter decay.

Fluoridation of water is an effective public health measure to prevent dental diseases. Fluoride is incorporated into the developing enamel of new teeth, strengthening them against the effects of acid and protecting them from the inside. The cost to fix a single cavity is greater than the cost of providing fluoridation to an individual for a lifetime. Today, most U.S. city water systems add fluoride to drinking water.

Disease at a Glance 23.1 summarizes features of dental caries.

DISEASE AT A GLANCE 23.1

Dental Caries

Cause Bacteria, particularly *Streptococcus mutans* (Gram-positive coccus).

Virulence factors Fimbriae and other adhesins; formation of biofilm called plaque and tartar; metabolizes sucrose into dextran; produces acid from sugars.

Portal of entry Ingestion.

Signs and symptoms Demineralization of tooth resulting in pits and holes; tooth pain.

Incubation period Acid production begins immediately.

Susceptibility Everyone, especially those with a high-sucrose diet.

Treatment Physical removal of plaque and tartar; filling of pits and holes.

Prevention Avoidance of foods containing sucrose; regular toothbrushing and flossing; fluoridation.

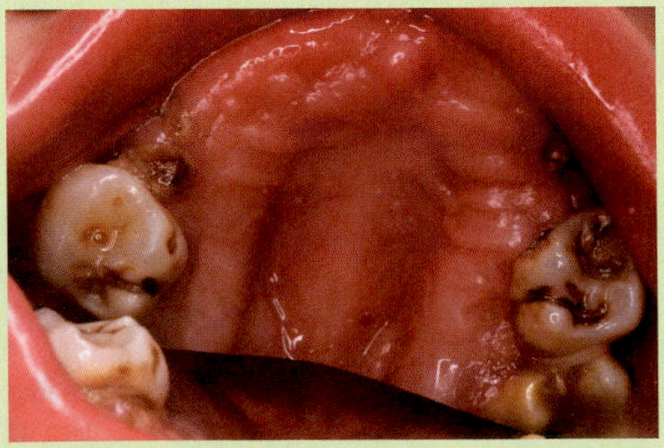

Peptic Ulcer Disease

LEARNING | OUTCOMES

23.7 Describe the effect of *Helicobacter pylori* on the lining of the human stomach.

23.8 Describe the virulence factors of *H. pylori* that allow it to colonize the stomach and to survive phagocytosis.

23.9 Describe the manifestations, treatment, and prevention of ulcers.

Peptic[4] **ulcers** are erosions of the linings of either the stomach *(gastric ulcer)* or duodenum of the small intestine *(duodenal ulcer)*. Ulcers that pierce the stomach or intestine are referred to as *perforations*. At one time, physicians considered peptic ulcers to be the result of drinking too much alcohol, smoking, eating the wrong foods, stress, or worry. We now know that most ulcers are actually due to the invasive activity of a bacterium.

Signs and Symptoms

Abdominal pain is the major symptom of ulcers; shock, in which the cardiovascular system fails to deliver enough blood to vital organs, is usually the major sign of a perforation. Some patients have nausea, vomiting (with or without blood, which looks like coffee grounds in the vomitus), weight loss, chest pain, or black, tarlike stools. Left untreated, ulcers can lead to a variety of complications, including internal bleeding and bowel obstruction—both of which constitute medical emergencies requiring immediate medical intervention.

Pathogen and Virulence Factors

Helicobacter pylori (hel´ī-kō-bak´ter pī´lō-rē)—a Gram-negative, slightly helical, highly motile bacterium—causes most peptic ulcers (see Disease at a Glance 23.2 on p. 722). *H. pylori* possesses numerous virulence factors that enable it to colonize the human stomach: a protein that inhibits acid production by stomach cells; flagella that enable the bacterium to burrow through the mucus lining the stomach; adhesins that facilitate binding to gastric cells; enzymes that inhibit phagocytic killing; and an enzyme—urease. Urease degrades urea, present in gastric juice, to produce highly alkaline ammonia, which neutralizes stomach acid.

Pathogenesis

FIGURE 23.4 illustrates the formation of a peptic ulcer. *H. pylori* (protected by urease) burrows through the stomach's protective mucus layer to reach the underlying epithelial cells **1**, where the bacterium attaches to the cells' cytoplasmic membranes and multiplies. A variety of factors—the triggering of inflammation by bacterial toxins and perhaps the destruction of mucus-producing cells—causes the mucus layer to become thin **2**, allowing acidic gastric juice to digest the stomach lining. Once gastric juice has perforated the epithelial layer, *H. pylori* gains access to the underlying muscle tissue and blood vessels **3**. Bacteria

[4]From Greek *pepto,* meaning "to digest," referring to the stomach enzyme pepsin.

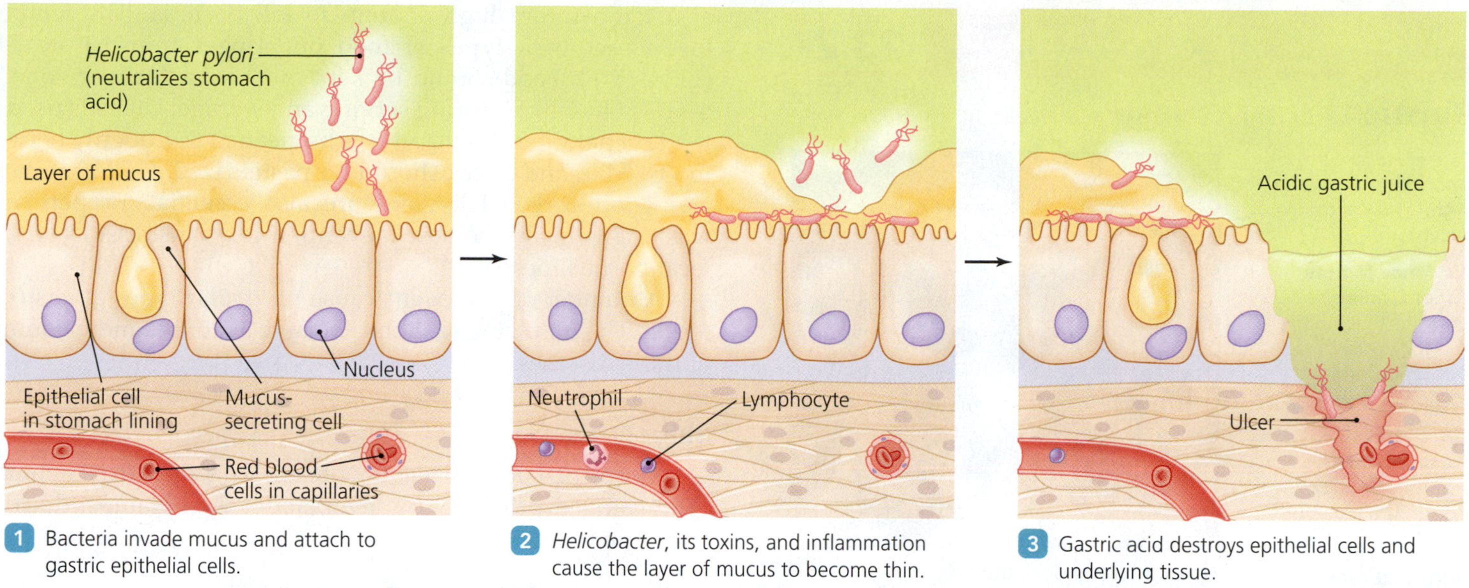

▲ **FIGURE 23.4** **The role of *Helicobacter pylori* in the formation of ulcers.**

that are phagocytized by leukocytes survive through the actions of catalase and superoxide dismutase, enzymes that neutralize part of the phagocytes' killing mechanism.

Epidemiology

About 456,000 people develop peptic ulcers annually in the United States. Not all strains of *H. pylori* appear to cause ulcers. Colonization of the stomach suggests a fecal-oral path of infection and studies have shown that *H. pylori* in human or cat feces on the hands, in well water, or on fomites infects humans.

Risk factors include use of aspirin, ibuprofen, or other nonsteroidal anti-inflammatory medications; excessive alcohol consumption; smoking cigarettes or using other tobacco products; and a family history of ulcers. Emotional stress does not cause ulcers but can worsen symptoms and make treatment more difficult.

Diagnosis, Treatment, and Prevention

An upper GI series of X-ray films following ingestion of barium reveals the presence of ulcers. Laboratory technicians find *H. pylori* in Gram-stained smears of clinical specimens, and a positive urease test indicates the presence of *H. pylori* in specimens from the stomach within one to two hours of culturing.

Treatment with one or more antimicrobial drugs—such as amoxicillin, clarithromycin, or tetracycline—in combination with drugs that inhibit acid production generally resolves most ulcers in six to eight weeks. Unfortunately, killing all the *Helicobacter* in a patient's stomach significantly increases the chance that the patient will develop esophageal cancer. The way in which *Helicobacter* might reduce this cancer is not known.

Recurrence of ulcers can be prevented by continuation of acid-blocking medications after antibacterial therapy has been completed. Surgery may be required in cases of excessive perforation or failure of medicinal interventions. Preventing infection involves avoidance of fecal-oral transmission of bacteria and lifestyle changes to eliminate other risk factors.

Disease at a Glance 23.2 on p. 722 summarizes the major features of peptic ulcers.

Bacterial Gastroenteritis

LEARNING | **OUTCOMES**

23.10 Compare and contrast the virulence factors of six bacteria that cause gastroenteritis.

23.11 Describe common prevention methods for avoiding gastroenteritis.

Bacterial gastroenteritis is an inflammation of the stomach or intestines caused by the presence of bacteria. Gastroenteritis occurs worldwide (the general incidence of bacterial gastroenteritis is roughly one in every 1000 people) but is most often associated with poorly prepared foods, contaminated washing or drinking water, and communities with poor living conditions. Institutional settings are prone to outbreaks in developed countries. Travel to areas with poor sanitation can lead to infection and export of disease.

General Features

Manifestations of gastroenteritis are usually similar regardless of the causative agent: some cases are asymptomatic or involve only mild diarrhea, but most patients have nausea, vomiting, diarrhea, loss of appetite, abdominal pain, and cramps. Some patients experience malaise and fever. In rare cases, the infection spreads beyond causing the initial gastrointestinal disease, resulting in kidney failure or anemia. In a severe and painful type of gastroenteritis known as **dysentery**, stools are loose, frequent, and contain mucus and blood.

Physicians diagnose most cases of bacterial gastroenteritis based on signs and symptoms, though victims with mild manifestations rarely seek medical treatment. Historically, laboratory

DISEASE AT A GLANCE 23.2

Peptic Ulcer Disease

Cause *Helicobacter pylori* (Gram-negative, slightly helical motile bacterium).

Virulence factors Flagella, adhesins, enzymes that inhibit phagocytosis, protein that blocks stomach acid production, urease.

Portal of entry Probably through the mouth as a result of fecal contamination.

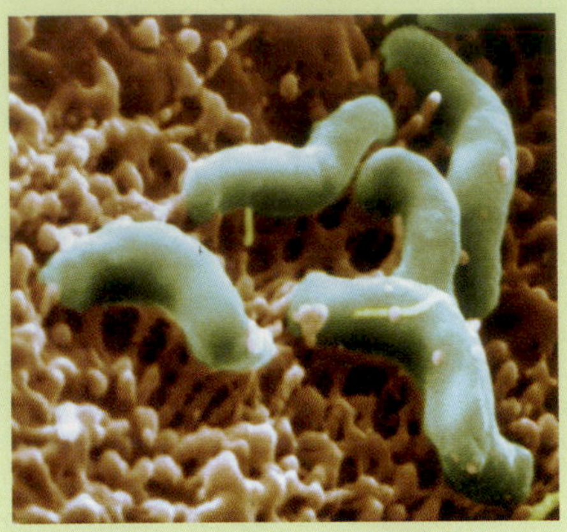

Helicobacter pylori, the cause of peptic ulcers

Signs and symptoms Abdominal pain 30 minutes to several hours after eating or skipping a meal, pain that is relieved with antacids or milk, heartburn, indigestion, nausea, vomiting of blood, weight loss, fatigue, and dark or bloody stools.

Incubation period Variable.

Susceptibility Anyone who becomes colonized with *H. pylori*, particularly people with a family history of ulcers, who use tobacco, aspirin, ibuprofen, or other nonsteroidal anti-inflammatory medications, or who consume excessive alcohol.

Treatment Antimicrobial drugs given in conjunction with acid-blocking drugs.

Prevention Good personal hygiene, adequate sanitation and proper food handling to decrease fecal-oral transmission, and lifestyle changes to reduce risk, including dietary changes to reduce stomach acid imbalances and reducing consumption of alcohol, tobacco, and aspirin-like pain medication.

scientists identified the causative bacterium from stool cultures (usually several are needed to ensure capture of the agent), fecal smears, or analysis of suspect food. In 2013, the U.S. Food and Drug Administration (FDA) approved the xTAG Gastrointestinal Pathogen Panel (xTAG GPP), a nucleic acid test that can identify seven bacterial, two viral, and two protozoan causes of gastroenteritis from a single stool sample.

Treatment of gastrointestinal diseases involves replacement of fluids and electrolytes lost to diarrhea and vomiting. In most cases, fluid replacement can be self-administered by drinking water and over-the-counter electrolyte solutions (sports drinks). Some patients need medication to suppress nausea so that fluids can be taken; in rare cases, intravenous fluid is required. Antidiarrheal drugs may prolong symptoms by allowing the organisms to remain in the intestines. Major symptoms of gastroenteritis generally disappear within hours or days, and recovery from dehydration may take up to a week.

Generally, prevention involves proper handling, storage, and preparation of food. Food should be thoroughly cleaned before consumption or use in cooking. All cooked foods, particularly ground meats, should be thoroughly cooked at a temperature high enough to kill bacteria. The temperature varies with types of food but ranges from 63° to 74°C (145–165°F). Food should be kept out of the "danger zone" of 4–60°C (40–140°F) during serving and storage. Utensils used for food preparation should always be cleaned thoroughly between foods. Milk and juices should be pasteurized. Good sanitation and good personal hygiene are also essential. Proper handwashing is likely the best prevention for transmission of these fecal-borne illnesses.

A number of pathogens cause particular types of gastroenteritis. The following sections examine six of the more common of these diseases.

Shigellosis

One form of bacterial gastroenteritis is **shigellosis**, which is characterized primarily by fever, abdominal cramps, diarrhea, and sometimes by a bloody stool.

Pathogens and Virulence Factors *Shigella* (shē-gel´ă) is a Gram-negative, nonmotile bacillus. Four different species of *Shigella* cause shigellosis: *S. dysenteriae* (dis-en-te´rē-ī), *S. flexneri* (fleks´ner-ē), *S. boydii* (boy´dē-ē), and *S. sonnei* (sōn´ne-ē). *S. sonnei* is the most common species isolated in industrialized nations, whereas *S. flexneri* is the predominant species in developing countries. Shigellosis is much more common outside the United States than in this country.

All four *Shigella* species produce *type III secretion systems* and diarrhea-producing *enterotoxins*. Type III secretion systems are complex structures composed of 20 different polypeptides that span both membranes of the bacterial cell. They insert into a host cell's cytoplasmic membrane, forming a channel through which bacterial proteins are introduced into the host cell. Enterotoxins are so named because they bind to surface proteins on epithelial cells lining the intestines, triggering the loss of electrolytes and water in a manner similar to cholera toxin. *S. dysenteriae* secretes **Shiga toxin**, which is an exotoxin that stops protein synthesis in a host's cells, resulting in a more severe form of shigellosis with a mortality rate as high as 20%.

Pathogenesis and Epidemiology Initially, *Shigella* colonizes cells of the small intestine, causing diarrhea; the main events in shigellosis, however, begin once the bacterium invades cells of the large intestine and end about seven days later. **FIGURE 23.5** illustrates the pathogenesis of *Shigella:* The pathogen attaches to epithelial cells in the large intestine 1, stimulating endocytosis of the bacterium 2, which then multiplies within the cell's cytosol 3. *Shigella* polymerizes the host's actin fibers, propelling itself out of the host cell and into adjacent cells 4, in the process evading the host's immune system. As the bacterium kills host cells, abscesses form in the mucosa 5; any bacteria that enter the blood from a ruptured abscess are quickly phagocytized and destroyed, making bacteremia a rare complication of shigellosis 6.

The World Health Organization (WHO) estimates about 90 million cases of shigellosis occur each year worldwide, with about 108,000 fatalities.

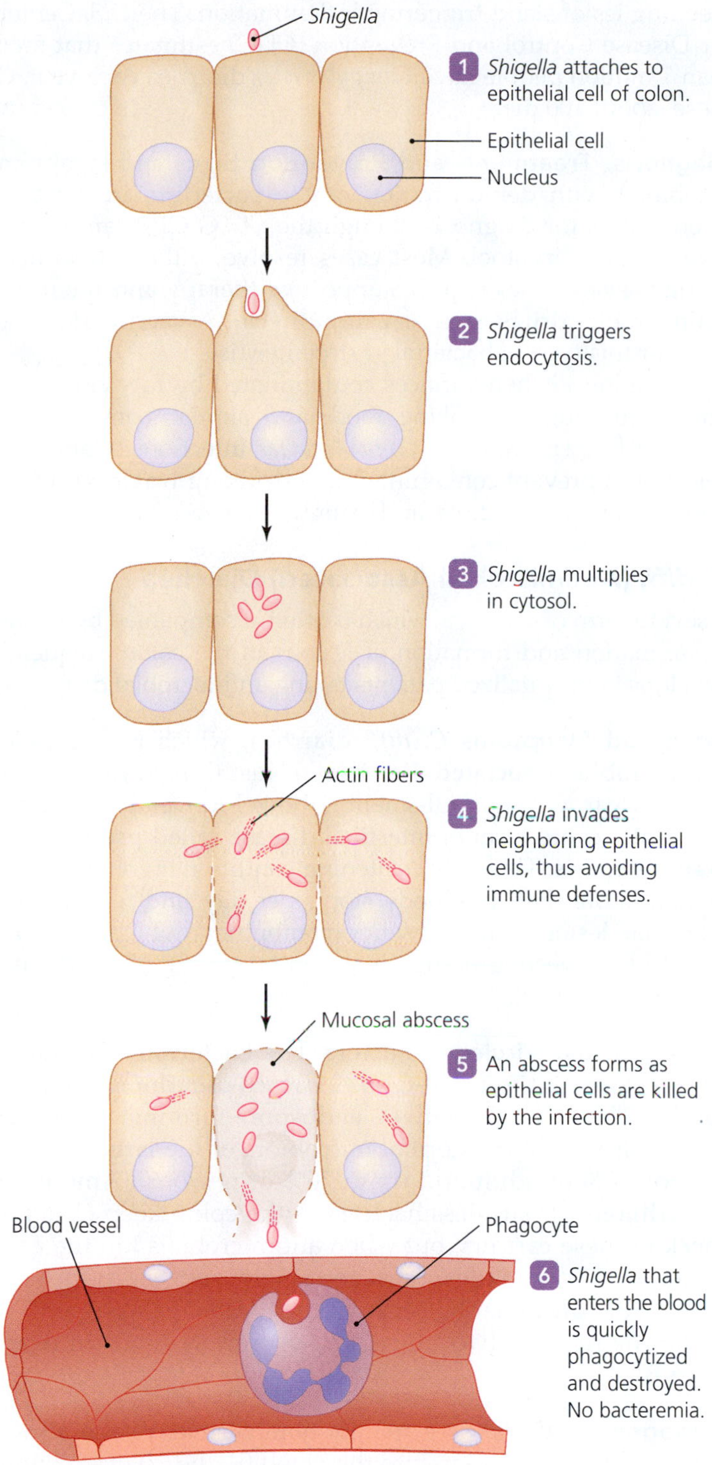
Shigella
1 Shigella attaches to epithelial cell of colon.
Epithelial cell
Nucleus
2 Shigella triggers endocytosis.
3 Shigella multiplies in cytosol.
Actin fibers
4 Shigella invades neighboring epithelial cells, thus avoiding immune defenses.
Mucosal abscess
5 An abscess forms as epithelial cells are killed by the infection.
Blood vessel
Phagocyte
6 Shigella that enters the blood is quickly phagocytized and destroyed. No bacteremia.

▲ **FIGURE 23.5 The events in shigellosis.**

Diagnosis, Treatment, and Prevention Physicians can diagnose shigellosis based on symptoms and the presence of *Shigella* in the stool. The newly approved nucleic acid test—xTAG Gastrointestinal Pathogen Panel—is one way that laboratory scientists can identify *Shigella* in a stool sample.

The diarrhea of shigellosis is usually self-limiting, so treatment involves supportive care. Physicians may prescribe an antimicrobial, such as ceftriaxone, which can shorten the duration of disease and reduce the spread of *Shigella* to close contacts of the patient. Resistance to antimicrobials among *Shigella* strains is widespread and increasing worldwide.

A recently developed, live, attenuated vaccine against *S. flexneri* has been successful in preventing the dysentery caused by this species, although the participants in the study still experienced mild diarrhea and fever. Researchers are working to perfect the vaccine so that it will not cause signs or symptoms.

Traveler's Diarrhea

Escherichia coli (esh-ĕ-rik´ē-ă kō-lī) is the most common and important of the bacteria causing diarrhea in travelers—thus its common name, *traveler's diarrhea.*

Pathogen and Virulence Factors *E. coli* is one of a group of colon-dwelling bacteria called *coliforms* (kŏ´li-formz). These bacteria are aerobic or facultatively anaerobic, Gram-negative bacilli that ferment lactose to form gas within 48 hours of being placed in a lactose broth at 35°C. Besides living in the intestinal tracts of animals and humans, coliforms can survive in soil and on plants and decaying vegetation, but their presence in water has historically been considered indicative of poor sewage treatment.

Scientists have identified numerous so-called O, H, and K antigens used to describe different strains of *E. coli.* A few antigens, such as O157, O111, H8, and H7, are associated with virulence. Virulent strains have genes (located on transmissible plasmids) for fimbriae, adhesins, and a variety of toxins that enable these strains to colonize human tissue and cause disease.

Two of the more dangerous of the toxins are the *Shiga-like toxins* of *E. coli* O157:H7; these toxins inhibit protein synthesis, kill cells, and can cause kidney failure, resulting in death. *E. coli* O157:H7 also produces a *type III secretion system.* One set of secreted proteins disrupts a host cell's metabolism; another set becomes lodged in a cell's cytoplasmic membrane and forms receptors for the attachment of additional *E. coli* O157:H7 bacteria. Such attachment apparently enables this strain of *E. coli* to displace normal harmless strains.

Pathogenesis and Epidemiology Generally, *E. coli* diarrhea appears 24–72 hours after consumption of the bacterium. Diarrhea is mediated by enterotoxins delivered via a type III secretion system. Strains that produce enterotoxins are common in developing countries and are important causes of pediatric diarrhea.

Shiga-like toxin attaches to the surfaces of neutrophils and is spread by them throughout the body, causing widespread death of host cells and tissues. Antimicrobial drugs induce *E. coli* O157:H7 to increase its production of Shiga-like toxin, worsening the disease. Investigators have found *E. coli* O157:H7 in almost 50% of beef carcasses in the United States. Despite its prevalence, only about 1 in 10 million deaths in the United States is attributed to consuming *E. coli* O157:H7 in ground beef.

Diagnosis, Treatment, and Prevention Physicians diagnose traveler's diarrhea based on signs and symptoms in patients returning from trips, though many cases are so mild that patients do not seek medical care. The xTAG Gastrointestinal Pathogen Panel can identify three pathogenic strains of *E. coli*.

Treatment involves replacing lost fluid and electrolytes. Antidiarrheal drugs prolong the symptoms by delaying expulsion of the bacterium from the digestive tract. Patients should avoid dairy products until diarrhea is over (generally after two to three days), because dairy foods can aggravate the symptoms resulting from temporary lactose intolerance in cases of *E. coli* gastroenteritis.

Antimicrobials—doxycycline, trimethoprim-sulfamethoxazole—treat bacterial gastroenteritis. No vaccine against *Escherichia* is available.

Campylobacter Diarrhea

Campylobacter is responsible for more cases of diarrhea that send people to doctors in the United States than any other bacterium.

Pathogen and Virulence Factors *Campylobacter jejuni*[5] (kam´-pi-lō-bak´ter jē-jū´nē), the causative agent of *Campylobacter* diarrhea, is a Gram-negative, slightly curved bacterium with polar flagella **(FIGURE 23.6)**. Scientists do not fully understand the virulence of *C. jejuni*, but the bacterium possesses adhesins, cytotoxins, and endotoxin (lipid A). The bacterium survives inside cells after being endocytized. Interestingly, nonmotile mutants of *C. jejuni* are avirulent.

Pathogenesis and Epidemiology One study found *Campylobacter* in 81% of chickens, which are the primary source of human infections. The virulence factors of *Campylobacter* enable colonization and invasion of the jejunum, ileum, and colon, producing bleeding lesions and triggering inflammation. The U.S. Centers for Disease Control and Prevention (CDC) estimates that more than 1 million people have *Campylobacter* diarrhea each year. Of these, about 100 die.

[5]From Greek *kampylos*, meaning "curved," and *jejunum*, the middle portion of the small intestine.

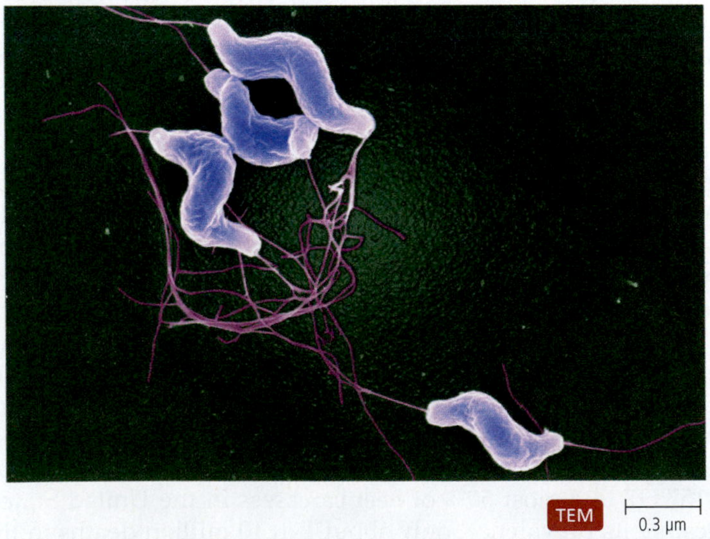

▲ **FIGURE 23.6** ***Campylobacter jejuni*, the most common cause of bacterial gastroenteritis in the United States.**

Diagnosis, Treatment, and Prevention Signs and symptoms combined with demonstration of the bacterium in the stool often suffice for diagnosis, though the xTAG GPP can identify *Campylobacter* in stool. Most cases resolve without treatment, though severe cases require supportive therapy and the use of antimicrobial drugs such as azithromycin. A vaccine does not exist for this type of bacterial gastroenteritis.

Washing kitchen surfaces contaminated by raw chicken or turkey, thoroughly cooking food, and similar commonsense culinary hygiene reduces *Campylobacter* infections. Communities should prevent contamination of drinking water with feces from stockyards, feedlots, and slaughterhouses.

C. diff. (Antimicrobial Associated) Diarrhea

A severe form of diarrhea, which is often accompanied by intense inflammation and formation of lesions in the colon, frequently develops in hospitalized patients taking antimicrobial drugs.

Signs and Symptoms ***C. diff.* diarrhea**, which is also called **antimicrobial associated diarrhea**, or just *C. diff.*, ranges from 5 to 10 clear, watery, foul-smelling bowel movements per day to a most severe form of intestinal disease called **pseudomembranous colitis**. This life-threatening condition involves inflammation, more than 10 bloody stools per day, and formation of intestinal lesions called pseudomembranes, which are composed of connective tissue, dying leukocytes, and dead colon cells **(FIGURE 23.7)**.

Pathogen and Virulence Factors The bacterium *Clostridium difficile* (klos-trid´ē-ŭm di-fi´sil-ē) causes pseudomembranous colitis. This Gram-positive, endospore-forming, anaerobic bacillus is part of the normal microbiota of the large intestine in about 5% of adult and up to 70% of newborn Americans. The trillions of harmless bacteria in the colon keep *C. diff.* in check in these carriers, but when antimicrobials kill the good bacteria, *C. diff.* endospores germinate and reproduce out of control, presumably because they have no competition. *C. difficile* releases two toxins known simply as toxin A and toxin B.

Pathogenesis *C. diff.* is generally noninvasive; that is, it does not move from the colon into the blood. Rather, it stays within the lumen of the colon, producing its two toxins. Toxin A breaks down the junctions holding cells of the colon's mucous membrane together, which triggers inflammation and allows loss of fluid. Toxin B kills the colon's cells outright and induces the formation of lesions that fuse into the characteristic pseudomembrane. Each toxin enhances the action of the other toxin by mechanisms that are not understood.

Epidemiology Pseudomembranous colitis is a by-product of modern medicine; until the widespread use of antimicrobial drugs, the condition was very rare. The CDC estimates that at

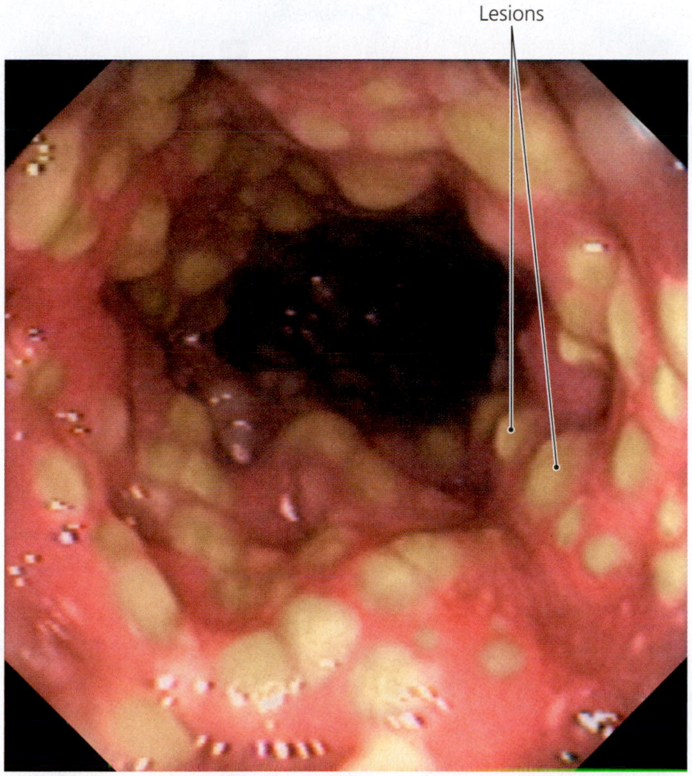

▲ **FIGURE 23.7 Pseudomembranous colitis.** Caused by *Clostridium difficile*, this can be a life-threatening inflammation of the colon.

minimum, more than 500,000 cases occur in the United States, with about 15,000 deaths, each year. Patients shed *Clostridium* in their feces, which can then infect hospital staff and other patients. About 20% of hospital patients carry *C. diff.*

Almost any antimicrobial can trigger the disease, but long-term use of newer, more powerful drugs as well as the simultaneous use of several drugs are two conditions that are more likely to be problematic. The elderly, burn patients, the immunocompromised, patients who have had a previous case of pseudomembranous colitis, and patients with kidney failure or who are recovering from abdominal surgery are particularly susceptible to pseudomembranous colitis.

Diagnosis, Prevention, and Treatment Clinical lab scientists test the stool for the presence of the two toxins or use the xTAG GPP to demonstrate *C. diff.* in the stool. This is sufficient for diagnosis in about 75% of patients. In the rest of cases, colonoscopy reveals the yellowish lesions of the pseudomembranes.

Prevention entails avoiding unnecessary use, especially prolonged use, of antimicrobial drugs. Hospital staff must follow excellent hygiene practices, especially with patients with diarrhea. The endospores of *Clostridium* are resistant to most disinfectants; therefore, handwashing and wearing gloves are the cornerstones of prevention. Early diagnosis allows treatment to begin immediately, reducing the chance of severe colitis.

Generally, cessation of the causative antimicrobial may be all that is necessary to restore normal microbiota. Treatment for moderately severe cases entails the use of oral metronidazole or vancomycin. Antidiarrheal drugs should be avoided, because diarrhea is actually beneficial in that it dilutes and eliminates bacterial cells and their toxins. Eating probiotics, such as *Lactobacillus*, may recolonize the colon with beneficial species that compete with *C. difficile*, though studies of probiotics' efficacy have shown mixed results. Physicians have successfully treated patients who have had recurrent *C. diff.* diarrhea with "fecal transplants." In this procedure, health care workers inject fluid containing fecal material, collected from a close relative or spouse of the patient, into the patient's colon through an enema or nasopharyngeal tube. The injected bacteria reestablish a normal microbiota in the recipient that successfully competes with *C. diff.* and eliminates further disease.

Disease at a Glance 23.3 summarizes the features of diarrhea caused by *Escherichia*, *Shigella*, *Campylobacter*, and *Clostridium difficile*.

Salmonellosis and Typhoid Fever

Salmonella cause two disease conditions: **typhoid fever** and a form of gastroenteritis called **salmonellosis**.

DISEASE AT A GLANCE 23.3

Bacterial Diarrhea

Causes Primarily *Escherichia coli*, *Shigella* spp., and *Campylobacter jejuni* (Gram-negative bacilli), and *Clostridium difficile* (Gram-positive, anaerobic, endospore-forming bacillus).

Virulence factors Adhesins, enterotoxins, Shiga toxin, ability to evade phagocytosis, ability to move between adjoining cells *(Shigella)*, Shiga-like toxin *(Escherichia)*, endotoxin (Gram-negative bacteria), endospore (Clostridium).

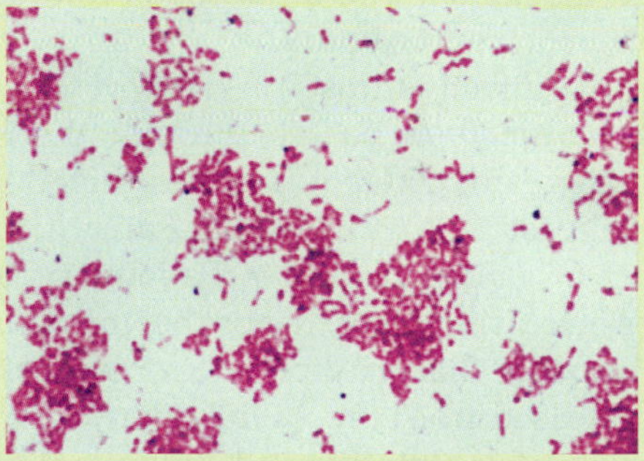

Escherichia coli LM 15 μm

Portal of entry Ingestion via fecal-oral route; *C. diff.* normally lives in 5% of Americans.

Signs and symptoms Abdominal cramps, bloody stools, nausea, vomiting, and diarrhea; *C. diff.* can additionally cause pseudomembrane formation.

Incubation period A few hours to days or weeks.

Susceptibility All humans are susceptible, but Gram-negative infection is generally worse in children and immunocompromised individuals; *C. diff.* is more prevalent in patients undergoing antimicrobial treatment.

Treatment Rehydration via fluid and electrolyte replacement; antimicrobials may shorten the duration.

Prevention Eat only thoroughly cooked meat products and pasteurized milk and juices, serve food piping hot, and practice good hygiene.

Pathogen and Virulence Factors *Salmonella* (sal´mŏ-nel´ă) is a genus of motile, Gram-negative, peritrichous bacilli that live in the intestines of virtually all vertebrates, especially reptiles, and are eliminated in their feces. *Salmonella* is not part of the normal microbiota of humans. Scientists have identified more than 2000 unique serotypes (strains) of *Salmonella,* though analysis of DNA sequences has revealed that all belong to a single species—*S. enterica* (en-ter´i-kă). However, many researchers and medical personnel continue to use historical names. Serotypes Typhi (tī´fē) and Paratyphi (par´a-tī´fē) (formerly *S. typhi* and *S. paratyphi*) cause typhoid fever, a disease unique to humans. Serotypes Enteritidis (en-ter-it´id-iss) and Typhimurium (tī´fē-mur-ē-ŭm) cause most U.S. cases of human salmonellosis.

Virulent serotypes of *Salmonella* tolerate the acidic condition of the stomach, passing into the intestine, where they attach via specific adhesins. They use type III secretion systems to introduce toxins into a host's cells. These toxins disrupt mitochondria, inhibit phagocytosis, rearrange the cytoskeletons of eukaryotic cells, or induce apoptosis.

Pathogenesis and Epidemiology Humans acquire typhoid fever via consumption of food or water contaminated with feces from a carrier of *S. enterica* serotypes Typhi or Paratyphi. The carrier may remain asymptomatic. People often acquire serotypes causing salmonellosis by eating or cooking with contaminated eggs. About one-third of chicken eggs carry *Salmonella,* even those laid by asymptomatic chickens. The bacterium in feces covers some eggs when they are laid; additionally, eggs may harbor *Salmonella* internally, having been produced by chickens with infected ovaries. Salmonellae released during the cracking of an egg on a kitchen counter and then inoculated into other foods can reproduce into millions of cells in just a few hours.

An infective dose of serotype Typhi is about 1000–10,000 cells. *Salmonella* passes through the intestinal cells into the bloodstream, where it is phagocytized. These defensive cells do not digest the pathogen but carry it to the liver, spleen, bone marrow, and gallbladder. Patients typically experience gradually increasing fever, headache, muscle pains, malaise, and loss of appetite that may persist for a week or more. Bacteria may be released from the gallbladder to reinfect the intestines, producing gastroenteritis and abdominal pain, followed by a recurrence of bacteremia. In some patients, the bacterium ulcerates and perforates the intestinal wall, allowing bacteria from the intestinal tract to enter the abdominal cavity, which causes *peritonitis* (inflammation of the peritoneum). Typhoid fever may last four weeks, and without treatment 12–30% of patients die.

FIGURE 23.8 depicts the events of typhoid fever and salmonellosis. After *Salmonella* passes through the stomach, it attaches to cells lining the small intestine 1 and inserts toxins to induce endocytosis 2. The pathogen reproduces within phagocytic vesicles 3, eventually killing the host cells 4 and inducing signs and symptoms of salmonellosis—fever, abdominal cramps, and diarrhea. Cells of some strains, most notably serotype Typhi, can subsequently enter the blood 5. Phagocytic cells then carry *Salmonella* through the blood to the liver, spleen, bone marrow, and gallbladder where, having resisted

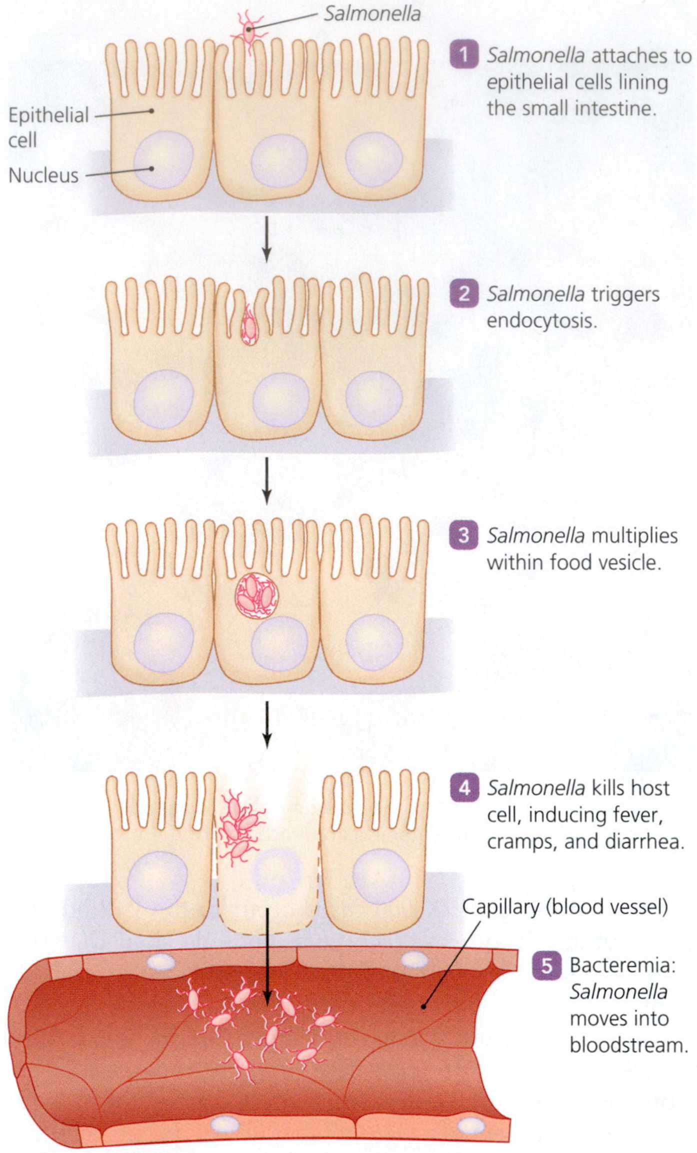

▲ **FIGURE 23.8 The events in salmonellosis.** *How is the multiplication of* Shigella *within host cells different from multiplication of* Salmonella *in host cells?*

Figure 23.8 Shigella multiplies in the cytosol of colon cells, Salmonella in phagocytic vesicles of cells of the small intestine.

phagocytic killing, they can establish semipermanent infection, particularly in the gallbladder. Carriers can remain infected for years even with treatment.

Typhoid fever is most common among poor people, particularly in the Southern Hemisphere. It is rare in industrialized countries.

Diagnosis, Treatment, and Prevention Diagnosis is made by finding *Salmonella* in the stools of patients, such as through the xTAG GPP nucleic acid test. Typhoid fever patients typically have a sustained fever as high as 40°C (104°F) accompanied by weakness, abdominal pain, headache, and loss of appetite.

Salmonellosis is generally self-limiting within a week. Health care providers replace lost fluids and electrolytes. Typhoid fever is treated with antimicrobial drugs, such as azithromycin or ciprofloxacin.

Prevention centers on good hygiene, especially in the kitchen. The CDC recommends "Boil it, cook it, peel it, or forget it"; that is, boil drinking water (though this will not protect from many other types of gastroenteritis), and avoid uncooked foods, except fruits or vegetables that can be peeled. People should wear gloves whenever handling pet reptiles and when cleaning their cages to avoid infection.

Disease at a Glance 23.4 summarizes the features of typhoid fever and salmonellosis.

Cholera

Epidemiologists have identified seven pandemics of **cholera** since 1817. A current epidemic, centering on India, may spread worldwide to become an eighth pandemic.

Pathogen and Virulence Factors *Vibrio cholerae* (vib´rē-ō kol´er-ī)—a slightly curved, Gram-negative bacillus with polar flagella (see Disease at a Glance 23.5 on p. 729)—causes cholera. The genus *Vibrio* is composed of species that occur naturally in estuarine and marine environments worldwide, preferring warm, salty, and alkaline water, and often in association with shellfish. *V. cholerae* is the only species that can survive in both salt- and freshwater. In saltwater it survives by forming biofilms, which are not infective, but in freshwater the biofilms fall apart, and single *Vibrio* cells become motile and infective.

A strain known as O1 El Tor is responsible for the pandemics, but a new strain, *V. cholerae* O139 Bengal, which arose in India in 1992, is spreading across Asia. This strain is the first non-O1 strain capable of causing epidemic disease. Other strains of *V. cholerae* do not produce epidemic cholera, only mild gastroenteritis.

Recent research indicates that the environment within the human body activates some *Vibrio* genes, making a bacterium in the body more virulent than its counterpart in water. Scientists hypothesize that such gene activation may explain the rapid,

DISEASE AT A GLANCE 23.4

Salmonellosis and Typhoid Fever

1. *Salmonella* is ingested in contaminated water or food, particularly chicken eggs.
2. Bacterium passes through the stomach, attaches to cells lining the small intestine, and induces endocytosis.
3. The pathogen eventually kills host cells.
4. This triggers fever, abdominal cramps, and diarrhea.
5. Cells of serotype Typhi can subsequently enter the blood, where they are phagocytized but not digested.
6. Phagocytes carry *Salmonella* serotype Typhi to the liver, spleen, bone marrow, and gallbladder.
7. *Salmonella* serotype Typhi can establish semipermanent infection in the gallbladder. Carriers can remain infected for years even with treatment.
8. *Salmonella* is shed in feces.

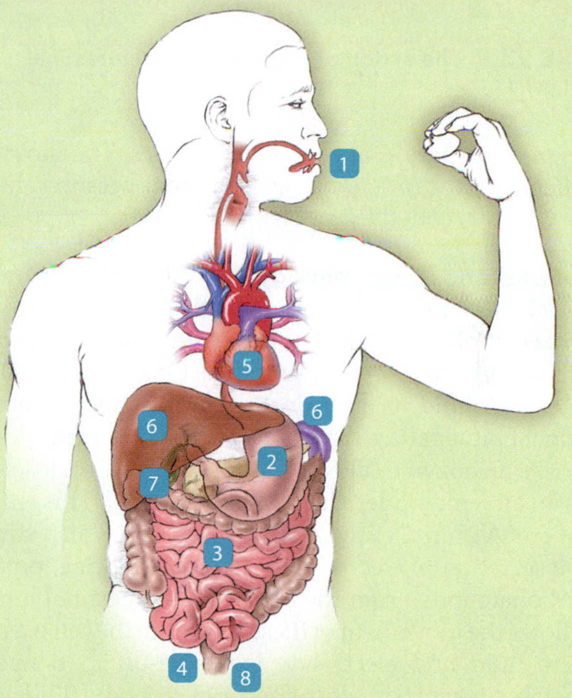

Cause Serotypes Typhi or Paratyphi of *Salmonella enterica* (Gram-negative, peritrichous bacillus) cause typhoid fever; serotypes Enteritidis and Typhimurium commonly cause salmonellosis.

Virulence factors More than 2000 serotypes; ability to tolerate low pH; adhesins; type III secretion system; toxins that disrupt mitochondria, inhibit phagocytosis, rearrange cytoskeletons of eukaryotic cells, or induce apoptosis.

Portal of entry Mouth and mucous membranes of the intestine by fecal-oral transmission; this involves ingestion of food or water contaminated with sewage from a carrier or food directly handled by an asymptomatic carrier.

Signs and symptoms Gradually increasing fever, headache, muscle pains, malaise, and loss of appetite that may persist for a week or more; "rose spot" rash may appear on lower chest and abdomen. Gastrointestinal symptoms common to other forms of bacterial gastroenteritis may occur. With typhoid fever, life-threatening complications are possible, including intestinal hemorrhage, perforation, kidney failure, or peritonitis (inflammation of the peritoneum).

Incubation period 8–48 hours.

Susceptibility Travel to countries lacking adequate sanitation; contact with asymptomatic carriers.

Treatment Fluid and electrolyte replacement are indicated for salmonellosis; antimicrobials are used against typhoid fever. Carriers may require removal of the gallbladder to end carrier status.

Prevention Proper and adequate sanitation and food handling, immunization for travelers to endemic areas, and preventing carriers from working as food handlers.

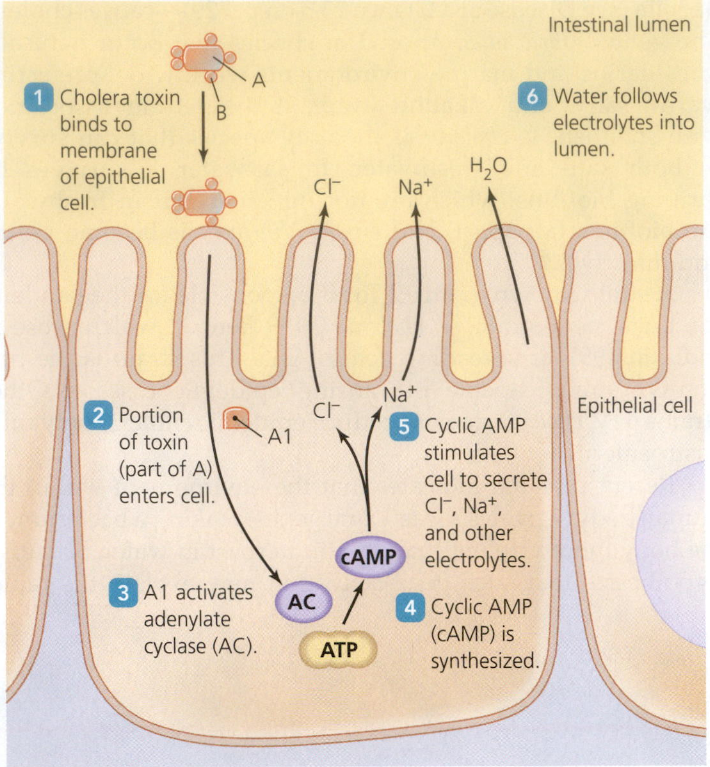

▲ **FIGURE 23.9 The action of cholera toxin in intestinal epithelial cells.**

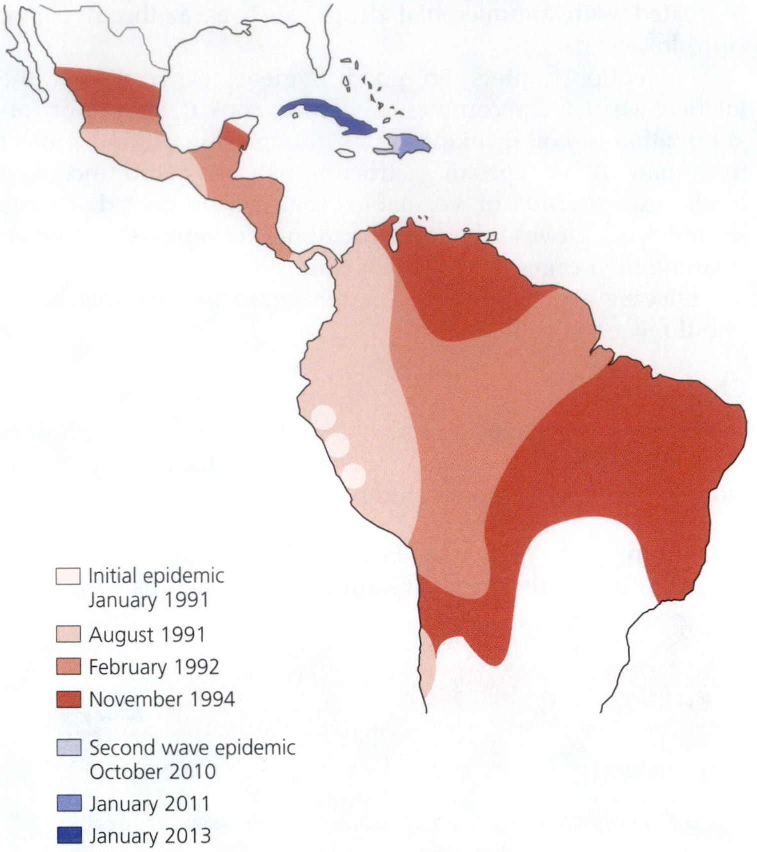

▲ **FIGURE 23.10 Cholera pandemic.** Cholera caused by *Vibrio cholerae* O1 El Tor, which started in Indonesia, spread throughout Latin America in the first half of the 1990s, and was reintroduced into Haiti in 2010. *Why is the cholera pandemic unlikely to establish itself in the United States?*

Figure 23.10 *The United States employs sewage treatment to prevent contamination of drinking water.*

almost explosive, nature of cholera epidemics. The most important virulence factor of *V. cholerae* is a potent poison called *cholera toxin,* which is coded by a plasmid.

Pathogenesis and Epidemiology FIGURE 23.9 illustrates the action of cholera toxin in producing the severe diarrhea that characterizes cholera. Cholera toxin is an A-B toxin composed of one A subunit and five B subunits. One of the B subunits binds to a glycolipid receptor in the cytoplasmic membrane of an intestinal epithelial cell, resulting in cleavage of the A subunit 1. Part of A, an enzyme called A1, enters the cell's cytosol 2 and activates an enzyme—adenylate cyclase (AC) 3. This enzyme in turn converts ATP into cyclic AMP (cAMP) 4, which stimulates the active secretion of electrolytes (sodium, chlorine, potassium, and bicarbonate ions) from the cell into the intestinal lumen 5. Water follows the movement of these ions from the cell via osmosis 6. Severe fluid and electrolyte losses result in dehydration, thirst, metabolic acidosis (decreased pH of body fluids) due to loss of bicarbonate ions, hypokalemia,[6] and hypovolemic shock caused by reduced blood volume in the body. These conditions can produce muscle cramping, lethargy (tiredness), sunken eyes, heartbeat irregularities, kidney failure, coma, and death.

Epidemiologists documented the spread of the seventh pandemic around the world, beginning in Indonesia in 1961. It reached South America, which had not seen cholera for a century, in 1991. FIGURE 23.10 illustrates the progression of cholera throughout Latin America in the first half of the 1990s. Over a million people reported symptoms, and researchers documented over 6300 deaths in Latin America alone. By 2002, the pandemic had subsided in South America, with only 23 cases reported all year. Cholera was apparently reintroduced in 2010 into Haiti by relief workers from Asia. From there, it spread into the Dominican Republic and Cuba.

Diagnosis, Treatment, and Prevention Physicians in endemic areas diagnose cholera based on its manifestations, particularly so-called "rice-water stool," which is watery, colorless, odorless, and flecked with mucus, which looks like bits of rice. In addition to supportive care, physicians may prescribe doxycycline for cholera; the drug reduces the production of cholera toxin.

Researchers have developed a vaccine, available in the United States, against the O1 El Tor strain of *V. cholerae,* but its protective value is unfortunately short lived; there is no vaccine for the O139 strain. Antimicrobial prophylaxis for those who travel to endemic areas has not proven effective. Fortunately, because the infective dose for *V. cholerae* is high, proper hygiene generally makes immunization and prophylaxis unnecessary.

[6]From Greek *hypo,* meaning "under"; Latin *kalium,* meaning "potassium"; and Greek *haima,* meaning "blood."

DISEASE AT A GLANCE 23.5

Cholera

Cause *Vibrio cholerae* (comma-shaped Gram-negative bacillus).

Virulence factors Type III secretion system, enterotoxins; Shiga toxin.

Portal of entry Ingestion of contaminated water or raw/undercooked seafood.

Signs and symptoms Sudden onset of "rice-water" diarrhea, dehydration (dry skin, excessive thirst, rapid yet diminished pulse, lethargy, sunken eyes), abdominal cramps, nausea, and vomiting. Death can occur within hours; the mortality rate is 25–50% in untreated patients, reduced to 1% with treatment.

Incubation period Generally two to three days, although an infected person may show symptoms in a few hours in some cases.

Susceptibility Humans living in endemic areas, especially in poverty-stricken areas; children tend to be affected more than adults are.

Treatment Fluid and electrolyte replacement and administration of a tetracycline.

Prevention When in endemic areas, boil water, eat only cooked food (especially seafood), avoid raw vegetables and fruit, and wash hands freqently.

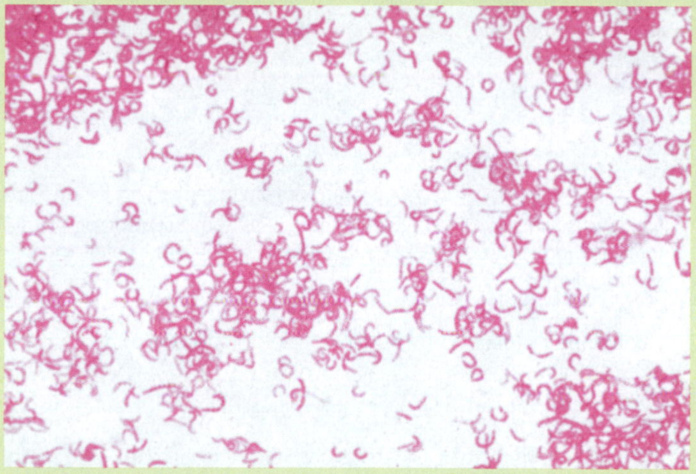

Vibrio cholerae, the causative agent of cholera

CLINICAL CASE STUDY

When "Health Food" Isn't

In a single day, two female students and one male student sought treatment at a university health clinic, complaining of acute diarrhea, nausea, and vomiting. No blood was found in their stools. One of the women was found to have a urinary tract infection. All three had eaten lunch at a nearby health food store the previous day. The man had a turkey sandwich with tomato, sprouts, pickles, and sunflower seeds. One woman had a pocket sandwich with turkey, sprouts, and mandarin oranges; the other woman had the lunch special, described in the menu as a "delightful garden salad of fresh organic lettuces, sprouts, tomatoes, and cucumbers with zesty raspberry vinaigrette dressing." All had bottled water to drink.

1. Which of the foods is the most likely source of the infections?
2. What media would you use to culture and isolate enteric contaminants in the food?
3. Which enteric bacteria could cause these symptoms?
4. How did the woman likely acquire the urinary tract infection?
5. What is the likely treatment of choice?
6. What steps can the food store's manager and the students take to reduce the chance of subsequent infections?

Disease at a Glance 23.5 summarizes and amplifies the major features of cholera.

TABLE 23.1 on p. 730 compares and contrasts the more common forms of bacterial gastroenteritis. **Clinical Case Study: When "Health Food" Isn't** examines cases of bacterial gastroenteritis.

Bacterial Food Poisoning (Intoxication)

LEARNING **OUTCOMES**

23.12 Distinguish between intoxication and gastroenteritis.

23.13 Describe the virulence factors of *Staphylococcus* that allow it to cause microbial intoxication.

23.14 Describe methods to prevent food poisoning.

Food poisoning is a rather broad term used collectively to refer to consuming either pathogens or their toxins. Infections of the gastrointestinal tract are more appropriately referred to as gastroenteritis. In this section, we focus on **bacterial intoxications** *(toxifications),* which are food poisonings caused by toxins—the microbe itself is either not present or not the immediate problem.

First, we examine staphylococcal food poisoning as a model. (Chapter 20 considers *botulism,* which is a form of intoxication that affects the nervous system.)

Signs and Symptoms

General symptoms of bacterial intoxication include nausea, vomiting, diarrhea, abdominal cramping, discomfort, bloating, loss of appetite, and fever. Symptoms may range from mild to severe. Some types of intoxications also produce weakness, headache, and difficulty in breathing. Symptoms differ to some extent depending on the toxins present and can be confused with bacterial or viral gastroenteritis. Dehydration resulting from fluid loss in diarrhea may become significant, but most cases, exemplified by staphylococcal food poisoning, are self-limiting and last no more than 24 hours.

TABLE 23.1 Common Forms of Bacterial Gastroenteritis

Disease	Pathogen (Minimum Infectious Dose)	Source of Infection	Incubation Period	Distinguishing Manifestations	U.S. Annual Incidence	Complications
Shigellosis	*Shigella dysenteriae*, *S. flexneri*, *S. boydii*, *S. sonnei* (200 cells)	Self-inoculation from fecally contaminated hands, secondarily through consumption of fecally contaminated foods; direct person-to-person spread	1–7 days	Purulent (containing mucus and pus) bloody stools, crampy rectal pain, fever, vomiting, and nausea lasting 2–3 days	14,000 cases	Severe dehydration; febrile seizures, confusion, and other neurological complications may appear in children
Traveler's diarrhea	*Escherichia coli* (unknown)	Fecally contaminated food or water	24–72 hours	Nausea, vomiting, and diarrheal symptoms lasting 1–3 days	Unknown, as reporting is not required; estimated >80,000 cases	Dehydration
***E. coli* O157:H7 infection**	*E. coli* strain O157:H7 (10 cells)	Fecally contaminated milk, fruit juice, or ground beef	24–72 hours	Bloody diarrhea, fatal hemorrhagic colitis, hemolytic uremic syndrome—destruction of erythrocytes and kidney failure	2000–3000 cases	Death
***Campylobacter* diarrhea**	*Campylobacter jejuni* (500 cells)	Zoonotic from domestic poultry, dogs, cats, rabbits, pigs, cattle, and minks through consumption of food, milk, or water contaminated with animal feces; close contact with infected humans	2–5 days	10 or more bowel movements per day lasting 2–5 days; blood may be present in diarrhea	More than 1 million cases estimated	Sepsis, arthritis, Guillain-Barré syndrome (temporary nerve paralysis), death
***C. diff* (antimicrobial associated) diarrhea**	*Clostridium difficile* (unknown)	5% of Americans carry *C. diff.* normally; 20% of hospital patients are infected	48 hours to six weeks	Numerous, watery, foul-smelling stools; pseudomembranes	Estimated 500,000 cases	Pseudomembranous colitis, death
Salmonellosis	*Salmonella enterica* serotypes Enteritidis and Typhimurium (>10^6 cells)	Zoonotic from domestic poultry through consumption of fecally contaminated meat or eggs, or consumption of inadequately pasteurized contaminated milk; close contact with infected reptiles; contact with human carriers	8–48 hours	Nonbloody diarrhea, nausea, vomiting, fever, headache, and pain lasting 1–2 weeks; rash of tiny rose spots may appear on the skin	50,000 cases	Dehydration
Typhoid fever	*Salmonella enterica* serotypes Typhi and Paratyphi (>10^6 cells)	Primarily contaminated water	8–48 hours	High fever (40°C), headache, muscle and stomach pain, malaise, loss of appetite, rose-colored spots	300–400 cases	Intestinal perforation, hemorrhaging, kidney failure, peritonitis, and death
Cholera	*Vibrio cholerae* (>10^8 cells)	Fecally contaminated food or water	48–72 hours	Rice-water stool (watery, colorless, odorless stools flecked with mucus) lasting 2–3 days; patients may lose up to 1 L of fluid per hour	0–8 cases	Death can occur within 48 hours of symptom onset if untreated (25–50% mortality rate)

Pathogens and Virulence Factors

Toxins of *Staphylococcus aureus,* which is a normal member of the microbiota of the skin and upper respiratory system, cause staphylococcal food poisoning. This is a common type of bacterial food poisoning. Food preparers frequently introduce *Staphylococcus* from their bodies into foods during cooking. The bacterium is salt tolerant and grows particularly well in foods at room temperature, where it produces toxins. Foods commonly associated with staphylococcal food poisoning include processed meats, custard pastries, potato salad, and ice cream.

S. aureus has several virulence factors, but important ones in cases of food poisoning are five enterotoxins. These proteins (designated A through E) stimulate intestinal muscle contractions, trigger nausea, and cause intense vomiting. The enterotoxins are heat stable, remaining functional at 100°C for up to 30 minutes, which means they are not usually inactivated by warming or reheating food.

Pathogenesis and Epidemiology

Bacterial intoxication can affect a single individual or hundreds of people at once. Outbreaks are usually associated with picnics, school cafeterias, or large social functions where food stands unrefrigerated or where food preparation is less than optimal. It takes several hours at room temperature or higher for *Staphylococcus* to grow and secrete toxins. *Staphylococcus* does not change the appearance or taste of food.

Because most cases of staphylococcal food poisoning are self-limiting and relatively mild, the number of cases is unknown—by the time a patient would see a doctor, the symptoms are gone.

Diagnosis, Treatment, and Prevention

Diagnosis is generally based on signs, symptoms, and patient history. Tests may be done on samples from vomit, blood, stool, or any leftover food deemed suspicious. Stool cultures positive for *S. aureus* are indicative of staphylococcal food poisoning, but other examinations are often inconclusive. Replacement of fluids and electrolytes is the only treatment and can be self-administered. Good hygiene and proper food handling reduce incidence.

Disease at a Glance 23.6 summarizes features of staphylococcal intoxication.

We have examined bacterial diseases and intoxication of the digestive system. The next section considers viral diseases of this system.

TELL ME WHY

Why is the elimination of sucrose sugar from the diet not enough to prevent the formation of all dental caries?

Viral Diseases of the Digestive System

Several viruses cause diseases of the digestive system almost indistinguishable from those caused by bacteria. Among these are various types of gastroenteritis. Viruses also cause far more severe digestive system diseases, including hepatitis.

We begin by considering *oral herpes*—a disease of the oral cavity and thus of the beginning of the digestive system.

DISEASE AT A GLANCE 23.6

Staphylococcal Intoxication (Food Poisoning)

Cause *Staphylococcus aureus* (facultatively anaerobic, Gram-positive cells arranged in clusters).

Virulence factors Heat-stable eterotoxins, salt tolerance.

Portal of entry Toxin crosses mucous membranes of the intestinal tract following consumption of contaminated food; *Staphylococcus* is not directly involved in the disease.

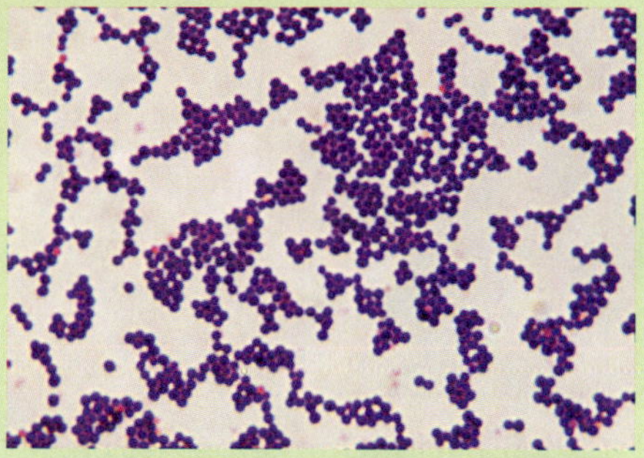

Staphylococcus aureus LM 5 μm

Signs and symptoms Nausea, vomiting, diarrhea, cramping, discomfort, bloating, loss of appetite, and fever; all lasting 24 hours or less.

Incubation period Four to six hours.

Susceptibility Everyone is susceptible because the organism is a normal member of the microbiota, but intoxication results only when inoculated food is improperly refrigerated or undercooked prior to consumption.

Treatment Self-administered replacement of fluids and electrolytes.

Prevention Thorough handwashing before and after handling foods, cleaning of utensils between use on different foods, and prompt refrigeration of leftovers all decrease risk of staphylococcal food poisoning.

Oral Herpes

LEARNING **OUTCOMES**

23.15 Describe the appearance of oral herpes infection.

23.16 Describe methods of prevention and options for treatment for oral herpes.

The family *Herpesviridae* contains a large group of linear dsDNA viruses with enveloped polyhedral capsids. Many of these herpesviruses infect humans, and their high infection rates make them among the more prevalent DNA viral pathogens. Oral herpes is among the most common of their infections.

Signs and Symptoms

Painful, itchy, creeping skin lesions on the lips, called **fever blisters** or **cold sores,** characterize **oral herpes**[7] **(FIGURE 23.11).**

---

[7]From Greek *herpo,* meaning "to creep."

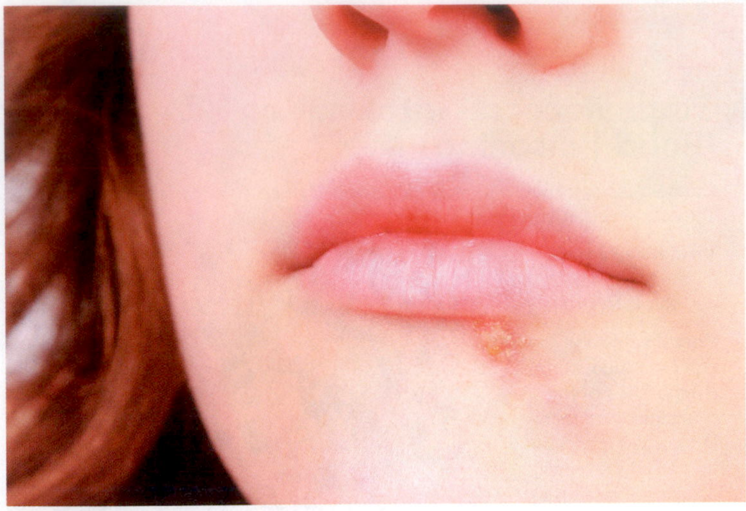

▲ **FIGURE 23.11** **Oral herpes lesion.**

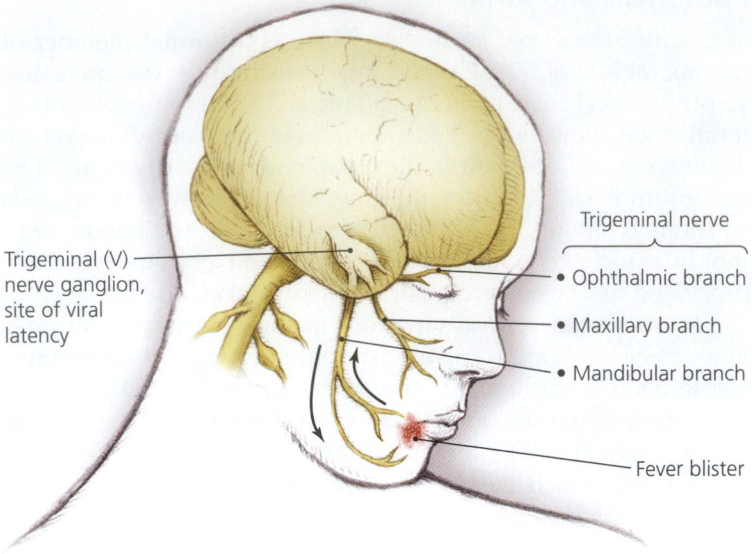

▲ **FIGURE 23.12** **Latency and reactivation of oral herpesviruses.** Viruses from the initial infection travel up the maxillary or mandibular branches of the trigeminal nerve to become dormant in the ganglion. Immune suppression allows the viruses to travel back down the nerve to cause oral herpes lesions.

Fever blisters appear one to two weeks after exposure to an infected individual. Initial infections may be accompanied by flulike symptoms such as malaise, fever, and muscle pain. The fluid-filled lesions eventually break, crust over, and fall off (within 7–10 days) to reveal pink healing skin. Subsequent lesions are generally milder.

Severe infections in which the lesions extend into the oral cavity, called *herpetic gingivostomatitis,* are most often seen in young patients and in patients with lowered immune function due to disease, chemotherapy, or radiation treatment. Young adults with sore throats resulting from other viral infections may develop *herpetic pharyngitis,* in which the pharynx becomes infected and inflamed. Immunosuppressed individuals may develop *herpes esophagitis,* characterized by extremely painful and difficult swallowing, fever, and sometimes chills. (Chapters 19 and 24 consider herpes skin infections and genital herpes, respectively.)

Pathogen and Pathogenesis

Human herpesvirus 1 (HHV-1; formerly called herpes simplex virus 1) causes most cases of oral herpes; HHV-2, which usually infests the genitalia, can also infect the oral cavity. After entering the body through cracks or cuts in mucous membranes, herpesviruses reproduce in epithelial cells near the site of infection, triggering inflammation and cell death and resulting in painful, localized lesions on the skin 2–12 days after infection. By causing infected cells to fuse with uninfected neighboring cells to form a structure called a *syncytium,* herpes virions spread from cell to cell, avoiding the host's immune system. Lesions usually last two to three weeks.

As illustrated in **FIGURE 23.12**, HHV-1 eventually establishes latent infections in the trigeminal nerve ganglion[8] by entering sensory nerve cells and being carried by cytoplasmic flow to the ganglion. Latent viruses may reactivate later in life when the immune system is suppressed by emotional stress, fever, trauma, sunlight, menstruation, or disease. Reactivated viruses travel down the nerve to produce recurrent lesions as often as every two weeks. Recurrent lesions are rarely as severe as the initial lesions because of immunological memory.

Epidemiology

HHV-1 accounts for 90% of all cold sores; HHV-2, the normal cause of genital herpes, causes the other 10% of cold sores. HHV-1 is transmitted by close contact with infected individuals who have active lesions. Primary HHV-1 infections typically occur via casual contact during childhood, and usually produce no signs or symptoms; in fact, HHV-1 has asymptomatically infected about 80% of children by age two.

Diagnosis, Treatment, and Prevention

Oral herpes is generally diagnosed by observation of the characteristic recurring lesions. Microscopic examination of infected tissue reveals syncytia. Positive diagnosis is achieved by immunoassay that demonstrates the presence of viral antigen. Topical creams containing penciclovir or acyclovir limit the duration of the lesions and reduce viral shedding, but they are not a cure and do not eliminate latency in the trigeminal ganglia.

Washing with soap and water may minimize spread of the virus, but prevention depends on avoiding direct contact with infected individuals. Patients with active lesions are more likely to spread the disease, but asymptomatic carriers still shed viruses and are contagious. Contaminated fomites, such as razor blades, toothbrushes, towels, and dishes, also spread herpesviruses. Promiscuous oral sex should be avoided to prevent transfer of HHV-1 to the genitalia. Patients should not touch their own lesions, because they may spread viruses into their eyes or onto other areas of the skin. Further, broken lesions may provide a portal, leading to secondary bacterial infections.

[8]A ganglion is a collection of nerve cell bodies containing their nuclei.

CLINICAL CASE STUDY

The Case of the Lactovegetarians

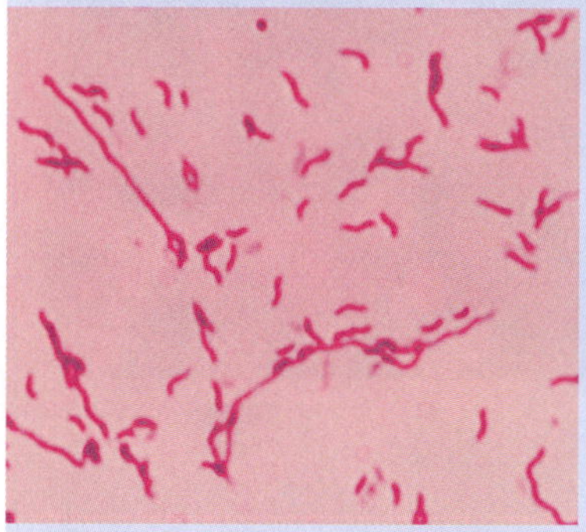

Two patients—a woman and her husband, ages 23 and 22, respectively—arrive at the health clinic one morning. They report having had severe abdominal cramps, grossly bloody diarrhea, nausea, and fever for 48 hours. Cultures of stool samples grown under microaerophilic, capneic conditions contain comma-shaped, Gram-negative bacilli (see the photo). Both the patients are lactovegetarians and report being part of a "cow leasing" program at a local dairy in which patrons lease part of a cow's milk production so they can drink natural, whole, raw milk. The couple devised the program so that they and several neighbors could circumvent state regulations prohibiting the sale of unpasteurized milk. Investigators obtained and cultured a milk sample from the dairy's bulk milk tank. The cultures contained the bacterium pictured.

1. What is the pathogen?
2. How did the couple become infected?
3. Are the couple's colleagues at work at risk of acquiring an infection from the couple?
4. What other foods that are common sources of this bacterium can be ruled out in this case?

Reference: *MMWR* 51:548–549. 2002.

DISEASE AT A GLANCE 23.7

Mumps

Cause Mumps virus (enveloped, helical, unsegmented, –ssRNA virus of genus *Rubulavirus*).

Virulence factors Adhesins, intracellular replication cycle.

Portal of entry Mucous membranes of the upper respiratory tract.

Signs and symptoms Parotitis (swelling of the parotid salivary glands), face pain, fever, headache, and sore throat are the most common symptoms. Some infections may be asymptomatic.

Incubation period 12–24 days.

Susceptibility Unimmunized individuals are at risk.

Treatment Comfort care only, including hot or cold packs, soft foods, fluids, and warm-water gargles.

Prevention MMR vaccine.

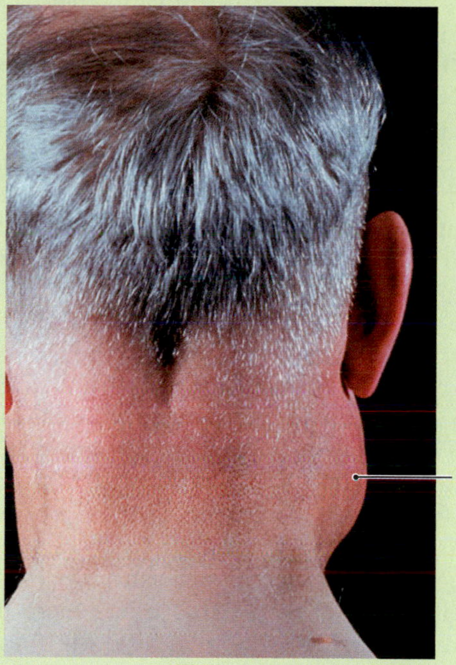

Mumps

Mumps

LEARNING | OUTCOME

23.17 Describe the cause, manifestations, and prevention of mumps.

Mumps, in which viruses infect the largest salivary glands located on each side of the face (see Disease at a Glance 23.7), was once among the more common of childhood diseases. Today, mumps is nearly nonexistent in developed nations as a result of effective childhood immunization. Epidemics of mumps in the late winter and early spring still occur in countries that lack immunization programs.

Humans are the only natural host for the mumps virus, which is a –ssRNA virus in the genus *Rubulavirus*. Mumps virus infects unimmunized children between the ages of 2 and 12 who are exposed to an infected person or to fomites carrying contaminated saliva. The virus enters via the respiratory tract, multiplies, invades the blood, and can infect many organs in addition to the salivary glands. Some patients suffer inflammation of the testes (resulting in sterility) meninges, or pancreas, and rarely mumps virus causes deafness.

Treatment for mumps is supportive care and pain medication. A recovered patient has effective, lifelong immunity. **Disease at a Glance 23.7** summarizes the features of mumps.

Viral Gastroenteritis

LEARNING | OUTCOMES

23.18 List the three viral causes of gastroenteritis.

23.19 Describe the diagnosis, treatment, and prevention of viral gastroenteritis.

Bacteria are not the only microbes that infect the digestive tract to produce gastroenteritis—many viruses can, too. Viral gastroenteritis, however, is generally less severe than bacterial forms of the disease.

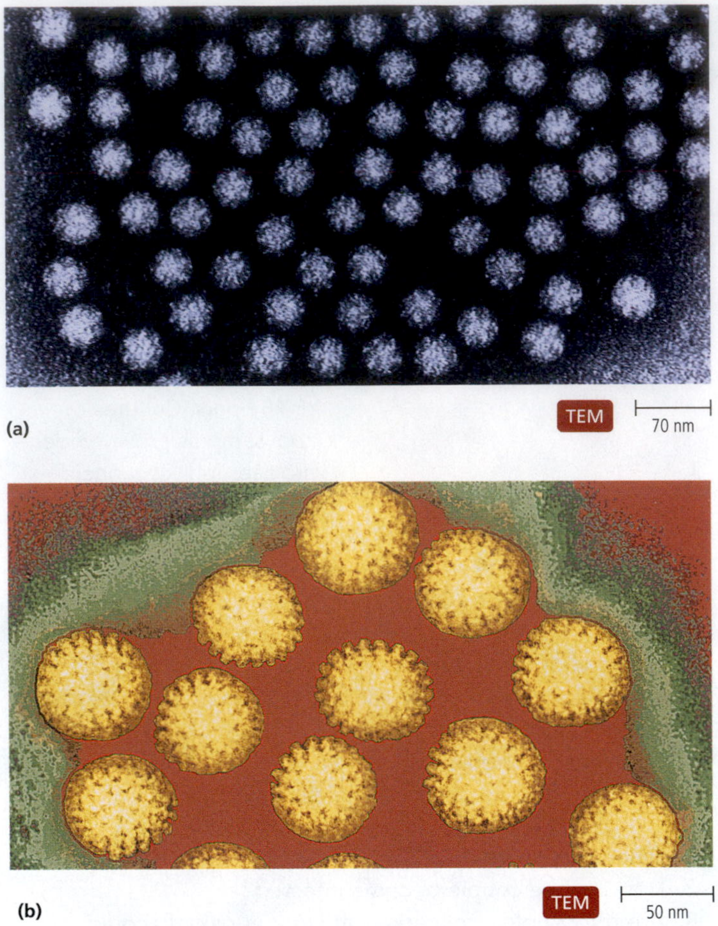

▲ **FIGURE 23.13 Some viruses causing gastroenteritis.** **(a)** Caliciviruses, such as noroviruses, and astroviruses have naked "star-shaped" capsids. **(b)** The wheel-like appearance of rotaviruses, from which they get their name.

Signs and Symptoms

The general manifestations of **viral gastroenteritis** (sometimes mistakenly called "stomach flu") are the same as for bacterial gastroenteritis—abdominal pain and cramping, diarrhea, nausea, and vomiting. Additional signs and symptoms may include fever, chills, clammy skin, weight loss, or lack of appetite. Dehydration is the most common complication. Symptoms generally appear within 24 hours of consuming contaminated food and resolve within 12–60 hours. Vomiting, bloody stool, life-threatening diarrhea, and dysentery may occur with viral gastroenteritis.

Pathogens and Pathogenesis

Common viral agents of gastroenteritis are caliciviruses, astroviruses, and rotaviruses. Caliciviruses (kal´i-sē-vī´rŭs-ez) and astroviruses (as´trō-vī-rŭs-ez) are two +ssRNA viruses that cause acute gastroenteritis. Both are small, naked, star-shaped, and have polyhedral capsids **(FIGURE 23.13a)** that enter the body through the digestive tract by consumption of contaminated food or water. The most studied of the caliciviruses are **noroviruses**, discovered in the stools of victims during an epidemic of diarrhea in Norwalk, Ohio, from which their name comes.

Rotaviruses (rō´ta-vī-rŭs-ez), members of the dsRNA *Reoviridae,* are almost spherical and have glycoprotein spikes **(FIGURE 23.13b)** that act as attachment molecules and trigger endocytosis. Rotaviruses are naked, though during replication they acquire and then lose envelopes. Their transmission is via the fecal-oral route from contaminated food or water. Infected children may pass as many as 100 trillion virions per gram of stool.

All three of these viruses—caliciviruses, astroviruses, and rotaviruses—infect cells lining the intestinal tract where they undergo lytic replication. As epithelial cells die, the normal function of the intestinal tract is lost. Infections are generally self-limiting—after the virus has destroyed the epithelial layer, replacement epithelial cells grow, and function is restored.

Epidemiology

Cases of viral gastroenteritis are more frequent in winter, being facilitated by close living conditions. Noroviruses cause 90% of nonbacterial gastrointestinal infections (about 10% of all cases of gastroenteritis) worldwide and have caused outbreaks of gastroenteritis in day care centers, schools, hospitals, nursing homes, restaurants, and in recent years, numerous epidemics on cruise ships. Generally, noroviruses infect adults and school-aged children.

Rotaviruses cause infantile gastroenteritis and account for approximately 50% of all cases of diarrhea in children requiring hospitalization because of fluid and electrolyte loss (up to 100,000 hospitalizations and 100 deaths per year in the United States). In developing countries, rotaviruses annually kill an estimated 600,000 children—about 5% of all childhood deaths **(FIGURE 23.14)**.

Diagnosis, Treatment, and Prevention

Serological tests performed on stool samples can distinguish among surface antigens of caliciviruses, astroviruses, and rotaviruses. The xTAG Gastrointestinal Pathogen Panel can identify noroviruses and rotavirus strain A. There is no specific treatment for any of these infections except support and replacement of lost fluid and electrolytes. Antidiarrheal medications may only prolong symptoms, because diarrhea tends to clear the viruses from the system.

Prevention involves adequate sewage treatment, purification of water supplies, frequent handwashing, good personal hygiene, and disinfection of contaminated surfaces and fomites.

Attenuated oral vaccines against rotaviruses exist and safely protect against up to 98% of severe rotaviral diarrhea requiring hospitalization. Depending on which of two vaccines, pediatricians administer two or three doses beginning at two months of age.

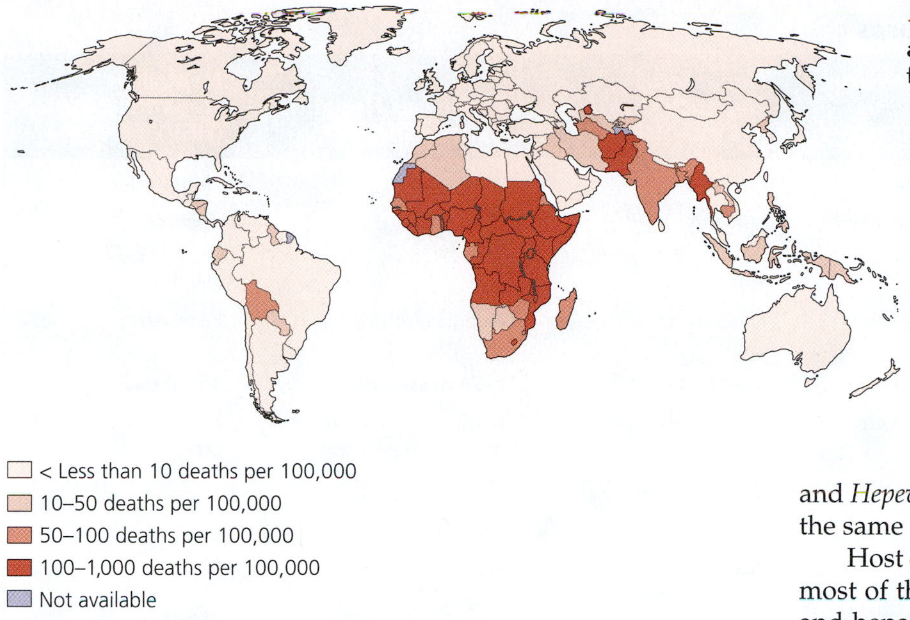

◀ **FIGURE 23.14 Deaths from rotaviral diarrhea are most common in developing countries.** (Data are from 2008.)

Viral Hepatitis

LEARNING OUTCOMES

23.20 Describe the three primary forms of viral hepatitis, the agents that cause them, and ways to prevent infection by each.

23.21 List the common risk factors for hepatitis.

Hepatitis (hep-ă-tī′tis) is inflammation of the liver produced by autoimmune disease, alcohol or drug abuse, genetic disorders, or microbial infection. Three viruses are responsible for most virally caused forms of the disease.

Signs and Symptoms

The liver has many functions, including synthesis of blood-clotting factors, storing glucose and other nutrients, assisting in the digestion of lipids, and removing wastes from the blood. When a viral infection damages the liver, all of these functions are disturbed, though signs and symptoms may not occur until years after initial infection. Manifestations may include yellowing of the skin and eyes called *jaundice* (jawn′dis) (see Disease at a Glance 23.8 on p. 738), abdominal pain and distention, dark urine, light-colored stools, loss of appetite, nausea, vomiting, fatigue, fever, and weight loss. Patients may become slow-acting and eventually go into a coma because of the accumulation of wastes in the blood. Complications from chronic (long-term) infection are serious and life threatening, including permanent liver damage (cirrhosis), liver failure, or liver cancer.

Pathogens and Pathogenesis

Five viruses cause hepatitis; they are *Hepatovirus Hepatitis A virus, Orthohepadnavirus Hepatitis B virus, Hepacivirus Hepatitis C virus, Deltavirus Hepatitis delta virus* (also called delta agent), and *Hepevirus Hepatitis E virus.* The taxonomic specific epithet is the same as the common name for each of these viruses.

Host cellular immune responses that kill infected cells cause most of the liver damage seen in hepatitis patients. Hepatitis A and hepatitis E viruses are usually cleared during the immune responses, but the other hepatitis viruses typically remain, resulting in chronic infection.

Hepatitis A virus (HAV) can survive on surfaces such as countertops and cutting boards for days, resists common household disinfectants such as chlorine bleach, and is transmitted in fecally contaminated food or water. Patients release virions in their feces and are infective even without developing symptoms. Hepatitis A, also called *infectious hepatitis,* is typically a mild condition with 99% of patients recovering fully.

Hepatitis B virus (HBV) replicates in liver cells and is released by exocytosis, rather than cell lysis, so infected cells serve as a source for the continual release of virions, resulting in billions of virions per milliliter of blood, which gives the disease its common name—*serum hepatitis.* Virions in the blood are shed into saliva, semen, and vaginal secretions, such that sexual transmission, particularly via anal intercourse, is the most common mode of transmission. The virus is also transmitted via contaminated needles, razors, toothbrushes, or contact with blood or open sores of an infected person. Babies of infected mothers can become infected during childbirth. The carrier state is age related—newborns are much more likely to remain chronically infected than are individuals infected as adults.

Strong medical evidence shows an association between HBV and hepatic (liver) cancer. Hepatic cancer is common in geographic areas with a high prevalence of HBV infection, and chronic carriers of HBV are 200 times more likely to develop hepatic cancer than noncarriers are. Furthermore, the HBV genome has been found integrated into hepatic cancer cells, and these same cells typically express HBV antigens. It is possible that integration of the virus activates oncogenes or suppresses oncogene repressor genes. Another theory is that repair and cell growth in response to liver damage proceeds out of control, resulting in cancer. Simultaneous infection *(coinfection)* with HDV increases the likelihood of severe liver damage.

TABLE 23.2 Comparison of Hepatitis Viruses

Feature	*Hepatovirus* *Hepatitis A virus* (HAV)	*Orthohepadnavirus* *Hepatitis B virus* (HBV)	*Hepacivirus* *Hepatitis C virus* (HCV)	*Deltavirus* *Hepatitis delta virus* (HDV)	*Hepevirus* *Hepatitis E virus* (HEV)
Virus family	*Picornaviridae*	*Hepadnaviridae*	*Flaviviridae*	*Arenaviridae*	*Hepeviridae*
Genome	+ssRNA	Partly ssDNA, partly dsDNA	+ssRNA	−ssRNA	+ssRNA
Envelope present?	No	Yes	Yes	Yes	No
Transmission	Fecal-oral	Needles; sex; blood and fluids	Needles; sex	Needles; sex	Fecal-oral
Incubation period	15–45 days	70–100 days	42–49 days	7–24 days	15–60 days
Severity (mortality rate)	Mild (<0.5%)	Occasionally severe (15–25%)	Usually subclinical (0.5–4%)	Requires simultaneous hepatitis B infection to replicate; together severity may be very high (10–20%)	Mild (1–3%; pregnant women 15–25%)
Chronic carrier state?	No	Yes	Yes	No	No
Common name of disease	Infectious hepatitis	Serum hepatitis	Non-A, non-B hepatitis; chronic hepatitis	Hepatitis delta	Enteric hepatitis
Other disease associations	—	Hepatic cancer	Hepatic cancer	Cirrhosis	—

Hepatitis C virus (HCV) is an enveloped RNA virus that lacks proofreading ability as it replicates, resulting in many genetic strains within a patient. This may explain the body's inability to clear HCV infections—as adaptive immunity clears one strain, another evolves to take its place.

Hepatitis delta virus (HDV) is unique in that it does not carry genes for a capsid; instead, it utilizes hepatitis B capsomeres. Thus, hepatitis delta virus can spread only from cells that also carry hepatitis B virus. Patients coinfected with both HBV and delta virus at the same time typically have a more severe acute disease than chronic HBV patients who are infected with delta agent at a later time—a condition called a *superinfection.*

Hepatitis E virus (HEV) is a naked RNA virus that replicates in human liver cells and that can infect monkeys, apes, swine, and rodents. Infection in humans appears to come from humans only.

TABLE 23.2 compares and contrasts the features of hepatitis viruses.

Epidemiology

Immunizations against hepatitis A and B viruses have reduced the number of cases significantly—from an estimated 312,000 annual cases of hepatitis A in 1990 to 1402 cases in 2012. Cases of hepatitis B declined 96% from 1990 to 2012.

Hepatitis C virus infects over 180 million people worldwide, including about 4 million U.S. residents. The virus kills 10,000 to 15,000 Americans each year, more than HIV. The virus is transmitted sexually and via contaminated needles. Over 80% of infected people remain chronically infected (the common name of the disease is *chronic hepatitis*), and 70% suffer serious liver damage, many requiring liver transplants. In the past, blood transfusions accounted for many cases of hepatitis C, but testing blood for HCV has considerably reduced the risk of infection by this means.

Most outbreaks of hepatitis E are associated with fecally contaminated drinking water, so the common name of the disease is *enteric hepatitis.* The disease is more common in parts of the world with hot climates; hepatitis E is rare in temperate climates. Cases of enteric hepatitis in the United States usually involve immigrants or travelers infected in an endemic region, such as Mexico, northern Africa, or the developing nations of Asia.

Diagnosis, Treatment, and Prevention

Initial diagnosis of hepatitis may involve observation of the presence of jaundice, an enlarged liver, or fluid in the abdomen. Laboratory tests include serological studies of body fluids to detect viral antigens or antibodies against hepatitis viruses, liver function tests, or liver biopsy to determine the extent of liver damage.

Microscopists can observe specific HBV proteins in body fluids: so-called Dane particles, spherical particles, and filamentous particles **(FIGURE 23.15)**. Dane particles are complete, infectious virions, whereas spherical and filamentous particles are "empty" viral surface antigens that serve as a decoy in the host—the binding of antibody to empty capsids reduces antibody response against Dane particles.

Treatment of hepatitis is rapidly developing as research progresses; in any case, treatment involves rest and reducing inflammation; there are no cures. Immunoglobulin against the viruses given immediately after exposure offers some protection.

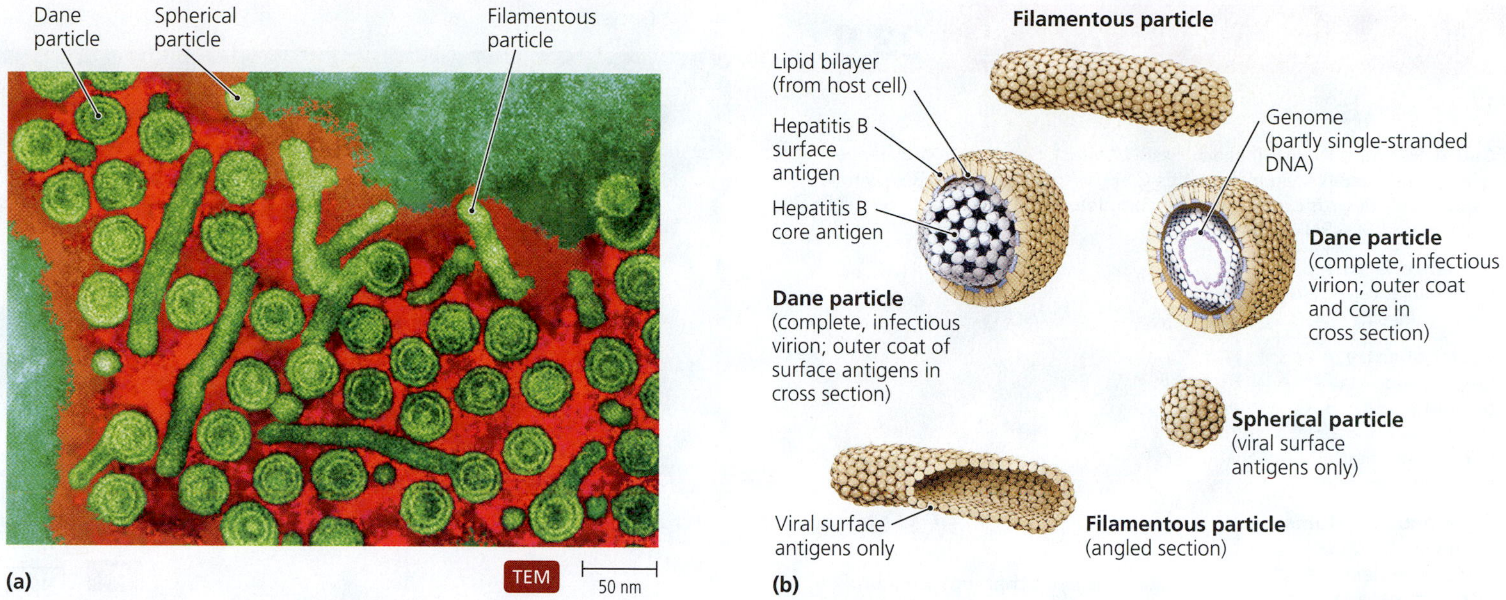

▲ **FIGURE 23.15 Three types of viral protein particles produced by hepatitis B viruses.** Dane particles are complete virions, whereas filamentous particles and spherical particles are capsomeres that have assembled without genomes. **(a)** Micrograph. **(b)** Artist's rendition.

Alpha interferon or nucleotide analogs, such as adefovir dipivoxil or lamivudine, help in 40% of cases of HBV infection. Alpha interferon, the nucleotide analog ribavirin, and the protease inhibitors telaprivir and boceprivir reduce the amount of hepatitis C virus in the blood to undetectable levels in 70% of patients—the closest thing to a cure yet developed for this chronic disease.

Prevention generally involves avoiding exposure. To reduce infections by HAV and HEV, wash hands often and avoid undercooked foods or contaminated water in endemic areas. For HBV, HCV, and HDV, avoid sharing needles for drugs, tattooing, or piercing. Heath care workers should be cautious with needles and other sharp instruments. All blood and blood products should be screened before use. Abstinence is the only sure way to prevent sexually transmitted infection; condoms can reduce risk but not eliminate it. The CDC recommends that all baby boomers (people born from 1945 through 1965) get tested for hepatitis C so that infected individuals can begin treatment.

Two vaccines are available against hepatitis A; each is administered in two doses separated by 6–12 months. Immunization is recommended for children between the ages of one and two years, adults who travel to high-risk areas (South America, Africa, and Asia other than Japan), men who have sex with men, and intravenous drug users.

The vaccine against HBV is given over a six-month period, resulting in protection against the virus in 95% of individuals. Immunity against HBV lasts for at least 15 years, and probably for life. Immunization is recommended for everyone.

There are no vaccines against hepatitis C, delta, or E viruses, though scientists have identified a protein-digesting enzyme required for HCV development and are working to develop a protease inhibitor.

Disease at a Glance 23.8 on p. 738 summarizes the features of viral hepatitis.

TELL ME WHY

In areas of poor sanitation, which form of hepatitis would you expect to be most common—infectious hepatitis, serum hepatitis, or chronic hepatitis? Why?

Protozoan Diseases of the Intestinal Tract

Protozoa as a whole are the most significant human pathogens worldwide, though relatively few of them cause infections of the gastrointestinal tract. Here we examine three protozoan parasites of the GI tract.

Giardiasis

LEARNING | **OUTCOMES**

23.22 Describe the disease giardiasis, including its cause, epidemiology, and treatment.

23.23 List methods to prevent giardiasis infections during recreational activities and in places such as day care centers.

Giardiasis (jē-ar-dī´ă-sis) is one of the more common waterborne gastrointestinal diseases in the United States. The disease is caused by *Giardia intestinalis* (jē-ar´dē-ă in-tes´ti-năl´is, formerly called *G. lamblia,* lăm´lē-a). It is often asymptomatic

DISEASE AT A GLANCE 23.8

Hepatitis

Cause Hepatitis A virus (naked, +ssRNA virus), hepatitis B virus (enveloped, partly dsDNA virus), hepatitis C virus (enveloped, +ssRNA virus), hepatitis delta virus (delta agent) (incomplete –ssRNA virus), hepatitis E virus (naked, +ssRNA virus).

Virulence factors Intracellular replication cycle, adhesins.

Portal of entry A and E—ingestion via fecal-oral route; B, C, delta—parenteral (blood- and fluid-borne) transmission.

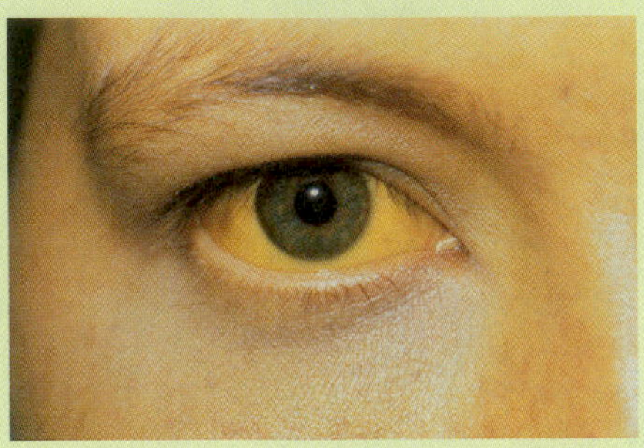

Yellow skin and eyes characterize jaundice.

Signs and symptoms Jaundice, fatigue, abdominal pain, loss of appetite, nausea, and diarrhea. Additional symptoms: A—fever; B—vomiting and joint pain; C—dark urine; delta—vomiting and dark urine, E—vomiting and dark urine.

Incubation period A—15 to 45 days; B—70 to 100 days; C—42 to 49 days; delta—7 to 24 days; E—15 to 60 days.

Susceptibility Adults generally display more symptoms than children do and are at greater risk because of behavioral activities, though children are more at risk for chronic infections and complications thereof.

Treatment For all—supportive therapy and rest; also: A—anti-HAV immunoglobulin; B—alpha interferon, adefovir dipivoxil, and lamivudine; C—alpha interferon, telaprivir, boceprivir, and ribavirin; delta—control of HBV coinfection.

Prevention Practice good hygiene and drink sterilized water, especially when traveling in areas endemic for hepatitis A and E. Limit activities where body fluids may be acquired (such as sexual intercourse, sharing needles) to help prevent transmission of hepatitis B, C, and delta. Vaccines against hepatitis A virus and hepatitis B virus are available.

but can cause one to two weeks of significant gastrointestinal distress. **Disease in Depth: Giardiasis** on pp. 740–741 examines *Giardia* and its disease more fully.

Cryptosporidiosis

LEARNING | **OUTCOMES**

23.24 Describe the cause and symptoms of cryptosporidiosis.
23.25 Describe methods to prevent the transmission of cryptosporidiosis.

In 1993, a water plant in Milwaukee, Wisconsin, malfunctioned for two weeks. Within days, over 403,000 people developed **cryptosporidiosis** (krip´tō-spō-rid-ē-ō´sis)—a zoonotic disease once thought to be limited to animals—and 100 died.

Signs and Symptoms

Severe watery diarrhea several times a day lasting about two weeks is a common manifestation of cryptosporidiosis. Headache, muscular pain, cramping, nausea, fatigue, and severe fluid and weight loss accompany the diarrhea. Life-threatening malabsorption, hepatitis, and pancreatitis can complicate the disease.

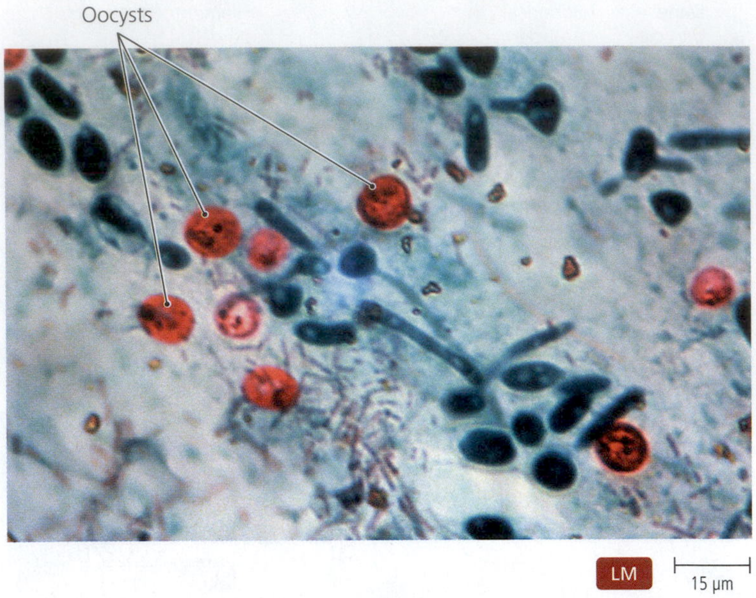

▲ **FIGURE 23.16 Oocysts of *Cryptosporidium parvum* in feces.** An acid-fast staining procedure colors the oocysts red.

Pathogen and Pathogenesis

The protozoan *Cryptosporidium parvum* (krip´tō-spō-rid´ē-ŭm par´vŭm) causes cryptosporidiosis. The infectious form of this parasite is a banana-shaped, motile *sporozoite* that has an apical complex of organelles specialized for penetrating host cells; thus, it is an *apicomplexan.* Sporozoites form thick-shelled *oocysts,* which are the infective stage, inside cells **(FIGURE 23.16)**. With a complex series of cell divisions and stages, oocysts eventually develop four internal sporozoites, which escape to continue the life cycle.

Infection most commonly results from drinking water contaminated with oocysts, but direct fecal-oral transmission resulting from poor hygienic practices also occurs, particularly in day care facilities. Scientists do not understand the pathogenicity of *Cryptosporidium,* but some think that destruction of intestinal cells and the subsequent inflammatory response trigger loss of electrolytes and water and decrease absorption of nutrients. In healthy adults, the infection clears spontaneously with time, often up to a month.

Epidemiology

About 30% of people living in developing nations carry *Cryptosporidium* asymptomatically. It is estimated that most natural waterways in the United States are contaminated with oocysts introduced in livestock wastes. U.S. physicians diagnose about 7000 cases of cryptosporidiosis each year. People with AIDS, transplant patients, and other immunocompromised individuals are at higher risk for severe disease. In HIV-positive individuals, chronic cryptosporidiosis is one of the life-threatening indicator diseases that reveals the clinical stage of AIDS.

EMERGING DISEASE CASE STUDY

Norovirus Gastroenteritis

Zack, Tran, Roy, and Justin were not happy roommates; in fact, things could not be much worse for the college friends. Each of the men had stomach cramps, fever, chills, muscle aches, extreme tiredness, nausea, and, most distressing, horrible diarrhea and persistent vomiting. They had been fighting over the toilet, and whoever wasn't in the bathroom often had his head in a trashcan. None of the four had left their suite for two days, being unable to venture more than a dozen feet. The friends had never experienced such an attack of gastroenteritis. *Norovirus* had arrived in the dorm.

Norovirus gastroenteritis afflicts people living in close quarters: prisoners, nursing home residents, vacationers on cruise ships, and students in college dormitories. People spread the hardy virus on contaminated hands and fomites due to poor personal hygiene. Students often blame food services for gastrointestinal distress, but noroviruses rarely travel in food, though they are often found in contaminated water.

The four roommates recovered as their bodies eliminated viruses from their digestive tracts. They also learned the value of handwashing, disinfecting bathrooms, and keeping the dorm room sanitized. For more about *Norovirus*, see p. 734.

1. Why is vigorously rubbing the hands with hot water and soap for at least 30 seconds necessary to limit the spread of noroviruses?
2. Alcohol disrupts lipids. Why isn't hand antiseptic effective against noroviruses?
3. How can noroviruses spread via laundromats?

Diagnosis, Treatment, and Prevention

The presence of oocysts in feces is diagnostic of the disease. Additionally, the xTAG GPP can identify nucleic acid patterns of *Cryptosporidium* in stool. Health care workers provide supportive care, primarily in the form of fluid and electrolyte replacement. Nitazoxanide shortens the duration of diarrhea. Prevention involves good hygiene, avoiding contaminated water or food, and avoiding fecal exposure during sex.

Amebiasis

LEARNING **OUTCOMES**

23.26 Describe the three forms of amebiasis seen in humans.
23.27 Describe methods for preventing amebiasis.

Amoebae (ă-mē´bē) are protozoa with no truly defined shape that use pseudopodia to move and acquire food. They are abundant throughout the world in water and moist soil.

Signs and Symptoms

Depending on the health of the host and the virulence of the particular infecting strain, three types of **amebiasis** (ă-mē-bī´-ă-sis) occur. (Note the disease and its cause are spelled differently.) The least severe form, *luminal amebiasis,* occurs in otherwise healthy individuals and is asymptomatic. Invasive *amebic dysentery* is a more serious form of infection characterized by severe diarrhea, colitis (inflammation of the colon), appendicitis, ulceration of the intestinal mucosa, bloody mucus-containing stools, and pain. In the most serious disease—*invasive extraintestinal amebiasis*—potentially fatal lesions of dead and dying intestinal cells form in the liver, lungs, spleen, kidneys, or brain.

Pathogen, Virulence Factors, and Pathogenesis

Entamoeba histolytica (ent-ă-mē´bă his-tō-li´ti-kă) causes all forms of amebiasis. A motile trophozoite (see Disease at a Glance 23.9 on p. 743) develops into an infective, resistant, chitin-shelled cyst. Virulent strains of *Entamoeba* produce adhesion proteins, proteases, proteins that create ion channels in host membranes, and other small proteins that appear to have toxic effects on cells and facilitate invasion. Avirulent strains of *Entamoeba* do not produce these four types of proteins and so remain in the lumen of the intestine.

Infection begins with ingestion of thick-shelled cysts, which pass successfully through the acid of the stomach and excyst in the small intestine to release trophozoites. These migrate to the large intestine and multiply by binary fission, producing any signs and symptoms in one to four weeks. Trophozoites use pseudopodia to attach to specific receptors on the intestinal lining, where they feed. Both trophozoites and cysts are shed into the environment in feces, but trophozoites die.

Trophozoites in the lumen of the intestine do little damage and typically produce no symptoms, but when trophozoites invade the peritoneal cavity and bloodstream, amebic dysentery

DISEASE IN DEPTH

GIARDIASIS

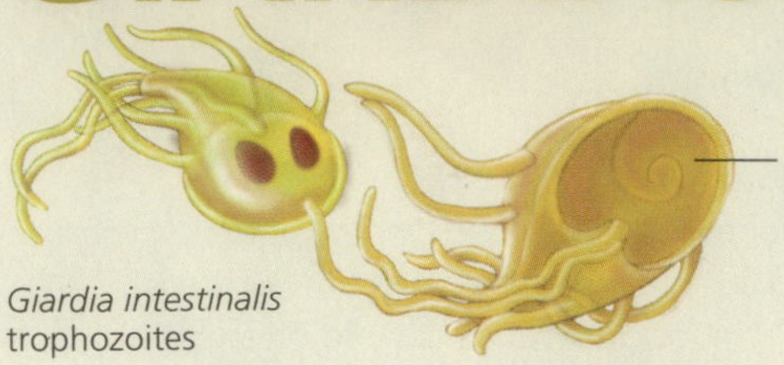

Ventral adhesive disk

Giardia intestinalis trophozoites

SIGNS AND SYMPTOMS

Giardiasis is often asymptomatic but can cause significant gastrointestinal distress, including severe, greasy, frothy, fatty diarrhea; abdominal pain; flatus; nausea; vomiting; loss of appetite; ineffective absorption of nutrients; and low-grade fever. Patients' stools usually have a "rotten egg" smell of hydrogen sulfide. Acute disease generally lasts one to four weeks following an incubation period of about two weeks. Giardiasis can be a chronic condition, often in animals.

Giardiasis is one of the more common waterborne gastrointestinal diseases in the United States. The *Giardia* life cycle is shown below:

GIARDIA LIFE CYCLE

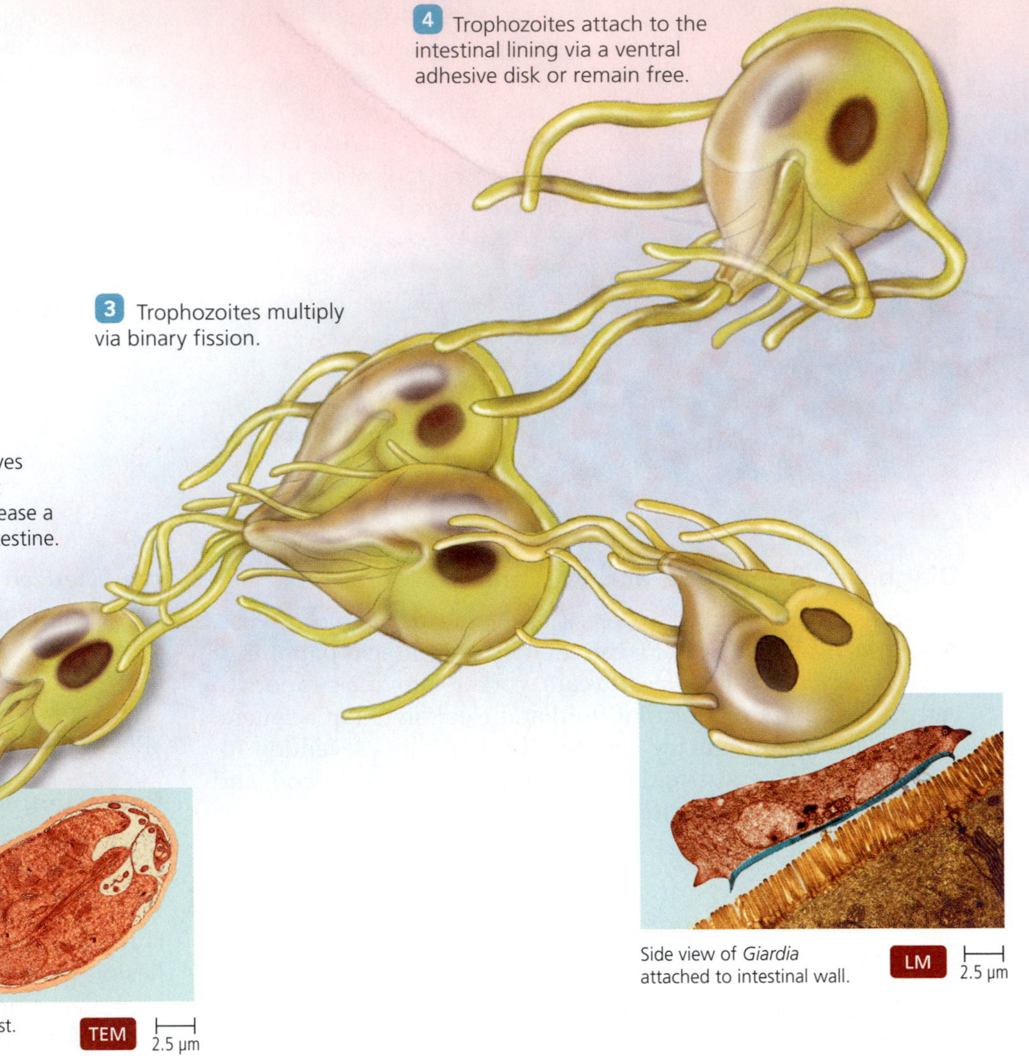

4 Trophozoites attach to the intestinal lining via a ventral adhesive disk or remain free.
3 Trophozoites multiply via binary fission.

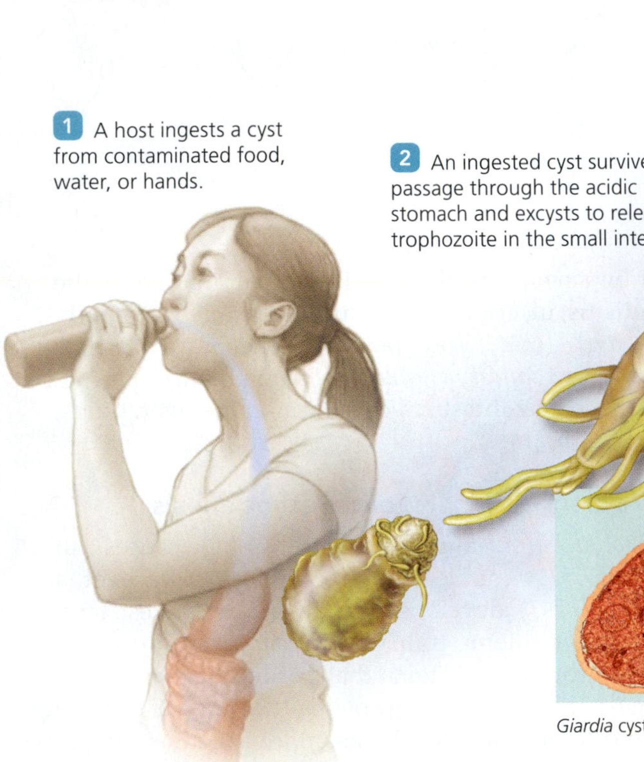

1 A host ingests a cyst from contaminated food, water, or hands.
2 An ingested cyst survives passage through the acidic stomach and excysts to release a trophozoite in the small intestine.
TEM
2.5 µm
LM
2.5 µm

Giardia cyst.

Side view of *Giardia* attached to intestinal wall.

INVESTIGATE IT!

Why is the eradication of giardiasis in the United States unlikely?

Scan this code to visit the Centers for Disease Control and Prevention website to investigate giardiasis. Then go to MasteringMicrobiology to record your research findings.

EPIDEMIOLOGY

Giardiasis occurs in both developed and developing countries; recent trends have shown increases in the number of cases worldwide. In the U.S., backpacking, camping, swimming, and contact with certain animals may increase the risk for giardiasis. Data from the CDC MMWR report.

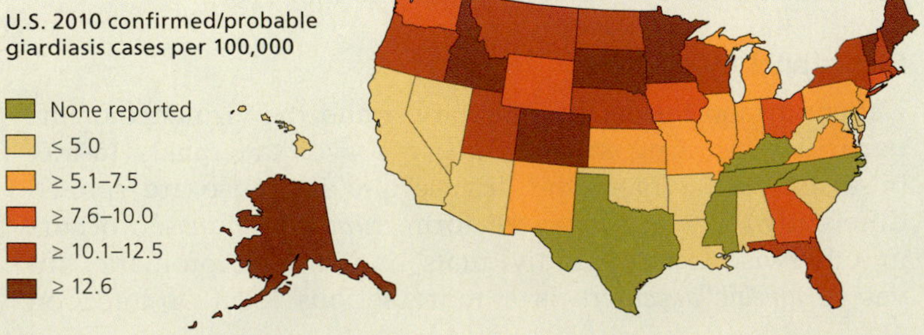

U.S. 2010 confirmed/probable giardiasis cases per 100,000
None reported
≤ 5.0
≥ 5.1–7.5
≥ 7.6–10.0
≥ 10.1–12.5
≥ 12.6

PATHOGEN

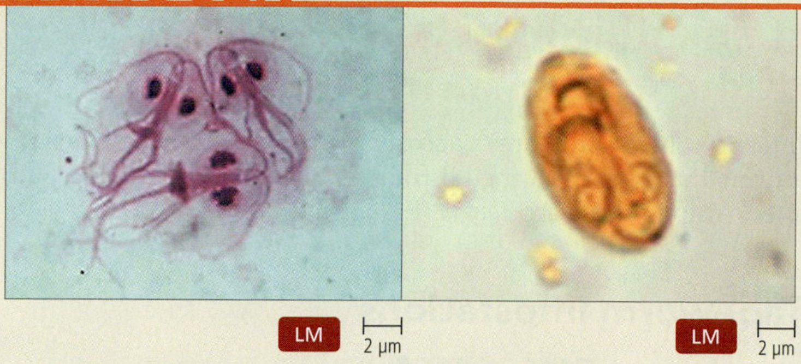
LM
2 μm
LM
2 μm

A diplomonad (two-nuclei) flagellate named *Giardia intestinalis* (formerly called *G. lamblia*) causes giardiasis. The protozoan has two forms—a motile feeding trophozoite shown on the left and a dormant cyst shown on the right, which has a tough shell composed of the polysaccharide chitin. The cyst is resistant to chlorine, heat, drying, and stomach acid.

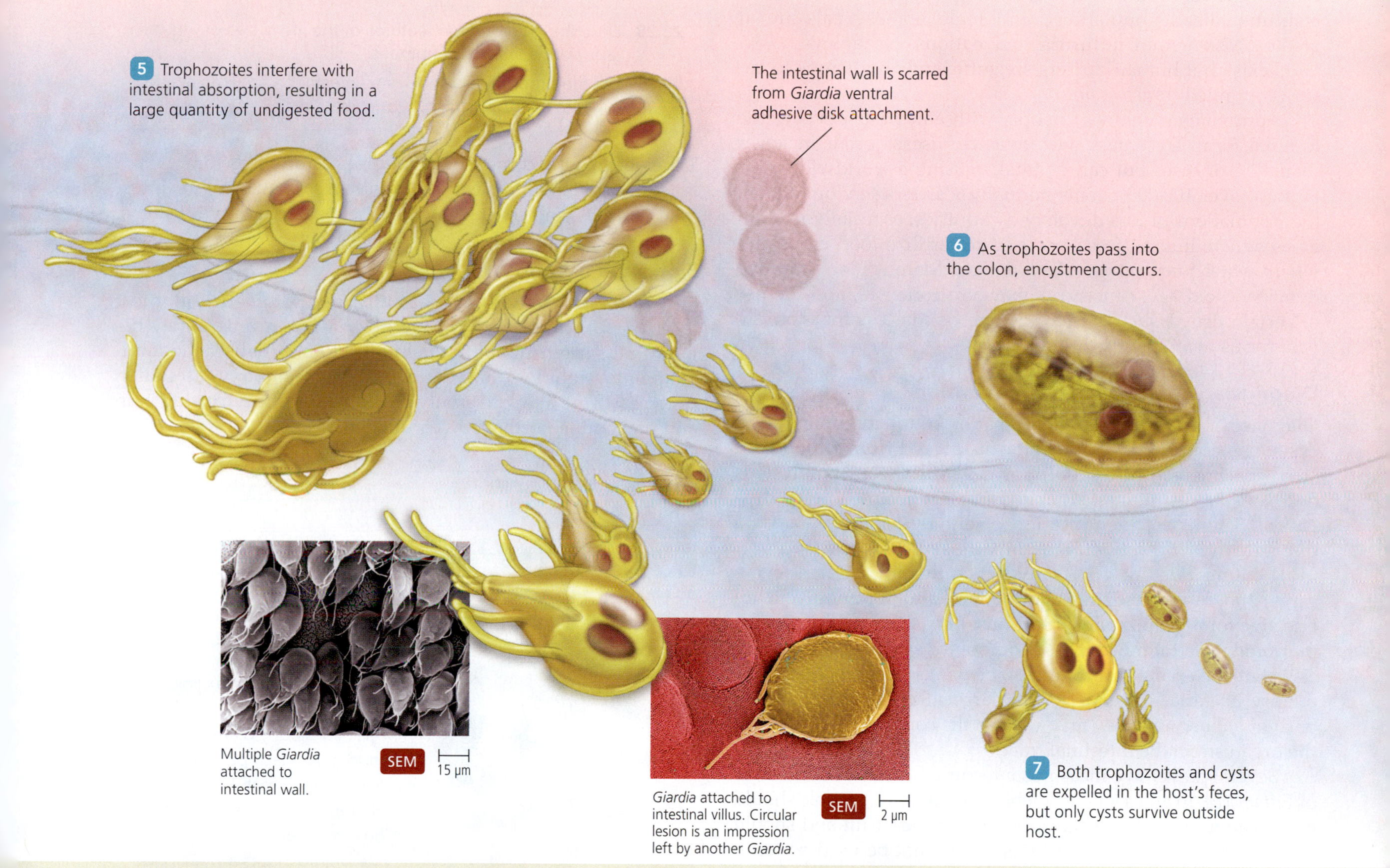
5 Trophozoites interfere with intestinal absorption, resulting in a large quantity of undigested food.
The intestinal wall is scarred from *Giardia* ventral adhesive disk attachment.
6 As trophozoites pass into the colon, encystment occurs.
Multiple *Giardia* attached to intestinal wall.
SEM
15 μm
Giardia attached to intestinal villus. Circular lesion is an impression left by another *Giardia*.
SEM
2 μm
7 Both trophozoites and cysts are expelled in the host's feces, but only cysts survive outside host.

DIAGNOSIS AND TREATMENT

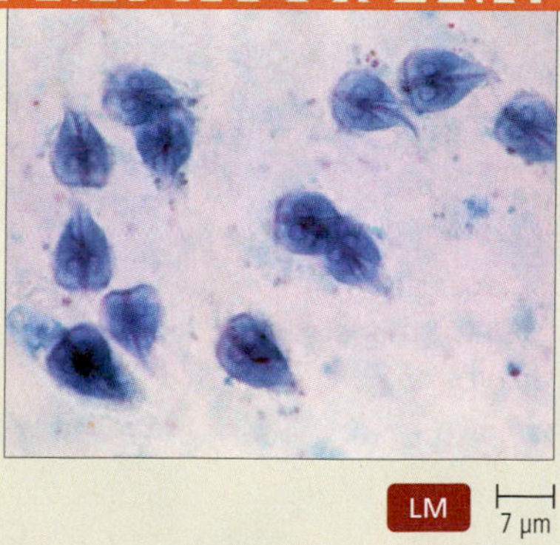
LM
7 μm

Diagnosis of giardiasis relies on microscopic examination of stool specimens to reveal oval cysts, or the xTAG GPP test for the presence of *Giardia* DNA. Some infections resolve spontaneously without treatment, but physicians can prescribe metronidazole for adults and furazolidone for children when diarrhea is present. Cure rates are generally high (80%), but drug resistance can lead to difficulties.

PREVENTION

Water must be filtered to prevent infection where *Giardia* is endemic. When hiking, humans and their pets shouldn't drink unfiltered stream or river water. In day care, excellent hygiene practices and the separation of feeding and diaper-changing areas are essential to avoid transmission.

or invasive extraintestinal amebiasis occur. The difference in severity is due to the production of differing virulence factors by various strains.

Epidemiology

E. histolytica is carried asymptomatically in the digestive tracts of roughly 10% of the world's human population. Infection arises following consumption of contaminated water or food, ingestion from contaminated hands, or during oral-anal intercourse. Cockroaches and houseflies can facilitate the spread of cysts under conditions of overcrowding. Travelers, immigrants, institutionalized populations, and male homosexuals are at greatest risk within industrialized nations. No animal reservoirs exist, but human carriers are sufficiently numerous to ensure continued transmission.

Approximately 90% of people with amebiasis develop the luminal form. Dysentery and invasive disease occur in fewer than 10% of cases but can be fatal, causing a worldwide mortality of over 100,000 people annually. Carriers predominate in the populations of less developed countries, especially in rural areas, where human feces are used to fertilize food crops and where water purification is inadequate. Malnutrition, immune deficiency, old age, cancer, pregnancy, alcoholism, and the use of certain drugs, such as steroids, are risk factors for more severe forms of amebiasis.

Diagnosis, Treatment, and Prevention

Diagnosis is based on the identification of microscopic, round cysts or pleomorphic trophozoites recovered from either fresh stool specimens or intestinal biopsies. Microscopic analysis is necessary to distinguish amebic dysentery from bacterial dysentery. Serological identification of antigens may be used to distinguish *E. histolytica* from nonpathogenic amoebae.

Treatment involves oral rehydration therapy (vital in severe cases) and antiamebic drugs. Iodoquinol and paromomycin are effective for asymptomatic infections. Physicians may prescribe metronidazole followed by iodoquinol for symptomatic amebiasis. Antibacterial agents may also be prescribed to prevent secondary bacterial infections. Antidiarrheal medications should be avoided because they may worsen the condition by retaining the organism in the intestinal tract.

Several preventive measures interrupt the transmission cycle of *Entamoeba*. In areas where amebiasis is common, people should avoid eating uncooked vegetables or unpeeled fruit and should drink bottled water. Human feces should not be used as fertilizer. Effective processing of water requires chemical treatment, filtration, or extensive boiling. Good personal hygiene and safer sexual practices can reduce transmission during intimate contact.

Disease at a Glance 23.9 summarizes the features of amebiasis.

TELL ME WHY

Why does the visually distinctive appearance of *Giardia* trophozoites improve the success of medical treatment of giardiasis as compared to that for amebic infections?

Helminthic Infestations of the Intestinal Tract

Intestinal helminths are macroscopic, multicellular, eukaryotic worms that can infest the GI tract as non-disease-causing parasites. Among these are *Taenia*, *Enterobius*, and *Anisakis*.

Tapeworm Infestations

LEARNING | **OUTCOMES**

23.28 Describe the common features of the life cycles of tapeworms that infest humans.

23.29 List the predominant modes of infestation for *Taenia*, and suggest measures to prevent infestation.

Tapeworm is the common name for a **cestode** (ses´tōd), which is a flat, segmented, parasitic helminth. All tapeworms are intestinal parasites and completely lack their own digestive systems.

Signs and Symptoms

Tapeworm infestations are usually asymptomatic; in most cases, people do not know they carry a worm unless they begin to pass segments of the helminth. Rarely, nausea, abdominal pain, weight loss, and diarrhea may accompany infestation or long worms may physically block the intestine, causing pain and preventing normal bowel function.

CLINICAL CASE STUDY

Painful Dysentery

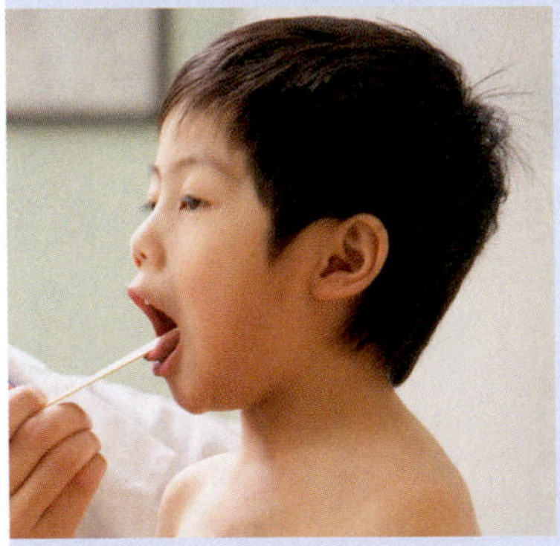

A 43-year-old immigrant from Thailand reports to a clinic with his 9-year-old son, who has had bloody diarrhea with mucus, stomach pain, and a fever for the past 24 hours. He reports that his son was healthy when they arrived in the United States a week prior to the start of his symptoms. Round protozoan cysts are subsequently microscopically identified in the boy's stool.

1. What is the diagnosis and causative agent?
2. What other diseases are caused by this organism?
3. Describe the other diseases caused by this organism.
4. How did the patient likely acquire this disease?
5. What is the best course of treatment?

DISEASE AT A GLANCE 23.9

Amebiasis

1. Person consumes cysts of *Entamoeba histolytica*, usually from contaminated water.
2. Excystment in small intestine releases trophozoites.
3. Trophozoites multiply in large intestine and attach to intestinal lining, causing luminal amebiasis.
4. They may invade the peritoneum to cause amebic dysentery.
5. They may invade the bloodstream to be carried throughout the body.
6. They cause invasive extraintestinal amebiasis when they infect the liver, lungs, spleen, kidneys, or brain, and they may cause death.
7. Cysts are shed in stool.

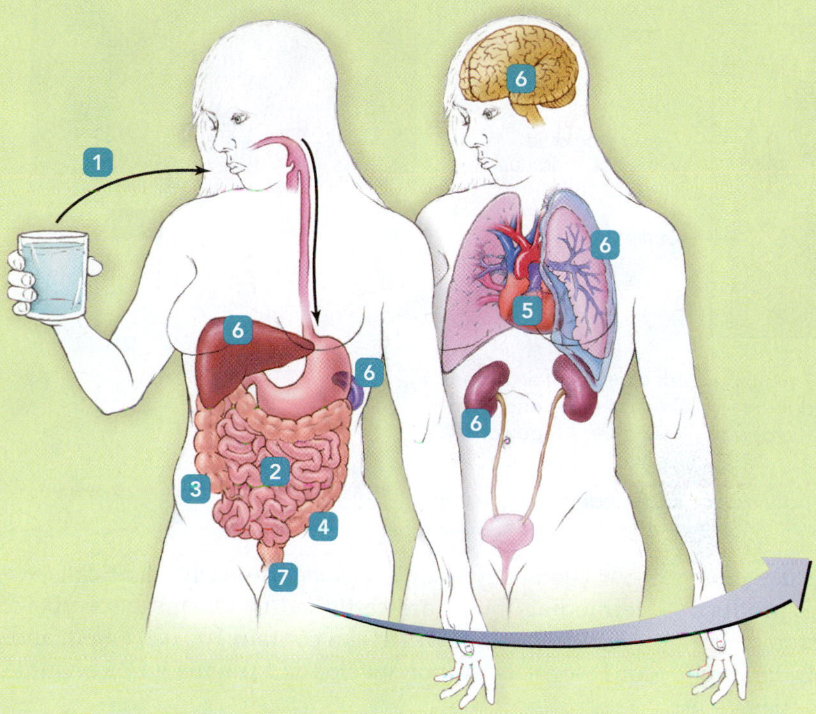

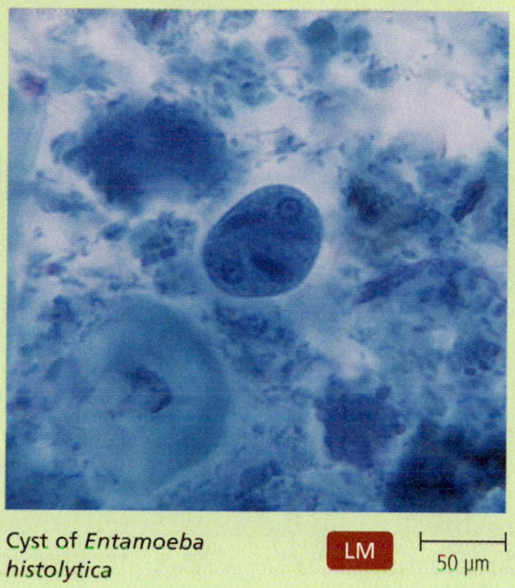

Cyst of *Entamoeba histolytica*

Cause *Entamoeba histolytica.*

Virulence factors Resistant cyst, adhesion proteins, proteases, ion channel proteins, toxins.

Portal of entry Oral.

Signs and symptoms Luminal amebiasis is asymptomatic. Invasive amebic dysentery involves severe diarrhea, colitis, appendicitis, ulceration of the intestinal mucosa, bloody and mucus-containing stools, and pain. In invasive extraintestinal amebiasis, potentially fatal necrotic lesions form in the liver, lungs, spleen, kidneys, or brain.

Incubation period 6–20 days.

Susceptibility People who live in developing nations that have poor sanitation; travelers, immigrants, institutionalized populations. Within industrialized nations, male homosexuals are at greatest risk.

Treatment Oral rehydration; iodoquinol or paromomycin for asymptomatic infections; metronidazole followed by iodoquinol for symptomatic amebiasis.

Prevention Avoid drinking contaminated water or eating foods washed or irrigated with contaminated water; do not use human feces as fertilizer; avoid oral-fecal contact during sex.

Pathogens

The common human tapeworms are *Taenia saginata* (te´ne-a sa-ji-na´ta), called the beef tapeworm, and *Taenia solium* (so´lī-um), called the pork tapeworm; the common names come from the fact that the worms develop during part of their lives in cattle and swine, respectively. **FIGURE 23.17a** illustrates a tapeworm body plan. The outer surface of the tapeworm is a *cuticle* ("skin"), through which a tapeworm steals nutrients from its host by absorption. The **scolex** (skō´leks) is a small attachment organ that possesses suckers and/or hooks to attach the worm to host tissue to prevent dislodgment **(FIGURE 23.17b)**. Behind the scolex is the neck region from which body segments called **proglottids** (prō-glot´idz) originate. New proglottids grow continuously from the neck, displacing older ones, which move farther from the neck to form a chain, or *strobila*. Tapeworms can attain lengths greater than 4 meters.

Proglottids mature as they are pushed away from the neck, producing both male and female internal reproductive organs. Each proglottid is monoecious[9] and may fertilize other proglottids of the same or different tapeworm. After fertilization, proglottids furthest from the neck become *gravid*[10] (full of fertilized eggs), break off the chain, and pass out of the intestine with

[9]From Greek *mono*, meaning "one," and *oikos*, meaning "house"; monoecious means that an organism contains both types of sex organs.

[10]From Latin *gravidus*, meaning "heavy," that is, "pregnant."

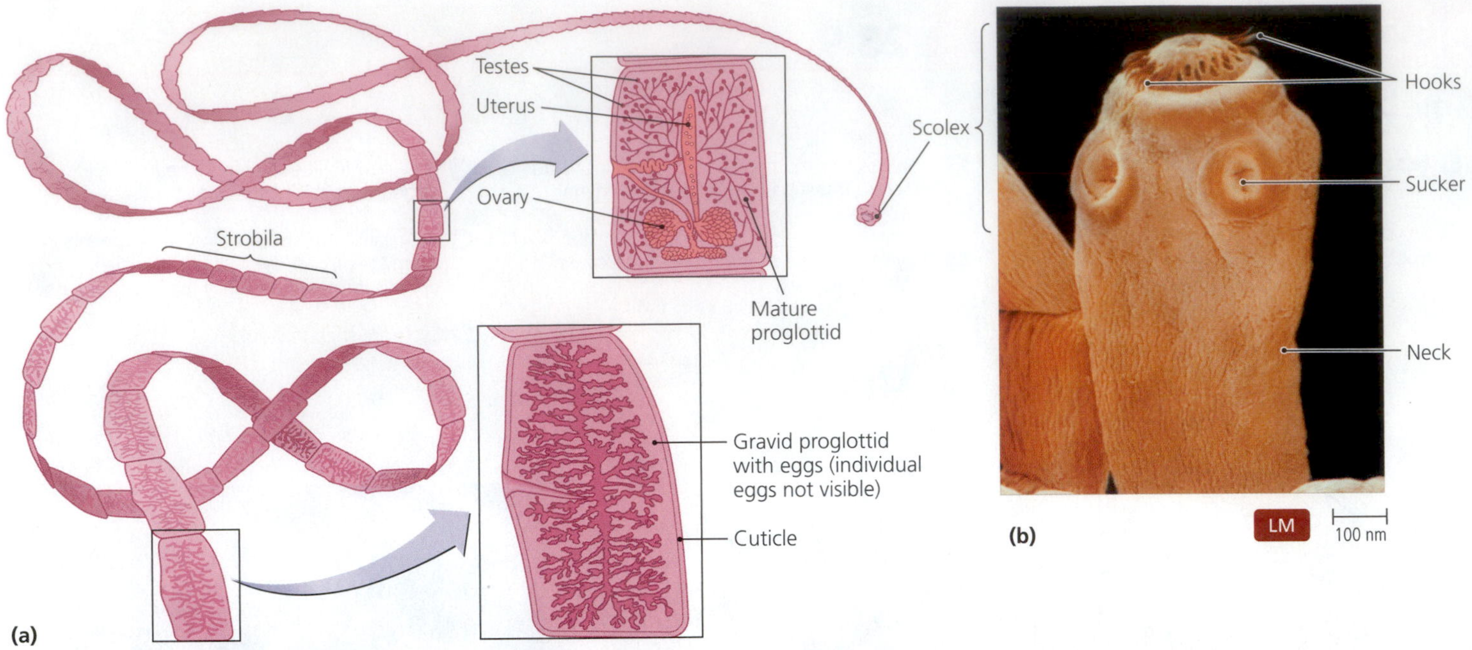

▲ **FIGURE 23.17 Features of tapeworm morphology.** Each tapeworm consists of an organ of attachment called a scolex (which in this case bears hooks in addition to suckers), a neck, and a long chain of segments called proglottids. **(a)** Artist's rendition. **(b)** Micrograph. *How does a tapeworm acquire food?*

Figure 23.17 *Tapeworms absorb food through their cuticles.*

feces. In a few cases, proglottids rupture within the intestine, releasing eggs directly into the feces. Some tapeworms produce proglottids large enough to be obviously visible in stools. Additionally, proglottids from the beef tapeworm are motile, providing a memorable experience for the person who passes a strobila.

Each tapeworm spends part of its life in a *definitive* or *primary host,* in which the sexual stage of the helminth develops, and part of its life in an *intermediate* or *secondary host.* **FIGURE 23.18** illustrates the life cycle of *Taenia solium* in its hosts—a human and a pig. Gravid proglottids and eggs enter the environment in feces from an infested human **1** and are consumed by a pig **2**. The eggs hatch into larvae in the pig's intestines and penetrate through the intestinal wall **3**. Larvae migrate to other tissues **4**, often muscle, where they develop into immature forms called **cysticerci** (sis´ti-ser-sī). Humans become infested by consuming raw or undercooked meat containing cysticerci **5**. Cysticerci excyst in the human intestine **6**, attach to the intestinal wall, and mature into new adult tapeworms **7**. *T. solium* adults average 1000 proglottids in length. Mature worms eventually shed gravid proglottids, thus completing the cycle. Infested human hosts pass approximately six proglottids per day. Each proglottid contains about 50,000 eggs.

Humans can become intermediate hosts of *T. solium* when they ingest eggs or gravid proglottids. Larvae released from the eggs invade human muscle or brain tissue and encyst. Encystment in muscle tissue generally results in no symptoms, but if cysticerci form in the brain, seizures and other neurological problems can occur.

The life cycle of *T. saginata* is nearly identical, except the intermediate hosts are cattle, adult worms have 1000–2000 proglottids (each of which can contain 100,000 eggs), and cysticerci of *T. saginata* cannot develop in humans who consume the eggs.

Epidemiology

Taenia species live worldwide in areas where beef and pork are food. The highest prevalence of human infestation is in poor rural areas with inadequate sewage treatment and regions where humans and livestock live in close proximity. Infestations are rare in the United States, though cases of cysticercosis are more prevalent because of immigration from endemic countries.

Diagnosis, Treatment, and Prevention

Clinicians can observe proglottids (or sometimes eggs) in fecal samples at least three months after infestation, which is the basis of diagnosis. Examination of the scolex is required to differentiate between *T. solium* and *T. saginata*.

Treatment with a single oral dose of niclosamide or praziquantel generally eliminates intestinal infestation. Rarely, surgery may be necessary to remove the tapeworms from the intestinal tract and to open blockages.

Thoroughly cooking or freezing meat is the easiest method of prevention. Because cysticerci are readily visible in meat, giving it a "mealy" look, inspecting meat either before shipment to market or before purchase can reduce the rate of infestation. Good sewage treatment to prevent human feces from entering the intermediate hosts' food also breaks a human tapeworm's life cycle.

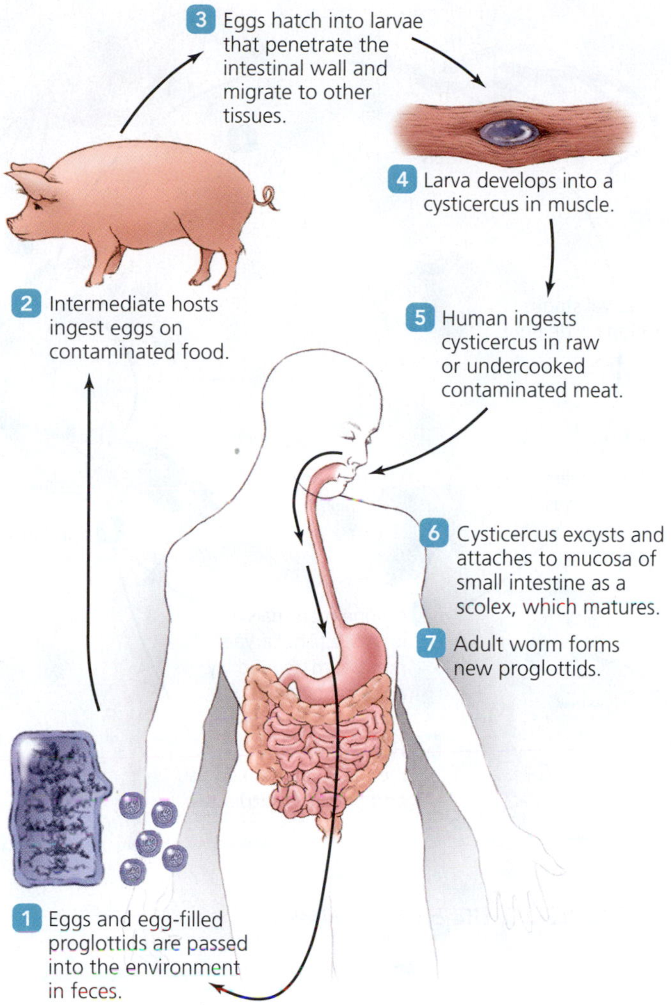
3 Eggs hatch into larvae that penetrate the intestinal wall and migrate to other tissues.
4 Larva develops into a cysticercus in muscle.
2 Intermediate hosts ingest eggs on contaminated food.
5 Human ingests cysticercus in raw or undercooked contaminated meat.
6 Cysticercus excysts and attaches to mucosa of small intestine as a scolex, which matures.
7 Adult worm forms new proglottids.
1 Eggs and egg-filled proglottids are passed into the environment in feces.

▲ **FIGURE 23.18** **Life cycle of *Taenia solium*.**

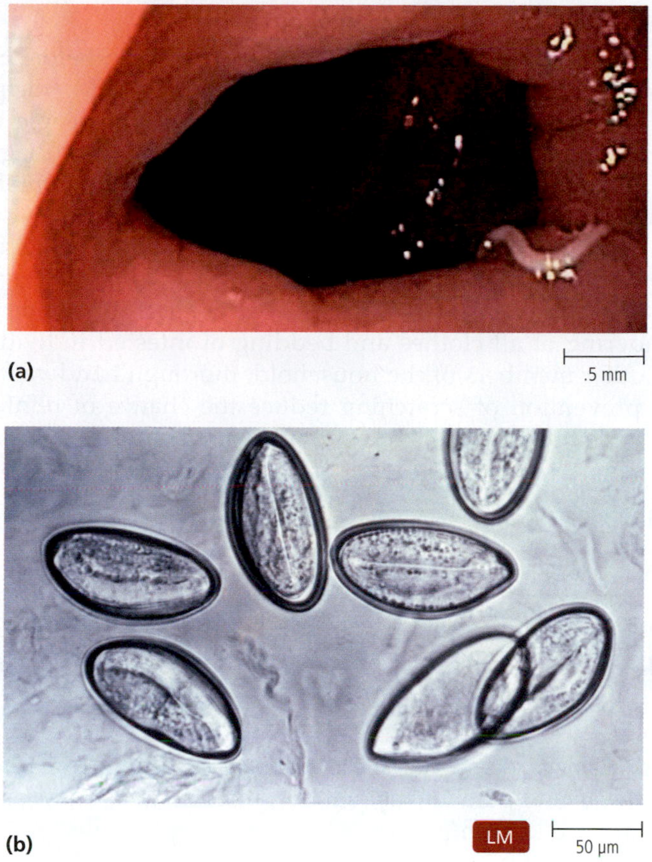
(a)
.5 mm
(b)
LM
50 μm

▲ **FIGURE 23.19** **Nematodes.** **(a)** Shown here is an adult female pinworm, *Enterobius vermicularis*, in a human colon. **(b)** Pinworm eggs.

Pinworm Infestations

LEARNING | OUTCOMES

23.30 Describe the common characteristics of nematodes.

23.31 Describe methods for preventing pinworm infestations.

Pinworm, *Enterobius vermicularis* (en-ter´ō´bī-ŭs ver-mi-kū-lar´is), commonly infests the intestines of children.

Signs and Symptoms

One-third of all pinworm infestations are asymptomatic. Symptomatic infestations involve intense perianal itching, irritability and sleep disturbance due to itching, decreased appetite, and possibly weight loss.

Pathogen and Infestation

Enterobius is a **nematode** (nem´ă-tōd)—a long, thin, unsegmented cylindrical helminth tapering to points at each end **(FIGURE 23.19a)**. All nematodes possess complete digestive tracts, have a protective cuticle, and are dioecious,[11] with females being larger than males. Nematodes as a group have variable reproductive strategies that make them highly successful parasites of almost all vertebrate animals, including humans. Humans are the only known host for pinworms.

After mating in the colon, female pinworms migrate at night to the anus, where they deposit eggs perianally (around the anus) before returning to the colon. Scratching dislodges eggs onto clothes or bedding, where they dry, become aerosolized, and settle in water or on food, which is consumed. Alternatively, scratching leaves eggs on the skin and under the fingernails, such that infested individuals can continually reinfest themselves by ingesting the eggs on their hands. Adult worms mature in several weeks and live in the intestinal tract for approximately two months.

Epidemiology

E. vermicularis infests about 500 million people worldwide, particularly in temperate climates, in school-aged children, and in conditions of overcrowding. *Enterobius* is the most common parasitic worm found in the United States, affecting 40 million Americans.

[11]From Greek *di*, meaning "two," and *oikos*, meaning "house"; dioecious worms have separate sexes.

Diagnosis, Treatment, and Prevention

Microscopy is used for diagnosis: in the morning, before bathing or defecation, transparent sticky tape is applied to the perianal area to collect the readily identifiable microscopic eggs (FIGURE 23.19b). Adult worms, if recovered, are also diagnostic. Treatment with albendazole, mebendazole, or pyrantel pamoate, followed by a second treatment two weeks later to kill any newly acquired worms, is usually successful. In some cases, an entire household must be treated.

Strict personal hygiene prevents reinfestation. Thorough laundering of all clothes and bedding of infested individuals and other members of the household, thorough handwashing, and prevention of scratching reduce the chance of reinfestation and spread of the worms to others.

Anisakiasis

LEARNING **OUTCOMES**

23.32 Describe the life cycle of *Anisakis*.

23.33 Explain how humans become infected, and describe the possible signs and symptoms of anisakiasis.

Anisakiasis (an-a-se-kī´a-sis) is a disease resulting from infestation by several parasitic nematodes, most notably, *Anisakis simplex* (an-a-se´kis sim´pleks).

Signs and Symptoms

Infested patients are typically asymptomatic, but some feel the larval stage of the worm moving around in their mouths or throats after eating infected meat. Some patients experience sudden and violent abdominal pain, nausea, vomiting, fever, and sometimes intestinal hemorrhaging. Following recovery from a first initial infestation, a few patients become sensitized and have acute, IgE-mediated allergic reactions to subsequent infestations or even exposure to dead worms in food. Allergic patients may develop a widespread rash, and some cough up adult worms—a distressing sign indeed. Very few allergic patients develop life-threatening anaphylactic shock.

Pathogen and Infestation

Anisakis simplex—the most common cause of anisakiasis—is a parasite of marine animals. The nematode has a complex life cycle with several larval stages (FIGURE 23.20). Marine mammals excrete the nematode's eggs into the ocean 1. Two larval stages develop one after the other in the egg 2, which subsequently hatches in seawater. The emergent second-stage larvae swim in the ocean 3, and small shrimplike crustaceans called krill eat them 4. In krill, the larvae develop into third-stage larvae (L3), which are infective to fish. Fish eat the krill, and the infective larvae invade from the fish's stomach into the fish's tissues 5. Other fish eat the infected fish, and the L3 larvae infect each predator fish in turn 6. In this manner, *Anisakis* L3 larvae are maintained in the fish population. Eventually, a marine mammal, such as a porpoise or whale, eats the infected fish. The larvae mature into adult worms that mate. The female worm lays eggs that are excreted in fish feces 7a. Humans who eat infected fish become accidental hosts to *Anisakis*, which matures in human and attaches to the esophagus, stomach, or intestine, producing anisakiasis and hypersensitivity reactions 7b. Humans are dead-end hosts, because they don't shed an abundance of eggs into seawater.

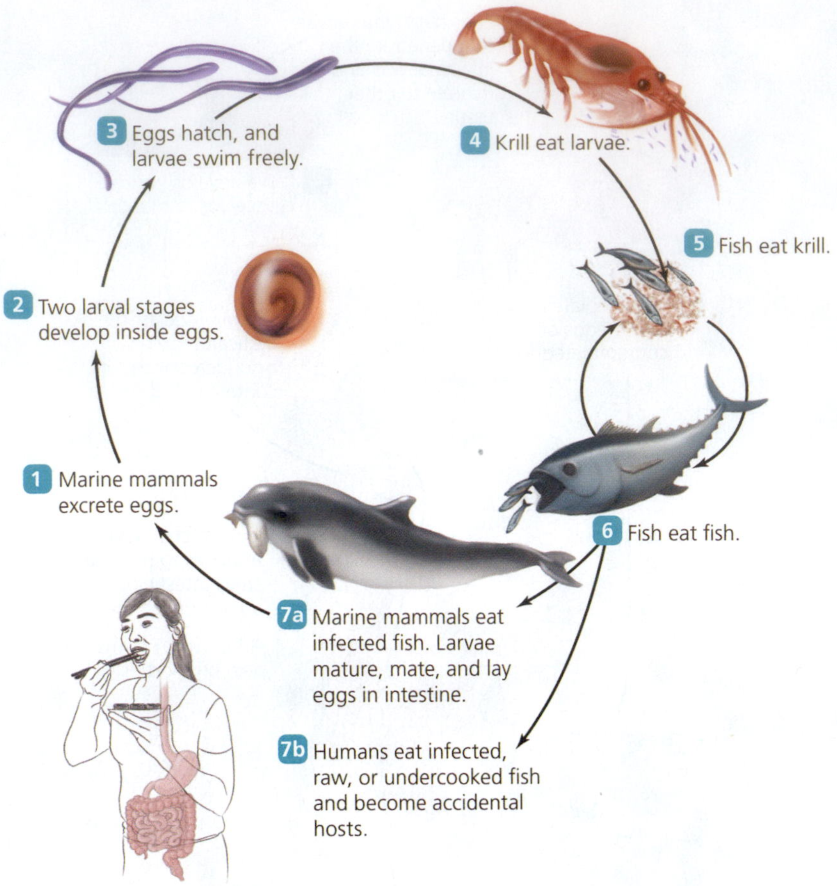

3 Eggs hatch, and larvae swim freely.
4 Krill eat larvae.
5 Fish eat krill.
2 Two larval stages develop inside eggs.
1 Marine mammals excrete eggs.
6 Fish eat fish.
7a Marine mammals eat infected fish. Larvae mature, mate, and lay eggs in intestine.
7b Humans eat infected, raw, or undercooked fish and become accidental hosts.

▲ **FIGURE 23.20 Life cycle of *Anisakis*.**

Epidemiology

Physicians report about 20,000 cases of anisakiasis worldwide annually.

Diagnosis, Treatment, and Prevention

Diagnosis is generally made by endoscopy in which a fiber-optic camera is inserted rectally into the voided intestinal tract. The worms can be seen crawling in the lumen or burrowing into the intestinal wall. Treatment involves removing the worms from the intestine with small forceps attached to a cable. Prevention of anisakiasis involves avoiding raw and undercooked marine fish.

TELL ME WHY

Tapeworm infestation of an intermediate host requires consumption of eggs in fecally contaminated food. Why is it possible for humans to become accidental intermediate hosts for *T. solium*?

MICRO IN THE CLINIC FOLLOW-UP

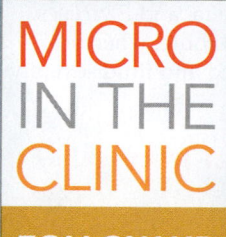

Trouble at the Rec Center

Andrea's entire family visits the doctor. They describe the diarrhea, provide a sample from one of the children's diapers, and explain that many of their neighbors have similar problems. The physician orders a series of three separate stool specimens, taken on three different days, for each family member. The stool specimens reveal that the entire family has *Giardia intestinalis*, a waterborne parasite that can reside in lakes, streams, and community swimming pools. The parasites are typically introduced in swimming pools by fecal matter—not unusual when young children are swimming. The doctor prescribes medication for the family members, tells them to drink lots of fluids, and warns Andrea to watch for dehydration in the children if the diarrhea does not subside quickly.

Water samples from the lake do not contain *Giardia*, but the parasites are in the wading pool, which is discovered to have a faulty chlorine dispenser (chlorine kills most *Giardia*) and an outdated filtration system. The rec center temporarily closes the wading pool until these things are fixed so as to prevent future outbreaks. In the meantime, Andrea and her family continue to enjoy socializing with their neighbors at the lake. However, Andrea warns her children not to swallow any of the lake water while swimming, and the whole family showers thoroughly after each swim.

1. **What environmental and biological characteristics of *Giardia intestinalis* contribute to widespread infections?**
2. **Why does *Giardia* produce the characteristic signs of frothy diarrhea, bloating, and gas?**

MM Check your answers to Micro in the Clinic Follow-Up questions in the MasteringMicrobiology Study Area.

CHAPTER SUMMARY

Structures of the Digestive System (pp. 716–717)

1. The digestive system is composed of the gastrointestinal (GI) tract and the accessory digestive organs.
2. The GI tract is composed of the mouth with its teeth, the esophagus, stomach, small intestine, large intestine, and anus. The primary function of the GI tract is to digest and absorb nutrients.
3. Digestive enzymes enter the GI tract from accessory digestive organs—the liver, gallbladder, salivary glands, and pancreas. The liver also aids in removal of wastes and poisons from the blood.

Normal Microbiota of the Digestive System (p. 717)

1. Normal microbiota colonize the digestive system except for some accessory organs and the esophagus, stomach, and duodenum.
2. **Viridans streptococci** are the most common normal microbiota of the mouth. *Bacteroides, Lactobacillus, Escherichia* and other enterobacteria, and *Candida* predominate in the lower small intestine and colon.
3. Intestinal normal microbiota provide some vitamins to the host. They also produce flatus and convert some substances to toxins and carcinogens. Their primary benefit is to inhibit pathogens by **microbial antagonism**.

Bacterial Diseases of the Digestive System (pp. 718–731)

1. Dental **caries**, or tooth decay, is the second most common infection of humans. Plaque and acid produced by mouth bacteria, usually *Streptococcus mutans* (a viridans streptococcus) and *Lactobacillus*, destroy the enamel of teeth to create cavities, which can lead to tooth loss.

2. Caries, plaque, and tartar when untreated can lead to a **periodontal disease** called **gingivitis**, which is inflammation of the gums. Severe periodontal disease manifests as increased inflammation, loss of bone, and tooth loss.
3. *Helicobacter pylori* causes **peptic ulcers**. The bacterium burrows through the mucus lining of the stomach, allowing stomach acid to produce painful erosion of underlying tissues.
4. **Bacterial gastroenteritis**, caused by many types of bacteria, is characterized by nausea, vomiting, diarrhea, cramping, and often fever. **Dysentery**, **shigellosis**, traveler's diarrhea, *Campylobacter* gastroenteritis, salmonellosis, and **cholera** are examples.
5. *E. coli* O157:H7 produces Shiga-like toxin as well as other enterotoxins to cause sometimes fatal gastroenteritis. Shiga-like toxin is related to the **Shiga toxin** of *Shigella,* several species of which produce gastroenteritis called shigellosis.
6. *Clostridium difficile* can cause ***C. diff.*** **diarrhea (antimicrobial associated diarrhea)**, which results when antimicrobial drugs kill normal microbiota, allowing *C. diff.* to predominate in the colon. Most severe cases—called **pseudomembranous colitis**—are lesions that fuse to make a false membrane of dead and dying cells and connective tissue.
7. Infections with *Salmonella* can result in **salmonellosis** or **typhoid fever**, the latter a more severe form of infection that can be fatal. *Salmonella* produce many virulence factors that aid in colonization of the host.
8. *Vibrio cholerae* produces **cholera toxin**, which stimulates the secretion of excess amounts of electrolytes and water from an infected cell, leading to severe dehydration.
9. Bacterial food poisoning includes GI tract infection (more properly called bacterial gastroenteritis) and **bacterial intoxication**, in which a toxin is present and active, while the bacteria are not.
10. Staphylococcal food poisoning is a common intoxication produced by enterotoxins of *Staphylococcus aureus*—a normal member of the microbiota.

Viral Diseases of the Digestive System (pp. 731–737)

1. **Fever blisters** and **cold sores** are alternative names for **oral herpes** infections. These recurrent lesions on and around the lips are usually caused by lifelong infection with human herpesvirus 1 (HHV-1).
2. **Mumps** is a relatively benign infection with the mumps virus of the parotid salivary glands of children. MMR vaccine has almost eliminated mumps from the United States.
3. **Viral gastroenteritis** displays symptoms similar to bacterial gastroenteritis except that in all cases the viral form is milder. Caliciviruses (especially **noroviruses**), astroviruses, and rotaviruses are the leading causes of viral gastroenteritis.
4. Many things can cause **hepatitis**—inflammation of the liver—including five viruses: hepatitis A, B, C, delta, and E viruses. Hepatitis A and E spread via the fecal-oral route, whereas hepatitis B, C, and delta are spread in blood and fluid sexually and via contaminated needles.

Protozoan Diseases of the Intestinal Tract (pp. 737–742)

1. **Giardiasis** is a severe, foul-smelling, watery diarrhea caused by *Giardia intestinalis,* a common resident of waterways throughout the United States.
2. **Cryptosporidiosis**—caused by *Cryptosporidium parvum*—is another intestinal disease with signs and symptoms that include diarrhea, cramping, and nausea.
3. Infection with *Entamoeba histolytica* can result in one of three forms of **amebiasis**: luminal amebiasis, which is generally asymptomatic; amebic dysentery, a significant diarrheal disease; and invasive extraintestinal amebiasis, a potentially fatal disease in which the **amoebae** invade the body and form necrotic lesions in various organs.

Helminthic Infestations of the Intestinal Tract (pp. 742–746)

1. **Tapeworm (cestode)** infestations are relatively rare in the United States as a result of inspection of meats, good sewage treatment, and freezing. Beef tapeworm *(Taenia saginata)* and pork tapeworm *(T. solium)* are acquired by eating meat containing tapeworm cysts; the worms then mature in the human intestine.
2. The tapeworm body plan includes an attaching **scolex**, a neck region, and a strobila (chain) of **proglottids** (body segments). Monoecious proglottids are fertilized to become gravid with eggs and shed into the environment where they are consumed by intermediate hosts. Eggs hatch in these animals and eventually form **cysticerci** (cysts) in muscle tissue.
3. **Pinworm** *(Enterobius vermicularis)*, a type of **nematode**, is a common helminthic parasite in the United States. Male and female worms mate in the human intestinal tract, and the female crawls out the anus to lay eggs. This leads to the intense itching characteristic of this type of infestation.
4. *Anisakis simplex,* a nematode, causes **anisakiasis**, a gastrointestinal disease manifesting with abdominal pain, vomiting, diarrhea, fever, and sometimes bleeding. Rarely, a patient develops an allergic reaction to *Anisakis.*

QUESTIONS FOR REVIEW

Answers to the Questions for Review (except Short Answer questions) begin on p. A-1.

Multiple Choice

1. Which of the following is *not* part of the gastrointestinal tract?
   a. stomach
   b. colon
   c. liver
   d. mouth
2. The major portion of food digestion occurs in the
   a. mouth.
   b. stomach.
   c. small intestine.
   d. large intestine.
3. The majority of microbes composing the normal microbiota of the GI tract are in the genus
   a. *Bacteroides.*
   b. *Escherichia.*
   c. *Clostridium.*
   d. *Staphylococcus.*
4. Diets high in sugars and starches increase the risk of tooth decay because they
   a. are acidic and destroy tooth enamel.
   b. are acidic and destroy the gingiva.
   c. are converted to acids by bacteria; the acid then destroys tooth enamel.
   d. are converted to acids by bacteria; the acid then destroys the gingiva.
5. Which of the following is a virulence factor important to *Helicobacter pylori* during the formation of ulcers?
   a. type III secretion system
   b. capsule formation
   c. flagella
   d. spore formation
6. Urease helps in the production of ulcers by *H. pylori* by
   a. increasing acid production by cells lining the stomach.
   b. neutralizing acid produced by cells lining the stomach.
   c. degrading mucus-producing cells.
   d. preventing destruction of *H. pylori* following phagocytosis.
7. Which causative agent of bacterial gastroenteritis is more commonly seen by doctors in the United States?
   a. *Campylobacter jejuni*
   b. *Vibrio cholerae*
   c. *Salmonella* serotype Typhi
   d. *Shigella* spp.
8. Oral fever blisters are most often caused by
   a. HBV.
   b. HCV.
   c. HHV-1.
   d. HHV-2.
9. Most of the symptoms associated with hepatitis are due to
   a. destruction of liver cells by virus.
   b. cellular immune reactions against infected liver cells.
   c. exotoxin production.
   d. endotoxin production.
10. One of the more common waterborne gastrointestinal diseases seen in the United States is
    a. amebiasis.
    b. *E. coli* O157:H7.
    c. giardiasis.
    d. salmonellosis.
11. Which one of the following diseases is an indicator disease signaling that an HIV-positive individual has progressed to AIDS?
    a. amebiasis
    b. cryptosporidiosis
    c. pinworm
    d. shigellosis
12. *Taenia saginata* is a
    a. bacterium.
    b. protozoan.
    c. cestode.
    d. nematode.
13. Which microbial group contributes the greatest number of causative agents to overall digestive tract diseases in the United States?
    a. bacteria
    b. helminths
    c. protozoa
    d. viruses
14. Typhoid fever is caused by a genus of bacteria that also causes
    a. cholera.
    b. cryptosporidiosis.
    c. mumps.
    d. salmonellosis.
15. An anxious mother consults you because her child has frothy, greasy diarrhea. After learning the child had returned from an overnight camping trip, you advise her that the child is likely infected with
    a. *Escherichia.*
    b. *Salmonella.*
    c. *Giardia.*
    d. *Helicobacter.*

Modified True/False

Indicate whether each statement is true or false. If the statement is false, change the underlined word or phrase to make the statement true.

1. _____ Plaque leads to the formation of dental caries <u>but does not</u> contribute to periodontal disease.
2. _____ <u>Stress</u> contributes to the formation of peptic ulcers.
3. _____ *Campylobacter jejuni* and *E. coli* are both causative agents of gastroenteritis, but only *E. coli* is an example of a <u>coliform</u>.
4. _____ Proper treatment of most forms of gastroenteritis <u>does not require</u> firm laboratory confirmation and identification of the causative agent.
5. _____ Viral gastroenteritis is generally <u>more severe</u> than bacterial gastroenteritis.

6. _____ Hepatic cancer is associated strongly with HBV infection.
7. _____ Giardiasis is a common bacterial infection of the digestive tract.
8. _____ Cryptosporidiosis has been proven to affect humans as well as livestock.
9. _____ Vaccines are available against hepatitis B and hepatitis C.
10. _____ *Vibrio cholerae* forms biofilms in seawater but not in freshwater.

Fill in the Blanks

1. Teeth are composed of three main layers, a hard layer called _______________ overlying the softer _______________ and the inner _______________.
2. The most common bacterium involved in cavity formation is _______________.
3. Peptic ulcers collectively include _______________ ulcers of the stomach and _______________ ulcers of the intestine.
4. _______________ toxin, produced by *Shigella dysenteriae,* is similar to the _______________ toxin produced by *E. coli* O157:H7.
5. "Cold sores" are alternatively known as _______________.
6. Swelling of the parotid glands is a major sign of _______________.
7. One of the primary symptoms of hepatitis is _______________, a yellowing of the skin and eyes.
8. Discovering oval cysts that have two nuclei is diagnostic for the disease _______________.
9. The B subunit of cholera toxin binds to an intestinal _______________ cell, and a portion of the A subunit acts as an enzyme that activates _______________.
10. The attachment organ of a cestode is its _______________.

Matching

Match the disease on the left with the causative pathogen on the right. Each answer is used only once.

1. _____ Viral gastroenteritis	A. *Shigella sonnei*
2. _____ Cholera	B. *Enterobius vermicularis*
3. _____ Typhoid fever	C. *Helicobacter pylori*
4. _____ Pinworm	D. *Entamoeba histolytica*
5. _____ Ulcers	E. *Vibrio* sp.
6. _____ Tapeworm	F. *Staphylococcus aureus*
7. _____ Bacterial gastroenteritis	G. Viridans streptococci
8. _____ Periodontal disease	H. *Taenia solium*
9. _____ Amebiasis	I. Rotaviruses
10. _____ Food intoxication	J. *Salmonella enterica* serotype Typhi

Short Answer

1. What role do normal microbiota play in protecting the GI tract from colonization by pathogens?
2. Describe how gingivitis arises from untreated dental cavities.
3. How do antacids help alleviate the symptoms of peptic ulcers?
4. Describe the process by which cholera toxin leads to severe fluid and electrolyte loss during diarrhea.
5. Why is poor sanitation such a big factor in the continued occurrence of gastrointestinal disease?
6. What is the difference between gastroenteritis and intoxication?
7. How does hepatitis B decoy the immune system?
8. Explain why it is unlikely that *Giardia intestinalis* will be eradicated from the environment.
9. What is the genetic difference between virulent and avirulent strains of *Entamoeba*?
10. The xTAG Gastrointestinal Pathogen Panel (xTag GPP) identifies 11 causes of gastroenteritis. What are the 11?

VISUALIZE IT!

1. Describe the events of tooth decay, shown below.

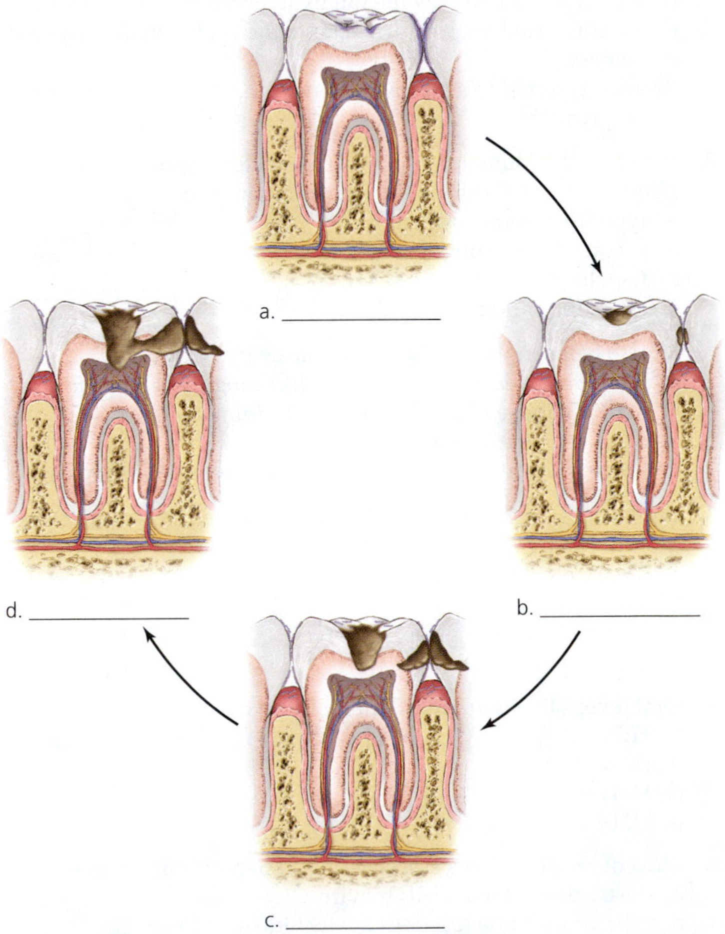

2. Label a Dane particle, a filamentous particle, and a spherical particle. How many of each are wholly or partially visible in the photo? Put a "D" on the particles that contain DNA and an "R" on the particles that contain RNA.

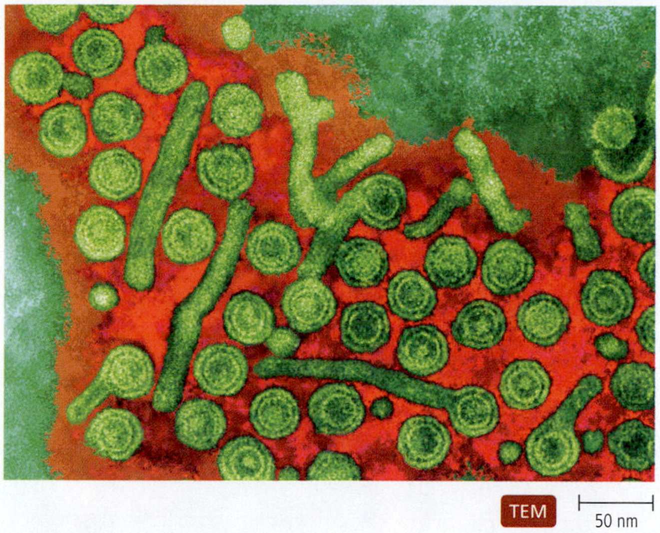

CRITICAL THINKING

1. Prescription antacids are useful for treating peptic ulcers and allowing the stomach lining time to heal. Antacids, however, can actually increase the risk of contracting other bacterial and viral gastrointestinal diseases. How?
2. Why is it necessary to establish childhood immunization for hepatitis B virus in order to attempt to eliminate hepatic cancer? Why would it not be enough to vaccinate only those who have been exposed to HBV?
3. Most gastrointestinal tract diseases are ultimately self-limiting and nonfatal, so long as fluids can be replaced. How do these two general observations help explain why such diseases so commonly affect humans?
4. Infections with HBV and HCV usually take years, even decades, before visible signs of hepatitis manifest themselves. Epidemiologically, how does this influence our ability to track such diseases in a population and prevent transmission?
5. Why did soldiers living in battlefield trenches in Europe during World War I frequently suffer from acute necrotizing ulcerative gingivitis (trench mouth)?
6. A respiratory therapy student is puzzled about mumps. Her textbook covers the disease in the chapter on the digestive system, but her teacher insists the mumps virus is respiratory. Explain why both sources are accurate.
7. No vaccine against hepatitis delta virus exists, and researchers are not seeking one—there would be no market for it. Why?
8. How could infection of the accessory digestive organs affect health?
9. How can eating spicy foods further antagonize tissue damaged by *Helicobacter pylori*?
10. A pharmacist in a small town of 4300 observes hundreds of people buying antidiarrheal medicines in one week. Obviously, there is an epidemic of gastroenteritis! What utilities and city services will epidemiologists examine? Explain your reasoning.
11. Rank the forms of gastroenteritis discussed in Table 23.1 on p. 730 in terms of severity, with the least severe first and the most severe last (consider this with regard to the most normal course of infection).
12. Based on your understanding of the human GI tract, explain why a disease such as herpes esophagitis would be limited to individuals who are immunocompromised or immunosuppressed.
13. Why and when should parents have their children immunized against mumps?
14. What social and environmental conditions contribute to the much higher rate of rotavirus infections outside the United States?

CONCEPT MAPPING

Using the following terms, draw a concept map that describes viral hepatitis. You may use some terms more than once. For a sample concept map, see p. 94. Or, complete this and other concept maps online by going to the MasteringMicrobiology Study Area.

Acute
Blood and body fluids (x2)
Blood test for antibodies (IgM)
Carrier state
Chronic
Chronic disease (x2)
Cirrhosis of the liver
Fecal-oral route
Hepatitis A
Hepatitis A vaccine
Hepatitis A virus (HAV)
Hepatitis B
Hepatitis B vaccine
Hepatitis B virus (HBV)
Hepatitis C
Hepatitis C virus (HCV)
Lamivudine and alpha interferon
Liver cancer
Mild
PCR or blood test for antibodies
Ribavirin and alpha interferon
Viral hepatitis

24 Microbial Diseases of the Urinary and Reproductive Systems

MICRO IN THE CLINIC

Picture-Perfect Romance?

Jordan and Lisa's whirlwind romance began six months ago with a blind date, followed by long walks along the lake, candlelit dinners, and a shared love of classic films. Before they got married, the couple underwent screening tests for sexually transmitted diseases (STDs). The results came back negative for both of them. Lisa was taking birth control pills, so they decided not to use condoms.

One morning, Jordan feels a burning sensation when he urinates. He then notices small red bumps on his penis that hurt when he touches them. In the next few days, the lesions enlarge into blisters. Jordan feels achy and tired. He also notices the lymph nodes in his groin are swollen.

Alarmed, Jordan visits an urgent care clinic. After examining him, the physician assistant uses a swab to sample the fluid in one of the lesions. The swab is sent to a lab for testing. Jordan can't believe what is happening. He and Lisa had been dating exclusively before they were married and had tested negative for STDs.

What will Jordan's test reveal? Is it possible that he has an STD? Turn to the end of the chapter (p. 777) to find out.

MM **Explore More:** Test your readiness and apply your knowledge with dynamic learning tools at MasteringMicrobiology.

Structures of the Urinary and Reproductive Systems

The urinary and reproductive systems in females are anatomically distinct. In males, the two systems share a portion of the "plumbing."

Structures of the Urinary System

LEARNING OUTCOME

24.1 Describe the function and structures of the urinary system.

Two kidneys, connected by ureters to the urinary bladder, and a urethra make up the urinary system in both females and males **(FIGURE 24.1)**. The *kidneys* remove wastes from the blood, excreting them in urine, which travels down the *ureters* to the *urinary bladder.* The bladder stores urine until there is an opportunity to eliminate it through the *urethra.* The urethra may be a portal for the entrance of microorganisms, especially in females because a female urethra is shorter (about 4 cm long, exiting near the opening of the vagina) than a male urethra (about 20 cm long, exiting at the tip of the penis). The urethra in males also functions as part of the reproductive system.

Each bean-shaped kidney has a mass of about 150 g and is roughly 11 cm × 6 cm × 3 cm in size—approximately the mass and size of a double deck of playing cards. A frontal section through a kidney reveals three distinct regions **(FIGURE 24.1b)**. A tough fibrous *renal*[1] *capsule* covers the outer surface. Beneath the capsule are an outer light-colored *renal cortex* and an inner, darker *renal medulla.* Conical structures called *renal pyramids* make up most of the renal medulla. The apices of the pyramids point into a hollow, flattened, funnel-like *pelvis,* which collects urine and empties it into the ureter. The ureter exits from the concave, medial surface of the kidney. Blood vessels also enter and exit the kidney at this point.

Each renal pyramid contains *nephrons,*[2] which are the actual functional units of the kidneys; they filter blood to form urine. A kidney has about 1.25 million nephrons, each composed of a ball of blood capillaries, called a *glomerulus,* and a tube with three distinct regions. A *glomerular (Bowman's) capsule* surrounds the glomerulus. Near the glomerulus, the tube loops back and forth and is called the *proximal convoluted tubule.* The tube then descends into the medulla as a U-shaped *nephron loop* (*loop of Henle,*[3] hen-lē´) and returns as the *distal convoluted tubule* to touch the glomerular capsule. The distal convoluted tubule empties urine into another tube called a *collecting duct.* The combined length of all the tubules in a single kidney is about 145 km (90 miles)!

Blood flows into the glomeruli through *afferent arterioles* and out through *efferent arterioles;* other arterioles twist around the convoluted tubules and the loops of Henle. Nephrons filter and cleanse blood flowing through the capillaries, removing wastes to produce urine, which flows into the thousands of collecting ducts and from there via the renal pelvis into the ureter.

[1]From Latin *ren,* meaning "kidney."

[2]From Greek *nephros,* meaning "kidney."

[3]Named for German anatomist Friedrich Henle.

Structures of the Reproductive Systems

LEARNING OUTCOME

24.2 Describe the anatomies of the male and female reproductive systems.

The female reproductive system consists of two *ovaries,* two *uterine tubes* (Fallopian tubes), a *uterus* (womb), a mucous-membrane-lined *vagina* (birth canal), and the external genitalia, which include the *clitoris,* two sets of *labia* (Latin, meaning "lips"), and the opening of the vagina **(FIGURE 24.1c)**. The ovaries produce haploid ova (eggs) before birth and typically release one ovum (egg) per month beginning at puberty. Cilia lining the uterine tubes sweep ova toward the uterus, which develops a blood-rich, thickened wall in anticipation of pregnancy.

When a sperm cell fuses with an ovum—a process called *fertilization*—the resulting *zygote* develops into an *embryo* that implants itself into the uterine wall, forms a *placenta,* and develops into a *fetus.* The placenta is a structure that places fetal blood vessels in close proximity to the mother's blood vessels, allowing the fetus to absorb nutrients and oxygen and eliminate wastes; the two bloodstreams never mix. A fully developed fetus passes through the *cervix* (neck of the uterus) into the birth canal (vagina) to be born.

When an ovum remains unfertilized, it and the lining of the uterus degenerate and pass out the vagina during *menstruation* in a monthly cycle. Ovarian hormones control the cyclic release of ova and the development and sloughing of the uterine lining.

Microorganisms can invade the female reproductive tract via the moist mucous membrane of the vagina, especially during sexual intercourse; however, the normal microbiota help maintain a vaginal pH of about 4.5, inhibiting the growth of pathogens.

Male reproductive anatomy consists of two *testes* (testicles) located in an external pouch called the *scrotum,* a system of *ducts, accessory glands,* and the *penis* **(FIGURE 24.1d)**. Beginning at puberty, each testis produces millions of sperm cells that mature and are stored in an *epididymis*—a coiled tube about 7 m long located above and behind each testis. Sperm pass from the epididymis via a tube, the *ductus deferens,* which joins the urethra near its exit from the urinary bladder inside the *prostate gland*—one of the male accessory glands. The prostate gland and other accessory glands add fluids to the sperm, forming *semen,* which passes from the penis. A hood of skin, the *foreskin,*

▶ **FIGURE 24.1 Urinary and reproductive systems. (a)** Structures of the urinary and reproductive systems. **(b)** Anatomy of a kidney. Each kidney contains about 1.25 million nephrons (filtration units), which are each composed of blood vessels, a glomerulus, and tubules. **(c)** Sectioned view of female genital and urinary tracts. **(d)** Sectioned view of male genitourinary tract.

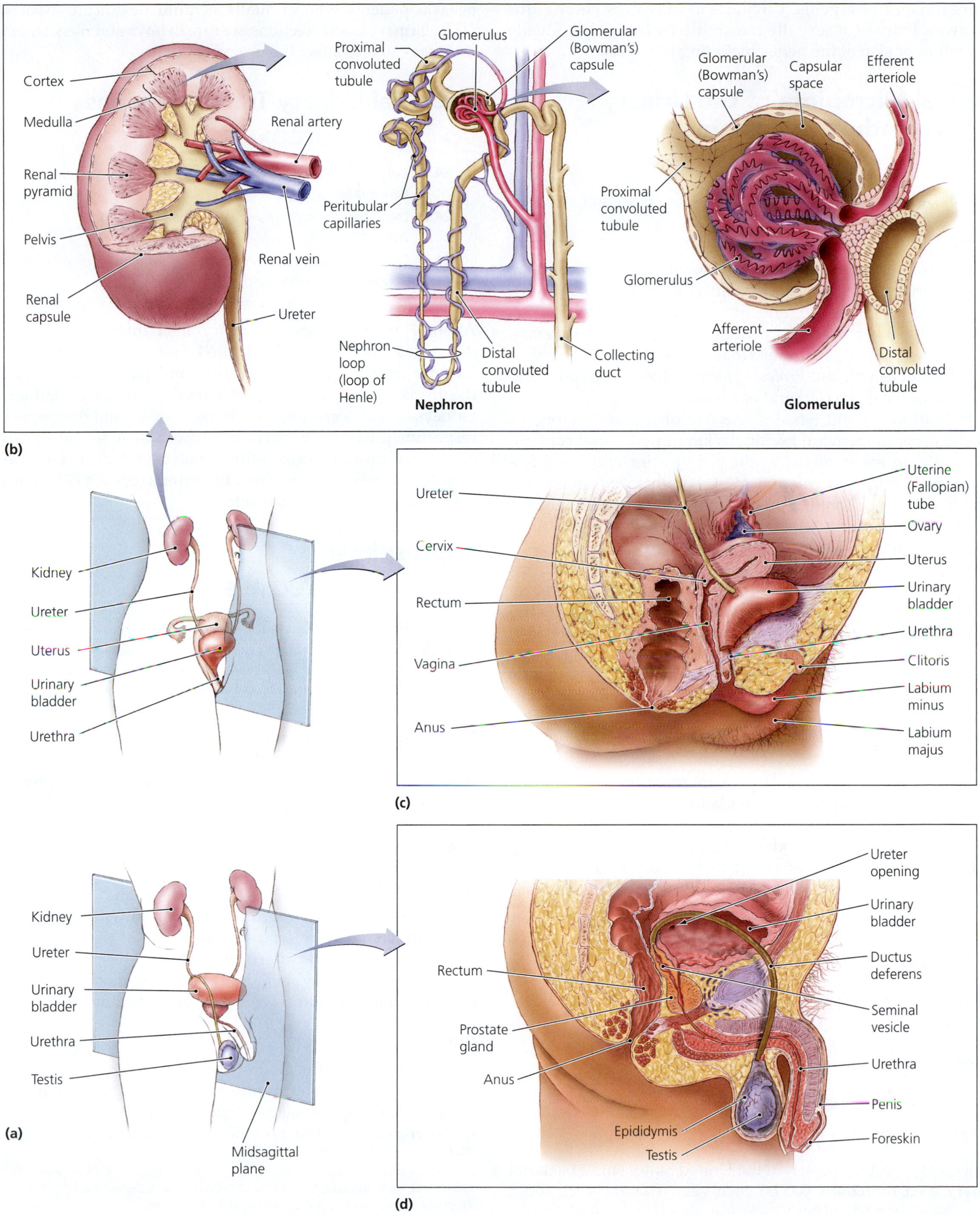
Cortex
Medulla
Renal pyramid
Pelvis
Renal capsule
Renal artery
Renal vein
Ureter
Proximal convoluted tubule
Glomerulus
Glomerular (Bowman's) capsule
Peritubular capillaries
Nephron loop (loop of Henle)
Distal convoluted tubule
Collecting duct
Nephron
Glomerular (Bowman's) capsule
Capsular space
Efferent arteriole
Proximal convoluted tubule
Glomerulus
Afferent arteriole
Distal convoluted tubule
Glomerulus
(b)
Kidney
Ureter
Uterus
Urinary bladder
Urethra
Ureter
Cervix
Rectum
Vagina
Anus
Uterine (Fallopian) tube
Ovary
Uterus
Urinary bladder
Urethra
Clitoris
Labium minus
Labium majus
(c)
Kidney
Ureter
Urinary bladder
Urethra
Testis
(a)
Midsagittal plane
Rectum
Prostate gland
Anus
Epididymis
Testis
Ureter opening
Urinary bladder
Ductus deferens
Seminal vesicle
Urethra
Penis
Foreskin
(d)

covers the end of a penis. Circumcision involves cutting this skin away. Urinary or sexually transmitted microbes can invade the urethra or skin of the penis, including the foreskin.

Normal Microbiota of the Urinary and Reproductive Systems

LEARNING | OUTCOME

24.3 Describe the normal microbiota of the urinary and reproductive systems.

The urethra normally supports the growth of some microbiota, chiefly avirulent species of *Lactobacillus* (lak-tō-bă-sil´ŭs), *Staphylococcus* (staf´i-lō-kok´us), and *Streptococcus* (strep-tō-kok´ŭs). Occasionally other bacteria such as species of *Mycobacterium* (mī´kō-bak-tēr´ē-ŭm), *Bacteroides* (bac-ter-oy´dēz), *Fusobacterium* (fū-sō-bak-tēr´ē-ŭm), and *Peptostreptococcus* (pep´tō-strep-tō-kok´ŭs) colonize the distal (far) end of a urethra.

In both males and females, the rest of the urinary organs and the urine in them are axenic (lacking in microbial contaminants) due to the normally acidic pH of urine and the flushing action of urination. Microbiota of a urethra do contaminate urine during urination; for this reason, normally voided urine contains some bacteria, whereas urine collected directly from a urinary bladder via a catheter is typically sterile.

A vagina is home to a wide assortment of microorganisms that vary with the levels of hormones, particularly estrogen. When estrogen levels rise, such as occurs at puberty, cells lining a vagina produce *glycogen,* a polysaccharide that lactobacilli convert into lactic acid. Acidity inhibits the growth of many opportunistic pathogens. Thus, prepubescent girls, who have little circulating estrogen, are more susceptible to vaginal infections. Likewise, as estrogen levels fall and rise during the menstrual cycle, some women cycle between periods of infection and periods of health.

In both males and females, microorganisms infecting the urethra can rarely move into the bladder, up the ureters, and infect the kidneys. Opportunistic and sexually transmitted microbes also infect the reproductive systems. In the following sections, we examine diseases of the urinary and reproductive systems, beginning with bacterial diseases of the urinary system.

TELL ME WHY

Why are newborn girls less likely to contract vaginal infections than are three-year-olds?

Bacterial Diseases of the Urinary System

Millions of girls and women in the United States suffer bacterial **urinary tract infections (UTIs)** each year. This includes about 600,000 patients who annually acquire healthcare associated UTIs. Urinary tract infections are rare in boys and men. Systemic diseases can also affect the urinary system.

Bacterial Urinary Tract Infections

LEARNING | OUTCOMES

24.4 List four bacteria that can infect the urinary tract and cause disease.

24.5 Describe the features of urethritis, cystitis, and pyelonephritis.

24.6 Describe five ways women can decrease the chance of acquiring a urinary tract infection.

About 7 million cases of urinary tract infections (UTIs) are diagnosed annually in the United States, mostly in women. Bacteria may trigger inflammation and pain in any or all of the urinary tract, including the urethra, urinary bladder, or kidneys—conditions called *urethritis, cystitis,*[4] and *pyelonephritis,*[5] respectively. Urinary bacteria can infect a male sexual accessory organ, the prostate—a condition called *prostatitis.* **Disease in Depth: Bacterial Urinary Tract Infections** (pp. 758–759) considers these infections in more detail.

Leptospirosis

LEARNING | OUTCOMES

24.7 Define *zoonosis.*

24.8 Describe the features of leptospirosis, including its cause, manifestations, pathogenesis, epidemiology, diagnosis, treatment, and prevention.

Not all infections of the urinary system originate from fecal contamination. **Leptospirosis** (lep´tō-spī-rō´sis) is a **zoonosis** (zō-ō-nō´sis)—a disease primarily seen in animals that spreads to humans. The causative agent enters the body through breaks in the skin or mucous membranes and spreads to the urinary system from the blood.

Signs and Symptoms

An abrupt fever, myalgia (muscle pain), muscle stiffness, and headache characterize leptospirosis. Half of patients develop nausea, vomiting, and diarrhea; one-third have a dry cough. Leptospirosis is rarely fatal, but when it is, mortality is due to kidney and liver failure, meningitis, or respiratory distress.

Pathogen

Over 200 strains of a Gram-negative spirochete cause leptospirosis. Taxonomists currently hypothesize that all the strains belong to a single species—*Leptospira interrogans* (lep´tō-spī´ră in-ter´ră-ganz). The specific epithet *interrogans* alludes to

[4]From Greek *kystis,* meaning "bladder," and *itis,* meaning "inflammation."
[5]From Greek *pyel,* meaning "vat" (referring to the renal pelvis).

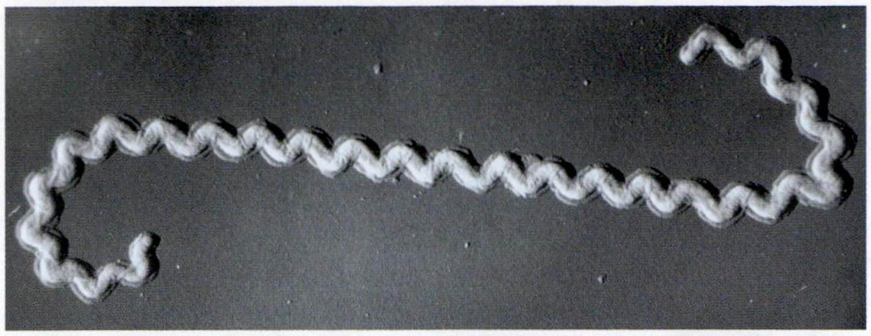

▲ **FIGURE 24.2** ***Leptospira interrogans.*** This spirochete, which causes leptospirosis, appears hooked like a question mark.

the fact that the end of the spirochete is hooked in a manner reminiscent of a question mark **(FIGURE 24.2)**. This thin (0.1 µm diameter) aerobic pathogen is highly motile by means of two axial filaments, each of which is anchored at one end. Clinicians and researchers grow *Leptospira* on special media enriched with bovine serum albumin or rabbit serum.

Virulent strains of *Leptospira* make adhesins that attach to human cells, are motile, are chemotaxic toward hemoglobin, and are evidently able to evade antibody-complement activity. The exact nature and actions of these virulence factors require further study.

L. interrogans normally lives in many wild and domestic animals—in particular, rats, raccoons, foxes, dogs, horses, cattle, and pigs—in which it grows asymptomatically in the kidney tubules. The spirochete can also survive in streams, rivers, and lakes.

Pathogenesis

Humans develop leptospirosis 2–26 days following direct contact with the urine of infected animals or with the spirochetes in animal-urine-contaminated streams, lakes, or moist soil, all environments in which the organisms can remain viable for six weeks or more. Person-to-person spread has not been observed.

Because *Leptospira* is thin and highly motile, it can penetrate intact mucous membranes or enter through invisible cuts and scrapes in the skin. It corkscrews its way through these tissues and then travels via the bloodstream throughout the body, including the central nervous system, damaging cells lining small blood vessels and triggering fever and the other signs and symptoms. Eventually, the bacteremia resolves, and the spirochetes live only in the kidneys. As the disease progresses, the spirochetes are excreted in urine.

Epidemiology

Leptospirosis occurs throughout the world. Farmers, ranchers, veterinarians, butchers, and people whose recreation takes them to potentially contaminated water are most at risk. National reporting in the United States ceased in part because the disease is rare in the U.S.; a total of only 89 cases occurred in 1993 and 1994, the last years that the disease was nationally reportable.

Diagnosis, Treatment, and Prevention

Leptospira does not stain well with Gram stain; therefore, a specific antibody test revealing the presence of the spirochete in blood or urine specimens is the preferred method of diagnosis. Intravenous penicillin G treats severe infections; oral doxycycline, ampicillin, or amoxicillin are drugs of choice for less severe cases. The most effective way to limit the spread of *Leptospira* is to refrain from wading in, swimming in, or consuming water contaminated with animal urine. Rodent control is also important, but eradication is impractical because the spirochete has many animal reservoirs. An effective vaccine is available for livestock and pets.

Disease at a Glance 24.1 on p. 760 summarizes the features of leptospirosis.

Streptococcal Acute Glomerulonephritis

LEARNING **OUTCOME**

24.9 Describe the manifestations and cause of streptococcal acute glomerulonephritis.

For an undetermined reason, the body does not remove from circulation antibodies bound to the antigens of some strains of group A *Streptococcus* (see p. 564). Instead, antibody-antigen complexes accumulate in the glomeruli of the kidneys, triggering inflammation of the glomeruli and nephrons, a condition called **glomerulonephritis** (glō-mār´yū-lō-nef-rī´tis). This obstructs blood flow through the kidneys and leads to hypertension (high blood pressure) and low urine output. The patient's urine often contains blood and proteins. Young patients usually recover fully from glomerulonephritis, but progressive and irreversible kidney damage may occur in adults.

We have examined bacterial diseases of the urinary system. In the following section, we examine diseases of the reproductive systems that are not regularly transmitted sexually, that is, *nonvenereal*[6] diseases.

TELL ME WHY

Why does insertion of a urinary catheter increase the likelihood of cystitis?

Nonvenereal Diseases of the Reproductive Systems

In previous sections, we considered two nonvenereal diseases of the urinary systems of males and females. Now, we consider three nonvenereal diseases of bacteria and a fungus in females.

[6]*Venereal* comes from the name Venus, the Roman goddess of sexual love.

BACTERIAL URINARY TRACT INFECTIONS

Escherichia coli and other bacteria

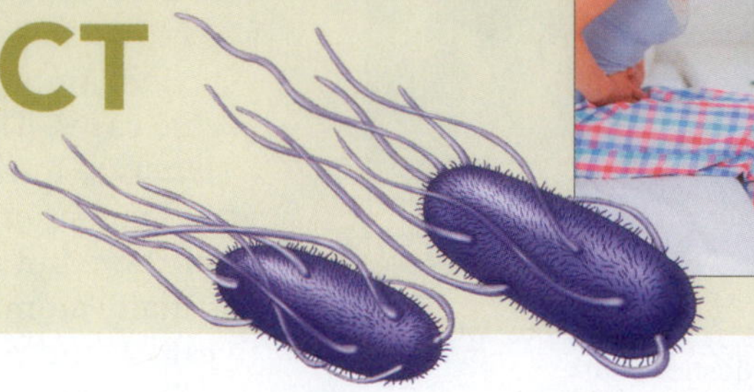

SIGNS AND SYMPTOMS

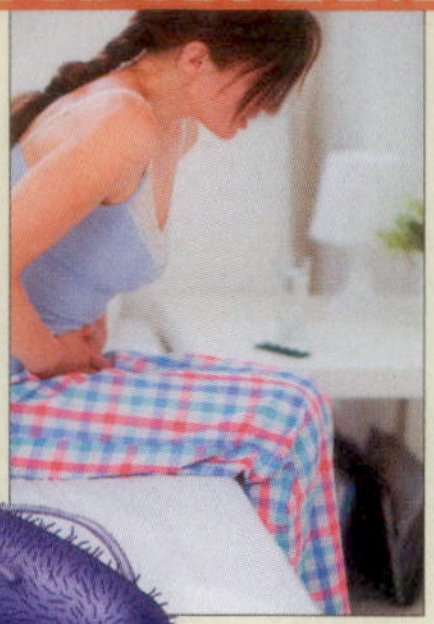

Patients with mild urethritis, cystitis, or prostatitis may have only a slight fever or no symptoms, but most of these diseases involve frequent, urgent, and painful urination—a condition called dysuria. The urine may be cloudy and bloody and have a strong, foul odor. Mental confusion, which results when the bacteria spread from the urinary system to the blood (bacteremia), is typically seen in the elderly.

Millions of people in the United States, mostly girls and women, suffer bacterial urinary tract infections (UTIs) each year. This includes about 600,000 patients who annually acquire healthcare associated UTIs. Urinary tract infections are rare in boys and men. Bacteria may trigger inflammation and pain in any or all of the urinary tract, including the urethra, urinary bladder, or kidneys—conditions called urethritis, cystitis, and pyelonephritis.

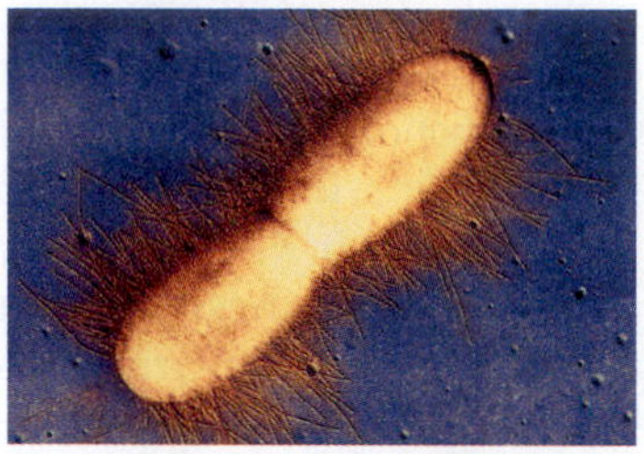

E. coli in early stage of binary fission. TEM 1 μm

PATHOGENESIS

In the Urethra (Urethritis)

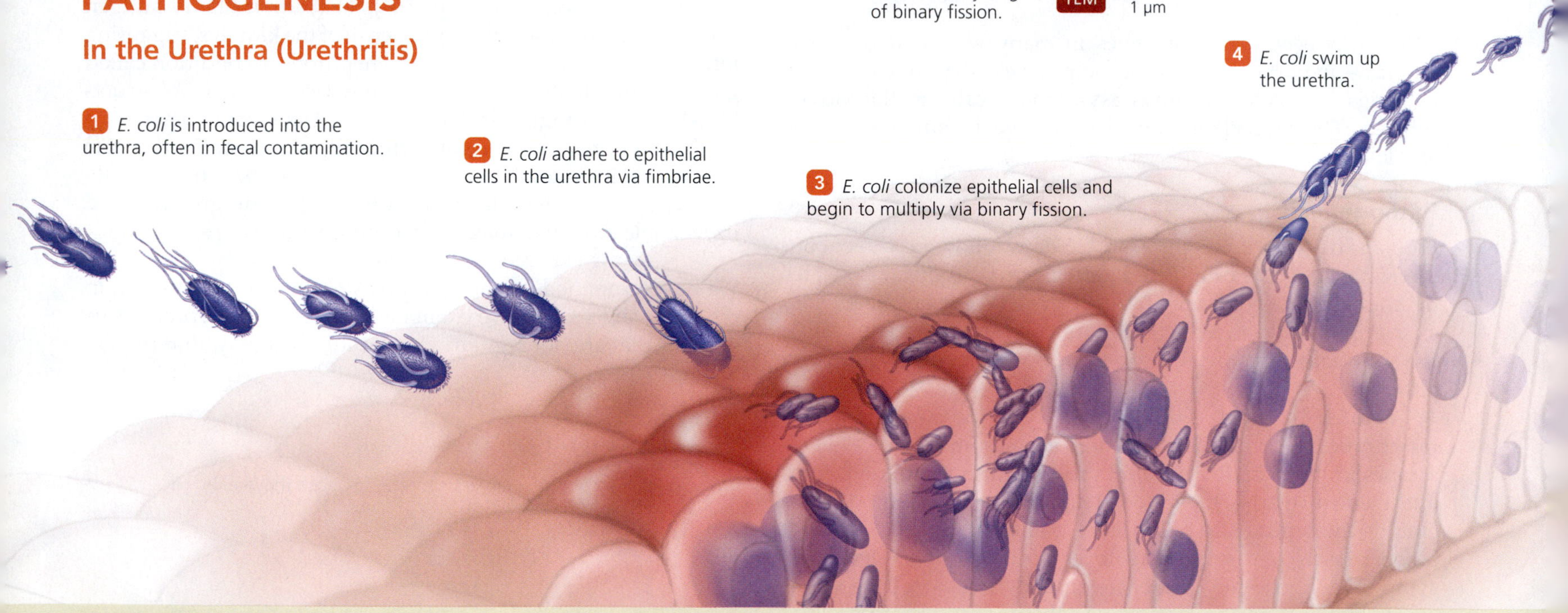
1 *E. coli* is introduced into the urethra, often in fecal contamination.
2 *E. coli* adhere to epithelial cells in the urethra via fimbriae.
3 *E. coli* colonize epithelial cells and begin to multiply via binary fission.
4 *E. coli* swim up the urethra.

INVESTIGATE IT!

Why are UTIs the most common of healthcare associated infections?

Scan this code to visit the World Health Organization website to investigate UTIs. Then go to MasteringMicrobiology to record your research findings.

EPIDEMIOLOGY

UTIs are more common in females than males because female urethras are shorter and closer to the anus. Diabetics, nursing home patients, elderly men who have trouble emptying their bladders because of prostate enlargement, patients with urinary catheters, women who use diaphragms for birth control, and people who do not drink adequate fluids are also at risk for UTIs.

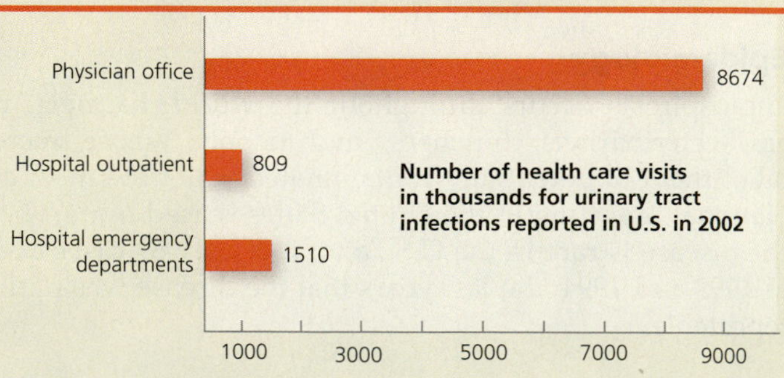
Physician office
8674
Hospital outpatient
809
Hospital emergency departments
1510
Number of health care visits in thousands for urinary tract infections reported in U.S. in 2002
1000
3000
5000
7000
9000

PATHOGEN

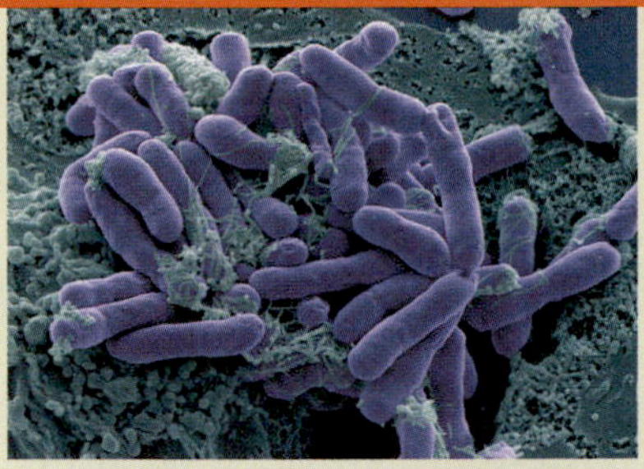

SEM 3 µm

Enteric bacteria—Gram-negative bacteria that are part of the intestinal microbiota—are the most common cause of urinary tract infections. *Escherichia coli* causes about 70% of cases; other bacteria from the intestinal tract, such as *Proteus* and *Klebsiella,* cause about 10% of UTIs. Nonintestinal bacteria, such as *Pseudomonas* and *Staphylococcus,* occasionally cause UTIs.

VIRULENCE FACTORS

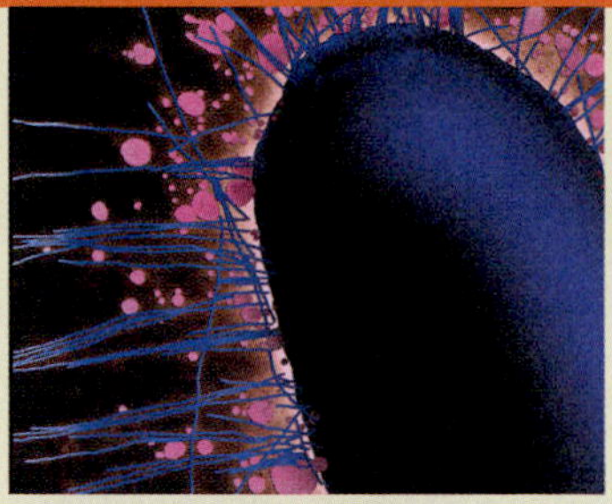

TEM 1 µm

E. coli, *Proteus*, and *Pseudomonas* have flagella that can propel the bacteria up the urinary tract. Strains of *E. coli* that infect the bladder have attachment fimbriae that bind specifically to epithelial cells lining the urinary bladder. Using fimbriae, these strains of *E. coli* move into bladder cells by an unknown mechanism and form biofilm-like aggregations within bladder cells' cytosol and avoid the body's defenses.

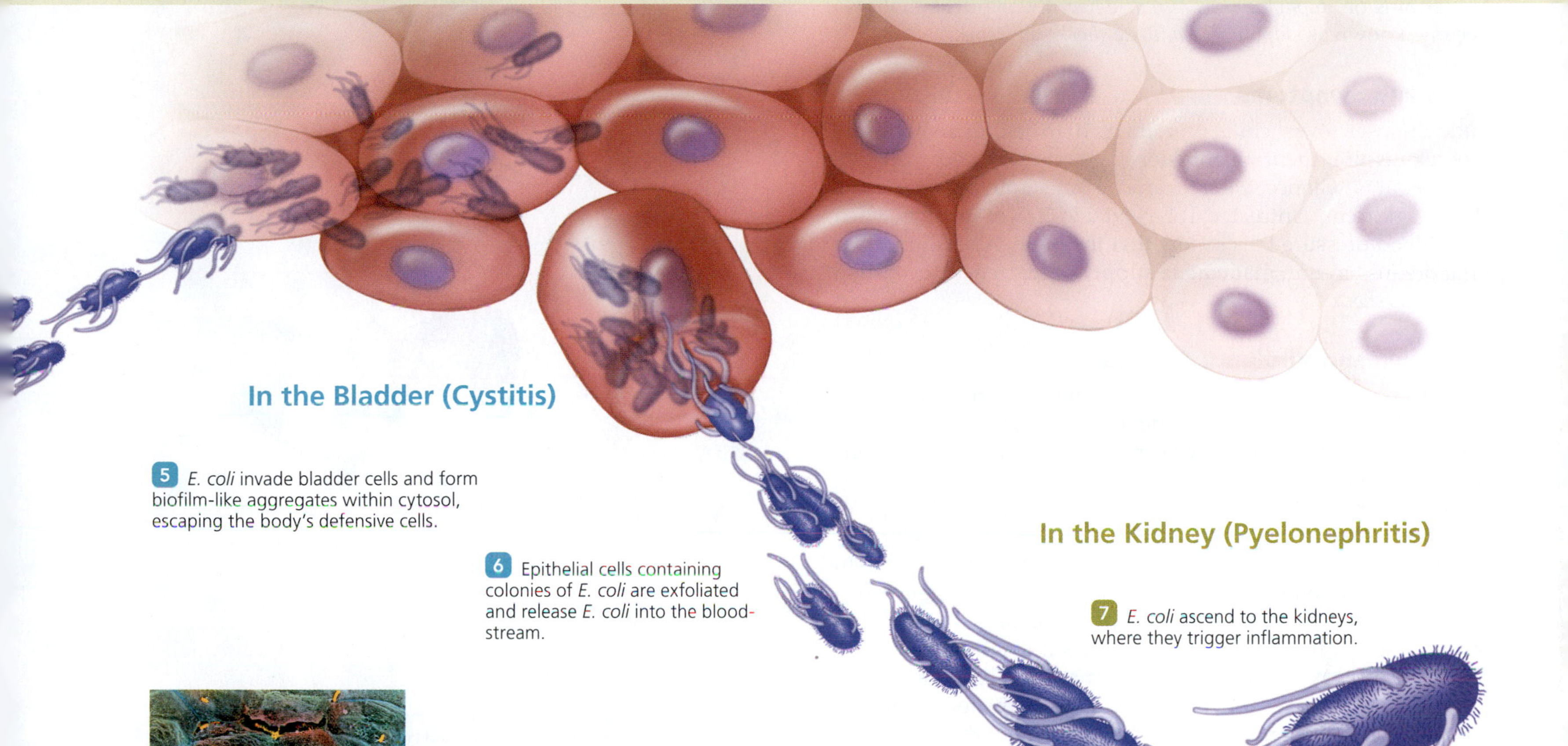

In the Bladder (Cystitis)

5 *E. coli* invade bladder cells and form biofilm-like aggregates within cytosol, escaping the body's defensive cells.

6 Epithelial cells containing colonies of *E. coli* are exfoliated and release *E. coli* into the bloodstream.

In the Kidney (Pyelonephritis)

7 *E. coli* ascend to the kidneys, where they trigger inflammation.

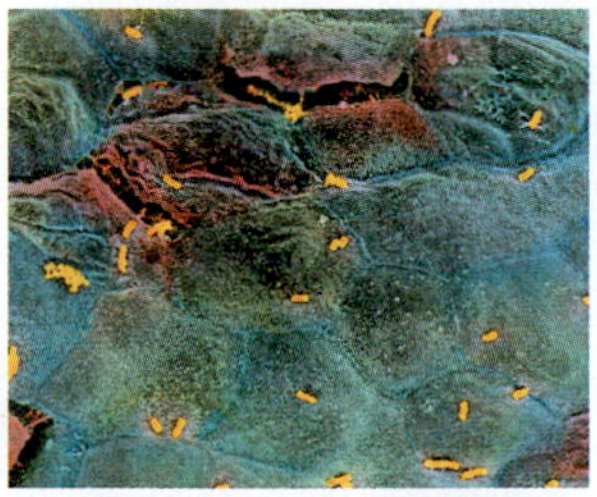

E. coli infection of bladder cells. SEM 10 µm

DIAGNOSIS

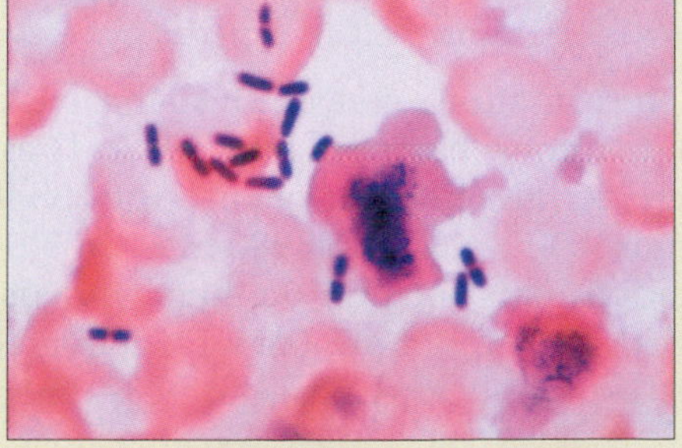

LM 4 µm

Urinalysis of patients with the signs and symptoms of urinary tract infections reveals leukocytes, erythrocytes, and the causative bacterium, usually *E. coli.*

TREATMENT AND PREVENTION

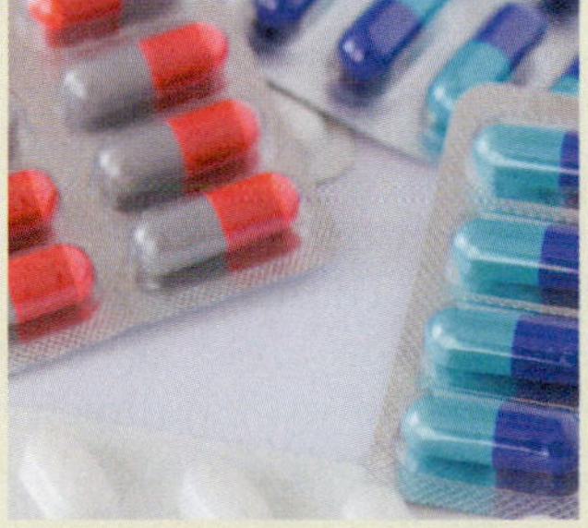

Mild UTIs resolve on their own without treatment, though antimicrobial drugs, such as cephalosporins, sulfonamides, and semisynthetic penicillins, can prevent the spread of infection to the kidneys and blood. Wiping from front to back after defecation to avoid dragging bacteria from the anus into the urethra is the most important step females can take to prevent UTIs.

Staphylococcal Toxic Shock Syndrome

LEARNING OUTCOME

24.10 Describe the features of staphylococcal toxic shock syndrome.

Physicians first described **toxic shock syndrome (TSS)** in the late 1920s, but in the 1980s the condition became epidemic in the United States among menstruating women who used superabsorbent tampons. Once epidemiologists determined that certain strains of *Staphylococcus aureus* cause the syndrome, it also became known as *staphylococcal toxic shock syndrome.*

Signs and Symptoms

Sudden-onset fever, chills, vomiting, diarrhea, extremely low blood pressure, mental confusion, and a severe red rash (see Disease at a Glance 24.2) characterize staphylococcal toxic shock syndrome. Untreated TSS is fatal to 50% of patients when their blood pressure falls so low that the brain, heart, and other vital organs have an inadequate supply of oxygen—a condition known as *shock.*

Pathogen and Virulence Factors

Staphylococcus is normally part of the skin and mucous membrane microbiota. Strains of *S. aureus* that cause staphylococcal toxic shock syndrome produce certain *exotoxins*—soluble toxins released from cells—called *toxic shock syndrome toxins (TSSTs).* These exotoxins bind simultaneously to major histocompatibility complex II molecules on antigen-presenting cells and T cell receptors on T cells; however, they bind to sites other than the normal antigen-binding sites of these immune system molecules. When a toxin molecule binds two defensive cells together in this fashion, it activates the T cell. As a result of the action of many such toxin molecules, many more T cells are activated than in a cellular immune response. These activated T cells release an overabundance of cytokines that trigger the manifestations of TSS.

TSST-1 causes 75% of cases of staphylococcal toxic shock syndrome. Other TSSTs, which are also *enterotoxins* (exotoxins that cause gastrointestinal distress when ingested), cause 25% of cases.

Pathogenesis and Epidemiology

When strains of *Staphylococcus* that produce toxic shock syndrome toxin grow in the vagina or in a wound, the toxin is absorbed into the blood, triggering TSS. In the investigation of the 1980 epidemic among menstruating women, researchers discovered that *S. aureus* grows exceedingly well on superabsorbent tampons, especially when a blood-soaked tampon remained in place for a prolonged period.

TSS occurs in males and females, but most cases have been identified in menstruating females. The U.S. Food and Drug Administration (FDA) ordered certain types of superabsorbent tampons withdrawn from the market in 1980, directed a reduction in the absorbency of all tampons in 1982, and mandated packaging educational information concerning the risks of TSS in every box of tampons beginning in 1982. These government directives reduced the number of cases of TSS in the United States from almost 1200 cases in 1980 to 61 in 2012 **(FIGURE 24.3)**. At-risk individuals include women who use tampons, vaginal sponges, or diaphragms; newly delivered mothers; surgery patients, especially after nasal surgery where

DISEASE AT A GLANCE 24.1

Leptospirosis

1. *Leptospira* in urine-contaminated water enters the body through mucous membranes or skin abrasion.
2. The spirochetes infect the blood (bacteremia).
3. *Leptospira* infects liver, central nervous system, kidneys, and other organs.
4. In most patients, the infection becomes localized in the kidneys, which can be severely or fatally damaged.
5. Patient sheds *Leptospira* in urine.

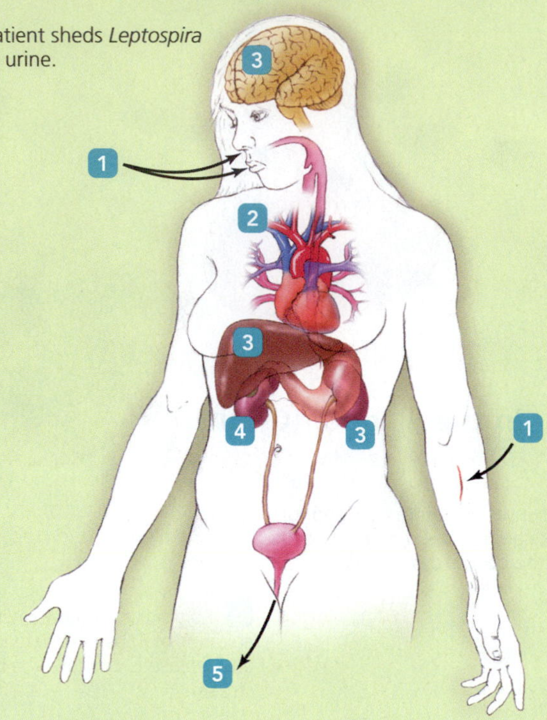

Cause *Leptospira interrogans* (aerobic spirochete).
Virulence factors Adhesins, motility, chemotaxis toward hemoglobin, ability to evade actions of antibodies and complement.
Portal of entry Contact with urine of infected animals or contaminated water or soil.
Signs and symptoms Myalgia, headache, abdominal pain, nausea, vomiting, and fever.
Incubation period Two days to four weeks.
Susceptibility All humans are susceptible regardless of age or sex.
Treatment Oral doxycycline, chloramphenicol, erythromycin, or intravenous ampicillin for more severe cases.
Prevention Use rodent controls and avoid water contaminated with animal urine. A vaccine is available for pets and livestock.

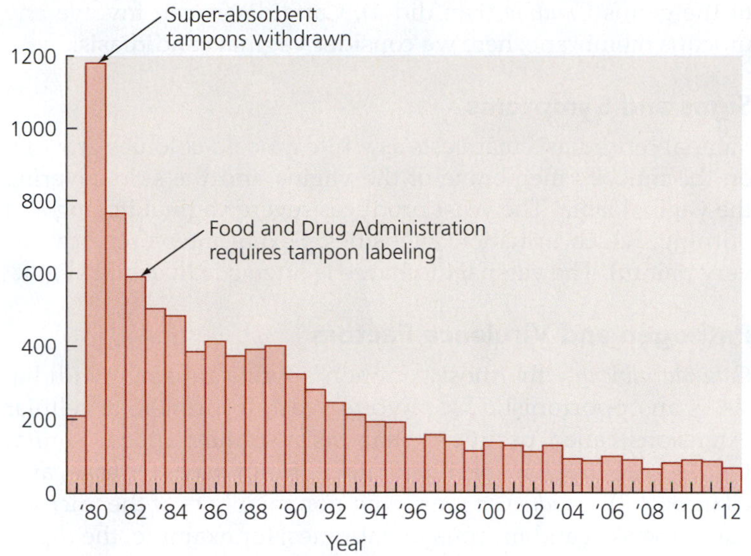

▲ **FIGURE 24.3 The incidence of staphylococcal toxic shock syndrome in the United States, 1979–2012.**

packing is employed to reduce bleeding; and anyone with an infection of *S. aureus*.

Diagnosis, Treatment, and Prevention

Physicians typically diagnose TSS based upon its signs and symptoms, particularly in menstruating women. In some cases, cultures of blood samples grow colonies of *S. aureus*.

TSS is a medical emergency; patients experiencing a sudden onset of fever or rash, particularly during menstruation or following surgery, should seek immediate treatment. Treatment involves removal of foreign material (tampon, sponge, diaphragm, nasal packing) and drainage of infected wounds. Supportive measures include intravenous fluid to support falling blood pressure and dialysis when the kidneys are affected. Physicians administer nafcillin, oxacillin, cephalosporin, or vancomycin to reduce the number of bacterial cells and prescribe anti-TSST immunoglobulin to counteract the toxin.

Women can reduce the risk of TSS by avoiding the use of tampons, vaginal sponges, and birth control diaphragms. Using less absorbent tampons and changing tampons frequently are also beneficial.

Disease at a Glance 24.2 summarizes the features of staphylococcal toxic shock syndrome.

Bacterial Vaginosis

LEARNING **OUTCOME**

24.11 Describe the characteristics of bacterial vaginosis.

Bacteria can infect the warm, moist vaginal lining, causing bacterial **vaginosis**. The condition does not involve inflammation; therefore, the disease is "vaginosis" rather than "vaginitis."

DISEASE AT A GLANCE 24.2

Toxic Shock Syndrome

Cause Exotoxin-producing strains of *Staphylococcus aureus* (Gram-positive coccus).

Virulence factors Exotoxins, enterotoxins.

Portal of entry *Staphylococcus* either grows in the vagina or enters the body through a wound, grows, and produces toxin that enters the bloodstream.

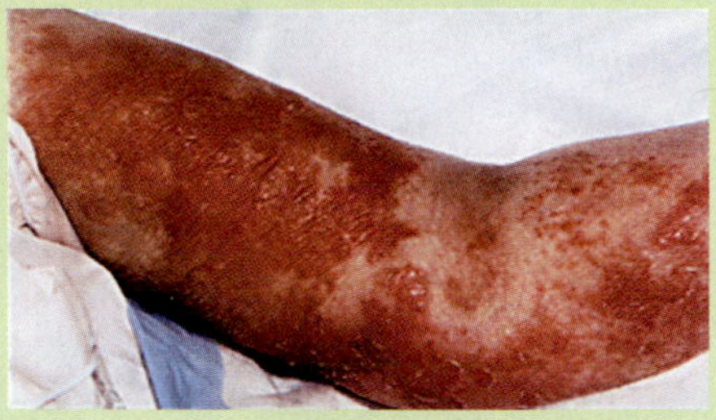

Red rash of staphylococcal toxic shock syndrome

Signs and symptoms Sudden onset of high fever, vomiting, rash, extremely low blood pressure, and sore throat.

Incubation period Two to three days.

Susceptibility Menstruating women who use highly absorbent tampons for extended periods, newly delivered mothers, and surgery patients are at higher risk.

Treatment Supportive therapy is extremely important; administration of vancomycin and antistaphylococcal immunoglobulin.

Prevention Avoid vaginal insertions such as highly absorbent tampons, vaginal sponges, or diaphragms, or use them intermittently and for shorter periods.

Signs and Symptoms

A homogenous white vaginal discharge with a "fishy" odor characterizes bacterial vaginosis. Some itching and irritation of the vaginal opening may also occur. Up to 50% of women with bacterial vaginosis report no symptoms.

Pathogens

Bacterial vaginosis results when the normal lactobacilli of the vagina are replaced with a large number of facultatively or obligate anaerobic bacteria such as Gram-positive *Gardnerella vaginalis* (gărd´ner-el´ă va-ji-nă´lis) and *Mycoplasma hominis* (mī´kō-plaz-mă ho´mi-nis). Scientists do not understand the underlying causes of this change in microbiota.

Pathogenesis and Epidemiology

Scientists do not know the exact cause or causes of bacterial vaginosis. However, a decline in the number of lactobacilli populating the vagina results in a pH in the vagina higher than the normal 4.5. This either promotes or allows the growth of the bacteria associated with bacterial vaginosis. Bacterial vaginosis is associated with having multiple sexual partners and vaginal douching.

Diagnosis, Treatment, and Prevention

Physicians diagnose bacterial vaginosis by its signs, including odor, discharge characteristics, and vaginal pH greater than 4.5. The presence of so-called clue cells, which are vaginal epithelial

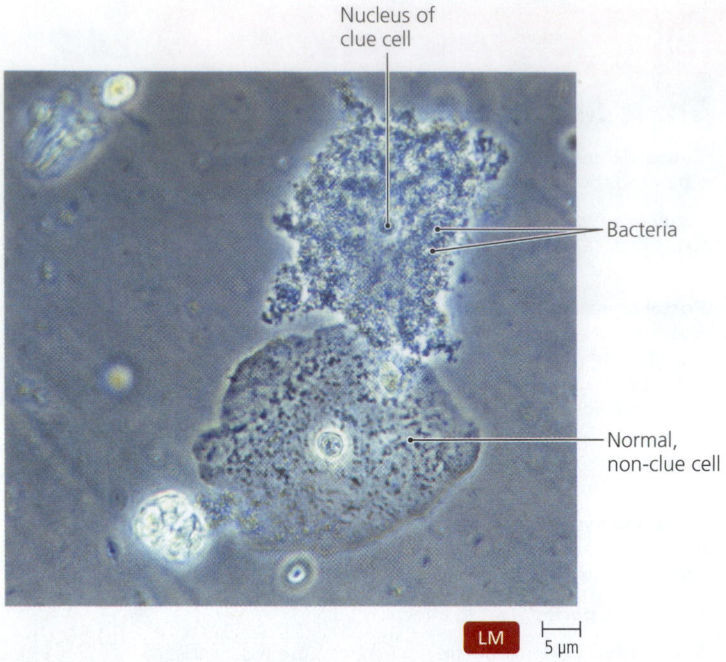

▲ **FIGURE 24.4 Clue cell.** Clue cells are vaginal epithelial cells covered with bacteria that suggest a diagnosis of bacterial vaginosis. *Name one bacterial species that can be found on a clue cell.*

Figure 24.4 *Gardnerella vaginalis is frequently the bacterium covering clue cells.*

cells completely covered with bacteria, supports the diagnosis of bacterial vaginosis **(FIGURE 24.4)**.

The U.S. Centers for Disease Control and Prevention (CDC) recommends treatment with oral or vaginal metronidazole or clindamycin. Researchers are studying the possibility of reestablishing normal vaginal microbiota and pH by using lactobacilli vaginal suppositories or by eating live cultures of *Lactobacillus.*

Since physicians and researchers do not fully understand the development of bacterial vaginosis, there are no absolute preventive measures, but sexual abstinence and refraining from douching are beneficial.

Bacteria are not the only nonvenereal pathogens of the reproductive system; fungi, particularly yeasts, can also infect reproductive organs. In the next section, we examine the most common fungal infection of reproductive tracts.

Vaginal Candidiasis

LEARNING | **OUTCOMES**

24.12 Describe *Candida albicans.*

24.13 Describe the features of vaginal candidiasis.

24.14 Explain why women taking antibacterial drugs are at risk for vaginal candidiasis.

Candidiasis (kan-di-dī´ă-sis) describes any of a variety of opportunistic yeast infections and diseases caused by various species of the genus *Candida* (kan´did-ă). Candidiasis can involve any mucous membrane; here we consider vaginal candidiasis.

Signs and Symptoms

Vaginal candidiasis manifests as white mucoid colonies growing on the mucous membrane of the vagina and the skin covering the vaginal labia. The yeast produces severe vaginal itching and burning, which urination intensifies. Sexual intercourse can be very painful. The vaginal discharge is often curdlike and slight.

Pathogen and Virulence Factors

Candida albicans, the most common species causing candidiasis, is an opportunistic ascomycete yeast. It forms long cellular extensions called **pseudohyphae** because they appear similar to the true hyphae of filamentous fungi (see Disease at a Glance 24.3). *Candida* spp. are common members of the microbiota of the skin and mucous membranes; for example, the digestive and reproductive tracts of 40–80% of all healthy individuals harbor the yeasts.

Pathogenesis and Epidemiology

C. albicans normally lives in the vagina in competition with lactobacilli and other bacteria. If the vaginal pH becomes more alkaline than usual or if normal bacterial populations are reduced by antibiotics, *Candida* can multiply rapidly, triggering inflammation and other manifestations of candidiasis. In immunocompromised individuals only, especially AIDS patients, *Candida* can become systemic.

The CDC estimates that 75% of women will experience candidiasis at least once; almost 50% will have two or more episodes. *Candida* is one of the few fungi that can be transmitted between individuals. From its site as a normal inhabitant of the female reproductive tract, for example, it can pass to babies during childbirth and, rarely, to males during sexual contact.

Predisposing factors for the development of candidiasis include cancer, invasive hospital procedures, antibacterial treatments (which inhibit normal bacteria that compete with *Candida*), diabetes, severe burns, intravenous drug abuse, and AIDS—nearly 100% of AIDS patients will develop candidiasis in their reproductive or digestive tracts.

Diagnosis, Treatment, and Prevention

Stained preparations of vaginal discharge reveal clusters of budding yeast and branching pseudohyphae, which are diagnostic in conjunction with symptoms.

Physicians prescribe topical azole cream or oral fluconazole to treat vaginal candidiasis. Azole creams are oil based; consequently, they may weaken latex condoms. Prevention involves maintaining the normal vaginal microbiota by avoiding excessive use of antibacterial drugs. Vaginal candidiasis is acquired rarely via sexual contact; therefore, treatment of sexual partners is not usually recommended.

Disease at a Glance 24.3 discusses other features of candidiasis. **Beneficial Microbes: Pharmacists of the Future?** examines a novel way to deliver antimicrobial drugs directly to the vagina to prevent infections by *Candida* and other pathogens.

DISEASE AT A GLANCE 24.3

Candidiasis

Cause *Candida* spp., particularly *C. albicans* (opportunistic fungus).

Virulence factors Normally a member of vaginal microbiota, resistant to antibacterial drugs.

Portal of entry Mucous membrane, normally part of microbiota of mouth, digestive tract, and vagina.

Signs and symptoms Oral: white plaques in mouth, tongue, and palate. Cutaneous: macular red rash in skin folds. Genital: white curdlike discharge, burning, redness, and painful intercourse. Invasive: fever, chills, and additional symptoms depending on the organ(s) affected.

Incubation period Generally 7–10 days.

Susceptibility Women taking antibacterial drugs, which inhibit or eliminate normal bacteria that compete with *Candida.* Immunocompromised individuals, especially AIDS patients.

Treatment Antifungal agents such as clotrimazole, miconazole, fluconazole, or terconazole in noninvasive disease; amphotericin B intravenously for invasive cases.

Prevention Avoid persistent moisture in genital area; for example, do not wear wet bathing suits for prolonged periods. Prophylactic treatment in AIDS patients may include oral fluconazole.

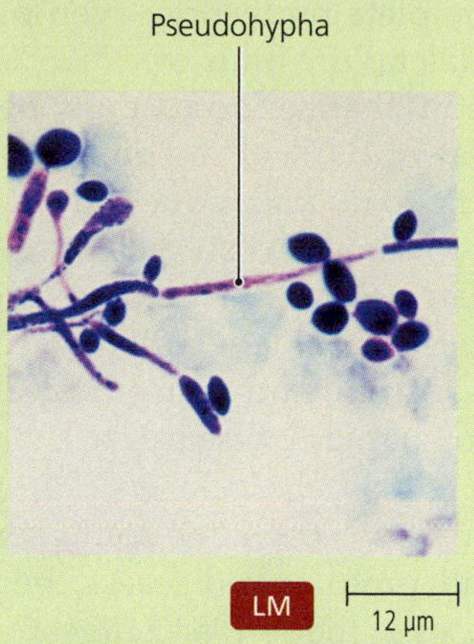

Candida albicans, a common cause of vaginal candidiasis, forms pseudohyphae and Gram stains violet.

TELL ME WHY

Why does *Candida albicans*, which is a member of the normal microbiota, sometimes cause disease?

Sexually Transmitted Infections (STIs) and Diseases (STDs)

LEARNING OUTCOMES

24.15 Discuss the incidence of STDs worldwide and in the United States.

24.16 Explain why adolescents are at greater risk than adults from STDs and the consequences of STDs.

24.17 Discuss prevention of STIs.

24.18 Describe the characteristics of pelvic inflammatory disease.

Sexual activity transmits microbes between partners. When the microbes are potential pathogens, such transmission is known as a **sexually transmitted infection (STI)**. Diseases resulting from STIs are **sexually transmitted diseases (STDs)**, which are also known as *venereal diseases.* A pandemic of STDs has developed over the last 50 years. The World Health Organization (WHO) estimates that 333 million new cases of STDs occur worldwide each year, though 50–90% of STDs worldwide, including half of the STDs in the United States, remain unreported or undiagnosed. Some epidemiologists estimate that sexually transmitted viruses infect one in five Americans,

BENEFICIAL MICROBES

Pharmacists of the Future?

Your microbiota—the trillions of microbes that live in and on your body—play essential roles in maintaining your health. As they grow and reproduce, they utilize nutrients that would otherwise go to pathogens; they excrete vitamins, aid in digestion, and change the pH of their environment, all of which benefit us. Now, scientists are on the verge of enhancing the protective role of the microbiota by giving them genes that code for chemicals that inhibit pathogens without being deleterious to the normal microbiota or to us.

In one experiment, researchers inserted a gene for an antibody against a yeast, *Candida*, into the genome of a benign species of *Streptococcus* bacteria—a member of the microbiota. Mice with candidiasis that received the recombinant *Streptococcus* recovered twice as fast as mice that received fluconazole—a standard antifungal medication. Since *Streptococcus* is normally in the vagina, it can reproduce there, maintaining a constant level of protective chemicals without requiring reinoculation and without harming the mice.

Other researchers have altered strains of *Lactobacillus*—a common probiotic—to express cyanovirin, which is a protein that successfully inhibits HIV. The idea is to colonize the vaginas of women at risk for infection with HIV with this altered but otherwise normal member of the vaginal microbiota. The modified bacterium would act as a potent first line of defense against infection.

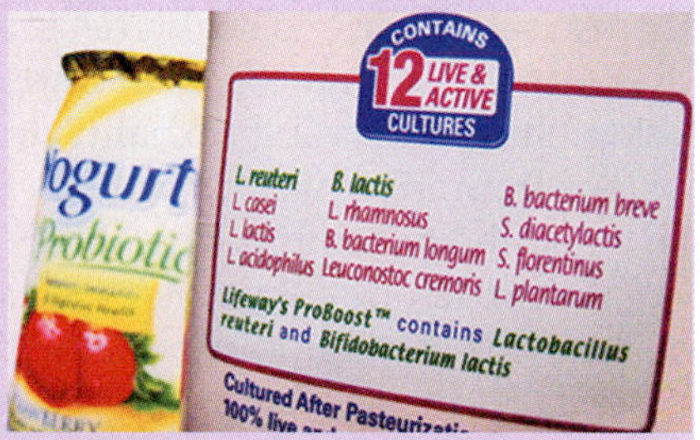

Benefits of using microbiota to deliver antimicrobial drugs are numerous. Such a drug would be relatively less expensive to manufacture—bacteria at the site would be the "factory"—and delivery of the drug would be automatic, not requiring a prescription to be filled, delivered, or administered more than once. Further, the concentration of drug at a site could be maintained without depending on a patient to keep a set schedule.

Perhaps your future pharmacist will be a microbe that has been tailored to deliver antimicrobial chemicals precisely to sites of infection.

DISEASE AT A GLANCE 24.4

Pelvic Inflammatory Disease (PID)

Cause *Neisseria gonorrhoeae* (Gram-negative diplococcus) or *Chlamydia trachomatis* (obligate intracellular bacterium); *Mycoplasma hominis* (pleomorphic bacterium) in rare instances.

Virulence factors
Neisseria—capsule, fimbriae, lipid A; *Chlamydia*—small size, intracellular lifestyle; *Mycoplasma*—small size, adhesins.

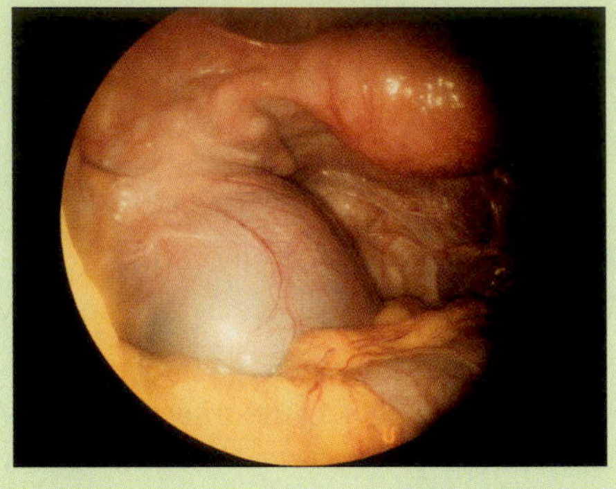

Portal of entry Mucous membrane of vagina, sexual transmission, and then migration deep into the uterus or uterine tubes.

Signs and symptoms Inflammation, fever, abdominal pain. Left untreated, PID can lead to ectopic pregnancies and sterility.

Incubation period Months to years after initial infection.

Susceptibility Women with untreated gonorrhea or chlamydial infections, especially those under age 20.

Treatment Antimicrobial drugs such as ofloxacin, metronidazole, doxycycline, and/or ceftriaxone.

Prevention Abstinence or mutual monogamy with an uninfected partner; early treatment of all STDs. Condoms provide some protection from *Neisseria* infection but may increase chance of chlamydial infections.

including one-fourth of teenaged girls. About half of all STDs affect people under age 25.

Young people often experiment with sex, and many have an it-cannot-happen-to-me attitude, which leads to exposing themselves to serious health risks. Additionally, embarrassment over contracting an STD can impair normal psychological development concerning sexuality. In all patients, the presence of lesions from STDs is a known risk factor for the transmission of HIV and the development of AIDS.

A female adolescent is especially at risk for STDs in that her cervical lining is especially prone to bacterial invasion. A sexually active 15-year-old girl has a 12.5% chance of developing **pelvic inflammatory disease (PID)**, which is a general term for inflammation and pain in the uterus, uterine tubes, or ovaries; at age 24 the probability has decreased to 1.25%. Further, STDs are more likely to be asymptomatic in younger women, so that these patients do not seek treatment and suffer horrendous long-term consequences, including development of birth defects in their babies, *ectopic*[7] *pregnancy* (implantation of a fetus outside the womb), miscarriage, sterility, or the development of cervical cancer. **Disease at a Glance 24.4** examines features of pelvic inflammatory disease.

[7]From Greek *ektos*, meaning "outside," and *topos*, meaning "place."

With the exception of rape, every STI and STD can be prevented by practicing sexual abstinence or completely faithful mutual monogamy. Latex and polyurethane condoms reduce the risk of contracting many STIs, but condoms don't provide complete protection; even properly used condoms have an annual failure rate of 17–25%. Further, epidemiological studies show that condoms must be used properly and consistently—that is, 100% of the time during every sex act—or they provide little to no risk reduction for most STDs.

Numerous studies have shown that circumcision reduces a man's chance of getting several sexually transmitted diseases. These include genital warts, AIDs, syphilis, genital herpes, chancroid, and gonorrhea. Circumcision also reduces transmission of many diseases to the men's sexual partners.

Here, we consider some sexually transmitted bacterial, viral, and protozoan diseases, beginning with bacterial STDs. (Diseases of the skin and digestive system, such as scabies infection, giardiasis, shigellosis, and hepatitis, can be transmitted sexually.) These infections are considered in the chapters covering the skin and digestive system.

TELL ME WHY

Why have STIs and STDs become pandemic over the past 50 years?

Bacterial STDs

Bacterial infections are among the more familiar STDs. Bacteria involved in STDs survive poorly on dry, cool, inanimate objects such as toilet seats; therefore, transmission is via sexual intercourse only. We begin with an examination of gonorrhea.

Gonorrhea

LEARNING **OUTCOMES**

24.19 Describe the origin of the name *gonorrhea*.

24.20 Compare and contrast the manifestations of gonorrhea in men and women.

24.21 Discuss the difficulties researchers face in developing an effective vaccine against *Neisseria gonorrhoeae*.

Physicians have known about the sexually transmitted disease **gonorrhea** (gon-ō-rē´ă) for centuries, although the disease was often confused with syphilis until the 19th century. In the 2nd century A.D., the Roman physician Claudius Galen named gonorrhea for what he thought was its cause—an excess of semen (gonorrhea means "flow of seed" in Greek). The disease is sometimes called "clap" from the archaic French word *clapoir*, meaning brothel.

Signs and Symptoms

In men, gonorrhea is usually insufferably symptomatic—acute inflammation typically occurs two to five days after infection in the urethra of the penis, causing extremely painful urination and a purulent (pus-filled) discharge. Rarely, the bacterium invades the prostate or epididymis, where the formation of scar tissue can render the man infertile.

In contrast, gonorrhea in women is often asymptomatic—50–80% of infected women have no symptoms or obvious signs of infection for years. It is only when they try to become pregnant that damage to the uterine tubes becomes obvious. About 25% of women with gonorrhea suffer pelvic inflammatory disease.

Pathogen and Virulence Factors

Neisseria gonorrhoeae (nī-se´rē-ă go-nor-rē´ī), also known as the *gonococcus,* causes gonorrhea. This Gram-negative bacterium usually forms pairs of cells that have fimbriae (see Disease at a Glance 24.5 on p. 766), polysaccharide capsules, and a major cell wall antigen called *lipooligosaccharide* (lip´ō-ol´ī-gō-sak´a-rīd, LOS), composed of lipid A (endotoxin) and sugar molecules. Cells that lack these three structural features are avirulent. The cocci also protect themselves from the immune system by secreting a protease enzyme that breaks down secretory IgA in mucus.

Pathogenesis

Gonococci adhere, via their fimbriae and capsules, to epithelial cells of the mucous membranes lining much of the genital, urinary, and digestive tracts of humans. As few as 100 pairs of cells can cause disease. The bacterium does not attach to cells lining the vagina; instead, the bacteria most commonly infect the cervix of the uterus. Via fimbriae, *N. gonorrhoeae* can attach to sperm cells, as they swim past, which allows the bacterium to hitchhike to the uterine tubes and beyond, producing pelvic inflammatory disease. Chronic infections can lead to scarring of the tubes, resulting in pregnancies outside the uterus or sterility.

Gonococcal infections of organs outside the reproductive tracts also occur. Because a woman's urethral opening is close to her vaginal opening, gonococci can infect the urethra during sexual intercourse. Anal intercourse can lead to *proctitis* (inflammation of the rectum), and oral sexual intercourse allows infection of the pharynx or gums, resulting in pharyngitis and gingivitis, respectively. These conditions are most commonly seen in men who have sex with men. Phagocytized bacteria survive and multiply within neutrophils, traveling within these leukocytes throughout the body. In very rare cases, gonococci travel to the joints, meninges, or heart, causing arthritis, meningitis, and endocarditis, respectively.

Babies delivered vaginally by infected mothers can be infected in their eyes, resulting in inflammation of the cornea, *ophthalmia neonatorum* (inflammation of the conjunctiva in newborns), or blindness.

Epidemiology

Gonorrhea occurs in humans only. The number of cases among civilians in the U.S. has been declining over the past three decades (**FIGURE 24.5a**). Most civilian cases occur in adolescents, particularly among those in several southeastern states who have multiple sexual partners (**FIGURE 24.5b**). Infection is four

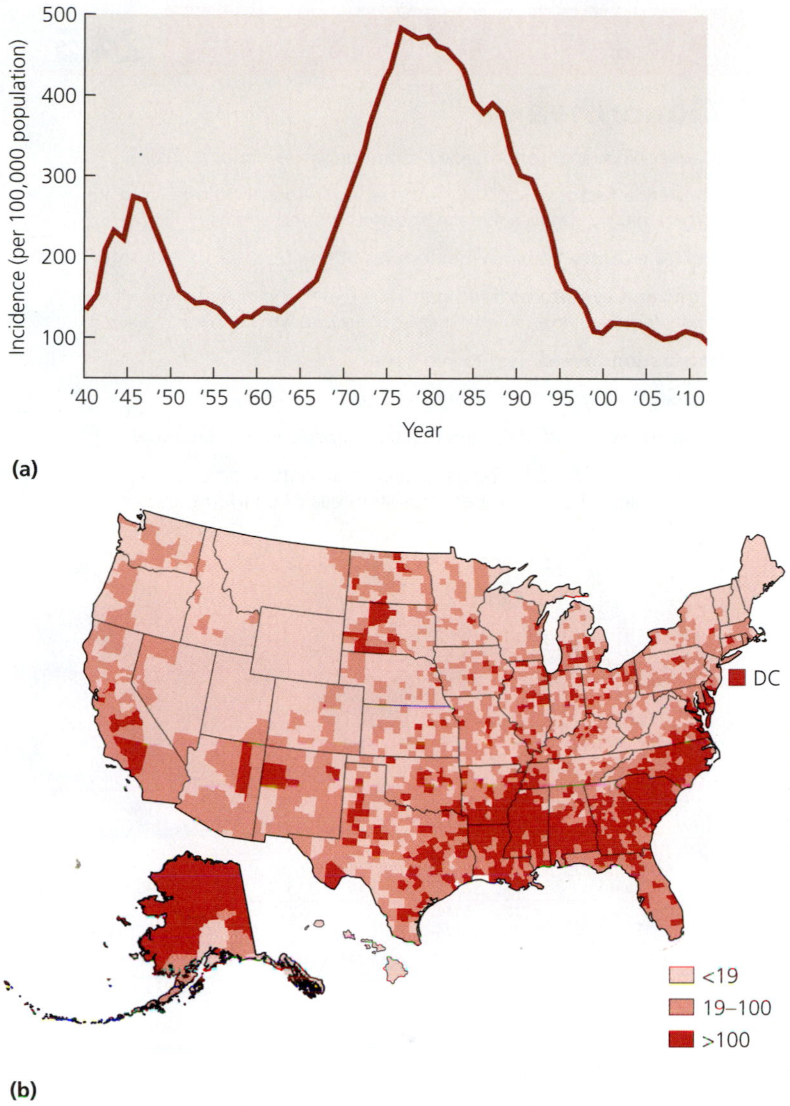

▲ **FIGURE 24.5 The incidence of civilian gonorrhea in the United States.** **(a)** Incidence 1941–2012. **(b)** The geographic distribution of cases reported in 2011 by county per 100,000 population.

times more common among blacks than nonblacks and slightly more common among males than females.

An individual's risk of infection increases with increasing frequency of sexual encounters. Women have a 50% chance of becoming infected during a single sexual encounter with an infected man, whereas men have only a 20% chance of infection during a single encounter with an infected woman. Infection of elementary-school-aged children with *N. gonorrhoeae* is strong evidence of sexual abuse by an adult.

Diagnosis, Treatment, and Prevention

The presence of Gram-negative diplococci in pus from an inflamed penis is sufficient for a diagnosis of symptomatic gonorrhea in men. For diagnosis of asymptomatic cases of gonorrhea in men and women, commercially available genetic probes provide direct, accurate, rapid detection of *N. gonorrhoeae* in clinical specimens.

DISEASE AT A GLANCE 24.5

Gonorrhea

Cause *Neisseria gonorrhoeae* (Gram-negative diplococcus).

Virulence factors Capsule, fimbriae, lipooligosaccharide (lipid A, sugar), highly variable surface antigens among strains.

Portal of entry Mucous membranes of genitalia; sexual transmission.

Signs and symptoms Men generally experience painful urination and pus-filled discharge. Women are generally asymptomatic.

Incubation period Two to five days.

Susceptibility Sexually active individuals.

Treatment Broad-spectrum oral cephalosporin or quinolones.

Prevention Effective: Sexual abstinence, mutual monogamy. Somewhat effective: Proper, consistent use of condoms.

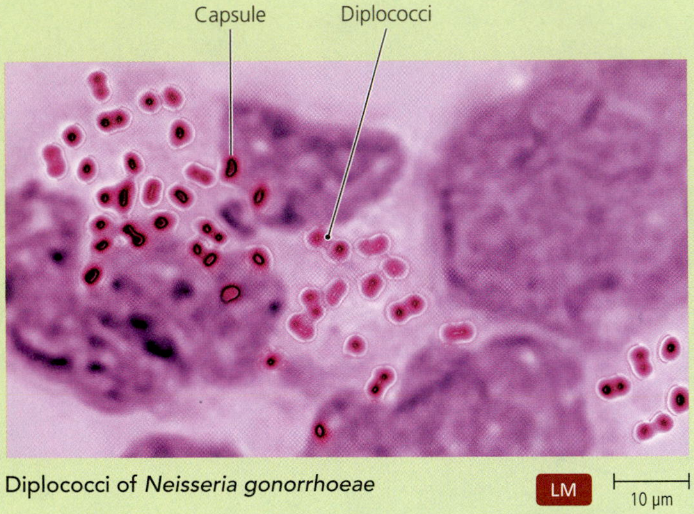

Diplococci of *Neisseria gonorrhoeae*

The treatment of gonorrhea is now complicated by the worldwide spread of gonococcal strains that are resistant to penicillin, tetracycline, erythromycin, aminoglycosides, and fluoroquinolones. Currently, the CDC recommends dual therapy with both a broad-spectrum oral cephalosporin (such as cefixime) and either azithromycin or doxycycline to treat gonorrhea.

There is no long-term specific immunity against *N. gonorrhoeae,* and thus people can contract gonorrhea multiple times. This lack of immunity is explained in part by the highly variable surface antigens in this bacterium—immunity against one strain often provides no protection against other strains. The existence of many different strains has prevented the development of an effective vaccine.

Routine administration of antimicrobial agents to newborns' eyes successfully prevents ophthalmic disease. Otherwise, chemical prophylaxis is ineffective in preventing disease. In fact, the use of antimicrobials to prevent genital disease may select for hardier resistant strains, worsening the situation. Preventive strategies, in order of effectiveness, are sexual abstinence, monogamy with a faithful partner, and the 100% consistent and proper use of condoms. Efforts to stem the spread of gonorrhea focus on education to change sexual behavior, aggressive detection, and the screening of all sexual contacts of carriers.

Disease at a Glance 24.5 summarizes the features of gonorrhea.

Syphilis

LEARNING **OUTCOMES**

24.22 Describe the four phases of untreated syphilis and the treatment for each.

24.23 Describe the cause, epidemiology, and prevention of syphilis.

Europeans first recognized **syphilis** in 1495, leading some epidemiologists to hypothesize that Spanish explorers brought syphilis to Europe. Another hypothesis is that the causative bacterium evolved from a less-pathogenic strain endemic to North Africa and that its spread throughout Europe was coincidental to Europeans' explorations of the New World. In any case, this sexually transmitted disease continues to afflict millions worldwide.

Signs and Symptoms

There are four phases of syphilis:

- *Primary:* A small, painless, reddened, hard lesion called a **chancre** (shang´ker) forms at the site of infection 10–21 days following exposure **(FIGURE 24.6a)**. Although chancres typically form on the external genitalia, about 20% form in the mouth, around the anus, or on the fingers, lips, or nipples. Chancres are often unobserved, especially in women, in whom these lesions frequently form on the cervix. Chancres last three to six weeks.
- *Secondary:* As the disease progresses, the patient suffers sore throat, headache, mild fever, malaise, myalgia, lymphadenopathy (diseased lymph nodes), and a widespread rash **(FIGURE 24.6b)** that can include the palms and the soles of the feet. Although this rash does not itch or hurt, it can persist for months.
- *Latent:* This phase has no symptoms and may last a decade or longer.
- *Tertiary:* Years later, some untreated patients experience dementia, blindness, paralysis, heart failure, or syphilitic **gummas** (gŭm´az), which are rubbery, swollen lesions that can occur in bones, in nervous tissue, or on the skin **(FIGURE 24.6c)**. Gummas rarely develop in patients in countries where antimicrobial drugs are available.

Pathogen and Virulence Factors

Treponema pallidum (trep-ō-nē´mă pal´li-dŭm) causes syphilis. The helical cells of this spirochete are so narrow (0.1 µm) that they are difficult to see by regular light microscopy in

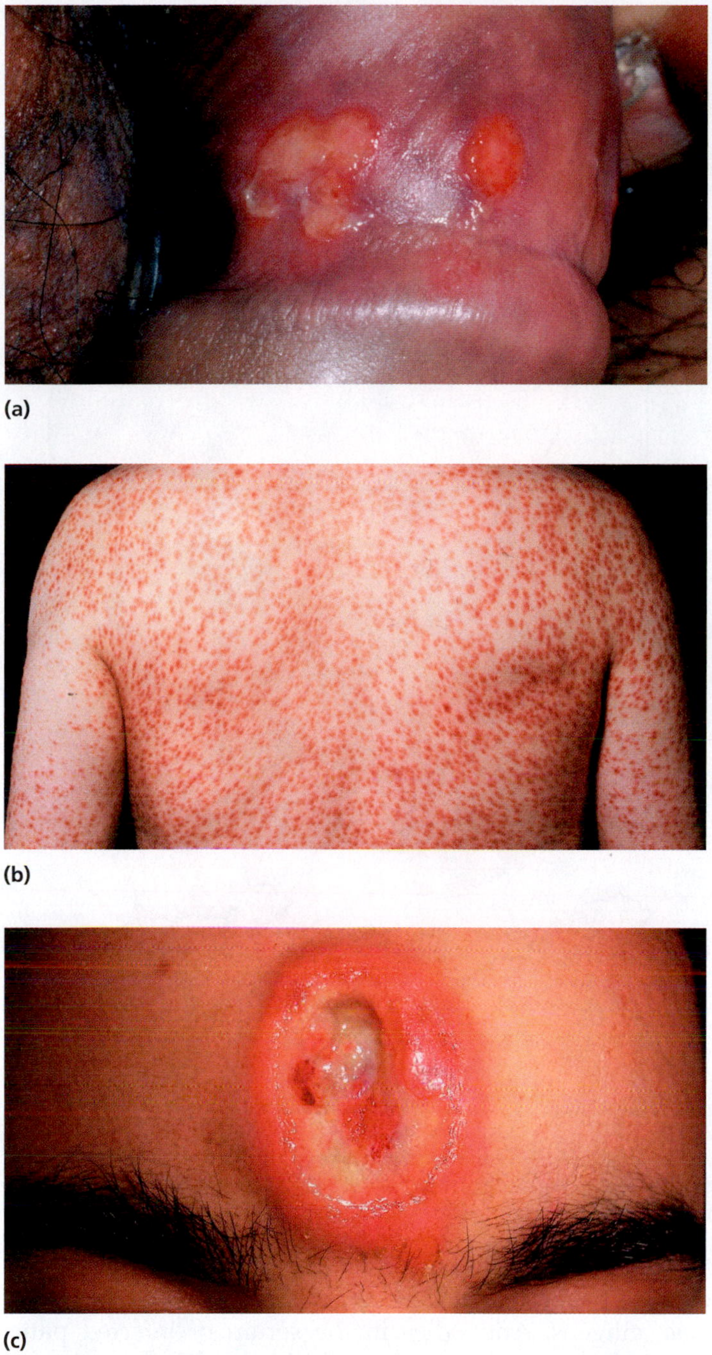

▲ **FIGURE 24.6 The lesions of syphilis.** **(a)** A chancre, a hardened and painless lesion of primary syphilis that forms at the site of infection, here the shaft of a penis. **(b)** Widespread rash characteristic of secondary syphilis. **(c)** A gumma of tertiary syphilis. This painful rubbery lesion often occurs on the skin or bones. *Why are gummas rare in industrialized countries?*

Figure 24.6 Antimicrobial drugs, available in industrialized countries, effectively treat syphilis, halting the progression of the disease.

Gram stained specimens. Therefore, scientists use either phase-contrast microscopy, dark-field microscopy, or special stains to increase contrast (see Disease at a Glance 24.6 on p. 769).

The bacterium lives naturally only in humans, and because heat, disinfectants, soaps, drying, the concentration of oxygen in the air, and pH changes destroy it, *Treponema* cannot survive in the environment. In fact, scientists have not successfully cultured *T. pallidum* in cell-free media, though they have coaxed it to multiply in rabbits, monkeys, and rabbit epithelial cell cultures. In these laboratory conditions, it multiplies slowly (binary fission occurs once every 30 hours), and only for a few generations.

Scientists have had difficulty identifying virulence factors of *T. pallidum*. They have used recombinant DNA techniques to insert genes from *Treponema* into *Escherichia coli* (esh-ĕrik´ē-ă kō´lī) and have then isolated the genes' proteins. Some of these proteins apparently enable *Treponema* to adhere to human cells. Virulent strains also produce hyaluronidase, which may enable *Treponema* to infiltrate intercellular spaces. The bacterium has a glycocalyx, which may protect it from phagocytosis by leukocytes.

CLINICAL CASE STUDY

A Painful Problem

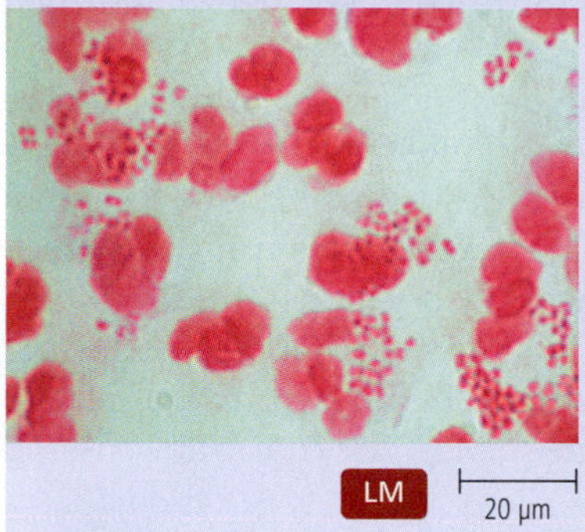

A 20-year-old man reports to his physician that he has experienced painful urination, as though he were urinating molten metal. He has also noticed a puslike discharge from his penis. A stained smear of the discharge is shown in the photo.

The man reports having been sexually active with two or three women in the previous six months. Because his partners reported being "absolutely sure" that they carried no sexually transmitted diseases, he had not used a condom.

1. What disease does this patient most likely have? What are the medical and common names for this disease?
2. How did the patient acquire the disease?
3. What could explain the lack of any history of sexually transmitted disease in his sexual partners?
4. What is the likely treatment?
5. Is the patient immune to future infections with this bacterium?

Pathogenesis

Because of its fastidiousness and sensitivity, *T. pallidum* must grow in humans and is transmitted almost solely via sexual contact, usually during the early stage of infection, when the spirochetes are most numerous. The risk of infection from a single, unprotected sexual contact with an infected partner is 10–30%. *T. pallidum* can also be spread from mother to fetus and rarely through blood transfusion. It cannot be spread by fomites such as toilet seats, eating utensils, or clothing.

Untreated syphilis has four phases: primary, secondary, latent, and tertiary syphilis. In **primary syphilis,** a chancre forms at the site of infection. The center of the chancre fills with serum that is extremely infectious because of the presence of millions of spirochetes. Chancres remain for three to six weeks and then disappear without scarring.

In about a third of cases, the disappearance of the chancre is the end of the disease. However, in most infections *Treponema* has invaded the bloodstream and spreads throughout the body to cause the symptoms and signs of **secondary syphilis,** including rash. Like the primary chancre, rash lesions are filled with spirochetes and are extremely contagious. People, including health care workers, are infected when fluid from the lesions enters breaks in the skin, though such nonsexual transmission of syphilis is rare.

After several weeks or months, the rash gradually disappears, and the patient enters a stage called **latent syphilis.** This is a clinically inactive phase of the disease. The majority of cases do not advance beyond this point, especially in developed countries where antimicrobial drugs are in use.

Latency may last 10 or more years, after which about a third of the originally infected patients proceed to **tertiary syphilis.** This phase is not associated with the direct effects of *Treponema,* but rather with severe complications resulting from inflammation and a hyperimmune response against the pathogen. Gummas (painful, rubbery lesions, often on skin or bones), destruction of cardiovascular or central nervous system tissue, personality changes, insanity, and blindness are manifestations of tertiary syphilis.

Congenital syphilis results when *Treponema* crosses the placenta from an infected mother to her fetus. Transmission to the fetus from a mother experiencing primary or secondary syphilis often results in the death of the fetus. If transmission occurs while the mother is in the latent phase of the disease, the fetus often suffers mental retardation and malformation of many organs. After birth, newborns with latent infections usually exhibit a widespread rash like that of secondary syphilis at some time during their first two years of life.

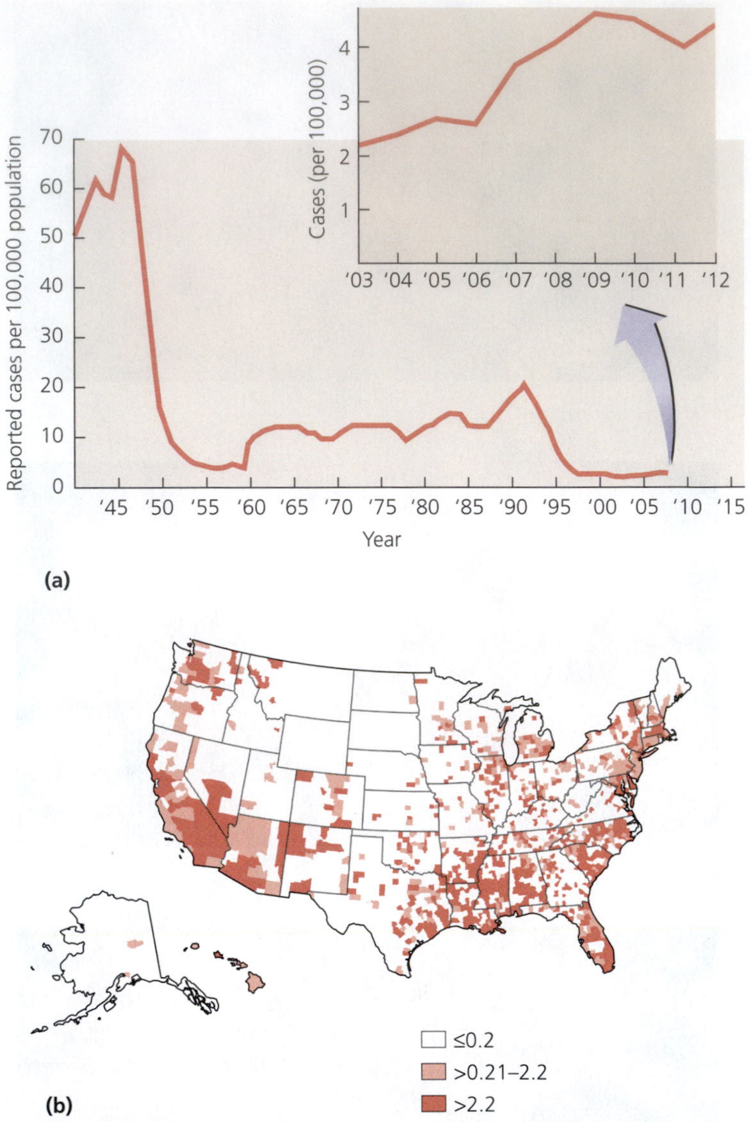

▲ **FIGURE 24.7 The incidence of syphilis in the United States.** **(a)** Nationwide incidence of syphilis, 1941–2012. **(b)** Reported incidence of syphilis by county per 100,000 population, 2011.

Epidemiology

Syphilis occurs worldwide; WHO estimates that syphilis afflicts more than 12 million new victims each year. The discovery and development of antimicrobial drugs has greatly reduced the number of cases in the United States **(FIGURE 24.7a)**, though the disease remains endemic among sex workers, men who have sex with men, and users of illegal drugs. It occurs throughout the United States.

The goal of the CDC to achieve an incidence of fewer than 0.2 cases of syphilis per 100,000 population in every county in the United States by 2010 was not met **(FIGURE 24.7b)**.

Diagnosis, Treatment, and Prevention

The diagnosis of primary, secondary, and congenital syphilis is relatively easy and rapid using specific antibody tests against antigens of *Treponema pallidum*. One such test is *MHA-TP* (*microhemagglutination assay against* T. pallidum), which uses red blood cells that have been artificially coated with *Treponema* antigens. Antibodies in the serum of infected patients will agglutinate (clump) the red blood cells—a positive test for exposure to *T. pallidum*. Spirochetes can be observed (see Disease at a Glance 24.6) in fresh discharge from lesions, but only when microscopic observations of clinical samples are made immediately—*Treponema* rarely survives transport to a laboratory. Nonpathogenic spirochetes, which are a normal part of the oral microbiota, can yield false positive results, so clinical specimens from the mouth cannot be tested for syphilis. Tertiary syphilis is extremely difficult to diagnose because it mimics many other diseases, because few (if any) spirochetes are present, and because the signs and symptoms may occur years apart and seem unrelated to one another.

Penicillin G is the drug of choice for treating primary, secondary, latent, and congenital syphilis, but it does not work for tertiary syphilis, because this phase is a hyperimmune response,

DISEASE AT A GLANCE 24.6

Syphilis

Cause *Treponema pallidum* (spirochete).

Virulence factors Glycocalyx, adhesins, hyaluronidase.

Portal of entry Mucous membranes of genitalia, sexual or congenital transmission.

Signs and symptoms Primary syphilis: hard, nonpainful, genital chancre that disappears after three to six weeks. Secondary syphilis: sore throat, headache, mild fever, malaise, myalgia, and rash that lasts several weeks or months. Latent syphilis: generally asymptomatic; lasts for decades. Tertiary syphilis: dementia, blindness, paralysis, and gumma lesions.

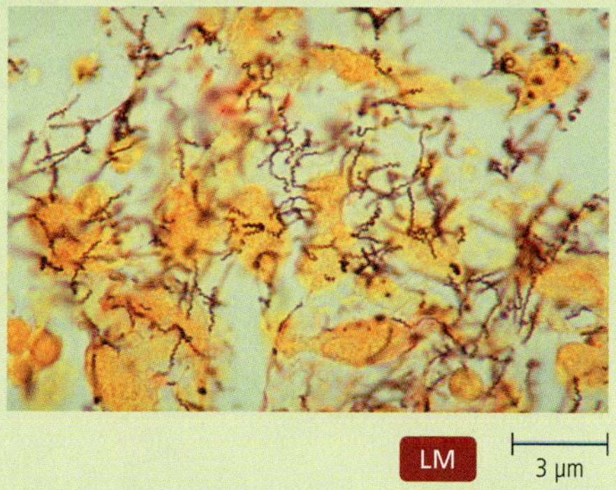

Treponema pallidum, the cause of syphilis. A special silver stain makes the thin spirochetes visible.

Incubation period 10–90 days (average 21 days) for the primary form of the disease.

Susceptibility Sexually active individuals and fetuses whose mothers are infected.

Treatment Penicillin.

Prevention Sexual abstinence, mutual monogamy, or use of condoms.

not an active infection. There is no proven alternative for patients who are allergic to penicillin.

A vaccine against syphilis is not available, so abstinence, faithful mutual monogamy, or consistent and proper condom usage are the primary ways to avoid contracting syphilis. All sexual partners of syphilis patients must be treated with prophylactic penicillin G to prevent the spread of syphilis.

Disease at a Glance 24.6 summarizes the characteristics of syphilis.

Chlamydial Infections

LEARNING | **OUTCOMES**

24.24 Describe lymphogranuloma venereum, including its manifestations, cause, pathogenesis, epidemiology, treatment, and prevention.

24.25 Explain how sexually inactive children may become infected with *Chlamydia trachomatis*.

The most common sexually transmitted bacterium is *Chlamydia trachomatis* (kla-mid´ē-ă tra-kō´ma-tis), which causes several sexually transmitted and congenital diseases, including *lymphogranuloma venereum* and *trachoma*.

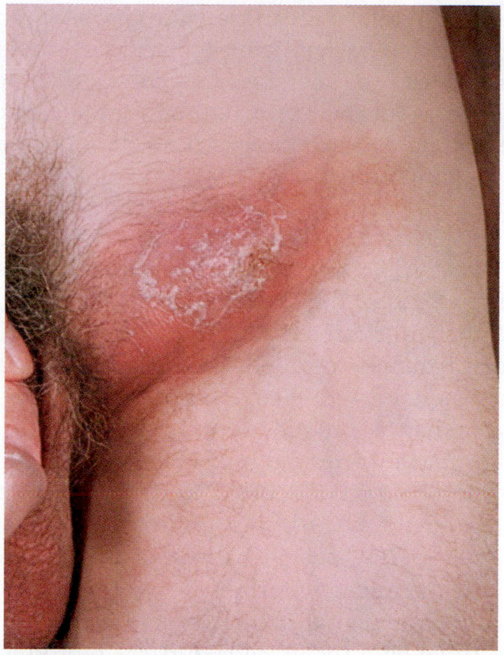

▲ **FIGURE 24.8 An advanced case of lymphogranuloma venereum in a man.** *Which microorganism causes lymphogranuloma venereum?*

Figure 24.8 *Chlamydia trachomatis* causes *lymphogranuloma venereum*.

Signs and Symptoms

About 85% of chlamydial genital tract infections in women are asymptomatic; in contrast, more than 75% of infected men show signs and symptoms similar to those of gonorrhea—urethritis, painful urination, and pus discharge from the penis. Chlamydial infection can also cause **epididymitis** (inflammation of the epididymis) or **orchitis**[8] (inflammation of a testis), either of which can lead to sterility.

When babies are infected at birth, an eye disease called **trachoma** develops. Trachoma is the leading cause of nontraumatic blindness in humans worldwide.

A severe form of chlamydial STD is **lymphogranuloma venereum**[9] (lim´fō-gran-ū-lō´mă ve-ne´rē-um), which is characterized by a transient genital lesion at the site of infection on the penis, urethra, scrotum, vagina, cervix, or external female genitalia. This is followed by development of a **bubo** (**FIGURE 24.8**)—a painfully inflamed lymph node—in the groin, fever, chills, anorexia, and muscle pain. Buboes may enlarge to the point that they rupture, producing draining sores.

Pathogens and Virulence Factors

Bacteria of the genus *Chlamydia* are nonmotile and grow and multiply only within vesicles inside host cells. Scientists once considered chlamydias to be viruses because of their small size, and intracellular lifestyle. However, chlamydias are cellular and possess DNA, RNA, and functional 70S ribosomes.

[8]From Greek *orchis*, meaning "testis," and *itis*, meaning "inflammation."
[9]From Latin *lympha*, meaning "clear water"; Latin *granulum*, meaning "a small grain"; Greek *oma*, meaning "tumor" (swelling); and Latin *Venus*, the goddess of sexual love.

Two membranes, similar to those of a typical Gram-negative bacterium, surround each chlamydial cell, but there is no peptidoglycan between the membranes—chlamydias do not have cell walls. Because of their unusual features and unique rRNA nucleotide sequences, taxonomists now classify chlamydias in their own phylum: Chlamydiae.

Chlamydia trachomatis has a very limited host range. With one exceptional strain, which may eventually be classified as a separate species, all strains are pathogens of humans, infecting the conjunctivas of the eyes, lungs, urinary tract, or genital tract. More than 10 strains of *C. trachomatis* cause STDs in humans; three of these—so-called *LGV strains*—trigger lymphogranuloma venereum. Other strains cause trachoma.

Chlamydia has a unique developmental cycle involving two cellular morphologies, both of which can occur within the phagosome of a host cell **(FIGURE 24.9a)**: tiny (0.2–0.4 µm) cocci called *elementary bodies (EBs)* and larger (0.6–1.5 µm) pleomorphic *initial bodies (IBs)*, which have also been called *reticulate bodies*. EBs are the infective forms; they are relatively dormant, resistant to environmental extremes, and can survive outside cells. IBs survive only inside cells and are reproductive rather than infective.

In the life cycle of *Chlamydia* **(FIGURE 24.9b)**, once an EB attaches to a host cell 1, it enters by triggering endocytosis 2. Once inside a vesicle, the EB converts into an IB 3, which then divides rapidly to form many IBs 4. About 21 hours after infection, IBs within the vesicle, which is now called an *inclusion body*, begin converting back to EBs 5, and about 19 hours after that, the EBs are released from the host cell via exocytosis 6, becoming available to infect new cells and completing the life cycle.

Pathogenesis

C. trachomatis enters the body through scrapes or cuts and infects cells of the conjunctiva or cells lining the mucous membranes of the trachea, bronchi, urethra, uterus, uterine tubes, anus, or rectum. Typically, the lesion at a site of infection is overlooked because it is small, painless, and heals rapidly. Headache, muscle pain, and fever may occur during this stage of the disease.

Proctitis[10] may occur in men or women because of lymphatic spread of *Chlamydia* from the genitalia or urethra to the rectum. About 15% of the cases of proctitis in homosexual men result from the spread of *C. trachomatis* via anal intercourse.

LGV strains of *Chlamydia* infect the lymph nodes in the groin, producing buboes. In a few patients, lymphogranuloma venereum proceeds to a third stage characterized by genital sores, constriction of the urethra, and severe genital swelling. Arthritis may also occur during this third stage, particularly in young white males for an unknown reason.

[10]From Greek *proktos*, meaning "rectum," and *itis*, meaning "inflammation."

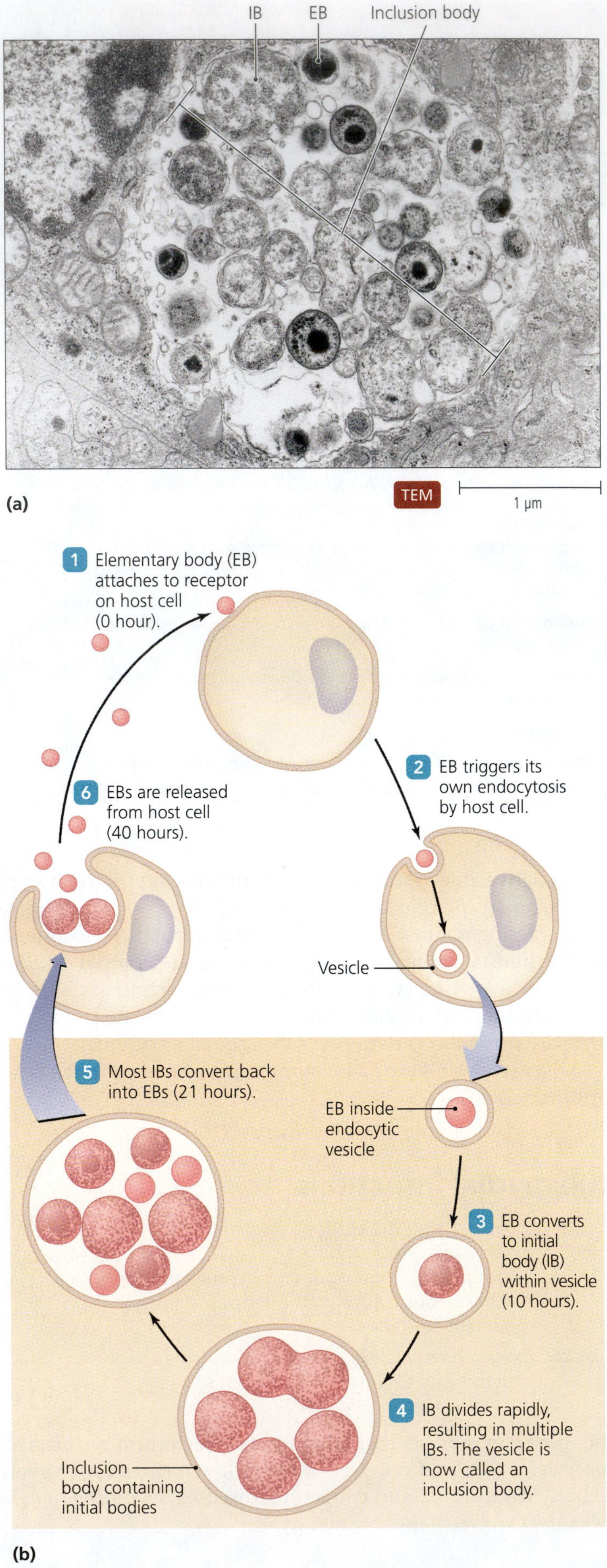
IB
EB
Inclusion body
(a)
TEM
1 µm
1 Elementary body (EB) attaches to receptor on host cell (0 hour).
2 EB triggers its own endocytosis by host cell.
Vesicle
EB inside endocytic vesicle
3 EB converts to initial body (IB) within vesicle (10 hours).
4 IB divides rapidly, resulting in multiple IBs. The vesicle is now called an inclusion body.
Inclusion body containing initial bodies
5 Most IBs convert back into EBs (21 hours).
6 EBs are released from host cell (40 hours).
(b)

▶ **FIGURE 24.9 The developmental forms and life cycle of *Chlamydia*.** **(a)** Elementary bodies and initial bodies within a host cell's vesicle. **(b)** The life cycle of *Chlamydia*. Times in parentheses refer to hours since infection.

The clinical manifestations of chlamydial infection, including pelvic inflammatory disease, result from the destruction of infected cells at the site of infection and from the inflammatory response this destruction stimulates. Reinfection in the same site by the same or a similar strain triggers a vigorous hypersensitive immune response that can result in blindness, sterility, or sexual dysfunction. Chlamydial infection in adolescence increases the risk of developing cervical cancer later in life.

Trachoma strains of *C. trachomatis* multiply in cells of the conjunctiva, killing them and triggering a copious, purulent (pus-filled) discharge that causes the conjunctiva to become scarred. Such scarring in turn causes the patient's eyelids to turn inward, such that the eyelashes scratch, irritate, and scar the cornea, which triggers an invasion of blood vessels into this normally clear surface of the eye. A scarred cornea becomes filled with blood vessels and is no longer transparent. The eventual result is blindness.

Epidemiology

Infection with *C. trachomatis* is the most common of the reportable sexually transmitted diseases in the United States—over 1,330,000 cases in 2012—but epidemiologists estimate that another 3–4 million asymptomatic cases go unreported annually. Sexually transmitted chlamydial infections are most prevalent among women under the age of 20 because they are physiologically more susceptible to infection. LGV strains of *Chlamydia* are found primarily in the tropics.

Researchers further estimate that over 500 million people worldwide, particularly children, contract eye infections with *C. trachomatis*. The bacterium infects children as they pass through the birth canal. The pathogen may also be transmitted from eye to eye via droplets, hands, contaminated fomites, or flies. Chlamydial eye infections are endemic in crowded, poor communities where people have poor personal hygiene, inadequate sanitation, and inferior medical care, particularly in the Middle East, North Africa, and India.

Diagnosis, Treatment, and Prevention

Historically, diagnosis of chlamydial infection involved demonstration of the bacterium inside cells from specimens obtained from the urethra or vagina by inserting, rotating, and then removing a sterile swab. Alternatively, clinicians amplified the number of bacterial cells by inoculating the specimen into a culture of susceptible human cells. They could then demonstrate the presence of *Chlamydia* in the cell culture by means of specific fluorescent antibodies or nucleic acid probes. Today, demonstration of chlamydial DNA in a specimen following PCR amplification of DNA is diagnostic for chlamydial infection.

Physicians prescribe doxycycline or erythromycin to eliminate genital infections in adults. Trachoma strains of *C. trachomatis* infecting the eyes of newborns are initially treated with erythromycin cream and then with oral erythromycin for 14 days. Surgical correction of eyelid deformities may prevent the scratching, scarring, and blindness that typically result from eye infections.

CLINICAL CASE STUDY

A Sick Mother-to-Be

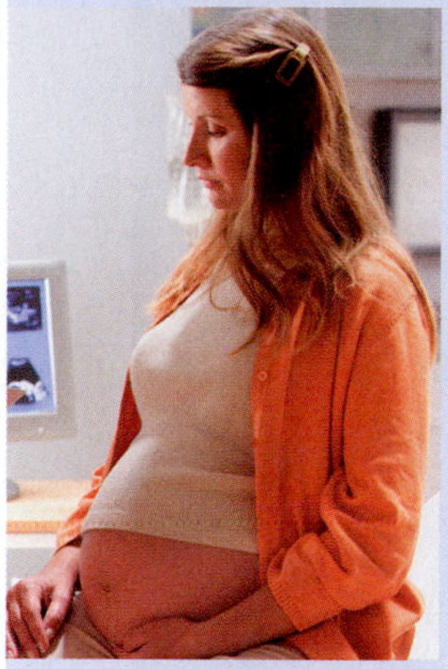

Preparing for delivery, an obstetrician notices genital sores and severe swelling of the labia on a pregnant 24-year-old female. Upon questioning the woman, the doctor discovers that she previously had swollen and painful lymph nodes near her groin but did not seek treatment because she assumed it was related to the pregnancy.

1. What is the disease and causative agent?
2. What risk is there, if any, to the unborn child?
3. How should the doctor treat the woman? The newborn?
4. Left untreated, what symptoms, if any, would the newborn experience?
5. Left untreated, how might the disease progress in the woman?

As with all STIs and STDs, abstinence or faithful mutual monogamy are the only preventives. Blindness can be prevented only by prompt treatment with antibacterial agents and prevention of reinfection. Unfortunately, chlamydial infections are often asymptomatic and frequently occur among populations that have limited access to medical care.

Chancroid

LEARNING | OUTCOMES

24.26 Describe the pathogenesis, epidemiology, diagnosis, treatment, and prevention of chancroid.

24.27 Contrast the appearance and symptoms of chancroid lesions with those of syphilis.

Chancroid (shang'kroyd) is another underdiagnosed and unreported bacterial STD that is present in the United States, though rare.

Signs and Symptoms

Chancroid is characterized by soft, painful genital ulcers called *soft chancres* that are 4–50 millimeters in diameter (**FIGURE 24.10**). These lesions, which form at the site of infection, are similar in appearance to primarily syphilitic lesions, but syphilis lesions are painless and hard. The base of a soft chancre is yellowish gray in color and bleeds easily when scraped.

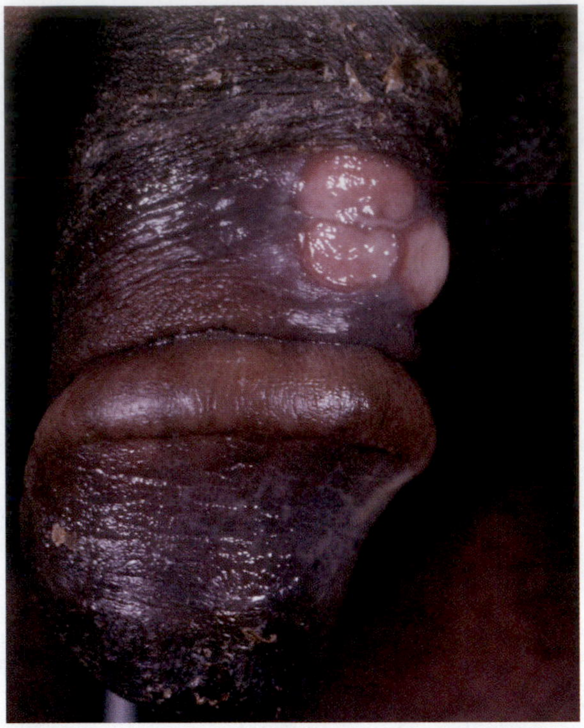

▲ **FIGURE 24.10 Soft chancres of chancroid.** *Haemophilus ducreyi* causes the painful, soft, genital lesions. *What do soft chancres have in common with primary syphilitic lesions?*

Figure 24.10 Both soft chancres and syphilitic lesions are located on genitalia and are manifestations of STDs.

Men typically have one ulcer; women may have four or more ulcers or be asymptomatic. A chancroid ulcer may block the opening of the urethra, causing pain during urination. This is the most common symptom in women.

Pathogen and Virulence Factors

Haemophilus ducreyi (hē-mof´i-lŭs doo-krā´ē) causes chancroid. This small, coccobacillus, Gram-negative bacterium requires heme and NAD^+, which can be derived from blood, for growth. As a result, *H. ducreyi* is an obligate parasite, colonizing humans. It produces a toxin that kills human epithelial cells.

Pathogenesis and Epidemiology

After incubating for 1–14 days, bacterial toxin kills cells at the site of infection, typically the labia of women and the foreskin or head of the penis in males—uncircumcised men are three times more likely to be infected than are circumcised men. Infection generates a small bump that develops into the characteristic ulcer within 24 hours. In half of patients, the bacterium may spread into lymph nodes in the groin, producing purulent buboes.

Although prevalent in Africa and Asia, chancroid is rare in Europe and the Americas, execpt the Caribbean islands. Most patients in the United States are infected during foreign travel. *H. ducreyi* spreads only via sexual intercourse. Chancroid resolves spontaneously, typically leaving only small scars, but because it is a risk factor for contracting HIV, the condition should be treated.

CLINICAL CASE STUDY

A Case of Genital Sores

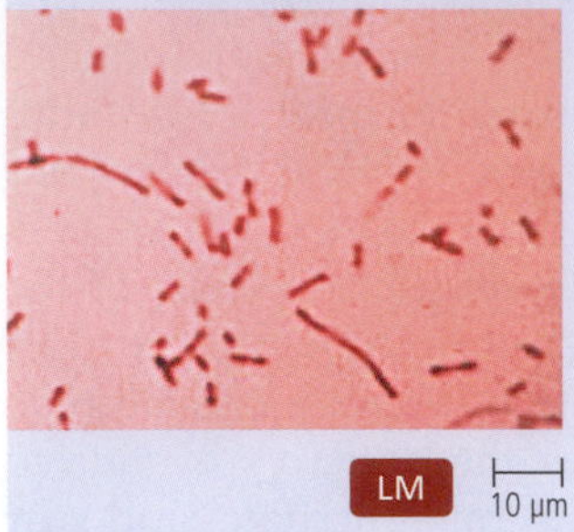

A 20-year-old sexually active college student reports to his campus clinic complaining of genital sores. The ulcers are soft, painful, and pus-filled. Upon questioning, the student reveals his current girlfriend is an Asian exchange student. The campus lab initially has difficulty identifying the causative agent, but finally is able to isolate a Gram-negative coccobacillus on blood agar.

1. What is the causative agent?
2. What is the mechanism of pathogenesis?
3. What may be the reason his partner has not sought treatment?
4. Why did the doctor ask about the nationality of the patient's partner?
5. What treatment is the doctor likely to recommend?

Diagnosis, Treatment, and Prevention

Physicians diagnose chancroid on the basis of the soft, painful lesions and presence of buboes. Stained specimens of infected tissue reveal the bacterium.

Effective antimicrobial drugs are azithromycin, erythromycin, ceftriaxone, and ciprofloxacin. Buboes need to be drained with a needle or local surgery.

Abstinence and mutual monogamy between uninfected partners are the only sure preventions. Condoms, when used consistently, provide some protection from infection.

TELL ME WHY

Why is erythromycin substituted for tetracycline in treatment of chlamydial infections in children? Why are penicillins and cephalosporins useless against *Chlamydia*?

Viral STDs

Viruses are the most common causes of sexually transmitted diseases, including AIDS, herpes, and genital warts. The following sections examine herpes and genital warts. (Chapter 18 examines AIDS as a syndrome resulting from the collapse of the immune system.)

Genital Herpes

LEARNING | OUTCOMES

24.28 Discuss the features of genital herpes, including manifestations, treatment, and prevention.

24.29 List conditions that may reactivate latent genital herpes.

The name *herpes,* which is derived from a Greek word meaning "to creep," is descriptive of the slowly spreading skin lesions that are characteristic of herpesvirus infections. (Chapter 23 examines oral herpes; here we consider genital herpes.)

Signs and Symptoms

Genital herpes manifests as numerous small blisters on the genitals, around the rectum, or on adjacent areas of skin. The blisters, which are filled with clear, straw-colored fluid, eventually break and become painful ulcers. Patients often experience fever, malaise (feeling of general discomfort or uneasiness), myalgia, and decreased appetite.

Pathogen and Virulence Factors

Two enveloped, polyhedral, double-stranded DNA viruses called *human herpesviruses* (HHV, genus *Simplexvirus*) cause genital herpes. (These viruses were known previously as herpes simplex viruses, abbreviated HSV.) About 85% of cases involve *HHV-2,* which has been called "below-the-waist herpesvirus." *HHV-1* ("oral herpesvirus" or "cold sore virus") causes the remaining cases of genital herpes, usually following oral-genital sexual intercourse. HHV-2 can also cause oral lesions. (Chapter 23 discusses oral herpes in more detail.) Herpesviruses can become latent in nerve cells, hiding the viruses from immune surveillance.

Pathogenesis

The membrane of a herpesvirus attaches to and fuses with a host cell's membrane, depositing the viral capsid and genome in the cytosol. The viral genome enters the cell's nucleus, where new viruses are synthesized and assembled. Virions acquire envelopes from the nuclear membrane and exit the cell via exocytosis or cell lysis. There is an incubation period of about five days between infection and manifestations of disease.

The fluid-filled blisters form as herpesvirus kills epithelial cells at a site of infection. When the blisters rupture, they release millions of herpesviruses and turn into painful ulcers.

Infected people shed herpesviruses in mucous secretions even in the absence of lesions. In fact, asymptomatic patients may spread genital herpesviruses more frequently than patients with ulcers.

Nerve cell nuclei are grouped in structures called *ganglia* **(FIGURE 24.11)**. Some viruses remain latent as circular DNA within ganglia, becoming active and triggering recurrent lesions intermittently for years at sites innervated by the infected nerve. The patient may experience pain, tingling, burning, itching, or skin sensitivity at sites where

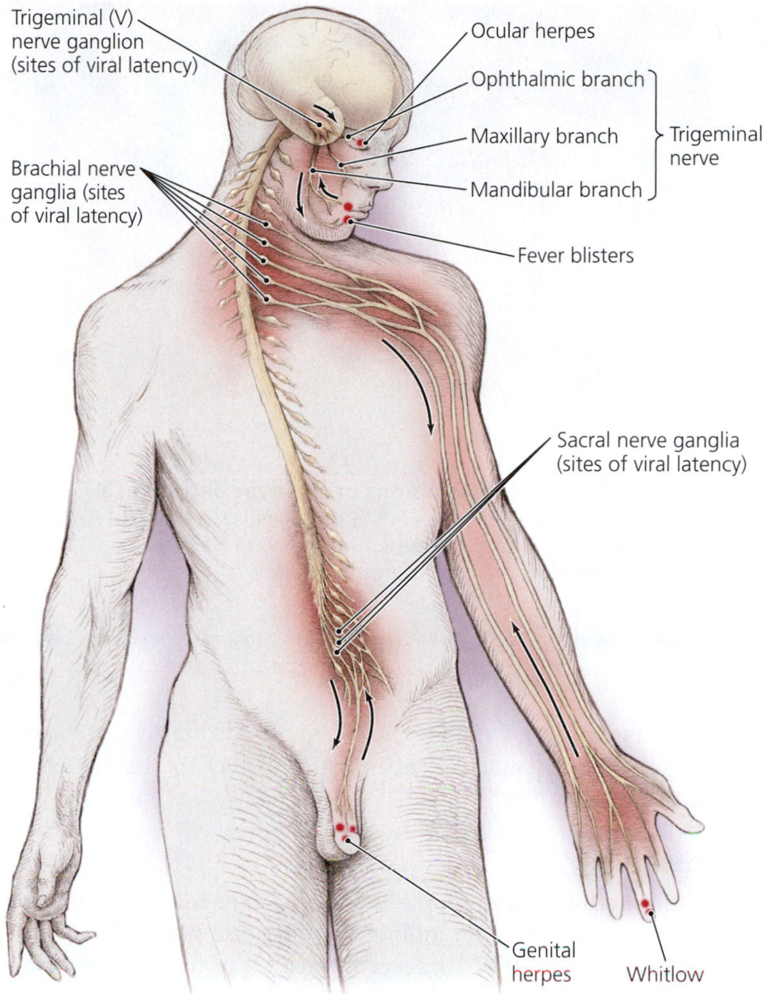

▲ **FIGURE 24.11 Sites of events in genital herpesvirus infections.** Infections occur when herpesviruses invade the lips, the genitalia, or broken skin. Viruses may remain latent for years in the trigeminal, brachial, or sacral ganglia before traveling down nerve cells to cause recurrent symptoms in the lips, genitalia, fingers, or eyes. *What factors may trigger the reactivation of latent herpesviruses and the recurrence of symptoms?*

Figure 24.11 *Emotional stress, fever, trauma, sunlight, menstruation, mechanical irritation, fatigue, or AIDS may trigger recurrent symptoms.*

new blisters will form. These may be far removed from the infected nucleus, such as the eye **(FIGURE 24.12a)**. Scientists do not fully understand what causes viral latency or triggers recurrence, but the latter may be induced by emotional or physical stress, such as fever, trauma, sunlight, menstruation, mechanical irritation, fatigue, or the immunosuppression of AIDS.

Herpesviruses spread by other than sexual means: Babies exposed to the initial genital herpes lesions as they pass through the birth canal can be infected. Herpesviruses can permanently disable or kill these young patients. Recurrent lesions, however, are not as harmful to newborns, presumably because maternal antibodies provide protection. Patients shedding herpesviruses from bursting blisters can infect their own or another person's esophagus, eyes, or skin. A blister on

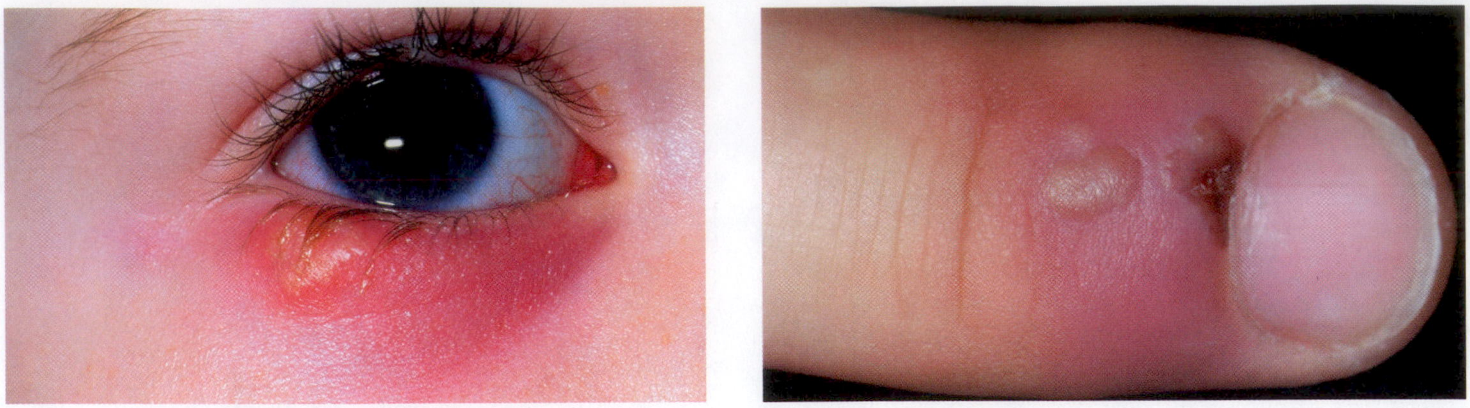

(a) (b)

▲ **FIGURE 24.12 Herpes lesions of the eyes and skin.** **(a)** Ophthalmic or ocular herpes results from reactivation of latent viruses in the trigeminal ganglion. **(b)** A whitlow on a health provider's finger, resulting from the entry of herpesvirus through a break in the skin of the finger. *How can health care providers protect themselves from a whitlow?*

Figure 24.12 *Health care workers should always wear gloves when handling patients who have lesions.*

a finger is called *whitlow* **(FIGURE 24.12b)**. Health care workers should always take precautions to protect themselves from ocular herpes and whitlow.

Epidemiology

An estimated 4 billion people worldwide are infected with herpesviruses, and about 86 million have genital herpes lesions. In the United States, HHV-1 infects about 15% of the populace, and HHV-2 infects 40%. Twice these numbers of people have antibodies against the viruses, indicating that they have been exposed.

Genital herpes quadruples the risk of HIV infection, presumably because herpes lesions provide a path for HIV to enter the body.

Diagnosis, Treatment, and Prevention

The characteristic lesions are indicative of genital herpes. Detection of herpesvirus DNA by PCR in fluid from a blister confirms the diagnosis.

As with many viral diseases, there is no cure for genital herpes, though treatment can relieve and lessen the manifestations. Daily oral dose of acyclovir or other antiviral agents over a 6- to 12-month period can be effective in preventing recurrent genital herpes lesions and in reducing the spread of herpes infections to sexual partners. Many obstetricians prescribe acyclovir during the final weeks of pregnancy for women with a history of genital herpes. (Chapter 10 considers the actions of these drugs.) Sexual abstinence and faithful monogamy between uninfected partners are the only sure ways of preventing genital infections.

Circumcised males are at a lower risk for herpes infections, as are their female sexual partners. Herpes lesions in women are usually on the external genitalia, so use of a condom during sexual intercourse provides little protection to their partners. People with active lesions should not have sex until the crusts have disappeared. Even in the absence of lesions, condoms should be used for every sexual encounter because asymptomatic individuals still shed viruses on their skin. Valacyclovir—a nucleotide analog that interferes with viral replication more than it does that of cells—reduces the risk of viral spread. Pregnant women with herpes infections, even if asymptomatic, should inform their doctors. To protect the baby, delivery should be by cesarean section (C-section) if genital lesions are present at the time of birth.

Genital Warts

LEARNING | **OUTCOMES**

24.30 Describe the characteristics and prevention of genital warts.

24.31 Describe the connection between genital warts and cervical cancer.

Warts (papillomas) are generally harmless growths of the epithelium of the skin on the face, trunk, hands, feet, elbows, knees, or genitalia. Here we examine **genital warts.**

Signs and Symptoms

Genital warts appear on the genitalia, surrounding skin, or in the anus and rectum. They range in size from almost undetectable small bumps to giant, cauliflower-like growths called **condylomata acuminata**[11] (kon-di-lō′mah-tă ă-kyū-mi-nah′tă) **(FIGURE 24.13)**. Besides being visually distressing, genital warts may be painful or itchy, and, rarely, they may bleed or increase vaginal discharge.

[11] From Greek *kondyloma*, meaning "knob," and Latin *acuminatus*, meaning "pointed."

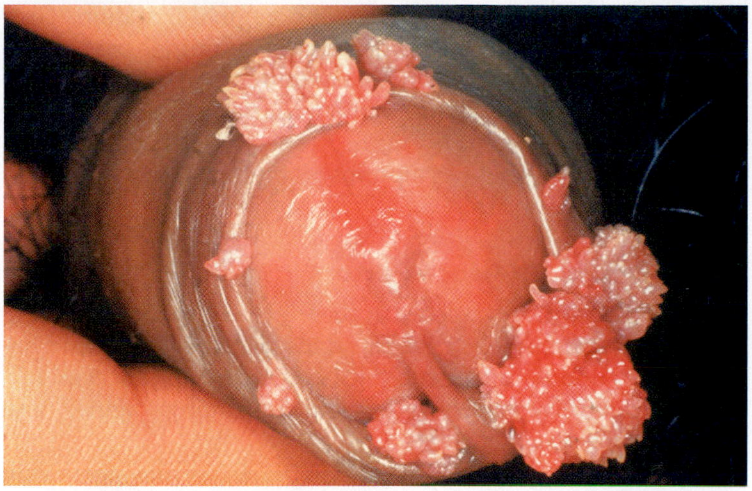

▲ **FIGURE 24.13 Genital warts.** Condylomata acuminata on a penis. Such cauliflower-like warts are considered precancerous. Genital warts can also occur on the external genitalia of women, on the lining of the vagina, on the penis, or on the skin around the anus in either sex.

Pathogen

About 30 varieties of double-stranded DNA, icosahedral naked viruses in the genus *Papillomavirus* cause genital warts. These viruses are called *human papillomaviruses (HPVs).* Half of the HPVs can permanently integrate into human chromosomes, where two viral proteins called E6 and E7 can trigger cancer, especially when they infect a patient who is also infected with herpesvirus.

Pathogenesis and Epidemiology

HPVs invade the skin or mucous membranes of the penis (particularly when it is uncircumcised), vagina, or anus during sexual intercourse. The incubation time from infection to the development of a wart is usually three to four months. Although all warts can be painful and are often unsightly, genital warts are more distressing because they cause nearly all cervical cancers as well as some vaginal, oral, penile, and anal cancers.

Genital warts have been the most common sexually transmitted disease in the United States, mostly in young adults. Sexually active adolescents, uncircumcised men, and the sexual partners of these individuals have a greater risk of contracting genital warts. Research suggests that if all men were circumcised, the worldwide incidence of cervical cancer would be reduced at least 43%.

Diagnosis, Treatment, and Prevention

Diagnosis of warts is a simple matter of observation. DNA probes can determine the exact strain of HPV involved. Cytotoxic T cells eventually recognize and destroy cells infected with HPV; therefore, warts disappear over time. Over-the-counter medications are often effective in removing warts from skin, but physicians recommend that such removal medications never be used on genital warts, in part because irritating chemicals may increase the transmission of other sexually transmitted diseases. Physicians can remove genital warts with surgery, freezing, burning, laser, or the use of caustic chemicals, though with any of these methods, viruses may remain in surrounding tissue, and warts may return.

Effective treatment of genital cancers depends on early diagnosis resulting from thorough visual inspection of the genitalia in both sexes, and in women by a *Papanicolaou* (pa´pa-nē´ ko-low) *smear (Pap smear)* to screen for cervical cancer. Treatment involves radiation or chemical therapy directed against reproducing tumor cells. Advanced cases of genital cancer necessitate removal of the entire diseased organ.

Prevention of genital warts is possible by abstinence or mutual monogamy with an uninfected partner. Condoms may reduce the risk of infection with HPV.

Using recombinant DNA techniques, scientists have created a vaccine that successfully prevents infection by the four most common strains of sexually transmitted papillomaviruses, including the strains associated with most cervical cancers. The CDC recommends three doses of HPV vaccine for all adolescents and adults younger than 50 years of age.

Disease at a Glance 24.7 on p. 776 summarizes the features of genital warts.

TELL ME WHY

Why are DNA viruses, such as herpesviruses and papillomaviruses, more likely to cause recurrent diseases and cancers than are RNA viruses?

CLINICAL CASE STUDY

A Very Sick Man

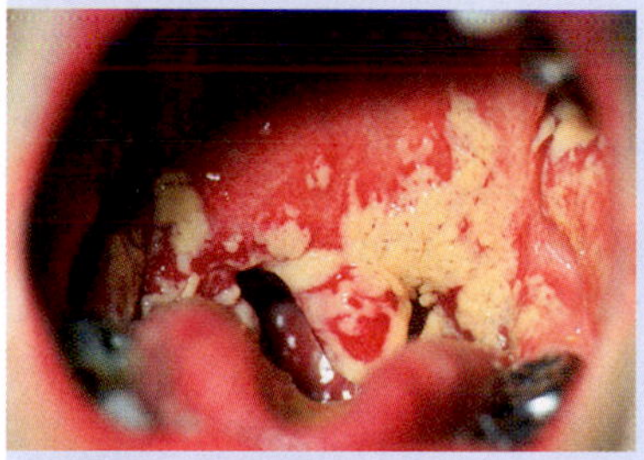

Thrush

A 25-year-old homosexual man was admitted to the hospital with oral candidiasis (thrush), diarrhea, unexplained weight loss, herpes lesions, and pneumonia. Cultures of pulmonary fluid revealed the presence of *Pneumocystis jirovecii.* The man admitted to frequently paying for sex.

1. Based on the man's symptoms, what diagnosis can be made?
2. What laboratory tests could confirm a diagnosis in this case?
3. How did the man most likely acquire the infection?
4. What changes to the man's immune system allowed the opportunistic infections of *Candida* (thrush) and *Pneumocystis* (pneumonia) to arise?
5. What precautions should be taken with the patient's blood?

DISEASE AT A GLANCE 24.7

Genital Warts

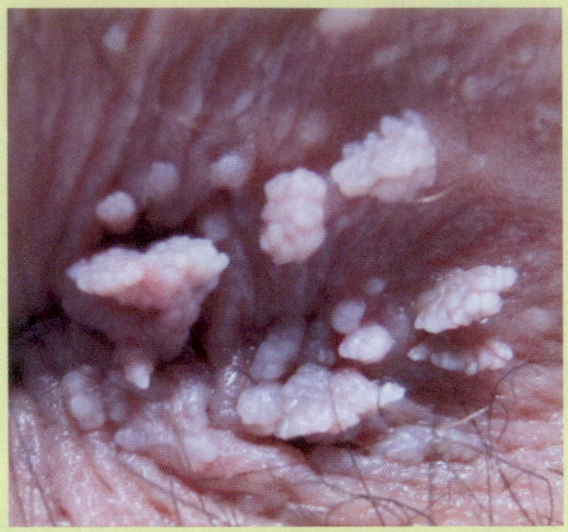

Cause Human papillomavirus (HPV, *Papillomavirus,* naked, double-stranded DNA virus).

Virulence factors Intracellular replication cycle, integration into human chromosome, ability of viral proteins E6 and E7 to trigger cancer.

Portal of entry Mucous membrane or skin of genitalia; direct contact with infected individuals or infected fomites.

Signs and symptoms Soft, small bumps to giant, cauliflower-like growths on the genitals.

Incubation period Three to four months.

Susceptibility Sexually active individuals, especially adolescents, uncircumcised men, and their sexual partners.

Treatment Removal of wart via surgery, freezing, cauterization, laser, or chemicals.

Prevention Abstinence and mutual monogamy.

Protozoan STDs

Sexual activity can transmit several protozoa, including the transfer of intestinal parasites such as *Giardia* and *Cryptosporidium* during oral-anal intercourse; however, this is not the usual way these protozoa are infective. This section examines *Trichomonas,* a protozoan that normally infects reproductive systems.

Trichomoniasis

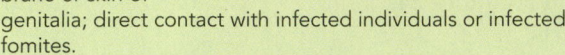

LEARNING | OUTCOME

24.32 Describe the features of trichomoniasis, including signs, symptoms, cause, pathogenesis, treatment, and prevention.

Trichomonads are flagellated protozoan parasites of animals and humans. Only one species is pathogenic, *Trichomonas vaginalis* (trik-ō-mō´nas va-jin-al´is), which causes an STD called **trichomoniasis** (trik´ō-mō-nī´a-sis) or *trich* (trik).

Signs and Symptoms

Trichomoniasis is usually symptomatic in women and asymptomatic in men. Infected women typically have a foul-smelling, yellowish-green vaginal discharge and vaginal irritation. They may also have lesions in the genitalia, abdominal pain, painful urination, and pain during sexual intercourse. In a few infected men, *T. vaginalis* causes inflammation of the urethra or prostate—urethritis and prostatitis, respectively—which causes painful urination. These symptoms in men might be confused with those of gonorrhea.

Pathogen and Virulence Factors

Trichomonas is a leaf-shaped, parabasalid protozoan, averaging 13 μm in length and about half as wide. It has five anterior flagella and an undulating membrane (see Disease at a Glance 24.8). The protozoan lives solely in humans, inhabiting vaginas, urethras, and prostate glands.

Trichomonas reproduces at pH 5 to pH 6. Normal lactobacilli keep the pH of the vagina at 4.0 to 4.5; thus, the vaginal microbiota help keep *Trichomonas* in check.

Pathogenesis and Epidemiology

T. vaginalis does not produce cysts and cannot survive long in the relatively high oxygen concentration and dry atmosphere outside the body; therefore, infection is nearly always via sexual intercourse. However, epidemiologists have documented a few cases of *Trichomonas* transferring to patients on fomites such as damp washcloths. Presence of *Trichomonas* in newborns indicates that the flagellate can infect babies during their passage through the birth canal.

For over 60 years researchers have puzzled over the exact method of *T. vaginalis* pathogenicity. Virulence factors include the ability to adhere to human cells, enzymes that break down proteins, the ability to hemolyze blood, and production of a cell-detaching protein, which disrupts tissue integrity.

Health departments report about 7.5 million new cases of trichomoniasis each year in the United States, and WHO estimates there are about 170 million new cases worldwide each year. The disease is the most common curable STD in women. People with multiple sex partners or other venereal diseases have the highest risk of contracting trichomoniasis. The disease increases patients' susceptibility to infection by HIV and may increase the likelihood that they will pass HIV to their sexual partners.

Diagnosis, Treatment, and Prevention

Diagnosis usually relies on identification of motile *Trichomonas* in secretions from the vagina, urethra, or prostate. The protozoan is found in only about two-thirds of specimens from people who have it, so fluorescent antibody staining of smears or culturing the parasite provides confirmation of a diagnosis.

Treatment is with a single, 2-gram dose of metronidazole or tinidazole taken orally. A physician must treat the male partner at the same time or the woman can be reinfected. No long-term immunity develops against *T. vaginalis.*

Prevention depends on refraining from sexual intercourse with infected persons. Babies can be protected by treating their mothers with metronidazole before they give birth.

Disease at a Glance 24.8 summarizes the features of trichomoniasis.

DISEASE AT A GLANCE 24.8

Trichomoniasis

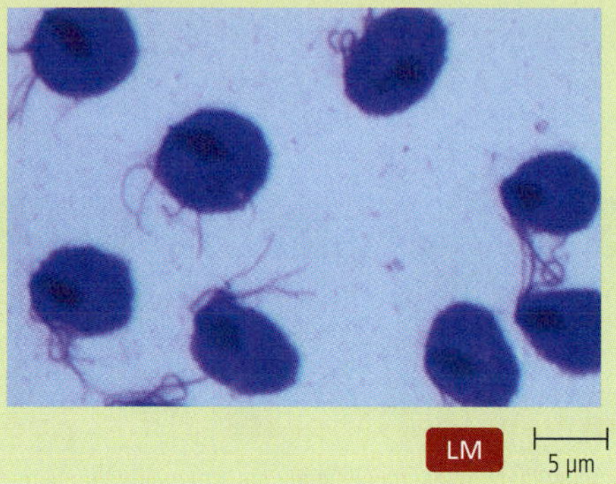

Cause *Trichomonas vaginalis,* a single-celled, flagellated protozoan in the taxon Parabasala.

Virulence factors Adhesins, proteolytic enzymes, hemolysis, cell-detaching factor.

Portal of entry Mucous membrane of genital tracts.

Signs and symptoms In females: malodorous, yellow-green vaginal discharge, vaginal irritation, possibly spot bleeding, dysuria; in males: usually asymptomatic, though it may cause urethritis and prostatitis.

Incubation period 4–28 days.

Susceptibility Sexually active individuals, more commonly heterosexual than homosexual; uncircumcised males may be more at risk than are circumcised males; in pregnant women, trichomoniasis can result in low birth weight and premature birth of baby.

Treatment Metronidazole or tinidazole.

Prevention Abstinence, faithful monogamy of both partners, consistent and accurate condom usage.

TELL ME WHY

Why might a course of antibacterial drugs trigger a case of trichomoniasis in a female patient?

MICRO IN THE CLINIC FOLLOW-UP

Picture-Perfect Romance?

Jordan is shocked when the test confirms a diagnosis of genital herpes, caused by *human herpesvirus 2* (HHV-2). The physician assistant reassures Jordan that although the infection cannot be cured, drugs such as acyclovir can diminish the frequencies of lesion outbreaks. The clinician tells Jordan that he should avoid sexual contact when an outbreak of lesions appears.

After Jordan's diagnosis, Lisa immediately gets tested for herpes as well. Her test, too, comes back positive for HHV-2. She learns that she was likely infected from a previous sexual relationship but experienced no obvious symptoms other than an occasional mild infection that she mistook for a yeast infection. Lisa periodically sheds the virus from her genital tract, but this was undetectable by either her or Jordan.

The couple learns that a standard STD screening test does not include a test for HHV-2. While the couple experiences an emotional rollercoaster after their shared diagnosis, they learn that having HHV-2 is not a hopeless condition. Even though the infection is irreversible, its symptoms are clinically manageable.

1. **Having genital herpes has been found to increase the risk of HIV infection fourfold. Why is this?**
2. **A couple wants to practice "safe sex" by using condoms whenever the woman has herpes lesions. Will condoms protect the man from becoming infected with human herpesvirus?**

MM Check your answers to Micro in the Clinic Follow-Up questions in the MasteringMicrobiology Study Area.

CHAPTER SUMMARY

Structures of the Urinary and Reproductive Systems (pp. 754–756)

1. The male and female urinary systems each consist of two kidneys composed of millions of nephrons that filter blood to form urine, two ureters that carry urine to a urinary bladder for temporary storage, and a urethra, which carries urine out of the body. The urethra may be a portal for the entrance of microorganisms.
2. The adult female reproductive system consists of two ovaries, which produce haploid ova, one of which is released monthly. Ova are swept through a uterine tube toward the blood-rich uterus, which is prepared for pregnancy.
3. If a sperm cell (haploid) fertilizes the ovum (also haploid), the resulting diploid zygote develops into an embryo that implants in the uterine wall and further develops into a fetus who passes through the cervix and the birth canal to be born approximately nine months from fertilization.
4. When an ovum is unfertilized, it and the lining of the uterus pass out of the vagina during menstruation. The vaginal opening can be a portal for microorganisms, especially during sexual intercourse.
5. The adult male reproductive system consists of two sets of these structures: a testis (located in an external sac, the scrotum), an epididymis, and a ductus deferens. Sperm formed by a testis are stored in an epididymis and pass through a ductus deferens, into the single prostate gland, where the left and right ducts join the urethra, which passes through the prostate gland (where fluid is added to the sperm to form semen) and penis. Sexually transmitted microbes can enter the body through the urethra or skin of the penis.

Bacterial Diseases of the Urinary System (pp. 756–757)

1. Bacteria may trigger inflammation in the urethra (urethritis), urinary bladder (cystitis), prostate (prostatitis), or kidneys (pyelonephritis). These are all types of **urinary tract infections (UTIs)**.
2. Mild urethritis, cystitis, or prostatitis cause problems with urination and produce fever. Pyelonephritis causes pain, higher fever, vomiting, and fatigue; if untreated, it can be fatal.
3. Enteric (fecal) bacteria most often cause UTIs.
4. The spirochete *Leptospira interrogans*, which is in the urine of infected animals, causes **leptospirosis**, a **zoonotic** disease (disease of animals acquired by humans), which results in bacteremia and infected kidneys.
5. When some strains of group A *Streptococcus* infect adults, antibody-antigen complexes accumulate in the glomeruli of the kidneys to trigger **glomerulonephritis**, inflammation of the glomeruli and nephrons. Streptococcal acute glomerulonephritis is a progressive and irreversible kidney disease in adults.

Nonvenereal Diseases of the Reproductive Systems (pp. 757–763)

1. Certain strains of *Staphylococcus aureus* cause staphylococcal **toxic shock syndrome (TSS)**, a potentially fatal condition observed historically in menstruating females who use super-absorbent tampons. Toxins produced by *S. aureus* in the vagina or in a wound stimulate T cells to release excessive cytokines, which triggers TSS.
2. Facultative or obligate anaerobic bacteria infecting the vaginal lining may cause bacterial **vaginosis**, a noninflammatory condition associated with odiferous vaginal discharge. Risk factors include having multiple sexual partners and vaginal douching. Douching reduces the normal population of lactobacilli in the vagina, elevating the pH and allowing pathogenic bacteria to grow.
3. Following changes in vaginal pH or in normal microbiota, opportunistic *Candida albicans* (yeast) causes vaginal **candidiasis** characterized by curdlike vaginal discharge, itching, and burning.
4. *Candida* forms diagnostic cellular extensions called **pseudohyphae**.

Sexually Transmitted Infections (STIs) and Diseases (STDs) (pp. 763–764)

1. **Sexually transmitted infections (STIs)** involve microbes shared by sexual partners. When they cause disease, such disease is a **sexually transmitted disease (STD)**. STDs are at epidemic proportions.
2. **Pelvic inflammatory disease (PID)** is a general term for inflammation and pain in the uterus, uterine tubes, or ovaries. Untreated PID can lead to ectopic pregnancies and sterility.

Bacterial STDs (pp. 764–772)

1. **Gonorrhea** is caused by *Neisseria gonorrhoeae,* which adheres to the epithelial cells of mucous membranes, especially in the genital, urinary, and digestive tracts, and can cause painful urination and pus-filled discharge from the penis in men, uterine tube damage or pelvic inflammatory disease in women, and eye damage in newborns infected during birth. Many women are asymptomatic. When the bacterium travels inside leukocytes throughout the body, other organs are affected.
2. The spirochete *Treponema pallidum* causes **syphilis**, an STD with four phases.
3. During **primary syphilis,** a small, hard, extremely infectious **chancre** (lesion) fills with spirochetes at the site of infection and remains for several weeks.

4. When *Treponema* spreads via the bloodstream, **secondary syphilis** results. It is characterized by malaise and a long-lasting contagious skin rash.
5. **Latent syphilis** is a clinically inactive phase, which may last decades.
6. **Tertiary syphilis** is associated with severe complications resulting from inflammation and a hyperimmune response. Untreated patients can experience dementia, blindness, paralysis, heart failure, and syphilitic **gummas** (swollen lesions on skin or other organs).
7. **Congenital syphilis** is a disease of a fetus following transmission in the womb of an infected mother. The results may be death, mental retardation, rash, or the malformation of organs.
8. *Chlamydia trachomatis* is the most common sexually transmitted bacterium; infections are often asymptomatic in women and produce symptoms similar to gonorrhea in men.
9. Chlamydial infection can also cause inflammation of an epididymis **(epididymitis)**, inflammation of a testis **(orchitis)**, eye disease in babies at birth **(trachoma)**, and **lymphogranuloma venereum**, which is a genital lesion followed by the development of an inflamed lymph node (a **bubo**).
10. *Haemophilus ducreyi* causes **chancroid**, which is characterized by soft, painful, genital ulcers called soft chancres.

Viral STDs (pp. 772–776)

1. Human herpesviruses HHV-2 ("below-the-waist" virus) and HHV-1 (the "cold-sore" virus) cause **genital herpes,** which is characterized by genital blisters full of virions. Some viruses remain latent in nerve ganglia and cause recurrent manifestations.
2. Herpesviruses also spread by nonsexual means, such as during birth or via contact with herpes blisters on the skin—a condition called whitlow.
3. *Papillomavirus* causes **genital warts,** which range in size from barely detectable to a cauliflower-like growth called **condylomata acuminata**. Some human papillomaviruses cause cancers.

Protozoan STDs (pp. 776–777)

1. Intestinal protozoa may be transferred by certain sexual practices.
2. The protozoan *Trichomonas vaginalis* causes **trichomoniasis**.
3. Trichomoniasis is symptomatic in women (an odiferous, yellowish green vaginal discharge and vaginal irritation) but usually asymptomatic in men.

QUESTIONS FOR REVIEW

Answers to the Questions for Review (except Short Answer questions) begin on p. A-1.

Multiple Choice

1. The functional unit of a kidney is the
   a. pelvis.
   b. nephron.
   c. glomerulus.
   d. renal pyramid.
2. Which of the following sequences most accurately describes the passage of sperm through the reproductive tract?
   a. testis, epididymis, ductus deferens, urethra, penis
   b. testis, ductus deferens, epididymis, prostate, urethra
   c. scrotum, ductus deferens, epididymis, urethra, prepuce
   d. vas deferens, epididymis, prostate, urethra, penis
3. Bacterial infection of the vagina that does *not* involve inflammation is called
   a. bacterial virulence.
   b. bacterial vaginitis.
   c. bacterial vaginosis.
   d. bacterial bulbosis.
4. Select the true statement about gummas.
   a. Gummas are small, painless, reddened, hard lesions.
   b. Gummas are diseased lymph nodes, which may persist for months.
   c. Gummas typically form during the primary phase of syphilis.
   d. Gummas are rubbery, painfully swollen lesions.
5. Which stage of syphilis is characterized by chancres filled with contagious spirochetes?
   a. primary
   b. secondary
   c. latent
   d. tertiary
6. Treatment of chlamydial infections involves
   a. erythromycin cream.
   b. tetracycline.
   c. surgical correction of eyelid deformities.
   d. all of the above.
7. In which stage of syphilis is penicillin ineffective?
   a. primary syphilis
   b. secondary syphilis
   c. tertiary syphilis
   d. Penicillin is ineffective for all of the above.
8. Which of the following is the most common of these sexually transmitted diseases in the United States?
   a. syphilis
   b. gonorrhea
   c. chlamydial infection
   d. AIDS
9. A foul-smelling, yellowish green vaginal discharge characterizes
   a. gonorrhea.
   b. trichomoniasis.
   c. chlamydial infection.
   d. syphilis.
10. Which of the following statements concerning properly used condoms is *false?*
    a. Condoms reduce the spread of genital warts.
    b. Condoms reduce the incidence of gonorrhea.
    c. Condoms may increase the likelihood of chlamydial infection.
    d. Condoms are effective in reducing the transmission of syphilis.

Matching

Match the term with its description.

1. _____ Dysuria
2. _____ Bacteremia
3. _____ Zoonosis
4. _____ Ophthalmia neonatorum
5. _____ Proctitis
6. _____ Pyelonephritis
7. _____ Orchitis
8. _____ Gummas
9. _____ Myalgia
10. _____ Cystitis
11. _____ Whitlow
12. _____ Prostatitis
13. _____ Condylomata acuminata
14. _____ PID
15. _____ Venereal disease

A. Inflammation of a kidney
B. Inflammation of the prostate
C. Inflammation of the urinary bladder
D. Inflammation of a testis
E. Inflammation of the rectum
F. Frequent, urgent, and painful urination
G. Muscle pain
H. Inflammation of the conjunctiva in newborns
I. Disease spread from animals to humans
J. Rubbery, painfully swollen lesions in bones, nervous tissue, or on the skin
K. Large cauliflower-like growths on the genitalia
L. General term for an STD
M. Infection of skin by herpesviruses
N. Bacterial invasion of the bloodstream
O. Inflammation of uterus, uterine tubes, and ovaries

Match the organism with the disease it causes.

1. _____ *Leptospira*
2. _____ *Haemophilus*
3. _____ *Chlamydia*
4. _____ *Treponema*
5. _____ *Neisseria*
6. _____ *Staphylococcus*
7. _____ *Candida*
8. _____ HHV-1
9. _____ HHV-2
10. _____ *Papillomavirus*

A. Cold sores
B. Toxic shock syndrome
C. Genital warts
D. Syphilis
E. Chancroid
F. Leptospirosis
G. Lymphogranuloma venereum
H. Genital herpes
I. Gonorrhea
J. Yeast infection

Fill in the Blanks

1. Circumcision involves cutting away the _______________ of the penis.
2. The majority (70%) of urinary tract infections are caused by the intestinal bacterium _______________.
3. When the normal lactobacilli of the vagina are replaced by *Gardnerella vaginalis* or *Mycoplasma hominis*, the resulting infection is called _______________.
4. Three structural virulence factors of a virulent strain of *Neisseria gonorrhoeae* are _______________, _______________, and _______________.
5. Trachoma, the leading cause of nontraumatic blindness in humans worldwide, is caused by the sexually transmitted bacterium _______________.
6. The most common curable protozoan STD in women is _______________.

Modified True/False

Change the underlined word or phrase in all false statements to make the statement true.

1. _____ Toxic shock syndrome <u>is</u> a sexually transmitted disease.
2. _____ Nearly all AIDS patients develop candidiasis in their <u>reproductive</u> or <u>digestive tracts</u>.
3. _____ Gonorrhea is frequently transmitted via <u>toilet seats</u>.
4. _____ The symptoms of gonorrhea in <u>women</u> are more obvious and painful than in <u>men</u>.
5. _____ There is <u>no</u> cure for genital herpes.
6. _____ The bacterium that causes syphilis is transmitted via <u>contact with toilet seats, soiled clothing, or other fomites</u>.
7. _____ Evidence suggests that male <u>circumcision</u> reduces the risk of cervical cancer in female sexual partners.
8. _____ Over-the-counter wart removal medications <u>are recommended</u> for genital warts.
9. _____ There <u>is</u> a vaccine for some strains of cancer-causing papillomavirus.
10. _____ A person infected with herpesvirus but lacking blisters <u>frequently</u> spreads genital herpes in mucous secretions.
11. _____ Although usually sexually transmitted, trichomoniasis <u>can be</u> transferred via damp fomites.

Short Answer

1. The normal microbiota of the female reproductive tract help maintain a vaginal pH of about 4.5. Why is this important?
2. If the urinary organs and the urine in them are usually axenic, why does normally voided urine contain bacteria?
3. What structure found in *Neisseria gonorrhoeae* allows the bacterium to attach to sperm cells and travel up the uterine tubes to produce pelvic inflammatory disease?
4. The Smith family gets cold sores regularly. Suggest an explanation for why this problem occurs.
5. Why have scientists had problems identifying virulence factors of *Treponema pallidum*?
6. Why do physicians place antimicrobial agents in babies' eyes at birth?
7. It has been observed that prepubescent girls are more susceptible to vaginal infections than adult women. Explain this observation, mentioning glycogen, lactic acid, pH, and estrogen in your discussion.
8. Although vaginal douching is a cleansing procedure, it can actually promote infection. Explain.
9. At a small farm, the children regularly played in a pond that was frequented by livestock. The children were unusually ill one year and complained of stomachaches, vomiting, headaches, and fever. What might be the cause?
10. Why is there no effective vaccine against gonorrhea?
11. Why is tertiary syphilis difficult to diagnose?
12. Describe and sketch the developmental cycle of *Chlamydia*.
13. Compare and contrast chancroid lesions with those of primary syphilis and genital herpes.
14. Contrast vaginosis with vaginitis.

VISUALIZE IT!

1. Label the following stages and structures of the chlamydia life cycle: *elementary body, endocytosis, vesicle, host cell, inclusion body, initial body.* Indicate how many hours typically transpire at each lettered step.

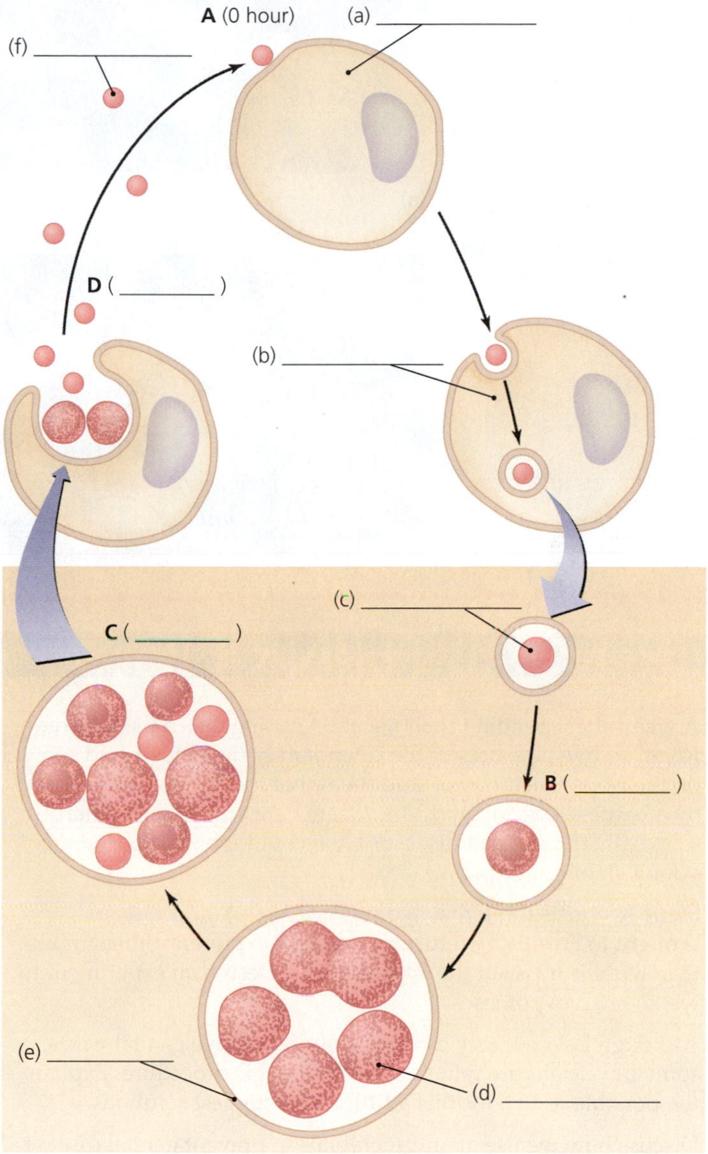

2. Name the following pathogens of the urinary/reproductive systems.

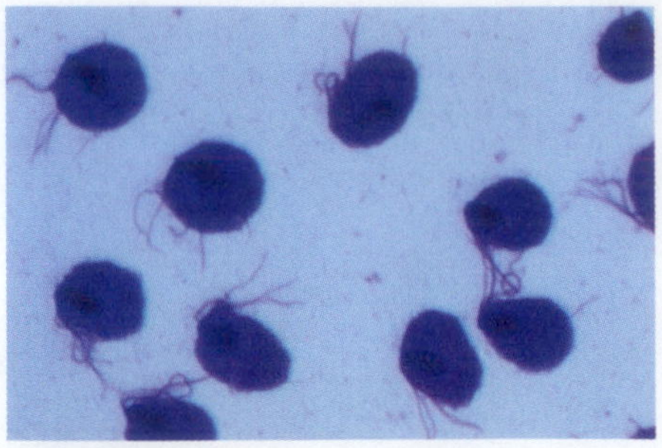

(a)

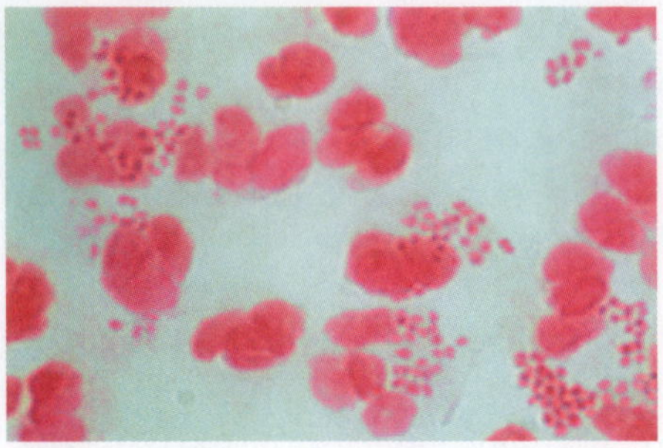

(b)

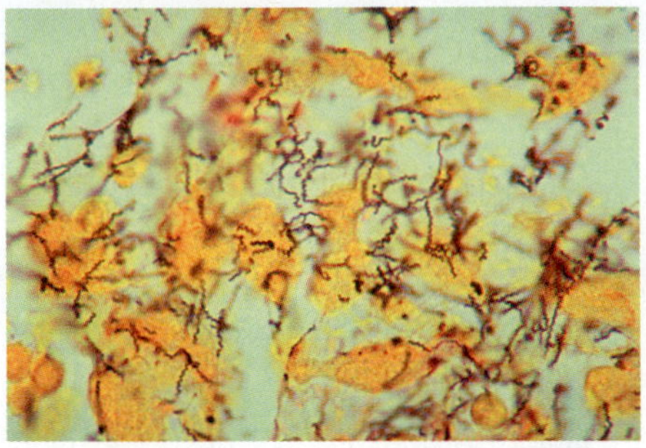

(c)

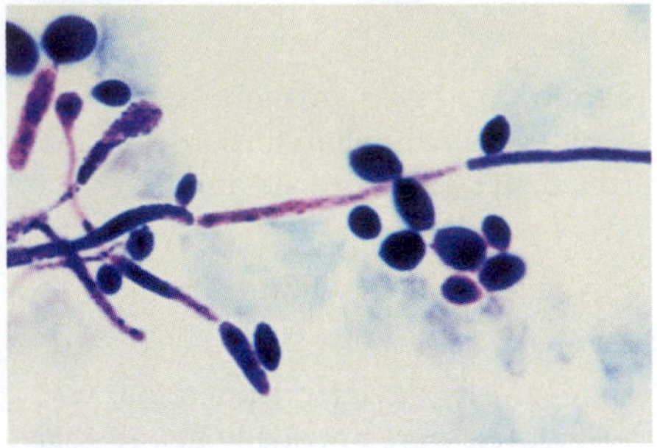

(d)

CRITICAL THINKING

1. A week after spending their vacation rafting down the Colorado River, all five members of the Chen family developed cold sores on their lips. Their doctor told them that the lesions were caused by a herpesvirus. Mr. and Mrs. Chen were stunned; isn't herpes a sexually transmitted disease? How could it have affected their young children? Explain.
2. Some scientists think that syphilis is a New World disease brought to Europe by returning Spanish explorers; others think that syphilis traveled the opposite way. Design an experiment to test the two hypotheses.
3. Although laser surgery can successfully remove genital warts, some physicians are reluctant to choose this procedure. Explain the possible reason for this attitude and suggest solutions.
4. Discuss how the use of antimicrobials to prevent genital disease may have negative effects.
5. After a patient complains that his eyes are extremely sensitive to light and feel gritty, his doctor informs him that he has ocular herpes. What causes ocular herpes? Which human herpesvirus, type 1 or type 2, is more likely to cause ocular herpes? Why?
6. Fifteen-year-old Dolores was embarrassed to talk with her mother about the pain she felt in her "private parts," but she was worried she might have gotten a disease, despite Nick's assurance that she was his "first." Her mother insisted that Dolores see a doctor, who discovered cervical lesions caused by a virus associated with cancer. What sexually transmitted virus is involved? How does the physician treat the lesions? How could Dolores have protected herself?
7. Explain why some natural-health advocates promote ingesting living *Lactobacillus* cultures as a way to prevent bacterial vaginosis. Why is the practice likely to fail?
8. Why does treatment of strep throat with an antibacterial drug increase the chance for vaginal candidiasis?
9. How could Koch's postulates be applied to prove that HPV causes cervical cancer?

CONCEPT MAPPING

Using the following terms, draw a concept map that describes syphilis. For a sample concept map, see p. 94. Or, complete this and other concept maps online by going to the MasteringMicrobiology Study Area.

Body-wide rash
Cardiovascular syphilis
Cell-free media
Chancre
Congenital syphilis
Dark-field microscopy
Gummas
Infect fetus in pregnant woman
Lasts for years
Latent stage
Neurosyphilis
Never progress
Penicillin
Primary syphilis
Secondary syphilis
Serological tests
Spirochete
Tertiary syphilis
Treponema pallidum

25 Applied and Environmental Microbiology

MICRO IN THE CLINIC

Food Poisoning at a Five-Star Restaurant?

As the master chef at a well-respected restaurant, Dan loves his work and takes it seriously. He understands the importance of sanitation in food preparation and is very strict about training his staff properly to prevent food contamination. He doesn't allow people who are ill to even enter his kitchen.

One day Dan gets a phone call from a long-time customer, who says he and his wife developed diarrhea after dining at Dan's restaurant. Dan reassures him and offers him a free entrée the next time they come in. The next day, another customer calls to complain of becoming ill. Dan begins to get a little worried, but he reviews safety guidelines with his staff and genuinely cannot find anything wrong.

Two nights later, Dan starts feeling nauseated and develops severe diarrhea. He has abdominal cramps, fever, and chills, and his muscles ache. At the emergency room, the physician is interested to discover that Dan is a chef at a well-known restaurant. It turns out other patients with similar symptoms became ill after eating at his restaurant. Dan soon becomes the focus of attention from both the infection control nurse and public health officials.

The public health inspectors visit Dan's spotless gourmet kitchen to take samples. Dan is baffled and worried. He not only is feeling sick himself but also is concerned about his customers' health and the reputation of the business he worked so hard to build.

Will Dan and his customers recover? What will the public health inspectors find in Dan's kitchen? Turn to the end of the chapter (p. 814) to find out.

MM **Explore More:** Test your readiness and apply your knowledge with dynamic learning tools at MasteringMicrobiology.

The metabolic activities of microorganisms shape much of our environment, and microbial reactions are essential to life on Earth. Bacteria and fungi in particular are capable of many metabolic processes that are useful to humans. In this chapter we examine some of the leading areas of applied microbiology as well as topics in the field of environmental microbiology. (Chapter 8 discusses some aspects of recombinant DNA technology that give organisms enhanced or new functions to make them more valuable in industry, agriculture, or medicine.)

Applied microbiology, the commercial use of microorganisms, encompasses two distinct fields: **food microbiology**, which includes the use of microorganisms in food production and the prevention of food spoilage and food-related illnesses, and **industrial microbiology**, which involves both the application of microbes to industrial manufacturing processes and solutions to environmental, health, and agricultural problems. **Environmental microbiology** explores where microorganisms are found in nature and how their activities affect other organisms (including humans) and the environment itself. We begin by considering food microbiology.

Food Microbiology

Microorganisms are involved in producing many of our favorite foods and beverages, from bread to wine to yogurt. Indeed, fermented foods are among some of the oldest foods known and are culturally very diverse. The characteristic flavors, aromas, and consistencies of such foods result from the presence of acids or sugars made by microbes during fermentation. Besides conferring taste and aroma, microbial metabolism also acts as a preservative, destroys many pathogenic microbes and toxins, and, in some cases, adds nutritional value in the form of vitamins or other nutrients. In the following sections, we explore how our knowledge of microbial growth and metabolism enables us to use microbes in food production and to control microbial activity that results in food spoilage.

The Roles of Microorganisms in Food Production

LEARNING | OUTCOME

25.1 Describe how microbial metabolism can be manipulated for food production.

Bread

Saccharomyces cerevisiae (sak-ă-rō-mī´sēz se-ri-vis´ē-ī) metabolizes sugars to leaven[1] bread. Bakers add the yeast to flour, salt, and other ingredients to make dough, which is kneaded to introduce oxygen. The dough rises when metabolic reactions release CO_2, producing expanding pockets within the dough. Ethanol produced by fermentation evaporates during baking. Sourdough bread is made using starter cultures consisting of yeast and lactic acid bacteria. Lactic acid produced by the bacteria gives sourdough its characteristic taste.

Biochemists use the word *fermentation* to refer to the partial oxidation of sugars to release energy using organic molecules as electron acceptors (see Chapter 5). In food microbiology, however, **fermentation** may refer to any desirable changes that occur to a food or beverage as a result of microbial growth. In contrast, **spoilage** denotes unwanted change to a food that occurs from undesirable metabolic reactions, the growth of pathogens, or the presence of unwanted microorganisms.

Fermentative microbes naturally occur on grains, fruits, and vegetables. In antiquity, people relied on these naturally occurring microbes to produce fermented foods and drinks. However, the same microbes are not always present on a food from harvest to harvest, yielding varying results. Most modern commercial food and beverage production relies on **starter cultures** composed of known microorganisms that perform specific fermentations consistently. In addition to an initial starter culture, *secondary cultures* may be added to further modify the flavor or aroma of foods. For example, the same starter culture is used to initiate formation of both Swiss cheese and blue cheese; the two cheeses differ because different secondary cultures are used in their production.

Fermented Vegetables

People around the world ferment many types of vegetables. Most of these vegetable products are the result of the actions of lactic acid bacteria, such as *Streptococcus* (strep-tō-kok´ŭs), *Leuconostoc* (loo´kō-nos-tŏk), *Lactobacillus* (lak´tō-bă-sil´ŭs), or *Lactococcus* (lak-tō-kok´ŭs), which specifically produce lactic acid during fermentation. Lactic acid acidifies the food and produces a "sour" flavor.

Food products derived from the fermentation of cabbage include kimchi (kim-chē´) from Korea and sauerkraut. Soy sauce is made by the fermentation of soybeans and wheat by lactobacilli, yeast *(S. cerevisiae)*, and the fungus *Aspergillus oryzae* (as-per-jil´ŭs o´ri-zī). Chocolate is derived from the fermentation f cacao seeds. Coffee production relies on natural fermentation to release the coffee bean from the outer layers of the coffee berry so that the bean can then be dried and roasted.

Pickles are another common fermented food. While people often equate "pickles" with cucumbers, other foods, such as beets and eggs, can also be pickled. *Pickling* refers to the process of preserving or flavoring foods with brine (saturated salt water) or acid. Microbial fermentation can be the source of the acid—as it is with dill pickles. Various spices can be added to the pickling solution to enhance taste. Because pickled foods are acidic, few pathogenic microorganisms survive, making pickling an excellent preservation method.

Fermented foods are not made for human consumption alone. *Silage*, a product used as animal feed on many farms, is made by the natural fermentation of potatoes, corn, grass, grain stalks, or other types of green foliage. The vegetation is cut up and stored in silos that are closed to air. Under such moist, anaerobic conditions, the vegetation in the silos

[1]From Latin *levare*, meaning "to raise."

ferments, producing many organic compounds that make the silage aromatic and tasty to livestock and more easily digested by them. Some farmers produce silage from baled hay and grasses wrapped in plastic.

Fermented Meat Products

Unless dried, cured, smoked, or fermented in some way, meat has a tendency to spoil rapidly. The combination of fermentation and drying or smoking has been used for centuries to preserve meats, particularly pork and beef. Dry sausages, such as salami and pepperoni, are made by grinding meat, mixing in various spices and starter cultures, allowing the mixture to ferment in the cold, and then stuffing it into casings. Fermented fish—common in Asian countries—are prepared by grinding fish in brine. Naturally occurring microbes ferment the mixture. The solids are removed and pressed to form a fish paste, while the liquid portion is drained off and mixed with flavoring agents to form fish sauce.

Fermented Dairy Products

Milk fermentation relies, for the most part, on the activities of lactic acid bacteria. Milk in an udder is sterile, but because milking introduces microorganisms, we pasteurize "raw" milk. The metabolic products of starter cultures give fermented milk products their characteristic textures and aromas.

Buttermilk is made from fat-free milk (skim milk) by adding a starter culture of *Lactococcus lactis* (lak´tis) subspecies *cremoris* (kre-mōr´is) and *Leuconostoc citrovorum* (sit-rō-vō´rum). Yogurt production utilizes starter cultures of *Streptococcus thermophilus* (ther-mo´fil-us) and *Lactobacillus bulgaricus* (bul-gā´ri-kŭs). Yogurt manufacturers mix pasteurized milk, milk solids, sweeteners, and other ingredients to a uniform consistency and then add the starter culture. They ferment the mix at 43°C and then cool it to stop microbial activity. They may add flavors prior to packaging.

Cheeses can be hard or soft and mild or sharp. Regardless of the end product, the cheese-making process typically begins with pasteurized milk **(FIGURE 25.1)**. A pound of cheese requires about 5 gallons of milk. Cultures of *Lactococcus lactis* coagulate protein in milk to form *curds* (solids) and *whey* (liquids). For cheeses sold as soon as they are made (called *unripened cheeses*, such as cottage cheese), the curd is removed, cut into small pieces, and packaged. Alternatively, the curdling process can be accelerated by the addition of the enzyme *rennin*, a type of protease (protein-digesting enzyme).

Other cheeses (so-called ripened cheeses) are aged until they have the desired texture or taste. Hard cheeses, such as Parmesan and cheddar, are pressed such that little water remains, while soft cheeses retain enough moisture to make them spreadable. Once the curds are pressed, the cheese is aged for months to years while continued microbial activity imparts characteristic smells and tastes. Cheese producers utilize numerous different microorganisms and their enzymes along with various ripening regimens and culture conditions to make numerous, distinct types of cheeses.

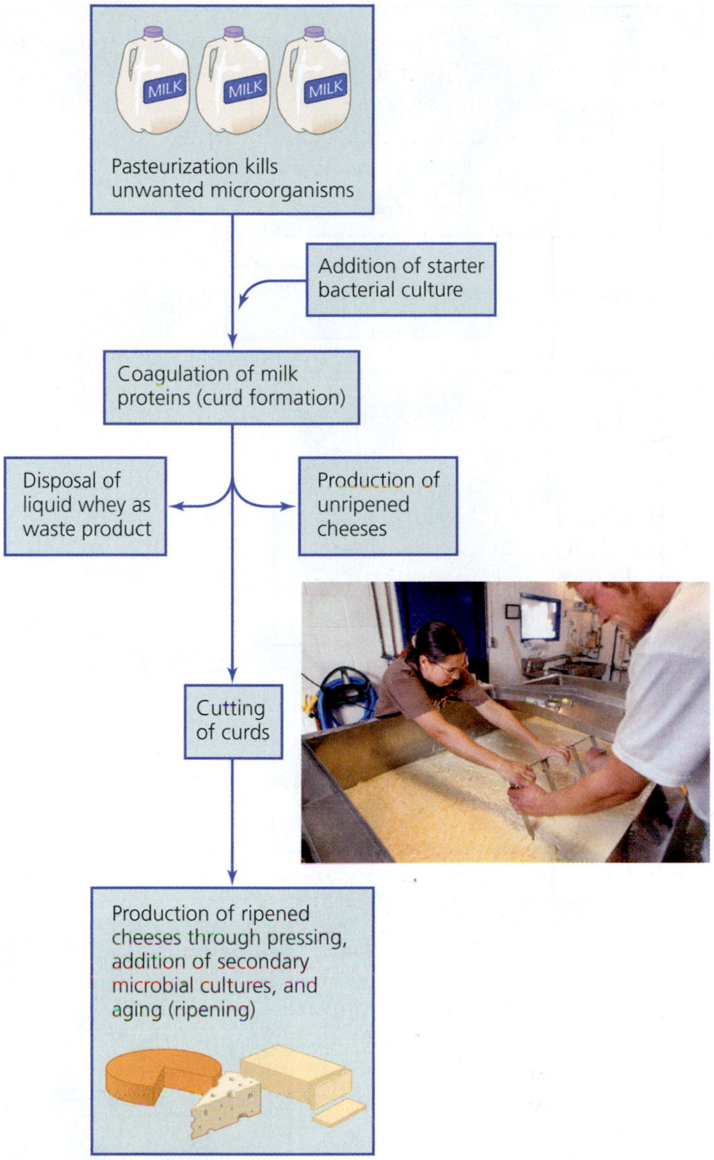

▲ **FIGURE 25.1 The cheese-making process.** The commercial production of all cheeses begins with the pasteurization of milk; the wide variety of cheeses available on the market results from the nature of subsequent processing steps. *Why do cheeses that are aged longer have more acidic, "sharper" tastes?*

Figure 25.1 *Fermentation during aging results in continued production of acid; the longer the cheese is aged, the more acid produced and the "sharper" the taste.*

Products of Alcoholic Fermentation

Alcoholic fermentation is the process by which various microorganisms convert simple sugars such as glucose into ethanol (drinking alcohol) and carbon dioxide (CO_2). These two products of fermentation take on different importance for different foods and beverages. As is the case for the production of fermented dairy products, manufacturers use specific starter cultures in the large-scale commercial applications of alcohol fermentation. In the next sections we consider the role of fermentation in the production of wine, spirits, beer, sake, and vinegar.

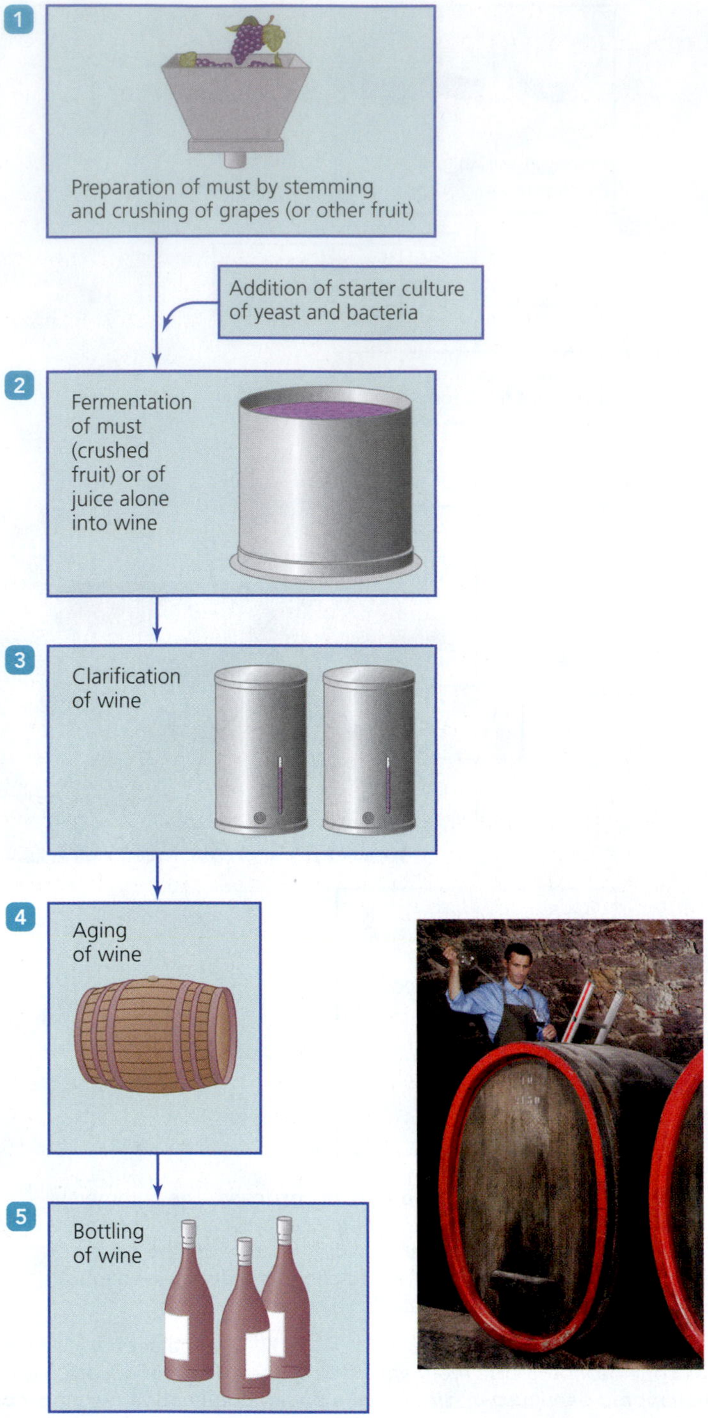

▲ **FIGURE 25.2 The wine-making process.** Red wine is produced by fermenting must (fruit solids and juice) from dark grape varieties, whereas white wine results from the fermentation of juice alone from either dark or light grape varieties.

Wine and Spirits Generally, wine production proceeds as follows (**FIGURE 25.2**):

1. *Preparation of must.* Winemakers crush the fruit and remove the stems to form *must* (fruit solids and juices). They make red wines from the entire must of dark grapes, whereas they make white wines using only the juice of either dark or light grapes. Weather conditions and soil composition affect the sugar content of fruit, which in turn greatly affects the quality of the wine.
2. *Fermentation.* A thin film of bacteria and yeast naturally covers grapes and other fruit, and in some wineries these natural microbes produce the wine. However, in most commercial operations, vintners add sulfur dioxide (SO_2) to retard growth of naturally occurring bacteria and then add starter cultures of yeast and bacteria to ensure that wine of a consistent quality is generated from year to year. *Saccharomyces* ferments sugar in fruit juice to alcohol to make wine. *Leuconostoc* removes acids naturally present in grapes.
3. *Clarification.* Filtration or settling removes solids—a process called clarification.
4. *Aging.* Wine is aged in wooden barrels. Continued yeast metabolism and the leaching of compounds from the wood add taste and aroma.
5. *Bottling.* The wine is then bottled and distributed.

Dry wines are those in which all the sugar has fermented, whereas sweet or dessert wines retain some sugar. Most table wines have a final alcohol content of 10% to 12%, but fortified wines have an alcohol content of about 20% because distilled spirits (discussed next) are added. Sparkling wine is produced by the addition of sugar and a second fermentation that produces CO_2 inside the bottle.

Distilled spirits are made in a process similar to wine making in that *Saccharomyces* ferments fruits, grains, or vegetables. The difference is that during the aging process, the alcohol is concentrated by distillation. In this process, the liquid is heated to drive off the alcohol, which is then condensed and added back. The net result is an increase in the alcohol content (100-proof spirits are 50% alcohol). Brandies are made from fruit juices, whiskeys are made from cereal grains, and vodka is often made from potatoes.

Beer and Sake Beer is made from barley in the following steps (**FIGURE 25.3**):

1. *Malting.* Barley is moistened and germinated, a process that produces catabolic enzymes, which convert starch into sugars, primarily maltose. Drying halts germination, and the barley is crushed to produce *malt.*
2. *Mashing.* Malt and additional sources of carbohydrate called adjuncts (such as starch, sugar, rice, sorghum, or corn) are combined with warm water, allowing enzymes released during malting to generate more sugars. The sugary liquid is called *wort.*
3. *Preparation for fermentation.* The brewer removes the spent grain from the wort and adds *hops*—the dried flower-bearing parts of the vinelike hops plant. The wort mixture is boiled to halt enzymatic activity, extract flavors from the hops, kill most microorganisms, and concentrate the mixture.
4. *Fermentation.* A starter culture of *Saccharomyces* is added to the wort. Most beers—called lagers—are made with the bottom-fermenting yeast *S. carlsbergensis* (karlz-bur-jen´sis). Other beers—called ales—are produced with the

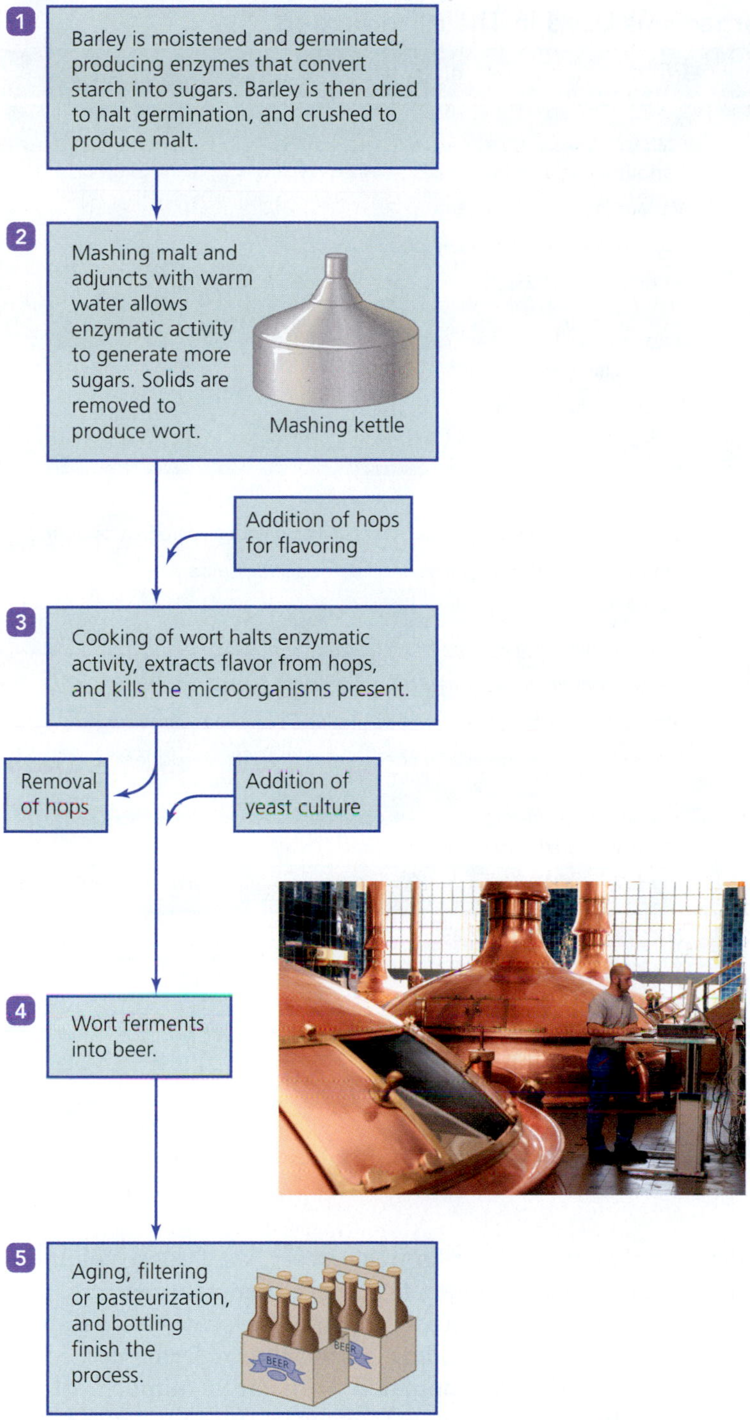

1
Barley is moistened and germinated, producing enzymes that convert starch into sugars. Barley is then dried to halt germination, and crushed to produce malt.
2
Mashing malt and adjuncts with warm water allows enzymatic activity to generate more sugars. Solids are removed to produce wort.
Mashing kettle
Addition of hops for flavoring
3
Cooking of wort halts enzymatic activity, extracts flavor from hops, and kills the microorganisms present.
Removal of hops
Addition of yeast culture
4
Wort ferments into beer.
5
Aging, filtering or pasteurization, and bottling finish the process.
BEER
BEER

▲ **FIGURE 25.3 The beer-brewing process.** The type of beer produced (ale, lager) is determined largely by the type of yeast used (top fermenting or bottom fermenting, respectively).

top-fermenting yeast *S. cerevisiae* (so called because the yeast floats in the vat).

5 *Aging.* The beer is aged, pasteurized or filtered, and bottled or canned. Beer has a considerably lower alcohol content (about 4%) than either wine or distilled spirits.

Sake (rice beer) is made from cooked rice in which the starch is first converted to sugar (by the fungus *Aspergillus oryzae*) and the sugar then is fermented by *Saccharomyces*. Sake has an alcohol content of about 14%.

Vinegar Vinegar[2] is produced when ethanol resulting from the fermentation of a fruit, grain, or vegetable is oxidized to acetic acid. *Acetic acid bacteria* such as *Acetobacter* (a-sē´tō-bak-ter) or *Gluconobacter* (gloo-kon´ō-bak-ter) perform this secondary metabolic conversion, generating about 4% acetic acid. Different flavors of vinegar are derived from different starting materials. Cider vinegar, for example, is made from apples.

TABLE 25.1 on p. 788 summarizes types of fermented foods along with the organisms involved in their production.

The Causes and Prevention of Food Spoilage

LEARNING **OUTCOMES**

25.2 Describe how food characteristics and the presence of microorganisms in food can lead to food spoilage.

25.3 List several methods for preventing food spoilage.

Whereas many chemical reactions involving food are desirable because they enhance taste, aroma, or preservation, other reactions are not. Spoilage of food involves adverse changes in nutritive value, taste, or appearance. Spoilage leads not only to economic loss but also potentially to illness and even death. In this section we examine the causes of food spoilage before considering ways to prevent it.

Causes of Food Spoilage

Foods may spoil because of inherent properties of the food itself. Such *intrinsic factors* include nutrient content, water activity, pH, and physical structure of the food as well as the competitive activities of microbial populations. Alternatively, food spoilage results from *extrinsic factors* that have nothing to do with the food itself but instead with the way it is processed or handled.

Intrinsic Factors in Food Spoilage The nutritional composition of food determines both the types of microbes present and whether they will grow. Some foods contain natural antimicrobial agents, such as the benzoic acid in cranberries, and thus are not prone to spoilage. In contrast, *fortified foods*—those enriched in vitamins and minerals to improve the health of humans—may inadvertently facilitate growth of microorganisms by providing more nutrients.

Water activity refers to water that is not bound physically by solutes or to surfaces and is thus available to microbes. The water activity of pure water is set at 1.0, and most microbes require environments with a water activity of at least 0.90. Moist foods, such as fresh meat, with water activities near 0.90 support microbial growth, whereas dry foods (e.g., uncooked pasta) with water activities near zero do not support microbial growth. Food processors reduce water activity by drying foods or by adding salts or sugars. Thus, even though jam is moist, its sugar

[2]From French *vinaigre*, meaning "sour wine."

TABLE 25.1 Some Fermented Foods and the Microorganisms Used in Their Production

Food	Starting Material	Representative Culture Microorganisms
Fermented Vegetables		
Sauerkraut/kimchi	Cabbage	Various lactic acid bacteria
Pickle	Cucumbers, peppers, beets	Various lactic acid bacteria
Soy sauce	Soybeans and wheat	*Aspergillus oryzae* and *Lactobacillus* spp.
Miso	Rice and soybeans or rice and other grains	*A. oryzae*, *Lactobacillus* spp., and *Torulopsis etchellsii*
Fermented Meat Products		
Dry salami	Pork, beef, chicken	Various lactic acid bacteria
Fish sauce/paste	Ground fish	Various naturally occurring bacteria
Fermented Dairy Products		
Milks		
Buttermilk	Pasteurized skim milk	*Lactococcus lactis* subspecies *cremoris* and *Leuconostoc citrovorum*
Yogurt	Pasteurized skim milk	*Streptococcus thermophilus* and *Lactobacillus bulgaricus*
Cheeses		
Cottage cheese	Pasteurized milk	*L. lactis*, including subspecies *cremoris*
Hard cheese (e.g., cheddar)	Milk curd	Starter culture as in cottage cheese without further additions
Soft cheese (e.g., Camembert)[a]	Milk curd	Starter culture as in cottage cheese plus *Penicillium camemberti*
Mold-ripened (e.g., Roquefort)	Milk curd	Starter culture as in cottage cheese plus *Penicillium roqueforti*
Animal Feed		
Silage	Corn, grains, vegetation	Various naturally occurring bacteria
Alcoholic Fermentations		
Wine	Grapes	*Saccharomyces cerevisiae*
Distilled spirits	Fruits, vegetables, grains	*S. cerevisiae*
Beer	Barley	*S. cerevisiae* or *S. carlsbergensis*
Sake	Cooked rice	*A. oryzae* and *Saccharomyces* spp.
Vinegar	Fruits, vegetables, grains	*S. cerevisiae* and *Acetobacter* or *Gluconobacter*
Bread	Flour, salt, etc.	*S. cerevisiae*

[a]In addition to being a soft cheese, Camembert is also a mold-ripened cheese.

content is very high, and its water activity is too low to support much microbial growth.

Acidity of food can either be an intrinsic chemical property of the food, as in the case of citrus fruit, or result from fermentation, as with dill pickles. In either case, a pH below 5.0 typically reduces microbial growth except by a few molds or lactic acid bacteria. A pH closer to neutral (7.0) supports a wider range of microbial growth. Since the pH of most foods is close to neutral, pH is generally not a deterrent to spoilage microorganisms.

Physical structure is a visible intrinsic factor. Rinds or thick skins usually protect fruits and vegetables, and eggs have shells. These coverings are dry and nutritionally poor, and thus they support little microbial growth. If the outer covering is broken or cut, however, microbes can reach the moist interior of the food and cause it to rot.

Ground meat has more surface area and more oxygen and may have bacteria mixed within it during the grinding process. Thus, ground meat supports microbial growth and spoils faster than whole cuts of meat. In uncut meat, the largest volume of the meat is anaerobic and not exposed to microorganisms.

To some extent, microbial competition can also be an intrinsic factor in food spoilage, especially in fermented foods. Fermented foods are populated with large numbers of fermentative bacteria but few pathogens because the latter do not readily grow in the environment produced by the fermenters. Furthermore, pathogens that require the rich nutrient levels found in the human body are not typically capable of growing on "lower-nutrient" foods.

TABLE 25.2 summarizes the effects of intrinsic factors on food spoilage.

Extrinsic Factors in Food Spoilage Extrinsic factors, including how food is processed, handled, and stored, also govern food spoilage. Microorganisms can enter food in a variety of ways; some are introduced accidentally during harvesting from soil or contaminated water, and commercial food processing

TABLE 25.2 Factors Affecting Food Spoilage

	Foods at Greatest Risk of Spoilage	Food at Least Risk of Spoilage
Intrinsic Factors		
Nutritional composition	Chemically rich or fortified foods (steak, bread, milk)	Chemically limited foods (flour, cereals, grains)
Water activity	Moist foods (meat, milk)	Dry foods or those with low water activity (pasta, jam)
pH	Foods with neutral pH (bread)	Foods with low pH (orange juice, pickles)
Physical structure	Foods without rinds, skins, or shells; ground meat	Foods with rinds, skins, or shells; intact foods
Microbial competition	Foods that lack natural microbe populations (ground beef)	Foods with resident microbial populations (pickles)
Extrinsic Factors		
Degree of processing	Unprocessed foods (raw milk, fruit)	Processed foods (pasteurized foods)
Amount of preservatives	Foods without preservatives (meats, natural foods)	Foods with either naturally occurring or added preservatives (garlic, spices, sulfur dioxide)
Storage temperature	Foods left in warm conditions	Foods kept cold
Storage packaging	Foods stored exposed	Foods wrapped or sealed

introduces others. Mechanical vectors, such as flies, also deposit microbes onto food. However, most microbial contaminants leading to spoilage can be traced to improper handling by consumers. Methods to prevent the introduction of microbes into food or to limit their growth will be presented shortly.

Classifying Foods in Terms of Potential for Spoilage

Based on the previous considerations, foods can be grouped into three broad categories, depending on their likelihood of spoilage. *Perishable* foods, such as milk, tend to be nutrient rich, moist, and unprotected by coverings. They need to be kept cold, or they spoil relatively quickly (within days). *Semi-perishable* foods, such as tomato sauce, can be stored in sealed containers for months without spoiling as long as they are not opened. Once opened, however, they may spoil within several weeks. *Nonperishable* foods are usually dry foods, such as pasta, or canned goods that can be stored almost indefinitely without spoiling. Nonperishable foods are typically either nutritionally poor, dried, fermented, or preserved (discussed shortly).

Prevention of Food Spoilage

Food spoilage can begin during production, processing, packaging, or handling. Spoilage results in significant economic losses to producers—in the form of lost productivity, food recalls, and even the loss of jobs resulting from the shutdown of unsafe food processing facilities—but also to consumers in the form of medical expenses and time lost from work. In the next sections, we examine some of the many techniques used to prevent food spoilage: food processing methods, the use of preservatives, and attention to temperature and other storage conditions.

Food Processing Methods *Industrial canning* is a major food packaging methodology for preserving foods **(FIGURE 25.4)**. After food is prepared by washing, sorting, and processing, it is packaged into cans or jars and subjected to high heat (115°C, 239°F) under pressure for a given amount of time, depending on the food. This temperature is not that of a sterlizing autoclave in a microbiology laboratory, so the food must be heated longer. Autoclave temperature (121°C, 250°F) would severely change the food's taste, texture, and nutritiousness. The heating is followed by rapid cooling. The heat kills vegetative mesophilic bacteria and destroys endospores formed by *Bacillus* (ba-sil´lŭs) and *Clostridium* (klos-trid´ē-ŭm). Although canning results in the elimination of most contaminating microorganisms, it does not sterilize food; hyperthermophilic microbes remain, but because they cannot grow at room temperature, they do not pose a threat.

▲ **FIGURE 25.4 Industrial canning.** Food is processed, put into cans or jars, and heated under pressure for a length of time and a temperature sufficient to kill most microbes, including endospore-forming bacteria. *Why isn't the worker wearing a mask and gloves while handling open cans of food?*

Figure 25.4 *The food is about to be steam sterilized, so any introduced microbes will be killed.*

Spoilage can occur if the food is underprocessed or if mesophilic microbes contaminate the can during cooling or thereafter. The two most frequent contaminants found in canned goods are *Clostridium* spp. and coliforms, organisms that grow well anaerobically in low-acid foods. Contamination with the resistant endospores of *Clostridium* is especially problematic because if the cans or jars are not heated sufficiently, the endospores germinate, grow, and release toxins inside the sealed container.

Pasteurization is less rigorous than canning and is used primarily with beer, wine, and dairy products because it does not degrade the flavor of these more delicate foods. By heating foods only enough to kill mesophilic, non-endospore-forming microbes (including most pathogens), pasteurization lowers the overall number of microbes, but because some microbes survive, pasteurized foods will spoil without refrigeration.

Moisture is essential for microbial growth; therefore, desiccation (drying) is an excellent food preservation technique. Drying greatly reduces or eliminates microbial growth, but it does not kill bacterial endospores. Although some fruits are still dried by being left in the sun, today most commercially dried foods are processed in ovens or heated drums that evaporate the water from the foods.

Lyophilization (lī-of´i-li-zā´shŭn), or *freeze drying,* involves freezing foods and then using a vacuum to draw off the ice crystals. Freeze-dried foods such as soups and sauces can be reconstituted by mixing with water.

Gamma radiation, produced primarily by the isotope cobalt-60, penetrates fruits, vegetables, and meats, including fish, poultry, and beef, to cause irreparable and fatal damage to the DNA of microbes. Such ionizing *irradiation* is controversial because some consumers believe erroneously that irradiated foods become radioactive, are less nutritious, or contain toxins. Nevertheless, because irradiation can achieve complete sterilization, it is used routinely to preserve some foods, such as spices.

UV light is not ionizing and does not penetrate very far. It is used to treat packing and cooling water used in industrial canning and to treat work surfaces and utensils in industrial meat processing plants.

In *aseptic packaging,* paper or plastic containers that cannot withstand the rigors of canning or pasteurization are sterilized with hot peroxide solutions, UV light, or superheated steam. Then a conventionally processed food is added to the aseptic package, which is sealed without the need for additional processing.

The Use of Preservatives Humans have preserved foods with salt or sugar throughout history. Both chemicals draw water by osmosis out of foods and microbes alike, killing microbes on the food and retarding the growth of any subsequent microbial contaminants. Bacon is an example of a high-salt food; jellies are examples of high-sugar foods.

Whereas salt and sugar act by removing water, some natural preservatives actively inhibit microbial enzymes or disrupt cytoplasmic membranes. For example, garlic contains *allicin,* which inhibits enzyme function. Benzoic acid, produced naturally by cranberries, also interferes with enzymatic function. Cloves, cinnamon, oregano, and thyme (and, to some degree, sage and rosemary) produce oils that interfere with the functions of membranes of microorganisms.

As we have seen, fermentation preserves some foods by producing an acidic environment that is inhospitable to most microbes. For other foods such as meats, the use of wood smoke during the drying process introduces growth inhibitors that help preserve the food. Other naturally occurring and synthetic chemicals can be purposely added to foods as preservatives. Acceptable preservatives are harmless and do not alter the taste or appearance of the food to any great extent. Organic acids, such as benzoic acid, sorbic acid, and propionic acid, are commonly used in beverages, dressings, baked goods, and a variety of other foods. Gases, such as sulfur dioxide and ethylene oxide, are used to preserve dried fruit, spices, and nuts. All such chemical preservatives inhibit some aspect of microbial metabolism, but many do not actually kill microbes; in other words, they are *germistatic* rather than *germicidal.* Some chemicals work better against bacteria than against fungi and vice versa. Benzoic acid, for example, has a largely antifungal function and does not affect the growth of many bacteria.

Attention to Temperature During Processing and Storage In general, higher temperatures are desirable during food processing and preparation to prevent food spoilage, whereas lower temperatures are desirable for food storage. High temperatures, such as those of pasteurization, canning, or cooking, kill potential pathogens because proteins and enzymes become irreversibly denatured. However, heat does not inactivate many toxins; for example, botulism toxin may remain in cooked foods even when the bacteria that produced it are dead.

Unlike heat, cold rarely kills microbes but instead merely retards their growth by slowing metabolism. Even freezing fails to kill all microorganisms and may only lower the level of microbial contamination enough to reduce the likelihood of food poisoning after frozen foods have been thawed.

To prevent food spoilage, foods should be prepared at sufficiently high temperatures and then stored under conditions that do not facilitate microbial growth. Wrapping leftovers or putting them in containers enables foods to be stored away from exposure to air or contact that could result in contamination. Storing foods in the cold of a refrigerator or freezer combines the protection of packaging with growth-inhibiting temperatures.

One microbe for which cold storage does not suppress growth is *Listeria monocytogenes* (lis-tēr´ē-ă mo-nō-sī-tah´je-nēz), the causative agent of listeriosis and a common environmental bacterium, which is prevalent in certain dairy products, such as soft cheeses. This bacterium grows quite well under refrigeration; therefore, it is best to prevent its entry into foods.

TABLE 25.3 Most Common Bacterial and Protozoan Agents of Foodborne Illnesses

Organism	Affected Food Products	Comments
Campylobacter jejuni	Raw and undercooked meats; raw milk; untreated water	Most common cause of diarrhea of all foodborne agents
Clostridium botulinum	Home-prepared foods	Produces a neurotoxin
Escherichia coli O157:H7	Meat; raw milk	Produces an enterotoxin
Listeria monocytogenes	Dairy products; raw and undercooked meats; seafood; produce	Common in soils and water, contamination from these sources occurs easily; grows at refrigerator temperature
Salmonella spp.	Raw and undercooked eggs; meat; dairy products; fruits and vegetables	Second most common cause of foodborne illness in the United States
Shigella spp.	Salads; milk and other dairy products; water	Third most common cause of foodborne illness in the United States
Staphylococcus aureus	Cooked high-protein foods	Produces a potent toxin that is not destroyed by cooking
Toxoplasma gondii	Meat (pork in particular)	Parasitic protozoan
Vibrio vulnificus	Raw and undercooked seafood	Causes primary septicemia (bacteria in the blood)
Yersinia enterocolitica	Pork; dairy products; produce	Causes generalized diarrhea and severe cramping that mimics appendicitis; grows at refrigerator temperature

Foodborne Illnesses

LEARNING | OUTCOME

25.4 Discuss the basic types of illnesses caused by food spoilage or food contamination and describe how they can be avoided.

Spoiled food, if consumed, can result in illness, but not all foodborne illnesses result from actual food spoilage. Foodborne illnesses may also result from the consumption of harmful microbes or their products in food.

Foodborne illnesses *(food poisoning)* can be divided into two types: **food infections**, caused by the consumption of living microorganisms, and **food intoxications**, caused by the consumption of microbial toxins instead of the microbes themselves. Typical signs and symptoms of food poisoning are generally the same regardless of the cause and include nausea, vomiting, diarrhea, fever, fatigue, and muscle cramps. Symptoms occur within 2–48 hours after ingestion, and the effects of the illness can linger for days. Most outbreaks of food poisoning are *common-source epidemics,* meaning that a single food source is responsible for many individual cases of illness.

The U.S. Centers for Disease Control and Prevention (CDC) estimates that about 48 million cases of food poisoning occur each year. Of these, about 128,000 people require hospitalization and 3000 die. Researchers identify the microbes involved in only about 14 million of the total cases. The U.S. Department of Agriculture estimates the economic cost of food poisoning—due to loss of productivity, medical expenses, and death—at roughly $5 billion to $10 billion per year.

More than 250 different foodborne diseases have been described. **TABLE 25.3** lists 10 common causes of foodborne illness and their sources. Except for the protozoan *Toxoplasma gondii* (tok-sō-plaz´mă gon´dē-ē), all are bacterial agents. A few fungi (*Aspergillus* and *Penicillium*) and viruses (e.g., hepatitis A, noroviruses) also cause food poisoning.

TELL ME WHY

Even though microbes are naturally present in raw milk, raw milk is not generally used for the production of cheeses. Why not?

Industrial Microbiology

The potential uses of microorganisms for producing valuable compounds, as environmental sensors, and in the genetic modification of plants and animals makes industrial microbiology one of the more important fields of study within the microbiological sciences. In the sections that follow we will examine the use of microbes in industrial fermentation, in the production of several industrial products, in the treatment of water and wastewater, and in the disposal and cleanup of biological wastes.

The Roles of Microbes in Industrial Fermentations

LEARNING | OUTCOME

25.5 Describe the role of genetically manipulated microorganisms in industrial and agricultural processes and the basics of industrial-scale fermentation.

In industry, the word *fermentation* is used differently than it is used in food microbiology or in the study of metabolism. *Industrial fermentations* involve the large-scale growth of particular microbes for producing beneficial compounds, such as amino acids and vitamins. Temperature, aeration, and pH are all regulated to

▲ **FIGURE 25.5 Fermentation vats.** Such large containers are used for growing microorganisms in the vast quantities needed for the large-scale production of many industrial, agricultural, and medical products. *Why are industrial fermentation vats made of stainless steel instead of wood?*

Figure 25.5 *Fermentation vats are made of stainless steel so that they can be cleaned more easily to avoid contamination of the product being produced.*

maintain optimal microbial growth conditions. Generally, industrial fermentations start with the cheapest growth medium available, often the waste product from another process (such as whey from cheese production). Because of its scale, industrial fermentation is performed in huge vats that can be readily filled, emptied, and sterilized **(FIGURE 25.5)**. Vats are typically made from stainless steel so that they can be cleaned and sterilized more easily.

There are two types of industrial fermentation. In *batch production,* organisms ferment their substrate until it is exhausted, and then the end product is harvested all at once. In *continuous flow production,* the vat is continuously fed new medium while wastes and product are continuously removed. For this setup to work, the organism must secrete its product into the surrounding medium.

Industrial products are produced as either primary or secondary metabolites of the microorganisms. *Primary metabolites,* such as ethanol, are produced during active growth and metabolism because they are either required for reproduction or by-products of active metabolism. *Secondary metabolites,* such as penicillin, are produced after the culture has moved from log phase of growth and entered stationary phase, during which time the substances produced are not immediately needed for growth.

Recombinant DNA techniques are used for the production of *recombinant microorganisms.* (Chapter 8 examines these techniques and some of their uses.) Most genetically modified organisms used in industry have been specifically designed to produce a stable, high-yield output of desirable chemicals or to perform certain novel functions.

Industrial Products of Microorganisms

LEARNING **OUTCOME**

25.6 List some of the various commercial products produced by microorganisms.

Microorganisms, particularly bacteria, metabolically produce an incredible array of industrially useful chemicals, including enzymes, food additives and supplements, dyes, plastics, and fuels. Recombinant organisms add to this diversity by producing pharmaceuticals, such as human insulin, which are not normally manufactured by microbial cells. In the sections that follow we consider a variety of industrial products produced by microbes, including enzymes, alternative fuels, pharmaceuticals, pesticides, agricultural products, biosensors, and bioreporters.

Enzymes and Other Industrial Products

Enzymes are among the more important products made by microbes. Most are naturally occurring enzymes for which industrial uses have been devised. For example, amylase, produced by *Aspergillus oryzae,* is used as a spot remover. Pectinase, obtained from species of *Clostridium,* enzymatically releases cellulose fibers from flax, which are then made into linen. Proteases from a variety of microbes are used in meat tenderizers, spot removers, and cheese production. Streptokinases and hyaluronidase are used in medicine to dissolve blood clots and enhance the absorption of injected fluids, respectively. Microbes also make the enzymatic tools of recombinant DNA technology, including restriction enzymes, ligases, and polymerases.

Some products made naturally by microorganisms are useful to humans as food additives and food supplements. *Food additives* generally enhance a food in some way, such as by improving color or taste, whereas *food supplements* make up for nutritional deficiencies. Amino acids and vitamins are two important microbial products used as supplements. Vitamins are added directly to foods or sold as vitamin tablets. Amino acids are either sold in tablet form or combined to make new compounds, such as the sweetener aspartame (made from the amino acids phenylalanine and aspartic acid). Other organic acids, such as citric acid, gluconic acid, and acetic acid (vinegar), are also microbially produced to be used in food manufacturing. Citric acid is used as an antioxidant in foods, while gluconic acid is used medically to facilitate calcium uptake.

Other industrial products made by microbes include dyes, such as indigo, which makes denim jeans blue (see **Highlight: Making Blue Jeans "Green"**), and cellulose fibers used in woven fabrics. Microbially produced biodegradable plastics can replace nonbiodegradable, petroleum-based plastics. This is possible because many bacteria produce a carbon-based storage molecule called polyhydroxyalkanoate (PHA), a polymer with a structure similar to petroleum-based plastics. Even though such biodegradable plastics are expensive to manufacture, they have been commercially available since the early 1990s.

HIGHLIGHT

Making Blue Jeans "Green"

Most blue denim today is colored with an indigo dye that has been synthetically created via a petroleum-based process because extracting the dye from indigo plants is labor intensive and expensive. Environmentalists are concerned that the potentially toxic by-products of the coal- and oil-dependent process pose risks to the environment.

To address such concerns, scientists have been investigating an alternative method of producing indigo using *E. coli*, which naturally produces the amino acid tryptophan, with a chemical structure very similar to that of indigo. Since 1983, scientists have known how to take advantage of this similarity and genetically alter *E. coli* so that the bacterium produces indigo instead. The problem? Bacterial synthesis of indigo was both slow and resulted in an unwanted red pigment that remained visible when the dye was used on denim.

These problems may soon be resolved. Scientists have succeeded in modifying the *E. coli* genome so that the red pigment is no longer produced. They have also boosted bacterial production of indigo by 60%. Though more work is needed to make the process economical, these results provide hope for environmentally sensitive blue jeans in the future.

Alternative Fuels

Photosynthetic microorganisms use the energy in sunlight to convert CO_2 into carbohydrates that can be used as fuels—so-called biofuels. Other microorganisms convert biomass (organic materials such as plants or animal wastes) into renewable biofuels, including ethanol, methane, and hydrogen.

Ethanol, which is made during alcoholic fermentation and is the simplest alternative biofuel to synthesize, can be mixed with gasoline to make *gasohol*, which is used in existing cars. Currently, in the United States, food crops such as corn are fermented to ethanol, but production of alternative fuel from nonfood crops or crop wastes—for example, inedible switchgrass or corn stalks—may use less water, fertilizer, and food supplies.

All microorganisms synthesize hydrocarbons as part of their normal metabolism, but only some microbes produce hydrocarbons that could be useful fuels. The colonial alga *Botryococcus braunii* (bot´rē-ō-kok´ŭs brow´nē-ē), for example, produces hydrocarbons that account for 30% of its dry weight. The technology to harvest such hydrocarbons does not yet exist, but one day this alga may be an important source of fuel. One exception is the harvesting of methane gas, which can be collected and piped through natural gas lines to be used for cooking and heating. The largest potential source of methane is from landfills, where methanogens anaerobically convert wastes into methane via anaerobic respiration **(FIGURE 25.6)**. Some communities already use methane from landfills to produce energy. Methane can also be used as a fuel in properly equipped cars.

Although a variety of microbes release hydrogen as part of their normal metabolism, the hydrogen-producing metabolic pathway does not cost effectively produce fuel. An economically efficient method for the production of hydrogen fuel using sunlight and photosynthetic microbes would have widespread applications and enormous appeal, but much research is needed to develop technologies to produce hydrogen as a viable energy source.

▲ **FIGURE 25.6 Collecting methane gas released from a landfill.** Such microbially produced methane could be an important alternative fuel source. *What are some potential benefits of renewable energy produced by microbes?*

Figure 25.6 *Microbial fuels may be easier to produce and less harmful to the environment.*

Pharmaceuticals

Foremost among the pharmaceutical substances microorganisms produce are antimicrobial drugs. About 6000 antimicrobial substances have been described since penicillin was first produced during World War II. Of these, 100 or so have current medical applications. The bacteria *Streptomyces* (strep-tō-mī´sēz) and *Bacillus* and the fungus *Penicillium* synthesize the majority of useful antimicrobials, though recombinant DNA techniques enable the modification of other microbes to produce new versions of old drugs.

TABLE 25.4 Some Products of Recombinant DNA Technology Used in Medicine

Product	Modified Cell	Uses of Product
Interferons	*Escherichia coli, Saccharomyces cerevisiae*	To treat cancer, multiple sclerosis, chronic granulomatous disease, hepatitis, and warts
Interleukins	*E. coli*	To enhance immunity
Tumor necrosis factor	*E. coli*	In cancer therapy
Erythropoietin	Mammalian cell culture	To stimulate red blood cell formation; to treat anemia
Tissue plasminogen activating factor	Mammalian cell culture	To dissolve blood clots
Human insulin	*E. coli*	For diabetes therapy
Taxol	*E. coli*	In ovarian cancer therapy
Factor VIII	Mammalian cell culture	In hemophilia therapy
Macrophage colony stimulating factor	*E. coli, S. cerevisiae*	To stimulate bone marrow to produce more white blood cells; to counteract side effects of cancer treatment
Relaxin	*E. coli*	To ease childbirth
Human growth hormone	*E. coli*	To correct childhood deficiency of growth hormone
Hepatitis B vaccine	Carried on a plasmid of *S. cerevisiae*	To stimulate immunity against hepatitis B virus

In addition to antimicrobials, genetically modified microbes are producing hormones and other cell regulators. **TABLE 25.4** lists some products of recombinant DNA technology used in medicine.

Pesticides and Agricultural Products

Farmers use microbes and their products in a variety of agricultural applications, particularly with regard to crop management. *Bacillus thuringiensis* (thur-in-jē-en´sis) is one of the more widely used organisms because during sporulation it produces *Bt toxin*, which, when digested by the caterpillars of such pests as gypsy moths, diamondback moths, and tomato hornworms, destroys the lining of the insect's gut wall, eventually killing it. Bt toxin isolated from the bacterium can be spread as a dust on plants, but with recombinant DNA technology the Bt gene can be added to a plant's genome. In this way plants such as corn, soybeans, and cotton protect themselves by manufacturing their own Bt toxin. There is little evidence of insect resistance to this Bt toxin despite the fact that large tracts of farmland are devoted to raising Bt-toxin-expressing plants. Some people are leery of the prospect of transgenic foods derived from these plants, whereas others contend that the genetic manipulation of crops is crucial if we are to continue to feed the world's population. In any case, one thing is clear: The genetic manipulation of plants is big business.

Pseudomonas syringae (soo-dō-mō´nas sĕr´in-jī) is another example of a bacterium with agricultural applications. This bacterium produces a protein that serves in the formation of ice crystals, but the protein is not essential to the survival of the organism, and scientists can remove its gene. When sprayed on crops such as strawberries, the strain lacking the gene, known as *P. syringae* ice$^-$ ("ice-minus"), inhibits the formation of ice, protecting the plants from freeze damage.

TABLE 25.5 lists selected industrial products produced by microorganisms and some of their uses.

Biosensors and Bioreporters

Among the relatively new applications of microorganisms to solve environmental problems are biosensors and bioreporters.

TABLE 25.5 Selected Industrial Products Produced by Microorganisms

Products Made	Use of Product
Enzymes	
Amylase, proteases	Spot removers
Streptokinase	Breakdown of blood clots
Restriction enzymes, ligases, polymerases	Molecular biology, recombinant DNA technology
Food Additives/Supplements	
Amino acids, vitamins	Health supplements
Citric acid	Antioxidant
Sorbic acid, lysozyme	Food preservatives
Other Industrial Products	
Indigo	Dye used in manufacturing clothes
Plastics	Biodegradable substitutes for petroleum-based plastics
Alternative Fuels	
Ethanol	Used in gasohol
Methane	Burned to generate heat and electricity
Hydrogen, hydrocarbons	Potential fuels
Pharmaceuticals	
Antimicrobial drugs	Treatment of bacterial infections
Insulin, human growth hormone	Replacement hormones
Taxol	Cancer treatment
Pesticides and Agricultural Products	
Bt toxin	Insecticide

Biosensors are devices that combine bacteria or microbial products (such as enzymes) with electronic measuring devices to detect other bacteria, bacterial products, or chemical compounds in the environment. **Bioreporters** are somewhat simpler sensors that are composed of microbes (again usually bacteria) with innate signaling capabilities, such as the ability to glow in the presence of biological or chemical compounds.

Currently, biosensors and bioreporters are used to detect the presence of environmental pollutants (e.g., petroleum) and to monitor efforts to remove harmful substances; they may also be useful in detecting bioterrorist attacks. Because bacteria are very sensitive to their environments, they can detect compounds in very small amounts. Biosensors and bioreporters could serve as early warning systems to give officials more time to respond by quickly detecting the metabolic waste products of weaponized biological agents.

Water Treatment

LEARNING OUTCOMES

- **25.7** Contrast water pollution and water contamination.
- **25.8** Describe two waterborne illnesses.
- **25.9** Explain how water for drinking and wastewater are treated to make them safe and usable.

Water Pollution

Water becomes polluted in three basic ways: *physically,* through the presence of particulate matter; *chemically,* from the presence of inorganic and organic compounds, usually derived from industrial activities or agricultural runoff; or *biologically,* through an overabundance of organisms or the presence of nonnative microorganisms. Many pollutants are not readily visible.

Although physical and chemical pollutants are important, of greater concern is biological contamination of water with human pathogens. They can cause significant human diseases, which can be prevented only through water treatment.

Waterborne Illnesses

Consuming contaminated water, either as drinking water or in water added to foods, can result in a variety of bacterial, viral, or protozoan diseases. Fungal water contaminants do not cause diseases.

Each year contaminated drinking water results in roughly 3 billion to 5 billion episodes of diarrheal disease worldwide, including over 3 million deaths among children ages five years or younger. Intoxication can also occur from the presence of microbial toxins in the water.

Water treatment removes most waterborne pathogens, so waterborne illnesses are generally rare in the United States as compared to countries with inadequate water treatment facilities. Outbreaks that do occur in the United States are *point-source infections,* in which a single source of contaminated water leads to illness in individuals that consume the water.

In some marine environments, eutrophic blooms of marine dinoflagellates, such as *Gonyaulax* (gon-ē-aw´laks) and *Gambierdiscus* (gam´bē-er-dis-kŭs), chemically pollute water with their toxins. These blooms are one cause of *red tides* because of the color often imparted to the water by the huge number of dinoflagellates. The toxins are absorbed and concentrated by shellfish, and human intoxication results from food consumption rather than water consumption.

TABLE 25.6 Selected Waterborne Agents and the Diseases They Cause

Organism	Disease
Bacteria	
Campylobacter jejuni	Acute gastroenteritis
Escherichia coli	Acute gastroenteritis
Salmonella spp.	Salmonellosis, typhoid fever
Shigella spp.	Shigellosis (bacterial dysentery)
Vibrio cholerae	Cholera
Viruses	
Hepatitis A virus	Infectious hepatitis
Norovirus	Acute gastroenteritis
Poliovirus	Poliomyelitis
Eukaryotic Parasites	
Cryptosporidium parvum	Cryptosporidiosis
Entamoeba histolytica	Amebic dysentery
Giardia intestinalis	Giardiasis
Schistosoma spp.	Schistosomiasis
Toxin Producers	
Gambierdiscus (ciguatoxin)	Fish poisoning
Gonyaulax (saxitoxin)	Paralytic shellfish poisoning

TABLE 25.6 lists some common human waterborne pathogens and toxins. **Emerging Disease Case Study: Attack in the Lake** on p. 796 considers a pathogen that has recently acquired more attention.

Clean water is vital for people and their activities. In the following sections we consider the treatment of drinking water and the treatment of wastewater (sewage). In both cases, treatment is designed to remove microorganisms, chemicals, and other pollutants to prevent human illness.

Treatment of Drinking Water

Potable water is water that is considered safe to drink, but the term *potable* (pō´tăbl) does not imply that the water is devoid of all microorganisms and chemicals. Rather, it implies that the levels of microorganisms or chemicals in the water are low enough that they are not a health concern. Water that is not potable is **polluted;** that is, it contains organisms or chemicals in excess of acceptable values.

The permissible levels of microbes and chemicals in potable water varies from state to state. Nationally, the U.S.

EMERGING DISEASE CASE STUDY

Attack in the Lake

Maria was more than worried—she was terrified. Her 10-year-old son wasn't acting normally. He was incoherent, and his eyes were rolling. It was as if he looked right through her, eyes unfocused and staring.

How life can change. Just two weeks ago the family had been enjoying their summer vacation at the lake. Everyone had been so happy, splashing, swimming, and diving, even though the lake had been warmer than normal. "Almost like a bathtub," her husband had said. Still, it had been a grand week, the opposite of this, the second week since they returned. This week seemed to have stretched to an eternity.

First, Manuel had complained of a headache, then he developed a fever, and on the third day the nausea and vomiting had started—he threw up 20 times by Maria's estimation. However, she could handle these things; after all, the flu or a "stomach bug" was not that uncommon to a mother of five older children. But now, Manuel was unable to move his head. He'd had a seizure and had been hallucinating. What was happening? Would it never end?

The doctors at the emergency room suspected bacterial meningitis and started an aggressive course of antibiotics; however, Manuel was not going to get better. Maria's worry was about to turn to anguish. Her son's brain was being consumed by *Naegleria fowleri*.

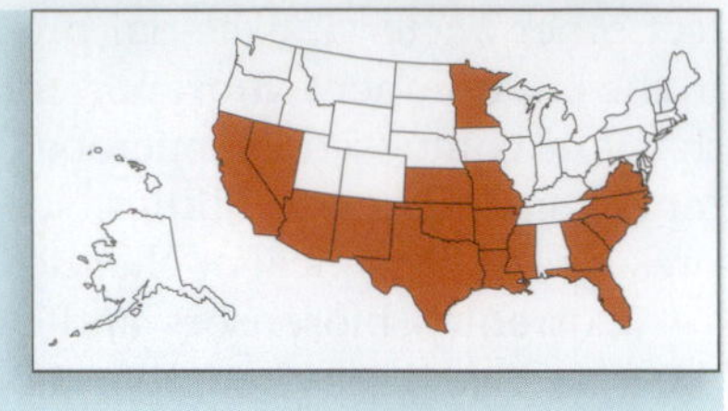

Naegleria is an amoeba that lives in warm freshwater, where it feeds on bacteria, grows, and divides. It normally doesn't bother people. But, when people jump or dive into water where *Naegleria* makes its home, the amoeba can enter the nose, penetrate the nasal mucous membrane, and crawl up the olfactory nerves to the brain, where it causes primary amebic meningoencephalitis (PAM). PAM results in part from the body's efforts to ward off the infection with inflammation; the brain swells, producing headaches and other neurological symptoms. No antimicrobial drugs are known to kill *Naegleria* in the brain. Finally, the patient lapses into coma and usually dies within one to two weeks of infection; only three survivors have been verified in hundreds of cases. Scientists have not been able to determine why some people are infected, while millions of others swim and play in warm lakes without consequence.

Epidemiologists have been unable to develop water safety standards to protect against *Naegleria*: they haven't developed a method to accurately measure the number of amoebae in the water, determine the concentration of *Naegleria* required to initiate an infection, or design preventive measures other than avoiding recreation in warm lakes and wearing nose clips while swimming. Fortunately, PAM is rare; only about 125 cases have occurred in the United States since 1962, though more cases are being reported now than in the past.

For Maria and her family, this emerging disease is not rare enough.

Environmental Protection Agency (EPA) requires that drinking water have a count of 0 (zero) coliform bacteria per 100 ml of water and that recreational waters have no more than 200 coliforms per 100 ml of water. Recall that coliforms are intestinal bacteria such as *Escherichia coli* (esh-ĕ-rik′ē-ă kō′lī). The presence of coliforms in water indicates fecal contamination and thus an increased likelihood that disease-causing microbes are present.

The treatment of drinking water can be divided into four stages **(FIGURE 25.7)**:

1. **Sedimentation**. During sedimentation, water from city drain pipes is pumped into holding tanks, where particulate materials (sand, silt, and organic material) settle.
2. **Flocculation**. The partially clarified water is then pumped into a secondary tank for *flocculation*, in which alum (aluminum potassium sulfate) added to the water joins with suspended particles and microorganisms to form large aggregates, called *flocs*. The flocs also settle to the bottom of the tank. The water above the sediment is then pumped into a different tank for filtration.
3. **Filtration**. In this stage, the number of microbes is reduced by about 90% in one of several ways. One method uses sand and other materials to which microbes adhere and form biofilms that trap and remove other microbes. *Slow sand filters* are composed of a one-meter layer of fine sand or diatomaceous earth and are used in smaller cities or towns to process 3 million gallons per acre of filter per day. Large cities use *rapid sand filters* that contain larger particles and gravel and can process 200 million gallons per acre per day. Both types of filters are cleaned by

(a)

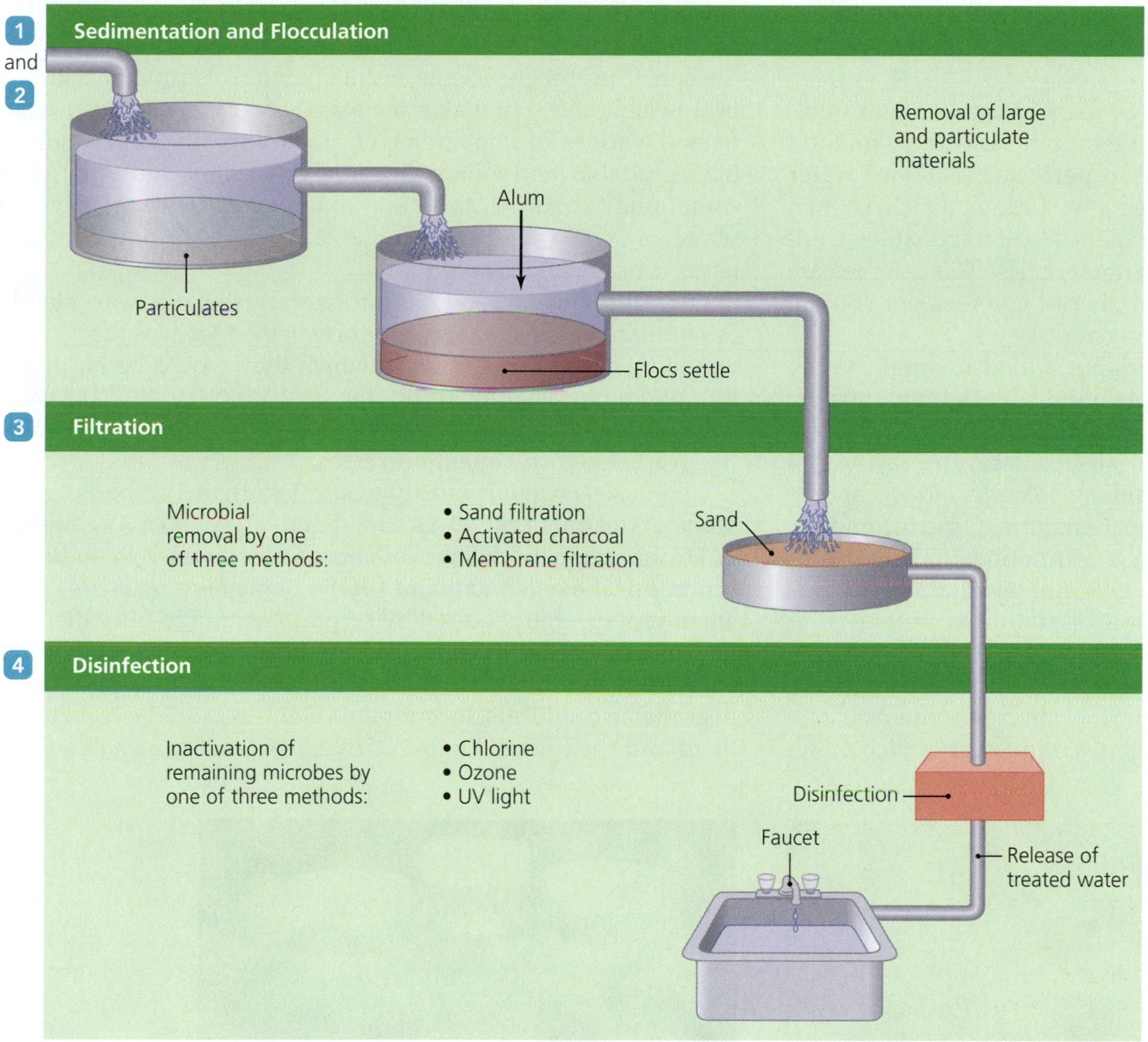

(b)

◀ **FIGURE 25.7 The treatment of drinking water. (a)** A water treatment facility. **(b)** The stages of water treatment: sedimentation, flocculation, filtration, and disinfection. *Why is it that chemical treatment cannot destroy most viruses?*

Figure 25.7 Most chemicals are designed either to inhibit some aspect of active metabolism or to damage cellular structures; therefore, they do not damage acellular and nonmetabolizing viruses.

back-flushing with water. Two other filtration methods are *membrane filtration,* which uses a filter with a pore size of 0.2 µm, and filtration with *activated charcoal,* which provides the added benefit of removing some organic chemicals from the water.

4 **Disinfection**. In this stage, ozone, UV light, or chlorination is used to kill most microorganisms prior to release of the water for public consumption. Many European communities use ultraviolet light. Chlorine treatment is widely used in the United States in part because it is least expensive.

Chlorine gas, an oxidizing agent, is thought to kill bacteria, algae, fungi, and protozoa by denaturing their proteins within approximately 30 minutes of treatment. Chlorine levels must be constantly adjusted to reflect estimates of *microbial load,* the number of microbes in a unit of water—a higher load requires more chlorination. Chlorination does not kill all microbes: Most viruses are not inactivated by chlorine, and bacterial endospores and protozoan cysts are generally unaffected by any chemical treatment. Only mechanical filtration can completely remove all viruses, endospores, and cysts.

Water Quality Testing

Water quality testing is a technique that uses the presence of certain **indicator organisms** to indicate the possible presence of pathogens in water. Because the majority of waterborne illnesses are caused by fecal contamination, the presence of *E. coli* and other coliforms in water indicates a probability that pathogens are present as well. *E. coli* is consistently prevalent in human waste as long as (if not longer than) most pathogens and is easily detected.

Several testing methods can be used to assess water quality. One testing method is the membrane filtration method **(FIGURE 25.8a)**, which is simple to perform: A 100-ml water sample is poured through a fine membrane, which is then placed on selective media agar plates and incubated. Coliforms colonies exhibit unique characteristics. The colonies are then counted and reported as number of colonies per 100 ml. This method gives a total coliform count.

In another test, water samples are added to small bottles containing both *ONPG* (*o*-nitrophenyl-β-D-galactopyranoside) and *MUG* (4-methylumbelliferyl-β-D-glucuronide) as sole nutrients. Most coliforms produce β-galactosidase, an enzyme that reacts with ONPG to produce a yellow color, but the fecal coliform *E. coli* produces an additional enzyme, β-glucuronidase, which reacts with MUG to form a compound that fluoresces blue when exposed to long-wave UV light **(FIGURE 25.8b)**. This test allows for the rapid detection of coliforms but, as with MPN, does not give an actual number.

The presence of viruses and particular bacterial strains cannot be determined with these tests; their presence must be confirmed by genetic "fingerprinting" techniques, in which water samples are collected and enriched to cultivate the organisms present. The DNA content of the enriched sample is then genetically screened for the identification of potential pathogens.

Governments are currently reconsidering the use of coliform tests to indicate fecal contamination because some coliforms grow naturally on plants even when there is little fecal contamination, giving a false-positive result for fecal pollution. Regulators are considering replacing coliform tests with genetic assays that would specifically indicate the presence of *E. coli*.

Treatment of Wastewater

Sewage, or **wastewater**, is typically defined as any water that leaves homes or businesses after being used for washing or flushed from toilets. (Some municipalities also include industrial water and rainwater as wastewater.) Wastewater contains a variety of contaminants, including suspended solids, biodegradable and nonbiodegradable organic and inorganic compounds, toxic metals, and pathogens. The objective of wastewater treatment is to remove or reduce these contaminants to acceptable levels.

At one time, "raw" (unprocessed) sewage was simply dumped into the nearest river or ocean; the idea was that wastes would be diluted to a point at which they would be harmless. Burgeoning populations and the realization that waterways were becoming increasingly polluted led to greater use of effective wastewater treatment processes.

Because sewage is mostly water (less than 1% solids), most sewage treatment involves the removal of microorganisms. A key concept in the processing of wastewater is reducing **biochemical oxygen demand (BOD)**, which is a measure of the amount of oxygen required by aerobic bacteria to fully metabolize organic wastes in water. This amount is proportional to the amount of waste in the water; the higher the concentration of degradable chemicals, the more oxygen is required to catabolize them and the higher the BOD. Effective wastewater treatment

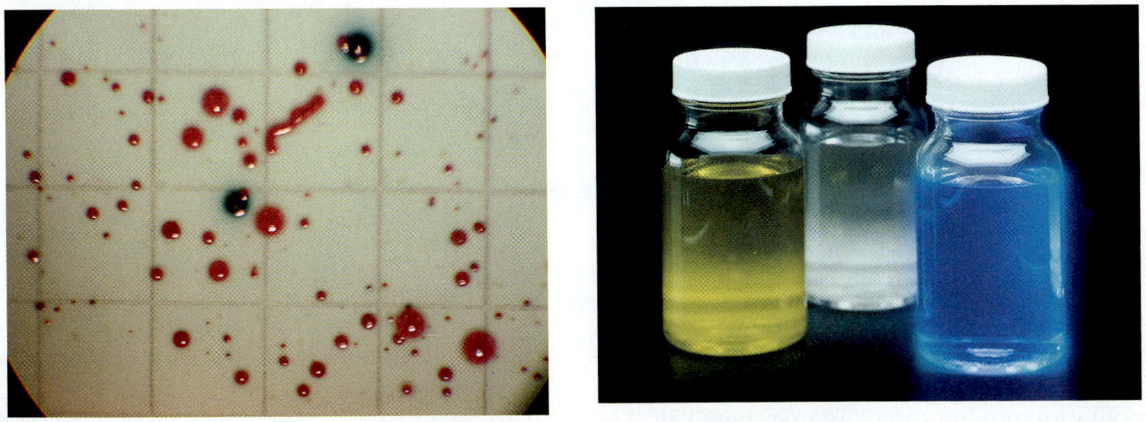

(a) (b)

▲ **FIGURE 25.8 Two water quality tests.** **(a)** Membrane filtration. The grid on the membrane makes it easier to count the colonies of fecal coliforms, which are green on this selective medium. **(b)** The ONPG and MUG tests. The yellow color of the ONPG bottle indicates the presence of coliforms, whereas the blue fluorescence in the MUG bottle indicates the presence of the coliform *E. coli*; the clear bottle is the negative control.

reduces the BOD to levels too low to support microbial growth, thus reducing the likelihood that pathogens will survive.

The following sections consider wastewater treatments of various types: the traditional sewage treatment used in municipal systems, treatments used in rural areas, a treatment used for agricultural wastes, and the use of artificial wetlands.

Municipal Wastewater Treatment Today, people living in larger U.S. towns and cities are usually connected to municipal sewer systems—pipes that collect wastewater and deliver it to sewage treatment plants for processing. Traditional sewage treatment consists of four phases **(FIGURE 25.9)**:

1 **Primary treatment.** Wastewater is pumped into settling tanks, where lightweight solids, grease, and floating particles are skimmed off and heavier materials settle onto the bottom as **sludge**. After alum is added as a flocculating agent, the sludge is removed, and the partially clarified water is further treated. Primary treatment removes 25% to 35% of the BOD in the water.

(a)

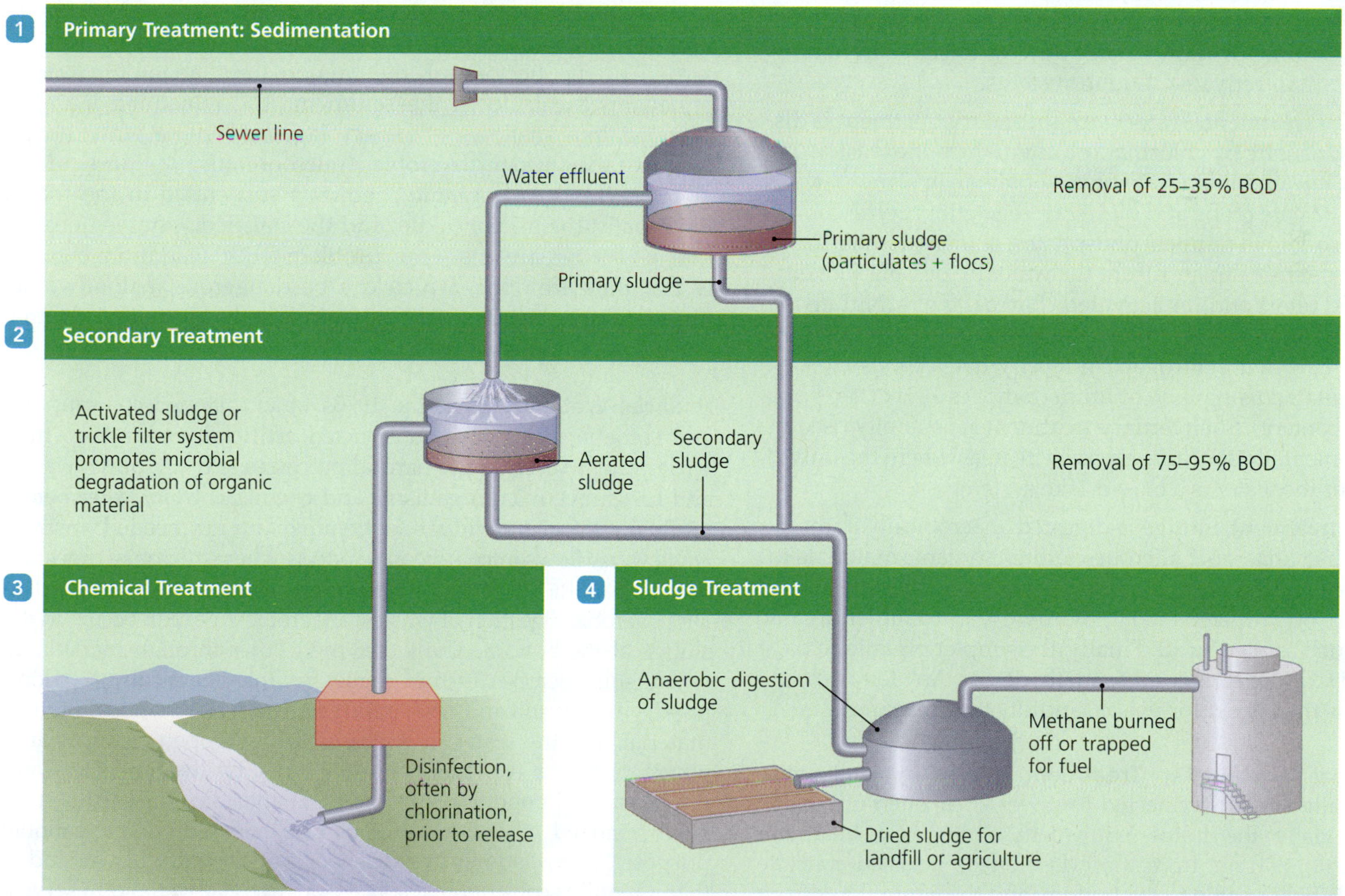

(b)

◀ **FIGURE 25.9 Traditional sewage treatment. (a)** A municipal wastewater treatment facility. **(b)** The traditional sewage treatment process. Microbial digestion during secondary treatment removes most of the biochemical oxygen demand (BOD) before the water is chemically treated and released; dried sludge is recycled as landfill.

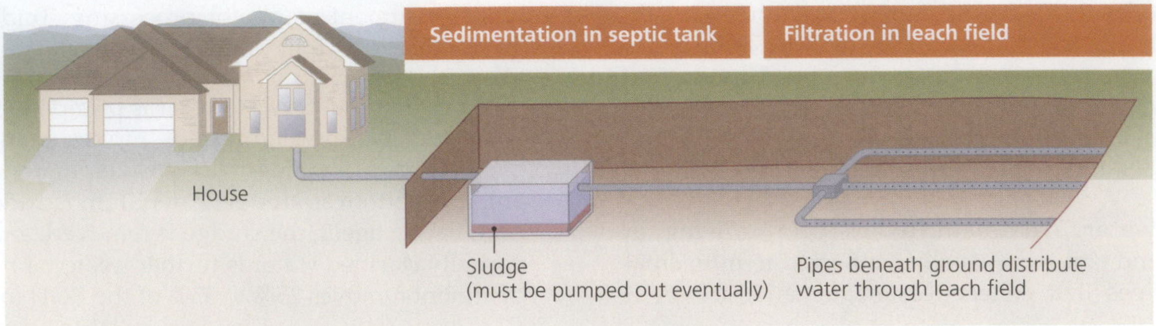

▲ **FIGURE 25.10 A home septic system.** After wastewater from the house enters the septic tank, solids settle out as sludge, and the effluent liquid is filtered by the soil in the leach fields.

2 **Secondary treatment.** The biological activity in this phase reduces the BOD to 5 % to 25% of the original. Most pathogenic microorganisms are also removed. The water is aerated to facilitate the growth of aerobic microbes that oxidize dissolved organic chemicals to CO_2 and H_2O. In an *activated sludge system,* aerated water is seeded with primary sludge containing a high concentration of metabolizing bacteria; flocculation also occurs during this step. Any remaining solid material settles and is added to the sludge from primary treatment. The combined sludge is pumped into anaerobic holding tanks. Some smaller communities accomplish secondary treatment using a *trickle filter system,* which is similar to the slow sand filters used in treating drinking water but less effective in removing BOD than activated sludge systems.

3 **Chemical treatment.** Water from secondary treatment is disinfected, usually by chlorination, after which the wastewater is either released into rivers or the ocean or, in some states, used to irrigate crops and highway vegetation. Some communities remove nitrates, phosphates, and any remaining BOD or microorganisms from the water by passing it over fine sand filters and/or activated charcoal filters. Nitrate is converted to ammonia and discharged into the air (removes roughly 50% of the nitrogen content), whereas phosphorus is precipitated using lime or alum (removes 70% of the phosphorus content). Such tertiary treatment is generally used in environmentally sensitive areas or in areas where the only outlet for the water is a closed-lake system.

4 **Sludge treatment.** Sludge is digested anaerobically in three steps: First, anaerobic microbes ferment organic materials to produce CO_2 and organic acids. Second, microbes metabolize these organic acids to H_2, more CO_2, and simpler organic acids, such as acetic acid. Finally, the simpler organic acids, H_2, and CO_2 are converted to methane gas. Any leftover sludge is then dried for use as landfill or fertilizer.

Nonmunicipal Wastewater Treatment Houses in rural areas, which typically are not connected to sewer lines, often use **septic tanks**, essentially the home equivalent of primary treatment **(FIGURE 25.10)**. Sewage from the house enters a sealed concrete holding tank. Solids settle to the bottom, and the liquid flows from the top of the tank into an underground *leach field* that acts as a filter. Sludge in the tank and organic chemicals in the water are digested by microorganisms. However, because the tank is sealed, it must occasionally be pumped out to remove sludge buildup. **Cesspools** are similar to septic tanks except that they are not sealed. As wastes enter a system of porous concrete rings buried underground, the water is released into the surrounding soil; solid wastes accumulate at the bottom and are digested by anaerobic microbes.

Treatment of Agricultural Wastes Farmers and ranchers often use **oxidation lagoons** to treat animal waste from livestock raised in feedlots—a penned area where animals are fattened for market. Oxidation lagoons accomplish the equivalent of primary and secondary sewage treatment. Wastes are pumped into deep lagoons and left to sit for up to three months, sludge settles to the bottom of the lagoon, and anaerobic microorganisms break down the sediment. The remaining liquid is pumped into shallow, secondary lagoons, where wave action aerates the water. Aerobic microorganisms, particularly algae, break down organic chemicals suspended in the water. Eventually, the microbes die, and the clarified water is released into rivers or streams. One problem with oxidation lagoons is that they are open, which can be dangerous if floodwaters inundate the lagoons and spread largely untreated animal wastes over a wide area.

Artificial Wetlands Since the 1970s, small planned communities and some factories have constructed **artificial wetlands** to treat wastewater. Wetlands use natural processes to break down wastes and to remove microorganisms and chemicals from water before its final release. Individual septic tanks are not needed; instead, wastewater flows into successive ponds where microbial digestion occurs **(FIGURE 25.11)**. The first pond in the series is aerated to allow aerobic digestion of wastes; anaerobic digestion occurs in the sludge at the bottom. The water then flows through marshland, where soil microbes further digest organic chemicals. A second pond, which is still and contains algae, removes additional organic material, and the water then passes through open meadowland, where grasses and plants trap pollutants. By the time the water reaches a final pond, most of the BOD and microorganisms have been removed, and the water can be released for recreational purposes or irrigation. One drawback of an artificial wetland is that it requires considerable space—an artificial wetland to serve a small community can cover 50 acres of land or more.

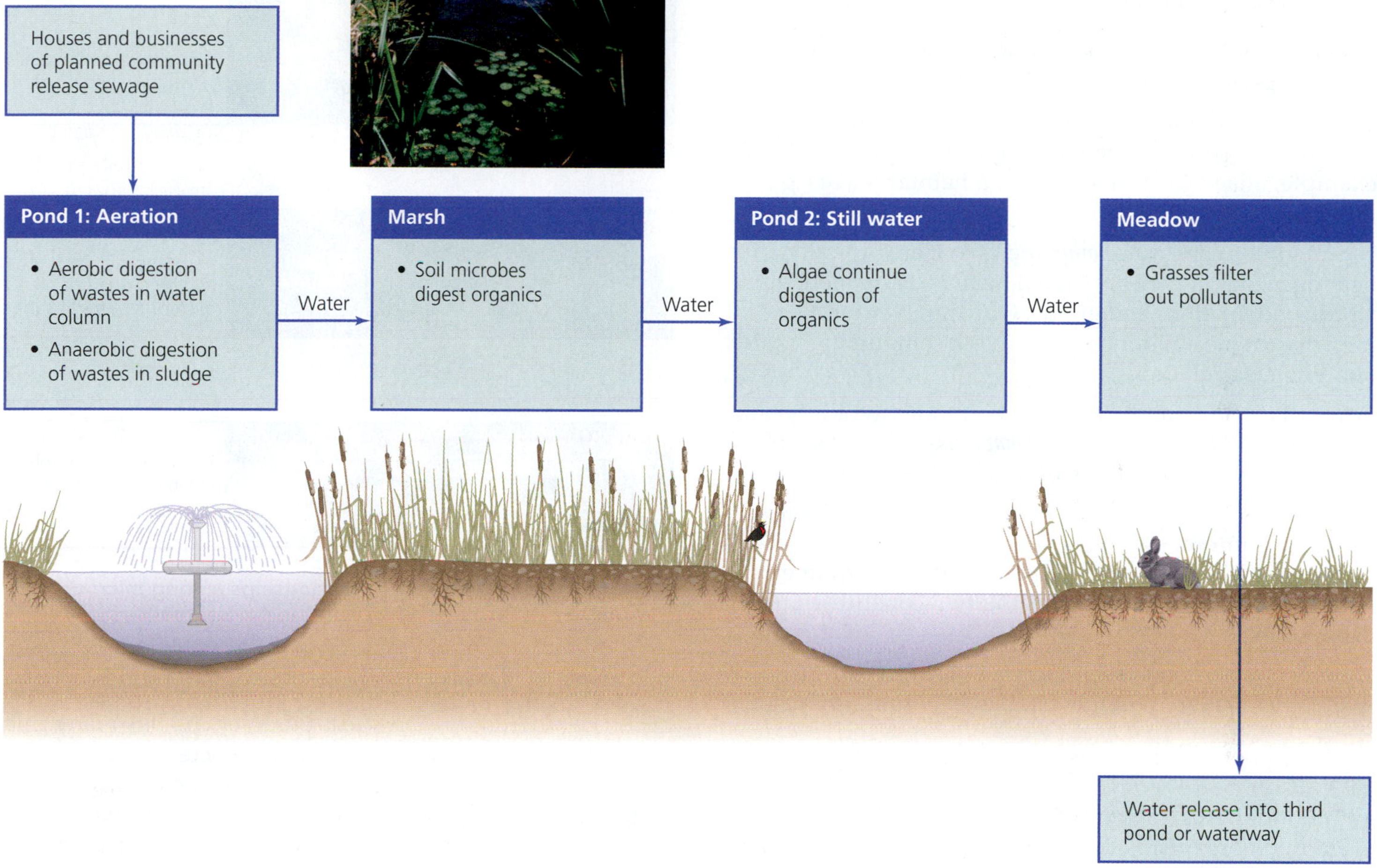

◀ **FIGURE 25.11 Wastewater treatment in an artificial wetland.** The majority of the BOD is removed by microbial action in the first pond; natural filtration by plants and the soil removes pollutants and the rest of the BOD: *Why aren't artificial wetlands feasible for major metropolitan areas?*

Figure 25.11 *Major metropolitan areas produce too much sewage and have too little available space for artificial wetlands.*

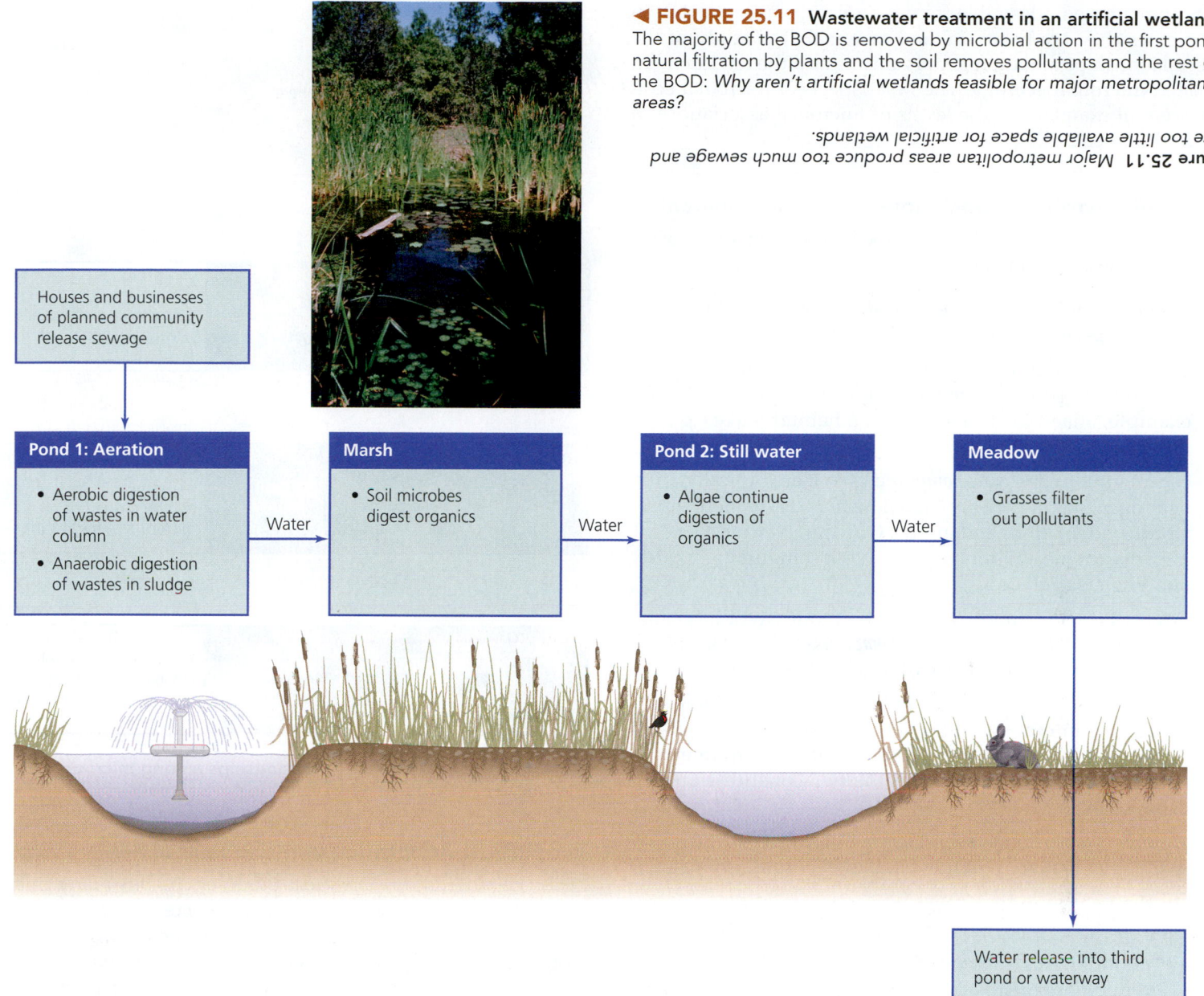

TELL ME WHY

In a polluted lake, the microbes reproduced prolifically, died, and then sank to the bottom, "feeding" anaerobes in the sediment. Even though the surface water looks clear, why is it still unsafe to drink the lake water?

Environmental Microbiology

Environmental microbiology is the study of microorganisms as they occur in their natural **habitats**—the physical localities in which organisms are found. Because of their vast metabolic capacities and adaptability, microbes flourish in every habitat on Earth, from Antarctic ice to boiling hot springs to bedrock.

In the following sections, we will explore the roles of microbes in the cycling of chemical elements in soil and aquatic habitats. First, however, we turn our attention to the relationships among microbes and between microbes and their habitats.

Microbial Ecology

LEARNING | **OUTCOMES**

25.10 Define the terms used to describe microbial relationships within the environment.

25.11 Explain the influences of competition, antagonism, and cooperation on microbial survival.

Microorganisms use a variety of energy sources and grow under a variety of conditions. They adapt to changing conditions, compete with other organisms for scarce resources, and change their habitat in many ways. In some cases, their effects on the environment are undesirable from a human perspective, but in most cases they are beneficial and essential. The study of the interrelationships among microorganisms and the environment is called **microbial ecology**. The first aspects of microbial ecology we will examine are the levels of microbial associations in the environment.

Levels of Microbial Associations in the Environment

A variety of terms are used to describe levels of microbial associations in the environment:

- Similar organisms produce a *population* (all the members of a single species) through reproduction.
- Populations of microorganisms performing metabolically related processes make up groups called *guilds*. For example, anaerobic fermenters in a habitat make up a guild.
- Sets of guilds constitute *communities* that are typically quite heterogeneous; a variety of relationships exist among the various populations and guilds. Only rarely—and usually only in extreme environments—does a community consist of a single population.
- Populations and guilds within a community typically reside in their own distinct *microhabitats*—specific small spaces where conditions are optimal for survival.
- Groups of microhabitats form habitats in which microorganisms interact with larger organisms and the environment. Together, the organisms, the environment, and the relationships between the two constitute an **ecosystem**.
- All of the ecosystems taken together constitute the *biosphere*, that region of Earth inhabited by living organisms.

To illustrate these concepts, consider a plot of garden soil (**FIGURE 25.12**). Soil is composed of tiny particles of rock and organic material, each of which forms a microhabitat for populations of microorganisms. Populations are distributed among the soil particles according to the available light, moisture, and nutrients. Populations of photosynthetic microbes, for example, live near the top and form a photosynthetic guild, which provides nutrients to other guilds living deeper within the soil. A soil community is composed of all of these populations and guilds, whose activities are a major factor affecting soil quality. Garden soil is just one type of soil occupying the soil ecosystem, which in turn is just one ecosystem within the biosphere.

Ecologists use the term **biodiversity** to refer to the number of species living within a given ecosystem, whereas the term **biomass** refers to the mass of all organisms in an ecosystem. By either measure, microorganisms are the most abundant of all living things: The number of species of microbes exceeds the number of larger organisms, and the biomass of microorganisms living throughout the biosphere—including places that are uninhabitable by plants, animals, or humans—is enormous.

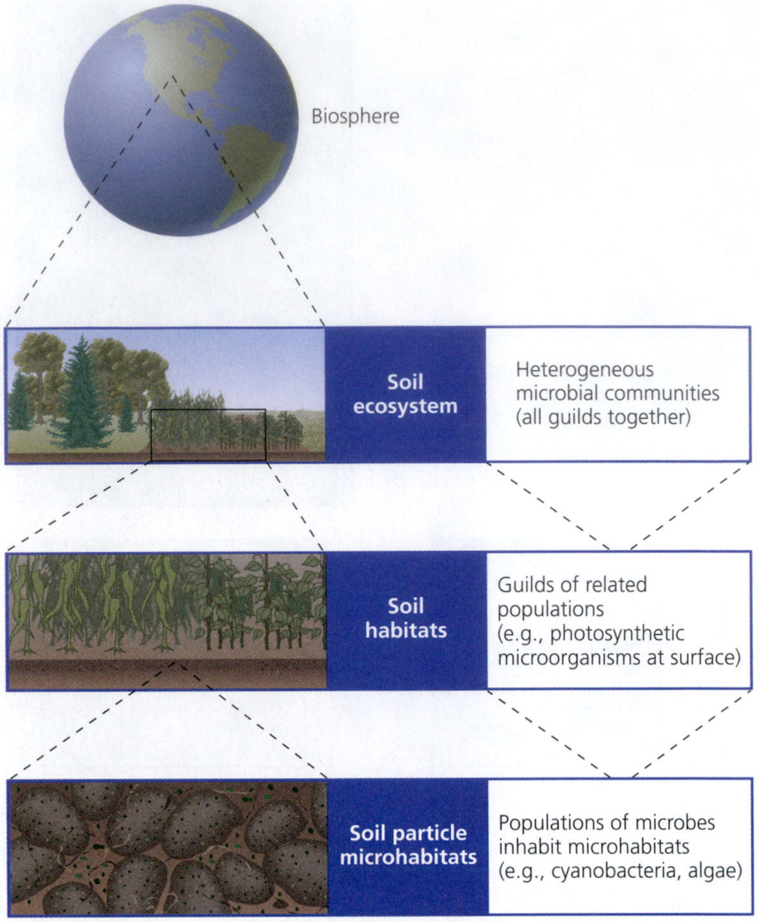

▲ **FIGURE 25.12 The basic relationships among microorganisms and between microorganisms and the environment.**

The Role of Adaptation in Microbial Survival

Lush ecosystems support great biodiversity, but many microorganisms live in harsh environments, where nutrient levels are low enough to limit growth. The harsher the environment, the more specially adapted a microbe must be to survive. Some environments cycle between periods of excess and depletion, and the microbes that live in them must be capable of adapting to such constantly varying conditions. *Extremophiles*—microbes adapted to extremely harsh temperatures, pH levels, and salt concentrations—are so adapted to extreme conditions that they cannot survive anywhere else.

Biodiversity can be held in check by *competition*. The best-adapted microorganisms have traits that provide them advantages in nutrient uptake, reproduction, response to environmental changes, or some other factor. *Antagonism*, in which a microbe makes some product that actively inhibits the growth of another, may also occur.

Although competition may limit biodiversity, rarely does a microorganism outcompete all other rivals; diversity is the norm, not the exception. Microbes use the waste products of other microbes for their own metabolism. Moreover, one microbe's metabolic activities sometimes make the environment more favorable for other microbes. Biofilms are examples of microbial cooperation.

The relationships among microorganisms and their environment constantly change; each microbe adapts to subtle changes in its environment. Most habitats support successions of microbial populations within a community. By studying how these successions occur, we learn much about the microbes' effects on the environment.

Bioremediation

LEARNING | OUTCOME

25.12 Describe the process of bioremediation.

Each year Americans produce more than 150 million tons of solid wastes, accumulated from household, industrial, medical, and agricultural sources; most of it ends up in landfills. A landfill is essentially a large, open pit into which wastes are dumped, compacted, and buried. Soil microbes anaerobically break down biodegradable wastes; methanogens degrade organic molecules to methane. When a landfill is full, it is covered with soil and revegetated.

To prevent leaching of potentially hazardous materials into the soil and groundwater, a landfill pit is lined with clay or plastic, and sand and drainage pipes lining the bottom filter out small particulates and some microorganisms. Unfortunately, chemicals and hazardous compounds might still leak from landfills. Some of these substances are potentially carcinogenic (benzene, phenol, petroleum products), others are toxic (gasoline, lead), and still others are *recalcitrant* (resistant to decay, degradation, or reclamation by natural means).

The reason that one synthetic molecule is biodegradable while another is recalcitrant relates to chemical structure; sometimes a subtle variation—perhaps in a single atom or bond—can be enough to make a difference. Most recalcitrant molecules resist microbial degradation because the microbes lack enzymes capable of degrading them; after all, until very recently synthetic compounds didn't even exist.

Bioremediation is the use of organisms, particularly microorganisms, to clean up toxic, hazardous, or recalcitrant compounds by degrading them to harmless compounds. Although most naturally occurring organic compounds are eventually degraded by microorganisms, synthetic compounds are not always easily removed from the environment. Petroleum, pesticides, munitions, herbicides, and industrial chemicals that accumulate in soil and water are degraded only slowly by naturally existing microbes.

A widely known example of bioremediation is the microbial degradation of complex hydrocarbons released in an oil spill (see **Beneficial Microbes: Oil-Eating Microbes to the Rescue in the Gulf**). Cleanup crews can enhance bioremediation by applying nitrogen and phosphorus fertilizers.

The Problem of Acid Mine Drainage

LEARNING | OUTCOME

25.13 Discuss the problem of acid mine drainage.

Acid mine drainage is a serious environmental problem resulting from the exposure of certain metal ores to oxygen and microbial action **(FIGURE 25.13)**. Coal deposits are often found associated with reduced metal compounds such as pyrite (FeS_2). Strip-mining for coal exposes pyrite to oxygen in the air, which oxidizes the iron, and bacteria such as *Thiobacillus* (thī-ō-bă-sil´ŭs) oxidize the sulfur. Rainwater then leaches the oxidized compounds from the soil to form sulfuric acid (H_2SO_4) and iron

BENEFICIAL MICROBES

Oil-Eating Microbes to the Rescue in the Gulf

In April 2010, the worst oil spill in U.S. history occurred when an oil rig explosion led to the release of millions of gallons of oil into the Gulf of Mexico. Environmentalists, government officials, oil industry executives, and residents of the Gulf states immediately began efforts to minimize damage to the marine and wildlife habitats caused by this oil spill.

By August, the cleanup crews had removed only about 2% of the oil from the Gulf. The huge oil slick, however, had seemingly disappeared. Where did all the oil go? While some of the oil simply evaporated, scientists determined that a significant portion of the oil was devoured by oil-eating fungi and bacteria, such as *Alcanivorax*. *Alacanivorax* is a bacterium that is native to the Gulf region and is named for its voracious appetite for alkanes, which are a major component of petroleum. *Alcanivorax*, along with other native species of oil-eating bacteria and fungi, metabolized much of the oil, converting it into carbon dioxide and water. Scientists were heartened to see that oil-eating microbes grew and reproduced faster in the warm waters of the Gulf of Mexico than they did in the cool waters off the Alaskan coast where the *Exxon Valdez* spill occurred in 1989.

The long-term effects of the oil spill on the shores of the Gulf are not yet known, though the ecosystem has proved to be surprisingly resilient, thanks in part to the bioremediation efforts of oil-eating microbes.

▲ **FIGURE 25.13 The effects of acid mine drainage.** Upon exposure to air, iron in water leached from mine tailings is oxidized to Fe^{3+}; the activity of iron-sulfur bacteria reduces the pH of the water to a level that kills plants and animals.

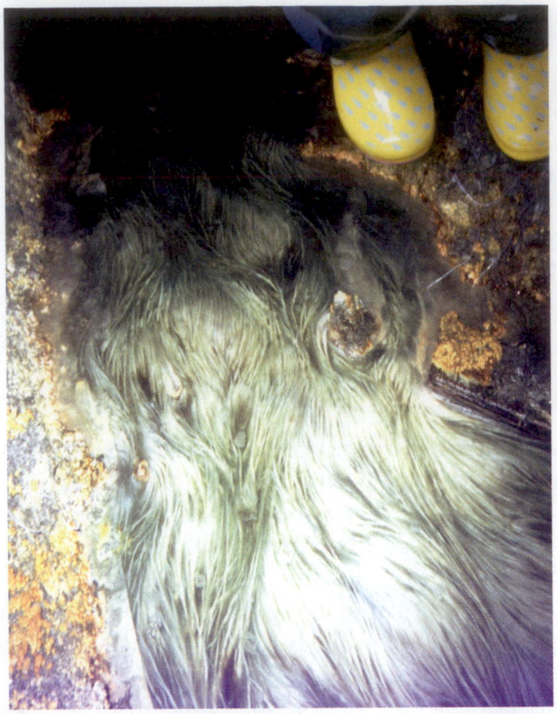

▲ **FIGURE 25.14 An acid-loving microbe.** The filamentous archaeon *Ferroplasma acidarmanus* growing as long filaments in acid mine runoff (pH -3.5) in California. The ore from this mine is rich in iron.

hydroxide [$Fe(OH)_3$], which are carried into streams and rivers, reducing the pH enough (pH 2.5–4.5) to kill fish, plants, and other organisms. Such acidic water is also unfit for human consumption or recreational use. The EPA requires that strip mines be reburied as soon as possible to halt the processes.

Underground mining operations pose similar (but somewhat less severe) problems resulting from runoff from mine tailings, which are the low-grade ores remaining after the extraction of higher-grade ores. Typically, subsurface mines are backfilled as richer veins of minerals are depleted, thus limiting the exposure of iron sulfides to oxygen.

Although acid mine drainage is generally devastating to the environment, some microbes—mostly archaea—actually flourish in acidic conditions: One unique archaeal species, found in mine drainage in California, is *Ferroplasma acidarmanus* (fe´rō-plaz-ma a-sid´ar-mă-nŭs; **FIGURE 25.14**), which lives in a pH near zero and obtains its energy from oxidizing pyrite in mine sediments. Such an organism is just one example of the incredible diversity of microorganisms that colonize every habitat on Earth.

The Roles of Microorganisms in Biogeochemical Cycles

LEARNING | **OUTCOMES**

25.14 Contrast the processes by which microorganisms cycle carbon, nitrogen, sulfur, phosphorus, and trace metals.

25.15 Explain the work of microorganisms in the carbon cycle.

25.16 Contrast the actions of microbes involved in nitrogen fixation, nitrification, ammonification, denitrification, and anammox reactions.

25.17 Describe the reduction and oxidation of sulfur by microbes.

25.18 Explain the use of microbes in biomining.

Six chemical elements make up most macromolecules—the building blocks of cells. These are hydrogen, oxygen, carbon, nitrogen, sulfur, and phosphorus.

Most elements are tied up in chemical and physical forms unavailable to organisms. The release of many elements from rock, for example, requires years of degradation by rain, wind, and microbes, so these processes contribute little to the day-to-day availability of nutrients for living things. As a consequence, the actions of organisms in recycling elements are the major components of **biogeochemical cycles**—the processes by which organisms convert elements from one form to another, typically between oxidized and reduced forms. Microbes are the primary biological components of the biogeochemical cycles.

Biogeochemical cycling essentially entails three processes: (1) *production,* in which organisms called producers convert inorganic compounds into the organic compounds of biomass; (2) *consumption,* in which organisms called consumers feed on producers and other consumers, converting organic molecules to other organic molecules; and (3) *decomposition,* in which organisms called decomposers convert organic molecules in dead organisms back into inorganic compounds. As with all cycles, balance is critical. When one part of a cycle becomes skewed relative to other parts, a cycle becomes inefficient, often with detrimental effects.

The following sections examine the biogeochemical cycles for carbon, nitrogen, sulfur, phosphorus, and some trace metals.

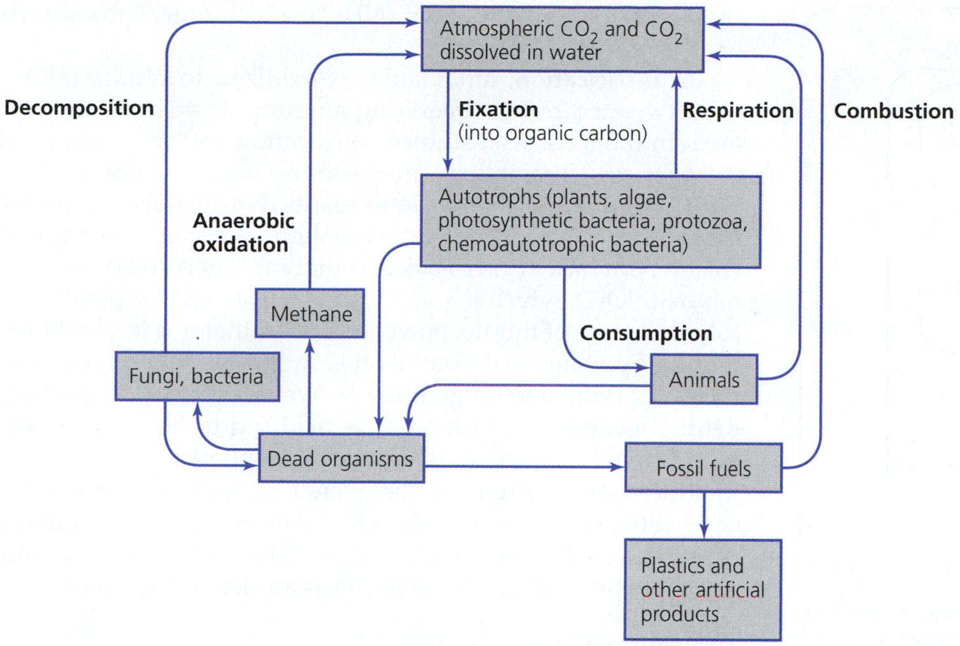

FIGURE 25.15 Simplified carbon cycle. The most mobile form of carbon in the cycle is CO_2; this inorganic molecule is fixed by autotrophs, which incorporate its carbon into organic molecules.

The Carbon Cycle

Carbon is the fundamental element of all organic chemicals. The continual cycling of carbon in the form of organic molecules constitutes the majority of the **carbon cycle** (**FIGURE 25.15**). Carbon in rocks and sediments has a very low *turnover rate* (rate of conversion to other forms)—this form of carbon is incorporated into organic chemicals only slowly over long periods of time.

The start of the carbon cycle is autotrophy. Photoautotrophic *primary producers*—cyanobacteria, green and purple sulfur bacteria, green and purple nonsulfur bacteria, algae, photosynthetic protozoa, and plants—convert CO_2 to organic molecules via carbon fixation in the Calvin-Benson cycle (see Figure 5.28). Photoautotrophs are restricted to the surfaces of soil and water systems because phototrophs require light. Chemoautotrophs can also fix carbon but acquire energy from H_2S or other inorganic molecules; therefore, chemoautotrophs are found in a greater variety of habitats. However, they do not fix as much carbon as photoautotrophs and are not as important in the carbon cycle.

Heterotrophs catabolize some organic molecules for energy in respiration, resulting in the release of CO_2. Other organic molecules made by autotrophs are subsequently incorporated into the tissues of heterotrophs. There they remain until the organism dies, and decomposers catabolize the organic materials, releasing CO_2—the reverse of autotrophy. Waste products are also broken down by decomposers.

The release of CO_2 starts the cycle over again as primary producers fix CO_2 once more into organic material. A rough balance exists between CO_2 fixation and CO_2 release.

Scientists are concerned by a growing imbalance in the carbon cycle due to an overabundance of CO_2 in the atmosphere. The burning of fossil fuels and wood sends tons of CO_2 into the atmosphere each year. Furthermore, in waterlogged soils, sewage treatment plants, landfills, and the digestive systems of ruminants, methanogens actively release methane gas (CH_4), which can be photochemically transformed into CO (carbon monoxide) and CO_2. Methane-oxidizing bacteria, living in conjunction with methanogens, can also directly convert methane to CO_2. Worldwide, the rate of CO_2 production exceeds the rate at which it is being incorporated into organic material.

Carbon dioxide and methane are called "greenhouse gases" because their presence in the atmosphere prevents the escape of some infrared radiation into space, redirecting heat back to Earth, much as the glass panes of a greenhouse trap heat. Such global warming can cause climate change.

The Nitrogen Cycle

Nitrogen is an important nutritional element required by organisms as a component of proteins, nucleic acids, and other compounds. Most nitrogen in the environment is in the atmosphere as dinitrogen gas (N_2), which is unusable by most organisms. The majority of organisms acquire nitrogen as part of organic molecules or from soluble inorganic nitrogen compounds found in limited quantities in soil and water. These include nitrate, nitrite, and ammonia. Microbes cycle nitrogen atoms from dead organic materials and animal wastes to these soluble forms of nitrogen. They also cycle nitrogen between the biosphere and the atmosphere. **FIGURE 25.16** summarizes the basic processes involved in the **nitrogen cycle**—nitrogen fixation, ammonification, nitrification, denitrification, and anammox reactions.

Nitrogen fixation, a process whereby gaseous nitrogen (N_2) is reduced to ammonia (NH_3), is an energy-expensive process in which the extremely stable nitrogen-nitrogen triple bond is broken in a reaction catalyzed by the enzyme *nitrogenase*. A limited number of prokaryotes—but no eukaryotes—fix nitrogen.

Nitrogenase functions only in the complete absence of oxygen, a condition that presents no problem for anaerobes. Aerobic nitrogen fixers, on the other hand, must protect nitrogenase from oxygen. They can do this in a number of ways.

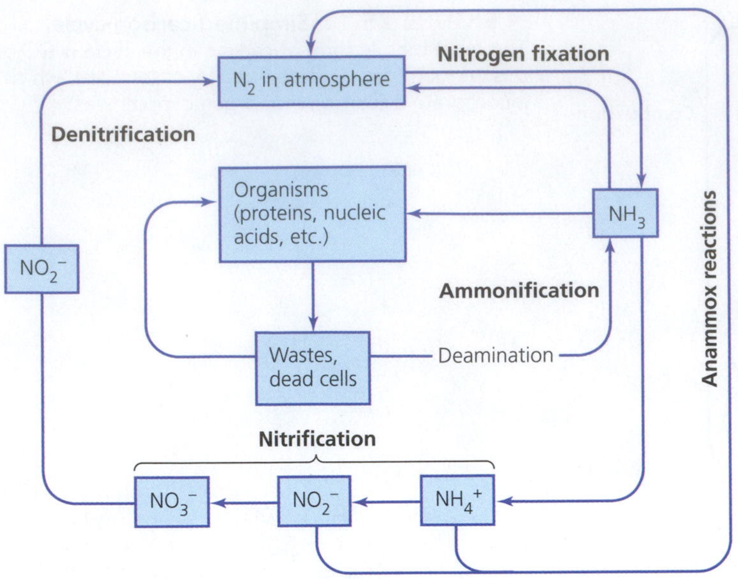

▲ **FIGURE 25.16 Simplified nitrogen cycle.** Even though nitrogen gas (N_2) is the most common form of nitrogen in the environment, most organisms cannot use it; to become available, nitrogen from the atmosphere must be incorporated into organic compounds by the very few species of nitrogen-fixing organisms.

For example, some aerobes use oxygen at such a high rate that it does not diffuse into the interior of the cell where nitrogenase is sequestered. Some cyanobacteria form thick-walled, nonphotosynthetic cells called *heterocysts* to protect nitrogenase from oxygen in the environment as well as from the oxygen generated during photosynthesis (see Figure 11.14a). Other cyanobacteria fix nitrogen only at night, when oxygen-producing photosynthesis does not occur, thereby separating nitrogen fixation from photosynthesis in time rather than in space.

Nitrogen fixers may be free living or symbiotic. Among the free-living nitrogen fixers are aerobic species of *Azotobacter* (ā-zō-tō-bak´ter), anaerobic species of *Bacillus* and *Clostridium*, and deep-sea-dwelling communities of archaea and bacteria. When they die, these microbes release fixed nitrogen into the soil or water, where it becomes available to other organisms, notably plants. None of the free-living genera are directly associated with the plants they fertilize.

Symbiotic nitrogen fixers, in contrast, live in direct association with plants, forming *root nodules* on legumes (e.g., peas and peanuts; see Figure 11.20). The predominant genus of nitrogen-fixing, symbiotic bacteria is *Rhizobium* (rī-zō´bē-ŭm). In the symbiosis, the plants provide nutrients and an anaerobic environment in the nodules, while the bacteria provide usable nitrogen to the plant. Farmers do not have to apply nitrogen fertilizers to legume crops because nitrogen fixers provide enough.

Bacteria and fungi in the soil decompose wastes and dead organisms, disassembling proteins into their constituent amino acids, which then undergo *deamination* (removal of their amino groups). Amino groups are converted to ammonia (NH_3)—a process called **ammonification**. In dry or alkaline soils, NH_3 escapes as a gas into the atmosphere, but in moist soils, NH_3 is converted to ammonium ion (NH_4^+), which organisms absorb, or ammonium is oxidized.

In **nitrification**, ammonium is oxidized to nitrate (NO_3^-) via a two-step process requiring autotrophic archaea and bacteria. In the most well-studied nitrification pathway, species of the bacteria *Nitrosomonas* (nī-trō-sō-mō´nas) convert NH_4^+ to nitrite (NO_2^-), which is toxic to plants. Fortunately, *Nitrosomonas* spp. are usually found in association with species of the bacterium *Nitrobacter* (nī-trō-bak´ter) that rapidly convert nitrite to nitrate (NO_3^-), which is soluble and can be used by plants. The soluble nature of nitrate, however, means that it is leached from the soil by water and accumulates in groundwater, lakes, and rivers. Certain microorganisms in waterlogged soils perform **denitrification**, in which NO_3^- is oxidized to N_2 by anaerobic respiration. N_2 gas escapes into the atmosphere.

Scientists have recently discovered an important new aspect of the nitrogen cycle—*anaerobic ammonium oxidation,* or **anammox**. Anammox prokaryotes oxidize 30% to 50% of the world's ammonium into nitrogen gas using nitrite as an electron acceptor.

The Sulfur Cycle

The **sulfur cycle** involves moving sulfur between several oxidation states **(FIGURE 25.17)**. Bacteria decompose dead organisms releasing sulfur-containing amino acids into the environment. Sulfur released from amino acids is converted to its most reduced form, hydrogen sulfide (H_2S), by microorganisms via a process called sulfur *dissimilation.* H_2S is oxidized to elemental sulfur (S^0) and then to sulfate (SO_4^{2-}) under various conditions and by various organisms, including nonphotosynthetic autotrophs such as *Thiobacillus* and *Beggiatoa* (bej´jē-a-tō´a) photoautotrophic green and purple sulfur bacteria. Sulfate is the most readily usable form of sulfur for plants and algae, which animals then eat. Anaerobic respiration by the bacterium *Desulfovibrio* (dē´sul-fō-vib´rē-ō) reduces SO_4^{2-} back to H_2S. Thus, the two major inorganic constituents of the sulfur cycle are H_2S and SO_4^{2-}.

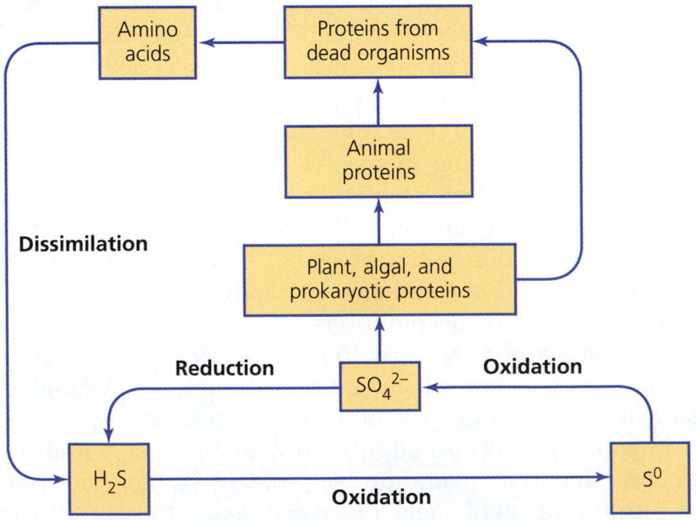

▲ **FIGURE 25.17 Simplified sulfur cycle.** The two main constituents of this cycle are hydrogen sulfide (H_2S) and sulfate (SO_4^{2-}), the fully reduced and oxidized forms of sulfur, respectively.

The Phosphorus Cycle

Unlike nitrogen and sulfur, phosphorus undergoes little change in oxidation state in the environment. Phosphorus usually exists in the environment though often in a metabolically limited amount. Organisms typically use phosphorus in phosphate ion (PO_4^{3-}). The **phosphorus cycle** involves the movement of phosphorus from insoluble to soluble forms available for uptake by organisms and the conversion of phosphorus from organic to inorganic forms by pH-dependent processes. No gaseous form exists to be lost to the atmosphere, but dissolved phosphates do accumulate in water, particularly the oceans, and organic forms of phosphorus are deposited in surface soils following the decomposition of dead animals and plants.

Too much phosphorus can be a problem in a habitat; for example, agricultural fertilizers rich in phosphate are easily leached from fields by rain. The resulting runoff into rivers and lakes can result in **eutrophication** (yū-trō´fi-kā´shŭn)—the overgrowth of microorganisms (particularly algae and cyanobacteria) in nutrient-rich waters. Such overgrowth, called a *bloom*, depletes oxygen from the water, killing aerobic organisms, such as fish. Anaerobic organisms then take over the water system, leading to an increased production of H_2S and the release of foul odors. When excess phosphate (and nitrogen) are removed, such a water system recovers over time.

The Cycling of Metals

Metal ions, including Fe^{2+}, Zn^{2+}, Cu^{2+}, Cd^{2+}, and Mg^{2+}, are important microbial nutrients. Though they are needed only in trace amounts, they can nonetheless be limiting factors in the growth of organisms. Many metals—including the most important trace metal, iron—are present in the environment in poorly soluble forms in rocks, soils, and sediments and are generally unavailable for uptake by organisms. The cycling of metal ions involves primarily a transition from an insoluble to a soluble form, allowing them to be used by organisms and to move through the environment.

Generally, oxidized metal ions (those having fewer electrons) dissolve more readily than reduced ions of the some metal. Miners have successfully used **biomining**—a process in which microbes (typically archaea) oxidize copper, gold, uranium, or other metals so that the metals dissolve in water. The miners can then extract the mineral-laden water and reduce the ions, which causes the metals to come out of solution where the miner can collect them.

Biogeochemical cycles are sustained by microorganisms, most of which live in soil. Next we examine aspects of soil microbes and their habitats.

Soil Microbiology

LEARNING | **OUTCOMES**

25.19 Identify five factors affecting microbial abundance in soils.
25.20 Describe several human and plant diseases caused by soil microbes.

Soil microbiology examines the roles played by organisms living in soil. They rarely cause human disease, though plant pathogens are prevalent in soils and are agriculturally and economically important.

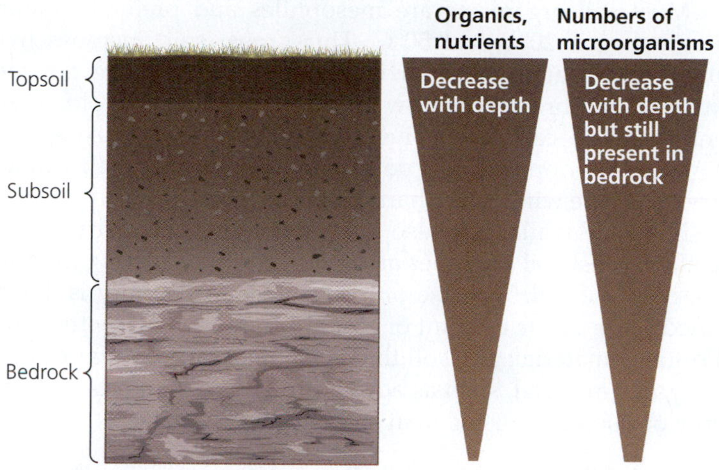

▲ **FIGURE 25.18 The soil layers and the distributions of nutrients and microorganisms within them.** Although topsoils in general are richer in nutrients and microbes than are subsoils, the nutrient and microbial content of topsoils is highly variable.

The Nature of Soils

Soil arises both from the weathering of rocks and through the actions of microorganisms, which produce wastes and organic materials needed to support more complex life forms, such as plants. Soil is composed of two major layers (**FIGURE 25.18**): *topsoil*, which is rich in *humus* (organic chemicals); and *subsoil*, which is composed primarily of inorganic materials. Soil overlies bedrock, which contains little organic material. Most microorganisms are found in topsoil, where the richness of the organic deposits sustain a large biomass. Topsoil itself, however, is highly heterogeneous, and therefore different kinds and amounts of microbes are found in various soils around the world.

Factors Affecting Microbial Abundance in Soils

Several environmental factors influence the density and the composition of the microbial population within a soil, including the amount of water, oxygen content, acidity, temperature, and the availability of nutrients.

Moisture and oxygen content are closely linked in soils. Moisture is essential for microbial survival; microbes exhibit lower metabolic activity, are present in lower numbers, and are less diverse in dry soils than in moist soils. Because oxygen dissolves poorly in water, moist soils have a lower oxygen content than drier soils. When soil is waterlogged, microbial diversity declines and anaerobes predominate, even at the surface. Weather patterns also affect oxygen content, as the presence or absence of rain water determines moisture and thus dissolved oxygen.

The pH of a soil determines in part whether it is rich in bacteria or rich in fungi. Highly acidic and highly basic soils favor fungi over bacteria, though fungi typically prefer acidic conditions. Bacteria dominate when soil pH is closer to 7.

Most soil organisms are mesophiles and prefer temperatures between 20°C and 50°C. Thus, most soil microbes live quite well in areas where winters and summers are not too extreme. Psychrophiles grow only in consistently cold environments and cannot survive in soils that experience spring thawing; the opposite is true for thermophiles, which cannot survive where winters are harsh.

Nutrient availability also affects microbial diversity in soil habitats. Most soil microbes are heterotrophic, utilizing organic matter in the soil. The size of a microbial community is determined more by the amount of organic material than by the kind of organic material: any soil that has a relatively constant input of organic material, such as agricultural land, supports a wider array of microorganisms than soil that is more barren.

Microbial Populations in Soils

Because of the variety of soils, microbial populations differ tremendously from soil to soil and even within the same soil over the course of a season. Bacteria are numerous and diverse inhabitants of soil and are found in all soil layers, where they often form biofilms. Archaea are present in soils, but the inability to culture many of them has limited our ability to study them. The fungi are also a populous group of soil microorganisms. Free-living and symbiotic fungi are found only in topsoil, where they can form gigantic mycelia that cover acres. Viruses are active within soil microorganisms; they are rarely found free.

Some algae and protozoa also live in soils. Soil algae live on or near the surface because as photoautotrophs they require light. Protozoa are mobile and move through the soil, grazing on other microbes. For the most part, protozoa require oxygen and remain in the topsoil. Neither algae nor protozoa can withstand dramatic environmental changes or the introduction of pollutants.

Wherever present, microbes perform a variety of necessary functions. They cycle nitrogen, sulfur, phosphorus, and other elements, converting them into usable forms. Microbes degrade dead organisms and their wastes, and some can clean up industrial pollutants. Further, microbes produce an incredible variety of compounds that have potential human uses. Researchers search natural populations of microbes to discover species that produce valuable chemicals or have useful biological functions.

Soilborne Diseases of Humans and Plants

Although the majority of soil microorganisms are harmless, there are exceptions. Soilborne infections of humans generally result from either direct contact with, ingestion of, or inhalation of microorganisms deposited in soil in animal or human feces or urine. In some cases the microbes live and replicate in the soil, but in most cases soil is simply a vehicle for moving the pathogen from one host to another. The majority of soilborne disease agents are fungal or bacterial. Few soil protozoa or viruses cause disease.

Soil pathogens include the bacterium *Bacillus anthracis* (an-thrā´sis), the causative agent of anthrax, which produces endospores shed from the skins of infected livestock. Endospores may remain dormant in soil for decades or centuries. Disturbing the soil can lead to infection if endospores enter cuts or abrasions on the skin (cutaneous anthrax) or are inhaled into the lungs (inhalation anthrax).

Histoplasma capsulatum (his-tō-plaz´mă kap-soo-lā´tŭm) is a fungus that causes histoplasmosis—a serious respiratory tract infection. *Histoplasma* grows in soil and is also deposited there as spores in the droppings of infected birds and bats. The spores can be inhaled by humans when contaminated soil is disturbed.

Hantavirus (han´tā-vī-rŭs) pulmonary syndrome is a life-threatening, viral respiratory disease acquired via the inhalation of soil contaminated by mouse droppings and urine containing *Hantavirus* (see Emerging Disease Case Study on p. 441). *Hantavirus* has been found throughout North America.

Soil contains many more plant pathogens than human pathogens. Microbial plant infections are generally characterized by one or more of the following signs: necrosis (rot), cankers/lesions, wilt (droopiness), blight (loss of foliage), galls (tumors), growth aberrations (too much or too little), or bleaching (loss of chlorophyll). Bacteria, fungi, and viruses all cause diseases in plants and spread either as airborne spores, through roots or wounds, or by insects.

TABLE 25.7 lists selected bacterial, fungal, and viral soilborne diseases of humans and plants.

TABLE 25.7 Selected Soilborne Diseases of Humans and Plants

Microorganism	Host	Disease
Bacteria		
Bacillus anthracis	Humans	Anthrax
Clostridium tetani	Humans	Tetanus
Agrobacterium tumefaciens	Plants	Crown gall disease
Ralstonia solanacearum	Plants	Potato wilt
Streptomyces scabies	Plants	Potato scab
Fungi		
Histoplasma capsulatum	Humans	Histoplasmosis
Blastomyces dermatitidis	Humans	Blastomycosis
Coccidioides immitis	Humans	Coccidioidomycosis
Polymyxa spp.	Plants	Root rot in cereals
Fusarium oxysporum	Plants	Root rot in many plants
Phytophthora cinnamomi	Plants	Potato blight; root rot in many plants
Viruses		
Hantavirus	Humans	*Hantavirus* pulmonary syndrome
Tobacco mosaic virus	Plants	Necrotic spots in various plants
Soilborne wheat mosaic virus	Plants	Mosaic disease in winter wheat and barley

Aquatic Microbiology

LEARNING | OUTCOME

25.21 Compare the characteristics and microbial populations of freshwater and marine ecosystems.

Aquatic microbiologists study microorganisms living in freshwater and marine environments. Compared to soil habitats, water ecosystems support fewer microbes overall because nutrients are diluted. Many organisms that live in aquatic systems exist in biofilms attached to surfaces. Biofilms allow aquatic organisms to concentrate enough nutrients to sustain growth; without forming biofilms, they likely would starve.

Types of Aquatic Habitats

Aquatic habitats are divided primarily into freshwater and marine systems. *Freshwater* systems, which are characterized by low salt content (about 0.05%), include groundwater, water from deep wells and springs, and surface water in the form of lakes, streams, rivers, shallow wells, and springs. *Marine* environments, characterized by a salt content of about 3.5%, encompass the open ocean and coastal waters, such as bays, estuaries, and lagoons.

Natural aquatic systems can be greatly affected by the release of so-called *domestic water,* which is water resulting from the treatment of sewage and industrial waste. Domestic water released into the environment affects water chemistry and the microorganisms living in the water. Changing levels of chemicals cause increases or decreases in microbial numbers. Furthermore, faulty treatment of sewage leads to contamination of natural water systems with pathogenic microorganisms.

Freshwater Ecosystems Microorganisms become distributed vertically within lake systems according to oxygen availability, light intensity, and temperature. Surface waters are high in oxygen, well lighted, and warmer than deeper waters. In large lakes, wave action continually mixes nutrients, oxygen, and organisms, which allows efficient utilization of resources. In stagnant waters, oxygen is readily depleted, resulting in more anaerobic metabolism and poorer water quality.

Scientists observe four zones in deep lakes **(FIGURE 25.19a)**: The **littoral zone** (li´ter-al) is the area along the shoreline where

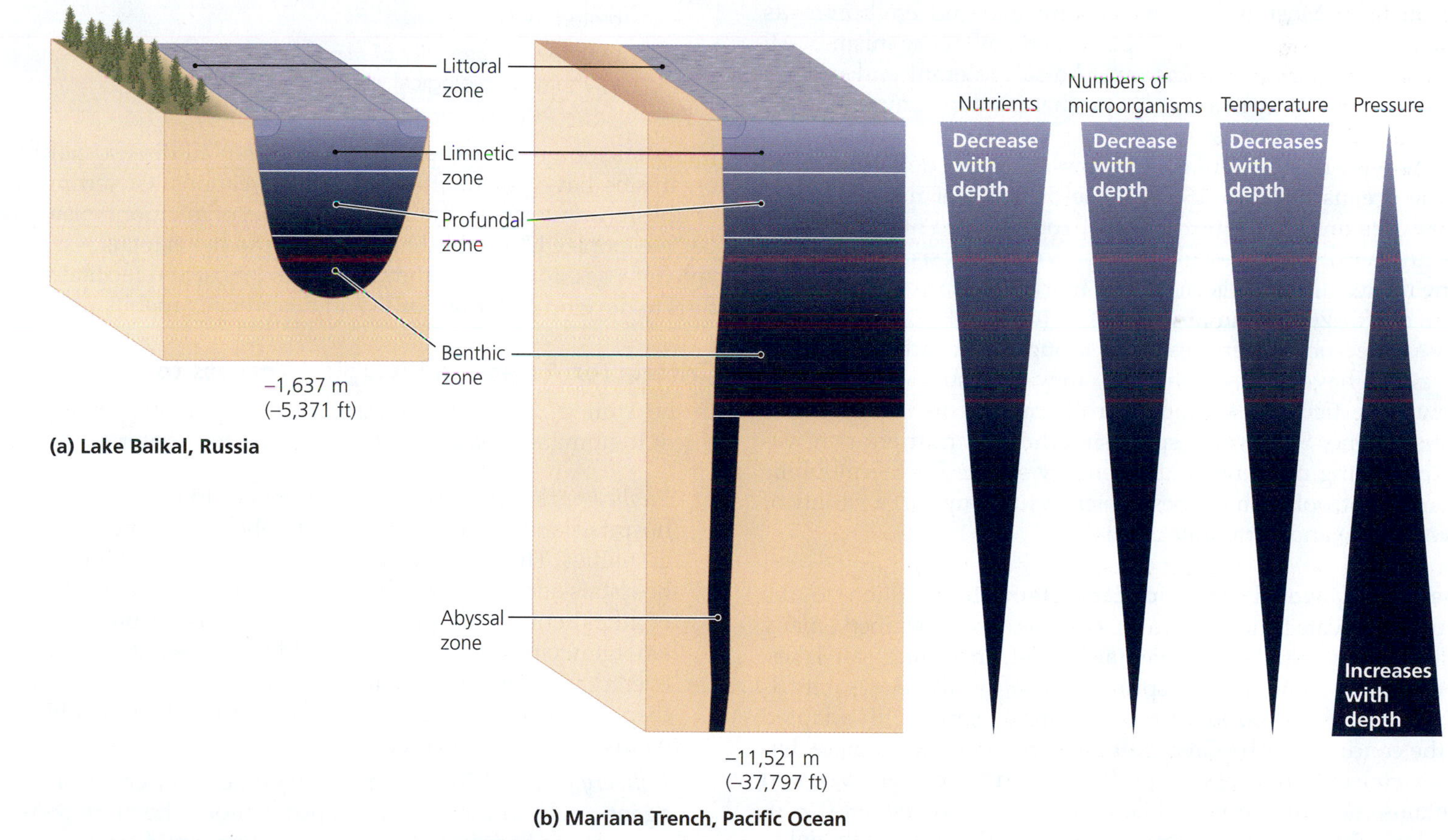

▲ **FIGURE 25.19 Vertical zonation in deep bodies of water. (a)** Lake Baikal, Russia, the world's deepest freshwater lake, and **(b)** the Mariana Trench in the Pacific Ocean, the deepest place in the ocean. Both deep lakes and oceans can be divided into zones that vary with respect to light penetration, concentration of nutrients, temperature, and pressure—and thus the types and abundance of microorganisms. *Why would a bacterium from the bottom of Lake Baikal die in the Mariana Trench?*

Figure 25.19 *A bacterium from the bottom of Lake Baikal could not survive the even greater pressure and the salinity of the seawater deep in the Mariana Trench.*

nutrients enter the lake. The littoral zone is shallow, and light penetrates it; most microbes live here. The **limnetic zone** (lim-net´ik) is the upper layer of water away from the shore. Photoautotrophs reside here and in the littoral zone. The **profundal zone** (prō-fun´dal) is the deeper water beneath the limnetic zone. It has a lower oxygen content and more diffuse light than the previous two zones. Some photosynthetic organisms, such as purple and green sulfur bacteria, perform anaerobic photosynthesis here. Beneath this is the **benthic zone** (bĕn´thik), which encompasses deeper lake water and sediments. Anaerobic bacteria in the sediments produce H_2S, which is used by organisms nearer the surface.

In contrast to lakes, streams and rivers are more uniform because organisms and nutrients are swept along and mixed. Biofilms are particularly important in moving waterways, and the majority of organisms live toward the edges, where surfaces are available, currents are less severe, and organic materials enter the water.

Marine Ecosystems Marine ecosystems are typically nutrient poor, dark, cold, and subject to great pressure. Photoautotrophic prokaryotes, diatoms, dinoflagellates, and algae are found near the surface. Most marine waters are extreme environments inhabited only by highly specialized microorganisms. All microbes in marine systems must be salt tolerant and possess highly efficient nutrient-uptake mechanisms to compensate for the scarcity of nutrients.

As with freshwater lake systems, scientists delineate zones in the oceans **(FIGURE 25.19b)**. The majority of marine microorganisms are found in the littoral zone, where nutrient levels are high and light is available for photosynthesis. The benthic zone makes up the majority of the marine environment. Oceans have a fifth zone—the **abyssal zone** (a-bis´sal), which encompasses deep ocean trenches. Even though the benthic and abyssal zones have sparse nutrients, they still support microbial growth, particularly around **hydrothermal vents** located in the abyssal zone. Such vents spew superheated, nutrient-rich water, providing nutrients and an energy source for thermophilic chemoautotrophic anaerobes, which in turn support a variety of invertebrate and vertebrate animals.

Specialized Novel Aquatic Ecosystems In addition to the two broad categories of water systems just described, many distinctive aquatic ecosystems also exist, including salt lakes, iron springs, and sulfur springs. Each of these systems is inhabited by specialized microorganisms that are highly adapted to the conditions. The Great Salt Lake in Utah, for example, has a salt concentration of 5% to 7%, depending on its level, and contains the extreme halophile *Halobacterium salinarium* (hā´lō-bak-tēr´ē-ŭm sal-ē-nar´ē-ŭm), an archaeon that thrives in highly saline water.

TELL ME WHY

A blogger stated that microorganisms are dangerous and should be avoided in all cases. Why is this idea wrong?

Biological Warfare and Bioterrorism

The properties of microorganisms that allow scientists to manipulate organisms to make advancements in medicine, food production, and industry also enable humans to fashion microbes into *biological weapons*, which can be directed at people, livestock, or crops. **Bioterrorism**, the use of microbes or their toxins to terrorize human populations, is a topic of major concern in today's world. A topic of growing concern is **agroterrorism**—the use of microbes to terrorize humans by destroying the food supply. International treaties and laws of the United States, Great Britain, and other countries prohibit the use of biological weapons.

Assessing Microorganisms as Potential Agents of Warfare or Terror

LEARNING OUTCOMES

25.22 Identify the criteria used to assess microorganisms for potential use as biological weapons or agents of bioterrorism.

25.23 List the characteristics of microbes that make them threats as agents of biological warfare and bioterrorism.

Many microorganisms cause disease, but not all disease-causing organisms have potential as biological weapons. Governments establish criteria for evaluating the potential of microorganisms to be "weaponized." Establishing such criteria helps focus research and defense efforts where they are needed most and facilitates efforts to develop better response and deterrence capabilities.

Criteria for Assessing Biological Threats to Humans

In the United States, the assessment of a potential biological threat to humans is based on the following four criteria:

- *Public health impact.* This criterion relates to the ability of hospitals and clinics to deal effectively with numerous casualties. The more casualties, the more difficult it is for hospitals and clinics to effectively respond to the needs of all patients. If an agent causes numerous serious cases, emergency response systems could be overwhelmed and even cease to function. If an agent is highly lethal, proper disposal of bodies might become difficult, which would contribute to the spread of disease.
- *Delivery potential.* This criterion evaluates how easily an agent can be introduced into a population. The more people that can be infected at one time, the more devastating a primary attack. If an introduced agent spreads on its own through a population, secondary and tertiary waves of illness will augment the number of casualties. Also assessed as part of delivery potential are ease of mass production (the easier an agent is to produce in quantity, the greater its potential threat), availability (the more prevalent it is in the environment, the easier it is to obtain and weaponize),

and environmental stability (the longer it remains infective once released, the greater the threat).

- *Public perception.* This criterion evaluates the effect of public fear on the ability of response personnel to control a disease outbreak following an attack. Agents with high mortality, few treatment options, and no vaccine instill greater fear into a populace, making quarantine (isolating infected and sick patients from the rest of the population) difficult to enforce. The resulting chaos could dramatically decrease the ability of response personnel to control disease transmission and treat patients.
- *Public health preparedness.* This criterion assesses existing response measures and attempts to identify improvements needed in the health care infrastructure to prepare for a biological attack. Diagnosis and recognition involve proper surveillance and training medical personnel to ascertain whether an attack has occurred. Once an attack has been confirmed, predetermined responses are required to reduce confusion and allow rapid control of the situation. Public health preparedness also involves funding for research and development of new vaccines, treatments, and diagnostic capabilities.

When assessing a threat, each potential bioterrorist agent is given a score for each of the criteria. Agents with the highest total scores are considered the most serious threats.

Criteria for Assessing Biological Threats to Livestock and Poultry

The criteria used to evaluate biological threats to livestock and poultry are very similar to those used to evaluate potential threats to humans and include agricultural impact, delivery potential, and plausible deniability.

Infectious agents prove the most devastating for agricultural livestock kept in large herds or flocks. Many animal pathogens exist naturally in soil or are already endemic in livestock and poultry, so they can be easily obtained and readily grown in large quantities. The highly infectious nature of some agents means that by the time a disease is recognized, much of a herd is infected already, so all must be destroyed in an attempt to control the outbreak. Some pathogens persist even after the animals are destroyed. Many animal diseases are spread either by contact or by inhalation, thus making attack easier for a terrorist. Though highly contagious among animals, most are not infectious to humans, making them "safe" for terrorists to handle.

Criteria for Assessing Biological Threats to Agricultural Crops

Plant diseases are generally not as contagious as animal or human diseases. Threats to crops are evaluated on predicted extent of crop loss, delivery and dissemination potential, and containment potential.

Plant pathogens that either cause severe crop loss or produce toxins are considered the greatest threats. Such agents already exist in the environment and can be readily obtained; however, plant pathogens are not as easily mass-produced as animal agents. Plant pathogens that can be spread systematically through fields by natural means, such as through contaminated soil or by insects, could remain in the environment even after destruction of the target crop. Successive plantings in contaminated fields could result in continued crop loss for as long as the agent persists. Because the causes of many plant diseases are not easily diagnosed, widespread dissemination of such a pathogen could occur before response measures were instituted. Economic losses would be staggering, particularly given that embargos on affected crops would likely remain in place for years after an attack.

Known Microbial Threats

In the following sections we briefly discuss some of the microorganisms currently considered threats as agents of bioterrorism. Note that as threat assessments become more refined and technology advances, the list of known bioterrorist agents is likely to change.

Human Pathogens

The U.S. government categorizes biological agents that could be used against humans into three categories (**TABLE 25.8** on p. 812). *Category A agents* are those with the greatest potentials as weapons. *Category B agents* have some potential as weapons but for various reasons are not as dangerous as category A agents (most lack the potential to cause mass casualties). *Category C agents* are potential threats; not enough is currently known about them to determine their true potential as weapons.

Smallpox currently tops the list of bioterrorist threats. Fortunately, it is difficult for would-be terrorists to acquire viral samples for propagation; it also takes a high degree of skill, in addition to specific containment facilities, to work with smallpox virus. An effective vaccine is available, and the vaccine is effective when administered soon after infection.

Animal Pathogens

Biological agents against animals are also divided into categories, with category A agents being the most dangerous. Whereas many potential agents are spread via inhalation, others are spread by insect vectors, making them less likely to be used as weapons. Some agents infect wild animal populations in addition to livestock, potentially amplifying any outbreak that might occur.

Foot-and-mouth disease virus, the most dangerous of potential agroterrorism agents, affects all wild and domestic cloven-hoofed animals. The virus is spread by aerosols and by direct or indirect contact. Humans can transport it from herd to herd on their person, on farm equipment, or through the movement of animals between auctions and farms. Any appearance of foot-and-mouth disease on a farm requires destruction of entire herds, complete disinfection of all areas occupied by the herd, and disposal of all animals by burning or burial. A vaccine exists but not in sufficient quantities to protect all animals.

Plant Pathogens

Many plant pathogens exist, but the categorization of plant pathogens as terrorist agents lags behind similar efforts for

TABLE 25.8 Bioterrorist Threats to Humans

Disease	Agent	Natural Source
Category A Threats: Highest Priority		
Smallpox	Variola major (*Orthopoxvirus*)	None
Anthrax	*Bacillus anthracis* (bacterium)	Soil
Plague	*Yersinia pestis* (bacterium)	Small rodents
Botulism	*Clostridium botulinum* toxin (bacterial)	Soil
Tularemia	*Francisella tularensis* (bacterium)	Wild animals
Viral hemorrhagic fevers	Filoviruses and arenaviruses	Unknown in most cases
Category B Threats: Moderate		
Q fever (fever + flulike syndrome)	*Coxiella burnetii* (bacterium)	Sheep, goats, cattle
Brucellosis (severe flulike syndrome)	*Brucella* spp. (bacteria)	Livestock
Glanders (pulmonary syndrome)	*Burkholderia mallei* (bacterium)	Horses
Melioidosis (severe pulmonary syndrome)	*Burkholderia pseudomallei* (bacterium)	Horses, livestock, rodents, soil
Viral encephalitis	Alphaviruses	Rodents, birds
Typhus fever	*Rickettsia prowazekii* (bacterium)	Humans
Toxins (especially of *Clostridium perfringens* and *Staphylococcus*)	Various bacteria	Various
Psittacosis (pneumonia-like syndrome)	*Chlamydophila psittaci* (bacterium)	Birds
Food safety threats (including *Salmonella* spp., *Escherichia coli* O157:H7, *Shigella*)	Various bacteria and viruses	Soil or animals
Water safety threats (e.g., *Vibrio cholerae*, *Cryptosporidium parvum*)	Various bacteria and viruses	Water
Category C Threats: Low Risk		
Nipah virus (encephalitis)	*Henipavirus Nipah virus*	Bats
Hantavirus pulmonary syndrome	*Hantavirus*	Rodents, soil

humans and animals. Most potential agents are fungi whose dissemination could easily result in contamination of soils, which would be difficult to neutralize. Attacks against grains, corn, rice, and potatoes are considered the most dangerous, because they would have significant negative impacts on national economies and food supplies. All of the agents are naturally present, so detecting the difference between a natural outbreak and an intentional attack would be difficult.

Defense Against Bioterrorism

No defense can completely prevent a carefully planned biological attack against any group or nation. However, much can be done to limit the impact of an attack. The key is coupling *surveillance*—the active diagnosis and tracking of human, animal, and plant diseases—with *effective response protocols.*

Because category A human biological agents are not common, the appearance of more than a few scattered cases is highly suggestive that an attack of some kind has occurred. This is one reason the diseases of category A are *reportable;* that is, they must be reported to state departments of health whenever they occur. Active monitoring of reportable diseases allows epidemiologists to quickly determine when unusual outbreaks are occurring. Once an unexpected pattern is seen, diagnostic confirmation can be obtained and, if an attack is deemed to have occurred, appropriate responses implemented (**FIGURE 25.20**). Such responses may entail forced quarantine, distribution of antimicrobial drugs, or mass vaccination.

Agroterrorism has become more of a concern with the realization that very little security protects the nation's agricultural enterprises. Livestock and poultry are routinely moved around the country without being tested for disease and without being quarantined prior to introduction into new herds or flocks. Infected animals could therefore spread disease as they pass from facility to facility. Compounding the problem is the fact that farms, ranches, auction houses, livestock shows, and irrigation facilities are all open to the public and impose few security measures to prevent purposeful infection of animals. It has been suggested that a step in defending against agroterrorism would be to restrict public access to such facilities. Additionally, effective screening of imported animals and plants would help prevent the introduction of foreign pathogens into the United States. Further, better diagnostic techniques, vaccines, and treatments need to be developed for animal pathogens.

HIGHLIGHT

Could Bioterrorists Manufacture Viruses from Scratch?

Researchers at the State University of New York at Stony Brook have managed an alarming achievement: They synthesized a fully functional poliovirus from materials that can be readily obtained from any of a number of molecular biology supply companies. After they pieced together sequences of RNA to form a full-length poliovirus genome, they successfully replicated and translated this material in cell-free extracts in test tubes. The resulting nucleic acids and proteins were then able to assemble spontaneously into fully infectious viral agents. The scientists began their work from genetic blueprints that exist in the public domain—in published journal articles and on Internet databases.

The ability to manufacture an infectious agent from scratch using preexisting, published knowledge is an unsettling development. Terrorists may be able to manufacture their own agents in similar fashion—rather than needing to steal agents from research facilities or isolate them from natural sources.

As a result of studies like that at Stony Brook, an ethical debate has arisen over whether such research should be pursued—and, if so, whether the details of such research should be published. Some argue that the pursuit and publication of such research unwittingly aids would-be terrorists; others argue that the dissemination of information is necessary for the effective sharing of research within the scientific community and for science to progress. What do you think?

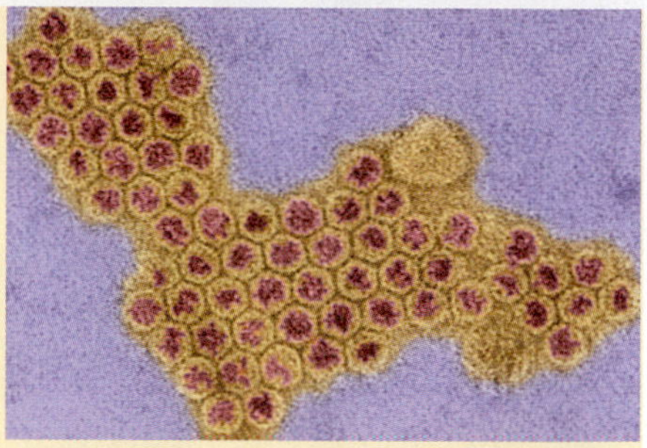

Polioviruses TEM

▲ **FIGURE 25.20 One aspect of the response to a bioterrorist attack.** Shown here are biohazard-suited personnel conducting an investigation at the time of the delivery of anthrax-containing mail in Washington, D.C., in October 2001.

The Roles of Recombinant Genetic Technology in Bioterrorism

Recombinant genetic technology could be used to create new biological threats or modify existing ones so that, for example, vaccines against them no longer work. Traits of various agents could be combined to create novel agents for which no immunity exists in the population.

In addition to the manipulation of existing threat agents, the techniques of recombinant DNA technology enable the synthesis of agents from scratch. Terrorists could, in theory, make their own microbes. To learn about the production of a completely manufactured poliovirus, see **Highlight: Could Bioterrorists Manufacture Viruses from Scratch?** Poliovirus is a relatively simple virus. There is no guarantee that such a process would work for more complex agents such as smallpox.

The techniques of recombinant genetic technology may also be used to thwart bioterrorism. Scientists can identify unique genetic sequences—"fingerprints" or signatures—of recombinants, which may aid in tracking biological agents and determining their source. Genetic techniques may also help in developing vaccines and treatments, and recombinant DNA technology could be used to create pathogen-resistant crops.

TELL ME WHY

Compare human, animal, and plant pathogens that could be used as biological agents in terms of environmental survivability. Why are animal and plant pathogens more common in environmental reservoirs than are human pathogens?

MICRO IN THE CLINIC FOLLOW-UP

Food Poisoning at a Five-Star Restaurant?

The public health officials culture *Salmonella enterica* serotype Enteritidis from the raw slivered almonds that Dan features in several popular dishes. Twenty-four cases of salmonellosis had been reported from nine states, all associated with almonds from one company in California. Dan's almonds come from the same supplier.

The connection with almonds initially perplexes researchers. Typically people contract salmonellosis by eating raw meat or eggs contaminated by infected poultry, reptile, or human feces. Almonds grow on trees and have shells, so how could they be contaminated with *Salmonella*? To make matter worse, almonds are California's largest crop, generating over $1.5 billion annually and supplying over 80% of the world's almonds. The outbreak has serious financial implications, so researchers must be cautious in assigning a source. The Centers for Disease Control and Prevention issues a worldwide recall of raw almonds. Researchers cannot identify a definitive source but suspect that reptile or rodent feces may have contaminated the almond-processing area in the company's facility.

Typically, over 40,000 cases of salmonellosis are reported annually. This represents only about 5% of actual cases because often the symptoms are mild and resolve in five to seven days, and patients don't seek treatment.

Dan recovers in less than a week, and so do his customers. However, several older people, infants, and patients with suppressed immune systems die. Dan is relieved that he is not at fault, but he vows never to use raw almonds again. Instead he switches to roasted almonds on his premier dishes, and his customers love them.

1. **Describe typical sources of *Salmonella* food infections.**
2. **Would the illness people suffered be considered a food infection or a food intoxication? Explain why.**

MM Check your answers to Micro in the Clinic Follow-Up questions in the MasteringMicrobiology Study Area.

Explore the Invisible

Visit the **MasteringMicrobiology Study Area** to challenge your understanding with practice tests, animation quizzes, and clinical case studies!

MasteringMicrobiology®

CHAPTER SUMMARY

Food Microbiology (pp. 784–791)

1. The commercial use of microorganisms is referred to as **applied microbiology** and includes two distinct fields: food microbiology and industrial microbiology.
2. **Food microbiology** involves the use of microorganisms in food production and the prevention of foodborne illnesses. In this context, **fermentations** involve desirable changes to food; **spoilage** involves undesirable changes to food.
3. Food fermentations involve the use of **starter cultures**—known organisms that carry out specific and reproducible fermentation reactions. For most of the great variety of fermented vegetables, meats, and dairy products, starter cultures of lactic acid bacteria are used. The acid produced results in "sour" flavors.
4. Alcoholic fermentations, usually performed by yeasts, convert sugars to ethanol and carbon dioxide. Alcoholic fermentation is used in the production of wine, distilled spirits, beer, vinegar, and bread.
5. Intrinsic factors of food spoilage are properties of the food itself, such as moisture content and physical structure, that determine how susceptible a food is to spoilage. Extrinsic factors of food spoilage include ways in which the food is handled.
6. Industrial processes preserve food via canning, pasteurization, drying, freeze drying **(lyophilization)**, irradiation, and aseptic packaging techniques. Natural and artificial preservatives are added to some foods to inhibit microbial growth. In stores and at home, foods should be properly stored in appropriate containers, cold foods should be kept cold, foods should be cooked

thoroughly, and leftovers should be refrigerated to reduce spoilage.

7. Food poisoning is a general term that describes instances of **food infections** (illnesses due to the consumption of living microbes) or **food intoxications** (illnesses due to the consumption of microbial toxins). Food poisoning frequently follows poor food handling.

Industrial Microbiology (pp. 791–801)

1. **Industrial microbiology** is concerned with the use of microorganisms for the production of commercially valuable materials. Such industrial fermentations synthesize desired products and can use genetically modified microbes.
2. Batch production is the growth of organisms followed by harvesting of the entire culture and its products. Continuous flow production involves the constant addition of nutrients to a culture and the removal of the products formed. Products may be either primary metabolites (produced during active growth) or secondary metabolites (produced during the stationary phase).
3. Microorganisms produce a variety of useful products, including enzymes, dyes, alternative fuels, plastics, pharmaceuticals, pesticides, biosensors, and bioreporters. Alternative fuels (biofuels) can be produced from products of photosynthesis or by fermentation of biomass into fuel. **Biosensors** combine microbes and electronics to detect microbial activity in the environment. **Bioreporters** use microbes alone as sensors.
4. **Potable** drinking water is derived via water treatment, which involves the removal of microbes and of organic and inorganic contaminants from water. **Polluted** water contains organisms or chemicals at unsafe levels. Some diseases result from consumption of polluted water or of food harvested from polluted water.
5. Water treatment involves four steps: **sedimentation**, **flocculation**, **filtration**, and **disinfection**. Sedimentation removes large materials. In flocculation, alum combines with suspended materials to make them precipitate. Filtration, using either slow or rapid sand filters, removes microorganisms and chemicals. Disinfection, usually chlorination, kills most microbes that remain after filtration. Water is potable following treatment if it has zero coliforms **(indicator organisms)** per 100 ml of water, as determined by one of several testing methods (MPN, membrane filtration, ONPG/MUG test).
6. **Wastewater** (sewage) refers to water used for washing or flushed from toilets. Wastewater treatment involves the removal of solids, organic chemicals, and microorganisms. **BOD (biochemical oxygen demand)** is a measure of the amount of oxygen required to fully metabolize organic wastes.
7. Municipal wastewater treatment involves four phases. Primary treatment entails sedimentation of large materials (primary **sludge**) and flocculation. Secondary treatment involves sedimentation of secondary sludge as well as the removal of microorganisms and organic material using activated sludge systems or trickle filters. In the third phase, effluent water is chemically treated (chlorinated) and released. In the fourth phase, primary and secondary sludge is digested and dried to produce landfill. Methane gas can also be recovered during the processing of sludge.
8. **Septic tanks** and **cesspools** are home equivalents of municipal wastewater treatment. After wastewater leaves the home, it is deposited in underground tanks. Sludge settles, and the water is released into the soil, where natural processes remove organic chemicals and microorganisms.
9. **Oxidation lagoons** are used by farmers and ranchers to process animal wastes. Waste is pumped into successive lagoons, where wastes are digested by microorganisms prior to the release of the water into natural water systems.
10. In **artificial wetlands**—found in some planned communities and industrial sites—ponds, marshes, and meadowland remove organic compounds, chemicals, and microorganisms from sewage as the water moves through them.

Environmental Microbiology (pp. 801–810)

1. Microorganisms live in microhabitats within larger **habitats**, or physical localities, in the environment. Organisms and habitats together form **ecosystems**. Single cells give rise to populations, populations performing similar functions form guilds, and many guilds together form a community of organisms living in a habitat. The study of the interactions of microorganisms among themselves and with their environment is **microbial ecology**, which is a part of **environmental microbiology** along with studies of microbial habitats.
2. All the ecosystems on Earth form the biosphere. **Biodiversity** describes the number of species living in a given ecosystem, whereas **biomass** refers to the quantity of all these species.
3. Microbes compete for the scarce resources that characterize the majority of habitats. Some microbes actively oppose the growth of other microbes (antagonism), but many microbes cooperate, forming complex biofilms.
4. **Bioremediation** is the use of microorganisms to metabolize toxins in the environment to reclaim soils and waterways. Industrial products are either biodegradable or recalcitrant (resistant to degradation by natural means). Recombinant DNA technology enables scientists to create microbes to degrade some recalcitrant chemicals.
5. Acid mine drainage is an environmental problem in places where ores contain iron. Microbial action on leached iron in water from mines results in the production of acid and ferric iron deposits that acidify water, which is destructive to most plant and animal life.
6. **Biogeochemical cycling** involves the movement of elements and nutrients from unusable forms to usable forms by the activities of microorganisms. These processes involve production of new biomass, consumption of existing biomass, and decomposition of dead biomass for reuse in the cycle. The four major biogeochemical cycles are the **carbon, nitrogen, sulfur**, and **phosphorus cycles**. The cycling of trace metals is also important.
7. In the **carbon cycle**, CO_2 is fixed by photoautotrophs and chemoautotrophs into organic molecules, which are used by other organisms. Organic carbon is converted back to CO_2 in aerobic respiration, by decomposition, and by combustion.
8. In the **nitrogen cycle**, nitrogen gas in the atmosphere is converted to ammonia via a process called **nitrogen fixation**. Ammonia may be converted to nitrate via a two-step process called **nitrification**. Organisms also use ammonia and nitrate to make nitrogenous compounds. Such compounds in wastes and dead cells are converted back to ammonia via **ammonification**. Nitrate can be converted to nitrogen gas by **denitrification**. **Anammox** prokaryotes oxidize ammonium anaerobically into nitrogen.
9. In the **sulfur cycle**, sulfur moves between several inorganic oxidation states (primarily H_2S, SO_4^{2-}, and S^0) and proteins.
10. The **phosphorus cycle** involves the conversion of PO_4^{3-} among organic and inorganic forms.

11. **Biomining** is a process that uses microorganisms (usually archaea) to oxidize metals in rocks to make a soluble ion. Mineral-laden water is collected and subjected to a reducing regimen, and the metal ions solidify.

12. **Eutrophication**—the overgrowth of microorganisms in aquatic systems—can result from the presence of excess nitrogen and phosphorus, which act as fertilizers. The overgrowth of microbes depletes the oxygen in the water, resulting in the death of fish and other animals.

13. Soil microbiology is the study of the roles of microbes in the ground. Soils are fairly diverse and differ greatly in nutrients, water content, pH, oxygen content, and temperature. Microbes inhabit topsoil in high numbers and are less abundant in deeper rock and sediments. Pathogenic microorganisms found in soil can be acquired through contact, but often disease follows the consumption of contaminated soil.

14. Aquatic habitats include freshwater and marine water systems. Scientists recognize four zones in freshwater based on temperature, light, and nutrient levels: the nutrient-rich **littoral zone** along the shore, the sunlit **limnetic zone** at and near the surface, the **profundal zone** just below the limnetic zone, and the **benthic zone** on the bottom, which is devoid of light and nutrients. Marine environments have the same four zones plus an **abyssal zone** (below the benthic zone), which is virtually devoid of life except around **hydrothermal vents**.

15. Microorganisms living in aquatic environments typically form biofilms to better accumulate nutrients that are limiting in most nonpolluted water systems.

Biological Warfare and Bioterrorism (pp. 810–813)

1. Of the relatively few microorganisms that can cause disease in humans, animals, and plants, some might be used to purposely infect individuals and are thus potential agents of biological warfare and **bioterrorism**. It is illegal to deploy such weapons. **Agroterrorism** is the deliberate infection of livestock or crops.

2. The degree to which an organism is considered a biological threat depends on several criteria concerning public health impact, dissemination or delivery potential, public perception, and public health preparedness.

3. Human, animal, and plant pathogens are categorized by threat level. Category A agents have the greatest potential to be used for bioterrorism, and category C refers to agents whose threat potential needs further study.

4. Defense against bioterrorism begins with surveillance—the reporting and monitoring required for effective response to biological attacks. To limit the impact of any attack that may occur, diagnoses must be reliable, and efficient control measures must be in place.

5. Recombinant genetic technology could potentially lead to the development of novel agents or the modification of existing agents to make them more difficult to control in the event of an attack. Such technology could also lead to potential vaccines, cures, or pathogen-resistant crops.

QUESTIONS FOR REVIEW

Answers to the Questions for Review (except Short Answer questions) begin on p. A-1.

Multiple Choice

1. Food fermentations do all of the following *except*
   a. give foods a characteristic taste.
   b. lower the risk of food spoilage.
   c. sterilize foods.
   d. increase the shelf life of the food.

2. Commercially produced beers and wines are usually fermented with the aid of
   a. naturally occurring bacteria.
   b. naturally occurring yeast.
   c. specific cultured bacteria.
   d. specific cultured yeast.

3. Which of the following lists foods in order, from perishable to nonperishable?
   a. pasta, cheese, fruit, uncooked ground beef
   b. pasta, fruit, uncooked ground beef, cheese
   c. uncooked ground beef, fruit, cheese, pasta
   d. uncooked ground beef, fruit, pasta, cheese

4. Which of the following would be the best growth medium to use for industrial fermentations?
   a. corn
   b. synthetic medium made by hand
   c. whey from cheese production
   d. brewing mash

5. Biodegradable plastics are made from which of the following microbial metabolites?
   a. sludge c. BOD
   b. PHA d. alum

6. Strains of the bacterium *Pseudomonas syringae* have been identified as being capable of
   a. producing plastics.
   b. producing alternative fuels.
   c. fermenting foods.
   d. preventing ice formation.

7. Which of the following is added during water or sewage treatment to promote flocculation?
   a. sludge c. BOD
   b. PHA d. alum

8. During chemical treatment of drinking water and wastewater, which of the following microbes is least likely to be inactivated or killed?
   a. algae c. fungal spores
   b. viruses d. bacteria

9. In which step is most of the organic content of sewage removed?
   a. primary treatment
   b. secondary treatment
   c. tertiary treatment
   d. sludge treatment

10. Microbial communities are composed of
    a. single, pure populations.
    b. all organisms in a locale.
    c. mixed populations of organisms.
    d. a biosphere.

11. In the environment, nutrients are generally
    a. limiting.
    b. present in excess.
    c. stable.
    d. artificially induced.

12. Most chemical elements exist in the environment as
    a. usable forms in soil and rock.
    b. usable forms in water.
    c. unusable forms in soil and rock.
    d. unusable forms in water.

13. In the carbon cycle, microbes
    a. convert CO_2 into organic material for consumption.
    b. convert CO_2 into inorganic material for storage.
    c. convert fossil fuels into usable organic compounds.
    d. convert oxygen into water as a by-product of photosynthesis.

14. Nitrification
    a. converts organic nitrogen to NH_3.
    b. converts NH_3 to NH_4^+.
    c. converts NH_4^+ to NO_3^-.
    d. converts NO_3^- to N_2.

15. In aquatic environments, most microbial life is found in the
    a. littoral zone.
    b. limnetic zone.
    c. profundal zone.
    d. benthic zone.
    e. abyssal zone.

16. Which of the following diseases is *not* caused by category A biological weapons agents?
    a. smallpox
    b. plague
    c. Q fever
    d. tularemia

17. Of the following characteristics, which would contribute most to making a microorganism an effective biological warfare agent?
    a. is readily available in the environment
    b. can be spread by contact after original dissemination
    c. cannot be treated well outside of a hospital
    d. is easily identified by symptoms

18. Anammox reactions are
    a. anaerobic and part of nitrogen cycling.
    b. anaerobic and part of carbon cycling.
    c. aerobic and part of sulfur cycling.
    d. aerobic and part of metal ion oxidation.

19. Industrial fermentation
    a. always involves alcohol production.
    b. involves the large-scale production of any beneficial compound.
    c. refers to the oxidation of sugars using organic electron acceptors.
    d. is any desirable change to food by microbial metabolism.

20. Lyophilization in food preservation is by
    a. cell lysis.
    b. gamma radiation.
    c. rapid heating.
    d. freeze drying.

Matching

Match each term with its correct definition.

1. _____ Organisms whose presence in water indicates contamination from feces
2. _____ Compound produced by a bacterium that kills insects
3. _____ Refers to water that is fit to drink
4. _____ Community of organisms surrounded by polysaccharides and attached to surfaces
5. _____ Used in the processing of animal wastes; mimics primary and secondary wastewater treatment
6. _____ Refers to compounds that are resistant to microbial degradation
7. _____ Water that is not bound by solutes
8. _____ Refers to quantity of all organisms present in an environment
9. _____ Process whereby pollutants accumulate to high levels in waterways, causing overgrowth and anaerobic conditions
10. _____ Process of reducing nitrogen from the atmosphere
11. _____ The process whereby organisms actively inhibit the growth of other organisms
12. _____ Undesirable fermentation reactions in food leading to poor taste, smell, or appearance
13. _____ Brief heating of foods during processing
14. _____ Descriptor of the level of organic material present in wastewater
15. _____ Fermentative products produced by microorganisms during stationary phase
16. _____ Process involving oxidation and then reduction of metals

A. Spoilage
B. Water activity
C. Coliforms
D. Pasteurization
E. Secondary metabolites
F. Bt toxin
G. Potable
H. BOD
I. Oxidation lagoon
J. Recalcitrant
K. Biomass
L. Antagonism
M. Biomining
N. Nitrogen fixation
O. Eutrophication
P. Biofilm

Modified True/False

Indicate whether each of the following statements is true or false. Rewrite the underlined phrase to make a false statement true.

1. _____ The fermentation of dairy products relies on <u>mixed acid</u> fermentation.
2. _____ Sauerkraut production involves the <u>alcoholic</u> fermentation of cabbage.
3. _____ Pasteurization kills <u>mesophilic</u> microorganisms except endospore formers.
4. _____ Methane is a gas produced by microbial metabolism that can be used directly as a <u>fuel source</u>.
5. _____ The treatment of drinking water and sewage involves <u>similar</u> processes.
6. _____ <u>Recalcitrant</u> molecules can be degraded by naturally occurring microorganisms.
7. _____ Biofilms of microorganisms form in <u>aquatic</u> environments only.
8. _____ <u>Cooperation</u> is common among microorganisms living in microhabitats.
9. _____ Aquatic microorganisms are <u>more</u> prevalent near the surface than at the bottom of waterways.
10. _____ <u>Abyssal</u> organisms are found near shores of oceans.

Fill in the Blanks

1. Intrinsic factors affecting food spoilage are properties of ________________ rather than ________________.
2. Leaving foods out at room temperature ________________ the likelihood of food spoilage.
3. The two types of industrial fermentation equipment are designed for ________________ production or ________________ production.
4. Potable water is allowed to have ________________ coliforms per 100 ml of water tested.
5. Leaching of compounds from mine tailings often results in the oxidation of two elements: ________________ and ________________.
6. Biogeochemical cycling involves three primary steps: ________________, ________________, and ________________.
7. Nitrogen exists primarily as ________________ in the environment.
8. Phosphorus exists primarily as ________________ in the environment.
9. A ________________ is a device composed of microbes and electronics used to detect other microbes or their products.
10. ________________ is the amount of oxygen required by aerobic organisms to fully metabolize organic waste in water.

VISUALIZE IT!

1. Label the general phases in the carbon cycle.

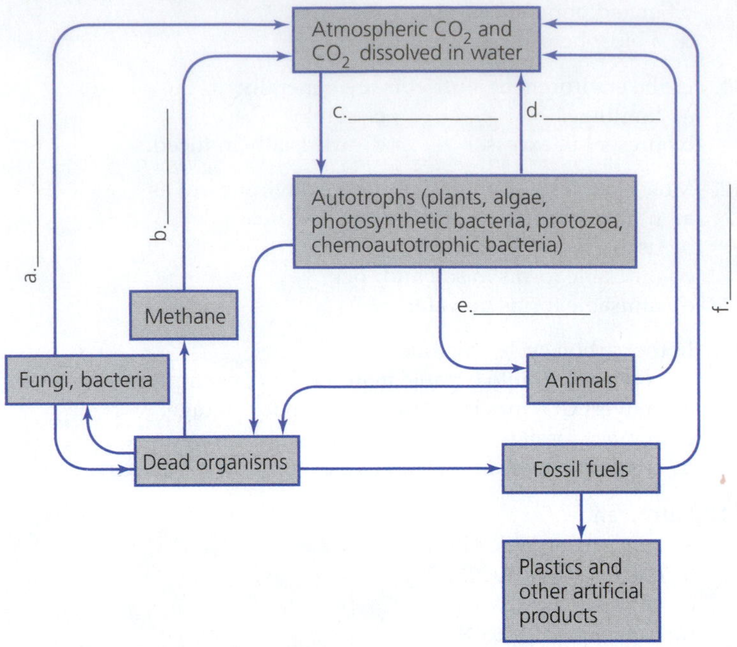

2. Label the processes of the nitrogen cycle.

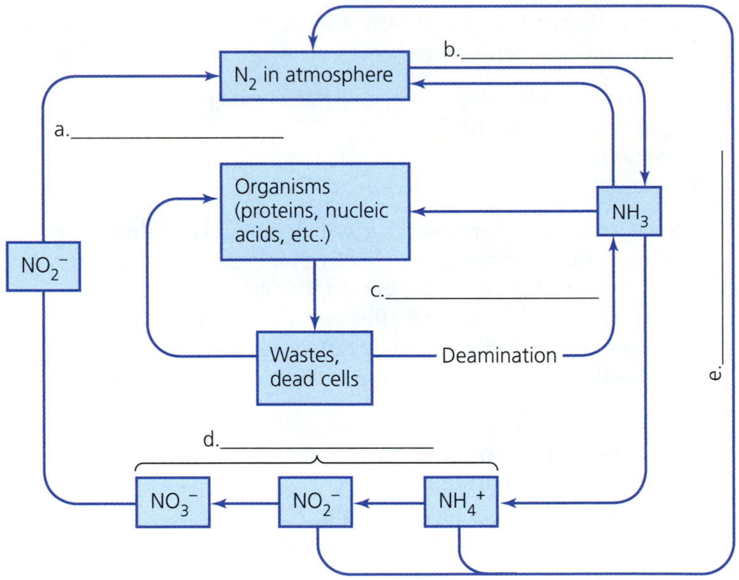

CRITICAL THINKING

1. Why does the application of recombinant DNA technology to food production have the potential to enhance not only food quality but also food output? Given that it has the potential to feed more people, why are some people opposed to genetic modification of foods?
2. Given what you know about microbial nutrition and metabolism, explain why it is technically more difficult to achieve high yields of a secondary metabolite than of a primary metabolite.
3. Compare the types of alternative fuels that could be produced by microbes. Based on starting materials, which would provide the most renewable energy?
4. One way that farmers are attempting to prevent the development of widespread insect resistance to Bt toxin is by planting non-Bt-producing crops in fields adjacent to Bt-containing crops. Insects can infest both fields, but the insects eating the non-Bt plants should survive at a higher rate than those in the Bt-producing field. Why would this prevent the dissemination of resistance to Bt toxin among insects?
5. Even though water and wastewater undergo essentially the same forms of treatment, treated wastewater usually is not put into the water system from which drinking water is derived. Why not?
6. Take a critical look at the garbage in all of your wastebaskets. How much of the material present could possibly be degraded by microbes? How much could be recycled either at a recycling center or in compost? What is left if the degradable or recyclable materials were removed?
7. Given the amount of pollutants and disease-causing microbes that are in soil and water, why don't we see higher incidences of soilborne and waterborne illnesses?
8. Explain why influenzaviruses could be potentially devastating biological weapons.
9. Explain why sake—sometimes called rice wine—would be more accurately described as "rice beer."
10. Inexpensive bulk wines and wines that have been left exposed to air for too long often acquire a "vinegary" smell or taste. Explain why this is so.
11. Describe the intrinsic and extrinsic factors that would contribute to the spoilage of butter, an apple, and a steak.
12. Suggest why neither the original food involved nor the organism responsible can ever be identified in more than 80% of food poisoning cases.
13. Which process would you expect to yield more product from the same-size vats over the same amount of time: batch production or continuous flow production? Why?
14. Explain how biosensors might be used for water quality testing. Would biosensors give more accurate identifications of the microorganisms present in water than an ONPG or MUG test? Is such specificity truly necessary?
15. Microbes are found mostly in topsoil, but some are found miles deep in bedrock. Nutritionally, how do deeply buried microbes survive?

CONCEPT MAPPING

Using the following terms, draw a concept map that describes microbial roles in food production. For a sample concept map, see p. 94. Or, complete this and other concept maps online by going to the MasteringMicrobiology Study Area.

Acetic acid in vinegar
Acetobacter
Alcohol in beer
Alcohol in wine
***Aspergillus* spp. and *Lactobacillus* spp.**
Bread to rise
Fermentation
Gluconobacter
Lactobacillus bulgaricus
Malt from grains
***Saccharomyces cerevisiae* (yeast)**
Soy sauce
Soybeans and wheat
Streptococcus thermophilus
Sugars in bread dough
Sugars in fruit juice
Sugars in milk
Yogurt

Answers to Questions for Review

Answers to Multiple Choice, Fill in the Blanks, Matching, True/False, Visualize It!, and Concept Mapping questions are listed here. Answers to Short Answer and Critical Thinking questions are available for instructors only in the Instructor's Manual that accompanies this text.

CHAPTER 1

Multiple Choice

1. a; **2.** c; **3.** d; **4.** a; **5.** c; **6.** d; **7.** a; **8.** b; **9.** d; **10.** d

Fill in the Blanks

1. Martinus Beijerinck and Sergei Winogradsky; **2.** Louis Pasteur and Eduard Buchner; **3.** Paul Ehrlich; **4.** Edward Jenner; **5.** John Snow; **6.** Robert Koch; **7.** John Snow; **8.** Louis Pasteur; **9.** Louis Pasteur

Visualize It

1. 1. cilium; 2. flagellum; 3. pseudopod; 4. nucleus; **2.** Microles were only on dust in neck's bend.

Matching

1. J; **2.** H; **3.** C; **4.** C, H, K; **5.** B; **6.** A; **7.** C; **8.** E; **9.** D; **10.** D; **11.** I; **12.** L

CHAPTER 2

Multiple Choice

1. c; **2.** d; **3.** c; **4.** c; **5.** b; **6.** c; **7.** a; **8.** a; **9.** a; **10.** c

Fill in the Blanks

1. valence; **2.** nonpolar covalent; **3.** ATP; **4.** fat, wax, amylase (starch), glycogen; **5.** functional groups; **6.** hydrolysis; **7.** exothermic; **8.** products; **9.** pH; **10.** ribose

Visualize It!

1. Primary structure: light blue strands. Secondary structure: green α-helices, gray β-pleated sheets. The entire molecule represents tertiary structure; **2.** See Figure 2.21, every angle formed by lines without a lettered designation represents a carbon atom. Purple box = amino group, orange box = carboxyl group, blue box = side group

CHAPTER 3

Multiple Choice

1. b; **2.** b; **3.** c; **4.** c; **5.** c; **6.** d; **7.** a; **8.** a; **9.** a; **10.** b; **11.** c; **12.** d; **13.** d; **14.** b; **15.** c

Matching

1. D Glycocalyx; B, H, I Flagella; F Axial filaments; H Cilia; A Fimbriae; C, G Pili; E Hami; **2.** A Ribosome; D Cytoskeleton; F Centriole; E Nucleus; I Mitochondrion; G Chloroplast; C Endoplasmic reticulum; H Golgi body; B Peroxisome

Visualize It!

1. a. cytoplasm—contains metabolic chemicals; b. nucleoid—site of DNA (genes); c. glycocalyx—adhesion; d. cell wall—protects against osmotic forces; e. inclusions—stored chemicals; f. flagellum—motility; g. cytoplasmic membrane—controls import and export; h. nucleolus—site of RNA synthesis; i. cilium—motility; j. 80S ribosomes—make proteins; k. nuclear envelope—bounds DNA (genes); l. mitochondrion—makes ATP (energy source); m. centriole—plays a role in cell division; n. Golgi body—packages secretions; o. rough endoplasmic reticulum (RER)—transports proteins; p. smooth endoplasmic reticulum (SER)—lipid synthesis; q. cytoskeleton—helps maintain cell shape; **2.** a. axial filament (endoflagella); b. peritrichous; c. polar (tuft); d. polar (single); **3.** Chemical A enters the cell via facilitated diffusion through a protein channel, which is likely specific for chemical A. Chemical B does not have such a route. At some concentration of chemical A, all the protein channels are filled, such that the rate of diffusion cannot increase. The cell could increase the rate of diffusion of chemical A by inserting more of the channel protein into the membrane. The rate of diffusion of chemical B could be increased by removing chemical B from the cytoplasm at an increased rate. This would have the effect of increasing its concentration gradient.

Concept Mapping

1. Gram-positive cell wall; **2.** Teichoic acids; **3.** Gram-negative cell wall; **4.** Periplasm; **5.** Peptidoglycan; **6.** Glycan chains; **7.** *N*-acetylglucosamine; **8.** Lipopolysaccharide (LPS); **9.** Porin; **10.** Lipid A

CHAPTER 4

Multiple Choice

1. c; **2.** d; **3.** c; **4.** d; **5.** d; **6.** d; **7.** b; **8.** a; **9.** a; **10.** d

Fill in the Blanks

1. 600X; **2.** heat fixation; **3.** increases, increases, more; **4.** Contrast; **5.** negatively

Visualize It!

1. a. scanning electron; b. bright-field light; c. phase-contrast light; d. fluorescent light; e. transmission electron; f. differential interference contrast (Nomarski); **2.** See Figure 4.4.

CHAPTER 5

Multiple Choice

1. c; **2.** a; **3.** c; **4.** a; **5.** a; **6.** c; **7.** b; **8.** d; **9.** a; **10.** c; **11.** d; **12.** d; **13.** a; **14.** a; **15.** a; **16.** c; **17.** a; **18.** c; **19.** a; **20.** c

Matching

1. C; **2.** B; **3.** E; **4.** A

Fill in the Blanks

1. the original reaction center chlorophyll; **2.** 2; **3.** pentose phosphate, Entner-Doudoroff; **4.** The Krebs cycle; **5.** Oxygen or ½ O_2; **6.** NO_3^-, SO_4^{2-}; CO_3^{2-}; **7.** inorganic; **8.**

Category of Enzyme	Description
Hydrolase	Catabolizes substrate by adding water
Isomerase	Rearranges atoms
Ligase/polymerase	Joins two molecules together
Transferase	Moves functional groups
Oxidoreductase	Adds or removes electrons
Lyase	Splits large molecules

9. chemiosmosis; **10.** NAD^+, FAD

Visualize It!

1. See Figure 5.12; **2.** a. Glycolysis (cytosol); b. Electron transport chains (cristae); c. Krebs cycle (matrix)

CHAPTER 6

Multiple Choice

1. b; **2.** c; **3.** b; **4.** a; **5.** b; **6.** b; **7.** d; **8.** a; **9.** c; **10.** a; **11.** b; **12.** d; **13.** c; **14.** a; **15.** a

Fill in the Blanks

1. electrons; **2.** singlet; **3.** nitrogen; **4.** Growth factors; **5.** minimum growth temperature; **6.** osmotic; **7.** halophiles; **8.** Carotenoid; **9.** fixation; **10.** streak plate; **11.** inorganic

Visualize It!

1. See Figure 6.3; **2.** Beta-hemolysis

CHAPTER 7

Multiple Choice

1. a; **2.** c; **3.** d; **4.** c; **5.** d; **6.** a; **7.** a; **8.** d; **9.** a; **10.** c; **11.** c; **12.** d; **13.** a; **14.** d; **15.** c; **16.** b; **17.** d; **18.** c; **19.** b; **20.** c; **21.** b; **22.** b; **23.** a; **24.** d; **25.** b

Fill in the Blanks

1. initiation of transcription, elongation of the RNA transcript, termination of transcription; **2.** codon; **3.** silence, missense, nonsense; **4.** frameshift; **5.** promoter, operator, a series of genes; **6.** inducible; **7.** semiconservative; **8.** transformation, transduction, bacterial conjugation; **9.** Transposons; **10.** Crossing over; **11.** Transfer; **12.** small interfering, micro

Visualize It!

1. a. replication fork; b. stabilizing proteins; c. nucleotide (triphosphate); d. leading strand; e. helicase; f. primase; g. DNA polymerase III; h. RNA primer; i. Okazaki fragments; j. DNA polymerase I; k. lagging strand; l. DNA ligase; **2.** GC base pairing is more stable than AT pairing because GC base pairs are held together by three hydrogen bonds. Therefore, the portion of the pictured molecule that appears as a single, thick structure has more GC base pairs than does the portion that has separated into two strands, forming a loop; **3.** These drugs are analogs of the nucleotide bases in DNA. When incorporated into a strand of DNA, the drugs prevent the replication of new DNA strands because they lack an —OH on the 3′ carbon of the sugar needed for continued elongation.

CHAPTER 8

Multiple Choice

1. d; **2.** b; **3.** c; **4.** d; **5.** a; **6.** c; **7.** c; **8.** b; **9.** a; **10.** d

Modified True/False

1. *cut DNA at specific sites*; **2.** True; **3.** *Electrophoresis*; **4.** True; **5.** *Southern blotting*

Visualize It!

1. Step 1, denaturation: 94°C; step 2, priming: DNA primer, deoxyribonucleotide triphosphates, DNA polymerase, cool to 65°C; step 3, extension: same reagents as step 2, 72°C; step 4, repeat; **2.** None of the DNA fingerprints exactly match the standard, but DNA from patients 167,179,165, 173, and 317 is very close; they likely have the disease. Patient 177 may be infected.

CHAPTER 9

Multiple Choice
1. a; **2.** b; **3.** d; **4.** d; **5.** d; **6.** a; **7.** d; **8.** d; **9.** c; **10.** a; **11.** d; **12.** d; **13.** a; **14.** c; **15.** c; **16.** b; **17.** a; **18.** d; **19.** d; **20.** a

Visualize It!
1. a. 1 min; b. 2.5 minutes; **2.** Ions of copper and zinc from the brass have antimicrobial effects.

CHAPTER 10

Multiple Choice
1. d; **2.** a; **3.** a; **4.** c; **5.** d; **6.** c; **7.** a; **8.** d; **9.** a; **10.** d

Visualize It!
1. See Figure 10.4; **2.** Etest; antimicrobial sensitivity and minimum inhibitory concentration; increasing the drug concentration would result in a zone of inhibition wider and longer than shown. Your oval should extend to at least .75.

CHAPTER 11

Modified True/False
1. asexually; **2.** vibrio; **3.** True; **4.** lack; **5.** rRNA; **6.** True; **7.** True; **8.** True; **9.** *Epulopiscium;* **10.** True

Matching
1. E; **2.** B; **3.** C; **4.** D; **5.** L; **6.** K; **7.** H; **8.** Q; **9.** I; **10.** N; **11.** P; **12.** O; **13.** M; **14.** F; **15.** A

Multiple Choice
1. c; **2.** a; **3.** c; **4.** d; **5.** a; **6.** d; **7.** a; **8.** b; **9.** c; **10.** d

Visualize It!
1. See Figure 11.1; **2.** (a) central (b) subterminal

CHAPTER 12

Multiple Choice
1. a; **2.** d; **3.** c; **4.** b; **5.** a; **6.** d; **7.** b; **8.** c; **9.** a; **10.** c; **11.** c; **12.** a; **13.** d; **14.** a; **15.** d

Matching
First section: **1.** E; **2.** B; **3.** D; **4.** C; **5.** A
Second section: **1.** A; **2.** F; **3.** E; **4.** C; **5.** D; **6.** B
Third section: **1.** C; **2.** E; **3.** B; **4.** D; **5.** A

Visualize It!
1. a. ascospore, sexual; b. basidiospore, sexual; c. conidio, asexual; d. sporangiospore, asexual; **2.** See Figure 12.19.

Fill in the Blanks
1. protozoology; **2.** mycology; **3.** phycology; **4.** mycoses; **5.** radiolarians

CHAPTER 13

Multiple Choice
1. c; **2.** c; **3.** a; **4.** a; **5.** b; **6.** c; **7.** a; **8.** d; **9.** d; **10.** b

Matching
1. H; **2.** G; **3.** C; **4.** B; **5.** D; **6.** E; **7.** F; **8.** A; **9.** J; **10.** I

Visualize It
1. See Figure 13.8; **2.** See Figure 13.5.

CHAPTER 14

Multiple Choice
1. a; **2.** b; **3.** b; **4.** a; **5.** d; **6.** a; **7.** d; **8.** c; **9.** a; **10.** d; **11.** d; **12.** b; **13.** c; **14.** d; **15.** b

Fill in the Blanks
1. pathogen; **2.** asymptomatic or subclinical; **3.** etiology; **4.** epidemiology; **5.** zoonoses; **6.** fomites; **7.** Nosocomial; **8.** prevalence; **9.** biological; **10.** lipid A

Visualize It!
1. a. Endemic (or sporadic); b. Epidemic; c. Pandemic; **2.** The red epidemic affected more people during the first 3 days. The red epidemic has a short incubation time and is much more contagious; as a result, the susceptible population was infected in a short time. The blue epidemic has a longer incubation time than the red, resulting in a longer time until cases show up and a longer time over which new infections are occurring.

CHAPTER 15

Multiple Choice
1. d; **2.** d; **3.** b; **4.** a; **5.** d; **6.** c; **7.** d; **8.** b; **9.** a; **10.** d

Modified True/False
1. dead; **2.** True; **3.** True; **4.** True; **5.** monocytes; **6.** True; **7.** pathogen; **8.** ingestion; **9.** phagolysosomes; **10.** pathogen's cytoplasmic membrane; **11.** inflammation; **12.** True; **13.** True; **14.** antimicrobial peptides; **15.** True

Matching
First section: **1.** B; **2.** B; **3.** B; **4.** B; **5.** A; **6.** B; **7.** A; **8.** B; **9.** A; **10.** A; **11.** A; **12.** B; **13.** A; **14.** C; **15.** A
Second section: **1.** J; **2.** E; **3.** G; **4.** D; **5.** C; **6.** H; **7.** A; **8.** B; **9.** F; **10.** I

Visualize It!
1. See Figure 15.6; **2.** See Figure 15.5

CHAPTER 16

Multiple Choice
1. b; **2.** e; **3.** e; **4.** b; **5.** d; **6.** a; **7.** c; **8.** d; **9.** e; **10.** a

Modified True/False
1. antigen-presenting cells; **2.** True; **3.** cytotoxic; **4.** Plasma cells; **5.** antibody

Matching
1. D Plasma cell; C Cytotoxic T cell; B Th2 cell; A Dendritic cell; **2.** D Artificially acquired passive immunotherapy; A Naturally acquired active immunity; B Naturally acquired passive immunity; C Artificially acquired active immunity

Visualize It!
1. See Figure 16.10 and 16.11; **2.** MHC I and MHC II are found on the cytoplasmic membrane; pseudopods are the thin, finger-like extension of the cell; vesicles are the white structures within the cell.

CHAPTER 17

Multiple Choice
1. d; **2.** c; **3.** b; **4.** c; **5.** e; **6.** e; **7.** e; **8.** a; **9.** b; **10.** e; **11.** a; **12.** d; **13.** d; **14.** d; **15.** c

True/False
1. False; **2.** False; **3.** True; **4.** True; **5.** False

Matching
1. D; **2.** C; **3.** A; **4.** B; **5.** C; **6.** A

Visualize It!
1. See Figure 17.13; **2.** Patients 5, 6, and 11 are most likely uninfected; patients 1 and 7 may be uninfected. The other patients have likely been infected.

CHAPTER 18

Multiple Choice
1. e; **2.** c; **3.** d; **4.** c; **5.** c; **6.** e; **7.** b; **8.** d; **9.** c; **10.** e; **11.** b; **12.** d; **13.** a; **14.** d; **15.** d

Modified True/False
1. Histamine, kinins, and/or proteases are; **2.** True; **3.** red blood; **4.** type IV; **5.** allograft or xenograft

Matching
1. A; **2.** D; **3.** E; **4.** D; **5.** E; **6.** D; **7.** A; **8.** C; **9.** A; **10.** A

Visualize It!
1. See Figure 18.13; **2.** See Figure 18.16; **3.** a. systemic lupus erythematosus (SLE), b. positive tuberculin test, c. rheumatoid arthritis, d. urticaria (hives)

CHAPTER 19

Multiple Choice
1. c; **2.** c; **3.** b; **4.** d; **5.** b; **6.** d; **7.** a; **8.** d; **9.** c; **10.** a; **11.** d; **12.** d; **13.** a; **14.** c; **15.** c

Matching
1. C; **2.** A; **3.** F; **4.** N; **5.** B; **6.** D; **7.** E; **8.** G; **9.** K; **10.** H; **11.** I; **12.** M

Fill in the Blanks
1. microbiota; **2.** exfoliative; **3.** impetigo; **4.** necrotizing fasciitis; **5.** pimples, sties, furuncles, carbuncles; **6.** *Dermacentor;* **7.** erythema infectiosum; **8.** Papillomaviruses; **9.** gas gangrene; **10.** bacteria, especially *Propionibacterium*

True/False
1. False; **2.** False; **3.** False; **4.** True; **5.** False

Visualize It!
1. See Figure 19.10; **2.** See Figure 19.12; **3.** a. chickenpox, b. shingles, c. oral herpes, d. whitlow

CHAPTER 20

Multiple Choice
1. d; **2.** b; **3.** a; **4.** c; **5.** b; **6.** b; **7.** a; **8.** b; **9.** d; **10.** d; **11.** c; **12.** c; **13.** d; **14.** a; **15.** b

Fill in the Blanks
1. synaptic cleft; **2.** polysaccharide capsule; **3.** *Streptococcus agalactiae;* **4.** foodborne, infant, and wound; **5.** *Clostridium botulinum;* **6.** cryptococcal meningitis; **7.** rabies; **8.** peripheral; **9.** dendrites; **10.** petechiae; **11.** trachoma; **12.** conjunctiva; **13.** sties; **14.** *Neisseria gonorrhoeae;* **15.** group B

Visualize It!
1. See Figure 20.6b; **2.** The needle is inserted through the skin (a), past the vertebrae (b), through the dura mater and subdural space (c) and arachnoid mater (d) to withdraw CSF from the subarachnoid space (e).

CHAPTER 21

Multiple Choice
1. d; **2.** a; **3.** b; **4.** b; **5.** a; **6.** c; **7.** a; **8.** c; **9.** d; **10.** b; **11.** a; **12.** d; **13.** a; **14.** b; **15.** a; **16.** a; **17.** b; **18.** c; **19.** b; **20.** b; **21.** a; **22.** c; **23.** b; **24.** d; **25.** c

Matching
First section: **1.** J; **2.** D; **3.** A; **4.** G; **5.** B; **6.** F; **7.** C; **8.** E; **9.** H; **10.** I
Second section: **1.** A; **2.** D; **3.** B; **4.** C; **5.** E

Fill in the Blanks
1. pulmonary semilunar; **2.** right atrioventricular; **3.** capillaries; **4.** disseminated intravascular coagulation; **5.** brucellosis; **6.** erythrocytic, exoerythrocytic, sporogonic; **7.** *Toxoplasma gondii;* **8.** Chagas'; **9.** flea; **10.** petechiae

Visualize It!
1. See Disease in Depth, pp. 662–663; **2.** The eggs and newly hatched spring larvae are not infected with *Borrelia*. Every other stage in the life of the tick and all the animals could be infected.

CHAPTER 22

Multiple Choice

1. b; **2.** b; **3.** a; **4.** b; **5.** b; **6.** c; **7.** a; **8.** d; **9.** a; **10.** b

Fill in the Blanks

1. streptococcal pharyngitis; **2.** diphtheria; **3.** syncytia; **4.** amphotericin B; **5.** Reye's

Modified True/False

1. does not produce; **2.** the susceptibility of AIDS patients; **3.** drift; **4.** exposure to; **5.** True

Visualize It!

1. See Figure 22.15; **2.** a. *Mycobacterium tuberculosis*, b. *Klebsiella pneumoniae*, c. *Corynebacterium diphtheriae*, d. *Streptococcus pneumoniae*

CHAPTER 23

Multiple Choice

1. c; **2.** c; **3.** a; **4.** c; **5.** c; **6.** b; **7.** a; **8.** c; **9.** b; **10.** c; **11.** b; **12.** c; **13.** a; **14.** d; **15.** c

Modified True/False

1. and does; **2.** *H. pylori;* **3.** True; **4.** True; **5.** less severe; **6.** True; **7.** protozoan; **8.** True; **9.** hepatitis A; **10.** True

Fill in the Blanks

1. enamel, dentin, pulp; **2.** *Streptococcus mutans* (viridans streptococcus); **3.** gastric, duodenal; **4.** Shiga, Shiga-like; **5.** Oral herpes or fever blisters; **6.** mumps; **7.** jaundice; **8.** giardiasis; **9.** epithelial, adenylate cyclase; **10.** scolex

Matching

1. I; **2.** E; **3.** J; **4.** B; **5.** C; **6.** H; **7.** A; **8.** G; **9.** D; **10.** F

Visualize It!

1. a. plaque (biofilm) formation, b. bacteria ferment sugar, producing acid, c. acid destroys enamel, d. acid destroys dentin; **2.** See Figure 23.15; 43 Dane particles, 12 filamentous particles, 10 spherical particles; Dane particles have DNA, the other two particles are empty; none of the particles contain RNA (HBV is a DNA virus).

CHAPTER 24

Multiple Choice

1. b; **2.** a; **3.** c; **4.** d; **5.** a; **6.** d; **7.** c; **8.** c; **9.** b; **10.** a

Matching

First section: **1.** F; **2.** N; **3.** I; **4.** H; **5.** E; **6.** A; **7.** D; **8.** J; **9.** G; **10.** C; **11.** M; **12.** B; **13.** K; **14.** O; **15.** L
Second section: **1.** F; **2.** E; **3.** G; **4.** D; **5.** I; **6.** B; **7.** J; **8.** A (sometimes H); **9.** H (sometimes A); **10.** C

Fill in the Blanks

1. prepuce; **2.** *Escherichia coli;* **3.** bacterial vaginosis; **4.** fimbriae, polysaccharide capsules, and lipooligosaccharide; **5.** *Chlamydia trachomatis;* **6.** trichomoniasis

Modified True/False

1. is not; **2.** True; **3.** sex; **4.** men/women; **5.** True; **6.** sexually; **7.** True; **8.** are not recommended; **9.** True; **10.** True; **11.** True

Visualize It!

1. See Figure 24.9; **2.** a. *Trichomonas vaginalis*, b. *Neisseria gonorrhoeae*, c. *Treponema pallidum*, d. *Candida albicans*

CHAPTER 25

Multiple Choice

1. c; **2.** d; **3.** c; **4.** c; **5.** b; **6.** d; **7.** d; **8.** b; **9.** b; **10.** c; **11.** a; **12.** c; **13.** a; **14.** c; **15.** a; **16.** c; **17.** b; **18.** a; **19.** b; **20.** d

Matching

1. C; **2.** F; **3.** G; **4.** P; **5.** I; **6.** J; **7.** B; **8.** K; **9.** O; **10.** N; **11.** L; **12.** A; **13.** D; **14.** H; **15.** E; **16.** M

Modified True/False

1. lactic acid; **2.** lactic acid; **3.** True; **4.** True; **5.** True; **6.** Biodegradable; **7.** True; **8.** Competition; **9.** True; **10.** Littoral

Fill in the Blanks

1. the food, processing or handling; **2.** increases; **3.** batch production, continuous flow production; **4.** zero; **5.** iron, sulfur; **6.** production, consumption, decomposition; **7.** dinitrogen gas (N_2); **8.** phosphate ion (PO_4^{3-}); **9.** biosensor; **10.** BOD (biochemical oxygen demand)

Visualize It!

1. See Figure 25.15; **2.** See Figure 25.16.

Appendix A Metabolic Pathways

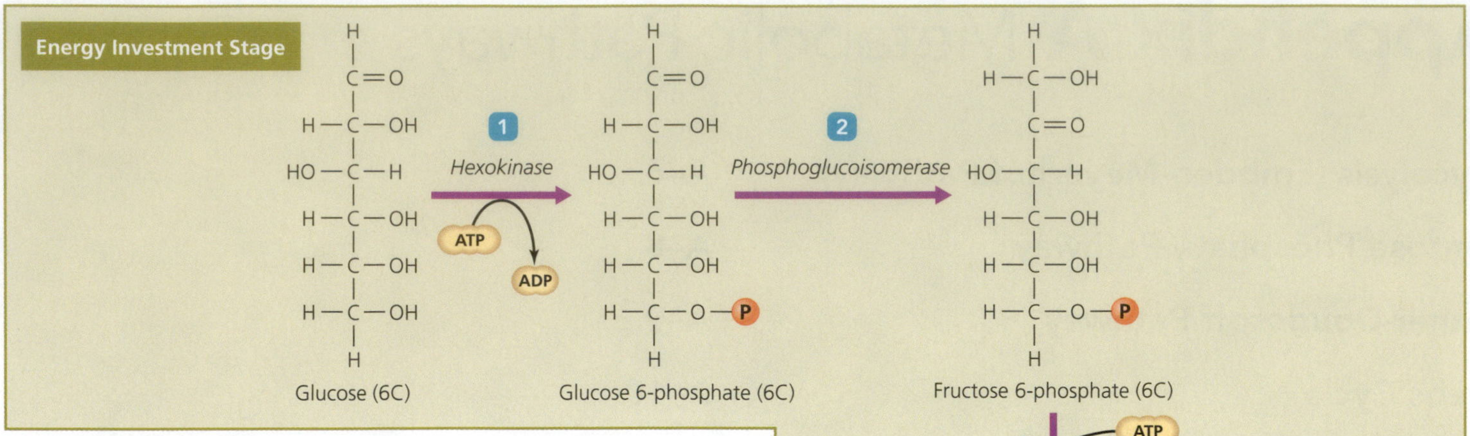

GLYCOLYSIS (Embden-Meyerhof Pathway):

Energy Investment Stage

Step 1. *Hexokinase* transfers phosphate from ATP to carbon 6 of glucose, forming glucose 6-phosphate—a charged molecule. Since the cytoplasmic membrane is impermeable to ions, phosphorylated glucose cannot diffuse out of a cell.

Step 2. *Phosphoglucoisomerase* rearranges the atoms of glucose to form an isomer—fructose 6-phosphate.

Step 3. *Phosphofructokinase* invests more energy by adding another phosphate group from ATP to form fructose 1,6-bisphosphate.

Lysis Stage

Step 4. *Aldolase* (fructose 1,6-bisphosphate aldolase) splits six-carbon fructose 1,6-bisphosphate into three-carbon glyceraldehyde 3-phosphate (G3P) and three-carbon dihydroxyacetone phosphate (DHAP). It is this cleavage that gives glycolysis its name.

Step 5. *Triose phosphate isomerase* rearranges the atoms of DHAP to form another molecule of glyceraldehyde 3-phosphate. From this point, every step of glycolysis occurs twice—once for each of the three-carbon molecules.

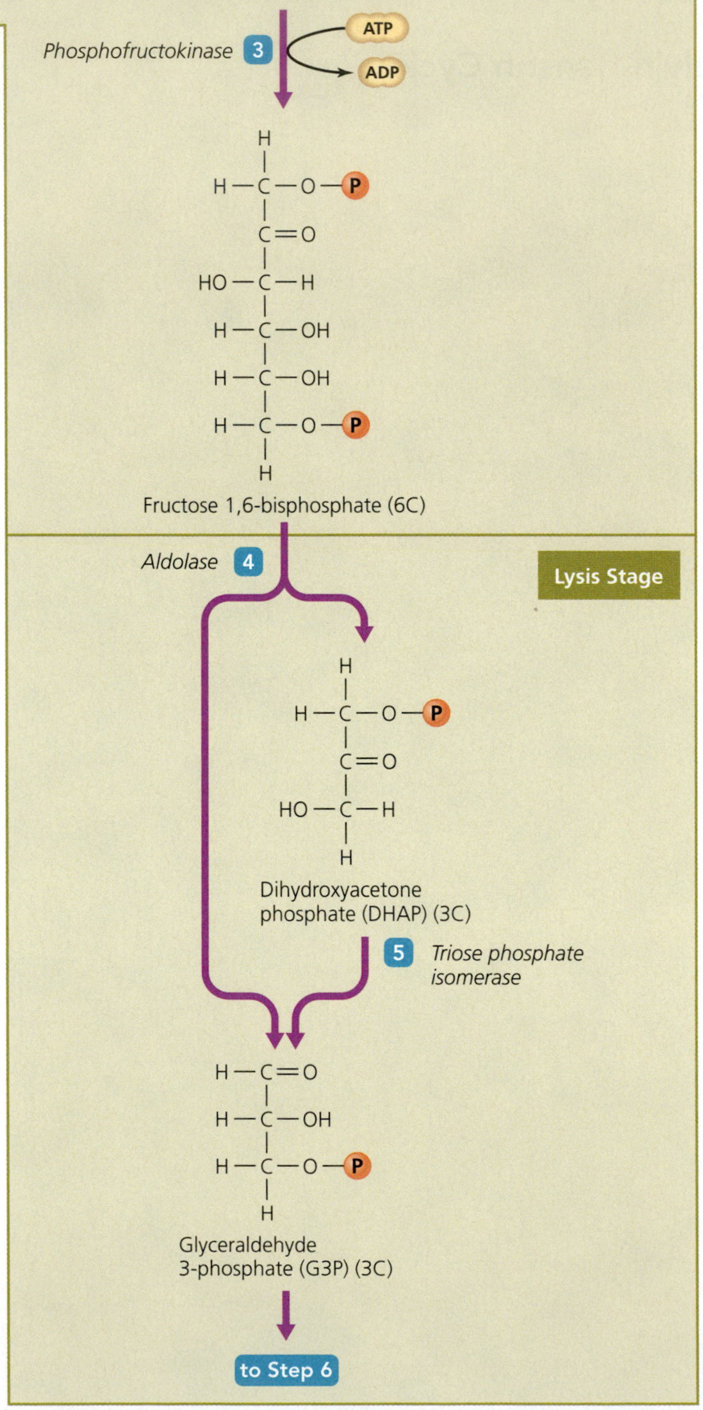

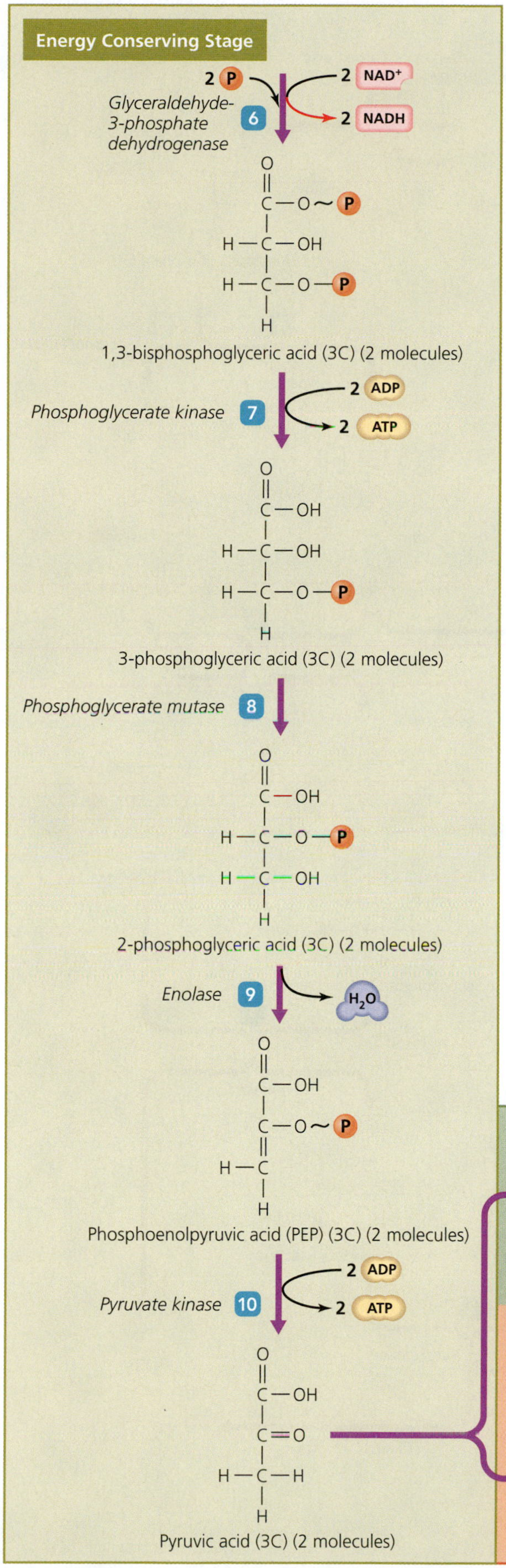

Energy Conserving Stage

Step 6. *Glyceraldehyde-3-phosphate dehydrogenase* catalyzes a reaction with two parts: (a) It oxidizes G3P, transferring electrons (and hydrogen) to NAD^+ to form NADH and (b) it adds inorganic phosphate from the cytosol to G3P with a high-energy bond to form 1,3-bisphosphoglyceric acid. This two-part reaction is among the more important in glycolysis because it generates a molecule of NADH for each molecule of G3P and because it creates the first high-energy intermediate.

Step 7. *Phosphoglycerate kinase* conserves the energy in the high-energy bonds in two molecules of ATP, yielding 3-phosphoglyceric acid molecules.

Step 8. *Phosphoglycerate mutase* rearranges the atoms to form 2-phosphoglyceric acid.

Step 9. *Enolase* removes a molecule of water from each substrate, forming a double bond and a high-energy bond with phosphate.

Step 10. *Pyruvate kinase* ends glycolysis by transferring energy to ATP, forming pyruvic acid. In the final analysis, two molecules of ATP are invested to yield four molecules of ATP—a net gain of two molecules of ATP—and two molecules of NADH. Pyruvic acid then undergoes respiration (when there is an inorganic, or rarely an extracellularly derived organic, final electron acceptor) or fermentation (when the final electron acceptor is an organic molecule from the cell).

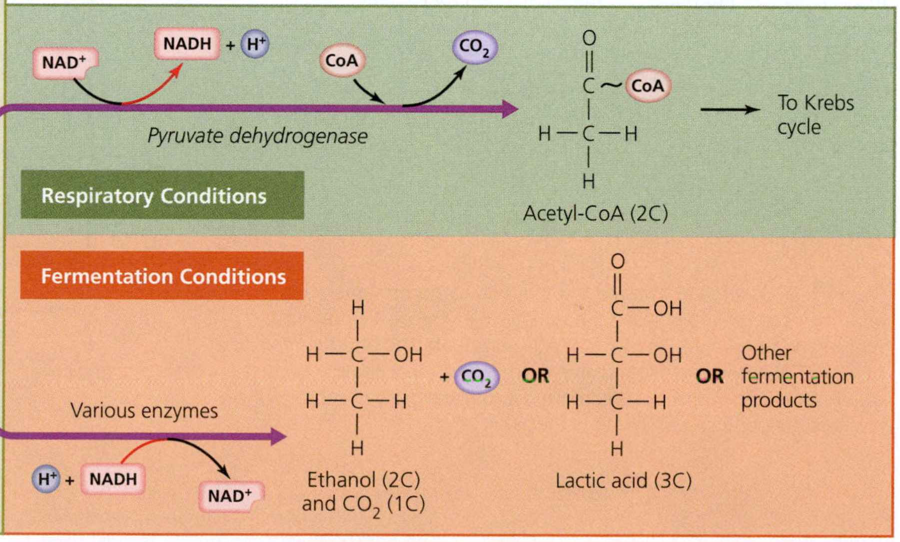

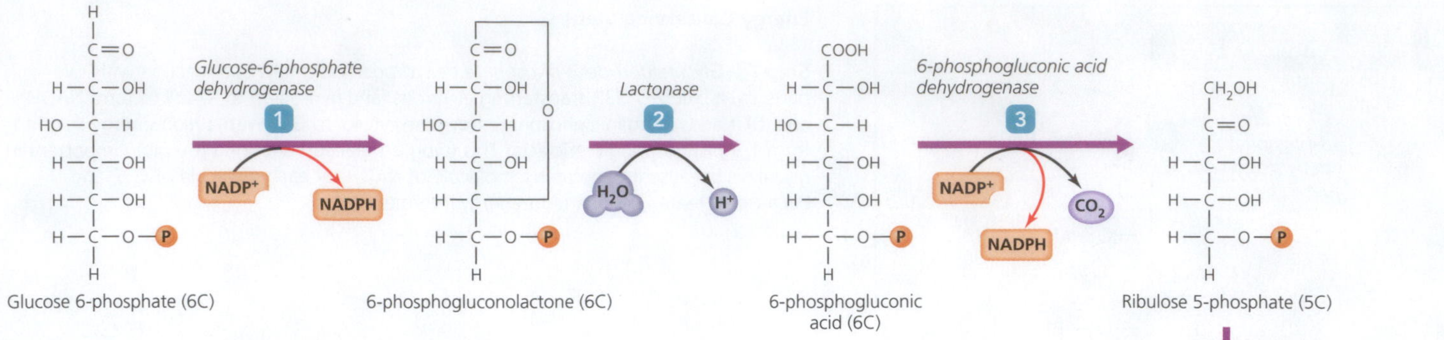

PENTOSE PHOSPHATE PATHWAY

Step 1. *Glucose-6-phosphate dehydrogenase* oxidizes glucose 6-phosphate to 6-phosphogluconolactone by transferring hydrogen to $NADP^+$, forming NADPH.

Step 2. *Lactonase* adds hydroxide from a water molecule, forming 6-phosphogluconic acid.

Step 3. *6-phosphogluconic acid dehydrogenase* oxidizes this intermediate (transferring hydrogen to another molecule of NADPH) and removes a molecule of carbon dioxide to form ribulose 5-phosphate. This five-carbon phosphorylated sugar is used in the synthesis of nucleotides, certain amino acids, and glucose (via photosynthesis).

Step 4. *Phosphopentose epimerase* converts some molecules of ribulose 5-phosphate to xylulose 5-phosphate.

Step 5. Simultaneously, *phosphopentose isomerase* converts other molecules of ribulose 5-phosphate to ribose 5-phosphate.

Step 6. *Transketolase* catalyzes a reaction in which a two-carbon fragment from xylulose 5-phosphate is transferred to ribose 5-phosphate, forming seven-carbon sedoheptulose 7-phosphate and three-carbon glyceraldehyde 3-phosphate (G3P).

Step 7. *Transaldolase* then transfers a three-carbon fragment from sedoheptulose 7-phosphate to G3P, yielding four-carbon erythrose 4-phosphate and forming six-carbon fructose 6-phosphate.

Step 8. *Transketolase* transfers another two-carbon fragment from another molecule of xylulose 5-phosphate to erythrose 4-phosphate, forming another molecule of fructose 6-phosphate and another molecule of G3P. G3P enters glycolysis at step 6; fructose 6-phosphate can enter glycolysis at step 1 or may be converted into glucose 6-phosphate, which reenters the pentose phosphate pathway.

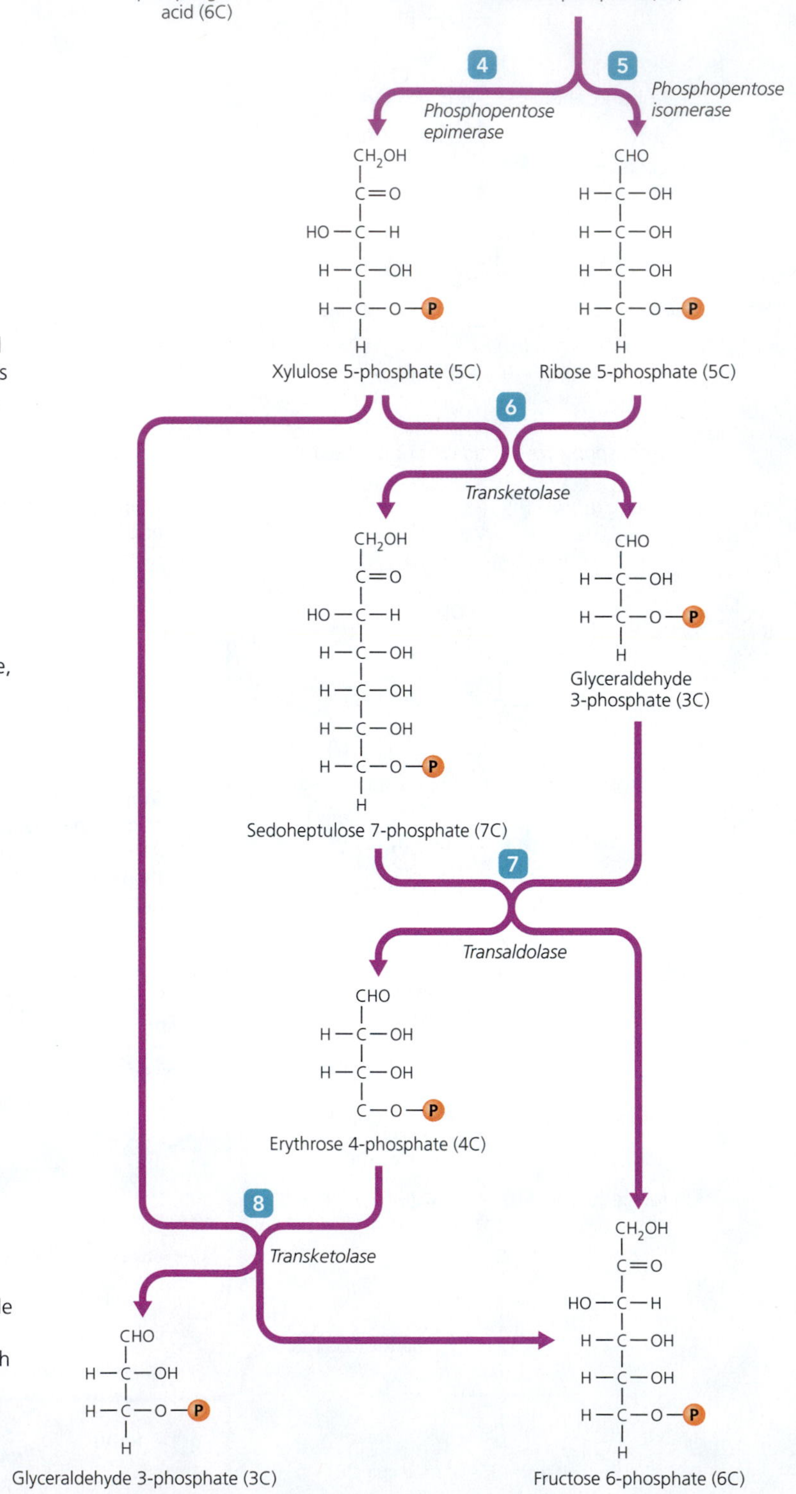

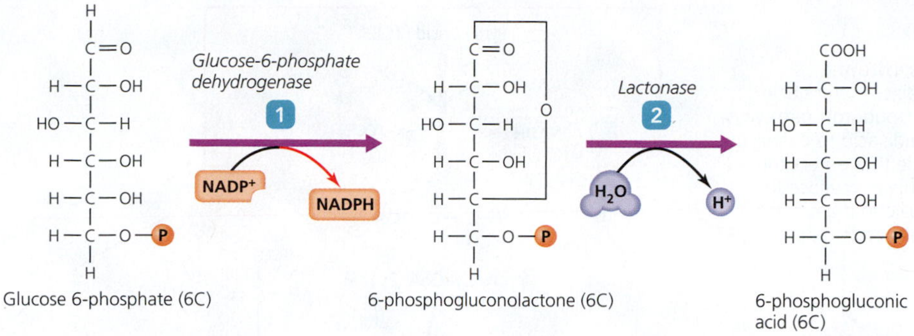

ENTNER-DOUDOROFF PATHWAY

The first two steps of the Entner-Doudoroff pathway are the same as the first two steps of the pentose phosphate pathway:

Step 1. *Glucose-6-phosphate dehydrogenase* oxidizes glucose 6-phosphate to 6-phosphogluconolactone by transferring hydrogen to $NADP^+$, forming NADPH.

Step 2. As in step 2 of the pentose phosphate pathway, *lactonase* adds hydroxide from a water molecule, forming 6-phosphogluconic acid.

Step 3. *6-phosphogluconic acid dehydrase* removes a molecule of water to form a six-carbon molecule. (The enzyme in this step should not be confused with the similarly named 6-phosphogluconic acid dehydrogenase of the pentose phosphate pathway.)

Step 4. *Aldolase* splits 2-keto-3-deoxy-6-phosphogluconic acid into two three-carbon compounds: pyruvic acid and glyceraldehyde 3-phosphate (G3P).

Step 5. G3P is converted to pyruvic acid by steps 6 through 10 of Embden-Meyerhof glycolysis.

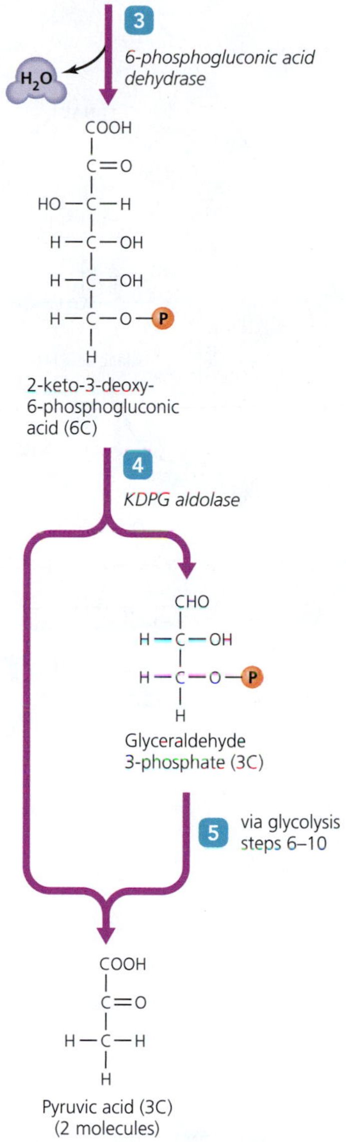

KREBS CYCLE

In a complex reaction with several parts, pyruvate dehydrogenase decarboxylates and oxidizes pyruvic acid (from the Embden-Meyerhof or Entner-Doudoroff pathways) and adds the remaining two-carbon acetic acid to coenzyme A, forming acetyl-CoA. It is this molecule that enters the Krebs cycle. Recall that for every molecule of glucose that enters glycolysis, two molecules of pyruvic acid are produced, necessitating two sets of Krebs cycle reactions.

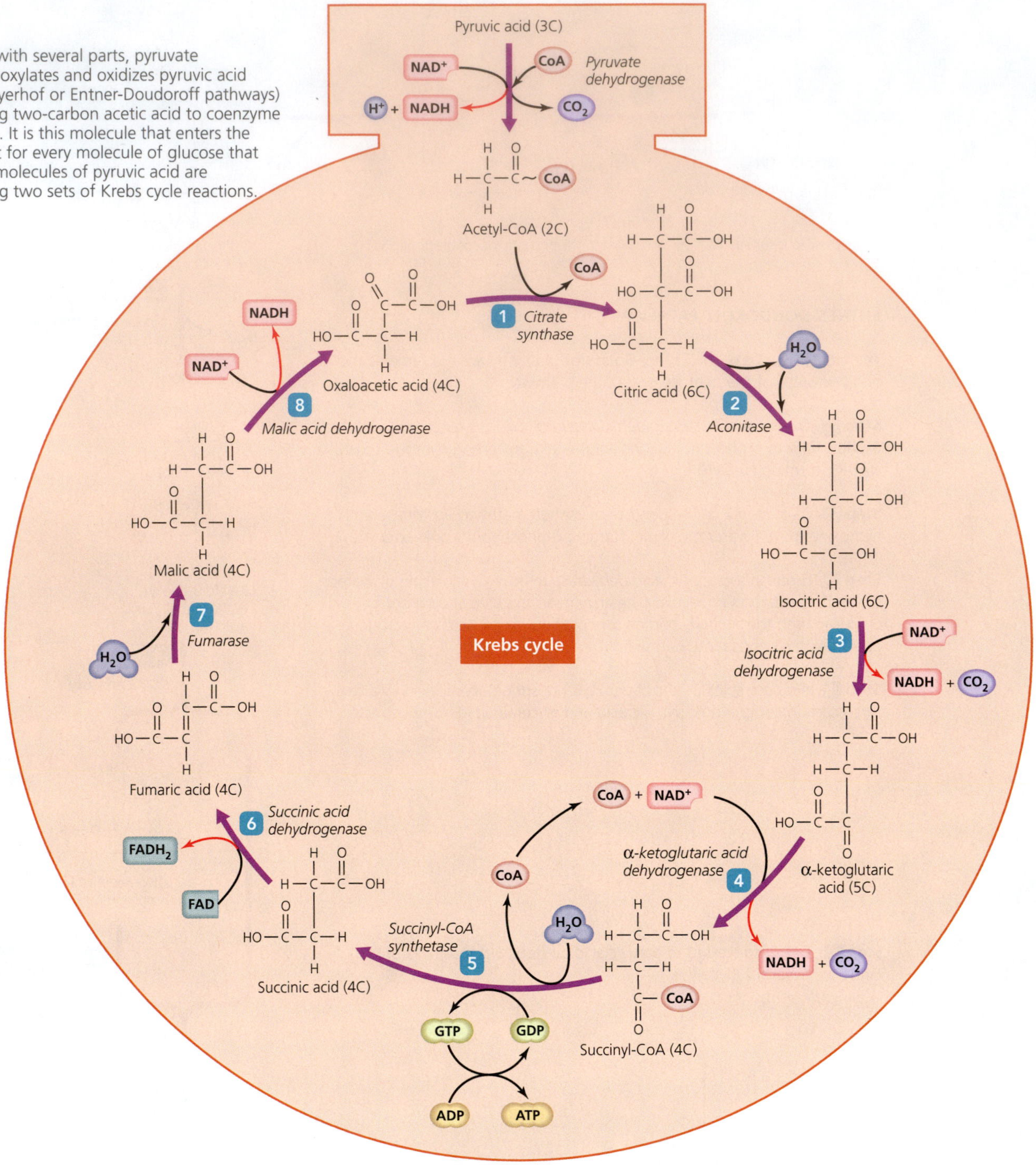

Step 1. *Citrate synthase* adds the two carbons of acetic acid from acetyl-CoA to oxaloacetic acid, forming citric acid. This step gives the Krebs cycle its alternate name—the citric acid cycle.

Step 2. *Aconitase* forms an isomer, isocitric acid, by removing a molecule of water and then adding another molecule of water back.

Step 3. *Isocitric acid dehydrogenase* oxidizes and decarboxylates six-carbon isocitric acid to five-carbon α-ketoglutaric acid; a molecule of NADH is formed in the process.

Step 4. *α-ketoglutaric acid dehydrogenase* oxidizes and decarboxylates α-ketoglutaric acid to form a four-carbon fragment that is attached to coenzyme A, forming succinyl-CoA.

Step 5. *Succinyl-CoA synthetase* forms four-carbon succinic acid and simultaneously phosphorylates GDP to form GTP. The latter molecule subsequently phosphorylates ADP to ATP.

Step 6. *Succinic acid dehydrogenase* oxidizes succinic acid to four-carbon fumaric acid, forming a molecule of $FADH_2$.

Step 7. *Fumarase* rearranges the four carbons to form malic acid.

Step 8. *Malic acid dehydrogenase* oxidizes malic acid to reform oxaloacetic acid, completing the cycle and forming another molecule of NADH.

NADH and $FADH_2$ carry electrons to an electron transport chain.

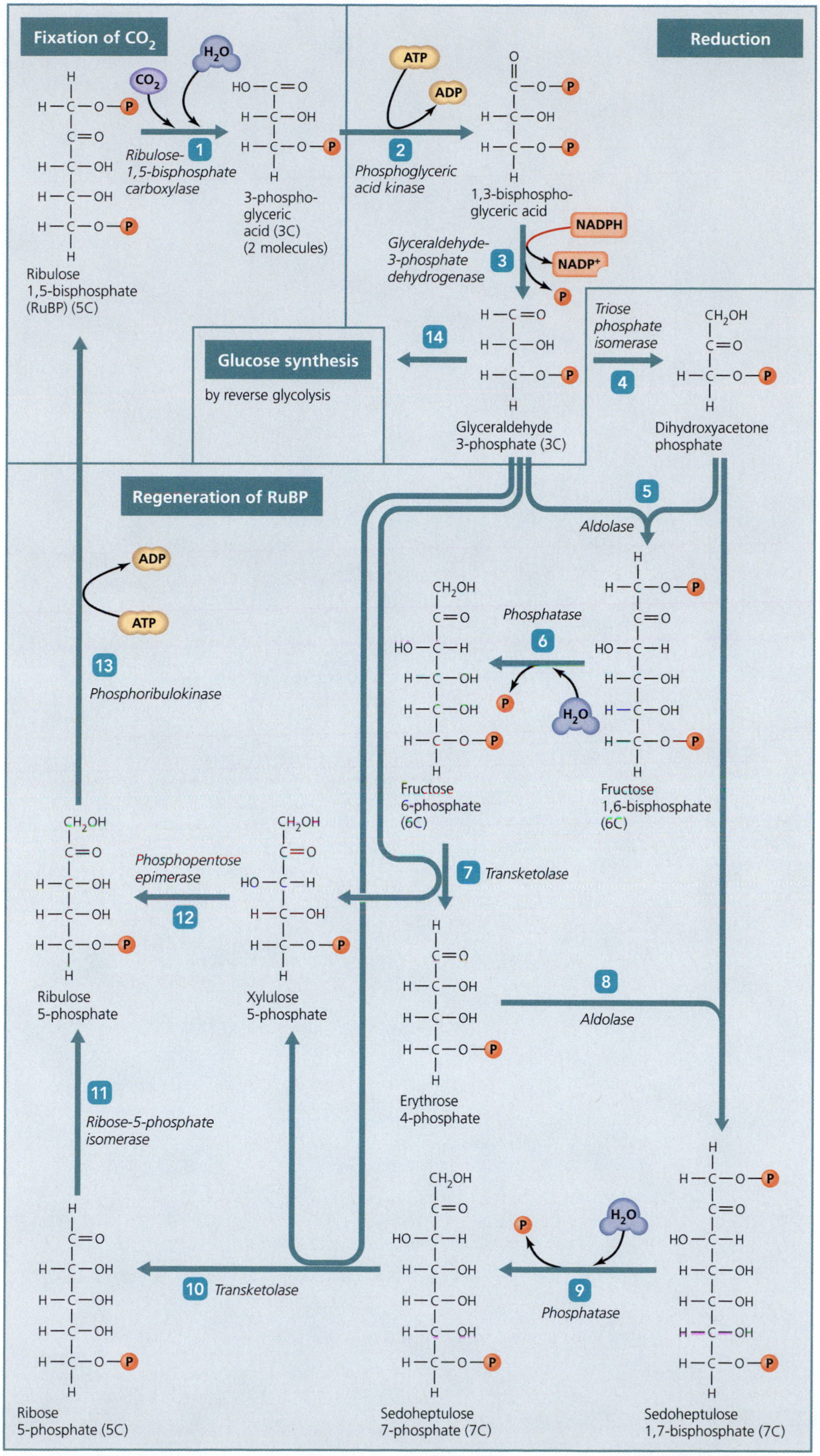

CALVIN-BENSON CYCLE

Fixation of CO_2

Step 1. *Ribulose-1,5-bisphosphate carboxylase* adds CO_2 and H_2O to five-carbon ribulose 1,5-bisphosphate (RuBP), forming a six-carbon intermediate (not shown) that is immediately split into two, three-carbon molecules of 3-phosphoglyceric acid.

Reduction

Step 2. *Phosphoglyceric acid kinase* phosphorylates phosphoglyceric acid at the expense of ATP to form 1,3-bisphosphoglyceric acid.

Step 3. *Glyceraldehyde-3-phosphate dehydrogenase* reduces (using NADPH) and removes one phosphate from 1,3-bisphosphoglyceric acid, forming glyceraldehyde 3-phosphate (G3P).

Regeneration of RuBP

Step 4. *Triose phosphate isomerase* changes some molecules of G3P to dihydroxyacetone phosphate (DHAP).

Step 5. *Aldolase* combines a molecule of G3P and a molecule of DHAP to form fructose 1,6-bisphosphate.

Step 6. *Phosphatase* adds water and removes a phosphate group, forming fructose 6-phosphate.

Step 7. *Transketolase* combines fructose 6-phosphate with a molecule of G3P (from step 3) to form five-carbon xylulose 5-phosphate and four-carbon erythrose 4-phosphate.

Step 8. *Aldolase* rearranges the carbons of erythrose 4-phosphate and of a molecule of DHAP (from step 4), forming sedoheptulose 1,7-bisphosphate.

Step 9. *Phosphatase* adds water and removes a phosphate group to form sedoheptulose 7-phosphate.

Step 10. *Transketolase* combines this molecule with a molecule of G3P (from step 3) to form five-carbon ribose 5-phosphate and five-carbon xylulose 5-phosphate.

Step 11. *Ribose-5-phosphate isomerase* rearranges the atoms of ribose 5-phosphate to form its isomer—ribulose 5-phosphate.

Step 12. *Phosphatase epimerase* similarly rearranges the atoms of xylulose 5-phosphate, forming another molecule of ribulose 5-phosphate.

Step 13. *Phosphoribulokinase* phosphorylates ribulose 5-phosphate to regenerate ribulose bisphosphate (RuBP).

Step 14. For every three molecules of CO_2 that enter the Calvin-Benson cycle (at step 1), one molecule of G3P leaves the cycle to be used for the synthesis of glucose via the reversal of the early reactions of glycolysis.

Appendix B Some Mathematical Considerations in Microbiology

Scientific Notation

Scientific notation (also called exponential notation) is a mathematical device developed to express numbers, particularly very small and very large numbers, in a manner that is convenient and readable. For example, 1.234×10^{-8} is much less cumbersome than 0.00000001234. Similarly, 5.67×10^{17} is more easily read than 567,000,000,000,000,000.

A number expressed in scientific notation is composed of a *coefficient*, which is always a number with only one whole number digit to the left of the decimal place, times 10 to some *exponential power*. In the examples above, the coefficients are 1.234 and 5.67, respectively. The exponents are −8 and 17, respectively.

To write a number in scientific notation, move the decimal point either right or left so that there is only one nonzero number to the left of it. For example,

454 becomes 4.54

In this case we moved the decimal point (which is understood to be at the end of a whole number) two places to the left. Since we moved two places left, the exponent will be positive 2, and the scientific notation is

$$4.54 \times 10^2$$

Similarly, 4,500,264.75 becomes 4.50026475×10^6 since we moved the decimal six places to the left.

As you would expect, when we have to move the decimal point to the right, as in 0.00562 (which becomes 5.62), then the exponent will be a negative number, in this case −3; therefore, $0.00562 = 5.62 \times 10^{-3}$.

Logarithms

A **logarithm** is the number of times a number, called the *base number*, must be multiplied by itself to get a certain number. For example, the number 10 must be multiplied by itself three times to get 1000 ($10 \times 10 \times 10$); therefore, the logarithm of 1000 in base 10 is 3. Similarly, the logarithm of 100,000 is 5. Though any number can be used as the base number, in microbiology, base 10 is commonly used. Microbiologists use base 10 logarithms to express pH values and the size of bacterial populations in culture.

When a number is expressed in scientific notation, the logarithm is the exponent when the coefficient is exactly 1. When the coefficient is a number other than 1, the logarithm must be calculated using the logarithm function on a calculator.

Generation Time

When bacteria and other microbes reproduce by binary fission, the number in the population doubles with each division cycle (generation). Such growth is called *logarithmic* (or *exponential*) *growth.*

The number in a population can be calculated as 2 (because each cell produces two offspring) multiplied by itself as many times as there are generations. In other words, the number of cells in a population arising from a single individual is expressed mathematically as

$$2^{\text{number of generations}}$$

A cell dividing for five generations would produce a population of 32 ($2^5 = 2 \times 2 \times 2 \times 2 \times 2 = 32$). When the population begins with more than one individual, then the final population equals

$$\text{original number of cells} \times 2^{\text{number of generations}}$$

For example, seven cells reproducing for five generations would produce a population of 224 cells ($7 \times 2^5 = 7 \times 2 \times 2 \times 2 \times 2 \times 2 = 224$).

In most cases, microbiologists are not concerned with the number of generations required to produce a given population, but they do want to know the **generation time**; that is, the time required for a bacterial cell to grow and divide. This is also the time required for a population of cells to double in number. To calculate generation time, scientists must first calculate how many generations have been produced, knowing the beginning and ending population sizes. When population sizes are converted to logarithms, the number of generations is calculated as

$$\text{number of generations} = \frac{\text{logarithm (log) number of cells at the end of reproduction} - \text{log number of cells initially}}{\log 2}$$

Log 2 is used in the formula because each cell produces two offspring each time it divides. Log 2 = 0.301.

The number of generations is used to calculate generation time:

$$\text{generation time (min/generation)} = \frac{60 \text{ minutes} \times \text{number of hours}}{\text{number of generations}}$$

For example, if 100 bacteria multiply to produce a population of 3.28×10^6 cells in 7 hours, then the generation time for this bacterium is calculated as follows:

$$\text{number of generations} = \frac{\text{logarithm (log) number of cells at the end of reproduction} - \text{log number of cells initially}}{\log 2}$$

$$= \frac{\log(3.28 \times 10^6) - \log(100)}{\log 2}$$

$$= 15 \text{ generations}$$

$$\text{generation time (min/generation)} = \frac{60 \text{ minutes} \times \text{hours}}{\text{number of generations}}$$

$$= \frac{60 \text{ min} \times 7 \text{ h}}{15}$$

$$= 28 \text{ minutes/generation}$$

Glossary

A site In a ribosome, a binding site that accommodates tRNA delivering an amino acid.

Abscess An isolated site of infection such as a pimple, boil, or pustule.

Abyssal zone In marine habitats, the zone of water beneath the benthic zone, virtually devoid of life except around hydrothermal vents.

Acellular Noncellular.

Acetyl-CoA Combination of two-carbon acetate and coenzyme A.

Acid Compound that dissociates into one or more hydrogen ions and one or more anions.

Acid-fast bacilli (AFB) *(acid-fast fast rod, AFR)* Bacilli that retain stain during decolorization by acid-alcohol, particularly species of *Mycobacterium*.

Acid-fast rod (AFR) Bacilli that retain stain during decolorization by acid-alcohol, particularly species of *Mycobacterium*.

Acid-fast stain In microscopy, a differential stain used to penetrate waxy cell walls.

Acidic dye In microscopy, an anionic chromophore used to stain alkaline structures. Works most effectively in acidic environments.

Acidophile Microorganism requiring acidic pH.

Acne Skin disorder characterized by presence of whiteheads, blackheads, and, in severe cases, cysts; typically caused by infection with *Propionibacterium acnes*.

Acquired immunodeficiency syndrome (AIDS) The presence of several opportunistic or rare infections along with infection by human immunodeficiency virus (HIV) or a severe decrease in the number of CD4 cells ($<200/\mu L$) along with a positive test showing the presence of HIV.

Acquired (secondary) immunodeficiency diseases Any of a group of immunodeficiency diseases that develop in older children, adults, and the elderly as a direct consequence of some other recognized cause, such as infectious disease.

Actinomycetes High G + C Gram-positive bacteria that form branching filaments and produce spores, thus resembling fungi.

Actinomycosis Disease caused by *Actinomyces*; characterized by formation of multiple, interconnected abscesses in skin or mucous membrane.

Activation energy The amount of energy needed to trigger a chemical reaction.

Active site Functional site of an enzyme, the shape of which is complementary to the shape of the substrate.

Active transport The movement of a substance against its electrochemical gradient via carrier proteins and requiring cell energy from ATP.

Acute anaphylaxis Condition in which the release of inflammatory mediators overwhelms the body's coping mechanisms.

Acute disease Any disease that develops rapidly but lasts only a short time, whether it resolves in convalescence or death.

Acute inflammation Type of inflammation that develops quickly, is short lived, and is usually beneficial.

Adaptive immunity Resistance against pathogens that acts more effectively upon subsequent infections with the same pathogen.

Adenine Ring-shaped nitrogenous base found in nucleotides of DNA and RNA.

Adenosine triphosphate (ATP) The primary short-term, recyclable energy molecule fueling cellular reactions.

Adherence Process by which phagocytes attach to microorganisms through the binding of complementary chemicals on the cytoplasmic membranes.

Adhesins Molecules that attach pathogens to their target cells.

Adhesion The attachment of microorganisms to host cells.

Adhesion factors A variety of structures or attachment proteins by which microorganisms attach to host cells.

Adjuvant Chemical added to a vaccine to increase its ability to stimulate active immunity.

Aerobe An organism that uses oxygen as a final electron acceptor.

Aerobic respiration Type of cellular respiration requiring oxygen atoms as final electron acceptors.

Aerosol A cloud of water droplets, which travels more than 1 meter in airborne transmission and less than 1 meter in droplet transmission.

Aerotolerant anaerobe Microorganism which prefers anaerobic conditions but can tolerate exposure to low levels of oxygen.

Aflatoxin Carcinogenic mycotoxin produced by *Aspergillus*.

African sleeping sickness Potentially fatal disease caused by a bite from a tsetse fly carrying *Trypanosoma brucei* and characterized by formation of a lesion at the site of the bite, followed by parasitemia and central nervous system invasion.

Agar Gel-like polysaccharide isolated from red algae and used as thickening agent.

Agglutination Aggregation (clumping) caused when antibodies bind to two antigens, perhaps hindering the activity of pathogenic microorganisms and increasing the chance that they will be phagocytized.

Agglutination test In serology, a procedure in which antiserum is mixed with a sample that potentially contains its target antigen.

Agranulocyte Type of leukocyte having a uniform cytoplasm lacking large granules.

Agroterrorism The use of microbes to terrorize humans by destroying the livestock and crops that constitute their food supply.

Airborne transmission Spread of pathogens to the respiratory mucous membranes of a new host via the air or in droplets carried more than 1 meter.

Alcohol Intermediate-level disinfectant that denatures proteins and disrupts cell membranes.

Aldehyde Compound containing terminal —CHO groups; used as a high-level disinfectant because it cross-links organic functional groups in proteins and nucleic acids.

Algae Eukaryotic unicellular or multicellular photosynthetic organisms with simple reproductive structures.

Alginic acid Cell wall polysaccharide of brown algae.

Alkalinophile Microorganism requiring alkaline pH environments.

Allergen An antigen that stimulates an allergic response.

Allergic contact dermatitis Type of delayed hypersensitivity reaction in which chemically modified skin proteins trigger a cell-mediated immune response.

Allergy An immediate hypersensitivity response against an antigen.

Allograft Type of graft in which tissues are transplanted from a donor to a genetically dissimilar recipient of the same species.

Allylamines Class of antifungal drugs that disrupt cytoplasmic membranes.

Alpha interferons (IFN-α) Interferons secreted by virally infected monocytes, macrophages, and some lymphocytes within hours after infection.

Alphaproteobacteria Class of aerobic Gram-negative bacteria in the phylum Proteobacteria capable of growing at very low nutrient levels.

Alternation of generations In algae, method of sexual reproduction in which diploid bodies alternate with haploid bodies.

Alveolar macrophage Fixed macrophage of the lungs.

Alveolates Protozoa with small membrane-bound cavities called alveoli beneath their cell surfaces.

Amebiasis A mild to severe dysentery that, if invasive, can cause the formation of lesions in the liver, lungs, brain, and other organs; caused by infection with *Entamoeba histolytica*.

Ames test Method for screening mutagens that is commonly used to identify potential carcinogens.

Amination Reaction involving the addition of an amine group to a metabolite to make an amino acid.

Amino acid A monomer of polypeptides.

Aminoglycoside Antimicrobial agent that inhibits protein synthesis by changing the shape of the 30S ribosomal subunit.

Ammonification Process by which microorganisms disassemble proteins in soil wastes into amino acids, which are then converted to ammonia.

Amoebae Protozoa that move and feed by pseudopods.

Amphibolic reaction A reversible metabolic reaction; that is, a reaction that can be catabolic or anabolic.

Anabolism All of the synthesis reactions in an organism taken together.

Anaerobe An organism that cannot tolerate oxygen.

Anaerobic respiration Type of cellular respiration not requiring oxygen atoms as final electron acceptors.

Analytical epidemiology Detailed investigation of a disease, including analysis of data to determine the probable cause, mode of transmission, and possible means of prevention.

Anammox Anaerobic ammonium oxidation, which is one aspect of the nitrogen cycle.

Anaphase Third stage of mitosis, during which sister chromatids separate and move to opposite poles of the spindle to form chromosomes. Also used for the comparable stage of meiosis.

Anaphylactic shock Condition in which the release of inflammatory mediators overwhelms the body's coping mechanisms, causing suffocation, edema, smooth muscle contraction, and often death.

Anaplasmosis *(human granulocytic anaplasmosis, HGA)* Tickborne disease caused by a rickettsia, *Anaplasma phagocytophilum*, manifesting with flu-like signs and symptoms. Formerly called human granulocytic ehrlichiosis.

Anion A negatively charged ion.

Anisakiasis Gastrointestinal disease caused by fish nematode parasite *Anisakis*, which may also manifest as an allergic reaction to the worm.

Anthrax Gastrointestinal, cutaneous, or pulmonary disease that is usually fatal without aggressive treatment; caused by ingestion, inoculation, or inhalation of spores of *Bacillus anthracis*.

Antibiotic Antimicrobial agent that is produced naturally by an organism.

Antibody *(immunoglobulin)* Proteinaceous antigen-binding molecule secreted by plasma cells.

Antibody-dependent cellular cytotoxicity (ADCC) Process whereby natural killer lymphocytes (NK cells) lyse cells covered with antibodies.

Antibody immune response (humoral immune response) The immune response centered around B lymphocytes and immunoglobulins.

Anticodon Portion of tRNA molecule that is complementary to a codon on mRNA.

Antigen Molecule that triggers a specific immune response.

Antigen-binding site Site formed by the variable regions of a heavy and light chain of an antibody.

Antigen-presenting cell (APC) Dendritic cells, macrophages, and B cells, which process antigens and activate cells of the immune system.

Antigenic determinant *(epitope)* The three-dimensional shape of a region of an antigen that is recognized by the immune system.

Antigenic drift Phenomenon that occurs every 2 to 3 years when a single strain of influenzavirus mutates within a local population.

Antigenic shift Major antigenic change that occurs on average every 10 years and results from the reassortment of genomes from different influenzavirus strains within host cells.

Antihistamines Drugs that specifically neutralize histamine.

Antimicrobial Any compound used to treat infectious disease; may also function as intermediate-level disinfectant.

Antimicrobial agent Chemotherapeutic agent used to treat microbial infection.

Antimicrobial associated diarrhea Watery stools resulting from elimination of normal intestinal microbiota by antimicrobial drug therapy; most severe form is pseudomembranous colitis caused by *Clostridium difficile*.

Antimicrobial enzyme Enzyme that acts against microbes.

Antimicrobial peptide *(defensin)* Chain of about 20 to 50 amino acids that acts against microorganisms.

Antiretroviral therapy (ART) A cocktail of antiviral drugs including nucleotide analogs, integrase inhibitors, protease inhibitors, and reverse transcriptase inhibitors.

Antisense nucleic acid RNA or single-stranded DNA with a nucleotide sequence complementary to a molecule of mRNA; used to control translation of polypeptide.

Antisense RNA RNA with a nucleotide sequence complementary to a molecule of mRNA; used to control translation of polypeptide.

Antisepsis The inhibition or killing of microorganisms on skin or tissue by the use of a chemical antiseptic.

Antiseptic Chemical used to inhibit or kill microorganisms on skin or tissue.

Antiserum In serology, blood fluid containing antibodies that bind to the antigens that triggered their production.

Antitoxin Antibodies formed by the host that bind to and protect against toxins.

Antivenin (antivenom) Antitoxin used to treat snake bites.

Antiviral proteins (AVPs) Proteins triggered by alpha and beta interferons that prevent viral replication.

Apicomplexans In protozoan taxonomy, group of pathogenic alveolate protozoa characterized by the complex of special intracellular organelles located at the apices of the infective stages of these microbes.

Apoenzyme The protein portion of protein enzymes that is inactive unless bound to one or more cofactors.

Apoptosis Programmed cell suicide.

Applied microbiology Branch of microbiology studying the commercial use of microorganisms in industry and foods.

Arachnid Group of arthropods distinguished by the presence of eight legs, such as spiders, ticks, and mites.

Arboviral encephalitis Inflammation of the brain and/or meninges caused by viruses transmitted by bloodsucking arthropods.

Arboviruses Viruses that are transmitted by arthropods; they include members of several viral families.

Archaea (*Archaeon*, sing.) In Woese's taxonomy, domain that includes all prokaryotic cells having archaeal rRNA sequences.

Archaezoa Protozoa that lack mitochondria, Golgi bodies, chloroplasts, and peroxisomes.

Arenaviruses Group of segmented, negative ssRNA viruses that cause zoonotic diseases.

ART (antiretroviral therapy) A cocktail of antiviral drugs including nucleotide analogs, integrase inhibitors, protease inhibitors, and reverse transcriptase inhibitors.

Arthropod Animal with a segmented body, hard exoskeleton, and jointed legs, including arachnids and insects.

Arthropod vector Animals with segmented bodies, external skeletons, and jointed legs that carry pathogens; vectors include ticks, mites, fleas, flies, and true bugs.

Artificial wetlands Use of successive ponds, marshes, and meadowlands to remove wastes from sewage as the water moves through the wetlands.

Artificially acquired active immunity Type of immunity that occurs when the body receives antigens by injection, as with vaccinations, and mounts a specific immune response.

Artificially acquired passive immunotherapy Treatment in which patient receives via injection preformed antibodies in antitoxins or antisera, which can destroy fast-acting and potentially fatal antigens, such as rattlesnake venom.

Ascariasis Symptomatic but not typically fatal disease caused by infection with the nematode *Ascaris lumbricoides*.

Ascomycota Division of fungi characterized by the formation of haploid ascospores within sacs called asci.

Ascospore Haploid germinating structure of ascomycetes.

Ascus Sac in which haploid ascospores are formed and from which they are released in fungi of the division Ascomycota.

Aseptate Lacking cross walls.

Aseptic Characteristic of an environment or procedure that is free of contamination by pathogens.

Aspergillosis Term for several localized and invasive diseases caused by infection with *Aspergillus* species.

Assembly In virology, fourth stage of the lytic replication cycle, in which new virions are assembled in the host cell.

Asthma Hypersensitivity reaction affecting the lungs and characterized by bronchial constriction and excessive mucus production.

Astroviruses Group of small, round enteric viruses that cause diarrhea, typically in children.

Asymptomatic *(subclinical)* Characteristic of disease that may go unnoticed because of absence of symptoms, even though clinical tests may reveal signs of disease.

Atom The smallest chemical unit of matter.

Atomic force microscope (AFM) Type of probe microscope that uses a pointed probe to traverse the surface of a specimen. A laser beam detects vertical movements of the probe, which a computer translates to reveal the atomic topography of the specimen.

Atomic mass *(atomic weight)* The sum of the masses of the protons, neutrons, and electrons in an atom.

Atomic number The number of protons in the nucleus of an atom.

ATP synthase (ATPase) Enzyme that phosphorylates ATP in oxidative phosphorylation and photophosphorylation.

Attachment In virology, first stage of the lytic replication cycle, in which the virion attaches to the host cell.

Attenuated vaccine Inoculum in which pathogens are weakened so that, theoretically, they no longer cause disease; residual virulence can be a problem.

Attenuation The process of reducing vaccine virulence.

Autoantigens Antigens on the surface of normal body cells.

Autoclave Device that uses steam heat under pressure to sterilize chemicals and objects that can tolerate moist heat.
Autograft Type of graft in which tissues are moved to a different location within the same patient.
Autoimmune disease Any of a group of diseases that result when an individual begins to make autoantibodies or cytotoxic T cells against normal body components.
Autoimmune hemolytic anemia Disease resulting when an individual produces antibodies against his or her own red blood cells.
Avirulent Harmless.
Axenic Having only one organism present.
Axial filament In cell morphology, structure composed of rotating endoflagella that allows a spirochete to "corkscrew" through its medium.
Azoles Class of antifungal drugs that disrupt cytoplasmic membranes.

B cell B lymphocyte.
B cell receptor (BCR) Antibody integral to the cytoplasmic membrane and expressed by B lymphocytes.
B lymphocyte *(B cell)* Lymphocyte that arises and matures in the red bone marrow in adults and is found primarily in the spleen, lymph nodes, red bone marrow, and Peyer's patches of the intestines and that secretes antibodies.
Bacillus Rod-shaped prokaryotic cell.
Bacitracin Antimicrobial that blocks NAG and NAM secretion from the cytoplasm, prompting cell lysis.
Bacteremia The presence of bacteria in the blood; often caused by infection with *Staphylococcus aureus* or *Streptococcus pneumoniae*.
Bacteria Prokaryotic microorganisms typically having cell walls composed of peptidoglycan. In Woese's taxonomy, domain which includes all prokaryotic cells having bacterial rRNA sequences.
Bacterial gastroenteritis Inflammation of the mucous membrane of the stomach and intestines, caused by a bacterial pathogen.
Bacterial intoxication *(toxification)* Food poisoning caused by bacterial toxin.
Bacteriophage *(phage)* Virus that infects and usually destroys bacterial cells.
Bacteriorhodopsin Purple protein synthesized by *Halobacterium* that absorbs light energy to synthesize ATP.
Bacteroid Diverse group of Gram-negative microbes that have similar rRNA nucleotide sequences, such as *Bacteroides*.
Balantidiasis A mild gastrointestinal illness caused by *Balantidium coli*.
Barophile Microorganism requiring the extreme hydrostatic pressure found at great depth below the surface of water.
Base Molecule that binds with hydrogen ions when dissolved in water.
Base pair (bp) A complementary arrangement of nucleotides in a strand of DNA or RNA. For example, in both DNA and RNA, guanine and cytosine pair.
Base-excision repair Mechanism by which enzymes excise a section of a DNA strand containing an error, and then DNA polymerase fills in the gap.
Basic dye In microscopy, a cationic chromophore used to stain acidic structures. Works most effectively in alkaline environments.
Basidiocarp Fruiting body of basidiomycetes; includes mushrooms, puffballs, stinkhorns, jelly fungi, bird's nest fungi, and bracket fungi.
Basidiomycota Division of fungi characterized by production of basidiospores and basidiocarps.
Basophil Type of granulocyte that stains blue with the basic dye methylene blue.
Bejel Childhood disease caused by the spirochete *Treponema pallidum endemicum;* characterized by rubbery oral lesions.
Benign tumor Mass of neoplastic cells that remains in one place and is not generally harmful.
Benthic zone Bottom zone of freshwater or marine water, devoid of light and with scarce nutrients.
Beta interferons (IFN-β) Interferons secreted by virally infected fibroblasts within hours after infection.
Beta-lactam Antimicrobial whose functional portion is composed of beta-lactam rings, which inhibit peptidoglycan formation by irreversibly binding to the enzymes that cross-link NAM subunits.
Beta-lactamase Bacterial enzyme that breaks the beta-lactam rings of penicillin and similar molecules, rendering them inactive.
Beta-oxidation A catabolic process in which enzymes split pairs of hydrogenated carbon atoms from a fatty acid and join them to coenzyme A to form acetyl-CoA.
Betaproteobacteria Class of diverse Gram-negative bacteria in the phylum Proteobacteria capable of growing at very low nutrient levels.
Binary fission The most common method of asexual reproduction of prokaryotes, in which the parental cell disappears with the formation of progeny.
Binomial nomenclature The classification method used in the Linnaean system of taxonomy, which assigns each species both a genus name and a specific epithet.
Biochemical oxygen demand (BOD) A measure of the amount of oxygen aerobic bacteria require to metabolize organic wastes in water.
Biochemistry Branch of chemistry that studies the chemical reactions of living things.
Biodiversity The number of species living in a given ecosystem.
Biofilm A slimy community of microbes growing on a surface.
Biogeochemical cycling The movement of elements and nutrients from unusable forms to usable forms because of the activities of microorganisms.
Biological vector Biting arthropod or other animal that transmits pathogens and serves as host for the multiplication of the pathogen during some stage of the pathogen's life cycle.
Biomass The quantity of all organisms in a given ecosystem.
Biomining The use of microbes to convert metals to soluble forms that can be more easily extracted.
Bioremediation The use of microorganisms to metabolize toxins in the environment to reclaim soils and waterways.
Bioreporter Type of biosensor composed of microbes with innate signaling capabilities.
Biosensor Device that combines bacteria or microbial products such as enzymes with electronic measuring devices to detect other bacteria, bacterial products, or chemical compounds in the environment.
Biosphere The region of Earth inhabited by living organisms.
Biotechnology Branch of microbiology in which microbes are manipulated to manufacture useful products.
Bioterrorism The use of microbes or their toxins to terrorize human populations.
Blastomycosis Pulmonary disease found in the southeastern United States, caused by infection with *Blastomyces dermatitidis*.
Blood group antigens The surface molecules of red blood cells.
Bodily fluid transmission Spread of pathogenic microorganisms via blood, urine, saliva, or other bodily fluids.
Botulism Potentially fatal intoxication with botulism toxin; three types include foodborne botulism, infant botulism, and wound botulism.
Bradykinin Peptide chain of nine amino acids that is a potent mediator of inflammation.
Broad-spectrum drug Antimicrobial that works against many different kinds of pathogens.
Bronchitis Inflammation of the bronchi.
Broth A liquid, nutrient-rich medium used for cultivating microorganisms.
Broth dilution test Test for determining the minimum inhibitory concentration in which a standardized amount of bacteria is added to serial dilutions of antimicrobial agents in tubes or wells containing broth.
Brucellosis Disease caused by *Brucella;* usually asymptomatic or mild, though it can result in sterility or abortion in animals.
Bruton-type agammaglobulinemia An inherited disease in which affected babies cannot make immunoglobulins and experience recurrent bacterial infections.
Bt toxin (Bt) Insecticidal poison produced by *Bacillus thuringiensis* bacteria.
Bubo Swollen inflamed lymph node.
Bubonic plague Severe systemic disease, fatal if untreated in 50% of patients, and characterized by fever, tissue necrosis, and the presence of buboes; caused by infection with *Yersinia pestis*.
Budding In prokaryotes and yeasts, reproductive process in which an outgrowth of the parent cell receives a copy of the genetic material, enlarges, and detaches. In virology, extrusion of enveloped virions through the host's cell membrane.
Buffer A substance, such as a protein, that prevents drastic changes in pH.
Bulbar poliomyelitis Infection of the brain stem and medulla resulting in paralysis of muscles in the limbs or respiratory system; caused by infection with poliovirus.
Bunyaviruses Group of zoonotic pathogens that have a segmented genome of three –ssRNA molecules and are transmitted to humans via arthropods.
Burkitt's lymphoma Infectious cancer of the jaw caused by infection with Epstein-Barr virus.

Caliciviruses Group of small, round enteric viruses that cause diarrhea, nausea, and vomiting.

Calor Heat.
Calvin-Benson cycle Stage of photosynthesis in which atmospheric carbon dioxide is fixed and reduced to produce glucose.
Cancer Disease characterized by the presence of one or more malignant tumors.
Candidiasis Term for several opportunistic diseases caused by infection with *Candida* species.
Candin Antifungal drug that inhibits cell wall synthesis.
Capnophile Microorganism that grows best with high levels of carbon dioxide in addition to low levels of oxygen.
Capsid A protein coat surrounding the nucleic acid core of a virion.
Capsomere A proteinaceous subunit of a capsid.
Capsule Glycocalyx composed of repeating units of organic chemicals firmly attached to the cell surface.
Capsule stain *(negative stain)* In microscopy, a staining technique used primarily to reveal bacterial capsules and involving application of an acidic dye that leaves the specimen colorless and the background stained.
Carbohydrate Organic macromolecule consisting of atoms of carbon, hydrogen, and oxygen.
Carbon cycle Biogeochemical cycle in which carbon is cycled in the form of organic molecules.
Carbon fixation The attachment of atmospheric carbon dioxide to ribulose 1,5-bisphosphate (RuBP).
Carbuncle The coalescence of several *furuncles* extending deep into underlying tissues; caused by infection with *Staphylococcus aureus*.
Carcinogen Chemical capable of causing cancer.
Caries *(cavities)* Tooth decay; caused by viridans streptococci and other bacteria.
Carotenoid Plant pigment that acts as an antioxidant.
Carrageenan Gel-like polysaccharide isolated from red algae and used as thickening agent.
Carrier In human pathology, continuous asymptomatic human source of infection.
Catabolism All of the decomposition reactions in an organism taken together.
Catarrhal phase In pertussis, initial phase lasting 1 to 2 weeks and characterized by signs and symptoms resembling those of a common cold.
Cation A positively charged ion.
Cat scratch disease Common and occasionally serious infection in children; characterized by fever and malaise plus localized swelling; caused by infection with *Bartonella henselae*.
CD4 Distinguishing cytoplasmic membrane protein of helper T cells, which is the initial binding site of HIV.
CD4 cell *(helper T cell, Th cell)* In cell-mediated immune response, a type of cell characterized by CD4 cell-surface glycoprotein; regulates the activity of B cells and cytotoxic T cells.
CD8 Distinguishing cytoplasmic membrane protein of cytotoxic T cells.
CD8 cell *(cytotoxic T cell, Tc cell)* In cell-mediated immune response, type of cell characterized by CD8 cell-surface glycoprotein; secretes perforins and granzymes that destroy infected or abnormal body cells.
CD95 pathway In cell-mediated cytotoxicity, pathway involving CD95 protein that triggers apoptosis of infected cells.
***C. diff.* diarrhea** Watery stools resulting from elimination of normal intestinal microbiota by antimicrobial drug therapy, which allows *Clostridium difficile* endospores to germinate.
Cell culture Cells isolated from an organism and grown on the surface of a medium or in broth. Viruses can be grown in a cell culture.
Cell wall In most cells, structural boundary composed of polysaccharide or protein chains that provides shape and support against osmotic pressure.
Cell-mediated immune response Immune response used by T cells to fight intracellular pathogens and abnormal body cells.
Cellular respiration Metabolic process that involves the complete oxidation of substrate molecules and production of ATP via a series of redox reactions.
Cellular slime mold Individual haploid myxamoeba that phagocytizes bacteria, yeasts, dung, and decaying vegetation.
Central dogma In genetics, fundamental description of protein synthesis that states that genetic information is transferred from DNA to RNA to polypeptides, which function alone or in conjunction as proteins.
Centrioles Nonmembranous organelles in animal cells that appear to function in the formation of flagella and cilia and in cell division.
Centrosome Region of a cell containing centrioles.
Cesspool Home equivalent of primary wastewater treatment in which wastes enter a series of porous concrete rings buried underground and are digested by microbes.
Cestodes *(tapeworms)* Group of helminths that are long, flat, and segmented and lack digestive systems.
Chagas' disease Potentially fatal disease caused by a bite from a kissing bug carrying *Trypanosoma cruzi* and characterized by the formation of swellings at the site of the bite, followed by fever, swollen lymph nodes, myocarditis, organ enlargement, and eventually congestive heart failure.
Chancre Painless red lesion that appears at the site of infection with *Treponema pallidum*, the agent of syphilis.
Chancroid Soft, painful, venereal ulcer at site of infection by *Haemophilus ducreyi*.
Chemical bond An interaction between atoms in which electrons are either shared or transferred in such a way as to fill their valence shells.
Chemical fixation In microscopy, a technique that uses methyl alcohol or formalin to attach a smear to a slide.
Chemical reaction The making or breaking of a chemical bond.
Chemiosmosis Use of ion gradients to generate ATP.
Chemoautotroph Microorganism that uses carbon dioxide as a carbon source and catabolizes organic molecules for energy.
Chemoheterotroph Microorganism that uses organic compounds for both energy and carbon.
Chemokine An immune system cytokine that signals leukocytes to rush to the site of inflammation or infection and activate other leukocytes.
Chemostat A continuous culture device controlled by adding fresh medium at the same rate old medium is removed.
Chemotactic factors Chemicals, such as peptides derived from complement and cytokines, that attract cells.
Chemotaxis Cell movement that occurs in response to chemical stimulus.
Chemotherapeutic agent Chemical used to treat disease.
Chemotherapy A branch of medical microbiology in which chemicals are studied for their potential to destroy pathogenic microorganisms.
Chickenpox *(varicella)* Highly infectious disease characterized by fever, malaise, and skin lesions, caused by infection with varicella-zoster virus.
Chitin Strong, flexible nitrogenous polysaccharide found in fungal cell walls and in the exoskeletons of insects and other arthropods.
Chlamydia Any of the small Gram-negative pathogenic cocci that grow and reproduce within the cells of mammals, birds, and a few invertebrates and that spread as elementary bodies.
Chloramphenicol Antimicrobial drug that blocks the enzymatic site of the 50S ribosomal subunit, inhibiting polypeptide synthesis.
Chlorophyll Pigment molecule that captures light energy for use in photosynthesis.
Chlorophyta Green-pigmented division of algae that have chlorophylls *a* and *b*, store sugar and starch as food reserves, and have rRNA sequences similar to plants. Considered the progenitors of plants.
Chloroplast Light-harvesting organelle found in photosynthetic eukaryotes.
Cholera Disease contracted through the ingestion of food and water contaminated with *Vibrio cholerae* and characterized by vomiting and watery diarrhea.
Cholera toxin Exotoxin produced by *Vibrio cholerae* that causes the movement of water out of the intestinal epithelium.
Chromatin Threadlike mass of DNA and associated histone proteins that becomes visible during mitosis as chromosomes.
Chromatin fiber An association of nucleosomes and proteins found within the chromosomes of eukaryotic cells.
Chromoblastomycosis Cutaneous and subcutaneous disease characterized by lesions that can spread internally; caused by traumatic introduction of ascomycete fungi into the skin.
Chromosome A molecule of DNA associated with protein. In prokaryotes, typically circular and localized in a region of the cytosol called the nucleoid. In eukaryotes, chromosomes are threadlike and are most visible during mitosis and meiosis.
Chronic disease Any disease that develops slowly, usually with less severe symptoms, and is continual or recurrent.
Chronic granulomatous disease Primary immunodeficiency disease in which children have recurrent infections characterized by the development of large masses of inflammatory cells in lymph nodes, lungs, bones, and skin.
Chronic inflammation Type of inflammation that develops slowly, lasts a long time, and can cause damage (even death) to tissues resulting in disease.
Chrysophyta Division of algae including the golden algae, yellow-green algae, and diatoms.

Cilia Short, hairlike, rhythmically motile projections of some eukaryotic cells.
Ciliate In protozoan taxonomy, group of alveolate protozoa characterized by the presence of cilia in their trophozoite stages.
Class Taxonomical grouping of similar orders of organisms.
Class switching The process in which a plasma cell changes the type of antibody F_c region (stem) that it synthesizes and secretes.
Clinical laboratory scientist An expert in health care-related microbiological laboratory procedures who has at least a bachelor degree.
Clinical specimen Sample of human material, such as feces or blood, that is examined or tested for the presence of microorganisms.
Clonal deletion Process by which cells with receptors that respond to autoantigens are selectively killed via apoptosis.
Clonal expansion In immunology, the reproduction of activated lymphocytes.
Clonal selection In antibody immunity, recognition and activation only of B lymphocytes with BCRs complementary to a specific antigenic determinant.
Coccidioidomycosis Pulmonary disease found in the southwestern United States, caused by infection with *Coccidioides immitis*.
Coccobacillus A prokaryotic cell intermediate in shape between a sphere and a rod, such as an elongated coccus.
Coccus Spherical prokaryotic cell.
Codon Triplet of mRNA nucleotides that codes for specific amino acids. For example, AAA is a codon for lysine.
Coenocyte Multinucleate cell resulting from repeated mitosis but postponed or absent cytokinesis.
Coenzyme Organic cofactor.
Cofactor Inorganic ions or organic molecules that are essential for enzyme action.
Coinfection Condition in which a patient is infected simultaneously with hepatitis B and D viruses.
Cold enrichment Incubation of a specimen in a refrigerator to enhance the growth of cold-tolerant species.
Cold sore Common name for oral lesion of a herpesvirus.
Coliforms Enteric Gram-negative bacteria that ferment lactose to gas and are found in the intestinal tracts of animals and humans.
Colony Visible population of microorganisms living in one place; an aggregation of cells arising from a single parent cell.
Colony-forming unit (CFU) A single cell or group of related cells that produce a colony.
Colorado tick fever Zoonosis caused by *Coltivirus* that is typically characterized by mild fever and chills.
Combination vaccine Inoculum composed of antigens from several pathogens that are administered simultaneously.
Commensalism Symbiotic relationship in which one member benefits without significantly affecting the other.
Communicable disease Any infectious disease that comes either directly or indirectly from another host.
Competence Ability of a cell to take up DNA from the environment.
Competitive inhibitor Inhibitory substance that blocks enzyme activity by blocking active sites.
Complement fixation test A complex assay used to determine the presence of specific antibodies in serum.
Complement system (complement) Set of blood plasma proteins that act as chemotactic attractants, trigger inflammation and fever, and ultimately effect the destruction of foreign cells.
Complementary DNA (cDNA) DNA synthesized from an mRNA template using reverse transcriptase.
Complex medium Culturing medium that contains nutrients released by the partial digestion of yeast, beef, soy, or other proteins; thus, the exact chemical composition is unknown.
Complex transposon Transposon containing genes not connected with transposition.
Compound A molecule containing atoms of more than one element.
Compound microscope Microscope using a series of lenses for magnification.
Concentration gradient The difference in concentration of a chemical on the two sides of a membrane. Also called a *chemical gradient*.
Condenser lens In a compound microscope, a lens that directs light through the specimen as well as one or more mirrors that deflect the light's path.
Condyloma acuminata Large, cauliflower-like genital warts caused by infection with a papillomavirus.
Confocal microscope Type of light microscope that uses ultraviolet lasers to illuminate fluorescent chemicals in a single plane of the specimen.
Congenital syphilis Disease characterized by mental retardation, organ malformation, and, in some cases, death of the fetus of a woman infected with *Treponema pallidum*, the agent of syphilis.
Conjugation In genetics: method of horizontal gene transfer in which a bacterium containing a fertility plasmid forms a conjugation pilus that attaches and transfers plasmid genes to a recipient; in reproduction of ciliates: coupling of mating cells.
Conjugation pilus Proteinaceous, rodlike structure extending from the surface of a cell; mediates conjugation.
Conjunctivitis *(pinkeye)* Inflammation of the lining of an eyelid.
Consumption Tuberculosis; refers to wasting away of a body affected with TB at several sites.
Contact immunity Immunity conferred to an unvaccinated individual following contact with an individual vaccinated with an attenuated vaccine.
Contagious disease A communicable disease that is easily transmitted from a reservoir or patient.
Contamination The presence of microorganisms in or on the body or other site.
Continuous cell culture Type of cell culture created from tumor cells.
Contrast The difference in visual intensity between two objects, or between an object and its background.
Convalescence In the infectious disease process, final stage during which the patient recovers from the illness, and tissues and systems are repaired and return to normal.
Convalescent phase In pertussis, final phase lasting 3 to 4 weeks during which the ciliated lining of the trachea grows back and frequency of coughing spells diminishes, but secondary bacterial infections may ensue.
Cord factor A cell-wall component of pathogenic *Mycobacterium tuberculosis* that produces strands of daughter cells, inhibits migration of neutrophils, and is toxic to body cells.
Coronaviruses Group of enveloped ssRNA viruses that cause colds as well as severe acute respiratory syndrome.
Coronavirus respiratory syndrome Severe respiratory disease caused by coronaviruses SARS virus and MERS virus.
Corticosteroids Another name for immunosuppressive agents that suppress the action of T cells.
Counterstain In a Gram stain, red stain that provides contrasting color to the primary stain, causing Gram-negative cells to appear pink.
Covalent bond The sharing of a pair of electrons by two atoms.
Coxsackieviruses Group of enteroviruses which cause a variety of diseases in humans, ranging from mild fever and colds to myocarditis and heart failure.
Crenation Shriveling of a cell caused by osmosis in a hypertonic environment.
Cristae Folds within the inner membrane of a mitochondrion that increase its surface area.
Cross resistance Phenomenon in which resistance to one antimicrobial drug confers resistance to similar drugs.
Crossing over Process in which portions of homologous chromosomes are recombined during the formation of gametes.
Croup Inflammation and swelling of the larynx, trachea, and bronchi and a "seal bark" cough, often caused by infection with a parainfluenza or rarely by other respiratory viruses.
Cryptococcosis Disease caused by the dimorphic fungus *Cryptococcus;* typically manifests as meningitis.
Cryptosporidiosis (Cryptosporidium *enteritis*) A gastrointestinal disease caused by infection with *Cryptosporidium parvum*; in humans, characterized by diarrhea, fluid loss, and weight loss; may be fatal in HIV-positive patients.
***Cryptosporidium* enteritis** *(cryptosporidiosis)* A gastrointestinal disease caused by infection with *Cryptosporidium parvum*; in humans, characterized by diarrhea, fluid loss, and weight loss; may be fatal in HIV-positive patients.
Culture Act of cultivating microorganisms or the microorganisms that are cultivated.
Cuticle Outer protective "skin" of a nematode.
Cyanobacteria Gram-negative photosystem bacteria that vary greatly in shape, size, and method of reproduction.
Cyclic photophosphorylation Return of electrons to the original reaction center of a photosystem after passing down an electron transport chain.
Cycloserine Semisynthetic antibiotic used to treat infections with Gram-positive bacteria.
Cyclosporine Immunosuppressive drug that inhibits action of activated T cells.
Cyst In protozoan morphology, the hardy resting stage characterized by a thick capsule and a low metabolic rate.

Cysticercus Immature tapeworm, usually in muscle of intermediate host.
Cytokines Proteins secreted by many types of cells that regulate adaptive immune responses.
Cytokinesis Division of a cell's cytoplasm.
Cytoplasm General term used to describe the semiliquid, gelatinous material inside a cell.
Cytoplasmic membrane Membrane surrounding all cells, and composed of a fluid mosaic of phospholipids and proteins.
Cytosine Ring-shaped nitrogenous base found in nucleotides of DNA and RNA.
Cytoskeleton Internal network of fibers contributing to the basic shape of eukaryotic and rod-shaped prokaryotic cells.
Cytosol The liquid portion of the cytoplasm.
Cytotoxic drugs Group of drugs that inhibit cells.
Cytotoxic T cell *(Tc cell, CD8 cell)* In cell-mediated immune response, type of cell characterized by CD8 cell-surface glycoprotein; secretes perforins and granzymes that destroy infected or abnormal body cells.

Dark-field microscope Microscope used for studying pale or small specimens; deflects light rays so that they miss the objective lens.
Dark repair Mechanism by which enzymes cut damaged DNA sections from a molecule, creating a gap that is repaired by DNA polymerase and DNA ligase.
Deamination Process in which amine groups are split from amino acids.
Death phase Phase in a growth curve in which the organisms are dying more quickly than they are being replaced by new organisms.
Decimal reduction time (D) The time required to destroy 90% of the microbes in a sample.
Decline In the infectious disease process, period in which the body gradually returns to normal as the patient's immune response and any medical treatments vanquish the pathogens.
Decolorizing agent In a stain, a solution that washes the primary stain away.
Decomposition reaction A chemical reaction in which the bonds of larger molecules are broken to form smaller atoms, ions, and molecules.
Deep-freezing Long-term storage of cultures at temperatures ranging from −50° C to −95° C.
Deeply branching bacteria Prokaryotic autotrophs with rRNA sequences and growth characteristics thought to be similar to those of earliest bacteria.
Defensins *(antimicrobial peptides)* Small peptide chains that act against a broad range of pathogens.
Defined medium *(synthetic medium)* Culturing medium of which the exact chemical composition is known.
Definitive host In the life cycle of parasites, host in which mature and sometimes sexual forms of the parasite are present and usually reproducing.
Degerming The removal of microbes from a surface by scrubbing.
Dehydration synthesis Type of synthesis reaction in which two smaller molecules are joined together by a covalent bond, and a water molecule is formed.
Delayed hypersensitivity reaction *(type IV hypersensitivity)* T cell-mediated inflammatory reaction that takes 24 to 72 hours to reach maximal intensity.
Deletion Type of mutation in which a nucleotide base pair is deleted.
Deltaproteobacteria Group of Proteobacteria that includes *Desulfovibrio*, *Bdellovibrio*, and myxobacteria.
Denaturation Process by which a protein's three-dimensional structure is altered, eliminating function.
Dendritic cells Cells of the epidermis and mucous membranes that devour pathogens.
Dengue fever Self-limiting but extremely painful disease caused by a flavivirus transmitted by *Aedes* mosquitoes.
Dengue hemorrhagic fever Potentially fatal disease involving a hyperimmune response to reinfection with dengue virus and causing ruptured blood vessels, internal bleeding, and shock.
Denitrification The conversion of nitrate into nitrogen gas by anaerobic respiration.
Deoxyribonucleic acid (DNA) Nucleic acid consisting of nucleotides made up of phosphate, a deoxyribose pentose sugar, and an arrangement of the bases adenine, guanine, cytosine, and thymine.
Dermatophyte Fungus that normally lives on skin, nails, or hair.
Dermatophytoses Any of a variety of superficial skin, nail, and hair infections caused by dermatophytes.
Dermis The layer of the skin deep to the epidermis and containing hair follicles, glands, and nerve endings.
Descriptive epidemiology The careful recording of data concerning a disease.
Desiccation Inhibition of microbial growth by drying.
Detergent Positively charged organic surfactant.
Deuteromycetes Informal grouping of fungi having no known sexual stage.
Diapedesis *(emigration)* Process whereby leukocytes leave intact blood vessels by squeezing between lining cells.
Diatom Type of alga in the division Chrysophyta; has cell walls made of silica arranged in nesting halves called frustules.
Dichotomous key Method of identifying organisms in which information is arranged in paired statements, only one of which applies to any particular organism.
Differential interference contrast microscope Type of phase microscope that uses prisms to split light beams, giving images a three-dimensional appearance.
Differential medium Culturing medium formulated such that either the presence of visible changes in the medium or differences in the appearances of colonies help microbiologists differentiate among kinds of bacteria growing on the medium.
Differential stain In microscopy, a stain using more than one dye so that different structures can be distinguished. The Gram stain is the most commonly used.
Differential white blood cell count Lab technique that indicates the relative numbers of leukocytes.
Diffusion The net movement of a chemical down its concentration gradient.
Diffusion susceptibility test *(Kirby-Bauer test)* Simple, inexpensive test widely used to reveal which drug is most effective against a particular pathogen. Procedure involves inoculating a Petri plate uniformly with a standardized amount of the pathogen in question and arranging on the plate disks soaked in the drugs to be tested.
DiGeorge syndrome Failure of the thymus to develop, and thus, absence of T cells.
Dimorphic Having two forms; for example, dimorphic fungi have both yeastlike and moldlike bodies.
Dinoflagellate In protozoan taxonomy, group of unicellular, flagellated, alveolate protozoa characterized by photosynthetic pigments.
Dioecious Male and female sex organs are in separate individuals.
Diphtheria Mild to potentially fatal respiratory disease caused by diphtheria toxin following infection with *Corynebacterium diphtheriae*.
Diphtheroids Generally nonpathogenic pleomorphic bacilli named for the similarity of their appearance to *Corynebacterium diphtheriae*.
Diplococcus A pair of cocci.
Diploid A nucleus with two copies of each chromosome.
Diploid cell culture Type of cell culture created from embryonic animal, plant, or human cells that have been isolated and provided appropriate growth conditions.
Dipstick immunochromatographic assay Rapid modification of ELISA test in which an antigen solution flows through a porous strip, encountering labeled antibody; used for pregnancy testing and for rapid identification of infectious agents.
Direct antibody test Immune test allowing direct observation of the presence of antigen.
Direct contact transmission Spread of pathogens from one host to another involving body contact between the hosts.
Direct fluorescent antibody test Immune test allowing direct observation of the presence of antigen in a tissue sample flooded with labeled antibody.
Disaccharide Carbohydrate consisting of two monosaccharide molecules joined together.
Disease Any adverse internal condition severe enough to interfere with normal body functioning.
Disease process Definite sequence of events following contamination and infection.
Disinfectant Physical or chemical agent used to inhibit or destroy microorganisms on inanimate objects.
Disinfection The use of physical or chemical agents to inhibit or destroy microorganisms on inanimate objects. In water treatment, ozone, UV light, or chlorination kill most microorganisms.
Disseminated intravascular coagulation (DIC) The formation of blood clots within blood vessels throughout the body; triggered by lipid A.
DNA fingerprinting *(genetic fingerprinting)* Technique that identifies unique sequences of DNA to determine paternity; connect blood, semen, or skin cells to suspects in criminal investigations; or identify pathogens.
DNA microarray Numerous distinct ssDNA molecules bound to a substrate and used to probe for complementary sequences.
Dolor Pain.
Domain Any of three basic types of cell groupings distinguished by Carl Woese, containing the Linnaean taxon of *kingdoms*.

Donor cell In horizontal gene transfer, a cell that contributes part of its genome to a recipient.
Droplet transmission Spread of pathogens from one host to another via aerosols, which exit the body during exhaling, coughing, and sneezing and travel less than 1 meter.
Dysentery Disease characterized by severe diarrhea often with stools containing blood and mucus.
Dyspnea Difficulty in breathing.
Dysuria Painful urination.

E site In translation, site at which tRNA exits from the ribosome.
Eastern equine encephalitis (EEE) Potentially fatal infection of the brain caused by a togavirus.
Ebola virus Virus of Africa causing a type of hemorrhagic fever fatal in 90% of cases.
Echinocandin Antifungal drug that inhibits cell wall synthesis.
Echoviruses *(enteric cytopathic human orphan viruses)* Group of enteroviruses that cause viral meningitis and colds.
Ecosystem All of the organisms living in a particular habitat and the relationships between the two.
Efflux pump Transmembrane pump that removes antimicrobial drugs from a cell or from the periplasm.
Ehrlichiosis *(human monocytic ehrlichiosis, HME)* Tick-borne disease caused by a rickettsia, *Ehrlichia chaffeensis*, manifesting with flulike signs and symptoms. Also previously used to refer to human granulocytic ehrlichiosis, now called *anaplasmosis*.
Electrical gradient Voltage across a membrane created by the electrical charges of the chemicals on either side.
Electrochemical gradient The chemical and electrical gradients across a cell membrane.
Electrolyte Any hydrated cation or anion; can conduct electricity through a solution.
Electron A negatively charged subatomic particle.
Electron transport chain Series of redox reactions that pass electrons from one membrane-bound carrier to another and then to a final electron acceptor.
Electronegativity The attraction of an atom for electrons.
Elek test Immunodiffusion assay used to detect the presence of diphtheria toxin in a fluid sample.
Element Matter that is composed of a single type of atom.
Elephantiasis Enlargement and hardening of tissues, especially in the lower extremities, where lymph has accumulated following infection with *Wuchereria bancrofti*.
Empyema In patients with staphylococcal pneumonia, the presence of pus in the alveoli of the lungs.
Encephalitis Inflammation of the brain.
Encystment In the life cycle of protozoa, stage in which cysts form in host tissues.
Endemic In epidemiology, a disease that occurs at a relatively stable frequency within a given area or population.
Endemic typhus *(murine typhus)* Disease transmitted by fleas and characterized by high fever, headache, chills, muscle pain, and nausea; caused by infection with *Rickettsia typhi*.
Endocarditis Potentially fatal inflammation of the endocardium; typically caused by infection with *Staphylococcus aureus* or *Streptococcus pneumoniae*.
Endocytosis Active transport process, used by some eukaryotic cells, in which pseudopods surround a substance and move it into the cell.
Endoflagellum A special flagellum of spirochetes that spirals tightly around a cell rather than protruding from it.
Endogenous antigen Antigen produced by microbes that multiply inside the cells of the body.
Endogenous healthcare associated infection An infection arising within a patient from opportunistic pathogens.
Endoplasmic reticulum (ER) Netlike arrangement of hollow tubules continuous with the outer membrane of the nuclear envelope and functioning as a transport system.
Endosome A sac formed during endocytosis containing the endocytized substance.
Endospore Environmentally resistant structure produced by the transformation of a vegetative cell of the Gram-positive genera *Bacillus* or *Clostridium*.
Endosymbiotic theory Proposal that eukaryotes were formed from the phagocytosis of small prokaryotes by larger prokaryotes, forming organelles.
Endothermic reaction Any chemical reaction that requires energy.
Endotoxin *(lipid A)* Potentially fatal toxin released from the lipopolysaccharide layer of the outer membrane of the cell wall of dead and dying Gram-negative bacteria.
Enrichment culture Technique used to enhance the growth of less abundant microorganisms by using a selective medium.
Enterobacteriaceae *(enteric bacteria)* Family of oxidase-negative Gram-negative bacteria, which can be pathogenic.
Enteroviruses Group of picornaviruses that are transmitted via the fecal-oral route but cause disease in any of a variety of target organs.
Entner-Doudoroff pathway Series of reactions that catabolize glucose to pyruvic acid using different enzymes from those used in either glycolysis or the pentose phosphate pathway.
Entry In virology, second stage of the lytic replication cycle, in which the virion or its genome enters the host cell.
Envelope In virology, membrane surrounding the viral capsid.
Environmental microbiology Branch of microbiology studying the role of microorganisms in soils, water, and other habitats.
Environmental specimen Sample of material taken from such sources as ponds, soil, or air and tested for the presence of microorganisms.
Enzyme An organic catalyst.
Enzyme immunoassay (EIA), enzyme-linked immunosorbent assay (ELISA) A family of simple immune tests that use enzymatic products as a label and that can be readily automated and read by machine.
Eosinophil Type of granulocyte that stains red to orange with the acidic dye eosin.
Eosinophilia An abnormal blood condition in which the number of eosinophils is greater than normal.
Epidemic In epidemiology, a disease that occurs at a greater than normal frequency for a given area or population.
Epidemiology Study of the occurrence, distribution, and spread of disease in humans.
Epidermis The outermost layer of the skin.
Epididymitis Inflammation of the epididymis.
Epitope *(antigenic determinant)* The three-dimensional shape of a region of an antigen that is recognized by the immune system.
Epsilonproteobacteria Group of Gram-negative rods, vibrios, and spiraled bacteria in the phylum Proteobacteria.
Erysipelas Impetigo spreading to lymph nodes, accompanied by pain and inflammation and caused by infection with group A *Streptococcus*.
Erythema infectiosum *(fifth disease)* Harmless red rash occurring in children and caused by infection with B19 virus.
Erythrocyte Red blood cell.
Erythrocytic cycle In the life cycle of *Plasmodium*, stage during which merozoites infect and cause lysis of erythrocytes.
Eschar Black, swollen, crusty, painless skin ulcer of anthrax.
Etest Test for determining minimum inhibitory concentration; a plastic strip containing a gradient of the antimicrobial agent being tested is placed on a plate inoculated with the pathogen of interest.
Ethambutol Antimicrobial drug that disrupts formation of arabinogalactan-mycolic acid by mycobacteria.
Etiology The study of the causation of disease.
Euglenids Protozoa that store food as paramylon, lack cell walls, and have eyespots used in positive phototaxis.
Eukarya In Woese's taxonomy, domain that includes all eukaryotic cells.
Eukaryote Any organism made up of cells containing a nucleus composed of genetic material surrounded by a distinct membrane. Classification includes animals, plants, algae, fungi, and protozoa.
Eutrophication The overgrowth of microorganisms in aquatic systems.
Evolution Changes in the genetic makeup of a population leading to the production of new varieties.
Exchange reaction Type of chemical reaction in which atoms are moved from one molecule to another by means of the breaking and forming of covalent bonds.
Excystment In the life cycle of protozoa, stage following ingestion by the host, in which cysts become trophozoites.
Exfoliative toxins Toxins of certain strains of *Staphylococcus aureus* that break down desmosomes in the skin, causing the outer layers of skin to slough off.
Exocytosis Active transport process, used by some eukaryotic cells, in which vesicles fuse with the cytoplasmic membrane and export their substances from the cell.
Exoerythrocytic phase In the life cycle of *Plasmodium*, stage during which infected mosquito injects sporozoites into the blood.
Exogenous antigen Antigen produced by microorganisms that multiply outside the cells of the body.
Exogenous healthcare associated infection An infection caused by pathogens acquired from the health care environment.
Exon Coding sequence of mRNA. Exons are connected to produce a functional mRNA molecule.

Exothermic reaction Any chemical reaction that releases energy.

Exotoxin Toxin secreted by a pathogenic microorganism into its environment.

Experimental epidemiology The testing of hypotheses resulting from analytical epidemiology concerning the cause of a disease.

Exponential (logarithmic) growth Increase in size of a microbial population in which the number of cells doubles in a fixed interval of time.

Extensively drug-resistant (XDR) tuberculosis Tuberculosis caused by extensively drug-resistant *Mycobacterium*.

Extremophile Microbe that requires extreme conditions of temperature, pH, and/or salinity to survive.

F (fertility) plasmid *(F factor)* Small, circular, extrachromosomal molecule of DNA coding for conjugation pili. Bacterial cells that contain an F plasmid are called F^+ cells and serve as donors during conjugation.

F_C region The stem region of an antibody.

Facilitated diffusion Movement of substances across a cell membrane via protein channels.

Facultative anaerobe Microorganism that can live with or without oxygen.

Family Taxonomical grouping of similar genera of organisms.

Fats Compounds composed of three fatty acid molecules linked to a molecule of glycerol.

Fecal-oral infection Spread of pathogenic microorganisms in feces to the mouth, such as results from drinking sewage-contaminated water.

Fecal transplant Injection of feces into a patient, typically via an enema tube; used to restore normal microbiota of the colon especially in recurrent cases with *Clostridium difficile* infection.

Feedback inhibition *(negative feedback)* Method of controlling the action of enzymes in which the end product of a series of reactions inhibits an enzyme in an earlier part of the pathway.

Fermentation In metabolism, the partial oxidation of sugar to release energy using an endogenous organic molecule rather than an electron transport chain as the final electron acceptor. In food microbiology, any desirable change to food or beverage induced by microbes.

Fever Body temperature above 37°C.

Fever blisters *(cold sores)* Painful, itchy lesions on the lips; characteristic of infection with *human herpesvirus 1*.

Filariasis Infection of the lymphatic system caused by a filarial nematode.

Filtration The passage of air or liquid through a material that traps and removes microbes. In water treatment, a process in which microbial biofilms on sand particles trap and remove other microbes.

Fimbriae Sticky, proteinaceous extensions of some bacterial cells that function to adhere cells to one another and to environmental surfaces.

Firmicutes Phylum of bacteria that includes clostridia, mycoplasmas, and low G + C Gram-positive bacilli and cocci.

Flagellates Group of protozoa that possess at least one long flagellum, generally used for movement.

Flagellum A long, whiplike structure protruding from a cell.

Flavin adenine dinucleotide (FAD) Important vitamin-derived electron carrier molecule.

Flea Vertically flattened, bloodsucking, wingless insect; vector of some pathogens.

Flocculation Process in water treatment in which alum (aluminum ammonium sulfate) added to the water forms sediments with particles and microorganisms.

Fluid mosaic model Model describing the arrangement and motion of the proteins within the cytoplasmic membrane.

Fluorescent microscope Type of light microscope that uses an ultraviolet light source to fluoresce objects.

Fly Insect with transparent wings that are not hidden or covered, including mosquitoes; vectors for many pathogens.

Folliculitis Infection of a hair follicle by *Staphylococcus aureus*.

Fomes (pl. fomites) Objects inadvertently used to transfer pathogens to new hosts, such as a glass or towel.

Food infection Type of food poisoning in whch living organisms are consumed.

Food intoxication Type of food poisoning resulting from consumption of microbial toxin.

Food microbiology The use of microorganisms in food production and the prevention of foodborne illnesses.

Food vesicle Sac formed during endocytosis of a solid; also called an endosome or phagosome.

Foodborne transmission Spread of pathogenic microorganisms in or on foods that are poorly processed, undercooked, or improperly refrigerated.

Foraminifera Type of armored marine amoeba.

Formed elements Cells and cell fragments suspended in blood plasma.

Frameshift mutation Type of mutation in which nucleotide triplets subsequent to an insertion or deletion are displaced, creating new sequences of codons that result in vastly altered polypeptide sequences.

Functional group An arrangement of atoms common to all members of a class of organic molecules, such as the amine group found in all amino acids.

Fungi Eukaryotic organisms that have cell walls and obtain food from other organisms.

Furuncle A large, painful, nodular extension of folliculitis into surrounding tissue; may be caused by infection with *Staphylococcus aureus*.

Gametocyte In sexual reproduction of protozoa, cell that can fuse with another gametocyte to form a diploid zygote.

Gamma interferons (IFN-γ) Interferon produced by T lymphocytes and NK lymphocytes; activates macrophages and neutrophils days after an infection.

Gammaproteobacteria Largest and most diverse class of Proteobacteria, including purple sulfur bacteria, methane oxidizers, pseudomonads, and others.

Gas gangrene Death of muscle and connective tissues accompanied by gaseous waste, caused by *Clostridium perfringens*.

Gaseous agent High-level disinfecting gas used to sterilize heat-sensitive equipment and large objects.

Gastroenteritis Inflammation of the mucous membrane of the stomach and intestines.

GC content The percentage of a cell's DNA bases that are guanine and cytosine.

Gel electrophoresis Technique used in recombinant DNA technology to separate molecules by size, shape, and electrical charge.

Gene A specific sequence of nucleotides that codes for a polypeptide or an RNA molecule.

Gene library Collection of bacterial or phage clones, each of which carries a fragment of an organism's genome.

Gene therapy The use of recombinant DNA technology to insert a missing gene or repair a defective gene in human cells.

Genera Plural of genus.

Generation time Time required for a cell to grow and divide.

Genetic engineering The manipulation of genes via recombinant DNA technology for practical applications.

Genetic fingerprinting *(DNA fingerprinting)* Technique that identifies unique sequences of DNA to determine paternity; connect blood, semen, or skin cells to suspects in criminal investigations; or identify pathogens.

Genetic mapping Application of recombinant DNA technology in which genes are located on a nucleic acid molecule.

Genetic recombination The exchange of segments, typically genes, between two DNA molecules.

Genetic screening Procedure by which laboratory tests are used to screen patient and fetal DNA for mutant genes.

Genetics The study of inheritance and heritable traits as expressed in an organism's genetic material.

Genome The sum of all the genetic material in a cell or virus.

Genomics The sequencing, analysis, and comparison of genomes.

Genotype Actual set of genes in an organism's genome.

Genus Taxonomical grouping of similar species of organisms.

Germ theory of disease Hypothesis formulated by Pasteur in 1857 that microorganisms are responsible for disease.

German measles *(rubella)* Disease caused by infection with *Rubivirus* resulting in characteristic rash lasting about 3 days; mild in children but potentially teratogenic to fetuses of infected women.

Giardiasis A mild to severe gastrointestinal illness caused by ingestion of cysts of *Giardia intestinalis*.

Gingivitis Inflammation of the gums.

Glomerulonephritis Deposition of immune complexes in the walls of the glomeruli—networks of minute blood vessels in the kidneys—that may result in kidney failure; typically caused by infection with group A *Streptococcus*.

Glucocorticoids *(corticosteroids)* Immunosuppressive agents, including prednisone and methylprednisolone, that suppress the response of T cells.

Glycocalyx (pl. **glycocalyces**) Sticky external sheath of prokaryotic and eukaryotic cells.

Glycolysis *(Embden-Meyerhof pathway)* First step in the catabolism of glucose via respiration and fermentation.

Goblet cells Mucus-secreting cells in the epithelium of mucous membranes.
Golgi body In eukaryotic cells, a series of flattened, hollow sacs surrounded by phospholipid bilayers and functioning to package large molecules for export in secretory vesicles.
Gonorrhea A sexually transmitted disease caused by infection with *Neisseria gonorrhoeae*.
gp41 Antigenic HIV glycoprotein that promotes fusion of the viral envelope with a target cell.
gp120 Antigenic glycoprotein that is the primary attachment molecule of HIV.
Graft Tissue or organ transplanted to a new site.
Graft rejection Rejection of donated tissue or organs by a transplant recipient.
Graft-versus-host disease Disease resulting when donated bone marrow cells mount an immune response against the recipient's cells.
Gram-negative cell Generally, a prokaryotic cell having a wall composed of a thin layer of wall material, an external membrane, and a periplasmic space between; appears pink after the Gram-staining procedures.
Gram-positive cell Prokaryotic cell having a thick wall; in bacteria, composed of a thick layer of peptidoglycan containing teichoic acids; Gram-positive cells retain the crystal violet dye used in the Gram-staining procedure, appearing purple.
Gram stain Technique for staining microbial samples by applying a series of dyes that leave some microbes purple and others pink. Developed by Christian Gram in 1884.
Granulocyte Type of leukocyte having large granules in the cytoplasm.
Granzyme Protein molecule in the cytoplasm of cytotoxic T cells that causes an infected cell to undergo apoptosis.
Graves' disease Production of autoantibodies that stimulate excessive production of thyroid hormone and growth of the thyroid gland.
Gross mutation Major change in the nucleotide sequence of DNA resulting from inversions, duplications, transpositions, or large insertions or deletions of nucleotides.
Group A *Streptococcus* *(S. pyogenes)* A Gram-positive coccus that produces protein M and a hyaluronic acid capsule, both of which contribute to the pathogenicity of the species.
Group B *Streptococcus* *(S. agalactiae)* A Gram-positive coccus that normally resides in the lower GI, genital, and urinary tracts but can cause disease in newborns.
Group translocation Active process, occurring in some prokaryotes, by which a substance being actively transported across a cell membrane is chemically changed during transport.
Growth An increase in size; in bacteriology, an increase in population.
Growth curve Graph that plots the number in a population over time.
Growth factor Organic chemical, such as a vitamin, required in very small amounts for metabolism. In immunology, an immune system cytokine that stimulates stem cells to divide, ensuring that the body is supplied with sufficient leukocytes of all types.
Guanine Ring-shaped nitrogenous base found in nucleotides of DNA and RNA.
Gumma Lesion that occurs in bones, in nervous tissue, or on skin in patients with tertiary syphilis.

HAART (highly active antiretroviral therapy) A cocktail of antiviral drugs including nucleoside analogs, integrase inhibitors, protease inhibitors, and reverse transcriptase inhibitors.
Habitat The physical localities in which organisms are found.
Halogen One of the four very reactive, nonmetallic chemical elements: iodine, chlorine, bromine, and fluorine. Used in disinfectants and antiseptics.
Halophile Microorganism requiring a saline environment (greater than 9% NaCl).
Hamus Proteinaceous, filamentous, helical extension of some archaeal cells that functions to attach the cells to one another and environmental surfaces.
Hansen's disease *(leprosy)* Disease caused by infection with *Mycobacterium leprae* that produces either a nonprogressive tuberculoid form or a progressive lepromatous form that destroys tissues, including facial features, digits, and other structures.
***Hantavirus* pulmonary syndrome (HPS)** Rapid, severe, and often fatal pneumonia caused by infection with *Hantavirus*.
Hantaviruses Group of bunyaviruses that are transmitted to humans via inhalation of virions in dried deer-mouse excreta and that cause *Hantavirus* pulmonary syndrome.
Haploid A nucleus with a single copy of each chromosome.
Haustoria Modified hyphae that penetrate the tissue of the host to withdraw nutrients.
Hay fever Allergic reaction localized to the upper respiratory tract and characterized by nasal discharge, sneezing, itchy throat and eyes, and excessive tear production.
Healthcare associated disease (nosocomial) A disease acquired in a health care facility.
Healthcare associated infection (HAI) (nosocomial) An infection acquired in a health care facility.
Heat fixation In microscopy, a technique that uses the heat from a flame to attach a smear to a slide.
Heavy-metal ions The ions of high-molecular-weight metals, such as arsenic, that are used as antimicrobial agents because they denature proteins. They have largely been replaced because they are also toxic to human cells.
Helminths Multicellular eukaryotic worms, some of which are parasitic.
Helper T cell *(Th cell, CD4 cell)* In cell-mediated immune response, a type of cell characterized by CD4 cell-surface glycoprotein; regulates the activity of B cells and cytotoxic T cells.
Hemagglutinin (HA) Component of glycoprotein spikes in the lipid envelope of influenzaviruses that help them attach to pulmonary epithelial cells.
Hemolytic disease of the newborn Disease that results when antibodies made by an Rh-negative woman cross the placenta and destroy the red blood cells of an Rh-positive fetus.
Hemorrhagic fever Viral syndrome characterized by fever, bleeding in the skin and mucous membranes, low blood pressure, and shock.
HEPA (high-efficiency particulate air) filter Filters built into biological safety cabinets; they prevent exposure to microbes by maintaining a barrier of moving filtered air across the cabinet's openings.
Hepatitis Inflammation of the liver.
Hepatitis A Inflammation of the liver resulting from infection with hepatitis A virus.
Hepatitis B Inflammatory condition of the liver caused by infection with hepatitis B virus and characterized by jaundice.
Hepatitis C Chronic inflammation of the liver that can cause permanent liver damage or hepatic cancer; caused by infection with hepatitis C virus.
Hepatitis D Inflammation of liver that can cause severe liver damage or hepatic cancer; caused by infection with hepatitis D virus.
Hepatitis E *(enteric hepatitis)* Inflammation of the liver, which is fatal in 20% of infected pregnant women; caused by hepatitis E virus.
Herd immunity Protection against illness provided to a population when a pathogen cannot spread because the majority of the group are resistant to the pathogen.
Herpangina Lesions of the mouth and pharynx caused by coxsackie A virus; resemble those of herpesvirus.
Herpes Painful, itchy skin lesions caused by herpesviruses.
Herpes zoster *(shingles)* Extremely painful skin rash caused by reactivation of latent varicella-zoster virus.
Heterocyst Thick-walled nonphotosynthetic cell of cyanobacteria; reduces nitrogen.
Hfr (high frequency of recombination) cell Cell containing an F plasmid that is integrated into the prokaryotic chromosome. Hfr cells form pili and transfer cellular genes more frequently than normal F^+ cells.
Highly active antiretroviral therapy (HAART) A cocktail of antiviral drugs including nucleotide analogs, integrase inhibitors, protease inhibitors, and reverse transcriptase inhibitors.
Histamine Inflammatory chemical released from damaged cells that causes vasodilation of capillaries.
Histone Globular protein found in eukaryotic and archaeal chromosomes.
Histoplasmosis Pulmonary, cutaneous, ocular, or systemic disease found in the Ohio River valley and caused by infection with *Histoplasma capsulatum*.
Holoenzyme The combination of an apoenzyme and its cofactors.
Horizontal (lateral) gene transfer Process in which a donor cell contributes part of its genome to a recipient cell, which may be a different species or genus from the donor.
Host In symbiosis, member of a parasitic relationship that supports the parasite.
Human herpesviruses Group of viruses of humans that cause skin lesions, which are often creeping; diseases include herpes, chickenpox, mononucleosis, and roseola.
Human immunodeficiency viruses Retroviruses that destroy the immune system.
Human T-lymphotropic viruses Group of oncogenic retroviruses associated with cancer of lymphocytes.

Humoral immune response (antibody immune response) The immune response centered around B lymphocytes and immunoglobulins.
Hybridomas Tumor cells created by fusing antibody-secreting plasma cells with cancerous plasma cells called *myelomas*.
Hydatid Fluid-filled.
Hydatid disease Potentially fatal disease caused by infection with the canine tapeworm *Echinococcus granulosus* and characterized by the presence of fluid-filled cysts in the liver or other tissues.
Hydrogen bond The electrical attraction between a partially charged hydrogen atom and a full or partial negative charge on a different region of the same molecule or another molecule. Hydrogen bonds confer unique properties to water molecules.
Hydrolysis A decomposition reaction in which a covalent bond is broken and the ionic components of water are added to the products.
Hydrophilic Attracted to water.
Hydrophobia Literally, a fear of water; symptom caused by painful swallowing characteristic of rabies infection.
Hydrophobic Insoluble in water.
Hydrothermal vent Vent in marine abyssal zone that spews superheated, nutrient-rich water.
Hydroxyl radical Most reactive of the toxic forms of oxygen.
Hypersensitivity Any immune response against a foreign antigen that is exaggerated beyond the norm.
Hypersensitivity pneumonitis A form of pneumonia.
Hyperthermophile Microorganism requiring temperatures above 80°C.
Hypertonic Characteristic of a solution having a higher concentration of solutes than another.
Hyphae Long, branched, tubular filaments in the bodies of molds.
Hypotonic Characteristic of a solution having a lower concentration of solutes than another.

Iatrogenic infections A subset of nosocomial infections that are the direct result of a medical procedure or treatment, such as surgery.
Illness In the infectious disease process, the most severe stage, in which signs and symptoms are most evident.
Immune complexes Antigen-antibody complexes.
Immune thrombocytopenic purpura Disease resulting when drugs bound to platelets bind antibodies and complement, causing the platelets to lyse.
Immunization Administration of an antigenic inoculum to stimulate an adaptive immune response and immunological memory.
Immunoblot *(western blot)* Variation of an ELISA test that can detect the presence of proteins, such as antibodies against multiple antigens.
Immunochromatographic assay Immune test in which antigen molecules form visible immune complexes with antibodies labeled with a colored substance.
Immunodiffusion An immune test in which antibodies and antigens diffuse from separate wells in agar to form a line of precipitate.
Immunofiltration assay Rapid modification of ELISA test using membrane filters rather than plates.
Immunoglobulin (Ig) *(antibody)* Proteinaceous antigen-binding molecule secreted by plasma cells.
Immunoglobulin A (IgA) The antibody class most commonly associated with various body secretions, including tears and milk. IgA pairs with a secretory component to form *secretory IgA*.
Immunoglobulin D (IgD) A membrane-bound antibody molecule found in some animals as a B cell receptor.
Immunoglobulin E (IgE) Signal antibody molecule that triggers the inflammatory response, particularly in allergic reactions and infections by parasitic worms.
Immunoglobulin G (IgG) The predominant antibody class found in the bloodstream and the primary defender against invading bacteria.
Immunoglobulin M (IgM) The second most common antibody class and the predominant antibody produced first during a primary humoral immune response.
Immunological synapse Interface between cells of the immune system that involves cell-to-cell signaling.
Immunology Study of the body's specific defenses against pathogens.
Immunophilins Immunosuppressive drugs, such as cyclosporine, that inhibit T cell function.
Immunotherapy Administration of antibodies (passive immunization) or dilute antigen so as to provide immunological protection against antigens.
Impetigo Presence of red, pus-filled vesicles on the face and limbs of children; caused by infection with *Staphylococcus aureus* or *Streptococcus pyogenes*.
Inactivated polio vaccine (IPV) Inoculum developed by Jonas Salk in 1955 for vaccination against poliovirus.
Inactivated vaccine Inoculum containing either whole agents or subunits and often adjuvants.
Incidence In epidemiology, the number of new cases of a disease in a given area or population during a given period of time.
Inclusion Deposited substance such as a lipid, gas vesicle, or magnetite stored within the cytosol of a cell.
Inclusion bodies In the life cycle of chlamydias, eukaryotic phagosomes full of chlamydial reticulate bodies.
Incubation period Stage in infectious disease process between infection and occurrence of the first symptoms or signs of disease. In a laboratory culture, the period between adding a sample to a plate and the development of colonies.
Index case In epidemiology, the first instance of the disease in a given area or population.
Indicator organisms Fecal microbe found in the environment that reveals potential contamination by feces.
Indirect contact transmission Spread of pathogens from one host to another via inanimate objects called *fomites*.
Indirect fluorescent antibody test Immune test allowing observation through a fluorescence microscope of the presence of antigen in a tissue sample flooded with labeled antibody.
Indirect selection *(negative selection)* Process by which auxotrophic mutants are isolated and cultured.
Induced-fit model Description of way in which an enzyme changes its shape slightly after binding to its substrate so as to bind it more tightly.
Inducible operon Type of operon that is not normally transcribed and must be activated by inducers.
Induction In virology, excision of a prophage from the host chromosome, at which point the prophage reenters the lytic phase.
Industrial microbiology Branch of microbiology in which microbes are manipulated to manufacture useful products.
Infection Successful invasion of the body by a pathogenic microorganism.
Infection control Branch of microbiology studying the prevention and control of infectious disease.
Infectious mononucleosis *(mono)* Disease characterized by sore throat, fever, fatigue, and enlargement of the spleen and liver; caused by infection with Epstein-Barr virus.
Influenza (flu) Infectious disease caused by two species of orthomyxoviruses and characterized by fever, malaise, headache, and myalgia; certain strains can be fatal.
Innate immunity Resistance to pathogens conferred by barriers, chemicals, cells, and processes that remain unchanged upon subsequent infections with the same pathogens.
Inoculum Sample of microorganisms.
Inorganic chemical Molecule lacking carbon.
Insect Arthropod with three distinct body divisions: head, thorax, abdomen; vectors for some helminthic, protozoan, bacterial, and viral pathogens.
Insertion Type of mutation in which a base pair is inserted into a genetic sequence.
Insertion sequence (IS) A simple transposon consisting of no more than two inverted repeats and a gene that encodes the enzyme transposase.
Integrase Enzyme carried by the virions of HIV that allows integration into a human chromosome.
Interferons (IFNs) Protein molecules that inhibit the spread of viral infections.
Interleukins (ILs) Immune system cytokines that signal among leukocytes.
Intermediate host In the life cycle of parasites, host in which immature forms of the parasite are present and undergoing various stages of maturation.
Intoxication (bacterial) Food poisoning caused by bacterial toxin.
Intron Noncoding sequence of mRNA that is removed to make functional mRNA.
In-use test Method of evaluating the effectiveness of a disinfectant or antiseptic which tests efficacy under specific, real-life conditions.
Inverted repeat (IR) Palindromic sequence found at each end of a transposon.
Ion An atom or group of atoms that has either a full negative charge or a full positive charge.
Ionic bond A type of bond formed from the attraction of opposite electrical charges. Electrons are not shared.
Ionizing radiation Form of radiation with wavelengths shorter than 1 nm that are energetic enough to create ions by ejecting electrons from atoms.
Ischemia Local anemia due to interruption of blood supply by mechanical blockage.
Isograft Type of graft in which tissues are moved between genetically identical individuals (identical twins).

Isoniazid (INH) Antimicrobial drug that disrupts formation of arabinogalactan-mycolic acid by mycobacteria.
Isotonic Characteristic of a solution having the same concentration of solutes and water as another.
Isotopes Atoms of a given element that differ only in the number of neutrons they contain.

Jaundice Yellowing of skin and eyes due to accumulation of bilirubin in the blood.

Kelsey-Sykes capacity test Standard assessment approved by the European Union to determine the ability of a given chemical to inhibit microbial growth.
Keratitis Inflammation of the cornea.
Kinetoplastid Euglenozoan protozoan with a single large mitochondrion that contains an apical region of mitochondrial DNA called a *kinetoplast*.
Kingdom Taxonomical grouping of similar phyla of organisms.
Kinins Powerful inflammatory chemicals released by mast cells.
Kissing bug Blood-eating insect of family Reduviidae that seemingly prefers oral blood vessels.
Koch's postulates A series of steps, elucidated by Robert Koch, that must be taken to prove the cause of any infectious disease.
Koplik's spots Mouth lesions characteristic of measles.
Korarchaeota Phylum of archaea; known only from environmental RNA samples.
Krebs cycle Series of eight enzymatically catalyzed reactions that transfer stored energy from acetyl-CoA to coenzymes NAD^+ and FAD.

Lag phase Phase in a growth curve in which the organisms are adjusting to their environment.
Lagging strand Daughter strand of DNA synthesized in short segments that are later joined. Synthesis of the lagging strand always moves away from the replication fork and lags behind synthesis of the leading strand.
Laryngitis Inflammation of the larynx.
Latency In virology, process by which an animal virus, sometimes not incorporated into the chromosomes of the cell, remains inactive in the cell, possibly for years.
Latent disease Any disease in which a pathogen remains inactive for a long period of time before becoming active.
Latent virus *(provirus)* An animal virus that remains inactive in a host cell.
Lateral (horizontal) gene transfer Process in which a donor cell contributes part of its genome to a recipient cell, which may be a different species or genus from the donor.
Leading strand Daughter strand of DNA synthesized continuously toward the replication fork as a single long chain of nucleotides.
Legionnaires' disease *(legionellosis)* Severe pneumonia caused by infection with a *Legionella* species, usually *L. pneumophila*.
Leishmaniasis Any of three clinical syndromes caused by a bite from a sand fly carrying *Leishmania* and ranging from painless skin ulcers to disfiguring lesions to visceral leishmaniasis, which is systemic and fatal in 95% of untreated cases.
Lepromin test Assay utilizing antigens of *Mycobacterium leprae* used in diagnosis of leprosy.
Leprosy *(Hansen's disease)* Disease caused by infection with *Mycobacterium leprae* that produces either a nonprogressive tuberculoid form or a progressive lepromatous form that destroys tissues, including facial features, digits, and other structures.
Leptospirosis Zoonotic disease contracted by humans upon exposure to infected animals; characterized by pain, headache, and liver and kidney disease; caused by infection with *Leptospira interrogans*.
Leukocyte White blood cell.
Leukopenia Decrease in the number of white blood cells in the blood.
Leukotrienes Inflammatory chemicals released from damaged cells that increase vascular permeability.
Lichen Organism composed of a fungus living in partnership with photosynthetic microbes, either green algae or cyanobacteria.
Light-dependent reaction Reaction of photosynthesis requiring light.
Light-independent reaction Reaction of photosynthesis not requiring light and synthesizing glucose from carbon dioxide and water.
Light repair Mechanism by which prokaryotic DNA photolyase breaks the bonds between adjoining pyrimidine nucleotides, restoring the original DNA sequence.
Limnetic zone Sunlit, upper layer of freshwater or marine water away from the shore.
Lincosamides Antimicrobial drugs that bind to the 50S subunit of bacterial ribosomes, preventing ribosomal movement.
Lipid Any of a diverse group of organic macromolecules not composed of monomers and insoluble in water.
Lipid A The lipid component of lipopolysaccharide, which is released from dead Gram-negative bacterial cells and can trigger shock and other symptoms in human hosts.
Lipopolysaccharide (LPS) Molecule composed of lipid A and polysaccharide found in the external membrane of Gram-negative cell walls.
Listeriosis Disease caused by *Listeria monocytogenes* and usually manifesting as meningitis and bacteremia.
Lithotroph Microorganism that acquires electrons from inorganic sources.
Littoral zone Shoreline zone of freshwater or marine water.
Log phase Phase in a growth curve in which the population is most actively growing.
Logarithmic (exponential) growth Increase in size of a microbial population in which the number of cells doubles in a fixed interval of time.
Louse (pl. *lice*) Sucking or biting parasitic insects that vector some bacterial pathogens.
Louse-borne relapsing fever Disease caused by *Borrelia recurrentis* transmitted between humans by the body louse *Pediculus humanus*.
Lumbar puncture (spinal tap) Collection of cerebrospinal fluid from the lower vertebral column for diagnostic purposes.
Lyme disease Disease carried by ticks infected with *Borrelia burgdorferi* and characterized by a "bull's-eye" rash, neurologic and cardiac dysfunction, and severe arthritis.
Lymph Fluid found in lymphatic vessels that is similar in composition to blood serum and intercellular fluid.
Lymph nodes Organs that monitor the composition of lymph.
Lymphangitis Condition in which inflamed lymphatic vessels become visible as red streaks under the skin.
Lymphatic system Body system composed of lymphatic vessels and lymphoid tissues and organs.
Lymphatic vessels Tubes that conduct lymph.
Lymphocyte Type of small agranulocyte which originates in the red bone marrow and has nuclei that nearly fill the cell.
Lymphocytic choriomeningitis (LCM) Zoonosis caused by an arenavirus; characterized by flulike symptoms and rarely by meningitis.
Lymphogranuloma venereum Sexually transmitted disease caused by infection with *Chlamydia trachomatis* and leading in some cases to proctitis or, in women, pelvic inflammatory disease.
Lyophilization Removal of water from a frozen culture or other substance by means of vacuum pressure. Used for the long-term preservation of cells and foods.
Lysogenic conversion Change in phenotype due to insertion of a lysogenic bacteriophage into a bacterial chromosome.
Lysogenic phage Bacteriophage that does not immediately kill its host cell.
Lysogenic replication cycle (lysogeny) Process of viral replication in which a bacteriophage enters a bacterial cell, inserts into the DNA of the host, and remains inactive. The phage is then replicated every time the host cell replicates its chromosome. Later, the phage may leave the chromosome.
Lysosome Vesicle in animal cells that contains digestive enzymes.
Lysozyme Antibacterial protein secreted in sweat.
Lytic replication cycle Process of viral replication consisting of five stages ending with lysis of and release of new virions from the host cell.

Macrolide Antimicrobial agent that inhibits protein synthesis by inhibiting the ribosomal 50S subunits.
Macrophage Mature form of monocyte, which is a phagocyte of bacteria, fungi, spores, and dust, as well as dead cells.
Macule Any flat, reddened skin lesion; characteristic of early infection with a poxvirus.
Magnification The apparent increase in size of an object viewed via microscopy.
Major histocompatibility complex (MHC) A cluster of genes, located on each copy of chromosome 6 in humans, that codes for membrane-bound glycoproteins called major histocompatibility antigens.
Malaise Feeling of general discomfort.
Malaria A mild to potentially fatal disease caused by a bite from an *Anopheles* mosquito carrying any of four species of *Plasmodium*; characterized by fever, chills, hemorrhage, and potential destruction of brain tissue.
Malignant tumor Mass of neoplastic cells that can invade neighboring tissues and may metastasize to cause tumors in distant organs or tissues.

Marburg virus Filamentous virus causing a type of hemorrhagic fever and fatal in 25% of cases.
Margination Process by which leukocytes stick to the walls of blood vessels at the site of infection.
Mast cells Specialized cells located in connective tissue that release histamine when they are exposed to complement.
Matter Anything that takes up space and has mass.
MDR TB (multi-drug-resistant TB) Tuberculosis caused by *Mycobacterium* resistant to at least isoniazid and rifampin.
Measles *(rubeola)* Contagious disease characterized by fever, sore throat, headache, dry cough, conjunctivitis, and lesions called Koplik's spots; caused by infection with *Morbillivirus*.
Mechanical vector Housefly, cockroach, or other animal that passively carries pathogens to new hosts on its feet or other body parts and is not infected by the pathogens it carries.
Medical technologist An expert in health care-related microbiological laboratory procedures.
Medium A collection of nutrients used for cultivating microorganisms.
Meiosis Nuclear division of diploid eukaryotic cells resulting in four haploid nuclei.
Membrane attack complexes (MACs) The end products of the complement cascade, which form circular holes in a pathogen's membrane.
Membrane filters Thin circles of nitrocellulose or plastic containing specific pore sizes, some small enough to trap viruses.
Membrane filtration Direct method of estimating population size in which a large sample is poured through a filter small enough to trap cells.
Membrane raft In a eukaryotic membrane, a distinct assemblage of lipids and proteins that remains together as a functional group.
Memory B cell B lymphocyte that migrates to lymphoid tissues to await a subsequent encounter with antigen previously encountered.
Memory response The rapid and enhanced immune response to a subsequent encounter with a familiar antigen.
Memory T cell Type of T cell that persists in lymphoid tissues for months or years awaiting subsequent contact with an antigenic determinant matching its TCR, at which point it produces cytotoxic T cells.
Meningitis Inflammation of the meninges, which can be caused by bacteria, viruses, fungi, or protozoa.
Meningoencephalitis Inflammation of the brain and of its meninges.
Mesophile Microorganism requiring temperatures ranging from 20°C to about 40°C.
Messenger RNA (mRNA) Form of ribonucleic acid that carries genetic information from DNA to a ribosome.
Metabolism The sum of all chemical reactions, both anabolic and catabolic, within an organism.
Metachromatic granules Inclusions of *Corynebacteria* that store phosphate and stain differently from the rest of the cytoplasm.
Metaphase Second stage of mitosis, during which chromosomes line up and attach to microtubules of the spindle. Also used for the comparable stage of meiosis.
Metastasis The spreading of malignant cancer cells to nonadjacent organs and tissues, where they produce new tumors.
Methane oxidizer Any Gram-negative bacterium that utilizes methane both as a carbon and as an energy source.
Methanogen Obligate anaerobe that produces methane gas.
Methicillin-resistant Staphylococcus aureus (MRSA) Strain of *S. aureus* that is resistant to many common antimicrobial drugs and has emerged as a major nosocomial problem.
Methylation Process in which a cell adds a methyl group to one or two bases that are part of specific nucleotide sequences.
Microaerophile Microorganism that requires low levels of oxygen.
Microbe An organism or virus too small to be seen without a microscope.
Microbial antagonism (microbial competition) Normal condition in which established microbiota use up available nutrients and space, reducing the ability of arriving pathogens to colonize.
Microbial death Permanent loss of reproductive capacity of a microorganism.
Microbial death rate A measurement of the efficacy of an antimicrobial agent.
Microbial ecology The study of the interactions of microorganisms among themselves and their environment.
Microbiota The group of microbes that normally inhabit the surfaces of the body without causing disease.
Microglia Fixed macrophages of the nervous system.
Micrograph A photograph of a microscopic image.
Microorganism An organism too small to be seen without a microscope.
MicroRNA (miRNA) Short (about 21-nucleotide) RNA molecule that binds to complementary segment of messenger RNA (mRNA), preventing translation.
Microscopy The use of light or electrons to magnify objects.
Microsporidia Unicellular, intracellular, parasitic fungi previously classified as protozoa.
Minimum bactericidal concentration (MBC) test An extension of the MIC test in which samples taken from clear MIC tubes are transferred to plates containing a drug-free growth medium and monitored for bacterial replication.
Minimum inhibitory concentration (MIC) The smallest amount of a drug that will inhibit a pathogen.
Mismatch repair Mechanism by which enzymes scan newly synthesized, nonmethylated DNA for mismatched bases, remove them, and replace them.
Missense mutation A substitution in a nucleotide sequence resulting in a codon that specifies a different amino acid: What is transcribed makes sense but not the right sense.
Mite Minute arachnid, which vectors *Orientia*, the agent of scrub typhus.
Mitochondria Spherical to elongated structures found in most eukaryotic cells that produce most of the ATP in the cell.
Mitosis Nuclear division of a eukaryotic cell resulting in two nuclei with the same ploidy as the original.
Mold A typically multicellular fungus that grows as long filaments called *hyphae* and reproduces by means of spores.
Molecular biology Branch of biology combining aspects of biochemistry, cell biology, and genetics to explain cell function at the molecular level, particularly via the use of genome sequencing.
Molecular mimicry Process in which microorganisms with epitopes similar to self-antigens trigger autoimmune tissue damage.
Molecule Two or more atoms held together by chemical bonds.
Molluscum contagiosum Skin disease caused by *Molluscipoxvirus;* characterized by smooth, waxy papules.
Monoclonal antibodies Identical antibodies secreted by a cell line originating from a single plasma cell.
Monocyte Type of agranulocyte that has slightly lobed nuclei.
Monoecious One individual contains both male and female organs.
Monomer A subunit of a macromolecule, such as a protein.
Monosaccharide *(simple sugar)* A monomer of carbohydrate, such as a molecule of glucose.
Morbidity Any change from a state of health.
Mordant In microscopy, a substance that binds to a dye and makes it less soluble.
Mosquito Type of fly with bloodsucking females; vector for many pathogens.
Most probable number (MPN) method Statistical estimation of the size of a microbial population based upon the dilution of a sample required to eliminate microbial growth.
Multi-drug-resistant (MDR) tuberculosis Tuberculosis caused by *Mycobacterium* resistant to at least isoniazid and rifampin.
Multiple drug resistance Lack of sensitivity to three or more antimicrobials by so-called superbugs.
Multiple sclerosis (MS) Autoimmune disease in which cytotoxic T cells attack and destroy the myelin sheath that insulates neurons.
Mumps Disease caused by infection with the mumps virus and characterized by fever, parotitis, pain in swallowing, and, in some cases, meningitis or deafness.
Mutagen Physical or chemical agent that introduces a mutation.
Mutant A cell with an unrepaired genetic mutation, or any of its descendants.
Mutation In genetics, a permanent change in the nucleotide base sequence of a genome.
Mutualism Symbiotic relationship in which both members benefit from their interaction.
Myalgia Muscle pain.
Mycelium Tangled mass of hyphae.
Mycetismus Mushroom poisoning.
Mycetoma Destructive, tumorlike infection of the skin, fascia, and/or bones of the hands or feet caused by mycelial fungi of several genera in the division Ascomycota.
Mycolic acid Long carbon-chain waxy lipid found in the walls of cells in the genus *Mycobacterium* that makes them resistant to desiccation and staining with water-based dyes.
Mycology The scientific study of fungi.

Mycoplasmas Class of low G + C bacteria that lack cytochromes, enzymes of the Krebs cycle, and cell walls and are pleomorphic.
Mycosis Fungal disease.
Mycotoxicosis Poisoning caused by eating food contaminated with fungal toxins.
Mycotoxins Secondary metabolites produced by fungi and toxic to humans.
Myxobacteria Gram-negative, aerobic, soil-dwelling bacteria with a unique life cycle including a stage of differentiation into fruiting bodies containing resistant myxospores.

Nanoarchaeum Small archaeon genus that possibly represents a fourth phylum of archaea; known only from environmental RNA samples.
Narrow-spectrum drug Antimicrobial that works against only a few kinds of pathogens.
Natural killer (NK) lymphocyte Type of defensive leukocyte of innate immunity that secretes toxins onto the surfaces of virally infected cells and neoplasms.
Naturally acquired active immunity Type of immunity that occurs when the body responds to exposure to antigens by mounting specific immune responses.
Naturally acquired passive immunity Type of immunity that occurs when a fetus, newborn, or child receives antibodies across the placenta or within breast milk.
Necrosis Death of a tissue or organ.
Necrotizing fasciitis Potentially fatal condition marked by toxemia, organ failure, and destruction of muscle and fat tissue following infection with group A *Streptococcus*.
Negative feedback *(feedback inhibition)* Method of controlling the action of enzymes in which the end-product of a series of reactions inhibits an enzyme in an earlier part of the pathway.
Negative selection *(indirect selection)* Process by which auxotrophic mutants are isolated and cultured.
Negative stain *(capsule stain)* In microscopy, a staining technique used primarily to reveal bacterial capsules and involving application of an acidic dye that leaves the specimen colorless and the background stained.
Negative-strand RNA (−RNA) Viral single-stranded RNA transcribed from the +ssRNA genome by viral RNA polymerase.
Negri bodies Aggregates of virions in the brains of rabies patients.
Nematodes Group of round, unsegmented helminths with pointed ends that have a complete digestive tract.
Neoplasia Uncontrolled cell division in a multicellular animal.
Nephelometry Automated method that measures the cloudiness of a solution by quantifying the amount of light it scatters.
Neuraminidase (NA) Component of glycoprotein spikes in the lipid envelope of influenzaviruses that provides access to cell surfaces by hydrolyzing mucus in the lungs.
Neutralization Antibody function in which the action of a toxin or attachment of a pathogen is blocked.
Neutralization test Immune test that measures the ability of antibodies to neutralize the biological activity of pathogens and toxins.
Neutron An uncharged subatomic particle.
Neutrophil Type of granulocyte that stains lilac with a mixture of acidic and basic dyes.
Neutrophile Microorganism requiring neutral pH.
Nicotinamide adenine dinucleotide (NAD^+) Important vitamin-derived electron carrier molecule.
Nicotinamide adenine dinucleotide phosphate ($NADP^+$) Important vitamin-derived electron carrier molecule.
Nitrification The process by which bacteria convert reduced nitrogen compounds such as ammonia into nitrate, which is more available to plants.
Nitrifying bacteria Chemoautotrophic bacteria that derive electrons from the oxidation of nitrogenous compounds.
Nitrogen cycle Biogeochemical cycle involving nitrogen fixation, ammonification, nitrification, denitrification, and anammox reactions.
Nitrogen fixation The conversion of atmospheric nitrogen to ammonia.
NOD protein In innate immunity, intracellular receptor for microbial component.
Noncommunicable disease An infectious disease that arises from outside of hosts or from normal microbiota.
Noncompetitive inhibitor Inhibitory substance that blocks enzyme activity by binding to an allosteric site on the enzyme other than the active site.
Noncyclic photophosphorylation The production of ATP by noncyclic electron flow.
Nonionizing radiation Electromagnetic radiation with a wavelength greater than 1 nm.
Nonliving reservoir of infection Soil, water, food, or inanimate object that is a continuous source of infection.
Nonpolar covalent bond Type of chemical bond in which there is equal sharing of electrons between atoms with similar electronegativities.
Nonsense mutation A substitution in a nucleotide sequence that causes an amino acid codon to be replaced by a stop codon.
Normal microbiota Microorganisms that colonize the surfaces of the human body without normally causing disease. They may be resident or transient.
Noroviruses Group of caliciviruses that cause diarrhea.
Nosocomial disease *(healthcare associated disease)* A disease acquired in a health care facility.
Nosocomial infection *(healthcare associated infection)* An infection acquired in a health care facility.
Nuclear envelope Double membrane composed of phospholipid bilayers surrounding a cell nucleus.
Nuclear pores Spaces in the nuclear envelope that function to control the transport of substances through it.
Nucleoid Region of the prokaryotic cytosol containing the cell's chromosome(s).
Nucleolus Specialized region in a cell nucleus where RNA is synthesized.
Nucleoplasm The semiliquid matrix of a cell nucleus.
Nucleoside Component of a nucleotide consisting of a nitrogenous base and a five-carbon sugar.
Nucleoside analog Chemical with a structure similar to a natural nucleoside.
Nucleosome Bead of DNA bound to histone in a eukaryotic chromosome.
Nucleotide Monomer of a nucleic acid, which is composed of a nucleoside and a phosphate.
Nucleotide analog Compound structurally similar to a normal nucleotide that can be incorporated into DNA; may result in mismatched base pairing.
Nucleus Spherical to ovoid membranous organelle containing a eukaryotic cell's primary genetic material.
Numerical aperture Measure of the ability of a lens to gather light.
Nutrient Any chemical, such as carbon, hydrogen, and so on, required for growth of microbial populations.

Objective lens In microscopy, the lens immediately above the object being magnified.
Obligate aerobe Microorganism that requires oxygen as the final electron acceptor of the electron transport chain.
Obligate anaerobe Microorganism that cannot tolerate oxygen and uses a final electron acceptor other than oxygen.
Obligate halophile Microorganism requiring high osmotic pressure.
Occult septicemia The condition of an unidentified bacterial pathogen being present in the blood and causing signs of illness.
Ocular herpes *(ophthalmic herpes)* Disorder characterized by conjunctivitis, a gritty feeling in the eye, and pain and characteristic of latent *human herpesvirus 1*.
Ocular lens In microscopy, the lens closest to the eyes. May be single *(monocular)* or paired *(binocular)*.
Operator Regulatory element in an operon where repressor protein binds to stop transcription.
Operon A series of genes, a promoter, and often an operator sequence controlled by one regulatory gene. The operon model explains gene regulation in prokaryotes.
Opportunistic pathogens Microorganisms that cause disease when the immune system is suppressed, when microbial antagonism is reduced, or when introduced into an abnormal area of the body.
Opsonin Antimicrobial protein that enhances phagocytosis.
Opsonization The coating of pathogens by proteins called *opsonins*, making them more vulnerable to phagocytes.
Optimum growth temperature Temperature at which a microorganism's metabolic activities produce the highest growth rate.
Oral herpes Painful, itchy skin lesions around the mouth and lips, caused by herpesvirus 1 or 2.
Oral polio vaccine (OPV) Inoculum developed by Albert Sabin in 1961 for vaccination against poliovirus.
Orchitis Inflammation of a testis.
Order Taxonomical grouping of similar families of organisms.
Organelle Cellular structure that acts as a tiny organ to carry out one or more cell functions.
Organic compounds Molecules that contain both carbon and hydrogen atoms.
Organotroph Microorganism that acquires electrons from organic sources.

Ornithosis *(parrot fever)* A respiratory disease of birds that can be transmitted to humans and is caused by infection with *Chlamydophila psittaci*.
Orphan virus A virus that has not been specifically linked to any particular disease.
Osmosis The diffusion of water molecules across a semipermeable membrane.
Osmotic pressure The pressure exerted across a selectively permeable membrane by the solutes in a solution on one side of the membrane. The osmotic pressure exerted by high-salt or high-sugar solutions can be used to inhibit microbial growth in certain foods.
Osteomyelitis Inflammation of the bone marrow and surrounding bone; often caused by infection with *Staphylococcus*.
Otitis media Inflammation of the middle ear, often caused by *Streptococcus pneumoniae*.
Oxazolidinone Antibacterial drug that inhibits initiation of polypeptide synthesis in Gram-positive bacteria.
Oxidase test Chemical test for presence of cytochrome oxidase in a cell.
Oxidation lagoons Successive wastewaster treatment areas (lagoons) used by farmers and ranchers to treat animal wastes and from which water is released into natural water systems.
Oxidation-reduction reaction *(redox reaction)* Any metabolic reaction involving the transfer of electrons from an electron donor to an electron acceptor. Reactions in which electrons are accepted are called *reduction* reactions, whereas reactions in which electrons are donated are *oxidation* reactions.
Oxidative phosphorylation The use of energy from redox reactions to attach inorganic phosphate to ADP.
Oxidizing agent Antimicrobial agent that releases oxygen radicals.

P site In a ribosome, a binding site that holds a tRNA and the growing polypeptide.
Palisade In cell morphology, a folded arrangement of bacilli.
Pandemic In epidemiology, the occurrence of an epidemic on more than one continent simultaneously.
Papilloma *(wart)* Benign growth of the epithelium of the skin or mucous membranes.
Papule Any raised, reddened skin lesion that progresses from a macule; characteristic of infection with a poxvirus.
Parabasalid Group of single-celled, animal-like microorganisms that contain a Golgi-like parabasal body.
Paracoccidioidomycosis Pulmonary disease found from southern Mexico to South America caused by infection with *Paracoccidioides brasiliensis*.
Parainfluenzaviruses Group of enveloped, negative ssRNA viruses that cause respiratory disease, particularly in children.
Parasite A microbe that derives benefit from its host while harming it or even killing it.
Parasitism Symbiotic relationship in which one organism derives benefit while harming or even killing its host.
Parasitology The study of parasites.
Parenteral route A means by which pathogenic microorganisms can be deposited directly into deep tissues of the body, as in puncture wounds and hypodermic injections.
Paroxysmal phase In pertussis, second phase lasting 2 to 4 weeks and characterized by exhausting coughing spells.
Parvoviruses Group of extremely small, pathogenic ssDNA viruses.
Passive immunotherapy *(passive immunization)* Delivery of preformed antibodies against pathogens to patients.
Pasteurellaceae Family of gammaproteobacteria, two genera of which—*Pasteurella* and *Haemophilus*—are pathogenic.
Pasteurization The use of heat to kill pathogens and reduce the number of spoilage microorganisms in food and beverages.
Pathogen A microorganism capable of causing disease.
Pathogen-associated molecular patterns (PAMPs) Molecules that are shared by a variety of microbes, are absent in humans, and trigger immune responses.
Pathogenicity A microorganism's ability to cause disease.
Pelvic inflammatory disease (PID) Infection of the uterus and uterine tubes; may be caused by infection with any of several bacteria.
Pentose phosphate pathway Enzymatic formation of phosphorylated pentose sugars from glucose 6-phosphate.
Peptic ulcer Erosion of the mucous membrane of the stomach or duodenum, usually caused by infection with *Helicobacter pylori*.
Peptide bond A covalent bond between amino acids in proteins.
Peptidoglycan Large, interconnected polysaccharide composed of chains of two alternating sugars and crossbridges of amino acids. Main component of bacterial cell walls.
Perforin Protein molecule in the cytoplasm of cytotoxic T cells which forms channels (perforations) in an infected cell's membrane.
Periodontal disease Inflammation and infection of the tissues surrounding and supporting the teeth.
Periplasmic space In Gram-negative cells, the space between the cell membrane and the outer membrane containing peptidoglycan and periplasm.
Peritrichous Term used to describe a cell having flagella covering the cell surface.
Peroxide anion Toxic form of oxygen which is detoxified by catalase or peroxidase.
Peroxisome Vesicle found in all eukaryotic cells that degrades poisonous metabolic wastes.
Pertussis *(whooping cough)* Pediatric disease characterized by development of copious mucus, loss of tracheal cilia, and deep "whooping" cough; caused by infection with *Bordetella pertussis*.
Petechiae Subcutaneous hemorrhages.
Petri plate Dish filled with solid medium used in culturing microorganisms.
pH scale A logarithmic scale used for measuring the concentration of hydrogen ions in a solution.
Phaeohyphomycosis Cutaneous and subcutaneous disease characterized by lesions that can spread internally; caused by traumatic introduction of ascomycetes into the skin.
Phaeophyta Brown-pigmented division of algae having cell walls composed of cellulose and alginic acid, a thickening agent.
Phage *(bacteriophage)* Virus that infects and usually destroys bacterial cells.
Phage typing Method of classifying microorganisms in which unknown bacteria are identified by observing plaques.
Phagocytes Cells, often leukocytes, that are capable of phagocytosis.
Phagocytosis Type of endocytosis in which solids are moved into the cell.
Phagolysosome Digestive vesicle formed by the fusing of a lysosome with a phagosome.
Phagosome A sac formed by a phagocyte's pseudopods; an intracellular food vesicle.
Pharyngitis *(strep throat)* Inflammation of the throat, often caused by infection with group A *Streptococcus*.
Phase microscope Type of microscope used to examine living microorganisms or fragile specimens.
Phase-contrast microscope Type of phase microscope that produces sharply defined images in which fine structures can be seen in living cells.
Phenol coefficient Method of evaluating the effectiveness of a disinfectant or antiseptic that compares the agent's efficacy to that of phenol.
Phenolic Compound derived from phenol molecules that have been chemically modified to denature proteins and disrupt cell membranes in a wide variety of pathogens.
Phenotype The physical features and functional traits of an organism expressed by genes in the genotype.
Phospholipid Phosphate-containing lipid made up of molecules with two fatty acid chains.
Phospholipid bilayer Two-layered structure of a cell's membranes.
Phosphorus cycle Biogeochemical cycle in which phosphorus is cycled between oxidation states.
Photoautotroph Microorganism that requires light energy and uses carbon dioxide as a carbon source.
Photoheterotroph Microorganism that requires light energy and gains nutrients via catabolism of organic compounds.
Photophosphorylation The use of energy from light to attach inorganic phosphate to ADP.
Photosynthesis Process in which light energy is captured by chlorophylls and transferred to ATP and metabolites.
Photosystem Network of light-absorbing chlorophyll molecules and other pigments held within a protein matrix on thylakoids.
Phototaxis Cell movement that occurs in response to light stimulus.
Phycoerythrin Red accessory pigment of photosynthesis in red algae.
Phycology Study of algae.
Phylum Taxonomical grouping of similar classes of organisms.
Picornaviruses Family of viruses that contain positive single-stranded RNA with naked polyhedral capsids; many are human pathogens.
Piedra Firm, irregular nodules on hair shafts caused by aggregates of fungal hyphae and spores.
Pilus *(conjugation pilus)* A tubule involved in bacterial conjugation.

Pinocytosis Type of endocytosis in which liquids are moved into the cell.
Pinta Childhood disease caused by the spirochete *Treponema carateum;* characterized by hard, pus-filled lesions.
Pinworm Common name of *Enterobius vermicularis*, whose adult female has a tail like a straight pin.
Pityriasis versicolor Condition characterized by depigmented or hyperpigmented patches of scaly skin resulting from infection with *Malassezia furfur*.
Plague Disease caused by *Yersinia pestis*, often manifesting with enlarged lymph nodes (bubonic plague) or with severe pulmonary distress (pneumonic plague).
Plaque In phage typing, the clear region within the bacterial lawn where growth is inhibited by bacteriophages.
Plaque assay Technique for estimating phage numbers in which each plaque corresponds to a single phage in the original bacterium/virus mixture.
Plasma The liquid portion of blood.
Plasma cells B cells that are actively fighting against exogenous antigens and secreting antibodies.
Plasmid A small, circular molecule of DNA that replicates independently of the chromosome. Each carries genes for its own replication and often for one or more nonessential functions such as resistance to antibiotics.
Plasmodial slime mold *(acellular slime mold)* Streaming, coenocytic, colorful filaments of cytoplasm that phagocytize organic debris and bacteria.
Platelet Cell fragments involved in blood clotting.
Platelet activating factor (PAF) Cytokine that is a potent trigger for blood coagulation.
Pleomorphic In cell morphology, term used to describe a variably shaped prokaryotic cell.
Pneumococcal pneumonia Inflammation of the lungs caused by *Streptococcus pneumoniae*—the pneumococcus.
Pneumococcus Common name of *Streptococcus pneumoniae*.
***Pneumocystis* pneumonia (PCP)** Debilitating fungal pneumonia that is a leading cause of death in AIDS patients and is caused by opportunistic infection with *Pneumocystis jirovecii*.
Pneumonia Inflammation of the lungs; typically caused by infection with *Streptococcus pneumoniae*.
Pneumonic plague Fever and severe respiratory distress caused by infection of the lungs with *Yersinia pestis*; fatal if untreated in nearly 100% of cases.
Point mutation A genetic mutation affecting only one or a few base pairs in a genome. Point mutations include substitutions, insertions, and deletions.
Polar In cell morphology, pertaining to either end of a cell, such as polar flagella.
Polar covalent bond Type of bond in which there is unequal sharing of electrons between atoms with opposite electrical charges.
Poliomyelitis *(polio)* Infection of varied degrees of severity from asymptomatic to crippling and caused by infection with poliovirus.
Polluted Containing microorganisms or chemicals in excess of acceptable values.
Polyenes Group of antimicrobial drugs such as amphotericin B that disrupt the cytoplasmic membrane of targeted cells by becoming incorporated into the membrane and damaging its integrity.
Polymer Repeating chains of covalently linked monomers found in macromolecules.
Polymerase chain reaction (PCR) Technique of recombinant DNA technology that allows researchers to produce a large number of identical DNA molecules *in vitro*.
Polyomavirus A cancer-causing virus.
Polysaccharide Carbohydrate polymer composed of several to thousands of covalently linked monosaccharides.
Polyunsaturated fat Triglyceride with several double bonds between adjacent carbon atoms in its fatty acids.
Portal of entry Entrance site of pathogenic microorganisms, including the skin, mucous membranes, and placenta.
Portal of exit Exit site of pathogenic microorganisms, including the nose, mouth, and urethra.
Positive selection Process by which mutants are selected by eliminating wild-type phenotypes.
Positive-strand RNA ($^{+}$RNA) Viral single-stranded RNA that can act directly as mRNA.
Postpolio syndrome Crippling deterioration of muscle function, likely due to aging-related aggravation of nerve damage by poliovirus.
Potable Fit to drink.
Pour-plate Method of culturing microorganisms in which colony-forming units are separated from one another using a series of dilutions.
Pox *(pocks; pustule)* Any raised, pus-filled skin lesion; characteristic of infection with a poxvirus.
Precursor metabolite Any of 12 molecules typically generated by a catabolic pathway and essential to the synthesis of organic macromolecules in cells.
Prevalence In epidemiology, the total number of cases of a disease in a given area or population during a given period of time.
Primary amebic meningoencephalopathy Often fatal inflammation of the brain characterized by headache, vomiting, fever, and destruction of neurological tissue; caused by infection with *Naegleria* or *Acanthamoeba*.
Primary atypical pneumonia So-called *walking pneumonia* characterized by mild respiratory symptoms that last for several weeks; caused by infection with *Mycoplasma pneumoniae*.
Primary immunodeficiency diseases Any of a group of diseases detectable near birth and resulting from a genetic or developmental defect.
Primary response The slow and limited immune response to a first encounter with an unfamiliar antigen.
Primary stain In staining, the initial dye, which colors all cells.
Prion Proteinaceous infectious particle that lacks nucleic acids and replicates by converting similar normal proteins into new prions.
Probe Nucleic acid molecule with a specific nucleotide sequence that has been labeled with a radioactive or fluorescent chemical so that its location can be detected.
Prodromal period In the infectious disease process, the short stage of generalized, mild symptoms that precedes illness.
Products The atoms, ions, or molecules that remain after a chemical reaction is complete.
Profundal zone Zone of freshwater or marine water beneath the limnetic zone and above the benthic zone.
Proglottids Body segments of a tapeworm, produced continuously as long as the worm remains attached to its host.
Progressive multifocal leukoencephalopathy (PML) Progressive, fatal disease in which JC virus (a polyomavirus) kills cells of the central nervous system.
Prokaryote Any unicellular microorganism that lacks a nucleus. Classification includes bacteria and archaea.
Promoter Region of DNA where transcription begins.
Prophage An inactive bacteriophage, which is inserted into a host's chromosome.
Prophase First stage of mitosis, during which DNA condenses into chromatids and the spindle apparatus forms. Also used for the comparable stage of meiosis.
Prostaglandins Inflammatory chemicals released from damaged cells that increase vascular permeability.
Protease Enzyme secreted by microorganisms that digests proteins into amino acids outside a microbe's cell wall; in inflammatory reactions, chemicals released by mast cells that activate the complement system; in virology, an internal viral enzyme that makes HIV virulent.
Protein A complex macromolecule consisting of carbon, hydrogen, oxygen, nitrogen, and sulfur that is important to many cell functions.
Proteobacteria Phylum of prokaryotes that includes five classes (designated alpha, beta, gamma, delta, and epsilon) of Gram-negative bacteria sharing common 16S rRNA nucleotide sequences.
Proton A positively charged subatomic particle, which is also the nucleus of a hydrogen atom.
Proton gradient Electrochemical gradient of hydrogen ions across a membrane.
Protozoa Single-celled eukaryotes that lack a cell wall and are similar to animals in their nutritional needs and structure.
Provirus *(latent virus)* Inactive virus in an animal cell.
Pseudohyphae Long cellular extension of the yeast *Candida* that look like the filamentous hyphae of molds.
Pseudomembranous colitis Inflammation of the colon, which is covered by a membrane consisting of connective tissue and dead and dying cells; condition is last state of antimicrobial-associated diarrhea.
Pseudomonad Any Gram-negative, aerobic, rod-shaped bacterium in the class Gamma proteobacteria that catabolizes carbohydrates by the Entner-Doudoroff and pentose phosphate pathways.
Pseudopods Movable extensions of the cytoplasm and membrane of some eukaryotic cells.
Psychrophile Microorganism requiring cold temperatures (below 20°C).
Pure culture *(axenic culture)* Culture containing cells of only one species.
Purple sulfur bacteria Group of gammaproteobacteria, which are obligate anaerobes and oxidize hydrogen sulfide to sulfur.

Pustule *(pox)* Any raised, pus-filled skin lesion; characteristic of infection with a poxvirus.
Pyoderma Any confined, pus-producing lesion on the exposed skin of the face, arms, or legs; often caused by infection with group A *Streptococcus*.
Pyrimidine dimer Mutation in which adjacent pyrimidine bases covalently bond to one another; caused by nonionizing radiation in the form of ultraviolet light.
Pyrogen Chemical that triggers the hypothalamic "thermostat" to reset at a higher temperature, inducing fever.

Q fever Fever caused by the bacterium *Coxiella burnetii;* cause was questionable (thus "Q") for many years.
Quaternary ammonium compound (quat) Detergent antimicrobial that is harmless to humans.
Quorum sensing Process by which bacteria measure their density in an environment by utilizing signal and receptor molecules.

Rabies Neuromuscular disease characterized by hydrophobia, seizures, hallucinations, and paralysis; fatal if untreated. Caused by infection with the rabies virus.
Radial immunodiffusion Type of immunodiffusion test in which an antigen solution is allowed to diffuse into agar containing specific concentrations of antibodies, causing a ring of precipitate to form.
Radiation The release of high-speed subatomic particles or waves of electromagnetic energy from atoms.
Radiolaria Amoebae with threadlike pseudopods and silica shells.
Reactants The atoms, ions, or molecules that exist at the beginning of a chemical reaction.
Reaction center chlorophyll In a photosystem, a chlorophyll molecule in which electrons excited by light energy are passed to an acceptor molecule that is the initial carrier of an electron transport chain.
Recipient cell In horizontal gene transfer, a cell that receives part of the genome of a donor cell.
Recombinant Cell or DNA molecule resulting from genetic recombination between donated and recipient nucleotide sequences.
Recombinant DNA technology Type of biotechnology in which scientists change the genotypes and phenotypes of organisms.
Recombinant vaccine Vaccine produced using recombinant genetic technology.
Red measles *(rubeola, measles)* Contagious disease characterized by fever, sore throat, headache, dry cough, conjunctivitis, and lesions called Koplik's spots caused by infection with *Morbillivirus*.
Red tide Abundance of red-pigmented dinoflagellates in marine water.
Redox reaction *(oxidation-reduction reaction)* Any metabolic reaction involving the transfer of electrons from an electron donor to an electron acceptor. Reactions in which electrons are accepted are called *reduction* reactions, whereas reactions in which electrons are donated are *oxidation* reactions.
Reducing medium Special culturing medium containing compounds that combine with free oxygen and remove it from the medium.
Regulatory RNA Form of ribonucleic acid used to control gene expression.
Regulatory T cell *(Tr cell, suppressor T cell)* Thymus-matured lymphocyte that serves to repress adaptive immune responses and prevent autoimmune diseases.
Release In virology, final stage of the lytic replication cycle, in which the new virions are released from the host cell, which lyses.
Reoviruses Group of naked, segmented, dsRNA viruses that cause respiratory and gastrointestinal disease.
Repressible operon Type of operon that is continually transcribed until deactivated by repressors.
Reproduction An increase in number.
Reservoir of infection Living or nonliving continuous source of infectious disease.
Resolution The ability to distinguish between objects that are close together.
Respiratory syncytial virus (RSV) infection Disease caused by a virus in genus *Pneumovirus* that causes fusion of cells in the lungs and difficulty in breathing.
Responsiveness An ability to respond to environmental stimuli.
Restriction enzyme Enzyme that cuts DNA at specific nucleotide sequences and is used to produce recombinant DNA molecules.
Reticulate bodies Noninfectious stage in the life cycle of chlamydias.
Retrovirus Any +ssRNA virus that uses the enzyme reverse transcriptase carried within its capsid to transcribe DNA from its RNA.
Reverse transcriptase Complex enzyme that allows retroviruses to make dsDNA from RNA templates; used in recombinant DNA technology to make cDNA.
Revolving nosepiece Portion of a compound microscope on which several objective lenses are mounted.
Rh antigens Cytoplasmic membrane proteins common to the red blood cells of 85% of humans as well as rhesus monkeys.
Rheumatic fever A complication of untreated group A streptococcal pharyngitis in which inflammation leads to damage of the heart valves and muscle.
Rheumatoid arthritis (RA) A crippling, systemic autoimmune disease resulting from a type III hypersensitivity reaction in which antibody complexes are deposited in the joints, causing inflammation.
Rhinosinusitis Inflammation of lining of nose and sinuses.
Rhinoviruses Group of picornaviruses that cause upper respiratory tract infection and the "common cold."
Rhodophyta Red algae, generally containing the pigment phycoerythrin, the storage molecule floridean starch, and cell walls of agar or carrageenan.
Ribonucleic acid (RNA) Nucleic acid consisting of nucleotides made up of phosphate, a ribose pentose sugar, and an arrangement of the bases adenine, guanine, cytosine, and uracil.
Ribosomal RNA (rRNA) Form of ribonucleic acid that, together with polypeptides, makes up the structure of ribosomes.
Ribosome Nonmembranous organelle found in prokaryotes and eukaryotes that is composed of protein and ribosomal RNA and functions to make polypeptides.
Riboswitch RNA molecule that changes shape in response to shifts in environmental conditions, which results in genetic regulation.
Ribozyme RNA molecule functioning as an enzyme.
Rickettsias Group of extremely small, Gram-negative, obligate intracellular parasites that appear almost wall-less.
RNA polymerase Enzyme that synthesizes RNA by linking RNA nucleotides that are complementary to genetic sequences in DNA.
RNA primer RNA molecule used by DNA polymerase or reverse transcriptase as a starting point for DNA synthesis.
Rocky Mountain spotted fever (RMSF) Serious illness caused by infection with *Rickettsia rickettsii* transmitted by ticks and characterized by rash, malaise, petechiae, encephalitis, and death in 5% of cases.
Roseola Endemic illness of children characterized by an abrupt fever, sore throat, enlarged lymph nodes, and faint pink rash; caused by infection with human herpesvirus 6.
Rotaviruses Group of reoviruses that cause a potentially fatal infantile gastroenteritis.
Rough endoplasmic reticulum (RER) Type of endoplasmic reticulum that has ribosomes adhering to its outer surface; these produce proteins for transport throughout the cell.
R plasmid Extrachromosomal piece of DNA containing genes for resistance to antimicrobial drugs.
Rubella *(German measles)* Disease caused by infection with *Rubivirus* resulting in characteristic rash lasting about 3 days; mild in children but potentially teratogenic to fetuses of infected women.
Rubeola *(measles, red measles)* Contagious disease characterized by fever, sore throat, headache, dry cough, conjunctivitis, and lesions called Koplik's spots, caused by infection with *Morbillivirus*.
Rubor Redness.

Salmonellosis A serious diarrheal disease resulting from consumption of food contaminated with the enteric bacterium *Salmonella*.
Salt A crystalline compound formed by ionic bonding of metallic with nonmetallic elements.
Sanitization The process of disinfecting surfaces and utensils used by the public.
Saprobe Fungus that absorbs nutrients from dead organisms.
Sarcina A cuboidal packet of cocci.
Satellite virus A virus, such as hepatitis D virus, that requires glycoproteins coded by another virus to complete its replication cycle.
Saturated fatty acid A long-chain, organic acid in which all but the terminal carbon atoms are covalently linked to two hydrogen atoms.
Scabies Skin disease caused by a burrowing mite.
Scalded skin syndrome Reddening and blistering of the skin caused by infection with *Staphylococcus aureus*.
Scanning electron microscope (SEM) Type of electron microscope that uses magnetic fields within a vacuum tube to scan a beam of electrons across a specimen's metal-coated surface.
Scanning tunneling microscope (STM) Type of probe microscope in which a metallic probe passes

slightly above the surface of the specimen, revealing surface details at the atomic level.

Scarlet fever Diffuse rash and sloughing of skin caused by infection with group A *Streptococcus*.

Schaeffer-Fulton endospore stain In microscopy, staining technique that uses heat to drive a malachite green primary stain into an endospore.

Schistosomiasis A potentially fatal disease caused by infection with a blood fluke in the genus *Schistosoma*; may cause tissue damage in the liver, lungs, brain, or other organs.

Schizogony Special type of asexual reproduction in which the protozoan *Plasmodium* undergoes multiple mitoses to form a multinucleate schizont.

Schizont Multinucleate body that undergoes cytokinesis to release several cells.

Scientific method Process by which scientists attempt to prove or disprove hypotheses through observations of the outcomes of carefully controlled experiments.

Scolex Small attachment organ that possesses suckers and/or hooks used to attach a tapeworm to host tissues.

Sebum Oily substance secreted by the sebaceous glands of the skin that lowers pH.

Secondary immune response Enhanced immune response following a second contact with an antigen.

Secretory IgA The combination of IgA and a secretory component, found in tears, mucous membrane secretions, and breast milk, where it agglutinates and neutralizes antigens.

Secretory vesicle In eukaryotic cells, vesicles containing secretions packaged by the Golgi body that fuse to the cytoplasmic membrane and then release their contents outside the cell via exocytosis.

Sedimentation Settling of particulate matter; the first step in treating water for drinking.

Segmented genome Genetic material consisting of more than one molecule of nucleic acid, used particularly for viruses.

Selective medium Culturing medium containing substances that either favor the growth of particular microorganisms or inhibit the growth of unwanted ones.

Selective toxicity Principle by which an effective antimicrobial agent must be more toxic to a pathogen than to the pathogen's host.

Selectively permeable In cell physiology, characteristic of a membrane that allows some substances to cross while preventing the crossing of others.

Semisynthetic antimicrobial Antimicrobial that has been chemically altered.

Septate Characterized by the presence of cross walls.

Septic shock Extremely low blood pressure resulting from dilation of blood vessels triggered by bacteria or bacterial toxins.

Septic tank The home equivalent of primary wastewater treatment, consisting of a sealed concrete holding tank in which solids settle to the bottom and the effluent flows into a leach field that acts as a filter.

Septicemia *(sepsis)* The condition of pathogens being present in the blood and causing signs of illness.

Serial dilution A stepwise dilution of a liquid culture in which the dilution factor at each step is constant.

Serology The study and use of immunological tests to diagnose and treat disease or identify antibodies or antigens.

Serum Blood plasma with clotting factors removed.

Serum sickness Type III hypersensitivity resulting from antibodies directed against antisera.

Severe acute respiratory syndrome (SARS) Manifestation of infection by a coronavirus called SARS virus.

Severe combined immunodeficiency disease (SCID) Primary immunodeficiency disease in children that affects both T cells and B cells and causes recurrent infections.

Sexually transmitted disease (STD) Disease resulting from a sexually transmitted infection.

Sexually transmitted infection (STI) Invasion of a pathogen into the body resulting from sexual activity.

Shiga toxin Exotoxin secreted by *Shigella dysenteriae* that stops protein synthesis in host cells.

Shigellosis A severe form of dysentery caused by any of four species of *Shigella*.

Shine-Dalgarno sequence The sequence of nucleotides in a molecule of mRNA where the smaller ribosomal subunit initiates translation. The sequence is named for its discoverers.

Shingles *(herpes zoster)* Extremely painful skin rash caused by reactivation of latent varicella-zoster virus.

Shock Severe disturbance of blood circulation resulting in insufficient delivery of oxygen to vital organs.

Siderophore An iron-binding molecule released by some bacteria and fungi.

Signs In pathology, objective manifestations of a disease that can be observed or measured by others.

Silent mutation Mutation produced by base-pair substitution that does not change the amino acid sequence, because of the redundancy of the genetic code.

Simple microscope Microscope containing a single magnifying lens.

Simple stain In microscopy, a stain composed of a single dye such as crystal violet.

Singlet oxygen Toxic form of oxygen, neutralized by pigments called carotenoids.

Sinusitis Inflammation of the nasal sinuses; typically caused by *Streptococcus pneumoniae*.

Slant tube (slant) Test tube containing agar media that solidified while the tube was resting at an angle.

Slime layer Loose, water-soluble glycocalyx.

Slime mold Eukaryotic microbe resembling a filamentous fungus but lacking a cell wall and phagocytizing rather than absorbing nutrients.

Sludge After primary treatment of wastewater, the heavy material remaining at the bottom of settling tanks.

Small interfering RNA (siRNA) RNA molecule complementary to a portion of a molecule of mRNA, tRNA, or a gene, rendering the target ineffective.

Smallpox Infectious disease eradicated in nature by 1980 and characterized by high fever, malaise, delirium, pustules, and death in about 20% of untreated cases.

Smear In microscopy, the thin film of organisms on the slide.

Smooth endoplasmic reticulum (SER) Type of endoplasmic reticulum that lacks ribosomes and plays a role in lipid synthesis and transport.

Snapping division A variation of binary fission in Gram-positive prokaryotes in which the parent cell's outer wall tears apart with a snapping movement to create the daughter cells.

SOS response Mechanism by which prokaryotic cells with extensive DNA damage use a variety of processes to induce DNA polymerase to copy the damaged DNA.

Southern blot Technique used in recombinant DNA technology that allows researchers to stabilize specific DNA sequences from an electrophoresis gel and then localize them using DNA dyes or probes.

Species Taxonomic category of organisms that can successfully interbreed.

Species resistance Property that protects a type of organism from infection by pathogens of other, very different organisms.

Specific epithet In taxonomy, latter portion of the descriptive name of a species.

Specific immunity The ability of a vertebrate to recognize and defend against distinct species or strains of invaders.

Spectrum of action The number of different kinds of pathogens a drug acts against.

Spinal tap *(lumbar puncture)* Collection of cerebrospinal fluid from the lower vertebral column for diagnostic purposes.

Spiral In cell morphology, a spiral-shaped prokaryotic cell.

Spirillus A stiff spiral-shaped prokaryotic cell.

Spirochetes Group of helical, Gram-negative bacteria with axial filaments that cause the organism to corkscrew, enabling it to burrow into a host's tissues.

Spliceosome Protein-RNA complex that removes introns from eukaryotic RNA.

Spoilage Any unwanted change to a food.

Spontaneous generation The theory that living organisms can arise from nonliving matter.

Sporadic In epidemiology, a disease that occurs in only a few scattered cases within a given area or population during a given period of time.

Spore Reproductive cell of actinomycetes and fungi.

Sporogonic phase In the life cycle of *Plasmodium*, stage during which sporozoites are produced in the mosquito's digestive tract, and migrate into the mosquito's salivary glands.

Sporotrichosis Subcutaneous infection usually limited to the arms and legs; lesions form around the site of infection with *Sporothrix schenckii*.

Spotted fever rickettsiosis Serious illness caused by infection with a rickettsial bacterium transmitted by ticks and characterized by rash, malaise, petechiae, and encephalitis, and death in 5% of cases.

Spp. Abbreviation used to indicate several species of a genus.

Staining Coloring microscopy specimens with stains called *dyes*.

Staphylococcal scalded skin syndrome (SSSS) Disease caused by exfoliative toxin of *Staphylococcus aureus* in which epidermis peels off.

Staphylococcal toxic-shock syndrome Potentially fatal syndrome characterized by fever, vomiting, red rash, low blood pressure, and loss of sheets of skin, usually caused by systemic infection with strains of *Staphylococcus* that produce toxic shock syndrome toxins.

Staphylococcus A cluster of cocci.

Starter culture Group of known microorganisms that carry out specific and reproducible fermentation reactions.

Stationary phase Phase in a growth curve in which new organisms are being produced at the same rate at which older organisms are dying.

Stem cells Generative cells capable of dividing to form daughter cells of a variety of types.

Sterile Free of microbial contamination.

Sterilization The eradication of all organisms, including bacterial endospores and viruses, although not prions, in or on an object.

Steroid Lipids consisting of four fused carbon rings attached to various side chains and functional groups.

Streak-plate Method of culturing microorganisms in which a sterile inoculating loop is used to spread an inoculum across the surface of a solid medium in Petri dishes.

Streptococcal pharyngitis *(strep throat)* Inflammation of the throat, often caused by group A *Streptococcus*.

Streptococcal toxic-shock syndrome (STSS) Shock produced by toxins of *Streptococcus* with manifestations similar to toxic-shock syndrome of *Staphylococcus*.

Streptococcus A chain of cocci.

Streptogramins Antimicrobial drugs that bind to the 50S ribosomal subunit and prevent ribosome movement along messenger RNA.

Structural analog Chemical that competes with a structurally similar molecule.

Sty Inflamed bacterial infection of the base of an eyelid.

Subacute disease Any disease that has a duration and severity that lies somewhere between acute and chronic.

Subacute sclerosing panencephalitis (SSPE) Slow, progressive disease of the central nervous system that results in memory loss, muscle spasms, and death several years after infection with a defective measles virus.

Subclinical *(asymptomatic)* Characteristic of disease that may go unnoticed because of absence of symptoms, even though clinical tests may reveal signs of disease.

Substitution Type of mutation in which a nucleotide base pair is replaced.

Substrate The molecule upon which an enzyme acts.

Substrate-level phosphorylation The transfer of phosphate to ADP from another phosphorylated organic compound.

Subunit vaccine Type of vaccine developed using recombinant DNA technology that exposes the recipient's immune system to a pathogen's antigens but not the pathogen itself.

Sulfonamide Antimetabolic drug that is a structural analog of para-aminobenzoic acid (PABA).

Sulfur cycle Biogeochemical cycle in which sulfur is cycled between oxidation states.

Superficial mycoses Fungal infections of the surface of the skin.

Superinfection Condition in which a patient infected with hepatitis B virus is subsequently infected with hepatitis D virus.

Superoxide radical Toxic form of oxygen that is detoxified by superoxide dismutase.

Surfactant Chemical that acts to reduce the surface tension of solvents such as water by decreasing the attraction among solvent molecules.

Symbiosis A continuum of close associations between two or more organisms that ranges from mutually beneficial to associations in which one member damages the other member.

Symptoms Subjective characteristics of a disease that can be felt by the patient alone.

Synapse In immunology, the interface between cells of the immune system that involves cell-to-cell signaling.

Syncytium Giant, multinucleated cytoplasmic mass formed by fusion of a virally infected cell to its neighbors.

Syndrome A group of symptoms, signs, and diseases that collectively characterizes a particular abnormal condition.

Synergism Interplay between drugs that results in efficacy that exceeds the efficacy of either drug alone.

Synthesis In virology, the production of new viral proteins and nucleic acids using the metabolic machinery of the host cell; third stage of lytic replication cycle.

Synthesis reaction A chemical reaction involving the formation of larger, more complex molecules.

Synthetic drug Antimicrobial that has been completely synthesized in a laboratory.

Synthetic medium *(defined medium)* Culturing medium of which the exact chemical composition is known.

Syphilis Sexually transmitted disease caused by infection with *Treponema pallidum*.

Systemic diseases Diseases caused by microbes spread via the blood and lymph that affect other body systems.

Systemic lupus erythematosus *(lupus)* A systemic autoimmune disease in which the individual produces autoantibodies against numerous antigens, including nucleic acids.

T cell T lymphocyte.

Tc cell *(cytotoxic T cell, CD8 cell)* In cell-mediated immune response, type of cell characterized by CD8 cell-surface glycoprotein; secretes perforins and granzymes that destroy infected or abnormal body cells.

Th cell *(helper T cell, CD4 cell)* In cell-mediated immune response, a type of cell characterized by CD4 cell-surface glycoprotein; regulates the activity of B cells and cytotoxic T cells.

Tr cell *(regulatory T cell, suppressor T cell)* Thymus-matured lymphocyte that serves to repress adaptive immune responses and prevent autoimmune diseases.

T cell receptor (TCR) Antigen receptor generated in the cytoplasmic membrane of T lymphocytes.

T lymphocyte *(T cell)* Lymphocyte that matures in the thymus and acts primarily against endogenous antigens in cell-mediated immune responses.

T-dependent antibody immunity Adaptive immune response resulting in immunoglobulin production that requires the action of a specific helper T cell (Th2).

T-dependent antigens Molecules that stimulate an immune response only with the involvement of a helper T cell.

T-independent antibody immunity Adaptive immune response resulting in immunoglobulin production following cross-linking of BCRs on numerous B cells and lacking involvement of helper T cells.

T-independent antigens Large molecules with repeating subunits that trigger an antibody immune response without the activation of T cells.

Tapeworms *(cestodes)* Group of long, flat, and segmented helminths.

Taxa Nonoverlapping groups of organisms sorted on the basis of mutual similarities.

Taxis Cell movement that occurs as a positive or negative response to light or chemicals.

Taxonomic system A system for naming and grouping similar organisms together.

Taxonomy The science of classifying and naming organisms.

Telophase Final stage of mitosis, during which nuclear envelopes form around the daughter nuclei. Also used for the comparable stage of meiosis.

Temperate phage *(lysogenic phage)* Bacteriophage that does not immediately kill its host cell.

Teratogenic Characterized by an ability to cause birth defects.

Terminator Region of DNA where transcription ends.

Tetanospasmin Neurotoxin of *Clostridium tetani* that blocks the release of inhibitory neurotransmitters in the central nervous system.

Tetanus Potentially fatal infection with *Clostridium tetani*, which produces tetanospasmin, a potent neurotoxin.

Tetracycline Antimicrobial agent that inhibits protein synthesis by blocking the tRNA docking site.

Tetrad In genetics: two chromosomes, which are each made up of two DNA molecules, physically associated together during prophase I and metaphase I of meiosis; in cellular arrangements four cocci remaining attached following cell division.

Thermal death point The lowest temperature that kills all cells in a broth in 10 minutes.

Thermal death time The time it takes to completely sterilize a particular volume of liquid at a set temperature.

Thermophile Microorganism requiring temperatures above 45°C.

Thrombocytopenia Decrease in the number of platelets in the blood.

Thylakoid In photosynthetic cells, portion of cellular membrane containing light-absorbing photosystems.

Thymine Ring-shaped nitrogenous base found in nucleotides of DNA.

Tick Bloodsucking arachnid, which vectors a number of bacterial and viral pathogens.

Tickborne relapsing fever Disease caused by *Borrelia* spp. transmitted between humans by soft ticks.

Tincture Solution of antimicrobial chemical in alcohol.

Titer In serology, a measure of the level of antibody in blood serum, determined by titration and expressed as a ratio reflecting the dilution.

Titration Serial dilution of blood serum to test for agglutination activity.
Toll-like receptors (TLRs) Integral membrane proteins that bind to specific microbial chemicals.
Total magnification A multiple of the magnification achieved by the objective and ocular lenses of a compound microscope.
Toxemia Presence in the blood of poisons called *toxins*.
Toxic shock syndrome (nonstreptococcal; TSS) Potentially fatal condition characterized by fever, vomiting, red rash, low blood pressure, and loss of sheets of skin, caused by systemic infection with strains of *Staphylococcus*.
Toxin Chemical that either harms tissues or triggers host immune responses that cause damage.
Toxoid vaccine Inoculum using modified toxins to stimulate antibody-mediated immunity.
Toxoplasmosis A disease affecting animals and caused by infection with *Toxoplasma gondii*. In humans, characterized by mild, febrile symptoms, but may be fatal in AIDS patients, and transplacental transmission may result in miscarriage, stillbirth, or severe birth defects.
Trace element Element required in very small amounts for microbial metabolism.
Trachoma Serious eye disease caused by *Chlamydia trachomatis*.
Transamination Reaction involving transfer of an amine group from one amino acid to another.
Transcription Process in which the genetic code from DNA is copied as RNA nucleotide sequences.
Transducing phage Virus that transfers bacterial DNA from one bacterium to another.
Transduction Method of horizontal gene transfer in which DNA is transferred from one cell to another via a replicating virus.
Transfer RNA (tRNA) Form of ribonucleic acid that carries amino acids to the ribosome.
Transformation Method of horizontal gene transfer in which a recipient cell takes up DNA from the environment.
Transgenic organism Plant or animal that has been genetically altered by the inclusion of genes from other organisms.
Translation Process in which the sequence of genetic information carried by mRNA is used by ribosomes to construct polypeptides with specific amino acid sequences.
Transmission electron microscope (TEM) Type of electron microscope which generates a beam of electrons that passes through the specimen and produces an image on a fluorescent screen.
Transport medium A special type of medium used to move clinical specimens from one location to another while preserving the relative abundance of organisms and preventing contamination of the specimen or environment.
Transposition Mutation in which a genetic segment is transferred to a new position through the action of a DNA segment called a transposon.
Transposon Segment of DNA found in most prokaryotes, eukaryotes, and viruses that codes for the enzyme transposase and can move from one location in a DNA molecule to another location in the same or a different molecule.
Trematodes *(flukes)* Group of helminths that are flat, leaf-shaped, have incomplete digestive systems, and have oral and ventral suckers.
Trench fever A disease common among World War I soldiers; caused by the bacterium *Bartonella quintana*.
Trichomoniasis Inflammation of the genitalia caused by *Trichomonas vaginalis*.
Trophozoite The motile feeding stage of a protozoa.
Tubercle Hard pulmonary nodule resulting from infection with mycobacteria.
Tuberculin response Type of delayed hypersensitivity reaction in which the skin of an individual exposed to tuberculosis or tuberculosis vaccine reacts to a subcutaneous injection of tuberculin.
Tuberculin skin test Test for a delayed hypersensitivity reaction to a subcutaneous injection of tuberculin.
Tuberculosis (TB) A respiratory disease caused by infection with *Mycobacterium tuberculosis*; its disseminated form can result in wasting away of the body and death.
Tularemia Zoonotic disease causing fever, chills, malaise, and fatigue, and caused by infection with *Francisella tularensis*.
Tumor In the pathology of cancer, a mass of neoplastic cells. In inflammation, a symptom of swelling *(edema)*.
Tumor necrosis factor (TNF) An immune system cytokine secreted by macrophages and T cells to kill tumor cells and to regulate immune responses and inflammation.
Turbidimetry Automated method that measures the cloudiness of a solution by passing light through it.
Type 1 diabetes mellitus Immunological attack on the islets of Langerhans cells in the pancreas resulting in the inability to produce the hormone insulin.
Type III secretion systems Complex proteinaceous structure that inserts into target cells, forming a channel for the secretion of bacterial toxin or enzymes.
Typhoid fever Fever, headache, and malaise produced by infection with *Salmonella enterica* serotypes Typhi and Paratyphi; severe infections may cause peritonitis.
Typhus *(epidemic typhus, murine typhus, scrub typhus)* A group of diseases caused by rickettsias transmitted by arthropod vectors.

Uncoating In animal viruses, the removal of a viral capsid within a host cell.
Unsaturated fatty acid A long-chain, organic acid with at least one double bond between adjacent carbon atoms, and thus at least one carbon atom bound to only a single hydrogen atom.
Uracil Ring-shaped nitrogenous base found in nucleotides of RNA.
Urticaria Hives.
Use-dilution test Method of evaluating the effectiveness of a disinfectant or antiseptic against specific microbes in which the most effective agent is the one that entirely prevents microbial growth at the highest dilution.

Vaccination Active immunization; specifically against smallpox.
Vaccine The inoculum used in active immunization.
Vacuole General term for membranous sac that stores or carries a substance in a cell.
Vaginosis Noninflammatory infection of the vagina.
Valence The combining capacity of an atom.
Vancomycin Antimicrobial drug that disrupts formation of Gram-positive bacterial cell walls by interfering with alanine-alanine crossbridges linking *N*-acetylglucosamine subunits.
Vancomycin-resistant *Staphylococcus aureus* (VRSA) Strain of *S. aureus* that is resistant to vancomycin and usually resistant to many common antimicrobial drugs as well.
Variant Creutzfeldt-Jakob disease (vCJD) Dementia caused by a prion that destroys brain tissue such that the brain appears spongelike—full of holes.
Varicella *(chickenpox)* Highly infectious disease characterized by fever, malaise, and skin lesions, and caused by infection with varicella-zoster virus.
Varicella-zoster virus (VZV) Virus that causes chickenpox (varicella) and shingles (herpes zoster).
Variola Common name for the smallpox virus.
Variola major Variant of smallpox virus, which causes severe disease with a mortality rate of 20% or higher.
Variola minor Variant of smallpox virus, which causes less severe disease and mortality rate of less than 1%.
Vector In genetics and recombinant DNA technology, nucleic acid molecule such as a viral genome, transposon, or plasmid that is used to deliver a gene into a cell. In epidemiology, an animal (typically an arthropod) that transmits disease from one host to another.
Vegetations Bulky masses of platelets and clotting proteins that surround and bury the bacteria involved in endocarditis.
Vehicle transmission Spread of pathogens via air, drinking water, and food, as well as bodily fluids being handled outside the body.
Venezuelan equine encephalitis (VEE) Potentially fatal infection of the brain caused by a togavirus.
Vesicle General term for membranous sac that stores or carries a substance in a cell; in human pathology, any raised skin lesion filled with clear fluid.
Viable plate count Estimation of the size of a microbial population based upon the number of colonies formed when diluted samples are plated onto agar media.
Vibrio A slightly curved rod-shaped prokaryotic cell.
Viral gastroenteritis Inflammation of the mucous membrane of the stomach and intestines, caused by a viral pathogen.
Viral hemagglutination inhibition test Immune test commonly used to detect antibodies against influenza, measles, and other viruses that naturally agglutinate red blood cells.
Viral neutralization Test of serum for presence of antibodies against a particular virus in which test serum is mixed with the virus, and then the mixture is added to a cell culture. Survival of the cells indicates antibodies in the serum neutralized the viruses.
Viremia Viral infection of the blood.

Viridans streptococci Group of alpha-hemolytic streptococci, which produce a green pigment when grown on blood media and normally inhabit the mouth and throat, as well as the GI, genital, and urinary tracts.
Virion A virus outside of a cell, consisting of a proteinaceous capsid surrounding a nucleic acid core.
Viroid Extremely small, circular piece of RNA that is infectious and pathogenic in plants.
Virulence A measure of pathogenicity.
Virulence factors Enzymes, toxins, and other factors that affect the relative ability of a pathogen to infect and cause disease.
Virus Tiny infectious acellular agent with nucleic acid surrounded by proteinaceous capsomeres that form a covering called a capsid.
Viviparity Process by which live offspring are produced in the body of a mother.

Wandering macrophage Type of macrophage that leaves the blood via diapedesis to travel to distant sites of infection.
Warts *(papillomas)* Benign epithelial growths caused by papillomaviruses.
Wastewater *(sewage)* Any water that leaves homes or businesses after being used for washing or flushed from toilets.
Water mold Eukaryotic microbe resembling a filamentous fungus but having tubular cristae in their mitochondria, cell walls of cellulose, two flagella, and true diploid bodies.
Waterborne transmission Spread of pathogenic microorganisms via water.
Wavelength The distance between two corresponding points of a wave.
Wax Alcohol-containing lipid made up of molecules with one fatty acid chain.
Western blot test *(immunoblot)* Variation of an ELISA test that can detect the presence of antibodies against multiple antigens; used to verify the presence of antibodies against HIV in the serum of individuals who have tested positive by ELISA.
Western equine encephalitis (WEE) Potentially fatal infection of the brain caused by a togavirus.
Whitlow Inflamed blister that may result from infection with *human herpesvirus 1* or HHV-2 via a cut or break in the skin.
Whooping cough *(pertussis)* Pediatric disease characterized by development of copious mucus, loss of tracheal cilia, and deep "whooping" cough; caused by infection with *Bordetella pertussis*.
Wild-type cell A cell normally found in nature (in the wild); a nonmutant.
Wound Trauma to body's tissue.

XDR-TB Tuberculosis caused by extensively drug-resistant *Mycobacterium*.
Xenodiagnosis Method of diagnosing Chagas' disease in which an uninfected *Triatoma* vector is allowed to feed on a patient. Subsequent presence of trypanosomes in the bug's gut indicates the patient is infected.
Xenograft Type of graft in which tissues are transplanted between individuals of different species.
Xenotransplant Technique involving recombinant DNA technology in which human genes are inserted into animals to produce cells, tissues, or organs that are then introduced into the human body.

Yaws Large, destructive, pain-free lesions of the skin, bones, and lymph nodes caused by *Treponema pallidum pertenue*.
Yeast A unicellular, typically oval or round fungus that usually reproduces asexually by budding.
Yellow fever Often fatal hemorrhagic disease contracted through a mosquito bite carrying a flavivirus.

Zone of inhibition In a diffusion susceptibility test, a clear area surrounding the drug-soaked disk where the microbe does not grow.
Zoonoses Diseases that are naturally spread from their usual animal host to humans.
Zygomycoses Opportunistic fungal infections caused by various genera of fungi classified in the division Zygomycota.
Zygomycota Division of fungi including coenocytic molds called zygomycetes. Most are saprobes.
Zygosporangium Thick, black, rough-walled sexual structure of zygomycetes that can withstand desiccation and other harsh environmental conditions.
Zygospores Haploid spores formed from the surviving nuclei within zygosporangia.
Zygote In sexual reproduction, diploid cell formed by the union of gametes.

Credits

Illustration Credits

All illustrations have been rendered by **Precision Graphics** unless noted otherwise below. All illustrations are copyright Pearson Education.

CHAPTER 1 **1.10, 1.12, 1.14:** J.B. Woolsey Associates, LLC.

CHAPTER 2 **2.13, 2.14:** J.B. Woolsey Associates, LLC.

CHAPTER 3 **3.2, 3.3, 3.6, 3.8, 3.14–3.16, 3.31, 3.34–3.37, 3.40, 3.41:** Kenneth Probst/Precision Graphics. **3.4, 3.32:** Darwen Hennings/Kenneth Probst/Precision Graphics. **3.19:** J.B. Woolsey Associates, LLC. **3.39:** Darwen Hennings.

CHAPTER 4 **4.2, 4.11, 4.12:** J.B. Woolsey Associates, LLC/Precision Graphics. **4.4, 4.6, 4.7:** J.B. Woolsey Associates, LLC. **4.22, 4.27:** Darwen Hennings/Precision Graphics.

CHAPTER 5 **5.18, 5.25–5.27:** Kenneth Probst/Precision Graphics.

CHAPTER 6 **6.3, 6.9, 6.10, 6.22, 6.23, 6.25, 6.26:** J.B. Woolsey Associates, LLC. **6.24:** J.B. Woolsey Associates, LLC/Precision Graphics.

CHAPTER 7 **7.29:** J.B. Woolsey Associates, LLC/Precision Graphics. **7.30, 7.31, 7.33:** J.B. Woolsey Associates, LLC.

CHAPTER 9 **9.8, 9.11:** J.B. Woolsey Associates, LLC/Precision Graphics. **9.10:** J.B. Woolsey Associates, LLC.

CHAPTER 10 **10.2:** Kenneth Probst/Precision Graphics. **10.12:** J.B. Woolsey Associates, LLC.

CHAPTER 11 **11.1, 11.6, 11.21, 11.26:** Kenneth Probst/Precision Graphics.

CHAPTER 12 **12.6, 12.8, 12.14, 12.19, 12.22, 12.24, 12.25, 12.27:** Kenneth Probst/Precision Graphics.

CHAPTER 13 **13.4, 13.6–13.8, 13.11, 13.12, 13.14, 13.18:** Kenneth Probst/Precision Graphics.

CHAPTER 14 **14.3, 14.4, 14.11, Table 14.2:** Kenneth Probst/Precision Graphics. **14.7:** J.B. Woolsey Associates, LLC.

CHAPTER 15 **15.2–15.4, 15.13–15.15:** Kenneth Probst/Precision Graphics.

CHAPTER 16 **16.2:** Kenneth Probst/Precision Graphics.

CHAPTER 17 **17.4, 17.7:** J.B. Woolsey Associates, LLC/Precision Graphics. **17.8, 17.9:** J.B. Woolsey Associates, LLC.

CHAPTER 18 **18.5, 18.8:** Cassio Lynm/Precision Graphics. **18.13:** J.B. Woolsey Associates, LLC. **18.15, 18.17, 18.18:** Kenneth Probst/Precision Graphics.

CHAPTER 19 **19.1, 19.10, 19.12:** Kenneth Probst. **19.6:** Cassio Lynm.

CHAPTER 20 **20.1, 20.17:** Kenneth Probst. **20.2, 20.10, 20.16:** Kenneth Probst/Precision Graphics. **20.8:** Cassio Lynm/Precision Graphics. **20.14:** Darwen Hennings.

CHAPTER 21 **21.1, 21.6, 21.21:** Kenneth Probst. **21.9, 21.16:** Kenneth Probst/Precision Graphics.

CHAPTER 22 **22.1, 22.10, 22.13, 22.14, Disease at a Glance 22.1:** Kenneth Probst. **22.11:** Kenneth Probst/Precision Graphics.

CHAPTER 23 **23.1, 23.2, 23.12, 23.15, 23.18, Disease at a Glance 23.4, Disease at a Glance 23.9:** Kenneth Probst/Precision Graphics.

CHAPTER 24 **24.1, 24.11:** Kenneth Probst. **Disease at a Glance 24.1, Disease at a Glance 24.5:** Kenneth Probst/Precision Graphics.

Photo Credits

FRONTMATTER: **Endsheet:** All photos Science Source; p. iv. Photo by Jeremy Bauman, copyright Robert Bauman.

CHAPTER 1 **Opener:** Graham Bell/Corbis. **1.1:** Pfizer. **1.2:** Alan Shinn. **1.3:** Richard Robinson, Pearson Education. **1.4:** L. Brent Selinger, Pearson Education. **1.5a:** Jeremy Burgess/Science Source. **1.5b:** Steve Gschmeissner/Science Source. **1.6a:** M I Walker/Photoshot. **1.6b:** M I Walker/Science Source. **1.6c:** Bruce J Russell/Biomedia Associates. **1.7a:** M I Walker/Science Source. **1.7b:** Jan Hinsch/Science Source. **1.8:** Sinclair Stammers/Science Source. **1.9:** Lee D Simon/Science Source. **Beneficial Microbes 1.1:** Marc Vermeirsc/iStockphoto. **Highlight 1.1:** Chung Sung-Jun/Getty Images. **1.11:** Images from the History of Medicine. **1.15:** National Library of Medicine. **1.16:** Kirk Hartwein. **1.17:** Science Source. **Clinical Case Study 1.1:** National Library of Medicine. **1.18:** National Library of Medicine. **1.20:** Christine Case. **Clinical Case Study 1:** Graham Bell/Corbis. **Emerging Diseases 1.1.1:** Keith Weller/United States Department of Agriculture. **Questions for Review 1.1:** M I Walker/Science Source. **Questions for Review 1.2:** Bruce J Russell/BioMedia Associates. **Questions for Review 1.3:** M I Walker/Photoshot.

CHAPTER 2 **Opener:** Poppy Berry/Corbis. **2.12b:** Felix Büscher/AGE Fotostock. **Beneficial Microbes 2.1:** Frank van den Bergh/iStockphoto.com. **Clinical Case Study 2.1:** Jonathan LittleJohn/Alamy. **Clinical Case Study 2:** Poppy Berry/Corbis.

CHAPTER 3 **Opener:** Jim Cummins/Getty Images. **3.1a:** Gopal Murti/Science Source. **3.1b:** M I Walker/Science Source. **3.3:** Don W. Fawcett/Science Source. **3.5a:** Kari Louatmea/Science Source. **3.5b:** Bergey's Manual Trust. **3.7a:** Biophoto Associates/Science Source. **3.7b:** Eye of Science/Science Source. **3.7c:** American Phytopathological Society. **3.8:** ASM/Science Source. **3.10:** Thomas Deerinck/Science Source. **3.11:** David Scharf/Alamy. **Highlight 3.1:** Public Library of Science. **3.23:** National Research Council of Canada. **Beneficial Microbes 3.1:** Josh Reynolds/AP Images. **3.25:** Rut Carballido-Lopez. **3.26 bottom:** Wiley-Blackwell Publishing Ltd. C. Moissl, R. Rachel, A. Briegel, H. Engelhardt, R. Huber, Mol Microbiol. 2005 Apr; 56(2):361-70; Figs. 1 and 2. **3.26 top:** C. Moissl, R. Rachel, A. Briegel, H. Engelhardt, R. Huber, Mol Microbiol. 2005 Apr; 56(2):361-70; Figs. 1 and 2. **3.27a:** Springer-Verlag GmbH & Co. **3.27b:** Dennis Searcy. **3.27c:** Mike Dyall-Smith. **3.28:** Donald L. Ferry. **3.29:** Don W Fawcett/Science Source. **3.30a:** Donald L. Ferry. **3.30b:** Donald L. Ferry. **3.30c:** Donald L Ferry. **3.30d:** Donald L Ferry. **3.31a:** SPL/Science Source. **3.31b:** Steve Gschmeissner/Science Source. **3.33:** Jennifer Waters/Science Source. **3.34a:** Conly Rieder. **3.34b:** Don W. Fawcett/Science Source. **3.35a:** Don W. Fawcett/Science Source. **3.35b:** Don W. Fawcett/Science Source. **3.35c:** Don W. Fawcett/Science Source. **3.36a:** Don W. Fawcett/Science Source. **3.36b:** Wellcome Images. **3.37a:** Don W. Fawcett/Science Source. **3.37b:** Biophoto Associates/Science Source. **3.38:** Biophoto Associates/Science Source. **3.40a:** Don W. Fawcett/Science Source. **3.40b:** Don W. Fawcett/Science Source. **3.41:** Electron micrograph by Wm. P. Wergin, courtesy of E. H. Newcomb,/University of Wisconsin. **Clinical Case Study 3:** Jim Cummins/Getty Images. **Questions for Review 3.1.2:** Don W. Fawcett/Science Source. **Questions for Review 3.2.1a:** ASM/Science Source. **Questions for Review 3.2.1b:** Biophoto Associates/Science Source. **Questions for Review 3.2.1c:** American Phytopathological Society. **Questions for Review 3.2.1d:** Eye of Science/Science Source.

CHAPTER 4 **Opener:** FSG/AGE Fotostock. **4.4a:** Charles D Winters. **4.8a:** Rich Robison, Pearson Education. **4.8b:** Rich Robison, Pearson Education. **4.8c:** Rich Robison, Pearson Education. **4.8d:** Rich Robison, Pearson Education. **4.9a:** Rich Robison, Pearson Education. **4.9b:** Rich Robison, Pearson Education. **4.10b:** CDC. **Highlight 4.1:** Bio-Rad Laboratories Inc. **4.11b:** Seelevel.com. **4.11c:** University of Texas Health Science Center at San Antonio. **4.13a:** Steve Gschmeissner/Photo Researchers. **4.13b:** Eye of Science/Science Source. **4.13c:** Andrew Syred/Science Source. **4.13d:** Eye of Science/Science Source. **4.14a:** Veeco Instruments, Inc. **4.14b:** Veeco Instruments, Inc. **Table 4.2.1:** Rich Robison, Pearson Education. **Table 4.2.2:** Rich Robison, Pearson Education. **Table 4.2.3:** Rich Robison, Pearson Education. **Table 4.2.4:** Rich Robison, Pearson Education. **Table 4.2.5:** Rich Robison, Pearson Education. **Table 4.2.6:** Bio-Rad Laboratories, Inc. **Table 4.2.7:** Stanley C. Holt, University of Texas Health Science Center. **Table 4.2.8:** Steve Gschmeissner/Science Source. **Table 4.2.9:** Veeco Instruments, Inc. **Table 4.2.10:** Veeco Instruments, Inc. **4.16a:** Elisabeth Pierson, Pearson Education. **4.16b:** Elisabeth Pierson, Pearson Education. **Table 4.17.1:** Rich Robison, Pearson Education. **Table 4.17.2:** Rich Robison, Pearson Education. **Table 4.17.3:** Rich Robison, Pearson Education. **Table 4.17.4:** Rich Robison, Pearson Education. **4.18:** Rich Robison, Pearson Education. **4.19:** Rich Robison, Pearson Education. **4.20:** P Bim/Custom Medical Stock Photo. **4.21:** L. Brent Selinger, Pearson Education. **Beneficial Microbes 4.1:** Yasunori Tanji, Tokyo, Institute of Technology, Department of Biotechnology. **Table 4.3.1:** Rich Robison, Pearson Education. **Table 4.3.2:** Rich Robison, Pearson Education. **Table 4.3.3:** Rich Robison, Pearson Education. **Table 4.3.4:** Rich Robison, Pearson Education. **Table 4.3.5:** P Bim/Custom Medical Stock Photo. **Table 4.3.6:** L. Brent Selinger, Pearson Education. **4.23a:** L. Brent Selinger, Pearson Education. **4.23b:** L. Brent Selinger, Pearson Education. **4.24:** Dade Behring Inc./Siemens Corporation. **4.25a:** L. Brent Selinger, Pearson Education. **4.26:** The Microbial Diseases Laboratory, Berkeley, California. **Emerging Diseases 4.1.1:** Bill Olmsted, The Janesville Gazette/AP images. **Questions for Review 4.1.1:** Eye On Science/Science Source. **Questions for Review 4.1.2:** John Durham/Science Source. **Questions for Review 4.1.3:** Rich Robison, Pearson Education. **Questions for Review 4.1.4:** CDC. **Questions for Review 4.1.5:** M Wurtz/Science Source. **Questions for Review 4.1.6:** M I Walker/Science Source.

CHAPTER 5 **Opener:** Dex Image/Alamy Images. **Beneficial Microbes 5.1:** Don Bendickson/Shutterstock. **Highlight 5.1.1:** Frederick R McConnaughey/Science Source. **5.25b:** Science Source. **Highlight 5.2:** AGE Fotostock. **Clinical Case Study 5:** Dex Images/Alamy. **Questions for Review 5.2** Don W. Fawcett/Science Source.

CHAPTER 6 **Opener:** UpperCut Images/Alamy Images. **6.2:** Brenda Wellmeyer, North Harris College Biology Dept., Houston TX. htpp://science.nhmccd.edu/biol/bwellmeyer.html. **Highlight 6.1:** J.W. Schaefer/Fotolia. **6.4b:** L. Brent Selinger, Pearson Education. **6.6a:** Doug Allen. **6.6b:** Wayne P. Armstrong. **Beneficial Microbes 6.1:** Savannah River National Laboratory. **Clinical Case Study 6.1:** Dr. Martin S Spiller, DMD. **6.8b:** L. Brent Selinger, Pearson Education. **6.9b:** L. Brent Selinger, Pearson Education. **6.10b:** Kirk Hartwein. **6.11:** L. Brent Selinger, Pearson Education. **6.12 left:** L. Brent Selinger, Pearson Education. **6.12 right:** L. Brent Selinger, Pearson Education. **6.13:** Rich Robison, Pearson Education. **6.14:** L. Brent Selinger, Pearson Education. **6.15abc:** L. Brent Selinger, Pearson Education. **6.17b:** Lee D Simon/Science Source. **Clinical Case Study 6.2:** CDC. **6.24b:** Biotechnology& Bioengineering 2008; 99/3; 634-43 Reproduced with permission of Wiley Inc. **6.24c:** L. Brent Selinger, Pearson Education. **6.26a:** Richard Megna/Fundamental Photographs. **6.26b:** TOPAC. **Clinical Case Study 6:** UpperCut Images/Alamy. **Questions for Review 6.2** Hans N.

CHAPTER 7 **Opener:** Woman with tattoo/Alamy Images. **7.2a:** Klaus Boller/Science Source. **7.2b:** David Dressler. **7.3a:**

Barbara Hamkalo. **7.3b:** Reproduced by permission from "Cell." Copyright (c) 2002 by Elsevier Science Ltd. Image courtesy of J. R. Paulsen and U. K. Laemmli. **7.3c:** G F Bahr/Armed Forces Institute of Pathology. **7.3d:** G F Bahr/Armed Forces Institute of Pathology. **Beneficial Microbes 7.1:** Ed Austin, Herb Jones/National Park Service. **7.19b:** Kiseleva, Elena. **Emerging Diseases 7.1:** Yuri Arcurs/Shutterstock. **Highlight 7.1:** Tomas Prokop/iStockphoto. **7.35a:** Veeco Instruments, Inc. **Clinical Case Study 7.1:** Mike Miller/Center for Disease Control.

CHAPTER 8 **Opener:** Janine Wiedel Photolibrary/Alamy Images. **Highlight 8.1:** James Gathany/Centers for Disease Control. **8.6b:** Science Source. **8.8b:** Laurence Game, CSC, IC Microarray Centre, London. **8.9d:** University of California Irvine University of California, Irvine Transgenic Mouse Facility. **8.10:** AdvanDx. **8.12:** D Parker/Science Source. **Highlight 8.2:** Werli Francois/Alamy. **8.13:** Gonsalves, Dennis. **Questions for Review 8.2** Centers for Disease Control and Prevention (CDC).

CHAPTER 9 **Opener:** Alex Segre/Alamy Images. **Emerging Diseases 9.1.1:** P. Garg & G. N. Rao/International Centre for Eye Health (ICEH). **9.4:** Ramon Flick, Ph.D. **9.7a:** Photofusion Picture Library/Alamy. **9.9:** Jonathan Blair/National Geographic Image Collection. **9.10b:** Science Source. **Highlight 9.1:** Ok nazarenko/Shutterstock. **9.12:** Richard Megna/Fundamental Photographs. **9.14:** Richard Megna/Fundamental Photographs. **Beneficial Microbes 9.1:** Scimat/Science Source. **9.16:** Robert Bauman. **Highlight 9.2:** Jiri Hera/AGE Fotostock America Inc. **Questions for Review 9.2** Rob Reed.

CHAPTER 10 **Opener:** Charlie Schuck/AGE Fotostock America, Inc. **10.1:** Rich Robison, Pearson Education. **Clinical Case Study 10.1:** P Marazzi/Science Source. **Highlight 10.1:** Fotosearch/Publitek, Inc. **10.9:** L. Brent Selinger, Pearson Education. **10.10:** Benjamin Tanner, PhD. **10.11:** Tim Pietzcker, Universitatsklinikum Ulm, Germany. **10.14a:** Com4. **10.14b:** Beth Yarbrough. **Beneficial Microbes 10.1:** Scimat/Science Source. **Emerging Diseases 10.1.1:** Toledo-Lucas County Health Department. **10.17:** Eddy Vercauteren. **Clinical Case Study 10.3:** Ariel Skelley/Getty Images USA, Inc. Courtesy of the Centers for Disease Control (1991). **Questions for Review 10.2:** Microrao, JJMMC, Davangere, Karnataka, India.

CHAPTER 11 **Opener:** Alamy. **11.1a:** SPL/Science Source. **11.1b:** David M. Phillips/Science Source. **11.1c:** Juergen Berger/Science Source. **11.1d:** AMI IMAGE/Science Source. **11.1e:** Andrew Syred/Science Source. **11.1f:** Janice Haney Carr, CDC. **11.1g:** Dr. Hans-Peter Klenk. **11.1h:** Bergey's Manual Trust. **11.2a:** Dr. Gary Kaiser. **11.2a:** Rich Robison, Pearson Education. **11.2b:** Rich Robison, Pearson Education. **11.4b:** Terry A Krulwich. **11.5:** Kim Findlay, John Innes Centre, Norfolk, UK. **11.7:** Esther R. Angert. **11.8a:** Mike Dyall-Smith, Dept. of Microbiology and Immunology, University of Melbourne, Australia. **11.8b:** Eastman Kodak Company. **11.8c:** L. Brent Selinger, Pearson Education. **11.8d:** L. Brent Selinger, Pearson Education. **11.8e:** L. Brent Selinger, Pearson Education. **11.9a:** William A. Clark/Centers for Disease Control and Prevention (CDC). **11.9b:** William A. Clark/Centers for Disease Control and Prevention (CDC). **11.9c:** Centers for Disease Control and Prevention (CDC). **11.9d:** William A. Clark, Centers for Disease Control and Prevention (CDC). **11.11a:** American Association for the Advancement of Science (AAAS)/K. Kashefi, D. R. Lovley. **11.11b:** Dr. Reinhard Rachel, University of Regensburg, Germany. **Beneficial Microbes 11.1:** Rick Gomez/AGE Fotostock. **11.12:** Sascha Burkard/Shutterstock. **11.13:** Nancy Nehring/iStockphoto. **11.14a:** MCC-NIES Microbial Culture Collection National Institute for Environmental Studies. **11.14b:** Yuuji Tsukii, Hosei University, Tokyo, Japan. **11.14c:** Wayne Lanier. **11.15:** David J Patterson. **11.21:** Alamy. **Highlight 11.1:** Merlin D. Tuttle, Bat Conservation International, www.batcon.org. **11.16:** F. Thiaucourt, CIRAD, Montpellier, France. **11.17:** SciMAT/Science Source. **Highlight 11.2:** Alessio Orrù/Fotolia. **11.18:** David Berd/Centers for Disease Control and Prevention (CDC). **11.19:** Prosthecate bacterium. **11.20:** Seelevel.com. **Beneficial Microbes 11.2:** Yves Brun. **11.23a:** David J Patterson. **11.23b:** University of Vienna. **11.24:** Dr. Tony Brain/Science Source. **Emerging Diseases 11.1.1:** Alamy. **11.25b:** Alfred Pasieka/Getty Images. **11.26b:** Ronald Garcia and Rolf Müller, Institute for Pharmaceutical Research Saarland. **Clinical Case Study 11:** Alamy. **Questions for Review 11.2a:** Rich Robison, Pearson Education. **Questions for Review 11.2b:** Rich Robison, Pearson Education.

CHAPTER 12 **Opener:** Blend Images/Alamy. **12.7:** Patrick Keeling. **12.2a:** Dartmouth College Electron Microscope Facility. **12.2b:** Michael V. Danilchik, Program in Molecular and Cellular Biosciences, Oregon Health Sciences University. **12.2c:** SciMAT/Science Source. **12.5:** Walter Dawn, Science Source. **12.8b:** Nicholas Despo, Biology and Life Sciences Department, Thiel College, Greenville, PA. **12.9:** Guy Brugerolle, Universitad Clearmont Ferrand. **12.10:** Meckes/Ottowas/Eye on Science/Science Source. **12.11b:** Biophoto Associates/Science Source. **12.12:** Georg Rosenfeldt, **12.13:** Eye of Science/Science Source. **12.14a:** Patrick W. Grace/Science Source. **12.14b:** Eye of Science/Science Source. **12.15a:** L. Brent Selinger, Pearson Education. **12.15b:** L. Brent Selinger, Pearson Education. **12.15c:** David Scharf/Science Source. **12.15d:** Christine Case/Skyline College. **12.16:** Jeremy Burgess/Science Source. **12.17:** George L. Barron Ph.D., D.Sc. **12.18a:** Lucile K George, Center for Disease Control and Prevention. **12.18b:** Buckman Laboratories International Inc. **12.18c:** David Scharf/Science Source. **12.21:** Guillaume Gouilloux/iStockphoto. **12.22:** Nino Santamaria. **12.31:** Steve Gschmeissner/Science Source. **12.23a:** Phyzome. **12.23b:** Paul J Fusco/Science Source. **12.24:** Merton F. Brown and Harold G. Brotzman from the APS Slide collection. **12.25 (left):** Eye of Science/Science Source. **12.26:** Fred Rhoades. **Beneficial Microbes 12.1:** Pixelmania/Fotolia. **Emerging Diseases 12.1.1:** The American Journal of Managed Care. **12.28:** D. P. Wilson/Science Source. **12.29:** David Patterson Courtesy of the Micro*scope website. **12.30:** Ethan Daniels/Shutterstock. **12.32:** Fred Rhoades. **12.33a:** Kent Wood/Science Source. **12.33b:** Department of Medical Technology, Faculty of Health Sciences, Kobe University School of Medicine. **12.33c:** Centers for Disease Control and Prevention. **12.33d:** Vincent S Smith. **12.33e:** Kim Taylor/Nature Picture Library. **12.33f:** Food and Environmental Hygiene Department. **12.33g:** Marcelo de Campos Pereira, The Veterinary Parasitology Images Gallery. **Clinical Case Study 12:** Blend Images/Alamy. **Questions for Review 12.1a:** Nino Santamaria. **Questions for Review 12.1b:** Merton F. Brown. **Questions for Review 12.1c:** David Scharf/Science Source. **Questions for Review 12.1d:** Lucile K. George, Centers for Disease Control and Prevention.

CHAPTER 13 **Opener:** Colin Jones/Impact/AGE Fotostock. **Beneficial Microbes 13.1:** NASA. **Clinical Case Study 13:** Colin Jones/Impact/AGE Fotostock. **13.1b:** David M. Phillips/Science Source. **13.2:** Huntington Potter. **13.3a:** Nigel Cattlin/Science Source. **13.3b:** Kevin M. Rosso Ph.D. **13.3c:** Eye of Science/Science Source. **13.5a:** Robley C. Williams, Jr., Vanderbilt University. **13.5b:** Biophoto Associate/Science Source. **13.5c:** Abergel Chantal, Information Genomique et Structurale UMR7256 CNRS, Aix-Marseille Université. **13.5d:** Frederick A. Murphy, University of Texas Medical Branch, Galveston. **13.6a:** Department of Microbiology, Biozentrum/Science Source. **13.7:** Klaus Boller/Science Source. **13.10:** M. Wurtz/Science Source. **Highlight 13.1:** Regalle Asuncion/AP Images. **Emerging Diseases 13.1:** James Gathany, CDC. **13.17:** Madboy, CC-BY-SA. **13.19:** OriGen Biomedical. **13.20:** Theodor O Diener. **13.21:** European and Mediterranean Plant Protection Organization (EPPO/OEPP). **13.23:** Enric Vidal - DVM, PhD. PRIOCAT laboratory, CReSA (UAB-IRTA). **Clinical Case Study 13.1:** Yuri Arcurs/Alamy. **Questions for Review 13.1a:** "Abergel Chantal, Information Genomique et Structurale UMR7256 CNRS, Aix-Marseille Université". **Questions for Review 13.1b:** Frederick A. Murphy, University of Texas Medical Branch, Galveston. **Questions for Review 13.1c:** Robley C. Williams, Jr., Vanderbilt University. **Questions for Review 13.1d:** Biophoto Associate/Science Source. **Beneficial Microbes 13.2:** Dr. Gerhard Wanner.

CHAPTER 14 **Opener:** Monalyn Gracia/AGE Fotostock. **14.1:** jeridu/iStockphoto. **14.2:** Tony Brain/Science Source. **Beneficial Microbes 14.1:** Ralph E Berry. **Clinical Case Study 14.1:** Corbis Images. **14.5b:** J. William Costerton, Montana State University. **14.12:** Andrew Davidhazy, Photo Arts and Sciences at Rochester Institute of Technology. **14.6:** David Scharf/Science Source. **14.10:** National Library of Medicine. **14.13:** Akintunde Akinleye/Reuters. **Clinical Case Study 14.2:** GlowImages/Alamy. **Clinical Case Study 14.3:** Vincent P Walter, Pearson Education. **Emerging Diseases 14.1.1:** David Cappaert. **Clinical Case Study 14:** Monalyn Gracia/AGE Fotostock.

CHAPTER 15 **Opener:** Blend Images/Alamy. **15.5a:** Gwen V Childs. **15.5a-2:** Gwen V Childs. **15.5b-1:** J. O. Ballard, M.D., Professor of Medicine & Pathology, Penn State University College of Medicine. **15.6:** Alison K Criss PhD. **15.1:** Eye of Science/Science Source. **Beneficial Microbes 15.1:** Science Photo Library/Science Source. **15.5a-3:** Michael Ross/Science Source. **15.5b-2:** J. O. Ballard, M.D., Professor of Medicine & Pathology, Penn State University College of Medicine. **15.10:** Sucharit Bhakdi. **Clinical Case Study 15:** Blend Images/Alamy. **Questions for Review 15.1:** Alison K Criss PhD. **Questions for Review 15.2a:** James O. Ballard, Pearson Education. **Questions for Review 15.2b:** Michael Ross/Science Source. **Questions for Review 15.2c:** Gwen V Childs. **Questions for Review 15.2d:** James O. Ballard, Pearson Education. **Questions for Review 15.2e:** Gwen V Childs.

CHAPTER 16 **Opener:** Fuse/Thinkstock. **16.1:** Biophoto Associates/Science Source. **16.6:** Cécile Chalouni, Genentech. **Highlight 16.1:** NIBSC/Science Source. **16.11b:** Tim Evans/Science Source. **Table 16.4.4:** Barbara Rice, Centers for Disease Control and Prevention (CDC). **Highlight 16.2:** Andrejs Liepins/Science Source. **16.17:** Steve Gschmeissner/Science Source. **Emerging Diseases 16.1.1:** Fullerene. **Clinical Case Study 16:** Fuse/Thinkstock. **Table 16.4.1:** PBWPIX/Alamy. **Table 16.4.2:** Diane Macdonald/Getty Images. **Table 16.4.3:** Supri/Reuters. **Questions for Review 16.2:** David M Phillips/Science Source.

CHAPTER 17 **Opener:** Blend Images/Alamy. **Highlight 17.1:** Barry Dowsett & David A.J. Tyrrell. **17.3:** Centers for Disease Control and Prevention. **Beneficial Microbes 17.1:** Centers for Disease Control and Prevention. **17.8b:** Karen E Petersen. **17.10:** Libero Ajello, Centers for Disease Control and Prevention (CDC). **17.11b:** Maxine Jalbert, Centers for Disease Control and Prevention (CDC). **17.13b:** L. Brent Selinger, Pearson Education. **Clinical Case Study 17:** Blend Images/Alamy. **Questions for Review 17.2:** "From: Seroprevalence of simian immunodeficiency virus in wild and captive born Sykes' monkeys (Cercopithecus mitis) in Kenya. Retrovirology 2004, 1:34. BioMed Central."

CHAPTER 18 **Highlight 18.1:** Robert Bauman. **18.2a:** Microfield Scientific/Science Source. **18.2b:** David Scharf/Science Source. **18.2c:** Andrew Syred/Science Source. **18.3:** BSIP/Science Source. **18.4:** Medical-on-Line/Alamy. **Highlight 18.2:** Masson/Shutterstock. **18.9:** P. Marazzi/Science Source. **18.10:** Riccardo Rondinone. **18.11:** B. Lopez Oblare (www.fotogeriatria.net). **Clinical Case Study 18.1:** Scott Camazine/Science Source. **18.12:** Seelevel.com. **18.14a:** Virat Sirisanthana, Department of Pediatrics, Chiang Mai University, Thailand. **18.14b:** A. Ramey/PhotoEdit. **Clinical Case Study 18:** Ricky John Molloy/Thinkstock. **Clinical Case Study 18.2:** SPL/Science Source. **Questions for Review 18.1a:** Riccardo Rondinone. **Questions for Review 18.1b:** B. Lopez Oblare (www.fotogeriatria.net). **Questions for Review 18.1c:** P. Marazzi/Science Source. **Questions for Review 18.1d:** BSIP/Science Source. **Highlight 18.3:** NASA/Johnson Space Center.

CHAPTER 19 **Opener:** John F. Wilson/Science Source. **19.2:** David McCarthy/Science Source. **Emerging Diseases 19.1.1:** Fernando Caro, Dermatologist, Lima, Perú. **19.3:** Used by permission of the Massachusetts Medical Society. **19.4:** BSIP/Science Source. **19.5:** P. Marazzi/Science Source. **19.7:** Tal Eidlitz-Markus, M.D., and Avraham Zeharia, M.D. New England Journal of Medicine 2006; 354:e17April 27, 2006. **19.8:** Leonard J. Morse. **Disease in Depth 19.2:** National Library of Medicine. **Disease in Depth 19.3:** Eye of Science/Science Source. **Disease in Depth 19.4:** Eye of Science/Science Source. **Disease in Depth 19.5:** Thomas Volk. **Beneficial Microbes 19.1:** Zuno Burstein, Dermatologia Sanitaria, Instituto de Medicina Tropical Daniel A. Carrion, Universitad Nacional Mayor de San Marcos, Lima, Peru. **Disease at a Glance 19.1:** Medical-on-Line/Alamy. **Disease at a Glance 19.2a:** Tom Murray/BugGuide.net. **Disease at a Glance 19.3:** Greater Southern Area Health Service, NSW Australia. **Clinical Case Study 19.1:** Angela Hampton/Alamy. **Disease at a Glance 19.4:** Science Photo Library/Science Source. **19.11:** BSIP/Science Source. **Disease at a Glance 19.5:** AGE Fotostock. **Disease at a Glance 19.6:** Scott Camazine/Alamy. **19.13a:** DermNet/www.dermnet.com. **19.13b:** Jankurnelius/Alamy. **19.13c:** SPL/Science Source. **19.13d:** P. Marazzi/Science

Source. **Clinical Case Study 19.2:** www.dermnet.com. **Disease at a Glance 19.7a:** Historical Pictures Service/Custom Medical Stock Photo. **Disease at a Glance 19.7b:** Dr. P. Marazzi/Science Source. **Questions for Review 19.3a:** Historical Pictures Service/Custom Medical Stock Photo. **Questions for Review 19.3b:** Dr. P. Marazzi/Science Source. **Disease at a Glance 19.8:** Custom Medical Stock Photo. **19.15:** P. Marazzi/Science Source. **Clinical Case Study 19.3:** Peter Sherrard/Getty. **19.17:** Michelle Vilasineekul of Veronicav photography. **Disease at a Glance 19.9:** mediacolor's/Alamy. **19.18:** www.dermnet.com. **19.19:** John Hadfield/Science Source. **19.20:** SPL/Science Source. **Clinical Case Study 19.4:** John Watney/Science Source. **19.21:** R.A.Schwartz/E.Baran/eMedicine/Medscape. **19.22:** From: Prise en charge des mycétomes en Afrique de l'Ouest. Develoux M., Dieng M. T., Kane A. and Ndiaye B. Bull Soc Pathol Exot, 2003;96(5):376.382. **19.23:** Custom Medical Stock Photo. **19.24:** Zuno Burstein, Dermatologia Sanitaria, Instituto de Medicina Tropical Daniel A. Carrion, Universitad Nacional Mayor de San Marcos, Lima, Peru. **Clinical Case Study 19.5:** "Ancient Leishmaniasis in a Highland Desert of Northern Chile." C. Matheson, L. Iachetta, A. Llagostera, O. Appenzeller. PLoS One. Costa MA, 2009 Sep 10;4(9):e6983. **19.25:** Oliver Meckes/Science Source. **Emerging Diseases 19.2:** Henk Bentlage/Shutterstock. **Clinical Case Study 19:** John F. Wilson/Science Source. **Questions for Review 19.3c:** BSIP/Science Source. **Questions for Review 19.3d:** Scott Camazine/Alamy.

CHAPTER 20 **Opener:** Custom Medical Stock Photo, Inc. **Emerging Diseases 20.1.1:** Microstock Man/Fotolia. **20.3:** AMI Images/Science Source. **20.4:** VEM/Science Source. **Disease in Depth 20.1:** B. BOISSONNET/Alamy. **Disease in Depth 20.2:** Julie A. Theriot and Timothy J. Mitchison. **Disease in Depth 20.3:** Lorry Eason/Digital Vision/Getty Images. **Disease in Depth 20.4:** Balasubr Swaminathan and Peggy Hayes, CDC. **Disease in Depth 20.5:** Pr. Courtieu/Science Source. **Disease in Depth 20.6:** L.T. Tilney, P.S. Connelly & D.A. Portnoy. **Disease in Depth 20.7:** Biophoto Associates/Science Source. **Disease in Depth 20.8:** Maren Caruso/Digital Vision/Getty Images. **Highlight 20.1b:** Kaltenbach LS et al. **20.5:** Rupak De Chowdhur/Thomson Reuters. **Highlight 20.1a:** CC-BY-SA image: Balaji. **Disease at a Glance 20.1:** CDC. **20.7:** CDC. **Clinical Case Study 20.1:** Hercules Robinson/Alamy. **Disease at a Glance 20.2:** Rich Robison, Pearson Education. **Beneficial Microbes 20.1:** Eric Florentin/Shutterstock. **20.9:** Corbis. **Disease at a Glance 20.3:** Joe Pereira (JoeGoaUk). **Clinical Case Study 20.2:** Remus Moise/Shutterstock. **20.12:** Biophoto Associates/Science Source. **Disease at a Glance 20.4:** Frederick A. Murphy, University of Texas Medical Branch, Galveston. **20.13:** Marion Sourisseau, Marie-Christine PrCvost, Olivier Schwartz, Institut Pasteur Paris. **20.14:** USGS National Wildlife Health Center. **Emerging Diseases 20.2:** CC-BY-SA photo: Richard Bartz, Munich Makro Freak. **Disease at a Glance 20.5a:** Jim Gathany, CDC. **Disease at a Glance 20.6:** Edwin P. Ewing Jr., CDC. **Clinical Case Study 20.3:** U.S. Geological Survey. **Clinical Case Study 20.4:** Centers for Disease Control and Prevention. **Disease at a Glance 20.7:** Ralph Eagle Jr./Science Source. **20.18:** Western Ophthalmic Hospital/Science Source. **Clinical Case Study 20.5:** Michael Abbey/Science Source. **Clinical Case Study 20:** Custom Medical Stock Photo, Inc.

CHAPTER 21 **Opener:** Westend61/AGE Fotostock. **21.2:** Science Source. **21.3:** CC-BY-SA photo: Hektor. **Beneficial Microbes 21.1:** SEM by Daniel Kadouri and George O'Toole, colorized by Russell Monds, Dartmouth Medical School. **Disease at a Glance 21.1:** Centers for Disease Control and Prevention (CDC). **Clinical Case Study 21.1:** L. Brent Selinger, Pearson Education. **21.5:** Brook Yockey, Centers for Disease Control and Prevention (CDC). **Disease at a Glance 21.2:** Centers for Disease Control and Prevention (CDC). **21.7:** Jim Marshall/The Image Works. **Disease at a Glance 21.3:** Jack Poland, Centers for Disease Control and Prevention (CDC). **21.8:** Sean E. Connolly/Stony Brook University. **Clinical Case Study 21.2:** CAPT Russell E. Enscore, Centers for Disease Control and Prevention (CDC). **21.9a:** American Lyme Disease Foundation. **Clinical Case Study 21.3:** Jill Fromer/iStockphoto. **21.13a:** Science Photo Library/Science Source. **21.13b:** M.A. Ansary/Science Source. **21.13c:** Robert Bauman. **21.13d:** National Institute of Health. **21.14:** Centers for Disease Control and Prevention (CDC). **21.15:** USDA/Science Photo Library/Science Source. **21.17:** Science Source. **Beneficial Microbes 21.2:** © 2011 Fischer et al. **Clinical Case Study 21.4:** Getty Images. **Highlight 21.1:** Dinodia Photos/Alamy. **Disease in Depth 21.1:** Sue Ford/Science Source. **Disease in Depth 21.2:** Gopal Murti/Science Source. **Disease in Depth 21.3:** Omikron/Science Source. **Disease in Depth 21.4:** Steven Glenn, Laboratory & Consultation Division,. **Disease in Depth 21.5:** AFP/Getty Images. **Disease in Depth 21.6:** Kletr/Fotolia. **Disease in Depth 21.7:** London School of Hygiene & Tropical Medicine/Science Source. **Disease in Depth 21.8:** London School of Hygiene & Tropical Medicine/Science Source. **Disease in Depth 21.9:** Mauro Fermariello/Science Source. **Disease in Depth 21.10:** Irene Abdou/Newscom. **Disease at a Glance 21.6:** Centers for Disease Control and Prevention. **Disease at a Glance 21.7:** Centers for Disease Control and Prevention (CDC). **Clinical Case Study 21.5:** Dmitry Melnikov/Shutterstock. **Disease at a Glance 21.8:** Centers for Disease Control and Prevention (CDC). **Emerging Diseases 21.1:** CC-BY-SA: User:Cornellier. **Clinical Case Study 21:** Westend61/AGE Fotostock. **Highlight 21.1a:** Balaji.

CHAPTER 22 **22.0a:** Getty Images. **22.0b:** Blend Images/Shutterstock. **Disease at a Glance 22.1:** Scott Camazine/Alamy. **22.2:** CC-BY-SA3.0: Dileepunnikri. **22.3:** "Infection Control", book edited by Silpi Basak, ISBN 978-953-51-1145-0, Published: May 29, 2013 under CC-BY-SA3.0 license, InTech. Figure 2 in Chapter 4, Ocular Infections Caused by Corynebacterium Species, by Hiroshi Eguchi. **22.4:** allesalltag/Alamy. **22.5:** Andrew Syred/Science Source. **22.6:** John W. Rippon, University of Chicago. **22.7:** Alan Rodwell. **22.8:** M.I. Walker/Alamy. **22.9:** Centers for Disease Control and Prevention (CDC). **Disease at a Glance 22.2:** Eye of Science/Science Source. **Disease in Depth 22.1:** Denis Meyer/imagebroker/Alamy. **Disease in Depth 22.2:** SPL/Science Source. **Disease in Depth 22.3:** National Institute of Allergy and Infectious Diseases. **Disease in Depth 22.4:** Biophoto Associates/Science Source. **Disease in Depth 22.5:** Biophoto Associates/Science Source. **Disease in Depth 22.6:** CNRI/Science Source. **Disease in Depth 22.7:** CNRI/Science Source. **Disease in Depth 22.8:** P. Marazzi/Science Source. **Disease in Depth 22.9:** Santibhavank P/Shutterstock. **Disease in Depth 22.10:** Life in View/Science Source. **Disease at a Glance 22.3:** NIBSC/Science Photo Library/Science Source. **22.10:** Otis Historical Archives, NMHM. **Clinical Case Study 22.1:** Ariel Skelley/Blend Images/Getty Images. **Emerging Diseases 22.1.1:** iStockphoto.com. **22.13:** Fahad Shadeed/Reuters. **Clinical Case Study 22.2:** Mark Henley/AGE Fotostock America Inc. **22_DG04:** NIAID. **Disease at a Glance 22.5:** graficart.net/Alamy. **Disease at a Glance 22.6:** James Gathany, Centers for Disease Control and Prevention (CDC). **Clinical Case Study 22.3:** St Bartholomew's Hospital, London/Science Source. **22.17:** "Dematology Photo Quiz." M. Miller. Resident and Staff Physician. May 2006; (52)5; Fig. 1. **22.18:** Science Source. **22.19:** John W. Rippon, University of Chicago. **Emerging Diseases 22.1.1:** iStockphoto.com. **Disease at a Glance 22.7a:** Centers for Disease Control and Prevention (CDC). **Disease at a Glance 22.7b:** Libero Ajello. **Clinical Case Study 22a:** Getty Images. **Clinical Case Study 22b:** Blend Images/Shutterstock. **Questions for Review 22.2a:** Biophoto Associates/Science Source. **Questions for Review 22.2b:** M.I. Walker/Alamy. **Questions for Review 22.2c:** From: "Infection Control", book edited by Silpi Basak, ISBN 978-953-51-1145-0, Published: May 29, 2013 under CC-BY-SA3.0 license, InTech. Figure 2 in Chapter 4, Ocular Infections Caused by Corynebacterium Species, by Hiroshi Eguchi. **Questions for Review 22.2d:** John Durham/Science Source. **Disease in depth 22.3:** NIAID. **22.5:** Andrew Syred/Science Source.

CHAPTER 23 **Opener:** Daniel Dempster Photography/Alamy. **Beneficial Microbes 23.1:** Andrew Syred/Science Source. **Disease at a Glance 23.1:** Custom Medical Stock Photo. **23.6:** Gary Gaugler/Science Source. **23.7:** David M. Martin/Science Source. **Disease at a Glance 23.3:** M. Peres/Custom Medical Stock Photography. **Clinical Case Study 23.1:** ene/Shutterstock. **Disease at a Glance 23.5:** Liverpool School of Tropical Medicine courtesy of A Stich, Wellcome Images. **Disease at a Glance 23.6:** Eye of Science/Science Source. **23.11:** Zara/Science Source. **Clinical Case Study 23.2:** Robert Weaver, Centers for Disease Control and Prevention (CDC). **Disease at a Glance 23.7:** Custom Medical Stock Photo. **23.13a:** Moredun Animal Health Ltd./Science Source. **23.13b:** Linda Stannard/Science Source. **23.15a:** Linda Stannard/Science Source. **23.16:** Science Source. **Disease at a Glance 23.8:** Garry Watsib/Science Source. **Emerging Diseases 23.1:** Piotr Marcinski/Fotolia. **Disease in Depth 23.1:** Cristina Pedrazzini/Science Source. **Disease in Depth 23.2:** DPD Image Library/Centers for Disease Control and Prevention Image Library. **Disease in Depth 23.3:** DPD Image Library/Centers for Disease Control and Prevention Public Health Image Library. **Disease in Depth 23.4:** Stan Erlandsen and Dennis Feely, Centers for Disease Control and Prevention. **Disease in Depth 23.5:** SPL/Science Source. **Disease in Depth 23.6:** Stan Erlandsen/Centers for Disease Control and Prevention Public Health Image Library. **Disease in Depth 23.7:** Dr. Stan Erlandsen, Centers for Disease Control and Prevention (CDC). **Disease in Depth 23.8:** Michael Abbey/Science Source. **Disease in Depth 23.9:** David P. Lewis/Shutterstock. **Clinical Case Study 23.3:** Veer/Corbis. **23DG09:** Centers for Disease Control and Prevention. **23.17b:** Andrew Syred/Science Source. **23.19a:** Gastrolabs/Science Source. **23.19b:** Centers for Disease Control and Prevention (CDC). **Clinical Case Study 23:** Daniel Dempster Photography/Alamy. **Questions for Review 23.2:** Linda Stannard/Science Source.

CHAPTER 24 **Opener:** Moment/Getty Images. **24.2:** Australian Research Council Centre of Excellence in Structural and Functional Mi. **24.4:** Centers for Disease Control and Prevention Public Health Image Library. **Disease in Depth 24.1:** Alliance/Shutterstock. **Disease in Depth 24.2:** CNRI/Science Source. **Disease in Depth 24.3:** Steve Gschmeissner/Science Source. **Disease in Depth 24.4:** SERCOMI/Science Source. **Disease in Depth 24.5:** Pietro M. Motta et al./Science Source. **Disease in Depth 24.6:** Michael Abbey/Science Source. **Disease in Depth 24.7:** Sasa Komlen/Shutterstock. **Disease at a Glance 24.2:** American Family Physician/Family Practice Management. **Beneficial Microbes 24.1:** Maureen Spuhler. **Disease at a Glance 24.3:** John Durham/Science Source. **Disease at a Glance 24.4:** Custom Medical Stock Photo, Inc. **Disease at a Glance 24.5:** James Cavallini/BSIP SA/Alamy. **24.6a:** BSIP/Photoshot Holdings Ltd. **24.6b:** Custom Medical Stock Photo. **24.6c:** © The Regents of the University of California, Davis campus. Originally published in Dermatology Online Journal. Solitary frontal ulcer: A syphilitic gumma. Andrade et al. Dermatology Online Journal 16 (9): 5 Fig. 1. All Rights Reserved. Used with permission. **Clinical Case Study 24.1:** Norman Jacobs, Centers for Disease Control and Prevention (CDC). **24.8:** Milton Reisch/Corbis Images. **Disease at a Glance 24.6:** Edwin P. Ewing, Centers for Disease Control and Prevention (CDC). **24.9a:** Science Photo Library/Science Source. **Clinical Case Study 24.2:** Blend Images/Veer/Corbis Images. **24.10:** M. A. Ansary/Science Source. **24.12a:** P. Marazzi/Science Source. **24.12b:** P. Marazzi/Science Source. **24.13:** Biophoto/Science Source. **Clinical Case Study 24.3:** Centers for Disease Control and Prevention (CDC). **24DG07:** CC-BY-SA: Colorectal. **Clinical Case Study 24:** Moment/Getty Images. **Disease at a Glance 24.8:** J.L. Carson/Custom Medical Stock Photo. **Questions for Review 24.2a:** J.L. Carson/Custom Medical Stock Photo. **Questions for Review 24.2b:** Norman Jacobs, Centers for Disease Control and Prevention (CDC). **Questions for Review 24.2c:** Edwin P. Ewing, Centers for Disease Control and Prevention (CDC). **Questions for Review 24.2d:** John Durham/Science Source.

CHAPTER 25 **Opener:** Tetra Images/Alamy. **25.1:** AP Images/Joe Hermitt. **25.2:** Paul Hudson/AGE Fotostock America Inc. **25.3:** Arnd Wiegmann/Reuters. **25.4:** EPA Photo/Attila Kisbenedek. **25.5:** Novo Nordisk. **25.6:** Hector Mata/Reuters. **Highlight 25.1:** Seelevel.com. **Emerging Diseases 25.1.1:** Alex Skelly/Flickr/Getty Images. **25.8a:** Central Virginia Governor's School for Science and Technology. **25.8b:** IDEXX Laboratories. **25.9:** JRC/Alamy. **25.11:** Ross Coleman/AP Images. **Beneficial Microbes 25.1:** U.S. Coast Guard. **25.13:** CSIRO Land and Water. **25.14:** K. J. Edwards/Woods Hole Oceanographic Institution. **25.14 inset:** Radhoose/Shutterstock. **25.20:** Kenneth Lambert/AP Images. **Highlight 25.2:** Dr. Klaus Boller/Science Source. **Clinical Case Study 25:** Tetra Images/Alamy.

Subject Index

NOTE: A *t* following a page number indicates tabular material, an *f* following a page number indicates a figure, and a *b* following a page number indicates a boxed feature.

SUBJECT INDEX

SUBJECT INDEX

SUBJECT INDEX

SUBJECT INDEX

SUBJECT INDEX

SUBJECT INDEX

SUBJECT INDEX

SUBJECT INDEX

SUBJECT INDEX

SUBJECT INDEX

SUBJECT INDEX

PRONUNCIATIONS OF SELECTED ORGANISMS AND VIRUSES

Absidia (ab-sid´ē-ă)
Acanthamoeba (ă-kan-thă-mē´bă)
Acetobacter (a-sē´tō-bak-ter)
Acinetobacter (as-i-nē´tō-bak´ter)
Acremonium (ak´rē-mō´nē-ŭm)
Actinomyces israelii (ak´ti-nō-mī´sēz is-rā´el-ē-ē)
Agaricus (a-găr´i-kus)
Agrobacterium tumefaciens (ag´rō-bak-tēr´ē-um tū´me-fāsh-enz)
Alternaria (al-ter-nā´rē-ă)
Amanita muscaria (am-ă-nī´tă mus-ka´rē-ă)
Amanita phalloides (am-ă-nī´tă fal-ōy´dēz)
Amoeba (am-ē´bă)
Amycolatopsis orientalis (am-ē-kō´la-top-sis o-rē-en-tal´is)
Anaplasma phagocytophilum (an-ă-plaz´mă fag-ō-sī-to´fil-ŭm)
Ancylostoma duodenale (an-si-los´tō-mă doo´ō-de-nā-lē)
Aquaspirillum magnetotacticum (ă-kwă-spī´ril-ŭm mag-ne-tō-tak´ti-kŭm)
Aquifex (ăk´wē-feks)
Ascaris lumbricoides (as´kă-ris lŭm´bri-koy´dēz)
Aspergillus oryzae (as-per-jil´ŭs o´ri-zī)
Azomonas (ā-zō-mō´nas)
Azospirillum (ā-zō-spī´ril-ŭm)
Azotobacter (ā-zō-tō-bak´ter)

Bacillus anthracis (ba-sil´ŭs an-thrā´sis)
Bacillus cereus (ba-sil´ŭs se´rē-ŭs)
Bacillus licheniformis (ba-sil´ŭs lī-ken-i-for´mis)
Bacillus polymyxa (ba-sil´ŭs po-lē-miks´ă)
Bacillus popilliae (ba-sil´ŭs pop-pil´ē-ī)
Bacillus sphaericus (ba-sil´ŭs sfe´ri-kŭs)
Bacillus stearothermophilus (ba-sil´ŭs ste-rō-ther-ma´fil-ŭs)
Bacillus subtilis (ba-sil´ŭs sŭt´i-lis)
Bacillus thuringiensis (ba-sil´ŭs thur-in-jē-en´sis)
Bacteroides fragilis (bak-ter-oy´dēz fra´ji-lis)
Balantidium coli (bal-an-tid´ē-ŭm kō´lī)
Bartonella bacilliformis (bar-tō-nel´ă ba-sil´li-for´mis)
Bartonella henselae (bar-tō-nel´ă hen´sel-ī)
Bartonella quintana (bar-tō-nel´ă kwin´ta-nă)
Bdellovibrio (del-lō-vib´rē-ō)
Beggiatoa (bej´jē-a-tō´ă)
Blastomyces dermatitidis (blas-tō-mī´sēz der-mă-tit´i-dis)
Bordetella pertussis (bōr-dĕ-tel´ă per-tus´is)
Borrelia burgdorferi (bō-rē´lē-ă burg-dōr´fer-ē)
Borrelia recurrentis (bō-rē´lē-ă re-kur-ren´tis)
Botryococcus braunii (bot´rē-ō-kok´ŭs brow´nē-ē)
Brucella abortus (broo-sel´lă a-bort´us)
Brucella canis (broo-sel´lă kā´nis)
Brucella melitensis (broo-sel´lă me-li-ten´sis)
Brucella suis (broo-sel´lă soo´is)
Burkholderia cepacia (burk-hol-der´ē-ă se-pā´se-ă)
Burkholderia pseudomallei (burk-hol-der´ē-ă soo-dō-mal´e-ē)

Campylobacter jejuni (kam´pi-lō-bak´ter jē-jū´nē)
Candida albicans (kan´did-ă al´bi-kanz)
Carsonella ruddii (kar-son-el´ă rŭd´ē-ē)
Caulobacter (kaw´lō-bak-ter)
Chlamydia trachomatis (kla-mid´ē-ă tra-kō´ma-tis)
Chlamydophila pneumoniae (kla-mē-dof´ī-lă noo-mō´nē-ī)
Chlamydophila psittaci (kla-mē-dof´ī-lă sit´ă-sē)
Chondrus crispus (kon´drŭs krisp´ŭs)
Chromatium buderi (krō-ma´tē-ŭm bū´de-rē)
Citrobacter (sit´rō-bak-ter)
Cladophialophora carrionii (klă-dŏf´ē-ă-lof´ŏ-rā kar-rē-on´ē-ē)
Claviceps purpurea (klav´i-seps poor-poo´rē´ă)
Clostridium botulinum (klos-trid´ē-ŭm bo-tū-lī´num)
Clostridium difficile (klos-trid´ē-ŭm di´fe-sēl)
Clostridium perfringens (klos-trid´ē-ŭm per-frin´jens)
Clostridium tetani (klos-trid´ē-ŭm te´tan-ē)
Coccidioides immitis (kok-sid-ē-oy´dēz im´mi-tis)
Codium (kō´dē-ŭm)
Coltivirus (kol´tē-vī´rŭs)
Cortinarius gentilis (kōr´ti-nar-ē-us jen´til-is)
Corynebacterium diphtheriae (kŏ-rī´nē-bak-tēr´ē-ŭm dif-thi´rē-ī)
Coxiella burnetii (kok-sē-el´ă ber-ne´tē-ē)
Cryptococcus neoformans (krip-tō-kok´ŭs nē-ō-for´manz)
Cryptosporidium parvum (krip-tō-spō-rid´ē-ŭm par´vŭm)
Cyclospora cayetanensis (sī-klō-spōr´ă kī´ē-tan-en´sis)
Cytomegalovirus (sī-tō-meg´ă-lō-vī´rŭs)
Cytophaga (sī-tof´ă-gă)

Deinococcus radiodurans (dī-nō-kok´ŭs rā-dē-ō-dur´anz)
Desulfovibrio (dē´sul-fō-vib´rē-ō)
Dictyostelium (dik-tē-ō-stē´lē-um)
Didinium (dī-di´nē-ŭm)
Diplococcus pneumoniae (dip´lō-kok´ŭs nū-mō´nē-ī)

Echinococcus granulosus (ĕ-kī´nō-kok´ŭs gra-nū-lō´sŭs)
Edwardsiella (ed´ward-sē-el´ă)
Ehrlichia chaffeensis (er-lik´ē-ă chaf-ē-en´sis)
Encephalatizoon intestinalis (en-sef-a-lat-e´zō-an in-tes´ti-năl´is)
Entamoeba histolytica (ent-ă-mē´bă his-tō-li´ti-kă)
Enterobacter (en´ter-ō-bak´ter)
Enterobius vermicularis (en-ter-ō´bī-ŭs ver-mi-kū-lar´is)
Enterococcus faecalis (en´ter-ō-kok´ŭs fē-kă´lis)

Enterococcus faecium (en´ter-ō-kok´ŭs fē-sē´ŭm)
Epidermophyton floccosum (ep´i-der-mof´i-ton flŏk´ō-sŭm)
Epulopiscium fishelsoni (ep´yoo-lō-pis´sē-ŭm fish-el-sō´nē)
Escherichia coli (esh-ĕ-rik´ē-ă kō´lī)
Euglena granulata (yū-glēn´ă gran-yū-lă´tă)
Eupenicillium (yū-pen-i-sil´ē-ŭm)
Exophiala (ek-sō-fī´ă-lă)

Fasciola gigantica (fa-sē´ō-lă ji-gan´ti-kă)
Fasciola hepatica (fa-sē´ō-lă he-pa´ti-kă)
Ferroplasma acidarmanus (fe´rō-plaz-ma a-sid´ar-mă-nŭs)
Fonsecaea compacta (fon-sē-sē´ă kom-pak´ta)
Fonsecaea pedrosoi (fon-sē-sē´ă pe-drō´sō-ē)
Francisella tularensis (fran´si-sel´ă too-lă-ren´sis)
Fusarium (fū-zā´rē-ŭm)

Gambierdiscus (gam´bē-er-dis-kŭs)
Gardnerella vaginalis (gărd´ner-el´ă va-ji-nă´lis)
Gelidium (jel-li´dē-ŭm)
Geogemma barossii (jē´ō-jem-ă ba-rōs´ē-ē)
Giardia intestinalis (jē-ar´dē-ă in-tes´ti-năl´is)
Gluconobacter (gloo-kon´ō-bak-ter)
Gonyaulax (gon-ē-aw´laks)
Gymnodinium (jīm-nō-din´ē-ŭm)
Gyromitra esculenta (gī-rō-mē´tră es-kū-len´tă)

Haemophilus ducreyi (hē-mof´i-lŭs doo-krā´ē)
Haemophilus influenzae (hē-mof´i-lŭs in-flu-en´zī)
Hafnia (haf´nē-ă)
Halobacterium salinarium (hă´lō-bak-tēr´ē-ŭm sal-ē-nar´ē-ŭm)
Hantavirus (han´tă-vī-rŭs)
Helicobacter pylori (hel´ĭ-kō-bak´ter pī´lō-rē)
Histoplasma capsulatum (his-tō-plaz´mă kap-soo-lā´tŭm)

Izziella abbottiae (iz-ē-el´lă ab´ot-tē-ī)

Klebsiella pneumoniae (kleb-sē-el´ă nū-mō´nē-ī)

Lactobacillus bulgaricus (lak´tō-bă-sil´ŭs bul-gā´ri-kŭs)
Lactococcus lactis (lak-tō-kok´ŭs lak´tis)
Legionella pneumophila (lē-jŭ-nel´lă noo-mō´fi-lă)
Leishmania (lēsh-man´ē-ă)
Leptospira interrogans (lep´tō-spī´ră in-ter´ră-ganz)
Leuconostoc citrovorum (loo´kō-nos-tŏk sit-rō-vō´rum)
Listeria monocytogenes (lis-tēr´ē-ă mo-nō-sī-tah´je-nēz)
Lyssavirus (lis´ă-vī-rŭs)

Madurella (mad´ū-rel´ă)
Malassezia furfur (mal-ă-sē´zē-ă fur´fur)

Methanobacterium (meth´a-nō-bak-tēr´ē-ŭm)
Methanopyrus (meth´a-nō-pī´rŭs)
Micavibrio (mī-kă-vib´rē-ō)
Microsporidium (mī-krō-spor-i´dē-ŭm)
Microsporum (mī-kros´po-rŭm)
Moraxella catarrhalis (mōr´ak-sel´ă kă-tah´răl-is)
Morganella (mōr´gan-el´ă)
Mucor (mū´kōr)
Mycobacterium avium-intracellulare
(mī´kō-bak-tēr´ē-ŭm ā´vē-ŭm in´tra-sel-yu-la´rē)
Mycobacterium bovis (mī´kō-bak-tēr´ē-ŭm bō´vis)
Mycobacterium leprae (mī´kō-bak-tēr´ē-ŭm lep´rī)
Mycobacterium tuberculosis
(mī´kō-bak-tēr´ē-ŭm too-ber-kyū-lō´sis)
Mycoplasma genitalium (mī´kō-plaz-mă jen-ē-tal´ē-ŭm)
Mycoplasma hominis (mī´kō-plaz-mă ho´mi-nis)
Mycoplasma pneumoniae (mī´kō-plaz-mă nū-mō´nē-ī)

Naegleria (nā-glē´rē-ă)
Necator americanus (nē-kā´tor ă-mer-i-ka´nus)
Neisseria gonorrhoeae (nī-se´rē-ă go-nor-rē´ī)
Neisseria meningitidis (nī-se´rē-ă me-nin-ji´ti-dis)
Neurospora crassa (noo-ros´pōr-ă kras´ă)
Nitrobacter (nī-trō-bak´ter)
Nitrosomonas (nī-trō-sō-mō´nas)
Nocardia asteroides (nō-kar´dē-ă as-ter-oy´dēz)
Nosema (nō-sē´mă)

Orientia tsutsugamushi (ōr-ē-en´tē-ă tsoo-tsoo-gă-mū´shē)
Orthopoxvirus variola (ōr-thō-poks´vī-rŭs vă-rī´ō-lă)

Paracoccidioides brasiliensis
(par´ă-kok-sid-ē-oy´dēz bră-sil-ē-en´sis)
Paramecium (par-ă-mē´sē-ŭm)
Pasteurella haemolytica (pas-ter-el´ă hē-mō-lit´i-kă)
Pasteurella multocida (pas-ter-el´ă mul-tŏ´si-da)
Penicillium chrysogenum (pen-i-sil´ē-ŭm krī-so´jĕn-ŭm)
Penicillium marneffei (pen-i-sil´ē-ŭm mar-nef-ē´ī)
Penicillium roqueforti (pen-i-sil´ē-ŭm rok´for-tē)
Pfiesteria (fes-tēr´ē-ă)
Phialophora verrucosa (fī-ă-lof´ŏ-ră ver-ū-kō´să)
Physarum (fī-sar´um)
Phytophthora infestans (fī-tof´tho-ră in-fes´tanz)
Piedraia hortae (pī-drā´ă hōr´tī)
Plasmodium falciparum (plaz-mō´dē-ŭm fal-sip´ar-ŭm)
Plasmodium knowlesi (plaz-mō´dē-ŭm nō-les´ē)
Plasmodium malariae (plaz-mō´dē-ŭm mă-lār´ē-ī)
Plasmodium ovale (plaz-mō´dē-ŭm ō-vă´lē)
Plasmodium vivax (plaz-mō´dē-ŭm vī´vaks)